Evolutionary events: Procaryotes, Protists, Fungi, and Plants	Evolutionary events: Animals
Gymnosperms on the rise; adaptive radiation of orchids Modern distribution of angiosperms and gymnosperms	Neolithic to modern civilization *Homo erectus, Homo sapiens* (Neanderthal), *Homo sapiens sapiens,* modern mammals
	Large extinction of earlier mammals
Grasslands spread	First hominids (Ramapithecus), numerous grazing mammals
	All of today's mammal families
Angiosperms and gymnosperms dominate	Earliest cetaceans, all of today's mammal orders formed
Most present-day angiosperm families develop	Age of Mammals begins; modern invertebrates appear
Extinction of dominant phytoplankton (coccolithophorids)	Massive extinction: nearly all dinosaurs and 70% of all animal species
Rise of angiosperms	Dinosaurs reach peak and rapidly decline; birds persist
Earliest angiosperms (restricted to higher tropical elevations), conifers, ferns, ginkgos dominant	First birds, teleost fishes, modern crustaceans; Age of Reptiles begins
Gymnosperms (conifers, cycads, ginkgos), ferns dominant plant form	First dinosaurs, earliest mammallike reptiles
Extinction of many dominant life forms	Most Paleozoic invertebrates (including trilobites) extinct
First gymnosperms; coal age forests (tall trees: lycophytes, horsetails, ferns, seed ferns)	Earliest reptiles, first winged insects, Age of Amphibians
Large terrestrial plants, first seeds, first vascular plants (psilopsids, lycophytes, sphenophytes)	First amphibians, sharklike fishes, bony fishes, lung fishes, mandibulate arthropods
Algae give way to terrestrial plants; Green, red, and brown algae	Age of Fishes: lobe-finned fishes, jawed fishes; first terrestrial invertebrates, coral reef building; first vertebrates: jawless fishes, chelicerate arthropods
Algal forms dominate	Marine invertebrates dominate, trilobites abound
Multicellular life (algae, fungi?)	Late Pre-cambrian: first evidence of multicellular animals: soft-bodied coelenterates and other marine invertebrates, burrowing wormlike animals
First eucaryotes (probable): green algae, other protists	
Photosynthetic anaerobes (procaryotes)	
Origin of life: first bacterialike procaryotes (anaerobic heterotrophs)	
Organic synthesis	

BIOLOGY
THE SCIENCE OF LIFE

ROBERT A. WALLACE
University of Florida

JACK L. KING
University of California, Santa Barbara

GERALD P. SANDERS
Science writer and consultant,
Goodyear Publishing Company

Goodyear Publishing Company, Inc.
Santa Monica, California

This book is dedicated

To King and Sanders
R.A.W.

To Ethel
J. L. K.

To my wife Mary and our clan for their faith and support, but especially to my daughter and chief literary critic, Cheri, who usually takes me seriously
G.P.S.

Library of Congress Cataloging in Publication Data

Wallace, Robert Ardell, 1938–
Biology the science of life.

Includes bibliographies and index.
1. Biology. I. King, Jack L., 1934– joint author. II. Sanders, Gerald P., joint author. III. Title.
QH308.2.W34 574 80–23154
ISBN 0–87620–083–8

Copyright © 1981 by Goodyear Publishing Company, Inc.
Santa Monica, California 90401

All rights reserved. No part of this book may be reproduced in any form or by any means without permission in writing from the publisher.

Current printing (last digit):

10 9 8 7 6 5 4 3 2 1

ISBN: 0-87620-083-8
Y–0838–6

Printed in the United States of America

Editorial Production Manager: Susan Smith
Art Director: Bob Hollander
Design: J. Paul Kirouac, A Good Thing, Inc.
Layout: Merilyn Yee Loo, A Good Thing, Inc.
Illustrations: Eric G. Hieber, E. H. Technical Services
Pencil Rendering: Kathleen Borowik
Photo Research: Mary-Lynn Riesmeyer and Karen Salzman, Photo Researchers, Inc.
Composition: Typothetae Book Composition

COVER PHOTO: *Like begets like in the basic miracle of life: the circular DNA of a bacterial plasmid is captured here as it opens out into identical daughter molecules.*

PREFACE

Someone once said that *two* people getting together to write a book is like *three* people getting together to have a baby. What can we say about *three* people getting together to write a book? We don't know how it is with other writing teams, but we have found our association over these last four years to be surprisingly pleasant, well-balanced, friendly, and mutually helpful. Surprisingly, because we had to merge, or submerge, three strong egos to a common end. Each of us has had to see our own precious words rewritten and our ideas recast by our coauthors, time and again, and have learned first to accept and then to appreciate the process. There is scarcely a sentence here that any one of us can lay unambiguous claim to. Although each of us has his own area of expertise in biology, there is not a chapter that does not bear our common stamp. We wouldn't have it any other way.

Among us, Jerry Sanders, whose own field of research is physiology and zoology, has perhaps the broadest knowledge of general biology and the most extensive teaching experience. Jerry is an organizer and an initiator, and was responsible for the first outlines of most of our chapters, as well as the art manuscript and the self-teaching material that is included at the end of each chapter. We've depended heavily on him to provide the foundation on which to build, especially in the inevitably many areas of biology in which no one of us had any special claim of sophistication.

Jack King's research expertise is in genetics, biochemistry, and evolutionary theory. His best-known work has to do with the mechanisms by which DNA and protein molecules slowly change over millions of years of evolution. In the creation of our book, he made a special effort to keep up with the constant flood of new research, particularly in the rapidly evolving field of molecular biology. Jack tended to get carried away with the excitement of it all, and we've had to throw out a lot of his exuberant ramblings, but we've done our best to retain his enthusiasm.

Bob Wallace's research expertise is in ecology and animal behavior. He has written several well-known books on animal behavior as well as BIOLOGY THE WORLD OF LIFE, an extremely popular textbook in general biology. In addition to his scientific qualifications, Bob is our best wordsmith and humorist. He also has the ability to make complex material seem simple. Biology is interesting; it is full of good stories; we wanted that to come across. Our chief storyteller and poet has contributed the most to the flavor and sustained *interestingness* of our book. The final rewrite, the polishing of phrases, have primarily been his responsibility. We must admit, though, that Bob tends to gleefully get out of hand. You should see some of the things we had to excise!

It's been quite an experience for all of us, the years of researching and of writing, only to have our words and ideas torn apart by a battery of expert reviewers as well as our coauthors; and then rebuilding, winnowing, deleting, rethinking, and doing the whole thing over yet one more time in the hope of creating the best biology book ever, always peering nervously at the long row of quite excellent competing texts, always aware of the constant thunder of current research that steadily reconstructs the science of biology even as we record it. For the final months we filled the Goodyear library with our own esoteric reference books, lived on Nippy Charlie tuna sandwiches, and exhausted ourselves working together in what seemed, to our delicate constitutions, like a sweatshop atmosphere, constantly working at details, details, details, while trying to maintain enough perspective to keep the big picture in view. None of us knew just how hard it would really be.

On the other side of the coin, it was beneficial, fun, and even exhilarating to work through the new ideas in biology, to discover to our frequent dismay that some things we had believed were dead wrong, to try to devise new ways to teach the material, to see if traditional information couldn't be handled a bit differently to make it more easily understood. We were surprised at how little we know about wombats, at the incredible clarity of Darwin's thinking, and at the amazing discrepancies between authoritative biology books. We also learned a lot of botany. Our close working association also enabled us to display to each other our wit and our talent for speaking in accents. So the task proved to be enormous, but the rewards were varied and worth the effort.

The greatest reward, of course, will be the acceptance of our efforts by the students who will use the book. The format is rather straightforward. After setting the stage historically with the remarkable Mr. Darwin, we proceed from small molecules to large systems. We've tried to be comprehensive—the facts are all there—but we've also emphasized the science involved, the experiments behind the stated conclusions, how we know what we know. Just as important are the general principles, the grand concepts such as natural selection and surface–volume phenomena. The essays, as you will see, stand alone and are simply asides that we thought anyone might find interesting or helpful; but the organization of the text itself reflects the needs of a comprehensive biology course, at the majors' level, with information and concepts to be mastered (though we'd prefer to think it reads like a novel).

A familiar complaint against general biology texts is that they are unfair to botany. We've taken great care to include a firm introduction to botany as well as zoology and microbiology. We hope that our genetics chapters and our presentation of the diversity of life are each strong enough to stand alone. We scoured the most recent literature to uncover information that people might find relevant in a number of contexts; the arguments surrounding recombinant DNA, brain chemistry, and sociobiology are important to everyone, and the more widely such concepts are understood the better.

It is a humbling experience to reread any biology textbook that is more than, say, a decade old. Many of the "facts" are still true, but are couched in what seem now to be hazy and fudging terms. And some of the most positively stated ideas are plainly wrong. It is both sad and hopeful to note that our textbook undoubtedly awaits the same fate, as new facts and new ways of looking at things are unveiled. Thus we have included at the ends of chapters some questions for which definitive answers do not exist. We hope they engender discussion and stimulate thinking; learning what is not known can be an even greater challenge than learning what is known.

We enjoyed the writing of this book and our working together. Perhaps because of that enjoyment and the general levity of our association, we occasionally felt free to cite our own opinions on some issue, but we hope such instances are clearly discernable and that they will provide a clear target for dissenting opinions. You may also notice that our writing style or phraseology may be personal or unusual in one or two instances. This is because, to each of us, biology is not only fascinating, but fun. It is after all the study of life, or living things, and one can't help but get occasionally giddy at the prospect of learning a little more. We offer few or no apologies.

A Special Note

The authors would like to make what may be an unusual statement. We love our production editor Susan Smith. It has been a constant delight to work with someone with such diligence, endurance, humor, intelligence, and graciousness.

We have all again benefited from the insight and camaraderie of our editor Clay J. Stratton. He not only conceived of the project, but offered invaluable guidance and suggestions at virtually every step of its development. We have gained enormous respect for his insight into what goes into the making of useful texts. In a real sense, this can be called his book.

We could expound on the expertise of others at Goodyear Publishing as well, such as Gerald Rafferty, one of the most knowledgable people in the business, we think; Bob Hollander, a professional nail-biter who measures everything carefully and then fully commits himself; and of course we must thank Al Goodyear who gave us a dry place to work with no heavy lifting.

ACKNOWLEDGMENTS

The authors and publisher would like to express their appreciation for the opinions, advice, criticism and support of a number of biologists who have greatly assisted in the development of the book. Their collective influence helped to provide a clear perspective for the book and greatly aided in the development of pedagogical techniques in teaching the basic principles of biology.

Harvey Alexander, Prince Georges Community College, Largo, Maryland
Ross Arnett, Jr., Siena College, Loudonville, New York
Vernon Avila, San Diego State University, San Diego, California
Mary Barber, California State University-Northridge, Northridge, California
Patricia Baker, University of California, Berkeley, California
Marvin Barnum, Florissant Valley College, St. Louis, Missouri
George Becker, Metropolitan State College, Denver, Colorado
William E. Barstow, University of Georgia, Athens, Georgia
Robert O. Bland, College of St. Thomas, St. Paul, Minnesota
Relis Brown, West Chester State College, West Chester, Pennsylvania
Richard K. Boohar, University of Nebraska, Lincoln, Nebraska
Frank Bushnell, St. Petersburg College, St. Petersburg, Florida
John Caruso, University of Cincinnati, Cincinnati, Ohio
Robert Catlett, University of Colorado, Colorado Springs, Colorado
Richard Collins, Louisiana State University, Baton Rouge, Louisiana
Philip Creighton, Towson State University, Towson, Maryland
Clara Dixon, Albion College, Albion, Michigan
David Dixon, Los Angeles Valley College, Van Nuys, California
Richard Dodge, Cerro Coso College, Ridgecrest, California
James Ebert, Pembroke State University, Pembroke, North Carolina
Michael G. Emsley, George Mason University, Fairfax, Virginia
Rich Goldberg, Temple University, Philadelphia, Pennsylvania
Phylis Gross, California State University-Hayward, Hayward, California
Michael Grant, University of Colorado, Boulder, Colorado
Robert D. Griffin, City College of San Francisco, San Francisco, California
Gayle Harper, University of Delaware, Newark, Delaware
Mildred Harry, University of Houston, Houston, Texas
Robert Hehman, University of Cincinnati, Cincinnati, Ohio
Robert Hellwig, Meredith College, Raleigh, North Carolina
Robert Henn, Sinclair Community College, Dayton, Ohio
John C. Hooker, Houston Baptist University, Houston, Texas
J. Houghton, Los Angeles Pierce College, Woodland Hills, California
David P. Husband, University of South Carolina, Columbia, South Carolina

Irwin R. Isquith, Fairleigh-Dickinson University, Teaneck, New Jersey
E. M. Ingersoll, Miami University, Oxford, Ohio
Richard J. Jensen, Wright State University, Dayton, Ohio
Norman Kerr, University of Minnesota, St. Paul, Minnesota
Steven Klug, Trenton State College, Trenton, New Jersey
Charles Kentz, University of Toledo, Toledo, Ohio
Fred Landa, Virginia Commonwealth University, Richmond, Virginia
David Lapin, Fairleigh-Dickinson University, Teaneck, New Jersey
Joseph Laufersweiler, University of Dayton, Dayton, Ohio
Lawrence Levine, Wayne State University, Detroit, Michigan
Charles K. Levy, Boston University, Boston, Massachusetts
Valerie A. Liston, University of Minnesota, St. Paul, Minnesota
Victor Lotrich, University of Delaware, Newark, Delaware
John Luchesi, University of North Carolina, Chapel Hill, North Carolina
Charles F. Lytle, North Carolina State University, Raleigh, North Carolina
James D. Mauseth, University of Texas, Austin, Texas
John Mecom, Richland College, Dallas Texas
Stephen Murphy, California State University, Fullerton, California
Marilyn Neulieb, St. Laurence University, Canton, New York
Charles O'Rear, East Carolina University, Greenville, North Carolina
Ivan Palmbladt, Utah State University, Logan, Utah
Linda Pesek, Long Island University, Greenvale, New York
Dottie Plott, Temple University, Philadelphia, Pennsylvania
Gerald G. Robinson, University of South Florida, Tampa, Florida
Robert Romans, Bowling Green University, Bowling Green, Ohio
Carol Ross, University of Delaware, Newark, Delaware
James Rooney, University of Missouri, Columbia, Missouri
Charles Schexnayder, Louisiana State University, Baton Rouge, Louisiana
Robert Schumacher, Kean College of New Jersey, Union, New Jersey
William R. Sigmund, Slippery Rock State College, Slippery Rock, Pennsylvania
Irwin Spear, University of Texas, Austin, Texas
Thomas A. Steyaert, Diablo Valley College, Pleasant Hill, California
Richard A. Strohman, University of California, Berkeley, California
Ralph Sulerud, Augsburg College, Minneapolis, Minnesota
Charles Swanson, Wayne State University, Detroit, Michigan
C. D. Therrian, Pennsylvania State University, University Park, Pennsylvania
Steve Williams, University of New Mexico, Albuquerque, New Mexico
Clyde Wilson, Laney College, Oakland, California
Weldon Witters, Ohio University, Athens, Ohio
Clarance Wolfe, Northern Virginia Community College, Annandale, Virginia
Richard D. Worthington, University of Texas, El Paso, Texas
Joe Wood, University of Missouri, Columbia, Missouri
Robert L. Wright, Jr., Highline College, Midway, Washington
Richard Wyatt, Wake Forest University, Winston-Salem, North Carolina
William Volker, Thomas More College, Ft. Mitchell, Kentucky
Charles Yokum, University of Michigan, Ann Arbor, Michigan

CONTENTS

PART ONE
Molecules to Cells 1

Chapter 1
Mr. Darwin and the Meaning of Life 2
1.1 Observation, Data, and an Idea 2
A Ratcatcher Goes to Sea, or Where Are the Rabbits? 2
An Explanation About Rabbits and Oceans 4
Darwin's Finches and Speciation 5
How to Go Against the Grain 7
Going Home to Applause and Illness 10
1.2 The Idea Becomes a Theory 13
The Theory of Natural Selection 13
A Mechanism of Natural Selection 16
How Species Are Formed 16
Why Are There Mosquitos? 18
Testing the Hypothesis 18
Group Selection 20
Reducing and Synthesizing 21
Application of Ideas 22
Key Words 23
Key Ideas 23
Review Questions 24

Chapter 2
Small Molecules 25
2.1 Basic Chemical Structure and Bonding 25
Elements, Atoms, and Molecules 25
Atomic Structure 26
Electron Orbitals and Electron Shells 28
Electron Pairing 28
Filled and Unfilled Shells 29
The Noble Elements 30
How an Atom Can Fill Its Outer Electron Shell 30
Ions and the Ionic Bond 30
Chemical Bonds and Molecular Diagrams 31
The Covalent Bond: Sharing Electron Pairs 32
Methane 32
The Hydrogen Bond and What, After All, Is Water? 36
Water Is a Polar Molecule 36
Gas, Liquids, Crystals, and Hydrogen Bonds 36
Water Forms Hydration Shells Around Ions 36
Water Is a Powerful Solvent 36
Water Is Wet 37
Essay 2.1 Acids and Bases 40
Nitrogen 41
Essay 2.2 Carbon Dioxide and the Bicarbonate Ion 43

Phosphorous and Phosphates 45
Photo Essay: Signs of Life 46
Sulfur 48
A Few Key Functional Side Groups 48
Chemical Reactions, Heat of Activation, and the Formation of Water 48
A System Goes to Its Lowest Energy State 49
Not All Covalent Bonds Are Equally Strong 49
Catalysts Lower the Heat of Activation 50
Application of Ideas 51
Key Words 51
Key Ideas 51
Review Questions 54

Chapter 3
Large Molecules 56
3.1 The Carbohydrates 56
Monosaccharides, Disaccharides, and Polysaccharides 56
Essay 3.1 Reading Structural Formulas 58
Starches 59
Amylose 59
Amylopectin 60
Glycogen 60
Primary and Secondary Structure of Starches 60
Polysaccharide Synthesis 61
Cellulose 61
Other Structural Polysaccharides of Plants 63
Chitin 63
Mucopolysaccharides 63
3.2 The Proteins 64
The Structure of Amino Acids 64
The Twenty Amino Acids and Their R Groups 65
Conjugated Proteins and Prosthetic Groups 67
Essay 3.2 Protein Structure 68
3.3 Biologically Active Proteins 69
Enzymatic Activity 69
Essay 3.3 How a Hydrolyzing Enzyme Works 70
Enzyme Kinetics 71
Effects of Temperature and Acidity 72
Control of Enzyme Activity: Negative Feedback and Allostery 72
Other Binding Proteins 73
Structural Proteins 74
Peptides and Peptide Hormones 75
3.4 The Lipids 75
Triglycerides 75
Saturated, Unsaturated, and Polyunsaturated Fats 77
Phospholipids 78
Application of Ideas 79
Key Words 80
Key Ideas 80
Review Questions 82

Chapter 4
Cells 84
4.1 The Units of Life 85
Cell Theory 85
Why Are There Cells? 85
Membranes and the Origin of Cells 85
Architecture of the Cell 86
Cell Diversity 86
The Cells of Very Different Organisms 86
Why Are Cells So Small? 87
The Surface–Volume Hypothesis 87
Essay 4.1 Exceptions to the Cell Theory 91
Essay 4.2 Small Units of Linear Measurement in the Metric System 93
4.2 Cell Barriers and Transport Mechanisms 95
The Cell Structures 95
The Plant Cell Wall 95
The Cell Membrane 95
Photo Essay: Looking at Cells 96
Essay 4.3 Ultracentrifugation 102
Diffusion, Facilitated Diffusion, Active Transport, and Osmosis 103
A Closer Look at Diffusion 103
Facilitated Diffusion and Permeases 104
Active Transport Across Cell Membranes 104
Osmosis 105
Osmosis in Living Cells 106
Junctions Between Cells 108

4.3 Structures Within Cells 109
Cell Organelles 109
The Nucleus 109
The Endoplasmic Reticulum 111
The Golgi Complex 111
Lysosomes 112
Peroxisomes 113
Plastids 114
Chloroplasts 114
Mitochondria 115
Vacuoles 116
Centrioles, Cilia, Flagella, and Basal Bodies 117
Cilia and Flagella 117
Microtubules 119
Application of Ideas 120
Key Words 120
Key Ideas 120
Review Questions 124

Chapter 5
Photosynthesis: Energy Captured 125
5.1 Energy, Chemiosmotic ATP Production, and Electron Carriers 125
Adenosine Triphosphate (ATP) 126
ATP Production by Chemiosmotic Phosphorylation 129
Essay 5.1 The Laws of Thermodynamics 131
Oxidation and Reduction 132
Essay 5.2 Peter Mitchell and the Overthrow of a Textbook-enshrined Scientific Dogma 133
NAD, $NADH_2$, NADP, and $NADPH_2$: Important Soluble Electron Carriers of the Cell 135
5.2 Light and the Light Reactions 136
Energy Enters the Ecosystem 136
The Physics of Sunlight 137
The Absorption Spectrum 137
The Chloroplast 140
An Overview of the Chemistry of Photosynthesis 140
The Photosynthetic Unit 140
Two Photosystems 141
The Role of Electrons in Energy Transfer 142
Receiving a Photon 142
The Z Diagram 143
The Oxidation of Water 144
The Electron Pathway 144
The Electron Pathway—P700 145
Summarizing the Light Reactions 145
5.3 The Light-Independent Reaction 145
Glucose Production in C3 and C4 Plants 145
The Calvin Cycle in C3 Plants 146
What's Wrong with the Calvin Cycle in C3 Plants? 148
Photosynthesis in C4 Plants 148
A Summary of Photosynthesis 151
Application of Ideas 152
Molecules of Energy Transfer 152
Soluble Electron Carriers 152
Molecules of the Light-Independent Reactions 152
Key Words 153
Key Ideas 153
Review Questions 156

Chapter 6
Respiration 158
6.1 Anaerobic Respiration 160
Glycolysis: Degrading Glucose Without Oxygen 160
Fermentation 164
Glycolysis in Muscle Tissue 165
Oxygen Debt 166
The Metabolism of Starch and Glycogen 166
Essay 6.1 A Marathon Runner Meets the "Wall" 168
6.2 Aerobic Respiration 168
Pyruvate to Acetyl CoA 168
The Citric Acid Cycle 169
An Overview of the Citric Acid Cycle 169
The Electron Transport System in Mitochondria 172
The ATP Balance Sheet in the Respiration of Glucose 174
Essay 6.2 The Energy Balance Sheet for the Respiration of Glucose 175
Application of Ideas 176
Key Words 176
Key Ideas 176
Molecules of Glycolysis and Fermentation 177
Molecules of the Citric Acid Cycle 177
Review Questions 178
Suggested Reading for Part One 179

PART TWO
Molecular Biology and Heredity 181

Chapter 7
The Central Dogma 182
7.1 DNA Structure and Replication 182
What Is the Central Dogma? 182
What Is a Nucleic Acid? 183
DNA Bonding 184
The Stability of DNA 187
DNA Repair 188
Essay 7.1 DNA Replication in Eucaryotes and Procaryotes 189
Essay 7.2 Exceptions to the Central Dogma 190
7.2 How We Know 192
A Brief History of the Discovery of the Genetic Material 192
Transformation 192
Hershey and Chase 193
Photo Essay: The Phage Virus 194
Essay 7.3 Radioactive Tracing 196
Essay 7.4 Meselson and Stahl 198
Chargaff's Rule 201
Watson and Crick and the Molecular Model of DNA 201
7.3 DNA, RNA, and Transcription 202
RNA: The Other Nucleic Acid 202
RNA Structure 203
RNA Synthesis 203
Varieties of RNA: Physical and Functional Classes 205
Ribosomal RNA 205
Messenger RNA and the Genetic Code 205
Transfer RNA 207
7.4 RNA, Translation, and Mutation 210
How Proteins Are Made 210
Initiation 210
Elongation 211
Polypeptide Chain Termination 214
Polyribosomes 214
Free and Bound Ribosomes 215
Mutations in DNA and Resulting Misreadings of mRNA 216
Effects of DNA Mutations at the Level of Protein Synthesis 216
Application of Ideas 217
Key Words 218
Key Ideas 218
Review Questions 222

Chapter 8
The Life of a Cell 223
8.1 The Nucleus and the Cell Cycle 223
Nuclear Structure 224
DNA, Chromatin, and Chromosomes 225
Nuclear RNA 226
Cycle or Die 227
The Stages of the Cycle 228
8.2 Mitosis 229
Mitosis and Cell Division 229
Details of Mitosis 230
Chromosome Condensation (Prophase) 230
Chromatids, Chromatin, and Centromere 233
The Spindle 234
Essay 8.1 Karyotyping 236
What Is the Centriole For? 237
Spindle Fibers 238
The Spindle in Action: Metaphase and Anaphase 238
Metaphase 238
Essay 8.2 How the Spindle Works 239
A Final Look at Mitosis and Cell Division 242

8.3 Meiosis and Sex Cells 243
Meiosis 243
Homologous Chromosomes 243
Where Meiosis Takes Place in Higher Organisms 244
Premeiotic Interphase 245
Meiosis I: The First Meiotic Division 245
Meiotic prophase: The First Stages 245
Metaphase I 248
Anaphase I 248
A Closer Look at Meiotic Prophase: Crossing Over 248
Telophase I and Meiotic Interphase 250
Meiosis II 250
Prophase II 250
Metaphase II and Anaphase II 250
Telophase II 250
Summing Up Meiosis 250
Meiosis in Females 251
Essay 8.3 When Meiosis Goes Wrong 256
Why Meiosis? 258
Application of Ideas 258
Key Words 259
Key Ideas 259
Review Questions 264

Chapter 9
Classical Genetics 265
9.1 Mendel and His Laws 265
When Darwin Met Mendel (Almost) 265
The Principle of Dominance 268
Mendel's First Law: The Segregation of Alternate Alleles 271
Mendel's Second Law: Independent Assortment 273
Mendel's Testcross 276
Backcross to the Recessive Parent 276
Backcross to the Double Dominant Parent 277
The Testcross 277
The Decline and Rise of Mendelian Genetics 277
9.2 Mendelian Genetics in the 20th Century 278
What Causes Dominance and Recessivity? 278
Other Dominance Relationships 279
Partial Dominance 279
Lethal-Recessive Dominant Alleles 280
Codominance 281
Overdominance 282
Multiple Alleles 282
Blood Groups 282
The RH Blood Group 283
Essay 9.1 Dominance Relationships Depend on How the Phenotype Is Classified: Sickle-cell Hemoglobin 284
Transplants 287
Gene Interactions and Modified Mendelian Ratios 287
Application of Ideas 289
Key Words 290
Key Ideas 290
Review Questions 295

Chapter 10
Genes, Chromosomes, and Sex 297
10.1 The Genetics of Sex 297
Sutton and Morgan 297
Sex Linkage in Humans 300
Color Blindness 301
Other Sex-Linked Conditions 301
Sex-Linked Dominants 301
Essay 10.1 The Disease of Royalty 302
10.2 Locating the Genes 304
Linkage, Crossing Over, and Chromosome Mapping 304
Mapping Genes 306
Bands and Puffs 309
10.3 Variations on Mendel 309
Gene Expression 309
Environmental Interactions 309
Modifier Genes 309
Variable Expressivity 309
Incomplete Penetrance 310
Sex-Limited and Sex-Influenced Effects 310
Variable Age of Onset 310
All Together Now 312
Continuous Traits and Polygenic Inheritance 312
Application of Ideas 315
Key Words 316
Key Ideas 316
Review Questions 319

Chapter 11
Evolution 321
11.1 Genes and Gene Frequencies 321

The Germ Plasm Theory 322
Gemmules and Pangenesis: Darwin's Big Mistake 322
Weismann and the Continuity of the Germ Plasm 322
Frequencies 323
The Population Genetics of Asexual Organisms 323
Antibiotic Resistance 324
Selective Advantage and Selection Coefficient 324
Selecting for Two Characteristics at Once 325
Asexually Reproducing Organisms 325
Sexually Reproducing Organisms 325
11.2 Population Genetics 326
Evolution in Diploid Organisms 326
Essay 11.1 The Story of the Peppered Moth 327
Genetic Variability in Diploid Populations 328
The Maintenance of Genetic Diversity 328
Short-Term Evolutionary Change in Diploid Populations 329
Selection for Going Up or Down 330
Natural Selection Compared with Artificial Selection 331
Small Babies and Large Babies—An Intermediate Optimum 332
Frequency-Dependent Selection 334
Geographic Differences in Selection, with Migration 334
The Rules of Bergmann, Allen, and Groger 335
Geographic Variation in the Human Species 337
Neutralists Versus Selectionists 337
Gene Changes in Evolution 338
Population Bottlenecks 338
The Balance Between Mutation and Selection 338
The Genetic Future of *Homo sapiens* 340
11.3 The Castle-Hardy-Weinberg Law 340
The Implications of the Castle-Hardy-Weinberg Distribution 344
An Algebraic Equivalent 342
Violating Assumptions 344
What Good Is the Castle-Hardy-Weinberg Distribution? 344
Application of Ideas 345
Key Words 345
Key Ideas 345
Review Questions 350

Chapter 12
Microbial Genetics 351
12.1 Genes and Metabolic Pathways 351
Inborn Errors of Metabolism 351
One Gene, One Enzyme 352
Beadle and Tatum 353
12.2 Genetics of Bacteria and Viruses 356
Recovering and Counting Mutations in Bacteria 356
Recovering Recombinants 357
The Life and Times of the Bacteriophage 357
Counting Viruses, or Holes in the Lawn 358
Bacterial Recombination 358
Transformation 358
Transduction 360
Plasmid Transfer 360
F^+ Episomes 360
Essay 12.1 Counting Bacteria 362
An Example of How Bacterial Genetics Is Done 363
The Great Kitchen Blender Experiment 365
12.3 The Operon 366
Gene Organization and Control in Bacteria 366
Background: Inducible Enzymes in Escherichia 366
Sexduction 366
Mutants of the Lactose Operon 366
Jacob's and Monod's Experiment 367
Dominance Relationships at the Operator Locus 367
Dominance Relationships at the i Locus 367
Jacob's and Monod's Conclusions: The Operon Model 368
The Control Mutants Explained 368
The "Invention" of Messenger RNA 368
Repressible Enzymes and Other Operons 370
Repressible Enzymes 370
Operons in Higher Organisms: Are There Any? 370

12.4 Recombinant DNA 372
What Have Molecular Geneticists Been Up To Recently? 372
Gene Splicing and Cloning, and the Recombinant DNA Debate 372
How Gene Splicing Is Done 372
What Recombinant DNA and Gene Cloning Can Do: The Threat and the Promise 373
Essay 12.2 Crippling a Microbe 376
Application of Ideas 377
Key Words 377
Key Ideas 377
Review Questions 382
Suggested Reading for Part Two 383

PART THREE
Microorganisms 385

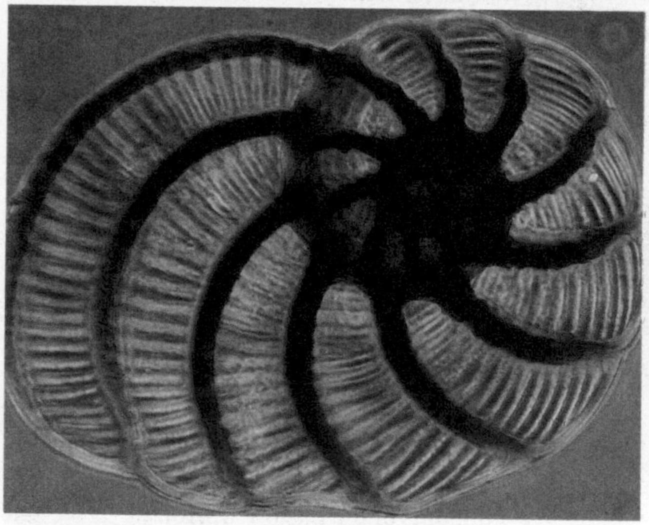

Chapter 13
Systematics and Taxonomy: Testing Evolutionary Hypotheses 386
13.1 The Species Problem 386
What Is a Species? 386
The Species Concept in Plants 388
Allotetraploids: Instant Species 389
Type Specimens and Biometrics 390
Numerical Taxonomy and the OTU 391
How Species Are Named 391
13.2 Systematics 392
Monophyletic, Polyphyletic, and Paraphyletic Groups 394
Essay 13.1 Systematics and Taxonomy: Are They Really Science? 396
Who Is Related to Whom? 396
Characteristics of a Taxon Can Be Described 396
The Degree of Similarity Between Two Taxa Can Be Described 397
Closely Related Taxa Will Be Similar Morphologically and Biochemically 397
The Degree of Similarity Between Taxa Decreases Over Time 397
Characteristics Once Lost in Evolution Tend Not to Be Regained 399
Homology Usually Can Be Distinguished from Analogy 399
Formal Construction of Trees 399
Essay 13.2 T. D. Lysenko 401
The Two-Kingdom Scheme and the Five-Kingdom Scheme 402
The Two-Kingdom Scheme 402
The Five-Kingdom Scheme 402
Application of Ideas 405
Key Words 405
Key Ideas 405
Review Questions 408

Chapter 14
The Procaryotes 409
The Bacteria 410
The Bacterial Cell 410
Bacterial Reproduction and Spore Formation 411
Essay 14.1 Tracing the Relationships of the Procaryotes 413
Essay 14.2 Bacterial Villains 415
Taxonomy of the Bacteria 415
Eubacteriae (True Bacteria) 415
Biochemical Adaptations 416
Myxobacteria 417
Chlamydobacteria (Mycelial Bacteria) 417
Spirochaetae 418
Mycoplasmae 418
The Cyanophytes (Blue-Green Algae) 419
The Eucaryotes: A Different Matter 421
The Symbiosis Hypothesis 421

Application of Ideas 424
Key Words 425
Key Ideas 425
Review Questions 427

Chapter 15
The Protista 428
The Algal Protists 428
Pyrrophyta (Dinoflagellates) 429
Euglenophyta (Euglenoids) 430
Chrysophyta (Diatoms and Their Relatives, the Yellow-Green Algae and the Golden-Brown Algae) 430
The Animallike Protists 432
Mastigophora (the Flagellates) 432
Sarcodina (Amebas) 433
Sporozoa (Malarial Parasites) 434
Ciliophora (Ciliates) 437
Application of Ideas 440
Key Words 440
Key Ideas 440
Review Questions 442

Chapter 16
The Fungi and Multicellularity 444
16.1 The Fungi 444
What Are the Fungi? 444
Subkingdom Gymnomycota: The Slime Molds 446
Subkingdom Dimastigomycota: The Oomycetes (Water Molds) 447
Subkingdom Eumycota: The "True Fungi" 449
The Chytridomycetes (Chytrids) 449
The Zygomycetes (Conjugating Molds) 449
Ascomycetes (Sac Fungi) 450
Basidiomycetes (Club Fungi) 454
The Fungi Imperfecti 456
Another Look at Fungal Ancestors 456
16.2 Multicellularity 456
The Origin of Multicellularity 457
Multicellular Plants 457
Multicellular Animals 458
Application of Ideas 458
Key Words 459
Key Ideas 459
Review Questions 461
Suggested Reading for Part Three 462

PART FOUR
Plants 463

Chapter 17
Evolutionary Patterns in Plants 463
17.1 The Nonvascular Plants 464
Division Rhodophyta: The Red Algae 465
Division Phaeophyta: The Brown Algae 466
Division Charophyta: The Stoneworts 467
Division Chlorophyta: The Green Algae 467
Division Bryophyta 472
Evolutionary Relationships 473
Class Musci (Mosses) 474
Class Hepaticae (Liverworts) 475
Class Anthocerotae (Hornworts) 477
17.2 The Rise of Vascular Plants 478
Division Tracheophyta 478
Subdivision Psilophyta 479
Subdivision Lycophyta (Club Mosses) 479
Subdivision Sphenophyta (Horsetails) 480
Subdivision Pterophyta (Ferns) 481
Subdivision Spermophyta 482

17.3 The Emergence of Seed Plants 484
The Gymnosperms 484
The Gingko 485
The Cycads 485
The Conifers 485
The Angiosperms: The Rise of Flowering Plants 486
Essay 17.1 Alternations of Generations in the Seed Plants 488
Angiosperms and Changing Dinosaurs 489
The Angiosperms Today 491
Origin and Phylogenetic Relationships 491
Application of Ideas 497
Key Words 497
Key Ideas 497
Review Questions 501

Chapter 18
Transport Systems in Plants 503
18.1 Vascular Systems and the Transport Problem 504
Vascular Plants and Nonvascular Plants 504
Water and Food Transport in Mosses 504
Transport and Support Structure in Vascular Plants 505
The Root 505
Root Types 506
The Root Tip 507
Root Vascular System 507
The Stem of a Woody Plant 510
Phloem 511
Xylem 511
The Leaf 512
18.2 Mechanisms of Transport 514
Water Transport in Plants 514
Root Pressure 515
The Transpiration, Cohesion, Tension Hypothesis 516
Pulling Water 516
Transpiration Pull 516
Food Transport 516
Essay 18.1 Testing the Transpiration, Cohesion, Tension Theory 517
Why Diffusion Is Not the Answer 518
Streaming Cytoplasm 519
Gaseous Exchange in Vascular Plants 519
Guard Cells 520
Application of Ideas 522
Key Words 522
Key Ideas 522
Review Questions 525

Chapter 19
Mineral Nutrition and Metabolism in Plants 527
Nutrient Requirements 529
The Macronutrients and Their Cycles 530
Nitrogen 530
Essay 19.1 Why Trees Don't Urinate on Dogs 533
The Phosphorous and Calcium Cycles 534
The Carbon Cycle 535
The Oxygen and Water Cycles 535
Potassium 536
Sulfates 537
Magnesium 537
Micronutrients 537
Essay 19.2 The Carbon Cycle and the Greenhouse Effect: A Destabilized Equilibrium 538
Transport of Mineral Nutrients Within the Plant 538
Metabolic Oddities 539
The Digestive Process in Plants 539
Application of Ideas 540
Key Words 540
Key Ideas 540
Review Questions 543

Chapter 20
Chemical Regulation in Plants 544
The Experiments of Boysen-Jensen and Pall 545
Auxin and Tropisms 546
Phototropism 546
Geotropism 547
Other Plant Growth Hormones 549
Gibberellins 549
Cytokinins 550
Ethylene 552
Abscisic Acid 552
Applications of Plant Hormones to Agriculture and Horticulture 553

Application of Ideas 553
Key Words 554
Key Ideas 554
Review Questions 555

Chapter 21
Plant Reproduction and Development 557
21.1 Asexual Reproduction in Plants 557
Vegetative Propagation 557
Grafting 558
Essay 21.1 Asexual Reproduction: The Long Run 559
Apomixis: Asexual Reproduction with Seeds 559
21.2 Sexual Reproduction in Flowering Plants 560
The Initiation of Flowering 562
Photoperiodicity 562
Transmitting the Stimulus 564
Pineapples and Auxin 565
Meanwhile, Back in the Ovary: Megasporogenesis 565
Similar Events in the Anthers: Microsporogenesis 566
Pollination and Fertilization 567
Seed Development in Flowering Plants 568
Photo Essay: The Flower 570
Further Seed Development in Dicots 574
Further Seed Development in Monocots 575
Seed Dormancy 575
Seed Dispersal 576
21.3 Germination, Growth, and Development 577
Germination and Primary Growth 577
The Development of the Plant 578
Growth and Development in the Root 578
Growth of the Stem 579
Growth and Development Within the Stem 580
The Problem of Differentiation 583
Information from the Study of Plant Cells 584
Application of Ideas 584
Key Words 584
Key Ideas 584
Review Questions 588
Suggested Reading for Part Four 589

PART FIVE
Animals 591

Chapter 22
Animal Diversity—The Nonchordates 592
22.1 Origins and Relationships 592
What Is an Animal? 592
Metazoan Origins and Phylogeny 593
Ernst Haeckle, Jovan Hadzi, and the Animal Tree 593
Trees and Trees 595
22.2 Animals Without a Coelom 597
Phylum Porifera: The Sponges 597
Phylum Coelenterata 598
Class Hydrozoa 599
Class Scyphozoa 599
Class Anthozoa 600
Phylum Ctenophora 601
Phylum Platyhelminthes: The Flatworms 602
Class Turbellaria 602
Class Trematoda 603
Class Cestoda 603
Phylum Nemertinea: The Nemerteans, or the Proboscis Worms 604
Phylum Aschelminthes 604
Class Nematoda 604
Class Rotifera 607
Other Aschelminthes 608

22.3 Animals with a Coelom 609
The Lopophorate Phyla 609
Phylum Annelida: The Segmented Worms 609
Class Oligochaeta: The Earthworms (Lymbricoides) 609
Class Hirudinea: The Leeches 611
Class Polychaeta 611
Phylum Mollusca 612
Class Amphineura: The Chitons 613
Class Gastropoda: The Snails 614
Class Pelecypoda: The Bivalves 614
Class Cephalopoda: The Squid and Octopus 614
Phylum Onychophora 615
Phylum Arthropoda 617
Subphylum Chelicerata 617
Class Arachnida 617
Subphylum Mandibulata 618
Class Crustacea 618
Classes Chilopoda and Diplopoda 620
Class Insecta 620
Phylum Echinodermata: The Spiny-Skinned Animals 624
Class Asteroida: The Sea Stars 625
Application of Ideas 627
Key Words 628
Key Ideas 628
Review Questions 633

Chapter 23
Animal Diversity—The Chordates 634
23.1 The Invertebrate Chordates 634
Phylum Hemichordata 634
Phylum Chordata 635
Subphylum Urochordata 635
Subphylum Cephalochordata 636
23.2 The Vertebrates 637
Subphylum Vertebrata: Animals with Backbones 637
Class Agnatha: Jawless Fishes 637
Class Placodermi: Extinct Jawed Fishes 638
Class Chondrichthyes: Sharks, Rays, Skates, and Chimera 639
Class Osteichthyes: The Bony Fishes 640
Class Amphibia: The Amphibians 642
Class Reptilia: The Reptiles 643
Class Aves: The Birds 646
Class Mammalia: The Mammals 649
23.3 Primates and the Evolution of Man 655
The Primates 655
Human Evolution 658
The Divergence of the Human Line 659
Australopithecus africanus 659
Australopithecus robustus 659
Homo habilis 659
Homo erectus 660
Homo sapiens: Neanderthals and Us 661
Application of Ideas 663
Key Words 663
Key Ideas 663
Review Questions 667

Chapter 24
Animal Support and Locomotion 669
24.1 Invertebrate Form and Movement 670
Exoskeletons 671
Insect Wings and Their Muscles 672
Mollusk Shells—A Type of Exoskeleton 672
Endoskeletons 674
Sponges 674
Echinoderms 674
24.2 Vertebrate Form 675
Vertebrate Supporting Structures 675
Bone 675
Essay 24.1 The Development and Structure of Bone 676
The Axial Skeleton 677
The Skull 677
The Vertebral Column 678
The Appendicular Skeleton 681
The Pectoral and Pelvic Girdles 681
Limbs 681
The Human Appendicular Skeleton 683
Essay 24.2 Evolution of the Mammalian Skull 684
24.3 Vertebrate Movement 686
Muscles and How They Move 686
Types of Muscle 686
Smooth Muscles 687
Cardiac Muscles 688
Skeletal Muscles 688
Muscle Antagonism 690
The Ultrastructure of Skeletal Muscle 690
Calcium and the Biochemical Mechanism of Muscle Contraction 693

Application of Ideas 695
Key Words 695
Key Ideas 695
Review Questions 699

Chapter 25
Digestion and Nutrition 701
25.2 Digestion 701
Digestion in Invertebrates 701
The Simpler Invertebrates 701
More Complex Invertebrates 702
Pseudocoeloms, Coeloms, and Hemocoels 703
Earthworms and Insects 704
Foraging and Digestive Structures in Fish, Amphibians, Reptiles, and Birds 707
Fish 707
Amphibians, Reptiles, and Birds 708
Foraging and Digestive Structures in the Mammals 710
Digestion in the Herbivores—Ruminants 710
The Digestive System of Humans 711
The Oral Cavity and Esophogus 712
The Stomach 714
Small Intestine 716
Liver 717
Pancreas 717
Large Intestine 717
25.2 The Chemistry of Digestion 718
Chemical Digestion and Absorption 718
Carbohydrate Digestion 718
Fat Digestion 720
Protein Digestion 720
Nucleic Acid Digestion 721
Integration and Control of the Digestive Process 721
25.3 Nutrition 722
What Happens to Nutrients? 722
Carbohydrates 723
Fats 723
Protein 724
Vitamins and Minerals 725
Application of Ideas 726
Key Words 726
Key Ideas 726
Review Questions 730

Chapter 26
Respiration and Transport 732
26.1 Evolution of Gas Exchange Surfaces 733
The Body Interface 733
Expanding the Interface: Trachae 734
Expanding the Interface: Branchiae 734
Complex Interfaces: Gills 734
Gill Baskets 736
Gills in Fishes 737
Complex Interfaces: Lungs 738
Evolution of the Vertebrate Lung 738
The Vertebrate Lung 739
The Anatomy of the Human Respiratory System 741
26.2 Gas Exchange and Respiration Control 742
The Breathing Movements 742
Partial Pressure 743
The Exchange of Gases in the Alveoli 743
The Control of Respiration 745
26.3 Evolution of Circulatory Systems 746
Transport in Multicellular Animals 746
Circulation in Vertebrates 748
Circulation in Fishes 748
Circulation in Amphibians and Reptiles 749
The Four-Chambered Heart 751
26.4 Human Circulation and Blood 751
The Human Circulatory System 751
Circulation Through the Heart 751
Control of the Heart 752
The Pacemaker 754
The Heart Beat 754
Circuits in the Human Circulatory System 756
Pulmonary Circuit 756
Hepatic Portal Circuit 757
Renal Circuit 757
General Body Circuit 758
The Role of the Capillaries 758
The Blood and Lymph 760
The Red Blood Cell 761
The White Blood Cell 761
Platelets 761
Plasms 763
The Lymphatic System 763
Application of Ideas 764
Key Words 765

Key Ideas 765
Review Questions 770

Chapter 27
Chemical Messengers in Animals 771
27.1 The Actions of Hormones 771
What Are Hormones? 771
Hormones and the Invertebrates 772
The Molting Activity in Crustaceans 772
Hormonal Activity in Insects 774
Hormones and the Vertebrates 776
The Pituitary 775
Control by the Hypothalamus 780
Hormones of the Anterior Pituitary 780
Hormones of the Posterior Pituitary 782
The Thyroid 783
Thyroxin and Triiodothyronine 783
Calcitonin 784
The Parathyroid Glands and Parathormone 785
The Pancreas and the Islets of Langerhans 785
The Adrenal Glands 787
The Ovaries and Testes 788
The Pineal body 789
27.2 How Hormones Work 790
The Molecular Biology of Hormone Activity 790
The Second Messenger 790
Developmental Hormones, Mitogens, and Growth Factors 790
Other Second Messengers? 791
Second Messengers as Gene Activators 792
Steriod Hormones; Direct Action at the DNA Level 792
Application of Ideas 793
Key Words 793
Key Ideas 793
Review Questions 796

Chapter 28
Homeostasis 798
28.1 Feedback Systems and Temperature Regulation 798
A Closer Look at Negative Feedback Loops 798
Thermoregulation 800
Why Thermoregulate? 800
The Countercurrent Heat Exchange Mechanism, or How Tuna Swim Fast 801
Regularity Behaviorally 802
Thermoregulation in the Homeotherms 805
Removal of Heat Energy 805
Conserving Heat 806
The Internal Source of Thermoregulation 807
A Closer Look at How the Hypothalamus Works 807
28.2 Osmoregulation and Excretion 808
Producing Nitrogen Wastes 808
The Osmotic Environment 809
The Marine Environment 809
Strategies of Marine Vertebrates 810
The Freshwater Environment 811
Vertebrates of the Freshwater Environment 811
The Terrestrial Environment 812
Invertebrates of the Terrestrial Environment 812
Vertebrates of the Terrestrial Environment 813
The Human Excretory System 813
Control of Nephron Function 816
Essay 28.1 Gout 817
28.3 The Immune System 819
How Are Specific Antibodies Produced? 820
The Germ-Line Hypothesis 821
The Clonal Selection Theory 821
Recognizing Self 821
Clonal Selection 821
Immunological Memory 822
Application of Ideas 823
Key Words 823
Key Ideas 823
Review Questions 828

Chapter 29
Animal Reproduction 830
29.1 Sex and Reproduction in Animals 830
Copulation and Viviparity 831
Reproduction and the Survival Principle 833

Reproduction in the Terrestrial Environment 833
Reproduction in Terrestrial Arthropods 833
Mating in Spiders 834
Reproduction in Reptiles and Birds 835
Reproduction in Mammals 837
29.2 Human Reproduction 837
The Human Reproductive System 837
The Male Reproductive System 837
Erectile Tissues 839
Sperm Production 839
The Female Reproductive System 840
External Genitalia 840
Internal Genitalia 841
Hormonal Control of Human Reproduction 844
Male Hormonal Action 844
Female Hormonal Action 845
29.3 Human Sexual Behavior 848
The Sex Act 848
Excitement 848
Plateau 849
Orgasm 849
Resolution 850
Masturbation 850
What Does Copulation Accomplish? 850
Fertilization 852
Contraception 852
Natural Methods 853
Mechanical Devices 854
Hormonal and Chemical Birth Control: The Pill 855
Surgical Intervention 856
Application of Ideas 858
Key Words 858
Key Ideas 858
Review Questions 862

Chapter 30
Development and Differentiation 864
30.1 The Gametes and Fertilization 865
The Gametes 865
The Sperm 865
The Egg 865
Fertilization 866
The Future of a Zygote 867
The First Cleavage 871
The Blastula and Gastrulation 872
Gastrulation in Birds 873
Gastrulation in Mammals 873
30.2 Morphogenesis: The Development of Form 875
The Fate of the Germ Layers 876
Embryonic Organizers and the Induction Theory 878
The Cell Nucleus in Differentiation 880
Nerve Cells and Developmental Patterns 882
Sprouting Factor and Antisprouting Factor 883
An Experimental Investigation of Nerve Specificity in Development 883
Physical Support and Life Support of Vertebrate Embryos 884
The Self-Contained Egg 885
Life Support of the Mammalia Embryo 886
30.3 Human Development and Birth 887
The Emergence of the Human Form 887
The First Trimester 888
Circulatory System 888
Respiratory and Digestive Systems 888
Essay 30.1 The Theory of Recapitulation 890
Limbs 891
Excretory System 891
Reproductive System 892
Prenatal Sex Hormones and the Development of Behavior 892
The Second and Third Trimesters 893
Birth 894
The Stages of Birth 894
Essay 30.2 Steroid Hormones and Development 895
Miscarriage (Spontaneous Abortion) 896
Physiological Changes in the Newborn Baby 896
Lactation 897
Application of Ideas 898
Key Words 898
Key Ideas 898
Review Questions 902
Suggested Reading for Part Five 903

PART SIX

Behavior 905

Chapter 31
The Nervous System 906
31.1 Neurons and Nervous Systems 906
What Neurons Are 906
How Neurons Work 906
The Resting Neuron 907
The Action Potential 908
The Synaptic Junction 908
The Neurotransmitters 910
Fast Neurons and the Schwann Cells 911
Integration of Transmissions 910
The Central Nervous System 913
Cephalization 913
The Coelenterates 914
Flatworms 914
Annelids 914
Mollusks 914
Arthropods 915
The Echinoderms 916
The Brains of Vertebrates 916
The Human Brain 917
The Medulla 917
The Cerebellum and Pons 917
The Thalamus and Reticular System 917
The Hypothalamus 919
The Cerebrum 919
Hemispheres and Lobes 920
The Autonomic Nervous System 923
31.2 The Senses 923
Perceptual Systems 923
Thermoreceptors 923
Heat Detection in Invertebrates 923
Heat Detection in Vertebrates 923
Tactile Receptors 924
Touch in Invertebrates 924
Touch in Vertebrates 924
Auditory Receptors 925
Hearing in Invertebrates 925
Hearing in Vertebrates 927
Visual Receptors 927
Seeing in Invertebrates 929
Seeing in Vertebrates 930
Chemoreceptors 933
Chemical Detection in Invertebrates 933
Chemical Detection in Vertebrates 934
Proprioception 936
Proprioception in Invertebrates 936
Proprioception in Vertebrates 936
Application of Ideas 937
Key Words 937
Key Ideas 937
Review Questions 942

Chapter 32
The Mechanisms and Development of Behavior 944
32.1 Instinct and Learning 944
Ethology and Comparative Psychology 944
Instinct 945
Essay 32.1 History of the Instinct Idea 946
Fixed Action Patterns and Orientation 946
Appetitive and Consummatory Behavior 949
Releasers 950
Action-Specific Energy and Vacuum Behavior 950
The Behavioral Hierarchy 951
Learning 952
Reward and Punishment 952
Social Reinforcement in Humans 954
Four Types of Learning 954
Habituation 954
Classical Conditioning 955
Operant Conditioning 956
Autonomic Learning 957
32.2 Memory 958
Memory and Learning 958
The Advantage of Forgetting 959
The Consolidation Hypothesis and Octopus Memory 959
The Mammalian Memory 960
How Instinct and Learning Can Interact 962

Application of Ideas 963
Key Words 963
Key Ideas 963
Review Questions 966

Chapter 33
The Adaptiveness of Behavior 967
Navigation and Orientation 967
Orientation 968
Homing Pigeons 970
Communication 970
Visual Communication 971
Sound Communication 972
Chemical Communication 973
Why Communicate? 974
Species Recognition 974
Individual Recognition 975
Aggression 975
Fighting 975
Is Aggression Instinctive? 977
Human Aggression 978
Territories 978
Cooperation 981
Altruism 982
Essay 33.1 Sociobiology 984
Application of Ideas 985
Key Words 985
Key Ideas 985
Review Questions 988
Suggested Reading for Part Six 989

PART SEVEN
Ecology 991

Chapter 34
Biomes, Communities, and Life-Support Systems 992
34.1 The Biomes 993
The Biosphere 993
The Distribution of Life: Terrestrial Environment 993
Biomes 995
The Desert Biome 995
Desert Plants 996
Desert Animals 997
The Grassland Biome 999
Tropical Rain Forest 1001
Chaparral or Mediterranean Scrub Forest 1003
The Temperate Deciduous Forest 1003
The Taiga 1005
The Tundra 1007
The Water Environment 1009
Coastal Communities 1011
Productivity in the Oceans 1012
The Fresh Waters 1013
Freshwater Lakes and Ponds 1013
34.2 Ecosystems 1013
Interaction in Ecosystems 1013
Abiotic Factors 1014
Biotic Factors 1014
Life-Support Systems 1016
Energy Flow Through Ecosystems 1016
Nutrient Cycles 1016
Measuring Ecosystem Productivity 1017
Essay 34.1 An Unusual Community 1017
The Atmosphere 1019
Ecological Succession 1020
Eutrophication in the Freshwater Community 1021
Nutrient Pollution 1021
Application of Ideas 1022
Key Words 1022
Key Ideas 1022
Review Questions 1027

Chapter 35
Populations 1028
35.1 Population Growth and Reproductive Strategies 1028
The Ecological Niche 1028
Population Changes 1028
Population Growth 1029
Carrying Capacity 1030

Reproduction Rate 1030
How Many Offspring Are Produced 1031
Tapeworms 1031
Chimpanzee 1032
Birds 1033
Humans 1033
35.2 Interactions in Populations 1035
Death 1035
Programmed Death 1035
Death: The Long View 1037
Controlling Populations Through Mortality 1037
Abiotic Control 1038
Density-Independent Mortality 1038
Biotic Control 1038
Density-Dependent Mortality 1038
Competition Within a Species for a Finite Resource 1039
Territoriality and Dominance Hierarchies in Intraspecific Competition 1039
Self-Poisoning Through Toxic Waste 1039
Disease and Parasitism 1041
Competition Between Members of Different Species 1043
Predation 1044
Essay 35.1 The Greenpeace Effort 1045
A Closeup of the Wolf as a Predator 1046
The "Balance of Nature" 1049
Emigration 1050
35.3 Human Populations: Growth and Limits 1051
Human Populations 1051
Early Humans 1051
Human Growth Rates 1052
Population Changes Since the 17th Century 1052
After 1850 1053
Age Profiles 1054
The Population Boom and Bust 1054
Essay 35.2 Birth Rates and Death Rates, Increases and Doubling Times 1056
Application of Ideas 1056
Key Words 1057
Key Ideas 1057
Review Questions 1060
Suggested Reading for Part Seven 1061

Appendix A The Classification of Organisms 1062
Appendix B A Biological Lexicon 1068
Glossary 1071
Illustration Acknowledgments 1109
Index 1112

Part One

Molecules to Cells

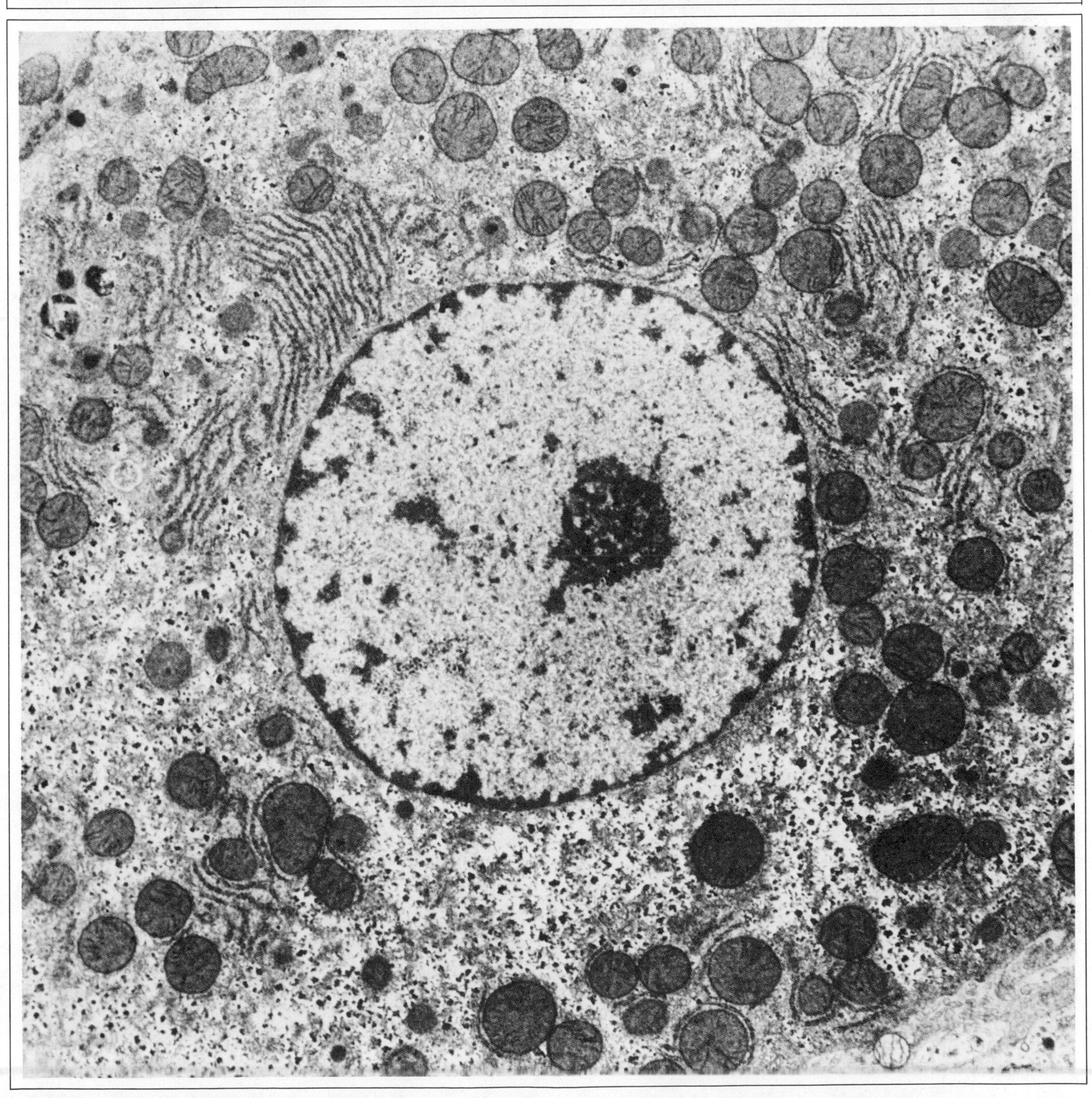

Chapter 1

Mr. Darwin and the Meaning of Life

Something was wrong. He couldn't put his finger on it, but he knew it just the same. The feeling had nagged at him before, but he had dismissed it. After all, it was hard to be pensive while standing on those grassy hills with the fresh winds of an exotic land brushing across his face.

The young naturalist was having the time of his life. The good ship *Beagle* lay at anchor off the coast of Argentina, and he himself was 200 miles inland, glad to be ashore, crossing the Argentine pampas on horseback. His only companions were a handful of rough, joking gauchos. He admired their superb horsemanship and their strong code of honor. They must also have admired his abilities. After all, he was their equal in horsemanship, having grown up in the English upper-class tradition of horses, dogs, and hunting. And he possessed not only an unusual physical strength but a vigor and endurance that was surprising in one reared under such sheltered conditions.

History doesn't record what the gauchos thought of the Englishman. What were they to think of a very young foreigner who insisted on climbing every mountain peak he came across, just to be the first Englishman to enjoy the view? Who spent days and sometimes nights feverishly digging huge bones out of canyon walls, who picked and examined every new flower he saw, who filled notebook after notebook with indecipherable jottings, who filled his pockets with rocks, who chased every animal, who never stopped asking questions?

1.1 Observation, Data, and an Idea

A Ratcatcher Goes to Sea, or Where Are the Rabbits?

Life had not always been a joyous adventure for Charles Darwin. His father and his grandfather had been among the wealthiest and most famous physicians in England. Charles had been expected to follow in their footsteps and, at the age of sixteen, was sent off to medical school. But he found that he became ill at the sight of blood, and he nearly fainted upon witnessing his first operation. He saved himself that disgrace only by rushing from the room. So, even in his teens Charles was already something of a failure, a dropout. To his family's despair he appeared to be interested only in his aristocratic hobbies of hunting, riding, and collecting beetles and rocks.

He worshiped his autocratic father, Dr. Robert Darwin, who was a commanding 6 feet 2 inches and 328 pounds of authority. The Darwin family said that the doctor's return home each evening was like the tide coming in. He was often especially tough on young Charles. At one point Charles was mortified to be told by his father that "You care for nothing but shooting, dogs, and ratcatching, and you will be a disgrace to yourself and all your family."

So Charles tried again—this time law. But he had no aptitude for it and was soon shuttled into training for the clergy. He was duly enrolled in divinity school at Cambridge, where he promptly showed almost as little aptitude for divinity as for medicine. His curriculum included Classics, which he loathed, and mathematics, which he couldn't understand. He once wrote a friend about his trouble with mathematics and said, "I stick fast in the mud at the bottom and there I shall remain." Still, his college experience had its pleasant aspects—he was a prominent member of the Glutton Club, a group of party-loving souls who often ended their evenings with drinks and a few hands of *vingt-et-un* (blackjack). And he kept up with his insect and rock collecting. At Cambridge he found a friend in one of his teachers, the Reverend John Henslow, who was a botanist as well as a clergyman.

Often the two would take long walks in the countryside around Cambridge and discuss the natural history of the area. Darwin was even known among some as "the one who walks with Henslow."

When at last Charles surprised himself and his family by passing his final examinations at Christ's College, he came home shouting "I'm through! I'm through!" No more school. He was ecstatic. Of course, he was now expected to enter the clergy, but, in truth, he had his private doubts about the whole matter. He spent his first postgraduate summer happily "geologizing" around the English countryside, away from difficult decisions and away from his family.

When Darwin returned home in the fall he found a letter from Henslow waiting. Henslow had been offered an appointment as naturalist on a British naval survey ship that was to sail around the world. Mrs. Henslow, however, had become so disconsolate at the idea of her husband being gone so long that he had reluctantly refused the offer. He recommended that young Darwin go in his place. Darwin, of course, would have accepted in an instant. But there were some problems.

Charles' father would have none of it. Darwin was disheartened, but finally his father (who wanted to be fair to his son) told him, "If you can find any man of common sense who would advise you to go, I will give my consent." Young Darwin thought this unlikely, and was prepared to decline the offer once and for all when his uncle Josiah Wedgwood (of pottery fame) said that he thought it was a splendid idea for his favorite nephew. The two of them together persuaded Robert to hold to his word.

On September 5, 1831, Darwin was summoned to London to be interviewed by the captain of the *Beagle*, Robert Fitzroy. Darwin, himself only twenty-two, was only a year younger than the captain, who nevertheless had already distinguished himself as a seaman of remarkable abilities. There was an initial awkwardness—Fitzroy thought that the shape of Darwin's nose indicated a weak character. But soon the two young men found that they got along quite well. Fitzroy's doubts about Darwin's nose evaporated and Darwin was accepted as the *Beagle*'s naturalist. In fact, Darwin shared quarters with the captain himself. There was no salary, and Darwin had to pay for his own room and board throughout the voyage.

Fitzroy's major mission was to chart the waters off South America. But the captain's private mission—his personal passion—was to find evidence that would establish once and for all the literal truth of the Genesis account of the creation of the world. For that he needed a naturalist, and this amiable young divinity student, who was hungry for adventure, seemed just right for the job.

Now, as Darwin gazed across the lush, endless-seeming, rolling grassland of the strange continent of South America, a question was forming in his mind. He knew that there was something askew, something not quite right. What was it? Why was he troubled? There was certainly no apparent reason to feel uneasy. The warm breeze gently smoothed the unkept grass, the sky was clear and blue, and his confidence and physical strength were high. But he sensed that, in the panorama that unfolded before him, something was wrong. Then it came.

There were no rabbits.

No rabbits. The phrase could have been engraved in the consciousness of Western civilization along with $E = mc^2$ and *E Pluribus Unum.* No rabbits; such an innocuous phrase. But this, perhaps, was the beginning.

In any case, Darwin knew rabbit country when he saw it, and this place was obviously a rabbit heaven. There was grass for rabbits to eat and bushes for rabbits to hide in and dirt for rabbits to dig in—but there were no rabbits. And Darwin wondered why.

Where rabbits should have been there were some strange little animals that hopped across the trail from time to time. They had long legs like rabbits, and large ears, and did many things that rabbits did, but they were clearly not rabbits. They looked more like guinea pigs (Figure 1.1). So where were the rabbits?

1.1 While rabbits are not native to South America, their niche is filled to some extent by the Patagonian hare, or mara, as it is known. The mara, *Dolichotis patagonum,* is placed with the Rodentia, along with squirrels and mice, so it is actually not closely related to rabbits. The mara's similarity to rabbits is considered to be an example of convergent evolution, where similar niches in geographically isolated regions are filled by animals that have similar physical features.

On one level Darwin knew perfectly well why there were no rabbits. There were no rabbits on the Argentine pampas because he was in South America, and rabbits don't live there. That kind of answer usually satisfies almost everyone, but in fact, it is no answer at all. If the question *Why aren't there any rabbits in South America?* had been pressed, another naturalist might have answered that South America was really not the proper place for rabbits, that the land couldn't support them. But Darwin knew rabbit country when he saw it. Perhaps he continued to mull over the question as he idly poked a stick into the campfire that night. The answer was slow in coming, but at last a simplistic answer formed in his mind.

Perhaps there are no rabbits in South America, he thought, *because rabbits can't swim across the Atlantic Ocean.*

That question and its startling answer were to change the world forever.

An Explanation About Rabbits and Oceans

If Darwin's tentative answer doesn't seem earthshaking, or even particularly interesting, perhaps it is because more than a century and a half have passed since Darwin looked across the lonely pampas. And the world has already changed. Of course, rabbits cannot swim across the ocean. But that answer leads to others, and still others, and eventually to the whole logic of evolution. It means that animals and plants are where they are not because they were put there, but because their ancestors either made the journey or themselves originated there. It leads on.

There are rabbits in Europe, Africa, and Asia, presumably because there were no ocean barriers to prevent their ancestors from moving from one place to another. But rabbits are different from one place to the next. Chinese rabbits are not the same as British rabbits. Were both species derived from the same stock? We would say yes, but such a response was nearly unthinkable in Darwin's time, when species were considered to be *by definition* unique, unmixing, and unchanging.

Why were the rabbitlike rodents of the pampas so recognizably similar to other South American rodents, such as guinea pigs? The conventional wisdom was that South American rodents were uniquely adapted (or designed) for life in South America. But in Darwin's mind, a newer, different answer began to take shape: South American rodents are similar to each other *because of common descent.*

It wasn't just the rodents and the rabbits that were stirring Darwin. He had dug up and reconstructed the bones of several extinct mammals, including a gigantic armadillo and some even larger, giant ground sloths that were very much like the hippopotamus. Sloths are still found only in South and Central America, but the present-day species are small creatures. The extinct giants were clearly similar to—in fact, Darwin had to say, were apparently *related* to—the small burrowing armadillos and the tree-dwelling sloths that were still around. So he had bones of animals that no longer existed and no evidence that modern animals had always roamed the earth in their present form.

And why were armadillos and sloths found only in Central and South America? Could it be due to a peculiarly South American environment to which the armadillo and sloth bodies were uniquely adapted? Darwin thought not. He speculated that the African hippopotamus itself would probably do rather well in South America, if it had some way of getting there.

Actually, Darwin found evidence that a number of giant South American mammals had become extinct. He speculated that they were driven to extinction by competition from invading North American species, in particular antelope and bison. He knew that English plants and animals that had been introduced into New Zealand were rapidly displacing native species. But why hadn't the South American extinctions occurred earlier? Why had the North American animals taken so long to go south? Darwin thought that perhaps the Isthmus of Panama had formerly been under water and that the animals had only advanced southward when a land bridge was formed. This was an excellent guess.

Darwin had taken with him, on the *Beagle,* a copy of the newly published first volume of his friend Charles Lyell's revolutionary book, *Principles of Geology.* By the time he had crossed the Atlantic, Darwin was a convert to the new geology. While in South America he received the second volume. Lyell had some rather startling new things to say about the physical evolution of the world. He said that the world was much older than anyone had imagined; that over long periods of time, continents and mountains raised slowly out of the sea, and that they just as slowly subsided again or were washed away. Most importantly, Lyell claimed that the very forces that had so changed the earth in the past were still at work, and that the world was still changing.

Darwin's own observations on South American geology seemed to confirm Lyell's position at every hand. In his adventurous climbing of the Andes, he had found clam shells at 10,000 feet. Below them, near an ancient seashore at 8000 feet, he found a petrified pine forest that had clearly once lain beneath the sea, because it, too, was covered and interspersed with seashells. In fact, the *Beagle* had arrived in Peru just after a strong local earthquake had destroyed several cities and had *raised the ground level by 2 feet.*

Darwin kept the most revolutionary of his thoughts to himself. Questioning the prevailing doctrine of the fixity of species, he was to say later, was like admitting to a murder. And he was full of doubts. He had been a Christian and had believed in the Bible, but the evidence seemed to contradict the creation account in every detail.

His cabinmate and erstwhile friend, Captain Fitzroy, had no such doubts. To him the bones of extinct giants merely proved the account of the flood, if one simply allowed that perhaps Noah hadn't been able to round up all the animals. If there were no rabbits in South America, it was because rabbits did not belong in South America. There was a very good reason for everything. Fitzroy believed in laws and rules, and he knew what the laws and rules were. Darwin, on the other hand, was blessed (or cursed) with an ever-inquiring mind and was always ready to consider an alternative hypothesis. The hypothesis that was forming now was a dangerous one, perhaps the most dangerous hypothesis of all: *The entire fabric of our understanding of the meaning of life was wrong and had always been wrong.*

The Copernican revolution had been a great blow to humanity's exhalted conception of itself, for it removed the earth from the center of the universe and placed humanity on an insignificant island in space. And, in a sense, Lyell's geology was just as revolutionary: It placed us on an insignificant island in time. But the greatest reorganization of humanity's conception of its own place in nature was yet to come, and it was to stem from that young man so laboriously poring over Lyell's writings. Even Darwin himself could not have been prepared for what was to come.

His thoughts began to come together on one momentous part of his 5 year trip. This was when the *Beagle,* on a dead run from the coast of South America, reached a peculiar little group of islands and clattered its anchor into the lee waters of an apparently insignificant island called Chatham by the English. Chatham was one of the Galapagos Islands, recently formed volcanic islands that form a small archipelago some 600 miles off the coast of Ecuador and lying astride the equator. Physically the islands were quite unlike anything Darwin had seen on the mainland: black, bleak, dry, and hot. And isolated—there were relatively few species of plants and animals. In fact, there were no mammals other than those brought by European ships, and only a few species of birds other than sea birds.

The plants and animals that did inhabit the Galapagos tended toward the bizarre. Darwin found giant tortoises there—a different species on each island. They were dimwitted and slow, and for centuries sailors from passing ships had captured the helpless reptiles. There were several species of iguanas, including the unique Galapagos marine iguana, a strange creature that lives on coastal rocks and dives into the ocean to graze on seaweed. And there were several species of drab little finches—about which we will say more later.

The Galapagos archipelago was, as Darwin wrote later, a little world to itself. And yet Darwin noticed that there was something familiar about the plants and animals, too. He observed that they seemed to have a distinctly South American character. Though most of the species were unique to the islands, Darwin had seen species like them only recently. They were similar to South American species and, as Darwin was to learn, unlike those of Europe, Asia, North America, or Australia. Why should this be?

The usual explanation just wouldn't do. There was nothing in the physical environment that suggested that the islands were somehow appropriate for South American creatures. Indeed, the environment was almost identical to that of the Cape Verde Islands, volcanic islands where the *Beagle* had tied up for nearly a month early in its voyage. The Cape Verde Islands, however, lay off the coast of Africa, and its species were typically African.

Because rabbits can't swim across the Atlantic . . .

Asian, Australian, and North American species could not cross the Pacific Ocean to settle on these volcanic islands. But perhaps, from time to time, a few seeds, a few reptiles on floating logs, and a male and female finch might have come by chance across the 600 miles from the South American mainland. And finding an empty island, free from competition, they could have stayed and propagated.

Perhaps the most revolutionary contribution of Darwin's slowly building hypothesis was this element of chance. Sometimes this has been called *historicity,* the notion that things are the way they are today because of events that have occurred in the past. It is important to note that none of the world views before Darwin had any place for chance or historicity. In seeking to find meaning and reason for the structure of the world, people had always rejected chance and had searched for higher *causes* and more noble *purposes.* So a kind of finch had, by chance, found its way to the Galapagos; it could have been a swift or a pigeon or a turkey, but it was a finch. And Darwin found a variety of little finches fluttering about the islands.

Darwin's Finches and Speciation

The finches Darwin observed on his historic visit are perhaps not so spectacular as the large lizards; however, Darwin's keen eye noticed something very remarkable about these thirteen species of small birds. They are all 4–8 inches long, and both sexes are drab-colored browns and grays. There are six species of ground finches on the islands. Of these, three live on the ground, each eating seeds of a different size, as evidenced by their different-sized bills. One eats cac-

1.2 Darwin's finches. The darker birds on the ground and standing on the low cactus are ground finches (9–14); those in the tree are tree finches (1–8). Are there similarities within each group? Can you account for this? What conclusions can you draw about the diets of different species by looking at the size and shape of their bills? The different species presumably arose from a single stock that gradually spread across the Galapagos Islands. The finches have a strong attachment for their home areas and are reluctant to fly across water. Once a group reached an island, it was likely to remain there and effectively be isolated, so it was left to follow its own evolutionary pathway. Eventually, groups differed so much that they could not interbreed.

The tree finches are: 1, *Camarhynchus pallidus* (the woodpecker finch); 2, *C. heliobates;* 3, *C. psittacula;* 4, *C. pauper;* 5, *C. parvulus;* 6, *C. crassirostris;* 7, *Certhidea olivacea* (the warblerlike finch); and 8, *Pinaroloxias inornata* (the Cocos Island finch). The ground finches are: 9, *Geospiza magnirostris;* 10, *G. fortis;* 11, *G. fuliginosa;* 12, *G. difficilis;* 13, *G. conirostris;* and 14, *G. scandens.*

tus as well as seeds, and two eat mainly cactus, the one with the long bill specializing on prickly pears.

There are also six species of tree finches on the Galapagos. One of these has a parrotlike beak and eats seeds and fruit. Four have bills well-suited for eating insects, and each species specializes in insects of a different type. But the strangest of all is the woodpecker finch. This little bird has a bill like a woodpecker, but it lacks the long protrusible tongue that the woodpecker uses to probe deep into crevices. So instead it pries grubs out of holes by using a long twig or cactus spine (Figure 1.2).

The thirteenth species of finch confined to the Galapagos has diverged markedly from the other finches. In fact, by external appearance and behavior it should not be considered a finch at all, but a warbler. Only the examination of internal structure unmasks it as an aberrant finch. Another finch lives isolated on one of the earth's most miserable places, the dense, mosquito-ridden Cocos Island.

So why have these little birds become so significant in the biological world? Their fame is due to Darwin's suspicion, supported by subsequent years of careful research by others, that the birds are all descended from the same stock (Figure 1.3). The essential idea is that long ago the islands were colonized by a species of ground-dwelling finch from the mainland. The colonizers were probably blown out to sea by some weather turbulence, and in time they became established on one or more of the islands. As their numbers increased, they could easily have spread throughout the islands. However, the rate of interbreeding among the birds of the various islands was low enough to ensure virtual isolation of each population. In time, then, as each small population was subjected to the specific and rigorous conditions of its particular island, changes would have appeared in the population through the process of *natural selection.* These changes, of course, would have been reflected in the appearance and behavior of the species.

1.3 The hypothetical relationship of Darwin's finches as they evolved from a common stock. The numbers refer to the birds in Figure 1.2.

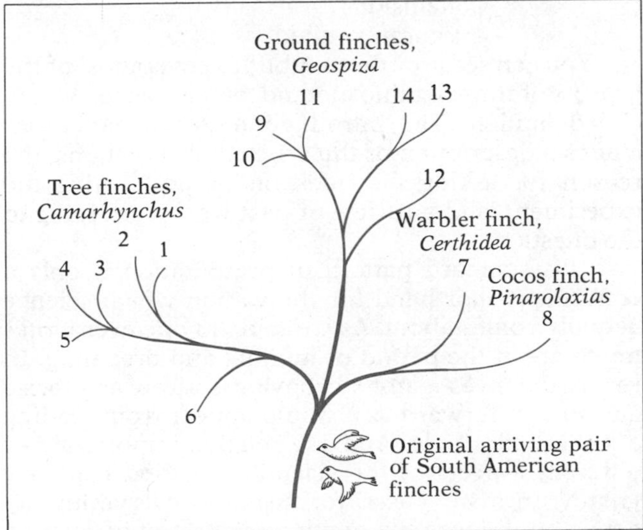

Under such conditions the population on each island would ultimately have become so different that they could no longer interbreed. The result would be *speciation,* or the formation of new species. It has been roughly calculated that each population would have to have been isolated for about 10,000 years in order to form a new species. We don't know how precise the time measurements are, but we do know that nature is in no hurry. And the fact that the colonizers became established was merely a fortunate interaction of characteristics between the colonizers and their new habitat.

Darwin reasoned that new species could not arise unless different populations of the ancestral species became geographically isolated. His analysis seems to have held up, and nearly all modern biologists go along with the idea that species can only arise in separated groups. Usually it is not a matter of one large species dividing into two large groups, but of a fairly widespread species spinning off little peripheral isolates, which then become distinct species. The usual fate of such a spinoff species is extinction, and occasionally, its demise is due to its parent group. After a period of adaptation and change, the derived species may reinvade its ancestral homeland; if it is still similar to the ancestral form, the two species will be in direct competition, and the derived form may die out. But it may conceivably have become superior to its ancestors in its ability to survive, and it may then force the parent group into extinction, instead. However, if the two species have developed sufficiently different adaptations, they may be able to coexist.

How to Go Against the Grain

But back to our story: Darwin and Fitzroy were now engaging in lively, if not heated, discussions about the nature and origins of life. Fitzroy would probably have been chagrined to think he was helping Darwin form his heretical ideas by providing a sounding board. Darwin probably was hesitant to say what he was really beginning to think, but he was really beginning to think that life does, in fact, change—that living species are modifications of earlier and different species. Other people, including Darwin's own grandfather, had said the same thing. But their musings were unsupported. Darwin was slowly gathering the buttressing material.

The hypothesis that Darwin was formulating was to be called *evolution,* or *descent with modification.* As his ideas gained substance, Darwin began to try to apply them to what he saw around him, and he now saw evidence for evolution everywhere; he came to believe that evolution was a fact. He knew, however, that the fact of evolution did not itself make a complete hypothesis. In a sense it was just an observation. Facts and observations are very important in science, but they are not science. What Darwin needed now was a *mechanism of evolution.*

The *scientific method* has been defined many ways, and yet it always eludes definition. Essentially, it is the process of establishing new facts and understanding mechanisms; but there is no set algorithm, no prescribed set of directions, for accomplishing these things. Still, we recognize the scientific process when we see it. The process usually begins and ends with observations, or facts if you will, about the real world. But, between the first observation and the last is interposed a fair amount of human mental activity. So, in a sense, science is the interaction of the human mind with the facts of nature. But where does it begin?

There are the initial observations, which are presumably not understood. Darwin, for example, observed that the land birds of the Galapagos are unique species with strong physical similarities to the finches of mainland South America. But what was he to do with such information?

What happens next is the least known aspect of science, and perhaps the most important. It begins with mulling over the facts and wondering, especially wondering *why.* Almost no one writes down very much at this stage, which may be why it is so little understood. Perhaps most scientists are wondering and mulling over facts most of the time, and most of the time nothing comes of it. Sometimes, though, wondering leads to speculation. Speculations are fine, bold, and usually untestable ideas. Sometimes these are even written down. We could no more do without speculation than we could do without facts. But even facts coupled with speculation do not add up to science. Darwin's grandfather, Erasmus Darwin, had speculated in print about evolution; for this he earned only scorn from his grandson. (Charles Darwin hotly denied ever having been influenced by his grandfather's speculations, but historians have their doubts about this.)

At some stage the investigator must formulate a hypothetical mechanism to explain the facts. A *hypothesis* is an assertion that is potentially provable or disprovable. The hypotheses of others may be used, because science is often a community effort. They can arrive fully formed, like Venus on a seashell, or in rough outline form so that modifications and refinements can be made to fit the new observations. At some point, however, if a scientific hypothesis is to be useful, it must make *predictions* and it must be *testable.* This is the crucial stage and it involves another observation—the observation that tests the prediction. One must seek some new fact that has not been observed before, but would be expected only if the hypothesis were true. In most cases the prediction is tested in an *experiment.* The investigator deliberately sets up carefully specified conditions under which certain observations or results can be expected according to the hypothesis being tested. The experiment must be set up in such a way that there cannot be alternate explanations for the observation; frequently, this involves running *controls* in which some crucial factor is altered or left out. Figure 1.4 shows an example of a controlled experiment.

The results of the experiment must then be interpreted in the investigator's *conclusions,* which is a declaration of what is believed to be the salient features of the case. The conclusions can be tentative or firm, depending on the investigator's confidence in the strength of the evidence. Throughout the entire procedure the investigator must constantly review what others have said and written, because someone else may have expressed the same ideas and made the same observations, but tested them and explained them in a different way.

Finally, publication is an important part of the scientific method. Discovering something and keeping it to oneself is not in keeping with the idea of modern science. The usual way to get the message out is to publish an article in a scientific journal. Usually the article is broken into several parts:

Introduction
Methods and materials
Results
Discussion
Conclusions
Literature cited

You can see how these subtitles cover most of the aspects of the scientific method, at least according to our definition. The introduction, for instance, includes a description of the original observations, the reason for asking the question or performing the experiment, and a review of past work pertaining to the question.

This standard pattern of presentation is only a convenient shorthand for the way in which science actually comes about. As we said, no one ever writes much about the period of mulling and dreaming. In reality the investigator's behavior is rarely as precise and straightforward as it would appear from reading a published article. And the relative importance of different aspects of the scientific method can vary greatly. In many cases, for instance, everything depends on the ingenuity with which a test is devised, and the progress of science is often paced by the invention of new techniques and instruments. In other cases, the crucial step may be simply the asking of a question that hadn't been asked before, or the offering of a persuasive explanation that hadn't been thought of by anyone else.

We've said that a hypothesis is tested in an experiment in most cases. It is important to note that not all experiments in biology are performed in laboratories. There have been many excellent field experiments (for instance, mark-and-recapture experiments in which animals are trapped, marked, released, and later recaptured, which provide information about such things as their movement, growth, or longevity). Hypotheses also can be tested without experiments at all. An ecologist might predict that a certain relationship will be found in nature and then test the prediction by going out into the field to look for it. But it is still necessary to first make the prediction and then test it.

The rejection of alternate hypotheses to explain an observation involves more than conducting controlled experiments. The discussion in a published article must include all the competing explanations that might be offered, as well as those that have been published by others, and it must show why these are inadequate or unnecessary or, at least, not as convincing as the interpretation being presented. Often, the simplest and most troublesome alternative explanation of the experimental results is that observations were due simply to a chance occurrence. An entire field of science—statistics—is devoted primarily to the problem of rejecting the possibility that results are due to chance.

A decisive ingredient of the scientific method, but one that is not usually published, is the scientist's own motivation, his or her need to know and or tell. The very best scientists are terribly curious, of course, but they also tend to be driving, egocentric, prideful enthusiasts who may covet fame, or titles and prizes, or the adulation of students. Perhaps the strongest and most common driving force of scientific investigation is the individual's very human need for the acceptance, approval, and admiration of the peer group—in this case, other scientists.

Darwin was no exception. When the *Beagle* arrived at Ascension Island near the conclusion of its voyage, Darwin found a letter from his sisters waiting for him. In it they mentioned that Adam Sedgwick, an English scientist, had told Robert Darwin that his son Charles would take a place among the leading scientific men, on the strength of the letters and specimens that Henslow had already received from the *Beagle*. Darwin wrote, "After reading this letter I clambered over the mountains of Ascension with a bounding step and made the volcanic rocks resound under my geological hammer!"

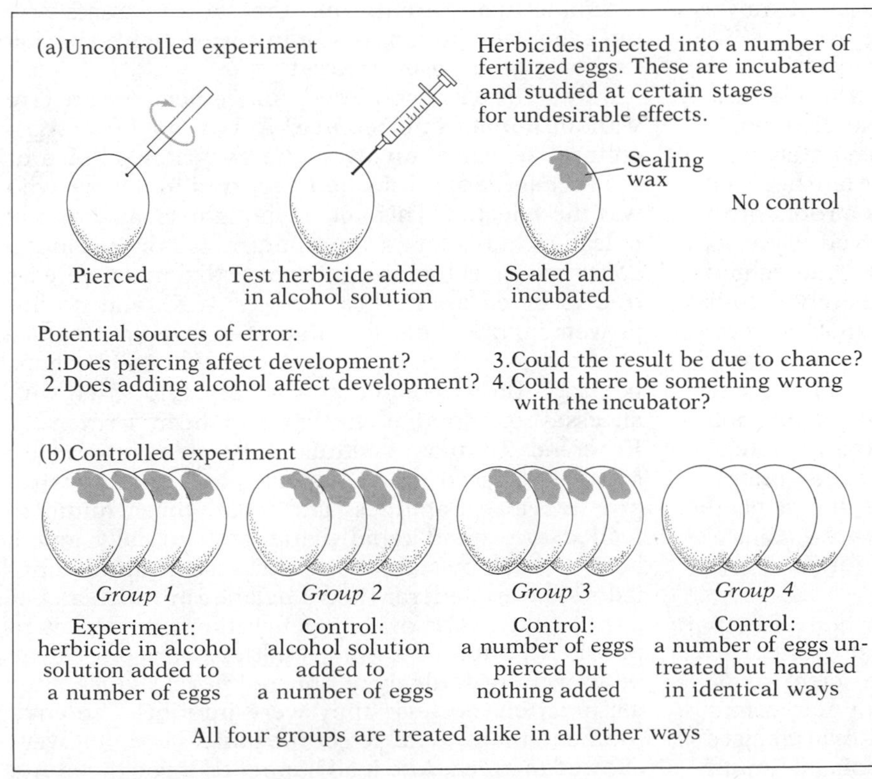
(a) Uncontrolled experiment
Herbicides injected into a number of fertilized eggs. These are incubated and studied at certain stages for undesirable effects.
Sealing wax
No control
Pierced
Test herbicide added in alcohol solution
Sealed and incubated
Potential sources of error:
1. Does piercing affect development?
2. Does adding alcohol affect development?
3. Could the result be due to chance?
4. Could there be something wrong with the incubator?
(b) Controlled experiment
Group 1
Experiment: herbicide in alcohol solution added to a number of eggs
Group 2
Control: alcohol solution added to a number of eggs
Group 3
Control: a number of eggs pierced but nothing added
Group 4
Control: a number of eggs untreated but handled in identical ways
All four groups are treated alike in all other ways

1.4 The experiment is intended to determine the effect of an herbicide (plant killer) on bird development prior to its use on plants. This would be one of many tests of the herbicide prior to its widespread application on foliage. The first procedure (a) includes piercing the fertile egg (under sterile conditions) and injecting a specific concentration of the herbicide into the air space below the shell. The herbicide must be dissolved in alcohol. Following incubation, the embryo is studied for defects at a desired stage.

As you may have surmised, the experiment contains several sources of potential error. First, if developmental abnormalities do occur, they might have been caused by the procedure, or perhaps the addition of fluid itself had an effect. In any case, it is difficult to draw a meaningful conclusion from one embryo. To correct these errors a controlled experiment must be devised.

In the controlled experiment, four large groups of fertile eggs are prepared. The first is treated in the manner described above. The second is treated identically, except that the injection is a *placebo*, a prepared alcohol solution that is identical except for the presence of the herbicide. A third group has been pierced and sealed, and a fourth group has no special treatment, but is incubated under identical conditions. Now the results of the four can be compared, and any difference attributed to the presence of herbicide.

Going Home to Applause and Illness

For Darwin, the key event of the voyage was the visit to the Galapagos. He was, from then on, convinced about the reality of evolution. He was especially impressed by the observation that the many small islands of the archipelago had different assemblages of species, and often their own unique species, even though the islands were all rather similar. This again suggested historicity, the idea that the animals and plants on each island were there because of past accidents of introduction; and it suggested that the different island species had evolved from one another. But he was still groping for the mechanism of evolution. Like any devoted observer of nature, he marvelled at how different organisms were so exquisitely adapted to their needs. And of one thing he was sure: *Any explanation of evolution that did not explain the adaptedness of species was no explanation at all.*

After the Galapagos the rest of the voyage was somewhat anticlimactic. The *Beagle* made for home as quickly as possible, which in this case meant continuing westward around the world. It called at Tahiti, New Zealand, Tasmania, Australia, Mauritius, South Africa, St. Helena, and Ascension. From there, the *Beagle* proceeded across the South Atlantic again to Brazil before returning to England. As always when the *Beagle* was under sail, Darwin was wracked with seasickness. In his letters home he wrote, "I loathe, I abhor the sea . . ." "Oh, the degree to which I long to be once again living quietly with not one single novel object near me."

By the time the *Beagle* entered English waters once again it was late in 1836, and Darwin was twenty-seven years old. When the boat tied up, a grateful Darwin leaped ashore and immediately took a carriage home. But since he arrived at a late hour, he decided not to rouse his family and took a room at an inn nearby. The next morning he received a joyous greeting from them all but he was especially delighted when his dog greeted him and immediately set off down the trail on which they had last enjoyed their morning walk 5 years before.

Darwin did indeed take his place among the scientific men. His observations on geology and zoology were published and he was revealed as a keen observer of nature, undoubtedly one of the greatest natural historians of all time. And he was a good storyteller. His journal, *The Voyage of the Beagle,* became successful popular reading in England and remains a classic today.

Darwin took lodgings in London and began rounds of scientific meetings and social events. He became good friends with many of the scientists of the day, including his idol Lyell. The popular young naturalist was elected a Fellow of the Royal Society on January 24, 1837, and married his cousin Emma Wedgwood 5 days later. By December of that year he had a son. He was accepted and admired by the leading scientists of England. He was a success.

Even his family life was enviable. Darwin was a patient and loving father who was very affectionate with all his children. His wife was a relaxed and unaffected woman who never cared much for Darwin's work, and it all came to be something of a joke between them.

Through all of this he had kept his ideas on evolution to himself. In 1837 he began to keep a journal titled *Transmutation of Species* and was soon making entries referring to "my theory." Reading that journal today, and the journal of the voyage of the *Beagle,* is like reading a detective story after you already know whodunnit—the suspense is in watching the detective sift through clues for the right answer. In his journal entries Darwin began to toy with the idea of natural selection, and it is fascinating to watch the idea develop.

Keep in mind that Darwin was a country boy. He knew a lot about agricultural practices, and he was particularly impressed with livestock breeding. Any good farmer knows that a breed can be improved by *selection,* simply by selecting the best individuals of each generation for breeding (Figure 1.5). By longstanding folk tradition, the best of the breed (whether cattle, fowl, dog, or cucumber) was honored annually at country fairs and was chosen for propagation. It's a simple matter: like begets like, offspring tend to resemble their parents, and the "best" parents produce the "best" offspring. In this way, highly specialized breeds had been created.

But Darwin wondered, did selection operate without human interference? And, if so, how? Agricultural selection involved the conscious choice of the breeder. So, if selection occurred in nature, who was the selector? This line of thought seemed at first to lead right back to a supernatural factor. Without a conscious selector, the inferior individuals were as free to breed as the best of the breed, and no improvement or change (or adaptation) would occur.

So, was conscious selection a requisite? Perhaps not. . . . Darwin read and was quite impressed with an essay on population that had been written by Reverend Thomas Malthus three decades earlier. Malthus stressed that all species had enormous reproductive capabilities and that their numbers tended to expand rapidly and geometrically unless held in check by starvation or disease. Natural populations, he argued, reached a balance in which all but a few of the young of each generation were forced to perish. All but a few. So, if Malthus was correct, the inferior individuals were not free to reproduce after all, precisely because they were inferior! The environment, then, could select the individuals that were allowed to breed. And it all happened through natural means.

1.5 The most recent ancestor of the modern horse was probably similar to Przewalski's horse (left), a hardy little animal of the Asian steppes. By selective breeding we have been able to produce forms as diverse as the Clydesdale (center) and the miniature horse (right). The breeding program must be rigorously maintained, however, because populations of horses left to their own devices will quickly revert from highly specialized stock to animals like the feral mustang.

By 1838 Darwin's journal showed that he had solved the major riddle of evolution. The mechanism of evolution, he said, was *natural selection.*

A penciled manuscript in 1842, and a somewhat lengthier one in 1844 laid out the entire theory of the origin of species through natural selection. Darwin did not publish it, however. He knew that the idea would arouse fierce resistance and that he would have to back up every part of his idea with facts and experimental evidence. So he set about preparing to defend his argument.

But then, Darwin's robust health began mysteriously to fail. He vomited frequently and complained of headache, nausea, fatigue, and heart palpitations. Because he tired quickly, he stopped seeing anyone, even his valued scientific colleagues. He resigned as secretary of the Geological Society. His father bought him a country house, and his wife became his nursemaid. Through this he continued to look completely healthy. In fact, his doctors, who could find no direct evidence of organic disease, suggested none to subtly that he suffered from hypochondria. Nevertheless, Darwin spent the rest of his long life as an invalid. He continued his work, but he did not begin until 9 o'clock each day and he stopped at lunchtime. His mind was a vigorous as ever, but his strength was gone.

The probable cause of Darwin's illness was not discovered until long after his death. Darwin was curious about everything. In August, 1835, he "experienced an attack of the Benchuga, the great black bug of the Pampas. . . . One, which I caught at Iquique, was very empty; being placed on the table and though surrounded by people, if a finger was presented, its sucker was withdrawn, and the bold insect began to draw blood. It was curious to watch the change in the size of the insect's body in less than ten minutes. There was no pain felt. This one meal kept the insect fat for four months; in a fortnight, however, it was ready, if allowed, to suck more blood" (*Voyage of the Beagle*). It turns out that the blood-sucking Benchuga (now known as the barbeiros), *Panstrongylus megistus,* is the carrier of a protozoan, *Trypanosoma cruzi,* that causes Chagas' disease. The Benchuga has the peculiar habit of defecating on the open wounds it has produced, and thus the protozoan enters the victim's bloodstream. After a latency period of some years, the parasites usually invade the heart and intestines, causing weakness, intestinal distress, and heart palpitations.

When he was able to work, Darwin set about compiling information to back up his theory. He even suggested further tests. But Darwin was an experimental biologist, too, and he devised a number of ingenious laboratory tests. One prediction of his theory was that the inhabitants of oceanic islands were descended from immigrants from the mainland; but how did these creatures make the journey? After all, seawater is deadly to many land species. Darwin had observed that all true oceanic islands lack mammals other than, significantly, bats. Also completely absent from these islands were newts, salamanders, frogs, and toads. And yet islands did not seem to be unsuitable to such species; "indeed it seems that islands are peculiarly well-fitted for these animals, for frogs have been introduced into Madeira, the Azores, and Mauritius, and have multiplied so as to become a nuisance. But as these animals and their spawn are known to be immediately killed by seawater, on my view we can see that there would be great difficulty in their transportal across the sea. . . ." So Darwin began a set of small but significant experiments. In his words from the *Origin of Species:*

> Until I tried, with Mr. Berkeley's aid a few experiments, it was not even known how far seeds could resist the injurious action of sea-water. To my surprise I found that out of 87 kinds, 64 germinated after an immersion of 28 days, and a few survived an immersion of 137 days. For convenience sake I chiefly tried

small seeds, without the capsule or fruit; and as these all sank in a few days, they could not be floated across wide spaces of the sea, whether or not they were injured by the salt-water. Afterwards I tried some larger fruits and capsules, and some of these floated for a long time. . . . I was led to dry the stems of 94 plants with ripe fruit, and to place them on sea water. The majority sank quickly, but some floated much longer; for instance, ripe hazel-nuts floated for 90 days and afterwards when planted they germinated; an asparagus plant with ripe berries floated for 85 days. . . . Altogether out of the 94 dried plants, 18 floated above 28 days, and some of the 18 floated for a much longer period. . . . Peas and vetches are killed by even a few days' immersion in sea-water, but some taken out of the crop of a dead pigeon, which had floated on artificial salt-water for 30 days, to my surprise nearly all germinated. . . . In the course of two months, I picked up in my garden 12 kinds of seeds out of the excrement of small birds, and some of them, which I tried, germinated. . . .

It has long been known what enormous ranges many fresh-water and even marsh-species have, both over continents and to the most remote oceanic islands. . . . I have before mentioned that earth occasionally adheres in some quantity to the feet and beaks of birds. Wading birds . . . are the greatest wanderers, and are occasionally found on the most remote and barren islands in the open ocean; they would not be likely to alight on the surface of the sea, so that dirt would not be washed off their feet. . . . I have tried several little experiments, but will here give only the most striking case: I took in February three tablespoonfuls of mud from the edge of a little pond. This mud when dry weighed only 6-3/4 ounces. I kept it covered up in my study for six months, pulling up and counting each plant as it grew. The plants were of many kinds, and were altogether 537 in number; and yet the viscid mud was all contained in a breakfast cup! . . . I suspended a duck's feet, which might represent those of a bird sleeping in a natural pond, in an aquarium, where many ova of fresh-water snails were hatching. I found that numbers of the extremely minute and just-hatched snails crawled onto the feet, and clung to them so firmly that when taken out of the water they could not be jarred off, though at a somewhat more advanced age they would voluntarily drop off. These just-hatched molluscs, though aquatic in their nature, survived on the duck's feet, in damp air, from twelve to twenty hours; and in this length of time a duck or heron might fly at least six or seven hundred miles, and would be sure to alight on a pool or rivulet if blown across the sea to an oceanic island. . . .

Almost all oceanic islands, even the most isolated and smallest, are inhabited by land-snails, generally by endemic species but sometimes by species found elsewhere. Now it is notorious that land-snails are very easily killed by salt; their eggs, at least such as I have tried, sink in sea-water and are killed by it. Yet there must be, according to my view, some unknown but highly efficient means for their transportal. Would the just-hatched young occasionally crawl on and adhere to the feet of birds roosting on the ground, and thus get transported? [probably not]. It occurred to me that land-snails, when hybernating and having a membranous diaphragm over the mouth of the shell, might be floated in chinks of drifted timber across moderately wide arms of the sea. And I found that several species did in this state withstand uninjured an immersion in sea-water during seven days. One of these shells was *Helix pomatia* (the common garden snail), and after it had again hybernated I put it in sea-water for twenty days, and it perfectly recovered and crawled away: but more experiments are wanted on this head.

Working only a few hours a day, Darwin published large works on the systematics, or natural classification, of barnacles (still a major reference work), on his theory of the origin of coral islands, and on geology. He showed his early manuscript on natural selection to only one close friend, the botanist Hooker, who remained doubtful.

1.6 In November 1859, Charles Darwin laid his ideas open to public scrutiny with the first printing of *Origin of Species by Means of Natural Selection.* The book was a sellout! As the printers rushed to restock, the Western world began to marshall forces as the scientific community and the rest of the intellectual world polarized into camps that still exist today.

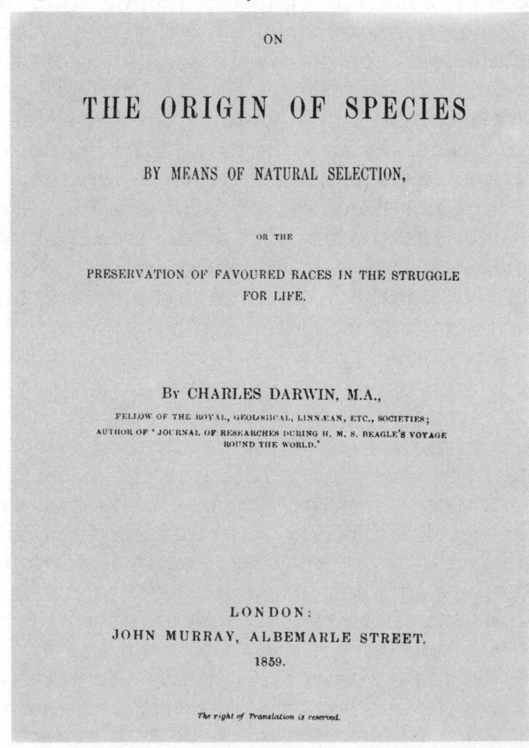

ON

THE ORIGIN OF SPECIES

BY MEANS OF NATURAL SELECTION,

OR THE

PRESERVATION OF FAVOURED RACES IN THE STRUGGLE FOR LIFE.

BY CHARLES DARWIN, M.A.,

FELLOW OF THE ROYAL, GEOLOGICAL, LINNÆAN, ETC., SOCIETIES;
AUTHOR OF 'JOURNAL OF RESEARCHES DURING H. M. S. BEAGLE'S VOYAGE ROUND THE WORLD.'

LONDON:
JOHN MURRAY, ALBEMARLE STREET.
1859.

The right of Translation is reserved.

Twenty years passed, and finally in 1856 Darwin began writing his major work, to be called *Natural Selection.* It was planned as an enormous, six-volume monograph. But in 1858 his work on the manuscript was suddenly interrupted. Unexpectedly, another manuscript—a very brief one—arrived in the mail from A. R. Wallace, a young naturalist working in Malaya. Wallace asked politely whether Darwin would care to make any comments on the manuscript. The article was a short but well-written statement about evolution and natural selection—Darwin's ideas had been quite independently deduced by someone else! Darwin was mortified. In a letter to his friend Hooker he lamented, "So all my originality, whatever it may amount to, will be smashed. . . . Do you not think that this sketch ties my hands? . . . I would far rather burn my whole book, than that he or any other man should think that I had behaved in a paltry spirit." Darwin had been scooped; it is not an uncommon experience in science. Would his own work now be thought to have been based on the ideas of another? Darwin's friends persuaded him to allow his previously unpublished 1842 summary and the text of a long letter describing his ideas to be presented together with Wallace's paper before the Linnean Society of London. They were presented in July and were published in August 1858. Now Darwin went furiously to work and, putting aside the idea of a huge, definitive monograph (which never was written), he quickly finished an "abstract" of it: the famous *Origin of Species,* published finally in 1859. The first edition sold out on the first day (Figure 1.6).

1.2 The Idea Becomes a Theory

The Theory of Natural Selection

Among Darwin's many gifts was an ability to see simple and obvious things that had escaped notice and to give them simple and obvious explanations that no one had thought of. At least they seem simple and obvious to us, living in a post-Darwinian age. But again, much of science involves clever people showing us things that seem so inherently apparent, once explained, that it seems like we must have known it all along. Our reactions must be somewhat irritating to the original scientists.

There has been a longstanding argument over just how much of an intellectual achievement the theory of natural selection really was. Some biologists and historians of science have maintained, in all seriousness, that the idea is so simple as to be trivial—a *tautology,* in fact. A tautology is an obviously true statement, which by its nature cannot be contradicted. "My father is a man" is a tautology. The tautology ascribed to Darwinian natural selection is that the fittest survive and fitness is defined by survivability. If the two terms define each other, then there is really no statement at all—certainly nothing that can be tested. Of course, the circle can be broken if we first devise a way to measure fitness other than survivability and then see if our "fit" individuals survive better than the others, in much the same way an animal breeder would.

Other people maintain that the concept of natural selection makes no sense and cannot be true. You'll have to judge for yourself. But let's let Darwin speak for himself. In the first chapter of his book, *Origin of Species,* Darwin sought to establish that animals and plants under domestication are extremely variable, that the variation is heritable, and that the many domestic varieties have risen, under artificial selection, from wild ancestors. In the second chapter, he drew together what evidence he could find to show that plants and animals in nature were variable too, and that species differences were only one aspect of variation. In the third chapter, he discussed Malthus' idea of the *struggle for existence:* the idea that the natural reproductive capacities of living things greatly outreach the ability of the environment to support them, so that most organisms must perish by starvation or disease before reproducing. In the fourth chapter, he introduced natural selection:

> How will the struggle for existence . . . act in regard to variation? Can the principle of selection, which we have seen is so potent in the hands of man, apply in nature? I think we shall see that it can act most effectively. Let it be borne in mind in what an endless number of strange peculiarities our domestic productions, and, in a lesser degree, those under nature, vary; and how strong the hereditary tendency is. Under domestication, it may truly be said that the whole organisation becomes in some degree plastic. . . . Can it, then be thought improbable, seeing that variations useful to man have undoubtedly occurred, that other variations useful in some way to each being in the great and complex battle of life, should sometimes occur in the course of thousands of generations? If such do occur, can we doubt (remembering that many more individuals are born than can possibly survive) that individuals having any advantage, however slight, over others, would have the best chance of surviving and of procreating their own kind? On the other hand, we may feel sure that any variation in the least degree injurious would be rigidly destroyed. This

preservation of favourable variations and the rejection of injurious variations, I call Natural Selection. Variations neither useful nor injurious would not be affected by natural selection, and would be left a fluctuating element, as perhaps we see in the species called polymorphic. . . .

It may be said that natural selection is daily and hourly scrutinising, throughout the world, every variation, even the slightest; rejecting that which is bad, preserving and adding up all that is good; silently and insensibly working, whenever and wherever opportunity offers, at the improvement of each organic being in relation to its organic and inorganic conditions of life. We see nothing of these slow changes in progress, until the hand of time has marked the long lapse of ages.

Although natural selection can act only through and for the good of each being, yet characters and structures which we are apt to consider as of very trifling importance may thus be acted on. When we see leaf-eating insects green, and bark-feeders mottled-grey; the alpine ptarmigan white in winter, the red-grouse of the colour of heather, and the black-grouse that of peaty earth, we must believe that these tints are of service to these birds and insects in preserving them from danger. Grouse, if not destroyed at some period of their lives, would increase in countless numbers; they are known to suffer largely from birds of prey; and hawks are guided by eyesight to their prey. . . .

In social animals, natural selection will adapt the structure of each individual for the benefit of the community, if each in consequence profits by the selected change. What natural selection cannot do is to modify the structure of one species, without giving it any advantage, for the good of another species. And though statements to this effect may be found in works of natural history, I cannot find one case which will bear investigation.

In order to make it clear how, as I believe, natural selection acts, I must beg permission to give one or two imaginary illustrations. Let us take the case of a wolf, which preys on various animals, securing some by craft, some by strength, and some by fleetness. And let us suppose that the fleetest prey, a deer for instance, had from any change in the country increased in numbers, or that other prey had decreased in numbers, during that season of the year when the wolf is hardest pressed for food. I can under such circumstances see no reason to doubt that the swiftest and slimmest wolves would have the best chance of surviving, and so be preserved or selected—provided always that they retained strength to master their prey. I can see no more reason to doubt this, than that a man can improve the fleetness of his greyhounds by careful and methodical selection, or by that unconscious selection which results from each man trying to keep the best dogs without any thought of trying to modify the breed.

Even without any change in the proportional numbers of animals on which our wolf preyed, a cub might be born with an innate tendency to pursue certain kinds of prey. Nor can this be thought to be very improbable, for we often observe great differences in the natural tendencies of our domestic animals; one cat, for instance, taking to catching rats, another mice. . . . The tendency to catch rats rather than mice is known to be inherited. Now, if any slight innate change of habit or of structure benefitted an individual wolf, it would have the best chance of surviving and leaving offspring. Some of its young would probably inherit the same habits or structure, and by the repetition of this process a new variety might be formed which would either supplant or coexist with the parent form of wolf. Or, again, the wolves inhabiting a mountainous district, and those frequenting the lowlands, would naturally be forced to hunt different prey; and from the continued preservation of the individuals best fitted for the two sites, two varieties might be formed. These varieties would cross and blend where they met, but to this subject of intercrossing we shall soon have to return. I may add that, according to Mr. Pierce, there are two varieties of wolf inhabiting the Catskill Mountains of the United States, one with a light, greyhound-like form, which pursues deer, and the other more bulky, with shorter legs, which more frequently attacks the shepherd's flocks. . . .

Thus I can understand how a flower and a bee might slowly become, either simultaneously or one after the other, modified and adapted in the most perfect manner to each other, by the continued preservation of individuals presenting mutual and slightly favourable deviations of structure. . . .

Natural selection can act only by the preservation of infinitesimally small inherited modifications, each profitable to the preserved being. And as modern geology has almost banished such views as the excavation of a great valley by a single diluvial wave, so will natural selection, if it be a true principle, banish the belief of the continued creation of new organic beings, or of any great and sudden modification in their structure. . . .

Isolation, also, is an important element in the process of natural selection. In a confined or isolated area, if it is not very large, the organic and inorganic conditions of life will generally be in a great degree uniform, so that natural selection will tend to modify all the individuals of a varying species throughout the area in the same manner. Intercrosses with individuals of the same species which otherwise would have inhabited the surrounding and differently circumstanced districts would be prevented. But isolation probably acts more efficiently in checking the immigration of better adapted species after any physical change, such as of climate. Thus new places in the economy of the country are left open for the original inhabitants to struggle for, and become adapted to, through modifications in their structure and constitution. Lastly, isolation, by checking competition, will give time for any new variety to be slowly improved, and this may be of importance in the production of new species. If, however, an isolated area be very small, the total number of individuals supported on it will necessarily be small, and fewness of individuals will greatly retard the production of new species through natural selection, by decreasing the chance of the appearance of favourable variations.

If we turn to nature to test the truth of these remarks, and look at any small isolated area, such as an oceanic island, although the total number of species inhabiting it will be found to be small, yet of those species a very large proportion are endemic—that is, have been produced there, and nowhere else. Hence an oceanic island at first sight seems to have been highly favourable for the production of new species. But we may thus greatly deceive ourselves, for we ought to make the comparison within equal times, and this we are incapable of doing.

Although I do not doubt that isolation is of considerable importance in the production of new species, on the whole I am inclined to believe that largeness of area is of more importance, especially in the production of species which will prove capable of enduring for a long period and of spreading widely. Throughout a great and open area, not only will there be a better chance of favourable variations arising from the large numbers of individuals of the same species, but the conditions of life will be infinitely complex from the large number of already existing species. And if some of these many species become modified and improved, others will have to be improved in a corresponding degree or they will be exterminated. Each new form, also, as soon as it has been much improved, will be able to spread over the open and continuous area, and will thus come into competition with many others. Hence more new places will be formed in the economy of nature, and the competition to fill them will be more severe, in a large than in a small and isolated area. Moreover, great areas, though now continuous, owing to oscillations of level will often have recently existed in a broken condition, so that the good effects of isolation will generally have occurred. Finally, I conclude that, although small isolated areas probably have been in some respects highly favourable for the production of new species, yet the course of modification will generally have been more rapid in large areas. And what is more important, the new forms produced in large areas, which already have been victorious over many competitors, will be those that will spread most widely, will give rise to most new varieties and species, and will thus play an important part in the changing history of the organic world. . . .

That natural selection will always act with extreme slowness, I fully admit. Its action depends on there being places in the economy of nature which can be better occupied if the inhabitants of the country undergo modifications of some kind. The existence of such [available] places will often depend on physical changes, which are generally very slow, and on the immigration of better forms having been checked. But the action of natural selection will probably still oftener depend on [changes in some of the species], the mutual relations of many of the other inhabitants thus being disturbed. Nothing can be effected, unless favourable variations occur, and variation itself is apparently always a very slow process. The entire process will often be greatly retarded by free intercrossing. Many will exclaim that these several causes are amply sufficient wholly to stop the action of natural selection. I do not believe so. On the other hand, I do believe . . . that this very slow, intermittent action of natural selection accords perfectly well with what geology tells us of the rate and manner at which the inhabitants of this world have changed.

Slow though the process of selection may be, if feeble man can do so much by his powers of artificial selection, I can see no limit to the amount of change, to the beauty and infinite complexity of the coadaptations between all organic beings, one with another and with their physical conditions of life, which may be effected in the long course of time by nature's power of selection.

This lengthy excerpt from Darwin's fourth chapter not only shows the author's own concept of what natural selection was and how it worked, but also gives one a sense of how he developed the analogy with artificial selection and how he tied natural selection in with geology, biogeography, and some very sophisticated basic ecology. By an organism's "place in the economy of nature," for instance, Darwin was clearly referring to what modern ecologists call a *niche.* Elsewhere in the *Origin* Darwin argues that only one species can occupy a "place in the economy of nature" at any one time and place, because if two species are too similarly adapted, one will drive the other to extinction through competition. Ecologists call this idea the *competitive exclusion principle* or *Gause's law,* after the 20th century Russian ecologist who rediscovered it. Darwin's ideas about continents being the principal source of species, and particularly of those species "capable of enduring and spreading," has been fully substantiated by modern ecological and evolutionary investigations.

The quoted passage also highlights Darwin's awareness of one of the difficulties with the theory—a problem that was to trouble him for the rest of his life: the fact that intercrossing, according to the blending theory of inheritance that was then accepted, would inevitably dilute and dissipate any progress made through natural selection. We'll return to this problem in Chapter 11.

A Mechanism of Natural Selection

The theory of evolution, or the origin of species by means of natural selection, was a large, encompassing concept that had many interdependent parts. Natural selection was perhaps the key to Darwin's argument, and it is this idea that is most closely associated with his name. But it was only part of the concept, as we shall see. It explained how species adapt to changes in their environment, but keep in mind that natural selection itself did not explain the origin of heritable variation or how new species arose.

So let's put some of this into context now. Perhaps before we consider the central problem of the origin of species, we should summarize how natural selection works.

We begin with variation among individuals in a population. For natural selection to work, there must be spontaneous, newly arising, random variation among individuals within a species. And this variation must be at least partly due to heredity.

Then, there must be competition between members of the same species, so that some individuals leave more offspring than do other individuals. The result of this is that any heritable variation that decreases the individual's reproductive output will tend not to be passed on to the next generation and, in time, is likely to disappear from the population.

Conversely, any heritable trait that increases the individual's reproductive output will be passed on and will tend to increase in the population until the entire species has been changed, unless conditions should change so that the trait loses its advantage.

Also, any new random variation is most likely to be favorable when there has been a change in the physical or biological environment, because if the organism is already adapted to a set of conditions, random changes are not likely to be helpful.

So, then, natural selection has been called *the survival of the fittest.* But who are the fittest? Those who survive, of course. So we are left with the tautology that natural selection is the survival of those who survive. Those who are fittest (meaning those who are reproductively most successful) have the most offspring—again, by definition. This is the sort of circularity that gives some people problems with the concept of natural selection. But in fact the process depends on quite a number of subtle, important conditions, such as the necessity that variation be both heritable and capable of arising spontaneously. In addition, there is the more difficult requirement that the units of inheritance must not be capable of being diluted and blended by sexual reproduction.

Darwin never pretended to understand the spontaneous origin of variation, which we know now to be due to chemical changes in DNA. He failed to understand the mechanism of inheritance, which his younger contemporary Gregor Mendel was to elucidate in 1865. Still, he made a good case for natural selection. But speciation was a more difficult problem for him. Let's see what he was up against.

How Species Are Formed

Darwin showed, rather convincingly, how natural selection could change a species. But how could it create new species? In order to explain the origin of species, Darwin would have to show how one species could split into two species, a process called *speciation.* It seemed to him that natural selection would be involved, but he knew that there had to be more to it than that.

This was a particularly demanding task because of the strongly held notion that species were definite, unique, and unchanging. After all, each species has a name, which in itself gives it a special kind of reality in people's minds. People in the most primitive societies have names for all the species they interact with, and they know when two plants or animals belong to the same species and when they do not. Species are separate entities *by definition,* and again, truth by definition is a very special (and tenuous) kind of truth. Cats and dogs do not mate; a cat is a cat and a dog is a dog; how can they be related? No wonder Darwin wrestled with his idea for 22 years before publishing it!

(A traveling religious fundamentalist, visiting college campuses, recently put the matter this way: "A fruit fly is a fruit fly. You can give it a mutation, and it is still a fruit fly. You can give it ten mutations or a million mutations, and it's still a fruit fly." But is it? If something is changed, is it still the same? How would you respond to this fundamentalist's statement?)

As we have seen in the passage from the *Origin of Species,* Darwin rejected the idea of "the creation of new organic beings, or of any great and sudden modification in their structure," opting instead for imperceptible, minute, and gradual changes. But this makes our task of searching for the origin of species more difficult, because any two species have to be substantially different from one another in order to be recognized as species. So if all change is gradual, just when does one species become two?

Darwin's answer to the challenge of the "fixity of species" is rather startling. *He simply denied that species exist!* In an extensive review, he listed the difficulties that zoologists and botanists had always had in deciding what was a species and what was not. Even then, it seemed that no two taxonomists could agree; what were three species to one taxonomist were three varieties to another, subspecies to a third, and something else to a fourth. While there is generally little difficulty in determining whether two individuals collected from the same locality belong to the same or to different species, how about similar individuals collected from different localities? Darwin argued that there was no basic difference between "varieties" and "species," except in degree (varieties were more similar), and that no one could say in every case which groups were more similar than others. German shepherds and Pekingese are varieties of dogs and wolves are often considered to be another species, but the German shepherd and the wolf are clearly more similar than the shepherd and the Pekingese.

The competitive exclusion principle explains why two varieties of a natural species cannot coexist in one locality but can coexist in separate localities. In one locality, two varieties would merge through interbreeding or one would outcompete the other. Thus, in any one place there is a sharp discontinuity between each species and the next, not by definition but because intermediate types are eliminated. However, if two groups are separated, they can remain only slightly different with no problems at all. Darwin also argued that, in fact, all geographically separated populations will diverge, principally by natural selection (adapting to different environments) but also by chance (randomly changing in different directions). His point was that variety, race, subspecies, species, genus, family, and so on, are not clear, real divisions but merely degrees on a continuum of divergence.

Speciation, Darwin believed, was not an event at all but only part of a slow process. He said it required a geographic barrier between two or more populations to prevent mixing and interbreeding. Once a population was isolated, it would continue to change and adapt through natural selection. But each separate population would change and adapt in different ways to different environments (Figure 1.7). A slight change would create the amount of difference that scientists would distinguish as varieties. The accumulation of more change would form what we would be willing to call a subspecies. More change, and some people would call the populations different species, while others would not be so sure. More change, and everyone would agree that the two populations were different species, belonging to the same genus. After a sufficient length of time and divergence under natural selection, scientists would begin to argue about whether the two species belonged to the same genus. So, said Darwin, there is nothing very special about the species level of population divergence.

There is one criterion of a species that had already been widely applied well before Darwin's time. Everyone agreed that if individuals of the different populations could mate and produce fertile offspring, they were of the same species. If they could not, they were of a different species. This criterion is still usually accepted. But Darwin attacked even this head on. He wondered why individuals of two species could not interbreed successfully. Was it a special rule of nature? Were species special after all? Darwin thought not and said that it was just the accidental result of the accumulation of many small physiological and biochemical changes in the two diverging groups. As each group became adapted to its own conditions, the two physiologies would no longer be compatible and the mating would fail.

Darwin pointed out that there were many different ways in which individuals of two populations could fail to mate successfully. Some species simply refuse to mate, while others may mate but produce no offspring. In other cases, offspring are produced, but they are sterile (Figure 1.8). Sometimes the male of one species could mate successfully with the female of the other species, but not vice versa. Sometimes species B could mate successfully with species A and

1.7 The Abert squirrel (a) and Kaibab squirrel (b) are two distinct species that live on opposite sides of the Grand Canyon. It is believed that they were once one population that was divided as the great chasm developed. The two populations followed their own paths of evolution and now are quite different and normally unable to interbreed.

1.8 This unusual member of the cat family is a *tiglon*, produced in a cross between a female lion and a male tiger. The reciprocal would be called a *liger*. This hybrid is never seen in the wild, since African lions and Siberian tigers would be unlikely to meet. If they did, their relationship would be unpredictable and would most likely not involve mating. Tiglons are probably sterile.

species C, but A could not mate with C. Sometimes two groups that were obviously separate species—not at all closely related—could produce hybrid offspring in captivity, but were not found to do so in the wild. So at which point has speciation occurred? At no point, Darwin argued, because, again, speciation is just an arbitrary stage of a process, and the process is the divergence of isolated populations under natural selection.

The matter is still argued by evolutionary theorists and systematists today, and Darwin's position is by no means the majority viewpoint. Many, perhaps most, evolutionary theorists believe that there is something special about species, and that a species is an organic, integrated entity. There is no argument, however, about the importance of the existence of some stage at which interbreeding is no longer possible. From this stage on, two separated populations are capable of coming back together again without merging.

Why Are There Mosquitos?

The theory of evolution through natural selection was immediately attractive to biologists because of its enormous explanatory power. Familiar and ancient observations, as well as more recent observations in natural history and paleontology, were seen for the first time to make sense, and even philosophical questions were brought under a new light. The explanatory power of the theory was its strength but also, paradoxically, its weakness. It has the embarrassment of a hypothesis that has the potential to explain everything and thus is nearly impossible to falsify.

To illustrate, let's consider some of the basic observations that could be explained by evolution through natural selection:

1. The marvelous adaptations of living organisms to their environments
2. The hierarchical ordering of natural species; groups are related to groups in such a way that all creatures readily lend themselves to being ordered in a continually splitting, treelike diagram
3. The geographical distribution of species
4. The persistence of well-adapted but primitive forms of life; the varying tempo of evolutionary change as seen in the fossil record, with periods of rapid change alternating with periods of relative stability; recurrent adaptive radiations of groups
5. The relationship between humans and animals

In a philosophical sense, the purpose of life and existence was given a new meaning. Organisms do not exist for each other or for any purpose outside themselves and their immediate offspring; but each one has the purpose, or rather behaves as if it had the purpose, of survival and procreation—because any organism that lacks such objectively purposive behavior will perish without descendants. The idea is simple: today's organisms are the offspring of the best reproducers of previous generations. Of course, the idea of such selfish purposefulness seemed morally repugnant to many of Darwin's Victorian contemporaries and is equally repugnant to many people today. But it explained something that had bothered people for millennia, and that is: why such things as mosquitos, bedbugs, rattlesnakes, tapeworms, and diseases exist in the world.

Testing the Hypothesis

Even Thomas Huxley, Darwin's most pugnacious defender in the 19th century, felt that the hypothesis would remain untested and unproven until someone directly observed the experimental creation of species that would be unable to breed. Darwin did not believe such a test was possible. One of the key ingredients in his theory is time—a great deal of time—and a hypothesis cannot be legitimately tested with observations that were made before the hypoth-

Cold-blooded gopher snakes (predator) and small rodents (prey)

Warm-blooded owls (predator) and small rodents (prey)

Warm-blooded predatory dinosaurs and herbivorous dinosaurs (prey)

1.9 Because a cold-blooded gopher snake is metabolically sluggish, it needs to consume few prey and thus tends to have a reduced effect on prey species. A given population of small, rodent prey can thus support a large number of predatory snakes. The more metabolically active owl must consume several small rodents each night, so there are many fewer owls than small rodents. In the dinosaur fossil records, the predatory dinosaurs are greatly outnumbered by their prey, which indicates that the dinosaurs probably had high metabolic rates typical of warm-blooded animals. There is currently a great debate over whether dinosaurs were warm-blooded.

esis was formed, so the power of the theory to explain ancient observations and relationships did not provide a valid proof. It is not enough to have explanatory power; a hypothesis must have predictive power if it is to be tested.

So, can hypotheses about the past be tested? Yes. Predictive power means only the power to predict the outcome of observations before they are made. For instance, hypotheses and predictions in paleontology can be made and then tested by digging up bones or analyzing bones that have been previously dug up. In one example, it has recently been hypothesized that predatory dinosaurs may have been warm-blooded. If this were true, one could predict that there would be far fewer predators than prey. On the other hand, if the predators were cold-blooded, predators and prey should have occurred in more nearly equal numbers, based on energy considerations and observations on modern ecological relationships (Figure 1.9). Surveys of data from past digs—which amounted to new observations—tested this hypothesis and found that indeed the prey greatly outnumbered their dinosaur predators.

Such experimental tests can be made of past evolutionary relationships, but it is not easy. The fossil record is spotty and ambiguous, especially when it comes to tracing transitional forms. Perhaps the best large-scale test of the theory of evolution has come with the advent of protein-sequencing techniques in the last 15–20 years. It was predicted, on the basis of evolutionary theory, that species that appeared to be closely related on the basis of morphology would prove to have closely similar proteins, and that the sequences of proteins would diverge in the same way the species were presumed to have diverged. Not too surprisingly, this prediction turned out to be true—but it would not have been predicted if the theory of evolution had been wrong, so in that sense the theory had been tested. It was found that the molecular structure of even a short molecule such as cytochrome *c* (104 protein subunits) reflects the evolutionary relationships of the organisms in which it is found. Figure 1.10 gives a tree of evolutionary relationships that is based entirely on a computer analysis of this one protein.

But keep in mind that speciation and evolution are not the same thing. The theory of evolutionary origins of species by repeated branching is one thing; the theory of natural selection is another, and the latter is exceedingly difficult to test experimentally. Of course, one can simulate natural selection in the laboratory. Darwin himself observed, "To keep up a mixed stock of even such extremely close varieties as the variously coloured sweet-peas, they must be each year harvested separately, and the seed then mixed in due proportions, otherwise the weaker kinds will steadily decrease in numbers and disappear" (*Origin of Species*). An experimental test of natural selection under more or less natural conditions is described in Chapter 11.

The difficulty with the hypothesis of natural selection is not that it predicts too little, but that it explains too much. As soon as it became apparent that natural selection explained the adaptations of organisms to their environment, natural selection began to be used in a lazy way to explain every kind of biological phenomenon observed. To explain anything, it began to seem that the speculating biologist

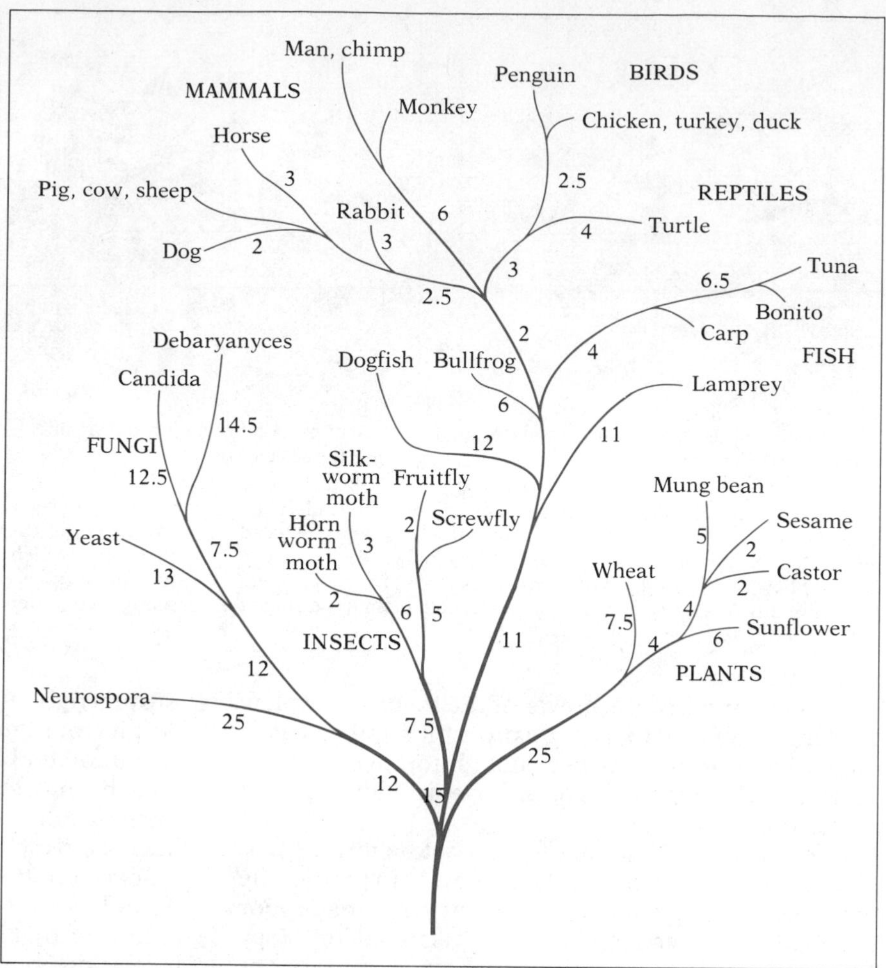

1.10 A computerized phylogenetic tree showing evolutionary relationships based on differences in a protein called cytochrome *c*. Essentially, the organisms are arranged according to differences in the protein. The numbers on the chart are a measure of those differences.

need not prove anything new, but needed merely to dream up some halfway plausible way in which the phenomenon might be of some benefit to the organism. And if the imagination failed in this, it seemed adequate to state that the benefit of the phenomenon to the individual was not obvious, but that surely there must be one. For instance, if the Indian rhinoceros has one horn and the African rhinoceros has two, it might be argued that there must be something about the two environments that makes this arrangement the best one for all concerned (Figure 1.11).

This extravagant faith in the power of natural selection leads to the benign views that everything is always for the best and there is a reason for everything. Such a viewpoint elevates natural selection to the status of a new, all-powerful deity, and such faith is inimical to science. It is certainly a contrast to Darwin's own concern with the role of chance and the constraints of reality and historicity in evolution. Natural selection is a powerful phenomenon, but it is limited. The adaptations of organisms are marvelous, but they are never perfect, just as evolutionary change is always opportunistic and never predictable.

Group Selection

It is interesting to read Darwin's discussion of the evolution of social interactions and group behavior—behavior that benefits others of the same species. Darwin carefully inserted the provision that such behavior will evolve if it also benefits the individual doing the behaving. One of the chief implications of the theory of natural selection is that adaptation is based on the effect a character (or behavior) has on the fitness of the individual or the individual's offspring, and not on its effect on the species as a whole. Darwin was emphatic on this point in his later writings. But the point is often lost, and it is not uncommon to read that certain adaptations have evolved "for the good of the species." In almost all cases, a careful investigation shows that the character or behavior that appears to be for the good of the species is, in fact, also beneficial to the individual and presumably has evolved on this latter basis. Even sterile social insects such as worker bees, which have a fitness of zero by definition, work not for the good of the species but for the survival and procreation of their fertile mothers, sisters, and brothers, who carry

1.11 Why does the African rhinoceros have two horns and the Indian rhinoceros have only one? Remember that any explanation would have to be testable. What explanations come to mind? Can they be tested? Is your hypothesis a tautology? Is it possible that there is no selective advantage of one horn over two? Or perhaps there is no longer a selective advantage. You won't find many biologists willing to say much about questions like this.

the workers' only hope of leaving their kinds of descendants. We will have more to say about all this later.

Reducing and Synthesizing

Darwin went on to publish other major theoretical works and continued his simple but first-rate experimentation. For example, he discovered plant hormones, as we'll see in Chapter 20. *The Expression of the Emotions in Man and Animals* (1872) was the foundation of the modern sciences of ethology and comparative animal behavior. *The Formation of Vegetable Mould through the Action of Worms* (1881) established the importance of earthworms in soil ecology. Because of his work on orchids, climbing plants, and insectivorous plants, modern botanists claim Darwin as a fellow botanist.

But it is on the *Origin of Species* and a related work, *The Descent of Man,* that Darwin's reputation is based, and it is the idea of natural selection that has become the central concept of the science of biology. You could go so far as to say that nothing in the rest of this book can be properly understood without an understanding of natural selection.

Of course, Darwin's ideas were immediately and bitterly controversial. And they continue to resist resolution. Unlike the once-controversial ideas of Newton and Einstein, his writings remain controversial. Almost all biologists believe in the central notions of natural selection and evolution, but, as we have seen, they still differ greatly among themselves on such questions as the meaning of species, the process of speciation, why different species can't mate, and whether most evolutionary change comes in small, continuous increments or in larger and less regular leaps. Darwin, through his works, remains an active participant in these controversies.

But Darwin's methods of scientific investigation are not the only methods. His view was wide-ranging and global. His gifts were common sense pushed to the extreme and an unwillingness to accept easy or conventional answers.

In truth, most of the scientific progress in biology in this century has not been through such grand conceptual breakthroughs, but through a different tradition in science, *reductionism.* The reductionist tradition is based on the controlled experiment, an experiment in which only one variable at a time is allowed to change and all others are kept rigorously constant. The data from such experiments tends to be in the form of numbers, often plotted in graphical form. Cause and effect are determined, whenever possible, by eliminating all competing explanations until one is left. The reductionist tradition looks for small answers—the smaller the better—to small questions, and seeks to determine mechanisms (not reasons) for observed phenomena. It is the tradition of physics and chemistry applied to biology. We see its most successful biological applications in biochemistry and molecular biology.

Of course, reductionism does not define modern science. There are also *synthesists* stalking the halls. Synthesists seek underlying order in new ways. They show that a number of seemingly unrelated observations are related after all. Synthesists are the emotional descendants of Darwin—clearly interested in forming grand rules and always looking at the big picture. For everything to work well, the reductionist approach and the synthesist approach should be interrelated. The synthesist scientist, after all, uses the accumulated observations and mechanisms that the reductionist scientist has produced to give new explanations of order and function. The synthesist's vision can then be validated only through additional experimentation by the reductionist. A recent example of this interaction, in the area of biochemistry and cellular physiology, is the chemiosmotic theory of Peter Mitchell (see Chapter 5). Mitchell used the known pieces to construct a larger, and previously unimagined, system that explained many previously puzzling questions about the synthesis of high energy compounds in cells.

Some scientific questions are not subject to rigidly controlled experiments, at least not rigidly controlled by the criterion of careful laboratory work. There are still scientists who must work with whole plants and animals, or with ecosystems and populations. There is no way, for instance, that a baboon can be brought into a laboratory so scientists can learn how it behaves in the wild; the behavior of wild baboons can be understood only by studies on wild baboons. Now, it may be necessary to catch a wild baboon and take some of its blood for a laboratory assay of its hormone levels, since hormones influence baboon behavior. An experimenter might even want to inject something into the baboon just to see what happens next. But in any case, a hypothesis must be formulated and some way of testing it must be devised.

Darwin, as we know, was much more of a synthesist than a reductionist. He brought together his own observations and those of hundreds of others to build and test a grand theory. There were great reductionist scientists among his younger contemporaries, however. Within Darwin's lifetime, Mendel unraveled the mathematical principles of inheritance in higher organisms by discovering the gene; and Meischner extracted and purified DNA, the very stuff of genetic material.

1.12 Thomas Huxley carried Darwin's message into the public arena, a process which Darwin himself found highly distasteful and perhaps fearsome. Huxley, an awesome debator and orator, was in his element and soon earned the title of "Darwin's Bulldog." Most famous of the great debates was that between Huxley and Bishop Samuel Wilberforce in 1860, which Huxley won by popular acclaim. At one point when Wilberforce turned and asked Huxley on which side of his family he was descended from the apes, Huxley muttered, "The Lord hath delivered him into my hands." When he took the podium he replied that he would rather be descended from an ape than be related to a man who would not use his God-given gifts of reason.

We'll close this chapter with a very famous quote by the brilliant debater Thomas Huxley, Darwin's contemporary and perhaps his best-known supporter (Figure 1.12). When the *Times* of London sent him a copy of the *Origin of Species* to review, he is said to have exclaimed of himself, "How extremely stupid not to have thought of that!"

Application of Ideas

1. In explaining scientific and other intellectual achievement, someone once said, "Chance favors the prepared mind." What does this mean to you, and how do you define "prepared"? As you answer, think of the background and experiences of Charles Darwin.

2. Develop an organizational diagram that illustrates how science might work. Include each of the intellectual and technological aspects of science and clearly distinguish these.

3. Distinguish between the idea of evolution and the idea of natural selection; between natural selection and adaptation.

4. One important aspect of the progress of scientific theory is the introduction of new technology: with inventions and new techniques, it is possible to test hypotheses that could not have been tested before. What new techniques or inventions, if any, might have played a role in the theoretical syntheses of Darwin and Wallace?

5. An antievolutionist derides the *Origin of Species,* saying that speciation is an "unscientific" idea, since it is a hypothetical event that has never been observed. How would you argue against such an assertion?

6. Are humans in modern society subject to natural selection, or have we managed to thwart the process as it occurs in other populations of organisms? How might humans living in advanced nations interfere with natural selection? Are new selective forces at work under these conditions?

Key Words

adaptation
artificial selection

biogeography
blending theory
botanist

character
common descent
control
controlled experiment

ecology (ecologist)
equilibrium
ethology
evolution
extinction

fossil

gene
geology

hybrid
hypothesis

insectivore
interbreed

mammary gland
mechanism
mutation

natural selection
niche

organism

paleontology
placenta
placental mammal
population
predator
prey

reductionism (reductionist)

scientific method
selection
speciation
species
statistics
synthesism (synthesist)
systematics

tautology
taxonomist
theory

Key Ideas

An Explanation About Rabbits and Oceans

1. The concept of *evolution* began to emerge in Darwin's thinking as he studied the animals of South America.
2. Darwin was able to find puzzling examples of *fossil* mammals in South America that had living descendants in Central America. He made guesses about the Isthmus of Panama having been of recent origin.
3. Lyell's studies of geological change gave Darwin both the geological framework and the time periods he needed to reinforce his growing concept of *evolution.*
4. Darwin's own thinking underwent a revolution as he began looking for alternatives to special creation.
5. Darwin's comparisons of the Galapagos plants and animals with those on the mainland added more fuel to his emerging ideas.

Darwin's Finches and Speciation

1. Thirteen *species* of finches inhabit the Galapagos. Some are ground foragers, others feed in trees, and one is a tool-user. All differ from mainland *species.*
2. Darwin's conclusion was that the finches descended from one mainland group.
3. One of the requirements for *speciation* to occur is isolation.
4. When newly diverged *species* again occupy the same range one will die out because of similar *ecological* requirements, unless sufficient evolutionary changes have occurred.

How to Go Against the Grain

1. As his *hypothesis* emerged, Darwin was aware that he could not yet suggest a *mechanism,* or driving force, to complete the idea.
2. The *scientific method,* although not entirely defined, begins with observations. These are followed by questions that ask, "Why?". A period of speculation may follow, which should lead to the formulation of a testable *hypothesis.* From *hypotheses* come predictions and then experimentation or simply more observation to test the predictions.
3. The problem of chance occurrences being responsible for experimental results has led to the application of *statistics.*

Going Home to Applause and Illness

1. Darwin's observations and collections were sent home far ahead of him. As a result, his fame as a collector and observer preceded his arrival.
2. As he reviewed his notes and collections, Darwin began to formulate his *hypothesis* on *natural selection.*
3. His knowledge of *artificial selection* was vitally important in arriving at the idea of *natural selection.*
4. The essays of Malthus, an economist, led him to arrive at competition and survival of the fittest as a part of *natural selection.*
5. *Natural selection* differs from *artificial selection* in that the rigors of the environment rather than man does the selecting.
6. Darwin turned to experimentation as he sought evidence for *natural selection.* He studied seed survival under varying conditions in an effort to understand island colonizations.
7. Darwin's decision to publish his thesis was hastened by the arrival of Alfred Wallace's manuscript in which the same conclusions about *natural selection* were reached.
8. Many of Darwin's observations and conclusions have become part of modern *ecology.*

The Theory of Natural Selection

A Mechanism of Natural Selection

1. *Natural selection* as a *mechanism* has several requirements:
   a. Variations in populations must arise spontaneously and randomly and must be inheritable.

b. Competition must exist within a *species.*
c. Any variation that decreases an individual's competitive ability will affect its reproductive output.
d. Conversely, a variation that is advantageous will increase an individual's reproductive output, thus increasing the number of offspring with that variation.
e. Random *mutations* (changes) will most likely be favorable when the physical or biological environment changes.

2. Darwin did not understand the source of variation, nor did he have a good grasp of elementary genetics.

How Species Are Formed

1. Explaining *speciation* was difficult for Darwin because of the accepted idea that *species* were fixed. In addition, he had the same problems that we have today in defining *species* and he preferred to deny the existence of species as such.
2. An older definition of *species,* which is often used today, is that two organisms are of the same *species* if they can successfully breed and produce fertile offspring.
3. Darwin believed that *speciation* was just an arbitrary stage in the process of *evolution.*

Why Are There Mosquitos?

A danger in the theory of *evolution* through *natural selection* is its very broad explanatory power. It does, however, offer an approach to many previously unexplained phenomena.

Testing the Hypothesis

1. It is possible to test the *natural selection hypothesis* because it has predictive power.
2. More recently, evolutionary *theory* has been tested and supported by the comparative study of amino acid sequences in proteins of different organisms.
3. Taxonomic schemes can be based upon these determinations.

Reducing and Synthesizing

1. *Reductionists* typically carry out controlled experiments, studying the details of a process.
2. *Synthesists* look for grander schemes and answers, utilizing the accumulated observations of the *reductionist.*
3. Although Darwin often worked as a *reductionist,* he is best known as a *synthesist* for his *theory* of *evolution* through *natural selection.*

Review Questions

1. What was the common explanation of varying but similar *species* of plants and animals in pre-Darwinian time? (p. 4)
2. How did Darwin explain the uniqueness of South American animals? (p. 4)
3. List some key ideas provided by Charles Lyell that helped Darwin formulate his *hypothesis* of *evolution.* (pp. 4–5)
4. Summarize the *evolutionary* lesson provided by the Galapagos finches. (p. 7)
5. Outline the steps commonly used by scientists in what is called "the *scientific method.*" (p. 8)
6. What is a *hypothesis?* What is the characteristic of a good *hypothesis?* (p. 8)
7. What was the "*mechanism*" Darwin groped for as he formulated his *theory* of *evolution?* (p. 10)
8. In the years immediately following his return from the *Beagle* voyage, Darwin became a successful scientist. What aspect of the *scientific method* had he achieved? What aspects had he yet to develop? (pp. 10–11)
9. What is *artificial selection?* How does this idea differ from *natural selection?* Give examples. (pp. 10–11)
10. Write a brief summary of Darwin's thinking on *natural selection.* What was it and how did it work? (pp. 13–15)
11. How does *speciation* occur? (p. 17)
12. What general rule about *species* is still used today in differentiating one *species* from another? (p. 17)
13. List the five observations that can be explained by applying the concept of *evolution* through natural selection. (p. 18)
14. Is *evolution* a testable *hypothesis?* Give an example with your answer. (pp. 18–19)

Chapter 2 Small Molecules

In my hunt for the secret of life, I started research in histology. Unsatisfied by the information that cellular morphology could give me about life, I turned to physiology. Finding physiology too complex I took up pharmacology. Still finding the situation too complicated I turned to bacteriology. But bacteria were even too complex, so I descended to the molecular level, studying chemistry and physical chemistry. After twenty years' work, I was led to conclude that to understand life we have to descend to the electronic level, and to the world of wave mechanics. But electrons are just electrons, and have no life at all. Evidently on the way I lost life; it had run out between my fingers.

ALBERT SZENT-GYÖRGI
Personal Reminiscences

Szent-Györgi's wry comment on his search for the secret of life makes some important points that may be humbling to a species that prides itself on its great intellect. He has described the reductionist's nightmare. As he notes, the "meaning of life" will never be grasped as a pure, crystalline gem of truth—that would be too simple! Life's many meanings lie somewhere in the very complexity that Szent-Györgi tried to discard. But at the same time, the processes of life depend ultimately on the behavior of lifeless molecules, atoms, and electrons spinning mindlessly in space. Thus, in order to make sense of the many things we do know about life, we must know about its components. So we find ourselves immersed, from time to time, in biochemistry.

Of course, biology can be "done" without biochemistry. Darwin, for example, had very little knowledge of the subject, and even now, the woods are full of biologists with a hatred of beakers, bases, and balanced equations but a love of the outdoors. However, perhaps their love would even increase if they had a greater appreciaiion for the complexities of photosynthesis. It cannot detract from the beauty of a delicately veined leaf to know that it can make food from carbon dioxide and water.

It may be true that some biochemists must periodically be convinced of the existence of the platypus, but only a biochemist can tell us how a fat little hummingbird is able to fly nonstop across the Gulf of Mexico. So in this chapter we'll be dealing with the biochemist's world. We will begin with some basics about atoms, inorganic molecules, and small organic molecules. At some point we'll traverse the lifeless world of chemicals to begin discussing life processes—but we're not sure where that line is.

2.1 Basic Chemical Structure and Bonding

Elements, Atoms, and Molecules

There are ninety-two naturally occurring *elements* in the universe. *An element is a substance that cannot be separated into simpler substances by purely chemical means.* Of these ninety-two elements, only six—sulfur, phosphorus, oxygen, nitrogen, carbon, and hydrogen—make up 99% of living matter. People have been known to use the word SPONCH (from the initial letters of their names) to remember these six elements. Although most of living matter is made up of SPONCH, life depends on other elements also. For instance, there could be no photosynthesis without magnesium; and while a pine tree can live without sodium, you cannot (Figure 2.1), because sodium is necessary to the functioning of your nerves and muscles. Table 2.1 lists the six SPONCH elements plus sixteen other elements that have significant roles in biology, along with some examples of their biological roles. Most of the remaining seventy natural elements are rare and not of much interest to us as biologists, although some of the heavy metal elements (gold, silver, osmium, and mercury) are used as laboratory and photographic agents.

An atom is the smallest indivisible unit of an element. To be more precise, the atom itself can be

2.1 While plants have no known need for sodium, it is a highly important element for animal life.

divided into smaller parts, as we'll see shortly; but if it is, the atom loses its special properties. In other words, it no longer constitutes the same element, because the properties of an element derive from the properties of the atoms that make it up.

Two or more atoms can be joined together into a discrete unit called a *molecule.* The atoms of a molecule can be of the same kind or of different kinds; that is, they can be the same or different elements. A molecule of oxygen consists of two oxygen atoms bound tightly together to form molecular oxygen, which is symbolized by the chemical formula O_2. The oxygen of the air is molecular oxygen. Hydrogen gas consists of molecular hydrogen, H_2. And you are undoubtedly aware that two atoms of hydrogen and one atom of oxygen combine to form one molecule of water, H_2O. However, chemistry gets a bit more difficult from there. Water, by the way, is a *compound. A compound is any pure molecular substance in which each molecule contains atoms of two or more different elements in specific proportions.*

Atomic Structure

Atoms are made up of subatomic particles. Something like a hundred distinct kinds of subatomic particles have been described, but most of them are short-lived and play no known role in biology. As a biologist, for instance, you don't really need to know that *hadrons* are made up of *quarks,* or that quarks come in four *flavors* (up, down, sideways, and strange by some accounts, or up, down, strangeness, and charm by others; or chocolate, vanilla, strawberry, and butter pecan if you prefer) and three *colors* (unnamed). Biology has its own color, charm, and strangeness.

The three stable subatomic particles that make up atoms are *neutrons, protons,* and *electrons,* and these alone concern us. Protons and neutrons are about equal to each other in mass, and are much heavier than electrons. Thus, protons and neutrons make up most of the mass of the universe. Protons have a positive (+) electrostatic charge and electrons have an equally strong negative (−) electrostatic charge; neutrons, on the other hand, have no electrostatic charge (they are neutral). Unlike charges attract and like charges repel, so protons and electrons are electrostatically attracted to one another.

Another kind of force, called the *strong nuclear force,* holds clusters of neutrons and protons together. An atom consists of an extremely small, incredibly dense cluster of neutrons and protons surrounded by orbiting electrons. The small, dense cluster of neutrons and protons is called the atomic *nucleus.* There are relatively small numbers of protons and neutrons in any atomic nucleus. The smallest atomic nucleus—that of the hydrogen atom —consists of a single proton and no neutrons. The largest naturally occurring atomic nucleus—that of uranium-238—consists of 92 protons and 146 neutrons. The number of neutrons plus protons determines the *atomic mass* (or *atomic weight*) of an atom. Atomic mass is measured in *daltons,* and an atom of hydrogen weighs 1 dalton. An atom of uranium-238 weighs just about 238 times as much as an atom of hydrogen, so it weighs 238 daltons (Figure 2.2). The *molecular mass* (or *molecular weight*) of a molecule is the sum of the atomic masses of the atoms that make up the molecule.

The number of protons in the nucleus determines the chemical properties of the atom, and hence determines what element the atom will be. The number of protons in an atom is called the *atomic number,* and so each element has its own atomic number. For instance, the atomic numbers of the six SPONCH elements are 16, 15, 8, 7, 6, and 1, respectively. (Notice that the acronym SPONCH lists the six elements in order of decreasing atomic number.)

2.2 The lightest and heaviest atomic nuclei. Uranium-238 has a massive nucleus when compared to that of hydrogen. The nucleus of uranium has 92 protons (+) and 146 neutrons. Hydrogen has just one proton and no neutrons (in its most common form). This tells us that the atomic mass of uranium is 238 times that of hydrogen (approximately).

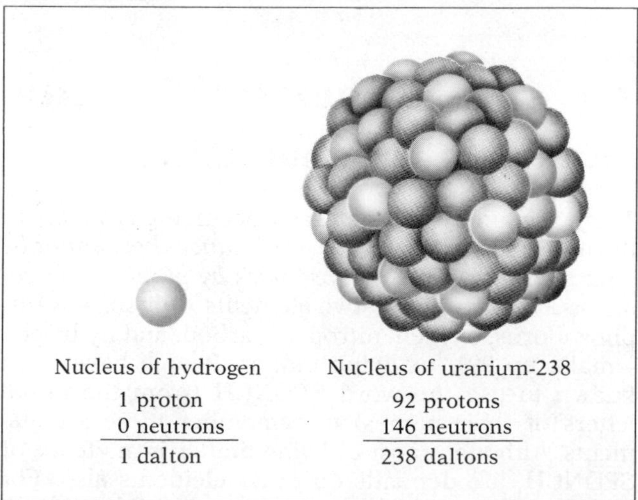

Table 2.1 Elements essential to the processes of life

Element	Symbol	Atomic number	Atomic mass	Example of role in biology
Calcium	Ca	20	40.1	Bone; muscle contraction
Carbon	C	6	12.0	Constituent (backbone) of organic molecules
Chlorine	Cl	17	35.5	HCl in digestion and photosynthesis
Cobalt	Co	27	58.9	Part of vitamin B_{12}
Copper	Cu	29	63.5	Part of oxygen-carrying pigment of mollusk blood
Fluorine	F	9	19.0	Necessary for normal tooth enamel development
Hydrogen	H	1	1.0	Part of water and of all organic molecules
Iodine	I	53	126.9	Part of thyroxine (a hormone)
Iron	Fe	26	55.8	Hemoglobin, oxygen-carrying pigment of many animals
Magnesium	Mg	12	24.3	Part of chlorophyll, the photosynthetic pigment; essential to some enzyme action
Manganese	Mn	25	54.9	Essential to some enzyme action
Molybdenum	Mo	42	95.9	Essential to some enzyme action
Nitrogen	N	7	14.0	Constituent of all proteins and nucleic acids
Oxygen	O	8	16.0	Molecular oxygen in respiration; constituent of water and nearly all organic molecules
Phosphorus	P	15	31.0	High energy bond of ATP
Potassium	K	19	39.1	Generation of nerve impulses
Selenium	Se	34	79.0	Essential to the workings of many enzymes
Silicon	Si	14	28.1	Diatom shells; glass sponge exoskeleton; grass leaves
Sodium	Na	11	23.0	Salt balance; nerve conduction
Sulfur	S	16	32.1	Constituent of most proteins
Vanadium	V	23	50.9	Oxygen transport in tunicates
Zinc	Zn	30	65.4	Essential to the workings of the alcohol oxidizing enzyme

Whereas the number of protons is fixed for any element, the number of neutrons is more free to vary—within limits. This means that different atoms of the same element may have different atomic masses. For instance, uranium-235 has only 143 neutrons instead of 146, so it is 3 daltons lighter than uranium-238. Atoms of the same element that have different atomic masses are called *isotopes* of the element.

Some combinations of neutrons and protons are stable, but others are internally unstable and tend to break down spontaneously, with the release of various subatomic particles and radiation. These unstable atoms are called *radioactive isotopes.* Although radioactive isotopes have no significant role for natural living things, they are extremely useful tools in the study of biology in the laboratory, as we'll see in later chapters. And radioactive isotopes have other applications as well. For instance, uranium-235, which is much rarer than uranium-238, is less stable internally and thus is highly prized by government officials for its applications in the manufacture of atomic bombs and in nuclear reactors.

Electron Orbitals and Electron Shells

Because of the attraction between protons and electrons, atoms tend to have the same number of electrons as protons. We have to use the weasel words "tend to" here, because there are many occasions when the numbers of protons and electrons of an atom don't exactly balance, as we'll see soon enough. But in the electrostatically balanced atom, the number of electrons is equal to the atomic number.

Electrons move, or occur, in definite regions or paths outside the atomic nucleus. The paths are called *orbitals* because of the way they were once depicted in drawings, namely as flat circles moving around the nucleus much as the planets move in planetary orbitals around the sun. There are many different ways of drawing electron orbitals, but none of them is really correct. For instance, one common way of drawing a picture of an atom is to show the electron orbitals as "probability clouds," depicted as shaded areas in which the density of the shading is supposed to be proportional to the probability that an electron will be exactly at that place at any one point in time. Unfortunately, such diagrams can be misleading, because, as we shall see, the shapes of the electron orbitals change when atoms join to form molecules.

Figure 2.3 shows the electron density clouds of helium, which has an atomic number of 2 and two electrons, and of neon, which has an atomic number of 10 and ten electrons. The two electrons of helium do not travel in flat circles, but can be found anywhere in a small, spherical orbital, fairly close to the nucleus. Two of the ten electrons of neon are also found in the same kind of small, spherical orbital, close to the nucleus, but the other eight electrons are found in four odd, dumbbell-shaped orbitals that stick out at right angles to one another. Note that the electrons travel in pairs, so eight electrons are found in the four orbitals—one spherical and three dumbbell-shaped. We'll soon see just how critical this arrangement is.

We'll use another common way of drawing an atom that conveys the maximum amount of useful information and still allows us to show the fine structure of simple molecules involving several atoms. The atoms are drawn as flat, two-dimensional structures, although, of course, real atoms exist in three dimensions. The nuclei are shown as small circles labeled with the number of protons and neutrons. The electron shells are depicted as concentric rings, which actually represent energy levels and are not to be confused with orbital paths (Figure 2.4). Although the pairs of electrons in the second electron shell of atoms in real molecules are actually found moving in four oddly shaped orbitals that together form a complex three-dimensional form (see Figure 2.3), in the flat diagrams they are shown as static, paired black dots placed at the top, bottom, left, and right sides of the circle. Figure 2.4 shows diagrams of helium and neon atoms; compare these drawings with the electron cloud diagrams in Figure 2.3.

Electron Pairing

Each of the five orbitals of neon contains two electrons. This is because electrons prefer to travel around in pairs. By "prefer" we mean that electrons are usually found in pairs, and that it requires a considerable amount of energy to pry apart two

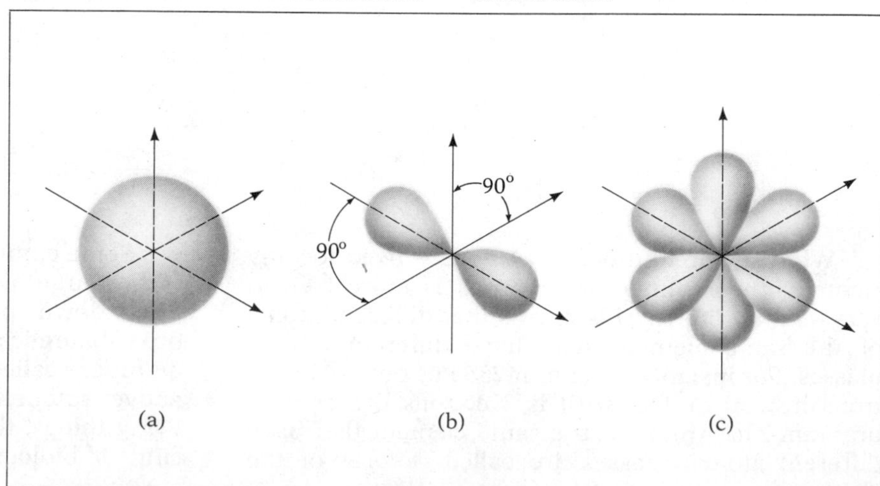

2.3 One way of depicting electron orbitals is with electron density clouds. The dots represent the relative probability that an electron will be at a certain place at any given moment.

The two electrons of helium occupy a single, spherical cloud (a), which constitutes the innermost electron shell. All heavier elements also have just two electrons in this innermost shell. For instance, neon with its ten electrons has, in addition to the innermost shell, four pairs of electrons in a second, outer shell. Each pair of electrons occupies a specific orbital. The first two electrons of neon's outer shell occupy another, somewhat larger sphere, much like that shown in (a). Each of the remaining three pairs occupies a strange, dumbbell-shaped orbital. A schematic, isolated dumbbell-shaped orbital is shown in (b). Either of its two electrons may be found anywhere within the dumbbell at any moment. The three dumbbell-shaped orbitals of neon's outer shell are arranged at right angles to each other (c). All elements heavier than neon have ten electrons in the same orbitals in the same first and second shells, but carry more electrons in yet larger electron shells.

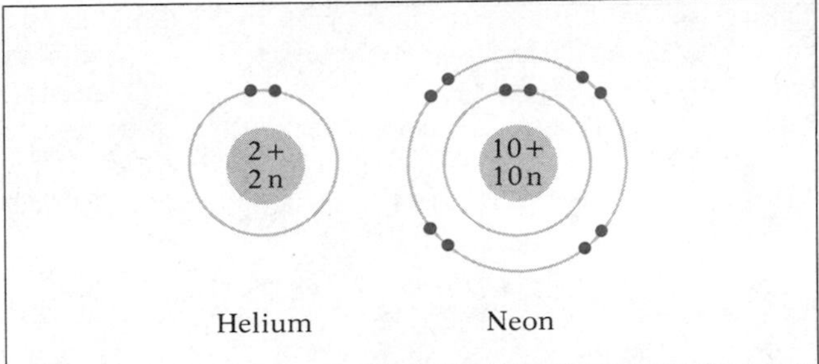

2.4 Representing atoms as flat concentric circles of electrons surrounding the atomic nucleus is not consistent with what we know about atomic structure, but it is sometimes useful to convey other types of information. Note that the electrons in each orbital have been depicted as pairs and that the pairs are separated from each other as far as possible. The concept of electron pairs is important in explaining how molecules are formed, which is one of the reasons for depicting the atoms in this manner.

electrons that are sharing an orbital. Correspondingly, two impaired electrons will release a considerable amount of energy if they are allowed to form a pair. The tendency of electrons to form pairs is one of the important forces in atomic and molecular structure and in chemical interactions.

Physicists say that the two electrons of a pair have "opposite spins." They then quickly add that they don't mean that electrons spin like tops but that they have an otherwise unexplainable property *called* spin. Similarly, each electron is said to act like a tiny bar magnet, with a north pole and a south pole. One way of explaining the energetic pairing of electrons is to picture one electron upside down relative to the other, north pole to south pole. Physicists will admit that this is also a kind of metaphor, but then they'll insist that people have to use metaphors, analogies, and peculiar diagrams in order to have any understanding of subatomic structure at all.

Filled and Unfilled Shells

We've noted the *strong nuclear force,* which is important in atomic physics and in the stability of isotopes. The strong nuclear force is not considered to be a *chemical* force, however. There are three important forces, or *energetic tendencies,* that determine most of what happens in chemistry. Conflicts between these forces are responsible for most of the properties of chemical interactions. One important characteristic of atoms is *electrostatic attraction* and *repulsion,* which is involved in the fairly strong tendency of atoms and molecules to balance positive and negative charges. A second important energetic characteristic is the tendency of orbiting electrons to travel in pairs. Now we'll consider a third energetic tendency that is probably the most important of all these competing forces when it comes to chemical reactions and molecular structure. This is the *tendency of atoms to establish completely full outer electron shells.* What do we mean by "completely full outer electron shells"? Here, we have to resort to a dogmatic description, sometimes called the "octet rule." The innermost electron shell has room for just two electrons—one electron pair. The next electron shell is filled when it has eight electrons—four electron pairs. In somewhat larger atoms the third shell, like the second, has room for four pairs of electrons. So counting outward, the first three shells are filled when they have two, eight, and eight electrons, respectively. In this chapter we'll be concerned only with those first three electron shells.

Smaller atoms, like those of SPONCH, always fill their electron shells progressively, completing the first before beginning the second, and then filling the second before beginning the third (Figure 2.5). Thus, if a SPONCH atom does not have enough electrons to fill all its electron shells, it is the *outer shell* that is partially unfilled. But, in fact, the atom

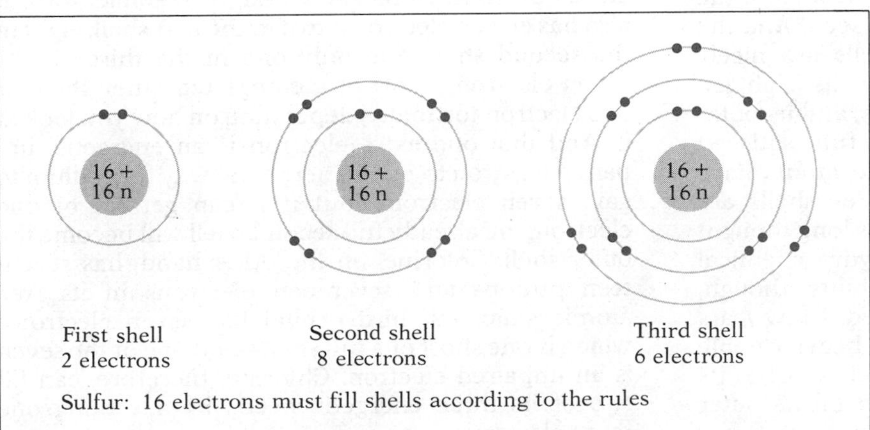

2.5 The electron shell model of sulfur is shown filling its shells according to the rules. Note the positioning of the electron pairs and also take note of the number of electrons in each shell. Atoms fill their shells from the inside out, and the maximum number of electrons in each shell (in the lighter elements) is 2, 8, 8. Only the outer shell remains unfilled, making it a good place for leftovers. This unfilled outer shell is what makes the elements interesting. You might say that chemistry is really the study of outer-shell electrons.

usually finds a way to fill that shell—*and that is what makes chemical bonds and allows chemical reactions.*

Obviously, the different energetic tendencies, or needs, of the atom are sometimes in conflict. As we have seen, one tendency is to balance proton and electron charges. But when the protons and electrons are balanced, the outer shell is frequently left unfilled. For example, oxygen has eight protons and thus is electrostatically balanced when it has eight electrons in its shells. But after the first shell is filled, there are only six electrons left to fill the second, or outer, shell, where eight is the stable number. Therefore, the outer shell of an electrostatically balanced oxygen atom is short two electrons. But if the oxygen atom were to fill its outer shell with electrons, it would have ten electrons and only eight protons, and would thus have lost the balance between protons and electrons. How are these competing energetic tendencies accommodated? We'll discuss this after we've looked more closely at a simpler class of atoms, the noble elements.

The Noble Elements

A few elements have no problems maintaining both balanced charges and full orbitals. For example, helium has two protons and two electrons, so it is electrostatically balanced. Its two electrons are paired; and its first electron shell, which is also its outer shell, is filled; so that requirement is satisfied too. As a consequence, helium doesn't form molecules with other atoms and isn't involved in chemical reactions, because chemical reactions and molecular bonds are the result of atoms seeking to accommodate their conflicting "needs" or energetic tendencies. Helium, with no conflicting needs, would have nothing to gain from interacting with another element or molecule. It seems that with atoms, as with people, the active ones are usually somehow unsatisfied.

Similarly, neon has ten electrons balancing the ten protons of its nucleus, as we have seen. And the first and second (outer) electron shells are nicely filled. Argon (atomic number 18) has its eighteen protons balanced by eighteen electrons, and its outer shell, the third shell in this case, is also fully satisfied (Figure 2.6). Elements in which single atoms have both balanced charges and filled outer shells are known as *noble elements* because it was long thought that they were above engaging in tawdry chemical reactions. (As often happens with nobility, though, some are more active than was suspected. A few years ago it was discovered that some of the heavier noble elements can react with fluorine, which satisfies its own energetically very strong need to fill its outer

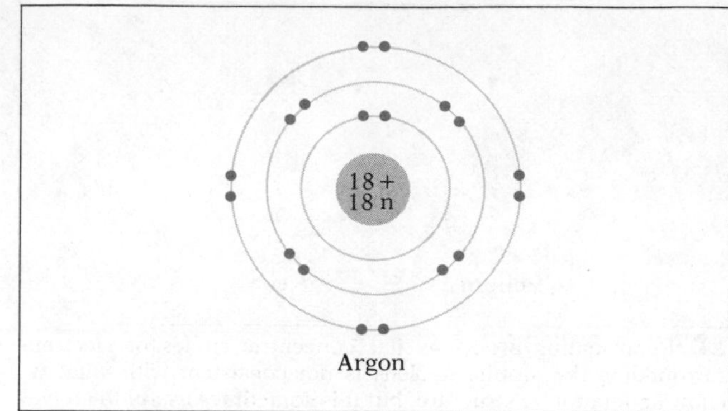

2.6 Electron shell diagram of argon, showing filled first, second, and third (outer) shells. This element is very stable because it has no unfilled (or unsatisfied) shell.

shell by forming bonds with electrons from the outer orbits of the heavy noble elements xenon or krypton, with atomic numbers of 54 and 36, respectively.)

How an Atom Can Fill Its Outer Electron Shell

An atom can fill its outer shell in one of only three ways: It can gain an electron from another atom, filling up the hole in its outer shell; it can lose one or two electrons to another atom, stripping its original outer shell bare and leaving a new, full outer shell on a lower energy level; or it can share one or more electron pairs with another atom. In all these cases, two or more atoms are involved, and thus these are all *chemical reactions.*

Ions and the Ionic Bond

Let's consider some examples. Sodium (Na) and chlorine (Cl) have complementary needs. Sodium has eleven protons. In its elemental, or metallic, form it also has eleven electrons: two in the first shell, eight in the second shell, and only one in the third shell—seven electrons short of a completed outer shell, or one electron too many, depending on how you look at it. And that one extra electron is an energetic, unpaired electron to boot. There is no way for sodium to gain seven electrons, but if it can get rid of one electron, the already full second shell will become the outer shell. Chlorine, on the other hand, has seventeen protons and seventeen electrons in its free atomic state. Its third orbital has seven electrons, which is one short of a full shell—and one of the seven is an unpaired electron. Chlorine, therefore, can fill two of its three energetic needs by accepting one more electron.

Because of their special properties, atomic sodium and atomic chlorine react together very swiftly, as an electron is passed from one to the other. Sodium is the *electron donor* and chlorine is the *electron acceptor.* Each element, in its pure state, is a deadly chemical, and when placed together they react explosively—but then the products of the reaction cool down to form table salt, NaCl (Figure 2.7).

The sodium now has only ten electrons to balance its eleven protons, so it takes on a net positive charge of +1. Chlorine now has eighteen electrons as compared to its seventeen protons, so it takes on a net negative charge of −1. The opposite charges attract, and sodium joins chlorine to form sodium chloride. Note that the name of the negatively charged atom has changed to chloride (although sodium is sodium, charged or not). Any charged atom or molecule, such as sodium (Na^+) or chloride (Cl^-), is an *ion.* In fact, we define an ion as any charged atom or molecule.

2.7 Sodium, at the left, has one energetic outer-shell electron, while chlorine, at the right, has seven. The outer shells of both atoms can be filled if sodium loses its lone electron and chlorine gains it. Of course, when this occurs, sodium will have a net positive charge, while chlorine will now be negative. The sodium and chlorine are then drawn together since negatively and positively charged atoms, or *ions* as they are better known, attract each other, forming what is known as an ionic bond.

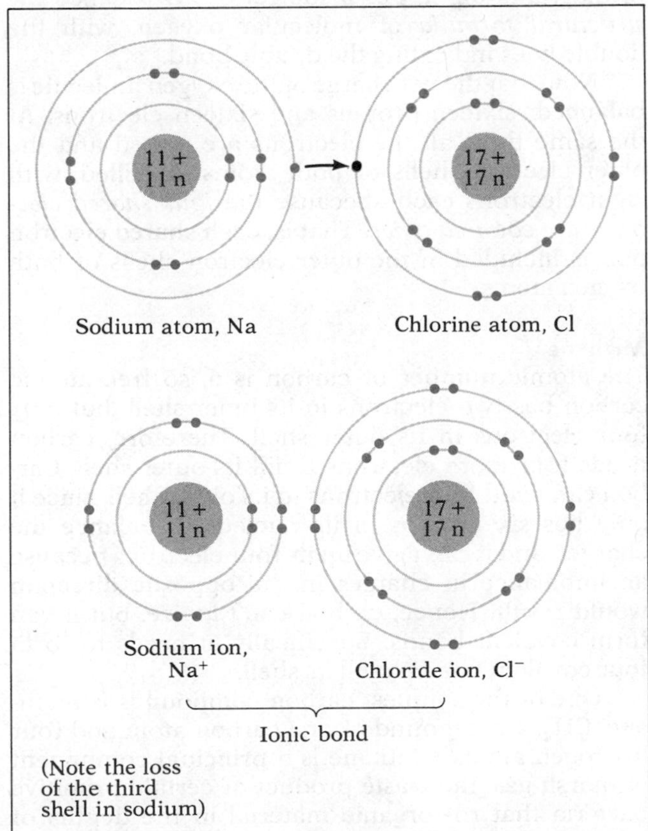

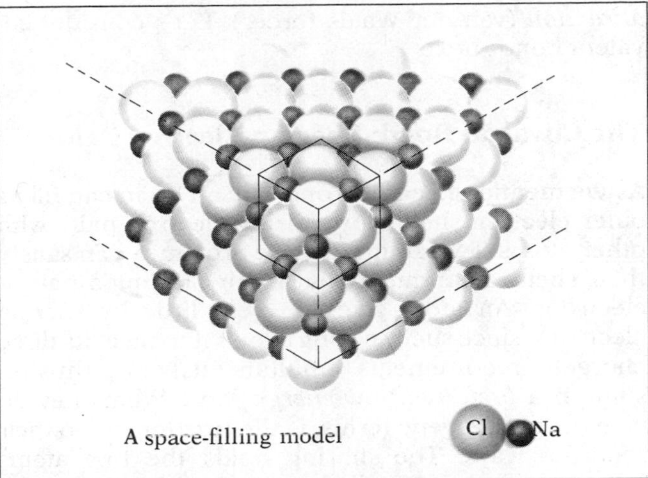

2.8 The sodium chloride crystal. Sodium and chlorine form ionic bonds, but true discrete molecules do not form. Because of their electrostatic charges, the sodium and chloride ions attract each other, accumulating into crystalline formations that consist of alternating sodium and chloride ions. Crystals are of indefinite size and can vary from invisible to enormous. Regardless of the size, the geometry of a salt crystal is definite, as shown here.

The electrostatic attraction between the positively charged ion and the negatively charged ion constitutes one kind of *chemical bond.* A bond is any force that tends to hold atoms together, and there are several kinds of chemical bonds. The bond in sodium chloride is called an *ionic bond* or *salt linkage,* and is defined as a bond formed by electrostatic attraction after the complete transfer of an electron from the donor to the acceptor.

The ionic bond, it turns out, is fairly weak. We don't speak of sodium chloride as a molecule because it occurs in *crystals.* These are massive accumulations of sodium chloride held together by the ionic bonds, or salt linkages, between the sodium and chloride ions (Figure 2.8). If the sodium chloride crystal is placed in water, even the thermal activity of the water and the simple competing attractions of the water molecules for charged ions will be enough to break the crystal into its component ions; thus, we say that the salt *dissolves.* The net charge of the *solution* formed must stay approximately balanced—you can't get a beaker full of dissolved, positively charged sodium ions without including approximately as many negatively charged ions. Incidentally, salt water is an excellent conductor of electricity because of its mobile charged ions.

Chemical Bonds and Molecular Diagrams

The ionic bond is one of four kinds of chemical bonds that we will consider in this chapter. The other three are *covalent bonds, hydrogen bonds,* and *nonpolar attraction* (van der Waals forces). Let's consider covalent bonds next.

The Covalent Bond: Sharing Electron Pairs

As we mentioned earlier, one way an atom can fill its outer electron shell is by sharing electron pairs with other atoms. Two atoms in close proximity can satisfy their shell requirements by simply sharing a pair of electrons. An atom gives up very little by sharing electrons, since such sharing allows it to meet its three energetic requirements simultaneously and thus results in a *preferred, low-energy state.* What they do give up, in energetic terms, is the freedom to go their separate ways. The sharing holds the two atoms together in what is called a *covalent bond,* thereby forming a molecule. Consider the simplest covalent molecule, molecular hydrogen (H_2). It consists of two hydrogen atoms, each comprised of a nucleus with one proton and a single electron. Since electrons prefer to travel together, the two hydrogen atoms can pool their electrons to make a pair. If they pool their electrons and then share the pair, they not only pair two isolated electrons, but they simultaneously satisfy the requirement of shell filling for both atoms (two in the first, or innermost, shell) and maintain the charge balance between protons and electrons. So we have a molecule of hydrogen, as seen in Figure 2.9. The covalent bond produced in such a way is a rather powerful bond, not easily disrupted.

Another relatively simple molecule is molecular oxygen (O_2). The two oxygen atoms are held together by sharing two electron pairs, and thus two covalent bonds are formed involving a total of four electrons. Two covalent bonds between the same two atoms is called, not surprisingly, a *double bond.* The idealized, free, uncharged oxygen atom has only six electrons in its outer shell. By sharing two of its own electrons and two electrons of another oxygen atom, it is able to fill its outer electron shell (Figure 2.10). Molecular oxygen is written O_2, but there are other shorthand ways of diagramming it. For instance, O=O is called the *structural formula* of molecular oxygen, with the double lines indicating the double bond.

Note that the net charge of the oxygen molecule is balanced: sixteen protons and sixteen electrons. At the same time, all the electrons are paired and the outer electron shells of both atoms are filled, with eight electrons each—because the *four shared electrons are counted twice.* That is, each shared electron pair is included in the outer electron shells of both oxygen atoms.

2.9 In covalent bonding two or more atoms form a molecule as outer-shell electrons are shared. The simplest example is hydrogen gas (H_2). Covalent bonds are considerably stronger than ionic bonds, because as the checklist indicates, covalent bonding satisfies all the individual requirements of both partners in the molecule.

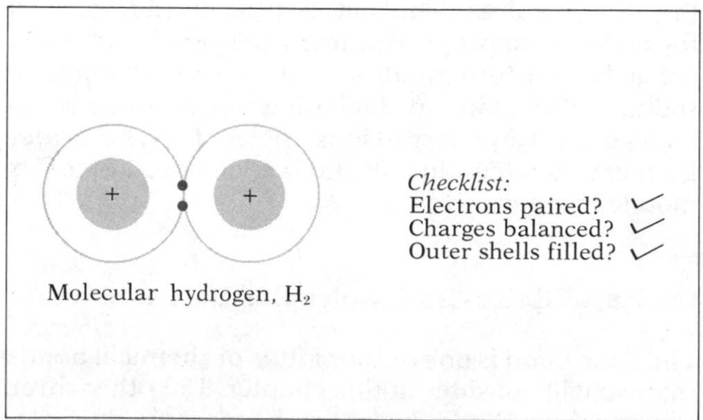

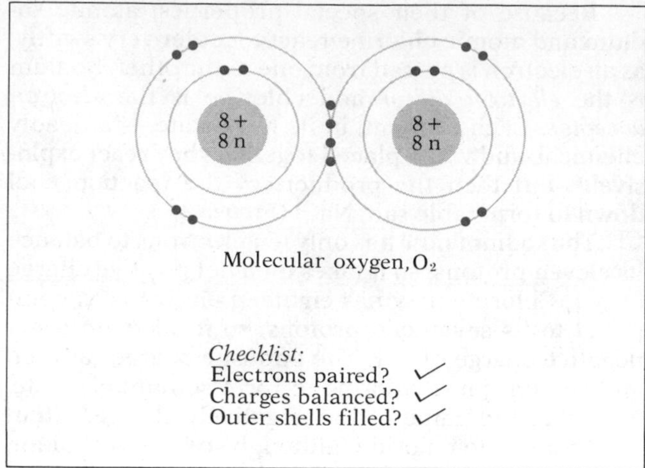

2.10 Two atoms of oxygen share two pairs of outer-shell electrons to form a molecule of O_2. In this manner, both oxygen atoms have achieved saturation of their outer shells. Sharing two pairs of electrons is not uncommon in molecules and is referred to as a double bond, symbolized as O=O. The checklist indicates that the requirements of both atoms have been satisfied.

Methane

The atomic number of carbon is 6, so free atomic carbon has two electrons in its inner shell, but only four electrons in its outer shell. Therefore, carbon needs four more electrons to fill its outer shell. Carbon can't add four electrons to its outer shell, since it only has six protons in its nucleus to balance the charges, and it can't give up its four electrons because an imbalance in charges in the opposite direction would result. Hence, carbon can't ionize, but it can form covalent bonds. Specifically, it needs to form four covalent bonds to fill its shell.

One of the simplest carbon compounds is methane, CH_4, a compound of one carbon atom and four hydrogen atoms. Methane is a principal component of marsh gas, the waste product of certain primitive bacteria that rot organic material in the depths of bogs where there is no oxygen. More importantly, perhaps, methane makes up most of the natural gas that the world seems to be running out of.

Each of the four hydrogens of the methane molecule forms a covalent bond with the carbon, sharing its own one electron and one of the outer four electrons of the carbon. That way each of the four hydrogens fills its outer shell with two paired electrons, (shared, of course), and the eight paired, shared electrons also fill the outer shell of the carbon atom. The usual way of writing the formula for methane is CH_4, but the structural formula of methane is written like this:

```
    H
    |
H — C — H
    |
    H
```

The single lines indicate single covalent bonds, but remember that each covalent bond involves two shared electrons. The electron shell diagram is shown in Figure 2.11.

Now we come upon a serious problem with this kind of flat diagram. We already know that the electron orbitals of the second shell don't really lie in a plane, but in three dimensions. Furthermore, as it turns out, the one spherical orbital and the three dumbbell-shaped orbitals of the free atom disappear when the atom forms covalent bonds. The four pairs of electrons in the outer shell of a covalently bonded carbon form four fully equivalent orbitals, each consisting of a single pear-shaped lobe. The four lobes stick out in four directions, as spread apart from one another as possible; they do not lie on a plane, but form a three-dimensional structure rather like a stack of four oranges. This is called a tetrahedronal arrangement, after the *tetrahedron*, a geometrical solid with four equal faces and four corners.

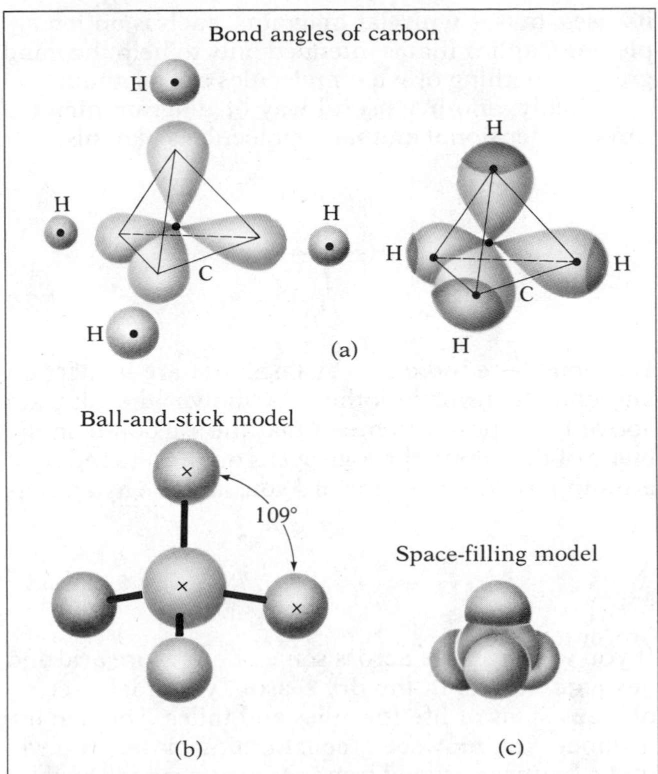

2.12 Methane (CH_4). The one carbon nucleus lies at the very center. Four hydrogen nuclei form the corners of an imaginary tetrahedron (a, dotted lines). Four pear-shaped orbitals, each containing one electron from the carbon's outer shell and one electron from a hydrogen, radiate from the carbon nucleus and envelop the hydrogen nuclei. Figures (b) and (c) depict the same molecule in the ball-and-stick model and the space-filling model respectively.

2.11 The electron shell diagram of methane (CH_4) illustrates how the four outer-shell electrons of carbon interact with those of four hydrogen atoms. The sharing of these electrons forms the four covalent bonds of CH_4.

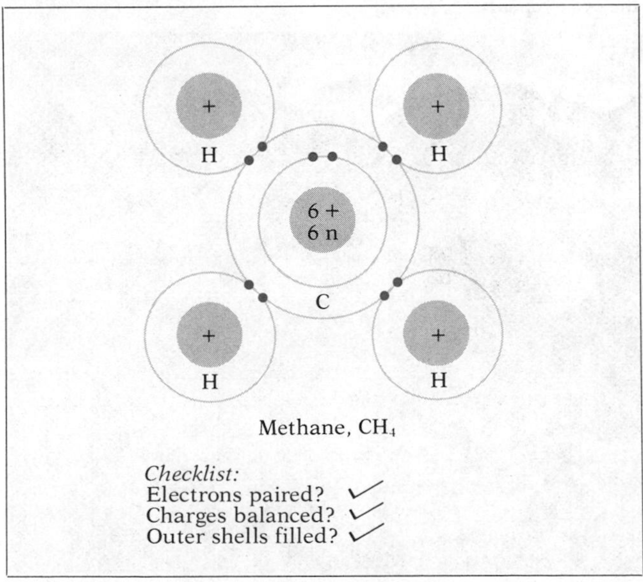

In methane, the nuclei of the four hydrogens lie embedded within the four lobelike electron orbitals of the carbon. This means that each hydrogen nucleus is exactly the same distance from each of the other three hydrogen nuclei, and that the four hydrogens are as far apart as possible from each other while being as close as possible to the carbon in the center. In fact, the four hydrogen nuclei form the four corners of a tetrahedron. If imaginary lines are drawn from the carbon to each of the four hydrogens, the angle between any two of them—the "bond angle"—is slightly more than 109°.

Now, tetrahedrons aren't easy to show on paper, but Figure 2.12 indicates what one looks like. It also shows several different "models" or ways of representing the methane molecule in three dimensions: the electron cloud model, the ball-and-stick model, and the space-filling model. Each kind of drawing has its uses, but as with flat diagrams, each is an incomplete metaphor that is intended only to help the mind grasp something of what molecules are all about.

Finally, another useful way of diagramming the three-dimensional methane molecule is like this:

$$\begin{array}{c} H \\ | \\ H-C-H \\ | \\ H \end{array}$$

The bonds are indicated by lines that are heavier on one end than on the other. As shown, the diagram above is supposed to mean that the carbon is in the plane of the paper, the hydrogens on the left and right are supposed to come toward you, and the hydrogens on the top and bottom are supposed to extend down below the plane of the paper.

Actually, for our purposes, methane itself really isn't all that important, except as a good example of a small, three-dimensional molecule. But the different ways of drawing it are important, because you'll be seeing them repeatedly in their various forms. Also, the tetrahedonal arrangement and the 109° bond angle are very important to us, as we'll see.

Carbon is extremely varied in the ways in which it can join with itself and with other elements to form molecules of all sizes. The chemistry of life is virtually the chemistry of carbon. But before we go on to describe any more about carbon chemistry, there is one molecule that needs to be considered in some detail, one that is equally important to life: simple water.

2.2 Water

If you were to walk across some of our more arid and desolate deserts in the dry season, you might see no obvious signs of life for miles and miles. Then, in the distance, you may see green. Coming closer, you will find a few cottonwood trees and scattered tall bushes. You know why the plants have appeared. You might not be able to see it, but you know that it's there, beneath the surface at least: water. Where there's life, there's water. And just about any place where there is water there is life.

Driving back through the same lifeless parts of the desert a week or two after a good rain, even there you may find signs of life in abundance. Hillsides that had stood only as rocky monuments covered with sand are now covered with a profusion of small, riotous flowers. Tiny plants have sprung up from long-dormant seeds and have, in a very short time, burst into bloom. They will be pollinated, make seeds, and die, probably before the next rain (Figure 2.13).

On a grander scale, life began in water, and the

2.13 Nowhere is the significance of water more dramatically evident than in a desert. Long periods of drought and dry, searing winds leave only the best adapted plants to dot the landscape. Everything else lies dormant awaiting the seasonal rains. When water is available, the desert bursts into a riot of color as the annuals grow, flower, and drop their seeds, often in a matter of a few short weeks. Soon the bleak desolated condition will return as the cycle repeats.

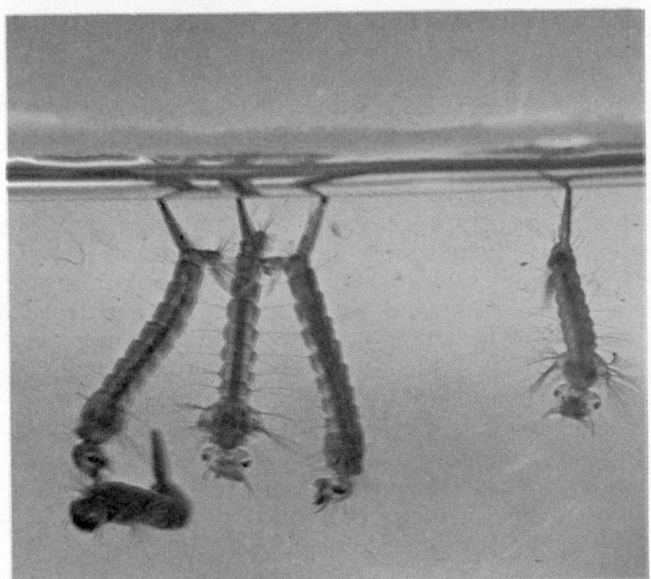

2.14 For many species, the embryonic or larval forms are particularly high in water content. The early stages are often the most vulnerable for living things, thus emphasizing the critical role of water in the processes of life.

association endures. Hope of discovering life on Mars faded when we found that water was virtually absent on our celestial neighbor. After all, most of life's chemical reactions occur in aqueous solution; the cells of most living organisms are about 80–90% water (Figure 2.14). In fact, you are about two-thirds water.

So, water is the essence of life, and on our unique planet it is plentiful. Three-fourths of the earth's surface is water and an incredible amount of water ebbs and flows through the atmosphere. Figure 2.15 shows the unending cycle of water between atmosphere and oceans. The cycle is not simply a physical entity but involves the organisms of the earth as well. Plants, animals, and the many other forms of life use water, incorporating it for a time in their cells as they go about their chemical activities. Eventually, all this water is returned to the cycle of the atmosphere and oceans as the organisms respire or die. In later chapters, we will see the importance of water to life in many different ways, but for now, let's consider some of the molecular peculiarities of this vital, life-sustaining substance.

In the water molecule, two hydrogen atoms are

2.15 The water cycle is a constant sequence involving evaporation and precipitation. Living organisms are a part of the cycle as plants absorb water through their roots and release it from their leaves in what is known as transpiration. Some of this water is used to build the molecules of life. All organisms carry on respiration, again freeing the water molecule as a waste product. In the death and decay process, microorganisms complete the breakdown of molecules, returning water to the cycle.

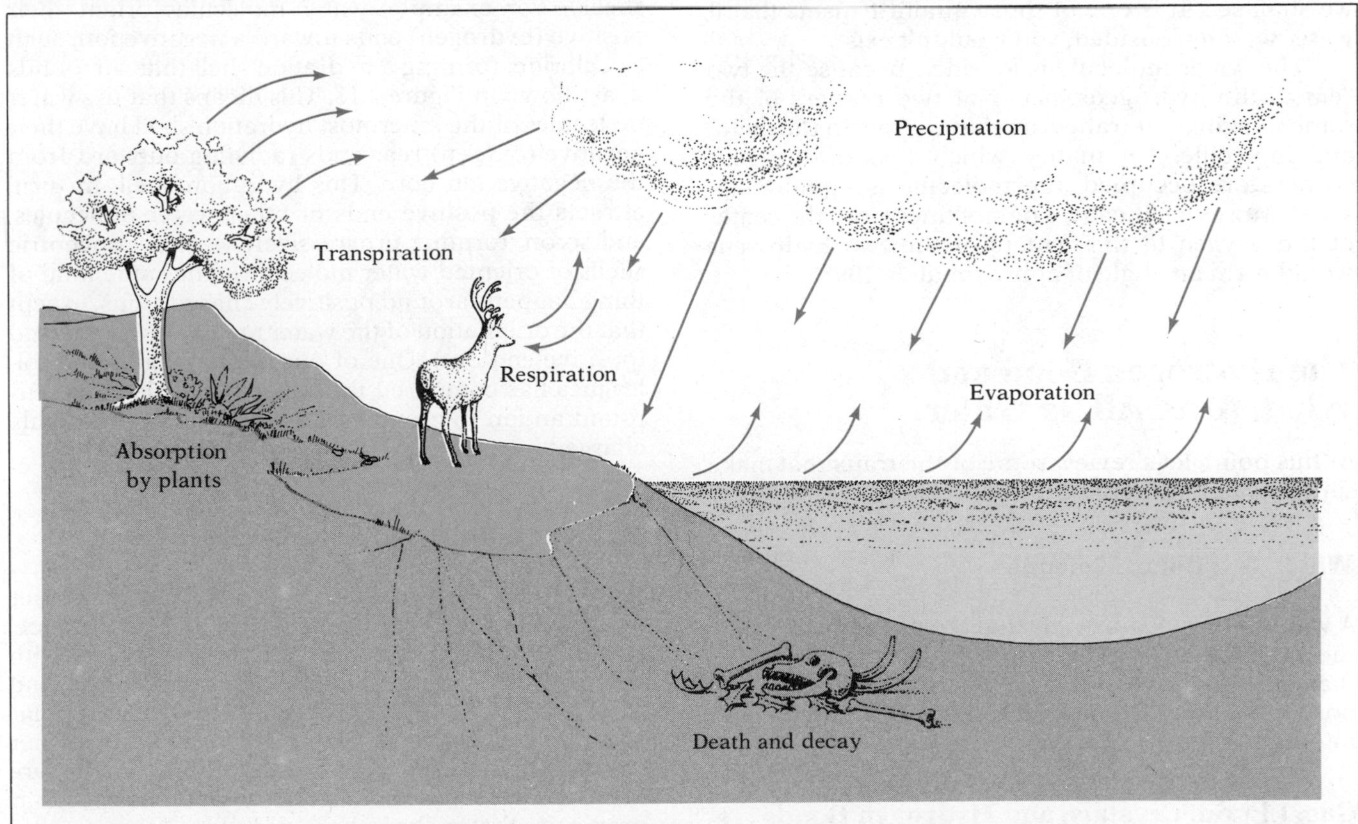

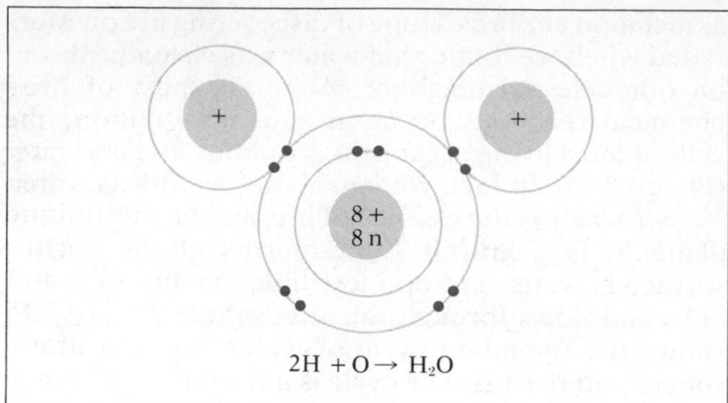

2.16 In water molecules, the electrical charge is unevenly distributed. The hydrogen end is positive, while the oxygen end is negative. Thus, water is a polar molecule and it readily dissolves other polar substances. Organisms evolving on the earth have become inseparably adapted to the peculiarities of water, making use of its characteristics in countless ways.

covalently bonded to one oxygen atom. That is, the two hydrogens share electron pairs in two of the four electron-pair orbitals of the oxygen's outer shell. Figure 2.16 shows an electron shell diagram of water. You will immediately note that the water molecule is lopsided—that is, it is *polar.* The diagram reminds some people of Mickey Mouse. (Now try looking at it without thinking of Mickey Mouse). This configuration is important to the magical properties of water, as we shall see. It is one of those quaint truisms that if water weren't lopsided, you wouldn't exist.

The water molecule is lopsided because the two "ears," the hydrogens, occur at two corners of the same imaginary tetrahedron that we saw in the methane molecule. No matter which two of the four corners are occupied, the molecule is equally lopsided. More specifically, a pair of lines from the center of the oxygen to the centers of the two hydrogens would form an angle of approximately 109°.

The Hydrogen Bond and What, After All, Is Water?

At this point, let's review some of the traits that make simple water so incredibly important.

Water Is a Polar Molecule

A water molecule has ends that are not alike. Because the two hydrogen atoms are displaced to one side, that side has two protons which give it a slightly positive charge. The opposite side has a corresponding slight negative charge.

Gas, Liquid, Crystals, and Hydrogen Bonds

Because water molecules are polar, with positively charged and negatively charged ends, they readily interact with one another, each clinging to the tail of the next, as it were. The slight attraction between the positively charged hydrogen end of one molecule and the negative rear end of the oxygen atom of the next molecule tends to hold two water molecules together, even if just for an incredibly brief instant. This forms the third biologically important category of molecular bonds, the *hydrogen bond.* Fortunately, the weak bond is easily broken by just a little heat. Above 0°C (32°F), water molecules move rather freely past one another, making and breaking the weak hydrogen bonds with dazzling speed. At below freezing temperatures, of course, a solid crystal of ice forms. The cooling removes the source of energy that breaks hydrogen bonds, and when they finally stop breaking so fast, ice appears. Ice, by the way, does not melt easily. It takes a surprisingly large amount of energy to turn the ice back into liquid water. Figure 2.17 reviews the three states of water.

Water Forms Hydration Shells Around Ions

We've already mentioned that salts such as sodium chloride dissociate into ions when dissolved in water. This dissociation is aided by the tendency of water to form *hydration shells* around charged molecules and atoms (don't confuse hydration shells with electron shells). For example, water molecules orient their positive (hydrogen) ends toward a negative ion, such as chloride, forming a hydration shell that surrounds it, as shown in Figure 2.18. This means that the water molecules of the innermost hydration shell have their negative (oxygen) rear ends radiating outward from the negative ion core. This hydration shell, in turn, attracts the positive ends of other water molecules, and so on, forming progressively weaker concentric shells of oriented water molecules. The same kind of thing happens around positively charged ions, except that the orientation of the water molecules is reversed (positive end out). One of our more imaginative colleagues has compared the water molecules that surround an ion to groupies clustering around a highly charged rock star.

Water Is a Powerful Solvent

Water will form hydration shells around any other polar molecule, even if it isn't an ion. For instance, sugars have slightly polar hydroxyl side groups with which water can build loose hydrogen bonds and form hydration shells. This keeps the somewhat polar sugar molecules from clumping together; in other words, sugar dissolves, and hydration shells are formed. So, because of its polarity, water has peculiar properties that make it the world's best solvent.

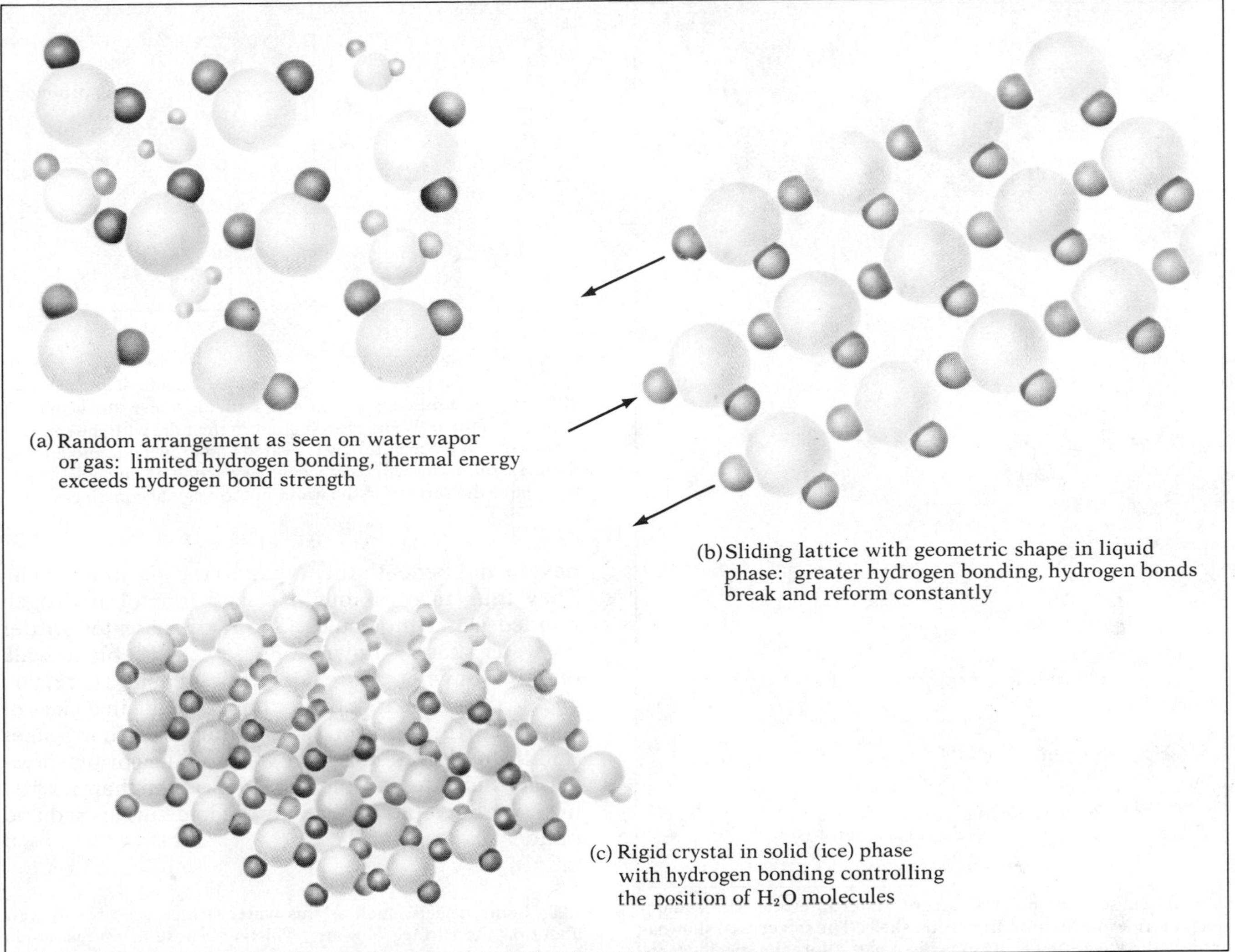
(a) Random arrangement as seen on water vapor or gas: limited hydrogen bonding, thermal energy exceeds hydrogen bond strength
(b) Sliding lattice with geometric shape in liquid phase: greater hydrogen bonding, hydrogen bonds break and reform constantly
(c) Rigid crystal in solid (ice) phase with hydrogen bonding controlling the position of H_2O molecules

2.17 The three states of water. Because of their polar structure, water molecules attract each other, forming weak hydrogen bonds. The degree of attraction depends upon the physical state of water. (a) In its gaseous form (above 100°C, or 212°F), thermal energy is high and there is little organization. Movement is rapid and molecular distances are too great for hydrogen bonds to hold. (b) In its liquid state (between 0 and 100°C, or 32 and 212°F), the molecules of water move much less (as thermal energy decreases), and hydrogen bonds form much more frequently. The association is still loose, however, resulting in sliding rows, or lattices, of molecules with some suggestion of geometric form. Because of this regular pattern, water has been called a "liquid crystal." (c) As thermal energy decreases and water reaches its freezing point, molecules become more rigidly held by hydrogen bonds. Eventually, crystals with definite geometric shapes will form. Now, the only molecular movement is vibrational. Note the relationships of the positive and negative ends of the molecule.

Water Is Wet

Water tends to get things wet. But what does this really mean? It means it forms hydrogen bonds with the surface molecules of solid objects, except, of course, with objects made of oily or waxy substances that are composed entirely of nonpolar molecules. This tendency to wet surfaces is due to liquid *adhesion*—adhesion is the attraction between two dissimilar substances. *Cohesion* is the attraction between similar substances; the hydrogen bonds between water molecules give water a considerable cohesion.

The tendency of water to adhere to and spread over solid surfaces gives it many of its special properties. If a thin glass tube is lowered into a beaker of water, the water wets the inside of the tube and, as it does so, a column of water rises in the tube until it is higher than the water level in the beaker. If glass tubes of different diameters are put into the same beaker, the water will rise higher in the tube with the smallest bore (Figure 2.19). This is called *capillary action,* and it is due in part to the adhesion of water to glass and in part to the cohesion of water to itself.

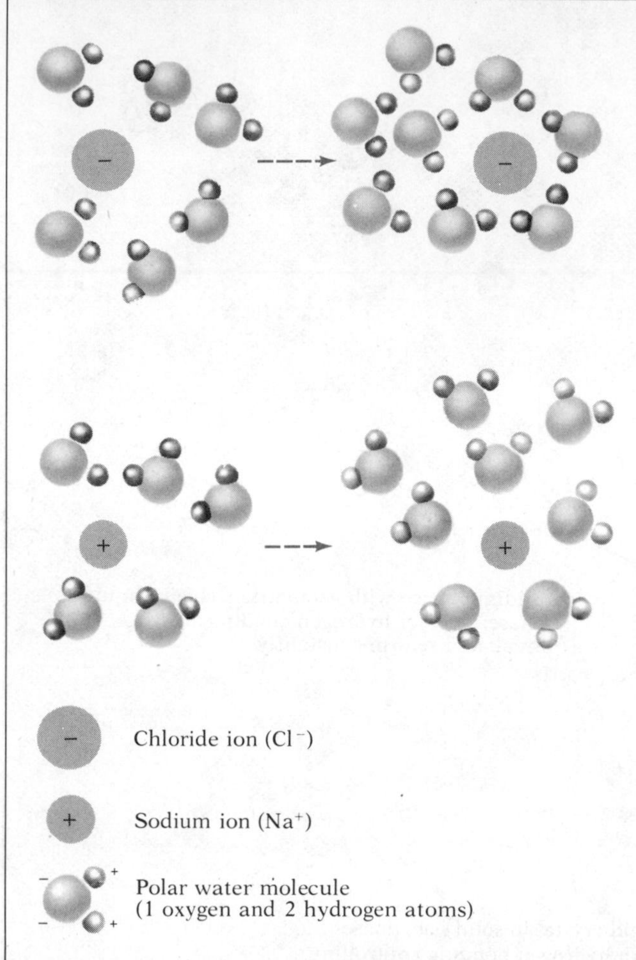

2.18 Because of its polar structure, water interacts with sodium and chloride ions to form hydration shells. The salt crystal shown is rapidly disintegrating as its ions dissociate. Note the specific manner in which the positive and negative ends of water orient to the negative and positive ions of chlorine and sodium. Many ionic compounds react in this manner with water.

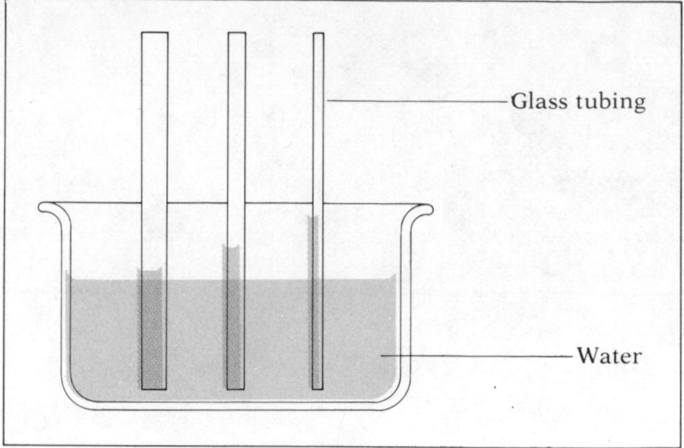

2.19 If glass tubes are placed into standing water, the water will rise up into them. Water rises highest in the tube with the smallest diameter. If a piece of tubing is heated and drawn out into hairlike thinness, the water will rise by capillary action as much as several feet above the surface of the water in the original container.

We'll have more to say about capillary action when we talk about how water rises in plants.

Similar to capillary action, but on a finer scale, is *imbibition.* This is the movement of water into porous substances, such as wood or gelatin. The substances swell as the water moves in, and in fact, the swelling can generate a startling force. Seeds can split their tough coats by the force of imbibition. And it has even been theorized that the great stones used in the construction of the Egyptian pyramids were quarried by driving wooden pegs into holes in the rock face and then soaking the pegs with water.

Water Has High Surface Tension

Surface tension is one aspect of the cohesion of water. Where air and water meet, the water molecules at the interface have a much greater attraction to the water next to and beneath them than to the air above them. They thus form a tough, elastic film of hydrogen bonded water molecules. The familiar water strider (or water skater) shown in Figure 2.20 is able to walk on this film without breaking it. With a little care, you can do the same trick by floating a needle in a glass of water. Surface tension is also the force that makes rain form discrete drops of water. Thus spring showers are delicate and pleasurable, rather than terrifying, as they would be if water fell in disorganized and ponderous masses.

2.20 Some insects, such as this water strider, are able to walk about on the surface of water. This is because at the water–air interface, water molecules have a stronger attraction for each other than for the air molecules. A close look at the appendages of such insects usually reveals that they are quite long and tipped with featherlike hairs, often coated with a nonwetting substance. Their weight is distributed on a large surface area because of their long legs, so their feet merely form indentations in the water surface.

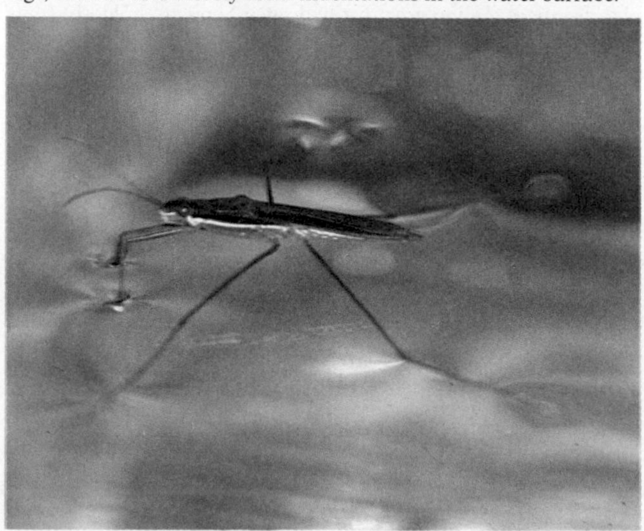

2.3 The Molecules and Ions Interact

Nonpolar Attraction, or van der Waals Forces

So far we've looked at three kinds of molecular bonds: covalent bonds, ionic bonds, and hydrogen bonds. A fourth chemical bonding force, much weaker than the others but still important, is a generalized and nonspecific attraction between molecules. It is seen most strongly in the clumping and weak binding of nonpolar (uncharged) molecules, and is sometimes called *nonpolar attraction* or *hydrophobic bonding*. The more proper term is *van der Waals forces*.

Uncharged molecules prefer to draw together as closely as possible, so long as their outer electron shell orbitals don't overlap. The result is as if nonpolar molecules were ever-so-slightly sticky. It is because of nonpolar attraction that oil globules form when you try to mix chicken fat with water, and that grease spots dissolve in gasoline. The nonpolar fat molecules tend to cling together and form globules, preventing any association with the polar water molecules. But organic solvents, such as gasoline, salad oil, and paint thinner, are liquids consisting of nonpolar molecules. They will easily surround and dissolve other nonpolar molecules such as grease. As we'll see later, side groups of large molecules such as protein can also be nonpolar, and van der Waals forces along with covalent bonding, ionic bonding, and hydrogen bonds are the intramolecular bonding forces that hold enzymes and other protein molecules in their extremely specific looped, tangled, and folded active shapes. Nonpolar attraction is even more important in the structure and function of cell membranes, as we'll see in later chapters.

Hydrophobia and Mayonnaise

Hydrophobia means "fear of water." Nonpolar molecules (or parts of molecules) are called *hydrophobic*, because they prefer to melt into each other rather than mix with water (as in the chicken fat and water example described above). Some molecules, however, are polar on one end and hydrophobic on the other, and therefore they can serve as links between water and fat. For instance, soaps and detergents are nonpolar (fat) on one end and charged on the other, so they can disperse nonpolar molecules in dishwater. Very small droplets of oil or grease are surrounded and infiltrated by the detergent molecules and are thus kept in solution and prevented from coalescing. Lecithin, a substance found in large quantities in egg yolk, is a natural detergent that is very useful in forming a bridge between dilute vinegar (which is polar) and salad oil (which is nonpolar). Beat well and the gel that results from your mixture of egg yolk, vinegar, and water is called mayonnaise.

Carbon Compounds

Now let's consider that enormous group of compounds that contain carbon. Molecules containing carbon, along with hydrogen and perhaps one or more other elements, are said to be *organic* molecules.

We have already mentioned the peculiar properties of carbon that make it so essential to life as we know it. Recall that the magic of carbon lies in the four electrons of its outer electron shell. We've already noted that since the atomic number of carbon is 6, and since two electrons are buried in the inner electron shell, the free carbon atom has four hungry, unpaired electrons in its outer shell. Carbon, remember, cannot ionize, since ionization would create too great a charge imbalance, and so it is doomed forever to seek electrons.

Thus, carbon can form four single covalent bonds, as methane does. But it can also form double bonds, as carbon dioxide does. Carbon dioxide contains two sets of double bonds, with the carbon in the middle and the double-bonded oxygens on either side: $O{=}C{=}O$. Carbon can even form *triple* bonds, as in acetylene gas: $H{-}C{\equiv}C{-}H$; and in hydrogen cyanide, $H{-}C{\equiv}N$. Most importantly, carbon can form bonds with other carbon atoms, and thus form long chains, rings, and other complex structures. There seems to be no limit to how large an organic molecule can be. A single giant DNA molecule, as we will see later on, can contain up to 50 billion atoms. These are aptly called *macromolecules*.

Carbon Backbones

Carbon can form covalent carbon–carbon bonds directly with one, two, three, or four other immediately adjacent carbon atoms. In many biological molecules, carbon atoms form long chains of atoms, which are sometimes called *carbon backbones* (because various side groups hang off the chains like ribs on a backbone).

Hydrocarbons

A *hydrocarbon* is a compound that consists solely of carbon and hydrogen. We've already looked at the simplest hydrocarbon—methane. Butane is a familiar hydrocarbon that is distilled from petroleum and

Essay 2.1 Acids and Bases

Although water usually exists as a neutral, uncharged molecule, a very tiny fraction of the molecules in a drop of water will briefly and spontaneously dissociate into a *hydrogen ion* (H^+) and a *hydroxide ion* (OH^-). That is, water molecules are continually breaking apart into ions, and the ions are continually rejoining to make the neutral molecule again. In pure water, at any one instant in time, something like 1 molecule in 550 million will be dissociated into a pair of positively and negatively charged ions. An instant later, they will be back together again, but another approximately 1 in 550 million water molecules will dissociate in the meantime.

The *molar concentration* of hydrogen ions in pure water is 0.0000001 *mole per liter.* A *mole* of any substance is the weight *in grams* that equals the molecular mass in daltons of one molecule. Thus, a mole of hydrogen ions weighs 1 gram, and the concentration of hydrogen ions by weight in pure water is 0.0000001 *gram per liter.* The number 0.0000001, or the digit "1" seven places to the right of the decimal point, can also be written as 10^{-7}, which is the *scientific notation* equivalent of 1 divided by 10 million. So the molar concentration of hydrogen ions in pure water is 10^{-7} mole per liter, and the same concentration is also 10^{-7} gram per liter.

Actually, the hydrogen ion (H^+) is a common but convenient fiction. A real hydrogen ion would be a naked proton, and while such things exist, they are rarely found dissolved in water. In reality, the hydrogen nucleus from a dissociating water molecule becomes bound up with another water molecule to make a *hydronium ion* (H_3O^+). Keep in mind that whenever we refer to a hydrogen ion, or write it as H^+, we mean a hydronium ion, H_3O^+. The dissociation and reassociation reaction may be symbolized as follows:

$$2\,H_2O \leftharpoondown\!\rightharpoondown OH^- + H_3O^+$$

where the longer arrow pointing left indicates that most molecules are in the H_2O form at any one time. In the standard convenient fiction, of course, the same reaction would be written this way:

$$H_2O \leftharpoondown\!\rightharpoondown OH^- + H^+$$

Some substances, when in water, release hydrogen ions in measurable quantities. We call solutions of such substances *acids.* Stomach acid, for instance, is dissolved hydrochloric acid, HCl. Pure HCl is a gas, but when it is dissolved in water, HCl ionizes to become paired H^+ and Cl^- (hydrogen ions and chloride ions).

Just how strong, or *acid,* an acid solution is depends on the concentration of the hydrogen ions. The molecular weight of hydrochloric acid is 36 daltons (1 for the hydrogen, 35 for the chloride). So 1 mole of pure HCl gas weighs 36 grams, and 36 grams of HCl dissolved in 1 liter of water would create a concentration of 1 mole of HCl per liter, a pH of −1.

Actually, since the HCl ionizes completely, the concentration of *chloride ions* would be 1 mole per liter and the concentration of *hydrogen ions* would also be 1 mole per liter—a concentration of hydrogen ions that is 10 million times greater than that of pure water. That's a rather strong acid.

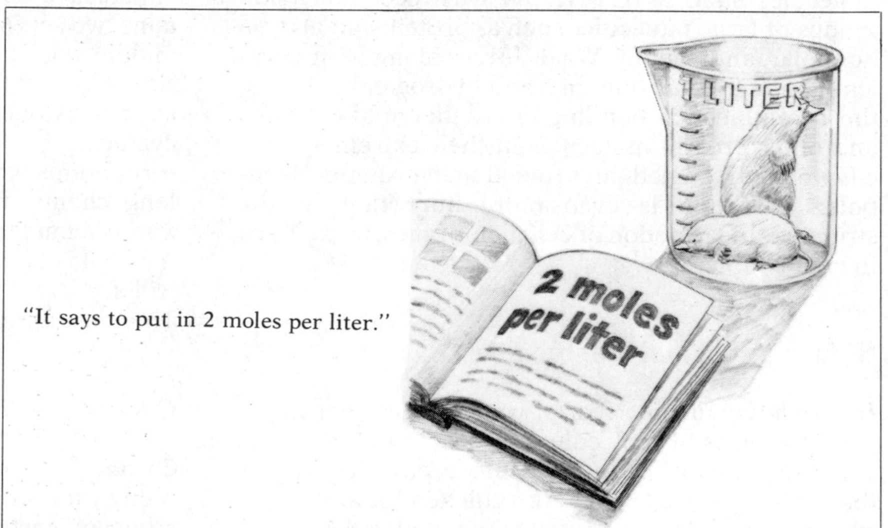

"It says to put in 2 moles per liter."

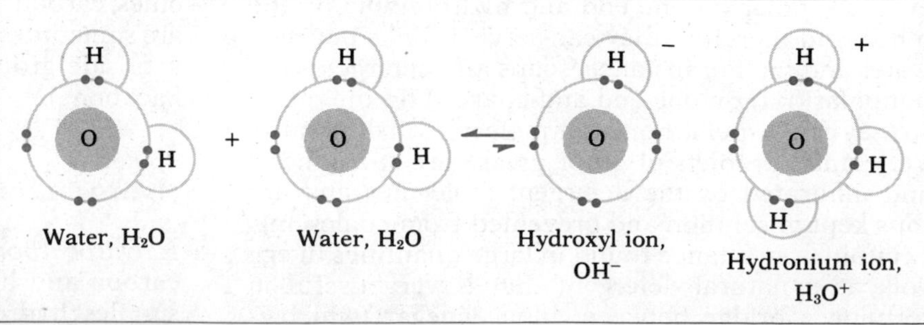

A shorthand notation for the strength of acid solutions, or for acidity in general, is the *pH scale.* This scale uses scientific notation to express the H^+ concentration in a solution. Thus, if the acidity of some rather tart orange juice is 0.01 mole of H^+ ions per liter, which is the same thing as 10^{-2} mole of H^+ per liter, that orange juice has a pH of 2. We just leave out the "ten to the minus" part of the number. On the pH scale, then, pure water has a pH of 7. The table gives the pH values of various substances. Note that the smaller the pH value, the more acid the solution.

A *base* is a substance that releases hydroxide ions (OH^-) when dissolved in water. Lye, or sodium hydroxide (NaOH), is a familiar example of a strong base. Ammonia is another base—although it contains no hydroxide ion itself, it reacts with water to form an ammonium ion and a hydroxide ion:

$$\underset{\text{Ammonia}}{NH_3} + \underset{\text{Water}}{H_2O} \rightleftharpoons \underset{\text{Ammonium ion}}{NH_4^+} + \underset{\text{Hydroxide ion}}{OH^-}$$

A solution can be acidic, neutral, or basic, but it cannot be acidic and basic at the same time. This is because hydroxide ions and hydrogen ions join spontaneously to form water. In pure water, which is neutral, the concentrations of hydrogen ions and hydroxide ions are both 10^{-7} mole per liter. But when the concentration of hydrogen ions rises to 10^{-1} mole per liter, as in gastric juice, the molar concentration of hydroxide ions falls to 10^{-13}, which is a very small number. In general, the exponents of the H^+ and OH^- ion concentrations in a solution always add up to -14. Thus, seawater, which is slightly basic, has a hydroxide ion concentration of 10^{-6} and a hydrogen ion concentration of 10^{-8}: $(-6) + (-8) = -14$. The seawater, then, has a pH of 8.0.

With certain exceptions (stomach contents, citrus fruit), most life processes go on at pH values between 6 and 8. Human blood has a pH of approximately 7.4, and the cell contents of most organisms have similar nearly neutral pH values.

Table of pH Values

Molar concentration of H^+ ions	pH	Example	Molar concentration of OH^- ions
$10 = 10^{+1}$	−1	Nitric acid	10^{-15}
$0.1 = 10^{-1}$	1	Gastric juice	10^{-13}
$0.01 = 10^{-2}$	2	Coca-Cola	10^{-12}
$0.001 = 10^{-3}$	3	Vinegar	10^{-11}
$0.000001 = 10^{-6}$	6	Saliva	10^{-8}
$0.0000001 = 10^{-7}$	7	Pure water	10^{-7}
$0.00000001 = 10^{-8}$	8	Seawater	10^{-6}
$0.000000000000001 = 10^{-15}$	15	Drain opener	10^{+1}

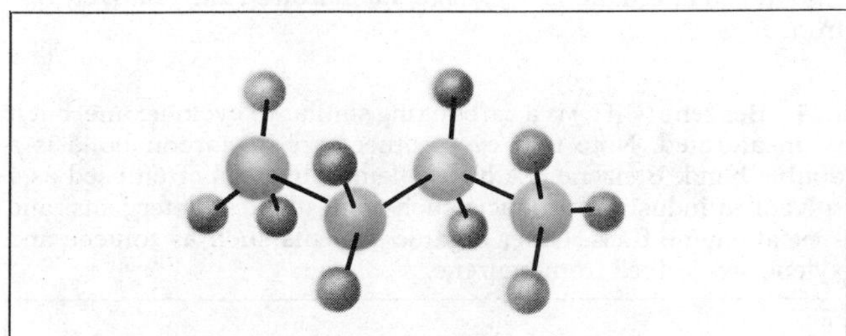

2.21 The ball-and-stick model of butane reveals its geometry. Note that butane's four carbon atoms are not in a straight line, but form angles. In addition, the shape of the chain can be changed by rotating the position of the carbon atoms on their bonds.

used in cigarette lighters. It has four carbons, as seen here and in Figure 2.21.

```
    H   H   H   H
    |   |   |   |
H — C — C — C — C — H
    |   |   |   |
    H   H   H   H
```

Note that the ball-and-stick model of butane shows that the carbon backbone is not really straight. The single covalent bonds are still forming tetrahedrons with angles of about 109°. Since single covalent bonds operate like little swivels, the carbon atoms can rotate freely about the axis of the chain.

But carbon can form more than chains. It can branch (Figure 2.22) and form rings (Figure 2.23).

The carbon–carbon linkages described above are single covalent bonds, but carbon–carbon linkages can also be double or triple bonds (if two adjacent carbons share four or six electrons, respectively). Double and triple bonds don't swivel, and thus these molecules are more rigid than molecules with only single bonds. Compare benzene, C_6H_6 (Figure 2.24), with cyclohexane, C_6H_{12} (Figure 2.23). In cyclohexane, all bonds are single bonds and each carbon has as many hydrogens as it can handle—this compound is therefore said to be *saturated.* In benzene, the carbons form double bonds between each other instead of with more hydrogens—this compound is said to be *unsaturated* with regard to hydrogen.

Nitrogen

Later on we're going to become rather involved with proteins and nucleic acids, which are both classes of large, nitrogen-containing organic molecules. So at

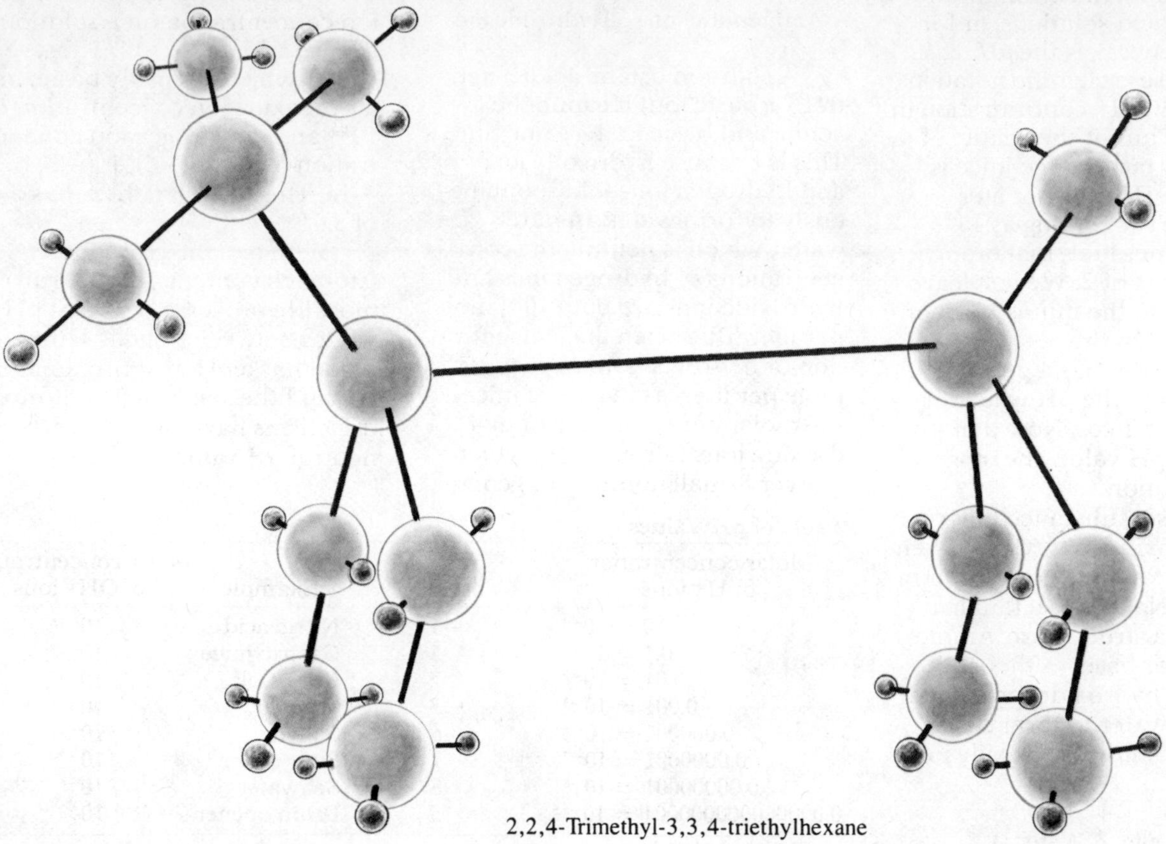

2,2,4-Trimethyl-3,3,4-triethylhexane

2.22 An organic chemist could probably draw the branched molecule 2,2,4-trimethyl-3,3,4-triethylhexane and end up with a dog. Such a molecule may have many different names—common, commercial, and scientific—but the name given follows the rules of the International Union of Pure and Applied Chemistry (IUPAC), an organization that names organic compounds. This provides chemists with an international language, standardized and relatively free from error.

2.23 Hydrocarbons occur in symmetrical rings, as seen in the hydrocarbon cyclohexane, C_6H_{12}. This compound is a useful solvent, but a poor, smoky fuel. More importantly, cyclohexane is a good starting molecule for producing other ring hydrocarbons.

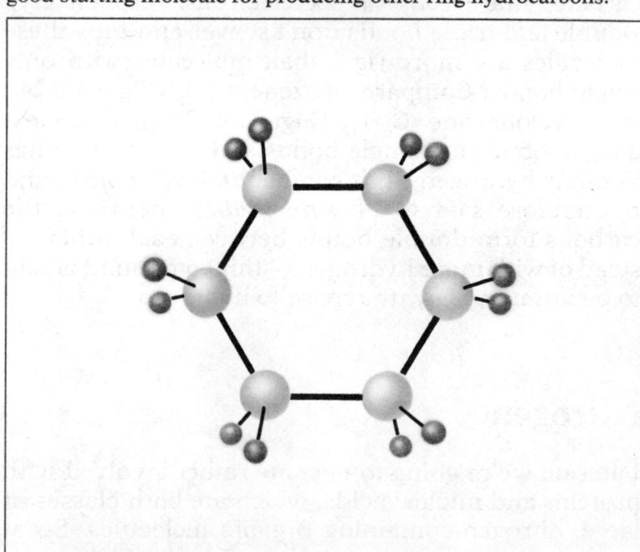

2.24 Benzene (C_6H_6) is a carbon ring similar to cyclohexane, but it is unsaturated. Note that every other carbon–carbon bond is a double bond. Benzene is a highly flammable fuel often used as a solvent in industrial products such as pesticides, detergents, and special engine fuels. Other organic solvents such as toluene and xylene are derived from benzene.

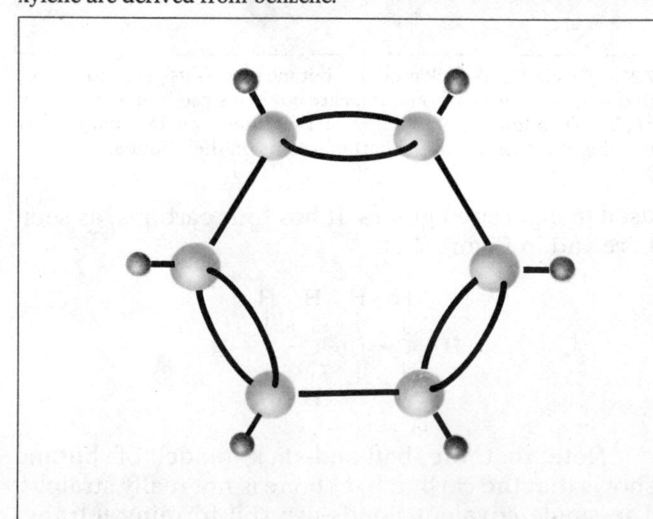

Essay 2.2 Carbon Dioxide and the Bicarbonate Ion

Carbon dioxide (CO_2) is an interesting and ubiquitous substance. It makes soft drinks and champagne tingle, but where does it come from? In champagne and beer, the carbon dioxide is produced as a waste product of yeast metabolism. (Another waste product of yeast is alcohol, of course.) Your own body is loaded with CO_2, which is a waste product of your own metabolism. The carbon dioxide in soft drinks comes from steel cylinders.

There is a tiny amount of CO_2 in the air—not much, about $\frac{1}{3}$% of air by weight, but this small amount is the only source of carbon for plants. Where does the CO_2 in the air come from? Some of it comes directly from the metabolism of living things, some of it comes from forest fires, and some of it comes from human energy-producing processes such as burning fossil fuels.

But what is CO_2? Carbon dioxide is a symmetrical, linear molecule, which can be written as O=C=O.

Carbon dioxide dissolves easily in water, and most of the CO_2 in beer and soft drinks occurs as CO_2 in simple solution. But CO_2 also reacts to some extent with the water to form *carbonic acid,* which is a weak acid that further dissociates into a *bicarbonate ion* and a *hydrogen ion:*

$$\underset{\text{Carbon dioxide}}{CO_2} + \underset{\text{Water}}{H_2O} \rightleftharpoons \underset{\text{Carbonic acid}}{H_2CO_3} \rightleftharpoons \underset{\text{Hydrogen ion}}{H^+} + \underset{\text{Bicarbonate ion}}{HCO_3^-}$$

The production of hydrogen ions is what makes CO_2 solutions acid (tart).

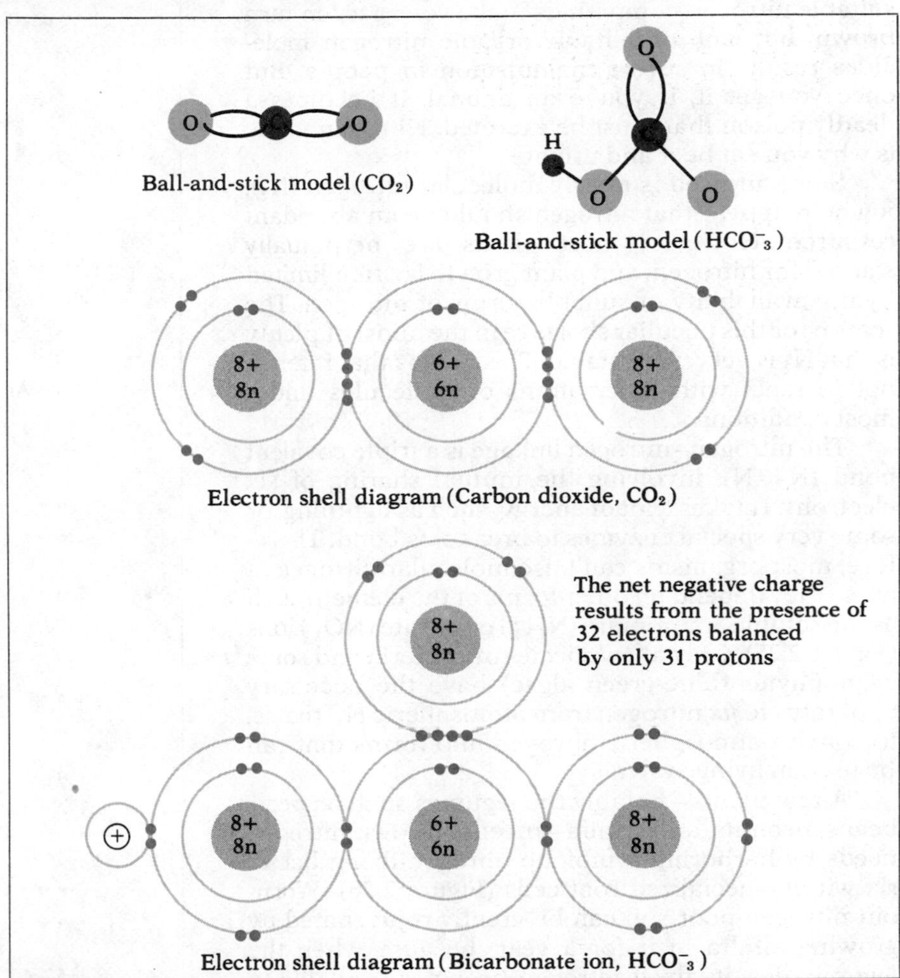

Ball-and-stick model (CO_2)

Ball-and-stick model (HCO_3^-)

Electron shell diagram (Carbon dioxide, CO_2)

Electron shell diagram (Bicarbonate ion, HCO_3^-)

Sodium bicarbonate is a sodium salt of the bicarbonate ion. When you dissolve a spoonful of "bicarb" in a glass of water, some of the bicarbonate ions combine with H^+ ions from the water to re-form into neutral carbonic acid, with the release of hydroxide ions:

$$\underset{\text{Water}}{H_2O} + \underset{\text{Bicarbonate ion}}{HCO_3^-} \rightleftharpoons \underset{\text{Carbonic acid}}{H_2CO_3} + \underset{\text{Hydroxide ion}}{OH^-}$$

The reaction tends strongly to go to the left, but the hydroxide ions produced in the reaction make sodium bicarbonate a weak base (with a high pH). In your stomach, sodium bicarbonate neutralizes some stomach acid (HCl) and releases CO_2, which takes the form of an unseemly belch (shown by the "up" arrow in the equation below). Note that the sodium and chloride ions don't actually enter into the reaction.

$$\underbrace{Na^+ + HCO_3^-}_{\text{Sodium bicarbonate}} + \underbrace{H^+ + Cl^-}_{\text{Hydrochloric acid}} \longrightarrow \underset{\text{Sodium ion}}{Na^+} + \underset{\text{Chloride ion}}{Cl^-} + \underset{\text{Water}}{H_2O} + \underset{\text{Carbon dioxide gas}}{CO_2\uparrow}$$

this point, let's learn a few basic points about nitrogen. This is a peculiar element because it is so common (78% of air is nitrogen) and yet it is difficult to obtain in usable forms. Not only does a lack of suitable nitrogen compounds make your garden turn brown, but lack of suitable organic nitrogen molecules results in severe malnutrition in people. But once you get it, if you're an animal, it becomes a deadly poison that must be excreted. Nitrogen, then, is why you eat beef and urinate.

Since air itself is mostly molecular nitrogen (N_2), it would appear that nitrogen should be an abundant resource. However, most plants are perpetually starved for nitrogen, and plant growth is often limited by the availability of suitable forms of nitrogen. The reason for this peculiar shortage in the midst of plenty is that N_2 is very, very stable. This means that it tends not to react with other atoms or molecules under most conditions.

The nitrogen–nitrogen linkage is a triple covalent bond (N≡N), involving the mutual sharing of six electrons. It takes a lot of energy, such as lightning, or some very special enzymes to break this bond. Therefore, most organisms can't use molecular nitrogen at all, so they depend on other forms of the element such as the soluble ammonium (NH_4^+) or nitrate (NO_3^-) ions (Figure 2.25). Only a few species of bacteria and some cyanophytes (blue-green algae) have the necessary apparatus to *fix* nitrogen from atmospheric N_2, that is, to convert atmospheric nitrogen into forms that can be used in living systems.

A few plants—notably the legumes such as peas, beans, peanuts, and alfalfa—meet their own nitrogen needs by harboring symbiotic nitrogen-fixing bacteria within specialized root cells (Figure 2.26). Worn-out nitrogen-poor soil can be greatly rejuvenated by growing alfalfa in it for a year, because when the legume dies, its fixed nitrogen becomes available to other plants.

Animal wastes are rich in available nitrogen and once were widely used as fertilizer; nowadays, however, nitrate and ammonium fertilizers are mostly made in chemical factories. These synthetic fertilizers are fairly inexpensive and so they can be liberally applied to the soil. The crops thrive—but the annual runoff of the excess chemicals can cause severe problems in water systems, as we'll see later. We'll also see how lightning can fix atmospheric nitrogen in the form of nitrous oxides, and how these readily convert to nitrates and other soluble and available forms. In addition, automobile engines can fix atmospheric nitrogen, converting it into nitrous oxides, which are brown, like the skies over large cities. Thus we see that nitrogen can enter biological systems through a number of routes.

Nitrogen is an essential component of proteins, nucleic acids, some carbohydrates, some lipids (fats), and a number of other biological molecules. We'll discuss these classes of compounds in Chapter 3.

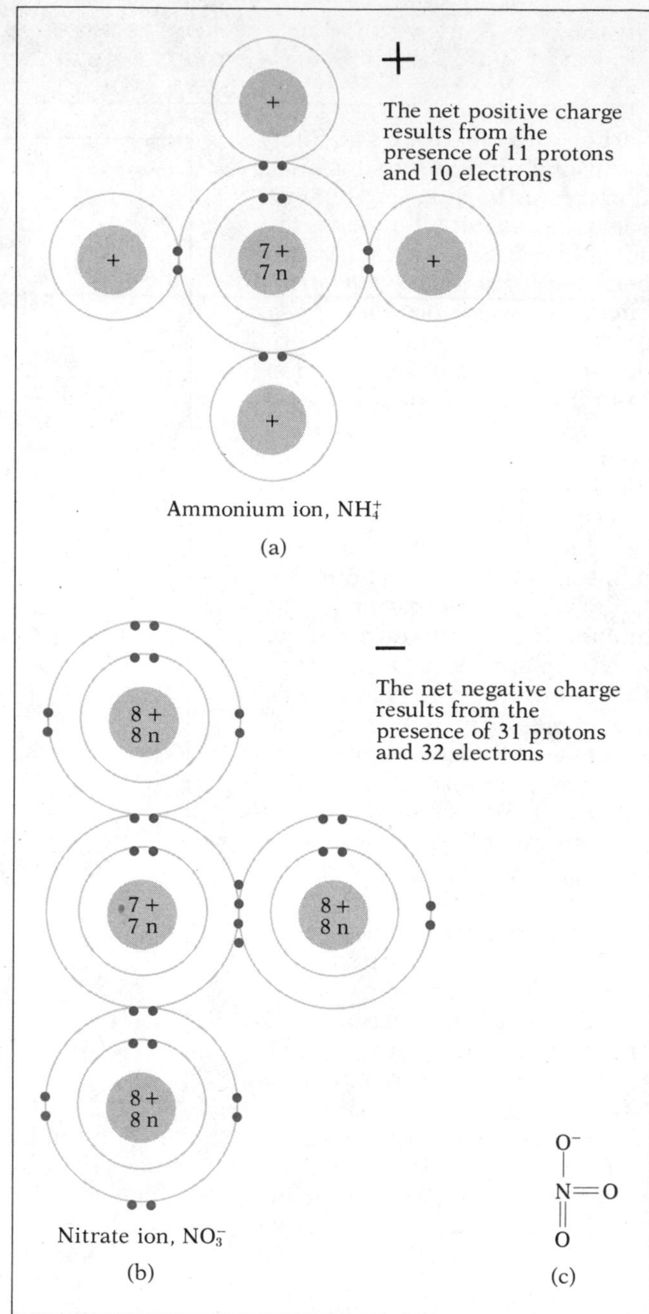

2.25 Ions of nitrogen. Most of the earth's organisms depend on ions of nitrogen such as the soluble ammonium (NH_4^+) (a) and nitrate (NO_3^-) ions (b), since they cannot use atmospheric nitrogen. These ions are both found in the ionic compound ammonium nitrate (NH_4NO_3), which is a crystalline substance and a valuable fertilizer.

Note that one of the three oxygens in the nitrate ion is double-bonded to the nitrogen atom. Although the remaining two oxygens form single bonds with the nitrogen, so that each carries a negative charge, one of these charges is balanced by a positive charge on the nitrogen atom. This gives the group as a whole a net charge of −1. A standard convenient fiction is to depict the nitrate as containing two double bonds (c).

2.26 Root nodules and nitrogen fixers. Root nodules are commonly found in leguminous plants. They contain populations of nitrogen-fixing bacteria which live comfortably in a mutual relationship with the plant. Using atmospheric nitrogen (N_2), the bacteria produce nitrogen compounds that the plant uses to produce proteins and nucleic acids. The nitrogen-rich plants can be used directly or, as is often the case, they can be plowed under so that they will decay in the soil, releasing the nitrogen compounds for use by another crop.

The atomic number of nitrogen is 7, so, since there are two electrons in its inner shell, its outer shell has five proton-balancing electrons. Three more electrons are needed to fill the outer shell. Nitrogen usually forms three covalent bonds, as in molecular nitrogen ($N{\equiv}N$) and in ammonia (NH_3). In some molecules, however, nitrogen forms four covalent bonds, as it does in the ammonium ion (NH_4^+). Four-bonded (*quaternary*) nitrogen always carries a positive charge.

Phosphorus and Phosphates

Phosphorus is another of the essential SPONCH elements. In biological systems it is always combined with oxygen as a *phosphate.* A phosphate can be an ion or can be combined with a larger molecule to form a *phosphate group* (Figure 2.27). In biological ranges of pH, phosphate ions exist as HPO_4^{--} or as $H_2PO_4^-$ (that is, with one or two hydrogen atoms and correspondingly, with two or one negative charges).

In phosphate, the four oxygens are tightly bound to the phosphorus atoms, forming the four corners of a tetrahedron with the phosphorus inside. Since the

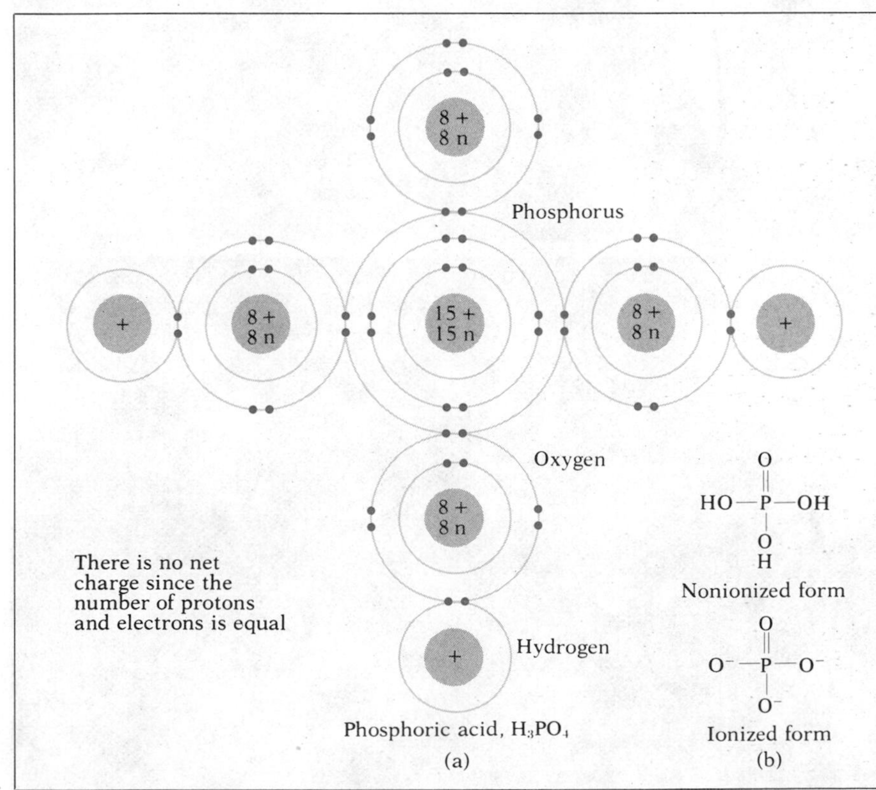

2.27 (a) The electron shell model of H_3PO_4, phosphoric acid (or simply phosphate). Note that the oxygen shown at the top of this diagram has a single bond to the phosphorus. Ordinarily, a single-bonded oxygen would have a negative charge, but in phosphate the negative charge is balanced by the positive charge of the phosphorus atom itself. To simplify things, another standard convenient fiction is to show the phosphorus atom as having a double bond to one of the oxygens (b). This seems to imply five pairs of electrons in the outer shell of the phosphorus, but don't let that worry you; it's a little white lie.

Signs of Life

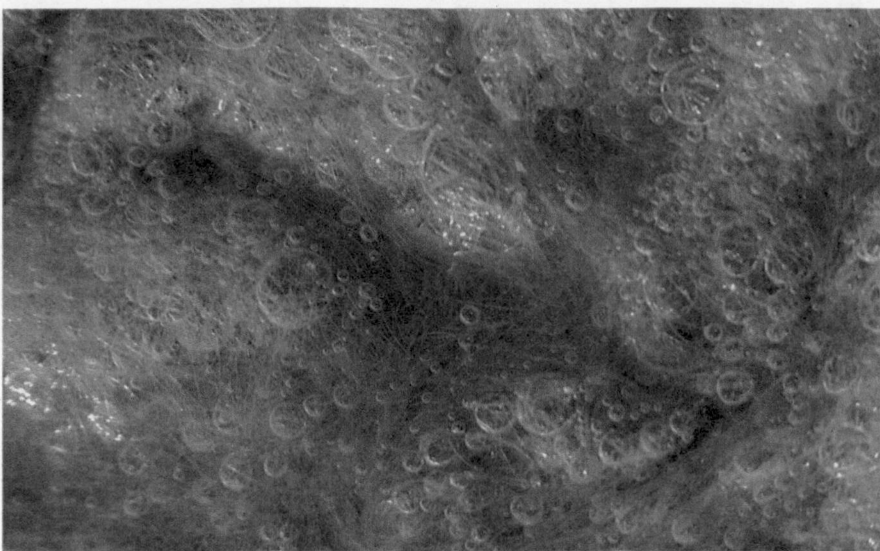

Life is notoriously hard to define, but we know that it has certain properties. For example, life requires energy in order to remain organized on an essentially disruptive planet. Many plants receive energy from the sun and produce oxygen as a by-product, shown here in bubbles.

Living things reproduce in a variety of ways. Some species care for their young, such as this water bug carrying her eggs.

Living things are sensitive to some degree, and some organisms are extraordinarily responsive to stimuli, such as this frog.

It is important that living things adapt to their environment. Those that fail may be doomed to extinction. The sea otter is remarkably adaptive and has learned to swim on its back, cracking shellfish on a rock that it carries on its stomach. It is presently threatened only by abalone fishermen who fear their profits are being reduced by this intelligent and inventive animal.

Life must exist within rather narrow limits and many forms have devised remarkable ways of regulating their internal environments. Bees may huddle to keep warm, and in hot weather may deposit water on the hive and then fan their wings vigorously, thus producing cooling air currents.

Organization is the catch-word of life. Even seemingly simple organisms such as these diatoms are highly organized, both structurally and chemically.

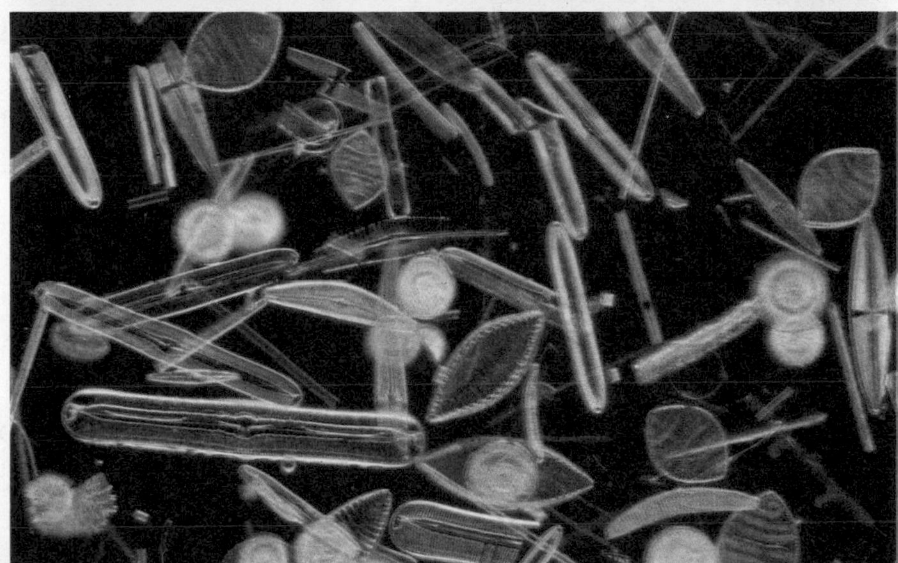

Living things must change from generation to generation as the environment changes. Short-term evolution has been demonstrated in the peppered moth. As trees in the English countryside became darkened by soot during the early industrial revolution, the dark form of the moth came to predominate since it became less visible to birds. With better antipollution devices on smokestacks, the English are witnessing a return of the lighter form of the moth.

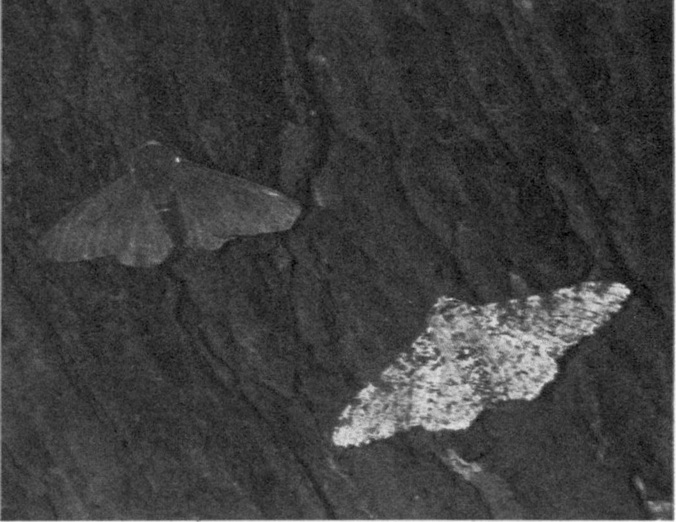

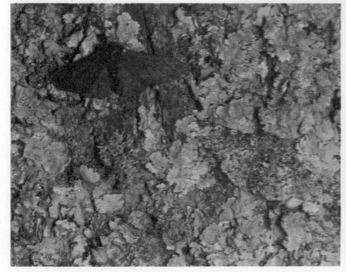

whole group seems to behave like one large atom, it is often given a symbol of its own, a P in a circle: so Ⓟ means phosphate.

When a bond is broken, the energy that previously held the atoms together becomes available to do work. Now it happens that the bond between two phosphate groups has some unusual properties. It takes a large amount of energy to form a phosphate–phosphate covalent bond. And breaking the bond releases the same large amount of energy. All living organisms use these high-energy phosphate bonds as a means of storing energy and shuffling it around to where it is needed.

Phosphates form the polar or charged end of phospholipids, which will be discussed in Chapter 3. These compounds are a basic subunit of all biological membranes. Phosphate is also used in calcium phosphate to form hard, inorganic crystals that are important in building bones and teeth.

Sulfur

Sulfur, which is the only SPONCH element we haven't considered so far, has several minor roles. It appears in the sulfhydryl side group (—SH) of proteins, which serves as a kind of "snap" or "hook" that can link molecules together. Two sulfhydryl groups form a reversible sulfhydryl bridge when their hydrogens are removed with the aid of enzymes:

$$\mathrm{R{-}SH + HS{-}R \rightleftarrows R{-}S{-}S{-}R + H_2}$$

A Few Key Functional Side Groups

A *functional group* is a part of a molecule that occurs fairly often in different kinds of molecules. These functional groups have characteristic makeups and appear frequently enough to be given names. In addition, a functional group will behave pretty much the same, no matter what kind of molecule it's attached to. A side group, of course, is a group that sticks out from the main body of the molecule. Some important functional side groups are:

R—OH	Hydroxyl group
$\mathrm{R{-}NH_2}$	Amino group
R—COOH	Carboxyl group
R—SH	Sulfhydryl group
R=O	Oxygen group
R╲NH╱R	Imino group
$\mathrm{R{-}CH_3}$	Methyl group
$\mathrm{R{-}H_2PO_4}$	Phosphate or phosphatyl group

The phosphate group has already been discussed, but we're including it here so we can generalize about functional side groups. The short lines to the left of the symbols indicate covalent linkages to the remainder (main body) of the molecule. The remainder is symbolized by R.

Some of the functional side groups listed above will ionize when a certain pH is reached. Let's list these, a little more diagrammatically and as they occur in more acid and in more basic solutions:

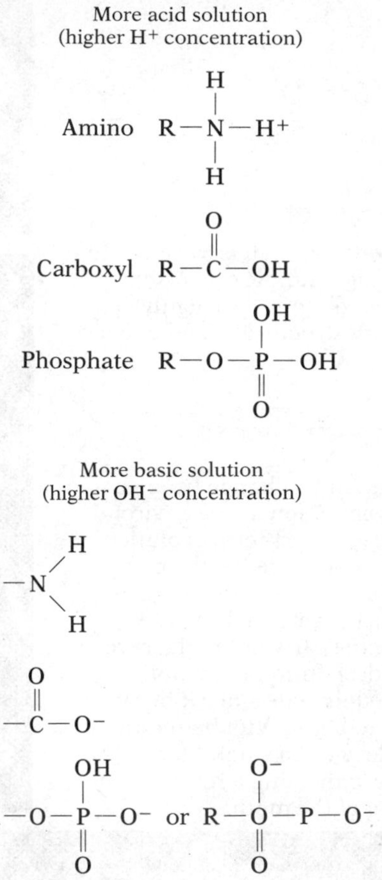

More acid solution
(higher H^+ concentration)

```
               H
               |
Amino      R—N—H+
               |
               H

               O
               ‖
Carboxyl   R—C—OH

                  OH
                  |
Phosphate  R—O—P—OH
                  ‖
                  O
```

More basic solution
(higher OH^- concentration)

```
      H
     /
R—N
     \
      H

    O
    ‖
R—C—O⁻

      OH                O⁻
      |                 |
R—O—P—O⁻   or   R—O—P—O⁻
      ‖                 ‖
      O                 O
```

At the normal pH range of living systems (pH 6–8), all these side groups are ionized (charged).

Chemical Reactions, Heat of Activation, and the Formation of Water

The *Hindenberg* (Figure 2.28), the most famous of the giant zeppelins, was filled with molecular hydrogen gas—unfortunately. Back in the thirties, the Third Reich intended to impress America with its bloated airship—and it succeeded. It indeed impressed

2.28 The hydrogen-filled dirigible, the *Hindenberg*, exploded in 1937. Hydrogen is rarely encountered in its molecular form (H_2), because it reacts readily with many other chemicals and is nearly always found combined with something. Hydrogen production is always risky because of its reactivity. The smallest spark or flame can provide the impetus needed for a rapid chain reaction such as the one shown here. We will never know what created the spark responsible for the *Hindenberg* disaster, but today, the few remaining lighter-than-air craft are filled with heavier, but unreactive, helium.

America. Somehow the dirigible caught fire while landing in New Jersey, and the hydrogen gas combined with the molecular oxygen of the atmosphere to produce water—fast!

But why did this happen? Why was the water formed so explosively? It was a matter of energy states. Water (H_2O) is in a lower energy state than an equivalent amount of H_2 and O_2. Thus, a mixture of molecular oxygen and hydrogen contains greater chemical potential energy than does water, and when the *Hindenberg* burned, the chemical mix quickly went to a lower energy state, producing low-energy water and releasing the excess energy as heat and light.

The molecular reaction of a tragedy may be written as

$$2\,H_2 + O_2 \rightarrow 2\,H_2O + \text{Energy!}$$

Note that the arrow has only one head because this is a highly directional kind of chemical reaction. The water won't easily change back to its components, hydrogen and oxygen.

So we see that oxygen and hydrogen will combine explosively to produce energy and water (actually water vapor). But don't try to impress a friend by mixing a little oxygen with hydrogen before his very eyes. He will be totally unimpressed because nothing will happen. Molecular hydrogen and molecular oxygen will just sit there with their outer orbitals nicely filled. What now? Perhaps a spark of intuition.

It turns out that with a tiny spark the mixture will blow his hat off. But why? Why didn't the mixture blow up without the spark? The reason why molecular oxygen and molecular hydrogen don't react with each other at room temperature is that the atoms must first be forced apart in order to rejoin violently. And they are forced apart by the spark that provides an intense input of energy.

Why would molecular oxygen and molecular hydrogen want to combine, anyway? We've seen that molecular oxygen and molecular hydrogen have already fulfilled their three basic energetic requirements of paired electrons, filled outer electron shells, and balanced net charge. Why should they combine to make water molecules?

A System Goes to Its Lowest Energy State

The answer is not easy. One way of stating it, as we've seen, is that $2\,H_2O$ is in a *lower energy state* than a mix of $2\,H_2 + O_2$, and that a *system tends to go to its lowest energy state*, giving off the energy difference in the form of heat and light. But this is really a description, not an explanation. And we haven't yet explained why a spark is necessary to get the reaction started. Let's try approaching the question a little differently.

Not All Covalent Bonds Are Equally Strong

The covalent bond between oxygen and hydrogen in water is stronger than the double covalent bonds between two oxygens or the bond between two hydrogens. By "stronger," we mean that it takes *more energy* to pull the water molecule apart than it does to pull the oxygen or hydrogen molecules apart.

Molecular bonds can be broken by heat. In a gas, for instance, most of the heat energy of the system is found in the *kinetic energy* of the rapidly moving molecules. (Kinetic energy is the energy of motion, or of momentum.) Those oxygens and hydrogens are bumping into each other at a great rate. Occasionally, the energy of a collision will force the molecules apart. For instance:

$$H_2 + \text{Energy} \rightarrow 2\,H$$

where H· represents the free atomic hydrogen atom (the dot represents the unpaired electron). Free atomic hydrogen is in a highly energetic state. Not only is its electron unpaired, but its outer electron shell is unfilled. How did the hydrogen get in that energetic state? The energy came from the thermal energy of the collision that broke the covalent bond. The above reaction actually *absorbs* energy. A reaction that absorbs energy is said to be *endergonic* (Figure 2.29). Although the thermal energy is "absorbed," it of course does not disappear. In its new form, the energy is said to be *potential chemical* energy. It is *potential* (stored) energy because it is available for other transformations.

In pure hydrogen gas, the above reaction would simply reverse itself when two free hydrogen atoms meet:

$$2\,H\cdot \longrightarrow H_2 + \text{Energy}$$

This reaction *produces* energy. A reaction that produces energy is said to be *exergonic.* A reversible reaction can be written with the usual double arrows:

$$2\,H\cdot \rightleftharpoons H_2 + \text{Energy}$$

2.29 A spontaneous endergonic reaction: the dissociation of a hydrogen molecule into two hydrogen atoms. An endergonic reaction is one that absorbs energy. Molecules in a gas move about because the temperature of the gas has given each molecule a certain amount of thermal kinetic energy. The molecules frequently collide. In most cases, such collisions result in the type of deflection common in billiards, with no net change in thermal kinetic energy. Occasionally, the energetic collision of two molecules will force the dissociation of one of the colliding hydrogen molecules, transforming it from a lower energy state (one hydrogen molecule) to a higher energy state (two hydrogen atoms). As thermal kinetic energy is transformed into chemical potential energy, there is a local decrease in the temperature. If the two hydrogen atoms should recombine (an exergonic reaction), their chemical potential energy would again be released as thermal kinetic energy (heat).

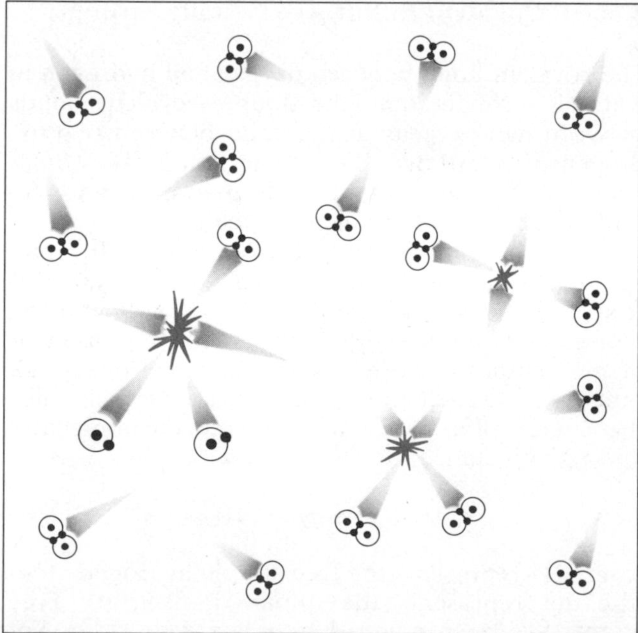

where the longer arrow indicates that, at equilibrium, most of the hydrogen will be in the H_2 form. The rate of bond breakage and reforming depends on the temperature of the gas, which is to say, it depends on the amount of thermal energy in the system. But, at equilibrium, there will be no net gain or loss in energy. Similarly, random thermal motion in molecular oxygen gas also breaks covalent bonds, which then reform spontaneously. Again, an equilibrium is reached and there is no net gain or loss in the total amount of energy.

But when the two gases are mixed, free atomic hydrogen and free atomic oxygen can combine to make a molecule of water. A lot of energy is given off in this reaction. (Actually, the formation of water doesn't occur in one step but involves several short-lived intermediate compounds, most of which are highly reactive.) The *overall net reaction* is

$$O_2 + 2\,H_2 \rightarrow 2\,H_2O + \text{Energy}$$

The important factor is the large net output of energy in the overall reaction, which is thus highly exergonic and essentially irreversible.

The reaction doesn't proceed at room temperature because the rate of spontaneous O_2 and H_2 bond breaking is too slow, and the heat produced by the few water molecules that are formed is dissipated too quickly. With the addition of concentrated energy from a spark or a flame, however, the relatively small input of energy necessary to break the H_2 and O_2 bonds is repaid many times over the output of energy from the formation of the strong bonds of the water molecule. This sets off a *chain reaction;* that is, it starts a cascade, an avalanche of reactions—or, we might say, the system catches fire. The energy required to break the original bonds and make the reaction go is called the *heat of activation.*

Catalysts Lower the Heat of Activation

Hydrogen and oxygen will combine to form water at room temperature if a *catalyst* is present. Catalysts provide shortcuts between the higher and lower energy states of a molecule, greatly lowering the heat of activation required to get things started. For instance, hydrogen and oxygen will combine readily, even at room temperature, if powdered platinum is present. Actually, the hydrogen first combines with the platinum and then with the oxygen, leaving the platinum in its original state (by definition, catalysts themselves are not changed by the reactions they initiate). Almost all new cars in the United States have platinum catalyst devices in their exhaust systems so that smog-producing unburned gasoline and other hydrocarbons can be combined with oxygen to form harmless carbon dioxide and water vapor. In biological systems, catalysts are called *enzymes.* We'll have much more to say about enzymes later.

So in this chapter we have discussed some small molecules, their interactions, and their parts. We'll get into more complicated molecules and their reactions in the next chapter. Of course, physicists would probably mumble that what we've been saying is all wrong—that elemental particles like protons and neutrons are made of strange strawberry quarks and other esoterica. And chemists may cluck over all the quantum mechanics we've left out. There is no end to physics or to higher chemistry, and the principles are indeed interesting, but these things can wait until we get a little better grip on the basics of chemistry as it applies to biology.

Why are we so concerned about chemistry in a book about biology? Why does so much of the rest of this book depend upon what we've discussed here? It's because the mechanisms of life's systems are so inextricably tied to the behavior of atoms and molecules that to understand anything about the nature of life, we must understand something about life's components, including our atoms, with their own blind, insistent energetic needs. From here we go to a discussion of larger molecules.

Application of Ideas

1. Reread Szent-Györgi's lament that opens this chapter. Actually Szent-Györgi is an extremely successful scientist, a Nobel Prize winner, and the discoverer of many of life's secrets (i.e., structure of vitamin C and much of the biochemistry of respiration). In what sense, then, did "life run out between his fingers"?

2. Assume you are part of a space probe assigned the task of looking for signs of life, as we know it, on distant planets. Make a list of the compounds and elements you would search for and explain why each of these would be vital to any conclusion you might make.

3. Elements like hydrogen, sodium, and chlorine are rarely found in their elemental form. Explain why this is true. From a periodic table of the elements, list other elements with the same chemical characteristics and try to determine whether these are ever found in their elemental form. (The organization of the periodic table will tell you what the others are.)

4. It is interesting to compare silicon with carbon, since they both have four electrons in their outer shells. Silicon and carbon can form long chains that are called silicones, which are structurally similar to hydrocarbons. In addition, silicon combines with oxygen (SiO_2) to form crystals of silica or quartz and readily combines with fluorine to form highly soluble SiF_4. This versatility has not escaped the attention of science fiction writers who have described extraterrestrial life based on silicon instead of carbon. What would such life forms be like? What might these beings have to say about the "special properties" of silicon and fluorine?

5. It could be argued that all of the reactions going on in living organisms are exergonic in their overall scheme of energy transfers. Support this argument.

Key Words

acid
adhesion
amino
ammonia
ammonium
aqueous
atom
atomic mass (weight)
atomic nucleus
atomic number

base
bicarbonate ion
biochemistry

capillary action
carbon dioxide (CO_2)
carboxyl

Key Ideas

Elements, Atoms, and Molecules

1. An *element* is a substance that cannot be separated into simpler substances by purely chemical means.
2. *Atoms* are the smallest indivisible units of *elements.*
3. A *molecule* is two or more *atoms* joined by a chemical *bond.*
4. A *compound* is a *molecule* containing two or more different *elements* in specific proportions.

Atomic Structure

1. *Atoms* consist of positively charged *protons,* neutral *neutrons,* and negatively charged *electrons.*
2. *Protons* and *neutrons,* also called *hadrons,* make up the central nucleus of an *atom. Electrons* are minute particles in motion about the nucleus.
3. The combined mass of the *protons* and *neutrons* (*hadrons*) is the *atomic mass* or weight of an *element.*
4. *Molecular mass* or weight of a *molecule* is the combined number of *hadrons* in that *molecule.*

catalyst
chain reaction
chemical bond
chemical reaction
cohesion
compound
covalent bond

dalton
dissociate
DNA
double bond

electron
electron acceptor
electron donor
electron shell
electrostatic
element
endergonic
energy
entropy
enzyme
exergonic

fossil fuel
free radical
functional group

gamma radiation

hadron
half-life
heat of activation
high-energy bond
hydration shell
hydrocarbon
hydrogen bond
hydrogen ion
hydronium ion
hydrophobic
hydroxyl
hydroxyl ion

imbibition
inorganic
ion
ionic bond
ionization
ionizing radiation
isotope

kinetic energy

macromolecule
metabolism
mole
molecule
molecular mass (weight)

neutron
nitrate

5. The *atomic number* of an *element* is the number of *protons* in its *atoms.*

6. *Isotopes* are *atoms* of an *element* with a different *atomic weight* than usual (more or fewer *neutrons*).

7. *Radioactive isotopes* are those which are unstable, emitting *energy* and/or matter.

Electron Orbitals and Electron Shells

1. *Electrons* move in definite paths (orbitals or *energy levels*) located at varying distances from the nucleus.

2. *Electron shells* can be visualized as concentric spheres each containing a specific number of *electrons.*

Electron Pairing

1. In most instances pairs of *electrons* with opposite spins travel together.

2. Where *electrons* are paired, there is considerable *energy* holding them together.

Filled and Unfilled Shells

1. *Atoms* have three chemically important energetic tendencies:
   a. *Atoms* tend to balance positive and negative charges.
   b. Orbiting *electrons* tend to travel in pairs.
   c. *Atoms* tend to fill their outer *shells.*

2. The shell-filling rule for the lighter *elements* (*SPONCH*) is 2-8-8. *Elements* with unfilled outer *shells* are chemically active and find ways to fill that *shell.*

The Noble Elements

The *noble elements,* including helium, neon, and argon have *saturated* outer *shells* and are not reactive.

How an Atom Can Fill Its Outer Electron Shell

Unsaturated *atoms* fill their outer *shells* in three ways:
a. They can lose outer *electrons* to another *atom.*
b. They can gain outer *electrons* from other *atoms.*
c. They can share outer *electrons* with other *atoms.*

Ions and the Ionic Bond

1. In their interaction, sodium, an *electron donor,* passes its outer *electron* to chlorine, an *electron acceptor.* The resulting unbalanced charges on each *atom* changes them to *ions.*

2. The attraction between negative and positive *ions* is known as an *ionic bond* or *salt linkage.*

3. *Ionic bonds* are weak and easily break, freeing the *ions,* when the substance is in water.

The Covalent Bond: Sharing Electron Pairs

1. In *covalent bonding, atoms* share *electrons,* reaching a preferred low-energy state, thus forming a stronger *bond* than the *ionic bond.*

2. Sharing satisfies the outer-shell-filling tendency of *atoms.*

3. Two pairs of *electrons* are often shared, forming *double bonds.*

Methane

1. Because carbon has two pairs of *electrons* in its outer *shell* it tends to form *covalent bonds* very readily.

2. Methane gas is a simple *compound* where carbon shares its four outer *electrons* with those of four hydrogen *atoms.* The methane molecule takes on the geometry of a *tetrahedron.*

Water

1. Water is essential to life, since organisms are composed largely of water and nearly all of life's chemical reactions occur in water.

2. In the formation of water, two hydrogens are bonded toward one side of the oxygen, producing a lopsided effect.

nitrogen-fixing
noble element
nonpolar attraction
nucleic acid

organic
oxidation-reduction

phosphate group
phospholipid
pH scale
polar
probability clouds
proton

quark

radiation
radioactive
radioactive isotope
reagent

salt linkage
shell
SPONCH
strong nuclear force
structural formula
subatomic
sulfhydryl
surface tension

thermal
tetrahedron

van der Waals forces

X-ray

The Hydrogen Bond and What, After All, Is Water?

Water Is a Polar Molecule
Water is polar because the hydrogen end of water has a slight positive charge, while the opposite end is negative.

Gas, Liquid, Crystals, and Hydrogen Bonds
1. Water molecules attract each other, hydrogen to oxygen, forming *hydrogen bonds.*
2. Below 0°C, the *hydrogen bonding* becomes more fixed, forming a rigid crystal known as ice.

Water Forms Hydration Shells Around Ions
Negative and positive *ions* attract the *polar* water *molecules* around themselves, forming *hydration shells.*

Water Is a Powerful Solvent
1. Water forms *hydration shells* around any *polar* substance, making it an excellent solvent.
2. The *adhesive* and *cohesive* action that accounts for water's rise in tubing is known as *capillary action.*
3. On a smaller scale, known as *imbibition,* water moves through porous substances.

Water Is Wet
Water has the tendency to adhere to, and spread over, solid surfaces through two mechanisms:
a. *Adhesion:* attraction to other substances
b. *Cohesion:* attraction to other water molecules

Water Has High Surface Tension
Through *cohesion,* water forms a tough surface film where it meets an air interface.

Essay 2.1 Acids and Bases
1. A minute proportion of water *molecules* form *hydrogen* (*hydronium*) and *hydroxyl ions* spontaneously.
2. When *hydrogen* (*hydronium*) *ions* outnumber *hydroxyl ions* in water, *acid* is formed.
3. The strength of *acids* (comparative number of *hydrogen ions* to *hydroxyl ions*) is measured on a *pH scale* from −1 through 6.
4. When *hydroxyl ions* outnumber *hydrogen ions* in water, bases form. The strength of a *base* is measured on a *pH scale* from 8 through 14.
5. Pure water is neutral with a *pH* value of 7.
6. Most chemical reactions in living organisms occur at pH 7.

Nonpolar Attraction, or van der Waals Forces
The tendency for *nonpolar molecules* to clump together in a weak binding manner is called *van der Waals forces.*

Carbon Compounds
1. *Organic compounds* contain carbon and are formed primarily in living organisms.
2. Carbon can form single, double, and even triple *bonds.*
3. A great variety of *molecules* with carbon backbones is possible.

Hydrocarbons
Hydrocarbons consist solely of carbon and hydrogen and are capable of forming simple chains, branched chains, and rings.

Nitrogen
1. Nitrogen is always a constituent of proteins and *nucleic acids* and is found in some lipids and carbohydrates.

2. Although the atmosphere is 78% molecular nitrogen, N_2 is unusable to nearly all forms of life. *Nitrogen fixing* organisms can convert N_2 to useful nitrogen *ions* such as *ammonium* and *nitrate.*

3. With five *electrons* in its outer *shell,* nitrogen is capable of forming three or four *covalent bonds* with other *elements.*

Essay 2.2 Carbon Dioxide and the Bicarbonate Ion

1. *Carbon dioxide* is a common respiratory waste of cells. It is also a by-product of combustion.

2. In water, CO_2 forms carbonic *acid,* which dissociates reversibly into *hydrogen* and *bicarbonate ions.*

Phosphorus and Phosphates

1. Phosphorus is utilized by organisms in the form of *phosphate ion* (HPO_4^{--} or $H_2PO_4^-$).

2. Bonds between adjacent *phosphates,* when broken, yield larger than usual quantities of *energy* and are known as *high-energy bonds.* Organisms use *phosphate* bonds to store and release *energy.*

3. *Phosphates* join with lipids to form *phospholipids,* important constituents of living membranes.

Sulfur

Sulfur is common in proteins, where it forms *sulfhydryl* bridges.

A Few Key Functional Side Groups

Functional groups are unique portions of *molecules* that *ionize* in water and interact with other substances in cells. They include $R{-}CH_3$, $R{-}SH$, $R{-}NH_2$, $R{-}OH$, $R{-}COOH$, $R{-}H_2PO_4$, and $R{=}O$.

Chemical Reactions, Heat of Activation, and the Formation of Water

1. In *exergonic* reactions, the reactants go from a higher to a lower energy state.
2. In *endergonic* reactions, reactants go from a lower to a higher energy state.
3. Energy of activation is the energy required to cause a chemical reaction.

A System Goes to Its Lowest Energy State

In *chemical reactions,* the tendency is for systems to go to their lowest energy states.

Not All Covalent Bonds Are Equally Strong

1. In the reaction between H_2 and O_2 to form water, the *covalent bonds* of water are stronger than those of the two gases.

2. When enough energy is available to a mixture of H_2 and O_2, a *chain reaction* is set off, resulting in the formation of water and the release of energy.

Catalysts Lower the Heat of Activation

1. *Catalysts* are substances that lower the *heat of activation* requirement.
2. Although *catalysts* interact with substances, they emerge unchanged.
3. *Catalysts* in living systems are called *enzymes.*

Review Questions

1. What are the apparent difficulties in defining life as expressed by Szent-Györgi? (p. 25)

2. Write the names, symbols, and *atomic numbers* of the chemical elements abbreviated in the term *SPONCH.* (pp. 25, 27)

3. Define the term *element* and list several examples. (p. 25)

4. Clearly distinguish between the terms *element, molecule,* and *compound.* (pp. 25–26)

5. Name and describe the three stable *subatomic* particles. Which of these comprise the nucleus? (p. 26)

6. Explain how the *atomic number* and the *atomic mass* of the *elements* are derived. (pp. 26–27)

7. In what way do *isotopes* of the same *element* differ? What is a *radioactive isotope?* (pp. 27–28)
8. Describe the phenomenon of *electron* pairing. (pp. 28–29)
9. How is the tendency of *atoms* to fill their outer *electron shells* important in chemical reactions? (p. 29)
10. State the *shell-filling rule* by using potassium (*atomic number* 19) as an example. (pp. 29–30)
11. Using argon as an example, explain the structure and behavior of the *"noble elements."* (p. 30)
12. Using potassium and chlorine (*atomic numbers* 19 and 17) as examples, define *ionic bonding* and show how it actually occurs between these two *elements.* (pp. 30–31)
13. Describe in detail what would happen to the compound formed in number 14 if it were placed in water. (pp. 31–32)
14. Using two *atoms* of nitrogen (*atomic number* 7) as examples, illustrate and explain *covalent bonding.* (p. 32)
15. Draw a *molecule* of ethane (C_2H_6) three ways, showing its chemical formula, its *structural formula,* and *space-filling* configuration. (Use methane for information.) (p. 33)
16. Using water as an example, explain what *hydrogen bonds* are. (p. 36)
17. Explain the behavior of water *molecules* and their *hydrogen bonds* at temperatures below 0°C, from O to 100°C, and above 100°C. (p. 36)
18. With a simple drawing, show how positive and negative *ions* form *hydration shells* around water *molecules.* (p. 36)
19. Describe six important characteristics of water. (pp. 36–38)
20. Define the terms *acid* and *base.* What determines the strength of each? (Essay 2.1)
21. List examples of substances with the following *pH* ranges: 7, 1, 8, 15, 2. What does the *pH* number actually refer to? (p. 41)
22. Using an example, describe the characteristics of *molecules* under the influence of *van der Waals forces.* (p. 39)
23. What is the usual behavior of *nonpolar molecules* in water? How can these *molecules* be dispersed? (p. 39)
24. For practice in visualizing carbon chains, draw *structural formulas* for the following: ethane (C_2H_6), propane (C_3H_8), pentane (C_5H_{12}), octane (C_8H_{18}). (pp. 39–40)
25. With all of the atmospheric nitrogen available, explain why plants are almost always in short supply of the *element.* (pp. 41, 44)
26. List three uses plants make of nitrogen. (p. 44)
27. Draw the *structural formula* for phosphate and list three uses for this *compound* in living things. (pp. 45, 48)
28. Identify the following *functional side groups:* $R-CH_3$, $R-SH$, $R-NH_2$, $R-OH$, $R-COOH$, $R-H_2PO_4$, $R=O$. (p. 48)
29. Review the *energy* states of molecular hydrogen, molecular oxygen, and water and explain the explosiveness of the reaction between the two gases. (p. 49)
30. Why would an *exergonic* reaction tend to be irreversible? (p. 50)
31. Define *heat of activation.* (p. 50)
32. What is the general role of a *catalyst* in chemical reactions? (p. 50)

Chapter 3
Large Molecules

Life is not simple. How many times we've heard that, usually from weary or worried people trying to live out theirs and hoping their problems are only part of a larger, unpleasant phenomenon. But even those of us who don't focus on life's problems would have to agree that it's complex. It's also complex in other ways—in its origins and mechanisms. Later on we'll say something more about guessing at origins, but here let's learn the fundamentals about the complex building blocks of a complex phenomenon.

In a sense, we can apply what we learned about small molecules to the study of large ones. Although large molecules can be dazzlingly complex, they are usually *polymers*—that is, of many parts. And those parts are small molecules.

In this chapter we'll look at the structures and functions of three major classes of large molecules: the big carbohydrates (called polysaccharides), proteins, and lipids. A fourth major class—the nucleic acids—is dealt with in Chapter 7. Actually, our categorizing may be unwarranted, because the four classes aren't really very distinct. Along the way we'll run into molecules that are part protein and part carbohydrate (glycoproteins and mucoproteins), some that are part protein and part lipid (lipoproteins), some that are combinations of protein and nucleic acid (nucleoproteins), and some that are part lipid and part carbohydrate (glycolipids).

3.1 The Carbohydrates

Monosaccharides, Disaccharides, and Polysaccharides

Carbohydrates are familiar to us as the sugars and starches in our diets. But, as we will see, they are important in other ways as well. Most carbohydrates have the empirical formula $(CH_2O)_n$. (Empirical formulas are reduced to their simplest terms.) The n tells us that any multiple of CH_2O can be found. Actually, the simplest common carbohydrates are $(CH_2O)_3$, or three-carbon compounds, but we will be primarily interested here in six-carbon carbohydrates and the way they are linked together to form the polymers of big molecules. Let's begin by looking at the organization of carbohydrates.

Like most large molecules in biology, carbohydrates are composed of repeated units of structure, covalently bonded together in a polymer. The basic unit is one of the *simple sugars*, or *monosaccharides*, as they are called. (*Mono-* means one, and *saccharide* means sugar.) There are many simple sugars, but the most familiar is glucose.

The next most complex carbohydrate is a disaccharide that is formed from two simple sugars. The most familiar of them is *sucrose*, or table sugar. Then, of course, more complex molecules can be formed from longer chains of simple sugars until we reach the *polysaccharides*, which may contain hundreds or thousands of simple sugars covalently linked into chains. Starch is one such polysaccharide. Others are glycogen (animal starch), cellulose, and chitin. Cellulose is a structural polysaccharide in plant cell walls, and chitin is the structural element of insect skeletons. Let's have a closer look now by first examining the structure of monosaccharides.

A *monosaccharide* consists of a short chain of carbon atoms (most commonly five or six carbon atoms), with nearly every carbon having a hydroxyl side group as well as a hydrogen side group (Figure 3.1). Note that the empirical formula $(CH_2O)_n$ for carbohydrates suggests one oxygen for every carbon. Usually, one carbon at the end of each simple sugar forms a double bond with its oxygen, producing an *aldehyde* group ($-CHO$). Simple sugars, we know,

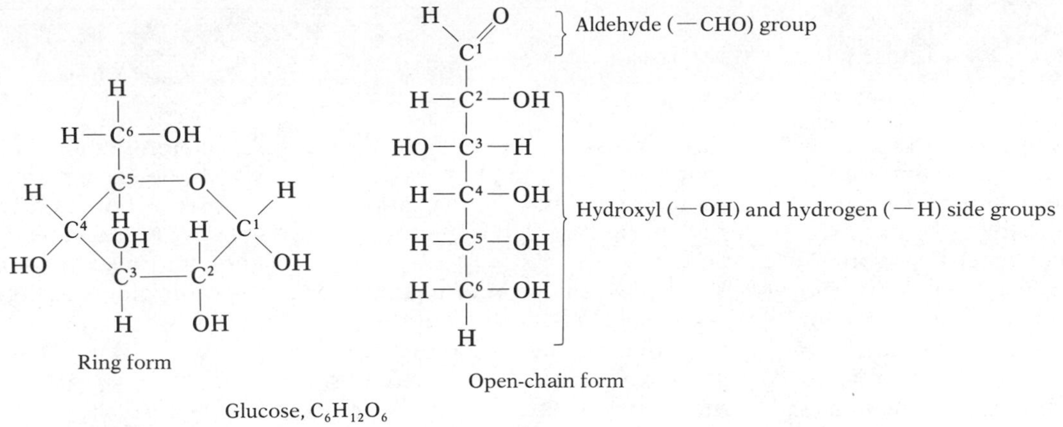

Aldehyde (—CHO) group

Hydroxyl (—OH) and hydrogen (—H) side groups

Ring form

Open-chain form

Glucose, $C_6H_{12}O_6$

3.1 Glucose is a simple sugar. Its chemical formula, $C_6H_{12}O_6$, can also be written as $(CH_2O)_6$, following the general formula for carbohydrates. Note the presence of hydroxyl and hydrogen groups and the aldehyde group, which is typical of many sugars. The numbers 1–6 identify the different carbons in the molecule.

can exist as rings or open chains, but in polysaccharides they are all in rings.

As we said, the most common and most important monosaccharide is *glucose,* the six-carbon sugar that is not only fundamentally involved in energy metabolism and photosynthesis, but is the building block from which starch, glycogen, and cellulose are all built. It is also one of the two subunits of sucrose. Glucose can be called blood sugar, corn sugar, or grape sugar, depending on its source. But wherever it comes from, it can be injected directly into the veins since it is also a digestive product of complex polysaccharides. (Physicians who do so often prefer its older name: "dextrose.")

Since sucrose is formed from glucose and fructose (Figure 3.2), it is a disaccharide. Technically, it is a twelve-carbon sugar consisting of two six-carbon sugars (glucose and fructose) linked together in a *dehydration linkage.*

Can you see from the figure what is meant by the term *dehydration linkage?* Two —OH side groups combine (R—OH + HO—R) to form an oxygen linkage (R—O—R) with the liberation of the hydrogens and oxygen as H_2O.

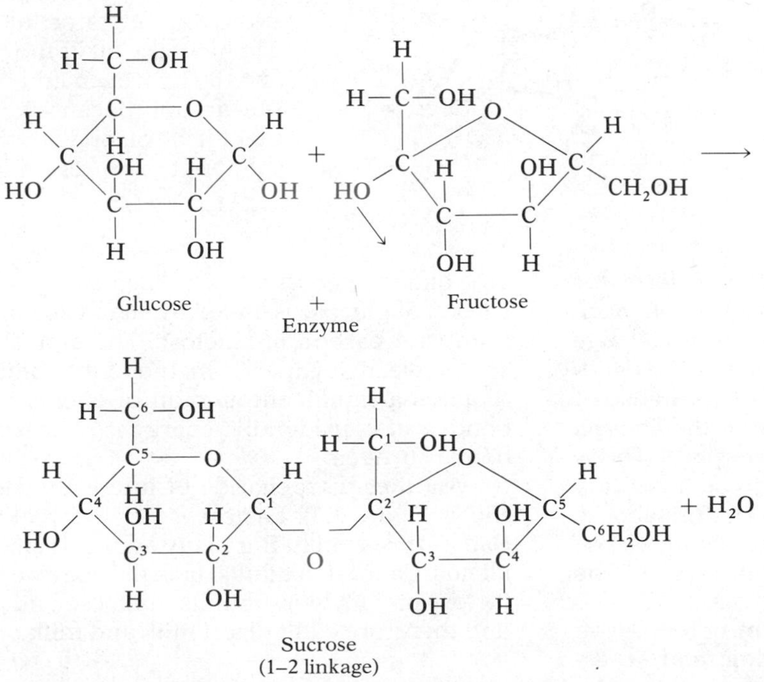

Glucose + Enzyme + Fructose →

Sucrose (1–2 linkage) + H_2O

3.2 Monosaccharides can be enzymatically joined to form disaccharides (double sugars). The diagram illustrates the manner in which chemical bonds join simple sugars to form the double sugar sucrose. In the presence of specific synthesizing enzymes, one —OH group combines with the hydrogen from an adjacent —OH group and a molecule of water is produced. The electrons of the remaining oxygen then pair up with those of the stripped carbon to form an oxygen bond. The process of enzymatic water removal, known as *dehydration,* is very common in the chemical activity of cells.

Essay 3.1 Reading Structural Formulas

A troublesome problem for many people when they are first introduced to the structural formulas of organic chemistry is the variable and inconsistent ways in which they are written. Why, they wonder, can't chemists pick some standard system of structural formulas and stick to it? The reason is, each of the systems used has its own strengths and weaknesses, and none will do for all purposes. In some cases, very fine detail is needed; for instance, if it is important to show where the electrons are, the electron shell diagram

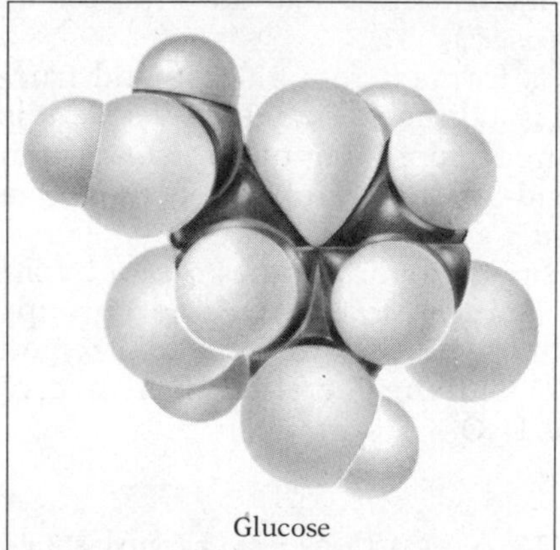
Glucose

may be appropriate. But most of the time one doesn't need to know where the electrons are.

As larger and larger molecules are considered, the detail has to be suppressed more and more so that the important features can be seen. Side groups can be simplified, or perhaps left out altogether.

Examples may be helpful at this point, so consider some of the ways that can be used to represent glucose. There is the space-filling model below left. Then there are two types of complete structural formulas with tapered bonds to indicate the three-dimensional structure—a ring form and a "chair" form:

A skeleton structural formula, in which the carbons that form the ring are left out and hydrogen side groups are indicated by short lines, may also be used to depict the same glucose ring:

And then there is the bare bones structural formula, showing only the ring, the oxygen in the ring, and the terminal hydroxyl group:

The last configuration is often used when glucose is incorporated into a polysaccharide, such as starch or cellulose. You can mentally fill in the carbons and the hydrogen and hydroxyl side groups, because you know that the hexagon with the tail is supposed to represent glucose.

In your gut (the biologists' indelicate term for gastrointestinal tract), the sucrose is broken back down into glucose and fructose by *hydrolytic cleavage* (*hydro*, water; *lysis*, rupture), which is just the opposite of a dehydration linkage. With the assistance of an enzyme, a water molecule is added to the linkage, breaking the sucrose into its component parts. This is necessary since your gut can't absorb sucrose very efficiently and the cells of your body can't metabolize it. (If you gorge yourself on table sugar, some sucrose will get into your bloodstream, but all of it will be excreted in your urine.)

The carbons of each sugar are numbered as you learned earlier when we drew the dog, and, in accordance with the system, the linkage between glucose and fructose is a 1–2 linkage, since the number 1 carbon of glucose is linked by an oxygen bridge to the number 2 carbon of fructose. The dehydration linkage between sugars is a high-energy bond; that is, it requires a significant amount of energy to form the bond, and considerable energy is released when the bond is broken.

Another disaccharide of interest is *lactose* (milk sugar, Figure 3.3). Lactose is not as sweet as sucrose. Babies thrive on it. But many adults (including nearly all non-Caucasian adults) lack the enzyme that breaks lactose into its two subunits—glucose and galactose—and therefore can't digest milk and milk products.

Galactose

Glucose

Lactose, $C_{12}H_{22}O_{11}$

3.3 Lactose, or milk sugar, consists of one molecule of glucose and one molecule of galactose. Note the slight difference between the two. These sugars are covalently bonded by dehydration (losing H_2O), yielding lactose and water. The formula of lactose is identical to that of sucrose, but the molecular structures of the two compounds and their chemical behaviors are quite different. As you can see, the glucose molecule was inverted when the bond formed. This is necessary because of the angles between the two —OH groups involved.

Starches

Starches form a large part of our diet, perhaps too large a part. Potatoes, bread, and Fritos are high-calorie foods largely because of their high starch content. The starches are important reservoirs of food, which are readily converted to energy. Amylose and amylopectin are the two most common food reserves of higher plants, and glycogen is a food reserve of animals. All three starches are made entirely of glucose subunits.

Amylose

Amylose is the simplest starch (Figure 3.4). Its molecules consist of unbranched chains of hundreds of glucose subunits, with 1–4 dehydration linkages between successive *monomers* (single units). Potato starch is about 20% amylose. In the gut or in a sprouting potato, breakdown (digestion) occurs in three steps: one enzyme breaks amylose into fragments of varying size, attacking the starch at random points; a second enzyme works on the ends of the fragments, cleaving off two glucose units at a time as disaccharides (glucose-1,4-glucose, or *mannose*); and a third enzyme cleaves the disaccharide into glucose monomers.

3.4 Polymers of glucose are exceedingly important to life. The polysaccharide amylose is a storage starch found in plants. It consists of hundreds of glucose molecules linked together at their number 1 and number 4 carbons. Amylose is insoluble in water and will form a helical structure as shown in part (b) as a result of its repulsion of water. Cooking amylose (boiling potatoes) fragments the chain, which becomes more soluble and easier to digest. The grains shown here have been photographed through polarizing filters, which reveal the regular structure of the starch molecules.

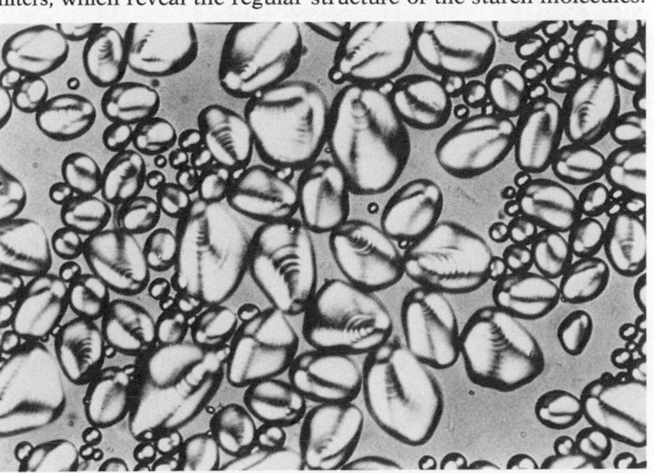

(a) Unbranched chain

(b) Helical structure

Amylopectin

The other 80% of potato starch is *amylopectin,* which is a large molecule consisting of short 1–4 glucose chains *cross-linked* with occasional 1–6 and 1–3 linkages between chains (Figure 3.5). The individual straight chains are twenty to thirty glucose units long. The amylopectin in rice starch is made up of about 80 to 100 such chains cross-linked into a huge mesh, with one or two cross-links per chain. The same enzymes that cleave amylose also break up amylopectin, but yet another enzyme is required to cleave the 1–6 and 1–3 linkages.

Glycogen

Glycogen is the reserve polysaccharide of animals; it is usually a much larger molecule than amylose or amylopectin, and is a highly branching structure. The predominant linkage is 1–4, but there are many 1–6 linkages as well. Glycogen is a temporary, short-term storage unit in animal cells, particularly prevalent in the liver and muscles of vertebrates (Figure 3.6).

When glycogen is metabolized within cells, the dehydration linkages are enzymatically cleaved by *phosphorylase.* The cleavage is called a *phosphorolytic cleavage* or *phosphorolysis.* The enzyme adds inorganic phosphate to the molecule instead of water (as in hydrolytic cleavage), and the product is not glucose but glucose-1-phosphate. This little trick saves the high-energy of the glucose-1,4-glucose bond as a high-energy phosphate bond, thus preparing the molecule for the vital energy-yielding reactions to follow.

Primary and Secondary Structure of Starches

The straight- and branched-chain structures described above are called the *primary structures* of the starches. The primary structure of a *macromolecule* (very large molecule) results from covalent bonding alone, and usually can be shown as a two-dimensional figure. Primary structure ignores folding of the molecule that may be due to salt linkages, van der Waals forces, or hydrogen bonds. When we referred to amylose as an unbranched chain, we had the primary structure in mind. In reality, amylose is a three-dimensional structure that tends to loop and form folds and helices, as shown in part (b) of Figure 3.4. Amylose forms helices because of hydrogen bonding between subunits along its length. These helices make up the *secondary structure* of the molecule. Whereas we are fairly confident about the secondary structure of amylose, the secondary structure of amylopectin and glycogen are less well understood.

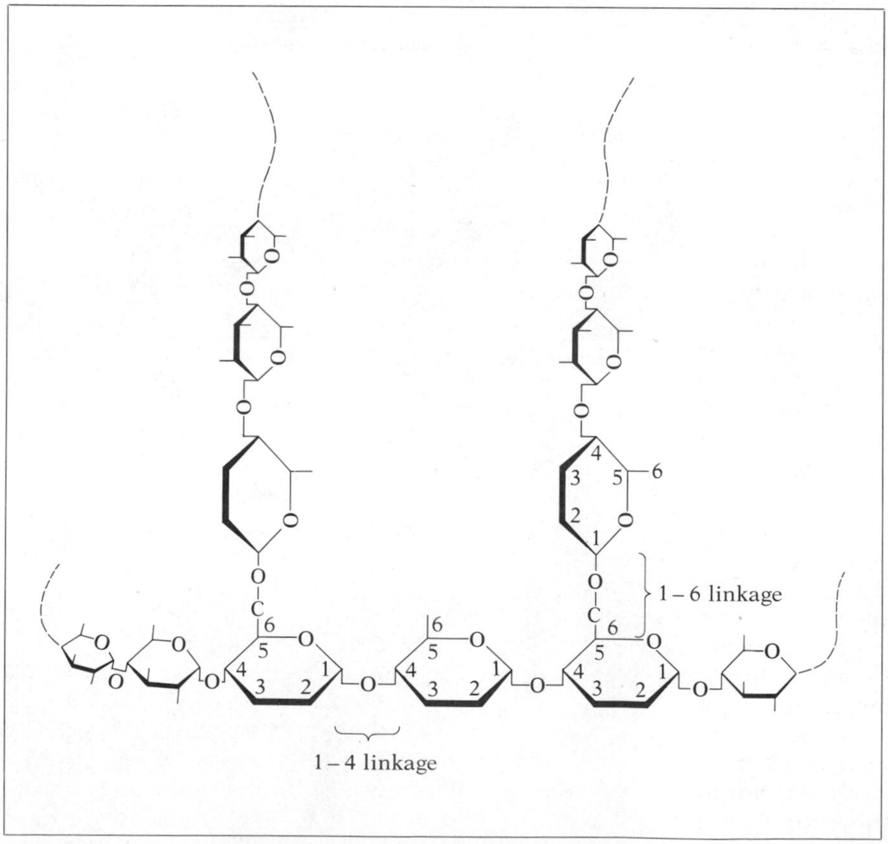

3.5 Amylopectin is a cross-linked polysaccharide. The short, helical chains consist of glucose subunits, joined at carbons 1 and 4 by enzymatic dehydration. The chains are joined by 1–3 and 1–6 linkages, which complicate digestion, since additional enzymes are required to break these linkages. Both amylose and amylopectin are excellent sources of glucose for animals.

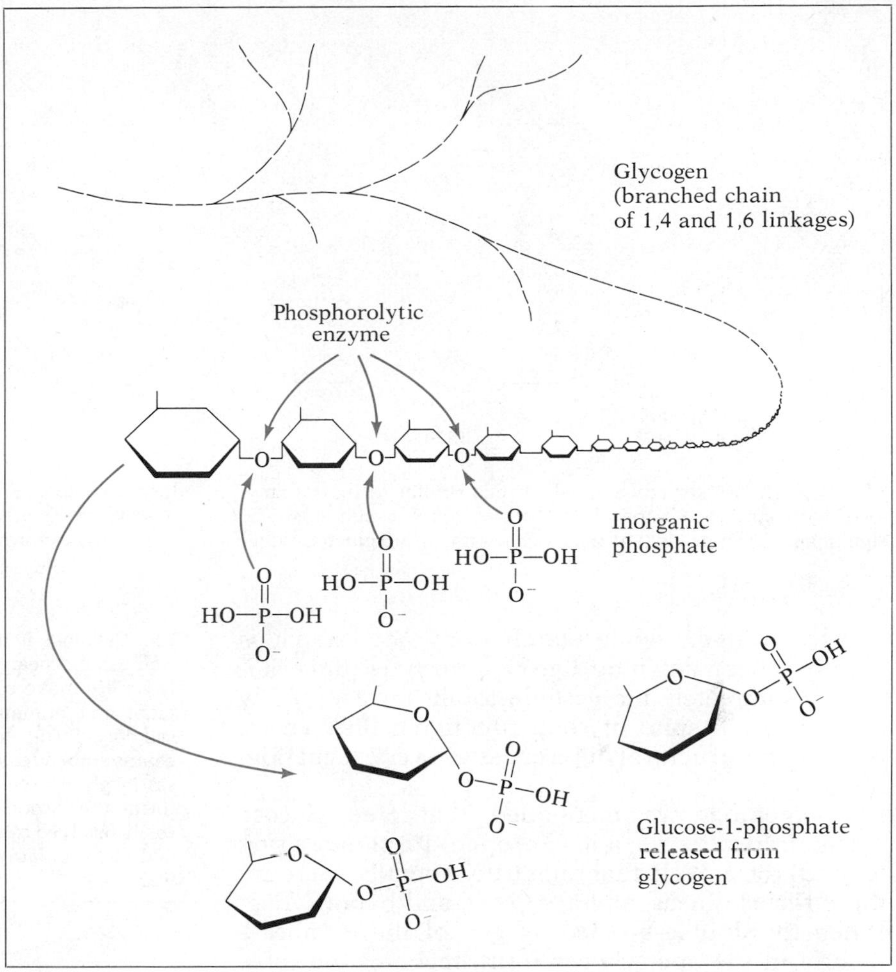

3.6 Animals, using glucose from plants, synthesize their own starch, the polysaccharide glycogen. Glycogen is similar to amylopectin with 1–4 linkages in the straight chains and 1–6 linkages where branching occurs. It is, however, more highly branched, heavier, and more compact than its plant counterpart. Glycogen is stored in skeletal muscle and in the liver cells, where it forms large granules. Storage is very temporary, however, since glycogen is rapidly converted to glucose-1-phosphate and transported throughout the animal body to cells where energy demands must be met. In the enzymatic conversion, phosphorolytic enzymes, rather than hydrolytic enzymes, are employed. The phosphorolytic enzyme utilizes inorganic phosphate in cleaving the 1–4 linkages between glucose subunits. The product is glucose-1-phosphate, which is valuable to cells as an energy source.

Polysaccharide Synthesis

The first step in polysaccharide synthesis from monosaccharide subunits is to come up with the energy needed to bind the subunits together. The energy comes from the high-energy phosphate bond of adenosine triphosphate (ATP). The ATP donates a phosphorus, with its high-energy bond, to glucose to form glucose-1-phosphate (all controlled by enzymes, of course). Then, in a reversal of phosphorolysis and under the control of other enzymes, the glucose-1-phosphates are joined by removing the phosphate while retaining the high-energy bond. At this point you may begin to think we were misleading you when we described the linkage between glucose subunits as a dehydration linkage, since in practice it is a *dephosphorylation* linkage. Perhaps we were, but we can say in our defense that the *net reaction* involves the loss of an oxygen and two hydrogens from the glucose subunits.

Polysaccharides are very variable molecules in the sense that the exact structure and molecular weight varies from one molecule to the next, even within a single cell. Because they are so variable, they can't be described precisely, and we assume that their synthesis is therefore not very rigidly controlled (as is the incredibly precise protein synthesis).

Structural Polysaccharides

An example of a *structural polysaccharide* is cellulose. Whereas plant and animal starches humbly function as glucose reservoirs, cellulose and other structural polysaccharides have a quite different function: they hold cells or even whole organisms, together. Chitin is another structural polysaccharide.

Cellulose

Cellulose is a linear polymer of glucose subunits put together with 1–4 dehydration linkages. If that sounds familiar, it's because amylose is also a linear polymer of glucose subunits put together with 1–4 dehydration linkages. But of course the starch amylose and cellulose are very different. Starch is fairly soluble, while cellulose is not; cellulose has great

Amylose

Alpha glucose

Cellulose

Beta glucose

3.7 The primary structure of cellulose is similar to that of amylose, with straight chains of 1–4 linked glucose subunits. The significant difference is that the glucose units are β-glucose rather than α-glucose. Since the position of the number 1-OH groups is upward, every other β-glucose subunit is inverted as the dehydration enzymes form the linkages.

tensile strength, while starch does not; starch is readily broken down by digestive enzymes, but cellulose is completely indigestible to all but a very few organisms. In spite of such functional differences, though, the structural differences are very slight (Figure 3.7).

Remember we mentioned that free glucose occurs in two forms, a linear (open) form and a ring (closed) form. Well, that's not true. Actually, there are three forms: linear, alpha (α) ring, and beta (β) ring. When the double-bonded oxygen of the number 1 carbon in the open, linear form becomes part of a hydroxyl group, there are two possible positions that the hydroxyl can take: pointing down (α-glucose) or pointing up (β-glucose), as shown in Figure 3.7. When pure glucose is in solution, about 34% of the molecules are in the alpha ring form, 65% in the beta ring form, and a fraction of a percent in the open, linear form.

In cellulose, then, glucose units are linked by what is called a *β-glycoside bond,* which is strong and very difficult to break. Relatively few organisms produce the enzyme *cellulase* necessary to break such a linkage. Garden snails can digest cellulose (perhaps you've seen the holes they eat in leaves and damp newspapers). Fungi and some bacteria have cellulases. Some animals don't produce any cellulase themselves, but harbor symbiotic intestinal protozoa that can break down cellulose and are themselves ultimately digested: cows and termites are the best examples. As for you, you can eat all the cellulose you want and it will just go right through. In fact, it may go through pretty fast because it abrades the gut lining and causes lubricating fluids to be released. Whole-bran cereals are very rich in cellulose.

The strength of cellulose lies in its organization, as seen in cell walls. As Figure 3.8 indicates, cellulose

3.8 Cellulose is a structural polysaccharide that forms a tough wall around nearly every cell in the plant body. Its strength and flexibility make cellulose an extremely valuable material to both plant and human consumer. (a) Unlike other polysaccharides, cellulose forms lengthy *microfibrils.* These consist of cellulose chains embedded in cementlike substances including pectin, extensin (a glycoprotein) hemicellulose, and lignin. This combination forms the "wood" of woody plants. (b) The microfibrils can be easily resolved by the electron microscope. Rows of microfibrils are laid down in laminated form, which explains the strength of cell walls.

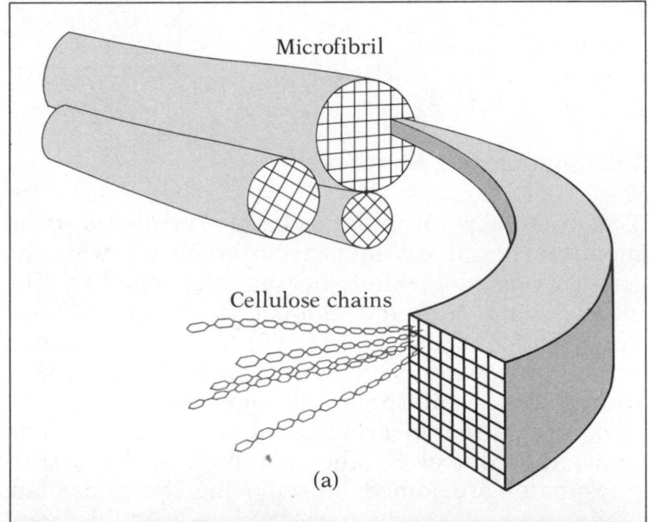

(a)

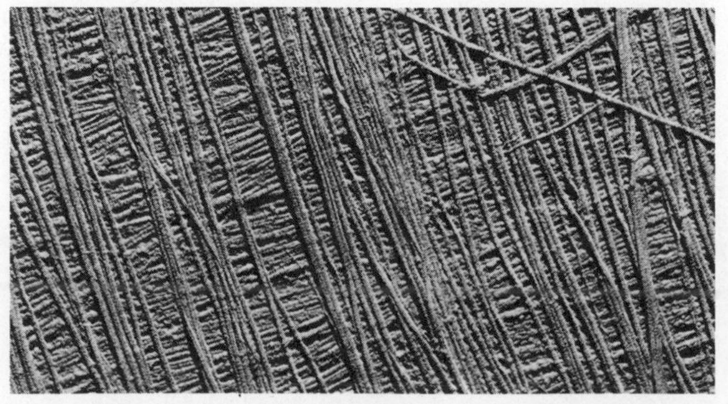
(b)

chains are organized into larger and larger units like the strands of a cable. These dense units are clearly visible in cell walls as seen with the electron microscope. Although cellulose is insoluble and strong, it is also flexible and even soft. After all, facial tissues are almost pure cellulose but it isn't like wiping your nose on a block of wood. Wood we know, is hard. The hardness of wood, though, comes from another class of large molecules, the *lignins,* which are superimposed on a cellulose matrix (Figure 3.8). Lignins are not polysaccharides but are in a class of their own. Wood processors, such as paper mills, have devoted a great deal of time in trying to find a use for lignin produced as a by-product but they've been only partially successful. For example, it can be decomposed into wood alcohol.

Other Structural Polysaccharides of Plants

In plant cell walls, cellulose fibers are imbedded in a gluey matrix of *hemicelluloses* and *pectins.* The hemicelluloses, despite the name, are not structurally related to cellulose but are polymers of some of the less common five carbon sugars. *Pectin* is the substance that gives jelly its consistency. The strong cell walls of higher plants can be compared to iron-reinforced concrete, or resin-impregnated fiberglass, or any other materials that owe their strength to rods or fibers imbedded in an amorphous matrix.

Other structural polysaccharides are found in various seaweeds (e.g., algin, agar, carrageen), from which they are extracted and sold to the food industry for such esoteric uses as thickeners in milk shakes. Also, many bacteria secrete slimy protective coats of polysaccharide material.

Chitin

Chitin is a principal constituent of the *exoskeletons* (shells) of insects and other arthropods, including lobsters and crabs (Figure 3.9). Chitin itself is rather soft and leathery, but can become very hard when impregnated with calcium carbonate. It is a structural polysaccharide similar in many ways to cellulose, except that instead of glucose, the basic unit is *N*-acetyl glucosamine. Notice that this carbohydrate contains nitrogen as well as carbon, oxygen, and hydrogen. [The *N*-acetyl part of the name of this compound means that an acetyl group ($-COCH_3$) is attached to the main structure through a nitrogen atom.] Chitin is indigestible to most animals, and you might keep in mind that eating too many insects can cause diarrhea.

Mucopolysaccharides

Another important group of nitrogen-containing polysaccharides of animals includes the *mucopolysaccharides,* which, as you might expect from the name, are a bit slimy. *Hyaluronic acid* is a mucopolysaccharide with an acid (carboxyl) side group. It shows up in quite a variety of places, such as in the jellylike substance filling the eye, in the lubricating fluid of joints, in the umbilical cord, and as a cementing substance holding the skin to the body. A closely related mucopolysaccharide is *chondroitin,* the principal component of cartilage and a significant part of adult bone as well as the cornea of the eye. Hyaluronic acid is made of repeating units of glucuronic acid and *N*-acetyl glucosamine; chondroitin is made of glucuronic acid and *N*-acetyl galactosamine.

N-acetyl glucosamine

Chitin

3.9 The lobster exoskeleton is composed of the complex carbohydrate chitin, a polymer of *N*-acetyl glucosamine. Arthropods commonly use this polymer as skeletal material, but in the lobster and other marine forms, additional firmness is provided by calcium carbonate.

3.2 The Proteins

Proteins are incredibly important molecules. And their importance is increasing as we add 70 million people each year to a hungry planet. As we will see, proteins are very large and complex, but they are made of smaller kinds of molecules that have been well-described. We will see that nitrogen groups are essential parts of proteins and that protein is extremely important to such things as brain development and the regulation of bodily processes. But these same nitrogen-containing groups form deadly poisons in animal bodies, and elaborate mechanisms have been developed within the body to eliminate the nitrogens animals try so hard to find in the first place. Let's see if we can put all this into some sort of perspective.

We've seen that carbohydrates function both as food reserves and as structural molecules. Proteins function in the same ways, but they have other uses as well. For example, they may take the form of enzymes and hormones. We'll say more about these later.

First, what *is* a protein? A *protein* is composed of one or more *polypeptides,* and may also contain saccharides, lipids, or other small molecules. Obviously, another definition is in order: A polypeptide is a linear polymer made up of a chain of *amino acids* linked together by dehydration linkages known as *peptide bonds.* The distinction between polypeptide and protein is not always clear, by the way. Many polypeptides are proteins and all proteins contain polypeptides. Usually, the term *protein* is reserved for a naturally occurring, biologically active molecule, whereas a polypeptide can be any linear chain of amino acids.

Amino acids are the building blocks of proteins. There are just twenty naturally occurring amino acids, and each of them has different and important properties. Twenty seems a small number when we consider how complex proteins are, but we might remember that the *Rise and Fall of the Third Reich* was written with only twenty-six letters. So let's have a closer look at these remarkable molecules and see if we can generalize about their structures.

The Structure of Amino Acids

First of all, each of the twenty amino acids has the structure shown in Figure 3.10. It is in the R portion of the molecule that the twenty amino acids differ from one another, as we will consider below.

The R group is always attached to a carbon atom. Attached to the same carbon atom are three other groups: an amino group, which is basic (in other words, it has a positive charge); a carboxyl group, which is acid (that is, it has a negative charge); and a hydrogen. In Figure 3.10, the amino group and the carboxyl group are shown in the ionized state, which is the actual condition of amino acids in solution under normal physiological conditions. An amino acid is sometimes said to be a *zwitterion* (in German this means "half-breed ion"), since it is both negatively and positively charged.

To make a polypeptide, the carboxyl group of one amino acid is linked, through a dehydration linkage, to the amino group of the next amino acid; the carboxyl group of the second amino acid is linked to the amino group of the third; and so on down the line. Each carboxyl–amino linkage is a peptide bond. The complete polypeptide has an *N*-terminal, or amino end (the first amino acid in the chain) and a *C*-terminal, or carboxyl end (the last amino acid). Polypeptides are traditionally written from left to right with the amino (basic) end on the left and the carboxyl (acid) end on the right (Figure 3.11).

The amino acid sequence of a protein is rigidly determined (as we'll see later), and specific proteins are therefore far more homogeneous than are polysaccharides. Proteins have precise, three-dimensional configurations determined by the amino acid sequence. (Although, in solution, a protein molecule may be somewhat flexible and, in enzymatic activity, may have "moving parts.") Proteins tend to be either long and stringy (fibrous proteins) or folded into nearly spherical balls, or globs (globulins).

Chemists continue to work long and hard to unravel the structure of proteins. The first thing they want to know is, what is the sequence of amino acids (the *primary* structure)? Then, how tight is the coil of

3.10 The basic structure of each amino acid includes a short carbon chain with at least one amino and one carboxyl (acid) group attached. The amino acids differ from each other in the kinds of side groups (R) attached. The simplest amino acid, glycine, has one hydrogen atom at this location. The illustration shows a general amino acid in two ways: nonionized and ionized. In the ionized state (whenever water is present, which is almost always), the carboxyl group releases a proton, which tells us it is an acid group. The amino end collects a proton, thus acting as a base.

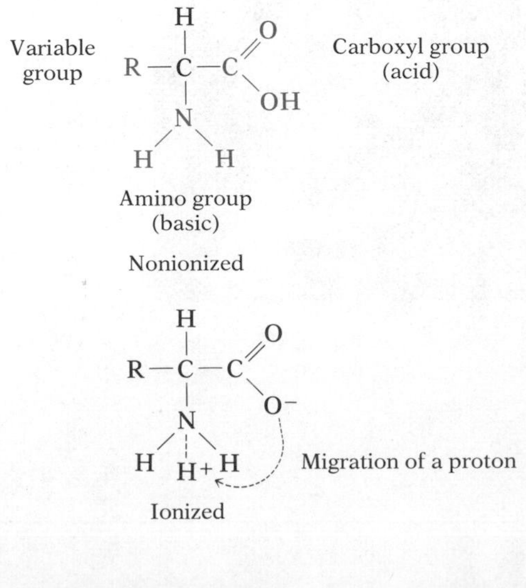

Variable group

Carboxyl group (acid)

Amino group (basic)

Nonionized

Migration of a proton

Ionized

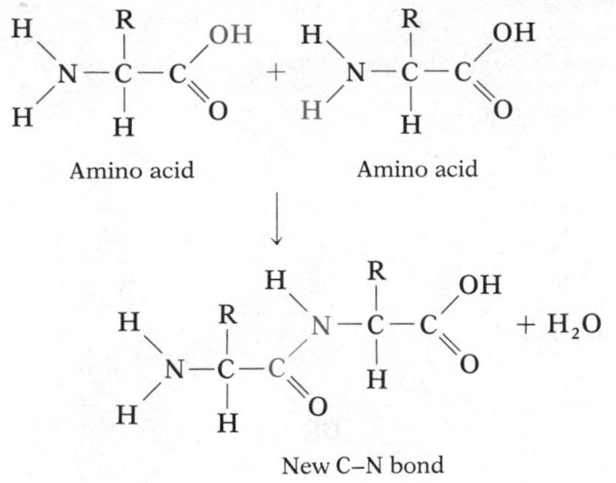

Amino acid + Amino acid

↓

+ H_2O

New C–N bond

3.11 Amino acids are joined through peptide bonds in the simplest or primary level of protein organization. Peptide bonds always occur between the carboxyl and amino groups of adjacent amino acids. A polypeptide is formed when many amino acids have joined in this manner.

the chain (the *secondary* structure)? And finally, how does the chain loop back on itself again and again to produce the rounded glob that is the functional protein (the *tertiary* structure)? The *quaternary* structure refers to the way in which separate protein molecules are joined together. Many proteins are *dimers* (two parts), while others, such as hemoglobin, may be *tetramers* (four parts).

To summarize, the forces at work in determining the shape of a protein include (a) the sulfhydryl covalent bonds between cysteine groups, (b) hydrogen bonding, (c) electrostatic or ionic interaction, and (d) hydrophobic interaction (see Essay 3.2). Each of these forces depends on the presence of specific amino acids in positions where the active groups are close together. In other words, the primary structure of a protein or polypeptide ultimately determines the kinds of folding a protein will undergo to assume its final shape.

The Twenty Amino Acids and Their R Groups

Table 3.1 lists all twenty amino acids and shows their R group structures. The R groups of the amino acids determine the physical and chemical properties of the proteins in several ways. First, such general properties of the protein as solubility and electrical charge depend on the net effect of all the side groups. Second, a few of the amino acid side groups of an enzyme participate directly in some enzymatic reactions. Third, the interactions between the different amino acid R groups of a given protein—especially such interactions as van der Waals forces, ionic bonding, hydrogen bonding, and even covalent bonding —determine the three-dimensional shape of the protein (Essay 3.2).

Proteins can become nonfunctional, such as by *denaturation*. The denaturation may be reversible or permanent. For example, proteins often may be reversibly denatured by adding urea or certain other small molecules to the solution. The urea competes for the hydrogen bonds that normally link some of the protein's side groups. When denatured in urea, most proteins lose their normal shape and all their enzymatic functions. However, if the urea is removed, the proteins usually regain their former shape and their functions.

Proteins may be irreversibly denatured by agents such as heat that break the usual covalent bonds and cause others to form. To illustrate how irreversible such processes are, try to unfry an egg!

The R groups interact in the four ways shown in Essay 3.2. A positively charged side group in one part of a protein molecule will be strongly attracted to a negatively charged side group in another part of the molecule, tending to bend the molecule into a particular configuration. This is an *ionic bond.* There are three amino acids (arginine, histidine, and lysine) that have additional (secondary) amino groups and are thus always positively charged at physiological pH (pH 6–8). And there are two other amino acids that have secondary carboxyl groups (aspartic acid and glutamic acid) so that they are negatively charged at physiological pH.

Among the remaining fifteen amino acids, a number are polar because their R groups contain hydroxyls or other functional groups capable of forming hydrogen bonds. Thus, they are *hydrophilic* (water-loving). These amino acids are asparagine, glutamine, glycine, serine, threonine, and tyrosine. The five ionic amino acids are also hydrophilic. Each of these has other special characteristics as well.

Of the remaining nine amino acids, eight are *hydrophobic* (water-hating). Just as oil and water separate, hydrophobic and hydrophilic side groups of a protein will not join. The hydrophobic side groups are attracted to one another by van der Waals forces. Most globular proteins have hydrophobic centers (where the hydrophobic amino acid R groups get together) and hydrophilic exteriors (which form hydration shells in an aqueous medium). The eight hydrophobic amino acids are alanine, isoleucine, leucine, methionine, phenylalanine, proline, tryptophan, and valine.

That leaves just one amino acid, cysteine, and this one has a very special kind of interaction. Cysteine has a *sulfhydryl* side group (—SH). Under mildly oxidizing conditions (which we'll consider later), two sulfhydryl groups will react to make a sulfhydryl bridge:

$$R{-}SH + HS{-}R + \tfrac{1}{2}O_2 \rightarrow R{-}S{-}S{-}R + H_2O$$

Sulfhydryl bridges formed between two cysteines in different parts of a protein molecule provide covalent

Table 3.1 The twenty amino acids* commonly found in proteins

Amino acids with nonpolar R groups

Alanine (Ala)

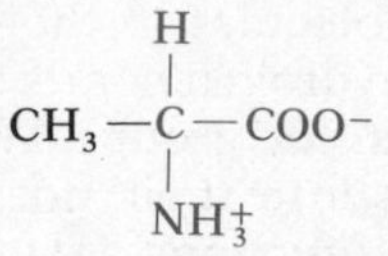

$$CH_3-\overset{H}{\underset{NH_3^+}{C}}-COO^-$$

Isoleucine (Ile)

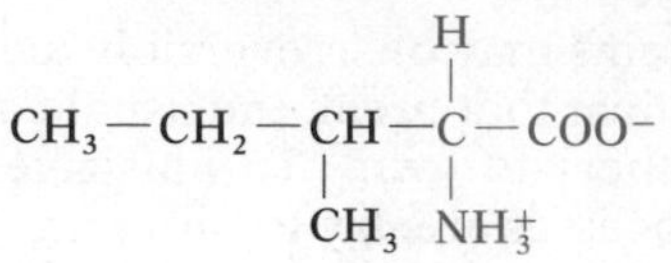

$$CH_3-CH_2-\underset{CH_3}{CH}-\overset{H}{\underset{NH_3^+}{C}}-COO^-$$

Leucine (Leu)

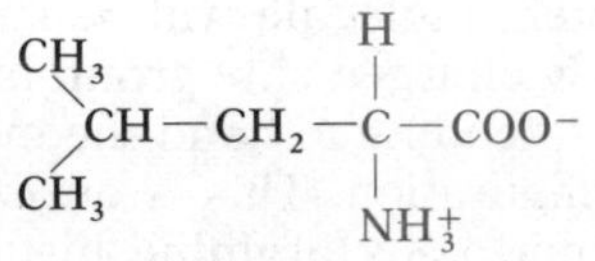

$$\overset{CH_3}{\underset{CH_3}{CH}}-CH_2-\overset{H}{\underset{NH_3^+}{C}}-COO^-$$

Methionine (Met)

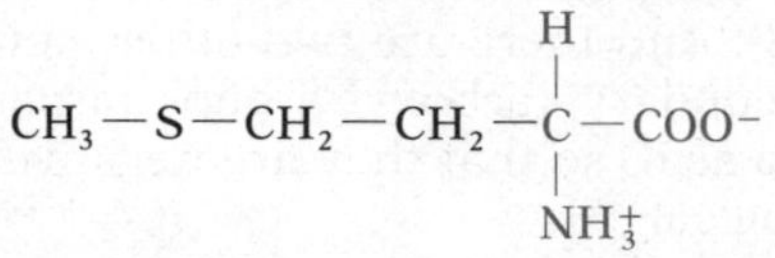

$$CH_3-S-CH_2-CH_2-\overset{H}{\underset{NH_3^+}{C}}-COO^-$$

Phenylalanine (Phe)

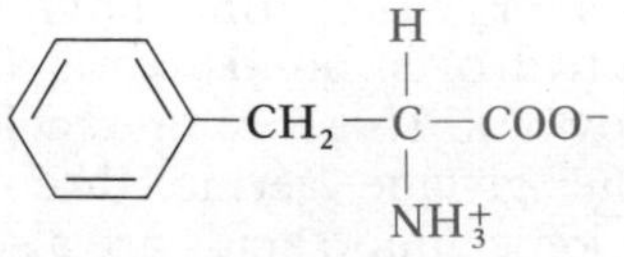

$$C_6H_5-CH_2-\overset{H}{\underset{NH_3^+}{C}}-COO^-$$

Proline (Pro)

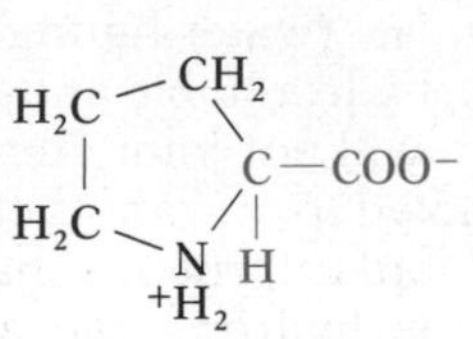

$$\begin{array}{l} H_2C-CH_2 \\ \;|\qquad\quad \backslash \\ \;|\qquad\quad\; C-COO^- \\ H_2C-\underset{+H_2}{N}-\!\!/\,|\,H \end{array}$$

Tryptophan (Trp)

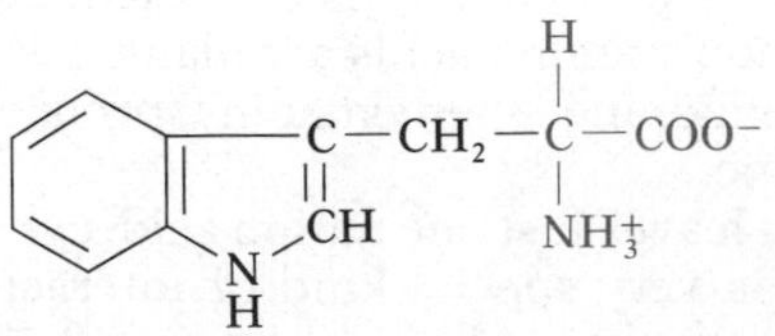

$$C_6H_4\text{-}\underset{\underset{CH}{\|}}{C}-CH_2-\overset{H}{\underset{NH_3^+}{C}}-COO^- \quad (\text{ring closed through } \underset{H}{N})$$

Valine (Val)

$$\overset{CH_3}{\underset{CH_3}{CH}}-\overset{H}{\underset{NH_3^+}{C}}-COO^-$$

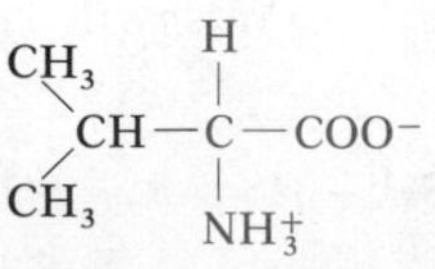

Amino acids with uncharged polar R groups

Asparagine (Asn)

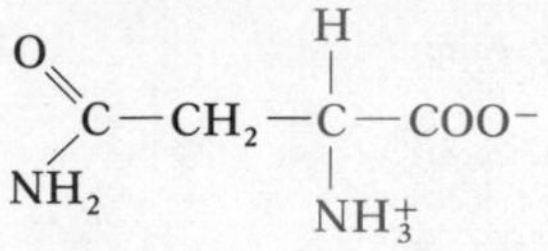

$$\overset{O}{\underset{NH_2}{C}}-CH_2-\overset{H}{\underset{NH_3^+}{C}}-COO^-$$

Cysteine (Cys)

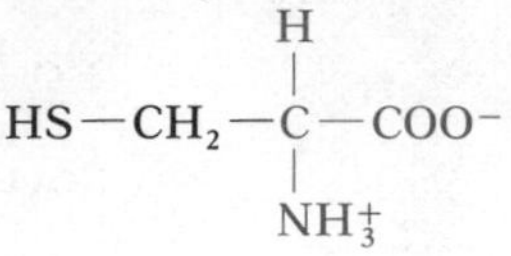

$$HS-CH_2-\overset{H}{\underset{NH_3^+}{C}}-COO^-$$

Glutamine (Gln)

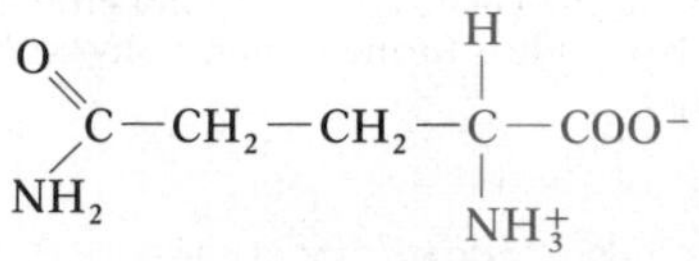

$$\overset{O}{\underset{NH_2}{C}}-CH_2-CH_2-\overset{H}{\underset{NH_3^+}{C}}-COO^-$$

Glycine (Gly)

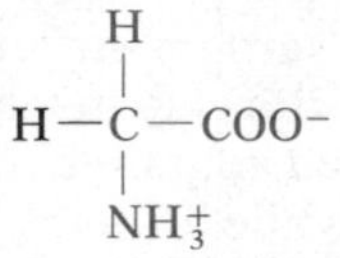

$$H-\overset{H}{\underset{NH_3^+}{C}}-COO^-$$

Serine (Ser)

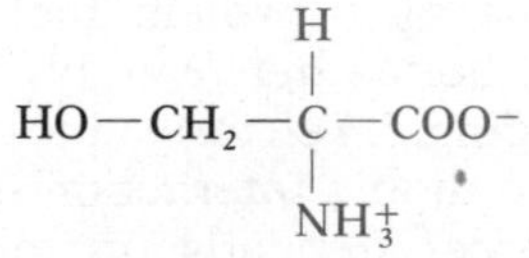

$$HO-CH_2-\overset{H}{\underset{NH_3^+}{C}}-COO^-$$

Threonine (Thr)

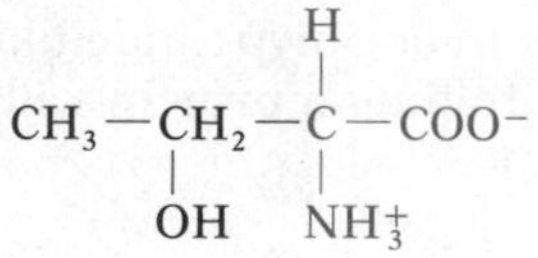

$$CH_3-\underset{OH}{CH_2}-\overset{H}{\underset{NH_3^+}{C}}-COO^-$$

Tyrosine (Tyr)

$$HO-C_6H_4-CH_2-\overset{H}{\underset{NH_3^+}{C}}-COO^-$$

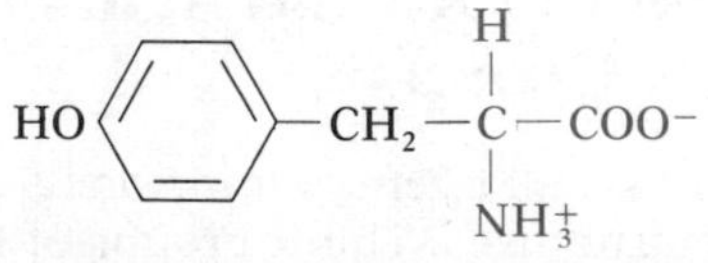

*The portion of the amino acid that is common to all is colored. Note that some of the amino acids contain more than one amino or acid group, giving them greater basic or acid qualities than the others. Cysteine contains a sulfur-hydrogen group at its *R*-terminal. This has special importance in determining the shapes of proteins. The abbreviations given in parentheses are used for convenience in writing protein formulas.

Table 3.1 *(continued)*

Amino acids with acid R groups (negatively charged at pH 6.0)

Aspartic acid (Asp)

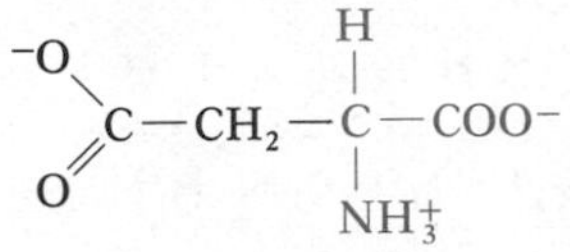

$$^{-}O(O{=})C-CH_2-\underset{NH_3^+}{\overset{H}{C}}-COO^-$$

Glutamic acid (Glu)

$$^{-}O(O{=})C-CH_2-CH_2-\underset{NH_3^+}{\overset{H}{C}}-COO^-$$

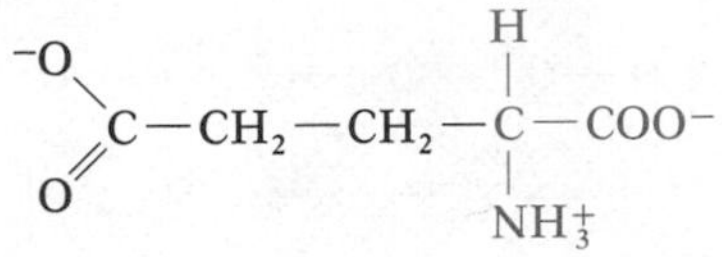

Amino acids with basic R groups (positively charged at pH 6.0)

Arginine (Arg)

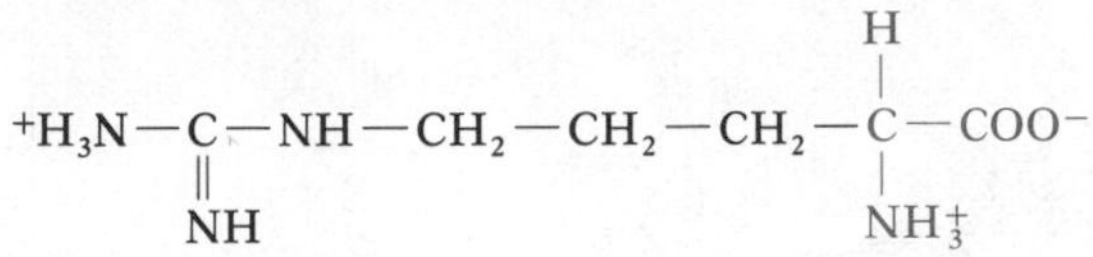

$$^{+}H_3N-\underset{NH}{\overset{}{C}}-NH-CH_2-CH_2-CH_2-\underset{NH_3^+}{\overset{H}{C}}-COO^-$$

Histidine (His)

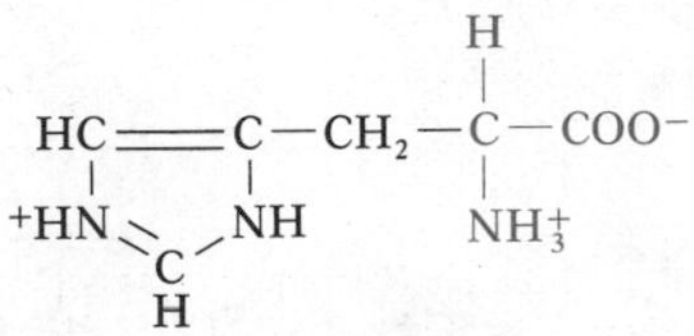

$$HC{=}C(-NH-\underset{H}{C}{=}N^{+}H-)-CH_2-\underset{NH_3^+}{\overset{H}{C}}-COO^-$$

Lysine (Lys)

$$^{+}H_3N-CH_2-CH_2-CH_2-CH_2-\underset{NH_3^+}{\overset{H}{C}}-COO^-$$

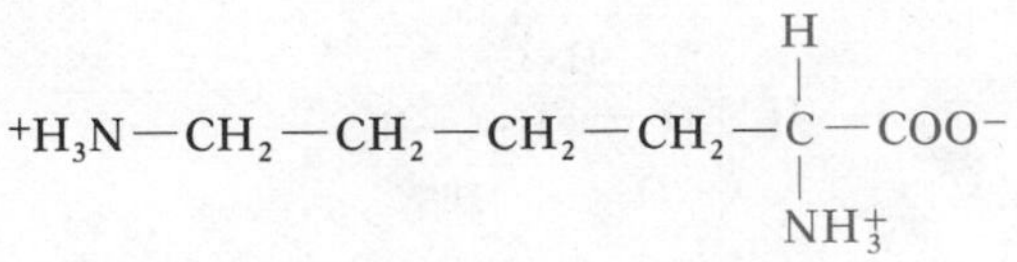

cross-links that hold the looping and winding molecule together. *Keratins,* the structural proteins of skin and hair, are very rich in sulfhydryl cross-linkages. Sulfhydryl bonds can be broken by some of the concoctions found in home-permanent kits. They act by reversibly denaturing hair molecules, allowing them to assume new, and hopefully more aesthetic, shapes.

Conjugated Proteins and Prosthetic Groups

A protein that consists only of amino acids, in one or more polypeptides, is called a *simple* protein. *Conjugated* proteins have something extra, either covalently bonded to the amino acid side groups or otherwise bound to the polypeptides. The something extra is called the *prosthetic group* (it might help to remember that a wooden leg, a glass eye, or a dental bridge is called a *prosthesis,* or added part). Among the conjugated proteins are nucleoproteins (here, the prosthetic group is nucleic acid), glycoproteins (prosthetic group is carbohydrate), chromoproteins (*chromo* refers to color, so the prosthetic group is a pigment), lipoproteins (prosthetic group is fat-soluble), and others. The hemoglobin molecule is a conjugated protein, because each subunit contains a *heme group* which, in turn, contains an iron atom (Figure 3.12). Since heme is deeply colored, hemoglobin is a chromoprotein, although pigmentation has nothing whatever to do with its function. Blood just *happens* to be red.

The prosthetic group of a protein often determines function. The catalytic activities of many enzymes, for instance, are dependent on *coenzymes,* which are small prosthetic groups attached to the enzyme. Neither the coenzyme by itself nor the remainder of the enzymatic protein alone has any catalytic activity. Many coenzymes have parts that cannot be synthesized by animals, so they must be included in the diet. These are called *vitamins.* No vitamins, no enzyme activity.

3.12 Hemoglobin, the respiratory pigment of many animals, is a conjugated protein consisting of four polypeptides (as shown in Essay 3.2). Each of the polypeptides contains a prosthetic group known as a *heme.* The heme groups contain iron and have the ability to bind with oxygen, forming *oxyhemoglobin.* The association of oxygen and hemoglobin is reversible, with oxygen released as blood circulates in oxygen-deficient tissues.

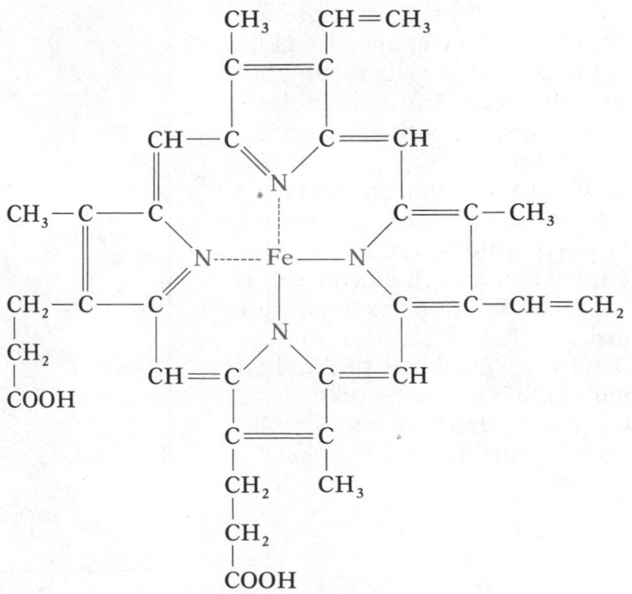

CH_3 $CH{=}CH_3$; $CH-C$, $C{=}CH$; CH_3-C-C, $C{=}C-CH_3$; N ---- Fe — N; CH_2-C-C, $C{=}C-CH{=}CH_2$; CH_2, $COOH$; $CH{=}C$, $C{=}CH$; $C{=}C$; CH_2 CH_3; CH_2; $COOH$

Heme group

Essay 3.2 Protein Structure

(A) The *primary structure* of a polypeptide or protein is the simple linear strand of amino acids attached together by peptide bonds. The number, kind, and order of amino acids is very specific for each protein. The number of amino acids incorporated into the primary structure varies greatly in the many different proteins.

(B) The *secondary structure* of proteins is commonly an alpha helix as shown. The helical folds are brought about chiefly through the attraction of R-terminal oxygen and hydrogen atoms, which form hydrogen bonds throughout the helix.

(C) In its *tertiary structure,* the alpha helix is itself coiled and folded into a very specific organization. This is commonly seen in globular proteins. The forces involved are complex and include several kinds of interactions among the amino acids. The clearest are the sulfhydryl bridges between cysteines, which actually form covalent bonds as shown in the model. In addition, hydrogen bonding, electrostatic attraction, and van der Waals forces are involved. The final shape of a globular protein is very significant, since this type of protein includes the biologically active enzymes.

(D) Some of the globular proteins assume a final *quaternary structure* level. Here, two proteins (or polypeptides) are interacting to form a *dimer.* Quaternary forces are the same as those at the tertiary level. The most familiar quaternary protein is probably the hemoglobin of red blood cells, which includes four interacting protein units (it is a tetramer).

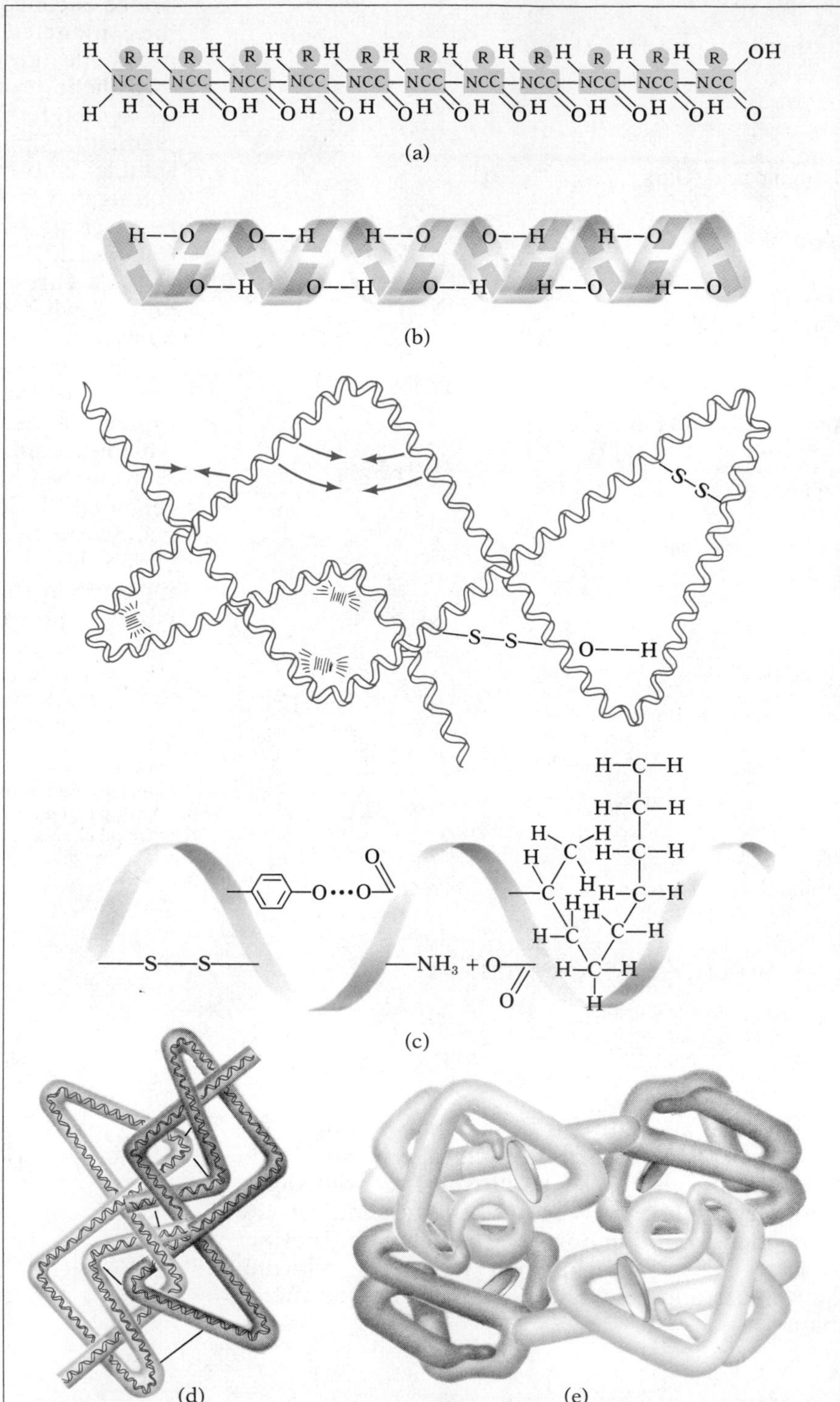

3.3 Biologically Active Proteins

Enzymatic Activity

In Chapter 2, we discussed the preferred energy states of atoms and molecules and the tendency of a mixture of molecules in a closed system to go to the lowest possible total energy state. For instance, a mixture of hydrogen (H_2) and oxygen (O_2) is not in its lowest energy state, but tends to react to form the lower energy state, H_2O, with the release of energy in the form of heat and light. However, in order for the reaction to proceed, an initial heat of activation, such as a spark, must be provided. The reaction itself then provides enough heat to be self-sustaining. Alternatively, a catalyst could be provided (Figure 3.13). In the case of H_2 and O_2, powdered platinum works well, as described in Chapter 2. A catalyst enables a chemical mixture to pass from one net energy state to another, lower net energy state, and (by definition) undergoes no net change itself. Enzymes are biological catalysts in that they function in living systems.

Proteins make efficient enzyme catalysts because their shapes can be defined so well. A crucial part of any enzyme is its *active site*, which is usually a groove or depression on the surface of the protein molecule, sometimes containing a prosthetic group. The shape of the active site, and the configuration of ionic bonding, hydrogen bonding, and van der Waals forces present in the active site, cause the enzyme to interact specifically with a certain *substrate*. The complex of enzyme and substrate is called, appropriately enough, the *enzyme–substrate complex*, or ES.

Enzymes work their wonders in different ways. For example, it is thought that most enzyme active sites are not rigid, but that they can move a bit to help the enzyme attach to its specific substrate. When the enzyme and substrate are positioned correctly, one or more molecules of the substrate may react directly (but reversibly) with an amino acid side group, or with the enzyme's prosthetic group. Or the enzyme may act only to bring two reactive substrate groups close together so that they can react on their own. In any case, when the enzyme dissociates from the substrate, the substrate or molecules are changed to a lower total energy state and can be called the *product* (P). The reaction is written:

$$\underset{\text{Substrate}}{S} + \underset{\text{Enzyme}}{E} \rightarrow \underset{\text{Enzyme–substrate}}{ES} \rightarrow \underset{\text{Product}}{P} + \underset{\text{Enzyme}}{E}$$

Since the enzyme comes out unscathed, the net reaction is:

$$S \rightarrow P$$

It should be noted that two or more substrates may be involved and that often there are two or more products. This is true because enzymes may act not only to join two molecules, but to break complex arrangements into their component parts. For instance, in the enzymatic hydrolysis of sucrose, the substrates (the substances entering the reaction) are sucrose and water, and the products are fructose and glucose:

$$\underbrace{\text{Sucrose} + H_2O}_{S} + \underset{E}{\text{Enzyme}} \rightarrow ES \rightarrow \underbrace{\text{Glucose} + \text{Fructose}}_{P} + \underset{E}{\text{Enzyme}}$$

(In the phosphorolysis of sucrose, the substrates are sucrose and phosphate, and the products are fructose and glucose-1-phosphate.)

In theory, all enzyme-catalyzed reactions are reversible. The theory states that if a high enough concentration of the product(s) is provided, the enzyme will change some of the product back into substrate. Although it is true that some enzymatic

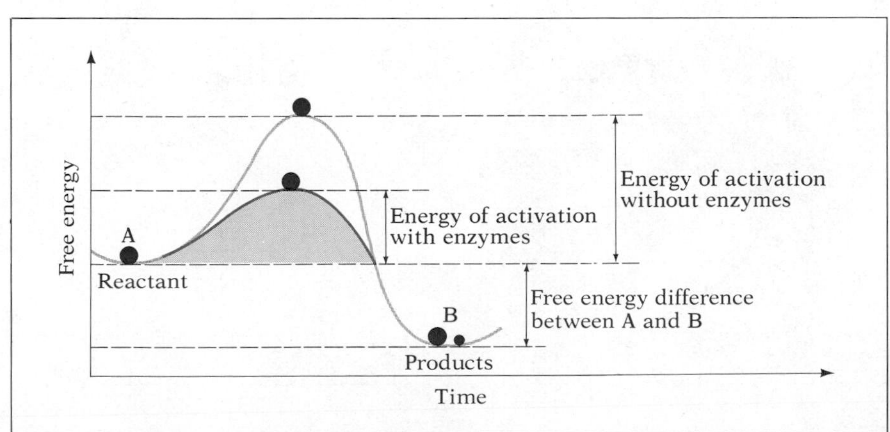

3.13 The graph compares the activation energy requirement for a reaction with and without an enzyme. The reactant (A) is stable at ordinary temperatures. To form the product (B), the activation energy barrier must be overcome. Without an enzyme, this would usually mean a considerable amount of heat must be applied, as shown by the upper curve. Enzymes, in ways that are not entirely known, provide the activation energy for biological reactions at considerably lower temperatures, as shown in the bottom curve. Thus, the energy barrier is much lower for an enzyme-catalyzed reaction.

Essay 3.3 How a Hydrolyzing Enzyme Works

An enzyme "recognizes" and attaches to a specific substrate. The substrate generally fits into an opening in the enzyme much as a missing piece might fit into a three-dimensional jigsaw puzzle. The substrate molecule is held in place by hydrogen bonds and van der Waals forces. The enzyme itself is not inactive, but a complex and dynamic structure. In most enzymes the substrate fits into a flexible *cleft* that opens to accept the substrate, closes down on it to perform the catalytic reaction, and opens again to eject the products of the reaction. The energy for this movement may come from random thermal energy or from the energy of the reaction itself. Notice that the initial fit between the enzyme and the substrate is not precise; this puts a mechanical stress on both molecules, leaving certain places particularly vulnerable to chemical interaction.

Here (a) depicts the initial fitting of the substrate into enzyme cleft. The bond to be broken occurs between two parts of the substrate, R_1 and R_2. Note that one of the amino acid side groups of the enzyme projects a hydrogen into the active center. This hydrogen will participate in the enzymatic reaction, a reaction that is essentially a hydrolytic cleavage. (The side group of the enzyme is shown as a dark spot.)

The amino acid side group of the enzyme moves to form an energetically favorable, but temporary bond with the substrate (b). It is said to "attack" the bond between R_1 and R_2, breaking it. The hydrogen from the active center of the enzyme moves to R_1, and the enzyme is now covalently bonded, through its side group, with R_2. R_1-H is one product of the reaction, which we can call P_1. Because of conformational changes in the enzyme, P_1 is no longer bound and it diffuses away as a water molecule diffuses in. The water molecule dissociates into a hydroxyl and a hydrogen (c). The hydrogen replaces the hydrogen that was originally a part of the enzyme active center, opening up the temporary enzyme–substrate bond, and the hydroxyl becomes attached to the R_2 subunit. The R_2-OH compound, which is the second product of the reaction (or P_2), has little affinity for the enzyme and drifts away. The enzyme is now restored to its original state. Since each of the intermediate states is energetically more favorable than the state preceding it, the system is now in its lowest available energy state.

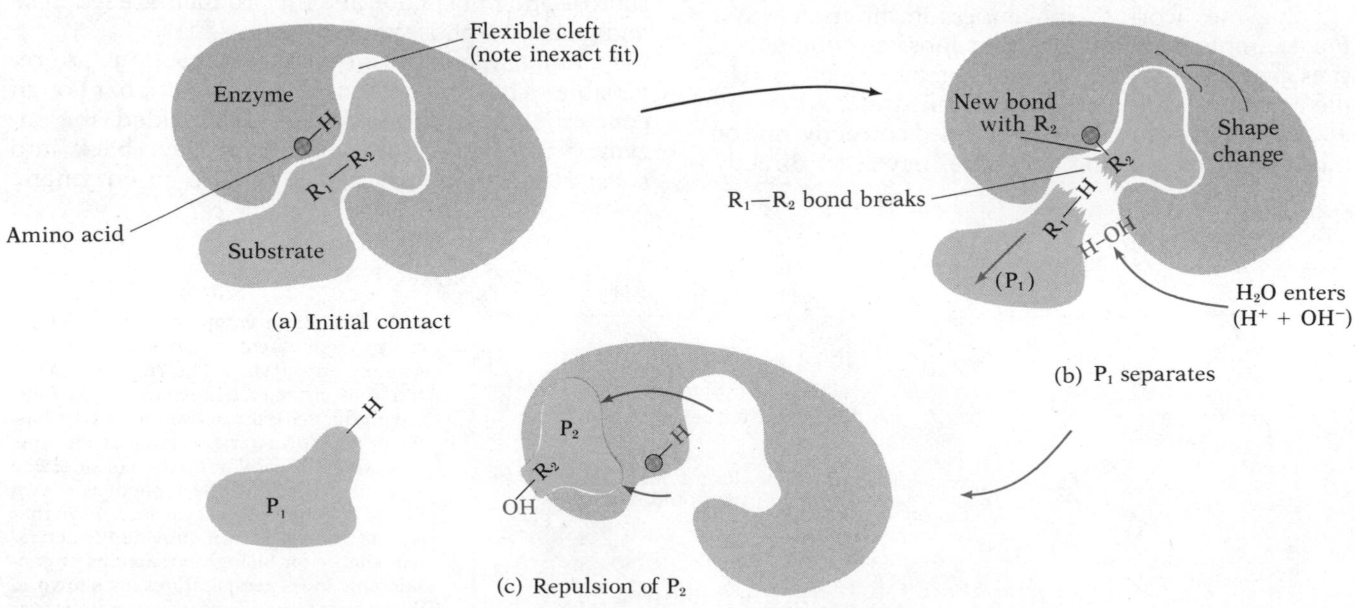

(a) Initial contact

(b) P_1 separates

(c) Repulsion of P_2

reactions actually are reversible, in fact, most are extremely one-directional and the conditions of the theory are seldom met. Problems arise, for instance, because it may be physically impossible to get a high enough concentration of the product or to measure the results of a reversed reaction, or the enzyme may become denatured in the process.

Another theory states that catalysts only speed up reactions that would happen anyway, although at a much slower rate. Whatever the technical merits of this theory, it has rather little practical application: a mixture of adenosine triphosphate (ATP) and glucose will never of itself produce measurable amounts of glucose-1-phosphate, though this is a simple enough enzymatic reaction. (Left to itself, in fact, a mixture of ATP and glucose would slowly decompose in a variety of other ways.)

You may have noticed another theoretical problem in all this. If enzymes always bring about lower energy states, how is it that some enzymatic reactions appear to be exactly the opposite of other enzymatic reactions? For instance, why is glucose enzymatically converted into starch, and at other times starch is enzymatically converted into glucose?

It turns out that there are several answers. To begin with, in most cases of synthesis and degradation, the reactions turn out not to be the same when all substrates and products are taken into account. For instance, in the case of starch synthesis, another coupled reaction takes place, the degradation of ATP to adenosine diphosphate (ADP) plus phosphate. The total set of reactions is this:

$$\text{ATP} + \text{Glucose} + \underset{\text{Enzyme}}{E_1} \rightarrow \text{ADP} + \text{Glucose-1-phosphate} + E_1$$

$$\text{Glucose-1-phosphate} + \underset{\text{Enzyme}}{E_2} \rightarrow \text{Starch} + \text{Phosphate} + E_2$$

The net reaction is this:

$$\text{Glucose} + \text{ATP} \rightarrow \text{Starch} + \text{ADP} + \text{Phosphate}$$

Although the synthesis of starch from glucose is endergonic (energy-requiring, or going from a lower energy state to a higher energy state), the hydrolysis of ATP into ADP and phosphate is exergonic (energy-releasing, or going from a higher energy state to a lower energy state). The two reactions are coupled, which, in this case, means that the energy provided by ATP is used in starch synthesis. The net reaction is exergonic (from a total higher energy state to a total lower energy state). In this example, two enzymes are involved sequentially, but often two coupled reactions are mediated simultaneously by a single enzyme (Figure 3.14).

In spite of all we've said here, however, sometimes a catalyzed reaction is actually reversible under normal conditions. For instance, in a reaction similar to the one above, glucose-1-phosphate is enzymatically converted into glycogen when the cellular pH is relatively high, and glycogen is broken back down into glucose-1-phosphate by the same enzyme when the pH is lowered a bit. Of course, it takes work to change the pH of a cell. Whether glycogen will be synthesized or degraded depends also, of course, on the concentration of glucose-1-phosphate in the cell. When glucose is being rapidly metabolized, the glucose-1-phosphate level drops, and the equilibrium shifts toward net breakdown of glycogen. Later, free glucose (transported in the blood) will be converted to glucose-1-phosphate with the sacrifice of a molecule of ATP for each glucose unit, and glycogen storage polymers will again be built up.

Enzyme Kinetics

Now let's take a look at the rate of enzyme-catalyzed chemical reactions and how these rates change in response to relative enzyme and substrate concentrations, to environmental changes, and over time as the reactions proceed. We are also interested in how the organism controls these enzymatic reactions.

Enzyme kinetics are most easily studied when either the substrate or the end product is a colored compound. Then the reaction can be followed as it occurs within the chamber of a colorimeter or a photospectrometer (color-measuring devices). Thus the disappearance or buildup of the colored compound can be monitered and recorded. If neither the substrate nor the end product is colored, a *coupled reaction* involving a dye that changes color in the presence of the end product may sometimes be used.

A graph of a typical reaction is shown in Figure 3.15(a). At first, a large amount of product is formed at a constant rate; then the amount of product formed during a given period of time suddenly levels off as the excess substrate becomes used up. The slope of the initial straight portion of the line in Figure 3.15(a) is called the *initial velocity* of the reaction under conditions where there is a specific amount of enzyme and a considerable excess of substrate. The substrate saturates the enzyme, and the rate depends on how fast the enzyme–substrate complex dissociates into enzyme plus product. This can be very fast indeed—some enzymes can break down many millions of molecules of substrate each second. Figure 3.15(b) shows the effect on the velocity of varying the substrate concentration.

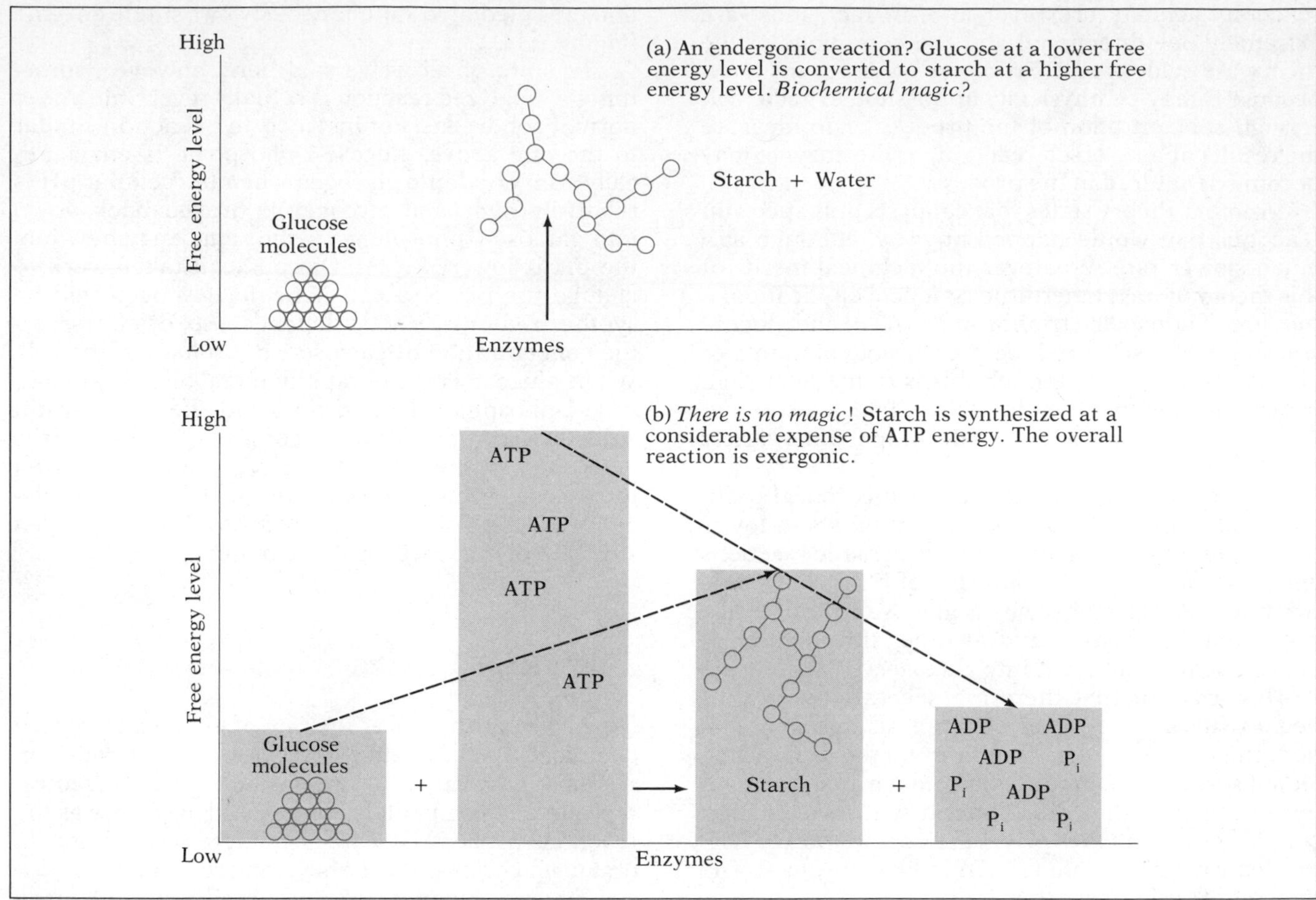

3.14 (a) At first glance it seems obvious that the conversion of simple glucose into complex starch is an endergonic reaction. In other words, through biochemical "magic," glucose at a lower free energy level is converted to starch at a higher free energy level. (b) However, look at the overall process of starch synthesis. We see that in the overall reaction ATP is used. The ATP is at a higher free energy level than starch and is degraded to ADP. Therefore, in terms of the entire event, starch synthesis is exergonic. The long view of this tells us that life goes on at the expense of some energy source with a free energy level which is higher than the sum total of free energy for all living organisms. What is that source?

Effects of Temperature and Acidity

Inorganic chemical reactions go faster if there is an input of energy in the form of heat. As a rule of thumb, an increase of 10°C doubles the rate of a chemical reaction. This general relationship holds pretty well for enzyme-catalyzed reactions also—but only up to a point. Then the (initial) rate of reaction actually decreases with increasing temperature. Apparently, high temperatures begin loosening hydrogen bonds, destroying the three-dimensional shape of the enzyme molecule and hence its catalytic activity. Further increases in temperature, in fact, may irreversibly denature the enzyme.

Each enzyme has an optimal temperature—that is, the temperature at which the initial velocity is greatest. Not surprisingly, the optimal temperature is often close to the normal temperature of the organism. Studies of some North American creek-bed fish ("suckers") revealed that certain enzymes of one species were slightly different in fish of the southern and northern parts of the species' range and that the optimal temperature of each enzyme form was appropriate to the summer temperature of the two habitats.

Acidity (pH) also affects enzyme activity. Again, enzymes are generally most efficient at conditions under which they normally operate. Pepsin, a stomach enzyme that helps break down proteins, works best at the typically acid pH 2 of the stomach, and trypsin, an intestinal enzyme with the same function, works best at the slightly basic pH 8 of the intestine.

Control of Enzyme Activity: Negative Feedback and Allostery

One of the most direct ways to control enzyme activity is through *negative feedback inhibition.* Negative feedback inhibition is a common phenomenon in the

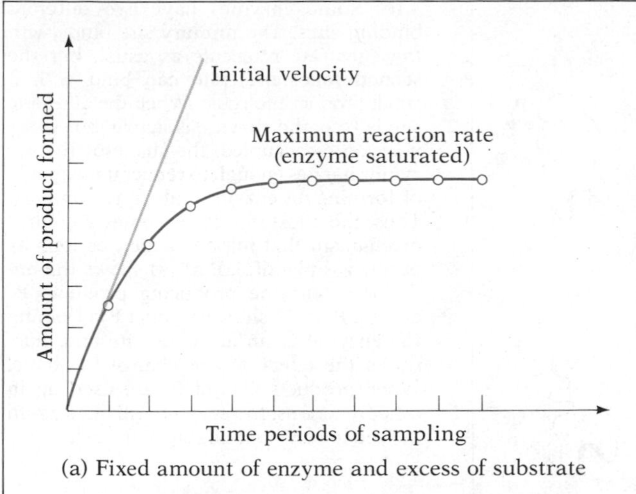

(a) Fixed amount of enzyme and excess of substrate

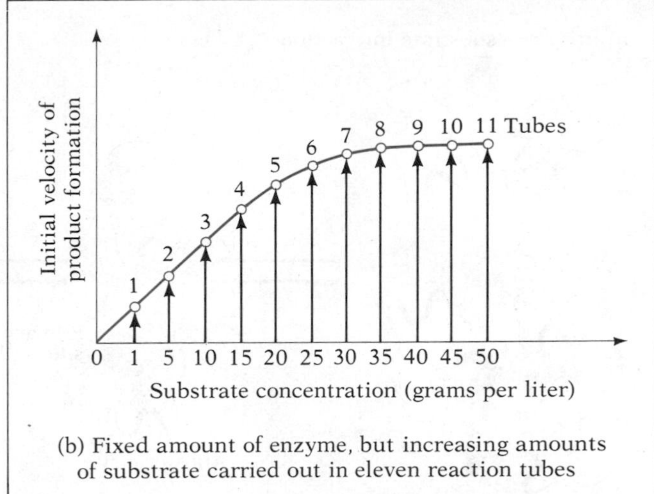

(b) Fixed amount of enzyme, but increasing amounts of substrate carried out in eleven reaction tubes

3.15 (a) The graph represents the amount of product formed when a fixed amount of enzyme and an excess of substrate is placed in a single reaction tube. The tube is checked at regular time intervals for the amount of product formed (perhaps in a colorimeter, which measures color change). As the graph indicates, the amount of product formed per unit time increases at first, but the reaction soon reaches a constant, maximum. We can conclude that the enzyme molecules become totally saturated—their active sites are all occupied—and that the rate of new product formation depends on the time each enzyme requires to complete its action.

(b) In this experiment eleven reaction tubes are prepared and each is checked after the same time interval. All tubes contain the same amount of enzyme, but the concentration of substrate in each varies, with the lowest concentration in tube 1 and the highest in tube 11. The points on the graph represent the initial velocity at the different substrate concentrations. As the graph indicates, increasing the substrate concentration raises the initial velocity, but only up to a point. The data indicate that increasing the quantity of substrate improves the chances of enzyme–substrate complexes forming. This is logical, since their formation relies on random movement and chance collision. At the right on the graph, however, the initial velocity stabilizes. This tells us that the enzyme molecules are totally saturated with those of substrate and no further increase in reaction velocity is possible. What might happen in tubes 8–11 if their temperatures were raised?

regulation of living systems. Essentially, as a product is formed, it hinders the mechanism that produced it, until the product falls off to such a degree that the mechanism is released and more product is formed.

Consider this reaction:

$$S + E \rightleftharpoons ES \rightleftharpoons EP \rightleftharpoons P + E$$

where E is the enzyme, S is the substrate, P is the product, and ES and EP are enzyme–substrate and enzyme–product complexes, respectively. A substrate (S) and free enzyme (E) must be available for the reaction to proceed. If there is very much P around, and if P binds E, then most of the enzyme will be tied up in EP complexes, and the rate of reaction will slow down considerably. This is just as well, since there is already a high concentration of product. As the product is used up, the breakdown of the substrate can again proceed. Any enzyme that is capable of being inhibited by its own product is called a *regulatory enzyme.*

Sometimes inhibition of the end product is more subtle. An enzyme may have a second binding site, called an *allosteric site,* that affects its activity. When a small molecule becomes bound at the allosteric site, the shape of the enzyme changes and it can no longer perform its usual task. Here, the inhibiting molecule need not be the direct product of the enzymatic reaction. Most commonly, the enzyme catalyzing the first reaction in a multistep biochemical pathway will be allosterically inhibited by the end product of the last reaction (Figure 3.16).

Other Binding Proteins

Proteins also perform a diversity of nonenzymatic, nonstructural functions. Many of these are related to protein's peculiar ability to form specific structures and shapes that enable them to bind to other substances. Hemoglobin, for instance, is a carrier protein; it binds oxygen in the lungs and releases it where it is needed in the body, and it binds carbon dioxide in the tissues and releases it in the lungs. There are a number of other circulating binding proteins, such as *transferin* (which binds iron) and *haptoglobin* (which binds hemoglobin). Egg white has a protein, *avidin,* which binds tightly to *biotin,* an important vitamin; fungi that attempt to invade eggs die of biotin vitamin deficiency.

One of the most critical functions of proteins is their ability to bind to other molecules as part of the antibody response. Some proteins are synthesized by white blood cells for the express purpose of binding

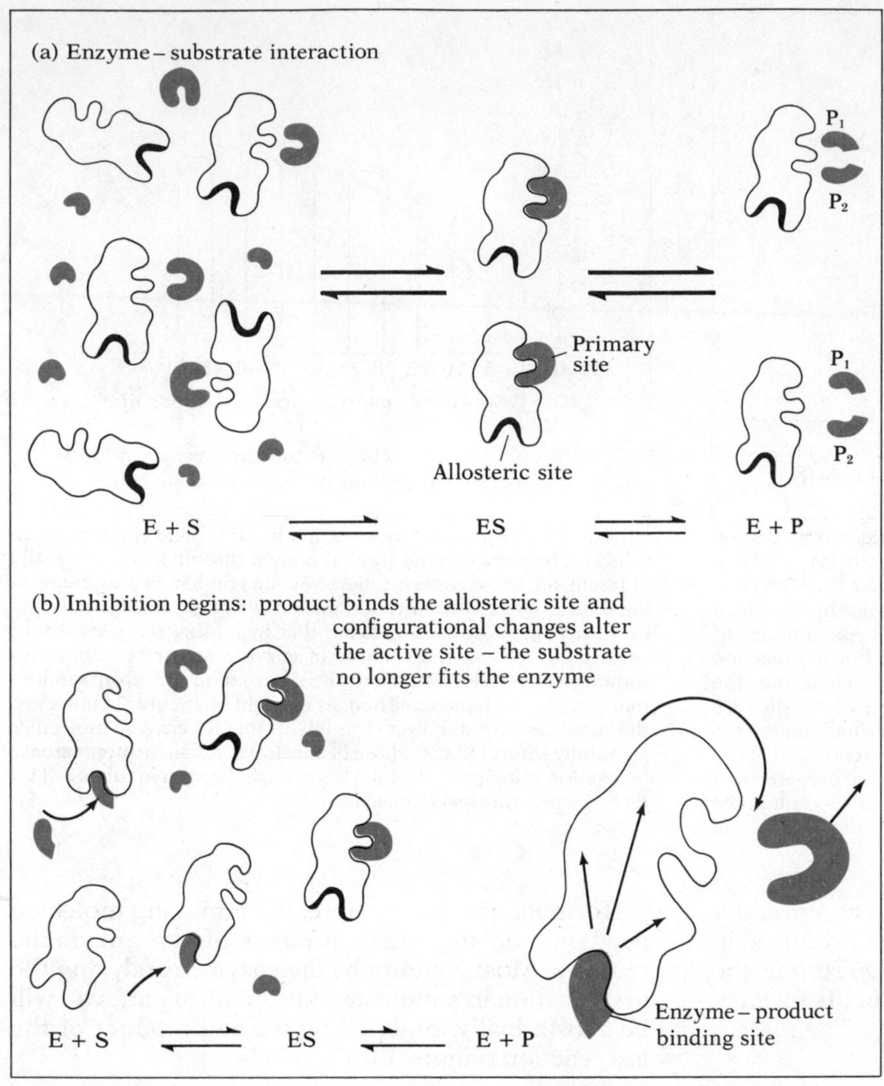

3.16 Some enzymes have two different binding sites. The primary site binds with the substrate molecule as usual, but the second (allosteric) site can bind with a small product molecule. When the allosteric site is free, the enzyme is active, but when it becomes occupied, the shape of the enzyme changes enough to render it incapable of forming an enzyme–substrate complex. Thus, the allosteric site becomes a control mechanism that inhibits action as long as product is plentiful. Part (a) shows the uninhibited enzyme producing products P_1 and P_2. Part (b) shows product P_2 rejoining the enzyme at an allosteric site and illustrates the effect of the change in shape. When products P_1 and P_2 are used up in other reactions, the enzyme will once again be free to carry out its role in the cell.

with and ultimately destroying or inactivating foreign molecules or particles. (See the discussion of immune responses in Chapter 28.)

We should add that there is good evidence that cells recognize each other during embryonic development through interactions of binding proteins on their surfaces. In addition, binding proteins on cell surfaces lend peculiarities to cells that enable us to smell and taste certain molecules.

Structural Proteins

Structural proteins may be *intracellular* (inside cells) or *extracellular* (outside cells). Common extracellular proteins in vertebrates are *collagen* and *elastin.* These are long fibrous molecules that clump together to make larger fibers. Collagen, the principal component of connective tissue such as tendons, ligaments, and muscle coverings, makes up about 25% of the protein in humans.

Elastin, another connective tissue protein, gets its name from its remarkable ability to stretch. Elastin fibers give elasticity to connective tissue, including that of your ears and skin. Ears are very rich in elastin, so if you reach over and tug on your neighbor's ear, it will snap right back to its original form (we hope). You can also use pinching as an age test. If you pinch the skin on the back of your hand and it snaps back, you're young. If a ridge stands there, you're old. This is because elastin loses its elastic properties in time. The bagginess in the face and neck that makes old people so happy is due to the same phenomenon.

Keratin is the protein of hair and of the outer layer of skin, as well as that of feathers, claws, nails, horns, antlers, and scales. In contrast to elastin or collagen, it is basically an intracellular protein. It is deposited as cells fill themselves with keratin, then dry up and die, leaving the keratin behind.

Peptides and Peptide Hormones

A short polypeptide—one with fewer than, say, thirty amino acid subunits—is usually just called a *peptide.* A peptide can be a degradation fragment of a larger protein, but some peptides are complete functional units themselves.

Amino acids are usually transported in the blood in the form of peptides. Normally, free amino acids are found only in the blood vessels between the gut and the liver. Proteins are broken down into amino acids in the gut and sent to the liver to be linked together as peptides before they enter the general circulation. Once these circulatory peptides reach cells in which active protein synthesis is occurring, they are again broken down into their constituent amino acids, which can then be resynthesized into structural proteins, enzymes, or other polypeptides.

An important class of very small peptides can also be found in the circulatory system. These are the *peptide hormones,* and there are many different kinds. The special properties of the hormones will be discussed later.

Some of these peptide hormones are small indeed. The hormone *oxytocin,* which stimulates milk production and contraction of the uterus following childbirth, is only nine amino acids long:

NH_2-Cys-Tyr-Ile-Gln-Asn-Cys-Pro-Leu-Gly-COOH

(The abbreviations used in this formula are those listed in Table 3.1.) The antidiuretic hormone, which causes a decrease in urine flow, differs only in the third and eighth positions:

NH_2-Cys-Tyr-Phe-Gln-Asn-Cys-Pro-Arg-Gly-COOH

Other peptide hormones are only a little larger: melanotropin has thirteen amino acids, gastrin has seventeen, corticotropin (ACTH) has thirty-nine, and insulin (with two chains) has fifty-one.

Actually, fifty-one amino acids is long enough for insulin to be called a polypeptide rather than a peptide (but the distinction is a matter of judgment). Lactogen, prolactin, and growth hormone have 190, 198, and 190 amino acid subunits, respectively, which makes them perfectly respectable proteins.

3.4 The Lipids

The *lipids* are a diverse group of molecules, defined by their solubility rather than by their structure. Lipids may be small molecules, large molecules, monomers, polymers, energy storage molecules, structural molecules, hormones, lubricants, prosthetic groups, or emulsifiers. Lipids are organic molecules that tend to be fat-soluble rather than water-soluble. They are generally nonpolar and tend to dissolve in organic solvents, such as gasoline, chloroform, paint thinner, and salad oil, which are rich in van der Waals forces. Whereas carbohydrates and proteins are hydrophilic, lipids are hydrophobic. It may seem odd that an important class of organic compounds is defined solely on the basis of a solubility characteristic, but it does appear that fat-solubility gives all lipids, no matter how different they are structurally, some common characteristics. For example, lipids tend to dissolve into one another and may often be found together in the same part of the cell.

Important groups within the category are fats, oils, sterols, waxes, and phospholipids. Most familiar to us are the fats (such as, beef tallow) and the oils (such as, Wesson, cod liver oil). Both are *triglycerides*—that is, compounds consisting of three *fatty acid* chains attached to a molecule of glycerol (see below). The difference between a fat and an oil is that a fat is solid at room temperature and an oil is liquid at room temperature. But both fats and oils tend to be liquid at the normal temperature of the living organism in which they occur. Because of this ambiguity, and to avoid confusion with petroleum oil and with other meanings of the word *fat,* biologists prefer to use the term *triglycerides.*

Triglycerides

Glycerol, formerly known as *glycerine,* is a small molecule. It has three carbons and three hydroxyl side groups:

```
  H
  |
 HC—OH
  |
 HC—OH
  |
 HC—OH
  |
  H
```

Glycerol

Linked to the glycerol in a triglyceride are three *fatty acid* molecules. A fatty acid molecule consists of a hydrocarbon chain (that is, a chain consisting of a backbone of carbon atoms with only hydrogens as side groups) with a carboxyl group at one end:

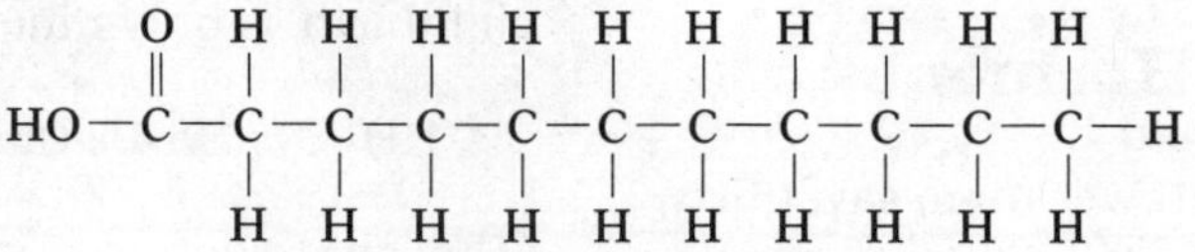

```
      O  H  H  H  H  H  H  H  H  H  H
      ‖  |  |  |  |  |  |  |  |  |  |
  HO—C—C—C—C—C—C—C—C—C—C—C—H
         |  |  |  |  |  |  |  |  |  |
         H  H  H  H  H  H  H  H  H  H
```

Caproic acid

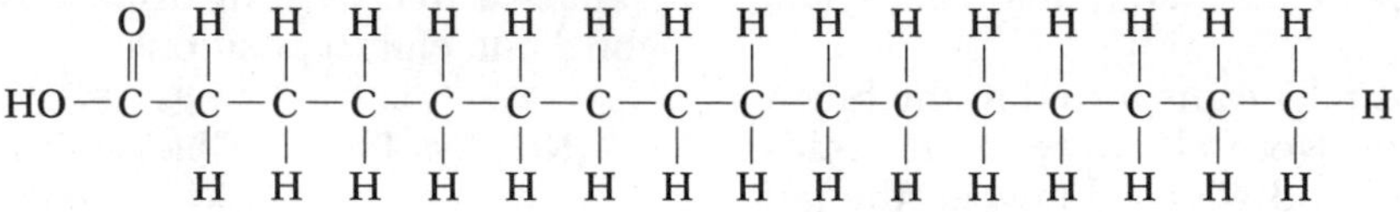

```
      O  H  H  H  H  H  H  H  H  H  H  H  H  H  H  H
      ‖  |  |  |  |  |  |  |  |  |  |  |  |  |  |  |
  HO—C—C—C—C—C—C—C—C—C—C—C—C—C—C—C—C—H
         |  |  |  |  |  |  |  |  |  |  |  |  |  |  |
         H  H  H  H  H  H  H  H  H  H  H  H  H  H  H
```

Palmitic Acid

```
      O  H  H  H  H  H  H  H     H  H     H  H     H  H  H
      ‖  |  |  |  |  |  |  |     |  |     |  |     |  |  |
  HO—C—C—C—C—C—C—C—C—C=C—C—C=C—C—C=C—C—C—H
         |  |  |  |  |  |  |        |  |     |  |        |  |
         H  H  H  H  H  H  H        H  H     H  H        H  H
```

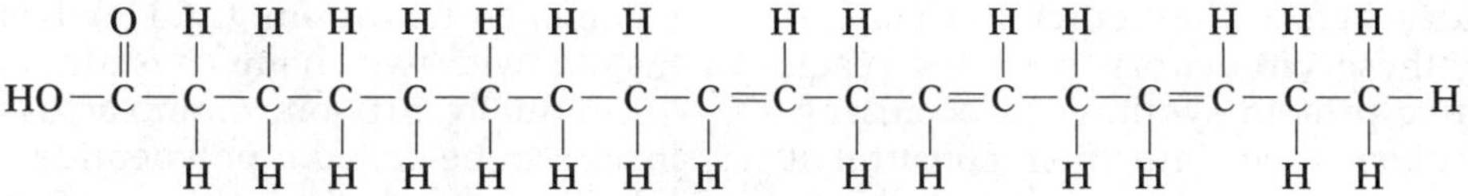

Linolenic Acid

The synthesis of a triglyceride from glycerol and three fatty acids is a dehydration process aided by enzymes:

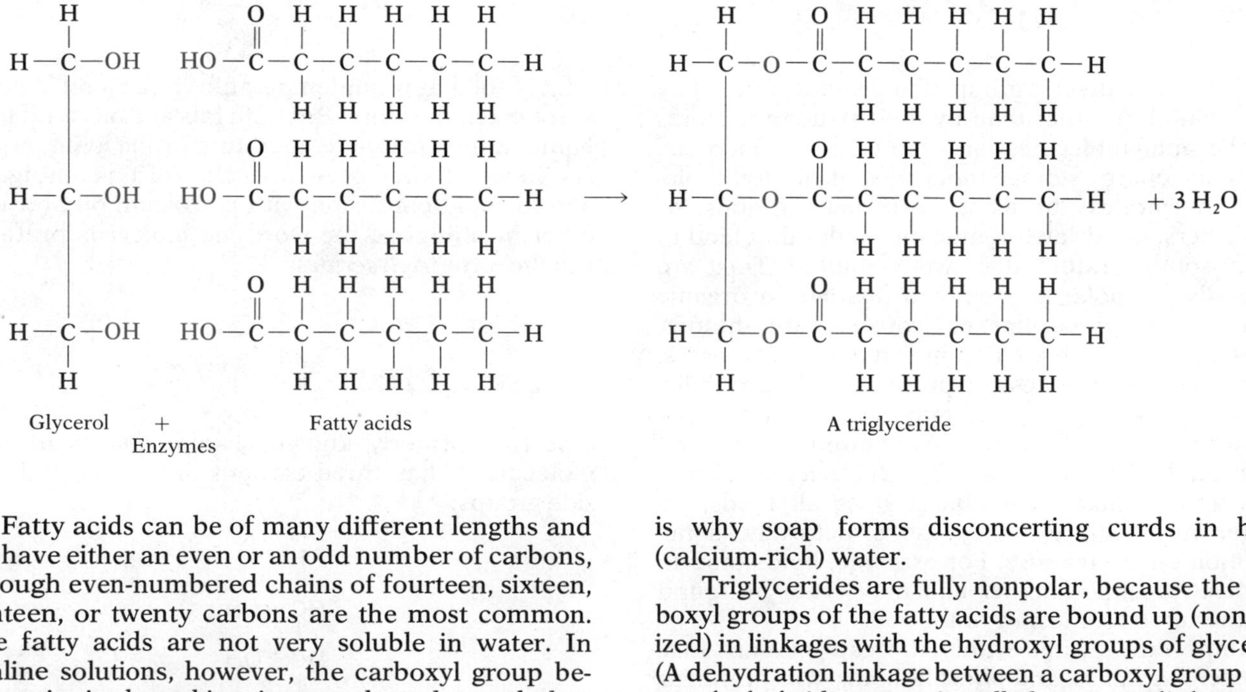

```
     H           O  H  H  H  H  H                 H      O  H  H  H  H  H
     |           ‖  |  |  |  |  |                 |      ‖  |  |  |  |  |
  H—C—OH    HO—C—C—C—C—C—C—H             H—C—O—C—C—C—C—C—C—H
     |              |  |  |  |  |                 |         |  |  |  |  |
     |              H  H  H  H  H                 |         H  H  H  H  H
     |           O  H  H  H  H  H                 |      O  H  H  H  H  H
     |           ‖  |  |  |  |  |                 |      ‖  |  |  |  |  |
  H—C—OH    HO—C—C—C—C—C—C—H    ⟶    H—C—O—C—C—C—C—C—C—H   + 3 H2O
     |              |  |  |  |  |                 |         |  |  |  |  |
     |              H  H  H  H  H                 |         H  H  H  H  H
     |           O  H  H  H  H  H                 |      O  H  H  H  H  H
     |           ‖  |  |  |  |  |                 |      ‖  |  |  |  |  |
  H—C—OH    HO—C—C—C—C—C—C—H             H—C—O—C—C—C—C—C—C—H
     |              |  |  |  |  |                 |         |  |  |  |  |
     H              H  H  H  H  H                 H         H  H  H  H  H

  Glycerol   +      Fatty acids                       A triglyceride
         Enzymes
```

Fatty acids can be of many different lengths and can have either an even or an odd number of carbons, although even-numbered chains of fourteen, sixteen, eighteen, or twenty carbons are the most common. Free fatty acids are not very soluble in water. In alkaline solutions, however, the carboxyl group becomes ionized, making it strongly polar and thus soluble. Sodium and potassium salts of fatty acids are known as "soap" and are soluble in water. Calcium salts of fatty acids are not soluble in water, which is why soap forms disconcerting curds in hard (calcium-rich) water.

Triglycerides are fully nonpolar, because the carboxyl groups of the fatty acids are bound up (nonionized) in linkages with the hydroxyl groups of glycerol. (A dehydration linkage between a carboxyl group and an *alcohol* side group is called an *ester linkage.* An alcohol is technically any substance with the formula R—OH. For example, the ethanol of spirits is $CH_3—CH_2—OH$. Thus, glycerol is an alcohol.)

Saturated, Unsaturated, and Polyunsaturated Fats

So let's talk about *saturated fats.* The obvious question here is, saturated with what? The answer is hydrogen.

The carbon backbone of the hydrocarbon tails of fatty acids is linked primarily with carbon–carbon single bonds (one pair of shared electrons between successive carbons). But some fatty acids have double bonds between one or more pairs of successive carbons in place of side bonds to hydrogen. Therefore, they have fewer hydrogens (Figure 3.17). Such fatty acids, and the fats containing them, are said to be *unsaturated,* because it is possible to add hydrogen to them (with the aid of a platinum catalyst and a little pressure). By adding hydrogens to an unsaturated triglyceride, one can produce, for example, hydrogenated vegetable oil. When all the double bonds are gone, the molecule won't accept any more hydrogen, and it is then a saturated fat.

So a fat with carbon–carbon double bonds is unsaturated, and one with more than one such double bond is *polyunsaturated.* Common fatty acids with carbon–carbon double bonds are oleic acid (vegetable oil) and linolenic acid (linseed oil). Highly saturated fats tend to have rather high melting points (tend to be solid fats), while polyunsaturated fats tend to have low melting points (tend to be oils). Simple vegetable oils are largely unsaturated (which is why old-fashioned peanut butter separates) and hydrogenated vegetable oils are hard fats (which is why new-fashioned peanut butter doesn't).

The membranes of cells are rich in unsaturated fatty acids. But the chemical processes of human metabolism can't form double bonds, and so it is necessary for us to eat unsaturated fats, although the amount needed is small.

Triglycerides are concentrated energy storage compounds. When oxidized, they yield about twice as much potential energy per gram dry weight as carbohydrates or protein. The difference is even more marked for calories per gram wet weight, since carbohydrates and protein absorb water and fat doesn't.

Because fat is so energy-rich, it is an ideal way to store energy. Plants store triglycerides in their seeds,

3.17 Stearic acid, found in animal fat (sold as lard), has the maximum number of hydrogen atoms, as do all saturated fats. Linolenic acid, found in linseed oil, has two double bonds in the carbon chain, and, as you can see by the gaps in the rows of hydrogen, is unsaturated.

```
    H     O
    |     ‖  H  H  H  H  H  H  H  H  H  H  H  H  H  H  H  H  H
H — C — O — C — C—C—C—C—C—C—C—C—C—C—C—C—C—C—C—C—C — H
    |        H  H  H  H  H  H  H  H  H  H  H  H  H  H  H  H  H
    |     O
    |     ‖  H  H  H  H
H — C — O — C  C—C—C—C—   —   —   —   —
    |        H  H  H  H
    |     O
    |     ‖  H  H
H — C — O — C—C—C—   —   —   —   —
    |        H  H
    H

              Stearic acid

    H     O
    |     ‖  H  H  H  H  H  H  H  H  H  H  H     H  H  H  H  H
H — C — O — C—C—C—C—C—C—C—C—C=C—C—C=C—C—C—C—C—CH
    |        H  H  H  H  H  H  H  H     H     H  H  H  H  H  H
    |     O
    |     ‖  H  H
H — C — O — C—C—C —   —   —   —
    |        H  H
    |     O
    |     ‖  H
H — C — O — C—C—   —   —   —
    |        H
    H

              Linolenic acid
```

and many kinds of animals lay on fat in anticipation of lean seasons or migration. Humans also store fat under the skin and around internal organs, although agriculture and storage techniques have largely exempted us from the rigors of seasonal food depletion.

Stored fats also serve other purposes, such as insulation and flotation. The blubber of sea mammals serves both purposes. Storage fat is used as padding in your fingers and in your bottom, and makes a significant contribution to the curvaciousness of women during their reproductive years. (See discussions of the development of secondary sexual characters in Chapter 29.)

Phospholipids

The *phospholipids* are structurally closely related to the triglycerides. In both molecules, two fatty acids (one saturated and one unsaturated) are linked to a backbone of glycerol by ester linkages. In place of the third fatty acid found in triglycerides, however, phospholipids have a phosphate group, which is commonly linked to still other organic molecules (variable groups) as shown in Figure 3.18. The phosphate or phosphate compound is hydrophilic, so phospholipids have detergent properties. In other words, they reduce the surface tension of water wherever it meets air or another liquid. Phospholipids can occasionally be found as food storage compounds (for example, lecithin in egg yolk), but they are more important in their roles as major components of cell membranes. A cell membrane, as we will see, consists of two layers of lipid, primarily phospholipid and glycolipid, along with a number of dispersed proteins. It is believed that the hydrophilic phosphate ends associate with the protein, while the hydrophobic fatty acid tails dissolve into one another, giving an essentially liquid, oily core to the membrane.

In vertebrates, nerves are often sheathed in many layers of cell membrane, and therefore nervous tissue—especially in the brain—has a high phospholipid content. So perhaps "fathead" is an unintentional compliment. In addition to the simple glycerol phospholipids already mentioned, brain and other ner-

3.18 Phospholipids are an essential component of cell membranes. They differ from triglycerides in several important ways. Phospholipids have only two fatty acid tails, one saturated and the other unsaturated. In place of the third fatty acid of triglycerides is a phosphate group, which is often attached to one of several organic molecules we have indicated as variable groups. The phosphate heads are polar, so they interact with water, which is also polar. The heads are therefore hydrophilic, while the fatty acid tails (without charges) are hydrophobic.

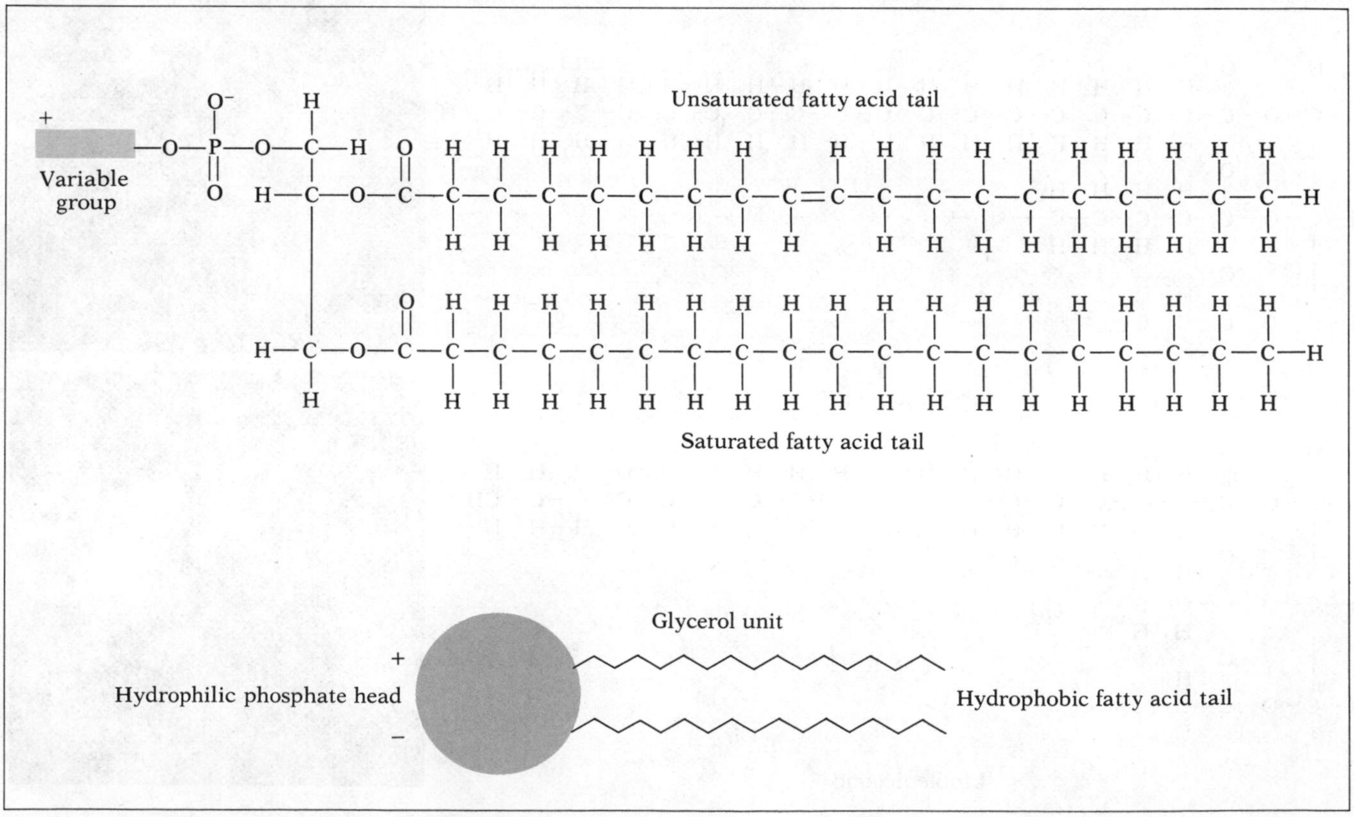

vous tissue has a wide array of marvelously named and complex phospholipids and glycolipids: sphingomyelin, dihydrosphingosine, cerebrosides, gangliosides, and many more.

Waxes, like neutral fats, are esters of fatty acids. However, the alcohol involved is not glycerol, but a long-chain alcohol with a single hydroxyl group. Beeswax consists largely of an ester of palmitic acid (sixteen carbons) with a straight-chain alcohol called myricyl alcohol (thirty carbons). Because of its hydrophobic (water-repellent) properties, wax is found in a variety of living things, especially those that need to save water. Insects generally have waxy cuticles. And so do plants: the waxes of leaves, fruit skins, and flower petals contain very long fatty acids (up to thirty-six carbons), both as free fatty acids and as esters of long-chain alcohols. Many marine invertebrates contain various waxes, and so, for some reason, do your ears. Waxes are generally much harder and even more hydrophobic than fats—two reasons why commercial wax is used as a protectant. Heads would turn if you announced that you were going to fat your car.

Steroids are really not structurally similar to fatty acid lipids, but since they are either partly or wholly hydrophobic, they qualify to be called lipids. All steroids contain a peculiar core consisting of four interlocking rings. The differences between various steroids depend on the variation in the side chains. Some steroids are highly hydrophobic, some less so. (By the way, steroids with —OH groups are also called *sterols*.) Let's consider a few of the places steroids are found.

Lanolin (or *lanosterol*) is a greasy, almost waxy substance that is commercially refined from sheep's wool. Small amounts of lanolin are found in your own skin and hair, and this helps keep them flexible.

Cholesterol is a substance we've all heard of because various food manufacturers have assaulted us over the airways with the message that their products don't contain any. Why should we care? Because cholesterol has been accused of coating the lining of our blood vessels and thereby reducing their diameter and increasing blood pressure the way one might increase water pressure by placing the thumb over the end of a garden hose (Figure 3.19). Cholesterol may be found in your skin as well as your arteries. But cholesterol is not all bad. In fact, it has a number of vital functions. For example, it is a major constituent of cell membranes. Also, bile acids, which are necessary for fat digestion, are modified cholesterol. When irradiated with ultraviolet light, cholesterol becomes vitamin D, which is necessary for normal bone growth and maintenance. Various tissues, such as your liver, can make their own cholesterol (more than you could possibly need), but you normally cannot make vitamin D without sunshine (ultraviolet light). You may have a greater appreciation for cholesterol when you realize that it can also be modified to make the various steroid hormones, such as the sex hormones.

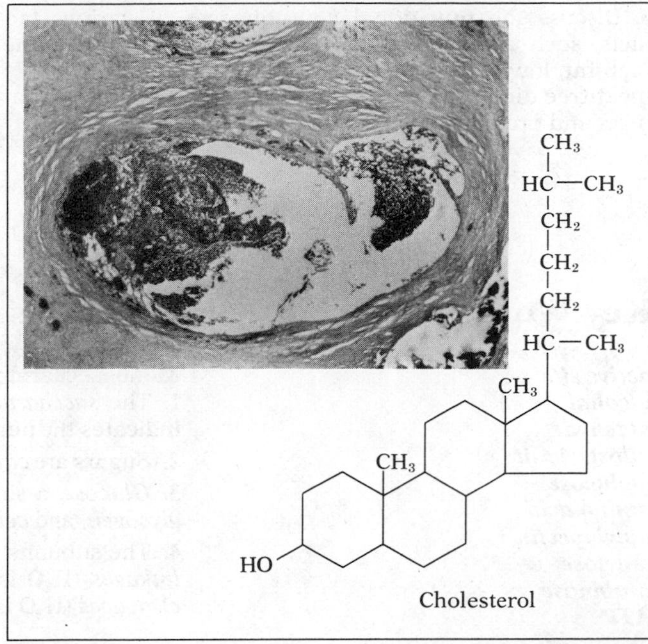

3.19 Steroids are complex four-ring molecules that have some lipid characteristics. One important steroid is cholesterol, which has been implicated in the disease atherosclerosis. The arterial linings form abnormal regions, thickenings and rough spots that may break away and clog vessels. In addition, the roughened lining may cause blood clotting within the circulatory system. Both situations are decidedly dangerous.

Application of Ideas

1. Nucleic acids and proteins are considered to be "informational molecules," but carbohydrates are not, even though they may be extremely large and complex. Why do carbohydrates fail to qualify? Could lipids be "informational molecules"?

2. We readily use enzymes to transform glucose and oxygen into water, carbon dioxide, and energy. Since all enzymatic reactions are reversible, can we make oxygen and glucose from water, carbon dioxide, and energy? (You may want to add more information after having studied Chapters 5 and 6.)

3. Discuss the nutritional consequences of various fad diets, such as the all-banana diet, the fat-free diet, the high-fat, low-carbohydrate diet, an all-meat diet, a totally meat-free diet, a brown rice diet, and the Lynn diet (vitamins and predigested protein only).

4. The statement has been made that all reactions proceed from a higher net energy state to a lower net energy state. In view of this, how can one account for the complex chemical organization of organisms? They are obviously at a higher energy state than the molecules from which they are constructed.

Key Words

active site
alcohol
aldehyde
allosteric site
α-glucose
amino acid
amylopectin
amylose
arabinose
ATP

β-glycoside
biochemical pathway

calorie
carbohydrate
cellulase
cellulose
chitin
cholesterol
chondroitin
chromoprotein
coenzyme
conjugated protein
C-terminal
cuticle

dehydration linkage
denaturation
dextrose
disaccharide

ester
ethanol

fat
fatty acid
fibrous protein
fructose

galactose
globulin
glucose
glucose-1-phosphate
glucoronic acid
glucosamine
glycerol
glycogen
glycolipid
glycoprotein

Key Ideas

Monosaccharides, Disaccharides, and Polysaccharides
1. The *saccharides,* or sugars, are all composed of simple sugars. The prefix indicates the number of individual sugar molecules contained.

2. Sugars are *carbohydrates* with the empirical formula $(CH_2O)_n$.

3. *Glucose,* a simple sugar, is the subunit of many *polymers* including *starch, glycogen,* and *cellulose.*

4. The subunits of *disaccharides* and *polysaccharides* are joined by *dehydration linkages* (H_2O is removed) when synthesized and broken down by *hydrolytic cleavages* (H_2O is added.)

Starches
1. *Amylose* is a simple plant *starch* consisting of hundreds of *glucose* molecules with 1–4 linkages.

2. *Amylopectin,* another plant starch, is highly branched, with both 1–4 and 1–6 linkages.

3. *Glycogen* is animal *starch* and contains both 1–4 and 1–6 linkages.

4. When *glycogen* is broken down into glucose, *phosphate* is added in an energy-conserving step called *phosphorolytic cleavage.*

Primary and Secondary Structure of Starches
Starches commonly take on a *secondary structure* when the chain or branched chains take the form of a helix or otherwise fold back on themselves.

Polysaccharide Synthesis
ATP energy is used to form *glucose* polymers.

Structural Polysaccharides
Cellulose and *chitin* are *polysaccharides* that form structural parts in organisms.

Cellulose
1. *Cellulose* is very strong, insoluble, and indigestible to most organisms.

2. *Cellulose* is composed of *β*-glycoside bonds, with every other *glucose* upside down relative to its neighbor.

3. The structure of *cellulose* owes its great strength to its organization into larger and larger *fibrous* units.

Other Structural Polysaccharides of Plants
Additional *polysaccharides* are *hemicellulose* and *pectin.*

Chitin
Chitin, the *polysaccharide* of arthropod exoskeletons, is composed of *N*-acetyl *glucosamine* units rather than *glucose.*

Mucopolysaccharides
The *mucopolysaccharides* include *hyaluronic acid* and *chondroitin,* which occur in connective tissue of the vertebrate body.

The Proteins
1. *Proteins* are used for energy, in structure, and as active molecules such as enzymes and *hormones.*

haptoglobin
heme group
hemicellulose
hemoglobin
hormone
hyaluronic acid
hydrolytic cleavage

inhibitor
initial velocity

keratin

lactose
lanolin
lecithin
lignin
linolenic acid
lipid
lipoprotein

mannose
micelle
monomer
monosaccharide
mucopolysaccharide

N-terminal
nucleic acid

oil
oleic acid

palmitic acid
pectin
peptide
peptide bond
phospholipid
phosphorolytic cleavage
phosphorylase
polymer
polypeptide
polysaccharide
polyunsaturated fat
primary structure
product
prosthetic group
protein

quaternary

regulatory enzyme

saturated fat
secondary structure
starch
steroid
substrate
sucrose

tertiary structure
tetramer
triglyceride

2. *Proteins* consist of amino acids arranged into *polypeptides.*

The Structure of Amino Acids

1. *Amino acids* contain two or more carbon atoms with at least one amino group and one carboxyl group.

2. When *amino acids* form *polypeptides, dehydration linkages* occur between carboxyl groups and *amino* groups.

3. The number, kind, and arrangement of *amino acids* are highly specific for each *protein.*

4. *Proteins* may have as many as four levels of organization: *primary, secondary, tertiary,* and *quarternary.* The first involves the arrangement of *amino acids,* the second and third, spiraling and folding, and the fourth, the union of two or more individual *proteins.*

The Twenty Amino Acids and Their R Groups

1. The individual R or side groups of *amino acids* are important to protein shape.

2. The forces involved in *protein* shape are covalent bonding, hydrogen bonding, van der Waals forces, and ionic bonding.

3. *Denaturation* (change in shape) of *proteins* can be caused by chemical and physical agents.

Conjugated Proteins and Prosthetic Groups

1. Conjugated proteins contain *prosthetic groups* such as heme that give the protein special properties.

2. Coenzymes are *prosthetic groups* that assist in the enzymes' function.

Enzymatic Activity

1. Enzymes are catalytic proteins with *active sites* that interact with the *substrate.*

2. *Enzymes* may interact directly with the *substrate,* may expose the *substrate* to a *prosthetic group,* or may bring two *substrate* molecules together to interact.

3. Enzymes emerge unchanged from reactions. The reactions are generally reversible.

4. When the total energy is considered, enzymatic reactions bring *substrate* to a lower net energy state.

Enzyme Kinetics

1. The rate of an enzyme-catalyzed reaction depends on the relative concentration of enzyme and *substrate.*

2. When all the enzyme molecules available are engaged, the rate of product production is determined by the time required for each enzyme-*substrate* reaction.

Effects of Temperature and Acidity

1. With every 10°C increase, the rate of enzyme action is doubled (until *denaturation*).

2. Specific enzymes have their own optimal temperature range.

3. pH has an effect on enzymes. Most operate near pH 7.

Control of Enzyme Activity: Negative Feedback and Allostery

1. *Regulatory enzymes* are capable of being inhibited by the presence of excess product.

2. Where teams of enzymes are involved, the first may have an *allosteric site* that can be joined by the product of a later step, altering the enzyme's shape and causing its inactivation.

Other Binding Proteins

Some *proteins* bind to substances. Examples are the oxygen-binding property of *hemoglobin* and binding *proteins* in the *immune system.*

unsaturated fat

vitamin

wax

xylose

zwitterion

Structural Proteins
Structural *proteins* are useful in connective tissue (which support animal bodies) and as coverings such as feathers, skin, hair, and scales.

Peptides and Peptide Hormones
Some *hormones* are *peptides* or *polypeptides*.

The Lipids
1. *Lipids* are *fat*-soluble, hydrophobic molecules, used for energy, membranes, hormones, food storage, and so forth.
2. The main *lipids* include *fats, oils* (*triglycerides*), *sterols, waxes,* and *phospholipids.*

Triglycerides
1. *Triglycerides* consist of one molecule of *glycerol* and three of *fatty acid.* They are nonpolar and not very soluble in water.
2. *Fatty acids* are hydrophilic at their carboxyl end and hydrophobic at their carbon chain end. They tend to form micelles in water.

Saturated, Unsaturated, and Polyunsaturated Fats
1. *Saturated fats* are those that have no carbon–carbon double bonds and hold all of the hydrogen linkages possible.
2. *Unsaturated fats* have carbon–carbon double bonds and could hold more hydrogen. *Polyunsaturated fats* are those that have more than one carbon–carbon double bond.
3. *Saturated fats* tend to be more solid (lard), while *unsaturated fats* are usually *oils* (corn oil).
4. Gram for gram, *fats* have twice the *calories* of *carbohydrates* and *proteins.* Animals store excess fats around internal organs and under the skin.

Phospholipids
1. In *phospholipids* two *fatty acids* and a phosphate group are linked to a glycerol backbone. Their *glycerol* heads are polar because of ionization in the phosphate group. The tails are hydrophobic. They act as a water barrier in the cell membrane.
2. *Waxes* differ from fatty acids in that they lack *glycerol* and are more hydrophobic. They serve as waterproofing in many plants and animals.
3. *Steroids* are hydrophobic, but otherwise are unlike other lipids. They contain four interlocking carbon rings with various side groups.
4. *Hormones, lanolin,* and *chloresterol* are common steroids.
5. *Cholesterol* is a constituent of cell membranes and *bile.* When modified it becomes vitamin D.
6. Animals have always relied on flavors and taste to fulfill their dietary requirements, but recently humans have begun to trick themselves by using so many artificial flavors in useless junk foods.
7. In modern industrialized societies, people often eat too much *protein.* While *protein* itself isn't dangerous, the accompanying *saturated* animal *fat* is.

Review Questions

1. List the four major classes of large molecules. What organization do they have in common? (p. 56)
2. List several ways in which *glucose* is used by organisms. (p. 57)
3. Name the subunits of *sucrose* and explain how they are linked together. (p. 57)
4. What does a 1–2 linkage mean? (p. 58)
5. Explain the difference between *amylose* and *amylopectin* in terms of the appearance of the molecule and the *glucose* linkages employed. (pp. 59–60)
6. Where is *glycogen* produced? Explain its structure. (p. 60)
7. Explain the difference between *hydrolytic cleavage* and *phosphorolytic cleavage.* What is the advantage of the latter? (pp. 58, 60)

8. Describe the secondary structure of *amylose*. What forces are involved in producing this level of organization? (p. 60)
9. List three characteristics of *cellulose* not shared with other *polysaccharides.* (pp. 61–62)
10. Compare the arrangement of *glucose* units in *amylose* with that of *cellulose.* (pp. 59, 61)
11. What structural characteristic of *cellulose* gives it its great strength? (pp. 62–63)
12. In addition to the strength of cellulose itself, how are plant cell walls further strengthened? (p. 63)
13. What is the role of *chitin* and how does it differ chemically from *cellulose?* (p. 63)
14. What is the difference between *proteins* and *polypeptides?* (p. 64)
15. Describe the general features of an *amino acid.* (p. 64)
16. Explain how *amino acids* are linked together in a *polypeptide.* (p. 64)
17. How does the *primary structure* of one *polypeptide* or *protein* differ from that of another? How many different *polypeptides* are possible? (p. 64)
18. Define the *secondary, tertiary,* and *quaternary* structure of *proteins.* (pp. 64–65)
19. List four forces that determine the higher levels of structure in *proteins.* (p. 65)
20. What characteristic of some *amino acids* enables them to engage in hydrogen bonding? (p. 65)
21. What is the special role of cysteine in *protein?* What kinds of bonds does it produce? (p. 65)
22. Many *proteins* are not "pure," but are conjugated. Explain this and give two examples. (p. 67)
23. List several ways in which enzymes "work their wonders." (p. 69)
24. If enzymes bring their *substrates* to a lower net energy state, how can one explain how *glucose* can be enzymatically converted to more complex *starch,* which is certainly at a higher energy level? (p. 71)
25. Draw a graph representing the rate of *product* production in an enzymatic reaction where *substrate* is in excess. (p. 71)
26. Describe the role of heat in an enzymatic reaction. Why would heat effect the rate of *product* formation? (p. 72)
27. Draw a graph representing the pH range of 1–14 and plot the activity of pepsin and trypsin. (p. 72)
28. Explain how a *regulatory enzyme* works. Of what value is this kind of control to the organism? (p. 73)
29. What is the difference between control in a *regulatory enzyme* and *allosteric* control? (p. 73)
30. Name three structural *proteins* and state their functions. (p. 74)
31. Describe a characteristic that all *lipids* have in common. (p. 75)
32. List the five major types of *lipids* and their natural sources. (pp. 75–79)
33. If all *triglycerides* are composed of basically the same parts, explain how so many *fats* and *oils* are formed by plants and animals. (p. 76)
34. *Fatty acids* are polar molecules, but *triglycerides* are not. Explain why this is so. (p. 76)
35. Explain the difference between *saturated, unsaturated,* and *polyunsaturated fats.* (p. 77)
36. Compare the energy value in the three major classes of foods. (p. 77)
37. How do *phospholipids* arrange themselves in a cell membrane and how does this produce a water barrier? (p. 78)
38. How do *waxes* differ from *triglycerides* and *phospholipids?* (p. 79)
39. What functions do *waxes* serve in plants and animals? Give examples. (p. 79)
40. Describe the basic structure of *steroids.* How do the steroids differ? (p. 79)
41. List three *steroids* and state their functions (p. 79)
42. What is the apparent danger of too much meat in the diet? (p. 79)
43. Our food is replete with artificial flavoring today. How might this effect our getting a complete diet? (p. 79)

Chapter 4
Cells

Robert Hooke had just been appointed Curator of Experiments for the prestigious Royal Society of London, and he knew he had to come up with something good for the next weekly meeting. He was only too aware that the elite of British science would be there, and he wanted to present a demonstration that would enlighten, entertain, and impress them. It would not be an easy task.

Hooke had an idea. Perhaps he would use an exciting new technology of the 17th century—lenses, cast and ground of glass. The world was buzzing with talk of lenses. With a pair of lenses in a frame the nearly blind were able to see—a miracle come true. Old men who had not been able to read for years had their books and letters restored to them. Earlier in the century, Galileo had pointed a lens to the sky and had started an intellectual revolution. The human eye has a voracious appetite and Hooke knew it. But Hooke was, himself, interested in a new use of the lens—to look at very small things. In fact, he had built his own microscope, one of the first in the world (Figure 4.1).

So for a scientific demonstration, Hooke thought of using the microscope to try to see why corks float. Cork was a mystery. It appears to be solid, yet it floats. Perhaps it is not so solid after all, Hooke thought. He decided to take a look.

Hooke trained his microscope on a cork, and what do you think he saw? Nothing. It turns out that microscopes do not work very well on reflected light. Perhaps he'd better try again. So he took out his pen knife and cut a very thin sliver of cork and shined a bright light *through* it. Then, when he peered through the microscope, he did see something. And it puzzled him. Hooke wrote at the time that the cork seemed to be composed of "little boxes." The little boxes, he surmised, were full of air, and that's why corks float. He called the little boxes *cells* because they reminded him of the rows of monk's cells in a monastery, and a new scientific field was born.

4.1 Hooke's primitive microscope consisted of two convex lenses at either end of a body tube some 6 inches in length. Focusing was done by twisting the body tube along its spiral threads. The subjects were simply stuck on a pin attached to the base of the microscope. The light source was a flame and its light was focused by a lens. The cork cells shown here are from a drawing by Hooke.

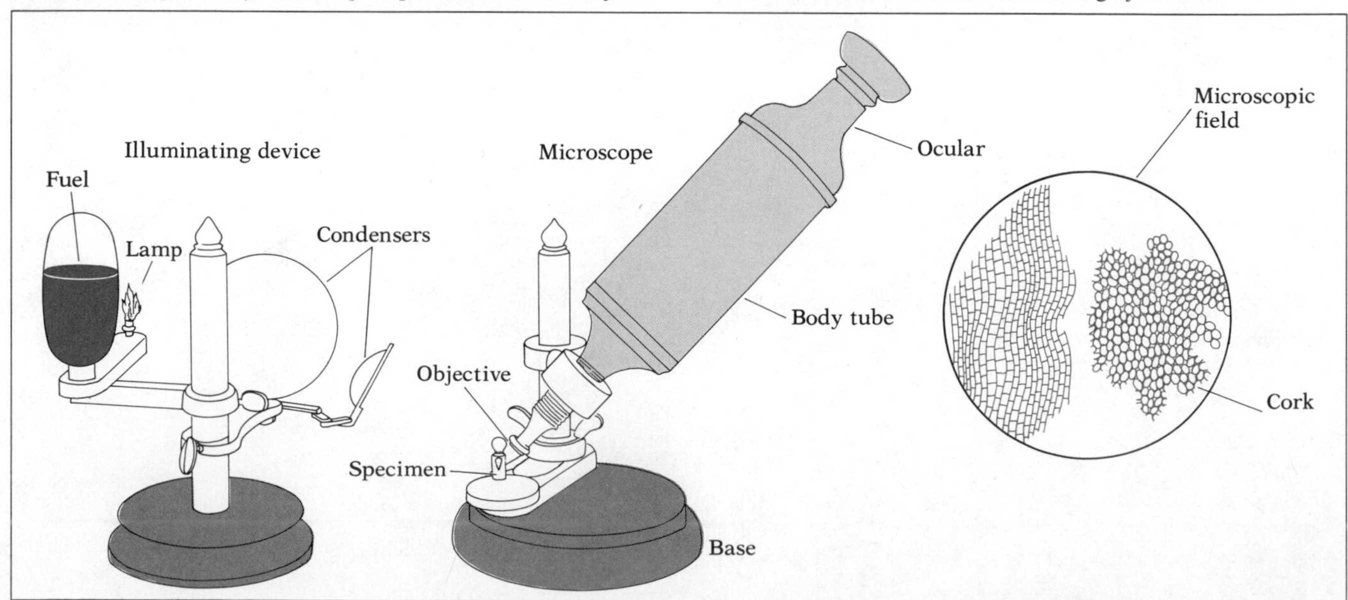

4.1 The Units of Life

Cell Theory

The group that week was pleased by Hooke's demonstration, but a full century would pass before the scientific world would grasp the meaning and importance of Hooke's cells. One of the first people to move on the idea was the German naturalist Lorenza Oken, who stated in 1805 what was to become known as the *cell theory* when he wrote, "All organic beings originate from and consist of vesicles or cells." But in spite of this rather clear and encompassing statement, the formulation of the cell theory is usually credited to two other Germans, the botanist Matthias Jakob Schleiden and the zoologist Theodor Schwann. In 1839, they published a conclusion that they had reached more or less simultaneously; they said that all living things, from oak trees to squid, and from worms to tigers, were composed of cells. Either they were better at public relations than Oken, or perhaps the world was simply more ready to listen in 1839, but Schleiden and Schwann got the credit. Fifty years later a fourth German, Rudolf Virchow, had boiled down the doctrine to "*omnia cellula e cellula*": all cells from cells.

We know now that not only must cells come from cells, but virtually every living thing is composed of cells. (However, exceptions do exist, as we shall see.) Once Hooke had described his little boxes, the act of microscopy blossomed and nothing—literally nothing—proved sacred. People were peering through their hand-made microscopes at everything. Of course, they went over the human body with a fine-tooth comb, looking at and into everything. What they found were not just the dead cell walls that Hooke had described, but variable, changing cells of all descriptions. They could see some of these easily, but others needed to be stained before they were clearly visible. Living cells, they found, were full of a very active, changing, shifting fluid—a puzzling situation, indeed. In time, cell researchers would find that the cells of every creature are special—differing not only from one place to the next within the individual, but from one organism to the next even within a given species.

Let's go on, now, with some of the ramifications of the cell theory.

Why Are There Cells?

The question is, since there are small organisms comprised of only one cell, why aren't there large single-celled organisms? Why isn't there an 800 lb ameba lurking on the bottom of the old swimming hole? Why aren't there single-celled oaks and elephants? Why are large organisms composed of tiny cells? The answer is twofold. First, cells permit specialization. This means that different kinds of cells come to perform certain tasks particularly well. And second, by breaking large organisms up into small units, the *membrane* area is vastly increased.

Membranes enclose cells and are regulatory structures, allowing only certain things to cross them. Thus, by increasing the membrane area, there is greater control or regulation over what is brought into the organism and what is excluded. This property is exceedingly important in any such tightly regulated structure as a living organism. Cells must keep their metabolic products inside and must shield those delicate processes of life from disruptive elements in the environment.

Membranes and the Origin of Cells

Clearly, one of the big questions about the diversity of life is just how cells got started, how they came to be. A clue to the origin of cells is the tendency of lipids to form membranes spontaneously when mixed with water (Figure 4.2). Lipids, as we will see, are a key constituent of membranes. In fact, if certain lipids (called phospholipids) from cell membranes are purified and added to water, they will spontaneously form into membranes that look very much like the cell membranes from which the lipids were extracted.

4.2 A micelle. When in water, phospholipids group together in clusters in a way that resembles the cell membrane's lipid component. Typically, a cluster will consist of an outer covering of hydrophilic, polar heads (the phosphate heads shown in Figure 3.18). The core of a cluster contains the hydrophobic fatty acid tails, which are nonpolar. In fact, the structure of the cell membrane was first suggested by the behavior of phospholipids in water.

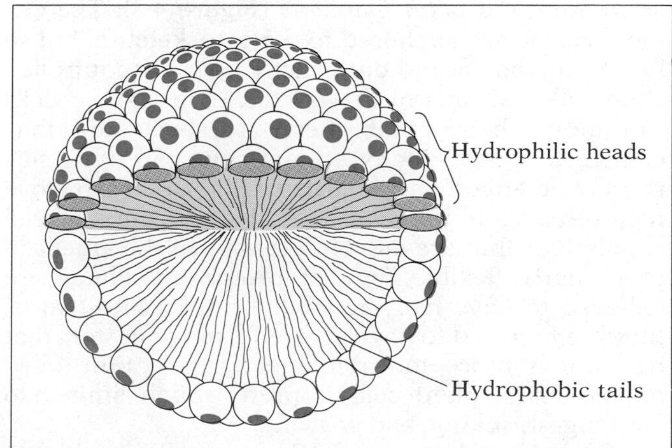

Upon close examination, one finds that the phospholipids and water form onionlike, layered structures of lipids, about the size of living cells. These consist of concentric layers of lipid membranes. The hydrophobic fatty acid tails of the phospholipids face toward each other, and the charged phosphate heads face toward the water. This is particularly interesting because just such an arrangement is found in the membranes of real cells.

Architecture of the Cell

Perhaps then, live, regulatory membranes were inevitable once certain cells in that distant primordial soup began to accumulate phospholipids. The simple membranes that resulted would have added to the stability of those cells so that they might enjoy a small measure of competitive success over any cell-like droplets that lacked phospholipids. In time, those early membranes might have been improved to the extent that they could have earned that venerable term, "living."

Whereas cell membranes are now considered living structures, bacteria and the cells of plants are enclosed in yet another structure outside their membranes—a *dead* structure. *Cell walls* are stiff extracellular constructions that impart a definite size and shape to these kinds of cells (Figure 4.3). Plant cell walls, which are what Hooke saw, are made primarily of the polysaccharides cellulose and pectin (Chapter 3). The cell membranes are pushed tight up against the insides of the cell walls by hydrostatic pressure. The pressure is due to *osmosis,* a process that we'll be discussing later in this chapter.

Because animal cells lack walls they are generally much less rigidly confined than are plant cells, but they are also restricted, and not just by membranes. The size and shape of most animal cells are determined by an internal *cytoskeleton* composed of *microtubules* and *actin filaments* (Figure 4.3). The cell membranes are anchored to the cytoskeleton, but so far no one has found out exactly how. By controlled chemical reactions of the cytoskeleton, animal cells can move their membranes extensively and thus change in shape. Many animal cells can even surround and engulf small particles. The ability to move membranes and to engulf particles are primitive cell capabilities that are most impressive in the ameba, a particularly flexible type of protozoan. Plants are believed to have had this ability but lost it—that is, they are believed to have evolved from ancestors that had much more movable membranes. Controlled movement in plant cells is pretty much limited to swelling, shrinking, and growing.

So, again, what is a cell? So far, we have said that a cell is the structural unit of plant and animal life; that it is enclosed in a phospholipid membrane; that the cell membrane encloses and protects the metabolic processes of the cell, preventing metabolites from diffusing away; and that the cell's shape, size, and ability to move are not determined by the membrane but by other special structures, notably by a cell wall or a cytoskeleton.

Cell Diversity

When one surveys the vast range of cell types, they seem so diverse that one wonders if they have any common traits at all. It turns out, however, that their differences are not so great—that they all have certain features in common, as shown in Figure 4.3.

But first, let's take a look at how different a few plant and animal cells can be (Figure 4.4). Keep in mind that each is specialized for a particular task.

The Cells of Very Different Organisms

Procaryotes are in the kingdom Monera and are comprised of the two major groups bacteria and cyanophytes (Figure 4.5). While these are cellular organisms, they lack some of the features common to the more complex plant and animal cells (*eucaryotes*). They are also generally much smaller than the eucaryotes (although cyanophytes are larger than bacteria). Procaryotes lack not only distinct nuclei (the nuclear material is not enclosed by a membrane) but most other internal membranes as well. (Nuclei, as we will see, contain genetic material and direct much of the cell's activity.) (However, the cyanophytes have extensive internal membranes, the *thylakoids,* which are involved in photosynthesis.) Because there is no nuclear membrane, there is no physical separation between the sites of gene activity and protein syntheses. Eucaryote cells, in contrast, are highly compartmentalized by internal membranes and have many diverse *membrane-bound organelles* ("little organs"), such as *mitochondria* (Figure 4.3). In eucaryotes, the transcription of messenger RNA from DNA sequences takes place inside the nucleus, and the translation of protein sequences from messenger RNA takes place on the outside (we'll discuss this further in Chapter 7).

It has been recently established that procaryotes are enclosed within two separate envelopes or cell membranes. Between the two procaryote cell membranes is a *periplasmic space.* Eucaryotes have single cell membranes.

Key differences among the procaryote, plant, and animal cell types are presented in Table 4.1. Note that the table includes some features that we will not discuss in this chapter. We won't get to such matters

Table 4.1 Key Differences among the Cells of Procaryotes, Plants, and Animals

Feature	Procaryote cell	Higher plant cell	Animal cell
Cell membrane	External only (two, separated by periplasmic space)	External and internal	External and internal
Supporting structure	Polysaccharide cell wall	Polysaccharide cell wall	Protein cytoskeleton
Nuclear membrane	Absent	Present	Present
Chromosomes	Single, circular, DNA only	Multiple, linear, complexed with protein	Multiple, linear, complexed with protein
Membrane-bound organelles	Absent	Many, including mitochondria, large vacuoles, and chloroplasts	Many, including mitochondria, lysosomes
Endoplasmic reticulum	Absent	Present	Present
Ribosomes	Smaller, free	Larger, some membrane-bound	Larger, some membrane-bound
Cell division	Fission	Mitosis	Mitosis
Sexual recombination	Only by plasmid-mediated transfer	Meiosis and fertilization	Meiosis and fertilization
Flagella or cilia (when present)	Solid, rotating	Never present[a]	Hollow, membrane-bound, nine plus two microtubule pattern
Ability to engulf solid matter	Absent	Absent[a]	Present, pseudopods and extensive movable membranes
Centrioles and asters	None	Absent[a]	Present

[a]Although absent in higher plants, these features are found in more primitive plants. Apparently they have been lost in the course of evolutionary change.

as mitosis, meiosis, and sex until later. But we can tell you now that sex in procaryotes is bizarre, incomplete, and infrequent.

Why Are Cells So Small?

How big are cells? Not very. Most cells are far too small to see with the naked eye. But, of course, there are exceptions; chicken egg is technically a cell, and a frog egg is somewhat more convincingly so. *Neurons* (nerve cells) may be 1 meter or more long (such as those that run down a giraffe's leg), but they are still too thin to be seen. However, apart from such specialized cells, eucaryote cells fall within a surprisingly small size range. Almost none are smaller than 10 μm in diameter, and only a few are larger than 100 μm. Plant cells tend to be somewhat larger than animal cells, but perhaps only because they contain large, water-filled internal cavities or *vacuoles;* the average amount of cytoplasm is about the same in plant and animal cells. Bacteria, as we will see, are much smaller than either, seldom exceeding a couple of micrometers in diameter.

So one wonders, why are eucaryote cells so similar in size? We can never give a definitive answer to that kind of question, especially when we already know that there are exceptions, but we can at least get some perspective on the problem.

One approach is to chart the growth of a cell under optimal conditions. Cells in a laboratory-grown culture can be measured or weighed at regular intervals, or the rate at which they incorporate radioactive nutrients into their bodies can be measured. People who measure these things have found that a cell will increase in size (or weight) until its volume is doubled; then it divides into two cells, each of which repeats the cycle. It is interesting to note that the rate of growth is not at all constant. One might think that a large cell would grow faster than a small cell; that is just what you would expect if the addition of new cellular material were proportional to the amount of cytoplasm, or even if it were proportional to the area of the cell membrane. But it just doesn't happen. Quite to the contrary; the cell grows very fast when it is smallest, immediately after a cell division. Then its rate of growth becomes slower and slower. It finally stops altogether some time before cell division begins. Apparently, there is something about being small that encourages growth. And for most cells, there is a maximum size beyond which they are unable to grow.

The Surface–Volume Hypothesis

The generally accepted explanation for this growth phenomenon is the hypothesis that normal growth and metabolism require a minimum area of exterior cell membrane for a given amount of cytoplasm. This

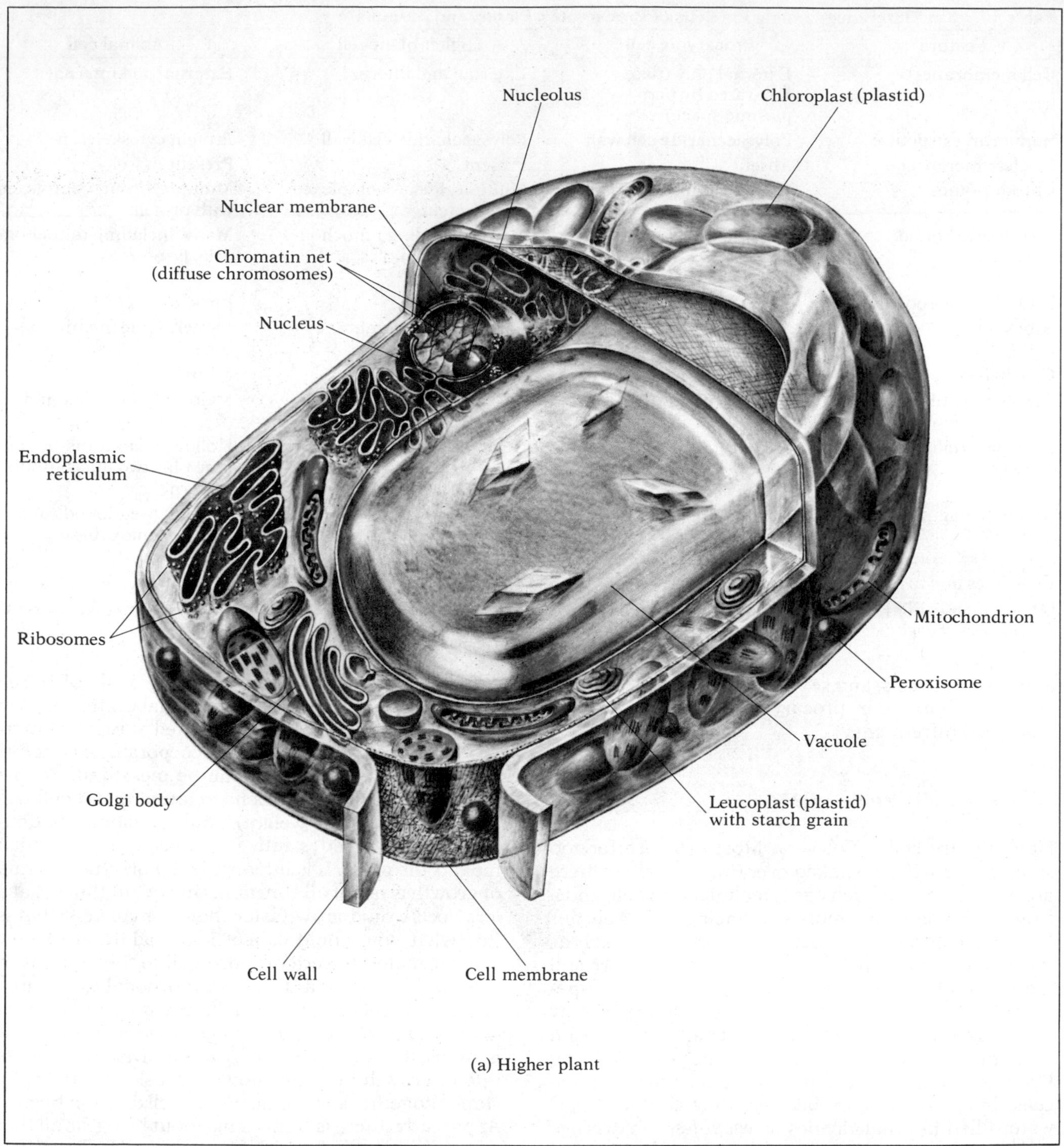

(a) Higher plant

4.3 Actually no one cell of either plant (a) or animal (b) shows all the characteristics shown in these composite drawings. These are both *eucaryote* cells, which means each has a *nucleus*—the more or less spherical body with the double membrane, inside of which the genetic material (here barely seen as a diffuse *chromatin net*) and the dense *nucleolus* are separated from the rest of the cell. Inside its cell membrane is a semifluid mass called *cytoplasm* in which there are numerous *inclusion bodies* and internal membranes. Both plant and animal cells have *mitochondria* in which food molecules are oxidized, *Golgi bodies* in which manufactured cell materials are collected, *ribosomes* on which proteins are synthesized, and an *endoplasmic reticulum* of variable internal membranes that communicate with both the nucleus and the cell membrane.

But plant cells and animal cells are not alike. One of the most prominent features of the plant cell is its huge *vacuoles* filled with *cell sap*, a clear, often pigmented fluid. (Vacuoles occur in animal cells less frequently and are usually small.) Plant cells are encased in a thick, semi-rigid *cell wall*. The animal cell, in contrast, owes its more variable shape to a shifting, dynamic fibrous *cytoskeleton* of

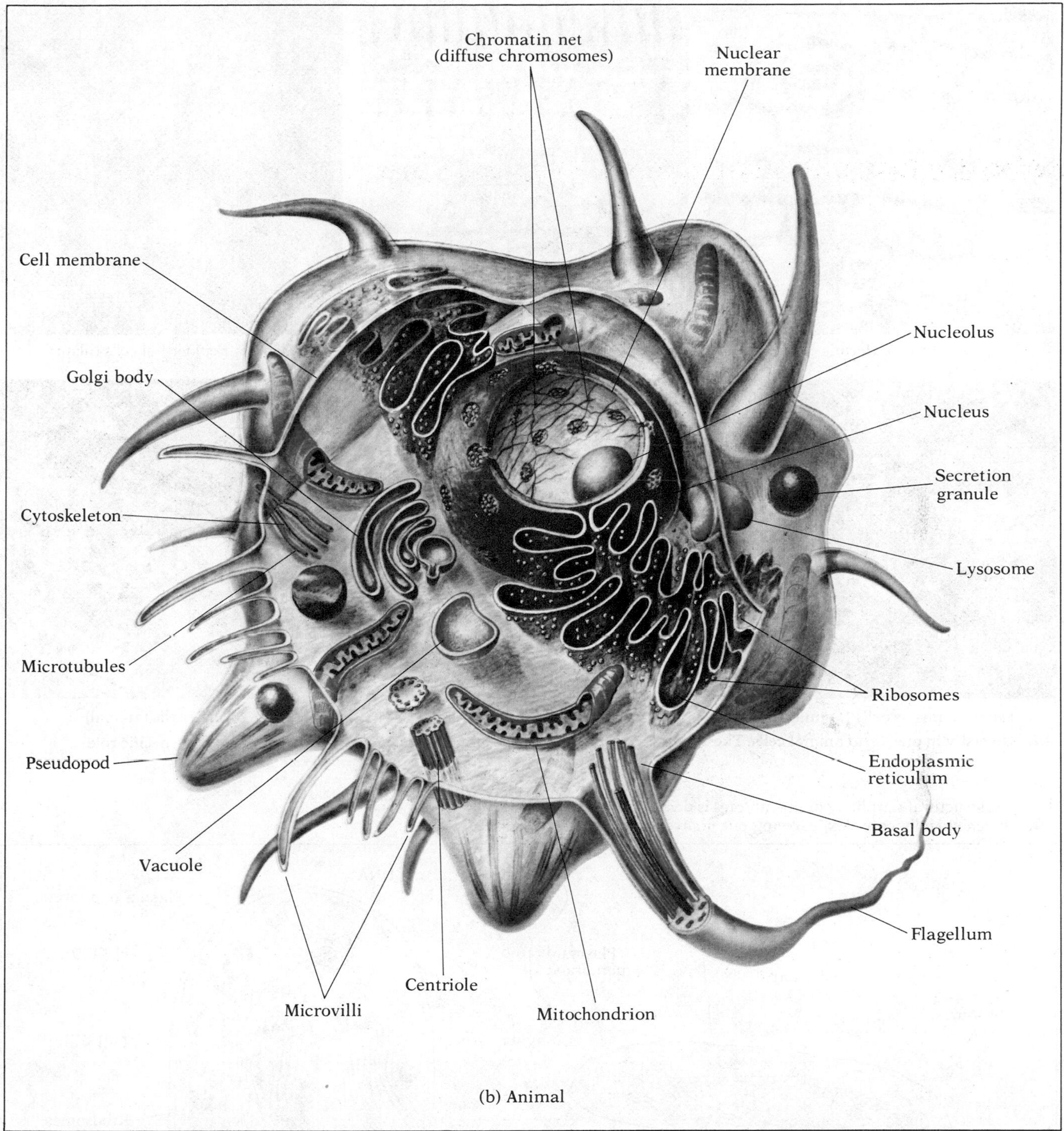

(b) Animal

microfilaments and *microtubules.* (Plant cells may have microfilaments and microtubules at some stages of growth.) Typical animal cell inclusions are *secretion granules* and the complex *centriole;* typical plant cell inclusions are the *plastids,* including the photosynthesizing *chloroplasts* and the *leucoplasts,* in which starch grains are stored; and membrane-bound *peroxisomes* that seem to be involved in protecting the cell from toxic by-products of photosynthesis. The cell surface of the animal cell may be thrown up into absorbing *microvilli* or grasping *pseudopods;* some animal cells have motile, complex *flagella* or *cilia* with their associated *basal bodies.* Higher plants lack cilia or flagella, although ferns and lower plants have motile sperm cells with flagella. Organelles found exclusively in animal cells are *lysosomes* and *secretion granules,* small membrane-bound sacs in which powerful enzymes or other materials are stored. By definition, plant cells do not have lysosomes, although some plant cell vacuoles may perform the same functions.

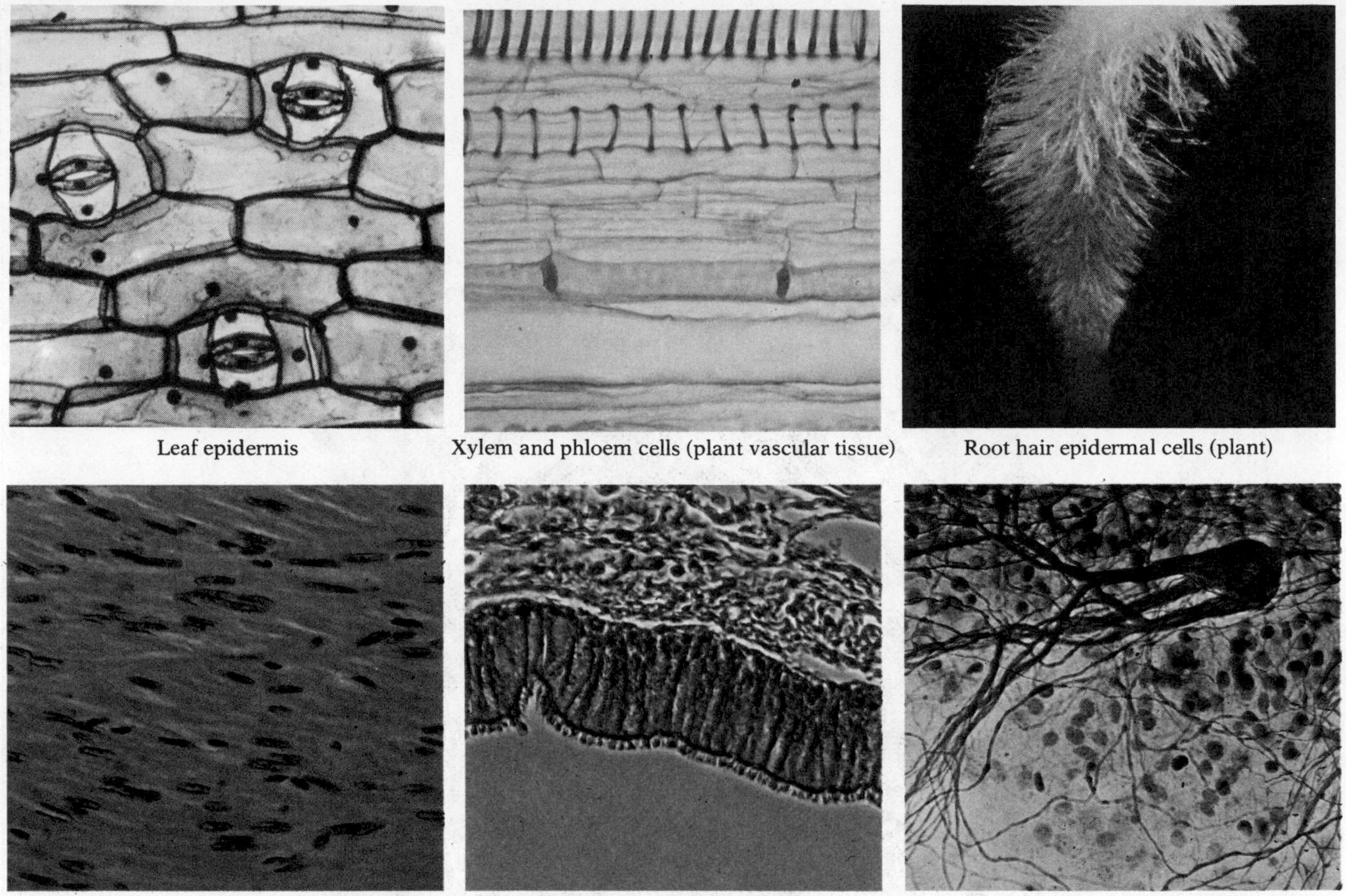

Leaf epidermis | Xylem and phloem cells (plant vascular tissue) | Root hair epidermal cells (plant)

Smooth muscle cells (animal) | Ciliated epithelium (animal) | Nerve cells (animal)

4.4 Diversity in plant and animal cells. These cells are from multicellular organisms, and each is well-adapted to a specific role.

4.5 The structural simplicity of procaryotes is deceptive. They are often biochemically complex, carrying out many of the chemical functions of eucaryotes.

(a) Bacterium (b) Cyanophyte

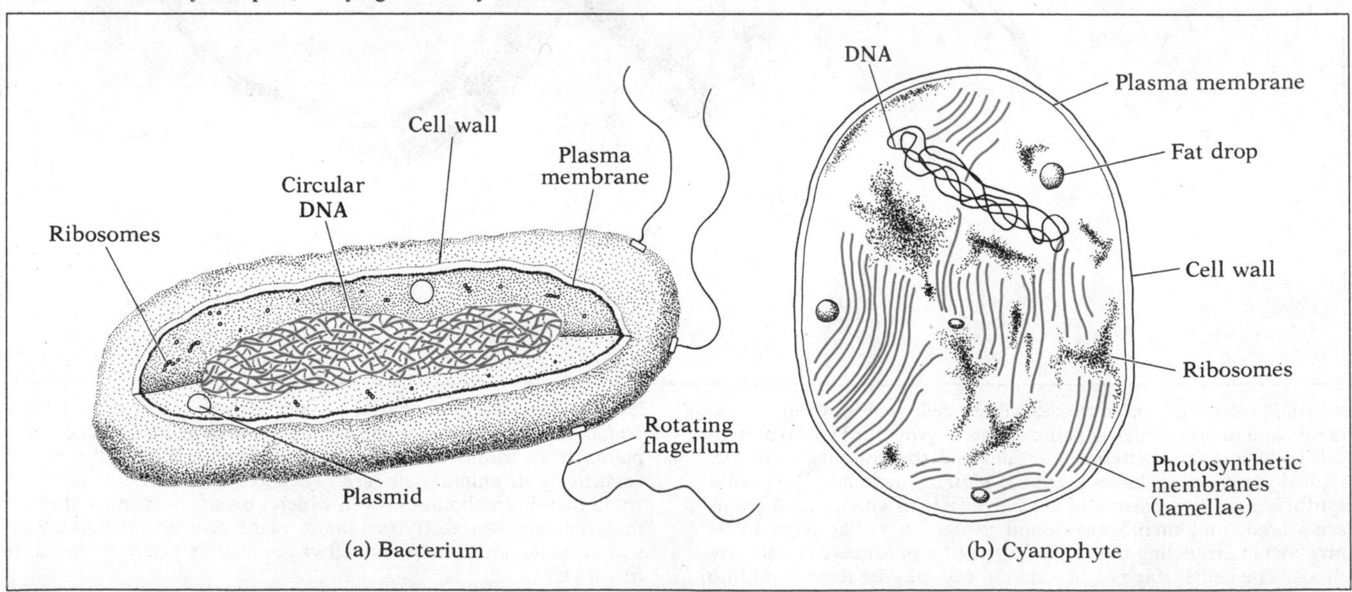

Essay 4.1 Exceptions to the Cell Theory

The notion that all living things are made up of cells is a generalization; it is by no means an immutable law. The cell theory, or cell doctrine, was formulated with multicellular organisms in mind—plants and animals. But even plants and animals are not necessarily composed entirely of cells; your tendons and ligaments, for instance, are extracellular materials (materials existing outside cells) that are secreted by cells. Are tendons and ligaments alive? It depends on your definition. On the other hand, hair and the outer layer of your skin consist of the remains of specialized cells that have died—that is, cells that were definitely alive once but have lost the ability to metabolize or to respond to any kind of stimulation.

Vertebrate skeletal muscle, which is certainly living, metabolizing tissue, also does not fit into the cell theory very comfortably. Skeletal muscle consist of *multinucleate* (having many nuclei) fibers, which can be quite large relative to single cells. There are no cell membranes separating the many nuclei. In the plant world, coconut milk (the milky fluid that sloshes around in the cavity of young coconuts) is a rather strange multinucleate tissue; the cell nuclei float freely in the fluid.

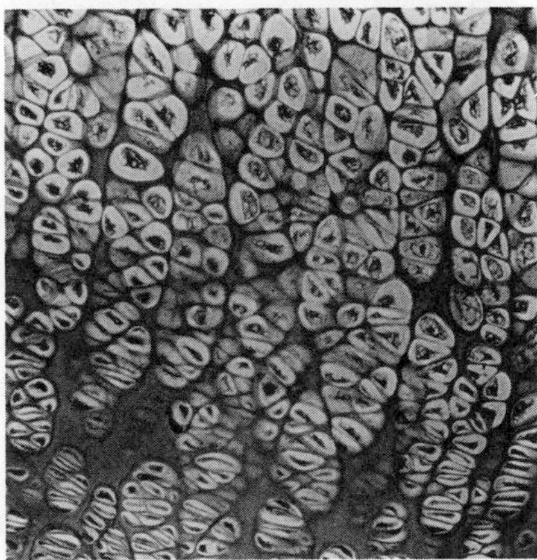
Cartilage

Muscle fibers

What does the cell doctrine have to say about *protozoa* and simple *algae?* The *paramecium,* a fairly common ciliated protozoan, is large enough to be seen with the naked eye (barely); it is a complex organism with many specialized parts, but it has no cellular boundaries between these parts. The whole organism is enclosed by a plasma membrane, however. Some biologists consider paramecia and other protozoa to be *acellular* (without cells), while others prefer to think of them as being *unicellular* (consisting of a single cell). This may be yet another instance of arguing over definitions, but clearly the cell theory is more comfortable with one cell than none.

Fungi, for the most part, don't fit the cell theory at all. The bodies of most fungi are long multinucleate tubes or *hyphae.* In some fungi the hyphae are divided into small compartments by cross-walls, but the cross-walls usually have big holes in them and the contents of the hyphae (*cytoplasm*) are continuous. Furthermore, *omnia cellula e cellula* doesn't quite work out, because the continuous cytoplasm does become pinched off into regular uninucleate cells, namely the *reproductive spores.*

A group of funguslike protozoa, ungraciously named the *acellular slime molds,* are normal amebalike cells during one stage of their life cycle; but at another stage, the little amebas flow together into a large, gooey mass of multinucleate living material which then proceeds to move around and over its substrate—usually a log or stump—like a 2 ft wide ameba. Later, this multinucleate mass will again bud into uninucleate cells.

Life cycles that alternate unicellular and multinucleate stages have evolved independently at least five times. This pattern is found in fungi and the unrelated slime molds, as we have seen, and also in certain green algae (Acetabularia), in some ciliate proto-

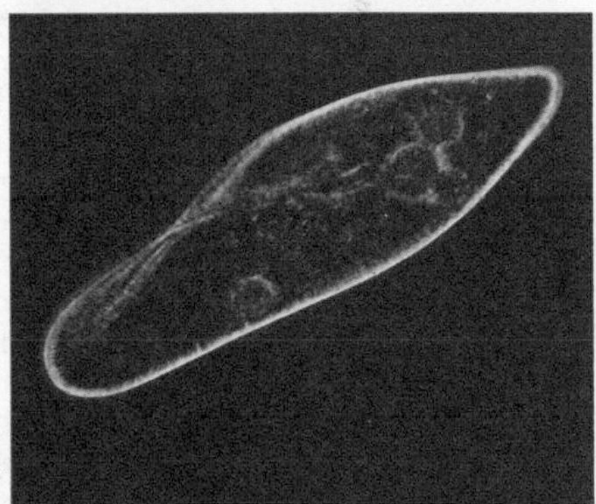
Paramecium

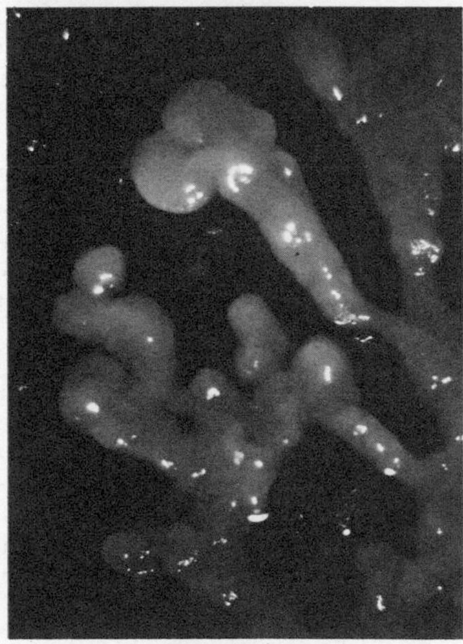
Slime mold

zoa, and even in certain kinds of bacteria (the bacteriomycetes).

Viruses are a special challenge. It is another favorite argument over definitions whether or not viruses are alive. One thing is certain, however: alive or not, they aren't cells. The infective virus particles have no lipid membrane or cytoplasm; they are just nucleic acid packaged into little boxes of protein, with some molecular mechanism of attachment to their prospective hosts and sometimes carrying a special enzyme or two. The best we can say for the cell theory when it comes to viruses is that the viruses can't metabolize or reproduce unless they are inside some other organism's cell.

Keep in mind that, in spite of these exceptions, the cell theory holds for most living things most of the time.

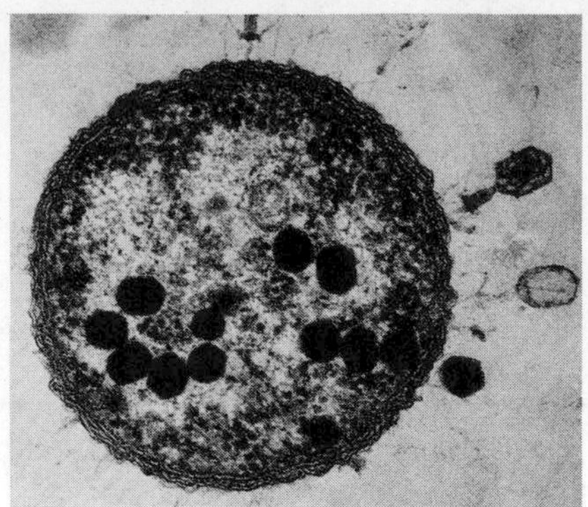

Viruses attacking a bacterium

---

can be called the *surface–volume hypothesis,* and it puts close restraints on how large cells can be. Let's take a look at why this should be.

Cells, especially growing cells, require nutrients and oxygen, and they must rid themselves of waste products. The nutrients and oxygen must move into the cell across the cell membrane and must then diffuse through the cytoplasm to sites of synthesis. For protein synthesis, the messenger RNA must diffuse out from the nucleus to these same sites. Since diffusion is a random process, the distances must not be too great, which is to say, the sites of synthesis in the cytoplasm must be fairly close to the surface of the cell, and also fairly close to the nucleus. Because materials entering the cell must join with materials leaving the nucleus in order for protein synthesis to occur, cells must remain small enough to ensure a high probability that these molecules will be able to get together.

There is another problem here, as well. As a cell increases in size, the volume grows faster than does the surface area (the membrane). Thus, in large cells there may not be enough membrane area to accommodate the cytoplasmic mass. At some critical point, then, growth stops. This point is reached very soon by bacteria. It has been argued that eucaryote cells can become much larger than bacteria because they possess internal membranes and compartments, but there is a limit even for eucaryotes. Plant cells, particularly those with large internal vacuoles, can stretch the limit a bit further by augmenting internal diffusion with a peculiar and puzzling process called *cell streaming,* in which the cytoplasm whirls around the cell in more or less definite channels.

The surface–volume hypothesis is explained mathematically by the observation that as the radius of a body, such as a sphere, is doubled, the *surface* area is *squared* while the *volume* is *cubed.* This relationship holds not just for spheres, but for any solid body with a well-defined shape (Figure 4.6). To keep the surface–volume ratio at a functional level, the cells must be small.

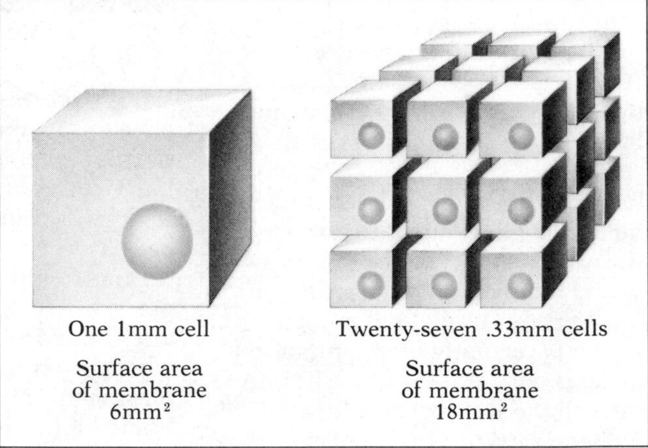

4.6 The cell membrane is vital to the transport of materials in and out of the cell. The total surface area determines how well the membrane can support the transport requirements. The membrane of the smaller cell has less cell volume to service per square unit of area, so it can operate more efficiently than the membrane of the larger cell. At some point in growth, biologists believe, the increased workload on the membrane of a metabolically active cell becomes intolerable. For most cells the answer appears to be cell division.

Then what about the exceptionally large cells we have been considering? The rules of surface–volume ratios still hold, so they must have a way of coping with the problem. For example, a nerve cell can grow to be very large but it does not maintain a well-defined shape; essentially it grows only in one dimension, and all parts of the cytoplasm remain very close to the surface. In fact, a long nerve cell has about the same surface–volume ratio as a short one. Some cells in the kidney have highly convoluted borders, as do cells of the small intestine (Figure 4.7). This development greatly increases the surface area of these cells whose primary function involves moving material across their membranes. The multinucleate cells of fungi occur in the form of long slender threads. Multinucleate muscle cells, as well as heart muscle cells, also take the form of slender tubes so that every part of the cell is close to the surface. The slime mold,

Essay 4.2 Small Units of Linear Measurement in the Metric System

The United States alone continues to use the peculiar English system of weights and measures in commerce and everyday life, but even in America the times are changing, as witnessed by a sure index: Ford and General Motors are shifting their production lines to the metric system. And men already have captured our attention by giving women's measurements (never their own) in metric units. Track meets now regularly feature the 100-meter dash rather than the 100-yard dash, and the 1500-meter "metric mile" is displacing the traditional mile. (The mile will probably hang on as a special event because of its sentimental value.)

The sciences, including biology, have always used the metric system. Even if you did a good job of memorizing the metric system in grade school, you may well come across some unfamiliar units of measurement in your study of cell structure. You will not be alone. Trained biologists have been having the same problem in recent years, because the international metric system itself has been changing. It is becoming even more regular, by decisions of international councils that have authority over such matters. The unit names in the metric system consist, usually, of a root word and a standard prefix. The root word gives the basic unit for a type of measurement: meter for linear measurement, liter for volume, gram for weight, and so on. The standard prefixes are added to these basic word roots to form larger or smaller secondary units. Each prefix indicates an increase or decrease by a specific positive or negative power of ten. Thus, for instance, *micro-* means one one-millionth; so a microliter is one one-millionth of a liter (10^{-6} liter), a microgram is one one-millionth of a gram (10^{-6} gram), a micrometer is one one-millionth of a meter

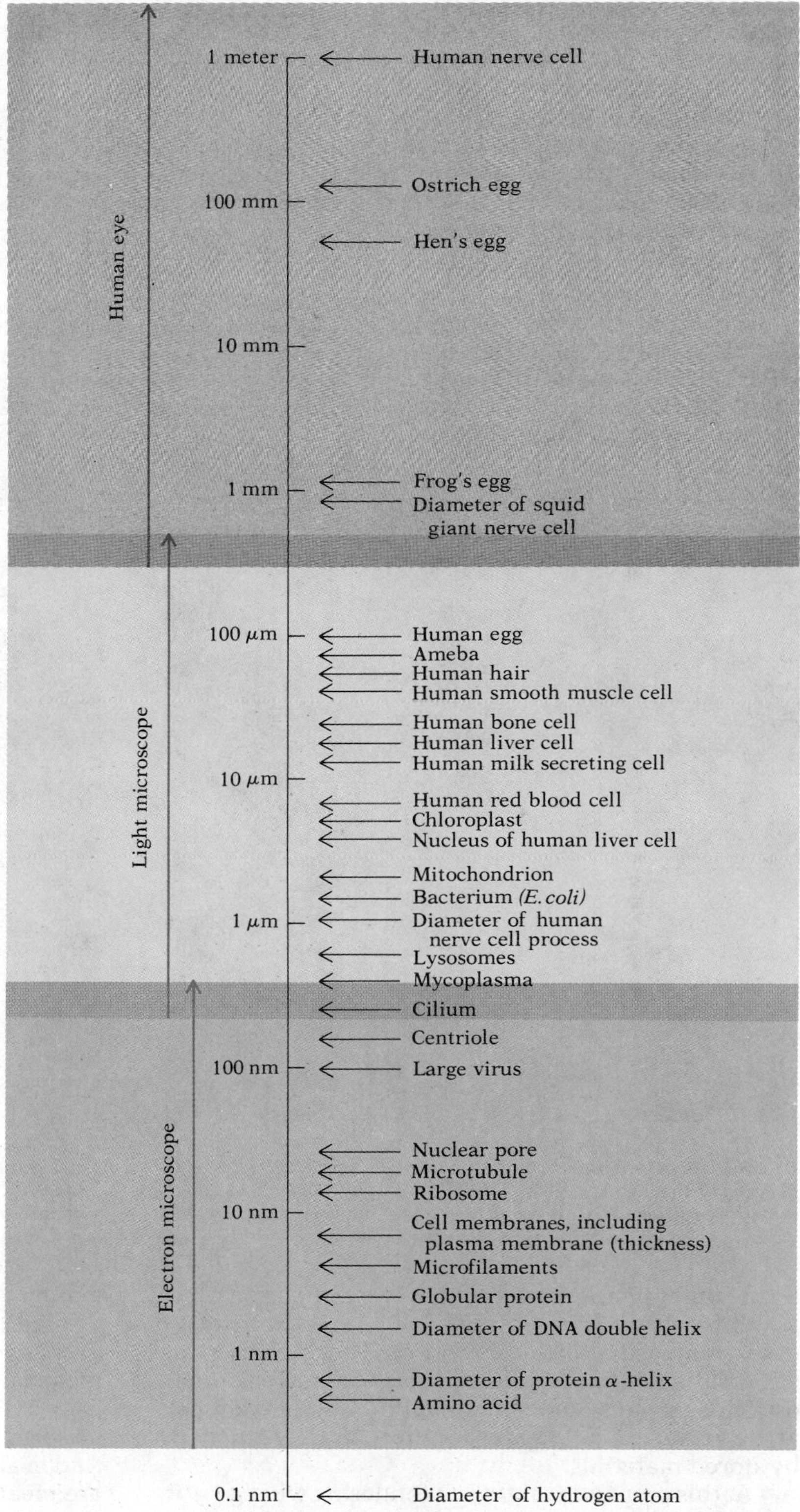

(10^{-6} meter), and a micromole is one one-millionth of a mole (10^{-6} mole). Another regularity that is being imposed on the metric system is this: each named unit of measurement is one one-thousandth (10^{-3}) of the next larger named unit. The table shows how this works out in practice for units of linear measurement. Note that the familiar centimeter, one one-hundredth of a meter, violates this regularity. Perhaps the centimeter will eventually be phased out of official use, too.

The basic unit of linear measurement is the meter. Other lengths relate directly to it, such as the kilometer (about 0.6 mile) and the centimeter (about 0.35 inch).

Units of Linear Measure

Unit name	Symbol	Portion of a meter	Equivalent
kilometer	km	10^3 m	1000 m
meter	m	1 m	0.001 km
centimeter	cm	10^{-2} m	0.01 m
millimeter	mm	10^{-3} m	0.001 m
micrometer	μm	10^{-6} m	0.001 mm
nanometer	nm	10^{-9} m	0.001 μm

In addition to the above units, which are now official, three other metric units of linear measurement are in the process of being phased out. These units are still common in scientific books and articles, so we'll add them to our table:

Unit name	Symbol	Portion of a meter	Equivalent
micron	μ	10^{-6} m	1 μm
millimicron	mμ	10^{-9} m	1 nm
angstrom	Å	10^{-10} m	0.1 nm

The change from the name micron to micrometer means the symbol has changed from μ to μm. The change from millimicron (mμ) to nanometer (nm) also means a new name and a new symbol for 10^{-9} meter. The angstrom (Å) is being dropped altogether, to be replaced by 0.1 nanometer.

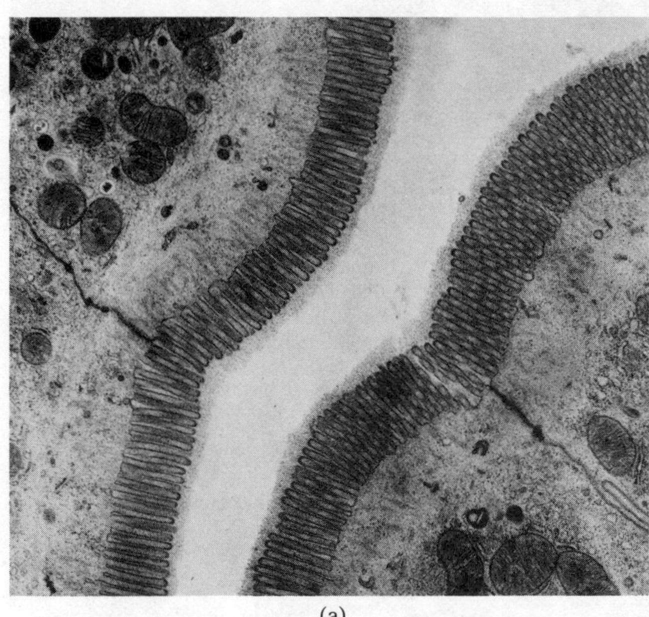

(a)

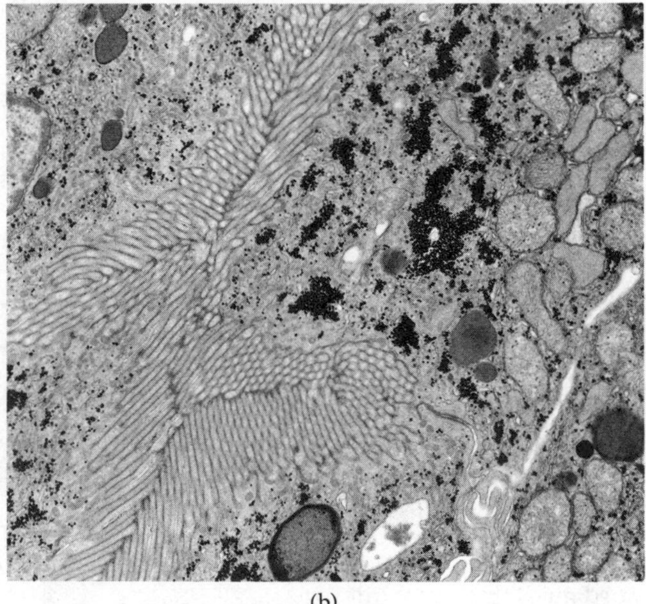

(b)

4.7 Surface-to-volume strategies in two cell types. (a) The epithelial cells of the intestine absorb small molecules through a surface that is greatly expanded by the presence of innumerable *microvilli,* small projections of membrane supported by bundles of microtubules (×50,000). (b) Kidney tubule cells also increase their area with microvilli, which fill the *lumen* (cavity) of the tubule.

on the other hand, spreads out in two dimensions, but it still forms a very thin mass, so oxygen and waste gases can easily diffuse in and out. Large, spherical egg cells, such as the yolk cell of a chicken egg or an ostrich egg, have almost no metabolic activity except at the very surface—the rest of the volume is taken up by stored materials.

At this point you might be wondering what good it does to divide a mass of tissue into a lot of little cells, if they just lay piled up in a mass. After all, a cell in the middle of the heap would still be just as far as ever from its source of nutrients and oxygen.

The answer is rather simple: Cells do not normally lie piled up in a mass, at least not in metabolically active tissues. For one thing, there are thin but important spaces between even closely packed cells. And in all large multicellular plants and animals there are means of carrying oxygen and nutrients to these metabolically active cells. Leaves, for instance, have air spaces within them that communicate to the outside by way of small pores. Insects have more or less the same kind of system, with air-filled internal passages that communicate directly with the outside air and with the surfaces of most of their cells. Both leaves and insects, and larger plants and animals in general, have vascular systems that bring water and nutrients to or near cell surfaces. There are seldom more than two cell layers between any part of the organism and a source of cell nutrients and oxygen. Thus, animals and plants that lack circulatory systems must remain small. In fact, as it turns out, their metabolically active tissues are usually only about two cell layers thick.

So that's why cells are as small as they are; so that they can maintain a favorable surface–volume ratio, and so that the sites of metabolic activity will not be too far from a source of oxygen and nutrients.

Why aren't cells smaller? We don't know.

4.2 Cell Barriers and Transport Mechanisms

The Cell Structures

Now that we have some idea about what cells are and why they evolved, let's see what's in them. But first, let's take a closer look at cell walls.

The Plant Cell Wall

As we've already noted, plant cells are surrounded by fairly rigid, nonliving outer walls composed primarily of cellulose and other polysaccharides (see Figures 3.8 and 4.3). Microfibrils of cellulose chains are arranged in layers. Each layer lies at an angle with respect to the one below it. The result is a laminated, very strong but porous covering over the cell. Once impregnated with hardening substances (pectin), the cell wall maintains the shape of the cell and has great strength and resiliency. We see this strength in wood and other plant fibers. The wall is secreted by the cell, and the cell is able to determine the shape of the wall in ways we don't entirely understand.

Actually, newly formed cell walls can't be too rigid, because each time a plant cell divides to form two cells, they must be able to grow to the size of the original cell. As the cell divides, a new cell wall appears, effectively separating the two new cells. Walls of adjacent cells are cemented together to form a layer known as the *middle lamella* (see Chapter 21).

As the plant cells mature and differentiate into specialized tissues, other substances are deposited into the cellulose and pectin matrix. In the stems of woody plants, for instance, *lignin*, a complex chemical that is technically an alcohol, is continually secreted into the cellulose, forming a hard, thick, decay-resistant wall. *Suberin*, a waxy substance, is secreted into the outer layer of stem cells, forming a protective, waterproof corky layer. Waterproofing of the upper surface of leaves, on the other hand, is accomplished by the secretion of *cutin*, a waxy substance added to the epidermal cell wall.

Plants that are not used as food are usually useful to humans because of the materials in their cell walls. Wood is not only a building material, but is a principal fuel in much of the world; cotton, linen, and all sorts of other fibers are useful because of their cellulose. We have only recently, and with great effort, managed to find economic uses for lignin other than to make wood alcohol, but delignified wood pulp is what this book is printed on. And it certainly wouldn't do to have a picnic under a limp oak.

The Cell Membrane

We've already mentioned this vital cell envelope, and vital is the right word because if anything is alive, the cell membrane is. Basically, it keeps the insides of the cell in and the outside out, and it lets through the materials that the cell must exchange with its environment. As we've noted, the membrane is composed of a double layer of phospholipids (Figure 4.2). This peculiar double membrane is somehow able to accept freely the movement of some substances through it, while utterly rejecting others and actively forcing some molecules in or out.

The cell membrane, then, is a critical structure, but for a long time some people refused to believe it existed. They can't be blamed because it is so thin that its existence, as well as its probable structure, was postulated entirely on circumstantial evidence until the advent of the transmission electron microscope (TEM). Before we take a closer look at the structure of the cell membrane as it is now understood, let's consider that evidence. It all has to do with what moves across the cell membrane and what doesn't.

Movement of molecules across a cell membrane can be divided into three types: *simple diffusion*, *facilitated diffusion*, and *active transport*.

Diffusion is the random movement of molecules in a fluid or gas. Actually, only the movement of individual molecules is random; the net movement of large numbers of molecules is not random at all, but is regular and predictable and follows well-known laws.

Looking at Cells

The Light Microscope

The compound lens *light microscope* had achieved near technical perfection by the end of the 19th century. Of course, there have been some embellishments since, but the essentials were all there. Not only had the machine itself been developed, along with the theoretical optics to make it work, but the equally important techniques of tissue preparation and staining had been worked out. As a result of these early technical breakthroughs, almost all cell structures that can be seen with a light microscope now had been named and catalogued before your grandfather was born.

The limits to what can be seen with any type of microscope are described as the instrument's *resolving power.* Resolving power is a measure of how close two points can be and still be distinguished as being separate. The resolving power of a normal, unaided human eye at close range is about 0.1 mm, or about the distance separating the dots in a good halftone print. Vision is variable at this range, however, and some people are capable of making much finer distinctions than others. The resolving power of the unaided eye is at its maximum when the object is about a foot away and diminishes with greater distances. With even a very simple microscope or a hand lens of good quality, you have no trouble making out the dots in even the best printed halftone. Thus, the resolving power of a light microscope is greater than that of the eye.

There is a theoretical limit to the resolving power of even the finest light microscope, and many important cellular structures are well below their resolving limits. Optical physicists have proven that no system can resolve points that are closer together than half the wavelength of the light used to view them. Amazingly, the fine microscopes of the turn of the century approached this theoretical limit very closely; and that is why there haven't been significant improvements since. The limitation is that of the nature of light itself.

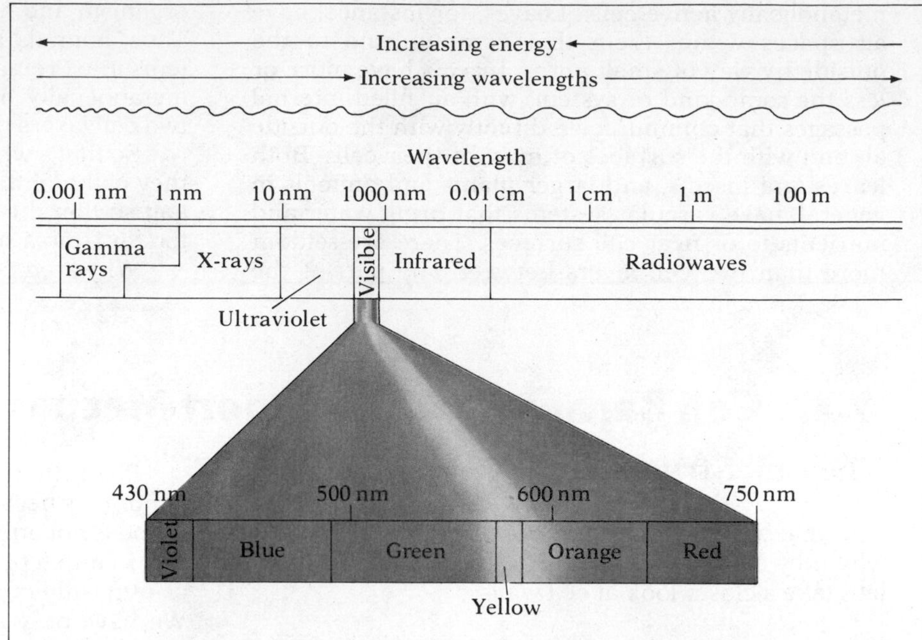

The electromagnetic spectrum. Our eyes are barraged by electromagnetic radiation of a variety of wavelengths, but we see only that portion with wavelengths between 430 nm and 750 nm. Within that range, light of different wavelengths is discerned as color. The wavelengths in the visible spectrum also restrict microscope resolving power, since objects closer together than 250 nm will interfere with light passing between them and the separation will not be visible. The 250 nm is just about half the average wavelength in the visible spectrum.

The visible spectrum includes light with wavelengths between 750 nm (red) and 430 nm (violet). The theoretical and actual resolving power of the best light microscopes is therefore approximately 250 nm (0.25 μm, 0.00025 mm). This is about 400 times the resolving power of the human eye. It is possible to magnify images more than 400 diameters, but doing so doesn't improve the quality of the image in the least. Even photographic enlargement cannot overcome the limits of resolving power.

A compound microscope consists basically of a tube with lenses at both ends. The *objective lens,* itself made up of a number of thinner lenses glued together, is brought very close to the object being viewed. The *ocular lens,* also made up of several thinner lenses, is brought close to the eye.

A third set of lenses is found in all but toy microscopes. This is the *condenser,* and it lies beneath the specimen, which is to say between the specimen and the source of light. This lens condenses the light; if an image is to be blown up some 400 diameters, the light passing through the specimen is thinned out $(400)^2 =$ 160,000 times. If the resulting image is not to be too dim, there has to be a lot of light passing through the specimen. More importantly, the condenser also focuses the light in a favorable way such that the rays of light pass through in a wide variety of angles and directions, but only those angles and directions that will pass into the objective lens. This greatly improves the quality of the image. Finally, the condenser contains an *iris diaphragm* that cuts out stray light that would cloud the image.

Fixing, Preparation, and Staining of Specimens for the Light Microscope

Small things tend to be transparent. The light just passes through without being absorbed. So special stains must be used to distinguish cellular structures; these are pigmented molecules that have strong affinities to one or another of the constituents of the material being looked at. For instance, basic stains have a natural affinity for acids and therefore tend to stain chromosomes, which are rich in nucleic acids.

Stains may be of a general nature, simply darkening the cytoplasm or nuclei of cells; or they may be highly selective for specific structures or molecules. A great deal of literature in the area of stain technology has been amassed over the past 150 years. Selective staining has been invaluable to the field of *cytology* (the study of cells) and *histology* (the study of tissues), and many of the old techniques are still in use.

In order to have enough light passing through the specimen, and in order for all of it to be in focus, the specimen must be very thin. Small, loose cells (such as bacteria) can be seen whole. They are killed, *fixed* (treated with chemicals that coagulate their cytoplasm into a solid gel), affixed to a glass slide, stained, and viewed. Somewhat larger cells can be squashed flat beneath a cover slip. Most biological material, however, must be cut into very thin slices, or *sectioned*, as Hooke found out with his cork. The procedure is rather complicated.

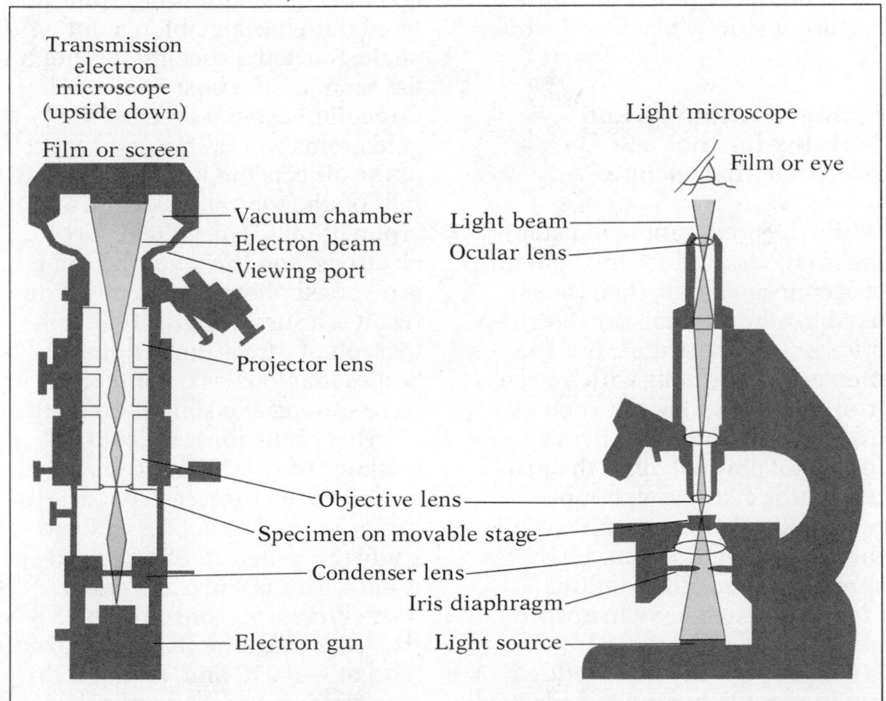

The light microscope and the transmission electron microscope work on very similar principles. In both cases a focused beam—photons or electrons—is transmitted through a very thin slice of the specimen and is absorbed differentially by different structures of stains. The emerging beam is then refocused to produce an image, either on a fluorescent screen, which can then be seen by the viewer (electron microscope), directly into the eye (light microscope), or onto film (either system). The light microscope lenses are glass, and the electron microscope lenses are magnetic fields. Here the electron microscope diagram is shown upside down to emphasize the correspondence between the two systems.

The Transmission Electron Microscope

The electron microscope has enormous powers of resolution, since the object is flooded with electrons rather than light waves and the wavelength of electrons is much smaller than the shortest light wave.

"Transmission" means that the electrons are shone through the specimen, rather than being reflected off its surface. The *transmission electron microscope* (TEM) is one answer to the problem of overcoming the resolving power limitations of the light microscope, and in general it is used only when one wishes to observe finer detail than can be seen with the light microscope. The principle upon which the electron microscope works is relatively simple, although the actual technology required is not. In place of light, the electron microscope uses electrons shot from an electron gun similar to the electron gun in the picture tube of a television set. Electrons move in transverse waves, just as do the photons of light, but the wavelength of moving electrons is very much shorter than that of visible light. This is the main reason that much smaller structures can be resolved.

Watching an electron microscope in action is a truly awe-inspiring experience. As the microscopist twists knobs that move the tiny specimen, vividly detailed, mysterious scenes move rapidly across a viewing screen about 8 inches in diameter. The microscopist is looking for something—perhaps a particularly favorable view of some special feature seen previously. But surprises come up too, and you can hardly wait to see what will appear next. We urge you to try to talk your way into an electron microscope laboratory some time, just to be a tourist in the incredible world of ultrastructure.

In the basic electron microscope system, electrons are emit-

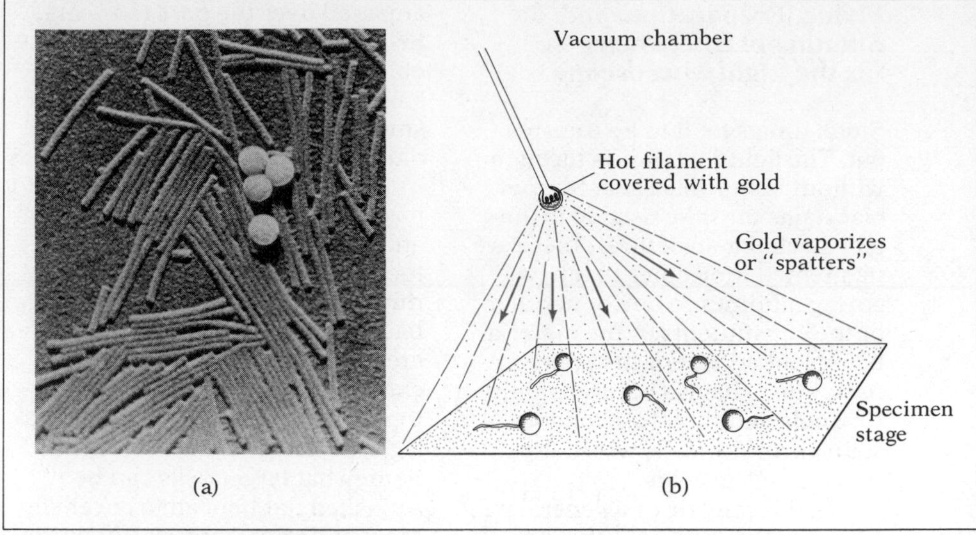

Heavy metal shadow casting. The objects in this photo are viruses. The moonscape appearance is attributed to heavy metal shadow casting. Molecules of heavy metals are sprayed at an angle and thus deposited on one side of each virus. The result is a distinct shadowing that helps emphasize the shapes. This is a positive print of the original electronmicrograph, so the heavy metal deposits are made to look light—strikingly like reflected sunlight. (b) Shadow casting is done by heating and vaporizing a small dab of gold or other heavy metal in a vacuum chamber. The gold "spatters" over the specimen.

ted from a cathode ray gun at the top of a chamber. A vacuum is maintained in the chamber to avoid the problem of the electrons being scattered by molecules of gases. The electron beams are focused by an electromagnetic condenser, which is only distantly related to the condenser of a light microscope. The negative charges of moving electrons make it possible for their paths to be directed by magnets in somewhat the same way light rays are directed by glass lenses. As the electrons pass through the specimen some are absorbed by the nuclei of heavier atoms and others pass through unhindered. Those that get through pass between additional electromagnetic lenses that spread them out. The electrons are finally focused onto a screen coated with phosphorescent material, or onto special photographic emulsions. The pattern created shows the image of the electron-absorbing parts of the specimen. Different magnifications are achieved by altering the strength of the magnetic lenses.

We have seen that the wavelengths of visible light occur around the range of 430–750 nm. By comparison, the electrons in the electron microscope have an average wavelength of only 0.05 nm. Theoretically, this should produce a resolving power of about 0.025 nm, but at present the best that can be accomplished is a resolving power of 0.5 nm. Compared with the 250 nm resolving power of the best light microscopes, the electron microscope is 500 times more powerful, but its picture is strictly black and white.

Fixing, Sectioning, and Staining for the Electron Microscope

While the preparation and staining methods used for the light microscope are harsh, then those used for the transmission electron microscope are brutal. The specimen is fixed, usually with an electron-dense metallic salt such as osmium tetroxide. The heavy metal not only solidifies the protein of the cytoplasm, but preferentially combines with it. Thus, the metal both stains and fixes the specimen, since the osmium atoms will later serve to absorb electrons.

The specimen is embedded not in paraffin but in hard plastic, which is cut with specially prepared broken glass. It is cut into exceedingly thin slices. The specimen has to be very, very thin, because electrons do not pass through solid material very well.

Small, nonsectioned material, such as loose viruses, chromosomes, or bacterial flagella, are prepared somewhat differently. They are first spread onto a thin film of protein that is supported by a wire screen. In a vacuum chamber, gold atoms are sputtered onto the specimen at an angle. Since the specimen is not a flat section but consists of small three-dimensional objects, the gold atoms will fall on one aspect of the objects the way evening sun falls on the tops and sides of a group of hills. The gold absorbs electrons, and the resulting image is reversed photographically. The result is a surprisingly lifelike photograph of three-dimensional bodies that look as if someone were shining a flashlight on them.

One of the most specialized methods of specimen preparation is of particular interest, because it has been used so successfully to study the structure of cell membranes. This is a process called *freeze-fracturing*, or sometimes *freeze-etching*. The tissue is frozen hard at −100°C and, at that temperature in a vacuum, is broken apart. At −100°C, the aqueous portions of the cell are frozen much harder than the thin lipid sheet that is sandwiched in the cell membranes, and these form natural weaknesses in the tissue. As a result, cell membranes may be split apart and opened up, exposing their inner structure from two aspects—the half of the mem-

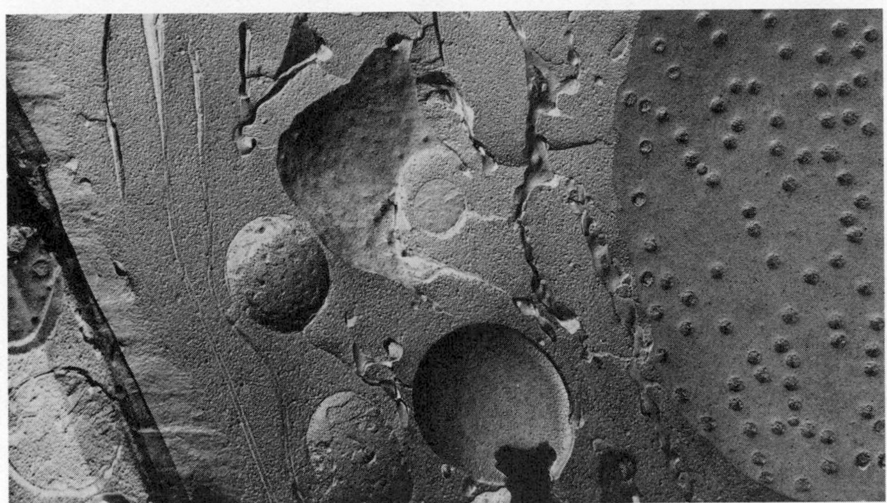

The freeze-fracture technique. The TEM photograph shown here was made of a specimen prepared by freeze-fracture. This technique produces fracture lines within membranes which reveal their inner structure. Here, the nuclear membrane has been cleaved, revealing its pores. This answers some of the older questions about transport in and out of the nucleus.

brane that faces out is broken apart from the half that faces in. Wherever water-binding materials, such as membrane proteins, are included within the layers of the ruptured membrane, indentations show on one inner side of the membrane and bulges show on the other inner side. The specimen is dried, and metallic salts are sputtered over the surface, forming a solid coating of varying thickness. The tissue itself is removed by acid treatment, leaving the metallic replica intact; and the replica is what is put into the machine.

Scanning Electron Microscope

The *scanning electron microscope* (SEM) was developed in the 1940s, but it was not perfected until the 1960s. While the resolving power of this instrument is far less than the standard electron microscope, it has the distinct advantage of producing the illusion of three-dimensional images, with unusually great depth of field. This capability makes it a unique tool for revealing surface features of a specimen.

The scanning electron microscope (SEM) is one of the newer innovations of electron microscopy. It involves more electronic wizardry than its predecessor, the TEM. A thin scanning beam of electrons is passed back and forth across an object, causing it to scatter and reflect the electrons. The reflected electrons are captured on a positively charged plate, giving rise to a small electric current. This current is amplified and fed into a cathode ray tube whose own scanning beam is synchronized with the scanning beam in the specimen chamber. The image appears on a television screen. The specimen is mounted on a movable mechanical axis, so its position can be changed for varying views.

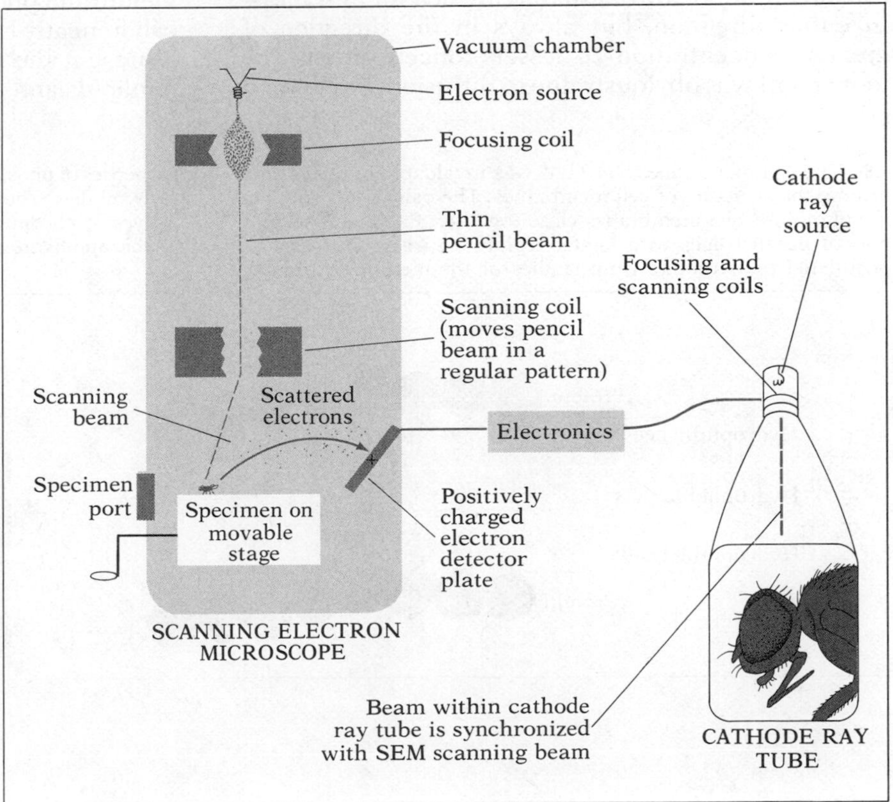

Put simply, the net movement is from regions of higher concentration to regions of lower concentration. If you drop some food coloring into one end of an aquarium and leave it, the color will diffuse away from the place you dropped it, until the entire aquarium is uniformly colored. And then net diffusion stops, although in fact the molecules of food coloring will be moving about as freely as ever, and in a random manner.

Some kinds of molecules are essentially ignored by the cell membrane and are allowed to diffuse in or out according to differences in their concentrations in just such a way, if perhaps a bit slower. In 1940, J. F. Danielli, a British biologist who specialized in lipid biochemistry, observed that fat-soluble molecules diffused through the membranes of living cells, while polar molecules generally did not. He concluded that the cell membrane must be largely made up of lipids, or material in which lipids could dissolve. In fact, at this point he had already postulated the phospholipid bilayer structure, but other observations suggested to Danielli that the cell membrane must have additional components as well.

Different kinds of molecules, Danielli found, flow at different rates across the membrane, but the rate of flow could be predicted for most molecules on the basis of their lipid solubility, polarity, and size. For example, very small polar molecules, such as water, could cross the cell membrane faster than larger molecules (but slower than their diffusion rate in the absence of a barrier). However, there were important exceptions to the general rule. Some large, polar molecules, such as glucose and certain amino acids, traveled quite readily across living cell membranes—in either direction, but always in the direction of greater concentration to lesser concentration. The movement was obviously due to diffusion, but it was a special type. The cell membrane simply appeared to be no barrier at all to these biological molecules. Danielli called the phenomenon *facilitated diffusion.* He postulated that the cell membrane, although primarily a phospholipid bilayer in structure, had protein-lined pores through which certain polar molecules were allowed to pass freely. He also postulated that globular proteins were somehow associated with the inner and outer surfaces of the membrane. His whole hypothetical structure became known as the *Danielli membrane model* (Figure 4.8). As we'll see, the hypothesis wasn't far off base.

The third type of movement across the membrane is *active transport.* Here the membrane, or something in the membrane, takes a more active role than merely preventing or allowing molecules to move through. Active transport has two important characteristics: 1. It allows the movement of dissolved molecules against the gradient—that is, from regions of lower concentration to regions of higher concentration. 2. It requires energy. (As in nearly all cell activities that require energy, the energy for active transport comes from the breakdown of ATP, a universal energy-carrying molecule that we'll be discussing in more detail in the next chapter.)

The development of the electron microscope supported most of the essentials of Danielli's model, which, keep in mind, had been based almost exclusively on the circumstantial evidence of diffusion studies. Membranes look like two dark lines separated by a clear line (see Figure 4.8). The clear line, it turns out, is about 5.0 nm wide, which happens to be twice the average length of the hydrocarbon tails of membrane phospholipids, so two layers of such tails can fit neatly back-to-back. The dark lines apparently represent the phosphorus-rich polar heads of phospholipids and the associated proteins.

4.8 The cell membrane. The TEM was invaluable in finally confirming the structure of cell membranes. The pair of dark lines in the photo are two membranes close together. Each consists of a pair of parallel lines with lighter regions between. Danielli first postulated the structure from studies of the membrane-forming properties of phospholipids. The drawing is adapted from one of his early models. The model consists of two layers of protein with a bilayer of phospholipids between. Note the orientation of hydrophobic and hydrophilic ends of each phospholipid molecule.

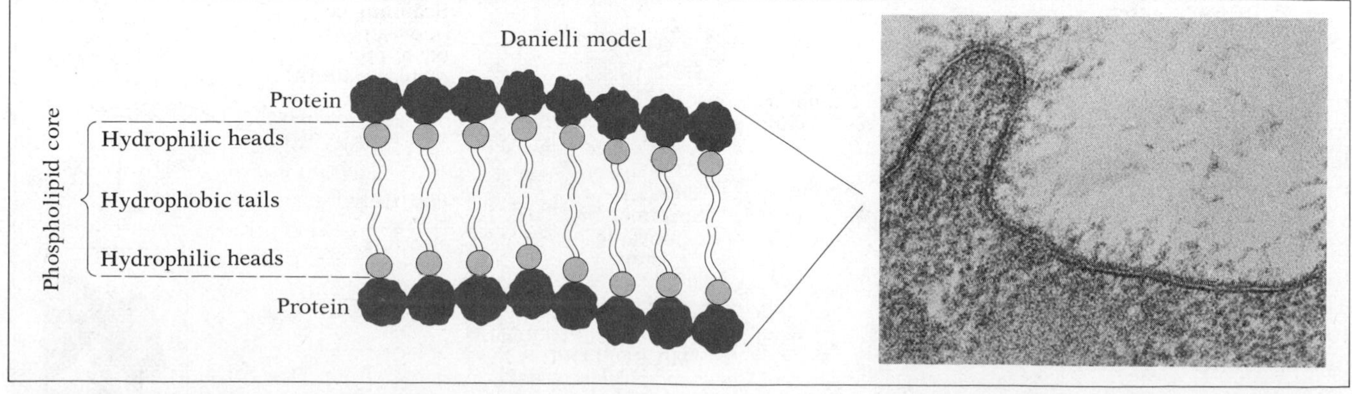

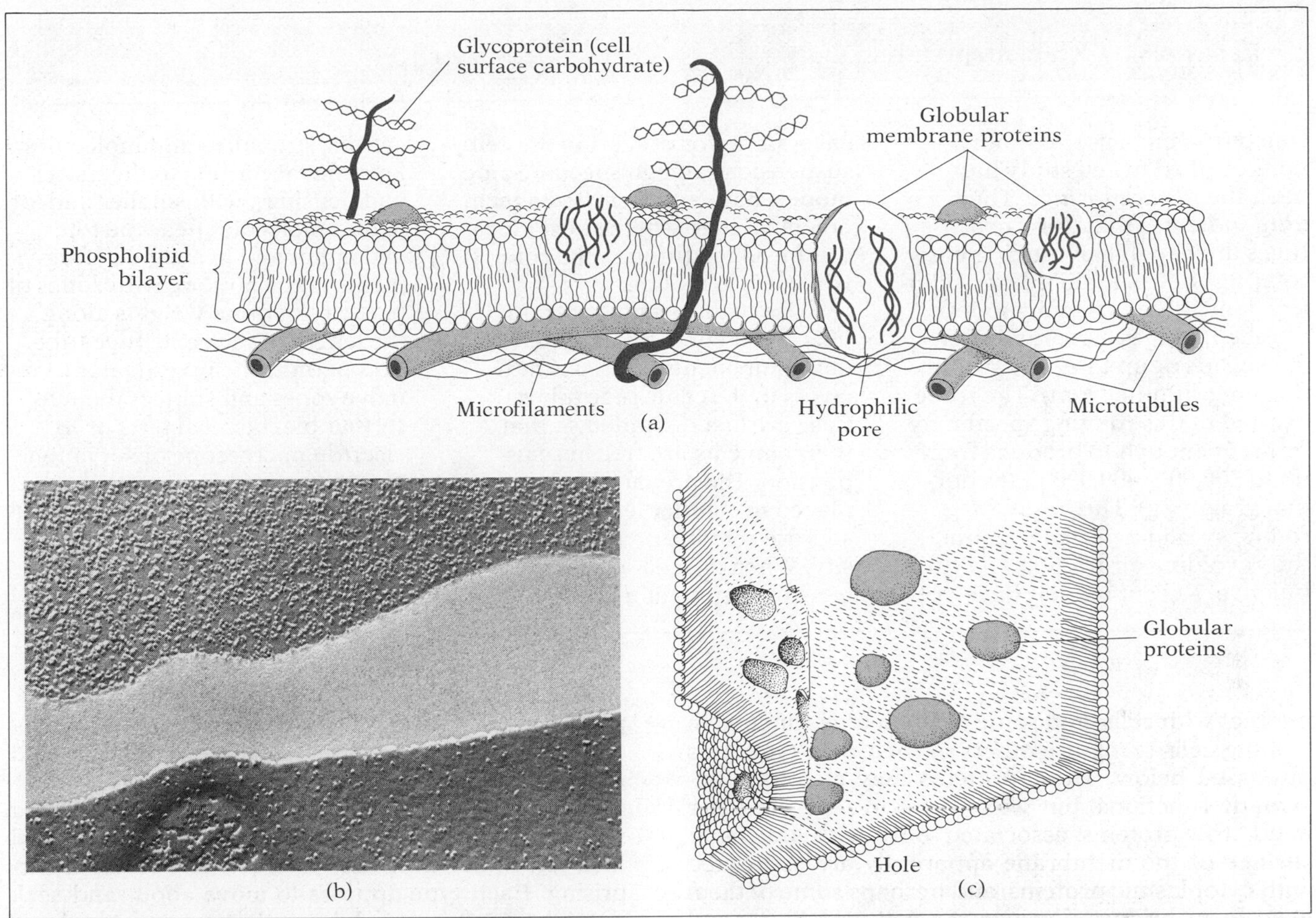

4.9 (a) The fluid mosaic model of the cell membrane reveals its best-known components. The basic structure includes the well-substantiated phospholipid bilayer with the hydrophilic heads on the outside and the double hydrophobic tails pointing inward. Numerous globular proteins are embedded throughout the membrane. Hydrophilic pores occur in some. Glycoproteins project their branched heads above the surface and attach to subsurface proteins below. Other subsurface structures include the microfilaments, and microtubules attach below, forming the cellular cytoskeleton.

(b) Freeze-fracture studies help substantiate the globular protein component of the membrane. These are clearly seen in the illustration where the upper lipid layer is torn away. Note the holes left by the transmembranal proteins as the lipid layer is raised (c).

Figure 4.9 shows a diagram of what we now believe the membrane structure to be. This is called the *fluid mosaic* model, or the *Singer model* (after S. J. Singer, who developed it). The phospholipids are shown as little (polar) spheres of glycerol, phosphorus, and perhaps choline; the polar heads have two hydrocarbon tails each, and these point inward. This arrangement forms a water-resistant barrier that excludes molecules that are not lipid-soluble (most water-soluble molecules—recall that the polar heads of the phospholipids are hydrophilic, but their tails are hydrophobic).

J. F. Danielli had proposed protein-lined pores through which facilitated diffusion and active transport could occur, but it seems that he was wrong on this point. Instead, large globular proteins are embedded right in the membrane, and these proteins are believed to have channels through which water-soluble materials can be passed. Some of the amino acids of which proteins are composed have hydrophobic lipid-soluble side groups. Membrane proteins are believed to be rich in such amino acids, and these serve to anchor them into the phospholipid bilayer.

Freeze-fracture preparations (Figure 4.9) have helped to substantiate this picture. When frozen cells are split apart right between the tails of phospholipids, the membrane proteins remain as large globs on one of the two surfaces. The other surface is pitted with indentations where the proteins had been.

In addition, smaller proteins may lie embedded in only the inner or outer membrane layer, and there are some that only lie against one of the membrane layers, but that do not pass through it all. Apparently, some of the proteins associated only with the outer membrane layer are the sites where hormones can

Essay 4.3 Ultracentrifugation

One of the most useful innovations applied to cell study has been the *ultracentrifuge.* The term *ultra* indicates that these centrifuges are capable of much greater rotating speeds than ordinary clinical centrifuges. In fact, modern ultracentrifuges can attain rotating speeds of up to 70,000 revolutions per minute (rpm). The force exerted at this rotating speed may be great enough to produce forces up to 300,000–400,000 times normal gravity (g). The *heads,* or *rotors,* spin in a partial vacuum, thus avoiding air friction and turbulence. Recent models have optical systems for observing the cell suspensions at high speeds. Some models have even reached a technological state where substances can be added through the rotor while it is in motion.

Use of ultracentrifugation permits researchers to separate various components of tissues into layers in the centrifuge tubes. Cells are first disrupted so that their contents are free in a suspension. The preparation is placed in another fluid, such as a sucrose solution of a known density. When the cell–sucrose suspension is spun at ultra high speeds, structures and molecules layer out according to their sizes and densities, with smaller and/or lighter molecules near the top, and larger and heavier cell fragments at the bottom, with zones of particles of other weights along the length of the centrifuge tube. This permits the investigator to remove zones and subject them to further biochemical tests or to electron microscope observation. Needless to say, having a relatively pure quantity of some cell organelle greatly simplifies the work of the cell biologist.

---

interact with cells, and some of them may function in helping cells recognize others of their own type, as is discussed below. These proteins may also have enzymatic functions, but we're not sure. On the other hand, the proteins associated only with the inner surface of the membrane apparently are associated with cytoplasmic proteins, and perhaps some of them even form bridges between the cell membrane and the microtubules and actin fibers that make up the cytoskeleton. This is all a bit conjectural at present. The model is new and scientists are still playing with the idea.

In any case, our present conception of cell membrane structure is primarily an elaboration on Danielli's basic model. The lipid bilayer is conceived of as behaving like a fluid, within which various proteins float in a kind of mosaic. Thus, the name *fluid mosaic.*

The membrane has a third component in addition to lipid and protein: carbohydrate (Figure 4.9). Some carbohydrates are joined with certain membrane proteins, forming glycoproteins. Others are associated with the outer cell membrane in more general (and less understood) ways. We do know, however, that the familiar A and B blood group types are due to cell-surface carbohydrates.

Not part of the membrane, but physically attached to its inner surface is the *cytoskeleton,* a network of *microtubules* and *microfilaments* that give the animal cell its shape and its ability to change that shape. We'll have more to say about microtubules, microfilaments, and the cytoskeleton later.

We should stress again that cells apparently recognize one another on the basis of membrane structure. The clue substances are both carbohydrates and proteins. Cell recognition is important in many processes, including embryological development and immune responses. In one rather interesting demonstration, cells from different tissues can be mixed together on the appropriate nutrient medium and then left to see how they interact. The result is surprising. Each type appears to move about and seek out others of its kind in such a way as to form individual masses of specific tissue type, presumably because each type of cell membrane has specific proteins and receptor molecules that react to these proteins. Cell-surface proteins (called *antigens*) also enable protective white blood cells to recognize and destroy most cancerous cells. And you undoubtedly have heard about the difficulties of rejection in organ and tissue transplants. Foreign tissues are rejected because their surface proteins and carbohydrates are not recognized by protective white blood cells, so they are destroyed just as any invader of the body would be.

This brief history of the study of cell membranes provides a good example of how science progresses. The Danielli model was a first step in understanding this important cell structure. It answered some questions and provided a target for disbelieving sharpshooters. More observations were made, new techniques were employed, and the model was improved to bring it into line with the newer data. The fact is, our newer models also leave many questions unanswered, so we can expect it to be changed again. Under the best of conditions, science progresses this way. It is unusual for a model or a hypothesis to remain unscathed for very long if people are interested in it. The model builders today are looking for even more versatile concepts of the membrane to explain new information regarding the membrane's role in the life of the cell.

Diffusion, Facilitated Diffusion, Active Transport, and Osmosis

Now let's take a closer look at diffusion, facilitated diffusion, and active transport, the processes introduced in our discussion of Danielli's model. First, we can divide the ways in which materials move across membranes into *active transport* and *passive transport* (which includes simple diffusion and facilitated diffusion). Essentially, the only difference between the two is their sources of energy. In the case of passive transport, the energy comes from the thermal energy of the cell's environment. Active transport, however, requires work, and it costs something in terms of the depletion of the cell's energy reserves.

A Closer Look at Diffusion

The molecules of any liquid or gas are moving constantly and randomly, forever bumping into each other and taking new paths. The kinetic energy of molecules constitutes thermal energy. The warmer the gas or liquid is, the faster the molecules move. The movement of individual molecules is random, but if there is an initial difference in the concentration of any particular kind of molecule, the net movement of that type of molecule is always from regions of higher concentration to regions of lower concentration, until the concentrations are equalized. Once this happens, the random movement of individual molecules continues, but there will be no further net movement in any direction.

Everyone who has survived an elementary inorganic chemistry course knows what happens when you uncork a container of hydrogen sulfide in a classroom. You can stay and feign innocence or you can leave quietly. The next person entering the room may cast a suspicious eye about until he realizes the probable source of the odor. He may then locate the container, if he's interested (which isn't likely) by tracing the odor along an increasing gradient to its source. How does the odor spread through the room? And why is the bottle harder to find, the longer it has been uncorked?

The crass soul who uncorked the bottle was behaving *as if* he were familiar with the laws of diffusion, as would the bottle-seeker. But each likely would have had difficulty explaining why the chemical acted this way in moving across the room. Actually, the original system—the corked bottle of hydrogen sulfide—is relatively ordered. In addition, the *free energy* of the system is great, because the bottle contains a higher concentration of molecules than the environment (Figure 4.10). When the cork is removed, the molecules can move out and through the room blindly bumping into walls, nasal receptors, and each other. Some molecules may cross the room twice and then return into the bottle; but that isn't

4.10 Diffusion of a gas. A container of hydrogen sulfide (H_2S) has been uncorked at one corner of a room. The container contains a higher concentration of molecules than the room. Thus, the gas has great free energy. Molecular motion is in all directions, but the greatest movement in any direction is toward areas where there are fewer molecules of H_2S. As movement continues, a concentration gradient is established, with the greatest concentration still in the container. Finally, a state of equilibrium is reached when the distribution of molecules is random throughout the bottle and the room. The free energy level of the system is at a new low. Molecular movement continues, but no net directional movement occurs. If someone opens the door to the room, the process will repeat itself as molecules begin to drift down the hall. The trend toward a low-energy state even in this dispersed system will continue as a diffusion gradient forms in the hall. When does diffusion end? That depends on the limitations of the system. Theoretically, it could be the universe!

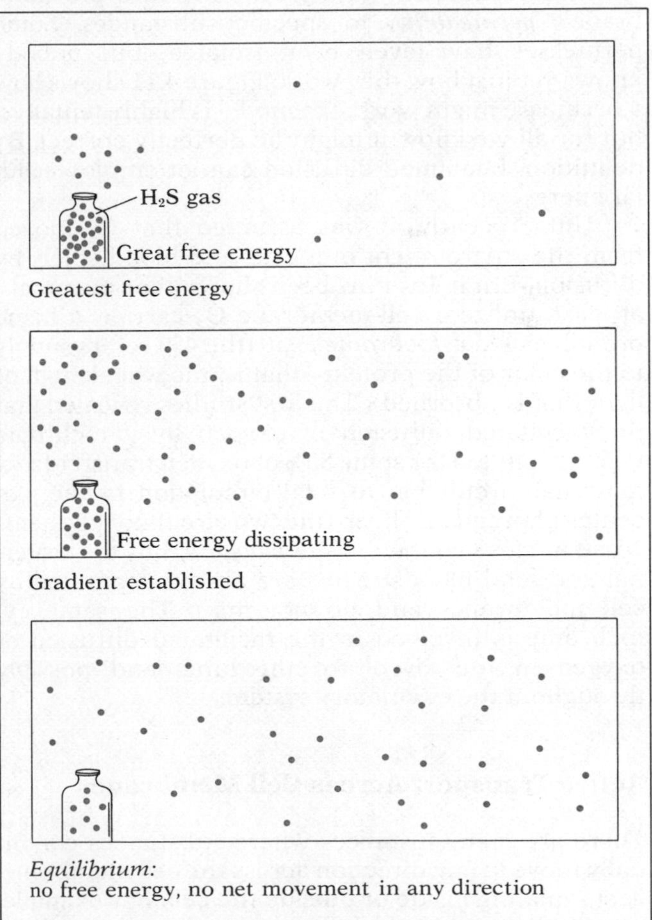

very likely. During the spreading process, the molecules take different random paths, and when the first hydrogen sulfide molecules have crossed the room, most of the rest are still jostling around somewhere near the bottle. The result is a temporary concentration gradient between the source and the walls of the room. Eventually, the molecules (and the odor) will be randomly dispersed in the room. Diffusion, then, is a random process by which molecules move under their own thermal energy away from their place of higher concentration.

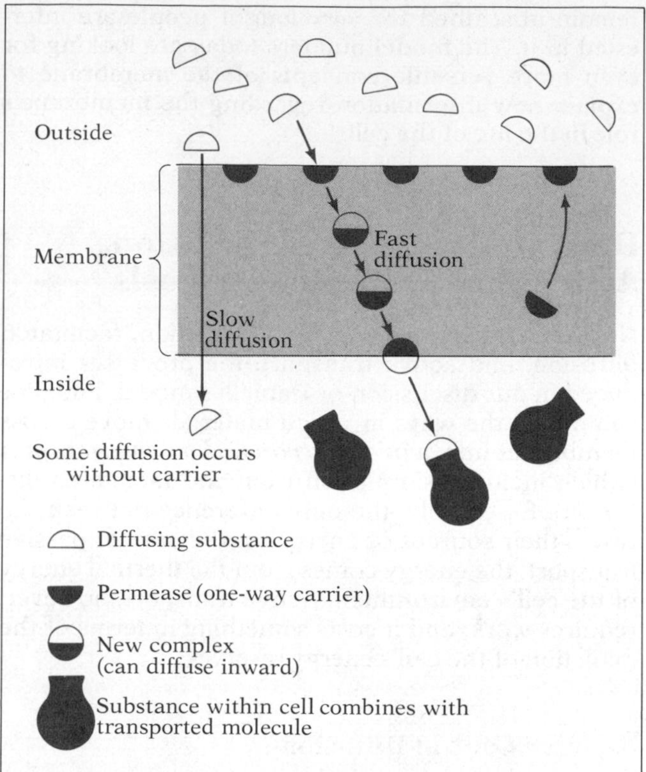

4.11 Facilitated diffusion can be observed and measured, but it continues to defy explanation. This conceptual model attempts to relate what is known about facilitated diffusion to what is known about the cell membrane. The rules of the game are simple: (1) The molecules must be capable of diffusing through the membrane without help, but they can be made to diffuse faster. (2) No cell energy source, such as ATP, can be used. In the situation illustrated, carrier molecules are hypothesized. These form a necessarily weak bond with the diffusing substance. The combination permits the carrier to diffuse through the membrane, where its cargo separates from its tenuous connection with the carrier. The carrier diffuses back to the cell surface and makes new contacts. Facilitated diffusion works in both directions across the membrane, so there is no net transport when the concentration inside the cell equals the concentration in the surrounding medium.

Facilitated Diffusion and Permeases

Facilitated diffusion is similar to simple diffusion in that the energy involved is thermal energy, and the net movement of molecules is always from regions of higher concentration to regions of lower concentration. It differs in that only certain kinds of molecules can cross cell membranes by facilitated diffusion, and these cross much more readily than do similar-sized, similarly charged competing molecules. Certain proteins called *permeases* are believed to be embedded in the membrane, and these greatly increase the membrane's *permeability* to specific substances. Some permeases have even been isolated, but nobody knows yet just how they work. Figure 4.11 shows how a permease might work; the model is highly tentative, but for all we know it might be perfectly correct. By definition, facilitated diffusion cannot employ cellular energy.

Until recently, it was assumed that O_2 moved from the environment into cells and blood solely by diffusion. But it has now been shown that mammals, at least, utilize a cell-membrane O_2 carrier, a heme protein called *cytochrome P450* (the 450 refers simply to the color of the protein—that is, the wavelength of light that is absorbed). The first studies revealed that the facilitated diffusion of oxygen by cytochrome P450 accounted for some 80% of oxygen transfer from maternal circulation to fetal circulation in the placentas of pregnant sheep (the two circulatory systems come in close contact in the *placenta,* but the maternal and fetal bloodstreams are always separated by cell membranes and do not mix). The same cytochrome is involved in the facilitated diffusion of oxygen in the alveoli of the lung, and possibly throughout the circulatory system.

Active Transport Across Cell Membranes

There are many instances where substances continually move in one direction across the cell membrane, accumulating inside or outside the cell. For example, mammalian red blood cells continually move sodium ions to the outside and carry potassium ions in, working constantly against the gradient. In these cases, passive transport cannot be involved. Similarly, many marine fish secrete sodium from the blood vessels of their gills, even though the sea is much saltier than their blood. Moving molecules against a gradient is like rolling boulders uphill; work has to be done, and energy has to be expended. This is why it is called active transport.

As with facilitated diffusion, a specific protein carrier is involved in pumping molecules such as sodium across a membrane against a gradient. In what is called a *sodium pump,* the carrier protein in red blood cells appears to exchange sodium ions for potassium ions. No one is sure how it works, but

Figure 4.12 shows what might be going on. We do know that each time the sodium pump pumps out a sodium ion, the universal cellular source of energy—ATP—is used.

There are other forms of active transport that involve extensive membrane movement. One of these was first observed in the feeding activity of amebas. It involves buckling the cell membrane inward, forming a depression and eventually a vacuole as the depression is pinched off from the surface. The resulting vacuole is a membrane-lined sac, in which the inner surface of the vacuole membrane is derived from the outer surface of the cell membrane. Material that was originally *outside* the cell now lies *inside* the vacuole inside the cell. The cell will eventually digest the contents within the vacuole, and the products of digestion will pass, by active transport, through the vacuole membrane into the cell's cytoplasm (Figure 4.13). The process is called *endocytosis.* Endocytosis is the general term covering the whole process. If the forming vacuole engulfs visible solid material, the process is called *phagocytosis;* if it engulfs only dissolved materials, such as proteins, the process is called *pinocytosis.*

4.12 Active transport—the sodium pump. Through an energy-requiring system, sodium (Na^+) and potassium (K^+) ions cross the cell membrane against the diffusion gradient. No one knows how this is done, but several hypotheses have been suggested. One idea involves a sodium pump. Such a pump would account for the peculiar concentration of potassium and active exclusion of sodium in some cells. In this scenario, a carrier protein attaches to a sodium ion at the membrane's inner surface (D). The carrier then moves through the membrane (E) and deposits the ion outside the cell (F). The release of sodium requires the presence of ATP since this is an energetic step. There the carrier picks up a potassium ion (A) and returns to the inner surface (B). Note that a diffusion gradient exists which, were it not for the active process, would lead eventually to an equilibrium of Na^+ and K^+ ions on either side of the membrane.

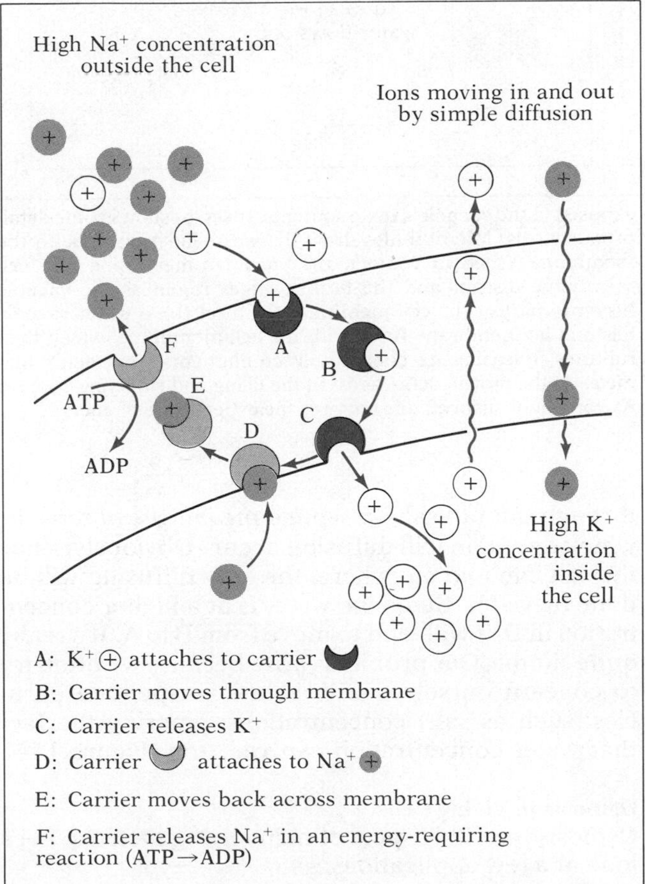
High Na⁺ concentration outside the cell
Ions moving in and out by simple diffusion
F
A
B
E
C
D
ATP
ADP
High K⁺ concentration inside the cell
A: K⁺ ⊕ attaches to carrier
B: Carrier moves through membrane
C: Carrier releases K⁺
D: Carrier attaches to Na⁺
E: Carrier moves back across membrane
F: Carrier releases Na⁺ in an energy-requiring reaction (ATP→ADP)

An equally interesting but opposite process is *exocytosis,* the expulsion of material from the cell (Figure 4.13). In this case, a vacuole containing material to be expelled travels to the cell membrane and fuses with it. The fusion is a complex and little understood process. From outside the cell membrane, the point of fusion appears first as a rosette, or little circle of bumps. When the cell membrane and vacuole membrane have fused, the site of contact somehow opens up and the vacuole contents are deposited to the outside as the vacuole membrane turns inside out. In addition to enabling such organisms as the ameba to eliminate digestive wastes, exocytosis is also commonly used by secretory cells. For example, endocrine glands secrete their powerful hormones in just this way.

Even water, apparently, can be moved actively across cell membranes—but not as individual molecules. Amebas and other freshwater protozoa often have *water vacuoles* that are specialized for the exocytosis of water. The process is only poorly understood. Apparently, bundles of microfilaments first pull on the water vacuoles to enlarge them, thus creating a lower pressure within the vacuole. Water then flows in because of the pressure gradient. Then other microfilaments contract and the pressure is suddenly reversed, and the water vacuole is squeezed as water empties through a pore to the exterior (Figure 4.13). Both phases require the expenditure of ATP. Active transport of water inward appears to occur in plant roots, at least sometimes, but the mechanism is not known.

Osmosis

Water itself diffuses, of course, just like any other substance. And like any other substance, it diffuses freely from regions of greater concentration to regions of lower concentration. However, the consequences of diffusing water are so far-reaching and can be so contrary to our notions of common sense that physiologists have given a special name to the process, *osmosis.*

Suppose wc havc two containers, A and B, with 1 liter of pure water in each. If we dissolve 10 grams of

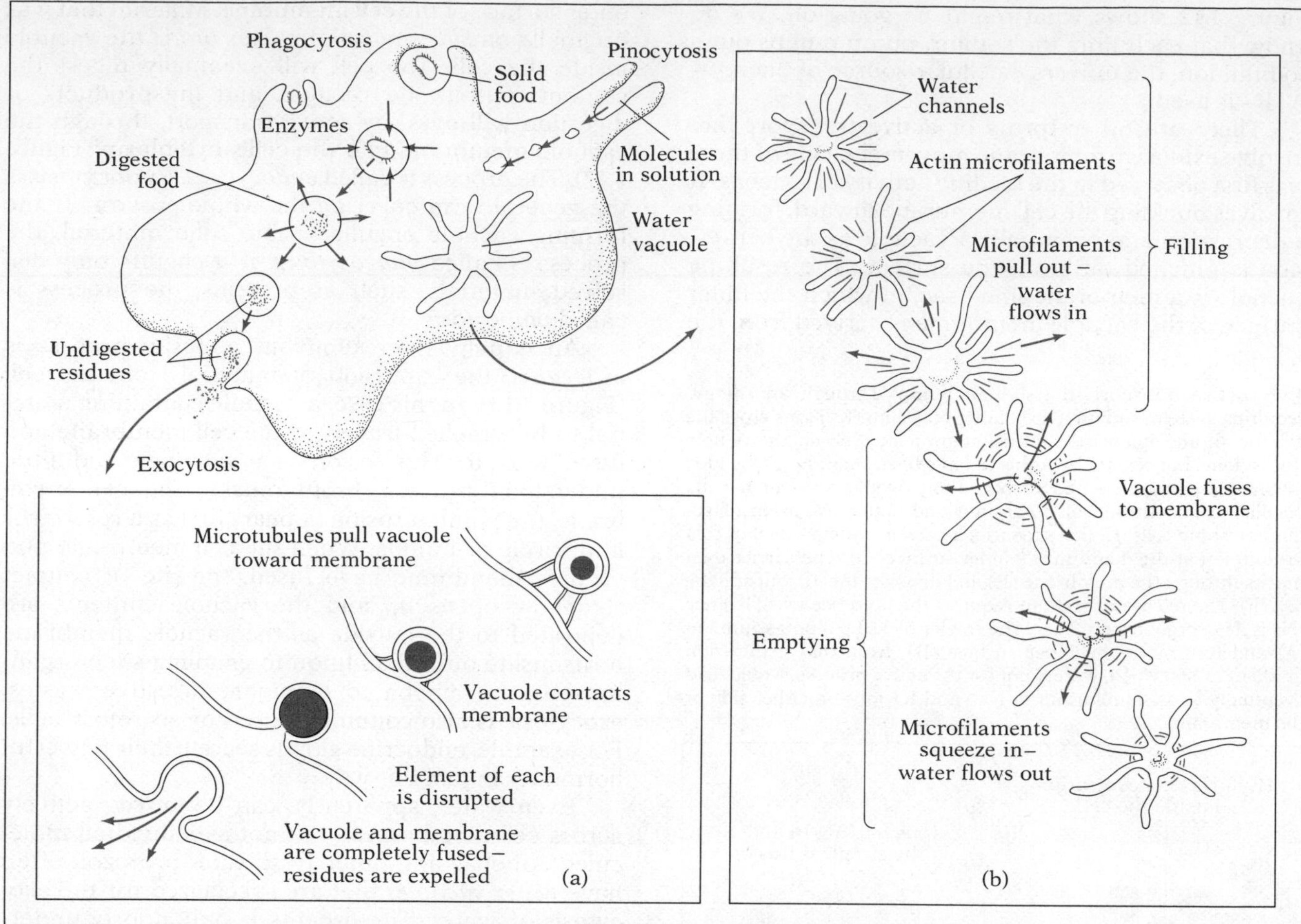

4.13 Active transport and vacuoles. A very busy ameba is demonstrating both endocytosis (phagocytosis and pinocytosis) and exocytosis in all their variations. At the upper region, it is engulfing a small ciliate protozoan by phagocytosis (cell eating). At the right, a channel has surrounded a solution of large molecules in a process called pinocytosis. Eventually, both processes will create a vacuole from the cell membrane. At the lower side of the ameba, the undigested residue from a food vacuole is being expelled by exocytosis. Actually, the food vacuoles merge or fuse their membranes with that of the cell, so there is never a time when the cytoplasm is exposed to the vacuole's toxic contents. Inset (a) shows some detail of exocytosis. Microtubules draw the waste-laden vacuole to the membrane. Once in contact, the vacuolar membrane and cell membrane disrupt and the broken edges rejoin as the vacuole becomes part of the cell membrane. In inset (b), a water vacuole fills and its membrane fuses with the cell membrane, which then ruptures, releasing its contents. Microfilaments, contractile filaments of the protein actin, assist in the filling and emptying process. As you might suspect, this requires the expenditure of energy.

salt in container A, which container has the higher concentration of salt? Container A, of course. It has a concentration of approximately 1% salt by weight, while container B has no salt. That was easy, but for some reason we often have trouble understanding the converse. If the salt concentration is higher in A, then the water concentration is higher in B. Obviously, if the solution in container A is approximately 1% salt, it is only 99% water, whereas the solution in container B is 100% water. So the concentration of water is greater in container B.

Now suppose that containers A and B are separated only by a membrane that will let water pass through but not salt—a *semipermeable membrane.* In which direction will diffusion occur? Obviously, since the salt can't go anywhere, the only diffusing will be done by water. Since the water is at a higher concentration in B, it will tend to move from B to A. It's really quite simple. Our problems arise because we often try to concern ourselves with solute (suspended particles, such as salt) concentrations, ignoring the fact that water concentration can vary too (Figure 4.14).

Osmosis in Living Cells

Osmosis is a vital aspect of many living systems. Let's look at a few applications.

The produce man at Safeway sprays his lettuce

and celery with cold water to keep them from wilting and to restore their crispness. Why? What makes lettuce and celery crisp in the first place? Celery, lettuce, and carrots are crisp because their cells are stiff; their cells are stiff because inside their confining cell walls they are under considerable hydrostatic pressure. It is the same kind of hydrostatic pressure that makes the water rise in Figure 4.14(b). The cells are like inflated tires. Pressure that stiffens an organism in this way is called *turgor pressure.*

If plants such as celery, lettuce, or carrots lose water through evaporation, the cells shrink. The turgor pressure within the cells diminishes, and the plants wilt. If the wilted plants have not died, they can be revived by spraying them or by placing them in water. The water on the surfaces of the celery and lettuce is at a much higher concentration (of water) than that inside the cytoplasm and vacuoles of the plant cells. It therefore diffuses inward until the turgor pressure reaches the point at which it balances the diffusion of water. If the produce man sprayed the lettuce with a solution of, say, salt, in which the number of solute molecules for any volume equaled the number in the cytoplasm of the lettuce cells, then nothing would happen. The water concentration would be the same inside and outside the cells, and the spray would be called *isotonic* to the lettuce cytoplasm. The pure water contains less solute and is therefore called *hypotonic.* If he sprayed them with a strong salt solution, it would be called a *hypertonic* solution. What would happen if he used a hypertonic solution? Water would leave the lettuce and it would

4.14 Osmosis is the diffusion of water through a semipermeable membrane. (a) In this situation, the diffusion of water is complete, because the membrane moves or expands, thus overcoming the resistance of hydrostatic pressure. (b) In this situation, the membrane cannot move or expand. Water diffusing along its gradient will end its net movement when it reaches equilibrium. This happens when the hydrostatic pressure on the right produces an equal movement of water molecules in both directions. Note that the membrane is *impermeable* to salt.

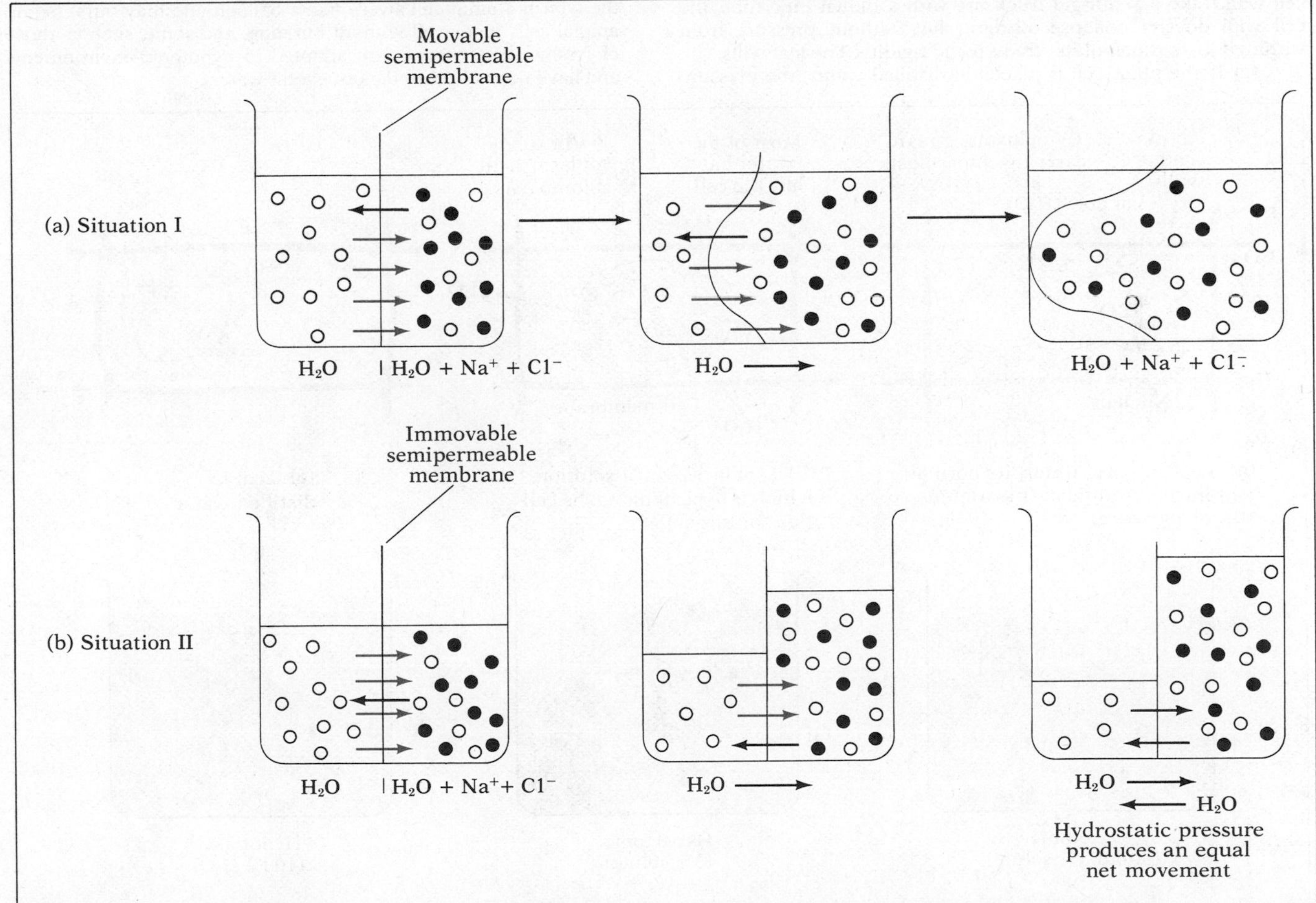

wilt even further (Figure 4.15). Like plant cells, animal cells react to their osmotic environment by swelling or shrinking (Figure 4.15). But because they lack semirigid cell walls, animal cells are more seriously affected. Distilled water can be lethal to an animal cell.

The pressure that builds up because of osmosis is called *osmotic pressure.* Physiologists also like to refer to the *osmotic potential* of a fluid. This is the hydrostatic pressure that would occur in the fluid if it were kept rigidly confined and separated from pure water by a semipermeable membrane. Obviously, the higher the concentration of solute molecules, the greater the osmotic potential of any solution.

Junctions Between Cells

Recent years have seen an upsurge of interest in membrane research, both on the biochemical and the morphological level. Junctions and channels of several kinds have been found between the closely packed cells of multicellular tissues. *Desmosomes* are cytoskeleton-associated organelles that link the membranes of adjacent cells together, something like little rivets. "Gap junctions" through membranes make the cytoplasm of adjacent cells effectively continuous, enabling the tissue to respond to the environment in a coordinated way. But the gap junction can quickly close, effectively sealing off any injured cells (Figure 4.16).

4.15 Diffusion of water in and out of cells. The cell contains two large central water vacuoles. (a) In its normal environment the cell remains turgid, with its membrane-bound cytoplasm pressed against the cell wall like an inflated innertube in a truck tire.

(b) Placing the leaf in a 3% salt solution subjects it to a hypertonic environment and water, now in greater concentration inside the cell, will diffuse out. The result is that the vacuole empties and the cell membrane pulls free from the semirigid cell wall. Like a semirigid truck tire with a blown innertube, the cell wall doesn't collapse outright; but without pressure from within, it loses much of its strength and rigidity. The leaf wilts.

(c) If the plant cell is placed in distilled water, the pressure within it will rise as water flows inward. But the cell will swell only slightly, like a moderately overinflated truck tire. The cell wall is sufficiently strong to prevent the cell from bursting.

(d) Within a multicellular organism, the animal cell is bathed with isotonic tissue fluid, and its shape is determined by its flexible, internal cytoskeleton. (e) If placed in (hypertonic) 3% saline, the entire animal cell shrinks, much as the plant cell protoplast shrinks within its cell walls. (f) In (hypotonic) distilled water, the typical animal cell swells like a balloon and may burst (some animal cells just swell without bursting, and some, such as those of freshwater protozoa, are adapted to hypotonic environments and have ways of pumping out excess water).

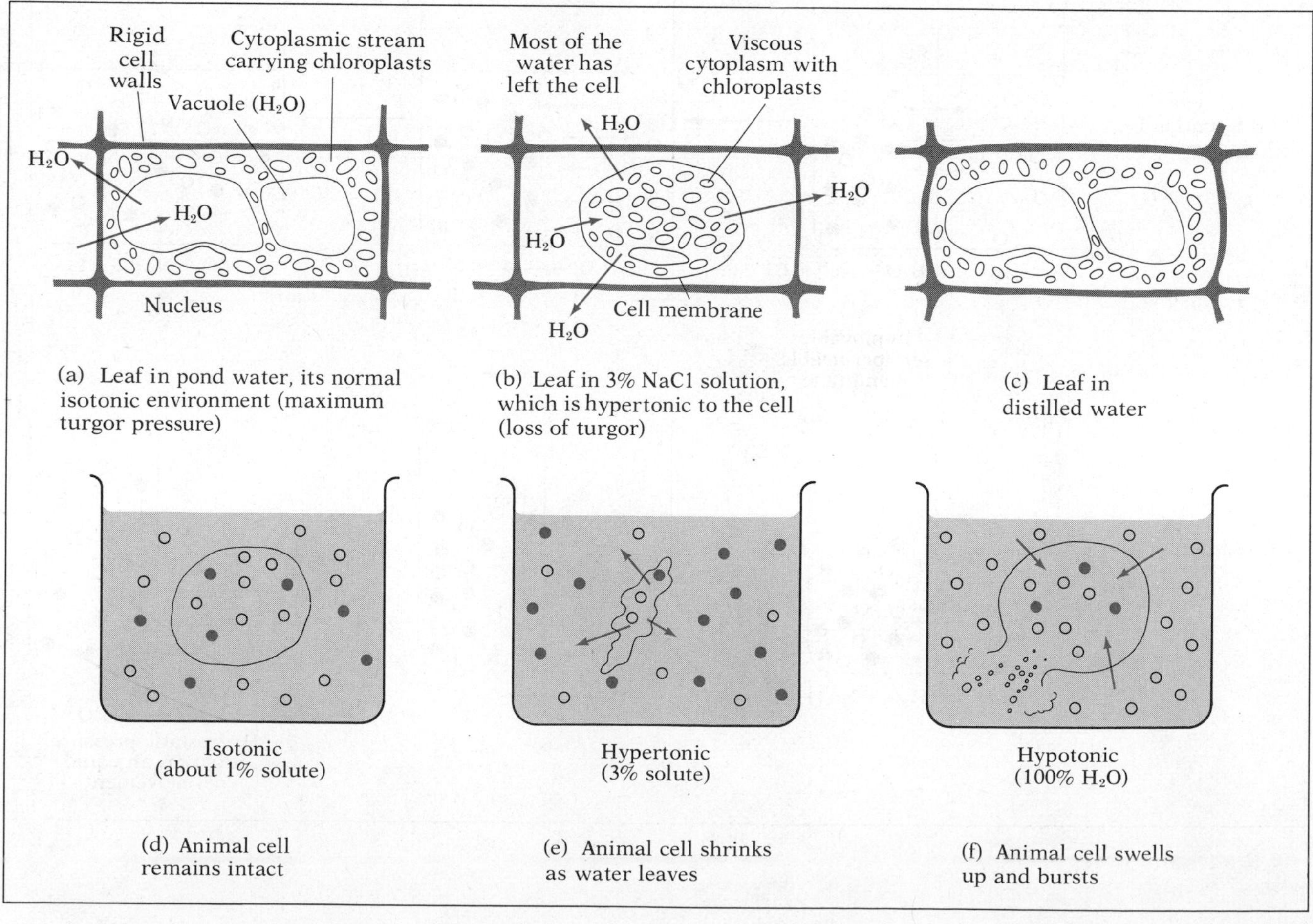

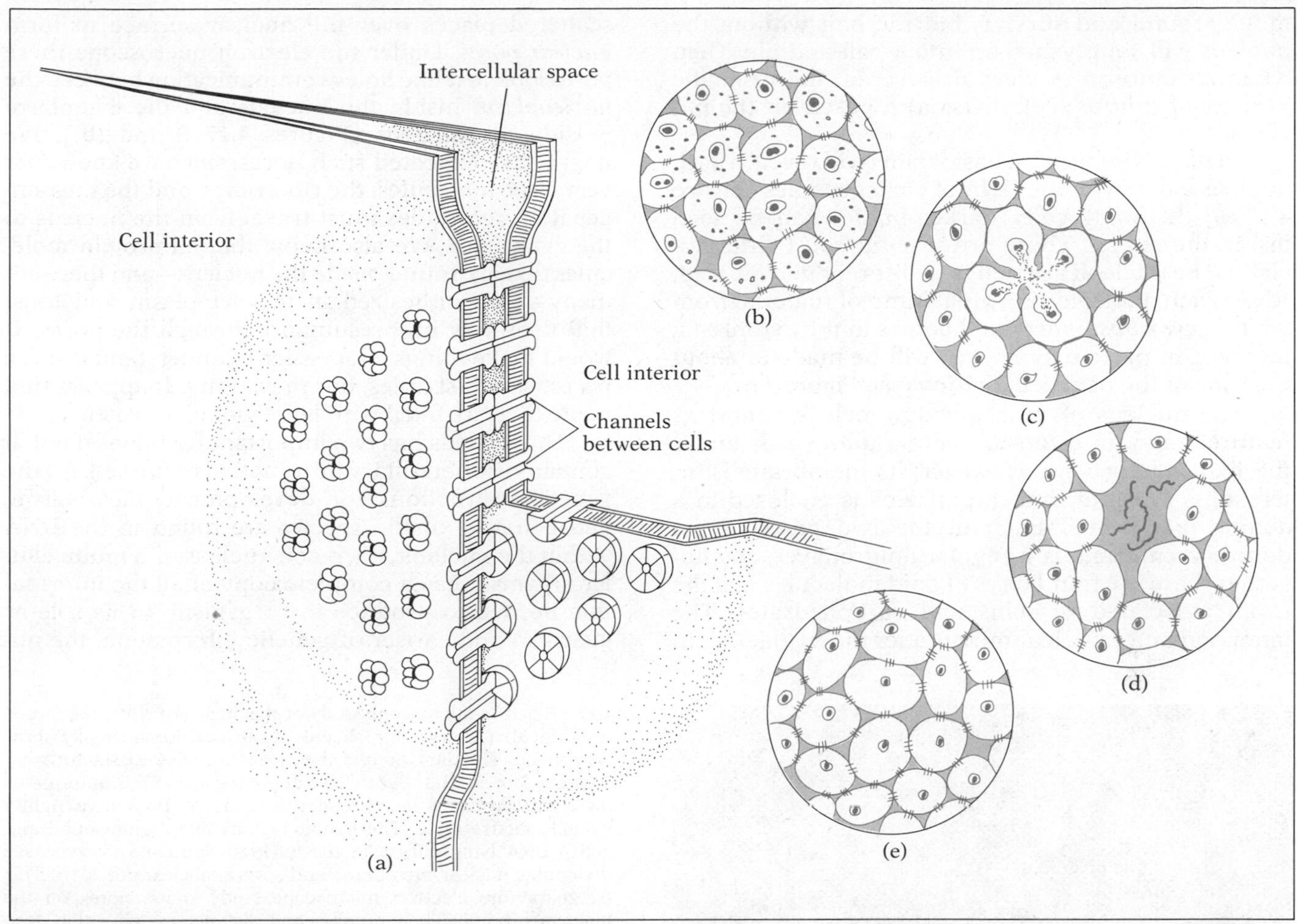

4.16 (a) Model of a *gap junction,* an organelle that allows the *direct* exchange of nutrients and intracellular hormones through channels that pass between cells. Each channel is created by a pair of "pipes," each pipe consisting of six dumbbell-shaped protein subunits. (b) Gap junctions make the cytoplasm of many multicellular tissues effectively continuous. (c) If some cells are injured, however, the gap junctions quickly seal off the wound, possibly in response to materials released by rupturing cytoplasmic lysosomes. (d) The injured cells undergo autolysis (self-digestion) by the released lysosomal enzymes. (e) Contact between the surviving cells is reestablished within 30 min. Mitosis and cell division may follow (not shown).

4.3 Structures Within Cells

Cell Organelles

An *organelle* is any specialized structure of the cell, especially one for which a specific function can be assigned. As we mentioned earlier, the name means "little organ," and is based on the analogy that cells have their organelles just as complex multicellular organisms have their livers, kidneys, stomachs, and other organs.

The Nucleus

The *nucleus* is the most prominent organelle of the cell. It is also the most easily stained, so it was described early in the history of cell study. In fact, the word, which means "kernel," was first used to describe cell nuclei in 1831, about the time of the emergence of the cell theory. Nuclei have since been found in virtually every type of eucaryotic cell. In this chapter we'll consider the nucleus only briefly, because it figures so prominently in our discussion of the life of a cell (Chapter 8).

The nucleus is where the *chromosomes* are. And chromosomes contain the organism's *genes,* which control its heredity. People didn't always know that, of course, but it was realized quite early that the nucleus was somehow important. It was found that if an ameba is cut in half, the half with the nucleus will move around and survive, but the half without the nucleus will simply curl up into a ball and die. Then when chromosomes were discovered, the dramatic process of *mitosis* (cell division) focused on the nucleus.

In nondividing cells, basic stains will reveal a dark nucleus laden with the DNA of chromosomes [Figure 4.17(a)]. In many cells, darker bodies can be seen inside the nuclei. These are the *nucleoli* ("little nuclei"). The nucleoli are rich in a different nucleic acid, RNA; each nucleolus is a great lump of material from which new *ribosomes* (small bodies in the cytoplasm, involved in protein synthesis) will be made at about the time of the next cell division (see Figure 4.3).

The nucleus of an undividing cell is almost as featureless in an electron micrograph as it is under the light microscope. However, its membrane is interesting. We find that the nucleus is enclosed in a *double* membrane. Each of the two parts of the double membrane is a regular lipid bilayer, so altogether there are four layers of lipid molecules and the usual associated proteins and carbohydrates. The inner and outer nuclear membranes pinch together in scattered places over the nuclear surface to form *nuclear pores.* Under the electron microscope these pores look just like holes communicating between the *nucleoplasm* inside the nucleus and the cytoplasm outside the nucleus [Figures 4.17(a) and (b)]. We might have expected such pores, since we know that very large molecules, the ribosomes, and the messenger RNA molecules must travel from the nucleus to the cytoplasm. We also know that all protein molecules that are found inside the nucleus—and there are many—are synthesized in the cytoplasm and must find their way in, presumably through the pores. It would make things a lot easier to understand if these pores were just holes, but they aren't. It appears that in life they are filled with large globular proteins.

The nucleus has two important functions. First, it contains the hereditary information of the cell. All the genetic instructions for development, metabolism, and behavior of the species are found in the DNA within the nucleus. Each cell nucleus of a multicellular organism has a complete copy of all the information needed to produce the organism. In its role as guardian and carrier of genetic information, the nu-

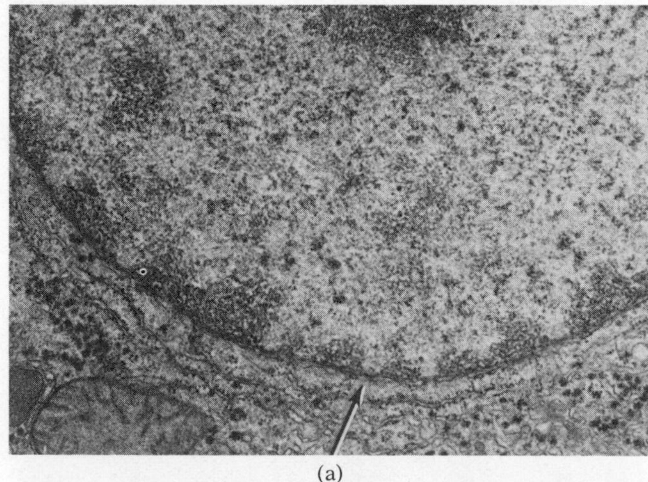

(a)

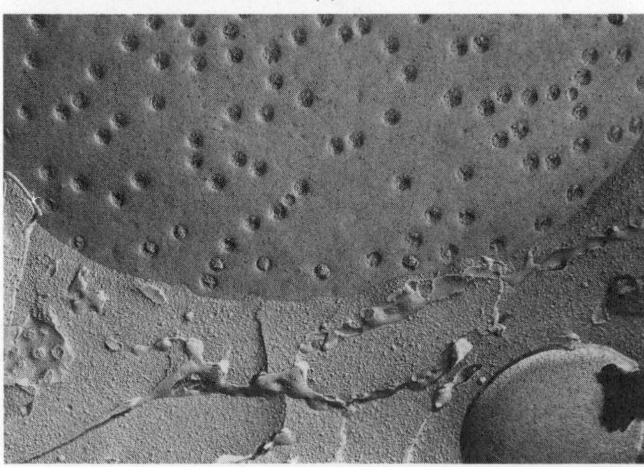

(b)

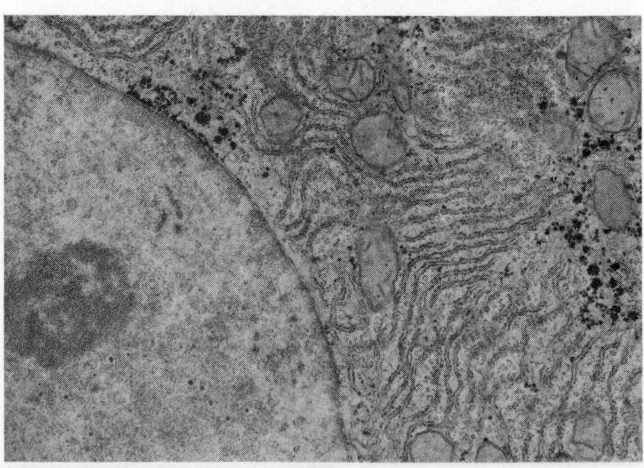

(c)

4.17 Electron microscope study of the nucleus. (a) We are forced to admit that a very thin slice of a nucleus doesn't really show very much. The dark area is the *nucleolus,* where RNA for ribosomes is stored. The slightly less dense regions in this amorphous mass are probably heavy concentrations of DNA in a tightly wound inactive state. The lighter regions may be unwound and active DNA lying diffuse in the nucleus. You can, however, see the double nuclear membrane and several nuclear pores. (b) This freeze-fracture electron microscope study shows pores on the inner membrane. Their number and distance matches that seen in the thin section shown in part (a). These are not believed to be simple holes, but may be filled with hydrated proteins. (c) This remarkable electron microscope photograph strongly suggests a connection between the endoplasmic reticulum and the outer nuclear membrane. Some observers interpret this to mean that they are continuous channels. Others believe that the endoplasmic reticulum actually forms from protrusions in the nuclear membrane. In any case, it is generally agreed that there is a close relationship between the nucleus and the endoplasmic reticulum.

cleus has the capability of preparing exact duplicates of this information for transmission to new generations of cells.

Second, the nucleus has the machinery for controlling various cellular activities. It does this by directing the synthesis of enzymes that are present in the cytoplasm through mechanisms we will discuss in Chapter 8.

The Endoplasmic Reticulum

The *endoplasmic reticulum* (ER) was unknown in the days when we were restricted to light microscopes. In fact, in those days the contents of cells were thought to be formless, soupy "protoplasm." But electron micrographs reveal a complex membrane system that takes up a large part of the cytoplasm of eucaryote cells, especially those cells that are engaged in significant protein synthesis. Although the ER system was first seen in the mid-1940s, it wasn't until 1953 that Keith Porter of the Rockefeller Institute first suggested the name. A year later, Porter and George Palade suggested that the reticulum was a very dynamic, ever-changing structure. Close study of serial sections reveal the ER to be a system of broad sheets forming channels and that the ER membrane is often continuous with both the cell membrane and the outer membrane of the nucleus. This suggests that the ER may be involved in the transportation of materials between the cell's environment and its nucleus [Figure 4.17(c)].

The membranes of the endoplasmic reticulum have the same basic structure as the cell membrane. In some preparations, the membranes appear to balloon out, forming vesicles that then pinch off to become closed sacs. Apparently, most of the saclike vesicles of the cell, including lysosomes and vacuoles (which we'll describe shortly) are formed in this way (although, as we have seen, some vacuoles are also formed by a folding in of the cell membrane).

The endoplasmic reticulum appears in two main forms: rough and smooth. The *rough endoplasmic reticulum* receives its name from the appearance of dense granules that tightly adhere to one side of the membrane, making it look a little like coarse sandpaper (Figure 4.18). Channels are formed by two sheets of this "sandpaper" lying side-by-side with their rough surfaces out. The granules are ribosomes. (It may sound strange, but those little grains can *read*, as we will see later.) Rough endoplasmic reticulum is seen most commonly in cells that manufacture proteins to be secreted outside the cell.

Smooth endoplasmic reticulum is primarily found in cells that synthesize, secrete, and/or store carbohydrates, steroid hormones, lipids, or other nonprotein products. These cells have ribosomes, but they are usually not bound to membranes. Instead, the ribosomes float freely in the cytoplasm. We find a lot

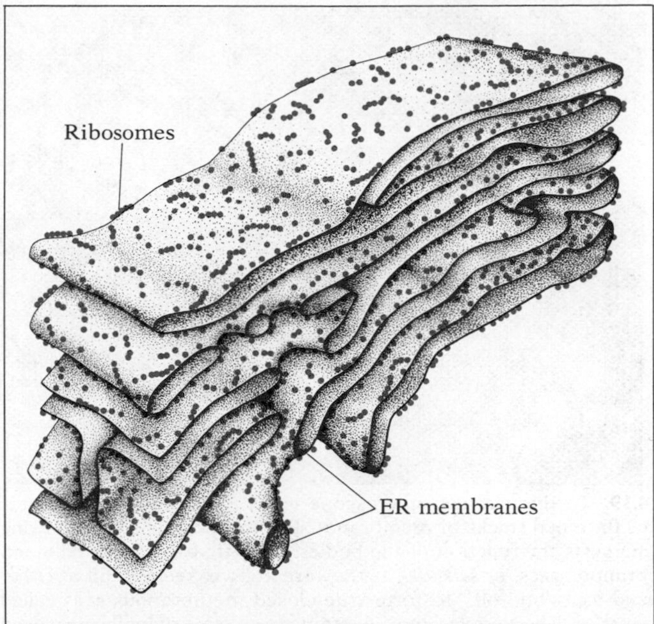

4.18 The rough endoplasmic reticulum (ER) consists of parallel rows of membranes surrounding deep channels. The three-dimensional view suggests the immense amount of folding in the ER. The small round bodies along the ER membranes are ribosomes. It is safe to conclude that a cell containing rough ER is highly involved in protein synthesis and probably secretes proteins out of the cell as fast as they formed.

of smooth endoplasmic reticulum in cells of the testis, oil glands of the skin, and some endocrine gland cells. It is currently believed that the enzymes of lipid and steroid synthesis are found in the smooth membranes.

The Golgi Complex

In 1898, Camillo Golgi, an Italian cytologist, discovered that when he treated cells with silver salts, certain peculiar bodies showed up in the cytoplasm. The "reticular apparatuses" he described had never been noticed before, and they didn't show up with other stains. But when other workers used Golgi's silver treatment, they found the bodies in a variety of secretory cells. Because the silver treatment was considered drastic, and because the peculiar bodies were never seen in living cells, cytologists argued for the next 50 years over whether *Golgi bodies* were really cell structures or just artifacts caused by the silver treatment.

Again, the electron microscope came to the rescue; Golgi bodies were real (Figure 4.19). It was found that they had a characteristic and identifiable structure no matter what kind of cell they were found in. In every case, the Golgi apparatus appeared as a group of flattened, baglike membranous sacs lying close to

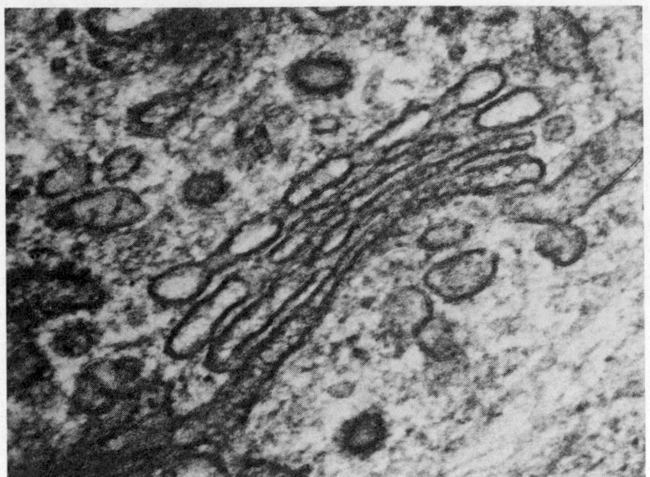

4.19 In this electron microscope view, the Golgi bodies appear as flattened stacks of membranes. The three-dimensional drawing suggests the function of the bodies. Note that the flattened membranous sacs, or *saccules* as they are known, seem to fill at either end and "bud off" to form true closed membranous sacs called *vesicles*. The vesicles may contain products originally produced in the ER, but modified and packaged in the Golgi bodies.

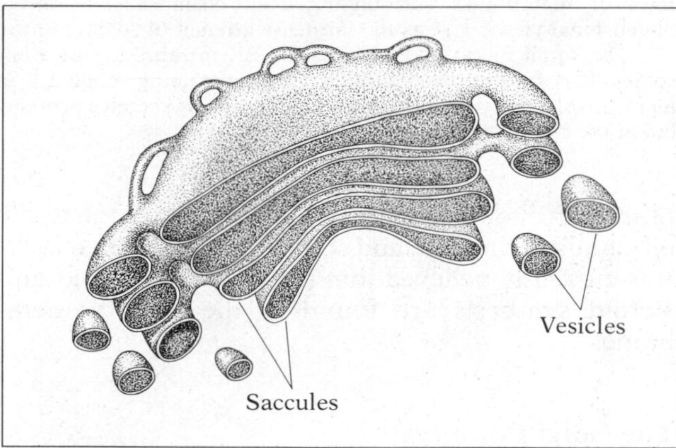

the nucleus and roughly parallel to each other. These were given a name, *saccules*, and the debate shifted to focus on their role. What do saccules do?

The answer was a long time coming, but now it is generally accepted that the Golgi apparatus is derived from the endoplasmic reticulum and that it serves as some sort of packaging center for the cell. Such things as enzymes, proteins, and carbohydrates are collected in these bodies and packaged into closed membranous sacs, called *vesicles*. Thus, they are effectively isolated from the rest of the cell.

In plants, Golgi bodies are thought to be directly involved in cell division and growth. They seem to release complex carbohydrates into the developing cell membrane, which then deposits them into the cell wall. The apparatus may also carry on some enzymatic activity, at least in some cells. Its close proximity to the nucleus in both plant and animal cells suggests some role in synthesis and metabolism (Figure 4.3).

Lysosomes

Lysosomes are membrane-bound sacs that are roughly spherical (Figures 4.3 and 4.20). But their insignificant appearance belies the rather startling role they play in the history of a cell. Lysosomes are bags of powerful hydrolytic enzymes, packaged and synthesized by the Golgi apparatus. If these digestive enzymes were floating free in the cell's cytoplasm, the cell itself would be quickly digested. Christian de Duve, who first described lysosomes, called them "suicide bags." His poetic fancy was not entirely unwarranted, since lysosomes sometimes actually serve the function of destroying the cells that bear them. This is not necessarily a treacherous act. Sometimes the destruction of cells is a normal part of embryonic development or is otherwise beneficial to the organism. For instance, lysosomes might destroy a superfluous cell that was not functioning well, or one in a part of the body that is undergoing reduction as part of a developmental process, such as the tissue between the fingers in a developing hand. Then the lysosomes open up and do their thing, and phagocytes clean up what's left.

Lysosomes also have specific functions in healthy, active cells. After phagocytosis—say, after the ingestion of a particle by an ameba, or after the ingestion of an invading bacterium by one of the

4.20 Lysosomes are storage bodies for powerful hydrolytic enzymes and hence their membranes are exceptionally strong. Their major function is in destroying or digesting. Their targets may include aging organelles within the cell or even the entire cell. They may also serve as storage places for unwanted and potentially dangerous substances that cannot be excreted.

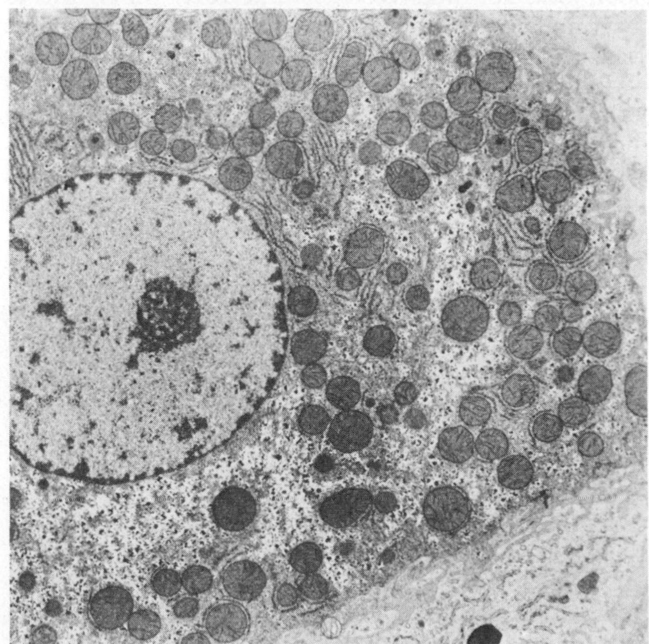

body's own white blood cells—a lysosome will fuse with the digestive vacuole and empty its contents into the vacuole, aiding in the digestion of whatever has been captured.

Obviously, such dangerous chemicals must normally be kept from spilling into the cell's interior, so the membranes of lysosomes are very tough. In fact, this toughness led to the discovery of the organelles. Certain organelles, researchers found, resisted all but the most vigorous attempts to rupture them. They could always be found intact after the cell was ravaged by a blender. When the mysterious particles were investigated with the electron microscope, de Duve found them to be composed of little baglike bodies. They were, in fact, lysosomes. Later, lysosomes were found in nearly all types of animal cells. Interestingly, there is some disagreement about whether they exist in plant cells. To a certain extent, it appears that different groups of people look at plant cells and animal cells, so similar organelles are sometimes known by different names. It is possible that some of the *plastids* found in plant cells are not much different from the lysosomes found in animal cells.

In recent years, evidence has been accumulating that points to a role of lysosomes in the regular turnover of cell components. It seems that lysosomes destroy various cell organelles at a fairly constant rate, presumably so that those parts can be replaced and the cell can remain forever young and healthy. For example, in some tissues mitochondria (which we'll discuss later) are replaced about every 10 days. Electron micrographs have occasionally shown a mitochondrion, or the half-digested remnants of a mitochondrion, enclosed in what seems to be a lysosome (meanwhile, new mitochondria are produced by the division of old mitochondria). Lysosomes may thus be important in the process of aging—a process that may not concern you now as much as it will later on.

Finally, there are a number of genetic disorders in which some essential lysosomal enzyme is altered or absent. The unreacted substrates of these enzymes accumulate in the cell, and so are called *storage diseases.* Most such diseases are fatal in the first years of human life. *Tay-Sachs disease,* a recessive genetic disease that is relatively common among infants of European Jewish descent, is one such storage disease.

Figure 4.21 summarizes some of the interrelationships that are believed to exist among the membrane-bound organelles of the cell.

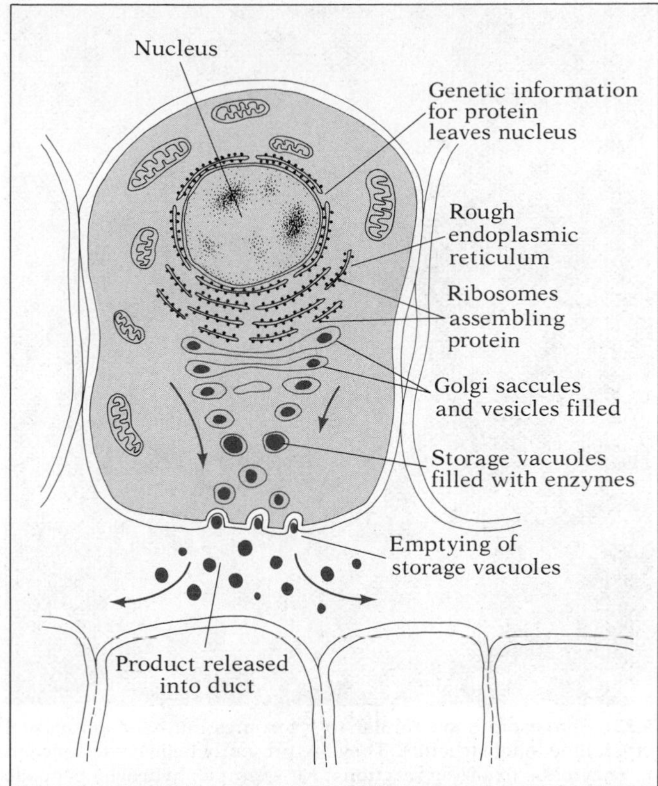

4.21 Secretory action in an exocrine cell: a suggestion for the relationship among rough endoplasmic reticulum, the Golgi complex, and the secretion of enzymes. The cell could be any secretory cell such as those from the pancreas or other gland of a mammal. Direct photographic evidence is meager, but the scenario suggests that protein synthesized by ribosomes first enters the membrane spaces of the ER. As the ER fills, it forms Golgi saccules in which the protein is modified into enzymes. As the saccules fill, vesicles bud off and become membrane-bound storage vacuoles. Finally, the vacuoles fuse with the cell membrane and dump their products into a channel, or duct, which will carry the products to the site of their action.

Peroxisomes

The *peroxisomes* are membrane-bound bodies that often lie near mitochondria or chloroplasts (both of these types of organelles are discussed below). Peroxisomes are found in a great variety of organisms, including plants, animals, and protists. In animals, they are most common in liver and kidney cells. Peroxisomes appear as very dense bodies with a unique crystalline core of tiny tubes, which causes the organelle to stand out unmistakably when it shows up in electron micrographs (Figure 4.22).

The peroxisomes contain enzymes, and perhaps they are a special class of lysosomes. The principal enzyme of the liver and kidney peroxisomes is *catalase,* which helps break down hydrogen peroxide into water and oxygen:

$$2\,H_2O_2 \xrightarrow{\text{Enzyme}} 2\,H_2O + O_2$$

Hydrogen peroxide is a fairly dangerous, highly reactive oxidizing chemical, and the role of the peroxisome here is probably protective.

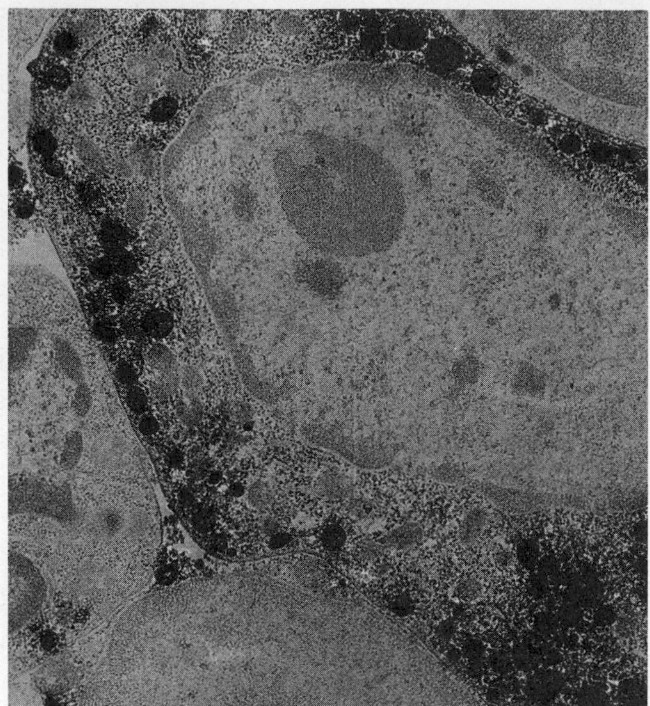

4.22 Peroxisomes are similar to lysosomes but have an unusual crystalline inner structure. They are presently believed to function in enzymatic oxidizing reactions; for example, hydrogen peroxide is oxidized to water and oxygen, rendering the peroxide harmless to the cell. The enzyme used here is catalase, which is found abundantly in liver tissue.

Plastids

Intracellular, membrane-bound inclusions in plant cells are called *plastids.* A wide range of organelles falls into this category. By definition, then, there are no plastids in animal cells. (Unfortunately, in bacteria, some completely unrelated genetic parasites are also called plastids— we mention this now in hopes that you won't be confused later.)

The most important plastid is the *chloroplast,* which was reported as early as 1702 although it wasn't named until 1883. But there are many other kinds of plastids, and plant cytologists have given them a variety of names, including *leucoplast* and *chromoplast.* Some botanists believe that all proper plastids are related, and that each type can develop from a single primitive type known as a *proplastid.* In addition, under certain conditions, they say, one specialized type can change into another type. Later we'll consider the idea that chloroplasts are the descendants of a photosynthetic parasite that once invaded the eucaryote cell, an intriguing notion.

The major plastids have double membranes (like the nucleus), and they also have their own packet of DNA. So here's a case of some DNA residing outside the nucleus. Chloroplasts and most other plastids contain a closed circle of DNA—this is exactly the same form of DNA that occurs in procaryotes.

Leucoplasts, the white plastids, are present in nearly all plant cells, but they can be most readily seen in leaf epidermal cells, onion or apple storage cells, or a variety of other white plant tissues specialized for storage. The principal role of leucoplasts is the storage of starch after it is formed from glucose. Leucoplasts appear to be nothing more than chloroplasts in which photosynthetic activity is temporarily suppressed in favor of starch storage. Chromoplasts (colored forms) are named for the pigments they contain; they impart many of the bright colors (other than green) that we see in plants, such as in their flowers. When chloroplasts are present, the chromoplasts are obscured. In ripening fruit, and in the dying leaves of autumn, chloroplasts lose their bright green color and the more persistent reds and yellows of the chromoplasts predominate. The colored pigments of chromoplasts include orange carotenes, yellow xanthophylls, and various red pigments.

Chloroplasts

Chloroplasts will be discussed in greater detail when we get into photosynthesis in the next chapter. For now we'll just examine their structure and general function.

Chloroplasts function in that critical and mysterious process called photosynthesis: the capture of the energy of light and its transformation into usable chemical reserves, specifically carbohydrates. *Chloroplast* means "green form," and the green comes from the photosynthetic pigment *chlorophyll.* Chlorophyll absorbs red and blue light, reflecting the green segment of the light spectrum. Cyanophytes also contain chlorophyll pigments, but in these procaryotes the chlorophyll is not organized into chloroplasts.

Inside their double membranes, the chloroplasts consist primarily of layers of flattened, membranous disks known as *thylakoids.* Each stack of disks is known as a *granum* (plural, *grana*). Many grana are distributed through the chloroplast in orderly rows. Each granum is connected to its neighbors by membranous extensions of the thylakoids. The surrounding matrix is called the *stroma* (Figure 4.23).

In Chapter 5, we'll consider the biochemistry of photosynthesis in detail, but we would like to make one important point about it here. Although we can identify many of the enzymes and particular biochemical steps involved in the transformation of light energy to chemical energy, we also know that this mysterious process depends on the very precise structure of the components of the chloroplast. If chloroplasts are broken open, photosynthesis ceases abruptly.

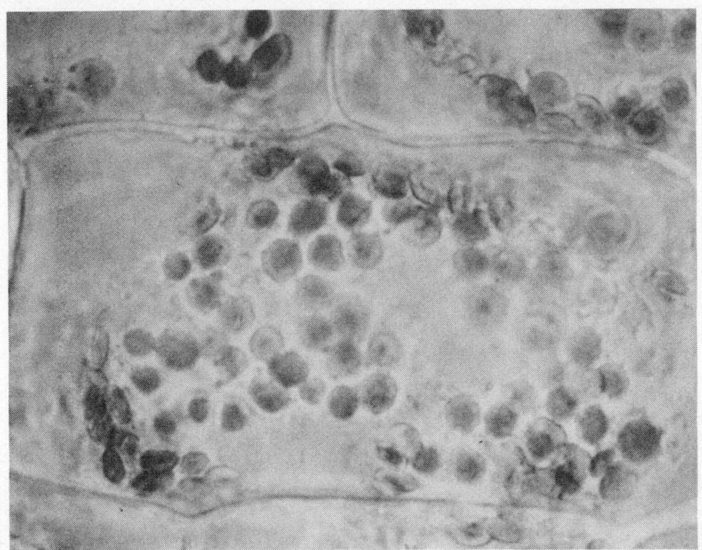

(a) Plant Cell

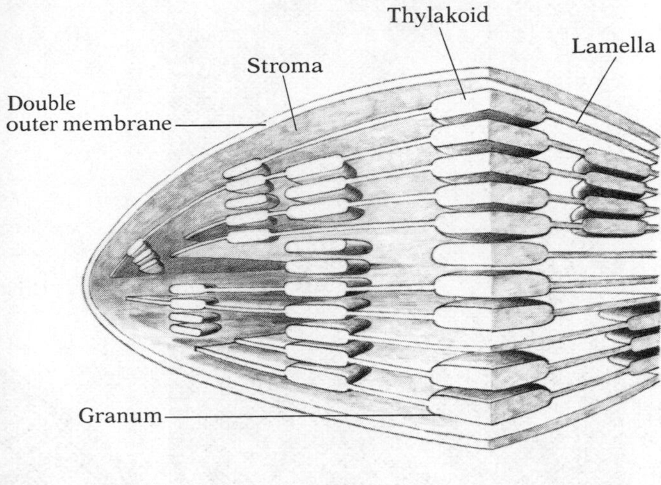

(c) Cross-section

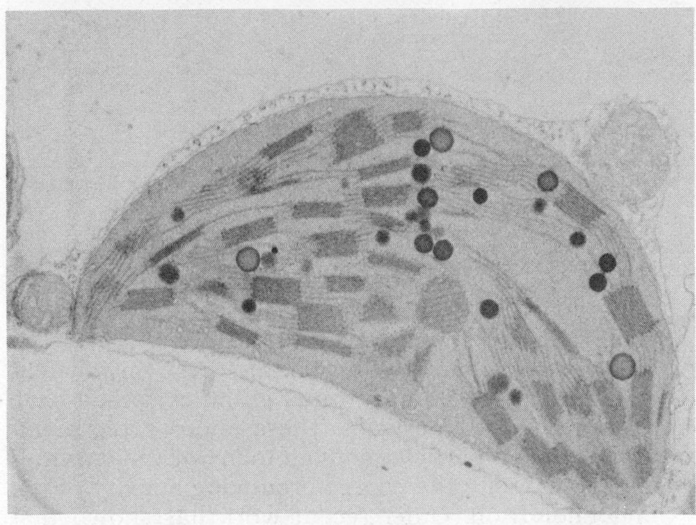

(b) Chloroplast

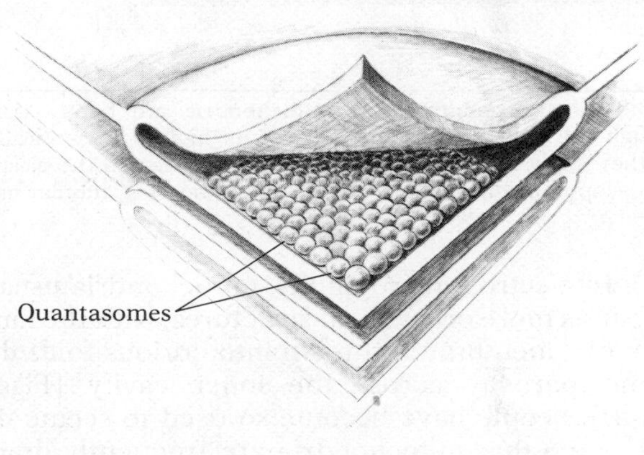

(d) Close-up View

4.23 Chloroplasts. (a) The chloroplasts in an intact plant cell appear as numerous minute, dark spheres. (b) At low electron microscope magnification the inner structure becomes visible. The dark, neat stacks are made up of thylakoids, each stack forming a granum. The thylakoid membranes extend from one granum to another. The clear substance suspending the grana is called the stroma. The dense spheres are grains of starch that have been produced by photosynthesis. (c) This three-dimensional drawing shows a cross-sectional view of a chloroplast, with its stacks of disklike thylakoids. (d) The central region of the thylakoid consists of a fine meshwork of *quantasomes,* which are the sites of the primary events of photosynthesis. The intricate structure of each chloroplast must be intact for photosynthesis to be completed.

Mitochondria

Mitochondria (singular, *mitochondrion*) are complex energy-producing organelles found in every eucaryote cell. Like chloroplasts, mitochondria (a) are enclosed in double membranes; (b) have their own circular DNA; (c) have their own ribosomes, transfer RNA, and other machinery of protein synthesis; and (d) are widely believed to be the descendants of a once-independent procaryote ancestor, engulfed by the eucaryote host cell more than a billion years ago. We'll give full attention to this controversial hypothesis of mitochondrial origins later, but for now, let's have a look at the structure and function of present-day mitochondria.

To some extent, chloroplasts and mitochondria do exactly opposite things. Put simply, chloroplasts turn energy into carbon compounds and oxygen, while mitochondria turn carbon compounds and oxygen into energy. We'll deal with both these important biochemical pathways in Chapters 5 and 6.

Mitochondria are considerably smaller than chloroplasts, especially in cross-section. In part, this is because chloroplasts tend to be more or less spherical, whereas mitochondria are usually long (Figure

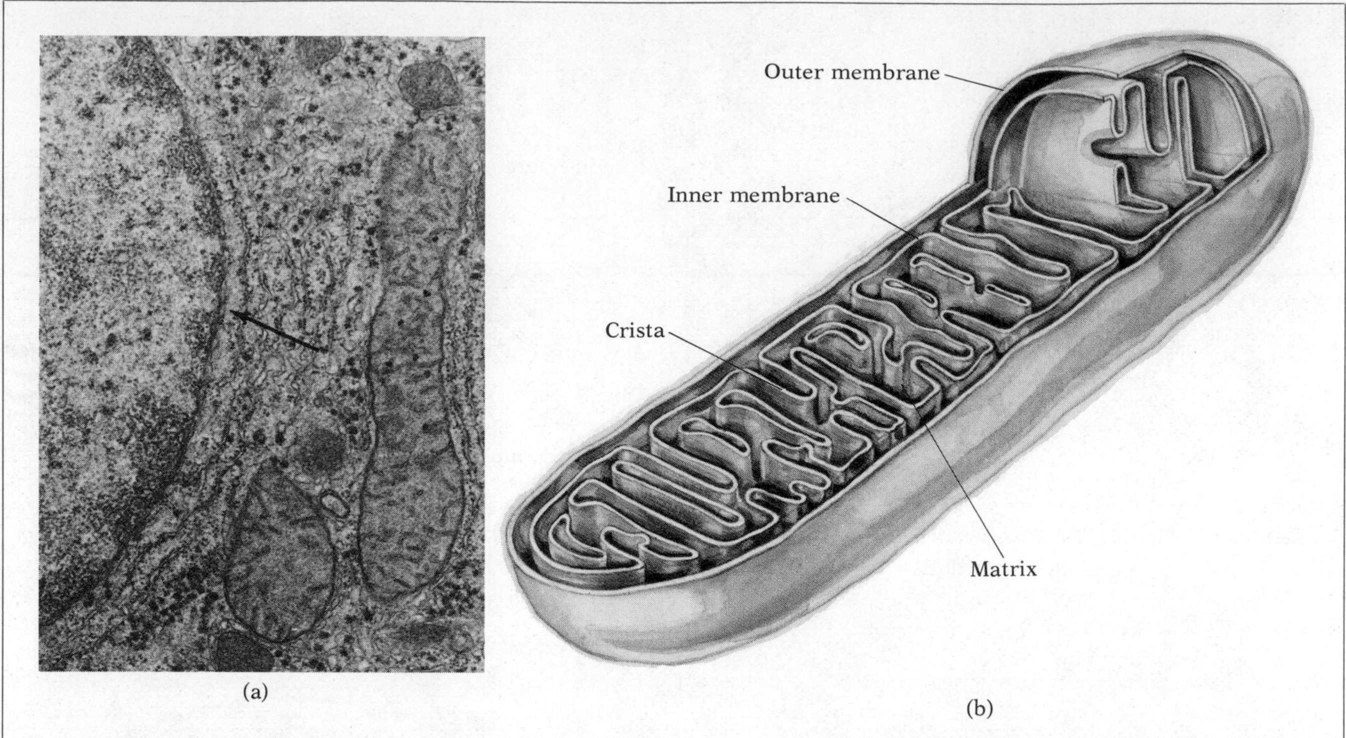

4.24 The mitochondrion. (a) Mitochondria are barely visible through the light microscope even at its highest magnification, but they show up clearly with low magnification in the electron microscope. In thin sections such as this, the long, tubular mitochondria appear as irregular elipses and circles. Each is surrounded by two membranes. The inner one is folded repeatedly to form inner shelves known as cristae. (b) The artist's reconstruction of a mitochondrion clearly reveals its intricate membranous structure.

4.24). In electron micrographs, mitochondria usually appear as more or less oval structures, with the inner of its two membranes thrown into curious folds that extend partway across the inner cavity [Figure 4.24(b)]. People have become so used to seeing this oval shape that mitochondria are frequently drawn schematically as being stubby oval or football-shaped organelles. In fact, they are more like long tubes; but when random, very thin cross-sections are cut through a lot of tubes, most of the cross-sections will be short ovals or elipses. Mitochondria are not easy to see in whole cells under the light microscope, even when specifically stained. When they are seen, they look rather like tiny needles or threads.

Although mitochondria are found in all eucaryotic cells, there are more in some cells than in others. The number varies in proportion to the cell's metabolic level—that is, in proportion to the rate at which the cell uses up oxygen. This is because mitochondria are specialists in oxygen metabolism, or respiration. For example, a single hard-working liver cell may contain as many as 1000 mitochondria, but few can be found in a fat storage cell. As you might expect, mitochondria are abundant in muscle cells.

The folds of the inner mitochondrial membrane are known as *cristae.* The cristae greatly increase the inner surface area of the organelle. Most of the biochemical work is done on the cristae. Special electron microscope techniques reveal that these tiny shelves are covered with small, round granules, which have been named F1 particles. These microscopic membrane-bound particles contain many of the enzymes associated with the energy-producing activity of the mitochondrion. Other recent work has shown that oxidative phosphorylation, the key process in the generation of ATP, depends on the passage of electrons through the membrane. In mitochondria, as in chloroplasts, the structure is critical to the function, and activity ceases if the membrane is ruptured.

Vacuoles

The term *vacuole* is rather general. It refers merely to any membrane-bound body with little or no inner structure. Vacuoles generally hold something, but their contents vary widely, depending on the cell and the organism.

Plant cells generally have more and larger vacuoles than do animal cells. In many types of plant cells, the vacuole dominates the central part of the cell, crowding all other elements against the cell wall

[Figure 4.15(a)]. Such vacuoles are filled with a fluid, the *cell sap,* which is primarily stored water with various substances in solution or suspension. The solutes may include atmospheric gases, inorganic salts, organic acids, sugars, pigments, or other materials. Sometimes the vacuoles are filled with water-soluble blue or purple pigments (the *anthocyanins*), which give color to blue and purple flowers.

The large, watery vacuoles of plants serve another curious purpose. While all animals have elaborate excretory systems for getting rid of cellular and metabolic wastes, plants lack such systems. Instead, metabolic wastes and other poisons are sequestered in the central vacuoles, frequently forming crystals. Some plants store large amounts of waste poisons in their vacuoles, and it has been suggested that this serves as a protective device against herbivores. Of course, if a herbivore should eat the plant and crush the cell, it will get not only food, but stored poisons.

Centrioles, Cilia, Flagella, and Basal Bodies

Centrioles are found in the cells of animals, most protists, fungi, and lower plants, but apparently higher plants have lost their centrioles in the course of evolution.

In those organisms that do have centrioles, the little organelle appears to play an important, but not well-understood, role in cell division. Under the light microscope, the centriole appears as a tiny dot, very close to the nucleus. It is made more prominent by an array of short rays projecting in all directions from it. During cell division, these rays become more prominent, and form the two *asters* and the *spindle* of the mitotic apparatus (Chapter 8).

The electron microscope reveals the centriole to be quite a complex structure. First, each centriole is really composed of two parts, two identical cylindrical structures that appear to lie at right angles to each other (Figure 4.25). Each cylinder (or *half-centriole*), in turn, is made up of nine sets of triplet tubules formed in a pinwheel arrangement, surrounding an amorphous center. Before cell division, the two half-centrioles separate and move to opposite sides of the nuclear membrane; then each cylinder sprouts a new half-centriole from its side so that a complete centriole is formed. We don't have a clue as to how they do it.

Cilia and Flagella

Cilia and *flagella* are fine, hairlike, movable organelles found on the surfaces of some cells. Actually, although cilia and flagella appear to be outside the cell, they really are not; the cell membrane protrudes to cover each cilium. Thus, we can consider them to be outpocketings of the cell proper.

Cilia and flagella are structurally almost identical. They differ only in length, the number per cell, and

4.25 Centrioles. A thin section seen by the electron microscope across the central axis of one of the paired bodies of a centriole. Note the nine sets of triplet microtubules. Each microtubule consists of protein spheres arranged in a spiral form to produce the tubular unit. (See Figure 4.28 for a closeup view of a microtubule.) Centrioles, like the closely related cilia and flagella, are not found in the higher plants at all. In other eucaryotes, however, they are part of the mitotic apparatus.

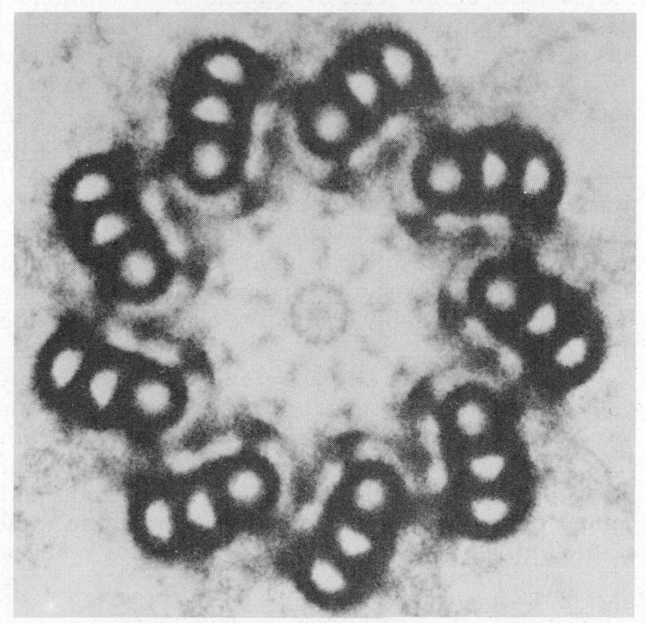

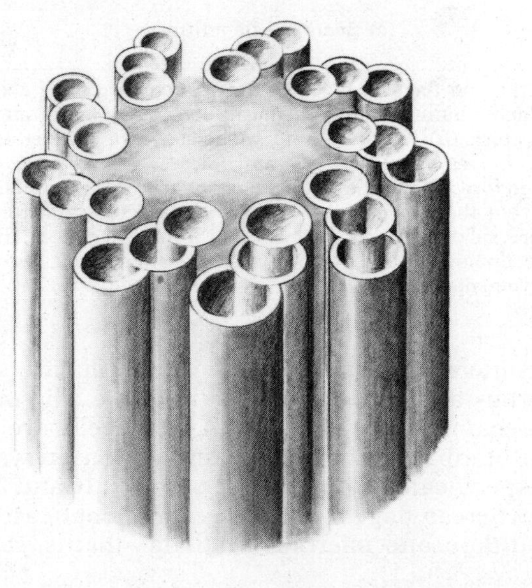

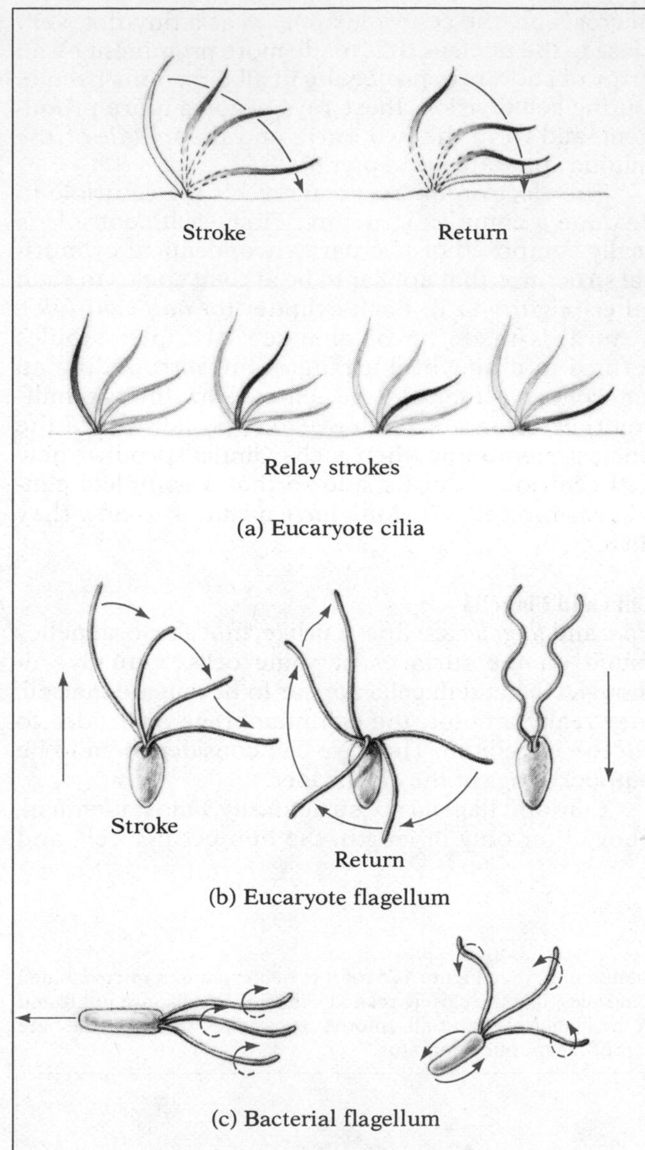

4.26 Cilia and flagella. (a) Note the sequential rowing action of cilia. Each cilium is extended during its power stroke and bent on the return. (b) The flagellum may be used in a number of ways to direct movement. As the diagram indicates, a different sequence is used in forward motion than in backing up. (c) Bacterial flagella differ from those of eucaryotes in both structure and function. They are solid filaments that actually rotate in place. Through the direction of rotation and positioning of the flagella, two different movements are possible.

their pattern of motion (Figure 4.26). Cilia are short, numerous and move like oars (note the cilia on the paramecium shown in Essay 4.1). Flagella are long, few in number, and move by undulation (in waves). More specifically, most cilia are between 10 and 20 μm long, whereas flagella may, in exceptional cases, be several thousand micrometers long—that is, several millimeters long. The flagellum on a *Drosophila* (fruitfly) sperm, for instance, may be longer than the fly itself!

Both cilia and flagella may either move the cell through some surrounding fluid or move some surrounding fluid past the surface of a stationary cell. For example, a sperm cell swims by undulations of its flagellum, a paramecium moves by the coordinated beating of rows of cilia, and the cilia of the cells that line the passages of your lungs (unless you have permanently paralyzed them with tobacco smoke) sweep a cleansing film of mucus upward to your pharynx.

Many protists, as well as the sperm of algae and some lower plants, have two flagella positioned on their front ends (because flagella can pull a cell as well as push it). Some groups of organisms lack cilia or flagella. These include some arthropods (such as insects), many fungi, red algae (a kind of seaweed), and all higher plants.

Procaryotes have flagella, but these are built on an altogether different plan and are not related to eucaryote flagella [Figure 4.26(c)]. Procaryote flagella are solid crystals of protein that stick out through holes in the cell membrane; they neither beat nor undulate, but spin like propellers.

Under the electron microscope, the internal structure of eucaryote cilia and flagella is similar to that of centrioles. While the half-centriole is made up of a circle of nine triplets of microtubules, both cilia and flagella have a "nine plus two" structure: a circle of nine pairs of microtubules, with two single microtubules in the middle (Figure 4.27). Later, we'll consider how this particular structure assists in the movement of cilia and flagella.

Each cilium and flagellum terminates beneath the surface of the cell at a *basal body,* which is structurally almost identical to a half-centriole. (This has not escaped the attention of biologists who have tried to explain the origin of cilia.) The nine pairs of microtubules of the cilium or flagellum, as they join the basal body, become nine triplet microtubules; the two single microtubules just end blindly. It's worth noting that all the groups of organisms that lack centrioles also lack cilia and flagella, although the reverse is not necessarily true. Incidentally, the basal bodies of cilia are usually joined by interconnecting fibers that presumably enable the cilia to coordinate their movements.

Cilia and flagella also apparently have some sort of primitive cellular sensory function. Many of our own sensory receptors have obviously evolved from cilia. Amazingly, the nine plus two arrangement of microtubules can still be seen in (a) the rods and cones of the retina, (b) the olfactory fibers of the nasal epithelium, and (c) the sensory hairs of the cochlea

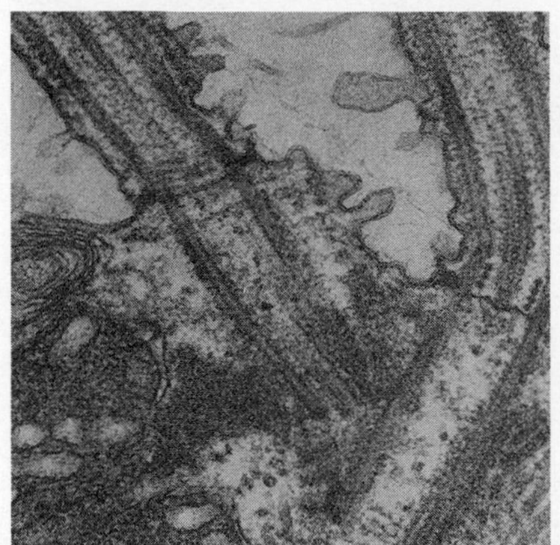

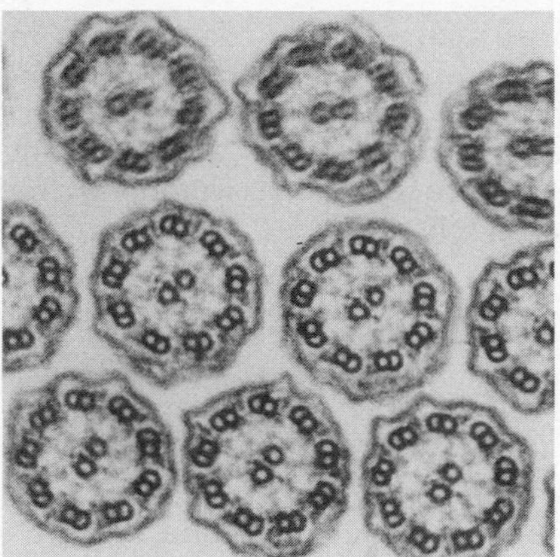

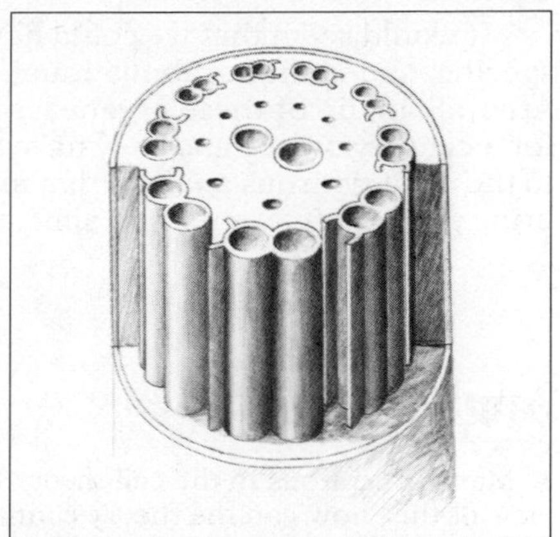

4.27 These two electron microscope views permit us to reconstruct the three-dimensional view of a cilium. The longitudinal view reveals the lengthy microtubules that extend to the basal body, while the cross-section reveals the arrangement around the cilia. Note the nine plus 2 arrangement of microtubule pairs.

and semicircular canals of the internal ear. It seems, then, that we see, smell, hear, and balance ourselves with highly modified cilia.

Microtubules

You may have noticed that in our study of the parts of a cell we've been unable to keep the categories from overlapping. Thus, we wrote about lysosomes destroying mitochondria before we told you what a mitochondrion was, and we have had several occasions to bring in microtubules before we got to this section. That's because the different parts of a cell are so closely interrelated. We have seen that microtubules are integral in the cytoskeleton of the cell, and that they also make up the internal structure of centrioles, cilia, flagella, and basal bodies. We have mentioned asters and the mitotic spindle in passing, but we didn't tell you that those structures are also made largely of microtubules.

It is generally accepted that all *microtubules* are made of a common protein, *tubulin.* But it is uncertain whether there is more than one variety of tubulin in a given cell; for that matter, it has not been proven that the microtubules of cilia and flagella are biochemically the same as the microtubules found in the cytoskeleton. However, for all microtubules, the name implies the structure: they are all small, hollow tubes. Single microtubules are invisible under the light microscope, although when they are tightly aligned as they are in the asters and the mitotic spindle, they can be seen. Under the electron microscope, their hollow tubular structure is evident. When microtubules are cut longitudinally, they appear as close, straight parallel lines. In cross-section, they are tiny circles. With special techniques, these tiny circles can be seen to be rings of about thirteen much tinier spheres (Figure 4.28). Apparently, each microtubule is made of a spiral arrangement of spherical bodies of tubulin protein.

4.28 Microtubules occur in many types of cells and apparently serve as a kind of cellular skeleton. They also aid movement of cells and within cells. With the electron microscope they appear as long rods, which, in cross-section, are found to be hollow. The tubes consist of spherical units of the globular protein tubulin.

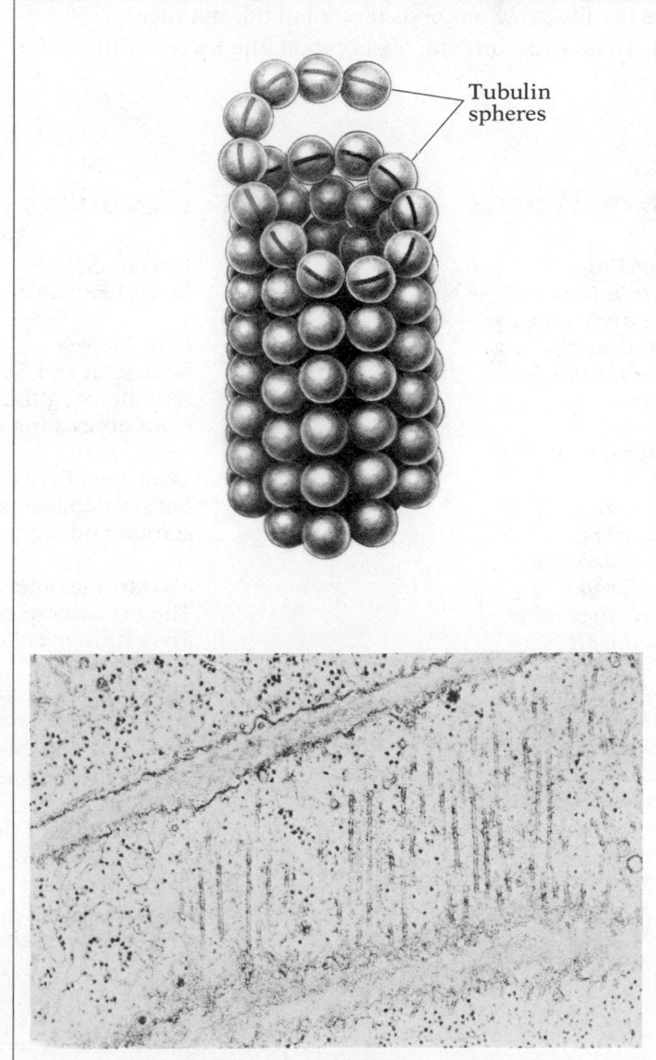

It would seem that we could have been a bit more specific and definite about much of this material. After all, many of these organelles have been known for a century or two and they have all been subjected to the most rigorous analysis that modern science can bring to bear. Yet we remain almost completely in the dark about many of these seemingly simple structures. It's almost as though Newton had been miraculously given a pocket calculator. It wouldn't have taken him long to be able to use it, but his exuberance would have faded before another question: How does the darned thing work?

Application of Ideas

1. Many exceptions to the cell theory have been noted. In view of this, how can the theory continue to be important and instructive? If you think it is not, offer reasons for your conclusions.
2. It was once thought that frogs could not survive in distilled water until someone actually kept some frogs in a pan of distilled water for months with no ill effects. What, theoretically, should have happened to the frogs? Why do you think it didn't?
3. Older methods of preserving food included drying, smoking, salting, or sugar curing. Offer a physiological explanation for the failure of spoilage bacteria and molds to readily grow on foods treated in this manner.
4. In an experimental observation, the leaves of the water plant, *Anacharis,* are treated in several ways and observed under the light microscope. The cells are normally rectangular in shape, with a large central vacuole and a thin membrane-bound layer of cytoplasm containing chloroplasts between the vacuole and cell wall. Compare and explain each microscopic observation.
   a. Leaves in pond water show no change.
   b. Leaves in distilled (pure) water show no change.
   c. Leaves in 1% and 2% glucose solution show no change, but in 3% glucose, the membrane-surrounded chloroplasts and cytoplasm have formed a tight sphere in the cell center.
   d. Leaves in a 1% NaCl solution show no change, but those in a 1.5% NaCl solution resemble those in the 3% glucose solution.

Key Words

acellular
actin filament
active transport
ameba
anthocyanin
aster

basal body

carotene
catalase
cell division
cell sap
cell streaming
cell wall
centriole
chloroplast
chromoplast
chromosome
cilia
compound lens
compound microscope
crista
cyanobacteria
cytochrome P450
cytoplasm
cytoskeleton

Key Ideas

Introduction
Robert Hooke used the term *cell* to describe the structure of cork.

Cell Theory
Schleiden and Schwann established the cell theory in 1805 when they theorized that all living things were composed of cells. Virchow added that all cells come from preexisting cells.

Why Are There Cells?
Larger organisms are always multicellular. Multicellularity permits cell specialization and increases *membrane* surface.

Membranes and the Origin of Cells
The behavior of phospholipids in water suggests how cells may have originated. They form *micelles,* or spheres with hydrophobic centers.

Essay 4.1 Exceptions to the Cell Theory
1. Numerous exceptions to the cell theory are found. Some are seen in the fungi, slime molds, protozoans, viruses, muscle tissue, and coconut endosperm (milk).
2. The theory generally holds for multicellular plants and animals.

Architecture of the Cell
1. *Cell walls* are produced by bacteria, blue-green algae, algal protists, and plants.
2. Animals and many protists lack *cell walls* and are generally more flexible. They utilize a *cytoskeleton* of *microtubules* and *microfilaments.*

Cell Diversity
Cells are diverse in size and shape, but have many factors in common. Cell differences generally reflect specialized functions.

desmosome
diffusion
digestive vacuole

electron microscope
endocytosis
endoplasmic reticulum (ER)
eucaryote
exocytosis

F1 particle
facilitated diffusion
flagellum
fluid mosaic model
free energy
freeze-fracture

gap junction
Golgi body (Golgi complex)

hydrostatic pressure
hypertonic
hypha
hypotonic

isotonic

leucoplast
lysosome

meiosis
membrane
messenger RNA
metabolic
micelle
microfilament
micrometer
microtubule
middle lamella
mitochondrion
mitosis
mitotic apparatus
multinucleate

nuclear membrane
nuclear pore
nucleolus
nucleus

organelle
osmosis
osmotic potential
osmotic pressure

passive transport
permeability
peroxisome
phagocytosis
photosynthesis
pinocytosis
plasma membrane

The Cells of Very Different Organisms

Cells can be organized into two very different types, the *procaryotes* and the *eucaryotes.* The first includes bacteria and blue-green algae, the second, all the other organisms.

Why Are Cells So Small?

1. Most cells are extremely small, less than 100 μm in length or diameter.

2. Cell growth continues until volume is approximately doubled, then they divide. Growth is fastest immediately after division.

The Surface–Volume Hypothesis

1. Cells remain small because of the surface–volume relationship. As size or volume increases, the surface area lags behind, increasing more slowly. Some cells overcome the surface–volume problem by folding the outer *membrane,* increasing inner *membranes,* cytoplasmic streaming, and forming large central *vacuoles.*

2. Larger organisms use transport systems (blood vessels or other vascular tissue) to support large masses of cells.

The Cell Structures

The Plant Cell Wall

1. The cellulose plant *cell wall* maintains the shape of the cell and gives it strength and flexibility.

2. As the wall develops, cellulose fibers are laid down in several directions and are impregnated with pectin. A middle lamella connects one *cell wall* with adjacent ones.

3. Various tissues have *cell walls* impregnated with lignin, a hardener, and suberin and cutin, waxy, waterproofing materials.

The Cell Membrane

1. Cell membranes contain a phospholipid bilayer and act as controlling barriers to keep some substances in or out.

2. Substances enter and leave cells through the processes of *diffusion, facilitated diffusion,* and *active transport.*

3. *Diffusion* is the random movement of molecules in a liquid or gas with a net movement from regions of higher concentration to regions of lower concentration.

4. The ease with which fat soluble molecules cross *membranes* suggested the phospholipid bilayer to Danielli.

5. Danielli notes that some molecules (such as glucose) travel faster than others across the *membrane,* but follow the general rules of *diffusion.* He called the transport process *facilitated diffusion.*

6. Danielli postulated the presence of proteins, which he believed formed an outer layer on each side of the *membrane.*

7. *Electron microscope* studies supported Danielli's model.

8. More recently it has been revealed that some proteins lie on the surface, but others pass all the way through. Some are believed to contain pores that conduct water-soluble substances.

9. Some surface proteins are now believed to interact with messengers such as hormones. Other proteins are recognition molecules that reject foreign substances and work in cell fusion to form tissues.

10. Glycoproteins are believed to have *antigenic* properties.

11. The current concept of cell membrane structure is known as the *fluid mosaic model.*

Diffusion, Facilitated Diffusion, Active Transport, and Osmosis

A Closer Look at Diffusion

1. *Diffusion* involves the random movement of molecules through *thermal energy.* Although individual molecular movement is random, net movement is toward areas of lower concentration.

plastid
procaryote
proplastid
protist
protoplasm

replication
ribosome
rough endoplasmic reticulum

scanning electron microscope (SEM)
semipermeable
smooth endoplasmic reticulum
sodium pump
solute
spindle
stroma

thylakoid
transfer RNA
transmission electron microscope (TEM)
tubulin
turgor pressure

unicellular

vacuole
vesicle

xanthophyll

2. Net movement continues until the system of molecules is randomized.

3. Unequally distributed concentrations of molecules represent systems with great *free energy*. In their net movement these molecules obey the rule of such systems and dissipate that energy.

Facilitated Diffusion and Permeases

Special proteins in the *membrane*, the *permeases*, are believed to assist molecules in crossing the *membrane* during *facilitated diffusion*.

Active Transport Across Cell Membranes

1. Studies of red blood cell membranes reveal that sodium and potassium ions move against the *diffusion* gradient.

2. A *sodium pump* has been hypothesized to explain *active transport*. Carrier molecules, using ATP energy, move ions across the *membrane*.

3. *Endocytosis* is the engulfing of substances by the cell *membrane*. It includes *phagocytosis*, taking in solids, and *pinocytosis*, taking in dissolved substances. Substances taken in are contained in *vacuoles* that form from the cell *membrane*.

4. *Exoctyosis* is the expulsion of material from *vacuoles*. These fuse to the cell *membrane* and empty the contents outside.

5. Using elements of the *cytoskeleton*, some organisms actively pump water out of their interior.

Osmosis

1. *Osmosis* is the *diffusion* of water through a *semipermeable membrane*.

2. When two different *solute* concentrations are separated by a *membrane*, water will diffuse into the higher *solute* (lower water) concentration.

3. Plant cells maintain *turgor pressure* through the inward movement of water, through osmosis.

4. The three terms *isotonic*, *hypotonic*, and *hypertonic* describe concentrations of *solute* on the outside of a cell *membrane*.

5. *Isotonic* indicates the *solute* concentration is the same on either side.

6. A *hypotonic* solution outside a cell has a lower concentration of *solute* than the solution inside. (Distilled water outside)

7. A *hypertonic* solution outside a cell has a higher concentration of *solute* than the solution inside. (Salt water outside)

8. *Osmotic potential* increases with the concentration of *solute* molecules.

Junctions Between Cells

1. *Desmosomes* link the *membranes* of adjacent cells together.

2. *Gap junctions* allow the direct exchange of nutrients and intracellular hormones through channels that pass between cells.

Cell Organelles

Organelles are structures of specific function within the cell.

The Nucleus

1. A membrane-bound *nucleus* is found in all *eucaryotic* cells.

2. The *nucleus* contains a matrix of DNA, and one or more RNA rich nucleoli.

3. The *nucleus* has a double *membrane* with many pores that permit the passage of RNA and protein.

4. The *nucleus* contains the hereditary information of the organism and is capable of duplicating this information for reproduction.

5. The nucleus controls cell activities through enzyme synthesis.

The Endoplasmic Reticulum

1. The *endoplasmic reticulum* (ER) is a system of broad membranes that extend through the *cytoplasm* of *eucaryotic* cells. It is believed to connect the cell *membrane* with the *nuclear membrane*.

2. The *ER* forms saclike *lyosomes* and *vacuoles*.

3. *Rough endoplasmic reticulum* contains minute *ribosomes* along its surface that function in protein synthesis.

4. *Smooth endoplasmic reticulum* lacks granules and is found where carbohydrates, steroid hormones, and lipids are stored or synthesized.

The Golgi Complex
The *Golgi bodies* appear as flattened saccules near the *nucleus.* They store and isolate many substances and are active in plant cell wall formation, providing complex carbohydrates.

Lysosomes

1. *Lysosomes* are membrane-bound sacs containing powerful hydrolytic enzymes. These destroy injured cells and cell organelles and function in intracellular digestion.
2. *Plastids* may carry out the lysosome's function in plants that lack them.
3. Absence of a lysosomal enzyme may cause storage diseases.

Peroxisomes

1. *Peroxisomes* are membrane-bound bodies containing a crystalline core of tiny tubes.
2. Liver and kidney *peroxisomes* are known to contain the enzyme *catalase,* which protects the cells by destroying hydrogen peroxide.

Plastids

1. *Plastids* are membrane-bound bodies in plant cells.
2. Several types of *plastids* are believed to develop from *proplastids.* The major types have double membranes and DNA.
3. *Chloroplasts* are green *plastids* that carry on *photosynthesis.*
4. *Leucoplasts* are starch storage bodies.
5. *Chromoplasts* arc *chloroplasts* that lose their green color and take on other pigments (e.g., autumn leaves and fruit).

Chloroplasts

1. *Chloroplasts* contain *chlorophyll,* the photosynthetic pigment.
2. *Chloroplasts* contain flattened, membranous disks, the *thylakoids.* These are arranged in stacks or *grana,* which are surrounded by a clear *stroma.*

Mitochondria

1. Mitochondria are ATP-producing structures found in eucaryotes. They contain their own DNA and ribosomes, and may have originated as independent organisms.
2. Mitochondria are long and slender with an outer and inner membrane. The inner membrane is folded into cristae, which are studded with F1 particles.

Vacuoles

1. *Vacuoles* are membrane-bound storage bodies with little or no inner structure.
2. Many plant cells have very large central *vacuoles* filled with water, various solutes, and metabolic wastes. Pigmented *vacuoles* give blue flowers their color.

Centrioles, Cilia, Flagella, and Basal Bodies

1. *Centrioles* are self-replicating bodies found in most eucaryotic cells (except those of higher plants). They function in cell division.
2. Each *centriole* is composed of two cylindrical structures lying at right angles to each other. Each cylinder consists of nine sets of three microtubules.

Cilia and Flagella

1. *Cilia* and *flagella* are hairlike structures surrounded by the cell *membrane.* They are identical except in length and movement. They function in moving the organism or moving substances past the cell.
2. *Procaryotes* have *flagella,* but these are protein crystals that protrude from holes and spin on their axes.

3. In cross section, eucaryotic cilia and flagella show a ring of nine pairs of *microtubules* and two single *microtubules* in the center. The *basal bodies* of *cilia* are often joined by fibers that coordinate their movement.

4. Organisms that lack *centrioles* lack *cilia* and *flagella* (the reverse is not true, however).

5. Cells of certain mammalian sensory structures (visual, olfactory, hearing, and balance) show the basic 9+2 structure, and may be derived from sensory cilia.

Microtubules

1. *Microtubules* are small hollow tubes composed of tiny spheres of the protein, *tubulin.*

2. In addition to their role in *cilia, flagella,* and *centrioles, microtubules* are part of the cell *cytoskeleton.*

Review Questions

1. Summarize Robert Hooke's contributions to the field of cell biology. (p. 84)
2. State the cell theory in the terms of Schleiden and Schwann. (p. 85)
3. What are two advantages of multicellularity over single cellularity? (p. 85)
4. What is important about sufficient *membrane* area? (p. 85)
5. Summarize the functions of the cell *membranes.* (p. 86)
6. List four features found only in *eucaryotes.* (Table 4.1)
7. List four differences between plant and animal cells. (Table 4.1)
8. Why is the loss of surface in the nuclear and cell *membranes* important as cells reach maximum size? (p. 92)
9. How do plants overcome the apparent surface–volume imperative? (p. 92)
10. Do nerve cells with long processes violate the surface–volume restriction? Explain. (p. 92)
11. Describe the manner in which plant *cell walls* are laid down. (p. 95)
12. What kind of indirect evidence led Danielli to suggest the bilayer of phospholipids? (p. 100)
13. Danielli is credited with the term *facilitated diffusion.* What observation was he trying to explain? (p. 100)
14. Make a drawing that illustrates the modern view of the cell *membrane,* the *fluid mosaic model.* (p. 101)
15. In addition to transport, what other functions do the surface proteins serve in the cell membrane? (p. 102)
16. Using the Danielli model as an example, discuss the value to science of model and hypothesis-building. (p. 102)
17. What is the essential difference between *active* and *passive transport?* (p. 103)
18. How long does the movement of molecules go on during *diffusion?* How long does *net* movement of molecules go on? (p. 103)
19. From the behavior of highly concentrated molecules in the hydrogen sulfide bottle, what can you generally predict about systems with great *free energy?* (pp. 103–104)
20. Describe what happens in the two types of *endocytosis.* How would you know that this is *active transport?* (p. 105)
21. What is the role of the *cytoskeleton* in *exocytosis?* (p. 105)
22. What would happen to a cell that contained 1% *solute* if it were placed in a solution containing 3% *solute?* (pp. 105–106)
23. Explain how *osmosis* assists in keeping leaves and soft stems erect. What is the pressure called? (pp. 106–107)
24. What would you expect to happen to animal cells suspended in an *isotonic* solution? A *hypotonic* solution? (p. 107)
25. What determines the *osmotic potential* of a solution? (p. 108)
26. What two characteristics of the *nucleus* helped make it one of the first cell *organelles* to be recognized? (p. 109)
27. Describe the contents of the *nucleus.* (p. 109)
28. Describe two unusual characteristics of the *nuclear membrane* and relate these to the molecules that pass through. (pp. 110–111)
29. How does the function of *rough endoplasmic reticulum* differ from that of *smooth?* (p. 111)
30. Describe the function ascribed to the *Golgi bodies.* What is a special function they serve in the plant cell? (pp. 111–112)
31. Explain what a "storage disease" is and what this has to do with *lysosomes.* (p. 113)
32. What characteristics of a *chloroplast* suggest it may have arisen as an independent organism? (p. 114)
33. Describe the inner structure of a *chloroplast.* (p. 114)
34. What characteristics of *mitochondria* suggest that they may have originated as invading or engulfed organisms? (p. 115)
35. Compare the function of *mitochondria* with that of *chloroplasts.* (p. 115)
36. What is the general role of a *vacuole?* What is their special role in plants? (pp. 116–117)
37. What structure and function do *centrioles, cilia,* and *flagella* have in common? (p. 117)
38. How does the fine structure of the *cilium* and *flagellum* differ from that of the *centriole?* (pp. 117–118)
39. Describe two functions of *cilia.* (p. 118)

Chapter 5

Photosynthesis: Energy Captured

It is somehow paradoxical that humans are so attracted to places like high mountain passes or distant peaks. One of the greatest of human experiences must be to step across the grassy hillocks of the continental divide or to stand on the wind-swept knolls of Haines Junction, Alaska. And, for some reason, we are impressed with the polar landscape.

Why are we so taken with flowered mountain passes, with snowy peaks looming in the distance? We think of them as beautiful and we often long to be among them, but they are places that few of us could survive for long. They still, as well as stir, our hearts. *Energy* is the reason.

Take the mountain pass, for instance. This is a cold place and we are thermoregulated, or warm-blooded, animals. Our cells can only operate within rather narrow temperature limits. So in a cold place, we must produce metabolic heat in order to keep our cells at a working temperature. In other words, fuel must be used to produce the energy or heat. That fuel, of course, is food. But on alpine slopes, where is the food? We are ill-equipped to derive sustenance from such places. We would soon exhaust any food types we could find and the energy we derived from them would hardly be worth the effort. Thus, whereas many of us think we would like to live in such places, it would be impossible unless materials and stored energy were imported from some more hospitable place (Figure 5.1).

5.1 The plants and animals clinging to life in this barren environment provide an uncomfortable reminder of our own tentative existence. It is sometimes helpful to consider the delicate links between organisms and the physical environment.

5.1 Energy, Chemiosmotic ATP Production, and Electron Carriers

This is an essentially hostile world in that our molecules tend to move continuously toward a disorganized state. In order to keep our molecules organized so that the complex activities of life can go on, energy must be expended. That energy comes from food, and our food sources sustain us with great reluctance. Leaves turn tough and unnutritious as fast as they can so that they will not be utilized by some energy-starved animal. Rabbits run away so that the energy they have stored in their bodies will not become available to us. Yet we must have the leaf and the meat. Humans scramble over the earth's surface, availing themselves of as much of its energy as possible, even at the expense of their fellow humans. One can safely say that many of the woes of the world are due to the relentless search for energy—energy that can be used to mold the world so that it is more compatible with cellular needs.

In Chapter 3, we introduced the notion of free energy, and the second law of thermodynamics, the law that says we must use energy to reduce entropy or die. *Free energy* is sometimes defined as *negative entropy.* Entropy is a measure of the *randomness* of a system, and free energy in a system can be measured by the degree of order in that system. Another way to look at free energy is that it is improbable. If you poured a bag of marbles on a slope, a few would bounce uphill and come to rest at a higher level than the others. It is improbable that any one marble will reach a higher level, but once the few marbles that do make it uphill are there, they will possess *potential energy,* or the ability to do work. A falling marble will have *kinetic energy,* or energy in motion.

The *second law of thermodynamics* says that entropy (disorder) is always increasing within any closed system. (A closed system is one in which neither matter nor energy enter or leave.) Another way of stating it is that the free energy of a system is always decreasing. But in a closed system the energy doesn't disappear; so where does it go? It becomes *heat.* Heat, after all, is a form of energy; but *if the heat is evenly distributed* (randomly dissipated), there is no way to organize it into potential energy again. The energy is there, but it is not free energy. There is no way it can be harnessed for work.

Now, you may think that the second law of thermodynamics represents a rather pessimistic view of the universe, and you wouldn't be alone. Many people have reached the same conclusion. After all, it's not pleasant to believe that things always become more disorganized. We abhor disorganization (even those who pride themselves on being able to function in a disorganized manner). Perhaps our distaste for disorganization stems from some primal knowledge that it spells death. Fortunately, there is an out. Remember, the law applies to *closed* systems—and the earth is not a closed system. The earth receives a constant input of sunlight. Sunlight is free energy in every sense—it doesn't cost us anything! So, it turns out that the second law of thermodynamics will have to wait for another 10 billion years or so, when the sun is scheduled to burn out. There is no way that we can, in any sense, identify with the earth and the things on it by that time, so we tend not to care what will happen when entropy takes over the planet.

So, life is a joke on the somber second law of thermodynamics. Living things are incredibly highly organized, and over their lifetimes they build up sizable stores of chemical (potential) energy. In fact, in the much longer run, evolution leads to ever-increasing complexity in life forms. Our 7 billion nucleotides of DNA information in every cell represent quite a store of negative entropy (or free energy, or organization, or potential energy).

Life can do this because it is a backwater, an eddy in the stream of energy dissipation. On our planet, light is captured and made to do tricks, most notably the trick of building ordered structures and energy reserves. Free energy is used to resist and reverse the inexorable trend toward increasing entropy. Of course, no organism can remain structured forever. The individual ordered structures and energy reserves will dissipate soon enough, but as long as the solar energy input continues, there will be new organisms to replace the old. In this chapter, we'll be concerned with how the raw energy of sunlight is put through the hoops. We'll see that life forms have evolved amazingly efficient ways of squeezing every bit of usable work out of the energy they find.

Adenosine Triphosphate (ATP)

We'll begin our look at cellular energetics with a discussion of a molecule that can accurately be described as a cellular workhorse. It is the coin of energy transactions in the cell; it is that famous molecule called *adenosine triphosphate,* or *ATP.* You've seen it before, but now we need to look at some of the specifics of its structure and its role in energy exchanges.

Adenosine triphosphate is used by all forms of life as the nearly universal molecule of energy transfer. If energy is produced by some cellular processes—photosynthesis or respiration (see Chapter 6)—it is stored by ATP. If energy is needed for some cellular process, it is provided by ATP. Energy may be stored in less bulky forms, such as fats or carbohydrates, but it is put into these storage molecules by ATP, and it is taken out of storage by ATP.

The ATP molecule consists of an *adenine* base, a five-carbon ring sugar named *ribose,* and *three phosphates* (Figure 5.2). The adenine is the same molecule that occurs in DNA and RNA and it is one of the four nitrogenous bases that make up the letters of the genetic code. The combination of adenine, ribose, and phosphate is found in ATP and in RNA, and we will soon encounter it in NAD and NADP, two other engery carriers.

The three phosphates are bonded together in a row. The chemical bonds between the phosphates are *high-energy bonds,* which means that an unusually large amount of energy must be used to make the bonds, and that energy becomes available when the bonds are broken. Most of the energy is stored in the bond between the second and third phosphates.

Two related molecules are *adenosine diphosphate* (*ADP*) and *adenosine monophosphate* (*AMP*). As you might gather from their names, these differ from ATP only in that ADP has two terminal phosphates and AMP has but one. The reaction in which energy is transferred usually involves changing ATP to ADP, with the release of an inorganic phosphate ion (PO_4^{3-}) and energy:

$$ATP + H_2O \rightleftharpoons ADP + P_i + Energy$$

5.2 Adenosine triphosphate (ATP) is a triphosphate nucleotide, identical to that found in RNA. It consists of adenine covalently bonded to the number 1 carbon of ribose. Ribose, in turn, is covalently bonded at its number 5 carbon to three phosphates. The last two contain high-energy bonds which, upon cleaving, yield about 8000 calories per mole.

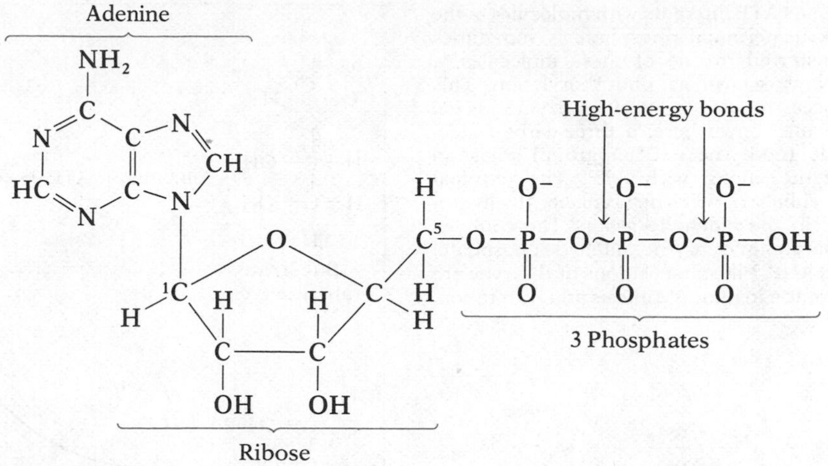

If we let A represent the adenine and ribose, together forming a *nucleoside*, let P represent the phosphate group (which includes a phosphorus atom and several oxygens); let P_i represent the released inorganic phosphate ion; let – represent an ordinary chemical bond; and let ~ represent a high-energy bond; this can be rewritten as

$$A{-}P{\sim}P{\sim}P + H_2O \rightleftharpoons A{-}P{\sim}P + P_i + \text{Energy}$$

Notice that this is a reversible reaction often portrayed as a cycle (Figure 5.3). If energy is put *into* the reaction, ATP is generated from ADP and inorganic phosphate. Of course, enzymes and other reactants are always involved; and most of the time no one bothers to write down the fact that water goes into the reaction. What happened to the water? Its atoms are needed to replace those lost when phosphate is moved around. If you remove a phosphate from ATP, some unpaired electrons will appear. These will form covalent bonds with oxygen and hydrogen in water.

5.3 Adenosine triphosphate (ATP) cycle. The ATP molecule stores energy for the cell's activities. Energy is released when the bond holding the third phosphate is broken, releasing the phosphate (P_i) and changing ATP to ADP. The ADP is reconverted to ATP during photosynthesis, and in mitochondrial respiration.

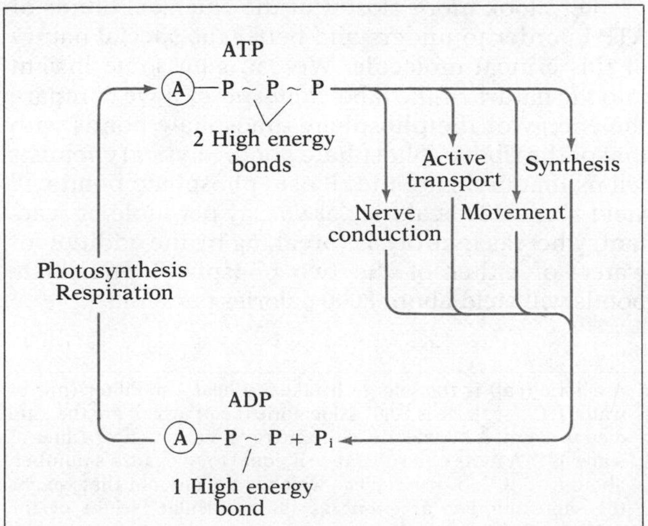

Sometimes when ATP is broken down, the inorganic phosphate is not released into the medium, but instead, is transferred to some substrate (high-energy bond and all). Such an event is called a *phosphorylation.* If we let R represent a substrate, the reaction might be written

$$ATP + R \rightleftharpoons ADP + RP_i$$

or

$$A{-}P{\sim}P{\sim}P + R \rightleftharpoons A{-}P{\sim}P + R{\sim}P_i$$

This reaction requires at least one enzyme (Figure 5.4). For some purposes, the phosphorylation reaction is written this way:

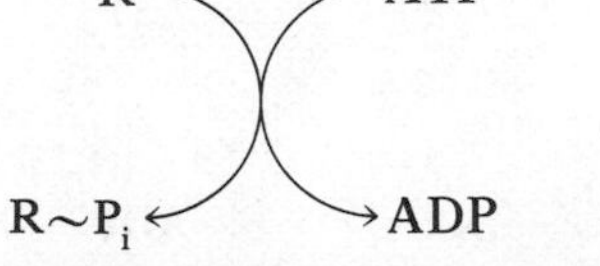

It isn't very often that ATP is changed to AMP, but in some cases it happens:

$$ATP + H_2O \rightleftharpoons AMP + P_iP_i + \text{Energy}$$

or

$$A{-}P{\sim}P{\sim}P + H_2O \rightleftharpoons A{-}P + P_i{\sim}P_i + \text{Energy}$$

The inorganic double phosphate (P_iP_i), with its high-energy bond, is known as *pyrophosphate.*

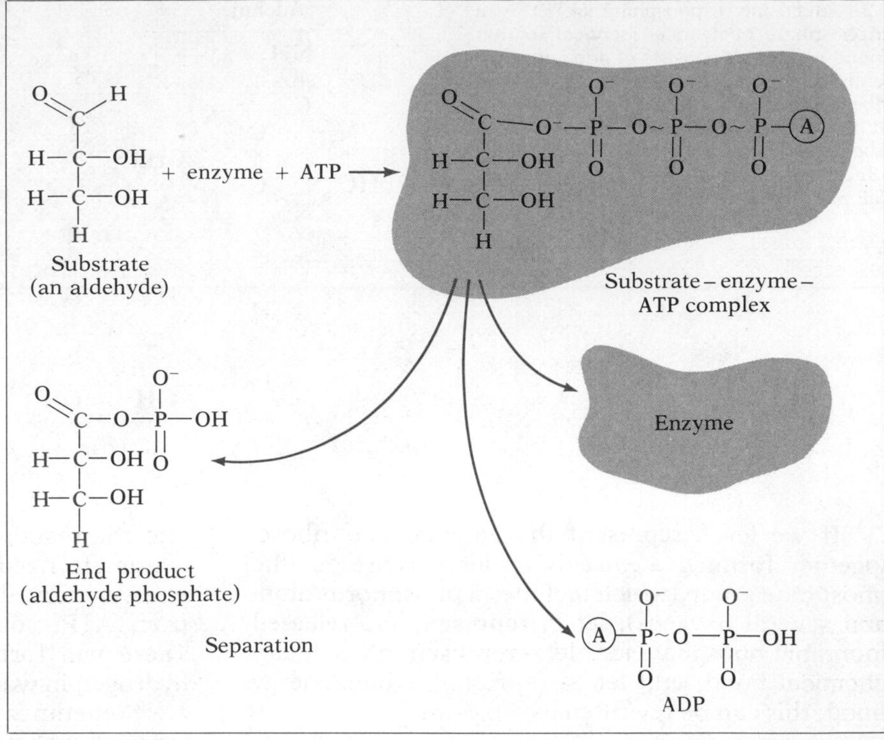

5.4 As ATP interacts with molecules in the cell, its terminal phosphate is sometimes transferred to one of these molecules, a process known as phosphorylation. This process requires a specific enzyme. In the scheme shown here, a three-carbon aldehyde (note the —CHO group) joins an enzyme, along with ATP. The terminal phosphate of ATP then replaces the hydrogen in the aldehyde group. The complex separates into enzyme, aldehyde phosphate, and ADP. Phosphorylations of this type are common in photosynthesis and respiration.

In many synthetic (which means "building") reactions, ATP supplies the energy to put two molecules together. This involves a complex binding of an enzyme and three substrates, one of which is ATP. If we let X and Y be the other two substrates, then the coordinated reaction is

$$\text{ATP} + \text{X} + \text{Y} + \text{Enzyme} \rightleftharpoons \text{ATP–X–Y–Enzyme complex} \rightleftharpoons \text{ADP} + \text{P}_i + \text{XY}$$

Sometimes we write such coordinated reactions this way:

X + Y → XY ; ATP → ADP + P_i

Similarly, if the breaking apart of a bond releases energy that is captured by ADP, the coordinated reaction might go like this:

XY → X + Y ; ADP + P_i → ATP

We'll see plenty of reactions like these in the pages to come. And we can see what happens to the ADP and inorganic phosphate after they have been separated by breaking down ATP. It turns out that they are rejoined to form ATP by a reaction that requires energy. This cyclic behavior of ATP (see Figure 5.2) can be shown by a diagram like this, where W, X, Y, and Z are four different molecular compounds:

X + Y → XY ; ATP → ADP + P_i ; ADP + P_i → ATP ; WZ → W + Z

Let's look more closely at the chemical bonds of ATP in order to understand better the special nature of this critical molecule. We can gain some insight into the nature of the label *high-energy* if we compare the energy of the phosphate–phosphate bonds with that of the ribose–phosphate bond. Physical chemists tell us that breaking the ribose–phosphate bond will yield about 2200 calories of energy per mole of reactant, whereas hydrolysis (breaking by the addition of water) of either of the two phosphate–phosphate bonds will yield about 8000 calories per mole.*

* A calorie (cal) is the energy it takes to heat 1 milliliter (ml) of water 1°C. A calorie is 0.001 kilocalorie (kcal), which are the kind dieters count. A kilocalorie is the energy it takes to heat 1 liter of water 1°C. A mole of a substance is equal to Avogadro's number, about 6×10^{23}, of molecules—which is the amount that weighs the same number of *grams* as the molecular weight of the compound in daltons.

As to why some chemical bonds require and release more energy than others, we can offer only a very rough explanation. Apparently, it is because the oxygens of the phosphate group draw electrons away from the phosphorus atom, rendering it positive. The charged condition produces a repulsion between phosphorus atoms. Significant bond energy is required to bring the reluctant phosphorus atoms together, and this energy is released when they are broken apart.

But what we're interested in here is the production of energy. So where does the energy—the 8000 cal per mole of ATP—go when ATP is broken down? Physical chemists measure it as heat output, but, as we have seen, it can also be transferred to other chemical compounds or it can be used for work in other ways. This, in fact, is what the idea of useful energy is all about. So, basically we're interested in where the energy goes. Some of the energy will always be converted to heat, because all energy transfers are inefficient to some degree. But we're interested in the energy that is transferred to chemical compounds that serve either as structural material for the organism or as centers for energy storage. Of course, this stored chemical energy can't remain stored forever. It will be broken down eventually, even if it means waiting until the individual dies. We can predict that sooner or later all the energy entering a living system will dissipate as useless heat because of the inexorable tendency for things to become random. It takes work to overcome this tendency, and a living organism can maintain an ordered existence only by continually bringing new energy into its system. If you're a plant, that energy can come directly from sunlight. If you're not, the energy must come indirectly from sunlight as you eat plants or plant eaters. We are children of the sun, every one of us.

ATP Production by Chemiosmotic Phosphorylation

Only recently has one very special kind of reaction involving ATP been described. The theory of *chemiosmotic phosphorylation* was proposed by Peter Mitchell in 1961. Mitchell, it seems, had his own peculiar way of looking at things. While other biochemists were trying to understand ATP synthesis by analyzing each specific component of its manufacture, Mitchell took a broader view and looked at ATP production in terms of intact chloroplasts and mitochondria. His first clue was the basic observation that neither chloroplasts nor mitochondria can make ATP unless they are physically intact. So Mitchell devised a scheme in which ATP synthesis was dependent on intact, closed membrane systems and membrane-separated cell compartments.

In Mitchell's system, hydrogen ions tend to build up on one side of a membrane of an organelle (Figure 5.5). This results in a pH differential and a difference in the net electrical charge across the membrane. Just as hydrogen ions build up differentially across the membrane, so do hydroxyl (OH^-) ions, at least in mitochondria and bacteria. A concentration differential, as we noted in our discussion of osmosis, represents potential free energy in a system (Figure 4.14). Now, it takes work to build up and maintain this hydrogen ion (and hydroxyl ion) differential, and creating the differential is half of the chemiosmotic story. The other half involves tapping the energy produced by the concentration differential to make ATP out of ADP and phosphate.

How can such a differential produce usable energy? In a number of ways. For example, perhaps you are aware that adding a strong base (high OH^- concentration) to a strong acid (high H^+ concentration) always results in the release of energy—sometimes explosively—in the form of heat. So when hydrogen ions flow back across the membrane to combine with hydroxyl ions and form water, some of the free energy of that reaction can be captured and used.

5.5 System (a) shows the hydrogen and hydroxyl ions at equilibrium on either side of a membrane. Free energy is low, as it would be in any random system. In (b), the molecular pumps have been at work, actively transporting ions. Now there are two systems of high energy separated by the membrane. This potential energy can be tapped if the ions are permitted to cross in a controlled manner as they are in the chloroplast and mitochondrion.

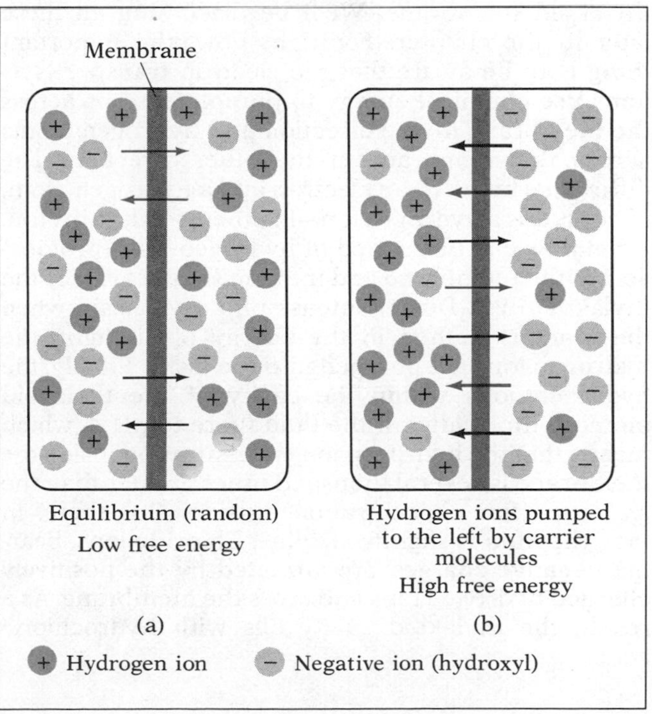

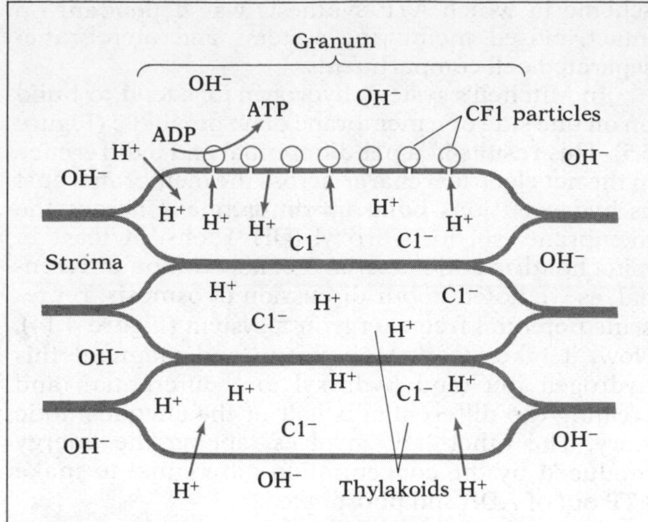

5.6 As the membranes do their work, the thylakoids of the chloroplast fill with hydrogen ions. In the presence of chloride ions, these form pools of hydrochloric acid. Outside, in the stroma, a dilute alkaline condition exists. Such a system—acid on one side, base on the other—represents great free or potential energy. The controlled admission of H^+ through the CF1 particles into the stroma provides a way of tapping the potential energy for ATP production.

Let's take a closer look, now, first considering chemiosmosis in the chloroplast. Within the chloroplasts are membrane-bound, flattened, disklike vesicles called *thylakoids* (see Figures 4.23 and 5.6). Photosynthetic pigments and their associated enzymes and *electron carriers* are all embedded in the thylakoid membrane. We'll be discussing all these later in the chapter. For right now, the important thing is to be aware that the electron transport systems use chemical energy to pump electrons across the membrane in one direction and hydrogen atoms across the membrane in the other direction. The difference between an electron and a hydrogen atom, of course, is a hydrogen ion—in other words, a proton. A net flow can be created of hydrogen ions (protons) across the membrane and into the small cavity of the thylakoid disk. During intense photosynthesis, when the plant is bathed in the energy of sunlight, the hydrogen ions are pumped into the disks. Finally, the hydrogen ions within the cavity of the thylakoid increase the acidity of the fluid there to pH 4, which means that the hydrogen ion concentration inside the membrane is several thousand times greater than the hydrogen ion concentration outside. Out there, in fact, the fluid is slightly alkaline. Chloride ions, bearing negative charges, are attracted by the positively charged hydrogen ions and cross the membrane. As a result, the thylakoid cavity fills with hydrochloric acid.

The hydrogen ions inside can't drift back out through the membrane proper, but they do have special channels through which they can leave. At the end of each channel, on the outside of the thylakoid membrane, is a knob of protein called the *CF1 particle* (C for chloroplast; the corresponding particle on the mitochondrial membrane is called the F1 particle) (Figure 5.7). The CF1 particle, which is on the end of a short stalk, is a complex group of enzymes, and it is here that the actual phosphorylation of ADP to ATP takes place. As we said, it takes energy to make ATP from ADP, and in the CF1 particle this energy comes from the protons that originated from inside the thylakoid disk. Scientists estimate that three hydrogen ions must pass through the CF1 particle in order to produce one high-energy bond of ATP. Little is known at present as to exactly how this enzymatic reaction takes place, but it is known to be reversible. Given an excess of ATP, the CF1 particles will use the terminal high-energy bond to push hydrogen ions back inside, against the gradient.

We could write the net reaction this way:

$$3\,H^+ + 3\,OH^- + ADP + P_i \rightleftharpoons 3\,H_2O + ATP$$

However, one of the limitations of writing something like chemiosmotic phosphorylation as a simple chemical reaction is that there is no way to indicate the very important concentration differentials or the membrane that separates them.

This, then, is what happens as chloroplasts make ATP. But almost exactly the same situation occurs in mitochondria and on the plasma membrane of aerobic bacteria. Again, biochemical reactions take place within the membrane itself that serve to pump hydrogen ions from one side of the membrane to the other. In these cases, however, the hydrogen ions build up outside the central cavity of the mitochon-

5.7 Coupled phosphorylation: the free energy of the H^+–OH^- gradient is harnessed to form high-energy phosphate bonds. One ATP is produced in a coupled reaction as $3\,H^+$ are permitted to cross the membrane into the alkaline exterior. Coupling events in the CF1 particle are not clearly understood, but are known to involve ADP, P_i, H^+, OH^-, and an enzyme complex.

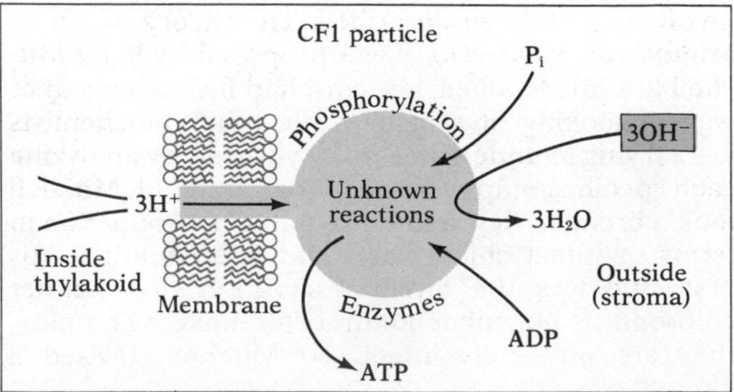

Essay 5.1 The Laws of Thermodynamics

There are some basic ideas in science that appear to hold up no matter how often they are tested. We call them *laws* and we like to think that these natural laws can't be broken under any circumstances. Biology, because of the almost infinite variety of life, is actually rather short of laws. But living systems are made of matter and chemistry, and the laws of physics and chemistry are as inviolable in living systems as anywhere else. One set of physical laws that biologists are always coming across have been codified as the *laws of thermodynamics.* There are four. The first two are the ones you hear more about.

The first law of thermodynamics: Energy can neither be created nor destroyed. This is also called the Law of Conservation of Energy. The total amount of energy in a closed system—when energy is neither entering nor leaving—remains constant. Energy can, of course, change its form, and it does so in a bewildering variety of ways. Energy in a can of gasoline, when combined with oxygen inside your lawnmower can be transformed into heat, noise, and mechanical motion. It is changed in a more subtle way, too. The highly organized system in which long hydrocarbon molecules are separated from the oxygen-rich atmosphere is transformed into a more disorganized form in which small molecules of carbon dioxide and water vapor spew aimlessly into the atmosphere. Eventually the lawnmower's noise and the mechanical motion stop, their energy transformed into heat and disorganization, or randomness. The heat itself dissipates (spreads) into the air. *Entropy* is the technical name for this randomness. The first law of thermodynamics has also been stated in more familiar ways: There is no such thing as a free lunch, or you can't get something for nothing.

The second law of thermodynamics: Entropy always increases. In a closed system, matter becomes increasingly disorganized and heat becomes uniformly distributed. Another way of stating the second law is to say that *a closed system changes only to a lower energy state.* This way of expressing it may seem more intuitive to us, but "lower energy state" means simply a state in which more of the energy is in the form of entropy and random, disorganized motion and less is in other forms. This may seem to be a technicality, but it has extremely far-reaching implications. Every conversion of energy from one form (such as chemical energy in gasoline and oxygen) to another form (such as mechanical motion) takes place *only* if the second state of the system is a lower energy state; this in turn means that some of the energy must be lost as entropy and random motion (the dissipation of heat). *No transfer of energy is 100% efficient.* If it were, the transfer wouldn't take place. The first two laws of thermodynamics, then, are Not only is there no free lunch, but You can't even break even.

Heat is a form of energy, of course. Heat can be used to generate mechanical work or other forms of energy, but *only* while the closed system is divided into regions of different temperatures. The ability to transform heat into other forms of energy, then, depends on the organization of the system. Once heat is uniformly distributed, it is unavailable for further energy transformations. Any energy that is available for further transformations is called the *free energy* of the system. The second law says that the free energy of a system always decreases.

The third law of thermodynamics states that there is no entropy in a system with a temperature of absolute zero. It's a technical statement that allows physical chemists to quantify entropy in terms of absolute temperature and specific heat. But that's of no great interest to us here.

The next and last law isn't called the fourth law, but instead, *the zeroth law of thermodynamics.* It says simply that if A and B are the same temperature, and B and C are the same temperature, then A and C are the same temperature. If that weren't true, none of the other laws would be true; perhaps that's why the zeroth law, once someone thought of it, was moved up ahead of the first one.

---

drion or outside the plasma membrane of the bacterium, and a corresponding increase in hydroxyl ions (forming a high, alkaline pH) builds up inside. Thus, the concentration differentials are reversed (Figure 5.8). Again, small stalked particles, here simply called *F1 particles,* are attached to the membranes. However, in bacteria and mitochondria, the particles are inside the membrane and point inward—the exact opposite of what we see in chloroplast thylakoids—and therefore any ATP that is generated is deposited inside the membrane. It seems, then, as far as chemiosmosis is concerned, mitochondria and thylakoid disks appear to be inside-out relative to each other. Another difference is that in the mitochondria, only

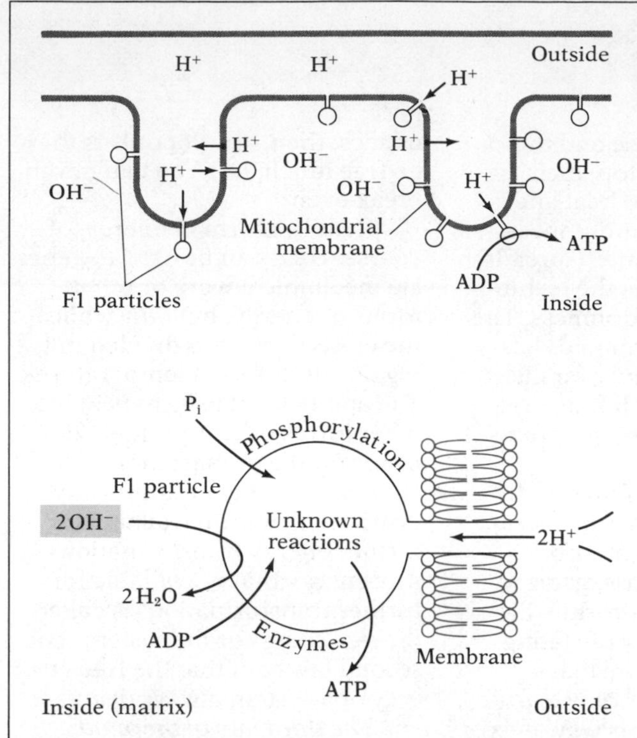

5.8 The chemiosmotic system in the mitochondrion is similar in principle to that of the chloroplast, but the accumulation of H^+ is spatially reversed. In the mitochondrion, hydrogen ions are pumped outside the central cavity (stroma). Coupled phosphorylation of one ATP occurs as two H^+ reenter the stroma through pores leading to the F1 particle. Compare this with Figures 5.6 and 5.7.

two hydrogen ions have to cross the membrane to produce one molecule of ATP.

So to summarize, the first half of the chemiosmotic system involves energetically expensive active transport to create the gradient. In the second half of the system, the ions are allowed to run freely with the gradient. As they pass through these peculiar stalked particles, part of their free energy is tapped to make ATP. It's rather as if the CF1 and F1 particles charged a toll for the privilege of crossing through them past the membrane.

The chemiosmotic hydrogen ion differential can be used for other things than ATP production. For example, mitochondria use the resulting gradient as an energy source for their active transport needs, among other things (ADP and phosphate are pumped into their inner compartments while ATP is pumped out). Bacteria also fuel some of their active transport systems with the hydrogen ion differential, and rotating bacterial flagella derive their energy directly from the flow of chemiosmotic hydrogen ions.

Before the chemiosmotic phosphorylation hypothesis was widely acepted, biochemists thought that much more specific coordinated reactions were responsible for the production of ATP in chloroplasts and mitochondria. There was a great deal of scientific disagreement over the exact number of ATP molecules produced by each reaction in photosynthesis or in oxidative respiration. One of the implications of chemiosmotic phosphorylation is that we should not expect to find an exact number of ATPs produced for each step in the overall process.

Oxidation and Reduction

If you borrow some iron nails and then leave them out in the back yard, the iron becomes oxidized—it rusts, the metallic iron combines with molecular oxygen: You should then apologize to the owner and say:

$$4\,Fe + 3\,O_2 \rightarrow 2\,Fe_2O_3$$

Similarly, if you burn a piece of paper, it too becomes oxidized, but much more rapidly. The cellulose of the paper combines with atmospheric oxygen:

$$\underset{\text{Cellulose}}{[CH_2O]_n} + n(O_2) \rightarrow n\underset{\text{Carbon dioxide}}{(CO_2)} + n(H_2O) + \text{Energy}$$

One might therefore conclude that *oxidation* means "combining with oxygen," but one would be wrong. To biochemists, *oxidation* means "the removal of electrons or hydrogen." That may not make much sense just yet, so we'd better go on.

In the case of the rusting nail, two things are happening. Iron is changing from its metallic state to its ferric (Fe^{3+}) state, and oxygen is changing from its molecular state to its charged state:

(1) $$4\,Fe \rightarrow 4\,Fe^{3+} + 12\,e^-$$

(2) $$3\,O_2 + 12\,e^- \rightarrow 6\,O^{2-}$$

where e^- stands for an electron. In other words, electrons are removed from iron and added to oxygen. And, another way of saying this is that iron is oxidized. As it happens here, oxygen is the electron acceptor, but that doesn't matter; any electron acceptor would do. For instance, what happens when iron filings are added to hydrochloric acid?

$$\underset{\text{Iron}}{2\,Fe} + \underbrace{6\,H^+ + 6\,Cl^-}_{\text{Hydrochloric acid (ionized)}} \rightarrow \underset{\text{Iron ion}}{2\,Fe^{3+}} + \underset{\text{Chloride ion}}{6\,Cl^-} + \underset{\text{Hydrogen}}{3\,H_2}$$

In this case, iron is again oxidized, but here, hydrogen is the electron acceptor (the chloride ions

Essay 5.2 Peter Mitchell and the Overthrow of a Textbook-enshrined Scientific Dogma

A new scientific truth does not triumph by convincing its opponents and making them see the light, but rather because its opponents eventually die, and a new generation grows up that is familiar with it.

MAX PLANCK
Scientific Autobiography

A whole generation was reared on a number of accepted "basic facts" about the mitochondrial electron transport system. Many of the electron carriers could be isolated and characterized biochemically. By assuming differences in their abilities to gain and lose electrons, we were able to visualize a kind of biochemical bucket brigade, passing electrons (and/or hydrogens) from one carrier to the next. "It does not matter whether the electron pairs are transported as naked electrons or as hydrogens," said the textbooks, "because the hydrogens readily dissociate into hydrogen ions and electrons, or reassociate as hydrogens again, and there are plenty of hydrogen ions in the medium." Biochemists worked out the individual biochemical reactions of each step. Unfortunately, biochemists traditionally worked with reacting molecules in free solution. And they ran into a few problems.

For example, although carriers floating free in solution (no longer bound to the mitochondrial membrane) would readily perform the individual transfers, no ATP would be made. For ATP production the mitochondrion must remain intact. Most scientists shrugged their shoulders at this inconvenient and mysterious fact, and continued to work out the details of the carrier-to-carrier reactions. Furthermore, some ingenious experiments seemed to indicate just where the ATP was generated in the intact electron transport chain. The experiments involved poisoning specific reactions (in intact mitochondria) with enzyme inhibitors, and noting how ATP production was affected.

The idea that the individual reactions were involved in "coupled phosphorylations" directly involving both ATP and specific enzyme carriers was never thought to be a hypothesis; it was considered an accomplished fact of science. And it had a certain strong appeal to the scientific (or human) love of precision. The energy relationships were just right; and with three sites of coupled phosphorylation along the electron transport pathway, it was readily calculated that every molecule of reduced nicotinamide adenine dinucleotide ($NADH_2$) that was oxidized yielded exactly three ATPs from ADP. The total energy yield of the oxidation of one molecule of glucose—including glycolysis, the citric acid cycle (see Chapter 6), and the electron transport pathway—was precisely 36 ATPs. "Energy yield budgets" based on this analysis filled even basic biology textbooks with rather precise numbers for students to memorize. The electron pathway of photosynthesis was similarly dissected.

But then in 1961, Peter Mitchell (at the Glynn Research Laboratories in England) presented a radically different idea, which he called the *chemiosmosis hypothesis.* He suggested that ATP synthesis in the mitochondrion depends on a hydrogen ion gradient across the mitochondrial membrane and that the electron transport system operates according to the physical and spatial relationships of its elements in the membrane. Mitchell challenged the notion that hydrogens and electrons were freely interconvertible. Indeed, he saw the difference in free electrons and hydrogen-bound electrons as crucial to his new idea. Mitchell hypothesized that the key steps of "coupled ATP synthesis" were in fact the steps at which hydrogen ions were pumped across the mitochondrial membrane. These electron transport reactions were involved with ATP formation, all right, but he saw the coupling as being indirect and imprecise.

This holistic way of looking at the problem was quite new and seemed bizarre to biochemists steeped in the established tradition of analyzing isolated chemical reactions of enzymes in artificial solution. Actually, Mitchell himself had originally only been interested in what had seemed to be a different problem, namely the active transport of hydrogen ions across membranes using ATP as an energy source. But then it occurred to him that the H^+ ion pump reaction might be reversible, and that ATP could be generated by a flow of hydrogen ions *back* across the membrane.

At first there was not much of a controversy; Mitchell was simply not taken very seriously. But he was more than just a theoretician. He also knew his way around the laboratory, and he had formed a very fortunate relationship with a particularly gifted colleague, Jennifer Moyle. Together they set out to do the experiments that would test the hypothesis. Over a period of 15 years they built up a body of experimental evidence that made the chemiosmotic interpretation increasingly plausible and the traditional isolated molecular "coupled phosphorylation" idea increasingly unlikely. Along the way, it became apparent that the chemiosmotic hypothesis could explain photosynthetic generation of ATP in chloroplasts as well as oxidative generation of ATP in mitochondria.

Unlike most scientists, Mitchell and Moyle now work at home—specifically, they do their work in a modern research laboratory that Mitchell built in his own home, which happens to be a medieval castle in Cornwall. The aristocratic Mitchell supports his laboratory through the unlikely combination of government research grants and proceeds from his dairy herd.

Other investigators became interested in Mitchell's idea and contributed to its scientific validation. A recent experiment strongly supported the chemiosmosis hypothesis. It consisted of isolating small, closed vessels made up of inner mitochondrial membrane. In the process of disrupting the mitochondria and isolating the vessels, the membranes were turned inside out, with the F1 particles on the outside—just as the CF1 particles are on the outside in living chloroplast membranes (thylakoids). The small vessels were soaked in an acid solution until their contents became acid. Then they were quickly separated and plunged into a basic medium. As the hydrogen ions passed back out through the tiny, isolated membranes, ATP synthesis could be measured. The experiment was repeated after the membrane vessels had been treated in a way that stripped off the F1 particles (as could be verified in electron micrographs). This time no ATP was generated.

Scientists may often be hostile to heretical new ideas. Perhaps it is just as well; otherwise, science would be constantly inundated by half-baked notions. But scientists are generally receptive to well-done research, and the chemiosmosis hypothesis has gradually won increasing acceptance. There are now very few biochemists who would not agree that it seems to be the most plausible explanation of mitochondrial and chloroplast function so far. Peter Mitchell was awarded the Nobel Prize in chemistry in 1978.

---

take no part in the reaction). The iron, then, loses electrons and is thus oxidized; the electron acceptor is said to be *reduced,* which is the opposite of being oxidized. In this reaction, hydrogen is reduced. In the previous reactions, oxygen was reduced. It works out so that whenever one substance is oxidized, another must be simultaneously reduced; all these reactions are *oxidation–reduction* reactions (Figure 5.9).

Now we must introduce a wrinkle in oxidation–reduction reactions that often bothers students and has also misled sophisticated biochemists. It is this: Oxidation–reduction reactions do not take place only by the subtraction or addition of electrons. The addition of entire hydrogen atoms is also reduction and the removal of entire hydrogen atoms is also oxidation. In other words, hydrogen atoms (H) and hydrogen molecules (H_2) are the full equivalent of electrons when it comes to oxidation–reduction reactions. The rationale, at least in aqueous systems, is that there are always plenty of hydrogen ions available, and electrons (e^-) and hydrogen ions (H^+) can freely associate and dissociate:

$$2\,e^- + 2\,H^+ \rightleftharpoons H_2$$

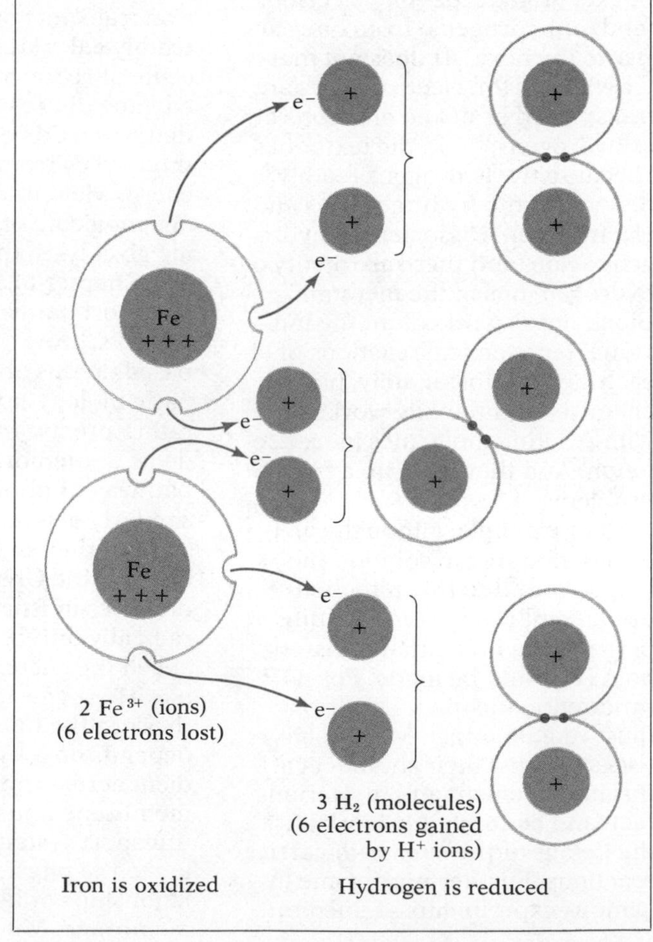

5.9 An inorganic oxidation–reduction reaction. In the oxidation of iron by hydrochloric acid, each iron atom loses three electrons to the six hydrogen ions of the acid. The hydrogen ions are thus reduced, since they have received electrons. (The chloride ions do not become involved in this reaction.) Oxidation can be defined as the loss of electrons and reduction as the gain. Oxidation and reduction, always coupled processes, are an integral part of the chemistry of photosynthesis and respiration.

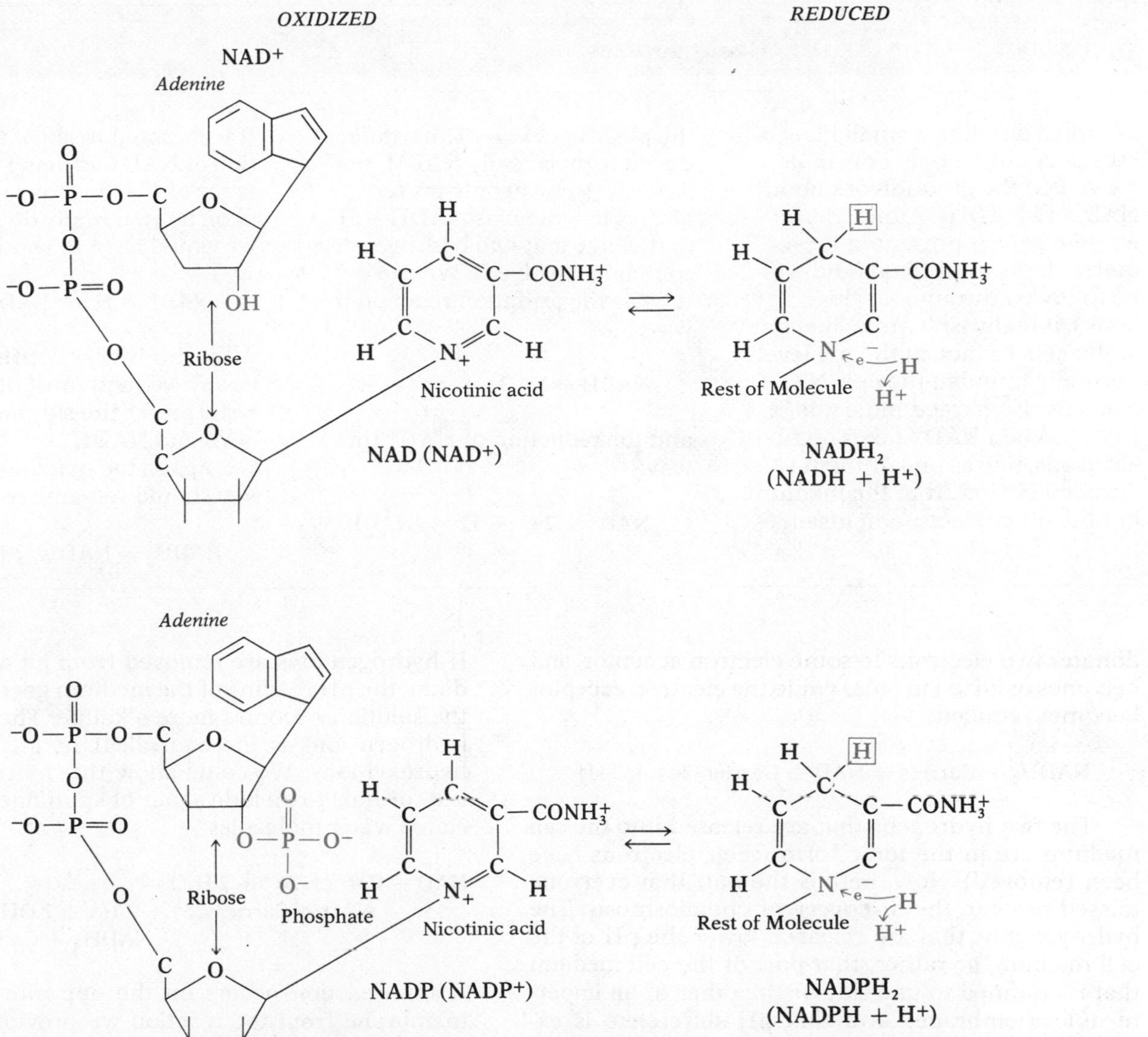

5.10 The electron and hydrogen carriers NAD and NADP. These molecules are nearly identical, except that NADP includes a phosphate group attached to the number 2 carbon of ribose. The active group of both NAD and NADP is nicotinic acid. Both the oxidized forms (NAD and NADP) and the reduced forms ($NADH_2$ and $NADPH_2$) are shown. Two hydrogen atoms are involved, with one proton and two electrons joining the nicotinic acid ring. The second proton is released into the aqueous medium.

NAD, $NADH_2$, NADP, and $NADPH_2$: Important Soluble Electron Carriers of the Cell

Many oxidation–reduction reactions in cells are mediated by specific electron carriers. Notable among these are two pairs: NAD (nicotinamide adenine dinucleotide) and $NADH_2$ (one pair); and NADP (nicotinamide adenine dinucleotide phosphate) and $NADPH_2$ (the other pair). In each pair, the first molecule is the oxidized form and the second molecule is the reduced form. NAD and NADP differ only by one phosphate group, but they participate in quite different cellular reactions. As we'll see, NAD is used more in oxidative respiration, and NADP is used more in photosynthesis. The structures of the two pairs of electron carriers are shown in Figure 5.10. Both molecules contain our old friend, adenine. They also contain nicotinic acid, which is derived from vitamin B_6 and not, as you might have supposed, from tobacco smoke. The behavior of NAD and NADP is similar, so let's follow one, NAD, through its reactions.

The oxidation of $NADH_2$ to NAD can be summarized as follows:

$$NADH_2 \rightleftharpoons NAD + 2\,H^+ + 2e^-$$

This is what chemists call a *half-reaction,* because, in fact, the electrons have to go somewhere—namely, to a different electron carrier. In other words, $NADH_2$

Essay 5.3 How $NADH_2$ Really Ionizes

We must admit to a small inaccuracy, a common one, now that we've laid the groundwork about NAD and $NADH_2$. Although the way we have represented these molecules is traditional (and will be followed throughout this book), it really isn't quite like that in the cell. In fact, at the pH levels normally found in the cell, NAD is a positively charged molecule, NAD^+. When NAD^+ takes on two electrons, it uses one of them to attract a H^+ ion from the medium, but the other electron neutralizes the positive charge. Thus, the reduced form is really NADH, not $NADH_2$ (some people prefer always to write it as $NADH + H^+$ to indicate that two hydrogens are somehow involved). We can rewrite the oxidation reaction this way:

$$NADH \rightleftharpoons NAD^+ + 2\,e^- + H^+$$

and the reduction of NAD^+ this way:

$$NAD^+ + 2\,e^- + H^+ \rightleftharpoons NADH$$

If we wanted to show the reduction of NAD^+ as being the acceptance of two hydrogen atoms (which often really does happen), we would have to show it like this:

$$NAD^+ + H_2 \rightleftharpoons NADH + H^+$$

All of this is very confusing, which is why we (and most other biologists) prefer the simple fiction of NAD and $NADH_2$.

Again, the oxidation reaction you should remember is,

$$NADH_2 \rightleftharpoons NAD + 2\,H^+ + 2\,e^-$$

---

donates two electrons to some electron acceptor, and becomes oxidized to NAD while the electron acceptor becomes reduced:

$$NADH_2 + \text{Carrier} \rightleftharpoons NAD + \text{Carrier}\cdot 2e^- + 2\,H^+$$

The two hydrogens that are released into the cell medium are in the ionic form (their electrons have been removed). Now, here is the part that everyone missed prior to the discovery of chemiosmosis: The hydrogen ions that are released lower the pH of the cell medium, or rather, that part of the cell medium that is confined to one side or the other of an impermeable membrane. And this pH difference is extremely important in chemiosmotic phosphorylation, as we have seen.

In the reverse reaction, NAD accepts two electrons from an electron donor and two hydrogen ions from dissociated water in the medium:

$$NAD + \text{Carrier}\cdot 2e^- + 2\,H^+ \rightarrow NADH_2 + \text{Carrier}$$

If hydrogen ions are removed from an aqueous medium, the pH reading of the medium goes up—that is, the solution becomes more alkaline. The removal of hydrogen ions is the equivalent of the addition of hydroxyl ions. We could show this by rewriting the last formula to include a pair of spontaneously dissociated water molecules:

$$NAD + \text{Carrier}\cdot 2e^- + 2\,H_2O \rightarrow$$
$$NAD + \text{Carrier}\cdot 2e^- + 2\,H^+ + 2\,OH^- \rightarrow$$
$$NADH_2 + \text{Carrier} + 2\,OH^-$$

If this reaction occurs on the opposite side of the membrane from the reaction we previously considered, the pH gradient across the membrane is again increased. And that pH gradient, as we have seen, can be used to generate ATP. Part of the energy released in either of these oxidation–reduction reactions, then, can be captured in the high-energy bonds of ATP.

5.2 Light and the Light Reactions

Energy Enters the Ecosystem

Apart from a few rare bacterial *chemotrophs* (Chapter 14) that live on the energy of inorganic chemicals from the bowels of the earth, all energy entering the *biosphere* (the realm of life) comes in as sunlight and is captured in the process of photosynthesis. Even the energy in fossil fuels comes from photosynthesis. No number of superlatives or platitudes can fully describe the significance of photosynthesis to life on earth; as we have stressed, it's our only edge on entropy. Knowledge of photosynthesis and its obvious applications to agriculture is increasingly critical to humanity—if the earth is to continue to support our burgeoning population.

For our purposes, photosynthesis is an appropriate place to begin the study of the pathways of energy. We'll first consider how energy is stored and then how it is used by living things. We'll start with some important preliminary concepts from the physical sciences, including a brief review of a few ideas from chemistry and one or two from physics.

The Physics of Sunlight

This small planet is bathed in radiation from other celestial bodies. Part of that radiation reaches us as visible light. Visible light is part of a continuum known as the *electromagnetic spectrum.* The entire spectrum includes (in ascending order of energy) radiowaves, microwaves, infrared radiation, visible light, ultraviolet radiation, X-rays, and gamma rays (page 96). The spectrum includes very long waves at the low-energy end and extremely short waves at the high-energy end. Visible light is in the middle. Visible light is interesting to biologists because it is capable of being absorbed by molecules in such a way as to bring about changes in energy levels in organic molecules. In fact, visible light is visible because it interacts with visual pigments in our eyes, and it is of interest to us here because it interacts with chlorophyll and other pigments of photosynthesis. Less energetic waves are useless because they are dissipated into thermal energy, and more energetic waves are too powerful for life forms to utilize since they tend to disrupt molecules. In other words, infrared warms us and ultraviolet burns us, and we can see what's in between. Of course, there are no really sharp dividing lines; both reds and violets become dully visible as they enter our range of sensitivity. Also, different wavelengths are interesting to different organisms. For instance, bees can see ultraviolet radiation but not red.

Light actually takes the form of tiny particles or packets of waves. These particles are called *photons,* and the amount of energy per photon is related to its wavelength. The more energy, the shorter the wavelength. Light of different wavelengths interacts differently with color-sensitive cells in our eyes and thus we are able to perceive colors. The colors of the rainbow, or visible spectrum, are related to their wavelengths and energy according to Table 5.1.

Table 5.1

Color (subjective)	Wavelength (nm)	Energy (cal/einstein[a])
Deep red	750	37,800
Red	650	43,480
Yellow	590	48,060
Green	540	52,970
Blue	490	57,880
Violet	430	71,800

[a]One einstein is a mole of photons (6×10^{23} photons).

The Absorption Spectrum

Photosynthesis in plants occurs in the chloroplasts, and the first step is the absorption of a photon of light by a molecule of *chlorophyll.* Chlorophyll absorbs light of many different wavelengths, but it absorbs some wavelengths more readily than others. If we measure the absorption of light by chlorophyll as a function of the wavelength, we obtain what is called an *absorption spectrum.* The absorption spectrum of the type of chlorophyll called *chlorophyll a* is given in Figure 5.11(c). As you can see, there are two peaks of light absorption, with a trough—a relatively insensitive area—between. The first peak is in the violet, shorter wavelength, region close to the upper limit of visual sensitivity (about 450 nm). The second is in the longer wavelength, red end of the spectrum (about 675 nm). Green light passes right through chlorophyll and is reflected from surrounding tissues, which is why the pigment itself looks green to us.

For photosynthesis to occur, chlorophyll *a* has to be *activated,* or raised to a higher energy state. There are two ways in which this can occur. One is that chlorophyll *a* can intercept a photon of light directly. The other way is for a photon of light energy to be captured by some other, *accessory* pigment, which then transfers its energy to chlorophyll *a.* The collection of accessory pigments is said to serve as an *antenna,* like a television antenna, in that it intercepts radiant energy and passes it on. One such pigment is *chlorophyll b,* which is molecularly almost identical to chlorophyll *a* and has a very similar absorption spectrum as seen in Figure 5.11(c).

A variety of other pigments, related to each other but structurally unrelated to the chlorophylls, are the *carotenoids.* The carotenoids absorb green and blue light, transmitting red or orange. This orange color makes their name easy to remember, because the carotenoids are named after a vegetable in which they occur in high concentration, namely the carrot (Figure 5.12). Actually, Figure 5.12 is not a carrot, but the absorption spectrum of a mixture of carotenoids.

Somewhat similar to the absorption spectrum, but calculated in an entirely different way, is the *action spectrum* (Figure 5.13). The absorption spectrum indicates which wavelengths of light are *absorbed,* but we don't know whether this absorbed light is doing any *work.* So the action spectrum is a measure of the amount of photosynthesis actually accomplished as a function of wavelength. The data graphed in Figure 5.13 were obtained from measurements of oxygen production by plants grown in light of different wavelengths. Oxygen production is directly proportional to the intensity of photosynthetic activity. Even though carotenoids were present in the chloroplasts of these plants, the ones grown in green

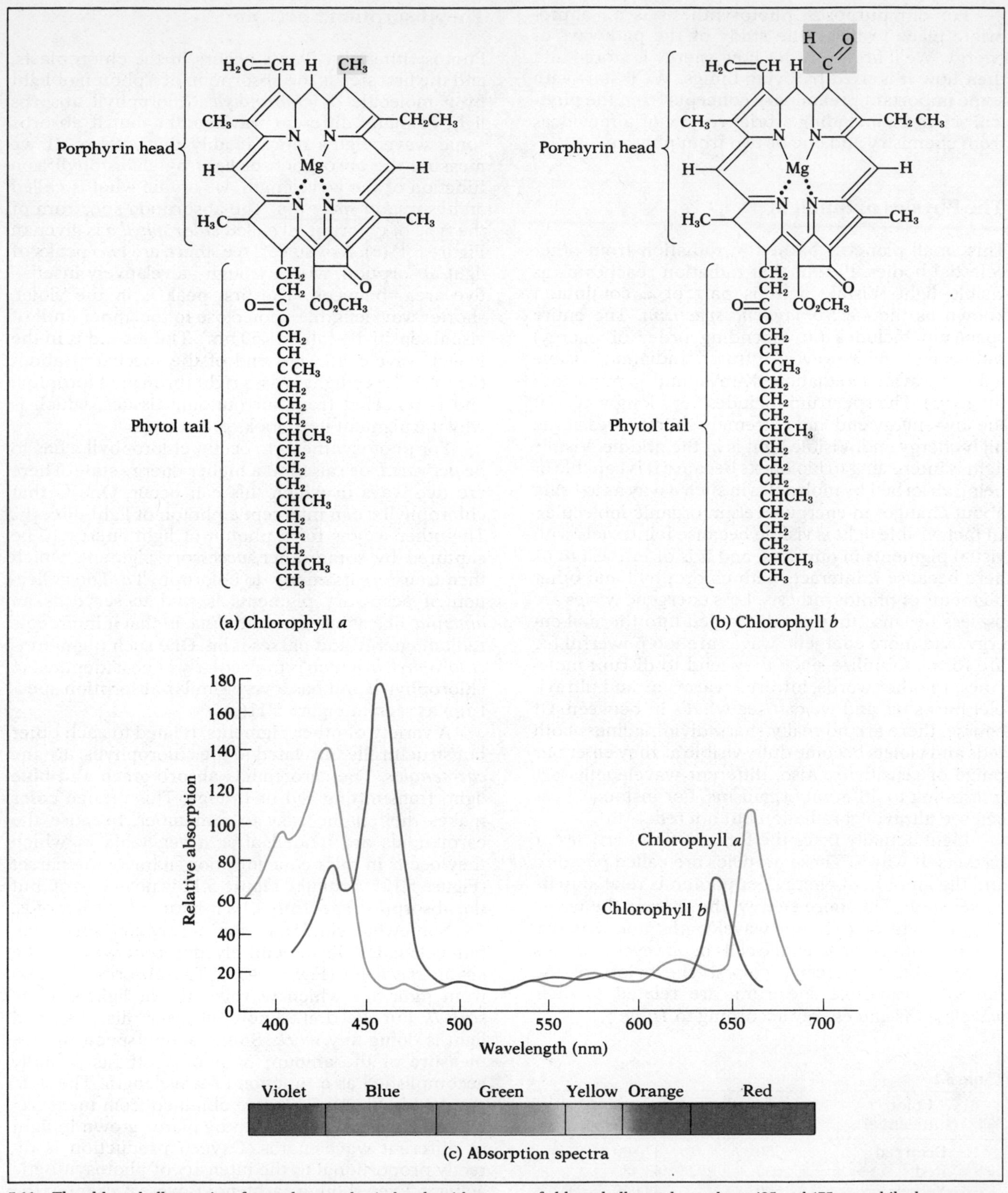

5.11 The chlorophylls consist of complex porphyrin heads with a lengthy hydrocarbon or phytol tail. Chlorophyll *a* differs only slightly (shaded area) from chlorophyll *b*. The absorption spectrum of chlorophyll *a* peaks at about 425 and 675 nm, while the spectrum of chlorophyll *b* peaks at about 460 and 660 nm. Absorption peaks represent energy absorbed and used in photosynthesis.

5.12 Two common carotenoids, β-carotene and lutein, are shown here. The carotenoids are accessory pigments to the chlorophylls. They absorb light only in the blue-green range. Their function is believed to be that of a light-absorbing antenna, shunting their energy into more active regions of chlorophyll.

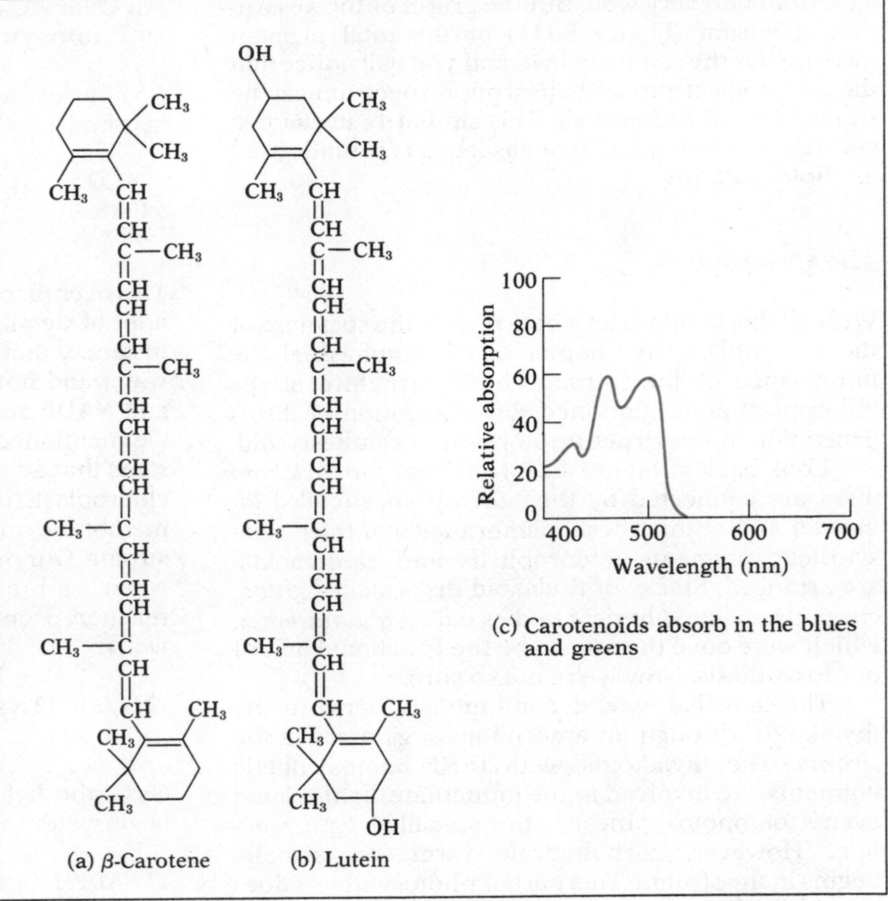

(c) Carotenoids absorb in the blues and greens

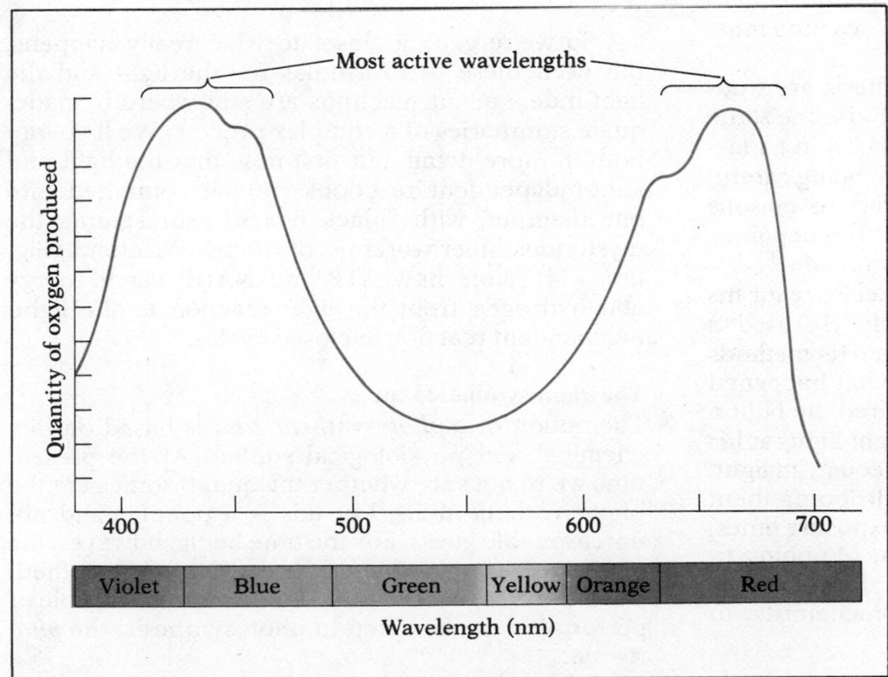

5.13 This photosynthetic action spectrum was constructed from measured volumes of oxygen produced by plants grown under different colored lights. Oxygen production is proportional to the intensity of photosynthetic activity. The two peaks indicate that plants grown in the violet-blue range and those grown in the red do best. Notice that when the action spectrum graph is superimposed over the absorption (Figure 5.11c) spectrum graph the fit is rather good (indicating what?).

light didn't do very well. But the graph of the absorption spectrum (Figure 5.11c) of the total pigment mixture has the same dip in it, and you will notice that the action spectrum and absorption spectrum can be superimposed rather well. This similarity in the two curves verifies that the light absorbed is actually used in photosynthesis.

The Chloroplast

With all this in mind, let's now review the structure of the chloroplast. In Chapter 4, we emphasized the importance of the intricate inner structure of the chloroplast and mentioned that disruption or disorganization of this structure stops photosynthesis cold.

Look back at Figure 4.23. Note that the *thylakoid* disks are connected by their membranes (called *lamellae*). It is within these membranes that the photosynthetic pigments (chlorophylls and carotenoids) are arranged. Stacks of thylakoid disks, called *grana,* contain smaller spherical bodies called *quantasomes,* which were once thought to be the functional unit of photosynthesis. Now we're not so sure.

The lamellae extend continuously between the thylakoids, through an amorphous region called the *stroma.* The thylakoids, with their photosynthetic pigments, are involved in the immediate, light-related events of photosyntheses—the so-called *light reaction.* However, carbohydrate formation actually begins in the stroma. This part of photosynthesis does not involve light directly and so is traditionally termed the *dark reaction.* It is also sometimes called the *light-independent reaction,* since some people tend to get confused and assume that the dark reaction must occur at night.

When the pathways of photosynthesis are studied, the usual procedure is to carefully isolate some chloroplasts (usually spinach chloroplasts) in an ice-cold, isotonic, pH-buffered suspension, being careful not to disrupt their structure. Then various poisons are added that inhibit the pathway at different points, and the intermediate compounds are studied.

Melvin Calvin, who received a Nobel Prize for his work, employed a different approach. He used a unicellular, chloroplast-rich green alga. His methods were not gentle. In order to find out what happened immediately after photons were captured, he built a device that would flash an array of bright lights at his ice-cold algae, bathing them instantaneously in light. Then he quickly killed the algae by dropping them into hot acetone. Using different light exposure times, he was then able to collect the products of photosynthesis at steps along the way. Through this and other work we have been able to relate the biochemistry to chloroplast structure.

An Overview of the Chemistry of Photosynthesis

So now let's look at the traditional formula for photosynthesis:

$$\underset{\text{Carbon dioxide}}{6\,CO_2} + 12\,H_2O + \text{Light} \rightarrow \underset{\text{Glucose}}{C_6H_{12}O_6} + 6\,H_2O + 6\,O_2$$

This overall formula, which is relatively ancient, hides a lot of detail. There's no telling by looking at it, for instance, that the liberated oxygen comes from the water and not from the carbon dioxide, or that ATP and NADP are intimately involved in the process. As we mentioned earlier, photosynthesis occurs in two steps that actually take place in different parts of the chloroplast: the light reaction within the thylakoid membrane and the light-independent reaction in the stroma. Our overall formula becomes much less exact when we break it into its two parts. First, the light reaction occurs, which begins with the *photolysis of water:*

$$12\,H_2O + 12\,\text{NADP} + \text{(approx) } 18\,\text{ADP} + 18\,P_i \rightarrow 6\,O_2 + 12\,\text{NADPH}_2 + \text{(approx) } 18\,\text{ATP}$$

Next, the light-independent reaction occurs, which begins with the *reduction of carbon dioxide:*

$$12\,\text{NADPH}_2 + 18\,\text{ATP} + 6\,CO_2 \rightarrow \underset{\text{Glucose}}{C_6H_{12}O_6} + 6\,H_2O + 12\,\text{NADP} + 18\,\text{ADP} + 18\,P_i$$

So we're getting closer to what really happens, but even these two formulas for the light and the light-independent reactions are still woefully inadequate summaries of a complex process. We'll go into both in more detail, but first note that the light and light-independent reactions can be combined into one diagram, with "black boxes" representing the mysterious inner workings of the two reactions (Figure 5.14). Note how ATP and NADP carry energy and hydrogen from the light reaction to the light-independent reaction in closed cycles.

The Photosynthetic Unit

The notion of a *photosynthetic unit* is based on biochemical and physiological studies. At the present time we're not sure whether the quantasomes are the photosynthetic units, but this is a popular and not unreasonable guess. For the time being, however, the photosynthetic unit has to be considered hypothetical. If it exists, it is the physical unit that is capable of performing the first step in photosynthesis, the *photoevent.*

In the photoevent (the light-dependent reaction), photons are absorbed and water is split into oxygen and hydrogen or, equivalently, into oxygen, hydrogen ions, and electrons. Oxygen is released as O_2; thus, the

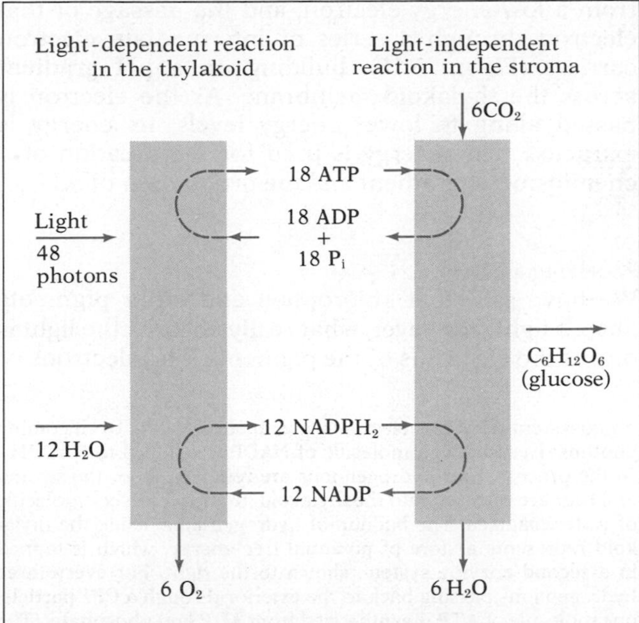

5.14 A "black box" view of photosynthesis. The basic formula for photosynthesis indicates the reactants and products in this complex process, but it doesn't reveal the mechanisms. Here, the reactants and products are arranged with regard to the two major mechanisms, the light-dependent and light-independent reactions. Biologists utilizing the skills and technology of physical scientists have been at work unraveling the mysteries in these "black boxes" for many long years.

photoevent is the capture of light and the splitting of *one* molecule of water into *two* hydrogen ions, *two* electrons, and one-half of an oxygen molecule. This fundamental chemical reaction of photosynthesis, you will note, involves no carbon compounds. It is called the *photolysis of water,* which means the breaking apart of water by light. The hydrogens and electrons that are liberated are later passed on to the light-independent reaction, and the oxygen is released into the air as a waste product.

Recent research indicates that each photosynthetic unit contains many pigment molecules. Most of the pigments act as *light antennae.* Their function is to absorb light energy and to transfer it from one to another, bucket-brigade fashion, until the energy reaches *reaction centers,* where the rest of the photoevent occurs.

We should add that each photosynthetic unit contains at least two reaction centers. In addition, the unit has an *electron transport system,* a series of carrier molecules that pass electrons from one to another through a series of reactions and eventually through the thylakoid membrane (Figure 5.15).

Two Photosystems

A photosynthetic unit contains two different *photosystems.* Each photosystem contains a slightly different kind of antenna and reaction center. One photosystem, for example, has an antenna rich in carotenols (a variety of carotene) and chlorophylls *a* and *b*, with chlorophyll *b* in greater abundance. The absorption spectrum of this photosystem peaks at 680 nm, and we can call it photosystem 680 (P680). The other photosystem has an antenna consisting of carotenes as well as both kinds of chlorophyll, and has an absorption peak at 700 nm, so we'll call it photosystem 700 (P700).

Both photosystems are necessary to complete a photoevent. Each must capture its own photon in order for one molecule of water to be split. In the transport of electrons, the two photosystems act sequentially, with P680 acting first followed by P700. (Incidentally, in some books the two systems are known as *photosystem II* and *photosystem I,* respectively.)

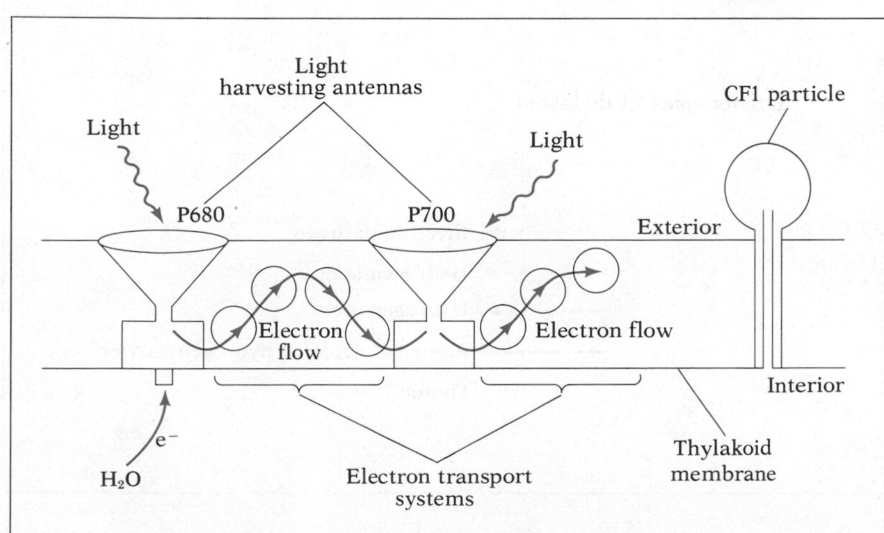

5.15 No one knows what a photosynthetic unit (a photounit) looks like, but biologists think they are getting close. The photounit is the array of pigments, accessories, and carrier molecules necessary to complete a photoevent. This scenario, which is not entirely imaginary, shows what the photounit may look like within the thylakoid membrane. Its elements are pigment systems P680 and P700, each with its light-gathering antenna and electron–hydrogen transport system. The spatial arrangement of the photounit's elements is essential to the functioning of the system. Each part must be in its place from the initial photon capturing stage to the final pumping of hydrogen ions into the stroma.

The Role of Electrons in Energy Transfer

It is no simple task to transfer photon energy into stored chemical energy. It is an uphill fight to turn water and carbon dioxide, which are very stable molecules with no free energy, into carbohydrate and oxygen, which have a lot of free energy in the form of potential chemical energy. It is important to keep in mind that the transformation of energy from light to molecules involves a stepwise transferring of electrons. We shall see that the initial steps of photosynthesis involve the activation of two molecules of chlorophyll, the creation of a high-energy electron from a low-energy electron, and the passage of that electron through a series of intermediate electron carriers (Figure 5.16), building up a pH gradient across the thylakoid membrane. As the electron is passed along to lower energy levels, its energy is extracted. This energy is used for the creation of a chemiosmotic gradient and the production of ATP.

Receiving a Photon

We have said that chlorophyll and other pigments absorb light. However, what really absorbs the light is one of the electrons of the pigments. The electrons of

5.16 Photosynthetic electron transport. The molecules of the photosynthetic light reaction are all embedded in the membranes of the thylakoids. Some face only the interior space of the thylakoid; some face only the exterior, the stroma, of the chloroplast; and some electron carriers extend through the membrane and face both compartments. The interior of the thylakoid serves as a reservoir of hydrochloric acid. A molecule of water is oxidized by protein Z, which releases two hydrogen ions and passes two electrons to photosystem 680. Photons strike the antenna of photosystem 680, exciting electrons which proceed to pass through a series of carrier molecules, each with a lower reducing potential. After entering photosystem 700, the electrons are again excited by incoming photons. Eventually, a molecule of NADP is reduced to $NADPH_2$. In the process, four hydrogen ions are removed from the stroma and four are released into the thylakoid interior for every molecule of water oxidized. The buildup of hydrogen ions inside the thylakoid represents a store of potential free energy, which is tapped in a second enzyme system, shown to the right. For every three hydrogen ions passing back to the exterior through a CF1 particle, one molecule of ATP is synthesized from ADP and phosphate. (The abbreviations are the same as in Figure 5.17.)

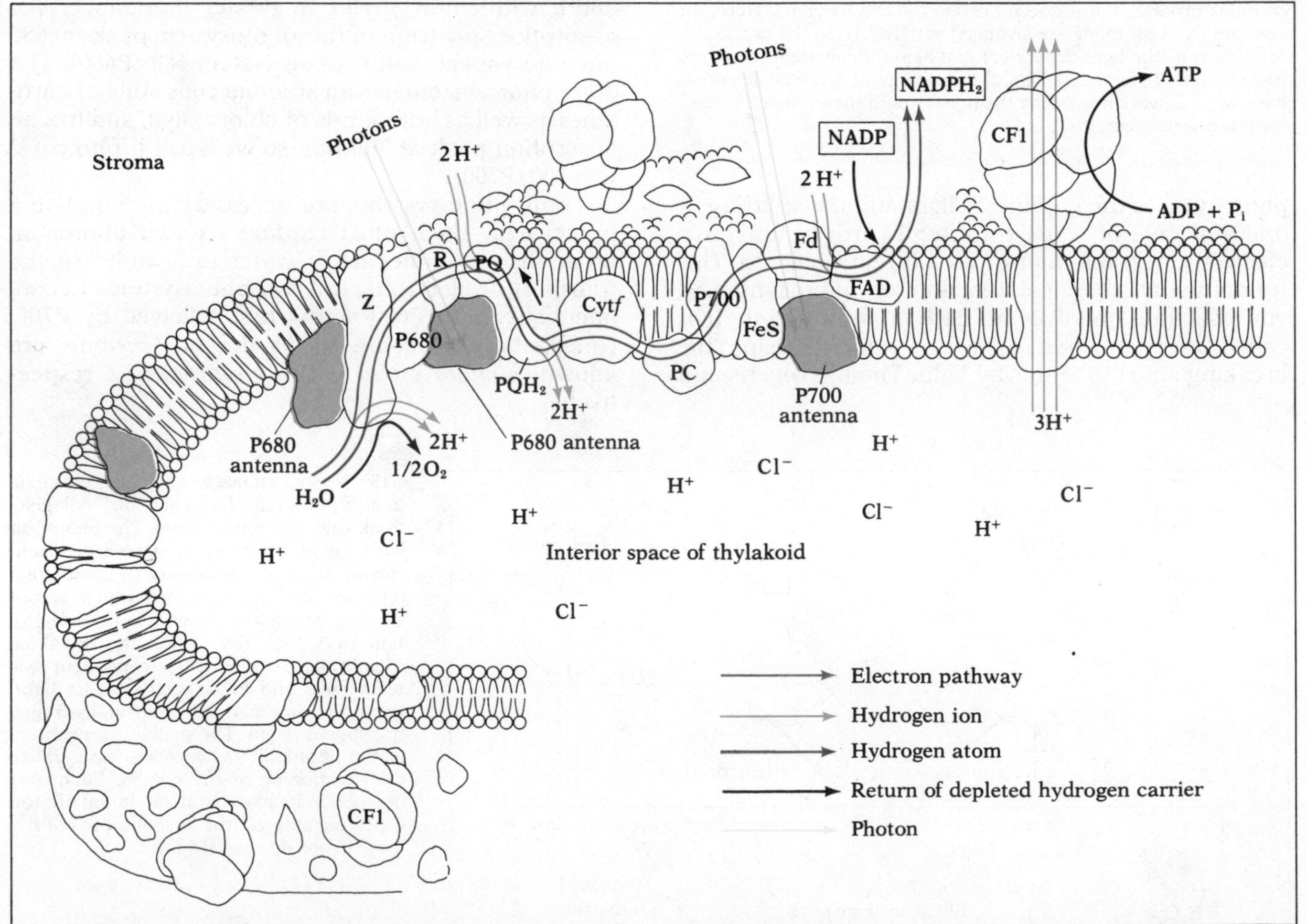

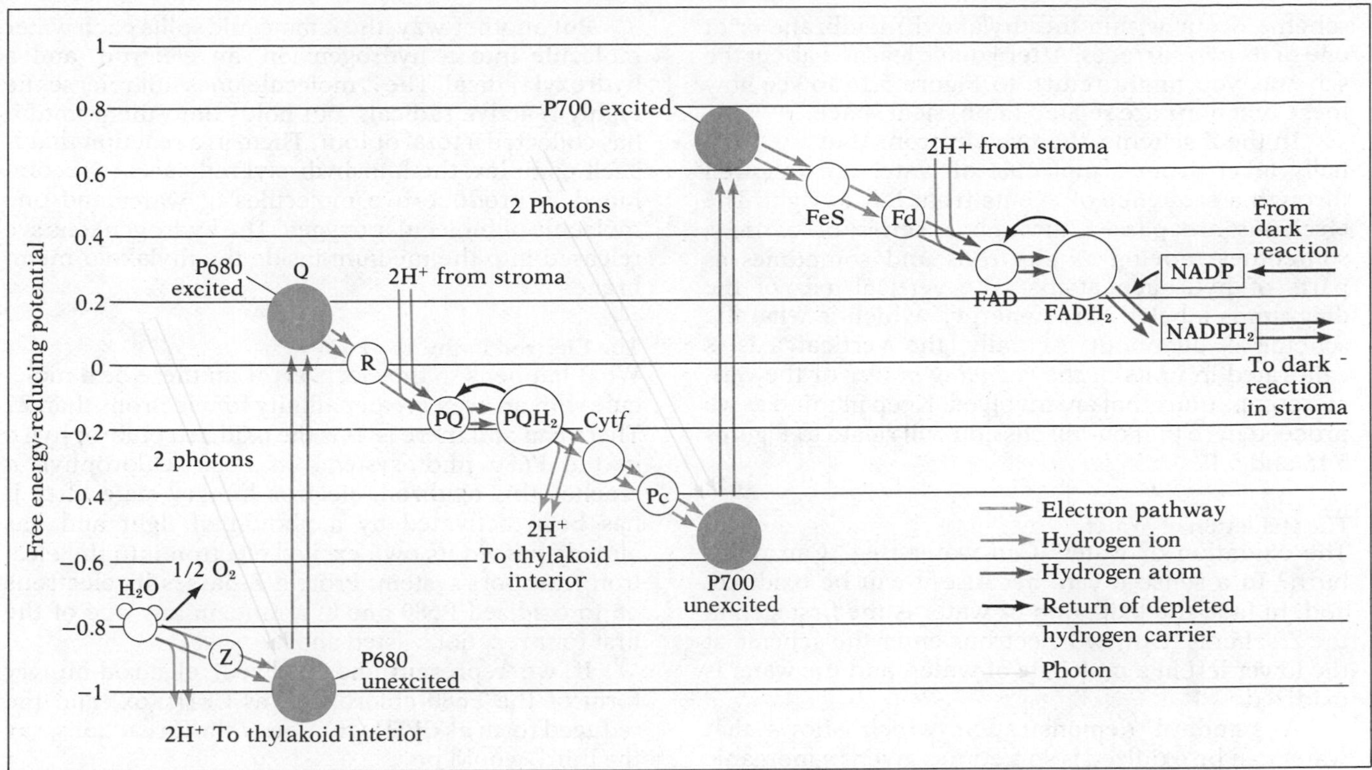

5.17 The Z scheme diagram. In the photolysis of water, two hydrogen ions are released into the interior space of the thylakoid, and two electrons enter the electron transport chain. Here, the paths of the two electrons are followed from left to right through a series of carriers. The vertical dimension represents the reducing potential of the various compounds, which is the equivalent of free energy in a system in which oxygen or oxidized substrates are available. Electrons tend to flow toward the most oxidized (least reduced) compounds. At two points, however, the electrons are boosted by the energy of captured photons to more excited states with increased reducing potential: once in photosystem 680 and once in photosystem 700. Note that a total of four hydrogen ions are released into the interior of the thylakoid, and a total of four hydrogen ions are removed from the stroma of the chloroplast for each molecule of water undergoing photolysis; and at the end of the reaction, two molecules of NADP are reduced to $NADPH_2$. The successive electron carriers, from left to right, are water; protein Z; chlorophyll *a* bound in P680 (which, in its excited state, is called Q); R; plastoquinone (PQ); reduced PQ (PQH_2); cytochrome *f* (cyt *f*); plastocyanin (PC); chlorophyll *a* bound in P700 (which in its excited state is called X); an iron and sulfur protein (FeS); ferredoxin (Fd); an enzyme complexed with flavin adenine dinucleotide (FAD); reduced FAD ($FADH_2$); and nicotinamide adenine dinucleotide phosphate (NADP and $NADPH_2$).

many substances are capable of absorbing light energy. When an electron absorbs a photon, the energy level of that electron is increased by a specific amount, according to the physics of quantum mechanics. There are a number of things that can happen next, some of which are important in photosynthesis and some of which are not. One of the important things is that the energy of excitation can pass from molecule to molecule, without the actual passing of electrons. This is what happens when light energy captured by one of the antenna pigments is passed to the chlorophyll *a* of the photosystem. Another thing that can happen that is important in photosynthesis is that the activated electron can actually leave its molecule, taking its energy of activation with it. This is what happens to the activated electron of the P680 photosystem.

When an electron leaves a molecule, that molecule is said to be oxidized. Of course, the electron has to go somewhere, and the chemical that accepts the electron (the electron acceptor) is then said to be reduced. So light energy activates the photosystem and *greatly increases its reducing power* because the energetic electron is so much more likely to leave the photosystem molecule and reduce something.

The Z Diagram

A *photoevent* is the series of reactions that are involved in the photolysis of a single molecule of water. Therefore, a photoevent can be considered to be the unit event of the light-dependent reaction of photosynthesis. The entire light-dependent reaction following the photolysis of a single water molecule is depicted schematically in Figure 5.17. This particular way of showing things is known as the *Z scheme.* Don't try to relate the Z diagram itself directly to chloroplast structure. All the events depicted in the Z scheme occur within the thylakoid membrane or at one of its two surfaces. After you've learned about the scheme, you might return to Figure 5.16 to see how these reactions are related in physical space.

In the Z scheme, the two electrons that are originally taken from a molecule of water are followed through a sequence of events from left to right. The electrons are passed through a series of carriers, sometimes moving as electrons and sometimes as parts of hydrogen atoms. The vertical axis of the diagram is labeled "Free energy," which is what the scheme is all about. Actually, the vertical axis is calibrated in units of the *reducing power* of the various compounds that are involved. Keep in mind as we proceed here that our discussion will relate to Figures 5.16 and 5.17.

The Oxidation of Water

The oxidation of water? Can water rust? Can water burn? In a sense it can, because it can be oxidized, and, in fact, the oxidation of water is the first step in the Z scheme. Our two electrons enter the scheme at the lower left in a molecule of water, and the water is oxidized.

A standard demonstration which shows that water can be oxidized is sometimes given in inorganic chemistry classes. A bell jar is filled with fluorine gas (F_2). Fluorine has a very strong affinity for electrons, and is a stronger oxidizing agent than oxygen itself. A dish of water is ignited with a spark, and in the fluorine atmosphere the water burns with a steady flame. The reaction removes hydrogens from the water, leaving behind molecular oxygen as the hydrogens combine with the fluorine to form hydrofluoric acid (HF, which incidentally ruins the bell jar). And the removal of hydrogens from a substance is oxidation, remember?

Water is oxidized in photosynthesis not by fluorine, but by a large protein called Z, which has some rather amazing properties. The actual reaction is not nearly as simple as the overall reaction seems to indicate, so let's look at some of the details.

As you may have guessed, "$\frac{1}{2}$ O_2" is another convenient fiction. The initial reaction is not really the splitting of one molecule of water to produce two hydrogens and half a molecule of oxygen. It is really a series of five sequential reactions in which the Z protein reduces four molecules of water to produce one molecule of oxygen, four electrons, four hydrogen ions, and two reconstituted water molecules:

(1) $$H_2O + Z \rightarrow Z\cdot OH + e^- + H^+$$

(2) $$H_2O + Z\cdot OH \rightarrow Z\cdot 2\,OH + e^- + H^+$$

(3) $$H_2O + Z\cdot 2\,OH \rightarrow Z\cdot 3\,OH + e^- + H^+$$

(4) $$H_2O + Z\cdot 3\,OH \rightarrow Z\cdot 4\,OH + e^- + H^+$$

(5) $$Z\cdot 4\,OH \rightarrow Z + 2\,H_2O + O_2$$

$$\text{Net reaction: } 2\,H_2O \rightarrow 4\,H^+ + 4\,e^- + O_2$$

Put another way, the Z molecule splits each water molecule into a hydrogen ion, an electron, and a hydroxyl radical. The Z molecule does not release the highly reactive radicals, but holds onto them until it has collected a total of four. Then, in a reaction that is itself complex, the four hydroxyl radicals are recombined to produce two molecules of water and one molecule of molecular oxygen. The hydrogen ions are released into the medium inside the thylakoid membrane.

The Electron Pathway—P680

What happens to the electrons? Can there be a molecule with an even greater affinity for electrons than Z? There can and there is. It is the oxidized chlorophyll *a* in the P680 photosystem. In P680, chlorophyll *a* reaches this oxidized, electron-hungry state after it has been activated by a photon of light and has already passed its own excited electron into the electron transport system. Protein Z passes its electrons on to oxidized P680 one at a time, in any one of the first four reactions listed above.

If we represent the oxidized, electron-hungry form of the P680 chlorophyll as CPHL(ox) and the reduced form as CPHL(re), one of the Z reactions, say the third, would be

$$H_2O + Z\cdot 2\ OH + CHPL(ox) \rightarrow H^+ + Z\cdot 3OH + CPHL(re)$$

and so on for the other Z reactions. The reduced P680 chlorophyll is then ready to receive a photon.

As soon as the P680 photosystem has received a photon (or the equivalent energy from its antenna), it jumps considerably in reducing power and potential energy. This is the big vertical jump shown on the left in Figure 5.17. The activated electron then moves from the P680 photosystem to the P680 electron transport system. Again, we'll follow the electrons along in pairs, although in some steps (as we've seen) they move individually—for instance, the entire P680 photosystem contains only a single molecule of specialized chlorophyll *a,* which can handle only one electron and one photon at a time.

One of the earliest carriers in the electron transport system is plastoquinone, a small molecule that is soluble in the membrane. Plastoquinone is actually a hydrogen carrier rather than an electron carrier. It picks up two electrons from the photosystem and two hydrogen ions from the medium outside the thylakoid. The reduced plastoquinone (PQH_2) molecule then diffuses over to the inner side of the membrane, where it passes its electrons to another electron carrier in the system *and releases its two hydrogens, as hydrogen ions, into the inner space* of the thylakoid.

There are a number of different electron and

hydrogen carriers in the P680 electron transport system. Some of them are known and named, and others are so far only hypothetical. Except for plastoquinone, each of them appears to be firmly embedded in specific placements in one or both surfaces of the lipid bilayer membrane. The electrons are said to "fall" down the series of carriers, since each carrier in the system has somewhat less reducing power than the one before it. In the course of the fall, as we have just seen, hydrogen ions are taken up from the chloroplast medium outside the thylakoid and hydrogen ions are released into the fluid medium inside the thylakoid. The electron transport system, then, is really a kind of *proton pump.*

The Electron Pathway—P700

At the end of the P680 electron transport chain, our two electrons have lost much of their energy, but they still have more reducing power than when they started out. At the bottom of the P680 "staircase," the electrons (one at a time again) reduce the oxidized chlorophyll of P700, and much of the process is then repeated. But the whole P700 system, which uses somewhat different carriers, operates at a higher level of reducing power than the P680 system, as you can see in Figure 5.17. The electrons that enter P700 from P680 must be boosted in energy a second time, each one using the energy of a new photon—the second big jump in the diagram. So, by the end of the entire two-stage system, each electron will have been boosted to a higher energy state twice, once by each of two photons. The photoevent, then, involves four photons altogether, two for each of its two electrons.

In P700, the two electrons again fall down a staircase of electron carriers, until they reach the final electron carrier at the end of the chain, which is the small NADP molecule. This final reduction occurs on the outer surface of the thylakoid, and as the NADP picks up the two photoevent electrons from the P700 electron transport system, it also picks up two hydrogen ions from the chloroplast medium outside the thylakoid. Since the original splitting of water produced two new hydrogen ions inside the thylakoid, this final electron transfer represents another net gain of two hydrogen ions inside the thylakoid, in addition to the two that were pumped across the membrane by the P680 electron transport system. *Each photoevent, then, increases the chemiosmotic gradient by four hydrogen ions, which can be cashed in later for ATP.*

The pumping of hydrogen ions into the thylakoid makes for a strong charge gradient across the membrane. Negatively charged chloride ions are attracted to the positive charges inside, and the membrane lets them pass in, where they become hydrochloric acid. It's easier to get in than out, and the membrane does not let the hydrogen ions out except at the complex "toll gates" we discussed earlier—the CF1 particles on the outer surface of the thylakoid membrane. The CF1 particles are complexes of enzymes that use the potential energy of the hydrogen ion gradient to make ATP. Apparently, one ATP molecule is generated for every three hydrogen ions that pass out through the CF1 particles. So, because the photoevent pumps four hydrogen ions across the membrane, it is responsible for the production of four-thirds molecules of ATP in addition to one molecule of $NADPH_2$. The ATP and $NADPH_2$ will both participate in the next part of photosynthesis, which is the light-independent reaction.

Summarizing the Light Reactions

Before we move on to the light-independant reaction, let's review the light reaction. The unit photoevent involves the photolysis of one molecule of water and absorbs four photons to produce half a molecule of oxygen, four-thirds molecules of ATP, and one molecule of $NADPH_2$. It begins with the oxidation of water by an elaborate protein called Z, which releases the oxygen of water as molecular oxygen, releases the hydrogen of water as hydrogen ions, and passes two electrons from each water molecule on to the photosystem of P680. In a two-step operation involving two photosystems (P680 and P700), the electrons are each boosted twice to reach high levels of energy and eventually are used to reduce a molecule of NADP to $NADPH_2$. For each molecule of water that is split, a net of four hydrogen ions are pumped into the cavity of the thylakoid. The resulting chemiosmotic gradient is tapped by specialized structures on the outer surface of the thylakoid to produce four-thirds molecules of ATP. And so, in the light reaction, the chloroplast gains some usable energy and a source of molecular hydrogen with which to reduce carbon dioxide.

5.3 The Light-Independent Reaction

Glucose Production in C3 and C4 Plants

The light-independent reaction takes place in the stroma immediately after the light reaction unfolds in the thylakoid. It is the second half of photosynthesis and the part in which carbohydrates are actually made. The overall net equation for the light-independent reaction is:

$$6\,CO_2 + 12\,NADPH_2 + 18\,ATP \rightarrow$$

$$\underset{\text{Glucose}}{C_6H_{12}O_6} + 12\,NADP + 18\,ADP + 18\,P_i + 6\,H_2O$$

Again, the overall formula hides a multitude of reactions—some we know about and some we don't. Basically, the process involves incorporating, or *fixing*, inorganic carbon dioxide into some kind of organic molecule. Just how this happens depends on what kind of plant is involved, because there are two quite different ways of fixing carbon dioxide in higher plants. In one kind of plant, the first organic molecule in which newly fixed carbon can be detected is a three-carbon compound, *3-phosphoglycerate* (shown in Figure 5.18). In other plants, the first molecule in which newly fixed carbon occurs is a four-carbon compound, *oxaloacetate*. The two kinds of plants are called C3 and C4 plants, respectively. The C4 plants actually have both the C3 and C4 pathways, but C3 plants have only the one way of fixing carbon dioxide. We've only recently learned about the C4 pathway, but the C3 pathway has been known for some time, and is called the *Calvin cycle* after its discoverer. So let's first look at this cycle, and then consider the more complex C4.

The Calvin Cycle in C3 Plants

Since we're considering a cycle, our description could theoretically start anywhere. So let's pick up the story just before carbon dioxide is fixed. Notice that the Calvin cycle is summarized in Figure 5.18. The first reaction starts with ATP and a five-carbon, phos-

5.18 The Calvin cycle. The light-independent reactions that comprise the Calvin cycle utilize the ATP and $NADPH_2$ generated in the light-dependent reactions, plus carbon dioxide from the atmosphere. There are six key steps involved in the abbreviated version shown here: ① Ru-P is phosphorylated by ATP. ② Carbon is added to Ru-DP, ③ forming an unstable intermediate that is cleaved into two molecules of 3-PG. ④ Another phosphorylation converts each 3-PG to DPGA. ⑤ Finally, $NADPH_2$ enters, reducing the two DPGA molecules to PGAL, which is a three-carbon carbohydrate. So, carbohydrate synthesis has been completed. The ADP remaining from the phosphorylation and the oxidized NADP will be recycled through the light-dependent reactions.

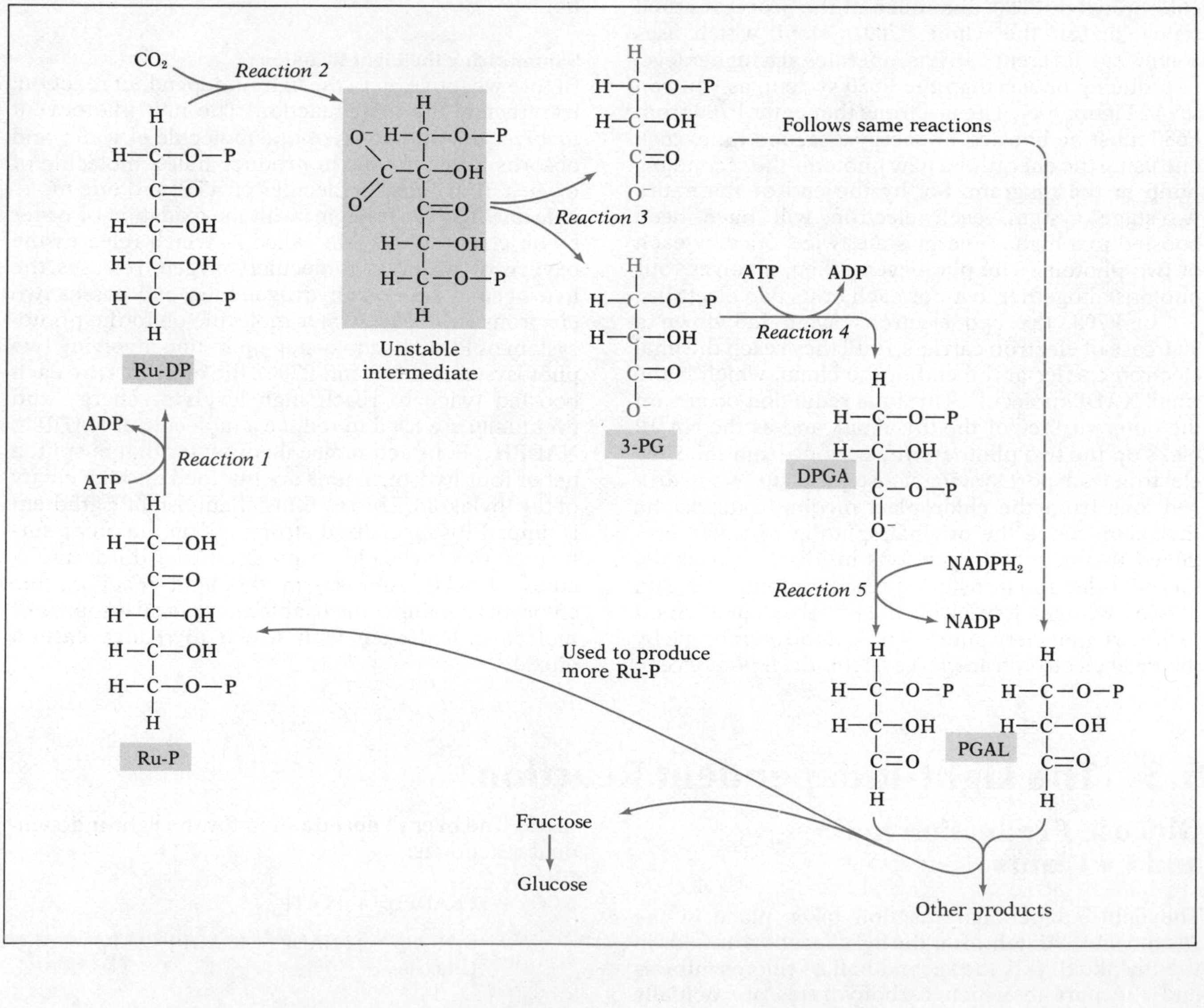

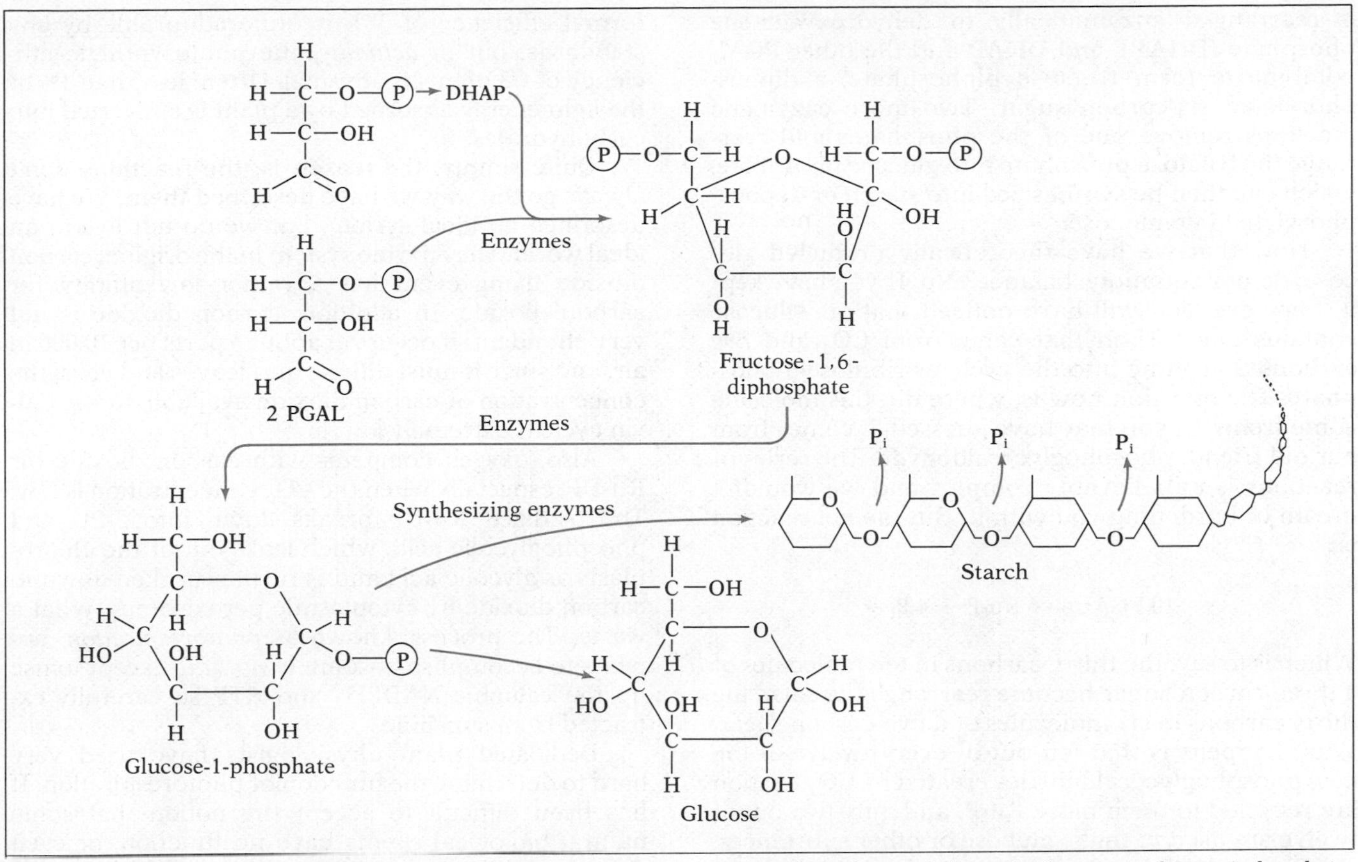

5.19 Under some conditions, PGAL is involved in glucose production. In such cases, one molecule of PGAL is converted to DHAP, which is then joined enzymatically to the other molecule of PGAL to form fructose diphosphate. Later, other enzymes remove a phosphate and change the fructose to glucose-1-phosphate. From its phosphorylated form, glucose can be converted to starch or dephosphorylated into simple glucose.

phorylated sugar, ribulose-5-phosphate (Ru-P). In this reaction, a molecule of Ru-P reacts with one molecule of ATP to form a molecule called ribulose 1,5-diphosphate (Ru-DP). (Refer to numbers 1, 2, 3, etc., in Figure 5.18 as you proceed.)

① Ru-P + ATP → Ru-DP + ADP

Then, Ru-DP is the substrate for the next reaction, the one in which carbon dioxide is fixed. In this step, the carbon dioxide is added to Ru-DP to form an unstable, six-carbon diphosphorylated sugar, which breaks down into two three-carbon compounds, 3-phosphoglycerate (3-PG):

②and③ Ru-DP + CO_2 →

[Unstable six-carbon intermediate] → 2 3-PG

So, here we have carbon dioxide incorporated into an organic molecule, an important step.

Next, ATP adds its terminal phosphate to each molecule of 3-PG to form a diphosphoryl three-carbon compound, 1,3-diphosphoglycerate (DPGA):

④ 2(3-PG + ATP → DPGA + ADP)

Then the $NADPH_2$ enters the cycle and reduces each DPGA. The reaction products are NADP, inorganic phosphate, and two molecules of a new three-carbon compound, phosphoglyceraldehyde (PGAL):

⑤ 2(DPGA + $NADPH_2$ → PGAL + NADP + P_i)

This is all obviously getting a bit complex, so let's take a look at a simplified version in Figure 5.18.

Note that now a carbohydrate has been produced. After all, phosphoglyceraldehyde is technically a sugar. It's not glucose yet, but it is a molecule from which glucose can be synthesized. Actually, it can produce not only glucose, but also maltose, starch, cellulose, fatty acids, amino acids, and other molecules. Notice that, at this point, we have used up the products of the light reaction, and so we could consider phosphoglyceraldehyde to be the end product of photosynthesis.

But we're interested in glucose production, and we don't have any glucose yet. So where does it come from? Actually, glucose is quite easily made from the two molecules of phosphoglyceraldehyde that we produced in photosynthesis (Figure 5.19). One PGAL is rearranged enzymatically to dehydroxyacetone phosphate (DHAP), and DHAP and the other PGAL combine to form fructose diphosphate, a diphosphorylated six-carbon sugar. Two more enzymatic reactions remove one of the phosphates and rearrange the fructose phosphate into glucose phosphate, which can then be synthesized into starch or dephosphorylated into glucose.

Now that we have theoretically produced glucose, do our equations balance? No. If you have kept a sharp eye, you will have noticed that our glucose contains *one* carbon that came from CO_2 and *five* carbons that came into the cycle as ribulose-5-phosphate. The question now is, where did this molecule come from? As you may have guessed, it comes from our old friend, phosphoglyceraldehyde. The series of reactions is unbelievably complex and we wouldn't dream of burdening you with it. But the net reaction is:

$$10\ \text{PGAL} \rightarrow 6\ \text{Ru-P} + 4\ P_i$$

Which is to say, the thirty carbons in ten molecules of a three-carbon sugar become rearranged to form the thirty carbons in six molecules of a five-carbon sugar. What happens is that ten out of every twelve of the new phosphoglyceraldehydes created by CO_2 fixation are recycled to form more Ru-P, and only two out of twelve are used to make glucose or other substances.

Remember that, according to the overall equation, six carbon dioxides enter at one point and one glucose leaves at another point. On the average, a carbon atom has to complete six turns of the cycle before it comes out as glucose. The overall formula for the Calvin cycle can be summarized this way:

$$6\ \text{Ru-P} + 6\ CO_2 + 12\ \text{NADPH}_2 + 18\ \text{ATP} \rightarrow \underset{\text{Glucose}}{C_6H_{12}O_6} + 12\ \text{NADP} + 18\ \text{ADP} + 18\ P_i + 6\ \text{Ru-P} + 6\ H_2O$$

The six ribulose-5-phosphates are shown on both sides of the equation to emphasize that the six on the left are consumed in the reaction, and the six on the right are created anew.

What's Wrong with the Calvin Cycle in C3 Plants?

The picture of the light-independent reaction in C3 plants, as we have presented it, is one of marvelous efficiency. After all, everything balances (perhaps with a few ATPs left over). It turns out that the thermal efficiency of the process is 38%, which means that about 38% of the energy in the original photons is eventually captured in the chemical bonds of glucose and the rest is lost as heat. Unfortunately, real life interjects itself to ruffle our formulas once again. A formal efficiency of 38% is quite admirable by any standards, but *in actuality,* the photosynthetic efficiency of C3 plants is abysmal. Often, less than 1% of the light energy absorbed by a plant is converted into carbohydrates.

Quite simply, the reason is, the reactions don't always go the way we have described them. We have described an ideal system, but we do not live in an ideal world. The enzyme system in the original carbon dioxide fixing event has a rather low affinity for carbon dioxide. In addition, carbon dioxide is not very abundant; it occurs at about 3 parts per 10,000 in air, and since it must diffuse into leaves and cells, the concentration of carbon dioxide available to the Calvin cycle is extremely low.

Also, oxygen competes with carbon dioxide for Ru-DP, especially when the CO_2 concentration is low. The oxidized Ru-DP breaks down into 3-PG and phosphoglycolic acid, which leaks out of the chloroplasts as glycolic acid and is further broken down to carbon dioxide by cytoplasmic peroxisomes. What a waste! The process, known as *photorespiration,* appears to accomplish absolutely *nothing* except to use up the valuable NADPH_2 and ATP so carefully extracted from sunshine.

Dedicated plant physiologists have tried very hard to determine the function of photorespiration. It has been difficult to accept the notion that some natural biological events have no function, or even that life processes are not always very efficient. So far, however, no one has been able to explain this paradoxical process.

Photosynthesis in C4 Plants

In the 1960s, there appeared some interesting evidence that 3-phosphoglycerate is not always the first organic molecule into which CO_2 is incorporated. It seemed that in maize, sugar cane, and a number of other plants (Figure 5.20), the initial fixation is in *oxaloacetate,* which is then reduced by NADPH to *malate*:

$$\text{Phosphoenolpyruvate} + CO_2 \rightarrow \text{Oxaloacetate} + P_i$$

$$\text{Oxaloacetate} + \text{NADPH}_2 \rightarrow \text{Malate} + \text{NADP}$$

This was surprising, and the next finding was just as unexpected. The malate diffuses out of the chloroplast and even out of the cell in which it was formed. It winds its way into a different kind of cell and there enters another cycle. The leaves of C4 plants, it seems, contain two different kinds of photosynthetic cells, which differ both structurally and biochemically. Note the internal structure of a leaf (Figure 5.21). The *bundle-sheath cells* surround the vascular system of the leaf, and the *mesophyll cells* are loosely arranged around the bundle-sheath cells. In C4 plants, the bundle-sheath cells have larger, greener chloroplasts,

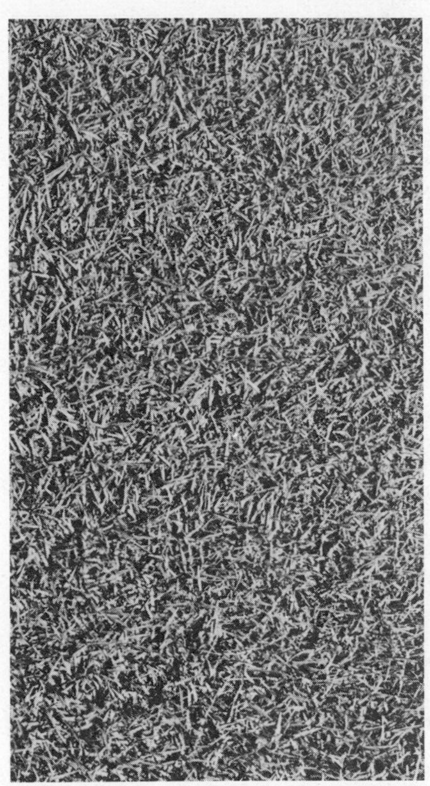

(a)

(b)

5.20 At least ten families and over one hundred genera include some C4 plants. The three examples here are (a) bermuda grass, (b) sugar cane, and (c) corn. Although these are all grasses, many

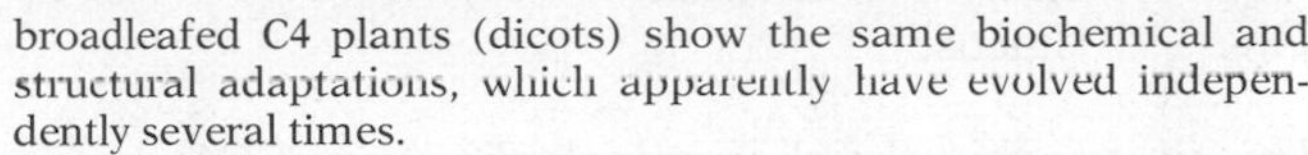

broadleafed C4 plants (dicots) show the same biochemical and structural adaptations, which apparently have evolved independently several times.

(c)

5.21 Some of the special features of C4 plants are easily seen in bermuda grass (*Cynodon dactylon*). The bundle-sheath cells are prominent with their large chloroplasts. Compare these to the mesophyll cell chloroplasts. The central group of cells comprise the vein with xylem, phloem, and supporting cells. The arrangement of mesophyll and bundle-sheath cells in C4 plants is an integral part of their special biochemistry. Carbon dioxide passes through the stroma and diffuses into a mesophyll cell, where it joins phosphoenolpyruvate to form oxaloacetate. Malate is then formed and transferred to the bundle-sheath cells, where carbon dioxide is released and its concentration grows. It is in the bundle-sheath cell where the carbon dioxide joins the Calvin cycle to be incorporated into carbohydrates.

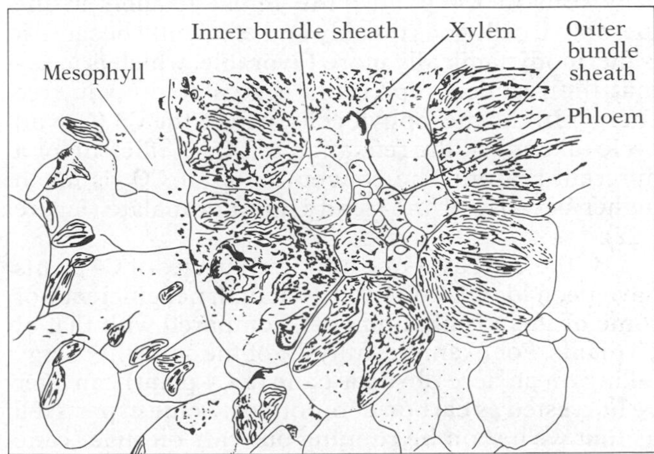

along with many more mitochondria and other organelles. The two types of cells are in contact, and the C4 cycle shuttles intermediate products between them. The C4 fixation that we have described occurs in the mesophyll cells. The four-carbon malate then enters the chloroplasts of the bundle-sheath cells. What happens then is also rather unexpected. The malate interacts with NADP to regenerate $NADPH_2$, but in the process it gives up the newly fixed CO_2:

$$\text{Malate} + \text{NADP} \rightarrow \text{Pyruvate} + CO_2 + NADPH_2$$

The pyruvate then diffuses back into the mesophyll cells, where it is regenerated as phosphoenolpyruvate by ATP. The net reaction is the hydrolysis of ATP to ADP, and the transfer of one molecule of carbon dioxide and two electrons from the mesophyll cell to the bundle-sheath cell. In the bundle-sheath cell, the carbon dioxide and the $NADPH_2$ are fed into the regular C3 (or Calvin) cycle.

The overall formula for the C4 light-independent reaction is the same as that for the C3 light reaction, except that twelve more ATPs are used up for each glucose molecule:

$$6\,CO_2 + 12\,NADPH_2 + 30\,ATP \rightarrow$$

$$\underset{\text{Glucose}}{C_6H_{12}O_6} + 12\,NADP + 6\,H_2O + 30\,ADP + 30\,P_i$$

This means that the *theoretical efficiency* of C4 photosynthesis is less than that of C3 photosynthesis. But in reality C4 photosynthesis is usually much more efficient. The reason is that the extra ATPs aren't wasted; instead, their energy is used to move CO_2 and electrons from one cell to another, and this, it turns out, is critical.

One of the problems with C3 photosynthesis, as we have seen, is that the concentration of CO_2 is too low for the system to operate with full efficiency. In part, this is because the first fixation reaction between CO_2 and Ru-DP is not very favorable thermodynamically. But in C4 plants, the initial fixation in the mesophyll cells goes quickly and efficiently because it is thermodynamically more favorable, which is to say that the reaction represents a greater drop in free energy. In the mesophyll cells, where the C3 (Calvin) cycle takes over, the refixation of CO_2 is efficient for a different reason: The concentration of CO_2 is much higher because of the steady input of malate (Figure 5.22).

By the way, over one hundred species of C4 plants have been identified. And the actual, net efficiency of some of them is phenomenal, compared with that of C3 plants. For example, fully 8% of the sunlight energy falling on an acre of sugar cane (a C4 plant) can later be harvested as carbon compounds. Some experts tell us that we'll soon be running our cars on sugar cane products, and, in fact, they're already doing it in Brazil.

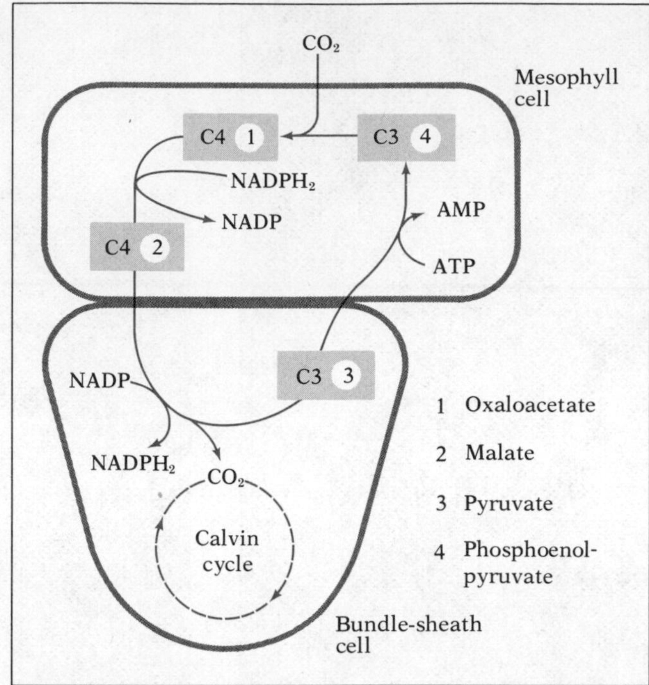

5.22 The purpose of this special C4 pathway is to concentrate carbon dioxide in the vicinity of the carboxylating enzyme of the Calvin cycle. This makes it possible for plants to incorporate carbon dioxide rapidly, improving their photosynthetic efficiency. The major step is the addition of CO_2 to phosphoenolpyruvate in the mesophyll cells, forming oxaloacetate. This is then converted to malate through reduction by $NADPH_2$. The malate moves to the bundle-sheath cells, where it is oxidized and decarboxylated, forming pyruvate. The decarboxylation makes CO_2 available to the Calvin cycle going on in the bundle-sheath cells. Pyruvate is then changed to phosphoenolpyruvate utilizing ATP. The cycle can then repeat.

In general, plant species from temperate climates tend to be C3 plants and tropical and desert plants tend to be C4 plants. Considering this relationship, we would surmise that tropical and desert plants have plenty of photons and must turn their efforts to the fully efficient capture and use of that rare resource, carbon dioxide. In plants from temperate climates, the input of light energy may be the limiting factor, and the more photon-efficient C3 system will do. However, that's just some second-guessing. It may simply be that the C4 system is always more efficient, but that it has evolved relatively recently in a limited number of lines and not all plants have the option of trying it.

Now, the C4 plants are better adapted to the tropics and to hot deserts than the C3 plants. This has led to some recent investigations in Death Valley. A desert shrub from this region, the salt bush, genus *Atriplex*, has both C3 and C4 species. These have been successfully compared under controlled conditions. *Atriplex rosea*, a C4 species, is able to carry on photosynthesis readily in the most intense light, while its C3 cousin, *A. patula,* fails. *Atriplex patula* cannot incorporate CO_2 under intense light, while *A. rosea* succeeds by utilizing every last scrap of CO_2. The stomata of *A. rosea* are nearly closed under intense light conditions and admit meager amounts of CO_2, but its efficiency in utilizing what is available makes up for the deficiency. Being able to photosynthesize with the stomatal openings restricted presents the plant with a distinct advantage. Open stoma, especially during the most intense heat of the day, will accelerate transpiration and subsequent water loss. This situation is intolerable to any desert plant where the next rainfall may be months away. Most plants survive by drastically slowing down their metabolic activities. But slow activity means slow growth (typical of desert perennials). Think of the advantage of the phosphoenolpyruvate cycle to C4 plants.

Incidentally, the efficiency of C4 plants has not escaped the attention of agricultural scientists. If fertile hybrids of C3 crops and C4 plants could be established, the result might be important in accelerating crop yields. But that's another story.

We should add that in case you're wondering where the bundle-sheath cells get the ATP needed to run the Calvin cycle, we're wondering too. However, all cells can use their energy stores to generate ATP through glycolysis and respiration if they have to.

A Summary of Photosynthesis

I. *Photosynthesis is the capture and transfer of light energy into the chemical bond energy of carbohydrates.*

A. In plants, the photosynthetic process takes place in chloroplasts and is intimately related to chloroplast structure.

B. The overall or general formula for photosynthesis in plants is

$$6\,CO_2 + 12\,H_2O + 48\text{ Photons} \rightarrow \underset{\text{Glucose}}{C_6H_{12}O_6} + 6\,H_2O + 6\,O_2$$

II. Plant photosynthesis can be divided into two parts: the light reaction, which takes place in the thylakoids of the chloroplast; and the light-independent reaction, which takes place in the stroma of the chloroplast.

A. The light reaction

1. Electrons of chlorophyll are excited by photons and are moved through an electron transport system to $NADH_2$.
2. A chemiosmotic pH gradient results in the production of ATP.
3. Two photosystems are involved, and each electron is boosted in energy level sequentially by two photons.
4. In one of the photosystems, the electron-depleted chlorophyll is reduced again by electrons taken from the oxidation of water and O_2 is released.
5. The overall formula for *one photoevent* of the light reaction is

$$H_2O + 2\,ADP + 2\,P_i + NADP + 4\text{ Photons} \rightarrow 2\,ATP + NADH_2 + \tfrac{1}{2}O_2$$

B. The light-independent reaction

1. The light-independent reaction is the fixation of carbon dioxide, using ATP and $NADPH_2$.
2. Two different cycles occur in different plants. The Calvin cycle, in which the initial product is a three-carbon compound, is found in all plants. The C4 cycle, in which the initial product is a four-carbon compound, is found only in some. Plants with the C4 cycle are called C4 plants, and those without it are called C3 plants.
3. The Calvin cycle combines carbon dioxide with a five-carbon compound to yield two three-carbon compounds. Ten out of twelve of these three-carbon compounds must be recycled to form the five-carbon compound, and the other two out of twelve (as phosphoglyceraldehyde) can be used to build any one of many compounds, including glucose. The overall formula for the Calvin cycle is

$$6\text{ Ru-P} + 6\,CO_2 + 18\,ATP + 12\,NADPH_2 \rightarrow \underset{\text{Glucose}}{C_6H_{12}O_6} + 6\,H_2O + 18\,ADP + 18\,P_i + 12\,NADP + 6\text{ Ru-P}$$

4. The C4 cycle involves two cell types. In the mesophyll cells, CO_2 is fixed into a four-carbon compound with the expenditure of ATP and $NADH_2$. A four-carbon product then diffuses to the bundle-sheath cells, where $NADH_2$ is regenerated and CO_2 is released into the Calvin cycle. The overall net formula is the same as for the Calvin cycle except that an additional 12 ATPs per glucose are involved in the net movement of CO_2 and hydrogen from one cell type to the other.
5. Although the theoretical photon efficiency of photosynthesis is higher in C3 plants, the realized efficiency of photosynthesis may be much higher in C4 plants, because they are more efficient in the use of water and carbon dioxide.

We have seen, then, that life on this particular planet depends on capturing the energy radiating from its nearest star. Energy is essential to living things because there is a strong tendency for systems to become randomized or disorganized. Life is highly organized and needs energy to maintain that level of organization. Perhaps on other worlds, the energy to sustain life comes from different sources, such as heat emanating from the planet's core. But here, most of our energy comes from our sun and can only be captured by organisms that are often rudely or cavalierly treated by the other species. Nonetheless, without plants quietly and thanklessly turning solar energy into food, this would be a far, far different place.

Application of Ideas

1. At one time, many inventors attempted to design "perpetual motion machines." All failed. The reasons for this consistent failure were not understood until the laws of thermodynamics were formulated. Explain how they apply to the problem.
2. Is there such a thing as an endergonic half-reaction? Does the idea of a half-reaction help clarify what is meant by endergonic and exergonic reactions? Explain.
3. When asked to define life, a thoughtful student once wrote that life is an interruption in entropy. Is this true? Comment on the statement, using the laws of thermodynamics as a basis.
4. Reviewing what has been presented about chemiosmosis and the membrane differential, use a rechargeable storage battery to develop an analogy to this energy-storing and releasing mechanism. In what ways are they roughly similar, and where does the analogy fail?
5. Earlier in this century, plant physiologists attempted to determine the precise role of water and carbon dioxide in photosynthesis by using the radioactive isotopes, carbon-14, oxygen-18, and hydrogen-3. Using your knowledge of photosynthesis, suggest how these isotopes might be used in such determinations, and what you might expect in the results.

Molecules of Energy Transfer

adenine
adenosine diphosphate (ADP)
adenosine monophosphate (AMP)
adenosine triphosphate (ATP)
inorganic phosphate (P_i)
nucleoside
phosphate (P)
pyrophosphate
ribose

Soluble Electron Carriers

nicotinamide adenine dinucleotide (NAD)
oxidized form: NAD
reduced form: $NADH_2$

nicotinamide adenine dinucleotide phosphate (NADP)
oxidized form: NADP
reduced form: $NADPH_2$

Molecules of the Light-Independent Reactions

DHAP
1,3-diphosphoglycerate (DPGA) (1,3-diphosphoglyceric acid)
fructose diphosphate
fructose phosphate
glucose phosphate
glycolate (glycolic acid)
malate (malic acid)
oxaloacetate (oxaloacetic acid)
phosphoglyceraldehyde (PGAL)
3-phosphoglycerate (3-PG) (3-phosphoglyceric acid)
pyruvate (pyruvic acid)
ribulose 1,5-diphosphate (Ru-DP)
ribulose-5-phosphate (Ru-P)

Key Words

absorption spectrum
accessory pigment
action spectrum

bundle-sheath cell

Calvin cycle
carotenoid
carrier molecule
CF1 particle
chemiosmotic
chemotroph
chlorophyll a
chlorophyll b
concentration differential
cytochrome c
cytochrome f
C3 plants
C4 plants

dark reaction

electromagnetic spectrum
electron carrier
electron transport chain
electron transport system
entropy

genetic code

hydroxyl radical

kinetic energy

light antenna
light-independent reaction
light reaction

mesophyll
microwave

negative entropy
nicotinic acid
nitrogenous base

oxidation
oxidative respiration

P680
P700
pH differential
photoevent
photolysis
photon
photorespiration
photosynthesis
photosynthetic pigment
photosynthetic unit
photosystem

Key Ideas

Introduction

1. All living organisms must continually find energy sources to prevent their own molecular disorganization from occurring.
2. The *second law of thermodynamics* reminds us that systems tend to move towards randomness or to a state of *entropy*.
3. Free energy can be defined as *negative entropy*.
4. *Entropy* is also a measure of the randomness of a system.
5. Matter raised to a higher energy level has *potential energy*.
6. Free energy is often defined as the ability to do *work*.
7. When systems move towards a lower net energy level the energy of motion is referred to as *kinetic energy*.
8. Energy in a system commonly dissipates in the form of heat.
9. The earth is not a closed system, but receives energy from the sun and other sources in space. This outside supply of energy supports highly organized living things.

Adenosine Triphosphate (ATP)

1. *ATP* is the nearly universal molecule of energy transfer.
2. Energy stored in proteins, fats, and carbohydrates must be transferred to *ATP* bonds before it is used by cells.
3. *ATP* is composed of the *nitrogen base adenine*, the sugar *ribose*, and three *phosphates*.
4. The high-energy bonds of *ATP* are in the two *phosphate* to *phosphate* (*pyrophosphate*) linkages.
5. When cellular work is done, energy is usually transferred from the terminal *phosphate* bond, changing *ATP* to *ADP*.
6. The *ATP* to *ADP* reaction is reversible. If energy is available, inorganic *phosphate* and *ADP* can produce *ATP*.
7. Phosphorylation is the transfer of *phosphate* from one molecule to another; usually ATP is either the donor or the recipient.
8. When two high-energy phosphates of *ATP* are removed, *AMP* is formed.
9. Synthetic reactions involve *ATP*, enzymes, and substrate.
10. High-energy bonds contain about four times the energy of low-energy bonds.
11. *Pyrophosphate* bonds in *ATP* contain excess energy because of strong repulsive force within the molecule.
12. We can predict that sooner or later all energy entering a system will dissipate in the form of heat.

ATP Production by Chemiosmotic Phosphorylation

1. The theory of *chemiosmotic phosphorylation* represents newly accepted thinking on the *phosphorylation* of *ADP*.
2. Clues to understanding *chemiosmotic phosphorylation* came from the knowledge that the membrane systems of chloroplasts and mitochondria had to be intact for the phosphorylation to work.
3. The basic idea of Mitchell's hypothesis is that *pH differentials* build up on either side of the membranes creating systems of great free energy.
4. In the presence of sunlight, hydrogen ions are actively pumped into the thylakoid disks creating an acid environment.
5. The energy reservoir is tapped by permitting H^+ to pass back across the membrane through pores in *CF1 particles*.
6. *CF1 particles* (and F1 particles in mitochondria) contain enzyme systems capable of directing *phosphorylation* of *ADP*.
7. The chemiosmotic gradient in mitochondria is essentially opposite to that in the chloroplast. Hydrogen ions accumulate outside the inner membrane.

photosystem I
photosystem II
plastoquinone
potential energy
proton pump

quantasome
quantum mechanics

reaction center
reducing power
reduction
respiration

stoma (stomata)
stroma

thermal efficiency
thermodynamics

ultraviolet light

visible light

X-ray

Z protein
Z scheme

Oxidation and Reduction

1. *Oxidation* is the removal of electrons or hydrogen.
2. *Reduction* is the addition of electrons or hydrogen.
3. Substances that lose electrons or hydrogen are said to be *oxidized,* while those that gain are said to be *reduced.* These reactions are always coupled.
4. *Electron transport chains* are a series of *carrier molecules* that pass electrons along in an *oxidation-reduction* series. They also transport hydrogen ions and molecular hydrogen.

NAD, $NADH_2$, NADP, and $NADPH_2$: Important Soluble Electron Carriers of the Cell

NAD and *NADP* are common soluble *electron carriers* in *respiration* and *photosynthesis,* respectively. When these carriers are oxidized, they yield two hydrogen ions plus two electrons. In the chemiosmotic system, the hydrogen ions are pumped across the membrane where they accumulate, forming the *pH differential.*

Energy Enters the Ecosystem

The Physics of Sunlight

1. The *electromagnetic spectrum* includes biologically important *visible light.*
2. Light energy takes the form of *photons* whose energy is inversely related to their *wavelength.*
3. The *visible spectrum* is a *wavelength* continuum from 750 nm (deep red) to 430 nm (violet).

The Absorption Spectrum

1. Light energy absorbed by chlorophyll is primarily in the violet and red regions.
2. Light energy is absorbed directly by *chlorophyll a* and by accessory pigments *chlorophyll b* and carotenoids, which act as a light-harvesting *antenna.*
3. *Action spectra,* when superimposed over *absorption spectra,* show that light absorbed is actually light used in *photosynthesis.*

The Chloroplast

The *light reactions* occur in the thylakoids, while the *light-independent reactions* occur in the *stroma,* which is outside of the thylakoids but inside the chloroplast.

An Overview of the Chemistry of Photosynthesis

1. The overall formula for *photosynthesis* is:

$$6\ CO_2 + 12\ H_2O + \text{Light} \rightarrow C_6H_{12}O_6\ (\text{glucose}) + 6\ H_2O + 6\ O_2$$

2. Considering the *light reactions* only, their overall formula is:

$$12\ H_2O + 12\ \text{NADP} + (\text{approx})\ 18\ \text{ADP} + 18\ \text{P} \rightarrow 6\ O_2 + 12\ NADPH_2 \quad (\text{approx})\ 18\ \text{ATP}$$

3. Considering the *light-independent reactions* only:

$$12\ NADPH_2 + 18\ \text{ATP} + 6\ CO_2 \rightarrow C_6H_{12}O_6\ (\text{glucose}) + 6\ H_2O + 12\ \text{NADP} + 18\ \text{ADP} + 18\ \text{P}$$

4. The *photosynthetic pigments* are believed to occur in granular *quantasomes.*

The Photosynthetic Unit

1. The *photosynthetic unit* is the physical unit capable of performing a *photoevent.*
2. The *photoevent* is the capture of light energy, splitting a water molecule (photolysis) into 2 H^+, 2 electrons, and half an O_2 molecule.
3. Energy from light *antennae* does its work in *reaction centers.*
4. A *photosynthetic unit* contains two light *antennae,* two *reaction centers,* and two *electron transport systems.*

Two Photosystems

Each photosynthetic unit contains two *photosystems.* These absorb light primarily at *680* and *700* respectively. Both *photosystems* cooperate to complete a *photoevent.*

The Role of Electrons in Energy Transfer

Light energy is initially transferred to electrons. These pass through *electron transport systems* and their energy is used to build up the pH gradient.

Receiving a Photon

Upon absorbing light energy, chlorophyll *a* molecules emit electrons, passing them to electron acceptors.

The Z Diagram

The *Z scheme* is an energy diagram that depicts the path of electrons during a *photoevent.* The levels in the *Z scheme* represent different free-energy states.

The Oxidation of Water

1. The *oxidation* of water is the first step of the *Z scheme.*
2. *Oxidation* is accomplished by the *Z protein* in five steps with the net reaction:

$$2\,H_2O \rightarrow 4\,H^+ + 4\,e^- + O_2$$

The Electron Pathway—P680

1. Once electrons are released from water, they reduce *chlorophyll a* in the *P680 photosystem.* Once reduced, *P680* can again absorb a *photon* and release another electron to the *electron carriers.*
2. Electrons from *chlorophyll a* go to carrier *plastoquinone,* a small molecule in the membrane. The electrons follow a zig-zag course through the carriers. Hydrogen ions are picked up outside the thylakoid and are eventually passed to the inside where the pH differential increases. The energy of the excited electrons is used, then, to pump protons across the membrane.

The Electron Pathway—P700

1. Electrons emerging from the *P680* pathway reduce the *P700 photosystem.*
2. *Photons* absorbed by *P700* boost electrons a second time.
3. *P700* electrons fall through the *P700* carriers to *NADP,* where they are joined by H^+ from the *stroma.*
4. Each *photoevent* increases the chemiosmotic gradient by four hydrogen ions, two during the *P680* pathway, and two during the *P700* pathway.
5. As hydrogen ions are permitted to pass through the *CF1 particles, ATP* is produced. One *photoevent* provides four hydrogen ions. Each *ADP phosphorylation* requires three. Therefore one *photoevent* produces four-thirds molecules of *ATP.*

Glucose Production in C3 and C4 Plants

1. In the *light-independent reactions,* a carbohydrate is produced by complex synthetic reactions in the *stroma.*
2. The overall equation for the *light-independent reactions* is:

$$6\,CO_2 + 12\,NADPH_2 + 18\,ATP \rightarrow C_6H_{12}O_6 + 12\,NADP + 18\,ADP + 18\,P + 6H_2O$$

3. The process involves fixing inorganic carbon dioxide into an organic molecule.
4. In *C3 plants* the newly fixed carbon can be detected in *3-phosphoglycerate* (3-PG) a three-carbon compound.
5. In *C4 plants* the newly fixed carbon is detected as *oxaloacetate,* a four-carbon compound.

The Calvin Cycle in C3 Plants

1. *ATP phosphorylates Ru-P* to form *Ru-DP.*
2. Carbon dioxide enters to join *Ru-DP,* forming an unstable six-carbon sugar that breaks down into two *3-PGs.*
3. *ATP* reacts with the two *3-PGs* to form two *DPGAs.*

4. *$NADPH_2$* reacts with the two *DPGAs* to form two *PGALs* (phosphoglyceraldehyde): end of *light-independent reactions.*
5. *PGAL* can be used to form carbohydrates, fatty acids, and amino acids.
6. Two *PGALs* are used in forming *fructose diphosphate,* from which *glucose phosphate* can be produced and dephosphorylated into glucose proper.
7. To provide enough *PGAL* to produce glucose and to keep the *Calvin cycle* going:
    a. Five-sixths or ten out of every twelve new *PGALs* formed are recycled to form more *Ru-P.*
    b. One-sixth or two out of every twelve new *PGALs* formed are used to make a molecule of glucose.
8. To revise the summarizing formula for the *light-independent reactions* (above), add six *Ru-P* to each side of the equation.

What's Wrong with the Calvin Cycle in C3 Plants?

1. When *photosynthesis* is at peak efficiency, 38% of the absorbed energy can later be found in the chemical bonds of glucose.
2. The efficiency can drop as low as 1% under certain conditions.
3. When CO_2 availability is low, *photorespiration* occurs. *Ru-DP* is wastefully broken down to CO_2.

Photosynthesis in C4 Plants

1. *C4 photosynthesis* is reviewed in Figure 5.22.
2. Carbon dioxide is fixed in the *mesophyll* cells but not into *3-PG.* It is used to convert *phosphoenolpyruvate* to *malate,* a four-carbon compound.
3. *Malate* leaves the *mesophyll* cells, entering *bundle-sheath cells,* where it reduces *NADP* and loses CO_2.
4. *Malate* is converted to *pyruvate,* giving off CO_2 to the *C3* photosynthetic activity.
5. *Pyruvate* then enters the *mesophyll* cells where it is changed to *phosphoenolpyruvate,* which combines with CO_2 to form *malate* once again.
6. *C4 photosynthesis* provides more CO_2 for the *Calvin cycle* than would otherwise be available.
7. *C4 plants* occur most commonly in hot desert regions where they can resist water loss and still capture enough CO_2 for efficient photosynthesis.

A Summary of Photosynthesis

1. *Photosynthesis* is the capture and transfer of light energy into the chemical bond energy of carbohydrates.
2. Plant *photosynthesis* can be divided into two parts: the *light reaction,* which takes place in the thylakoids of the chloroplast; and the *light-independent reaction,* which takes place in the *stroma* of the chloroplast.

Review Questions

1. What does the phrase "life is a joke on the somber *Second Law of Thermodynamics*" mean? (pp. 125–126)
2. Using the terms, *ATP*, P_i, *ADP*, *photosynthesis, respiration,* and biological *work,* construct a simple *ATP* cycle with the components in place. (p. 126)
3. Compare the structure of *AMP, ADP,* and *ATP* in terms of their chemical constituents and high-energy bonds. (pp. 126–127)
4. List the components necessary in a synthetic reaction employing *ATP.* Show these in a simple formula. (pp. 127–128)
5. Compare the bond energy in a *pyrophosphate* bond with that between *phosphate* and *ribose* in the *ATP* molecule. (What does *calories per mole* mean?) (p. 128)
6. Why is there more energy in a *pyrophosphate* bond than other bonds? (p. 129)
7. In very general terms, how does the explanation of *photosynthesis* introduced by Mitchell differ from that of earlier workers? (p. 129, Essay 5.2)
8. How might a high H^+ concentration on one side of a membrane and a high OH^- concentration on the other provide energy for the *phosphorylation* of *ADP?* (p. 129)
9. Explain the role of *CF1 particles* in *ATP* synthesis. (p. 130)

10. Compare chemiosmotic phosphorylation in the mitochondria with that of the chloroplast. (pp. 130, 132)
11. Explain the common reaction between sodium and chlorine (Chapter 2) using the terms *oxidized, reduced, oxidation, reduction* oxidizing agent, and reducing agent. (p. 132)
12. Write *NAD* and *NADP* in their *reduced* and *oxidized* forms. What is their general role in energetic reactions in the cell? (p. 135)
13. What is the ultimate source of energy for living things on the earth? Are there exceptions? Explain. (p. 136)
14. List the parts of the *electromagnetic spectrum.* Which parts are biologically useful and helpful? (p. 137)
15. What is a *photon*? How do we describe its energy? (p. 137)
16. List the primary colors and their wavelengths. Describe their energies from one end of the *visible spectrum* to the other. (p. 137)
17. Compare an *absorption spectrum* for *chlorophyll a* with an *action spectrum* (Figure 5.11). What does the comparison tell us? (p. 137)
18. What is the role of the accessory pigments (*chlorophyll b* and the carotenes) in light absorption? (p. 137)
19. What is the rationale for using oxygen production as a measure of photosynthetic activity? (p. 140)
20. Where in the chloroplasts do the *light reactions* and the *light-independent reactions* occur? (p. 140)
21. What is the general (net) equation for *photosynthesis?* (p. 140)
22. Following the equation for the *light reactions,* explain what happens to water (in general terms). (p. 140)
23. Following the equation for the *light-independent reactions,* explain (again, generally) what happens to carbon dioxide. (p. 140)
24. Summarize the *photoevent.* (pp. 140–141)
25. What is the physical content of a *photosynthetic unit?* (p. 141)
26. Name the *photosystems.* What is their relationship to the *photoevent?* (p. 141)
27. The apparent link between sunlight and chemical bonds in glucose is the electron. Amplify this comment from the introduction on page 142.
28. Following the *Z scheme* from left to right, summarize what is happening to electrons. Begin by explaining what the vertical scale at the left means (just trace the electron pathway). (p. 143 and Figure 5.17)
29. Looking in more detail at *photolysis,* explain the fate of four molecules of water. (p. 144)
30. The text tells you that *chlorophyll a* has even greater affinity for electrons than *protein Z.* This means that *chlorophyll a* has been previously oxidized. Explain how this happened. (p. 144)
31. The *electron carrier*'s role is passing electrons. How does this relate to the buildup of hydrogen ions in the lumen of the thylakoid? (p. 144)
32. Why is *plastoquinone*'s mobility so important to the *chemiosmotic* process? (p. 144)
33. Using your answers from 31 and 32, explain how the *electron transport system* amounts to a "*proton pump.*" (p. 145)
34. What happens to electrons from *P680* when they enter *P700?* Where do they end up after their second "uphill" trip? (p. 145)
35. Summarize the effect of each *photoevent* to the chemiosmotic gradient. (p. 145)
36. Reviewing the role of *CF1 particles* again, summarize the *ATP* gain in one *photoevent.* What else is gained? (p. 145)
37. Write a word or chemical formula for the *light-independent reactions.* (p. 145)
38. How is *ATP* first used in the *Calvin cycle?* (p. 147)
39. Explain how CO_2 enters the *Calvin cycle.* (p. 147)
40. How is *ATP* used again in the cycle? (p. 147)
41. Explain the role of $NADPH_2$ in the cycle. What is the product? (p. 147)
42. How does PGAL finally become glucose? What other products are produced from PGAL? (pp. 147–148)
43. Compare the theoretical efficiency with the real efficiency of *C3 plants.* (p. 148)
44. How does *photorespiration* affect the efficiency of *C3 photosynthesis?* (p. 148)
45. Explain what happens to CO_2 entering the *mesophyll cells* of a *C4 plant.* (p. 148)
46. In simple terms, what do *mesophyll cells* supply to *bundle-sheath cells?* (p. 148)
47. What does the *C4 plant* gain by the shuttling of molecules between *mesophyll* and *bundle-sheath cells?* What problems does this overcome? (p. 150)
48. How does the *C4* system better adapt plants to desert conditions? (pp. 150–151)

Chapter 6
Respiration

We've seen that life can only exist through intense and precise efforts to keep its molecules organized. As soon as the tendency to remain organized ceases, so does life. A corpse is a once-organized entity, gradually becoming disorganized as it decays, until finally, its molecules have no more to do with each other than they do any other molecules. At this point they are behaving randomly and organization no longer exists at any level. And the corpse is gone.

We have just seen how living things manage to capture the energy of sunlight and use that energy, in part, to keep their molecules organized. So now let's take a look at what living things do with the energy they have acquired.

Energy is used for movement, active transport, and biosynthesis, and in living things, it is usually transferred by the molecule called ATP. Thus, cells need a constant supply of ATP. Some cells of green plants can get ATP directly from photosynthesis, but that works only in the daytime and doesn't work, for example, in roots. Animals can't photosynthesize at all, but they still need ATP. Both plants and animals, then, get their needed ATP by utilizing organic molecules that have been made from photosynthesis. The process of turning the chemical energy of cellular fuels (carbohydrates, fats, and proteins) into the high-energy bonds of ATP is called *respiration.*

You should keep clearly in mind the simple differences in cellular respiration and photosynthesis. Photosynthesis begins with water, carbon dioxide, and energy and can result in any of several different end products. On the other hand, cellular respiration can begin with any of many different stored molecules and always ends up with carbon dioxide, water, and energy. In either case, a good example of the overall reaction is perhaps most easily demonstrated by the metabolism of glucose.

Compare the overall net formula for the photosynthesis of glucose:

$$6\,CO_2 + 6\,H_2O + \text{Energy} \rightarrow \underset{\text{Glucose}}{C_6H_{12}O_6} + 6\,O_2$$

to the overall net formula for the respiration of glucose:

$$\underset{\text{Glucose}}{C_6H_{12}O_6} + 6\,O_2 \rightarrow 6\,CO_2 + 6\,H_2O + \text{Energy}$$

The formulas appear to be the same, with only the direction of the reaction reversed. Actually, the energy input in the photosynthesis equation is in the form of photons, and the energy output of respiration is in the form of heat and the high-energy bonds of ATP.

Energetically, the fact that the two reactions are reversed is highly significant. Photosynthesis is essentially an *uphill* process, with the final chemical products (oxygen and glucose) possessing much more free energy than the starting chemical materials (water and carbon dioxide). This is accomplished at the expense of the energy of photons. The second process, cell respiration, is a *downhill* reaction, in which the glucose and oxygen are allowed to recombine and fall to the substantially lower free-energy state of carbon dioxide and water—with some of the energy saved in the high-energy bonds of ATP and the rest lost as heat. Figure 6.1 illustrates the relationship between the two processes in terms of energy and molecular organization.

Keep in mind that we are interested not so much in the free energy of individual molecules as that of entire systems. For example, pure oxygen (O_2) has no free energy and neither does pure hydrogen (H_2). As separate systems, each is already in its lowest energy state and we really aren't very interested in them. But a *mixture* of the two has a lot of free energy. And a mixture of oxygen and glucose is a system with a lot of free energy—energy that can be tapped by cellular respiration.

But there's a twist to this. Whereas the oxygen alone has no free energy, pure glucose in the absence of oxygen does represent a store of available free energy. This is because the carbon, hydrogen, and oxygen atoms of the glucose molecule are highly organized and are not in their lowest possible energy state. They can be rearranged, with a net release of energy.

Because free energy is available in pure glucose, even in the absence of oxygen, many organisms can get their energy needs by breaking glucose down into simpler molecules *anaerobically* (that is, in the absence of oxygen). The organism that is best known for this trick is yeast, which in the absence of oxygen gets its energy needs by breaking down glucose into alcohol and carbon dioxide. The process is a form of respiration called *fermentation* and is the basis of the distilleries industry. The net formula for fermentation is this:

$$\underset{\text{Glucose}}{C_6H_{12}O_6} \rightarrow \underset{\text{Alcohol}}{2\,C_2H_6O} + 2\,CO_2 + \text{Energy (as ATP)}$$

The reaction proceeds as written because the alcohol and carbon dioxide mixture has a lower level of organization and free energy than does pure glucose (Figure 6.2).

However, the free energy of pure glucose is only a small fraction of the free energy of a mixture of glucose and oxygen. In *aerobic* (oxygen-using) organisms, most of the energy of respiration comes from the oxidation of the components of glucose. This is why any organism that lives in your gut, or deep in the sediment of a stagnant pond, is not likely to be very

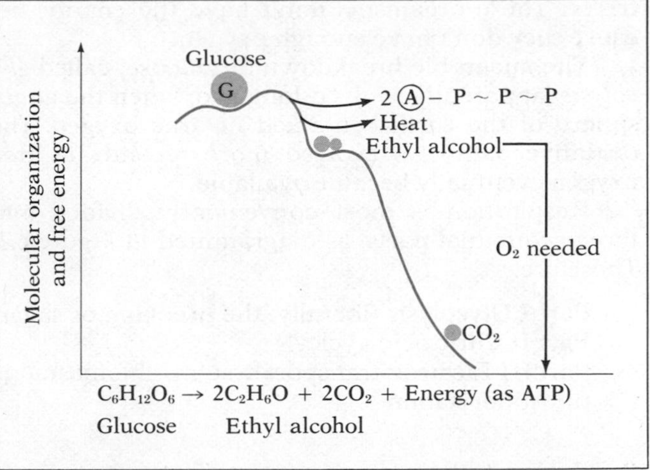

6.2 The energy hill and fermentation. Here, only the right half of the energy hill is shown. This time anaerobic respiration is diagrammed. As yeasts respire anaerobically, only a small amount of energy is transferred to ATP and released as heat. The products are carbon dioxide, which has little free energy, and ethyl alcohol, which has a great deal, since it comes to rest only part way down the hill where it will remain until oxygen is available.

6.1 The relationship between photosynthesis and respiration in terms of energy and molecular organization is shown here. In the left half of the graph, photosynthesis is portrayed as an uphill (endothermic) process. Carbon dioxide and water, with little free energy and organization, are reorganized by using solar energy to form glucose, which is high in free energy and organization. At the right, respiration begins with molecules of high free energy and proceeds downhill (exothermic), with the transfer of some energy into ATP and much energy released as heat. The products are low-energy carbon dioxide and water, again with little free energy and very simple organization. The small bump in the curve at the right of glucose represents the energy of activation needed to get stable glucose to react (see Chapter 3).

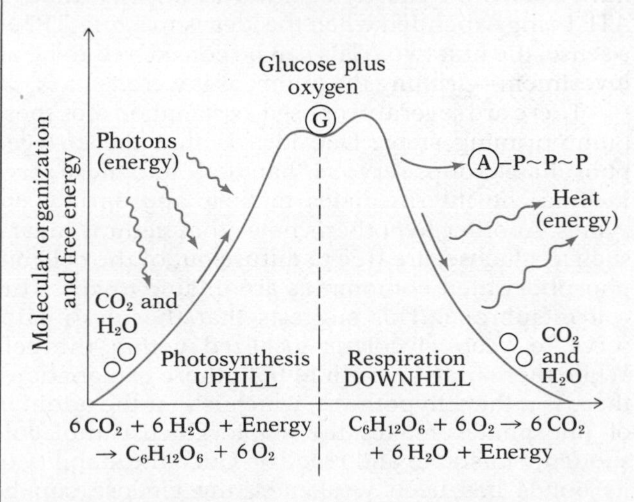

6.3 Respiration may be divided into three sequential parts: glycolysis; citric acid, or Krebs, cycle; and electron transport/chemiosmotic phosphorylation. Part III includes most of the actual ATP production.

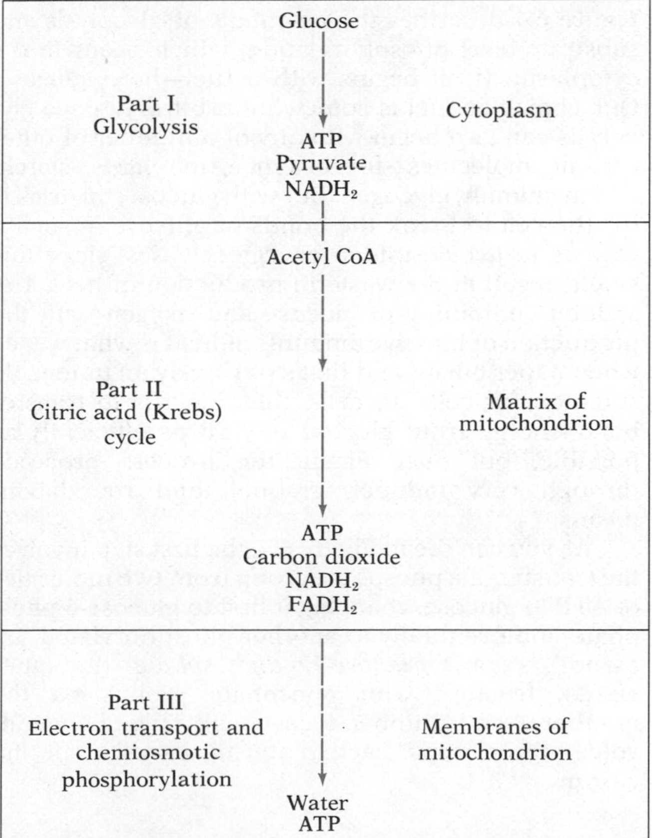

frisky. These organisms don't have the energy, because they don't have enough oxygen.

The anaerobic breakdown of glucose, called *glycolysis,* apparently evolved long ago, when the atmosphere of the earth contained no free oxygen. The oxidative pathways evolved more recently as free oxygen eventually became available.

Respiration is most conveniently divided into three sequential parts, as diagrammed in Figure 6.3. These are:

Part I Glycolysis (literally, the breaking of sugar)
Part II Citric acid cycle
Part III Electron transport system/chemiosmotic phosphorylation

Part I, glycolysis, occurs in the cytoplasm. Parts II and III take place in the mitochondria.

There are two different ways in which ATP is produced by metabolizing glucose. One is chemiosmotic phosphorylation, which we have already encountered in photosynthesis. Because it is intimately involved with oxygen utilization, chemiosmotic phosphorylation in the mitochondrion is also known as *oxidative phosphorylation.* As previously noted, one of the differences between chemiosmosis in the chloroplast and the mitochondrion is that the direction of the pH gradient is reversed: hydrogen ions accumulate *outside* the inner cavities of the mitochondrion and hydroxyl ions accumulate *inside.*

The other kind of ATP formation occurs during glycolysis. It happens in the cytoplasm, and does not involve oxygen or membranes. It is called *substrate-level* phosphorylation, which means that ADP receives its terminal high-energy phosphate on or near the substrate molecule. While there are relatively few organisms still around that cannot tolerate oxygen, all aerobic organisms, including plants and humans, are still capable of some measure of anaerobic respiration.

6.1 Anaerobic Respiration

Glycolysis: Degrading Glucose Without Oxygen

Figure 6.4 describes the essentials of glycolysis and substrate-level phosphorylation, which occur in the cytoplasm. It all begins with a fuel—here, glucose. Our choice of fuel is somewhat arbitrary, since glycolysis can also begin with any of a number of other organic molecules—for instance, in plants, starch, and in animals, glycogen. But with glucose, the trick is for the cell to break the bonds of glucose in such a way as to avoid sudden energy releases, since this would result in the wasteful production of heat. The sudden combining of glucose and oxygen with the production of massive amounts of heat is what we see when paper burns, and that's obviously an untenable situation for cells. In cells, the object is to transfer bond energy from glucose to ATP as efficiently as possible, but once again, the process proceeds through very indirect, gradual, and roundabout means.

As you can see in Figure 6.4, the first step involves the transfer of a phosphate group from two molecules of ATP to glucose, changing it first to glucose-6-phosphate and eventually to another phosphorylated six-carbon sugar, *fructose-1,6-diphosphate* (in other words, fructose with phosphate groups on the number 1 and number 6 carbons). This change involves three steps, each requiring its own specific enzyme.

$$① \text{ Glucose} + \text{ATP} \xrightarrow{\text{Enzyme}} \text{Glucose-6-phosphate} + \text{ADP}$$

$$② \text{ Glucose-6-phosphate} \xrightarrow{\text{Enzyme}} \text{Fructose-6-phosphate}$$

$$③ \text{ Fructose-6-phosphate} + \text{ATP} \xrightarrow{\text{Enzyme}} \text{Fructose-1,6-diphosphate} + \text{ADP}$$

[*Note:* Numbers in circles refer to Figure 6.4.]

At this point, the cell has used up two ATP molecules, degrading them to ADP. This is a bit paradoxical since, if the object of respiration is to build ATP from ADP and phosphate, things seem to have started off exactly backwards. So why, then, is ATP being expended when the idea is to gain ATP? In a sense, the first two ATPs can be considered to be an investment—priming the pump, as it were.

There are several proposed explanations for these pump-priming steps. One idea is that the charged phosphate groups serve as "handles," and these serve to bind the intermediates to their appropriate enzymes. Another hypothesis notes that neutral sugars, such as glucose, are free to diffuse out of the cell, but phosphorylated compounds are unable to cross the cell membrane. This suggests that the phosphates serve to keep glycolysis localized within the cell. While there may be truth to both these explanations, there is a third hypothesis, which is that the addition of phosphates makes the stable glucose molecule suddenly unstable and reactive. Once the stability of its bonds has been weakened, the glucose can be broken apart, and at the right moment the phos-

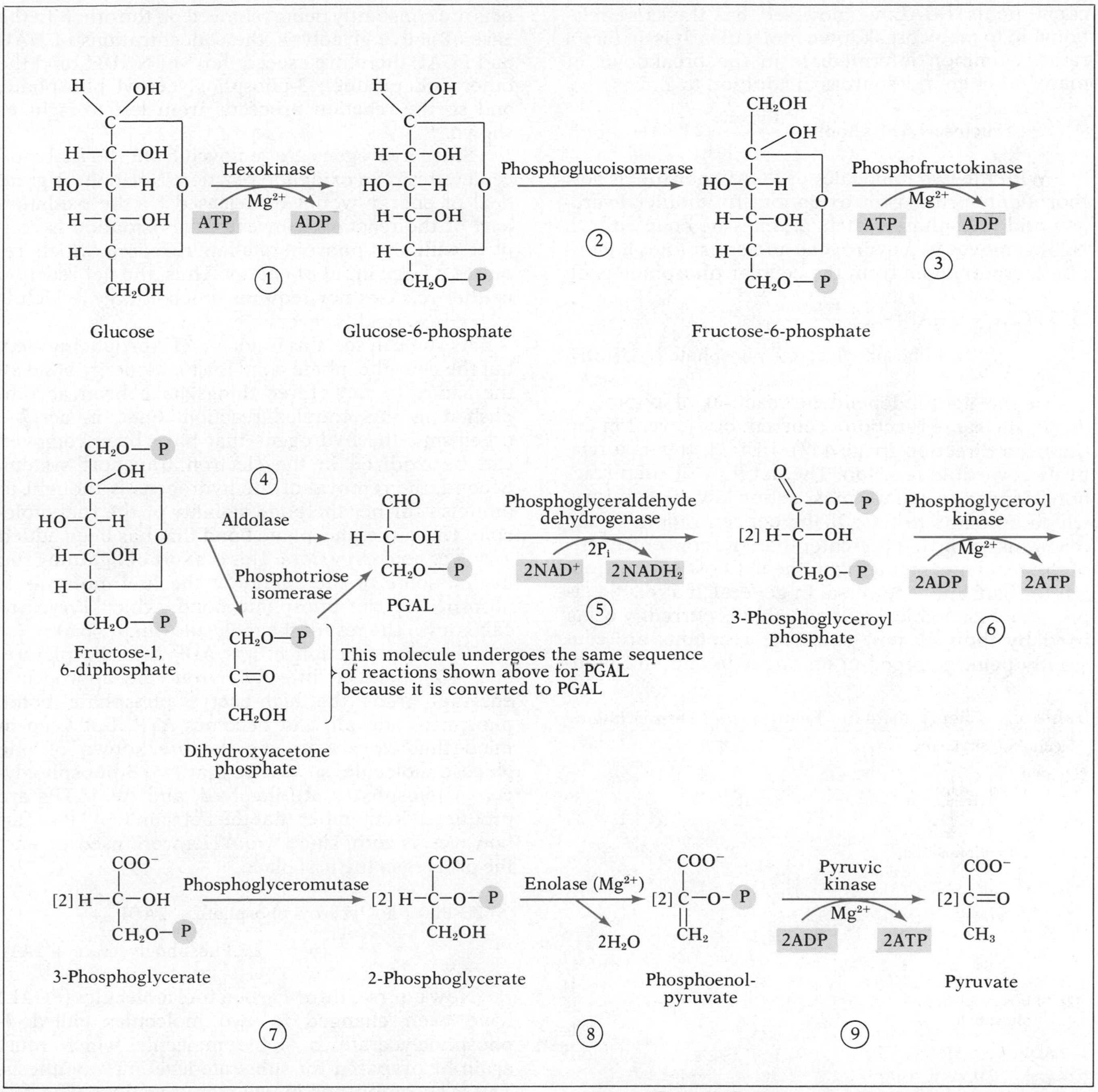

Glucose
Hexokinase
Mg^{2+}
ATP ADP
①
Glucose-6-phosphate
Phosphoglucoisomerase
②
Fructose-6-phosphate
Phosphofructokinase
Mg^{2+}
ATP ADP
③
Fructose-1, 6-diphosphate
Aldolase
④
PGAL
Phosphotriose isomerase
Dihydroxyacetone phosphate
This molecule undergoes the same sequence of reactions shown above for PGAL because it is converted to PGAL
Phosphoglyceraldehyde dehydrogenase
$2P_i$
$2NAD^+$ $2NADH_2$
⑤
3-Phosphoglyceroyl phosphate
Phosphoglyceroyl kinase
Mg^{2+}
2ADP 2ATP
⑥
3-Phosphoglycerate
Phosphoglyceromutase
⑦
2-Phosphoglycerate
Enolase (Mg^{2+})
$2H_2O$
⑧
Phosphoenol-pyruvate
Pyruvic kinase
Mg^{2+}
2ADP 2ATP
⑨
Pyruvate

6.4 In glycolysis, the first stage of respiration, phosphorylation of ADP, occurs at the substrate level. This means that the bond energy in the fuel molecule (such as glucose) is used directly to form the terminal high-energy phosphate bond of ATP. There is no electron transport system or chemiosmotic gradient involved. There is a total of nine major events, each with its own specific enzyme. Two molecules of ATP must be invested to begin the reactions ① and ③. These are returned further along in an oxidation–phosphorylation reaction ⑤, followed by ATP synthesis ⑥. Adenosine triphosphate is again produced further along ⑨. In all, glycolysis represents a net gain of two ATPs. The final reaction yields two molecules of pyruvate, each containing a considerable amount of potential energy.

phates can be removed along with some of the chemical bond energy that was invested in them.

The fructose-1,6-diphosphate produced is at once cleaved into two three-carbon compounds. Through a second reaction, these are transformed into two molecules of phosphoglyceraldehyde (PGAL). These enzyme-mediated reactions do not require the input of additional energy. If you'll think back to our discussion of the Calvin cycle in the light-independent reaction of photosynthesis, you might remember that we discussed how PGAL was a basic molecule that could be incorporated into any of many different cell constituents. PGAL, we now see, has the same relationship to many breakdown molecules. It is, in fact, a rather common intermediate in the breakdown of many other energy sources in addition to glucose.

④ Fructose-1,6-disphophate $\xrightarrow{\text{Enzyme}}$ 2 PGAL

With the two molecules of PGAL, our fuel is now thoroughly primed for oxidation. In another hydrogen and phosphate switch, an enzyme coupled with NAD removes two hydrogens and substitutes a phosphate group taken from the cellular phosphate pool:

⑤ 2 PGAL + 2NAD + $2P_i \xrightarrow{\text{Enzyme}}$

2 3-Phosphoglyceroyl phosphate + $2NADH_2$

In the light-independent reaction of photosynthesis, the same reaction occurred, but it went in the opposite direction (page 147). In fact, this is a relatively reversible reaction. The net flow of such reactions depends upon the mass action law of chemistry, which amounts to this: If the concentrations of the reactants on the left is greater than the concentrations of the reactants on the right, the net flow will be from left to right, and vice versa. In general, if a reaction is part of a biochemical pathway that is currently being used by the cell, new substrate reactants are constantly being pumped in on one side and end products are constantly being removed on the other. In the case of active glycolysis, the concentrations of NAD and PGAL therefore exceed those of $NADH_2$ and the other end product, 3-phosphoglyceroyl phosphate, and so the reaction proceeds from left to right as shown.

Since hydrogens are removed from the fuel molecules, this is an oxidation reaction. Normally, a great deal of energy would be released by the oxidation part of the reaction. However, the oxidation is coupled with the phosphorylation reaction, which requires a large input of energy. Thus, the net reaction neither releases nor requires much energy, which is why it is so readily reversible.

As you can see, this is not an ATP-producing step, but the new phosphate bond is a high-energy bond all the same. In fact, three things have been accomplished in this coupled reaction. First, in aerobic organisms, the hydrogens that have been removed can be oxidized in the electron transport system. Second, the removal of the hydrogens is thought to diminish further the bond stability of the fuel molecule. Third, the phosphate bond that has been added serves as an *energy store.* This is a sort of lightning rod in that some of the energy of the fuel molecule is absorbed in the phosphate bond, which serves to calm down the rest of the molecule (Figure 6.5).

The next reaction brings ADP into the picture, and ADP (with a little help from the appropriate enzyme) grabs the high-energy phosphate bond, phosphate and all, and becomes ATP. But keep in mind that we are tracing the breakdown of one glucose molecule, so we see that two 3-phosphoglyceroyl phosphates are involved, and two ATPs are produced. Remember that the net gain in ATP so far, however, is zero, since two ATPs were used to start the process in the first place.

⑥ 2 3-Phosphoglyceroyl phosphate + 2ADP $\xrightarrow{\text{Enzyme}}$

2 3-Phosphoglycerate + 2ATP

Now our two three-carbon fuel molecules (PGAL) have been changed to two molecules called 3-phosphoglycerate, a stable molecule which must again be prepared for substrate-level ATP synthesis. This is accomplished in two more reactions. The first reaction moves the phosphate from its end carbon to the center carbon, where it becomes a potential energy store. The second event, a dehydration reaction, removes two hydrogens and an oxygen. So, a second disruption in energy distribution has been accomplished, with a shift of energy to the energy store. The phosphate bond, which had previously been a relatively low-energy bond, becomes a high-energy bond, ripe for picking.

⑦ 2 3-Phosphoglycerate $\xrightarrow{\text{Enzyme}}$ 2 2-Phosphoglycerate

Table 6.1 Energy Yield in Glycolysis and Fermentation

Glucose (680 kcal)		
2 ATP → 2 ADP (Reactions 1 and 2)	−16 kcal	
Fructose-1,6-diphosphate		
(2) PGAL		
← P_i		
(2) 3-Phosphoglyceroyl phosphate		
2 ADP → 2 ATP (Reaction 6)		+16 kcal
(2) Phosphoenolpyruvate		
2 ADP → 2 ATP (Reaction 9)		+16 kcal
(2) Pyruvate		
Total	−16 kcal	+32 kcal
Net gain	2 ATP	16 kcal
Percentage efficiency	$\frac{16}{680}$ =	.024 or 2.4%

6.5 What is an energy store? How a stable molecule is prepared for an energy transfer and the role of phosphate in this process. The stable molecule is first oxidized (by removing one of its hydrogens). This creates a highly unstable situation in which the molecule would normally react with some other substances in an undirected manner. But, instead, the molecule is phosphorylated. The presence of phosphate produces an energy store into which excess energy can be drawn. Here, the phosphate is eventually removed and used to change ADP to ATP.

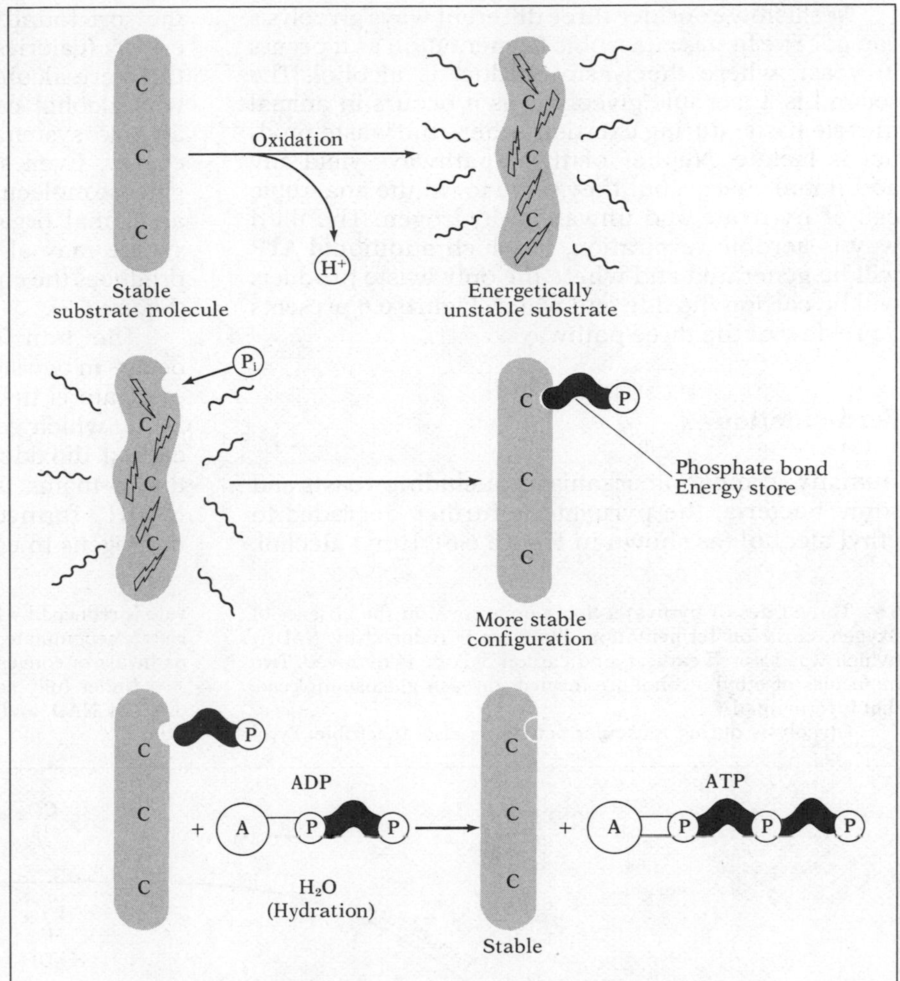

Another enzyme transfers the high-energy bond and phosphate to ATP, leaving our two fuel molecules as *pyruvate* ions.

⑧ 2 2-Phosphoglycerate $\xrightarrow{\text{Enzyme}}$ 2 Phosphoenolpyruvate + $2H_2O$

⑨ 2 Phosphoenolpyruvate + 2ADP $\xrightarrow{\text{Enzyme}}$ 2Pyruvate + 2ATP

Now, we can see that ATP synthesis has returned the original ATP investment and produced a gain of two ATPs through glycolysis. It is here, by the way, that fermentation and glycolysis begin to differ from one another. But before we look at these differences, let's see how much energy the cell has gained so far (Table 6.1).

The high-energy bonds of ATP have a caloric value of 8 kcal/mole. Ignoring the $NADH_2$ and pyruvate for the moment, the net yield of glycolysis of 1 mole of glucose to the pyruvate stage is 2 moles of ATP, worth 16 kcal. Now, 1 mole of glucose weighs 186 grams (or about 2.5 ounces), and if it is burned in oxygen, it can release 680 kcal. On the molecular level, then, the energy stored in the two ATP high-energy bonds represents about 2.4% of the oxidizable energy of a molecule of glucose (that is, the glycolysis is 2.4% efficient). This is a low yield compared with aerobic respiration, but if the cell has no oxygen, it must do the best it can. Why would an organism utilize such an inefficient energy-producing process? As we mentioned earlier, anaerobic organisms can live in places where other organisms cannot, for instance, in mud, in draft beer, or in improperly sterilized canned goods.

We'll now consider three different ways glycolysis can go. The first is anaerobic fermentation as it occurs in yeast, where the waste product is alcohol. The second is anaerobic glycolysis as it occurs in animal muscle tissue during exertion, where the waste product is lactate. Neither of these pathways yield any additional energy, but they serve to rid the anaerobic cell of pyruvate and unwanted hydrogen. The third way is aerobic respiration, in which additional ATP will be generated and where the only waste products will be carbon dioxide and water. Figure 6.6 presents a preview of the three pathways.

Fermentation

In many fermenting organisms, including yeasts and some bacteria, the pyruvate is further degraded to ethyl alcohol (as shown in Figure 6.6). Ethyl alcohol, the sort found in spirits, still has a lot of potential energy (calories) in an aerobic system, which is why there are alcohol stoves and alcohol lamps and also why alcohol can be fattening. However, in an anaerobic system, alcohol is nearly depleted of free energy. Even so, some microorganisms can milk a glucose molecule even further, giving off methane gas as a final organic waste. Hydrogen and hydrogen sulfide may also be released. The hydrogen sulfide produces the characteristic smell of swamps and bogs (Figure 6.7).

The transformation of pyruvate into alcohol occurs in two steps, each with its specific enzyme. The pyruvate is first acted upon by a *decarboxylase* enzyme, which removes the carboxylic acid group as carbon dioxide. The nearly depleted fuel molecule that remains is known as *acetaldehyde*. Next, the $NADH_2$ formed earlier in glycolysis adds its two hydrogens to each molecule of acetaldehyde, reduc-

6.6 Three fates of pyruvate. Some organisms, in the absence of oxygen, carry on fermentation. Pyruvate is reduced by $NADH_2$ (which was formed earlier), and carbon dioxide is removed. Two molecules of ethyl alcohol are formed for each glucose molecule that is fermented.

Glycolysis during muscular activity is also anaerobic. Pyruvate is reduced by $NADH_2$ to lactate. Although large quantities of lactate accumulate, eventually all of it will be sent through aerobic pathways or converted back to glucose.

Under fully aerobic conditions in muscles, pyruvate is oxidized by NAD, and CO_2 is removed, producing two-carbon acetyl CoA.

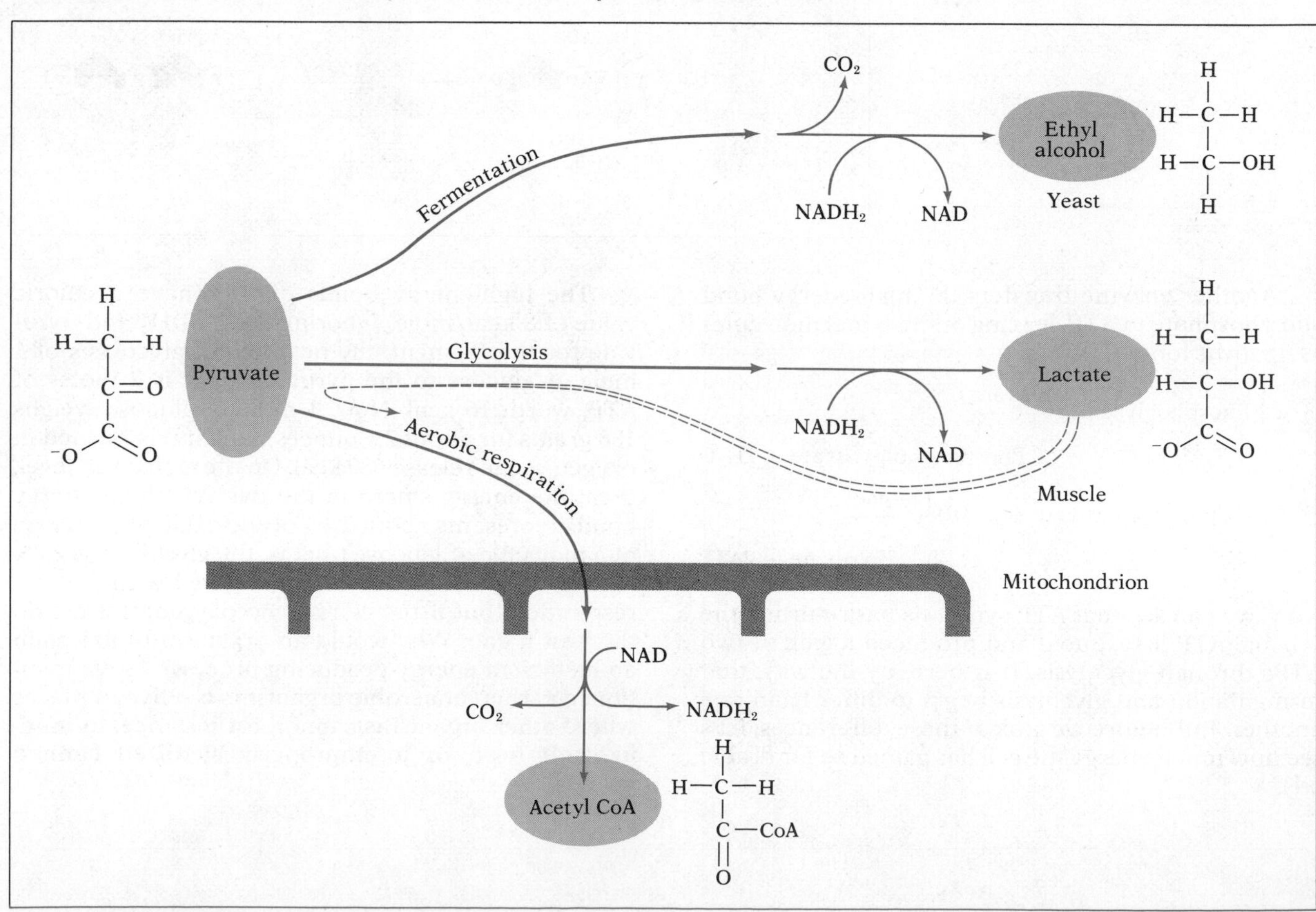

6.7 Marshes, bogs, and swamps are stagnant places that are notably short on oxygen. They are the habitat of anaerobic organisms whose presence is easily detected by the presence of foul-smelling gases.

ing it to ethyl alcohol and regenerating NAD. This step is important to fermentation, because the NAD supply must be recycled in order for glycolysis to continue. In the absence of oxygen, the hydrogen of $NADH_2$ is not a fuel but a waste product. The final step of fermentation, then, is the combining of two waste products, hydrogen and acetaldehyde, to form one waste product, alcohol (Figure 6.8).

6.8 In total contrast to the anaerobic swamp, where nature has her way, is this scene in a commercial winery. Under the strictest sanitary conditions, an anaerobic state is maintained during the production of wine. Pure cultures of yeast, used in the conversion of carbohydrate to ethyl alcohol, are introduced after pasteurization.

Yeasts in general are facultative anaerobes, which means that they are opportunistic in that they can switch their metabolism from full oxidative respiration to anaerobic fermentation and vice versa, depending upon conditions. The aerobic respiration of yeast, by the way, is biochemically identical to human aerobic respiration.

Glycolysis in Muscle Tissue

During heavy exertion, a great deal of ATP can be used up in a hurry. Muscle cells normally operate aerobically, but in larger animals there is no way for the circulatory system to bring in oxygen fast enough to replace the ATP through oxidative respiration, so muscle cells under heavy exertion have two backup systems. The first backup system is a store of high-energy bonds in a molecule that is abundant in muscles, *creatine phosphate.*

```
      High-energy
         bond
                        H
     H    |             |
     O    |  H   H—C—H   H        OH
     |    ↓  |      |    |       /
   O=P  ~  N—C—N————C—C
     |       ||          |       \\
     O       N           H        O
     H       |
             H
```

Creatine phosphate
(Phosphocreatine)

Creatine phosphate is a much more compact molecule than ATP and is less reactive, and the cell can store quite a lot of quick energy in this form. Creatine phosphate doesn't provide muscle contraction energy directly, but it can transfer its own high-energy bond to ADP to regenerate ATP:

$$\text{Creatine phosphate} + \text{ADP} \rightarrow \text{ATP} + \text{Creatine}$$

The reaction goes readily in either direction, but when the reserves of ATP are low and the concentration of ADP is high, the mass action law dictates that the net flow of the reaction will be from left to right.

As the creatine phosphate is gradually used up, the muscle tissue falls back to yet another quick energy source, anaerobic glycolysis. The pathway to pyruvate is just as we have described it above. As before, the $NADH_2$ has to be rapidly recycled to NAD. This time, however, $NADH_2$ is combined directly with pyruvate to form lactate instead of alcohol, and no carbon dioxide is released (Figure 6.6).

Lactate accumulates rapidly during intense muscular activity. Animals can remove the lactate in two ways. Some of it may remain in the muscles until the amount of oxygen being brought in by the circulation exceeds the muscle's current needs, in which case the reaction is reversed and the pyruvate proceeds into the oxidative respiration pathway. There it will combine with oxygen to produce ATP and, eventually, carbon dioxide and water. In addition, some of the lactate is washed away by the circulatory system and carried to the liver where, once enough oxygen is available, it will be metabolized back into glucose. Figure 6.9 summarizes these events.

Oxygen Debt

After a period of heavy exertion, the muscle tissues in humans and other vertebrates will be depleted of creatine phosphate and both the liver and the muscle will be loaded with lactate. Perhaps you have experienced this condition; fatigue hurts, and most people try to avoid it, although a few madmen like marathon runners claim to find it exhilarating. The final stage is sheer exhaustion. When the activity stops, it takes a period of time and a large amount of oxygen and ATP for the lactate to be metabolized and for the creatine to be regenerated as creatine phosphate. During this time you can expect to find yourself continuing to breathe hard, taking in as much oxygen as the lungs can handle. The state of oxygen and creatine phosphate depletion is known as the *oxygen debt,* which can be measured in the amount of extra oxygen needed to restore the system to its preexertion equilibrium. How long it takes to repay the oxygen debt obviously depends on a person's physical condition. *Physical conditioning,* in turn, involves increasing respiratory and circulatory capacity as well as increasing the storage capacity and possibly the mass of the muscles themselves. Conditioned runners have enlarged lung capacities, increased capillary beds, hearts that pump more blood with each stroke, and an increased ability to utilize oxygen. Some may have a high tolerance for pain—at least they act as though they do.

The Metabolism of Starch and Glycogen

Starch and glycogen are polymers of glucose in which the glucose units are held together by 1–4 linkages (Chapter 3). The energy in this type of bond is rela-

6.9 The energy reserves of muscle. (a) In sustained muscular activity, the ATP reserves are depleted in the first few minutes. (b) To refurbish the supply, a much larger reservoir of creatine phosphate is tapped. Its high-energy phosphate is transferred to ADP, producing more ATP. As activity continues, even the creatine phosphate becomes depleted. To restore high-energy bonds to creatine, ATP must be produced in the mitochondria, which is a lengthy process. (c) Meanwhile, the muscle switches to the fast but limited process of anaerobic glycolysis for its ATP supply. Here, glucose is converted to lactate which will soon accumulate in muscle cells. Then the inevitable "oxygen debt" sets in. (d) Recovery requires ATP input from aerobic respiration in the mitochondria. Some of the lactate is used for this. ATP from the oxidation of lactate restores the ATP and creatine phosphate reservoirs. During this recovery period, runners breathe heavily as their lactate accumulation is removed. When all three reservoirs are replenished, recovery is complete.

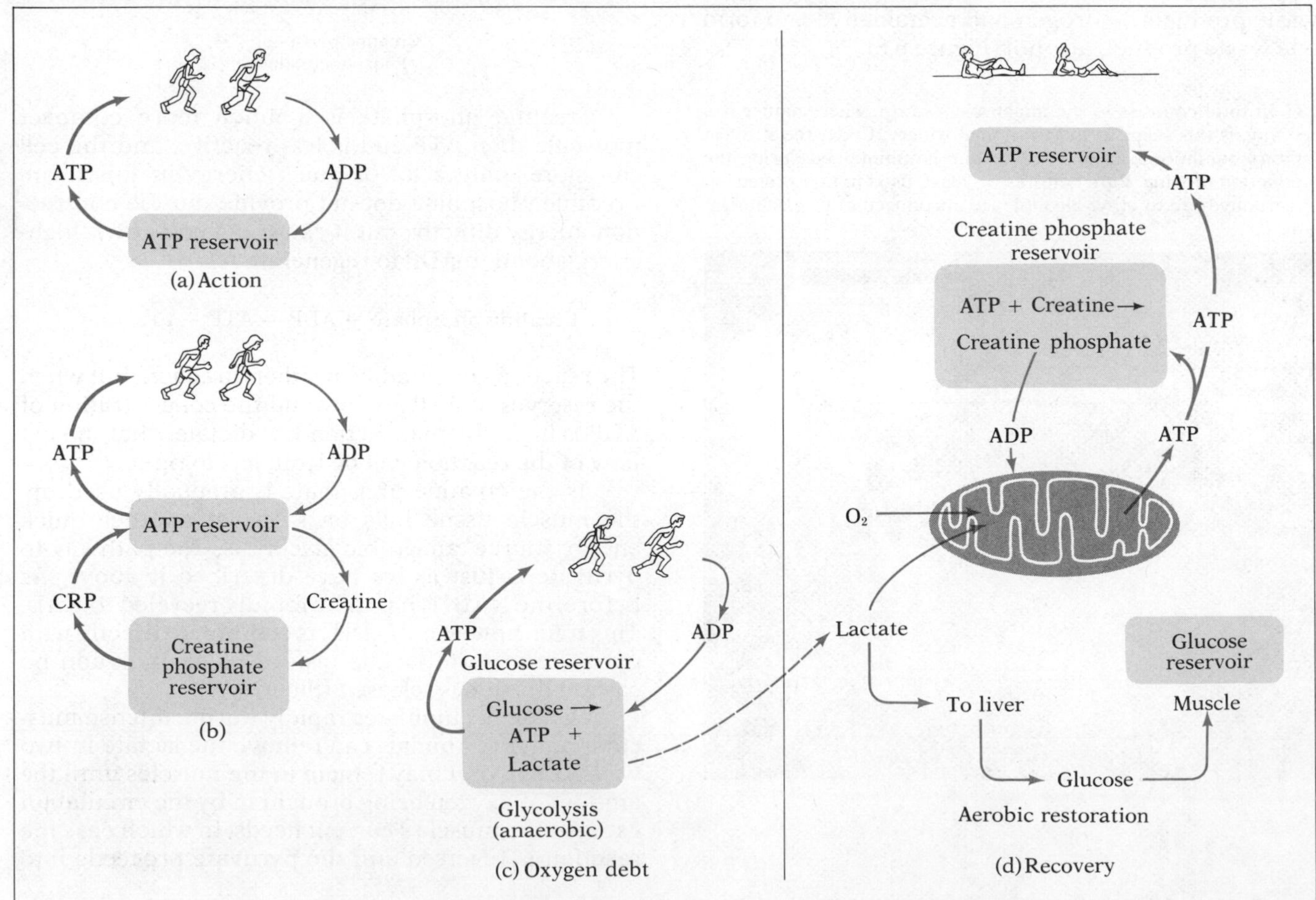

6.10 Long-chain polysaccharides can be broken down into their simple components in preparation for glycolysis by a special process that results in higher yields of ATP than usual. Instead of breaking the linkages by adding H_2O (hydrolysis), the 1–4 linkages are broken by phosphorolysis. Here, the enzyme phosphorylase cleaves the glucose units and at the same time adds inorganic phosphate to each number 1 carbon. No ATP is required. As a result, the first step in glycolysis is bypassed and no ATP is invested until step 3 in Figure 6.4 (the production of fructose-1,6-diphosphate). The net yield at the end of glycolysis is then three ATPs instead of the usual two.

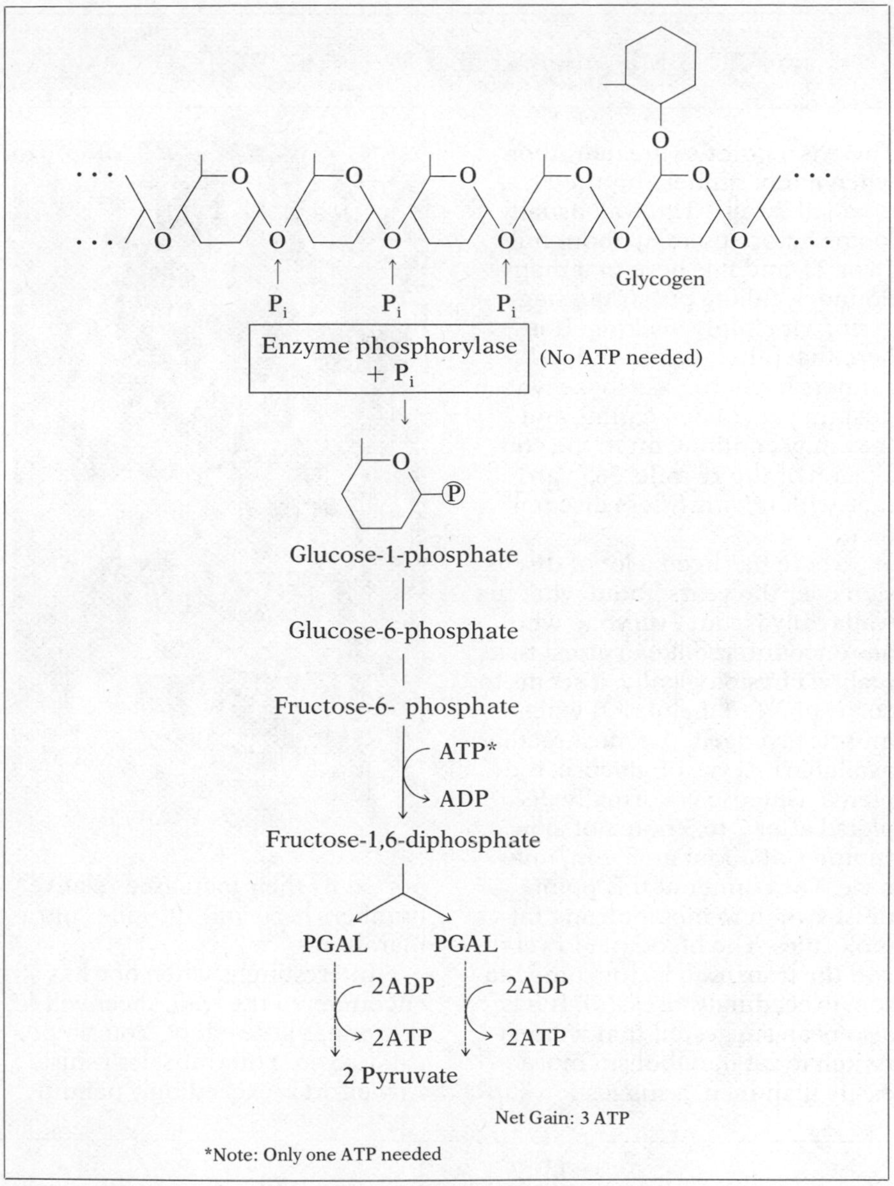

tively small, but the 1–4 linkage means that these molecules contain more energy than does free glucose. In fact, the carbohydrate energy stores of cells are usually in the form of starch or glycogen, and few cells contain very much free glucose at any one time. The polymers are broken down into single glucose units by *phosphorolysis,* in which the 1–4 bond is split by an enzyme that substitutes a phosphate ion on the number 1 carbon. The resulting breakdown product of both starch and glycogen is glucose-1-phosphate, which can be changed to glucose-6-phosphate by another enzyme. While none of these reactions require ATP or any other energy input, the phosphorylation of glucose itself, as we have seen (Figure 6.4), requires the expenditure of ATP. As a result, the net gain in the anaerobic glycolysis of starch and glycogen is three ATPs per glucose subunit rather than just two (Figure 6.10).

Hard-working muscle tissue is not the only animal tissue in which anaerobic glycolysis takes place. Other cell types can also use the lactate pathway. Red blood cells, for instance, lack nuclei and mitochondria, and respire anaerobically. Many tumors and cancers metabolize glucose anaerobically, throwing off large amounts of lactate into the blood stream. The reason for this shift in respiratory pathways is not known, but many cancer researchers have suggested that it is of basic importance in understanding cancer. One prominent hypothesis is that the shift to anaerobic metabolism comes first, and that this change causes the tissue to be cancerous.

Essay 6.1 A Marathon Runner Meets the "Wall"

There is a point in the marathon when many runners hit the dreaded "wall." The wall usually looms before them at about mile 18 or 20, and it is here that many flounder, falling prostrate, staggering, or simply walking. It is here that other, better-trained runners begin to pass those whose dust they had been eating, and they may continue on to the conclusion of the 26 mile, 385 yard race with *relatively* less discomfort.

There has been a lot of discussion over the years about what the wall really is, but everyone who has encountered it can attest to its reality. Physiologically, it seems to correspond to the period when muscle glycogen, the most readily available reserve of glucose, is depleted. Glucose is normally depleted after 2 to 3 hours of slow running, at about an 8 min/mile pace. The runner at this point must switch to metabolizing fat molecules. The blood pH lowers and the transition is, for some reason, exceedingly stressful. It has also been suggested that women switch to fat metabolism more easily than men, a suggestion supported by their increased relative numbers in 50 and 100 mile "ultramarathons."

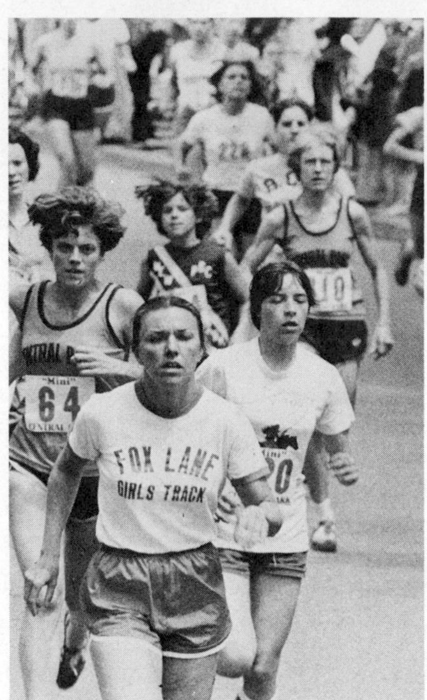

Interestingly, when one has encountered the wall, sheer willpower has little effect. You simply can't go on. Your muscles refuse. Any effort is exceedingly painful and fatiguing. In addition, blood sugar levels drop and you may suffer the psychological depression associated with hypoglycemia.

Fluid loss can only be replaced by fluid intake, particularly of isotonic fluids such as Gatorade. Tomato juice also quickly replaces the potassium lost in sweat. Experienced runners drink en route, but avoid salt tablets, since they may distort the salt/water balance if not taken with a precise quantity of water.

Marathoners have been able to beat the wall by training so hard that it no longer exists for them. This training includes long runs of 20 miles or more during which muscle glycogen is repeatedly depleted and then restored in even greater abundance during the following week. They have also been able to move the wall back by a depletion run 1 week before the race followed by 3 days of low-carbohydrate dieting and 3 days of carbohydrate "loading," when it is theorized that the deprived muscles overcompensate and store glycogen in large quantities.

6.2 Aerobic Respiration

Pyruvate to Acetyl CoA

The third possible route in the glycolysis pathway is the most important one in all eucaryote cells. This is the one that involves *oxidative phosphorylation.* The products of fermentation are still high in oxidizable energy. The aerobic metabolism of glucose yields far more energy than anaerobic metabolism. Compared with the two or three molecules of ATP yielded by anaerobic glycolysis, aerobic respiration yields anywhere from 21 to 36 ATPs per glucose molecule.

As you recall, we left anaerobic glycolysis with pyruvate and $NADH_2$, products formed in the cytoplasm. The site of activity then changes since the remainder of respiration takes place within the mitochondrion.

The $NADH_2$ cannot enter the inner chambers of the mitochondrion, but it can pass its electrons to a shuttle carrier on the surface of the inner membrane and build up the supply of electrons inside in this way. The pyruvate, however, can enter the mitochondrion, where it is altered so that it can take part in the citric acid cycle. This alteration turns out to be a fairly complex process. The overall reaction is

$$\text{2 Pyruvate} + \text{2 NAD} + \text{2 Coenzyme A} \rightarrow \text{2 Acetyl CoA} + 2\,NADH_2 + 2\,CO_2$$

Coenzyme A (CoA) is covalently linked by a high-energy bond to what's left of our fuel molecule, which has now been reduced from a three-carbon pyruvate to a two-carbon acetyl, to form *acetyl CoA* (Figure 6.11).

Acetyl CoA is a vital intermediate in the utilization of cellular fuels other than glucose. We have mentioned that marathon runners eventually burn fat. In other metabolic pathways, amino acids (from proteins) and glycerol (from fats) are also broken down into acetyl CoA.

The Citric Acid Cycle

The *citric acid cycle* (or *Krebs cycle*) was first postulated in 1937 by H. A. Krebs. Legend has it that his paper, which has become a classic in biochemistry, was rejected by the disbelieving editors of the first journal he submitted it to.

The citric acid cycle occurs in the inner matrix of the mitochondrion. As soon as acetyl CoA enters this inner chamber, it reacts with a four-carbon molecule known as *oxaloacetate* and an enzyme called *citric synthetase.* The high-energy bond between coenzyme A and the acetyl group causes the reaction to go quickly, with the release of a considerable amount of free energy. The coenzyme A is simply released to return through the inner mitochondrial membrane, but the acetyl group combines with the oxaloacetate to form a new six-carbon compound, citric acid—the same citric acid found in lemons. However, the interior of the mitochondrion is alkaline, so the citric acid is neutralized to form *citrate.* One molecule of water is used up in the citric synthesis reaction.

6.11 Upon entering the mitochondrion, pyruvate must first be converted to acetyl CoA before it can join the citric acid cycle. The pyruvate is joined to an enzyme complex, decarboxylated, reduced by NAD, and finally joined by a high-energy bond to coenzyme A. For each molecule of glucose entering respiration, two pyruvate molecules are formed and these are converted to two molecules of acetyl CoA.

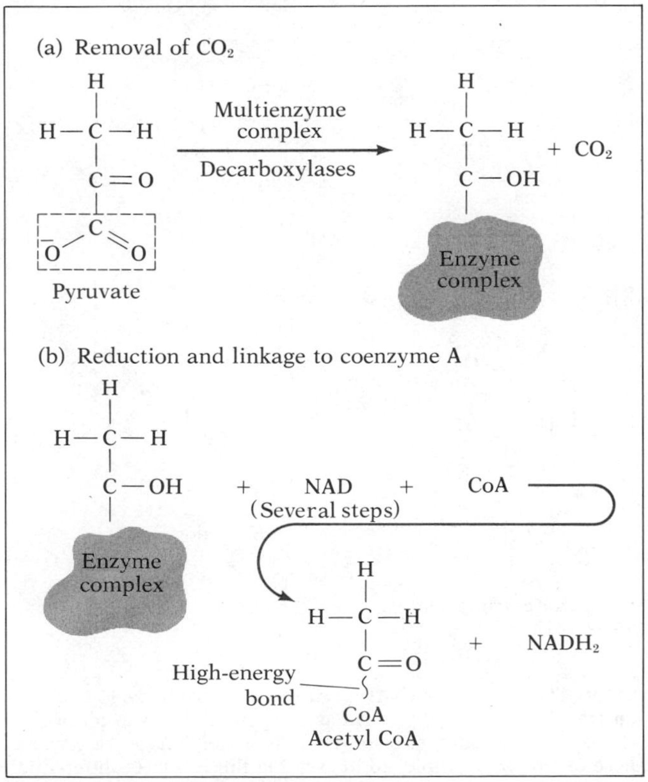

(a) Removal of CO_2

Pyruvate

Multienzyme complex

Decarboxylases

$+ CO_2$

Enzyme complex

(b) Reduction and linkage to coenzyme A

Enzyme complex

$+$ NAD $+$ CoA

(Several steps)

$+ NADH_2$

High-energy bond

CoA

Acetyl CoA

An Overview of the Citric Acid Cycle

With the union of oxaloacetate and acetyl CoA to form citrate and free CoA we have formally entered the citric acid cycle. Figure 6.12 illustrates the entire cycle, showing the series of events as our fuel molecule fragment is acted upon sequentially by a battery of enzymes. The numbers in the following discussion refer to the numbers in the diagram. As you can see, in the course of the cycle, a six-carbon intermediate is broken down to a five-carbon compound, with the release of a carbon dioxide (between steps 3 and 4). The five-carbon intermediate is altered a bit (step 4) and then broken down to a four-carbon compound, with the release of another useless carbon dioxide. In each cycle, two carbons leave as carbon dioxide and two enter as an acetyl group. Eventually, the four-carbon compound is changed to oxaloacetate (step 9), and the cycle continues.

The energy of the acetyl group, you see, is milked from the cycle in repeated removals of hydrogens and electrons. These are transferred to NAD and another hydrogen carrier, flavine adenine dinucleotide (FAD), reducing them to $NADH_2$ and $FADH_2$ (steps 4, 5, 7, and 9). Every hydrogen and electron removal is another oxidation of the fuel molecule, so the glucose is very gradually "burned." At one point along the way, an additional bit of energy is picked off in a high-energy substrate-level phosphorylation (step 6). Notice also that three water molecules enter the cycle at various points, including the water that enters into the synthesis of citrate; one water molecule leaves the reaction (steps 1, 2, 3, and 8). The net reaction of one turn of the cycle, leaving out the various cofactors (such as NAD and FAD), is

$$\underset{\text{Acetic acid}}{C_2H_4O_2} + 3\,H_2O \rightarrow H_2O + 2\,CO_2 + 8\,H$$

Some of the hydrogens that are sent on to the electron transport system don't come from the original fuel molecule itself, but from the water that enters the reaction.

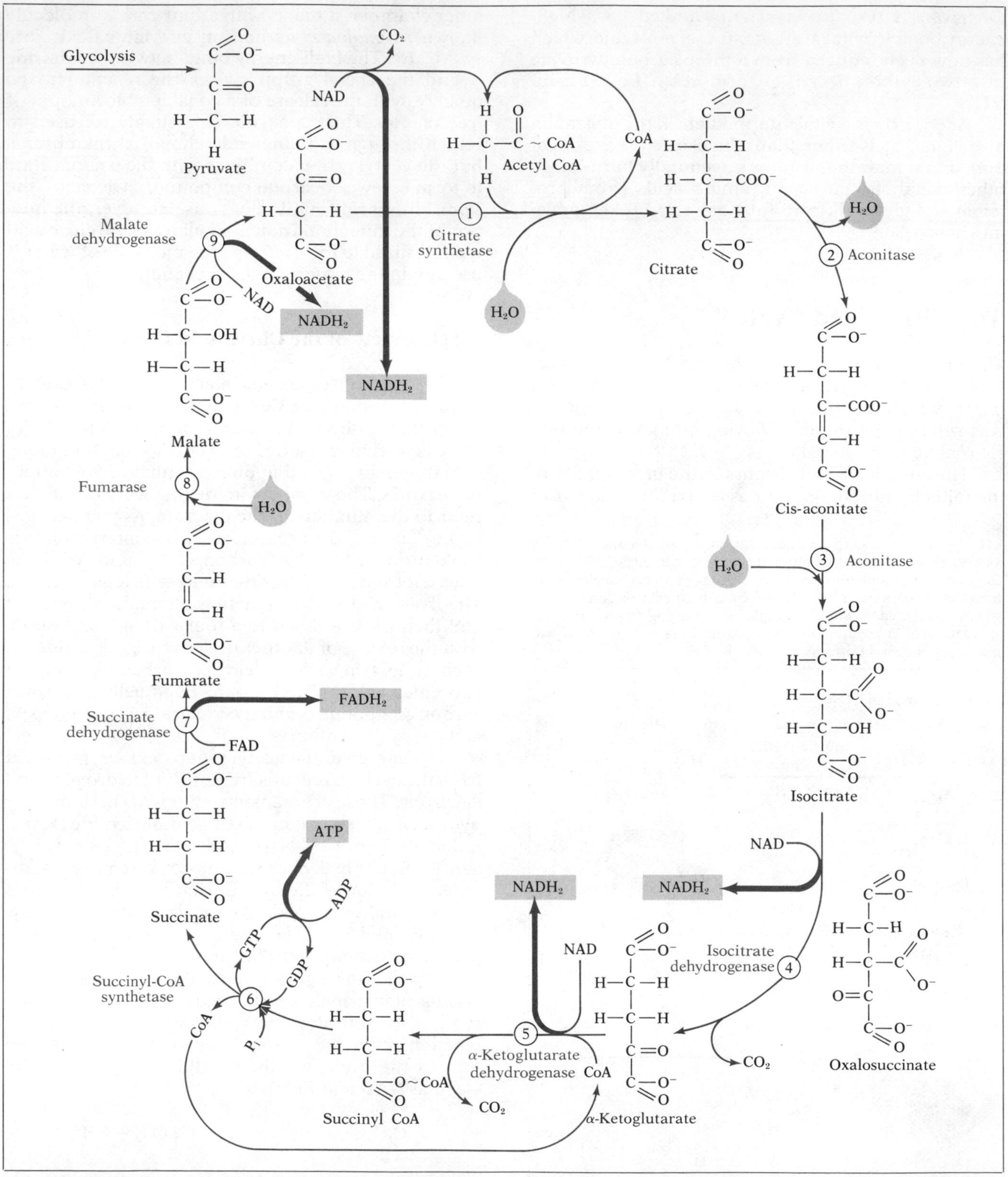

6.12 The citric acid cycle, which occurs in the mitochondrial matrix, has nine major enzymatic steps. Each turn of the cycle begins and ends with oxaloacetate. First, oxaloacetate joins with acetyl CoA and the CoA is released, thereby forming citric acid. As each enzyme does its job in the succeeding steps, molecules change, CO_2 is removed, NAD performs its oxidations, H_2O enters and leaves, and a GTP is formed. The main object in all of this is to use the molecules as a source of hydrogen ions and electrons. These can then be employed in creating the H^+–OH^- differential in the chemiosmotic production of ATP.

If we include the cofactors, one turn of the cycle can be written as:

$$\text{Acetyl CoA} + 3\,\text{NAD} + \text{FAD} + 3\,H_2O + \underset{\text{Guanosine diphosphate}}{\text{GPD}} + P_i \rightarrow \text{CoA} + 3\,\text{NADH}_2 + \text{FADH}_2 + \underset{\text{Guanosine triphosphate}}{\text{GTP}} + H_2O + 2\,CO_2$$

Of course, if we want to continue following both of the acetyl CoA molecules that result from the breakdown of one glucose molecule, the terms in the formula have to be doubled.

In Figure 6.12, we have included the names of all the enzymes and intermediates and have given the structural formulas as well, to fill you in on the details of what's happening. In our discussion, however, the names of these substances are secondary. We are more interested in following the events. So, now for a quick run through the cycle.

Let's proceed clockwise from step 1, which we have discussed already, on through steps 2 and 3, in which another water molecule is added and the citric acid molecule is subtly rearranged, losing a water molecule (and being renamed), to step 4. In step 4, isocitrate is oxidized to oxalosuccinic acid, as NAD is reduced to $NADH_2$. Oxalosuccinate, while still joined to the isocitrate dehydrogenase enzyme, loses a carbon dioxide. The remaining five-carbon compound, called α-ketoglutaric acid, is then released. Let's review the first four steps now, since a lot has happened:

① $\text{Acetyl CoA} + \text{Oxaloacetate} + H_2O \xrightarrow{\text{Enzyme}} \text{Citrate} + \text{CoA}$

② $\text{Citrate} \xrightarrow{\text{Enzyme}} \text{Cis-aconitate} + H_2O$

③ $\text{Cis-aconitate} + H_2O \xrightarrow{\text{Enzyme}} \text{Isocitrate}$

④ $\text{Isocitrate} + \text{NAD} \xrightarrow{\text{Enzyme}} \underset{\text{(bound to enzyme)}}{\text{Oxalosuccinate}} + \text{NADH}_2 \xrightarrow{\text{Enzyme}} \alpha\text{-Ketoglutarate} + CO_2$

The enzyme alphaketoglutarate dehydrogenase then enters the picture and performs a dazzling display of chemical virtuosity. Essentially, what happens is, NAD is reduced, carbon dioxide is released, and the remaining four-carbon skeleton is covalently linked to coenzyme A as succinyl CoA.

⑤ $\alpha\text{-Ketoglutarate} + \text{CoA} + \text{NAD} \xrightarrow{\text{Enzyme}} \text{Succinyl CoA} + CO_2 + \text{NADH}_2$

In the next reaction (step 6 in Figure 6.12), the high-energy bond is picked up by guanosine diphosphate (GDP) to form guanosine triphosphate (GTP), as coenzyme A is released. We don't know why the cell uses GDP here instead of ADP, but it does. In any case, the high-energy bond of GTP is readily transferred to ADP:

⑥ $\text{Succinyl CoA} + P_i + \text{GDP} \xrightarrow{\text{Enzyme}} \text{Succinate} + \text{CoA} + \text{GTP}$

$$\text{GTP} + \text{ADP} \rightleftharpoons \text{GDP} + \text{ATP}$$

Proceeding along to step 7, we find another oxidation (or dehydrogenation) reaction. But instead of the hydrogen being passed to NAD, as we have come to expect, coenzyme FAD does the job.

⑦ $\text{Succinate} + \text{FAD} \xrightarrow{\text{Enzyme}} \text{Fumarate} + \text{FADH}_2$

In step 8, the four-carbon skeleton adds another water to become malate. Malate dehydrogenase then oxidizes the malate (step 9) and produces our old friends oxaloacetate, to start the cycle again. As you see, NAD is again reduced to $NADH_2$ in the process.

⑧ $\text{Fumarate} + H_2O \xrightarrow{\text{Enzyme}} \text{Malate}$

⑨ $\text{Malate} + \text{NAD} \xrightarrow{\text{Enzyme}} \text{Oxaloacetate} + \text{NADH}_2$

Let's stand back now and take a look at the damage. We have made a shambles of our glucose molecule, having oxidized it completely to carbon dioxide. And what do we have to show for it? We have accumulated a net of only four ATPs, two from glycolysis and two (via GTP) from the citric acid cycle. This may not seem quite right, since the citric acid cycle was touted as the major energy producer. So far it's no better than inefficient glycolysis. But keep in mind that we have accumulated a lot of hydrogens in the form of $FADH_2$ and $NADH_2$—actually, twenty-

Table 6.2 Adding Up Hydrogen

The total hydrogen gained per glucose from all of respiration is 24:

Glycolysis (steps)	2 $NADH_2$
Pyruvate to acetyl CoA	2 $NADH_2$
Citric acid cycle	
Step 4	2 $NADH_2$
Step 5	2 $NADH_2$
Step 7	2 $FADH_2$
Step 9	2 $NADH_2$
Total	12 $NADH_2$ or 24 hydrogens

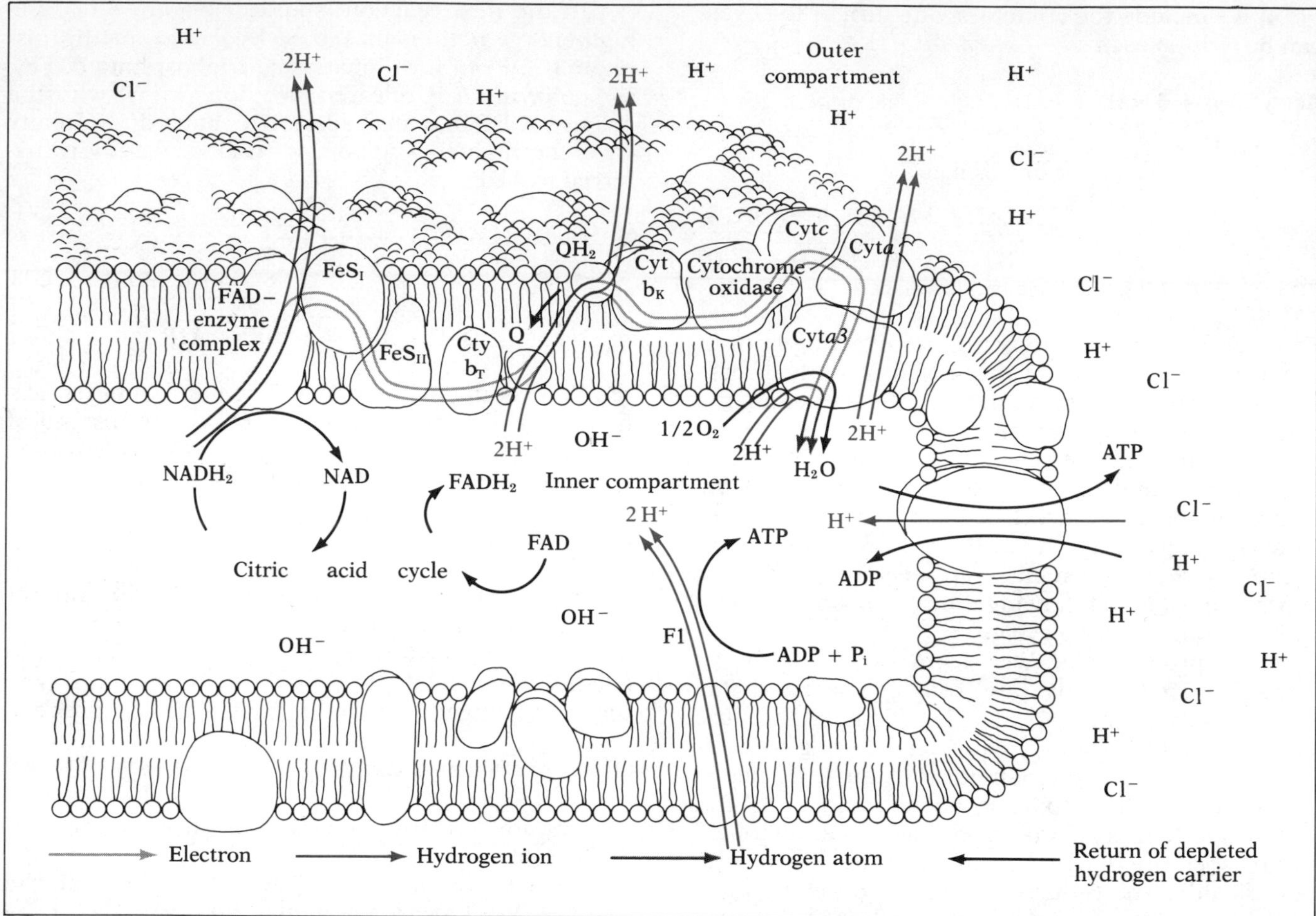

6.13 Electron transport in mitochondria. The enzymes of the citric acid cycle are in the matrix of the inner compartment of the mitochondrion; the enzymes and electron carriers of the electron transport system are embedded in the membrane of the crista. The electron transport system begins with the transfer of two hydrogens from $NADH_2$ to the FAD–enzyme complex. The hydrogens are moved across the membrane, where they are dissociated into electrons and hydrogen ions. The hydrogen ions are released into the outer compartment of the mitochondrion, and the two electrons flow through the following series of electron-carrier proteins: iron–sulfur protein I (FeS_I), iron–sulfur protein II (FeS_{II}), and cytochrome b_T. The small hydrogen carrier Q takes on the two electrons and two hydrogen ions (from the inner compartment) to become QH_2. QH_2 diffuses across the membrane to release the two hydrogen ions, and passes the electrons through the next series of electron-carrier proteins: cytochrome b_K, cytochrome oxidase, cytochrome *c*, cytochrome *a*, and cytochrome *a*3. Finally, the two electrons reduce an oxygen atom to water, removing two hydrogen ions from the interior compartment in the process. Hydrogen ions are also pumped across the membrane at one other point, probably at the cytochrome *a*–*a*3 complex as shown here. Altogether, the oxidation of $NADH_2$ to NAD and water results in the net transport of six hydrogen ions across the inner mitochondrial membrane. The $FADH_2$ from the citric acid cycle reacts with ubiquinone (Q), by-passing the first proton transport and resulting in the net flow of only four hydrogen ions across the membrane. In either case, the resulting H^+–OH^- gradient represents a store of potential free energy, which is tapped in several ways. In the F1 particle, two hydrogen ions are admitted into the mitochondrial matrix and the energy is used to phosphorylate ADP to ATP. Another proton-powered active transport system uses one hydrogen ion to pump out an ATP molecule and simultaneously bring in an ADP molecule.

four hydrogens altogether. That's a pretty good trick if you come to think about it, since glucose has only twelve hydrogens to start with. As we have seen, the other twelve hydrogens come from the six molecules of water that enter the cycle as the two acetyl CoA molecules (from glucose) go through the citric acid cycle (Table 6.2). These hydrogens are going to have a great deal to do with making more ATP.

The Electron Transport System in Mitochondria

We've already been through an electron transport system in the thylakoid membrane of the chloroplast. The electron transport chain of the mitochondrion has many similar features, which should not be too surprising. Both systems apparently evolved from a common ancestor long ago. You may wish to compare Figure 5.16 with Figure 6.13.

As in the case of the chloroplast, all the electron carriers are embedded in the membrane; here, they are attached to the inner membrane of the mitochondrion. As you saw in Chapter 4, the mitochondrion is a cell organelle with two membranes. The outer membrane is simple in structure, is highly permeable, and has the form of a closed sac. The inner membrane, however, is highly convoluted and forms extensive folds or shelflike cristae that reach into the interior of the organelle. The inner membrane is impermeable to hydrogen ions and many other substances. It is, essentially, a protein membrane, in which large proteins are bound together by phospholipids—more or less the way bricks in a wall are bound together by mortar (see Figure 4.24). The enzymes and electron carriers of the electron transport system are tightly bound within the inner membrane. Stalked F1 particles, which appear to be identical to CF1 particles, project inward from the surface of the inner membrane. This is the opposite of the arrangement of CF1 particles, which are on the outer surface of the thylakoid membrane.

There are other differences in the electron transport system of chloroplasts and mitochondria, of course. In particular, the photosynthetic electron transport chain *ended* with the electrons being passed from the last membrane-bound carrier to NADP; the mitochondrial electron transport system, on the other hand, *begins* with electrons being passed from $NADH_2$ to the first membrane-bound carrier in the chain. In addition, the flow in the Z scheme of photosynthesis *began* with the splitting of water into hydrogen ions, electrons, and molecular oxygen; in the mitochondrial system, the flow *ends* with the combining of hydrogen ions, electrons, and molecular oxygen into water. So mitochondrial and chloroplast electron transport systems are not only inside out but also backwards relative to one another.

In the chloroplast transport chain, the electrons started on one side of the membrane (inside the thylakoid) and finished on the other side (the stroma). Actually, each electron made three trips across the membrane (Figure 5.16). In contrast, the electrons of the mitochondrial chain start on the inside and end on the inside, crossing the inner mitochondrial membrane no fewer than six times, or three round trips. In each round trip, the electrons travel out as hydrogen atoms in one kind of carrier and return as electrons in a different kind of carrier. Each time a hydrogen carrier reaches the outside surface, it liberates hydrogen ions into the external medium, but the electrons are passed to an electron carrier in the membrane. Then each time an electron carrier reaches the inner surface, the electrons are passed to a hydrogen carrier, which combines them with hydrogen ions taken from the internal medium. For each electron that makes three round trips, the electron transport system pumps three hydrogen ions across the membrane.

Of course, the electrons and hydrogens follow a thermodynamically predictable pathway, and the free energy of each carrier state is less than that of the one before. When the electrons and hydrogens are finally allowed to combine with oxygen, they are almost depleted of energy—most of it has been lost as heat or transferred into the free energy of the chemiosmotic differential (buildup of hydrogen ions on one side of a membrane). We can depict the stepwise decline in oxidizable free energy in a falling-down-the-steps diagram as shown in Figure 6.14.

So, the cycle usually begins inside the membrane with the coenzyme $NADH_2$ which reduces a complex comprised of FAD and protein. The FAD–enzyme complex, a large molecule that spans the membrane, releases two hydrogen ions across the membrane and into the outer compartment of the mitochondrion. The two electrons pass to the first of a pair of iron–sulfur proteins. Back at the interior side of the membrane, the iron–sulfur proteins pass electrons to cytochrome b_T, another protein. The electrons pick up hydrogen ions from the inner medium to become hydrogen atoms once again, in the process converting a small molecule called Q to its reduced form QH_2. QH_2 diffuses to the outer surface of the membrane, where it releases the hydrogen ions and passes the two electrons to another electron carrier, cytochrome b_K, which in turn passes them to cytochrome oxidase. The final carriers are cytochromes *c*, *a*, and *a*3. Finally, cytochrome *a*3 passes two electrons to an oxygen atom, and another two hydrogen ions are removed from the inner medium to make a water molecule. Another two hydrogen ions are pumped across the membrane at some as yet unknown point, probably at the cytochrome *a*–*a*3 complex as shown in Figure 6.13. Figures 6.13 and 6.14 summarize the mitochondrial electron transport system.

So now we have answered the question about the role of oxygen in oxidative respiration. It enters the scene at the very end of the process, collecting the energy-poor electrons after they have been shamelessly milked of their reducing power. One question that has bothered people for decades, and which remains unanswered, is just how the oxygen molecule finally enters the picture. One theory is that the cytochromes work in pairs, or possibly in quartets, and that the final reaction is between one O_2 molecule, four electrons, and four hydrogen ions to make two molecules of water. Another possibility, for which there is some evidence, is that every second reaction involves two hydrogen ions and two electrons, which break up an oxygen molecule into two highly reactive, toxic hydroxyl radicals ($OH\cdot$); the other reactions then supply two more hydrogen ions and two more electrons to turn the free radicals harmlessly into water. For various reasons, no one is really satisfied with either hypothesis.

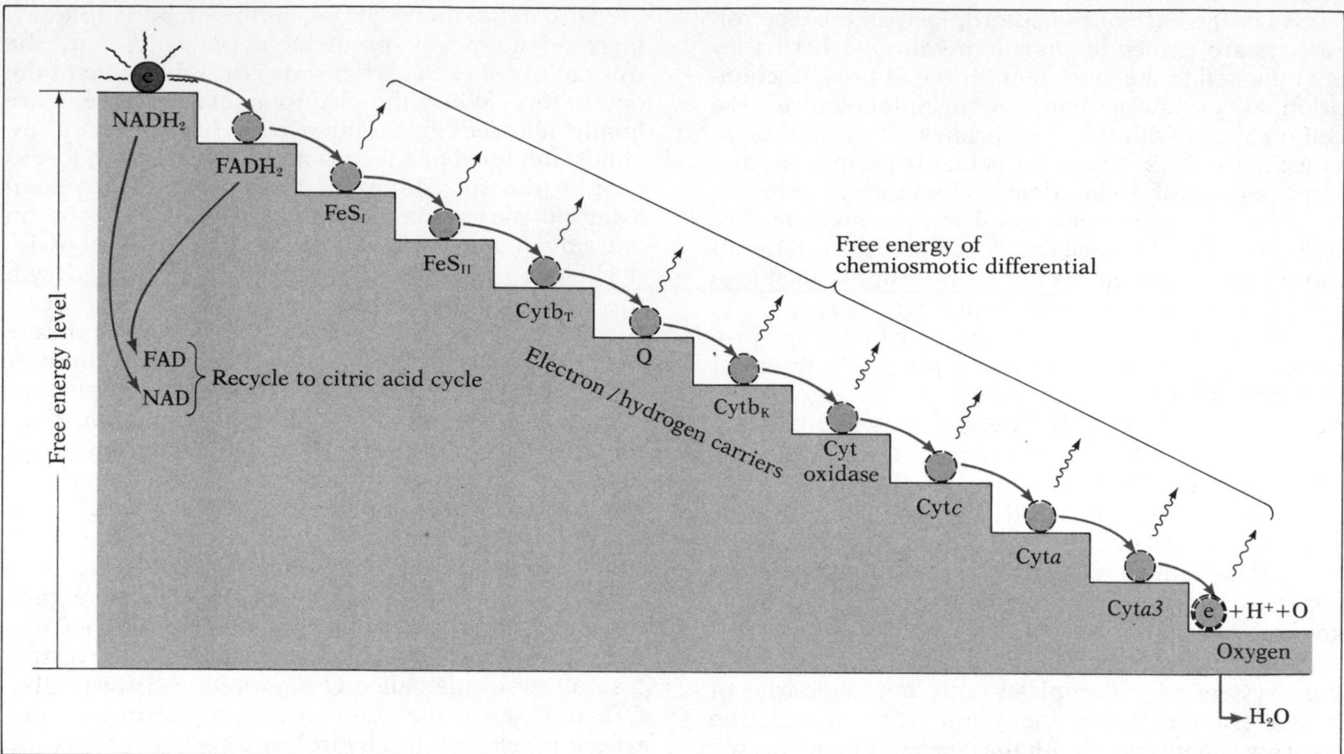

6.14 The passage of electrons through the electron transport system has been compared to a marble bouncing down stairs. As the electrons pass from one carrier to another, their energy level is decreased. In the staircase analogy, the marble loses its potential energy in the form of heat as it hits each step. Similarly, some of the electron's energy is lost as heat, but much of it is transferred into the free energy of the chemiosmotic differential. This is accomplished as hydrogen ions are transported across the membrane of the crista.

Let's now return to the falling-down-the-stairs diagram in Figure 6.14. There are several important points to make about the system. First, most of the electrons from the citric acid cycle are carried as $NADH_2$, so they can start at the beginning of the chain, and each electron will serve to pump three hydrogen ions out against the gradient. One of the reactions in the citric acid cycle, however, passed a pair of hydrogens directly to a lower-energy FAD coenzyme. The $FADH_2$ can't start at the beginning of the cycle, but must pass its hydrogens to the cycle further down the line. These lower-energy hydrogens miss out on the first active transport reaction and will contribute only two hydrogen ions each to the chemiosmotic gradient.

There is another source of reducing power (free energy) for the electron transport system. Recall from the discussion of glycolysis that each glucose molecule which is broken down produces, among other things, two cytoplasmic $NADH_2$ molecules. The $NADH_2$ itself cannot enter the mitochondrion. In another minor miracle of nature, however, a special "shuttle" reduces $NADH_2$ to NAD and carries the electrons across the inner mitochondrial membrane, where they reduce FAD. The $FADH_2$-bound hydrogens can then enter the transport chain a little downstream from the beginning and can account for another contribution of two hydrogen ions apiece to the chemiosmotic gradient.

The ATP Balance Sheet in the Respiration of Glucose

One of the favorite activities of biochemists is to try to balance the books in a complicated series of reactions such as we have seen. It's the traditional way to sum up respiration, so we'll have a go at it. Nowadays, however, we are less sure of our numbers. Remember, the development of the chemiosmotic view of chloroplast and mitochondrion function has made any exact accountings impossible and has put something of a damper on any tit-for-tat view of the process. Even if it were possible to determine exactly how many protons were transported across the membrane for each reaction, and how many protons were passed back across the membrane to produce each ATP, we'd still have problems calculating precise numbers. For one thing, laboratory measure-

Essay 6.2 The Energy Balance Sheet for the Respiration of Glucose

In aerobic eucaryotes, including humans, the three parts of glucose respiration are usually well-integrated. When they are working together, the end products of glycolysis include two molecules of pyruvate, a three-carbon compound. Before the pyruvate enters the citric acid cycle, however, it is broken down into acetyl CoA (a two-carbon compound attached to a coenzyme, read as "acetyl coenzyme A"). The three parts of respiration can then be summarized in these equations.

Glycolysis:

$$\text{Glucose} + 2\,\text{ADP} + 2\,\text{P}_i + 2\,\text{NAD} \rightarrow 2\,\text{Pyruvate} + 2\,\text{NADH}_2 + 2\,\text{H}_2\text{O} + 2\,\text{ATP}$$

Pyruvate to acetyl CoA:

$$2\,\text{Pyruvate} + 2\,\text{NAD} + 2\,\text{Coenzyme A} \rightarrow 2\,\text{Acetyl CoA} + \text{NADH}_2 + 2\,\text{CO}_2$$

Citric acid (Krebs) cycle:

$$2\,\text{Acetyl CoA} + 6\,\text{NAD} + 2\,\text{FAD} + 2\,\text{GDP} + 2\,\text{P}_i + 6\,\text{H}_2\text{O} \rightarrow 2\,\text{Coenzyme A} + 4\,\text{CO}_2 + 6\,\text{NADH}_2 + 2\,\text{FADH}_2 + 2\,\text{GTP} + 2\,\text{H}_2\text{O}$$

and through substrate-level phosphorylation:

$$2\,\text{GTP} + 2\,\text{ADP} \rightarrow 2\,\text{GDP} + 2\,\text{ATP}$$

Electron transport system:

$$10\,\text{NADH}_2 + 2\,\text{FADH}_2 + 32\,\text{ADP} + 32\,\text{P}_i + 6\,\text{O}_2 \rightarrow 10\,\text{NAD} + 2\,\text{FAD} + 32\,\text{ATP} + 44\,\text{H}_2\text{O}$$

Net Reaction:

$$\text{Glucose} + 6\,\text{O}_2 + 36\,\text{ADP} + 36\,\text{P}_i + 6\,\text{H}_2\text{O} \rightarrow 6\,\text{CO}_2 + 48\,\text{H}_2\text{O} + 36\,\text{ATP}$$

---

ments indicate that the membranes leak. For another, some of the free energy of the hydrogen ion gradient is siphoned off for uses other than ATP formation. For instance, some of the hydrogen ion gradient energy is used to power active transport across the mitochondrial membrane.

The latest data indicate that chemiosmotic phosphorylation requires the passage of only two hydrogen ions through the F1 particle for each ATP generated (Figure 5.8). This fits well with older calculations, which indicated that the mitochondrial electron transport system generated ATPs at three different steps along the way. We now believe that the three key steps in the electron transport system each involve the transport of two hydrogen ions across the membrane rather than directly coupled phosphorylation as such. The good fit may be more apparent than real, however. The older data indicated that only three steps were involved and each step was related to ATP synthesis. At that time, everyone assumed that each step involved the synthesis of exactly one ATP molecule. We have noted in our consideration of photosynthesis, however, that each pair of hydrogen ions pumped across the thylakoid membrane accounts for the synthesis of only two-thirds of an ATP molecule in that system.

According to thermodynamic measurements, there isn't quite enough energy in transferring a pair of hydrogen ions to account for one ATP synthesis, if all reactants are present in equimolar concentrations. Apparently, the reaction goes only because there is a high concentration of ADP within the inner cavity of the mitochondrion. This ADP concentration is there only because of a proton-powered active transport exchange shuttle that brings in an ADP and a hydrogen ion while simultaneously removing an ATP. So it appears that each ATP synthesis, in the long run, requires the movement of three hydrogen ions across the membrane after all—one to bring in ADP and two to phosphorylate it.

If we assume that three hydrogen ions are involved, the total number of ATPs produced is twenty-one. On the other hand, if we assume that only two hydrogen ions are involved in each chemiosmotic phosphorylation, the grand total is thirty-six ATPs per molecule of glucose respired. Actually, it is more likely that the efficiency of respiration is not constant, but varies according to a number of environmental factors. Laboratory experimentation has led to many quite different direct estimates. We can assume only that the average yield is somewhere between twenty-one and thirty-six ATPs produced per glucose molecule oxidized.

About 680 kcal are released in the complete oxidation of 1 mole of glucose. The caloric value of the high-energy bond in ATP is about 8 kcal/mole. The 21–36 ATPs produced by the metabolism of glucose represent a range in efficiency of 25–43%. Even the lower value is quite respectable by any standards, and you can be certain that your mitochondria are more efficient than your car.

We have seen that food undergoes extremely delicate, complex, and rigorous changes in living systems. The energy that has been stored there by plants is gradually released to do the work required in order to keep the molecules of life organized.

Application of Ideas

1. What does a runner mean by "going anaerobic"? Who depends more on anaerobic respiration, a sprinter or a marathon runner? Explain your answers.
2. Is anaerobic respiration really inefficient? Consider the question in terms of the free energy actually available in an anaerobic environment. Also consider the answer in terms of whether or not anaerobic organisms are "successful" organisms.

Key Words

acetyl group
aerobic
anaerobic

biochemical pathway

citric acid cycle (Krebs cycle)
chemiosmotic phosphorylation
cofactor
creatine phosphate
cytochrome

electron transport system
energy store

FAD, $FADH_2$
fermentation
F1 particle

glycolysis

hypoglycemia

kilocalorie

mass action law

NAD, $NADH_2$

oxidative pathway
oxidative phosphorylation
oxygen debt

phosphorolysis

respiration

substrate-level phosphorylation

Key Ideas

Introduction

1. Energy is required to maintain organization and resist entropy. In death, an organism's molecules become disorganized, eventually behaving independently as entropy occurs.
2. ATP energy is required for all cellular activity, including movement, active transport, and biosynthesis.
3. *Respiration* is the process of turning chemical bond energy in fuels into high-energy bonds of ATP.
4. The general formula for *respiration* is the opposite of photosynthesis:

$$C_6H_{12}O_6 + 6\ O_2 \rightarrow 6\ CO_2 + 6\ H_2O + \text{Energy}$$

5. Photosynthesis is essentially an uphill process, ending with molecules containing greater free energy.
6. *Respiration* is a downhill process, ending with molecules containing less free energy. Some energy is conserved in ATP, but much is lost as heat as ATP is produced.
7. Molecular oxygen has no free energy of its own. Glucose, because of its molecular complexity, does have some free energy by itself. The mixture of oxygen and glucose, however, has a great deal of free energy.
8. Organisms such as yeasts can obtain energy from glucose *anaerobically,* through *fermentation.*
9. The net formula for *fermentation* is:

$$C_6H_{12}O_6 \rightarrow 2\ C_2H_6O\ \text{(alcohol)} + 2\ CO_2 + \text{Energy (ATP)}$$

10. Alcohol is at a lower level of organization and free energy than glucose.
11. *Anaerobic respiration, fermentation,* and *glycolysis* probably evolved first, *aerobic respiration* later.
12. The two major processes of ATP synthesis are *substrate-level phosphorylation* and *chemiosmotic phosphorylation.*

Glycolysis: Degrading Glucose Without Oxygen

1. *Substrate-level phosphorylation* occurs during *glycolysis,* a long roundabout process whose object is to transfer chemical bond energy to ATP without sudden and uncontrolled heat losses.
2. In the initial reactions of *glycolysis,* glucose is phosphorylated by ATP. Enzymes change glucose-6-phosphate to fructose-6-phosphate, and a second ATP is added forming *fructose-1,6-diphosphate.*
3. Adding phosphate may aid in forming enzyme–substrate complexes, may prevent diffusion out of the cell, and may make glucose more reactive.
4. Many of the *glycolytic* reactions are reversible, following the *mass action law.* They proceed to the right when there is abundant substrate and the products are removed or used up.
5. *Fructose-1,6-diphosphate* is then cleaved into two three-carbon *PGALs,* important intermediates in both *respiration* and photosynthesis.
6. The *PGALs* are at once oxidized by *NAD,* which becomes *$NADH_2$,* and two *inorganic phosphates* are added, replacing the lost hydrogen. PGAL becomes *3-phosphoglyceroyl phosphate.*

Molecules of Glycolysis and Fermentation

acetaldehyde
ethyl alcohol
fructose-1,6-diphosphate
phosphoglyceraldehyde (PGAL)
glycogen
lactate
3-phosphoglycerate (3-PG)
1,3-diphosphoglycerate
pyruvate

Molecules of the Citric Acid Cycle

acetate
acetyl CoA
α-ketoglutarate
α-ketoglutarate dehydrogenase
citrate
citric synthase
coenzyme A (CoA)
fumarate
guanosine diphosphate (GDP)
guanosine triphosphate (GTP)
isocitrate
malate
malate dehydrogenase
oxaloacetate
oxalosuccinate
succinate
succinyl CoA

7. Next, two phosphates are transferred to 2 ADPs, producing ATP (no net ATP gain since two were used initially). The remaining molecules are *3-phosphoglycerate (3-PG).*

8. The two *3-PGs* are dehydrated and changed to *2-phosphoglycerate.* Two ADPs again react yielding two ATPs (net gain of two) and the *3-PGs* become *pyruvate.*

9. The net gain in *glycolysis* is two ATP, or sixteen kcal, representing only 2.4% of the energy available in glucose in the presence of oxygen.

Fermentation

1. Some organisms can survive on this limited efficiency.

2. The products of *glycolysis* follow three main pathways, depending on the organism and the availability of oxygen:
   a. Alcoholic *fermentation: Pyruvate* to *ethyl alcohol* or methane
   b. Cellular *anaerobic respiration: Pyruvate* to *lactate*
   c. Oxidative (*aerobic*) *respiration: Pyruvate* to CO_2 and H_2O

3. In *fermentation,* carbon is removed from *pyruvate,* forming CO_2. Next, the molecule is reduced by *NADH*$_2$ to *ethyl alcohol.*

4. Yeasts are fermenters that can switch to *aerobic respiration.*

Glycolysis in Muscle Tissue

1. During heavy exertion ATP reserves are backed up by *creatine phosphate,* which reconstitutes ADP to ATP.

2. Following *CP* exhaustion, muscles switch to *glycolysis. Pyruvate* is reduced by *NADH*$_2$ to *lactate* (*lactic acid*).

3. *Lactic acid* is either used in oxidative *respiration* and/or sent to the liver to be converted back into glucose.

Oxygen Debt

1. After exertion, *lactic acid* accumulates in muscles and liver. *Oxygen debt* is the amount of extra oxygen needed to restore the resting condition.

2. Physical conditioning increases the heart's capacity and rate, thus more oxygen reaches the muscles.

3. Marathon runners are believed to utilize stored fats as energy sources. They also improve their ability to store *glycogen.*

The Metabolism of Starch and Glycogen

1. Cells store carbohydrate in the form of starch (plants) or *glycogen* (animals).

2. The breakdown into glucose through phosphorolysis (using inorganic phosphate) produces glucose-1-phosphate without the need of ATP.

3. This conversion prepares glucose for *glycolysis* ahead of time and conserves one ATP that would have to be used.

Pyruvate to Acetyl CoA

1. During *oxidative phosphorylation,* 21–36 ATPs can be produced.

2. *NADH*$_2$ produced in *glycolysis* can send its electrons into the mitochondrion.

3. *Pyruvate* enters the mitochondrion where it is enzymatically converted to *acetyl CoA,* accompanied by reduction of *NAD* to *NADH*$_2$ and the release of CO_2. (Fatty acids and some amino acids can also be converted to *acetyl CoA.*)

The Citric Acid Cycle

1. The *citric acid cycle* occurs in the matrix within the inner membrane of the mitochondrion.

2. Upon entering the cycle, *acetyl CoA* joins with *oxaloacetic acid,* forming *citrate* (citric acid) and releasing the *coenzyme A (CoA).* It is from this acid that the cycle receives its name.

An Overview of the Citric Acid Cycle

1. As reactions of the cycle occur, the six-carbon *citrate* is altered and changed, carbons are removed as CO_2, water enters and leaves, and the last product is *oxaloacetate,* where the cycle began.

2. As the events occur, hydrogen is removed several times, reducing *NAD* and a similar carrier, *FAD*, to *$NADH_2$* and *$FADH_2$*. One GTP is produced directly for each turn of the cycle and is readily converted to ATP.
3. The major steps of the cycle can be summarized as follows:

① Acetyl CoA + Oxaloacetate + $H_2O \rightarrow$ Citrate + CoA
② ③ Citrate $\rightarrow \rightarrow$ Isocitrate
④ Isocitrate $\rightarrow \rightarrow$ α-Ketoglutarate + CO_2 + $NADH_2$
⑤ ⑥ α-Ketoglutarate $\rightarrow \rightarrow \rightarrow$ Succinate + ATP + $NADH_2$
⑦ Succinate $\rightarrow$ Fumarate + $FADH_2$
⑧ ⑨ Fumarate + $H_2O \rightarrow$ Malate $\rightarrow$ Oxaloacetate + $NADH_2$

4. From two *acetyl CoAs* (representing one glucose), two ATPs were formed directly, and twenty hydrogens in the form of *$NADH_2$* and *$FADH_2$* are made available to the electron transport system (twenty-four, if we add two $NADH_2$ from glycolysis.)

The Electron Transport System in Mitochondria
1. Mitochondrial organization differs from chloroplast organization in the following ways:
   a. The *F1 particles* project inward from the inner membrane.
   b. The mitochondrial electron transport chain begins with *$NADH_2$* passing electrons to the first carrier.
   c. The electron flow ends with H^+ and electrons joining oxygen to form water.
   d. Electrons in the mitochondrion start on the inside and cross the inner membrane six times, compared with three transits in photosynthesis. Protons are picked up inside and delivered outside the inner membrane.
2. Two $NADH_2$ electrons enter at the beginning of the electron transport chain and are responsible for pumping six H^+ against the gradient (three per electron).
3. Two $FADH_2$ electrons enter the *electron transport system* at a lower energy level and are responsible for pumping only four H^+ against the gradient.
4. *$NADH_2$* from glycolysis passes its electrons across the outer mitochondrial membrane to *FAD*, from which they enter the *electron transport system* at the *$FADH_2$* level.

The ATP Balance Sheet in the Respiration of Glucose
1. *Chemiosmotic phosphorylation* requires the passage of two H^+ through the *F1 particle* for each ATP generated.
2. Since a proton is required to pump ADP into the matrix, the actual requirement for one ATP is three H^+ ions (protons).
3. The average yield in *chemiosmotic phosphorylation* is 21–36 ATPs per glucose.
4. In terms of percentage efficiency, the total gain in *aerobic respiration* is 25–43%.

Review Questions

1. List three significant uses for ATP energy. (p. 158)
2. Write the net formulas for photosynthesis and respiration and then describe the obvious differences. (p. 158)
3. Explain the statement, "photosynthesis is described as an uphill process, while *respiration* is downhill." (p. 158)
4. Define *anaerobic respiration*. (p. 159)
5. Compare the free energy of *ethyl alcohol* with that of glucose and comment on the efficiency of organisms that carry on fermentation for their ATP. (pp. 159–160)
6. List the three parts of respiration and explain where in the cell each takes place. (p. 160)

7. List the two ways in which ATP is synthesized and explain the difference. (p. 160)
8. What is the *anaerobic* process called when it occurs in animals like ourselves? How does it differ from *anaerobic respiration* in microorganisms such as yeast? (p. 160)
9. Why is it necessary for cells to carry out *respiration* in an indirect, gradual, roundabout manner? (p. 160)
10. The initial steps in *glycolysis* require two phosphorylation steps when glucose is used. Summarize the three hypothetical explanations for why this is necessary. (pp. 160–161)
11. Write the word formula for the first *oxidation* step in *glycolysis* and explain what has happened. (p. 162)
12. Briefly explain how a reversible reaction can be made to go in one direction only. (p. 162)
13. List three things that have been accomplished during the first oxidation step of glycolysis. (p. 162)
14. Describe the first ATP-producing step in *glycolysis* and explain why there is no actual net gain in ATP. (p. 162)
15. Describe the second ATP-producing step in *glycolysis.* How much of an ATP gain occurs here for each glucose molecule used? (pp. 162–163)
16. List in *kilocalories* (kcal) the net gain in ATP energy during *glycolysis* and explain how this figure is derived. (p. 163)
17. Describe the three main pathways followed by pyruvate. What determines which pathway? (p. 164)
18. What is the fate of $NADH_2$ in the following:
    a. When *ethyl alcohol* is produced (p. 164)
    b. When *lactic acid* (*lactate*) is produced (pp. 165–166)
    c. When *aerobic respiration* proceeds directly after *glycolysis* (pp. 165–166)
19. List the four sources of energy for muscle activity in the order in which they are used. (pp. 165–166)
20. Define *oxygen debt* and explain how this debt is eventually paid. (p. 166)
21. Explain why physical conditioning shortens the period of *oxygen debt.* (p. 166, Essay 6.1)
22. Discuss the advantage of the *phosphorolysis* of *glycogen* or starch to glycolytic efficiency. (p. 167)
23. Describe the conversion of *pyruvic acid* to *acetyl CoA.* (p. 168)
24. Draw the mitochondrion and identify: outer membrane, inner membrane, *F1 particles,* and *electron transport systems.* (p. 173)
25. Summarize the function of the *citric acid cycle.* List the important products of its enzymatic reactions (not the names of the acids). (p. 169)
26. Write word formulas for each step in the *citric acid cycle* that reduces *NAD* or *FAD.* (p. 171)
27. In which step is ATP produced in a *substrate-level phosphorylation?* Explain the role of *GDP* (*guanosine diphosphate*). (p. 171)
28. Compare the following aspects of chloroplast and mitochondrial structure and function (p. 173)
    a. Position of the *CF1* and *F1 particles*
    b. Buildup of H^+ in the chemiosmotic differential
    c. Direction of electron flow between NADP, NAD, and the electron carriers
    d. The role of water and oxygen in the two pathways
    e. The path of electrons across the membranes
29. What is the precise role of oxygen in cellular *respiration?* (In other words, give a chemical explanation for why we breathe in oxygen.) (p. 173)
30. Compare the ability of $NADH_2$ and $FADH_2$ to add protons to the chemiosmotic gradient. (p. 174)
31. Why is it no longer feasible to determine the exact number of ATPs produced for each mole of glucose consumed in cell *respiration?* (pp. 174–175)
32. What is the current estimate of the number of H^+ (protons) required to pass through an *F1 particle* for each ATP produced? (p. 175)
33. In terms of percentage efficiency of aerobic respiration, what is the ballpark estimate and how is this figure determined? (p. 175)

Suggested Reading

Anderson, C. M., F. H. Zucker, and T. A. Steitz. 1979. "Space Filling Models of Kinase Clefts and Conformation Changes." *Science* 204:375. A graphic look at how an enzyme opens and closes its active center during enzymatic activity.

Bassham, J. A. 1962. "The Path of Carbon in Photosynthesis." *Scientific American,* June. The Calvin cycle illustrated.

Calvin, M., ed. 1973. *Organic Chemistry of Life: Readings from Scientific American.* W. H. Freeman, San Francisco. A collection of *Scientific American* reprints, featuring their superb technical illustrations and written for the intelligent lay person.

Darwin, C. 1859. *On the Origin of Species through Natural Selection.* A facsimile of the first edition. Harvard University Press, Cambridge, Mass. We strongly urge all serious biology students to take the time to read it as it was written; Darwin makes much better reading than any of the innumerable books that have since been written about him. Although the writing style is early Victorian, the ideas are fresh and modern and the spirit is infectious.

de Beer, G. 1965. *Charles Darwin: A Scientific Biography.* Doubleday, New York. A sober but intelligent account of Darwin's life and work, with a strong emphasis on the evolution of his scientific thought.

———. 1974. "Darwin, Charles" in *The New Encyclopedia Brittanica,* 15th ed., 5:492. An excellent and readable account of Darwin's life and ideas.

Dickerson, R. E., and I. Geis. 1969. *The Structure and Action of Proteins.* Harper and Row, New York. After more than a decade, this remains the best source book on protein structure. A supplement is available that gives three-dimensional stereograms of important proteins.

Folsome, C. E., ed. 1979. *Life: Origin and Evolution.* A *Scientific American* book. W. H. Freeman, San Francisco.

Govindjee and R. Govindjee. 1974. *"Primary Events of Photosynthesis." Scientific American,* December.

Hall, D. L., P. D. Tessner, and A. M. Diamond. 1978. "Planck's Principle." *Science* 202:717. Do younger scientists accept new scientific ideas—in this case, the theory of natural selection—more readily than older scientists?

Hanawalt, P. C., and R. H. Haynes, eds. 1973. *The Chemical Basis of Life: An Introduction to Molecular and Cell Biology.* A *Scientific American* book. W. H. Freeman, San Francisco.

Hinkle, P. C., and R. E. McCarty. 1978. "How Cells Make ATP." *Scientific American,* March. A thoroughly convincing presentation of the chemiosmosis hypothesis and the evidence for it in both photosynthetic and oxidative phosphorylation. We have drawn heavily on this article in preparing chapters 5 and 6.

Ifftt, J. B., and J. E. Hearst, eds. 1974. *General Chemistry.* A *Scientific American* book. W. H. Freeman, San Francisco.

Jensen, W. A. 1970. *The Plant Cell.* 2d ed. Wadsworth, Belmont, Calif.

Kennedy, D., ed. 1974. *Cellular and Organismal Biology.* A *Scientific American* book. W. H. Freeman, San Francisco.

King-Hele, D. 1974. "Erasmus Darwin, Master of Many Crafts." *Nature* 247:87. Charles' wonderful grandfather, poet, inventor, and essayist, made some amazingly clear and sensible statements about evolution and sexual selection. Unlike Charles, he never bothered to collect the facts that would back them up.

Kozlowski, T. T. 1964. *Water Metabolism in Plants.* Harper and Row, New York.

Laüger, P. 1972. "Carrier-mediated Ion Transport." *Science* 178:24. Experimental work on the transport of potassium ions by valinomycin, and a general consideration of how ion carriers function across cell membranes.

Lehninger, A. L. 1975. *Biochemistry.* 2d ed. Worth, New York. The current standard of excellence in biochemistry texts.

Margaria, R. 1972. "The Sources of Muscular Energy." *Scientific American,* March.

Margulis, L. 1971. "Symbiosis and Evolution." *Scientific American,* August. Still the best presentation of her once startling theory of the origin of eucaryotes.

Margulis, L., L. To, and D. Chase. 1978. "Microtubules in Prokaryotes." *Science* 200:1118. Margulis found in 1978 just what she had predicted in 1968: some spirochaete bacteria have tubulin-based microtubules.

Moorehead, Alan. 1969. *Darwin and the Beagle.* Harper and Row, New York. We think this book is the most thoroughly enjoyable account of Darwin's seminal years aboard the H.M.S. *Beagle.* Lavishly illustrated.

Porter, E. 1971. *Galápagos.* Ballantine, New York. All evolutionary biologists dream of making the trip to the Galápagos—and some of them make it. A different world.

Satir, P. 1974. "How Cilia Move." *Scientific American,* October. An intriguing but still speculative model of how sliding microtubules might be responsible for the movement of cilia and flagella.

Schopf, J. W. 1978. "The Evolution of the Earliest Cells." *Scientific American,* September. Biochemical pathways clearly demonstrate that basic life processes were evolved in an anaerobic world, and that steps involving free molecular oxygen have been "tacked on" to previously evolved systems.

A *Scientific American* book. 1978. *Evolution.* W. H. Freeman, San Francisco. The September 1978 special issue on evolution, reissued as a book. Superior articles include R. E. Dickerson, "Chemical Evolution and the Origin of Life"; J. Maynard Smith, "The Evolution of Behavior"; J. W. Valentine, "Multicellular Plants and Animals"; R. M. May, "Ecological Systems"; and J. W. Schopf, "The Earliest Cells."

Singer, S. J., and G. Nicolson. 1972. "The Fluid Mosaic Model of the Structure of Cell Membranes." *Science* 175: 720. The original description of the now-accepted fluid mosaic model.

Stephenson, W. K. 1978. *Concepts in Cell Biology.* John Wiley, New York. A very good study aid with a succinct, self-teaching approach.

Stent, G. S. 1972. "Prematurity and Uniqueness in Scientific Discovery." *Scientific American,* September. A valuable essay on scientific creativity by an established creative scientist.

Van Valen, L. 1971. "The History and Stability of Atmospheric Oxygen." *Science* 171:439. Although oxygen-producing photosynthesis evolved prior to the change from a reducing atmosphere to an oxygen-rich atmosphere, Van Valen argues that the oxygen atmosphere was almost certainly not created by biological forces, but by long-term spontaneous dissociation of water molecules in the ionosphere and the subsequent loss of hydrogen ions to space.

Weissmann, G., and R. Clairborne, eds. 1975. *Cell Membranes: Biochemistry, Cell Biology and Pathology.* HP Publishing Company, New York. Superbly illustrated articles on all aspects of cell membranes, incorporating the latest research findings. Notable are E. Racker on the inner mitochondrial membrane and the chemiosmosis hypothesis; W. R. Lowenstein and G. D. Pappas on cell-to-cell communication via membrane junctions; and J. F. Danielli on the Danielli lipid bilayer concept as it has evolved over the years.

White, E. H. 1964. *Chemical Background for the Biological Sciences.* Prentice-Hall, Englewood Cliffs, N.J.

Whittingham, C. P. 1977. *Photosynthesis.* Carolina Biological Supply Company, Burlington, N.C.

Zelitch, I. 1971. *Photosynthesis, Photorespiration,* and *Plant Productivity.* Academic Press, New York. What is accomplished by the apparently wasteful process of photorespiration?

Part Two

Molecular Biology and Heredity

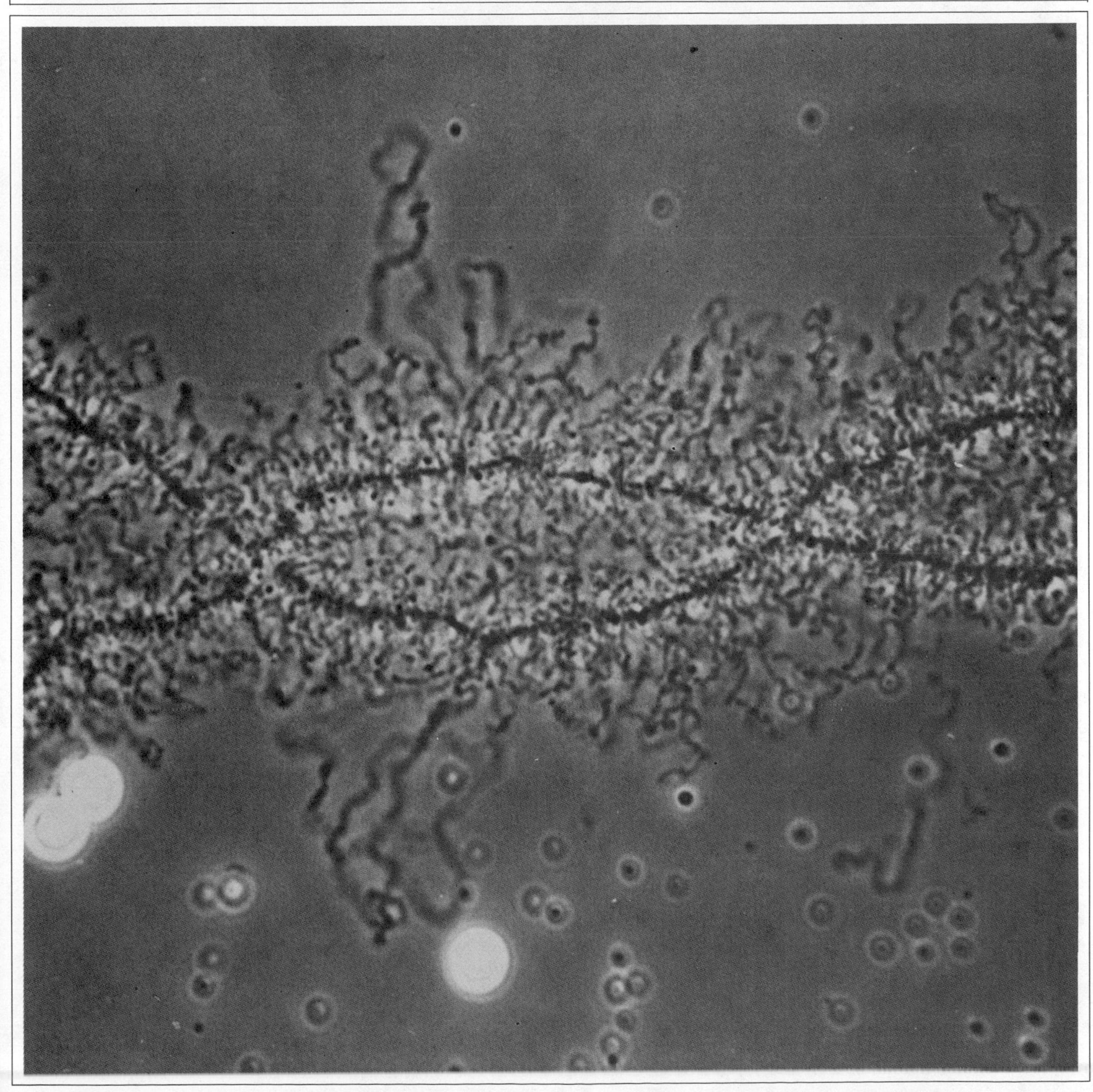

Chapter 7
The Central Dogma

Not very long ago, as recently as the early 1950s, the basic processes of *genetics* were not only unknown but were widely considered to be so mysterious and complex as to be quite literally beyond human comprehension. Scientists had argued about whether the gene itself had a chemical composition and had generally agreed that it did. But what genes were, how they were put together, and how they worked seemed hopelessly beyond grasp. There had been some good evidence in the 1940s that the genetic material was *deoxyribonucleic acid* (*DNA*), but still most biologists—if they allowed themselves an opinion at all—would have said that genes were probably complex structures made of protein. They knew that enzymes were proteins, and they believed that genes must be much more complex even than enzymes. And at that time, enzymes themselves seemed to be so enormous and complex that their own structure could never be known. Thus how could we ever hope to understand genes? After all, it was said, genes not only had to direct all the life processes of the cell and the development of the organism, but also somehow had to be able to make exact copies of themselves every cell generation.

The more pessimistic souls must have been shocked when two vigorous young researchers, James Watson and Francis Crick, showed that the structure of genes and the mechanism of gene function and replication were essentially not so difficult to understand after all. It has turned out to be one of the basic stories of biology. In fact, we'll review essentially everything Watson and Crick found out in this chapter. It makes a compelling story, and one that is pleasingly logical and understandable when it is approached in a step-by-step manner. So let's take it one step at a time, keeping in mind that many scientists of the 1950s found it difficult to accept what we're about to say. The mechanical simplicity of the ideas almost threatened to take away the central mystery of life.

7.1 DNA Structure and Replication

What Is the Central Dogma?

Since this is a science textbook, you may be surprised to see the word *dogma* in the titles of this chapter and section. *Dogma* usually refers to something that can't be questioned and is generally reserved for matters of religion and politics. In a sense, the only proper dogma in science is that there is no dogma, and anything can be questioned. The term *central dogma* was originally meant as a joke of sorts, initiated by Crick. But the name stuck.

The reasons serious scientists accept the term "dogma" are, first, the simplicity of the statements; second, the virtually universal acceptance of them by biologists; and, most of all, the absence of any real evidence that these are the only ways in which biological information is preserved and passed on. That's really the dogmatic part.

The central dogma completed the intellectual process begun by Mendel and gave our concept of life (including human life) a rigorous chemical and physical basis. Of course, there are still important details to be worked out, many smaller mysteries remain to be solved, and some questions, perhaps, may never be answered. But *the essence of this life process is understood.*

So what is the central dogma? Simply this:

1. All DNA is copied from other DNA (a process called *replication*).
2. All RNA is copied from DNA (*transcription*).
3. All proteins are copied from RNA in such a way that three sequential RNA nucleotide bases specify one amino acid in a protein chain, using a genetic code that is the same for all organisms (*translation*).

As it happens scientists quickly found exceptions to the three rules above, but, as far as we know, the exceptions occur only in certain viruses:

4. For some viruses, viral RNA is copied from viral RNA (*RNA replication*).
5. For other viruses, viral DNA is copied from viral RNA (*DNA-dependent RNA synthesis* or *reverse transcription*).

These principles are presented in Figure 7.1.

Hardly the stuff, it would seem, for an earth-shattering intellectual revolution. But the effects are far-reaching. Let us not dwell on those effects at this point. For now, be aware that the essence of the dogma is this: *All genetic information* (here, "information" refers to how certain structures direct other structures) *is contained in linear sequences of nucleic acid bases* (how the bases are arranged in a line), *is retained when these sequences are copied into other nucleic acid sequences, and can be expressed in the synthesis of proteins with specific linear amino acid sequences* (the principle of collinearity). This statement may be befuddling at present, but our task here is to unfuddle it so that it makes easy reading.

7.1 The core of the central dogma includes *replication, transcription,* and *translation.* ① DNA carrying the genetic instructions of a species is copied faithfully through generations, assuring survival through the process of *replication.* ② Within the active cells of an organism, the genetic instructions are copied in RNA in a highly organized and selective manner. This is *transcription.* ③ Several types of RNA, in concert with the ribosomes, assemble proteins according to the three-letter coding system transcribed from DNA. Converting the genetic code into proteins is known as *translation.* ④ and ⑤ But the central dogma must be amended to accommodate viruses. Some viruses contain RNA but no DNA. Therefore, RNA is replicated by RNA. Other viruses have reversed the roles of RNA and DNA. Their RNA is the permanent genetic code and it is transcribed into DNA. Viruses present the only known exceptions to the central dogma, however.

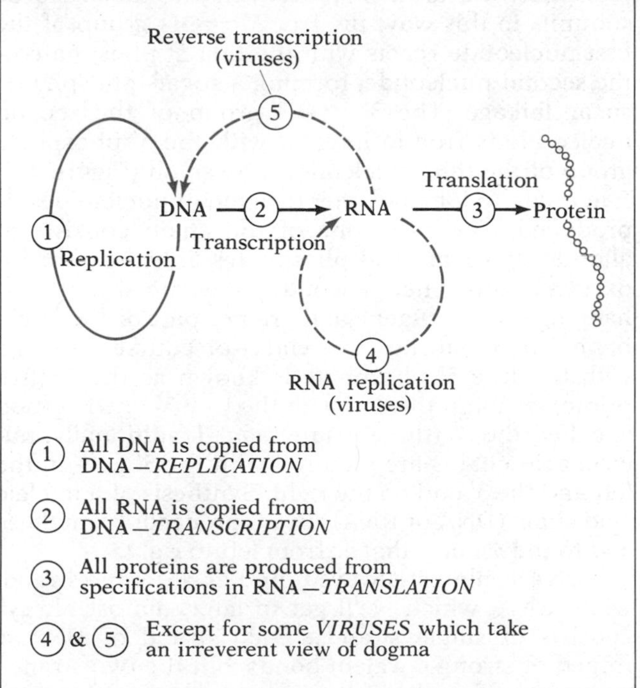
Reverse transcription
(viruses)
5
DNA
2
RNA
Translation
3
Protein
1
Replication
Transcription
4
RNA replication
(viruses)
1 All DNA is copied from DNA—*REPLICATION*
2 All RNA is copied from DNA—*TRANSCRIPTION*
3 All proteins are produced from specifications in RNA—*TRANSLATION*
4 & 5 Except for some *VIRUSES* which take an irreverent view of dogma

Figure 7.1 shows that DNA makes RNA, which makes protein; but DNA also makes other DNA, and RNA makes other RNA as well as DNA. If this doesn't make sense at the moment, that's OK. It's just a reference to return to as we proceed. Let's first take a closer look at some of these molecules and see how they work.

What Is a Nucleic Acid?

We have already indicated that there are two kinds of nucleic acid, DNA (deoxyribonucleic acid) and RNA (ribonucleic acid). As the "deoxy" implies, there is one less oxygen atom in each unit of DNA than in RNA. Let's talk about DNA first, since it essentially has one structure whereas RNA has several.

The basic unit of DNA is called a *deoxynucleotide,* or more frequently just a *nucleotide.* The nucleotides can exist as free-floating molecules in the cell fluids, and when they are strung together into long polymers, they form DNA *strands.*

The free-floating nucleotides are usually found as triphosphates, meaning that they have an appendage of three phosphates strung together. In Chapters 5 and 6, we described ATP as the universal energy currency in the cell. This is because the two phosphate–phosphate bonds in ATP are *high-energy bonds.* ATP is a ribose nucleotide, not a deoxynucleotide, but the free-floating deoxynucleotides also have high-energy bonds. In addition to the phosphate subunits, the deoxynucleotide consists of a sugar (*deoxyribose*) and any one of four different *nucleotide bases.*

Deoxyribose is a five-carbon sugar, exactly like the more common sugar ribose except that one of the oxygens has been removed. In other words, on one of the five carbons a —OH side group is replaced by a —H. Deoxyribose attaches to one of the nucleotide bases at its number 1′ carbon.

H
O
H—5′C—H
O
OH
4′C—H
C1′
H H
H
3′C
C2′
O
H
H In ribose, this is —OH

Deoxyribose

The nucleotide bases are also called *nucleic acid* bases, *nitrogenous* or *nitrogen* bases (the latter because they consist of nitrogen-containing rings). Two of the bases, *thymine* and *cytosine,* each consist of a single six-cornered ring. They are called *pyrimidines.* The other two, *adenine* and *guanine,* each consist of a five- and a six-cornered ring connected together. Adenine and guanine are *purines.* The shaded NH groups are the sites of reaction with the deoxyribose. (The bigger molecules, you notice, have the shorter name.)

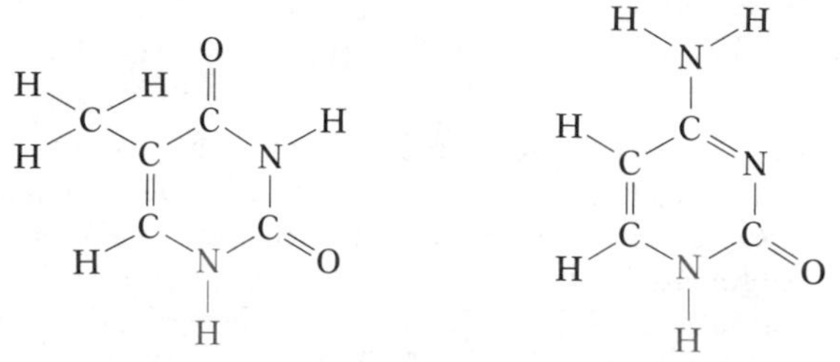

Thymine (T)

Cytosine (C)

Pyrimidines
(single ring)

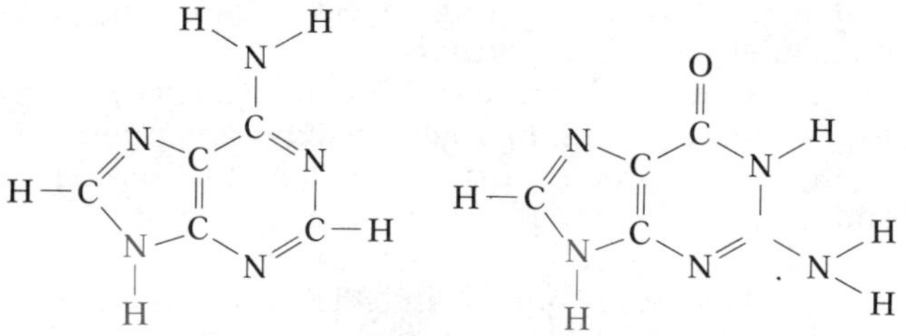

Adenine (A)

Guanine (G)

Purines
(double ring)

Adenine, guanine, thymine, and cytosine are also known simply by the letters A, G, T, and C, respectively. The corresponding nucleotide triphosphates are ATP, GTP, TTP, and CTP. (*Adeno-* means gland; adenine and thymine were first isolated from the thymus gland. *Cyto-* means cell, and, as you may have guessed, cytosine is present in all cells. Guanine was first isolated from guano, a rather unromantic undertaking.)

The five carbons of deoxyribose are traditionally numbered as 1′ through 5′ (read "one prime through five prime"). In a free-floating nucleotide triphosphate, the string of three phosphates is covalently linked through the —O— side group of the 5′ carbon, and the nucleotide base is covalently linked by one of its nitrogens directly to the 1′ carbon.

So the nucleotide consists of phosphates, sugar, and a nucleotide base linked together. The 1′ carbon of deoxyribose is also linked, through an oxygen, to the 4′ carbon, making the sugar a pentagonal ring. The 2′ carbon is the one without any —OH side groups. This leaves only the 3′ carbon with a reactive —OH side group in the free-floating nucleotide. This information will come in handy later.

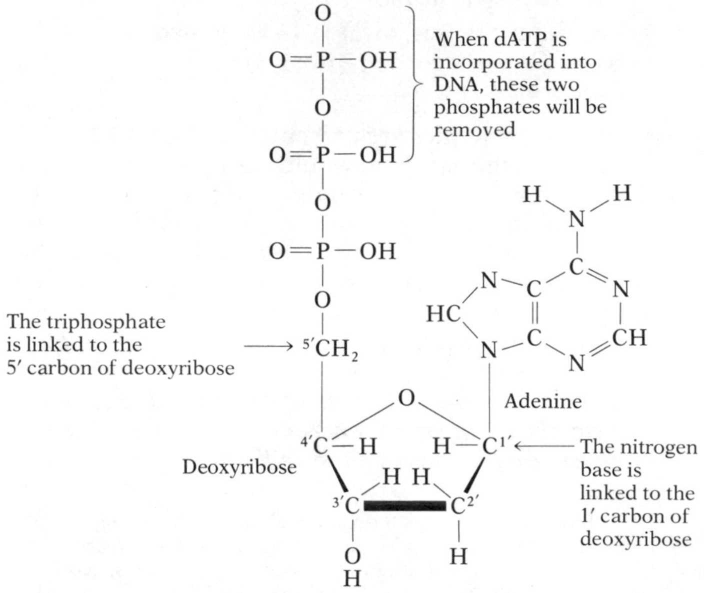

When dATP is incorporated into DNA, these two phosphates will be removed

The triphosphate is linked to the 5′ carbon of deoxyribose

Adenine

Deoxyribose

The nitrogen base is linked to the 1′ carbon of deoxyribose

Deoxyadenosine triphosphate (dATP)

DNA Bonding

In the synthesis of DNA chains, the high-energy bond between two of the phosphates is utilized to form a strong covalent bond between nucleotides. In the process, two of the phosphate groups are liberated, leaving one phosphate attached. This remaining phosphate forms a link between two deoxyribose subunits in this way: the free 3′ —OH group of the first nucleotide reacts with the first 5′ phosphate of the second nucleotide, forming a sugar–phosphate–sugar linkage. The 3′ —OH group of the second nucleotide is free to interact with the 5′ phosphate group of the third nucleotide, and so on (Figure 7.2). Eventually, a long polymer (a chain of nucleotides) is produced. The backbone of the chain consists of alternating sugars and phosphates linked by shared oxygen atoms. The nucleotide bases are side groups hanging off the sugars, and are not part of the backbone. The chain has two ends—of course—the one with the free 5′ phosphate is known as the *5′* (five prime) *end,* and the end with the free 3′ —OH group is called the *3′* (three prime) *end.* Traditionally, nucleic acid chains are pictured with the 5′ end on the left and the 3′ end on the right. Synthesis of a nucleic acid chain (DNA or RNA) always proceeds from the 5′ end to the 3′ end—that is, from left to right.

DNA is almost always found as a double strand, while RNA, which we'll get to later, almost always consists of single strands. Each strand of DNA is forged of strong covalent bonds, but the two strands

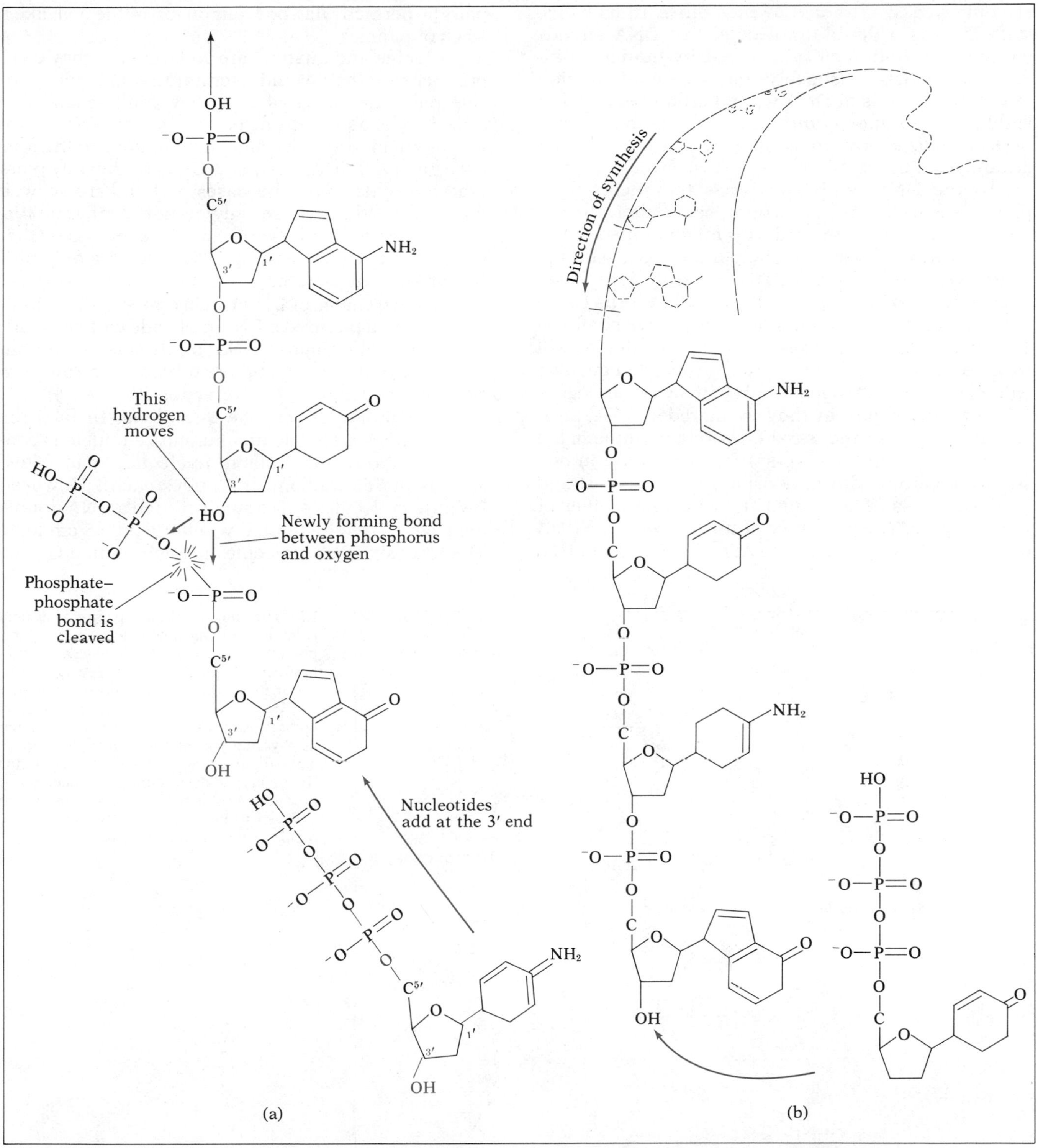

7.2 DNA bonding. (a) Nucleotides are added to a growing DNA chain by DNA polymerase. Although not shown here, there is always a preexisting strand which serves as a template or guide for new strands. As each nucleotide is added, two phosphates are cleaved from the nucleotide triphosphate, and the hydrogen from the —OH group of the 3′ sugar carbon is transferred to the liberated diphosphate molecule. Newly added nucleotides are subsequently attached at the 3′ carbon of the deoxyribose, forming 3′-5′ phosphate linkages as the strand grows. (b) The single strand of DNA grows in length to become an incredibly long and thin polymer. All four nucleotide bases must be present in great abundance to accommodate nucleic acid synthesis. During this synthetic period of the cell cycle, nearly all the cell's resources are devoted to this immense task.

are only weakly attracted to each other. In its native state DNA is a double molecule: two DNA strands wrapped around each other, held by numerous individually weak hydrogen bonds. The configuration of the two strands of DNA wound around each other is that of the famous *double helix.* (A double helix is best visualized by imagining two wires twisted around each other.)

In the DNA double helix, as the sugar–phosphate—sugar chains curl around each other, the nucleotide bases point inward. It is much as if someone had twisted a ladder, and the pairs of bases are the rungs. Watson and Crick determined that the bases pair up in a curious fashion, but one that turns out to have critical functional implications, as we shall see. The structures of the bases are such that adenine will form two hydrogen bonds with thymine, and guanine will form three hydrogen bonds with cytosine (Figure 7.3). And this is the way they are paired in DNA. Each adenine in one strand is paired with a thymine in the other strand (and vice versa); each guanine in one strand is paired with a cytosine in the other strand (and vice versa). This results in the peculiar coiling of the molecule. The geometry of the inside of the DNA molecule is very precise. The A:T, T:A, C:G, and G:C pairs lie perfectly flat, one pair on top of the next like a stack of pennies.

Adenine and guanine are so large that they can't pair because they would overlap; thymine and cytosine pairs are so small that they can't reach each other; and side groups of pairs of adenine and cytosine or guanine and thymine don't match. So the four base pairs A:T, T:A, C:G, and G:C are the only pairs that can match up. The bases of DNA are so well matched partly because only certain configurations fit together and partly because the enzymes that synthesize DNA are very specific, allowing only particular kinds of patterns.

The two chains of DNA run in opposite directions. That is, if you pictured a DNA molecule vertically on a page, one of the chains would run from top to bottom in its 5′-to-3′ direction, and the other would run from bottom to top in its 5′-to-3′ direction (Figure 7.4).

DNA molecules can be quite long. In fact, the average DNA molecule in a human cell nucleus consists of about 160 million nucleotide pairs (two strands of 160 million nucleotides each)! However, because molecules are so small, if such a DNA double helix were stretched out, it would be only 5 cm long. The forty-six DNA molecules in each human G1 cell

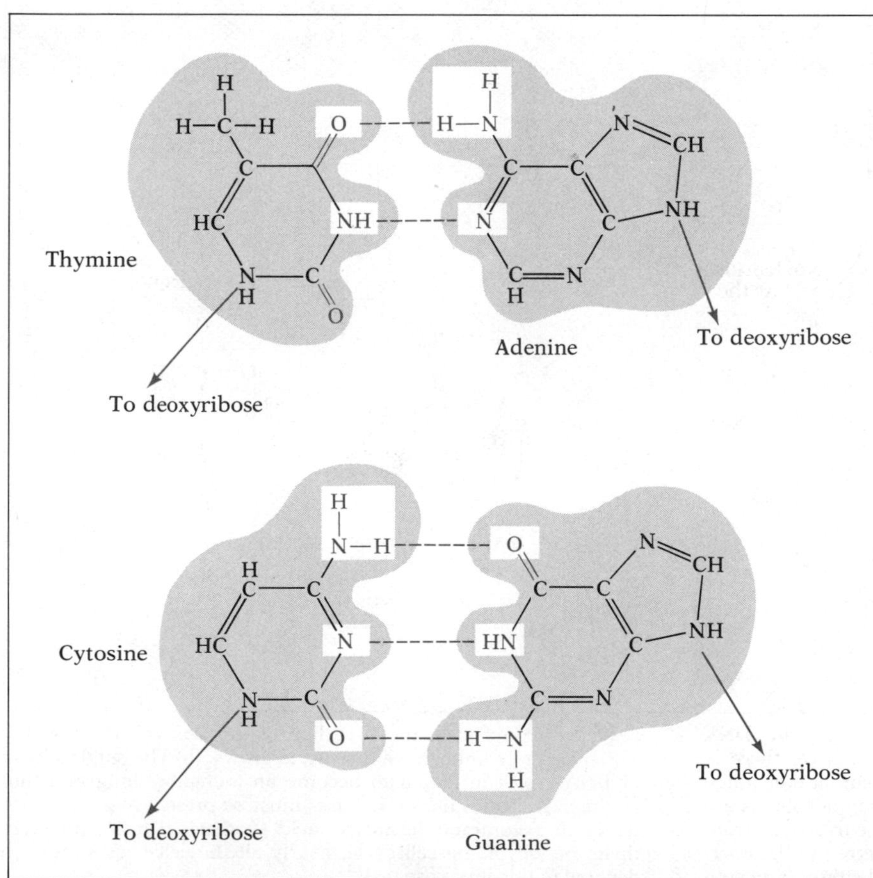

7.3 The concept of base pairing is critical to understanding the central dogma. Simply put, pairing in DNA always occurs between A and T and between G and C. You can see here how hydrogen bonding will occur only when the base pairing is correct. Any time a mistake occurs, such as thymine pairing with guanine, there is a misfit. Bonding will not readily occur. (In reality, it does sometimes occur, but produces problems. Mispaired bases are believed to be spontaneously removed by repair systems in the nucleus.)

7.4 In fully assembled DNA, one strand appears upside down to the other.

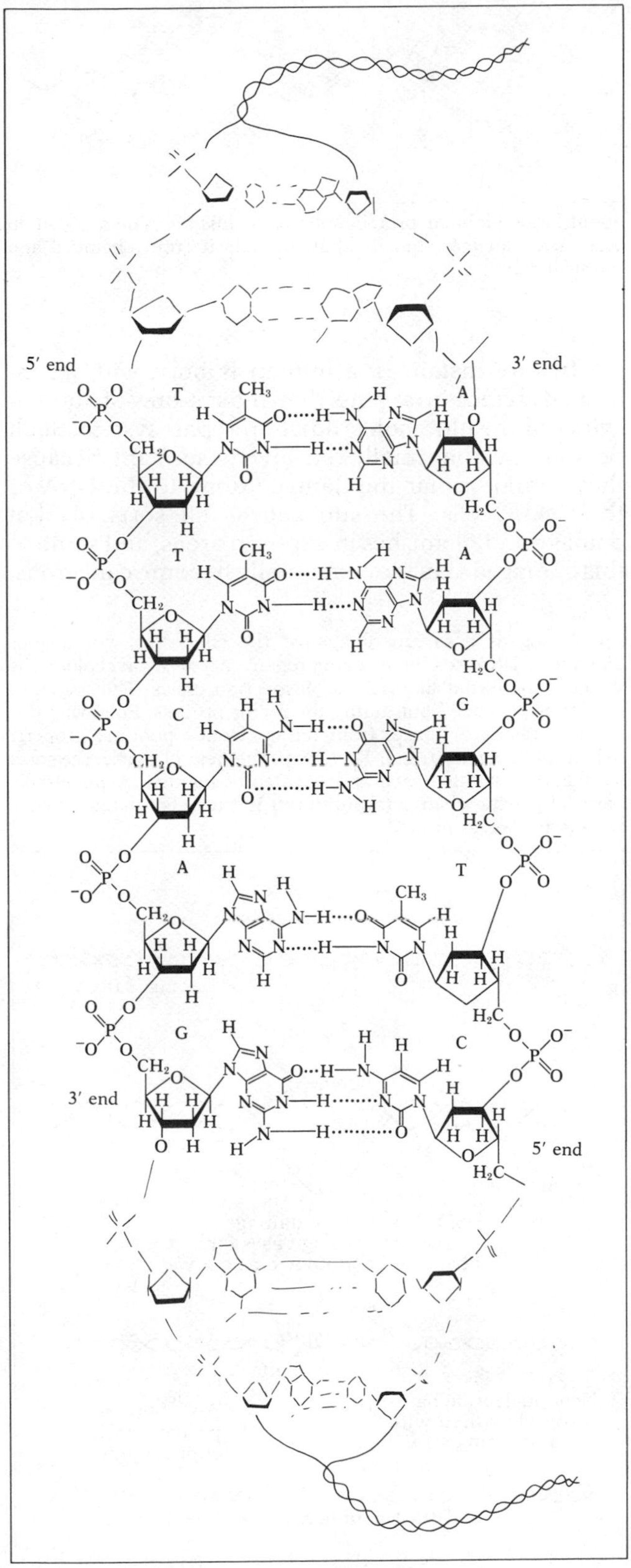

nucleus, placed end to end, would measure almost exactly 2 m. One might argue that that's pretty big to be contained in a microscopic cell. It's all a matter of perspective.

The Stability of DNA

The DNA molecule is well-adapted to its function as a repository of genetic information. For one thing, it is a relatively stable molecule. On the outside it shows only the sugar–phosphate backbone; and the deoxyribose sugar has all its potentially reactive side groups already bonded. Most of the potentially reactive side groups of the nucleotide bases are immobilized by hydrogen bonds. In any case, the nucleotide bases are safely tucked away inside, and pretty well immobilized just by the geometric tightness of the molecule. So DNA does not readily react with the various chemicals and the occasional highly reactive free radicals that occur from time to time in the nucleus.

All molecules, including DNA, are subject to some spontaneous denaturation, which is to say that chemical bonds are sometimes altered in a random way. Such a change can inactivate an enzyme, for instance. Spontaneous changes in DNA are potentially far more dangerous than spontaneous changes in other molecules. Such changes alter the genes themselves, and therefore are *mutations.* Mutations in the tissue that produces sperm or eggs can result in seriously ill (or dead) descendants; mutations in the other cells of the body can cause cell death or, less frequently, cancer. Obviously, then, it's best for so critical a molecule as DNA to be relatively resistant to spontaneous changes.

DNA in higher organisms is well protected from chemical damage. It is tightly wound around special proteins called *histones* (Figure 7.5). The tightly binding histone proteins have strong positive charges and form salt linkages with DNA by attaching to the negatively charged phosphates. Histones also bind DNA by specific folding and shape interactions. The histones, then, further immobilize DNA and protect it from thermal and chemical damage.

The most important protection that DNA has against random mutation, however, is biological rather than strictly chemical. It depends upon the specific A:T and G:C pairing that we have discussed. The biological information that must be preserved lies in the *specific order* of the base pairs. That information is redundant (present more than once) in the DNA molecule, in the sense that the specific order of bases in one chain uniquely determines the specific order of bases in the other chain. The redundancy helps to maintain the integrity of the information. This means that if one of the two chains is accidentally altered (say, by ultraviolet radiation), the organ-

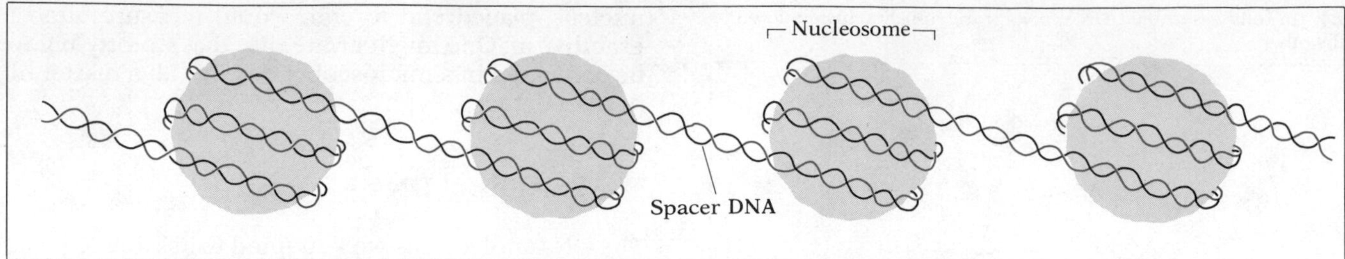

7.5 Because of its critical role, DNA, once formed, must be protected. The double helix with all its bases turned inward is one level of protection. In addition, DNA in eucaryotes is tightly wound about large, globular protein spheres of histone. The association gives DNA an inert quality that protects it from chemical and physical damage.

ism throws out the damaged chain and makes a new, perfectly good one by using the remaining intact partner as a template, following the A:T and G:C format.

DNA Repair

Several different, complex *DNA repair systems* are present in the cells of every organism. Some systems are specific for the kind of DNA damage caused by ultraviolet radiation, and other systems are more general. This is how they work: A complex of enzymes roves up and down the DNA, checking it out, as it were. The enzymes are very sensitive to any alterations in the standard Watson–Crick double helix. If any bases are incorrectly paired (say, a guanine paired with a thymine), or are linked together improperly, or have been altered to some form other than the standard four bases, the repair enzyme complex goes to work (Figure 7.6). A chunk of perhaps a couple of hundred nucleotides is cut out of the offending chain, leaving the intact partner briefly naked. Then the other enzymes of the repair system construct a new length of DNA, following the rules we just discussed—putting an adenine nucleotide opposite every thymine, a cytosine opposite every guanine, a thymine opposite every adenine, and a guanine opposite every cytosine—until a proper DNA chain is synthesized. The new piece is sutured into the chain being repaired by yet another enzyme, *ligase.* In this way, all but the most gross damage to DNA is fixed.

Actually, then, the term *mutation* should be used to refer not to any random changes in the DNA molecule, but to the very small proportion of those random changes that cannot be repaired, or that are repaired incorrectly. Even DNA repair complexes make mistakes sometimes. Still, experiments with microorganisms and fruit flies indicate that something like 99.9% of all damage to DNA is completely repaired by the repair enzymes. Repair enzymes can even reconstruct chromosomes (and other DNA molecules) that have completely broken apart.

In rare instances, a human is born with an inherited defect in a gene that makes one of the enzymes of the ultraviolet radiation repair system. Such persons are abnormally sensitive to sunlight, because they cannot repair the damage done to the DNA of their skin cells. The sun causes all sorts of skin damage and blemishes in exposed areas, and some of this damaged skin may eventually become cancerous.

7.6 Acting as a safeguard against the effects of spontaneous damage to DNA, teams of roving repair enzymes travel along the strands, somewhat like railroad inspection crews. The damaged portion is "clipped" out during the repair process. Following this bit of ultrasurgery, bases from the nucleotide pool are properly paired into a short strand. Then, with the help of ligase, the short strands are fit into place and covalently bonded. A potentially harmful genetic change (a mutation) is thus eliminated and the DNA is as good as new.

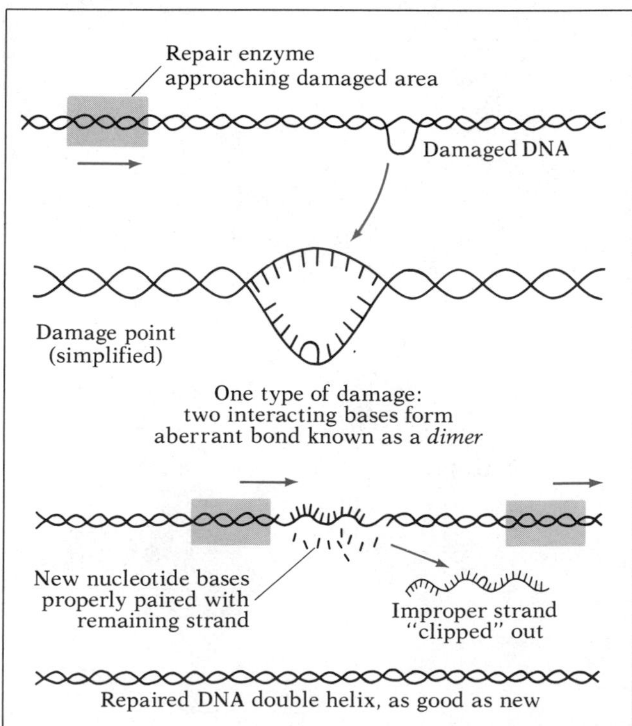

Essay 7.1 DNA Replication in Eucaryotes and Procaryotes

In both procaryotes and eucaryotes, DNA replication begins with the two sides of the twisted ladder of DNA pulling apart at some point along the length of the strand and unwinding, forming a "bubble." Each end of the bubble, where the two single strands of DNA emerge from the double-stranded DNA, is called a *replication fork.* Newly synthesized DNA strands are formed enzymatically on both of the unwound single strands, following the Watson–Crick pairing rules. As replication proceeds, further unwinding moves the replication forks in opposite directions, causing the bubble to become larger (Figure 7.7).

In eucaryotes, each of the numerous chromosomes contains one very long, linear DNA molecule prior to DNA replication. At the initiation of DNA replication, hundreds of bubbles are formed along the length of each long DNA molecule. As the bubbles enlarge, the replication forks of different bubbles run into each other, and the bubbles merge. One replication fork runs off each end of the molecule, and eventually two long, linear DNA molecules exist where there was one before.

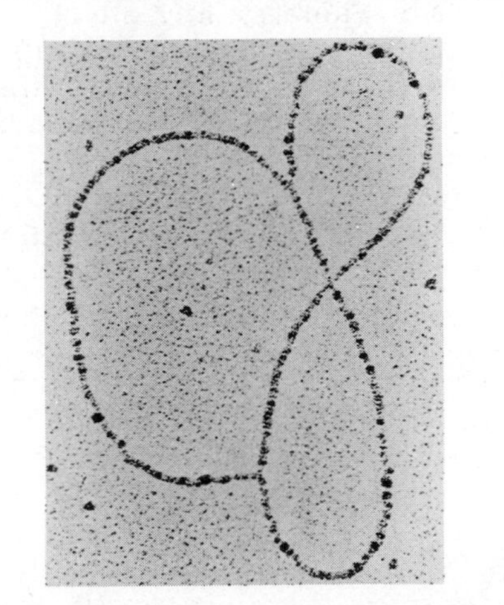

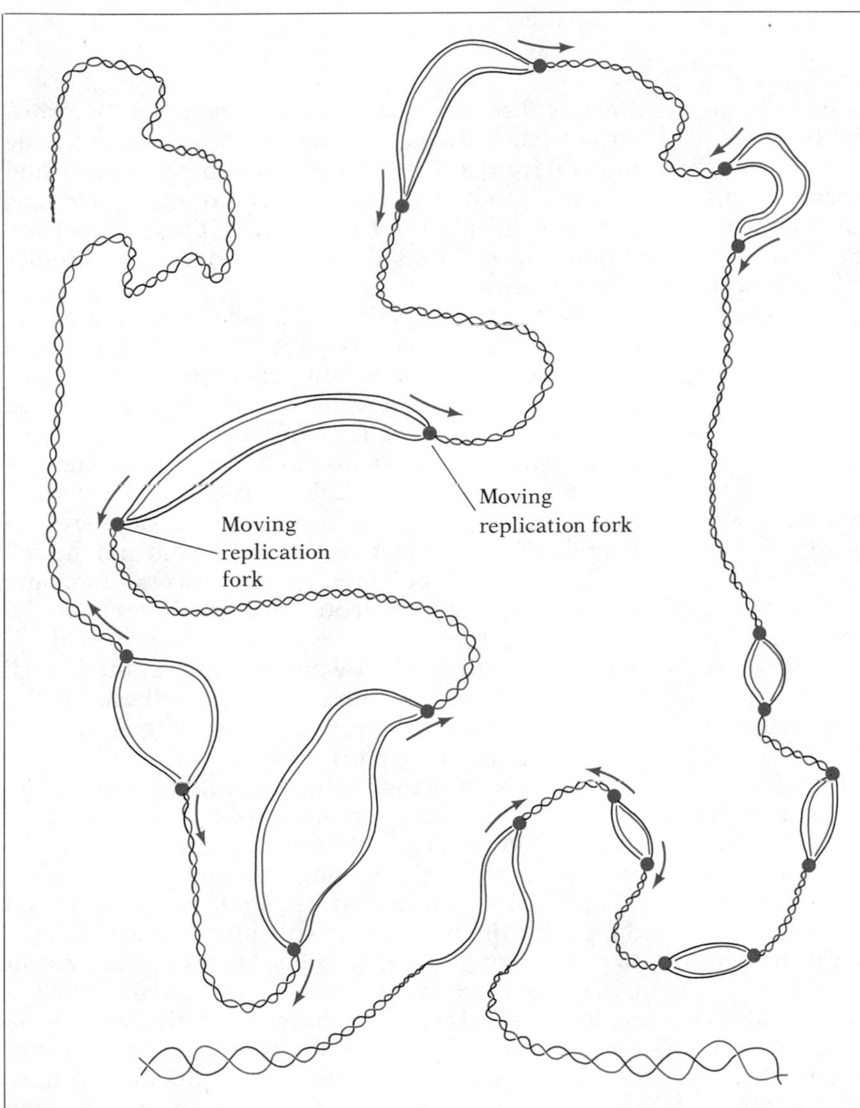

In bacteria and cyanophytes, most or all of the DNA is in a single circular molecule that is not nearly as long as a eucaryote DNA molecule (you have about 1400 times more DNA per cell than an average bacterium). There is no end and no beginning of a circle, of course, but replication in the common intestinal bacterium, *Escherichia coli,* always begins at a single, specific initiation point. Again unwinding proteins form a bubble of single-stranded DNA, and the unpaired strands are replicated with new Watson and Crick strands in the usual way. The two replication forks—and in procaryotes there are only two—travel in opposite directions all the way around the circular DNA molecule, eventually meeting each other at a specific termination point halfway around the circle. Above, a smaller *E. coli* chromosome (plasmid) is caught in the act of replicating.

Essay 7.2 Exceptions to the Central Dogma

It is a bit embarrassing, after being so dogmatic, that there appear to be some exceptions to the universality of the central dogma (in addition to the exceptions found in rules 4 and 5). For instance, *centrioles* are small bodies found in cells, and they appear to arise only from other centrioles. Also, there is some evidence that the whiplike *flagella* possessed by some cells and small structures called *ciliary basal bodies* arise only from basal bodies or from centrioles. In certain weird soil bacteria, short, circular polypeptides with unusual amino acids are produced without the direction of RNA. (However, the synthesizing enzymes, themselves, are made in the usual way.) Certain specific enzyme proteins make chemical alterations of selected nucleotide bases in at least one class of nucleic acids, the transfer RNAs. That's a kind of information transfer, of course, from a protein to a nucleic acid. Finally, replication and transcription don't occur without enzymes, and enzymes are proteins. One could argue that in some sense information transfer in these cases does not quite fit the central dogma. Dogma or not, however, in biology the basic scheme of gene action and protein synthesis is certainly central.

Persons diagnosed for this genetic disease, *xeroderma pigmentosum*, must protect themselves by becoming night people and staying indoors all day.

Inherited defects in the other DNA repair systems are even more serious. Affected persons may be characterized by very poor growth (due to cell death), anemia, a high incidence of chromosome breakage in their cells, and a high probability of dying of leukemia or some other form of cancer.

DNA Synthesis and Replication

Synthesis means making something and *replication* means making an exact copy of something. With very few exceptions, DNA synthesis and DNA replication are exactly the same thing, because DNA molecules are made only by copying other DNA molecules. It was realized very early that one of the properties that genetic material must have is the ability to replicate, because when a cell divides, a complete set of chromosomes has to be passed down to each new cell, and any reproducing animal must pass along a full set of chromosomes to its offspring. Watson and Crick were aware that the specific pairing of DNA nucleotides suggested how DNA itself might be replicated. "It has not escaped our notice that the specific pairing we have postulated immediately suggests a possible copying mechanism for the genetic material," they said. And the suggestion was soon shown to be correct.

The very special way in which nucleotide bases pair in DNA suggests the analogy of a positive and a negative photographic film. The same information is present in both; you can reproduce a positive from a negative and a negative from a positive. Similarly, DNA replication enzymes can produce a "Watson" strand (actually, either one of the strands—a little in-joke) from a "Crick" strand (the other strand), and a "Crick" strand from a "Watson" strand; in so doing, a new Watson strand and a new Crick strand are produced, and thus there are two DNA molecules where there was one before.

We must confess that we don't really know all the details of DNA replication, such as what initiates the process or how DNA is duplicated only once in cell division, but we do know most of the essentials. These are summarized in Figure 7.7. We are aware that, first, an *unwinding protein* unwinds the two strands of DNA so that there are regions, each a few hundred nucleotides long, in which the two strands are unpaired, with their nucleotides sticking out into the cell fluid. Then DNA polymerases go to work, matching free-floating deoxynucleotide triphosphates from the cell fluid to the bases presented by the opened-up, single-stranded DNA. Always working in the 5′-to-3′ direction of the new chain being synthesized, the enzymes condense reactive 5′ phosphates of free nucleotide triphosphates with the free 3′ —OH groups of the growing chain. As the nucleotides of the unwound segments find new partners, unwinding protein moves along the strand, unwinding more DNA and creating a moving *replication fork* [Figure 7.7(a)]. Since the original strands run in opposite directions, the two newly replicated strands are synthesized in opposite directions. One is synthesized in the same direction as is traveled by the moving replication fork, and the other is synthesized in the opposite direction. Interestingly, both new chains are first laid down in short sections of a couple of hundred nucleotides known as *Okazaki fragments.* Then the enzyme ligase joins together the ends of the fragments [Figure 7.7(b)].

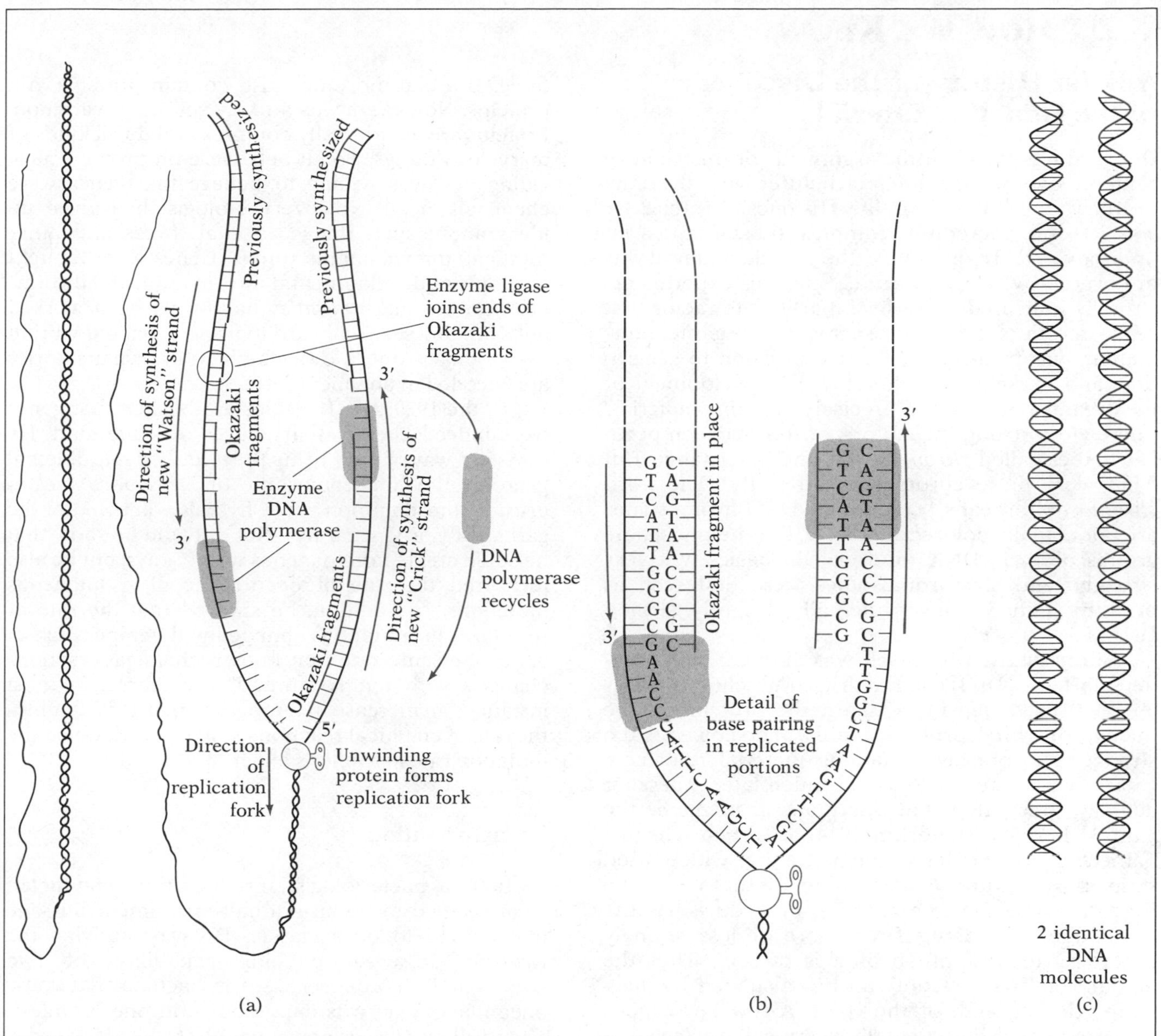

7.7 DNA replication. (a) Prior to the onset of replication, DNA is wound tightly in its helical configuration. The signal "start replicating" will begin the process. The precise nature of this signal is not known. Replication begins when the double helix is opened by unwinding proteins. (b) As unwinding proceeds, nucleotides are paired along the preexisting strands. The entire *replication fork* moves down the old DNA double helix, unwinding it as it goes. Short sections of single-stranded DNA, called *Okazaki fragments,* are assembled by the enzyme DNA polymerase and then joined together by the enzyme ligase. Replication of the separated original strands occurs simultaneously but in opposite directions, always following the 5′-to-3′ direction. As the replication activity proceeds, each new double strand assumes the usual double helix configuration. The DNA polymerase enzyme dissociates and returns to the region of the moving replication fork. (c) The two newly produced DNA molecules will be identical. Note that each replicated molecule contains either an original Watson strand or an original Crick strand, both fully intact. This is known as *semiconservative replication* and is much easier to describe than to prove.

It is apparent that the processes of DNA repair and DNA synthesis have a lot in common. In fact, they even appear to utilize many of the same enzymes, including ligase. And both synthesis and repair involve the matching of free nucleotides against an intact DNA single strand (or chain). DNA synthesis is termed *semiconservative.* This means that each *new* double helix is made up of *one new strand* and *one old strand* (one of the two strands unwound from the old double helix). Thus, the redundancy of information that enhances repair processes also makes accurate replication possible.

7.2 How We Know

A Brief History of the Discovery of the Genetic Material

One of the great scientific arguments of this century centered over what cellular structures were the units of inheritance. Was it DNA? Histones? Protein? Or was it some incredibly complex substance as yet undiscovered? In one sense, the genetic material was discovered by Gregor Mendel, since his experiments strongly indicated a physical, particulate factor that replicated in cell division, segregated in gamete formation, was transmitted from generation to generation, and in some way directed the development of the organism. But what, precisely, was this material? There were strong indications of ties between peculiar bodies called *chromosomes* and inheritance. But what were these chromosomes exactly? The word *chromosome* means "colored body." Chromosomes are not actually colored in nature, but the phosphate groups of their DNA bind certain basic dyes very strongly, and so chromosomes were first seen as brightly stained objects—actually, brightly stained DNA.

Surprisingly, DNA itself was discovered in Mendel's lifetime. In 1869, Friedrich Miescher, a Swiss chemist, used pepsin to digest the proteins of the nucleus of white corpuscles (pus), and showed that a strange, phosphorus-containing material remained. In a private letter, discovered much later, Miescher actually speculated that this material might be the stuff of heredity. Later, in 1914, a German chemist named Robert Feulgen invented a still widely used staining procedure (*Feulgen staining*) that is specific for DNA. The Feulgen staining procedure has the advantage of staining DNA more or less strongly according to how much of it is present. Thus the amount of DNA present can be calculated by measuring the strength of the color. A few key experiments revealed that virtually every cell nucleus in a given plant or animal has the same amount of DNA, except for gametes (eggs and sperm), which have *half* that amount. Could it be that DNA is the genetic material? Of course, you already know it is, but it was a great puzzle to an earlier generation. Most biologists couldn't bring themselves to take DNA seriously, for a few very convincing reasons. In the first place, the structure of DNA is very simple: just four different nucleotides are present. How could anything so simple be the physical basis of anything so wonderful as the gene? How could four nucleotides produce the wondrous variations of life? In the second place, DNA didn't seem to *do* anything. It just sat there, some scientists said, probably holding the chromosome together, or making it acid, or doing some trivial thing. But chromosomes also contain proteins. Ah, proteins. Now *there* was a likely source of variation. Proteins are wonderfully complex and do all kinds of marvelous things. So all bets were on proteins, providing one was willing to believe that genes were chemicals at all. Not every biologist had given up ideas about such things as vital forces and other mystical, unexplainable things. Genes were thought of simply as developmental information, and the idea of *informational molecules* hadn't yet been worked out. If this all seems absurd to us now, remember that even today no one knows whether our own memories are encoded in chemical form.

In the 1920s, H. J. Muller established that genes were indeed chemical structures of some kind. He devised a way of measuring the *mutation rate* of lethal genes in the X chromosome of *Drosophila melanogaster,* the common fruit fly (also known as the garbage fly or vinegar fly). Not only did he show that he could create mutant genes with X-rays, but he also measured the rate of occurrence of spontaneous mutations. In addition, he showed that the rate of spontaneous mutations changed with temperature at about the same rate that known chemical reactions changed with temperature. For example, in most instances an increase in temperature of 10°C doubles the rate of chemical reactions, and it also doubled the mutation rate in Muller's experiments.

Transformation

In 1928, a bacteriologist, Fred Griffith, conducted what seemed to be an oddball experiment but one that proved to be a classic. He was studying the *virulence* (disease-producing capability) of two strains of *Pneumonococcus,* the bacteria that cause pneumonia. One was dangerous and one harmless. The virulent (disease-producing) strain produced a smooth, gummy polysaccharide that seemed to protect it from the host's defenses; the harmless strain (a laboratory curiosity) did not. When grown on agar in Petri dishes, the virulent strain produced "smooth" colonies; the harmless strain lacked the proper enzymes to coat themselves and produced "rough" colonies.

When Griffith injected smooth strain bacteria into mice, the mice died. When he injected rough strain bacteria into mice, the mice did not die. He then killed some smooth strain bacteria by heating them, and injected their bacterial corpses into more mice. The mice did not die. So far, all he had shown was that the smooth polysaccharide itself didn't kill the mice when the bacteria were dead. But then Griffith mixed dead smooth strain bacteria with live rough strain bacteria—both of which had proved to be harmless—and injected the mixture into still more mice. These mice came down with severe pneumonia and died. At this point you might pause and make your own best guess about what was happening. Did the chemical remains of the smooth strain bacteria help the rough strain do its dirty work? To further confuse things, mouse autopsies showed that the dead mice were full of virulent, living, smooth bacteria! Where did they come from?

Griffith thought that perhaps he had erred in his experimental technique. So he repeated the experiment with great care, again and again. The results were clearly not due to accidental contamination; the dead smooth strain *Pneumonococci* were dead, alright. As Sherlock Holmes said, when you have eliminated the impossible, whatever is left must be the truth, no matter how improbable. It didn't seem likely, but apparently the living rough strain *Pneumonococcus* had somehow been *transformed* by material from the dead smooth strain and had become smooth (Figure 7.8).

Others improved on the experiment, trying to discover what was behind these results. They found that transformation could occur in test tubes, as well as in living mouse hosts. Various materials from the dead smooth bacteria were isolated and purified and then injected to see if they were the mysterious *transforming substance.* Not until 1944 was it demonstrated that pure DNA extracted from smooth strain *Pneumonococcus* could transform rough strain bacteria, giving them the ability to synthesize the necessary enzymes for making the smooth polysaccharide coat. After years of research, we now know that the harmless rough cells actually take in pieces of smooth-cell DNA. With a low but measurable frequency, the repair enzymes of the rough cell suture the deadly smooth strain DNA fragments into place over the rough strain's defunct gene. And the rough bacteria are then able to synthesize new enzymes and become smooth.

Anyway, by 1944 it was clear that DNA was the genetic material. Clear, that is, to a relatively few younger and more open-minded microbiologists. Most workers were still far from convinced, even after some thirty different instances of bacterial transformation by purified DNA had been published.

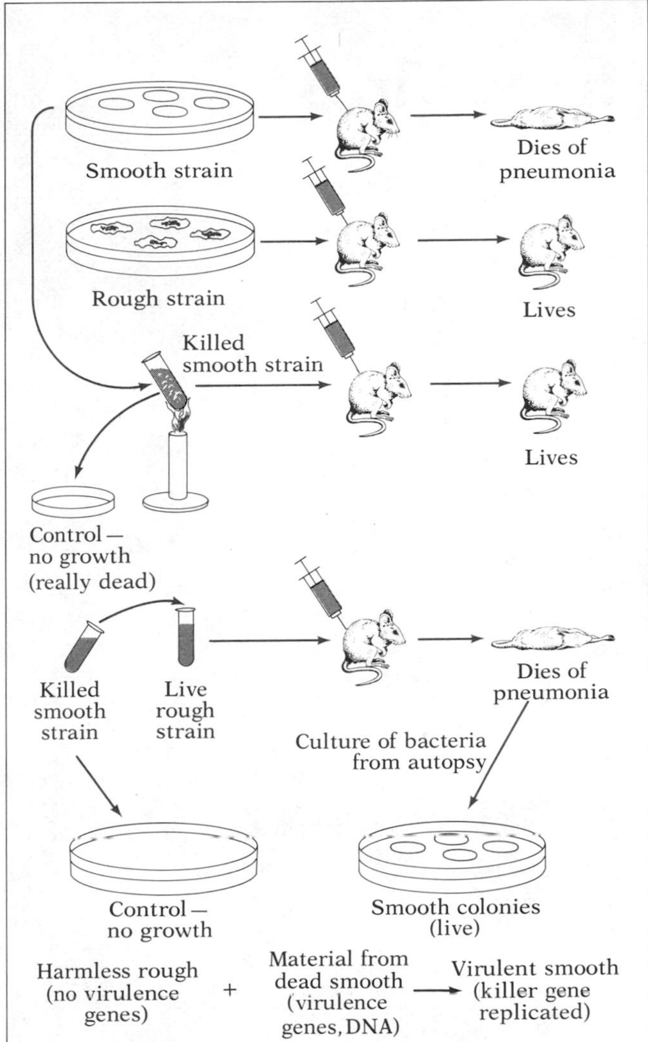

7.8 Transformation (1928). Twenty-five years before the central dogma was finally resolved, Griffith's experiments clearly laid the groundwork for the idea that DNA was the genetic material. Griffith worked with two strains of *Pneumonococcus,* the agent of bacterial pneumonia. His smooth strain was virulent, while his rough strain was rendered harmless by mouse body defenses. Griffith found that some substance or factor from heat-killed smooth *Pneumonococci* could be transferred to the rough strain, transforming the rough form to the smooth form and rendering the bacteria virulent. He had no idea why this happened, but today we know that bits of chromosome (DNA) were being taken in by the rough bacteria and copied during replication. Descendants of the bacteria had incorporated smooth DNA genes and become smooth and deadly.

Hershey and Chase

Alfred Hershey and Margaret Chase performed another classic experiment that, in retrospect at least, firmly established that DNA was the genetic material. They used a new tool in genetic analysis, the *bacteriophage* (called *phage* for short). If you have ever wondered whether a germ can get sick, you will be glad to learn that it can. It can become infected with phage (a kind of virus) and may even die.

The phage Hershey and Chase worked with destroys *Escherichia coli,* the common, rod-shaped bacterium that lives harmlessly in your intestine. The phage they used consists of a DNA chromosome contained in a body made of protein. The whole thing resembles a moon lander or a hypodermic needle. When the phage touches down, tail first, on the surface of its bacterial host, it makes a small hole in the bacterial wall and shoots its DNA inside. There,

The Phage Virus

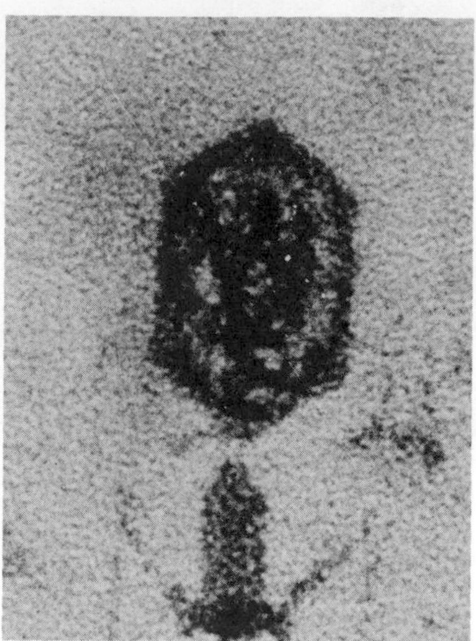

The bacteriophage or phage consists of a head and a narrow tail ending in several tail fibers. The head contains the viral DNA protected by a protein covering. The tail is a hollow cylinder composed of contractile proteins. It terminates in a spiked end plate from which the fibers (long, thin, spidery legs) emerge. The phage so described is a member of the T-even bacteriophages that infect the common colon bacterium, *Escherichia coli.* The term *T-even* refers simply to the even-numbered members of a series which were numbered as they were discovered.

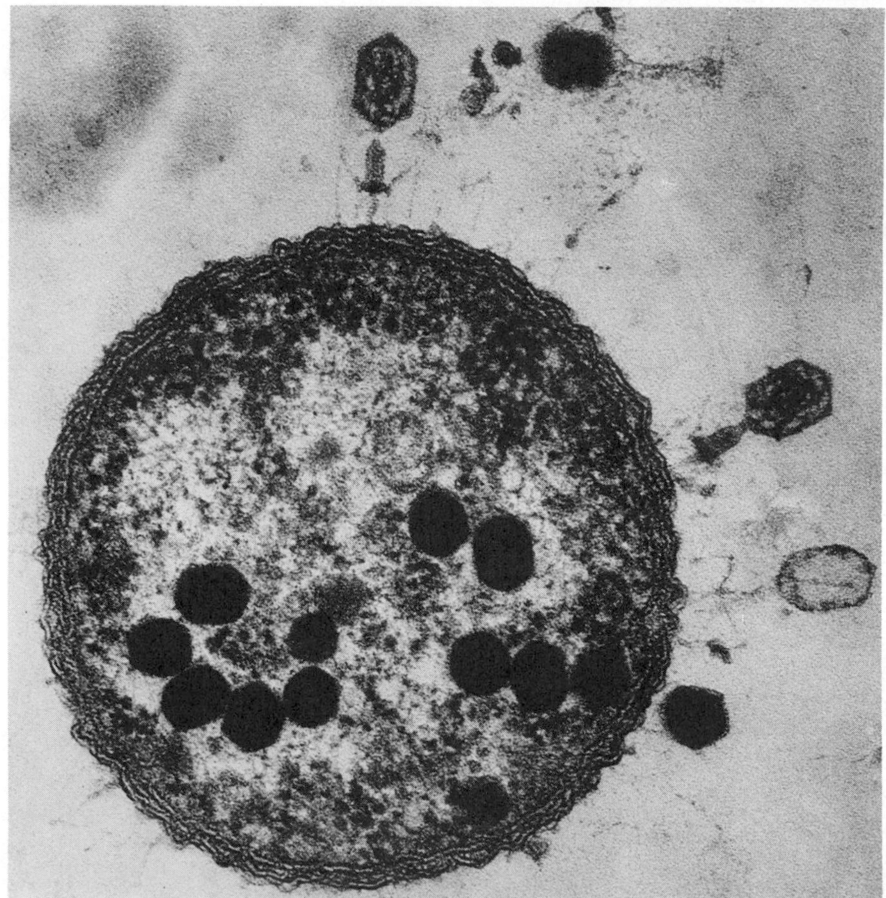

This low-magnification electron micrograph view shows the T-even phage particles attached to *E. coli* and close by. Once inside the host, the viral DNA carries out an insidious program of genetic takeover. The host DNA is destroyed, and the host transcription and translation enzymes are put to work making phage mRNA and proteins. This electron micrograph view shows the new phage coats being assembled in the bacterial host. When all is ready, the viral DNA undergoes repeated rounds of replication, finally equipping the protein shells with viral chromosomes.

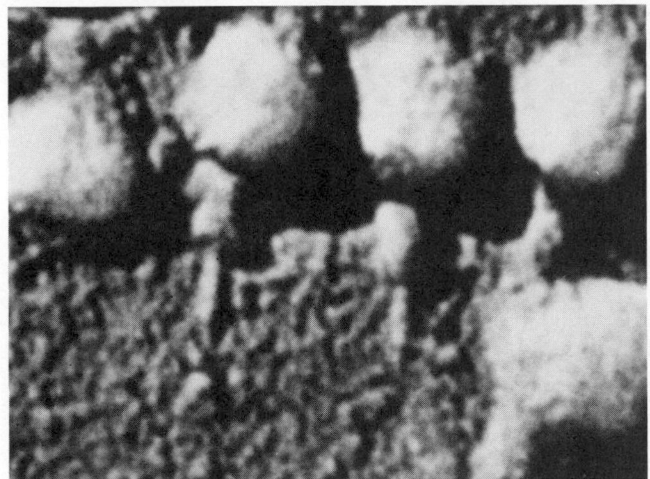

At much higher magnification, the phage particles are shown injecting their chromosomes into the host cell. When the particle attaches to the bacterial wall (just through random bumping around), its tail fibers bond to the wall. This only happens when the bacterium has the proper coat proteins so that the phage fibers can recognize them. Phage–host relationships are very specific! Once attached, the T-even virus makes a small hole in the host's wall and goes into action. The hollow tail contracts and the head collapses, much like a syringe, injecting the viral DNA into the host.

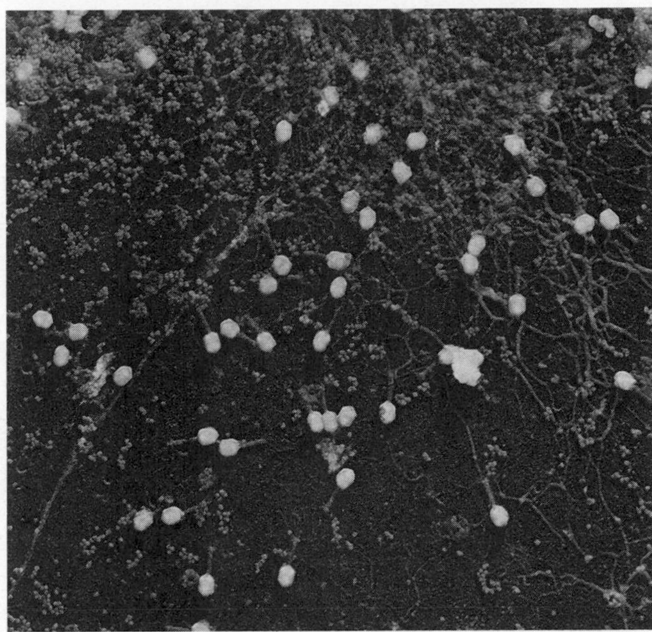

The final event in phage infection is cell *lysis.* The host has served its purpose and now bursts, releasing the newly created phage particles. The phage descendants are now ready for a new round of infection and reproduction.

The cycle of phage activity can take on a quieter aspect as well. At times, the phage DNA may not interfere at all with the host, but inserts into the host chromosome, where it simply replicates every time the host does. Its replication products are then passed on to the daughter cells. Its descendants will eventually be a part of an entire host population. At some signal—we're not sure what—the viral DNA breaks free of the DNA of one of the descendant hosts, and begins the rapid process that ends in host cell lysis and the release of infective virus particles.

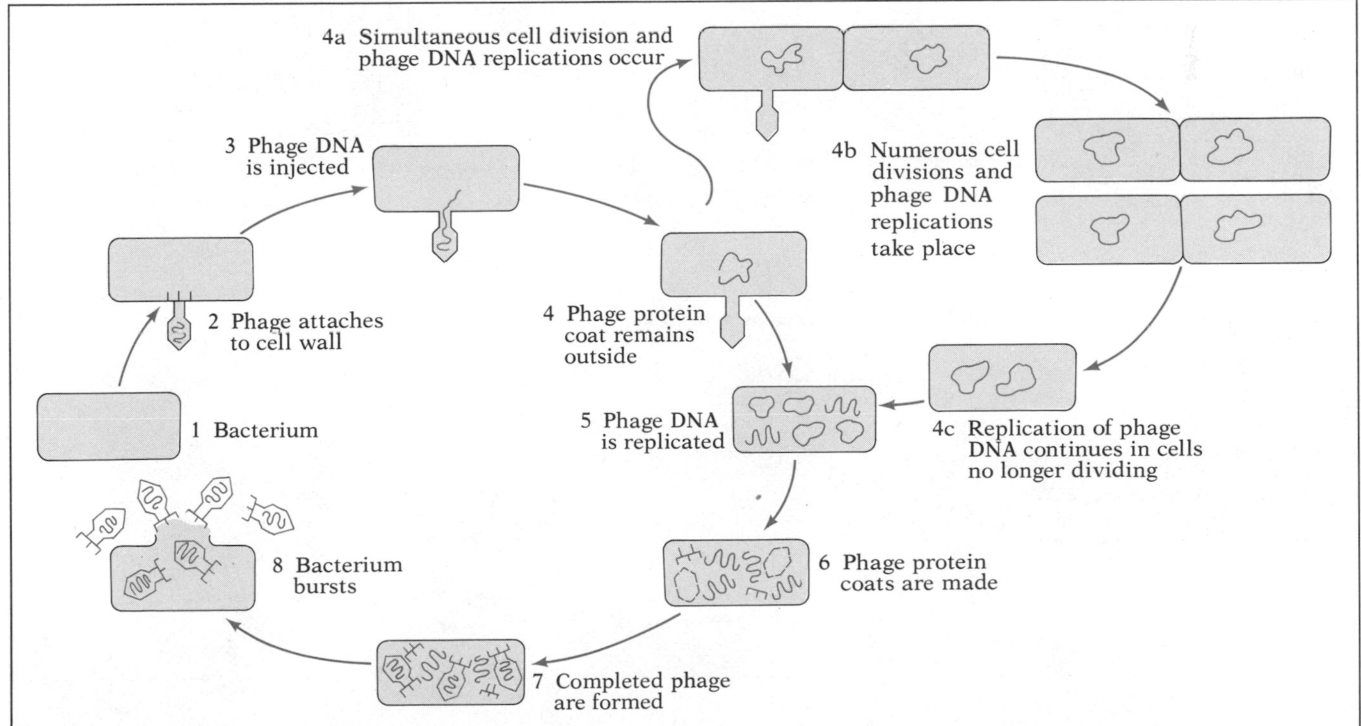
1 Bacterium
2 Phage attaches to cell wall
3 Phage DNA is injected
4 Phage protein coat remains outside
4a Simultaneous cell division and phage DNA replications occur
4b Numerous cell divisions and phage DNA replications take place
4c Replication of phage DNA continues in cells no longer dividing
5 Phage DNA is replicated
6 Phage protein coats are made
7 Completed phage are formed
8 Bacterium bursts

Essay 7.3 Radioactive Tracing

The nuclei of some atoms are unstable. Of the more than 320 *isotopes* that are known to exist in nature, about 60 are unstable, or *radioactive.* In addition, nuclear physicists have created about 200 more radioactive isotopes, called *radioisotopes.*

When a radioisotope *decays* (spontaneously decomposes), it releases radiation in some form: either an energetic subatomic particle (*alpha* or *beta ray*) or a highly energetic photon (*gamma ray*), or some combination of these. In the process of decaying, the radioisotope usually changes from one element to another. The new atom may also be radioactive, or it may be stable.

The time it takes for half of the atoms of any radioactive material to change is called the isotope's *half-life.* Half-lives can vary considerably. Most radioactive substances found in nature are extremely durable; some half-lives are billions of years. All potassium, for instance, is weakly radioactive, but it has an extremely long half-life. Most of the potassium formed at the beginning of time, about 14 billion years ago, is still here. However, radioisotopes that have been created in laboratories are usually much shorter-lived.

Such short-lived radioisotopes can be used as *tracers,* in order to determine how chemicals behave in living material. Since all isotopes of any material are chemi-

(1) Cells in culture flask are starved for nitrogen, halting all protein and nucleic acid synthesis.

(2) Medium containing needed nutrients and tritiated (radioactive) uracil is added to the cell culture. RNA synthesis begins at once, incorporating radioactive uracil into the macromolecule.

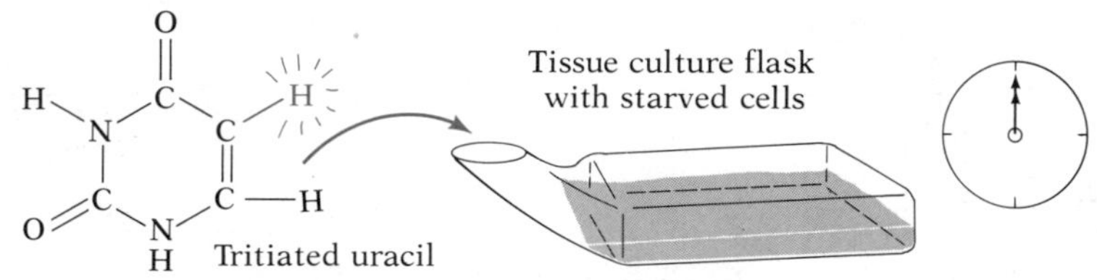

(3) After 5 minutes the cells are "washed" by pouring off the radioactive medium and flooding the cells repeatedly with fresh medium containing an excess of cold (not radioactive) uracil. The cold uracil treatment is known as a "chase."

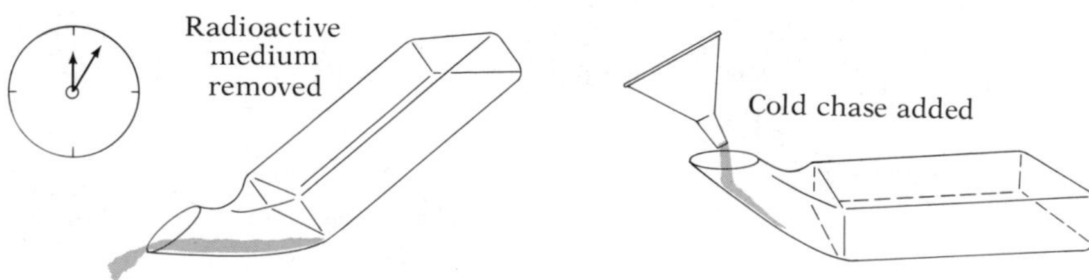

(4) At regular intervals, small samples of cells are removed from the culture flask.

(5) Each sample is spread on a microscope slide, killed, stained with microscope dyes for visibility, and covered with photographic emulsion.

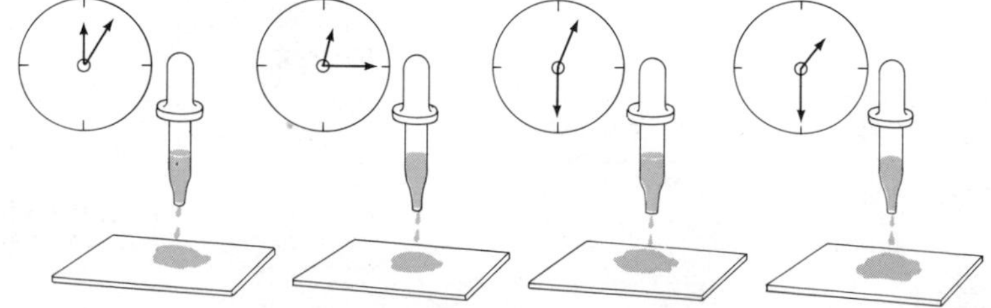

cally the same, radioisotopes can take the place of any normal biological isotope. For example, to trace the course of phosphorus in a plant, radioactive phosphorus can be mixed with ordinary phosphorus. Suppose a botanist wants to know *when* phosphorus from the roots reaches a leaf. Radioactive phosphorus can be made available to the plant roots and then a radiation counter can be placed on the leaf. When the radioisotope reaches the leaf, the counter will immediately register the radioactivity.

If the botanist wants to know more precisely *where* the phosphorus lodges in a leaf, the leaf may be placed on a photographic plate, and the disintegrating isotopes will reveal their whereabouts. On a much smaller scale, the same process—which is known as *autoradiography*—can locate specific chemicals within cells. Flattened cells, or flat sections of killed cells, are covered with a layer of photographic emulsion. The radioactive atoms within the labeled molecules disintegrate over a period of weeks or months, their tiny explosions leaving tracks in the photographic emulsion. The emulsion can then be developed photographically. Dark spots will appear over those parts of the cell that have incorporated the labeled molecules. Chemical analysis will further show just what molecules and what parts of molecules have incorporated the tracer.

Radioisotopes have also been used to detect abnormalities of certain organs in the human body. For example, the rate at which the thyroid gland accumulates iodine indicates the general health of the gland. If a patient is given radioactive iodine, the rate at which it accumulates in the thyroid gland can be measured; a very high rate may indicate a thyroid tumor.

(6) The microscope slide preparations are stored in a darkroom for several months, while the radioactive tritium decays and exposes the photographic emulsion.

(7) The photographic emulsions are developed in the darkroom. Dark spots appear wherever tritium had decayed—that is, directly over the places where the uracil is located.

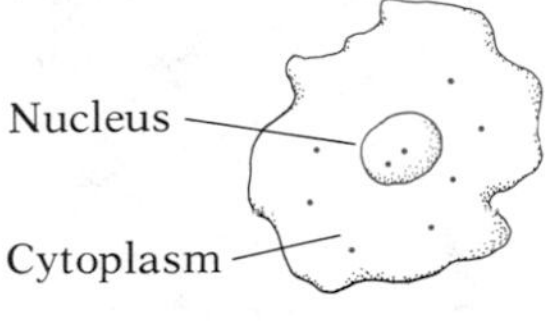

(a) Killed after 5 minutes.

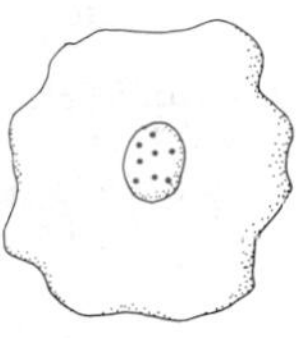

(b) Killed after 15 minutes.

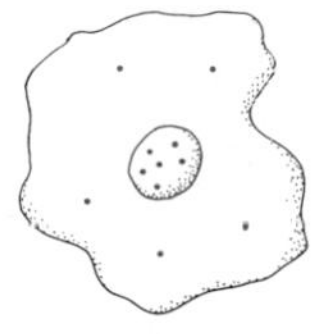

(c) Killed after 30 minutes.

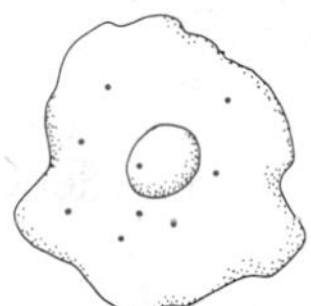

(d) Killed after 90 minutes.

(8) Interpretation of these results:

(a) After 5 minutes, radioactive uracil has entered the cell.

(b) After 15 minutes, uracil is restricted to the nucleus and is probably already incorporated into RNA.

(c) By 30 minutes, newly synthesized nuclear RNA has begun to leave the nucleus.

(d) By 90 minutes, virtually all of the RNA that was synthesized in the first few minutes has gone out into the cytoplasm.

Essay 7.4 Meselson and Stahl

Although the structure of DNA was clarified by Watson and Crick in 1953, the details of replication remained something of a mystery. Watson and Crick's information strongly suggested a *semiconservative* mechanism, in which the two strands of the double helix separate during replication, but each remains intact, acting as a template for the assembly of a new partner. It was not inconceivable, however, that DNA replication might be *conservative* or perhaps even *dispersive.* In the conservative replication hypothesis, the entire molecule would remain intact and another double strand of new material would be replicated alongside. In the disruptive replication hypothesis, the strand would be dismantled piece by piece and replication would occur along the pieces.

In 1957, Matthew Meselson and Franklin Stahl set out to determine how replication was accomplished. Their procedure utilized two modern tools of biology, isotope-labeled biological chemicals and the ultracentrifuge. Using special techniques for isolating DNA from bacteria, they were able to centrifuge it in a cesium chloride (CsCl) solution, which bands each molecule according to its specific gravity (density). Since bacteria are normally exposed only to nutrients containing common nitrogen-14, a ^{14}N reference line was available. In other words, DNA containing ^{14}N settled in the centrifuge tubes at a certain level. Next, a reference line was established for bacterial DNA extracted from cells that had been grown for many generations in culture media containing ^{15}N labeled nucleotides. Since ^{15}N is heavier than ^{14}N, the sedimentation value of the ^{15}N DNA was different; it settled at a somewhat lower position in the centrifuge tube. With these reference points established, the experiment could begin.

Bacteria were grown in a culture medium containing ^{15}N nucleotides for several generations, so that DNA containing ^{15}N would be present in the vast majority of bacterial chromosomes. Bacteria from this culture were then removed, washed, and resuspended in a culture medium containing only nucleotides with ^{14}N. ONE ROUND OF REPLICATION WAS PERMITTED. Cells from this stage were then removed, and the DNA was extracted and centrifuged in the CsCl gradient.

Predictions of competing hypotheses of DNA replication

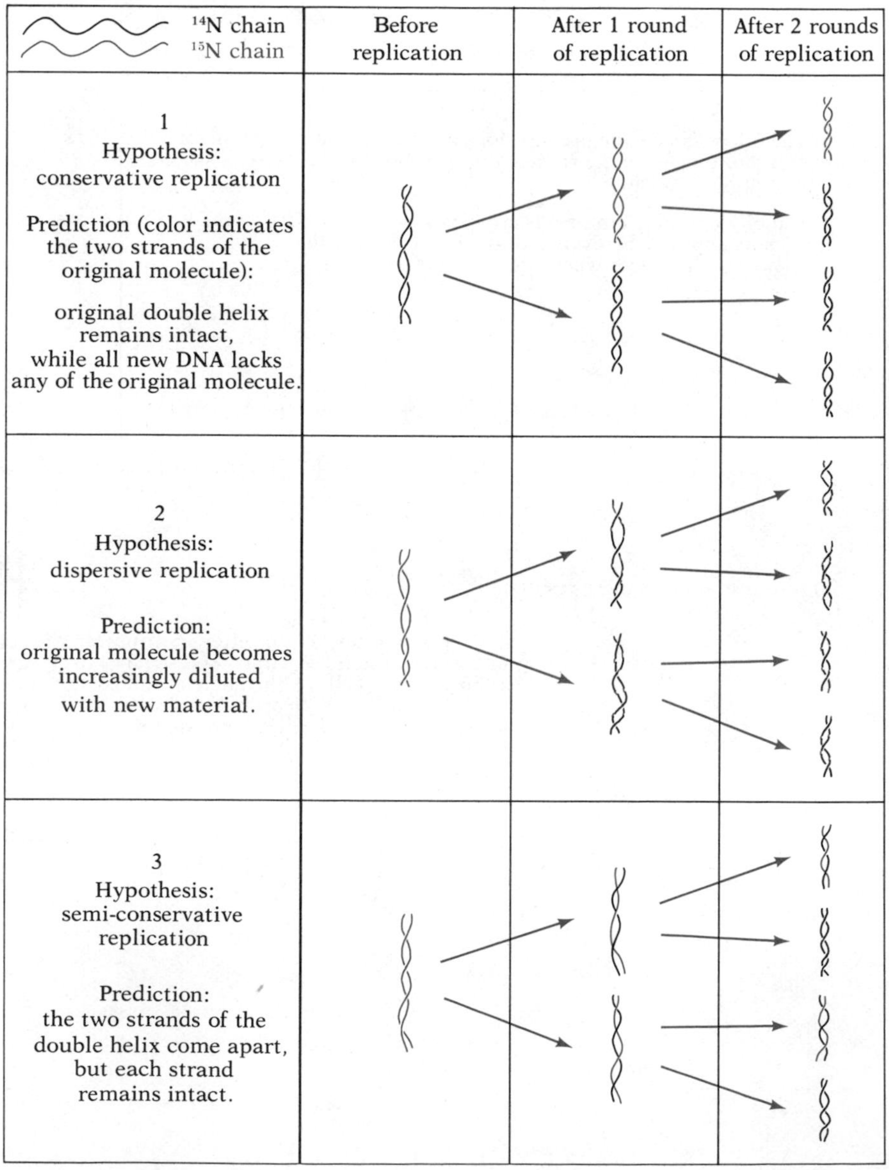

^{14}N chain / ^{15}N chain	Before replication	After 1 round of replication	After 2 rounds of replication
1 Hypothesis: conservative replication Prediction (color indicates the two strands of the original molecule): original double helix remains intact, while all new DNA lacks any of the original molecule.			
2 Hypothesis: dispersive replication Prediction: original molecule becomes increasingly diluted with new material.			
3 Hypothesis: semi-conservative replication Prediction: the two strands of the double helix come apart, but each strand remains intact.			

Results and interpretation

Time	Observation: Observed density gradient centrifuge bands — ^{14}N (light)	hybrid (intermediate)	^{15}N (heavy)	Quantitative results	Interpretation (^{15}N DNA strand / ^{14}N DNA strand)
Before replication			band	100% ^{15}N DNA	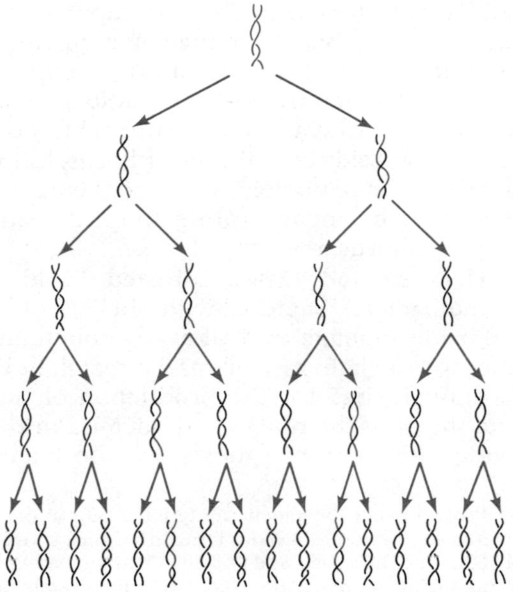
After 1 round		band		100% hybrid DNA	
After 2 rounds	band	band		1/2 hybrid DNA 1/2 ^{14}N DNA	
After 3 rounds	band	band		1/4 hybrid DNA 3/4 ^{14}N DNA	
After 4 rounds	band	band		⅛ hybrid DNA ⅞ ^{14}N DNA	

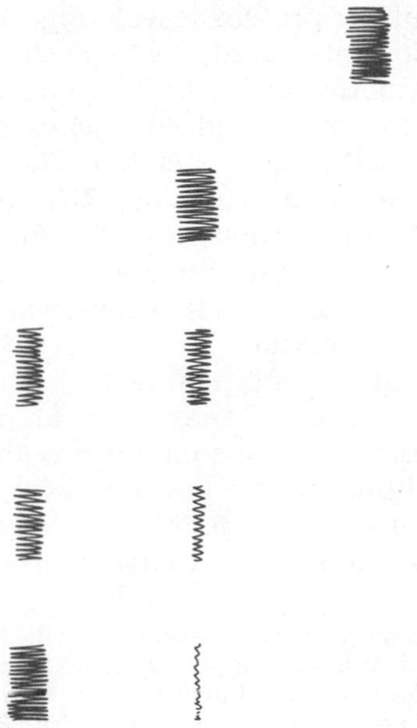

Prediction:

1. If DNA replication is *conservative,* then two regions of DNA will be seen in the centrifuge tube. One will contain only ^{14}N DNA, while the other will contain only ^{15}N DNA.
2. If replication is *dispersive,* all DNA after one round of replication would be halfway between the reference densities of ^{14}N and ^{15}N DNA.
3. If DNA replication is *semiconservative,* then only one region of DNA sediment will be found. This region will contain *hybrid* DNA, each molecule containing a ^{14}N strand and a ^{15}N strand. The sedimentation will occur halfway between the two reference lines established earlier.

In other words, one round of replication could not distinguish between the dispersive and semiconservative hypotheses. But other bacteria were allowed to continue into A SECOND ROUND OF REPLICATION. The cells were then removed, and the DNA was isolated and centrifuged in the CsCl gradient.

Prediction:

1. If DNA replication is *conservative,* then the same two sedimentation bands will appear as did in Prediction 1, but the ^{14}N band will contain 3 times as much DNA as the ^{15}N band.
2. If DNA replication is *dispersive,* all DNA after 2 rounds of replication would be uniform, namely 25% ^{15}N and 75% ^{14}N, and should appear as a single band appropriately spaced between the two reference points.
3. If DNA replication is *semiconservative,* then two equal sedimentation bands will appear. One will contain *hybrid* DNA and will form a band at the same location as did hybrid DNA in Prediction 2. The other will contain ^{14}N DNA only and will settle out at the ^{14}N reference line established earlier.

Results: The results, as diagrammed here, were clear. What Meselson and Stahl saw were the bands shown on the left. Their interpretation is shown on the right—only the semiconservative hypothesis could be supported. Note that the experiment was followed for four rounds of replication, just to be sure.

the phage DNA takes over the synthetic machinery of the host cell, causing it to make a hundred or so new viruses and then to rupture, releasing a myriad of newly constructed viruses.

The important point, with respect to the experiment that Hershey and Chase were going to perform, is that only the DNA of the phage enters the host cell; the protein portion of the phage stays outside. They didn't know this at the time, of course; they discovered it themselves through experimentation. They did this by growing bacteriophage on a medium containing radioactive sulfur (^{35}S) and radioactive phosphorus (^{32}P), and then tracking the radioactive materials. Here's how: Proteins contain sulfur, but no phosphorus; nucleic acids contain phosphorus, but no sulfur. So protein and nucleic acids conveniently labeled themselves by incorporating the radioactive sulfur and phosphorus, respectively.

Hershey and Chase infected "cold" (normal, nonradioactive) bacteria with their "hot" (radioactive) bacteriophage, and allowed enough time for the phage to attach themselves and inject their DNA—but not enough time for the production of new phage. Then they put the phage and bacteria into a kitchen blender. The empty "ghosts" of the bacteriophage were dislodged from the bacterial surface and could be separated by centrifugation. It turned out that all the radioactive sulfur was found in the empty bacteriophage ghosts; all the radioactive phosphorus was found inside the infected bacteria. Thus, Hershey and Chase proved that only the DNA enters the cell. And, more importantly, they showed that naked DNA alone has all the information necessary to enable the host to make new bacteriophage DNA, and new bacteriophage protein as well (Figure 7.9).

So, it appeared, DNA was the stuff of which genes were made—at least, the genes for smooth coats in *Pneumococcus* and the genes of certain bacterial viruses. Actually, Hershey and Chase were lucky, because some viruses use RNA as their genetic material. And in certain viruses, some or all of the protein does in fact enter the host cell. But at the very least, they demonstrated that DNA was *capable* of being the genetic material.

Still, no one had very much of an idea of how DNA worked. Genes were known to produce enzymes, which are complex proteins. The big question was, how could a molecule with only four different subunits determine the specificity of proteins, which are comprised of twenty different amino acids? The

7.9 Protein is not the genetic material (1950). In the early 1950s, Hershey and Chase followed a hunch that phage viruses injected only DNA into their hosts and that DNA had the genetic instructions for producing more viruses. At this time, people weren't sure whether it was the DNA or the protein that carried the instructions. The researchers grew phage viruses on radioactive sulfur (^{35}S) and phosphorus (^{32}P). After allowing the labeled phage particles to infect their hosts, but permitting no time for replication, Hershey and Chase succeeded in dislodging the empty shells of the phage. The rest is history. Only the radioactive ^{32}P entered the host cells, so DNA was almost certainly the genetic material. Another link in the central dogma had been forged.

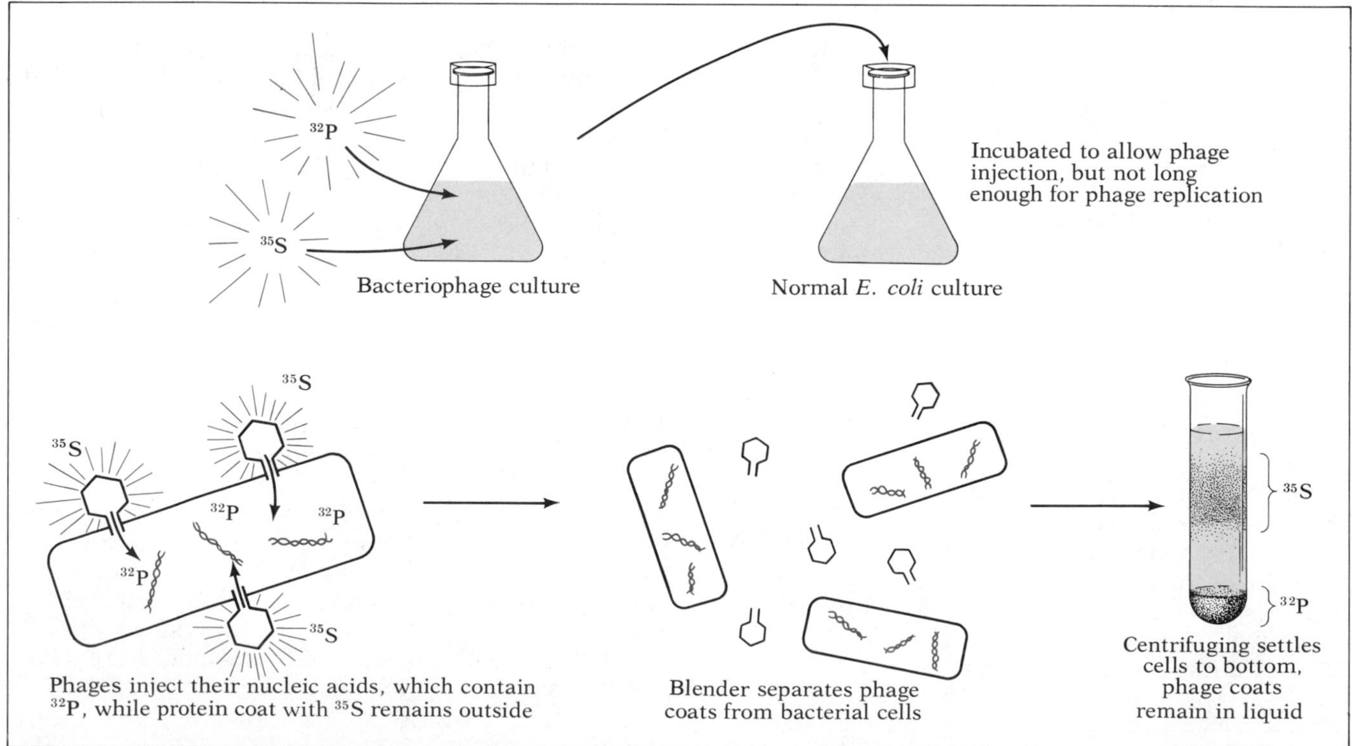
32P
35S
Bacteriophage culture
Incubated to allow phage injection, but not long enough for phage replication
Normal E. coli culture
35S
35S
32P
32P
32P
35S
Phages inject their nucleic acids, which contain 32P, while protein coat with 35S remains outside
Blender separates phage coats from bacterial cells
35S
32P
Centrifuging settles cells to bottom, phage coats remain in liquid

next thing to do was to look at the molecule as closely as possible, and from every possible angle.

Chargaff's Rule

For a long time it had been thought that the basic subunits A, T, C, and G occurred in equal frequencies. In fact, it was once believed that the basic unit of DNA was a simple, repeating tetranucleotide consisting of one each of the four different nucleotides.

But then in 1950 E. Chargaff showed that DNA from different sources had *different base ratios,* that is, different frequencies of the four subunits. *Escherichia coli,* for instance, had about 25% A, 25% T, 25% C, and 25% G—fitting the old ideas—but the DNA of humans and other mammals was about 21% C, 21% G, 29% A, and 29% T. As Chargaff looked at the DNA of more and more organisms, he found increasingly different ratios, but he also discovered one general rule: The number of A nucleotides in his samples was always equal to the number of T nucleotides, and the number of G nucleotides was always equal to the number of C nucleotides (Table 7.1). Or, in shorthand, A = T and G = C. A most interesting observation, of course. We know the reason for these equalities now, but to Chargaff they were only intriguing, mysterious observations. Nevertheless, they were key observations that enabled Watson and Crick to work out the structure and function of life's key molecule.

Table 7.1 Chargaff's Rule (1949–1953).[a]

	Base composition (mole percent)			
	A	T	G	C
Animals				
Human	30.9	29.4	19.9	19.8
Sheep	29.3	28.3	21.4	21.0
Hen	28.8	29.2	20.5	21.5
Turtle	29.7	27.9	22.0	21.3
Salmon	29.7	29.1	20.8	20.4
Sea urchin	32.8	32.1	17.7	17.3
Locust	29.3	29.3	20.5	20.7
Plants				
Wheat germ	27.3	27.1	22.7	22.8
Yeast	31.3	32.9	18.7	17.1
Aspergillus niger (mold)	25.0	24.9	25.1	25.0
Bacteria				
Esherichia coli	24.7	23.6	26.0	25.7
Staphylococcus aureus	30.8	29.2	21.0	19.0
Clostridium perfringens	36.9	36.3	14.0	12.8
Brucella abortus	21.0	21.1	29.0	28.9
Sarcina lutea	13.4	12.4	37.1	37.1
Bacteriophages				
T7	26.0	26.0	24.0	24.0
λ	21.3	22.9	28.6	27.2
ϕX174, single strand DNA[b]	24.6	32.7	24.1	18.5
ϕX174, replicative form	26.3	26.4	22.3	22.3

[a]By determining the composition of nitrogen bases in the DNA of a variety of organisms, Chargaff and his contemporaries were able to provide vital information as the central dogma emerged. Pay close attention to the relative quantities of A and T, and G and C here. (But note that the values are not exactly equal due to experimental error.)

[b]Note that this virus has single-stranded DNA, which does *not* follow Chargaff's rule. Why not?

Adapted from A. L. Lehninger, *Biochemistry,* 2d ed. (New York: Worth, 1975).

Watson and Crick and the Molecular Model of DNA

One of the key devices in cracking the problem of how DNA is constructed was a tool that physicists, metallurgists, geologists, and chemists use to explore the fine structure of crystals. The procedure is *X-ray crystallography,* and which involves shooting X-rays at a crystal and noting how the rays are bent (diffracted) by the regular, repeating molecular structures within the crystal. The closer together the regularly repeated structures, the greater the angle through which diffracted X-rays are bent. The pattern produced on a photographic plate consists of whorls and dots, with those farthest from the center of the plate representing the most closely spaced repeating structures (Figure 7.10). In this way, the relatively simple structures of inorganic crystals can be worked out.

But, organic chemicals can be crystallized, too. In recent years, the three-dimensional structure of many proteins and of some RNA molecules have been worked out. In Watson and Crick's time, however, only preliminary work on a few proteins had been done. People were just starting to look at DNA—among them Maurice Wilkins (who received a Nobel Prize with Watson and Crick) and Rosalind Franklin (who died before her work was fully appreciated). Their studies revealed a few intramolecular distances, such as the 0.34 nm that later turned out to be the distance between the stacked nucleotide base pairs, and the 3.4 nm that later turned out to be the distance of one complete turn of the double helix. Wilkins and Franklin also saw a pattern they recognized as indicating a helical molecule and evidence that the phosphates were on the outside of the molecule. Franklin even argued that there were probably two strands, not one or three. Wilkins and Franklin had formulated a pretty good idea of what the molecule was like, but hadn't really gotten into the specifics. That was left open to Watson and Crick. Their successful analysis of DNA structure opened "the golden era of molecular biology." Things happened fast after the publication of their short little paper in April of 1953. Incidentally, the order of names (Watson and Crick, rather that Crick and Watson) was decided on the toss of a coin.

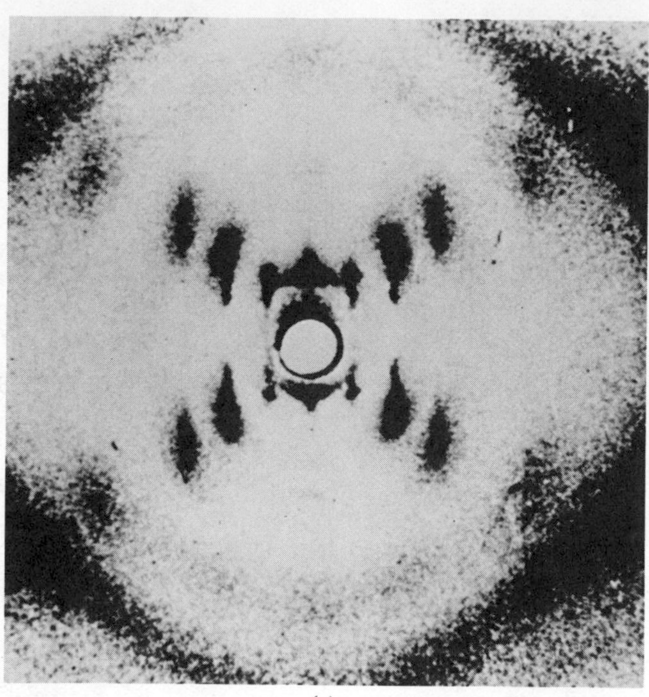

(a)

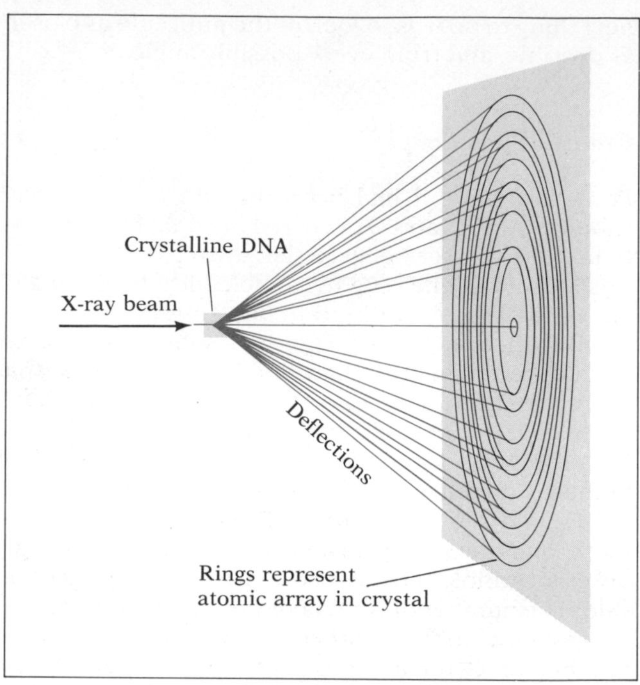

(b)

7.10 (a) Even to many biologists, the X-ray diffraction pattern of DNA might as well be a Rorschach ink blot used in detecting personality disorders. But to Watson, Crick, and others schooled in X-ray crystallography, this picture was a thing of beauty, actually suggesting the helical structure of DNA. The dark upper and lower patterns represent the dense packing of atoms in the nitrogen bases at the core of DNA, while the ×-shaped central pattern is interpreted as a helical structure. From patterns like this, scientists have deduced the structure of DNA, RNA, and many complex proteins.

(b) In X-ray diffraction, an X-ray beam is deflected by repeating structures in the regular crystalline array of closely packed molecules. The closer together two repeating structures are, the farther from the center the beam will be deflected. There are now computers that can be fed the raw data and will immediately draw three-dimensional pictures of the molecule, but keep in mind that Watson and Crick didn't even have a pocket calculator.

7.3 DNA, RNA, and Transcription

RNA: The Other Nucleic Acid

When it appeared that DNA was the material of which genes were made, and thus the controlling molecule of the life of the cell, new and puzzling questions were introduced. For example, when and where were DNA, RNA, and protein synthesized? Molecular biologists, a new breed, used radioactive tracking techniques, labeling molecules that would only be associated with one type of substance in question. For DNA, radioactive thymine was used, since only DNA incorporates thymine. For checking out protein synthesis, radioactive amino acids could be used. And RNA incorporates *uracil*, a nucleotide base that is never present in DNA. (If guanine was purified from guano, guess where uracil was first found.)

Cells can be cultured for a short time with radioactive thymine, uracil, or amino acids, and then examined by *autoradiography* to see where the labeled molecules went (see Essay 7.3). Autoradiography of eucaryote cells during the 1960s revealed consistent patterns of synthesis for DNA, protein, and RNA. First, DNA synthesis occurs only in the nucleus, and only during a specific phase of the cell cycle (the S phase, which stands for synthesis). Autoradiography indicated that all the synthesized DNA remains in the nucleus (these earlier studies did not pick up the relatively small amounts of mitochondrial DNA, or the somewhat larger amount of plastid DNA in plant cells). The amount of DNA is exactly doubled during synthesis; all cells in an organism, except for the gametes (eggs and sperm), have the same amount of DNA prior to synthesis and twice that amount after synthesis. In higher organisms, DNA occurs only in the chromosomes (again, excepting mitochondrial and chloroplast DNA).

Second, protein synthesis occurs only in the cytoplasm. Even nuclear protein is synthesized in the cytoplasm and then migrates into the cell nucleus.

Note that these two observations already present a problem. If DNA is the genetic material, it controls the specificity of protein synthesis; but, somehow, it must do so at a distance.

But let's go on with the story: RNA synthesis occurs only in the nucleus (again, early studies did not pick up the miniscule amount of RNA synthesis that occurs within mitochondria and plastids). Unlike DNA, however, RNA occurs in large amounts in the cytoplasm. Most of the movement of RNA from the nucleus to the cytoplasm occurs during mitosis (when, it turns out, newly synthesized ribosomes are released), but some RNA is continuously migrating from the nucleus to the cytoplasm.

We now know that DNA controls protein synthesis only through RNA intermediates, but it took a while for biologists to learn this and to learn how it was done. We'll drop the historical approach for a while and take a close look at what is now known about RNA and about its various roles in the life of the cell.

RNA Structure

First, let's briefly review the structure of RNA. We should begin by noting that RNA differs from DNA in a number of ways (Figure 7.11):

1. The pentose (five-carbon sugar) is a ribose instead of deoxyribose.
2. While both RNA and DNA contain adenine (A), guanine (G), and cytosine (C), the fourth nucleotide base differs in the two molecules. DNA contains thymine (T) and RNA contains a closely related base, uracil (U).
3. DNA almost always occurs as a double-stranded helix. RNA almost always occurs as a single-stranded molecule, which often has a complex *secondary* and *tertiary structure* (twisting and folding on itself).
4. DNA molecules are almost always much larger than RNA molecules—typically, a thousand to a million times larger.
5. DNA is generally more stable than RNA; it is more resistant to spontaneous and enzymatic breakdown. This is due in part to the double-stranded structure of DNA; but also, as we have seen, DNA is protected by its redundancy and its repair enzyme systems. RNA is more, reactive partly due to the additional reactive —OH side group of ribose. There apparently is no RNA repair system, although there are enzymes that alter RNA in complex but regular ways.
6. There are many different kinds and sizes of RNA, each with its own function.

RNA Synthesis

As we have seen, DNA synthesis is also called DNA replication, because the molecule being synthesized turns out to be the same as the molecule being copied.

DNA

To deoxyribose

① Pyrimidine base, thymine

RNA

To ribose

Pyrimidine base, uracil

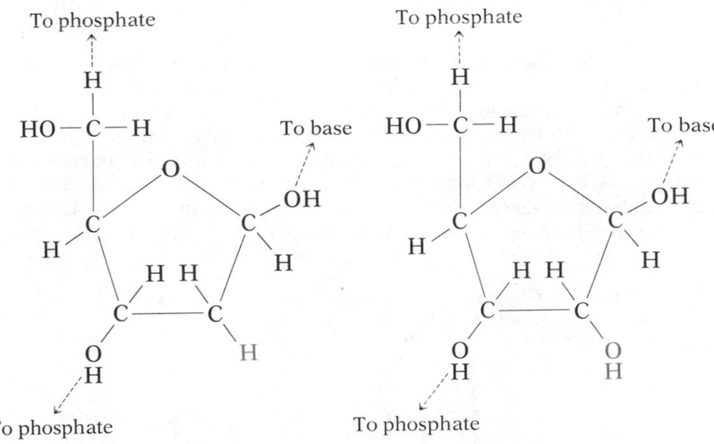

7.11 RNA nucleotide: identical to DNA, with two exceptions. In RNA, the nitrogen bases used include uracil as a substitute for the thymine that occurs in DNA. Except for hydrogen replacing a methyl group, these bases are identical, but as you must realize by now, minor changes make big differences in biochemistry. The second difference is in the pentose sugar used. Where deoxyribose is used in DNA, ribose is the pentose of RNA. Again, the difference is minor—the hydrogen on the number 2 carbon of deoxyribose is replaced by an —OH group in ribose.

In RNA synthesis, the molecule being copied is just one of the two strands of a DNA molecule; thus, the molecule being synthesized is different from the molecule being copied. So a different term, *transcription*, is used to describe this process; we say that an RNA molecule is *transcribed* from one of the two DNA strands. It is also useful, here, to be able to distinguish between the two strands of the parent DNA as the *transcribed strand* (which is copied) and the *nontranscribed strand* (which is not). The length of the DNA molecule on which a single RNA is transcribed is, in some sense, equivalent to a gene. Only specific portions of the long DNA molecule are transcribed.

We have seen that DNA replication is *semiconservative*. RNA transcription has many features in common with DNA replication, but it is not semiconservative; it is *conservative*, since the molecule being copied is conserved—that is, it is not changed by the process of RNA synthesis. The two DNA strands unwind, and through a process we do not yet understand, only one of the two strands is transcribed. Later, the original two DNA strands will wind up again.

The base pairing rules of RNA synthesis are very similar to those of DNA replication except, of course, that RNA uses uracil where DNA uses thymine. The RNA strand is synthesized in the usual 5′-to-3′ direction, using triphosphonucleotides (Figure 7.12) and an enzyme, *RNA synthetase*. For every C in the transcribed DNA strand, RNA synthetase puts in a G ribonucleotide; for every G, a C; and for every T, an A. Since it is RNA that is being made, the enzyme matches a U nucleotide against every A in the transcribed DNA strand. When the process is completed, the new RNA has the same order of bases as the appropriate stretch of the nontranscribed strand of DNA, with U substituting for T.

Many RNA molecules can be transcribed from different parts of the same DNA gene simultaneously, for as soon as the first few bases of a sequence have been copied, they are free to begin a new RNA strand. In Figure 7.13, a series of genes actively transcribing RNA is shown. The axes of the featherlike forms are the DNA genes covered with globular RNA synthetase enzymes. The side branches are the partially synthesized RNA molecules. The RNA strands are short near the beginning of the gene, and get longer and longer toward the end of the gene. In Figure 7.13, there are

7.12 Synthesis of RNA: transcription. (a) The tightly wound helix of inactive DNA is opened by unwinding proteins, and the hydrogen bonds of adjacent nitrogen bases are broken. The bases are then exposed for transcription. So far the activity is similar to replication. (b) RNA synthetase is the enzyme of importance in transcription. It travels along one of the opened strands (the other is apparently ignored in transcription) base pairing RNA nucleotides to the DNA template. (c) Cytosine pairs with guanine and uracil pairs with adenine (remember, uracil replaces thymine in RNA). Once the phosphate–sugar bonds have formed, the RNA polymer dislodges from DNA. It will perform its function as a single strand. (d) RNA is produced in copious amounts when a gene is being transcribed. As soon as one RNA synthetase has moved up the transcription strand, another will follow it, until several are being transcribed simultaneously. When transcription is complete, DNA will resume its inactive form.

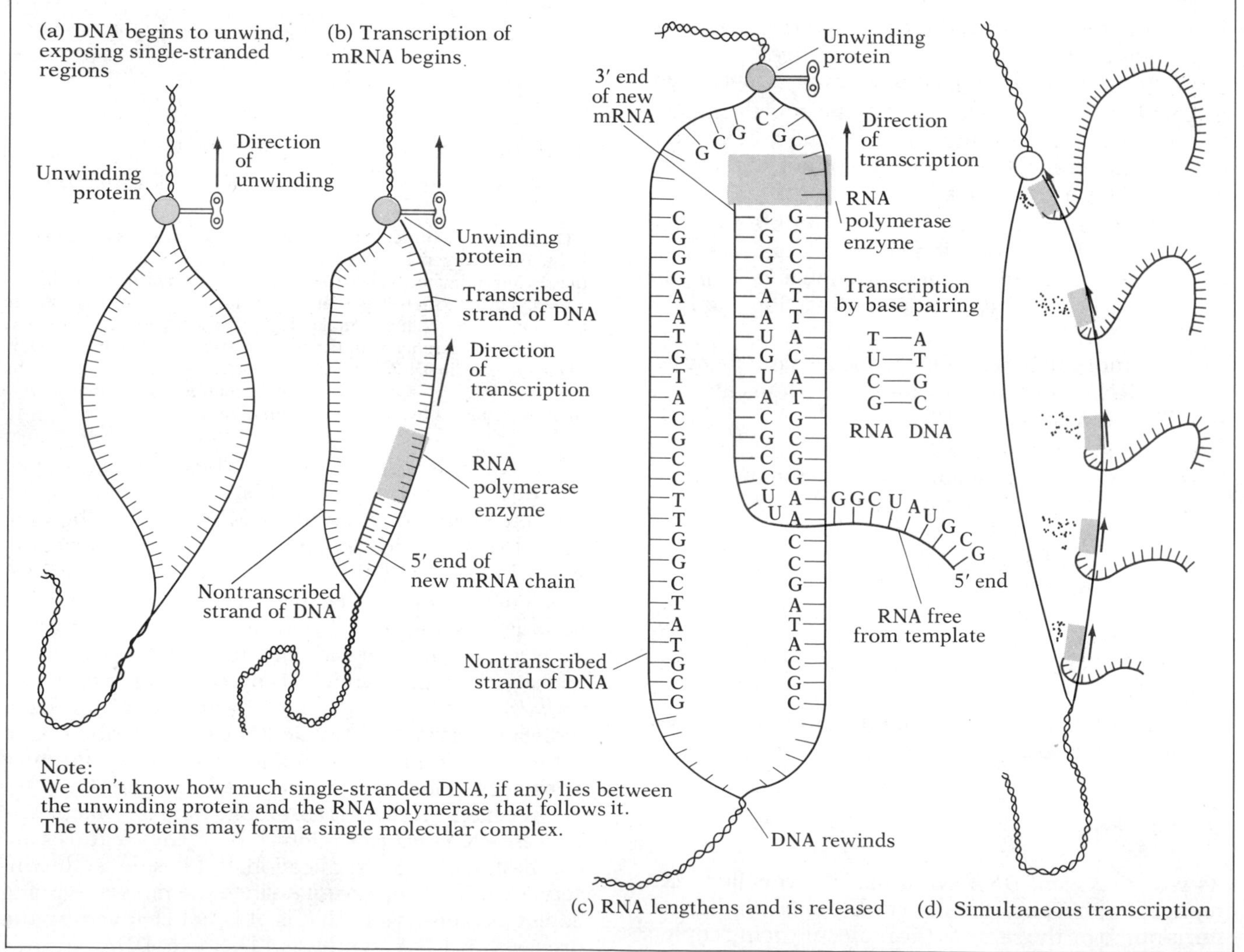

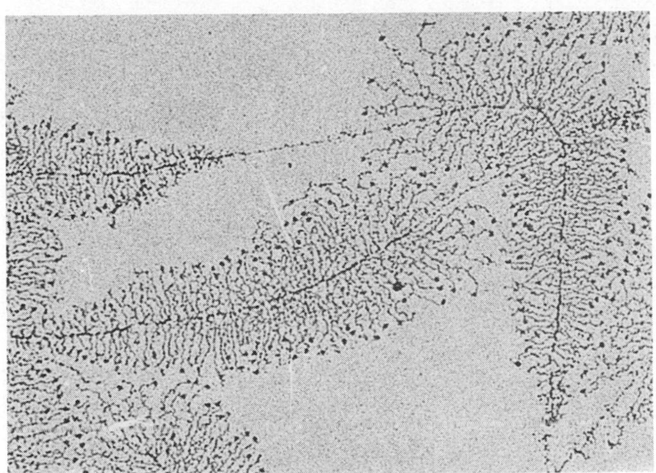

7.13 Here, transcription is occuring along the chromosomes of an amphibian egg cell. Each transcribing section of DNA has many RNAs emerging. You can follow the direction of transcription by determining the length of the strands. The finished strands are released at the 3′ end. The DNA regions not involved in transcription are called "spacer regions" because they don't seem to have any other function.

many active genes, all producing one specific type of RNA, namely *ribosomal RNA* (*rRNA*). Note that there are spaces between each gene where the DNA is not being transcribed. These are called *spacer* regions.

Varieties of RNA: Physical and Functional Classes

Let's enumerate, here, the kinds of RNA that are important to us:

1. Ribosomal RNA: rRNA
2. Messenger RNA: mRNA
3. Transfer RNA: tRNA
4. Heterogeneous nuclear RNA: HnRNA
5. Viral genetic RNA

We might note here that rRNA, mRNA, and tRNA all undergo "tailoring" in the nucleus. After the original transcripts, or *precursors* are synthesized, parts are cut out by specific enzymes. The cut ends are rejoined to make the *mature* rRNA, mRNA, or tRNA molecules.

Viral genetic RNA functions in information storage and replication, much like the genetic DNA of other organisms. Heterogeneous nuclear RNA occurs only in the nucleus and is variable in size and, most likely, in function. Included in the HnRNA are the precursors, or original transcripts, of other classes of RNA. We'll return to describe these mysterious HnRNA molecules later, after we've dealt with the classes of RNA that we think we understand better.

We should begin by saying that ribosomal RNA, messenger RNA, and transfer RNA are all devoted to only one task: the synthesis of proteins.

Ribosomal RNA

Ribosomal RNA is found in ribosomes. In fact, rRNA is the principal constituent of ribosomes if only weight is considered. Each ribosome is made up of two spheres, one somewhat larger than the other. Each spherical subunit contains its own large rRNA chain. The intact ribosome also contains two much smaller RNA chains, called 5S RNA and 7S RNA, as well as about one hundred different proteins. Ribosomes, then, are very large and complex molecules, consisting of many smaller molecules joined together by hydrogen bonds. They are large enough to be seen in electron micrographs (Figure 7.14), which show them as peppery dots.

It should be stressed that ribosomes are present wherever protein synthesis takes place—in fact, ribosomes are the only places where proteins are synthesized. A ribosome can be likened to a kind of workbench or to the drive and reading head of a tape recorder. The ribosome is where mRNA, tRNA, and the growing polypeptide chain of a protein all get together to create their magic: making proteins. The ribosome itself is not passive in this process. Its surface has three *attachment sites* (specifically shaped cavities): one site for a short, three-nucleotide stretch of mRNA and one site each for a new and a used tRNA as we'll see.

Messenger RNA and the Genetic Code

Messenger RNA carries the message—in this case, the specific information regarding the sequence of amino acids of the polypeptide to be produced. The linear sequence of each protein is encoded in the DNA of a specific gene on one of the chromosomes. But DNA never makes proteins directly—it can direct the synthesis only of RNA or of copies of itself. In higher organisms, in fact, protein synthesis occurs only in the cytoplasm, and DNA occurs only in the nucleus. So messenger RNA (mRNA) is the physical link between the gene and the protein. The mRNA molecule is synthesized on DNA and faithfully incorporates the information necessary to specify a protein. The information is written in the *genetic code.* Each code word, or *codon,* is made up of three adjacent nucleotides. The three nucleotides specify one of the twenty common amino acids (Table 7.2). The sequence GAG, for instance, specifies the amino acid glutamic acid.

Since there are four different RNA nucleotides that can occur in the first position of a codon, four different nucleotides that can occur in the second position, and four that can occur in the third, there are $4 \times 4 \times 4 = 64$ different codons. Three of these,

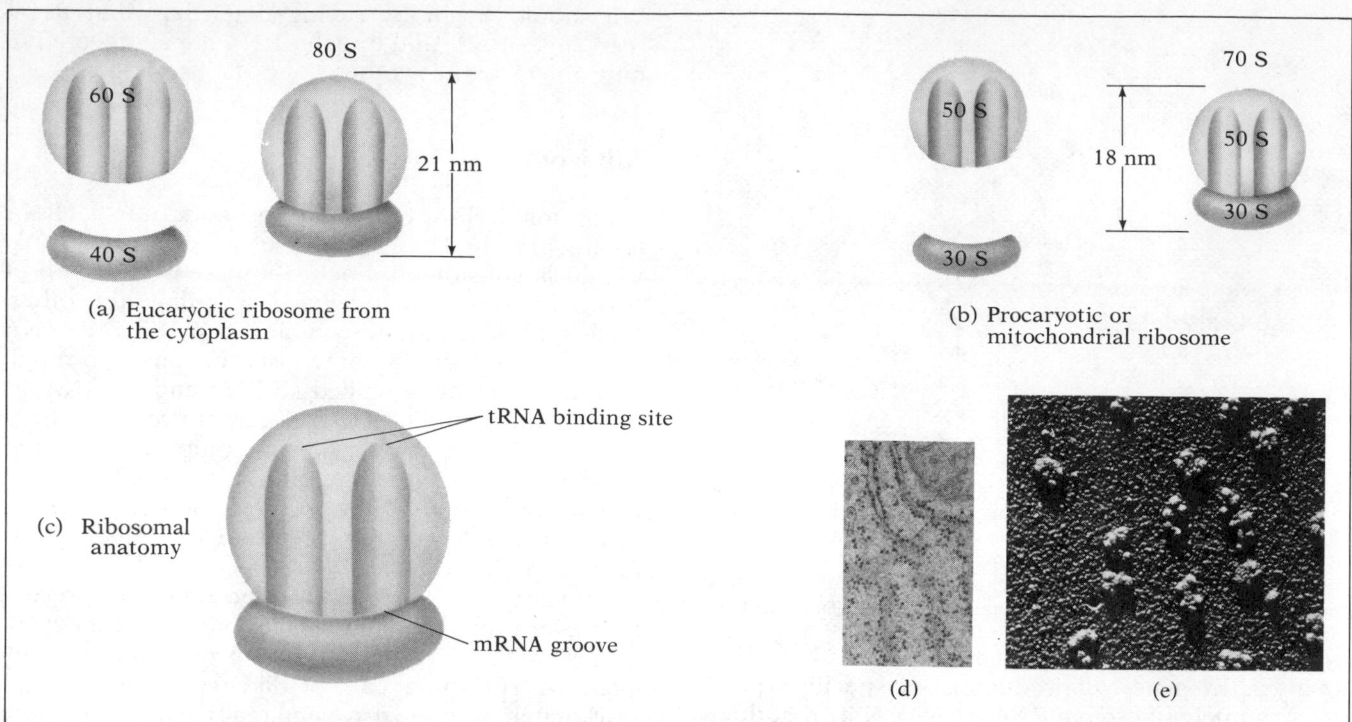

7.14 Ribosomes are the sites of translation. Translation is protein production following the instructions transcribed in mRNA. Ribosomes are composed of two subunits that join during protein synthesis. (a) Eucaryotic ribosomes, 80 S when intact, have a larger 60 S and a smaller 40 S subunit. (The S is the *Svedberg unit,* a measure of the speed of sedimentation. Note that one can't add the S values of the parts to get the S value of the whole.) (b) Bacterial and other procaryotic ribosomes (and those of mitochondria and chloroplasts) are smaller, 70 S, and can be separated into 50 S and 30 S subunits. (c) Ribosomes are composed chiefly of protein and rRNA. Each has three attachment sites identified here as tRNA binding sites and the mRNA groove. (d) Membrane-bound ribosomes (dots) on rough endoplasmic reticulum. (e) Electron micrograph of ribosomal particles.

Table 7.2 The Genetic Code

The genetic code can be described in terms of the codons in mRNA. The table is read in the following manner: There are sixty-four possible codons. The left-hand column contains the first letters of the codons. Across the top are the second letters, and at the right are the third letters. If you wanted to know which amino acid was coded by CAU, you would find C at the left, A at the top, and U at the right. Where the three letters intersect in the table, you will find CAU and the abbreviation His, which stands for the amino acid histidine.[a] Note that the code is *degenerate,* meaning that there is more than one codon for each amino acid (with two exceptions). UAA, UAG, and UGA do not code for amino acids; they signal STOP, and are known as *terminators.* And one more irregularity needs to be mentioned. AUG has two purposes—it codes for methionine and it also means START. Every mRNA begins with AUG, so it is an *initiator.*

FIRST LETTER	SECOND LETTER: U		C		A		G		THIRD LETTER
	U		C		A		G		
U	UUU	Phe	UCU	Ser	UAU	Tyr	UGU	Cys	U
	UUC		UCC		UAC		UGC		C
	UUA	Leu	UCA		UAA	STOP	UGA	STOP	A
	UUG		UCG		UAG	STOP	UGG	Trp	G
C	CUU	Leu	CCU	Pro	CAU	His	CGU	Arg	U
	CUC		CCC		CAC		CGC		C
	CUA		CCA		CAA	Gln	CGA		A
	CUG		CCG		CAG		CGG		G
A	AUU	Ile	ACU	Thr	AAU	Asn	AGU	Ser	U
	AUC		ACC		AAC		AGC		C
	AUA		ACA		AAA	Lys	AGA	Arg	A
	AUG	Met	ACG		AAG		AGG		G
G	GUU	Val	GCU	Ala	GAU	Asp	GGU	Gly	U
	GUC		GCC		GAC		GGC		C
	GUA		GCA		GAA	Glu	GGA		A
	GUG		GCG		GAG		GGG		G

[a]Amino acid abbreviations: alanine, Ala; arginine, Arg; asparagine, Asn; aspartic acid, Asp; cysteine, Cys; glutamic acid, Glu; glutamine, Gln; glycine, Gly; histidine, His; isoleucine, Ile; leucine, Leu; lysine, Lys; methionine, Met; phenylalanine, Phe; proline, Pro; serine, Ser; threonine, Thr; tryptophan, Trp; tyrosine, Tyr; valine, Val.

UAA, UAG, and UGA, are "stop" codons that specify the end of a protein, like the period at the end of a sentence. The remaining sixty-one codons specify the twenty amino acids. Obviously, there are more types of amino acid-specifying codons than there are types of amino acids, so most amino acids are coded by more than one codon. For instance, codons GGU, GGC, GGA, and GGG all code for one amino acid, glycine. These are *synonymous* codons. Because of the lack of a one-to-one relationship between codons and amino acids, the genetic code is said to be *degenerate.*

There is only one ambiguous codon, namely AUG. It can either specify the amino acid methionine in the middle of a protein, or serve as a signal that directs the enzymes of the cell to begin polypeptide synthesis. In the latter case, the AUG triplet is called an *initiator* codon. Apparently, how the cell reacts to an AUG codon depends on the folding (secondary structure) of the mRNA. Given this folding, there is really no ambiguity anywhere in messenger RNA.

A region of a nucleic acid that specifies a polypeptide sequence is called a *cistron* (Figure 7.15). The cistron of a mRNA molecule directs translation directly; the corresponding region of the DNA gene is also called a cistron, or, alternatively, a *structural gene* (because it determines protein structure).

7.15 Eucaryotic mRNA with cistron. In eucaryotes, the group of codons necessary to produce one polypeptide is called a *cistron.* Each cistron begins with AUG (the initiator codon) and ends with one or more of the three terminator codons: UAA, UAG, or UGA. At either end of the cistron are sequences of uncertain function. The 5′ end is protected by a *cap* and the 3′ end is protected by a string of about 200 adenine nucleotides, called the *poly-A tail.* Both the cap and the poly-A tail are added by enzymes in the nucleus.

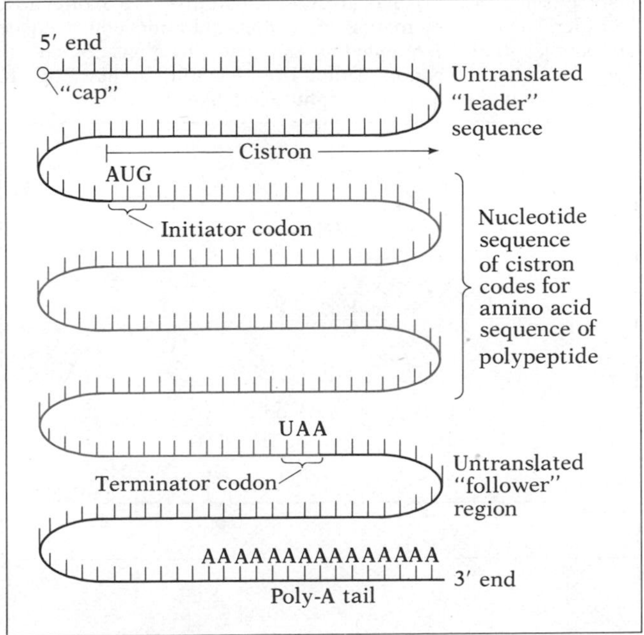

If you'll think back to the spacer regions in Figure 7.13, you'll remember that not all parts of the DNA molecule are transcribed into RNA. Some DNA regions are not transcribed at all; other regions may be transcribed into RNA, but then parts of the RNA are cut out before the mature messenger is complete. For unknown reasons, the cistrons of higher organisms occur in pieces scattered along a length of DNA. The base sequences between the pieces are called *intervening sequences* or *introns.* The entire sequence is transcribed, but the introns are enzymatically removed in the cell nucleus, and the remaining pieces are rejoined to form the mature messenger RNA.

Similarly, not all parts of the mRNA molecule are translated into protein. Only about half of the total length of a mRNA molecule consists of codons that specify amino acids, which is to say that only about half is written in the universal genetic code. The rest of the mRNA molecule is nevertheless faithfully copied from DNA. We don't know what function it has, or if it has any function at all.

Transfer RNA

The mRNA carries the coded message (a long sequence of nucleotides) to the ribosome, where it is decoded. The decoding determines what will be the sequence of amino acids in the developing polypeptide. But the ribosome itself cannot tell one codon from the next except, perhaps, for the initiator and terminator codons. Deciphering the codons, one at a time, is the job of the tRNA.

There are different tRNAs for every amino acid. Special enzymes recognize each particular type of tRNA and, at one point, link it to its own special amino acid with a high-energy, covalent linkage. Thus, one such enzyme (called a *phenylalanine tRNA charging enzyme*) might pick up a phenylalanine tRNA in one of its receptive sites, a phenylalanine amino acid in another site, and, with the aid of ATP, join the specific amino acid to its specific tRNA. In addition to the twenty classes of tRNA for the twenty amino acids, there is often further diversity within each class, because amino acids with several different codons may have overlapping sets of tRNA that recognize some but not all of their codons. And recognizing codons is what tRNA is all about.

Each tRNA molecule is a relatively short length of RNA, consisting of about ninety nucleotides [Figure 7.16(a)]. When it is first transcribed from DNA, the tRNA is somewhat longer, but before it can be active it undergoes some *posttranscriptional modifications.* Special enzymes cut segments off each end, a process called *tailoring.* Other enzymes make chemical modifications of some of the bases in special places on the different tRNAs, so that the completed molecule contains "exotic" RNA bases in addition to the usual four. These exotic RNA bases may serve in part to prevent enzymatic degradation. Yet another enzyme adds three more nucleotides to the 3′ end of each and every tRNA, so that all completed tRNAs end with the sequence —CCA. (Does it seem that this violates the central dogma? *Protein* enzymes here determine a *nucleic acid* sequence.)

The tRNA molecule is precisely coiled and loops back on itself to form a characteristic conformation. The folded tRNA has three loops and a stem [Figure 7.16(b)], and the whole molecule is held in a twisted, L-shaped configuration by hydrogen bonds, not only between its nucleotide bases but also between the bases and the free —OH side groups of the ribose units [Figure 7.16(c)]. Such contortions, of course, would be quite impossible for DNA, because deoxyribose lacks the free —OH group.

It has been speculated that early in the evolution of life, nucleic acids were not just information carriers but also did all the enzymatic dirty work of the first primitive organisms, and that, only gradually, did proteins take on the role of enzymes. Now, all enzymes are proteins—except, perhaps, that one can consider tRNA molecules and ribosomes to have crudely enzymelike functions.

So what can this tRNA do? It turns out that it can do three things:

1. One end—the *anticodon loop*—can recognize specific codons of mRNA.
2. The other end—the —*CCA stem*—is capable of being covalently bonded to an amino acid by a charging enzyme. Not surprisingly, the amino acid attached is always the amino acid that is coded for by the codon.
3. A portion of at least one side loop can make a specific Watson/Crick (that is, C:G, A:U, etc.) pairing with one of the RNA constituents of the ribosome. This portion of the loop is the same for all tRNAs, because they all need to connect to ribosomes in the same way.

The "recognition" between codon and anticodon is yet another example of the Watson/Crick type of pairing. That is, U in the codon pairs with A in the anticodon, C pairs with G, and so on. Thus, the anticodon AGC forms a Watson/Crick interaction with the codon GCU (which codes for the amino acid *alanine*, Figure 7.17). Remember that strands run in opposite directions in Watson/Crick pairing, so that in this case the A of the anticodon 5′-AGC-3′ pairs with the U of the codon 5′-GCU-3′, the anticodon G with the codon C, and the anticodon C with the codon G. It might seem to make sense to print anticodons in the opposite directions, so as to emphasize their pairing

7.16 Transfer RNA molecules contain about ninety nucleotides and are transcribed similarly to mRNA. (a) Upon leaving the nucleus, the *precursor* of a tRNA (that is, the original long RNA transcript before it becomes tRNA) is "tailored"—drastically modified by cytoplasmic enzymes. These modifications include clipping off the ends and short pieces out of the middle, leaving about ninety nucleotides, some of which are chemically modified from the U, C, A, and G of the original transcript. (b) In its modified form, tRNA contains three larger loops and one smaller (variable) loop. These have names: *D loop, T loop, anticodon loop,* and *variable loop.* The 3′ end always contains the three bases CCA and is the attachment site for amino acids. The *anticodon loop* contains three nitrogen bases which base-pair with codons on mRNA. (c) tRNA achieves its final form by folding from its cloverleaf shape into a twisted form resembling an inverted L. Note that the anticodon loop is still prominent. (d) The exact sequence of an alanine tRNA. The unfamiliar symbols stand for various modified bases. Every cell has at least one kind of tRNA for each of the twenty amino acids. Because of the degeneracy of the genetic code, most amino acids utilize several varieties of tRNA. In biochemical parlance, the base sequence is called *primary* structure; the helical base pairing is called *secondary* structure; and the final folding into a complex three-dimensional shape is called *tertiary* structure.

with codons, but the tradition is that all nucleic acid sequences are shown with the 5′ end on the left and the 3′ end on the right.

As we have noted, each type of tRNA has its own specific *charging* enzyme that interacts with the tRNA and with its specific amino acid to form a covalent linkage between the 3′ —OH group of the tRNA and the —COOH group of the amino acid. The charging enzyme presumably recognizes both partners by feel, and forcibly introduces them, using energy from ATP. Some of the details of tRNA charging are shown in Figure 7.18.

7.17 One of the highly specific regions of tRNA is the anticodon loop. Here, we see the anticodon 5′-AGC-3′ base pairing with the mRNA codon 5′-GCU-3′, a perfect match, which will code for alanine (see Table 7.2). The arrows indicate the 5′-to-3′ direction. In order for a proper match, the codon and anticodon must be antiparallel—that is, oriented in opposite directions. Since the diagram shows the codon reading from left to right, like English, the anticodon must read from right to left, like Hebrew.

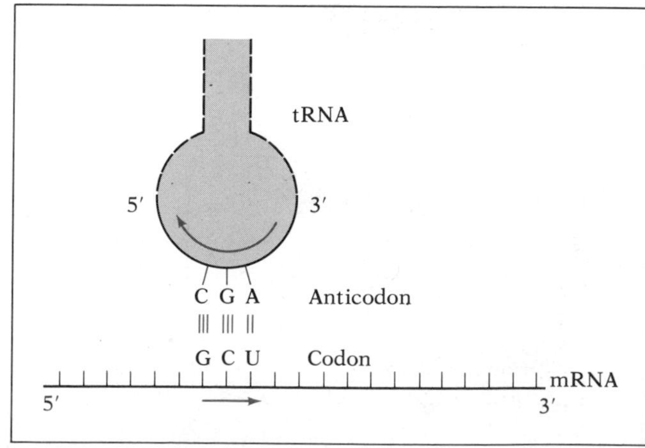

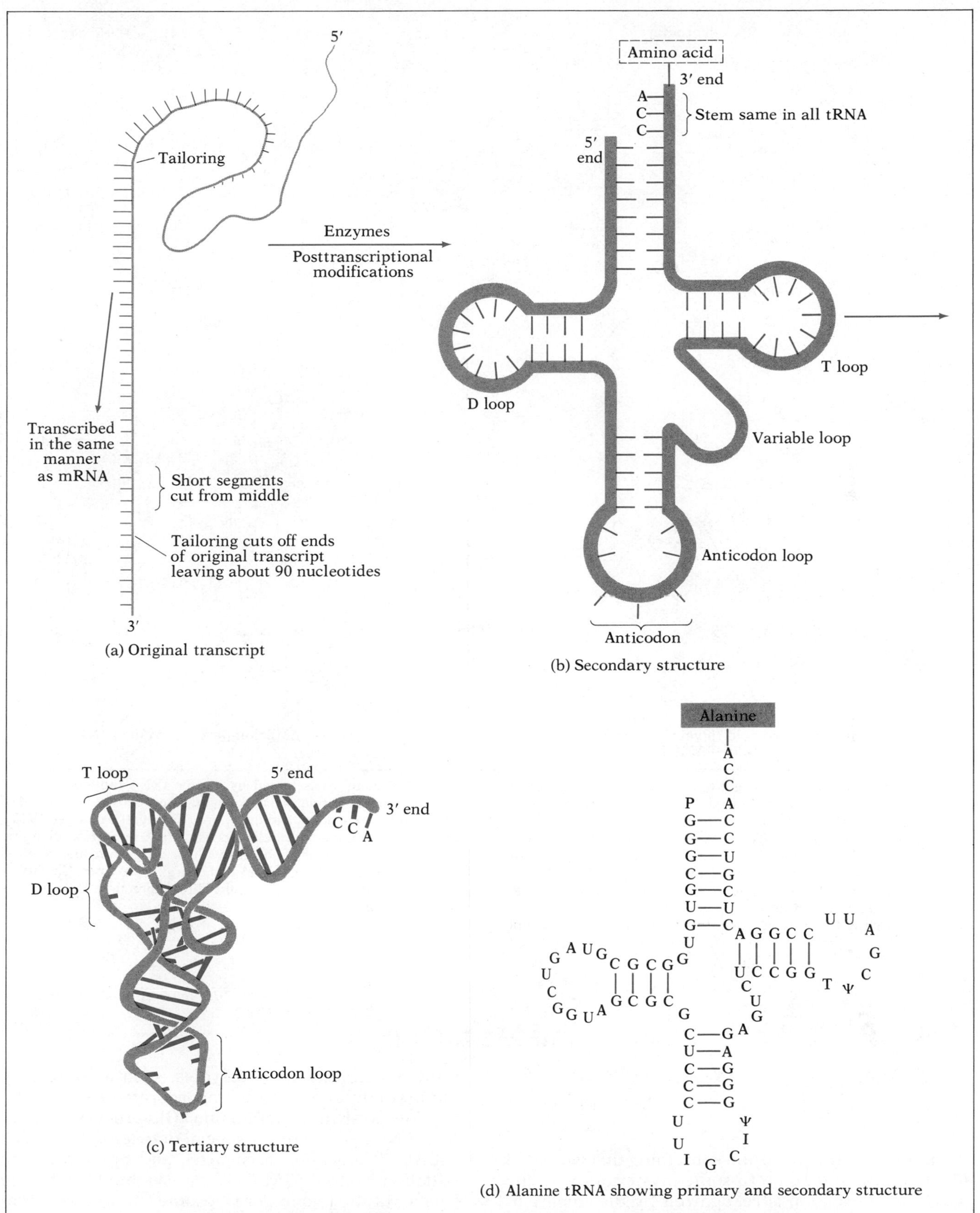

(a) Original transcript

(b) Secondary structure

(c) Tertiary structure

(d) Alanine tRNA showing primary and secondary structure

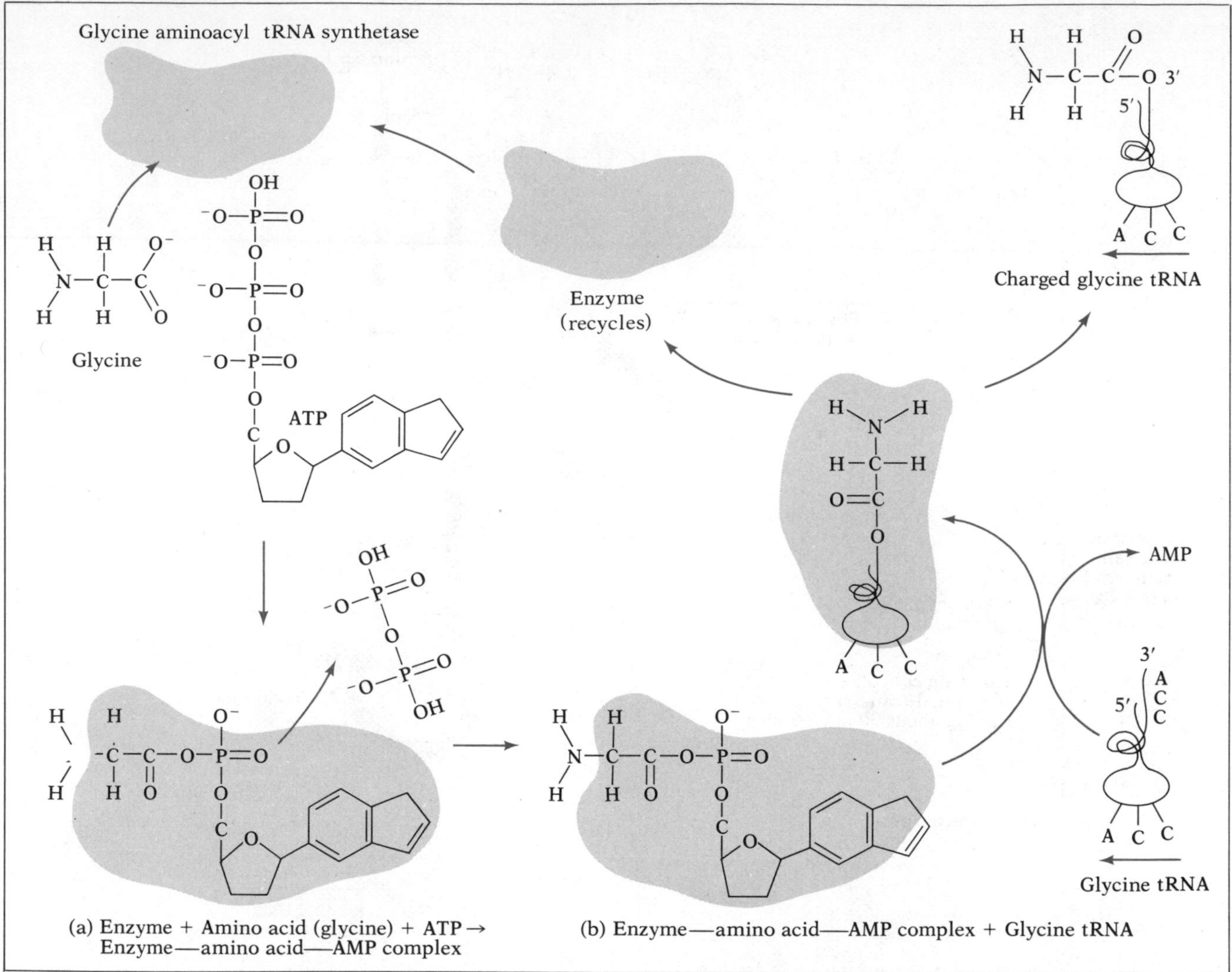

7.18 Glycine charging enzyme system. A vital link in the chain of events leading to translation is the charging of tRNA with its specific amino acid. This requires four factors: a specific charging enzyme, ATP, a specific tRNA, and the proper amino acid. Here we can follow the sequence in the charging of glycine tRNA. (a) The proper charging enzyme, glycine aminoacyl tRNA synthetase, and ATP join to form an active complex. The two high-energy phosphates are released and the high-energy bonds provide the necessary energy for the sequence. (b) Next, the enzyme–amino acid–AMP complex is joined by glycine tRNA (anticodon CCA), whereupon AMP is released and a bond forms between the 3′ end of tRNA and the carboxyl end of glycine. Finally, the charged tRNA, carrying its amino acid, is released ready for use, and the charging enzyme recycles. All of this occurs with dizzying speed and involves all the amino acids in the tRNA and charging enzymes. The arrow beneath the anticodon indicates the 5′-to-3′ direction. Here it reads right to left, like Arabic.

7.4 RNA, Translation, and Mutation

How Proteins Are Made

Initiation

We immediately come upon something that we would like to know, and don't. In some mysterious way, the AUG initiator codon is recognized by the smaller of the two bulbous ribosome units, and this begins the process of *initiation,* one of the most complex steps in protein synthesis. In eucaryotes, such as ourselves, the first amino acid to be incorporated into a polypeptide is always methionine. (In procaryotes, it's *N*-formyl methionine.) A special *initiator* tRNA with a 5′-CAU-3′ *anticodon* recognizes and pairs with the initiation codon 5′-AUG-3′. It also binds with the smaller of the two major ribosome subunits, as mentioned, in an energy-utilizing reaction involving at least three specific initiation proteins. Only after this *initiation complex* (three initiation proteins, the smaller ribosomal subunit, a charged methionine tRNA, and the mRNA initiator codon) is formed can the two subunits of the ribosome combine to form a functional, intact ribosome (Figure 7.19).

The methionine (look at it again) will form the *N-terminal end* of the growing polypeptide. Proteins are synthesized from the *N*-terminal end (*N*-terminus) to the *C*-terminal end (*C*-terminus), just as nucleic acids are synthesized from the 5′ end to the 3′ end. And protein sequences are traditionally written with the *N*-terminal amino acid on the left and the *C*-terminal on the right. *N* Terminal amino acids are called that because they have free (unlinked) H_2N- amino groups, and *C*-terminal amino acids are named after their free (unlinked) $-COOH$ carboxyl groups. Every amino acid has a primary H_2N- group and a primary $-COOH$ group, as you recall from Chapter 3; except for the *N*-terminal and the *C*-terminal, the primary amino and carboxyl groups are linked to form peptide bonds. The net reaction, leaving out the ATPs, phosphates, and tRNAs, is a dehydration bond:

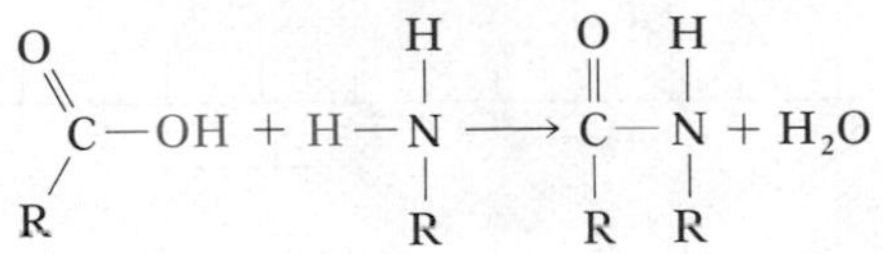

```
 O              H          O  H
  \\            |          ||  |
   C—OH + H—N  ——→  C—N + H2O
  /             |          |  |
 R              R          R  R
```

and the complete protein may be diagrammed:

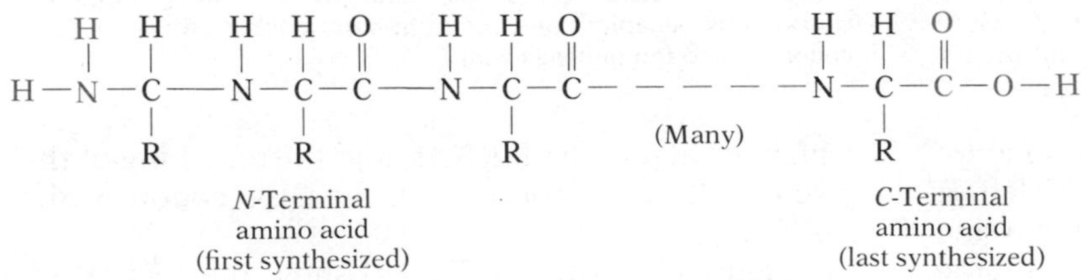

```
    H  H      H  H  O      H  H  O               H  H  O
    |  |      |  |  ||     |  |  ||              |  |  ||
H—N—C————N—C—C————N—C—C— — — — —N—C—C—O—H
       |         |              |       (Many)         |
       R         R              R                      R
      N-Terminal                               C-Terminal
      amino acid                               amino acid
   (first synthesized)                      (last synthesized)
```

where R represents the residue, or remainder of the amino acid, which is the part that is different for each of the twenty amino acids.

Often, the *N*-terminal H_2N- group isn't free at all, but is "capped" with a peptide type of linkage with formic acid (HCOOH) to produce an *N*-formyl terminus:

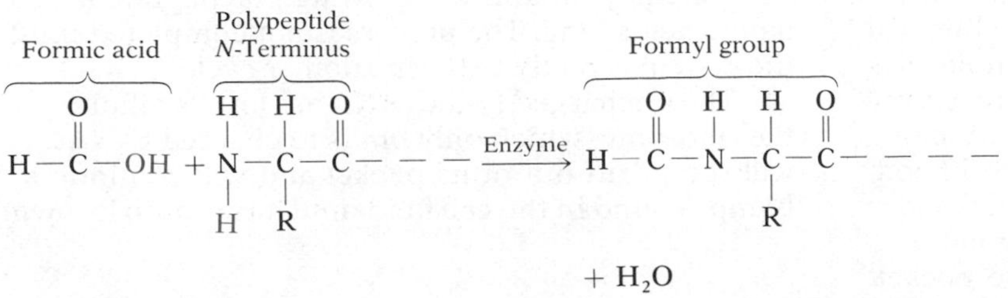

```
                 Polypeptide
  Formic acid    N-Terminus                 Formyl group
      O          H  H  O                   O  H  H  O
      ||         |  |  ||      Enzyme      ||  |  |  ||
   H—C—OH + N—C—C— — —  ——————→  H—C—N—C—C— — — —
                 |  |                              |
                 H  R                              R

                                 + H2O
```

Formation of an *N*-formyl terminus

In procaryotes, in fact (but not in our kinds of cells), the initiator RNA is charged not with methionine but with *N*-formyl methionine. We eucaryotes use plain old methionine. In either case, the methionine or *N*-formyl methionine is almost always immediately clipped off the newly completed protein by a specific enzyme. The methionine has served its purpose in the initiation process and is no longer needed. Other enzymes may or may not then add formic acid residues onto the new *N*-terminus, depending on the particular protein that is being formed.

Elongation

So that's how a protein starts, but how does it grow? How are additional amino acids added to the chain? Let's take a look at the process about halfway along. Polypeptides come in all lengths, but a length of 250 amino acids is about typical; so let's say that the first 125 amino acids have already been synthesized, and the 126th is about to be added on.

Figure 7.20 shows the situation. A length of mRNA lies in the groove between the two halves of the ribosome—nine bases reading —AAA—GGC—UUA—, which are three codons specifying lysine, glycine, and leucine, respectively. Let's say that these will be the 125th, 126th, and 127th amino acids in the protein. Lysine, the 125th, is already part of the chain. The ribosome has two large pockets, right next to each other, that are the tRNA attachment sites. The messenger RNA runs along the bottom of the two pockets, with codon AAA in one pocket and the next codon, GGC, lying in the bottom of the other pocket.

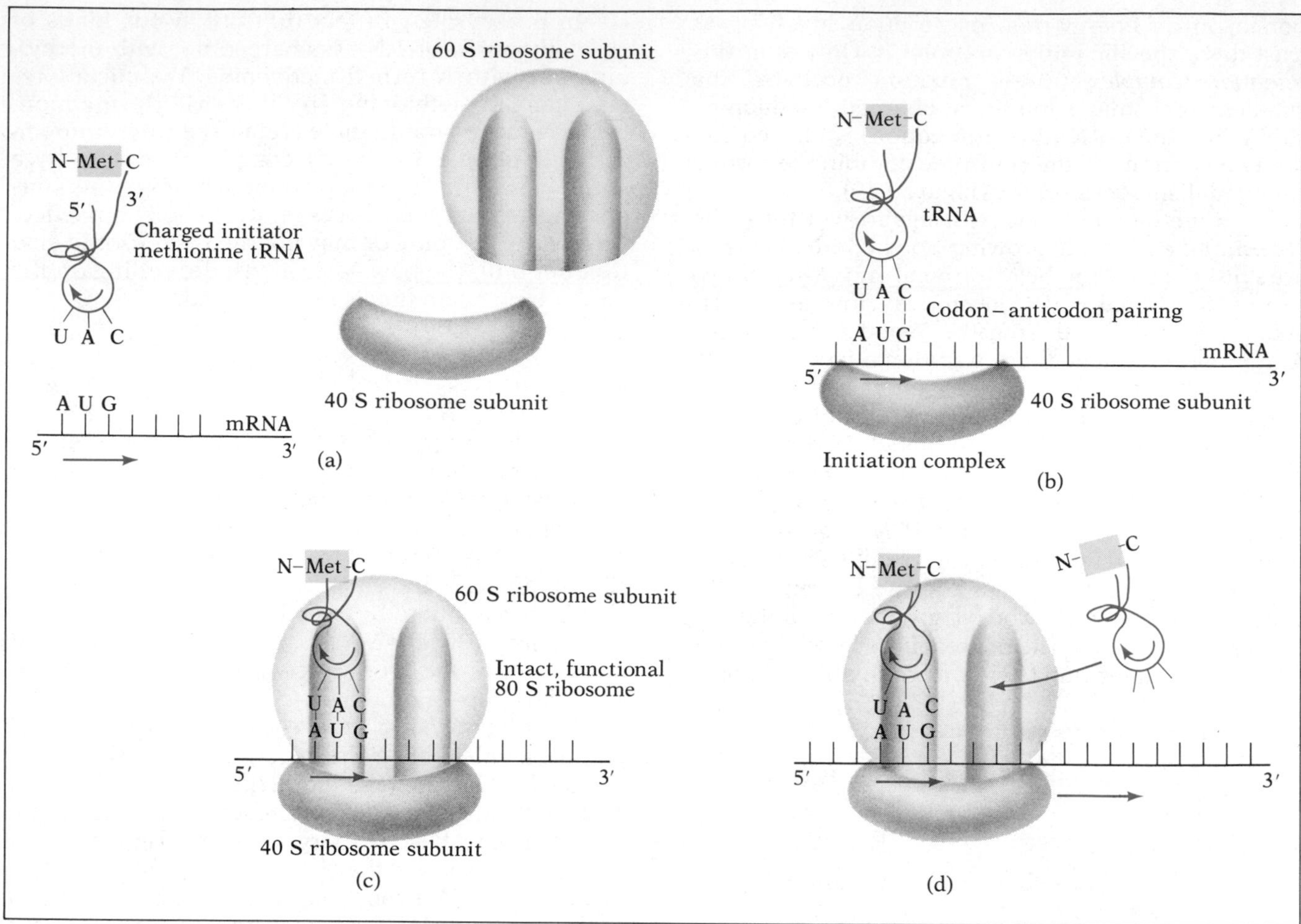

7.19 Part one of protein synthesis or translation: initiation. (a) The required elements are mRNA, the two ribosomal subunits, and charged methionine tRNA. (b) The initial event is base pairing between the mRNA initiator, AUG, and the anticodon UAC, which is only found on methionine tRNA. As the base pairing occurs, the smaller ribosomal subunit joins the RNAs. (c) Only then can the larger subunit move in and complete the protein-synthesizing complex. All is ready. (d) The left-hand pocket (binding site) of the ribosome is occupied, and the right-hand pocket awaits the next codon–anticodon pairing event.

At the particular moment [captured in Figure 7.20(a)], a lysine tRNA lies in the pocket on the left. It is still attached to the carboxyl group of its own lysine amino acid. Attached to the amino group of this same lysine, however, is a long string of 124 other amino acids floating out in the cell medium. So, as of this moment, the lysine is the temporary *C*-terminal end of a 125 amino acid polypeptide. The other end of the *nascent* (half-formed) polypeptide—the *C*-terminal end—is the methionine that started the process.

The other tRNA pocket is empty, except for the next mRNA codon that lies along the bottom of it. This situation lasts for a few microseconds, which is a short time for us but a fairly long time for a chemical reaction. During this time, all sorts of small molecules randomly bump in and out of the nearly empty pocket in the ribosome, including any tRNA molecules that might happen to be in the neighborhood. Sooner or later, a charged glycine tRNA will wander in, and it will fit so well that it sticks. The good fit is due to the combination of the shape of the pocket, which fits all charged tRNAs, and the matching of the glycine tRNA anticodon with the GGC codon of the mRNA.

Figures 7.20(b) and 7.20(c) show the two tRNA pockets in the ribosome filled with the lysine and glycine tRNAs. Part (c) shows that the entire 125 amino acids, ending with lysine, have been transferred from the stem of the lysine tRNA to the amino side group of the glycine, as a peptide bond is formed between the carboxyl group of the lysine and the amino group of the glycine (using the energy of the lysine tRNA high-energy bond). The lysine tRNA is now uncharged, and the growing polypeptide is 126 amino acids long. The process is not completed until the system is ready to begin another cycle.

The uncharged lysine tRNA has lost its affinity for the ribosome, which only binds to charged tRNAs. It will soon drift out of its pocket and will continue to bump around in the cell fluid until it runs into its own special charging enzyme, which will charge it with another lysine so that it can enter the process again.

Between Figure 7.20(c) and 7.20(d), the glycine tRNA has moved from one pocket to the other, bringing along with it both the growing polypeptide and the mRNA, which is still bound to its anticodon. This crucial step is called *translocation*. Translocation, you will notice, moves the ribosome three nucleotides to the right along the mRNA. As a consequence, the UUA codon now lies along the floor of the newly empty tRNA pocket. Soon a charged leucine tRNA will randomly bump into place, and the elongation process will proceed. This, then, is why we compared the ribosome with the reading head and

7.20 Part two of protein synthesis: elongation. (a) The illustration picks up the process midway through the synthesis of a polypeptide. Three code groups of mRNA are shown with the left one occupied by lysine tRNA. Its amino acid, lysine, has already covalently bonded to the growing polypeptide chain. (The previous occupant is shown tilting away for recycling.) The right pocket opposite GGC is empty, but moving in is glycine tRNA with the proper anticodon for a fit. (b) Now glysine tRNA has landed in the right-hand pocket. (c) The two amino acids will become covalently bonded. (d) Then the ribosome is ready to move down one codon (translocation). Following translocation, lysine tRNA is released to recycle, glycine tRNA occupies the left pocket, and the right pocket is again empty. Leucine tRNA, a newcomer, is about to drop into the empty pocket where its anticodon is a match for the mRNA codon UUA. Then bonding and translocation will occur once more as the elongation process continues. As far as it is known, all the highly organized activity in translation is brought about by totally random events. The artist has tried to capture the frenzied activity about a ribosome as numerous molecules strike and bounce away from the waiting ribosomal pocket. Eventually, the molecule to hit the pocket will be a charged tRNA with the right anticodon for a fit. Then the process continues.

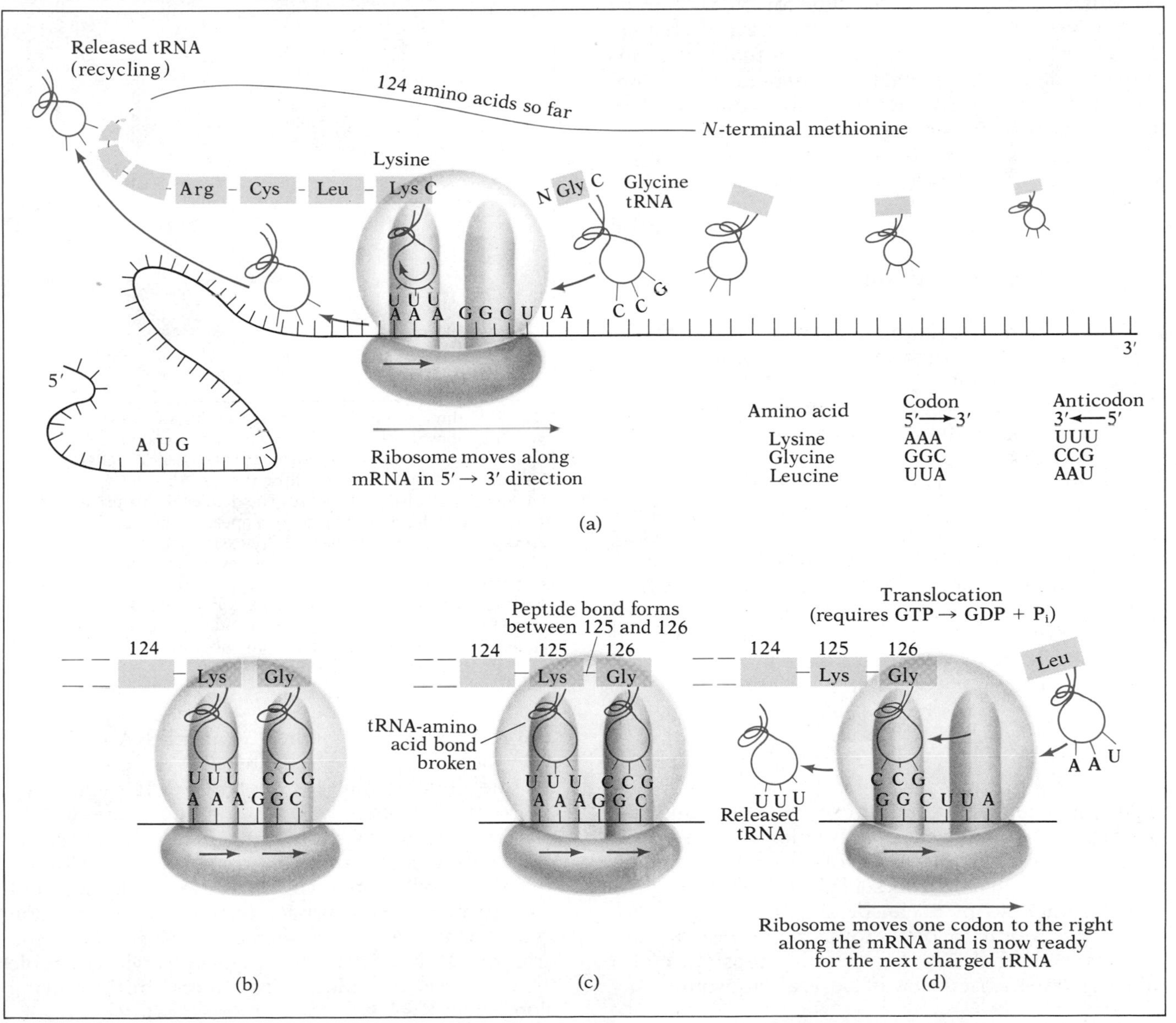

tape drive of a tape recorder: the ribosome not only "reads" the mRNA, but moves it along. (It is arbitrary, of course, whether we consider that the mRNA moves past the ribosome, or that the ribosome travels down the length of the mRNA.)

Translocation requires energy. Energy for cell reactions usually comes from ATP, but in this case, the process gets its energy from the high-energy bond of guanosine triphosphate (GTP).

The question most often asked when translation is explained is, how does everything know where to go? The scheme just described is beautiful but it prompts skepticism. The movement of the molecules appears to be totally random and there is every reason to believe it is. In other words, the base pairing of codon and anticodon is just a matter of chance contact. Of course, no combination but the correct one will work and all others will be rejected. One way of improving the odds is for a great deal of charged tRNA to be around and in motion. Actually, it seems that most chemical events in cells depend on random motion and fortuitous but predictable collision. Does this mean protein synthesis is a slow process? A ribosome of *E. coli* requires about 6 sec to produce a polypeptide! Our cells are comparatively sluggish so it takes us 2–3 min.

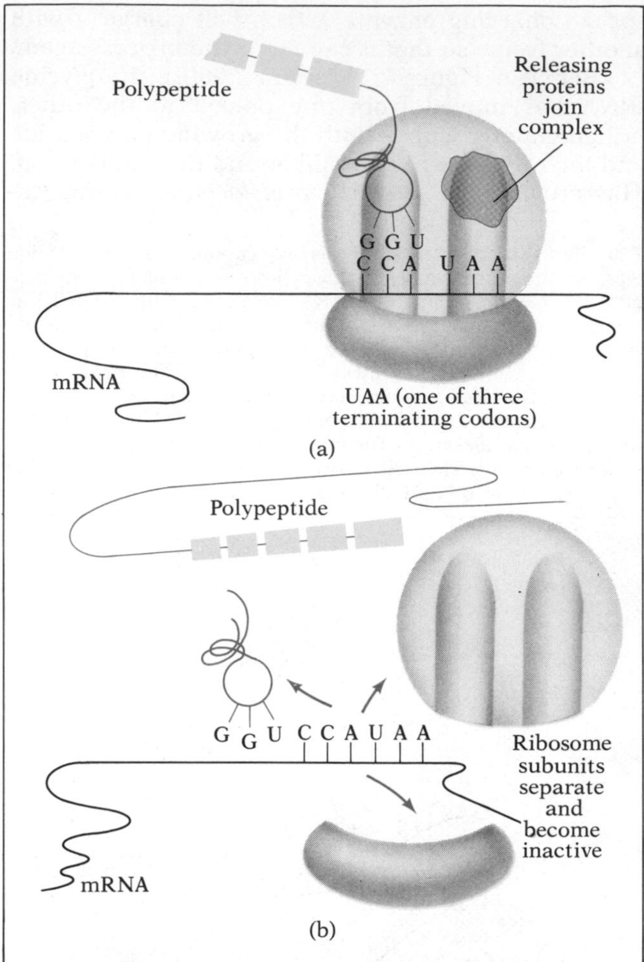

7.21 Part three of protein synthesis: termination. (a) In the first event, the ribosome has reached a terminator codon (UAA). There is no opposing anticodon, since no AUU-bearing tRNA exists. The amino acid just added will be the last. (b) The second event is a "derailing" of the ribosome. The details of this process aren't at all clear right now, but there appear to be special proteins involved that have been named "releasing factors" (this term is deliberately vague in anticipation of new information). Somehow, the presence of UAA triggers a change in the ribosome, and its subunits separate as the final tRNA is released. Our polypeptide moves away for final shaping into secondary, tertiary, and perhaps quaternary form, and the ribosomal subunits go back to "start" for another round of translation.

Polypeptide Chain Termination

Well before the ribosome reaches the end of the mRNA molecule, it runs into a chain *terminator,* or stop codon. There are three of these: UAA, UAG, and UGA. Sometimes there are double stops (e.g., UAAUAG), apparently just to be sure that the ribosome gets the message.

Normally, no tRNAs have anticodons that are complementary to any of the three stop codons. Instead, there are specific proteins that apparently occupy the tRNA site once a stop codon has been reached and they clog the works, grinding it to a halt. Then, yet another protein factor frees the *C*-terminal carboxyl group from the last tRNA, the completed protein is released, and the ribosome falls apart again into its two principal components (Figure 7.21).

Polyribosomes

High-speed centrifugation of crushed cells can separate cell contents into various fractions, according to the size and specific gravity of the solid particles. Ribosomes appear in two such fractions. Under the electron microscope it can be seen that single ribosomes are found in one group; *polyribosomes* are found in the other. The heavier polyribosomes (also called *polysomes*) consist of several ribosomes, say two to ten, bound together by an mRNA molecule. They look like little strings of beads (Figure 7.22). Why are they bound together in groups? It seems that different ribosomes are reading the same mRNA molecule, and are spaced along it at appropriate intervals—a minimum of twenty-five nucleotides apart. Each ribosome will travel the whole length of the cistron of the messenger, from the initiator codon to the terminator codon; then each will fall apart and drop off. Meanwhile, other ribosomes will assemble themselves at the initiator codon and begin moving along the mRNA.

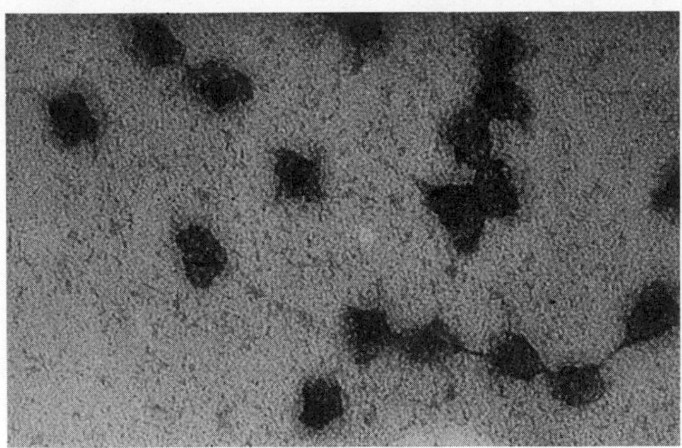

7.22 This electron microscope photograph provides evidence of the existence of polyribosomes (or polysomes). The amount of protein synthesis is greatly speeded when several ribosomes read the mRNA code at the same time. In *Escherichia coli*, polyribosomal translation is so rapid that ribosomes often move along a mRNA that is still being transcribed along the chromosome.

The several ribosomes reading the same mRNA are like a group of ancient Talmudic scholars all reading different parts of the same long scroll. While the last scholar to arrive begins with reading Genesis, the first scholar to arrive may be just finishing Deuteronomy, while others are working on Exodus, Leviticus, and Numbers. When the first scholar is finished, he can take a break, begin a new scroll, or go back to "In the beginning. . . ." Thus, the different ribosomes in a polysome can be producing different copies of the same protein simultaneously, each working on a different portion of the sequence.

Free and Bound Ribosomes

In Chapter 4, we noted that ribosomes occur under two different circumstances: free in the cytoplasm or bound to membranes. In higher organisms, *bound ribosomes* are attached to one side of the membranes of the rough endoplasmic reticulum; in fact, their pebbly appearance gives the rough endoplasmic reticulum its name (see Figure 4.18). In bacteria, there may be ribosomes bound to the inncr surface of the cell membrane itself.

Polysomes also form among the bound ribosomes. The polypeptides that are produced suffer a different fate from those produced by the free ribosomes: Whereas polypeptides produced by free ribosomes are just released into the cytoplasm, those produced on the bound ribosomes are somehow moved across the membrane and appear on the opposite side from the ribosomes and the mRNA (Figure 7.23).

The rough endoplasmic reticulum forms outpocketings, or vesicles, full of all sorts of poisonous or degradative proteins, which are thus isolated from the protein synthesis mechanism and the rest of the cell in general (see Chapter 4).

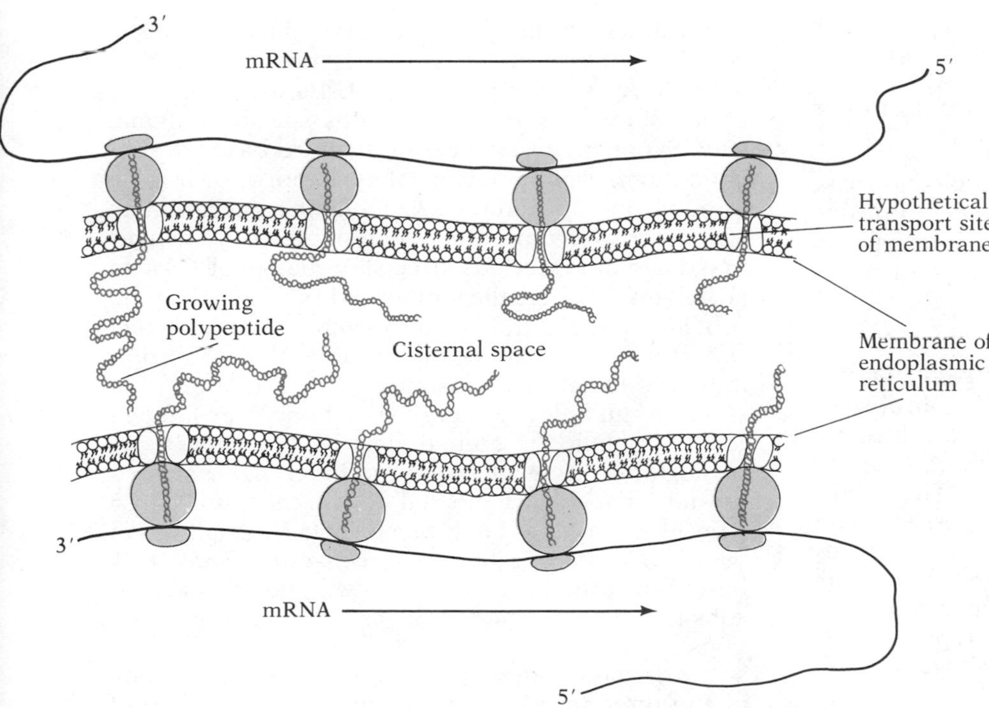

7.23 The activity of free ribosomes, as we have seen, is fairly easy to describe. However, large numbers of ribosomes appear to be bound to the membranes of the rough endoplasmic reticulum (see Figure 4.18). A problem with bound ribosomes concerns the fate of newly produced polypeptides. Studies of protein synthesis at the endoplasmic reticulum clearly indicate that somehow the polypeptides end up in the cisternae of the endoplasmic reticulum for packaging and delivery—often outside the cell. This hypothetical scheme offers an explanation, but it is still unproven. As the polypeptide chain grows, it may be directed into the endoplasmic reticulum through a special transport site in its membrane. The *N*-terminal end of the polypeptide may provide the key needed to open the transport site.

Another theory is that the *N*-terminal end of the growing polypeptide simply dissolves into any part of the membrane, then reappears on the other side as the protein grows. Apparently "bound" ribosomes themselves do not in fact adhere directly to the ER membrane; they are simply attached to the growing polypeptide, which in turn has an affinity for the membrane.

We don't know just how the proteins are moved across the membranes. Some investigators have drawn imaginative pictures of bound ribosomes with holes, through which the growing polypeptide chains are unceremoniously stuffed. More recently it has been argued that the *N*-terminal ends of the proteins have an affinity for the membranes, causing the growing chain to dissolve into, and then through, the membrane. In fact, it now appears that "bound ribosomes" are not bound directly to the membranes at all: they are merely ordinary ribosomes that are attached to growing polypeptide chains which, in turn, are dissolving into the membranes.

Mutations in DNA and Resulting Misreadings of mRNA

Once a ribosome has recognized an initiation codon, it continues to move along the mRNA three bases at a time and to construct polypeptides blindly until it hits a terminating codon. The DNA sequence (that codes for the mRNA sequence that codes for the polypeptide sequence) sometimes changes permanently—by being copied incorrectly, or by being denatured in place. Such an event is a *primary mutation*. As we have seen, the redundancy of information in the DNA double helix allows for efficient repair, and most primary mutations do no harm. But repair is not perfect, and occasionally a permanent alteration in the DNA occurs. It is these permanent alterations that are usually referred to as *mutations*. Mutations occur spontaneously, but can also be caused by chemicals or by ionizing radiation. Any substance that causes a mutation is called a *mutagen*.

At this point, it will be informative to look at some basic kinds of mutation. At the DNA level:

1. Base substitutions—the number of nucleotide base pairs in a length of DNA is unchanged, but one base pair is changed from one of the four types to another. This is the most common type of mutation.
2. Insertion—some extra bases are added.
3. Deletion—some bases are lost and the gap closed.
4. Chromosomal rearrangement—DNA is broken in two different places or two DNA molecules are simultaneously broken; then the *wrong ends* are joined together by repair enzymes. This is the unwanted result of normal repair processes; repair enzyme complexes join together the ends of any broken DNA molecules. Since single breaks (with two free ends) are much more common than double breaks (with four free ends), this repair usually results in putting the broken ends together correctly.
5. There are other kinds of mutation, such as whole chromosome loss or gain, but we will consider these later.
6. Most DNA (of higher organisms at least) does not code for polypeptide sequences. It has been estimated that only about 1% of human DNA is written in the genetic code. A much smaller proportion codes for non-mRNA with known functions, such as the rRNA and tRNA genes. Of the remaining 98 or 99% of human DNA, some proportion may form spacer DNA between genes; some, for all we know, may have no function at all ("junk" DNA); but it is usually assumed that most of the DNA has some kind of controlling function, such as regulating the transcription of the structural genes. We don't know how this control information is coded, or whether "coded" is the right word to use to describe it. We do know that some, and perhaps all, gene control involves complex sequence-specific binding interactions with controlling proteins. We can assume that permanent changes (mutations) in the controlling regions of DNA are important, but we do not know exactly how they might work or how much influence, if any, single base changes might have in the binding of controlling proteins. We understand much better the effects of changes in structural gene DNA (cistrons).

Effects of DNA Mutations at the Level of Protein Synthesis

1. Base substitutions. Single base changes alter a codon from one of the sixty-four possible types to another. A change from GAA to GAG would have no effect, since both of these codons specify glutamic acid. Synonymous changes in the third base are very common in evolution; since they do not appear to be weeded out by natural selection, they accumulate randomly over time. But other single base changes may have more serious effects. A change of GAA to GUA would change the amino acid somewhere in the protein from glutamic acid to valine. An *amino acid substitution* may have no measurable effect at all, or it may be lethal. For instance, the mutation that changes glutamic acid to valine in the sixth position in the beta chain of human hemoglobin causes the hemoglobin molecule to form long, sickle-shaped crystals that distort the red blood cells and cause painful, eventually fatal, blood clots in small blood vessels. This genetic disease is *sickle-cell anemia*. DNA base substitutions that cause polypeptide amino acid substitutions are sometimes called *missense* mutations.

If the new codon produced by a mutation is one of the three termination codons, the mutation (not surprisingly) is called a *chain-termination mutation*. Unless the terminating codon is very close to the far end of the structural gene, chain-termination mutations result in totally nonfunctional polypeptides. The majority of *lethal* mutations, at least in microorganisms, are of the chain-termination type.

Sometimes, the normal termination codon mutates to some other form, say UAA to GAA. When this happens, the ribosome puts in the newly coded amino acid and continues to read out amino acids from the (probably) random assemblage of mRNA to the far right of the normal termination sequence. Since $\frac{3}{64}$ of all random codons are chain terminators, the ribosome will eventually hit the combination that allows it to let go and fall apart. Mutations of this type are known for both the alpha chain and the beta chain of human hemoglobin, where they occur as inherited family defects. Twelve extra amino acids are tacked onto the *C*-terminal end of the abnormal alpha chain, and thirty-one extra amino acids are tacked onto the end of the abnormal beta chain. Persons who are heterozygous for these strange hemoglobin genes are not very healthy, but they seem to be able to survive. Initiator codon mutations aren't known, but they surely must occur.

2. Insertions and deletions. What would you expect to happen when extra nucleotides are inserted into a structural gene? If the number of extra nucleotides is exactly divisible by 3, the effect of inserting 3*n* nucleotides into a structural gene is to insert *n* amino acids into the relevant protein. *Deletion* of 3*n* nucleotides results in the deletion of *n* amino acids from the protein. Whether or not such insertions or deletions will completely wreck the function of the protein depends on where they occur.

Insertions and deletions don't have to come in threes, and usually don't. The effect of inserting or deleting a single nucleotide (or any number of nucleotides not exactly divisible by 3) is to cause a *reading-frame shift*. Ribosomes are not very intelligent: they continue to read mRNA nucleotides three at a time. Thus, if the normal sequence were:

```
            U
            ↓
5′ →       UGG           GAG        AAA      AAA         UUU          AAG  → 3′
      Tryptophan — Glutamic — Lysine — Lysine — Phenylalanine — Lysine
                       acid
```

and a U were inserted at the point indicated by the arrow, the sequence would be translated:

```
5′ →    UGU        GGA          GAA        AAA        AUU         UAA   G → 3′
      Cysteine — Glycine — Glutamic — Lysine — Isoleucine — STOP
                                acid
```

Not surprisingly, frame shift mutations are usually lethal.

Application of Ideas

1. Before the discovery of RNA-dependent DNA synthesis (reverse transcription), the central dogma had included the provision that DNA arises only from DNA. This newer knowledge was then incorporated into the dogma. Discuss how such a flat contradiction affects a theory. Is the theory invalidated?

2. It has recently been shown that UGA codes for tryptophan and CUA codes for threonine in the mitochondrial translation systems of yeasts and hamsters. What does that do to the "universal code"? Comment on the significance, if any, of these minor departures from the code.

3. In the prehistory of life, nucleic acids might have come before proteins, or proteins before nucleic acids. How could a purely protein organism reproduce or store information? (Can enzymes be self-perpetuating?) Is there some other way around the problem? (For example, suppose that the original self-replicating organism had a nucleic acid with dozens of different nucleotide bases, which were later reduced to the four used in protein synthesis.)

4. With the increased use of nuclear power, the subject of harmful ionizing radiation has once more become a public issue. Radiation is known to alter DNA in random ways. Its effect has two aspects, medical and genetic. Distinguish between the two in terms of what cells are involved and the potential results of both.

5. Proponents of nuclear energy are comparing the risk of a nuclear accident such as the one at Three Mile Island with the risk of airplane accidents, freeway travel, and petroleum fires. What, if any, is the flaw in this kind of argument? (You might want to look into the accumulative effects of radiation and consider *all* forms of life in your answer.)

Key Words

adenine
anticodon loop
attachment site
autoradiography

base substitution
bound ribosome

central dogma
chain-termination mutation
charging enzyme
chromosome
cistron
codon
colinearity
C-terminal
CTP
cytosine

degenerate code
deletion
deoxynucleotide
deoxyribonucleic acid (DNA)
deoxyribose
DNA polymerase
double helix

elongation

five-prime end (5′ end)

gamete
genetic code
GTP
guanine

heterogeneous nuclear RNA (hnRNA)
histone

initiation complex
initiator
insertion

ligase

messenger RNA (mRNA)
missense mutation
mutagen
mutation
mutation rate

nitrogenous base
nontranscribed strand
N-terminal
nucleic acid
nucleotide base
nucleotide triphosphate

Okazaki fragment

Key Ideas

What Is the Central Dogma?

1. The *central dogma* is the way in which genes function, giving us a rigorous chemical and physical basis for understanding life.
2. The *central dogma* dictates that:
   a. *DNA* is *replicated* from other *DNA*.
   b. *RNA* is *transcribed* from *DNA*.
   c. All proteins are *translated* from *RNA*.
3. The *genetic code* is believed to be universal (used by all organisms).
4. The principle of *colinearity* states that linear sequences of *nucleic acid* bases determine the linear sequence of amino acids in a protein.

What Is a Nucleic Acid?

1. *DNA* is *deoxyribonucleic acid* and *RNA* is *ribonucleic acid*.
2. *DNA* is composed of long polymers of *nucleotides*. These are the units of *nucleic acid* structure.
3. Each *DNA nucleotide* consists of phosphate, sugar (*deoxyribose*), and one of four *nucleotide bases*.
4. *Nucleotide* or *nitrogenous bases* are the *pyrimidines, thymine* and *cytosine*, and the *purines, guanine* and *adenine*.
5. Covalent bonds link the *nitrogenous base* to the sugar and the sugar to the phosphate.

DNA Bonding

1. As individual *nucleotide* triphosphates are assembled into the *DNA* polymer, two phosphates are split off. The remaining phosphate is linked to the sugar group of the next *nucleotide*.
2. The *DNA* polymer or chain is pictured with the free *5′ end* on the left and the free *3′* OH group on the right. Assembly proceeds from the *5′ end* to the *3′ end*.
3. A *DNA* molecule contains two chains of *nucleotides*, linked together by pairs of *nitrogen bases* that point inward. The bases are attracted together in pairs by weak hydrogen bonds.
4. The entire double strand is twisted into a *double helix*, with the phosphate–sugar strands outside and the bases pointing inward.
5. The *nitrogen bases* pair up in a specific manner, with *adenine* pairing with *thymine* and *guanine* pairing with *cytosine*.
6. The two chains of bases run in opposite directions with a *5′*-to-*3′* chain usually shown on the left (right side up) and a *5′*-to-*3′* chain on the right (upside down).
7. The average *DNA* molecule in humans contains about 160 million nucleotide pairs.

The Stability of DNA

1. *DNA* is relatively inactive in its *double helix*, since the bases are inside and there are no reactive groups outside.
2. In higher organisms, *DNA* is further protected by surrounding *histone* proteins and by redundancy in the strands. Each linear code is actually copied in the other chain. When *DNA repair* occurs, damaged strands are discarded and new segments paired up with the surviving strand.
3. Spontaneous changes in the *DNA* molecule are called *mutations* and may alter normal protein structure.
4. *Mutations* in sperm- and egg-producing cells produce abnormalities in offspring. Those in other cells can cause cell death or cancer.

DNA Repair

1. Repair is accomplished by complex repair systems containing enzymes that clip out damaged regions and add in correct sequences using the undamaged strand as a template.

phage (bacteriophage)
poly-A
polyribosome (polysome)
poly-U
posttranscriptional modification
primary mutation
purine
pyrimidine

radioactive precursor
radioactive tracing
reading-frame shift
replication
replication fork
ribonucleic acid (RNA)
ribosomal RNA (rRNA)
RNA synthetase

sickle-cell anemia
S phase
spontaneous mutation
structural gene

terminator
three-prime end (3′ end)
thymine
transcribed strand
transcription
transfer RNA (tRNA)
transformation
translation
translocation
TTP

unwinding protein
uracil

viral genetic RNA
virulence
virus

xeroderma pigmentosum
X-ray crystallography

2. *Xeroderma pigmentosum* is a skin condition involving repair systems that have suffered a *mutation,* leaving the victim without a way to repair ultraviolet light damage.

DNA Synthesis and Replication

1. The *DNA* molecule replicates itself for cell division in growth and gamete production.
2. *DNA* accomplishes *replication* by unwinding portions of the molecule and incorporating *nucleotides* according to base pairing rules. These are assembled along the *5′*-to-*3′* direction.
3. The unwound region is known as a *replication fork* and the new strands of nucleotides are added in *Okazaki fragments* using the enzyme *ligase* to connect them.
4. *Replication* is called *semiconservative,* which means that each new DNA molecule contains one old strand and one new strand.

Essay 7.1 DNA Replication in Eucaryotes and Procaryotes

In eucaryotes the *DNA* molecules are linear, and *replication forks* occur all along the molecule. Procaryotic *DNA* is circular, so one *replication fork* proceeds in each direction until the entire molecule has replicated.

A Brief History of the Discovery of the Genetic Material

1. Gregor Mendel's work first suggested that the genes were physical, particulate factors.
2. *Chromosomes* were long believed to be the hereditary factors.
3. In 1869 Miescher determined the presence of a phosphorus-containing material in the nucleus, and in 1914, *Feulgen* developed a specific *stain* for nucleic acids that was quantitative. Further experiments determined that the same quantity of *DNA* was present in each body cell and one half that amount was present in the sperm and egg.
4. In 1920, Muller proved that the genes were chemical structures. He created *mutations* using *X-rays* and also determined *spontaneous mutation rates.*

Transformation

1. In 1928 Griffith discovered that *virulence* could be transferred from one strain of bacteria to another.
2. In 1944 it was determined that the *transformation* factor in Griffith's experiments was *DNA*. Fragments of the dead *virulent* bacterial *DNA* were incorporated into the nonvirulent strain's *chromosome.*

Hershey and Chase

1. Using *bacteriophage* that contained radioactive sulfur and phosphorus, *Hershey and Chase* determined that only the *phage DNA* was necessary to produce more *phage* particles in a host cell. The possibility of protein being the genetic material was eliminated.
2. *DNA*'s simplicity was a serious problem. How could a molecule with only four different subunits specify complicated proteins?

Chargaff's Rule

Chargaff was able to determine that the amounts of the four different *nitrogen bases* differed in different organisms, but the ratios of *adenine* to *thymine* and of *cytosine* to *guanine* always remained the same.

Watson and Crick and the Molecular Model of DNA

1. X-ray diffraction studies of crystals of *DNA* by Wilkins and Franklin indicated some of the dimensions of the *DNA* molecule (such as distance between *nucleotide* pairs and length of one turn of the helix).
2. Using observations made by Wilkins, Franklin, and others, and data from Chargaff, Watson and Crick developed the first model of *DNA.* They published in 1953.

RNA: The Other Nucleic Acid

1. The search for answers about *RNA* included *radioactive tracing* techniques. *DNA, RNA,* and protein could be differentially labeled and their pathway or activity traced in the cell. The process, known as *autoradiography,* utilizes photographic emulsions that are exposed by radiation.

2. *Autoradiographic* studies helped establish that DNA remains in the nucleus and produces RNA, which migrates to the cytoplasm where protein synthesis occurs.

RNA Structure

RNA differs from DNA in the following ways:
1. *RNA* uses *ribose* sugar.
2. *Uracil* is used instead of *thymine.*
3. *RNA* is single stranded.
4. *RNA* is considerably shorter.
5. *RNA* is relatively unstable.
6. *RNA* occurs in many kinds and sizes.

RNA Synthesis

1. *RNA* is *transcribed* from one of the two *DNA* strands, occurring as *DNA* unwinds and enzymes assemble RNA *nucleotides* along the strand. Wherever an *adenine* occurs along the *DNA* strand, it pairs with *uracil.*

2. *Many RNA* molecules can be assembled at one time along a gene. In the electron microscope these appear as a featherlike assembly line.

Varieties of RNA: Physical and Functional Classes

The kinds of *RNA* are: *ribosomal, messenger, transfer, heterogeneous nuclear RNA,* and viral genetic *RNA.* The first three are involved in protein synthesis.

Ribosomal RNA

1. Ribosomes contain ribosomal *RNA* as well as about twenty different proteins. Ribosomes are present wherever protein synthesis occurs.

2. Messenger RNA, tRNA, and rRNA meet at the *ribosome* to translate the *genetic code* into protein.

Messenger RNA and the Genetic Code

1. *Messenger RNA* carries the DNA coding for assembling amino acids into a specific protein. The code is in *codons,* each consisting of groups of three *nucleotides.* Each *codon* calls for the positioning of one of the twenty amino acids.

2. There are sixty-four different ways of arranging four different *nucleotides,* thus there are sixty-four *codons* in the *genetic code.*

3. The code contains synonymous *codons.* More than one *codon* may call for the same amino acid. This is called *degeneracy.* One *codon* calls for methionine and is also the *initiator codon* for starting a message. No *codon* calls for more than one specific amino acid.

4. The portion of *messenger RNA* that actually specifies a polypeptide sequence is called a *cistron* or *structural gene.*

Transfer RNA

1. *Messenger RNA* does not recognize amino acids itself. This is the role of *transfer RNA.* Different *tRNA*s occur for each of the specific amino acids.

2. *Transfer RNA* molecules contain about ninety *nucleotides* and are tailored by enzymes to a specific shape. They contain both the means of identifying an amino acid and the means of placing it at the correct place on *messenger RNA.* The *anticodon loop* can recognize and interact with specific *codons* on *mRNA.* One end is capable of being joined to a specific amino acid by a *charging enzyme.*

3. *Codons* on *mRNA* and *anticodons* on *tRNA* base pair as amino acids are assembled.

4. Each type of *tRNA* has its own *charging enzyme,* which, along with ATP, forms the complex between an amino acid and its proper *tRNA.*

5. Protein synthesis occurs as charged *tRNA*s bring their amino acids to the *ribosome* where they meet *mRNA* also attached to the *ribosome.*

How Proteins Are Made

Initiation

1. The first amino acid to be brought into the *ribosome–mRNA* complex is always methionine. It is carried by the *tRNA* bearing the *anticodon* CAU, which base pairs on *mRNA* with the *initiator codon* AUG.

2. Methionine forms the *N*-terminal end of a polypeptide. The next amino acid will attach to its carboxyl group.

Elongation

1. Each *ribosome* has two pockets. Two *codons* of *mRNA* align with these pockets.

2. The *tRNA* carrying the last amino acid to attach to the growing polypeptide is in the left pocket. An incoming *tRNA* will land in the right-hand pocket where a *codon–anticodon* match is made.

3. As the entering amino acid is attached to the last one in the growing chain, *translocation* occurs. The left-hand *tRNA* breaks away, and the *tRNA* with its amino acid newly added to the growing chain moves into the left pocket. Another *tRNA* with its amino acid, base pairs into the right pocket and the process continues. Each time translocation occurs, the ribosome moves to the right, one codon at a time.

Polypeptide Chain Termination

1. Each *cistron* ends with one of the codons UAA, UAG, or UGA. These are the *terminator codons.*

2. When a *ribosome* reaches a *terminator codon,* the polypeptide is released and the ribosomal parts separate.

Polyribosomes

Polyribosomes consist of a number of ribosomes attached to the same *mRNA* and carrying on simultaneous *translation.* This increases the rate of protein synthesis.

Free and Bound Ribosomes

1. Bound *ribosomes* in higher organisms are attached to the endoplasmic reticulum. In bacteria they attach to the inner cell membrane.

2. Protein produced by bound ribosomes is believed to enter the lumen of the endoplasmic reticulum from which vesicles fill and pinch off.

Mutations in DNA and Resulting Misreadings of mRNA

1. Permanent changes in *DNA* sequences are known as *primary mutations.* Most are repaired or compensated for by the DNA redundancy factor.

2. *Mutations* may be caused by chemicals (mutagens) or ionizing radiation but some are simply copying errors, or are due to spontaneous thermal denaturation.

3. *DNA* level *mutations:*
   a. Base substitutions (most common).
   b. Insertion and deletion of bases.
   c. Chromosomal rearrangements—breaks rejoined in wrong position.
   d. Whole *chromosome* loss or gain.
   e. Most *DNA* is not involved in protein synthesis but probably serves some controlling function. How *mutations* in this *DNA* affect the cell is unknown.

Effects of DNA Mutations at the Level of Protein Synthesis

1. Specific effects:
   a. Single-base changes alter a *codon* from one of the sixty-four to another. If the change is in the third base there is no effect (see Table 7.2).
   b. Other single-base changes will code for a different amino acid in that position. Some are harmless, but some have a structural effect on protein. (e.g., sickle-cell hemoglobin).

c. When the change places a terminator *codon* somewhere in the *mRNA*, useless polypeptide fragments are formed.
d. When the change alters a normal *terminator*, extra amino acids are added to the polypeptide, making it abnormally long.
e. *Initiator codon mutations* are believed to occur but cannot be observed, since no polypeptide forms.
2. Single insertions or deletions create a *reading-frame shift.* All of the encoded message beyond the *mutation* changes and an aberrant protein is produced.

Review Questions

1. List the three principal ideas of the *central dogma.* (p. 182)
2. Summarize what is meant by the term *colinearity.* (p. 183)
3. List the components of a *nucleotide triphosphate.* (p. 184)
4. List the four kinds of *nucleotides* in *DNA*, grouping them according to whether they are *purines* or *pyrimidines.* (p. 184)
5. Explain the linkages that bond the components of a *nucleotide* together, using the numbering system applied to *deoxyribose.* (p. 184)
6. Explain the *5′-end, 3′-end* organization of a *DNA* strand. How is this related to the synthesis of the polymer? (pp. 184–185)
7. Describe the assembly of two chains of *nucleotides* into the *double helix* (include bonding, base pairing, and general geometry of the molecule). (p. 186)
8. How does the geometry of *DNA* protect it from reactive chemicals in its environment? (p. 187)
9. Summarize the protective function of *histones.* What are they? How do they interact with DNA and prevent mutations? (p. 187)
10. Through a characteristic known as redundancy, *DNA* resists permanent alteration. Explain how this operates. (pp. 187–188)
11. Describe how a *DNA repair system* works to correct spontaneous changes in *DNA*. (p. 188)
12. Explain how *xeroderma pigmentosum* relates to a *DNA repair system.* (pp. 188, 190)
13. What is the importance of *replication?* (p. 190)
14. Outline the process of *replication* using the terms *Okazaki fragments, DNA polymerase, replication fork, unwinding protein, semiconservative,* and *Watson* and *Crick strands.* (pp. 190–191)
15. Describe a procedural difference between eucaryotic and procaryotic *replication.* (Essay 7.1)
16. Describe two quantitative discoveries that were made possible by Feulgen's staining technique. (p. 192)
17. How were the studies of Muller instrumental in showing that the gene was a chemical structure? (p. 192)
18. What did Griffith's experiments contribute to our understanding of the gene? (pp. 192–193)
19. How did the experiments of Hersey and Chase help clarify the role of *DNA?* (pp. 193, 200)
20. From Chargaff's data, explain how the base-pairing rule could be derived. Does his data indicate universality in the rule? Explain. (p. 201, Table 7.1)
21. Meselson and Stahl were able to show that *DNA replication* was *semiconservative.* What was their evidence? (Essay 7.4)
22. Describe the actual contribution of Watson and Crick. (p. 201)
23. What were the radioactive precursors used in detecting the presence of DNA, protein, and RNA? (p. 202)
24. List the three discoveries about synthesis made possible by *autoradiography.* (pp. 202–203)
25. List the six characteristics that distinguish *RNA* from *DNA.* (p. 203)
26. Compare *transcription* with *replication* by pointing out how they are different. (p. 203)
27. Describe the appearance of a *DNA cistron* undergoing simultaneous *transcription.* Explain the different lengths of *RNA.* (p. 204)
28. List the five types of *RNA* and indicate which are directly involved in protein synthesis. (p. 205)
29. Describe the structure of a *ribosome.* (p. 205)
30. Explain how a single strand of *DNA* constitutes a genetic message. Use the term *codon.* (p. 205)
31. Using the *genetic code* table, find one ambiguous codon and two synonymous codons. How many examples of degeneracy can you find? (Table 7.2)
32. Describe the role of each of the important parts of a *tRNA* molecule. (p. 208)
33. Describe how a charging enzyme prepares a *tRNA* for protein synthesis. (pp. 207–208)
34. List the participants and events in the formation of an initiation complex. (p. 210)
35. With four simple drawings review the sequence of events as two new amino acids are added to a growing polypeptide chain. (pp. 211–213)
36. Define *translocation* and explain what the energy source is for this important event. (p. 213)
37. The events of protein synthesis are complex and involve many participating molecules. Explain how all of these things manage to happen at the right time and place. (pp. 212–213)
38. Prepare a simple drawing of polypeptides being produced along polyribosomes. No detail is necessary, but explain the different polypeptide lengths. (pp. 214–215)
39. List five kinds of *mutation* at the *DNA* level. (p. 216)
40. Explain why third-letter (base) changes have no effect on most codons. (Table 7.2 and p. 216)
41. Using sickle cell anemia as an example, describe the potential significance of a single *base substitution.* (p. 216)
42. What is the obvious effect of mutations that affect the *terminator* codons? (pp. 216–217)
43. What causes a *reading-frame shift?* How does a frame shift effect the polypeptide produced? (p. 217)

Chapter 8
The Life of a Cell

You are not the same person you were a few years ago. In fact, you aren't the same person you were a few seconds ago. This is not to say that the first sentence here is so profound that you will never be the same for having read it. It's just that even as you read, your cells are growing, dying, multiplying, and dividing. Some cells, of course, are more active in such processes than other cells. For example, the palms of your hands, and other parts of your body that receive excessive friction, must constantly replace worn cells. Other parts of your body, however, have very low cell turnover. And some, such as your skeletal muscles and nervous system, are never replaced in adults. Nonetheless, even as you sit here, your cells are at work quietly obeying their genetic overlords, many of them slowly and methodically reorganizing themselves and dividing again and again, duplicating themselves ever so precisely in their unending pageant until that day when it all ends.

And that brings us to another point: We will stress in this chapter that cells duplicate themselves. That is, they produce identical copies of themselves. But that isn't really true. Even cells that rise anew, fresh and vigorous after only a few seconds of existence, may be ever so slightly different from their parent cells. Some of these differences are accidental mutations, but in other cases the new cells are genetically programmed to be different, and perhaps less robust than the old cells that gave rise to them.

If you look in the mirror each morning and see a fresh, clear-eyed face—unworn and unwrinkled—you are young. Of course, this means that subconsciously you believe that you are destined to forever be one of the earth's much-vaunted "young people," and that the oldsters around you were obviously born old in spite of their denials. The oldsters, though, know better. They recall being young and they may also recall their dismay when they discovered that their cells were letting them down—literally (Figure 8.1). Why did the cells under their eyes lose their tone and begin to sag under the weight of new fat deposits? Why does skin on the back of their hands show ever larger brown splotches? Why do muscle cells contract more slowly now? Why are they changing? And so rapidly?

They are, of course, changing because their cells and supporting tissues are changing. We will have more to say about all this later. For now just keep in mind that the process of cell cycle and renewal, which we will be describing here, seems to offer the assurances that cells won't change from one cell generation to the next. But they will.

"The nucleus is the controlling center of the cell." This sentence, or something like it, is found in every biology text. Our job in this chapter and the next will be to explore this generalization, locate its flaws and its strengths, and discover as much as we can about just how the nucleus really interacts with the cytoplasm and the surrounding environment. But first, a little history is in order. The nucleus has received a great deal of attention since its discovery in the 19th century.

8.1 The Nucleus and the Cell Cycle

The nucleus is a very prominent structure in most stained cells. In fact, it's so prominent that people fiddling with primitive microscopes saw nuclei at least 175 years ago. Leeuwenhoek, himself, observed the nuclei of red blood cells in fish in one of his many crude observations, and the word *nucleus* appears in the literature as early as 1823. Discovery of the nucleus, however, is generally credited to a Mr. Brown, an early microscopist who also discovered that cell contents move or vibrate under random thermal collisions (*Brownian movement*). The proponents of the cell theory, Schleiden and Schwann, wrote about the nuclei in the cells they observed. So, the nucleus was there, everyone agreed, but what did it do?

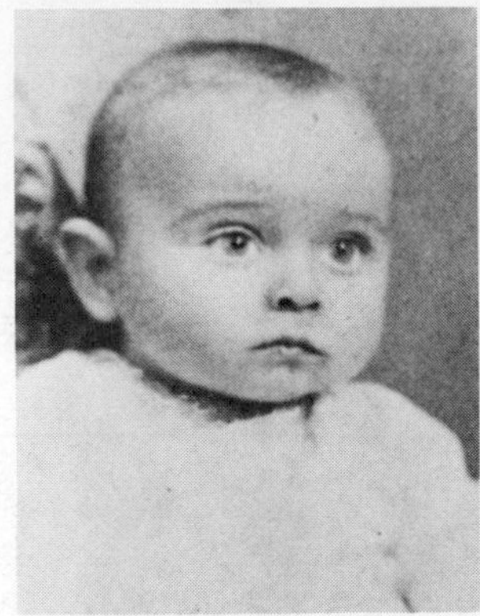

8.1 The aging process is not well-understood. Countless cell divisions, each an experience in precision, followed by unknown forces of differentiation and maturation produce the adult. Once the molding forces finish their processes, the vigorous cellular activity slows in most of the body. A steady decline begins as the years progress. Signs of the decline are unavoidable.

Scientists began to learn something about the role of the nucleus in cell division in the late 1800s, first observing plant cells, and later looking at animal tissues, particularly the cells of early fish and amphibian embryos. In fact, many of the terms we will use here were introduced by Walther Fleming back in 1882. So, by the turn of the century, many details of cell division had been worked out, and much of what we will describe here was already known by then. But there have been a few more recent developments, and new facts and perspectives keep coming in.

Nuclear Structure

We described the nuclear structure in Chapter 4, so we will be brief here. Recall that the fluid material of the nucleus, known generally as *nucleoplasm,* is surrounded by the familiar double-layered membrane, often portrayed as continuous with the endoplasmic reticulum. The nuclear membrane is riddled with numerous porelike structures that are particularly evident in freeze-fracture preparations (see Figure 4.17). In fact, the combined area of these *pores* is estimated to account for as much as 25% of the nuclear membrane in some cells. The pores, also known as *annuli,* permit the passage of molecules across the membrane, but they are not simple openings (although that is what they look like in electron micrographs). Presumably the holes are filled with hydrated proteins. The nuclear membrane is known to be highly selective in what it permits to pass. For example, we know that comparatively large mRNA molecules are assembled in the nucleoplasm and move readily into the cytoplasm to carry out their functions. Yet, much smaller molecules are not allowed to pass through the pores of the nuclear membrane. Therefore we must assume the nuclear membrane dictates what passes through it and what doesn't. Because of the great selectivity of the membrane, the nuclear materials are quite different from cytoplasmic materials. But there are times in the lives of eucaryotic cells when this is not true. During cell division in higher eucaryotes, the nuclear membrane appears to fade away into the cellular matrix, only to reappear later when the process has been completed. So, what does this do to the nuclear integrity? Apparently, at this time in the life of a cell, it is not so important that the pristine nucleoplasm be separated from the rough and tumble world of the cytoplasm. Some reasons for this will be suggested as we look more closely at the process of cell division.

One very obvious feature of the nucleoplasm is the *nucleolus* ("little kernel"; plural, *nucleoli*). One or more nucleoli are seen in stained sections of most nuclei. (Keep a close eye on the spelling here.) The nucleolus has given cytologists fits. One reason for this is that it appears to be both granular and fibrous in electron micrographs. We've only recently learned something about the function of this almost structureless, dense mass, but a lot of head scratching and chin stroking remains to be done. However, we do know that the nucleolus is the site of rRNA production and appears to be the source of ribosomes themselves.

DNA, Chromatin, and Chromosomes

The terminology used to describe the nuclear contents may get a bit confusing as we go along, so some definitions are in order. We discussed DNA earlier, but now we will be paying more attention to other terms such as *chromatin* and *chromosomes.* Chromatin is DNA that is complexed with proteins, including *histones* and other proteins known poetically as *nonhistone chromosomal proteins.* We'll have more to say about both these classes of proteins.

When cells are dividing, the chromatin is condensed or coiled into the much larger, highly visible bodies we call chromosomes. When the chromosomes "relax" (or, more properly, become *diffuse*), they form incredibly long threads (Figure 8.2). In fact, if a DNA molecule from the largest human chromosome were straightened out, it would be about twelve centimeters long. Diffuse DNA, it turns out, occupies most of the nucleoplasm most of the time. The DNA is bunched up as tiny, beadlike globules, each consisting of about 200 DNA base pairs wound around clusters consisting of eight molecules of those proteins, called histones. Between each of these beadlike globules, the DNA molecule continues as an open string of about fifty nucleotide pairs.

Each chromosome, then, consists of one long DNA molecule wrapped around globules of histone, and hundreds of nonhistone chromosomal proteins.

Chromatin occurs in two forms: *euchromatin* and *heterochromatin* ("true and beautiful chromatin" and "other chromatin"). The euchromatin contains nearly all the functional genes, at least nearly all the genes we have been able to map and identify. Between stages of cell division, much larger, irregular blobs of chromatin can be seen in the nucleus; this is the heterochromatin, and it comes in several varieties.

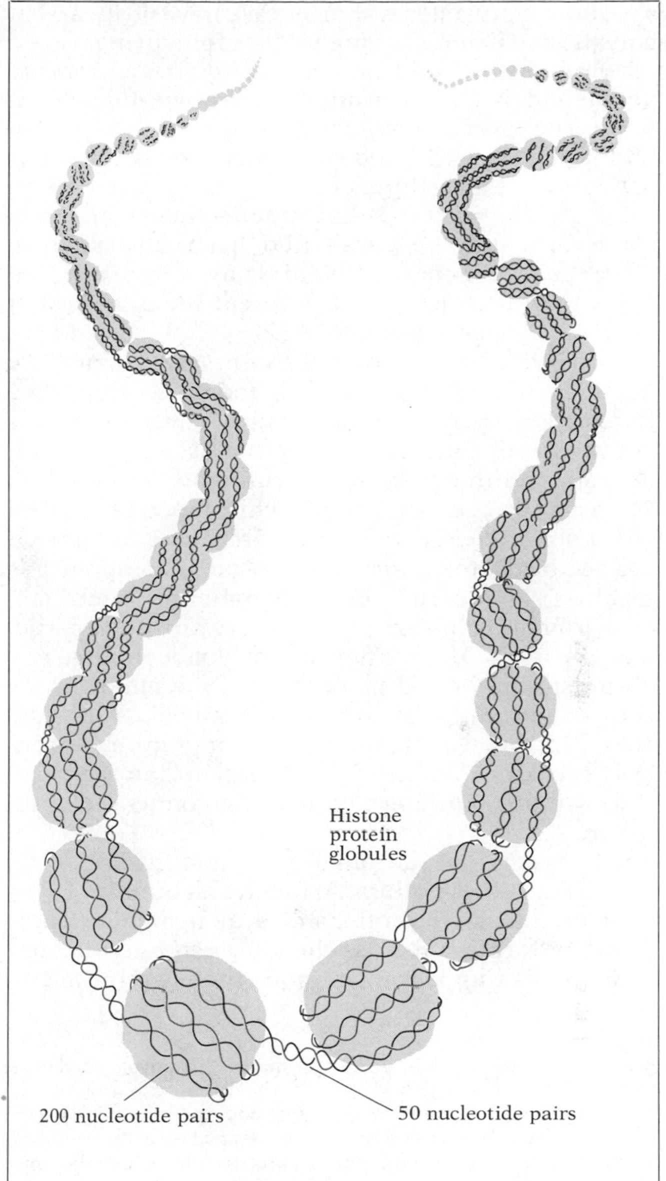

8.2 In eucaryotes, the DNA strand is wrapped around numerous globules of histones, creating a beadlike appearance. Each bead contains about 200 nucleotide pairs on its surface, while the strand between beads consists of fifty nucleotide pairs. The one long DNA molecule, its histone complex, and its associated nonhistone proteins constitute a single eucaryotic chromosome. The chemical substance of eucaryotic chromosomes in general is termed *chromatin.*

Constitutive heterochromatin, for example, largely consists of histone and thousands, or even millions, of tandem repeats of short and probably meaningless sequences of DNA nucleotides. It is sometimes called *centric heterochromatin* because it is concentrated on either side of the *centromere* (spindle fiber attachment) of each chromosome. (We will have more to say about centromeres shortly.) We have no idea what constitutive heterochromatin does for a living.

Actually, the difference between euchromatin and heterochromatin is not as clear as we might like it to be. The term *heterochromatin* applies to any chromatin that is condensed (and thus not metabolically active) at any given time between cell divisions. So the same genetic material that is active euchromatin in some cells may be condensed heterochromatin in other cells. When we identify any chromatin that appears different ways at different times, we call it *facultative heterochromatin.*

Consider two prominent examples of facultative heterochromatin: (1) In female mammals (including humans), only one of the two X chromosomes is active in any one cell; the other is inactive, condensed, and therefore heterochromatic (Figure 8.3). (2) In certain cells, such as the white blood cells called *polymorphonuclear leukocytes,* and the red blood cells of birds, the entire chromosomal component is inactive, condensed heterochromatin. Such heterochromatin appears to be inactive and of no particular use at all. As a matter of fact, you won't find any chromatin in the red blood cells of mammals. These cells have simply discarded their nuclei as they develop. In this way, mammals are one up (evolutionarily speaking) on birds in that mammalian red blood cells are not burdened by useless, condensed chromatin.

The mass of chromatin in nondividing cells, which resembles nothing so much as a bundle of yarn subjected to the attentions of a demented kitten, is sometimes referred to as the *chromatin net* because of its netlike appearance under the electron microscope. Parts of the chromatin net are thin and diffuse, while other parts are thicker and more tangled, but both are considered to be composed of euchromatin. It has been shown that only the DNA in the diffuse portions of the net shows functional gene activity—i.e., transcribing mRNA—at any one time. The relatively condensed portions are temporarily inactive.

8.3 Two examples of heterochromatin (inactive, condensed chromatin), (a) the Barr body and (b) the drumstick chromosome, are easily seen in human females. Barr bodies are found in buccal (mouth) smears where the mucosa is scraped away and stained. Drumsticks are seen in simple blood preparations where the white cells are stained and the polymorphonuclear leukocytes may be identified.

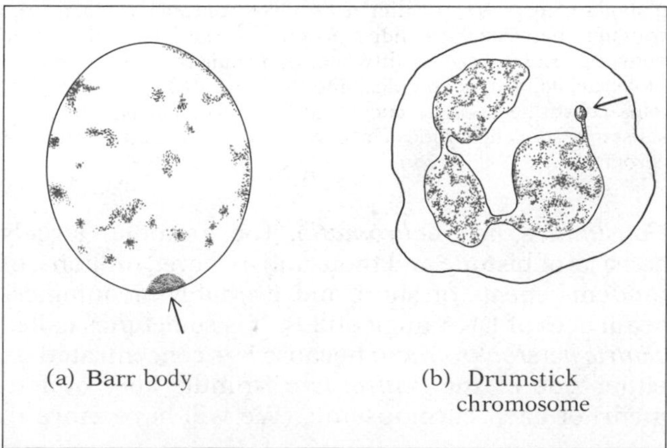

One of the central questions in genetics involves the control of the activity of chromosomes. Keep in mind that virtually every cell in the body has an identical set of chromosomes. So if chromosomes determine the behavior of a cell, dictating whether the cell will become muscle, blood, nerve, or whatever and if all the cells have the same kinds of chromosomes, how do cells specialize? How do they become different from each other? Obviously, all that DNA in any cell simply defines that cell's *potential.* The entire DNA complement can't be active. Some parts must operate while others are shut down.

It has been suggested that cells may regulate some of their activities in part through the use of agents that cause condensation and diffusion of different regions of chromatin. This would explain one way in which genes might act in a selective manner. There are certainly others. As far as regulation itself is concerned, however, it really pushes the question back one step. What regulates condensation and diffusion of chromatin?

And the operative parts of the DNA complex can change with time. Thus, we don't continue to grow all our lives. The parts of the chromosomes that are responsible for growth and development shut down at some point, others come into play and the aging process takes over. We can't reactivate those parts of the chromosomes. The first person who is able to make old cells young again will be rich and the idol of millions.

And another thing, why would a cell have inactive DNA at all? Well, except for the drastic measure taken by developing red blood cells, there simply seems to be no efficient way for the cell to get rid of genes that it doesn't need. And, as we have seen, the need may vary. Some genes are needed only at certain times in the life of the organism. So condensed DNA may simply be in mothballs, so to speak.

Nuclear RNA

In addition to chromatin, the nucleoplasm contains several kinds of RNA, including some tRNA, early stages of rRNA, and mRNA destined for the cytoplasm. There are also large amounts of long RNA molecules of unknown function, termed *heterogeneous nuclear RNA* (*hnRNA*). Heterogeneous nuclear RNA includes the initial transcripts (precursors) of mRNA, tRNA, and rRNA, but may also include RNAs of still unknown function. Most of it never leaves the nucleus and is very rapidly metabolized there.

The nucleoplasm also contains a potpourri of small molecules, including the ones that are the building blocks of DNA and RNA, various enzymes, a lot of ATP, and a number of inorganic ions. What this all adds up to is that the nucleus is a behaviorally prominent structure, surrounded by a functionally complex membrane. So, to sum up, within the nucleus we find one or more nucleoli (dark-staining bodies involved in ribosome production); pools of small molecules that are important in metabolism; a variety of RNA molecules, some with functions we know and some we don't; and most importantly, the chromosomes.

Different species, by the way, have widely different numbers of chromosomes. No one knows why (Table 8.1).

Cycle or Die

A cycle is endless, and the cell cycle is an endless repetition of *mitosis, cytokinesis,* growth, chromosomal replication, and more mitosis. We'll be more specific about the phases of the cycle later, but first we should point out one thing: Not all cells do it—cycle, that is. Although any cell has innumerable cell cycles in its past, it may have none in its future; some break out of the cycle. But when they do, they pay dearly. They die. In fact, many specialized cells are actually in a terminal condition. They will live in this state for a time (sometimes a very long time), fulfill some function, and pass on, to be replaced with newly differentiated cells that have broken out of their own cycles in order to specialize. Fingernail cells, for example, fill themselves with keratin (a tough protein) and die. In birds, reptiles, and fish, the nuclei of red blood cells are present, but they are permanently turned off, so these cells will die without undergoing another division. And, of course, the red blood cells of mammals, which have no nuclei, are living on borrowed time. Blood *stem cells* (Chapter 26), the cells from which other blood cells arise, continue to divide, but their daughters may have different fates. One daughter cell can differentiate into a red blood cell, perhaps into a *leucocyte* (white blood cell), while another continues to cycle as a dividing stem cell. A final example of noncycling, doomed cells are the gametes (the eggs and sperm), with the exception of the very few that unite to form a new individual. Of course, if they survive the incredible odds against them, they enter the events of growth, replication, and division once again.

Table 8.1 Chromosome numbers.[a]

Alligator	32
Ameba	50
Brown bat	44
Bullfrog	26
Carrot	18
Cat	32
Cattle	60
Chicken	78, 77
Chimpanzee	48
Corn	20
Dog	78
Dogfish	62
Earthworm	36
Eel	36
English holly	40
Fruit fly	8
Garden pea	14
Goldfish	94
Grasshopper	24
Guinea pig	64
Horse	64
House fly	12
Human being	46
Hydra	32
Lettuce	18
Lily	24
Magnolia	38, 76, 114
Marijuana	20
Onion	16, 32
Opossum	22
Penicillium	2
Pheasant	82, 81
Pigeon	80, 79
Planaria	16
Redwood	22
Rhesus monkey	42
Rose	14, 21, 28
Sand dollar	52
Sea urchin	40
Starfish	36
Tobacco	48
Turkey	82, 81
White ash	46, 92, 138
Whitefish	80
Yucca	60

[a]There is no apparent significance to chromosome number as far as biologists can determine. If you feel good about your 46 (see human being), check the turkey, ameba, cattle, tobacco, and yucca. Note the variation in some plant and animal species. Plants undergo spontaneous doubling and tripling of chromosome number. In chickens and turkeys the larger number signifies a male.

There are exceptions (biology is full of exceptions) to the dictum "cycle or die." Nerve cells specialize early and some remain alive through the lifetime of the individual. The same is true of skeletal muscle cells and many of the tissue cells of the body. Nerve and skeletal muscle cells formed in the vertebrate embryo, once differentiated, are no longer capable of cell division. So you're born with virtually all these cells you're ever going to get. Although such cells don't divide, they are capable of growth. So you have just as many biceps muscle cells as any weightlifter. Point that out to the next one you see. Such functioning, metabolizing cells that are nevertheless unable to divide are said to be *terminally differentiated.* In the end, of course, they too die.

Differentiation is not always irreversible. For example, some terminally differentiated white blood cells (those same polymorphonuclear leukocytes), which normally cease dividing, can be grown in a laboratory where they can be experimentally induced to undergo further mitosis. Also, certain cells in the reproductive system of rats can be induced to retrace their developmental route and dedifferentiate into an unspecialized, or more primitive, condition. Redifferentiation of these more primitive cells may or may not trace the same route to specialization. In other words, cells differentiate (specialize), then dedifferentiate (revert to an earlier stage), from which they can take a number of developmental routes, perhaps ending up as entirely different sorts of cells. Dedifferentiation is not just a laboratory trick; regeneration is not possible without dedifferentiation. If an arm of a salamander is amputated, the cells near the stub will revert to a more primitive condition and respecialize as the arm grows back. So what was once a cartilage cell could end up as a muscle cell.

Many embryonic cells undergo *programmed death.* But their death is part of an organism's development. For instance, the human embryonic hand starts as a paddlelike structure; separate fingers are formed by the deaths of cells in the spaces between the future fingers (Chapter 30). And other cells throughout the body must die to make channels for blood vessels.

We should point out that the number of terminally differentiated cells in any organism depends largely on the species involved. For example, in many adult insects, all cells are terminally differentiated except for those of the germ-line (gamete-forming) tissue. Plants have a high proportion of terminally differentiated cells also. To illustrate, the *phloem* cells of higher plants (tubelike cells that carry food down from the leaves, see Chapter 18), like red blood cells, lack nuclei (but they manage to remain metabolically active).

The Stages of the Cycle

At this point we can begin our discussion of the cell cycle proper. What stages do cycling cells undergo? First, the cell cycle is traditionally divided into four phases: M phase, G_1 phase, S phase, and G_2 phase (Figure 8.4). The M stands for *mitosis* (including cell division), S stands for *synthesis* of DNA, and G_1 and G_2 stand for *gap one* and *gap two,* respectively. (The terms may seem a bit primitive, but they come from an earlier age.)

Briefly, in M phase (mitosis), the chromosomes condense, the nuclear membrane and nucleoli disappear, the mitotic spindle forms, the two identical sister strands of each chromosome separate and go to opposite poles, two new nuclear membranes and two new sets of nucleoli form, and the cell divides into two daughter cells.

The remaining three phases are collectively called *interphase.* The chromosomes decondense as each new daughter cell enters G_1, the first stage of interphase. This is generally a very active period, the time when the cell synthesizes the enzymes and structural proteins necessary for cell growth. In this stage, each chromosome consists of only a single unreplicated DNA strand with its associated histones and nonhistone chromosomal proteins.

The S phase is the period in which the DNA and chromosomal proteins are replicated. This phase typically lasts a few hours, and can be distinguished in labeling experiments as the stage in which the cell nucleus will incorporate radioactive thymine.

The G_2 phase is simply the period between synthesis and mitosis. The proteins of the *mitotic spindle* are synthesized in this time, in preparation for the coming nuclear division. The mitotic spindle is an elaborate structure that is involved in chromosome movement during mitosis; it is constructed anew for each cell cycle, then dismantled after it has been used once (the name comes from a fancied resemblance to the spindle of a hand loom). As in G_1, the cell may be metabolically active and growing. In G_2, each chromosome consists of two strands, now called *chromatids.* The two chromatids of a G_2 chromosome are so tightly bound that they cannot be distinguished visually until they separate during mitosis.

The relative amount of time the cell spends in each stage varies greatly from one cell type to another. S phase and G_2 can be considered to be preparation for mitosis. Terminally differentiated cells, if they have nuclei at all, are usually frozen in the G_1 phase.

In addition to the mitotic cell cycle, there is the much more strange and complex meiotic cell cycle, but more about that later.

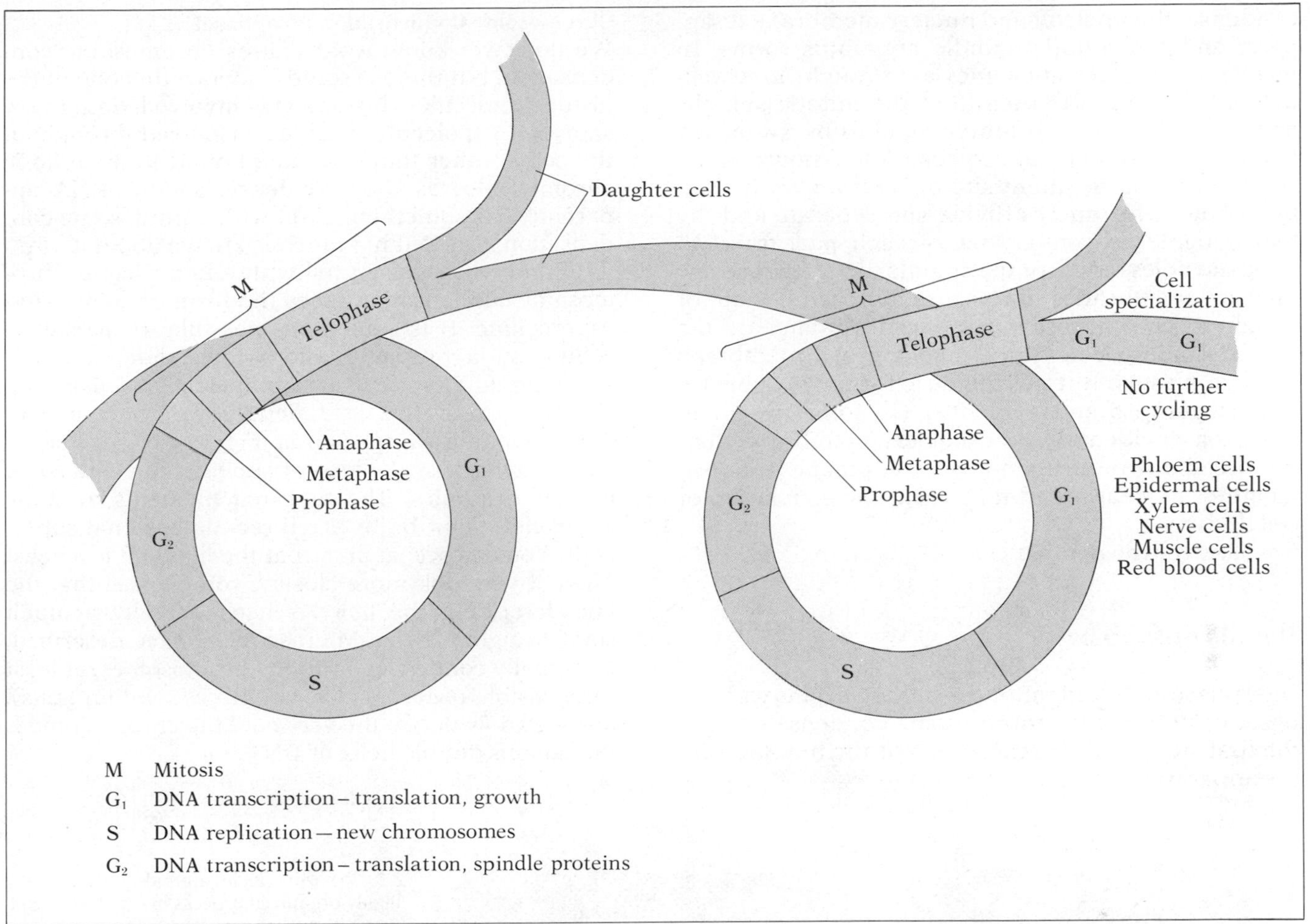

8.4 The cell cycle. This scheme represents several generations of cells which follow a cyclical series of events as they divide, grow or synthesize, replicate, and prepare to divide again. Each cycle is divided into four major parts which have been designated M, G_1, S, and G_2. The M stage represents mitosis and cell division. It includes four substages called prophase, metaphase, anaphase, and telophase. Mitosis is followed by G_1, where the cell actively synthesizes proteins, builds structure, and carries on all kinds of metabolic activity. The S stage is a period when many of the cell's resources are devoted to DNA replication. The G_2 stage is a time when the cell continues preparations for division, although it may still be metabolically active in other ways. At one point in the scheme, a cell becomes arrested in G_1. This cell has differentiated into a specialized type and will no longer cycle.

8.2 Mitosis

Mitosis and Cell Division

Before we go any farther, we should make it clear that *mitosis* refers to the duplication and division of the nucleus and of the chromosomes, while *cell division* (also called *cytokinesis*) refers to the actual division of one entire cell into two *daughter cells*.

In the interphase nucleus—G_1, S phase, and G_2—a lot happens biochemically, but we can't see anything very interesting under the light microscope. Presumably, this is why earlier microscopists called most of the cell cycle interphase, or, worse, "resting phase." The cells certainly aren't resting at this time. The electron microscope has revealed an unbelievably busy cytoplasm, roiling and seething, with bodies appearing and disappearing in a cauldron of activity. There appears to be almost total anarchy in the cytoplasm, but organization is the key to life and so we can be sure that all those processes are incredibly regular and specific, and that each molecule is in a tightly controlled transition or at least not interfering with those that are.

Mitosis (M phase) is itself divided into four subphases: *prophase, metaphase, anaphase,* and *telophase.* Briefly, in *prophase,* the chromosomes condense, the nucleoli and nuclear membrane disappear, and the mitotic spindle apparatus forms. In *metaphase,* the chromosomes are brought to a well-defined plane in the middle of the mitotic spindle, apparently by the microtubule spindle fibers attached to each chromosome at a specific site, known as the spindle fiber attachment site or *centromere.* In *anaphase,* the centromeres divide and separate and the two daughter chromosomes of each pair travel to opposite poles (ends) of the spindle. In *telophase,* new nuclear membranes form around each group of daughter chromosomes, the nucleoli appear, the chromosomes decondense, and the cell membrane and cytoplasm of the cell divide to form two daughter cells (Figure 8.5 and Table 8.2). (In some fungi and protozoa, incidentally, the nuclear membrane does not break down during mitosis, but pinches into two daughter nuclear membranes near the conclusion of cell division.)

Now for the details.

Details of Mitosis

Interphase ends and mitotic prophase begins with the onset of two events: chromosome condensation and the first steps in the organization of the mitotic spindle apparatus.

Chromosome Condensation (Prophase)

We don't yet know what causes chromosome condensation, but there is some evidence that one of the histone molecules (histone I) is involved. Some very short RNA molecules that are dispersed throughout the cell at other times become bound to the mitotic chromosomes as they condense, so this RNA apparently has something to do with chromosome condensation also. But not much is known about it.

From what we can tell with a microscope, chromosome condensation takes the form of coiling and supercoiling. It is sometimes possible to see supercoiling on a prepared microscope slide; a three-dimensional effect can be obtained by focusing up and down, but this is difficult to show in a two-dimensional drawing such as Figure 8.6. Let's see if we can get across the idea of supercoiling with some homely examples. The tungsten filaments of some large, clear light bulbs are three-dimensional supercoils. You can see, at first, that the filament is a helix. Then, if you look more closely, you can see that the line describing the helix is itself actually a much smaller, tighter helix. Microscopists have described, in partially condensed mitotic chromosomes, at least three visible orders (helix within helix within helix), and we know that at the very core the chromosome is the famous double helix of DNA.

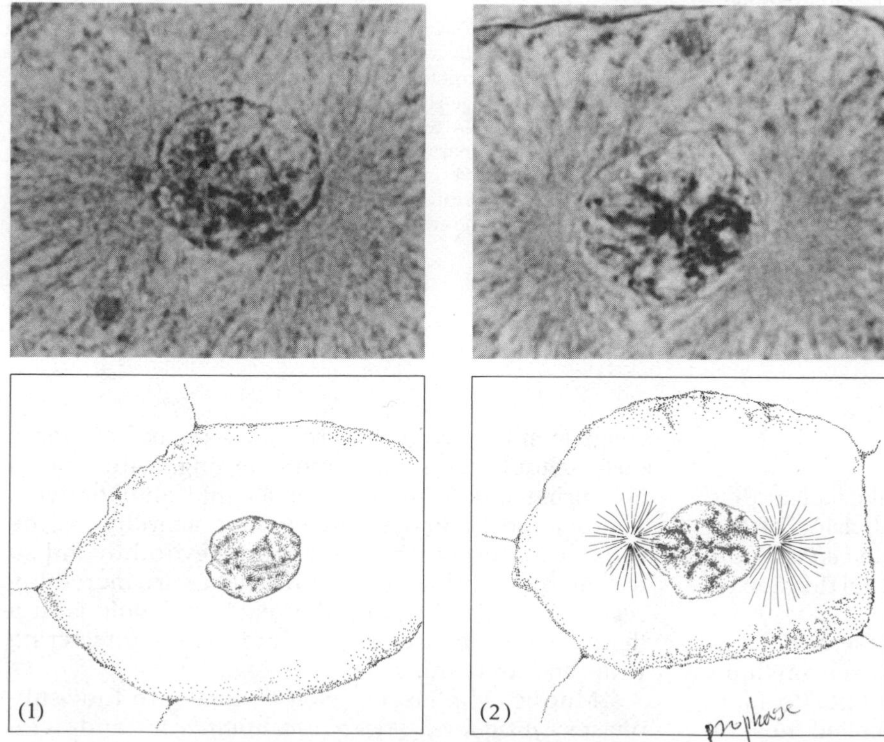

8.5 (a) Mitosis in animals. Mitosis is a highly organized process in which the products of replication, the chromatids, are separated at their centromeres and exactly and equally divided between the emerging daughter cells. The process is divided into five phases, each with its specific events, as shown here with light microscope photographs and drawings. The cells shown here are from the blastula of a whitefish.

(1) In interphase, the nucleus shows no hint of impending events, but a vast amount of biochemical activity precedes mitosis. Each DNA molecule replicates and gradually condenses into visible chromosomes. (2) Chromosomes begin to condense in early prophase.

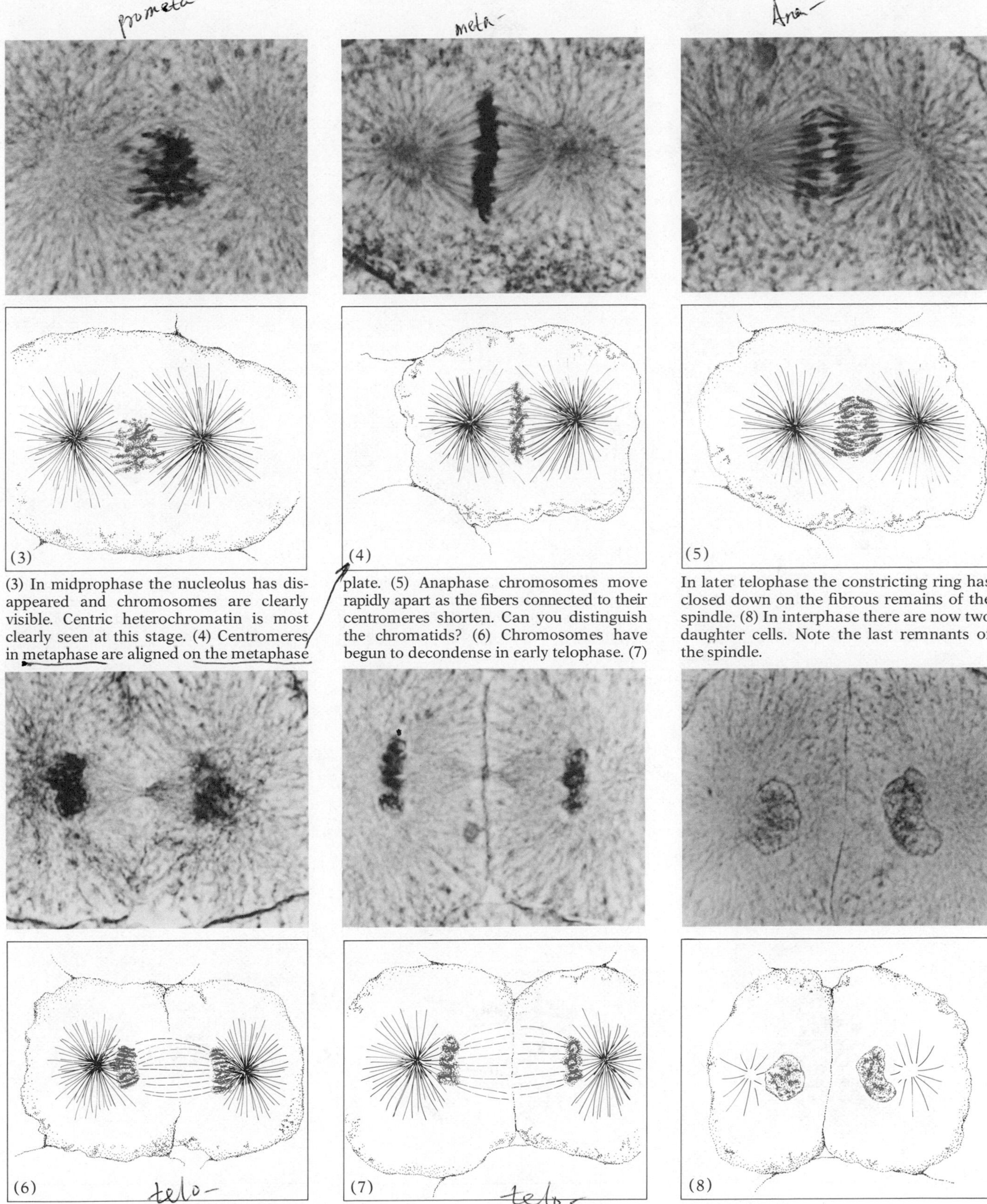

(3) In midprophase the nucleolus has disappeared and chromosomes are clearly visible. Centric heterochromatin is most clearly seen at this stage. (4) Centromeres in metaphase are aligned on the metaphase plate. (5) Anaphase chromosomes move rapidly apart as the fibers connected to their centromeres shorten. Can you distinguish the chromatids? (6) Chromosomes have begun to decondense in early telophase. (7) In later telophase the constricting ring has closed down on the fibrous remains of the spindle. (8) In interphase there are now two daughter cells. Note the last remnants of the spindle.

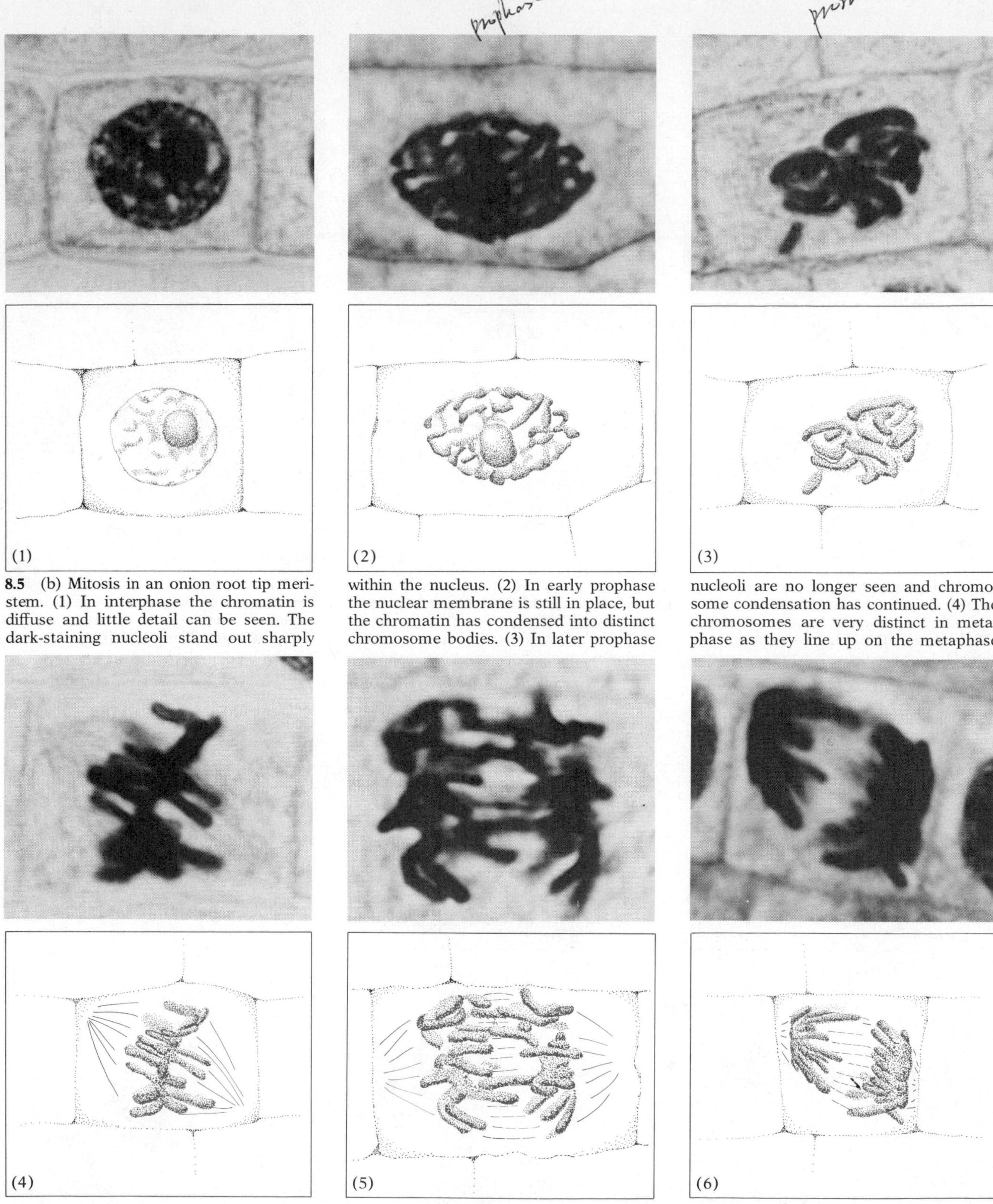

8.5 (b) Mitosis in an onion root tip meristem. (1) In interphase the chromatin is diffuse and little detail can be seen. The dark-staining nucleoli stand out sharply within the nucleus. (2) In early prophase the nuclear membrane is still in place, but the chromatin has condensed into distinct chromosome bodies. (3) In later prophase nucleoli are no longer seen and chromosome condensation has continued. (4) The chromosomes are very distinct in metaphase as they line up on the metaphase

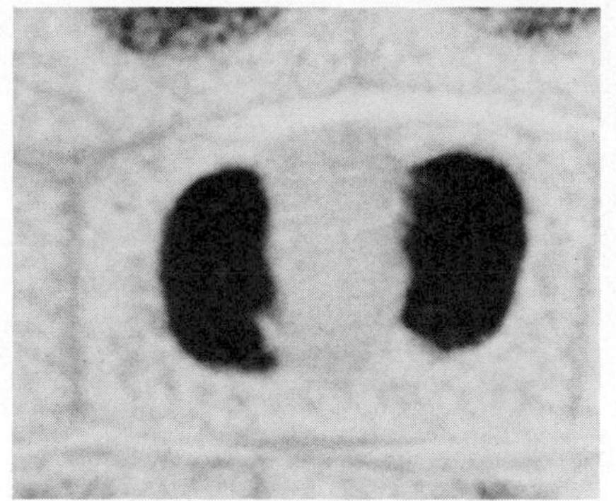

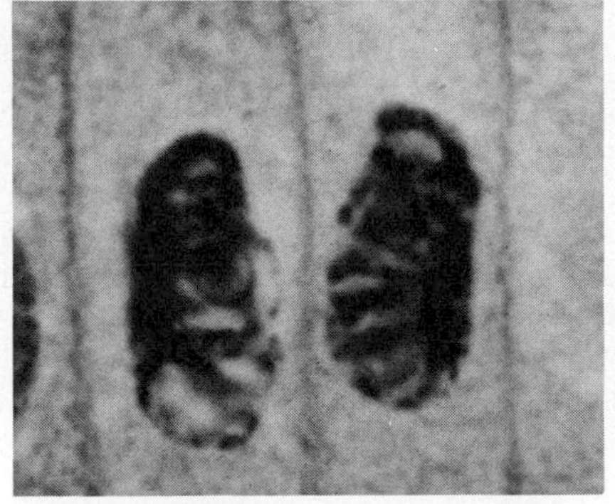

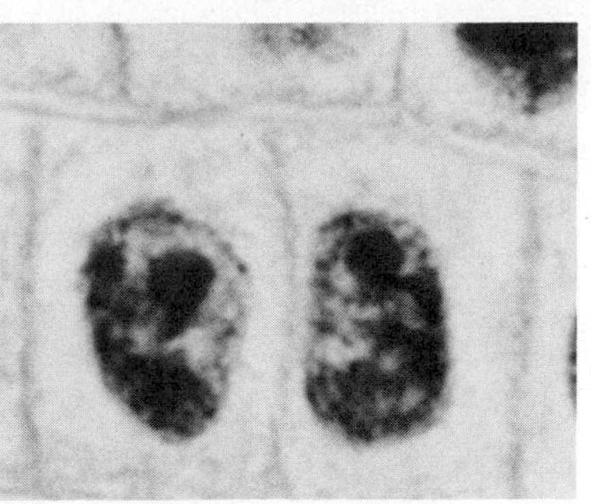

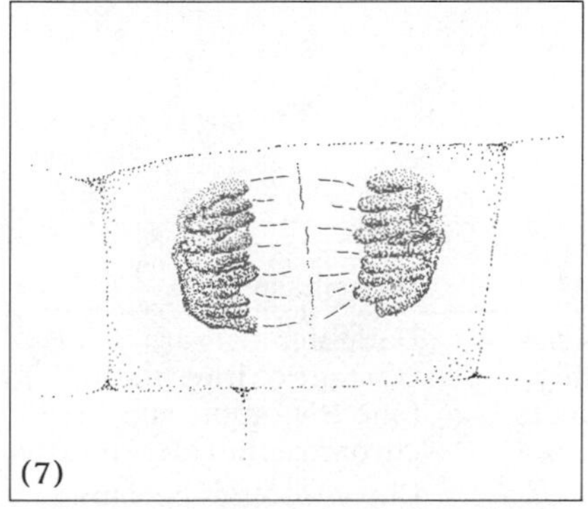

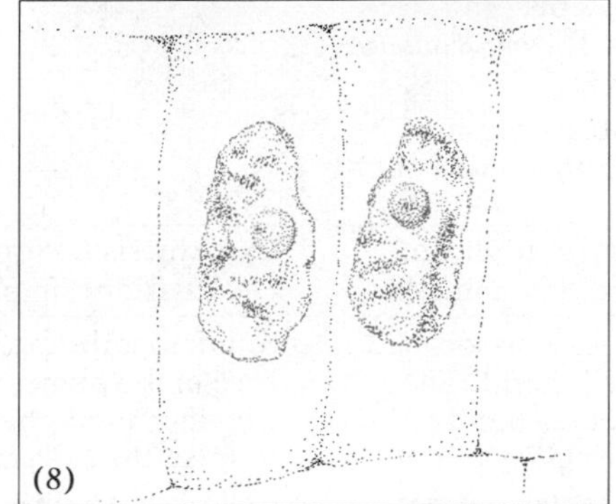

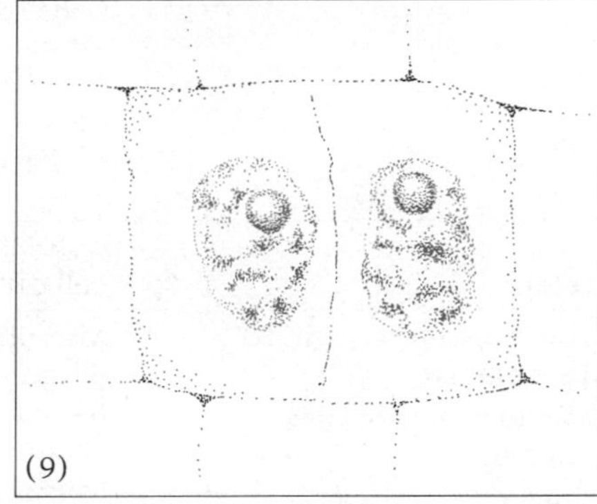

plate. This is a brief interlude and some of the centromeres have already divided. (5) Centromere division is complete in early anaphase, and (6) in later anaphase the daughter centromeres have been drawn to opposite sides of the cell. (7,8) In early telophase the individual chromosomes begin losing their identity as decondensation begins. The newly forming cell plate can be seen between the two nuclei. This is visible in (7). (8) The chromosomes continue their decondensation and nucleoli are just visible. Rigid cell walls have now formed. (9) The nuclear membrane has reformed and nucleoli are clearly visible.

For another example of superhelices, watch a kid wind up a rubberband-powered model airplane. The rubberband first forms a simple double helix. No kid is satisfied with this, however. He keeps winding. As the rubberband gets tighter, it suddenly throws loops into the coil and begins to form a supercoil. When the whole length of the rubberband is supercoiled into a thick cord, his imagination begins to run wild. He keeps winding. Finally a third order of coiling may begin. Supercoiled chromosomes are something like that, but there is no external force holding the ends, so the supercoiled chromosome shortens. The rubberband can't shorten, of course, so eventually it breaks, dashing the hopes of the kid and balling up into a tight mass of coiled rubber. The youngster at this point may take some consolation in the fact that he has produced interesting side pieces that stick out from the mass, each consisting of a length of rubber folded back and coiled around itself. Supercoiled chromosomes have a very similar appearance, as you can see in Figure 8.7.

So why do chromosomes condense in mitosis? ("Why" questions in biology usually mean, what good is it?) We have an intuitive answer, though it would be hard to prove. Chromosomes that are condensed into tight sausagelike bundles are simply much easier to move around the cell than the long, wispy strands of uncondensed chromosomes. You may well ask, then, why the chromosomes don't simply remain in their shortened, coiled form all the time, even through interphase? The answer to that seems to be that uncoiling is a prerequisite to action. Like many of us, chromosomes can't do their job when they're all wound up. Their job, of course, is to direct protein synthesis.

Chromatids, Chromatin, and Centromere

The single strand of chromatin of each G_1 chromosome has already doubled in the S phase, so each chromosome now consists of two strands of *chromatin*—here called *chromatids*—as it enters prophase. The doubled structure is still considered to be one chromosome (consisting of two chromatids) until anaphase, when it divides into two chromosomes—each with one *chromonema* (strand of chromatin). There is no such thing as a chromosome with only one chromatid, since the term *chromatid* is defined as one subunit of a doubled chromosome. And that's the only difference between *chromonema* and *chromatid.* All this is essentially a matter of semantics, and while

Table 8.2 Mitosis and Cell Division

Interphase	Prophase	Metaphase	Anaphase
Chromosomes are decondensed *S phase:* DNA and chromosomal proteins synthesized *Spindle proteins* and other mitotic proteins formed in G_2 phase Cell increases in volume	Chromosomes condense (*centric heterochromatin* first) Separate chromatids may become visible Asters and mitotic spindle begin to form *Nuclear membrane* breaks down *Nucleolus* disperses *Centromeres* become attached to the *centromeric spindle fibers* Chromosomes migrate toward the *metaphase plate*	Spindle is fully formed Chromosomes are aligned with their centromeres on the *metaphase plate* Centromeres begin to divide	Centromeres divide Daughter centromeres go to opposite poles Sister chromatids separate to become chromosomes Discontinuous spindle fibers shorten; the spindle as a whole elongates
Telophase	**Cytokinesis (cytoplasmic cell division)[a] in animals**	**Cytokinesis (cytoplasmic cell division)[a] in plants**	**Interphase**
Chromosomes begin to decondense Nuclear membranes reform Nucleoli reappear Spindle disappears Cytokinesis (cytoplasmic cell division) usually occurs	Microfilaments associated with the cell membrane form a circular band around the cell Microfilaments contract, pinching apart daughter cells	Small membrane vesicles fuse in the plane of the previous metaphase plate to form the *cell plate* Cell plate grows to separate daughter cells *Middle lamella* and *cell walls* are laid down	Daughter cells are in the G_1 stage of interphase (one DNA molecule per chromosome) after mitosis Chromosomes become diffuse

[a]Not a feature of mitosis proper.

it is clearly more important for you to understand what happens physically than to know what these things are called at the various stages, naming them carefully the first time through will make learning the processes easier. One trick to counting chromosomes is to count centromeres.

The two chromatids of the chromosome are not visibly distinguishable in early prophase, and, in the cells of many tissues, do not become distinct until metaphase or even early anaphase. In some cells, however, the two chromatids become quite distinct by late prophase. Each is supercoiled individually, and, by metaphase, the two chromatids are held together only by a structure known as the *centromere.* The centromere, as seen under the electron microscope, is a ringlike structure surrounding the thickened chromatids, rather like a napkin ring with two napkins in it. On the centromere, the actual sites of insertion of the spindle fibers appear to be thin, discrete plates of protein. The chromatids are somewhat constricted at the centromere. In stained preparations, the centromere appears under the light microscope as a prominent light-staining body separating the chromosome into "arms."

The centromere is made of protein, and it is considered to be a permanent part of the chromosome. It always appears at exactly the same place in any given chromosome, presumably under the direction of DNA. The constancy of the placement of the single, highly visible centromere is one of the most reliable characters used for identifying specific chromosomes (see Essay 8.1).

The Spindle

Before we start discussing this peculiar, transient, and almost ethereal structure, we should point out that flowering plants do things a little differently from other eucaryotic organisms. Because of this, we are temporarily excluding flowering plants from this part of our discussion.

The *spindle apparatus* is architecturally at the very core of mitotic movement. It consists primarily of two important proteins involved in all eucaryotic cell movement: *actin* and *tubulin.* (Elsewhere, actin is involved in muscle movement and in ameboid movement; tubulin is involved in the movement of cilia and flagellae. But they operate together in mitosis.) Tubulin is the more common mitotic component if you

Human chromosomes
(a)

Supercoil
(b)

Coil within coil
(c)

Further supercoiled chromosome strand
(d)

DNA wound around histone proteins to form nucleosomes
(e)

DNA double helix – the final level of coiling
(f)

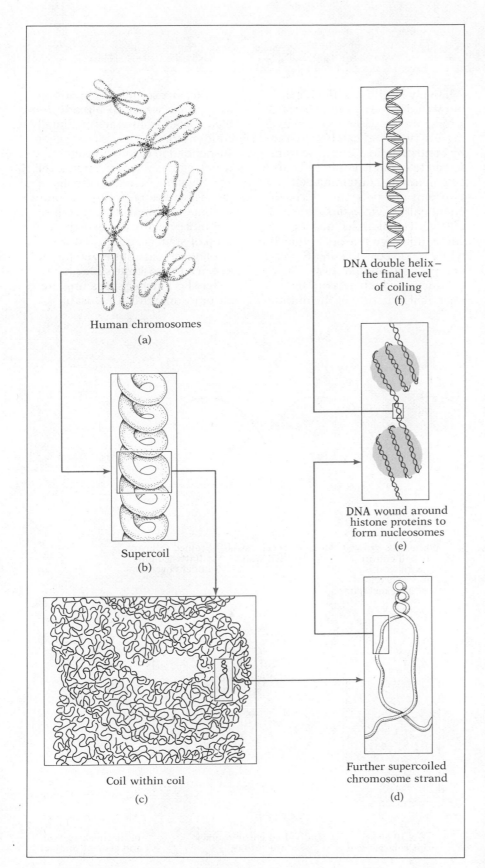

8.6 Supercoiling and the levels of chromosomal organization. The condensed chromosome, easily visible with proper staining and the light microscope, is a supercoil. It consists of many levels of organization, each a coil of its own. The human chromosomes in part (a) are shown as they appear through the light microscope. The coiling nature of the chromosome arms cannot be readily discerned in this view, but with the greater magnification in part (b), the arms are seen to consist of a thickened helix. (c) Examining a small segment of this supercoil reveals that it consists of an incredibly long thread of DNA coiled back and forth as it is packed into the supercoil. (d) Yet more coiling of the chromosome strand. Here a short length coils back on itself. (e) The nucleosomes: DNA coiled around histone molecules. Structure at this level of organization can just be resolved by the electron microscope, but its final helical organization cannot. The now famous DNA helix must be determined using techniques such as X-ray diffraction. (f) So, with greater, imaginary magnification, we see the last level of coiling, the double helix of DNA.

8.7 This view of the chromosome, magnified about 100,000× by the electron microscope, can be compared to the supercoiled rubberband. The protrusions sticking out of the sides are loops coiled back on themselves, and not broken pieces. This chromosome was photographed at metaphase. It still consists of two chromatids joined at the centromere. Centromeres are permanent regions on chromosomes where spindle fibers attach. The position of the centromere, in spite of the term itself, can be almost anywhere along the chromosome. The centromere will divide after metaphase and each chromatid will then be a fully qualified chromosome.

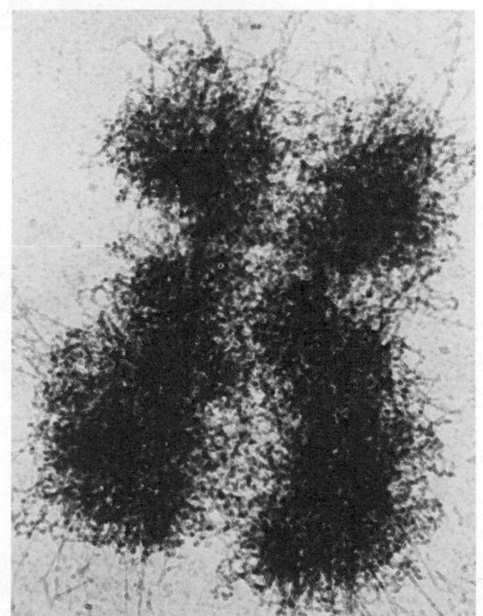

Essay 8.1 Karyotyping

A *karyotype* is graphic representation of the chromosomes of any organism, in which the chromosomes are systematically arranged according to size and shape. Each species has its particular karyotype, and so we know the number and kinds of chromosomes found in cabbages, fruit flies, and people. Karyotyping is done according to an established method.

Consider, for example, human karyotyping. (a) A blood sample is withdrawn and (b) a few drops are put into distilled water. The relatively fragile red blood cells burst under the resulting osmotic stress, but the white blood cells merely swell and can be separated by centrifugation. They are transferred to (c) an isotonic cell culture medium, containing cell nutrients and two plant-derived chemicals: *plant lectins* and *colchicine*. The plant lectins, for some unknown reason, induce the white blood cells to enter S phase and proceed on to mitosis. The colchicine, which acts on the chemical structure of the mitotic spindle, arrests all cells in mitotic metaphase, when the already doubled chromosomes are maximally condensed. (d) The cells are again suspended in distilled water, which again causes them to swell. This swelling increases the distances between the chromosomes so that individual chromosomes can more readily be seen. (e) A drop of stain is added, and the swollen cells are squeezed flat between a microscope slide and a cover slip, further spreading the chromosomes. (f) A favorable

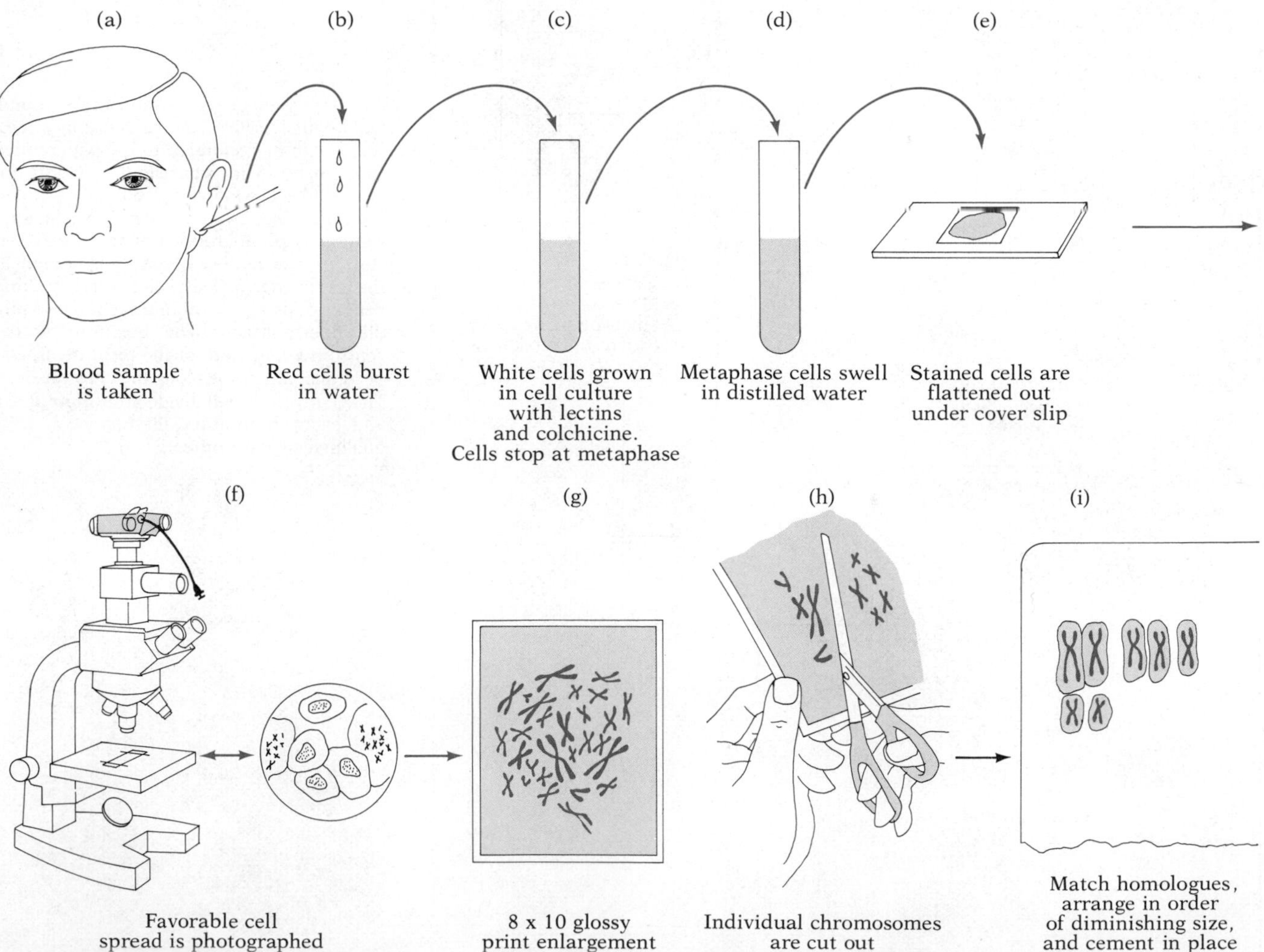
(a)
Blood sample is taken
(b)
Red cells burst in water
(c)
White cells grown in cell culture with lectins and colchicine. Cells stop at metaphase
(d)
Metaphase cells swell in distilled water
(e)
Stained cells are flattened out under cover slip
(f)
Favorable cell spread is photographed
(g)
8 x 10 glossy print enlargement
(h)
Individual chromosomes are cut out
(i)
Match homologues, arrange in order of diminishing size, and cement in place

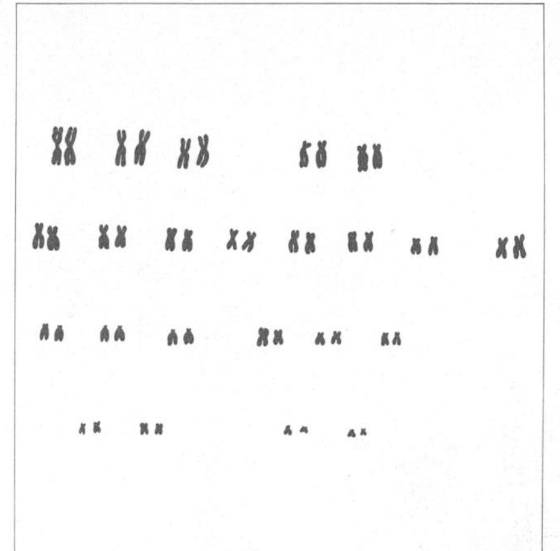

cell—one showing all the chromosomes spread out and separated from each other—is photographed through a light microscope. (g) The photograph is enlarged to an 8 × 10 glossy print. (h) Here comes the most technically sophisticated trick of all: The investigator uses scissors to cut out each chromosome from the photograph. (i) The cut-out images of the chromosomes are sorted by size and shape. The homologous pairs are matched up, and the chromosomes are laid out in one or several lines, from the largest to the smallest; and fastened down with rubber cement. This is the karyotype. (j) Once the karyotype has been constructed, the investigator can count the chromosomes and can identify individual ones. By comparing the karyotype with a karyotype of a normal individual, the investigator can detect even minor chromosomal abnormalities. (The X shape of the chromosomes is due to the fact that the two condensed chromatids of each chromosome have almost completely separated and are held together only by the centromere.)

consider weight. In fact, in the late G_2 phase, tubulin accounts for as much as 10% of the total protein of the cell. Then, somehow, these subunits assemble themselves into *microtubules* (see Figure 4.28), but only after some as yet unknown signal is given (Figure 8.8).

During interphase, the *centrioles* (see Figure 4.25) look like a pair of bright dots under the light microscope. But under the electron microscope each is seen to be a pair of complex short cylinders made up of microtubules.

At the onset of mitosis, the two centrioles, which have been situated fairly close to each other near the nucleus, separate and migrate to opposite sides of the nucleus. The position of the two centrioles apparently determines the orientation of the spindle. All we know, though, is that the poles of the spindle converge at the centrioles (Figure 8.9). As the nuclear membrane disappears, the spindle forms between the centrioles, and astral rays (comprised of microtubules) radiate outward from the centrioles.

8.8 Each fiber of the mitotic spindle is actually a hollow microtubule (see Figure 4.28). This high-magnification electron microscope view shows several microtubules attached to a chromosome. Each microtubule is composed of many spheres of the protein tubulin.

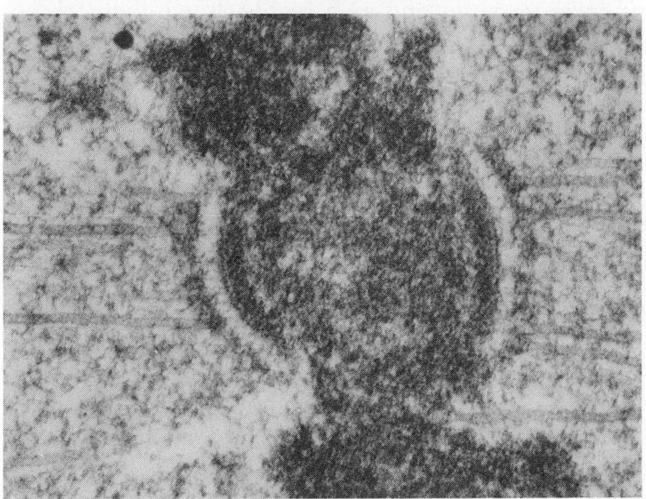

(Each centriole will later divide so that each daughter cell will have a pair when mitosis is over. Somehow, the two little cylinders come apart, and each causes a new one to form at right angles to itself. The result is two complete centrioles.)

What Is the Centriole For?

In recent years, the role of the centriole in organizing the spindle has been questioned. It has been noted, for example, that flowering plants, which lack centrioles, have perfectly good spindles. It has further been noted that all organisms that have flagellae and/or cilia also have centrioles, so someone came up with a counterhypothesis for centriolar function: The centrioles, it was suggested, are only "seeds" that give rise to basal bodies and flagellae. [Basal bodies are very similar to centrioles in that they both are comprised of a ring of nine groups of tubules. And these basal bodies are found in association with cilia and flagella (see Chapter 4).] Actually, there is some hard evidence that basal bodies are formed from centrioles. But then, why do the centrioles migrate to the future poles of the spindle? Perhaps it is merely a way to ensure that each daughter cell gets one.

It turned out that the new hypothesis was short-lived. In laboratory preparations of tubulin subunits taken from cells in early mitotic prophase, microtubules would not form unless intact centrioles were added. When the centrioles were added, beautiful *asters* (radiating clusters of microtubules) formed around them. So, it seems, centrioles do organize asters (and spindles), after all. It is now believed that even flowering plants have something like spindle organizers, but these happen to be amorphous

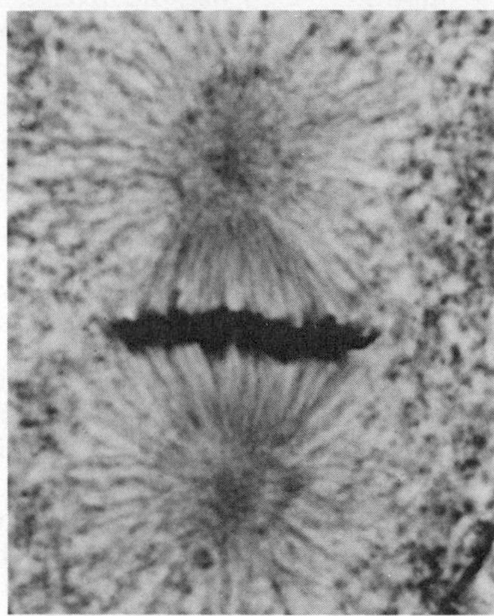

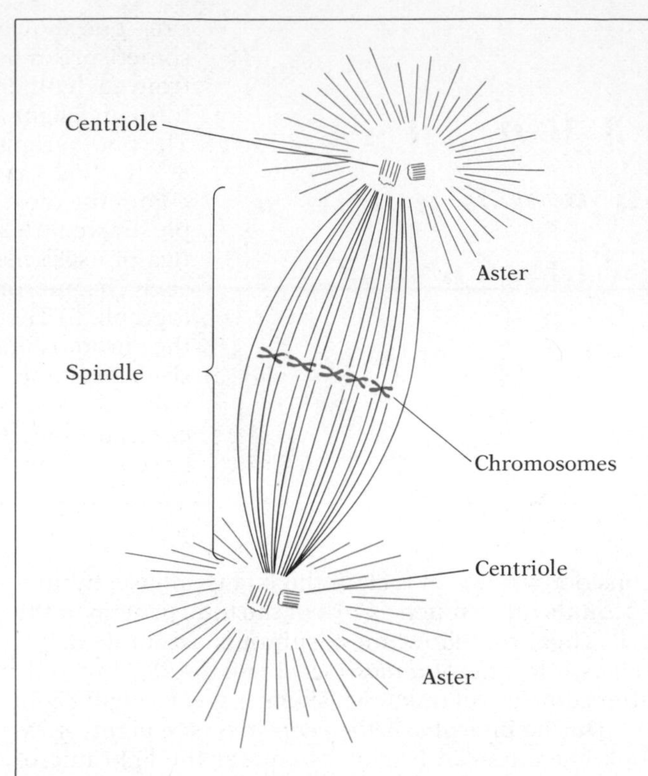

8.9 This photograph shows the mitotic apparatus in the blastula of a whitefish. Many animal cells form two centers of radiating microtubules about the centrioles. These are known as *asters.* At the center of the spindle are the chromosomes.

(shapeless) and nearly invisible under either light or electron microscopes.

Spindle Fibers

In addition to its actin microfilaments, the spindle is comprised of two visibly different kinds of microtubules, both called *spindle fibers. Centromeric* spindle fibers are attached at one end to the centromere of a chromosome and at the other end to a spindle pole (see Essay 8.2). Note that a number of spindle fibers are attached to each centromere, and each of these fibers terminates at the centriole.

The other type of spindle fibers has traditionally been called *continuous spindle fibers,* since it was believed that they were indeed continuous from pole to pole. This has turned out to be a misnomer, because in 1976 electron microscope studies revealed that each "continuous" fiber ends blindly somewhere short of the opposite pole. However, they do overlap considerably along the length of the spindle, extending well past the chromosomes.

But what about those astral rays? They too are made of tubulin and terminate at the spindle pole. However, they are not part of the spindle proper since they radiate away toward the periphery of the cell. We can only guess at what asters do. Perhaps nothing. Maybe they are just leftover tubulin that didn't get into the spindle, or perhaps they anchor the spindle to the cell membrane. We know, after all, that flowering plants do just fine without asters. But then, flowering plants have relatively rigid cell walls, so perhaps the asters serve to make the cytoplasm of animal cells temporarily rigid. Actually, the spindle, asters, centrioles, and chromosomes do form a rather rigid mitotic apparatus. Two researchers, Daniel Mazia and S. K. Dan, astonished almost everyone in the late 1950s by isolating the mitotic apparatus intact. This was surprising because the spindle and asters had always been thought to be extremely fragile. However, these researchers carefully washed away the cell membrane and the other loose cytoplasm with weak detergents and were surprised to find the mitotic apparatus remaining as a feeble, but intact, framework (Figure 8.9).

The Spindle in Action: Metaphase and Anaphase

Metaphase

With the formation of the spindle, the breakdown of the nuclear membrane, and the attachment of the centromeric spindle fibers to the condensed chromosomes, the mitotic apparatus is set to do its dance. The chromosomes are lined up on a plane called the *metaphase plate* that lies in the middle of the spindle and perpendicular to its axis. Actually, only the centromeres are in the plane of the metaphase plate; the arms of the chromosomes may hang off in all directions. Throughout this process the chromosomes are

Essay 8.2 How the Spindle Works

We know that the spindle serves to move the daughter chromosomes to opposite poles during anaphase. How it accomplishes this has been a popular subject of debate among biologists for half a century or more. We now have a pretty good idea of how it's done.

The basic observations are simple. During anaphase (the period of chromosome movement), the spindle drastically changes its shape in two ways: (1) the centromeric spindle fibers (the ones attached to the chromosomes) become shorter; and the spindle itself becomes longer, with the poles moving farther apart.

These two classic observations have given rise to a hoary old controversy over the forces at work during chromosome movement. The "pull" theory holds that chromosome movement is due to the shortening of the centromeric spindle fibers, which reel in the chromosomes like trout on a line. The "push" theory holds that the chromosomes are forced apart by the elongation of the spindle, which in turn is due to the elongation of the continuous spindle fibers; the centromeric spindle fibers simply bind each set of chromosomes to its proper pole. The argument, each side with its distinguished proponents, has continued for decades. But we do have some new information, gained from carefully measuring electron micrographs.

Of course, the partisans of both sides now feel that their long-argued positions have been fully vindicated. Yes, indeed! The centromeric spindle fibers do get shorter, apparently exerting some pull on the chromosomes. Yes, indeed! The spindle itself does get longer, pushing the poles apart. The ("continuous") spindle fibers, which are not continuous after all, do not themselves grow longer; instead, the two overlapping sets slide apart from one another, leaving a smaller and smaller region of overlap. It isn't known whether this sliding apart of two sets of fibers (microtubules in this case) is related to the sliding together of other kinds of filaments that occurs in muscle contraction (actin and myosin myofilaments), but it is known that actin microfil-

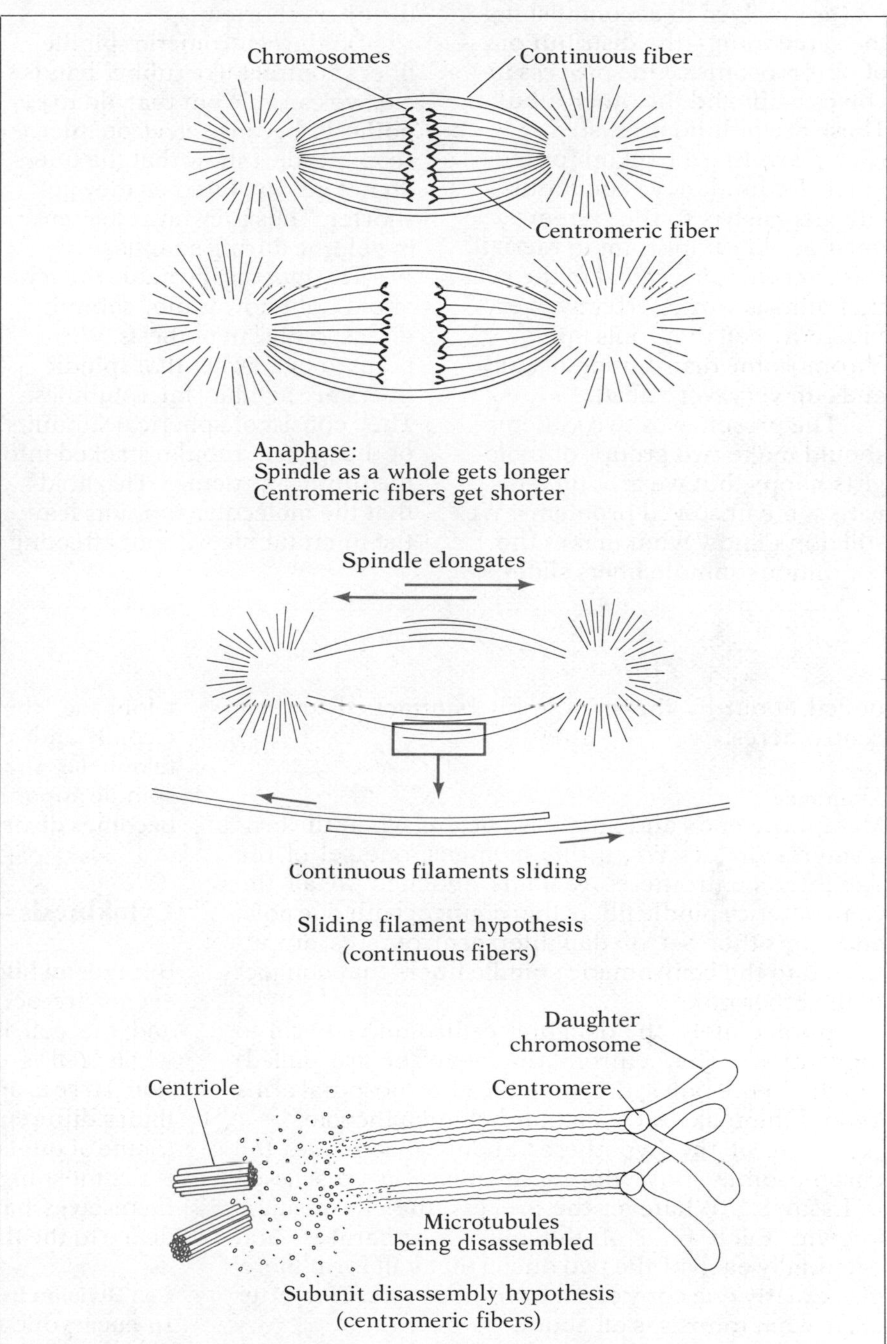

aments are also present in spindles.

It now seems likely that there are two entirely different processes involved in accomplishing the same thing—the distribution of chromosomes. One process involves actin and the other tubulin. These are redundant systems, each providing a backup for the other. Redundancy is undoubtedly expensive, so the system must be very important to merit it. It is exceedingly critical to the cell that mitosis work perfectly every time. Any cell that fouls up its chromosome distribution usually ends up very, very dead.

The presence of two systems should make two groups of biologists happy, but we are still left with some unsolved problems. We still don't know what makes the continuous spindle fibers slide apart, or even whether this sliding apart is the cause or the result of spindle elongation. We also don't know how the centromeric spindle fibers shorten.

Do the centromeric spindle fibers contract like rubber bands? No, we can rule out that old hypothesis, because electron microscope studies show that the fibers do not get any fatter as they get shorter. Thus they must decrease in volume during anaphase.

The most respectable theory afloat right now is the "subunit disassembly" hypothesis. We pointed out earlier that spindle fibers are actually microtubules. They consist of spherical subunits of the protein tubulin stacked into the tubular structure. This holds that the molecular subunits leave the microtubule without affecting the integrity of the part that remains, so that the microtubule grows shorter while somehow remaining attached at both ends. There are many variants of this hypothesis. According to some theorists, either the centromere, or the centriole, or both, have the ability to actively dissociate molecular subunits from the ends of the microtubules, while still keeping a grip on the fibers. This reminds us of one of the favorite games of the old frontier parties in which each of two players held one end of a long piece of licorice in his or her mouth; then they chewed their way to each other's lips. If you can imagine that the two players are the centriole and the centromere, and the licorice is a spindle fiber, you are in a position to envision a thoroughly preposterous analogy.

---

pulled about by the spindle fibers attached to their centromeres.

Anaphase

Metaphase ends and anaphase begins when the centromeres divide. When this happens, one set of the daughter centromeres remains attached to all the centromeric spindle fibers that connect with one pole, and the other set of daughter centromeres are attached to the centromeric spindle fibers that connect to the other pole.

Immediately, the daughter centromeres begin to move apart. The centromeres move (or are pulled) rapidly to the poles, dragging the chromosomal arms behind them like raccoon tails behind a bicycle.

Some of the hypotheses about what makes the chromosomes move during anaphase are discussed in Essay 8.2. Whatever the process, the end result is known: Each pole of the mitotic apparatus, and eventually each of the two nuclei that will form there, gets exactly one copy of each chromosome. This, after all, is what mitosis is all about.

Telophase

If you're following all this, you should be able to visualize the differences between telophase and prophase. Telophase is somewhat the reverse of prophase, and now there are two nuclei where there was one. In prophase, the chromosomes condense; in telophase, they decondense. In prophase, the nucleolus and the nuclear membrane disappear; in telophase, they reappear. In prophase, the mitotic spindle apparatus becomes organized; in telophase, it becomes disorganized.

Cytokinesis—Cell Division

But meanwhile, out in the cytoplasm, other dramatic events are occurring. The centrioles are replicating, and the cell itself is dividing. This division of the whole cell is called *cytokinesis,* or *cytoplasmic division.* Here is another place where flowering plants do things differently from every other kind of organism. (Some of our colleagues insist that the flowering plant is the most highly developed form of life. The plants themselves have expressed no opinion, lending credence to the theory.)

Cell Division in Animals

In eucaryotic organisms other than flowering plants and fungi, cell division generally begins with an equatorial furrow. This is an indentation in the surface of the cell, usually at about the plane of the metaphase plate. It looks as though someone were squeezing the cell with a tight but invisible band. Actually, the furrow is created by actin microfibrils in the cytoplasm just beneath the cell membrane. In most eucaryotes other than flowering plants and fungi, cell membranes generally have the power of movement thanks to these actin microfibrils. The furrow continues to deepen, to become a deep groove still being drawn in from below by the actin microfibrils. The remains of the spindle may be caught by the tightening ring of contracting microfilaments, but no matter, the constriction continues until eventually the cell is pinched in two (Figure 8.10).

Of course, there are exceptions to this general pattern. Some cells undergo a number of mitoses (nuclear divisions) without any cytoplasmic division, to become multinucleated cells. Other cells may normally undergo two or more rounds of chromosome replication without either nuclear or cytoplasmic division, and end up having multiples, greater than two, of each kind of chromosome (some of your liver cells, for instance).

Cell Division in Plants

But what about cell division in plants? Plant cell walls are not very flexible and are not given to bending and folding. They are capable of stretching, but this appears to be the result of an increase in hydrostatic pressure, coupled with selective weakening of appropriate portions of the cell wall. Cell walls, of course,

8.10 Cytokinesis in animal cells. Following the division of chromosomes and their migration to opposite sides of the cell, cytokinesis (or cytoplasmic division) begins (a). Typically, indentations form on each side and then deepen around the cell, forming a definite furrow (b). The furrow deepens and eventually the cytoplasm is pinched in two (c) and (d). The structures responsible for the formation and deepening of the cleavage furrow are the actin microfibrils which, in a manner not entirely understood, pull the cell membrane in until separation is complete.

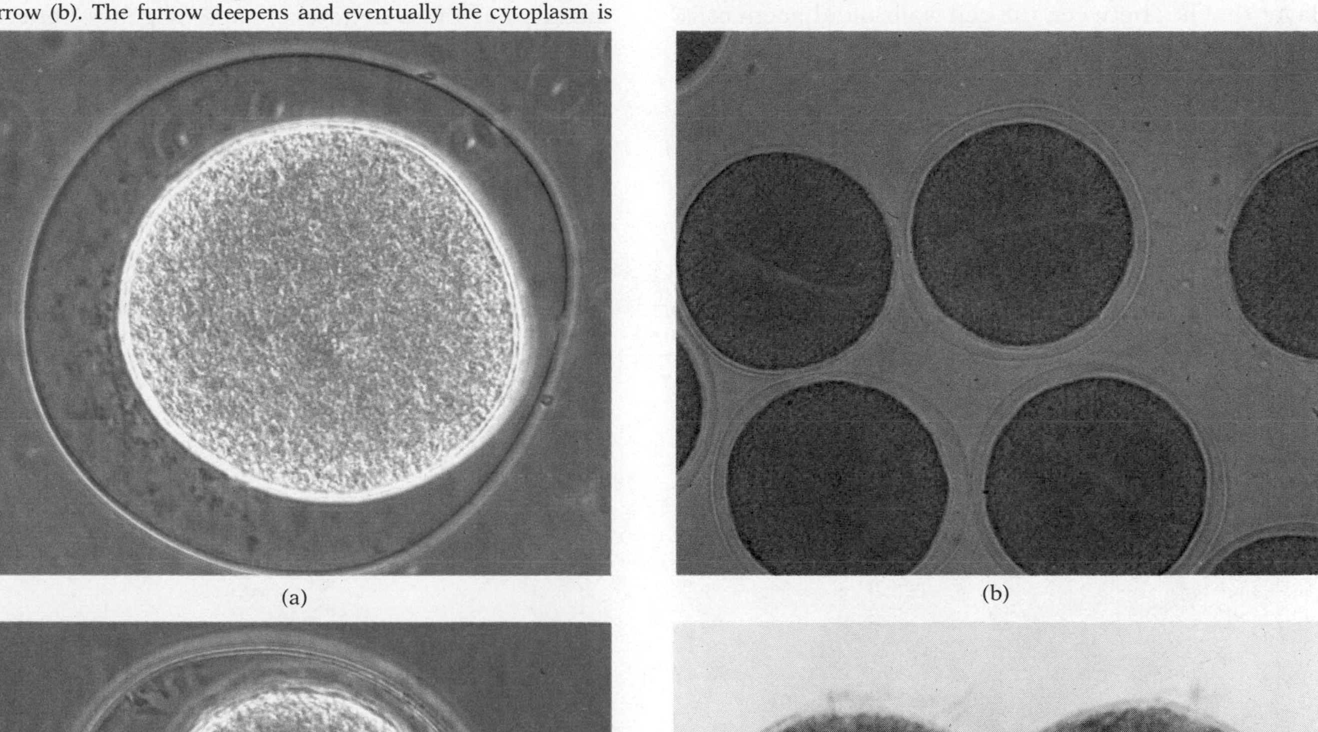

(a) (b)

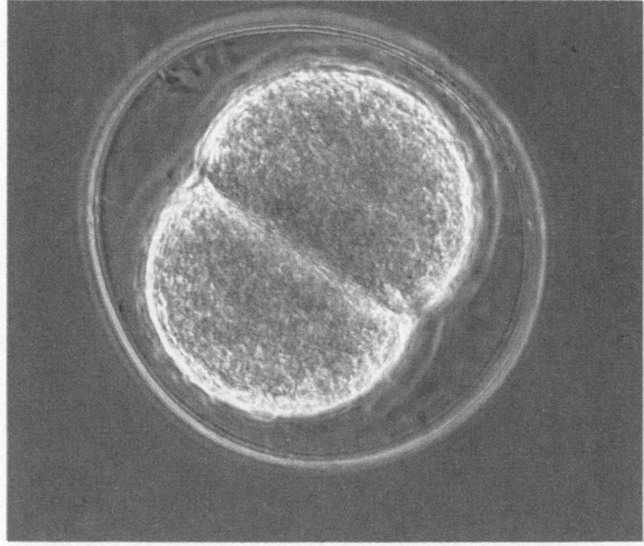

(c)

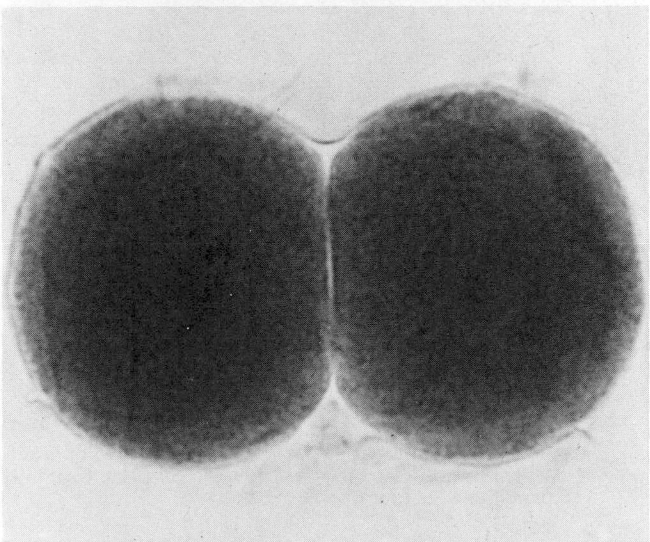

(d)

are laid down by the membrane, are metabolically inert, and are considered legally "dead."

Since plant cells have rather rigid cell walls and relatively immobile cell membranes, they have their own way of dividing. At about the time of telophase, little membranous vesicles of electron-dense material form in the plane of the metaphase plate. These continue to form (from Golgi bodies) and join up with one another until they make a more or less continuous double membrane filled with heavy material. When complete, this structure is called the *cell plate.* The cell plate begins to form in the middle of the cytoplasm and then grows outward to join the periphery of the cell—exactly the opposite of the animal cell division furrow, which begins at the periphery and extends toward the middle.

The cell plate becomes impregnated with pectin and ultimately forms the *middle lamella,* the gummy layer that lies between the cell walls of adjacent cells. The new cell membranes then grow in over the middle lamella, secreting new cell walls as they grow (Figure 8.11). The details of plant cell wall production are described in Chapter 21.

A Final Look at Mitosis and Cell Division

So, to summarize, cell replication consists of *mitosis,* which refers to the nuclear and chromosomal events, and *cytokinesis,* which refers to the cell membrane and cytoplasmic events. Mitotic cell division serves organisms in two major ways. First, for many single-celled organisms, mitotic cell division is a way of accomplishing rapid increases in population when growth conditions are good. For example, a single paramecium (a ciliate protozoan) introduced into a water medium containing bits of decaying plant matter will give rise to thousands of offspring in only a few days. This form of reproduction is asexual, which means that there is no exchange of genes between individuals. The colony will be made up of individuals with genes identical to those of the founder. Such a

8.11 Cytokinesis in plants. The cytoplasmic division following chromosome separation in plants is quite different from that in animals. Rigid cell walls preclude any form of cleavage from occurring. Instead, a new cell wall is constructed between daughter cells. In the sequence shown here, evidence of a cytoplasmic division is first seen as the formation of vesicles at the middle of the cell. These coalesce and extend across the cell, forming the *cell plate.* The microtubules constituting the spindle fibers apparently play a role in forming the plate, although the details are obscure. The first electron micrograph shows clusters of microtubules ending in the emerging vesicles. As the plate thickens, these begin to dwindle, as shown in the second electron micrograph. The completed plate is known as the *middle lamella,* and from this structure, the primary cell wall will emerge.

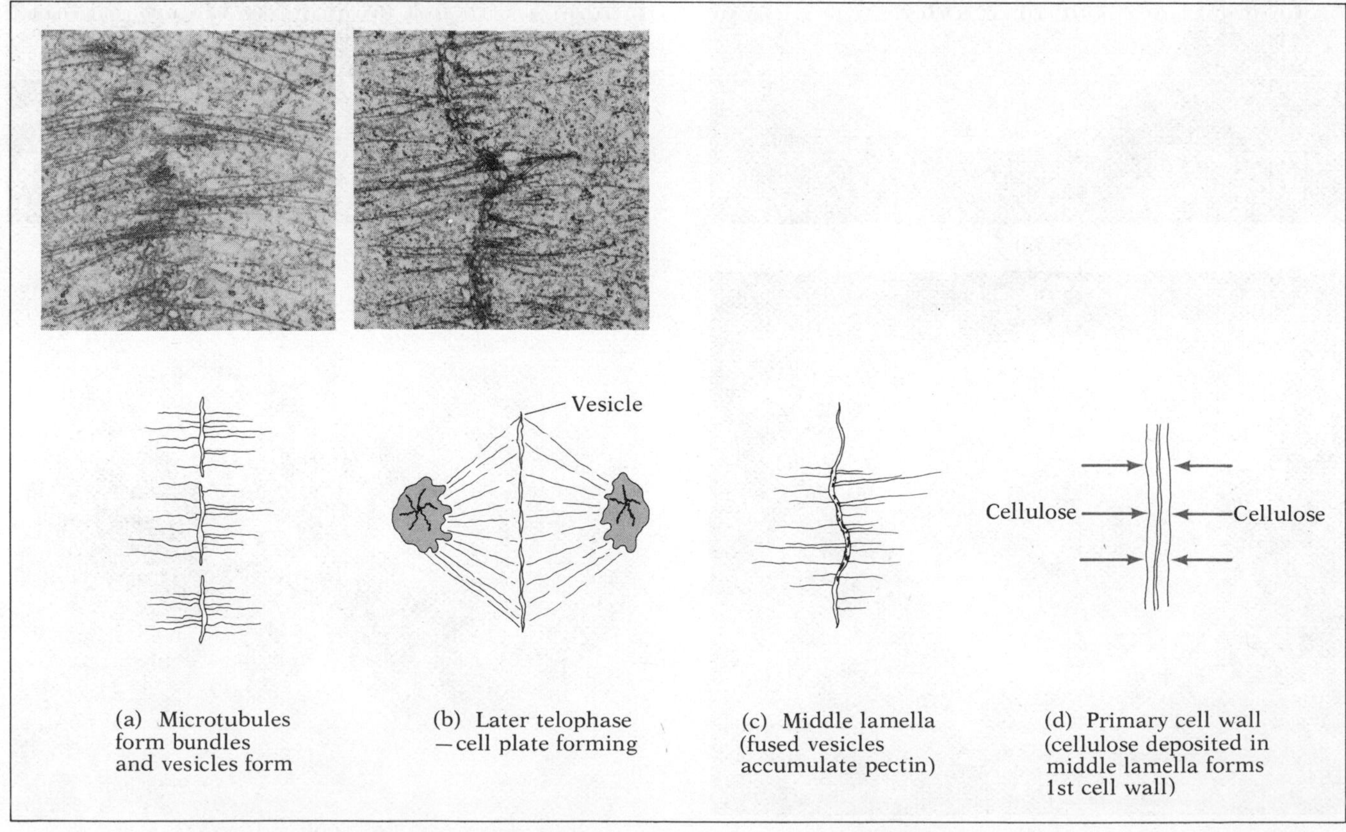
Vesicle
Cellulose
Cellulose
(a) Microtubules form bundles and vesicles form
(b) Later telophase —cell plate forming
(c) Middle lamella (fused vesicles accumulate pectin)
(d) Primary cell wall (cellulose deposited in middle lamella forms 1st cell wall)

group of genetically identical organisms is called a *clone.*

In multicellular organisms, mitotic cell division has several functions. Most of all, it allows an organism to grow to a respectable size while maintaining favorable surface-to-volume relationships (Chapter 4). Next, it allows for the specialization of cell types through cellular differentiation—remember that daughter cells are not necessarily identical. Some simple organisms reproduce through budding, as we described in Chapter 4. Other species can regenerate lost parts (e.g., salamanders and trees) or produce scar tissue (as do humans) through mitosis. Most of the vast energy expended in mitotic processes throughout the world of multicellular plants and animals, however, goes toward increasing the number of the cells of individual organisms: growth. Each mature plant or animal is the product of several to millions of mitotic events. We will return to the subject of development from fertilized egg to adult later in this book. For now, let's turn to a form of nuclear division that is intimately involved in the process of sexual reproduction.

8.3 Meiosis and Sex Cells

Meiosis

In higher organisms, both plants and animals, each individual is *diploid.* This means that every cell in the organism has a double set of genes and chromosomes: one set from a father and one set from a mother. When an individual reproduces sexually, a partner of the opposite sex must be involved. The two partners produce gametes (eggs and sperm if they are animals), which join to produce an offspring. Their offspring will receive one set of genes and chromosomes from each of the two parents.

So here are two related problems: First, if fertilization joins the genetic material of two parents, why is it that the amount of genetic material is not doubled in every generation? Second, if each parent in fact has two sets of genes and chromosomes, how is it that the offspring receives only one set from each parent?

Obviously, a solution to the second problem is also a solution to the first one. If each gamete enters into fertilization with only one set of genes and chromosomes instead of two, there is no doubling in each generation. The problem then becomes, how is the number of chromosomes halved—during gamete formation—from two complete sets per cell to one complete set per cell? The answer is, it's not easy. It requires a complex and orderly set of steps that is called *meiosis.* We can define meiosis as the process whereby a diploid (double) set of chromosomes is reduced to a *haploid* (single) set of chromosomes in a cell. In diploid organisms, the process guarantees that the chromosome number will remain stable from generation to generation and that each sexually reproduced offspring will get two complete sets of genetic instructions, each set being a random mix taken from the parents' own two sets of genetic instructions (Figure 8.12). Humans, for instance, have forty-six chromosomes in each cell—twenty-three from the father's sperm and twenty-three from the mother's egg. In meiosis and gamete formation, the forty-six chromosomes per cell will be reduced to twenty-three chromosomes per gamete. And each gamete will be different from every other gamete.

Homologous Chromosomes

The twenty-three chromosomes in each human gamete are all different from one another. That is, they contain different genes, which is to say they contain genetic instructions for different aspects of development and metabolism. In the human karyotype (Essay 8.1), if the cells have been properly stained, each of the twenty-three different types can be identified by its unique pattern of bands and constrictions. The complete set of genetic instructions is called the *genome* of the species.

The forty-six chromosomes in each human diploid cell, then, naturally come in twenty-three pairs—one member of each pair comes from the father's sperm and one from the mother's egg. The two members of each pair are called *homologues,* or *homologous chromosomes.* The two homologues are functionally equivalent in that they have the same kinds of genes, arranged in the same order. Thus, if the chromosome from the father's sperm has a gene that makes an enzyme involved in the development of pigment in the iris, and the gene is located on the chromosome at one particular place, then the homologous chromosome from the mother's egg will have a similar gene for an enzyme involved in the development of pigment in the iris and it will be located in the same position on the chromosome. The two homologues differ from each other only in the usually minor ways in which two individuals within a species might differ from one another. For instance, the chromosome from the father might carry a form of the eye color gene that tends to make eyes blue, and the homologous chromosome from the mother might

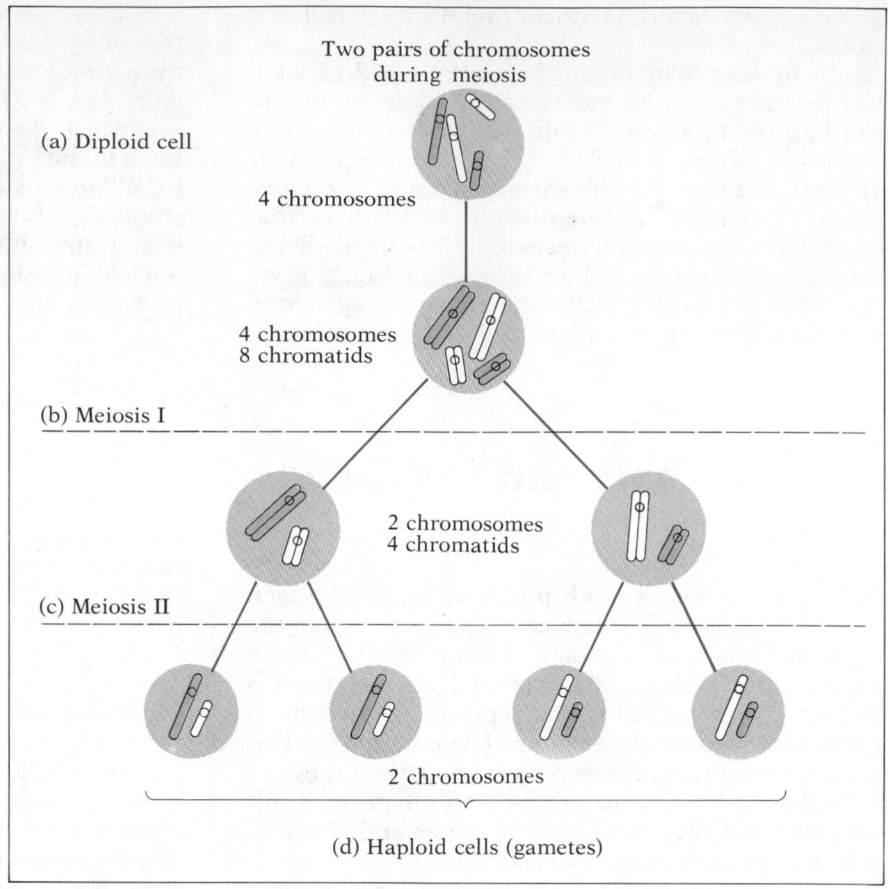

8.12 In this simplified scheme, a cell with two pairs of chromosomes enters meiosis. (a) In preparation, the cell goes into interphase, where each chromosome replicates. (b) The cell emerges at prophase with two pairs of chromosomes, but eight chromatids as in mitosis. (c) Next, the cell divides, but the centromeres do not. Thus, we have two cells, each with two half-pairs (but four chromatids). (d) Finally, with the second meiotic division, centromeres divide and the daughter cells now have two chromosomes per cell. Compare these to the cell at the top. These haploid cells will develop into gametes and remain haploid unless they become fertilized by another haploid cell. Fertilization produces a diploid fertilized egg.

carry a form of the gene that tends to make eyes green.

During meiosis, three important things happen to each pair of homologous chromosomes. First, the two homologues pair and are scrambled. These two processes are called *pairing* and *crossing over.* Briefly, what happens is that the homologue from the father lies next to the chromosome from the mother, and the two chromosomes fragment into shorter stretches of DNA and recombine to make new, unique chromosomes that belong neither to the individual's father nor to the mother, but are unique to the individual in which meiosis is taking place. We'll explain the scrambling process more clearly later on.

The third important thing that happens is that the scrambled chromosomes separate and go to different daughter cells, in such a way that each of the eventual daughter cells gets only one copy of a gene from each gene pair. In our blue eye/green eye example, some of the gametes would carry the blue eye form of the gene and some would carry the green eye form of the gene, but none of the gametes would carry both types. In order to get back to the paired condition again, the gamete will have to fuse with a different kind of gamete from some other source.

Where Meiosis Takes Place in Higher Organisms

In multicellular organisms, meiosis is restricted to the germinal tissues in specialized organs. In animals, these are the *gonads*—the ovaries of the female and the testes of the male. In flowering plants, meiosis occurs in flower parts—the ovary and the anthers.

The term *germinal* refers to reproduction, from the Latin word for seed. Germinal tissue is sometimes referred to as *germ-line* tissue, and the cells that might later give rise to gametes are said to be in the germ line. In plants, any actively dividing tissue, such as the meristem described in Chapter 21, can give rise to reproductive organs (flowers) and is therefore at least potentially germ-line tissue. In animals, however, a group of cells is set aside early in development, and germinal tissue will arise only from these cells. The germ line is the cellular line of descent through the *zygote* (fertilized egg), the early group of germinal cells that is set aside, the mature germinal tissue of the gonads, the meiotic cells, and finally the gametes—ready again to join in fertilization. The germ-line cells are, in a sense, potentially immortal. All other cells are *somatic cells* or *somatic tissue,* from the Latin word for body. These cells and all their cellular descendants will perish.

The gonads, of course, have their own supporting somatic tissue. The germinal tissue of the gonads is the *germinal epithelium* (Chapter 29) (*epithelium* is a general term for surface tissue, either internal or external). The germinal epithelium of the testes and the ovary, like any other tissue, undergoes repeated mitoses during development. There is nothing special about these mitoses.

In human females, mitosis in the germinal epithelium of the ovary ceases during fetal development. The remaining germ-line cells undergo most of the stages of meiosis, more or less simultaneously, during the final months of fetal development. When a baby girl is born, her germ-line cells, now referred to as *oocytes,* are well along in the process of gamete formation. They will stay in this condition, without finishing the final stages of meiosis, for about 10–12 years. Then they will be released, one or sometimes two at a time, about once a month for the next 35 years or so. Meiosis will not be completed unless the oocyte is fertilized. Although a woman may have several thousand oocytes, only about 400–500 will become mature, and of course most of these will not be fertilized.

In human males, the germinal epithelium of the testes forms the lining of long, convoluted tubes—the spermatic tubules (Chapter 29). After fetal development, both mitosis and meiosis are suppressed for perhaps 12 years. Then the germinal epithelium undergoes a type of assymetrical mitotic division that is common in epithelial tissues of all types. With each mitotic division, one of the daughter cells remains a germinal epithelium cell, capable of further mitoses, but the other daughter cell becomes differentiated as a *primary spermatocyte,* a specialized diploid cell ready to begin that once in a life-cycle event, meiosis. It will proceed through two divisions eventually forming four sperm cells. Mitosis and meiosis will continue to produce billions of sperm daily in the male germinal epithelium for the rest of the man's life.

Premeiotic Interphase

The *primary gametocyte* (also called a *primary spermatocyte* in mature males and a *primary oocyte* in fetal females) is actually through with mitosis for this life cycle, but it will undergo one more interphase before meiosis itself takes place. This interphase, called the *premeiotic interphase,* consists of the usual G_1 phase, S phase, and G_2 phase.

In many organisms, including humans, the premeiotic interphase also sees a strange and completely mysterious event called the *premeiotic condensation.* The chromosomes condense, as if they were about to undergo mitosis. And then *they decondense again,* and the cell reverts to the G_2 phase. No one knows what, if anything, the premeiotic condensation accomplishes, or whether it is important. Perhaps in some way it prepares the cell for meiosis.

The fact that the premeiotic interphase ends with the primary gametocyte in the G_2 phase means that when the cell enters the first stage of meiosis (*meiotic prophase*), its chromosomes are already replicated. Each chromosome will consist of two chromatids (ninety-two chromatids per cell all told in humans), just as in the prophase of mitosis.

Meiosis I: The First Meiotic Division

Meiotic Prophase: The First Stages

Meiotic prophase is terribly long and complex, compared with mitotic prophase. As in mitotic prophase, the chromosomes condense, but they take their time about it. Perhaps one reason it takes so long is that in the early stages, the chromosomes are shifting about in that murky soup of the nucleus, almost as though they were looking for something. That something turns out to be each other. Each chromosome must somehow find its chromosomal counterpart, the other member of its pair, its homologue.

How do homologues find each other? Chromosomes can't move under their own power. Even on the spindle they are pulled about by their centromeres, and the spindle is not present in prophase. Pairing can be seen to occur when homologous chromosomes touch each other in the right places. But first, there must be a lot of random groping. To speed things along, the entire nuclei of most organisms actually rotate during the early stages of meiosis, so that with time-lapse photography they look rather like dryer windows in a laundromat.

Ever since fertilization the two homologues of every chromosome pair have ignored each other, each going through mitosis after mitosis completely independent of the other, and interacting, if at all, only through the biochemical interactions of their gene products. But in early meiotic prophase all that changes, and suddenly chromosome number 13 begins to have an affinity for the other chromosome number 13, and number 7 for number 7, and so on. They have now changed so that when two homologues happen to touch in the same region, they hold. The result, over a period of time, is that the two homologues seem to come together like the two halves of a zipper. It turns out that there is a specialized structure, the *synaptinemal complex* (made up of protein and RNA), that bridges the two homologues in meiotic prophase, and it even *looks* like a zipper (see Figure 8.16). In photomicrographs "half zipped" homologues can be seen. This stage of meiotic prophase is called *zygotene,* meaning "yoked threads" (Figure 8.13).

The two homologous chromosomes are already partially condensed when the synaptinemal complex bridges them. Actually, as you already know, each chromosome consists of two chromatids, but they are not discernable at this stage. At this time, each partially condensed chromosome appears to be a series of irregularly spaced lumps strung together. The lumps are tightly coiled regions of various sizes, and are called *chromomeres.* The chromomeres appear in the same pattern in any two homologues. This is especially noticeable as the chromomeres match up in prophase.

Zygotene is not the only stage of meiotic prophase that has a specific name. There are a total of five stages. These are described here and in Figure 8.13.

1. Leptotene, when chromosomes first condense.
2. Zygotene, when homologous chromosomes pair.
3. Pachytene, when the paired chromosomes condense some more and crossing over occurs.

8.13 Stages of the meiotic prophase in humans. In oocytes, most of the stages of meiotic prophase are found only in fetal ovaries. Meiotic prophase in spermatocytes, however, occurs throughout the adult life of males. These remarkable photomicrographs were taken by Jean-Marie Lucciani of the Pasteur Institute, Paris, France. (a) Leptotene, human fetal oocyte. The chromosomes begin to condense. (b) Zygotene, human oocyte (from fetal material). The chromosomes are in the process of point-by-point pairing. Only the chromosomes are stained in this preparation. Note the precise alignment of chromomeres—the irregular bumps and swellings along the pairing maternal and paternal chromosomes. (c) Pachytene, human fetal oocyte. The paired chromosomes now consist of four aligned chromatids, two homologous pairs of chromatids. (d) Pachytene, human spermatocyte. The centromeres

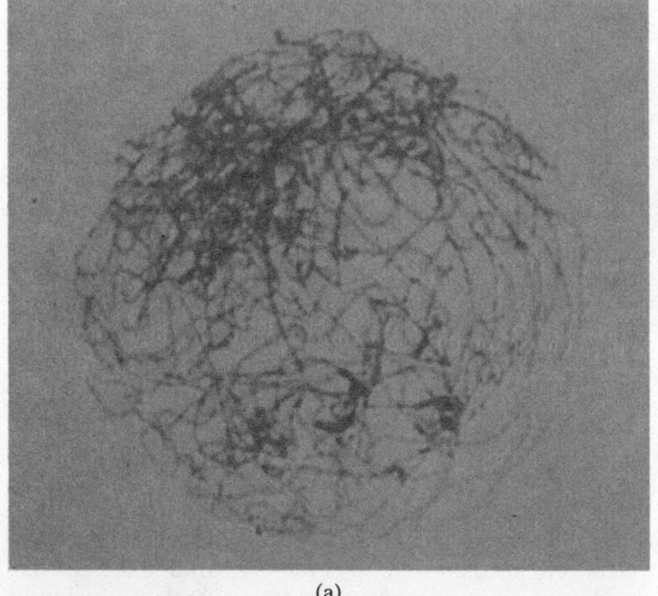
(a)

(b)

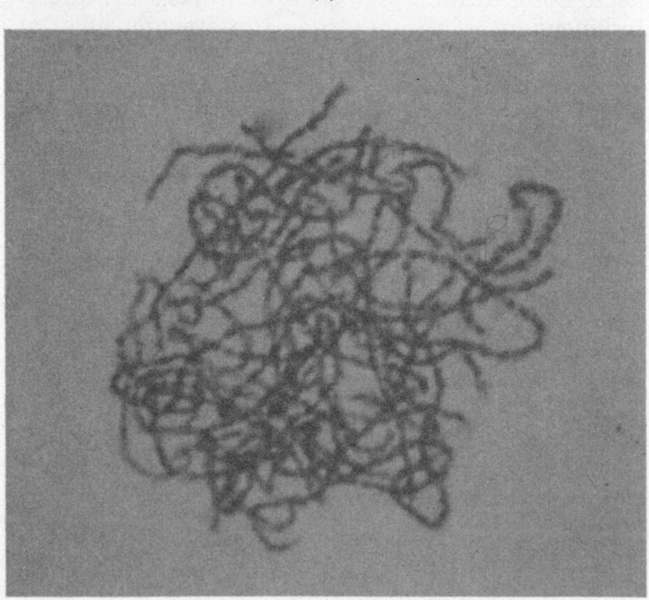
(c)

(d)

4. Diplotene, when the results of crossing over are first visible, and in which the four chromatids of each homologous pair can sometimes be seen. There may be some partial decondensation.
5. Diakinesis, in which the chromosomes are very, very condensed, and in which homologues come to be held together only at their ends.

After zygotene, the meiotic cell of a human has twenty-three separate genetic entities called *tetrads*, each consisting of a packet of four chromatids. (Pay careful attention to the phrasing here. You would do well to note these terms as we go along and try not to jump to conclusions.) Each tetrad, however, has two complete centromeres. So how many chromosomes are there during meiotic prophase? It is not considered polite to ask this question—even of a professional biologist—but if you recall that you can count chromosomes by counting centromeres, you will see that in humans there are still forty-six chromosomes throughout meiotic prophase.

appear as dark-staining spots. (e) Diplotene, human spermatocyte. The homologous chromosomes have begun to separate. Individual chromatids are not visible, but can you identify the chiasmata (places where homologous chromosomes have crossed over)? (f) Diplotene and zygotene, human fetal oocyte. Two nuclei are seen here. In diplotene, the chromosomes may partially decondense, so the nucleus that looks more spread out is the diplotene one. The violet spheres are primary and secondary nucleoli, seen only at this stage. The more condensed nucleus is in the zygotene stage; note that the ends of the chromosomes have not yet fully paired. (g) Diplotene, human oocyte. Here, the chromosomes are stained yellow and the nucleoli red, against a dark background. (h) Diakinesis, grasshopper testis (courtesy of James L. Walters). This is a stage intermediate between diplotene and metaphase I.

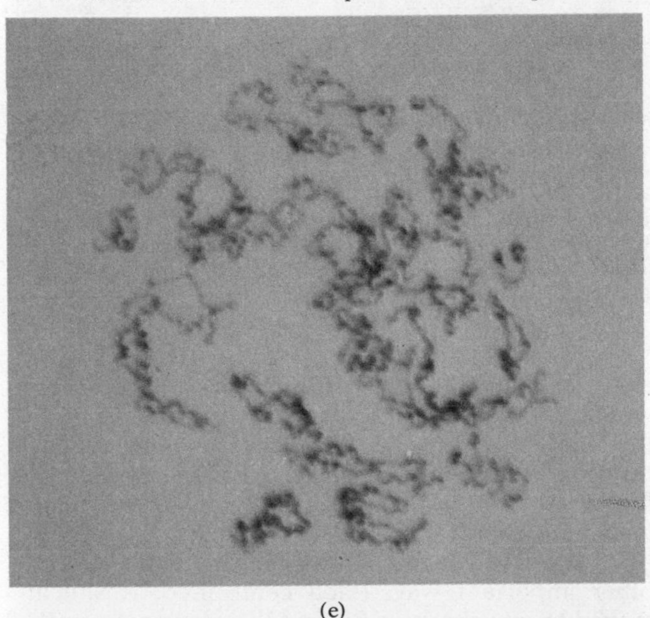

(e)

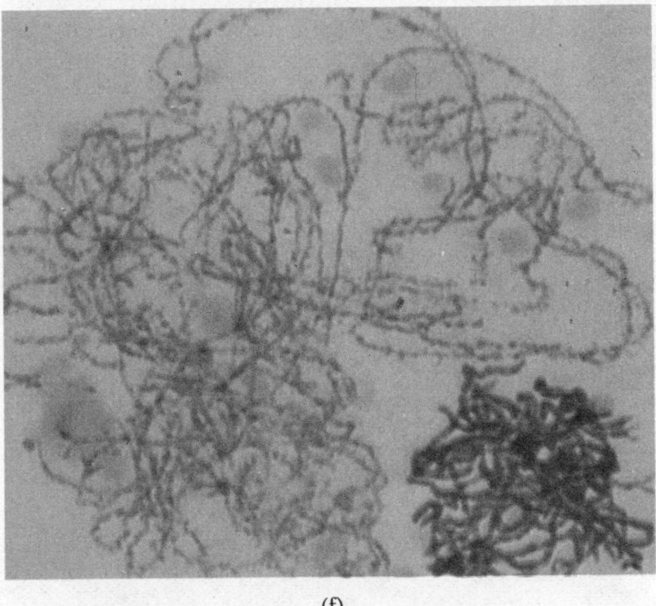

(f)

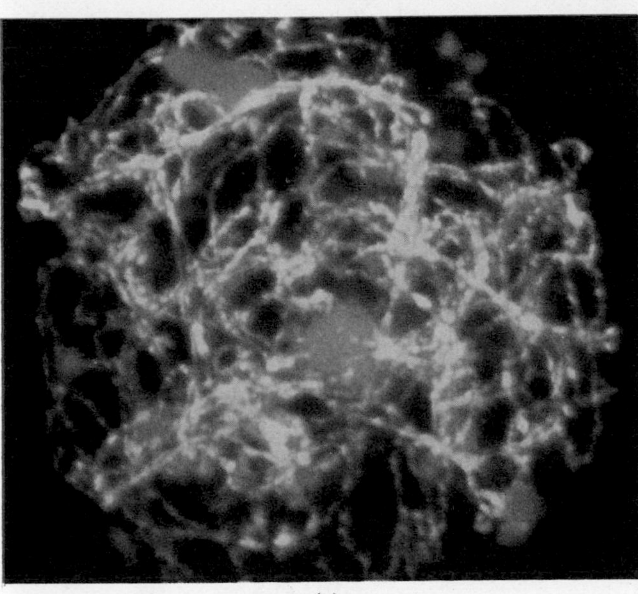

(g)

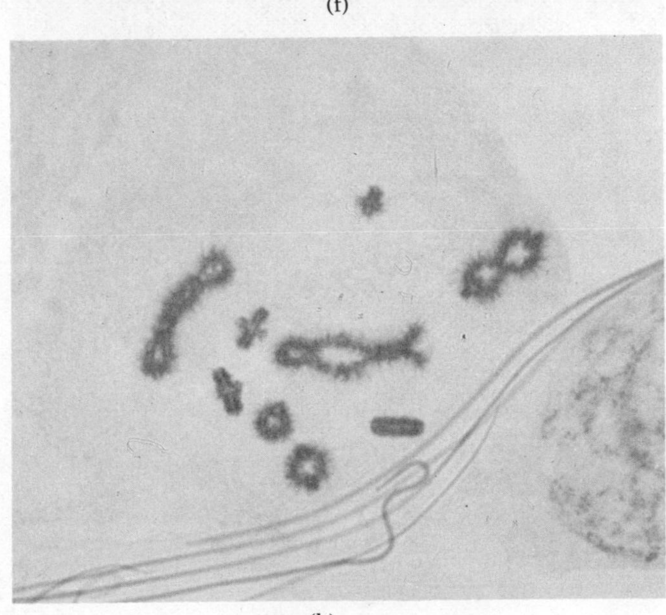

(h)

Metaphase I

That soon changes. The change takes place in two parts because meiosis actually consists of two cell divisions, complete with two metaphases, two anaphases, and two telophases. In the first metaphase (Figure 8.14), the tetrads are brought to the metaphase plate. Metaphase I is distinctly different from any mitotic metaphase. The two homologous centromeres appear to be separated from one another as far as possible, with only the ends of the chromosome arms touching. Furthermore, each centromere is attached to the spindle fibers of only one of the two poles.

Anaphase I

Anaphase I follows quickly (Figure 8.15), but as the genetic material separates in this stage, there is no division of the centromeres—just separation of homologues. As a result, half of the centromeres go to each pole, one from each pair of homologues. In humans, this means twenty-three chromosomes go to each pole. If we follow our rule that the number of chromosomes is equal to the number of centromeres, we can see that there has been a reduction (by half) of the number of chromosomes arriving in each daughter cell at the end of anaphase I. For this reason, the first meiotic division is sometimes known as the *reduction division.*

Anaphase I is unlike any other anaphase in another important way. Each centromere is still attached to *two* chromatids. And they are not even the same two chromatids that were attached to it at the beginning of prophase I. To see why that is so, we have to go back to where we left off our discussion of meiotic prophase—at the end of zygotene—and take a closer look at what has happened to those chromatids in pachytene, diplotene, and diakinesis.

8.14 Metaphase of meiosis I. When the chromosomes move to the metaphase plate in meiosis I, they align in a manner that indicates two distinct differences from any other metaphase. The repulsion of homologous chromosomes is very pronounced, as though the pairs were trying to get away from each other. They eventually touch only at the tips of the chromosome arms. In addition, they attach to the spindle fibers of only one pole each, which causes the homologous centromeres to pull outward from one another. This is different from metaphase of mitosis and meiosis II, where a pole-to-centromere spindle fiber attaches to each side of each centromere. In this preparation of metaphase I chromosomes of a human spermatocyte, the spindle and the centromeres cannot be seen. Not all the chiasmata have fully terminalized. The long figure near the center of the spread, without any chiasma, is formed by the X and Y chromosomes paired at their tips.

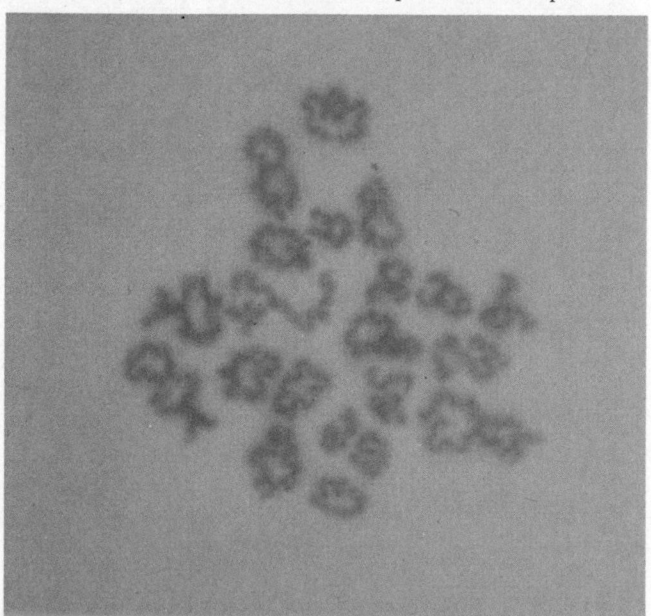

Metaphase, human spermatocyte

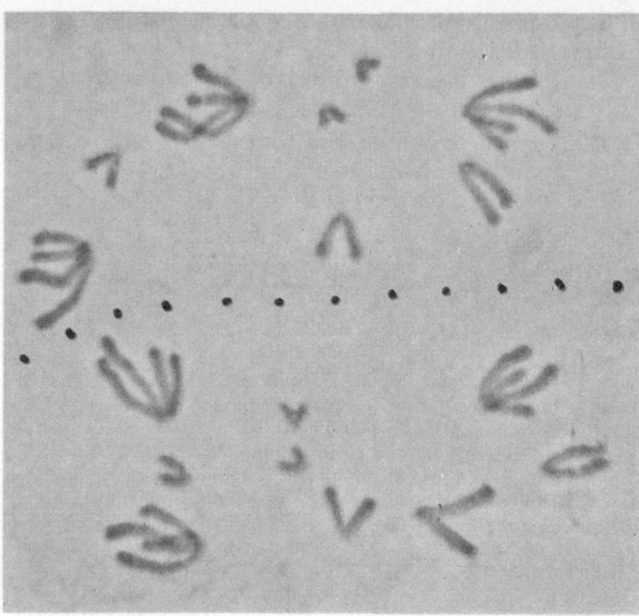

Anaphase, grasshopper testis

8.15 Anaphase of meiosis I proceeds without the division of centromeres, as is always seen in mitosis. The sister chromatids, still attached by their centromeres, have separated from their homologues. Along with the separation go all of the crossover products formed during prophase. Two homologous sets of grasshopper chromosomes have separated, as shown by the dotted line. The X chromosome (lower right) lacks a homologue in grasshopper males. Can you match the other pairs?

A Closer Look at Meiotic Prophase: Crossing Over

Recall that the cell enters prophase I with a maternal and a paternal chromosome making up each homologous pair, but that these become thoroughly scrambled in zygotene. In the next stage (pachytene), the four chromatids are scrambled, or *cross over.* It turns out that at any point along the DNA backbone of a chromatid, special enzymes can cause the chromatid to unwind so as to reveal two single strands of DNA. A complex series of events then occurs, which we won't describe here, except to say that at each crossover event, several thousand base pairs of the Watson strand of one chromatid pair with their counterparts on the Crick strand of one of the chromatids of the homologous chromosome. Then there are more breakages and the chromatids untangle themselves again. Meanwhile, other enzymes are busy repairing the breaks in the DNA strands.

The final result is that, at randomly located but exactly equivalent places along the chromatids, the DNA of one of the original four chromatids is covalently bonded with, and continuous with, the DNA of one of the other three chromatids. The process is reciprocal, so that at the point where one chromatid leaves off being paternal and begins being maternal, the opposite chromatid leaves off being maternal and begins being paternal, in that they join the centromere descended from the other parent. Each such breakage and rejoining event is called a *crossover* (Figure 8.16). In human cells in meiosis, there is an average of about ten crossovers for every meiotic tetrad. In other species, there may be fewer. But it clearly is no longer appropriate, after meiotic prophase, to refer to individual chromatids as maternal or paternal. The scrambling process produces entirely new chromatids and is an important source of variation in the population. This constantly renewed variation is crucial to the evolutionary process. The scrambling process, of course, is the way that both maternal and paternal genes end up on the same chromosome.

The results of crossing over are seen first in the diplotene stage. The *sister strands* (identical DNA

8.16 Crossing over and chiasma formation. (a) In pachytene, following the zipperlike pairing of homologous chromosomes in the previous zygotene stage, the maternal and paternal chromosomes are intimately associated and bound together by the synaptinemal complex. Each of the two chromosomes consists, in turn, of two sister chromatids, which are even more intimately associated and cannot be distinguished even under the electron microscope. Thus the whole complex consists of four strands. Breakage and reunion of the DNA strands in pachytene is known to occur in pachytene, although it cannot be seen under the microscope. (b) In a later stage, the homologous maternal and paternal chromosomes separate (where they can), but the sister-strand regions are still tightly held together as indicated. Wherever crossing over has occurred (physical rejoining of homologous chromatids), the chromosomes are still held together. The visible evidence of this exchange is a cross-shaped conformation called a *chiasma* (plural, *chiasmata*). Here two chiasmata are shown as they would appear in the diplotene stage. (c) If the four chromatids could be unwound and separated, and the regions of maternal and paternal origin indicated, they would look like this. Actually, the four chromatids will separate in later stages of meiosis. Any one exchange involves only two of the four chromatids, but ordinarily there are many exchanges between each pair of homologues and all four chromatids become scrambled.

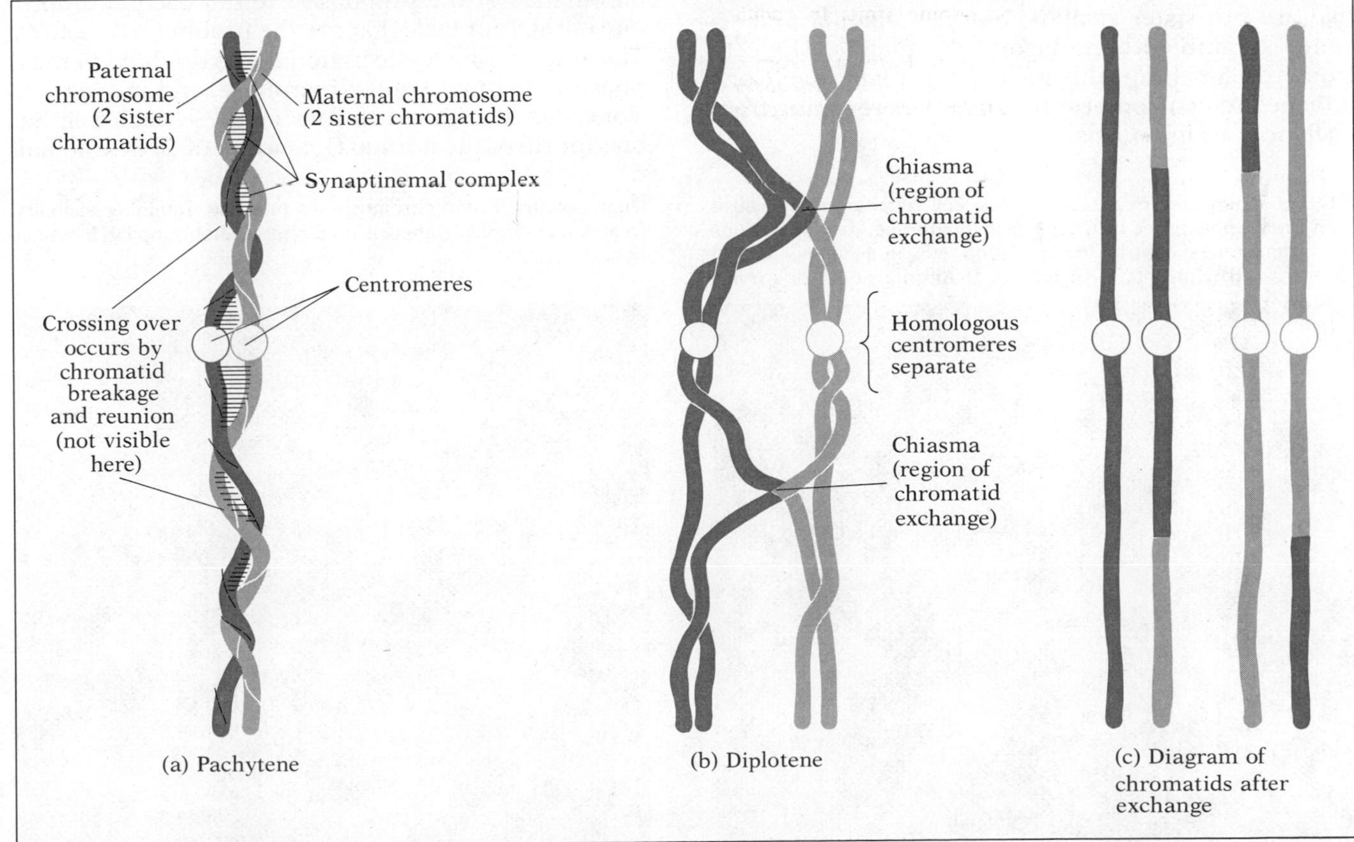

double helices produced by replication) of each of the two homologues are still held tightly together, but the homologues begin to separate in diplotene. Where a chromatid exchange has taken place, however, the homologues form a cross-shape configuration, called a *chiasma* (plural, *chiasmata*).

Telophase I and Meiotic Interphase

Anaphase I is followed by telophase I (Figure 8.17), complete with the reorganization of nuclei, decondensation of chromosomes, duplication of centrioles, and cytoplasmic division. The cell then enters its one and only *meiotic interphase.* (The previous interphase is known as the *premeiotic interphase,* since meiosis is considered to start with prophase I.)

Meiotic interphase is unlike other interphases in one very important detail. *There is no S phase* (in other words, no DNA replication). Meiotic interphase may be extremely brief or extremely long, and the chromosomes may or may not completely decondense, depending on the organism.

Meiosis II

Prophase II

Prophase II is rather like a mitotic prophase. The chromosomes condense and there are only two chromatids attached to each centromere. This time, the two are not sister strands, but two distinct and recently scrambled homologous chromatids. The centromeres head for the metaphase plate, each one attached to centromeric spindle fibers reaching from each pole, as in mitosis.

Metaphase II and Anaphase II

Then, in metaphase II, the centromeres, which failed to divide in metaphase I, finally divide. The reorganized chromatids now become full-fledged chromosomes. The second meiotic division is sometimes referred to as the "equational" division to contrast with the first or "reduction" division, but it is important to remember that the chromatids that have separated in anaphase II are not sisters and are not genetically identical. So, actually, both anaphase I and anaphase II are distinctly different from mitotic anaphase.

Telophase II

Anaphase II proceeds into telophase II as the nuclear membranes reform and the chromosomes decondense, but in this telophase the centrioles usually do not replicate. (Why do you suppose this is?)

Then the four resulting cells enter interphase, where they are destined to remain in G_1. DNA replication is over until fertilization triggers the continuation of the process (Figure 8.18).

Summing Up Meiosis

So here we are. The four newly scrambled, unique chromatids of the prophase I tetrad each end up in one of the four cells that are the products of meiosis. The amount of genetic material was doubled in the S phase of the premeiotic interphase, and in two divisions was distributed to four cells. So each cell has one-fourth of the normal G_2 amount of DNA, and half

8.17 (a) Telophase of meiosis I: The cell retreats from meiosis with the reappearance of the nuclear membrane, decondensation of chromosomes, centriolar duplication, etc., in a manner similar to mitosis. (b) Interphase of mitosis I: An interphase of varying length occurs. Some chromosomes may lag, retaining visibility. In meiotic interphase, there is no S phase—that is, no DNA replication.

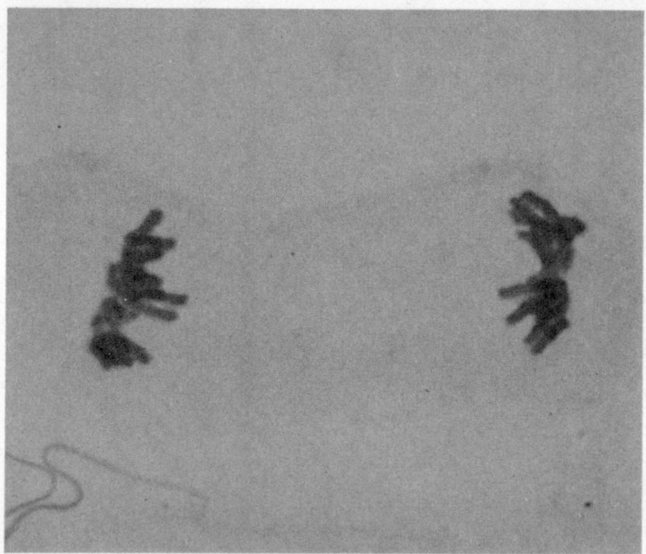

(a) Telophase I, grasshopper testis

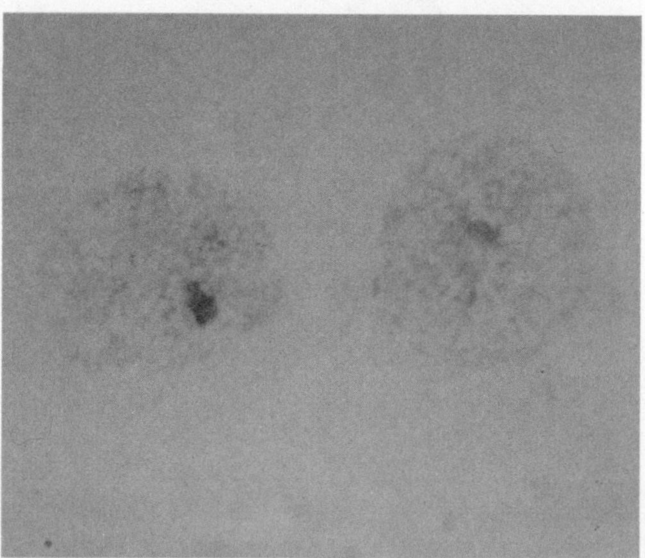

(b) Interphase, grasshopper testis

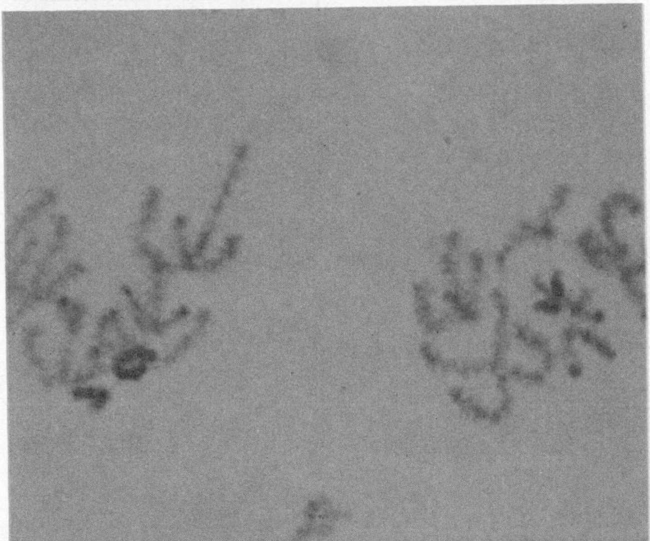

(a) Prophase II, grasshopper testis

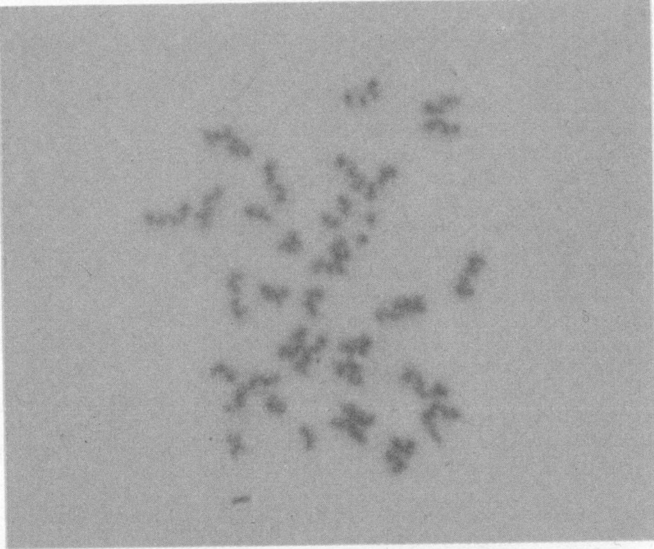

(b) Metaphase II, human spermatocyte

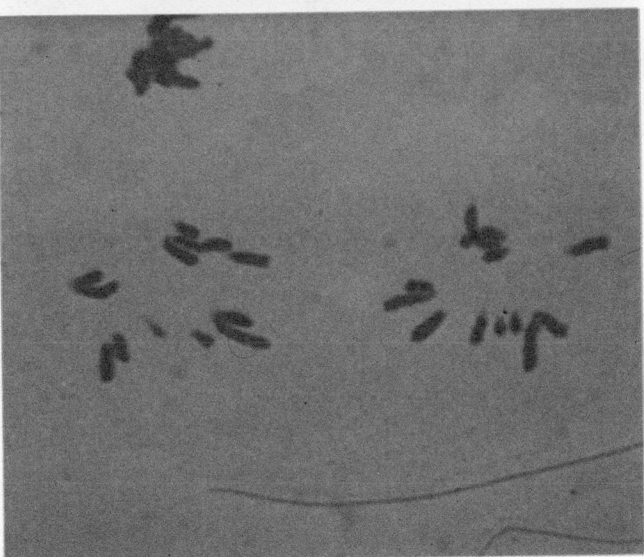

(c) Anaphase II, grasshopper testis

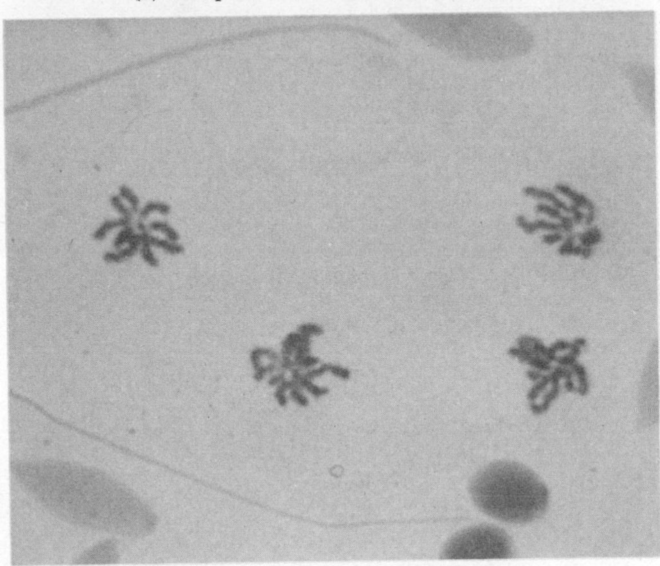

(d) Telophase II, grasshopper testis

8.18 Meiosis II is strikingly similar to mitosis. (a) Prophase II: The simple prophase of mitosis is recalled here. (b) Metaphase II: In this meiotic metaphase, the chromosomes, consisting of sister chromatids attached at their centromeres, align on the metaphase plate. Centromeric fibers extend from both sides of each centromere. (c) Anaphase II: Now the centromeres have separated. Each chromatid, now a chromosome, has separated from its formerly attached member and they migrate in opposite directions. (d) Telophase II: Each nucleus now contains a single chromosome from each of the original pair. The cells retreat from telophase in the usual manner. The four cells emerging from meiosis are haploid and will remain so until fertilization.

the normal G_1 amount of DNA. Each of these cells also has only one of each of the different types of chromosomes of the species (for example, twenty-three chromosomes in human gametocytes). Such cells are haploid (*haplo-* means "single" in Greek).

Notice that four haploid cells are produced. These four products of meiosis are called *gametocytes,* and they will undergo cellular differentiation to become gametes, at least in males. Meiosis in females, which we will get to in the next section, has a few additional wrinkles.

All these DNA duplications and cell divisions get confusing, so perhaps the summary in Table 8.3 will help. Figure 8.19 compares meiosis with mitosis.

Meiosis in Females

The chromosomes behave the same way in meiosis in males and females, so most of the above account applies equally to both sexes. But there are some differences. For example, timing is different in males and females (perhaps you've noticed that). And the cytoplasmic events are quite different in the sexes.

As you read a few pages back, meiosis is begun but is only partly completed in human females shortly before birth. All oocytes are frozen in the diplotene stage of meiotic prophase, just before metaphase I.

8.19 The differences between mitosis and meiosis become apparent when they are seen together. We have already described the longer and more elaborate prophase of meiosis I, so here we will simply point out that in mitosis, homologous chromosomes have no particular interest in each other and are arranged randomly in the nucleus, while meiotic homologues have paired up.

At metaphase, the alignments are also different. Mitotic chromosomes align randomly with the spindle fibers attached on each side. In meiotic metaphase I, attachments form on one side only.

At anaphase the mitotic centromeres divide and chromatids separate, while in anaphase I of meiosis they do not. Sister chromatids remain attached and homologous chromosomes, still doubled, move apart.

Mitosis is complete with telophase. The cell enters interphase and its DNA will be replicated in the S phase. In addition, the manner of division at anaphase assures that each daughter cell will have the same number of chromosomes as the cell which produced it. The meiotic daughters will not enter an S phase and no DNA replication occurs.

The meiotic event is only half finished. A second division will occur with centromeres now dividing and sister chromatids (now chromosomes) moving to opposite poles, just as happened in mitosis. Unlike mitosis, however, the four daughter cells will be haploid with exactly half the chromosome number of the mitotic daughter cells.

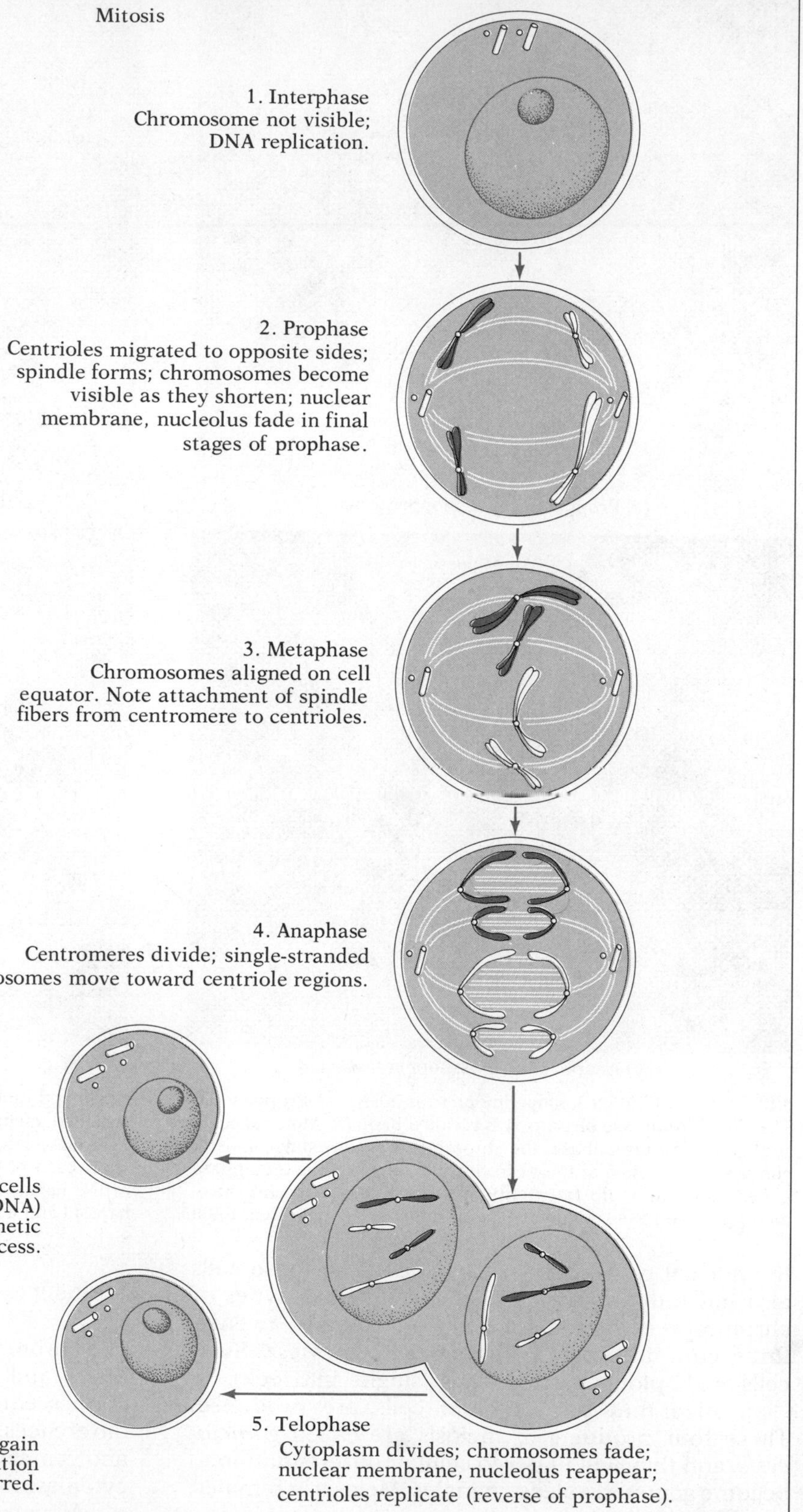

Mitosis

1. Interphase
Chromosome not visible;
DNA replication.

2. Prophase
Centrioles migrated to opposite sides;
spindle forms; chromosomes become
visible as they shorten; nuclear
membrane, nucleolus fade in final
stages of prophase.

3. Metaphase
Chromosomes aligned on cell
equator. Note attachment of spindle
fibers from centromere to centrioles.

4. Anaphase
Centromeres divide; single-stranded
chromosomes move toward centriole regions.

5. Telophase
Cytoplasm divides; chromosomes fade;
nuclear membrane, nucleolus reappear;
centrioles replicate (reverse of prophase).

6. Daughter cells
Two cells of identical genetic (DNA)
quality; continuity of genetic
information preserved by mitotic process.

These cells may divide again
after growth and DNA replication
has occurred.

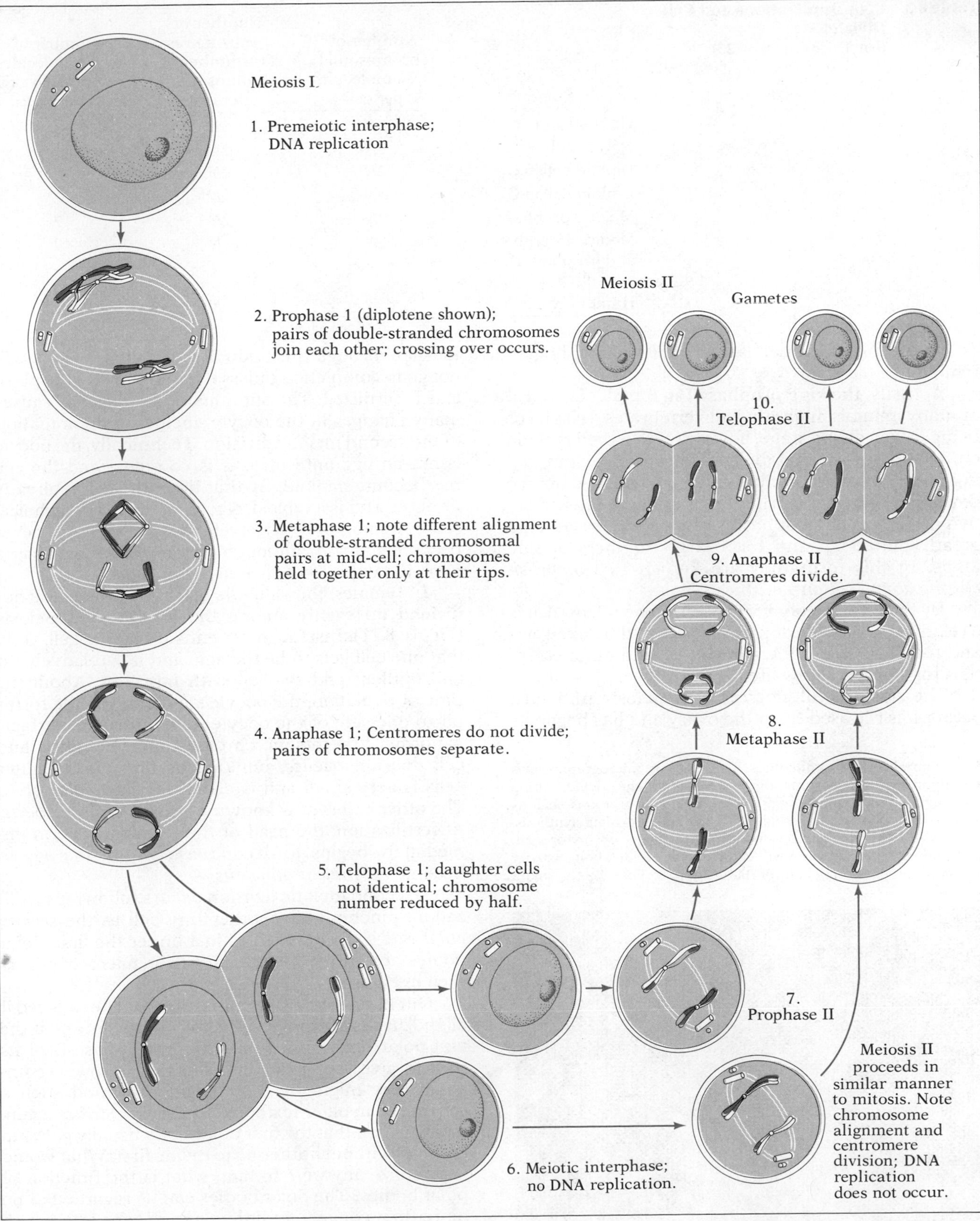
Meiosis I
1. Premeiotic interphase; DNA replication
2. Prophase 1 (diplotene shown); pairs of double-stranded chromosomes join each other; crossing over occurs.
3. Metaphase 1; note different alignment of double-stranded chromosomal pairs at mid-cell; chromosomes held together only at their tips.
4. Anaphase 1; Centromeres do not divide; pairs of chromosomes separate.
5. Telophase 1; daughter cells not identical; chromosome number reduced by half.
6. Meiotic interphase; no DNA replication.
Meiosis II proceeds in similar manner to mitosis. Note chromosome alignment and centromere division; DNA replication does not occur.
7. Prophase II
8. Metaphase II
9. Anaphase II Centromeres divide.
10. Telophase II
Meiosis II
Gametes

Table 8.3 DNA Duplications and Cell Division (for Humans $N = 23$)

Stage	Number of chromosomal DNA molecules per cell	Number of centromeres (= number of chromosomes) per cell	Number of DNA molecules attached to each centromere
Haploid gamete	N	N	1
New zygote	$2N$	$2N$	1
Diploid cell in G_1	$2N$	$2N$	1
Diploid cell in G_2	$4N$	$2N$	2
Meiotic prophase I	$4N$	$2N$	2
Meoitic interphase and prophase II, per cell	$2N$	N	2
Haploid gametocyte	N	N	1

So meiotic prophase can last 45 or 50 years in human females.

Actually, the long prophase I in females is typical of many animals, including all vertebrates. In spite of being frozen at one stage, however, the paired meiotic chromosomes are metabolically very active, producing large amounts of ribosomes and mRNA for the developing oocyte. During this period of intense activity, the female prophase I chromosomes become greatly puffed up and form large numbers of side loops. In this form they are known as *lampbrush chromosomes* (Figure 8.20).

By the time the oocyte is ready to be released, it is a relatively huge cell, actually visible to the naked eye, and full of yolk, mRNA, ribosomes, and other essentials for starting a new life.

The oocyte will not resume meiosis until just before it is released from the ovary and has begun its journey through the oviduct. Even then meiosis will not go to completion unless the oocyte meets a sperm and is fertilized. The stimulus of fertilization causes many changes in the oocyte, including the initiation of the second meiotic division. Technically, the oocyte is not an egg until meiosis is complete and the cell has become haploid. By that time, the cell is already fertilized and is a diploid zygote—so in a very technical sense, human oocytes never become eggs and never become haploid, although the fertilized egg has a haploid nucleus—the *female pronucleus.*

8.20 Lampbrush chromosomes. This electron micrograph shows activity in a meiotic chromosome from a diplotene oocyte. During the lengthy prophase, the egg's DNA will open in loops such as those shown here, and produce RNA for active protein synthesis. Some of the proteins will be stored in the massive egg cytoplasm, while others (the enzymes) will carry on varied metabolic activity related to preparing the egg for its future role. Note the chiasmata (arrows).

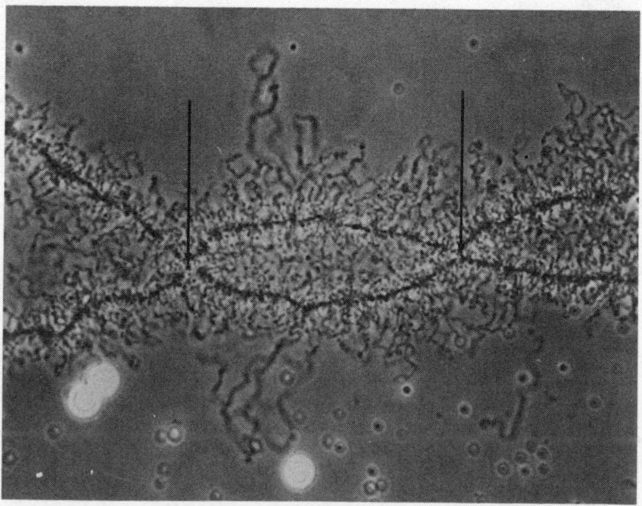

In females, the valuable cell constituents are not divided up evenly among the products of meiosis (Figure 8.21). Instead, most ends up in one cell. Only that one cell gets to be the egg, and it is relatively fat and opulent and swollen with nutrients. About the time of ovulation, the oocyte's meiotic spindle forms off to one side of the oocyte, just under the surface. A normal first meiotic chromosome separation and cell division occurs, but one of the two daughter cells is very small and is called the *first polar body.* The other cell is now known as the *secondary oocyte.* At fertilization, the head of the fertilizing sperm immediately begins to decondense within the egg to become the *male pronucleus.*

A second meiotic division occurs following fertilization, pinching off another tiny cell as the *second polar body,* which emerges just under the first. Polar bodies are clearly visible with light microscopes, as seen in Figure 8.22.

Our natural love of symmetry would be satisfied if all the divisions of meiosis were completed, and if the first polar body joined in with a metaphase II of its own to give four products of meiosis, however unequal they might be. You might even find such a scheme in an older textbook. But reality strikes again, and we learn that the first polar body usually gives up and dies without bothering to divide first. What would be the use, anyway? In fact, what is the function of polar bodies? The polar bodies can be regarded as no more than convenient little garbage cans into which

8.21 By the time a baby girl is born, her ovaries have produced all of the oocytes that they will ever form. Each oocyte begins meiosis and stops in diplotene of prophase I, remaining there until about the time of ovulation. Even then the egg must be penetrated by a sperm cell if meiosis is to go to completion. The cell is then arrested in metaphase II and will not complete meiosis unless fertilization occurs. Note in both anaphase I and II how meiosis occurs at the edge of the cell and the very unequal cleavage of cytokinesis. Meiosis I forms the first polar body. Following fertilization, a second polar body is produced directly beneath the first. Both polar bodies degenerate without further cell division. The sperm head decondenses to become the male pronucleus, which will fuse with the female pronucleus.

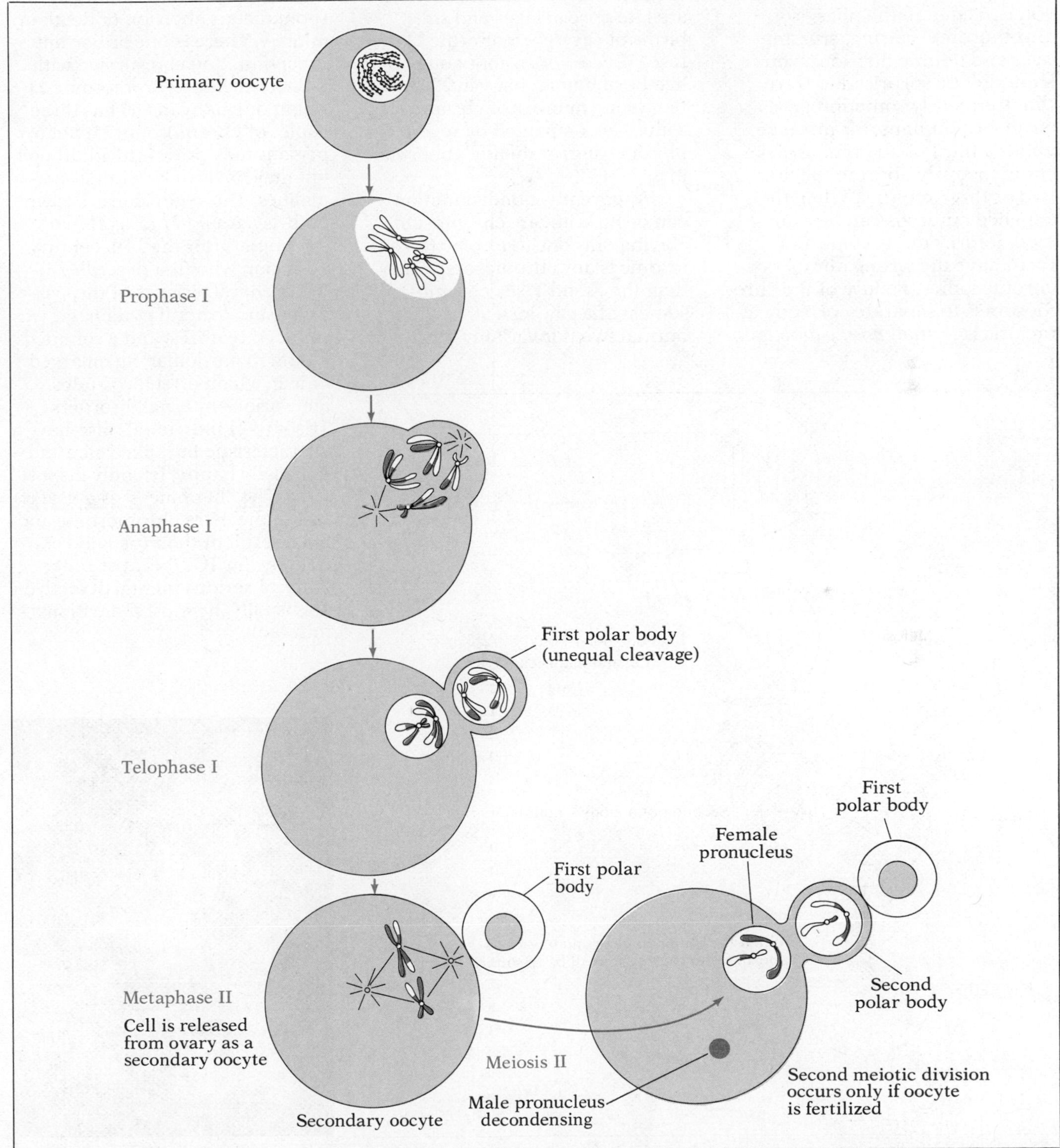

Essay 8.3 When Meiosis Goes Wrong

Meiosis is a much more complicated process than mitosis. When you consider all the phases of chromosome pairing, crossing over and double divisions, you shouldn't be surprised to learn that frequently something goes wrong. In humans, for instance, about a third of all pregnancies spontaneously abort in the first two or three months. When the expelled embryos can be examined, it turns out that most of them have the wrong number of chromosomes. Failure of the chromosomes to separate correctly at meiosis is termed *nondisjunction.*

But not all failures of meiosis result in early miscarriage. There are late miscarriages and stillbirths of severely malformed fetuses. Even worse, about one live-born human baby in 200 has the wrong number of chromosomes, accompanied by severe physical and/or mental abnormalities.

Apparently, nondisjunction can occur with any chromosome. Having only one *autosome* (an autosome is any chromosome other than the X and Y sex chromosomes) of a pair instead of the normal two is invariably fatal. Having three instead of two is almost always fatal, resulting in spontaneous abortion or death in infancy. There is one important exception. You can survive with three of the tiny chromosome 21. About one baby in 600 has three copies of chromosome 21. Such persons may grow into adulthood, but they have all kinds of abnormalities. The syndrome is known both as *trisomy-21* or as *Down's syndrome,* after the 19th century physician who first described it. Other characteristics of the syndrome are general pudginess, rounded features, and a rounded mouth in particular, an enlarged tongue which often protrudes, and various internal disorders. Trisomy-21 individuals also have a characteristic barklike voice and unusually happy, friendly dispositions. The "happiness" is a true effect of the extra chromosome and not a result of their (usually) extremely low IQs, because other kinds of serious mental defectives are usually, by most indications, miserable.

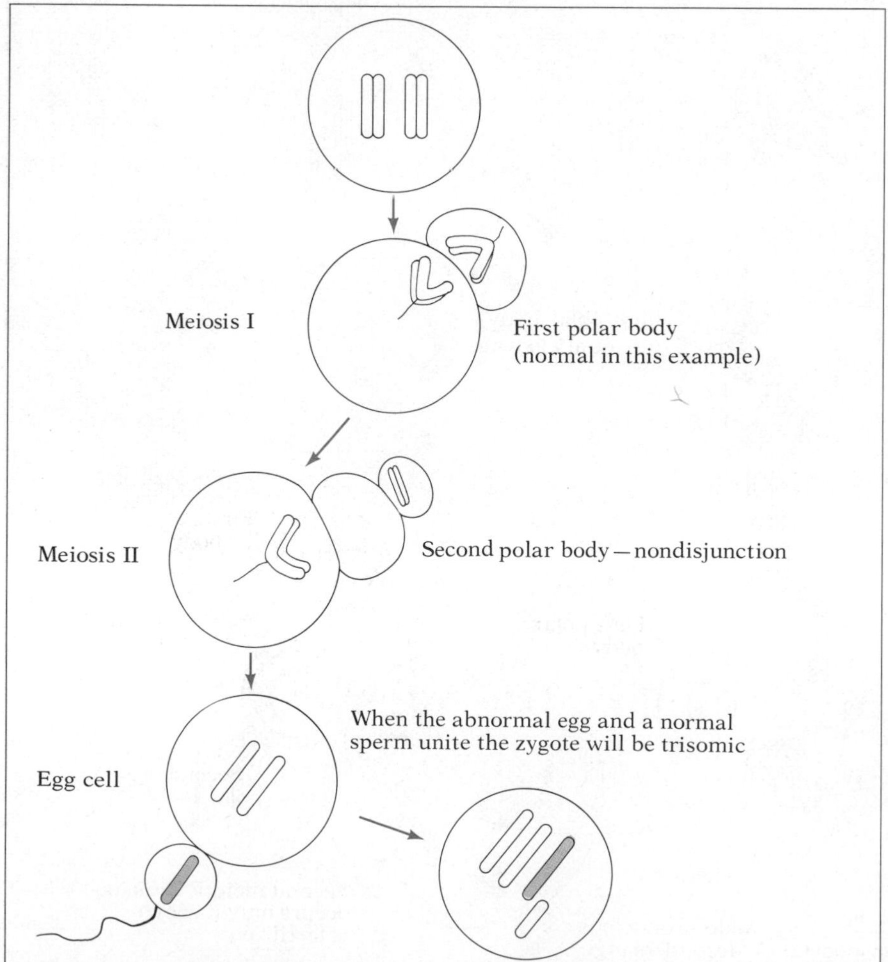

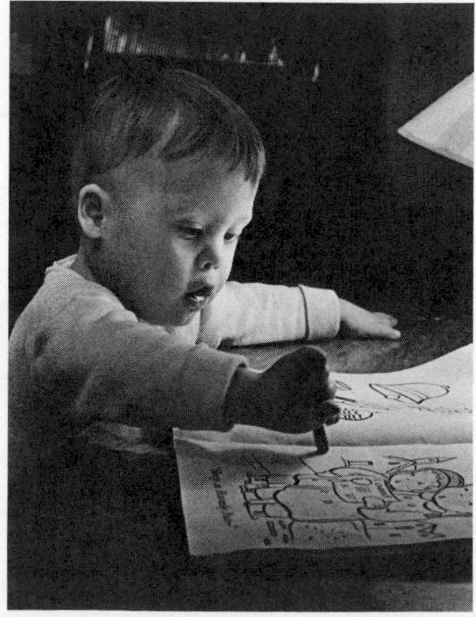

Trisomy-21 occurs most frequently among babies born to women over 35 years old, affecting up to 2% of such births. The age of the father apparently has a much smaller effect. We can guess that the much prolonged prophase I of the human oocyte might have something to do with this. Remember, the oocyte would have been arrested in this stage for 35 years.

Wrong numbers of X and/or Y chromosomes in humans result in live babies and abnormal adults. There are many varieties of sex chromosome conditions. The shorthand designations of the most common are as follows:

XX	Normal female (two X chromosomes)
XY	Normal male (one X, one Y)
XO	Turner's syndrome female (one X, no Y)
XXY	Klinefelter's syndrome male (2 Xs, 1Y)
XYY	Extra Y, or XYY, syndrome male
XXX	Trisomy-X, or XXX, female

In addition to the above, there are many more extreme situations, such as XXYY, XXXY, XXXYY, XXXX, XXXXX, XXXXYY, and so on, each syndrome having its own distinguishing characteristics. However, we can make four generalizations: First, one must have at least one X chromosome to live. Second, the presence of a Y causes the individual to develop as a male, and the absence of a Y causes the individual to develop as a female. Third (and this is probably why these syndromes are not fatal), all but one of the X chromosomes will condense into heterochromatin, and be visible as a Barr body when stained, so that XO females lack a Barr body, XXY males have one, XXX females have two, and XXXXXYY males have four, and so on. Fourth, the more sex chromosomes a person has, the taller he or she will be, so that XO females are tiny, while XXX, XXY, and XYY individuals are usually much taller than chromosomally normal men and women, and XXYY men are huge.

XO individuals are phenotypically female, but do not develop ovaries. They remain sexually immature as adults, unless given hormones. XXY males are tall and have small, imperfect testes and low levels of male hormones. They may have female-like breast development and somewhat feminine body contours. XXX females are tall and frequently sterile, but appear otherwise normal. XYY males appear normal except for their extreme height and for a tendency toward severe acne. They are also generally sterile. On the average, they have somewhat reduced IQs and, in common with other low -IQ groups, they average significantly increased criminal arrest records. At one time there was speculation that XYY males had so-called "genetic criminal tendencies," but further analysis showed that an XYY male is no more likely to be arrested than an XY or XXY male of the same IQ.

Actually, all three types of individuals with three sex chromosomes (XXX, XXY, and XYY), in addition to being taller than average, have somewhat depressed IQ scores, though the distributions overlap the population mean of 100. XXX females and XYY males cannot usually be diagnosed except by karyotyping, so these relatively innocuous conditions have not been well-studied; the trisomy-X condition is most often found when a patient complains that she seldom or never menstruates, but the condition of XYY men is usually not discovered unless some researcher happens to include them in a population survey. XXY males (Klinefelter's syndrome males), however, make up a substantial percentage of the institutionalized mentally retarded and are relatively easily diagnosed. Persons with four or more sex chromosomes, such as XXXY, are usually severely retarded.

The Frequency of Some Common Sex Chromosome Abnormalities in the United States

Syndrome	Number of chromosomes	Estimated number born in 1973	Estimated number living in the United States in 1974
Triple X (XXX)	47	3,000	200,000
Turner's (XO)	45	1,300	85,000
Klinefelter's (XXY)	47	8,000	530,000
Double Y (XYY)	47	12,000	850,000

Adapted from G. J. Stine, *Biosocial Genetics* (New York: Macmillan, 1977).

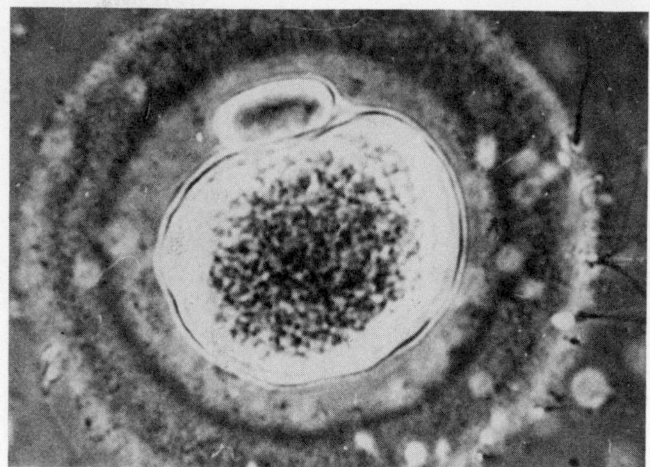

8.22 A living human egg at the moment of fertilization. One polar body is seen at top, and numerous sperm in contact with the egg are clearly visible.

an unwanted three-quarters of the genetic material of the primary oocyte can be dumped, leaving one big happy egg with its haploid complement of newly scrambled chromosomes and the lion's share of its predecessor's cytoplasm and nutrients.

Why Meiosis?

Meiosis, we now know, is a difficult and complicated process, and all too frequently something goes wrong (see Essay 8.3). So what's it all about? Why do organisms bother? Perhaps you haven't noticed that we have been dealing with sex all along. Meiosis, after all, is part of sex. So the next question is, what is sex good for in the long run? (You probably have an answer about the short run.)

But let's talk about the long run. In the long run we are all dead. But our genes live on. When we leave offspring (this "we" is used here to include all us sexual beings), we pass our reproductively "successful" genes on to our descendants. The problem is, our descendants may not find the world to be the same as the one *we* lived in; the weather changes, new diseases show up, and unexpected competitors eat our food. Obviously, no one particular combination of genes, including ours, will be adequate to meet all new situations. So, according to evolutionary theory, sexual reproduction comes to the rescue. Sexual recombination keeps reshuffling the genes of all the successful individuals in the population, so that virtually infinite possible combinations of genes will be produced. Most of the new combinations will actually be worse than the old ones. That's unfortunate, but in the long run it's the cost of success. The winning evolutionary strategy is to cover as many bets as possible by having highly variable offspring. Then, when the weather changes or the Creeping Purple Flu shows up, perhaps not all of your descendants will be wiped out, because some of them are likely to have the right combinations of genes to be able to live under these conditions.

We can see this principle at work in certain organisms that can reproduce either sexually or asexually. For instance, *Daphnia,* the little water flea, reproduces asexually for generation after generation, females producing only females. In fact, sometimes one can see an asexual female embryo within an asexual female embryo within an asexual adult! The little animals keep up this kind of reproduction as long as conditions are stable. But when their pond begins to dry up, or things otherwise get dangerous and unpredictable, they change strategies. They begin to produce both males and females. These go on to mate and reproduce sexually in the usual way. Thus, when hard times come, or in the next spring, or the next rain, a few of their highly variable offspring may have that particular combination of genes that will allow them to cope with their new environment.

There are two major take-home lessons from the story of meiosis. First, meiosis halves the chromosome number of eggs and sperm, making fertilization feasible. Second, meiosis provides a means of shuffling and reorganizing chromosomes, thus increasing genetic variation in offspring. This reorganization takes place in the prophase I scrambling of chromosomes through crossing over; in the random lining up of homologous chromosomes in metaphase I, so that maternal and paternal centromeres are randomly distributed to the two poles; in the random distribution of chromosomes into polar bodies; in the random ascendancy of one oocyte in the ovary to begin development while the rest remain frozen in prophase I; and in the chancey competition of billions of individually unique sperm for a single unique egg. Not to mention the somewhat random way two sexual individuals get together in the first place.

Application of Ideas

1. Single-celled organisms can die—they do die, and in enormous numbers. Potentially, however—at least in some sense—such organisms are immortal. Multicellular organisms that can reproduce by budding are also potentially immortal in the same sense. Discuss the evolution of sex and death.

2. In cycads and ginkgos, unlike other seed plants, the sperm are flagellated (this is evidently the primitive condition of seed plants). Centrioles are generally absent from

cycad sporophyte cells, but appear in cycad microgametocytes at about the time of meiosis, apparently function in the organization of meiotic spindles, and are present in the cells of the microgametophyte (pollen grain). The details of this are obscure. Discuss the relevance of this observation to the theory that basal bodies arise developmentally from centrioles, to the generalization that centrioles arise only from other centrioles, and to the notion that seed plants may have invisible cytoplasmic bodies that function as centrioles in spindle formation.

Key Words

actin
anaphase
annulus (pl., annuli)
asexual reproduction
aster
astral ray
autosome

Barr bodies
budding

cell division
cell plate
centric heterochromatin
centriole
centromere
centromeric spindle fiber
chromatid
chromatin
chromomere
chromonema (pl., chromonemata)
chromosomal replication
cleavage furrow
clone
constitutive heterochromatin
continuous spindle fiber
crossing over
cytokinesis

daughter cell
dedifferentiation
diakinesis
diploid (2N)
diplotene
drumstick

euchromatin

facultative heterochromatin
fertilization

gamete
gametocyte
genome
germinal epithelium
germ line
gonads
G_1 phase
G_2 phase

Key Ideas

Introduction

1. Cells of the body are in a constant state of change with aging, turnover, death, and division.

2. The presence of the nucleus has been known for 175 years and the term itself was used as early as 1823. Many details of *cell division* were described by 1882.

Nuclear Structure

1. The double-membraned nucleus contains numerous pores called *annuli,* which are protein filled and are very selective in permitting the passage of materials.

2. In many eucaryotes the cell membrane disappears during cell division.

3. The prominent nucleoli are believed to be the site of ribosomal RNA production.

DNA, Chromatin, and Chromosomes

1. DNA occurs in protein (histone) complexes known as *chromatin,* which condenses to form *chromosomes.*

2. Most of the time, DNA is diffuse, extended, and threadlike, with beads of histones along its length.

3. *Chromatin* occurs as genetically functional *euchromatin* and *heterochromatin.*

4. *Heterochromatin* may be *constitutive* (function unknown) or *facultative* (active in some cells). The latter includes *X chromosomes* seen in *Barr bodies* and *drumsticks.*

5. The thinner regions of the *chromatin net* are active in producing RNA. These differ in various cells and their selective activity is the basis for cell specialization.

Nuclear RNA

In addition to chromatin, *nucleoplasm* contains tRNA, mRNA, rRNA, hnRNA, and a variety of smaller molecules, enzymes, ATP, and inorganic ions.

Cycle or Die

1. Many cells cycle from division to division, but some mature, stop cycling and eventually die.

2. Nerve and muscle cells that are *terminally differentiated* are exceptions, since they continue to live after cycling stops.

3. In regeneration of tissue, some cells dedifferentiate and enter differentiation anew.

4. Differentiation of muscle and nerve cells begins early in the development of an embryo.

5. The death of some cells in the embryo is programmed as a part of development.

6. Insects and plants have a large proportion of *terminally differentiated* cells.

The Stages of the Cycle

1. There are four phases in the cell cycle:
   a. *M phase: mitosis* and *cell division*

haploid (N)
heterochromatin
homologue

interphase

karyotyping
Klinefelter's syndrome

lampbrush chromosome
leptotene

meiosis
meiotic interphase
metaphase
metaphase plate
microfibril
microtubule
middle lamella
mitosis
mitotic spindle
M phase

nondisjunction
nonhistone chromosomal protein
nucleoplasm

oocyte
ovary

pachytene
polar body
polymorphonuclear leukocyte
premeiotic condensation
premeiotic interphase
primary gametocyte
primary oocyte
primary spermatocyte
prophase

reduction division

secondary oocyte
sex chromosome
sexual recombination
sexual reproduction
somatic
S phase
spindle (spindle apparatus)
spindle fiber
supercoiling
synaptinemal complex

telophase
terminal differentiation
testes
tetrad
trisomy-21 (Down's syndrome)
trisomy-X (XXX)
tubulin
Turner's syndrome

X chromosome
XYY syndrome

b. G_1 *phase:* active synthesis of protein; chromosomes in form of one *chromonema*
c. *S phase:* DNA replication
d. G_2 *phase:* period between *S* and *mitosis;* chromosomes in form of 2 *chromonemata*

2. The second, third, and fourth phases are known as *interphase.*

Mitosis and Cell Division

1. *Mitosis* is the replication and division of the chromosomes.
2. *Cell division* (*cytokinesis*) is the division of the cytoplasm.
3. *Mitosis* is subdivided into four subphases:
   a. *Prophase:* migration of centrioles, condensation of chromosome, loss of *nucleoli* and nuclear membrane, and onset of *spindle*
   b. *Metaphase:* chromosomes move to middle of *spindle*
   c. *Anaphase: centromeres* divide, chromosomes separate
   d. *Telophase:* events of *prophase* reverse and cell divides

Details of Mitosis

Chromosome Condensation (Prophase)

1. Chromosome condensation may involve histone I and involves *supercoiling* into three visible orders, coil within coil within coil. At the lowest order is the DNA helix.
2. Highly condensed, coiled DNA may be easier to move about during *mitosis,* but must uncoil to be functional in *interphase.*

Chromatids, Chromatin, and Centromere

1. Chromosomes replicated in *S phase* are doubled, *forming chromatids* that are not usually visible until *metaphase. Chromatids* are held together by *centromeres,* seen as somewhat constricted regions of the chromosome.
2. The specific location of *centromeres* is reason to believe they are a permanent part of chromosomes. Their location is used as a means of identifying individual chromosomes.

The Spindle

1. Flowering plants lack *asters* and *centrioles* and do not form *cleavage furrows.*
2. The *spindle apparatus* consists of *actin* and *tubulin. Tubulin* is assembled into *microtubules* to form the *spindle* fibers.
3. Each centriole consists of two cylindrical bodies lying at right angles to each other. Each cylinder consists of microtubules grouped in threes to form the body of the cylinders.
4. *Centrioles* migrate to opposite poles at the start of *mitosis.* At the end of *mitosis, centrioles* replicate themselves.

What Is the Centriole For?

Experimental evidence indicates that *centrioles* give rise to *asters* and *spindles.* These may be present as amorphous bodies in flowering plants.

Spindle Fibers

1. *Spindle fibers* are *centromeric* (attached to centromeres) and continuous (*aster* to *aster*).
2. The function of *asters* is unknown, but they may aid in making cytoplasm rigid and acting as anchors for the *spindle.*
3. Both *spindles* and *asters* are rigid enough to be isolated from the cell.

The Spindle in Action: Metaphase and Anaphase

Metaphase

At the end of *prophase* chromosomes are moved about on the spindle until they line up on the metaphase plate.

Y chromosome

zygote
zygotene

Essay 8.2 How the Spindle Works

1. Two older theories (push and pull) have been explored. The first states that pole-to-pole fibers push apart, moving the chromosomes. The pull theory states that *centromeric fibers* get shorter, pulling the chromosomes.

2. The most respectable theory involves subunit disassembly. It states that *centromeres* or *centrioles*, or both, disassemble the *microtubules* as they pull the chromosomes along toward the poles. This hypothesis agrees with visual observations.

Anaphase

1. Anaphase begins with the division of the *centromeres*. Centromeric fibers, attached to each *chromatid*, separate them and the chromosomes migrate to opposite poles.

2. The end result is that each new cell will receive a complete set of chromosomes identical to the former cell.

Telophase

In *telophase:*

1. Chromosomes decondense.
2. Nucleoli and nuclear membrane reappear.
3. Mitotic apparatus disappears (disassembled).
4. *Centrioles* replicate.

Cytokinesis—Cell Division

Cell division is also known as *cytokinesis* or *cytoplasmic division*.

Cell Division in Animals

1. *Cell division* begins with *microfibrils* pulling the membrane into a *cleavage furrow* at the *metaphase plate*. The *furrow* deepens until the cytoplasm is pinched in two.

2. In some instances mitosis occurs without cell division (multinucleated cells and multiple chromosome replication).

Cell Division in Plants

1. The rigidity of cell walls in plants leads to a different process of *cell division*.

2. Following *mitosis*, a new *cell plate* is laid down between daughter nuclei. The plate becomes impregnated with *pectin* and new membranes grow on cither side.

A Final Look at Mitosis and Cell Division

The functions of *mitosis* and *cell division* are:

1. Reproduction in simpler forms of life, producing *clones*
2. Regeneration of lost parts and repair through scar tissue
3. Growth and development from fertilized egg to adult in multicellular forms of life

Meiosis

1. Higher organisms are diploid (2N); therefore *meiosis* solves the problem of potential chromosome doubling at fertilization. In meiosis chromosome numbers are halved as the gametes are produced.

2. Humans have forty-six chromosomes, but sperm and egg cells each contain 23. The full number is restored in fertilization.

Homologous Chromosomes

1. The twenty-three chromosomes in human *gametes* contain all of the genetic instructions for development and metabolism. The complete set is known as a *genome*.

2. Chromosomes come in pairs and members of each pair are called *homologues* or *homologous chromosomes*. One is the functional equivalent of the other.

3. Although each member of a *homologous* pair carries the same genetic instructions, variations in these instructions exist (blue eyes on one, brown eyes on other, but both are color genes).

4. *Meiosis* assures that each gamete will have a complete *genome.*

Where Meiosis Takes Place in Higher Organisms

1. *Meiosis* in higher animals occurs in the *germinal epithelium* of the *gonads* (*ovaries* and *testes*). In higher plants *meiosis* occurs in *ovaries* and *anthers.*

2. *Germ-line cells* are those that have the potential for producing *germ cells* (*gametes*). Cells not in the *germ line* are called *somatic cells.*

3. Boys produce *gametes* from puberty on. Girls produce all of their gametes (*oocytes*) before birth. *Oocytes* become arrested in prophase I of meiosis with one or two maturing each month. *Meiosis* resumes but will not be completed unless *fertilization* occurs.

4. In males the *germinal epithelium* is in the *spermatic tubules.* As *mitosis* occurs in this tissue one cell remains behind to divide mitotically again while the other becomes a *primary gametocyte* and will divide meiotically to form *gametes.*

Premeiotic Interphase

1. The *primary gametocyte* enters a *premeiotic interphase* with the usual cell cycle phases.

2. During *premeiotic interphase,* chromosomes condense and then decondense. When *premeiotic interphase* ends in humans each chromosome will have replicated, forming two *chromatids,* with a total of ninety-six *chromatids* per cell.

Meiosis I: The First Meiotic Division

Meiotic Prophase: The First Stages

1. In zygotene, homologous chromosomes only partially condensed attach or fuse through bridges of protein and RNA called the *synaptinemal complex.*

2. Stages of *meiotic prophase* include:
   a. *Leptotene:* condensation begins
   b. *Zygotene:* pairing of *homologues*
   c. *Pachytene:* further condensation and crossing over
   d. *Diplotene:* partial decondensation, *chromatids* and crossover effects visible
   e. *Diakinesis:* highly condensed, *homologues* attached only at ends

3. After *zygotene,* chromosomes are grouped into *tetrads.* (In humans there are twenty-three tetrads. Each has two *centromeres,* so the centromere count at this time is forty-six.)

Metaphase I

1. When chromosomes align on the *metaphase plate,* the *homologues* are still attached, but only at the tips of their arms.

2. *Centromeric fibers* attach to the *tetrad* from each side.

Anaphase I

1. Homologous chromosomes separate during anaphase I. Unlike mitosis, the centromeres holding chromatids together do not divide. At the end of anaphase each newly forming nucleus will contain twenty-three chromosomes, with twenty-three undivided centromeres.

2. *Anaphase* I differs from that of *mitosis* or *meiosis* II as follows:
   a. Each *centromere* is still attached to two *chromatids.*
   b. The *chromatids* have undergone a mixing during *crossover.*

A Closer Look at Meiotic Prophase: Crossing Over

During *pachytene, chromatids* unwind in places and Watson and Crick strands are exchanged between *homologous chromosomes.* Maternal and paternal *chromatids* therefore become intermingled, then break and rejoin.

Telophase I and Meiotic Interphase

1. *Telophase* includes the usual reorganization events and *cytoplasmic division.*
2. *Daughter cells* enter *interphase,* but there is no *S phase,* thus no chromosome replication.

Meiosis II

Prophase II

Prophase II resembles *mitosis* superficially. *Chromatids* still attached to *centromeres* receive *spindle fibers* from each pole and align on the *metaphase plate.*

Metaphase II and Anaphase II

The *centromeres* divide in *metaphase II.* The *chromatids* separate. Because of *crossing over,* the newly designated chromosomes are not identical and should never be referred to as "maternal" and "paternal."

Telophase II

Telophase includes *cytoplasmic division,* but *centrioles* do not replicate. The four *daughter cells* enter *interphase* but do not go past G_1. *They remain haploid.*

Meiosis in Females

1. *Meiosis* in male and female vertebrates differs primarily in the *cytoplasmic division* and the long *prophase* encountered by *oocytes.* During the long *prophase,* the chromosomes may be metabolically active. The active regions appear to puff out and are called *lampbrush chromosomes.*
2. *Cytoplasmic division* in *meiosis* I and *meiosis* II is highly unequal with most of the cytoplasm ending up in one *daughter cell.* The other, a tiny, deprived cell, called a *polar body,* never develops.

Why Meiosis?

1. *Meiosis* is an essential part of the sexual process, since it includes gene shuffling in crossing over and produces haploid gametes.
2. In *sexual recombination,* these shuffled genes are recombined in an organism and in a population of organisms with others, producing considerable variation in offspring. Variation in offspring means that at least a few can cope successfully with a changing environment.
3. Some organisms reproduce asexually while conditions are stable, but switch to *sexual reproduction* when conditions become variable or stressful.
4. *Meiosis* provides two levels of gene shuffling. One is *crossing over.* The other in the random arrangement and division of *homologues* at the *metaphase plate.*

Essay 8.3 When Meiosis Goes Wrong

1. Many spontaneous abortions can be traced to nondisjunction, mistakes in chromosome divisions during *meiosis.* In some cases, individuals with abnormal chromosome numbers survive. One human in 200 has this problem.
2. *Trisomy-21* or *Down's syndrome* is caused by the presence of an extra twenty-first chromosome. It includes mental and physical abnormalities.
3. *Trisomy-21* increases in children born to women over thirty-five, so an aging process in the *oocyte* is suspected.
4. Nondisjunction can affect the *sex chromosome* distribution. Most cases involve extra *X chromosomes,* but one involves the loss of one *X chromosome* and is designated XO, or *Turner's syndrome.* At least one *X* is required for survival.
5. Extra *X chromosomes* are common in both males (*Klinefelter's syndrome*) and females. These people can have fertility problems, are usually tall, and have limited IQs.
6. One abnormality in males, the *XYY syndrome,* is caused by an extra *Y chromosome.* These very tall individuals have lower than average IQs, which in turn leads to an increased risk of imprisonment.

Review Questions

1. How do the *annuli* aid in preserving the contents of the *nucleoplasm?* (p. 224)
2. Review what is known about the function of the nucleolus. (p. 225)
3. Distinguish between *euchromatin, heterochromatin, constitutive heterochromatin,* and *facultative heterochromatin.* Use a diagram if it helps. (pp. 225–226)
4. List common examples of *facultative heterochromatin.* What evidence suggests cells could get along without it? (p. 226)
5. How do we explain nonfunctioning DNA? (p. 226)
6. What is the fate of cells that do not continue cycling? List two exceptions in vertebrates. (p. 227)
7. Why are *gametes* considered noncycling cells for the most part? (p. 227)
8. Describe a situation where *terminally differentiated* cells dedifferentiate and then follow new directions of development. (p. 228)
9. List the phases of the cell cycle and briefly describe the events occurring in each. (p. 228)
10. Clearly distinguish between the terms *cytokinesis* and *mitosis.* (p. 229)
11. List the subphases of *mitosis* and briefly summarize the events in each. (pp. 229–230)
12. With the aid of a simple drawing describe the organization of *supercoils* in condensed DNA. (p. 230)
13. What is the advantage of condensation and *supercoiling* to a cell going through *mitosis* or *meiosis?* (p. 233)
14. When does a chromosome contain *chromatids* and when is a *chromatid* no longer a *chromatid?* (p. 233)
15. What is a *centromere?* What is the evidence that suggests they are a permanent part of a chromosome? (p. 234)
16. How does the mitotic apparatus differ between plants and animals? (p. 234)
17. Describe the structure of a *spindle fiber.* When are the elements of a *spindle fiber* produced? (pp. 234, 237)
18. Describe the structure of a *centriole* and its behavior during *prophase.* (p. 237)
19. What is the evidence that indicates the role of *centrioles* in *spindle* formation? (p. 238)
20. Describe the visual differences between *centromeric spindle fibers* and *continuous spindle fibers.* (p. 238)
21. Describe the arrangement of chromosomes on the *metaphase plate.* (p. 238)
22. Describe the logic of the *subunit disassembly* in overcoming the objections to the push and pull hypotheses. (Essay 8.2)
23. Describe the organized separation of *chromatids* in *anaphase* and how this can be explained by the specific attachment of *centromeric fibers.* (p. 240)
24. List the events of *telophase.* (p. 240)
25. Describe the role of *microfibrils* in *cytokinesis.* (pp. 240–241)
26. Using the terms *cell plate* and *middle lamella* describe cell division in the plant cell. (p. 242)
27. In what situations does *mitosis* and *cytokinesis* become a means of reproduction for the organism. (p. 242)
28. What are the two purposes of *cell division* in multicellular organisms? (pp. 242–243)
29. What basic problem in reproduction does *meiosis* solve in *diploid* organisms? (p. 243)
30. If a *diploid* organism had fourteen G_1 chromosomes, how many would be present in its *gametes?* In its *zygotes?* (p. 243)
31. Write a careful definition for *homologous chromosomes* and include a word about the genes on each. (pp. 243–244)
32. List the two important things that happen to *homologous chromosomes* in *meiosis.* (pp. 243–244)
33. Distinguish between *germinal tissue* and *somatic tissue.* (p. 244)
34. How does the functioning of *germinal epithelium* differ in human males and females? (p. 245)
35. Describe the peculiarity of epithelial tissue, which makes it possible for males to produce *gametes* throughout life. (p. 245)
36. *Premeiotic condensation* seems to be a false start on *meiosis.* Explain why this is the case. (p. 245)
37. Describe the special behavior of *homologous* chromosomes during *prophase* of *meiosis* I. (p. 245)
38. Describe the formation of the *synaptinemal complex* during *zygotene.* Include a visual and a molecular description. (p. 245)
39. List the stages of meiotic *prophase* and summarize the events in each. (pp. 246–247)
40. Describe the arrangement of *tetrads* during *metaphase I.* (p. 247)
41. Compare the spindle hookup and behavior of *centromeres* in *anaphase* of *meiosis I* with that of *mitosis.* (p. 248)
42. Explain the term *reduction division* as applied to *anaphase I* of *meiosis.* (p. 248)
43. Returning to the DNA molecule level, explain how *crossing over* occurs. (pp. 248–249)
44. How does *meiotic interphase* differ from mitotic interphase? (p. 250)
45. Compare the separation of chromosomes in *metaphase* II and *anaphase* II with that described in *metaphase* I and *anaphase* I (meiosis). (p. 250)
46. How many *gametes* result from each *meiosis* in male animals? In female animals? (pp. 251, 254)
47. What does a *lampbrush chromosome* indicate about metabolic activity in an *oocyte?* What is the purpose of this activity? (p. 254)
48. Why is it technically inaccurate to use the term, "human egg cell"? (p. 254)
49. Describe the unusual *cytokinesis* that occurs during *meiosis* in females. How does this affect the number of *gametes* from each meiotic event? (p. 254)
50. Describe the two ways in which meiosis increases the genetic variability of *gametes.* (p. 258)
51. Using a diagram to show nondisjunction, explain *trisomy-21.* (Essay 8.3)
52. What is the apparent relationship between maternal age and *trisomy-21* births? What explanation for the relationship can you offer? (Essay 8.3)
53. Using Xs and Ys, write the sex chromosomes for the following abnormalities (Essay 8.3):
    a. *Turner's syndrome*
    b. *Klinefelter's syndrome*
    c. *extra Y syndrome*
    d. *trisomy-X*
54. What are the *X* and *Y chromosome* requirements for producing a female? A male? (Essay 8.3)
55. Generally, what are the physical and mental effects of extra X chromosomes? (Essay 8.3)

Chapter 9 Classical Genetics

Occasionally you may hear someone just back from a trip to Mexico or Spain boasting of having "fought a bull." In fact, that very noun may come to mind when you learn that the "bullfight" was with a heifer (a young cow). At certain times of the year, guests are invited to the ranches where fighting bulls are bred to watch the testing of the young cows—the time when the ranchers determine which stock will be used for breeding. As part of the festivities, the guests may be invited to try their hand at caping a heifer. The ranchers seem to think that nothing is quite so festive as watching a gringo trying to stand his or her ground while avoiding an angry cow!

Professionally, the ranchers are interested in how cows fight because it is axiomatic in such circles that a fighting bull gets its strength from its father, but its courage from its mother. The origin of such beliefs is lost in tradition, but it is evident that people utilized the principles of genetics long before they knew anything about genes. In this country, you can find people who are totally ignorant of genetics and biology but possess great stores of native wisdom about how to breed show dogs or race horses.

Charles Darwin, as the son of a gentleman farmer, was well aware of the basic principle of inheritance, namely that offspring tend to resemble their parents. The offspring of heavier animals, he knew, tended toward heaviness; among cows, the daughters of high-yielding milk producers tended to produce more milk. The principle seemed clear enough to animal and plant breeders: certain traits tend to be transmitted from generation to generation.

9.1 Mendel and His Laws

When Darwin Met Mendel (Almost)

Darwin was probably the 19th century's greatest biologist, and perhaps its fourth best geneticist (after Mendel, Galton, and Weissmann). Even so, he was a terrible geneticist, and most of his ideas about heredity were wrong! Darwin's chief mistake was to accept the only intellectually respectable theory of his day, which was *blending inheritance.* It was believed that the "blood" or hereditary traits of both parents blended in the offspring, just as two colors of ink blend when they are mixed. The blending hypothesis appeared to work reasonably well for some traits—such as height or weight, where there is a continuous gradation of possible values—but it couldn't account for other observations. It's always surprising, in hindsight, to recognize the degree to which a strongly held belief will blind its proponents to obvious contradictions. The blending theory would predict that the offspring of a white horse and a black horse should always be gray, and that the original white or black should never reappear if gray horses continued to be bred. But it turns out that the offspring are not always gray, and black or white descendants do appear. Similarly, if a honey-colored cocker spaniel were crossed with a black one, the offspring would usually be black or honey, not some kind of golden black. Something obviously was wrong with the theory, but no one seemed to notice, or if they did, they tried to ignore it.

Actually, the blending theory doesn't really work even for continuous, graduated traits like height and weight. If it did, the offspring in every generation would always be less extreme than their parents for every trait. Every time a fat horse mated with a thin horse, the offspring should always have been an "average" horse. So the world, by now, should be full of average, gray horses and average everything else; but, in fact, some horses that eat about the same amounts are still fat and others are still thin. Horses are also still found in many different colors, and even expert breeders usually can't predict the color of a foal.

When Darwin finally got around to publishing his theory of natural selection, some of his sharper critics used the weakness of his genetic arguments against him. They pointed out, quite correctly, that natural selection will not work with blending inheritance, because any variation would be blended away. Try as he might, Darwin could not come up with an answer for them, and the tired old man, so sickened in his later years, began to backtrack. In later editions of the *Origin of Species* he began to give more and more weight to the possibility of the inheritance of acquired traits, and he came up with desperate explanations that were so bad that we won't bother you with them (but see Chapter 11).

But Darwin himself was a keen enough observer to be well aware of the fact that some parental traits could, in fact, disappear in the offspring and reappear in the following generation. Actually, while he was still working on the *Origin,* he did some very good experiments to try to solve this nagging puzzle. He performed almost exactly the same experiment that the Austrian monk, Gregor Mendel, was performing at just about the same time; and the two great scientists, unknown to each other, got almost exactly the same results. But only Mendel was able to understand what had happened. This almost incredible historical incident points up an aspect of Darwin's science that is sometimes overlooked: he was an experimenter of dazzling abilities.

In his set of experiments, Darwin crossed two *true-breeding* strains of snapdragons (true-breeding means that when individuals of a strain are crossed, the offspring are the same, generation after generation). The two strains differed in only one respect: One strain had abnormal, radially symmetrical flowers (what botanists called *peloric* flowers), and the other had the normal, bilaterally symmetrical snapdragon flowers (Figure 9.1). Of the progeny of this first cross, Darwin wrote: "I thus raised two great beds of seedlings, and not one was peloric." Darwin called this tendency of one trait in a cross to suppress another, *prepotence.* We would now call it *dominance.* We'll leave Darwin's genetics experiment now, but return to it later.

Even while Darwin was working on his snapdragon experiment, scientific history was being made across the Channel, deep in Europe. A self-effacing but incredibly bright and dedicated monk was crossing strains of garden peas.

The Abbot Gregor Johann Mendel (1822–1884), a member of an Augustinian order in Brunn, Moravia (now part of Czechoslovakia), is often depicted as a kindly old man of the cloth who, while puttering around his monestary garden, somehow stumbled onto important genetic laws. Again, our tendency to embellish the memory of already worthy individuals has led us astray. The real Mendel was remarkable

9.1 One of Charles Darwin's many excellent experiments included breeding snapdragons. In one observation he crossed true-breeding *peloric*-flowered snapdragons (a) with true-breeding regular-flowered snapdragons (b). He found that the peloric flower trait was lost in the progeny. In describing what we would call the dominance of one trait over another, he coined the term *prepotence.*

(a)

(b)

9.2 Gregor Mendel (1822–1884). Bringing together an innate curiosity, keen observational powers, and mathematical training, Mendel developed the basic laws of heredity. He was undoubtedly the first mathematical biologist. His work, published in 1866, went unnoticed for many years until it was rediscovered about the turn of the century. Once his ideas were understood, they opened the door to 20th century genetics.

enough in his own right. In many ways he seems more like a 20th century biologist, somehow displaced into the wrong century (Figure 9.2). His first published paper is a landmark in its clarity and a model modern experimental report. Many scientists feel that it is perhaps the best scientific paper ever written.

Early in his life Mendel began training himself, and he became a rather competent naturalist. To support himself during those years, he worked as a substitute high school science teacher. The professors at the school, noting his unusual abilities, suggested that he take the rigorous qualifying examination and become a regular member of the high school faculty. Mendel took the test and did reasonably well, but he failed to qualify. The standards for high school teachers were very rigorous. Mendel joined a monastic order.

His monkish superiors, confident of his abilities, sent him in 1851 to the University of Vienna for 2 years of concentrated study in science and mathematics. Here he learned about the infant science of statistics, which was to serve him well.

When Mendel returned to the monastery he began his plant breeding studies in earnest. He impressed his fellow monks with his intelligence and his vigor, and he applied these qualities to the study of plant breeding. He developed new varieties of fruits and vegetables, kept abreast of the latest developments in his field, joined the local science club, and became active in community affairs.

Mendel's Crosses

Mendel began to experiment in the effects of crossing different strains of the common garden pea. To begin with, Mendel based his research on a very carefully planned series of experiments and, more importantly, on a statistical analysis of the results. The use of mathematics to describe biological phenomena was a new concept. Clearly, Mendel's 2 years at the University of Vienna had not been wasted.

The careful planning that went into Mendel's work is reflected in his selection of the common garden pea as his experimental subject. There were several advantages in this choice. Pea plants were readily available and fairly easy to grow, and Mendel was able to purchase thirty-four true-breeding strains. These strains differed from each other in very pronounced ways, so that there could be no problem in identifying the results of a given experiment. Mendel chose seven different pairs of traits to work with:

1. Seed form—round or wrinkled
2. Color of seed contents—yellow or green
3. Color of seed coat—white or gray
4. Color of unripe seed pods—green or yellow
5. Shape of ripe seed pods—inflated or constricted between seeds
6. Length of stem—short (9–18 inches) or long (6–7 feet)
7. Position of flowers—axial (along the stem) or terminal (at the end of the stem)

Partly because of his mathematical training, he generally worked with quite large numbers, since he knew that large numbers yield more solid data. This was also because of his exceedingly fortunate choice of the pea; each pea in a pod is an essentially new plant, with its own genes and traits—or its own *genotype* and *phenotype.* (The total combination of an organism's genes is called its *genotype,* and the combination of its visible traits is called its *phenotype.*) So for the first two pairs of traits listed, the kinds of descendants the plant produces can be easily counted by simply examining the peas. The old description, "alike as two peas in a pod," then, is not a very good one. Mendel just had to look at the peas to observe the results of many of his crosses.

Mendel's approach, a novel one at that time, was to cross two true-breeding strains that differed in only one characteristic, such as seed color. Peas ordinarily self-fertilize, so to cross two strains, the pollen had to be transferred by hand. We would now call the original parent generation P_1 and designate the first-generation offspring F_1 (first filial) generation. When the F_1 plants are crossed with each other or are allowed to self-pollinate, the resulting offspring are called the F_2 generation, and so on.

The Principle of Dominance

Now, when Mendel crossed his original P_1 plants, he found of course that the characteristics of the two plants didn't blend, as prevailing theory said they should. When plants grown from round seeds were crossed with plants grown from wrinkled seeds, their F_1 offspring were not intermediate seeds. Instead, all of them were round seeds. Mendel termed the trait that appeared in the F_1 generation the *dominant* trait and the one that had failed to appear the *recessive* trait. But he was now left with a vexing question. What had happened to the recessive trait? It had been passed along through countless generations, so it couldn't have just disappeared.

Mendel then allowed his F_1 pea plants to pollinate themselves. And, in England, Darwin allowed his two great beds of snapdragon seedlings to pollinate themselves.

In his book, *Variation in Plants and Animals under Domestication,* Darwin recorded his results. Of the offspring of the F_1 snapdragons that bore flowers, he had found that 88 had normal flowers, 37 had abnormal (peloric) flowers, and 2 had a few flowers he honestly couldn't classify. So he could classify 125 of the F_2 plants, and could see that these fell into two discrete groups of unequal size. But where had the peloric flowers come from? Weren't they bred out in the F_1 generation? What about the mixing theory? The trait that had disappeared in the first generation had reappeared in the second. Now, here was something that had to be explained! Darwin was greatly puzzled but finally came up with a suitable explanation. He said the phenomenon was due to *latency.* He said that the latent character had "gained strength by the intermission of a generation." Apparently satisfied, he directed his attention to other things. Darwin went to his grave unaware of how close he had come to another landmark idea.

Mendel, however, stayed with the problem. In the offspring of the F_1 generation, which we call the F_2 generation, Mendel found that roughly one-fourth of the peas were wrinkled and that about three-fourths were round. He repeated the experiment with other pea strains and got about the same results. He crossed a yellow-pea strain with a green-pea strain. All of the F_1 peas were yellow, but in the F_2 generation about one-fourth of the peas were green again. The ratios did not escape the tenacious Mendel, determined to badger the problem until he would make some sense of it.

Darwin, when he recorded his own numbers, probably did not notice that $\frac{37}{125}$ is fairly close to one-fourth. If he had noticed, it would probably have had no more significance to him than one-third or one-fifth. Darwin apparently had a very different kind of mind than Mendel. Whereas Mendel was the first mathematical biologist and was familiar with statistics, he also knew how to isolate and work out small parts of great problems. Darwin's genius, on the other hand, was in the enormous breadth and scope of his ideas and the ability to fit seemingly unrelated details into a grand scheme.

Mendel continued and even expanded his experiments. He continued to use the trusty garden pea, in spite of the fact that his artificial crosses demanded a lot of surgical manipulation. Left to themselves, most pea plants will simply self-fertilize and produce plants just like themselves. To get a cross between two plants or two strains it was necessary to open the normally closed pea flower, remove the pollen-producing anthers (to prevent self-pollination), and apply the foreign pollen with a small paintbrush. But Mendel controlled his crosses, prevented accidental contamination, and allowed self-pollination only when it suited his needs.

One of Mendel's early observations was critical to his development of his genetic model. With regard to round and wrinkled peas, Mendel realized that there were two kinds of round peas: the true-breeding kind, like the original parent stock, which would grow into plants that would bear only round peas; and another type, which when grown and self-pollinated, would produce pods containing both round and wrinkled peas. Two kinds of round peas: one pure-breeding, one not. So, he wondered, were there also two kinds of wrinkled peas? He found that there were not. When wrinkled peas were cultivated and allowed to self-pollinate, they always bore only wrinkled peas. We now call the true-breeding peas *homozygous,* and the other kind *heterozygous.* Homozygous means that the organism bears only the genes for one trait; heterozygous, as we will see, means it bears related genes for different traits, no matter what the appearance of the organism.

It was impossible to tell the difference between the two kinds of round peas unless they were allowed to grow up and reproduce, an experiment called *progeny testing.* Then two seemingly identical round seeds would show just how different they really were. The kind we call homozygous could be found bearing only one type of pea on self-pollinating, while the heterozygous plant bore two types of peas. Mendel decided he needed an F_3 generation. He planted the round peas of the F_2 generation, waited another year for them to grow and bear pods, and then looked into the pods to determine whether the plants were true-breeding or not. He found that one-third of the F_2 round peas had been true-breeding (we would say homozygous) and that two-thirds had not (and thus, we would say, were heterozygous). More numbers! Darwin would have fallen asleep by this time.

Three-fourths of the F_2 generation had been round, and one-third of the round F_2 had been true-breeding. One-third of three-fourths is one-fourth. So, when he considered all the F_2 peas together, Mendel found the following (see Figure 9.3):

$\frac{1}{4}$ of the total F_2 were round and true-breeding (homozygous) }
$\frac{1}{2}$ of the total F_2 were round and not true-breeding (heterozygous) } $\frac{3}{4}$ round

$\frac{1}{4}$ of the total F_2 were wrinkled and true-breeding (homozygous) } $\frac{1}{4}$ wrinkled

The same thing worked for yellow and green peas. There were two kinds of yellow and one kind of green. Mendel looked at all seven characters (Table 9.1), and the same principle worked in each case; roughly the same ratios were generated. In every case, one form, which Mendel called the *dominant* trait, was present in 100% of the F_1 peas or pea plants. In every case, three-fourths of the F_2 peas or pea plants showed the dominant form of the character (we would say the dominant phenotype), and one-fourth of the F_2 peas or pea plants showed the contrasting phenotype, which Mendel called the *recessive* trait.

In every case, a breeding test revealed that, in actuality, one-fourth of the F_2 were the dominant form and true-breeding; one-half of the F_2 were the dominant form but not true-breeding; one-fourth of the F_2 were the recessive form and true-breeding.

Clearly, something determining the recessive form was passed down from the true-breeding recessive parental strain, through the hybrid F_1, to the true-breeding recessive F_2; but whatever it was, it was skipping the F_1 generation. Mendel, at that time the world's only mathematical biologist, thought he could use algebraic symbols to express his dilemma (Figure 9.4).

He let **A** represent the factor that determines the dominant form and let **a** represent the factor that determines the recessive form. The F_1 hybrid, and in fact all the non-true-breeding plants, must have both factors present, as represented by the symbol **Aa**. Since there are two parents, Mendel figured that in the hybrid, **A** comes from one parent and **a** from the other. Mendel determined experimentally that it didn't matter which parental strain bore the peas and which provided the pollen. He let the capital letters signify dominance and the lowercase letters signify recessivity, which merely means that the **Aa** individuals will look exactly the same as the **A**-bearing parental stock.

If the heterozygous plants get an **A** from one parent and an **a** from the other parent, and are symbolized **Aa**, it makes sense that the true-breeding dominant forms get two **A** factors—one from each parent—and can be symbolized **AA**. In the same way, the true-breeding recessive forms get **a** factors from both parents and can be symbolized **aa**. We can use **AA** to symbolize the dominant *homozygote,* (when the factor from each parent is identical), **aa** to symbolize the recessive homozygote, **Aa** to represent the heterozygote (when the factor from each parent is differ-

Table 9.1 Mendel's F_2 Generations. The dominant and recessive traits analyzed by Mendel are shown along with the results of F_1 and F_2 crosses. Note the large numbers he worked with. How does a large sample size help with the conclusions? How well do his numbers in the last two columns agree with what probability tells us to expect in the crosses? The proportion of the F_2 generation showing recessive form is in the last column.

Dominant form	Number in F_2 generation	Recessive form	Number in F_2 generation	Total examined	Ratio	Proportion of F_2 generation
Round seeds	5,474	Wrinkled seeds	1,850	7,324	2.96:1	0.253
Yellow seeds	6,022	Green seeds	2,001	8,023	3.01:1	0.249
Gray seed coats	705	White seed coats	224	929	3.15:1	0.241
Green pods	428	Yellow pods	152	580	2.82:1	0.262
Inflated pods	882	Constricted pods	299	1,181	2.95:1	0.253
Long stems	787	Short stems	277	1,064	2.84:1	0.260
Axial flowers (and fruit)	651	Terminal flowers (and fruit)	207	858	3.14:1	0.241

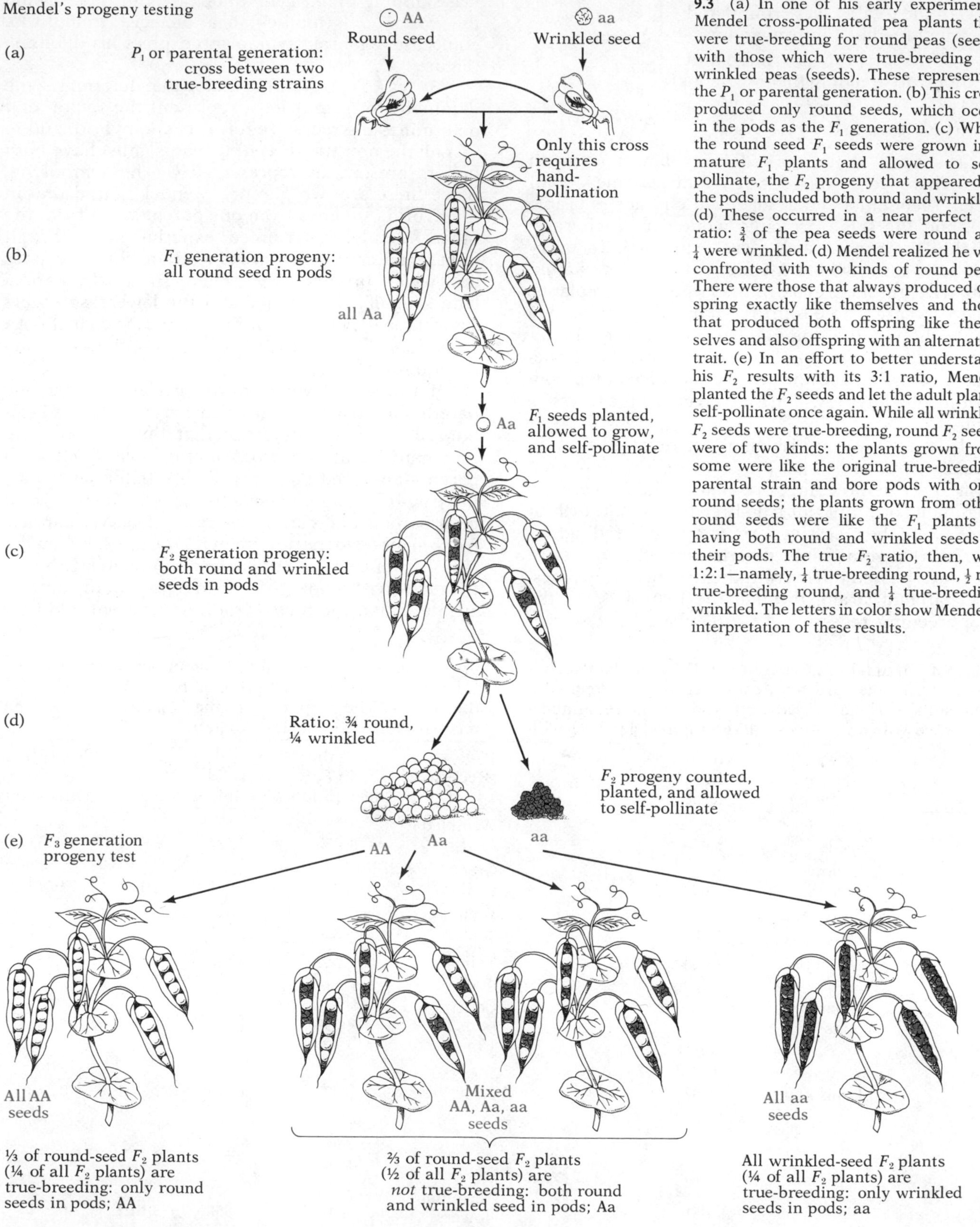
Mendel's progeny testing
AA
Round seed
aa
Wrinkled seed
(a)
P_1 or parental generation: cross between two true-breeding strains
Only this cross requires hand-pollination
(b)
F_1 generation progeny: all round seed in pods
all Aa
Aa
F_1 seeds planted, allowed to grow, and self-pollinate
(c)
F_2 generation progeny: both round and wrinkled seeds in pods
(d)
Ratio: ¾ round, ¼ wrinkled
F_2 progeny counted, planted, and allowed to self-pollinate
(e) F_3 generation progeny test
AA
Aa
aa
All AA seeds
Mixed AA, Aa, aa seeds
All aa seeds
⅓ of round-seed F_2 plants (¼ of all F_2 plants) are true-breeding: only round seeds in pods; AA
⅔ of round-seed F_2 plants (½ of all F_2 plants) are *not* true-breeding: both round and wrinkled seed in pods; Aa
All wrinkled-seed F_2 plants (¼ of all F_2 plants) are true-breeding: only wrinkled seeds in pods; aa

9.3 (a) In one of his early experiments, Mendel cross-pollinated pea plants that were true-breeding for round peas (seeds) with those which were true-breeding for wrinkled peas (seeds). These represented the P_1 or parental generation. (b) This cross produced only round seeds, which occur in the pods as the F_1 generation. (c) When the round seed F_1 seeds were grown into mature F_1 plants and allowed to self-pollinate, the F_2 progeny that appeared in the pods included both round and wrinkled. (d) These occurred in a near perfect 3:1 ratio: $\frac{3}{4}$ of the pea seeds were round and $\frac{1}{4}$ were wrinkled. (d) Mendel realized he was confronted with two kinds of round peas. There were those that always produced offspring exactly like themselves and those that produced both offspring like themselves and also offspring with an alternative trait. (e) In an effort to better understand his F_2 results with its 3:1 ratio, Mendel planted the F_2 seeds and let the adult plants self-pollinate once again. While all wrinkled F_2 seeds were true-breeding, round F_2 seeds were of two kinds: the plants grown from some were like the original true-breeding parental strain and bore pods with only round seeds; the plants grown from other round seeds were like the F_1 plants in having both round and wrinkled seeds in their pods. The true F_2 ratio, then, was 1:2:1—namely, $\frac{1}{4}$ true-breeding round, $\frac{1}{2}$ not true-breeding round, and $\frac{1}{4}$ true-breeding wrinkled. The letters in color show Mendel's interpretation of these results.

AA × **aa**	P_1 true-breeding strains
Aa	F_1 progeny
Aa × **Aa**	F_1 cross
AA : 2**Aa** : **aa**	F_2 progeny
¼ round (**AA**)	= homozygous round
½ round (**Aa**)	= heterozygous round
¼ wrinkled (**aa**)	= homozygous wrinkled

9.4 Mendel's symbolism. From his many trials Mendel knew that his heterozygous F_1 peas carried two factors responsible for producing the pea shape round or wrinkled. He represented these two factors as **A** and **a**. The capital **A** represented dominance, while the lowercase **a** represented recessiveness. Each individual in his crosses can be represented with a pair of letters. As you can see, the progeny of an F_2, represented this way, applies itself to algebraic factoring to obtain the genotypes of the F_1 that produced it! Likewise, you can begin with the F_1 and simply multiply the algebraic factors **Aa** × **Aa** together to predict the F_2 results.

ent), and **A–** to symbolize those plants with the dominant phenotype where the complete genotype is not known, (this usage, now common, differs slightly from Mendel's own).

Briefly, then, the basic F_2 ratio of Mendelian genetics is:

Cross: **Aa** × **Aa**

Offspring: $\frac{1}{4}$ **AA**, $\frac{1}{2}$ **Aa** } $\frac{3}{4}$ dominant phenotype, **A–**

$\frac{1}{4}$ **aa** } $\frac{1}{4}$ recessive phenotype, **aa**

Now Mendel was ready for some conclusions. First, each of his seven characters was controlled by transmissible factors that came in two forms, dominant and recessive. Second, every hybrid individual had two different copies of the factor controlling each character, one from each parent. Third, if the factors were the same (we would say if the individual were homozygous), the individual would be true-breeding. Fourth, if the factors were not the same (we would say if the individual were heterozygous), the dominant factor would determine the appearance of the plant (we would say: the dominant *allele*—the alternate form of a gene on two homologous chromosomes—would determine the phenotype). Mendel considered these conclusions preliminary.

Mendel's First Law: The Segregation of Alternate Alleles

When a heterozygote reproduces, its gametes (eggs and sperm in animals, or pollen and ovules in plants) *will be of two types in equal proportions.*

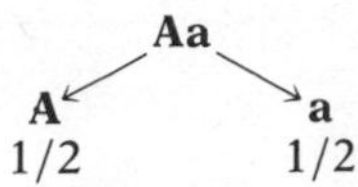

Now, how did he come up with that? He came up with it by his playing with the numbers his experiments were generating, and realizing what they meant. Perhaps it is more precise to say that Mendel's first law was a *model* that gave predictions that were consistent with his observations. Anyway, this is how he had it figured: When the heterozygous plant was allowed to pollinate itself, it would produce half **A** ovules and half **a** ovules. And it would also produce half **A** pollen and half **a** pollen. So, when it self-pollinated, what proportion of all the fertilized ovules should be **aa**?

Now the probability of any given egg being **a** is $\frac{1}{2}$. The probability that this egg should be fertilized by an **a** pollen is also $\frac{1}{2}$. It is one of the basic laws of probability, the so-called multiplicative law, that: *The probability of two independent events both occurring is equal to the product of their individual probabilities* (Figure 9.5). Since the event of the egg being **a** has no influence on the event of the pollen being **a**, or the other way around, the probability that both of these things will happen for any given pea is $\frac{1}{2} \times \frac{1}{2} = \frac{1}{4}$. (If the probability of your flunking your next exam is $\frac{1}{6}$ and the probability of rain on that day is $\frac{1}{5}$, the probability that both these things will happen is $\frac{1}{30}$, unless your test performance influences the weather, or vice versa.)

The same reasoning applies to the quarter of the F_2 progeny that are **AA**: $\frac{1}{2}$ chance of **A** pollen times $\frac{1}{2}$ chance of **A** ovule gives a $\frac{1}{4}$ chance of a zygote being **AA**.

But half of the F_2 peas were **Aa**. How do we account for that? There are two different ways that **Aa** zygotes are formed in an **Aa** × **Aa** cross. It's fairly obvious. Either an **a** pollen fertilizes an **A** egg, or an **A** pollen fertilizes an **a** egg. The probability of the first combination of events is $\frac{1}{2} \times \frac{1}{2} = \frac{1}{4}$, and the probability of the second combination of events is $\frac{1}{2} \times \frac{1}{2} = \frac{1}{4}$.

There is no other way of getting an **Aa** zygote from such a cross, and the two possibilities are *mutually exclusive* (they can't both happen to the same zygote). And Mendel had shown that the **Aa** heterozygotes were identical regardless of which parent contributed which factor.

Another basic law of probability, the so-called additive law, is: *The probability of either one or another of two mutually exclusive events occurring is equal to the sum of their individual probabilities.* If there is a $\frac{1}{2}$ probability of a coin coming up heads, and a $\frac{1}{2}$ probability of it coming up tails, then you can be 100% sure ($\frac{1}{2} + \frac{1}{2} = 1$) that it will come up either heads or tails.

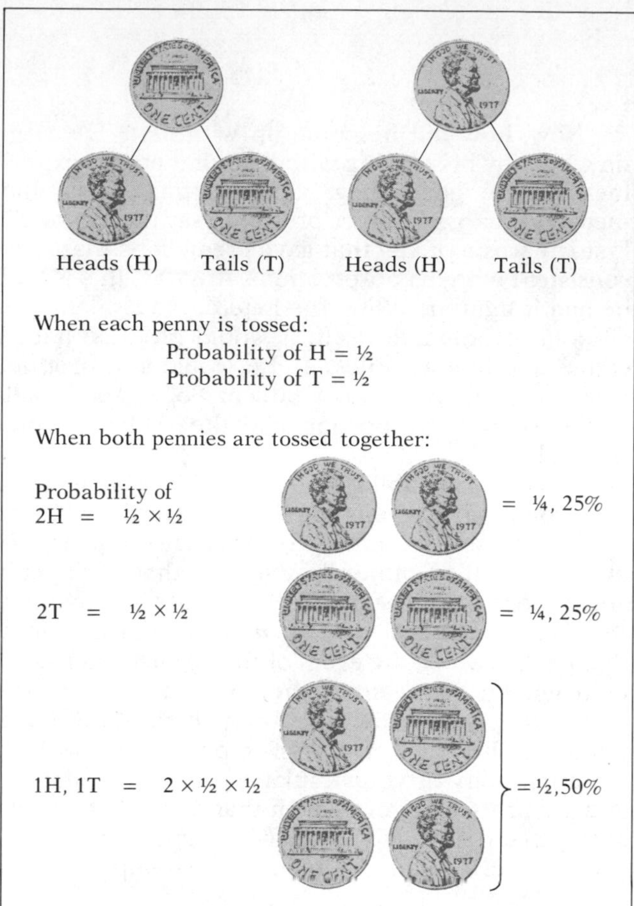

9.5 The multiplicative law. The probability of two independent events both occurring is equal to the product of their individual probabilities. The law can be tested with coins. In this scheme, two coins are used to represent two independent events. Toss two pennies 100 times. Record the results on a graph. Explain why the probability of getting one head and one tail is $2 \times \frac{1}{2} \times \frac{1}{2} = \frac{1}{2}$. What does this have to do with Mendel?

In this case, the two mutually exclusive events are (1) **a** pollen and **A** ovule, and (2) **A** pollen and **a** ovule. If either one or the other of these events occur, the zygote will be an **Aa** heterozygote. The probability of a zygote from such a cross being an **Aa** heterozygote is $\frac{1}{4} + \frac{1}{4} = \frac{1}{2}$.

Another example of the additive law in probability has already been used: the events "zygote is **AA**" and "zygote is **Aa**" are mutually exclusive. The probability of the first, in an **AA** × **Aa** cross, is $\frac{1}{4}$, and the probability of the second is $\frac{1}{2}$. Thus, the probability that an individual will have the dominant phenotype **A–** is $\frac{1}{4} + \frac{1}{2} = \frac{3}{4}$, which, of course, is what Mendel originally observed.

Early in the 20th century, these relationships were put into a graphic form by a fan of Mendel named Reginald Crandall Punnett. Figure 9.6 shows a Punnett square. Each little square represents the simultaneous occurrence of the event directly above it and to its left. Note that there are two **Aa** squares, which must be added to account for all the **Aa** individuals in the F_2 progeny.

By the way, we have called Mendel a mathematical biologist not just because he was trained both in mathematics and in biology, or because he was the first biologist to use statistical analysis in his work, but because of the way he arrived at his conclusions. So what is a mathematical biologist and how does he or she work? Usually he or she starts with some set of observations. In Mendel's case it was the dominance of one trait in the first generation and the reappearance of the recessive trait in some following generation. By some mental process involving both intuition and logic, the mathematical biologist then constructs a *model*. The model is an imaginary biological system, based on the smallest possible number of assumptions, and it is expected to yield numerical data consistent with past observations. New experiments are then performed to test further predictions of this model. If the new data don't fit the predictions, the model is discarded or adjusted to fit the new observations so that further experiments can be done. A model, then, is a biological hypothesis with mathematical predictions.

Models and hypotheses cannot be *proven* with experimental data. We can only say that the data are consistent with the model. Mendel did not *prove* his model of the segregation of alternate traits, but the

9.6 Punnett squares are often used as a means of keeping track of the factors being multiplied in a cross. In the simple cross illustrated here, the factors **A** and **a** are separated (as they are in meiosis) from each parent and placed on the squares. We derive the genotypes of the offspring where the factors intersect in the squares. In this example, each square represents one-fourth of the offspring.

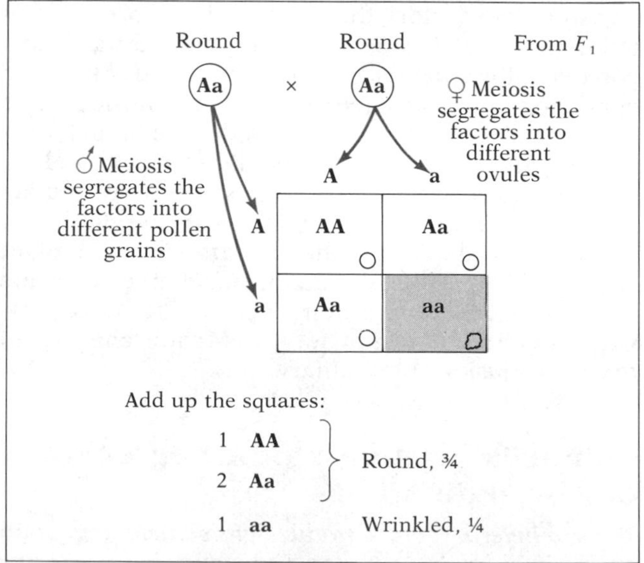

simplicity of the model and the excellence of the fit enabled him to make new predictions and test them. His success came very close to a proof, at least as far as he was concerned. However, others were unconvinced until after the discovery of chromosomes and meiosis. Have you noticed how well Mendel's findings fit with what you already know about meiosis? Imagine how elated Mendel would have been if meiosis had been discovered in his own lifetime!

Mendel's Second Law: Independent Assortment

We have made note of Mendel's success in breaking his problem down to its smallest parts, partly by studying only one character at a time. The next step, obviously, was to study two characters at a time. So he crossed a true-breeding strain that bore round, yellow peas with another true-breeding strain that bore wrinkled, green peas.

The F_1 offspring (which could be categorized and counted while still in the pod) were all round and yellow. We can symbolize this as follows:

$$\mathbf{AABB} \times \mathbf{aabb} \rightarrow \mathbf{AaBb},$$

where **A** and **a** are symbols for the two *alleles* of the round-wrinkled gene locus and **B** and **b** are symbols for the two *alleles* of the yellow-green gene locus. Although we are using Mendel's basic symbols and concepts, we will begin to introduce more modern terms. *Allele* means "a particular form of a gene at a locus." The term *locus* derives from our modern knowledge that each gene occupies a specific place, or *locus* on the chromosome (plural, *loci*). For Mendel's purposes, it was sufficient to suppose that different features of the plant were affected by *alternate factors.*

Mendel's first law (segregation of alternate traits in a heterozygote) refers only to alternate forms (alleles) of a gene that occupy a particular locus on two homologous chromosomes. Here **A** is the symbol for the dominant (round) allele of the round-wrinkled locus and **a** is the symbol for the recessive allele (wrinkled) at the round-wrinkled locus; **aa** is the genotype that gives rise to the not-round, or wrinkled, phenotype. The symbols **B** and **b** represent the dominant and recessive alleles of the yellow-green locus, and the genotypes **BB**, **Bb**, and **bb** give rise to yellow, yellow, and not-yellow (green) pea phenotypes, respectively. Again, **A–** will refer to the set of all individuals with the dominant **A** phenotype (whether they are **AA** or **Aa**) and **B–** will refer to all individuals with the dominant **B** phenotype (**BB** or **Bb**).

Now let's see what happened in the F_2 generation when Mendel crossed plants that were different in two ways (called a *dihybrid cross*) (Figure 9.7). First, the F_1 peas were uniformly round and yellow, as might be expected. In the F_2—the offspring of $F_1 \times F_1$ (**AaBb** × **AaBb**)—Mendel found and classi-

9.7 Independent assortment of two pairs of alleles. Mendel developed his second law by crossing pea plants for two traits simultaneously. The scheme here illustrates the crosses from P_1 through F_2, using the Punnett square to show the results of the F_1 self-pollinated cross. The traits being considered are round and wrinkled shape and yellow and green color, both characteristics of seeds. When true-breeding round-yellows **(AABB)** are crossed with true-breeding wrinkled-greens **(aabb)** (a), the F_1 is heterozygous round-yellow **(AaBb)**, since these are the dominant traits (b). Inbreeding the F_1 generation is a bit more complicated, because each F_1 individual produces four different types of gametes. The F_2 generation is diagrammed in the Punnett square (c). It is comprised of four distinct phenotypes, which include all the possible color and shape combinations. These occur in a 9:3:3:1 ratio (d). Mendel determined this ratio by counting and classifying 556 pea seed offspring.

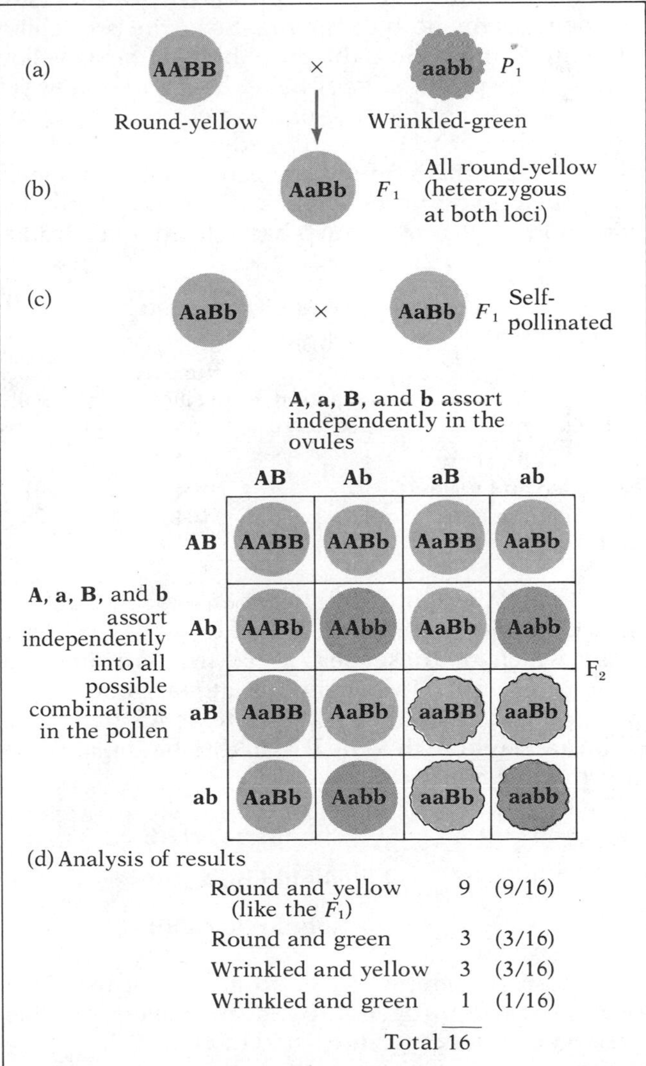

fied 556 peas. He was able to divide them into four groups:

315 round and yellow **A–B–**

101 wrinkled and yellow **aaB–**

108 round and green **A–bb**

32 wrinkled and green **aabb**

Note that 133 peas altogether were wrinkled and 140 peas altogether were green. In either case, this comes close to 139, which is one-fourth of 556.

So about one-fourth of the F_2 peas were wrinkled and one-fourth were green, while three-fourths were round and three-fourths were yellow, which demonstrates Mendel's first law. But the data indicated more than that. Mendel now suspected that the two pairs of contrasting characters were inherited independently. We have already mentioned the multiplicative law of probability, which gives the probability of two independent events both occurring. So, if the probability of being round is $\frac{3}{4}$ and the probability of being yellow is also $\frac{3}{4}$, the probability of being both round and yellow is $\frac{3}{4} \times \frac{3}{4} = \frac{9}{16}$. And $\frac{9}{16}$ of 556 peas is 312.75 peas. Mendel observed 315 round, yellow peas. Remarkably close! Researchers rarely reach such good agreement with their hypotheses. It has been suggested that somebody up there must have liked old Brother Mendel (Table 9.2).

Table 9.2 Mendel's Predictions and Results for F_2 Phenotype

Phenotype of F_2	Fraction predicted	Number predicted out of 556	Number actually observed
Round and yellow	$\frac{9}{16}$	312.75	315
Wrinkled and yellow	$\frac{3}{16}$	104.25	101
Round and green	$\frac{3}{16}$	104.25	108
Wrinkled and green	$\frac{1}{16}$	34.75	32

Mendel thought the fit between expected and observed numbers was impressive. He went on to determine which of these peas were true-breeding and which were not. Here again the fit was good. He expected one-fourth to be **AA**, half to be **Aa**, one fourth to be **aa**; one-fourth to be **BB**, half to be **Bb**, and one-fourth to be **bb**. Then:

$\frac{1}{4} \times \frac{1}{4} = \frac{1}{16}$ should be **AABB**

$\frac{1}{2} \times \frac{1}{4} = \frac{1}{8}$ should be **AaBB**

$\frac{1}{2} \times \frac{1}{2} = \frac{1}{4}$ should be **AaBb**

and so on. Mendel was able to get 529 of his 556 F_2 peas to bear F_3 progeny, so that he could score their genotypes. His results are listed in Table 9.3.

Table 9.3 Mendel's Predictions and Results for F_2 Genotype

Genotype of F_2	Fraction expected	Number expected according to hypothesis	Number actually observed
AABB	$\frac{1}{16}$	33	38
AAbb	$\frac{1}{16}$	33	35
aaBB	$\frac{1}{16}$	33	28
aabb	$\frac{1}{16}$	33	30
AABb	$\frac{1}{8}$	66	65
aaBb	$\frac{1}{8}$	66	68
AaBB	$\frac{1}{8}$	66	60
Aabb	$\frac{1}{8}$	66	67
AaBb	$\frac{1}{4}$	132	138

Again, Mendel felt that the numbers fit his model quite well. He tried combinations of other traits; he even tried three traits together. In each case, the different pairs of alternative traits behaved independently. We know why, of course: It is because maternal and paternal chromosome pairs line up and separate independently during meiosis (Figure 9.8). But Mendel had never heard of meiosis (except that in a sense, he discovered it). Anyway, in Mendel's words:

> The behavior of each pair of differing traits in a hybrid association is independent of all other differences between the two parental plants.

This is known as Mendel's second law, or the law of independent assortment. It can be restated:

> If an organism is heterozygous at two unlinked loci, each locus will assort independently of the other.

This is a modern form, based on Mendel's work as modified by subsequent findings. But how did the word *unlinked* get in there? It was added many years later. What Mendel didn't know was that the law of independent assortment works only for genes that are not on the same chromosome. If **A** and **B** are on the same chromosome, they are said to be linked, and Mendel's law of independent assortment just doesn't hold for such loci. So we might rewrite the law to say:

> When an organism is heterozygous at two different loci, the loci will assort independently, except when they don't.

But we'd better not. In any case, as it happened, none of the three gene loci that Mendel used in his multifactor crosses (round-wrinkled, yellow-green albumin, and grey-white seed coat) were linked. Perhaps Mendel was just lucky. In fact, Mendel had been

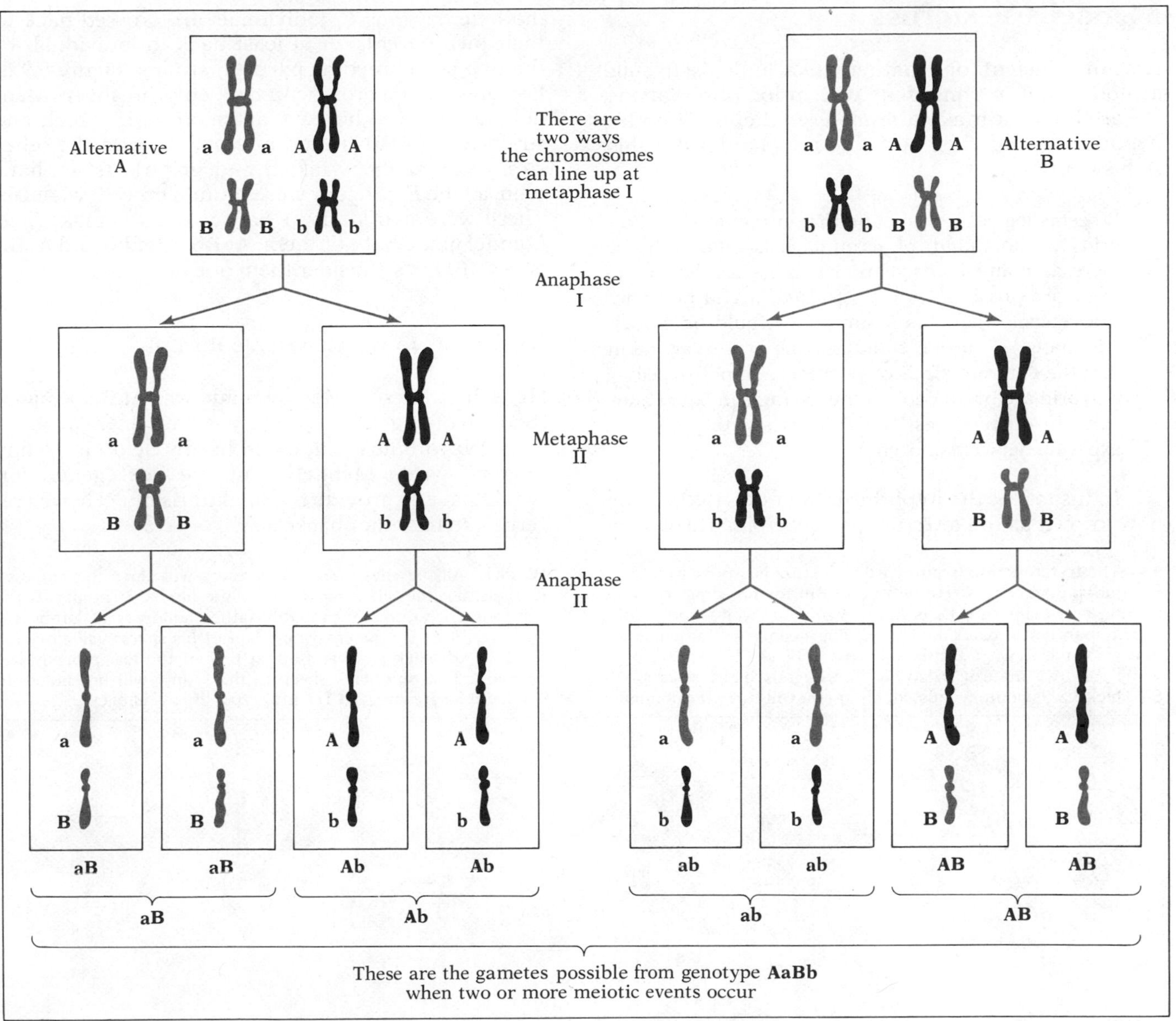

9.8 Independent assortment of chromosomes. When Mendel formulated his second law, the law of independent assortment, he was also describing the behavior of chromosomes in meiosis. Now we can describe the meiotic events that are the chromosomal basis for the second law. The cell entering meiosis with two pairs of genes on two different chromosome pairs has two alternatives at metaphase of meiosis I. The chromosome pairs can align themselves in either manner (A or B in the diagram), and either is a matter of chance. Of course, once a meiotic cell goes to either alternative, that's it for that cell. You have to have two or more cells going through meiosis to get both alternatives. In genetics one must count large numbers of individuals or gametes. In the large number of crosses Mendel carried out, either alternative occurred half the time. Following the alternatives as each continues through meiosis reveals that there are four equally likely combinations possible in the gametes. (If crossing over is taken into account, it is possible to get all four gamete types in a single meiosis. This doesn't change the all-over 1:1:1:1 ratio of the four combinations.)

very lucky.* Anyway, with a little different luck, he might have become completely confused and given up, or he might have discovered gene linkage!

*In 1936, some 36 years after the rediscovery of Mendel's paper in the early part of this century, his data were reanalyzed by R. A. Fisher, a noted statistician and geneticist. The data fit Mendel's model, alright. But it fit *better* than it should have by random chance; his data were literally too good to be true. Either Mendel had fudged his data, consciously or unconsciously, or he had presented only his best results and had left out other, less favorable experiments. Or he was absurdly lucky to an unlikely degree. You can take your choice. Perhaps it is worth noting that Mendel recorded several thousand pea and pea plant phenotypes, and, unlike many other researchers, and unlike Darwin with his snapdragons, never once found a pea he felt he couldn't classify—one way or another.

Mendel's Testcross

It is not sufficient for a mathematical biologist to build a model that is consistent with prior observations. He or she must make a *testable* prediction from the hypothesis—and then test it. And Mendel did this. In his words:

> It seems logical to conclude that in the ovaries of hybrids as many kinds of germinal cells—and in the anthers as many kinds of pollen cells—are formed as there are possibilities. . . . Indeed, it can be shown theoretically that this assumption would be entirely adequate . . . if one could assume at the same time that the different kinds of germinal and pollen cells of a hybrid are produced, on the average, in equal numbers. In order to test this hypothesis, the following experiments were chosen. . . .

In his test of the hypothesis, Mendel used what is now known as the *testcross* or *backcross.* This means the heterozygous F_1 individuals are crossed back to their own parents, or at least back to individuals of the two pure-breeding parental strains. Figure 9.9 illustrates such a cross with coat color in sheep. Mendel chose to use his two-factor cross, in which one strain was **AABB** (round and yellow) and the other strain was **aabb** (wrinkled and green). As we have seen, all the F_1 progeny were round and yellow, **AaBb**. There were two kinds of backcrosses to make, and Mendel made both of them: **AaBb** × **aabb** and **AaBb** × **AABB**. Let's consider them one at a time.

Backcross to the Recessive Parent

Here the test cross Mendel made was **AaBb** × **aabb**. (See top of page 277.)

"A favorable result could hardly be doubted any longer," wrote Mendel, "but the next generation would have to provide the final decision." He was referring to the other backcross.

9.9 The testcross. Sir Beauregard Thickfuzz is a prize ram by all ram-judging standards. He will be useful for breeding, but only as long as we are sure he is homozygous for white wool. White is dominant over black in sheep. The genotypes for color are **WW** = white, **Ww** = white, and **ww** = black. To test our prize white ram, we mate it to some homozygous black ewes. (a) If Sir Thickfuzz is homozygous, all his sperm will carry the dominant **W** allele. All the eggs of the black ewes will carry the recessive **w** allele, but the offspring will be white lambs. If he passes the test, our prize ram will probably father hundreds of lambs. (b) However, if he is a heterozygote, half of his sperm will carry the allele **w** for black wool and about half of the test progeny will be black. Just one black sheep in the family will be enough to condemn Sir Beauregard Thickfuzz to a life of celibacy!

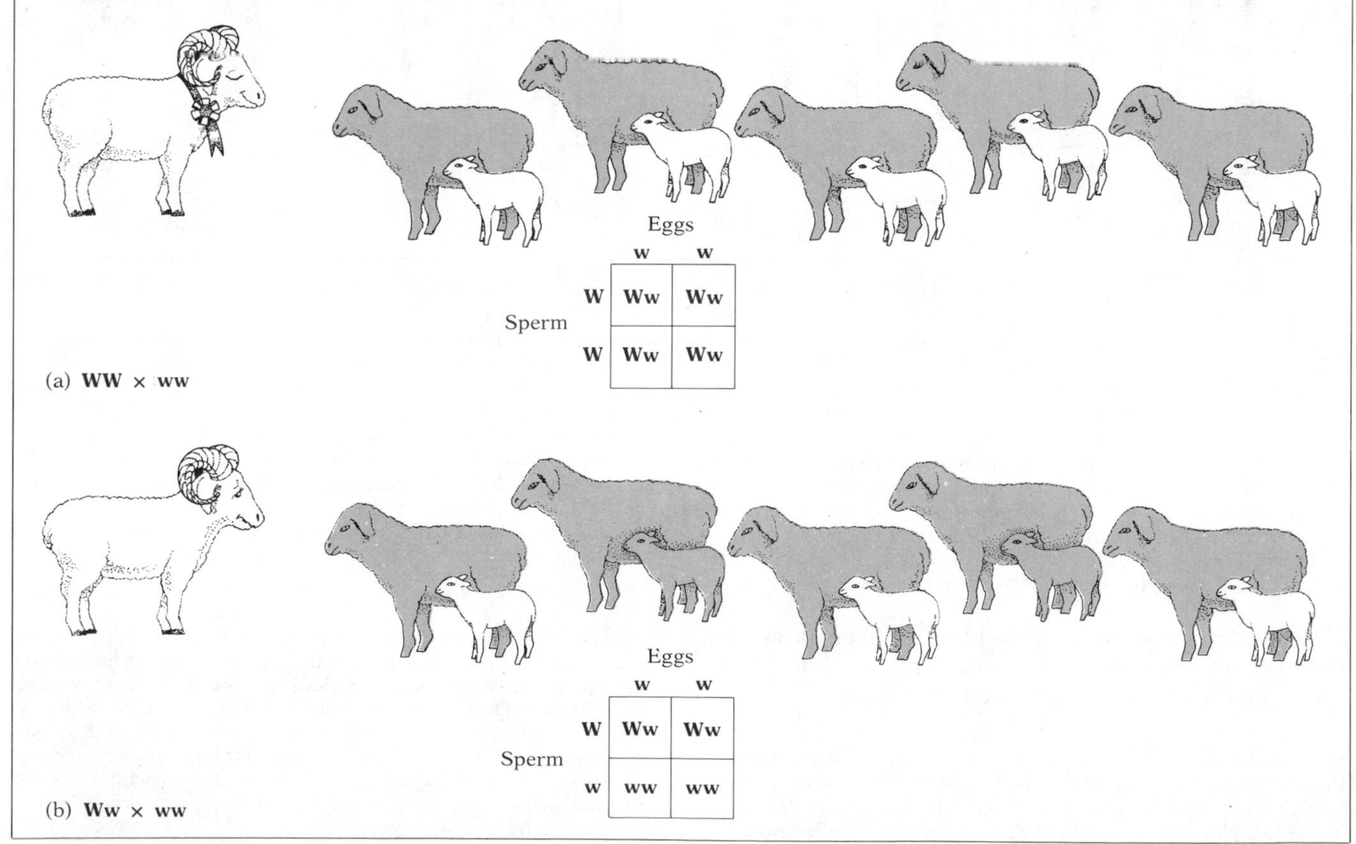

	Gametes from **AaBb** parent			
	AB 25%	**Ab** 25%	**aB** 25%	**ab** 25%
Gametes from **aabb** parent **ab** 100%	**AaBb**	**Aabb**	**aaBb**	**aabb**
Phenotypes	Round, yellow	Round, green	Wrinkled, yellow	Wrinkled, green
Expected (208)	52	52	52	52
Observed (208)	55	51	49	53

Backcross to the Double Dominant Parent

Here the cross was **AaBb** × **AABB**. We can analyze it the same way:

	Gametes expected from **AaBb** parent			
	AB 25%	**Ab** 25%	**aB** 25%	**ab** 25%
Gametes from **AABB** parent **AB** 100%	**AABB**	**AABb**	**AaBB**	**AaBb**
Phenotypes	Round, yellow	Round, yellow	Round, yellow	Round, yellow
Expected	100% round, yellow			
Observed	100% round, yellow			

The only phenotype expected among the progeny of such a cross was **A–B–**, and that was the only one found. However, there were four different *genotypes* that Mendel expected to occur in equal numbers (as shown in the table above), and these could be determined by progeny testing. Mendel planted his round, yellow peas, allowed them to grow up and self-pollinate, and then looked into their pods. Even without counting he could tell whether the plant was **AABB** (it would have only round, yellow peas in its pods) or **AaBb** (it would have some green and some wrinkled peas in its pods as well), or any of the other genotypes. That's why the next generation would have to provide the final decision. His results are given in Table 9.4.

Table 9.4 Progeny Test of Backcross to AABB

Found in pods	Expected	Observed
Round, yellow peas only (**AABB**)	44	45
Round, yellow and green (**AABb**)	44	42
Round and wrinkled, yellow (**AaBB**)	44	47
Round and wrinkled, yellow and green (**AaBb**)	44	43

Mendel had been able to predict the *genotypes* of his all-round, all-yellow backcross progeny. It was quite an achievement. He was able to use his theory to predict something new, and to test his prediction. It held.

The Testcross

Before going on, let's think for a second. What can a testcross tell us? Actually, several things. For instance, if we are not sure whether something that shows the dominant phenotype (**A–**) is homozygous (**AA**) or heterozygous (**Aa**), we can perform a testcross by mating it with a homozygous **aa** individual, as in Figure 9.9. If the unknown is homozygous **AA**, all its offspring will have the dominant phenotype; if it is heterozygous **Aa**, half will have the recessive phenotype.

AA × **aa** → 50% **AA**, 50% **Aa** (100% **A–**)
Aa × **aa** → 50% **Aa**, 50% **aa**

The testcross can also be used to determine whether two genes are linked. How? If the answer doesn't come to you, read on. In this case, a known double heterozygote is crossed with a test stock that is homozygous recessive at both gene loci. If there is independent assortment, the four kinds of offspring occur in equal proportions—why? The reason is:

AaBb × **aabb** → 25% **AaBb**, 25% **Aabb**, 25% **aaBb**, 25% **aabb**

However, if the two gene loci are close together on the same chromosome, some other distribution of progeny will occur, because the two loci will not assort independently, and the multiplicative law can't be used.

Somehow Mendel figured this out, and so he performed the appropriate test to see whether his differing traits assorted independently or not. As we have seen, they did.

The Decline and Rise of Mendelian Genetics

In 1865, after 7 years of experimentation (at the very time Darwin was beating his brains out over the enigma of heredity), Mendel presented his results to a meeting of the Brunn Natural Science Society. His audience of local science buffs just sat there politely, probably not understanding a word of what they were hearing. His single paper was published in the society's proceedings the following year, and was actually distributed rather widely. The learned scientists of the day were just as baffled and just as uninterested as Mendel's original audience. Eventually a German botanist included an abstract of Mendel's work in an enormous encyclopedia of plant breeding. Again, the world responded with silence. Apparently, no one had the foggiest notion of the importance of the Austrian monk's experiments and analysis; 1866 minds were just not ready for 20th century mathematical biology.

Historians have come up with a small, sad but remarkable piece of information. In Darwin's huge library, which is still intact, a one-page account of Mendel's pea work appears in that German encyclopedia of plant breeding. Some relatively obscure work is described on the facing page, and it is covered with extensive notes in Darwin's handwriting. The page describing Mendel's work is clean. Darwin must have seen the paper that would have explained his own snapdragon work and, more importantly, could have clarified his theory of natural selection, saving him years of agony and uncertainty. But even Darwin was not ready for mathematical biology, and he too failed to grasp Mendel's simple but profound ideas.

Mendel's work continued to be ignored until 1900, the year of the great Mendel revival. In that year three biologists in three different countries, each trying to work out the laws of inheritance, each searched through the old literature and came up with Mendel's paper. They immediately recognized its importance. Science had changed in 35 years. The 20th century had arrived, and the obscure monk became one of the most famous scientists of all time. But he had been dead for 16 years.

Later in the 20th century, Mendelian genetics was applied to the theory of natural selection, and Darwin's reputation, which had faded considerably, ascended to new heights. Darwin had been right all along, if only he had gotten his genetics straight. It is often too easy for those of us who have just had something explained to us to say, " . . . but of course." However, breaking new conceptual ground—even a little—can be stultifyingly difficult. Mendel's work and his reasoning, like Darwin's, are so obvious to us now that we often forget how formidable the monk's task was as he placed those first seed peas in the carefully cultivated soil of that monastary garden. You may well have found that even when Mendel's careful reasoning is all laid out, it can still be difficult to follow. If it's any consolation, you are not alone. Biologists, as well as their students, can usually be divided into two groups: Those who dote on abstract reasoning, who eagerly devour Mendel's work and find that solving genetics problems is more fun than crossword puzzles; and those who can't wait to see the last of Mendel and get on to the next topic.

9.2 Mendelian Genetics in the 20th Century

What Causes Dominance and Recessivity?

The various ways in which two alleles at one gene locus can affect the phenotype are called *dominance relationships.* Genes are informational molecules that direct the production of proteins and determine the course of development. In many organisms, including humans, they come in pairs, one member of each pair coming from the father and one coming from the mother. This double dose of information provides, among other things, a failsafe or backup system in case one of the genes isn't working properly.

But what happens when the gene from the father and the corresponding gene from the mother give conflicting information? Say, an allele for brown eyes comes from one parent and an allele for blue eyes comes from the other. What then? Actually, many things can happen, but one of the most common results is that the information from one allele will appear to be ignored. In that case, as we've seen, the other allele is dominant. That's what Darwin and Mendel saw in snapdragons and peas, respectively. *Dominant,* as you know by now, means that the phenotype of the *heterozygote* (the individual with the conflicting genetic instructions) will be exactly like that of one of the two homozygotes. The allele that is expressed no matter what its partner is, is called the *dominant allele,* and the allele that is suppressed in the heterozygote is called the *recessive allele.*

But so far we have given only observations and definitions, not explanations. What really happens when two different alleles occur together? How is one suppressed? And how does the organism "choose" between its two sets of information? Actually, there are several different ways that dominance can happen.

The most usual case is simply that the recessive allele isn't doing anything. For instance, many genes make enzymes; an allele that isn't making anything will result in a phenotype based on the absence of an enzyme's function. This is clearest in rare medical disorders, where the absence of an enzyme can have a severe and often lethal effect. In a relatively benign example, *albinos* lack an enzyme that is necessary to make melanin pigments. They didn't get a functioning gene from their father, and they didn't get one from their mother, so their cells cannot make the pigment. Such individuals are homozygous for recessive alleles, which in this case are alleles that aren't functioning. Heterozygotes for albinism or other enzyme deficiencies, on the other hand, have one working allele and one that doesn't work, and produce only half the usual amount of enzyme; but in most cases half the normal amount of enzyme is still enough to metabolize all the enzyme's substrate, and the phenotype of the heterozygote will be perfectly normal.

Many other genes appear to be involved in the control of developmental processes. We don't understand the genetic control of development nearly as well as we can understand the production of enzymes, but a lot of evidence indicates that *control proteins* are involved—proteins that interact with other genes as well as those that direct other cell processes. Again, half the normal dose of the gene product is usually enough to ensure normal development, and we once again have a situation where a nonfunctioning gene is fully recessive in the heterozygote.

Although we have been talking about recessive alleles that don't do anything, the same situation occurs when the recessive allele is simply doing less of something. Such alleles can still be masked by more active dominant partners.

Sometimes dominance is in the eye of the beholder. For instance, many mutant genes have been found that create bizarre changes in the eye color of *Drosophila* (see Figure 10.2). Almost all of them are recessive to the normal alleles, which are involved in the synthesis of one or both of two strong eye color pigments. For some of the gene loci involved, dominance is once again just a matter of half enough enzyme being enough to complete some necessary step. For other genes, however, the situation is different. Even in these the recessive mutant allele is not functioning; but here, half as much gene activity results in exactly half as much eye pigment being produced. This can be determined easily enough if the pigments are extracted and quantified. But in the living fly, the observer can't tell the difference between a fly with its full, normal amount of eye pigment and a heterozygote with only half as much pigment, and so the normal allele is recorded as being dominant. It's as though the three genotypes were represented by a jar full of India ink, a jar full of water, and a jar with a 50/50 mixture of ink and water. All anyone can see is that there are two jars with black fluid and one with clear fluid. There's an important lesson here: *Dominance depends on how the phenotypes are classified.*

In some cases the dominant and recessive interact in another way. The recessive allele is involved in some normal function, and the dominant allele makes some kind of gene product that prevents that function. For example, true-breeding white leghorn chickens are homozygous for an allele, **I**, that inhibits melanin (color) formation. That's why they're white. Other chicken breeds are homozygous for the alternative allele, **i**, and can be all sorts of colors. In a cross between a white leghorn chicken and another chicken, all the offspring are F_1 heterozygotes:

$$\mathbf{II} \times \mathbf{ii} \rightarrow \mathbf{Ii}$$

And, because the inhibitor allele is dominant, all the offspring of such a cross will be white also. A similar example of dominant color inhibition exists in sheep, where the black recessive allele shows up occasionally in spite of attempts to breed the gene out (Figure 9.9).

Other Dominance Relationships

Whatever its basis, full dominance is common; but it is not universal. There are other ways in which alleles with conflicting instructions can interact. All these different ways are called *dominance relationships.* Sometimes half enough gene activity is *not* enough to ensure normal development. Often, both alleles are "normal," and the conflicting instructions must be compromised. Sometimes, the two alleles do different things, and the organism is better off with these two different alleles than it is with two doses of either one.

Partial Dominance

Whenever the heterozygote is somewhere between the phenotypes of the two homozygotes, and not exactly like either one of them, we can call the relationship *partial dominance.* A classic example of partial dominance is found in snapdragons, in crosses between strains with red and white flowers. Here, the two alleles can be symbolized by **r** and **w**. When the homozygous **rr** red-flowered snapdragons are crossed with homozygous **ww** white-flowered snapdragons, the plants in the F_1 generation, the **rw** heterozygotes, all have pink flowers. If the pink-flowered F_1 plants are self-pollinated to produce an F_2 generation, we have an interesting Mendelian ratio. Instead of a 3:1 or 75:25 ratio, the F_2 snapdragons are 25% red, 50% pink, and 25% white. Since all three genotypes are easily distinguished, the F_2 phenotypic ratio is exactly the same as the F_2 genotypic ratio (Figure 9.10).

Sometimes the phenotype of the heterozygote is not simply a compromise between the two homozygous phenotypes, but has some unique characters of its own. Consider Roy Rogers' horse Trigger, for instance. Trigger was a beautiful palomino horse, with a golden coat and blonde mane and tail. (You can see for yourself. He is stuffed and mounted in the Roy Rogers & Dale Evans museum in Southern California.) All palomino horses are heterozygotes (Figure 9.11). Crosses between palomino horses yield brown, palo-

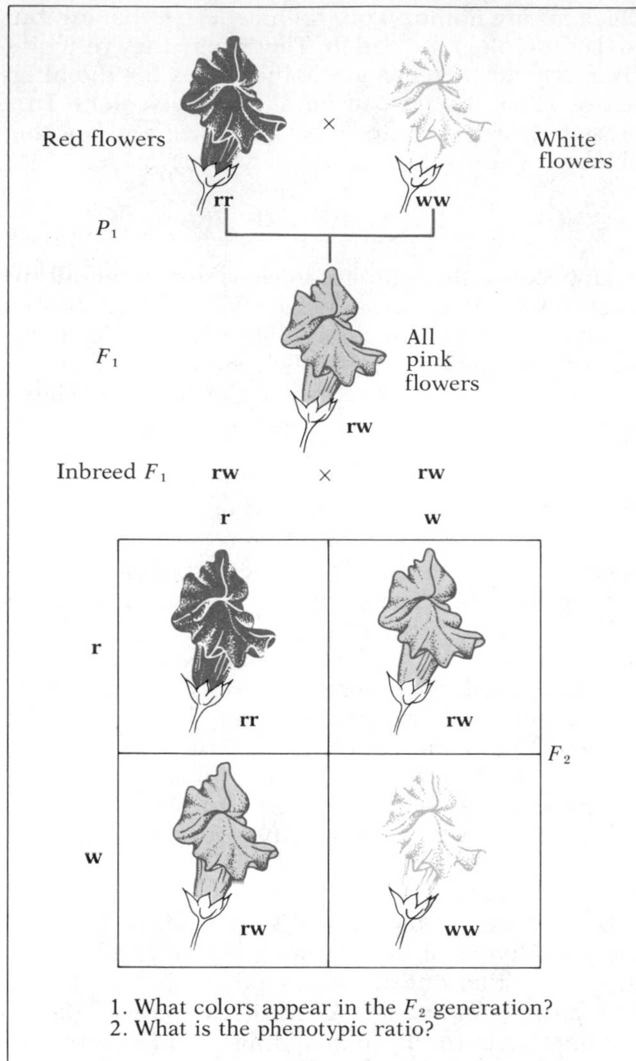

9.10 Snapdragons represent a good example of *partial dominance*, or *no dominance.* When white-flowered snapdragons are crossed with red, the offspring are pink, as shown in the first cross. In the second, pink heterozygotes have been crossed. The result is the appearance of red, pink, and white in the progeny. The ratio of the three colors is 1:2:1. Since we can identify the heterozygous progeny by their color, we have derived both the phenotypic and genotypic ratios. Is it possible to produce a true-breeding strain of pink snapdragons?

mino, and white foals in an approximate 25:50:25 ratio.

Lethal–Recessive Dominant Alleles

Here we are once again in the realm of abnormal, mutant alleles. *Lethal–recessive dominant* alleles, however, don't have to become homozygous to work their terrible ways. A single dose is enough to alter the phenotype substantially. A lethal–recessive dominant allele is recessive with regard to lethality, which means that homozygous individuals will die—usually when they are still tiny embryos. The heterozygote doesn't usually die, but it isn't normal, either.

Lethal–recessive dominant is an awkward phrase, and such alleles are usually called *rare dominants.* Although such alleles are indeed rare, they are dominant only in the sense that their visible effects are not recessive. A couple of examples will be more helpful than any definition.

Probably the most familiar rare dominant trait is *achondroplastic dwarfism,* or *achondroplasia.* The heterozygous carriers of the defective allele appear normal at birth, but they don't grow normally. The arm and leg bones, in particular, stay very short, and there are usually facial bone anomalies as well. Affected persons have nearly normal bodies and normal-sized heads, and are generally in good health and often extremely athletic, although they are conspicuously stunted because of their short limbs (Figure 9.12).

Achondroplastic dwarfs are usually fertile, although studies have shown that they tend not to marry. If they do marry, and their mates happen to be normal homozygotes, about half their children will be dwarfs like themselves (in keeping with Mendel's first law). Sometimes, two achondroplastic dwarfs will marry each other. Such matings might be expected to produce children who are homozygous for the dwarf allele; apparently, however, the double dose of the allele is usually not compatible with survival of the embryo.

Dominant traits, unlike recessive traits, do not skip generations (except for the special case of alleles with incomplete penetrance, which we will deal with later). An individual cannot carry the allele with-

9.11 A palomino horse with her foal. The beautiful golden color is the result of heterozygosity of alleles for brown and white coat color. Palomino horses do not breed true.

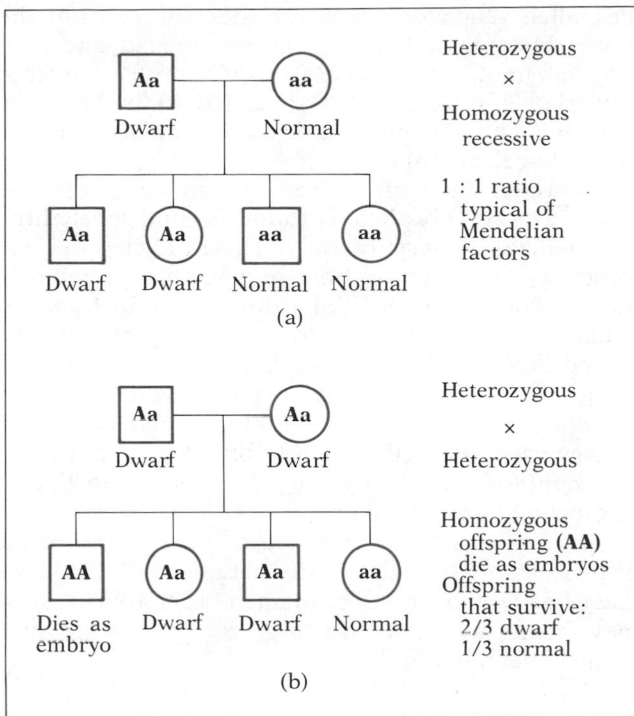

9.12 Achondroplasia is a dominant genetic trait in which an individual's arms and legs remain very short. Two pedigrees are shown here: (a) an achondroplastic dwarf **(Aa)** marries a normal person **(aa);** on the average, the couple can expect about half their children to be affected **Aa** heterozygotes. (b) Occasionally two dwarfs marry **(Aa × Aa).** One-fourth of the zygotes they produce will be **AA,** which apparently is lethal in embryonic development. Of their surviving children they can expect, on the average, that about two-thirds will be affected **Aa** heterozygotes. About one-third will be normal, and will carry no genetic tendency for the condition.

out showing its effects. So if you are normal but your mother and two brothers are dwarfs, you don't have to worry about passing the dominant gene on to your own children: you don't have it.

There are many less common or less noticeable dominant traits. In congenital night blindness, the affected carriers have no functional rods in their retinas; they have only cones. In one family pedigree, the trait has been traced through 300 years, with each affected person having either an affected mother or an affected father.

How do we know that these rare dominant alleles are lethal when they are homozygous? For most rare dominants in humans, we really don't know this. But similar conditions occur in other organisms where we can make the crucial genetic crosses, and in these organisms nearly all dominant visible abnormalities are also recessive lethals. All manx cats, for instance, are heterozygous for a peculiar allele that leaves them without tails. A mating of two manx cats produces a litter of (on the average) two-thirds manx cats and one-third alley cats. Actually, the ratio at the time of conception is one-fourth alley cat, one-half manx cat, and one-fourth embryonic lethal; but since we don't see the last category, the proportions in the litter become $\frac{1}{3}$, $\frac{2}{3}$, and 0, respectively.

In our earlier discussion of dominance, we noted that one functioning gene was usually enough to produce a normal phenotype. But for some gene loci, both copies of the gene have to be working in order for development to be normal. Then just one normal allele is not enough, and the heterozygote does not develop normally. The defective allele in such a case would be classified as a dominant. If, in addition, the complete absence of the normal allele were lethal, the defective allele would be classed as a lethal–recessive dominant.

Codominance

Sometimes one homozygote will show one phenotypic trait, the other homozygote will have a different phenotypic trait, and the heterozygote will show both traits. The condition is called *codominance.* The simplest and most dramatic case occurs in spotted cats. One gene locus produces spotting as a simple recessive trait. However, another gene locus determines the color of the spots. Homozygous **BB** cats have black spots. Homozygous **RR** cats have orange spots. Heterozygous **BR** cats have black spots *and* orange spots.

Codominance is actually rather rare among traits that are easily seen, but it is common among genetic traits that can be measured by biochemical and immunological tests (so called *in vitro* (in glass) phenotypes). Codominance is often encountered, for instance, in the genetics of blood groups. Let's consider one of these as another example.

In the MN blood group system, the two alleles are **M** and **N**. The three genotypes, then, are **MM**, **MN**, and **NN**. The in vitro phenotype is the agglutination of red blood cells with special reagents, called *antisera,* which contain antibodies from immunized rabbits. Two antisera are used, anti-**M** and anti-**N**. Here is how the three genotypes are determined:

Genotype	Reaction with anti-**M**	Reaction with anti-**N**
MM	+	−
MN	+	+
NN	−	+

Where, + means that the red blood cells agglutinate, or clump, and − means that they do not. Look at the reaction with anti-**M**. Considering these three reactions alone, it is clear that allele **M** is dominant. Now look at the reaction with anti-**N**. For this test, **N** is clearly dominant. When both antisera are used, alleles **M** and **N** are codominant.

Overdominance

The phenotype of the heterozygote does not always have to be within the range bounded by the phenotypes of the two corresponding homozygotes. For instance, we see *overdominance* if one homozygote is short, the other homozygote is tall, but the heterozygote is even taller than the tall homozygote. Overdominance with regard to Darwinian fitness (ability to survive and reproduce) is the most interesting case to biologists. When overdominance increases fitness, both of the alleles can become common in the population.

Pigeons have two common alleles for slightly different forms of a protein called *egg-white transferrin.* The alleles are simply codominant by the in vitro criterion of gel electrophoresis, in which the two protein variants can be characterized by their rate of movement in an electric field. But they are overdominant by other criterion, namely Darwinian fitness as measured by egg hatchability. One allele appears to give the egg a greater resistance to fungal infections, while the other allele appears to give greater resistance to bacterial infections. The two together give the egg a greater chance of surviving than either allele alone can provide, a clear case of overdominance.

Multiple Alleles

Fortunately, Mendel dealt with only two alleles at any gene locus, and most of our examples so far have dealt with only two alleles. However, there is a great deal of genetic variation among the normal, or wild-type, alleles. Often, there are many forms of a gene that can occur at a given gene locus, although any one individual can have only two—one from each parent. A Harvard research group recently did an intensive study of a randomly chosen enzyme locus in a North American fruit fly population. They uncovered thirty-seven different alleles at one locus, out of a sample of only 146 flies. As far as could be determined, all thirty-seven variants functioned normally although they produced proteins with different amino acid sequences.

In a much larger sample, even more alleles of human *beta hemoglobin* have been found. One variant at this locus is the **S** allele, which causes sickle-cell anemia in homozygotes, although it gives heterozygotes some protection against malaria (see Essay 9.1). Most people are homozygous for the **A** allele of the beta hemoglobin locus, even in Africa where the **S** sickle-cell allele is fairly common. In other parts of the world where malaria is a problem, other alleles of the hemoglobin locus may be common. For example, in parts of Asia, **C** and **E** alleles occur. Like the sickle-cell allele, each of these variants differs from the normal hemoglobin by a single amino acid, and each appears to offer heterozygotes some protection from malaria. Homozygous **CC** and **EE** individuals are somewhat anemic, but the effects are not as severe as in sickle-cell anemia.

Another allele of the hemoglobin locus is **th** (for *thalassemia*). This allele is found in moderately high frequencies in parts of Italy. The **th** allele does not produce an abnormal beta hemoglobin protein—in fact, it doesn't produce anything at all. In homozygous persons, it gives rise to a severe genetic disease called *thalassemia major.* Children with thalassemia major usually die in infancy. Heterozygotes have half the normal amount of hemoglobin, which makes them somewhat anemic (their condition is known as *thalassemia minor*) but apparently gives them some kind of protection against malaria.

In addition to these only moderately rare forms, nearly 200 other alleles of the beta hemoglobin locus have been reported. All of them are rare. In fact, most have been found only in a single family or sometimes in just one individual.

Blood Groups

Moving from the hemoglobin inside the red blood cell to the molecules on the surface of the same cell, we have already noted that people have different *blood types.* The most important and best known blood group system is the ABO system, and the corresponding four blood groups to which people belong are A, B, O, and AB. The blood types are determined by cell-surface *antigens* carried by persons of different genotypes. The term *antigen* refers to any substance that produces an immune response, usually by producing *antibodies.* Antibodies are proteins that are produced in response to any foreign substance, such as an antigen. The antigens of the ABO system happen to be polysaccharides. The red blood cells of a type A person, for instance, have A cell-surface antigens and the blood cells of a type B person do not, but instead have B cell-surface antigens. Type AB people have both A and B antigens, so the genes for A and B are codominant. The two alleles are usually written $\mathbf{I^A}$ and $\mathbf{I^B}$, in part so that it's clear whether one is talking about the allele that makes the antigen or about the antigen itself. The fourth blood type, or blood group, is type O, and type O individuals have neither the A antigen nor the B antigen on the surfaces of their red blood cells. A third allele, then, is $\mathbf{I^O}$; this allele is recessive to $\mathbf{I^A}$.

Genotype	Blood type (phenotype)
$\mathbf{I^AI^A}$	A
$\mathbf{I^AI^O}$	A
$\mathbf{I^OI^O}$	O

The allele I^O is also recessive to I^B (in a similar manner). This is a case of multiple alleles, then. The three alleles can combine into six different genotypes, but since I^AI^A and I^AI^O are both type A, and I^BI^B and I^BI^O are both type B, there are only four ABO blood groups.

The ABO genotypes have another, more indirect effect on the phenotype. The immune systems of all individuals, for reasons that are not entirely understood, always produce antibodies against the A and B antigens, whichever is lacking on the individuals' own cell surfaces. Thus, type A persons produce anti-B antibodies and type B persons produce anti-A antibodies; type O persons produce both kinds of antibodies, and typc AB pcrsons produce neither.

The antibodies are specific, two-headed binding proteins. By *two-headed* we mean that each antibody is able to bind to two of the specific molecules that it is directed against. The anti-B antibody, for instance, can bind tightly to the B molecule (antigen) on the surface of one red blood cell and at the same time bind tightly to another B molecule on the surface of another red blood cell. Since each red blood cell is covered with these antigens, the antibodies will cause susceptible cells to clump, or *agglutinate.* In fact, the agglutination reaction can be seen to occur in drops of blood on a glass slide, and this is how blood of different types is determined (Figure 9.13).

As a result of the A and B antibodies produced by persons of various genotypes, the blood types of donors and recipients have to be very carefully matched. Type A blood transfused into a type B person will be agglutinated by the antibodies of the recipient, forming fatal clots. In some cases, type O blood can be transfused into persons of other blood types, but even this is unwise, because the introduced type O blood brings in with it some anti-A and anti-B antibodies.

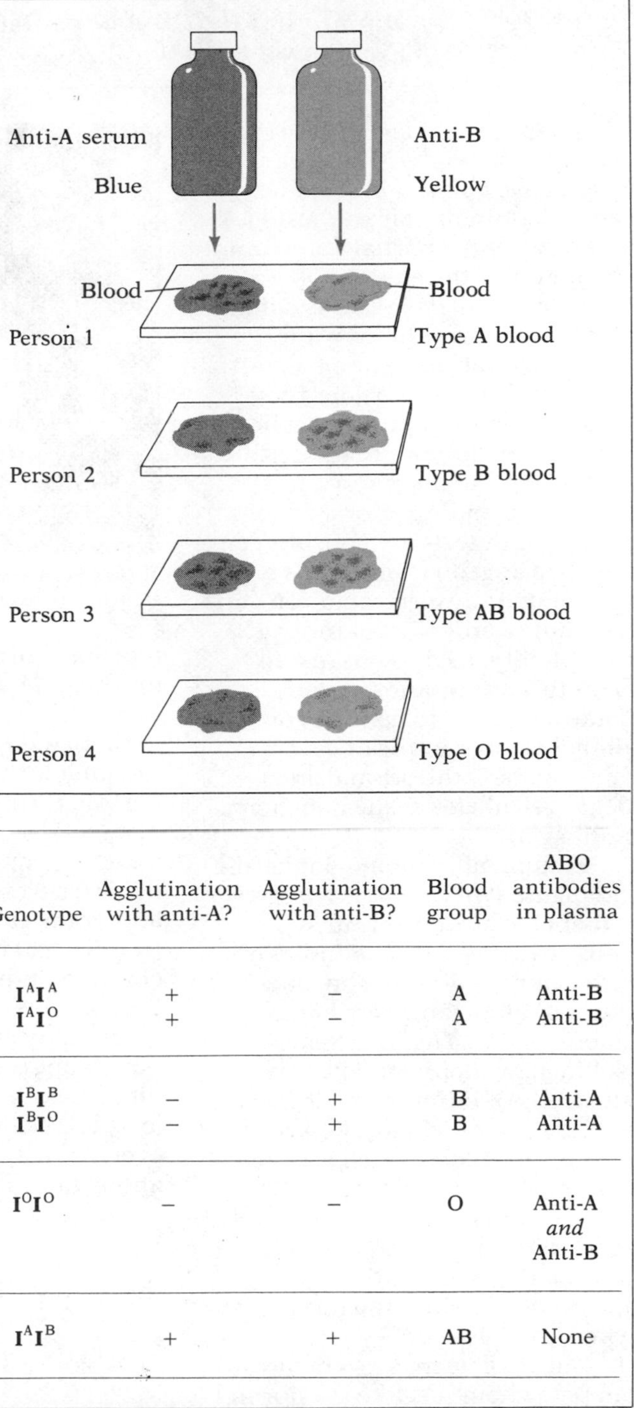

Genotype	Agglutination with anti-A?	Agglutination with anti-B?	Blood group	ABO antibodies in plasma
I^AI^A	+	–	A	Anti-B
I^AI^O	+	–	A	Anti-B
I^BI^B	–	+	B	Anti-A
I^BI^O	–	+	B	Anti-A
I^OI^O	–	–	O	Anti-A *and* Anti-B
I^AI^B	+	+	AB	None

9.13 ABO blood typing. The presence of A and B antigens in one's blood can be readily determined in a simple test using anti-A and anti-B test reagents. (These reagents are usually extracted from plants and are usually color-coded for convenience.) As the scheme shows, two drops of blood and two drops of reagent (one of anti-A and one of anti-B) are used. The reaction (or the lack of reaction) is then used to determine the blood type. A positive reaction, called agglutination, is characterized by a distinct graininess occurring in the mixture. The red cells eventually form a patchwork pattern in the serum. Where there has been a negative reaction (no agglutination), the blood remains homogeneous, with an even texture.

The Rh Blood Group

Other blood group loci also frequently have multiple alleles. For example, there is the Rh (from rhesus, the monkey in which it was found) blood group system, which has eight fairly common alleles and many rarer ones. You're probably aware that you're referring to the Rh system when you add "positive" or "negative" to your blood type. It's particularly important to have this information if you intend to use someone else's blood or bear children, but, here, we're interested in the genetics of the system.

As in the case of the ABO system, the different alleles can be identified according to how the blood types react to each other. One Rh antigen is particularly reactive with other types and the Rh blood groups can be divided into *Rh positive* and *Rh nega-*

Essay 9.1 Dominance Relationships Depend on How the Phenotype Is Classified: Sickle-Cell Hemoglobin

Dominance relationships are in the eye of the beholder. For instance, normal eye color in *Drosophila* can be a dominant if you just look at the eye, but a partial dominant if you measure the amount of pigment. Manx allele in cats is a dominant if you consider only the presence or absence of a tail, but a recessive if you consider the presence or absence of the whole cat. The **M** allele is a dominant by one agglutination test and a recessive by another, or a codominant if both tests are considered together. Pigeon transferrins are codominant by electrophoretic criteria but overdominant for egg hatchability. And so on. Just to drive this important principle home, we'd like to consider one all-inclusive example of everything, namely the normal and sickle-cell alleles of the beta hemoglobin gene.

Human adult hemoglobin (the red pigment of blood) is a large protein consisting of four subunits: two *alpha chains* and two *beta chains*. Different gene loci code for alpha and beta hemoglobin chains. The beta hemoglobin gene locus actually has hundreds of different alleles, most of them rare. We'll just consider the normal allele, which is usually written Hb^A, and the most common mutant allele, the sickle-cell allele, which is written Hb^S. We'll shorten the symbols down to **A** and **S** and consider the three genotypes **AA**, **AS**, and **SS**.

The **S** allele makes a hemoglobin beta chain just like the normal beta chain, except for 1 out of its 146 amino acids. The normal hemoglobin beta chain has a glutamic acid at position 6, and sickle-cell beta chain has a valine at that position.

The valine on the surface of the molecule causes it to form large, sickle-shaped crystals that deform the cell and reduce the oxygen-carrying capacity of the molecule. In addition, the **S** allele produces less hemoglobin—only about 40% as much as the **A** allele (although this varies from family to family). The homozygous **SS** person, then, has a severe anemic disease. His or her blood will also tend to form small clots that clog the arterioles of the muscles, heart, and brain, causing pain, heart trouble, and brain damage. The condition is usually lethal for **SS** homozygotes, but **AS** heterozygotes enjoy essentially normal health under normal modern conditions. So **A** appears, clinically at least to be dominant over **S**. However, the **AS** carrier may have only about 70% as much hemoglobin as a normal **AA** person, and is somewhat more subject to anemia when placed under stress. Which makes the **A** allele only *partially dominant* over the **S** allele.

The sickle-shaped cells occur in the blood of heterozygous carriers when a sample of that blood is deprived of oxygen and viewed under the microscope. Carriers can also be detected by gel electrophoresis, since the two proteins have different electrophoretic mobilities. **AA** individuals show a fast band on the electrophoretic apparatus; **SS** individuals show a slow band; and **AS** heterozygotes show both bands. That's codominance.

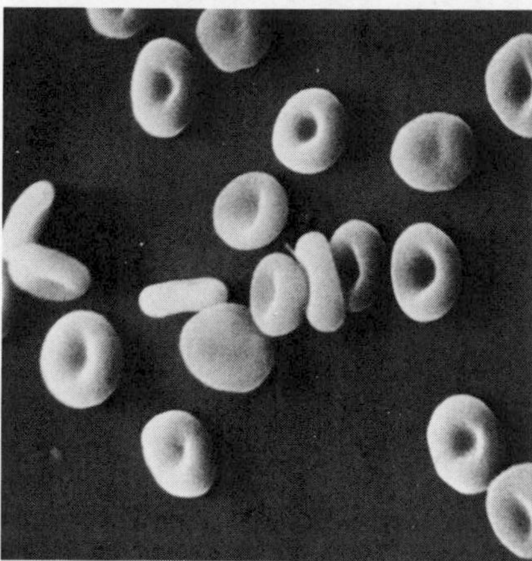

The severe changes in red blood cells are shown here as sickling occurs. The normal red blood cells are shown at the left for comparison.

Threonine — Proline — Glutamic acid — Glutamic acid

Normal hemoglobin (**A**) positions 4–7

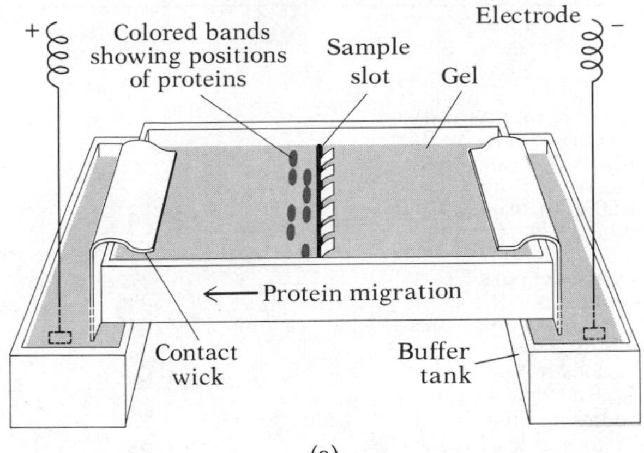

(a)

Phenotype	Genotype	Hemoglobin electrophoretic pattern (Origin → +)	Hemoglobin types present
Normal	**AA**		A
Sickle-cell trait	**AS**		S and A
Sickle-cell anemia	**SS**		S

(b)

Using gel electrophoresis (a), it is possible to compare the hemoglobin in the three genotypes in question, **AA**, **AS**, and **SS** (b). Because of the differences in amino acid composition, which are reflected in the charges on the protein, each of the hemoglobins migrates differently. Normal hemoglobin (**AA**) forms a single band and is seen to migrate the furthest. Homozygous (**SS**) sickle-cell hemoglobin also forms a single band, but lags behind the normal hemoglobin. The heterozygote (**AS**) hemoglobin produces two bands; one of these has the same mobility as the **AA** band and the other has the same mobility as the **SS** band. As far as gel electrophoresis is concerned, the two alleles are simply *codominant.*

The **S** allele is common in parts of Africa where malaria is endemic, and there it shows *overdominance* with the **A** allele with respect to Darwinian fitness. Where malaria is endemic, EVERYONE is bitten by malaria-carrying mosquitoes and EVERYONE contracts malaria. Heterozygotes (**AS**) are much less seriously affected by malaria because blood cells influenced in any way by sickling do not provide a suitable home for the malarial parasite. Thus, in these areas, heterozygotes survive better, live longer, and have more children than homozygous (**AA**) persons—in other words, there is overdominance for Darwinian fitness. This means that, in these areas, there is selection *for* the sickling allele so that in some places nearly 40% of the population are carriers for the sickle-cell gene. Few adults, however, are sickle-cell homozygotes, because **SS** persons don't live very long, especially under primitive conditions.

Is the sickle-cell allele a dominant allele, a recessive allele, a partial dominant allele, an overdominant allele, a codominant allele, or a lethal—recessive dominant allele? It is all of these, depending on how the phenotype is classified.

1. *Dominant:* The **S** allele is dominant for the trait *sickle-shaped cells under the microscope.*
2. *Recessive–lethal dominant:* The **S** allele is dominant for the above trait, and effectively a recessive–lethal under primitive conditions.
3. *Recessive:* The **S** allele is recessive for the severe anemic disease syndrome.
4. *Codominant:* The **S** and **A** alleles are codominant for their electrophoretic patterns in vitro.
5. *Partial dominant:* The **A** and **S** alleles are partial dominants for the amount of hemoglobin, the **AS** genotype being just halfway between **AA** and **SS.**
6. *Overdominant:* Under conditions of endemic malaria, the **A** and **S** alleles are overdominant with regard to Darwinian fitness. The fitness of the **AS** heterozygote exceeds that of either homozygote.

Our use of the **S** and **A** alleles as examples of dominance, codominance, partial dominance, and overdominance may have seemed confusing or even perverse, but it was deliberate, and for a good reason. The moral of the sickle-cell anemia tale is that dominance is, in fact, in the eye of the beholder—a function both of the alleles and of the particular aspect of their effects that one chooses to call *phenotype.*

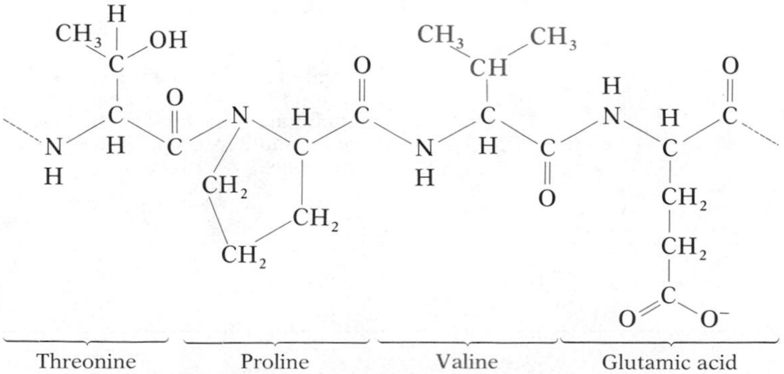

Threonine — Proline — Valine — Glutamic acid

Sickle-cell hemoglobin (**S**) positions 4–7

9.14

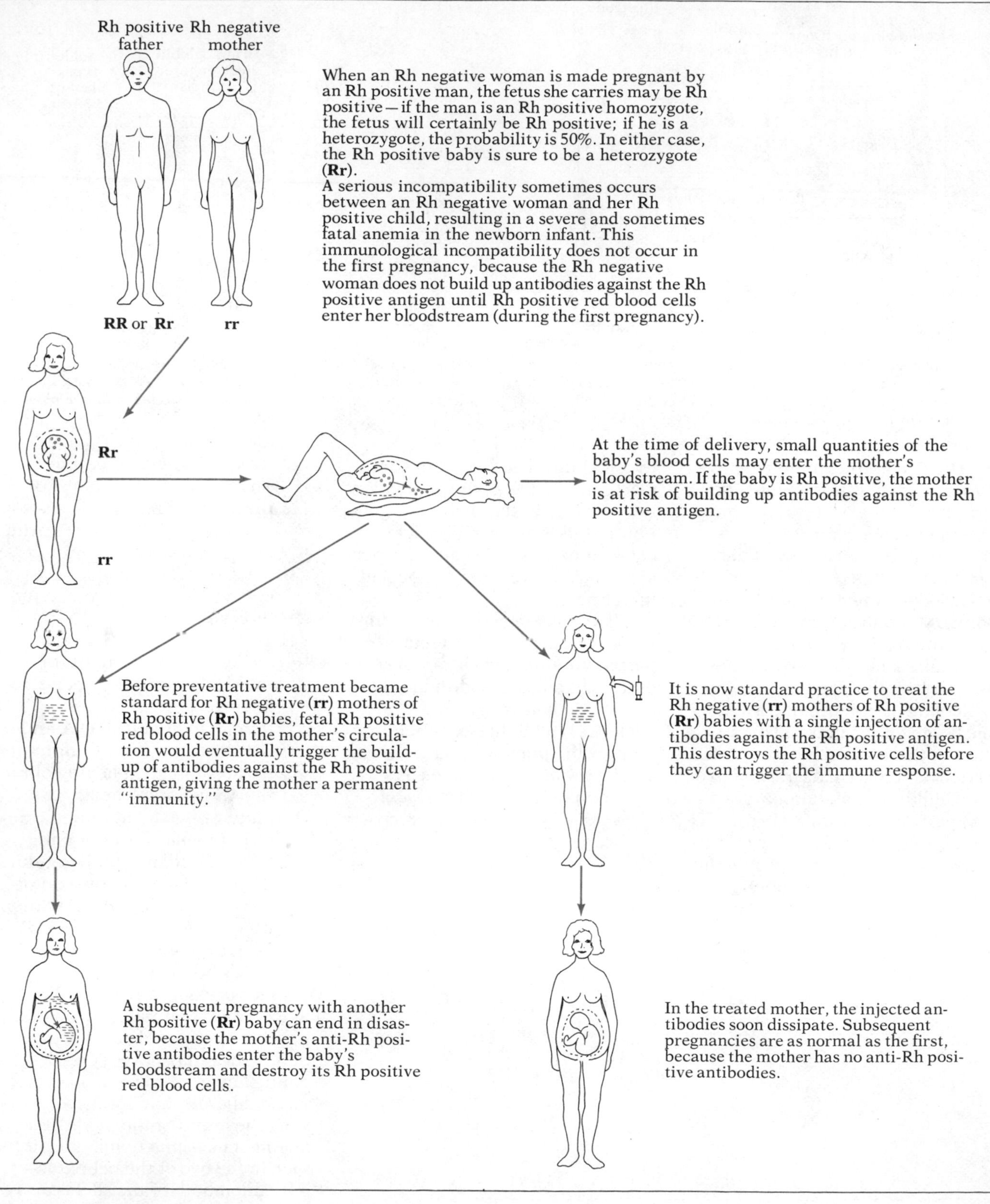

Rh positive father
Rh negative mother
When an Rh negative woman is made pregnant by an Rh positive man, the fetus she carries may be Rh positive—if the man is an Rh positive homozygote, the fetus will certainly be Rh positive; if he is a heterozygote, the probability is 50%. In either case, the Rh positive baby is sure to be a heterozygote (**Rr**).
A serious incompatibility sometimes occurs between an Rh negative woman and her Rh positive child, resulting in a severe and sometimes fatal anemia in the newborn infant. This immunological incompatibility does not occur in the first pregnancy, because the Rh negative woman does not build up antibodies against the Rh positive antigen until Rh positive red blood cells enter her bloodstream (during the first pregnancy).
RR or **Rr**
rr
Rr
rr
At the time of delivery, small quantities of the baby's blood cells may enter the mother's bloodstream. If the baby is Rh positive, the mother is at risk of building up antibodies against the Rh positive antigen.
Before preventative treatment became standard for Rh negative (**rr**) mothers of Rh positive (**Rr**) babies, fetal Rh positive red blood cells in the mother's circulation would eventually trigger the buildup of antibodies against the Rh positive antigen, giving the mother a permanent "immunity."
It is now standard practice to treat the Rh negative (**rr**) mothers of Rh positive (**Rr**) babies with a single injection of antibodies against the Rh positive antigen. This destroys the Rh positive cells before they can trigger the immune response.
A subsequent pregnancy with another Rh positive (**Rr**) baby can end in disaster, because the mother's anti-Rh positive antibodies enter the baby's bloodstream and destroy its Rh positive red blood cells.
In the treated mother, the injected antibodies soon dissipate. Subsequent pregnancies are as normal as the first, because the mother has no anti-Rh positive antibodies.

tive on the basis of the presence or absence of this one—the most potent—Rh antigen. Care is taken when blood transfusions are given to match the donor and recipient for both ABO and Rh blood groups. A mismatch in either case can have serious consequences, involving anything from malpractice awards to death.

People don't naturally have anti-Rh antibodies, but giving an Rh negative person an Rh positive blood transfusion would *immunize* him or her against the Rh positive factor, causing that person to *produce* antibodies and making a future transfusion risky. (Any large foreign molecule introduced into the bloodstream causes the immune system to start making specific antibodies against that molecule. If the system meets that molecule again, then it is ready to react against it; see Chapter 28.)

Let's see why Rh negative women seem to be so kindly disposed toward potential mates who are also Rh negative. (Figure 9.14). It is basically because Rh negative women sometimes build up antibodies against the Rh positive antigen once they have given birth to an Rh positive baby, since mixing of infant and maternal blood sometimes takes place during the birth process. Thus, Rh negative women can become "innoculated" against their Rh positive fetuses. This can cause trouble in subsequent pregnancies, because the mother's new antibodies can enter the blood of the next embryo and destroy its red blood cells (Figure 9.15).

But Rh negative women no longer have to search out Rh negative men. Those that have just given birth to an Rh positive baby are now routinely given injections of anti-Rh serum, which destroys any fetal red blood cells that have leaked into the mother's body before the mother's immune system starts building up antibodies against them. This prevents Rh incompatibility problems with any future pregnancies.

9.15 Erythroblastosis, the destruction of red blood cells in the fetus, is now uncommon in medically sophisticated societies. The child from which this blood sample was taken inherited its father's Rh positive phenotype. Its mother, an Rh negative person, was probably sensitized by a former pregnancy. Her anti-Rh antibodies found their way into the baby's bloodstream and began the systematic destruction of red cells shown here. Mass transfusions were necessary to save the baby.

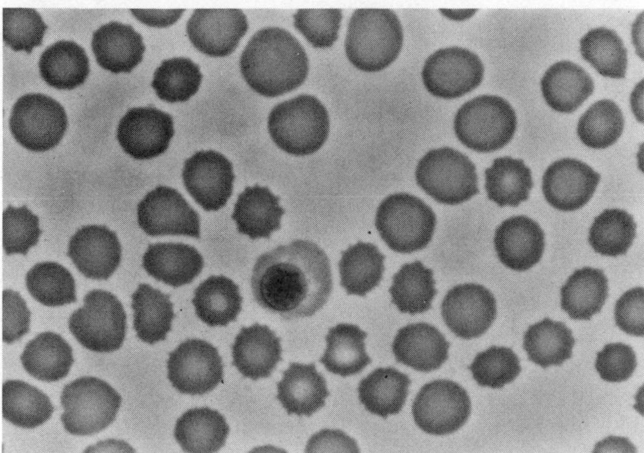

Transplants

Careful genotype matching must also be done in the case of heart and other organ transplants. Cells other than red blood cells have *histocompatibility antigens* on their surfaces; these are controlled principally by four closely linked *histocompatibility loci.* In spite of there being only four loci, matching a donor and a recipient is extremely difficult. This is because each histocompatibility gene locus has hundreds of alleles, and none of them are common. Virtually every person is unique for his or her cell-surface antigens, identical twins being the obvious exception, and immediate family members more likely to show similarities. So you are unique. But if you should ever need a heart transplant, you may wish you were not so special. One technique that facilitates transplantation is to chemically destroy the body's immune system. The side effects, however, are traumatic, and treated individuals are vulnerable to a host of invading disease organisms.

Gene Interactions and Modified Mendelian Ratios

You will recall that Mendel found that the F_2 of the round, yellow and wrinkled, green cross yielded a 9:3:3:1 ratio. After Mendel's work was rediscovered, his enthusiastic followers gleefully produced this ratio again and again. Their work is a textbook standard, and clearly indicates some of the ways in which genes at different loci can interact to produce different phenotypes.

This all worked out very nicely for Mendel and his followers because the round, wrinkled and yellow, green character pairs don't react with one another, but this is not true of all pairs of gene loci. For instance, it doesn't work for loci that regulate coat color in mice. In this case, we find that at one gene locus, **B**, is dominant to **b**, such that **BB** and **Bb** mice are black and **bb** mice are brown. A cross between a homozygous black (**BB**) mouse and a homozygous brown (**bb**) mouse will produce nothing but black heterozygotes (**Bb**). A cross between two heterozygous black mice produces an F_2 generation with three-fourths black mice and one-fourth brown mice. So far so good.

But at another gene locus, **C** is dominant to **c**, such that **CC** and **Cc** mice can make pigment (black or brown) and **cc** mice cannot, and are thus white albinos. The **C** allele, in effect, allows for coat color. So the two gene loci, at **B** and **C**, control two different steps in the biochemical pathway that produces the pigment normally present in mouse fur.

Now consider a mating between a true-breeding white mouse and a true-breeding brown mouse. What would you expect? You might *not* expect the litter to be entirely black. Mendel, however, wouldn't have been surprised. After all, here is one possibility in a mouse cross:

CCbb (brown) × **ccBB** (white) → **CcBb** (black)

So the black F_1 are heterozygous at two gene loci. The real surprise comes at the next cross, however. If you mate two such double heterozygotes, you will find that in the F_2 generation one-fourth are **cc** (white), regardless of what's happening at the **B** locus. Of the remaining colored mice, three-fourths are black and one-fourth are brown. But $\frac{3}{4} \times \frac{3}{4} = \frac{9}{16}$, as before, and $\frac{1}{4} \times \frac{3}{4} = \frac{3}{16}$. The phenotypic classes of the F_2 are: $\frac{9}{16}$ black; $\frac{3}{16}$ brown; and $\frac{4}{16}$ white (Figure 9.16). This is just the old 9:3:3:1 ratio with the last two terms combined (9:3:4), because once a mouse is white you can't tell whether it is brown or black. Perhaps we should say this another way. One cannot distinguish between **BBcc**, **Bbcc**, and **bbcc** without doing a progeny test. We say that the **Cc** locus is *epistatic* to the **Bb** locus. *Epistasis* refers to the masking of a trait determined by one pair of genes by the action of another pair of genes.

Similarly, one can combine two *different gene loci* that have *identical effects.* For instance, **pp** mice are also white, while **PP** and **Pp** mice are normally pigmented. These genes control yet another step in the pigment pathway. The F_1 offspring of **PPcc** and **ppCC** are **PpCc**, a genotype which gives normally pigmented mice (let's say black). In the F_2 generation, about half are white and about half are black. If we have large enough numbers, we may be able to show that the ratio is not really half black or half white, but nine-sixteenths black to seven-sixteenths white. This is the 9:3:3:1 ratio again, but this time the last three groups are lumped together:

$\frac{9}{16}$ **P–C–**	Black	$\frac{9}{16}$ black
$\frac{3}{16}$ **P–cc**	White	
$\frac{3}{16}$ **ppC–**	White	$\frac{7}{16}$ white
$\frac{1}{16}$ **ppcc**	White	

Or, put another way, once a mouse is white, it can't be any whiter, but as long as it has both a **P** allele and a **C** allele, it will produce the normal black coat color.

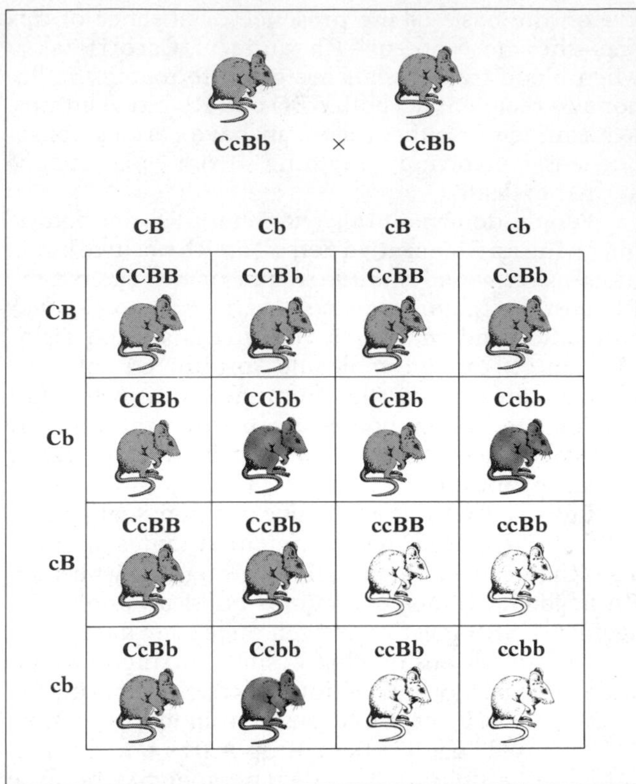

9.16 In this scheme we follow the path of two pairs of genes that influence the coat color of mice. The **B** (black) and **b** (brown) alleles occur at one locus, and the **C** (color) and **c** (albino) alleles occur at another locus. The two pairs assort independently in typical Mendelian fashion but produce strange ratios when heterozygous F_1 are inbred **(CcBb × CcBb).** The resulting ratio is 9:3:4 because of an epistatic interaction.

In white leghorn chickens, you will recall, the dominant **I** allele produces an inhibitor of melanin production. This means that **II** and **Ii** chickens are white and **ii** chickens are pigmented. But other strains of chickens may be white because of a homozygous recessive allele at another locus, **Cc**. A mating of two true-breeding white chickens then produces an F_1 generation **IiCc**, which is all white because of the dominant inhibitor gene. In the F_2 generation, we obtain the white to colored ratio 13:3 as follows:

$\frac{9}{16}$ **I–C–**	White
$\frac{3}{16}$ **I–cc**	White
$\frac{3}{16}$ **iiC–**	Colored
$\frac{1}{16}$ **iicc**	White

It should be apparent at this point that with a little ingenuity, it is possible to find examples of almost any lumping of two loci that will give modified 9:3:3:1 ratios. Mendel himself came up with one, which he reported in an appendix to his pea paper as a "partly successful" hybridization experiment with string beans. The two true-breeding bean species differed in a number of traits; in particular, one had white flowers, while the other had crimson flowers. The blossoms of the F_1 generation were less intensely colored than those of the crimson parental species, so this was a case of incomplete dominance. The F_1 plants were relatively sterile, but Mendel did manage to get thirty-one F_2 plants to germinate and eventually flower. Only one had white flowers; "the remaining 30 plants developed flower colors that represented several gradations, from crimson to pale violet." What was going on? If you know, you have an incredibly good grasp of genetics at this point.

In the case of the string beans, Mendel inferred that there were two more or less independent systems determining flower color. The crimson flower resulted from the presence of both pigment systems; and the white flower from the absence of both. The remaining intergradations, then, developed from partial presences and absences. If one considers only the presence versus absence of pigment (colored versus white), the modified ratio is 15:1. The observed occurrence of one out of thirty-one is well within the limits of error. (Mendel was careful to mention that his data were equally close to what one would expect if there were three pigment systems, in which case one F_2 plant out of sixty-four should have white flowers. Why is this true?)

Mendel gave some tables of the F_2 genotypes expected under this hypothesis. He called the parental strains $\mathbf{A_1A_1A_2A_2}$ (where two alleles, $\mathbf{A_1}$ and $\mathbf{A_2}$, code for crimson) and **aaaa** (white). The F_1, then, was $\mathbf{A_1aA_2a}$. So, in the F_2 we can expect to find 1:2:1 ratios for each of the two loci independently, and together:

$\frac{1}{16}$ $\mathbf{A_1A_1A_2A_2}$	$\frac{2}{16}$ $\mathbf{A_1aA_2A_2}$	$\frac{1}{16}$ $\mathbf{aaA_2A_2}$
$\frac{2}{16}$ $\mathbf{A_1A_1A_2a}$	$\frac{4}{16}$ $\mathbf{A_1aA_2a}$	$\frac{2}{16}$ $\mathbf{aaA_2a}$
$\frac{1}{16}$ $\mathbf{A_1A_1aa}$	$\frac{2}{16}$ $\mathbf{A_1aaa}$	$\frac{1}{16}$ $\mathbf{aaaa}$

Mendel did not specify which of the genotypes should be associated with which intermediate color, only that $\mathbf{A_1A_1A_2A_2}$ was crimson and **aaaa** was white. Infertility of the hybrids prohibited further analysis, much to Mendel's annoyance. But two things were suggested here that later were to become important: the 15:1 modified F_2 ratio and the Mendelian basis of the inheritance of *quantitative* (more or less) characters. We'll return to quantitative inheritance later.

Application of Ideas

1. Promiscuous Alice had five lovers in one week, became pregnant, and doesn't know who the father is. Alice received marriage proposals from her five former lovers, but she has decided that only the true father of her child can have her hand in marriage. Scientific Alice's blood type is B negative (i.e., has B antigen, no A antigen, no Rh^+ antigen). Her five suitors have volunteered blood samples of their own:

Henry	O negative
Arthur	A negative
William	B negative
Schuyler	A positive
Buddy	AB positive

The child is born before indecisive Alice can make up her mind, and proves to be O positive. Which of the five men could have been the child's father?

2. In one breed of chickens, *pea* comb **(PP, Pp)** is dominant to *single* comb **(pp)**. In another breed of chickens, *rose* comb **(RR, Rr)** is dominant to *single* comb **(rr)**. When a homozygous *pea* comb chicken **(PPrr)** is mated with a homozygous *rose* comb chicken **(ppRR)**, the F_1 progeny all have yet a fourth phenotype, *walnut* comb **(P–R–)**. What are the expected phenotypic ratios among the F_2 progeny of two F_1 *walnut* comb **(PpRr)** chickens? In the progeny of an F_1 *walnut* comb chicken **(PpRr)** crossed to a *single* comb **(pprr)** chicken?

3. In another breed of chickens, the **BB** chickens are *black*, **Bb** chickens are *blue*, and **bb** chickens are *white*. A rooster from a true-breeding strain of *black, single* comb chickens **(BBpp)** is mated with several hens from a true-breeding strain of *white, pea* comb chickens **(bbPP).** What is the expected phenotype of the F_1 offspring, and what kinds of chickens can be expected in the F_2 generation in what expected frequencies? Construct a Punnett square of the $F_1 \times F_1$ (i.e., **BbPp** × **BbPp**) cross.

4. Assume that crossing over does not occur in the region of each chromosome that is associated with the centromere. You have 46 such centromeric regions in each cell, 23 that are copies of DNA in your mother's egg and 23 that are copies of DNA in your father's sperm. Any gametes that you make will have just 23 centromeric regions.
   a. For a given gamete, what is the probability that all 23 centromeric regions will be copies of the same ones that you got from your mother?
   b. Considering maternal and paternal origin of centromeric regions alone, how many different kinds of gametes can you produce?
   c. It has been estimated that the average person is heterozygous for at least 2000 protein-forming genes. How many different kinds of gametes is it theoretically possible for each person to produce?

5. Hagiwara crossed two true-breeding strains of Japanese morning glories, both of which had blue flowers. The plants of the F_1 generation all had purple flowers. When Hagiwara crossed the purple-flowered F_1 back to one of the parental strains, the progeny of the backcross were approximately

50% blue-flowered and 50% purple-flowered. When the purple-flowered F_1 was crossed to the other parental strain, the results were the same. But when the F_1 purple-flowered progeny were crossed to each other, the F_2 offspring included purple-flowered, blue-flowered, and scarlet-flowered plants. Hagiwara concluded that two gene loci were involved, an **Aa** locus and a **Bb** locus. The scarlet-flowered plants, then, must have been **aabb.**

a. What were the genotypes of the parental strains, the F_1 hybrid, and the progeny of the two backcrosses?
b. All phenotypes and all genotypes appear in the F_2. List the phenotypes and the genotypes, with their expected ratios.
c. Below is a list of individual crosses between pairs of F_2 plants, together with the ratios of phenotypes found in the progeny. Determine the genotypes of the parents in each cross.

6. Baur crossed two true-breeding strains of *Antirrhinum majus*, the same snapdragon that Darwin had experimented with. One strain had regular (bilaterally symmetrical) white flowers, and the other had peloric (radially symmetrical) red flowers. The F_1 progeny, which had regular pink flowers, were crossed to each other to produce an F_2 generation. Here is Baur's data on the F_2, in numbers of plants of each flower phenotype:

regular pink	94
regular red	39
regular white	45
peloric pink	28
peloric red	15
peloric white	13

a. Explain the results.
b. For the observed total of 234 F_2 progeny, compute the *expected* number of each phenotype, and compare with the observed number.
c. Calculate the chi-square value: that is, the square of the difference between expected and observed numbers, divided by the expected number for each category; summed over the six categories.

Male parent	Female parent	Progeny: Blue	Purple	Scarlet
purple	scarlet	—	100%	—
purple	scarlet	50%	50%	—
purple	scarlet	25%	50%	25%
blue	blue	25%	50%	25%
blue	blue	75%	—	25%
blue	blue	100%	—	—
blue	blue	—	100%	—
blue	bluc	50%	50%	—

Key Words

ABO system
acondroplasia
additive law
agglutination
albino
allele
antibody
antigen
antisera

backcross
beta hemoglobin
biochemical pathway
blending inheritance
blood group

carrier
cell-surface antigen
codominance

Darwinian fitness
dihybrid cross
dominance

Key Ideas

When Darwin Met Mendel (Almost)

1. Darwin had little knowledge of genetics and subscribed to the idea of *blending inheritance.*
2. *Blending inheritance* presumes that traits from both parents "blend" or mix to produce traits in the offspring.
3. Critics of natural selection were quick to point out how blending would destroy the variations which were a key to Darwin's explanation of evolution.
4. Darwin carried out a cross between peloric and regular flowers in the snapdragon. He referred to *dominant* traits as prepotent.
5. Mendel's preparation for the work he was to undertake included concentrated study in science and mathematics.

Mendel's Crosses

1. Mendel used carefully planned experiments and applied statistical analysis to his data.
2. Mendel reported on experiments with seven different characteristics of garden peas. These included seed form, color of contents, color of coats, color of pods, shape of ripe pods, length of stem, and position of flowers.
3. His approach was to cross *true-breeding* strains, manipulating pollen by hand to avoid *self-pollination.*
4. He didn't use the symbols P_2, F_1, F_2, etc., to designate generations beginning with parental, but these are now traditional.

epistasis

filial generation (F_1, F_2, F_3)

gel electrophoresis
genotype

heterozygous
histocompatibility antigen
homozygous
hybrid

inhibitor allele
in vitro

latency
law of independent assortment
laws of probability
law of segregation
lethal–recessive dominance
linked gene
locus (loci)

MN blood group system
multiple alleles
multiplicative law
mutually exclusive

overdominance

P_1
partial dominance
phenotype
progeny
progeny testing

quantitative inheritance

rare dominant
recessive
Rh negative
Rh positive

self-fertilization
self-pollination
sickle-cell
statistical analysis

testcross
thalassemia
transferrin
true breeding

5. In crossing two *true-breeding* strains with different characteristics he did not observe blending, but did note that one characteristic, the *recessive* one, was absent in the F_1. He called the trait that did appear *dominant.*

6. Both Mendel and Darwin perceived the same results in their crosses. The *recessive* trait disappeared in the F_1 generation and reappeared in the F_2.

7. Darwin called this phenomenon *latency* and pursued the problem no further.

8. Mendel noted that the reappearance of a *recessive* trait occurred with a definite frequency in one-fourth of the F_2.

9. In peas, each pea in a pod has its own *genotype* and *phenotype. (Genotype:* total combination of an organism's genes; *phenotype:* combination of observed or measured traits, generally what is readily visible.)

10. Mendel accomplished *cross pollination* in peas by opening the flower, removing the anthers, and introducing pollen with a small brush. This prevented *self-pollination.*

11. He determined that for each dominant trait there were two kinds (*genotypes*), those that produced two kinds of offspring (*heterozygous*), and those that produced one (*homozygous*).

12. To determine whether an individual with a dominant trait is *heterozygous* or *homozygous* requires *progeny testing.*

13. Mendel carried out *progeny testing* by breeding an F_3 generation from F_2 round peas. He determined that one-third of the round F_2 peas were *true-breeding,* two-thirds were not. From this he determined that one-fourth of the total F_2 were round and *true-breeding;* one-half of the F_2 were round and not *true-breeding;* and one-fourth of the F_2 were wrinkled and *true-breeding.*

14. Using the seven traits mentioned earlier, Mendel tested his findings and they were consistent with the above data each time.

15. Mendel soon realized that his crosses could be represented algebraically. Thus the F_1 cross **Aa** × **Aa** can be stated (**A** + **a**) × (**A** + **a**) or simply $(\mathbf{A} + \mathbf{a})^2$. When multiplied this becomes $\mathbf{A}^2 + 2\mathbf{Aa} + \mathbf{a}^2$, which is the same as **AA** + 2**Aa** + **aa.** From this analysis he decided the following:

   a. Characters were controlled by factors in two forms, *dominant* and *recessive.*
   b. *Hybrids* had two different factors, one from each parent.
   c. If the factors were the same, the individual was true-breeding (*homozygous*).
   d. If the factors were not the same (*heterozygous*), the *dominant* factor would prevail (produce the *phenotype*).

16. From these data Mendel formulated his first law.

Mendel's First Law: The Segregation of Alternate Alleles

1. *Heterozygous* individuals produce two kinds of *gametes* in equal proportions:

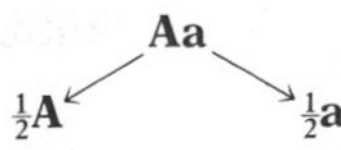

2. With these numerical factors known, the multiplicative law from the *laws of probability* can be applied to crosses.

3. "The probability of two independent events both occurring is equal to the *product* of their individual probabilities."

4. The F_1 cross **Aa** × **Aa** can be stated ($\frac{1}{2}$**A** + $\frac{1}{2}$**a**) × ($\frac{1}{2}$**A** + $\frac{1}{2}$**a**). Multiplying produces $\frac{1}{4}$**AA** + $\frac{1}{2}$**Aa** + $\frac{1}{4}$**aa.**

5. The $\frac{1}{2}$**Aa** is determined algebraically but can be explained genetically. The combination **A** + **a** can occur two ways, **A** + **a** or **a** + **A.**

6. The *additive law* is applied: The probability of either one or another of two mutually exclusive events occurring is equal to the *sum* of their individual probabilities.

7. The probability of an **A** pollen and an **a** ovule combining is one-fourth as is the probability of **a** + **A** and they are mutually exclusive. Therefore the probability of a *heterozygous* F_2 individual is $\frac{1}{4} + \frac{1}{4} = \frac{1}{2}$.

8. Punnett illustrated Mendel's principles using squares:

	A	a
A	AA	Aa
a	aA	aa

When summed up the results are

Genotypically: $\frac{1}{4}$**AA** + $\frac{1}{2}$**Aa** + $\frac{1}{4}$**aa**
Phenotypically: $\frac{3}{4}$ *Dominant*
$\frac{1}{4}$ *Recessive*

9. Mendel can be considered a mathematical biologist because he constructed a model that yielded numerical predictions consistent with observations. A model is a biological hypothesis with mathematical predictions. A convincing test of Mendel's model was the discovery of chromosomes and meiosis.

Mendel's Second Law: Independent Assortment

1. Mendel's next task involved studying two characters at the same time.
2. His cross involved two *alleles* of the round *locus*, **A** and **a**, and two *alleles* of the yellow *locus*, **B**, and **b**. *Locus* refers to a specific place on a chromosome (location of a gene).

AABB × **aabb** (P_1 cross)
all **AaBb** (F_1 offspring)
AaBb × **AaBb** (F_1, *dihybrid cross*)

3. For Mendel's results see p. 273. To predict the results consider the following:
   a. If the two traits are first considered separately you know that **Aa** × **Aa** produces $\frac{1}{4}$ **AA**, $\frac{1}{2}$ **Aa**, and $\frac{1}{4}$ **aa**. Likewise **Bb** × **Bb** produces $\frac{1}{4}$ **BB**, $\frac{1}{2}$ **Bb**, and $\frac{1}{4}$ **bb**.
   b. To predict the results when both are considered simultaneously follow the *multiplicative law* for all possible **A** and **B** combinations. These are as follows:

Genotype	Separate probabilities		Combined probabilities	Grouped into *phenotypes*
AABB	$\frac{1}{4} \times \frac{1}{4}$	=	$\frac{1}{16}$	
AaBB	$\frac{1}{2} \times \frac{1}{4}$	=	$\frac{1}{8}$	Round and yellow
AABb	$\frac{1}{4} \times \frac{1}{2}$	=	$\frac{1}{8}$	$\frac{9}{16}$
AaBb	$\frac{1}{2} \times \frac{1}{2}$	=	$\frac{1}{4}$	
aaBB	$\frac{1}{4} \times \frac{1}{4}$	=	$\frac{1}{16}$	Wrinkled and yellow
aaBb	$\frac{1}{4} \times \frac{1}{2}$	=	$\frac{1}{8}$	$\frac{3}{16}$
AAbb	$\frac{1}{4} \times \frac{1}{4}$	=	$\frac{1}{16}$	Round and Green
Aabb	$\frac{1}{2} \times \frac{1}{4}$	=	$\frac{1}{8}$	$\frac{3}{16}$
aabb	$\frac{1}{4} \times \frac{1}{4}$	=	$\frac{1}{16}$	Wrinkled and Green $\frac{1}{16}$
			$\frac{16}{16}$	$\frac{16}{16}$

4. Mendel's findings as seen in Table 9.2 are very consistent with this prediction.
5. The findings indicate more than this, however. The two characters, pea shape and color, are inherited independently. If they were not, the multiplicative law wouldn't have worked.
6. Today we know that all of the characters Mendel studied this way were located on different chromosomes. Because of the random way pairs of chromosomes align at metaphase and separate at anaphase, one pair of *homologues* separates independently of any other pair of *homologues*.
7. Mendel's second law can be restated: "If an organism is *heterozygous* at two unlinked *loci*, each *locus* will assort independently of the other."

Mendel's Testcross

1. Mendel used *testcrosses* or *backcrosses* to determine whether a *dominant* type was *homozygous* or *heterozygous*, and whether assortment was independent.

2. *Testcrosses* can be made by *backcrossing* suspected individuals to fully *recessive* and fully *dominant* individuals. These are called *backcrosses* because the P_1 is the source of the types needed.

3. Mendel's *testcrosses:*

AaBb × aabb and **AaBb × AABB**

4. In the first cross, if the test individual was *heterozygous,* the expected result would be: all four categories of round, wrinkled, yellow, and green in equal numbers.

5. If the test individual was partially or fully *homozygous,* or if assortment was not independent, other results would be found.

Backcross to the Double Dominant Parent

1. In the second cross, back to a *homozygous* dominant, all the offspring would be *dominant* round and yellow. To complete the test each offspring would have to be *progeny-tested.*

2. By *self-pollinating* the *heterozygous* offspring, Mendel determined by the results, the validity of his model.

The Testcross

Testcrossing back to *recessive* types can help determine two things:

1. It can determine whether the test subject is *homozygous* or *heterozygous.*
2. It can determine if genes are linked. When *testcrossed,* double *heterozygotes* will produce a 1:1:1:1 ratio in the offspring if there is independent assortment. If the gene pairs are linked, some other ratio will show up.

The Decline and Rise of Mendelian Genetics

Mendel's findings were not understood in his time, even by Darwin who, it is believed, might have glanced over them. His work was rediscovered about the turn of the century.

What Causes Dominance and Recessivity?

1. The double dose of genetic information provides organisms with a backup system in case of failure.

2. When the information is in conflict, one *allele* (*recessive*) is usually ignored, the other (*dominant*) is expressed.

3. *Recessive* may mean that a gene is simply not functioning. An example of this is *albinism* where a *recessive* cannot fulfill its role in producing the pigment melanin. Two alleles failing to produce pigment results in an albino individual.

4. *Recessive alleles* may be functioning, but to a lesser degree than *dominant* genes.

5. In some instances *dominance* is apparent visually, but on closer examination or measurement there is a difference between one gene functioning normally and two genes functioning normally. In fruit flies, for example, there may be less pigment produced by one gene than by two.

6. Some *dominant alleles* are also known as *inhibitor alleles,* since they prevent *recessive alleles* from expressing themselves. The *dominant* white of white leghorn chickens is an example.

Other Dominance Relationships

Four specific *dominance* relationships are: *partial dominance, lethal–recessive dominance, codominance,* and *overdominance.*

Partial Dominance

Partial dominance in snapdragons is seen when red and white are crossed. The offspring are pink. *Phenotypic* and *genotypic* ratios are the same ($\frac{1}{4}$:$\frac{1}{2}$:$\frac{1}{4}$).

Lethal–Recessive Dominant Alleles

Lethal–recessive dominance or simply *rare dominance* is seen in *achondroplasia* or *achondroplastic dwarfism. Homozygous* offspring die as embryos, while *heterozygotes* survive. The purely *dominant* does not get to be expressed, therefore the name *lethal–recessive dominance.*

Codominance

1. *Codominance* occurs when one *homozygote* expresses a trait differently from the other *homozygote*, but the *heterozygote* shows both traits. In cats, black spotting *or* orange spotting represent the two *homozygotes*. Black *and* orange spotting represents the *heterozygote*.

2. The blood groups **M** and **N** are another example of *codominance*. When tested with *antisera*, *agglutinations* show that each *genotype* can be biochemically identified. Both **M** and **N** are clearly expressed.

Overdominance

In *overdominance* the *heterozygote* expresses the *dominant* trait to an even greater degree than does the *homozygous dominant*. In pigeons, one *allele* helps fight off fungi while the other helps fight off bacteria. Together they increase hatchability or *Darwinian fitness*.

Essay 9:1 Dominance Relationships Depend on How the Phenotype Is Classified: Sickle-Cell Hemoglobin

1. Clinically, the normal gene (**A**) for *beta hemoglobin* appears to be *dominant* over the *sickle-cell* gene (**S**). (The *heterozygote* functions normally enough.)

2. Measuring the amoung of normal hemoglobin produced by the *heterozygote* (**AS**) tells us that it produces only 70% of the hemoglobin of the *homozygous dominant* **AA.** So *partial dominance* is indicated.

3. In Africa, the *heterozygote* is better able to resist malaria than the *homozygous dominant*, and of course the *homozygous recessive* is already ill. Therefore *overdominance* is indicated.

4. Electrophoretic patterns show that **A** and **S** genes are simply *codominant*.

5. For cell sickling the *recessive–lethal dominant* situation can be applied. The **S** *allele* behaves as a *dominant* since it creates the effect in spite of the presence of the **A** *allele* in the *heterozygote*.

Multiple Alleles

1. The term *multiple alleles* means that in a population there may be more than two alternative *alleles* for a character. **A, B, C,** and **D** may be alternative genes for the same character. Individuals in the population, however, still get their *alleles* in pairs.

2. The gene that translates into *beta hemoglobin* (one of the polypeptides of the hemoglobin tetramer) can have the **A** or **S** *allele*, but there are others. (Examples are the **C** and **E** *alleles* from Asia and the **th** allele from the Mediterranean region.)

3. It is possible for individuals to have a pair of *alleles* such as **S** and **th,** but it is rare and survival questionable.

4. Variants in the hemoglobin *alleles* usually produce amino acid substitutions, some more serious than others. (Recall from Chapter 7 what the base substitutions of DNA do.)

Blood Groups

1. The blood types of the *ABO system* are determined by *cell-surface antigens*. *Antigens* are large molecules that are capable of reacting with specific *antibodies*. In terms of *dominance*, A and B are *codominant* and both are *dominant* over O. A, B, and O represent multiple *alleles*.

2. From the *multiple alleles* A, B, and O there are six *genotypes* possible. From these six *genotypes* there are only four blood groups or *phenotypes*.

3. The immune systems of individuals produce either anti-A or anti-B *antibodies*. Type A persons produce anti-B, B persons produce anti-A, AB people produce neither, and O people produce both.

4. When *antibodies* meet opposing *antigens* on red cell surfaces they bind to these and form bridges to other *red* cells nearby. The massive binding of these cells is called *agglutination* or clumping.

5. Because of *antigen–antibody* reactions, blood transfusions have to be preceded by careful blood-matching tests.

The Rh Blood Group

1. *Rh* blood groups can be divided into Rh^+ and Rh^-. The Rh^+ *antigen* is very potent. In blood transfusions, this factor must be considered along with the *ABO* factors.

2. A and B antigens are common in most living organisms, and the related *antibodies* probably evolved as a defense against disease organisms.

3. *Rh antibodies* are only present when a person receives Rh^+ *antigens.* The presence of *antigen* induces an immune reaction and the *antibodies* are produced.

4. The immune reaction of Rh^- people to Rh^+ *antigens* extends into reproduction. Rh^+ fathers pass the *antigen*-producing gene to their offspring. When Rh^- mothers bear Rh^+ children there can be a mixing of maternal and fetal blood causing the mother to produce *antibodies.* Subsequent pregnancies are potentially dangerous, since the *antibodies* can enter the fetal blood and cause *agglutination.* The immune response can be clinically suppressed if the incompatibility is known.

Transplants

Organ and tissue transplants are difficult because of four *histocompatibility loci* that produce *histocompatibility antigens.* Because of the large number of *multiple alleles,* matching between people is extremely difficult and the immune system of the recipient must be suppressed to avoid graft rejection.

Gene Interactions and Modified Mendelian Ratios

1. Mendel's two factor crosses worked well because the pairs of genes involved did not interact with each other.

2. In many instances the 9:3:3:1 ratio Mendel observed is not produced when other *loci* or organisms are studied in similar crosses.

3. Sometimes pairs of *alleles* control different steps in producing the same trait. These are known as *epistatic* genes or *alleles.*

4. An example of *epistatic* genes at work is seen in mouse hair color. One pair of *alleles* produces brown or black, but another pair permits or prevents color at all. When the color preventer (a *recessive*) is *homozygous,* the mice are white.

5. In the *epistatic* cross **BbCc** × **BbCc,** the *phenotypic ratio* in the offspring turns out to be 9:3:4, instead of the classic 9:3:3:1. This is because every offspring with a **cc** combination is white.

6. In the *epistatic* cross **PpCc** × **PpCc,** two color preventors are at work and the *phenotypic ratio* in the offspring becomes 9:7. This is because either **pp** or **cc** showing up in the offspring produces white (or prevents color).

7. *Epistatic* effects are seen in white leghorn chickens and in sheep. Mendel observed an unusual ratio in string beans, 30 colored to 1 white. He noted that this was close to the 15:1 ratio expected if there were two pigment systems at work.

Review Questions

1. What is *blending inheritance?* How did Darwin's use of this concept open him to severe and justifiable criticism of his natural selection hypothesis? (pp. 265–266)

2. Describe Darwin's experiments in the heredity of snapdragons. What did he mean by prepotence? (p. 266)

3. What kind of preparation did Mendel have for mathematical biology? What other characteristics led to his success? (pp. 266–267)

4. List the pea characters Mendel chose for his crosses. (p. 267)

5. How did Mendel overcome the problem of *self-pollination* in his garden peas? (p. 267)

6. Describe Mendel's general procedure from P_1 to F_2. What kinds of crosses did he make? (pp. 267–268)

7. What important question confronted Mendel as he observed his F_1 *generation* in each cross? (p. 268)

8. Compare Darwin's and Mendel's handling of the reappearance of recessive traits in the F_2. (p. 268)

9. Distinguish between the terms *genotype* and *phenotype.* (p. 267)

10. Carry out one of Mendel's crosses from P_1 to P_2 (round × wrinkled) and verify his ratios in the F_2. State the *phenotypic* ratio of the F_2. (p. 268)

11. Distinguish between the terms *heterozygous* and *homozygous.* (p. 268)

12. Explain how Mendel used *progeny* testing to determine the *genotypes* of his F_2 peas. What did he learn about the F_2 round peas? (pp. 268–269)

13. What conclusions finally led Mendel to determine that the traits he studied were controlled by pairs of factors? (p. 269)

14. State Mendel's first law and show it symbolically using **Aa.** (p. 271)

15. Using the cross **Aa** × **Aa** and applying the *multiplicative law,* answer the following (p. 271):
   a. What is the probability of an **A** sperm fertilizing an **A** egg? Why?
   b. What is the probability of an offspring carrying the **Aa** combination? Explain carefully.
   c. What is the probability of an **a** sperm fertilizing an **A** egg? Why?
   d. How does the *additive law* apply to predicting the **Aa** offspring?

16. Carry out Mendel's two character cross from P_1 through F_2. The P_1 will be **AABB** × **aabb.** (pp. 271–272)

17. From the above cross write the *phenotypic* ratio of the F_2. (pp. 271–272)

18. State Mendel's second law and explain what it has to do with the results of the above cross. (p. 273)

19. Carry out the cross **AaBb** × **AaBb** as though the two pairs of *alleles* had their *loci* linked close together on the same pair of chromosomes. Why do your results differ from Mendel's? (p. 274)

20. Using diagrams of chromosomes to represent the cross **AaBb** × **AaBb,** show how *independent assortment* really works. You will have to review meiosis and look ahead to Chapter 10 unless you already understand this. Hint: There are two ways the homologous pairs of chromosomes can align on the metaphase plate. (pp. 273—274)

21. Show the two types of *testcrosses* or *backcrosses* Mendel made for the double *heterozygote* **AaBb.** What is the purpose of a *testcross?* (p. 276)

22. How would a *testcross* tell you whether a white ram carried a *recessive* gene for black? Prove your answer. (p. 276)

23. Why do you need to do *progeny testing* if your *testcross* was to a *homozygous dominant* individual? (p. 277)

24. Explain in terms of pigmentation how *dominance* works in *albinism.* (p. 278)

25. What is meant by the statement *dominance* depends on how the *phenotypes* are classified? Give several examples. (pp. 278–279)

26. Define and give an example of each of the following (pp. 279–282)
   a. *Partial dominance*
   b. *Lethal–recessive dominance (rare dominance)*
   c. *Codominance*
   d. *Overdominance*

27. What is an *antiserum?* How are the **M** and **N** *antisera* produced? (p. 281)

28. List the responses of the **MM, MN** and **NN** *blood groups* to anti-**M** and anti-**N** *antisera.* Use the terms *agglutination* and no *agglutination.* (p. 281)

29. How can the *alleles* controlling the protein *transferrin* be both *codominant* and *overdominant?* (p. 282)

30. Review the following aspects of *sickle-cell anemia* (Essay 9.1):
   a. Clinical symptoms
   b. Visual effect on red cells at low oxygen levels
   c. Transmission in families

31. How is *sickle-cell anemia* an example of (Essay 9.1):
   a. Simple *dominance*
   b. *Partial dominance*
   c. *Codominance*
   d. *Overdominance*

32. Carefully explain what is meant by *multiple alleles.* (p. 282)

33. List two combinations of *alleles* for the *beta hemoglobin locus* (other than **SS**), which would render a person very, very sick. (p. 282)

34. List both the blood *phenotypes* (types) and the *genotypes* possible in the *ABO blood system.* (pp. 282–283)

35. Why is blood type O considered to be a universal donor? (pp. 282–283)

36. What *antibodies* do people in each of the ABO blood groups produce? (pp. 282–283)

37. Using a diagram of red cells show how cell surface *antigens* and *antibodies* interact to produce the *agglutination* effect. (p. 283)

38. Review the problem of *Rh* incompatibility between Rh^+ males and Rh^- females. What are the clinical solutions? (pp. 283, 287)

39. What are *histocompatibility antigens* and what do they have to do with organ transplants? Why is it that identical twins can trade organs without problems? (p. 287)

40. Carry out the *epistatic* cross between two **BbCc** mice (as described in the text). From your results derive the 9:3:4 ratio and explain why the familiar 9:3:3:1 *phenotypic* ratio did not show up. (pp. 287–288)

41. Define the term *epistasis.* (p. 288)

42. In what instance did Mendel run across *epistasis* and how did he handle the problem? (pp. 288–289)

Chapter 10

Genes, Chromosomes, and Sex

The 19th century was a very comfortable one for the Victorian spirit before Darwin so reluctantly and abruptly jarred its sensibilities. The 20th century, however, descended abruptly on our small planet like a crazy quilt being tossed on a globe. The pattern of the crazy quilt was in strongly demarcated patches. This was indeed to be a patchy century. Technology began to blossom, even in the first few years, and small isolated groups of researchers worked feverishly to bring their own area of expertise to its culmination. Among the many advances in those wee years of the century was the incredibly rapid development of good light microscopes. In fact, we have not yet been able to improve much on the resolving powers of those instruments.

Not only were microscopes vastly improved, but stains were developed and perfected and people were seeing things that had never been seen before. Interest in the cell and its constituents was growing and the questions were becoming more sophisticated. Perhaps, though, in this proud age, one of the greatest discoveries was . . . the 19th century monk, Gregor Mendel.

Once his writings were brought to light, a chorus swelled, "Of course!" One could almost hear the heels of hands slapping against foreheads. Suddenly an army of experimenters emerged to duplicate old Mendel's work, and 9:3:3:1 became as familiar as other great numbers like 1776.

10.1 The Genetics of Sex

Sutton and Morgan

We were not to wait long for yet other advances. The next important one came from a bright young graduate student at Columbia University. His name was Walter Sutton and in 1902 he published a paper in which he attested that he could see a relationship between Mendelian inheritance and meiosis. Keep in mind that Mendel didn't know about meiosis and those studying it weren't sure how it fit into the big picture. The field was patchy, not tied together. One of the most difficult chores of scientists is to merge patches.

Sutton reasoned that, since there are only a small number of chromosomes in any one cell, and there are many hereditary factors, each chromosome must carry many genes. In a sense, he was predicting gene linkage well before it was actually observed.

Subsequent work on chromosome segregation further established the connection between Mendel's work and meiosis. Investigators found *heteromorphic* chromosome pairs in the meiotic cells of male grasshoppers. In heteromorphic chromosome pairs, the two homologous chromosomes can be told apart by some visible difference, such as an extra knob of chromatin on one end, or a dark or faint band. The X and Y chromosomes, which are clearly distinguishable in most species, are one prominent example of a heteromorphic chromosome pair. In the grasshopper testis there is just one X chromosome, and it could be seen to segregate to only half the gametes. This was a visual confirmation of Mendel's first law. When two different heteromorphic pairs were present in the same organism, they were seen to segregate independently in different meiotic cells, giving visible proof to Mendel's second law. Mendel drew his inferences from statistical analysis of numbers, but the cytologists could actually view *alternate segregation* and *independent assortment* in the meiotic process through their light microscopes.

In 1910, an innocuous little insect entered the world of genetics. Thomas Hunt Morgan, also at Columbia University, began a program of breeding experiments with the fruit fly, *Drosophila melanogaster.* This is the little brown fly you find quietly flying around your bananas.

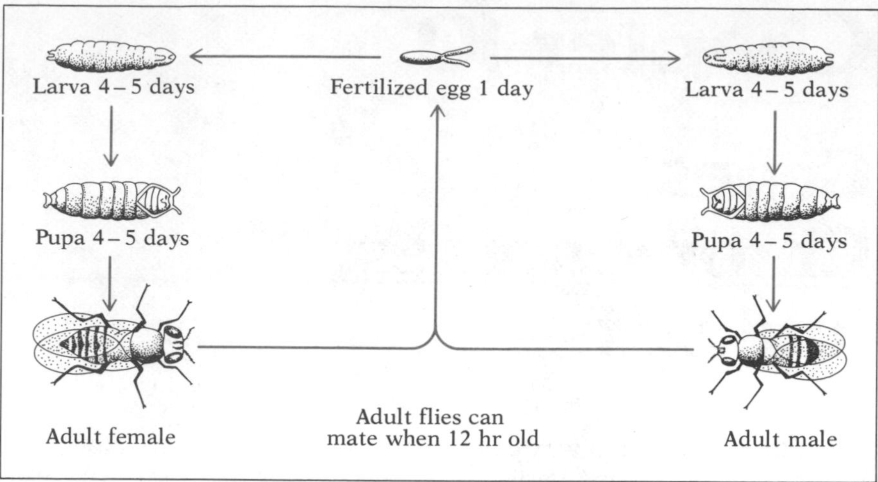

10.1 *Drosophila melanogaster* and its life cycle. The fruit or vinegar fly, *Drosophila melanogaster,* is a favorite subject of genetic and population experiments. Using this insect in experiments has many decided advantages. Its life cycle is short, it produces many offspring, it is cheap to raise, and it doesn't really bother anyone. *Drosophila* undergoes a complete metamorphosis in about 10 days from egg to emerging adult. Each female is capable of producing hundreds of offspring.

Drosophila turned out to have many advantages for genetic studies. They are easy to maintain in the laboratory; they mate readily; each female lays hundreds of eggs; the insects mature in 10 days (Figure 10.1); they can be anesthetized for easy inspection; they have only four pairs of chromosomes, and in certain organs of the larvae these become giant chromosomes on which specific loci are readily distinguishable under a low-powered microscope. In addition, mutants are readily recognized (Figure 10.2). People have now identified about 1000 gene loci, many with a number of known mutant alleles. *Drosophila melanogaster* has little economic importance, and except for occasionally floating belly-up in your cider or flying up your nose, they don't seem to bother people much. In fact, it has been said that God must have invented *Drosophila melanogaster* just for Thomas Hunt Morgan.

10.2 Some mutants of *Drosophila.* The wing and eye mutations shown here were created in the laboratory using X-radiation. In doing this, geneticists are able to create variants from the normal gene and subsequently study the genetic control of the trait. In addition, they have learned a good deal about the nature of mutation and radiation damage.

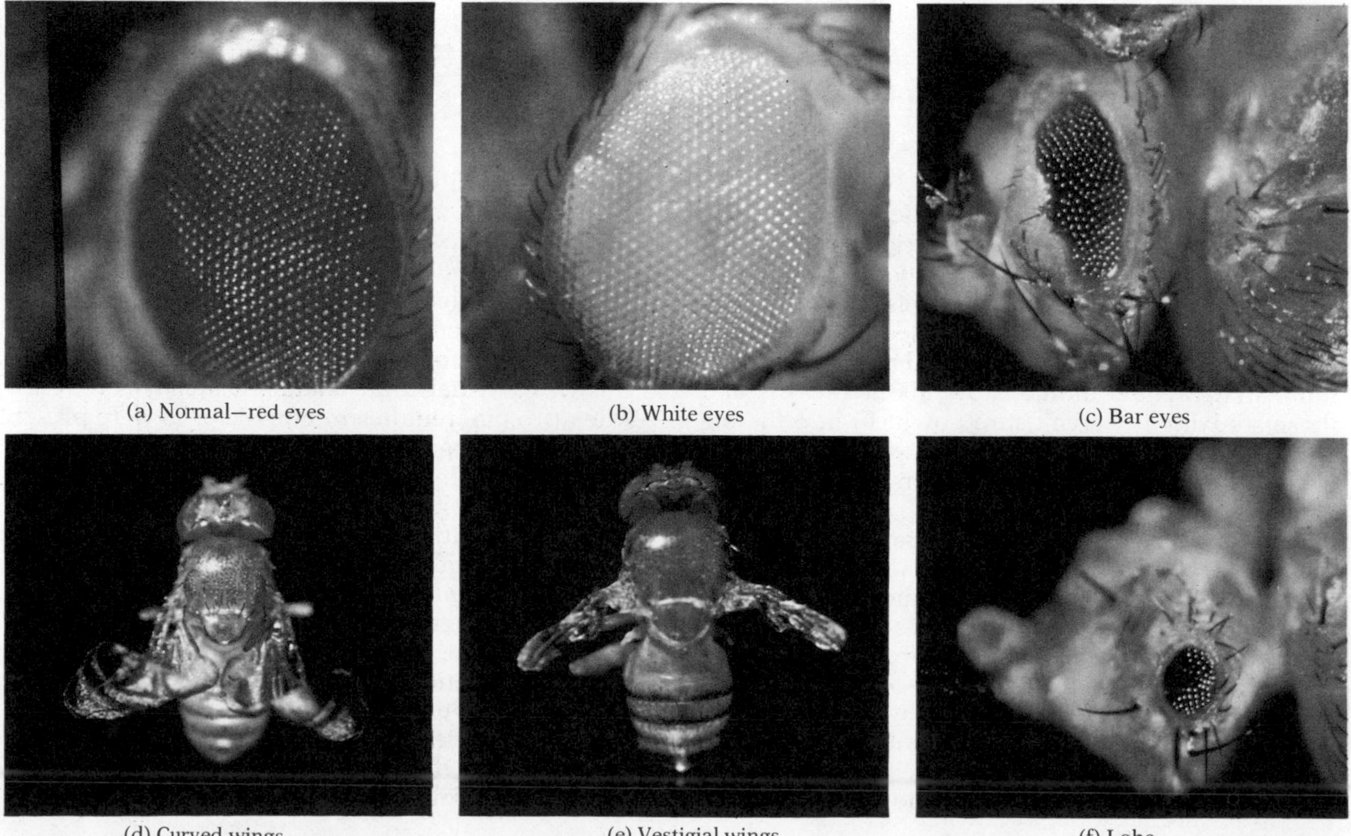

(a) Normal—red eyes (b) White eyes (c) Bar eyes

(d) Curved wings (e) Vestigial wings (f) Lobe

Now, in order to study inheritance one must have variation. But, at first, all Morgan's flies looked exactly alike, except that the males could be told from the females. However, as Morgan carefully scrutinized each new generation, he eventually turned up one variant—among a group of flies with normal brick-red eyes, Morgan found a single male with white eyes. He was convinced that this was due to a *mutation,* or spontaneous change in a gene (and he was right). Morgan decided to apply Mendel's newly appreciated techniques. He carefully nurtured his little white-eyed specimen and crossed it with several of its red-eyed virgin sisters. (In the laboratory, a fruit fly will mate with any other fruit fly of the opposite sex, if given no other choice. As it happens, female fruit flies, if given a choice, prefer to mate with strangers rather than their own brothers, but they will accept their brothers if no other males are around. Male fruit flies don't seem to care one way or another and will even attempt to mate with each other.) From these matings, all the F_1 were red-eyed, to the surprise of none of the new Mendelians. Obviously, *white eyes* was a recessive trait.

The experiments continued, and when the F_1 flies were mated with one another to produce an F_2 generation, sure enough, about one-fourth of the F_2 flies were white-eyed and about three-fourths were red-eyed. The actual numbers were not as close to this expected ratio as Morgan had hoped because, as it turned out, the white-eyed flies have a somewhat lower rate of survival than the normal flies. But there was something more disturbing about the F_2 flies. Every single white-eyed fly was a male! In fact, the F_2 ratio approximated:

$\frac{1}{4}$ red-eyed males: $\frac{1}{2}$ red-eyed females: $\frac{1}{4}$ white-eyed males.

At this point, you may have decided that, obviously, only males can be white-eyed. You would be wrong. Morgan discovered this for himself when he first did the testcross, mating his original, now geriatric, white-eyed male to its own red-eyed F_1 daughters. A simple testcross should have provided a 1:1 ratio of dominant to recessive, and this one did. In fact, the testcross offspring consisted of approximately equal numbers of red-eyed males, red-eyed females, white-eyed males, and white-eyed females. So females could have white eyes. But even more surprises were in store. When Morgan mated white-eyed females to red-eyed F_1 males, in what is called a *reciprocal testcross,* again, half of the offspring were red-eyed and half were white-eyed. But now every single one of the males were white-eyed, and every female had red eyes! (See Figure 10.3.)

If you have a penchant for puzzles, can spare a few minutes, and think you're smart, try to figure out how this could have happened. Keep in mind what you learned about sex chromosomes in the chapter on meiosis. Meanwhile, the rest of us will forge ahead.

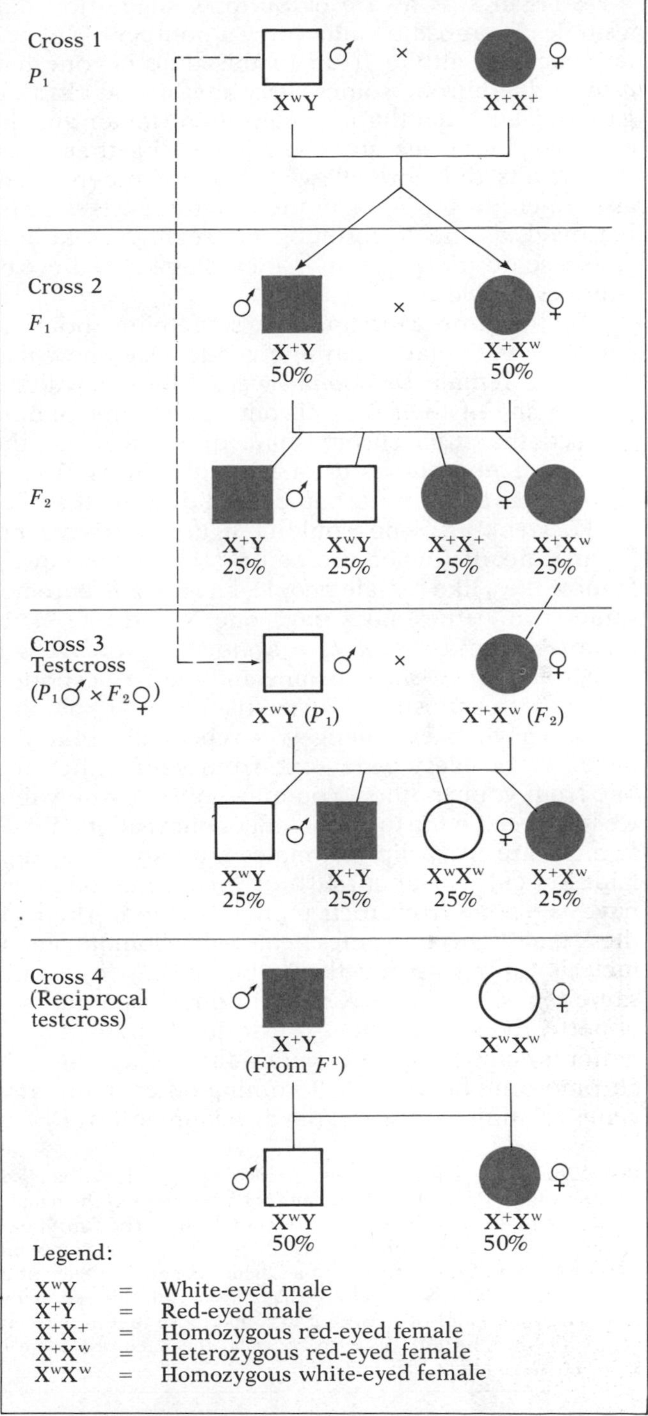

10.3 Morgan's crosses. In his P_1 cross Morgan mated his newly discovered white-eyed male with a normal red-eyed female, starting a revealing set of experimental crosses. All the F_1 offspring were red-eyed. Inbreeding the F_1 produced an F_2 that was three-fourths red-eyed and one-fourth white. But all the white-eyed flies were males.

The first testcross was between an F_2 red-eyed female and the P_1 white-eyed male. The offspring clearly showed that the F_2 female was heterozygous and that females could be white-eyed. Note the testcross results.

A reciprocal testcross was done using an F_1 red-eyed male and one of the white-eyed testcross females. This time white eyes appeared in all the male progeny, and only in males. From these crosses arose the concept of sex-linkage.

Morgan was aware of Sutton's suggestion that a single chromosome may carry a number of hereditary factors—although up to this time no one had *proven* that chromosomes carry anything at all. Morgan surmised that the sex-determining factor and the eye-color factor are somehow linked together, since these traits did not follow the law of independent assortment in the F_2 or in the testcrosses. Hence, he reasoned, as the X chromosome segregates at anaphase, so do the genes on it, including the red-eye or white-eye alleles.

By this time, Morgan knew something about the chromosomes that determine gender. He knew that male and female *Drosophila* were different with regard to one of their four chromosome pairs, and he guessed that this chromosome difference was the *cause* and *not the result* of sex differences. The X chromosome had been named and described earlier by H. Henking, who couldn't figure out what his finding meant (hence the name "X" for unknown). Female flies, like female people, have two X chromosomes, while the males have one X and one Y. At metaphase in *Drosophila melanogaster*, the X is a long, rod-shaped chromosome and the Y is a shorter, J-shaped chromosome (Figure 10.4).

We have been telling you repeatedly that you have two of every gene—one from your father and one from your mother. You may realize by now that we have been lying to about half of our readers. While females are truly diploid, males are only partly diploid. Actually, at about 10% of their gene loci, men have one gene from their mother—period. Although the X and Y chromosomes behave like homologues in meiosis (lining up together), they do not carry the same genes. While the X chromosome has thousands of perfectly normal genes—those determining growth patterns, enzymes, and so on—the stunted little Y chromosome bears almost nothing other than a few genes relating to male sexual development. In *Drosophila*, the Y chromosome carries about six genes, all having to do with male fertility, while the X chromosome has approximately 1000 genes.

10.4 *Drosophila* chromosomes. *Drosophila* has four pairs of chromosomes. Each of the four pairs is homologous in the female, but in the male only three pairs are homologues. The fourth pair consists of an X chromosome which is identical in shape and content to that of the female, and a Y which is not. The Y chromosome of fruit flies is actually J-shaped, as you can see in the illustration. It is almost devoid of loci, and those that are present deal with male fertility only. In contrast, the X chromosome is known to have about 1000 loci.

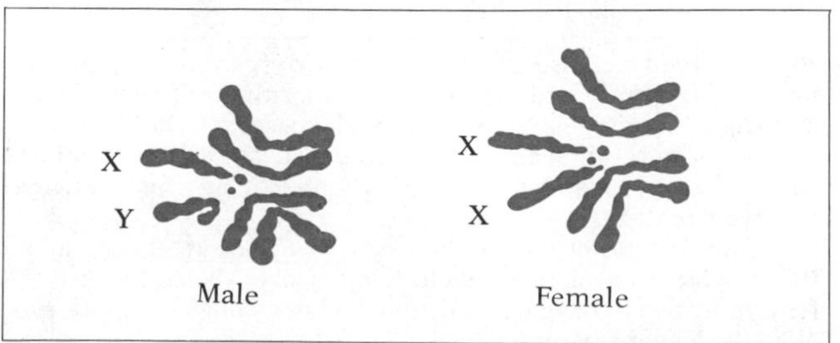

In genetic crosses, such as the white-eye versus red-eye experiment done by Morgan, any gene on an X chromosome doesn't have to interact with, or yield to, any homologous gene on the Y chromosome. So, the Y chromosome behaves as if it has a recessive allele for virtually all the X chromosome loci. Thus, a female must have a recessive allele on both of her two X chromosomes in order for a recessive to be expressed. But since a male has only one copy of any X-linked gene, it will express itself, whether it is recessive or dominant. For X-linked genes in males, the term *hemizygous* (*hemi* means half) is used instead of homozygous or heterozygous.

Thus, Morgan had discovered X-linked inheritance—a discovery with vast implications. But more importantly, at the same time, he had (a) proven that genes are on chromosomes and (b) discovered the general concept of linkage.

Sex Linkage in Humans

Although normal *Drosophila* and human females are XX and normal *Drosophila* and human males are XY, the physiological mechanism determining sex is somewhat different in the two species. In *Drosophila*, sex is determined by the number of Xs: two Xs, female; one X, male. Abnormal XXY flies are fully functional females, and XO flies (one X, no Y) are sterile males that look normal. In humans and other mammals, the *active* presence of a Y determines the development of testes, which in turn determines male development. Thus, XO humans are sterile, abnormal females, and XXY humans are sterile, abnormal males.

Even though human females have two X chromosomes in every cell nucleus, they only use one in any given cell (aside from those in the ovaries). The other X chromosome becomes inactive and *condensed* during embryonic development and remains in this compact condition through subsequent cell divisions.

Color Blindness

Color blindness in humans is usually caused by a recessive allele at either of two closely linked gene loci on the X chromosome. Human color vision depends on the differential sensitivity of three groups of receptors in the retina called *cones*. One group of cones is maximally sensitive to blue light, one to red light, and one to green light. Perception of other colors and of subtle hues depends on the relative stimulation of these three types of cones. Persons homozygous (or hemizygous) for a recessive allele at one of the two X-linked loci lack the cones that are most sensitive to green, and homozygotes for recessive alleles at the other X-linked locus lack the cones that are maximally sensitive to red. By the way, the locus controlling the blue-sensitive group is autosomal, and blue-insensitive color blindness is very rare. Both X-linked defects are called *red-green color blindness* (Figure 10.5) because neither type of color-blind people can distinguish between red and green. The defect was first described in a little boy who couldn't learn how to pick ripe cherries. He brought home a mix of red and green fruit.

About 8% of American men have one form or another of X-linked color blindness—about 6% are hemizygous for a recessive allele at one of the two loci and about 2% are hemizygous for a recessive allele at the other locus. Women are affected far less often—only about 0.4% of American women are red-green color blind. The reason for the sex difference is fairly simple. To be affected, a man need only receive one recessive allele from his mother. But an affected woman must receive a recessive allele from her mother, and one from her father as well. The chance of both of these things happening is much smaller than the chance of just one of them. Recall our probability discussion in Chapter 9. The likelihood of two independent events occurring is equal to the product of their individual occurrences.

A woman who is heterozygous for color blindness, shows no symptoms of the condition. Among her children, however, she can expect half her sons to be color blind and half her daughters to be carriers like herself—assuming that she marries a man with normal vision. What could she expect if she marries a color-blind man? Figure 10.6 shows the appearance of color blindness in one family.

10.5 A color-blindness test. You can test your color vision with this color plate. A person with normal vision sees the number *ninety-seven.* Actually, the test is more extensive than the one shown here, but if you flunked this test, you might want to take the long version.

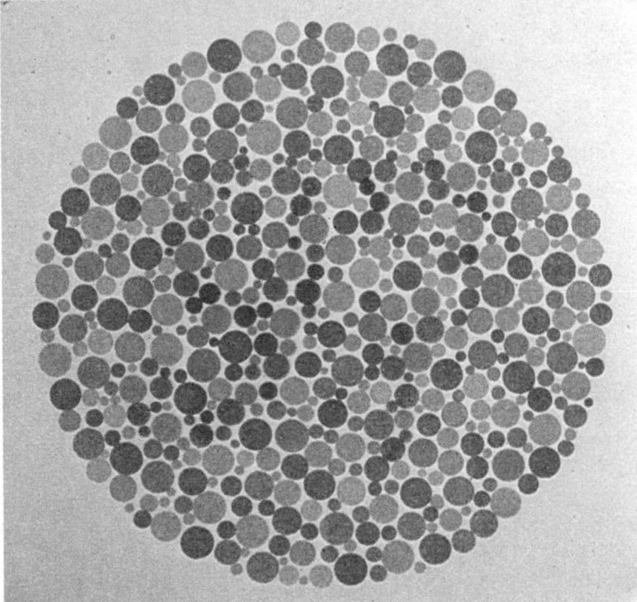

Somebody in the British Navy was aware of this situation well over a century ago, and set about developing "running lights" that even color-blind men could tell apart. The green (starboard side) had a touch of blue and the red (port side) had a touch of orange. When traffic lights were introduced on railroads and later on city streets, these readily recognizable hues were the logical choice and so we now see them daily.

Other Sex-Linked Conditions

Many sex-linked genetic conditions have had great impact on people's lives. Two of the most common, but also dramatic, of all genetic disorders are *muscular dystrophy* (muscle tissue breaks down in late childhood) and *hemophilia* (bleeder's disease, when the blood doesn't clot normally). The usual forms of both of these are sex-linked. There are actually two common forms of X-linked hemophilia, governed by different X-linked loci. Most hemophiliac males formerly bled to death in their youth. But in recent years modern medicine has allowed affected hemophiliacs to survive and reproduce, thanks to blood transfusions and to transfusions of a blood-derived substance know as *anti-hemophilic factor* that supplies the critical substance missing in hemophiliacs. So now we find adult hemophiliac males, a new situation in human history. In fact, even hemophiliac females occur occasionally (in spite of the fact that homozygous females must receive a recessive allele from each parent and hemophiliac fathers are still rare).

There is, as yet, no effective treatment for muscular dystrophy, so boys with this genetic disease still die before reaching adulthood. Muscular dystrophy has a delayed age of onset, so that hemizygous boys appear perfectly normal as infants and toddlers, only to begin wasting away some time during their elementary school years. Their heterozygous mothers appear to be unaffected. Interestingly, however, microscopic tissue samples of females heterozygous for muscular dystrophy show that clusters of muscle cells, accounting for about half the total number of muscle cells, have atrophied by adulthood. We assume that these are the muscle cells in which the normal X happened to become condensed. The women do not weaken because the remaining muscle cells expand in response to the increased load they must carry—another example of how dominance can work.

Sex-Linked Dominants

A few sex-linked mutant alleles are dominant, or are recessive–lethal dominants. *Dominant brown spotting* of the teeth, for instance, is passed from affected men to *all their daughters* and *none of their sons*—work that one out. Dominant brown teeth affects about twice as many women as men, since they have two chances of receiving the X chromosome with the

Essay 10.1 The Disease of Royalty

Because it was the practice of ruling monarchs to consolidate their empires through marriage alliances, hemophilia was transmitted throughout the royal families of Europe. Hemophilia is a sex-linked recessive condition in which the blood does not clot properly, so that any small injury can result in severe bleeding, and if the bleeding cannot be stopped, in death. Hence, it has sometimes been called the *bleeder's disease.*

Hemophilia has been traced back as far as Queen Victoria, who was born in 1819. One of her sons, Leopold, Duke of Albany, died of

the disease at the age of thirty-one. Apparently, at least two of Victoria's daughters were carriers, since several of their descendants were hemophilic. Hemophilia played an important historical role in Russia during the reign of Nikolas II, the last Czar. The Czarevich, Alexis, was hemophilic, and his mother, the Czarina, was convinced that the only one who could save her son's life was the monk Rasputin—known as the "mad monk." Through this hold over the reigning family, Rasputin became the real power behind the disintegrating throne.

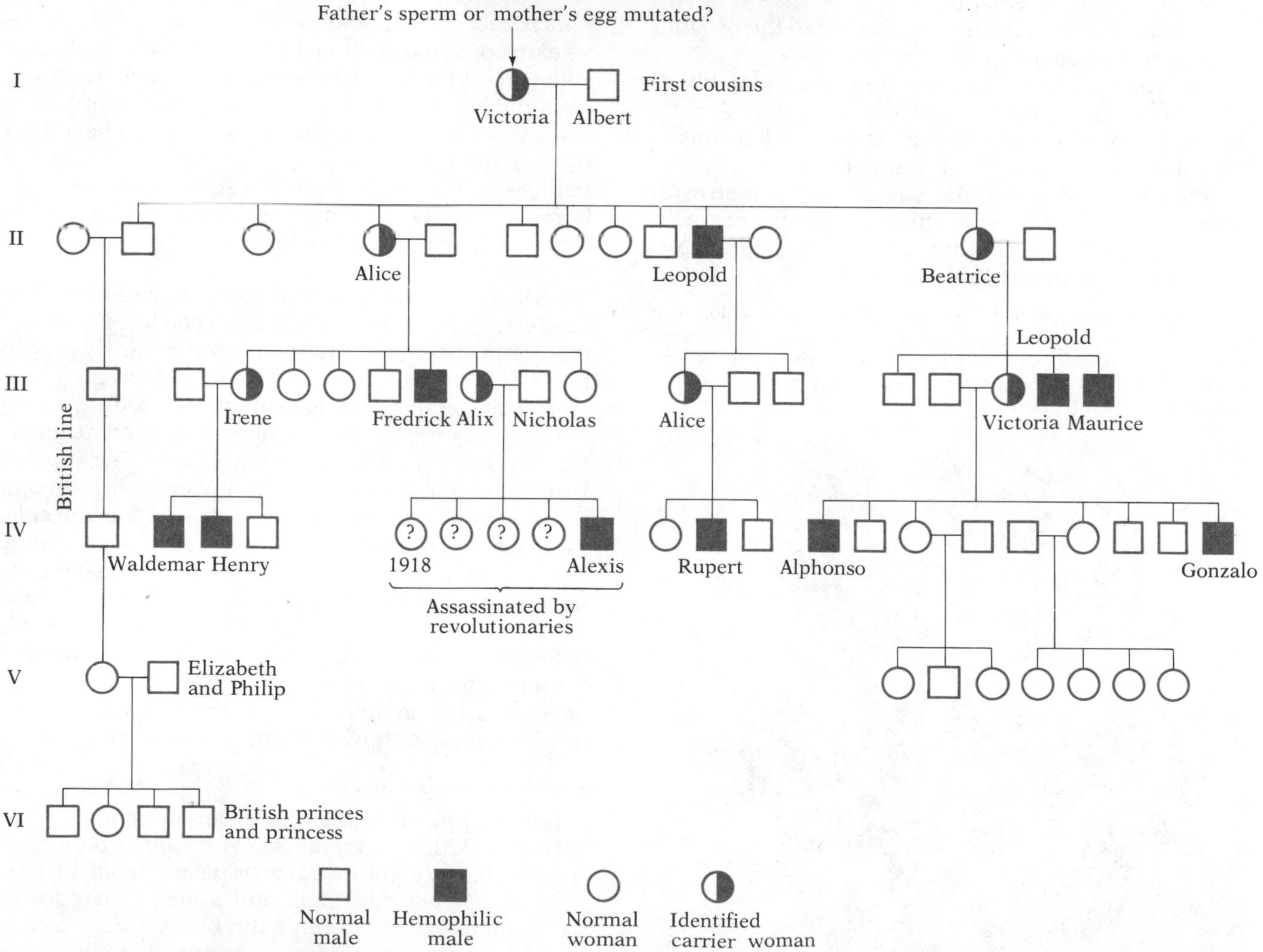

Father's sperm or mother's egg mutated?

I Victoria Albert First cousins

II Alice Leopold Beatrice

III Irene Fredrick Alix Nicholas Alice Leopold Victoria Maurice

British line

IV Waldemar Henry 1918 Alexis Rupert Alphonso Gonzalo

Assassinated by revolutionaries

V Elizabeth and Philip

VI British princes and princess

Normal male · Hemophilic male · Normal woman · Identified carrier woman

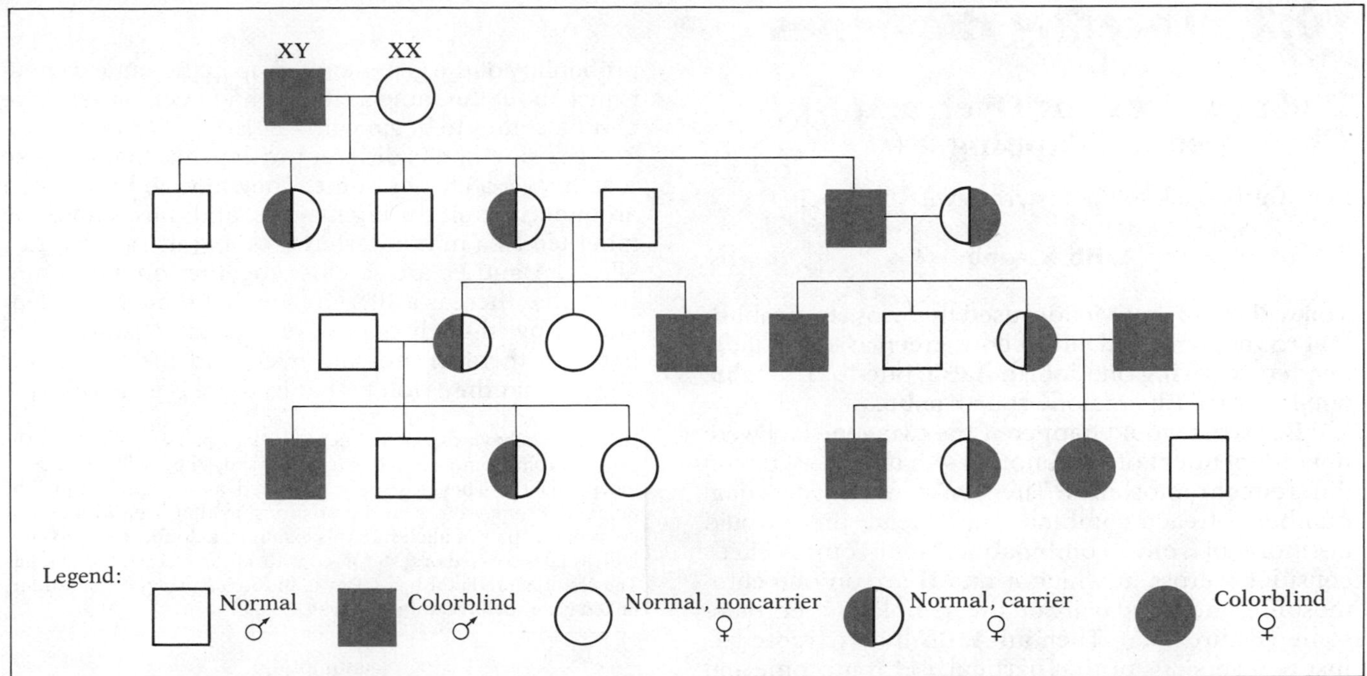

10.6 In this hypothetical family tree, color blindness can be traced back to the great-grandfather and great-grandmother, although the problem was intensified by the marriage of a color-blind man and a carrier woman as shown at the right. We did not complete the great-grandmother's circle because we want you to decide whether she was a carrier.

brown-spotting gene (Figure 10.7). Another curious X-linked genetic disease is known as the oral–facial–digital syndrome, which involves irregularitics of the mouth, face, fingers, and toes. This condition affects only women, who get it only from their mothers. Such women pass the condition on to half their daughters but to none of their sons. It turns out that as a group, affected women have twice as many daughters as sons. We can therefore surmise that the allele in the hemizygous state is an early lethal.

10.7 The family pedigree shown here is for the sex-linked dominant, brown spotting in the teeth. It is unusual in that it is expressed in twice as many women as men. As you can see by the chart, fathers pass the trait to their daughters, but not to their sons. Afflicted mothers can pass the trait to either with equal probability.

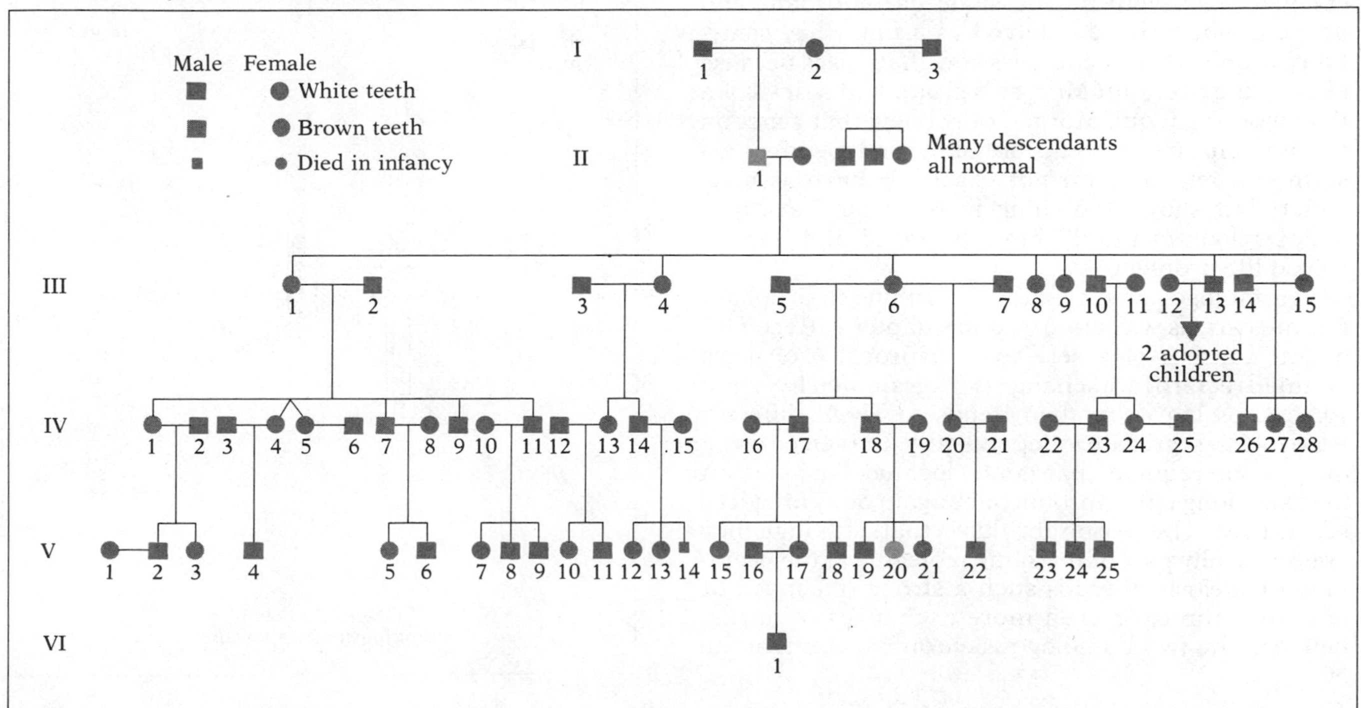

10.2 Locating the Genes

Linkage, Crossing Over, and Chromosome Mapping

Let's think back to the testcross with two loci:

AaBb × **aabb** → ?

You will recall that Mendel used this cross to establish that round, wrinkled and yellow, green assorted independently, giving one-fourth **AaBb**, one-fourth **Aabb**, one-fourth **aaBb**, and one-fourth **aabb**.

But what would happen if the two gene loci were not independent of one another—that is, were not on different chromosomes? Then we would not get equal numbers of each combination. Instead, there would be more of some combinations than others. Let's consider a cross in which **A** and **B** are on one chromosome, and **a** and **b** are on the homologous chromosome (Figure 10.8). Then an **AaBb** heterozygote has just two versions of that particular chromosome and can make only two kinds of gametes: **AB** and **ab**. In a testcross with a double recessive homozygote, **aabb**, only two kinds of offspring occur, **AaBb** and **aabb**, in equal numbers. Right?

We hope you didn't say right. But if you did, you erred in good company, because that is just what Morgan and his co-workers expected, knowing as they did about a large number of genes being associated with a small number of chromosomes. But things just didn't work out that way. To make the problem more difficult for Morgan, even when different gene loci were on the same chromosome, and hence ought to have behaved as a unit, they sometimes didn't. You, of course, know that this is because of crossing over, but Morgan's group had a devil of a time working it out. Morgan only knew that genes on a single chromosome do not show independent assortment, and they do not exactly behave as units either, but show something in between. Let's take a closer look at the dilemma he faced and how he solved the problem.

Homologous chromatids recombine in prophase I of meiosis, as was discussed in Chapter 8. Recombination consists of a series of reciprocal exchanges (a single reciprocal exchange is diagrammed in Figure 10.9). Crossing over, then, tends to allow different gene loci to segregate independently. In fact, crossing over is so frequent that genes located far apart on the same long chromosome once again obey Mendel's second law. The reason for this is that, although they are physically part of the same molecule at the beginning of meiosis, there is such a strong likelihood of one, two, three, or even more exchanges occurring between the two homologous chromosomes that the probability of the genes ending up in the same gamete is just about the same as if they had been on separate chromosomes to begin with—50%.

The situation is different with genes that lie close together on a chromosome. They tend to be shunted around as a unit during meiosis, and thus tend to be inherited as a unit in testcrosses. Let us say that two alleles, **A** and **B**, are so close together on a chromosome that there is a 10% chance that they will be separated by the chromosome breaking somewhere between them in crossing over, and there is a 90% chance that they won't. That is, there is a 10% chance

10.8 The behavior of linked genes. If two pairs of alleles on the same chromosome are *completely* linked, they will not assort independently. They will be transmitted as a unit, limiting the number of phenotypes in the offspring to that seen in a simple cross of one pair of alleles. In this example, a doubly heterozygous individual is crossed or testcrossed with a fully recessive individual. The result is predictably a 1:1 ratio in the offspring. But not many linked genes behave quite this way.

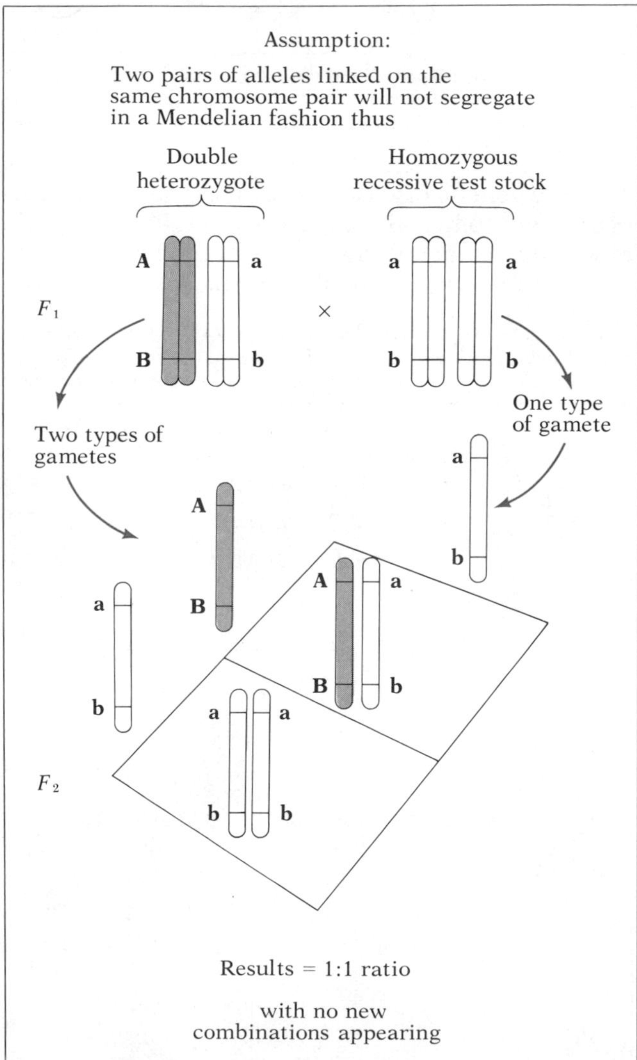

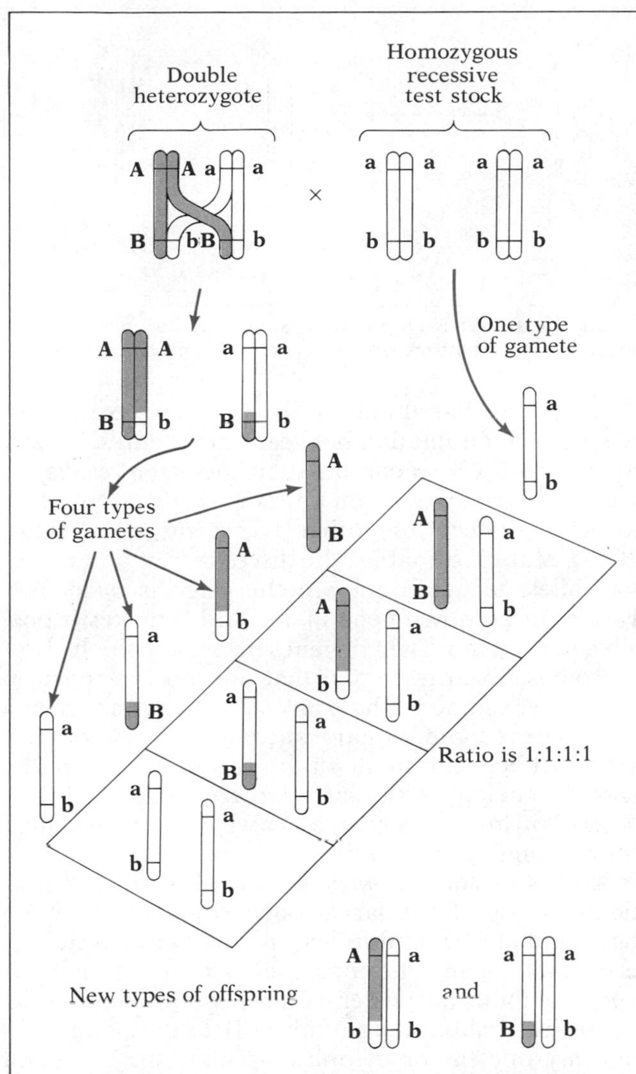

10.9 Continuing from Figure 10.8, we look at the same cross with a single crossover occurring. Note that with a single crossover, the types of gametes doubles. The result is the appearance of two new classes of offspring.

of recombination in the region between the two loci. Now, as an exercise, start with a double heterozygote, with one chromosome carrying **AB** and with its homologue carrying **ab**. You can see that if there were no crossing over at all, half the gametes would carry the **AB** chromosome and half would carry the **ab** chromosome, and none would carry **aB** or **Ab**. But only 90% of the chromosomes are *nonrecombinants.* Of the other *recombinant* chromosomes, half will be **aB** and half will be **Ab**. So we will expect the following kinds of gametes (see Figure 10.10):

Nonrecombinants, 90%: 45% **AB**
45% **ab**

Recombinants, 10%: 5% **Ab**
5% **aB**

In a testcross, all of the gametes from the double homozygote test stock parent would be **ab**, so the distribution of the testcross progeny should be the same as the distribution of the gametes:

45% **AaBb**, 45% **aabb** } Nonrecombinants = 90%

5% **Aabb**, 5% **aaBb** } Recombinants = 10%

10.10 We continue from Figure 10.9 with the same example, and this time limit the frequency of crossover to 10%, which means the noncrossover frequency is 90%. We end up with the same types of gametes as in Figure 10.9, but now each gamete has a percentage that must be considered. Also, recall that when we are using Punnett squares we are multiplying sperm × egg. When tabulated, the results do not indicate a 1:1:1:1 ratio at all, but actually a 9:9:1:1 ratio. (Divide each value in the offspring by the smallest, 5, to get this ratio.) Perhaps you can see what Morgan and his contemporaries were up against with ratios like this one to contend with.

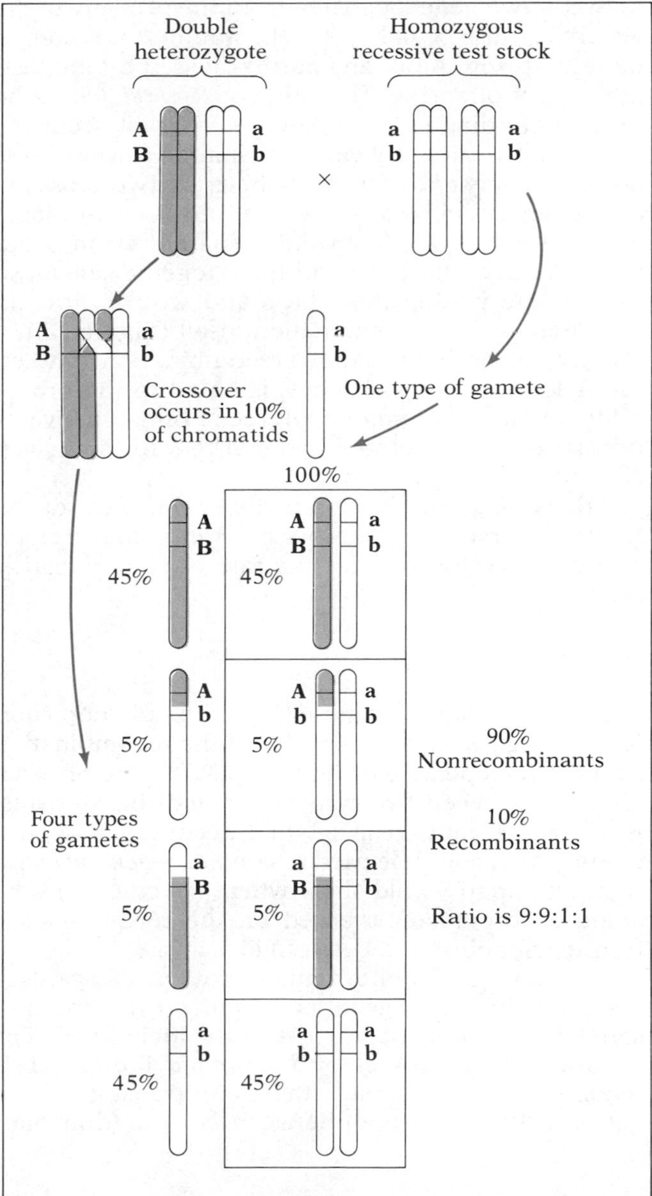

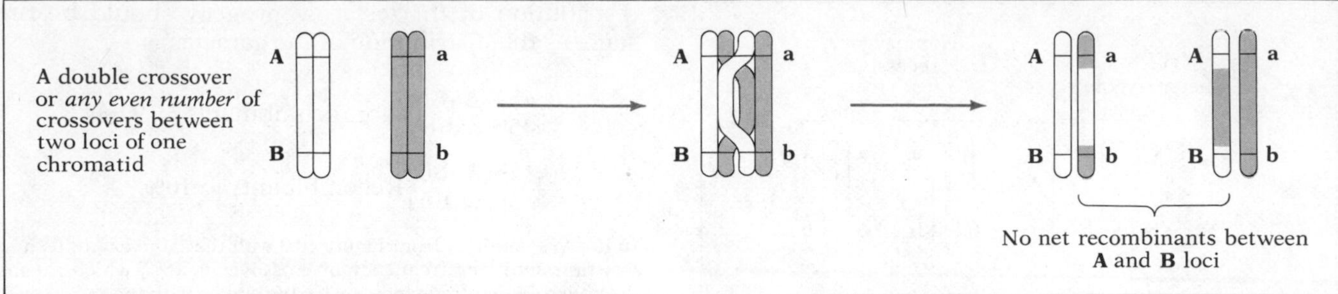
A double crossover
or *any even number* of
crossovers between
two loci of one
chromatid
A
B
a
b
A
B
a
b
A
B
a
b
A
B
a
b
No net recombinants between
A and B loci

10.11 Two-strand double crossover. If a crossover occurs between **A** and **B**, what happens if another crossover should occur between them? The answer is above. There will be no new recombinants in the gametes if both crossovers occur in the same two chromatids.

Therefore, the total frequency of recombination between the **A** locus and the **B** locus is 10%. In general, let R represent the frequency of recombination between two gene loci. Then from the progeny of the testcross **AaBb** × **aabb**, R is the number of recombinant offspring (**Aabb** and **aaBb**) divided by the total number of offspring. The value of R is *related* to the probability that a crossing-over event will occur between two loci on a given chromatid. However, it is *not exactly equal* to this probability. If two crossing-over events occurred between **A** and **B** on the same chromatid, then the first would separate **A** and **B** and the second would put them back together again, so the gamete getting that chromatid would show no evidence of any recombination at all (Figure 10.11). Then R, the probability of net recombination between the **A** locus and the **B** locus, is equal to the probability of an *odd number* of recombination events occurring between these two markers for any given chromatid.

Because each chiasma involves only two of the four chromatids, the number of *chiasmata per tetrad* in a given region is twice the rate of *recombination per chromatid.*

Mapping Genes

Soon after Morgan stumbled upon the phenomenon of crossing over and inferred that the recombination frequency depended at least in part on the physical distance between two gene loci, one of his students, A. H. Sturtevant, realized that recombination frequency data could be used to construct *genetic maps.* A genetic map would show where specific genes lie along a chromosome, based on the recombination frequencies observed between loci.

A map yields information, not only regarding the order in which gene loci occur on the chromosome, but the distances between the loci as well. For instance, if testcrosses of *Drosophila* indicate 13% recombination (meaning they are transmitted together 87% of the time) between *bar eye* (dominant trait for eye shape) and *garnet* (recessive bright red eye), 7% recombination between *garnet* and *scalloped wings,* and 6% recombination between *scalloped wings* and *bar eye,* we can infer that *scalloped wings* is located between the other two gene loci (Figure 10.12). Mapping enables the discoverer of a new mutant allele to determine whether the discovery is a variant of a known gene locus. And it makes it possible to determine which genes belong to which chromosomes. Even gene loci that are too far apart on one chromosome to show any significant linkage in a two-locus recombination test can be related to other gene loci between them on the chromosome, and be shown to belong to the same *linkage group.*

Following Sturtevant, other ways were developed for mapping genes. We have mentioned that the chromosomes of some *Drosophila* larval organs, in particular those of the larval salivary glands, are very large and are banded in very regular ways. Actually, each such "giant chromosome" is a tight bundle of about a thousand sister strands lying side-by-side, the result of about ten rounds of DNA doubling without any nuclear or cytoplasmic division. The giant salivary chromosomes are interesting in their own right—they seem to be necessary for the production of the huge amounts of saliva needed by the little maggots—but they are even more interesting to genet-

10.12 Gene mapping. By studying the crossovers between groups of genes, their relative locations on the chromosome can be determined and their distance apart in map units decided. Thus genetic maps of chromosomes can be prepared. In this example, the locus of scalloped wings was determined to be between *bar eye* and *garnet eye*, all mutant characters of *Drosophila.* Can you follow the logic used?

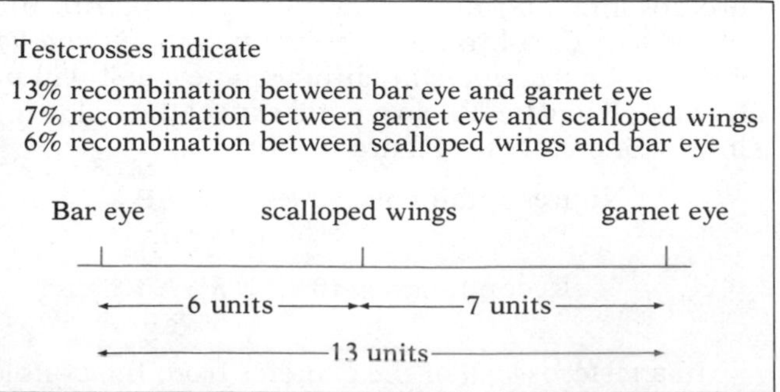
Testcrosses indicate
13% recombination between bar eye and garnet eye
7% recombination between garnet eye and scalloped wings
6% recombination between scalloped wings and bar eye
Bar eye
scalloped wings
garnet eye
6 units
7 units
13 units

icists because they provide a physical picture of each chromosome that can be related to its genetic content. One can draw physical chromosome maps (Figure 10.13) showing a total of about 5000 cross-bands of different widths that occur in recognizable patterns in the four chromosome pairs of *Drosophila.*

Individual gene loci can be correlated with individual giant chromosome bands through a process known as *deletion mapping.* In deletion mapping, chromosomes are used that have been broken apart by X-radiation. Any segments missing as a result can be identified under the microscope by noting which salivary chromosome bands aren't there. A chromosome with a small missing segment (deletion) behaves in genetic crosses as if it had recessive alleles at all the gene loci that normally occur in the missing bands. Flies with missing segments can then be crossed with flies carrying known mutant alleles, to find out which genes belong with which chromosomal bands. Such maps are called *deletion maps* or *cytological maps,* to distinguish them from the *recombination genetic maps* we have just described.

Comparisons between physical chromosome maps (based on direct observations of altered chromosomes) and recombination maps (inferred from counts of recombinant offspring) have turned up some interesting things. First, the different gene loci appear in exactly the same order in both kinds of maps—which is a relief. Second, the relative distances between genes are not the same on the two maps, which means that the frequency of crossing over is not the same all along the DNA molecule, but varies from place to place on the chromosome. For instance, the heterochromatin regions on either side of the centromere are actually quite large, but since almost no recombination takes place there, these regions appear to be short on the recombination map.

The third thing learned from comparing *Drosophila* physical chromosome maps and recombination maps is perhaps the most interesting. The great majority of gene loci that have been mapped do not affect things like the shape or color of eyes and bristles, but are *vital* genes that can have *recessive lethal* alleles that kill the fly before it can grow up and be counted. The wild-type alleles of the recessive lethal

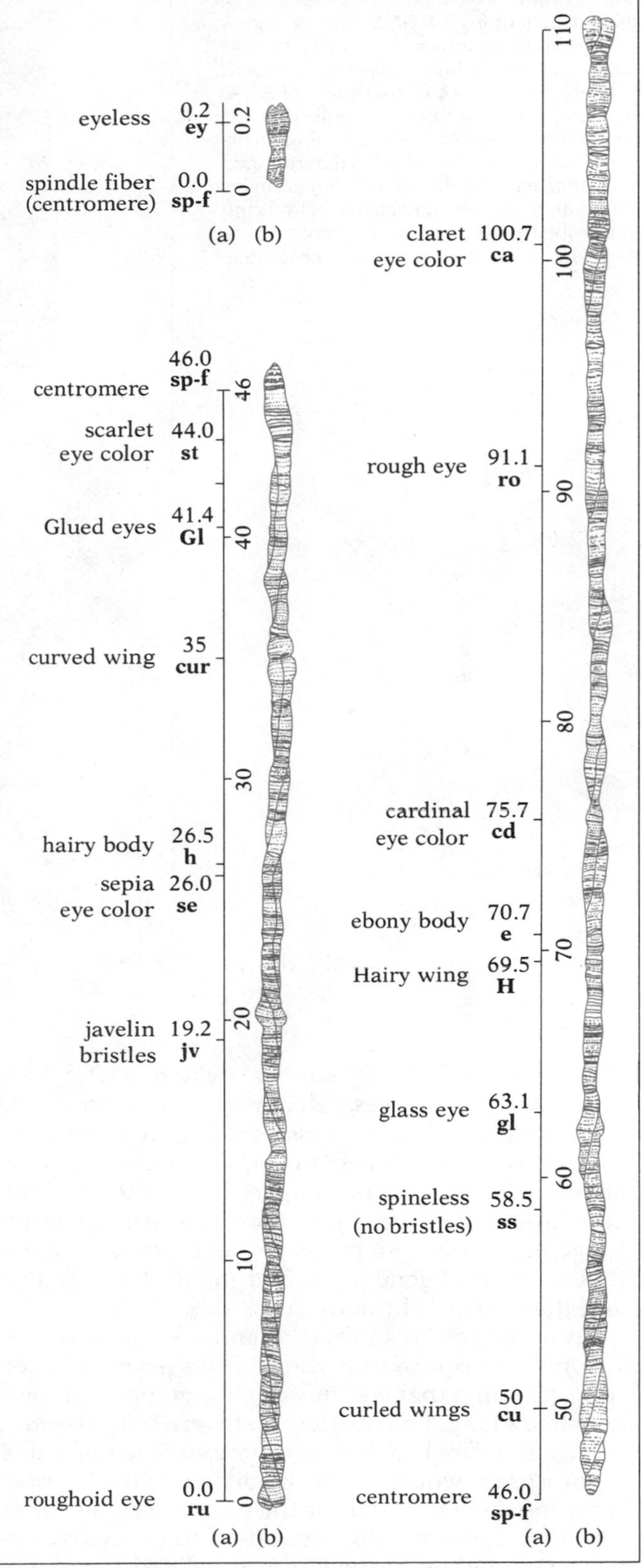

10.13 This illustration contains two kinds of maps of *Drosophila.* Parts labeled (a) are genetic maps. These were constructed by exhaustive studies of crossovers among the mutant loci. The numbers refer to the percentage of recombinations that occur between the loci, which are represented by the letter abbreviations. Parts labeled (b) are cytological maps that have been prepared from microscopic observations and microphotographs of the giant salivary gland chromosomes obtained from *Drosophila* larvae. The giant chromosomes represent numerous replications of DNA rather than a single strand. The bandings are very distinct, and geneticists have identified bandings and associated them with the actual loci for specific mutants. Shown here are the tiny fourth chromosome and the left and right arm of chromosome 3.

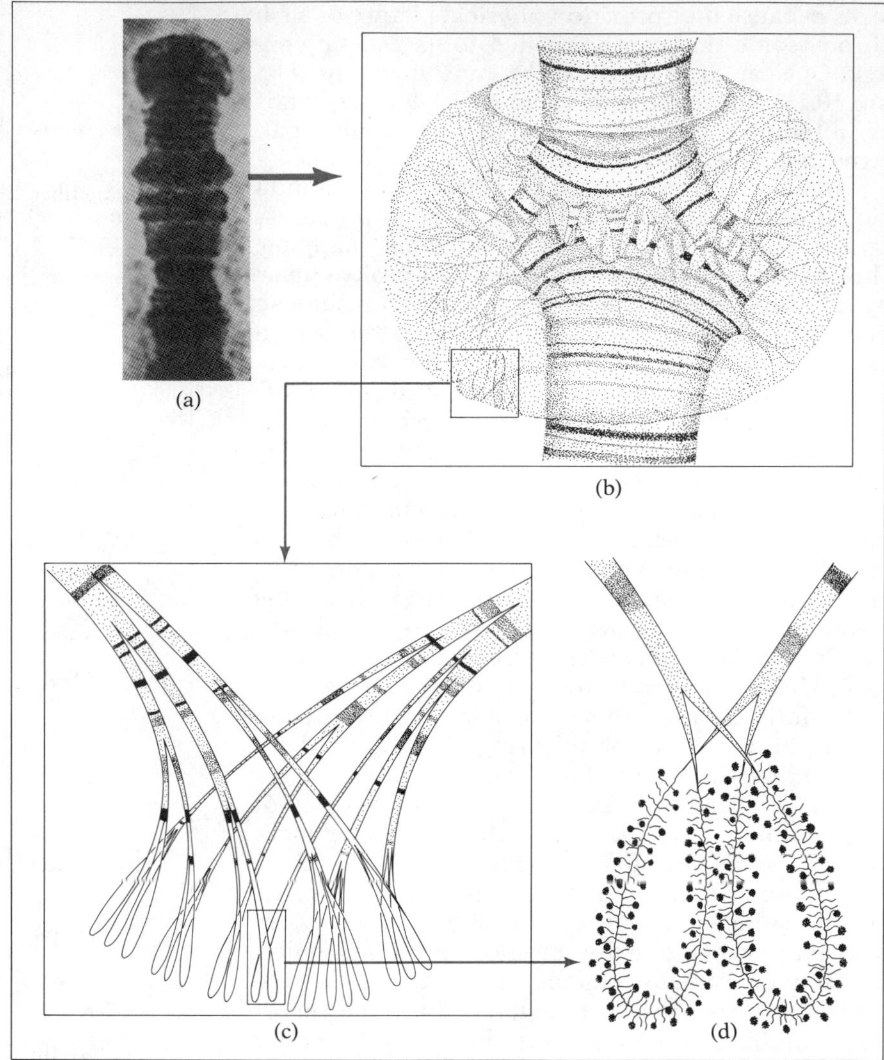

10.14 Chromosome puffs, a general loosening and looping of DNA from a region on the giant chromosomes, occur with regularity in fruit fly larvae. These are known to be regions of DNA transcription of RNA. (a) Microphotograph of one puff after experimental exposure to the molting hormone, ecdysone. (b–d) Puff is diagrammed in increasing magnifications. The granules are thought to be messenger RNA being transcribed. The autoradiograph made shows many clusters of radioactive uridine.

loci are necessary for fly survival, which is why these are called *vital genes.* Recessive lethals can be mapped like any other recessive gene, but there are a lot more of them. When the physical chromosome map of vital gene loci is compared with the recombination map of vital gene loci, there is a striking result: As a general rule, with rather few exceptions, there is exactly one vital gene in each of the dark bands that mark the salivary chromosomes.

In an intensive study of a limited region of one *Drosophila* chromosome, Burke Judd used a deleted X chromosome that was missing a segment just fourteen bands long. Female flies that were heterozygous for this test stock X chromosome and a normal X chromosome would survive, because the normal X chromosome covered for the deleted region. However, any recessive lethal mutation that occurred in the corresponding region of the undeleted X chromosome would not be covered and could be detected. After collecting hundreds of strains of flies heterozygous for recessive lethal mutations in this region, Judd made crosses that showed that there were exactly fourteen different vital gene loci involved: Every one of their hundreds of recessive lethal mutations were alleles of one or another of these vital gene loci. Deletion mapping further showed that each vital gene locus was in a different chromosome band.

This seemed to indicate that there is one vital gene in each chromosome band, and that when we look at a giant banded salivary gland chromosome, we are in effect looking at a linear array of genes. It has since been shown that some bands may contain two different vital genes, so the relationship is not absolute. However, using biochemical techniques, geneticists have estimated that there are about 5000 different kinds of mRNA in *Drosophila.* Assuming that most functioning genes produce a single kind of mRNA, we can conclude that there are about 5000 genes in the organism. This estimate is identical to estimates based on mutation rate studies and is also identical to the number of microscopically visible salivary chromosome bands. So, the evidence for the existence of about 5000 *Drosophila* genes is pretty good. Furthermore, the good match between the number of bands and the estimated number of vital genes helps to establish the believability both of the gene number estimate, and of the one-band–one-gene hypothesis.

Human cells in tissue culture have been studied the same way, and J. Bishop has calculated that such cells have about 35,000 different kinds of mRNA, most of them represented by only a few molecules. This estimate is in agreement with work done by J. L. King who had estimated an upper limit of about 40,000 vital genes in humans, based on mutation rate studies. You may be relieved to know that you have seven times as many genes as a fruit fly.

Bands and Puffs

The giant salivary chromosomes, bizarre as they are, are good functioning chromosomes, and some of their genes are busy transcribing mRNA, as one would expect. But in these cases we have an exceptional opportunity to leave our theorizing and simply watch the system working. Because we can identify specific bands, we can actually see genes doing their job. At specific stages in *Drosophila* larvae, highly localized and predictable bands of giant chromosomes loosen up and send great loops of chromosomal material into the nuclear sap. In such places, chromosomes are no longer dense and compact, but diffuse with tiny individual strands of DNA looking like the tangled line on an overly ambitious flycaster's reel. These areas are called *puffs* (Figure 10.14). Experimentation has shown that radioactive uracil becomes concentrated in these puffs, indicating that mRNA transcription is taking place. The largest puffs show the most RNA transcription activity. Then, as the larva proceeds into the next stage of development, the puffs recede, transcription ceases, the chromosomes rewind and settle down to lie quietly side-by-side, and the ordinary banding pattern once again appears in that place. Then the puffing begins in a new and different set of bands. As the larvae mature, they need different kinds of enzymes, and hence they need new RNA to organize them. Thus, one area puffs and then recedes as the strands of another unwind to expose the four critical bases that are the foundations of life—at least on this planet. These early observations, by the way, were among the first confirmations of the central dogma that chromosomal genes function by transcribing RNA.

10.3 Variations on Mendel

Gene Expression

We've already discussed several ways genes interact, such as in dominance relationships, and we know how genes can interact to upset our neat 9:3:3:1 ratio in the F_2 generation. But you should be aware that there are other factors that can influence the way genes are expressed, such as the environment in which the gene appears, genetic interaction, and chance.

Environmental Interactions

Environmental interactions may be very direct or very subtle. Let's consider a straightforward example: the Siamese cat. One of the enzymes in its pigmentation pathway is temperature-sensitive; it won't function when it is warm. As a result, pigmentation of the fur occurs primarily in the colder extremities of the cat: the ears, the tail, the feet, and the nose. While these parts are black or dark brown, the rest of the cat is cream or almost white, and that is why Siamese cats look like Siamese cats. If you keep your Siamese cat in the warm house, it may grow to be almost white, but if you put it out at night, it will get to be quite dark.

Modifier Genes

Sometimes the expression of genes at one locus is affected by the alleles present at another locus. In *Drosophila,* for instance, the mutant called *forked* (having misshapen bristles) may be modified by an allele at another gene locus. This modifier allele is called *suppressor of forked* and makes the homozygous forked fly appear to be normal. Another *Drosophila* mutant, called *prune* (purple eyes), may be modified by the *prune-killer* allele, which is lethal only in homozygous prune flies. The prune-killer has no other known effect. Other modifiers may have more subtle effects, such as those involved in expressivity and penetrance.

Variable Expressivity

Normal phenotypes are the result of fairly tight controls built into development systems, so that most traits don't vary much among normal individuals. The normal phenotype, then, can be assumed to be well-buffered from random effects of the environment. Mutant phenotypes, however, are much more variable since they are not so well-buffered. In other words, mutants may be greatly influenced by changes in the environment that would scarcely affect a normal individual. Some *Drosophila* mutants, for instance, are generally much more grotesque when grown at some temperatures than at others (Figure 10.15).

The stabilizing buffering system may also protect the organism from changes brought about by modifier genes. Mutants may be quite vulnerable to a range of such genes.

The example of the environmental effects in Siamese cats was also an instance of variable expressivity. And variation may also be due to modifying genes, or what is vaguely referred to as the "genetic background." Finally, some variation appears to be due neither to environmental nor genetic effects, but to random change in the developmental pathway that we might call "developmental chance."

Incomplete Penetrance

If a mutant phenotype is sufficiently variable, it may overlap the normal phenotype. Then an individual may have an abnormal genotype without showing it. For instance, a rare dominant in human genetics is *polydactyly,* the tendency to have extra fingers or toes. Persons carrying this dominant gene show variable expressivity in that all four extremities may be affected, or only the hands, or only the feet, or perhaps only one hand or one foot will have extra digits (Figure 10.16). Both hands and both feet of a given carrier individual have the same genes, and the same environment, but may differ in having normal or abnormal numbers of digits. This would seem to indicate that developmental chance is at work. Of course, this means that a person carrying the gene may just be lucky enough to have only five toes on each foot and only five fingers on each hand. In such a case, fortune has smiled four times and the person would be unaware of carrying this gene if it weren't for the fact that about half of his or her children have extra fingers and toes.

You may have picked up on the fact that Siamese cats may show incomplete penetrance. You may also have noted that we earlier claimed that dominant mutations do not skip generations, and that we did not tell the whole truth.

Sex-Limited and Sex-Influenced Effects

A dominant gene is known to be responsible for a rare type of cancer of the uterus. This is a *sex-limited* trait since, needless to say, it has minimal influence on men. On the other hand, the most common kind of middle-aged baldness is also due to a dominant gene, which doesn't usually affect women. It doesn't affect eunuchs, either—unless they have been given injections of male sex hormones. Some genetic conditions affect one sex more often, or more severely, than the other and are said to be *sex-influenced.* For instance, *pyloric stenosis,* a common and serious congenital malformation of the digestive tract, runs in families but affects five times as many boy babies as girl babies. And men are more frequently affected by stomach ulcers than are women.

Variable Age of Onset

Baldness and muscular dystrophy are two genetic conditions that we've already mentioned. They both have delayed, variable ages of onset. Muscular dystrophy can begin at very different ages even in affected brothers—people who have received the same abnormal X-linked allele from their mother. Another example is *Huntington's chorea,* which is a severe neuromotor disease due to a dominant gene. It does not begin to show its ultimately lethal effects until some time in adulthood (Figure 10.17). The average age of onset for Huntington's chorea is 40, but it can begin affecting men and women as early as age 15 or as late as age 60. Usually, the victim learns he or she

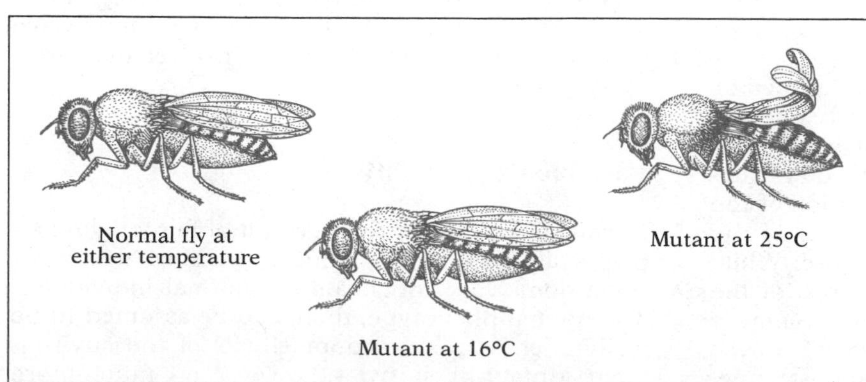

10.15 Curly wings and environmental temperature. The wings of fruit flies are normally straight under any environmental temperature in which the insect lives. One type of wing mutation, known as *curly wing,* causes the wings to curl up under warmer temperatures. As long as the flies are kept at about 16°C. the wings look normal. When the temperature gets higher—nearing room temperature—inherited wing defect reveals itself.

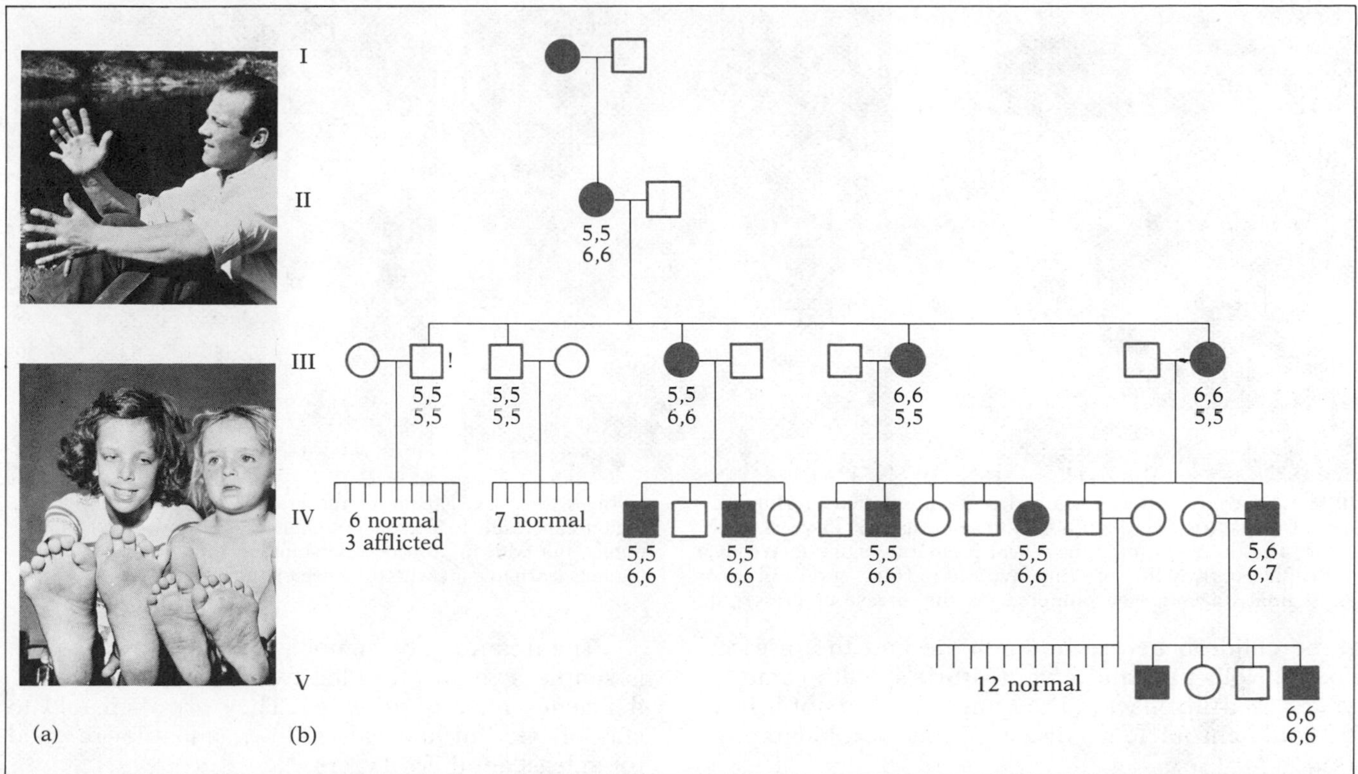

10.16 Polydactyly and incomplete penetrance. (a) The individuals shown here have inherited a dominant gene that expresses itself in the development of extra fingers and/or toes. Although the gene is dominant, its expression is somewhat incomplete. Some individuals carry the gene but do not express it. Their children may not be so lucky. The variables involved in polydactyly are not known. (b) In the family portrayed in the chart, the numbers accompanying the individuals represent the number of digits on the left and right hands (at the top) and left and right feet (at the bottom). The dominant gene expressed itself in either the hands or feet of every carrier except for the first male in the third generation. Although he was phenotypically normal, three of his children were afflicted–so he must have been a carrier. For some reason, unknown to geneticists, his dominant gene did not *penetrate* during his embryological development.

has the disease—and the defective gene—only after it is too late to make a conscious decision as to whether to have children. The disease begins as a personality disorder, and progresses through muscular shakiness, with symptoms similar to intoxication, on to complete paralysis and death in about 15 years. (If a carrier of the Huntington's chorea allele should die before he or she had become affected with the disease, it would be a case of incomplete penetrance.)

The best-known victim of this dominant allele was the folk-singer Woody Guthrie (Figure 10.18), who died of the disease in 1967. He and his wife had

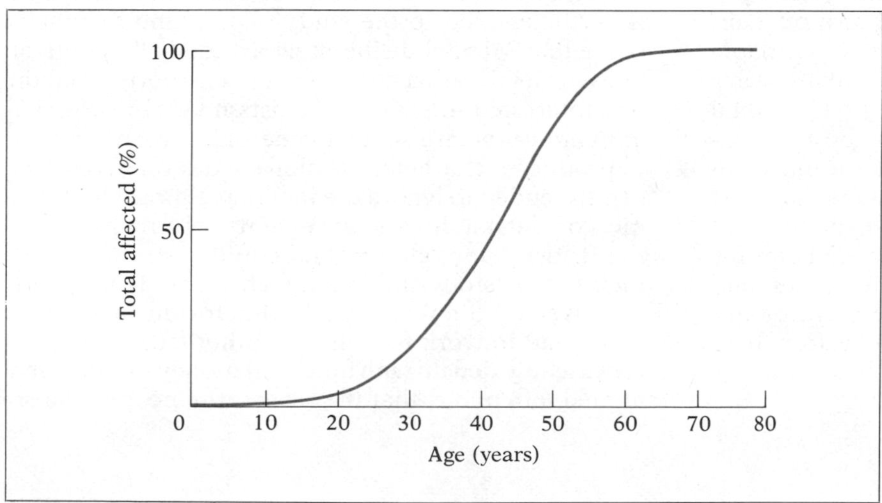

10.17 Huntington's chorea is a progressive nervous disorder that is genetic in origin. Its onset (appearance of symptoms) can occur at any time between ages 15 and 60, but most of its sufferers experience its onset between the ages of 30 and 50, as this graph reveals. Unfortunately, people can be well into their reproductive years before Huntington's chorea can be diagnosed, thus leaving the unfortunate legacy to another generation.

10.18 Woody Guthrie, the famed folk-singer, died of Huntington's Disease (or Huntington's chorea) at age 55, 13 years after onset. The three photographs reveal premature aging in Woody, but cannot portray the suffering involved in Huntington's Disease as its final stages are encountered. As the disease progresses, it involves the basal ganglia of the brain. Damage to cells in this region can result in personality changes and involuntary movements, but with improved understanding (and some love) most patients learn to cope with their personal misfortune.

three children before he knew he had the disease: Joady, Nora Lee, and Arlo. Guthrie's death came 13 years after the onset of the symptoms, but not before he could chronicle his disaster in his autobiography, *Bound for Glory.*

All Together Now

One final, all-purpose example gives an indication of how sex, environmental factors, and time can operate together to influence genes. The disease *hemochromatosis* is a recessive condition affecting iron uptake in the intestines. Normal people absorb iron from their diet only when they need it. But homozygotes for the defective hemochromatosis allele are unable to adjust their intake, and continue absorbing iron into their systems after they have had enough. The problem is, iron can be poisonous in large amounts. Symptoms of iron poisoning are diabetes, enlarged liver, lethargy, brownish skin, loss of armpit hair, and eventual death. The disease shows (1) variable expressivity, depending on sources of drinking water and on such things as whether or not the individual uses iron cookware; (2) incomplete penetrance, because many homozygotes can be shown to have abnormal patterns of iron uptake but show no overt symptoms; (3) a variable age of onset, usually about age 50 or later, since it apparently takes a long time for the iron to build up to toxic levels in the tissues; and (4) sex-limitation: only men have overt symptoms, presumably because women need more iron than men in the first place.

The treatment of hemochromatosis, by the way, is simple. Patients are bled twice weekly until iron-deficiency anemia develops. They are then told to stay off Geritol and calves' liver, and they're good for at least another 50 years.

Continuous Traits and Polygenic Inheritance

In Chapter 9, we expressed the opinion that Mendel was the greatest geneticist of the 19th century and that Darwin was the fourth greatest. That seems to leave a hole, doesn't it? Well the second best geneticist of the 19th century was probably Francis Galton, who just happened to be Darwin's younger cousin. Galton had a theory he wanted very much to prove: that "genius" was inherited and tended to run in families. He used his own family as an example, which doesn't say much for his modesty. His ideas, nevertheless, led to the study of the kind of inheritance that Mendel deliberately avoided—"matters of more or less" rather than discrete categories. Galton was interested in what geneticists now call *polygenic inheritance* (where several gene pairs work together to produce a character). Galton's work on hereditary genius tended to ignore the fact that almost all British men of outstanding achievement were members of the British upper classes, and could afford to devote their time to study and writing. This included Darwin, who lived off the Wedgwood china fortune and never really had to work for a living. Inherited genius was inextricably tied up with inherited money. (It has been argued, of course, that the reason the people had so much wealth was because they were smart. We doubt this.)

An even more serious trouble with genius, as far as Galton's studies were concerned, was that it was impossible to measure, except subjectively. He tried to develop quantitative ways of measuring mental characteristics, such as by measuring sensitivity or reaction time, but failed in this (Binet was later to succeed in quantifying mental characteristics with his IQ tests. Or was he? That question is still being hotly debated.) So Galton proceeded to measure other things that varied continuously, like height and weight. Mendel and Galton studied different kinds of things, in different ways, and both of them discovered a number of important principles of heredity.

In particular, Galton and his followers (such as the brilliant statistician, Karl Pearson) developed various *empirical laws.* Empirical laws are laws that simply describe a relationship, without offering an explanation of why the relationship works. If Mendel had established the 3:1 F_2 ratio without trying to explain why it occurred, or without testing his hypothesis, that would be an empirical law.

Galton's and Pearson's work was based on the almost universal *normal distribution* of continuous phenotypes, such as height or weight, in a *population.* On a graph, the normal distribution generates the familiar *bell-shaped curve,* with most of the population clustering around the *mean,* and a sharp drop-off tapering into two long tails on either end (Figure 10.19). The normal distribution is found in such diverse characters as hat, waist, and shoe sizes; height and weight of people, pigs, and pineapples; velocity of molecules in a gas; raw scores on midterms in a large class; length of life of light bulbs; barbs on the tongue of woodpeckers; and bristles on the belly of *Drosophila melanogaster.*

Once Mendel's work was rediscovered in this century, as we have seen, a stalwart band of devotees followed enthusiastically in the great man's experimental footsteps. They saw discrete differences everywhere and doggedly pursued the hallowed 3:1 ratio at every opportunity. But Galton and Pearson had their followers too, and to distinguish themselves from the Mendelians, they called their study *biometry* and themselves *biometricians.* Both the Mendelians and the biometricians claimed to understand heredity. And they HATED each other—in fact, it was probably the most intense feud in the history of biology. Neither side would even listen to the other. The Mendelians found 3:1 ratios everywhere. The biometricians found normal distributions everywhere. The biometricians could not understand why anyone would bother to generalize about the behavior of a few bizarre, abnormal, mutant traits. The Mendelians could not understand how the biometricians could be so stupidly involved in sterile empiricism offering no explanations. Journals controlled by one faction refused to publish papers favorable to the other faction. Even the prestigious journal, *Nature,* refused to publish the first account of Mendelian genetics in natural population, which is now known as the Castle–Hardy–

10.19 The normal distribution. (a) Polygenic genetic traits, when plotted by phenotypes gathered from large samples, tend to form curves approximating the idealized distribution shown here. (b) The accompanying photograph shows a good example. The World War I soldiers are arranged in rows according to height. While the curve thus formed at the top of each row is not perfectly symmetrical, there is an apparent cluster around a *mean,* with rapidly tapering tails.

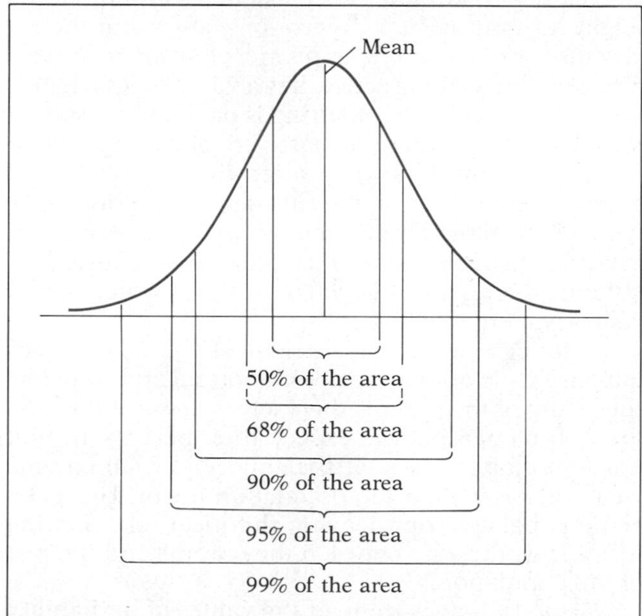

Weinberg law. (After reading the next chapter, ask yourself which side *Nature* was on.)

Finally, in 1918, the brilliant mathematician R. A. Fisher showed that all the biometricians' empirical laws followed quite logically from Mendel's laws of inheritance, as long as one assumed that the *continuous* (quantitative) traits were controlled by many loci, each contributing a small amount to the final phenotype, and with some variability accounted for by environmental effects. This is now called *polygenic inheritance,* to indicate that many (*poly-*) genes are involved. Actually, Mendel himself would have had no quarrel with the biometricians. He seemed to have understood the possibilities of polygenic inheritance and wrote about one such case in his discussion of the inheritance of flower color in green beans. After Fisher's paper, it would be nice to be able to say that the Mendelians and the biometricians made up in an orgy of contrite handshaking. Not so. In fact, the old antagonists went to their graves, each convinced that the others were nuts. As often happens in science, however, new generations of investigators were less rigid, and they were finally able to resolve the dispute, incorporating aspects of both ideas and developing an even better theory.

We don't have the time to delve deeply into modern biometry, so we will consider only the most basic of its tenets as they relate to inheritance. This has to do with the concept of heritability, which is traditionally symbolized h^2. In its simplest terms, offspring tend to resemble their parents, but this tendency is more pronounced for some traits than it is for others. Heritability is simply *a measure of the degree to which offspring tend to resemble their parents.* Galton found that the tendency was not absolute; sometimes offspring did not resemble their parents much at all. And, besides, what does "resemble" mean? This had to be defined.

The first step in studying a particular trait is to characterize each individual in terms of a deviation from the mean for that trait and for a particular population. For instance, if the mean height of men in a given population is 69 inches and Arthur is 72 inches tall, he is 3 inches above the mean. The average height of American women is 64 inches. At 67 inches each, Ed is a shorter-than-average man (2 inches below the mean) and Edna is a taller-than-average woman (3 inches above the mean). Separate means, and separate deviations from these means, must be calculated for men and for women. This helps to define what was meant by "resemble." The observation that offspring tend to resemble their parents can be reduced algebraically to:

$$E(Y') = h^2(X')$$

What does this mean? The X' stands for the average value of the two parents' deviations from the mean. The Y' is the offspring's deviation from the mean, keeping in mind that each deviation is measured from the appropriate mean of the two sexes where these are different. What about E? It means *expected value;* that is $E(Y')$ is the expected value of Y', which is the average value around which many actual values of Y' will be distributed. This will become clearer after we go through an example. As for h^2, that is an empirical value that measures the degree to which variation in some particular trait is inherited. For some traits, h^2 is very low and for others, it is very high. It has a theoretical maximum value of 1 (an entirely gene-determined trait) and a minimum value of 0 (an entirely environmental or chance trait). (The reason for h^2 traditionally having the form of a squared value is technical and need not concern us here.)

Now for an example. Recall that the population mean of height for adult American men is 69 inches and the mean for adult American women is 64 inches. The heritability (h^2) for human height in America is known from previous studies to be $h^2 = 0.75$ (these are all approximate, but realistic, values). A man is 74 inches tall and his wife is 67 inches tall. What is the average or expected height of their sons, when grown? Of their grown daughters?

The man's deviation is $+5$ and his wife's deviation is $+3$, so their mean deviation (X') is $+4$. The heritability of human height is $h^2 = 0.75$, so

$$E(Y') = h^2(X')$$
$$E(Y') = 0.75(4) = 3 \text{ inches}$$

It is expected that their daughters will grow up to be 3 inches taller than average women, or $64 + 3 = 67$ inches tall, and their sons will also grow up to be 3 inches taller than average men, or $69 + 3 = 72$ inches. In fact, the children of this particular pair of parents might turn out to be midgets—or giants—but the *expectation gives the best estimate of what to expect.* Since h^2 always, in practice, has a value of less than 1, the expected value of offspring is called *regression to the mean.* It says that the expected, or average, value of the offspring is always somewhere between the average value of the parents and the population mean. This seems counterintuitive to many people, but it is true just the same. Just remember that expectations are only average values around which actual values will fluctuate.

Although the average height of the offspring of tall parents is above the population mean but below the height of the parents themselves, *some* of the children of tall parents will exceed their parents' height. For that matter, some of the children of tall parents will be shorter than the population mean. These deviations balance out, leaving the *mean* of offspring just 0.75 of the way between the parents and the rest of the population.

Table 10.1 gives some of the values of heritability for various characters in farm animals.

Table 10.1

Cattle	h^2	Chickens	h^2
Birth weight	0.49	Body weight, 8 weeks	0.31
Gestation length	0.35	Shank length	0.50
Milk yield	0.43	Egg production	0.21
Conception rate	0.03	Egg weight	0.60
White spotting	0.95	Hatchability	0.16

This chapter has covered a lot of ground and, it must be admitted, genetics is not the easiest of topics to deal with, but it is increasingly important to understand its basic principles. Some people may be interested in genetics for very specific reasons, such as how to make the sandy soil of Georgia yield peanuts or how to breed guard dogs that can't be stolen. The population in general, however, is rapidly approaching the need to understand genetics for broader reasons. We want to know why those miracle crops haven't yet begun to feed this hungry planet, as once promised. We also want to know about the inheritance of our dispositions. Are we naturally aggressive and can we reduce the likelihood of our aggression in our children by choosing a docile mate? We want to know the answers to a host of questions, but we will continue to respond off the top of our heads unless we apply ourselves to learning how genetics really works.

But there is new ground to cover, so let's go on to learn about the genetics of populations.

Application of Ideas

1. Eleanor Perkins is phenotypically normal, but her family has its share of sex-linked genetic abnormalities. Her husband, Garvey, has the X-linked dominant allele for brown teeth, but is otherwise normal. Her brother Arthur and her son, Little Ned, both suffer from hemophilia A, the most common type of sex-linked bleeders' disease. Her father, Grandpa Bob, is not hemophilic, but he is color blind.
   a. Draw and label a pedigree diagram of this family.
   b. From this information alone, Eleanor knows she herself is a carrier. Which X-linked mutant gene or genes does Eleanor carry?
   c. Once Grandpa Bob came to breakfast wearing one red sock and one green sock, and couldn't understand why Eleanor, Garvey, and Arthur were all laughing at him. But when Little Ned didn't get the joke, Eleanor knew that her son was both hemophilic and color blind —and she suddenly realized what must have happened in one of her own oocytes, shortly before her birth. Explain.
   d. Eleanor and Garvey are expecting another child. If it is a daughter, what is the probability that their daughter will be hemophilic? Color blind? Have brown teeth?
   e. If their child is a son, what is the probability that he will be hemophilic? Color blind? Have brown teeth?
   f. What is the probability that such a son will be normal—that is, neither hemophilic, nor color blind, nor have brown teeth? (Note: this last question requires more information. Haldane and Smith calculated that the recombination frequency between the red/green color blindness and hemophila **A** loci is about 10%. List the expected frequencies of all possibilities for Eleanor's and Garvey's sons, and be sure they add up to one.)

2. Thomas Hunt Morgan, the discoverer of X linkage in fruit flies, later studied inheritance in a bug, *Euchistus*. The bug also proved to have XX females and XY males. In one species, *Euchistus variolarius*, all males have a black spot on the abdomen. In a closely related (and interfertile) species, *Euchistus servis*, the males are spotless, just as are the females of both species. A spotless female *E. variolarius* was mated with an equally spotless male *E. servis*. Among their numerous F_1 hybrid progeny, all the males had spots and the females were as spotless as ever. When these hybrid brothers and sisters were allowed to mate and reproduce, Morgan found 97 spotless females, 32 spotless males, and 84 spotted males in the F_2 generation.
   a. How many gene loci are involved in the spotting phenomenon?
   b. Is spotting a dominant or recessive trait?
   c. Is the spotting gene on an X chromosome, a Y chromosome, or on an autosome? Explain your answer.
   d. If the original cross had been between a spotted male *E. variolarius* and an unspotted female *E. servis*, what would have been seen in the F_1 and F_2 generations?

3. Many plant species have separate male and female sporophytes. Bauer and Shull took pollen from a narrow-leaved male *Lychnis alba* and dusted it on the flowers of a broad-leaved female plant. In the F_1, both males and females were all broad-leaved. The F_1 male and female plants were crossed to each other to produce an F_2 generation. All the F_2 females were broad-leaved, but there were both broad-leaved and narrow-leaved F_2 males.
   a. Why?
   b. Is breadth of leaf sex-linked or sex-limited?
   c. What would you expect would happen in the F_1 and F_2 of a cross between a male of a true-breeding broad-leaved strain and a female of a true-breeding narrow-leafed strain?

4. *Cannibis sativa*, as knowledgeable growers know, also has separate male and female sporophytes. Occasionally, however, male plants will bear a few runty flowers that are able to function as female flowers and set seed. McPhee grew some of these seeds to maturity, and found that they produced males and females in a 3:1 ratio.
   a. How do you explain these results?
   b. How would you test this hypothesis?

5. Recently, Vogt determined that sex in turtles and many other reptiles is not determined by chromosomes or even by genes, but is irreversibly determined by the temperature at which the eggs are incubated (efforts to save endangered turtle and tortoise species have resulted in years of wasted effort, as diligent but unaware workers inadvertently raised brood after brood of males under artificial conditions). According to Vogt, the reptile ancestor of birds and mammals probably had this kind of sex determination.
   a. How does this observation make it easier to understand the probable evolution of sex chromosomes in birds and mammals?
   b. What problems remain?

6. T. H. Morgan crossed a *white*-eyed, *yellow*-bodied *Drosophila* female with a *red*-eyed, *brown*-bodied male. In the F_1 progeny, the females were red-eyed and brown-bodied, while the males were white-eyed and yellow-bodied. Allowing these to cross to produce an F_2 generation, Morgan counted:

Eyes	Body	Sex	Numbers
white	yellow	male	474
white	yellow	female	543
red	brown	male	512
red	brown	female	647
white	brown	male	11
white	brown	female	6
red	yellow	male	5
red	yellow	female	7

a. What happened?
b. Of the total of 2205 F_2 flies, what proportion were white-eyed? Yellow-bodied? Male? Recombinant?
c. Do these values depart from the expected 50% of Mendel's first and second laws? Why?
d. What is the map distance between the *white* and *yellow* loci?

Key Words

anti-hemophilic factor

Barr body
biometry (biometrician)

carrier
chromosome puff
color blindness
crossing over
cytological map

deletion map
Drosophila melanogaster
drumstick

empirical laws

genetic background
genetic map
giant chromosome

hemizygous
hemochromatosis
hemophilia
heritability (h^2)
heteromorphic chromosomes
Huntington's chorea

incomplete penetrance

linkage group

modifier gene
muscular dystrophy

Key Ideas

Sutton and Morgan

1. Sutton proposed that the chromosomes carried the hereditary factors that had been hypothesized by Mendel.
2. *Heteromorphic* chromosome pairs such as the X and Y with identifiable differences had been detected, so researchers were able to identify and follow homologous chromosomes through meiosis.
3. Cytologists were able to *see* Mendel's laws in the meiotic process.
4. The use of *Drosophila melanogaster,* the fruit fly, was begun about 1910 by T. H. Morgan.
5. The advantages of using fruit flies in genetic research were:
   a. Ease of maintaining, readily mated
   b. Large output of offspring (hundreds per female)
   c. Short generation time (10 days)
   d. Only four pairs of chromosomes
   e. *Giant* larval *chromosomes*
   f. Little economic importance (not agricultural pests)
6. The first variation (mutant) Morgan discovered was white eyes.
7. Morgan began a series of matings with his white-eyed male. Through this study he discovered sex-linkage. His crosses were:
   a. White male × red female
   F_1 = All red
   b. F_1 red × F_1 red
   $F_2 = \frac{3}{4}$ red, $\frac{1}{4}$ white
   All white-eyed flies were males
   c. P_1 white-eyed male × F_1 red-eyed female (testcross)
   Offspring = $\frac{1}{2}$ red, $\frac{1}{2}$ white
   Red and white eyes evenly distributed in sexes
   d. Any red-eyed male × any white-eyed female (reciprocal testcross)
   Offspring = $\frac{1}{2}$ white, $\frac{1}{2}$ red
   All females, red-eyed; all males, white-eyed
8. The answer to these puzzling results is in the peculiar X and Y *sex chromosomes.*
9. Morgan surmised that the sex-determining factor and eye color genes were linked on the *X chromosome.*

nonrecombinant chromosome
normal distribution

oral–facial–digital syndrome

polydactyly

recombinant chromosome
recombination frequency
recombination map

sex chromosome
sex-influenced trait
sex-limited trait

variable age of onset
variable expressivity

X chromosome
X-linked inheritance

Y chromosome
Y-linked inheritance

10. He knew that male and female *Drosophila* had *X* and *Y* (J-shaped) *chromosomes* that were morphologically different.

11. Female flies have two *X chromosomes,* while males have one *X* and one *Y* (as do humans).

12. In *Drosophila, X chromosomes* are now known to carry about 1000 genes, while the *Y* has only 6 (fertility genes).

Sex Linkage in Humans

1. The influence of sex chromosomes differs in humans and *Drosophila* as follows:
   a. In humans, the presence of one *Y chromosome* produces maleness. The lack of a *Y chromosome* produces femaleness.
   b. In fruit flies, the presence of two or more *X chromosomes* produces femaleness. Only XY produces a normal male, XO is a sterile male.

2. Human females have two *X-chromosomes* in each cell, but one becomes condensed and functionless. The distribution of active and inactive *X chromosomes* is random in terms of paternal and maternal X's. The condensed *X chromosome* of women shows up in *Barr bodies* and *drumstick* chromatin.

Color Blindness

1. *Color blindness* in humans is caused by a recessive allele at either of two closely linked loci on the *X chromosome.*

2. Blue, red, and green sensitive cone cells are affected, but blue blindness is an autosomal abnormality.

3. About 8% of American males are afflicted by color blindness, which occurs in less than 1% of women.

4. Normal-visioned, *carrier* women can expect half their sones to be *color blind* and half their daughters to be *carriers.*

Other Sex-Linked Conditions

1. *Hemophilia,* or bleeder's disease, occurs in two forms and is *sex-linked.* With clinical treatment *hemophilic* males can live to reproduce.

2. *Muscular dystrophy* has a delayed onset, its symptoms beginning in early childhood. The symptoms show up in half the muscle cells of carrier women. This can be related to the random way one *X chromosome* or the other becomes condensed into heterochromatin.

Sex-Linked Dominants

1. Dominant brown spotting is a *recessive–lethal dominant.* The trait passes from fathers to all their daughters, and the frequency in females is twice that of males.

2. *Oral–facial–digital syndrome* is passed from mother to daughter. None of the sons have it. These women have twice as many daughters as they do sons.

Linkage, Crossing Over, and Chromosome Mapping

1. When the cross **AaBb** × **aabb** is carried out, the expected results are: $\frac{1}{4}$ **AaBb**, $\frac{1}{4}$ **Aabb**, $\frac{1}{4}$ **aaBb**, and $\frac{1}{4}$ **aabb.** These results occur when the two pairs of alleles are on different chromosomes and assort independently.

2. If the two pairs of alleles are linked on the same chromosome pair then one *might* expect the results to be: $\frac{1}{2}$ **AaBb** and $\frac{1}{2}$ **aabb.**

3. Actually these results are encountered infrequently. Because of *crossing over,* it is more likely that all four classes of offspring will be seen, but not necessarily in a 1:1:1:1 ratio. The gametes are shown below for a situation where *crossover* occurs in 10% of the chromatids (and does not occur in 90% of the chromatids). Chromosomes experiencing *crossing over* are called *recombinant chromosomes.*

Nonrecombinants, 90%:	45% **AB**
	45% **ab**
Recombinants, 10%:	5% **Ab**
	5% **ab**

4. When the cross is carried out the results are:

45% **AaBb**
45% **aabb**
5% **Aabb**
5% **aaBb**

5. Even numbers of *crossover* events do not show up in crosses because the second *crossover* puts the gene back where it was. Because only odd numbers of *crossovers* have a recombining effect, the probability of net recombination has a maximum value of 0.5. The probability of *crossover* is directly proportional to the distance between gene loci on the chromosome.

Mapping Genes

1. *Recombination frequencies* are used to construct *genetic maps.* Map distances are measured in percent of recombinations (successful *crossovers*).

2. *Genetic maps* show the relative location of genes, determine which chromosome a gene is on, and help determine whether new mutants are variants of known loci.

3. Studies of *Drosophila's giant chromosomes* are used to construct *cytological maps. Giant chromosomes* represent ten rounds of replication, side by side. They contain distinct cross banding that help identify chromosome pairs and loci on each chromosome.

4. *Deletion mapping* is used to identify specific loci. After X-ray exposure, various loci are destroyed and survivors are studied for the specific location of the damage.

5. Comparisons of *deletion maps (cytoligical maps)* with *recombination* or *genetic maps* verify the accuracy of the latter as far as gene order is concerned.

6. The distance between loci, however, are not the same. The differences can be explained in part by the presence of large *heterochromatic* regions around the centromere, which do not get involved in *crossover.*

7. *Vital gene* loci are identified by comparing the two maps. One *vital gene* exists in each of the dark bands of the *Drosophila* chromosomes. Studies of vital genes indicate that there is a total of 5000 genes in each *Drosophila* gamete. Studies of mRNA in cells of *Drosophila* verify the gene count since there are 5000 different kinds of mRNA molecules. Similar studies of human cells indicate there are 35,000 genes per cell.

Bands and Puffs

Activity in giant salivary gland chromosomes is shown by the appearance of *puffs* where the chromosome decondenses. Studies show that radioactive uracil becomes incorporated in these *chromosome puffs,* indicating RNA synthesis. During different stages of larval development, different bands become involved in puffing.

Gene Expression

Gene expression can be influenced by environmental conditions, genetic interaction, and also chance.

Environmental Interactions

In the Siamese cat, the hair-pigment-producing genes are inactive in warmer body parts.

Modifier Genes

Some genes suppress the action of others. A few of these modifiers are involved in penetrance and expressivity.

Variable Expressivity

The normal phenotypes are well buffered from random effects of the environment. Mutant phenotypes lack this buffering and become more grotesque under certain environmental conditions.

Incomplete Penetrance

1. Abnormal genotypes sometimes fail to express themselves.

2. In the dominant *polydactyly* condition, extra fingers and toes may not appear, or may appear in one hand or one foot or both.

Sex-Limited and Sex-Influenced Effects
Sex-limited traits only appear in one sex, while *sex-influenced traits* occur in both sexes, but more frequently in one.

Variable Age of Onset
Baldness, *muscular dystrophy,* and *Huntington's chorea* have delayed onset. The average age of onset for *Huntington's chorea,* a fatal neuromuscular disease, is 40, but there is considerable variation.

All Together Now
Hemochromatosis is an example of how sex, environmental factors, and time interact in influencing genes. The individual cannot control the uptake of iron. Excess iron causes severe to fatal symptoms. The disease is influenced by many factors, including the amount of iron in the diet, *sex limitation, incomplete penetrance,* and a *variable age of onset.*

Continuous Traits and Polygenic Inheritance
1. Francis Galton was interested in nondiscrete genetic traits. He tried, unsuccessfully, to measure genius and IQ. He and his followers developed *empirical laws* (laws that describe a relationship, but do not explain it). They observed the normal distribution of various characters (height, weight, etc.), and began the field of biometry, which was in total disagreement with Mendelian genetics.
2. In 1918, Fisher showed how the *biometricians' empirical laws* followed Mendel's laws. The key was that continuous traits were controlled by many loci.
3. A basic tenet of *biometry* is the concept of *heritability.* It is a measure of the degree to which offspring tend to resemble their parents.
4. The observation that offspring tend to resemble their parents is reduced algebraically to:

$$E(Y') = h^2(X')$$

X' = Average value of two parents' deviations from mean
Y' = Offspring's deviation from mean
E = Expected value
h^2 = Measures the degree to which variation in a trait is inherited (values ranges from 1 to 0)

5. The values of h^2 are applied to plant and animal breeding to predict yield commodities such as milk, eggs, corn, etc.

Review Questions

1. What were the two technologies which rapidly advanced cellular knowledge at the start of this century? (p. 297)
2. Sutton's hypothesis is one of the milestones in understanding particulate genetics. Review his contribution. (p. 297)
3. What are *heteromorphic chromosomes* and how did their discovery help in understanding meiosis? (p. 297)
4. List six important advantages in using *Drosophila* for research in genetics. (p. 298)
5. Carry out the crosses Morgan used with his newly discovered white-eyed male fruit fly. Use symbols or words to show the steps and summarize the results of each cross. (p. 299)
6. What was Morgan's thinking on the behavior of the white-eyed gene in his crosses? (p. 300)
7. Compare the *X* and *Y* chromosomes of *Drosophila* in the following aspects (pp. 298–300):
   a. Their microscopic appearance
   b. The gene complement of each
   c. The recessive behavior of the *Y*
8. With a diagram, show how males determine the sex of their offspring and why the probability of male and female offspring is equal. (p. 303)
9. Explain why males always inherit X-linked traits from their mothers and never their fathers. (pp. 300–303)
10. Compare the role of X and Y chromosomes in determining sex in humans and fruit flies. (pp. 299–303)
11. Explain what is meant by the statement, "Women are just as *hemizygous as men,*" and how does this relate to *Barr bodies* and *drumstick* chromatin? (p. 300)

12. What problems can a *color-blindness* carrier woman and her normal-visioned husband expect in their children? (p. 301)

13. Study Essay 10.1 and try to determine the following (p. 302):
    a. Explain why hemophilia is seen only in males. (Can females be hemophiliacs?)
    b. Victoria and Albert (top) had four sons. How did the laws of probability work out for their sons?
    c. Answer the above question for the sons of Alice (row 2).

14. What characteristic of *muscular dystrophy*-carrying females indicates that the gene is not really recessive on the tissue level? (p. 301)

15. Explain with diagrams why the *oral–facial–digital syndrome* never effects males. (p. 303)

16. Carry out the cross **AaBb** × **aabb** in three ways (p. 304):
    a. Where the pairs of alleles are on different chromosomes
    b. Where they are too closely linked on one chromosome for *crossing over* to occur
    c. Where they are linked on the same chromosome but there are 80% *nonrecombinants* and 20% *recombinants*

17. If the probability of a *chiasma* between two loci is .30 per *tetrad,* what is the recombination frequency *per chromatid?* (p. 306)

18. How does preparing a *genetic* or *recombination* map differ from preparing a *cytological* or *deletion* map? (pp. 306–307)

19. What four kinds of information are available from *genetic mapping?* (pp. 306–307)

20. Draw the map representing bar eye, garnet, and scalloped wings, using the data presented. (p. 306)

21. Explain briefly how *deletion mapping* is done. (p. 307)

22. In general, how well do *genetic* and *cytological maps* compare (p. 307):
    a. In terms of relative loci
    b. In terms of actual distances (explain this)

23. How many genes are present in *Drosophila* gametes? How do we know this? (pp. 308–309)

24. What is the estimated number of human genes and how did Bishop determine this? (p. 309)

25. What seems to control the expression of pigment genes in Siamese cats? What parts are usually pigmented? (p. 309)

26. What is genetic buffering? How does it differ between normal genes and mutants? (pp. 309–310)

27. Explain how *polydactyly* can be an example of *incomplete penetrance.* (p. 310)

28. How do *sex-limited traits* differ from *sex-influenced traits?* (p. 310)

29. What is *Huntington's chorea* and how does its expression differ from that of most simple dominants? (p. 310)

30. What is *hemochromatosis* and how does it illustrate *variable expressivity, incomplete penetrance, variable age of onset,* and *sex limitation?* (p. 312)

31. Compare the approach to heredity followed by *Mendelians* with that followed by *biometricians.* (pp. 312–314)

32. What are *empirical laws?* (p. 313)

33. Draw a curve representing a *normal distribution.* Where do we find measurements that fit *normal distributions?* (p. 313)

34. How did R. A. Fisher finally settle the controversy between the *Mendelians* and the *biometricians?* (p. 314)

35. Define the term *heritability.* What practical value does *heritability* have for people? (pp. 314–315)

36. If a man and wife are 70 inches and 65 inches tall respectively and the h^2 factor in their population is 0.68, what is the expected height of their sons? Of their daughters? (p. 314)

Chapter 11
Evolution

Evolution is one of the most pervasive and explanatory themes in modern biology. It is an intellectual fulcrum that has been used to pry loose countless gems of truth from the real world. In fact, it is so important that one wonders how biology could ever have been done without a clear understanding of the principles of evolution. Of course, as we know, much biology was done without it, and, as we also know, much of it was wrong. So let's take a close look at Charles Darwin's venerable old idea. We will see that some parts of it have weathered, aged, hardened, and "cured," while other parts have been changed, and new parts—parts that Darwin could never have imagined—have been added, as the concept of evolution has itself evolved. In particular, let's focus on how evolution actually works and how good the evidence is.

11.1 Genes and Gene Frequencies

Let's set the stage by noting that *individuals do not evolve,* at least not in the sense that we will consider it. *Populations evolve.* All the genes of any population at any given time comprise the *gene pool,* and the ratio of various alleles in that pool can change over time. As the ratio changes, evolution occurs. Evolution can proceed somewhat randomly as this or that gene accumulates by mere chance, or it can proceed under the directing influences of natural selection. In this case some genes are favored by the environment and help the organism to survive and reproduce its own kind of genes, including the favored ones. Other genes may not help their bearers to fit into the existing environment as well, or to reproduce as prolifically, and these genes and their bearers may dwindle into oblivion. So where do "good" or "bad" genes come from? From existing genes that have changed through mutation—permanent, random chemical change in the DNA molecule itself. A gene may also remain unchanged through generations but may change in its *survival value* if the environment changes.

To be more precise, evolution essentially involves a change in *allele frequencies:* The relative number of one form of the gene increases, and the relative number of a different form decreases. This progressive change in allele frequencies is a crucial stage in the evolutionary process that occurs between the origin of a new allele by chance mutation and the final replacement of the original form by the descendants of the newer form of the gene. The net result in the long term is that the group's DNA changes, and the change spreads, perhaps slowly, perhaps rapidly, through a local population or an entire species.

Darwin knew about evolutionary changes in physical appearances (what we now call phenotypes) and he knew also that these changes must reflect changes in the hereditary makeup. But whereas Darwin discovered the process of natural selection, he did not know about alleles, Mendel's "alternate factors." We've repeatedly made the point that an understanding of the mechanism of evolution required a meshing of Darwinian and Mendelian thinking, because Darwinian evolution ultimately depends on Mendelian genes. So, in this chapter, we'll explore the relationship between the two.

We should begin by noting that before Darwin's century ended, another new idea was formulated by August Weissmann—an idea that profoundly changed the way that people thought about heredity and made the rediscovery of Mendel possible. This was the idea of "the continuity of the germ plasm." No experiments or new observations were involved, and no mathematical analyses were done; Weissmann merely introduced a new way of thinking about heredity, but he was astoundingly convincing in his arguments.

The Germ Plasm Theory

"He has his father's eyes." The language of our folk awareness of heredity betrays a subtle fallacy that lay undiscovered in the thinking of Darwin and everyone else (except Mendel) until Weissmann's time. It's a silly fallacy, really, but an important one to think through. The child may inherit land or silverware from his parents, but not eyes. We now know that the child simply inherits a physical entity, DNA, that may make his eyes similar to his father's eyes. But this distinction was lost on earlier biologists.

The early biologists assumed that there was some direct physical connection, some emanation or influence, from the parent's *body* to the child's *body*. Since Leeuwenhoek's time it was realized that heredity was transferred through the egg and sperm, but still the unthought and unspoken assumption was that the parent's body itself—what we would call the phenotype—somehow influenced the form of the offspring. Until Darwin's time, in fact, no one had really given it very much thought at all.

Gemmules and Pangenesis: Darwin's Big Mistake

Darwin was different. He was not one to let unstated assumptions go by without working them out to their logical conclusions. Nor did he believe in mysterious influences, or emanations or magic of any kind. He was a complete materialist. Darwin knew that there must be some *physical* connection between the bodies of the parent and the child, between the father's eyes and the baby's eyes. At the same time, he knew that this physical connection had to be transmitted through the gametes. So far, so good. Then he made the greatest mistake of his scientific life.

In his theory of *pangenesis,* Darwin proposed that every organ of the adult body, possibly even every cell, produced tiny packets of information, perhaps embryonic copies of themselves. These tiny bits of hereditary information he called *gemmules.* According to the theory, gemmules from all over the body were somehow transported to the gonads and packed into the gametes. Darwin knew that a baby did not inherit his father's eyes, but proposed the next closest thing: the baby inherited gemmules that were tiny physical essences of his father's eyes, and were produced by his father's eyes. In the new organism, of course, the gemmules of both parents would be blended together and eventually would sprout like seeds to determine the character of each part of the body. It was a perfectly logical theory, although it was completely wrong. Remember, Darwin subscribed to the idea that use and disuse of organs influenced heredity. We now know that this idea was wrong, but it was widely believed in Darwin's time and may have given rise to the theory of pangenesis.

Weissmann and the Continuity of the Germ Plasm

Weissmann also believed that there must be a physical continuity between, say, the father's eyes and the son's eyes. However, the line of cause and effect that he proposed was somewhat different from that of Darwin's pangenesis (Figure 11.1). Weissman proposed that hereditary information was contained in the *germ plasm,* which was conceived as being an amorphous substance (because Mendel's unit factors hadn't been rediscovered yet). The germ plasm, in Weissmann's scheme, was continuous. Germ plasm arose only from germ plasm. The rest of the body, however, also arose from the germ plasm, and in its development, the body followed the hereditary information present in the germ plasm. Thus, the father's eyes did not *influence* the son's eyes in any way. If the father's eyes and the son's eyes were similar, it was because they were *both* influenced by a partly shared germ plasm.

Most animals, early in their embryonic development, actually do set aside the cells and tissues that will give rise to the germinal epithelium of the gonads. But if Weissmann had been a botanist, he might never have come up with his idea, because plants do not segregate specific germ-line tissue until late in development. In other words, with plants it is difficult to say which particular tissue will someday form gametes until the plant is already well-developed. Nevertheless, the distinction between a continuous line of hereditary information and its expression in development is essentially valid for both plants and animals. We know now that it depends on the fact that DNA has two functions: replication (duplicating itself) and transcription (making mRNA). Through replication, hereditary information is passed from cell to cell and organism to organism. Through transcription, this hereditary information is put into effect in the development of the phenotype. But the *phenotype itself never directly affects the hereditary material,* except insofar as the phenotype's interactions with the environment determine whether the hereditary material will be preserved or lost through natural selection.

Weissmann's theory allowed evolution to be understood in new terms. *Nothing evolves but the germ plasm.* In this view, giraffes do not evolve long necks, but giraffe populations do evolve and accumulate the genetic material that gives rise to long necks. The germ plasm is the physical link, the continuity between ancestors and descendants. Through the physical link to recent common ancestors, and through the union of gametes in fertilization, the germ plasm unites the individuals of a species. Natural selection preserves or eliminates variation in the germ plasm, depending on its effect on the developed phenotype of the organism carrying it.

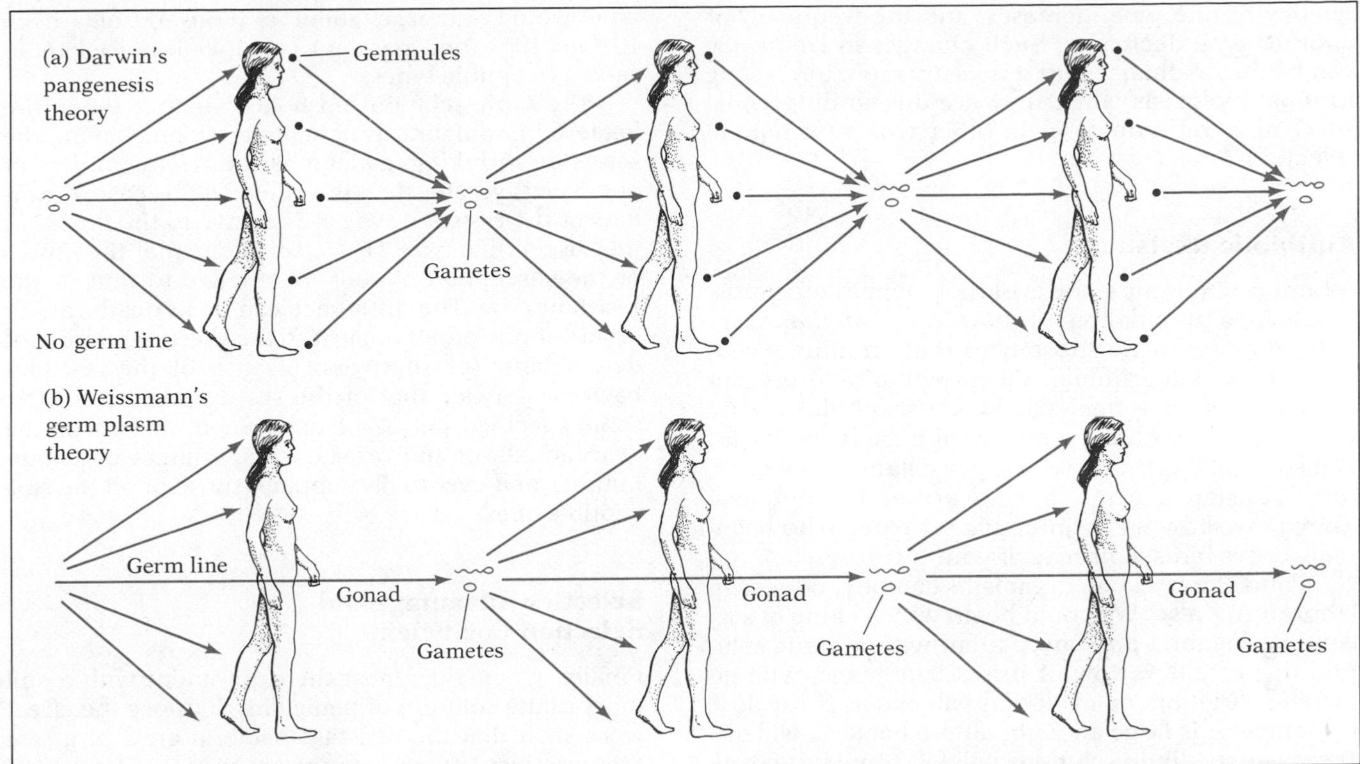

11.1 (a) Darwin hypothesized that offspring resembled their parents because the parents' phenotypes *directly* influenced the offspring phenotypes. Hereditary information was supposed to travel as "gemmules" from the various organs to the gonads to be packaged into gametes. In the next generation, this information directed development, and the cycle was repeated.

(b) Weissmann reasoned that the hereditary information was passed directly, through the gonads and the gametes, from generation to generation. Any physical resemblance between parents and offspring would be because the phenotypes of both individuals developed from partially shared hereditary information. The phenotype of the parent does not influence the hereditary information or the offspring's phenotype directly (although it may, through natural selection, determine whether or not that information is passed along at all).

Still, the idea was not quite accurate, because hereditary material is not an amorphous substance but a collection of discrete pieces of information. It is Mendel's genes, rather than Weissmann's plasm or Darwin's gemmules.

Frequencies

The notion of *frequency* is fundamental to population genetics. We will eventually be discussing *phenotype frequencies, genotype frequencies,* and *allele frequencies.* Here the term *frequency* has a special meaning, which has nothing to do with how frequently something happens or how often something occurs in time, such as when we refer, say, to the frequency of tornados. In population genetics, frequency is a *proportion,* namely a proportion of items of a particular kind in a more general class. Thus, if there are 10,000 registered voters in town, and 4300 are registered Republicans, the *frequency* of Republicans in this town is $4300/10{,}000 = 0.43$. Frequencies are always numbers between 0 and 1, because a frequency is always a fraction consisting of a part divided by the whole. Both the numerator and the denominator are counts of individual items, and any individual that appears in the numerator must also appear in the denominator. For instance, the 10,000 registered voters *includes* the 4300 registered Republicans.

The Population Genetics of Asexual Organisms

In asexual organisms, since the genetic material doesn't break down into unit factors or recombine, there is no practical population distinction between genes, genotypes, and phenotypes. There are just one or more kinds of individuals, and each kind has its distinctive genes, its distinctive genotype, and its associated phenotype.

If all individuals in a population are genetically identical, the population is a clone. No natural selection can take place in a clone. But if there are two or more types, each type will be present at a certain frequency. Evolutionary change occurs when the frequency of one type increases, and the frequency of another type decreases. Such changes in frequency can be due to chance, but it is vastly more interesting to most biologists when they are due to differential survival or reproduction—in other words, to natural selection.

Antibiotic Resistance

A familiar example is the evolution of antibiotic resistance in a population of bacteria. *Antibiotics* are natural or synthetic substances that, in minute concentrations, kill or inhibit the growth of microorganisms. For instance, penicillin kills susceptible bacteria by interfering with the synthesis of bacteria cell walls. But some bacteria are naturally resistant to penicillin, their resistance genes having arisen by mutation. They may show up by infecting someone who habitually doses himself or herself with the drug.

Penicillin-resistant organisms can be produced in laboratories also. We could begin with a clone of susceptible bacteria and cover them with a dilute solution of penicillin. Now, if it is a true clone, with no genetic variation, no evolution will occur. If the dose of penicillin is large enough, all the bacteria will die. Increasingly dilute solutions will kill some but not all of the bacteria, or will slow their growth.

But clones of genetically identical bacteria do not remain identical for long. Bacteria occur in large numbers, and we can predict that in a culture of any size there will be one or more spontaneous genetic mutants to the penicillin-resistant state. The *frequency of such mutants,* in the absence of penicillin, may be something like one resistant individual among a hundred million. We can symbolize the frequency by p, thus:

$$p = 1/100{,}000{,}000 = 0.00000001 = 10^{-8}$$

If we then kill all the susceptible bacteria with penicillin, perhaps only one or two resistant cells will remain, but the *frequency of the resistant type among the survivors* will have gone up to $p = 1$. That's one way of producing *fast* evolution! The descendants of the resistant bacteria will then fill an entire culture vessel in a short time.

If the penicillin is diluted, so that it just kills *some* of the susceptible bacteria, the change in frequency is not so great. Suppose that the penicillin kills half the susceptible bacteria. Then the frequency of the resistant type goes from one in a hundred million to approximately two in a hundred million—that is, from 10^{-8} to 2×10^{-8} (0.00000002). That's a large *relative* increase—a doubling—but the *absolute* increase is very small. However, if the difference in survival went on for ten generations, the frequency of the resistant type would increase about a thousandfold—from 10^{-8} to 10^{-5} (one resistant type per hundred thousand susceptible types).

The same relationship holds whether the entire bacterial population is actually growing, staying the same, or shrinking, as long as the *relative fitness* of one type is greater than the other. In this case, the *fitness* of the resistant type is 2 relative to the fitness of the susceptible type. Or we could say that the fitness of the susceptible type is 0.5 relative to that of the resistant type. The difference can be in death rate or in rate of reproduction, just so the average number of descendants (or survivors) of each of the resistant bacteria is twice that of the susceptible types after some specified period of time. Soon, of course, the descendants of the resistant bacterium would outnumber and eventually supplant those of all the susceptible ones.

Selective Advantage and Selection Coefficient

Finally, we might repeat the experiment with a still more dilute solution of penicillin. Suppose the effect were such that the resistant bacteria grew at a rate that was just 10% greater per generation. The rate of change in the frequency of the resistant type would be slower, but again the favored type would become relatively more common and the unfavored type would become relatively less common. If we made a graph of the frequency of the penicillin-resistant type over time, it would show a *sigmoid* or S-shaped curve (Figure 11.2).

Notice that the absolute change in frequency is initially very slow. It then increases, reaching a maximum *rate of change* at $p = 0.5$. After this point, it slows down again to a *fixation* level, where it very slowly (*asymptotically*) nears the value $p = 1.0$ (when every individual in the population is penicillin-resistant). The value of p theoretically gets closer

11.2 Increase in frequency of a superior new mutant strain in a population of asexual organisms. If a beneficial mutation occurs in a strain of asexual organisms, over a number of generations it will increase in frequency at the expense of the less favored type. A graph of the frequency as it changes through time takes the form of a smooth, S-shaped curve (sigmoid curve).

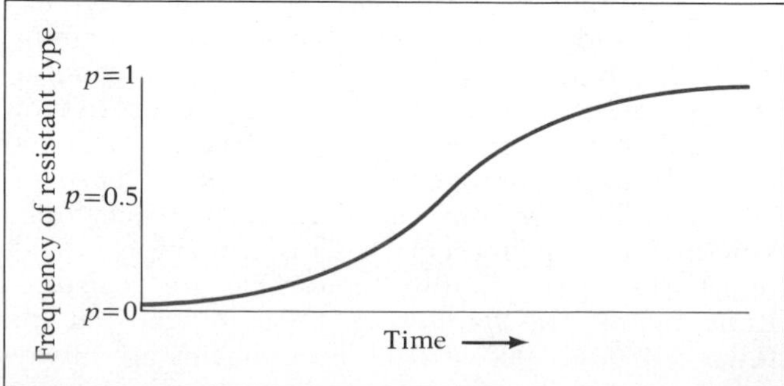

and closer to 1.0 without ever quite reaching it. In a real population, of course, infinitely small frequencies don't occur, and eventually the last surviving non-resistant type would bite the dust.

Selecting for Two Characteristics at Once

So far, we've considered changes in a single trait, which would be controlled by one gene locus. Now let's see what sort of interactions are possible when natural selection operates on two phenotypes, and two gene loci, at once.

Asexually Reproducing Organisms

In populations of asexual organisms, there may be selection going on simultaneously for two or more different factors. For instance, one new type might have a selective advantage because, let us say, it has a new mutant enzyme that enables it to metabolize arginine more efficiently. Another strain of the bacterium may have a selective advantage because a different gene gives it better resistance to damage from ultraviolet radiation. What happens now?

With asexual organisms, it's always a matter of "may the best strain win." If the strain with better ultraviolet resistance can outreproduce either of the other two strains (with and without the mutant enzyme), it will drive both of them to extinction—and the new and improved arginine enzyme would be lost. No matter, perhaps, because things like bacteria are so numerous that a very similar mutation will probably show up again in our hypothetical ultraviolet-resistant strain, and then it can take over. The rule in asexual organisms, however, is that only one new evolutionary change can be fixed at a time, so progressive evolutionary changes are *sequential* (Figure 11.3).

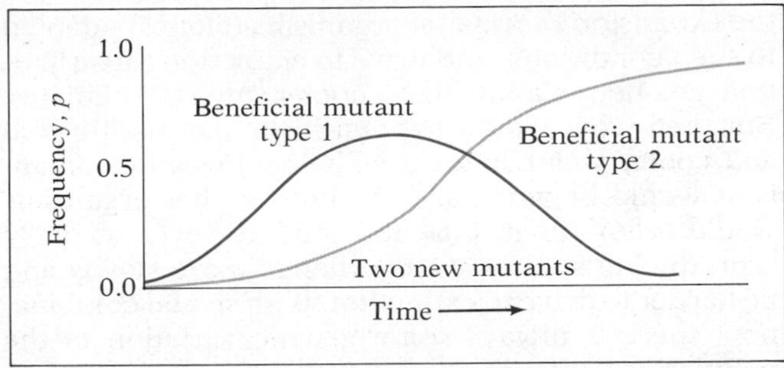

11.3 Increases in the frequencies of two superior new mutant strains in a population of asexual organisms. When two unrelated mutant strains occur in the same population of asexual organisms and both are superior to the original type, they will be in competition with each other. Eventually, only the one strain with the higher reproductive potential will survive (type 2), and the other potentially beneficial mutation will be lost (type 1).

Sexually Reproducing Organisms

In sexual organisms, two different favorable mutations can arise in separate individuals and go to fixation simultaneously. For instance, an ultraviolet-resistant mutation in one organism and a mutation with more efficient metabolism in another organism would not be competing against each other, as they were in the asexual population. Genetic recombination frees unrelated genes from competing with each other, and each gene locus behaves as if it were independent. Recombination will create individuals bearing both favorable mutations (Figure 11.4).

We should remind ourselves of a small but important point here. It is only in sexually-reproducing populations that unlinked genes behave as independent units in evolution. They can be separated by various "shuffling" processes discussed earlier and be brought together independently of each other. Gene A need not always appear with gene B. Natural selection can then mold many different aspects of an evolving organism at the same time, and every beneficial change that arises can be combined into the evolving phenotype. This is what *recombination* is all about. According to people who have given the question much thought, this is the chief advantage of sexual recombination and perhaps the reason that meiosis and sex evolved in the first place.

One claim for the evolutionary advantage of sex is that sexual organisms can evolve more rapidly, since many genes can be selected for simultaneously. Opponents of this view object that asexual organisms can evolve just as rapidly or even more rapidly, at least when only one trait is involved. And that, in fact, asexual organisms can evolve *too* rapidly, leading a line of asexual organisms down a blind alley. For instance, a succession of hot, dry years would favor

11.4 Increases in the frequencies of two superior new mutant genes in a population of sexually reproducing organisms. With sexual recombination, alleles at different loci are not in competition with one another. If two or more beneficial mutant alleles are increasing in the same population, recombination will allow the emergence of individuals that carry the best allele at each gene locus, and different beneficial mutations can go to fixation independently and simultaneously. This is one of the long-term advantages of sexual reproduction.

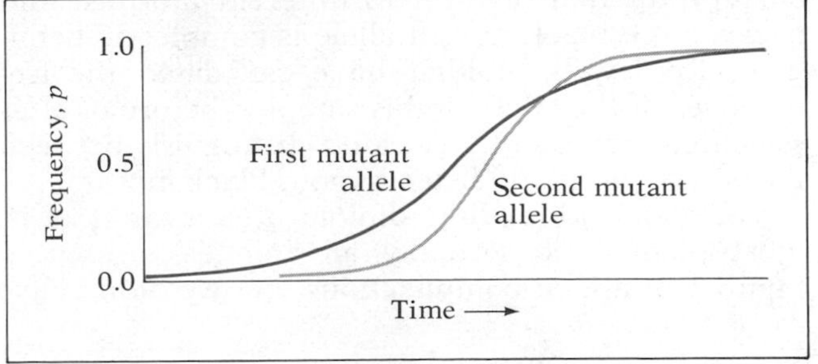

the expansion of asexual organisms (clones) adapted to such conditions, and drive to extinction those lines that are better adapted to cooler, moister climates. But then what would happen if the dry spell ended and cooler weather set in? The cool-weather organisms would be gone, and the hot-weather organisms would follow them into oblivion. However, sexually reproducing species would change more slowly and be harder to drive to extinction. With sexual combination, there is always some genetic variation in the population—unexpressed recessive genes in heterozygotes, perhaps, but there nonetheless. Thus, as the environment changes, a population with hidden variation is more likely to be able to track it.

The last advantage of sex that we will mention is related to heterozygosity. The heterozygosity permitted by sex could be a way of shedding harmful mutations (which occur constantly). With a constant reshuffling of genes, any harmful genes might ride along harmlessly, their ill effects being swamped by the good genes, until, as chance would have it, some hapless individual winds up with a heavy load of bad genes and dies. Thus, a high number of bad genes are removed from the population in one fell swoop.

A relevant study has been done on the short-term coevolution of a species of pine and the scale insect that infects it. Pines, like other plants, have a wide variety of chemical defenses against insect pests—but in pines, as in other plants, there is usually one or a few insect parasites that are able to overcome these barriers. In this case, the pines are much more long-lived than the insects. Thus, the short-lived insects undergo many generations on a single host. As the tree ages, so does the population of scale insects, with the best adapted parasites surviving. In time the pest population becomes increasingly adapted to its venerable host.

The extent of this adaptation can be demonstrated readily: Scale insects can be successfully transplanted from one part of a tree to another part, but they nearly all die if they are moved onto a different tree, even one of the same species. In fact, for the most part, the scale insects cannot infect the young seedlings that grow from the parent tree to which they are adapted. The reason is sexual recombination. The young seedlings have different genotypes from those of their parents, and thus they have new and different combinations of defenses against insect attack, defenses for which the insects aren't adapted. Most of the young pine seedlings are therefore relatively free of scale insects, but even on the hardiest hosts, a few invading insects manage to survive. Then, as the tree ages, descendants of the first colonists become increasingly resistant to their host's defenses through natural selection that favors new combinations of the insect's own counterdefenses. With such recombination, adaptation is fairly rapid. In this continuing relationship, sexual recombination is important to the scale insects. Thus, sex is important to both the tree and the insect.

The problem of why sex ever arose to begin with has plagued biologists for years. It has been pointed out, for example, that if the name of the "fitness" game is simply to leave one's kinds of genes, the best way would be to make replicas of one's self rather than to produce offspring that carried only half of one's genes. Obviously the advantages of reshuffling genes and distributing them among variable offspring would have to be enormous to offset the advantage of simply replicating one's self. Nevertheless, the question of what sex is really good for has never been satisfactorily answered. We will return to the question in Chapter 29. Now let's go on to the evolution of diploid organisms. Have a look at the story of the peppered moth in Essay 11.1, which serves as an introduction.

11.2 Population Genetics

Evolution in Diploid Organisms

The example of natural selection in British moths (Essay 11.1) is incomplete, because we have no data on reproduction of the two different morphs. And because the black morph allele is completely dominant, Kettlewell couldn't have calculated the frequencies of the alleles in his sample—before or after selection—because he couldn't distinguish between the homozygous and heterozygous black moths.

Sexual, outbreeding, diploid organisms (that is, most plants and virtually all animals) present a number of added complications that we didn't have to deal with in our brief discussions of the evolution of asexual organisms and sexual haploids. For instance, if an allele is recessive, it can "hide behind" a dominant allele in the heterozygous genotype. Keep in mind that if it doesn't affect the organism, the hidden allele is completely invulnerable to natural selection. There are other complications, too. If two alleles are involved, we have to deal with three genotypes—the heterozygotes as well as the homozygote—which gets us into all those confusing dominance relationships. And, as we'll see toward the end of this chapter, the frequencies of the three genotypes depend on such things as whether or not the organism in question tends to mate with its close neighbors, or mate with its own relatives, or self-pollinate, or whatever.

Essay 11.1 The Story of the Peppered Moth

Let's look at a case where natural selection has been observed operating on real organisms. *Biston betularia* is a British moth, commonly called the *peppered moth.* It occurs in two *morphs;* that is, it may have either of two distinct appearances, as shown in the photograph. One moth morph is light and mottled, or peppered, and the other moth morph is black. The British have a long tradition of butterfly and moth collecting, and records on the peppered moth go back two centuries. The black morph, whose color is controlled by a single dominant gene, originally showed up in 18th century collections as a rare, highly prized variant, or mutant. In the early stages of the industrial revolution (in the 1840s), the black form began to show up in greater frequencies in collections, especially near cities. The black morph continued to become more and more common in industrialized areas, until it greatly outnumbered the light peppered morph. In Manchester, England's industrial center, the dominant black morph achieved a frequency of 98%. Meanwhile, the light peppered morph remained the predominant form in rural areas.

The environment had changed and the species, through differential mortality and a change in allele frequencies, adapted to it. The environmental factor was soot from burning coal. Industrial England, as the 19th century proceeded, quietly submitted to its dark cloak of carbon. Meanwhile, the *Biston betularia* adapted. Bird predation is probably the species' principal cause of death. Over the long course of past evolutionary time, the moth had achieved a camouflaging coloration that blended well with the light, peppered appearance of lichen-overed tree trunks. But pollution killed the lichens and blackened the trees, making the light peppered morph highly visible and extremely vulnerable to predation. In industrialized areas, the black morph achieved a significant selective advantage, because it became the form that was less easily spotted by birds.

From a graph of the frequency increase of the black form in the historical data, J. B. S. Haldane (one of the founders of population genetics) calculated its relative fitness in an industrial environment to be twice that of the more conspicuous peppered form. But a British naturalist, H. B. D. Kettlewell, performed the crucial experiment. He released known numbers of marked black and light peppered moths in unpolluted woodlands and in polluted, soot-blackened woodlands. In each habitat, after a period of time had elapsed, he recaptured a portion of the released moths. Here are some of Kettlewell's mark-and-recapture data:

For the first set, released in the unpolluted woodland, almost exactly twice as many light forms survived as black forms. That's equivalent to a 100% advantage of the light type in a brief exposure to predation. The frequency of the black morph in this sample fell from $p = 0.488$ to $p' = 0.326$ (i.e., from 473 out of 969 to 30 out of 92).

In the second data set, selection was against the light peppered morph. Almost exactly twice as great a percentage of the favored black type survived. The frequency of the black morph rose from $p = 0.765$ to $p = 0.872$.

Incidentally, England has been doing pretty well of late in its battle against air pollution. The woodlands near the cities are once again becoming covered with lichens, and the soot is disappearing. As one might predict, the black morphs of *Biston betularia* are now declining in frequency.

Dorset, England—unpolluted woodland	Light peppered moths	Black moths
Marked and released	496	473
Recaptured later	62	30
Percentage recaptured	12.5	6.3
Birmingham, England—soot-blackened woodland		
Marked and released	137	447
Recaptured later	18	123
Percentage recovered	13.1	27.5

As a result, the algebra of natural selection in higher organisms can put the most enthusiastic biologist into a glassy-eyed stupor. So we'll skip the algebra and go right on to what can actually be observed about genetic variability in natural diploid populations.

Genetic Variability in Diploid Populations

There is a fundamental difference in the way in which natural selection works in higher organisms, and simple things like bacteria and unicellular algae. The big difference is due to the much greater amount of genetic variability in populations of diploid sexual organisms. Whereas a population of bacteria may often be essentially a clone of identical individuals, or at most a mix of a very few clones, in diploid populations every individual is unique. This genetic variability allows sexual diploid populations to respond very rapidly to natural selection.

We humans are quite aware of this variability in our own species, of course. We are incredibly sensitive to human differences, especially when it comes to facial features (Figure 11.5). And facial differences in humans are almost entirely genetic, as we are reminded when we are confronted with identical twins. It turns out that although individuals of other species might look very similar to us on casual observation, they are just as variable as we are (Figure 11.6). As we'll see when we consider animal communication in Chapter 33, other species are fully aware of these individual differences among their own kind, and, for all we know, they might think that all humans look exactly alike.

11.5 Human beings, like all large populations of sexually reproducing organisms, are extremely variable. No two individuals are at all alike, with the exception of single-egg twins.

11.6 Penguins, like humans, form large populations of sexually reproducing individuals. No two penguins are at all alike—at least to another penguin. Every Adélie penguin here knows its mate, its offspring, and all its nesting-ground neighbors.

Our uniqueness and variability are not limited to our faces; we are utterly unique, even down to our chemistry. This can be demonstrated by simple chemical classification of blood proteins and other biochemical differences. Consider the difficulties of organ transplant resulting from tissue incompatibility. Other evidence that shows that natural populations have a lot of genetic variability comes from experiments in which plants or animals are selected for special traits or characteristics. It seems that a breeder can choose to emphasize any aspect of the phenotype in a population. This proves that there must be genetic variability affecting that trait in the population at the outset, because neither natural selection nor artificial selection can work unless genetic variability exists.

It is no accident that Darwin began his book on evolution by discussing variation and differences between individuals, because this is the raw stuff of natural selection. Previous thinkers from Plato on had always thought of species as perfect, idealized types, and had considered the variation that they found in real plants and animals to be due to the imperfection of nature, a nuisance that was not of fundamental importance. But it turns out that genetic variation is a treasure. It is the species' main chance for coping with an unpredictable future.

The Maintenance of Genetic Diversity

What maintains all this genetic variability in diploid sexual species? We don't know. The problem is not that there is no answer, but that there are too many ideas and we don't know which of our current ones are important and which ones are simply wrong. A whole generation of population geneticists and evolutionary theorists are devoting their professional lives to just this question. Part of the responsibility for variability lies in dominance relationships, specifically the tendency of recessive effects to "go into hiding" whenever they are present in low frequencies. Another part of the answer is overdominance, where the heterozygous genotype has a greater reproductive fitness than either homozygote. We saw an example of overdominance in the sickle-cell and normal alleles of beta hemoglobin, where malaria is endemic. Since each of the two homozygous types are inferior to the heterozygote, natural selection keeps both alleles in the population in intermediate frequencies.

Short-Term Evolutionary Change in Diploid Populations

Darwin drew his idea of natural selection from what he knew about *artificial selection.* In a sense, then, artificial selection can mimic natural selection, at least in the short term. In fact, the results of artificial selection experiments can be startling, because it is so rapid and so effective. Considering how we have, in a very brief time, successfully created the vast array of domestic dogs that range from 200 lb behemoths to hairless, mouselike creatures, through artificial selection, a visitor from outer space might well conclude that *Homo sapiens* is, above all, a whimsical creature.

That visitor would have other evidence as well—such as tall chickens. Figure 11.7 shows the result of a famous experiment done in Berkeley, California, in which a population of chickens was selected for long legs. Note, by the way, that some of the long leg genes are associated with more general effects and also cause increases in neck and tail length. It seems evident that another century of artificial selection would suffice to produce chickens with legs like storks and necks like giraffes.

11.7 University of California geneticists have selected a line of chickens for longer and longer legs. The female on the left is a representative bird from the unselected control line; the female on the right is from the line selected for long legs. Artificial selection for this character has obviously been successful; in addition, there has been a coordinated response in neck, tail, and body length.

In another set of experiments, the British population geneticist D. S. Falconer selected for body weight in mice for twenty-three generations. A large group of mice were divided into eighteen separate groups. In six such groups, only the heaviest mice in each generation were allowed to breed. In six other groups only the lightest mice were allowed to breed. In the remaining six populations, which served as a control for environmental variation, mice to be bred were chosen by lot. The results are shown in Figure 11.8. Selection for greater or smaller body weight in mice is obviously successful. The jagged data of individual lines is due primarily to sampling error and to slight, subtle variations in the environment from generation to generation, but on the average all lines responded in a predictable and repeatable way. This graph, in fact, is quite typical of hundreds of graphs from similar experiments with mice, chickens, corn, and fruit flies. The response to artificial selection is immediate, it is substantial, and, at least for a while, it is essentially linear. After a dozen or two generations, however, the rate of phenotypic change gets much smaller, and eventually there may be no more response to selection for a while.

Not visible on the graph, but present in Falconer's experiment (and in virtually *all* artificial selection experiments) is the fact that each of the experimental populations got progressively weaker, had smaller and smaller litter sizes, and showed increasing infertility and higher infant mortality. In other words, their own natural fitness decreased under artificial selection, no matter whether the artificial selection was for larger or for smaller mice. This pattern is typical, seen again and again, no matter what phenotypic character is being selected for. In agriculture this problem presents a great risk to people who grow new miracle grains and other highly selected and inbred strains.

So just what is going on? First, with the mice and with most selection experiments, we are dealing with a *polygenic* character. Thus, selection is operating on many gene loci. Also, the fact that the initial response is strong and linear means that the experimenter is not dealing with rare alleles. You will recall that initial responses are very small when alleles are rare. In these cases, the alleles for variable body weight must already be present in the original population and must be *polymorphic*—with both lightweight and heavyweight alleles in intermediate frequencies.

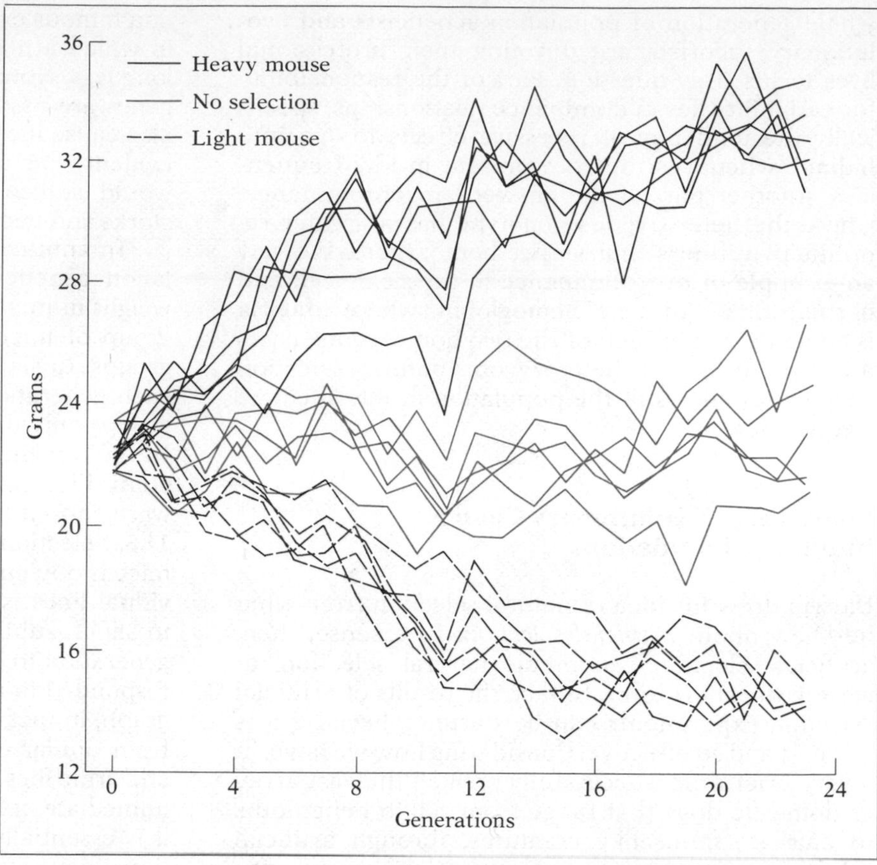

11.8 Eighteen populations of mice were involved in an artificial selection program in which body weight at 6 weeks was measured. The experiment was carried out for twenty-three generations. Six replicate populations were selected for increased body weight, six replicate populations served as unselected controls, and six replicate populations were selected for smaller body weight. Although there were random variations between replicates, the general trends are clear. Note that there is relatively much less progress in the selected lines in the last twelve generations, compared with the rather substantial progress made in the first twelve.

This is an example of the pervasive genetic variability that is found in diploid sexual populations, the same variability that makes life so difficult for analytical physiologists. In short-term artificial selection, and presumably in short-term natural selection as well, any changes must arise from variation *already present in the population.* By the end of the experiment, the low-line, intermediate and high-line populations still have the same alleles, but in different frequencies. The low line has accumulated a lot of lightweight alleles, while the high line has accumulated a lot of the heavyweight alternatives of these alleles. The intermediate line remains more variable than either the high or the low line, as shown in Figure 11.8.

But you're probably jumping up and down and screaming, "what causes the rate of selection response to slow down after ten or twenty generations?" Aren't you? Actually, a number of factors conspire to give this result. One might be that some of the alleles become fixed in the populations, or they may attain very high frequencies, so that we are seeing the high end of the sigmoid curve of Figure 11.2. In other words, the experimental lines run out of genetic variability, and without genetic variability there is no evolution. Another factor is the decreasing fitness of the selected lines. As litters get smaller and fewer, it is more and more difficult for the experimenter to pick parents that are notably heavier or lighter than their littermates. And frequently, the most extreme animals are the least fertile.

But why do the selected lines lose vigor and reproductive fitness? Again there are several reasons. One is simple: The delicate physiology of the animal is adjusted to a certain range of body weight, and outside that range there may be physiological problems of various sorts. Another reason is that some of the combinations of alleles that are selected for by the experimenter are simply harmful to the mouse. Each gene has effects on many systems, and a lightweight gene might well have an effect on, say, hormone production or thermoregulation.

Finally, by repeatedly choosing just the most extreme animals for mating, the experimenter inevitably begins to mate related animals, and the selected lines become increasingly inbred. Inbreeding of normally outbred organisms always causes drastic reductions in fitness, primarily because it allows harmful recessive alleles to become homozygous.

Selection for Going Up or Down

Figure 11.9 shows response to selection, in two directions, for geotaxis in *Drosophila.* Positive *geotaxis* is the tendency to move up. Jerry Hirsch invented an

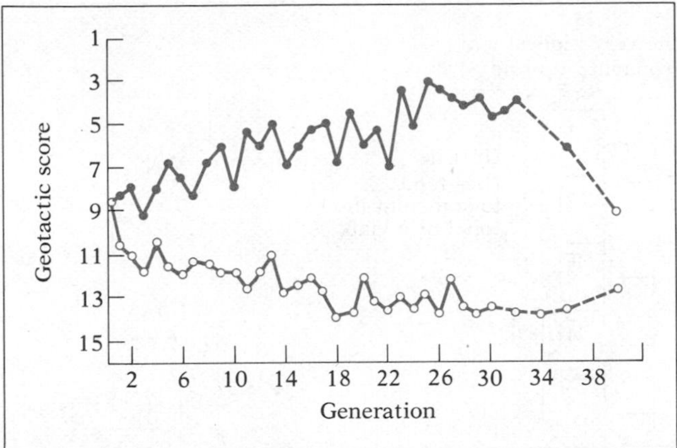

11.9 The results of selection for geotaxis in *Drosophila.* Open triangles indicate the progress of populations selected for positive geotaxis (down); solid triangles indicate the progress of populations selected for negative geotaxis (up). The dotted lines indicate the progress of the same populations after selection had been "relaxed." Note that the up line, in particular, rapidly lost its selected behavior.

ingenious maze that enabled experimenters to select flies for geotaxis (Figure 11.10). The maze consists of a series of interconnected chambers. As the fly enters each chamber, it finds only two exits: one straight up and one straight down. To get from one end of the maze to the other, the fly must make an up–down choice fourteen times. At the end of the maze are fifteen collection vials with fly food in them. The flies that end up in the top vial have to have made the up choice fourteen times in a row; the flies that end up in the bottom vial have to have made the down choice fourteen times in a row. The flies that go up seven times and down seven wind up in the middle vial (vial 8). The experimenter can simply put a whole population—hundreds of flies—in one end of the maze and collect the high scorers and low scorers from the top and bottom vials at the opposite end of the maze. In a selection experiment, only the high-scoring flies are kept in the up line, and only the low-scoring flies are kept in the down line.

In the original experimental runs, the flies started off by showing no particular preference—as a population, that is. But within the original population there were some flies that showed very slight tendencies to prefer going up and some flies that had equally slight tendencies to go down. This behavior is another polygenic trait; there are many gene loci that are variable for alternate up and down alleles in the *behavioral phenotype.* A selection response in both directions continued for twelve to sixteen generations. By that time, almost all the flies in the down line always ended up in the extreme down vials (numbers 14 and 15). The average of the up line was around vial number 4, which corresponds to three down choices and eleven up choices. As we said before, an experimenter can select successfully for just about any imaginable trait.

As usual for selection experiments, the fitness of the extreme up and extreme down lines declined markedly. While it is normal for a fly population to have a mix of both kinds of alleles, it is not normal to accumulate too many of one type and too few of another. We have no idea of what other things these alleles are doing, but they evidently are harmful in the wrong balance and combinations. After twenty generations, the researchers who ran this particular experiment stopped selecting the most extreme flies, and just let their populations mate and reproduce as they pleased. Every few generations the "relaxed" populations were tested. Natural selection had taken over, and the balance of up and down alleles returned toward normal in both populations (Figure 11.9, dotted lines).

Experiments like these show how easily natural populations are changed, and how rapid evolution can be when selection is very strong. But what does it tell us about natural selection?

Natural Selection Compared with Artificial Selection
To some extent, artificial selection experiments have given us misleading ideas about how natural selection works in the long-term evolution of species. In these experiments, all the genetic variation is present in the beginning population. We thus see only the (admittedly crucial) stage in which allele frequencies are changing within a population, which of course alters the average phenotype of members of the population. But one vital ingredient in long-term evolution is missing—the unpredictable appearance of totally new genes by random mutation. Mutation is a slow process; beneficial mutations, in particular, occur so rarely that they can't be expected to show up in the course of laboratory experiments on higher organisms. So artificial selection just simulates one part of the whole complex process of adaptive evolution.

Dogs and cats have comparable gene loci for the same *classes* of effects (such as those that produce coat color), but they don't have the same alleles in different frequencies; they have different alleles that arose in the past as beneficial (adaptive) mutations in the two lines of descent. In fact, although humans and chimpanzees, which are much more closely related than are cats and dogs, do have some identical alleles (as determined by protein analysis), most of their alleles are qualitatively different. Long-term evolution, then, does involve a stage at which alleles change in frequency in response to natural selection, but it also includes the origin of qualitatively new forms of genes. New mutations, the replacement of one allele by another, and the maintenance of the kind of genetic variability that allows for rapid-term response to changes in the environment, are all equally important genetic aspects of evolution.

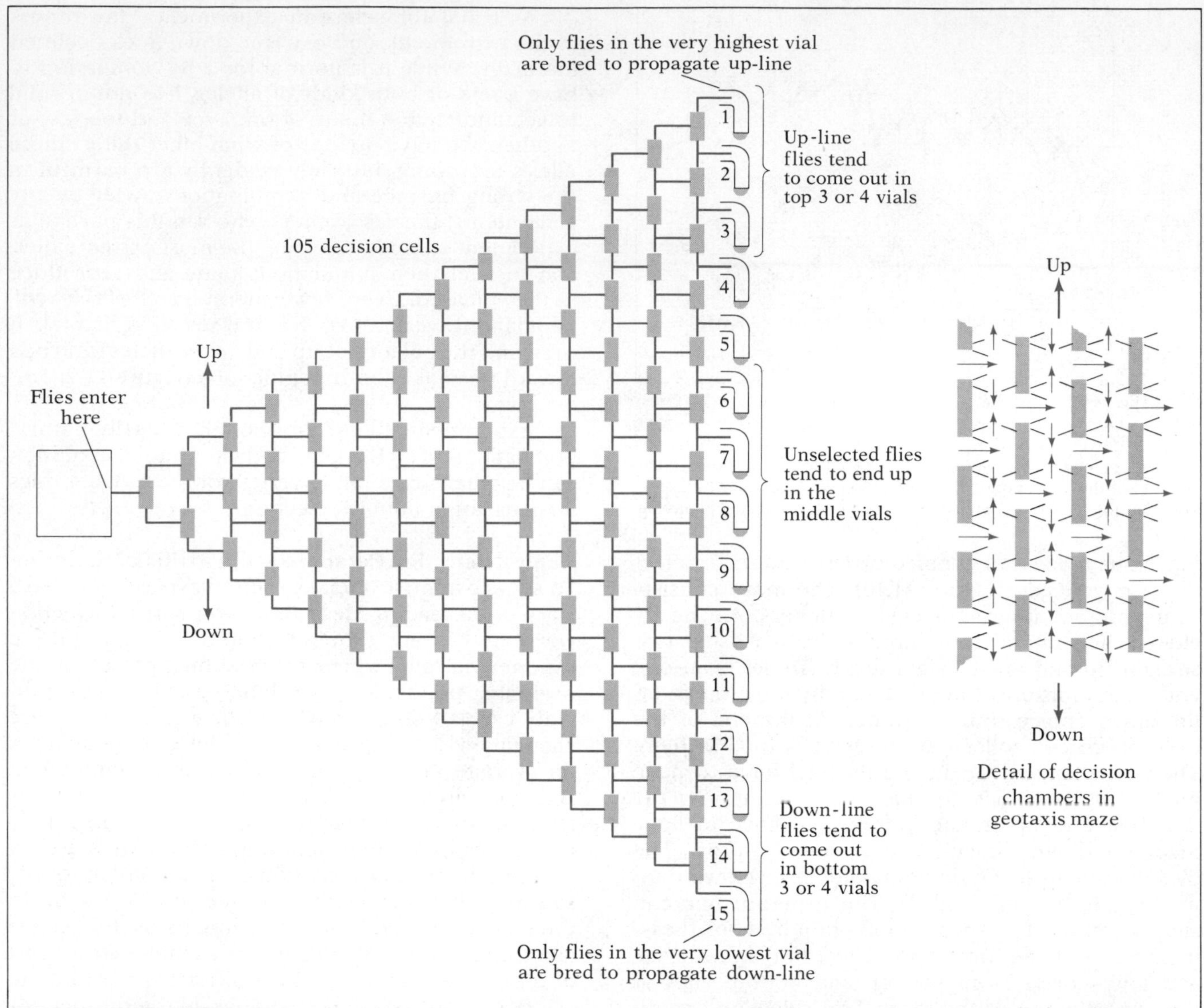
Only flies in the very highest vial
are bred to propagate up-line
Up-line
flies tend
to come out in
top 3 or 4 vials
105 decision cells
Up
Flies enter
here
Down
1
2
3
4
5
6
7
8
9
10
11
12
13
14
15
Unselected flies
tend to end up
in the
middle vials
Down-line
flies tend to
come out
in bottom
3 or 4 vials
Only flies in the very lowest vial
are bred to propagate down-line
Up
Down
Detail of decision
chambers in
geotaxis maze

11.10 Jerry Hirsch's amazing do-it-yourself *Drosophila* geotaxis maze. Flies are put into the maze at the far left. Crawling through a series of one-way, cone-shaped baffles, the flies repeatedly find themselves in decision chambers (see detail) in which there are two ways out: straight up or straight down. Every fly passes through fourteen such decision chambers before reaching one of the fifteen food-filled vials at the right. The top vial receives only those flies that have gone up fourteen times in a row; the bottom vial receives only the ones that have consistently moved down. By breeding flies with the most extreme behavior in each generation, the experimenter can create strains with strong tendencies for positive or negative geotaxis (see Figure 11.9).

Small Babies and Large Babies—An Intermediate Optimum

Generally, natural selection does not favor extremes. It usually favors the intermediate condition, just as we saw in the geotaxis experiment. In a sense, then, selection acts against both extremes. Perhaps the best-studied example of selection for an intermediate condition is related to birth weight in human babies. If we plot survival against birth weight, we find that small babies have relatively low rates of survival, a fact which is not too surprising. But we may be surprised to observe that large babies also have lower survival rates (Figure 11.11). The highest survival rate is for babies around 3.4 kg (7.4 lb). In this case, the optimal birth weight is almost exactly the average birth weight. Genes for large and small birth weight are both selected against. Similar situations are found for almost any "continuous" trait (any trait that can be measured on a continuous scale, like height or weight). In essence, the average tends to be the best. So "survival of the fittest" is usually "survival of

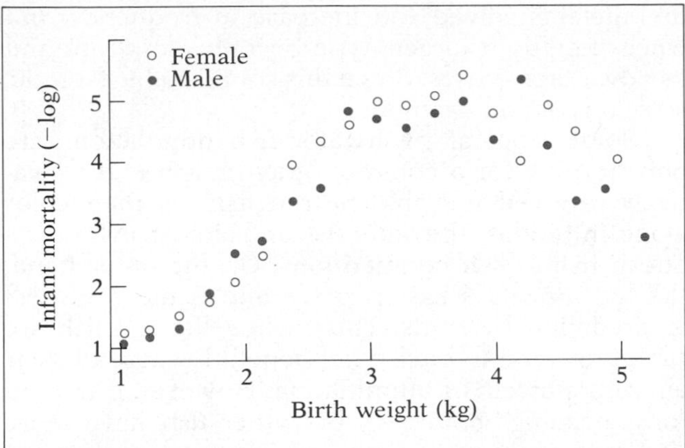

11.11 Stabilizing selection for human birth weight. The negative log of the probability of infant mortality is plotted for various birth weight categories. The optimal birth weight range appears to be between 2.7 and 3.6 kg (6 and 8 lb).

the most mediocre." You may have noticed that we are treading on the edge of circular reasoning, because natural selection will tend to *ensure* this condition. If larger individuals were better (yielded more offspring), then large-size genes would increase until a large size was the average, and the new extremes would once again be selected against. Directional and stabilizing selection are illustrated in Figures 11.12 and 11.13.

Incidentally, when Darwin spoke of the "fitness" of an individual, or of selection favoring the most "fit," he just meant that evolution favored the organism whose phenotype and behavior were most appropriate to its role, the one that most closely *fit* into its ecological niche or its "place in the economy of nature." Since Darwin's time, the idea of the "survival of the fittest" has worked its way into our language, and we now may speak of "feeling fit as a fiddle" or of "physical fitness," equating *fitness* for *health and vigor.* Darwin was closer to the mark. The *fittest* may be closest to the average, when the average phenotype is one that is already well-adapted to the species' needs.

The entire process can become quite complex, of course. Genes at different loci tend to balance each other. An allele for small birth weight at one locus may be favored if it occurs in the presence of an allele for large birth weight at a different locus. What's more, as the environment changes, the intermediate optimum also changes, and the genetic pool of the species must "track" this elusive optimum. In one dusty year, long-nose alleles may have an average selective advantage over short-nose alleles; but the next damp year, the direction of selection may be reversed, and short noses are in. As the gene pool continually adjusts to a changing environment, if the environmental change is cyclic, there will always be a lag while the gene pool changes. Then, before the adjustment is complete, a new genotype becomes favored. This is believed to be one way that high genetic variability is maintained in populations.

11.12 Directional selection favors phenotypes at one extreme of the distribution. In this example we consider the *past* evolution of the giraffe, an animal that browses on tree leaves. Along with each drawing is a population frequency distribution, which characterizes the mean and spread of the population with respect to an important giraffe character, height. (a) Among the antelopelike ancestors of the giraffe, height is variable, as is any other character (bell-shaped distribution). (b) The tallest individuals with the longest necks are best able to reach the foliage of trees on the African veldt; these individuals survive and/or reproduce more offspring than shorter animals. The average height of the surviving/reproducing individuals is greater than that of the population as a whole, as can be seen in the frequency distribution. (c) The offspring of the survivors tend to resemble their successful parents, although there is some regression to the mean. The average height increases over the course of one generation (exaggerated here). Over many generations, giraffes become taller and taller. (And so, incidentally, do the trees, as only the tallest trees escape defoliation by giraffes.)

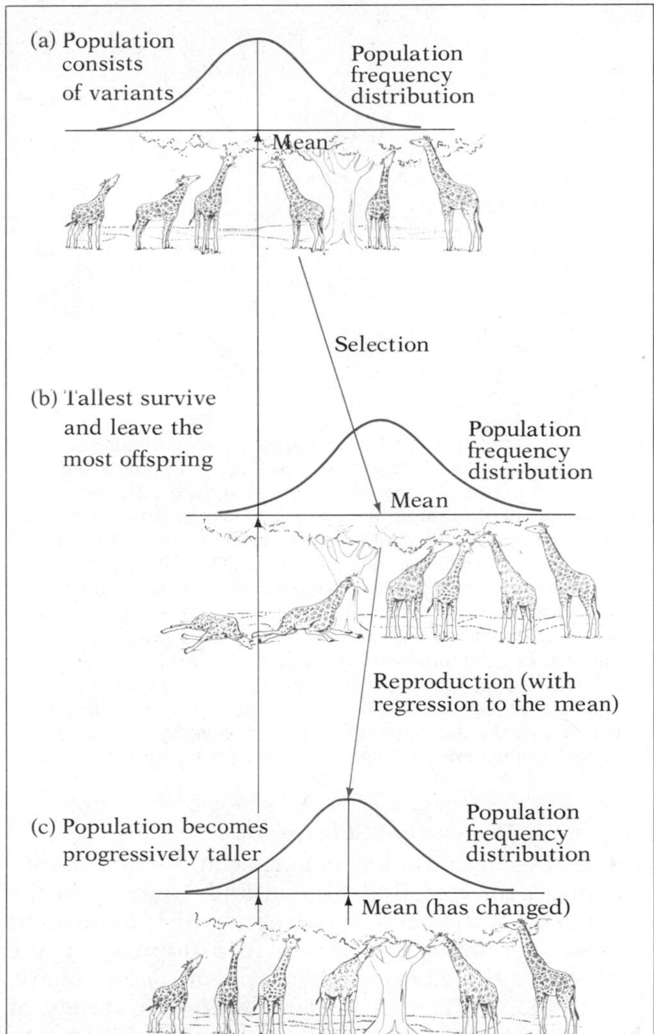

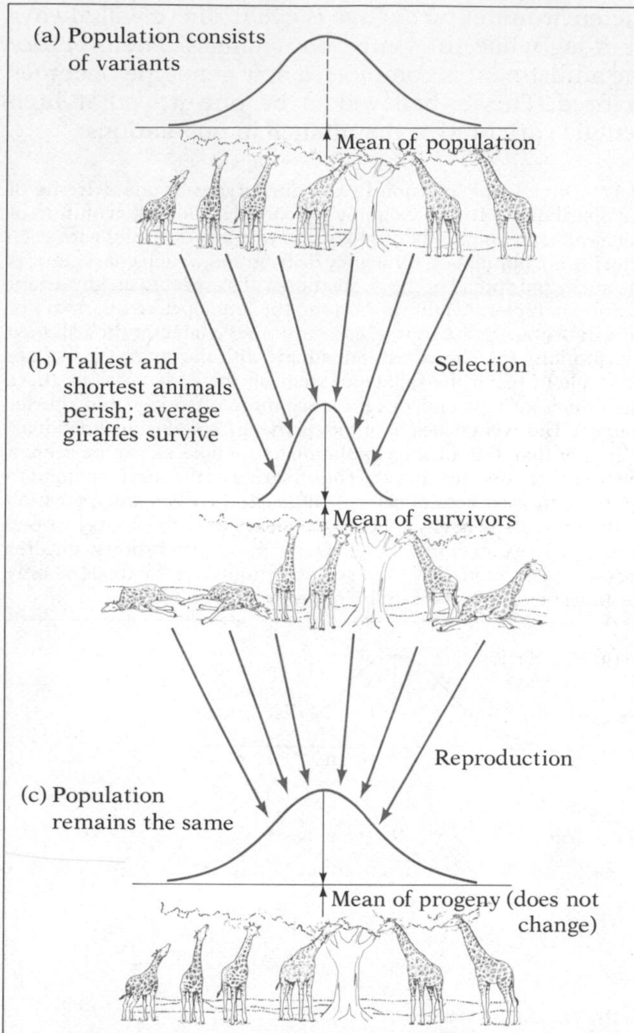
(a) Population consists of variants
Mean of population
(b) Tallest and shortest animals perish; average giraffes survive
Selection
Mean of survivors
Reproduction
(c) Population remains the same
Mean of progeny (does not change)

11.13 Stabilizing selection. Phenotypes are usually already well-adapted to the needs of the organism and are not under directional selection. (a) In this scheme, we assume that the giraffe population is already at its optimal tallness—on the average. But there is still some variation, with some giraffes being too short to browse well and some being too tall for their own good. (The tallest giraffes may have trouble drinking efficiently, or perhaps may be too tall for the trees, or may be subject to high blood pressure or enormous sore throats.) (b) The most successful giraffes are no longer the tallest individuals, but the most average individuals, and these leave the most offspring. (c) The population mean is not expected to show any further change under these conditions. Because of genetic recombination, the distribution (spread) of phenotypes also remains the same from generation to generation (curves a and c).

Frequency-Dependent Selection

Frequency-dependent selection occurs when the fitness of a genotype depends on its frequency in the population. If the success of a genotype is dependent on how frequently it appears in a population, the result can be a stable polymorphism (*poly,* many; *morph,* shape). That is, it can produce a variety of phenotypes that are distinctly different. This happens if a genotype has a net advantage when it is rare and a net disadvantage when it is more common. The problem is, net advantage when rare means that the alleles involved will increase in frequency; but when they do, the genotype becomes less rare and its advantage ceases. Since this sounds a bit esoteric, let's consider an example.

Some tropical freshwater fish populations are polymorphic for a common gray morph and a relatively rare red morph. The red fish, by their color alone, intimidate the other fish and almost always win out in fish-to-fish competitions. On the other hand, the red morph is easier to see and is more subject to predation by birds. Thus, when the red fish are rare, they tend to have a net benefit because of their amazing powers of intimidation. However, if the red form becomes common, the other fish have more frequent encounters with red competitors and get wise to the fact that those red fish aren't so tough after all. Or, the gray fish that are harder to intimidate produce more offspring than the more timid gray fish so that a braver gray morph evolves. Also, as red fish become more common, the birds that feed on them learn to spot them more easily. So each morph has its advantages, but their relative numbers are important. The red forms are kept at an *equilibrium frequency* at which the benefits of being red just balance the disadvantages (Figure 11.14).

Geographic Differences in Selection, with Migration

The different environments found in different parts of a species' range may favor different alleles. For instance, an enzyme that has a higher activity at a low temperature may be favored where the winters are severe, while an enzyme that is less easily denatured by high temperature may have a selective advantage in part of the range that is subject to blistering summers. If the two regions are separated, and if there is no migration between the two parts of the range, different alleles come to dominate as the two populations simply become "fixed" for alternate alleles. However, if individuals occasionally migrate from one part of the range to another, or if the ranges are continuous, there may be a *cline* in allele frequencies from one end of the range to the other. A cline is a simple gradient, or regular change, in the geographic variation of a given character. Logically enough, phenotypic clines are presumed to be the expression of genetic clines.

Examples of both kinds of geographic variation are found in two Australian birds, *Rhipidura* and *Seisura* (Figure 11.15). In both species, the birds are larger in the cooler southern part of their range. (Remember, in the land "down under," the south is colder.) *Rhipidura* has a continuous distribution, with considerable gene flow throughout; and it shows a gradual cline in size (and, presumably, in the frequency of the many genes involved in size). *Seisura,* which is divided into three discrete populations with-

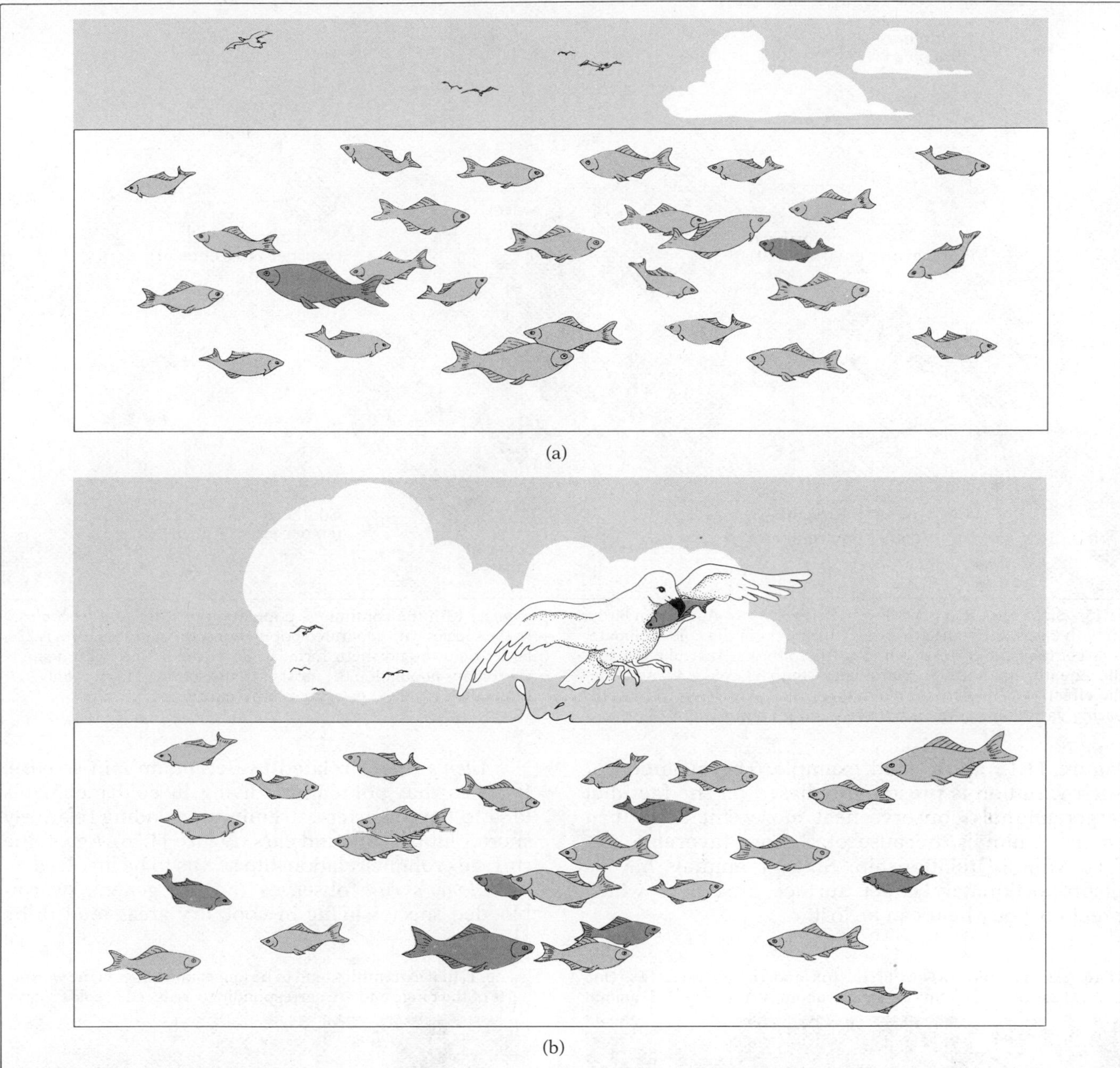

11.14 Frequency-dependent selection. (a) Among certain species of tropical fish, rare red individuals may have a selective advantage over the more common gray form because their bright coloration somehow tends to intimidate other fish of the same species. The red forms get the best breeding territories and produce more offspring. (b) If the red form becomes too common, it no longer has its advantages. Not only do birds learn to search for the easily spotted red fish, but the common gray fish learn not to be intimidated. In the end, a balance is struck between advantages and disadvantages, and the red forms stabilize in numbers and continue to persist at a low frequency.

out much gene flow between them, simply has three different size phenotypes. Significantly, the overall difference between extremes in *Seisura* is about twice as great as that seen in the species with regular gene flow throughout the population (*Rhipidura*). This seems to indicate that gene flow from other parts of the range can prevent a local population from reaching the phenotype that is best for that particular place.

The Rules of Bergmann, Allen, and Gloger

Which would you expect to be larger, a fox that lives in Georgia or one that lives in Minnesota? There is a general trend among the warm-blooded species for individuals to be larger in the cooler parts of the species range. This generalization is called *Bergmann's rule.* The Australian birds shown in

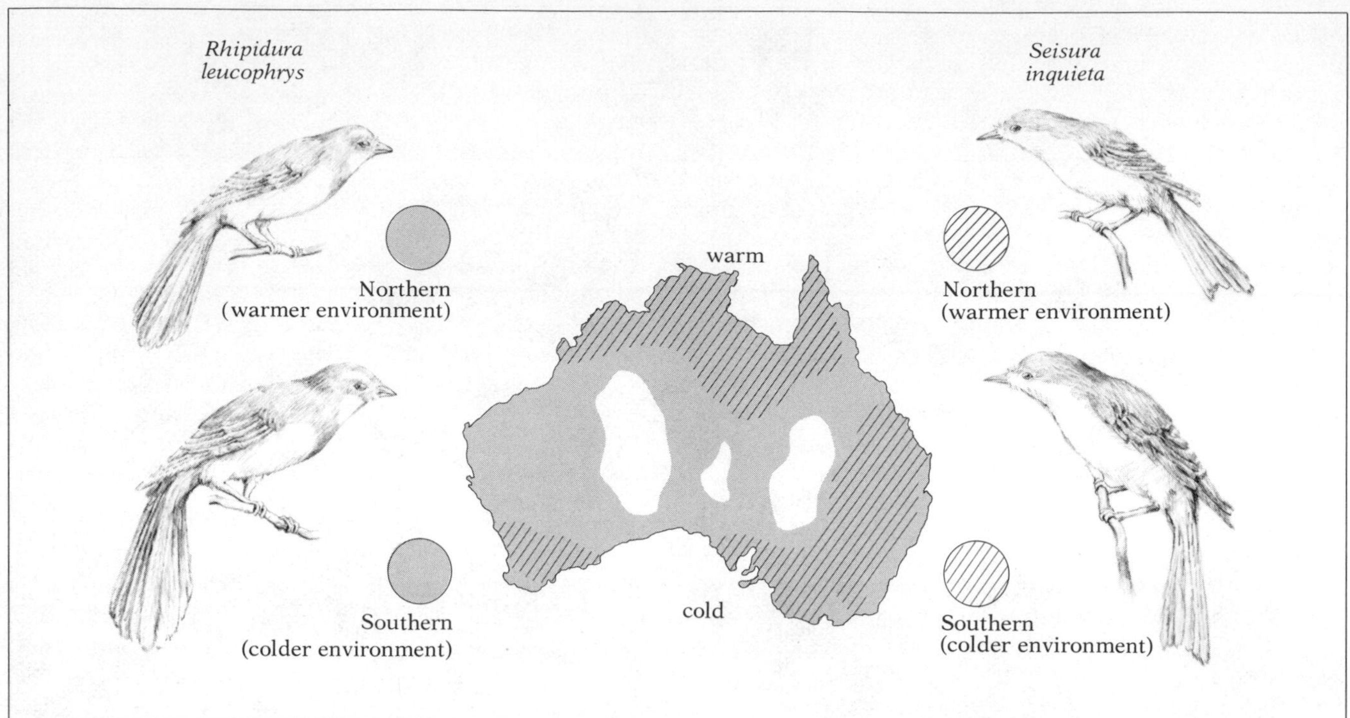

11.15 Size clines and gene flow in two groups of Australian birds. In both cases, the southern form, which lives in the colder climate, is larger than the corresponding northern form. (Australia is below the equator so north is warm and south is cold.) In addition, the effect of gene flow on phenotypic clines can be seen by comparing variation in the discrete (separated) populations of *Seisura inquieta* with the continuous population of *Rhipidura leucophrys.* In the species with separated populations, the southern form is 22% larger than the northern form, while in the species with a single continuous population the most extreme southern form is only 11% larger than the most extreme northern form.

Figure 11.15 are a good example. The evolution of such variation is presumably based on the fact that larger animals conserve heat more efficiently than smaller animals, because of a more favorable surface–volume relationship. Smaller animals have a disproportionately larger surface area over which precious body heat can be lost.

Allen's rule is related to Bergmann's, in a sense. It states that populations living in colder climates tend to have shorter extremities, including relatively shorter limbs, tails, and ears (Figure 11.16). Again, the surface–volume relationship seems to be involved.

Gloger's rule observes that, in general, warm-blooded species living in cool, dry areas tend to be

11.16 Allen's rule: (a) an arctic fox and (b) a desert fox. One generalization that can be made about warm-blooded animal groups is that extremities tend to be longer and larger in the warmer parts of the range and are correspondingly reduced in colder parts.

(a)

(b)

lighter colored than their relatives in warmer, moister regions. This may be in part because moister backgrounds are usually darker, but it may also have something to do with heat loss. A warm black body, after all, radiates more heat than one of a lighter color, especially at night.

These rules are based on simple observations and correlations. There are far too many exceptions to allow us to apply them as hard-and-fast laws, but there is one "accidental experiment" to draw on. In the last two centuries, the English sparrow has been introduced, deliberately or inadvertently, to cities all over the world, some tropical, some arctic. In this short period of time, the species has changed in different areas in just the ways that Bergmann, Allen, and Gloger would have predicted.

Geographic Variation in the Human Species
Humans, by the way, tend to follow Bergmann's, Allen's, and Gloger's rules reasonably well—or at least did prior to the great European expansions of the last 500 years. Equatorial peoples tend to be smaller and darker than natives of more temperate climates, but at the same time they often have proportionally longer limbs (Figure 11.17).

There are some exceptions to the generalization, notably the very small-bodied Lapps and the rather dark Eskimos. However, some human evolutionists believe that the relationship of human skin color to latitude has less to do with temperature regulation than with the amount of sunlight available, which is related to vitamin D synthesis. Whereas dark skin protects against sunburn, it also tends to prevent ultraviolet radiation from reaching the capillary beds where cholesterol is converted into this vital vitamin. Eskimos, we know, get all the vitamin D they need by eating fish liver and therefore do not need much sunlight. Evolution has apparently settled on a tradeoff of advantages that is different for each latitude.

Neutralists Versus Selectionists

In recent years there has been some controversy over how much natural variation is due to *neutral mutations and random drift.* The neutralists say that a lot of variation on the molecular level is simply incidental and unadaptive. For example, there are frequently several slightly different allelic forms of any enzyme in a population, differing from each other by only one or two amino acids. *Neutralists* hold that most of this structural variation has no effect on the function of the enzyme. They think that functionally equivalent alleles just happen by chance mutation and that the allele frequencies *drift* around meaninglessly. *Selectionists,* on the other hand, prefer to believe that virtually all variation is due to natural selection and has some adaptive basis, even if we don't happen to know what it is. Take note that the disagreement centers over normally *invisible traits,* not over obvious differences in phenotypes. Neither side maintains that any *visible* phenotypic variation is likely to be meaningless. The neutralist and selectionist hypotheses are restricted to the question of whether normally *invisible* details of molecular structure are always subject to differential natural selection.

11.17 Allen's rule in humans. The Nilotic peoples have inhabited equatorial East Africa for many thousands of years. Following Allen's rule, they have longer and thinner extremities than other populations of the same species (*Homo sapiens*). The warrior here is a Masai.

In *Origin of Species,* Darwin himself summed up the neutralist argument: "Variations neither useful nor injurious would not be affected by natural selection, and would be left either a fluctuating element, as perhaps we see in certain polymorphic species, or would ultimately becomes fixed. . . ."

It can be shown mathematically that a completely neutral mutation would usually disappear by chance, but that it does have a finite probability of spreading through the population and thus creating a molecular polymorphism. There is also a somewhat smaller chance that such a neutral mutation would drift to fixation in the species, effortlessly ousting its fully equivalent predecessor.

Horse beta hemoglobin and human beta hemoglobin are exactly the same at 129 out of 146 amino acid positions, but are different at the 17 remaining positions. Since the time of the last common ancestor of the two species, which was a little insectivore that lived about 80 million years ago, the beta hemoglobin gene has undergone seventeen evolutionary changes—seventeen occasions on which a new mutant has replaced an older allele, in one line of descent or the other. Has all this change been adaptive? Are these changes examples of the replacement of an inferior allele by a superior new one through natural selection? Or, have some or most of the changes been due to meaningless neutral mutations and chance alone? We don't know.

Gene Changes in Evolution

The remarkable fact that has emerged from molecular evolutionary studies is that the rate of change in proteins (an indication of genetic change) is nearly constant in different lines of descent. For instance, since the time of the last common ancestor of a human and a carp—an ancestor that was surely a bony fish—the rate of change in hemoglobin molecules has been the same in the carp line of descent and in the human line, about one amino acid change per protein every 7 million years. There are, of course, thousands of proteins, and other proteins change at different rates, but each protein type has its characteristic evolutionary pace.

Now, this is all a bit unexpected. One amino acid change every 7 million years? Yet species change astoundingly rapidly. Doesn't a great morphological change reflect a great genetic change? Perhaps not. First of all (as we will see), some genes control great blocks of other genes, and a single change in a controlling gene could therefore alter the expression of the rest. Also, many genes continue through generation after generation, unexpressed and unsummoned, until things change and they are suddenly needed. They then express themselves and the phenotype quickly changes.

As we saw in Chapter 1, in the fossil record, whole groups appear suddenly, evidence of a period of rapid specialization, change, and adaptive radiation. Then there may be extended periods in which no change is seen. Most species (according to what we can learn from fossils) survive for millions of years without any visible change, only to be replaced suddenly with new species or new groups of species.

Population Bottlenecks

We know that even harmful mutations—especially if they are recessive so that they are shielded from natural selection—can also increase by pure chance (genetic drift). This is especially true if the population goes through a *population bottleneck*, which is a time in which the population size is temporarily reduced, perhaps by a natural disaster. The few individuals who survive the bottleneck crisis may well carry some rare recessive genes. Then if the population begins to increase, the descendants of these few survivors will carry the same alleles. The "rare" recessive gene will now be much less rare (Figure 11.18). There are many examples of the *bottleneck effect* or *founder effect* in human populations as well as in other species. For example, most of the Afrikaaners (Boers) of South Africa are descended from about thirty Dutch families, and today the Afrikaaners have rather high frequencies of several recessive diseases that are rare in other populations. Closed religious communities in America, such as the Hutterites, the Dunkers, and the Amish, also have their share of special recessive diseases.

There is good evidence that the Ashkenazi Jewish population underwent a population bottleneck somewhere in Eastern Europe during the middle ages. One or more of the survivors of the bottleneck must have carried the recessive allele for Tay-Sachs disease, an enzyme deficiency that leads to blindness, idiocy, and early death in infant homozygotes. There are now millions of persons of Ashkenazi Jewish descent, and about one in twenty-seven is a heterozygous carrier of this tragic disease. Ashkenazi Jews also have relatively high frequencies of five other recessive genetic diseases, and a rather high incidence of diabetes (a polygenic trait); but the same group is relatively free of other recessive genetic diseases and has no higher overall incidence of genetic disease than does any other human population. A population bottleneck many increase the relative frequencies of some rare alleles, but at the same time it may eliminate other rare alleles altogether.

The Balance Between Mutation and Selection

We mentioned the way in which natural selection sometimes acts by causing a rare, beneficial mutation to increase in a population because of the advantages it confers. We must admit, however, that this is a rare event. After all, most mutations can't be expected to make a gene work better than it did before; the great majority will make the gene work less well. It's as if you raised the hood on your sports car and let your little brother randomly change something. He may have made precisely the adjustment needed to make the car run better. But the people who would risk it could probably hold their annual convention in a phone booth. The point is, most mutations are harmful. Still, they occur at a surprisingly high rate, and even harmful mutations have important effects on the variability of a population.

Usually, any mutant gene will be weeded out of the population by natural selection. A gene is harmful (by definition) if it lowers the reproductive fitness of its carriers. But keep in mind that this decrease in reproductive ability also tends to eliminate the mutant gene itself.

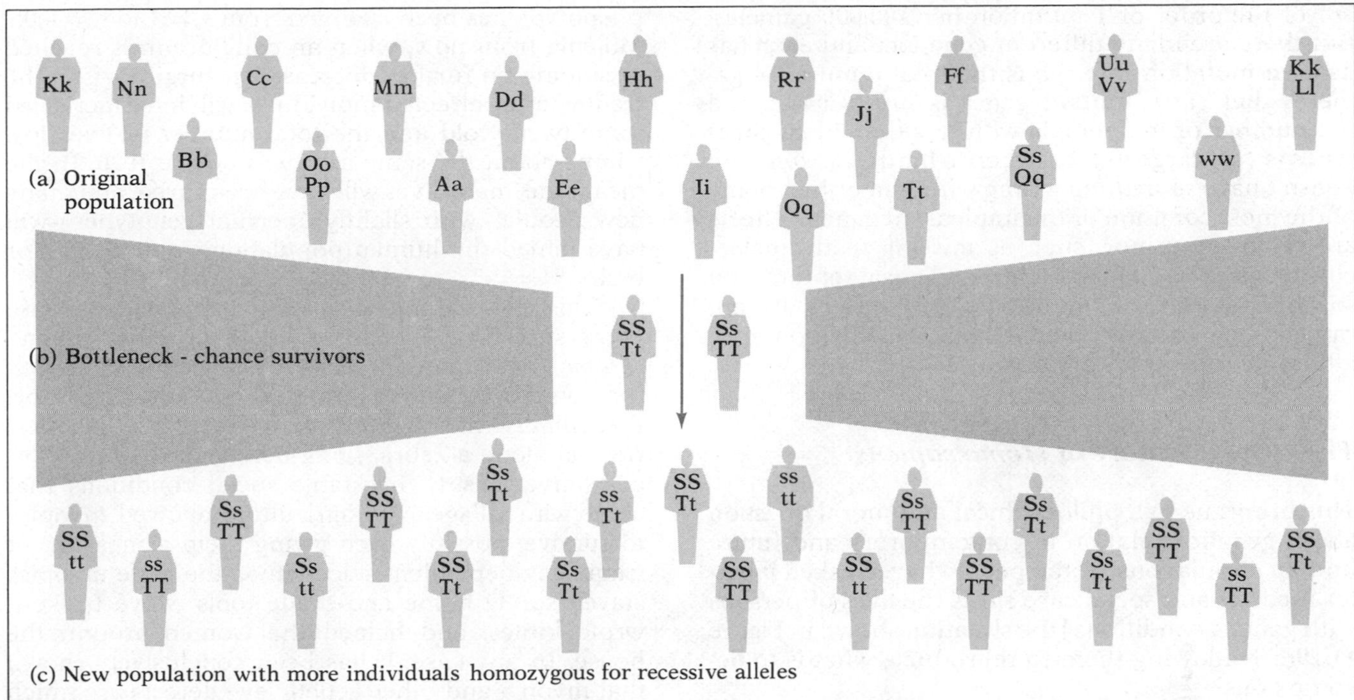

11.18 The bottleneck effect. (a) In any large, diploid population, most individuals will be heterozygous carriers for rare recessive alleles at several different gene loci. (b) At the time of a population bottleneck, only a relatively few individuals survive. The survivors will carry a random sample of the rare alleles that were present in the formerly large population. (c) After the bottleneck, the few survivors will produce large numbers of descendants. Many of these progeny will carry some of the same recessive alleles, which will no longer be rare. Numerous individuals will become homozygous for these alleles (shading). On the other hand, many rare recessive alleles from the original population will not occur at all in the new population.

In any population, any specific allele will mutate at one time or another, usually to a nonfunctional or harmful form. The proportion of gametes carrying new mutant alleles of a given locus is called the *mutation rate* and is symbolized by u. The amount by which the harmful genotype lowers the fitness of the average individual is called the *selection coefficient,* symbolized by s. The relative fitness of the affected genotype is defined as $1 - s$. For a lethal genotype, that is, a genotype with 0 fitness, the value of s is 1.

Without going through the mathematics, we'll just tell you that a recessive gene will rise in allele frequency until the genotype frequency of affected (homozygous) individuals is u/s, which is the mutation rate divided by the selection coefficient. So, if $u = 10^{-5}$ and $s = 0.10$, that is, if 1 out of 100,000 eggs and sperm carried a new mutant, and if the homozygote were 10% worse off than the average individual, eventually about 1 individual out of 10,000 would be born an affected homozygote. If the mutant gene were only a tenth as bad, so that it reduced the fitness by only 1%, then ten times as many individuals would be affected. For a harmful dominant gene, the allele frequency would rise until the proportion of affected genotypes in the population was $2u/s$.

In either case, the number of affected individuals is *directly* proportional to the mutation rate, but *inversely* proportional to the selection coefficient (Figure 11.19). This is because the more severe an allele is, the more quickly it is weeded out of the population. Thus, fewer generations of carriers are affected before the allele is eliminated. Typical mutation rates

11.19 The balance between mutation and selection: a schematic analogy. Water in the beaker represents mutant genes in the gene pool. Water enters the beaker at a constant rate (spontaneous mutation). Water flows out of the beaker (natural selection) at a rate that depends on the current level in the beaker. (a) At *mutational equilibrium,* the flow of mutant alleles into the gene pool through spontaneous mutation exactly equals the loss of mutant alleles through natural selection. (b) If the mutation rate is increased, perhaps because of radiation or other mutagens in the environment, the level of mutant alleles in the gene pool will rise until the outflow again equals the inflow. (c) Similarly, if natural selection against the mutant alleles is reduced, for instance, by improved medical treatment of affected individuals that allows them to reproduce, the level of mutant alleles in the gene pool will rise until a new equilibrium is reached between the inflow by mutation and the outflow through natural selection.

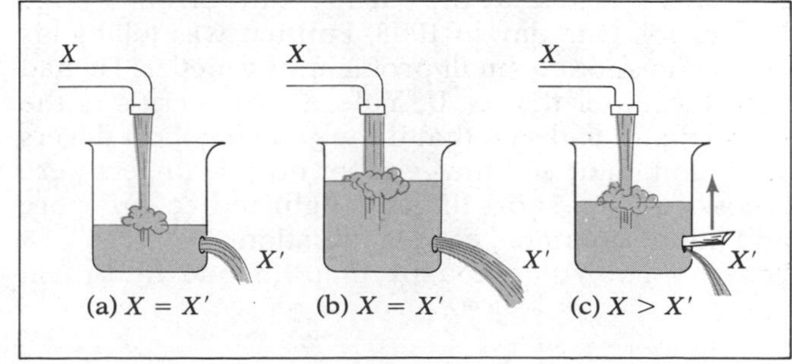

are of the order of 1 mutation per 100,000 gametes. But there are many different gene loci and each has its own mutation rate. Thus the total number of gametes that carry mutant genes is pretty large, and the number of individuals with less than ideal genotypes is very large indeed. There's hardly *anyone* who doesn't have *something* wrong with him or her. Some of the most common of the unpleasant genetic effects are relatively minor, such as missing teeth, malocclusion, near-sightedness, deviated nasal septum, and so on. But some of the more severe effects are not rare enough, such as phenylketonuria, schizophrenia, albinism, and hereditary deafness.

The Genetic Future of *Homo sapiens*

This brings us to a philosophical and moral question about genetic variation in contemporary and future human populations. Some people haved asked if better medical and social care saves the lives of persons with genetic conditions [the situation shown in Figure 11.19(c)], allowing them to reproduce, what is to become of us?

For instance, the genetic condition *pyloric stenosis,* an abnormal overgrowth of a stomach valve muscle in infants, was once invariably fatal. Since the 1920s, a simple surgical procedure has saved the lives of nearly all affected infants in developed countries. About half the offspring of such people are also affected, and there are just as many new mutants as ever, so the genes for the condition have increased in frequency. Where will we be in 10,000 years?

What has happened is that a genetic condition for "certain infant death by intestinal obstruction" has been transformed to a genetic condition for "simple abdominal surgery needed in infancy." Suppose that the risk of the condition, including possible misdiagnosis as well as surgical mishap, becomes as low as 5%. The selection coefficient for the affected phenotype has been changed from $s = 1$ to $s = 0.05$. Millenia from now, when an equilibrium is reached (assuming no further decreases in surgical risk), the frequency of affected individuals will have increased some twentyfold and the total number of lives lost will be about the same as it was before 1920. In the meantime, many lives will have been saved, and many new people—with slightly aberrant genotypes—will have joined the human population. Is that good or bad?

This sort of thing has already happened, because every species adapts to its changing environment. *Myopia* (nearsightedness) is a near-lethal condition for persons in a nomadic hunting–gathering tribe. Presumably, nearsighted aborigines can't find berries, let alone a zebra. Long before the invention of corrective lenses, the stable social conditions that came with villages and agriculture allowed myopics to survive, possibly even giving them a measure of frequency-dependent selection as the male myopics stayed safely home and made tools, wove baskets, wrote bibles, and helped the women around the house. In any case, it has been conclusively shown that myopia and other genetic eye defects are much more common among people with a long history of agriculture and urban civilization than among groups that have more recently given up the hunting–gathering life. For instance, American Indians and American blacks have far lower rates of vision defects than do Americans of European or Oriental ancestry.

So, a gene for sublethal blindness has become a gene for needing glasses—a minor nuisance—and the gene has increased in frequency accordingly. Part of the increase may have been due to subtle and indirect advantages of the myopia allele, rather than to repeated mutation alone, but in any case, the balance has certainly been shifted. Should we be alarmed by such evidence of genetic deterioration under civilization?

11.3 The Castle–Hardy–Weinberg Law

G. H. Hardy, an eminent mathematician, had few professional interests in common with R. C. Punnett, the young Mendelian geneticist, but they frequently met for lunch and tea at the faculty club of Cambridge University. One day in 1908, Punnett was telling his colleague about a small problem in genetics. He had heard a rumor that G. U. Yule, a strong critic of the Mendelians, had said that if the gene for short fingers were dominant and the gene for normal fingers were recessive, then short fingers ought to become more and more common each generation. Within a few generations, Yule thought that no one in Britain should have normal fingers at all. Punnett didn't think this argument was correct, but he couldn't explain why.

Hardy said he thought the problem was simple enough, and wrote a few equations on his napkin. He showed that, given any particular frequency of genes for normal fingers and genes for short fingers in a population, the relative numbers of people with normal fingers and people with short fingers ought to stay the same for generation after generation as long as there was no natural selection involved.

Punnett was excited, and wanted to have the idea published (on something besides a napkin) as soon as possible. But Hardy was reluctant. The idea was so simple and obvious, he felt, that he didn't want to have his name associated with it and risk his reputation as one of the great mathematical minds of the day. But Punnett prevailed, and the relationship between genotypes and phenotypes in populations quickly became known as *Hardy's law.* Hardy, who was indeed one of the great mathematical minds of the day, is now known almost solely for this modest contribution. Yule, incidentally, denied ever having said that dominant traits should increase from generation to generation, so this little incident in the history of science was based on a misunderstanding from the outset.

In Germany, Hardy's law was known as *Weinberg's* law, since it had been discovered by a German physician of that name and published within weeks of Hardy's short paper. Eventually the formula became known as the *Hardy–Weinberg law,* and then later as the *Castle–Hardy–Weinberg distribution,* in recognition of the belated discovery that an American, W. E. Castle, had published a neglected exposition of the relationship in 1903. The Castle–Hardy–Weinberg distribution is the basic starting point of the population genetics of diploid sexual species.

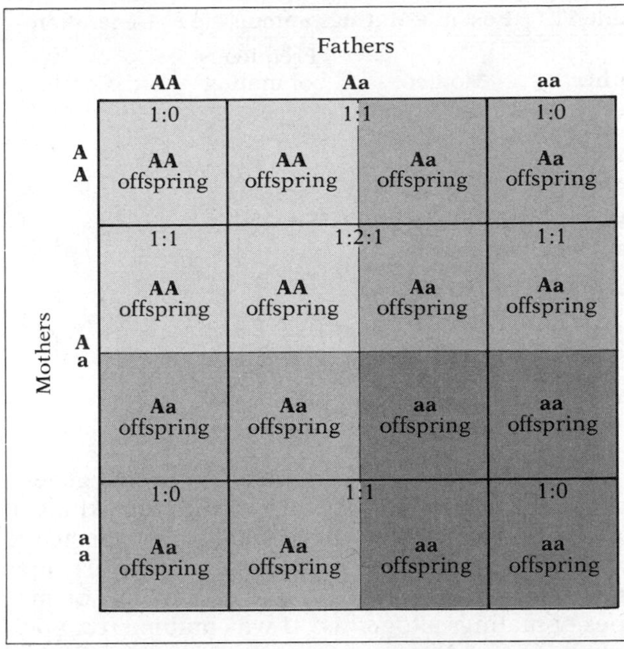

Fathers

Mothers	**AA**	**Aa**		**aa**
	1:0	1:1		1:0
A A	**AA** offspring	**AA** offspring	**Aa** offspring	**Aa** offspring
	1:1	1:2:1		1:1
A a	**AA** offspring	**AA** offspring	**Aa** offspring	**Aa** offspring
	Aa offspring	**Aa** offspring	**aa** offspring	**aa** offspring
	1:0	1:1		1:0
a a	**Aa** offspring	**Aa** offspring	**aa** offspring	**aa** offspring

11.20 Random mating between F_2 individuals to create an F_3 generation. In the F_2 generation of a Mendelian cross, both males and females occur in three genotypes **AA**, **Aa**, and **aa**, and the genotype frequencies are $\frac{1}{4}$, $\frac{1}{2}$, and $\frac{1}{4}$, respectively. If random mating occurs, there will be nine different kinds of mating pairs. Here, the nine kinds of mate pairs are represented by the rectangles bounded by heavy black lines. Within each type of mating, Mendelian genotypic ratios again occur: 1:0, 1:1 and 1:2:1, depending on the mating. Each area bounded by heavy lines, then, is the Punnett square for the appropriate cross. The F_3 offspring genotypes within each mating type are represented by different colored areas. Overall, the F_3 generation again has the same distribution of genotypes: $\frac{1}{4}$ **AA**, $\frac{1}{2}$ **Aa**, and $\frac{1}{4}$ **aa**.

The Implications of the Castle–Hardy–Weinberg Distribution

To rephrase the problem of the Mendelians, brown eyes are dominant over pale blue eyes in humans. So why doesn't everyone have brown eyes by now?

To approach such questions, we must begin by considering genes in populations. We have been using the term *population,* and you may have taken it to mean a group of individuals. You were not wrong, but at this point we can give it a more precise meaning. In biology, a population designates a group of interbreeding or potentially interbreeding individuals. With this definition in mind, let's now consider the ratios of different alleles for a specific characteristic, such as eye color, in a population.

First, imagine a population of only two individuals in which the male is homozygous for dominant trait **A** and the female is homozygous for recessive trait **a**. Now, we know that all their F_1 offspring will be heterozygous **Aa** for that characteristic. Assume that F_1 individuals mate to produce an F_2 generation. The Punnett square then shows us that in the F_2 generation three out of four individuals will show the dominant trait and only one will show the recessive trait. So it might appear that we are on the way to eliminating the recessive gene from the population. However, if we now plot the F_3, F_4, F_5, and so on, generations (Figure 11.20), we will find that the proportion of dominant and recessive genes in the population has not changed at all. In fact, if we think again of the F_1 and F_2 generations, we see that there is a 1:1 ratio of the two alleles even at these stages.

A population of two individuals is unrealistic, but it serves to point out how the phenomenon occurs in larger populations. We can show by means of Punnett squares that the frequency of alleles for any characteristic will remain unchanged in a population through any number of generations—unless this frequency is altered by some outside influence.

According to the Castle–Hardy–Weinberg law: In the absence of natural selection, mutation, or drift (forces that change gene frequencies in populations), when random mating is permitted, the frequencies of each genotype will tend to remain constant through the following generations.

Note the ratios of the combinations **AA, Aa,** and **aa** in the F_2 generation (in Figure 11.20), which is our first opportunity to see all the possible combinations. We see in the F_2 that one-fourth of the population is **AA,** one-half is **Aa,** and one-fourth is **aa.** Now,

Table 11.1 Possible Matings among the F_2 Generation

Father	Mother	Frequency of mating	Combined
AA ($\frac{1}{4}$)	**AA** ($\frac{1}{4}$)	$\frac{1}{16}$	$\frac{1}{16}$
AA ($\frac{1}{4}$)	**Aa** ($\frac{1}{2}$)	$\frac{1}{8}$	$\frac{1}{4}$
Aa ($\frac{1}{2}$)	**AA** ($\frac{1}{4}$)	$\frac{1}{8}$	
AA ($\frac{1}{4}$)	**aa** ($\frac{1}{4}$)	$\frac{1}{16}$	$\frac{1}{8}$
aa ($\frac{1}{4}$)	**AA** ($\frac{1}{4}$)	$\frac{1}{16}$	
Aa ($\frac{1}{2}$)	**Aa** ($\frac{1}{2}$)	$\frac{1}{4}$	$\frac{1}{4}$
Aa ($\frac{1}{2}$)	**aa** ($\frac{1}{4}$)	$\frac{1}{8}$	$\frac{1}{4}$
aa ($\frac{1}{4}$)	**Aa** ($\frac{1}{2}$)	$\frac{1}{8}$	
aa ($\frac{1}{4}$)	**aa** ($\frac{1}{4}$)	$\frac{1}{16}$	$\frac{1}{16}$

in order to find out what happens in the next generation, in Table 11.1 we list all the different kinds of matings and how often these should be expected to occur. With three different genotypes, there are three kinds of males and three kinds of females, or nine types of matings altogether. If we combine *reciprocal matings* (e.g., **AA** × **aa** and **aa** × **AA**), there are still six different types. Each type of mating can be expected to have certain kinds of offspring in the usual Mendelian ratios.

These six different types of matings, together with the proportions of the total offspring that will occur in each of the three offspring types are listed in Table 11.2. For instance, $\frac{1}{4}$ of the matings are **AA** × **Aa** (or **Aa** × **AA**), and here we would expect a 50:50 Mendelian ratio of **AA** and **Aa** children; so, among the F_3, a total of $\frac{1}{8}$ will be **AA** children and $\frac{1}{8}$ will be **Aa** children from this kind of mating.

From Table 11.2 and Figure 11.20, we can see that random mating—the assumption that the frequency of each type of mating is exactly the product of the genotype frequencies involved—produces an F_3 generation that is $\frac{1}{4}$ **AA,** $\frac{1}{2}$ **Aa,** and $\frac{1}{4}$ **aa,** just as in the F_2. Obviously, the F_4 and F_5 will also have the same genotypic ratios. We also have to assume that each type of family will have the same numbers of offspring. And that is why we continue to have blue eyes and other recessive traits in our population.

Table 11.2 Expected F_3 Generation

Mating	Frequency	Offspring expected: AA	Aa	aa
AA × **AA**	$\frac{1}{16}$	$\frac{1}{16}$	—	—
AA × **Aa**	$\frac{1}{4}$	$\frac{1}{8}$	$\frac{1}{8}$	—
AA × **aa**	$\frac{1}{8}$	—	$\frac{1}{8}$	—
Aa × **Aa**	$\frac{1}{4}$	$\frac{1}{16}$	$\frac{1}{8}$	$\frac{1}{16}$
Aa × **aa**	$\frac{1}{4}$	—	$\frac{1}{8}$	$\frac{1}{8}$
aa × **aa**	$\frac{1}{16}$	—	—	$\frac{1}{16}$
Total	1	$\frac{1}{4}$	$\frac{1}{2}$	$\frac{1}{4}$

An Algebraic Equivalent

In the example above, the two alleles start out at the same frequency—half **A** and half **a.** The Castle–Hardy–Weinberg distribution also maintains stable genotype frequencies in populations in which the alleles are at different frequencies. For those burning with a fierce love of mathematics, let's put it all into algebra. First, let p be the allele frequency of allele **A,** while q is the allele frequency of allele **a.** If there are only two alleles, $p + q = 1$.

The Castle–Hardy–Weinberg distribution says that the expected genotype frequency of **AA** is p^2, the expected genotype frequency of **Aa** is $2pq$, and the expected genotype frequency of **aa** is q^2. We can show that these frequencies will also be stable. As before, we list the six (combined) types of matings, the frequencies in which they should occur, and the distribution of offspring of each type of family (Table 11.3). The genotype frequencies of the offspring will be p^2, $2pq$, and q^2, just as in the parents' generation.

The Castle–Hardy–Weinberg distribution is easier to understand if we forget about random mating of diploid individuals and just consider the *random association of gametes* (Figure 11.21). It can be proven that this amounts to the same thing. After

11.21 The Castle–Hardy–Weinberg equilibrium. If there are two alleles, **A** and **a**, occurring in relative frequencies p and q, respectively, and mating is random, a proportion p of the sperm will carry the **A** allele, and p of the eggs will also carry the **A** allele. For each zygote formed, the probability that the egg and sperm will both carry **A** alleles is $p \times p$ or p^2. Similarly, the probability that both of the uniting gametes will carry **a** alleles is q^2. There are two ways that an **Aa** zygote can be formed: **A** sperm uniting with **a** egg and **a** sperm uniting with **A** egg. The total probability of one of these two events occurring is $pq + qp$, or $2pq$. (The arrow connects the two areas that represent the same **Aa** genotype.) The relative frequencies of the three kinds of genotypes in the population will be equal to the individual probabilities of each kind of event: p^2, $2pq$, and q^2 will be the frequencies of genotypes **AA**, **Aa**, and **aa**, respectively.

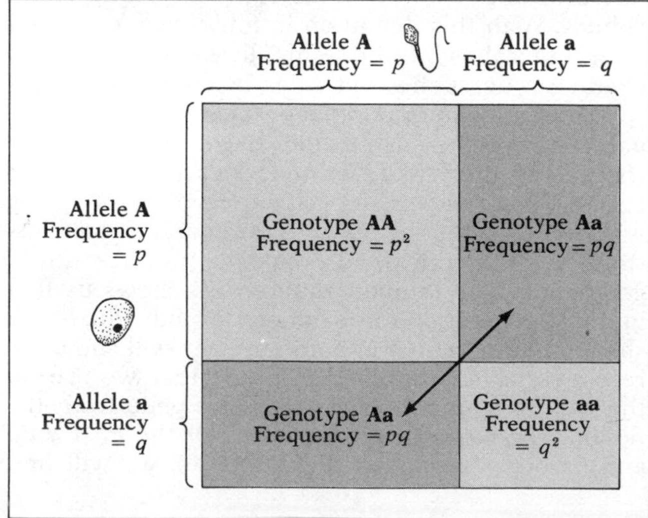

Table 11.3 Expected Allele Frequencies

Mating ♀ ♂	Mating frequency	Offspring frequencies AA	Aa	aa
AA × **AA**	p^4	p^4	—	—
{**AA** × **Aa**, **Aa** × **AA**}	$4p^3q$	$2p^3q$	$2p^3q$	—
{**AA** × **aa**, **aa** × **AA**}	$2p^2q^2$	—	$2p^2q^2$	—
AA × **Aa**	$4p^2q^2$	p^2q^2	$2p^2q^2$	p^2q^2
{**Aa** × **aa**, **aa** × **Aa**}	$4pq^3$	—	$2pq^3$	$2pq^3$
aa × **aa**	q^4	—	—	q^4
Total	$(p^2 + 2pq + q^2)^2 = 1$	$p^2(p^2 + 2pq + q^2) = p^2$	$2pq(p^2 + 2pq + q^2) = 2pq$	$q^2(p^2 + 2pq + q^2) = q^2$

all, when an egg and sperm meet—either in the open sea, as with sea urchins, or in the dark confines of an oviduct—the parents' diploid genotypes no longer really matter. All that matters is the haploid genotypes of the two gametes, and if there is random mating, the probabilities for each gamete will be p and q of being **A** or **a**, respectively.

The Castle–Hardy–Weinberg law has very specific and important implications. For example, if we know the prevalence of the trait of a recessive condition, for instance albinism (the absence of normal melanin pigment), in the population, we can predict, within limits, the probability that a couple will have an albino baby. Here's how this would work. Normal skin and eye pigment in humans is dominant over the albino condition **a**. The genotype **aa** occurs in about 1 of every 20,000 people. According to the Castle–Hardy–Weinberg equation, this frequency would be given by q^2, so the frequency of genotype **aa** is

$$q^2 = \frac{1}{20{,}000}$$

and the frequency of a single allele for this trait is thus

$$q = \sqrt{\frac{1}{20{,}000}} = \frac{1}{141}$$

The frequency of the dominant allele **A** would then be

$$p = 1 - q$$

or

$$p = 1 - \frac{1}{141} = \frac{140}{141}$$

The heterozygous condition **Aa** would, therefore, occur in the population with a frequency of

$$2pq = 2 \times \frac{140}{141} \times \frac{1}{140}$$

$$= \frac{1}{70}, \text{ or about } 1.4\%$$

Since 1.4% of 20,000 is 280, this means that about 280 people in every 20,000 will be carrying a recessive allele for albinism, while, as we have seen, only one is affected. Hence, in the absence of a family history of this characteristic in either parent, the chance that any couple will have an albino child is very slim.

Albinism has been a favorite textbook example for decades, in part because everyone is familiar with the striking phenotypes of affected homozygotes and has some idea of what is being discussed. We've just used it ourselves to show how the Castle–Hardy–Weinberg law can be applied. Now we'll show just how easily it can be misapplied.

One day, not too many years ago, an albino man and his albino wife produced a child who was normally pigmented (in fact, black). Their doctor, who remembered his introductory biology course, cheerfully informed the parents that their child was illegitimate, since two recessive homozygotes could not possibly have a child with the dominant phenotype. This point of view did not favorably impress the albino father, who fumed that any fool could tell that he and his wife were different kinds of albinos—for instance, she could not stay out in the sun for even 5 minutes without burning badly, while he could last a couple of hours out-of-doors.

A geneticist investigated the situation and found that the father's intuition was perfectly true. There are several different points in the pathway of melanin development where mutant alleles can block pigment formation. The man and his wife were recessive homozygotes at different loci—let us say that the father was **aaBB** and the mother was **AAbb,** in which case their child was **AaBb,** and thus (phenotypically) perfectly normal. The geneticist then developed a simple biochemical test that differentiated the two types of albino: If a few hairs were pulled out by the roots and incubated in a tyrosine-rich medium, one type of albino hair root produced dark pigment and the other type of albino hair root did not (tyrosine is an amino acid from which melanin is synthesized).

The hair-root test was performed on a sample of albinos from different families. About half responded to the test with pigment formation and about half did not. So there are *at least two* different gene loci that can produce the albino phenotype. Actually, there are probably quite a few, and some of these have several alleles—alleles that produce weakly functional enzymes, blue eyes and light yellow hair, as well as alleles that produce totally nonfunctional enzymes, pink eyes and white hair. Some albino alleles are dominant, including one dominant condition in which only the head of the affected individual is albino.

The Castle–Hardy–Weinberg expectations based on total frequencies of albinos actually gave incorrect estimates of the allele frequencies because the simplest assumption was violated: the assumption that a single gene locus was involved.

Violating Assumptions

What happens when the assumptions of the Castle–Hardy–Weinberg law are violated? In physics you might quite casually begin the study of some area of interest by making absurd assumptions: frictionless surfaces, falling bodies without air resistance, mass concentrated at a single point, and so on. Just because these things don't really happen doesn't mean that the formulas you derive aren't any good. Similarly, the Castle–Hardy–Weinberg expectations are based on a set of idealized circumstances that may seldom occur in nature, but the expectations are useful all the same. The logic of the random association of gametes depends on the following:

1. A single autosomal diploid locus with two alleles
2. Alternate segregation of alleles in heterozygotes (Mendel's first law)
3. Random sampling of the population in obtaining data
4. Equal allele frequencies in both sexes
5. No changes in allele or genotype frequencies due to differential viability
6. Equal fertility and mating success of all genotypes
7. No allele change through mutation
8. No change in allele frequencies due to immigration
9. Random mating, which includes:
   (a) No inbreeding (no increased probability of mating with relatives)
   (b) No assortative mating (no preference for similar or dissimilar mates)
   (c) All data collected from the same geographical region; no pooling of samples from different regions
   (d) No preferential mating within ethnic groups, or at least no pooling of data from different groups
10. No chance variation; infinite population and sample size

Of these stipulations, item 10 is the most blatantly unrealistic, since all populations and all samples are finite. In practice, then, one seldom gets, or really expects to get, the exact Castle–Hardy–Weinberg expectations.

What Good Is the Castle–Hardy–Weinberg Distribution?

By this time you might be wondering why we need to predict idealized expected genotype frequencies instead of simply going out and measuring real genotype frequencies. It turns out that there are three ways in which the Castle–Hardy–Weinberg distribution is useful.

1. *Population genetics theory and evolutionary theory:* Much of the theory of the genetics of populations is built on *hypothetical models* of what we think is going on. In many of these models, the Castle–Hardy–Weinberg acts as the ground floor on which everything else is built. (Algebraic models in biology are only useful, of course, only if they make predictions that can be tested.)
2. *The study of natural populations:* For many purposes, the Castle–Hardy–Weinberg predictions are the most useful when they don't occur. We've listed several things that can happen in natural populations that will give rise to departures from Castle–Hardy–Weinberg expectations. These departures themselves are interesting, and can sometimes tell us such things as how much inbreeding or natural selection may be going on.
3. *Estimating recessive allele frequencies when there is complete dominance:* This special use of the Castle–Hardy–Weinberg distribution depends on all the assumptions being at least nearly true. As we saw in the case of recessive albinism, it's very easy to reach incorrect conclusions when you don't know all the facts. An extension of this is using the Castle–Hardy–Weinberg distribution to test whether some specific condition *is* or *is not* a simple Mendelian trait. If the frequencies of supposed carriers and alleged homozygotes don't match expectations, or if the *ratios* in the offspring of certain types of matings don't match Mendelian laws, the investigator is forced to try another hypothesis.

So we have seen that the earth is a variable and changing place and that the tiny coiled chromosomes of living things are responsive (indirectly) to those differences. The environment, in a sense, winnows the genes, enhancing the survival of those that help the organism adjust to the environment and eliminating those that are detrimental. We have discovered many of the rules of natural selection and have been able to describe, rather precisely in some cases, how it operates. In other instances, though, tantalizing mysteries remain as mute encouragement to our continuing efforts.

Application of Ideas

1. In a herd of forty wild mustangs, Greg Meddlesome counted ten palominos, twenty-nine dark (brown or black) horses, and one white horse. (a) If the genotypes are **Aa**, **AA**, and **aa** respectively, what are the respective genotype frequencies? (b) What is the allele frequency of **a** in this group? (c) Is this group in approximate Castle–Hardy–Weinberg equilibrium? Calculate the Castle–Hardy–Weinberg expectations. (d) Greg noted that the herd had only one stallion, which happened by pure chance to be the white (**aa**) horse. The rest of the horses constituted a harem of females—not an unusual situation with groups of wild horses. What will be the approximate allele frequency of **a** in the next generation, assuming that the group remains isolated? (e) Would you expect the next generation to be in Castle–Hardy–Weinberg equilibrium? Why or why not?

2. Among Americans of European descent, about 70% find weak solutions of a particular chemical, phenylthiourea (also called phenylthiocarbamide or PTC), bitterly distasteful. The other 30% are unable to taste the chemical unless it is extremely concentrated. The principal cause for this difference between people is a single gene locus with two alleles. The ability to taste weak solutions of phenylthiourea is dominant; **TT** and **Tt** persons are tasters, **tt** persons are nontasters. (a) If 70% of the population have the dominant **T**–phenotype and 30% have the recessive **tt**–phenotype, what is the frequency of the nontaster (**t**) allele? (b) What is the frequency of the taster (**T**) allele? (c) What are the frequencies of the three genotypes **TT**, **Tt**, and **tt**? (d) What is the probability that a taster is homozygous?

Key Words

albinism
allele frequency
Allen's rule
antibiotic resistance
artificial selection

behavioral phenotype
Bergmann's rule
Biston betularia

Castle–Hardy–Weinberg law
cline
clone
continuous distribution

ecological niche
equilibrium frequency

fitness
fixation
founder effect
frequency
frequency-dependent selection

gemmules
gene flow
gene pool

Key Ideas

Introduction

1. Much of modern biology rests on the premise of evolution.
2. Individuals do not evolve; *populations* evolve through changes in the composition of the gene pool.
3. All the genes contained in a *population* constitute the *gene pool.*
4. Evolution can be a random process or can be guided by *natural selection.*
5. Some genes change through the processes of evolution; others remain unchanged over long periods of time.
6. Evolution can be defined or reduced to the definition of "a change in *allele frequencies.*"
7. When a change in a gene has moved to a high frequency in a population or species, evolution has occurred.
8. Our understanding of evolution required a melding of Darwinian and Mendelian thinking.
9. Weissmann's theory of "the continuity of the *germ plasm*" was a whole new way of looking at heredity that made Mendel's work more understandable.

The Germ Plasm Theory

1. We know that physical structures such as eyes, hair, disposition, and so forth are not inherited. Only the physical entity, DNA, is passed from one generation to another.
2. For years it was generally accepted that parts of the parent's body contributed somehow to the offspring.

genetic variability
genome
genotype frequency
geotaxis
germ plasm
Gloger's rule

intermediate optimum

mark and recapture
morph
mutant
mutation rate

natural selection
neutralist
neutral mutation

pangenesis
phenotype frequency
polymorphic
population
population bottleneck

random drift
relative fitness

selection coefficient
selectionist
selective advantage
sigmoid curve

Tay-Sachs disease

"use and disuse"

Gemmules and Pangenesis: Darwin's Big Mistake

In his theory of *pangenesis,* Darwin suggested that *gemmules* from each cell were gathered to produce the gametes. The idea of *pangenesis* may have originated in the accepted theory of *use and disuse.* Structures (perhaps their gemmules?) that were not used were lost in evolution.

Weissmann and the Continuity of the Germ Plasm

1. Weissmann proposed that hereditary information was contained in *germ plasm. Germ plasm* arose continuously from other *germ plasm* and in this manner passed through generations. It indirectly influenced the appearance of traits.
2. While the idea of *germ plasm* contributing to *germ plasm* is essentially correct in animals, it is much less obvious in plants, because *germ plasm* can arise from any undifferentiated plant tissues.
3. *Germ plasm* contributing to *germ plasm* describes the replication role of DNA.
4. The phenotype of an individual never directly affects the hereditary material (except when *natural selection* removes a phenotype from the population).
5. A conclusion from Weissmann's theory is that nothing evolves but the *germ plasm.*
6. Mendel's concept of particulate factors (genes), not *gemmules* or *germ plasm,* is the closest to the "central dogma."

Frequencies

In population genetics, *frequency* is a proportion (e.g., the proportion of items of one kind in a group containing more than one kind). Frequencies are expressed as a fraction of the whole, thus are always less than one.

The Population Genetics of Asexual Organisms

1. Since segregation and recombination do not occur, there is no population distinction between genes, genotypes, and phenotypes.
2. Evolution through natural selection cannot occur in pure clones, since any selective pressure will effect the whole population identically. But when asexual populations have several variations, natural selection will eliminate all but one of the variants.

Antibiotic Resistance

1. Antibiotics and microorganisms can be used to set up laboratory demonstrations of evolution in populations. An example is penicillin and susceptible bacteria (penicillin kills by inhibiting cell wall synthesis, thus stopping reproduction):
   a. If a *clone* population is subjected to a killing concentration, all bacteria will die.
   b. If in the *clone* population a *mutant* for penicillin resistance exists, all but that individual will die.
   c. At the start, in a population of 100 million, the *mutant's frequency, p,* was 10^{-8}. But after the death of the *nonmutants, p* will equal 1 (all).
   d. If the penicillin concentrate is such that only some of the bacteria will die, the *frequencies* for the two types change.
   e. The change in *frequencies* between the two will not be as drastic, but over many generations the *frequency* of the *mutant* will increase until it replaces the *nonmutant.*
   f. These results will occur as long as the *relative fitness* of one type is greater than the other.

Selective Advantage and Selection Coefficient

1. Graphing the *frequency* of the penicillin resistant *mutant* over time would produce a *sigmoid curve.*
2. The shape of the *sigmoid curve* tells us that the increase in the *mutant* type is slow at first, speeds up, and then slows as the *frequency* approaches 1.0.

Selecting for Two Characteristics at Once
Asexually Reproducing Organisms

When two factors are selected for simultaneously, only one factor persists, the other is lost. The rule is that in asexual organisms, only one new evolutionary change can be fixed at a time. Evolutionary changes are sequential.

Sexually Reproducing Organisms

1. In sexually reproducing organisms two factors do not compete against each other. They are freed from competition by genetic recombination.

2. The gene rather than the *genome* becomes the unit of evolution.

Essay 11.1 The Story of the Peppered Moth

1. A classic case of *natural selection* is found in *Biston betularia,* the peppered moth. Two traits, peppered and black, are of interest. The black moth, rare in rural areas, greatly increased in growing industrial regions where it was protected by sooty trees. The peppered type, however, was easy prey for birds. The opposite was true in nonindustrial regions.

2. Release and recapture of black and peppered moths in hostile environments clearly showed that predation was a factor. Black moths survived twice as well as peppered in soot-covered forests, and peppered survived twice as well in clean forests.

3. It is important to note that these data were gathered before the moths reproduced, so sexual reproduction was not a complicating factor.

Evolution in Diploid Organisms

Diploid organisms introduce many problems in the analysis of *frequency changes.* This is because of dominance and recessiveness and the fact that recessives can hide from *natural selection.* All of the genetic complexities must be accounted for.

Genetic Variability in Diploid Populations

1. *Genetic variability* allows sexual diploid populations to respond rapidly to *natural selection.*

2. Diploid individuals in a population are all genetically unique from their physical appearance down to their body chemistry.

3. Variation, as pointed out by Darwin, is the raw material of evolution and is the species' main chance for quickly coping with an unpredictable future.

The Maintenance of Genetic Diversity

Genetic diversity may be maintained through many mechanisms. Among these is the ability of recessiveness to stay hidden and factors such as overdominance, where the heterozygote is favored. In overdominance, dominant and recessive genes are retained indefinitely.

Short-Term Evolutionary Change in Diploid Populations

1. *Artificial selection* mimics *natural selection* in short-term effects. It works very successfully as shown in test experiments with white leghorn chickens and mice, but populations so derived show a decrease in viability in many ways. Decreased viability indicates that the trait selected carried with it other traits that decrease fitness. These were hidden in the unselected population where their effects were not as drastic.

2. The rapid response seen in *artificial selection* indicates that the alleles for the desired trait were not rare. Otherwise the response would have been much slower at first.

3. In short-term selection, changes must arise from variation already present in the population.

4. The rate of selection slows down eventually because *genetic variability* runs out and because of a decreasing fitness or loss of vigor.

Selection for Going Up or Down

1. Up or down experiments for fruit flies demonstrated the existence of a behavioral phenotype as selection was intensified. The usual results for artificial selection were also encountered:
   a. The general fitness of highly selected populations declined.
   b. The rate of selection response declined after several generations.

2. The up or down experiment also showed how easily natural populations could be changed when selection was strong. When allowed to mate indiscriminately, the whole population returned to its previous state.

Natural Selection Compared with Artificial Selection

1. *Artificial selection* experiments can be misleading because they are based on the presence of preexisting alleles and do not include *genetic variation* by random mutation, a key factor in evolution.

2. It was once thought that species differ because of different frequencies of the same allele. We now know that the alleles themselves are different.

3. Dobzhansky stated that "evolution is a change in the genetic constitution of a population."

Small Babies and Large Babies—An Intermediate Optimum

1. *Selection* appears to act against extreme phenotypes, favoring the intermediate condition. An example is birth weight, where the optimum (best survival) is the average weight.

2. *Selection* appears to generally favor the average or intermediate type, but as the environment changes, the intermediate or optimum condition also changes. The *gene pool* must follow or track these changes. Cyclic changes in the environment bring about high *genetic variability.*

Frequency-Dependent Selection

Selection may favor a genotype only as long as it remains rare in a population. For example, a few red fish in a population may startle competitors for mates, but too many red fish lose their startling effect and they also become easier for predators to spot. The red phenotype is kept at an *equilibrium frequency.*

Geographic Differences in Selection, with Migration

Certain favored alleles dominate the ends of ranges when there is no migration between. When migration does occur a cline of allele frequencies appears with phenotypes not clearly suited for either end.

The Rules of Bergmann, Allen, and Gloger

1. *Bergmann's rule* states that warm-blooded animals tend to be larger in cooler regions.

2. *Allen's rule* states that warm-blooded animals tend to have shorter limbs, tails, and ears in cooler regions.

3. *Gloger's rule* claims that warm-blooded animals in cool, dry areas tend to be lighter colored than those in warm, moist regions.

Geographic Variation in the Human Species

Before the European expansion, humans followed the Bergmann, Allen, and Gloger rules fairly closely. There appears to be a correlation between skin color and the amount of sunlight exposure. Light-skinned northerners can better absorb limited sunlight for vitamin D synthesis. Where sunlight is abundant, dark skin absorbs sufficient light for vitamin D synthesis, and also offers the advantage of protection from sunburn.

Neutralists Versus Selectionists

1. *Neutralists* subscribe to *random drift* of *allele frequencies.* These are brought about by *neutral mutations,* those that produce variation without affecting functions.

2. *Selectionists* believe that nearly all variation is due to *natural selection.*

3. Both accept the idea that visible variations are subject to, or are the result of, *selection.*

4. *Neutral mutations* have a chance of disappearing or becoming fully fixed.

5. There is no way to determine how much genetic change has been *random* or *selective.*

Gene Changes in Evolution

The rate of change (amino acid substitution) for specific proteins appears to be the same in different lines of descent. The fossil record does not support such slow change in visible form, but reveals the sudden appearances of new species.

Population Bottlenecks

Population bottlenecks occur when a population is temporarily small. By pure chance, there may be an increase in the frequency of rare recessives in the residual population. As the residual population grows, a *founder effect* may prevail. For example, recessive diseases may be more common. *Tay-Sachs disease* may be an example of *founder effect. Founder effect* may also cause the loss of undesirable alleles.

The Balance Between Mutation and Selection

1. Beneficial mutations are extremely rare. Most mutations have a harmful effect, and are selected out.

2. Any specific allele will mutate at one time or another. The *mutation rate* is the proportion of gametes carrying a new *mutant* at a gene locus.

3. The amount by which a harmful genotype lowers fitness is the *selection coefficient.*

4. Generalizations about *mutations:*
   a. Number of affected individuals is directly proportional to the *mutation rate* and inversely proportional to the *selection coefficient* (severe alleles are weeded out at a faster rate than those which are less severe).
   b. A typical *mutation rate* is 1 *mutation*/100,000 genes.
   c. Each locus has its own rate and the total gametes with *mutant* alleles is large.
   d. Fortunately more mutations produce minor effects.

The Genetic Future of *Homo sapiens*

The frequency of harmful alleles tends to increase when medical intervention permits the bearer to live and reproduce (lowers the *selection coefficient*). In human history, the more stable the social system, the less selection against formerly debilitating alleles.

The Castle–Hardy–Weinberg Law

The *Castle–Hardy Weinberg law* explains why recessive traits do not change *frequency* in a *population* (unless they are selected against).

The Implications of the Castle–Hardy–Weinberg Distribution

1. The biological definition of a *population* is: a group of interbreeding or potentially interbreeding individuals.

2. A *Castle–Hardy–Weinberg* population is an ideal or model concept and has stringent requirements.

3. One of the observations possible from a *Castle–Hardy–Weinberg* population is that the allele frequencies do not change because of dominance. This answers the question, Why don't recessives simply disappear in a population? See page 342 and Table 11.2.

An Algebraic Equivalent

1. All of the possible matings and their *frequencies* are shown on page 343. The results indicate that the *allele frequencies* remain the same after all possible matings are carried out.

2. From the *Castle–Hardy–Weinberg law,* it is possible to predict, within limits, the outcome of mating. This requires that the allele frequencies be determined. Allele frequencies can be determined by applying the formula $p + q = 1$. Here p is the frequency of the dominant allele, while q is the frequency of the recessive. Together they equal all the alleles of the population. To find q, the recessive individuals in a population (**aa**) are counted. The frequency of **a**, or q, is $\sqrt{\mathbf{aa}}$. When this is known, p can be readily determined by $p = 1 - q$. When both are known, then the frequency of the three genotypes can be determined.

$$\mathbf{AA} = p^2$$
$$\mathbf{Aa} = 2pq$$
$$\mathbf{aa} = q^2 \text{ (counted in the population)}$$

3. Knowing the frequency of recessive homozygotes can be used to predict allele frequencies only if the traits in question are controlled by a single gene locus.

Violating Assumptions

The *Castle–Hardy–Weinberg law* is based on idealized circumstances:

a. One pair of autosomal alleles
b. Alternate segregation of alleles in heterozygotes
c. Random sampling
d. Equal *allele frequency* in both sexes
e. No differential viability
f. Mating is totally random (without genetic preference)
g. No mutation
h. No immigration or emigration
i. Infinite population and sample size

What Good Is the Castle–Hardy–Weinberg Distribution?

1. It helps construct hypothetical models.
2. Failure of data to fit the *Castle–Hardy–Weinberg law* is instructive, since the cause of variation indicates things are happening to a population.
3. It is useful in determining whether a specific condition is or is not a simple Mendelian trait.

Review Questions

1. What is meant by the term, *gene pool?* (p. 321)
2. What causes the *gene pool* of a population to change? (p. 321)
3. Using the term, *allele frequencies,* define evolution and explain your definition. (p. 321)
4. Why was a melding of Darwinian and Mendelian thinking needed to provide an understanding of evolution? (p. 321)
5. How did Weissmann's *germ plasm* theory differ from *pangenesis* as presented by Darwin? (p. 322)
6. What did Mendel's discoveries lend to Weissmann's germ continuity theory? (p. 322)
7. Explain the phrase, "Nothing evolves but the *germ plasm.*" (p. 322)
8. Under what conditions is a *clone* subject to *natural selection?* (p. 323)
9. Review the experiment with penicillin and susceptible bacteria and explain how the p value went from 10^{-8} to 1. (p. 324)
10. Graphing the *frequency* of penicillin-resistant types over time typically produces a *sigmoid curve.* Explain the bottom, middle, and top of the curve. (p. 324)
11. Explain why progressive evolutionary changes are sequential in asexual organisms. (p. 325)
12. What aspect of evolution was left out of the *Biston betularia* experiment? How would including this factor have complicated the study? (pp. 327–328)
13. What is the importance of *genetic variability* to evolution? (p. 328)
14. Describe the chicken and mouse *artificial selection* experiments in terms of the following (p. 329):
    a. General success in affecting phenotypic change
    b. The rate at which phenotypic change occurs
    c. The effect on general health and viability
15. What is the evidence suggesting that fruit flies have a "behavioral" phenotype? (p. 331)
16. What important aspect of evolution is left out in *artificial selection* experiments? (pp. 330–331)
17. What does the study of birth weight and mortality tell us about the direction of *natural selection?* (p. 331)
18. What is meant by the statement, "the *gene pool* continually tracks a changing environment"? (p. 333)
19. What is the effect of a *cline* on the fixing of alternative alleles at opposite ends of a range? (p. 334)
20. Summarize the rules of *Bergmann, Allen,* and *Gloger.* (pp. 335–337)
21. How does the thinking of *neutralists* and *selectionists* differ? What do they agree on? (p. 337)
22. Describe the *founder effect* and give two examples (pp. 338–339)
23. Define the terms *mutation rate* and *selection coefficient.* (p. 339)
24. Why is it that we all carry many *mutant* genes? (pp. 339–340)
25. What determines how fast a *mutant* gene will be weeded out of a population? (pp. 339–340)
26. What effect does modern medicine seem to have on the frequency of dangerous *mutant* alleles? (p. 340)
27. To what basic question about dominant and recessive alleles does the *Castle–Hardy–Weinberg law* address itself? (pp. 341–342)
28. Review and summarize the proof of the *Castle–Hardy–Weinberg law.* (pp. 342–343)
29. Explain what is meant by the formula $p + q = 1$. (p. 342)
30. Explain how the frequency of a dominant and recessive gene are determined when the frequency of the recessive genotype is known. Use the *albinism* example if it helps, but explain what you are doing. (p. 343)
31. Assume that the recessive genotype **yy** appears in 36% of a population. Find the gene frequency of **y** and its alternative **Y**. Find the frequency of the homozygous dominants and of the heterozygotes. (p. 343)
32. What assumptions about alleles are necessary for the *Castle–Hardy–Weinberg law* to apply? (p. 342)
33. What are the assumptions about *allele frequency* changes? (p. 344)
34. What valuable information can we obtain from populations that depart from *Castle–Hardy–Weinberg* predictions? (p. 344)

Chapter 12

Microbial Genetics

High points in the history of science are of two types: those with immediate impact, such as Darwin's, and those with delayed impact, such as Mendel's. Such differences arise because of the patchwork design of science. Different areas may progress at different rates, and precisely how one rapidly advancing area may fit into the overall design is not always immediately clear. We're about to consider another of those findings that were unfortunately neglected until many years after they were reported.

12.1 Genes and Metabolic Pathways

Inborn Errors of Metabolism

The world was not ready for the first major contribution to the biochemistry of the gene. In 1908, A. E. Garrod, a physician influenced by the Mendelian revival, published a book called *Inborn Errors of Metabolism.* His subject was *Homo sapiens,* which, at the time, was an unusual experimental organism for genetics research. Garrod was interested in metabolic defects, breakdowns in the complicated biochemical processes of life. He searched for them in urine, because it was known that metabolic failure anywhere in a long chain of reactions could produce unusual substances in the urine. Of special interest to Garrod was *alkaptonuria,* a disease in which the urine contains *metabolites* (products of metabolism) called *alkaptones*—substances that happen to turn black upon oxidation, and so are easily revealed. Persons with alkaptonuria are usually detected as soon as their diapers start turning black. As the child grows older, black pigments begin to settle in cartilage and other tissues, blackening the ears and even the whites of the eyes. Other, more serious effects include arthritis, caused by the accumulation of the metabolite in the cartilage of the joints.

Garrod observed that the disease tended to be found in several brothers or sisters in a single family. By studying family histories, he correctly inferred that alkaptonuria and certain other inborn errors of metabolism were caused by Mendelian recessive alleles. The problem, he deduced, is caused by the absence of specific enzymes that are necessary for the long chains of biochemical reactions to occur. If an enzyme for a particular reaction is absent, no reactions can take place past the point where it normally enters the chain, and the substance that the enzyme acts on builds up.

Other inborn errors of metabolism produce albinism (a complete or partial lack of melanin pigment in the hair, skin, iris, and sometimes the retina) and phenylketonuria (which also has an effect on hair and skin pigmentation, but has a much more severe effect on mental development because of the accumulation of metabolites that are toxic to the nervous system). As it turns out, albinism, phenylketonuria, and alkaptonuria are due to defects in enzymes that act on the metabolism of phenylalanine and tyrosine (Figure 12.1).

Garrod showed that genes can act directly on biochemical metabolism through enzymes—something all of us can readily appreciate now. But his work, like Mendel's, was not influential in its own time, nor for long afterward.

The significance of Garrod's work was that he showed the Mendelian inheritance of enzyme metabolism. But he was ignored. The geneticists of the time were interested in "more important" things. Genes, they knew, were the determinants of *morphology:* wrinkled versus round peas, tall versus dwarf plants, red eyes versus white eyes, curly wings versus wild-type, variously shaped chicken's combs, and the like. Morphology—now, that was a mystery worth tackling!

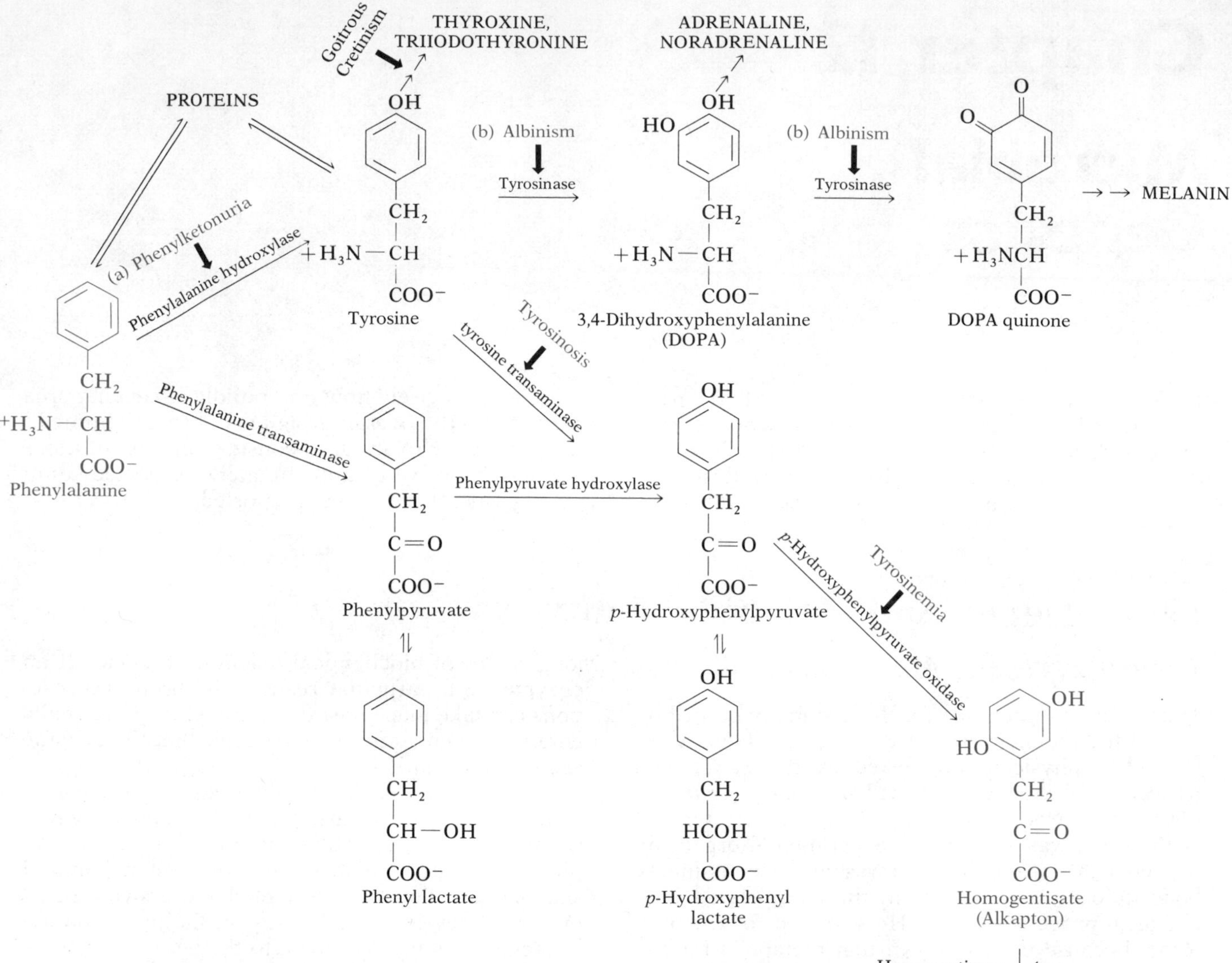

12.1 The normal metabolic pathways that utilize the amino acids tyrosine and phenylalanine. Seven genetic disorders, including two forms of albinism, phenylketonuria, and alkaptonuria, result from enzyme deficiencies in the pathways. An enzyme deficiency means that the DNA segment that codes for the enzyme has undergone a genetic mutation resulting in the absence of a key enzyme, the accumulation of some unmetabolized substrate, and the loss of important products further down the pathway. (a) Phenylketonuria, the accumulation of phenylalanine and loss of the subsequent product tyrosine, occurs when the enzyme *phenylalanine hydroxylase* is absent. (b) Albinism, the absence of the pigment melanin, occurs when the enzyme *tyrosinase* is absent. (c) Alkaptonuria, near the end of the pathway, occurs when the enzyme *homogentisase* is absent, leading to an accumulation of homogentisate (alkapton) in the body.

But, as it turned out, Garrod was on the right track. It is now taken for granted that genes direct the formation of enzymes and proteins, and that therefore the various genetic controls of morphology must operate through variation in proteins. However, that information really does not get us very far; ironically, we still know next to nothing about the genetic control of morphology. Also, while we know that genes can have powerful and specific effects on behavior, we're almost totally in the dark when it comes to the molecular basis of behavior. We have to be satisfied, for now, with what we *do* know.

One Gene, One Enzyme

"One gene, one enzyme," a slogan that was an electrifying bit of public relations in its day. It first caught on just before World War II, thanks to the Nobel Prize-winning work of George Beadle and Edward Tatum, who applied genetic analysis to biochemical

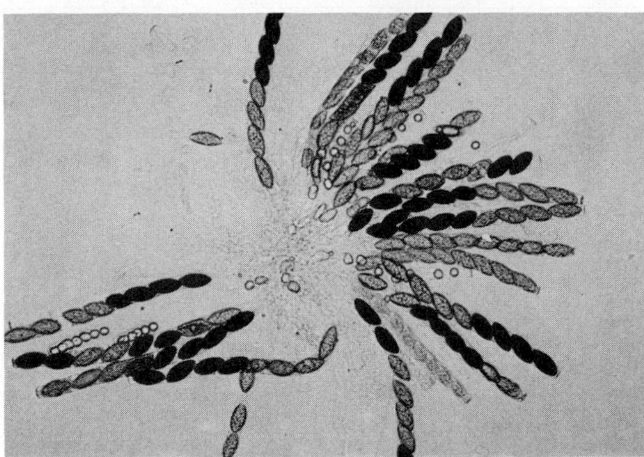

12.2 *Neurospora crassa.* Besides being a pesky contaminant in the baking industry and a delightful additive in the manufacture of *ontjom,* a Javanese delicacy made from peanuts, *Neurospora* helped lay the foundation for research in microbial genetics and molecular biology. Its usefulness to geneticists is due in part to a very long haploid stage in its life cycle. This means the researcher has to contend with only one gene for a trait rather than the allelic pair of *Drosophila* or garden peas.

pathways, imaginatively choosing a very unusual experimental organism: the fungus *Neurospora* (Figure 12.2).

Beadle had previously worked with Boris Ephrussi on the biochemical pathways of eye pigment synthesis in *Drosophila.* Their procedure had involved the incredibly laborious and delicate transplantation of developing organs from the larvae of one *Drosophila* strain to the abdominal cavities of larvae of another genetic strain. (Imagine that operation!) They then observed the transplanted organs as their larval host went through its various developmental stages. They finally threw in the towel, but before they gave up, they learned that each of the various *Drosophila* eye color mutants appeared to block the biochemical pathway of normal pigment synthesis at a specific step. Thus, they reasoned, eye color differences occurred because the chain of reactions was not complete. Intermediate products built up because they couldn't be metabolized. In the transplants, various intermediate products would diffuse between the host and transplant and that normal eye color would result as a missing enzyme product was supplied.

Thus Garrod's idea was rediscovered: Specific genes could block specific metabolic steps. Or, put another way, specific metabolic steps in biochemical pathways depended on the presence of functioning alleles.

Beadle and Tatum

Ephrussi went on to other things, but Beadle was entranced by the problem. In looking for something easier than *Drosophila* to work with, Beadle and his new partner, Edward Tatum, initiated a whole new era of genetic research: *microbial genetics* and *molecular biology.* They chose *Neurospora* because they found there was a way to screen millions of fungal spores in their searches for specific mutations.

Wild-type *Neurospora* (Figure 12.2) can grow in a *minimal medium* consisting of water, glucose, minerals, and one vitamin (biotin). Beadle and Tatum were interested in finding mutant strains that were unable to synthesize some nutrient and thus could not grow in this minimal medium. The *haploid spores* were first irradiated to increase the mutation rate. (What was the advantage of working with a haploid organism? A single mutation will alter the cell's only working copy of a gene, so Beadle and Tatum didn't have to worry about dominance interactions and they didn't have to inbreed to get homozygotes.) The spores were then allowed to grow in an aerated minimal-medium liquid broth. All the normal, wild-type spores would germinate, growing into long filaments called *hyphae.* Beadle and Tatum then strained the whole mess through cheesecloth. The hyphae that had developed from normal, unmutated spores stayed in the cheesecloth, but any abnormal spores that couldn't grow went through. The ungerminated spores were then spread on the surface of an agar gel of a minimal medium plus certain added metabolites, such as a specific amino acid. Some of the spores that had failed to germinate in minimal medium began to grow. For instance, some could grow on a minimal medium only if it were supplemented with arginine; others required pyrodoxine; and so on. Ordinary cells also need these metabolites, but could make their own. It was assumed that, for example in the case of the arginine, the enzyme necessary to change some compound to arginine was missing and when the cell was supplied with arginine, the biochemical pathway could continue (Figure 12.3).

Because *Neurospora* does have a sex life (it forms diploid zygotes), wild-type strains could be crossed with the additive-requiring mutant strains. The mating procedure had to take place on the supplemented medium. It is important to remember that *Neurospora* is haploid until it reproduces sexually. After sexual union, it soon goes into meiosis, again producing haploid spores. Upon germinating, each spore produces a fungal growth which is still haploid.

Beadle and Tatum found that the nutritional deficiencies were indeed straight Mendelian factors, since meiosis in the heterozygote produced haploid spores in the expected 1:1 ratio of wild-type and mutant (nutrient-requiring) types. For example, if a wild-type (+) was mated with an arginine mutant (**arg**), the short-lived diploid zygote would be a heterozygote, +/**arg**. With meiosis, two kinds of spores, + (wild type) and **arg** (mutant), would be produced in equal

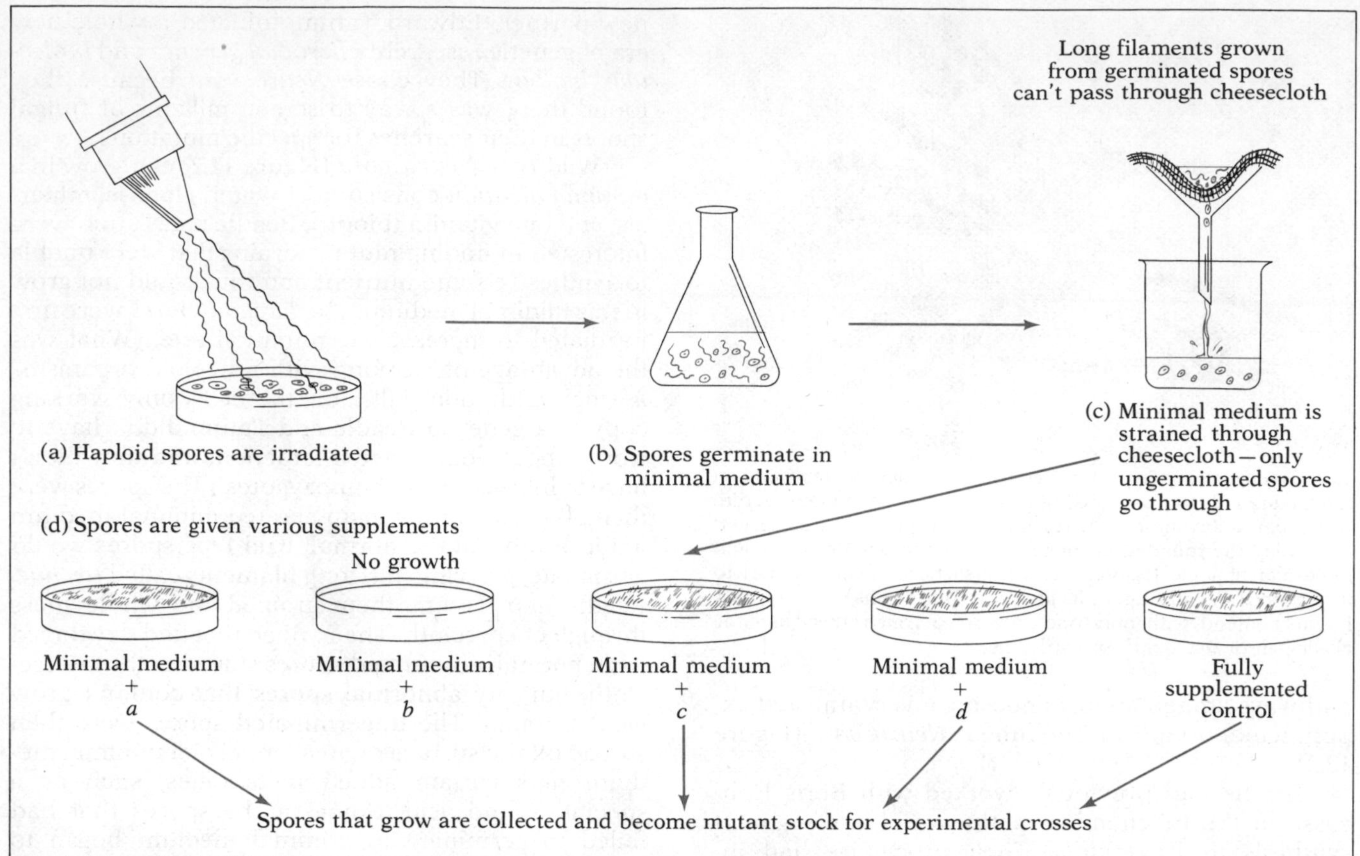
Long filaments grown from germinated spores can't pass through cheesecloth
(a) Haploid spores are irradiated
(b) Spores germinate in minimal medium
(c) Minimal medium is strained through cheesecloth—only ungerminated spores go through
(d) Spores are given various supplements
No growth
Minimal medium + a
Minimal medium + b
Minimal medium + c
Minimal medium + d
Fully supplemented control
Spores that grow are collected and become mutant stock for experimental crosses

12.3 Experimental procedures of Beadle and Tatum. (a) The procedure for producing and isolating nutritional mutants in *Neurospora* begins with irradiaton of haploid spores. This greatly increases the number of mutations. (b) The spores are then placed in a minimal medium, where only the wild-type grow into hyphal masses. (c) The ungerminated, mutant spores are then isolated by allowing them to pass through cheesecloth. (d) The ungerminated spores are then placed on agar that is supplemented in various ways with vitamins, amino acids, or other metabolites. Spores that germinate are classified according to which supplement they require.

numbers. Furthermore, genes for different deficiencies could be mapped through recombination in the same way that *Drosophila* genes had first been mapped a generation before (see Chapter 10).

Beadle and Tatum also crossed two mutant strains that both required the same added nutrient, and sometimes recovered wild-type (normal) progeny. For instance, they might have isolated two different arginine-requiring mutant strains. A cross between two such strains often resulted in progeny that were 75% arginine-requiring and 25% wild-type. Beadle and Tatum concluded that two different gene loci were involved, which they called **arg**-1 and **arg**-2.

Suppose that the wild-type *Neurospora* had functional alleles at both loci, for example, + +. Strain 1 might then have had a mutant, nonfunctional allele at the first gene locus, and its genotype would be **arg**-1 +, where **arg**-1 represents the mutant allele at the first arginine locus. Strain 2 might have a mutant, nonfunctional allele at the second gene locus, and its genotype would be + **arg**-2. When the cross was made, the diploid zygote would be **arg**-1 +/+ **arg**-2. Assuming that the two alleles would be unlinked, meiosis would form four different kinds of gametes in equal numbers: 25% **arg**-1 +, 25% + **arg**-2, 25% **arg**-1 **arg**-2, and 25% + +, in accordance with Mendel's law of independent segregation of alleles at two heterozygous loci. Since *both* wild-type alleles are needed for arginine synthesis, only one genotype, + + (25%), would be capable of surviving without added arginine (Figure 12.4).

Sometimes the two investigators would collect and cross two different arginine-requiring strains and get another result: All the progeny would still be arginine-requiring. In this case, they concluded that the different strains had received mutations at the same gene locus, let us say the **arg**-1 locus. Then the genotypes of the two mutant strains might be **arg**-1^x and **arg**-1^y where these are two different mutations that both destroy the function of the same gene. In the cross, the zygote **arg**-1^x/**arg**-1^y would follow

12.4
Segregation at first **arg** locus: 50:50 ratio of + to **arg**-1
Segregation at second **arg** locus: 50:50 ratio of + to **arg**-2
These are independent events. To find joint probabilities, probabilities of independent events can be multiplied.

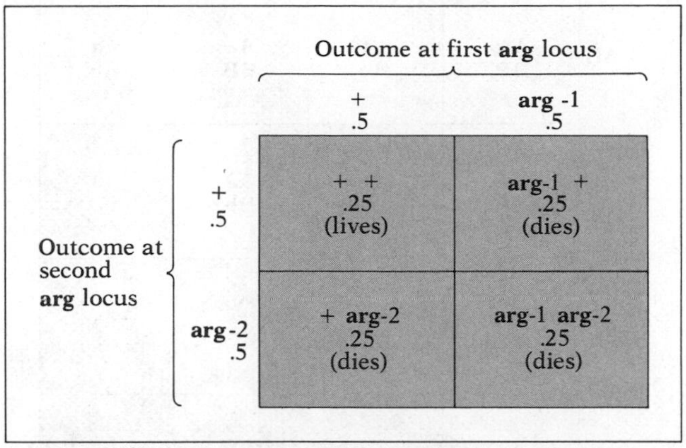

		Outcome at first **arg** locus	
		+ .5	**arg** -1 .5
Outcome at second **arg** locus	+ .5	+ + .25 (lives)	**arg**-1 + .25 (dies)
	arg -2 .5	+ **arg**-2 .25 (dies)	**arg**-1 **arg**-2 .25 (dies)

Independent segregation in a haploid organism segregating for two arginine-requiring mutants. The probability of getting the wild-type allele is .5 at each locus; the probability of getting wild-type alleles at both loci is .25, the product of independent probabilities.

Mendel's first law: 50% **arg**-1^x and 50% **arg**-1^y, neither of which can survive without added arginine.

By collecting a large number of arginine-requiring mutants and by making crosses between all possible pairs, Beadle and Tatum found that all their mutant strains fell into exactly three classes. Crosses between two strains within any one class never produced wild-type progeny, but any cross between any individual from one class with any individual from a different class always produced 25% wild-type progeny. Therefore, they reasoned, there were just three gene loci involved in arginine requirement. They named the gene loci **arg**-1, **arg**-2, and **arg**-3. The wild-type alleles were **arg**-1^+, **arg**-2^+, and **arg**-3^+, or simply +, +, and +. Mutant alleles were named **arg**-1^-, **arg**-2^-, and **arg**-3^-. (The use of "+" to indicate the wild-type allele of any gene locus is derived from *Drosophila* studies and is standard in microbial genetics.)

They found that the three groups had different biochemical properties. Mutants at the **arg**-1 locus could grow if arginine were added to the medium, of course, but these strains could survive and make their own arginine if provided with either of two other nutrients, *citrulline* or *ornithine.* Mutants at the **arg**-2 locus could make arginine from citrulline, but could not make it from ornithine. And the **arg**-3 mutants required arginine for survival and would accept no substitutes.

Beadle and Tatum's interpretation was this: Wild-type *Neurospora* synthesizes its own arginine in a biochemical pathway involving at least three enzymes, produced by the wild-type alleles of **arg**-1, **arg**-2, and **arg**-3. The first enzyme in the pathway is produced by the **arg**-1 gene. It combines two amino acid precursors (aspartic acid and glutamate semialdehyde) to produce the intermediate compound, ornithine. The second enzyme (the product of the **arg**-2 locus) catalyzes the reaction that changes ornithine into citrulline. The third enzyme (from the **arg**-3 locus) catalyzes the last step in the biosynthetic pathway, which is to change citrulline into arginine. In this simple, linear example, three enzymes are involved sequentially in a short biosynthetic pathway; the pathway doesn't produce its end product, arginine, if any one of the three enzymes are missing (Figure 12.5).

Other longer and more complex biochemical pathways were similarly subjected to genetic analysis. In every case, a single gene locus was responsible for a single enzyme. Thus emerged the slogan, "one gene, one enzyme," even though the function of DNA and the genetic code were not yet known and no one yet knew just *how* the wild-type gene actually made enzymes. But, in any case, among the things that could be better understood after the develop-

12.5 The arginine pathway. Each enzyme creates some change in its substrate molecule, and eventually arginine emerges from the pathway. A mutation in any of the loci can interrupt the sequence. (Note that RNA is required for transcription and translation of the enzyme in each step.)

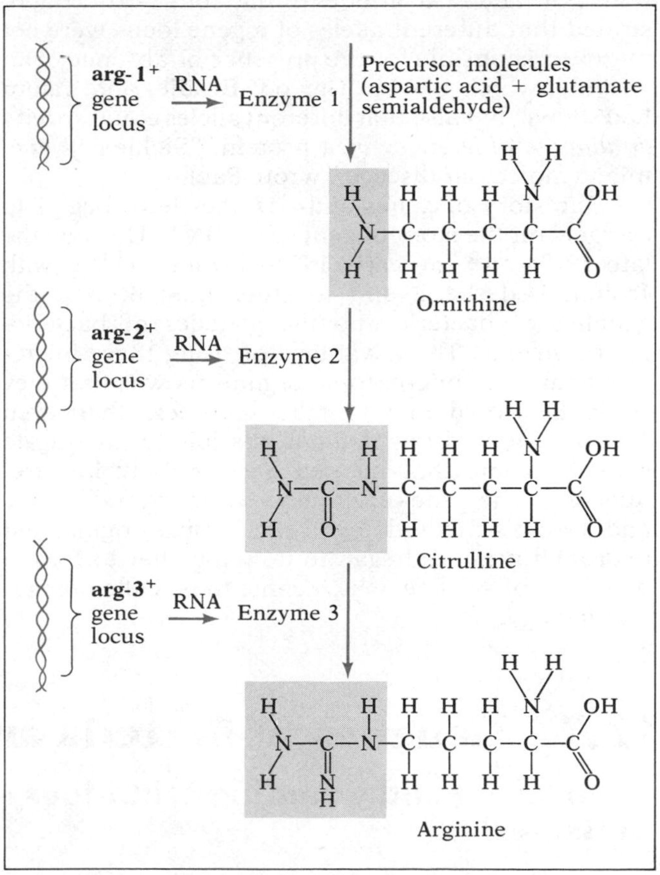

ment of the one gene, one enzyme hypothesis were:

1. The likely basis of dominance: one dose of an enzyme is enough, since most reactions are not limited by enzyme concentration
2. The basis of the Mendelian ratios with two interacting genes: for instance, a cross between two double heterozygotes for different white flower mutations in the flower pigment biosynthetic pathway gives a 9:7 ratio, as illustrated in Figure 12.6, because the pathway can be stopped at either of two steps (with a total probability of $\frac{7}{16}$)

At this point the sickle-cell hemoglobin allele entered the intellectual history of gene function. The sickle-cell case shows up often in the study of genetics, in part because of intensive efforts that have been made to understand this debilitating disease. Actually, however, sickle-cell anemia is just one of many autosomal recessive diseases, though it is a distressingly common one.

In 1949, Linus Pauling made an acute observation: In an electrophoretic apparatus, the rate of migration of the hemoglobin of homozygous sickle-cell anemics is slower than the hemoglobin of normal people, while known heterozygous carriers of the mutant gene have two different kinds of the protein: both slow and fast varieties. It had thus been demonstrated that different alleles of a gene locus were not merely responsible for the presence or absence of an enzyme, which is what Garrod, Beadle, and Tatum had shown, but also that different alleles could specify *qualitative differences* in a protein. "Sickle-cell anemia: a molecular disease," wrote Pauling.

The story of genes and enzymes here begins to merge with the story of genes and DNA. Through the late 1940s and the early 1950s Tatum worked with Joshua Lederberg, and together they utilized the genetics of bacteria and the genetics of bacteriophage viruses. The advantage of using these microscopic and submicroscopic organisms was that they could be raised in far higher numbers than even *Neurospora,* and this made it possible to investigate genetic events that happen at extremely low frequencies. The gene–enzyme hypothesis (and Beadle and Tatum's methodology), the central dogma, and microbial genetics began to flow together to form a new kind of science, which came to be called *molecular biology.*

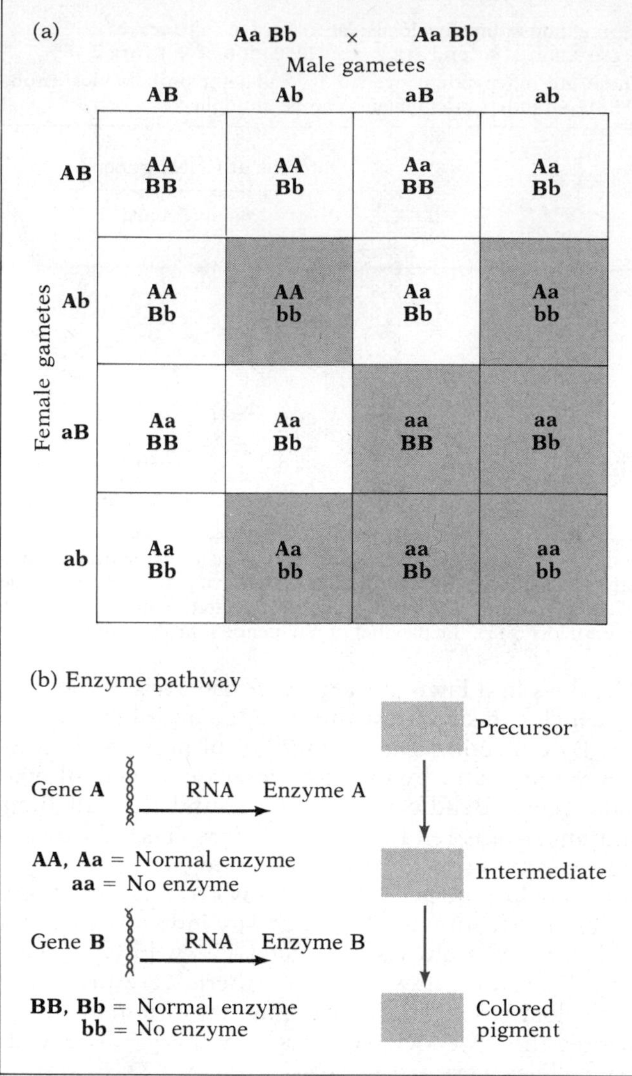

12.6 The one gene, one enzyme theory applied to unusual Mendelian ratios. (a) In a cross involving flower color where two pairs of alleles are involved, it is possible to get highly unusual ratios. This example is a cross between two doubly heterozygous flowers, **AaBb** × **AaBb**. The results are not the expected 9:3:3:1, but a strange 9:7 ratio. (b) The explanation is in the biochemical pathway that produces flower color. Gene **A** and gene **B** control two steps in color production. Therefore, any individual plant that is homozygous **aa** *or* **bb** will be unable to produce pigment. If you were to go through the Punnett square, you would discover the proportion of plants expected to produce white flowers (those with **aa** or **bb**).

12.2 Genetics of Bacteria and Viruses

Recovering and Counting Mutations in Bacteria

Genetic experiments in bacteria require collecting nutritional mutants that are unable to grow on minimal medium—just as Beadle and Tatum collected nutritional mutants of *Neurospora.* Different techniques have to be used, however. One trick is to use *penicillin* in the otherwise minimal medium.

Penicillin is a powerful antibiotic, and it kills cells by interfering with cell wall synthesis—the growing cell tries to divide, and ends up dead. In these experiments the experimenter puts a large, known quantity of wild-type bacteria into the minimal medium plus penicillin. Among the bacteria are a small but unknown quantity of nutritional mutants. As the normal cells metabolize, try to divide, and die, the mutants just sit there, unable to metabolize and uninterested in cell division. The penicillin doesn't kill them in this state. After a while, the experimenter can centrifuge them out (they are intact and heavier than the fragmented remnants of the dead bacteria) and then wash them carefully to get rid of the penicillin. The recovered cells are serially diluted and again spread on agar plates, this time on a Petri dish gel containing a supplemented medium. Now the mutant cells can grow—and they find themselves only with others of their kind. Nutritional mutant colonies can be counted in the manner described in Essay 12.1, so mutation rates can be calculated for any class of mutation. Another ingenious way of recovering mutations is *replica plating,* developed by Esther and Joshua Lederberg (Figure 12.7). Among other things, the replica-plate experiment proved for the first time that nutritional mutants occurred spontaneously on supplemented medium. Some critics had previously claimed that minimal medium itself had *caused* nutritional mutants in earlier experiments.

12.7 Replica plating is a method of locating nutritionally mutant bacteria. In this scheme, plate I is coated with a dilute mixture of nutritionally normal bacteria and thus far unidentified nutritional mutants. The medium in the plate is completely supplemented so that both types of bacteria can grow into discrete colonies that make a random pattern of dots on the agar. Then, a velvet-covered disk is pressed against the surface, picking up some bacteria from each of the colonies. The disk is then touched gently to the surface of plate II, which has been prepared with minimal medium. Only the normal bacteria can grow on plate II, so a comparison of the patterns of colonies on plates I and II reveals which original colonies are nutritionally mutant bacteria. The colonies that are missing on plate II are still present on plate I, and can be retrieved if desired. These are the nutritionally mutant bacteria.

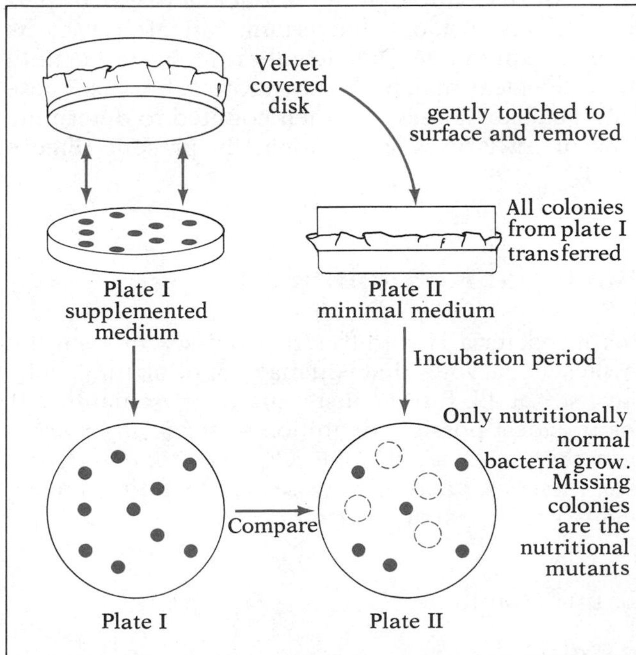

Recovering Recombinants

Different strains of bacteria are able to recombine in various ways that we'll discuss shortly. However, they have never mastered the process of meiosis, and so the recombination doesn't involve sex as we know it. What passes for sex in bacteria is varied, bizarre, and infrequent. So infrequent that the odds of a bacterium having sex is in the order of one in a million. In order to study such infrequent behavior, it's obviously necessary to take some shortcuts. Even with serial dilution techniques, no one wants to count a million bachelor bacteria before finding one in the act of mating.

Suppose you have two nutritionally mutant strains of *Escherichia coli;* one strain needs tryptophan (symbolized **trp**$^-$, **arg**$^+$) and another needs arginine (**trp**$^+$, **arg**$^-$). Neither can grow on minimal medium. To find that one-in-a-million recombinant, you can mix 50 million or so bacteria of each type, let them mate (or undergo whatever other kind of genetic recombination event they can manage), dilute them several times, and plate them out on minimal medium. The **trp**$^-$, **arg**$^+$ bacteria and the **trp**$^+$, **arg**$^-$ bacteria will both just sit there. Those of the two strains that have managed to recombine to produce **trp**$^+$, **arg**$^+$, however, will form colonies (Figure 12.8).

The Life and Times of the Bacteriophage

The phage, a bacterial virus, has already been mentioned with respect to Hershey and Chase's classic experiment, which showed that only the viral DNA entered the bacterial cell while the empty shell stayed outside (see Figure 7.9). Let's consider what happens once the viral DNA gets inside a bacterium.

First, it may immediately run into trouble with one of the bacterium's protective mechanisms. Many bacteria have cytoplasmic *restriction enzymes* that chop up foreign DNA sequences. Each restriction enzyme recognizes a specific, short DNA sequence (eight to twelve nucleotides long) and chops it in the middle. The bacterium has other enzymes that protect its own DNA. And not all restriction enzymes kill all phage. The bacterial host and its specific bacteriophage parasites evolve together, and the parasites manage to stay one jump ahead of their host—if they didn't, they wouldn't survive. The phage does this, for instance, by changing its DNA to remove any

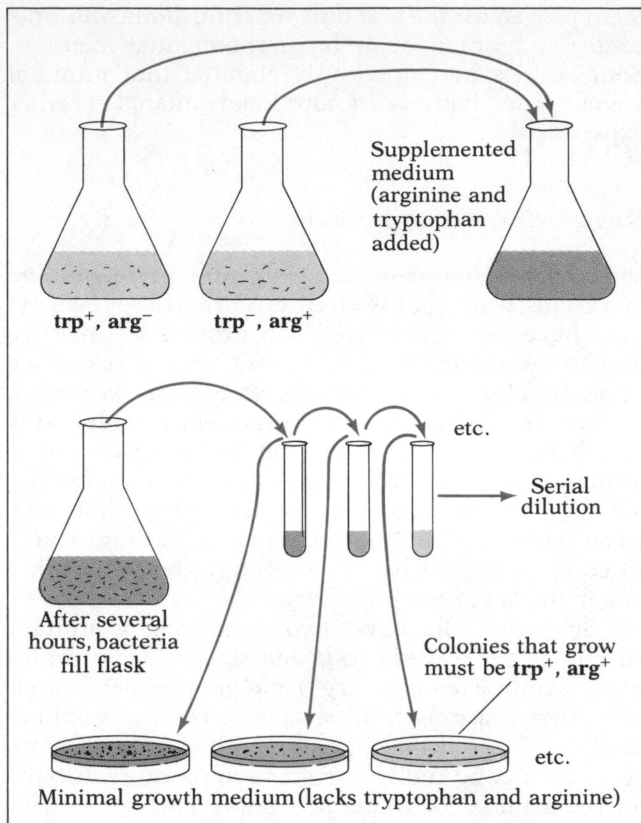

12.8 Biochemical and genetic evidence for sexual reproduction in bacteria was first gathered with the techniques shown in this scheme. Two strains, **trp**$^+$, **arg**$^-$ and **trp**$^-$, **arg**$^+$, which were selected through replica plating, were grown together in supplemented medium. Then, following serial dilution, samples were plated out on minimal medium that lacked tryptophan and arginine. If colonies grew, they represented genetic recombinants with the **trp**$^+$, **arg**$^+$ genotype.

occurrence of the fatal DNA sequence that the bacterium can recognize. The result of this and other defense mechanisms of both bacteria and phage is that every strain of bacteria is subject to infection by only a few bacteriophage strains, and every strain of bacteriophage is able to infect just a few host strains.

If the viral DNA gets past the restriction enzymes, it may encounter a friendlier host protein: *RNA polymerase.* RNA polymerase, unlike the restriction enzymes, isn't able to distinguish between its own DNA and foreign DNA. It reacts to the viral DNA as if it were its own, and transcribes mRNA from it—*viral* mRNA is produced. The viral mRNA is translated by the host's ribosomal machinery into viral proteins.

Then the virus turns tables on the host. Some of its own newly made enzymes can also distinguish between host DNA and viral DNA. These enzymes chop the host DNA into fragments, presumably so that the nucleotides can be recycled into viral DNA. Other viral enzymes make copies of the viral DNA; some even make more mRNA to speed up the process. The virus genes also make *coat proteins* and the leg-like "landing gear." The final act of the viral proteins is to *lyse* (rupture) the envelope of the host cell, now hardly more than a bag of virus particles. Lysis of an *E. coli* cell may release about a hundred new, infectious particles.

And that's the basic bacteriophage life cycle, at least for the group known as the T-even phages. There are also other important phage life cycles. Sometimes the phage doesn't immediately kill its host. Instead of destroying the host genome, the phage DNA may cut into and join the host DNA. The phage is said to *lysogenize* the cell, and the whole phenomenon is called *lysogeny.* The lysogenized viral DNA undergoes replication with the rest of the bacterial chromosome and is passed on to all the bacterium's descendants. While on the chromosome, it does very little except to put out a cell-surface protein that protects the cell from further attacks by the virus' close relatives. Then, in some future generation, the viral DNA may *excise* —that is, break away from the host DNA—and the cycle is taken up where it left off. Again the cell is lysed, releasing a hundred or so active virus particles (Figure 12.9).

Counting Viruses, or Holes in the Lawn

Viruses are much smaller than bacteria, so they can occur in even greater concentrations. The serial dilution technique is used to count them. Viruses do not make colonies on minimal medium; they grow only inside bacterial cells. So the viruses are diluted serially and then spread on agar plates that have been especially prepared with a *lawn*—a continuous surface layer—of susceptible bacteria. Each active virus particle infects just one bacterium, but after lysis its progeny spread to the neighboring bacteria, until there is a clear hole in the lawn for each virus. These holes, called *plaques,* are then counted to determine how many viruses were originally present (Figure 12.10).

Bacterial Recombination

Sex in bacteria is so different from sex in even the lowliest eucaryotes that you may not think they really have sex at all. But it's just a matter of semantics. If the broadest possible definition of sex is *any mechanism that combines into one cell, genetic material from two,* then yes, bacteria have sex. In fact, they have a highly varied sexual repertoire.

Transformation

The first kind of recombination to be observed in bacteria was *transformation* (Figure 7.8). It seems that bacteria can—with extremely low efficiency—take up naked DNA from other bacteria into their

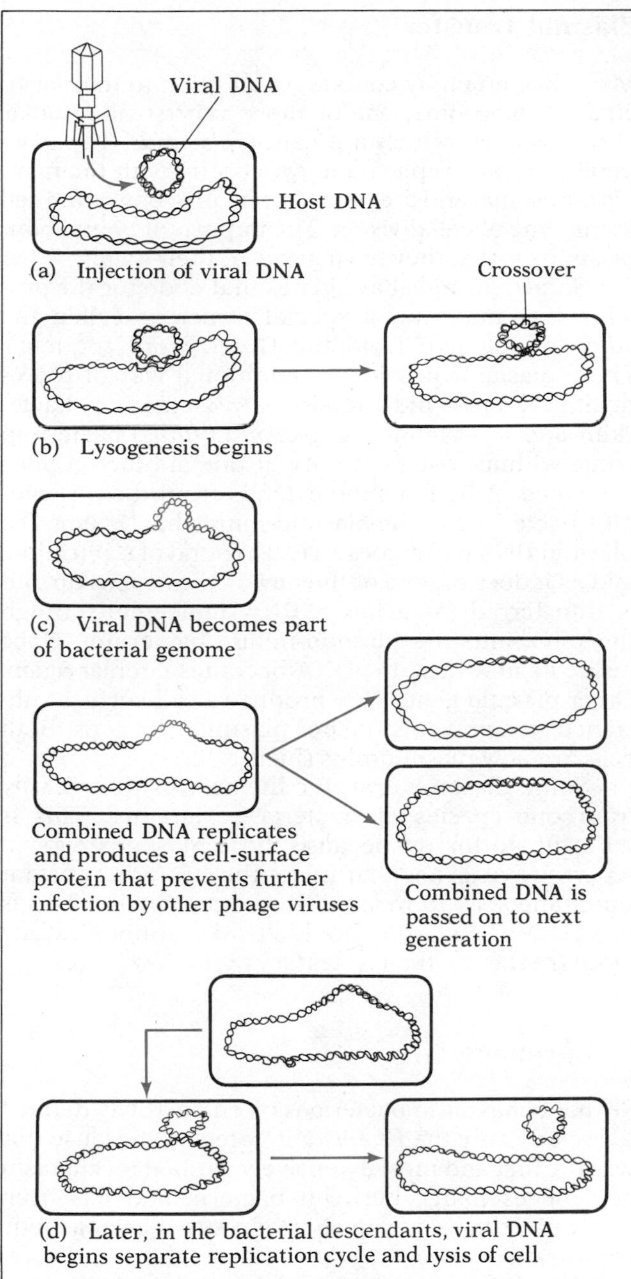
Viral DNA
Host DNA
(a) Injection of viral DNA
Crossover
(b) Lysogenesis begins
(c) Viral DNA becomes part of bacterial genome
Combined DNA replicates and produces a cell-surface protein that prevents further infection by other phage viruses
Combined DNA is passed on to next generation
(d) Later, in the bacterial descendants, viral DNA begins separate replication cycle and lysis of cell

12.9 Bacteriophage lysogenizing a bacterial cell. (a) In this series of events, a phage injects its DNA into the host cell. (b) The circular viral DNA joins the much larger circular bacterial DNA in an event that requires one crossover. (c) Eventually, the host will replicate both its own DNA and that of the viral segment. This may go on for generations. (d) Then, some unknown factor induces the phage DNA to excise and begin the events that end in phage production and cell lysis.

cells and incorporate it into their own genetic makeup. We know that the bacterial chromosome is circular, and presumably the fragment of DNA that is taken up is linear. Once the fragment is inside the cell, it can become incorporated into the bacterial chromosome by what amounts to double crossing over (Figure

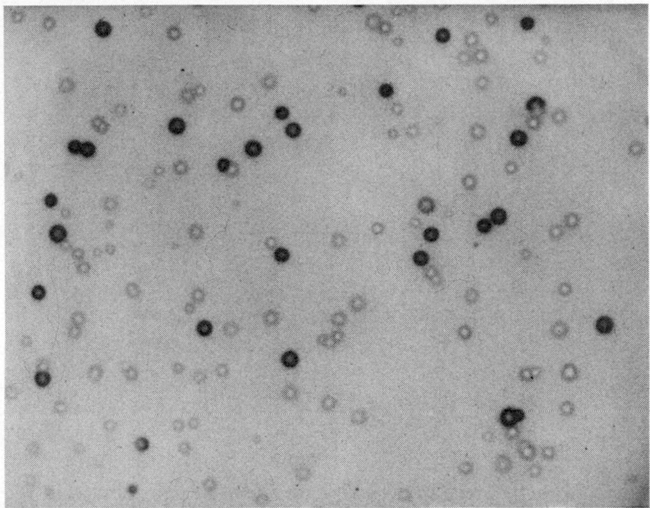

12.10 Viruses are counted by being serially diluted and spread over plates that have previously been prepared with a "lawn" of bacteria. After a sufficient period of incubation, plates such as the one shown here are examined for the presence of plaques (clear circles) in the lawn. Then, by counting the plaques and multiplying back through the dilution factors, the number of viruses per milliliter of liquid can be determined.

12.11). The fragment aligns with the corresponding section of the host's chromosome, and the two sections are exchanged. The leftover segment is then simply dismantled by host enzymes. The new addition is passed along to the cell's descendants when the cell divides.

12.11 In the transformation process, bacteria take in fragments of DNA from other bacteria. As the illustration shows, once inside, the fragment aligns itself along the opposing segment of the circular DNA and becomes incorporated through double crossing over. All the bacterium's descendants will then carry the new recombinant chromosome.

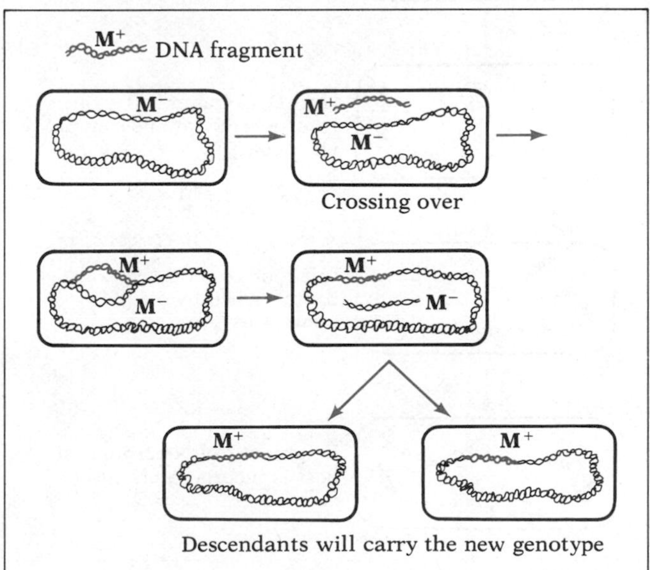
M+ DNA fragment
M−
M+
M−
Crossing over
M+
M−
M+
M−
M+
M+
Descendants will carry the new genotype

Transduction

Transformation is a natural process, but *transduction* is usually done in the laboratory. It involves the transfer of genetic material from one bacterium to another using a bacteriophage as the carrier. What happens is this: Sometimes, when the phage is inside the bacterium, the coat protein envelopes have already been made, and they are beginning to fill with DNA, there may still be fragments of the host's DNA floating about. Some of this bacterial DNA may be taken up in place of the viral DNA. The developing virus coat doesn't know the difference! After lysis, the viral particle with the bacterial DNA attacks a new host, injecting the piece of foreign bacterial DNA. Then recombination can take place. The protein body of the virus has been sexually used to get bacterial DNA from one bacterial strain into another (Figure 12.12).

12.12 Transduction is a sort of DNA hitchhiking transfer. During phage lysing, a segment of the host's fragmented DNA is incorporated into the viral coat. When that virus infects a new host, the segment of bacterial DNA that it carries is then added to the new host's DNA through recombination.

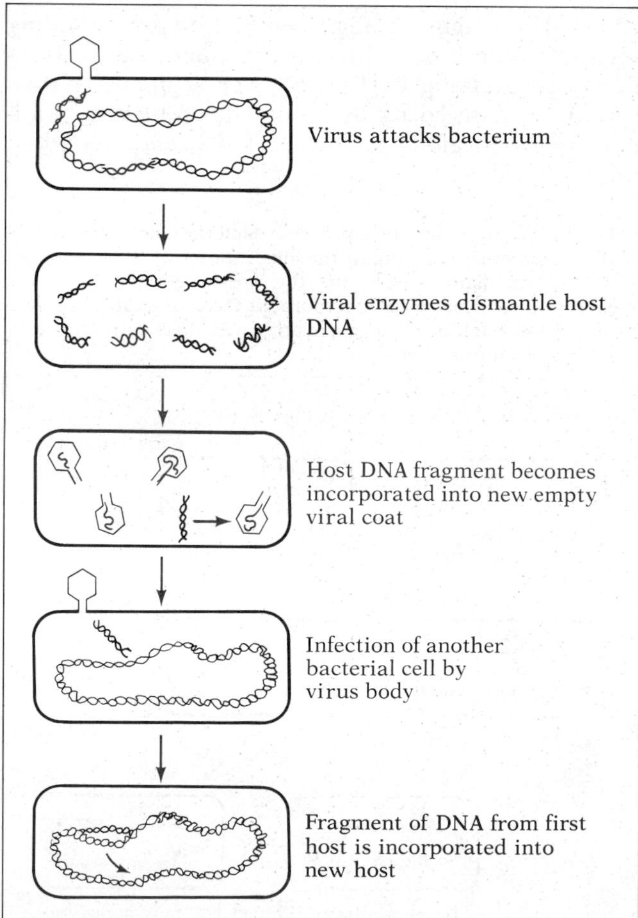
Virus attacks bacterium
Viral enzymes dismantle host DNA
Host DNA fragment becomes incorporated into new empty viral coat
Infection of another bacterial cell by virus body
Fragment of DNA from first host is incorporated into new host

Plasmid Transfer

Many bacteria may contain, in addition to their principal chromosome, one or more very small circular chromosomes, which are called *plasmids.* The plasmids generally replicate in synchrony with the main chromosome, and are passed on to progeny of the cell at the time of cell division. The important point about plasmids is that they have genes of their own.

Some plasmids have genes that code for the proteins that make up a special structure called the *pilus* (plural: *pili*; from the Greek word for hair). These plasmids provide an additional way of reproducing. If a plasmid-containing (*plasmid-plus*) bacterium and a plasmid-free (*plasmid-minus*) bacterium come within close proximity of one another, a pilus is formed: A hollow tube extends from the plasmid-plus bacterium to the plasmid-minus bacterium. The plasmid DNA undergoes an extra round of replication, and as it does so, one of the new copies opens up and is transferred (as a linear DNA molecule) through the pilus into the plasmid-minus bacterium. Once inside its new host, the DNA becomes circular again. Other plasmid genes now produce a cell-surface substance that prevents further plasmic invasions. Both cells are now plasmid-plus (Figure 12.13).

Some plasmids transfer themselves quite easily from one species of bacteria to another. This is unfortunate for us, because these plasmids carry a repertoire of genes that make the bacteria resistant simultaneously to many different antibiotics. This is one reason why you shouldn't take antibiotics any more frequently than necessary.

F^+ Episomes

Sexual behavior in bacteria is often narrowly defined to include only the *F^+ episome* category, because this was the first and most extensively studied mechanism of gene exchange between bacteria and has been historically important in our understanding of genetic organization in procaryotes. This is the type of recombination originally found by Joshua Lederberg and Edward Tatum. So why are we listing it fourth instead of first? Because, when all is said and done, the F^+ episome is just another plasmid, and F^+ sexual transfer in bacteria is in fact an aberration of plasmid transfer. But let's begin with a bit of history that shows why this one plasmid is so important to geneticists.

In 1946, Lederberg and Tatum performed a rather imaginative experiment: They simply mixed two nutritionally deficient strains of colon bacteria together to see if they could get any wild-type progeny. It was an unusual experiment because until then *everyone was convinced that bacteria were confirmed celibates.*

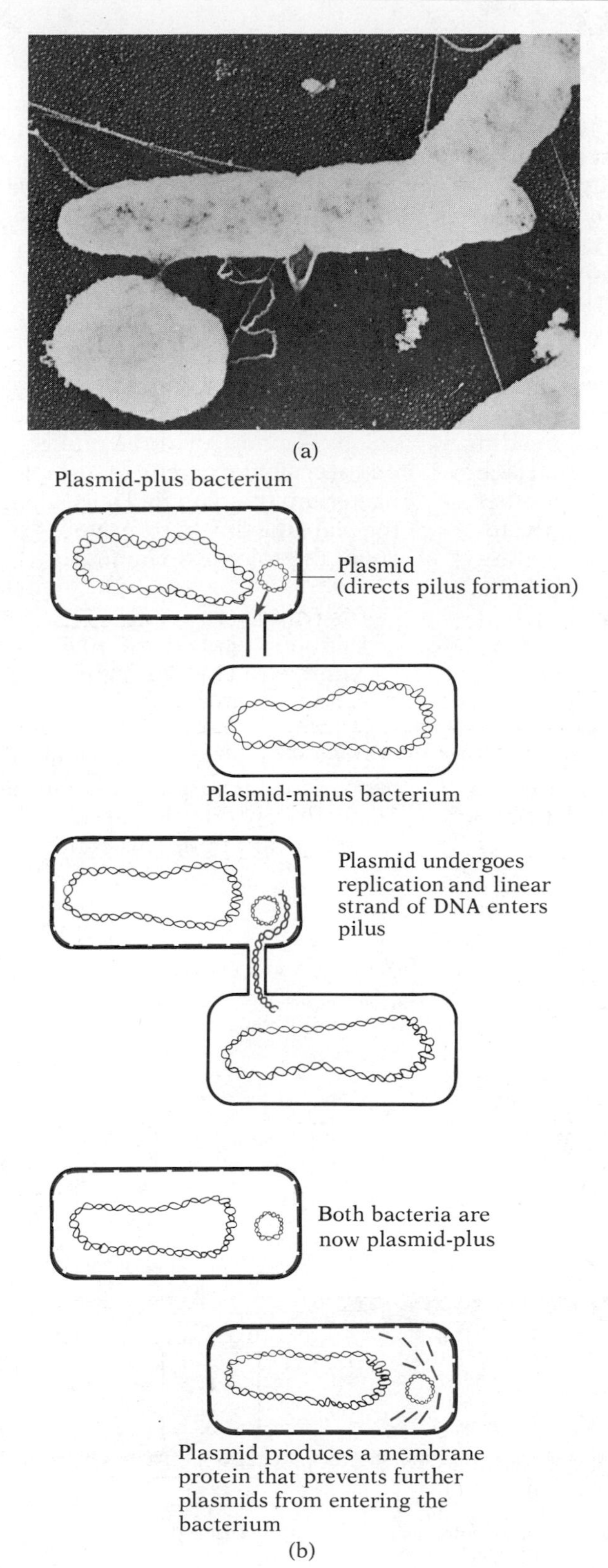
(a)
Plasmid-plus bacterium
Plasmid
(directs pilus formation)
Plasmid-minus bacterium
Plasmid undergoes replication and linear strand of DNA enters pilus
Both bacteria are now plasmid-plus
Plasmid produces a membrane protein that prevents further plasmids from entering the bacterium
(b)

12.13 The plasmid and bacterial recombination. (a) The slender tube between the two bacteria of different strains in this electron micrograph is known as a *pilus.* Its proteins are produced in a *plasmid-plus* bacterium by genes on the plasmid chromosome. The pilus is a plasmid's way of moving its genes from one bacterium to another. (b) Once the pilus forms, the plasmid undergoes a round of replication. The new plasmid chromosome passes through the pilus into the plasmid-minus bacterium. The arrival of a plasmid changes the recipient into a plasmid-plus bacterium, which is capable of repeating the process.

The two strains were:

Strain A: **met$^-$, bio$^-$, thr$^+$, leu$^+$, thi$^+$**

Strain B: **met$^+$, bio$^+$, thr$^-$, leu$^-$, thi$^-$**

where the minus means that the nutrient is required and the plus means that it is not, and the nutrients are methionine, biotin, threonine, leucine and thiamine, respectively. Neither strain could grow on minimal medium.

It turned out that about one cell in 10 million grew perfectly well on minimal medium, and therefore had the following *recombinant* genotype:

Recombinants: **met$^+$, bio$^+$, thr$^+$, leu$^+$, thi$^+$**

This was not the kind of experiment one could do with fruit flies, because it is too much work to count tens or hundreds of millions of insects. But it's easy to count millions or billions or trillions of bacteria using the serial dilution technique.

Lederberg and Tatum next tried to work out linkage relationships. They had a terrible time! Genes that appeared to be closely linked in one experiment were evidently distantly linked in another. After discarding any number of ingenious (but erroneous) hypotheses, they eventually concluded that the bacterial chromosome was circular. Imagine the problems anyone would have in trying to work out linkages with circular chromosomes!

Other investigators picked up Lederberg's and Tatum's techniques, and as new data came in, a strange and confusing picture began to emerge, piece by piece. For example, it was discovered that bacterial strains could be divided into two groups, which were promptly (perhaps too promptly) dubbed "male" and "female." Males mated only with females, and mixtures of two male strains or two female strains never produced recombinant progeny.

What distinguished male from female was that gene transfer was always in one direction, from the male to the female. But the surprising thing was that the male usually transferred only a few genes at a time. This is a very different situation from that of the truly sexual eucaryotes, where the zygote has an equal input of genes from both parents.

Essay 12.1 Counting Bacteria

It is often necessary to count bacteria. You might want to know, for instance, how many bacteria are present in a 1 ml sample. Let us say that in fact there are on the order of a billion (10^9), a rather large number—but the actual number is not known. How can you count a billion bacteria? The procedure is not as difficult as it sounds, if you are willing to accept a close estimate.

First, you set up a rack of perhaps a dozen test tubes all in a row. Each tube is numbered and you put 0.9 ml of sterile water into each one. You also prepare a dozen sterile pipettes, each able to hold 0.1 ml, and a dozen Petri dishes of nutrient agar. To count the bacteria in your sample, take a 0.1 ml aliquot from it and blow it out into the first test tube, using your first pipette. Stir in the *inoculum* (bacteria sample), and discard the pipette.

If the sample has 10^9 cells per milliliter, then the first test tube has 10^8 cells. You really don't know yet what the concentration is, but you do know that the concentration in the first test tube is 0.1 times that in the original sample (because you used one-tenth of the sample).

Now take the second pipette and transfer 0.1 ml of the new mixture to a Petri dish, spreading the fluid evenly over the surface of the agar gel. Transfer another 0.1 ml from the first test tube to the second test tube, stir in the mixture, and discard the pipette. Repeat the process until each of your twelve test tubes has been inoculated and all of your Petri dishes have been spread. The process is called *serial dilution*—each of the tubes, and each of the dishes, has one-tenth the concentration of bacteria of the previous one.

Now put the Petri dishes in an incubator overnight. When you check them, you find that the first five dishes are solid with bacteria. Throw them out. Dishes 9, 10, 11, and 12 are sterile—no bacteria at all; throw them out, too. That leaves dishes 6, 7, and 8. On the gel surface of each of these are small but easily visible *colonies,* little round spots of bacteria. Each colony is made up of the numerous overnight progeny of a single bacterium that you had spread on the plate the day before. Now count the colonies. You find:

Dish number	6	7	8
Dilution factor	10^{-6}	10^{-7}	10^{-8}
Number of colonies counted	73	6	2

The best count is in dish number 6. The other counts are consistent with it, within the limits of sampling variation. This count

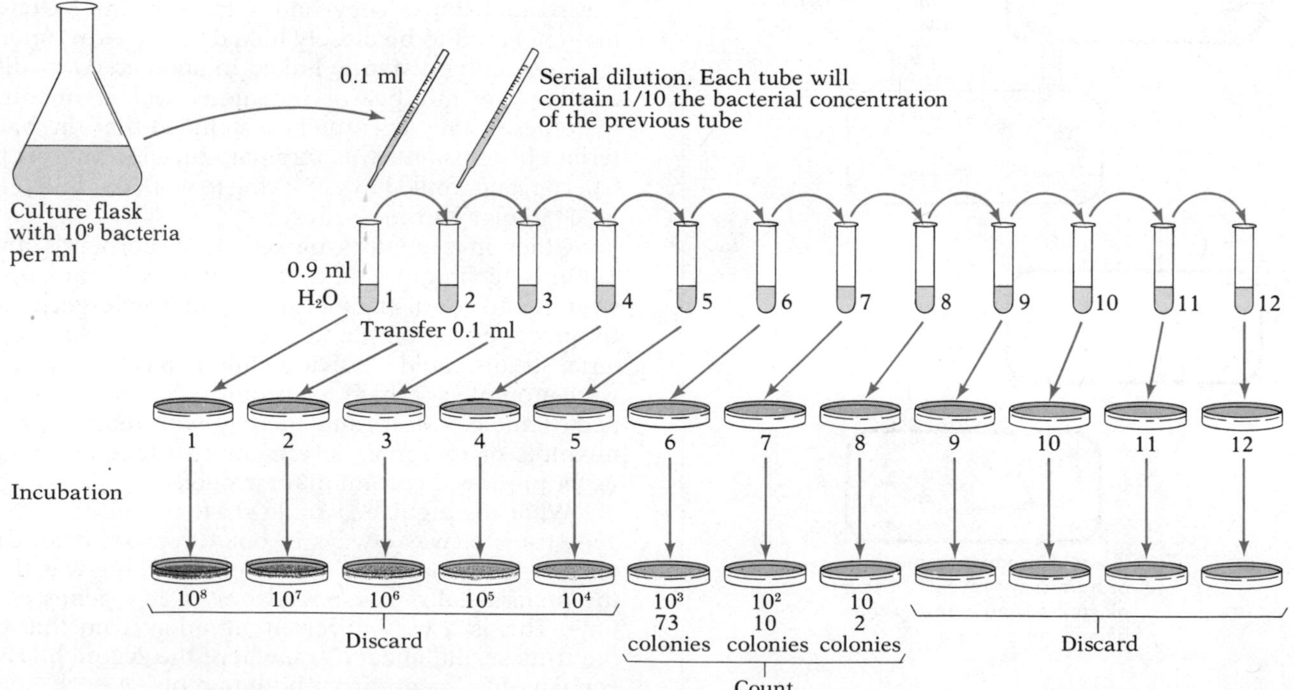

shows that there were 73 platable bacteria in a 0.1 ml subsample, or a concentration of approximately 730 bacteria per milliliter in tube 6. The dilution factor was 10^{-6}, which means that the concentration must be multiplied by 10^6 in order to find the concentration of the original sample: $730 \times 10^6 =$ 730,000,000 bacteria per milliliter in your original sample! This is usually written in the form 7.3×10^8. And remember that while this is an *approximation,* it is a pretty good one.

An Example of How Bacterial Genetics Is Done

How was it discovered that usually only a few genes from the male are transferred to the female? Let's look at a typical cross between two strains, as it was done in the early days. The fact that a combination of strains with different nutritional requirements can give rise to completely wild-type progeny doesn't tell us much, other than that recombination does take place. To get to the root of the matter, it was necessary to look at the transfer of *unselected* loci; that is, the inheritance of nutritional mutants under conditions where either the plus or the minus allele could survive. Such unselected genes were called *markers,* because they were simply used to mark places (loci) on the chromosomes in the linkage studies and did not affect whether or not the bacterium would be recovered. To be sure that recombination is taking place, a medium supplemented with only some of the nutrients is used. Then the recombinant progeny are looked at more closely, with replica plating, on other kinds of partially supplemented agar plates.

For example, suppose the original strains,

Strain A: **met⁻, bio⁻, thr⁺, leu⁺, thi⁺**

Strain B: **met⁺, bio⁺, thr⁻, leu⁻, thi⁻**

are mixed and plated on a medium supplemented with biotin, leucine, and thiamine and lacking methionine and threonine. Then all the surviving progeny have to be **met⁺, thr⁺**, indicating that a mating has taken place. But what are their genotypes with regard to the three genetic markers? There are actually eight different possible genotypes among the recombinant colonies (Table 12.1). To diagnose the genotypes and to count the number of colonies of each type, three replicate plates are made. Each replicate plate is supplemented with two nutrients and lacks a third. Linkage relationships can be worked out from the relative numbers of colonies of each of the eight types. For instance, if **leu⁺, thi⁺**, and **leu⁻, thi⁻** combinations were common, while **leu⁺, thi⁻** and **leu⁻, thi⁺** were both rare, one could conclude that the leucine locus and the thiamine locus were close together on the chromosome.

The results of experiments such as these also showed that the males contributed relatively little to the recombinant progeny. For instance, if strain A were male and strain B were female, most of the recombinant progeny in the above experiment would be **met⁺, bio⁺, thr⁺, leu⁻, thi⁻**—just like their strain B mothers for all unselected marker loci. (It doesn't help to know the numbers for *selected* loci, because all the progeny must be **met⁺** and **thr⁺**.)

Table 12.1 Eight Possible Genotypes among the Recombinant Colonies

Genotype	Growth on replicate plates[a]		
(all **met⁺, thr⁺**)	Lacking only biotin	Lacking only leucine	Lacking only thiamine
bio⁺, leu⁺, thi⁺	+	+	+
bio⁺, leu⁺, thi⁻	+	+	−
bio⁺, leu⁻, thi⁺	+	−	+
bio⁺, leu⁻, thi⁻	+	−	−
bio⁻, leu⁺, thi⁺	−	+	+
bio⁻, leu⁺, thi⁻	−	+	−
bio⁻, leu⁻, thi⁺	−	−	+
bio⁻, leu⁻, thi⁻	−	−	−

[a] + grows on medium; − does not grow on medium

What recombinant genotype would be the most common if strain A were female and strain B were male? In line with the above reasoning, the most common of the **met⁺, thr⁺** recombinants would be **bio⁻, leu⁺**, and **thi⁺**—just like their strain A mothers.

In the early 1950s, William Hayes came up with some unnerving experiments. In separate experiments, he treated either the male or the female strain with streptomycin before mixing them together. Streptomycin prevents cell division, but not bacterial conjugation. The streptomycin-treated females produced no viable progeny because they couldn't divide, but the streptomycin-treated males could and did leave progeny of their own by transferring some of their genes into the receptive, untreated females. In fact, this bizarre experiment was what was originally used to distinguish male from female strains.

Then Hayes showed that the "maleness" of the bacteria was catching! Male and female bacterial strains (with easily distinguishable genotypes) were mixed together and plated. The resulting colonies were then tested for maleness or femaleness. It turned out that the male bacteria were all still male, but now about a third of the bacteria of the previously all female strain were now male! Hayes thought that was a curious kind of sex. Later, it was discovered that maleness is due to a plasmid, and this plasmid is readily transferred from a male (plasmid-plus) to a female (plasmid-minus) bacterium through a pilus.

Once the new host receives the sex plasmid, it becomes a male (plasmid-plus).

Now, electron microscopes were not well enough developed in the 1950s for anyone to know about pili, but because both sexual recombination and contagious masculinity obviously required direct physical contact, Hayes had a pretty good idea of what was going on. Something, originally called the *fertility factor,* was apparently transferred from individual to individual. The male strains were renamed F^+, short for *fertility-positive,* and the female strains were renamed F^-, short for *fertility-negative*—although in fact both types were necessary for successful fertility (an obvious case of male chauvinism). Soon the fertility factor itself was identified as a circular piece of DNA and became known as the *F^+ episome.* It wasn't until many years later that it was realized that the F^+ episome was only one of a large class of plasmids.

The F^+ episome is one of many plasmids that ensure their own survival by inducing the formation of pili and transferring copies of themselves to bacteria. Like other plasmids, it also causes its host to produce a substance that makes it immune to infection from other F^+ episomes. The "contagiousness" of the F^+ condition, then, is just like any other parasitic contagion.

While the F^+ episome is normally a closed circle of DNA, there is one place on the circle that is capable of opening up, so that the episome temporarily becomes a linear structure. When this happens, as with other plasmids, there is a special round of DNA replication, during which one of the DNA copies crawls through the pilus into the F^- "hostess" bacterium, while the other copy stays behind. Once safely through the tunnel and into the F^- bacterium, the linear F^+ episome becomes a circle again, and the new host becomes infected with masculinity. It, too, is now able to form a pilus.

What about the transfer of bacterial genes? After all, that's what makes the F^+ episome so special in the first place. It turns out that while simple transfer is very common, in rare cases the F^+ episome may become inserted into the chromosome of its host. The plasmid has no preferred site of insertion, and can show up anywhere in the host's circular chromosome. After insertion, the F^+ DNA is replicated right along with the host DNA and can be passed on to all the bacterial offspring. So far, this sounds like phage lysogeny. But later, when the plasmid forms a pilus and attempts the transfer to an F^- bacteria, strange things happen. First, the F^+ episome opens up, as usual. But since it is now part of a larger circle, this converts the entire structure—host chromosome and all—into a linear DNA molecule. Then, when one end of the F^+ episome goes through the pilus, it starts to pull the entire host chromosome in after it. Because the chromosome is very long, it takes about 89 minutes for the whole thing to get through. In fact, the pilus usually breaks apart before transfer is completed. Since the recipient cell usually gets only a piece of the F^+ sequence, it isn't transformed into an F^+ bacterium, but the bacterial DNA that is brought in can recombine, by crossing over, with the DNA of the recipient ("female") bacterium (Figure 12.14).

It wasn't long before Hayes and others were able to isolate clones of individuals in which the F^+ episome had been integrated into the host chromosome. Then they didn't have to depend on rare, random insertions. Every bacterium in such a clone had the episome integrated in exactly the same place. These strains showed a potential for recombination that was 1000 times greater than that of the ordinary F^+ strains, because in these strains every bacterium had an integrated F^+ episome, while integration is a rare event

12.14 Transferring the integrated episome. In the rare cases where the F^+ episome joins the entire bacterial chromosome, the sex act changes. When the newly converted F^+ bacterium mates with an F^- type, the entire chromosome replicates, episome and all. The episome start through the pilus, dragging the lengthy bacterial chromosome behind it. Quite often, the pilus breaks before the entire transition is completed. Any portions that do make the transfer may be incorporated into the F^- host through crossing over and recombination. When this happens, the F^- remains an F^-.

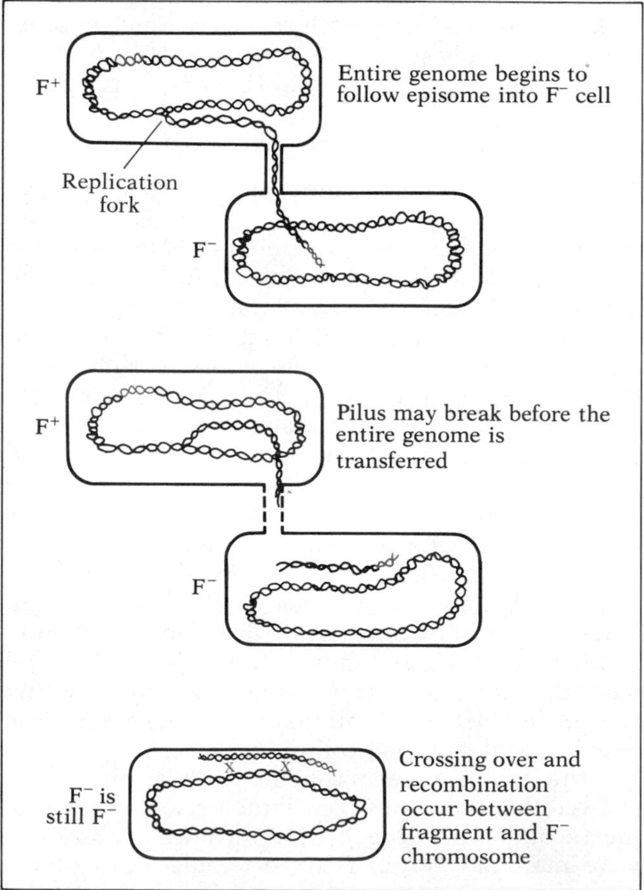
F+
Entire genome begins to follow episome into F− cell
Replication fork
F−
F+
Pilus may break before the entire genome is transferred
F−
F− is still F−
Crossing over and recombination occur between fragment and F− chromosome

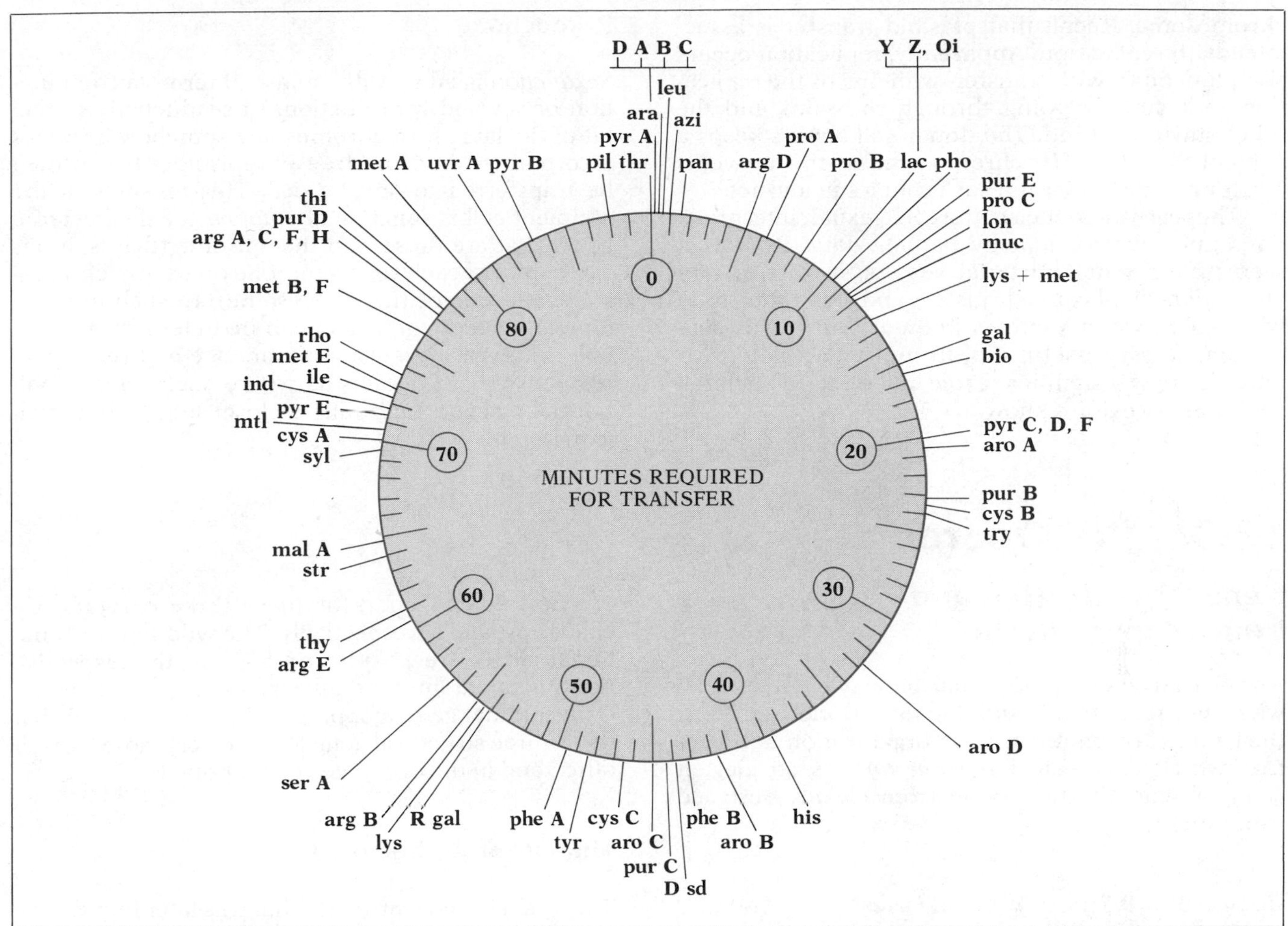

12.15 This genetic map of the circular chromosome of *Escherichia coli*, produced by recombinant studies of interrupted Hfr chromosomes, is divided into 89 parts, corresponding to the 89 min required for a complete chromosome transfer. Each gene locus is identified by code letters that represent the ability or inability to synthesize various substances. Note that many of the codes represent amino acids using the standard abbreviations.

in other F^+ strains. The new clones were called Hfr, for *high-frequency of recombination.*

The Great Kitchen Blender Experiment

Progress in molecular biology often follows the development of new technology. In this case, the technology was provided by Fred Waring, a popular band leader of the 1950s. When he was not leading his group, The Pennsylvanians, in song, he was busy inventing the Waring blender. The blender, of course, whips food into a mush, and in so doing, it must disrupt cells. The molecular biologists happened to need a good way to disrupt cells and confiscated the blenders from their kitchens.

In one of the first experiments using a blender, François Jacob and Elie Wollman broke up mating bacteria. They let Hfr and F^- bacteria mate for various lengths of time before pushing the button, interrupting the coitus.

As we've mentioned, it takes about 89 min for the Hfr chromosome to transfer to the F^- bacterium. If the transfer is interrupted after just a few minutes, some of the genetic markers would make it across while others remain behind. As the periods of mating were extended, more and more markers were transferred and could be recovered in the progeny, although at lower efficiencies because of spontaneous breakups of the bacterial couples. The last genetic markers come through, but with extremely low efficiency, toward the end of the 89 min.

Jacob and Wollman found that they could rapidly map Hfr bacterial chromosomes by using their blender. They found that on a standard Hfr chromosome the first marker passed through in about 2 min, and each subsequent marker was given a position appropriate to the time it entered the recipient cell. Figure 12.15 shows the genetic map of *Escherichia coli,* which is always given in minutes.

One question that might have occurred to you is whether the donor Hfr cell survives the transfer of its chromosome. Recall that plasmid transfer is associated with replication. Apparently, replication occurs simultaneously with transfer, with one of the replication fork strands going through the pilus and the other staying behind. The donor cell always keeps a copy of the entire Hfr chromosome, so the answer is No, for a male bacterium sex is not a suicidal act.

The early geneticists forced sexual interpretations upon their originally baffling data. By now it seems clear that "bacterial sex" is, in fact, a rare aberration of plasmid transfer, and has little to do with sex as we eukaryotes know it. But it was, and remains, a very useful tool in molecular biology. It may also play a significant role in long-term bacterial evolution—we don't know.

Sexduction

Sexduction is admittedly an awful term—a combination of sex and transduction. In sexduction, a small bit of the large host chromosome somehow becomes incorporated into the free F^+ episome. It can then be transferred to any F^- cell. This means that the recipient cell is functionally *diploid* for the bacterial genes that are present in the episome (that is, it has two copies of each such gene, one on its own chromosome and one of the F^+ episome). In such partially diploid bacteria, gene loci can be heterozygous, and one can even investigate dominance and recessivity relationships in organisms where such things don't normally occur. Let's take a closer look at one such experiment.

12.3 The Operon

Gene Organization and Control in Bacteria

A major advance in molecular biology came in 1961, when François Jacob and Jacques Monod unveiled their model of bacterial gene organization and control, which they called the *operon*. It's an elegant story of scientific deduction from cleanly designed experiments.

Background: Inducible Enzymes in Escherichia coli

Some of *E. coli*'s many enzymes are produced all the time. Microbiologists call the genes for these proteins *housekeeping genes*, perhaps because housework is something that has to be done continually. But the synthesis of other enzymes is under some kind of control, and the tiny bacterium can make them when it needs them and stop production when it doesn't.

For instance, if *E. coli* is grown on glucose, it doesn't need to make its own, so it doesn't synthesize enzymes that help it form glucose from other sugars. Specifically, it doesn't make the enzymes that metabolize lactose (milk sugar). If these same bacteria are put into a medium containing lactose, but not glucose, they will soon begin to produce lactose-metabolizing enzymes. These lactose-metabolizing enzymes are said to be *inducible*, because they can be induced by an appropriate substrate.

There are three enzymes that are used in the metabolism of lactose and other closely related sugars. These are *β-galactosidase, galactoside permease,* and *thiogalactoside transacetylase.* The structural genes, or *cistrons,* that code for these three enzymes are called **z**, **y**, and **a**, respectively. The wild-type or functional alleles are $\mathbf{z}^+$, $\mathbf{y}^+$, and $\mathbf{a}^+$, and the respective nonfunctional mutant alleles are $\mathbf{z}^-$, $\mathbf{y}^-$, and $\mathbf{a}^-$. Conventional gene mapping techniques showed that these three structural gene loci are very close to each other, and in the order we've mentioned.

Mutants of the Lactose Operon

The $\mathbf{z}^-$ allele had a property that was later found to be the key to the experimental work. Although $\mathbf{z}^-$ did not make a functional enzyme, it did make a protein. The mutation fouled up the sequence of the protein so that it had no enzymatic activity, but it could still be identified by antibody-binding analysis. The nonfunctional protein produced by the $\mathbf{z}^-$ allele was termed CRM, for *cross-reacting material* (because it reacted with rabbit antibodies induced by the functional enzyme).

In the wild-type bacteria, Jacob and Monod noted that all three enzymes appeared together or disappeared together in response to lactose. They hypothesized, and later proved, that the three cistrons were under a single control system. They were interested in gene control more than in carbohydrate metabolism, and began to search for mutants in the control system. They found two general types of gene-control mutants:

Constitutive mutants continuously manufactured great quantities of all three enzymes, whether lactose was present in the medium or not. If mutant gene control can be likened to a defective light switch, in constitutive mutants the control is permanently *broken on.*

Noninducible control mutants were just the opposite. Noninducible mutant bacteria ignored the inducer molecule lactose, and refused to make any of the three lactose-metabolizing enzymes under any circumstance. To follow the defective light switch analogy, these mutants permanently were *broken off.*

The control mutants could be mapped, too. Jacob and Monod collected many different constitutive mutants and many different noninducible mutants, and proceeded to map them all. Here they found something new: (1) Each category of control mutant was composed of two groups; each group mapped at a different site. For instance, one class of constitutive mutants mapped just to the left of the **z** cistron, and the other class of constitutive mutants mapped at another, fairly distant gene locus. (2) The constitutive and noninducible mutants mapped at the *same* two gene loci; that is, they were allelic, either at the locus close to **z** or the other distant locus.

Thus, there were three alleles at each control gene locus: wild-type, constitutive, and noninducible. The locus that was just to the left of **z** was named **o** (short for *operator*), and the other gene locus was named **i** (for *inducer*). The wild-type alleles were symbolized by $\mathbf{o}^+$ and $\mathbf{i}^+$, respectively. The constitutive mutant alleles were $\mathbf{o}^c$ and $\mathbf{i}^c$, and the noninducible mutants were $\mathbf{o}^n$ and $\mathbf{i}^n$. (This terminology is only slightly different from that used by Jacob and Monod in 1961.)

Jacob's and Monod's Experiment

Here is what they had learned so far: The three enzyme cistrons were genetically linked and were under a single control system. The control system had two elements. One element of the control system was immediately to the left of the structural genes, and the other was somewhere else on the bacterial chromosome. Mutants of either gene locus could occur in broken off or broken on states.

But just how did control work? In some way, the wild-type control system had to be sensitive to the presence of inducer molecules (lactose) in the medium or in the cytoplasm of the cell.

Jacob and Monod used the *dominance relationships* of the different mutant alleles to unravel their functions. Now, as you know, dominance relationships are supposed to be limited to diploid organisms, and *E. coli* is haploid. They got around this difficulty by using a technical trick that we've already mentioned: *sexduction.* They used a bizarre laboratory curiosity that has had an important place in the history of molecular genetics: the *F⁺ lac episome.* This F⁺ episome had somehow incorporated the entire segment of *E. coli* chromosome that contained the lactose-responsive genes **i, o, z, y,** and **a** (the two control loci and the three enzyme loci). Jacob and Monod could collect mutants of any of these gene loci on either the F⁺ lac episome or on the bacterial host chromosome. By allowing the episome to transfer to an F⁻ strain, they could produce any combination of alleles in a *partial diploid.*

Table 12.2 Six Possible Combinations—the o and z Loci

Episome	Host	β-Galactosidase phenotype ($\mathbf{z}^+$)	CRM phenotype ($\mathbf{z}^-$)
$\mathbf{o}^c, \mathbf{z}^+$	$\mathbf{o}^+, \mathbf{z}^-$	Constitutive	Normally inducible
$\mathbf{o}^+, \mathbf{z}^+$	$\mathbf{o}^c, \mathbf{z}^-$	Normally inducible	Constitutive
$\mathbf{o}^n, \mathbf{z}^+$	$\mathbf{o}^+, \mathbf{z}^-$	Noninducible	Normally inducible
$\mathbf{o}^+, \mathbf{z}^+$	$\mathbf{o}^n, \mathbf{z}^-$	Normally inducible	Noninducible
$\mathbf{o}^c, \mathbf{z}^+$	$\mathbf{o}^n, \mathbf{z}^-$	Constitutive	Noninducible
$\mathbf{o}^n, \mathbf{z}^+$	$\mathbf{o}^c, \mathbf{z}^-$	Noninducible	Constitutive

Dominance Relationships at the Operator Locus

In considering their crosses, we can ignore the **y** and **a** loci, since they behave exactly the same as the **z** locus. Also, it was shown that the alleles in question operated perfectly well regardless of whether they were on the episomal chromosome or on the host chromosome, so we won't have to look at completely reciprocal crosses. Jacob and Monod combined the three different operator (**o**) alleles with the two **z** alleles. Table 12.2 shows what they found, giving all six possible combinations. Although it didn't matter whether the alleles in question were on the host chromosome or on the episomal chromosome, it did matter whether the alleles that were interacting were on the same chromosome. Looking over these results, it is clear that the operator locus controls the **z** locus that is on the same piece of DNA, but that there is no interaction between the two different pieces of DNA. The operator locus is said to be *cis-dominant* with respect to the **z** locus.

Dominance Relationships at the i Locus

When the same kinds of experiments were done with the three different alleles of the **i** locus in combinations with the two alleles of the **z** locus, the results were different, as shown in Table 12.3. For interactions between the **i** locus and the **z** locus, it didn't matter whether the interacting loci were on the same piece of DNA or not. And the **i** alleles showed a clear

Table 12.3 Six Possible Combinations—the i and z Loci

Episome	Host	β-Galactosidase phenotype ($\mathbf{z}^+$)	CRM phenotype ($\mathbf{z}^-$)
$\mathbf{i}^c, \mathbf{z}^+$	$\mathbf{i}^+, \mathbf{z}^-$	Normally inducible	Normally inducible
$\mathbf{i}^+, \mathbf{z}^+$	$\mathbf{i}^c, \mathbf{z}^-$	Normally inducible	Normally inducible
$\mathbf{i}^n, \mathbf{z}^+$	$\mathbf{i}^+, \mathbf{z}^-$	Noninducible	Noninducible
$\mathbf{i}^+, \mathbf{z}^+$	$\mathbf{i}^n, \mathbf{z}^-$	Noninducible	Noninducible
$\mathbf{i}^n, \mathbf{z}^+$	$\mathbf{i}^c, \mathbf{z}^-$	Noninducible	Noninducible
$\mathbf{i}^c, \mathbf{z}^+$	$\mathbf{i}^n, \mathbf{z}^-$	Noninducible	Noninducible

dominance series, with $\mathbf{i}^n$ being dominant to $\mathbf{i}^+$, and $\mathbf{i}^+$ being dominant to $\mathbf{i}^c$. When the investigators combined different **i** alleles with different **o** alleles, the results were similar. The **i** alleles showed the same kinds of dominance relationships, and the **o** alleles continued to act only on the structural genes that were adjacent to them on the same chromosome.

The fact that the **i** locus on one chromosome could control the **z** locus on a different chromosome meant, to Jacob and Monod, that there must be some kind of diffusible molecule produced by **i** that could move through the cell to interact with **z.** They interpreted the dominance relationships for the **i** locus in the way in which dominance is usually understood, at least in molecular biology. Normally, a dominant allele is doing something that the recessive allele isn't doing. In the present case, the specific dominance relationships for all three **i** alleles seemed to imply that $\mathbf{i}^n$ was doing something when it shouldn't be (that is, when lactose was absent), and that $\mathbf{i}^c$ was not doing something when it should be (that is, when lactose was present). This meant that $\mathbf{i}^n$ was repressing enzyme synthesis in spite of the presence of lactose, and that $\mathbf{i}^c$ was failing to repress enzyme synthesis in spite of the absence of lactose.

What else can you tell from the crosses? Jacob and Monod thought they could tell a great deal.

Jacob's and Monod's Conclusions: The Operon Model

Here is the model of normal lactose operon function that was consistent with the above results:

1. The **i** locus produces a *repressor* molecule (later shown to be a protein).
2. The repressor molecule diffuses into the cytoplasm. There it may (or may not) encounter an *inducer* molecule—in this case, lactose. The repressor has a specific affinity for the much smaller inducer molecule, and will combine with it to make a *repressor–inducer complex.*
3. The repressor molecule also has an affinity for the *operator* locus, a stretch of DNA right at the beginning of the first of three cistrons. When the repressor molecule is complexed with the operator, mRNA cannot be transcribed and no enzyme is produced.
4. The repressor–inducer complex, on the other hand, has no affinity for the operator locus. There is a finite number of repressor molecules in the cell (later experiments showed that there are about ten repressor molecules per cell). When all the repressor molecules are complexed with lactose, the operator is unblocked and the genes are transcribed and translated. Thus, the presence of the inducer molecule brings about the production of specific enzymes by tying up the usual repression mechanism (Figure 12.16).

If your fancy turns to analogies, RNA polymerase is a train that must run along a track (DNA). The repressor protein is an elephant that has a propensity for sitting on a favorite spot on the track, except when there are peanuts available. Then the elephant lumbers away to eat peanuts. So the availability of peanuts controls whether the train can run or not.

The Control Mutants Explained

Now we can reconsider what was happening with the constitutive and noninducible mutants. The $\mathbf{i}^c$ allele produced no repressor protein, and this allowed the structural alleles to produce in the absence of lactose. The $\mathbf{i}^n$ allele, on the other hand, made a faulty repressor that either had no affinity for lactose, or that continued to bind with the operator in the presence of lactose. The $\mathbf{i}^n$ allele was dominant because it continued to repress even in the presence of normal $\mathbf{i}^+$ gene products.

The $\mathbf{o}^c$ allele was changed in such a way that it no longer bound the repressor protein. This allowed RNA polymerase to proceed. The $\mathbf{o}^n$ allele, on the other hand, blocked RNA polymerase no matter what.

The "Invention" of Messenger RNA

Jacob and Monod called their entire system—control loci together with structural gene loci—an *operon.* As we've seen, their model of the operon involved

12.16 The Jacob–Monod operon model. The production of the three inducible enzymes responsible for the metabolism of lactose is under the control of a system known as an *operon.* The operon consists of three principal parts, as shown in the diagram; these are the regulator gene, the operator gene, and the structural gene region (consisting of several cistrons). In the lac operon, the structural gene region includes cistrons **z**, **y**, and **a**, one for each of the three lactose-metabolizing enzymes.

(a) The production of enzymes is shut down. The repression of the structural genes is accomplished by an interaction between the regulator and operator genes. The regulator gene slowly but constantly produces a protein known as a *repressor,* which binds to the operator gene and renders it incapable of unwinding so that transcription of mRNA can begin.

(b) When lactose enters the cell it immediately acts as an *inducer,* by activating the production of the enzymes that use lactose as a substrate. To accomplish induction, it attracts and ties up the repressor protein. The operator then opens up and transcription of the entire operon region occurs with the eventual production of a single, long, polycistronic mRNA. Even while this mRNA is still being transcribed, ribosomes bind at initiator codons and begin translating the three enzymes. This process continues as long as lactose is available. Once all the lactose molecules have been metabolized, induction ceases. The regulator gene, methodically producing repressor protein, again succeeds in coating the operator gene. The structural genes become quiescent, no longer able to transcribe mRNA. The operon has shut down and will remain inactive until more lactose is present.

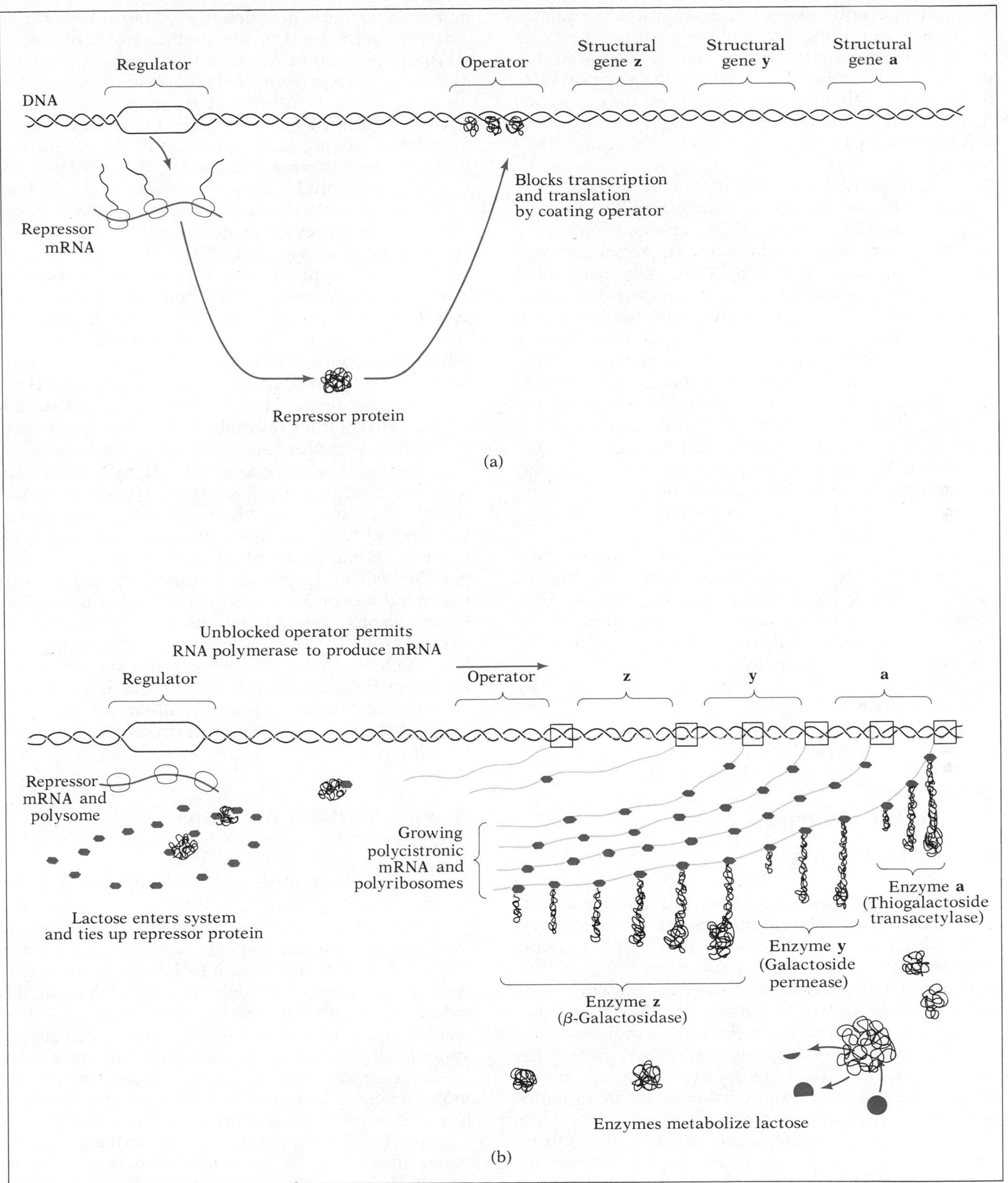
Regulator
Operator
Structural gene **z**
Structural gene **y**
Structural gene **a**
DNA
Blocks transcription and translation by coating operator
Repressor mRNA
Repressor protein
(a)
Unblocked operator permits RNA polymerase to produce mRNA
Regulator
Operator
z
y
a
Repressor mRNA and polysome
Growing polycistronic mRNA and polyribosomes
Lactose enters system and ties up repressor protein
Enzyme **a** (Thiogalactoside transacetylase)
Enzyme **y** (Galactoside permease)
Enzyme **z** (β-Galactosidase)
Enzymes metabolize lactose
(b)

blocking and unblocking of certain mRNA synthesis. There was one difficulty, however. *No one had ever heard of mRNA,* let alone isolated or even detected it. The French microbiologists had to hypothesize it, in order for their model to work. So they dreamed it up and then named it. Now, of course, we know that mRNA really exists and that it can be purified and measured.

Why was mRNA necessary for their scheme? It was because of the other cistrons in the operon. The operator locus, one might suppose, could control the expression of the **z** locus, because it was right on top of it. But how, short of magic, could it control the nonadjacent **y** and **a** cistrons? Jacob and Monod had to assume that only one gene product was made under the control of the operator locus, and that this one gene product then made three different enzymes. The gene product was mRNA. (In all fairness, we have to mention that other investigators had indicated there might be an RNA intermediate between DNA action and protein synthesis. Jacob and Monod's own experiments involved kinetics studies—the time relationships of induction, gene transfer, and protein synthesis—that also seemed to require some kind of intermediate.)

And this is often how science is done. In pursuit of one objective—here, the understanding of gene control—a consistent model is proposed. But in order to make the model work, certain things have to be hypothesized. If the model is correct (and the operon model certainly is), the invented "saving" hypothesis may turn out to be a fundamental intellectual discovery.

Repressible Enzymes and Other Operons

Repressible Enzymes

Many other operons have since been discovered and have been analyzed in detail. Some of them work almost exactly the same way the lactose operon does, but there are also many variations on that basic model. Most bacterial operons, however, are polycistronic; that is, there are usually several different cistrons on one bacterial mRNA molecule.

The lactose-metabolizing enzymes, you will recall, are *inducible.* They are not present in cells grown on glucose, but they can be induced by small molecules in the medium. In contrast, other enzymes are *repressible,* which is more or less the opposite of inducible. The synthesis of some proteins can be repressed by certain substances in the medium. For instance, bacteria that are grown on glucose and minerals regularly produce *tryptophan synthetase,* an enzyme that is used in the synthesis of tryptophan. Tryptophan is an amino acid that the cell must have if it is to make protein. If tryptophan itself is added to the medium, synthesis of tryptophan synthetase enzymes abruptly ceases (Figure 12.17). In this case, then, the end product of the synthetic pathway is a repressor of enzyme synthesis.

The tryptophan operon is similar to the lactose operon in all respects except one. In the tryptophan operon, the repressor protein itself has no affinity for the operator sequence in the absence of tryptophan; but the tryptophan–repressor protein complex does have an affinity for the operator sequence, and can block RNA polymerase. This is just the opposite of the situation with the lactose operon, which is why repressible operons have the opposite pattern of gene control from that of inducible operons. Here we have a different kind of elephant, one that sits on the railroad track only when it has a peanut to eat!

Both the tryptophan operon and the lactose operon are under *negative control*—and both involve the selective blocking of mRNA transcription by something in the cytoplasm, whether it is a repressor protein or a repressor protein–repressor molecule complex. Some bacterial operons, however, employ *positive control.* In this case something in the cytoplasm induces or increases mRNA transcription. And some operons use combinations of positive and negative controls. In addition, some operator sequences have multiple binding sites for repressor molecules that allow for graduated responses; that is, the switch doesn't necessarily have to be either "off" or "on," because the control system has (in effect) a dimmer switch.

Operons in Higher Organisms: Are There Any?

As soon as Jacob and Monod's revolutionary findings hit the scientific journals, other biologists began eagerly to look for operons in higher organisms. At first, everyone optimistically assumed that such a tidy control system must control genes in all organisms. Expectations ran high. This is it, thought the optimists. Alas, the eucaryotes didn't cooperate—they yielded little data suggesting that they shared such a system. All evidence now indicates that eucaryotes never have more than one cistron on any one mRNA molecule. So if *operon* is defined as a control system in which several cistrons are under the coordinated control of a single operator, the provisional answer to our question is "No, there are no operons in higher organisms."

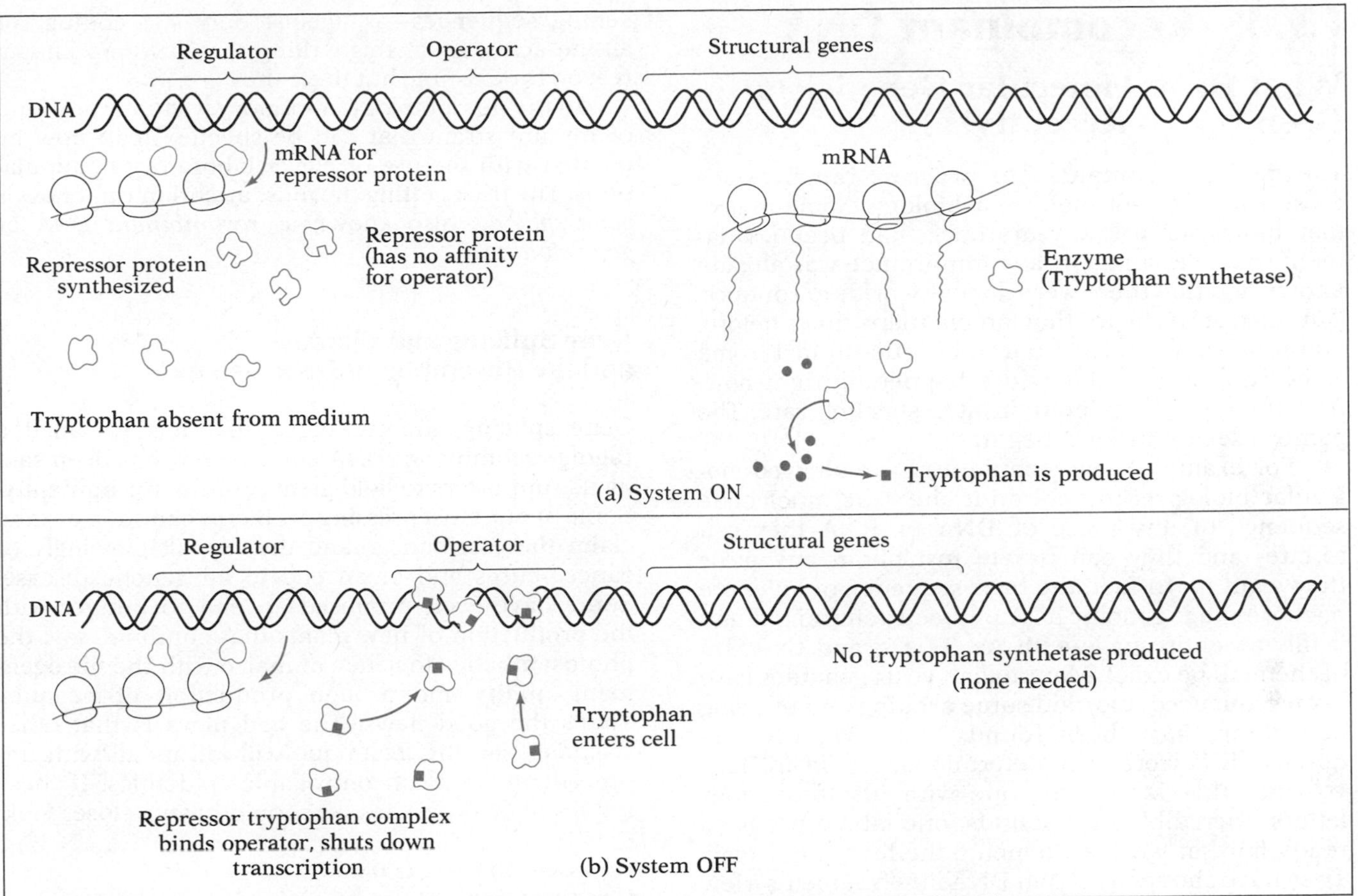

12.17 (a) The repressible operon works in the opposite way. In this system, cells grown continuously in glucose continually produce the enzyme tryptophan synthetase, which is essential in producing the amino acid tryptophan. Although the regulator gene produces the repressor protein, it cannot bind the operator gene in its present form. Therefore, the system remains turned on. (b) If the cells are fed the amino acid tryptophan, the system immediately shuts down. Tryptophan joins the repressor protein to form a tryptophan–repressor complex. This complex, in turn, binds to and inhibits the operator. This shuts down the structural genes that produce tryptophan synthetase. As you can see, this is a conserving process. Why produce the enzyme when its product is abundant?

For a long time there was no evidence for cytoplasmic controlling proteins either, and the initial bright optimism gradually faded to gloom. Part of the problem was simply that the enormous complexity of eucaryote cells made controlled experimentation very difficult.

In recent years, experimenters have found some control systems in higher organisms that are reminiscent of gene control in bacteria. For example, steroid hormones can control gene activity rather directly. The oviducts of baby chicks have been found to be responsive to estrogen, the female hormone; in the presence of estrogen, the chick oviduct will make albumin and other egg white proteins. The steroid molecules enter the cell, where they are bound by specific cytoplasmic proteins. These protein–steroid complexes then diffuse into the nucleus. There they are bound by other proteins that are tightly attached to the chromosomes. This complex interaction then initiates mRNA transcription. Apparently, this is positive control; only one kind of protein is made by any given mRNA (although the same hormone controls many different genes); but all in all, the system is not too different in principle from Jacob and Monod's operon. The complete control mechanism, of course, must be much more complex in higher organisms. Estrogen doesn't induce the synthesis of egg white proteins in other chick cells, those which lack the specific receptors. And to complicate the issue, the same hormone has very different effects in different tissues. Estrogen can suppress synthesis of some proteins as it induces the synthesis of others, while causing entire tissues to proliferate in other target organs. Estrogen also, as you probably know, affects behavior.

12.4 Recombinant DNA

What Have Molecular Geneticists Been Up To Recently?

The field has continued to progress rapidly since those early days of molecular biology. Some molecular biologists, a few years back, had begun to lament that everything of any importance was already known and that there were no new worlds to conquer. But science has heard that lament many times before, and it has never turned out to be true. In fact, some rather astonishing things are happening right now. And new questions pour in at a startling rate. The game, it seems, has just begun.

For example, new techniques have allowed molecular biologists to determine the exact nucleotide sequence of any piece of DNA or RNA they can isolate—and they can isolate just about any piece they want to. In fact, the base sequence of an entire bacteriophage genome has now been determined, and it fills an entire page with As, Ts, Cs, and Gs—5315 of them, to be exact. Mammalian virus genomes have been sequenced, too, and some exciting and puzzling new things have been found. Your own DNA sequence, if it were ever determined, would fill 1000 volumes the size of this one with just those four letters. Incredible as it sounds, one laboratory is already hard at work sequencing the human genome. They have chopped human DNA into fragments a few tens of thousands of nucleotides long, they have inserted the fragments into the DNA of a bacteriophage, and they are maintaining 3 million separate cultures of cloned human DNA. Such a collection is called a *DNA library.*

The new sequencing techniques have enabled geneticists to look more closely at the mechanism of operons. In fact, that's how they found the multiple repressor sites we've mentioned. Comparisons between the DNA sequences of different species have shed new light on evolutionary processes. For instance, the hemoglobin gene DNA sequences of humans and rabbits have been compared. It turns out that human and rabbit hemoglobin DNA sequences are similar but not identical, and most of the differences that have accumulated since our last common ancestor lived, about 75 million years ago, have been changes in the third positions of amino acid codons and in those mysterious, recently discovered, untranslated *gene inserts.*

Gene inserts are long stretches of DNA that occur (unexpectedly) in the middle of eucaryote structural genes. They are transcribed, but are cut out of mRNA and discarded in the *tailoring* process in the nucleus. In the human beta chain hemoglobin structural gene, for instance, there are 550 nucleotides in the intervening sequences—as against only 438 coding for amino acids. As of this writing they have no known reason for existing, but there they are.

In the microbiology laboratory, virtually any kind of mutant strain that can be specified can now be created with the use of clever laboratory manipulations. The trick getting the most attention right now is *gene splicing,* also known as *recombinant DNA* or *gene cloning.*

Gene Splicing and Cloning, and the Recombinant DNA Debate

Gene splicing, the technique that has spawned a raging recombinant DNA controversy, has been said by its supporters to hold great promise for humanity. Some of our most pressing problems can soon be met, claim these scientists, and they speak glowingly of cancer cures and of an end to all genetic disease. Some have even promised an end to hunger with the production of new plants that combine, say, the photosynthetic efficiency of maize with the nitrogen-fixing ability and protein production of peanuts. That's the good news. The bad news is that other scientists say the technique will kill us all with unprecedented and uncontrollable epidemics. If these are the stakes, perhaps we should have a closer look.

How Gene Splicing Is Done

Gene splicing uses *restriction enzymes,* which are a normal part of the bacterial cell's defenses against infection with viral DNA. Laboratory techniques have evolved to the point that these enzymes can be purified and sold by the bottle. There are many different ones available, each with the ability to recognize and cleave a different specific pattern in DNA sequences.

For example, a restriction enzyme called *Eco-1* (from *E. coli*) recognizes the following sequence of double-stranded DNA, and cuts it wherever it is found:

↓
-X-X-X-**C-T-T-A-A-G**-X-X-X
-X-X-X-**G-A-A-T-T-C**-X-X-X-
↑

where X stands for any unspecified nucleotide. Note that the two strands are not cut directly opposite each other, but that the cuts are offset (arrows). This leaves free ends of complementary, single-stranded DNA:

-X-X-X-**C-T-T-A-A**-   -**G**-X-X-X-
-X-X-X-**G**-   -**A-A-T-T-C**-X-X-X-

The free, single-stranded ends will recognize and base pair with one another, given the opportunity. For this reason they are called *sticky* ends. After pairing, the DNA can be *healed* again with the repair enzyme, *ligase*. Now comes the key part of the laboratory procedure. Since the restriction enzymes leave the same sticky ends on any DNA it cuts up, any two appropriately cut ends can be rejoined. The implications of this are enormous. The molecular biologist can use the restriction enzymes on DNA from two entirely different sources and link them together, then put them back into a bacterial host. Do you want a chicken–bacterium hybrid organism? It's been done. If you think this sounds amusing, you'd better ponder the implications very carefully.

The usual trick is to cut open the circular DNA of a bacteriophage or a bacterial plasmid, using just one cut, and splice in a fragment of DNA from some other organism. The healed plasmid, with the foreign DNA insert, is then grown in a bacterial host. As the host rapidly multiplies in its usual manner, it makes unlimited numbers of the new DNA fragment. Making many genetically identical copies of anything, using asexual processes, is called *cloning* and the set of identical copies is called a *clone*. This is how someone made those 3 million clones of human DNA we mentioned earlier.

The laboratory techniques are not difficult. Plasmid DNA rings are easily separated from *E. coli* DNA by centrifugation. After they have been opened up with the restriction enzyme, which (by careful selection of appropriate enzymes and plasmids) finds only one specific place on the ring to cut, they are mixed with enzyme-cut fragments of foreign DNA and the proper amount of ligase. Most of the plasmids will heal again without taking in a fragment, but some will first include a fragment of foreign DNA. These altered plasmids will be a little larger and heavier than the normal plasmids and can be separated from them by centrifugation. Now more *E. coli* hosts can be infected with the altered plasmids using another trick: The cell walls of the susceptible bacteria are dissolved away, leaving naked bacteria (now called *protoplasts*), which readily incorporate the altered plasmid DNA. The individually infected bacteria can then be spread onto Petri dishes and grown into cultures, or clones (Figure 12.18).

The plasmids—in enormous numbers—can be isolated again at any time, and the foreign DNA can be released by the same restriction enzymes that were used in the first place. In this way any amount of specific DNA desired can be made. Keep in mind that bacteria multiply rapidly and that each time they divide, a new copy of the DNA insert is formed.

And this is just where some people begin to get worried. It is possible to put a gene for botulin (the deadly botulism toxin) into *E. coli*, the bacterium that is already well adapted to live in your gut. What would happen if you swallowed it? Who would do such a thing, you ask? Well, possibly the same folks who brought you nerve gas, cruise missiles, and napalm. It is also possible to put the DNA of a cancer-causing virus into a plasmid host. It's been done. The frightening prospect is that some well-meaning scientist, perhaps trying to understand the causes of cancer, might not handle his or her laboratory cultures carefully enough. We shouldn't forget that the last two minor epidemics of smallpox in Europe—including the very last death ever to be caused by smallpox—were due to escaped laboratory strains. (The deadly smallpox virus, which only two decades ago affected millions of people, is now extinct except for laboratory cultures.)

What Recombinant DNA and Gene Cloning Can Do: The Threat and the Promise

You may still be wondering, how can a piece of DNA inserted into a bacterial plasmid be of much interest to anyone at all? Why are some people so happy about it and others so nervous? (Other than the fact that some people basically like technology and others basically fear it.) Let's begin with some of its potential benefits.

The principal benefit, in the minds of the proponents, is that scientists like themselves will be able to do more experiments and learn more things faster. That's nothing to scoff at, but there are also some more direct and more practical benefits on the horizon. One of the most dramatic is that we are now in the position to make human insulin inexpensively and in unlimited quantities. Insulin is a polypeptide hormone that is synthesized in the pancreas. Many people are unable to synthesize normal amounts of insulin, and suffer from diabetes, a severe disease. Before insulin was discovered, most diabetics faced certain death. For about the last 50 years, insulin extracted from the pancreas of cattle has met most human needs, but it is very expensive. Also, some diabetics develop allergies to animal insulin, since it is not quite the same as human insulin.

Molecular biologists have recently extracted and purified insulin mRNA from human pancreatic cells and then made double-stranded DNA copies of it, using a viral reverse transcriptase enzyme. Only an extremely small quantity of such DNA was recovered, and it was contaminated with copies of other forms of mRNA. After many attempts, a DNA copy of insulin mRNA was successfully incorporated into a plasmid. From this experiment, unlimited quantities of insulin DNA became available. The next trick was to get the bacteria to transcribe and translate it.

This was done with more gene splicing. A single copy of the purified insulin DNA was inserted into an *E. coli* operon, in fact a constitutive operon with its control mechanism permanently broken on. Along

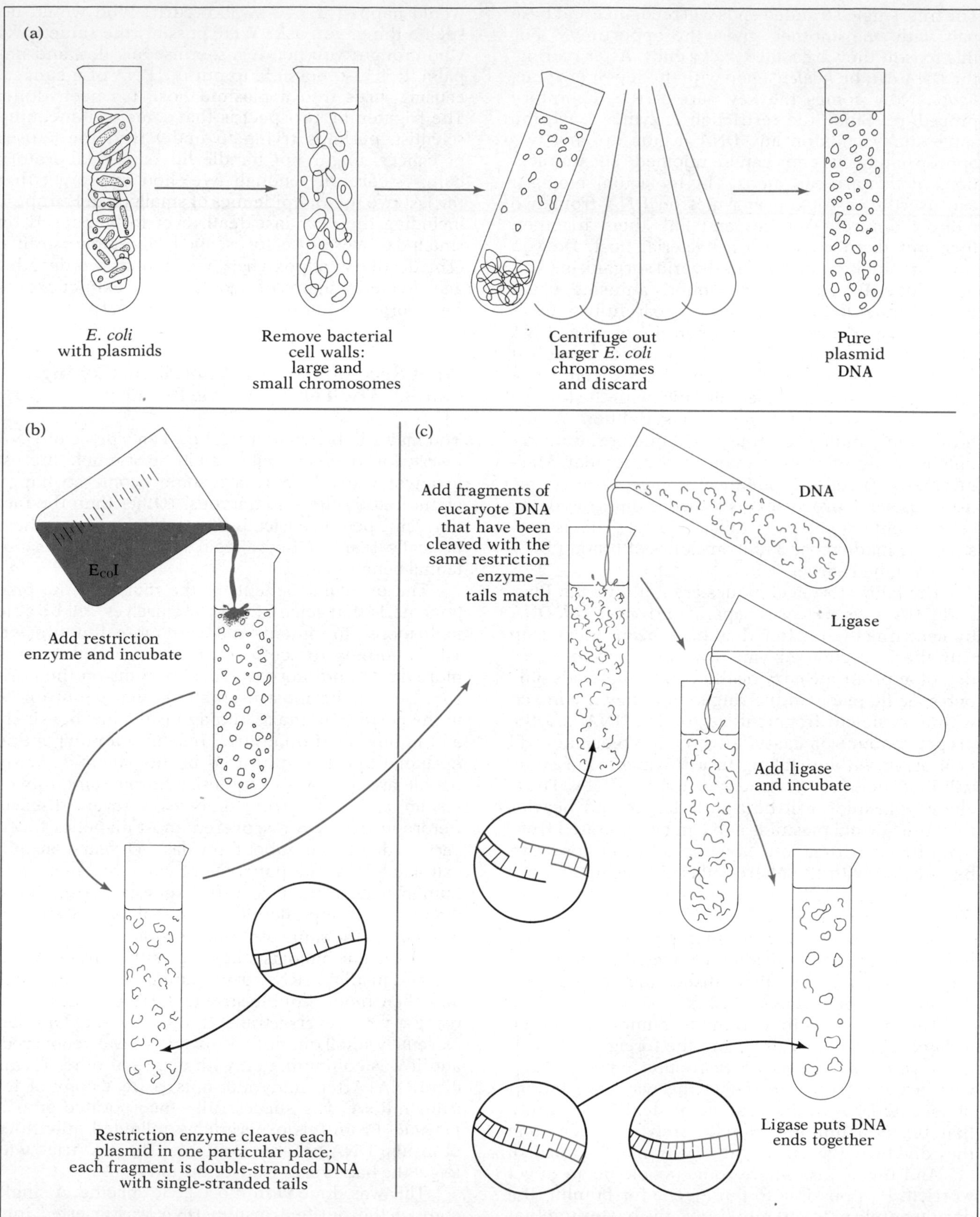
(a)
E. coli with plasmids
Remove bacterial cell walls: large and small chromosomes
Centrifuge out larger E. coli chromosomes and discard
Pure plasmid DNA
(b)
ECoI
Add restriction enzyme and incubate
Restriction enzyme cleaves each plasmid in one particular place; each fragment is double-stranded DNA with single-stranded tails
(c)
Add fragments of eucaryote DNA that have been cleaved with the same restriction enzyme—tails match
DNA
Ligase
Add ligase and incubate
Ligase puts DNA ends together

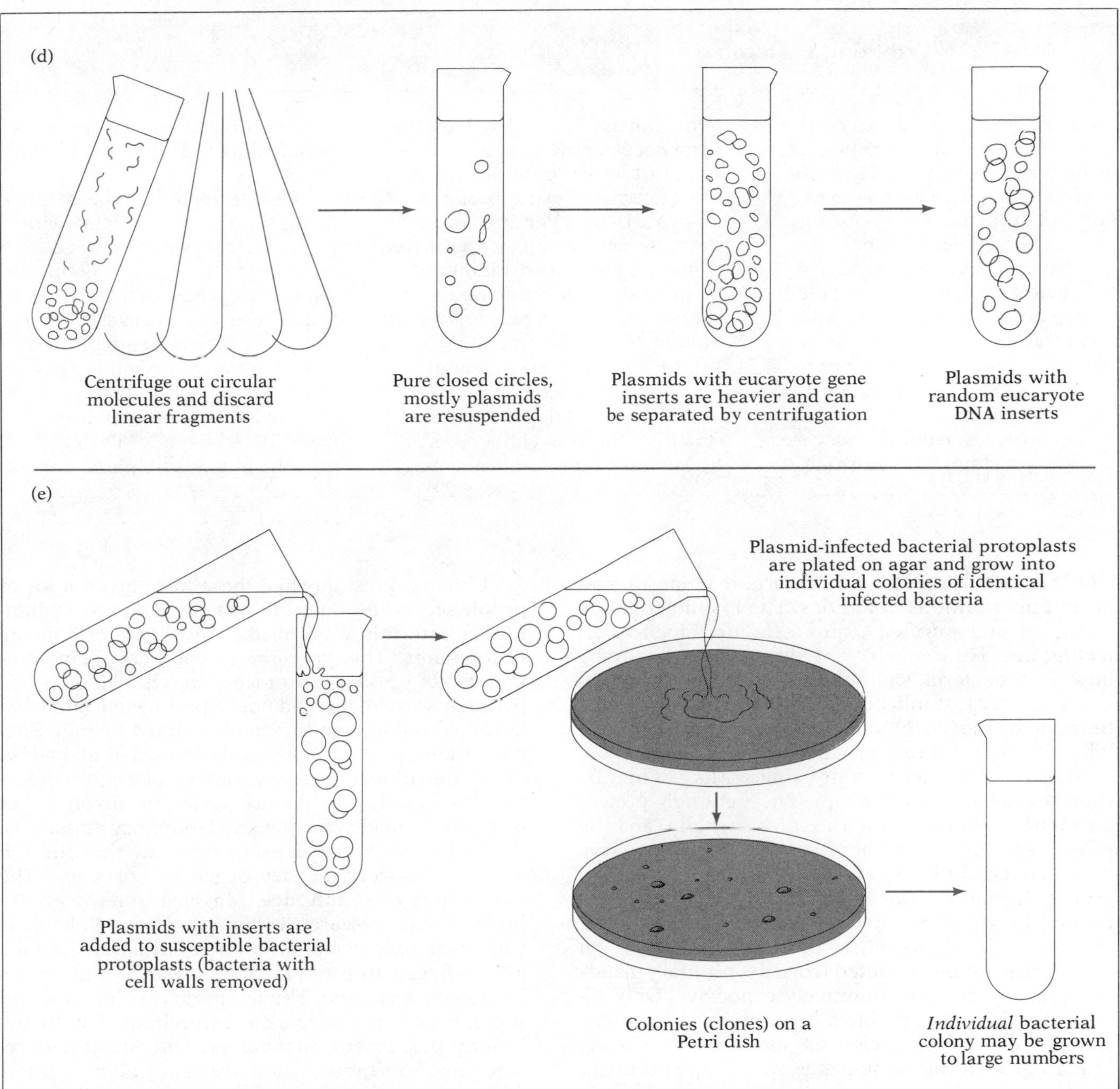
(d)
Centrifuge out circular molecules and discard linear fragments
Pure closed circles, mostly plasmids are resuspended
Plasmids with eucaryote gene inserts are heavier and can be separated by centrifugation
Plasmids with random eucaryote DNA inserts
(e)
Plasmids with inserts are added to susceptible bacterial protoplasts (bacteria with cell walls removed)
Plasmid-infected bacterial protoplasts are plated on agar and grow into individual colonies of identical infected bacteria
Colonies (clones) on a Petri dish
Individual bacterial colony may be grown to large numbers

12.18 Gene splicing permits researchers to insert segments of foreign DNA into bacteria, where it can transcribe mRNA and produce proteins in huge quantities. The techniques and skills required are not technically sophisticated by today's laboratory standards, although much of the procedure relies on chance events. With bacteria, the numbers of cells raised is unlimited, so the problems of chance are not serious, just tedious. (a) The procedure makes use of bacterial plasmids for the splicing of genes. These are obtained by removing the bacterial walls and differentially centrifuging the plasmids from other cell elements. (b) The relatively pure plasmid fractions are then cleaved with restriction enzyme (Eco I). (c) Eucaryote DNA is then cleaved with the same restriction enzymes, added to the cleaved plasmids, and the combination treated with ligase, the suturing enzyme. (d) The plasmids with successful splices are then separated from unspliced plasmids and linear fragments of DNA. What remains are many circular spliced plasmids, part bacterial and part eucaryote. (e) The bacterial–eucaryote plasmids are then added to bacterial protoplasts, taken in, and grown in large numbers on regular nutrient agar plates. Transfers can then be made to broth culture tubes where even larger numbers are grown. The eucaryote DNA can then be extracted, purified, and used as desired (e.g., sequenced, or inserted into a bacterial operon by additional trickery).

Essay 12.2 Crippling a Microbe

At one of the conferences on the risks of recombinant studies, geneticist Roy Curtiss III volunteered to produce a weakened mutant—one that could not survive outside the laboratory.

Given the go-ahead, he first produced an *E. coli* with a defective gene that prevented it from manufacturing its protective coat. The material that the gene normally made had to be provided artificially. But microbes mutate on their own, so soon some were back to producing the normal gene. Curtiss then deleted another gene necessary to coat production. But he was outfoxed by the crafty germs; they reproduced anyway. Dennis Pereira, a graduate student working with Curtiss, found that they were manufacturing a sticky substance called *colonic acid,* which acted as a kind of coat. So Curtiss and Pereira produced a microbe that couldn't make colanic acid. Finally, they had a germ that depended on scientists for its livelihood. As an unexpected bonus, this new bug was sensitive to ultraviolet radiation. Ordinary daylight would kill it.

One problem remained. Even dying *E. coli* can conjugate with normal *E. coli,* so an escaped germ might be able to pass its dangerous gene to a healthy colleague. Curtiss, however, altered the gene of the dependent bacteria that makes thymine. Thymine, therefore, had to be supplied and, without it, how could DNA be made? Perhaps now the bug was helpless enough.

with bacterial protein, the bacterium made human insulin in enormous quantities. The insulin could be extracted and purified using standard biochemical techniques. How much? The limits are set only by how fast bacteria can multiply—and that, as you know, is very fast indeed. So human insulin can now be made by the barrelful. That's not an abstract possibility; it's real and very practical.

Let's consider another possibility. *Midgets* may be short because they don't produce enough growth hormone. Some also don't mature sexually, and the growth centers of their bones never fuse. A few have been successfully treated with purified human growth hormone and, in fact, one patient grew to normal height after having been only 4 feet tall at the age of 35. However, at present, human growth hormone must be extracted from the pituitary glands of cadavers and is prohibitively expensive. Only experimental treatments have been done so far. Gene splicing techniques have been successful in cloning human growth hormone genes, and are expected to produce a ready supply of inexpensive growth hormone. In another generation, perhaps, no one will be any shorter than he or she wants to be. Does that seem to you like a blessing, or like technology has gone mad?

Gene splicing techniques are so powerful that, in fact, they have terrified a lot of people, including some of the scientists who had a hand in developing them. What if some fool actually developed that *E. coli* bacteria that could make botulin toxin? What if zany coaches started feeding growth hormones to their teams, and we ended up with an illicit "hormone culture" to go along with our drug culture? What, in fact, might happen that no one can even guess at now?

Chilling questions like these have given a lot of people second thoughts. In fact, not long ago a voluntary moratorium was called on further gene splicing experiments. This gave the researchers themselves time to set up safety protocols on what kinds of experiments could or could not be performed, and what kinds of containment precautions must be met. Special strains of *E. coli* were developed that had so many nutritional and environmental requirements that they could not possibly grow in anyone's gut or survive outside of expensive laboratory apparatus.

Of course, when nonscientists saw that the scientists were worried, they began to worry too. The city fathers of Cambridge, Massachusetts, voted to prohibit such research within city limits. Politicians with their usual smattering of information jumped in and began to press for laws. Perhaps the results have been fortunate. The temporary, voluntary, and self-imposed restrictions on research are rapidly becoming permanent, mandatory, and monitored restrictions. Some researchers are happy about it; most feel hobbled. To put it mildly, there are differences of opinion as to whether the regulation of gene recombination research is a good thing or a bad thing.

So far, however, nothing the least bit harmful has come of gene splicing experiments. Complete cancer virus DNA has already been spliced into *E. coli* plasmids; fortunately, huge doses of the carrier bacteria had no ill effect on laboratory animals. The abnormal plasmids, and the bacteria carrying them, can scarcely compete with nature's own. If you will recall the original transformation experiments, you'll remember that bacteria have always had the ability to incorporate fragments of DNA into their own chromosomes. Our guts have always been full of DNA fragments from plants, animals, and microorganisms. But perhaps we shouldn't be too complacent just because the potential for disaster has always been with us. After all, any number of other disasters have always been with us, but that doesn't make them any more enjoyable.

Whether gene splicing techniques will work for good or ill, the genie is out of the bottle now. It is certain that molecular biology will never be the same.

Application of Ideas

Recall that the *E. coli tryptophan synthetase* gene is in a repressible operon. That is, the organism produces the enzyme at all times, except when tryptophan is present in the medium. An investigator collects a number of *constitutive mutants* for tryptophan synthetase: strains that continue to produce the enzyme even when tryptophan is added to the medium.

a. In what part or parts of the gene control system might such mutants occur?

b. The same investigator also has isolated an F^+ episome that contains the entire wild-type tryptophan operon and control system. When the constitutive mutant cells are infected with the F^+ $tryp^+$ episome, they become partial diploids for the tryptophan operon. The investigator finds that the mutant strains then fall into two groups: in one group the constitutive mutant is dominant (the infected partial diploid cell continues to make enzyme constitutively in the presence of tryptophan); in others, the constitutive mutant is recessive (all enzyme production is repressed when the partial diploid cells are grown in tryptophan). Diagram the two kinds of partial diploid cells. In each diagram, show both complete operons: the mutant one of the bacterial chromosome, and the wild-type one on the F^+ episome. Label each diagram completely: show where within the chromosomal operon the constitutive mutation has occurred, indicate the nature of the repressor protein, and explain what happens in the presence of tryptophan.

c. What classes of *tryptophan synthetase negative* mutants—strains that show no enzyme activity even in the absense of tryptophan in the medium—might be recovered? How could these classes of tryptophan synthetase negative mutants be distinguished experimentally?

Key Words

agar gel
albinism
alkaptonuria
arginine
auxotroph

bacteriophage
biochemical pathway
biotin

cis-dominant
citrulline
constitutive mutant
cross-reacting material (CRM)

episome

fertility factor

galactosidase
galactoside permease
gene cloning
gene splicing
genetic marker

Key Ideas

Inborn Errors of Metabolism

1. The first interest in the biochemistry of the gene was shown by A. E. Garrod. In his work with *alkaptonuria* he was able to show that an enzyme deficiency was responsible and that the defect was genetically produced.
2. Garrod correctly believed that the responsible enzyme was one in a long chain of reactions. Its absence caused the buildup of an unfinished substrate.
3. Other disorders with similar causes are *albinism* and *phenylketonuria.* All three disorders occur in the metabolism of *phenylalanine* and *tyrosine.*
4. Garrod showed that the defects followed Mendelian laws, but his contemporaries were more interested in visible or morphological traits, so there was little interest.

One Gene, One Enzyme

Beadle and Ephrussi worked on *Drosophila* eye pigments, trying to clarify their biochemical pathways. They found that mutations block pathways, but were unable to clarify the steps.

Beadle and Tatum

1. Using the fungus *Neurospora,* Beadle and Tatum were responsible for ushering in a new era of *microbial genetics* and *molecular biology.*
2. The choice of *Neurospora* as an experimental organism offered many advantages:
   a. It grew on simple minimal media and offered the opportunity to study *nutritional mutants.*
   b. Because the organism is primarily haploid, the genetics was easier.

High frequency of recombination (Hfr)

inborn errors of metabolism
inducer
inducible enzyme
inducible operon
innoculum

lac operon
leucine
ligase
lysis
lysogeny

metabolic defect
methionine
microbial genetics
minimal medium
molecular biology

negative control
Neurospora
noninducible control mutant
nutritional mutant

operator
operon
ornithine

penicillin
Petri dish
phenylalanine
pilus (pili)
plaque
plasmid
positive control
protoplast

recombinant DNA
replica plating
repressible operon
repressor-inducer complex
repressor molecule
restriction enzyme
reverse transcriptase

serial diluton
sexduction
spore
supplemented medium

T-even phage
thiamine
thiogalactoside transacetylase
transduction
transformation
threonine
tryptophan
tryptophan synthetase
tyrosine

velvet-covered disk

wild-type

c. Because there is a brief sexual stage, crosses could be made and recombinants studied.

d. Large numbers of offspring, represented by *spores,* could be easily handled.

3. Beadle's and Tatum's initial experiments involved irradiating haploid *spores* and growing them on minimal medium.

4. Separating *nutritional mutants* was done by straining out the growing nonmutants.

5. *Nutritional mutants* were then analyzed by growing their *spores* on various supplemented media until the nutritional requirements were identified.

6. If a mutant could not produce a certain substance, such as *arginine,* it was known that that enzyme was missing or defective and that the gene responsible for its production had mutated. The spore would germinate and grow if *arginine* was added to the *minimal medium.*

7. By collecting various mutant strains, carrying out crosses, and studying the recombinant types, they were able to determine that the nutritional requirements were straight Mendelian factors.

8. For example, a failure to produce *arginine* could be caused by gene mutation in more than one place in the *biochemical pathway* leading to its production. If two mutants with different gene deficiencies were crossed, some of the progeny would be able to produce *arginine* and others would not.

9. In the results of this type of cross they found that 75% required *arginine* and 25% did not (the *wild type*).

10. They determined that arginine was produced in a three-step process. Each step utilized a different enzyme and each enzyme had its own gene. Gene mutations in any of the three loci produced a nutritionally deficient strain.

11. From this and other similar studies it was deduced that, *as a rule* one gene produced one enzyme. This discovery helped to explain the following:

a. The likely basis of dominance (one *wild-type* gene could produce enough enzyme).

b. The 9:7 ratio in some crosses was explained in terms of enzyme pathways.

12. In his studies of the sickle-cell hemoglobin, Pauling modified the "*one gene, one enzyme*" idea by noting that genes could specify qualitative differences in a protein.

13. Following the era of *Neurospora* research, bacterial and *bacteriophage* studies began. The gene–enzyme hypothesis, the central dogma, and *microbial genetics* merged to become the new science of *molecular biology.*

Recovering and Counting Mutations in Bacteria

1. *Nutritional mutants* in bacteria were studied in a manner similar to the *Neurospora* experiments.

2. Bacteria offer the advantage of enormous numbers of progeny in a short generation time. This permits the recovery of very low frequency mutants.

3. To separate mutants, irradiated bacteria are grown in *minimal media.* Penicillin is then added. It kills the nonmutants that grew in *minimal media.* The dead organisms are removed and the remainder are the mutants. These are then grown in various types of *supplemented media,* categorized and used for experiments.

Recovering Recombinants

1. Bacteria are procaryotes so they do not undergo meiosis, produce gametes, or reproduce sexually in the usual way.

2. Recombinations do occur, but are very rare. They are selected by mixing two different *nutritional mutant* strains and then looking for a few cells that will grow on minimal media.

The Life and Times of the Bacteriophage

1. The *bacteriophage* is a virus that infects bacteria.

2. Bacteria are somewhat protected by their *restriction enzymes,* which recognize some viral DNA and cleave it into fragments.

3. Some viral parasites escape *restriction enzymes* through variations in their DNA sequences, which are then not recognized.

4. When a *bacteriophage* gets by the host defenses, the bacterial RNA polymerase transcribes mRNA from viral DNA, which is then translated into viral proteins.

5. Next, the host-produced viral proteins (enzymes) fragment the host DNA and use its nucleotides to produce more viral DNA. The viral proteins are assembled into viral coatings. When many viruses have been produced this way the cell bursts (*lyses*) and a new infection cycle can start with other bacteria.

6. Alternatively, *bacteriophage* may lysogenize the bacterial cell. The viral DNA may join the bacterial DNA, undergo replication and be passed this way through many generations of bacteria before lysing occurs.

Counting Viruses, or Holes in the Lawn

Viruses can be counted. The procedure is to serially dilute them, plate them on an *agar plate* previously covered with bacteria, and count the empty spaces (plaques) where the viruses have done their work.

Bacterial Recombination

Sex is defined here as an act that combines the genetic material from two cells into one cell.

Transformation

Bacteria can take in and incorporate DNA from other bacteria.

Transduction

In the laboratory, *bacteriophage* sometimes take in bits of bacterial DNA that can then be transferred to other bacteria infected by the phage.

Plasmid Transfer

1. *Plasmids* are smaller circular chromosomes containing a separate genome.

2. One type of *plasmid* produces a *pilus*, or connecting tube between one bacterium and another.

3. *Plasmid-plus* strains produce *pili* and send a copy of the *pilus*-producing DNA into *plasmid-minus* strains, making them also *plasmid*-plus.

4. In some instances, the *plasmid* plus strain is also resistant to antibiotics and can transfer this resistance via *plasmid* DNA to other bacteria.

F^+ Episomes

1. The F^+ *episome* is another example of a plasmid.

2. By crossing strains of *nutritional mutants*, Lederberg and Tatum discovered that one cell in 10 million were recombinants that grew on *minimal media.*

3. From many failures in trying to determine linkage, they discovered that the bacterial chromosome was circular.

4. Further studies produced evidence that there were "male" and "female" mating types. The transfer of DNA was always in one direction.

An Example of How Bacterial Genetics Is Done

1. Evidence for the one way transfer of DNA came from studies of the transfer of unselected loci. This means that transfers were done under conditions where either the plus or the minus allele could survive.

2. Table 12.1 (p. 363) shows the various combinations tried in determining linkage groups.

3. F^+ *episomes* assure survival by inducing *pili* to form and transferring *episome* copies to other bacteria. They also confer immunity to the host so that other F^+ *episomes* cannot infect them.

4. During transmission, the *episome* opens to a linear form, replicates, and sends one copy through the *pilus.* Once in the new host it assumes its circular shape.

5. In rare instances, the F^+ *episome* joins the bacterial DNA circle, goes through replication, and is passed along to descendants.

6. Later, when the F^+ portion opens to its linear form to pass through the *pilus*, it begins to take the entire chromosome with it. When this happens, the long

chromosome can enter into crossing over with the new host chromosome. Strains of bacteria with such *integrated* F+ episomes were eventually isolated. Because they showed a high frequency of recombination, they became known as *Hfr* strains.

The Great Kitchen Blender Experiment

1. Using a blender, Jacob and Wollman were able to break up mating bacteria at selected times in the transfer of the linearized chromosome.
2. Since the chromosome required 89 minutes for the transfer, it could be broken at different places along its length.
3. From this procedure they were able to determine the order in which genes were arranged on the chromosome. From this information *genetic maps* of the bacterial DNA were constructed and the units are given in minutes.

Sexduction

In *sexduction* a part of the large host chromosome becomes incorporated into The F+ *episome.* The *episome* renders the bacterial host a *partial diploid.*

Gene Organization and Control in Bacteria

The Jacob and Monod model of gene organization and control was produced in 1961.

Background: Inducible Enzymes in Escherichia coli

1. The synthesis of some enzymes was known to be under some kind of control.
2. When *E. coli* are fed glucose they stop producing glucose synthesizing enzymes. On the other hand, if they are fed lactose, they produce enzymes that change lactose to glucose. Enzymes that can be induced to form in the presence or absence of a certain substrate are called *inducible* enzymes. Three enzymes were found in lactose metabolism.
3. The three enzymes and their gene designations are *β-galactosidase* (**z**), *galactoside permease* (**y**), and *thiogalactoside transacetylase* (**a**).
4. Their wild and mutant types are known as $\mathbf{z}^+$, $\mathbf{y}^+$, and $\mathbf{a}^+$, and $\mathbf{z}^-$, $\mathbf{y}^-$, and $\mathbf{a}^-$.

Mutants of the Lactose Operon

1. The $\mathbf{z}^-$ mutant produced a nonfunctional protein called CRM. Its presence was important to understanding how the enzyme control system worked.
2. All three enzymes were under a single control system.
3. Two types of mutant control systems were found:
   a. *Constitutive* produced enzyme all the time.
   b. *Noninducible* would not work at all.
4. Three alleles were determined for each gene locus, *wild-type, constitutive,* and *noninducible.* The two control loci were mapped: **o** (*operator*) and **i** (*inducer*). The mutants for these were named for the three types of alleles: $\mathbf{o}^+$, $\mathbf{o}^c$, $\mathbf{o}^n$, $\mathbf{i}^+$, $\mathbf{i}^c$. and $\mathbf{i}^n$. The *constitutive mutants* were $\mathbf{o}^c$ and $\mathbf{i}^c$, and $\mathbf{o}^n$ and $\mathbf{i}^n$ were the *noninducible mutants* ($\mathbf{o}^+$ and $\mathbf{i}^+$ were normal).

Jacob's and Monod's Experiment

Dominance relationships were used to determine how the control system functioned. Since this requires diploidy, *sexduction* was employed, introducing an F+ episome that bore the lactose regulating system.

Dominance Relationships at the Operator Locus

1. All combinations of the **o** locus and the **z** locus were studied (see Table 12.1 for results).
2. Results show that the **o** locus controlled the **z** locus, but only if it was on the same DNA segment.

Dominance Relationships at the i Locus

1. The same experiments were carried out with **i** and **z** alleles (see Table 12.2).
2. Results showed that **i** could interact with **z** regardless of whether it was on the same DNA strand.
3. The **o** locus affected only its own chromosome, while the **i** locus produced a diffusible substance (protein) that could work at the distant control system.

4. From these observations, Jacob and Monod developed the *operon* model.

Jacob's and Monod's Conclusions: The Operon Model

1. The *operon* model is an explanation of how enzyme-producing genes get turned on and off in response to the presence or absence of *inducer* substances in the cell.

2. The *operon* described here is the *lac operon,* named for lactose, which is the inducer for the production of three enzymes that metabolize lactose.

3. The functioning *operon* consists of five loci. These are:
   a. The regulator gene
   b. The *operator* gene
   c. The three enzyme-producing genes

4. The regulator gene produces mRNA, which translates into a blocking protein known as a *repressor*. In the absence of an *inducer* it blocks the *operator* gene.

5. While the operator gene is inactivated, the enzyme-producing loci cannot function and no enzyme is produced.

6. When lactose enters the cell it binds with the repressor protein, freeing the *operator* locus.

7. Once the operator locus is unblocked, it permits transcription of the three enzyme-producing genes to occur. Messenger RNA is produced and translated into the three lactose-metabolizing enzymes.

8. As soon as the lactose has been metabolized, the repressor protein becomes active, binding the operator. Once the *operator* gene is blocked, the enzyme-producing loci again become inactive.

The "Invention" of Messenger RNA

Jacob and Monod had to hypothesize mRNA (unknown at that time) since their *operon* model required the activity of an intermediate molecule between DNA and protein.

Repressible Enzymes and Other Operons

Repressible Enzymes

1. While the lactose-metabolizing enzymes are inducible, other enzymes are repressible. In repressible systems, the presence of certain substrates in the cell does not induce enzyme production, but, rather, stops it. An example is the enzyme *tryptophan synthetase.*

2. Bacteria grown on glucose and minerals produce this enzyme for the production of the amino acid *tryptophan.*

3. When the cells are fed *tryptophan*, enzyme production stops.

4. The *tryptophan operon* is similar to the *lac operon* except that the repressor protein will not bind the operator unless it first complexes with *tryptophan*, thus blocking transcription.

Operons in Higher Organisms: Are There Any?

1. The initial search for *operons* in eucaryotes failed to show their presence. Part of the problem was that the complexity of eucaryote cells made clear-cut experiments very difficult.

2. At least one system has been discovered in baby chicks. Estrogen induces oviduct cells to produce albumen.

What Have Molecular Geneticists Been Up To Recently?

1. *Microbial geneticists* have been able to determine sequences of DNA and RNA in entire bacteriophage and viral genomes. Work in sequencing the human genome is now progressing.

2. Knowing DNA sequences has shed new light on evolution. Evolutionary distances are now determined by comparing DNA sequences.

Gene Splicing and Cloning, and the Recombinant DNA Debate

How Gene Splicing Is Done

1. *Gene splicing* and recombination is a highly emotional and controversial issue. The technology involves the use of synthetic enzymes that cleave and suture DNA

segments at will. This permits the hybridization of desired combinations of DNA from many sources.

2. Inserts can be made in *plasmids* and from the huge numbers of bacteria that can be cultured, the products of these genes can be collected in large quantities. The procedure includes:
   a. *Plasmids* are removed by ultracentrifugation.
   b. The *plasmids* are then cleaved by restriction enzymes and foreign DNA segments added, using *ligase.*
   c. Bacterial cells, with their walls dissolved away, take in the hybrid *plasmids.* Huge clones with the hybrid DNA are produced.
   d. Hybrid DNA can then be inserted into a bacterial operon, and clones of the altered bacterium will produce a desired product.
3. There is considerable fear that improper techniques and a lack of precaution in recombination experiments could produce highly virulent bacteria. Once out of the lab, these new mutants could sweep through a population, out of control, according to some critics.

What Recombinant DNA and Gene Cloning Can Do: The Threat and the Promise

1. The benefits:
   a. More knowledge of genetics.
   b. Using new splicing techniques, huge quantities of hormones such as insulin and growth hormone could be produced cheaply. Other products would follow.
2. The problems:
   a. There is the real danger of creating bacterial strains that are extremely dangerous.
   b. There is the potential for producing deadly "germ warfare" organisms.

Review Questions

1. What was Garrod's evidence for deciding that *alkaptonuria* was an inherited defect? (p. 351)
2. What do *albinism, alkaptonuria,* (Figure 12.1) and *phenylketonuria* have in common? (p. 352)
3. What are some of the advantages of using *Neurospora* as an experimental organism? (pp. 353–354)
4. Explain Beadle and Tatum's procedure for isolating *nutritional mutants* of *Neurospora.* (pp. 353–354)
5. Show the cross between two different *arginine* mutants that proved to Beadle and Tatum that they were studying two different loci. (pp. 354–355)
6. What were Beadle and Tatum's final conclusions about the *arginine* pathway? Diagram the steps, enzymes, products, and genes. (p. 355)
7. List two problems that were better understood as a result of the *one gene, one enzyme* idea. (pp. 355–356)
8. Describe Beadle and Tatum's procedure for recovering *nutritional mutants* in bacteria. (p. 355)
9. How did the Lederbergs recover their *nutritional mutants?* (p. 357)
10. How do *restriction enzymes* protect bacteria from phage viruses? (p. 357)
11. Explain in detail how *bacteriophage* prepare for *lysis* in their hosts. (p. 358)
12. Describe the process of *lysogeny.* Of what advantage is this to the virus? (p. 358)
13. Explain the *plaque* system for counting viruses. (p. 358)
14. Give a definition of sex that covers bacteria as well as other organisms. (p. 359)
15. How does simple *transformation* work? (pp. 358–359)
16. Explain how *transduction* produces new recombinants. (p. 360)
17. What are *plasmids* and what special structure do they produce? (p. 360)
18. Explain the steps in the transfer of *plasmids* from one bacterium to another. (p. 360)
19. Diagram the strain A × strain B cross that led to Lederberg and Tatum's discovery of sex in bacteria. (p. 361)
20. Explain how Hayes came to the conclusion that "maleness was catching." (p. 363)
21. Describe the movement of DNA from an F^+ *episome* into another bacterium. (p. 364)
22. Explain how the kitchen blender was employed in mapping the chromosome of *E. coli.* (p. 365)
23. What is an *inducible enzyme?* How are such enzymes induced? (p. 366)
24. What is the difference between *constitutive* and *noninducible control mutants?* (pp. 366–367)
25. List the four loci that constitute the *lac operon* and give the function of each locus. (p. 368)
26. Explain the role of lactose as an *inducer* in the *lac operon.* (p. 368)
27. What causes the *lac operon* to shut down? (p. 368)

28. What factors in their *operon* theory led Jacob and Monod to hypothesize the existence of messenger RNA? (pp. 368–370)

29. How does a *repressible operon* differ from an *inducible operon*? (p. 370)

30. Explain how a knowledge of DNA sequences is shedding new light on evolution. (p. 372)

31. Explain how *restriction enzymes* can be used to produce selected fragments of DNA. (pp. 372–373)

32. Review the procedure for splicing genes into a *plasmid*. There are four important steps. (pp. 372–373)

33. List some of the undesirable possibilities of *gene splicing*. (p. 373)

34. How can *gene splicing* be used to produce hormones like insulin and human growth hormone? (p. 373)

35. Describe the protective measures taken to assure that recombinant *E. coli* will not "escape" from the research laboratories. (Essay 12.2) (p. 376)

Suggested Reading

Allison, A. C. 1956. "Sickle Cells and Evolution." *Scientific American,* August. The first demonstration of overdominance: the sickle-cell allele is actually beneficial to carriers in certain environments.

Avery, O. T., C. M. MacLeod, and M. McCarty. 1944. "Studies on the Chemical Nature of the Substance Inducing Transformation of Pneumococcal Types." *Journal of Experimental Medicine* 79:137. The classic experiment that proved to the world—at least in hindsight—that the genetic material is DNA.

Bishop, J. A., and Laurence M. Cook. 1975. "Moths, Melanism and Clean Air." *Scientific American,* January. How the peppered moth is readapting to the improving atmosphere of postindustrial England.

Campbell, A. M. 1976. "How Viruses Insert Their DNA into the DNA of the Host Cell." *Scientific American,* December.

Crick, F. H. C. 1962. "The Genetic Code." *Scientific American,* October.

———. 1966. "The Genetic Code III." *Scientific American,* October.

———. 1979. "Split Genes and RNA Splicing." *Science* 204: 264. The codiscoverer of the structure of DNA was active in the race to decipher the genetic code and has some cogent things to say about those mysterious newly discovered entities, gene inserts or introns.

Darnell, J. E. 1978. "Implications of RNA–RNA Splicing in the Evolution of Eukaryote Cells." *Science* 202:1257. Another thoughtful attempt, by one of the leaders in the field of eucaryote gene function, to make some sense out of those puzzling introns.

Darwin, C. 1859, 1966. *On the Origin of Species.* A facsimile of the first edition. Harvard University Press, Cambridge, Mass.

Dawkins, R. 1976. *The Selfish Gene.* Oxford Press, New York and Oxford. Dawkins is an enthusiastic and often persuasive writer who, in this popular paperback, puts modern evolutionary theory into clear and vivid language. His approach is logical and nonmathematical, with some emphasis on theories of the evolution of animal behavior.

de Beer, G. 1974. "Evolution." In *The New Encyclopedia Brittanica,* 15th ed. 7:7. A concentrated synopsis of evolutionary thought, with an extensive bibliography.

Dickerson, R. E. 1972. "The Structure and History of an Ancient Protein." *Scientific American,* April. How cytochrome C has evolved its present shape and amino acid sequence over the last 2 billion years.

Dobzhansky, T. 1963. "Evolutionary and Population Genetics." *Science* 142:3596.

———. 1970. *Genetics of the Evolutionary Process.* Columbia University Press, New York. A great compendium of observations and interpretations of the genetics of natural populations by an influential evolutionary geneticist.

Du Praw, E. J. 1970. *DNA and Chromosomes.* Holt, Rinehart and Winston, New York (paperback).

Friedman, T. 1971. "Prenatal Diagnosis of Genetic Diseases." *Scientific American,* November. A primer of transabdominal amniocentesis and a valuable discussion of the moral implications involved in the interruption of pregnancy.

Gilbert, L. E., and P. H. Raven. 1975. *Coevolution of Plants and Animals.* University of Texas Press, Austin, Tex. Such fascinating esoterica as flowers that mimic insects and insects that mimic flowers, as well as such important topics as the coevolution of plants and their herbivores.

Goodenough, U. 1978. *Genetics,* 2d ed. Holt, Rinehart and Winston, New York. Of the current general genetics texts, Ursula Goodenough's has the best coverage of recent developments in molecular genetics and is particularly strong in the molecular genetics of higher eucaryotes.

Grant, V. 1963. *The Origin of Adaptations.* Columbia University Press, New York. A balanced view of evolutionary adaptations in plants and animals.

Haldane, J. B. S. 1932. *The Causes of Evolution.* Cornell University Press, New York (paperback). All the difficult math is relegated to the appendix in this rather old—but very wise—discussion of evolutionary genetics. A must for advanced students of population biology.

Hamilton, W. D. 1967. "Extraordinary Sex Ratios." *Science* 156:477. Why have some species departed from the usual one-to-one ratio of males to females?

Jacob, F., and J. Monod. 1961. "Genetic Regulatory Mechanisms in the Synthesis of Proteins." *Journal of Molecular Biology* 33:318. A modern classic and an example of fine scientific writing and reason, this is an account of the revolutionary experiments that demonstrated the existence of operons, the operator, cytoplasmic regulating molecules, and messenger RNA.

Joravsky, D. 1970. *The Lysenko Affair.* Harvard University Press, Cambridge, Mass. Fascinating reading about the man whom biologists—especially geneticists—love to hate.

Kettlewell, H. B. D. 1956. "Further Selection Experiments on Industrial Melanisms in the Lepidoptera." *Heredity* 10: 287. Here you can read for yourself one of the most often cited experiments in modern evolutionary biology.

King, J. L., and T. H. Jukes. 1969. "Non-Darwinian Evolution." *Science* 164:788. The authors suggest, among other things, that most evolutionary changes on the molecular level may be meaningless noise, the result of mutation and random drift; that most DNA in higher organisms is not genetic material, and that no more than about 1% codes directly for proteins.

Kretchmer, N. 1972. "Lactose and Lactase." *Scientific American,* 227:70. Milk produces flatulence in adults of oriental or African ancestry because they lack the enzyme needed to break down milk sugar.

Lack, D. 1947. *Darwin's Finches.* Cambridge University Press, New York. Lack investigates the coevolution of competing species on different islands in the Galápagos archipelago.

Maynard Smith, J. "Group Selection and Kin Selection." *Nature* 201:1145.

McKusick, V. A. 1965. "The Royal Hemophilia." *Scientific American,* February. Queen Victoria was heterozygous for the X-linked recessive allele and through political marriages of her daughters managed to pass it on to all the leading royal families of Europe.

Mendel, G. 1965. "Experiments in Plant Hybridization (1865)." Translated by Eva Sherwood. In *The Origin of Genetics,* ed. C. Stern and E. Sherwood. W. H. Freeman, San Francisco. In addition to the full text of Mendel's classic paper, this volume includes some of Mendel's letters and minor works, the three "rediscovery" papers of 1900, and a fascinating exchange between R. A. Fisher and Sewall Wright on the question of whether or not Mendel dry-labbed the whole thing.

Meselson, M., and F. W. Stahl. 1958. "The Replication of DNA in *E. coli.*" *Proceedings of the National Academy of Sciences* (U.S.) 44: 671. The brilliant and influential experiment that proved that DNA unwinds and replicates semiconservatively, as foreseen by Crick. All biology students must study this work in detail at some time or another.

Miller, O. L. 1973. "The Visualization of Genes in Action." *Scientific American,* March. Some remarkable electron micrographs of transcription and translation, looking almost exactly like diagrams that had originally been made on biochemical evidence alone.

Mourant, A. E., et al. 1978. *The Genetics of the Jews.* Clarendon (Oxford University Press), New York and Oxford. Ashkenazi and Sephardic Jews form a genetically distinct racial group after all, according to blood group and enzyme polymorphisms. Basically Palestinian, they show surprisingly little evidence of past mixing with European groups but rather more (5–10%) negroid admixture, presumably from the time spent in slavery in Egypt.

Okazaki, R. T., et al. 1968. "Mechanism of DNA Chain Growth: Possible Discontinuity and Unusual Secondary Structure of Newly Synthesized Chains." *Proceedings of the National Academy of Sciences* (U.S.). On the Okazaki fragments.

Patterson, C. 1978. *Evolution.* Cornell University Press, Ithaca, N.Y. (paperback). This short textbook makes a useful supplement to a general biology course or stands on its own. It is a good secondary reference for Darwin's finches and industrial melanism.

Sanger, F., et al. 1977. "Nucleotide Sequence of Bacteriophage ϕX174 DNA." *Nature* 265:687. The first publication of the entire genome of any organism and a *tour de force* of molecular biology.

A *Scientific American* book. 1978. *Evolution.* W. H Freeman, San Francisco. Reprinted from the September 1978 special issue on evolution.

Shine, I., and S. Wrobel. 1976. *Thomas Hunt Morgan: Pioneer of Genetics.* University of Kentucky Press, Lexington, Ky. Interesting narrative and lively anecdotes of the early days of genetics in America.

Stebbins, G. L. 1971. *Processes of Organic Evolution.* Prentice-Hall, Englewood Cliffs, N.J. Somewhat dated, but still a superior textbook on evolution.

Strickberger, M. W. 1976. *Genetics,* 2d ed. Macmillan, New York. Of the current general genetics texts Strickberger's has the clearest treatment of classical Mendelian genetics. In addition, we feel that the 120 pages that are included on population genetics and quantitative genetics happen to constitute the best textbook in print on these difficult subjects.

Temple, S. 1977. "The Dodo and the Tambalacoque Tree." *Science* 197:885. On Mauritius there are some geriatric trees whose seeds normally germinate only when passed through the gizzard of a dodo. But the last dodo died in the 17th century.

Van Valen, L., and G. W. Mellin. 1967. "Selection in Natural Populations. 7. New York Babies." *Annals of Human Genetics* (London) 31:109. Among newborns, it's better to be average, because small and large babies are both at risk.

Watson, J. D. 1968. *The Double Helix.* Atheneum, New York. Deftly hidden in the narrative of this witty, often hilarious and picaresque account of the personal triumph of a young scientist and an old graduate student is a surprising amount of solid scientific information. Certainly the most enjoyable account of how "the scientific method" actually works in practice.

Watson, J. D., and F. H. C. Crick. 1953. "Molecular Structure of Nucleic Acids. A structure of deoxyribose nucleic acid." *Nature* 171:737. This is the one that started it all: the most influential single page in scientific history.

Wilson, E. O., and W. H. Bossert. 1971. *A Primer of Population Biology.* Sinauer, Sunderland, Mass. Students tell us that this self-teaching approach to the elementary mathematics of population genetics and population ecology is more helpful than most formal courses on the subjects.

Part Three

Microorganisms

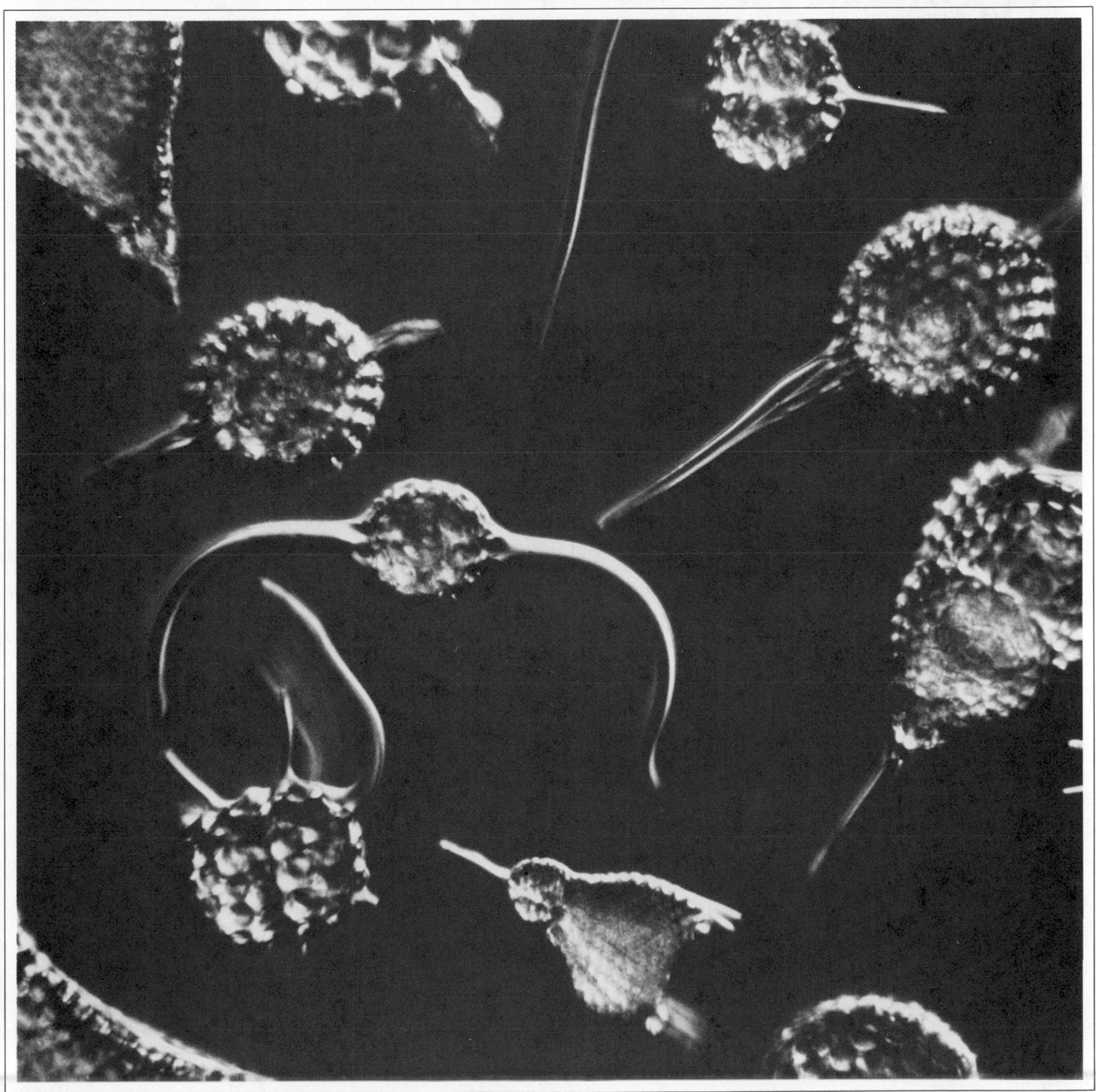

Chapter 13

Systematics and Taxonomy: Testing Evolutionary Hypotheses

There are 2 million species of living things on the earth.

There are 10 million species of living things on the earth.

One is true or neither is true. Both these figures can be found in textbooks, both written with solemn authority. Obviously, we can't learn much from such statements, but they do tell us two things. First, there are probably a lot of species. And, second, there is some disagreement over *what a species is.* This latter point is of greatest interest, because by now you'd think we would have had that ironed out.

13.1 The Species Problem

What Is a Species?

Often, a *species* is defined as a group of organisms that interbreed or that *could* interbreed. Obviously, if organisms are to interbreed, they must have certain things in common, such as their morphology, behavior, and internal chemistry.

In general, of course, this kind of definition works fine because it tells us that African golden jackals comprise a species. They breed among each other, but they show no sexual interest in the wildebeest or lions around them. Golden jackals, then, are one of the distinct *species* you might see on a safari.

The jackals are certainly a different species from the wildebeests and lions. But are they a different species from the timber wolf? That seems like an easy enough question. In the first place, jackals don't look at all like wolves. They are adapted to hot, dry, open country and timber wolves are adapted to cold, damp forests. Jackals live in Africa, and timber wolves live in northern Europe, Siberia and North America. They are indeed two different species.

But we'd best take another look. If we do, we learn that jackals can mate with domestic dogs, and the offspring are fertile. Furthermore, domestic dogs can and do interbreed with wolves and, for that matter, with coyotes and Australian dingos (Figure 13.1).

So we see that the criterion of "potential interbreeding," timber wolves, jackals, coyotes, and dingos would all be one big, highly variable species, because all of them will interbreed with domestic dogs.

But it's even more complicated than that. It turns out that there are three quite distinct species of jackals, the golden jackal (*Canis aureus*), the side-striped jackal (*C. adjustus*) and the black-backed jackal (*C. mesomelas*). And while they will all hybridize with dogs, they apparently don't hybridize with each other. In the Serengeti of Eastern Africa, the ranges of all three jackals overlap, and there they simply ignore each other. A highly territorial jackal of one species won't even bother one of a different species that wanders through its personal domain. On the other hand, the range of the coyote overlaps slightly with that of the now-rare red wolf (*Canis rufus*) in Texas; and there, wolf-coyote hybrids have been found in nature (Figure 13.2). But still wolves remain wolves, and coyotes remain coyotes, and the two are physically quite distinct animals. Aside from this and the promiscuity of man's best friend, then, the interbreeding situation among the species of the genus *Canis*—eight altogether—appears to be more one of unwillingness rather than inability. The eight species are similar enough *physiologically* to be able to produce fertile hybrid offspring if they should mate, and their similarity places them in a single genus, *Canis.* But they are true species, reproductively isolated in nature, partly because of geographical separation and partly because their behavioral tendencies do not encourage interbreeding.

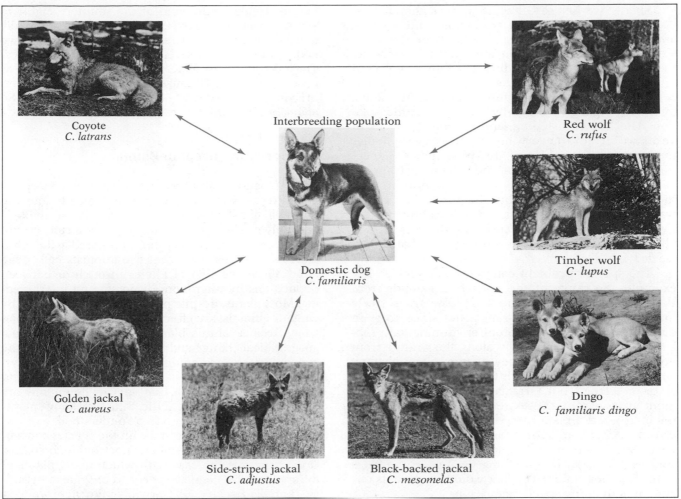

Coyote
C. latrans

Interbreeding population

Red wolf
C. rufus

Timber wolf
C. lupus

Domestic dog
C. familiaris

Golden jackal
C. aureus

Dingo
C. familiaris dingo

Side-striped jackal
C. adjustus

Black-backed jackal
C. mesomelas

13.1 The "dog" species: the genus *Canis*. Certain problems with the species concept are shown here. Each of the *Canis* species here is considered separate, but each is known to mate readily with the domestic dog. However, they generally do not mate with the others in the group, even if they are given the opportunity. If these matings were to occur, the offspring would presumably be fertile hybrids. So are they all members of one highly variable species? The arrows between recognized groups indicate well-documented interbreeding.

13.2 Hybrids in the dog group. (a) Interbreeding between wolves and domestic dogs is fairly common, even though they are recognized as separate species. The parents of these superb animals were a dog *(Canis familiaris)* and a timber wolf *(Canis lupus)*. (b) Hybrids sometimes occur between species in nature. In Texas, where the range of the red wolf (*Canis rufus*) overlaps with that of the coyote (*Canis latrans*), natural hybrids are common—as this 16 month old male.

Should we just say, then, that those animals that do interbreed successfully are of the same species, and let it go at that? It would certainly be the safest bet, but it's a bit arbitrary. Furthermore, it is so restrictive that it would be misleading. The timber wolves of Siberia and Alaska are identical, and no one has suggested that they are different species, but actual interbreeding is prevented by their geographic isolation. The large, glistening, and raucous grackles of Texas are, in almost every important way, very similar to the smaller grackles of the West Indies. Except for size, their physical, vocal, and behavioral traits are alike. They would probably manage to interbreed if they had the opportunity, but the physical barrier of water keeps them apart. It seems misleading to classify them as different species, but there is a total absence of evidence to the contrary. So what should we do?

The species problem can get much stickier. For example, the range of the grass frog extends from Vermont to Florida (Figure 13.3). The frogs freely interbreed all along their range, but since frogs in different regions face different environmental problems, they change gradually along their range from north to south. In fact, they change so much that, whereas a frog from North Carolina may look with favor on a frog from Virginia, Florida frogs and Vermont frogs—from the extremes of the range—will not interbreed. In laboratory forced crosses, the eggs do not hatch. So are Florida and Vermont frogs different species? As of now they are said to be of the same species, but what if something decimated the frogs in the middles states? The link would be gone. Is our system of defining species capricious?

For a natural experiment, we can turn to a group of salamanders in California. Their range extends over a circular route, and at the southern end, two variants overlap (Figure 13.4). They breed continually along their range, but where the variants from the two ends of the circle overlap, they do not interbreed. They do not look alike. They treat each other as different species. Because of the continuity of their breeding range, they are considered one species. So again, our definition gives us problems.

The Species Concept in Plants

The ambiguous situation of the species concept in grass frogs—where one population blends into the next, while distant populations are quite distinct—is probably due to the fact that frogs, as a rule, do not move very often or very far, compared with other animals. This restricts what evolutionists call "gene flow," the movement of alleles through or between populations by migration, dispersal and interbreeding. Most plants, despite pollen and seed dispersal by wind or animals, are effectively even less motile than frogs. Because of this, plants often show very gradual geographical change and there are ambiguities in the concept of species among botanists. Botanical systematists have learned to live with the situation, and tend to accept the idea that species definitions are often arbitrary and pragmatic. Still, they know, plants must be named if they are to be understood.

There are also problems with the species concept that are unique to plants and do not trouble zoologists. One of these is the ease with which many plants of different, well-recognized species continue to hybridize. Largely because of Darwin's influence, we tend to assume that species can arise only by branching off from existing species. This concept has problems

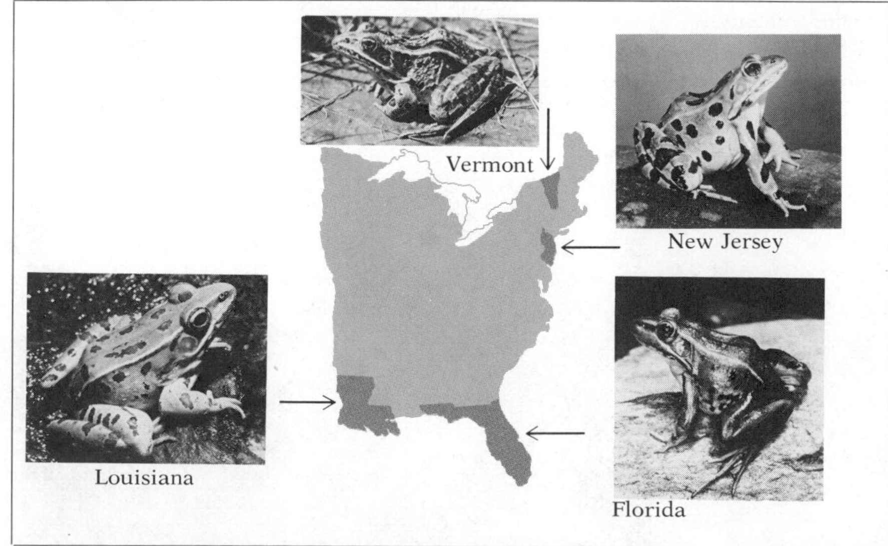

13.3 *Rana pipiens,* the grass frog. All of these frogs are in the same species of the genus *Rana,* although there are important differences between them. Neighboring populations can interbreed, forming a breeding continuum throughout the range. However, frogs from extreme ends of the range (Florida and Vermont) cannot interbreed. By some definition, then, the Florida and Vermont varieties are really two different species. No one seems anxious to correct this small embarrassment, though. Obviously, tradition often influences our classification of animals and plants.

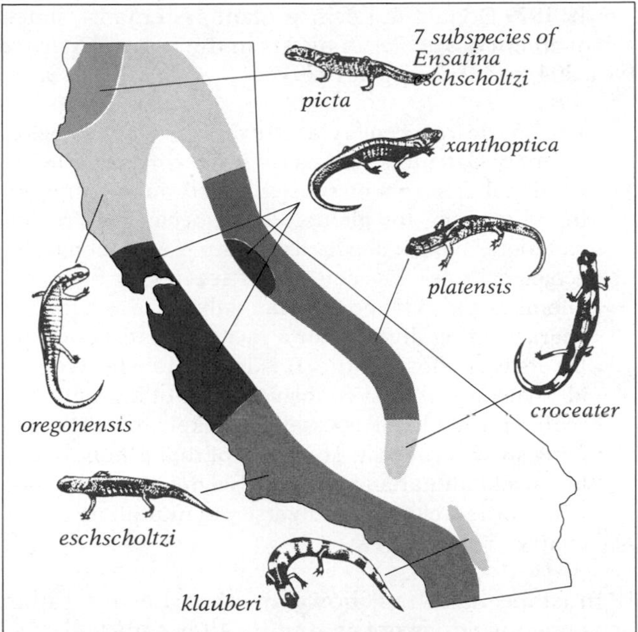

13.4 *Ensatina* and the species argument. Populations of the salamander, *Ensatina eschscholtzi* are found in California along the coast and in the inland mountains. Skin color and size varies more or less continuously along the range, which forms a circular shape, as shown on the map. In the southernmost end of the range, two variants overlap, sharing the same territory, and individuals of each type may live just a few yards from each other. These variants, as far as can be determined, do not interbreed. Yet studies made along the range of the salamander indicate that neighboring variants can and do interbreed. Have the southernmost varieties evolved into two species? So far there hasn't been much agreement on the answer.

enough of its own, since evolution is a continuous process while "species" is a discrete category. But in fact speciation doesn't always follow the line of bifurcation—one species becoming two—not in plants, anyway. In plants there are many, many examples of new species arising by the *fusion* of other species. One problem is that we don't know when or how often such evolutionary events might have happened.

Hybrids between animal species are usually infertile, if they survive at all; if fertile, the offspring are usually weak and unable to compete effectively, so hybridization is almost always an evolutionary dead end in the animal kingdom. But not in plants. There are many examples of human-mediated hybridizations, some of them representing the great agricultural advances of our time. And it is apparent that the same thing happens in nature. "Hybrid swarms," or populations of hybrid organisms, often exist in great numbers, especially at the interface between the ranges of the parent species. Live oaks (of the genus *Quercus*) hybridize so readily and so often that some people maintain that there is no such thing as species in the genus *Quercus*. Fossil records show that extensive hybridization has been going on in this genus for many millions of years, while the different forms (species) of *Quercus* have remained as distinct as ever. Zoologists find such situations extremely enigmatic.

Hybridization can only be considered successful in an evolutionary sense when at least some of the offspring are vigorous, fertile and able to contribute genetic material to the future. Whereas such hybrids may form new species, at other times the hybrid plants cross back to one of the parental strains, and their descendants are eventually absorbed into it. This is an important evolutionary event because it means that genetic material from one species enters the gene pool of another, a phenomenon called *introgression*. Introgression makes tracing evolutionary relationships very, very difficult. And tracing evolutionary relationships is one of the major jobs and chief tools of the systematist.

Most interspecific hybrids in plants are not very fertile, even if the hybrid plants themselves appear vigorous. One reason is the frequent failure of meiosis in hybrids. In meiosis, you will recall, the homologous chromosomes must pair and exchange parts. If the two chromosome sets are from different species, they may not pair effectively or they may not pair at all. Unbalanced gametes, and inviable zygotes, are the usual result. So great is the potential reproductive performance of most plants, however, that even if only a fraction of a percent of the gametes produce viable seeds, that small percentage may be enough to propagate the hybrid. For instance, if a hybrid is crossed back to one of its parent, a few of the hybrid gametes may have—by pure chance—just the right balance of chromosomes to match those of the parent. That gamete, then, with the gamete from the parental strain, will result in a viable seed. We should add that it's possible for a whole chromosome or part of one thus to be introduced into the parental species, where it just might be a big hit, spreading rapidly through the population as an important genetic advance. Modern plant breeders frequently use this technique deliberately to introduce some valuable characteristic—some fungal resistance factor, for instance—from one species into another.

Allotetraploids: Instant Species

Something even more dramatic can happen with a hybrid plant. Some of its cells can become *polyploid* through the failure of nuclear division after a mitotic chromosomal replication. When the chromosomes of such cells do not separate, the cells then have *four* complete sets of chromosomes, two sets from each parental species. It is possible that only part of the plant such as a flower or a branch might become polyploid this way. When meiosis occurs in such a cell, the chromosomes have no trouble in finding partners just like themselves. The (doubled) chromosomes of species **A** pair with their species **A** counterparts, and the (doubled) chromosomes of species **B** pair with their counterparts. If the plant is one that is capable of self-fertilization, it is an instant new species—called an *allotetraploid.* If parental species **A** had, say, 12 chromosomes, and parental species **B** had 14, the new species would have 26 chromosomes—complete diploid sets from both parents. Its offspring would in turn have 26 and so on as the descendants grew in number. They would constitute a unique and perfectly good species by every definition. Many plant species, both wild and domestic, are polyploid, most of them having arisen by species hybridization in just this way. Wheat is one example. Most wheat is actually an *allohexaploid*, a genetic combination of the full diploid chromosome complements of three entirely different Middle Eastern species of wild grass.

Polyploidization also occurs in animals, but rather rarely. There are polyploid lizards and salamanders, but they reproduce only asexually. Chromosomal sex determination probably rules out successful polyploidization, because polyploidy would disrupt the necessary meiotic pairing between X and Y chromosomes. (If there are two Xs and two Ys, the Xs will pair with Xs and the Ys with Ys instead of Xs with Ys.) On the other hand, there are many polyploid fish; in many groups of fish, unlike in most other animals, sex is not determined chromosomally—individual fish can even change sex as they get older.

Interestingly, the plant allotetraploids are not necessarily reproductively isolated from their parent species. A tetraploid plant may sometimes cross back to a diploid parent, producing a highly infertile triploid hybrid in which meiotic pairing is badly disrupted. A few gametes of this triploid hybrid, however, might just happen to get the right balance of chromosomes—just by chance—and produce viable seeds when crossed back to one or the other of the parental species. Thus a small but continuous amount of introgression may go on between members of a *polyploid species complex*: a number of related species, including diploids and polyploids, that continue to exchange chromosomes and genes. This has been documented in wild wheat, for instance.

Polyploidy may cloud the evolutionary relationships and species definitions among plants, but it presents an excellent mechanism for increasing the diversity of the plant kingdom. For reasons not fully understood, polyploid species often have much greater tolerance to harsh climatic conditions. As evidence, one study revealed that 26% of the vascular plants studied in tropical regions were polyploids, while 86% of all species in northern Greenland were polyploid. In more moderate climates, the percentage is intermediate. So it appears that polyploidy is more than a freakish occurrence in plants, but it is instead enormously advantageous to plants and many of them have availed themselves of its benefits.

In 1979 Donald A. Levin, a plant systematist, stated the problem of species in plants in the journal *Science* (vol. 204, p. 381, 27 April 1979) :

> The species concept is a central tenet of biological diversity. Attempts to describe diversity have led to empirical concepts of species based on assumptions found wanting for plants. Plant species lack reality, cohesion, independence, and simple evolutionary or ecological roles. The concept of species for plant taxonomists and evolutionists can only serve as a tool for characterizing diversity in a mentally satisfying way. Diversity is idiosyncratic. It is impossible to reconcile idiosyncrasy with preconceived ideas of diversity. The search for hidden likenesses is unlikely to yield a unifying species concept. The concept that is most operational and utilitarian for plants is a mental abstraction which orders clusters of diversity in multidimensional character space.

It must be admitted, however, that Levin's rather extreme position is not shared by all botanists.

Type Specimens and Biometrics

Our large museums are full of *type specimens.* These are the preserved remains of individuals, sometimes the first of their kind to be discovered, against which other animals and plants are measured in order to determine whether they fall into that species. If they are the same in critical points, they are said to be of the same species. If they diverge too widely, they are not. For a long time, those individuals that conformed closely to the type specimen were said to comprise a "good" species. Variants were not thought to be "good"—there was some question about their lineage.

Type specimens were common in Darwin's day. His geologist friend, Charles Lyell, recalled meeting a dealer in seashells who discarded any variant. He said his customers would reject even a beautiful shell if it were not of a "good" species.

We no longer rely so strongly on type specimens to indicate what the traits of a species "should" be. That aged process is yielding to more general approaches, including the mathematical assessment called *numerical taxonomy.* Numerical taxonomy is extremely complex and differs from the use of type specimens not only in the measurement of more traits (which yields more precise descriptions) but in the recognition of variation within species. Admitting variation, of course, means that one must deal with *populations*, since there is no single representative (or ideal) animal or plant. According to this philosophy, then, a species has no single phenotype but is a *population of variants*.

Numerical Taxonomy and the OTU

Numerical taxonomy is the statistician's approach to taxonomy. Numerical taxonomists take pains to reject the intuitive or "subjective" approach of traditional taxonomy, and will have nothing to do with "the species problem": even the word *species* is dropped, and every grouping is flatly called an *OTU*, or *operational taxonomic unit*. The ideal of numerical taxonomy is this: every individual collected is rated with regard to a large array of *character states*, with each rating being either 1 (present) or 0 (absent). The more character states that can be thought up, the better. For example:

five digits on front legs: present (1).
four digits on front legs: absent (0).
zygomatic arch wider than it is long: present (1).
anthocyanin pigments in inner petals: absent (0).

Of course, there is no benefit in considering any character state that is always present or always absent in the organisms being looked at; so *two eyes in head*, and *three digits on each extremity*, would be equally useless character states in a study of monkeys, all of which have the first and none of which have the second character state.

Ultimately each specimen in the numerical taxonomist's group of organisms to be studied is represented by a number, a long string of ones and zeroes. These numbers are then fed into a computer. In an elaborate statistical procedure, the individuals of various kinds are grouped into a series of clusters within clusters, where "cluster" simply means a group that shares in common a certain percentage of character states. Each cluster, and each larger cluster of clusters, is an OTU; the smallest clusters may or may not correspond to traditional species, and the larger clusters (comprised of groups of smaller clusters) may or may not correspond to families, orders, and so on. As a matter of fact, however, the "objective" clusters of numerical taxonomy and the "subjective" classifications of traditional taxonomy usually coincide quite nicely, which would seem to reinforce our faith in both procedures. Where there are differences between the two methods, they tend to show up just where you would expect: in those groups where traditional taxonomy has always had the most problems and the most apparent contradictions.

The numerical taxonomist has accomplished two things: first, due recognition is given to variability that occurs within species, which are now correctly seen as populations consisting *entirely* of variants. Each species—pardon, OTU—is seen as a variable population with a range of phenotypes and ecological adaptations, separable statistically from all other such groups. Note that this is a completely phenetic (descriptive) approach; in the cluster analysis, no inferences are made about past genetic relationships. For that matter, present genetic relationships (the hybridization criteria) are also ignored.

The other accomplishment, or claim, of numerical taxonomy has been the banishment of subjectivity and its replacement with complete objectivity. Some traditionalists doubt this, pointing out that the choice of character states depends on an intimate knowledge of what to look for in the groups concerned, and is entirely subjective.

At a recent scientific lecture a leading numerical taxonomist was describing the superiority and objectivity of his methods, giving as an example his recently completed analysis of the phenetic relationships of a group of mosquitos. A traditionalist critic asked how he was able to prevent the inclusion of a horse in his sample. "A horse?" "Yes," repeated the critic, "what keeps you from including a horse along with the mosquitos?"

"Well," said the numerical taxonomist, "I may be a statistician but I can tell a horse from a mosquito."

"Then you've just made your first subjective judgement," came the rejoinder.

How Species Are Named

Humans come in two types: those who like asparagus and those who don't. Or those who like biology and those who don't. The point is, it's only human to divide the world up, pigeonholing everything.

Pigeonholing living things has proved difficult, however, because people disagree on how large the hole is, if it exists at all, and what should go into it. We've already mentioned part of the problem: Living things vary. How much can an animal vary from the others before it goes into a different hole (that is, gets a different name)? Another problem is, we don't know enough about relationships. We need to know about these things, because in each hole are yet smaller holes and these have holes that are smaller yet. It's not too difficult to place most things in the largest holes. For example, if it has leaves and roots, we can safely place it in the great pigeonhole labeled "plants." If it has ears and it bites, it's definitely an "animal." But not all plants have leaves and not all animals have ears, so the story already grows complicated.

What we've done is divide living things into categories, and categories within categories. The largest categories are called *kingdoms*, and the smallest are *species* (or, in some cases, *subspecies* or *varieties*). As we proceed from kingdoms into smaller categories, the plants and animals in each category have more and more traits in common until they are so much alike that they can interbreed. These are the species. We'll discuss these categories shortly, but first we should say something about the *binomial system*.

Binomial means "two names." This simply refers to the way scientists label living things. Thus, humans are *Homo sapiens*. Our genus (or our *generic* name) is *Homo*. Our species (or specific) name is *Homo sapiens*. (Note that when species are called by their scientific names—generic plus specific names, both are italicized. Sometimes, however, the generic name can be abbreviated. For example, *H. sapiens* and *E. coli* are acceptable when the reader knows what the *H.* and *E.* stand for.)

The two-name system of classification was developed by Karl von Linne, who latinized his own name to Carolus Linnaeus (1707–1778). Linnaeus is responsible both for the idea of categories within categories and for naming things with two Latin names. Latin was chosen because it's a dead language, not commonly spoken anywhere outside a few academic halls or in religious ceremonies. So it isn't likely to change much. It's also the root language of a number of present-day languages and therefore latinized names can transcend language barriers.

In a monumental undertaking, Linnaeus singlehandedly classified and named many of the earth's creatures; but today, the rules for assigning names are rigidly enforced by international commissions. In fact, Charles Darwin sat on the very first such commission. There are now international nomenclature commissions for zoology, botany, bacteriology, and virology (not to mention enzymology, organic chemistry and just about any other branch of science where names are important).

13.2 Systematics

"King Philip Came Over from Greece, Singing Songs"

This nonsense about King Philip, or some ribald variant of it, has been memorized by generations of biology students, some of whom have grown old and died, others just old, some now writing textbooks and some, it must be allowed, who find after twenty years that it is the *only* part of biology they can still remember. It is actually simply a device for remembering the categories of living things (Figure 13.5):

Kingdom
Phylum
Class
Order
Family
Genus
Species
Subspecies

13.5 Classification of two mammals and a tree. The two animals whose classification from kingdom to species are listed here represent the largest and smallest members of a diverse class, the mammalia. The great blue whale grows to about 100 feet in length, can weigh over 150 tons, and devours 8 tons of food per day. The shrews weigh about 4 grams (0.14 oz). Relatively speaking, shrews eat far more than whales. In fact, a shrew eats the equivalent of its body weight each day in an effort to keep up with its record-breaking metabolic rate. Both these animals have hair, are warm-blooded, suckle their young, and nourish their embryos through a placenta. As it happens, neither species is divided into subspecies. The red maple is a flowering plant of considerable size and is rather closely related to those other species that are also called maples.

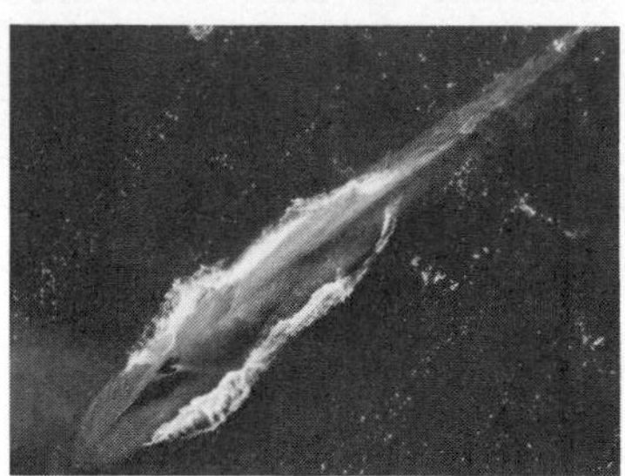

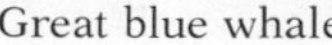

Great blue whale

Dwarf shrew

Red maple

Kingdom:	Animalia	Animalia	Kindgom:	Plantae
Phylum:	Chordata	Chordata	Division	Tracheophyta
Subphylum:	Vertebrata	Vertebrata	Subdivision	Pterophyta
Class:	Mammalia	Mammalia	Class	Angiospermae
			Subclass	Dicotyledoneae
Order:	Cetacea	Insectivora	Order	Sapindales
Family:	Mysticeti	Soricidae	Family	Aceraceae
Genus:	*Balenoptera*	*Suncus*	Genus	*Acer*
Species:	*B. physalus*	*S. etruscus*	Species	*A. rubrum*

(For plants and bacteria, the term *division* is often, but not always, used in place of *phylum*. The other categories remain the same. King David . . . ?)

In essence, these are just names applied to groups of organisms assumed to be related to one another. Each group, from subspecies to kingdom, is a *taxon* (pl. *taxa* from *taxis*, "to put in order"). Naming, of course, is necessary if we are to talk or write about living things. But we must keep in mind that names are human contrivances, created for human purposes, and have no other importance and no separate reality in themselves. At the same time, if a name is going to be useful, it shouldn't be totally arbitrary, but should reflect some kind of reality. Let's look at the "reality" of the taxonomist.

All earthly living things are related by descent, as we have noted when considering glycolysis and the genetic code. This means they are not only related to each other, but to each other's ancestors. Your distant ancestors had more in common with the ancestors of a crow than you do with today's bird. Darwin proposed an interesting analogy using a tree (Figure 13.6):

> The affinities of all the beings of the same class have sometimes been represented by a great tree. I believe this simile largely speaks the truth. The green and budding twigs may represent existing species; and those produced during each former year may represent the long succession of extinct species. At each period of growth all the growing twigs have tried to branch out on all sides, and to overtop and kill the surrounding twigs and branches, in the same manner as species and groups of species have tried to overmaster other species in the great battle for life. The limbs divided into great branches, and these into lesser and lesser branches, were themselves once, when the tree was small, budding twigs; and this connexion of the former and present buds by ramifying branches may well represent the classification of all extinct and living species in groups subordinate to groups. Of the many twigs which flourished when the tree was a mere bush, only two or three, now grown into great branches, yet survive and bear all the other branches; so with the species which lived during long-past geological periods, very few now have living and modified descendants. From the first growth of the tree, many a limb and branch has decayed and dropped off; and these lost branches of various sizes may represent those whole orders, families and genera which have now no living representatives, and which are known to us only from having been found in a fossil state. As we here and there see a thin straggling branch springing from a fork low down in a tree, and which by some chance has been favoured and is still alive on its summit, so we occasionally see an animal like the *Ornithorynchus* [platypus] or *Lepidosiren* [dugong], which in some small degree connects by its affinities two large branches of life, and which has apparently been saved from fatal competition by having inhabited a protected station. As buds give rise by growth to fresh buds, and these, if vigorous, branch out and overtop on all sides many a feebler branch, so by generation I believe it has been with the great Tree of Life, which fills with its dead and broken branches the crust of the earth, and covers the surface with its ever branching and beautiful ramifications.

13.6 Darwin's tree of life. The growing tips of twigs on Darwin's phylogenetic tree represented living species. These branched from recent ancestral types, which in turn diverged from more primitive ancestors. The trunk itself represented the original ancestral form of life. Dead twigs are extinct species, and some of the lower branches are entire extinct groups. The progressive division into trunks, limbs, and branches may correspond to higher taxonomic groupings. The concept is useful for illustrating relationships among various groups or species and is frequently used for this purpose.

Darwin saw contemporary species as the fresh growing tips of twigs, with their ancestors represented by boughs, who were the descendants of branches, and ever-larger limbs, finally returning to the original great-grandparent of them all—the common ancestor of all living things, the trunk.

Such trees are still used in order to depict *phylogeny* (*phylo*, kind; *geny*, origin of; so, "origin of kinds"). The tree results from virtually all species originating from the splitting and subsequent divergence of earlier species.

Instead of a tree, Darwin could have used the metaphor of boxes within boxes within boxes. The largest boxes would be the kingdoms; and each box would contain one or more smaller boxes until the smallest box, the species, is opened to reveal a plethora of individuals. Divisions below the species level, such a subspecies, race, variety, or local population, are of intense interest to the evolutionary and population biologist, but they generally don't interest the systematist or the taxonomist.

And what, then, is a systematist? The *systematist* is a biologist who tries to determine the evolutionary relationships of organisms. The systematist is interested in the tree and not the names. The *taxonomist*, on the other hand, keeps track of the names—especially the species' names. Much of the taxonomist's job consists of forming precise descriptions of new species, or of identifying unclassified specimens. Systematists, of course, are often taxonomists as well, but the two groups are quite distinct, as evidenced by some of their intense arguments.

When you think about it, there can be only one true phylogeny. History unfolded along one route. The problem is we have trouble tracing it. Our understanding of it changes, and people disagree about the evidence. It is the same problem that makes the study of classical history so vexing. Whom are we to believe? In biology, however, we have an additional problem and it centers over philosophy. "Lumpers" prefer fewer, larger taxa; "splitters," given the same relationships, prefer more numerous, smaller taxa (Figure 13.7).

Changes in taxonomy (naming) follow, or at least are supposed to follow, new discoveries in systematics (relationships). What was formerly thought to be one species of fiddler crab, for instance, has now been found to be three species with very similar appearances. It turns out that they are biochemically distinct and will not mate with one another. In another case, what was originally described as two distinct species has been found to be the larval and adult forms of a single species. At the level of higher taxa, groups that were thought to be closely related are now considered not to be—or vice versa.

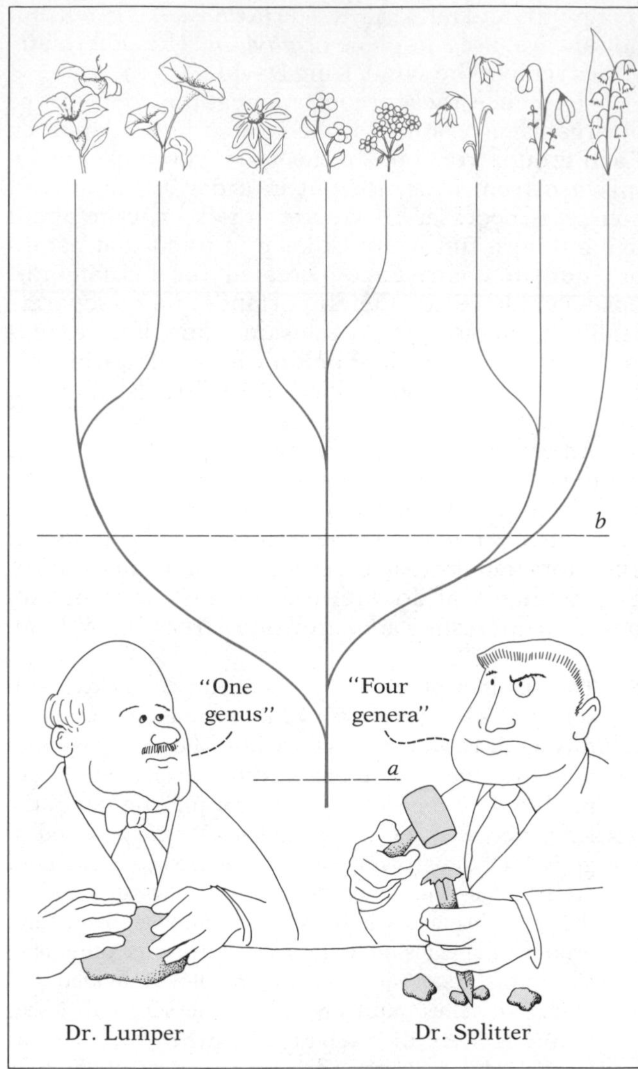

13.7 Both Dr. Lumper and Dr. Splitter agree that there are eight species of plants being classified. Dr. Lumper, however, places all eight in one genus, believing that the species all descended from a single genus that flourished at time *a*. Dr. Splitter disagrees, and calls for four genera. He concludes that the eight species descended from four distinct groups that already had been established at time *b*. By the looks of things, both may be right, depending on where one draws the line—literally. Arguments among lumpers and splitters are more common when the higher taxa are being considered. Thus, we encounter such taxons as *superfamily*, *infraclass*, and *suborder* as taxa are arranged and rearranged in ever-changing phylogenetic schemes.

Monophyletic, Polyphyletic, and Paraphyletic Groups

Things change so frequently that the doggerel about King Philip has lasted longer than the names of many of the taxa it celebrates. The names and groupings of organisms must always be considered provisional, subject to change based on new information.

In the ideal evolutionary taxonomy, every taxon would be represented by a part of a branching tree, and any branch could be, metaphorically, broken off at a single point and waved around as a single, discrete unit. In such a system, each taxon would have a single ancestral species that *would also* (if it were still alive) *fit the definition of the group*, and when any branch were broken off, there would be no members of this group left on the tree. *Every* living member of the taxon would be more closely related to every other member than *any* would be to the member of another taxon. Such an ideal group is said to be *monophyletic.*

For instance, the vertebrate class *Aves* (birds) is a monophyletic group (Figure 13.8). As far as we can tell, all birds have descended from some single species in the past, and that ancestral species was itself a true bird. Thus any two birds are more closely related to each other than either is to any other animal. Similarly, the mammals appear to constitute a monophyletic group. However, some dissenters theorize that the monotremes (egg-laying mammals, the platypus and the echidna) arose from reptilian ancestors independently of the other mammals. If, indeed, the last common ancestor of the monotremes and the other mammals should prove to have been a reptile, then we would have to say that the mammals constitute a *polyphyletic* group—a group with more than one point of origin. That is, they now have similar characteristics that cause us to label them as mammals, but they arose from different reptilian lines and independently evolved similar features.

13.8 The phylogenetic tree of terrestrial vertebrates. The branches bearing the mammals and birds illustrate what is meant by *monophyletic* groups. Each group has a common ancestor of its own that gave rise to no other groups (*a for the mammals, *b for the birds). The reptiles are an entirely different matter, as you can see by the multiple branching. The reptiles are a *paraphyletic* group: all are descended from a common ancestor that was also a reptile (*c); but this common ancestor also gave rise to birds and mammals, which are not reptiles. Today's amphibians represent a mystery, but may be a *polyphyletic* group with several fishlike ancestors.

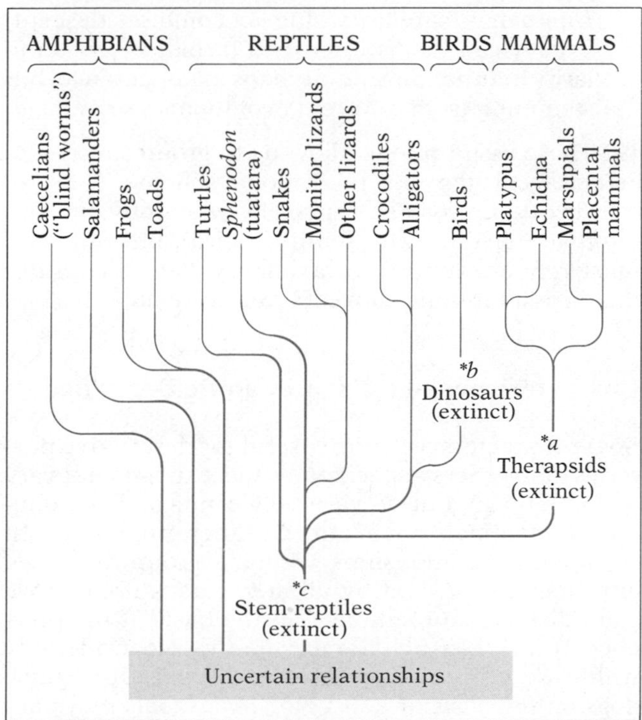

A clearer example of a polyphyletic group is the fungi. Some say the fungi are a division of the plant kingdom, and others consider it to comprise a unique kingdom of its own; but in either case, it is a single taxon. However, it is almost certainly a polyphyletic taxon. *Slime molds* are usually considered to be fungi, because of their appearance, their saprophytic habits and, above all, because of their stalked fruiting bodies. They certainly look like fungi, yet molecular studies clearly show that slime molds are much more closely related to certain protozoa than they are to other fungi. For that matter, two kinds of slime molds, the cellular and the acellular ones, are extremely similar to one another in habits and morphology but are probably not at all closely related. They are grouped together because of phenotypic convergence. They have similar traits; however, their similarities are not due to a shared common ancestry but to *convergent evolution* (see Fig. 13.10).

Rigid adherence to the ideal of monophyletic groupings clearly doesn't always work. Even if convenient polyphyletic groups could be broken up successfully, we would still need another kind of group, the *paraphyletic* taxon. In a paraphyletic group, every member is descended from a common ancestor that also belongs to the group, just as in any monophyletic taxon. However, a paraphyletic group is one that has itself given rise to other groups that have diverged to the extent that they no longer belong to the parent group.

A familiar paraphyletic group is the vertebrate class *Reptilia*, the reptiles. Reptiles have a number of features in common, and they are all derived from one ancestor that was also a reptile. But some reptilian ancestors gave rise to nonreptiles, namely, the birds in one case and the mammals in another. In Figure 13.8, note that one line of descent leads to the crocodilians, dinosaurs, and birds, while other, earlier departures lead to turtles, to mammals, and to snakes and lizards.

If we were to quantify "closeness" of evolutionary relatedness in terms of recentness of common ancestry—which makes sense—then birds are more closely related to alligators than they are to lizards. More to the point, alligators are more closely related to birds than they are to lizards, even though alligators *look* a lot more like lizards. Appearances can be deceiving.

While we're on the subject of lizards, they too form a (smaller) paraphyletic group. The monitor lizards are believed to be more closely related to snakes than they are to other lizards. On the other hand the "glass snake" is not really a snake at all, but a legless lizard.

Paraphyletic taxa are common in the plant kingdom. The two major groups among the flowering plants are the monocots (grasses, palms, lilies, orchids,

Essay 13.1 Systematics and Taxonomy: Are They Really Science?

One definition of *science* is "systematic knowledge of the physical or material world," so the fields of systematics and taxonomy are definitely science—in one sense at least. But our gut feeling about science is that ideally it is a process involving the generation and testing of *hypotheses.* And this involves making *testable predictions.* So, are systematics and taxonomy really sciences in this sense? Again, the answer is yes. Every conclusion a systematist draws about the probable evolutionary relationships of a group of organisms is, in essence, a hypothesis subject to further testing. The systematist predicts that any further investigations will support his or her hypothesis.

For example, a systematist examines the seed anatomy of three plant species A, B, and C, previously thought to be all closely related and thus classed as members of a single *genus.* He or she concludes that A and B are indeed dosely related, but that species C resembles A and B only superficially and is in fact much more distantly related—the *systematics* part of the work. In order for the *taxonomy* of this group to remain consistent with established principles, the systematist creates a new genus, and gives species C a new name—the taxonomy part of the work. The new classification, like all biological classification, is tentative. Once published, it becomes the new predictive hypothesis. It predicts, for instance, that an examination of protein structure in the three plants (perhaps by a different worker in the distant future) would show that A and B have relatively similar proteins, and the protein structure in species C would be less similar. A contrary finding could always prove the hypothesis to be wrong, of course, and this would require another renaming or possibly a revival of the original names and relationships.

It has sometimes been said that the theory of evolution is an untestable hypothesis. But it is being tested every day, in great detail, in the field and in the laboratories of systematic biologists.

---

and so on) and the dicots (beans, apples, magnolia, clover, geraniums, oaks, avocado, and many more). The two groups are named for the number of cotyledons in the plant embryo. The monocots are a monophyletic taxon. Although the origin of the monocots is obscure, it is believed that the monocots branched off from one group of the dicots, making the dicots a paraphyletic taxon. Which is just to say that some dicots are more closely related, in time at least, to the monocots than they are to certain other dicots. Among the nonflowering plants, the living gymnosperms (ginkgo, cycads, conifers) are also a paraphyletic group.

Who Is Related to Whom?

Phylogenetic trees are constructed to indicate how species and groups of species are believed to have diverged from one another in the past. How are such decisions made? One must begin with assumptions:

1. Characteristics of a taxon can be described.
2. The degree of similarity between two taxa can be described.
3. Closely related taxa will be similar morphologically and biochemically.
4. The degree of similarity between taxa decreases over time—generally, the greater the passage of time since common ancestry, the greater the difference.
5. Characters once lost in evolution tend not to be regained.
6. *Homology* (similarity due to common descent) usually can be distinguished from *analogy* (similarity in function, and perhaps in appearance, but stemming from different evolutionary origins).

These assumptions allow us to group species together according to some rational scheme. The process is often informal, especially when obvious and intuitive (anyone can identify a bird), but often we must rely on numerical taxonomy. Let us consider the six assumptions above in greater detail.

Characteristics of a Taxon Can Be Described

Some characters are more useful for descriptive purposes than others—specifically, those that do not vary within a taxon, but do vary between taxa. For example, all mammals have hair; no other group does. In considering relationships within a group, however, any character shared by all members is useless; we can't distinguish mammals on the basis of the presence of hair. Most characters that vary considerably within any group are also useless; we can't distinguish dogs on the basis of hair color, since coats may vary widely even within the same litter of puppies. In classical taxonomy, just as in numerical taxonomy, each character is considered to exist in one of two specific and identifiable *character states*. Traits that may vary gradually between individuals are generally useful only if they do not overlap between taxa; for example:

1. Leaves more than 4 cm long.
2. Leaves less than 4 cm long.

The leaves can vary in length as long as the categories do not overlap.

The Degree of Similarity Between Two Taxa Can Be Described

In numerical taxonomy, the degree of similarity between two taxa is reflected by the *number of characters shared*. In some schemes, certain characters may be considered more important (given more weight) than other characters, but generally, if characters are good enough to be used at all, they are all given the same weight.

Closely Related Taxa Will Be Similar Morphologically and Biochemically

This assumption simply states that newly diverged species will retain many of the same traits. This seems like sheer common sense, but departure from it almost destroyed the serious study of biology in the Soviet Union. A charlatan who gained favor with Stalin, T. D. Lysenko, claimed that oat seeds spontaneously appeared on wheat plants. The rise of dramatically new and different species *de novo* has occasionally been suggested throughout the history of biology, but generally, such beliefs are without foundation—evolution is usually a slow and conservative process. Even when new plant species arise suddenly by allotetraploidy, the two parent species are closely related to one another and the new allotetraploid species is recognizably similar to both.

The Degree of Similarity Between Taxa Decreases Over Time

Generally, the greater the passage of time since two branches diverged from a common ancestor, the more different they will be. Theoretically, species *can become more similar* over a time, at least with regard to a specific character, through convergent evolution. Convergent evolution is due to a similar response of two species to a similar environment. For example, Darwin noticed that the South American mara was very similar to the European rabbit (Figure 1.1) but must have developed rabbitlike traits independently. Convergent evolution is sometimes striking (Figures 13.9 and 13.10). But generally, species tend to become increasingly different, rather than more similar, and over time the number of differences continues to increase. The differences accumulate much faster at some times than others.

Rates of morphological evolution fluctuate among different species. Some lines of descent, such as birds, are characterized by relatively rapid changes; in others, such as horseshoe crabs, there is very little change over very long periods of time. The members of slowly evolving groups are often called *living fossils*, because they closely resemble their ancient ancestors.

13.9 Divergent and convergent evolution. This hypothetical phylogenetic tree describes both divergent and convergent evolution. From a common ancestor, two lines diverged far back in time. As the two types continued to evolve, individual species emerged in a diverging pattern. The distance between branches indicates the degree of divergence. In the center we see an example of convergence, as branches from each of the original lines come close together. The species represented as V and VI, although they may be totally separated geographically, have followed a similar course in their evolution and made similar morphological adaptations. They may both be flyers or swimmers, or they may have adapted to open grassland, but their similarities are striking. When their structural features have been examined closely, it may be found that they are merely analogous features. This means, for example, that they both have wings, but these develop differently, even to the point of being derived from different embryological origins (such as the wings of birds and insects).

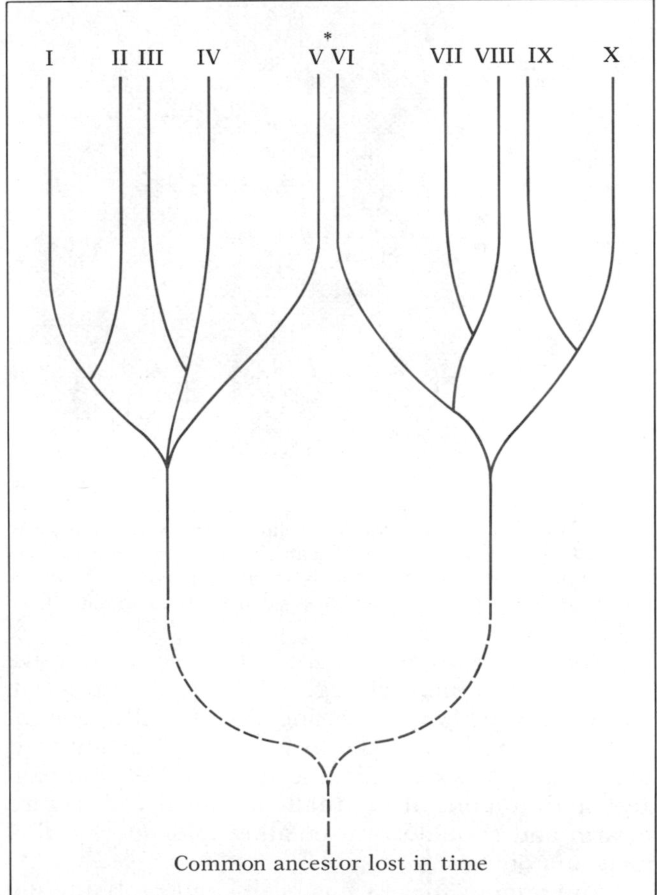

13.10 Examples of convergent evolution. (a) A tropical Asian cyprinid fish (above) and an African characin. (b) An Australian agamid lizard (above) and a North American iguanid. (c) An aye-aye (left), a primate from Madagascar, and a striped opossum from New Guinea, both of which use specialized middle fingers to extract insects from trees. (d) The eastern meadowlark of America (left) and the yellow-throated longclaw of Africa. (e) Unrelated cactuslike plants adapted to different desert environs.

Sometimes some morphological features evolve slowly while others change with surprising speed. For example, chimpanzees and humans differ considerably with regard to skulls, jaws, pelvises and feet; those characters have diverged rapidly. Yet chimpanzees and humans are virtually identical with regard to arm and shoulder musculature and internal thoracic anatomy.

The human species has been—and probably still is—the most rapidly evolving animal species known with regard to skeletal morphology and, especially, behavior (including walking, talking, and building automobiles). We are biochemical slowpokes, however. Humans and chimpanzees have identical hemoglobins, fibrinopeptides, cytochrome *c*, and lysozyme. On the molecular level, chimpanzees and humans are just about the most closely related two species known. We are biochemically far more similar than, say, horses and donkeys, or brown and black bears (Figure 13.11).

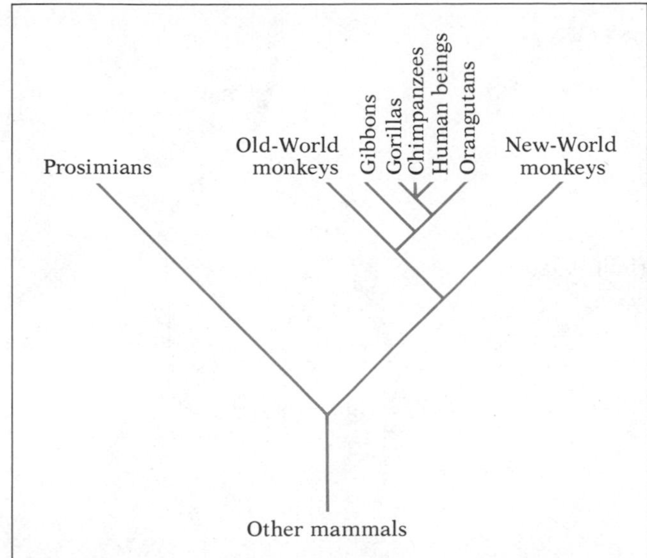

13.11 Evolutionary distance between humans and the other primates. The tree shown here was prepared from biochemical studies of the primates. The distances indicate relative biochemical similarity. It is still not known whether chimpanzees are more closely related to humans or to gorillas, hence the three species are shown as diverging from one point.

Characters Once Lost in Evolution Tend Not To Be Regained

This is known as *Dollo's law.* It is not a law in the strict sense, but a generalized observation. Of course, old folk tales mention "throwbacks" to ancestral states, but this is probably based on a failure to understand Mendelian inheritance of recessive characters. Once genetic information is gone, that's it. Louis Dollo based his diagnostic conclusion on consideration of fossils: Shell or skeletal features that disappear in certain geological strata rarely reappear in more recent rock. Keep in mind that an *analogous* feature may be reevolved; thus, dorsal fins reappear on some dolphins, but these fins are not *homologous* with the dorsal fins of the dolphin's fishy ancestor.

An interesting apparent exception to Dollo's law is the appearance of C4 photosynthesis in widely scattered, distantly related plant species. In all cases both the detailed biochemical pathway and the specialized C4 tissue anatomy are the same. At present this phenomenon defies explanation.

Homology Usually Can Be Distinguished from Analogy

This is a necessary assumption for the systematist. If a systematist can't make such distinctions, his or her work is meaningless.

The area of *comparative anatomy* is largely a study of homologies. One may begin by studying the segmented musculature of a tiny fish and then follow these segments through increasingly complex specializations, finally to arrive at mammalian musculature with a clearer idea of how such a system could have evolved. The task is not always easy, nor the conclusions immediately evident. Often only the fossil record retains the necessary clues. *Homologous* bones, muscles, and other organs have undergone extensive modifications in different lines. Homology of structures is suggested if there are intermediate forms between two existing species, or if the fossil record indicates that particular structures on two living forms have descended from a common ancestral structure. Homology is also indicated if the structures being considered in two species have a common embryonic origin. The bones of the forearm, for example, show a fairly clear homology (Figure 13.12). Even obscure homologies can be traced by these methods. The tiny bones of the middle ear are homologous with reptilian jaw bones, which in turn evolved from the bony supports of the gills of ancient fish. Muscle homologies in vertebrates can be inferred from intermediate stages deduced from fossil skeletal muscle attachments.

The problem of studying homology, of course, is complicated in those species in which we lack fossils. This is one reason the taxonomy of the tiny protists is so tenuous; we have no ancient fossil monuments of their ancestors.

Formal Construction of Trees

The animal and plant kingdoms are so vast and varied, we are often at a loss to describe the pedigree of a living species. The fossil record is simply incomplete, or absent. Therefore, constructing a tree is quite a task. If we simply lump living species, we end up with incorrect trees. For example, lizards and crocodiles would be positioned on the same limb (and, as we mentioned earlier, this wouldn't be correct). A somewhat better procedure is to construct all possible trees and choose the one that requires the smallest number of evolutionary changes. We could call this the *minimum mutation tree.* The problem is, nature does not necessarily minimize the number of necessary evolutionary changes. Natural selection can be a wastrel indeed. Another problem is, for a group of any size, the number of possible trees is astronomical—for only 15 living species, there are exactly 7,905,853,580,625 theoretically possible trees! For sixteen species there are 27 times that many. Even computers can't evaluate that many trees. Thus, taxonomists must take some shortcuts, such as starting with smaller trees and adding a branch at a time. Even then, it takes a high-speed computer to consider just the more plausible relationships. Figure 13.13 is based on such a computer analysis.

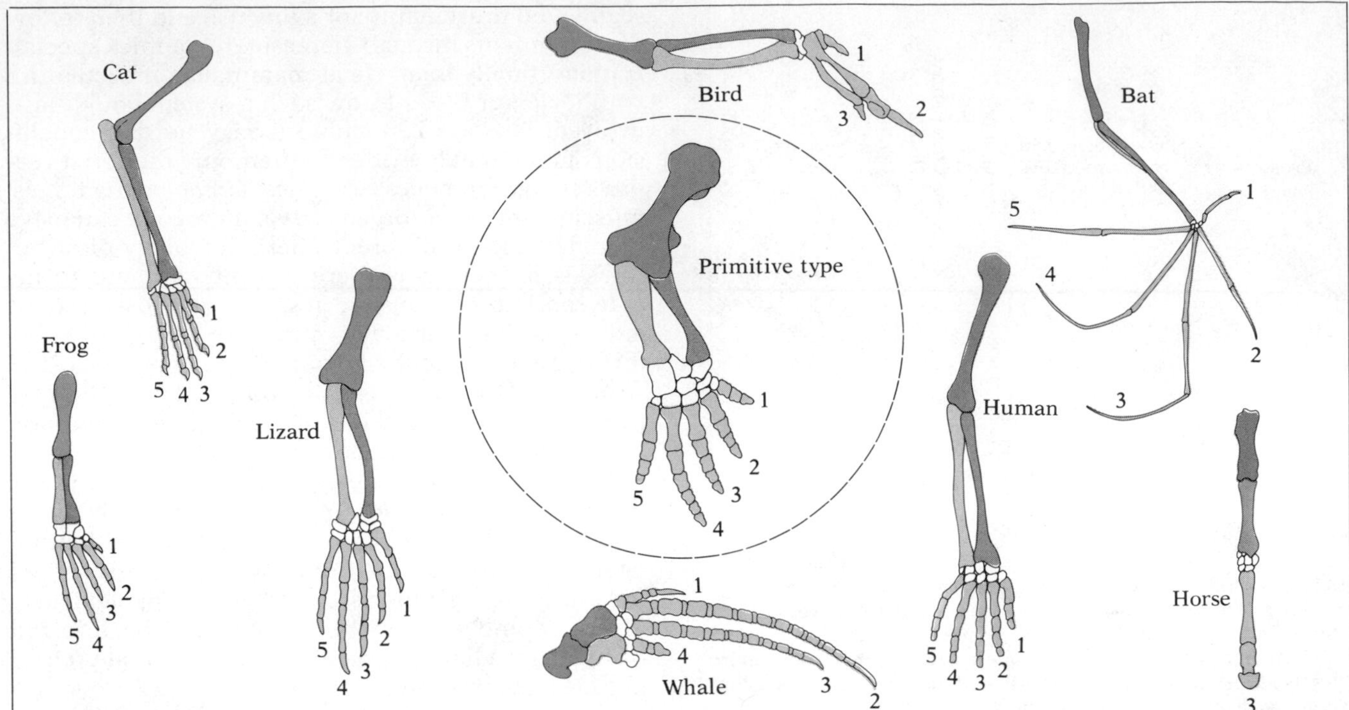

13.12 Homologous structures. The forelimbs of several representative vertebrates are compared here with those of the suggested primitive ancestral type. In each case, individual bones have undergone modification, although they can still be traced to the ancestor. Since they all have the same embryological origin and they can be traced to the ancestor, they are said to be *homologous*. The most dramatic changes can be seen in the horse, bird, and whale. Many individual bones have undergone reduction and, in some instances, fusion. Another interesting modification is seen in the greatly extended finger bones of the bat.

13.13 This computer-generated phylogenetic tree represents amino acid differences in the polypeptide sequence of respiratory pigment cytochrome *c*, which is found in all aerobic organisms. Each amino acid difference represents at least one DNA mutation that has been "accepted" in evolution. The estimated number of DNA changes (numbers on the tree) determine the evolutionary distance from one organism to another. In general, the tree derived from one short protein is consistent with other trees based on morphology or the fossil record, although the computer seems to think that the turtle is a bird.

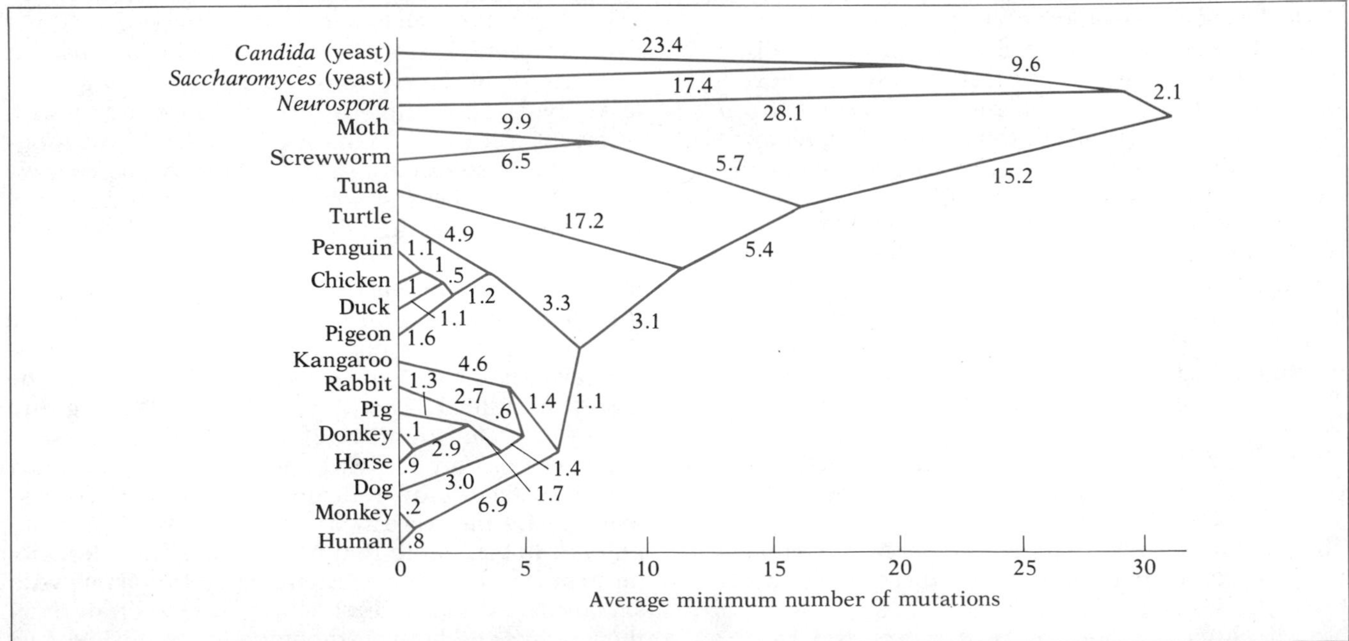

Essay 13.2 T. D. Lysenko

The story of Soviet biology in the period of 1937–1964 is among the most unfortunate in all of science. During this period, one man (not a good scientist, and apparently not even very bright) imposed his own crackpot, neo-Lamarkian ideas on the nation. He destroyed Soviet biological science, and set back Soviet agriculture by several decades. At the height of his reign, official Soviet publications regularly bleated denunciations of Gregor Mendel, T. H. Morgan, and anyone else so corrupt as to believe in genes.

The Soviet Union had been quite advanced in genetics in the 1920s and 1930s, but this man, T. D. Lysenko, was able to send the most prestigious geneticists to slave labor camps and to silence the rest. Lysenko is now regarded as an illiterate and fanatical charlatan who was allowed absolute dictatorship over Soviet biology and agriculture.

Lysenko's scientific credentials initially rested on his 1926 report of a finding (the true source is unknown) that under mild winter conditions, planting peas before a cotton crop buffered the cotton against the ravages of winter. This was a modest but useful finding which Lysenko, who did have a fine instinct for public relations, tried to sensationalize. He was ignored by his colleagues then, as well as when he later claimed credit for the "discovery" of *vernalization*, a process whereby the germination of winter wheat could be speeded by soaking the seeds in ice water. The process had been used by peasants for at least a century, but Lysenko claimed that *one such treatment would permanently change a winter wheat strain to a spring wheat.* The paper in which he made his announcement drew criticism because of its false claims to originality, its dependence on Lamarkian reasoning, and its aura of inflated importance. But in the Soviet Union at that time, only one opinion counted—that of Josef Stalin—and he was impressed.

Pravda carried the banner headline: "It is possible to transform winter into spring cereals: an achievement of Soviet Science." It seems that the idea that the environment could change the genetic condition of organisms was in line with Marxist dogma, which stresses the idea that a man was the creation of his environment. At the time, racism and claims of genetic superiority were rampant in Hitler's Germany; and Stalin, who often expressed his contempt for the traditional Russian character and "our slavish past," was trying to create what he termed "the new Soviet man." Lysenko's rejection of genetics, and his claims for the influence of the environement on heredity, fit well into Stalin's philosophical framework and political needs. Lysenko became Stalin's man, and Lysenkoism became the official science. In one sense, the dictator was being idealistic in his rejection of genetics, a fact we can't ignore if we want to understand the true horror of what was going on. The imposition, by brutal force, of "correct" ideology and wishful thinking on science in fact destroyed the very essence of science, and brought only farce and disaster.

Lysenko, as director of the Soviet Institute of Genetics, promoted more and more bizarre "findings." He and his growing band of opportunistic followers published "results" that showed, for instance, that wheat stalks sometimes produced oat grains, and that animal hybrids could be formed by injections of blood or plant hybrids by intraspecific grafts. Now, one of science's most cherished but vulnerable rules is that other scientists accept reported observations at face value; we must trust each other. But this was too much. It was clear to every scientist outside the Soviet Union that Lysenko was a fraud. In official communist publications, however, angry ranting at G. Mendel and T. H. Morgan continued unabated. For a while it looked bad for Lysenko when Stalin died in 1953, but Lysenko's grip on science, biology, and agriculture increased under the rule of Stalin's eventual successor, Nikita Khruschev.

I. A. Rapoport was among the most prestigious of the older Russian geneticists to have escaped the purges. He survived by dropping his genetics studies and moving into other areas of biology. Then one day in 1964, nearly 30 years after Lysenko's rise to power, Rapoport received an urgent call from the Kremlin. Could he write an encyclopedia article favorably describing Mendelian genetics? The startled Rapoport said that he certainly could. Could the manuscript be ready in 24 hours? Well, yes. Actually it took Rapoport 30 hours, but the next day, the first page of *Pravda* carried his detailed scientific article on the accomplishments of Gregor Mendel, complete with a laudatory banner headline. Ordinary readers may have been perplexed, but the more astute knew that this could mean only one thing: Nikita Khruschev had been toppled from power.

Lysenko and his by now numerous followers were also stripped of power, but they retained their academic ranks and the privileges that go with them. As of this writing, an aged Lysenko is probably still planning or dreaming of a return to authority.

The Two-Kingdom Scheme and the Five-Kingdom Scheme

It might be instructive to see how the living world has been divided according to two different ways of looking at things.

The Two-Kingdom Scheme

The notion of two kingdoms is based on ancient folk tradition, which recognized every living thing as either an animal or a plant (Figure 13.14a). According to the old line, everything that was photosynthetic, or did not actively ingest food, or had spores, had to be a plant. Thus *Euglena,* a photosynthetic flagellate, was considered a plant although it also moves around and eats things. (Some zoologists questioned this classification and called *Euglena* animals.) And fungi were considered to be plants, although they are not photosynthetic, because they do not move around or ingest anything. Furthermore, *Dictyostelium,* the cellular slime mold, was a plant; it is not photosynthetic, it moves around, it ingests food—but it forms spores. Bacteria were also placed in the plant kingdom, probably because plant taxonomy was a much more developed science than animal taxonomy was at the time, but also because bacteria have cell walls and don't ingest their food.

The Five-Kingdom Scheme

The old tradition of two kingdoms has largely given way, but there is yet no universal agreement on the major kingdoms of living organisms. At the present time one of the most widely accepted schemes is the one we follow in this book, the five-kingdom classification proposed by R. H. Whittaker (Figure 13.14e).

1. Procaryotes: bacteria and cyanophytes (blue-green algae). Whittaker himself called this kingdom the *monera.*
2. Protists: various simple, mostly single-celled eucaryotes, including both protozoa (animal-like protists) and *protophyta* (photosynthesizing algal protists).
3. Fungi
4. Plants: most algae, *nonvascular* and *vascular* plants (*vascular* refers to an internal transport system, or vessels)
5. Multicellular animals, including sponges (parazoa) and other animals (metazoa)

Keep in mind that the five-kingdom scheme is only provisional, and itself is a compromise between the ideal of an evolutionary phylogeny and the legitimate demands of tradition and convenience—not to mention some remaining pure and simple ignorance. As might be expected, arguments about the five-kingdom classification have already arisen. Sometimes all *green algae* are included in the plant kingdom (their pigments indicate that they are more closely related to vascular plants than are other protists) and sometimes they are not. While Whittaker included *brown and red algae* among the plants because they are multicellular photosynthetic organisms, Margulis and others have put them in with the protists (Figure 13.14d).

Sponges are usually included in the animal kingdom, because like all animals they are differentiated, multicellular organisms that ingest food. But there is good evidence that sponges arose from one of the protozoan groups entirely independently of the metazoa. This makes Whittaker's animal kingdom a polyphyletic group. For this reason, sponges are sometimes given a kingdom of their own, the *parazoa.*

The protists make up the most diverse eucaryote group. The evolutionary relationships of the various protists are not at all clear, and the kingdom is largely a grab-bag of small eucaryotes that don't readily fall into any other kingdom.

You may have noticed that the *viruses* do not appear anywhere in the five-kingdom scheme. One reason is that they fail to meet some standard definitions of "living." (What must one think about something that can crystallize?) Also, we don't know whether viruses are degenerate forms of more complex ancestors, or are elaborations of originally simpler forms derived from maverick genetic material of some ancient hosts. We don't even know if viruses are monophyletic or polyphyletic. All things consid-

13.14 The kingdoms of life. How do we divide the living organisms of the world into groups? There are two problems: first, although we know that all are related, we don't know all the details of the true phylogeny. Second, even if we did know the true evolutionary relationships, there would still be disagreement as to how the tree of life would be divided into kingdoms. Here we accept, provisionally, one speculative phylogeny of all life. Some of the details are pure guesswork: we don't really know, for instance, where the red algae fit in, and the true phylogeny of protists is unknown and complex. This phylogeny is divided into kingdoms according to five different schemes (color overlays). Green, plants; brown, animals; yellow, fungi; orange, protists; blue, monera (procaryotes). (a) A traditional two-kingdom scheme: everything that is not a multicellular animal (metazoan or sponge) is a plant. In other two-kingdom schemes, protozoa (nonphotosynthetic unicellular eucaryotes) are considered to be animals. (b) A three-kingdom scheme. Protozoa are added to the animals, and the procaryotes are given their own kingdom. (c) A four-kingdom scheme. The *protists* are now defined to include bacteria and cyanophytes. (d) Margulis' five-kingdom scheme, an adaptation of Whittaker's. Multicellular algae and slime molds become protists. (e) H. C. Whittaker's five-kingdom scheme, generally followed in this book. Many other schemes have been proposed; sometimes, for instance, cyanophytes are plants and bacteria are not. Recently Carl Woese proposed a novel three-kingdom scheme: eucaryotes, procaryotes, and methanogenic bacteria.

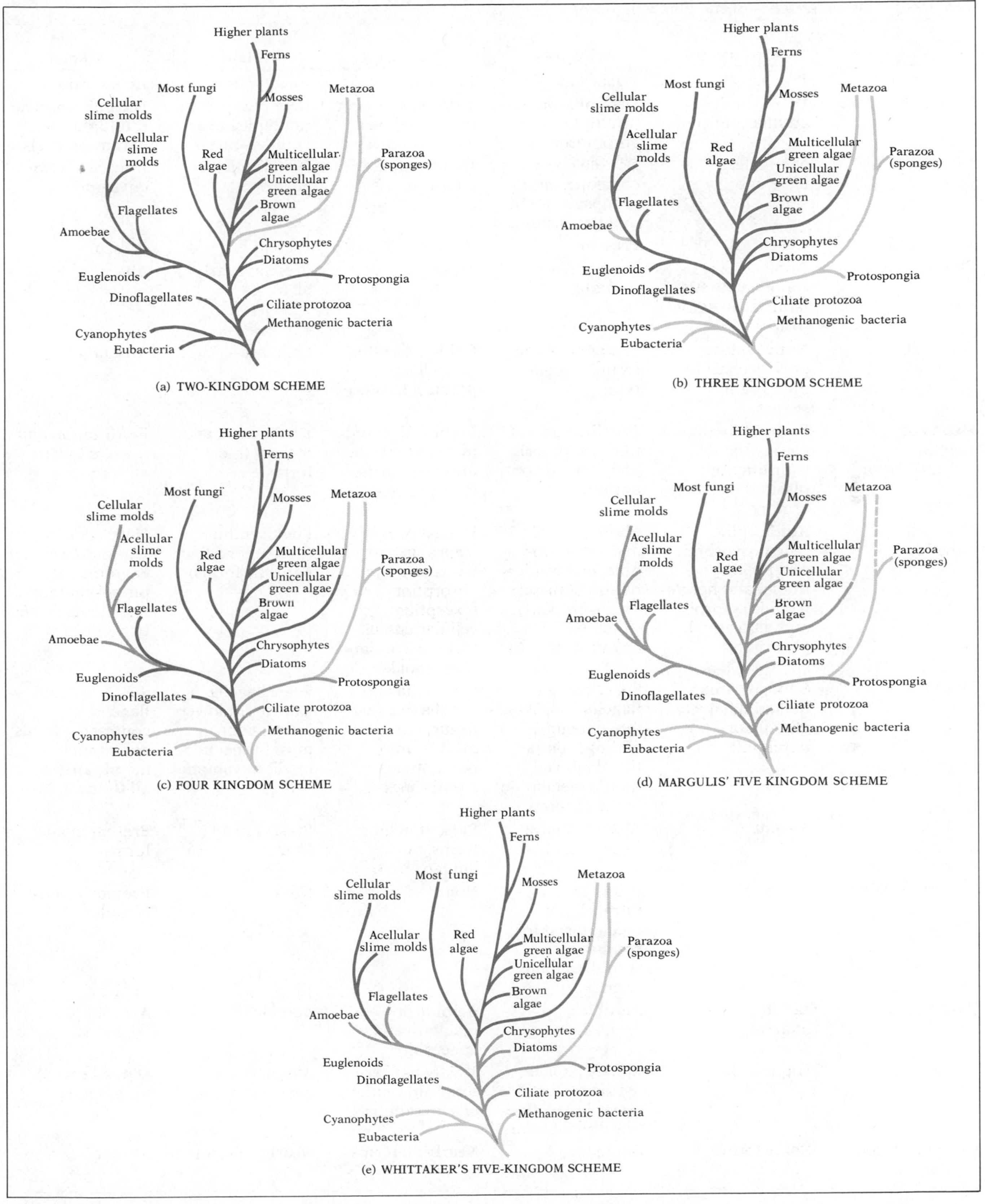

(a) TWO-KINGDOM SCHEME

(b) THREE KINGDOM SCHEME

(c) FOUR KINGDOM SCHEME

(d) MARGULIS' FIVE KINGDOM SCHEME

(e) WHITTAKER'S FIVE-KINGDOM SCHEME

Table 13.1 Characteristics of the Five Kingdoms

	Procaryotes	Protists	Fungi	Plants	Animals
Cell type	Procaryotic	Eucaryotic	Eucaryotic	Eucaryotic	Eucaryotic
Genetic Material	DNA (single circular molecule)	DNA plus protein (multiple linear chromosomes)	DNA plus protein (multiple linear chromosomes)	DNA plus protein (multiple linear chromosomes)	DNA plus protein (multiple linear chromosomes)
Nucleus	No nuclear envelope	Double nuclear envelope, most groups; single in dinoflagellates	Double nuclear envelope	Double nuclear envelope	Double nuclear envelope
Mitochondria	Absent	Present	Present	Present	Present
Chloroplasts	None (photosynthetic lamellae in some forms)	Present (some forms)	Absent	Present, nearly all forms	Absent
Cell wall	Noncellulose polysaccharide plus amino sugars	Present in some forms, various types	Chitin and other noncellulose polysaccharides	Cellulose	Absent
Means of genetic recombination	Plasmid-mediated conjugation; transduction; plasmid transfer, or none	Fertilization and meiosis; meiosis and conjugation; or none	Fertilization and meiosis; or none; dicaryon formed (some forms)	Fertilization and meiosis (most forms)	Fertilization and meiosis (nearly all forms)
Mode of nutrition	Autotrophic (chemosynthetic and photosynthetic) and heterotrophic (saprobic and parasitic) by absorption	Photosynthetic and heterotrophic, or a combination of these; by absorption or cellular engulfment	Heterotrophic (saprobic and parasitic), by absorption (exception: cellular engulfment in cellular slime molds)	Photosynthetic (except for a few parasitic forms)	Heterotrophic, by ingestion; cellular engulfment; photosynthetic symbionts hosted by some
Motility	Solid, rotating bacterial flagella; gliding; or nonmotile	9 + 2 cilia and flagella, amoeboid movement: gliding; contractile fibrils and motile membrane in many forms	9 + 2 cilia and flagella in a few forms, none in most forms; nonmotile membranes	9 + 2 flagella in gametes of lower forms; none in most forms; nonmotile membranes	9 + 2 cilia and flagella, contractile fibrils and motile membranes in all forms
Multicellularity and cell specialization	Absent	Absent in most forms	Present in most forms, but limited	Present in all forms	Present in all forms
Nervous system	None	Primitive intracellular mechanisms for conducting stimuli in some forms	None	None	Present, often complex
Respiration	Aerobic or anaerobic	Aerobic	Aerobic or facultatively anaerobic	Aerobic	Aerobic
Life cycle	Haploid, fission	Mostly haploid; fission or sexual recombination; a few diploid forms	Mostly haploid; some alternation of generations	Alternation of generations	Diploid except for gametes
Unicellular spore formation	Some forms	Some forms	Nearly all forms	Nearly all forms	Absent

ered, it seems wiser to bar the viruses membership in the club of life until we can dig up their birth certificates.

In any case, apart from the virus puzzle, the major division of the living world is between the procaryotes (or monera) and the remaining four (or more) kingdoms, collectively called the eucaryotes.

Scientists traditionally have faith in the future, and indeed new data are pouring in. Someday we may know the answers to all our vexing questions, and a universally accepted higher taxonomy may then be chisled in stone. And yet, perhaps not. It's possible that some of the events of our biological past will never be unraveled, and that these same vexing questions will remain forever as unweathered monuments to human limits.

Application of Ideas

1. The fossil record for nearly all forms of multicellular life on the earth begins rather abruptly in the Cambrian Period of the Paleozoic Era. This raises numerous questions and adds fuel to the creationist's arguments. Offer two different hypothetical explanations for the seemingly abrupt appearance of multicellular life.
2. Until recently, phylogenetic trees were hypotheses that could only be tested in terms of the characters contrived by the tree's originator. This is no longer true. Describe two important discoveries of this century which offer other ways for testing these hypotheses and explain how such testing is done.
3. In assigning organisms to the various taxons, a dichotomous key is commonly used. Find an example of a dichotomous key in one of the numerous sources in the library and explain in some detail how one is used.
4. Even though numerical taxonomy is a relatively recent innovation, Darwin subscribed (in essence) to the concept of the "operational taxonomic unit," a recently developed replacement for species. Explain how Darwin reflected this concept in his thinking.

Key Words

allotetraploid
analogous

binomial system

character state
class
comparative anatomy
convergent evolution
cyanophyte
cytochrome c

Dollo's law

embryology

family
fibrinopeptide

generic name
genus

homologous
homology

introgression

kingdom

Key Ideas

What Is a Species?

1. There is little agreement on the total number of *species* on the earth.
2. A general definition of *species* is any group of organisms that could interbreed. The definition works well in many situations but is very argumentative in others.
   a. Some well-defined *species* do interbreed under laboratory conditions.
   b. Some might interbreed if they were not separated by barriers.
   c. Some *species* interbreed along a cline, but not between the ends of the range, even under laboratory conditions.

The Species Concept in Plants

Defining species with plants is often more confusing than with animals. Being nonmotile, they tend to show considerable geographic change. Further, many plant species readily hybridize. While the hybrids are relatively infertile, there are so many offspring (seeds) produced that chance favors an occasional fertile one.

Allotetraploids: Instant Species

1. Hybridization is facilitated by polyploidy. The problem of infertility because of chromosome incompatibility is overcome by spontaneous doubling. Species created this way are called allotetraploids.
2. Polyploidy in plants may just be a common evolutionary mechanism. Polyploid plants are most numerous in harsh climates.

Type Specimens and Biometrics

1. *Type specimens* are traditionally chosen to represent a species, and variants have been treated suspiciously. *Type specimens,* however, are *human* contrivances and must be considered as such.

lumper
lysozyme
monophyletic
numerical taxonomy
operational taxonomic unit (OTU)
order
paraphyletic
parazoa
phylogeny
phylum
polyphyletic
polyploidy
population of variants
protist
race
specific name
species
splitter
subspecies
systematics
systematist
taxon
taxonomist
taxonomy
tetraploidy
tetrapod
type specimen
varieties

2. *Numerical taxonomy* utilizes the measurements of many variables in defining a *species*. According to numerical taxonomists, a species is a *population of variants*.

Numerical Taxonomy and the OTU

1. Numerical taxonomists have substituted the OTU (operational taxonomic unit) for species and other groupings. Numerical values are assigned to character states. These are clustered into OTUs.

2. Numerical taxonomy recognizes and deals with variability. It claims to have eliminated subjectivity in classifying organisms.

How Species Are Named

1. *Species* are named according to the *binomial system* of nomenclature. Each has a *generic name* (*genus*) and a *specific name* (*species*).

2. Taxonomic names are latinized. Latin is unchanging and forms the root words of many European languages.

3. Carolus Linnaeus originated the *binomial system*. Today commissions decide on *species* names.

Systematics

1. *Species* are organized into larger groups called *genera* (*genus*) which in turn form *families*. *Families* are grouped into *orders* and *orders* into *classes*. *Classes* form *phyla* (*phylum*) or *divisions* and finally, *phyla* and *divisions* are grouped to form the *kingdoms*.

2. Below the *species* level are the *subspecies*, *varieties*, and *races*. The latter two are least commonly used.

3. The various groupings in the hierarchical system is supposed to reveal the relationships between *species*.

4. Each group from *species* to *kingdom* is referred to as a *taxon*.

5. Charles Darwin proposed the analogy of the tree to show relationships between *taxa*. The tips of branches represented living groups or *species*, while the branches were lines of descent and ancestral types. The main trunk represents the common ancestor.

6. *Taxonomists* are responsible for naming new *species*, and identifying unclassified specimens.

7. *Systematists* try to determine evolutionary relatedness of organisms. (Individuals are often both *taxonomists* and *systematists*.)

8. There is a perpetual argument between two groups of *systematists* known as *lumpers* and *splitters*. The former favors large groupings while the latter tend to break large groupings down.

9. *Taxonomy* and *systematics* are constantly changing because of new information and new techniques for comparing organisms.

Monophyletic, Polyphyletic, and Paraphyletic Groups

1. *Taxonomy* has been changing at the *kingdom* level. Where there were previously two *kingdoms*, there are now five. There is considerable disagreement about this organization.

2. A monophyletic group is one that can be traced to a single common ancestor. Members of a monophyletic group are more closely related to each other than to members of any other group.

3. The birds are a *monophyletic* group, as (probably) are the mammals. The reptiles are a *paraphyletic* group, with at least three ancestral lines. Within the reptiles, the crocodilians are more closely related to birds than to other reptiles. The reptiles are retained in one class out of tradition and out of recognition of the many character states they still share.

4. Reptiles are paraphyletic as a class because they have given rise to groups (birds and mammals), which are no longer classified with the reptiles.

5. The dicot plants gave rise to the monocots, a taxonomically distinct group. The dicots are also considered to be paraphyletic.

Who Is Related to Whom?
The assumptions used in designing *phylogenetic* trees are as follows:

1. Characteristics of a *taxon* can be described. (Each character used must have two or more specific and identifiable character states.)
2. The degree of similarity between two *taxa* can be described. (This is reflected in the number of characters shared.)
3. Closely related *taxa* will be similar morphologically and biochemically. (Newly diverged species will retain many similar traits.)
4. The degree of similarity between two *taxa* decreases over time.
   a. As time passes, greater numbers of changes occur. An exception is *convergent evolution*, where two distinct *species* evolve in similar directions in response to a similar niche.
   b. The rate of change is highly variable, with long periods of slow change and sudden vast changes.
   c. Humans represent a morphologically rapidly changing *species*, but we are slow to change biochemically.
   d. Humans and chimpanzees are very close biochemically.
5. Characters once lost in evolution tend not to be regained. (This is *Dollo's law.* Once a gene is gone, it is gone. Actually, analogous structures can reevolve, but this does not violate *Dollo's law*.)
6. *Homology* can usually be distinguished from *analogy.*
   a. *Homologies* are similarities which arise from common descent. *Analogies* are similarities which stem from different evolutionary origins.
   b. *Homologous* organs show changes within a *taxon*. They can be determined through fossil and embryological study. Examples are the forelimbs of all the vertebrates.
   c. *Analogies* have no evolutionary commonality. They are coincidence, often cases of *convergent evolution.* An example is the wing of the insect and bird. Their source in the embryo is different and there is no common fossil ancestor to the two.

Formal Construction of Trees
The number of possible trees is astronomical, but it is possible to begin with smaller trees and add a branch at a time.

The Two-Kingdom and Five-Kingdom Schemes

The Two-Kingdom Scheme

1. The two-*kingdom* scheme is traditional. It reflects the subject matter of the two earliest divisions of biology, botany and zoology, and includes the plant and animal *kingdoms*.
2. Assignment to the plant *kingdom* was based on: photosynthesis, no active ingestion of food, production of spores, or the presence of cell walls.
3. All other organisms were classified as animals. There are many examples of whole groups of organisms that do not clearly fit in either *kingdom.*

The Five-Kingdom Scheme

1. The five-*kingdom* scheme includes
   a. Procaryotes—bacteria and cyanobacteria (blue-green algae) (single cells, no nucleus)
   b. Protists—mostly single-celled eucaryotes, protozoa, and some of the algae
   c. Fungi—mostly multicellular, with cell walls, nonphotosynthetic, life cycle mainly haploid
   d. Plants—multicellular photosynthesizers, nonmotile, with cell walls
   e. Animals—multicellular, motile, nonphotosynthetic, food ingesters
2. The five-kingdom scheme is far from perfect but does address itself to the problem of classifying protists, fungi, and procaryotes.
3. Problems still exist with the algae, some *protists,* sponges, and (to many botanists) the fungi.

4. The five-*kingdom* scheme should be considered a hypothesis from which predictions can be made and tested. This is being done through biochemical taxonomic studies.
5. Viruses have not been assigned to a *kingdom* in the five-kingdom scheme.

Review Questions

1. Review the traditional definition of *species* and comment on the difficulties experienced when it is applied. (pp. 386–390)
2. What special problem do the frogs of the eastern United States offer the *taxonomist?* (p. 390)
3. Define the *type specimen* and comment on the limitations of its usefulness to *taxonomists.* (p. 390)
4. How does the phrase *population of variants* help overcome the problem of defining species? (p. 390)
5. Support the idea that *taxonomy* and *systematics* are really sciences. Are they predictive? (Essay 13.1) (p. 396)
6. Explain the *binomial system* of classification, its origin, language, and organization. (p. 392)
7. Define the term *taxon.* (p. 393)
8. Explain the organization of Darwin's phylogenetic tree. (p. 393)
9. What is the advantage, if any, of the operational taxonomic unit? (p. 391)
10. Distinguish between the work of *taxonomists* and *systematists.* (p. 394)
11. How would the *lumper* and the *splitter* treat the California salamanders? (pp. 390–393)
12. Describe an application of biochemical taxonomy. (p. 394)
13. Using mammals as an example, explain what is meant by a *monophyletic* group. (p. 395)
14. What are the reasons why reptiles are considered a *paraphyletic* group? (p. 395)
15. Explain the rule about character states in describing a *taxon.* (p. 396)
16. How are similarities between taxa described in *numerical taxonomy?* (p. 396)
17. State the assumption about the morphology and biochemistry of newly diverged species. (p. 397)
18. Review the fallacies in the thinking of T. D. Lysenko. (Essay 13.2) (p. 401)
19. Compare man and the chimpanzee, morphologically and biochemically. (p. 398)
20. What is *Dollo's law?* Is *convergent evolution* a violation of this law? (p. 399)
21. Distinguish between *homologous* and *analogous* structures and present examples of each. (p. 399)
22. Why is the study of *embryology* important to the *taxonomist* and *systematist?* (p. 399)
23. Briefly review the problems of constructing phylogenetic trees. (pp. 399–400)
24. Describe the two-kingdom scheme of classification and, using examples, review its inadequacies. (p. 402)
25. List the *kingdoms* in the five-kingdom scheme and name a few types of organisms in each. (pp. 402–404)
26. Comment on the adequacy of the five-kingdom scheme, offering some examples of problems which still remain. (pp. 402–404)

Chapter 14
The Procaryotes

For some 2.7 billion years the world belonged to the procaryotes alone. The oldest evidence of life on earth, filamentous photosynthetic bacteria discovered in Australia in 1980, are 3.5 billion years old. There are good reasons to believe that some groups of anaerobic bacteria are even older. The extremely well-preserved fossil record of 2 billion years ago reveals hundreds of different forms of bacteria, about half of which appear to be identical to forms that are still living. In contrast, the first undisputed eucaryote fossils are only about 800 million years old. The eucaryotes—the "higher" organisms, which have cell nuclei—were latecomers to an already old and biologically diverse world. That was the world of the procaryotes—the bacteria and cyanophytes—which comprise the kingdom Monera.

As we mentioned in Chapter 13, the biggest taxonomic distinction in biology is between the procaryotes and the eucaryotes. Procaryotes, as we will see, aren't like the other forms of life—those you are most familiar with. By comparison, plants and animals are very similar to one another in genetic function, enzymes, and fine cell structure. The plants and animals are even relatively similar to fungi and protozoa. All these eucaryote groups obey Mendel's laws, and go through mitosis and meiosis in very similar ways. But the procaryotes don't; they're very different from us.

Because the bacteria and cyanophytes were already old and well-diversified long before the first nucleated organism showed up, they form a widely varying group today. It has proven extremely difficult to determine just how the different kinds of procaryotes might be related to one another evolutionarily, since their morphologies are so simple and since their common ancestors lived and died so very long ago. In spite of their diversity, however, the procaryotes do have enough characters in common so that they make a stable taxonomic group. Although some of the living types of bacteria are extremely ancient—and extremely different from other bacteria—we will follow the usual tradition and consider the kingdom Monera to be composed to two large groups: the cyanophytes (blue-green algae) and the bacteria.

By definition, since the monera are procaryotic, they lack a nuclear membrane, endoplasmic reticulum, mitochondria, Golgi bodies, and lysosomes. Procaryote cells are also smaller than eucaryotic cells. They are roughly $\frac{1}{10}$ as long and about $\frac{1}{1000}$ as large in volume as the smallest eucaryotic cells (Figure 14.1).

14.1 The size of things. This collection includes (a) phage virus (for comparison), (b) mycoplasma, (c) coccus bacterium, (d) bacillus, (e) spirochaete, and (f) coccoid blue-green alga. The diameter or length of each is given in micrometers. The bacteriophage is shown for comparison only, since it has really not been assigned to any kingdom. The mycoplasmas are the smallest organized cells.

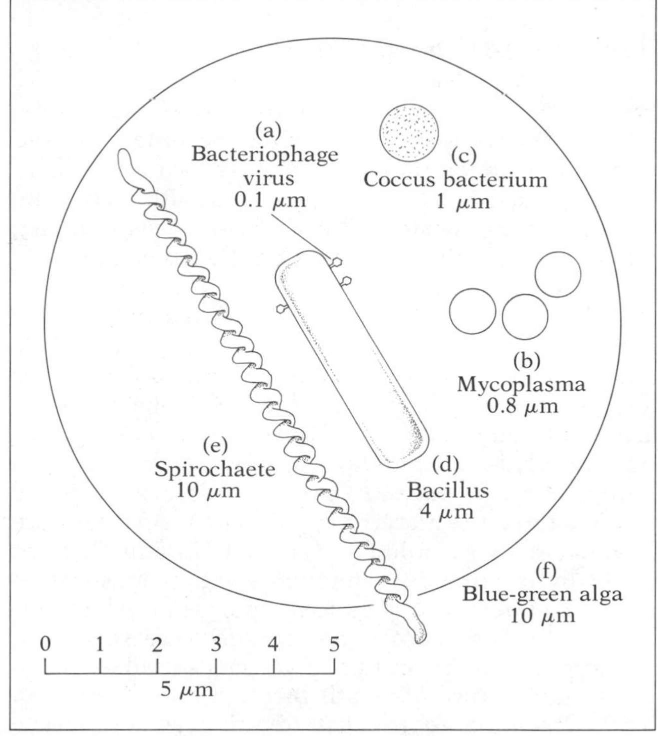

On the other hand, all procaryotic cells possess a cell membrane, or *plasmalemma*. And they nearly all have cell walls. (You will soon see how important these are, to them and to us.) However, the cell walls are extremely variable from one group to the next. Some species possess whiplike flagella, but these are not to be confused in any way with the flagella of eucaryotes. Many procaryotes also boast an array of tiny spikes called *pili* (singular, *pilus*). Some species protect themselves with a slimy jellylike coat (e.g., the slime that strangles diphtheria victims). At least some have a double cell membrane. Between the inner and outer cell membranes is a *periplasmic space*.

Inside, the genetic material lies clumped in the cytoplasm and consists of a single circular molecule of DNA. Most procaryotes lack internal membranes, except the cyanophytes (blue-green algae), which lack chloroplasts and hang their photosynthetic pigments on layers of internal membranes. We will discuss the cyanophytes in more detail shortly.

Where do these unusual creatures live? Practically everywhere. You can find them in your nose and swarming by the millions in the dark recesses of your bowels. They are quite succesful at living in boils and pimples, and they literally cover your skin. They live in mud, in hot springs (the blue-green algae lend the delicate hues to the springs at Yellowstone); and they flourish in frigid Antarctic waters. So let's take a closer look at these tough, ubiquitous creatures.

The Bacteria

We hear about bacteria almost daily, and almost all their press is bad. We have become obsessed with killing bacteria in everything from our teeth to our toilets. Most of us forget that although some bacteria are known to deal in disease (are *pathogenic*), others are essential to life. *Are* bacteria being unfairly maligned?

All bacteria are procaryotic, lacking inner membranous structures, including nuclear membranes, endoplasmic reticulum, mitochondria, Golgi bodies, and lysosomes. Generally, four divisions or phyla are recognized: *true bacteria, myxobacteria, mycelial bacteria,* and *spirochaetes*. Recently, a peculiar fifth group, the *mycoplasmas* has been described. Some true bacteria, the anaerobic methane fermenters, are biochemically so unlike other bacteria and all other living things that investigators have proposed that they might best be placed in a separate kingdom.

Part of the reason that bacteria are so reviled is that bacterial infections of varying sorts have been the leading cause of death through most of human history. Adding to the list of miseries they cause us are food spoilage, food poisoning, diseases of crop plants, and diseases of domestic animals (see Essay 14.2). Actually, although some bacteria have historically wreaked havoc on humans, most of them really present little threat to people today, at least where medical treatment and preventive measures are available. Part of our success in overcoming the problem of bacterial disease in the past three or four decades has been due to the development of antibiotics. Of far greater significance to human health, however, has been a 19th century idea attributable primarily to Louis Pasteur—the *germ theory of disease*. The idea was a momentous one. The idea that "germs" were responsible for disease caused people to gradually change their behavior in ways we simply take for granted today. Thus cooks wash their hands and no longer spit in the soup.

We must also acknowledge the development of large-scale public sanitation programs—food inspections, pasteurization of milk, sewage control, drinking-water sanitation, public immunization, and a variety of other measures. These, plus almost yearly advances in antibiotic therapy and aseptic medical procedures, have helped to curb the once-rampant spread of many communicable diseases. Because of such innovations, and, of course, because of improved methods of producing and preserving foods, human population growth has accelerated at an unprecedented rate.

The good side of microbial life is too easily ignored. Bacteria, along with fungi, provide invaluable service to organisms of the earth. We describe the details of much of this activity in Chapter 19, so we'll be brief here. Bacteria are *decomposers*, or *reducers*. That is, they secrete enzymes that help decompose all kinds of dead organisms—plants, animals, and even other microorganisms. Decomposition is the reducing of life's complex molecules into simpler components. Thus, vital molecules are permitted to recycle through the ecosystems of the earth. Without this activity, essential chemicals would become locked away in immutable corpses that would eventually cover our small planet. You may not be overjoyed at the thought of a rotting corpse, but think of the alternative!

In addition to their ecological roles, bacteria are used to commercial advantage. They are used in making yoghurt, some cheeses, buttermilk, and a host of other products. Fermenters are responsible for producing silage (cattle feed), aided by other bacteria in a succession of activities. And we owe the sharp flavors of sauerkraut, pickles and tea to still other bacteria.

The Bacterial Cell

Typically, a bacterium consists of *protoplast*, or bag of living material, surrounded by a cell wall (Figure 14.2). We still don't completely understand

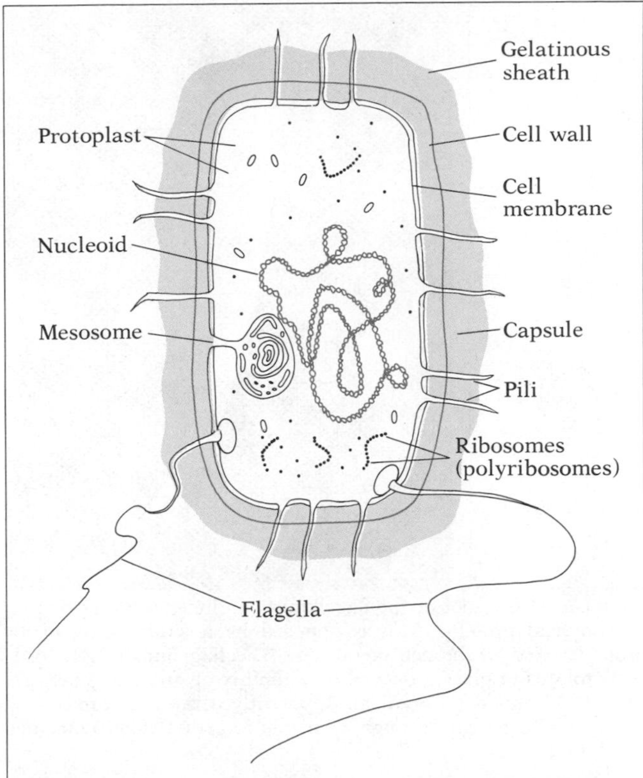

14.2 The bacterial cell shown here is a rod-shaped bacillus. Surrounding the cell is a jellylike capsule, and within this is a dense cell wall, which contains the protoplast. The protoplast includes a cell membrane. There are no membrane-bounded organelles like the ones found in eucaryotes. The nuclear area contains naked DNA in the form of a circular chromosome. Ribosomes are present, but are free-floating. The folded, membranous mesosome appears to have something to do with cell division (bacterial fission). Penetrating the cell wall are numerous spike-shaped pili. The nucleoid is one circular chromosome of naked DNA attached to a mesosome. The flagella are solid protein structures that actually rotate like propellors.

how the cell wall is constructed, but we know that it varies from one species to the next. Among its more common constituents are lipids, polysaccharides, and *amino sugars* (short amino acid chains containing two sugar groups). Some bacteria secrete tough capsules around themselves, while others produce slime.

The protoplast enclosed within the cell membrane is a very thick, but fluid, substance. Lying clumped in the protoplast is a long, coiled circular molecule of DNA. The protoplast also contains free-floating ribosomes. There is little to suggest any kind of membranous inner structure other than recently discovered structures called *mesosomes,* which appear to be inward extensions of the cell membrane. Their function is uncertain, but it has been suggested that mesosomes contain respiratory enzymes, or function in the production of cell walls, or help in DNA separation during fission. The last idea is supported by electron microscope photographs (Figure 14.3), which indicate that the mesosomes attach to the replicated strands of DNA prior to division.

Bacteria that can move commonly do so by propelling themselves by one or more flagella. These flagella are very different from eucaryotic flagella (see Figure 4.26). While eucaryote flagella occur as extensions of the cell membrane with the characteristic 9+2 organization of microtubules within, the bacterial flagella are solid polymers of a structural protein, *flagellin.* And whereas eucaryote flagella move by undulating waves, the helical bacterial flagella spin like propellors. At their bases, each bacterial flagellum passes through the cell wall and attaches to a tiny disk, which rotates like an electric motor and is, in fact, powered by a flow of protons across the chemiosmotic gradient (Figure 14.4).

The fact that bacterial flagella actually rotate was established in an interesting way. A nonmotile mutant of *E. coli* was found in which the flagella were straight, rather than spiral. The straight flagella didn't propel the bacterium, but they did rotate. This was observed by adding tiny visible particles which adhered to the straight flagella, and then watching them whirl around. By the way, no other life forms have rotating parts, as far as we know.

Bacterial Reproduction and Spore Formation

Bacteria reproduce through the primitive fission process, but as we discussed in Chapter 12, some bacteria have a form of sexual recombination. In

14.3 The bacterial mesosome. Mesosomes are saclike inward protrusions from the cell membrane as seen in this electron micrograph. Their precise function has not been clarified, although there is reason to believe they may be involved in the separation of DNA replicas during cell division.

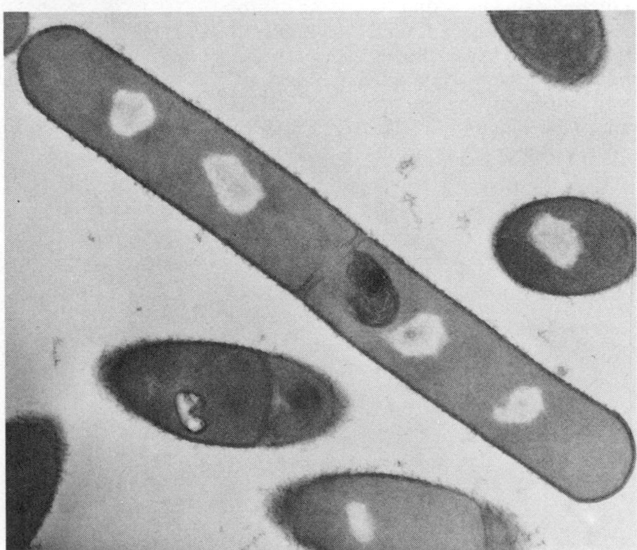

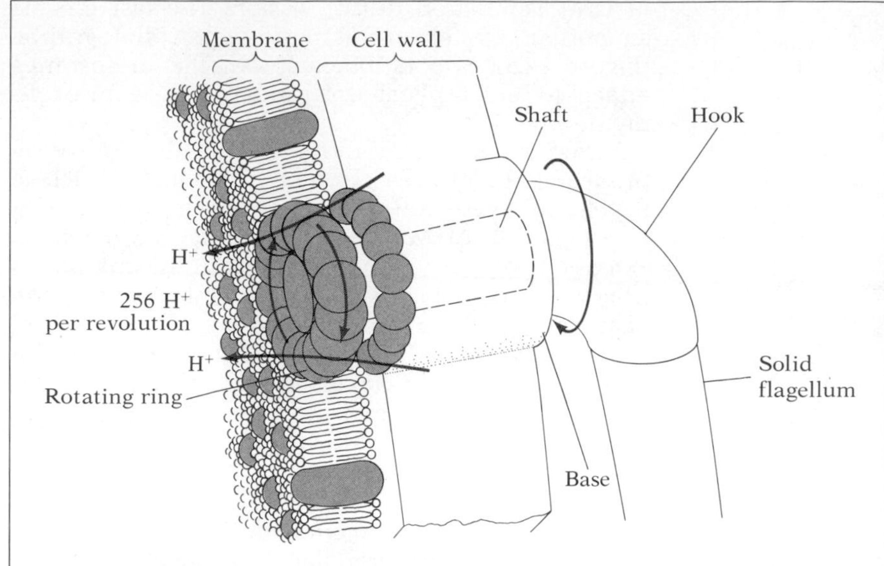

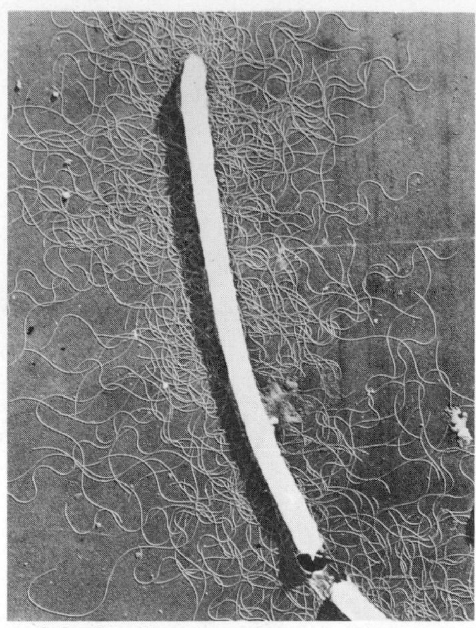

14.4 The bacterial flagellum is unlike that of any eucaryote. It represents the only known instance of a truly rotating part of an organism. The electron micrograph shows numerous flagella on a bacillus. Flagella are a common feature in water and soil bacteria. The flagellum does not have the internal microtubules seen in the eucaryote, but consists of a curved "hook" and a single, solid crystal of the protein flagellin. The flagellum is anchored in a ringlike base in the cell wall. Its shaft passes through the base. At the root of the flagellum is an inner rotating ring of 16 spherical proteins. This is opposed by a similar ring of 16 proteins fixed to the cell membrane. The flagellum ring of proteins rotates in place, powered by an influx of protons generated by the chemiosmotic gradient. Apparently it takes 16 protons to move the flagellum through $\frac{1}{16}$ revolution, or 256 protons per revolution.

fission, the circular chromosome replicates; then a deep groove forms in the cell wall, the DNA replicas separate, and the cell divides. Under the best of conditions, the process takes a mere 20 min. At this rate, 1 bacterium could generate 1000 tons of offspring in just 36 hr. Luckily, not every bacterium survives.

Some bacteria can form spores and thus survive conditions that would otherwise have killed them. Many soil bacteria, for example, have this ability. In general, spores are formed in response to unfavorable conditions. It is a most effective form of defense and it begins by the bacteria losing fluid and secreting an incredibly tough shell around its nucleoid. In such a state, it can survive almost anything in nature. It remains in a state of *dormancy*, or suspended animation, and can remain this way for many years. Some spores can even survive boiling for extended periods. Fortunately, none of the boilable bacteria are pathogens (Figure 14.5).

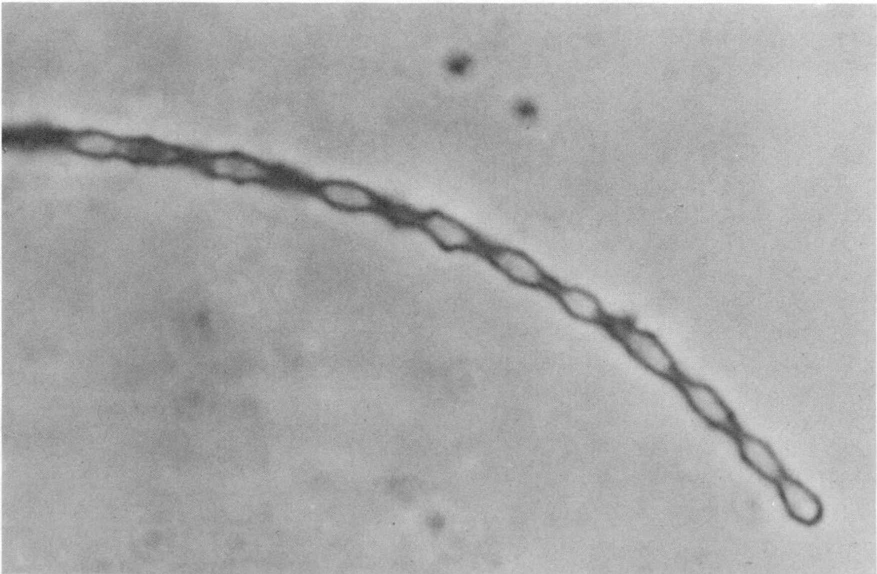

14.5 Bacterial spores consist of tough, highly resistant shells that protect the chromosome and small amount of cytoplasm within. They form during unfavorable conditions germinating when things get better. Numerous spores are seen in this chain of *Bacillus subtilis.* The remainder of the cell will soon disintegrate.

Essay 14.1 Tracing the Relationships of the Procaryotes

Comparative anatomy has been a great tool in determining the phylogenetic relationships of plants and animals. Unfortunately, it doesn't work very well for procaryotes. The principal reason is that procaryotes don't have very much anatomy to start with. There are a few basic forms—spheres, rods, and filaments—but no one is prepared to state that two bacteria are closely related just because they are both round. Some feeble attempts have been made in this direction: sheathed, filamentous cyanophytes (blue-green algae), for instance, are considered to be distinct from *coccoid* (spherical) cyanophytes. Bacteria that glide form a different group from those that have flagella; and branching, budding bacteria are considered to form a distinct group.

The different morphologies (shapes) of bacteria are extremely ancient. The best-preserved ancient fossils are the bacteria of the Gunflint chert which is an outcropping of 1.9 billion year old rock in the Canadian Shield. The fossils include branching forms, spherical forms, grapelike clusters, filaments, filaments within sheaths (like many modern cyanophytes), and rods. Some rare bacterial forms, found living today only in offbeat anaerobic environments such as the bottom of the black sea, are in fact indistinguishable from some of these 1.9 billion year old, extremely well-preserved fossils.

Almost as well preserved, and far more ancient fossil procaryotes were discovered in 1980 in 3.5 billion year old rocks collected near North Pole, a remote Australian mining town (see Figure 14.11). Stanley Awramik and William J. Schopf have described five distinct forms, including free rods, spheres, and both sheathed and unsheathed filaments. The organisms were found in large domelike, layered sedimentary structures, called *stromatolites,* that are similar to stromatolites formed today by inorganic deposits in shallow water mats of living photosynthetic cyanophytes.

One interesting old fossil, found in great numbers in the Gunflint chert and thus dating back to the early days of the planet, was shaped rather exactly like an umbrella, complete with ribs and a shaft with a lobed handle. No similar fossils are found in the succeeding 2 billion years of the fossil record, and this one could be dismissed as a long-extinct evolutionary oddity were it not for the fact that recently an identical organism was found—alive—in the soil beneath an ancient stable near Harlech Castle, Wales. This *Kakabekia umbellata* grows only in the presence of 5 to 10 molar ammonium hydroxide. No one knows where it was hiding for 2 billion years, but for the last seven centuries the soil in which it is now found has been continuously enriched with ammonia from horse manure and urine.

For any further understanding of the evolutionary relationships of the procaryotes, systematists must resort to subtler measures, which are mostly biochemical. Traditionally, bacteria have been classed as *gram-positive* and *gram-negative,* depending on whether or not they can be stained with a particular laboratory reagent. Also, bacteria that can utilize oxygen in metabolism (*aerobes*) are considered to be phylogenetically advanced over those that cannot (*anaerobes*), even though similar morphologies may be found in both groups. Variations in biochemical pathways may divide bacteria, and thus can also be used to suggest evolutionary pathways. *Photosynthetic* bacteria have long been considered distinct from *nonphotosynthetic* bacteria—although more recent work has shown the distinction to be shaky—and different photosynthetic pathways have been associated with specific major groups. For example, only the cyanophytes among the procaryotes can utilize water as a photosynthetic hydrogen source, and they are the only ones that produce oxygen as a waste product.

Recent advances in molecular biology promise to elucidate further the ancestry of the various procaryote groups, untangling the branching of phylogenetic divisions that must have occurred billions of years ago. Amino acid sequences of proteins and nucleotide sequences of nucleic acids show some clear relationships. Some molecular evolutionists, pointing out the great dissimilarity of rRNA of apparently primitive, anaerobic bacteria that live in unusual habitats such as hot springs, and the relative similarity between common aerobic bacteria (such as *Escherichia coli*) and eucaryotes, suggest that the procaryotes are such a diverse group that eucaryotes might well be considered to be a minor offshoot of one particular branch of the procaryote tree.

A few molecules are found in all eucaryotes and in many procaryotes. Richard Dickerson, of the California Institute of Technology, has compared the three-dimensional structures of the cytochromes, homologous molecules that are involved in the electron transport pathways of mitochondrial oxidative phosphorylation, chloroplast photosynthetic phosphorylation, cyanophyte photosynthesis, and both oxidative and photosynthetic phosphorylation of various bacteria. In some photosynthetic bacteria that are also capable of oxidative metabolism, the same electron transport system is used in both functions. Dickerson supports the symbiosis hypothesis of Lynn Margulis (discussed at the

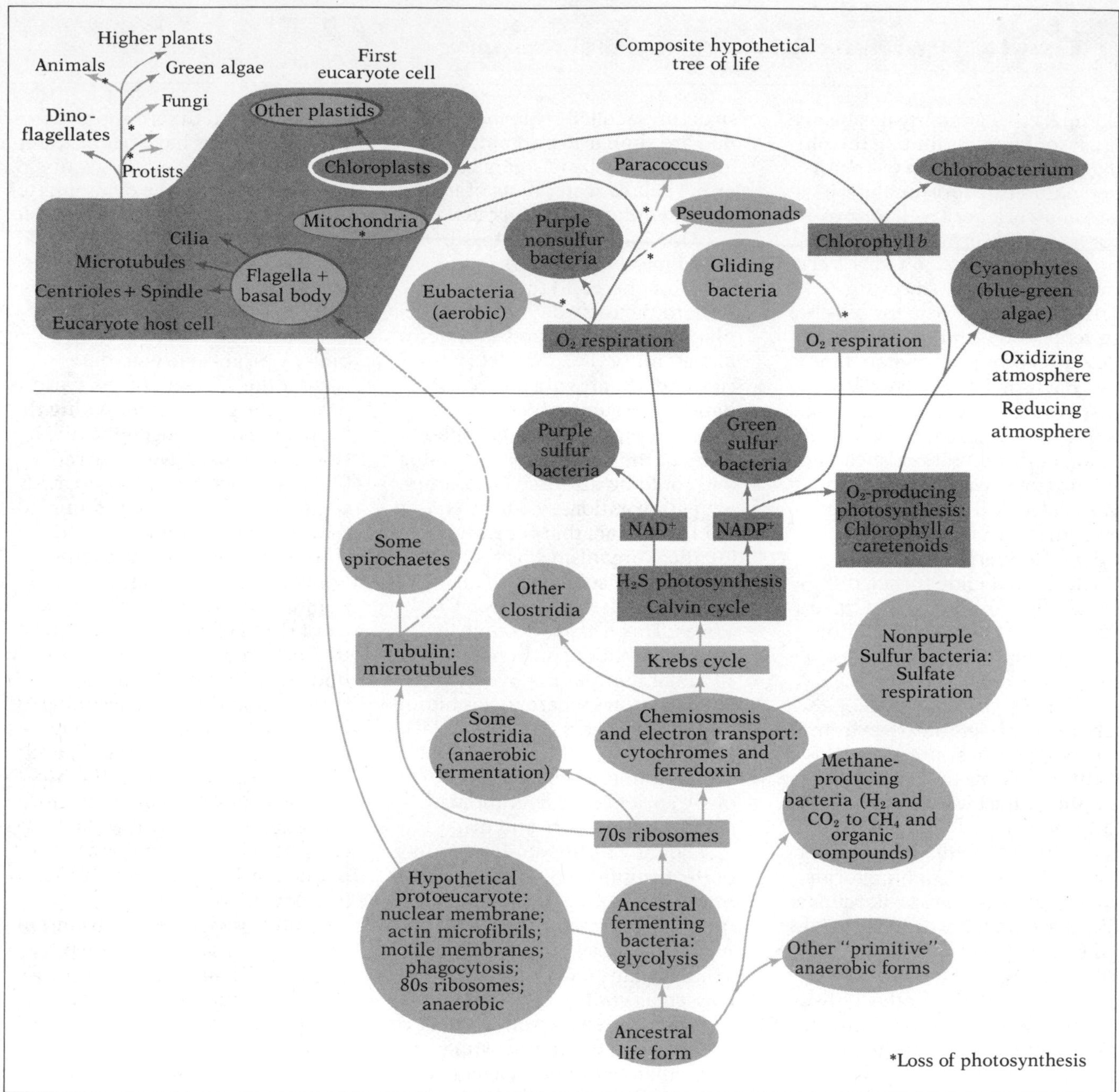

end of this chapter) concerning the origin of mitochondria (from purple nonsulfur bacteria) and chloroplasts (from relatives of cyanophytes), and has presented a tentative partial phylogeny of the aerobic procaryotes based on cytochrome structure. Margulis, herself, used shared biochemical pathways to present a tentative procaryote phylogeny, and M. O. Dayhoff has combined the data from cytochromes, ferredoxin (another pigment of the photosynthetic pathway), and 5S rRNA to compile a composite tree. Carl Woese has shown that methane-producing bacteria are only distantly related to all other procaryotes. Here we present a composite scheme relating these and other recent theories of the relationships between major procaryote groups. Evolution that is thought to have occurred before the origin of an oxygen-rich atmosphere is shown below the dotted line running across the center. Photosynthetic forms are indicated in color. You can pass over the complex details of this scheme, because it is only one of many to be proposed; all such phylogenies must be considered to be hypotheses subject to further testing.

Essay 14.2 Bacterial Villains

Writing about bacterial villains may seem to make as much sense as a treatise on criminality in prisons, but that's because we so often neglect beneficial bacteria (except when referring to cheese). We should keep in mind that we live in a sea of bacteria. We are each an island of bacteria. Everything we touch is loaded with them, or only temporarily cleansed. Because of their ubiquity and our general good health, we should assume that most of them are not harmful.

But some bacteria are harmful—or may become harmful, given the proper circumstances. For example, the *Staphylococcus* bacteria that covers your skin in grapelike clusters is usually harmless. But it mutates quickly and so that same bacteria can produce the dreaded "staph" infections that sometimes sweep through hospitals. Those same organisms may also be responsible for pimples and boils if they penetrate the skin or move into pores.

And bacteria can cause a host of other diseases of humans and other animals. We need only think of anthrax, diphtheria, gonorrhea, leprosy (perhaps the least contagious of bacterial diseases), pneumonia, syphilis, tetanus, tuberculosis, typhoid fever, and whooping cough. Not to mention the germs that cause bad breath. And cabbages must live in a world of black rot, just as roses fall to crown gall.

Bacteria can be harmful indirectly by causing food to spoil. But, here, our divisions may be considered fairly arbitrary. If we want to eat some food and it is rotten, the bacteria have spoiled it; if we don't intend to eat it, the bacteria are decomposing it. One kind of bacterium produces one of the most deadly poisons known. A spore-forming bacterium called *Clostridium botulinum* grows in foods that are deprived of oxygen, such as in contaminated canned goods. The fatality rate in botulism is high if untreated with antitoxin. Home canning is the biggest problem, but the poisons can be destroyed by boiling a few minutes.

Another dangerous spore former, *Clostridium tetani,* grows best in deep puncture wounds. We don't hear much about people dying of tetanus anymore because of the readily available vaccine. But the problem is always lurking in the background since there is no effective treatment once the infection takes hold.

Taxonomy of the Bacteria

Now let's look at the taxonomy of bacteria. We can begin with the phylum (or division),* Eubacteriae, a group which best exemplifies some of the things we have been discussing.

Eubacteriae (True Bacteria)

The true bacteria are small, even for bacteria. They are about 1 micrometer in diameter and one to several micrometers in length. The majority grow singly, although some species grow in colonies. Eubacteria commonly come in two shapes: *coccus* (spherical) and *bacillus* (rod-shaped).

The spherical cocci grow singly or in colonies. Some of the colonies, however, scarcely deserve the name since they are composed of only two cells (*diplococci*), while others form longer chains (*streptococci*), grapelike clusters (*staphylococci*), or cubes of four or eight cocci (*sarcina*) (Figure 14.6). The first three may be familiar to you since the gonorrhea organism is a diplococcus and you may have read about gonorrhea. Other diplococci cause pneumonia. And some species of *streptococcus* causes blood poisoning and "strep" throat. Scarlet fever, rheumatic fever, and nephritis (kidney inflammation) are indirect, allergic responses to strep throat. *Staphylococci* are well-known agents of skin infections, associated with boils, pimples, carbuncles, and abscesses. In spite of all this, however, most cocci are harmless *saprophytes* (bacteria which use nonliving food sources). A streptococcus organism is used in the manufacture of buttermilk.

The other form of Eubacteriae is the bacillus. Species of these rod-shaped cells are capable of causing anthrax, tetanus, botulism, diphtheria, dysentery, meningitis, whooping cough, plague, tuberculosis, and typhoid fever. Aside from that they're no problem at all.

In fact, they have their good side. They are of great importance in the process of decomposition and nitrogen fixation. No eucaryotes can utilize atmospheric nitrogen, and without bacterial and cyanophyte nitrogen fixation, the life forms on earth would be drastically different or cease altogether.

*The botanical term *division* applies here and elsewhere in this chapter, since bacteria were traditionally classified as plants. The distinction between phylum and division is moot, but bacteriologists as well as botanists have agreed on the use of the latter and it is specifically stipulated in "The International Code of Botanical Nomenclature."

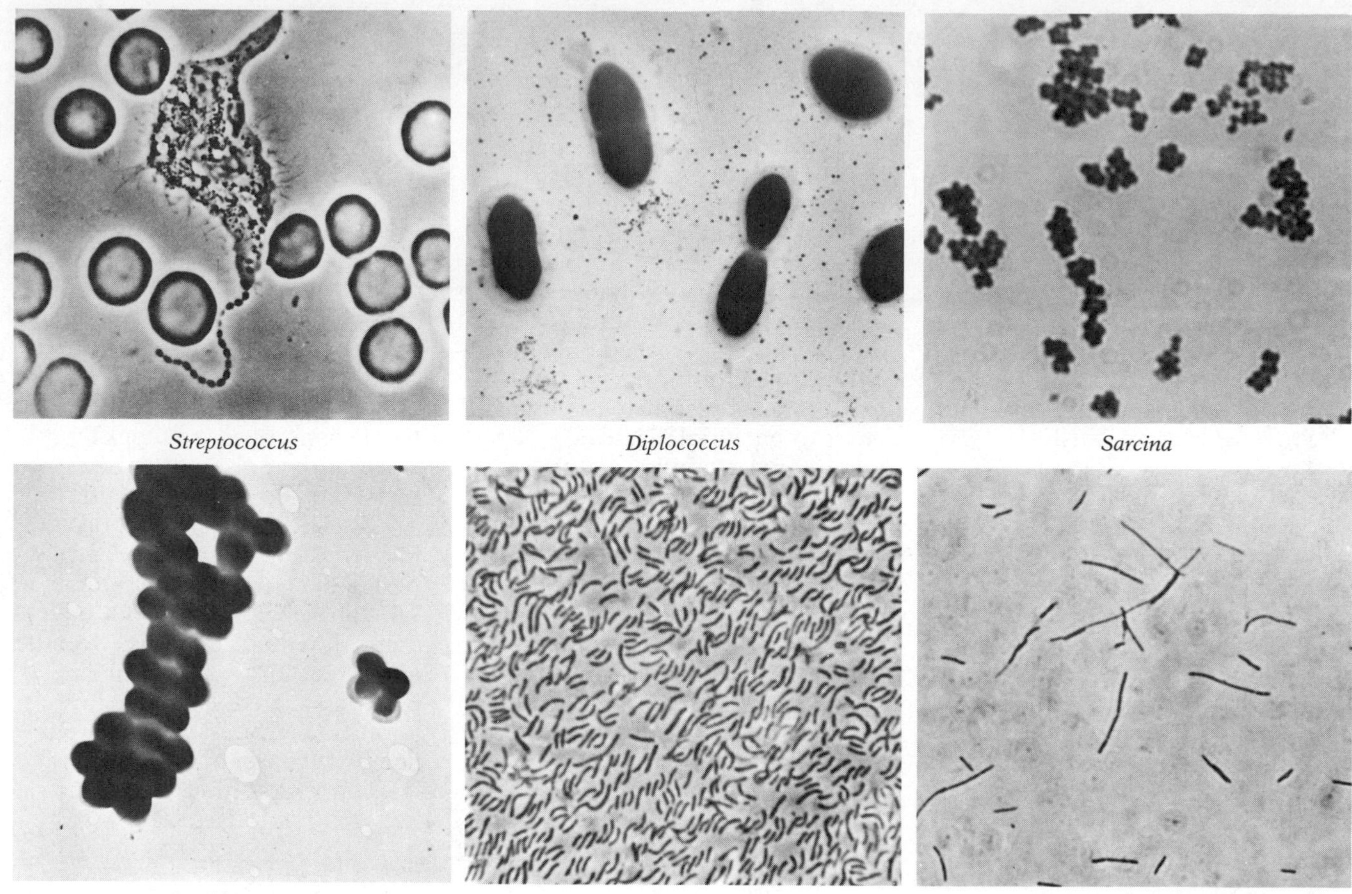

14.6 Form and arrangement in bacteria. Eubacteriae occur in the two basic forms shown here: the coccus and the bacillus. The coccus forms may have definite arrangements in colonies. Both their forms and arrangements are used to identify specific bacteria. For example, the diplococcus form and arrangement shown here is typical of *Diplococcus pneumoniae,* the agent of bacterial pneumonia. The *Streptococcus* above is being phagocytized by a white blood cell.

In addition, *Lactobacillus* is used in the preparation of cheese, yoghurt, sauerkraut, and pickles. *Lactobacillus* bought at the drugstore may also aid in your recovery from certain intestinal disorders, and many antibiotics are produced by bacteria. Silage requires the action of bacilli, and, of course, the famed *Escherichia coli* of your large intestine is a bacillus. *E. coli* helps provide you with, among other things, vitamin K and biotin (Chapter 25).

Biochemical Adaptations

Eubacteria demonstrate a wide range of biochemical adaptations. For example, many are aerobic, requiring oxygen for respiration in much the same way we do, and producing H_2O and CO_2 as they rid themselves of spent organic fuels. Others are *facultative anaerobes,* not requiring oxygen, and some, the *obligate anaerobes,* cannot even survive in the presence of oxygen.

Anaerobic bacteria may produce CO_2 from the breakdown of carboydrates, but must use elements other than oxygen to rid themselves of hydrogen. Some produce methane (CH_4, swamp gas). Some reduce sulfur, forming hydrogen sulfide. Their presence can be detected by a foul smell—sewers, for example, smell of hydrogen sulfide.

The earth's organisms, including bacteria, are placed in two broad categories, the heterotrophs and the autotrophs, according to their relationship with carbon and energy. Heterotrophic bacteria utilize organic carbon compounds (the carbohydrates, lipids, and amino acids) as energy sources and as molecular building blocks. So their biochemistry is like that of heterotrophic eucaryotes—(that is, animals). The bacterial autotrophs, however, don't necessarily behave like the eucaryote autotrophs—(that is, plants).

Bacterial *autotrophs* include *chemotrophic* bacteria and *phototrophic* bacteria. Both forms utilize carbon dioxide as a carbon source, but the energy for changing CO_2 to carbohydrate comes from different sources. *Chemotrophic* bacteria (or *chemosynthetic* bacteria, as they are also known) utilize inorganic energy sources such as hydrogen, sulfur, hydrogen sulfide, or ammonia. They leave behind the oxidized remains as wastes. A common example of a chemotroph is the bacterium *Nitrosomas,* an important component of the nitrogen cycle. *Nitrosomas* oxidizes soil ammonia in the presence of oxygen to produce nitrite (NO_2^-). Energy from the reactions is used by the bacterium in reducing CO_2 to carbohydrate. The ammonia-oxidizing reaction proceeds thus:

$$2\,NH_3 + 3\,O_2 \xrightarrow{\text{Enzyme}} 2\,NO_2^- + 2\,H_2O + 2\,H^+ + \text{Energy}$$

Much of the chemistry of this process remains unknown, but the overall carbon-fixing reaction is:

$$2\,H_2O + 4\,NH_3 + 6\,CO_2 \xrightarrow{\text{Enzyme}} \underset{\text{Carbohydrate}}{C_6H_{12}O_6} + 4\,NO_2^- + 4\,H^+$$

The *photosynthetic* bacteria, known as *purple bacteria*, are inhabitants of brackish and highly polluted waters. These phototrophs have been very useful in research on the photosynthetic process. They contain purplish *bacteriochlorophylls* which are somewhat different in their ring structures from the green chlorophylls of eucaryotes and cyanophytes. Unlike cyanophytes and eucaryotic photosynthesizers, the bacteria do not use water as a source of electrons. One group, the *purple sulfur bacteria* uses hydrogen sulfide as an electron donor, and uses light energy to oxidize the molecule to elemental sulfur. The overall reactions can be summarized as follows:

$$6\,CO_2 + 12\,H_2S \xrightarrow[\text{Chlorophyll}]{\text{Light}} \underset{\text{Carbohydrate}}{C_6H_{12}O_6} + 12\,S + 6\,H_2O$$

You can compare this to the overall formula for cyanophyte and eucaryotic photosynthesis by substituting water in the formula for H_2S, and getting $6\,O_2$ as a product instead of 12 S.

Another group of photosynthesizers, the *purple nonsulfur bacteria* are even more unusual in terms of their metabolism. The nonsulfur group has bacteriochlorophyll and utilizes light energy, but oddly, it must have organic compounds for a source of electrons to reduce carbon dioxide. They can use a variety of alcohols and are known to use succinate (one of the intermediates of the citric acid cycle) as an electron source. This organic carbon requirement tends to strain the meaning of the term *autotroph* somewhat. Obviously, these organisms cannot live by themselves but must exist as members of biological communities. Apparently purple nonsulfur bacteria have found a middle ground between autotroph and heterotroph.

Myxobacteria

The myxobacteria include the slime or gliding bacteria (Figure 14.7). The secreted slime forms a slippery layer on solid surfaces, allowing the bacteria to "creep" or "glide" over it. It has been proposed by some investigators that the mucus is secreted in a concentrated form, and that it expands as it takes up water; this expansion then pushes the organism along. (The eucaryotic diatoms apparently use this same manner of locomotion.)

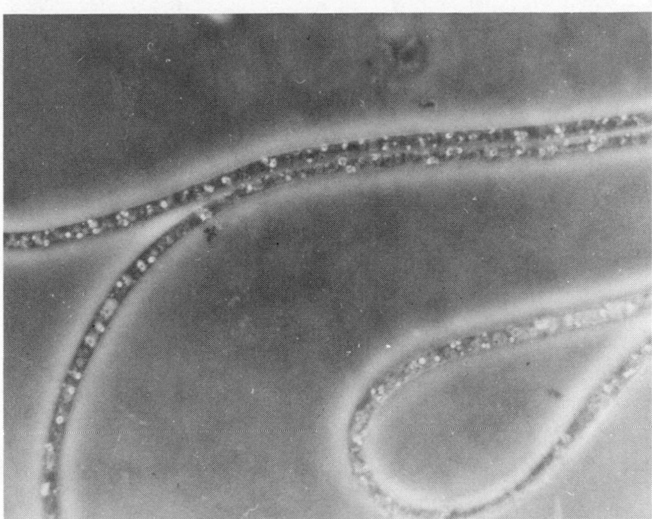

14.7 The Myxobacteria are more commonly known as the *slime bacteria* because of the layer of slippery material they secrete. These bacteria are unusual in that some produce fruiting bodies in which spores form. Because of this apparent specialization, they are believed to be multicellular, at least on a primitive level.

The reproductive behavior of these bacteria is interesting because it is so much like that of the eucaryotic slime molds (but it may admittedly fail to capture your imagination if you are romantically inclined). What happens is that individual creeping, slimy cells converge into a slippery mass. From this mass of slime, a multicellular *fruiting body* grows up, holding aloft a *cyst* in which many spores form. When the cysts rupture, the spores are released. Myxobacteria thus have a degree of cellular specialization entitling them to be considered truly multicellular organisms, lowly procaryotes though they are. Similar levels of cellular specialization are achieved also by cyanophytes.

Chlamydobacteria (Mycelial Bacteria)

The mycelial bacteria have characteristics which appear to be intermediate between true bacteria and fungi. *Mycelia* refers to very fine, cobwebby filaments. *Chlamydobacteria* are sheathed, filamentous bacteria that occur in colonies. The sheath is extracellular, secreted material in which the bacterial cells are lined up end-to-end (Figure 14.8). One species is common in polluted water; its filaments actually attach to the underside of the water surface. In its dispersal phase, swarms of motile, flagellated cells are released from the sheaths.

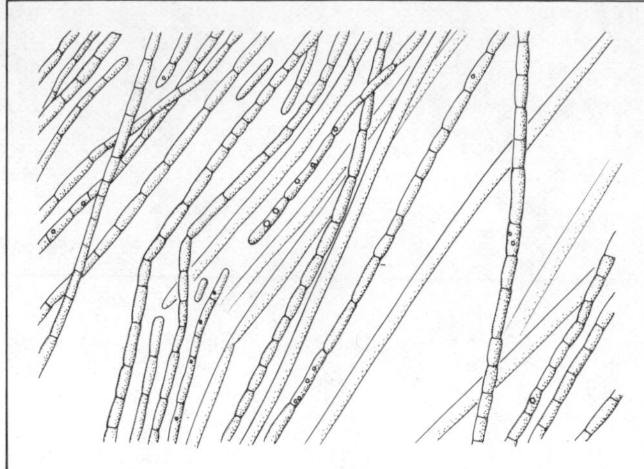

14.8 Chlamydobacteria form sheaths around themselves. The sheath generally attaches to some object in its watery surroundings or to the water surface itself. This particular organism is *Sphaerotilus,* a bacterium capable of oxidizing iron compounds.

One of the best known groups, the *Actinomycetes,* are bacteria with highly branched filaments, but without any suggestion of cross-walls. They may give rise to thin, clublike asexual reproductive structures which rise vertically from the mass. The swollen tips break off, each a potential new colony. The sporelike reproductive cells that are released may be flagellated, or nonmotile and windborne. *Actinomycetes* may also reproduce by fragmentation and fission. In these cases filaments become fragmented and then the fragments, themselves, fall apart and the bacteria exist independently as typical bacterial spheres or rods.

Chlamydobacteria are medically important as the source of several antibiotics, notably streptomycin, aureomycin, and actinomycin, produced by *Streptomyces, Aureomyces,* and *Actinomyces,* respectively.

Spirochaetae

The spirochaetes are corkscrew-shaped bacteria (Figure 14.9) that propel themselves by undulating movements. Among them are the largest known bacteria. A member of one genus, *Spirochaeta,* grows to a length of 500 μm, or half a millimeter! Spirochaetes are best known for the disastrous effects of one of their kind, a comparatively tiny one called *Treponema pallidum,* which is the causative agent of syphilis (see Essay 14.2). Prophylaxis (the use of condoms and germicides) and antibiotics reduced the incidence of syphilis considerably in the recent past, but it has again become an epidemic disease in the United States. Syphilis is particularly prevalent among people in their late teens, perhaps because they tend to be less informed and more optimistic than other people. Unfortunately, its really vile effects, which may include insanity, blindness, paralysis, and rampant death of tissues, typically do not show up for decades after the initial infection and are untreatable when they do occur. Many spirochaetes are obligate parasites (which means they cannot survive outside their hosts). *Treponema pallidum* cannot even survive for a few minutes on a towel or toilet seat, tales to the contrary notwithstanding.

Mycoplasmae

The mycoplasmas may be the smallest living things, some species being less than 0.16 μm in diameter. They are smaller than some viruses, and are intracellular parasites of humans and other animals. They apparently *lack a cell wall* and therefore have no definite shape. Mycoplasmas are completely resistant

14.9 The Spirochaetes. Members of this phylum are helical (corkscrew-shaped) and can become very long. The *Spirochaeta* shown here grows to 0.5 mm in length—long enough to be visible—but only 0.5 μm in diameter.

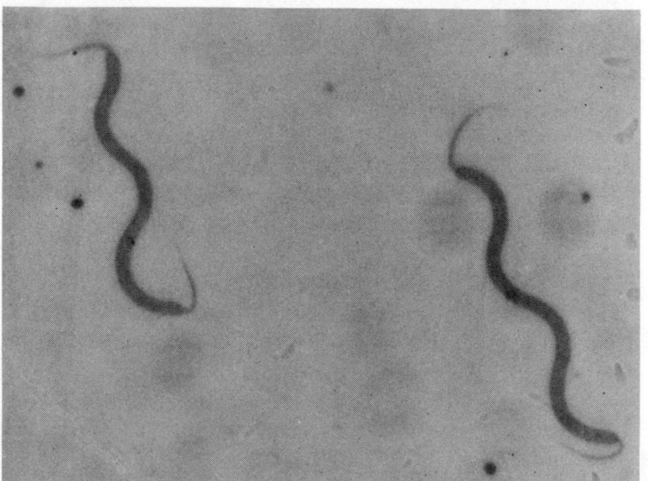

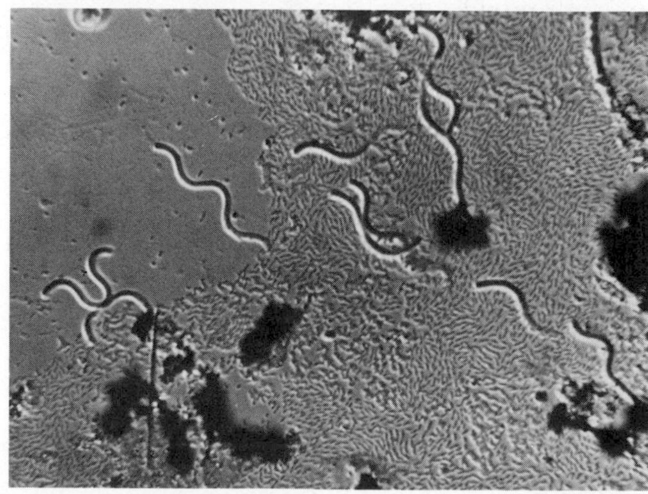

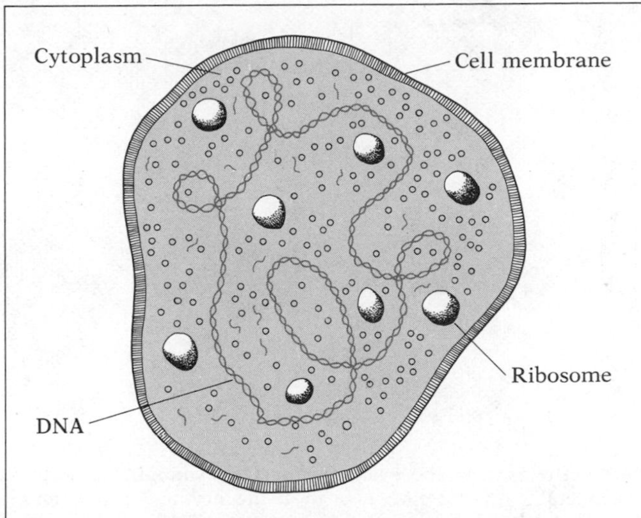

14.10 The mycoplasmas are the smallest bacteria and the smallest cells known. Their structure indicates that they are true cells, although they lack the cell wall found in other bacteria. It is extremely doubtful whether organisms smaller than the mycoplasmas exist, other than viruses. Anything smaller than the size of the mycoplasmas would not have the volume necessary to carry out the synthetic processes necessary for life.

to penicillin, presumably because penicillin normally kills by interfering with growth of bacterial cell walls. They sometimes form colonies resembling mycelia. Reproduction is by binary fission. One form is responsible for a relatively mild form of human pneumonia (Figure 14.10). Mycoplasmas, or something very much like them, have recently been found in plant cells.

The Cyanophytes (Blue-Green Algae)

The *cyanophytes* are sometimes called the *cyanobacteria*; until recently they were commonly known as *blue-green algae*. The group has an ancient lineage. In fact, some authorities believe they are the oldest form of life on earth. The fossil record seems to support this contention, since the oldest records of life on earth are found in *stromatolites* that may have been formed by *cyanophytes* 3.5 billion years ago (Figure 14.11).

Today, 2500 species of cyanophytes are known. We find them everywhere, including ponds, lakes, tidal flats, moist soil, swimming pools, and even around leaky faucets. They occur in the oceans but they are not common there. We mentioned earlier that they flourish in hot springs where temperatures reach 75°C (166°F). They can often be recognized by their blue-green color, but they also come in blacks, purples, browns, and reds.

Cyanophytes occur as single cells, as colonies of cells, or even as multicellular organisms. At least one group actually shows a degree of cell specialization. It has been demonstrated that food can be transferred from *vegetative cells* to *reproductive cells* through thin connecting strands. Similar connections exist between certain other cell types, such as those specialized for photosynthesis and others specialized for nitrogen fixation. This certainly represents a degree of cellular specialization and cooperation, and illustrates the problem of building hard-and-fast rules in biology.

Many cyanophytes occur in long chains of cells called *filaments*. The filaments may or may not be branched. In other instances, cyanophytes form gelatinous masses of no definite shape (Figure 14.12). The cyanophytes can move: filaments such as the common *Oscillatoria* rotate in a screwlike manner, while the gelatinous forms glide along in a mucuslike slime they produce.

The cells of cyanophytes show how surprisingly complex procaryotes can be (Figure 14.13). Unlike the bacteria, they have numerous internal membranes—the photosynthetic lamellae, which contain chlorophyll. They are photosynthetic autotrophs.

In other respects, however, the cyanophytes are typically procaryotic. They lack other membrane-bounded organelles, such as the endoplasmic reticulum, mitochondrion, or nucleus. Their DNA is found clumped in the central portion of the cell. The viruses that infect them closely resemble bacteriophages, and their ribosomes are bacterialike. The entire cell is surrounded by a cell wall comprised of tough carbohydrates. Outside this is the slimy coating, and often an extracellular sheath.

A few more points: In their photosynthetic activity, cyanophytes generate oxygen as do plants and true algae. Their respiration is aerobic, so they can

14.11 This is the oldest known fossil—a photosynthetic, sheathed, filamentous procaryote, possibly a cyanophyte or the ancestor of one. At 3.5 billion years old, it is about 80% as old as the earth itself. Note cross-walls (arrows).

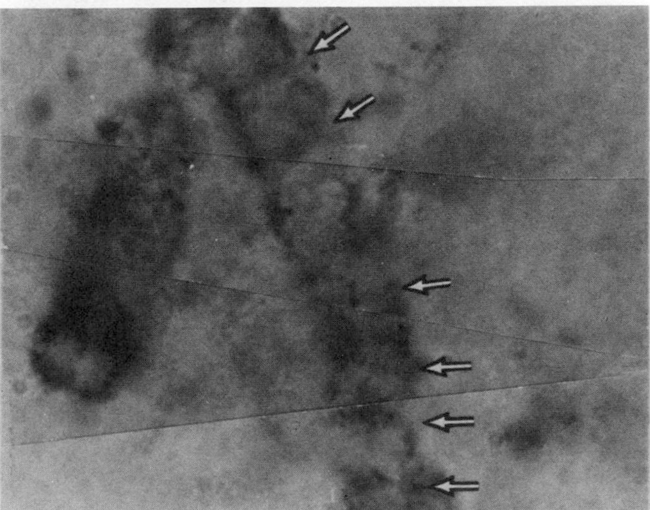

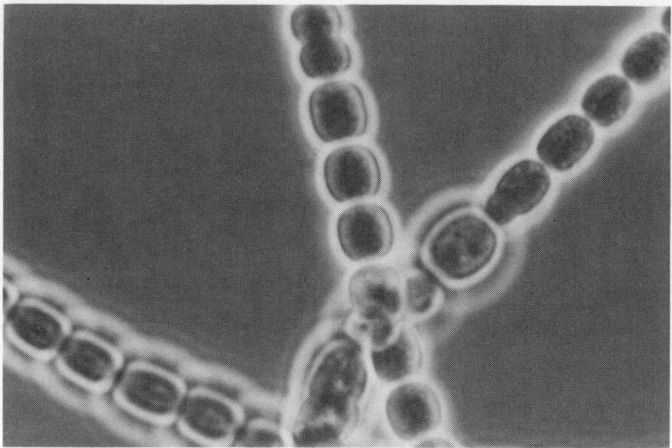

Nostoc

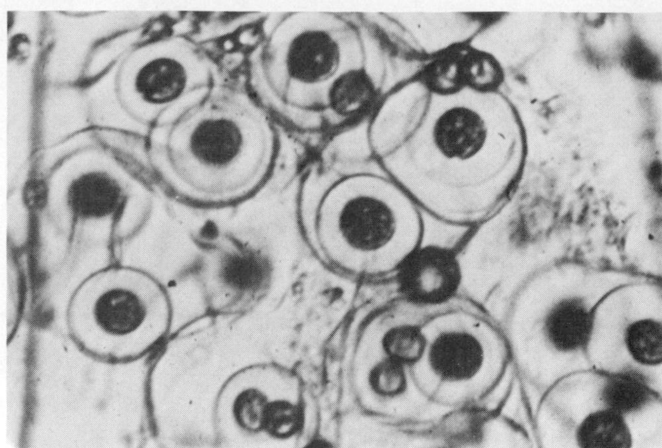

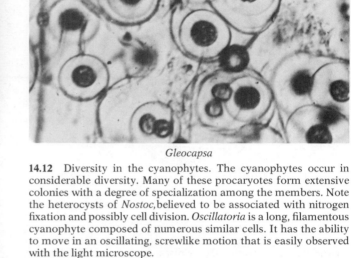

Gleocapsa

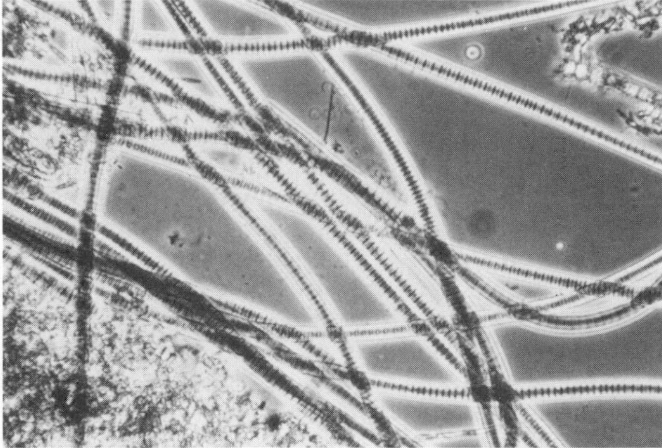

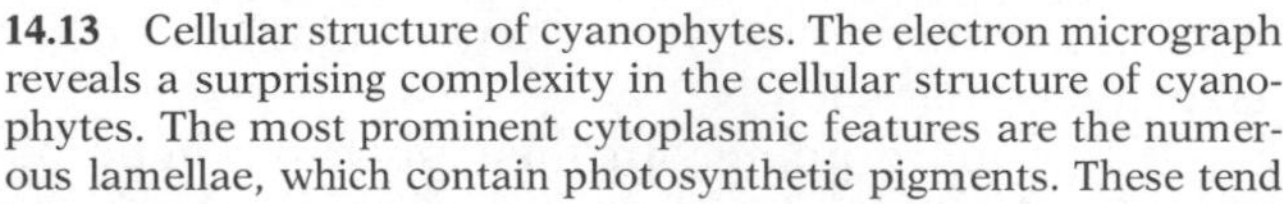

Oscillatoria

14.12 Diversity in the cyanophytes. The cyanophytes occur in considerable diversity. Many of these procaryotes form extensive colonies with a degree of specialization among the members. Note the heterocysts of *Nostoc*, believed to be associated with nitrogen fixation and possibly cell division. *Oscillatoria* is a long, filamentous cyanophyte composed of numerous similar cells. It has the ability to move in an oscillating, screwlike motion that is easily observed with the light microscope.

14.13 Cellular structure of cyanophytes. The electron micrograph reveals a surprising complexity in the cellular structure of cyanophytes. The most prominent cytoplasmic features are the numerous lamellae, which contain photosynthetic pigments. These tend to form a circular pattern around the outer cytoplasm. In the clearer central portion, the circular DNA appears to occur in scattered fibrils. There are also numerous ribosomes, storage granules, and odd-shaped structures of unknown function.

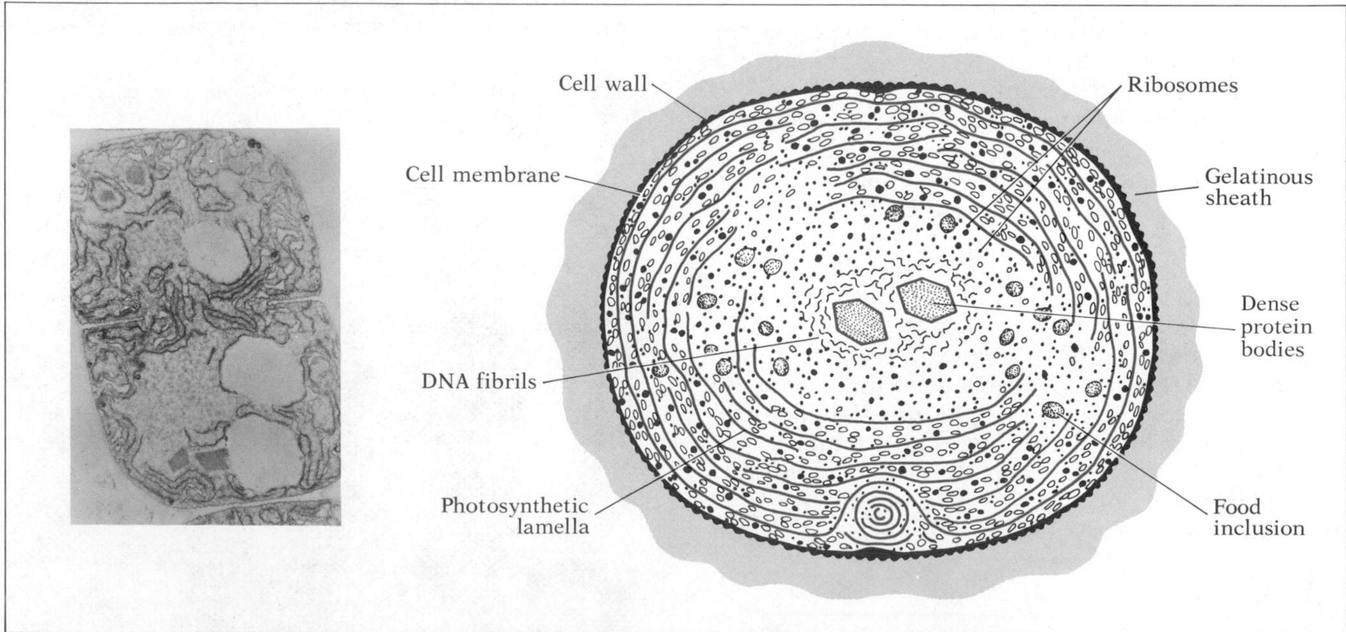

also consume oxygen. In fact, they can't live without it. Their photosynthetic pigments include chlorophyll *a* and beta carotene. In addition, they produce blue pigments called *phycobilins,* which contribute to their characteristic color. They store their own form of starch. Also, many cyanophytes are capable of nitrogen fixation and thus are important ecologically in the nitrogen cycle.

Reproduction in the cyanophytes is entirely asexual, most commonly by fragmentation. However, some species produce numerous endospores within large specialized cells, while other species produce both *heterocysts* (heavy-walled cells) that are the nitrogen fixers, and large, multinucleate reproductive cells.

The ability of some cyanophytes to fix nitrogen has not gone unnoticed in experimental agriculture. The nitrogen compounds produced by these simple organisms can become a valuable fertilizer. In India and Japan, scientists have been experimenting with the notion for many years. In Japan, 30% increases in rice yield have been made possible by growing certain cyanophytes in the paddies.

The Eucaryotes: A Different Matter

So the procaryotes, as we have seen, are an ancient lineage. It is a moving experience to enter a paleobiologist's laboratory and see these ancient procaryote remains. Such organisms have been on the planet for untold eons and are alive and well now. They were here and thriving when eucaryotes emerged. And then we eucaryotes managed to come up with the notion that the earth is somchow rightfully ours. One can only hope the procaryotes have their version of humor.

The origin of eucaryotes is shrouded in mystery. While no one doubts that the earliest eucaryotes must have been protists—relatively simple, unicellular forms—but there is no surviving fossil record to support this. The earliest undisputed fossil records of eucaryote life consist of worm tracks from the ancient sea floor. Within a relatively short time fossils of all the major animal phyla appear. L. Margulis has argued that this sudden appearance of complex eucaryotes was real, and is not simply an artifact of an incomplete fossil record. Once a major step has been taken, evolution can be very rapid indeed.

In any case, all life forms are related at some level, and it seems that the eucaryotes did, indeed, evolve from procaryotes. There are many ideas on how the required dramatic changes came about, but one hypothesis—the *symbiosis hypothesis*—is outstanding and convincing. Part of its plausibility stems from evidence from organisms that are with us today.

The Symbiosis Hypothesis

Lynn Margulis at Boston University has incorporated information from a wide variety of fields (no easy task) and has developed an ingenious and widely accepted hypothesis to explain the origin of the eucaryote cell. She maintains that the primitive eucaryote cell was derived by three separate events that involved the union of four separate procaryotic lines. One of these hypothetical lines (line A) is assumed to be extinct.

Line A, which she calls the protoeucaryote, had evolved the ability to move its cell membrane, and thus had the ability to engulf particles and to form digestive vacuoles and other internal membranous structures. It was the first predator. It was capable only of anaerobic respiration (glycolysis), but it may have had multiple chromosomes and a nuclear membrane. There is no such organism alive today.

Line B was an *aerobic* bacterium, not too dissimilar to *E. coli,* that was engulfed by the line A cells (Figure 14.14). Eventually, a *symbiotic* (mutually beneficial) relationship developed so that engulfed line B cells were not digested, but, instead, they began helping to break down and oxidizing other foodstuffs ingested by the larger cell. In time, the two became completely dependent on one another. Finally, line B cells lost the ability to live outside their hosts, and (Margulis hypothesizes) their descendants exist today as mitochondria. Over time, much of the mitochondrial genetic material was shifted to the host chromosomes, but, even today, mitochondria retain a complete functioning set of tRNAs, bacteria-

14.14 The union of procaryote lines A and B. Two hypothetical procaryotes, now extinct, included a protoeucaryote (line A), which was a simple anaerobe with the ability to phagocytize its food. One organism it fed upon was an aerobic bacterium from line B. Eventually, line B cells became incorporated, and a symbiotic relationship developed. They became interdependent, with the mitochondrialike bacteria carrying on aerobic respiration within the line A cells. The line B cells lost their ability to live outside their hosts. As the host reproduced, the line B cells did likewise, just as mitochondria do in eucaryotic cells today.

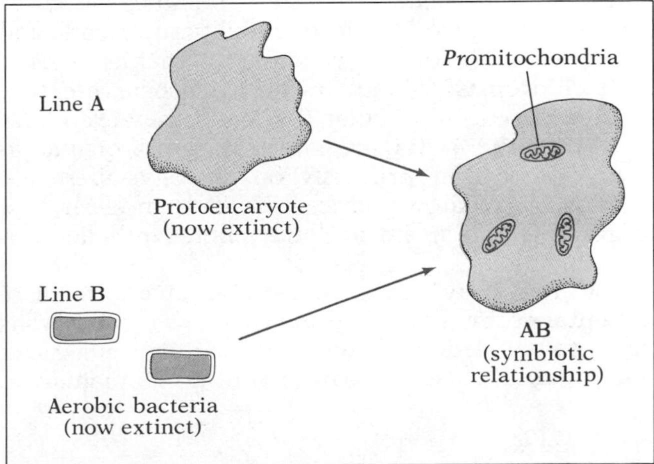

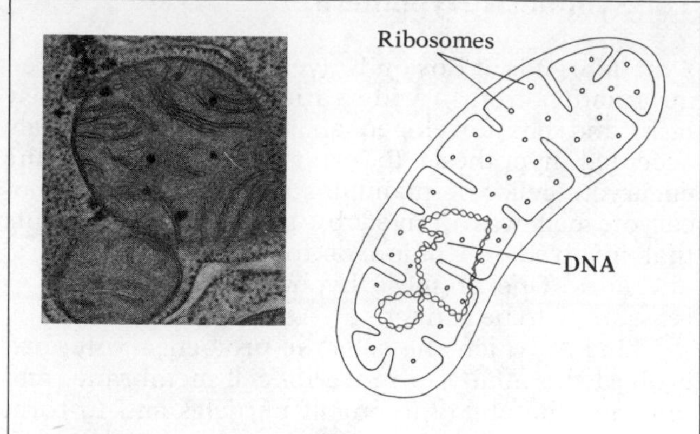

14.15 Mitochondria are known to contain protein-synthesizing ribosomes that are smaller than cytoplasmic ribosomes and about the size of those found in bacteria. In addition, like the chloroplasts, they contain their own DNA.

like ribosomes, and a bacterialike circular chromosome of nearly naked DNA (Figure 14.15). That, we are told, was the first symbiotic event.

The second symbiotic event perhaps involves a bit more faith. A new organism enters the picture: line C. According to Margulis, line C resembled a modern procaryote, the *spirochaete,* in that it was long, thin, and highly motile. Margulis assumed that it contained microtubules in a 9+2 arrangement. The line C organism first attached to the outer surface of the AB complex to become the first flagellum. It introduced the protein *tubulin,* which was to give rise to cilia, flagella, basal bodies, and centrioles (Figure 14.16). Acquisition of cilia and flagella gave further mobility to the evolving eucaryote cell. It also gave the eucaryote cell something else: the centrioles and the mitotic spindle. Thus the stage was set for mitosis.

The third symbiotic event was the acquisition of cells of line D. Margulis proposed that line D cells were simply cyanophytes. Of course, in none of these steps did it pay the host to be too voracious. The eucaryote cell able to engulf photosynthetic algae without digesting them acquired a reliable source of energy—and gave its guests mobility, nutrients, and protection. Part of the evidence for this last step is that chloroplasts retain their own procaryote type of ribosomes and circular DNA, and they even have their own tRNAs (Figure 14.17). Margulis' own argument was based primarily on analogy: there are hundreds of known cases in which modern host organisms take in intracellular photosynthetic symbionts.

Margulis' hypothesis did not receive immediate acceptance by all biologists. There were numerous holes to be filled. Evolutionist Peter Raven suggested that the real ancestral symbiont of green plants was

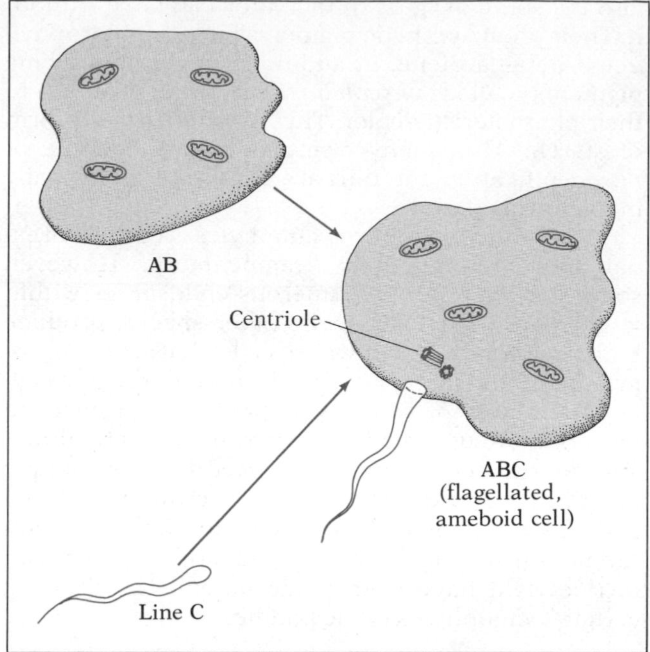

14.16 The union of procaryote lines AB and C. The new AB symbiont now took one step closer to a eucaryotic status by incorporating into its cytoplasm a cell from line C. These new cells brought with them microtubule proteins and the flagellate structure. In addition, it is hypothesized, the tubular centriolar structure arose from this new incorporation. Presumably, this made the mitotic process a future possibility. The new development improved the ability of the symbiont to move about, becoming a more efficient heterotroph.

14.17 The union of procaryote lines ABC and D. In the final step, the highly improved procaryote incorporated a cyanophyte cell with efficient photosynthesizing structures. By this step, the procaryote had several of the organelles of today's eucaryotes. These included mitochondria, contractile proteins, centrioles, cilia, and now chloroplasts. All that remained was for the membranous endoplasmic reticulum and the nuclear membrane to evolve and a modern eucaryote would emerge.

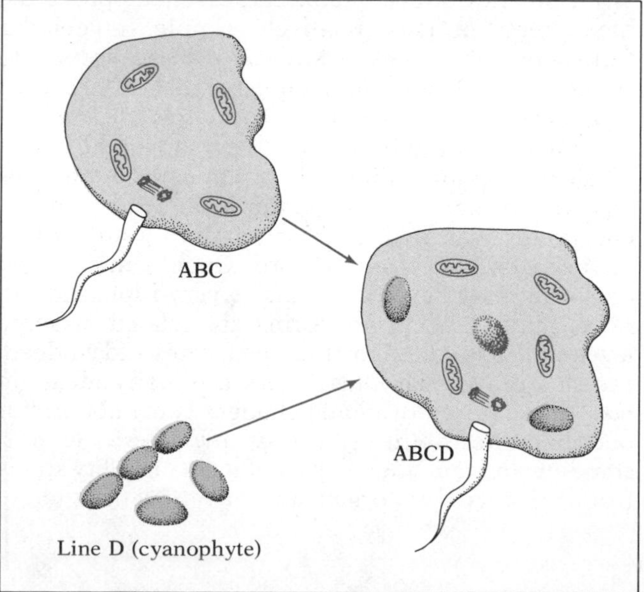

not a cyanophyte but a different photosynthetic procaryote, one which used chlorophylls *a* and *b* and carotenoids, but not phycobilins. He supposed that this hypothetical ancestral group was now extinct. He must have been happily surprised when a living procaryote that fit his description was soon found, in an unlikely place—living as a symbiont with a tunicate in Baja California (Figure 14.18). Thus, a new procaryote group, called the *chloroxybacteria* was promoted to the position of ancestor of the group from which the chloroplasts of green plants evolved.

So here we are with a workable—no, a nifty—theory of how eucaryotes came to be. Since one doesn't tamper with a working clock, do we hang the theory on the wall and look at it with a complacent smile? Or do we continue to question? We continue to question. And the first question is, how much confidence can we have in the theory? It is certainly not a far-fetched notion. Hereditary symbiosis is not uncommon in the living world. Time and again organisms have taken in useful guests. Corals and giant clams live off photosynthetic dinoflagellates that grow within their cells. Lichens, after all, are formed from a hereditary symbiosis between a green alga and a fungus. But the algae can be separated from their coral, clam or fungal hosts and grown independently, at least in the laboratory. Chloroplasts, mitochondria, and flagella cannot be cultured outside living cells.

14.18 Tunicate and chlorobacteria symbiont. It was recently discovered that a tunicate harbors a photosynthetic symbiont that contains typical eucaryotic chlorophylls (*a* and *b*, and carotenoids). The photosynthetic symbiont is now indicated as a possible ancestor of today's chloroplasts.

Margulis' case seems strongest for the origin of chloroplasts. Their internal structure and biochemistry are very similar to that of cyanophytes. Only cyanophytes, chloroxybacteria and plastids store starch. Chloroplasts arise only by division of other chloroplasts. Both chloroplast and mitochondrial DNA are susceptible to certain antibiotics that do not affect the eucaryote nucleus—further evidence of similarity between the inclusions and procaryotes. When the chloroplasts of *Euglena* have been destroyed by such antibiotics, the parent cell line can be kept growing indefinitely in a nutrient broth, but the chloroplasts never reappear.

The case for the bacterial origin of mitochondria is almost as good. True, mitochondria have internal membranes (cristae) unlike anything seen in bacteria, and they seem to have very few functional genes. But at least one mitochondrial protein, cytochrome *c*, is recognizably similar by shape and sequence—presumably by descent—to the cytochromes of certain bacteria, especially the photosynthetic nonsulfur purple bacteria. It now seems likely that the original mitochondrial symbiont was a photosynthetic symbiont, and that it later became restricted to respiratory tasks.

The case is weakest for the symbiotic origin of flagellar structures. Although centrioles and flagellar basal bodies arise only by fission, and are nearly identical in appearance, no flagellar DNA or protein-synthesizing machinery have been reported, and no free-living organisms with the characteristic 9+2 microtubule organization are known.

Margulis argued for the origin of flagella by symbiosis with an impressive analogy that is truly a parade of symbioses within symbioses. Try to follow this argument: Termites eat wood but have no enzymes capable of digesting cellulose. Instead, all termites harbor in their guts various symbiotic protozoa, which ingest and digest the wood particles. The termites eventually digest the bodies of excess protozoa, but the arrangement is nonetheless mutually beneficial—neither organism can exist without the other. Now, one such symbiotic protozoan, *Myxotricha paradoxa*, lives in the gut of certain Australian termites. In addition, the protozoan lives its own hereditary symbiosis with no less than *three* bacterial species! One bacterial species lives in the protozoan cytoplasma and aids in the digestion of wood. The other two bacteria live on the surface of *Myxotricha* and provide the protozoan with a unique form of locomotion. When *Myxotricha* was first described, it was thought to be just another multiflagel-

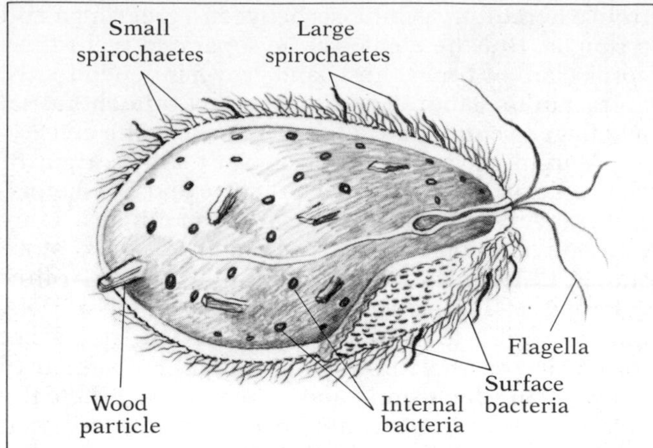

14.19 *Myxotricha paradoxa.* The protozoan symbiont of termites, *Myxotricha paradoxa*, illustrates that the type of complex symbiotic relationships suggested by Margulis can occur. In this system, a symbiont harbors its own symbionts. Part of the organism's mobility is provided by numerous spirochaetes, which affix themselves to its surface. In addition, it harbors a number of cellulose-digesting bacteria within its gut and additional bacteria living symbiotically on its surface.

lated protozoan with an unusually smooth manner of swimming. What were thought to be flagella turned out to be spirochaete bacteria, wriggling away furiously with their basal ends firmly embedded in the host's membrane. Furthermore, each spirochaete is associated with another symbiont, a nonmotile bacterium also attached to the host's surface (Figure 14.19). *Myxotricha paradoxa* has a few true flagella of its own as well.

Of course, analogies are *not* proof. But, still, the idea that a similar kind of symbiosis has occurred before (a billion or so years ago) seems a little more plausible in light of this odd five-species symbiotic complex, which *does* exist.

Margulis presented her theory in 1967. Subsequently, a massive amount of new biochemical evidence seemed to indicate that she was correct about the symbiotic origin of the mitochondria and the chloroplast—but failed to support her notion about the origin of microtubule structures from a symbiotic spirochaete. Then, in 1978, Margulis published new findings of her own: large spirochaetes (also from termite hindguts) with internal microtubulelike structures that cross-react antigenically with antibodies against eucaryote tubulin. This gave some biochemical support, though not proof, for the last part of Margulis' hypothesis.

The symbiosis hypothesis is a superb example of the synthesist aspect of science. It was a bold idea that brought together seemingly diverse observations into a major synthesis. More importantly, it provided testable hypotheses, which could then be addressed by reductionist methods: that is, it predicted that mitochondria should be biochemically similar to some bacteria, that chloroplasts should be biochemically related to cyanobacteria, and that cilia should share biochemical properties with spirochaetes. For instance, 5S ribosomal RNA is found in mitochondria, chloroplasts, eucaryote cytoplasm, bacteria, and cyanobacteria (but not in cilia): a computer analysis of such 5S RNA sequences produces exactly the relationships that are uniquely predicted by Margulis' hypothesis. The symbiotic origins of mitochondria and chloroplasts are now as thoroughly established as any "fact" in biology today. And we ourselves are descended from bacteria.

Application of Ideas

1. There are strong arguments against the current liberal use of antibiotics. Numerous kinds of bacteria live on and inside the body, often in an innocuous manner. Many are saprophytes, while others are marginal parasites, and still others are a constant threat as virulent parasites. What is wrong with using antibiotics as often as seems necessary? Why not kill all the bacteria possible?

2. Despite the great care taken in the commercial canning of foods, an occasional spoiled can appears. These can usually be detected by their bulging out at the ends. What causes the bulging? How can bacteria survive the canning process and grow in a sealed environment?

3. By gathering together pertinent information about bacteria, develop a table or chart that could be used in identifying individual groups or kinds. It is customary to use both physical (visual) and biochemical characteristics, including nutritional requirements or peculiarities.

4. Review the Margulis hypothesis with care and determine the major hypothesis. Then break it down into its several parts. Did Margulis make predictions from the hypothesis? If so, were they tested? Can the general hypothesis really be tested?

Key Words

actinomycin
aerobe
aseptic
aureomycin
autotroph

bacillus
bacteria
bacteriochlorophyll
binary fission
biotin

chemosynthetic
chemotrophic
Chlamydobacteria
chloroxybacteria
coccus
cyanobacteria
cyanophyte
cyst

decomposition
diplococcus

endospore
Eubacteriae

facultative anaerobe
filament
fruiting body

heterocyst
heterotroph

kingdom Monera

mesosome
monerans
mycelia
mycelial bacteria
mycoplasma
Myxobacteria

obligate anaerobe
obligate parasite

paleobiology
phototrophic
phycobilin
plasmalemma
protoeucaryote
protoplast
purple nonsulfur bacteria
purple sulfur bacteria

reducer

silage
spirillum

Key Ideas

The Procaryotes

1. Procaryote fossils are believed to be 3.5 billion years old, the oldest eucaryotes, about 800 million years.

2. Together, the *bacteria* and *cyanophytes* comprise the *kingdom Monera.* Monerans are genetically and morphologically different from the eucaryotes. They lack the nuclear membrane and most other membrane-bounded organelles and are $\frac{1}{10}$ the size, with $\frac{1}{1000}$ the volume of eucaryote cells.

3. Other characteristics are: a rotating flagella, pili, slime or capsule covering, circular, naked DNA, and unique cell wall chemistry.

4. *Monerans* live in all environments and in places where eucaryotes cannot survive.

The Bacteria

1. *Bacteria* are all procaryotic and include four major groups, the *true bacteria,* myxobacteria, the *mycelial bacteria,* and the *spirochaetes.* A fifth may be the *mycoplasmas.*

2. *Bacteria* get their bad reputation from their role as infectious parasites and their destruction of goods.

3. Our conquest of bacterial disease followed the germ theory of disease, antibiotics, public sanitation, and public health measures.

4. Ecologically, *bacteria* are vital for their activity as *reducers,* recycling essential nutrients.

The Bacterial Cell

1. The cell wall consists of lipids, polysaccharides, and amino sugars. It is often surrounded by capsules and slimy secretions.

2. The *protoplast* within contains circular DNA, free ribosomes, and sometimes *mesosomes.*

3. Bacterial flagella are solid protein polymers that spin in place.

Bacterial Reproduction and Spore Formation

1. Bacteria reproduce by simple *binary fission.* Sexual recombination occurs, but is rare.

2. There is no mitosis or meiosis. The circular chromosome replicates and the two replicas are separated by a dividing wall.

3. Many *bacteria* form spores. These are highly resistant to drying, heat, and chemical damage.

Eubacteriae (True Bacteria)

1. Eubacteriae are extremely small—1 μm × several μm.

2. There are three shapes: *coccus, bacillus, spirillum.*

3. There are three arrangements of *cocci: diplococcus, streptococcus, staphylococcus.*

4. *Among the cocci* are well-known pathogens.

5. The *bacilli* are also known as pathogens, but include important *reducers* or decomposers as well.

6. *Bacilli* are used for producing antibiotics, food flavoring, and silage, and one produces vitamin K and *biotin* in our intestine.

Biochemical Adaptations

1. Biochemically, bacteria include aerobic forms, facultative anaerobes, and obligate anaerobes.

2. *Heterotrophic* bacteria are similar to eucaryotes in some of their biochemistry.

3. *Autotrophic* bacteria are different from autotrophic eucaryotes. They include two groups, the chemotrophs and the phototrophs.
   a. *Chemotrophic* bacteria utilize hydrogen, sulfur, hydrogen sulfide, or ammonia.

spirochaete
spore
staphylococcus
streptococcus
streptomycin
symbiont
symbiosis hypothesis

true bacteria

b. One *chemotroph, Nitrosomas,* converts ammonia to nitrite and is an essential participant in the nitrogen cycle.

c. The *purple bacteria* are photosynthesizers using a purple bacteriochlorophyll.

d. *Purple sulfur bacteria* use H_2S for an electron supply in their photosynthetic activity.

e. *Purple nonsulfur bacteria* use organic compounds for a photosynthetic electron supply and represent a middle ground between *autotroph* and *heterotroph.*

Myxobacteria
These are the slime or gliding bacteria. They move on mucus secretions. When they reproduce, cells converge and produce *fruiting bodies,* which contain *cysts* with many *spores.* The *spores* are released and disseminated.

Chlamydobacteria (Mycelial Bacteria)
1. These *bacteria* are colonial, sheathed, and filamentous. *Mycelium* refers to the filamentous growth.
2. One group, the *Actinomycetes,* have branching filaments and produce club-like structures that break away and form new colonies.
3. The Actinomycetes are used as a source of antibiotics.

Spirochaetae
These are corkscrew-shaped, undulating *bacteria.* Some are the largest known bacterial cells. The most familiar member *Treponema pallidum* causes syphilis.

Mycoplasmae
These are the smallest cells (0.16 μm in diameter). There is no cell wall: They reproduce by *fission.*

The Cyanophytes (Blue-Green Algae)
The cyanophytes occur as single cells and colonies, and some are considered multicellular. They move by rotation or gliding along slime secretions. They have internal membranes similar to those of the eucaryotic chloroplast, which contain chlorophyll and beta carotene. Phycobilins impart their characteristic color. They are aerobic, generating oxygen during photosynthesis. Some are ecologically important nitrogen fixers. Cyanophytes reproduce by asexual fragmentation, and some produce endospores.

The Eucaryotes: A Different Matter
1. The earliest eucaryotes are believed to have been simple protists.
2. The *symbiosis hypothesis* describes how eucaryotes might have evolved from procaryotes.

The Symbiosis Hypothesis
1. Line A cells were simple anaerobes which became phagocytic and evolved chromosomes and a nuclear membrane.
2. Line AB cells evolved when line A cells engulfed aerobic bacteria of line B, which survived as symbionts and became today's mitochondria.
3. ABC cells evolved when line B formed an association with spirochaetes of line C, which had already produced tubulin. This gave rise to cilia, flagella, basal bodies, and centrioles. The ABC cells now had motility and a mitotic and meiotic apparatus.
4. ABCD cells evolved when photosynthetic cells of line D were engulfed. Like the aerobic cells, these became symbionts represented today by the chloroplast.

Related Ideas
1. Mitochondria and chloroplasts today have ribosomes and their own DNA.
2. Cytochrome *c* is similar in mitochondria and bacteria.
3. Symbionts like those described in the hypothesis exist today.
4. Some tunicates harbor photosynthetic symbionts.
5. The protist *Myxotricha* harbors spirochaetes that provide locomotion.
6. One spirochaete has been found to contain microtubules of tubulin.

Review Questions

1. *Cyanophytes* may be among the oldest living organisms. How old are they and how does this compare to the age of eucaryotes? (p. 409)
2. How do procaryotes differ morphologically from eucaryotes? (p. 409)
3. List several characteristics that are special to procaryotes. (p. 410)
4. What are some bacterial diseases and what measures have people taken to overcome their threat? (p. 410)
5. Why was Pasteur's contribution of primary importance in the war against bacterial disease and spoilage? (p. 410)
6. What are *reducers* and why is this activity of great ecological importance? (p. 410)
7. List several uses man has made of *bacteria.* (pp. 410–418)
8. Describe the structure of bacterial cell walls and bacterial *protoplasts.* (p. 411)
9. What are three hypothetical functions of the *mesosome?* (p. 411)
10. Compare the bacterial flagellum with that of the eucaryotes. (p. 411)
11. Describe the process of *binary fission* in *bacteria.* (p. 412)
12. What are *spores* and what is the advantage of *spore* formation to *bacteria?* (p. 412)
13. Prepare a simple diagram showing the three forms of bacterial cells and the three arrangements of *cocci.* Label each item. (p. 415)
14. List several diseases caused by *coccus* and *bacillus bacteria.* (p. 415)
15. Define the terms *aerobe, facultative anaerobe,* and *obligate anaerobe.* (p. 416)
16. Describe the ways in which *chemosynthetic bacteria* obtain energy. (pp. 416–417)
17. Write the chemical formulas that represent the oxidation of ammonia by *Nitrosomas.* (p. 417)
18. List two types of *purple bacteria* and describe how each obtains energy. (p. 417)
19. How does photosynthesis in *purple sulfur bacteria* differ from eucaryotic photosynthesis? (p. 417)
20. What characteristic of myxobacteria makes them appear multicellular? (p. 417)
21. Describe the special reproductive structures in *Actinomycetes.* (p. 418)
22. Describe the body form of spirochaetes and comment on their size. (p. 418)
23. Describe the *mycoplasmas.* (pp. 418–419)
24. What are the common habitats of blue-green algae? (p. 419)
25. What observation suggests that some *cyanobacteria* are multicellular? (p. 419)
26. Explain how the photosynthetic structures of blue-green algae resemble those of eucaryotes. (p. 419)
27. In what ways are the *cyanophytes* typical procaryotes? (p. 419)
28. How are the *cyanophytes* of great ecological importance? (p. 421)
29. What is the general theme of the *symbiosis hypothesis?* (p. 421)
30. Describe each of the four cell lines: What did each bring to the hypothetical primitive eucaryotic cell? (pp. 421–422)
31. What is the pecularity of chloroplasts and mitochondria that makes their origin as independent organisms a credible idea? (pp. 422–423)
32. How does *Myxotricha paradoxa* help with the problem of motility in the symbiosis hypothesis? (p. 423)
33. The hypothesis explains the presence of tubulin in the new eucaryote. Why is this, and what recent discovery helps with the problem? (p. 423)
34. Explain how the multilevel symbiosis in the termite lends credibility to the *symbiosis hypothesis.* (pp. 423–424)

Chapter 15
The Protista

We're not sure that there *is* a kingdom Protista. That is, we're not sure the organisms that fall into this kingdom are any more closely related to each other than they are to some organisms in other kingdoms. Nevertheless, they have certain features in common that enable us to lump them together. According to the five-kingdom scheme of H. L. Whittaker (see Figure 13.14), the Protista are those species that have the following traits:

1. They are unicellular (acellular) or simple colonial
2. They have the eucaryotic forms of cellular organization
3. They may have diverse nutritive modes, including photosynthesis, ingestion, and absorption (plus various combinations of these)
4. They may reproduce both asexually and sexually, by mitotic and (in most groups) meiotic processes
5. They have motility by eucaryotic cilia or flagella, or by other means (such as pseudopods); or they may be nonmotile

All these traits are shared with some members of other kingdoms. Only when all five traits are present does an organism qualify as a protist, and even then there will be some overlap with other kingdoms. At the moment, the protista is a veritable potpourri of organisms, many of which have been placed in the group by default rather than design.

If you look back at Figure 13.14, and review the entire five-kingdom scheme, you will see that the arrangement of the kingdoms indicates their relationships. The protists form a broad paraphyletic group underlying the three other eucaryote kingdoms. Groups of protists are located directly under the roots of the other kingdoms because those kingdoms all have protist ancestors. Thus, below the Plants are the plantlike protists—mainly algae. Below the Fungi are the protists that are most like fungi; and below the Animal kingdom are the protists that are believed to share ancestry with the animals. Admittedly, many of the protists appear to have evolved separately and are going their own evolutionary ways. And, of course, there is a lot of educated guessing going on here. Should the protists be divided up into several separate kingdoms? Maybe. Or maybe they should be parceled out among the three other existing eucaryote kingdoms. As you can see, there is always room for debate in the fields of taxonomy and systematics. But we've made our choices; now let's look at the protists themselves in more detail.

We can begin by dividing the protists into plantlike and animallike groups. The protists that resemble plants are called *algae,* and the protists that resemble animals are called *protozoans,* by tradition. This division is obviously broad, but it doesn't begin to cover all the cases. For example, there are the puzzling little *Euglena* (see Figure 15.3). They can be photosynthetic, like the algae, because they contain chlorophyll. But if they are grown in the dark so that they can't photosynthesize, their green area fades and they become heterotrophic, like any animal. They also have a very close relative called *Astasia,* which is morphologically almost identical, but which is nonphotosynthetic and *saprophytic* (meaning that it lives on dead matter). Biochemical studies show that *Euglena* is related to *Crithidia,* a parasite protozoan. So, what is *Euglena?* Because of their flagella, some taxonomists place both *Euglena* and *Astasia* in the protozoan phylum called the Mastigophora—the flagellates. *Euglena* and *Astasia* are markedly different from the other flagellates, however. For example, they don't reproduce sexually.

The task of sorting all this out must fall on the systematists of the future, because we need to know much more about evolutionary relationships. But in any case, let's keep in mind that our categorizing is tenuous and subject to change if new information should arrive.

The Algal Protists

There are about 10,000 named species of algal protists. As we mentioned, algae are photosynthetic plantlike eucaryotes, but they can't be called true plants because they are not multicellular (Whittaker puts multicellular algae in the plant kingdom). Algal protists make up the floating *phytoplankton* of the ocean, and produce more new living material every year than all other photosynthetic forms put together. In addition, many algal protists live as intracellular photosynthetic symbionts in corals and other animals.

Pyrrophyta (Dinoflagellates)

Visitors to tropical waters are often surprised and delighted as they are rowed back to their anchored ship in a dugout canoe after a rousing night ashore. Each time the paddle slides into the water, it seems to explode with tiny iridescent lights. Even the wake of the canoe is aglow. Objects tossed overboard leave a shimmering trail as they disappear into the briny depths. It seems like magic, but it is the magic of the pyrrophyta, the "fire plants."

The same little sea creatures that flash so brilliantly when disturbed are also responsible for another dramatic, and far more dangerous phenomenon: the dreaded *red tide.* When conditions are right, such as often occurs in warm and stagnant lagoons, certain dinoflagellates multiply to incredible densities. The species with reddish pigments may also contain a potent nerve poison, and the results of their periodic "blooms" spell death for enormous numbers of fish. The water turns a rusty or bloody color, and fish by the thousands appear belly up (Figure 15.1). Clams, oysters, and mussels are unaffected by the dinoflagellates, but they can accumulate the poisons, making their flesh toxic to humans. So avoid eating bivalves during the summer months, especially from waters where a dangerous condition is posted. Some people are even so allergic to the toxins that they can become ill even when swimming in affected waters.

15.1 The dead fish strewn on this beach have washed ashore after being killed in a red tide. This condition results from a sudden population bloom of dinoflagellates, some of which excrete highly toxic metabolic products. The dinoflagellates are always present, but they tend to increase seasonally due to enrichment of the waters by upwellings of nutrients from deep sediments.

So, what are these fascinating little creatures? Again, we're not sure. Figure 15.2 shows a variety of dinoflagellates. They have contractile vacuoles. They all have two flagella. Most have chloroplasts and chitinous cell walls. Some eat other organisms, others are parasites and yet others are rather formless photosynthetic symbionts in larger organisms. But most dinoflagellates are free-living and photosynthetic. Their large chromosomes are permanently condensed and remain attached to their nuclear membranes, much as bacterial nucleoids are attached to

15.2 Several common dinoflagellates. There is considerable variation in size and shape. *Gonyaulax* and *Gymnodinium* cause red tide.

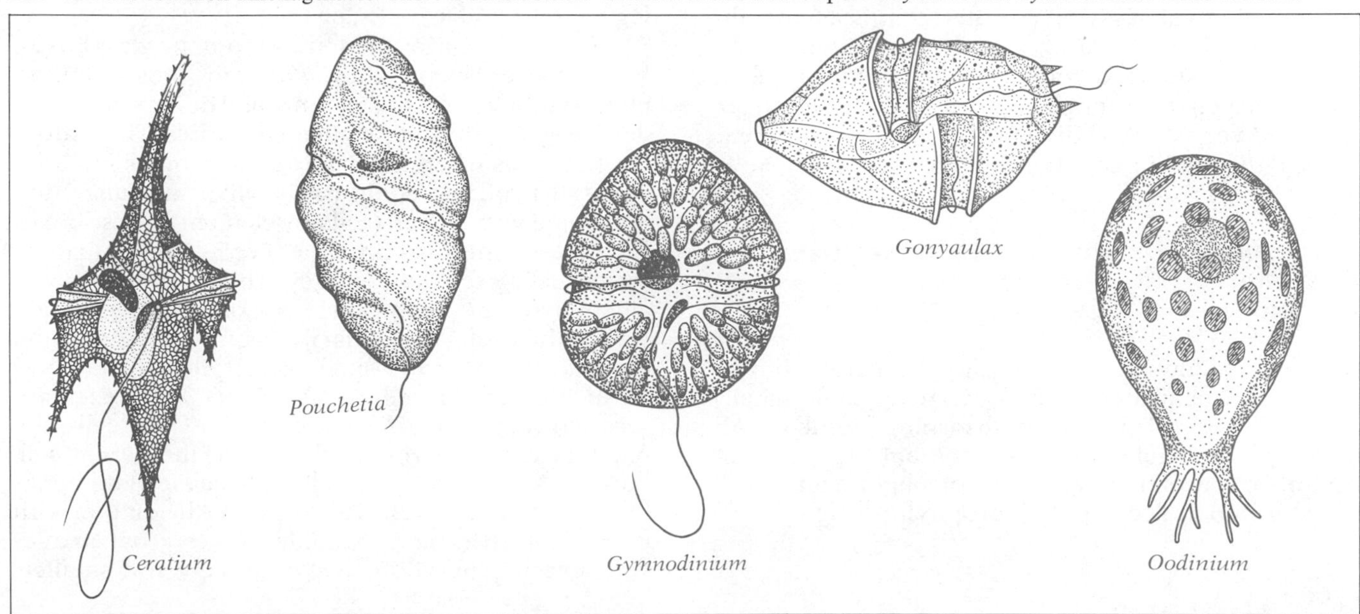

the bacterial cell membrane. The nuclear membranes themselves are unusual, being a single membrane, whereas all other eucaryotes have double nuclear membranes. The nuclear membrane does not break down during dinoflagellate mitosis, which has many features reminiscent of procaryote fission. Meiosis does occur, but it too is strange. Apparently it involves only one division, not two as in other eucaryotes. In general, dinoflagellates have many unique cellular features and seem to have diverged from all other eucaryotes very early in the evolution of life. Some specialists consider them to be intermediate to procaryotes and eucaryotes, and have called them *dinocaryotes.*

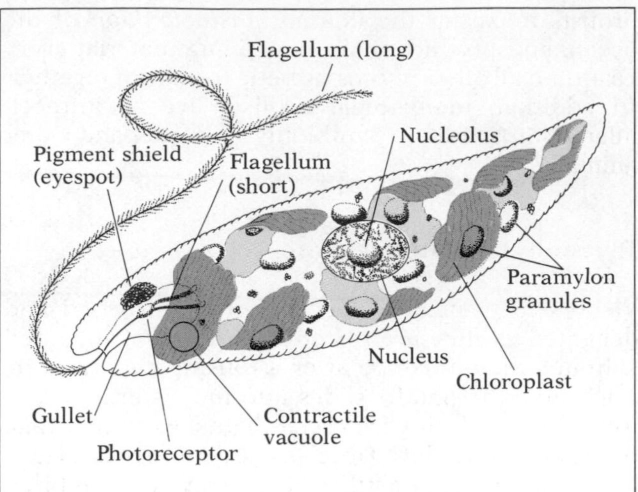

15.3 *Euglena gracilis,* a well-known euglenoid, is shown here. Note that its chloroplasts are more organized than the cells of the cyanophytes and are membrane-bounded. Attached to each chloroplast is a *paramylon body* where starch is stored. The large, membrane-bounded, circular structure at the center is the nucleus, which, with its nucleolus, is typically eucaryotic. At the base of the euglenoid is the flagellum, and enclosed in the flagellar membrane is the photoreceptor. The photoreceptor, along with its *pigment shield,* permits the protist to orient itself toward light, which it requires for photosynthesis. Because of the structural association between the photoreceptor and the flagellum and the fact that it is the flagellum that orients the protist toward light, the two are believed to interact in some way. The pigment shield, a reddish lipid, shades the photoreceptor from light when *Euglena* is in its oriented position.

Euglenophyta (Euglenoids)

We have mentioned *Euglena* before, primarily because it serves as a clear example of an organism that is difficult to classify. But it's also interesting for other reasons.

The name Euglenophyta comes from *eu-,* "true"; *-glene-,* "eye"; and *-phyton,* "plant." The visible "eyespot" of euglenoids (Figure 15.3) is really not an eye at all, but is a shield of red pigment lying next to a light-sensitive area. Euglenoids can therefore detect light and move toward it because of the arrangement of the shield. It works out so that when the sensitive area is not shielded, the organism moves. It thus continually approaches the light that it needs for its photosynthesis. Euglenoids can capture the sun's energy with other red pigments and are responsible for pink snow. They can also turn tidepools or salty desert lakes a blood red, as their numbers explode.

Euglenoids are unicellular, flagellated, asexual, and mostly photosynthetic. The most widespread and typical genus is *Euglena.* Euglenoids lack rigid cell walls, but tend to maintain a flasklike shape. A single long flagellum arises from an anterior *gullet,* a mouthlike depression. There is actually no evidence that *Euglena* ingests solid food particles with its gullet, but other genera of euglenoids are known to do so. A second very short, rudimentary flagellum is present in the gullet, but its function is not known.

Chrysophyta (Diatoms and Their Relatives, the Yellow-Green Algae and the Golden-Brown Algae)

Both the yellow-green algae and the golden-brown algae are freshwater species, deriving their names from the color of their light-absorbing pigments. All Chrysophyta ("golden plants") contain high concentrations of carotenes, which absorb light energy and transfer it to molecules of chlorophyll.

This is a diverse group. Some live in glass boxes; some are flagellated, others not; most are solitary, but a few genera produce filamentous colonies, and some are multinucleate.

Although they are photosynthetic, there are some ameboid species of golden-brown algae that can ingest solid food. Many yellow-green algae can be blown about by the wind and are often found growing on tree trunks, rocks, or soil.

Perhaps you are wondering about the glass boxes. These are best seen in the *diatoms,* the most common photosynthetic marine organism. The boxes are indeed very fragile and peculiar structures. The diatom shell consists of an inner box and an outer lid, both made of pectin (a colorless, jellylike substance) impregnated with silicon (a glassy element). These boxes resemble nothing so much as Petri dishes. Diatoms lack flagella except when they are reproducing sexually. When they reproduce asexually, they divide within their shell by binary fission, whereupon the two old halves of the shell each becomes the outer lid of a daughter cell, as new inner half-boxes are secreted (Figure 15.4). One of the daughter cells, the one that inherited the inside half of the parent cell, must be smaller than the other, because glass boxes don't stretch very well. This presents the diatom with a problem. After all, it wouldn't do to grow smaller with every generation. So, eventually the smallest

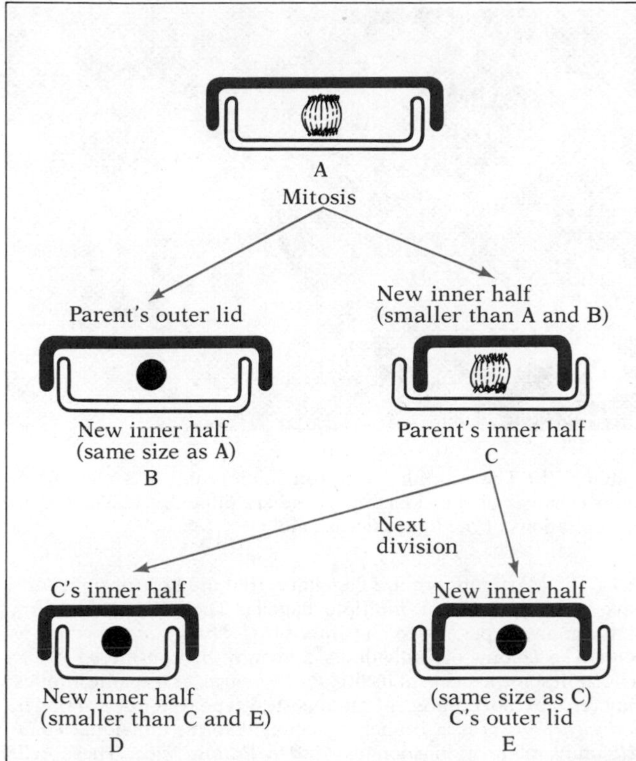

15.4 Asexual reproduction in diatoms. Cell division in the diatoms is complicated by the problem of what to do about dividing the glassy skeleton. Actually, however, the problem is solved rather simply. The two halves of the skeleton separate with the daughter cells. One new cell gets the outer lid of the parent and the other, the inner half (B, C). You might think that all that remains is for the two new diatoms to synthesize the missing halves. Unfortunately it doesn't work out that way. Diatoms only synthesize inner halves, as shown here. This means that the one receiving an inner half (C) produces another inner half, thereby becoming smaller than its parent cell. Then, when this smaller diatom divides, one of its offspring will be even smaller (D). This keeps happening until the diatoms reach a critical minimum size, and then they must enter a sexual process.

diatom undergoes meiosis in its box, but only one haploid gamete survives. It grows flagella, kicks off its box, and becomes the free and naked sexual form. It seeks out another form in the same state and joins with it to produce a diploid zygote, called an *auxospore* because it increases in size for a while (*auxo*, "increasing"). The auxospore then secretes a new glass house and the cycle begins again (Figure 15.5).

The walls of marine diatoms are highly ornamented and beautiful to the human eye. The intricate designs are produced by the arrangements of tiny holes through which gas and water are exchanged. Diatoms can be radially, biradially, or triradially symmetric (Figure 15.6).

Ocean floor sediments consist largely of diatom shells. Deposits of ancient diatom shells occur as *diatomaceous earth,* which is used in toothpaste, swimming pool filters, and insulating material. The

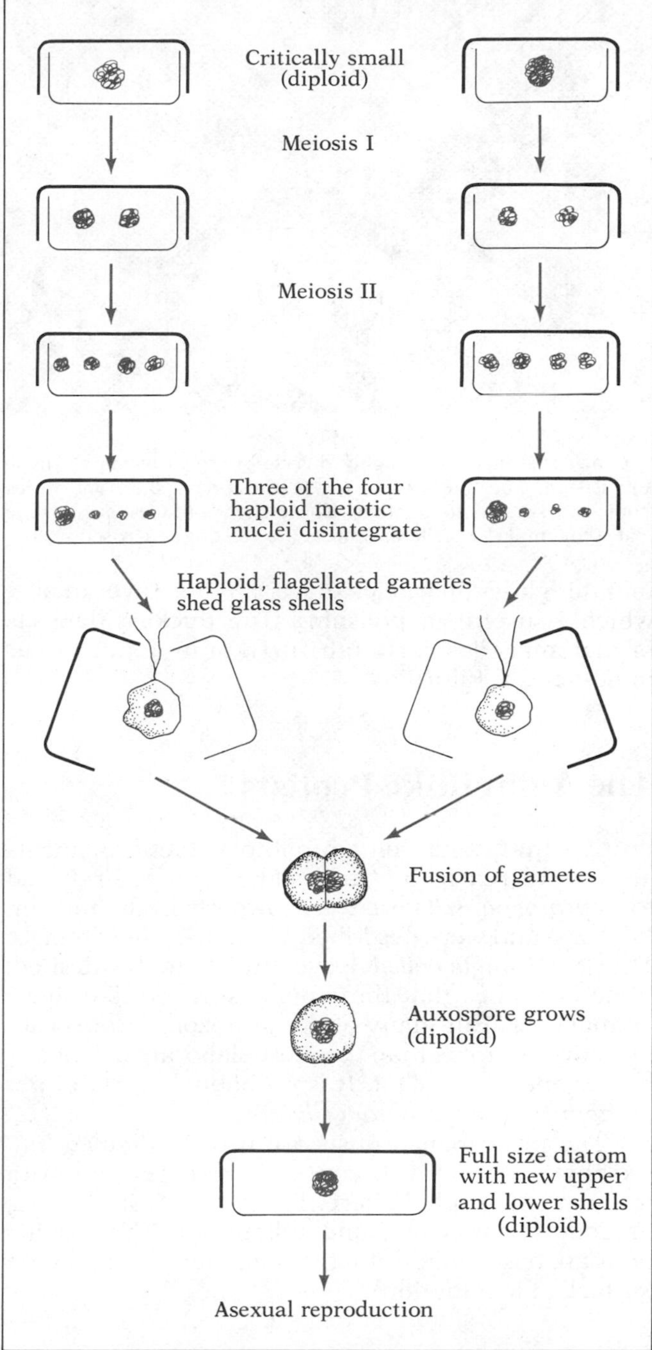

15.5 Sexual reproduction in diatoms. When the diatom reaches a minimum size, it enters a sexual cycle. Meiosis occurs, in typical eucaryote fashion, with four haploid nuclei forming after meiosis II. Three of these simply disintegrate, while the fourth becomes the haploid nucleus of the gamete. The flagellated gamete then slips out of its glass shell and joins an apparently identical gamete from another diatom that has undergone the same process. These fuse in fertilization, to become a diploid auxospore, which produces a new shell, large enough to go through many asexual divisions before the sexual process must be repeated.

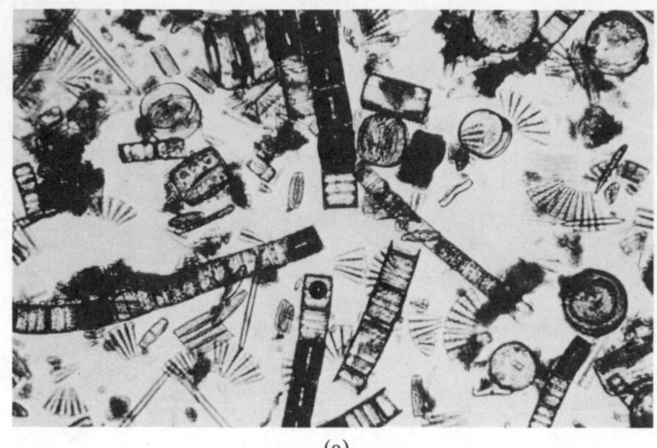

(a)

(b)

15.6 (a) Diatoms are a highly diverse group in terms of shape and design. The fine sculpturing of their glass skeletons varies from one species to another. The etched surfaces are so fine that they make excellent objects for testing microscope lens systems. (b) The scanning electron micrograph is a closeup of the two halves of the skeleton. These are difficult to discern with the limitations of the light microscope.

minute glassy particles have an abrasive quality which is useful in polishing. The thickest deposits of diatomaceous earth (about 1 km in depth) occur in Lompoc, California.

The Animallike Protists

Protists that don't contain photosynthetic pigments are considered to be "animallike" and are referred to as *protozoans.* These were formerly in the phylum Protozoa and were described as single-celled animals. The term *single-celled* leaves much to be desired, since it implies that these organisms are like single animal cells. But many of the protozoan protists are far more complex than the most elaborate cell of any higher animal or plant. It is probably better to refer to them as small and *noncellular.*

The protozoan protists are usually divided into six phyla, but we will concern ourselves only with the four major phyla here. These are Mastigophora, Sarcodina, Sporozoa, and Ciliophora. The distinctions are based mainly on their manner of locomotion (or lack of locomotion).

Mastigophora (the Flagellates)

The Mastigophora are the flagellated protozoans (Figure 15.7). Most members of this phylum propel themselves by these whiplike flagella, which occur singly, in pairs, or in greater numbers. The flagella have the 9 + 2 microtubule structure that is typical of eucaryotes. Movement through undulation of the flagella permits the flagellate protozoan to move efficiently in any direction (see Figure 4.26).

Flagellated protozoa feed in a variety of ways. They may hunt and capture prey or simply absorb

15.7 The Mastigophora are flagellates that are heterotrophic and have single, paired, or multiple flagella. These are only a few of the many types. Note the interesting *Protospongia,* which is actually a colony of individuals and may be considered multicellular in some aspects of its life. *Protospongia,* as its name implies, may (or may not) represent an ancestral type to today's sponges. *Codosiga botrytis* is a branching colonial form; individual *collar cells* may relate it to sponges and to *Protospongia.* These cells have vaselike screens, or collars, that surround a flagellum and are used in taking in food. *Trichomonas vaginalis,* also descriptively named, can inhabit the human vagina and create considerable discomfort. The multiflagellated *Trychonympha* is one of several flagellates that inhabit the termite's gut, digesting cellulose for the host in return for a continuous supply of food. A related organism was used to illustrate the symbiotic hypothesis of eucaryote evolution (Chapter 14).

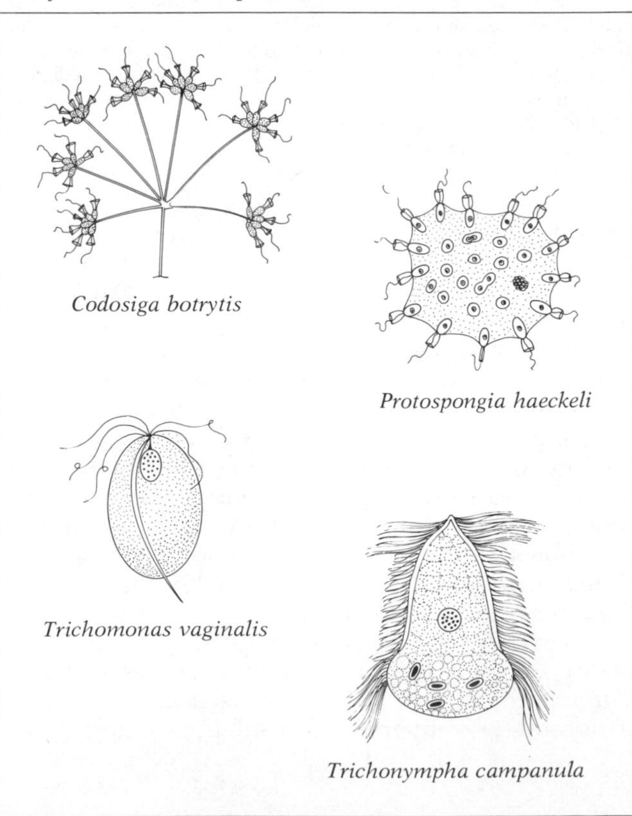

nutrients through their body covering. Reproduction is primarily by asexual cell division. Protozoans are eucaryotes, and so we expect to see nuclear division with the typical movement of chromosomes and action of the mitotic apparatus. And we can expect meiosis and sexual fusion; but our expectations aren't necessarily met. In fact, our knowledge of sexual reproduction in the flagellates is indeed scant, as it is with many other protozoans. We continually add new members to the roll of organisms with known sex lives. As the years go by, one group of protozoa or algae after another falls from the category "sexual fusion and meiosis have *never* been observed" to the category "sexual fusion and meiosis have *seldom* been observed."

One of the most notorious flagellates, *Trypanosoma gambiense,* is the causative agent of *African sleeping sickness.* The trypanosome is carried by the equally infamous *tsetse fly,* whose bite injects the parasite into the mammalian victim. Control is very difficult, since nearly all large mammals in tropical Africa harbor the parasite. Figure 15.8 depicts the life cycle of the sleeping sickness trypanosome.

Sarcodina (Amebas)

Imagine a toughened, brutal, well-trained army on the move into some primitive land. Then imagine that army brought to a dead halt—as soldiers drop their weapons, clutch at their bellies, and dart for the nearest trees. It has happened. Armies have literally been stopped by a tiny protozoan called *Entamoeba histolytica,* the carrier of amebic dysentery (Figure 15.9). The ameba is transmitted by infected human feces entering the food or water system. That, of course, sounds highly unlikely in civilized countries such as our own, but keep in mind that hepatitis is often transmitted the same way—and it's on the rise in the United States. The problem with amebic dysentery, though, is that it can be transmitted by people who carry the disease but show no symptoms.

These impressive amebas are just one kind of organism in the family Sarcodina. The sarcodines are the ameboid protozoans. They are best known for their plasticity. They have no definite form and can send out a temporary extension of the body wall, known as a *pseudopod* ("fake foot") whenever they

15.8 Life cycle of *Trypanosoma gambiense.* (a) The parasite *Trypanosoma gambiense* is shown here in a stained blood smear from an infected person. (b) In the drawing, the flagellum is shown to have a thin membranous flap. It moves with an undulating motion toward its pointed end, where a bit of free flagellum is seen. The spherical nucleus is positioned toward the central region. When a tsetse fly draws blood from an infected animal, the parasite enters the fly's body and reproduces in the midgut. The immature offspring then make their way into the salivary glands, where they become mature and capable of infecting a new host. When humans are bitten, the trypanosome reproduces in the blood and lymph glands and eventually enters the cerebrospinal fluid. The infection is known as *African sleeping sickness* (*trypanosomiasis*) and the symptoms are irregular fever, swollen lymph nodes, painful edema, and—as the parasites invade the central nervous system—tremors, headache, apathy, and convulsions. The person sleeps a great deal, eventually lapsing into coma and death. The entire course of the disease may run from weeks to months.

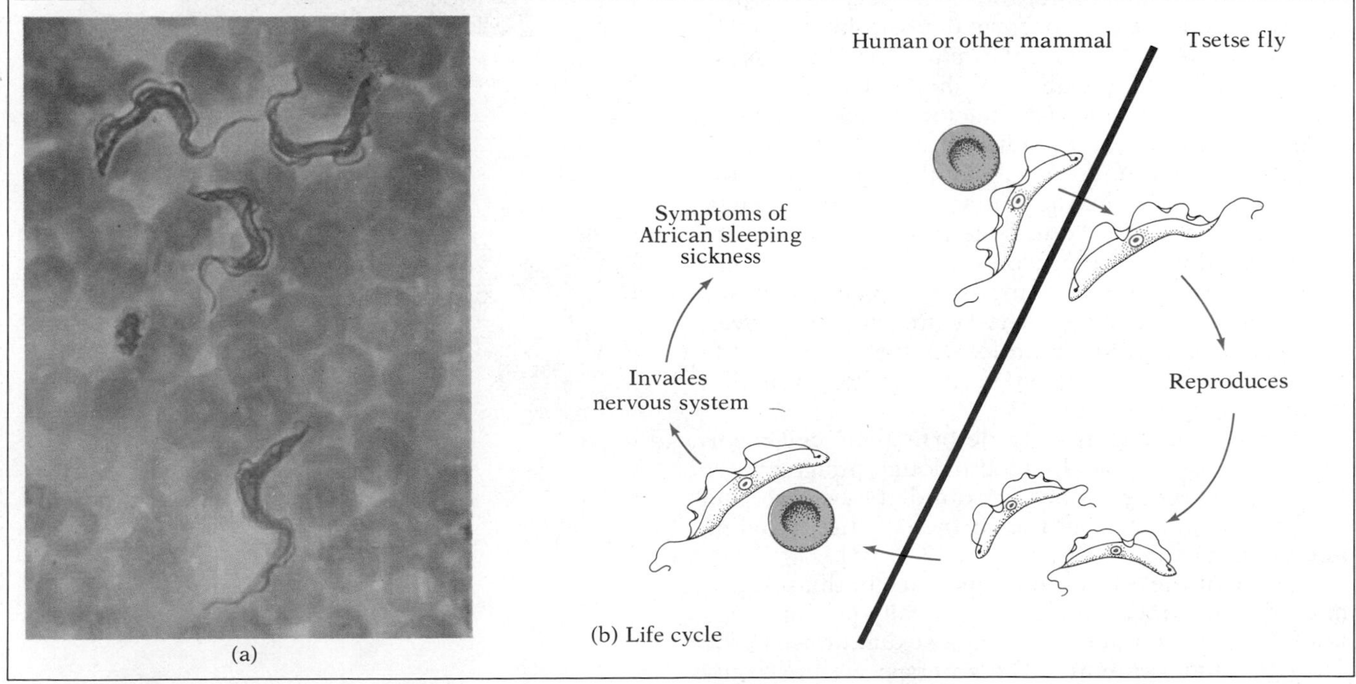

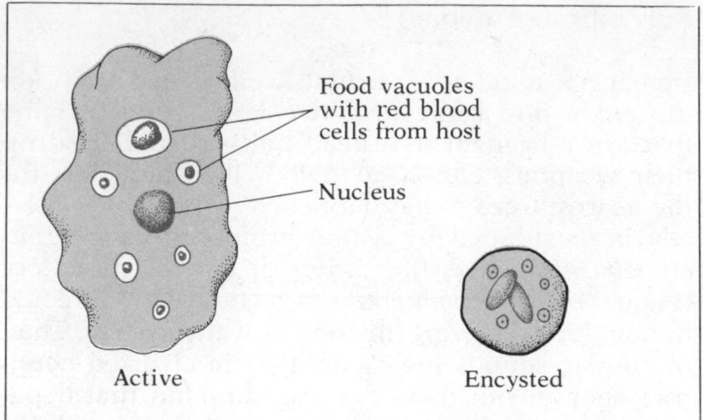

15.9 *Entamoeba histolytica,* the agent of ameboid dysentery, is shown here in both the active state and in its protective cyst. The spread of ameboid dysentery (*amebiasis*) occurs through contaminated water supplies, food, utensils, etc. The ameboid cyst ruptures in the intestine of the person who has eaten it and there, the ameba reproduces. The progeny then penetrate the wall of the colon, where they create ulcerous erosions.

wish to move. Basically, they simply shoot out an arm and then flow into it. Ameboid movement is accomplished by complex actions and dynamic reorganization of the actin microfilament cytoskeleton. Certain white blood cells are creeping through the tissues of your body in just such a way this very moment.

One group of sarcodines, the heliozoans, produces a hardened capsule of silica with long, slender microfilament spines (or *axonemes*) upon which pseudopods creep back and forth gathering food (Figure 15.10). Another group, the *radiolarians,* have been on earth a long time, and their silaceous corpses on the ageless ocean floor form deposits known as the *radiolarian ooze.* The *foraminiferans* also form skeletons. They produce elaborate shells of calcium carbonate. So the radiolarian skeleton is glassy and that of the foraminiferans is chalky. The "foram" shell is pitted with numerous pores through which the pseudopods move in and out as the organism feeds. Foraminiferans also contribute to the dense ooze of the ocean floor. In fact, some land areas that were under the sea are still composed of dense deposits of foraminiferan shells. The white cliffs of Dover, England, are a good example. Oil geologists use the presence of certain ancient foram bodies to predict where oil might be found.

Sarcodines can readily deform their cell membrane, and they typically feed through phagocytosis. Food particles or prey are surrounded by converging pseudopods, or stick to the surface of the sarcodine and are then engulfed.

Some of the Sarcodina reproduce both asexually and sexually, and some only asexually (as far as we know). So far, no one has seen a sexual phase in the most familiar sarcodine, the amebas; we have only seen them reproduce by mitosis and cell division. In the species that reproduce sexually, diploid cells undergo meiosis and form haploid sex cells, which are often flagellated. These can then join in fertilization to produce a diploid zygote that can then reproduce asexually for many generations. Many sarcodines can form protective cysts when conditions dictate, enabling them to withstand long periods of drying. Some Sarcodina are shown in Figure 15.11.

Sporozoa (Malarial Parasites)

From one point of view, the sporozoans may be among the most important of all protozoa to humans. That's because they can cause *malaria,* a disease that has been called humanity's greatest curse.

So what is a sporozoan? This group of protists is characterized by three significant features:

1. They form spores.

15.10 The heliozoan *Actinosphaerium* has a spherical body covered with long, fine spines (axonemes), which are surrounded by cytoplasm capable of a streaming action. Waves of cytoplasmic movement carry trapped food particles into the main body, where they may enter the cell by pinocytosis or phagocytosis. Food vacuoles in the cytoplasm are the sites of digestion. Contractile vacuoles aid in maintaining a suitable water balance.

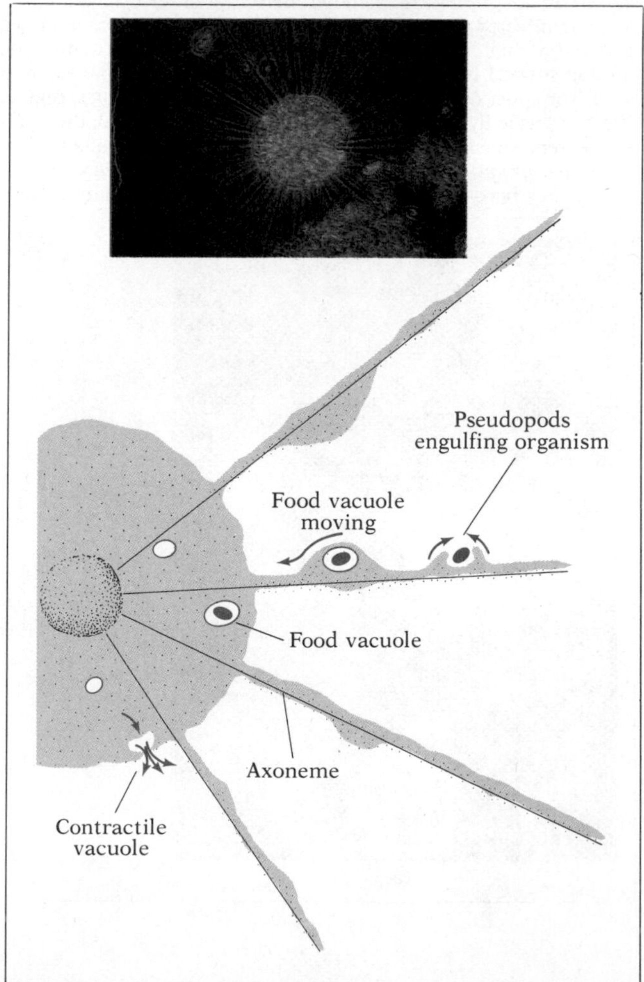

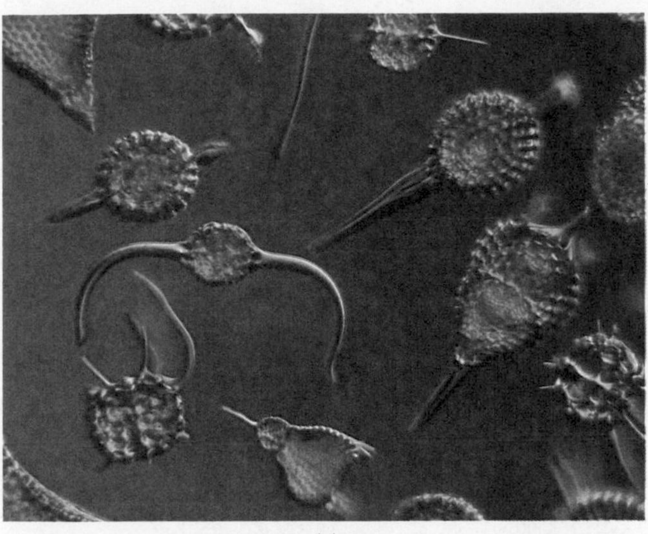

(a)

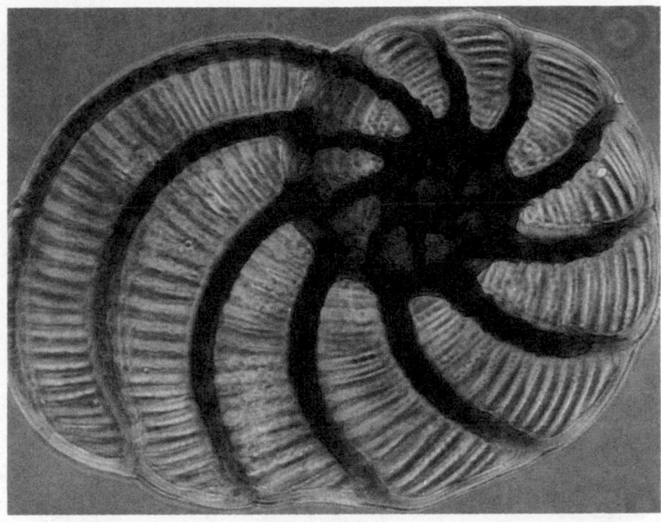

(b)

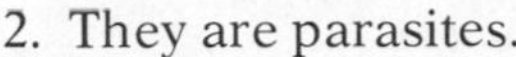

15.11 Sarcodines occur in great variety and may be very beautiful. (a) The radiolarians are typically spherical, with highly sculptured skeletons. (b) The foraminiferans tend to form spiral shapes reminiscent of tiny snails. The skeletons of marine sarcodines contribute to the thick ooze common on the ocean bottom.

2. They are parasites.
3. They have no special means of locomotion in the mature stages.

Sporozoans exhibit complex life cycles, including both sexual and asexual phases. They often complete different portions of their parasitic life cycle in separate hosts.

Malaria, of course, is spread from person to person by the *Anopheles* mosquito, a tiny insect harboring a tinier parasite called *Plasmodium.* The *Anopheles* is alive and well throughout the world (and is recognized by its habit of standing on its head when it feeds; see Figure 15.12). It requires constant vigilance to prevent malaria epidemics, even in the more developed regions such as the United States. The protozoan is most easily stopped by stopping the mosquito, and a number of methods have been devised to do just that. One of the most effective mosquito control agents has been DDT, but its use is highly controversial and has been banned in some places. Because of the dangers of DDT, our best tactics against malaria may be draining ponds, putting mosquito-eating fish in streams, or filling in ditches to eliminate the mosquito's breeding grounds.

The life cycle of *Plasmodium* is reviewed in Figure 15.13. As you can see, it reproduces in both the mosquito and the vertebrate host. The cycle begins with the injection of mosquito saliva containing,

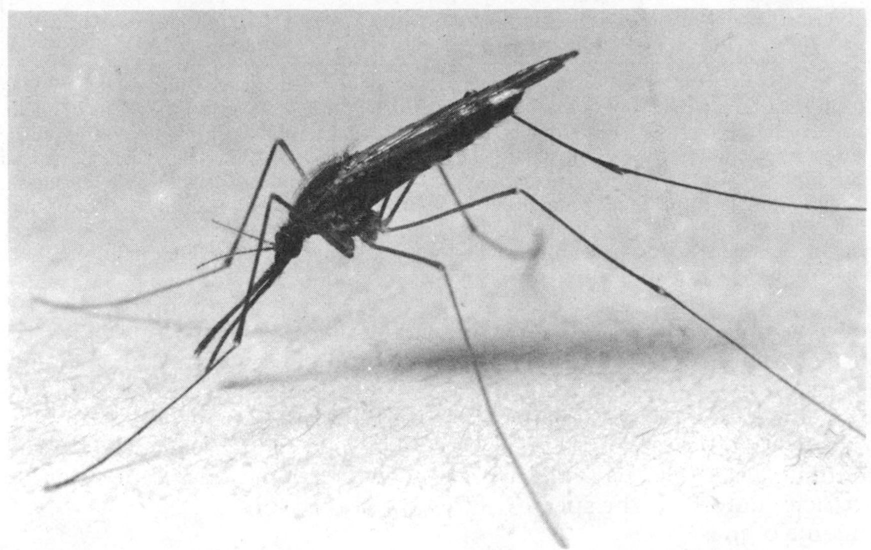

15.12 The mosquito quietly feeding on its host's blood is an *Anopheles* female. We know this because *Anopheles* feed with their tails in the air. If the tail were not raised in the air, it would probably be a mosquito from the genus *Culex,* which prefers to remain horizontal when feeding. We know, further, that it is a female, because males do not feed on blood. The *Anopheles* mosquito is a known carrier of the malarial parasite, *Plasmodium vivax,* at least where malaria is endemic. The mosquito injects the parasite into the bloodstream along with saliva containing an anticoagulant that prevents clots from forming.

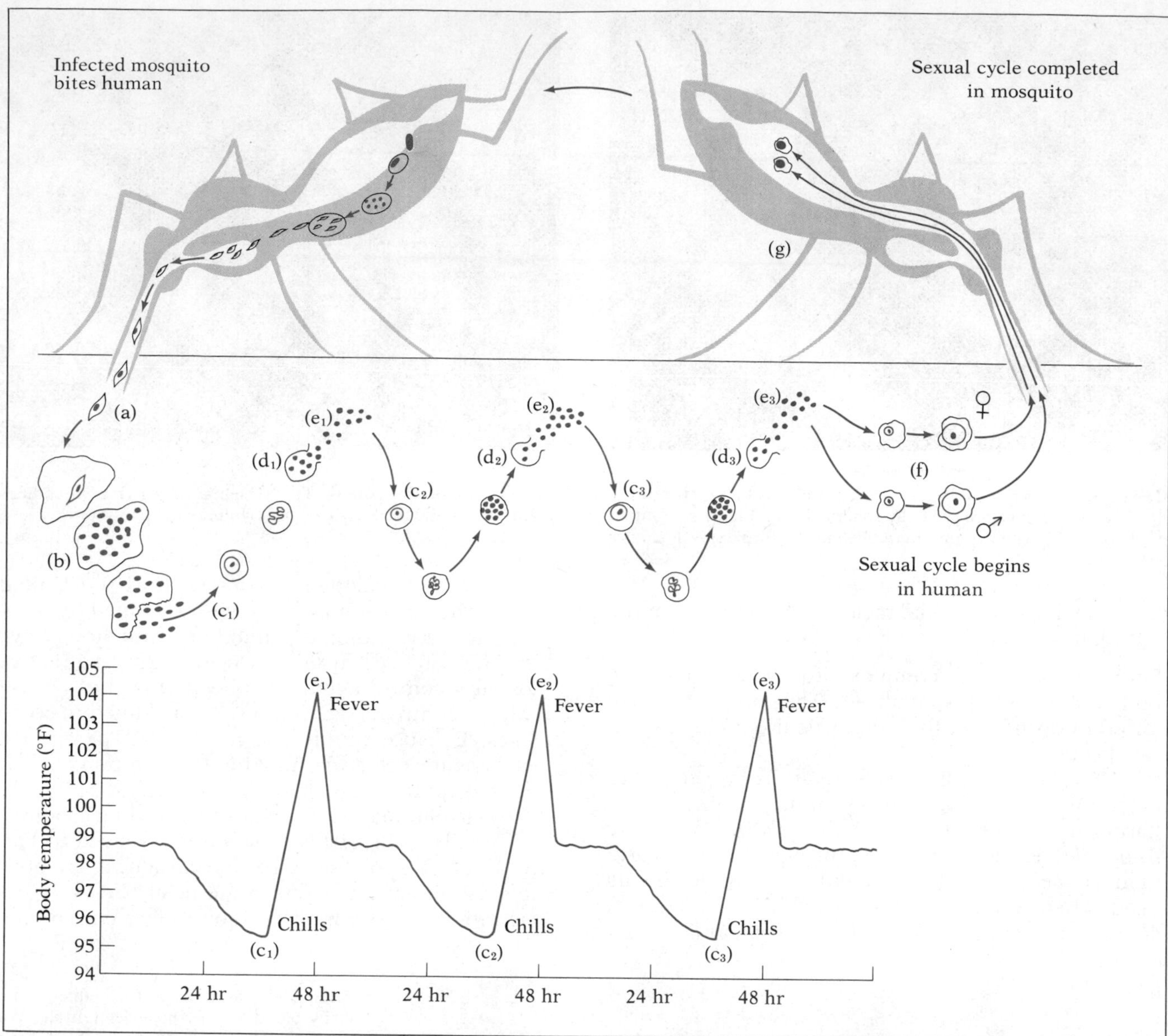

15.13 The life cycle of *Plasmodium vivax* takes place partly in the mosquito and partly in the *reservoir animal,* in this case a human. The periodic symptoms of malaria reflect the life cycle of the organism. (a) The cycle in a human begins with the injection of sporozoites into the bloodstream. (b) From there, they enter the liver and produce large numbers of merozoites. (c) These reenter the bloodstream and (d) invade red blood cells to reproduce once again. (e) The sudden, coordinated release of numerous parasites brings on a period of fever and chills. As the fever subsides, a new infection of red cells occurs, ultimately bringing about a new release of offspring. Then fever and chills occur again. (f) Eventually, gamonts (sexual forms) are produced. When these are taken into the intestine of a mosquito, they begin a sexual cycle. The gamonts release sperm and egg cells. After fertilization, the zygote develops in the intestinal wall of the mosquito. (g) Eventually, the maturing sporozoites venture into the salivary gland from where they move into the saliva that is injected into a new mammalian host.

along with an anticoagulant, protozoan *sporozoites* (the ameboid stage). From there, the parasite makes its way through the bloodstream and enters the liver cells, where it undergoes a quiet asexual phase. At this point, the host still suspects nothing. However, the parasite is reproducing rapidly. Eventually large numbers of the next phase, called *merozoites,* enter the bloodstream. These are the deadly agents of malaria. Their task is to invade the red blood cells, where they repeat their asexual activities. They reproduce inside red blood cells in synchrony, finally rupturing the cells and releasing poisons throughout the host's body. The poisons bring on the familiar fever and chills of malaria. The cycle repeats itself every 48 hr, 72 hr, or longer periods, depending on the species of *Plasmodium* involved.

Eventually, the *Plasmodium* enters a sexual phase. This occurs when some merozoites develop into potential gamete producers known as *gamonts.* There are two types—male and female. The gamonts are quite impotent within the vertebrate host. To produce gametes, they must first find their way into a mosquito. So, when an infected person is bitten, the mosquito sucks a few gamonts into its gut, where sperm and egg cells form and fertilization occurs. The zygotes then develop in the walls of the intestine. The fertilized cell, here known as an *oocyst,* divides asexually, producing large numbers of sporozoites which migrate to the salivary glands of the mosquito. When the mosquito bites, she (only the females bite) injects a little of her infected saliva.

The cycle can be broken by isolating an infected person in a mosquito-proof room. This is exactly what is done in developed countries, thus drastically lowering the probability of an epidemic. Unfortunately, when the seemingly simple precautions are attempted in underdeveloped regions of the world, endless economic and sociological complications arise. So in these areas people continue to resort to DDT. Malaria endemic regions are shown in Essay 9.1 in Chapter 9.

Ciliophora (Ciliates)

Ciliates are identified by having cilia at some stage in their life cycles. The cilia occur in rows, either longitudinal or spiral. Some of these highly diverse protozoa are undoubtedly the most complex single cells on earth (Figure 15.14). For example, their size ranges from about 10 μm to 3000 μm (approximately the same relative difference that exists between shrews and blue whales).

Cilia arise from *ciliary basal bodies* (see Figures 4.26 and 4.27). Cilia can beat, but it wouldn't do for each cilium to wave wildly, independent of the rest. Thus, they are highly coordinated by fibers connecting the basal bodies. When the cilia move, they

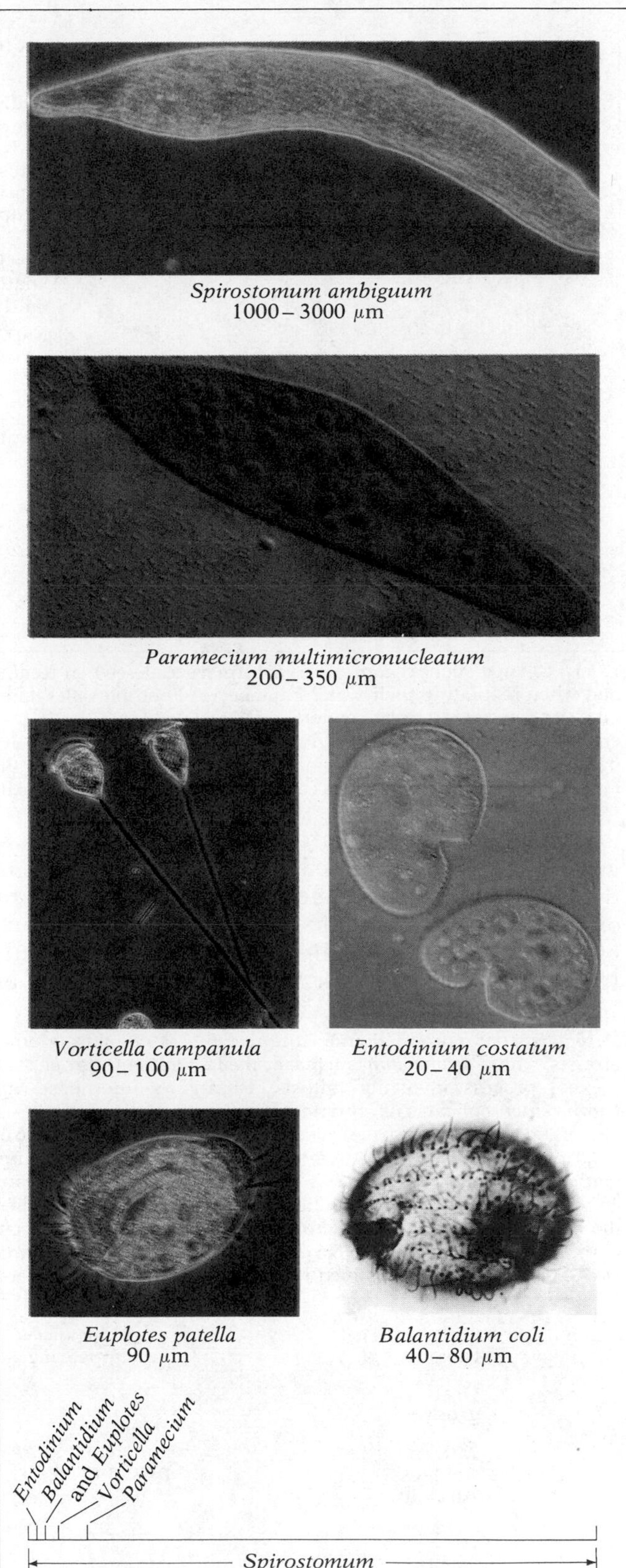

15.14 The ciliates are enormously varied in both size and structure. The largest is *Spirostomum,* which is 3000 μm (or 3 mm) in length, and easily visible to the unaided eye. The smallest ciliates are only $\frac{1}{300}$ the length of this giant. One of these, *Entodinium,* is shown here. Among the most structurally different is *Vorticella,* which is somewhat shaped like a funnel and is attached to a long contractile stalk. Its open end has a ring of cilia that creates a vortex, drawing in its food particles. A peculiar variation is *Euplotes,* with fused cilia forming spinelike projections; the ciliate uses these to scurry along the bottom of its watery environment or to row itself through water. *Balantidium,* one of the few parasitic ciliates, invades the colon of humans and has been known to cause ulceration of the entire lining. The main symptom in most cases is persistent diarrhea. *Paramecium* is one of the best-known ciliates, simply because it is so often used as a laboratory subject. Note the scale showing the comparative lengths of the ciliates described. The variation is somewhat in the order of variation in mammals, from shrews to whales.

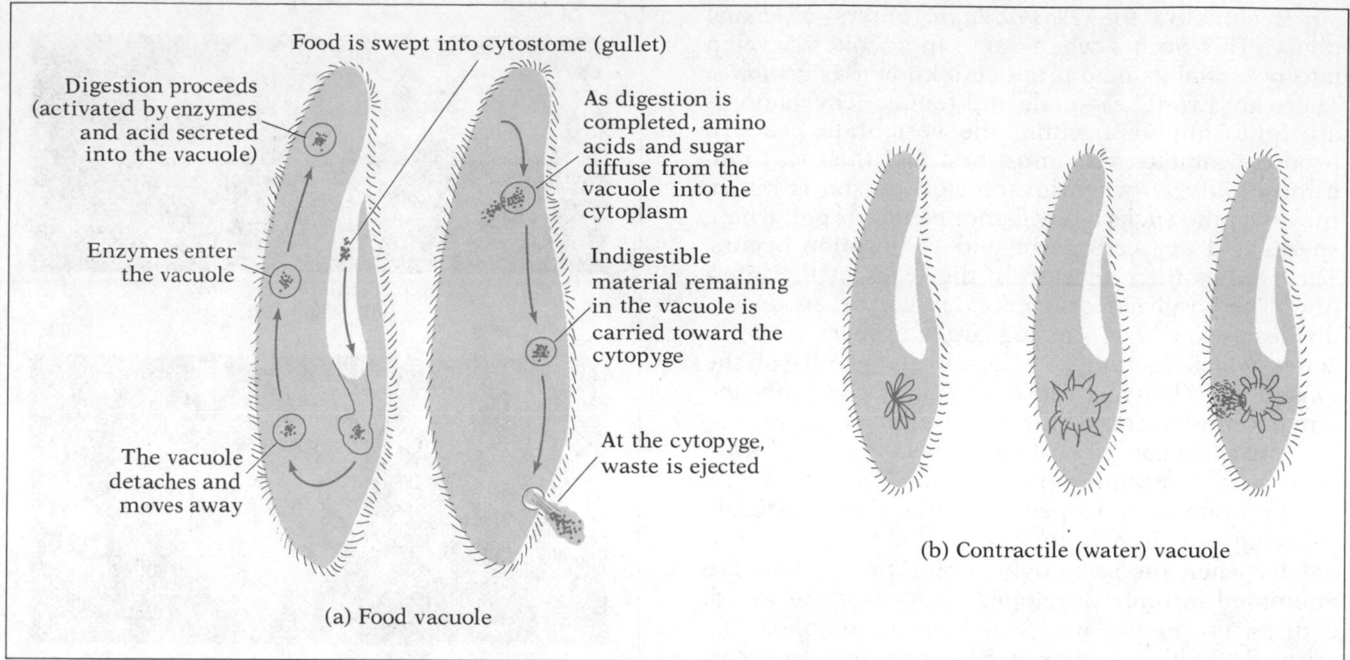
Food is swept into cytostome (gullet)
Digestion proceeds (activated by enzymes and acid secreted into the vacuole)
As digestion is completed, amino acids and sugar diffuse from the vacuole into the cytoplasm
Enzymes enter the vacuole
Indigestible material remaining in the vacuole is carried toward the cytopyge
The vacuole detaches and moves away
At the cytopyge, waste is ejected
(a) Food vacuole
(b) Contractile (water) vacuole

15.15 Ciliates such as *Paramecium* form vacuoles when feeding and when regulating their water balance. (a) Food particles taken into the *cytostome* (gullet), enter food vacuoles for digestion. As enzymes and hydrochloric acid digest food, the food vacuoles move by cytoplasmic streaming. Small food molecules diffuse into the cytoplasm. When absorption is completed, the vacuole empties its waste contents at a special region in the pellicle known as the cytopyge. (b) Water regulation is managed by the contractile vacuoles, which are sometimes found in pairs, one at either end of the organism. They periodically balloon out as they fill and then squeeze in, emptying their contents to the outside.

move in sequence (much like wheat bends before gusts of wind), and the ciliate is propelled along. In the *Paramecium,* one of the best-known genera, some of the cilia are fused into a kind of membrane and line the *cytostome,* which serves as a sort of mouth. In other genera, such as *Euplotes,* the cilia may be arranged in tufts, which help this group scramble over the sediment of its watery habitat, or enable it to dogpaddle through the water.

The body "wall" of ciliates is a tough but elastic *pellicle.* Because it is elastic, the ciliates can bend and squeeze and wriggle and contort and manage

15.16 Asexual reproduction in *Paramecium.* Like numerous other protists, the *Paramecium* increase their numbers through an asexual process involving mitosis. During its interphase, the *Paramecium* replicates its chromosomes in preparation for mitosis. Mitosis occurs, with the usual events, including the condensation of chromosomes, formation of a mitotic spindle, separation of chromatids, and nuclear division. Where this ciliate differs from many is in its nuclear structures. The nucleus is separated into two bodies, the *macronucleus* and the *micronucleus.* The macronucleus contains multiple strands of DNA which is referred to as *somatic DNA* because it is not destined to be preserved in future generations. The micronucleus is the germ nucleus, which contains the full diploid genetic information of the species, carefully preserved and passed on to future generations through the relatively exact processes of mitosis and meiosis. In asexual reproduction, only the micronucleus enters mitosis. The macronucleus simply divides by pinching in two. During cell division, the oral groove of *Paramecium* closes and cytoplasmic division occurs across that region. As division proceeds, new grooves begin to form. After the two resulting cells separate, all the cilia and other cytoplasmic organelles are reproduced.

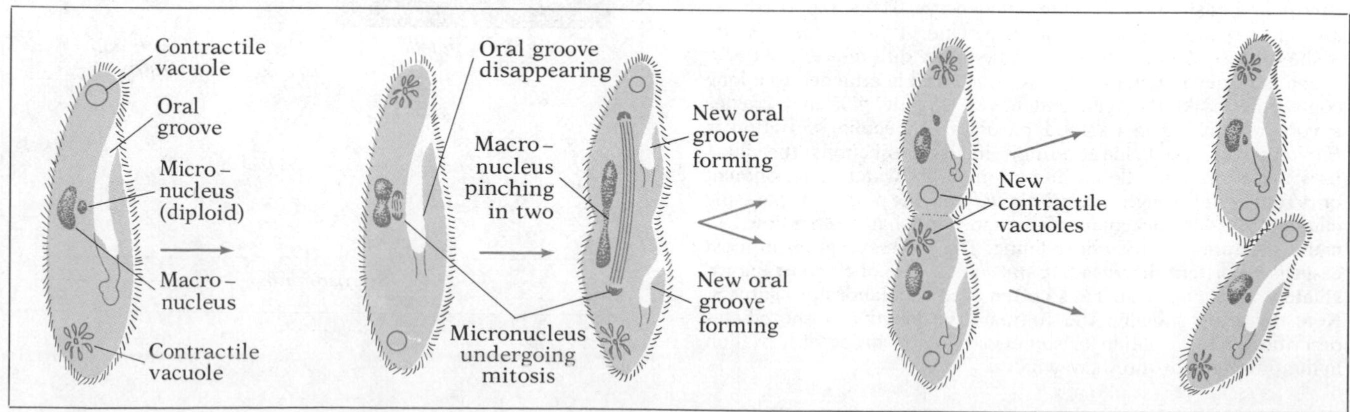
Contractile vacuole
Oral groove
Micro-nucleus (diploid)
Macro-nucleus
Contractile vacuole
Oral groove disappearing
Macro-nucleus pinching in two
Micronucleus undergoing mitosis
New oral groove forming
New oral groove forming
New contractile vacuoles

to get past or through all sorts of obstructions. The pellicle itself consists of an outer membrane with numerous membranous fluid-filled cavities underneath.

Ciliates also have a contractile vacuole (or vacuoles) with which they actively pump out excess water. Both the filling and emptying cycles appear to be due to ATP-powered contractions of cytoplasmic microfibrils. Whereas the contractile vacuole(s) may be fixed, the cytoplasm also contains *food vacuoles* that move in well-defined paths through the cytoplasm. The food vacuoles usually pinch off from the cytostome when food enters it and quickly fill with acids and digestive enzymes. If the cytostome is the protozoan's mouth, then the *cytopyge* may be considered to be its anus. The cytopyge appears periodically—always in the same place—when solid wastes are ready to be expelled (Figure 15.15).

Reproduction in ciliates is extremely complex and has some fascinating aspects. Actually, the process outwardly seems simple enough: Ciliates can reproduce asexually by mitosis and cell division (Figure 15.16). But that's the easy part. It's when they

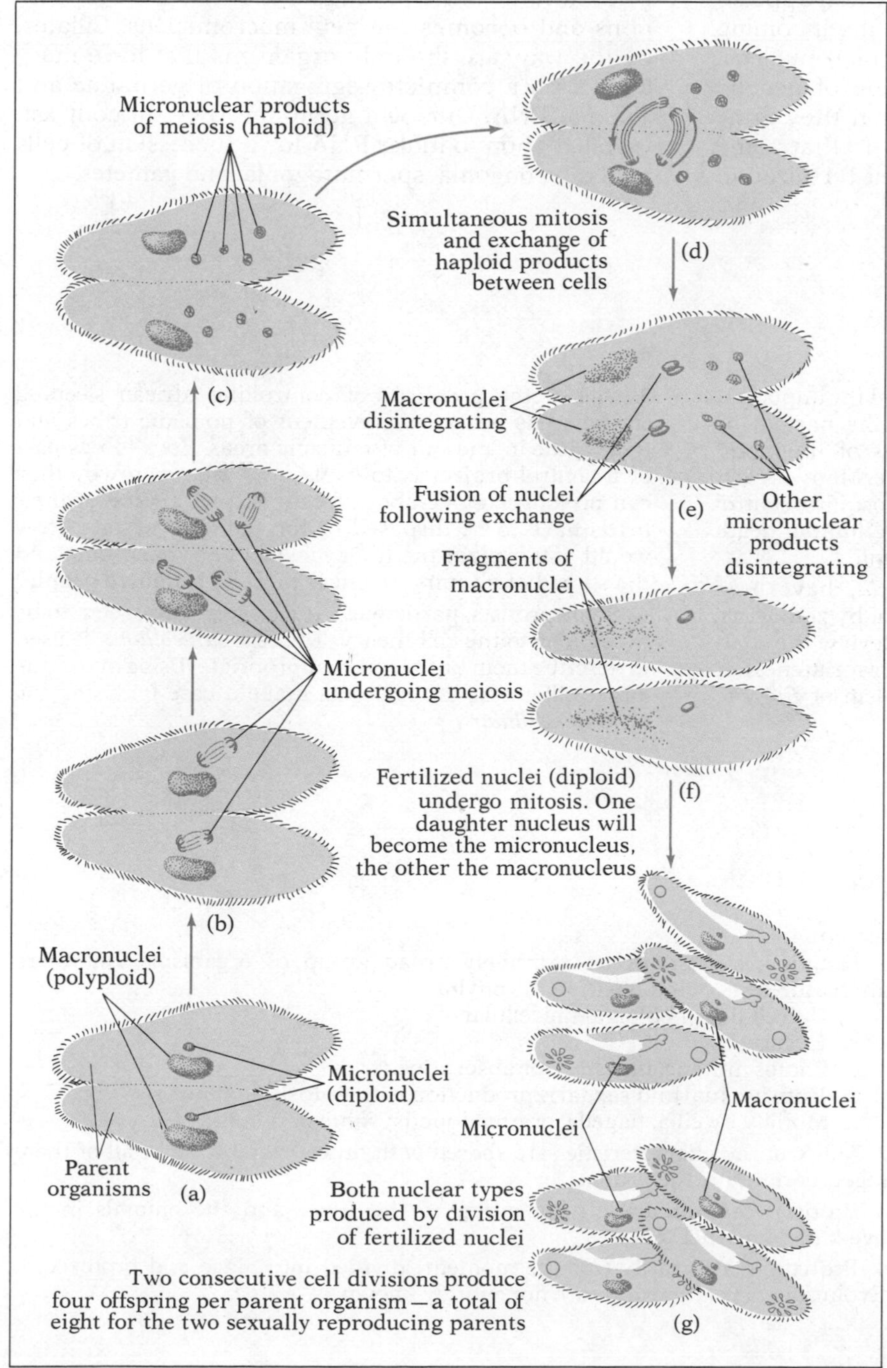

15.17 Sexual reproduction in *Paramecium*. The *Paramecium* reproduces sexually through a process of conjugation involving two individuals of opposite mating types. (a) These partners fuse together at their oral grooves and the process begins. (b) The diploid micronuclei undergo meiosis producing four haploid micronuclei in each partner. (c) Three of these disintegrate and the remaining one undergoes mitosis, forming two haploid micronuclei. These are the equivalent of gametes in other sexually reproducing organisms. (d) Following this, one of each pair of haploid micronuclei is exchanged across the cytoplasmic bridge. (e) In each partner, the exchanged micronucleus fuses with the one remaining to produce a new diploid micronucleus. At this time, the macronucleus is disassembled. (f) In each partner the new micronucleus undergoes mitosis without cell division. This is followed by several rounds of DNA replication in one of the daughter nuclei, to form the new large, polyploid macronucleus. The partners separate. (g) Following conjugation, each partner goes through two more rounds of mitosis, producing four offspring each.

become involved with sex that their lives become complicated. Sexual reproduction occurs through a process called *conjugation.* Any two ciliates must be of different mating types in order to conjugate, and there are eight types in all. They are not really sexes, although the analogy is clear. By definition, there are only two sexes, in the strict sense, among all the living things. Interestingly, no one can explain why there are only two sexes or why *Paramecium* needs eight mating types. In any case, in conjugation, meiosis in both partners produces several haploid *micronuclei* (see Figure 15.17). A pair of these are then exchanged through a cytoplasmic bridge that appears between the partners. In each partner, the incoming micronucleus fuses with a resident micronucleus. The result is a new diploid combination of genetic material in each of the conjugates when they draw apart. One puzzling aspect of all this is that some ciliates are capable of a form of self-fertilization wherein two haploid micronuclei fuse. We're not sure of the evolutionary advantage of such behavior.

The most interesting part of ciliate life is the fact that the ciliates keep their genetic material separated into a germ line (involved in reproduction) and a somatic line (involved in the maintenance of the individual). The germ-line DNA is confined to the diploid micronucleus. The somatic DNA, from which all RNA is transcribed, occurs in the polyploid macronucleus. After conjugation, the macronucleus disintegrates while the micronucleus undergoes mitosis, forming two daughter micronuclei. One of these enters the germ line, the other undergoes multiple DNA replications and becomes the new macronucleus. Ciliates, by the way, are the only organisms that have managed such a complete segregation of germ-line and somatic DNA. Our own germ-line DNA, in contrast, is called upon to make RNA for a succession of cells including oogonia, spermatogonia, and gametes.

Application of Ideas

1. The use of DDT has been greatly curtailed for important ecological reasons, yet it has proved in the past to be the greatest deterrent to malaria in parts of the world where other measures have been ineffective. Many organizations would like to see this form of mosquito control continued. Discuss the issue of preventing human death or disease versus protecting the environment.

2. Euglenoids, particularly *Euglena gracilis,* have been classified as a plant by botanists, an animal by zoologists, and now a protist in the newer schemes. Review the characteristics of *Euglena* and discuss the appropriateness of calling them protists. What does the problem of *Euglena* reveal about taxonomy and systematics?

3. One of the problems of controlling African sleeping sickness has been the movement of nomadic tribes and their cattle in and out of endemic areas. Your job as part of a control project is to explain to tribesmen why they can no longer move about freely and why some of their herds have to be disposed of for sanitary reasons. How would you explain the life cycle of the trypanosome and the spread of sleeping sickness to these primitive people?

4. Some protists, particularly the large ciliates, seem to be exceptions to the cell theory. The term *unicellular* is used to describe them but seems inappropriate. Using one of the larger ciliates as an example, make a case for using the term *noncellular.*

Key Words

absorption
African sleeping sickness
ameba
amebic dysentery
ameboid movement
Anopheles
auxospore
axoneme

biradial

chrysophyte
ciliary basal body
ciliate
conjugation

Key Ideas

Introduction

1. Kingdom Protista is an extremely broad group of organisms, but there are features in common within the phylum:
   a. Unicellular or colonial unicellular
   b. Eucaryotic
   c. Photosynthetic, ingestive, or absorptive
   d. Both asexual and sexual reproduction with mitosis and meiosis
   e. Motility by cilia, flagella, or pseudopods; some nonmotile

2. Many of the characteristics are shared with other kingdoms, but all of them together describe the protists.

3. Protists form an underlying group to the fungi and the animals in the five-kingdom scheme.

4. Protista is arbitrarily (for the moment) divided into algae and protozoans. Evolutionary relationships are not entirely known.

contractile vacuole
cyst
cytopyge
cytostome

DDT
diatom
diatomaceous earth
dinocaryote
dinoflagellate

Euglena
euglenoid
eyespot

flagellated protozoan
food vacuole
foraminiferan

gametangia
gamont
germ-line DNA
golden-brown algae
gullet

heliozoan

ingestion

macronucleus
malaria
merozoite
micronucleus

noncellular

oocyst

pectin
pellicle
phagocytosis
protozoa
pseudopod
pyrrophyte

radiolarian ooze
red tide

sarcodine
silicon
somatic-line DNA
sporangia
sporozoan
sporozoite

triradial
trypanosome
tsetse fly

yellow-green algae

The Algal Protists
Algae are photosynthetic, plantlike eucaryotes. The algal protists are not multicellular. Their spore-forming bodies are unicellular. Cell division is animallike.

Pyrrophyta (Dinoflagellates)
1. Many of the *dinoflagellates* are luminous. Some are responsible for *red tides.*
2. They have *contractile vacuoles,* move by flagella, lack chloroplasts and cell walls, and some feed on other organisms. Some are parasites and some are photosynthetic.
3. The chromosomes are permanently attached to the unique single nuclear membrane. Both mitosis and meiosis are aberrant by eucaryotic standards.
4. *Pyrrophytes* represent an early divergence from other eucaryotes.

Euglenophyta (Euglenoids)
1. *Euglenoids* live in snow, desert lakes, tide pools, etc. They sometimes turn snow or water red.
2. The red *eyespot* is a pigment shield that helps it orient toward light.
3. *Euglenoids* are unicellular, flagellated, asexual, and photosynthetic. Some ingest food.
4. Cell division follows mitosis and is longitudinal.

Chrysophyta (Diatoms and Their Relatives, the Yellow-Green Algae and the Golden-Brown Algae)
1. *Yellow-green* and *golden-brown algae* live in fresh water. Their color comes from carotenes.
2. Some have glass skeletons. Most are solitary, but some form filamentous colonies. All are photosynthetic, but some ingest food in amebalike fashion.
3. *Diatoms* have glass coverings with a large and smaller lid. During asexual division, each half-lid produces a new one. As a result they get progressively smaller. Then meiosis occurs, flagella grow, and the flagellated cells fuse in a sexual process. After growth, new glass walls are produced and the cycle repeats.
4. Marine *diatom* walls are highly ornamented and sculptured. Their walls have accumulated into great deposits of *diatomaceous earth.*

The Animallike Protists
1. The animallike protists, the protozoans, are sometimes placed in the animal kingdom.
2. Many are far too complex to be accurately called single-celled. *Noncellular* may be a better description.

Mastigophora (Flagellates)
1. Flagella are used in locomotion. They are efficient eucaryotic (9+2) structures. They feed by prey capture and by absorption.
2. Reproduction is asexual through cell division. Sexual reproduction with meiosis is seldom observed.
3. *Trypanosoma gambiense* is an important flagellate parasite and is the agent of *African sleeping sickness.* Its life cycle occurs in the tsetse fly and in a human or other mammalian host.

Sarcodina (Amebas)
1. The group is of great historical importance because of the parasite, *Entamoeba histolytica,* the agent of *amebic dysentery.*
2. The freshwater *heliozoans* and the marine *radiolarians* both secrete porous, silica-impregnated capsules, through which they extend long, thin pseudopods stiffened by rods of microtubules. These are used in feeding. *Radiolarian* skeletons form the *radiolarian ooze* of the ocean floor.
3. The *foraminiferans* produce calcium carbonate shells. *Pseudopods* move in and out through pores. Their chalky shells form thick deposits.
4. *Sarcodines* feed by *phagocytosis.* Reproduction is asexual and sexual. In sexual reproduction, flagellated haploid gametes form and join in fertilization. Many form protective *cysts* for surviving unfavorable conditions.

Sporozoa (Malarial Parasites)

1. The group is best known for its malarial parasite.
2. The *sporozoa* are nonmotile spore-forming parasites. Their life cycles are complex with sexual and asexual phases of many steps.
3. *Malaria* is caused by *Plasmodium*. It completes its life cycle in the mosquito and man.
4. The life cycle of *Plasmodium:*
   a. Injection of *sporozoites* into host by mosquitos
   b. Asexual reproduction in liver
   c. Invasion of red blood cells by *merozoites*
   d. Reproduction in red blood cells and onset of *malaria* symptoms
   e. *Gamont* formation and return to mosquito
   f. *Gamonts* produce gametes that fuse in fertilization in mosquito gut
   g. *Oocysts* form, divide sexually, and release many *sporozoites*
   h. *Sporozoites* enter salivary glands ready for next mammalian host
5. Malarial control by isolation of infected persons works well in developed countries, but not in most *malaria* endemic regions. *DDT* continues to be the most effective control agent.

Ciliophora (Ciliates)

1. Most complex of the protozoans, the *ciliates* all have cilia for locomotion and/or feeding. They are a large, diverse group.
2. Cilia arise from basal bodies and their movement is coordinated through interconnecting fibers.
3. *Paramecium* is the best known *ciliate*. It has a complex mouth with fused cilia. Cilia are also fused in *Euplotes*, permitting the protozoan to "walk" or paddle.
4. *Ciliates* have a complex outer covering, the *pellicle*.
5. They form *contractile vacuoles* for water control and *food vacuoles* for digestion. Wastes leave through the *cytopyge*.
6. *Ciliates* reproduce asexually by mitosis and cell division. Sexual reproduction is through conjugation.
7. Both *macronuclei* and *micronuclei* occur. *Macronuclei* contain *somatic-line DNA*, which transcribes all RNA. *Micronuclei* contain *germ-line DNA*, which replicates in reproduction.
8. During sexual reproduction, *conjugants* undergo meiosis in the *micronuclei*, and the paired *ciliates* exchange haploid *micronuclei*. Extra meiotic products disintegrate. Fusion of haploid nuclei completes the process.
9. Following *conjugation*, *macronuclei* disintegrate, and *micronuclei* undergo mitosis. One new *micronucleus* contains *germ-line DNA*, while the other undergoes multiple replication, forming the new *macronucleus*.

Review Questions

1. Review the taxonomic problems of phylum Protista. (Why is the phylum so conditional and what are some problems that remain to be worked out?) (p. 428)
2. List the characteristics that define phylum Protista (p. 428).
3. What are the characteristics of *Euglena* that show the broadness of phylum Protista? (p. 428)
4. List the main characteristics of the algae. (p. 428)
5. In what way are the *pyrrophytes* ecologically important? (p. 428)
6. In what way are the *dinoflagellates* a threat to humans? (p. 429)
7. Describe the role of the red *eyespot* of *Euglena*. (p. 429)
8. Should you avoid eating pink snow? Why or why not? (p. 429)
9. List several characteristics of the *chrysophytes*. (p. 430)
10. Why does the *diatom* have to periodically stop reproducing asexually and resort to sexual reproduction? (p. 430)
11. How does the skeleton of marine *diatoms* differ from those of *radiolarians?* (p. 430)
12. What is *diatomaceous earth* and how do we make use of it? (p. 430)
13. Why do some people consider the term *noncellular* a better description of many of the protozoans? (p. 430)

14. List the main characteristics of the Mastigophora. (p. 432)

15. How do flagellates reproduce? What is the status of sexual reproduction in the group? (pp. 432–433)

16. Review the cause and problem of *African sleeping sickness.* (p. 433)

17. Describe the main characteristics of the *Sarcodina.* (p. 433)

18. What structures do the *heliozoans* and *foraminiferans* have in common? How do they differ? (p. 434)

19. Review reproduction in the *Sarcodina.* (p. 435)

20. List three characteristics of the *Sporozoa.* (pp. 434–435)

21. List the main stages in the life cycle of *Plasmodium* and indicate where the stages occur. (p. 436)

22. Describe the common measures of *malaria* control. Which method seems to be the most effective? (p. 435)

23. Describe the arrangement and coordination of cilia in the ciliates. (p. 437)

24. List the types of vacuoles in *Paramecium* and explain their functions. (p. 438)

25. What are the primary roles of *macronuclei* and *micronuclei* in the *ciliates?* (p. 437)

26. How do *conjugants* prepare for sexual reproduction? Be specific. How is it actually accomplished? (p. 439)

27. Review the nuclear events that occur after *conjugation.* (p. 440)

Chapter 16

The Fungi and Multicellularity

You may have been surprised one day when walking in a damp woodland to suddenly come upon a large mushroom on the forest floor. It's especially exciting, perhaps, if you don't know much about mushrooms, because their size, shape, and even coloration may seem rather outlandish (Figure 16.1). They have a distinct sense of mystery about them, possibly because they are so often associated with dark, wet forests—pensive places. If the area is remote, the day is overcast, and you are alone, it is not hard to secretly believe in the "little people," if only for a moment.

If you have experienced a moment like this, you may not be pleased to hear that the mushroom is technically a fungus. A fungus! The word *fungus* is simply *not* associated with beauty and mystery. A fungus, to many people, is bad; it grows on things, it lurks on shower floors. You won't be any happier to learn that fungi, *all* fungi, are parasitic, or worse, they live off the dead. Nonetheless, mushrooms, along with yeast and a number of other kinds of organisms, are fungi. And fungi really can be beautiful and, furthermore, can be delicious. So let's have a closer look.

16.1 The circle of mushrooms seen here is probably the result of a fungal spore. Spores from some distant mushroom were carried by air currents to this ideal location, where dead vegetation, moisture, and warmth made growth possible. Beneath the surface is a vast fungal *mycelium* (body) which spread from a central point in the circle, finally encountering mycelia of another mating type and sending fruiting bodies upward around the perimeter. Each mushroom shown will in turn produce millions of spores that will have to find their own ideal environments.

16.1 The Fungi

What Are the Fungi?

As a group, the fungi are heterotrophs. They require an organic food source much the same as other heterotrophs, although some are able to use rather simple organic substances. The fungi feed by absorption as they secrete digestive enzymes into their surrounding substrate and take in the products of digestion. In this nutritive mode, the fungi are best known as *saprophytes* (that is, they live on dead matter; the *-phytes* part of this term harks back to their former taxonomic affiliation with the plants). Those that are not saprophytes are *parasites*—they obtain their organic requirements in a similar manner, but from living hosts. The majority of fungi species are parasitic. Many of the parasitic fungi are medically and economically quite important.

Traditionally, the fungi have been classified as plants, and, in some schemes, they are considered protists. However, in the five-kingdom scheme, you will notice that they are placed in their own kingdom (see Figure 13.14).

Many of the fungi are multicellular. Although they lack tissue organization in their somatic or vegetative structures, they often produce highly elaborate reproductive structures. These are often as complex as those of many plants and animals, but they are very different. Unlike the plants, the cytoplasm of fungi is sometimes organized, not into cells, but into an extensive, flowing *syncytium,* which is a multinucleate cytoplasm.

The fungi were once believed to have evolved from algal stock, but most biologists now believe they sprang from the colorless flagellates. This ancestry further separates the fungi from the plant line. Whether the fungi arose as a monophyletic or a polyphyletic group is still unsettled, but we will shortly review the arguments for both cases. For now, we might just note that there are enough significant differences between the fungi and the plants to warrant their separation. While the phyletic relationships of the fungi are a challenging problem to biologists, there are other aspects of this group that are equally fascinating.

One fungal group helped to change the course of history, at least for the Irish. If you are of Irish ancestry, you just might owe your American citizenship to the activity of a fungus. The heaviest period of Irish immigration to the United States occurred between 1843 and 1847 as a direct result of the famous *potato famine*. The loss of crops in those years brought about the death of 250,000 Irish people and the emigration of a million or more to America. The culprit responsible for this great exodus was *Phytophthora infestans,* better known as *late blight,* a parasitic fungus of potatoes. It just so happened that weather conditions in Ireland were ideal for the fungus during those 4 years.

On a less morbid note, some people may move for another reason—because of what they call itchy feet. But they might examine the basis of that itch since another fungus, *Trichophyton mentagrophytes,* causes athlete's foot.

The list of parasitic activities among the fungi is long and infamous and ranges from rose mildew to jock itch. There is, however, another side to the fungal world. Since they share with bacteria the role of reducer or decomposer, utilizing dead organisms as a source of organic nutrient, they aid in the cycling of important nutrients that would otherwise be tied up in corpses. We also make use of some fungi in such significant commercial activities as the manufacture of cheeses, antibiotics, linen, and, of course, bread and beer.

You will notice, as we consider each of the fungal phyla, that their names end in *-mycota* or *-mycetes.* These suffixes are derived from the Greek word for fungus, which is *myketos.* The root word is also our source for the term *mycelium,* which is the body of the fungus. The mycelium of many of the fungi is made up of threadlike masses of individual filaments known as *hyphae* (Figure 16.2). The hyphae may be composed of individual tubular cells with one nucleus each (*septate*), or they may be tubular but without cell walls separating the nuclei (nonseptate). Cell walls in some fungi are composed of *chitin* (the material of insect cuticles), although some groups produce *cellulose* (a common plant material). Whatever its shape, the mycelium is a feeding structure that secretes enzymes into its substrate and then absorbs the digestion products through the thin mycelial walls.

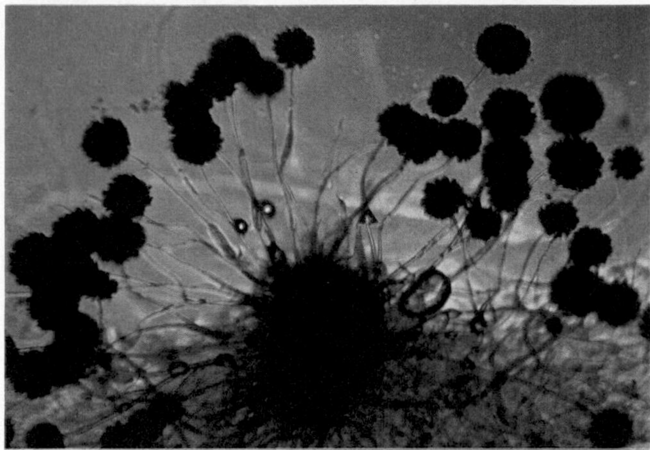

16.2 Most fungi grow by producing filamentous hyphae, fine tubes of cytoplasm that penetrate the substrate to obtain food. The entire mass of hyphae is known as the *mycelium*. When conditions favor vertical growths of hyphae, the *sporangiophores* rise up and produce *sporangia* at their tips, eventually releasing numerous haploid spores into the environment.

Typically, fungi often develop erect, spore-producing organs called *sporangia*, which make tiny spores by the millions. Spores are easily borne by wind or water. The presence of fungal spores in the air everywhere on earth means that your bread, oranges, leather goods, stored grains, fruits, and even your body are subject to fungal growth.

In the fungi, spores are not simply a way of surviving harsh conditions, as is the case in bacteria. Here, they are a means of asexual reproduction and dispersal. Fungi also reproduce sexually, often with complex life cycles. In parasitic forms, this may involve two unrelated hosts. We will discuss a few examples shortly.

The fungi are a diverse group, and so there are exceptions to practically every rule concerning them, and we will try to point these out as we go along. There are numerous schemes used to organize the fungi in order to show relationships. The reasoning behind the various schematic organizations is often complex and always open to argument. We will sidestep some of the taxonomic problems by restricting our discussion to three major groups of fungi, the *Gymnomycota* (slime molds), the *Dimastigomycota* (egg molds, water molds), and the *Eumycota* (true fungi). Each of these represents a subkingdom, which is somewhat distinct from the others by virtue of cellular structure and materials, mode of growth, and reproductive characteristics.

Subkingdom Gymnomycota: The Slime Molds

Perhaps we are off to a bad start with the poetically christened *slime molds.* Some biologists include these with the protists, but we will go along with the people who believe that they belong here. Actually, there are two distinct groups—the ascellular slime molds and the cellular slime molds—that may not be closely related even to each other. Anyway, the slime molds are truly unusual. Any description is reminiscent of the story of the blindmen describing an elephant—each touching a different part.

Acellular slime molds produce a slimy mass called a *plasmodium;* this has all the characteristics of a huge ameba. The masses, looking like refugees from a science-fiction movie, may be found creeping over the moist underside of rotting tree trunks. Although it moves like an enormous ameba, the mass turns out to be multinucleate, the result of many synchronous mitotic divisions. As it creeps along, it feeds by phagocytizing bits of organic matter. The mass shows some sensitivity and avoids obstacles and dry areas. It might seem that what we have described so far makes the slime mold a prime candidate for kingdom Protista or perhaps the Animal kingdom, but let's look a bit further and examine the sex life of these strange organisms (Figure 16.3).

At some time in the creeping stage of its life cycle, a slime mold will seek out a drier habitat (or perhaps its moist habitat will dry out). It is only then that it shows the characteristics of a fungus. The drying mass produces sporangia in the form of rounded bodies perched atop slender vertical props (*sporangiophores*). Each sporangium then undergoes a number of meiotic divisions, producing numerous haploid spores. The spores then emerge and are carried by air currents to settle in new places. If a spore lands in a suitable place, it begins to divide and produce an ameboid growth known as a *myxameba.* The myxameba, itself, may divide a number of times, producing many ameboid descendants. In some species, pairs of the offspring become flagellated *swarm cells* that can swim about and fuse together, thereby re-

16.3 Life cycle of an acellular slime mold. Acellular slime molds have essentially two phases to their life cycles. These are a creeping *plasmodium* phase and a fruiting stage in which sporangia rise up on short stalks and release spores into the environment. As you can see in this drawing, the creeping stage begins with the zygote, which forms a large *plasmodium* as the cytoplasm increases and nuclei reproduce mitotically, without cell division. This is the feeding stage. The plasmodium stage ends with the production of sporangia in which spores form by meiosis. The second half of the life cycle begins as spores are released and germinate into small, single-nucleate ameboid cells or flagellated cells, some of which form the gametes of the acellular slime mold. Gametes then fuse, forming zygotes, and the cycle repeats.

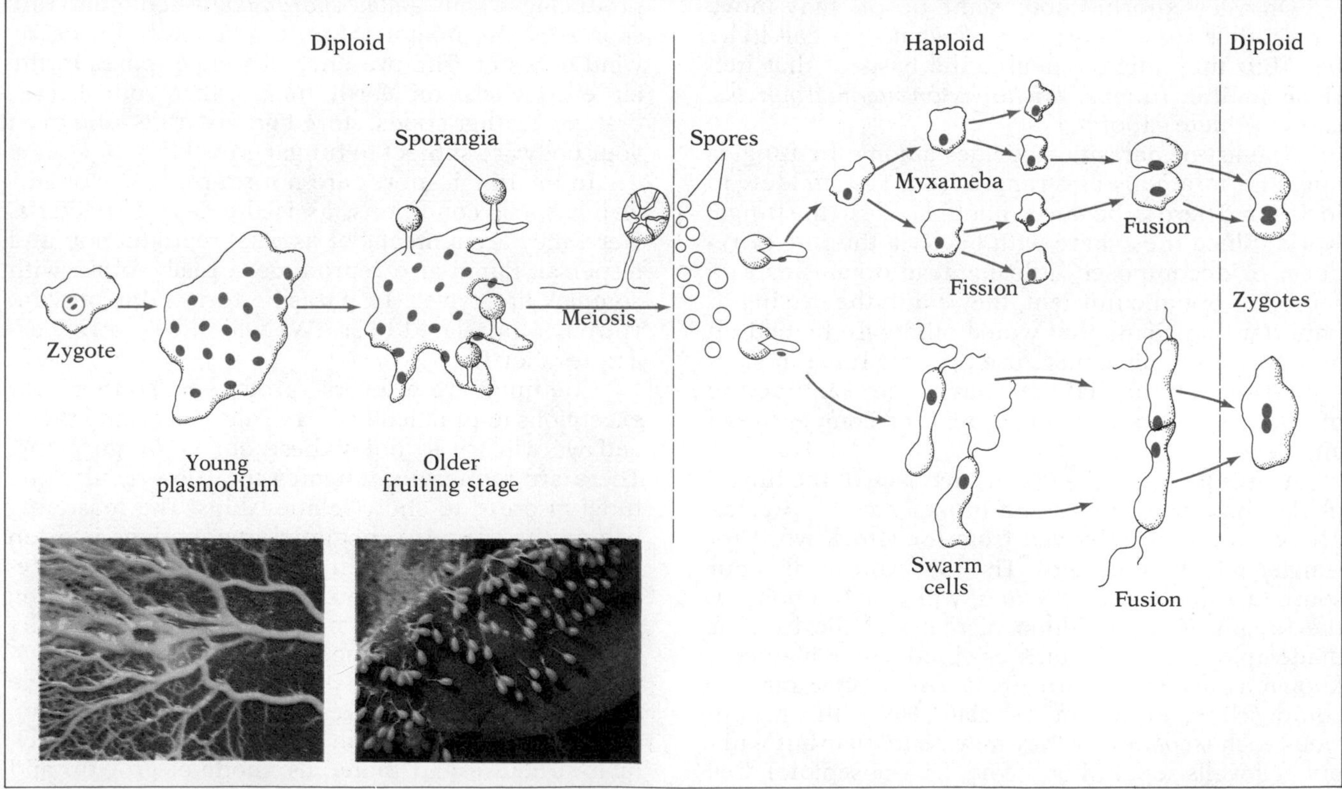

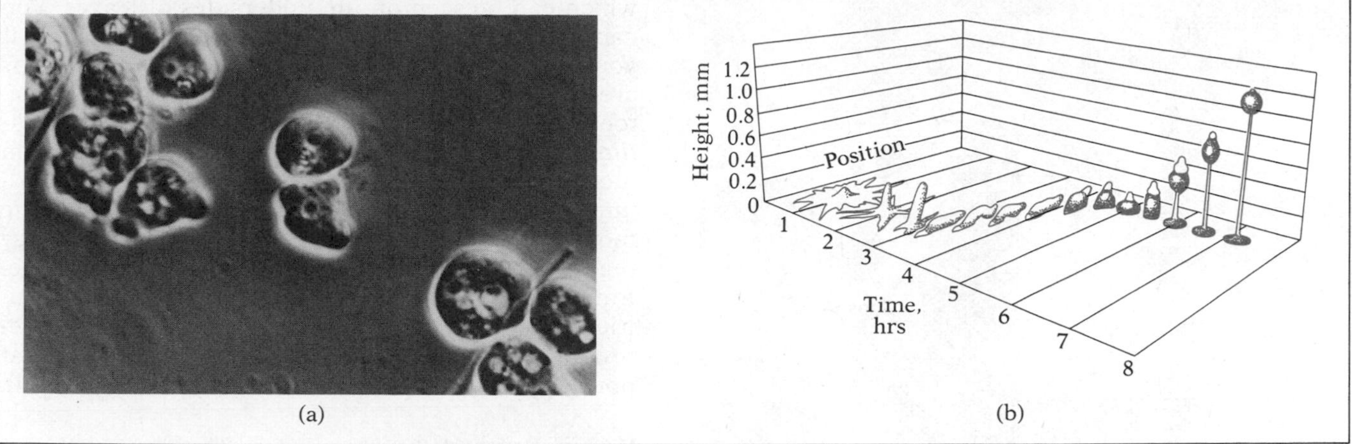

16.4 (a) Cellular slime molds live as independent, unicellular amebas, phagocytizing soil bacteria and reproducing asexually during the vegetative part of their life cycle. In the sexual phase, individual cells coalesce into a heap, which differentiates into a motile, nonfeeding "slug." (b) Eventually the slug stops crawling; a stalk appears, first as a bump which then lengthens, lifting a ball of cells to its tip. The ball (sporangium) develops spores as it rises higher with the further elongation of the stalk. Eventually the spores are released into the wind.

storing the diploid chromosome number. In any case, the ameboid cells are capable of undergoing numerous nuclear mitotic events, thereby producing the slimy multinucleate plasmodium once again.

The cellular slime molds never have a plasmodium stage. Throughout their entire *vegetative* (nonsexual) phase, they are free-living, independent amebas, not easily distinguishable from any other soil ameba, gathering food by ingestion (phagocytosis). They will divide mitotically and proliferate as amebas as long as the environment remains favorable. When their substrate dries out or becomes exhausted of nutrients, some of the amebas begin to exude a *pheromone* (a chemical messenger that is transmitted between individuals). In this case the pheromone is cyclic AMP (cAMP), and it causes the independent amebas to move toward one another until they have formed a heap. Then a truly amazing transformation takes place. The heap of cells—from diverse sources, mind you—becomes a primitive multicellular organism, complete with cell specialization. First it forms a *slug* that crawls around like, well, a slug. Then the slug forms a fruiting body, some of the former amebas becoming cells of the *base, stalk,* or *sporangium.* Only a few become reproductive cells (Figure 16.4).

Subkingdom Dimastigomycota: The Oomycetes (Water Molds)

The oomycetes are considered primitive because of two characteristics. First, some species produce a mycelium composed of very few cells. Where an extensive mycelium does develop, it is nonseptate or *coenocytic* (that is, having no cell walls) and lacks organization. Second, they commonly form flagellated spores, a decidedly protistan characteristic.

Most of its members are known as *water molds,* which, with a few exceptions, are not parasitic. The oomycetes receive their name from their sexual cycle in which large egg cells are produced inside a special structure called an *oogonium.* Many of the water molds are saprophytes, while a few are "weak" parasites. By weak, we mean that they can only invade living plants or animals that have been previously damaged by injuries or disease. You may have had trouble with some of these molds in your aquarium, since they often parasitize fish that have been wounded by other fish.

In their asexual cycle, the water molds produce sporangia. From these, flagellated *zoospores* emerge, which then swim about in search of a food source. When food is found, hyphae grow and a new colony develops [Figure 16.5(a)].

Sexual reproduction in water molds begins with the growth of unusually thick hyphae that produce egg cells in sherical structures called *oogonia.* Male gametes reach the egg cells in an unusual way. The hyphae growing near an oogonium send branches over the spherical body. Within these branches are the *antheridia,* the sperm-producing structures. The fingerlike branches become fertilization tubes by penetrating the egg cells, permitting the sperm to reach the eggs. Some of the events are clearly reminiscent of fertilization in higher plants, and the possibility that at least some oomycetes are closely related to higher plants has not been firmly ruled out [Figure 16.5(b)].

Many of the terrestrial species of the oomycetes, that is, oomycetes that are *not* water molds, are parasitic. This group includes *downy mildews* and *blights,*

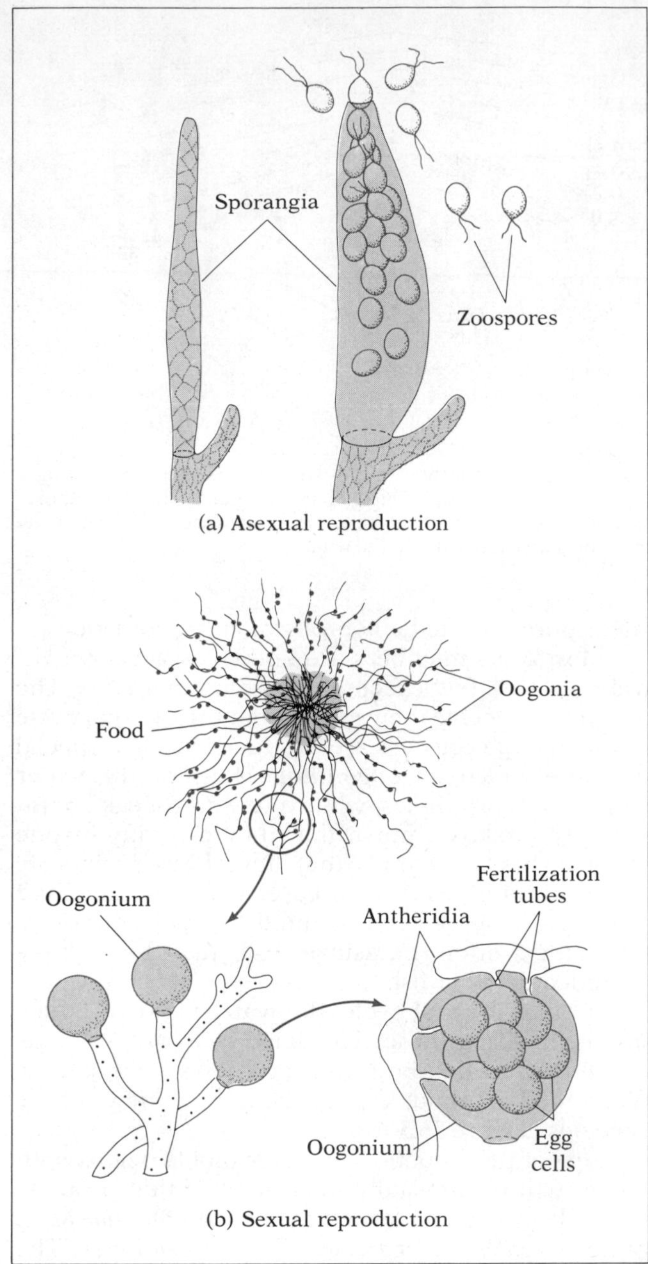

16.5 Asexual and sexual reproduction in the water molds. (a) Asexual reproduction in some of the water molds reveals certain protist tendencies in the group. Sporangia form at the tips of certain hyphae and produce numerous spores. These emerge in their watery environment as flagellated cells (*zoospores*). They eventually reach a new food source and produce the usual hyphal growth. Having flagellated spores is undoubtedly an adaptation for water dwelling, so the similarity of this asexual stage to flagellated spores in the protists may be a matter of convergent evolution. (b) The sexual reproductive activity of water molds begins with rounded thickenings (*oogonia*) forming in certain hyphae. Other hyphae send fingerlike projections (*antheridia*) around the enlarging structures. Eventually, the swelling becomes an egg-producing oogonium, forming clusters of haploid eggs within. The antheridia produce the fungal sperm cells. The sperm reach the eggs through growing fertilization tubes, as shown here. After fertilization, each diploid zygote will itself form a nonmotile *zygospore,* resistant case and all. Under proper conditions, it will germinate into another hyphal mass, and the cycle will repeat.

which can be seen on the undersides of leaves. You can expect them on your house or garden plants if you tend to overwater or undercultivate. Many crops are especially susceptible to infection, such as potatoes, beans, melons, sugar beets, and grains. *Phytophthora,* the villian of the Irish potato famine, grows on moist leaves by penetrating their pores with its mycelium. These natural openings on leaves permit the mold to grow throughout the air spaces of the spongy regions of the leaf. Portions of the mycelium penetrate the photosynthesizing cells, absorbing the nutrients as they are produced. Eventually, asexual reproductive shoots grow back out through the leaf pores, producing sporangia at their tips (Figure 16.6).

16.6 Infestation of the potato leaf by *Phytophthora infestans.* When any downy mildew or late blight infests a plant, it can be readily seen on the underside of the leaves. *Phytophthora* sends its mycelial mass throughout the spongy tissue of the leaf, where it can absorb the plants photosynthetic products. This eventually weakens and kills the plant, but not before *Phytophthora* completes its asexual cycle. Hyphae of *Phytophthora* emerge through the leaf pores, sending sporangiophores out into the air. At their tips, numerous sporangia develop to produce immense numbers of haploid spores, which can then spread to other plants to produce a new mycelium (as shown in the photograph).

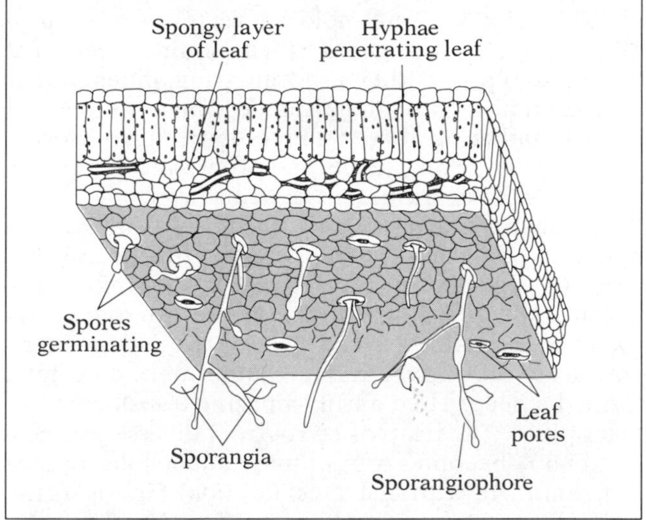

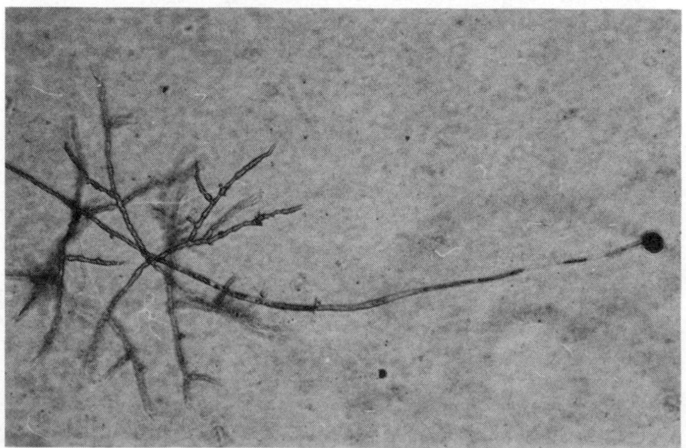

Subkingdom Eumycota: The "True Fungi"

The third subkingdom is comprised of four divisions (phyla): Chytridomycetes (the chytrids), Zygomycetes (the conjugating molds, bread molds), Ascomycetes (the sac fungi, including *Neurospora* and yeasts), and Basidiomycetes (the club fungi, including mushrooms and toadstools).

The Chytridomycetes (Chytrids)

While some chytrids have hyphae and flagellated spores and gametes, many others are unicellular, lacking an extensive mycelium. Terrestrial chytrids are parasitic, infecting other fungi or plants. Some aquatic forms are known to parasitize algae.

The Zygomycetes (Conjugating Molds)

The Zygomycetes, like the water molds, lack cell walls, and have hyphae that lack regular form in the mycelial mass. Most of the phylum is saprophytic. They are named after the *zygospore*, which is the result of sexual reproduction in this division.

The common bread mold, *Rhizopus stolonifer*, is bound to turn up in everyone's refrigerator at one time or another (Figure 16.7). But you shouldn't necessarily look for it on bread, because *Rhizopus* grows vigorously on many foods—and, besides, most of our bread is so full of preservatives that fungal growth is inhibited. Spores that land on a suitable medium will absorb water and germinate. Hyphae sprout in all directions, branching time and again until the mycelial mass has permeated the food. The favored conditions for the growth of *Rhizopus*, and many fungi, are moisture, darkness, warmth, and available nutrients. When the bread mold mycelium has grown for a time, it develops long horizontal hyphae known as *stolons*, which venture along the surface of the food, sending rootlike growths called *rhizoids* down into the food. This peculiar growth gives the food a fuzzy appearance.

You have undoubtedly noticed the tiny black spheres on moldy bread. These are the sporangia. They appear at the tips of vertical hyphae (the sporangiophores) when the mold is reproducing asexually. Numerous spores arise out of the sporangia. When released, the spores will be borne by the air to seek their fortune. Also, many of the spores fall right back onto the food that bore them and so it may become covered with *Rhizopus* overnight.

Rhizopus also has an interesting sex life (Figure 16.8). It begins with conjugation. When opposing mating types are grown in the same medium, specialized club-shaped hyphae are produced. There are no true males and females, so we refer to compatible mating types as *plus* (+) and *minus* (−). When the hyphae fuse, they form closed chambers known as *gametangia*. These chambers contain haploid nuclei from each parent (nuclei in the mycelium are generally haploid in fungi). The haploid nuclei fuse, forming a diploid zygote or *zygospore*. In typical fungal fashion, the diploid period is short. The tough, resistant zygospore may remain dormant for a considerable period, but under newly favorable conditions meiosis occurs within the zygospore, followed by degeneration of all but one haploid nucleus. The remaining haploid organism finally emerges and

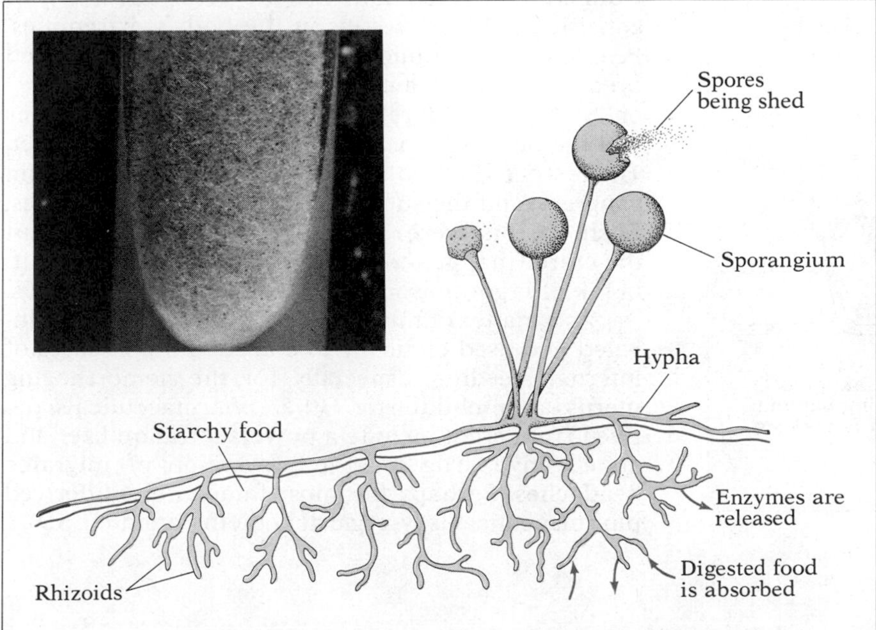

16.7 *Rhizopus*, the common black bread mold, grows well on any starchy medium. Its mycelium grows in a tangled mat throughout the medium and is not easily visible, but its sporangia, which are black when mature, form the large mass seen in the photograph. Like most saprophytic fungi, *Rhizopus* secretes enzymes from the hyphae into the food where digestion occurs extracellularly. The simplified products of digestion are then absorbed into the cells.

16.8 Sexual reproduction in *Rhizopus*. *Rhizopus* will enter into a sexual phase if opposing mating types are present in its medium. These strains are simply called *plus* (+) and *minus* (−). Branches from neighboring plus and minus hyphae send filaments together. When they touch, each produces a clublike end, which in turn forms cross-walls. Many mitotic divisions of the nucleus of each occurs and the two gamete-producing cells (*gametangia*) fuse, with nuclei joining to form a diploid zygospore. This diploid structure, contained in a highly resistant spore case, is then capable of beginning a new fungal growth. Meiosis occurs within the zygospore and all but one of the resulting haploid nuclei degenerate. When the zygospore germinates, a haploid hypha is produced, and the cycle begins again. An asexual sporangium may form, producing haploid spores by mitosis.

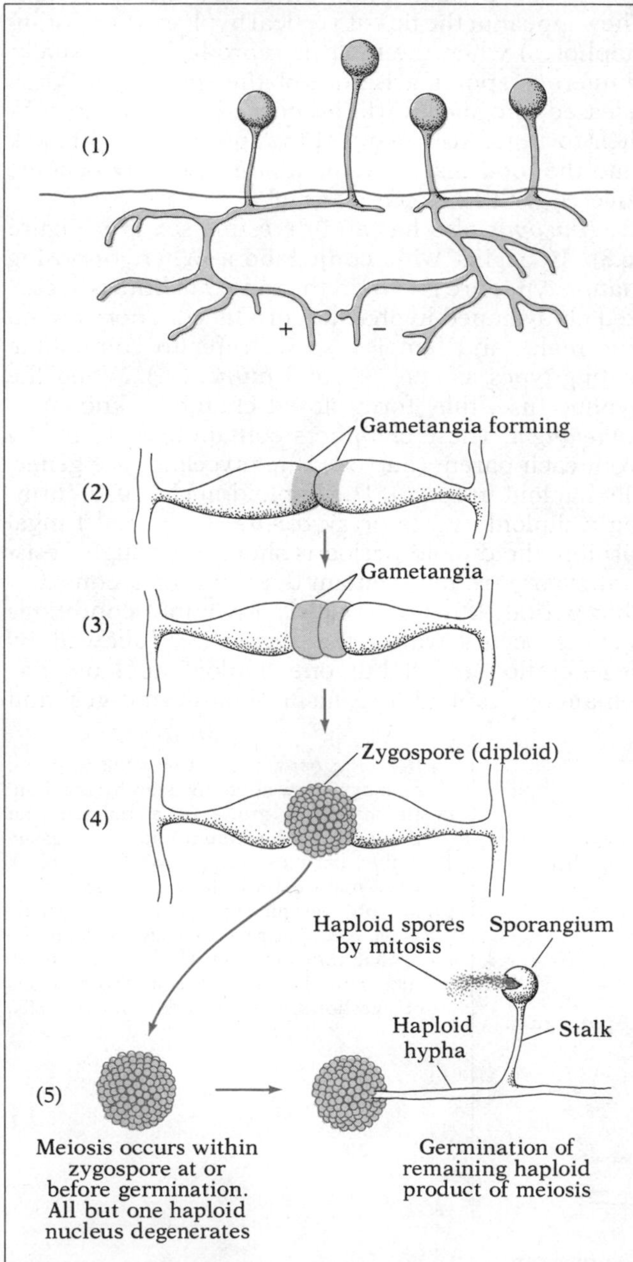

produces its own mycelial mass. The new organism will have a rearranged genome because of meiotic segregation. Keep in mind this general pattern: sexual reproduction is associated with times of environmental stress, with long periods of dormancy, and with the exploitation of newly favorable environments.

Perhaps the life cycles of these organisms is confusing when you're trying to work out the chromosome complement at different stages. Actually, though, there isn't much of a problem here. Just keep in mind that the zygospore is the only structure in the life cycle that we can consider diploid. The + and − mating types segregate in meiosis, so that the haploid hyphae of half of the germinating zygospores will be of one type and half will be the other.

Ascomycetes (Sac Fungi)

The divisions (phyla) Ascomycetes and Basidomycetes together are commonly called the *higher fungi*. This elevation in status is primarily due to their septate condition. This simply means that cell divisions are accompanied by the formation of cross-walls (*septa*) along the length of the hyphae. Generally, each resulting compartment has one nucleus; but the septa have large central openings through which the nuclei can pass.

The Ascomycetes include some familiar members such as the edible mushroomlike *Morchella*, powdery mildew, and the blue and green molds of citrus fruit. Yeasts are also considered to belong to the division. One member that holds rather ghastly implications for humans is the genus *Claviceps*, a parasite of rye and other grasses. Infected rye contains a mycelium known as *ergot*. Rye bread that has been produced from ergot-infected flour, has the strange property of constricting blood vessels in the body's extremities. People on a continuous diet of ergot contaminated bread suffer from a condition known as "Holy Fire" or "St. Anthony's Fire," named for both hallucinations and the burning sensation felt in the hands and feet. The restricted blood flow to these parts can result in gangrene and the subsequent loss of the extremities. Such horrors were common in past centuries, but they rarely happen today, although there was an outbreak of ergot poisoning in France in the late 1960s.

As a matter of medical interest, ergot is now cultivated and used clinically to control difficult cases of internal bleeding, especially for the hemorrhaging uterus after childbirth. Other pharmaceuticals derived from ergot include a powerful tranquilizer and an effective drug for the prevention of migraine headaches. Perhaps the most famous ergot-derived pharmaceutical is lysergic acid diethylamide (LSD).

In addition to producing cell walls, the Ascomycetes characteristically produce *conidiospores* during their asexual reproductive stages. Conidiospores are produced in long chains at the ends of specialized hyphae known as *conidiophores* (Figure 16.9).

There is great diversity in the sexual characteristics of Ascomycetes. The name *sac fungi* originates from the production of *ascospores* in a saclike container called the *ascus* (from which the phylum name is derived). A group of *asci* (the plural form of *ascus*) are usually found in a fruiting structure known as the *ascocarp*. The ascocarp occurs in a variety of shapes, but it is most commonly spherical or cup-shaped. What goes on inside the ascocarp is most unusual (Figure 16.10). The diploid ascospores are produced by a strange process involving delayed fertilization. When the hyphae of different mating types of sac fungi come into contact, one hyphae produces a large multinucleate swelling, an *ascogonium* (sort of an ovary). The other produces a multinucleate *antheridium* (a sperm-producing organ). These fungi are still designated + and − but they have apparently evolved differently than the + and − types we witnessed in the bread molds. In conjugation, each reproductive structure supplies many nuclei. The fungus is now a dicaryon (meaning it has "different nuclei"). The nuclei, however, do not immediately fuse. Binucleate hyphae emerge—each compartment containing one + and one − nucleus. The hyphae may contain many of these binucleate cells. The fusion of nuclei occurs after the ascocarp has formed and the binucleate asci have begun to mature. And then meiosis immediately ensues, completing the sexual process.

Thus, nuclear fusion in the sac fungi is *immediately* followed by meiosis. Four meiotic products undergo one round of mitosis to produce eight haploid ascospores. When the ascospores are fully mature, the ascus breaks open and the spores disperse. Each will be capable of producing new fungal growths. These growths remain haploid until another fertilization produces the brief diploid interlude.

Another well-known group of sac fungi lives with certain algae and, together, they form *lichens*. (Actually, some of the other fungi are also capable of this relationship.) This symbiotic association is mutualistic, which means that both parties benefit. One benefits more than the other, however, because the algal partner can get along without the fungus, but the fungus is dependent on the alga.

About 17,000 combinations of lichens are known. Ecologically, the lichens are important as soil builders, eroding rock surfaces and harboring bits of organic matter in their crusty bodies. The efforts of these hardy pioneers pave the way for a succession of organisms in many desolate, rocky environments. You have seen these organisms on rocks, although they may also be found on trees and on the soil.

16.9 Asexual reproduction in the sac fungi occurs when specialized hyphae, called *conidiophores*, undergo a change. The terminal cell rounds out and matures into a conidiospore, followed by the one below and so on, until a long chain of haploid spores forms. These will then break away and become airborne, finding their way to new food sources. This manner of spore formation occurs in two citrus molds, *Penicillium* and *Aspergillus*. Although these two molds are very similar to the naked eye, they can be distinguished clearly with the microscope. Note the branching conidiophores in *Penicillium*. Much of the taxonomy of the sac fungi is based on such differences.

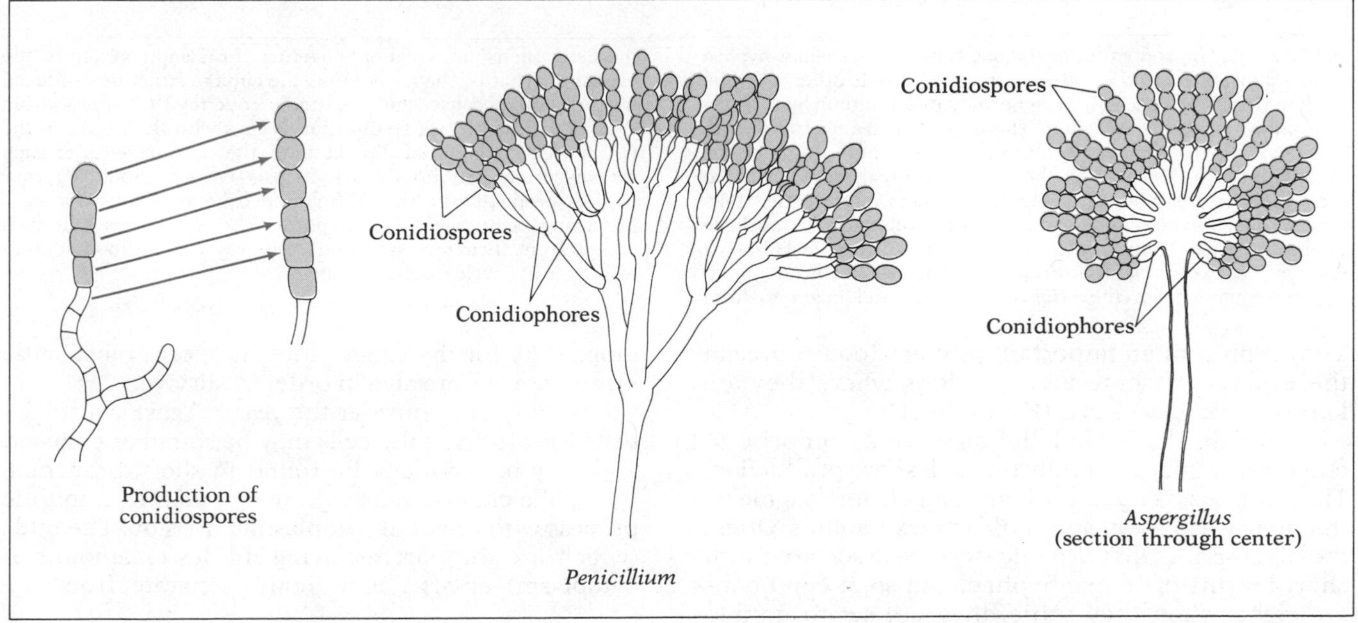

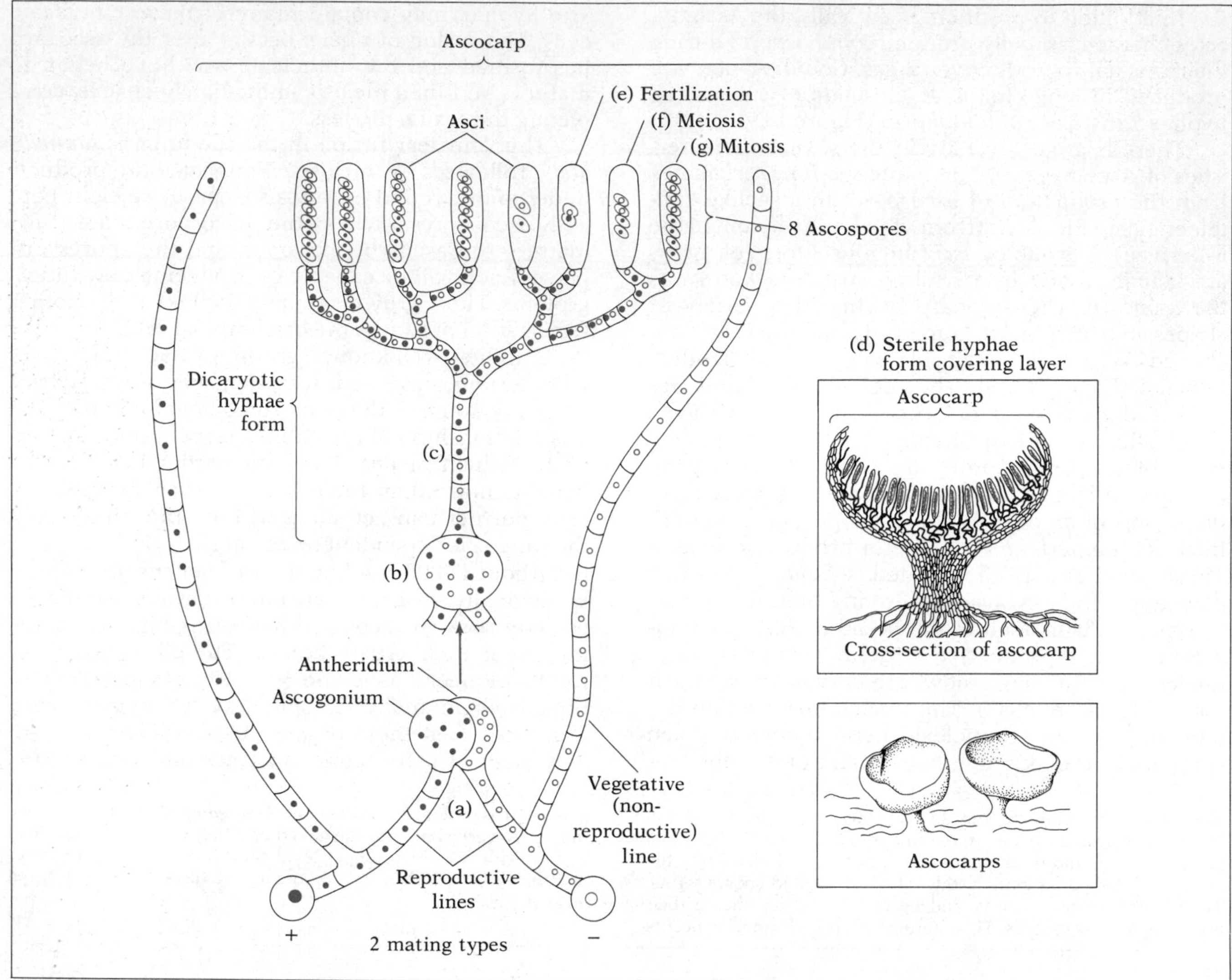

16.10 Sexual reproduction in the sac fungi occurs when hyphae from opposing + and − strains approach each other. (a) One hyphae (it can be either + or −) produces a multinucleate structure known as an *ascogonium*. The opposing strain produces an *antheridium*, which is also multinucleate, and grows against the ascogonium. (b) Eventually, the cell walls break down between the two structures and the nuclei within them intermingle. However, fertilization does not occur at this time. (c) Instead, binucleate hyphae emerge. Many branches are produced and each branch ends in a saclike *ascus* (plural: asci). (d) In the meantime, other hyphae emerge from the original branches and begin to form the *sterile* (nonspore-forming) covering of the entire structure, the *ascocarp*. (That is, the ascocarp is the cuplike structure produced by the combined asci and the sterile covering.) (e) Now, within each binucleate ascus, fertilization occurs with the fusion of the two nuclei. In this brief diploid state, the chromosomes of each mating strain join briefly. (f) Meiosis ensues, producing four haploid products. (g) This is followed by one round of mitosis, producing eight haploid ascospores. The ascospores will then escape from their sacs and find their way to new food sources where the life cycle can be repeated.

They represent an important primary food source in the ecology of some arctic regions where they are known as *reindeer moss* (Figure 16.11).

Both the fungi and the algae that comprise lichens reproduce asexually through spore production. They are also capable of growing from fragments that may wash away and settle in new habitats. Often, the algal spores that are released are associated with short lengths of fungal hyphae, and so, if conditions are right, when they settle, they are set to go into business. Reproduction through spores is a good deal more risky for the fungi, since its spores must settle on an appropriate alga in order to survive.

Finally, let's consider the *yeasts*. Yeast is a single-celled sac fungus; the cells may be round or ellipsoid and may occasionally be found in short, branched, hyphalike chains. Yeasts grow by *budding*, a mitotic process with unequal cytoplasmic division. The buds (which are the part receiving the lesser amount of cytoplasm) enlarge and finally separate from the

16.11 Lichens are a mutualistic symbiotic association of fungi and green algae. There are thousands of different known associations, most involving sac fungi. The lichens shown left have formed a crusty mat on the surfaces of rocks. They play an important part in the subsequent erosion process. Other lichens, much more luxuriant in their growth, are found in the Arctic tundra and are called *reindeer moss* (right).

parent cell. A complete organism grows from these tiny, dispersing bodies.

As you may now agree, seemingly simple fungi may not lead very simple lives. And, to drive the point home, consider the sex lives of the yeasts. Common bakers' yeast can be diploid or haploid, and either form can undergo asexual reproduction by budding (Figure 16.12). The diploid form can also undergo meiosis. The four haploid products of meiosis become separate *meiospores* inside the original cell wall of the parent, which thus becomes an ascus. (That, by the way, is why yeasts are classed as ascomycetes.)

Two of the four meiospores will be the plus (+) mating type and two will be the minus (−) mating type. They can undergo immediate fusion to regain the diploid state, or sexual fusion between the + and − cells can occur much later, at the yeasts' convenience. Generally, diploid strains do not un-

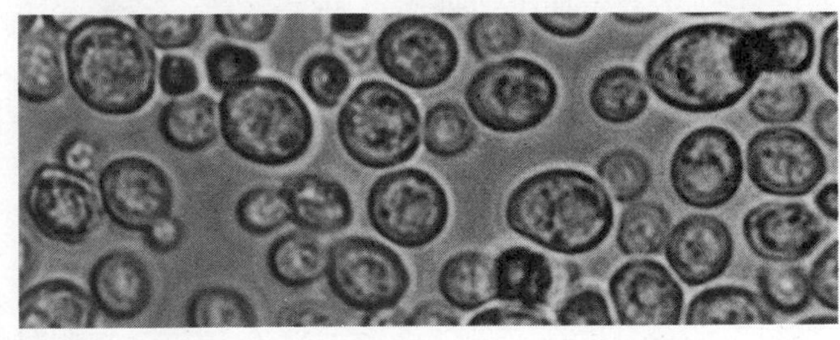

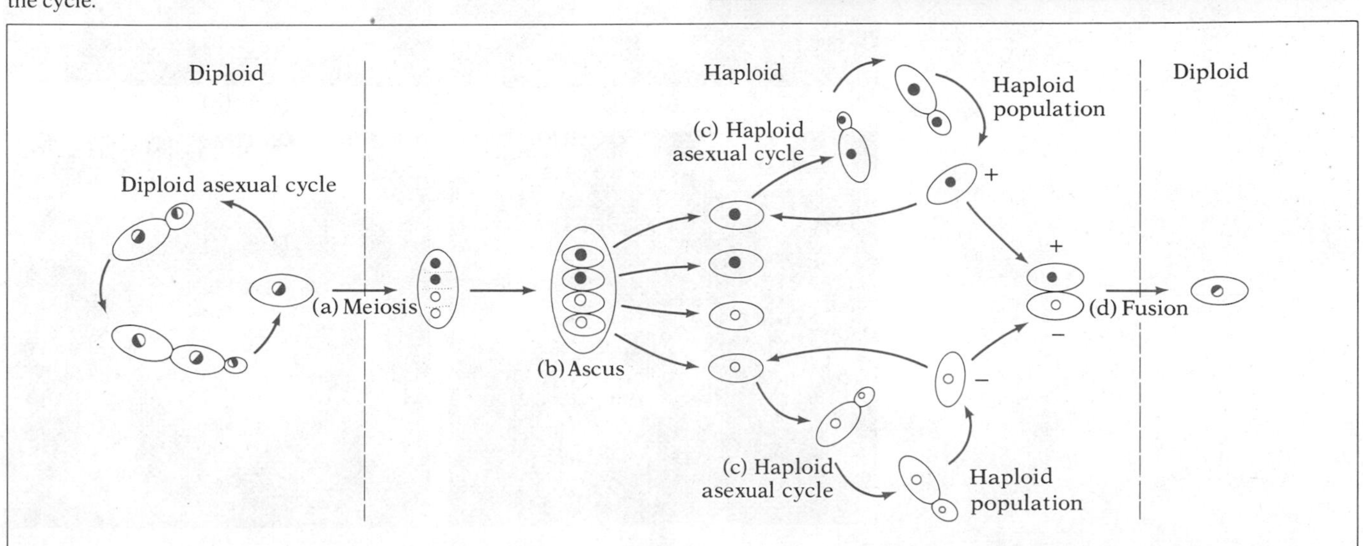

16.12 In their life cycle, yeasts have both a haploid state and a diploid state, which resemble each other rather closely and are without cellular specialization. (a) In the diploid state, a cell reproduces asexually by mitosis and budding (photo), or may undergo meiosis, (b) producing four haploid ascospores within the ascus. Two will be + and two will be −. (c) These ascospores emerge from the cell and carry out the usual yeast activities, including budding for a time, so that large haploid populations are produced. (d) Then if both + and − mating strains are present, a fusion will occur, returning the yeast to the diploid state of the cycle.

dergo meiosis when conditions are good. Meiosis is induced by starvation or desiccation.

Yeasts have mitochondria and when oxygen is available, they can oxidize sugar completely to carbon dioxide and water. But under anaerobic conditions, the citric acid cycle is shut down, and the yeasts derive their energy from anaerobic glycolysis. However, the waste product in this case is not lactate but *ethanol* (ethyl alcohol). Anaerobic glycolysis also produces carbon dioxide, which makes bread rise and beverages bubbly. The yeasts will continue to ferment away until their own alcoholic waste poisons them. This occurs when the concentration of alcohol reaches about 13%, the percentage of alcohol in wine. It is possible to get a higher concentration of alcohol, of course, but you need a still.

Not all yeasts are so well suited to human needs. Some yeasts can produce rather nasty vaginal infections, for instance. Yeast infections are most common in persons who have overdosed on antibiotics, thus killing off the natural bacterial flora that normally inhibit yeast growths.

Basidiomycetes (Club Fungi)

The Basidiomycetes are significant to us in a number of ways. Although the group includes such familiar groups as the mushrooms, toadstools, and shelf fungi, it also includes some devastating parasites. Among these are wheat rust and corn smut. Considering the world-wide dependence on wheat and corn, it is easy to appreciate the concern these parasites cause. In fact, the development of rust-resistant strains of wheat is one of the great success stories of applied genetics. Saprophytic members of the Basidiomycetes are also partly responsible for the decomposition of dead trees and litter on forest floors. (Figure 16.13).

Like the Ascomycetes, the Basidiomycetes are all septate, but, although they do reproduce sexually, they don't produce such distinct sexual structures as the asci. The phylum can be roughly divided into two groups that differ both nutritionally and structurally. We will use the saprophytic mushroom to represent one group and the parasitic wheat rust to represent the other.

The mycelial mass of the common mushroom is not seen above ground; it invariably lies deeply embedded in organic matter, breaking it down and using its stored nutrients. What we see growing on the forest floor is the spore-producing *basidiocarp*, which produces *basidiospores*. The mature basidiocarp develops a large number of *gills* on its underside. These contain numerous *basidia*, single cells that are the spore-producing structures (*basidio-*, "little pedestal"). Meiosis occurs in each basidum, and the

16.13 Fungi can be large and colorful as the mushroom and bracket fungus seen here. Many are saprophytes, but some, such as corn smut, are devastating parasites. While many mushrooms are edible, the innocent looking "death angel" is very poisonous. Nonedible mushrooms are commonly called toadstools. (a) Edible mushroom, of the genus *Lepiota*. (b) A bracket fungus. (c) Corn smut. (d) A poisonous mushroom from the genus *Amanita*.

four haploid products bud off as spores (Figure 16.14). Each mushroom is capable of producing millions to billions of spores.

The sexual phase of mushrooms occurs before the basidiospores develop. The basidiocarp, like the ascocarp, is a *dicaryon* with two nuclei in each cell. Cells containing two nuclei are found in the immature basidium. The nuclei eventually fuse in the basidium, making it briefly diploid. Then the basidial cells undergo meiosis, budding off four haploid cells that will eventually develop into basidiospores. As in most fungi, the diploid condition represents only the briefest interlude in an otherwise haploid life cycle.

We can't really leave the subject of mushrooms without a word about picking your own wild specimens for the table. People who do this are often referred to as "patients" and sometimes as "departed." Why is this so? You probably already know that some species of mushrooms (called toadstools) are deadly poisonous. But which? Even the experts often have difficulty distinguishing certain poisonous species from the grocery store variety. Your best bet is to obtain your dinner mushrooms from the same source as do experienced mycologists—the corner supermarket.

Now let's consider wheat rust, *Puccinia graminis*, a species that follows a life cycle involving two hosts. In addition to wheat, the parasite infects the common barberry plant, *Berberis vulgaris*. While in the wheat, the rust fungi produces spores of two types. One—the red spore—is capable of infecting other wheat plants, while the other—the black spore—is the rust's investment in future generations. Black spores survive the winter season and germinate in moist soil in

16.14 Life cycle of a mushroom. The mushroom seen growing above ground is only the *basidiocarp*, the reproductive structure. Below ground is an extensive mycelium produced by the germination of haploid spores some time past. Fusion of hyphae of different mating types produces a dicaryotic mycelium (a). In the sequence of events shown here, the basidiocarp of the death angel mushroom emerges from the ground and begins to elongate. At this time, the cap is rounded and the stalk is short. As growth proceeds, the stalk elongates and the cap opens up like an umbrella (b). Within the cap are a large number of *gills*, which house the *basidia* (c). The basidia produce the *basidiospores*. The number of spores produced in one basidiocarp is enormous, with some mushrooms dropping millions of spores daily. The basidiocarp is formed by binucleate hyphae, with fertilization occurring just before the spores are produced. At that time a brief diploid state is reached followed immediately by meiosis. Each haploid nucleus is taken into an immature spore; then the basidiospores develop and disperse.

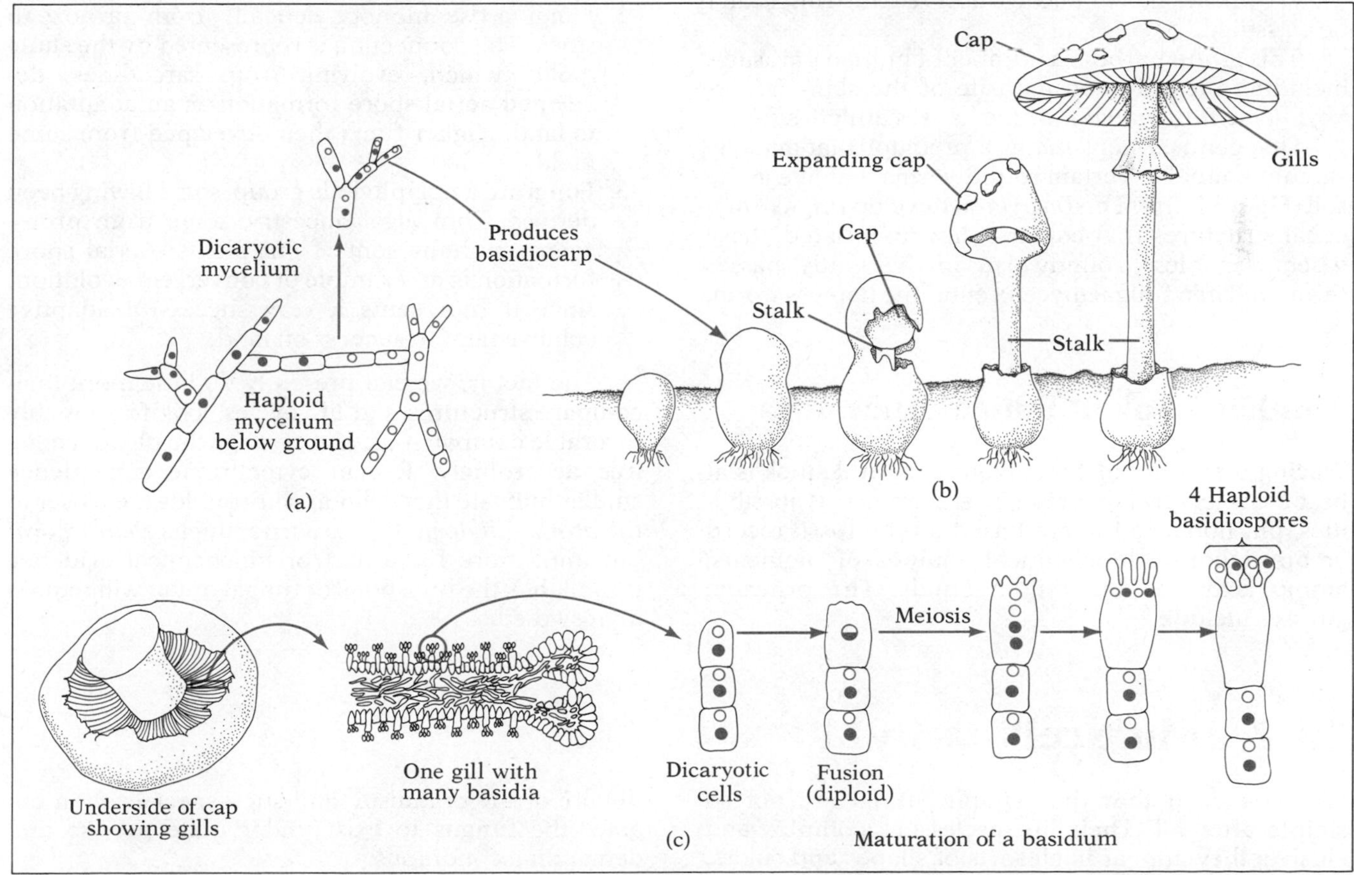

the spring. Upon emerging from its protective spore case, the black spore undergoes meiosis and forms four basidiospores. These then can infect the wild barberry.

It is only in the barberry leaf that the sexual phase occurs. Fertilization is followed by the production of binucleate spores. These, in turn, infect wheat. Wheat infected by *Puccinia graminis* is invariably weakened or killed. After the life cycle was discovered, it became common practice to break the infection cycle by burning contaminated fields and by systematically destroying barberry bushes.

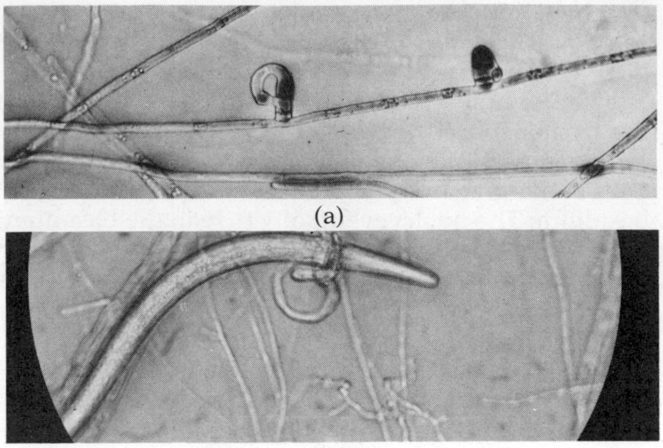

The Fungi Imperfecti

Now let's consider a group called Fungi Imperfecti. The name will first need explaining. *Imperfect* is a botanical term referring to a lack of sexual reproduction, and the approximately 24,000 named species of Fungi Imperfecti apparently have no sexual phase. Actually the Fungi Imperfecti require a lot of taxonomic work. Many will undoubtedly eventually be placed with the sac and club fungi. But a vital part of fungal classification is based on sexual stages, and if these phases haven't been observed, the fungi can't be classified.

This group harbors a number of human parasites, including the common parasite of the skin, *Trichophyton mentagrophytes,* which causes athlete's foot.

One genus, *Dactylaria,* is a predator that actually catches animals—certain roundworms that live in the soil (Figure 16.15). The *Dactylaria* develop ringlike mycelial structures that constrict like an inflated noose when a hapless roundworm inadvertently passes through. Then fungal mycelia enter the hapless worm.

(a)

(b)

16.15 (a) *Dactylaria,* a microscopic fungus, bears many snares (open loops). These swell rapidly on touch, trapping anything inside. (b) Gotcha! The nematode worm shown here is secured by a snare only a thousandth of an inch thick. Soon the prey will be invaded by rapidly growing fungal filaments.

Another Look at Fungal Ancestors

Tracing the origin of fungi from ancestral stock is at best a conjectural exercise. There are many possibilities, but none, so far, are based on the fossil record or an elaborate biochemical analysis of pigments, amino acids, or other compounds. The principal guesses include:

1. Fungi are monophyletic, derived from green algal stock. The derivation was accompanied by a loss of the photosynthetic pigments of algal forerunners, and the retention of cell walls.
2. Fungi arose monophyletically from protozoan stock. The connection is represented by the slime molds which, evolving from sarcodines, developed aerial spore formation as an adaptation to land. Higher fungi then developed from slime molds.
3. Fungi are a polyphyletic group, some having been derived from algal ancestry, some from protozoans, perhaps some from plants. Aerial spore formation is an example of convergent evolution, since it represents a very successful adaptive achievement to success on land.

The fact is, we can presently do little more than compare structures and life cycles, look for possibly favorable comparisons, and arrange the phylogenetic tree accordingly. Recent cytochrome *c* sequence studies indicate that cellular slime molds are closer to the protist *Euglena* than to a true fungus *Neurospora;* but until more fossil and/or biochemical evidence is available, the questions of fungal origin will remain unanswered.

16.2 Multicellularity

We have seen that the "simple" fungi are not so simple after all. Their life cycles are complex and elusive; they appear in all sorts of shapes and colors, and they are rather complex in structure. Most fungi are not only multicellular, but are a variety of cell types. Each cell type, of course, plays its own role in the life of the organism, and such specialization enables the fungus to exist under quite precise and demanding conditions.

Let's consider for a moment, the biological implications of multicellularity and cell specialization.

Why are so many successful species composed of many cells of different types? The answers may provide an important insight into the nature of living things and how they have adapted to such a complex planet.

The Origin of Multicellularity

If we begin by defining multicellularity, some aspects of its origin may be more easily clarified. *A multicellular organism is one that is composed of a number of cells that cooperatively carry out the functions of life.* This definition is meant to distinguish the multicellular organisms from aggregations and colonies of single-celled things, although the distinction is not always clear. For our purposes, multicellular organisms are made up of cells that cannot survive independently under natural conditions.

So, we are assuming that multicellularity began when aggregations of cells became interdependent. This would have happened, of course, when different cells became adapted for different roles. For example, some cells might produce or obtain food, others aid in its transport, and still others become motile, able to move the aggregate to a more favorable place. We might imagine that, finally, some cells came to specialize in reproduction. Other cell types then become terminally differentiated.

As you know, the organisms of the earth are divided into taxonomic groups according to shared characteristics. Two of these, the Monera and the Protista, are commonly referred to as unicellular. This description should exclude these taxa from membership in the multicellular club. However, the taxonomy is not that clear-cut. Some Monera and Protista are clearly multicellular. Further, some of the Fungi are unicellular.

Fungi bridge the gap between unicellular-colonial and multicellular organisms. While yeasts are unicellular, Fungi Imperfecti form hyphae that may be considered to be colonies of identical cells; they also form specialized dispersal cells (spores). The reproductive structures of Ascomycetes and Basidiomycetes may be quite complex, with cells specialized for stalk, cap, gills, and covering. The slime molds also have specialized, terminally differentiated tissues. The noose of the predatory fungus *Dactylaria* is one exceptional example of fungal cell specialization in a nonreproductive tissue.

But remember, for the most part there is little specialization in the fungal mycelium. Cell specialization is almost entirely reserved for the reproductive structures. The plants and animals have made the transition to somatic cell specialization—to the tissue, organ, and organ system levels of organization. We can't let this transition slip by unnoticed.

It seems obvious that when the necessary labors of life are doled out to specialized cells in a colony, then each will become better at performing its particular task and the result will be greater efficiency. It follows, then, that multicellular organisms must somehow be better off than the single-celled organisms. But this is not necessarily so. Single-celled organisms have survived very well through evolutionary history and haven't been displaced by multicellular life. We have emphasized that some of the protists, such as the ciliates, are highly complex and nicely adapted creatures. Their lineage has certainly continued without the drastic adaptive changes that have marked ours. And they have become successful without following the multicellular trend. So if single cells are so successful, why did multicellular life develop?

The answer has already been stated: in a word, it's *specialization*. This term has two connotations, however. Not only have cells become specialized and interdependent so that the success of one type depends on the success of others, but the relative numbers of each type, the direction of their specialization, and the absolute numbers of the cells (the size of the multicellular body) have permitted various kinds of living things to specialize by adapting to particular kinds of environments. For example, some multicellular bodies may have disproportionate numbers of contractile (muscle) cells, so they can survive under environmental conditions where strength is important. They may not have very many cells that are specialized to detect soundwaves, however, so, in parts of the environment where hearing is important, they must yield before other organisms that perhaps have fewer muscles but that can hear better. Thus, cellular specialization means that organisms may come to occupy different niches, and may branch out, specialize, and diversify, creating the staggering array of living things we see around us.

Multicellular Plants

We seem to like to talk about what is the dominant form of life on the earth. Probably one reason for this is that we think *we* are and it gives us a chance to say so. Of course, someone invariably tries to put us in our place by saying something about rats or cockroaches. However, if we were being objective about all this, we would have to seriously consider multicellular plants. Look around. Unless you live in New York City, plants are everywhere, in all sorts of places and in all kinds of shapes and sizes. Some are cute and live in pots, and others are huge and can crush cars, but there they are, from temperate areas to the tropics. Probably, the ancestors of the many kinds of plants covering the earth came, like our ancestors, from the sea.

Multicellularity in plants probably occurred in the sea before they invaded the land: many marine red, brown, and green algae are multicellular, although the level of cell specialization is comparatively low in these groups. We don't know how it happened or what the first land plants were like, but they probably lived a rather marginal existence on the damper edges of dry land, such as we see in some of today's mosses and liverworts. Maybe the first group was ancestral to today's mosses, perhaps evolved from a colonial green alga. However, mosses are not believed to have given rise to the higher plants we see today (those with flowering structures and vascular transport systems). In any case, they made it onto dry land where competition was undoubtedly lessened, but where conditions were harsh and variable, unbuffered by the water medium. Under such conditions, multicellularity became important because the problems of survival were great and the solutions demanded the kind of complexity that comes with cellular specialization.

The land invaders needed protective coverings, water transport systems, anchoring and absorbing structures, and protective reproductive structures for the embryo. Furthermore, some manner of competing for sunlight had to arise among the ground-hugging pioneers. The answer was to become taller and larger; thus, there was a need for supportive tissue. The key evolutionary "invention" was *lignin,* the complex substance that provided stiff conduction tubes that wouldn't collapse in dry weather, and which, by making wood hard and strong, made possible the growth of tall, erect stems. So today's land plants are structurally complex. Complexity, of course, paves the way for diversity, as we shall see in the coming chapters.

Multicellular Animals

Animals are clearly heterotrophs, feeding upon plants and each other. Most animals have nervous systems and respond to stimuli with highly organized movement. They must locate food and avoid predators. Increased size is one way to avoid being eaten, and often gives an advantage in competitions within species. With increased size comes the problem of unfavorable surface-to-volume ratios, and the solution, as we saw in Chapter 4, is multicellularity. And the complex behavior of animals required cell and tissue specialization. As we noted earlier, the earliest unambiguous eucaryotic fossils are already complex, highly differentiated animals. It appears as if all the major animal phyla evolved almost overnight, soon after the perfection of the eucaryote cell and meiotic sex. Note that "almost overnight" may mean 100 million years when one refers to events that may have taken place perhaps 700 to 600 million years ago.

The development of complex systems in animals was evolutionarily rapid, but of course did not really occur overnight. So, one wonders, how did it happen? What was the sequence? Our simpler contemporaries present some opportunity for insight into emerging patterns. For example, sponges have a simple multicellular organization. Not only are they multicellular, but they show some cellular specialization. So here is a rather primitive example of tissue organization. But do they have organs? There is room for argument here, but the lining of their body cavities may qualify. We must add a word of caution. We are not saying that sponges are a link to more complex organisms. Sponges, evolving from protist ancestors, went their own way (as did the mosses). These mindless filter feeders are not believed to represent a step in the main line of animal evolution. However, they may be similar in function to some distant group that did give rise to animals with distinct organs.

Coelenterates have gone a step further. Though simple in structure, they have become organized along the organ level for feeding. Their bodies, however, consist only of two organized layers of cells on either side of an inner, much less organized layer.

All true animals (metazoa) are held together by connective tissue based on *collagen,* an unusual extracellular protein. Collagen is unusual because one of its principle amino acids is *hydroxyproline,* which is not among the twenty directly coded by mRNA; so the synthesis of collagen involves extensive enzymatic modification. Before collagen evolved, large motile animals may have been impossible. But once the first such animals came into being, there was a premium on ever-increasing complexity and diversity. Many animals have evolved elaborate organ systems, including complex nervous systems with acute sensory devices. And basically, animals evolved in response to the need to eat and avoid being eaten.

So, from humble multicellular origins, complex plants and animals rose. That complexity will become more apparent when we look at plant and animal diversity and tour the insides of all sorts of organisms in following discussions.

Application of Ideas

1. The slime molds are a taxonomic enigma since they apparently bridge the kingdoms in their characteristics. They have been classified as plants, fungi, and protists. Review their characteristics and comment on the importance of such organisms in establishing evolutionary relationships.

2. The manner in which wheat rust can be controlled is an important example of intelligent ecological management. Describe this method and compare its environmental effects to those of agricultural pesticides or insecticides.

3. The fungi are still considered to be plants in some taxonomic schemes. In what ways are they plantlike? If they are plants, do they represent primitive plants or degenerate plants? Be sure to define your terms when you support one idea or another.

4. In explaining the evolution of the first land plants, we have to avoid using terms that credit the organisms with concious purpose or forethought. Present a hypothetical explanation for the colonization of land, but use terms like variation, adaptation, and natural selection.

Key Words

aggregation
amanitin
antheridia
asci (ascus)
ascogonium
ascospore
athlete's foot

basidiocarp
basidiospore
binucleate
blue mold
bread mold
budding

chytrid
club fungi
coenocytic
colony
conidia
conidiophore
conidiospore
corn smut

downy mildew

ergot

Fungi Imperfecti
fungus

gametangia
gill
green mold

heterocaryon
hyphae

late blight
lichen

mating type
meiospore
multinucleate
mushroom
mycelium
mycota
myxameba

nonseptate

Key Ideas

Introduction

Except for the *mushrooms* alongside their steaks, most people regard the *Fungi* with suspicion. The bad press of *Fungi* is associated with the large number of parasites of plants and animals and the *saprophytes* that invade our food and goods.

What Are the Fungi?

1. *Fungi* are heterotrophic, falling into *saprophytic* or parasitic categories. *Saprophytes* are reducers, feeding on dead organisms.
2. Taxonomic difficulties arise because:
   a. They have plant characteristics, but do not photosynthesize.
   b. They have protist characteristics, but are multicellular for the most part.
   c. They have animallike nutritive features, but do not share other animal characteristics.
   d. *Fungi* are almost unique in their sexual activites.
   e. The origin and relationships are only poorly understood.
3. *Fungi* have played a prominent role in history. The great *potato famine* caused by fungal *late blight,* brought a million or more Irish to the New World.
4. The *fungi,* like the bacteria, have a vital ecological role as reducers. This activity recycles minerals. They are commonly used in manufacturing foods and medicines.
5. The suffix, *-mycota,* means *fungi,* and the root word is also used in *mycelium,* thc fungal body. *Mycelia* take the form of threadlike *hyphae,* which are septate or nonseptate depending on the individual species. The hyphae secrete enzymes into food and absorb the digested residue.
6. *Fungi* are spore formers. Spores are carried by air and water and germinate under proper conditions to form *mycelia.* Spore formation is also a part of the sexual cycle. Sexual cycles in *fungi* are often highly complex.

Subkingdom Gymnomycota: The Slime Molds

1. *Slime molds* have a slimy, multinucleate, *myxameba* stage in which the organisms take on protistlike characteristics, feeding by phagocytosis and moving by streaming. The spore-forming stage of *slime molds* is fungallike. *Sporangia* form atop *sporangiospores.* Meiosis occurs, forming numerous haploid spores.
2. The *myxameba* can itself divide into a swarm of flagellated offspring, where fusion produces a diploid state. The multinucleate state occurs when numerous mitotic events occur without cytoplasmic division.

Subkingdom Dimastigomycota: The Oomycetes (Water Molds)

1. The *oomycetes* are considered primitive because they produce *nonseptate* cells and form flagellated spores.
2. Most are *saprophytes.* The *water molds* form large eggs from which their name is derived.
3. Most are weakly parasitic. The exception is a group known as *downy mildew,* which is the agent of *late blight. Downy mildews* penetrate leaf stoma and absorb the leaf nutrients. They venture out to produce spores.

oogonium
oomycete

powdery mildew

reindeer moss

sac fungi
saprophytic
septa
septate
shelf fungi
slime mold
smut
sporangium

toadstool
true fungi

water mold
wheat rust

yeast

zoospore
zygospore

4. The *oogonia*, egg-producing structures, become covered with fingerlike *antheridia*, which are sperm-producing structures. After fertilization a *zoospore* may form or the zygote may simply grow a new *mycelium*.

Subkingdom Eumycota: The "True Fungi"

The Chytridomycetes (Chytrids)
These are unicellular parasites of *fungi*, plants, or algae.

The Zygomycetes (Conjugating Molds)
1. These are *nonseptate*, lacking in the usual mycelial mass. Zygomycetes includes the familiar *bread molds*.

2. A typical example is *Rhizopus stolonifer*, common black *bread mold*. Its spores will germinate on a suitable medium and spread a growing *mycelium* into the food. Horizontal *hyphae* (stolons) send rootlike growths down, while sporangiophores rise up to produce spores.

3. Negative and positive mating types undergo conjugation, forming *gametangia* where they touch. Nuclei fuse, forming a *zygospore*, the only diploid stage. When it germinates it undergoes meiosis, forming a haploid *mycelium* typical of all *fungi*.

Ascomycetes (Sac Fungi)
1. *Sac fungi* and *club fungi* are known as higher *fungi*. Both have *septate mycelia*. *Sac fungi* include *mushrooms*, *powdery mildew*, and *blue* and *green molds*. One member, *Claviceps*, is responsible for *ergoted* rye, which will cause restricted blood flow when eaten. Pharmaceuticals derived from *ergot* include tranquilizers, hemorrhage control drugs, and LSD.

2. Asexual spores of *sac fungi*, known as *conidiospores*, are produced atop *conidiophores*.

3. In their sexual activity, *sac fungi* produce *ascospores* in cuplike *asci*. When mating types meet, an *ascogonium* forms on one, an *antheridium* on the other. Together they form *binucleate* cells. Fusion of nuclei occurs when the *ascus* forms, and is immediately followed by meiosis.

4. *Sac fungi* and certain algae form some of the many known lichen associations. Their relationship is mutualistic since both benefit.

5. *Lichens* are important soil builders, and some are an important food source for animals.

6. *Lichens* are reproduced asexually when algal spores and bits of fungal *hyphae* break away and grow again in a new location.

7. *Yeasts* are single-celled *sac fungi*. They grow asexually by *budding*, and strains can be diploid or haploid. Diploid *yeasts* undergo mitosis forming four *meiospores* inside the cell, which becomes known as an *ascus*. After meiosis *plus* and *minus mating* types segregate and can fuse, becoming diploid again. Either mating type can go on for a long period as a haploid cell, however. Under anaerobic conditions, *yeasts* become fermenters, producing ethanol and carbon dioxide.

Basidiomycetes (Club Fungi)
1. This group includes *mushrooms, toadstools, shelf fungi,* and parasitic *wheat rust* and *corn smut*. The latter two are of immense economic importance.

2. In *mushrooms*, the *mycelium* sends a spore-producing *basidiocarp* above ground. It produces haploid *basidiospores* meiotically all along the *gills* under the cap.

3. The sexual stage occurs just before *basidiospores* develop. Cell fusion produces binucleate cells. The nuclei fuse in a brief diploid state, followed immediately by meiosis.

4. Poisonous *mushrooms* (toadstools), such as the "death angel,"contain the toxic compound *amanitin*, which causes respiratory and circulatory failure.

5. Even experts have difficulty identifying toxic *mushrooms*.

6. The *wheat rust, Puccinia graminis*, follows a complex life cycle in two hosts. In wheat it produces red spores that infect other wheat hosts and black spores that overwinter and can infect only wild barberry.

7. Sexual reproduction occurs in the barberry and the resulting spores again infect the wheat. Wheat rust is controlled by eliminating barberry plants.

The Fungi Imperfecti

1. The *Fungi Imperfecti* have no known sexual stages. As they do become known, the individuals are usually reclassified into one of the other fungal groups.

2. The *Fungi Imperfecti* includes the skin parasite of *athlete's foot, Trichophyton mentagrophytes*.

Another Look at Fungal Ancestors

1. Three possible origins of *fungi* have been hypothesized.
   a. Monophyletic from green algal stock.
   b. Monophyletic from protozoan stock—the connection is the *slime molds*.
   c. Polyphyletic group, some from *algae*, others from protozoans. Aerial spore formation in this event would have to have evolved twice (coevolution).
2. Fossil and biochemical evidence may help clarify their relationships to other kingdoms.

The Origin of Multicellularity

1. A multicellular organism is one that is composed of a number of cells that cooperatively carry out the life functions. *Aggregations* and *colonies* are not included, but multicellularity may have *begun* this way. Multicellularity is not a clearly taxonomic category since all kingdoms have some multicellular forms. The *fungi* form a good bridge to multicellularity.

2. Multicellularity did not replace unicellularity since there are many examples of successful species retaining this organization. It did permit specialization, and this permitted organisms to fit new niches, leaving the old to the unicellular types.

Multicellular Plants

The origin of multicellular plants may have been the sea. From there they first explored the moister regions. Multicellularity was undoubtedly a prerequisite to the exploration of the land.

Multicellular Animals

1. Animals, unlike plants, attained complex multicellular organization in the sea, long before the transition to land.

2. The earliest eucaryotic fossils are those of multicellular animals, although these early animals must have arisen from yet earlier, unrecorded protist ancestors.

3. Multicellularity might have arisen in groups with the simple organization of sponges, although they are not believed to be in a direct line.

4. Greater complexity is seen in the coelenterates where organs are found.

5. Increasingly complex developments are seen in the other animal groups with the eventual emergence of organ systems.

Review Questions

1. List the characteristics of the *fungi.* (p. 444)

2. What was the historical importance of *Phytophthora infestans?* (p. 445)

3. Describe the vital ecological role of the *saprophytic fungi.* (p. 445)

4. What are the two principal roles of fungal spores? (p. 445)

5. Review the arguments about *slime mold* taxonomy. (p. 446)

6. Review the events in the life cycle of a *slime mold.* Which steps are protistlike and which are funguslike? (p. 446)

7. What is the source of the name *oomycetes?* (p. 447)

8. Describe the body form of the *water molds.* Include the terms *nonseptate* and *coenocytic.* (p. 447)

9. List an important example of *downy mildew* and review its method of infection. (pp. 447–448)

10. Describe the role of *oogonia* and *antheridia* in sexual reproduction in the *water molds.* (p. 448)

11. List the unusual features of the *chytrids.* (p. 449)

12. Describe the growth and feeding of the *bread mold Rhizopus stolonifer.* (p. 449)

13. How does asexual reproduction occur in the *bread molds*? (p. 449)
14. Describe the events of sexual reproduction in *bread molds*. (p. 449)
15. How do the hyphae differ in higher *fungi*? (p. 450)
16. List several important members of the *sac fungi*. (p. 450)
17. Describe the effect of *Claviceps*-infested rye when eaten. What is this infection called? (p. 450)
18. What are *asci* and *ascospores?* Describe their formation. (p. 451)
19. What is a *heterocaryon* and how is this important in the sexual cycle of *sac fungi*? (p. 451)
20. Describe the association of cells in *lichens*. How does asexual reproduction occur? (p. 451)
21. What is the ecological importance of the *lichens*? (p. 452)
22. How do asexual and sexual reproduction occur in *yeast*? (p. 452)
23. List common members of the *club fungi*. Which are parasites? (p. 454)
24. Using the terms *basidiocarp* and *basidiospore*, review the life cycle of a mushroom. Indicate which parts are diploid and which are haploid. (p. 454)
25. Construct a simple diagram showing the infection cycle of *wheat rust*. Include both wheat and wild barberry and the events that occur in each. (pp. 455–456)
26. What does *Fungi Imperfecti* mean? Why is sexuality so important in classifying fungi? (p. 456)
27. List the three possible origins of *fungi* and explain what evidence is needed to make decisions about the correct lineage. (p. 456)
28. Define the term *multicellularity* and distinguish this from *aggregations* and *colonies*. (pp. 456–457)
29. Why is multicellularity not a clear-cut taxonomic category? (p. 457)
30. Why hasn't multicellularity completely replaced unicellularity? (p. 457)
31. Do mosses represent an ancestor to today's multicellular plants? Explain. (pp. 457–458)
32. What special conditions on land make multicellularity adaptive? (p. 458)
33. What might have prompted the great diversification of animal life? (p. 458)
34. Are any animal groups in existence today considered to be the ancestors to more highly developed groups? Explain. (p. 458)

Suggested Reading

Alexopoulos, C. J. 1962. *Introduction to Micology*. John Wiley, New York. For enthusiasts of things fungal.

Banks, H. P. 1975. "Early Vascular Plants: Proof and Conjecture." *Bioscience* 25:730. On the origin of one of the five kingdoms.

Barghoorn, E. S. 1971. "The Oldest Fossils." *Scientific American,* May.

Dagley, S. 1975. "Microbial Degradation of Organic Compounds in the Biosphere." *American Scientist* 63:681. Microbes have ecology, too.

Dickerson, R. E. 1978. "Chemical Evolution and the Origin of Life." *Scientific American,* September. We think that this is the best no-nonsense review of this primarily speculative field.

Dodson, E. O. 1974. "Phylogeny." In *The New Encyclopedia Brittanica,* 15th ed. 14:376. Dodson includes a full historical account of the various one-, two-, three-, four- and five-kingdom schemes as well as an overview of phylogenetic relationships within the plant and animal kingdom.

Leedale, G. F. 1974. "How Many Are the Kingdoms of Organism?" *Taxon* 23:261.

Levin, D. A. 1979. "The Nature of Plant Species." *Science* 204:381. The so-called "biological species concept" quickly breaks down when plant species are considered.

Margulis, L. 1968. "Evolutionary Criteria in the Thallophytes: A Radical Alternative." *Science* 161:1020. A cogent synthesis of the problem of the origin of eucaryotes and the kingdoms of organisms.

———. 1971. "Symbiosis and Evolution." *Scientific American,* August.

Maynard Smith, J. 1971. "What Use Is Sex?" *Journal of Theoretical Biology* 30: 319. One major difference between procaryotes and eucaryotes is the invention, in the latter, of regular biparental reproduction. Being widespread and of ancient origin, it presumably serves some function, but what is it?

Miller, S. L. 1935. "Production of Some Organic Compounds under Possible Primitive Earth Conditions." *Journal of the American Chemical Society* 77:2351. The first *experimental* approach to the question of the origin of life on earth.

Parkinson, D., and J. S. Waid. 1960. *The Ecology of Soil Fungi.* Liverpool University Press, Liverpool, England.

Raven, P. H. 1970. "A Multiple Origin for Plastids and Mitochondria." *Science* 169:641.

Salle, A. J. 1979. *Fundamental Principles of Bacteriology.* Not for leisure reading, but if there is anything you want to know about bacteria you'll find it in here.

Schopf, J. W., and D. Z. Oehler. 1971. "How Old Are the Eukaryotes?" *Science* 193:47. Perhaps they are not as old as had been thought, but originated only a few hundred million years before the first oldest known fossil animals.

Silver, W. S., and J. R. Postgate. 1973. "Evolution of Asymbiotic Nitrogen Fixation." *Journal of Theoretical Biology* 40:1.

Stanier, R. Y., and M. Douderoff. 1973. *The Microbial World,* 3d ed. Prentice-Hall, Englewood Cliffs, N.J.

Whittaker, R. H. 1959. "On the Broad Classification of Organisms." *Quarterly Review of Biology* 34:210. Whittaker's influential revolt against the classic plant–animal dichotomy.

Part Four

Plants

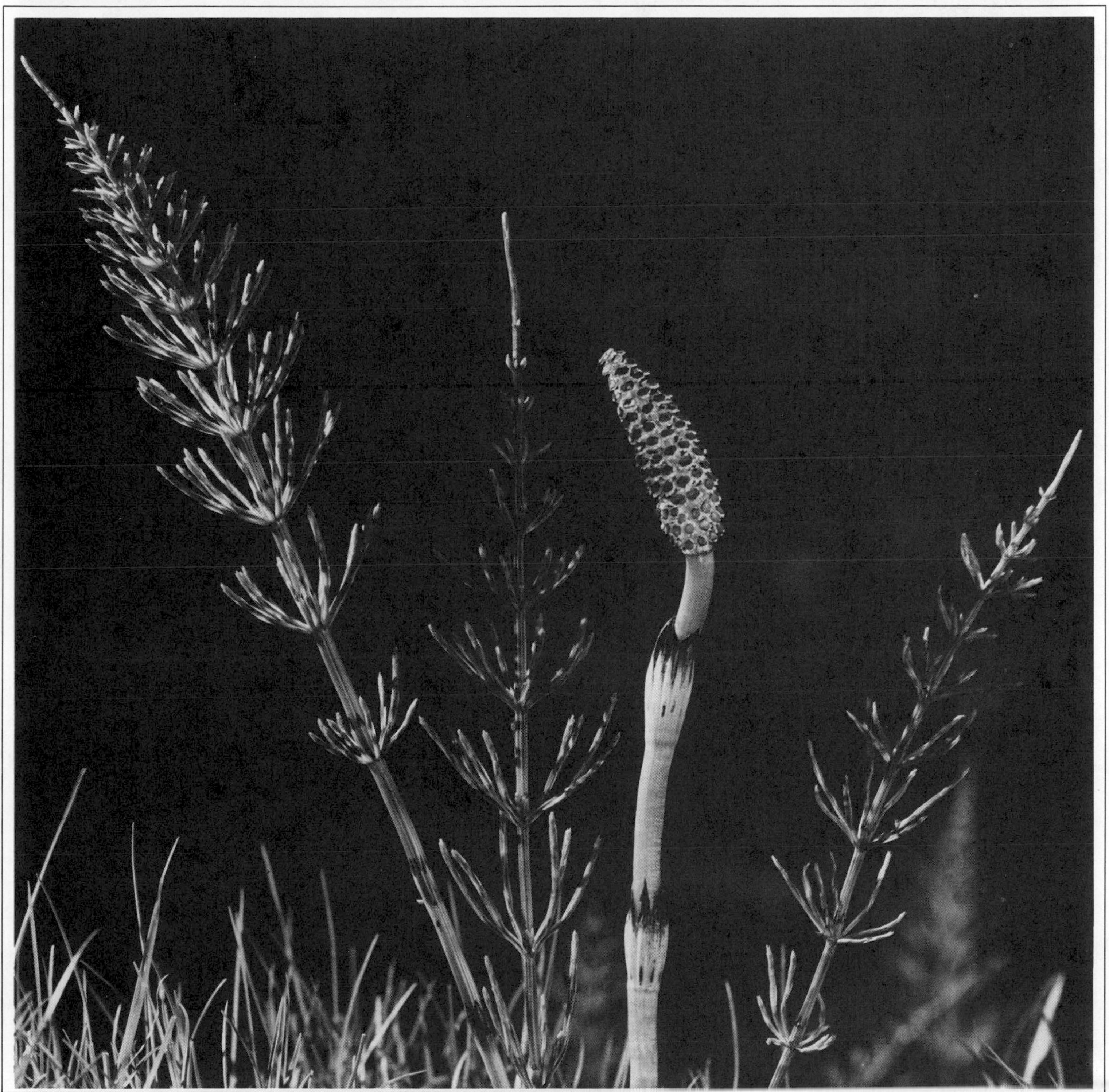

Chapter 17

Evolutionary Patterns in Plants

Our current definition of the word *plant* represents an uneasy truce among life scientists. In some cases we're sure, but in other instances we're not. No one has ever denied that a potato is a plant. And moss surely belongs to the Plant kingdom. But there has been quite a tussle over algae, and even now many biologists would be reluctant to expell any photosynthesizing eucaryotes from the Plant kingdom. In Chapter 13, we pointed out that cyanophytes (blue-green algae), fungi, slime molds, bacteria, and even protozoa have sometimes been included in the Plant kingdom. But we have chosen to follow the five-kingdom scheme (see Figure 13.14), which at least addresses the problem of what to do with such great diversity. We must remind ourselves, however, that even this scheme is provisional. As biochemical taxonomy proceeds to straighten out the phylogenetic relationships—if indeed it does—we'll very likely see yet another revolution in basic systematics. It is even possible that "plant" and "animal" will eventually be considered to be scientifically useless words, and will be returned, with thanks, to the public domain.

17.1 The Nonvascular Plants

For now, following the terms of the truce, we can confidently report that the Plant kingdom contains six distinct divisions* (phyla) grouped into three subkingdoms. These subkingdoms have been established according to the structure of the plants, their photosynthetic pigments, and their forms of starch storage. The first subkingdom, Rhodophycophyta, consists only of red algae, all organized into one division, Rhodophyta. The second, Phaeophycophyta, contains the brown algae, again in only one division, Phaeophyta. The third subkingdom, Euchlorophyta, contains all the rest of the plants, organized into four divisions, Charophyta, Chlorophyta, Bryophyta, and Tracheophyta. The euchlorophytes are a strange collection of types ranging from some very simple green algae to the largest of trees. The grouping, however, is not based on size or appearance, but on biochemistry. All euchlorophytes utilize chlorophylls *a* and *b* and carotenoids in photosynthesis and store carbohydrates in the form of amylose and amylopectin. As you will see, the red and brown algae are quite different in these aspects.

Some or all of the organisms in each of the subkingdoms are multicellular, which lifts the groups out of the kingdom Protista. Plants, then, are multicellular photosynthetic organisms; or, occasionally, single-celled photosynthetic organisms clearly related to multicellular plants.

The three subkingdoms apparently arose independently in dim prehistory from different protists. Because they separated so early, a pine tree, it seems, may be no more closely related to a kelp than it is to your uncle Max. Thus, the Plant kingdom is clearly a polyphyletic group, comprising remarkably unrelated groups.

Plants, nevertheless, do have common features other than photosynthesis. They produce cellulose walls and are nonmotile; most have specialized organs and tissues; and they are primarily sexual reproducers, with alternating haploid and diploid generations. When flagellated cells occur, each usually has two flagella on the anterior end. These should only be considered general or primitive characteristics of the kingdom, because some of the features have become lost or changed in several of the more recently evolved species. For instance, some true plants are nonphotosynthetic parasites; and in flowering plants, the haploid generation is greatly reduced (although it is still present) and flagellated cells are absent.

*The term *division* is the preferred term in botanical nomenclature, and is the correct term according to official botanical rules.

Division Rhodophyta: The Red Algae

The division *Rhodophyta* (*rhodo-*, "red;" *-phyta*, "plants") has about 300 species of seaweed, all called *red algae*. In spite of the name, most red algae are green or black. But all contain the pigments *phycocyanin* (meaning "algal blue-green") and *phycoerythrin* (meaning "algal red"), as well as chlorophyll *a*, and many have chlorophyll *d* (an oxidation product of chlorophyll *a*). The pigments of the red algae are strikingly similar to those of the procaryotic cyanophytes (blue-green algae). This fact has long troubled taxonomists looking for evolutionary relationships, since the red algae are in all other respects unabashedly eucaryotic. The puzzle has been resolved with the decision that the red algae are probably derived from an independent symbiotic union between an ancient, free-living, procaryotic cyanophyte and what was once an independent eucaryotic host. Part of the evidence is based on the fact that the chloroplasts of red algae, which contain all those photosynthetic pigments, still closely resemble cyanophytes in structure as well as in chemistry. Other plants, in contrast, have more highly specialized chloroplasts, thought to have evolved much earlier from a different photosynthetic symbiont, a chlorobacterium.

Flagella and cilia, typically found in some form or another in the rest of the eucaryotic divisions and phyla, are never found in the red algae. Even the sperm lack tails and must float passively to receptive female cells.

All red algae are aquatic, and nearly all of them are marine (Figure 17.1). You've undoubtedly seen red algae if you've ever looked into a tide pool. In fact, red algae comprise a large portion of what we commonly call *seaweed*. You might not recognize them as red algae, since they are usually green or black, and some are blue or violet, but those growing well below low tide may even be red. Actually, color can't even be used to identify species. Individuals of the same species often are found to be of different colors at different depths, possibly in an adaptive response to the changing spectra of light available for photosynthesis. Different pigments, of course, each tend to absorb certain wavelengths of light. For example, bright red algae have been found growing 200m below the surface of the sea, where any light that is present is highly penetrating shortwave blue or green—colors that are most readily absorbed by red pigment.

17.1 The plant shown here is a rhodophyte, more commonly known as red algae. Most red algae are small compared to the brown algae, measuring from a few centimeters to perhaps a meter in length. Most species are marine and are found in warm waters, usually attached to rocks by their *holdfasts*. The branched bodies are usually frilly and delicate, but some form the widened, flat blades seen here.

Marine red algae grow primarily on rocky coasts, where they attach firmly to the substratum by specialized structures known as *holdfasts*. Holdfasts are simply anchors, not roots. After all, algae grow in water and therefore have no need of a root system to extract water and minerals from the soil.

Red algae store their foods in the form of a starch called *floridean* and produce a number of other polysaccharides. One of these, *agar-agar* (or *agar* for short), is used by Indonesians to thicken soup and by biologists as a jellylike medium on which to grow bacteria. The bacteria cannot metabolize the agar, but use only the nutrients that have been added to it. Red algal polysaccharides cannot be metabolized by most microorganisms or by anything else. However, we eat them anyway. A species of red alga known as *Irish moss* is harvested in enormous quantities for its polysaccharide *carrageenin*, which gives a fake richness to such things as chocolate-flavored dairy drink and milkshakes. But although you're fooled into believing that you're ingesting something that you're not, there probably is little harm done. Indigestible though it is, it is "food from the sea," and, being nonfat and no-calorie, is probably better for most of us than real cream and sugar. Red algae blades—leaflike structures—are an important part of the Japanese diet; they are usually wrapped around a small portion of rice.

The red algae present us with an unusually clear example of *alternation of generations*, although all three subkingdoms exhibit this phenomenon. Alternation of generations (where a diploid, asexual generation is followed by a haploid, sexual generation) is clearly seen in the life cycle of *Polysiphonia*, a common genus of red alga (Figure 17.2). In this genus, the sexual haploid and the asexual diploid are both multicellular forms with *thalli* (branched bodies) that closely resemble one another; in fact, they can be distinguished only by inspecting the reproductive structures under the microscope (or by chromosome counts). Neither generation is dominant—that is, neither is decidedly larger nor more common than the other.

Now, don't be misled by talk of sex in plants. The terms *sexual* and *asexual* are used somewhat arbitrar-

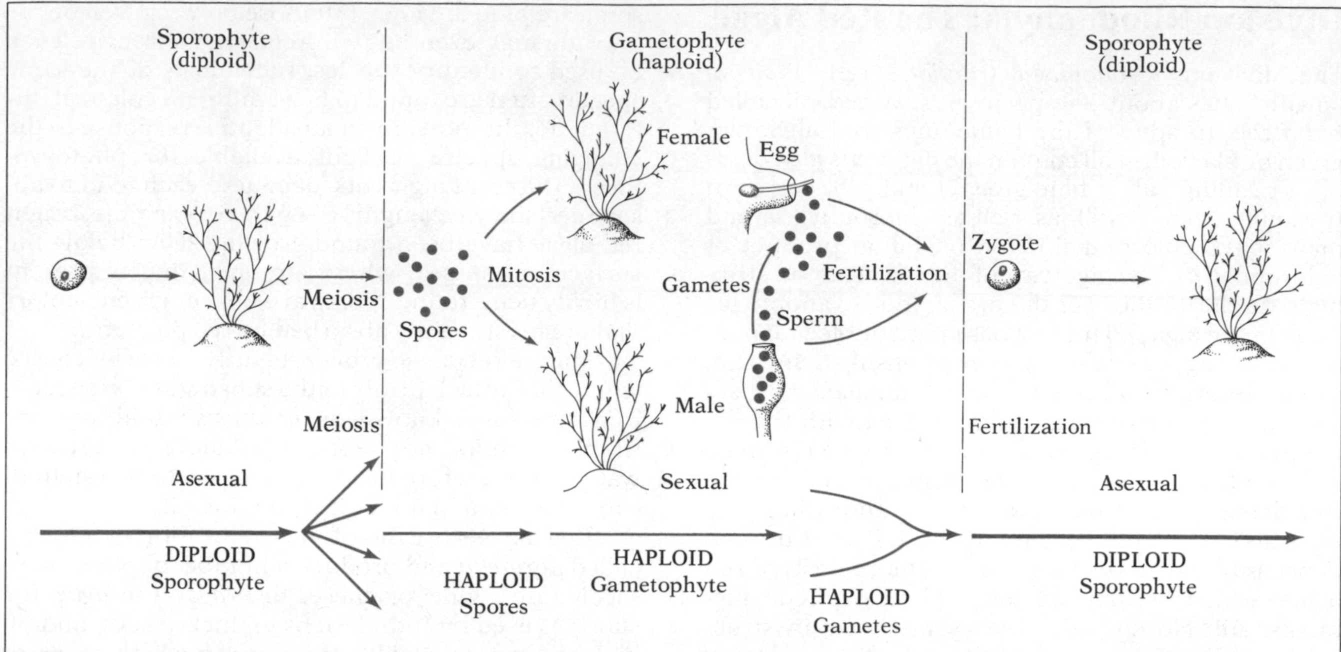

17.2 The life history of the red alga *Polysiphonia,* like that of many of the algae, consists of separate sporophyte and gametophyte generations. The *sporophyte is diploid* and begins with the fertilized egg cell. The egg cells mature and grow into the sporophyte. The sporophyte generation ends with meiosis and the production of *haploid spores,* marking the beginning of the *gametophyte* generation. In *Polysiphonia,* the gametophyte *thallus* is identical to that of the sporophyte, which is not always the case in other algae. Female gametophytes produce egg cells in a pouchlike structure that has a long slender filament protruding outward. The male gametophyte produces *nonmotile* sperm cells that are released into the water to be carried by currents. The sperm adhere to the filament, their nuclei penetrate this structure and then migrate down to fertilize the egg cells. Fertilization ends the haploid gametophyte generation, starting a new, diploid sporophyte generation.

ily when talking about plant generations. The diploid phase, called the *sporophyte* in all plant groups, is the asexual generation, but it is formed by the union of gametes into a zygote—definitely a sexual process. It is in the sporophyte that meiosis takes place, and meiosis, you will recall, is also quite definitely a sexual process. The problem is, meiosis in plants rarely produces gametes. Instead, it produces *spores,* reproductive cells that (usually) disperse, germinate, and grow by mitosis into multicellular haploid organisms, the *gametophytes.* The gametophytes then produce gametes, and these will join in fertilization to produce the sporophyte. So the gametophyte is considered to be the sexual generation (Figure 17.2).

In certain species of red algae, either the sporophyte or the gametophyte generation will be clearly dominant. In some species, the gametophyte generation is reduced to a single-cell stage, and eggs and sperm develop directly by meiosis, just as in animals.

The fossil history of the red algae is not very complete, but it seems likely that they shared the earth with the green and brown algae (as well as with the far older cyanophytes) as far back as the Cambrian period at the beginning of the Paleozoic era, some half billion years ago.

Division Phaeophyta: The Brown Algae

Phaeophyta is rather small, as plant divisions go, and contains only about 1000 named species. The *brown algae* are distinguished from all other living things by their characteristic brown pigment, *fucoxanthin.* Brown algae at least have the decency to be brown—all of them. All of them also store carbohydrates in the form of *laminarin* and *mannitol,* and they have characteristic structural polysaccharides as well (notably *algin,* an important constituent of commercial ice cream and frozen custards). Unlike the red algae and most advanced flowering plants, the brown algae have flagellated sperm and, strangely enough, some female reproductive cells are flagellated as well.

The brown algae live only in the ocean, particularly in cold coastal waters. An exception of sorts is the genus *Sargassum,* which is found in great masses in the warm water that flows through the middle of the Atlantic ocean, an area called the Sargasso Sea (Figure 17.3). Even *Sargassum,* however, is really a coastal seaweed. The plants in that floating island of seaweed have been uprooted by storms and have drifted out to sea. Though they blissfully continue to photosynthesize, grow, and release gametes, the zy-

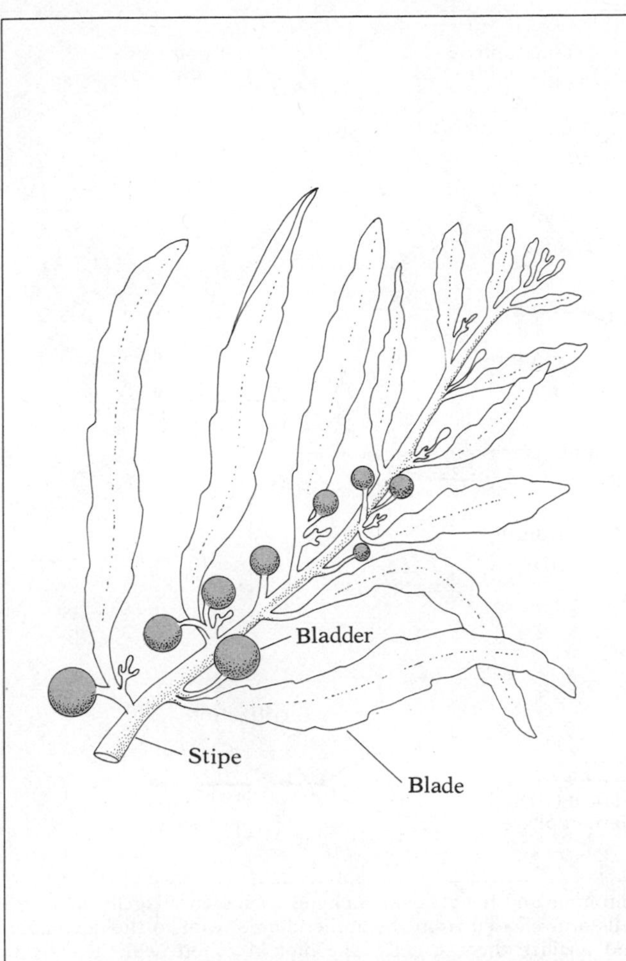

17.3 The brown alga *Sargassum* is a floating seaweed commonly found in the warm southern Atlantic waters. Their floating masses are reported to occupy an area of over two million square kilometers in what is known as the Sargasso Sea. You have probably read about sailing ships becalmed and trapped by *Sargassum,* but these stories are pure fiction. *Sargassum,* or *gulf weed,* as it is also known, forms branching *stipes* with flat, leaflike *blades.* The spherical objects attached to the stipes are bladders (floats) that help the seaweed remain near the surface where light is plentiful. The gulf weed is actually a coastal marine alga, but it constantly breaks away from its point of attachment to drift with the Atlantic currents.

gotes that are produced fall to the bottom of the ocean and join the marine food chain.

The brown algae come in all shapes and sizes and have a wide range of life histories. The life cycle of one of these, *Lamminaria,* is shown in Figure 17.4. Some brown algae are microscopic filaments, while others are enormous, such as the giant kelps. One of the largest kelps, *Nereocystis* (Figure 17.5), can grow to 30 m (100 ft) tall. The length of the kelps rivals the height of many terrestrial trees.

The kelps are structurally simple compared to flowering plants, but they appear to be the most advanced brown algae. For example, *Macrocystis,* another giant kelp, has highly specialized tissues and organs (Figure 17.5). These include the convoluted *holdfast;* the *stipes,* which are stemlike structures supporting the *blades;* the *blades* themselves, which resemble flattened leaves; and spherical, hollow *bladders* (or floats), which keep the photosynthetic cells of the plant near the surface. The brown alga *Postelsia* even has specialized conducting tissue that closely resembles that of some vascular plants. But in general, algae do not possess elaborate conducting systems.

Division Charophyta: The Stoneworts

The *stoneworts* are a remote and ancient alga. Their fossils are found in Silurian deposits, which date back to 425 million years ago. Today, this venerable and complex plant is a bottom dweller, found in sluggish streams, ponds, and lakes. The stoneworts hold the depth record for the euchlorophytes, flourishing at 140 m in Lake Tahoe, California. Only one-tenth of the surface light penetrates to this depth.

The best-known stonewort is *Chara,* a complex plant that shares such characteristics as *apical growth* (see Chapters 18 and 21) with higher plants. It further differs from other euchlorophytes by incorporating calcium salts into its cell walls (thus the name *stone*-wort), and producing complex reproductive structures. These odd plants long ago diverged from the ancestral green alga and followed a different evolutionary direction. But they changed slowly. As a matter of fact, they are not much different from their fossil ancestors.

Division Chlorophyta: The Green Algae

There are about 7000 named species of *green algae,* and whereas most of them are freshwater forms, there are a respectable number of marine species as well. A few terrestrial/airborne species appear on the surface of melting snow, on the moist sides of trees, or free in the soil. Although, for the most part, we will be considering multicellular plants in this chapter, the Chlorophyta is a group that includes both unicellular and multicellular species. Many single-celled green algae have become photosynthetic symbionts in lichens, ciliates, and invertebrates.

The green algae occur as single-celled flagellated or unflagellated forms, as chains or filaments, as inflated "fingers" (*Codium,* also called "dead man's fingers"), and as delicate flattened blades (*Ulva,* the delicate *sea lettuce* of tide pools). The group is so diverse that it is difficult to find a representative species, so we'll just arbitrarily choose a few examples.

Sporophyte (diploid)

Gametophyte (haploid)

Sporophyte (diploid)

Zygote

Meiosis

Spores

Female

Oogonium

Male

Antheridium

Gametes

Fertilization

Zygote

Meiosis

Fertilization

DIPLOID Sporophyte

HAPLOID Spores

HAPLOID Gametophyte

HAPLOID Gametes

DIPLOID Sporophyte

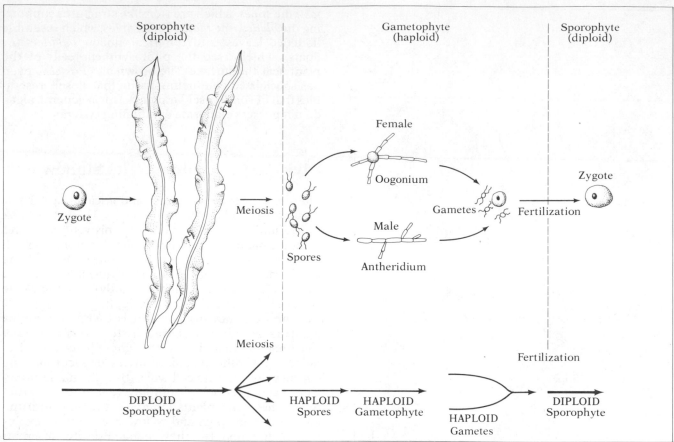

17.4 The life history of *Laminaria*, a brown alga, is characterized by a very dominant sporophyte generation and a brief gametophyte generation consisting of a small mass of cells that produce gametes. The biflagellated spores are produced in the blade and are released to swim about in the water. The spores develop into male and female gametophytes as shown. Flagellated sperm cells are released from the antheridium, swim to the oogonium, and fertilize the egg cells. The diploid zygote will then divide rapidly and develop into the mature sporophyte.

17.5 The giant kelps such as *Nereocystis* and *Macrocystis* thrive in cold oceans. (a) *Nereocystis* is a common sight along the beaches of the Pacific coast, particularly after storms, when its holdfast is torn from the ocean floor. Note the long whiplike stipe with its single large bladder from which the blades emerge. (b) *Macrocystis* differs considerably, with a highly branching stipe and smaller bladders at the base of each blade. Kelps like these occupy large submerged beds along the continental shelf, and form the habitat of many other marine species. Commercial kelp harvesting, which has long been practiced along the Pacific coast, must be carefully managed with its impact on the habitat in mind.

(a)

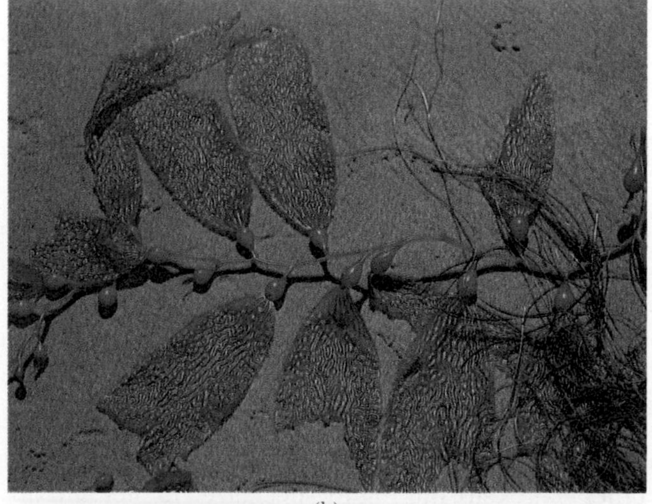

(b)

The most primitive of the green algae are the *single-celled* and *colony-forming* types. As far as anyone can figure, this group, although related to the multicellular green algae, is not descended from them and thus simply has never gotten past the protist level of organization.

Chlamydomonas (Figure 17.6) is a favorite organism of many biologists. The small water plant is easily grown, and its genetics and physiology have been studied in detail. The cells have two flagella, a single light-sensitive *red eyespot*, and one cup-shaped chloroplast. They are phototactic, and *Chlamydomonas* in a beaker of pond water will congregate on the side closest to the window. *Chlamydomonas* are usually seen in their haploid phase, that is, in the gametophyte generation. In this stage it reproduces asexually by mitosis and cell division and builds huge clone populations. It will not enter into a sexual cycle unless conditions are right.

The right conditions include the presence of opposite mating types (the plus and minus strains) as well as some form of environmental stress (such as absence of nitrogen compounds). When both mating types are present, the haploid *Chlamydomonas* produce *isogametes* mitotically. (Isogametes are gametes in which the two types are identical in appearance.) The two isogametes fuse by butting, head on, and entering a wildly spinning "nuptial dance" as their nuclei slowly fuse. Following the union, the zygote, now representing the sporophyte generation, forms a tough, resistant *zygospore*. The zygospore may remain quiescent for a considerable time or, if conditions are right, it may immediately enter into meiosis, producing four haploid *meiospores*. This begins a new gametophyte generation, which matures into the familiar haploid population.

In summary, *Chlamydomonas* populations are haploid, increasing their population size as meiospores go through repeated mitoses and cell divisions. Under special conditions, a sexual cycle occurs and isogametes are produced. Fusion of these isogametes restores the brief diploid sporophyte generation: only the zygote itself is diploid. (An organism which is haploid throughout its life cycle except for the diploid zygote is called a *haplont*. In haplonts, mitosis occurs only in haploid cells. In *diplonts*, all cells are diploid except gametes, and mitosis occurs only in diploid cells; you are a diplont.)

Another remarkable green alga is *Volvox* (Figure 17.7). Although it is not related to any truly multicel-

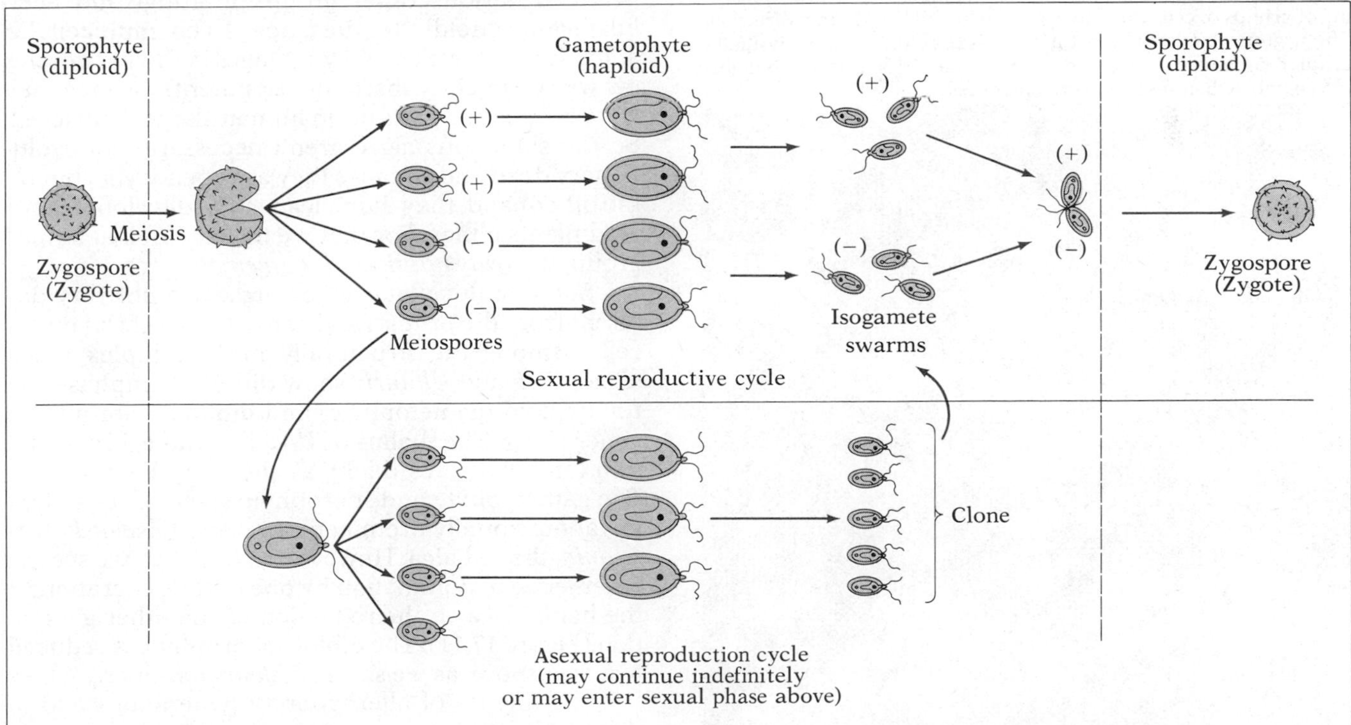

17.6 Life history of *Chlamydomonas*. The sporophyte is a brief interlude in an otherwise lengthy gametophyte history. Zygospores enter meiosis and produce flagellated meiospores. These can enter either a sexual phase, producing many isogametes, or an asexual phase, producing enormous cloned populations through repeated mitotic divisions. Sexual reproduction occurs when different (+ and −) mating strains meet. Flagellated isogametes fuse head-to-head, producing diploid zygospores once again. The haploid gametophyte generation is strikingly prominent in this alga, since the diploid sporophyte generation is represented only by the transient zygospore.

lular form, its history tantalizingly suggests what the earliest beginnings of multicellularity *might* have been like. *Volvox* is composed of a spherical colony of virtually identical cells, each one, in fact, very similar to an individual *Chlamydomonas*. So, in a sense, *Volvox* is a group of individuals in a sphere behaving as an organism. Or is the *sphere* really the individual? The seeming simplicity of this sort of question is deceptive. Philosophers of biology have been concerned with it for years.

Volvox is not only interesting because it suggests a way multicellularity could have arisen, but also because it shows a kind of rudimentary differentiation within the colony:

1. The sphere swims in an organized manner, with the flagellated cells in front pulling and those in the rear pushing.
2. In the sexual phase some cells become specialized for reproduction, producing a few large eggs or many smaller sperm.
3. In asexual reproduction, pockets of the sphere depress inward, and new spheres of cells pinch off and come to lie inside the parental sphere. The new, young spheres are inverted, however, with their flagella directed inward. The daughter volvox colony has to turn itself inside-out within the parent structure. Eventually it is released through a hole in the parent colony (Figure 17.8).

17.7 *Volvox,* a green alga, is a spherical colony of tiny, interconnected flagellated cells. There is enough differentiation in the colony to suggest a trace of tissue-level organization. Individuals provide the colony with motility in a coordinated manner. The cells are immersed in a gelatinous sheath and are interconnected by strands of cytoplasm. The larger concentrations of cells that are visible in the colony are sexually reproductive bodies, consisting of groups of cells that have become specialized in that direction. The argument over whether *Volvox* is a colony or a truly multicellular organism is still not settled, but it is certain that this alga is more than just an aggregate of cells.

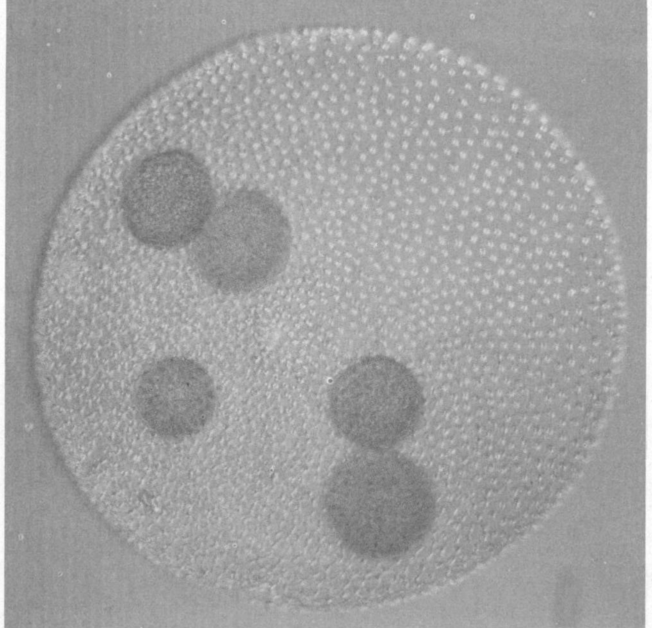

Green algae of another group (the *siphonous algae*) have taken off in a different evolutionary direction. Old botany texts once called it "an evolutionary dead end," but this is probably unfair, because they're still around—and doing quite well, in fact. Algae of this group aren't exactly multicellular, but they're not exactly unicellular either. In fact, they present a direct challenge to the famous cell doctrine of Schleiden and Schwann. Some individuals grow to a considerable size, often producing branched *thalli* (bodies). Although the nuclei undergo mitosis, the cytoplasm doesn't divide and cell walls are not laid down between the newly separated nuclei. The result is a multinucleate mass of cytoplasm (or *coenocyte*) within an enveloping cell membrane and cell wall. *Codium* is in this group. So is *Acetabularia*, a weird little plant that looks like a toadstool or parasol. Its single cell stands about 5–9 cm tall. Some *Acetabularia* are multinucleate coenocytes, just like *Codium*. But in other species, the multiple nuclei are not dispersed through the cytoplasm; instead, they clump into a single *compound nucleus* in the foot, so that the structure of the whole organism is very like that of an ordinary single cell (Figure 17.9).

Siphonous green algae are not the only coenocytic organisms. The same kind of organization appears in various other groups in fungi and acellular slime molds, for instance. Even multicellular animals may have coenocytic stages in their life cycles (as we see in early insect development), or they may have coenocytic parts (as in human skeletal muscles). So the siphonous algae aren't necessarily an evolutionary dead end because they are coenocytic, but on the other hand, they didn't lead to the development of land plants either. For that we have to look at a third group, the *truly multicellular green algae.*

Some multicellular green algae are not very different from the protists and consist of single chains of cells. Among the structurally more complex green algae, *Ulva* and *Ulothrix* show different emphases on the haploid (gametophyte) and diploid (sporophyte) generations. The thallus of *Ulva* is leaflike, broad and thin (only two cells thick), as shown in Figure 17.10. The gametophyte and sporophyte stages of this plant are about equally prominent, as were those of *Polysiphonia*, the red alga. However, in *Ulothrix*, we see the emergence of domination by one of the generations—the haploid—with the reduction of the other generation (Figure 17.11). The diploid sporophyte is reduced to a zygospore, as we saw in *Chlamydomonas*.

The themes of alternation of generations and of the tendency of one generation to become dominant continue in the evolution of the land plants. There is no direct evidence that the ancestors of the land plants were closely related to present-day multicellu-

17.8 The *Volvox* colony or organism reproduces both asexually and sexually. (a) In its asexual reproduction, individuals of the sphere grow dramatically in size; then they divide, grow, and divide again. At first, the cells are much larger than others in the sphere, but then they undergo a rapid division, soon reaching the size of the outer flagellated cells of the colony. As you can see, the result is a miniature colony, but with the flagella all pointed inward. Then, in a strange, animallike phenomenon, the colony inverts within the parent and finally reorients its flagella, forming a replica of the parent colony. Eventually, it passes out through a hole in the parent colony and is freed to live on its own. Large numbers of daughter colonies are often found still within the parent colony. (b) To reproduce sexually, flagellated cells of the sphere differentiate into sperm- and egg-producing structures. Many sperm are produced by a single cell, while single egg cells are produced by enlargement of single parental cells. Sperm are released and may then enter and fertilize the egg cells of another colony. With fertilization, a zygospore is produced, complete with roughened, resistant casing. *Volvox* colonies are haploid, so the first event following zygospore germination is for the zygote to undergo meiosis, producing haploid, flagellated spores. These will grow into new colonies by mitotic division.

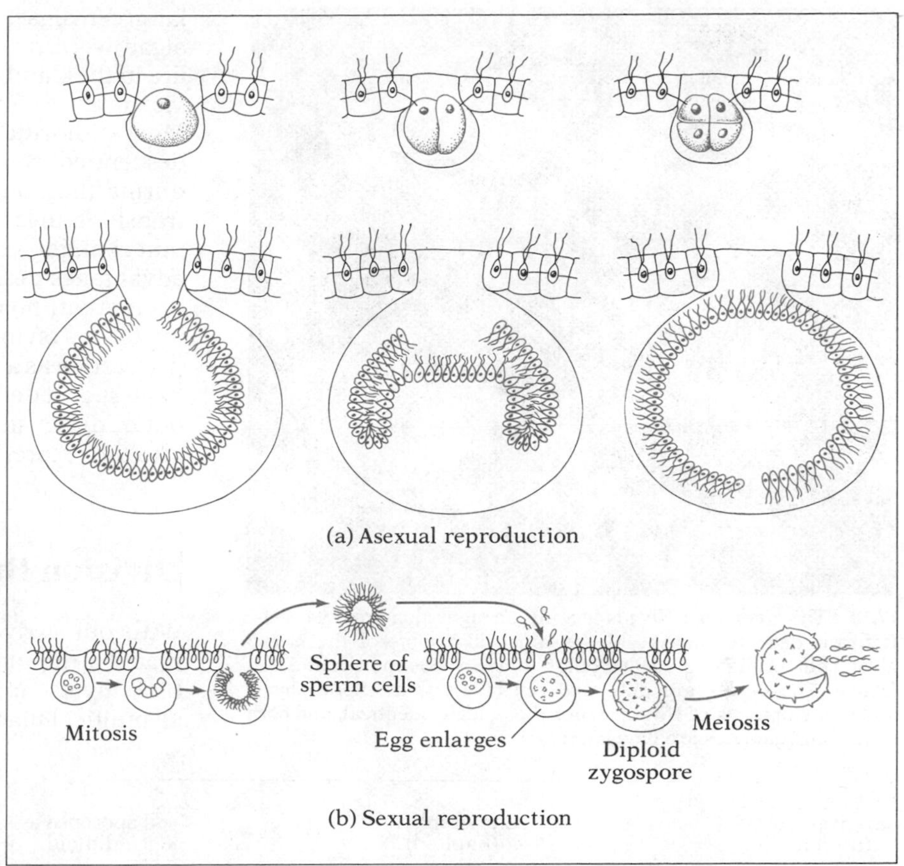

17.9 The siphonous algae, *Acetabularia* and *Codium*. (a) *Acetabularia*, romantically called the "mermaid's wineglass," is a most unusual organism, seemingly having gone off in its own evolutionary direction. This alga is actually "single-celled." *Acetabularia* grow as long as 9 cm, yet have only one (compound) nucleus throughout most of their life. The nucleus is found at the base of the stalk. These delicate organisms are found in warm, tropical waters, and can be seen in this continent along the Gulf of Mexico, Florida Bay, and in the waters of the overseas highway from the mainland to Key West. (b) *Codium* is not clearly multicellular, but is like a fungus in that cell walls are not formed after mitosis. In effect, its cytoplasm is continuous and multinucleated. *Codium* produces a branching fingerlike growth, from which it gets its common name, "dead man's fingers." It can often be found growing on the shells of mollusks.

(a)

(b)

17.10 The green alga *Ulva* is a common sight along rocky coasts. It forms delicate, leaflike blades much like some of the brown algae, but brilliant green in color, thus its common name, *sea lettuce.* The blades are only two cell layers thick. Sporophytes and gametophytes of *Ulva* are morphologically identical, and both spores and gametes are flagellated.

lar green algae, but it is a reasonable guess. The green algae were probably the dominant form of aquatic life in the Cambrian period and throughout the Ordovician and Silurian, too (600–425 million years ago). Most authorities believe that the first terrestrial plants descended from the filamentous green algae, possibly during the Silurian or Lower Devonian period. The transition to land involved many intricate and important changes—the brown algae never made it—but the advantages to the first successful invaders must have been great: no competition; no predation; and nothing to step on them, mow them, or spray them. But the dry earth was a dangerous place. These pioneers must have succeeded at first only in moist places and only out of direct sunlight. Sagebrush was still a long way into the future.

Division Bryophyta

With our arrival at the bryophytes we have surely reached the plant realm by anyone's taxonomy. The bryophytes—mosses, liverworts, and hornworts—are all multicellular plants with well-differentiated tissues

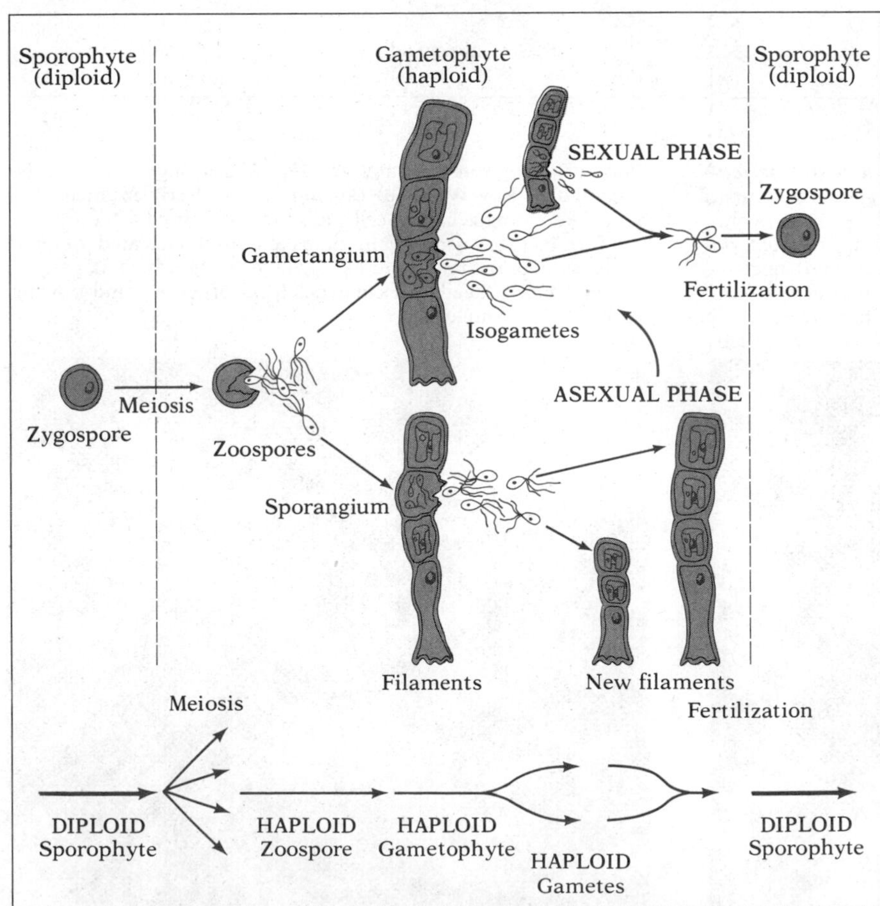

17.11 Alternation of generations in *Ulothrix.* Like *Chlamydomonas, Ulothrix* has an extremely abrupt sporophyte generation. The diploid zygospore enters meiosis, producing numerous zoospores. In an asexual phase, the zoospores produce filaments, which, in turn, develop sporangia and more motile, tetraflagellate spores, which settle to begin the growth of new filaments. In the sexual phase, some filaments develop gametangia, which produce biflagellate isogametes. Different mating types fuse to produce diploid zygospores once again.

that efficiently divide the tasks of life. Although all these species are terrestrial, they retain reproductive features that resemble those found in aquatic plants.

Bryophytes are readily distinguished from the more advanced plants in that they lack specialized vascular (circulatory) and supporting tissue. Thus, there are no tall mosses; they lie limply hugging the earth. Nevertheless, bryophytes have been very successful, if somewhat restricted, in their invasion of the land. There are, according to some estimates, as many as 23,000 named species of bryophytes, most of them mosses. The other bryophytes, the liverworts and hornworts, are much less common. You probably can't recall ever having seen a hornwort.

Evolutionary Relationships

Since the bryophytes represent a successful transition to life out of the water, we'll digress a bit to explore this major evolutionary step. This is a crucial transition and it seems that the bryophytes were the first to make it.

Whatever the situation, the movement onto land must have been extremely advantageous. The statement is really a truism since the survivors' progeny are with us—in fact, they *include* us. But since the land environment presented so many problems to its colonizers, things must have been pretty tough and competitive in the water in order for natural selection to favor the few stalwarts that first ventured into the uncomfortable terrestrial realm. In this new place they must have indeed been initially ill-adapted. Fish out of water, so to speak.

Consider the problems faced by an aquatic organisms on land. First, desiccation: How is one to keep from drying up? The skin covering must either be kept moist at all times or become waterproof—or, in some cases, the organism has to learn to live with periodic desiccation. Next, what about support? Buoyancy provides a great deal of support for aquatic organisms, but on land one realizes the gravity of the situation. Also, water is easy to travel through and it permits sperm and eggs to join each other with few problems. Water provides oxygen and minerals and washes away wastes. On land these advantages are lost. Nonetheless, they came ashore.

There is some disagreement about whether organisms made the changes gradually or abruptly, but most evidence suggests that they were probably gradual. Many primarily aquatic groups have members capable of living temporarily out of the water, or members capable of subsisting in the moist and damp places that are probably similar to the initial sites of invasion by land colonizers. However, as soon as these niches were filled, the pressure of competition increased again. Selection favored new variants that could venture into even drier areas.

Today's bryophytes are the descendants of plants that successfully made the transition. However, the bryophytes have clung to their ancestral reproductive mode. They have flagellated sperm that require at least a film of water to swim in. So, in spite of successfully invading the land, the bryophytes have simply not developed a means of getting the sperm to the egg without water. They have, however, evolved the appropriate structures for embryological development on land, including a protective shield (the *archegonium*), which surrounds the embryo and protects it from drying out.

Since the bryophyte is multicellular, it can devote various tissues to different tasks such as water absorption, anchorage, foliage support, and photosynthesis. But keep in mind that we are not referring to roots, stems, and leaves. Those are reserved for the vascular plants. However, the bryophytes produce structures that carry out the same functions. In the bryophytes, the analogue of the root is an elongated thread of cells that emerges from the thallus as a *rhizoid* (Figure 17.12). It anchors the plant in the soil and absorbs water and minerals. The "stem" is a complex cylinder of many cell layers, surrounded by a photosynthetic *epidermis* (skin). The epidermis surrounds a region of *cortical* (outer) *cells*, which in turn contains a *central cylinder* composed of tough, thick-walled cells, some of which vaguely resemble the supporting and conducting cells of more advanced vascular plants. Whether the central cylinder actually conducts, or whether it plays a simpler structural, supporting role, is not clear, and botanists disagree. Transport in bryophytes may be simply cell-to-cell diffusion with no specialized tissue involvement.

The "leaves" (actually *leaf scales*) make up most of the thallus; they lack the complexity of real leaves, but they serve the same function. Each leaf scale is composed of a flattened, ribbonlike array of photosynthetic cells. The epidermis (outer cell layer) is coated with *cutin*, a moisture barrier, in some species. The leaf scales in these bryophytes contain numerous pores similar to those of vascular plants.

How do the bryophytes fit into the evolution of vascular plants? The consensus among botanists is, they don't! It is believed that both bryophytes and vascular plants arose from common algal ancestors, each going its separate way. Most bryophytes became specialized in inhabiting the semiaquatic or moist environments, while vascular plants were unusually successful in drier places. In the alternation of generations, the gametophyte has become dominant in the bryophytes, and the reduced sporophyte is dependent upon it. In tracheophytes, as we shall see, the direction of evolution has been the reverse, with the progressive reduction of the gametophyte. Vascular plants, of course, have proved to be much more successful, as witnessed by the fact that they are all

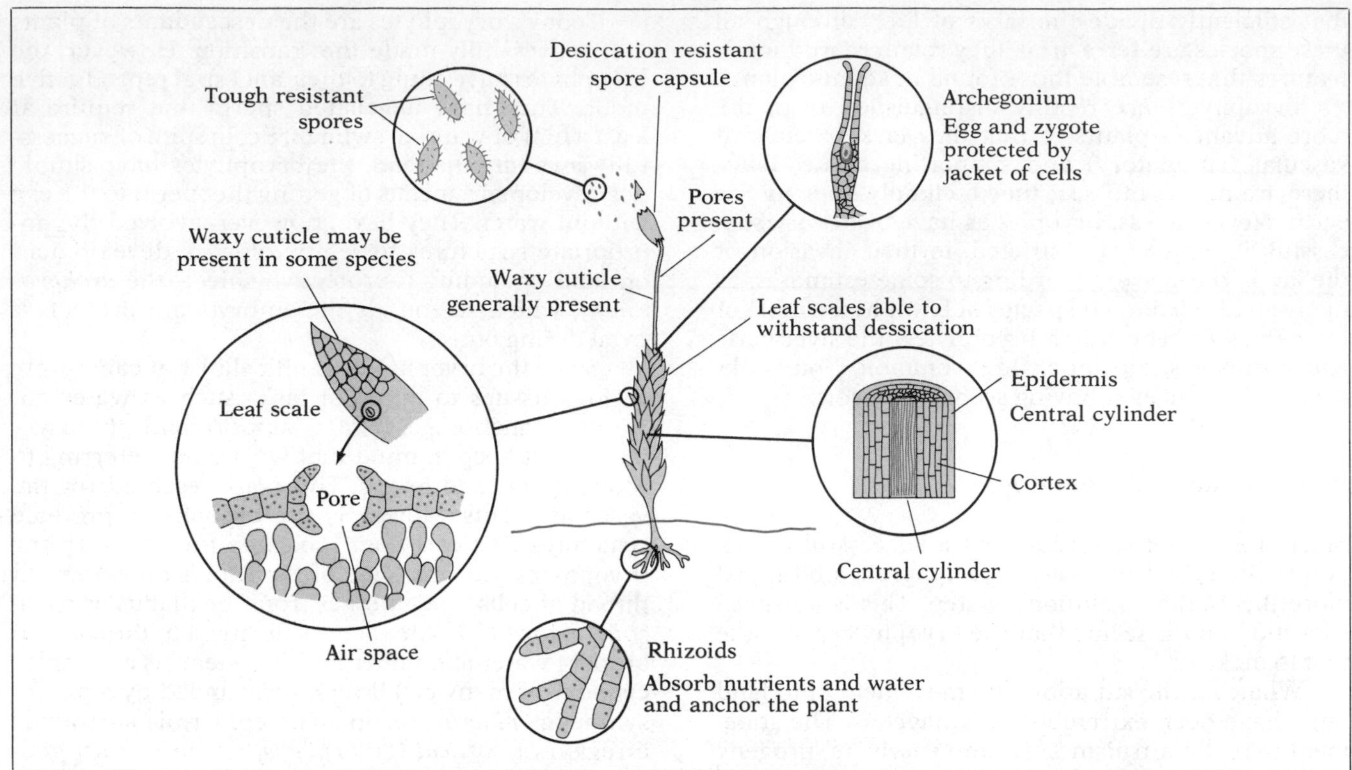

17.12 Bryophyte adaptations to land life. The major problems of adaptation to terrestrial life concern desiccation. The bryophytes have solved this difficult problem in a number of ways, as shown here. The multicellular condition permits many regions of the plant to specialize. Specialized structures, the *rhizoids*, penetrate the soil, absorbing water and minerals and holding the plant in place. Above ground, the leafy gametophyte exposes its chloroplasts to the sunlight. Most (but not all) moss gametophytes lack the kind of epidermal waterproof coating found in higher plants, and mosses are subject to desiccation. They solve this problem by being able to dry out almost completely, ceasing all activity, only to spring back to life with the next rain or dewfall. The small diploid sporophyte, however, usually has a waterproof coating that resists desiccation. The problem of admitting air to the photosynthesizing cells is solved by controlling its entrance through numerous pores surrounded by *guard cells.* Further, the embryo is protected from the drying atmosphere by development in the *archegonium,* surrounded by cells of the gametophyte. Although specialized food and water conducting cells have not evolved in the bryophytes, a central cylinder is seen. It provides support and may function in conducting water. When spores are formed, these are protected by their hardened and highly resistant cases. One problem they haven't solved is that of dry transport of sperm to egg. Bryophytes are dependent upon rainwater or dew in this vital aspect of their lives.

around us. The byrophytes of today are from ancient stock, having changed little since their origin. Our best guess is that the common ancestor of bryophytes and vascular plants resembled the green alga, *Fritschiella* (Figure 17.13).

The bryophytes are important parts of certain food chains. In addition, they play a major role in the natural aging of lakes and ponds. They grow in great masses in the shallow areas, gradually filling in the body of water and finally choking it out of existence.

Class Musci (Mosses)

The *mosses* are by far the most conspicuous members of the division. Like the other bryophytes, they grow best in moist conditions and are found everywhere growing in the soil, rocks, and trees. Their distribution is cosmopolitan and they thrive worldwide, wherever conditions permit. Sphagnum (peat moss) is used as a mulching or bedding material for lawns and decorative plants (and is also burned to provide the smoky flavor of Scotch whisky).

The mosses, like all the bryophytes, show a clear alternation of generations, with the gametophyte being dominant. Most of the structures described here are found in the gametophyte generation. The velvety green carpet we think of as "moss" is probably the haploid gametophyte generation (Figure 17.14). The gametophyte is leafy, often quite large, and may bear both male and female sexual structures. At the gametophyte's base are extensive rhizoids. The sporophyte in most species takes the form of a capsule (*sporangium*) borne on a lengthy stalk that grows out of the thallus (body) of the gametophyte. Within the sporangium, cells undergo meiosis, and the resulting haploid cells then multiply through mitosis, producing numerous spores. When the capsule is mature and filled with spores, its cap falls away and the tiny spores are scattered to the winds. The likelihood of landing in a suitable place is not very good, but the low odds are defeated by the great number of spores.

Once a haploid spore has established itself in a suitable place, it will absorb water and sprout into a threadlike structure, the *protonema*. The young protonema is very similar to certain filamentous green algae. If conditions are favorable, the protonema will bud in several places, finally producing mature gametophytes of separate sexes. Gametophytes will bear sperm-producing structures called *antheridia* and egg-producing structures called *archegonia*. Some species are *monoecious* (producing both egg and sperm in the same thallus), but the majority of species are *dioecious* (separate sexes, producing only one type of gamete within a given thallus).

When sufficient water is available, sperm emerge from the antheridium of one gametophyte and swim into the archegonium of another, where fertilization occurs. The event ushers in the diploid sporophyte generation, with the sporophyte growing as a parasite from the female gametophyte. Meiosis occurs in a new sporangium and, once again, haploid spores are produced.

Mosses—the dominant moss gametophytes, at least—do not resist desiccation in the way other land plants and most land animals do. When their habitat becomes dry, the mosses also dry out, and all biochemical activities ceases. Dry moss is not dead, however. With the next rainfall or heavy dew, the plant "comes to life" and resumes where it left off.

Interestingly, one kind of terrestrial animal does the same trick. *Tardigrades*, which are microscopic and somewhat obscure arthropods, are also capable of being desiccated (sometimes for years) and of then rehydrating successfully. *Tardigrades* are found in only one habitat: the surface of mosses.

Class Hepaticae (Liverworts)

Liverworts may sound like they were invented by J. R. Tolkien, but actually they are a fairly common if inconspicuous group of plants. They produce a rather simple leafy gametophyte which tends to be flat, re-

17.13 *Fritschiella* and similar green algae may be the ancestor of vascular plants and bryophytes. This illustration reveals certain aspects of its growth and form that make it an intriguing possibility. Note the branching filaments and the ability of cells to change their spindle orientation as division occurs. The inset shows the manner in which cells divide at the filament tips and further back. Changing the plane of division allows the filament to be thicker than it would be otherwise. In other words, a greater number of cells for specialization may be produced, and the structure is no longer strictly filamentous. This mode of growth is suggestive of how multicellarity might have started. In its earliest stages of growth, the moss gametophyte forms a *protonema* ("original thread") that greatly resembles mature *Fritschiella*.

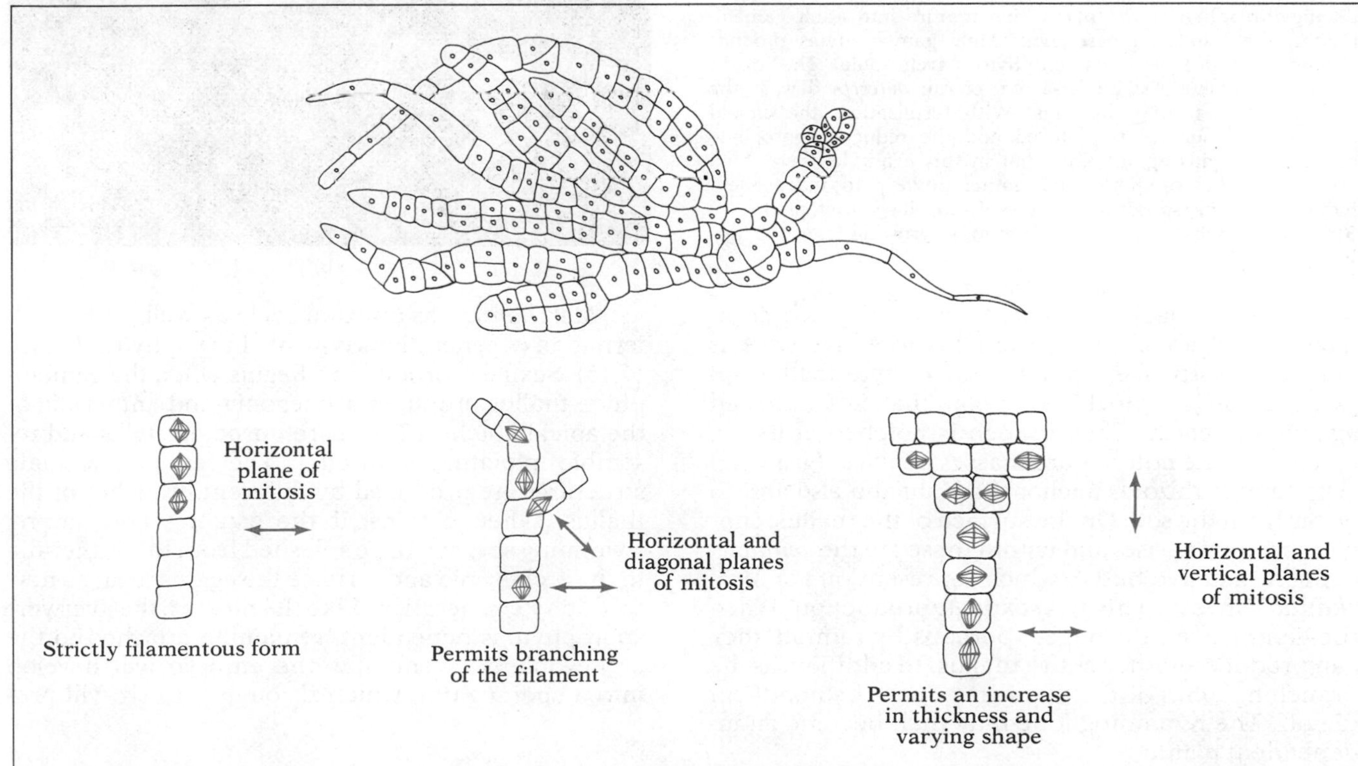

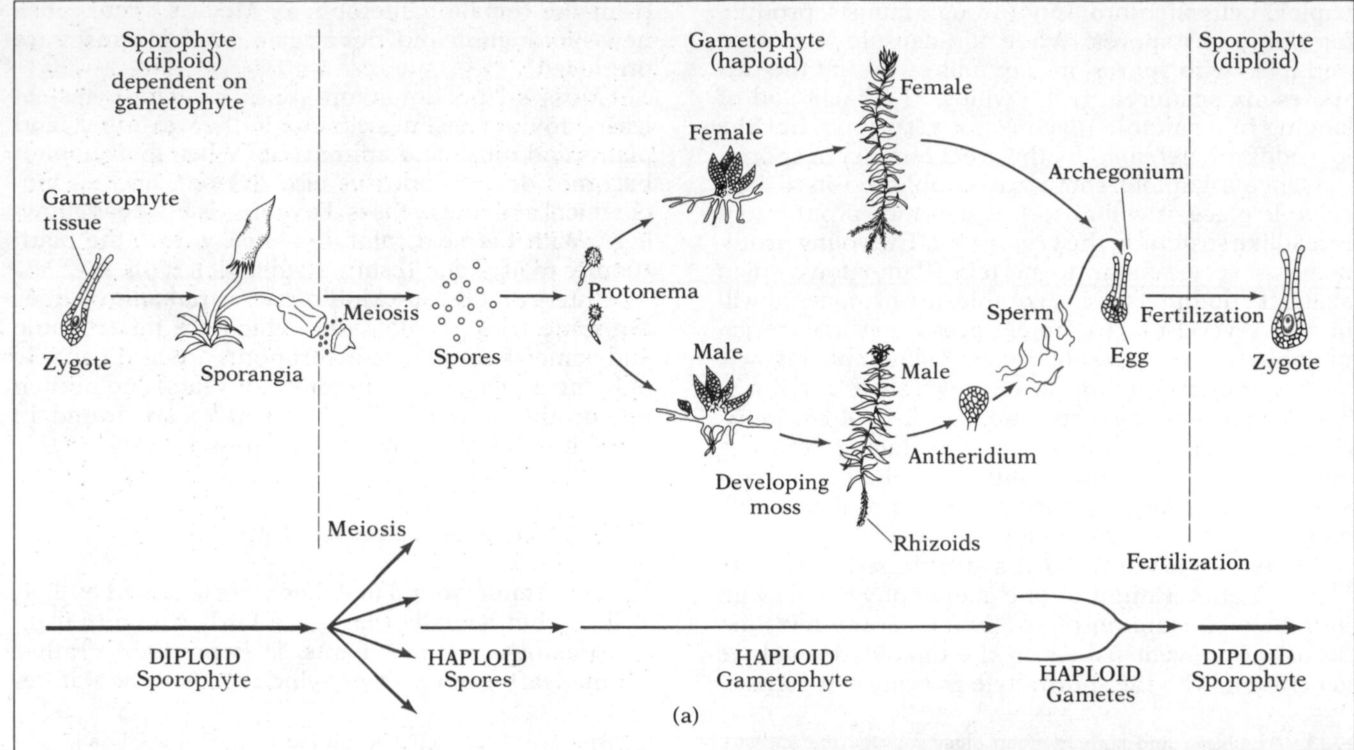

17.14 Alternation of generations in mosses. (a) Mosses typically have prominent gametophyte generations. The sporophyte is usually nonphotosynthetic, consisting of a stalk growing out of the gametophyte and topped by a sporangium where meiosis and spore formation occur. With meiosis and spore formation, the haploid gametophyte generation begins. Spores germinate, producing threadlike *protonema,* which mature into adult gametophytes, the familiar green plant. Male gametophytes produce antheridia and female gametophytes, archegonia. The sperm from the antheridia swim, in a film of rainwater or dew, to the archegonia, fertilizing the eggs. With fertilization, the diploid chromosome number is restored and the reduced sporophyte generation begins again. Note that in this example, there is no separation of sporophyte and gametophyte into independent individuals. The sporophyte grows from the gametophyte. (b) Stalked sporophytes of *Polytrichum* moss growing from the lowlying gametophytes.

sembling (for lack of a better analogy) green cornflakes. Just about everyone's favorite liverwort is *Marchantia.* Its deep green gametophyte thallus appears to be a branching ribbon that lies flattened against the earth. Each branch is notched at its tip, and within the notches are masses of *apical* (growing) cells. Simple rhizoids anchor the plant and also absorb water from the soil. On the surface of the thallus, cuplike structures arise, and within these are the *gemmae,* (small vegetative buds), which represent one of *Marchantia's* investments in asexual reproduction. When the gemmae are dislodged, perhaps by rainfall, they can produce another entire thallus. In addition, as the branching lobes of the thalli grow, cells behind them die off. The remaining lobe may then become an independent plant.

Marchantia has a sexual cycle as well, with an alternation of generations typical of bryophytes (Figure 17.15). Sexual reproduction begins when the gametophyte thallus produces archegonia and antheridia at the apical notches. These are borne on stalks and resemble miniature umbrellas. The female and male structures are produced by different branches of the thallus. When it rains, if the gametes are mature, swimming sperm will be splashed from the antheridia to the archegonia and fertilize the eggs, starting a new sporophyte generation. Like the mosses, the liverwort sporophyte is dependent, remaining attached to the archegonium. Eventually, the embryo will develop into a sporangium, which through meiosis will produce numerous haploid spores, ending the brief diploid sporophyte generation. Each spore, under suitable conditions, can produce the ribbonlike gametophyte thallus.

Class Anthocerotae (Hornworts)

The *hornworts* are a relatively minor group, with only 300 or so named species. They are structurally simpler than the liverworts. The gametophyte is a small, slightly lobed growth. Antheridia and archegonia are all produced directly in the thallus (Figure 17.16). Fertilized egg cells develop into an unusual diploid sporophyte, from which the name *hornwort* was derived. The sporophytes are long, slender growths that separate into two spore-producing "horns." Following meiosis, the haploid spores are carried away to begin a new gametophyte. The sporophyte is unusual in two other aspects of its growth. First, it can grow continuously, producing spores over a long period of time. This is made possible by a *meristematic* area of actively-dividing cells at its base. Second, under exceptional conditions, it can survive on its own, should the supporting gametophyte die off. It does this by producing its own water-conducting rhizoids. This

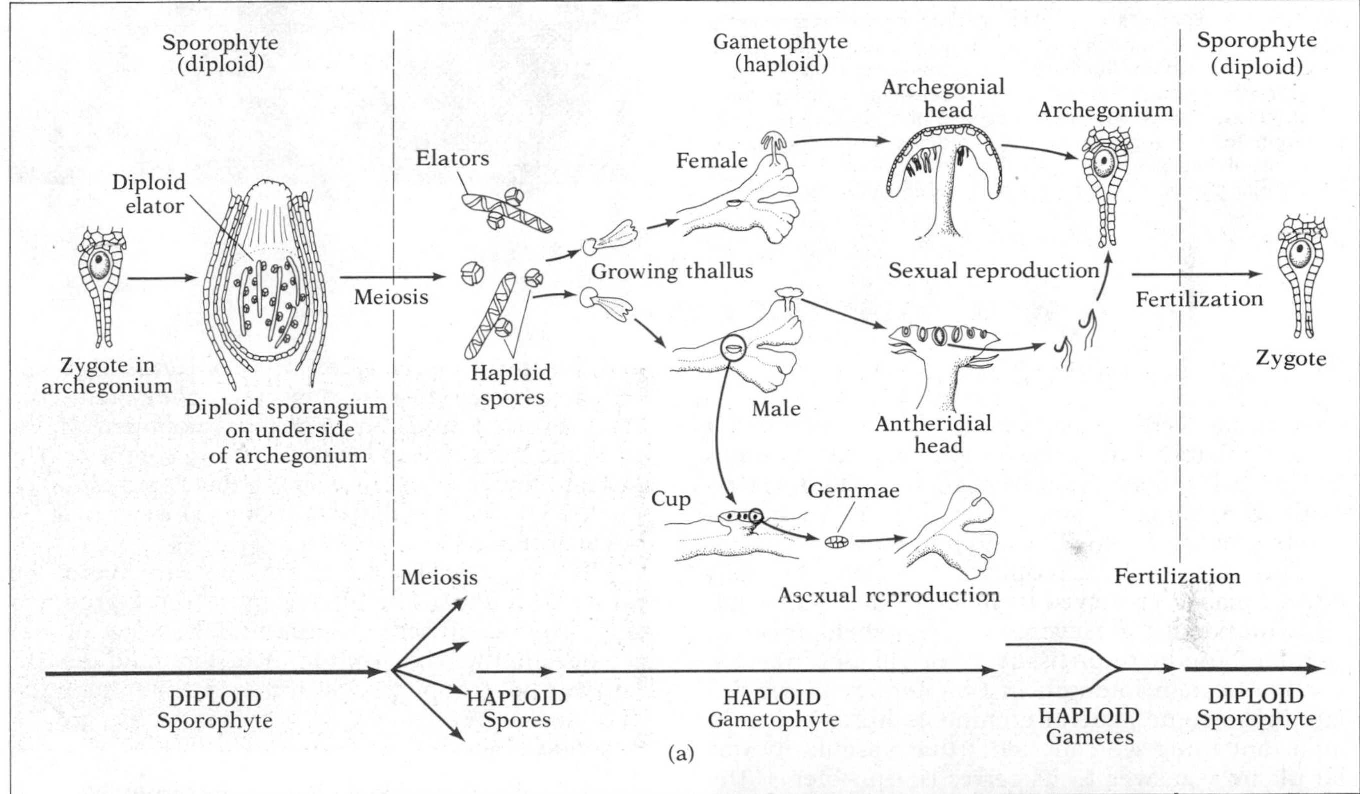

(b)

17.15 Alternation of generations in *Marchantia,* a liverwort. (a) The brief diploid sporophyte generation in *Marchantia* includes the zygote, which develops into a sporangium. This is housed in the archegonium of the gametophyte. When meiosis and spore formation occur, the sporophyte generation ends. With maturation, haploid spores are propelled from the sporangium by springlike *elators,* and germinate to produce the gametophyte thallus. *Marchantia* frequently undergoes an asexual cycle, producing new thalli from reproductive buds known as *gemmae.* Each gemma, when freed from its cup, can produce a regular thallus. Sexual reproduction involves the development of stalked *archegonial heads* (with archegonia on the underside) and *antheridial heads* (with antheridia on the top side), on separate thalli. Sperm from mature antheridia are splashed by rain to a neighboring archegonium and fertilize the egg cell within. This event begins the diploid sporophyte generation once again. (b) Liverworts. Archegonial heads, bearing female reproductive cells, look like small palms (lower left). Flattened antheridial heads are at the lower right.

may appear to be a minor point—a peculiarity of the hornwort—but it also suggests another, more important point. The hornwort is the only bryophyte that produces a semiindependent sporophyte. Yet all the higher (vascular) plants have independent sporophytes, which, as you will see, are the dominant part of their life cycles. Perhaps the hornwort sporophyte is similar to an evolutionary stage in the development of higher plants.

17.16 In *Anthoceros,* a hornwort bryophyte, the gametophyte is apparently dominant, but the sporophyte grows on a conspicuous stalk above the gametophyte thallus, as seen here. The sporophyte does not form the spore-producing cap found in the mosses. Instead, spores are continuously formed from an active cellular region below. Spores are released as they mature in the upper region. The life history of *Anthoceros* is unusual, also, in that the sporophyte can become independent under certain conditions. The sporophyte contains photosynthetic cells, and it may produce a footlike base with absorbing tissue. Upon the decay of the gametophyte, the sporophyte grows independently for a time. This unusual divergence places the hornwort closer, in one way, to the vascular plants, which always have independent sporophytes.

17.2 The Rise of Vascular Plants

Division Tracheophyta

If someone were to ask you to draw a plant, you would probably draw a tracheophyte. And you probably know far more about tracheophytes than you do about other kinds of plants. After all, trees are tracheophytes, and so are flowers and potatoes.

The division Tracheophyta (*tracheo-*, "tubes"; *-phyta,* "plants") received its name from its outstanding evolutionary achievement, the development of *vascular* (conducting) tissue. We will describe the vascular system in-depth in Chapter 18, so we will only make some general comments here. The most important thing we can say is that vascularity was the planet's answer to its terrestrial problems. The transition to terrestrial life was dependent on getting water up from the soil to the photosynthetic tissues. The theory is that, as plants ventured further and further from their ancestral environment, the sources of water became more remote from the leaf. The water accumulated in the soil, but the leaves had to continue to stretch skyward to capture the sun's energy. Root systems penetrated deeper and deeper to find soil water and the leaves grew taller or broader in response to increasing competition from other vascular plants.

Today, vascular plants inhabit nearly every part of the planet that protrudes above water, and a few that don't. As they radiated out over the landscape, they adapted to their environments in different ways and thus produced a wide variety of forms. Not only did they develop leaves, stems, and root systems, but some evolved *seeds.* With the development of the seed, the transition to land was nearly complete. The delicate finishing touch was the development of an unflagellated sperm that does not need water in order to get around.

The division Tracheophyta contains five major groups of plants. The phylogeny of these groups is somewhat confused at present, so some of our organization will be arbitrary. Keep in mind that the names and groupings will differ in different reference books. But here we will say that the five groups, or subdivisions are:

Psilophyta ("naked plants"); psilotum
Lycophyta ("spiderlike plants"); ground pine or club mosses
Sphenophyta ("wedge plants"); horse tails
Pterophyta ("winged plants"); ferns
Spermophyta ("seed plants"); conifers and flowering plants

The first three were once an important part of the earth's foliage, but their heyday is over, and they are now relatively minor groups. However, we will consider them anyway, because they are interesting and because they may represent steps in the evolution of the seed plants. Of the estimated 261,000 named species of vascular plants, the first three groups comprise only about 1000 species, and, of these, all but 28 are found in the subdivision Lycophyta. So, we will be

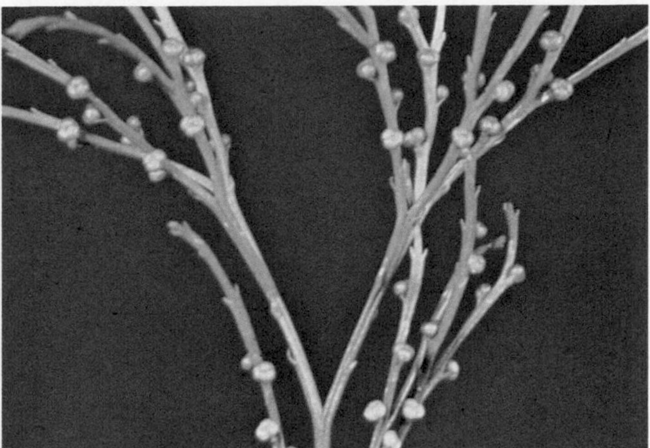

17.17 *Psilotum,* one of the more primitive vascular plants, produces a primitive vascular system in its stem. It grows erect, with a highly branched stem, which is quite an improvement over the ground-hugging bryophytes. The foliage seen here is the sporophyte generation of *Psilotum.* Spores are produced meiotically in the scattered sporangia. The haploid gametophyte that appears following the germination of the spore consists of a mass of tissue bearing both antheridia and archegonia. The flagellated sperm cells require water for transfer to the archegonia. When fertilization occurs, the zygote grows out of the gametophyte thallus into a mature sporophyte again. This mode of development is not far removed from that of bryophytes, but the mature sporophyte routinely becomes independent of its gametophyte parent.

looking at a very few survivors of another time, before the ferns and then the seed plants became the dominant forms of plant life on earth.

Subdivision Psilophyta

This remarkable persistent group of plants is represented today by two genera and only four species, found mainly in the tropics. The genus *Psilotum* (Figure 17.17) produces hairy rhizoids similar to those of the bryophytes, but no true roots. The stems are fairly well-developed, with simple, primitive vascular tissue. The stems bear primitive sporangia and are highly branched with tiny scalelike leaves. Both stem and leaf scales are photosynthetic and the sporophyte generation is dominant.

The psilophytes are well-represented in the fossil record of 300 million years ago. They may represent an early stage in the evolution of vascular plants, but we can't be sure because the fossils of other, more advanced plants date even further back. We have no way of knowing whether still older fossils of the suspected ancestral types exist. This is a common kind of problem when one is trying to figure out phylogeny on the basis of fossil records. Structurally, the psilophytes are a good candidate to represent the transition of green algae to terrestrial vascular plants.

Subdivision Lycophyta (Club Mosses)

We will concentrate on one genus, *Lycopodium* (Figure 17.18). You may have seen *Lycopodium* without knowing it, since they are often mistaken for pine seedlings and in some parts of the United States they are commonly known as *ground pine* or *club moss.* If you live in the eastern or northwestern states, you may have gathered them to make Christmas wreaths. The custom was once considered quaint but now it has endangered the plant and is illegal in some states.

How does one recognize a *Lycopodium?* First, it has true roots, stems, and leaves, but that doesn't narrow it down much. So let's take a closer look

17.18 Lycophytes, known as club mosses, are not mosses at all but vascular plants with true roots, stems, and leaves. Its roots emerge from numerous leafy horizontal stems that lie on the soil surface. The upright stems are sent up to produce and release spores from terminal strobili. Most lycophytes live in moist regions where water is available for sperm transport. A few have adapted to drier regions. An example is the so-called resurrection plant (*Salaginella*) whose range extends from Texas to Peru. During droughts, this lycophyte forms a brown mass of apparently dead tissue. When moisture returns, however, the resurrection plant uncoils its stems, producing a luxuriant green foliage.

at the roots. In the adult plant, they are only of the *adventitious* type; that is, they emerge from the stem. As we will see, this is not the case in higher plants.

The vascular system of *Lycopodium* is well-developed, extending from the adventitious roots through stem and leaves. The leaves, however, are small and usually arranged in a whorl around the stem. Some species produce *runner stems,* which eventually form a mat over the forest floor, with occasional roots and vertical stems springing from the runners.

The lycophyte forms spores in a terminal cone-like region known as the *strobilus.* As its generations alternate, the haploid spores germinate into minute gametophytes. Fertilization occurs when there is sufficient water over the gametophyte for the sperm to swim to the archegonium. The diploid sporophyte then emerges from the gametophyte to repeat the cycle.

The lycophytes were not always the humble, ground-hugging plant we see today. There are fossils of one order, the Lepidodendrales, which produced great treelike plants over 50 m in height and 2 m in diameter. In fact, the lycophytes were the dominant plants of the Devonian and Carboniferous forests (Figure 17.19). Their ghosts now haunt us—the corpses of these great plants became partly decomposed to form coal, gas, and crude oil, affecting modern global politics as we squabble over their remains.

While today's lycophytes are not seed plants, some of the extinct forms appear to represent the earliest seed producers.

The fossil record once more baffles us as we attempt to sift out phylogenetic relationships among the tracheophytes. In addition to the dominant lycophytes, the Paleozoic forests included sphenophytes (our next subdivision), ferns, seed ferns, and gymnosperms—some of these are definitely seed plants. So how do they all fit into an evolutionary pattern? We are left with the impression that there may have been several lines of evolution from the ancestral green algae. If we accept this idea, then the first three or four subdivisions of vascular plants are *cousins to the seed plants,* rather than their ancestors. Following this idea, the differences among the vascular plant groups of today may only represent various adaptations to niches that have persisted throughout time. Therefore, the so-called primitive survivors are few in number, compared to the so-called advanced seed plants only because their most favorable niches are fewer in number today. Actually, where the habitat is appropriate, the "primitive" vascular plants manage to survive well enough alongside the seed plants. Should the earth return to the conditions of the Paleozoic era, who knows what would become the dominant plant? Perhaps the hardy seed plants would become the minority again. Would some future biologist then consider them primitive, and the spore-forming, vascular plants, with their sperm and eggs, to be "advanced"?

17.19 A Paleozoic forest. The earliest forests became established in the Devonian period of the Paleozoic era. By the Carboniferous and Permian periods, huge forests of lycophytes, psilophytes, sphenophytes, and primitive seed ferns flourished on the earth. In this scene, some of the giants of these now insignificant plant groups are seen. Included are giant lycopsids, with five leaves and cones (upper left), seed ferns (left) and sphenopsids with whorled branches (right). The luxurious growth seen here is now represented on the earth by coal deposits, pools of crude oil, and natural gas far below the earth's surface.

Subdivision Sphenophyta (Horsetails)

This once-thriving group of plants is now represented by the sole genus *Equisetum* (Figure 17.20). All plants in this genus have leafless vegetative (nonreproductive) shoots with whorls of short, lateral branches at widely spaced nodes, and rather tall, straight, hollow reproductive shoots, which give rise to tiny scalelike, nonphotosynthetic leaves in circular intervals around the stem. Since the leaves are so reduced, the stems themselves are the principal photosynthetic organs. Vertical strands of vascular tissue are located around the hollow centers of both the vegetative and reproductive stems. In the primitive fashion, roots arise directly from stem nodes.

Equisetum (*equi-,* "horse"; *-setum,* "bristle") is a spore former, with a terminal strobilus on each repro-

17.20 *Equisetum,* the horsetail, represents another remnant from the Paleozoic era. Its growth often produces two types of shoots: one that is vegetative with whorls of photosynthetic branches, and another that is hollow, unbranched and bears a spore-producing strobilus at its tip (called a reproductive shoot or a fertile branch). The reproductive shoot has tiny scalelike leaves at regular nodes along its stem; the nonreproductive shoot has side branches but no leaves. Both belong to the dominant sporophyte generation. Spores are produced meiotically and when they germinate give rise to miniature male and female gametophytes, which produce antheridia and archegonia, respectively. Upon fertilization, the diploid sporophyte generation begins again, growing from the tiny, degenerating archegonium and quickly putting out its own roots.

ductive stem. The sporophyte is dominant and grows to be about 1 m tall, whereas the gametophyte is no larger than the head of a pin. In this species, also, the sperm must swim through a watery film to reach the egg.

You have probably seen sphenopsids growing as decorative plants. Commonly known as *horsetails,* they grow in dense matted beds in yards and gardens, but their natural habitat is along streams and in marshes of both tropical and temperate regions. So the descendants of those vast and unruly Paleozoic forests now tamely decorate suburban gardens.

Subdivision Pterophyta (Ferns)

Ferns are undoubtedly among the most enchanting of plants. The spring forest, dripping with cool rain, is accented by delicate ferns rising with tiny bowed heads from the damp floor. Later, the plants will lend an exotic touch to the woods as they stand full grown, their leaves splayed, as if they had been placed there as a decoration.

Ferns are interesting botanically because they have survived in great numbers and have adapted more easily to changing environments than have the other primitive vascular plants. Some 11,000 named species of ferns cover the globe; some even live in parched and arid regions, although they grow best under moist, shaded conditions (Figure 17.21).

The fern sporophyte (Figure 17.22) typically consists of a thick underground rhizome from which arise many fine roots. Single leaves emerge above ground as the familiar *fiddleheads,* each of which unfolds into a large and often frilly compound leaf (that is, a leaf subdivided into leaflets). The leaves come in all sorts of sizes and designs. All the ferns have well-developed vascular tissue. A few, such as the tree ferns, have tall stems supporting a leafy rosette.

The ferns alternate generations, with a decidedly dominant sporophyte. Spores are usually produced on the underside of the leaves in structures known as *sori,* each of which is actually a group of sporangia, sometimes hidden under a scalelike cover (Figure

17.21 The large ferns shown here are tree ferns. Ferns like these still grow in many regions of the earth, but they favor tropical and subtropical climates. Ferns are still a large and diverse group, perhaps because they have not been diminished by changing geological conditions as were the other primitive vascular plants.

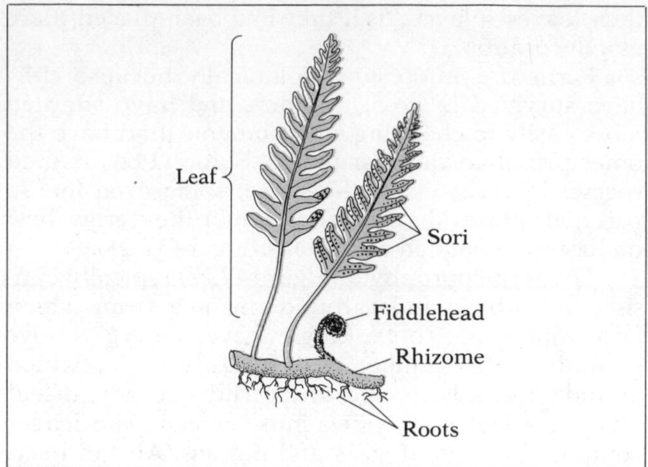

17.22 Anatomy of a fern: the fern thallus. Below the soil, an extensive underground rhizome sends fronds above ground to form the cluster shown here. The growths emerge from the rhizome as highly curled *fiddleheads,* which gradually open into leafy structures. In addition to fronds, the rhizome produces hairy root growths to absorb water and minerals. In many species, the underside of the frond bears *sori,* which are circular structures made up of sporangia.

17.23). The sporangia themselves may be rather complex. You have undoubtedly seen the sori on the underside of ferns. They are little black, red, or white dots in rows, and are often mistaken for an insect or fungal invasion. The mechanism for spore release may be fairly complex and involves the bursting of the *annulus* (the wall of the sporangium).

Figure 17.24 illustrates a typical alternation of generations in the fern. Haploid spores formed by meiosis in the sporangia are released into the breeze. When one settles into a moist, well-protected surface, it germinates into a delicate, heart-shaped gametophyte known as the *prothallium.* This photosynthetic plant produces true roots, and for a time, it lives as a completely independent, if seemingly insignificant, individual. Eventually, however, each prothallium produces antheridia and archegonia where haploid gametes are produced through mitosis. If at least a film of water is available, sperm cells from the antheridia will become active, swim to the archegonia, and fertilize mature egg cells. The diploid zygote, housed and protected for a brief time by the prothallium, develops into the large sporophyte plant.

17.23 Fern sporangia occur in a variety of forms and patterns, but the sorus shown here is common. Each sorus consists of a number of sporangia, commonly covered with a scalelike *indusium.* Spores are produced meiotically in the sporangia, and remain there until the *annulus* of the sporangium splits and spores are ejected. The spores become air-borne, traveling for many miles on even the gentlest of currents. Although we consider seed production to be an advanced way of ensuring the survival of offspring, spore production has many obvious advantages when it comes to spreading the species far and wide.

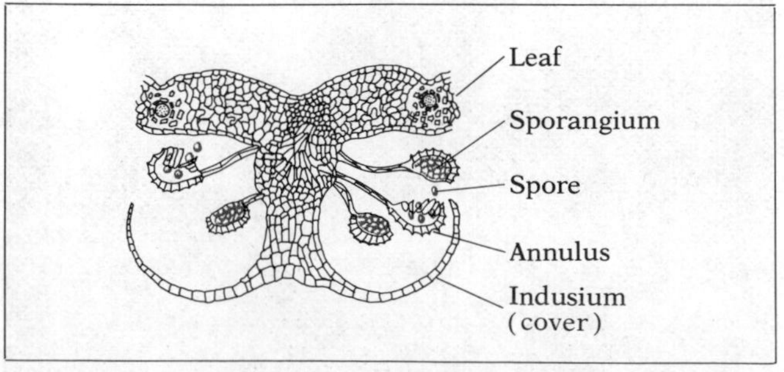

One might wonder whether ferns tend to self-fertilize by using such a system. After all, it usually isn't very far from an antheridium to the nearest archegonium on the wet surface of a single fern gametophyte. Actually, self-fertilization is not uncommon, but cross-fertilization is more desirable from a genetic point of view. Thus many ferns have evolved mechanisms for inhibiting self-fertilization. The first prothallium to mature produce a hormone known as *antheridogen* (which is similar to the gibberellic acid discussed in Chapter 20). This stimulates neighboring prothallia to develop antheridia, while inhibiting their own antheridial development. Thus the hormone's effect is to produce separate male and female prothalli, assuring cross-fertilization. In other species, sperm and eggs from one gametophyte are physiologically incompatible and only fertilizations involving different genotypes are successful.

The ferns were among the dominant plant species of the Paleozoic forest, along with the other primitive vascular plants of the day. Some of those ancient ferns had even developed seeds, but no representatives of these seed-producing groups have survived. The fossil seed ferns are an obvious candidate as an ancestral line to the seed plants, although there is little direct evidence to support the idea.

Subdivision Spermophyta

The primitive vascular plants and their less conspicuous contemporaries, the early seed plants, persisted through the Devonian, Carboniferous, and Permian periods. Toward the close of the Permian period, at the end of the Paleozoic era (about 225 million years ago), the plant life of the earth experienced a mysterious and dramatic change. It seems to be the same sort of drastic change that would mark the end of the dinosaurs some 100 million years later. Powerful geological movements produced a different landscape and climate than the world had ever seen. The earth had been a rather smooth globe and its surface had permitted countless warm, lowland, shallow seas to cover the planet. Rather quickly, though, things changed. Soggy lowlands were uplifted to form vast mountain ranges. The monotonous warm climate gave rise to a general cooling, even as the new up-

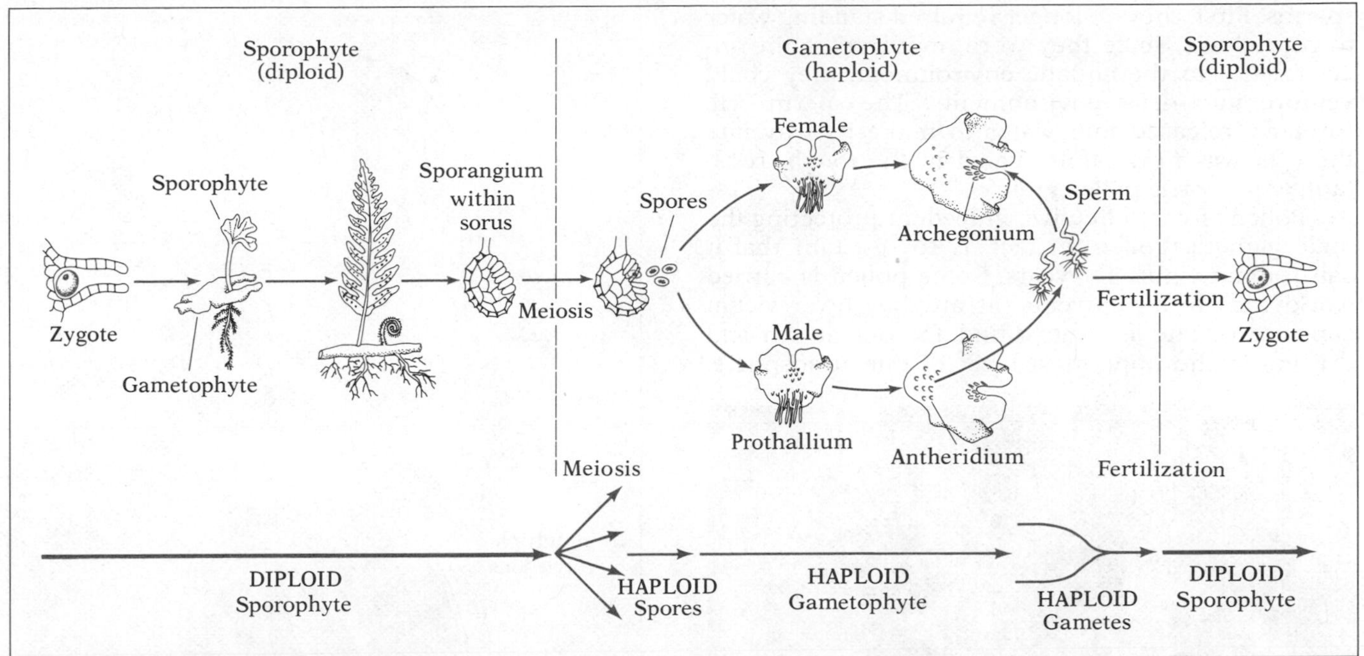

17.24 Alternation of generations in a fern. In ferns, as in other primitive vascular plants, the sporophyte is highly dominant and the gametophyte is a reduced but separate generation. The sporophyte produces spores meiotically in incredible numbers, suggesting a comparatively infinitesimal chance of success for any individual spore. Spores that succeed produce a heart-shaped gametophyte (*prothallium*) which bears antheridia and archegonia, the sperm- and egg-producing structures. When fertilization has been accomplished, the young diploid sporophyte begins its growth from the archegonium, soon producing its own roots and photosynthetic tissue as the prothallium withers and dies.

lands dried. Thus, the primitive giants of the late Paleozoic era perished, leaving only scattered remnants of the once prominent groups.

With the demise of the ancient forests, the competitive edge in the plant world passed to the inconspicuous seed plants. They were better prepared to survive on this new kind of earth. They grew; they changed; they invaded. They appeared everywhere over the new Mesozoic landscape. The earliest of these were the primitive *gymnosperms*, the forerunners of today's evergreens. The secret of their new-found success was their ability to draw water from deep in the dry surface of the earth and to conserve it in their stems and leaves. They also developed a different mode of reproduction—using *pollen* and *seeds*. Although a minute quantity of fluid is required for pollen germination, the sperm no longer had to swim from one plant to the next. The gymnosperms came to dominate their new environment (Figure 17.25).

Let's take a closer look at the drastic reproductive changes that occurred in these early gymno-

17.25 Early Mesozoic landscape. The dawn of the Mesozoic era saw changes in both the land and the plants and animals that inhabited it. There were moist areas, but much of the land at this time was dry and wind-swept, conditions to which gymnosperms are well adapted. The primitive vascular plants gave way to the new seed plants, principally gymnosperms. The great age of reptiles also began. The lycophytes, psilophytes, and the like were pushed aside as climatic conditions changed. The ferns probably persisted, mostly in the remaining moist lowlands. The new arrivals included primitive gymnosperms such as those shown here. The ginkgos and modern conifers were to arrive later. Foliage and seeds from these gymnosperm ancestors undoubtedly provided food for the emerging reptiles.

sperms. First, they no longer required standing water to reproduce. Since they were now free of the ancestral ties to the aquatic environment, they could venture into drier environments. The sperm cell, formerly released into water to make its way into the egg, was now safely sealed in the tough, resistant, wind-borne pollen grain.

Pollen is very effective, indeed, at protecting the male gamete. Pollen, in fact, is so resistant that it can remain viable for years. Some pollen is carried world-wide in air currents (as any hay fever victim can attest), and is even found far out at sea and drifting in the uppermost layer of our atmosphere.

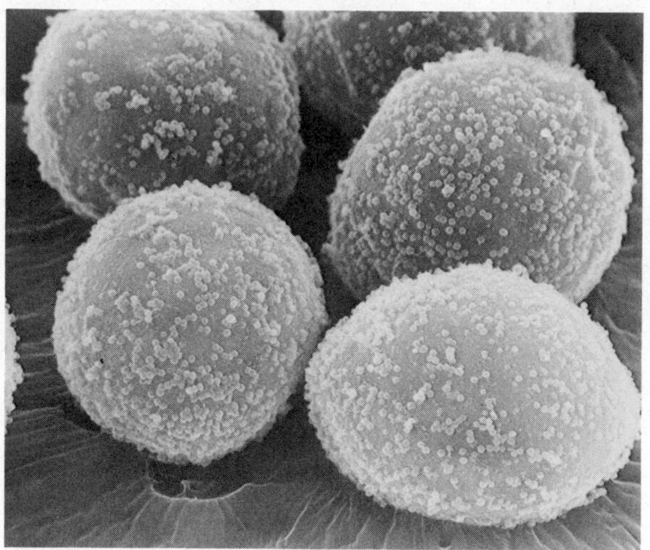

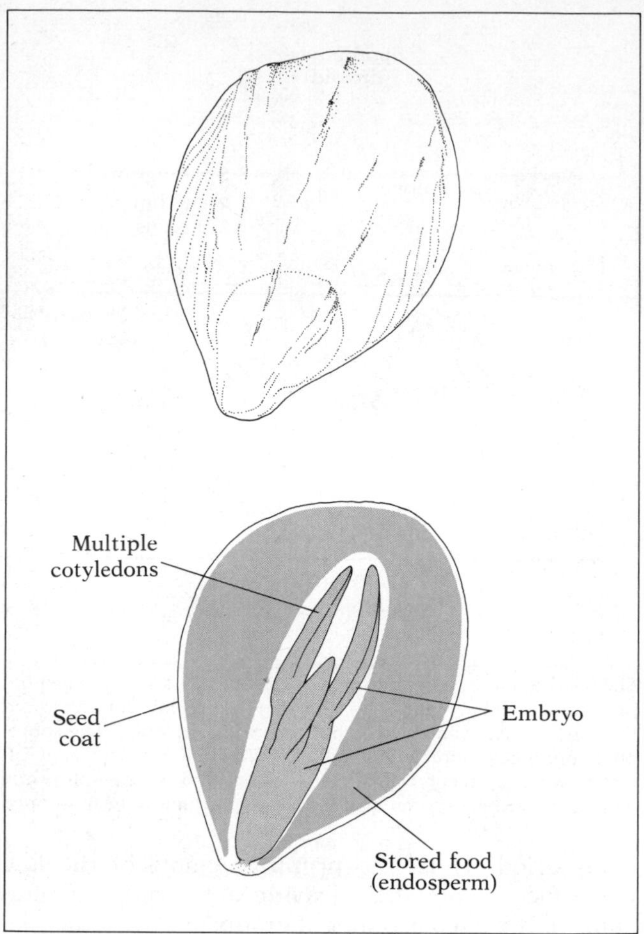

17.26 Two reproductive developments helped the gymnosperms adapt to the new conditions. The pollen grain carried the male gamete safely in a tough resistant case, and the embryo came to be protected in an equally tough container of its own, the seed. The pollen shown here in the SEM photograph is from the northern white cedar. The seed contains the *embryo* of the plant, some *stored food,* and a surrounding *seed coat,* which is hard and water-resistant. The seed of the pine is illustrated here. The protective seed coat makes it possible for the embryo to remain intact for considerable periods of time, until conditions are right for growth.

Thus, the pollen structure is protective and it travels easily.

Incidentally, pollen grains also make excellent fossils. The very specific shape and sculpturing of fossil pollen grains help to reconstruct the nature of the flora of bygone periods. In fact our limited knowledge of flowering plant origins relies heavily on the analysis of fossil pollen from soil and rocks. People who study fossil pollen are called *palynologists.*

A second reproductive development in this harsh new world was the *seed.* A seed consists of an *embryo* and *stored food* enclosed within a hardened *seed coat* (Figure 17.26). Most gymnosperms produce seeds in cones, without a surrounding fleshy fruit. (*Angiosperms,* the flowering seed plants, produce a fruit around the seed.)

As you may recall, seeds did not first appear with the gymnosperms. They were produced by a few ancient lycophytes and some of the ferns. In fact, members of one extinct group, the pteridospermophytes (a rather common Devonian group), produced pollen and seeds in typical gymnosperm fashion.

17.3 The Emergence of Seed Plants

The Gymnosperms

The name *gymnosperm* literally means "naked seed," which refers to plants that have seeds without fruit. The gymnosperms are divided into three taxonomic classes. Two of these, the *Ginkgos* and the Cycads, are not of great importance today. The third class, the Conifers, contains only a few hundred species, but these are present in extraordinary numbers and so the conifers share much of the earth with flowering plants.

The Ginkgo

The one species of *Ginkgo* that still exists is considered to be a living fossil. *Ginkgo biloba,* the *maidenhair tree,* is a native of China, but it is now cultivated in many other parts of the world (Figure 17.27). From its appearance, you would probably think it was an angiosperm, but its life cycle reveals that it is a gymnosperm.

Reproductively, the tree is intermediate between a primitive mode of fertilization and that of the gymnosperms. Male *Ginkgo* trees produce air-borne spores that germinate close to the ovules in female trees. The male gametophyte forms a pollen tube that grows toward the egg cell. On reaching the vicinity of the egg cell, it releases two motile, multiflagellated, swimming sperm cells. But, water is not required for them to get around. Fluids for the very short distance the sperm must travel are produced by the receptive sporophyte itself. Following fertilization, the female *Ginkgo* produces seeds in the manner of a typical gymnosperm. The trees were once known to Europeans only from fossils and were thought to be extinct, but living *Ginkgos* were subsequently found on the grounds of oriental temples. They are now common decorative plants throughout the world, and are often found lining city streets. If you see one, just keep in mind that *Ginkgos* covered great parts of the earth in the Carboniferous period!

The Cycads

The cycads had their day during the late Triassic period of the Mesozoic era, some 200 million years ago. We know from fossils that they were among the most common plants in those ancient forests. They may even have been an important part of the diet of giant reptiles of that era. Today, the hundred or so remaining species are found mainly in tropical regions (Figure 17.28). You may have seen cycads in museums and parks, and you probably thought they were palms. They weren't.

In the Cycads, pollen is produced in conelike sporophylls. Sexes are separate, and pollen is carried by the wind to the ovulate cones. The wind-borne pollen produces sperm cells that are flagellated, representing our last look at this primitive mode of sperm transport in plants.

The Conifers

The conifers (class Coniferinae) include nine families containing only about 300 named species, not a large number in spite of their great populations in the coniferous forests of the world. The most common and best known is Pinaceae, the pines. In addition to the familiar pines, this family includes firs, spruces, hemlocks, douglas firs, junipers, and larches.

The Conifers are the *cone bearers* whose corpses decorate your house at Christmas. Many of them are well-adapted to the cold climates of the earth, but some live in warm areas as well (Figure 17.29). They typically produce needles or scalelike leaves through-

17.27 The *Ginkgo,* to a casual observer, appears to be like almost any dicot angiosperm, broad leaves and all. However, it is a gymnosperm, albeit a strange one. The leaves of the *Ginkgo* are fan-shaped, divided slightly into two lobes. There are two sexes—male and female trees—and the ginkgo retains the flagellated sperm of its fernlike ancestors. The ginkgo, the only survivor of a once dominant order, has been maintained for millennia in Japanese and Chinese temple gardens. It is extinct in the wild. It is remarkably fungus-, smog-, and insect-resistant, and some living ginkgoes are 1000 years old.

17.28 Cycads belong to another ancient gymnosperm group. Their ancient fossil history can be traced to the Mesozoic era, when they were widespread. Today's survivors grow in the tropical and subtropical regions of most continents, and with care they can be grown as decorative plants in harsher climates. Their leaf form is very similar to that of the palms, at least to the casual observer. The sperm, like those of the ginkgo, are flagellated.

out the year, shedding the older leaves as they grow. For this reason, they are often referred to as *evergreens.* The vascular system, though highly developed, lacks certain vessels found in flowering plants.

Conifers typically bear *staminate* (male, pollen-bearing) and *ovulate* (female, ovule-bearing) cones. So a tree is bisexual, but a cone is male or female. Not surprisingly, each species has its own distinctive cones. Figure 17.30 illustrates the life cycle of the pine, a typical conifer.

The importance of conifers to animal life is well-known. Their seeds (such as pine nuts) serve as a food source for numerous herbivores (including humans), and their foliage serves as shelter for a host of species. We need not dwell on the economic importance of these plants. Most lumber used in the construction of homes comes from coniferous forests.

17.29 Coniferous forests are found chiefly in the cold regions of the earth. They form a continuous belt across North America, northern Europe, and the northern Soviet Union. Such forests also extend southward in mountain ranges, and some types of conifers thrive in more temperate regions.

The dominance of the gymnosperms and the dinosaurs were strangely intertwined. For some reason, they both fell from prominence at the end of the Mesozoic era. The conifers, of course, survived these changes, but their numbers fell off drastically. With the arrival of the Cenozoic era about 135 million years ago, they were replaced over much of the earth by the upstart flowering plants. With their greater adaptability to constantly fluctuating conditions, the flowering angiosperms fanned out over much of the planet in those distant days that marked the beginning of the Cenozoic. They abounded from the tropics to the temperate regions, leaving the colder regions near the poles to the surviving gymnosperms.

However, the most recent chapter in the saga of the interaction between gymnosperms and angiosperms is an interesting one. The gymnosperms are on the comeback trail! In the last few million years, extensive northern spruce and pine forests have displaced tracts of hardwood angiosperms. We appear to be in a period marked by rapid speciation and adaptive radiation of conifers. We don't understand why this is happening, but it makes one hesitant to label any organism as "primitive."

The Angiosperms: The Rise of the Flowering Plants

The astounding diversity of flowering plants serves to dramatize the seeming sameness of the conifers. This genetic flexibility—or ability to diversify—may be a vital clue in our attempt to understand the rise of angiosperms to dominance at the close of the Mesozoic era. But their rise also presents many unanswered or partially answered questions. When did flowering plants evolve? Were they inconspicuously sprinkled among the dominant primitive conifers, cycads, and ginkgos of the Mesozoic? Or were they absent altogether in those days? And particularly, what kind of conditions would have promoted the sudden explosion of flowering plants?

We really aren't prepared to answer these questions. Perhaps the question should be, if there *were* flowering plants during the Mesozoic, *where* were they? The answer, based on fragmentary evidence, seems to point to the drier upland regions of the earth (Figure 17.31). The lowland gymnosperms were in an ideal location for formation of fossils, whereas the upland regions were too dry to form fossils. This may be why we find few fossils of flowering plants from the Mesozoic, while fossil gymnosperms are common.

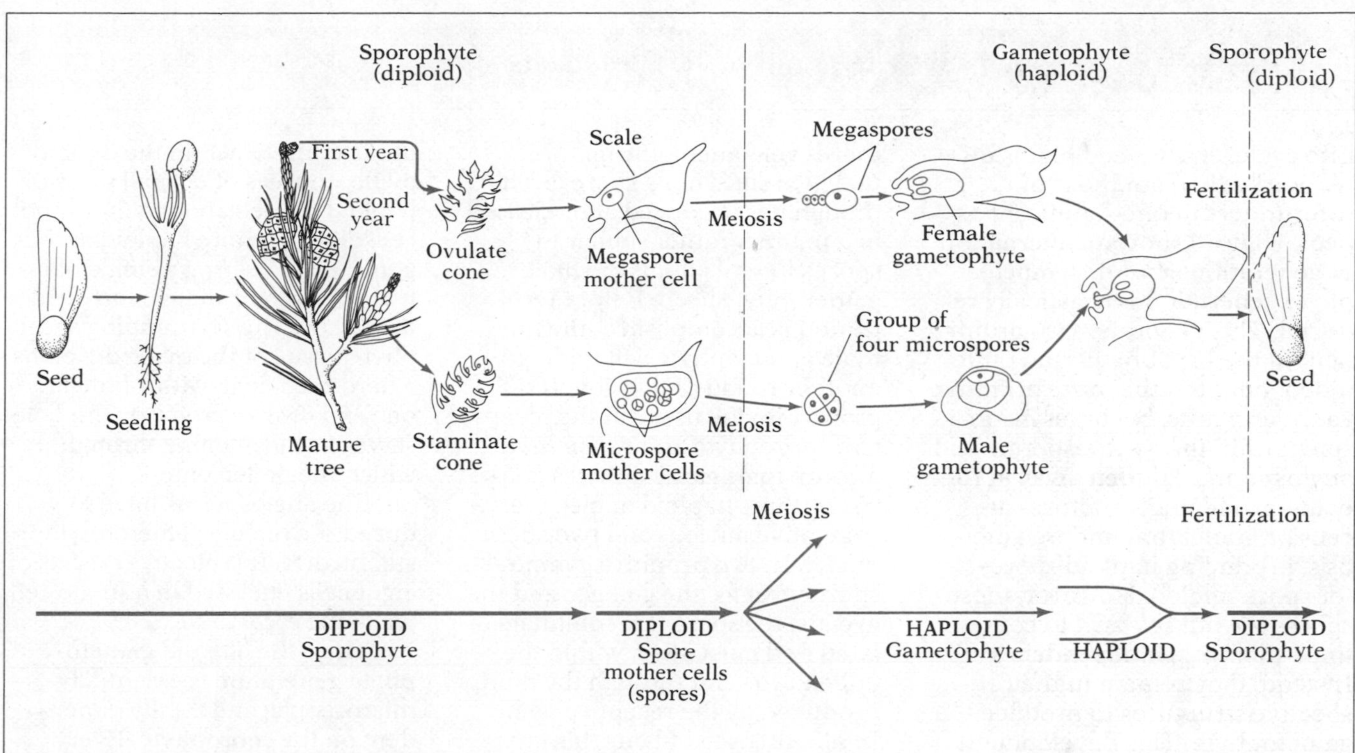

17.30 Alternation of generations in a conifer. With the exception of cells hidden in their reproductive structures, and an occasional yellow cloud of wind-borne pollen, all of the conifer we see is the sporophyte. Unlike other plants, the conifers and the angiosperms have no truly separate gametophyte generation. The trend toward reduced gametophytes is therefore nearly complete. Pines are typically monoecious, with both staminate and ovulate cones. Cells in the staminate cones, known as microspore mother cells, enter meiosis and form microspores. These microspores, enclosed in a winged pollen grain, give rise to gametes. In the ungerminated pollen there are two cells known as the generative and tube cell. The generative cell will give rise to two sperm nuclei just before fertilization, while the tube cell apparently controls the metabolic activity associated with the pollen tube growth. Cells in the ovulate cone, known as megaspore mother cells, enter meiosis and produce megaspores. These produce the female gametophyte, with two archegonia. When pollen, the male gametophyte generation, lands on the ovulate cone, a pollen tube grows into the tissue surrounding the egg cell within an archegonium. The two sperm cells penetrate the egg cytoplasm, with one fertilizing the cell and the other disintegrating. The embryo and surrounding seed of the pine may take as long as two years to mature and be ready for germination and growth.

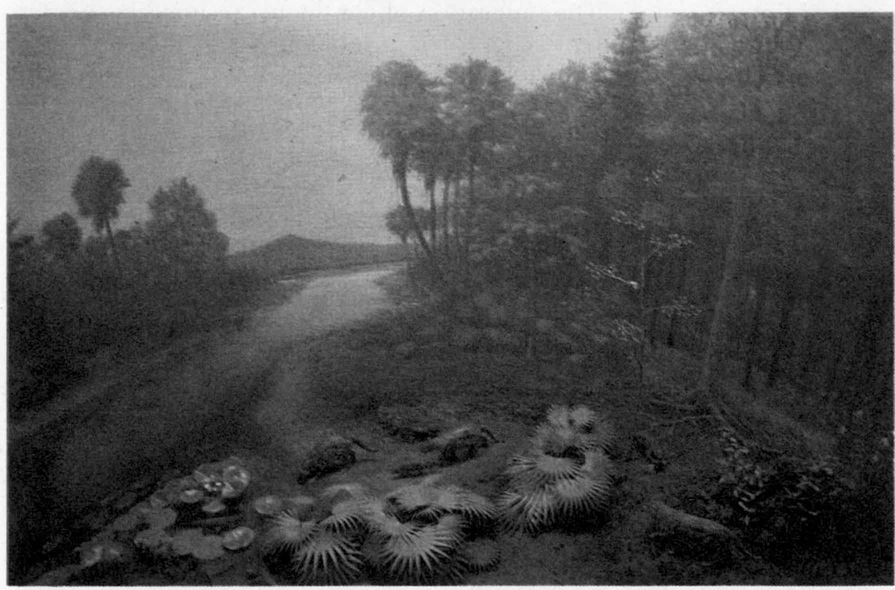

17.31 The scene from the late Mesozoic era shown here is an artist's reconstruction. The angiosperms (flowering plants) have established themselves and are becoming dominant. The flowered shrubs are similar to today's magnolias. Along with the angiosperms came the insects that were needed to pollinate them. Reptiles had given way to mammals, and birds were undergoing a rapid evolutionary radiation into many new niches. In another 60 million years the first hominids would join the descendants of these mammals.

Essay 17.1 Alternations of Generations in the Seed Plants

Life cycles in the seed plants differ markedly from those of their forerunners in one significant aspect. They all show an alternation of generations, but the gametophyte generation is drastically reduced. This generation, so prominent in the bryophytes and quite independent in the more primitive vascular plants, becomes inconspicuous in the gymnosperms and angiosperms. Hidden away in the ovule-producing structures are cells or nuclei that undergo meiosis, producing haploid spores (or spore nuclei). However, these spores *are not released* to continue their role independently. Instead, they remain in their respective structures to produce the gametophyte. This development occurs within the ovule, finally producing female gametic nuclei.

The male gametophyte begins as a single haploid cell, the *microspore*, a direct product of meiosis. Within its hardening cell wall, the nucleus undergoes one cell division to form a two-celled organism, the pollen grain. Because of this development, the mature pollen grain is not a spore, even though it is produced and released in a manner rather similar to spores in ferns but is the male gametophyte itself. Before fertilization is accomplished, the tiny male gametophyte will undergo another round of division. It grows into the tissue of the receptive sporophyte as a *pollen tube*. Mature male gametophytes consist of three haploid nuclei, the vegetative nucleus and two sperm nuclei. In two primitive gymnosperm groups, the ginkgos and the cycads, the sperm are multiflagellated and must swim within the pollen tube and through the fluid produced by the receptive ovule. In all other seed plants this last link to an aquatic past is lost. Sperm nuclei are transferred from male gametophyte to female gametophyte by direct cell-to-cell contact, which is accomplished by pollen tube growth.

In the gymnosperms, the female haploid gametophyte, developing entirely within the diploid ovule, consists of a small mass of haploid tissue—about a thousand cells. Two or more haploid archegonia, each bearing a single haploid egg, differentiate in one end this tissue. The surrounding sporophyte tissue of the ovule develops a hard seed coat, with a hole in one end (the *micropyle*). The hole serves as the opening through which the pollen enters.

The angiosperms have reduced the female gametophyte still further. It typically consists of eight cells enclosed in a single cell wall—the *embryo sac*.

Thus, the haploid gametophyte generation is essentially microscopic, and totally dependent on the sporophyte. Every seed plant you see has a similar cycle. So what we actually see, except for pollen, is the diploid sporophyte only. Grasses, pines, oaks, carrots, roses, lilies, and all the other 250,000 or so named species of seed plants share this condition.

Hypothetically, the sudden emergence of flowering plants can be explained on the basis of geological changes at the end of the Mesozoic, along with drastic changes in climate (an incredible average drop in temperature over the earth of perhaps 20°C) during the early Cenozoic. The changes in climate were brought about by two geological events. The first, as we mentioned, was a period of mountain building. The second was a series of events that geologists are only now beginning to understand. These events are called *plate tectonics* or *continental drift.* According to evidence from studies of the ocean floor, the major land masses of the earth began shifting at about the start of the Mesozoic era, some 200 million years ago (Figure 17.32). Most of the world's land mass broke apart, and the pieces drifted northward, so that by the Cenozoic the continents had roughly assumed their present positions.

The separation of continents resulted in the permanent geographic isolation of some animal groups, among them the lungfish of Africa, Australia, and South America; and the great flightless birds—the ostrich of Africa, the rhea of South America, and the moa and kiwi of New Zealand.

It is just these sorts of drastic changes on the earth's surface that would have marked the demise of many living things and provided countless opportunities for other organisms to expand and diversify.

Were climatic and geological events the only factors responsible for the change in plant dominance? Perhaps not. Perhaps the great diversity of flowering plants itself suggests another factor. Perhaps there is just greater genetic variability in flowering plants—greater genetic plasticity—than in many other living things. In other words, perhaps flowering plants can form new species faster than other organisms. There is some evidence that this is true. For example, the ability of angiosperms to form fertile hybrids is well-

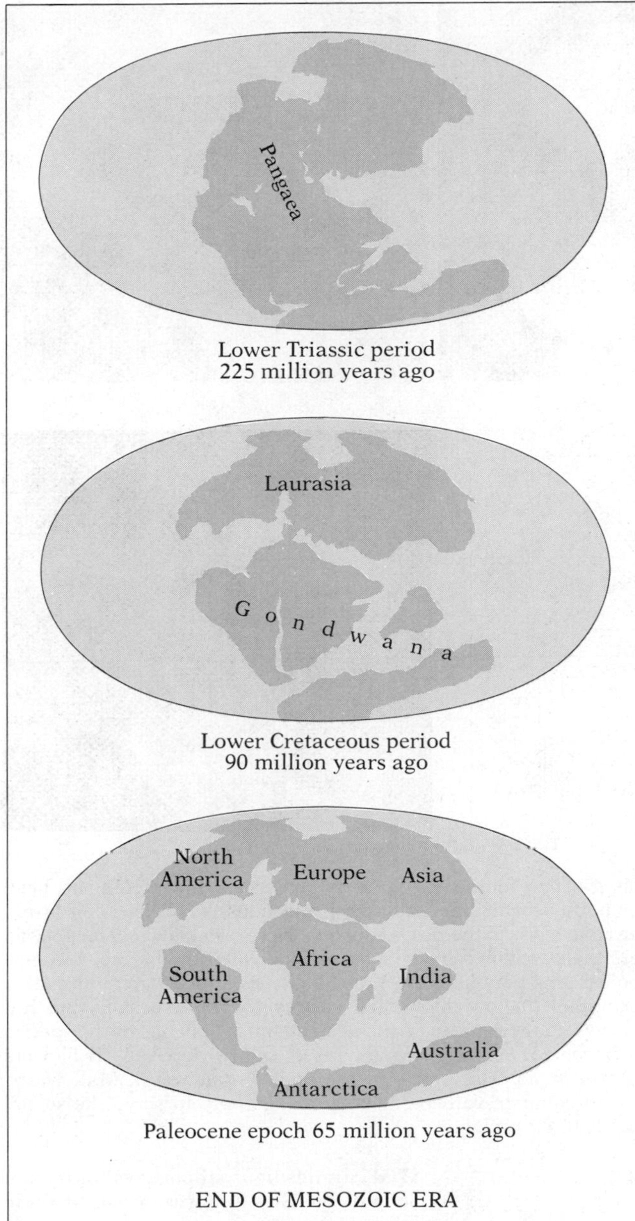

17.32 Continents adrift. Each geological era was ushered in by vast changes that included uplifting of major mountain ranges. These events changed the landscape and climate, and greatly influenced the habitat of the organisms that had flourished under the old conditions. Only the versatile ones survived to start new trends in evolution. A second great series of events, just beginning to be understood, was also quietly but irrevocably going on. The great land mass called Pangaea, of the Triassic period (some 225 million years ago), began to slip apart. By the lower Cretaceous period (about 90 million years ago), most of the continents had taken form, but North America and Eurasia formed one mass, called Laurasia, while South America and Africa, which were part of Gondwanaland, had nearly separated. Finally, by the Paleocene epoch (65 million years ago), only land bridges joined some of the continents. Along with their separations, some of the drifting continents had changed latitude significantly. Latitudinal changes bring on vast climatic differences, which added to the important geological events over the ages.

known. Hybrid flowering plants can even form "instant species" through *polyploidy,* a sudden doubling in chromosome number through irregularities in the meiotic process following hybridization (Chapter 13).

Perhaps the most important factor in the rise of the flowering plants was a partnership with the insects (Figure 17.33). Insects became pollen carriers as they visited the flowers in search of commodities. Once flying insects had evolved, attractive flowers were not far behind. This is a classic case of coevolution. Not only is insect pollination more efficient than wind pollination, it is much more selective (once an insect finds nectar in one flower, it will return only to that kind of flower). This makes it easier for new species to survive, and for many different species to coexist. With selective pollinations, a new species can survive even when its individuals are sparse and scattered. Far-flung wind-pollinated plants might die untouched by the pollen of distant neighbors.

Of course, not all flowering plants are pollinated by insects, but at one point probably all the ancestral angiosperms were. Even today, insect pollination is widespread and it appears to be the *primitive* condition. That is, it seems that the common ancestor of all present-day angiosperms was insect pollinated and that wind pollination in the grasses and other angiosperms is a more recent development.

The bright petals, attractive scent and sweet nectar of flowers are undoubtedly an evolutionary response to insect pollination. Interestingly, many familar flower scents mimic or duplicate the scent *pheromones* (chemicals used in signaling) of bees. Bees produced *geraniol* before geraniums did. Somewhere along the line, an ancestral plant "discovered" that it could attract bees by smelling like one.

Angiosperms and Changing Dinosaurs

So the mystery continues. Why did the nonflowering conifers and fernlike cycads give way before the flowering plants? R. T. Bakker, a paleontologist at Johns Hopkins University, has suggested that the dinosaurs may have had something to do with it. Bakker is the developer of the theory that the dinosaurs were not just giant, cold-blooded, sluggish lizards, but active, warm-blooded animals that dominated the earth far longer than we mammals have.

He suggests that until the beginning of the Cretaceous period, about 135 million years ago, the greatest plant eaters were the huge sauropods, such as the long-necked Brontosaurus and Brachiosaurus (Figure 17.34). Scientists had always believed that these animals were too massive to live on the land, but were forced to browse in lakes, supported by water. It was assumed that their long necks allowed them to reach above water for air and also to forage in lake bottoms. Bakker thinks that this is wrong. Re-

(a) (b) (c)

(d) (e) (f)

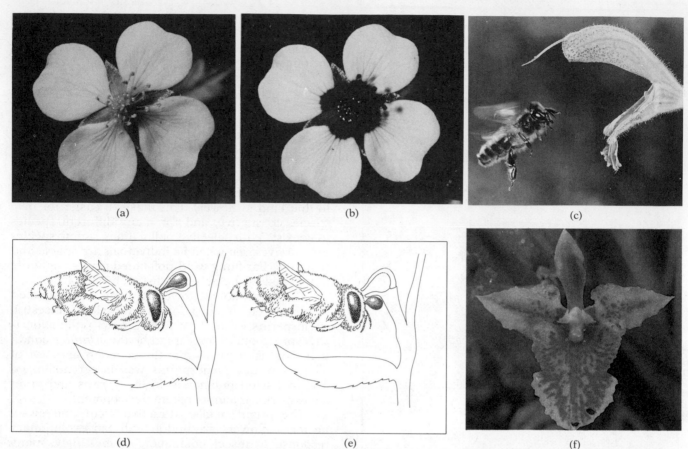

17.33 Insect pollination devices. Some plants have evolved elaborate devices for assuring pollination. Many flowers have markings known as *nectar guides*. These are pigment lines that are easily visible to insects and guide them into the nectaries. In some cases, the guides are not visible to us (a), but show up under ultraviolet light, which is visible to bees (b). Perhaps the most intricate and elaborate mechanism is found in orchids. One orchid, *Oncidium*, which has specialized anthers with very adhesive surfaces, imitates male bees, and transfer of the pollinium (pollen sac) occurs when an aggressive, territorial male bee is deceived into picking a fight with a flower (c). Note the pollinium on the bee forehead. When a bee enters the orchid, its head or body brushes the adhesive projections, which become firmly attached (d). As the bee backs out, these break away. The pollinia are then deposited in the next orchid visited by the bee, and new pollen sacs may be picked up (e). These devices are all excellent examples of how interacting species coevolve, usually, but not always, to their mutual benefit. Sometimes only one of the species is benefited. For example, the insect can be deceived. (f) Flowers of the orchid *Ophrys insectifera* mimic female wasps. Male wasps try to copulate with them but manage only to serve the sexual needs of the orchid.

17.34 In this unusual portrayal of the life style of the Brontosaurus, a pair of these lofty giants is seen at work grazing on the tender shoots of primitive conifers. According to one theory, these were terrestrial beasts and coexisted nicely with their food supply. The theory goes on to state that when angiosperms began to compete and outgrow their neighboring gymnosperms, the angiosperms were better able to withstand grazing in their younger stages of growth than were the gymnosperms. It is therefore concluded that the browsing dinosaurs may have been a deciding factor in the rise of angiosperms to dominance in the tropical and temperate regions of the earth. (Another theory has it that the giant reptiles were unable to adapt to a diet of angiosperms and died out as a result of the change in plant life.)

cent evidence suggests that the Brontosaurus and Brachiosaurus were fully terrestrial, and that they moved across the earth in great, lumbering herds, browsing from the tops of pine trees and sometimes rearing up on their enormous hind legs. They could have coexisted nicely with the conifers because their browsing would not have killed large trees and they would not have easily been able to eat tiny seedlings.

But when the Cretaceous period began, these high-browsing animals mysteriously died out and were replaced by smaller, low browsers of the Ornithischian type (Figure 17.35).

About 5–10 million years after the change in the dinosaurs, the insignificant little flowering plants began to make the earth bloom. The great coniferous forests began to dwindle, their southern borders continuously giving way. The flowering plants grow more rapidly than the conifers and they produce seeds earlier in their life cycles, always regenerating any lost leaves or branches. Bakker reasons that the low-browsing dinosaurs would have eaten a conifer before it was able to reproduce, but a flowering shrub or seedling would stand a better chance of forming seeds before being eaten. Thus, the flowering plants would have been able to escape devastation by these low browsers by quickly invading denuded ground and leaving their seeds before the small dinosaurs returned.

The Angiosperms Today

So the angiosperms have inherited the earth—for the time being. The flowering plants today constitute the vast majority of plant species. As a matter of fact, we really don't know how many there are, although estimates range from 200,000 to 250,000 named species. The point is, flowering plants occur in a dizzying array of diversity.

In spite of such diversity, the angiosperms can be divided taxonomically into two major groups or classes. These are the Monocotyledonae (monocots) and the Dicotyledonae (dicots). The first is in the minority, containing about 50,000 named species. The two are believed to represent different evolutionary lines, and they vary from each other in several important aspects as shown in Figure 17.36.

We won't dwell on the diversity of angiosperms here since this group is the subject of the next four chapters. Let's look instead at some evolutionary aspects of the flowering plants. Figure 17.37 presents a scheme of the alternation of generations in angiosperms. You might like to compare this cycle to the other cycles shown earlier in this chapter.

Origin and Phylogenetic Relationships

There are, of course, many taxonomic schemes that attempt to show evolutionary relationships among the angiosperms. There is some agreement, however, that the family Magnoliaceae, from the Ranales, is the paraphyletic ancestral group. Remember, there are two schools of thought on the origin of angiosperms. The first says that they existed through at least part of the Mesozoic, very inconspicuously and gradually taking advantage of their opportunities. The second postulates a rather sudden burst of plant species at the end of the Mesozoic resulting, in part, from rapid genetic changes. The fossil record isn't much help in resolving the issue. With the exception of a few scattered "probables" in the form of fossil parts and fossil pollen, the Mesozoic strata are devoid of angiosperm fossils. Such fossils begin to appear in the strata of the late Cretaceous period and in greater numbers in the early Cenozoic era. Most of our presumed plant relationships, therefore, are based on the study of today's plants, particularly

17.35 Ornithischian dinosaur. When the lofty gymnosperms went, so went the Brontosaurus, according to a recent theory. The Brontosaurus was replaced by new herbivores like this low-browsing Ornithischian dinosaur. These herbivores fed from the young shoots of the rapidly emerging angiosperms, which were better able to recover from the constant onslaught. Angiosperms produced flowers and seeds at a greater rate than their slower contemporaries, thus increasing their number and becoming the dominant form of plant life.

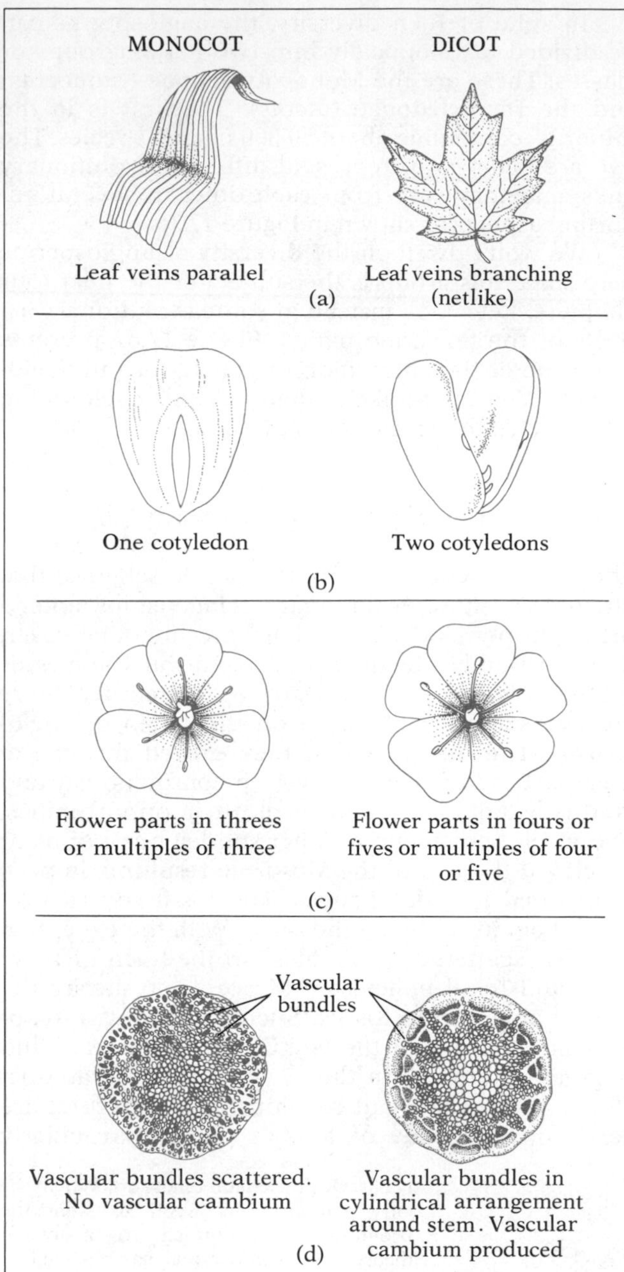

17.36 Key characteristics of monocots and dicots. Angiosperms are divided into monocots and dicots according to five major structural differences. (a) The most obvious difference is in the pattern of veins in the leaf, which is netlike in the dicot and parallel in the monocot (for example, the leaves of corn, a monocot, as compared to those of the sugar maple, a dicot). (b) The seeds also differ in that two *cotyledons* are seen in the dicots, while only one is present in the monocots (thus their names). The cotyledons of some dicots emerge from the soil and become photosynthetic. (c) The flowers of dicots have four or five floral parts, or multiples of these numbers (when the floral parts are divided). On the other hand, monocots have three floral parts, or multiples of three. (d) The vascular system of dicots is generally a neat ring of vascular bundles arranged in a circle around the stem. Vascular cambium is present. In the monocots, the vascular system occurs in scattered bundles, and vascular cambium is absent. While all these differences may not apply in every example, some of them will always occur, so it is relatively easy to make the distinction when studying a plant.

the comparison of their flowers. As an example, the Magnolia family is suggested as the ancestral type because of its primitive flower structure.

When botanists use the floral structure to determine taxonomic relationships, first they must decide what is primitive (original equipment) and what is advanced (new)—and that presents problems.

The concepts of *primitive* and *advanced* are always troublesome. Whole taxa can be considered primitive in the sense that they represent older, ancestral groups; we might consider ferns to be primitive compared with seed plants. However, every living organism has exactly as long an evolutionary history as every other living organism—some 3 billion years—even though some groups may have changed rapidly while others appear to have been marking time. "Primitive" taxa usually have members with advanced features. It's best never to consider any living organism, or even any current taxon, to be either primitive or advanced. Instead, individual features within a group may be primitive (present in the original founders of the group) or derived (evolved subsequently by only some members of the group; advanced). If only specific features are considered as primitive or derived, semantic and philosophical problems disappear, leaving only questions of fact. And we can see that every living organism has a mix of primitive and derived features.

When we look at the monocots and dicots, then, we should concentrate on the features that distinguish them. For each character we should ask, Which state is primitive? And, Which is derived?

The scattered fossil evidence from the early Cenozoic era at least gives us a starting point. From here, though, a number of evolutionary systems have been hypothesized. One of these, developed by C. E. Bessey, is often given most credence. Bessey's scheme differs from that of his leading opponent, Adolph Engler, in that it proposes that monocots are advanced over dicots. Engler believes the opposite is true. There is a corresponding disagreement as to which families are the most primitive. More succinctly, was the first angiosperm a monocot or a dicot? The arguments tend to be highly technical and are beyond the scope of our discussion, so we'll just discuss the most commonly accepted theory here.

Bessey has worked out what he considers to be "primitive" floral features. Any departure from these types represents an advancement or divergence from the ancestral type. Let's consider a member of the order Ranales, the buttercup (*Ranunculus*), to illustrate the primitive condition (Figure 17.38). (The magnolia flower itself is too confusing and difficult to deal with.) So according to Bessey, primitive characteristics are:

1. Floral parts are arranged in spirals.
2. *Carpels* (ovule-containing parts) are superior to

Sporophyte (diploid)

Gametophyte (haploid)

Sporophyte (diploid)

Gamete (embryo sac)

Meiosis

Megaspore mother cell

Fertilization

Zygote

Seed (embryo) germination

Zygote

Meiosis

Plant with flower

Sporophyte

Microspore mother cell

Gamete (pollen)

Sperm cells

Meiosis

Fertilization

DIPLOID Sporophyte

HAPLOID Gameto- phyte

DIPLOID Sporophyte

Megaspores and microspores

Gametes

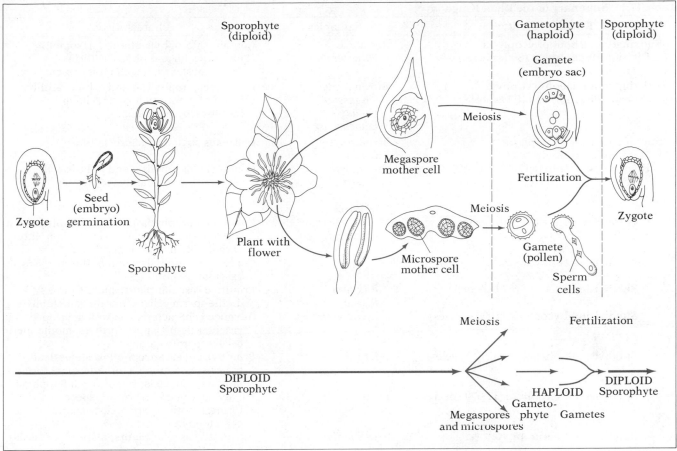

17.37 Alternation of generations in angiosperms. The sporophyte generation in angiosperms is the dominant part of the life cycle. The haploid female gametophyte generation is reduced to mere patches of tissue in the flower. Starting with a zygote tucked away safely in the ovary, an embryo emerges and is confined within the seed. Seeds germinate into seedlings that produce mature plants. When flowering occurs, the gametophyte generation is ushered in by meiotic events in the megaspore and microspore mother cells. The haploid gametophytes consist of the embryo sac and the pollen grain. When fertilization is accomplished, the sporophyte generation begins again.

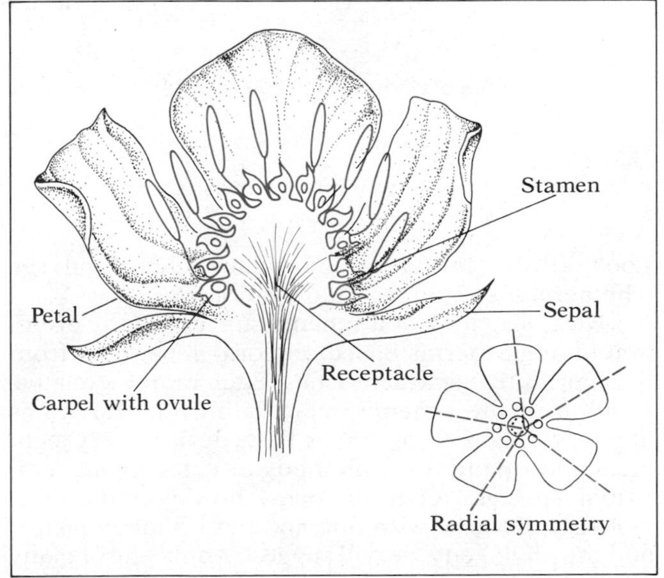

17.38 Floral structure of the buttercup, *Ranunculus.* The buttercup flower is believed to be representative of the primitive condition. Though it is not obvious, the arrangement of floral parts is in a spiral. The *carpels* of the buttercup are above the other floral parts. The carpels (with one ovule each) are all separate, as are the *stamens.* The *petals* are all separate and individual. Both male and female structures are present, along with petals and *sepals,* making the flower perfect and complete. Finally, the buttercup is radially symmetrical; that is, a slice through the plant at any vertical angle will produce identical halves. Any departure from the arrangement shown here is considered derived from the primitive condition.

(above) the *receptacle* (enlarged end of floral stem) and other floral parts, never inferior (below).

3. Carpels and stamens (pollen-producing parts) are numerous and not fused.
4. *Petals* are separate or completely divided, never fused.
5. The symmetry of the flower is radial (*regular*), not bilateral (*irregular*).
6. Flower is *perfect* (has both male and female parts) and complete (contains *sepals, petals, stamens,* and *carpels*—all the floral parts).

Table 17.1 Summary of the Plant Kingdom

Kingdom Plantae	Examples	Characteristics
Subkingdom Rhodophycophyta Division Rhodophyta (300 species)	Red algae *Polysiphonia*	Contain chlorophyll *a* (and *d* in some), phycocyanin and phycoerythrin, floridean starch; flagella not present
Subkingdom Phaeophycophyta Division Phaeophyta (1000 species)	Brown algae *Sargassum* *Fucus* *Ectocarpus*	Contain chlorophylls *a* and *c*, fucoxanthin, starch; anterior pair of flagella in motile cells; giant kelps
Subkingdom Euchlorophyta Division Charophyta	Stoneworts *Chara*	Cell walls contain calcium salts
Division Chlorophyta (7000 species)	Green algae *Volvox* *Ulva* *Ulothrix*	Contain chlorophylls *a* and *b*; little cell differentiation; mainly freshwater
Division Bryophyta (23,000 species)	Mosses, liverworts, hornworts	Contain chlorophylls *a* and *b*; all multicellular with considerable cell specialization; little vascular tissue developed
Division Tracheophyta		Contain chlorophylls *a* and *b*, starch; vascular tissue
Subdivision Psilophyta (4 species)	*Psilotum* *Rhyna*	Primitive vascular plant; spore formers; motile sperm cells; separate gametophyte
Subdivision Lycophyta (1000 species)	*Lycopodium*	Advanced characteristics; well developed vascular tissue; spore formers, motile sperm cells; separate gametophyte
Subdivision Sphenophyta (12 species)	*Equisetum*	Primitive appearance; hollow stems, leaf scales are nonphotosynthetic spore formers; motile sperm cells; separate gametophyte
Subdivision Pterophyta (11,000 species)	Ferns	Highly developed leaf frond; spore former; motile sperm cells; separate gametophyte
Subdivision Spermophyta	Seed Plants	Advanced vascular plants; all produce seeds and pollen
Gymnosperms		Advanced vascular plants; highly developed root systems; naked seeds
Ginkgo (1 species) Cycad (100 species)	*Ginkgo biloba* Cycad	Primitive conifers; motile sperm cells; naked seeds
Conifer (550 species)	Pines Spruces Hemlocks	Cone producers; nonmotile sperm cells
Angiosperms		Most advanced vascular plants; produce flowers; fruit surrounds seeds
Monocotyledonae (50,000 species)	Lily, grass, palm	Floral parts in threes and multiples of three; parallel veined leaves; one cotyledon
Dicotyledonae (200,000 species)	Rose, Bean, Oak	Floral parts in fours or fives; netlike veined leaves; two cotyledons

Look at the buttercup in Figure 17.38, and see whether it meets all the qualifications.

Now, let's look at a scheme suggesting the evolution of angiosperms based on floral departures from these primitive characteristics. Each family's relative position in the scheme depends on how far it has departed. Plant taxonomists have devised very technical and quantitative methods of categorizing individual species. As it happens, however, they still violently disagree with one another on many points, and the phylogeny we will suggest is only one of many competing hypotheses.

The scheme in Figure 17.39 suggests three main lines of evolution. At the left, the monocots diverge, with the lily family leading the way. The orchids, as you see, are the most advanced monocot group.

The dicot line culminates in the mint family, Labiatae. Botanists generally consider the mints to be as advanced as the orchids, but in their own direction. A separate continuous line produced the family Compositae, represented here by the daisy.

This scheme, as all others, will undoubtedly be revised again and again as new kinds of evidence turn up.

In addition to the evidence from flower parts, there is one feature of the stem that seems to indicate that monocots were derived from dicot ancestry. Woody dicots show a complex pattern of *secondary growth* produced by vascular cambium, in which the vascular bundles of the shoot are converted into concentric rings of vascular tissue, followed by increases in stem diameter. This pattern is also seen in gymnosperms, so it is presumed to be ancient and primitive.

17.39 Evolutionary scheme for the angiosperm families. According the C. E. Bessey's system of character study, the angiosperm families can be arranged in the manner shown. Only a few of the hundreds of angiosperm families are included here. As you can see, the scheme suggests that flowering plant families diverged from the ancestral paraphyletic family, Ranunculacae (magnolia, buttercup). Off to the left the monocots have diverged, with the orchids considered the most advanced from the primitive type. There is little argument here, considering their unique and highly specialized life style and their highly developed mechanisms for assuring pollination. Family Compositae, the largest angiosperm family, occupies the place of honor at the center. Chrysanthemums, daisies, dandelions, and some of the sagebrush species are among these highly evolved members. The division at the right terminates in Labiatae, the mints. Its members sound like the spice rack, including rosemary, thyme, sage, winter savory, and lavender.

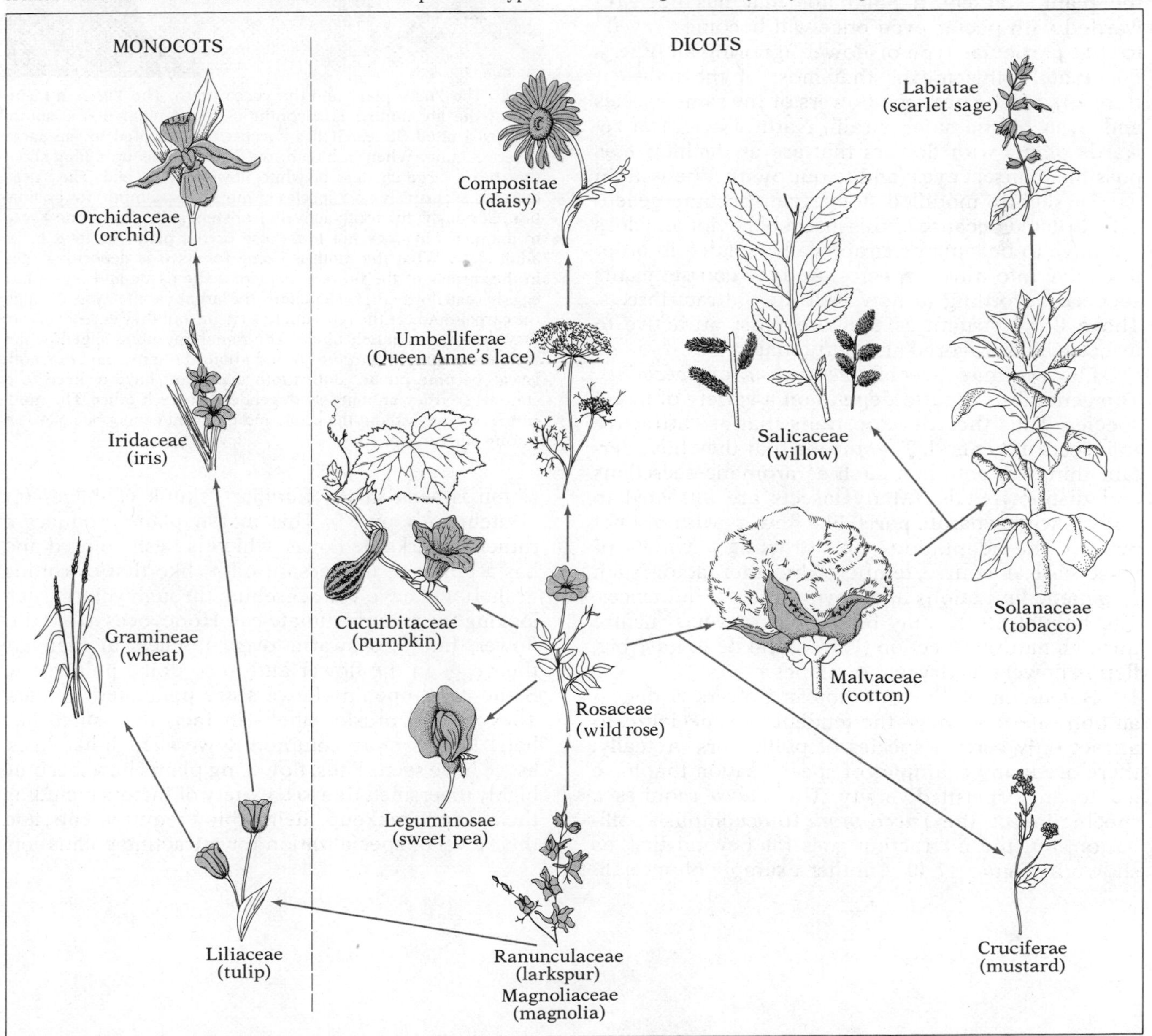

Some short-lived, herbaceous dicots lack secondary growth—and so do nearly all monocots, since they lack vascular cambium. Some monocots have evolved very different mechanisms of diffuse secondary growth. The inference is that monocots evolved from a herbaceous dicot that had already lost this complex pattern of secondary growth.

Now that we have seen a hypothetical evolutionary pathway for the flowering plants, we can return to our old question. Why is there so much diversity in the flowering plants? One answer involves the potential isolating mechanism of insect pollination. Bees and other pollinators have an important behavioral trait—important to them, and important to the plant. Namely, a pollinator that has been rewarded with nectar even once will become "keyed" to that particular type of flower, ignoring all others. Fortunately, this means that most of the pollen it transfers will be between flowers of the same species and even of the same strain. Natural selection rewards plants with flowers that are as distinctive as possible to insect eyes (and to our own). A new strain with a slightly modified flower can become genetically isolated because of this insect behavior and does not have to become geographically isolated in order to evolve into a new species. And, of course, plants succeed according to how well they attract insects. Those floral structures that are most attractive to insects will be favored and perpetuated.

Flowers may become *generalists* or *specialists.* The generalists tend to depend on a variety of insect species. Thus they develop traits that are attractive to insects in general. This means that they have certain things in common such as aromatic secretions and distinct petals. (Many insects are attracted to objects with separate parts, like flowers with distinct petals.) The adaptations for attracting a variety of insects are, of course, tempered by other factors such as genetic limitations and environmental influences. The habitat, itself, may help mold flower structure through natural selection (It wouldn't do to have big, floppy flowers on dry, windy slopes.)

But much of the variation in flowers is due to an opposite tendency—the tendency to specialize, to attract only certain species of pollinators. Actually, there are many examples of specialization that have led to angiosperm diversity. The *Yucca* requires a specific insect, the *Yucca moth* to accomplish pollination, but the interaction goes far beyond that, as shown in Figure 17.40. Another example of specialization is seen in the familiar "skunk cabbage" (or "Dutchman's pipe"). This marsh plant produces a rather remarkable flower which is flesh colored and has a pungent, unpleasant odor, like that of rotting flesh. It will have you searching through your garden looking for an unfortunate cat. Honeybees avoid this flower, but flies swarm over it. Flies, in fact, lay their eggs in the flower and in so doing pollinate it. Some night-opening flowers are pollinated by bats. They have a musky smell—in fact, they smell like bats! Just as more common flowers smell like bees, as we have seen. Thus, flowering plants have become highly diversified due to a variety of factors including their genetic makeup, their habitat requirements, and their level of specialization for attracting pollination.

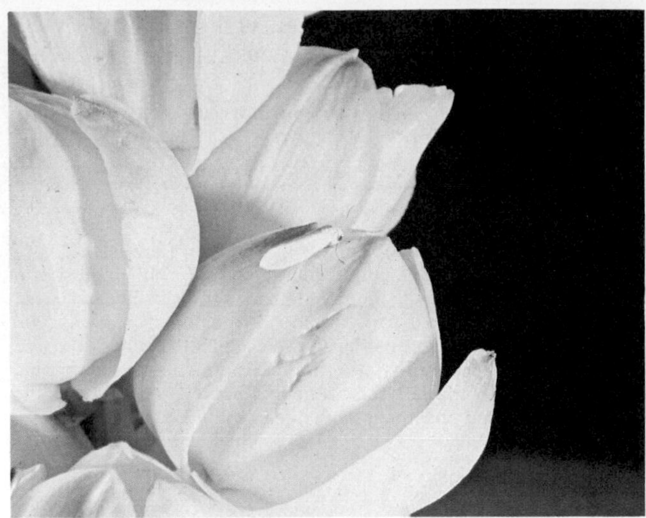

17.40 The *Yucca* plant and the yucca moth. The *Yucca,* a member of the lily family, is a common desert plant of the southwestern United States. It has successfully adapted to the harsh desert climate. When in bloom, the *Yucca* sends up a long shoot that bears large clusters of white flowers at its end. The *Yucca* depends exclusively on a species of moth (yucca moth) for pollination. Curiously, the moth actively transfers pollen from one *Yucca* to another, but does not make use of the plant for food in its adult state. What the moth is doing for itself is depositing eggs in the carpels of the flowers. It pierces the tissue and lays a few eggs in each flower. Upon hatching, the larvae eat their way through the carpels and eat the reproductive tissue, but they emerge before they destroy too much tissue. The remaining undamaged ovules form seeds and drop to the ground about the same time the moth larvae become pupae. Both moth and plant have evolved to a state where they are entirely dependent on each other. The moth cannot survive without the plant, and the plant cannot be pollinated by any other means.

Application of Ideas

1. Present an argument for dividing the plant kingdom, as defined in the five-kingdom scheme, into three kingdoms. What would the kingdoms contain and what would their ancestors have been?

2. Describe the conditions (climate, landscape, etc.) that would have to prevail for the Bryophytes to return to the prominence they once held. Which of the animal groups might persist under these conditions?

3. The fossil record clearly demonstrates the presence and even prominence of fernlike plants that produced seeds in the Devonian forests. Their significance to seed-producing gymnosperms and angiosperms is unknown. Present two hypotheses that might account for seeds occurring twice in the fossil records of plants. Support the one that sounds more probable.

4. Pollen is believed to have been a Mesozoic innovation. What, if any, significance would pollen have had if it had developed in the Devonian or Carboniferous periods of Paleozoic?

5. The involvement of animals in plant reproduction may have been a key factor in the rise and spread of angiosperms. Discuss this proposition and explore two aspects of angiosperm reproduction that involves animals today. How might these have influenced animal evolution?

Key Words

Acetabularia
adventitious root
alternation of generations
amylopectin
amylose
angiosperm
antheridium
antheridogen
archegonium

bladder
blade
brown algae
bryophyte

Cambrian
capsule
carboniferous
carpel
carrageenin
chlorophyte
complete flower
conducting tissue
cone
conifer
continental drift
cycad

Devonian
dicot
dioecious

fiddlehead
flowering plant
fucoxanthin

gametophyte
gemma
green algae
gymnosperm

Key Ideas

Introduction

1. Three subkingdoms—Rhodophycophyta, Phaeophycophyta, and Euchlorophyta—are established according to the photosynthetic pigments, starch storage, and structure.

2. Rhodophycophyta are the *red algae*, Phaeophycophyta, the *brown algae*, and Euchlorophyta, the rest of the plants, including the *green algae* and all land plants.

3. All euchlorophytes have chlorophylls *a* and *b* and carotenoids and they all store the carbohydrate *amylose.*

4. The Plant kingdom is polyphyletic with at least three origins.

5. Plant characteristics (some are lost in recent evolution and some do not apply to *green algae*):
   a. Photosynthesis, with chlorophyll
   b. Cellulose cell walls
   c. Nonmotile
   d. Specialized organs and tissues
   e. Primarily sexual reproducers
   f. Alternating haploid and diploid generations
   g. Multicellular

Division Rhodophyta: The Red Algae

1. There are 300 species of red algal *sea weeds.* All contain *phycocyanin* and *phycoerythrin,* chlorophyll *a*, chlorophyll *d.*

2. Their procaryotic ancestors are believed to be cyanophytes and primitive eucaryotes, which developed symbiotically.

3. *Red algae* are a coastal organism, attaching to the substratum by *holdfasts.* Their polysaccharides are used to produce *agar agar* and *carrageenin.*

4. *Red algae* undergo an *alternation of generations.* Alternation is from a diploid asexual generation to a sexual haploid generation, both in different individuals. The *sporophyte* generation produces spores by meiosis. These haploid spores produce the *gametophyte* generation. Haploid *gametophyte* individuals produce gametes that will join in fertilization, producing a diploid *sporophyte* again.

Division Phaeophyta: The Brown Algae

1. *Brown algae* are unique in their production of *fucoxanthin.* They store the carbohydrates *laminarin* and *mannitol.* They produce flagellated sperm and some produce flagellated eggs.

2. *Brown algae* inhabit cold coastal waters. *Sargassum* is a common coastal brown algae that drifts into a vast area known as the *Sargasso Sea.*

holdfast
hornwort
horsetail

isogamete

kelp

laminarian
leaf scale
liverwort
lycophyte

mannitol
Mesozoic
metaphyte
micropyle
monocot
monoecious

oogonium
ovule

Paleozoic
perfect flower
Permian
peristome
petal
pheromone
phloem
phycocyanin
phycoerythrin
pistilate
plate tectonics
pollen grain
pollination
polyploidy
prothallus
protonema
psilophyte
pterophyte

red algae
rhizoid
rhizome

sclerenchyma
seed coat
seed plant
sieve tube
Silurian
siphonous algae
sorus (sori)
spermophyte
sphenophyte
sporophyte
stamen
staminate
stipe
stonewort
strobilus

thallus
tracheophyte
Triassic

3. *Brown algae* vary from microscopic forms to the giant kelps that are the most complex. Some have specialized tissues and organs (*holdfasts, stipes, blades*). One species produces simple vascular tissue.

Division Charophyta: The Stoneworts

Stoneworts are ancient plants, found today in fresh water. *Chara,* one of the few species, is similar in several ways to higher plants. The name *stonewort* refers to calcium in the cell walls. This remote group followed its own evolutionary path having diverged from *green algae.*

Division Chlorophyta: The Green Algae

1. Some are marine *seaweeds,* but most are freshwater forms. A few, including lichen symbionts, grow on the land.

2. *Green algae* occur as single-celled, in filaments, and in *blades.* Single-celled *green algae* are both flagellated and nonflagellated.

3. The single-celled *Chlamydomonas* is used frequently as a laboratory organism. The most commonly seen are *gametophytes,* which reproduce asexually, forming clones. When mating strains are present they produce *isogametes,* which fuse in conjugation. *Zygospores* form and later enter meiosis, producing *meiospores,* which produce the *gametophyte* generation again.

4. *Volvox* forms spherical, flagellated colonies, suggesting a way that multicellularity might have begun. Movement in the colony is coordinated; cells specialize in reproduction, both sexual and asexual. In asexual reproduction, small spheres grow, invert themselves, and form new colonies within the parental colony.

5. *Siphonous green algae* form branched chains. They are coenocytic. *Acetabularia* is highly unusual, representing a single cell that is 5–9 cm tall. A compound nucleus is found in the "foot."

6. *Ulva* and *Ulothrix* are truly multicellular. In *Ulva* (*sea lettuce*), *gametophyte* and *sporophyte* generations are separate but identical in appearance.

7. In *Ulothrix,* the haploid *gametophyte* is prominent, and the *sporophyte* is greatly reduced, found only as a *zygospore.*

8. Terrestrial plants most likely evolved from filamentous *green algae,* which were dominant during *Silurian* and Lower *Devonian.*

Division Bryophyta

1. *Bryophytes* (*mosses, liverworts,* and *hornworts*) are clearly multicellular plants and many are terrestrial. All require water to reproduce. *Bryophytes* differ structurally from other terrestrial plants by the absence of conducting tissue. The division has made a very successful transition to land life with about 23,000 species, mostly *mosses.*

2. The problems confronting the early land plants included:
   a. Desiccation in dry air
   b. Support to replace bouyancy
   c. No medium for gamete transport

3. As semiaquatic pioneering filled this niche, adaptation to drier regions began.

4. Multicellularity permitted the *bryophytes* to form specialized tissue for water absorption, anchorage, foliage support, photosynthesis, and resistance to desiccation.

5. *Bryophytes* form a rootlike *rhizoid,* a *primitive,* stemlike supporting tissue without *xylem* or *phloem,* but with *leaflike scales.* The *leaf scales* are often pitted with pores for exchanging gases.

6. *Bryophytes* are ecologically important in soil formation and in their use as food by animals.

Evolutionary Relationships

Bryophytes are believed to have followed their own evolutionary pathway. Their algal ancestor may be represented today by the primitive *green algae, Fritschiella.*

Class Musci (Mosses)

1. *Mosses* are the largest class. They have a clear-cut *alternation of generations* with a prominent *gametophyte* (haploid) generation.

vascular tissue

xylem

zygospore

2. Leafy *gametophytes* produce male and female sexual structures, the *antheridium* and *archegonium,* at their tips. Sperm cells swim to the *archegonia* and fertilize egg cells, producing a diploid zygote and starting the *sporophyte* generation. These are dependent, growing from the *gametophyte* to produce spores through meiosis. Haploid spores start the gametophyte generation again. Successful spores germinate, producing a *protonema,* which develops into another leafy *gametophyte.*

Class Hepaticae (Liverworts)

1. *Liverwort gametophytes* tend to be ground-hugging with ribbonlike *thalli* that are often highly branched.

2. *Marchantia* produces *gemmae* that are cuplike asexual structures that can break off and grow on their own.

3. The *gametophyte* produces *antheridia* and *archegonia,* which are borne on aerial stalks. Upon fertilization, the *sporophyte* grows out of the *archegonium,* eventually producing spores through meiosis and restoring the *gametophyte* generation.

Class Anthocerotae (Hornworts)

Hornworts are a minor group. The dominating *gametophyte* is small and leafy. It produces *antheridia* and *archegonia,* and fertilization begins the *sporophyte* generation. The *sporophytes* are "hornlike." Unlike *sporophytes* in other *bryophytes,* they can survive on their own and grow continuously. This is a move away from the totally dependent *sporophyte* generation.

Division Tracheophyta

1. *Tracheophytes* have developed vascular tissue that enables plants to become large and live in drier regions of the land. They have true roots, stems, and leaves and many produce seeds. Most do not require water for fertilization.

2. The five major subdivisions of *tracheophytes* are Psilophyta, Lycophyta, Sphenophyta, Pterophyta, and Spermophyta. The first three are minor groups today.

Subdivision Psilophyta

1. Only four species of *psilophytes* remain. *Psilotum* produces *primitive* roots (*rhizomes*) and stems with simple vascular tissue. Leaves are actually *leaf scales* along the stem. Sporangia produce spores through meiosis and a highly reduced *gametophyte* generation develops from these.

2. Fossil records show the *psilophytes* to be at least 300 million years old. They may represent a step in the transition from algae to vascular plants.

Subdivision Lycophyta (Club Mosses)

1. *Lycophytes* have true roots, stems, and leaves (all with vascular tissue). Roots arise from underground stems.

2. The *sporophyte* generation is highly prominent, producing spores in a conelike *strobilus.* Minute *gametophytes* develop independently, producing *archegonia* and *antheridia.* When fertilization occurs, the diploid zygote matures into the adult *sporophyte.* Water is required for fertilization.

3. *Lycophytes* were treelike plants, dominant in *Paleozoic* (*Devonian*) and are now represented by coal deposits.

Subdivision Sphenophyta (Horsetails)

1. *Horsetails,* represented by one surviving genus, *Equisetum,* form a vegetative shoot with whorls of side branches and tall, straight, hollow reproductive shoots. The leaves are tiny and scalelike, and *primitive* roots arise from stem nodes.

2. The prominent *sporophyte* produces spores that develop into minute *gametophytes.* Swimming sperm persist in this group, and upon fertilization, the zygote develops into the *sporophyte.*

Subdivision Pterophyta (Ferns)

1. *Ferns* are still successful with 11,000 species.

2. The prominent *sporophyte* produces its leaves from underground *rhizomes,* which also produce fine roots. The leaf produces *sori,* which house *sporangia.*

Spores form meiotically and become airborne. Upon germination each forms a small, independent *gametophyte* called a *prothallium. Antheridia* and *archegonia* produce sperm and eggs. Water is required for the swimming sperm. Upon fertilization, the diploid zygote develops into the familiar leafy foliage of the *sporophyte.* Self-fertilization probably occurs in *prothallia* of some species, but at least one mechanism is known that prevents it in others.

3. *Ferns* were dominant in *Devonian* forests, and some produced seeds.

Subdivision Spermophyta

1. *Primitive* vascular plants persisted to the end of *Paleozoic,* when many were replaced by *seed plants.* The shift was accompanied (or caused) by great geological and climatic changes. *Seed plants* were well adapted for the climatic changes. Most do not require water for fertilization, since a new form of gamete transfer using pollen had evolved. In addition, *seed plants* were able to produce extensive root systems to reach distant water.

2. The earliest *seed plants* were probably the *gymnosperms,* although some ferns were seed producers.

3. Seeds consist of an embryo, stored food, and hardened *seed coats.* They are resistant to harsh conditions.

Essay 17.1 Alternation of Generations in the Seed Plants

1. *Seed plants* have an *alternation of generations,* but there are no independently growing *gametophyte* individuals. The sporophyte is highly dominant. Megaspores are formed meiotically, but they remain in the reproductive structures. Pollen is released as a free-floating microgametophyte, but it can germinate and grow only in receptive sporophyte tissues.

2. The female *gametophyte* in *gymnosperms* consists of a small mass of cells. Two or more haploid *archegonia,* each with an egg, differentiate in one end. The *sporophyte* tissue surrounding forms the *seed coat.* The sperm cell, traveling through a pollen tube, enters through the *micropyle.*

The Gymnosperms

Gymnosperms produce naked seeds without fruits. The *conifers* have a wide distribution today and are the most significant gymnosperms.

The Ginkgo

The only species is *Ginkgo biloba,* the maidenhair tree. Like some of the other tracheophytes, its sperm cells swim to the egg, but the plant produces its own fluid medium. They produce aerial spores.

The Cycads

Cycads were widely distributed in *Triassic,* but are more restricted today. They produce pollen and flagellated sperm cells.

The Conifers

1. Conifers are most common in northern regions and at high altitudes. They produce leaves year-round and bear their seeds in cones. Ovulate cones produce ovules while pollen is produced in staminate cones.

2. *Conifer* dominance and dinosaur dominance ended about the same time, with the arrival of *Cenozoic.* They were replaced in many places by the more diverse *flowering plants.* Today, the *gymnosperms* may be on the rise, as *conifers* move toward southern climes, replacing hardwoods.

The Angiosperms: The Rise of the Flowering Plants

1. The physical changes accompanying the rise of *flowering plants* included a drastic rise in temperature, the uplifting of great mountain ranges, and the shifting of land masses. The movement of the continents (continental drift) began some 200 million years ago, completing their movement to their present position by *Cenozoic.* The changes contributed to new climatic conditions.

2. There is reason to believe that *flowering plants* have greater genetic variability with more rapid speciation.

Angiosperms and Changing Dinosaurs

1. The Bakker theory suggests that dinosaurs were active, warm-blooded creatures and even the largest lived on land.

2. The large dinosaurs browsed on the tops of pines, leaving the seedlings alone. With the demise of the larger dinosaurs, low-browsing, smaller dinosaurs ate the seedlings. *Flowering plants* were better able to survive the low-browsers, and eventually replaced the faltering *conifers.*

The Angiosperms Today

1. *Angiosperms* today comprise up to 250,000 species. They occur in enormous variety. The two major classes are Monocotyledonae (*monocots*) and Dicotyledonae (*dicots*).
2. The *monocots* contain about 50,000 species and represent a divergent evolutionary line.
3. Differences between *monocots* and *dicots* are seen in seeds, flowers, leaves, vascular system, and vascular cambium.

Origin and Phylogenetic Relationships

1. Family Magnoliaceae, from order Ranales, represents the ancestral type. It is not known whether *angiosperms* existed quietly in *Mesozoic* and gradually rose to dominance in *Cenozoic,* or whether they originated at the end of *Mesozoic,* bursting into prominence at the start of *Cenozoic.* Mesozoic is devoid of angiosperm fossils. Therefore phylogenetic organization is based on comparative studies of today's plants rather than fossil evidence.
2. According to the Bassey hypothesis, primitive floral features are represented by *Ranunculus* from order Ranales. These primitive features are:
   a. Floral parts in spirals
   b. *Carpels* superior
   c. *Carpels* and *stamens* separated and numerous
   d. *Petals* never fused
   e. Regular and radial symmetry
   f. *Flowers perfect* and *complete*
3. Departures from the *primitive* state are rated numerically to determine evolutionary distance.
4. The analysis of plant families in terms of Besseyian measurements suggests three evolutionary lines (see Figure 17.39).
5. One reason for *angiosperm* diversity is the competition for insect *pollinators.* Insect behavior allows new plant species with distinctive flowers to become genetically isolated. Flowers become *generalists* or *specialists.* Orchids and *yucca* are clear examples of pollinator specialists, as is the *skunk cabbage,* which attracts flies.

Review Questions

1. What factors are used to distinguish the three subkingdoms of kingdom Plantae? (p. 464)
2. List six characteristics of plants. (p. 464)
3. List the pigments of the *red algae* and determine which of these indicate a relationship with the cyanophytes. (p. 465)
4. What is the environment of *red algae* and in which part are they most commonly found? (p. 465)
5. Using *Polysiphonia* as an example, draw a simple scheme showing an *alternation of generations.* (p. 465)
6. When did the *red algae* originate and what other alga shared this period of time? (p. 466)
7. List the pigment and polysaccharides found in the *brown algae.* (p. 466)
8. List two advanced *brown algae* and explain why they are believed to be advanced over other algae. (p. 467)
9. List the pigments and the carbohydrates of the euchlorophytes. (p. 467)
10. List the various habitats of the *green algae.* (p. 467)
11. Describe the three major forms taken by green algae. (p. 468)
12. Briefly review the alternation of generations in the single-celled *Chlamydomonas*. Be sure to indicate haploid and diploid stages. (p. 469)
13. Explain why *Volvox* is often considered as an example of how multicellular organisms arose. (p. 469)
14. Review the processes of sexual and asexual reproduction in *Volvox.* (p. 470)
15. What features of the *stoneworts* lead botanists to believe they may be on their own evolutionary pathway? (p. 470)
16. What are the special features of the *siphonous algae* and in what other kingdoms do we find their unusual cellular organization? (p. 470)

17. Compare the *sporophyte* and *gametophyte* generations in *Ulothrix* with those of *Ulva*. (p. 470)

18. What are the *bryophytes* and why is it safe to call them plants "in anyone's taxonomy"? (p. 472)

19. Explain the basic differences between *bryophytes* and *tracheophyte* (p. 473)

20. What were the major problems confronting plants as they emerged from a watery environment? (p. 473)

21. Explain how *bryophytes* solved the problems of the rigorous land environment. (p. 473)

22. Review with simple diagrams the *alternation of generations* seen in mosses. Be sure to indicate *sporophyte* and *gametophyte* generations and haploid and diploid states. (p. 472)

23. Describe the *thallus* of *Marchantia,* including the *gemmae, archegonia,* and *antheridia.* (p. 475)

24. Diagram the *alternation of generations* in *Marchantia.* (p. 477)

25. In what ways does the *sporophyte* of the *hornwort* indicate changes from other *bryophytes?* (p. 477)

26. What is vascular tissue and why was its development so important to the progress of land plants? (p. 478)

27. List the five phyla of vascular plants and give an example of each. Indicate which are important today. (p. 478)

28. What key evolutionary place do the *psilophytes* occupy? (p. 479)

29. List the characteristics of the *lycophytes.* (p. 479)

30. Compare the *lycophytes* of today with those of the *Paleozoic.* (p. 480)

31. Review the hypothesis about the five phyla of vascular plants being "cousins." (p. 480)

32. Describe the body form of the sphenophyte (horsetail). (p. 480)

33. Compare the present-day success of the *ferns* with that of the three *earlier* groups of *tracheophytes.* (p. 481)

34. Review the *alternation of generations* in *ferns.* (p. 482)

35. Explain the mechanisms for preventing self-fertilization in the *fern prothallus.* (p. 482)

36. Describe the geological and climatological changes that ushered in the *seed plants.* (p. 482)

37. List three important changes in *seed plants* that accounted for their success in the changing environment. (p. 483)

38. Compare the *alternation of generation* in *seed plants* with that in other plants. Explain the fate of the *gametophyte* in detail. (Essay 17.1) (p. 488)

39. In its sexual activity, the *Ginkgo* represents a transitional or intermediate type. Explain this. (p. 485)

40. List four characteristics of *conifers.* (p. 485)

41. List the geological and climatic changes that ushered in the *angiosperms.* (p. 486)

42. How does the unusual genetics of *flowering plants* suggest they may evolve faster than *conifers?* (p. 488)

43. What role did the insects play in the rapid evolution of *flowering plants?* (p. 488)

44. Review the Bakker theory of how the large dinosaurs and *conifers* faded away together. (p. 489)

45. List the differences between *monocots* and *dicots* in the following categories: seeds, flowers, leaves, stems. (Figure 17.36) (p. 492)

46. Explain the main problem in understanding *angiosperm* taxonomy. (p. 491)

47. According to Bessey, what are the characteristics of a *primitive* flower? (p. 492)

Chapter 18

Transport Systems in Plants

There are giants on earth. But, fortunately, the greatest of these are plants—towering invincible plants, great trees that have resisted every threat except the American need for lawn furniture. But the great redwoods are not the only giants. In their persistent struggle for survival, many of today's plants have become quite large. Why should this be? There are decided advantages to being tall if one is a forest plant, since there is intense competition for sunlight. The lofty giants are able to expose their photosynthetic tissue to unobstructed light, while shorter and smaller plants grow very slowly in the shade of the forest floor. Also, keep in mind that the large plants produce vast root systems that go deep into the soil, allowing them better access to water and minerals. Furthermore, tall trees can disperse their pollen and seeds farther than their shorter competitors.

Largeness, however, has its price. The needles of a giant redwood (Figure 18.1), perhaps 100 meters above the ground, constantly lose water to the atmosphere and must have that water replaced. But the water must come from a root system far below. Of course, root tissues have easy access to water, but if they are to grow and respire, they must be fed carbohydrates manufactured in the distant leaves. The plant must somehow bring food down to the roots, and it must bring water and minerals from the soil up to its leafy canopy. So, in the process of evolution, higher plants have met the cost of their large size by developing extensive vascular systems for the transport of water and nutrients.

But the cost of large size does not end there. The high forest canopy must be supported, even in periods of intense and unpredictable winds and heavy snows. Therefore, much of the plant tissue must be devoted to physical support. The forest giants, and other higher plants as well, solve both their support problem and their water transport problem with the same tissue—the *xylem,* or woody tissue of the plant. Wood is marvelously strong supporting material, particularly as it is arranged in the trunks and stems of trees. And the ability of plants to raise water several hundred feet above the earth is an unsurpassed engineering accomplishment. The achievement has long mystified the human engineering mind,

18.1 This towering California coast redwood *Sequoia sempervirens* is the world's tallest tree (107 m, or 350 ft). It is truly awe-inspiring to stand in one of the remaining groves of these giants. If you know about problems of water transport, the marvel of it all increases. Where stands have developed unmolested for hundreds of years, the coastal redwood competes for sunlight only with its fellows. Nothing else comes close. This tree is also awe-inspiring to lumbermen who stand in reverence at visions of board-feet and profits after taxes.

since the work is done with *no apparent expenditure of plant energy.*

On the other hand, the distribution of food molecules does require the expenditure of energy on the part of the plant, but it has its mysterious aspects as well. The tree does not simply send food molecules blindly from its leaves downward; it responds to the specific needs of the different tissues and shifts the food distribution according to those needs. It can even send food up from the roots; in spring, trees and bushes that have gone through the winter with bare, leafless branches transport stored food molecules up from their roots to the newly forming leaves and twigs.

18.1 Vascular Systems and the Transport Problem

Vascular Plants and Nonvascular Plants

In most of this chapter we will be describing the support and transport systems of the *tracheophytes,* or *vascular plants.* This group includes most of the familiar terrestrial plants; all the trees, grasses, shrubs, and other flowering plants; as well as all the ferns and horsetails. They are all vascular, which means that they have specialized systems for the internal transport of water and food molecules. But before we describe the structure and function of these systems, let's briefly review what goes on in a primitive plant that lacks the highly specialized conducting tissue of the higher plants. We'll find that some important principles of water and food transport are the same in vascular plants and their simpler relatives.

Water and Food Transport in Mosses

You might recall from Chapter 17 that the bryophytes, (the group that includes mosses, liverworts, and hornworts) are terrestrial plants that have not developed vascular systems. Perhaps you also recall that the mosses and their relatives are probably not directly related to higher plants, but have evolved independently from green alga ancestors.

The rootlike rhizoids of the bryophytes anchor them to their substrates and can help absorb water (and dissolved minerals) that will be utilized by other tissues. In turn, the rhizoids must receive food molecules from the photosynthetic tissues. But both the water and the food molecules can pass through the plant only by slow cell-to-cell transport, either by diffusion or by active transport (Figure 18.2).

Water conduction is usually no problem to a moss because it is usually wet anyway. Thus, water and dissolved minerals can flow into all its cells directly from the exterior by osmosis (see Figures 4.14 and 4.15). The net flow of water inward ceases only when the hydrostatic pressure inside increases to the extent that the osmotic pressure is balanced. The cells are by that time fairly stiff (turgid). Thus, osmosis not only acts as transport, it provides the turgor that helps support these small, low-lying plants.

Under drier conditions, the moss rapidly loses water, and with its turgor pressure lower, it wilts. However, the rhizoids and the lower parts of the plant may still be in contact with moisture in the soil. As the cell loses water and its contents become more concentrated, osmotic pressure increases and water flows into those cells that are in direct contact with the soil. Thus, they remain rigid and the plant suffers little in its anchorage.

Water that enters the rhizoids passes from cell to cell and will reach the leaflike photosynthetic organs of the moss. Eventually the flow of water entering the leaf cells just balances the flow of water leaving the cell surfaces by evaporation. This situation is described as a *steady state,* which is not quite the same as an *equilibrium.* An equilibrium occurs in a closed system—one in which neither matter nor energy enters or leaves—whereas a steady state is a property of an open system (Figure 18.3).

Plant physiologists note that the movement of water in and out of a cell is due to the balance between *osmotic potential* and *hydrostatic pressure.* Osmotic potential draws water in, hydrostatic pres-

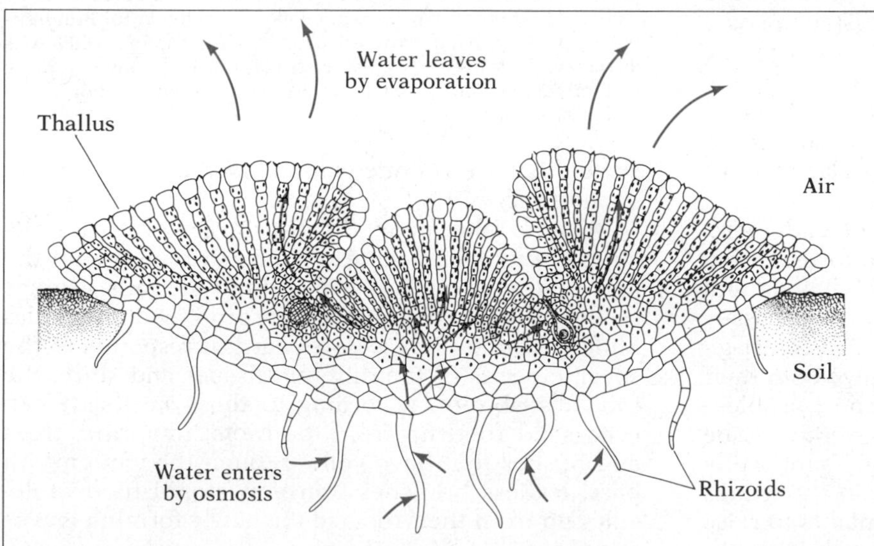

18.2 Water and ion transport in *Riccia.* In the small, ground-hugging bryophyte, *Riccia,* water is rather easily transported from the soil. There is no true vascular tissue, so the water simply travels by diffusing from cell to cell. When air is dry, there is a regular osmotic gradient from the water-rich cells of the rootlike rhizoids to the relatively dry upper surface. Since the plant is small, no cell is far from the source.

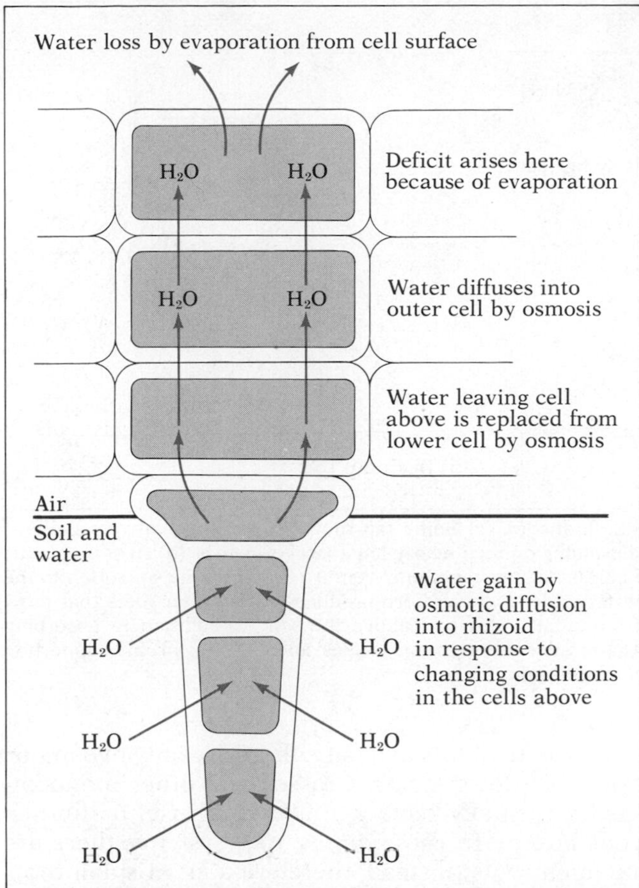

18.3 The bryophyte reaches a *steady state* when water lost is replaced by water gained in a constant "tug-of-war" among forces. Water loss here is represented by evaporation from the upper surface (through epidermal cells or from parenchymal cells and through pores in the epidermis). Deficits created by evaporation are quickly recovered as water diffuses into cells whose solute concentration is increased by the loss. The cell-by-cell transfer of water is supported by diffusion in the rhizoids, which increases and decreases with changing conditions in the plant. The net flow is due to an osmotic gradient.

sure pushes it out. The difference is called the *water potential:* the net tendency of water to flow into the cell. When the soil is moist and the air is dry, there is a continuous gradient in water potential from cell to cell.

The movement of water from cell to cell is always along a gradient from lower water potential to higher water potential and from higher hydrostatic pressure to lower hydrostatic pressure. It is important to emphasize that this movement is a *passive* diffusion process. The internal hydrostatic pressure of the cells is not due to biochemical work done by the plant (other than the work needed to maintain its semipermeable membranes) but is due to osmosis. It does, in fact, take energy to move the water, but the energy comes from evaporation at the surface of the plant. The *free energy of evaporation* is an indirect form of solar energy since the sun affects wind, humidity, and temperature.

The cell-to-cell diffusion process works reasonably well over short distances—but only over short distances. And that is why bryophytes never grow very tall, and why even low-lying mosses are so often subject to drying out.

The flow of stored food molecules in bryophytes is also a cell-to-cell process. Some active transport may be involved, but there are no specialized conducting tissues for food in the thallus.

Transport and Support Structure in Vascular Plants

Vascular plants have two separate parts to their vascular systems, although there may be interconnections between the two. One part is specialized for the conduction of water and solutes; it is made up of *xylem* tissue. Paradoxically, it does not begin to function until after it dies, when its old cell walls serve as conducting tissue, primarily in conducting water up from the roots. Transport in xylem is mostly or entirely passive. The xylem also serves as the principal supporting tissue of the plant. In fact, the xylem of mature trees is called *wood,* and xylem probably holds up your house.

The other part of the vascular system specializes in carrying food molecules, and its tissue is called *phloem.* Phloem cells are living—some of them lack nuclei, but they are living all the same. Transport in phloem may be either upward or downward. Under certain conditions, food molecules are transported downward from the photosynthesizing leaves to the stem and roots; but under other conditions, the food molecules are carried upward from the root and stem to the leaves. Transport in phloem is an active, energy-consuming process.

Before we consider the functions of the xylem and phloem in greater detail, let's take a look at how these tissues are distributed in higher plants.

The Root

Quite a variety of materials enter the plant through its roots. The water that enters contains nitrogen, sulfur, and phosphate, as well as other vital plant nutrients. The poet Joyce Kilmer said something about "A tree whose hungry mouth is prest/ Against the earth's sweet flowing breast." That breast flows with water and dissolved minerals.

The roots absorb water through the numerous tiny root hairs, just above the root meristems. The water taken in must balance the tremendous amount of water lost through the leaves. The delicate balance is not always maintained, since water may pass

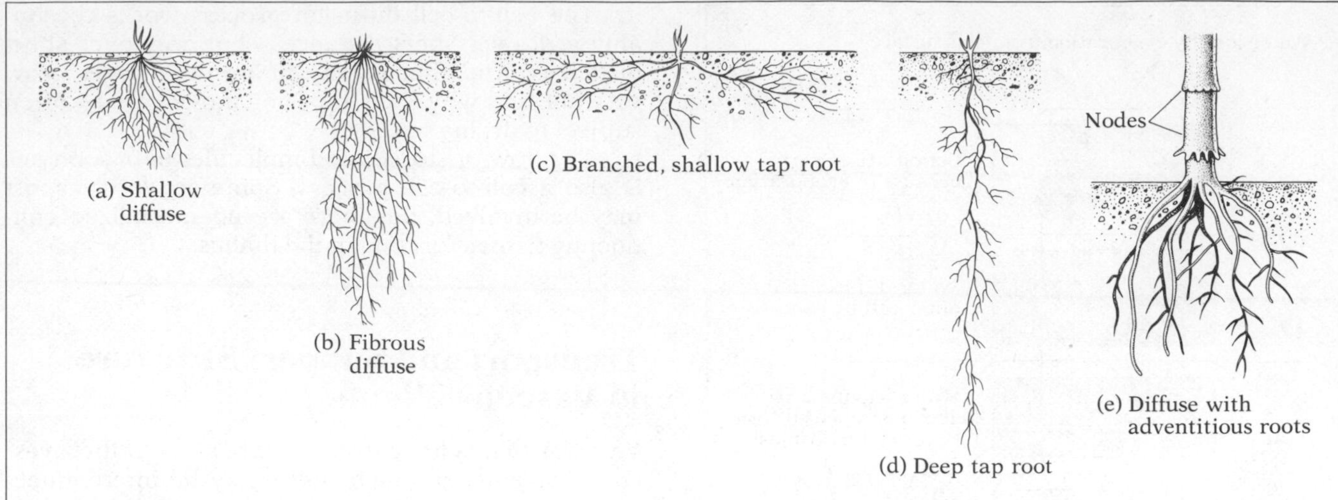

18.4 Root systems can be roughly divided into two main groups, *diffuse* and *tap*. The two varieties of diffuse roots shown here are actually different responses by two species to the problem of water supply. (a) The shallow diffuse root system at the left is efficient for absorbing surface water only, while (b) the fibrous diffuse root system at right reaches for deeper supplies. Both are highly branched and lack a single dominant root. The tap root has a central, vertical root and may or may not have extensive side branches. (c) Some tap root systems are relatively shallow, depending on their nearly horizontal branches. (d) In other plants, unbranched tap roots may venture deep into the soil, often to the water table itself. (e) Corn produces *adventitious* roots that grow from stem nodes just above the soil. In addition to absorbing water, adventitious roots act as anchors and props, supporting the heavy corn plant.

through roots and leaves at different rates that depend on environmental conditions. But in the long run, a balance must be achieved.

The relationship between leaf area and root hair surface area may have become painfully clear to you if you have tried to transplant a small tree. For one thing, the absorptive root hairs are concentrated at the ends of the roots, generally in the zone just under the edge of the plant's canopy. As a rule of thumb, there is just about as much of the plant below ground as there is above. If you consider the extent of a root system and how much is torn away while transplanting, the frequently discouraging results are not too surprising.

Absorption of water into the root hairs is largely a passive, osmotic process. There is some evidence, however, that active transport can be involved, at least indirectly. Plants can extract water from what appear to be the most unpromising soils. In recent years, Israeli agronomists have made astonishing progress at growing salt tolerant crops such as barley in desert soil watered with seawater.

Root Types

Competition for soil water and minerals is believed to be one of the deciding factors in how many plants an ecosystem can support. Thus we find a wide diversity of root types (Figure 18.4). Why? (This is an important ecological question.) Competition for water among plants is often intense and spatial variation in root systems can reduce that competition.

In spite of this diversity, there are only two major types of root systems. Grasses and other monocotyledons usually have a shallow, diffuse, or fibrous root system. In these *diffuse root* systems, there are no main roots; instead, there are a lot of small roots of about equal size. They branch continuously as they grow, and eventually they produce an enormous surface area (Figure 18.5). It has been estimated that a

18.5 The bluegrass root system reveals the extensiveness of root growth in grasses.

root system from *one* rye plant produced nearly 14 million root branches! The combined length was calculated at 624 km, and the surface area at 2.3 million square centimeters! Desert plants often have shallow diffuse systems so that they can take advantage of any brief showers that only dampen the upper part of the soil. The saguaro cactus, for example, has a root system that radiates out in all directions along the surface, like spokes on a wheel.

The alternative to the diffuse root is the *tap root* system. In this system, the plant produces a main root with few to many side branches. It is commonly, but not always, found in woody plants such as trees. (It is also found in carrots, beets, and radishes.) Most people, if they think of tap roots at all, think of a single, extremely deep vertical root. And such tap roots do exist. Where the single tap root grows deep, this system is especially efficient where water supplies lie well beneath the ground surface. A deep tap root also helps provide a firm anchor for a large tree. In many plants with tap root systems, however, shallow, horizontal branches may dominate (and may break your sidewalk).

Many plants produce adventitious roots that extend from stems or even leaves. Good examples are the aerial roots in ivy and the prop roots of corn, which help in both water absorption and support. The growth of adventitious roots from stems permits the propagation of plants from cuttings.

The Root Tip

Root tips are a good place to begin our study of the transport system in plants. This is the subterranean business end of a plant, as well as the growth point from which the vascular system of the root develops. The root tips contain the all-important *apical meristem* tissue, which is located just above the *root cap* (Figure 18.6). The apical meristem is a region of small, rapidly dividing cells. This tissue region contributes cells to the expanding and maturing vascular tissue above and to the root cap below. To continue this process throughout the lifetime of the plant, some cells of the meristem are set aside after division as a reserve of embryonic tissue. The reserve tissue will always be found at the same location in each root tip, since it is constantly pushed along as the cells above it elongate.

The epidermal tissue of each root tip, above the meristem, develops the root hairs. As shown in Figure 18.6, each hair is a long extension of an epidermal cell. All together, the multitudes of root hairs on numerous root tips account for a tremendous surface area. This area is essential, since it is the root hairs of the epidermis that absorb water and all chemical nutrients. This is just as true for radishes as it is for oaks.

Root Vascular System

The vascular system of the root begins in the region of the root hairs. Notice in Figure 18.6 that the vascular cells occupy a cylinder of tissue in the center of the root. The general region of vascular tissue is known as the *stele* (Figure 18.7).

The stele is surrounded by a large area of loosely arranged, thin-walled cells (known as the *cortex*) that extends from stele to epidermis (Figure 18.7). The cortex consists of parenchyma cells with numerous intercellular spaces. While the cortex is tissue primarily for storage rather than transport, it does provide an avenue for the movement of water and minerals from the root hairs to the stele.

The stele itself is surrounded by a very special ring of cells called the *endodermis.* We will discuss their specialness shortly. The *pericycle,* just inside the endodermis, also conducts water and minerals inward to the vascular tissue, but it has another function as well. The pericycle, a parenchyma tissue, has the capacity to produce secondary roots. These branch roots are started deep inside and, through numerous divisions and subsequent growth, push their way through the cortex, bursting through the epidermis of the primary root. Eventually each branch root will develop its own vascular tissue that will join that of the primary root.

In cross-section, the vascular tissue within the pericycle of a typical dicot looks like a four-armed star (Figure 18.7). The broad arms contain the *primary xylem,* which is made up of large, thick-walled, water-conducting cells. The *primary phloem* with its sieve-like end walls (food-conducting cells) are found between the arms of the star. Each group of phloem cells is surrounded by a half-circle of *vascular cambium* made up of small, dividing cells that will later produce xylem on one side and phloem on the other.

This primary growth arrangement becomes drastically reorganized in plants that are capable of secondary growth. With the exception of a few large annuals (e.g., sunflower), secondary tissue is restricted to the perennial plants (those capable of more than 1 year's growth). Essentially, secondary growth is growth in diameter, and can be defined as growth that is produced by vascular cambium. The change from primary to secondary growth patterns in roots will be discussed in Chapter 21.

In the walls of the endodermal cells, an interesting development occurs that influences the uptake of water. These cells become partially *suberized,* or waxed. (Suberin is a waterproof waxy substance). The wax is deposited in the cell walls of the endodermis, forming a semiclosed cylinder known as the *Casparian strip* (Figure 18.8). This strip prevents water from diffusing freely in and out of the vascular system along the otherwise water-permeable cellulose walls. Because of the Casparian strips, water must pass across cell membranes and then through cytoplasm of the endodermal cells in order to enter the conduct-

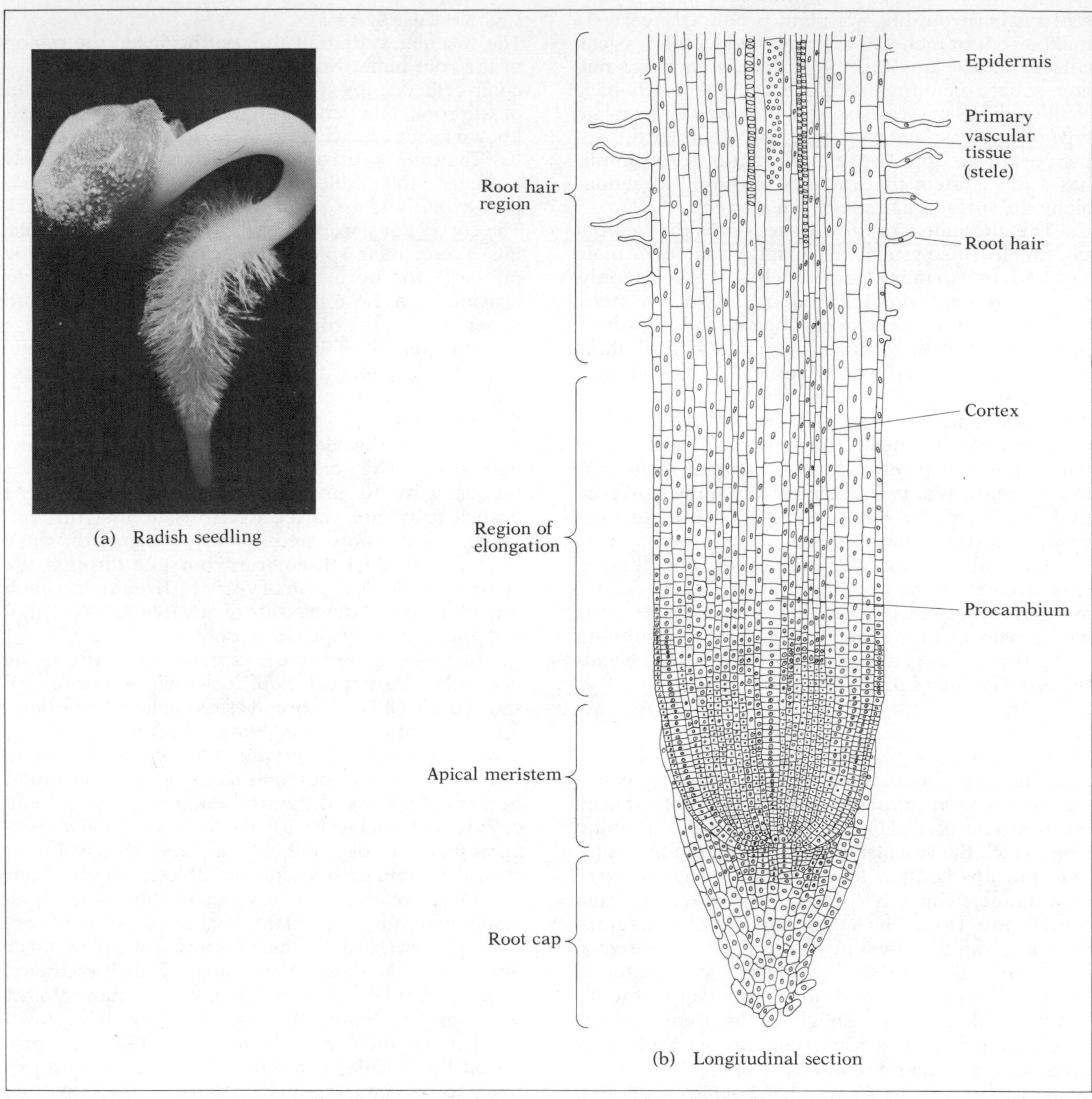

18.6 Root tip anatomy is most easily studied in the seedling before things get too complex. (a) Here, a radish has been germinated in a moist chamber to show the development of root hairs. (b) In the longitudinal section, the tip of the root reveals several distinct regions. First is the root cap, a sort of shield that protects the critical *apical meristem* tissue just above. The meristem is easily recognized by its tiny cells, many of which are visibly engaged in mitosis. Above the meristematic region, cells just produced begin to elongate. Signs of cellular differentiation can be seen in the center of this region of elongation; note the beginning of primary vascular tissue in the central core, or *stele,* surrounded by a cylinder of *cortex.* Just above the region of elongation is the *root hair* region. Root hairs are extensions of epidermal cells. They are actually the water- and mineral-absorbing structures of the plant. Cellular differentiation is more pronounced here.

ing system. The significance of this appears to be that it allows the plant to exercise some control of the water and minerals in the conducting system, since they must pass through the membrane and cytoplasm. This would not be true if the water simply passed in and out through a continuous line of nonselective, dead, cellulose walls. To the extent that active uptake of water is occurring in the roots, or when the soil becomes temporarily dry, the Casparian strip prevents this hard-won moisture from diffusing right back out.

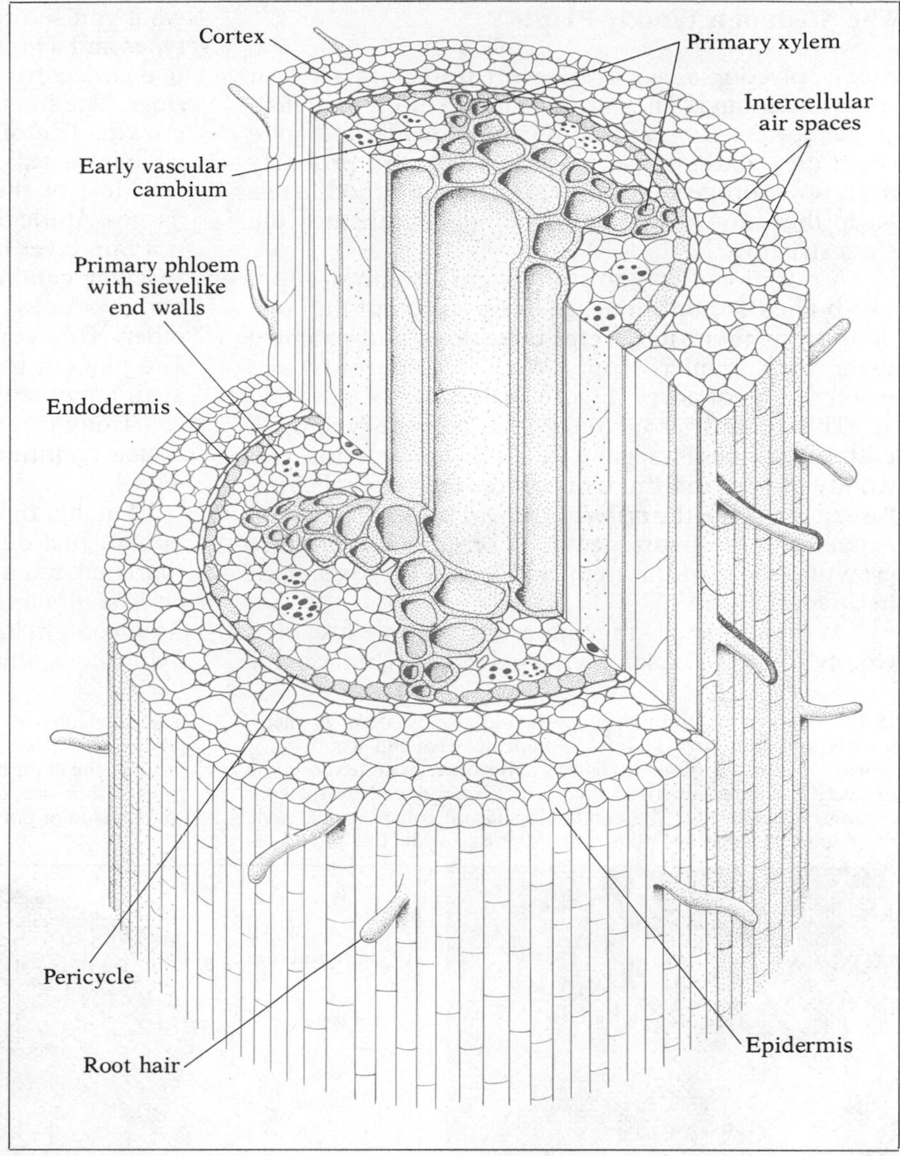

18.7 Conducting and supporting tissues begin to form in the young root. In this cross-section of the area just above the root tip, a loosely arranged *cortex* is visible under the epidermal layer. It extends to the *endodermis,* a layer of cells that encircles the *pericycle.* The cylindrical arrangement of endodermis and pericycle surrounds the conducting cells. The *primary xylem* is easily recognized as the large, thick-walled cells, which in this plant resemble a four-armed star. Between the arms of the star is the *primary phloem,* shown here with their sievelike end walls. The entire cylinder of conducting and supporting tissue is known as the *stele.*

18.8 The Casparian strip consists of partly suberized (waxy) cell walls in the endoderm. The suberized areas are believed to block the passage of water through the porous cell walls into the xylem. Thus, the water is routed inward through the cytoplasm of the endodermal cells, and cannot diffuse passively outward by way of the cellulose cell walls. This arrangement has raised speculation that active transport is involved in water movement.

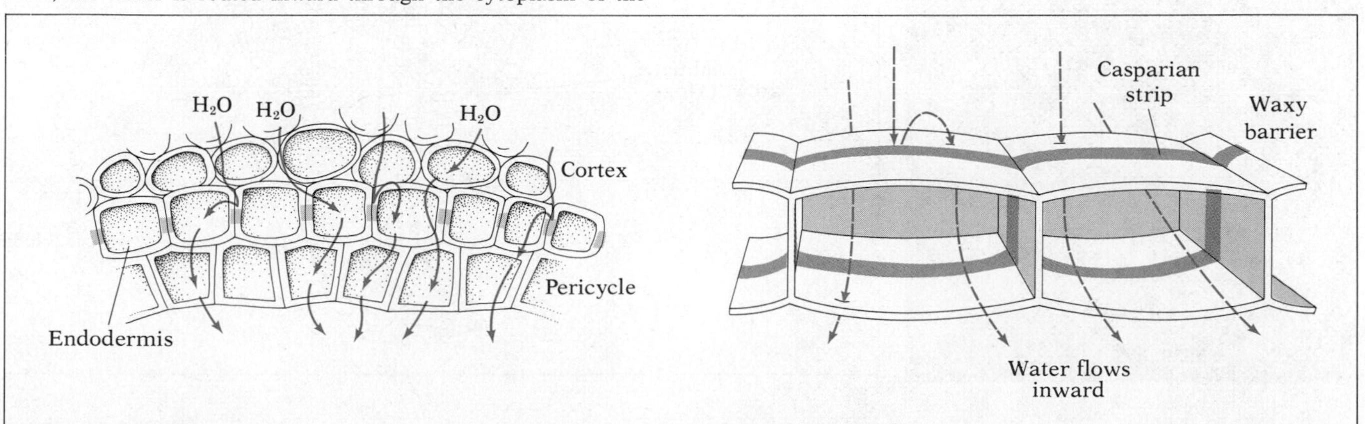

The Stem of a Woody Plant

Stems, of course, are avenues of transport between roots and foliage. They are also much more than that, since they produce and support the foliage, and quite often they act as storage regions. You can probably list a few storage stems that are common food items. Keep these four functions in mind as you study the stem structure.

A traditional way to study stem structure—from soft-bodied annuals to large trees—is to cut thin sections from them in several directions and examine them with a microscope. We'll consider first the woody trunk of a tree (Figure 18.9).

The pattern is actually the *secondary growth* pattern, which is seen only in gymnosperms and woody dicots; and there only after the original structure of the stem, the *primary growth* pattern, has been extensively reorganized. We'll consider the primary growth pattern of the stem, and how it is reorganized in Chapter 21.

At first glance, the apparent complexity of a woody stem can be somewhat perplexing, but most of what you see can be sorted out to a few basic tissue types and a great deal of repetition. Notice the repetitive concentric circles. These are, of course, annual rings. The inner two represent the first two season's growth. The outer ring is the final season's growth, which was still in progress when the stem was cut.

Most of the stem consists of xylem and related tissues. At the border between the xylem and phloem is a thin layer of undifferentiated cells known as the vascular cambium. It continues to undergo mitosis and produces phloem on one side and xylem on the other. This continues through the growing season. The phloem forms triangular blocks of cells, which are interrupted by the widening areas of phloem rays, consisting of parenchyma cells. These rays, you will notice, continue into the xylem like spokes in a wagon wheel.

Finally, three additional tissues need to be mentioned. Just outside the phloem is an area of *cortex* (parenchyma and collenchyma) followed by a ring of *cork cambium,* and beyond that is the *cork* itself. As you might have guessed, cork cambium consists of living cells that produce cork (what layman calls

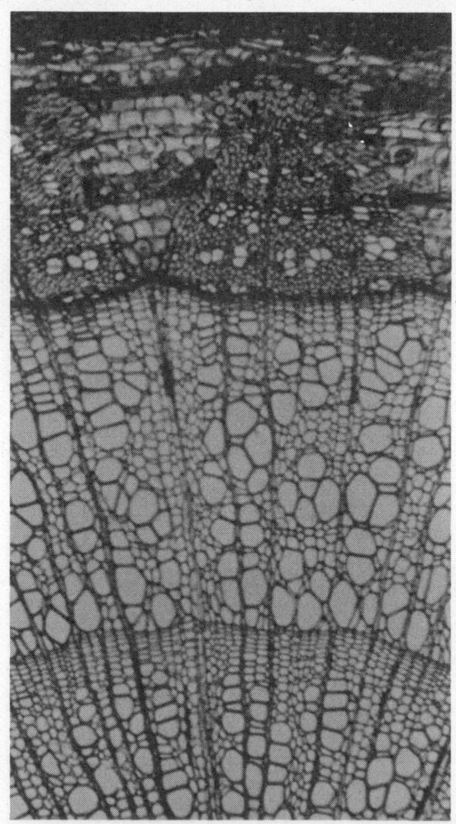

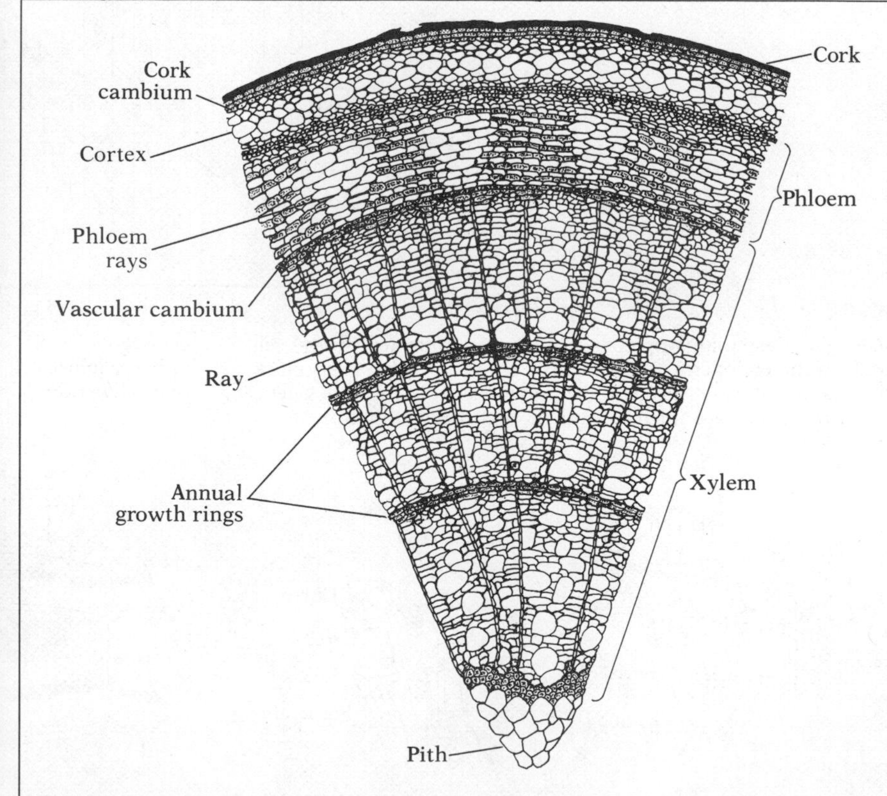

18.9 Cross-section through a 3 year old woody stem. A photomicrograph of a cross-section through a 3 year old woody stem reveals regions of different tissues. The outermost tissue is the protective cork, a suberized layer of cells produced by the cork cambium just inside. The conducting tissue consists of a circle of phloem divided by phloem rays. Just inside the phloem is a thin cylinder of vascular cambium which seasonally produces phloem and xylem. Within the vascular cambium, and making up most of the stem tissue, is the xylem, or "wood." The concentric rings within the xylem are annual rings. At the very center, a small region of pithy parenchyma remains.

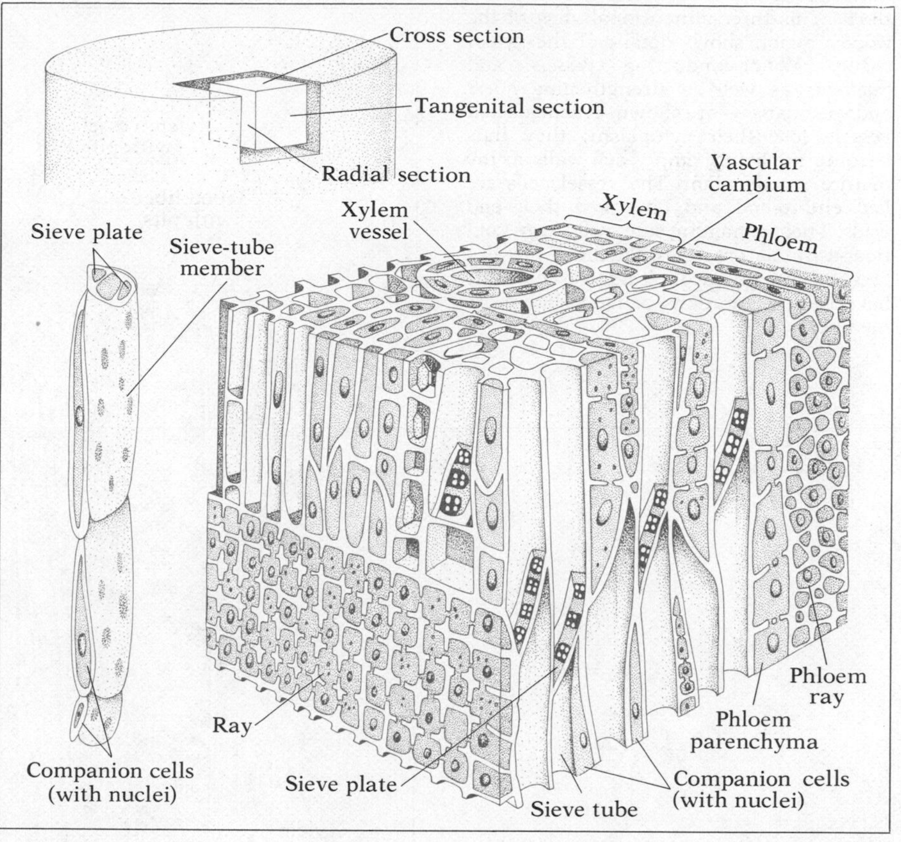

18.10 In this three-dimensional view of the stem, the region of phloem is shown in its functional detail. The cells in the background are xylem, and the large craterlike space in the center is the outline of a xylem vessel. Between the xylem and phloem regions is a layer of vascular cambium, the undifferentiated cells that seasonally produce xylem and phloem. Forward of the cambium is the phloem region. Included in the cell types of phloem are the *phloem rays, sieve tubes, companion cells,* and *phloem parenchyma.* The conducting members of the phloem tissue are the sieve tubes. Their perforated walls, the *sieve plates,* are clearly shown.

bark; to the botanist, *bark* is everything from the vascular cambium out). The cork itself waterproofs and protects the surface of the stem, thus keeping fluids where they are needed.

Phloem

Phloem lies just inside the cork of the stem, forming a ring (in cross-section; actually, a cylinder) just outside the vascular cambium. The ring is divided by rays that section the stem into a number of parts. These rays are thought to be primarily storage tissues, but they may provide lateral conduction of materials as well. The phloem tissue consists of *sieve-tube members* and *sieve tubes,* along with *companion cells, phloem fibers,* and *phloem parenchyma.* These cells represent *secondary growth*—that is, growth in the *diameter* of the stem—and are produced annually by *mitosis* in the vascular cambium.

So far, we have been considering phloem as it appears in cross-section. But perhaps a three-dimensional view would help if we are to understand phloem functionally (Figure 18.10). We are primarily interested in sieve-tube members here, since they are the food-conducting cells of the phloem. Sieve-tube members are long and narrow, arranged end-on-end. Each member ends with a *sieve plate* containing minute pores. Thus, the sieve-tube members make up sieve tubes that are interrupted by porous plates. (Incidentally, we have avoided the term *sieve cell* and have used *sieve-tube member* instead. Sieve cells refer to nucleated cells in gymnosperms, in which the attached companion cells are absent—just a technicality.)

Cytoplasm from adjacent cells is continuous through the sieve pores, so we can consider the fluid in sieve tubes to be a continuous protoplast. Strangely, the sieve-tube members lose their nuclei during development but retain active cytoplasm throughout their functional lives. The loss of nuclei must have some significance, but so far we don't know what it is. Companion cells extend along the walls of sieve tubes. We are unsure of their function, but we do know that they retain their nuclei. They may serve in a supporting, nutritive relationship with the leaderless sieve-tube members. We will get back to hypotheses about phloem function later.

Xylem

Within the vascular cambium lies the xylem, which consists of *tracheids, fibers, xylem parenchyma,* and, in flowering plants, *vessels* (Figure 18.11). The vessels and tracheids are water-conducting elements. Vessels are continuous tubes formed by nonliving, empty cell walls. As the xylem cells that form them die, their walls thicken with cellulose and lignin and their end walls become perforated or disappear. The cellulose

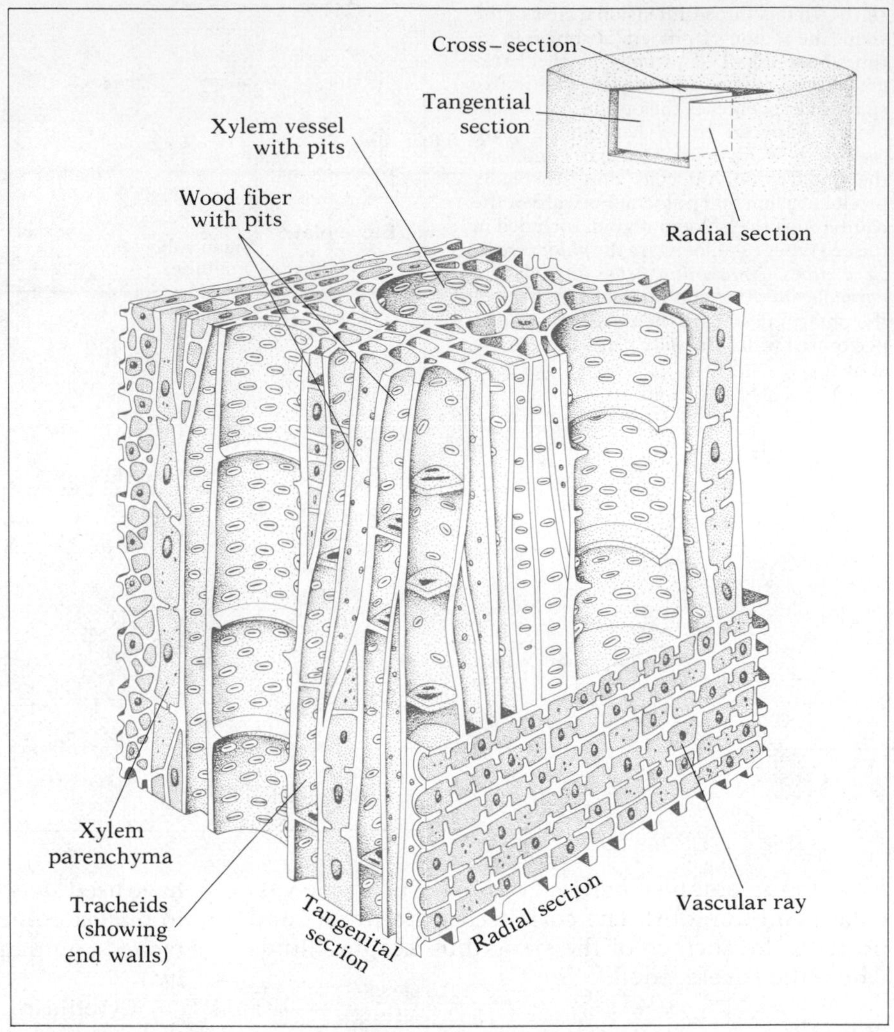

18.11 This three-dimensional view of the woody region shows details of the xylem tissue. Water-conducting *vessels* and *tracheids* as well as strengthening *fibers* and vascular *rays* are shown. Tracheids and vessels lose their cytoplasm; they thus become nonliving, empty cell walls in the mature woody plant. The vessel cells are laid end-to-end and have lost their end walls. Their remaining side walls form continuous miniature pipes. Tracheids are also laid end-to-end, but their angled end walls have thin perforations or ladderlike cross members.

may be deposited in any of a number of patterns. Thus, some *vessel elements* are ringed, others are spiraled, and still others are pitted along the sidewalls. Like the sieve tubes of phloem, xylem vessels are formed by long cells laid end-on-end.

Functioning tracheids, like vessel elements, are nonliving, empty cell walls. Most are heavily pitted, and the pits of adjacent cells are in contact. The end walls of the tracheids are also in contact, but are much more angular. In most gymnosperms and lower vascular plants, vessels are absent and water conduction in xylem is via tracheids only.

The water-conducting tissue of the stem in a woody plant consists of thousands of vessels and tracheids laid end-to-end. They form ultrathin pipelines that extend from root to leaf, carrying water and minerals to regions of food manufacture.

Vascular systems in the young shoots of woody dicots and in herbaceous plants are also composed of xylem and phloem. The arrangement of these tissues in the stem, however, is the *primary growth pattern,* which may differ considerably from what we have seen so far (Figure 18.12).

In monocots (such as grass, corn, orchids, and palm trees), and in the primary growth pattern of dicots, the vascular tissues do not form concentric, continuous rings. Instead, they are broken up into *vascular bundles.* Each small bundle contains a strand of xylem and a strand of phloem. The dicot vascular bundles, however, are arranged in a discontinuous ring, while monocot vascular bundles are scattered.

The Leaf

The vascular system of the leaf is simply an extension of the xylem and phloem of the stem. As these tissues enter the leaf they form one or more discrete vascular bundles. The bundle or bundles run parallel through the *petiole* (the stalklike portion of the leaf), and branch extensively in the *blade* (flat part) of the leaf. These branched vascular bundles are familiar to us as the *veins* of the leaf. In dicots, one enlarged vascular bundle is contained in the *midrib,* the large vein that

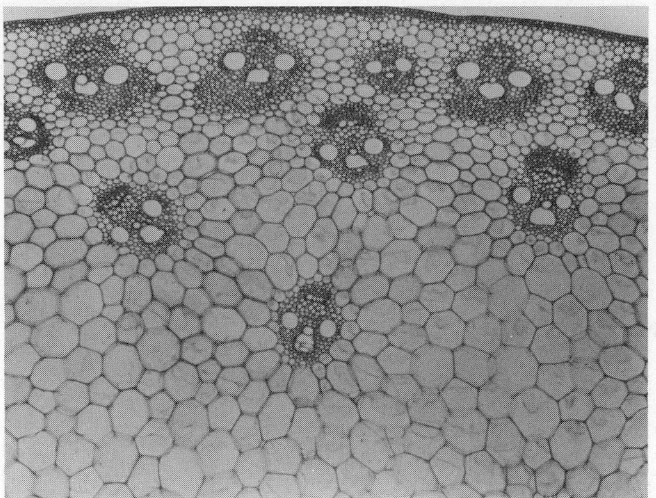

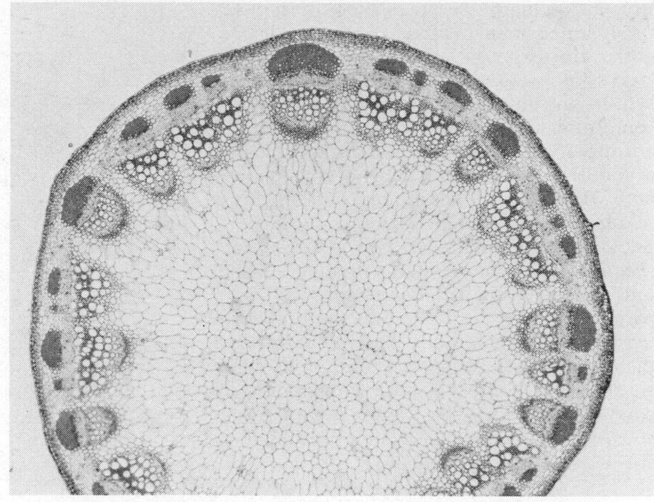

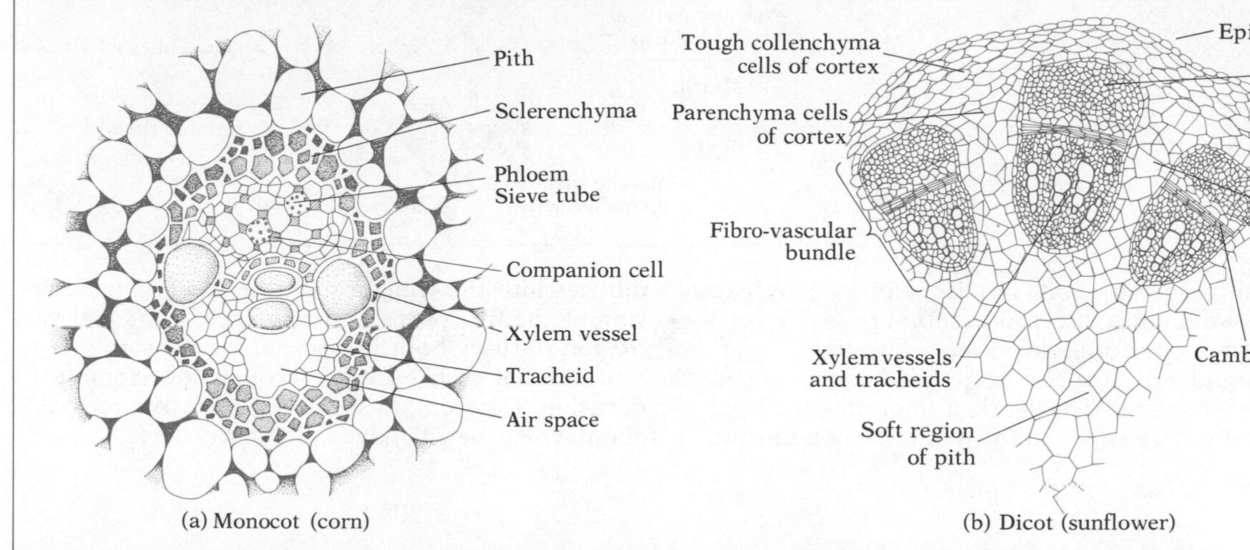

18.12 The vascular systems of herbacious stems: primary growth patterns. (a) A cross-section through a corn stem, showing the rather typical structure of monocots. The vascular system is found in bundles scattered through the pithy stem. The principal cells of each bundle include tough, fibrous sclerenchyma, water-conducting xylem, nutrient-conducting phloem and, in younger plants, bits of procambium. Monocots are incapable of secondary growth, and lack a cambium layer. (b) The stem of a herbaceous dicot, the sunflower, is similar to the young stem of a woody dicot. The vascular tissues are either in the form of a cylinder or in bundles as shown here. Note that the bundles contain elements similar to those of monocots, but they form a neater ring around the stem periphery.

runs down the center of the blade. Branches of this bundle form smaller and smaller veins, some rejoining repeatedly to complete the netlike venation that typifies this kind of leaf (see Figure 17.36). In monocots, the readily visible leaf vascular bundles appear as parallel veins, but these are interconnected with much smaller bundles that form a complex microscopic network.

In both types of leaves, the larger leaf veins are surrounded by parenchyma cells that contain few or no chloroplasts. The smaller veins may be enclosed by one or more layers of compactly arranged parenchyma that form a *bundle sheath* (Figure 18.13). The bundle sheath assures that no part of the vascular tissue is exposed to the air which lies in the intercellular spaces. All substances entering and leaving the leaf through the vascular tissue must pass through the enclosing sheath.

Extensions of the bundle sheath connect to the various rather generalized and undifferentiated *mesophyll* cells. The mesophyll cells contain numerous chloroplasts and represent the photosynthetic tissue of the leaf. Conduction of water and nutrients here must be by cell-to-cell transmission, much as in

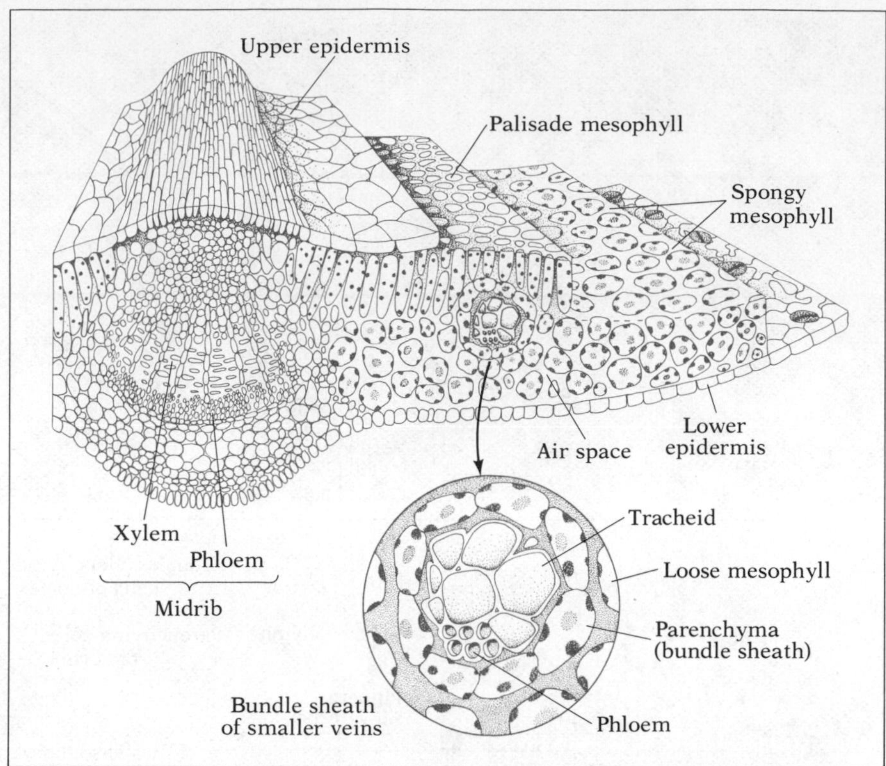

18.13 Conducting tissue of the leaf. In this three-dimensional cutaway of the leaf, notice the structure of the *midrib* and the layers of *mesophyll* tissue in the blade. Within the midrib is a vascular cylinder containing a lower region of phloem and an upper region of xylem. To the right is a smaller vein (a branching of the main one). Its structure is simpler, but regions of phloem and xylem are still apparent. Surrounding the smaller vein is a *bundle sheath* consisting of parenchyma which may or may not be photosynthetic. This parenchyma border surrounds the branching veins to their very ends, separating them from the loose, spongy mesophyll outside. The mesophyll is also important in conduction. Its moist recesses permit gaseous exchange throughout the mesophyll.

the primitive moss plant, but is facilitated by gap junctions—channels of cytoplasm that pass from cell to cell through perforations in the cell walls.

It should not surprise you that leaves contain extensive air spaces. These are, in fact, the principal avenues of gas transport in the plant. Carbon dioxide diffuses into the leaf through the *stomata* (singular, stoma), the tiny pores of the surface of the leaf, and oxygen diffuses both in and out by the same route. Water vapor also escapes through the stomata by diffusion. The stomata are surrounded by *guard cells,* about which we will have more to say later.

18.2 Mechanisms of Transport

Water Transport in Plants

Plants need a lot of water, as you know if you've ever been in charge of the family lawn. In the early 18th century, Stephen Hales (an inventor and plant physiologist) calculated that his sunflower plant "imbibed" and "perspired" seventeen times as much water as he himself did, pound-for-pound and hour-for-hour. Why do plants need so much water?

One reason is because most of the water taken up by plants rapidly evaporates. This process, in plants, is called *transpiration.* We don't know what good, if any, transpiration is to the plant. It may help to cool it on hot days, but it seems more likely that the water is lost as an unavoidable side effect of the plant's necessarily extensive exchanges with the air.

Just as animals must breathe air, thereby losing some water vapor from their lungs, plants must also "breathe." Animals are only after the oxygen in the air, but plants need both oxygen and carbon dioxide. During photosynthesis, plants have to pass enormous amounts of air through the tiny stomata of their leaves. And the plant's carbon dioxide, like the animal's oxygen, cannot move into the cells without first being dissolved in a film of water. Thus, the gas-exchange surfaces in the interior of the leaf, like an animal's lungs, must be kept wet. This, then, is where much of the water brought up from the roots goes, and a lot of it is lost through the stomata.

We know that the movement of water through the vascular system up to the leaves is a constant, pervasive process. During daylight, of course, a little water is consumed in photosynthesis, but this doesn't amount to much. Most of it, as we have noted, is lost by transpiration. Obviously, when soil water is no longer sufficient to offset these demands, the plant wilts. Wilting, which is a loss of turgor in the leaf and stem, is rarely seen in plants with deep root systems, but it may be a frequent occurrence in small herbs and seedlings. Wilting is probably just an unavoidable response, but at least it closes the stomata, which reduces further water loss.

If you're still not sure exactly how water moves through plants, don't feel too badly. We don't know either, and it's our book. In fact, the search for the mechanisms of water transport has intrigued botanists for over 250 years. Of course, water movement is rather easily visualized in the smaller plants such as mosses, ferns, and herbs. The real problem occurs when we try to explain the movement of water from deep in the ground up through the trunks of trees that are 20, 30, or even 100 m tall. Our best bet may be in reexamining the properties of water.

What causes water to move through the vessels of plants? Another way to state this question is: How can a column of water be lifted as high as the length of a football field?

Root Pressure

First, let's consider the possibility of *root pressure.* You may not realize it, but you have certainly seen one of the most dramatic results of root pressure. If you've ever gotten up early in the morning—or gotten home early in the morning—and walked across the lawn, you've probably noticed that early morning grass is wet, especially if it has been well watered. This is usually not dew; that is, it usually isn't the same kind of early morning water that glistens on your windshield. The drops of water on the grass blades have been pushed up by the roots of the grass through the plant's tiny vascular system. The blades of grass have special openings, near their tips, where excess water can escape. This loss of water in its *liquid* phase is known as *guttation,* not to be confused with transpiration, which involves the evaporation of water *vapor.* No one is sure how root pressure is developed, but it seems to be the result of an energy-requiring mechanism, a kind of "water pump."

Root pressure can be demonstrated by removing the stem from a small plant and attaching a tube filled with mercury in its place (Figure 18.14). A rise in a column of mercury is generally used as a measure of atmospheric pressure. With this apparatus it has been possible to demonstrate that roots can generate pressures of about 3–5 atmospheres. So the immediate question is: Is this sufficient to push water to the top of a tall tree? The answer is no. The weight of a 100 m column of water is at least two times too great to be supported by a root pressure of even 5 atmospheres. The weight of such a column ought to push water right out of the vascular cylinder, back into the earth. Yet the redwood cylinder stands there, with water constantly flowing upward and out of the leaves.

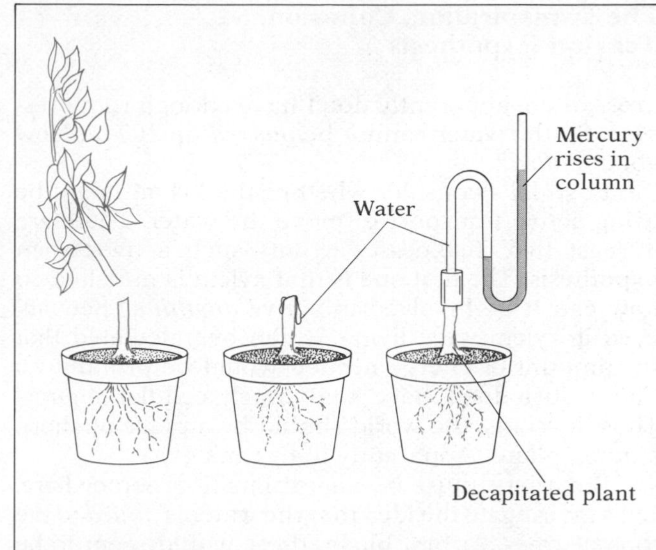

18.14 A device for measuring root pressure. The potted plant has been decapitated and the S-shaped tube is fastened to the stump. The subsequent movement of water into the plant lifts the column of mercury and the distance is measured. Such studies indicate that root pressure alone cannot account for the rise of water to the leaves of a very tall tree.

One more observation should dispel the notion of root pressure as a total explanation for water transport. The pressure inside trees is apparently less than the pressure outside. If the roots were generating great pressures, cutting into the xylem would produce a *spurt of water.* Actually, however, cutting produces a sucking action as air is drawn *in!* In summary, then, root pressure apparently plays a role in water transport, but it cannot be considered to be the major force.

Most plant physiologists now believe that root pressure, when it occurs, is only an indirect effect of the active transport of inorganic nutrients into the roots. We know that plants do expend energy to extract dissolved nitrates, potassium, and phosphates from the soil and to pump them into the xylem. Once inside, they produce an osmotic gradient, and, as you know, water moves toward regions where it is less concentrated—in this case, into the root.

This explanation doesn't seem to apply to that Israeli barley grown on seawater, however, so it's clear that something else must be involved. All we can say right now is that root pressure is one aspect of water transport that simply is not understood. Sometimes the scientists involved in studies of this phenomenon seem to be embarrassed and attempt to discount its importance. After all, we have remained ignorant about the matter for many decades. What can we say about a process that can produce as much as 5 atmospheres of root pressure in small plants that don't really seem to need it, and produces zero or negative root pressures in trees that would seem to need it most?

The Transpiration, Cohesion, Tension Hypothesis

Trees, then, apparently don't have enough root pressure. So the water cannot be *pushed* up 100 m. How about *pulled?*

First, let's consider whether the xylem might be using active transport to move the water. There are at least two weaknesses to any such active xylem hypothesis. The first one is that xylem is not alive, so how can it be involved in active *anything?* Second, even if xylem were living, it can be calculated that the amount of energy needed would be prohibitive. Plants just don't have that much cellular energy. Thus, it seems, the world should be a place of short, stunted plants. Apparently trees can't exist.

But there must be an explanation somewhere. Let's investigate the idea that the water is *pulled* to the tops of trees. At first blush, there would seem to be two difficulties with the "leaf pull" hypothesis: First, it is difficult to pull water. Second, we are still stuck with the need for enormous amounts of energy; leaves, though they are specialists in energy production, just don't have the resources.

Pulling Water

It seems that, in a sense, water *can* be pulled. The last time you drank water with a plastic straw, you expended energy to suck it into your mouth. As the column of air in the straw was decreased, water was apparently pulled along behind it. Would this work for plants?

It won't work. In sucking, you simply reduced the air pressure inside your mouth. The atmospheric pressure, which was then greater than the pressure inside the straw, *pushed* the water up. However, water can't be lifted this way over a height of 10 m because the weight of a column of water 10 m high already balances atmospheric pressure. So suction certainly won't work for our 100 m tree.

To try to find an answer that will work, let's consider two physical characteristics of water. One is the attraction of water molecules to surfaces such as glass. This attraction is called *adhesion.* The other characteristic is the attraction of water molecules to each other, which is called *cohesion.* Adhesion explains how water rises by itself in thin glass tubing (see Figure 2.19). Water adheres to the glass, tending to creep upward. The weight of the water column pulls downward at its center, thus forming a concave surface. The smaller the diameter of the tube, the higher the water will rise. If a number of glass tubes of different diameters are put into a beaker of water, the water will be seen to rise highest in the tube with the smallest diameter. This is called *capillary action.* Capillary action also causes cigarette lighter fluid to rise up a wick. And if you drape a towel over the edge of a full bathtub and leave it, capillary action will soak the towel and the bathroom floor.

So, will capillary action suffice to raise water to the top of a tall tree? The answer from the plant physiologists is a resounding, "maybe, but probably not."

Capillary action, however, is not the most popular hypothesis. Instead, plant physiologists have focused on cohesion, which gives water significant *tensile strength.* Tensile strength is the ability to be pulled without breaking—for example, steel wire has more tensile strength than cotton string, which has more tensile strength than wet spaghetti. Water would seem to have even less tensile strength than wet spaghetti. But the tensile strength of a column of a fluid depends on all sorts of complex factors. In general, the smaller the cross-section, the greater the relative tensile strength. By some extensive calculations, the tensile strength of a column of water in a tiny xylem vessel has been found to approach that of a steel wire of the same diameter!

Transpiration Pull

So capillary action allows water to rise in a tube, but it certainly wouldn't cause it to continue to flow out of the top. And tensile strength allows a long column of water to be pulled upward, but doesn't provide the pull. So what provides the pull?

Here the answer is less mysterious. The pull is provided by the *free energy of evaporation*—better known as *transpiration.* This free energy of evaporation is an indirect form of solar energy. Keep in mind that evaporation occurs only under nonequilibrium conditions (under equilibrium conditions, no evaporation would occur, because the atmosphere would be saturated with water vapor and there would be no wind. In the real world the atmosphere is usually not saturated, there is wind, and moisture evaporates).

So transpiration removes water from cells within the leaves, reducing their turgor pressure and increasing their osmotic potential. Then, through osmosis, water from the xylem tissues flows into the leaf cells. And this water pulls behind it an entire column of water. A partially depleted cell, by the way, can generate osmotic pressure equal to 12 atmospheres, which is enough pressure to lift, or pull, a column of water almost 130 m high. This is enough—finally, barely—to get water to the top of our giant redwood.

Food Transport in Plants

In 1671, just 6 years after he described the cells of a cork, Robert Hooke and a colleague, Robert Brotherton, *ringed* (girdled) a tree. That is, they removed a ring of bark around the trunk of the tree, including the soft, moist layer beneath the dead outer cork. The tree eventually died, but they observed that it

Essay 18.1 Testing the Transpiration, Cohesion, Tension Theory

The part of the hypothesis that remains problematical is the cohesion–tensile strength idea. The transpiration part is more easily demonstrated. Here we can apply the techniques used to demonstrate root pressure, cutting the stem of a plant (under water so as not to break the column) and connecting the upper stem and leaves to a glass tube of water. (If dye is added to the water, it will soon appear in the leaves and flowers: This is a trick that is sometimes used by florists to dye flowers odd and unnatural colors.) If the lower end of the water-filled glass tube is put into a dish of mercury, the column of mercury rises in the tube substantially, indicating a strong pull. Unfortunately, this demonstration doesn't mimic the real situation very well, because the mercury-in-the-tube apparatus depends on suction and atmospheric pressure, and it can't generate more than one atmosphere of "pull."

It also remains to be demonstrated that it is evaporation and not some other force that is involved in transpiration. A clever mechanical model seems to support this idea. The setup is the same as before, but this time the top of the glass tube is attached to a porous clay jug, neck down, instead of the crown of a plant. There is no air in the jug, only water. The wetted pores of the jug are too small to allow air in, but water can ooze out to the surface of the jug through capillary action. Moisture evaporates from the damp outer surface and the column of water rises, bringing behind it the column of mercury. The rise is much faster if a fan is used to blow air around the jug.

In spite of the objections to the transpiration pull model, such as the disconcerting lack of data, it is presently the preferred explanation. Its position is tenuous, however, as with all models or hypotheses. This one will continue to be subjected to relentless testing by the plant physiologists as their methods, equipment and imagination allow.

Two very clever experiments lend strong support to the transpiration, cohesion, tension theory. In one, a plant physiologist at Carnegie Institute, D. T. MacDougal, considered the enormous tension which must be placed on water columns in the xylem if the transpiration pull idea

(a) In a demonstration of the pulling force of transpiration, a fresh leafy branch is attached to a vertical glass tube that is filled with dyed water. The other end of the tube rests in a beaker of mercury. As the living branch transpires, dye appears in the leaves and water is drawn into the tube with a force that lifts the mercury up the column. Transpiration pull exists. Furthermore, energy is expended. But how?
(b) In an analogous, but completely nonliving system, a porous clay pot is substituted for the leafy branch. As water evaporates from the surface of the pot, more water is pulled in from the tube and the column of mercury rises. Transpiration does not require the expenditure of cellular energy. (And never mind the irrelevant fact that it is atmospheric pressure that really pushes the mercury up the tube in this experiment, just as it can push milkshake up a straw. In the vascular system of a real tree, if not in a glass tube, water really can be pulled as well as pushed.)
(c) Changes in the diameter of a tree over several days can be registered on a drum recording. D. T. MacDougal, testing the hypothesis that transpiration pull places enough tension on columns of water within the stem to compress the trunk, found that his intriguing data support the hypothesis. The 12 hour variations in trunk diameter correspond nicely with measured transpirational activity. Plants transpire most at midday and least at night. Thus the transpiration pull hypothesis is supported.

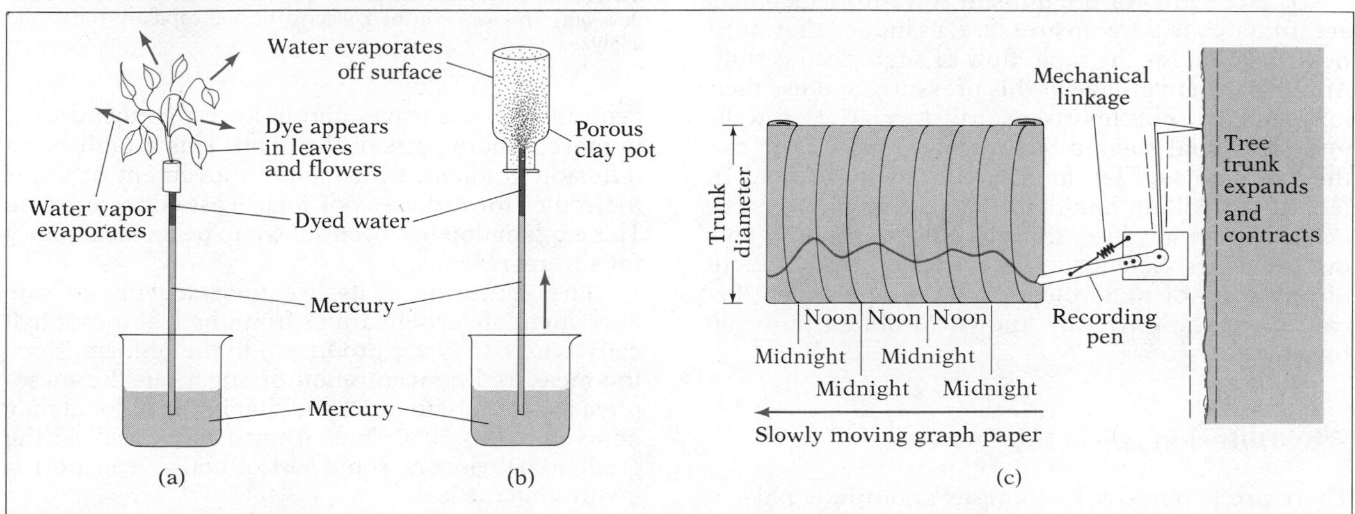

were correct. He reasoned that such a tension might effect the stem diameter, causing a slight decrease during peak activity. MacDougal tested this idea by devising an instrument which could measure minute changes in the stem. The instrument, a dendrometer (later called a dendrograph when a recording drum was added), did in fact detect the expected changes. Not only did the stem decrease in diameter, but it did so with regular periodicity. The fluctuations followed a day–night cycle. Plants transpire much less at night when evaporation rates are low. The measurements coincided with this very nicely. (You might have noted that the same data would also be consistent with the rival hypothesis of a sucking action in transpiration, that is, with the idea that transpiration flow occurs upward when pressure within the water column is reduced to less than atmospheric pressure. The reduction in trunk girth could be due to atmospheric pressure pushing inward. Of course, we know that suction alone won't work.)

continued to grow for some time, putting out new leaves and branches. They also noted something odd. The trunk of the tree increased in diameter above the ring *but not below it.* The roots did not grow, and in fact began to shrivel. Hooke and Brotherton concluded that the material a plant receives from the air is transported downward in the bark, while the material it receives from the roots is transported upward in the wood. Hooke's and Brotherton's conclusions have remained unaltered for 300 years. We now call the fleshy part of the bark, just beneath the dead outer part, the phloem, and the xylem is the wood that transports material upward from the roots.

So let's discuss the transport of nutrients in plants. Since we are interested in nutrients, and not water, we will focus on the phloem. Sugar and other nutrients move through the phloem in either direction, down toward the roots when the leaves are photosynthesizing and up toward the crown at other times, especially in the spring when new leaf buds are growing. It seems clear that some form of active transport is involved in the movement of sugar and other nutrients through the phloem.

The contents of the phloem sieve-tube members are under considerable pressure, as indeed they must be to account for the rapid flow of such viscous stuff. Aphids take advantage of this pressure, and use their long, hollow mouthparts to drill tiny artesian wells into individual sieve tubes (Figure 18.15). They can then sit back and let the nutrient-laden sap flow. In fact, if you kill an aphid and cut off its mouthparts, leaving them in place, the sap will continue to flow out for hours. Scientists sometimes do this—not out of any spirit of malevolence toward aphids, but because it is the only way known to obtain pure sap for analysis.

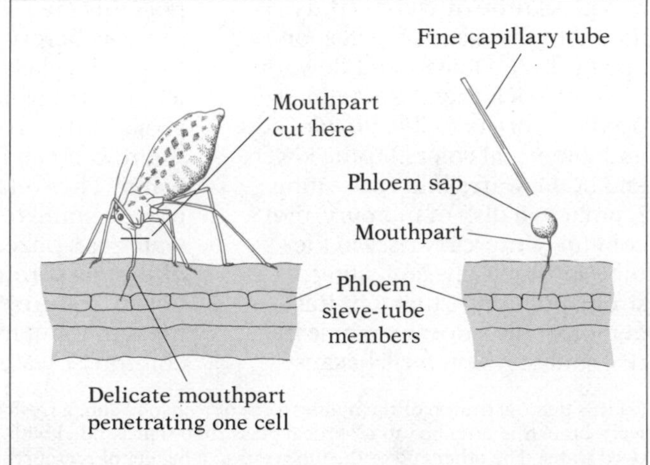

18.15 The aphid and phloem content analysis. Efforts to find out exactly what is transported in phloem sieve tubes and what mechanisms are involved have always been fraught with technical problems. An ingenious method of obtaining the pure sap involves the use of aphids, whose pipettelike mouthparts are used by the insect to suck out the contents of sieve tubes as they feed. When the aphid has pierced the phloem, it is anesthesized and the mouthpart is cut off in the head region. When the operation is successful, the mouthpart remains in place and sieve-tube fluids continue to flow out. This sap can be collected in fine capillary tubes and analyzed.

Why Diffusion Is Not the Answer

There are two major hypotheses about how phloem moves. The first was proposed in the 1930s by Ernest Munch, who suggested a diffusion process. In essence, Munch hypothesized that sugars were in highest concentration in the leaves and in lowest concentration in more remote parts of the plant. This establishes a diffusion gradient, with the net movement of sugar molecules toward areas of their least concentration. This explanation has been shown to be unsatisfactory for several reasons.

First, diffusion in itself cannot account for the movement of carbohydrates from the leaf mesophyll cells (where they are produced) to the phloem, since the measured concentration of sugars in the mesophyll is not more than 3% and that in the phloem may be as much as 30%. Thus, it must move *against* the gradient. Obviously, some sort of active transport is involved in the leaf.

Second, the sap flows too fast to be accounted for by diffusion or cell-to-cell osmotic pressure. In fact, it roars along at 3 m per hour. Calculations based on viscosity and sieve plate pore diameters, as well as actual measurements, indicate that pressures of several thousand pounds per square inch (hundreds of atmospheres) would be required to produce such movement. Although phloem sap is indeed under pressure, the pressure is nowhere near that great, and such force could not be produced by osmotic pressure.

The third objection is that different food molecules may show different patterns of movement. We have already noted that phloem transport can go in opposite directions at different times; more remarkably, different nutrients can go in opposite directions *at the same time.* Even different nutrients going in the same direction may move at very different rates. The diffusion notion, it seems, just won't hold water.

At least half a dozen alternative hypotheses have been proposed seriously. Some of them are very vague (but perhaps correct). For instance, several British physiologists have suggested that the sap either moves nutrients by active transport or somehow manages to reduce resistance to diffusion. All well and good, but *how?* There is no need to go into all the hypotheses here, but we do think one other major hypothesis, the *cyclosis* or *cytoplasmic streaming* theory, deserves some explanation.

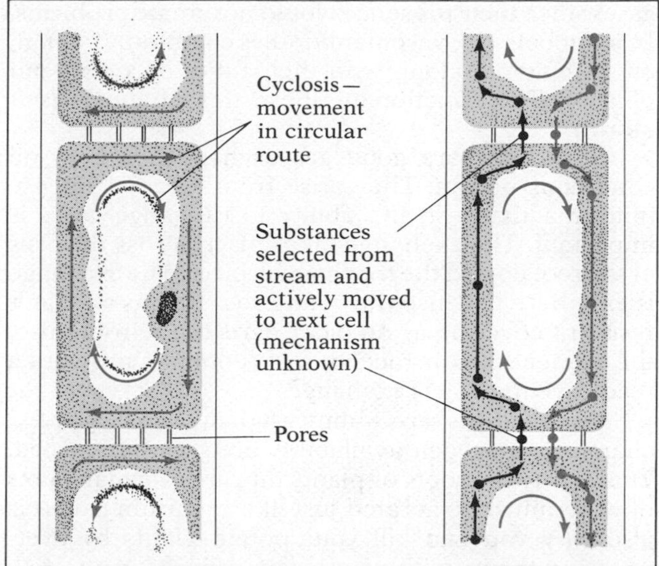

18.16 One theory of phloem transport deals with the phenomenon of *cyclosis,* or *cell streaming.* This action has been observed in cells other than phloem for many years. In the diagram shown here, cyclosis is shown to move substances in opposite directions at the same time, as proposed by John Field. Since the stream is composed of substances dissolved in living cytoplasm, the theory is tenable. Some kind of sorting process would be necessary to direct some substances downward and others upward. Perhaps a type of metabolic pump is employed. The idea requires considerable further study.

Streaming Cytoplasm

The *cyclosis theory* has been supported recently by English and American physiologists, but it was first proposed back in 1885 by the Dutch botanist Hugo de Vries (who was later to gain fame as one of Mendel's three rediscoverers). If you look at some plant cells under an ordinary light microscope, you will see tiny currents of cytoplasm cruising around and around (Figure 18.16). Apparently, the cell fluid is moved by the actin and tubulin cytoskeleton, although not very much is known about the phenomenon. In any case, the streaming is called *cyclosis,* and an English botanist, John Field, has reported seeing very rapid streaming in the phloem of wheat. So perhaps the food molecules in sap are carried along by these tiny rivers. The cyclic movement within cells could even account for bidirectional flow, provided that there is some way to pick up the proper nutrients at one end of the cell and deposit them at the other.

In spite of Field's reported observations, and the fact that cyclosis is commonly seen in many cells, most investigators have been unable to find cyclosis in the sap of mature phloem cells. Thus, the issue remains unresolved and the cytoplasmic streaming hypothesis is a tenuous one at best. In any case, active, energy-requiring mechanisms are undoubtedly involved, since only living and respiring phloem has the ability to move sap. Any movement by diffusion must be buttressed by a considerable amount of work by the plant.

Gaseous Exchange in Vascular Plants

Plants continually exchange gases with the environment while carrying out both photosynthesis and respiration. But, in general, plants don't deal with gases too well. For example, few plants have developed any way to move gases around within their own tissues. Instead, gas is transported by diffusion where thin-walled cells of the plant contact the air, such as in root hairs, pores along the stem (*lenticels*), and the stomata of the leaves. The oxygen–carbon dioxide exchange found in all metabolically active cells of the plant may be assisted by some form of active transport, but no one is sure. Of course, air bubbles must be kept out of the xylem, since bubbles break the cohesion and capillary action in these columns. If this should occur, water must move into adjacent columns through pits in the walls of the vessel (see Figure 18.11). Since the xylem itself is, for the most part, nonliving, it doesn't need to exchange gases through its walls. It is more likely that living phloem is more intimately involved in carrying dissolved gases since their presence would not create problems. In monocots, the vascular bundles commonly contain an air channel along with their tubes of xylem and phloem. The function of these tiny channels isn't known.

Root hairs are good gas exchangers when soil conditions permit. They arise from areas of intense mitotic activity, so the ability to exchange gases is important. The exchange takes place across the cells of the root tip and the root hair surfaces. The exchange is easy here because the young root tip has no water resistant covering as do older parts of the root. Also, the tremendous surface area of root hairs provides a ready avenue of gas exchange.

Where roots are submerged in water, gas exchange and oxygen availability are greatly retarded. This is why the roots of plants that are grown in tanks of water must be aerated just like aquarium fish. It is also why you can kill your potted plants by overwatering them: without gaseous oxygen in the soil, they simply drown. Some plants, however, have adapted to flooded, oxygen-poor soils. These include swamp plants and rice. It turns out that many of these plants have large hollow air channels in their stems (not to be confused with the much smaller channels associated with some vascular bundles). The "knees" of cypress trees reach above the stagnant waters and absorb air into their loosely arranged tissue. The air then diffuses down into the roots (Figure 18.17).

The lenticels, which occur in the stems of some plants, are a type of pore that allows gas exchange. These are eruptions in the cork—large enough to be readily seen in smooth bark—filled with loosely arranged parenchyma cells.

Guard Cells

Guard cells mark an important adaptation to the land environment. As we mentioned earlier, when aquatic plants made the evolutionary transition to land, they faced the obvious risk of drying out. Plants developed a watertight epidermis, and that helped, but obviously they couldn't afford to seal themselves off completely. In their continuing adaptation, the higher plants evolved guard cells. These help to close the *stomata* (pores) when conditions dictate (Figure 18.18). Guard cells are located mainly on the underside of leaves, but are also found on the green stems of many herbaceous plants and leafless plants such as cacti. Primitive guard cells are found in the waxy cuticle of moss sporophytes.

Guard cells regulate the size of the stomatal openings by swelling or shrinking. A pair of guard cells can press together and close the stoma or they can pull apart to allow gas in and out of the plant. They open and close in response to their own internal turgor pressure. In general, the guard cells increase their turgor (swell) and thus open during daylight periods, and they close by decreasing their turgor (shrinking) at night. Thus, gas is allowed to pass through the stomata when the plant is engaged in photosynthetic activity.

What causes the cells to open and close? At first glance (Figure 18.18), it might appear that an increase in turgor should close the stomatal opening. But it doesn't. The key to its action lies in the inner walls, which are much thicker than the rest of the walls. Increased turgor, we see, results in a general rounding of the guard cells. The thinner parts of the cell walls stretch more than the thicker part. This, in turn, bends the thicker walls into a concave shape, increasthe size of the stoma. A loss of turgor permits the semirigid walls to return to their former shape, closing the stoma.

(In cactuses and other plants that are adapted to extremely dry climates, the opening and closing cycle

18.17 The bald cypress (*Taxodium distichum*), with its roots submerged in stagnant water, has adapted to its oxygen-poor environment by producing "knees." These root extensions rise well above the surface of the water and absorb air through their spongy tissues. The air can then diffuse into actively growing root regions.

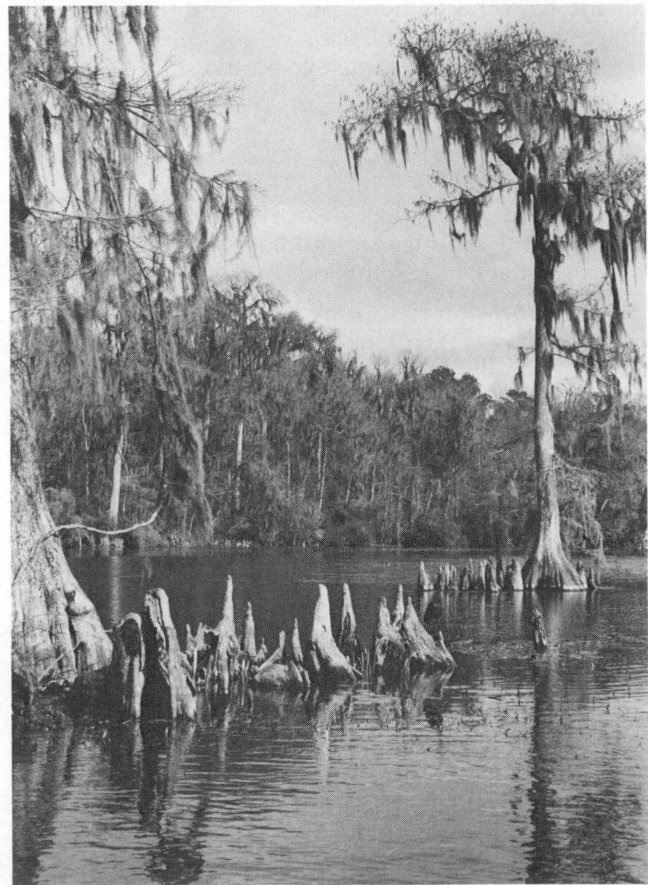

of the stomata is reversed. In these plants, the stomata are closed in the daytime and opened at night. This helps to prevent water loss; but it means that access to atmospheric carbon dioxide is cut off during photosynthesis, when it is needed. To get around this problem, carbon dioxide is fixed at night. So, the plants have to use their own starchy energy reserves at night to provide the necessary $NADH_2$ and ATP to fix atmospheric CO_2 into organic compounds. In the daylight, the carbon dioxide is released within the cell and subsequently recycled back into glucose, and starch, using energy from photosynthesis.)

How does turgor increase and decrease in guard cells? We don't know. Unfortunately, all the data are not in, and much of what has been reported is conflicting. For a long time a major clue seemed to be the presence of chloroplasts in the guard cells, since they are usually lacking in other epidermal cells. The reasoning was as follows: Chloroplasts carry on photosynthesis, and photosynthesis results in the production of sugars. The presence of sugar molecules increases the solute concentration in the guard cell, setting up an osmotic gradient between the cell and the surrounding watery environment. Hypothetically, then, water should enter the cell by diffusion and cause an increase in turgor. This, in turn, would open the stomata. Then, when the sun goes down, the processes slow down, sugar is metabolized, water diffuses out, turgor decreases, and the stomata close. It's a rather neat explanation. It means that the plant conserves water at night when CO_2 isn't needed for photosynthesis.

But it's wrong. As usual, some restless soul came along and carried out one experiment too many. The sugar–osmosis hypothesis didn't hold water, as it were. It turned out that observed changes in turgor occur much too fast to be explained by the production of sugar solutes in photosynthesis.

A second, related hypothesis was that the depletion of CO_2 by photosynthesis raises the pH of the guard cell, permitting pH-sensitive enzymes to hydrolyze starches stored in the cell sugars. This would quickly increase the solute concentration, thus greatly hastening the osmotic process. Unfortunately, there are guard cells that don't contain any starch—and they also open and close!

So we have a third hypothesis. Plant physiologists have found that plants in direct light show an increase

(a)

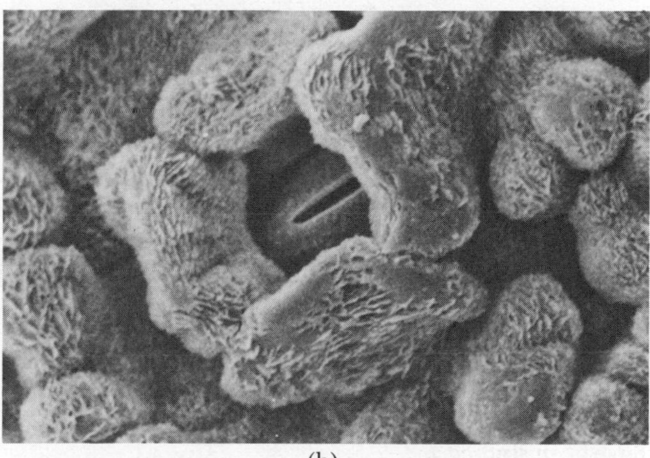

(b)

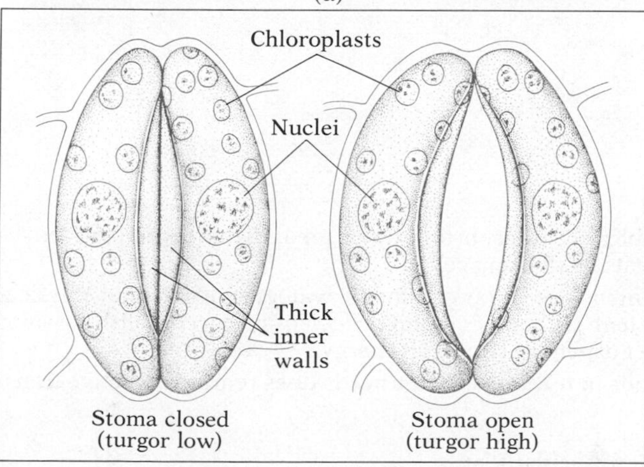

(c)

18.18 Guard cells must allow gases to pass and yet prevent excessive water loss. These cells are distributed over the epidermis of leaves and green stems. (a) The light micrograph shows two open guard cells in the epidermis of a leaf. (b) The scanning electron micrograph reveals a three-dimensional view. (c) The guard cells in the drawing look like twin sausages, with their curved shapes surrounding the stoma. Note the presence of chloroplasts in the cells. They are not found in any other epidermal cells. Note also that the inner walls of each guard cell are thicker than the rest of the cell wall. These thickened walls apparently resist stretching, and when turgor is lost, they close the stomal opening.

in the active transport of potassium ions into the guard cells. It turns out that only the blue part of the light spectrum is effective, because the process is triggered by a blue-sensitive pigment called *phytochrome.* This active transport of potassium increases the solute concentration in the cell, speeding the uptake of water, which subsequently opens the stomata (Figure 18.19). Unfortunately, the *potassium transport mechanism,* which is now well-substantiated, has not answered all the questions. The link between blue light and the guard cell opening may be direct, with phytochrome itself powering potassium transport; or it may be quite indirect, with the light reactions providing a coded signal for a potassium transport mechanism. The latter explanation is preferred at the moment, although the relationship between light reaction activity and K^+ pumping is not known.

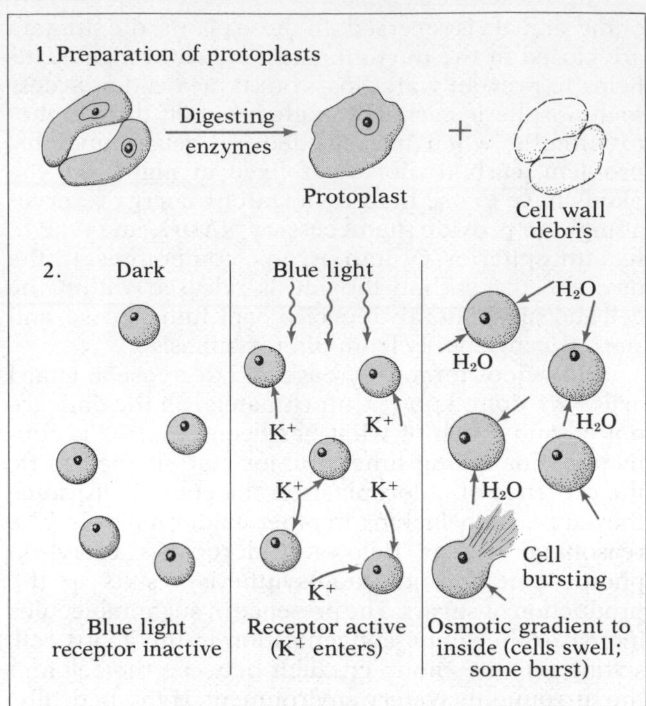

18.19 Recent studies of potassium transport in guard cells indicate that a process utilizing blue light is involved in the changes in turgor. Researchers have subjected guard cell protoplasts (cells without walls) to light and found that blue light (410 nm) stimulates the rapid uptake of K^+ followed by the osmotic influx of water, swelling the guard cell protoplast as much as 50%.

Application of Ideas

1. The formerly popular idea of *hydrotropism* that explained the growth of roots "toward" water has been discarded. Suggest a reasonable hypothesis that might explain how random growth and differential survival enable plant roots to "find" water.

2. An important characteristic of perennial plants is indeterminate growth. What is this and what characteristic of plant growth and development provides for indeterminate growth? Do any animals share this characteristic? (You may want to return to this question again when animals have been studied.)

3. Compare plants to animals in terms of system organization. Animals are organized into organ systems including, among others, the nervous, respiratory, digestive, excretory, vascular, dermal, immunological, and endocrine systems. Do plants have systems? If so, what are their systems and how do they compare to systems in animals? What animal systems are absent in plants and in what ways, if any, do plants compensate for the lack of these systems?

Key Words

adhesion
annual ring
apical meristem

bundle sheath

cambium
capillary action
Casparian strip
cohesion

Key Ideas

Introduction

1. Plants have the problem of transporting water great distances to their leaves and distributing nutrients to all living cells.

2. Tallness in plants intensifies the problems of water transport and physical support. The *woody* stem with its complex *xylem* tissue accomplishes water transport with no direct expenditure of ATP energy.

3. The transport of foods in meeting cellular needs does require an active transport process.

Vascular Plants and Nonvascular Plants

The vascular plants (*tracheophytes*) include the great majority of plants seen around us.

collenchyma
companion cell
cork
cork cambium
cortex
cyclosis

dendograph
diffuse root

elongation
embryonic tissue
endodermis
epidermis

fiber

guard cell
guttation

hydrostatic pressure

lenticel
lignin

maturation
meristematic tissue
mesophyll
midrib

osmotic gradient

parenchyma
pericycle
permanent tissue
petiole
phloem fiber
phloem parenchyma
phloem ray
primary phloem
primary xylem

root
root cap
root hair
root pressure
root tip apical meristem

sclerenchyma
secondary growth
secondary tissue
sieve plate
sieve tube
sieve-tube member
steady state
stele
stoma (stomata)
suberin

tap root
tensile strength
tracheid
tracheophyte
transpiration
transpiration-pull
turgid

Water and Food Transport in Mosses
1. Mosses and other bryophytes have no vascular tissue. Transport is through cell-by-cell diffusion and by active transport.
2. Water transport is through *osmosis.* Water maintains the soft plant in a state of *turgidity. Wilting* occurs with a loss of water. A *steady rate* is reached when water entering is balanced by water loss.

Transport and Support Structure in Vascular Plants
Water transport is carried out in nonliving *xylem,* while nutrients are actively transported in the *phloem,* a living tissue.

The Root
1. In addition to water, *roots* absorb mineral nutrients. The intake of water and minerals occurs across the *root hairs* of root tip *epidermal* cells. Water absorption by root hairs is a passive osmotic process.
2. The total root system, is about equal in size to the total foliage and stem.

Root Types
Two major types of *root* system, *diffuse* and *tap,* are found, with many variations on these. Grasses have enormous *diffuse root* systems, while *woody* plants typically produce *tap root systems,* with one main *root* and many branches. This helps in anchorage as well as in water transport.

The Root Tip
1. *Root tips* contain *apical meristem,* a region of *embryonic* cells. Below the *apical meristem* is a dense cellular *root cap,* while above is a region of *cell elongation* where cells lengthen, mature, and differentiate. Just above the *meristem, epidermal cells* produce extensive *root hairs,* whose function is absorption.
2. The *embryonic* region continually supplies cells above and below, while retaining some in an *embryonic* state. The *root tip* is pushed along through the soil by the *elongation* process.

Root Vascular System
1. The *vascular tissue* begins to appear in cells above the *root tip meristem* as maturation occurs. Further up, a cylinder of *vascular tissue,* the *stele* takes form. Between the *stele* and the *epidermis* a region of thin-walled *cortex* is found. The *stele* consists of an outer *endodermis,* a layer of *parenchyma, pericycle, primary xylem, phloem,* and *cambium.*
2. In *roots* capable of *secondary growth,* the *vascular system* becomes more complex. A *suberized* wall, the *Casparian strip,* forms in the *endodermis.* Its role is to force water to enter the *stele* through the endodermal cytoplasm. This is believed to set up a control system, preventing backflow.

The Stem of a Woody Plant
1. Stems function in foliage support, transport, and storage.
2. *Woody* stems consist of several types of tissue, arranged and functioning as follows:
   a. Central core: *xylem* and related tissue; conduction of water and support.
   b. *Vascular cambium: meristematic* tissue; produces *xylem* inside and *phloem* outside.
   c. *Phloem: phloem* and companion cells; conduct nutrients.
   d. *Phloem rays: parenchyma;* conduction, storage.
   e. *Cork cambium: meristematic tissue;* produces *cork.*
   f. *Cork: suberized* tissue; protects stem from water loss and parasites.

Phloem
1. *Phloem* is a tissue consisting of *sieve-tube members, companion cells, phloem fibers,* and *phloem parenchyma.*
2. *Sieve-tube members* are arranged end-to-end. Cytoplasm extends from one cell to another through sieve plates. The *sieve-tube members* are living but have no nucleus. In angiosperms they are accompanied by nucleated *companion cells.*

vascular bundle
vascular cambium
vascular plant
vascular system
vein
vessel

woody

xylem

Xylem

1. *Xylem* consists of *tracheids, fibers, xylem parenchyma,* and in angiosperms, *vessels.*
2. *Vessels* and *tracheids* conduct water. End walls of *vessels* are perforated or are absent. *Tracheids* are also pitted. The two form continuous microscopic tubes.
3. The *vascular system* in monocots and many herbaceous dicots consists of vascular bundles rather than rings of vascular tissue.

The Leaf

1. The stem *vascular system* extends into the leaf via the *petiole* and *midrib.* The *vascular system* forms *veins* that contain tough, fibrous *collenchyma* tissue.
2. *Veins* in dicots branch into a netlike arrangement, but are parallel in monocots. Smaller *veins* are surrounded by *parenchyma,* forming a *bundle sheath.* (These cells are important in C-4 photosynthesis.)
3. Leaves contain numerous air spaces for gas transport. Air enters and leaves through the *stomata,* which are surrounded by pairs of guard cells.

Water Transport in Plants

Water is unavoidably lost to plants through leaf evaporation, a process known as *transpiration.* The inner surfaces of leaves are kept moist to act as an absorbing surface for gases. Because of loss, water is in great demand. The problem of water replacement and its rise to the tops of tall trees has yet to be fully resolved. There are several hypotheses.

Root Pressure

1. Evidence of *root pressure* is seen in plants that ooze water through their leaves, a process known as *guttation.*
2. *Root pressure* alone cannot account for water rise, since there is not enough pressure to force water up through tall trees. Part of *root pressure,* the inward movement of water under force, is the active intake of minerals that can create an inward *osmotic gradient.*

The Transpiration, Cohesion, Tension Hypothesis

1. The consensus is that *xylem* does not utilize active transport.
2. The "pull" hypothesis has two requirements. A force to pull and the expenditure of great amounts of energy.

Pulling Water

1. Two forces, *adhesion* and *cohesion,* help explain the pull idea. *Adhesion* is the tendency for molecules of different substances to cling together (attract each other). *Cohesion* is the tendency for similar molecules (water) to cling or attract each other. (Cohesion in water is explained in Chapter 2.)
2. The two forces explain the rise of water in capillary tubing, and the phenomenon is called *capillary action.*
3. The water cohesive force in *xylem* is so great that the water column has the strength of steel wire of the same diameter.

Transpiration Pull

The pulling force on a column of water is provided in plants by the free energy of evaporation (*transpiration*). As water evaporates from the surface of cells, the *osmotic gradient* brings in more, and the movement creates the pull on the *xylem* columns. Each depleted cell can generate a pulling pressure equal to 12 atmospheres. This is enough for the tallest tree.

Essay 18.1 Testing the Transpiration, Cohesion, Tension Theory

Attempts to measure the pulling force:

1. The force of evaporation can also be demonstrated.
2. *Dendrographs* have established that the diameter of a tree decreases when *transpiration* rates are high, as the water column is "stretched."
3. Studies with thermocouples indicate a faster rate of rise in the upper parts of a tree in the early hours.

Food Transport in Plants

Observations by Robert Hooke long ago determined that sugars are transported in the *phloem* region. Sugar movement goes in both directions in *phloem,* depending on the needs of the plant. Active transport is indicated. Aphid mouthparts have been used to study the flow and content of *phloem* tubes.

Why Diffusion Is Not the Answer

1. The relative concentrations of sugars in plants indicate that diffusion cannot account for most of its transport.
2. In addition, sap flowing through *phloem* moves too fast to be explained by diffusion. Further, the movement of food molecules can go in either direction, in spite of the concentration gradients.

Streaming Cytoplasm

Cytoplasmic streaming (*cyclosis*) could explain the differential movement of food solutes, but how these are selected out of the moving stream remains to be explained.

Gaseous Exchange in Vascular Plants

Gas movement in plants depends on diffusion, probably through the *phloem.* Air channels are seen in the *vascular bundles* of monocots, and larger air spaces occur in some plants, particularly in land plants that are adapted to marshes. Cypress knees are an example.

Guard Cells

1. *Stomata* surrounded by pairs of *guard cells* permit gases to be exchanged, while preventing excessive water loss. *Guard cells* open and close in response to changing turgor pressure. Opening and closing in most plants cycles with day and night.
2. Increase and decrease in turgor pressure in *guard cells* has been explained in a number of hypotheses.
   a. *Guard cells* have chloroplasts. When photosynthesis occurs, sugars produced create an *osmotic gradient,* bringing in water. *Criticism: Turgor* changes occur too rapidly for this mechanism.
   b. Depletion of CO_2 raises *guard cell* pH, permitting enzymes to hydrolize starches, freeing sugars and increasing solute content. Water enters following the new gradient. *Criticism: Guard cells* that do not contain starch also open and close.
   c. In direct sunlight, plants increase the active transport of potassium ions into the *guard cells.* This increases the solute concentration and water enters. The light receptor mechanism is known but its action is unknown. *Criticism:* This new hypothesis has many unanswered questions.

Review Questions

1. Describe the transport problem created by competition for sunlight. (p. 503)
2. What two main purposes do the *xylem* and related tissues serve in the tree? (p. 503)
3. How do the nonvascular plants transport water and foods throughout their bodies? (p. 504)
4. All movement requires energy. Explain the sources of energy for maintaining a *steady state.* (p. 504)
5. Compare the *root* area with that of the foliage in a tree. (p. 505)
6. Describe and give examples of the two main types of *root* systems. (p. 506)
7. Prepare a simple drawing of a *root* tip and label the following: *apical meristem, root cap,* areas of *elongation* and *maturation, embryonic tissue,* and *root hairs.* (p. 507)
8. List the types of cells and tissues you would encounter in a mature region of a young *root,* starting from the outside. (p. 507)
9. Describe the structure and importance of the *Casparian strip.* (p. 507)
10. List the tissues found in a young *woody* stem and write a general function of each. (p. 510)
11. List the contents of *phloem* tissue. (p. 511)
12. Describe the structure of *sieve-tube* members. (p. 511)
13. List the components of *xylem* and describe the conducting cells in detail. (pp. 511–512)
14. How does the vascular tissue of gymnosperms differ from that of angiosperms? (p. 512)
15. Compare the arrangement of the *vascular system* in monocots and dicots. (pp. 512–513)
16. Describe the location and function of the *bundle sheath.* (p. 513)

17. What is the purpose of air spaces in the leaf? (pp. 513–514)

18. What is the relationship between *transpiration* and gas exchange? (p. 514)

19. What adaptation does *wilting* accomplish in the *guard cells* of a plant? (pp. 514–515)

20. According to plant physiologists, *root pressure* is related to the intake of mineral nutrients. Explain this. (p. 515)

21. Summarize the arguments that have been used to argue against the importance of *root pressure.* (p. 515)

22. List two reasons why active transport in *xylem* is not a reasonable explanation for the rise of water. (p. 516)

23. Explain capillary action in terms of *adhesion* and *cohesion.* (p. 516)

24. What strange-sounding characteristic of water makes the pull hypothesis tenable? (p. 516)

25. Summarize three lines of evidence that indicate that water is actually PULLED through the *xylem.* (Essay 18.1)

26. Explain why girdling a tree (as Hooke did) kills the *roots,* but permits growth to continue for a time above the cut. (pp. 516, 518)

27. Criticize the Munch hypothesis for sugar movement in the *phloem.* (p. 518)

28. What is *cyclosis* and what is its hypothetical basis? (p. 519)

29. What problem remains in explaining *phloem* transport through *cyclosis?* (p. 519)

30. List the areas where gas exchange is possible. (pp. 519–520)

31. What are the special adaptations made by plants that grow in marshes and swamps? (p. 520)

32. Explain the mechanism of *guard cell* opening. (p. 520)

33. Summarize the photosynthesis hypothesis for *guard cell* operation and review the criticism against it. (p. 521)

34. Criticize the pH shift hypothesis of *guard cell* operation. (p. 521)

35. Review the potassium transport mechanism of *guard cell* operation. (p. 522)

Chapter 19

Mineral Nutrition and Metabolism in Plants

A young willow sapling is cleaned of all debris, weighed, planted in a tub of carefully weighed soil, and protected and nurtured for 5 years. At the end of that time, the tree is removed from the soil, cleaned, and weighed again. The soil is also weighed again. It turns out that the willow has gained 164 lb. What about the soil? Would you expect it to weigh less? The same?

If you were an Aristotelian, you would probably say that the soil should weigh less. Furthermore, if Jan Baptist van Helmont had not performed this very experiment, he would probably have agreed with you. But he did the experiment, and it made him the world's first plant physiologist.

For a man of the 17th century, van Helmont was unusual in that he sought answers to questions about life processes from chemistry and physics, which of course were still in a rather primitive state. Van Helmont was also a mystic of sorts, who believed in "vital forces" and spontaneous generation of life. But through his experiments, he obtained some interesting results that gave his work a different aspect from that of his contemporaries, who dealt mainly in the realm of philosophy.

Now, what about the soil? It seems that van Helmont's soil lost only 2 oz in 5 years! So Aristotle was wrong; plants don't generate their bodies from the soil alone. Then how did the willow gain 164 lb? Where did this mass come from? Van Helmont seized upon the obvious: "It's the water!" he cried. And he was half right (Figure 19.1).

It wasn't until late in the next century that the mysteries of photosynthesis began to be unravelled. In fact, it was over a century after van Helmont's death that Joseph Priestley crossed the English channel to chat with Antoine Lavoisier about the strange gas that was emitted by animals and seemed to support plant life. It was too bad that Priestley didn't

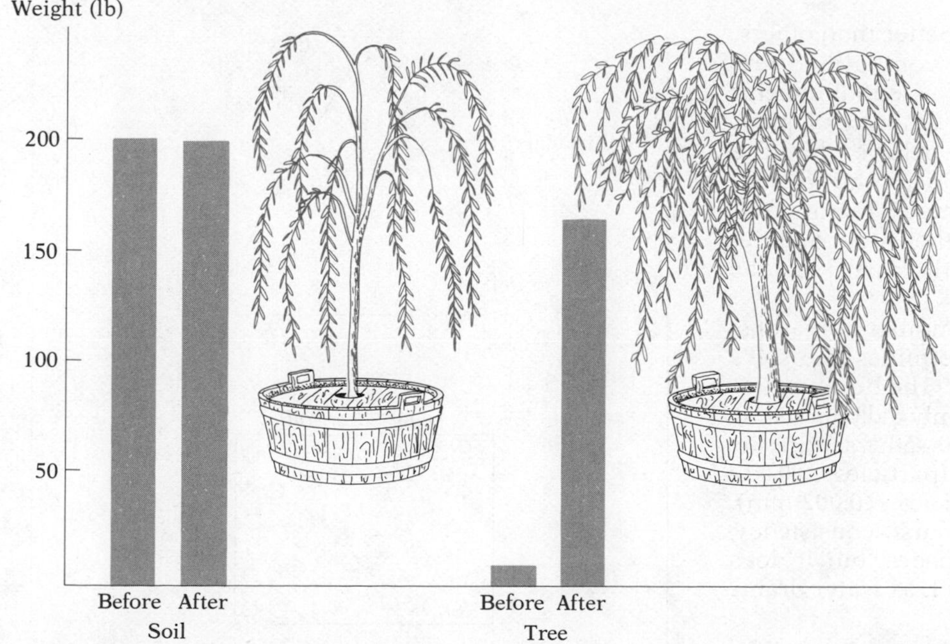

19.1 Van Helmont's experiment. "I dried 200 pounds of earth in a furnace and then put it into a large pot. The earth was then moistened with rainwater and a small willow tree was planted. At this time the willow tree weighed 5 pounds. The earth was kept moist with rainwater or distilled water. Nothing else was added. In order that no dust would get into the pot, and add to the weight of the earth, I put a cover on the pot, leaving a hole for the trunk of the tree.

"After five years the entire tree was removed and found to weight 169 pounds and 3 ounces. Since the tree had weighed 5 pounds at the start, the gain in weight was more than 164 pounds.

"I removed the earth from the pot, dried and then weighed it. It weighed 199 pounds and 14 ounces.

"Therefore, 164 pounds of wood, bark, and roots were formed out of water only." (Source: M. Gabriel and S. Fogel, *Great Experiences in Biology,* Prentice-Hall, 1955.)

know about the work of a Joseph Black, who had carefully described carbon dioxide 25 years earlier. Scientists of that time didn't have the advantages in communications we have today. There were very few scientific societies, and publications were rare.

The van Helmont experiment, aside from its historical interest, draws our attention to some significant points. First, a half-right hypothesis based upon experimental data is better than assumptions with no supporting data, even if the assumptions are ancient and revered. Second, and more specifically, the total mineral nutrient uptake in a plant body is amazingly meager in comparison to the quantity of hydrogen and carbon incorporated from the atmosphere during photosynthesis. But meager doesn't mean unessential. No matter what the availability of water and carbon dioxide, a seed cannot produce a full-grown plant without mineral nutrients.

Plant nutrition, to clarify our terms, includes two categories: One is *food*, which the plant manufactures and uses. This category includes proteins, carbohydrates, lipids, and other organic substances. The other category—the subject of this discussion—includes the *mineral nutrients*. These are inorganic substances, primarily ions, that the plant requires.

The nutritional aspects of plant physiology are vitally important. This becomes painfully obvious in the light of world food supply problems. We now face the prospect of feeding more people with less cultivated and productive land per capita than ever before. We're adding about 70 million people to the planet each year and just about all the land that can be economically cultivated is already under production. The point here is that the soil is important to plants other than as an anchor. It is a reservoir of water and minerals.

Some reservoirs, however, are better than others. Soils vary greatly in their mineral content and texture. A shortage or absence of any essential mineral nutrient means a limited type or amount of plant growth (Figure 19.2). Ecologists believe that the availability of soil nutrients comprises the *limiting factors* of plant communities. Climatic factors, soil type, and mineral availability together mold and define the great plant complexes of the earth, such as grasslands, deserts, and jungles.

Soil texture is important for a number of reasons. The texture, or particle size, determines the soil's ability to retain minerals and water. The best mixture for general plant growth is commonly called *loam*. It has a very specific consistency: 30–50% sand (particles >0.02 mm), 30–40% silt (particles >0.02–0.002 mm), and 10–25% clay (particles <0.002 mm). Sand and silt alone produces a coarse consistency, which is good for retaining air spaces, but it does not retain water or minerals. In fact, as water drains through sand, it leaches the minerals away. Clay, on the other hand, has excellent water- and mineral-binding qualities, but it is dense and does not form air spaces readily. Loam, produced by the proper combination of sand, silt, and clay, permits soil to retain organic matter, which becomes fixed in the clay. The presence of organic matter, then, greatly increases the soil's water-holding capacity (Figure 19.3).

19.2 This soil has been depleted of one or more of the essential soil nutrients by poor management. For the foreseeable future, growth of plants will be extremely limited here.

19.3 The texture of soil has a great deal to do with its ability to hold water and minerals. In porous sandy soil, minerals are quickly leached away by water percolating into the regions below. Clay, on the other hand, binds water and minerals, but is often deficient in oxygen. The best soils are a mixture of clay, silt, and sand. The soil triangle illustrates the size of the three soil particles. A correct mix produces loam, an ideal combination for most plant growth.

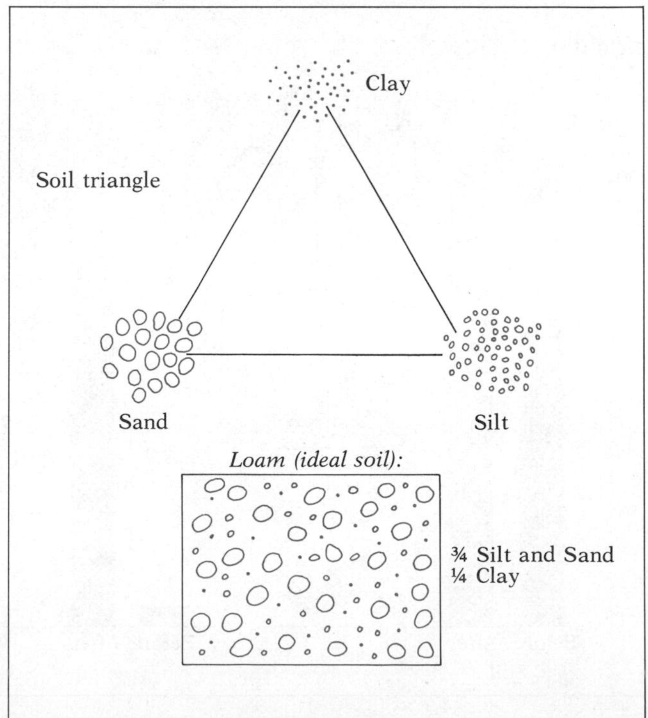

The organic matter of the soil is also the source of many mineral nutrients. As the organic matter decays, the nutrients are released and are recycled to new plants. The organic matter can come from any organism, including the remains of plants, animals, and microorganisms, and can be found in varying states of decomposition. Materials that are partway between decay and nutrient are known as *humus*. In addition to the organic matter, the soil contains great numbers of living organisms from bacteria to arthropods, all dying, defecating or digesting and all contributing to the production of humus and cycling of mineral nutrients.

The pH is an important factor in determining the fertility of a soil. The best pH range is 5–7. At a lower pH, valuable positive ions no longer adhere to negative clay and organic particles. They are replaced by hydrogen ions and tend to leach out of the soil (Figure 19.4).

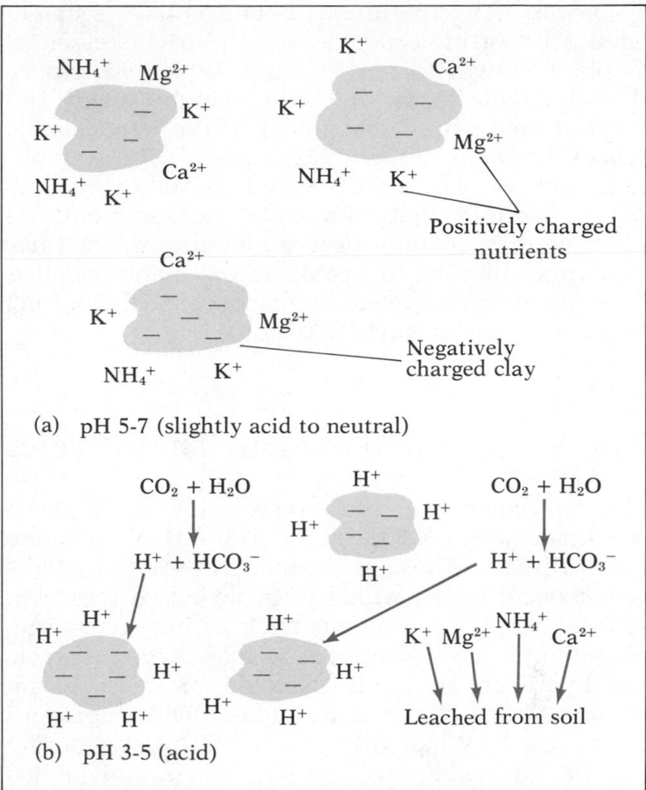

19.4 (a) Negatively charged clay particles attract valuable positive ions (such as ammonium, calcium, magnesium, and potassium). (b) These ions are displaced when carbon dioxide, released by plants from their respiratory activity, produces carbonic acid in the soil water, or when soils are otherwise made acid. Thus, the various positive ions, replaced on the clay particles by hydrogen ions, are leached out of the soil by moving water.

Nutrient Requirements

Plant scientists have used many approaches to determine the mineral requirements of plants. One method, for example, involves careful chemical analysis of the substances in the plant itself. In other words, it answers the question: What does the plant incorporate into its body? This is a somewhat indirect approach, however, since the compounds found in plants may have undergone profound chemical alterations as they were incorporated. And plants can incorporate materials they don't need or use.

A second approach is the use of deprivation experiments in which plants are grown in aerated distilled water of rigidly controlled purity with specific mineral nutrients added. Once the minimal nutrients for growth are known, it is easier to determine the particular effect of any mineral simply by withholding it from the solution. There are, of course, many different ways to organize these experiments. In some instances, it is sufficient to grow the plants from seeds, but where trace amounts of minerals are involved, it is necessary to remove plant embryos from the seeds, and begin their growth on a mineral-free medium, such as agar. When growth is sufficient, the shoots are moved into water. In this way, the minerals present in the storage region of the seed are eliminated. Figure 19.5 illustrates how one such study was done.

19.5 Plant mineral requirements are determined by growth under rigidly controlled conditions. The plants on the right in this experiment have been deprived of nitrate.

From such experiments botanists have accumulated a list of important mineral nutrients essential to plant growth that is given in Table 19.1. Notice that the table is divided into *macronutrients* and *micronutrients* (trace elements). These terms simply reflect the relative quantities required. For example, ions such as chloride (Cl^-) are required in such minute quantities that they defy measurement, yet photosynthesis cannot proceed without them. Trace elements often cause problems for plant scientists in trying to control experiments, because if you can't find it, you can't always keep it out.

The Macronutrients and Their Cycles

Many of the mineral nutrient requirements of plants are made available through cycles that are often quite intricate. These are usually referred to as *biogeochemical cycles*, which is an awkward term, but self-defining. Key elements such as nitrogen, sufur, phosphorus, and calcium, as well as carbon, oxygen, and hydrogen, are made available through constant recycling between the living and nonliving organisms of the earth. When these cycles fail, for whatever reason, the result can be infertile soil and limited plant growth.

The prime movers of mineral nutrient cycles are the microorganisms of the soil, which act as reducers or decomposers. When plants or animals die (or deposit wastes), the role of decomposer bacteria and fungi begins. These often act as teams, each using the wastes of another as its energy source. Decomposition is the first step in most cycles, but sometimes scavenger animals hasten the process somewhat by first feeding on dead organisms and producing simplified wastes for the decomposers to work on.

Eventually, the biologically useful energy in a dead organism's body is depleted. Very little is then left in the form of organized molecules. The remaining substances may be ions of nitrogen, phosphorus, calcium, sulfate, and so forth, ready for absorption into the roots of a plant. But, interestingly, even some of these simple ions contain enough energy to interest a few soil bacteria, so plants still have to compete for the simplified products of bacterial action, as we shall see.

Nitrogen

The most intricate biogeochemical cycle (and perhaps the most famous) is the *nitrogen cycle* (Figure 19.6). Nitrogen, you recall, is a key element of proteins and nucleic acids, so vigorous plant growth (as any gardener knows) requires a continuous nitrogen supply. In a balanced system, there is rarely a need for human intervention with chemical fertilizers. But, of course, modern, intensive agriculture isn't a naturally balanced system. Balanced systems

Table 19.1 Elements and Nutrients Essential to Plant Growth

Element	Taken in as	Examples of use
Macronutrients		
Calcium	Ca^{2+}	Cell walls, cell membrane activity
Carbon	CO_2 (atmosphere)	Proteins, lipids, carbohydrates
Hydrogen	H_2O (soil water)	Proteins, lipids, carbohydrates
Magnesium	Mg^{2+}	Porphyrin ring of chlorophyll
Nitrogen	NO_3^-	Amino acids, purines, pyrimidines
Oxygen	O_2	Cell respiration
Phosphorus	$H_2PO_4^-$	Nucleic acids, phospholipids, ATP
Potassium	K^+	Cell membrane activity
Sulfur	SO_4^{2-}	Proteins
Micronutrients		
Boron	BO_3^{3-}, $B_4O_7^{2-}$	Cell elongation, carbohydrate translocation
Chlorine	Cl^-	Exact role not clear but accumulates as HCl in chemiosmotic photosynthesis
Copper	Cu^{2+}	Enzyme activity
Iron	Fe^{2+}	Enzyme activity (synthesis of chlorophyll)
Manganese	Mn^{2+}	Enzyme activity
Molybdenum[a]	Mo^{3+}	Enzyme activity of nitrogen fixers
Zinc	Zn^{2+}	Auxin activity

[a]Indirectly important in the nitrogen cycle.

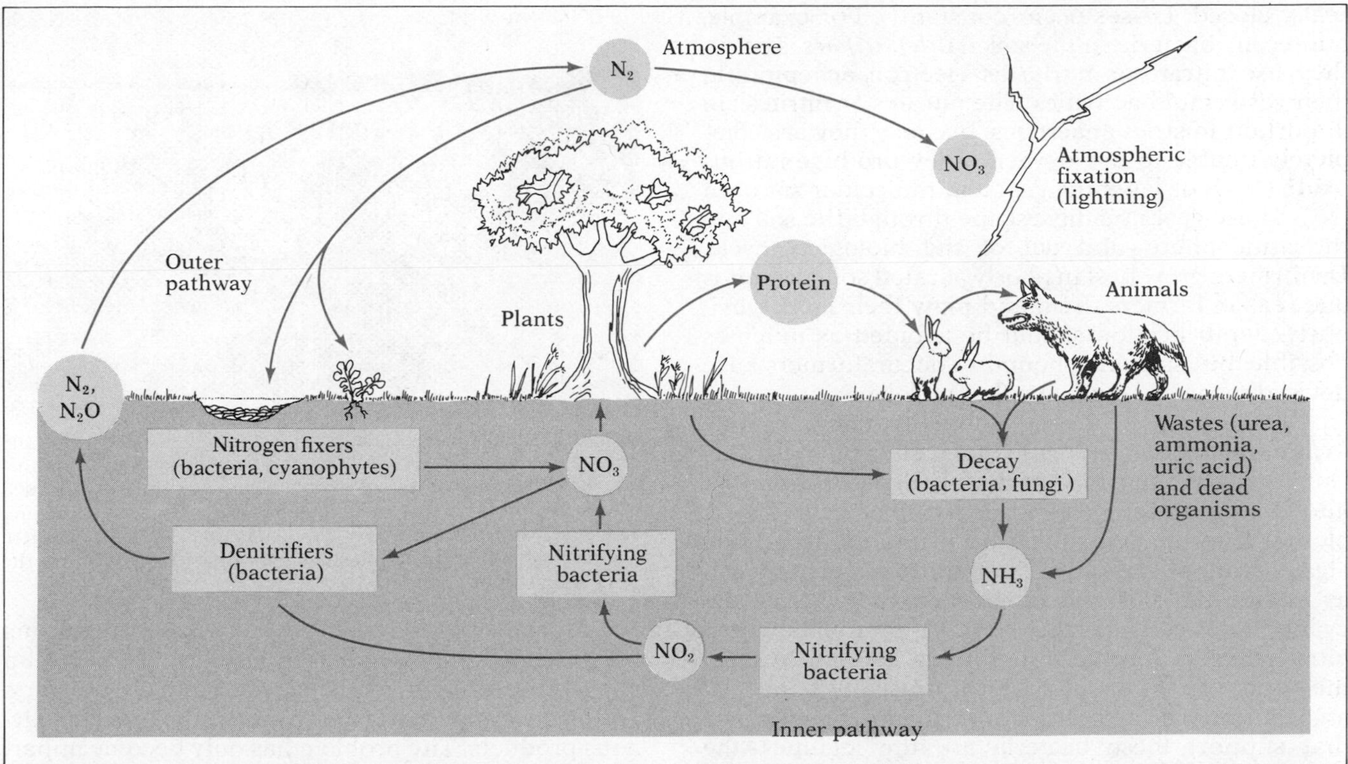

19.6 The nitrogen cycle. Nitrogen, a key element to life, follows an intricate cycle that includes an *inner pathway* among organisms and an *outer pathway* through the atmosphere. Plants absorb nitrogen in the form of nitrate from the soil, and incorporate it into large molecules such as proteins and nucleic acids, where it will remain through the plant's lifetime: Animals feeding on plants and each other pass the incorporated nitrogen along. As molecules are rearranged according to the needs of the organisms, some nitrogen is released to the cycle in the form of nitrogen wastes—ammonia, urea, and uric acid. The rest will reenter the cycle upon the animal's death. All dead organisms are acted upon by a series of microorganisms including fungi, protists, and bacteria. The products of one group's metabolism are used by the next. Eventually, ammonia, nitrite, and nitrate ions are freed into the soil and water. One group of bacteria, the anaerobic denitrifiers, can oxidize the nitrogen compounds to nitrogen gas and nitrous oxide. The subsequent loss of available nitrogen is, in the long-run, balanced by nitrogen-fixing cyanophytes and bacteria. These organisms convert nitrogen gas from the atmosphere to NH_3, which is then converted to nitrate and used by plants. And the cycle continues.

are possible only where cycling is unimpeded, and that requires mixed or rotated crops. Modern agriculturalists find the use of chemical fertilizers to be an easier and cheaper method of providing the plant's needs.

A cycle can begin anywhere, so we'll begin with plants absorbing nitrogen compounds from the soil and using them with the products of photosynthesis to form amino acids and nucleotide bases (adenine, guanine, etc.). These molecules, once formed, will sooner or later become available to herbivores and reducers. Herbivores digest their plant food, break down the proteins, and rearrange the amino acids and nucleotides to suit the needs of their own cells. The process is repeated from one feeding level to the next as herbivores are eaten by carnivores and carnivores are eaten by bigger carnivores. But even the mightiest carnivore must finally succumb to the silent activities of the lowly decomposers and is thus recycled through the soil to the plants.

But let's take a closer look at the processes involved. The reducer populations digest and metabolize the proteins of larger organisms. In the initial decay action, these microorganisms create the nitrogen-rich waste, ammonia (NH_3, which becomes the ammonium ion, NH_4^+, when dissolved in water). The process is known as *ammonification*. Ammonia is a major product of decay, along with other nitrogen compounds of disrepute, such as *putricene* and *cadaverine*—terms that need no definitions. Although these compounds have extremely unpleasant odors, many organisms find the chemicals not only quite acceptable, but even essential.

The next step is the *nitrifying process*. A number of kinds of bacteria use the hydrogen in the ammonia as well as other decay products for their energy. As they oxidize the hydrogen of ammonia, the nitrogen is released as nitrite ions (NO_2^-), and finally, other bacteria change them to nitrate ions (NO_3^-). So we end up with the all-important ion, nitrate, which is ready to reenter the root hairs of a plant and begin the cycle anew.

This all appears to form a rather closed and tidy cycle; seemingly, nothing is lost. But the cycle is not really closed. Losses occur constantly. For example, numerous bacteria in the soil are *denitrifiers*. That is, they use nitrate or nitrite as electron acceptors in their respiratory activities. The nitrates or nitrites are important to strict anaerobes, because they are completely unable to use oxygen. They produce nitrous oxide (N_2O, or laughing gas) and molecular nitrogen (N_2). These gases readily escape through the soil into the atmosphere—and out of the biological cycle. Denitrifiers grow best in poorly aerated soils, so this is one reason farmers drain and plow their land. Obviously, denitrification should be avoided as much as possible, but since it is bound to occur, farmers have devised means of replacing that nitrogen.

Nitrogen fixers are used. These organisms are able to manufacture ammonia so that it ends up in the soil. Only a few organisms can fix nitrogen and they fit into two broad categories: free-living types and symbionts. The free-living ones are primarily blue-green algae (cyanophytes). Where conditions permit (such as in rice paddies and other stagnant waters), the cyanophytes convert N_2 to NH_3. In the soil, however, nitrogen fixers are chiefly symbionts, bacteria living in the roots of certain plants in a mutually beneficial association. To farmers, the most important plants that support these bacteria are the legumes—the group that includes beans, peas, alfalfa, and clover. Interestingly, the primitive cycads and ginkgos (see Chapter 17) house nitrogen fixers as do our more modern alders and buckthorns.

As symbionts, the nitrogen-fixing bacteria live in nodules on the roots of these plants. As a colony of nitrogen fixers grows, the root responds by developing a nodule around the region. The bacteria benefit from a constant supply of food and water, while the plants get usable nitrogen. It has long been a common practice in farming to grow legumes in alternate years. The crop is then plowed under, increasing soil fertility for other crops.

Incidentally, nitrogen fixation fails when molybdenum is absent in soils. Although it is only needed in trace amounts, this element is absolutely essential. The Australian southern tablelands (Figure 19.7) were once an economic wasteland. Then someone discovered that a nearly total absence of molybdenum was the problem. It turns out that the nitrogen fixers need trace amounts of molybdenum to form *nitrogenase*, the enzyme responsible for converting N_2 to NH_3. The Australian soil fertility problem was solved by adding only 50–75 grams of molybdenum to each acre of land.

Nitrogen may also be fixed the hard way, by lightning, and more quietly, by the action of the sun on certain chemicals. However, nitrogen fixed through atmospheric action amounts to less than 10% of that fixed by organisms.

19.7 Tablelands of Australia. Until soil scientists diagnosed the problem, tablelands such as the area shown here were almost totally nonproductive. Plants simply would not grow well in such places. The problem was an almost total absence of molybdenum, an element which is absolutely required in the fixation of nitrogen. Here a fence separates the heavily grazed pasture land from the generating scrub.

A problem has rather recently arisen regarding our manipulation of nitrogen. We are not using up all our nitrate as you might expect. Quite the contrary. We are overloading the environments with nitrogenous products. The problem has only become apparent since we learned to synthesize ammonia through industrial means. The technique was developed by the Germans during World War I. Nitrates needed for the manufacture of gunpowder became critically short when Britain blockaded the German ports. The ingenious Fritz Haber solved the problem by producing synthetic nitrate. The process uses natural gas (methane, CH_4), high-pressure steam, and atmospheric nitrogen. In a multistep catalytic process, ammonia is produced. Of course, the technique requires a considerable energy input, but in spite of today's energy costs, the process is relatively economical, so synthetic fertilizers are widely used. They are applied liberally to the soil and thus enter the natural nitrogen cycle. If you add to this the nitrogen compounds produced by natural nitrogen fixation and by automobile exhaust (another major new source), the total amount of available nitrogen is astounding. C. C. Delwiche, at the University of California at Davis, has calculated that there is a net gain of 9 million metric tons of fixed nitrogen to the biosphere each year, representing 10% of the total fixed annually! The big question is, where is this surplus going? In California's central valley, which receives more fertilizer than any place on earth, the answer is: right into the water table and river systems, and from there into San Francisco Bay.

Now let's ask a few questions. How does excess nitrogen affect the water supplies in the soil? What is its effect on river systems and their estuarine life? What is the effect of the nitrogen load on the marine environment, keeping in mind the added nitrogen from sewage dumped into the ocean? One answer is

Essay 19.1 Why Trees Don't Urinate on Dogs

Perhaps you've had the pleasure of seeing a dog lift his leg and urinate on the base of a tree. And perhaps you've noticed that the tree rarely retaliates. In fact, you may be unable to recall ever seeing a tree urinate at all. Why do animals, but not plants, rid themselves of noxious nitrogens?

To begin with, there are fundamental differences in plant and animal metabolism. For instance, we know that plants make their own food but that they require minerals—including nitrogen-containing ones such as nitrates—from the environment, just as animals do. But animal nutrition is concerned primarily with such needs as essential amino acids, polyunsaturated fats, and vitamins. Plants, on the other hand, make their own; they need only minerals, water, carbon dioxide, and oxygen.

The distant primitive ancestors of animals could probably manufacture their own vitamins and amino acids, just as bacteria and most protists can now. But a billion years of eating organic foods, often rich in these items, has allowed the evolutionary loss of many biosynthetic pathways. The process still goes on: While most animals can make as much vitamin C as they need, other animals (including the primates) cannot, presumably because for millions of years their ancestors were able to find fruits rich in vitamin C. When these animals came out of the trees and started chasing rabbits, they found that their vitamin C machinery had, in effect, rusted into disrepair.

Animals and plants have different problems with regard to nitrogen. Most plants are perpetually starved for nitrogen, which explains the dramatic effect you get by fertilizing your lawn. To help meet this need, plants have evolved elaborate enzymatic mechanisms for recycling what nitrogen they do have. Thus, plant nitrogen compounds are rapidly recycled: the nucleic acids are broken down and recycled into proteins and vice versa, the nitrogens of one amino acid are readily transferred to another, nitrogenous plant alkaloids are readily synthesized and broken down, and so forth. Nitrogen is valuable to plants, and none of it goes to waste.

Animals generally lack this enzymatic dexterity. While we can shift nitrogens between certain amino acids, this ability is limited, and we have to acquire other amino acids in our diets—directly or indirectly from plant sources. In fact, we need a lot of nitrogenous products in our diet—humans require about 80–100 grams of protein a day. But because of our limited ability to recycle nitrogenous compounds, most of the nitrogen we eat quickly goes to waste. Our elaborate excretory systems are largely devoted to removing the toxic breakdown products of the nitrogenous compounds we eat. (The other principal function of the animal excretory system is water and salt balance.) Our metabolic inefficiency means we must continually eat protein merely to make up for the nitrogen we've just urinated away.

Do plants then have no toxic metabolic wastes? As a matter of fact, they often do, although by no means to the degree that animals of similar size do. Plants can manage their wastes by *sequestering* them—often in crystalline form, such as calcium oxylate—in cell vacuoles. In trees, some waste materials are deposited in the nonliving heartwood. Occasionally, waste materials (such as salts) are exuded from openings of the vascular system at the tips of leaves. If the metabolic wastes of a plant are poisonous, so much the better—from the plant's point of view! Grazing herbivores may well learn to avoid such plants. Some plant *alkaloids* (nitrogenous products such as morphine and caffeine) may be merely sequestered metabolic wastes, or perhaps they have evolved as mechanisms that make the plant less palatable to pests and herbivores. Some botanists have suggested, in all seriousness, that the best way for a plant to get rid of an unwelcome herbivore is to get it high on dope, whereupon its uninhibited behavior will quickly make it prey to some serious-minded predator.

---

provided by the appearance of *methemoglobinemia*, a blood condition that occurs in infants and children exposed to dangerous levels of nitrite (NO_2^-) in their water supply. Hemoglobin is converted to methemoglobin, losing much of its oxygen-carrying capacity. Curiously, methemoglobinemia occurs most frequently in the children of migrant farm workers. A second visible effect is the sporadic choking of waterways by algal blooms (Figure 19.8). Are there other effects? We really don't know. But obviously, we need to find out. It is quite conceivable that we can feed the world's burgeoning populations without calling the environment down around us. By coming to understand the nitrogen cycle better, perhaps we will

19.8 A dramatic increase in algal growth occurs when excess nitrate suddenly enriches a body of water. Conditions like this often result in the destruction of other precariously balanced life forms that inhabit the water. Masses of dead and decaying algae will support huge populations of bacteria, which, in turn, will deplete the oxygen supply. The result is the formation of an anaerobic water "desert," with only a few species of anaerobes and highly tolerant aerobes surviving.

find it vital and even economically feasible to keep our agricultural practices within the limits of natural systems.

The Phosphorus and Calcium Cycles

Now let's briefly consider two more cycles, those of phosphorus and calcium (Figure 19.9). Their cycles do not involve the atmosphere, but they do involve water. Phosphorus enters the roots of plants as soluble anions in the form of HPO_4^{2-}. Phosphates are required for the production of many familiar cellular substances, such as ADP and ATP, phospholipids, nucleic acids, and compounds that function in photosynthesis and respiration. Obviously, a shortage of phosphates will dramatically slow down the growth of plants by interfering with their energy transfer systems and with biosynthesis.

Calcium, another mineral nutrient, enters the root as the cation Ca^{2+}. It is critical to the proper functioning of cell membranes and many enzymatic reactions. A shortage of calcium can disrupt transport processes in the plant cells, causing the plant to die. Although it functions in transport systems, the ion itself is rather immobile once in place. When local shortages

19.9 Phosphorus and calcium move in a cycle between organism and environment, but unlike the nitrogen cycle, the atmosphere is not involved. The cycle begins with the uptake of calcium and phosphate ions by plants. Upon the death of a plant, reducers release the ions into the soil once again. Or, animals graze on the living plants and incorporate calcium and phosphate into their bodies. Upon the death of the animals, some of the ions are released by decomposers, but because skeletons and shells resist this action, they tend to accumulate in the soil and sediments. Deep marine sediments are periodically distributed by upswellings, which bring these minerals back into the cycle again.

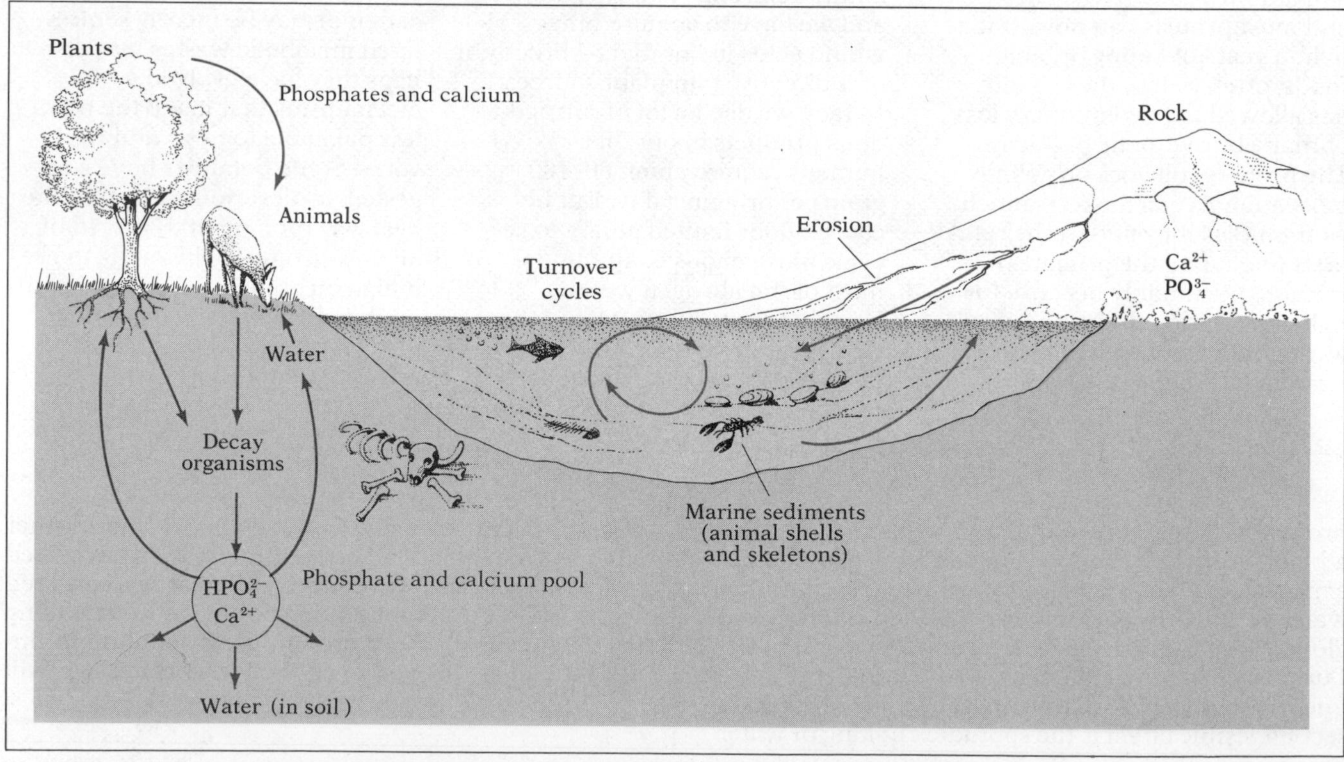

of other ions occur, they can simply be brought in from other parts of the plant. But not calcium, it must be taken in constantly by the roots.

The cycles of these elements are rather straightforward. Phosphates, for example, are released by the action of decomposers. Since they are soluble, they may be recycled at once. Calcium, on the other hand, may cycle more slowly, since it can remain in skeletons for years. When a vertebrate dies, calcium from its bones is slowly leached back into the soil. In freshwater biomes, phosphorus and calcium may lie bound in the bottom sediments for long periods of time until something agitates those murky depths. In marine biomes, occasional upwellings bring the sediments and dissolved ions to the surface, where they reenter the cycle via the phytoplankton.

Calcium that is fixed in the exoskeletons of marine plankton may form thick chalky deposits on the ocean floor. Eventually, these deposits become limestone and then marble, which are forms of calcium carbonate. Increases in pH will dissolve limestone and even marble. There is a constant equilibrium in the ocean between dissolved and precipitated calcium carbonate. Calcium deficiency is never a problem for marine organisms.

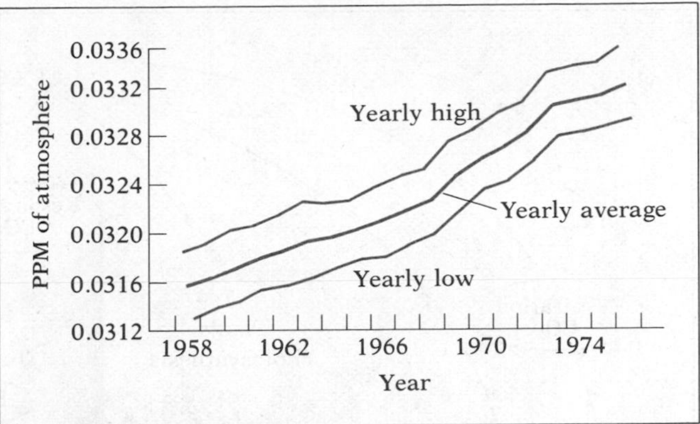

19.10 Studies of carbon dioxide content in the atmosphere have been monitored at the Mauna Loa Observatory in Hawaii for many years. The trend is obviously rising, and the rate of its rise is disconcerting. The increase in atmospheric carbon dioxide is primarily attributed to the increasing rate of fossil fuel expenditure by humans. Everything that is burned produces carbon dioxide. The yearly highs and lows reflect variations in carbon fixed during photosynthesis. Since many of the plants are seasonal, the winter months in the Northern Hemisphere are represented by highs in free CO_2. These predictable changes help prove the accuracy and sensitivity of the measuring procedures at the observatory.

The Carbon Cycle

You already are aware that life is based in carbon. Since it is a key element, its very availability may determine the size of populations. Carbon is mainly available as the atmospheric gas, carbon dioxide (CO_2). Plants can draw on this supply as they manufacture the carbon compounds of life, mainly through photosynthesis. Paradoxically, it is present in such small proportions in the atmosphere (less than 0.04%) that it can almost be called a rare gas. The quantity is admittedly small when compared to other gases, but it still represents an enormous amount in absolute terms. As you will see, the quantity must also remain constant if delicate environmental balances are to be maintained.

In addition to the carbon found in the atmosphere, the earth has a sizable reservoir in the form of carbonate rock (such as limestone) and fossil fuels (oil, coal, peat). However, this reservoir is being steadily altered by human and geological processes. Since the industrial revolution began, CO_2 from fossil fuels has been released into the atmosphere in steadily increasing amounts (Figure 19.10). Meteorologists estimate 5–6 billion metric tons of CO_2 from fossil fuels are released into the air annually! Fortunately, as we shall see, about two-thirds of this amount is quickly removed by the oceans and by photosynthesizers. (The ocean is a bigger dumping ground than you may have thought.) But let's return to cycling, our subject here, and find out what happens to CO_2.

In this cycle, carbon is shunted around among populations of plants, animals, and microorganisms. As you can see from Figure 19.11, CO_2 enters the cycle from the atmosphere through photosynthesis. Plants make CO_2 through respiration, just as animals do, so some goes directly from the plant back to the atmosphere. Animals eat the plants, ingesting the carbon which the plants have incorporated into their bodies, and in so doing, they release more of the carbon as CO_2 as they carry on their own cell respiration. Finally, the plants and animals meet their fate and are greeted by the decomposers, which eventually return the last molecules of CO_2 to the atmosphere. Thus, a carbon atom that perhaps was once a part of the great Leonardo da Vinci wanders into an open stoma of a plant.

This, then, is the main carbon cycle. But there are other factors at work. The earth's solid reservoirs of carbonate deposits are made available by the erosion of carbonate rock and by volcanic activity at a rather constant rate. This reservoir is partially replenished as carbon compounds locked in the skeletons of dead animals enter the sediment. Thus, there is a rough equilibrium established—or there would be were it not for the carbon we add to the atmosphere by burning fossil fuels (see Essay 19.2).

The Oxygen and Water Cycles

We won't say much about these cycles in this chapter. The plants' constant need for water has already been discussed, and the photolysis of water in photosynthesis and the reforming of water in respiration are

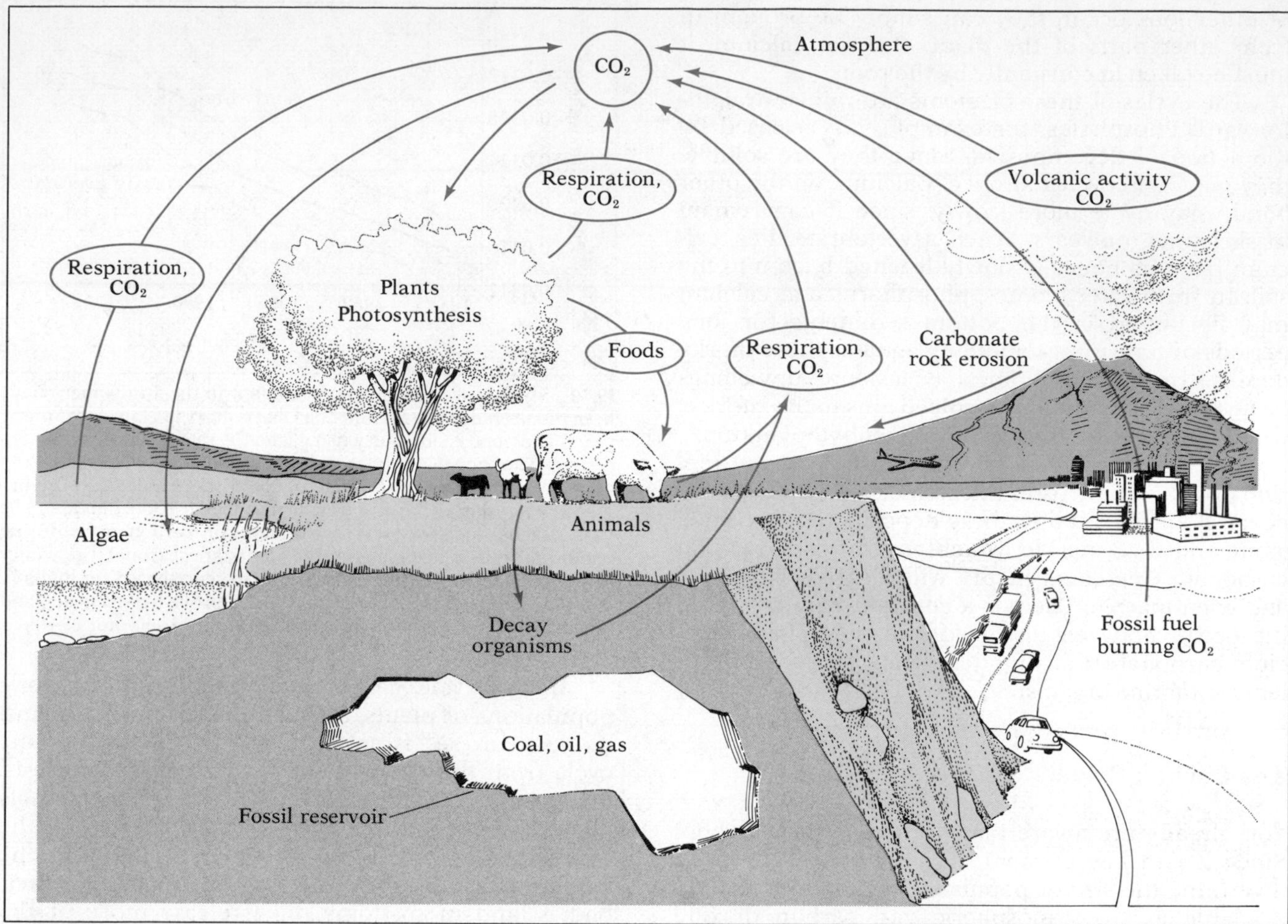

19.11 The carbon cycle. The atmosphere and ocean are the earth's greatest reservoirs of carbon. However, a sizable amount resides in carbonate rock and fossil fuels. The cycling of carbon begins with CO_2 entering plants during photosynthesis and becoming incorporated in carbohydrates. Plants then carry on respiration, releasing some carbon back into the atmosphere and soil as carbon dioxide. Some of the plant carbohydrates enter animals and other heterotrophs, where respiration returns more CO_2 to the atmosphere. The complete cycle requires the action of microorganisms, which utilize the carbon compounds of dead plants and animals in their own respiration, and then release CO_2. A recent addition to this cycle is one more significant source of CO_2—the burning of fossil fuels by humans.

also familiar to you (see the diagram of the water cycle in Figure 2.15).

You probably learned in the third grade that plants provide the oxygen that animals breathe. You may not have realized then that the plants don't care a fig whether animals breathe or not, but respiration and the breakdown of water by plants are certainly key events in the biological part of the oxygen cycle. So it doesn't matter that plants don't care about animals—as long as they behave as though they do!

Geologists are presently divided over the hypothesis that the earth's oxygen primarily came from photosynthesis. Some maintain that photolysis of water vapor high in the atmosphere, with the subsequent loss of protons into outer space, was and is a more important source of molecular oxygen.

Potassium

The word *potassium* comes from *pot ash*, a reference to the material in which potassium hydroxide salts were originally discovered. Plants tend to absorb potassium from the soil and to accumulate large concentrations of it in their tissues, so the bodies of plants are rich in this element. Plants need potassium in larger quantities than any inorganic ion besides nitrate, although we really don't know why. Probably, the most important function of potassium is in regulating the osmotic balance of cells. Plants that are deficient in potassium tend to accumulate small molecules such as amino acids and simple sugars in their cytoplasm, and may show abnormally low turgor pressure.

Plants don't use potassium as a building block for other molecules, perhaps because it doesn't readily form covalent bonds, so it remains in its ionic form as the cation K^+. The ion is very mobile, and it moves rapidly through the plant, accumulating in meristematic regions and other places with abundant cytoplasm and intense biochemical activity. Many enzymes require potassium, including those that make chlorophyll. Plants grown on sandy soils may become potassium-deficient. This deficiency causes slow growth, pale, chlorophyll-deficient leaves, and a generally wilted appearance.

Potassium ions are involved in the opening and closing of the guard cells around stomata, which is a result of an osmotic event. Plant movement in general utilizes osmotic mechanisms, and potassium is probably always involved. Potassium may also be involved in the active transport of other ions.

Sulfates

Generally, plants take in sulfur in the form of sulfates. The sulfur is used primarily in the production of two amino acids, cysteine and methionine:

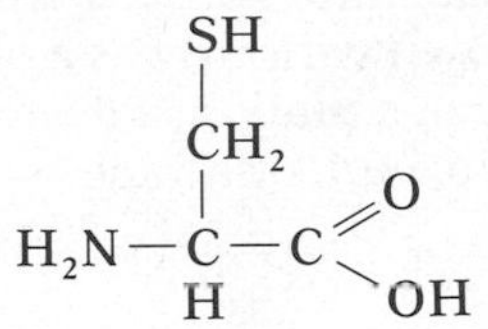

$$\begin{array}{c} SH \\ | \\ CH_2 \\ | \\ H_2N-C-C\overset{\displaystyle O}{\underset{\displaystyle OH}{\lessgtr}} \\ H \end{array}$$

Cysteine

$$\begin{array}{c} CH_3 \\ | \\ S \\ | \\ CH_2 \\ | \\ CH_2 \\ | \\ H_2N-C-C\overset{\displaystyle O}{\underset{\displaystyle OH}{\lessgtr}} \\ H \end{array}$$

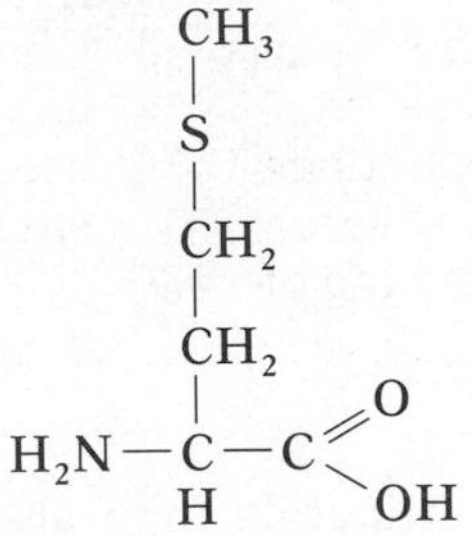

Methionine

You might wonder why sulfur is so important if it is only used in two amino acids. If you recall from Chapter 3, cysteine is important in the formation of sulfhydryl bridges between polypeptides as proteins are assembled. These bridges result in the tertiary folding of the molecule, and this folding is important to the biological activity of many proteins, especially enzymes.

Magnesium

The last macronutrient on our list is magnesium, which is transported as the cation Mg^{2+} in a variety of combinations. Every chlorophyll molecule contains a porphyrin ring with magnesium in its center. Magnesium-deprived plants produce yellow foliage, reflecting a lack of chlorophyll (a condition called *chlorosis*). Obviously, photosynthesis is retarded in these plants. In addition to its essential role in the chlorophyll molecule, ionic magnesium (like potassium) acts as an enzyme activator. It is particularly important in energy-producing reactions that involve phosphates.

Micronutrients

Micronutrients are required only in minute quantities. They are often measured in parts per million (ppm), but the health of the plant depends on their presence nonetheless. From the list of micronutrients given in Table 19.1, zinc, iron, manganese, copper, and molybdenum are all important to the activity of enzymes.

Iron, for example, is essential to the cytochromes of electron transport systems of both photosynthesis and respiration. In addition, iron is required for the manufacture of chlorophyll, although it is not incorporated into the molecule. Lack of iron therefore produces a chlorosis similar to that caused by magnesium deficiency.

Zinc apparently plays a role in stem elongation and leaf growth. Its action is not firmly established, but it is thought to have some connection to the production of auxin, a growth-promoting plant hormone.

Manganese deficiency results in leaf chlorosis, producing a mottled appearance. It is important in the activation of enzymes of the citric acid cycle and in the photosynthetic breakdown of water.

Copper is essential to plants, in much the same manner as manganese. It activates enzymes, but in electron transport systems in both photosynthesis and respiration.

Earlier, we reviewed the importance of *molybdenum* to the nitrogen-fixation process. As we mentioned, it is incorporated into the enzyme nitrogenase, which functions in the conversion of atmospheric nitrogen to ammonia.

The exact role of *boron* is unknown at the present time, but it has been established through deprivation experiments that a boron deficiency interferes with stem and shoot elongation and flowering, and its absence can cause other abnormalities. It is further involved in the breakdown and translocation of carbohydrates.

Chlorine is a micronutrient, but, because of its ubiquity is never in short supply outside of experimental situations. Chloride ions accumulate inside the thylakoid disks as *hydrochloric acid* (HCl) in the chemiosmotic process of photosynthesis.

We will end our discussion of micronutrients with *cobalt*. This element is not in itself required by plants, but they take it in. In regions where it is lacking, grazing animals do poorly and develop vitamin B_{12} deficiency. It turns out that the source of B_{12} in ruminants is the microorganisms in their gut, and these organisms fail when cobalt is absent. The web of life is indeed intricate.

Interestingly, if any of the micronutrients are too abundant, they may interfere with plant growth. For example, copper compounds are well-known algae

Essay 19.2 The Carbon Cycle and the Greenhouse Effect: A Destabilized Equilibrium

In recent years, atmospheric scientists have become increasingly concerned with some aspects of the carbon cycle. The enormous amounts of carbon dioxide released by the burning of fossil fuels, and the even greater amounts released by the clearing of forests, have not all been absorbed by the ocean. The concentration of carbon dioxide in the atmosphere has been increasing, and by some measures it is up substantially over what it was 150 years ago. What does this mean to us?

It means trouble. One of the physical characteristics of carbon dioxide gas is that it does not absorb or reflect short lightwaves, but it does absorb and reflect the longer infrared lightwaves. Thus, it does not deflect solar energy entering the biosphere, but it does reflect back the radiant energy that emanates from the surface of the sunlit earth. So carbon dioxide is a factor in the balance between radiant energy coming from the sun and radiant energy that is reflected back into space. An increase in atmospheric carbon dioxide should cause an increase in the world's average temperature.

Greenhouses also work by letting in the short lightwaves and retaining the longer waves of heat energy, so the carbon dioxide effect is known as the *greenhouse effect.* So far, the actual effect seems to have been small, apparently because it is countered by a different geophysical cycle that would otherwise be cooling the earth.

Still, according to the experts, the temperature cycle will eventually reverse, and when it does we may see some really severe greenhouse effects. Warming of the earth would result in a warming of the ocean. If the ocean water gets warmer, the solubility of carbon dioxide will decrease. There are much greater reserves of dissolved carbon dioxide in the ocean water than there are in the atmosphere, and an ocean temperature rise of only 1 or 2°C would unload more carbon dioxide into the atmosphere. This in turn would increase the greenhouse effect and raise the world's temperature, which would cause the release of even more carbon dioxide. To make matters worse, decreases in the size of the polar ice caps would further decrease the amount of solar energy that is reflected back into space. This is what is known variously as a positive feedback situation, a destabilized equilibrium, or a vicious circle.

---

killers, and farmers have long known that a copper nail driven into a stump will retard shoot growth. Boron is used as a weed killer, and heavy doses of zinc salts are highly toxic to plants.

Transport of Mineral Nutrients Within the Plant

Some nutrients, such as calcium, are passively absorbed and carried through the xylem to become permanently fixed in plant tissues. However, most other mineral nutrients are highly mobile, and are shunted around to meet the metabolic needs of various parts of the plant. In fact, the plant may selectively incorporate whichever minerals it needs. For example, dissolved ions are preferentially absorbed by root hairs, using active transport. The ions can then be moved passively up the xylem aided by the pull of transpiration, but they can also be moved actively and in other directions by the phloem. As we saw in Chapter 18, different substances can move in opposite directions at the same time and even in the same phloem channels (see Figure 18.16).

Minerals may also move through interconnections between the xylem and the phloem. These interconnections, passing through the living cells of the vascular cambium, help regulate the flow of nutrients in the xylem columns. In particular, salts left behind as water is transpired would accumulate in the leaves and poison the plant if they weren't actively removed via these channels.

In *deciduous* plants (plants that seasonally drop their leaves), mineral nutrients are removed from the leaves just before the leaves are shed. Thus, potassium, sulfate, nitrate, magnesium, and iron are salvaged and recycled through the phloem into other tissues of the plant.

We don't know how plant tissues determine their mineral needs and then cause the proper chemicals to move selectively through the phloem to areas where they are needed. The selective process may be mediated by plant hormones. Perhaps, the mechanism is some kind of facilitated diffusion, with specific molecules following gradients that are set up by active transport through the membranes of the cells that are actually using the mineral nutrients. But these are just guesses.

Metabolic Oddities

In the inexorable search for energy and nutrient sources, a few plants have given up self-sufficiency in favor of a hemiparasitic or even a parasitic existence. Mistletoe (*Phoradendron*) is a hemiparasite. It grows on trees, sending its funguslike roots into the xylem of the host tree (by-passing the nutrient-laden phloem) to obtain water and minerals. However, it carries out photosynthesis for itself, taking advantage of its lofty exposure to the sun. On the other hand, dodder (*Cuscuta*, or "witches' hair") is a total parasite that utilizes its host's carbohydrate and water supply for its own purposes. Visitors to the chaparral of southern California can see its bright orange or yellowish tangle of stems on some of the local plants (Figure 19.12).

Perhaps the strangest nutritional oddities of all are the insectivorous plants, which Darwin studied extensively. These plants supplement their needs by capturing and digesting small insects. Examples include the sundew (*Drosera*), the Venus flytrap (*Dioneae*) and the pitcher plant (*Sarracenia*) (Figure 19.13). The most unusual is probably the Venus flytrap, since it actively captures its prey. These plants are adapted to bogs, where nitrogen compounds and other nutrients are leached away by water. Thus, trapped insects supplement the nutritional supplies of these photosynthesizers. Digestion and absorption occur gradually. Actually, their predatory habit is unnecessary when nitrogen supplies are added, but they go right on killing flies.

The Digestive Process in Plants

Can plants have indigestion? Or, better yet, can they have *digestion?* Unlike the animals, plants have no digestive system. They have no need for one, since

19.12 Dodder is a true parasite, utilizing both water and carbohydrate supplies of the host. Its golden weblike growth is a common sight in the chaparral of southern California.

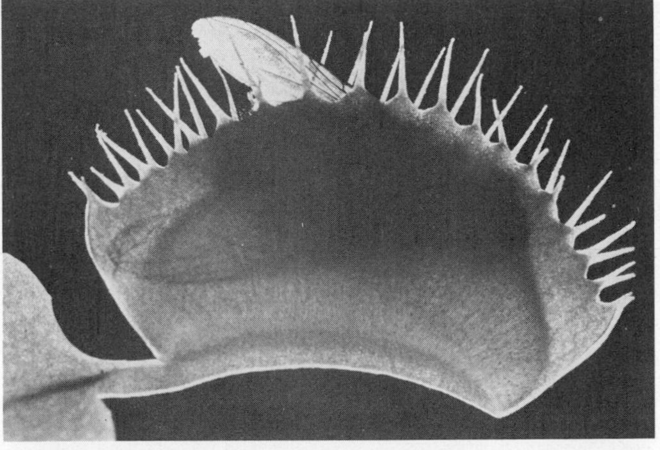

(a)

(b)

(c)

19.13 Insectivorous plants. A few plants are able to supplement their nitrogen supplies by capturing and digesting insects. (a) The Venus flytrap (*Dioneae*) actively snares insects in its highly modified, spiked leaves. (b) The pitcher plant (*Sarracenia*) has evolved a one-way passage into its trap. Insects venturing into the vaselike structure find their return blocked by downward pointing hairs. (c) The sundew (*Drosera*) makes use of a thick, sticky fluid that traps the insect. These are all plants that grow in nitrogen-poor swamps and bogs.

they manufacture their own food molecules. This is not to say that they don't carry on digestion, however. When they begin to use stored food, the first step is its enzymatic hydrolysis. Furthermore, plants must break down the same molecules that animals do: proteins, carbohydrates, and lipids.

Plants store foods chiefly in roots, stems, seeds, and fruits. In their stored form, the food molecules are relatively immobile. Consider starch. (You may recall that plants store glucose molecules in the form of starch.) When starches are to be redistributed, perhaps to rapidly growing regions, they must first be broken down into soluble glucose. These smaller molecules can then be shifted around. Proteins and lipids are also hydrolyzed by enzymes into their component parts, and those molecules are redistributed according to the needs of the plant.

Seeds are food storage centers for the plant embryo. The foods remain in their semidry storage form until water is absorbed. Then the seed germinates and begins its growth. While germination is occurring, digestive enzymes in the seed coat are produced and these move through the seed, breaking down large molecules, particularly carbohydrates, and making energy available for the early growth and development of the shoot.

Application of Ideas

1. Because of the CO_2 problem and the greenhouse effect, some environmentalists have predicted temperature rises on the earth as high as 2° to 3° on the average. This doesn't seem significant at its face value. How can a minor change in the average temperature become disastrous and what are some of the predicted effects?
2. Nitrogen and phosphorus are often considered to be the "limiting factors" of a specific region (i.e., lakes, forests, grasslands). Discuss man's capacity for altering these regions by increasing or decreasing the amount of nitrogen or phosphorus available. How can we do this and what are the effects?
3. The trap of the Venus flytrap is a modified leaf that closes rapidly when triggered by touch. The exact mechanism is unknown, but one hypothesis deals with rapid ion transport. Develop this idea into an explanation of how the trap might work. (So far, no one has located a nervous system in plants.)

Key Words

ammonia
ammonification
ammonium ion

biogeochemical cycle
bundle-sheath cell

carbonate rock
carbon-4 plant (C4)
carboxylase
cation
chlorosis

dentrification

fossil fuel

greenhouse effect

hemiparasite
humus

insectivorous plant

limiting factor
loam

Key Ideas

Introduction

1. The first experimenter on record in plant physiology was van Helmont. He concluded from his willow experiment that plant growth and weight increase was almost solely due to the water it took in. He could not have known about CO_2.
2. The requirement of CO_2 for growth was worked out by Priestley and Lavoisier. They recognized the reciprocal requirements of plants and animals for CO_2 and O_2.
3. As meager as the mineral requirements of plants are, it is absolutely essential that the minerals are available.
4. The mineral content of soil is listed among the *"limiting factors"* for plant growth (as defined by ecologists).
5. Soil texture is determined by the relative quantity of clay, silt, and sand. The ideal mixture is called *loam.*
6. Excessive acidity in soils produces infertility. Mineral ions, usually adhering to clay particles, are displaced by $H+$ and once freed, leach out of the soil.
7. Organic matter in soil is an important source of minerals. Its decomposition requires the action of many soil organisms.

Nutrient Requirements

1. Specific requirements are determined in deprivation experiments. By withholding each of the known minerals in a series of containers, their individual effects on plant growth become known.
2. The known mineral requirements have been organized into the *macronutrients* and the *micronutrients.* The latter are often needed in *trace* amounts only.

macronutrient
malic acid
mesophyll
methemoglobinemia
micronutrient
mineral nutrient

nitrate
nitrification
nitrite
nitrogenase
nitrogen fixation
nitrogen fixer
nitrous oxide
nodule

oxaloacetic acid

photorespiration
pyruvic acid

reducer

trace elements

The Macronutrients and Their Cycles

1. The availability of many of the essential *mineral nutrients* is assured by cycling. Many soil organisms (bacteria, fungi, protists, and some animals) are involved, but the principal *reducers* are bacteria and fungi.

2. The products of decomposition are mineral ions, CO_2, ammonia, and H_2O.

Nitrogen

1. Nitrogen is essential for producing protein, nucleic acids, coenzymes, vitamins, and hormones.

2. Plants utilize nitrogen ions in producing their own compounds. In turn animals and reducers work on the plant body. Eventually animals and their wastes become the energy source of reducers.

3. Recycling of nitrogen compounds in the soil includes the following steps.
   a. *Ammonification:* Bacteria metabolize amino acids producing *ammonia* and ammonium ions.
   b. *Nitrification:* Other bacteria oxidize *ammonia* to *nitrite. Nitrite* is then changed to *nitrate* by yet other bacteria.
   c. Incorporation: Plants use *nitrates* to produce amino acids.
   d. *Dentrification:* Some microorganisms convert *nitrite* or *nitrate* to *nitrous oxide* and molecular nitrogen. These escape to the atmosphere representing a loss.

4. New input into the nitrogen cycle occurs through *nitrogen fixation.* Molecular nitrogen is converted to *ammonia* by:
   a. Cyanophytes
   b. Nitrogen-fixing bacterial symbionts (root nodules of legumes)
   c. Lightning and sunlight (a minor amount)

5. An essential ion for *nitrogen fixing* is molybdenum, which is used by bacteria to produce the enzyme *nitrogenase.*

6. In agriculture, most of the nitrogen fertilizers are produced synthetically. Methane is reacted with nitrogen to produce *ammonia.* The application of huge amounts of commercial fertilizer has often resulted in excess nitrogen entering the environment.

7. Excess nitrogen compounds in the water table in central California presents a threat of *nitrite* poisoning in the water supply. Chiefly affected are infants who can develop *methemoglobinemia.*

8. A second threat occurs in the waterways where the nitrogen load creates choking algal blooms, drastically altering the distribution of life forms.

The Phosphorus and Calcium Cycles

1. Phosphorus and calcium ions cycle in water. The first is required for ATP, phospholipids, nucleic acids, and other compounds. The second is critical to enzymatic and membrane functions in plants.

2. Phosphorus and calcium are restored in simple cycles, released by decomposers as soluble ions. A constant loss occurs through sedimentation, but it is balanced by erosion and upwellings.

The Carbon Cycle

1. Carbon is essential for all molecule building. Plants require carbon dioxide for photosynthesis.

2. Reservoirs of carbon exist in the CO_2 of the atmosphere, in carbonate rock, dissolved in ocean water, and in fossil fuel (coal, oil, gas).

3. Atmospheric carbon enters the plant during photosynthesis. As carbohydrates are metabolized the plant gives off carbon dioxide. Animals and microorganisms give off CO_2 during respiration.

4. Additional CO_2 is provided by erosion and volcanic activity, while some is lost to the sediments in the skeletons and shells of animals.

Essay 19.2 The Carbon Cycle and the Greenhouse Effect: A Destabilized Equilibrium

Enormous amounts of CO_2 added to the atmosphere when fuels are burned, threaten to create a *greenhouse effect.* CO_2 in the atmosphere permits light energy to enter unhindered, but acts in an increasing manner to trap heat energy within. The threat is to the earth's atmospheric temperature, which, if elevated only a degree or two, could create drastic climatic changes.

The Oxygen and Water Cycles

1. Oxygen enters the atmosphere when plants carry on photolysis. Both plants and animals and other organisms as well use oxygen in aerobic respiration, producing water as a waste product.

2. So the water and oxygen are actually cycling between photosynthesis and respiration, with the atmosphere acting as a constantly filling and emptying reservoir.

3. Some authorities believe that the earth's oxygen came originally from photosynthesis, but others argue for its nonbiological formation in the atmosphere.

Potassium

Plants take in potassium ions, maintaining them in mobile *cation* form. Potassium ions are required for enzymatic activity, osmotic activity, and active transport.

Sulfates

Sulfur enters plants as the sulfate ion. It is required for producing cysteine and methionine, two bridge-forming amino acids that bond polypeptides into their specific protein shapes.

Magnesium

Magnesium enters the plant as a *cation.* It is required in each chlorophyll molecule and is essential as an enzyme activator.

Micronutrients

Micronutrients are often required in minute and *trace* amounts. These include:
1. Iron: cytochromes; chlorophyll production
2. Zinc: stem elongation
3. Manganese: enzyme activation (respiration, photolysis)
4. Copper: enzyme activation (respiration)
5. Molybdenum: for *nitrogenase* production
6. Boron: cell elongation and flowering
7. Chlorine: chemiosmotic photosynthesis
8. Cobalt: direct need unknown, but if absent in plants, herbivores will suffer vitamin B-12 deficiency

Transport of Mineral Nutrients Within the Plant

Many minerals in ion form are actively transported into root hairs. Ions move passively in xylem, but actively in phloem. Minerals pass between xylem and phloem through interconnections.

Metabolic Oddities

1. Some plants have evolved into parasites, using the host's carbohydrates, or into *hemiparasites,* using the host's water and mineral supply.

2. *Insectivorous plants* have evolved mechanisms for trapping insects and obtaining nitrogen compounds through digestion. These plants generally live in nitrogen-poor soils.

The Digestive Process in Plants

Plants lack a digestive system. Digestion occurs mainly in storage regions where starches and lipids accumulate. Seeds are important storage centers, where digestive enzymes hydrolyze foods for use by the growing embryo.

Review Questions

1. Explain the statement that van Helmont was "half right" in his conclusions from the willow tree experiment. (p. 527)
2. Did the minor change in the soil's weight in van Helmont's experiment mean soil was insignificant? Explain. (p. 528)
3. Describe the characteristics of a good quality *loam.* (p. 528)
4. Why is an acid soil usually devoid of *mineral nutrients?* (p. 529)
5. Describe the manner in which deprivation experiments are carried out in the laboratory. (p. 529)
6. Why is nitrogen essential for plant growth? (pp. 530–531)
7. Describe the steps that occur between death of an organism and the incorporation anew of the useful nitrogen compounds by plants. (p. 531)
8. What is *dentrification* and what is its effects on soil fertility? (p. 532)
9. List the organisms that are capable of *nitrogen fixation.* (p. 532)
10. In addition to fixation by organisms, what are two other natural means of adding nitrogen compounds to the soil? (p. 532)
11. Why is it that *nitrogen fixation* cannot occur in molybdenum-free soils? Do these exist anywhere? (p. 532)
12. What is an argument for and against the use of commercially prepared fertilizer? (p. 532)
13. How can excess *nitrites* endanger human life? (pp. 532–533)
14. Describe the cycling of phosphorus and calcium. (pp. 534–535)
15. List the commonly used ions of phosphorus and calcium and explain their use in plants. (p. 535)
16. Draw a simple scheme showing the cycling of carbon. (p. 535)
17. What are the carbon reservoirs of the earth and how does carbon get released from these in a usable form? (p. 535)
18. Explain the problem of the *greenhouse effect,* its source, action, and possible outcome. (Essay 19.2)
19. In a simple diagram, show how photosynthesis and respiration fit into oxygen and water cycling. Add carbon dioxide if you wish. (p. 536)
20. For what purposes do plants require potassium? (pp. 536–537)
21. What would you expect to happen in plants deprived of adequate levels of sulfate compounds? (p. 537)
22. List two important uses the plant makes of magnesium. (p. 537)
23. Review the transport mechanisms used in moving minerals (p. 538):
    a. Into the root
    b. Up into the leaves
    c. Across and around the plant
24. Describe two plants that live in a parasitic or hemiparasitic mode. (p. 539)
25. What are *insectivorous plants* and what is a biological explanation for their existence? (p. 539)
26. Where in plants does digestion occur? What is the role of digestion in the storage and mobilization of foods? (pp. 539–540)

Chapter 20

Chemical Regulation in Plants

Charles Darwin was intensely interested in the evolution of behavior, and he set about studying it with his usual uncanny insight and methodology. His publication, *The Expression of the Emotions in Man and the Animals* (1872), was to become the foundation of animal psychology. But the subject was incredibly complicated, so Darwin set about reducing his problem to its most basic elements. His first step was to try to analyze the simplest kind of behavior. And what could be more simple than the behavior of plants? So Darwin became the first plant psychologist and, perhaps, the last.

Of course, one is tempted to ask: Do plants *behave?* In fact, they do—if behavior is considered to be a response in the form of a movement.

Darwin began his studies with the most interesting of all plant behavior, that of the insect eaters, such as the Venus flytrap and the sundew (Figure 19.13). He described what triggers the plants and how they respond in exquisite detail in *Insectivorous Plants* (July 1875), but he was at a loss to explain just *how* these things were done. So he changed the focus of his research; he looked at the behavior of climbing plants.

The tips of climbing plants, he found, move about in circles while they grow, *as if* seeking something to twine around. If no object is encountered, the tip eventually grows straight upward. But if a twig or a beanpole is encountered, the slowly rotating growing tip spirals around it and tightens up. In this way, vines and climbing plants can reach great heights without having to invest in thick woody supporting structures such as those needed by trees (Figure 20.1).

Darwin also found that on a hot day a new shoot of a hop plant can revolve once in 2 hr 8 min. After twenty-seven such revolutions without encountering anything, the tip of the shoot will then be describing a circle 19 inches in diameter and, surprisingly, if one looked closely enough, one can actually see the movement.

Darwin described many variations of this basic pattern and interpreted them in terms of natural selection. He showed that the underlying mechanism was one of differential elongation of first one, and

20.1 Climbing plants are adapted to reaching upward to the sunlight at little metabolic expense. These plants don't invest their energy in producing thick supporting stems; instead, they make use of other plants or objects for support. As they grow, they send out slender, fast-growing shoots known as *tendrils*, which are sometimes modified leaves and in other instances are extensions of the stems. When the tendrils make contact with a solid object, they begin a coiling growth around the object. Once this has happened, new tendrils are produced and the process continues. Climbing plants are common in tropical forests where competition for sunlight favors these unusual patterns of stem growth.

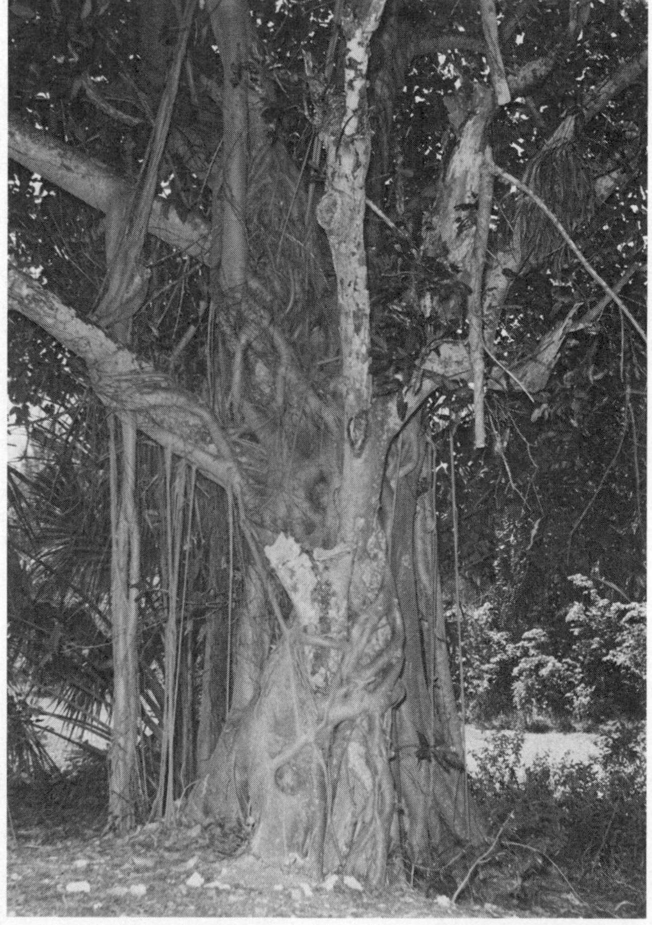

then another, part of the stem below the growing shoot, causing it to bend first one way and then another (*Climbing Plants,* September 1875).

Like any reductionist physiologist, Darwin sought to define further the mechanism for this movement by looking for ever simpler and more basic phenomena. He decided that all plants possess some ability to move their parts. "In accordance with the principles of evolution," he said, "it was impossible to account for climbing plants having been developed in so many widely different groups, unless all kinds of plants possess some slight power of movement of an analogous kind." He knew that, for instance, all plants bend their growing tips toward the light. He found that here, too, the bending was due to an unequal elongation of the stem, with the side opposite the light elongating at a slightly faster rate than the side facing the light. But what brought about this differential elongation?

Darwin, now an old man, worked on the problem with his son, Francis (Figure 20.2). Their experimental organism was canary grass. Like other grasses, in its early stages of embryonic development it produces a tubular sheath, the *coleoptile.* This emerges from the soil, surrounding the primary (first) leaf. They found that when the coleoptile is illuminated from one side, it bends toward the light. The bending does not occur at the tip, but in the elongating part of the leaf—well below the tip. Was light acting directly on the bending part? Apparently not. Darwin illuminated only the growing tip of the coleoptile, and it bent as before. With a small piece of foil, he covered the tip and exposed the rest of the plant to light. Nothing happened; it didn't curve. Darwin then knew that *something was happening in the growing tip that caused certain areas of the stem beneath to elongate.* In *The Power of Movement in Plants* (1880), Darwin ascribed this fundamental discovery to "some matter in the upper part which is acted upon by light, and which transmits its effect to the lower part. . . . When seedlings are freely exposed to a lateral light some influence is transmitted from the upper to the lower part, causing the latter to bend." Thus, Charles Darwin, working with his son, was the first to discover a *plant hormone.* Darwin was 71.

20.2 Charles Darwin and his son Francis were the first to report studies of the response of plants to light. They observed the growth of shoots of canary grass subjected to light coming from only one direction. Growth occurred as shown in (a). To determine where the photosensitive region was, they used foil caps to cover the growing tips (b). As the results in (b) indicate the part of the shoot most sensitive to light is its tip.

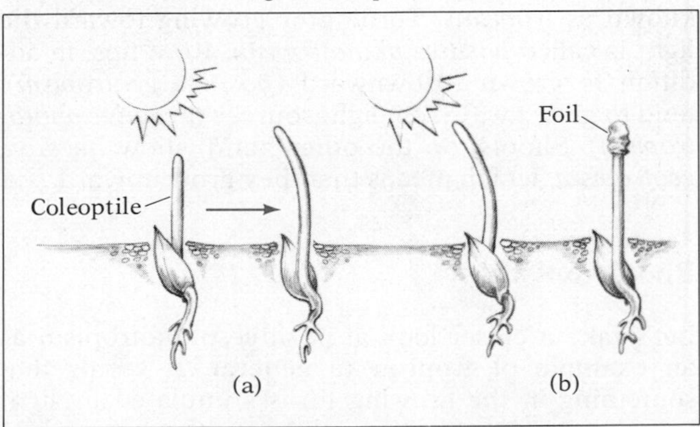

The Experiments of Boysen-Jensen and Paal

Early in this century two researchers named P. Boysen-Jensen and A. Paal extended Darwin's observations. Boysen-Jensen decapitated a coleoptile of an oat seedling a few millimeters from the tip, put a block of gelatin on the stump, replaced the tip, and illuminated it. The coleoptile bent as before, showing that Darwin's "influence" was something that could move through gelatin. Boysen-Jensen then cut off the tips of coleoptiles and replaced them over slivers of impervious mica. There was no bending. When he inserted mica into slits cut partway through coleoptiles and then illuminated them from different sides, he found that the active substance did not move down the illuminated side, but down the side away from the light (Figure 20.3). The bending, then, was due to the relative *lack* of a stimulus to elongation on the illuminated side of the coleoptile.

Paal decapitated oat coleoptiles and replaced the growing tips on one side only. The coleoptiles always bent away from that side, even in the dark. This clearly suggested that there was a material substance emanating from the tip, and that this substance stimulated the elongation of cells below the tip. The next logical step was to isolate this material and see precisely what it could do.

Any biochemical isolation requires an *assay system,* a way of measuring whether a substance is present, and in what concentrations. An assay system for the coleoptile growth substance was worked out by the Dutch botanist, F. W. Went. He used oat seedlings with agar gels instead of gelatin, and calculated hormone concentration by placing gels impregnated with different levels of hormone on one side of coleoptiles and then measuring the angle of bending. The greater the angle, the more hormone was present. Went then began to look for the plant hormone and found that the substance was widespread in nature.

Unknown to Boysen-Jensen, Paal, or Went, the active substance itself had been isolated from fermentations many years before (in 1885), by a pair of biochemists, E. and H. Salkowski. But the Salkowskis hadn't had a clue about its biological significance. The substance was not successfully isolated again until 1934, when it was finally purified from (of all

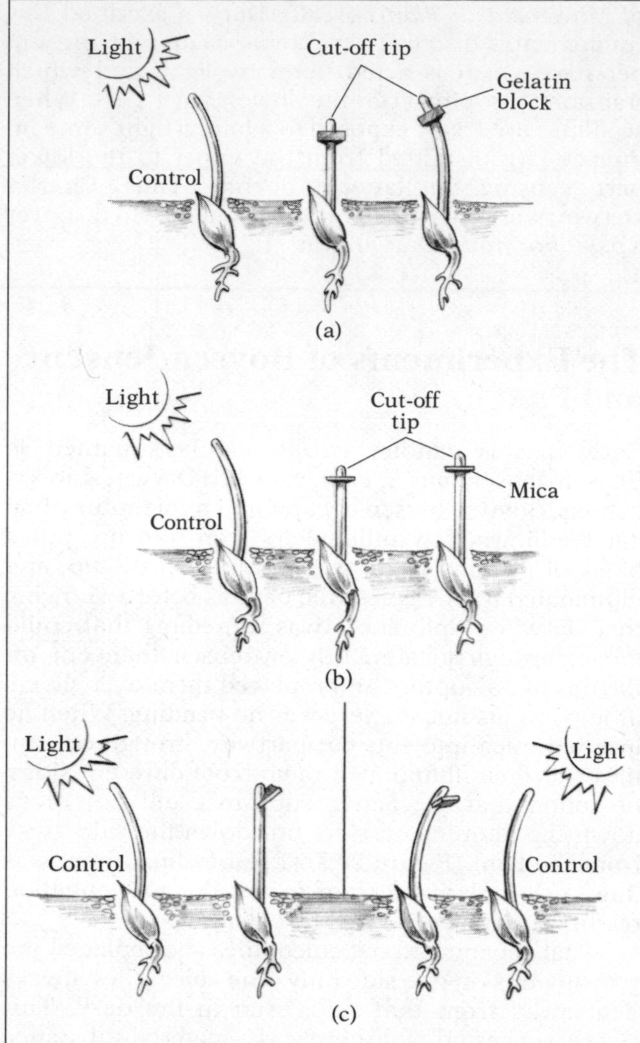

20.3 Boysen-Jensen elaborated upon Darwin's experiments by using surgical procedures. In one experiment (a) he decapitated seedling tips, placed a small gelatin block on the cut, and then replaced the tip. The plant bent toward the light indicating that the stimulating substance passed through the gelatin. In a second series (b) the gelatin was replaced by mica, an impervious substance. No curvature occurred, substantiating his results from the first series. In a third experiment (c), slivers of mica were inserted into cuts made halfway through the tip. The response to light indicated that the substance responsible for elongation passed down the unlighted side of the stem only.

things) human urine. The mysterious coleoptile growth substance, which is actually a widespread and fundamental plant growth hormone, is variously known as *auxin* or *indoleacetic acid* (IAA).

Auxin and Tropisms

The structure of auxin is very similar to the amino acid tryptophan, which is not surprising since tryptophan is its primary building block. It also bears a close similarity to serotonin, a neurohormone in the central nervous systems of animals, which also happens to be derived from tryptophan.

Indole — Acetic acid group

Auxin, or Indoleacetic acid (IAA)

Tryptophan

Auxin doesn't cause cells to proliferate by mitosis, but rather promotes cell enlargement, particularly by elongation. This cell lengthening occurs during the growth of stems after the tiny, tightly packed daughter cells are first formed in the mitotic area, the apical meristem. In horizontal stems, the lengthening occurs on the underside. In root tips, the same sort of growth relationships occur, but upside down of course, so that cells elongate behind and above the mitotic activity of the root meristem.

Although we still don't really know just how auxin works, it is currently believed that elongation requires a loosening of the closely bound filaments of cellulose at the ends of the vertical cell walls. This permits turgor pressure to expand the cells in those areas. In most cases, the actual cytoplasm of the cell doesn't increase appreciably; instead, the central vacuole enlarges. Interestingly, auxin clearly doesn't act directly on cell walls, but rather on the genetic mechanism of RNA transcription. Under experimental circumstances, auxin has been shown to induce RNA synthesis in cells. The RNA synthesis is a step, then, in the loosening of the cell walls, but we still don't know the mechanism.

It now appears that auxin is responsible for nearly all the specialized growth and movement patterns that so intrigued Darwin. These are collectively known as *tropisms*. Turning or growing toward the light is called *positive phototropism*. Root tips, in addition to growing downward (*positive geotropism*), tend to grow away from light sources (*negative phototropism*). Shoots, on the other hand, show *negative geotropism*, which means that they grow upward.

Phototropism

Let's take a closer look at positive phototropism as an example of tropism in general. It seems that something in the growing tip is stimulated by light and then perhaps reduces the manufacture and/or release of auxin, or causes the auxin to move laterally in the apical meristem, or possibly stimulates its degradation. (The initial photoreceptor is a red-light sensitive protein complex called *phytochrome,* about which we will have more to say later.) The unilluminated half of the growing tip still produces or receives auxin in quantity, and this moves down the side of the shoot, causing it to elongate on one side and thus bend (Figure 20.4). Why doesn't the auxin diffuse sideways as well as downward? We don't know.

The auxin migrating downward from the tip doesn't accumulate in the stem. In common with many hormones, auxin is short-lived. It is inactivated by specific enzymes in the lower parts of the plant. The inactivation process is necessary if the auxin's control is to be precise. In addition, excess auxin can actually inhibit growth.

Geotropism

The continued upward growth (negative geotropism) of the plant is a complex process because of the production of lateral branches. One particular growing tip, usually the original one, becomes dominant to any lateral growing tips. In many plants, particularly the conifers, a single dominant growing tip shows strong negative geotropism, while the growing tips of the lateral branches grow horizontally. This phenomenon, known as *apical dominance,* results in the familiar triangle shape of the evergreens (Figure 20.5).

Apical dominance also operates in another way. The apical meristem of the dominant growing tip, and of lateral growing tips as well, inhibits the sprouting of new branches. So what do you suppose is the way to produce a bushy shrub as opposed to a tall one? If an evergreen is "topped," the inhibitory effects of apical dominance are removed. The highest remaining lateral branches then begin to grow upward, seeming to vie with each other for apical dominance. Ultimately, one of the lateral branches will win out and be transformed into the main trunk of the tree.

The same principle works in broadleaved flowering trees, but the mechanics are more complex than in conifers. For one thing, flowering trees may have a single trunk up to a certain height and then branch out into several large limbs, with no one limb truly dominant. Some form of apical dominance may occur within each of these major limbs [Figure 20.6(a)].

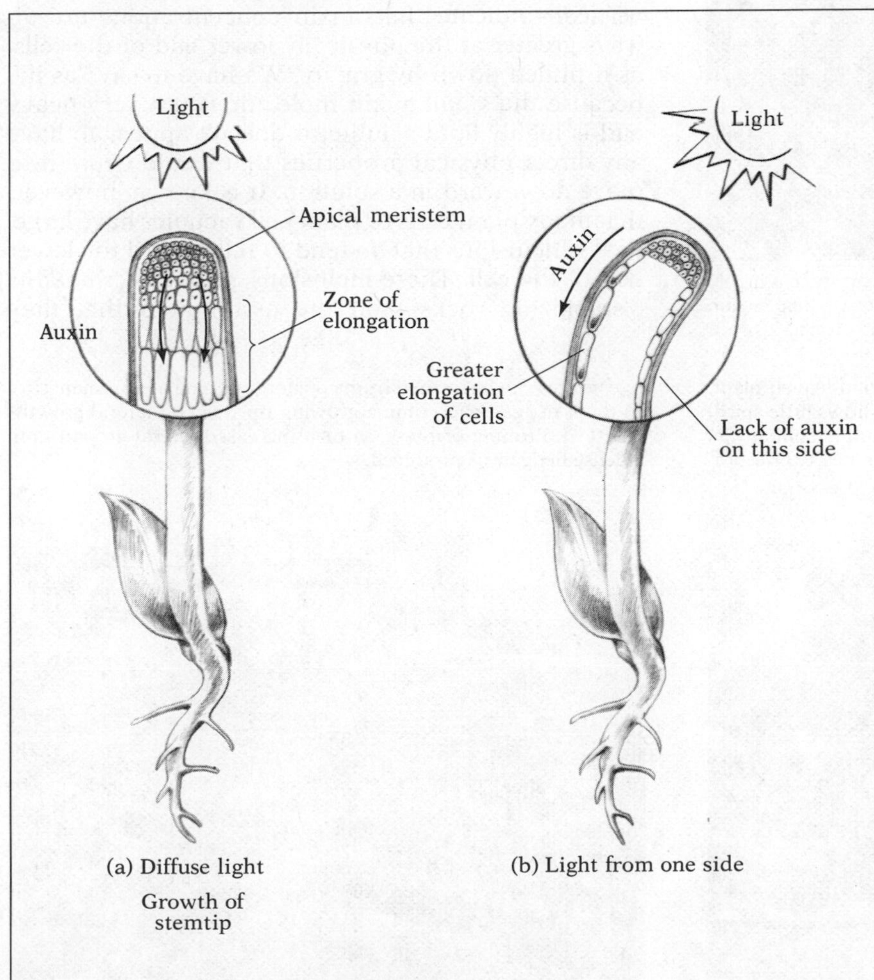

20.4 Auxin is believed to promote cell elongation in the growing tips of stems. The region known as the apical meristem continually divides, producing cells which can then elongate. (a) When light is multidirectional, elongation occurs around the entire stem. (b) But when light is available from one side only, just the cells on the unlighted side elongate. Studies show that auxin concentrates on the unlighted side.

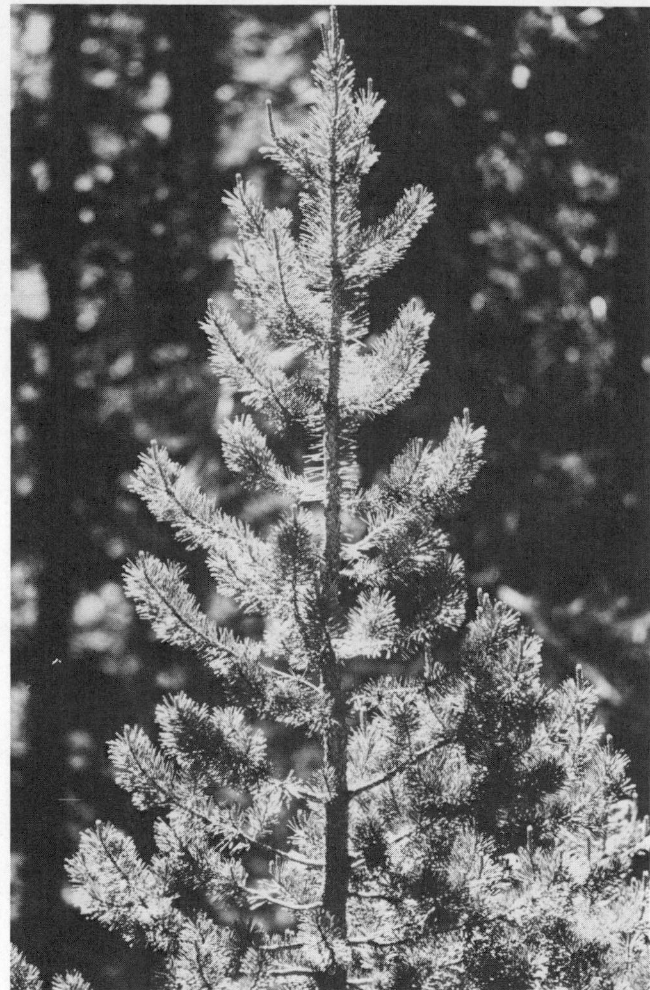

20.5 Apical dominance in the conifers. Conifers produce a highly symmetrical, conical growth form. Typically, there is one region of apical dominance at the top, which produces this form.

Continued removal of growing tips, as is done in pruning, increases the number of lateral shoots. This results in a decorative bushy plant growth [Figure 20.6(b)]. Pruning can also persuade a fruit tree to put more effort into fruit and less into wood.

One thing we can learn from all this is that plants, because of their many centers of auxin production and because of their ability to replace dominant growing tips, can survive extensive damage from, say, browsing animals, windstorms, and inept gardeners.

There are many other plant hormones, some of which we'll be dealing with soon. But before going on, we should mention that auxin, the factor that is involved in phototropism, is involved in geotropism too. If a young seedling is removed from the soil and placed in a horizontal position, it will bend its stem upward and its roots downward. Auxin assays show that, in the supine plant, the auxin itself moves downward as if pulled by gravity. This means that the concentration of auxin is much greater on the underside of the horizontal plant. It also means that auxin can move sideways with regard to the structure of the plant after all, as long as it is moving with respect to gravity.

Studies of auxin in living plants—horizontal or vertical—indicate that auxin concentrations are always greater at the physically lower end of the cells, as if pulled down by gravity. We have to say "as if," because the small auxin molecule is not very heavy and is highly lipid-soluble; it doesn't appear to have any direct physical properties that would cause it to move downward in a solution. It is known, however, that many plant cells or plant cell vacuoles have large, heavy inclusions that *do* tend to fall toward the lower end of the cell. These inclusions are called *statoliths* ("stabilizing rocks") on the assumption that they

20.6 Lack of single apical dominance in the broad-leaved plant. (a) The growth form of many broad-leaved trees shows little apical dominance. Rather, each of the large branching limbs contains its own dominant tip, which outgrows the lateral branches. (b) Decorative hedges can be formed from constant pruning of dominant tips. In these hedges, each time a growing tip was cut, lateral growths emerged. Pruning of these, in turn increased lateral growth until a dense hedge was produced.

(a)

(b)

serve some kind of sensory function in plants. Plant physiologists are presently looking for some kind of relationship between statolith movement and the active transport of auxin. Auxin is thought to be moved by active transport because it can move against its own concentration gradient at rates that are much too fast for any kind of diffusion. Metabolic inhibitors also block auxin movement downward in both upright plants and horizontal seedlings.

Now, returning to our example of the horizontal seedling, we note again that the shoot is in the process of bending upward while the root is bending downward, and that auxin is concentrated on the underside of the plant. So, let's see if we can construct an explanatory hypothesis. Perhaps root cells and shoot cells react in different ways to the same hormone; perhaps the same auxin that causes shoot cells to elongate inhibits the elongation of root cells. After all, it is well-established that different "target cells" in animal tissues can react in directly opposite ways to the same hormone.

A second, related, hypothesis suggests that there are three levels of response to auxin. It can be shown directly that excess auxin actually slows growth, as we've already noted. Different concentrations, then, can have different effects. As another possibility, perhaps root cells are simply much more sensitive to auxin than are shoot cells—in particular, they may be much more sensitive to inhibition by higher concentrations of auxin. Thus, the roots would grow down because the high auxin concentration on the underside would slow elongation while the upperside would still have enough auxin for rapid elongation. This is presently the most widely accepted explanation of the contrary action of auxin in shoots and roots. But you may have noticed that, actually, there is little real difference between the two points of view (Figure 20.7).

Some botanists are bothered by the lack of proof for either choice and prefer a simpler hypothesis: namely, that the root tip is producing some as yet unidentified auxinlike growth factor that moves upward. In fact, this was actually the suggestion of Charles and Francis Darwin in 1880, when they concluded "that it is the (root) tip alone which is acted upon (by gravity), and that this part transmits some influence to the adjoining parts, causing them to curve downward." The Darwin hypothesis has not yet been disproven. It has been shown, however, that growing root tips, like growing shoot tips, do produce auxin, which presumably moves upward.

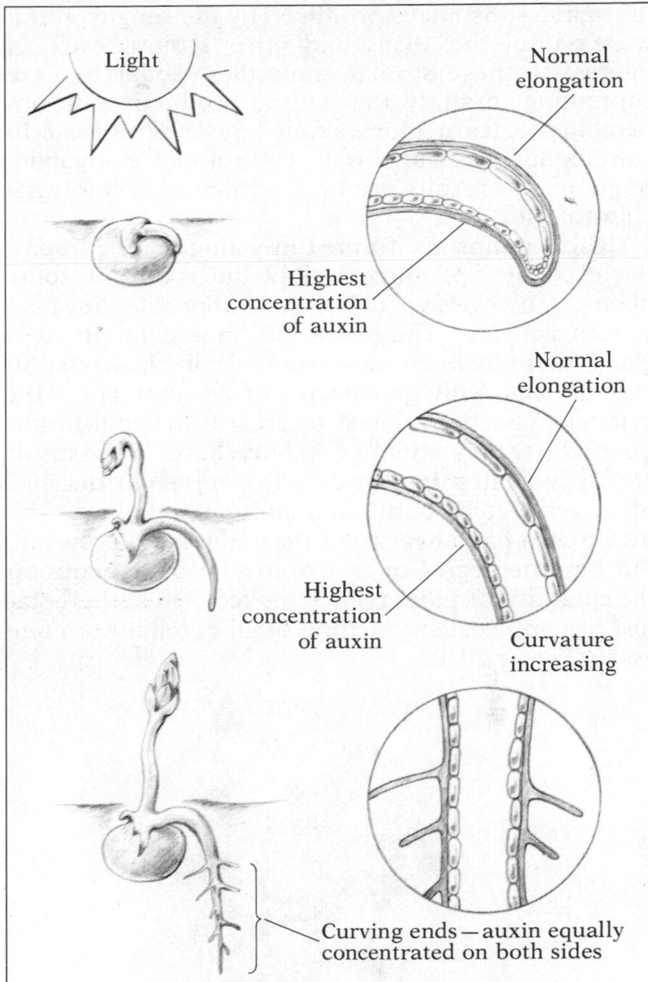

20.7 Auxin and root growth. Here, a pea seedling has been laid on its side following germination in the dark. As time passes, the young stem will become erect, growing upward, while the root will turn downward and penetrate the soil. Whereas phototropic responses are fairly well understood, geotropic responses by the root remain unexplained. One hypothesis for the opposite action of auxin in root growth is that the plant hormone inhibits growth in these tissues. While concentrations of auxin in a stem cause elongation, the root may react differently to the same concentrations so that elongation is inhibited. As shown here, the higher side of the root contains the lower concentration of auxin, but the root cells respond to this low concentration by elongating, while the cells on the underside (with higher concentrations) are inhibited. The root turns downward, penetrating the soil.

Other Plant Growth Hormones

A number of plant growth hormones have now been identified, and some of their stories are quite fascinating. Let's have a closer look at four of these.

Gibberellins

The *gibberellins,* a family of some forty molecules, received their name from the source of their discovery, *Gibberella fujilcuroi,* a fungus. The fungus, which once threatened rice harvests in Japan, causes the stem to elongate strangely and the failure of the plant to produce normal flowers. Between 1926 and 1935, Japanese botanists had already isolated and purified the active substances produced by the fungus. But it wasn't until the 1950s that other nations took an interest in these strange molecules. Today, we are continuing to study the natural production of gibberellins to learn more about how they operate in conjunction with auxins to control cell elongation. Some of the results are finding their way into agricultural use.

Gibberellins are formed in young leaves around the growing tip, and possibly the roots, of some plants. (However, we don't know what role they play in root activity.) The power of gibberellins in stem elongation has been most dramatically illustrated in experiments with genetic dwarfs. Dwarf corn, for instance, can be induced to grow to normal height after the application of gibberellins. This result clearly indicates that the dwarf corn is short because of a genetically controlled hormonal failure. The dwarf corn had always had the potential to grow tall. Further, the degree of growth in a dwarf depends on the quantity of gibberellins applied. Thus, the botanist has an excellent method of gibberellin hormone assay (Figure 20.8).

20.8 Gibberellins have a dramatic effect on stem growth, as seen in these five cabbage plants. The plants at the left were grown normally, while the ones at the right were treated with gibberellins. Note the development of the flower at the tips of the treated plants. Normally, domestic cabbage will flower in its second year.

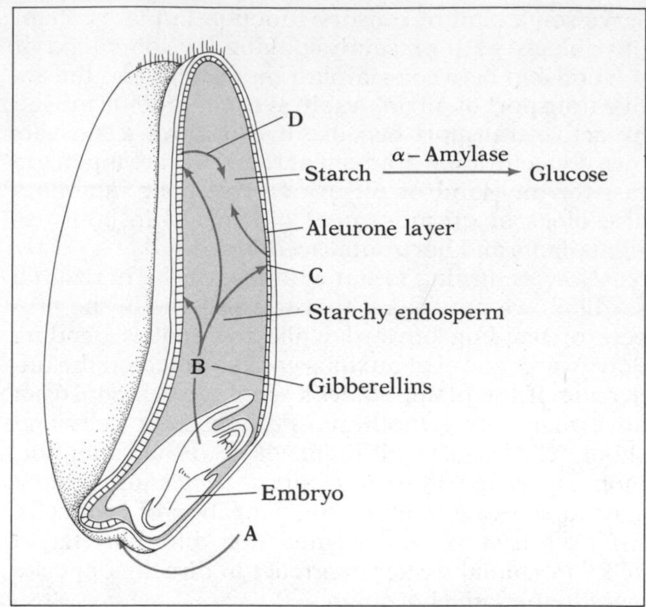

20.9 Gibberellins alter the conversion of starch to glucose in the germinating barley seed by the sequence of events seen here: (A) Water is absorbed by the seed. (B) The cotyledon secretes gibberellins. (C) Upon reaching the aleurone cell layer, gibberellin stimulates certain of these cells to produce α-amylase. (D) The release of α-amylase into the endosperm hydrolyzes starch to glucose, and thus the seedling has its supply of fuel for growth and development.

But gibberellins also have another role. They act as chemical messengers to stimulate the synthesis of an enzyme called *α-amylase* in some seeds such as barley and corn. As these grains germinate, the embryo secretes gibberellins, which move to the cell layer (*aleurone*) that surrounds the starchy endosperm. The cells of this layer, apparently stimulated by the hormonal messenger, begin to produce α-amylase, which then breaks down starch and makes glucose available to the growing plant (Figure 20.9). There may be many steps (or only a few) between the arrival of the hormone messenger and the transcription activity required for producing the enzyme. The fact is, we are just beginning to understand a little bit about how hormones alter genetic activity, as we will see.

Cytokinins

Most of what we know about this family of hormones springs from work begun in the 1950s, when it was found that plant growth could be influenced by something from corn kernels. In 1964, the first of the molecules, *zeatin,* was described; since then, three others have been identified.

CH_3
|
$C-CH_3$
||
CH
|
CH_2
|
NH
(purine ring: N, N, N, N–H)

Zeatin

Before zeatin was isolated from young sweet corn kernels, all we had was the circumstantial evidence that some plant materials stimulated cell division. Research scientists found, for example, that in order to grow cultures of plant tissues, they had to add coconut milk to the medium. What was so special about coconut milk and corn kernels? Biologists began to isolate chemicals found in both sources and the magical substance turned out to be *cytokinins.*

The first thing biologists found out about cytokinins was that they stimulated cell division in plants. Actually, though, cytokinins alone could not stimulate cell division. They had to be coupled with other plant hormones. Why? How did they work?

The rather startling answer came from work on tissue taken from tobacco plants. Botanists mixed cytokinins and auxins in different ratios and then grew tobacco pith tissue in the various media. It turned out that different ratios of the two elements produced entirely different kinds of growth. Specifically, tobacco pith grown in a higher auxin-to-cytokinin ratio madly produced roots. In other words, pith tissue differentiated into root tissue. But low auxin-to-cytokinin ratios changed pith into apical meristem, which then produced stems and leaves. When the substances were more nearly equal, the pith produced an undifferentiated kind of *callus* tissue, the kind that grows at the site of an injury in a plant (Figure 20.10).

The studies indicated that those undifferentiated and ordinary cells of the pith, the parenchyma cells, contain all the genetic information necessary to develop into a number of kinds of other plant cells, producing a variety of tissues and organs. Even more differentiated tissue, grown on the proper medium, would sprout into complete, normal plants. This knowledge is incredibly important. The value of these studies, once the techniques are perfected, staggers the imagination—entire individuals grow from a piece of tissue.

The third role of the cytokinins is associated with aging in plants. Aging, or *senescence,* occurs, for example, in leaves prior to their seasonal fall from the branches of deciduous trees. Leaves also age if they are picked while in the prime of life. Then, of course, they die. When picked leaves are treated with cytokinins, however, death comes far more slowly—aging is retarded. The chlorophyll does not disintegrate (so the leaves stay green), protein synthesis continues,

20.10 Cytokinins and differentiation. This unseemly mass of tissue is known as a *callus.* It consists of tobacco pith (parenchyma) and a combination of auxin and cytokinin. The specific ratio of this combination was correct for producing a stem shoot, which is seen emerging from the callus. By varying the ratio of auxin to cytokinin, the specific type of tissue produced can be controlled by the experimenter. The illustration shows the results of four different combinations of auxin and gibberellin.

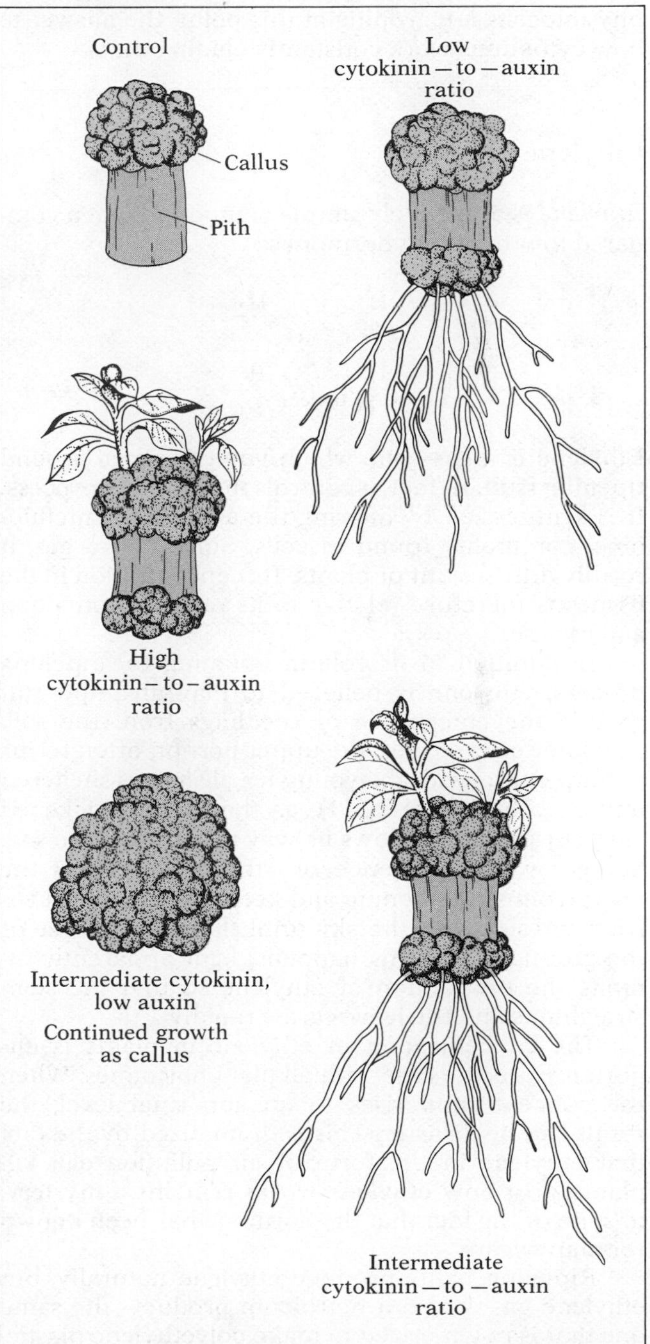

and carbohydrates do not break down. Synthetic cytokinins have been applied to harvested vegetable crops such as celery, broccoli, and other leafy foods in order to extend their storage life.

No one is quite sure how cytokinins work, but we have a few ideas about it. It is known, for example, that some of the tRNA molecules contain cytokinins as a functional part of their structure. It is possible that the addition of free cytokinins to a cell may facilitate some types of protein synthesis. On the other hand, cytokinins may actually act through tRNA in some other way. Frankly, however, plant physiologists are groping at this point, the answer to how cytokinins work constantly eluding them.

Ethylene

Ethylene is a relatively simple compound when compared to other plant hormones:

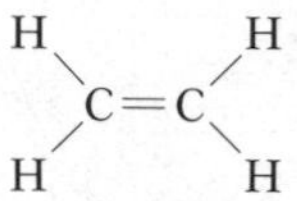
H
H
C=C
H
H

Ethylene

Ethylene is a gas—one which you can smell around ripening fruit. In fact, it controls the ripening process. It is synthesized by altering the amino acid methionine, commonly found in cells. Since it is a gas, it readily diffuses out of plants. Its concentration in the tissues is therefore relative to its rate of production and escape.

In addition to its role in initiating the ripening process, ethylene is believed to play an important part in the emergence of seedlings from the soil. As some seeds sprout, the upper portion often forms a sharp curve, with the young fragile leaves sheltered underneath (Figure 20.11), as the tough, thickened curve of the shoot plows its way up through the soil. Somehow, the presence of ethylene prevents the plant from straightening and keeps the fragile leaves from unfolding to the sky until the shoot is free of the ground. Once this happens, light apparently inhibits the production of ethylene so that the stem straightens and the leaves grow rapidly.

The concentration of ethylene in plants is important, as is the case with all plant hormones. When the concentration rises to an abnormal level, the result can be disaster. This is dramatized by the fact that ethylene in the form of air pollution can kill plants. Just how ethylene works remains a mystery, in spite of the fact that the hormone has been known for many years.

Ripening fruits produce ethylene naturally, but ethylene gas is also a petroleum product—the same one that is polymerized to make polyethylene plastic.

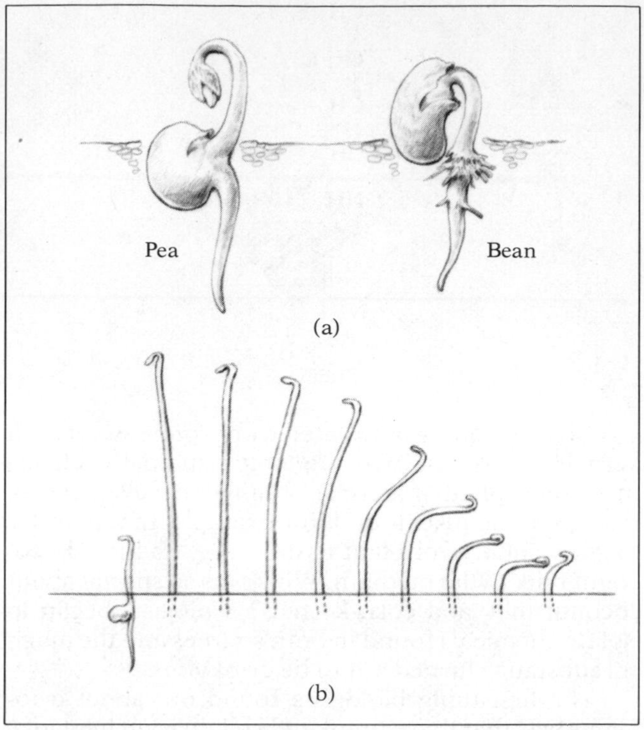
Pea
Bean
(a)
(b)

20.11 Ethylene is usually associated with fruit ripening, but it also has a role in shoot curvature. (a) Note the typical curvature of seedlings as they emerge from the soil. The curve, called the *epicotyl hook*, is thought to act as a protective "bumper." (b) This photograph shows pea seedlings that have been grown in increasing concentrations of ethylene gas. As the concentration is increased the curvature becomes greater.

The synthetic gas has long been used to initiate ripening of fruits in transit to the world's markets. In particular, the banana industry owes much of its success to this simple chemical. It is possible to ship unripened fruit to world markets in prime condition without great losses because of untimely ripening. The release of ethylene in warehouses full of green bananas, or in a ship's cargo hold will produce ripe, yellow bananas at just the time they reach the consumer.

Abscisic Acid

Our last example of a plant hormone is *abscisic acid.* Its molecular structure consists of a carbon ring with a short carbon chain containing a carboxyl (acid) group:

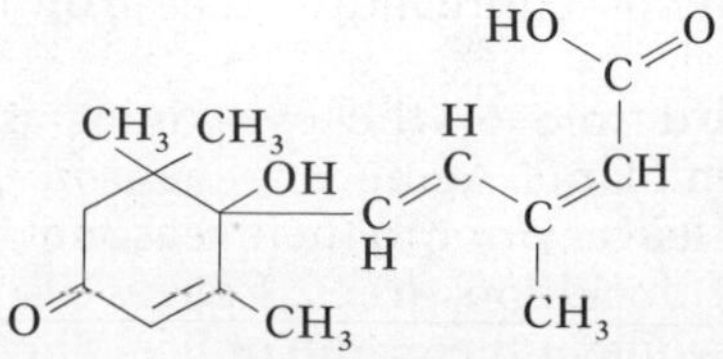
HO
O
C
CH3 CH3
H
OH
C
CH
C
C
H
O
CH3
CH3

Abscisic acid

The molecule is named for its principal function, the control of leaf *abscission,* or separation. The fall of leaves may be a seasonal event or it may follow accidental wind damage, animal browsing, or drought. The primary problem for the plant is to prevent water loss and parasite invasion at the place where the leaf was attached. You may recall that the vascular system of plants extends directly into the midrib of the leaf; therefore, its exposure could be hazardous. The leaf must seal off the wound as rapidly as possible. Each leaf attaches to the stem via a special layer of cells at the base of the petiole, known as the *abscission zone.* In a complex interaction with other plant hormones, especially auxin, abscisic acid causes the cells in the abscission zone to die and harden in preparation for the fall of the leaf. Thus, when the leaf finally falls, the plant is prepared (Figure 20.12).

It is generally believed that the abscisic acid also induces winter dormancy by inhibiting the growth-promoting hormones, auxin and gibberellin. The inhibition may be by suppressing mRNA production and is a general phenomenon produced throughout the plant. This suppression is obviously important to deciduous plants of the temperate regions. Only dormancy can save a plant which is exposed to winter's freezing temperatures and the water shortages common in frozen ground.

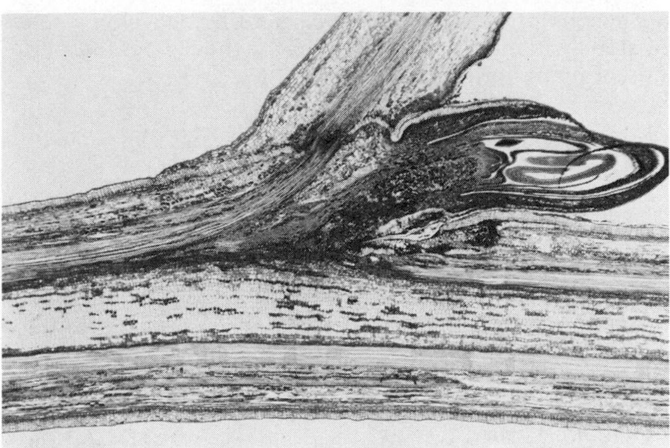

20.12 Abscisic acid and the leaf abscission layer. Before the leaf falls, the cells of the abscission zone (shown here) die and harden under the influence of abscisic acid. Eventually, the leaf will break away and fall. The dead abscission zone then seals the scar, averting water loss and any danger from infection.

Applications of Plant Hormones to Agriculture and Horticulture

Knowledge about plant hormones has obviously been extremely useful to people in a variety of ways. The investigation has led to the development of increasingly precise analytical techniques such as the hormone assay. The improved techniques have enabled us to regulate and understand (sometimes in that order) a wide range of hormones and to produce the sorts of plants we want. Sometimes, of course, we want those plants dead, such as some weeds and at one time, the trees of Vietnamese forests.

Since the discovery of auxins, chemists have developed an entire family of analogous compounds. Some of these have turned out to have even greater growth-promoting effects than the original auxin. One such compound that is a very active growth promoter is 2,4-dichlorophenoxyacetic acid, known mercifully as 2,4-D. While 2,4-D promotes growth at very *low* concentrations, it kills at higher concentrations. This compound is a very common and potent agent of weed control. Industrial ingenuity has made it possible to produce this compound by the ton. It is highly selective in that it is usually toxic to dicotyledons while sparing the monocotyledons (such as grasses—wheat, oats, etc.). This means you can kill the dandelions in your lawn with 2,4-D, while sparing the rye and bermuda grass. Actually, you *can* kill the monocots if you use a high concentration. It's just that broad-leaved plants such as weeds are more sensitive to the herbicide. The major importance of 2,4-D in weed control where grain crops are involved is obvious. (It beats pulling weeds in a Kansas wheat field.) In addition to its effect on broad-leaved weeds, 2,4-D also offers the advantage of rapid *biodegradability* (it rapidly loses its biological activity when exposed to the elements and organisms of the soil).

2,4-Dichlorophenoxyacetic acid is not uncontroversial, however. There are serious and ominous questions about its effects on the reproductive activity of mammals (it may cause birth defects), so its use in weed control in national forests and parks (as a form of fire prevention) has to be measured carefully in terms of risk to wildlife. The problem is, we create a weed problem by clearing the land, building roads, and otherwise disrupting the natural habitat. Then we solve the weed problem by applying unnatural controls, which in turn tend to create additional, often unexpected problems. We will have more to say on this subject later.

Application of Ideas

1. The use of specific auxins or their analogs is probably a great improvement over other means of chemical weed control (such as petroleum distillates). Suggest why this may be true. What might the advantages of hormonal control be when applied to insects?
2. Identifying and characterizing plant hormones must be

a rigorous process if it is to be scientific. Develop a series of steps you might apply as rules for the proper identification of plant hormones.

3. Employing modern technology and capability, suggest an experiment that could be used to better isolate and study the phenomenon of geotropism.

Key Words

abscisic acid
abscission zone
aleurone
analog
apical dominance
apical meristem
assay system
auxin

callus tissue
coleoptile
cytokinin

2,4-D

ethylene

genetic dwarf
geotropism
gibberellin

hormone

IAA
indoleacetic acid

negative geotropism
negative phototropism

pith
positive geotropism
positive phototropism
pruning

root meristem

senescence
statolith

tendril
tropism

zeatin

Key Ideas

Introduction

1. Charles Darwin described the behavior of climbing plants, observing that the underlying mechanism was differential elongation. In attempting to understand this phenomenon, he carried his studies into the natural bending of plants toward light.

2. He and his son Francis observed that the sheathlike *coleoptile,* particularly its tip, responded to light. He was able to show that the upper part transmits its effect to lower parts, resulting in elongation of cells on the unlighted side.

The Experiments of Boysen-Jensen and Paal

1. Boysen-Jensen's procedure was to cut off the *coleoptile* tip and cover it with a slab of gelatin. Placing the decapitated tip on the gelatin and illuminating it from one side produced the usual bending results. This told him that something was passing through the gelatin.

2. Boysen-Jensen used mica slivers inserted half way through the lighted and unlighted sides. In the later step, no elongation or bending occurred, indicating that something was passing down that side that caused elongation.

3. Paal used decapitated tips. By placing the tips on one side of the stem at a time, he showed that the tip produced a substance that passed into the stem causing elongation. Light inhibits the production of this substance.

4. Went later carried out assaying techniques to determine the concentration of the substance. He used agar instead of gelatin, impregnating the agar with different amounts of hormone and measuring the angle of bending.

5. The growth-promoting substance was finally identified as *indoleacetic acid (IAA),* or *auxin.*

Auxin and Tropisms

1. *IAA* is derived from tryptophan. Its action is to promote growth by cell elongation. Cell elongation commonly occurs below *apical meristem* in stem tips and above *apical meristem* in root tips.

2. Experiments indicate that auxin works at the gene level, causing transcription of mRNA to begin. The link between mRNA and cell wall loosening is not known.

Phototropism

1. *Auxin* is involved in nearly all *tropisms* including *phototropism* (toward or away from light) and *geotropism* (toward or away from gravity). Growth towards water is another matter, and it may be that roots only grow when there is water.

2. In many plants the growing tip is *apically dominant,* meaning that it elongates fastest, with elongation in subordinate tips somewhat suppressed. Removal of the tip removes inhibition in lateral branches.

Geotropism

1. *Auxin* appears to move toward gravity, but is too small to be affected directly. Heavy inclusions, or *statoliths,* also appear to fall to the lower end of the cells. There may be a relationship between *auxin* movement and *statoliths.*

2. In a horizontal seedling, the stem grows up, the root, down. There are three hypotheses to explain this different response.

   a. Root cells and stem cells may simply respond differently to the same hormone. This is not an unusual phenomenon in hormone activity.

b. Root cells may be more sensitive to *auxin*, thus the higher concentration on the underside may retard elongation there. It is well known that *auxin* in *high* concentrations retards elongation.
c. Root tips may produce a growth factor that moves *up* from the root tip.

Other Plant Growth Hormones

Gibberellins

1. The *gibberellins* were first discovered in a fungus that infested rice, causing abnormally rapid growth. They are also produced in growing tips and may work in concert with *auxin* to induce elongation.
2. *Gibberellins* also act as a chemical messenger in seeds. When secreted by the embryo, they stimulate the *aleurone* layer to produce α-amylase. The enzyme digests the starchy endosperm, producing glucose. They can be used to stimulate germination and promote stem thickening.

Cytokinins

1. *Cytokinins* together with *auxin* stimulate cell division. Different ratios of *cytokinin* to *auxin* produces different growth results. One study utilized pith cells from tobacco in a growth medium with the following results:
   a. High auxin—low *cytokinin*: roots produced
   b. Low auxin—high *cytokinin*: stems produced
   c. Equal concentrations: undifferentiated callus produced
2. *Cytokinins* are also known to delay *senescence* (aging). This knowledge has been applied to prolonging the storage life of vegetables.
3. The precise action of *cytokinins* is not understood, but they appear to have a close association with RNA.

Ethylene

1. *Ethylene* is a gas. It initiates ripening of fruit and plays a role in the emergence of seedlings from the soil.
2. It is believed to help maintain the sharp curve in some seedlings, which acts as a bumper while the seedling pushes up through the soil.
3. *Ethylene* is used to ripen fruit while in transit.

Abscisic Acid

This hormone prepares the deciduous plant for leaf fall. Interacting with *auxin* and other hormones, it induces winter dormancy. It may act by suppressing mRNA production as the plant shuts down for winter. It is often used to inhibit seed germination.

Applications of Plant Hormones to Agriculture and Horticulture

1. Much of our knowledge of plant hormones can be applied to plant control.
2. In addition, it has been possible to produce synthetic analogs to hormones that are very active. *2,4-Dichlorophenoxylacetic* acid (*2,4-D*) is an *analog* of *auxin*. It is used as a herbicide to kill dicot weeds.
3. 2,4-D may have a deleterious effect on mammal reproduction, so its use is controversial.

Review Questions

1. What observations in climbing plants prompted Darwin to experiment with grass seedlings? (p. 544)
2. Explain Darwin's procedure and state his conclusions in his grass seedling experiment. (pp. 544–545)
3. What is a *coleoptile* and what is its function? (p. 545)
4. With the aid of simple diagrams, explain how Boysen-Jensen experimented with *coleoptiles* and the growth phenomenon. (p. 545)
5. Describe Paal's contribution to the growth experiments. (p. 545)

6. Describe Went's procedure as he developed an *assay system*. (p. 545)
7. Where, in the plant, is *auxin* produced and what is its general effect on the plant? (p. 546)
8. What is the specific effect of *auxin* on the plant cell wall? (p. 546)
9. Describe the different kinds of *tropisms*. (p. 546)
10. What happens to *apical dominance* when a plant is pruned? (p. 547)
11. Explain the *statolith* hypothesis of *auxin* travel. (pp. 548–549)
12. Describe three hypotheses that attempt to explain why stems grow up and roots grow down. (p. 549)
13. Review the discovery of *gibberellins*. (pp. 549–550)
14. What appears to be the general role of *gibberellins* in plant growth and what is the evidence for this conclusion? (p. 550)
15. Explain how *gibberellin* acts as a chemical messenger in seed germination. (p. 550)
16. In what way do *cytokinins* affect plants? (pp. 550–551)
17. Review the evidence from tobacco pith experiments that suggests that *cytokinins* and *auxins* work in concert. (p. 551)
18. What role do *cytokinins* apparently play in the process of *senescence* and how can this information be put to use? (pp. 551–552)
19. Present a theoretical explanation of how *cytokinins* may actually work. (p. 552)
20. What effect does *ethylene* have on fruits? On seedlings? (p. 552)
21. How might *abscisic acid* prevent damage to deciduous plants when winter approaches? (pp. 552–553)
22. What is a molecular *analog*? How might these do the work of hormones? (p. 553)
23. What are the benefits and problems associated with the use of 2,4-D? (p. 553)

Chapter 21

Plant Reproduction and Development

In reproducing on dry land, plants have been faced with some very special problems. For example, most plants (other than a few water-borne species) don't move around much, so they've had to make do with what life has dealt them. An animal might move from a dry place to a wetter one, while a plant might solve the problem of dryness by reproducing madly whenever water is temporarily available. Some plants keep their offspring near, saturating an area with their own kind, while others literally scatter them to the four winds. Yet others take an intermediate route and simply cast their seeds a short way. Who would have guessed that as winged maple seeds drop from the trees, they first fall, whirling, straight down, but as they gather momentum, their aerodynamics dictate that they suddenly begin to spin sideways, sailing from under the sheltering leaves of the parent plant? So, in this chapter we will look at how plants reproduce and then how the offspring develop, keeping in mind the special problems of this kingdom.

21.1 Asexual Reproduction in Plants

Vegetative Propagation

In most seed plants, asexual reproduction takes the form of *vegetative propagation*. Here, the plants simply form new plants from portions of their own roots, stems, or leaves. This capability has been exploited so extensively in agriculture that we often forget that plants carry out the same process in nature. Actually, many plants routinely reproduce vegetatively, while others may initiate the process as a response to injury (Figure 21.1).

It is generally known that many plants can be propagated from *cuttings*. People sneak into other people's yards, snip a twig, and tiptoe into the shadows, their heads full of images of the splendid plant that will soon attract neighbors to their own yards. Sometimes it works and sometimes it doesn't, because not all plants can generate roots from a planted stem. In nature, however, broken branches may fall to the ground and take root, a fact we often overlook. Willows, for example, can reproduce this way. When people attempt to propagate plants from cuttings, they usually keep them in water until roots appear and then plant them. Sometimes, they add auxin to the water in which the plant is kept. Root growth is generally preceded by the appearance of *callus* tissue, which is an undifferentiated mass of thin-walled cells produced at the site of injury.

New plant growth can also emerge from what are known as *adventitious buds*. These are common on stems where the plant has been severely damaged or actually cut off. They are also found on the roots of some plants such as the silver leaf poplar and black locust. In these cases, hidden buds along underground roots begin to sprout, sending young stems upward and suffusing the ground beneath with tiny roots. They are often referred to, rather ungallantly, as *suckers*, and are chopped off because they draw on materials that might otherwise go into the productive parts of the parent tree.

In a few instances, shoots can emerge vegetatively by *leaf generation*, such as in *Bryophyllum* (Figure 21.1). If a bryophyllum leaf falls into a moist environment, it may produce a new plant. Home gardeners often propagate bryophyllum by floating the leaves on water until young shoots appear.

Some plants send *runners* or stolons along the ground; these are wispy stems that snake away from the parent plant until they send down roots at some distance from their origin and eventually develop into independent plants. The runner stems have intermittent nodes, where adventitious roots and shoots are produced. Hardly anyone grows strawberries from seeds, since it is such a simple matter to pinch off a runner and transplant the shoots arising from it.

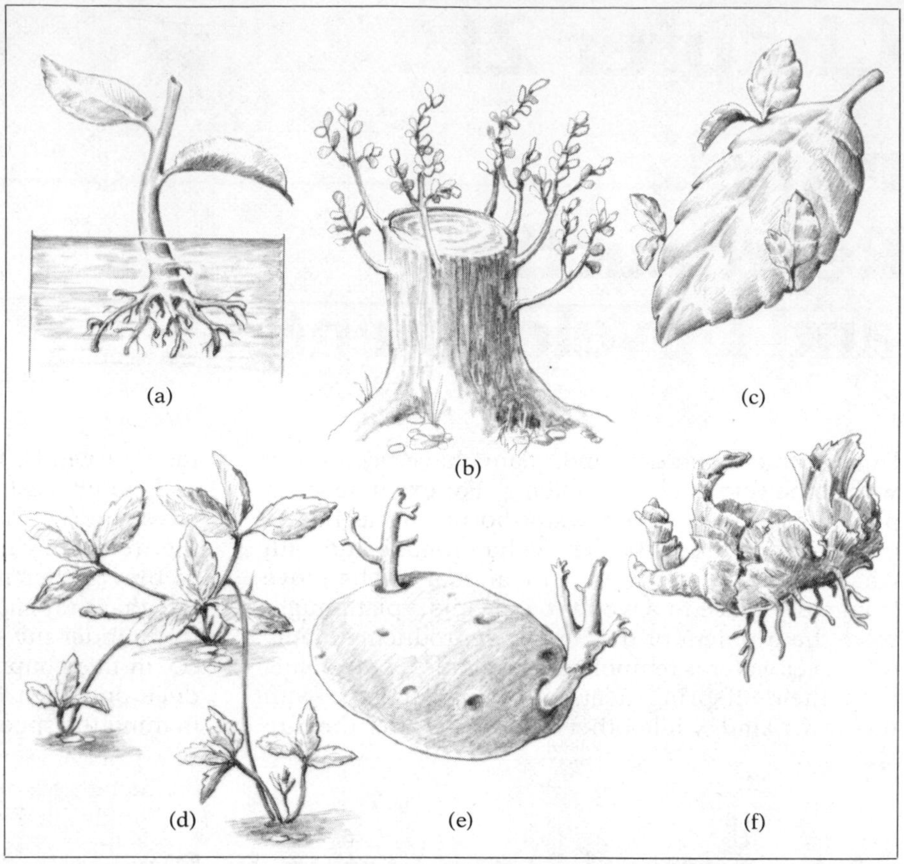

21.1 In the seed plants, asexual reproduction occurs chiefly from the growth of new plants from preexisting parts. Six ways in which asexual reproduction may occur in nature are shown here. All these have been applied to agricultural practices.

(a) *Cuttings.* Bits of stem tissue can form roots and continue the growth of the stem and leaves. Plants usually propagated from cuttings include bananas, ivy, geraniums, and many conifers.

(b) *Adventitious buds.* These buds emerge from stem tissue in plants such as poplars, the black locust, and raspberry. The buds produce shoots, often called *suckers.*

(c) *Leaf generation.* Leaves of *Bryophyllum* will produce roots and shoots in the notches of the leaf blade when they lie on a moist surface or in water.

(d) *Stolons,* or *runners.* Strawberries commonly send out runners above ground. They will produce roots and shoot growth some distance from the parent plant.

(e) *Tubers.* Tubers such as the Irish potato are actually parts of the stem. The familiar eyes are really lateral buds, capable of producing adventitious roots and stem shoots.

(f) *Rhizomes.* Rhizomes are fleshy underground stems. The parent plant sends out rhizomes in which the nodes are separated by lengthy internodes. Nodes are capable of root and stem shoot production, as we see in such plants as canna, bermuda grass, crabgrass, and ferns.

Tubers are also stems, but they are typically thick and are found underground. The potato is our best example. The potato doesn't look much like a stem, but careful examination reveals typical stem features. The most significant is the eye, which is actually a lateral bud, fully capable of producing stem and adventitious root growth. In fact, a new potato crop is usually produced by planting pieces of potato that include at least one eye.

Crab grass and bermuda grass can reproduce vegetatively through underground stems known as *rhizomes*. These stems produce nodes at various distances from each other. Each node sends roots down and stems up. In addition to their insidious method of spreading, these grasses are C4 plants with improved photosynthetic efficiency, so that propagating bermuda grass is easy. A cylindrical core cutter is used to obtain "plugs" of the grass, which can then be set into new ground.

Grafting

In agriculture and horticulture, *grafting* is a very effective means of propagation (Figure 21.2). Through grafting, older stem stocks can be used as recipients for young, vigorous shoots. In many cases, the root stock of one species or strain may have strong advantages—such as resistance to fungal parasites—while the stem stock of another related species or strain may have different advantages—perhaps greater fruit yield or more desirable fruit. Thus, English walnut stems are usually grafted to black walnut root stock. European varietal grape branches are usually grafted to the hardier American grape root stock, and the situation is similar with different kinds of avocados. In the case of the familiar navel orange, grafting onto the root stock of other orange tree strains, or growth from cuttings, are the only ways this seedless mutant can be propagated.

The concept of grafting is rather straightforward. Stems or shoots of plants are inserted into the recipient in a manner that brings the vascular cambium of both into contact. The cambium doesn't just grow together; instead, a callus develops. The callus then differentiates into a new region of phloem, xylem, and vascular cambium, and the union is completed.

Many kinds of strange combinations are possible. For example, someone once succeeded in grafting a tomato stem onto a potato stock. The result after a few months was a good growth of both tomatoes and potatoes, on the same plant!

So one wonders, what other combinations are possible? Probably a great variety, because plants don't show the powerful immune response that makes animal grafting (such as organ transplants)

Essay 21.1 Asexual Reproduction: The Long Run

If asexual reproduction is so successful, you may ask, why is it that plants don't develop it more fully and rely on it as a major means of increasing their populations? The answer lies in the mechanics of genetics and evolution. Any asexually reproduced individual is genetically identical to its parent. This, of course, is exactly why vegetative propagation is so valuable to agriculture and horticulture—we like to have one Irish potato be as similar as possible to the next. We have enough surprises in life without dealing with variable potatoes.

As long as we protect and nurture a plant strain, all may be well. But plants in nature are not always so lucky. As you know, populations that have no genetic variability face particular problems in a changing environment. How can a homogeneous population change with a changing environment?

It is interesting that while completely asexual plant species are very common, none of these species appears to have been around very long in evolutionary terms, and none of them can be expected to hold out in the evolutionary long run. Most apomictic plant species are closely related to sexual species and are presumed to have been derived from sexually reproducing ancestors in the relatively recent past.

---

so difficult. Grafting between closely related species seems to produce few problems. One variety of apple tree, for instance, can accept grafts from many other varieties. Furthermore, plums, apricots, and nectarines, can be grafted onto peach trees, and grape varieties can be readily interchanged.

Apomixis: Asexual Reproduction with Seeds

Sexual reproduction in higher plants is always through seeds, but, interestingly, not all seeds are produced sexually. *Apomixis* is the production of seeds without the union of gametes. For example, dandelions can be pollinated but not fertilized. (You hardly ever hear about *this* at your parent's knee.) The megaspore mother cell simply skips meiosis and the seed embryo begins development without benefit of a father. Of course, the new embryo has exactly the same chromosomes and genetic constitution as the single parent sporophyte on which it is borne. Dandelions, then, essentially produce clones. The evolutionary implications of this drastic shift in reproduction are discussed in Essay 21.1. The seed develops and is dispersed in just the same ways that sexually produced seeds are dispersed—in the case of dandelions, the seeds have airy plumes and are dispersed by the wind.

One might wonder why such a plant, called an *apomict*, undergoes pollination at all. Actually, some

21.2 Grafting is a common method of propagating woody plants such as fruit trees and grape vines. The four usual techniques include (a) *budding,* (b) *cleft grafting,* (c) *whip grafting,* and (d) *bridge grafting.* All these use the same basic principle. Stem tissues from a desired stem stock (donor) is cut and inserted into a cut in the recipient root stock. In each case, the vascular cambium of the two is brought into contact and subsequently fuses. Bridge grafting isn't used very frequently but can save badly damaged trees. When successful, the bridges act as new supporting and conducting tissues, covering the damaged area.

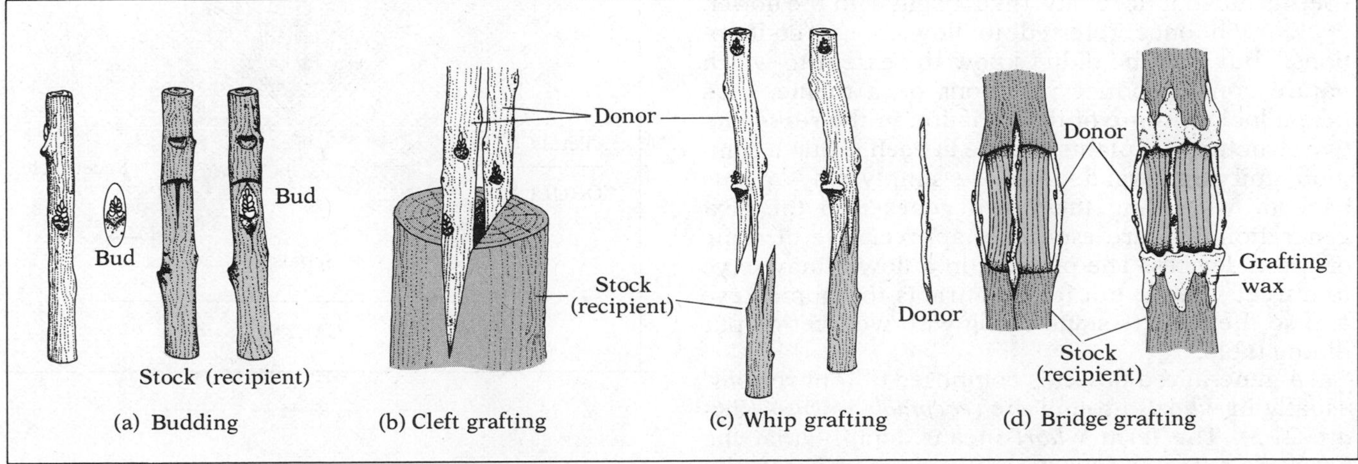

apomicts don't require pollination, but, of those that do, the pollen provides a male nucleus for the development of the triploid endosperm, even though it does not contribute to the genetic constitution of the diploid embryo itself. In other apomicts, the pollen contributes neither to the embryo nor to the endosperm, but the act of pollination somehow triggers the beginning of development. The triggering may be due to the fact that growing pollen tubes synthesize large amounts of auxin and perhaps other growth hormones, thus stimulating growth.

21.2 Sexual Reproduction in Flowering Plants

In the flowering plants—the angiosperms—the role of sexual reproduction falls on the flower. After pollination, the petals drop off, and part of the flower enlarges to become the seed-bearing *fruit.* But keep in mind that the term has a special botanical meaning. In everyday usage it usually refers to a sweet and juicy structure such as the familiar apple, grape, or banana. Botanically, however, the *fruit* is the ripened or mature ovary. Whereas it does commonly consist of a sweet fleshy structure, it may also include any other mature ovary such as the familiar string bean, cereal grain, pumpkin, or tomato. The term *fruit* also applies to such seed *capsules* as the walnut shell and sunflower-seed shell, as well as the plumed capsule that carries the dandelion seed.

The United States Supreme Court at one time was required to decide whether the tomato was a fruit or a vegetable. It seems that Congress had passed a 10% tariff on vegetables, but there was no tariff on fruit. We know, of course, that the tomato is a fruit, because it is a ripened ovary with seeds. But by that criterion, zucchini, string beans, and eggplant are fruit too. On the other hand any sensible person knows that all three are "vegetables." The nine learned judges, wise men all, took due note of proper botanical usage but decided that the tomato was not sweet enough to be a fruit, and was cooked for dinner often enough to be a vegetable.

The fruit, of course, bears the seed, which is the flowering plant's investment in its future. But any consideration of sexual reproduction in the angiosperms must of necessity, then, begin with the flower.

Darwin once referred to flowers as "contraptions," but even he didn't know the extent to which nature could produce variations on a theme. Let's take a look at some of the variation in the reproductive structures of plants. Notice in each group a common unifying thread: they are simply ways plants have of wheedling their own genes into the next generation. We are especially appreciative of some of these devices. The pigment in a flower may serve to attract insects, but it also attracts the human eye and so the flowers stand dying with wet feet on our dining table.

A generalized flower is composed of four regions, usually in *whorls* around the *receptacle* or *base* (Figure 21.3). The term *whorl* means, simply, that the parts are repeated in a circle. Each whorl is actually a modified leaf, but some are much more modified than others. The outermost whorl consists of the *sepals,* which are green and leaflike. They surround the flower bud before it opens. The whorl of sepals is collectively known as the *calyx.*

The second whorl of floral parts contains the *petals.* These are often large and colorful, but still somewhat leaflike. A whorl of petals is known as a *corolla.*

The third region, also a whorl, consists of the *stamens,* which are the male reproductive structures of the flower. They are also modified leaves, but they have changed so much that their ancestry is not apparent. Each stamen includes a slender stalk known as a *filament.* Atop the filament is the *anther* where pollen is produced. Taken together, all the stamens in the whorl are called the *androecium.* The term translates rather loosely as "house of man,"

21.3 A generalized flower. Flowers generally consist of four major parts: The *calyx* contains the leaflike *sepals.* The *corolla* contains the larger, often colorful *petals.* The *androecium* is a whorl of *stamens,* each of which consists of a *filament* and an *anther.* The *pistil* (or *gynoecium*) consists of the sticky *stigma,* the long *style,* and the *ovary.*

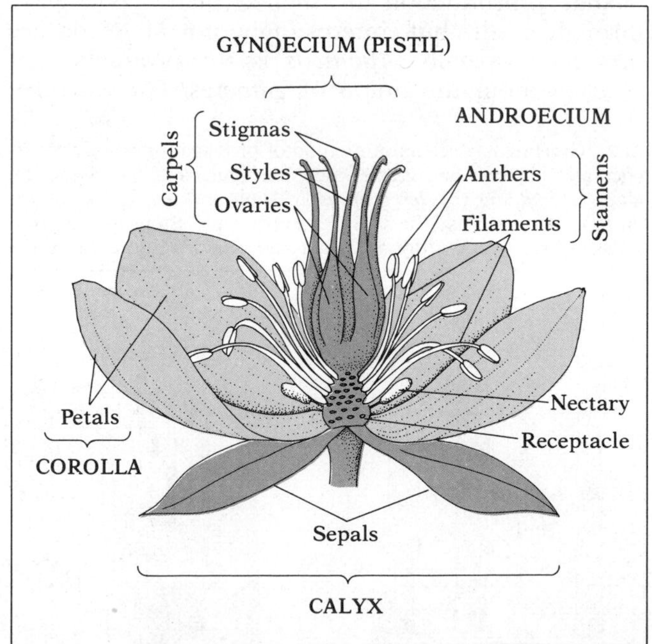

a rather anthropomorphic description of a male plant structure.

The fourth region of the flower, the pistil, is also a whorl, but it is often too modified to be recognizable as a leaf-derived structure unless one considers its embryonic development. Pistils consist of three parts: the carpel, style, and stigma. The carpel (or carpels, since they are often multiple) produces the ovules. The entire body is known as the ovary. Extending from the ovary is a stalk known as the *style*, which emerges from the ovary; and an enlargement, the *stigma*, at the tip of the style. The stigma is often hairy or sticky and is the pollen receptacle. The pistil comprises part of the plant's female tissues, but it is actually the ovary that produces the egg cell. Unfortunately, the matter is not as straightforward as one might wish, since the egg cell stage is preceded by both meiosis and mitosis in a somewhat complex series of events. The pistils are collectively or singly known as the *gynoecium*, a term you can translate for yourself.

In summary, the flower consists of a calyx (whorl of sepals), corolla (whorl of petals), androecium (whorl of stamens), and the gynoecium (whorl of carpels). The essential parts for reproduction are the last two. In fact, sepals and petals are known as *accessory parts*, since they play no direct role in sexual reproduction.

We mentioned earlier that flowers are often quite colorful and have fragrant odors. They also produce sugary nectar in structures at the base of the flower, known as *nectaries*. Color, fragrance, and nectar are all essential for flowers that depend upon insects or birds (like the hummingbird) for transferring pollen from one flower to another. There are variations on this theme, however, such as we see among the flowers in Figure 21.4. Keep in mind that humans are poor judges of flower color. For example, insects may see the ultraviolet waves emanating from what, to us, is a white flower. And most insects can't see red at all, so a beautiful red rose appears to them as black.

There is also a very large group of plants that are wind-pollinated. These include the grasses, such as grains and corn. It may seem strange that grasses produce flowers, but they are, in fact, angiosperms. In fact, the grass flowers are in some ways similar to those in insect-pollinated flowers, but they are not conspicuous, since there is no reason for their being so.

Now the question arises, since all angiosperms have flowers, when do they develop? And how? It is easy to find out when. The how is more difficult.

The Initiation of Flowering

Every flower has its time. Throughout much of the United States the early-flowering crocus is the famous herald of spring. As the days lengthen, species after species bursts into bloom—and in a rather predictable

21.4 *Rafflesia* flowers are the world's largest, up to 1 m in diameter. Their flowers have the odor of rotting meat and attract flies and other carrion insects.

sequence. Through the summer and into the fall, as the days again begin to grow shorter, yet new species will unfurl their colors. Hardly any other good news in this life is so dependable.

But one wonders, how does the crocus know that spring has arrived? And how does the ragweed know that fall is here? What physiological processes tell the tissues of the plant to start making flowers? The question is not a simple one. A lot of people would like to know. Any biologist who is interested in the control of morphology would like to have some idea of what causes the plant to turn from producing leaves and stems to flowers. If the process could be controlled, florists could have their product ready just in time for Mother's Day, but, more importantly, farmers could regulate the production of fruit and develop a more steady market. They would be interested not only in how to start the reproductive process, but how to inhibit it. After all, flowering or "going to seed" can destroy a carrot, celery, or lettuce crop.

There has been an intense interest in these questions for many years, but the answers are not all in by any means. For example, flowering seems to be under hormonal control, but no specific flowering hormone has ever been discovered. Auxin is at least indirectly involved, but in a negative way. The concentration of auxin drops in plants that are about to go into flower; furthermore, the application of auxin can sometimes be used to prevent or delay flowering.

Photoperiodicity

As to how a plant knows what the season is, we do have some answers. The critical factor is the length of the night. Of course, the length of night varies with the latitude and with the season. So the plant simply counts the hours—somehow. We don't know how. The length of daylight is not important, but when the period of darkness extends for a certain period, the plant "knows" that it is, say, April and behaves accordingly. The sensitivity to light–dark cycles is known as *photoperiodicity*.

Different kinds of plants vary in their response to changes in the length of night, and can be roughly organized into three categories according to those responses. Plants that begin the flowering process before the summer solstice, June 21, are called *long-day* plants because they flower only when the shortening nights reach a critical length, or brevity. Plants that do not begin to flower until after the summer solstice are called *short-day* plants, because after the solstice the days grow shorter and the nights longer; short-day plants respond to lengthening nights. The third category of flowering plants are the *indeterminate* or *day-neutral* plants, which appear to be indifferent to the light cycles. They may flower continuously or in response to other stimuli (Figure 21.5).

How was it determined that it was the duration of *darkness* that triggers flowering, and not the duration of *daylight?* Actually, the demonstration is rather simple and it can be done in laboratories, where the light and dark periods can be regulated. Under such conditions it was found that the period of light can be changed without causing any flowering changes, but if the length of darkness is tampered with, the plant responds. More dramatically, in plants exposed to normal outdoor cycles, a single relatively brief exposure to intense light in the middle of the night can fool a long-day (short-night) plant into behaving as if the short nights of spring had arrived. Such a plant can then be made to bloom in any season. The same trick, if done every night, can prevent a short-day (long-night) plant from flowering at its normal late summer or autumn time, since it will react by behaving as if the nights were still too short for such activity.

This bit of academic tinkering was immediately put into practical use by chrysanthemum growers, who for many years had extended the chrysanthemum season into early winter by artificially lengthening the days. They had been doing this by keeping bright lights turned on for several hours after sunset. But now they found they needed only to turn on the lights for a few minutes each night in order to get the same results. Similarly, but much more important economically, sugar cane growers can stall flowering in their fields by turning floodlights on at night. Sugar-laden sap that might otherwise be used to produce useless flowers can then continue to accumulate in the stems.

The cocklebur has proved to be an ideal organism for studying the initiation of flowering. It is so hardy that researchers have been able to alter it drastically and be sure it will survive. The cocklebur is a short-day (long-night) plant, and will put out homely little green flowers if exposed just once to a period of darkness longer than $8\frac{1}{2}$ hr.

Armed with this knowledge, plant physiologists began to toy with the plant. One of the first things they found was that the stem and even the incipient flowers were insensitive to the midnight flash effect; only the *leaves* were sensitive.

The scientists also found that varying the wavelength of the lights they used had curious results (Figure 21.6). The most effective light for establishing the photoperiod was found to be in the red or orange-red region. Far-red light, with its longer wavelength, had the opposite effect; that is, a flash of far-red

21.5 Variation in the flowering response. The three plants shown here have a specific photoperiod requirement for flowering. (a) The chrysanthemum is a short-day plant that produces flowers in the autumn. Flowering can be stalled if the plant receives an hour of light during the night. (b) Poinsettias are also short-day plants and will bloom when the night is longer than 13 hr. Their blossoming can be easily controlled by illuminating the greenhouses in which they are grown; thus, poinsettias can be brought into bloom just in time for Christmas. (c) The henbane requires a long daylight (short night) period of time, usually in excess of 12 hr, for flowering.

(a)

(b)

(c)

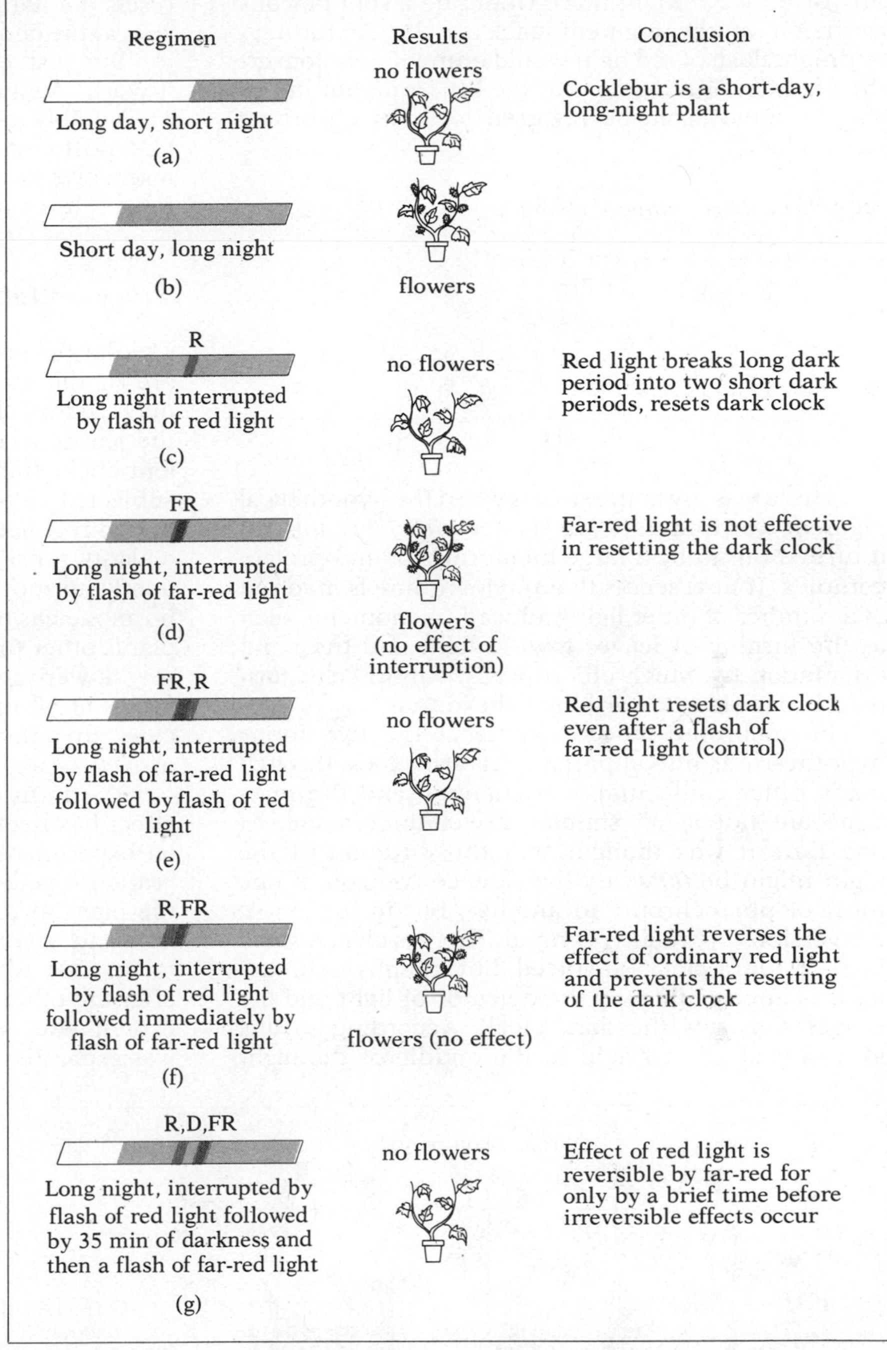

Regimen
Results
Conclusion
no flowers
Long day, short night
(a)
Cocklebur is a short-day, long-night plant
Short day, long night
(b)
flowers
R
no flowers
Long night interrupted by flash of red light
(c)
Red light breaks long dark period into two short dark periods, resets dark clock
FR
Long night, interrupted by flash of far-red light
(d)
flowers (no effect of interruption)
Far-red light is not effective in resetting the dark clock
FR,R
no flowers
Long night, interrupted by flash of far-red light followed by flash of red light
(e)
Red light resets dark clock even after a flash of far-red light (control)
R,FR
Long night, interrupted by flash of red light followed immediately by flash of far-red light
(f)
flowers (no effect)
Far-red light reverses the effect of ordinary red light and prevents the resetting of the dark clock
R,D,FR
no flowers
Long night, interrupted by flash of red light followed by 35 min of darkness and then a flash of far-red light
(g)
Effect of red light is reversible by far-red for only by a brief time before irreversible effects occur

21.6 Photoperiods in the cocklebur. (a, b) The cocklebur, a short-day plant, will flower when the nights become long. Its requirement of a specific dark period to initiate flowering is demonstrated at (c), where the period is interrupted by a flash of red light. (d) A similar flash of far-red light has no effect. (e) Preceding the red flash by a flash of far-red light apparently has no effect on the inhibiting effect of red. (f) But when the flashes are reversed, with far-red following red, the effect of red is reversed and flowering occurs. (g) The timing of these flashes is critical, since a lapse of 35 min or more between red and far-red flashes is apparently enough time for the effect of red light to reset the "dark clock," producing no flowering.

light actually reversed the effect of the midnight flash of either white or red light if it immediately followed either of them. However, if the far-red flash preceded the red flash, or was delayed for more than 35 minutes after the red flash, the red flash had its full effect. Curious, indeed.

The midnight flash experiments gave rise to the hypothesis that there was some receptive pigment involved. The receptive pigment could not be chlorophyll (which is abundant in the photosensitive leaves), because it responds to different wavelengths from those produced by the floodlights. So the scientists suggested that perhaps there is a pigment that occurs in two forms: Pr, the red-absorbing form; and Pfr, the far-red absorbing form. Absorption of light of the appropriate wavelength would change this hypothetical pigment from one form to another. During the day, the predominance of red light over far-red would convert all the pigment from the Pr form into the

Pfr form, but at night there would be a spontaneous reversion of the pigment back to the Pr form. A midnight flash of red light would immediately convert the pigment once again into the Pfr form, but in this state it could again be reversed to Pr by absorbing far-red light.

Hypothetical light-induced changes:

$$\mathrm{Pfr} \underset{\text{Red light}}{\overset{\text{Far-red light}}{\rightleftarrows}} \mathrm{Pr}$$

Spontaneous dark reaction:

$$\mathrm{Pfr} \rightarrow \mathrm{Pr}$$

There was joy in the streets when the hypothetical pigment was found. It was named *phytochrome*, and it turned out to be a large membrane-bound protein complex. It now seems that phytochrome is involved in a number of other light-induced phenomena, such as the turning of leaves toward light and the rapid orientation by which chloroplasts, within cells, turn their broad sides toward the light source.

Phytochrome indeed occurs in the two forms hypothesized, but apparently it also goes through many other conformational changes, and the reactions are not at all simple or well-understood. At one time it was thought that the duration of the night might be *timed* by the slow conversion of one form of phytochrome to another, but in fact these conversions are relatively rapid. There is clearly some kind of "dark clock" involved, but the phytochrome itself is apparently only the detector of light and the trigger that sets the dark clock. According to this idea, a flash of red light in the middle of the night resets the dark clock back to zero, just as it is set at zero at the end of a normal day.

But just what *is* the dark clock, and how does it work? Again, we must admit we don't know. It is presumably some kind of chemical reaction, but unlike most chemical reactions it is almost completely insensitive to temperature differences within normal ranges. It is a most peculiar phenomenon indeed.

Transmitting the Stimulus

The dark clock and its photoreceptor, phytochrome, are in the leaves. But the signal that actually induces flower development has to be delivered from the leaves to other parts of the plant. In the resilient cocklebur, an isolated, amputated leaf can be subjected to an appropriate flower-inducing photoperiod regimen, and subsequently grafted back onto a plant—causing the plant to flower. The signal is evidently hormonal, but, as we said, no flower-inducing hormone has been found. In some experiments using plants other than the cocklebur, it appears that there are flower-*repressing* hormones produced by the leaves at all times *except* when the dark clock indicates that the appropriate time for flowering has arrived.

The hormonelike transmission of the photoperiod effect has been demonstrated most convincingly by an experiment in which six cocklebur plants were grafted together in a row (Figure 21.7). One leaf of the plant at one end of the line was enclosed in a box and given the appropriate photoperiod treatment, after which *all six plants* put out flowers, one after the other, right down the line. In other grafting experiments, the cambium of the conjoined plants was separated by a sheet of paper, but the message

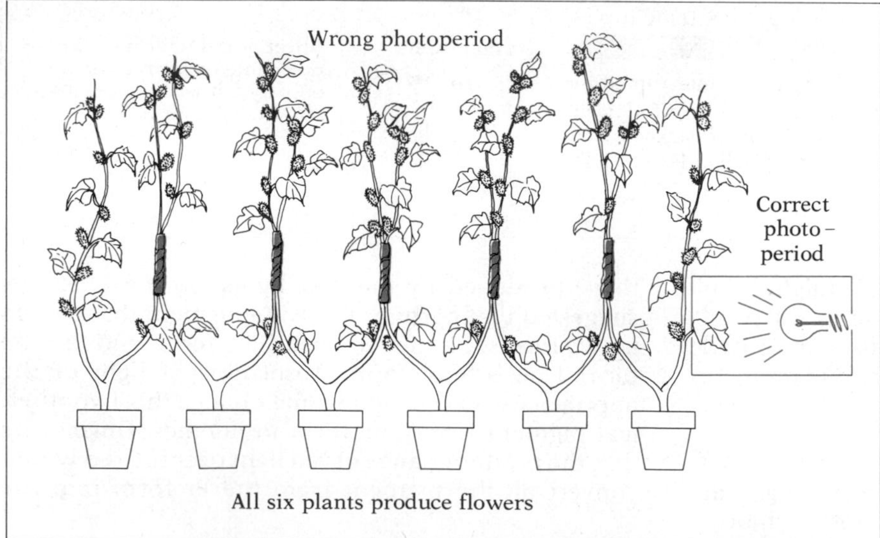

21.7 Of the many experiments that reveal the photoperiod effect this is the most dramatic. It establishes conclusively that the leaf is the photoreceptor. Only one leaf in this series of six grafted cocklebur plants is exposed to the proper photoperiod, yet all six plants produce flowers. It seems obvious that a hormone is being transferred from the exposed leaf to the other plants, but as yet no specific chemical messenger is known. There may be such an agent, or there may be several. The action of the agent(s) may be direct or target cells may be influenced to produce other messengers. For now, we can only conclude that the photoperiod effect is transferrable.

got through anyway, just as we would expect if the signal were hormonal.

The hypothetical flower-inducing hormone has somewhat presumptuously been named *florigen*. However, because it has proved to be so elusive, some plant physiologists believe that there is no special flower hormone at all, but that the message is conveyed by particular levels and combinations of auxin, gibberellins, and other known plant growth hormones.

Paradoxically, we are aware that auxin alone has an inhibitory effect on flowering, and that the application of auxin can sometimes prevent or retard the process. Since growing tips of shoots produce auxin in quantity, flowering can sometimes be induced or increased by simply cutting off the shoot tips. Rose growers have discovered this, and the clicking of pruning shears can be heard over the countryside at the beginning of the floral season.

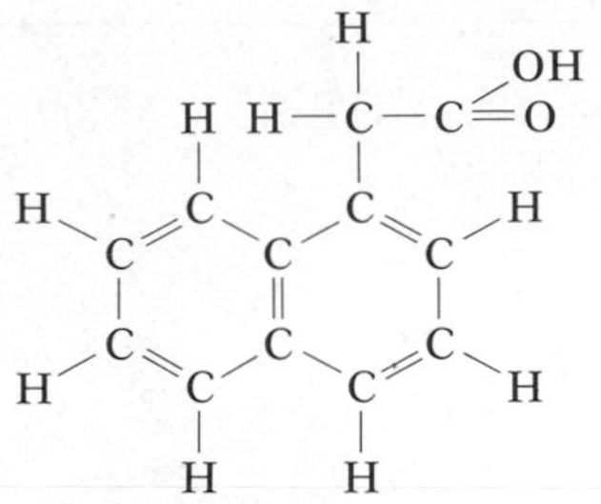

NAA

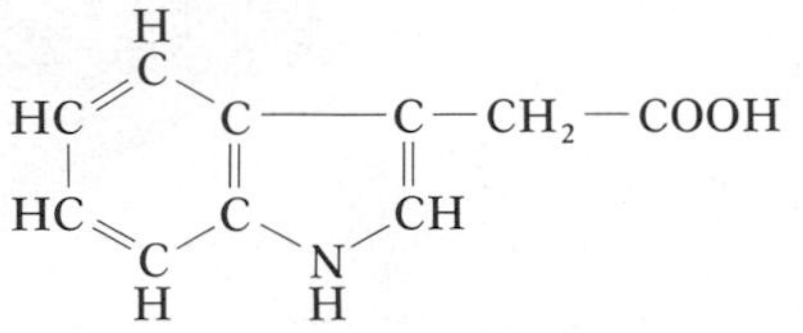

IAA

21.8 Naphthalene acetic acid (NAA), an artificial auxin. Compare its structure with natural auxin, indoleacetic acetic acid (IAA).

Pineapples and Auxin

Pineapples are different. In pineapples, auxin helps induce flowers. Pineapples are also tropical plants, living where there is little seasonal change in day length, and, not surprisingly, they are indifferent to photoperiods. They normally bloom and set fruit on irregular schedules of their own, all year long. In the past, pineapple pickers had to roam the fields daily, looking for ripe pineapples—an expensive and time-consuming process. Now, growers synchronize the flowering of whole fields by the application of artificial auxins, notably naphthalene acetic acid (Figure 21.8). The plants dutifully flower all at once, and, after an appropriate period of time, ripen together and can be efficiently harvested by machines. That's why pineapples are cheaper than they used to be. (Do you suppose we're paying any hidden costs by spraying artificial auxins into the environment?)

Meanwhile, Back in the Ovary: Megasporogenesis

Within the ovary of the flower, minute but complex multicellular organs have developed from the relatively undifferentiated tissue of what is called the flower's *placenta*. These organs are the *ovules*, and they consist of a stalk, a surrounding tissue called the *nucellus*, and one or two *integuments*, or skinlike protective coverings. At the exposed end, a small opening in the integuments is called the *micropyle*.

Within each ovule is a single large cell. This is the *megaspore mother cell*. Its role is reviewed in Figure 21.9. Like its male counterpart, the *microspore mother cell*, it will undergo meiosis; but it will ultimately produce only one haploid gametophyte (whereas the microspore mother cell will produce four, as we shall see.)

Meiosis in the megaspore mother cell involves the usual two divisions, so at the end of the process the cell has four haploid cells. Three of these distintegrate, leaving a very large cell with one small haploid nucleus. This cell is the *megaspore*. The megaspore's one remaining haploid nucleus undergoes three successive nuclear mitotic divisions, giving what is now called the *megagametophyte* its eight identical haploid nuclei. Note that mitosis of haploid nuclei (a process that doesn't normally occur in diploid animals) is a regular feature of the life cycle of plants. So, the gametophyte generation is greatly reduced in flowering plants, but it is still present.

The eight nuclei migrate to the two ends of the megagametophyte, and then two of the nuclei, one from each end, migrate back to the center of the cell. Finally, new cell plates are laid down and the cytoplasm is divided into separate cells. The division of cytoplasm is unequal: one cell, the largest one, ends up in the center with two nuclei. This binucleate cell, after fertilization, will give rise to the *endosperm* of the future seed, but for now it is known as the *endosperm mother cell*. The entire seven-celled structure (six haploid cells and one binucleate cell) is still a female gametophyte or megagametophyte, but it now has a special name of its own: the *embryo sac*. Of the six mononucleate haploid cells, only one is the *egg cell*, the single female gamete (or megagamete) of the megagametophyte. There will be no more nuclear divisions in the embryo sac until after fertilization. Both the egg cell and the endosperm mother cell will participate in that impressive event known as *double fertilization*.

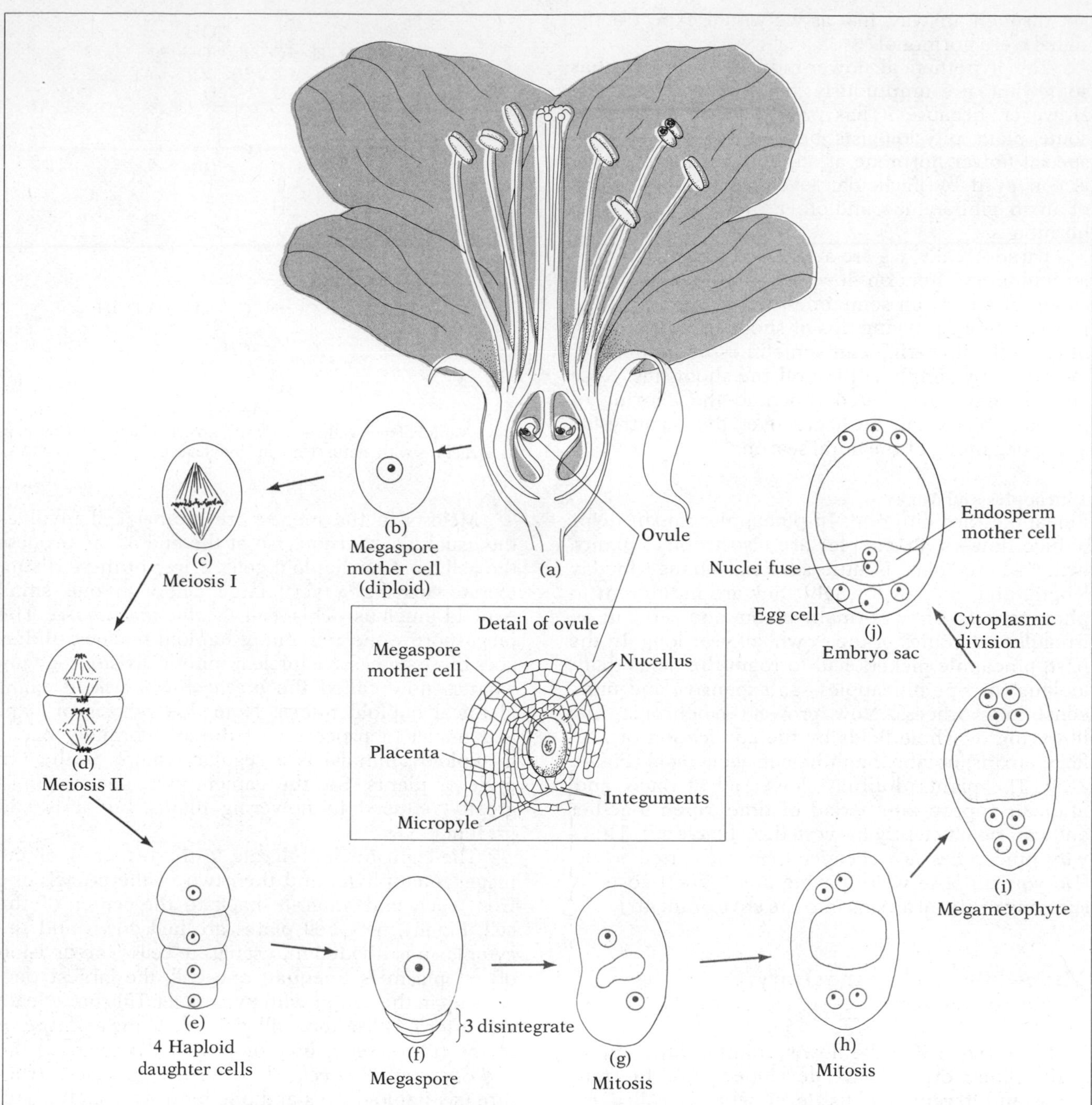

21.9 Megasporogenesis. (a) The female gametes develop within the ovules. The ovule, enlarged in the inset, contains the *megaspore mother cell* and its accompanying tissues, the nucellus, and enfolding integuments, all arising from the placenta. (b–e) The megaspore mother cell undergoes meiosis, producing four haploid *megaspores.* (f) Three of these disintegrate while the fourth enlarges. (g–i) The large haploid nucleus then undergoes mitosis three times and finally forms a *megagametophyte* with eight nuclei. These nuclear divisions are not accompanied by cell divisions. (j) Finally, the *embryo sac* matures and seven cells form. The largest is a binucleate *endosperm mother cell* (with two nuclei). At each end, three cells form. The center cell at one end becomes the *egg cell.* The binucleate endosperm mother cell and the egg cell will participate in fertilization.

Similar Events in the Anthers: Microsporogenesis

While the ovary has been preparing its embryo sac, similar, but less complex events have been occurring in the anthers, the site of *microsporogenesis* (Figure 21.10). Typically, the anthers contain four chambers known as *pollen sacs*. Within the pollen sacs are numerous *microspore mother cells* that will enter meiosis. By the end of meiosis, each microspore mother cell will have produced four *microspores.* The haploid nucleus of each microspore will then undergo

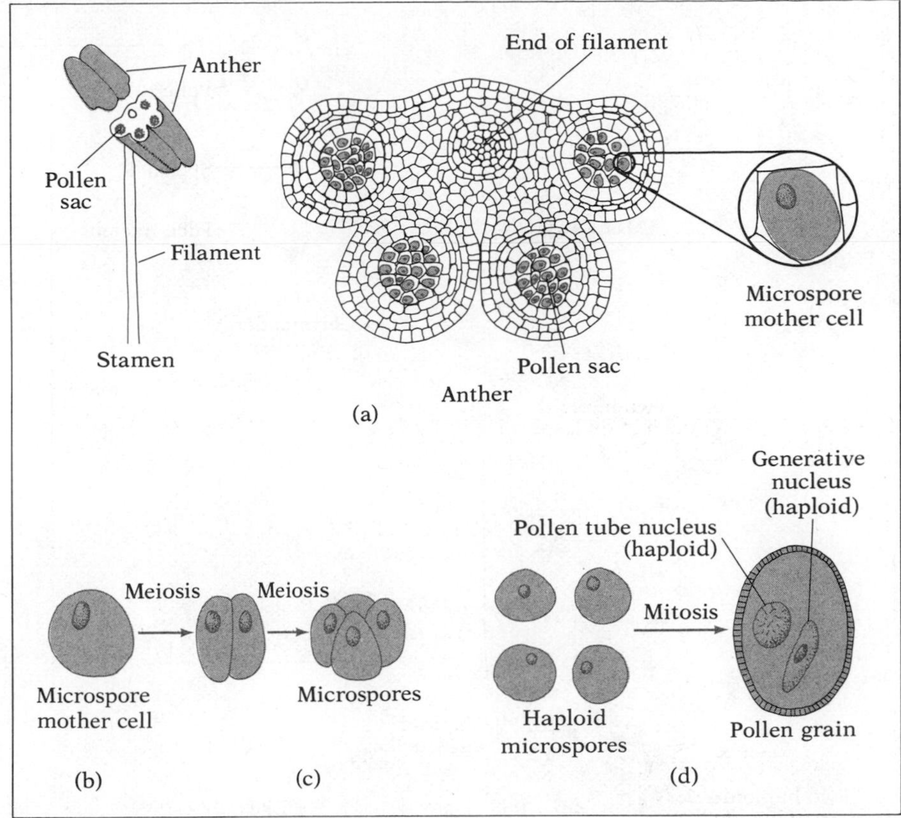

21.10 Microsporogenesis. The development of male gametes occurs within the anthers. (a) Within each anther are four *pollen sacs.* (b) Each pollen sac contains numerous *microspore mother cells,* one of which is shown enlarged. (c) In the development of pollen, each microspore mother cell enters meiosis and, at its completion, forms four haploid *microspores.* (d) Each microspore then divides mitotically, forming a *tube nucleus* and a *generative nucleus.* The microspore, meanwhile, matures into a *pollen grain,* awaiting its role in pollination and fertilization.

mitosis, producing two nuclei, a *generative nucleus* and a *tube nucleus*. Each of these is still haploid, since mitosis doesn't change the final chromosome number. When this mitotic event has been completed, each microspore produces a tough, resistant coating around itself and becomes a grain of *pollen*. No cellular activity will occur after this until the pollen finds its way to the stigma of a flower. At pollination the generative nucleus will form two sperm nuclei while the tube nucleus will direct the production of essential enzymes.

Pollination and Fertilization

Pollination technically occurs when pollen is deposited on a receptive stigma. The source of the pollen may be the male parts of the same flower, other flowers on the same plant, or some distant colleague. Of course, we are aware of the dangers of inbreeding and that self-fertilization is inbreeding of the severest sort, so some plants have ways of avoiding self-pollination, such as by the position of their anthers, or physiological incompatibilities. But, strangely enough, self-pollination is a regular event in certain other species.

When pollen germinates, it produces a *pollen tube,* which makes its way down through the stigma and into the long style (Figure 21.11). The pollen produces enzymes that actually digest the tissue ahead of the tube. As the tube growth proceeds, the nuclei stay near the growing tip and periodically appear to lay down new cell walls behind themselves as they advance. During this advance, the generative nucleus undergoes mitosis, producing two *sperm nuclei.*

Finally, the tube penetrates the ovule at that tiny pore called the micropyle. The two sperm nuclei then move from the tube and enter the embryo sac. One fertilizes the egg cell and a diploid zygote is formed. The fertilized egg, or zygote, will develop into the plant embryo. The second sperm penetrates the endosperm mother cell, where its chromosomes fuse with those of the two nuclei already present, producing a triploid nucleus. In many species, the triploid endosperm will form the starch food storage region of the seed, but in others it will simply disintegrate and be absorbed.

After fertilization, the flower begins to change. Deep in its ovary, the seed immediately starts to develop a large starch storage region and an embryo. Now the flower need no longer maintain its earlier

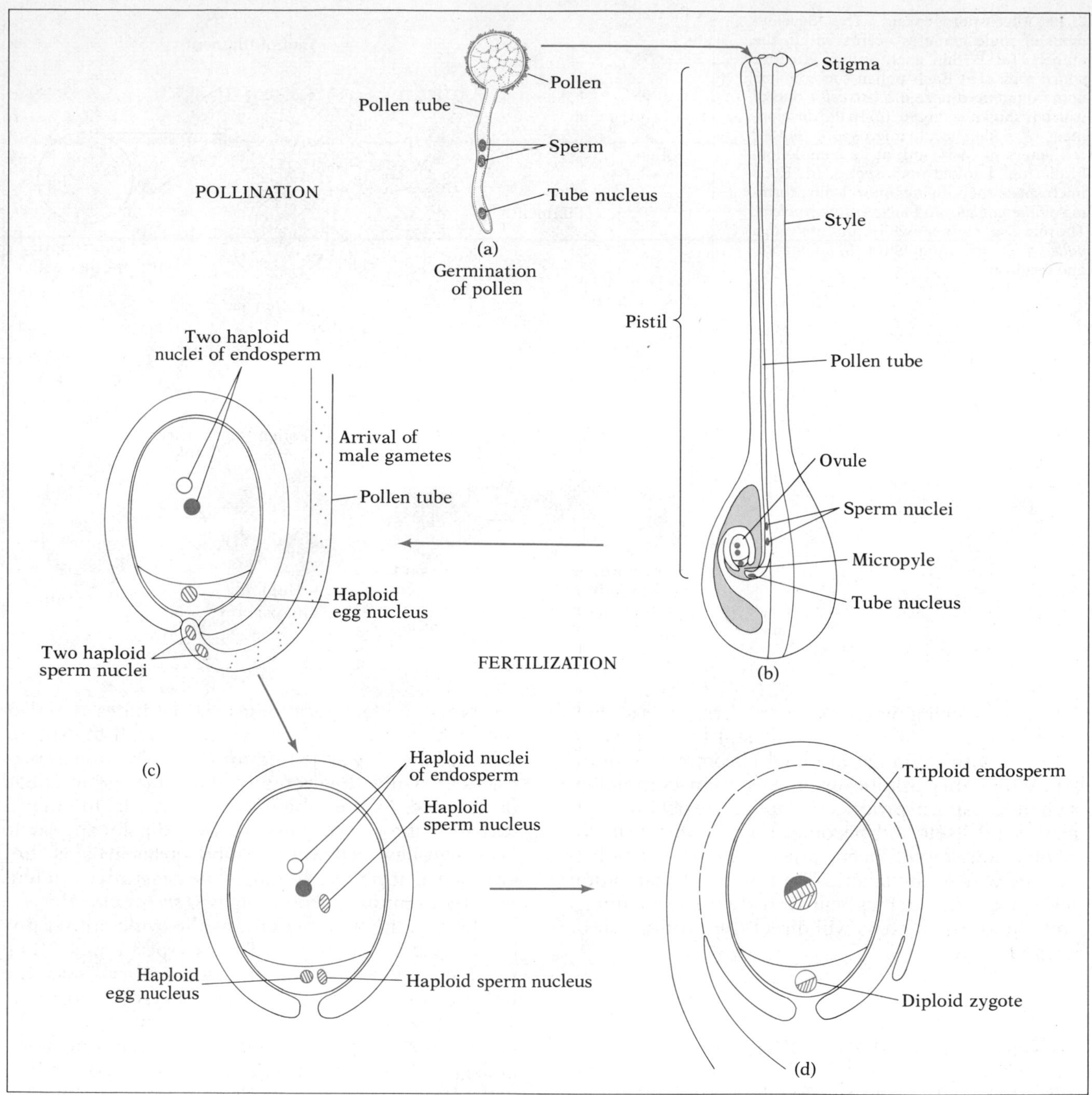

21.11 (a) Pollination begins when pollen lands on the stigma and produces a *pollen tube.* (b) As the pollen tube grows down through the style, the generative nucleus undergoes mitosis, producing two sperm nuclei. The tube nucleus then penetrates the ovule and the two sperm nuclei enter the embryo sac. (c) One sperm nucleus fertilizes the egg cell, while the other enters the endosperm mother cell and joins its two nuclei forming a triploid nucleus. (d) Fertilization itself is completed when the nuclei fuse.

firm beauty. Those parts that are not needed to sustain the seeds begin to wither and fall off. Then the ovary itself will swell, sometimes to enormous proportions. In fact, oranges, and watermelons are largely swollen ovaries. However, the vital activity is the development of the seed, and it is here that we will focus our attention now.

Seed Development in Flowering Plants

After fertilization, the new diploid zygote, the triploid endosperm cell, and all the associated maternal tissues begin the cell divisions and movements necessary to produce the plant *embryo* and its life support systems. The finished product—the *seed*—will consist of an embryo, some kind of food supply, and a protective *seed coat*.

The manner in which food is stored differs considerably in major groups of plants. In the monocots such as the grains (grasses), it is stored primarily in undifferentiated endosperm tissue. In these, the white or yellow starchy bulk of the kernel is triploid endosperm tissue, and the wheat germ and corn germ are the diploid embryos [Figure 21.12(a)].

The seed coats of wheat and corn are far more complex than they seem. They are, in part, maternal diploid tissue from the ovary wall, the pericarp and, in part, the specialized endosperm tissue called the *aleurone*. In the colorful "Indian corn" it is triploid aleurone that contains the vivid pigments. The wheat milling process separates the seed coats (*bran*) and embryos (*wheat germ*) from the endosperm (*flour*) and these are sold separately to the cattle feedlot and health food industries. (Wheat germ oil is highly nutritious, but it tends to spoil easily, which is why whole wheat flour does not keep well.)

21.12 Comparing corn (monocot) and bean (dicot) "seeds" can be misleading. While the familiar bean is actually a seed, the corn grain is a fruit that contains the equivalent of a seed. At the upper part of the illustration, corn and bean ovaries (fruits) are compared. (a) Each kernel of corn on an ear is a complete ovary (the strands of corn silk are the pistils). In corn and other monocots, the entire ovary becomes enclosed by a hardened covering formed from the maternal tissue called the *pericarp*. The endosperm becomes filled with unorganized stored starch. The embryo (germ) is seen at the lower left. (b) In the bean, the seed is one of several that has left the ovary. Each of these true seeds is surrounded by a coat derived from portions of the ovules. The embryo includes the miniature embryonic plant and two organized food storage regions, the cotyledons, which have absorbed all of the endosperm.

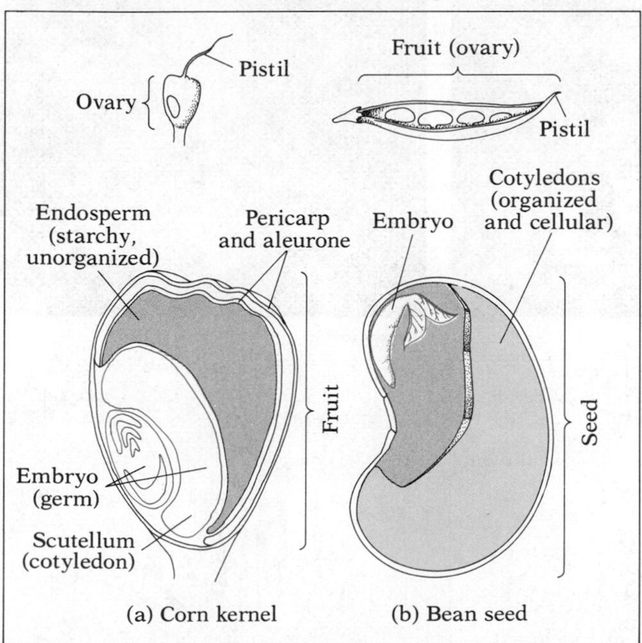

The corn germ and wheat germ contain the embryo of the plant and the single cotyledon that gives the monocots their name. The cotyledon or *scutellum*, as it is also known, will serve primarily as a digestive organ in corn; upon germination, it will absorb the starchy food reserves of the endosperm and transmit them to the growing seedling.

The seeds of some dicots also have endosperm storage reserves. In many other dicot species, however, the endosperm disappears early in the seed development, and the principal storage organs of the mature seeds are the two cotyledons that give the dicots their name [Figure 21.12(b)]. The cotyledons are modified leaf structures, and indeed they are sometimes known as *seed leaves*, but they are not true leaves.

The most familiar dicot seeds are probably peas, beans, and peanuts. All three belong to the same family, the *legumes*. The embryo takes up most of the dicot seed and technically consists of the two cotyledons and the miniature "plant." The new plant can be found where the two cotyledons are joined.

In other, less familiar dicot seeds, the two cotyledons may be quite leaflike. Often, they can be seen in pairs, emerging from a patch of moist soil as the first sign that a tiny seedling has germinated.

But let's get back to tracing seed development (Figure 21.13). We see that after fertilization, the ovule contains an embryo sac with a diploid zygote, a triploid endosperm cell, and several layers of surrounding and supporting diploid cells of the maternal tissue, all borne on a short stalk. The remaining haploid cells of the embryo sac are of no further interest to us. From here, the nuclei of the endosperm continue to divide until a multinucleate mass surrounds the zygote. This stage is carried to an extreme in the coconut, since "coconut milk" is nothing but a mass of white, fluid endosperm cytoplasm with free nuclei floating in it. Ultimately, in most seeds, cell walls are laid down around the endosperm nuclei.

The zygote itself does not usually begin to develop into an embryo until the endosperm has grown to a considerable size. Then the zygote begins a series of cell divisions and will eventually produce the embryo. The first few divisions produce a vertical row of cells. Then, at one end of the row, cell divisions occur so rapidly that a ball of cells is formed. It is this ball that will become the embryo. The cells at the other end of the row divide much more slowly, but they become large cells, swollen by the uptake of water. These great bloated cells produce a bulbous swelling called the *suspensor*, which anchors the embryo in place and orients it to its future food supply.

The developing seed continues to change as the embryo sends the first cells into the surrounding endosperm. Up to this point, development has been about the same in monocots and dicots, but now we see the fundamental difference between the two

The Flower

Variation in Flowers

The seemingly endless variety of flowers can be better organized following the botanist's classification. Two major groupings known as *complete* and *incomplete* separate those flowers with all major parts present (sepals, petals, stamens, pistils) from those with parts absent. A second division can be made with incomplete flowers. Those with sepals and/or petals absent but with both male and female reproductive parts are known as *incomplete-perfect* flowers.

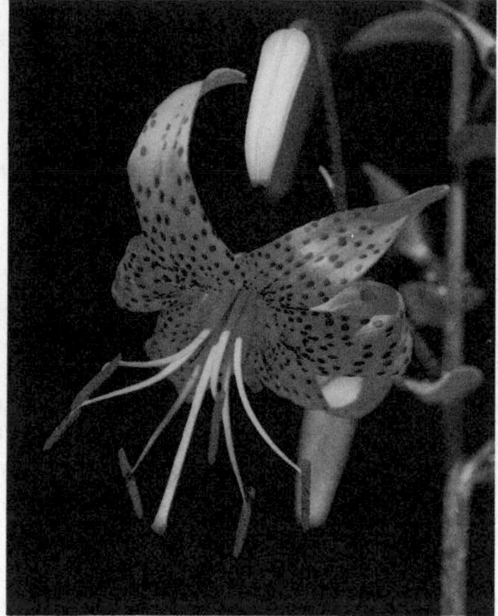
Complete flower: Tiger lily

Incomplete-perfect flower: Calla lily (Sepals and calyx absent)

Variations in Ovary and Floral Parts

Flowers can also be classified according to the position of floral parts in respect to the ovary. Three categories are recognized:

1. Hypogyny: floral parts originate below the ovary (often called a superior ovary).
2. Perigyny: floral parts originate at the ovary.
3. Epigyny: floral parts originate above the ovary (often called an inferior ovary).

Hypogyny: The tulip

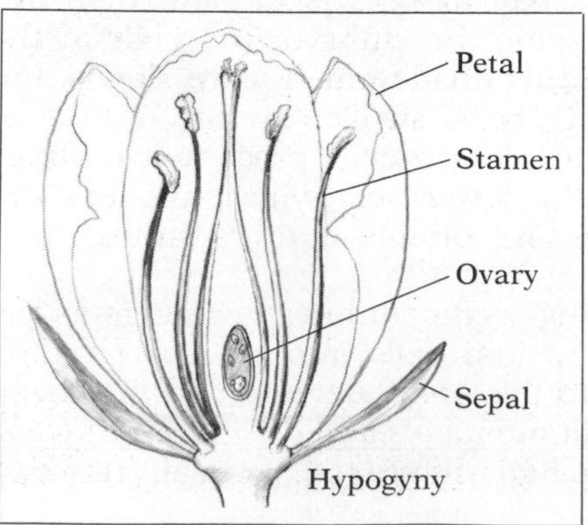

Perigyny: The cherry

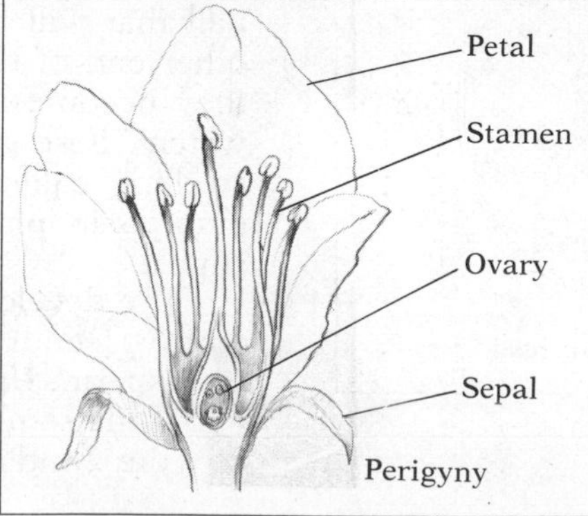

Epigyny: The daffodil

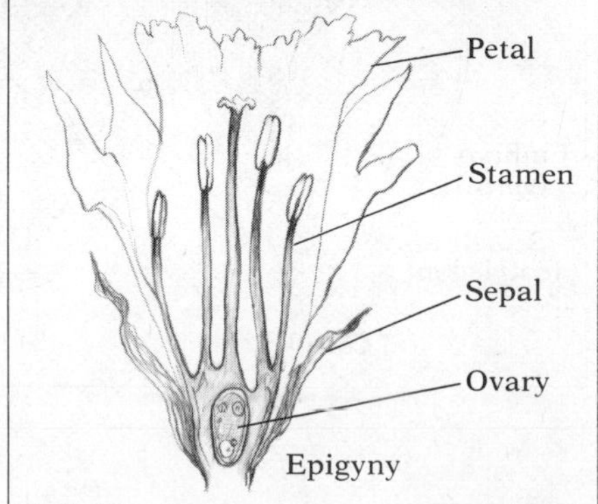

Flowers of the Grasses

There is no reason for grass flowers to develop color; they are almost always wind-pollinated. So grass flowers are usually a bit drab. The two examples shown here are oats and corn. In corn, the flowers are either male or female flowers. The male flower (the *tassel*) is at the top of the plant and is a conspicuous frilly structure. The female flower is the *ear* with its *corn silk*. Each filament of the silk is actually part of a pistil. At its base is an ovule. This means that each seed or kernel of corn is a separate offspring, fertilized independently of the others. If corn is allowed to grow without interference, each ear can express its own phenotypic ratio, which is dramatically seen in "Indian corn."

Flower of wild oat

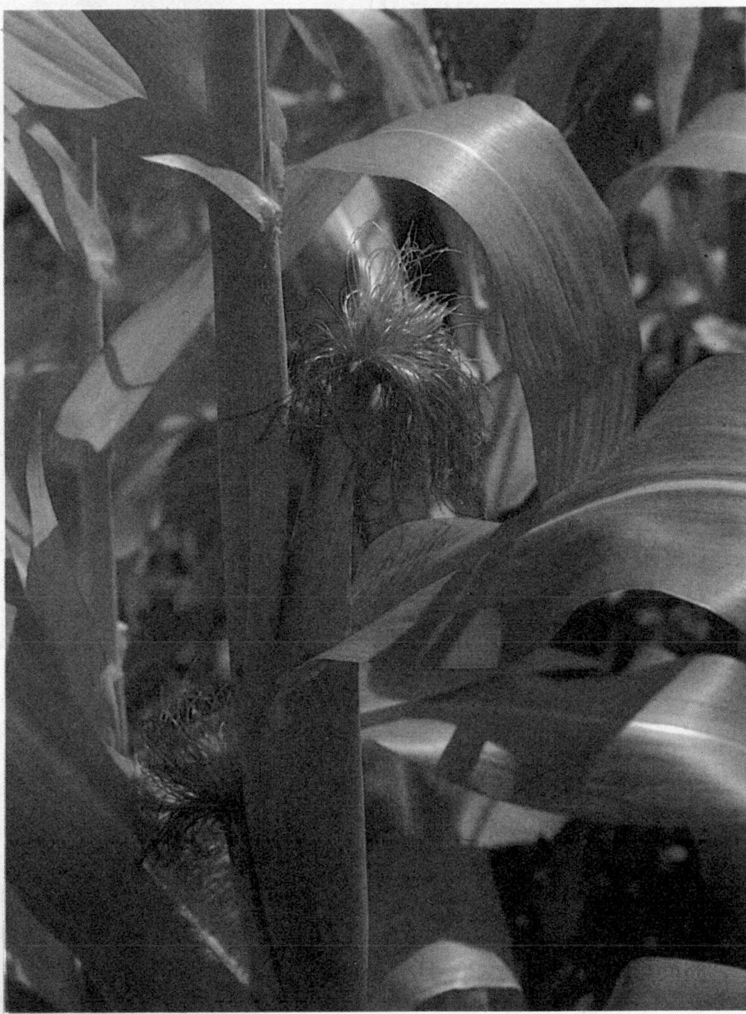

Flower of maize, or Indian corn

An ear of corn expressing a 9:3:4 ratio for color of kernel.

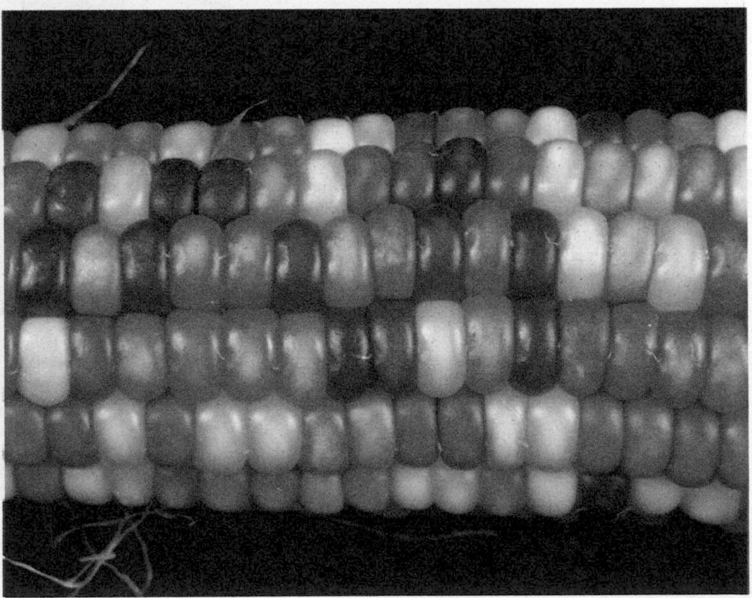

Fruits of Flowering Plants

Technically, a fruit is a ripened ovary. But there may be other floral parts closely associated with the ovary, thus adding to their great diversity. In spite of the apparent variety, however, there are only three basic types of fruits: *simple, aggregate,* and *multiple.* The type depends on the number of ovaries and flowers that make up the structure.

Simple fruits are derived from a single ovary. They are:

1. Legumes or pods. Beans and peas form pods that are the fleshy ovary surrounding the seeds. Each pod is a single *carpel* from one flower.
2. Capsules. The lily produces a simple fruit in the form of a capsule. Each capsule consists of three united *carpels.*
3. Drupes. Plums, cherries, and peaches are simple fruits of the rose family. A single seed is enclosed in a hard covering. The overlying fruit has a thick fleshy middle layer with a thin covering.
4. Berries. Surprisingly, tomatoes, watermelon, cucumbers, and citrus are *berries.* The berry develops from a compound ovary and contains seeds embedded in its fleshy region.

Aggregate fruits are derived from numerous separated carpels from a single flower. Blackberries and raspberries are aggregate fruits. Each consists of many simple fruits clumped together to form the familiar fruit.

Multiple fruits are formed from the single ovaries of many flowers joined together. This group includes pineapple, figs, and mulberries. The pineapple itself includes various accessory floral parts, which help to confuse the casual observer, causing him to wonder why pineapples are in this group.

Pea pod

Lily capsule

Peach

Watermelon

Pineapple

Blackberry

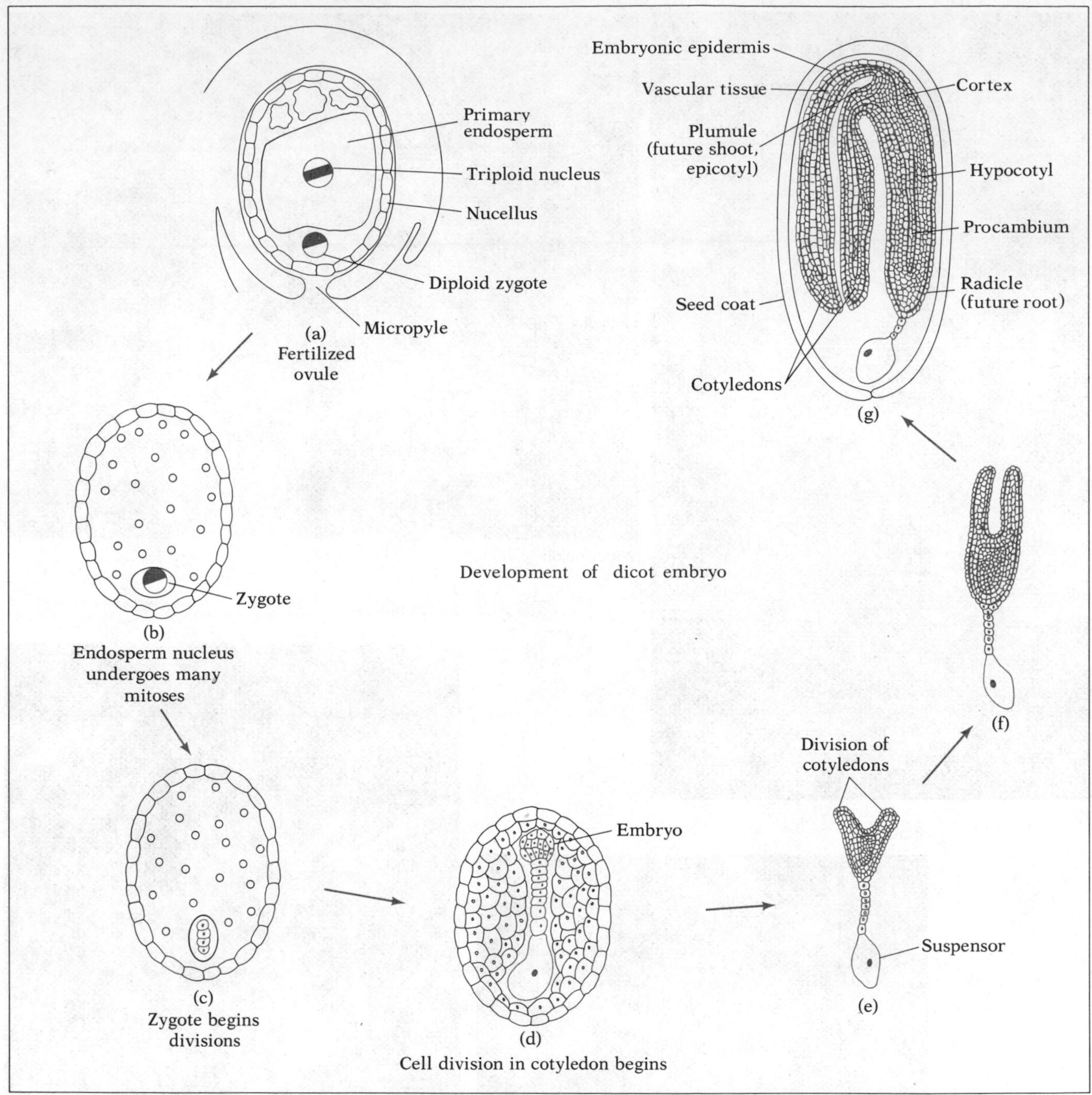

21.13 Seed and embryo development in the dicot. (a) The development of both the seed structures and the embryo in dicots begins with the fertilized ovule. (b) In the endosperm there is extensive mitotic activity, but without cell divisions, at least at first. (c) The zygote then divides several times, forming a chain of cells. (d) The embryo becomes a ball of cells connected to a bulbous suspensor. (e, f) As the embryo progresses, its ball of cells forms a two-pronged extension to produce the cotyledons. The advancing cotyledons then make a U-turn toward the base of the embryo. As they grow, the cotyledons absorb the nutrients from the endosperm. (g) Below the cotyledon branch, cells of the epicotyl produce the embryonic shoot. The completed embryo of a dicot consists of two cotyledons, an *epicotyl*, a *hypocotyl*, and a *radicle*, or embryonic root.

groups. In dicots, the advancing cells of the embryo form a two-pronged attack as they penetrate the endosperm, while in monocots, the cells invade in a single file. The reason is, the tissues that are pushing into the endosperm will form the cotyledon or cotyledons, as the case may be.

Further Seed Development in Dicots

In legumes and many other dicots, the cotyledons continue to grow, absorbing all, or nearly all, of the endosperm and filling most of the space of the growing embryo sac. As they grow, they turn, bringing the advancing cotyledons back toward the base of the seed.

Meanwhile, at the region near the suspensor, two groups of unobtrusive-appearing cells quietly begin to undergo a series of divisions. These divisions will produce the embryo. It contains two general regions: the *epicotyl*, at the cotyledon junction, and the *hypocotyl*, just below. The epicotyl contains at its tip a region of cells known as shoot apical meristem. These cells will contribute to new shoot growth throughout the life of the plant. In the embryo they produce the shoot or first embryonic leaves. The *radicle* contains a similar group of cells, the root apical meristem, that will contribute to root tip growth also throughout the plant's life. The radicle will develop into the primary root of the seedling as it grows.

As embryonic growth continues, a third region of meristematic cells forms midway between the first two. This is the *procambium*, which will eventually produce the vascular tissue of the plant—the xylem and phloem.

Further Seed Development in Monocots

The fruits of corn and other grasses are rather different from those of dicots. A cereal grain may look like a seed, but it is actually a hard fruit encasing a single large seed. In other words, each kernel of corn is a separate fruit. In these plants, the endosperm is not dissolved until after germination and, as we have seen, it remains the principal food storage organ. The single cotyledon never becomes a seed leaf. In fact, it doesn't even emerge from the grain or appear above the ground with the shoot. Instead, it becomes specialized as the scutellum, a digestive and absorptive organ.

In addition, the embryo consists of a radicle and an embryonic shoot, separated by a short length of undifferentiated stem tissue. The shoot apex is surrounded by partially developed embryonic leaves. This leafy region is itself surrounded by a closed, tubular protective structure known as the *coleoptile*. (You may recall from Chapter 20 that the coleoptile has been the favorite organ for the study of phototropic responses in plants, beginning with the pioneering work of Charles Darwin.) The coleoptile will first be visible as the shoot emerges from the soil. In a short time, the coleoptile will split open and the first leaf will emerge. The structures in typical mature monocot and dicot seeds are shown in Figure 21.14.

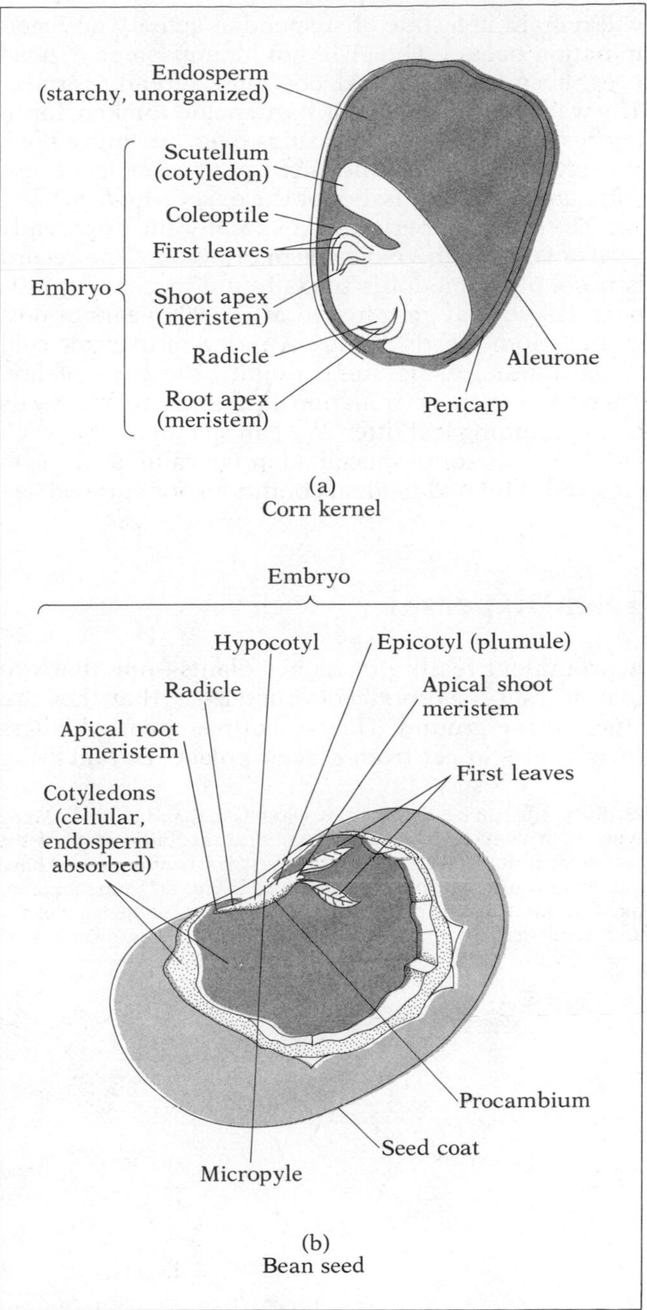

21.14 (a) Corn, a monocot, differs substantially in its embryonic organization from the dicot. Note that the endosperm is an unorganized region of food storage, surrounded by its aleurone layer. The corn's equivalent of the bean's cotyledon, known as the *scutellum*, is a single unpaired structure. It never undergoes the extensive development that the cotyledons of the dicot do and it does not emerge from the grain. The monocot embryo is surrounded at its apex by a unique structure known as the *coleoptile*. Within the coleoptile are the embryonic first leaves, produced by shoot apex meristem. Below is the radicle and root apex meristem.

(b) The common bean contains seed structures typically found in the dicots. The seed is covered with a hardened seed coat. At the concave side the micropyle still marks the point of sperm entry. Most of the seed is occupied by the paired cotyledons, which have absorbed all of the earlier endosperm. Other regions of the embryo include the epicotyl, hypocotyl, and radicle.

Seed Dormancy

With the completion of primary differentiation of the meristematic areas—a very busy period, indeed—many plant embryos simply become dormant. They will remain in a state of suspended activity until germination occurs. They will not lie unprotected, however, because their seed coats are already formed. The walls of the outer cells harden and thicken, forming resistant, protective casings over the entire seed. We have seen that, in the angiosperms, seeds are also surrounded by the tissues of the ovary (the *fruit*).

The dormant period varies greatly and apparently seeds can remain viable for many years. The record is probably some lotus seeds found in peat deposits near Tokyo that germinated after 2000 years of dormancy. Some seeds require a period of extreme cold to germinate, while some require the heat of fire. There are even germination inhibitors in the seeds or surrounding leaf litter. We can speculate that each variation has some special adaptive value to the species and is related to ideal conditions for survival.

Seed Dispersal

A prominent feature of higher plants—one that you may in fact have noticed yourself—is that they are stuck in the ground. This, of course, greatly hinders their ability to get from place to place. But all living things must be able to move about, or at least to send genetic emissaries in their place, and seeds are one kind of emissary. It is necessary to be able to get one's self, or one's genes, around, because after all, no place on earth is going to be a suitable habitat forever.

Dispersal is a particularly critical factor in the life cycles of *perennials,* plants that live from year to year. The problem is especially important in trees because undispersed seeds that germinate beneath their parent's canopy must compete with it for water, mineral resources, and, especially, sunlight. If a seed germinates close to the parent plant, the new seedlings may also be subject to heavy attack by insect predators and fungal root parasites attracted to the parent.

The primary natural seed carriers are water, wind, and animals. Perhaps the best-known example of a water-borne seed is the largest of all seeds, the coconut (Figure 21.15). Coconut palms have managed to invade even the most remote and tiniest atolls of the Pacific, and are found throughout the tropical and subtropical regions of the earth. Wind dissemination is common in the conifers, which often produce winged seeds. Winged seeds are able to spin in a strong wind, spiraling to new frontiers away from

21.15 Seeds can be dispersed by wind, water, and animals. Many types of fruit have evolved structures to take advantage of the particular traits of each of these carriers. The fruit in part (a) have plumes or wings that help in lofting them in breezes from the parent plants to some distant place. (b) The coconut has enough buouancy and water-resistance to enable it to be carried long distances by ocean currents. The cranesbill, foxtail, and burr clover fruit in part (c) are adapted for clinging to the fur of mammals. The cranesbill fruit, in addition, has the unique ability to screw itself into the ground. As humidity rises and falls, the spiral opens and closes and the seed turns. The seeds of berries are carried for a while in an animal's digestive tract and then ultimately deposited.

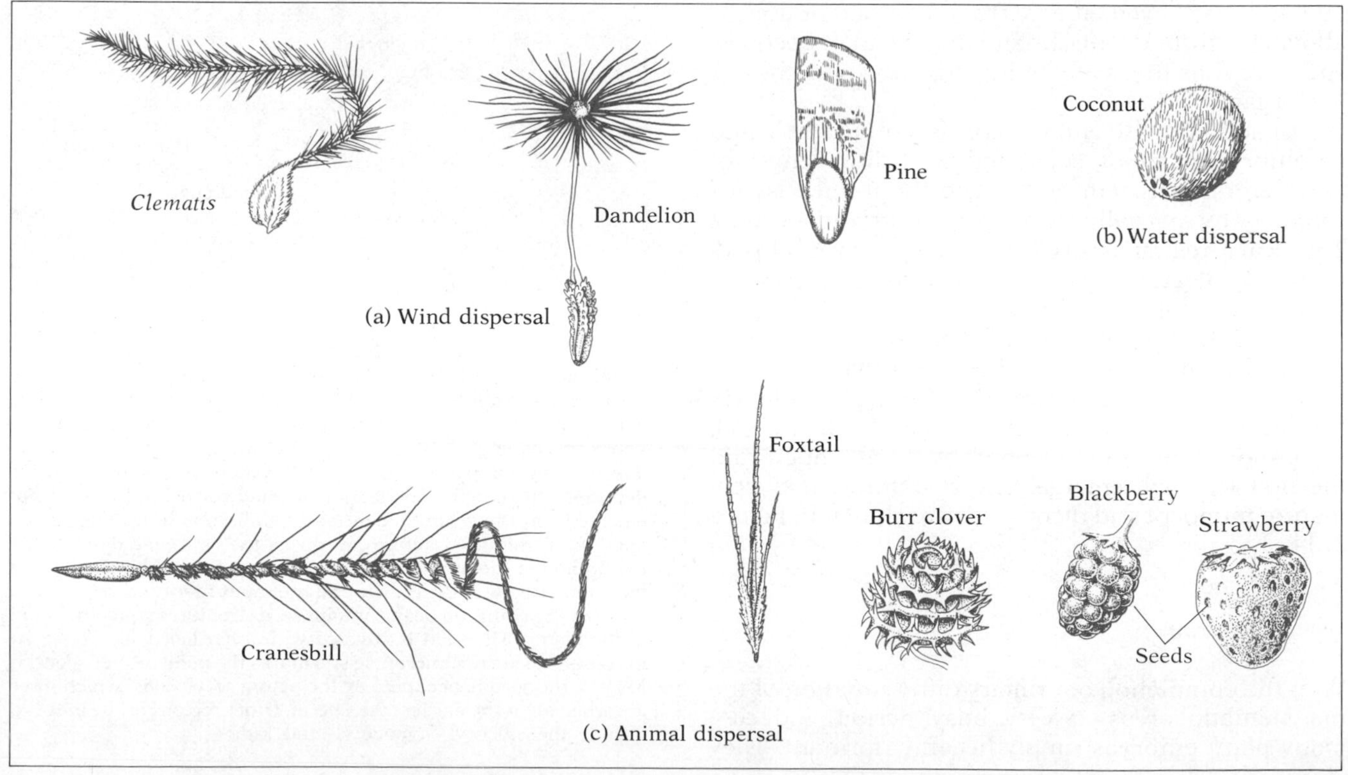

their parents. Dandelion seeds are carried in the wind on fine plumes. An odd adaptation to wind dispersal is the tumbleweed: the whole plant forms a huge, light ball that dries up, breaks off at its base and rolls across the prairie, scattering seeds as it tumbles, and giving rise to songs.

Seeds can be carried by animals in two ways: on the outside and on the inside. Those that are carried on the outside often have spiny, barbed fruit called *capsules* (the geranium and burr clover are familiar examples) or have barbs or sticky coats of their own. What dog owner hasn't had to remove a foxtail from the animal's paw?

Perhaps the most successful evolutionary strategy has been to tempt animals to eat indigestible seeds, which are thus carried for a while in the animal's digestive tract and then ultimately deposited, intact and ready to germinate, in a mound of excrement. This is what bright-colored, fleshy, sweet-tasting fruits and berries are all about—from the plant's point of view. Many such fruits contain powerful laxatives to help the process along. Sometimes, the seed itself is what attracts the dispersing animal. Watch birds eat up the tiny seeds at a feeder. Since seeds are small, a plant can produce many of them and, although they can be digested (else why would the birds eat them?) some small portion will escape the bird's digestive processes to germinate later in a nitrate- and phosphate-rich dropping. Some animals also store seeds for later use, and many an oak has grown up from an acorn that some absent-minded squirrel has misplaced and forgotten.

Interestingly, sometimes a seed won't germinate unless it has been subjected to an animal's digestive processes. On the island of Mauritius there are eleven huge trees of a very rare species. All of the trees are about 300 years old; there are no younger trees. Every year they put out a crop of huge seeds with incredibly thick seed coats. But the seeds never germinate. It seems that normal germination requires that the seed be passed through the crop of another native of the island, a bird called *Raphus solitarius*, better known as the dodo. But, thanks to us, the dodo has been extinct for just about 300 years, so the seeds just lie around waiting for the bird that will never come. (You may be pleased to learn, however, that recently, the naturalist who worked out this curious relationship managed to get several of the seeds to germinate by passing them through a turkey.)

21.3 Germination, Growth, and Development

Germination and Primary Growth

We've mentioned germination before, but perhaps now it is time to define it more precisely. Germination is a sort of reawakening of the dormant embryo. The length of dormancy varies considerably with different species of seed plants, but no matter when it ends, most plants have the same requirements for germination. They include an adequate supply of water, favorable temperatures, and available oxygen. The reawakening proceeds when water is absorbed into the dry interior of the seed.

One of the first tasks in germination of monocot grains is the mobilization of the large starchy endosperm. This is a roundabout process involving the gibberellin hormones (see Chapter 20). The targets of the gibberellins are the cells of the aleurone layer of the endosperm, just below the seed coats. As we mentioned in Chapter 20, these aleurone cells respond by producing α-amylase, a starch-digesting enzyme that helps form sugars that the embryo can use.

In the dicot, the early visible signs of germination include swelling of the seed and the splitting of the restrictive seed coats. If we were to open a germinating bean seed rather carefully, we would detect signs of growth in the embryo as well. On the concave side of the seed, the embryo itself can be readily seen (Figure 21.16).

As embryos develop the hypocotyl is the first part to emerge from the seed. It is pushed out of the seed as cells in the radicle rapidly divide and elongate. Again, the intake of water into cells is an important factor in elongation.

At the other end of the embryo, we can also see early signs of leaf development. The embryonic leaves, still white in color, are produced from cells in the shoot apical meristem. The young leaves are known as the *plumule* (the plumule sometimes develops before germination, as in the peanut). In beans, rapid division and elongation of cells occur in the hypocotyl just above the radicle. The hypocotyl elongates, elevating the two cotyledons and the epicotyl. The hypocotyl, as we saw earlier in Figure 20.11, will form a "bumper" as the embryo pushes its way through the soil. Once the loop is in the air, the plant will straighten, lifting both cotyledons and the epicotyl up toward the light. Then the epicotyl will emerge and begin intensive cell division and elongation, carrying the young leaves and stem away from the cotyledons, where the leaves will be exposed to sunlight.

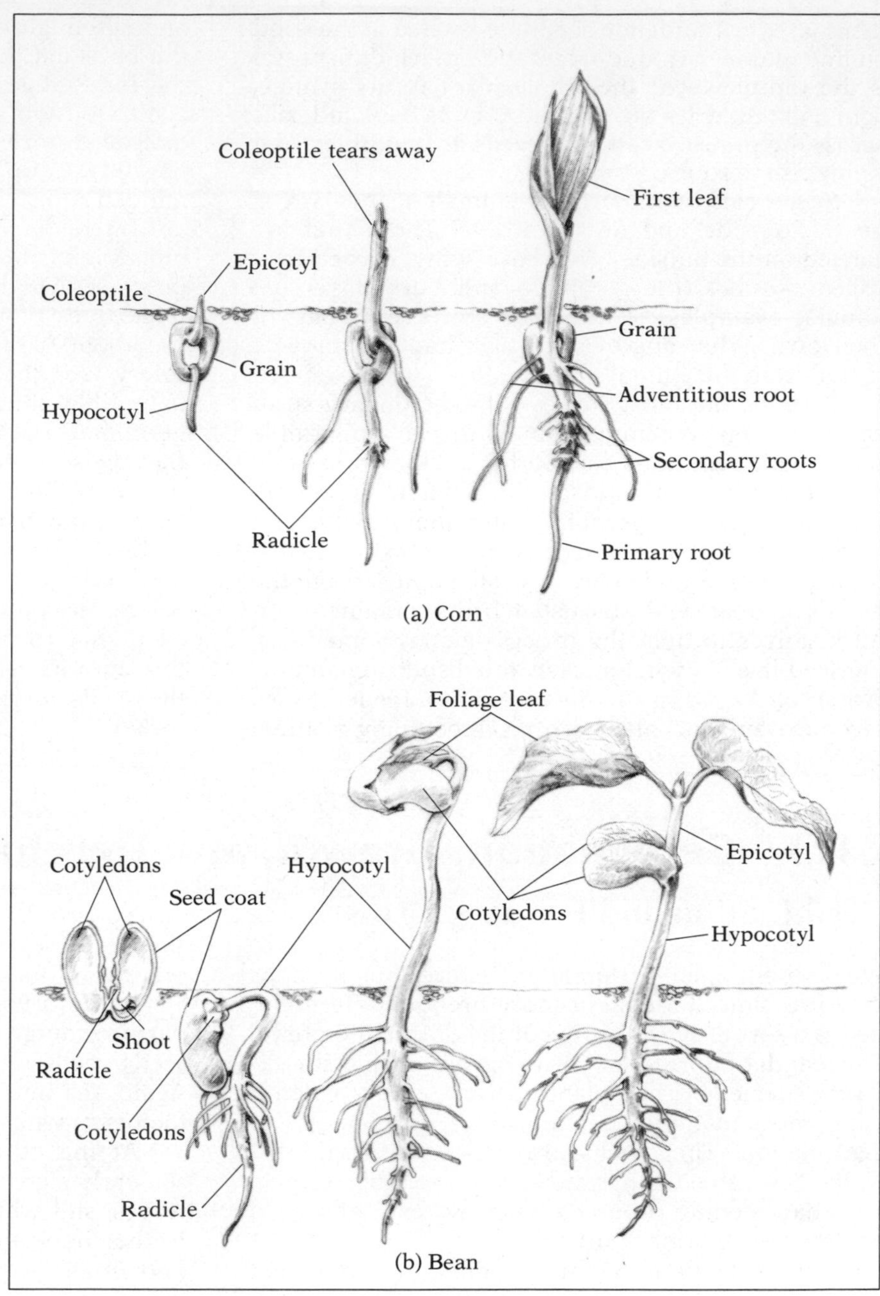

21.16 Differences in germination between (a) corn, a monocot, and (b) the bean, a dicot. The most obvious difference appears when the shoots emerge. Notice that the entire seed is lifted out of the ground in the bean as the curved hypocotyl elongates. In contrast, the corn grain and its remaining food reserves remain in the soil with only its coleoptile-covered shoot emerging. As you can also see, the production of first leaves occurs early in the monocot, while that in the dicot lags behind, as the photosynthetic cotyledons open and spread. Later, the cotyledons will wither and fall as their food reserves diminish.

The Development of the Plant

Even while germination is occurring, the parts of the plant standing above ground are already producing their first chloroplasts, causing the plant to turn from a pale to a deep green. The growing embryonic leaves will expose their green surfaces to energy-laden light. And, as the leaves develop in dicots, the cotyledons wither. In monocots, the starchy endosperm is exhausted. Those food stores are no longer needed by a photosynthesizing plant, so they fall away, leaving their telltale scars on the young shoot. The root also becomes greatly enlarged, each new root tip putting out a multitude of tiny root hairs, greatly expanding the vital water-absorbing area.

Growth and Development in the Root

In spite of the great differences in the events occurring above and below ground, if you were to view both the emerging root and shoot at a microscopic level, you would see very similar events. Root apical meristem produces cells that will form the root cap, which is a protective layer of undifferentiated cells that shields the delicate root as it pushes through the soil. Its abraded cells are constantly replaced as the root probes downward. Other cells of this meristem become the critical epidermal layer. The root hairs form as this outer layer develops. Behind this rapidly dividing meristematic region, the cells elongate, providing the "push" downward.

Elongation, as we saw in Chapter 4, is essentially a cell wall phenomenon. In the young cells above the root meristem, the primary wall is still thin and pliable. As turgor pressure causes the cell to elongate, the first layers of cellulose fibrils circling the cell change their orientation to diagonal and longitudinal. Any additional fibrils laid down during this process will be at right angles to the original fibers (Figure 21.17). Once the cell wall is completed and hardened, no further elongation is possible. We know from Chapter 20 that auxins are involved in cell elongation in the stem, but their role in the lengthening of the root is not so clear.

Once their elongation has been completed, the older cells will be left behind the growing root tip, and will become differentiated. In the central portion, cells form the *cortex,* the *endodermis,* the *pericycle,* and the *procambium.* Procambial cells, as mentioned earlier, will continue to differentiate into phloem and xylem for conducting sugars and water (Figure 21.18).

21.17 Plant cells elongate through turgor pressure. The elongation causes changes in the orientation of cellulose fibrils in the young primary wall. Subsequent layers of cellulose change their angles as elongation continues. However, once the wall becomes firm, no further lengthening is possible.

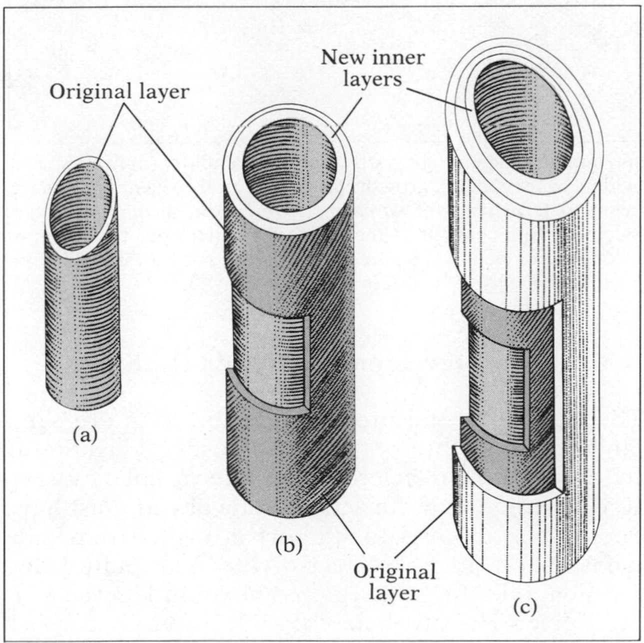

Growth of the Stem

The organization of the apical meristem of the shoot is somewhat more complex than that of the root. A rather small mound of tiny, actively dividing cells which constitutes the meristem is found at the very tip of the shoot. As cell division occurs, daughter cells at the lower side of the meristem undergo extensive elongation, which pushes the meristem upward. Below this region of elongation, cells that have completed the process undergo differentiation. In the center, thin-walled parenchyma cells will form the *pith.* Just outside the pith, we see primary xylem and phloem surrounded by cortex and finally epidermis. Since things get more complicated from here, a look at Figure 21.19 will help as we proceed.

In addition to the production of these tissues, the apical meristem of the stem produces *leaves, leaf primordia,* and *branch primordia.* The primordia are patches of embryonic tissue that give rise to these structures. As the meristem itself is pushed upward by the underlying region of elongating cells, primordia are left behind in an active region of potential leaf and stem production known as a *node.* The stem region between each successive node is very logically called an *internode.* The length of an internode depends on how much elongation has occurred and often marks the extent of growth in a season. Nodes and internodes are most easily seen in deciduous woody plants such as apple, peach, and pear trees (Figures 21.19, 21.20). These are seasonally growing plants that shed their leaves in the autumn. Since growth occurs in surges, each previous year's growth is marked by a nodal or bud-scale scar. In addition, when a leaf falls, it leaves a scar that remains visible for a time in the young branch. Above each leaf scar is a tiny bud which in the younger region will produce leaves in the spring. Some of these buds may produce flowers as well. Buds further down the stem at older regions typically produce branches.

The lateral branch buds in some woody plants are protected not only by the leaf petiole but also by *bud scales,* tiny hardened structures that are believed to be modified leaves. What is being protected is the small bit of soft, tender meristematic tissue inside the bud. It may remain dormant indefinitely. The bud is inhibited from sprouting into a lateral branch by auxin from the apical meristem. (This is the apical dominance phenomenon we described earlier). Under the proper conditions, depending on the plant species, the position of the bud relative to the apex, the environment, and the nutritional state of the plant, the lateral branch bud will throw off its dormancy and become a lateral branch. Or, if the apical meristem itself is removed, the lateral branch bud may develop a new apical meristem and continue the upward growth of the principal stem.

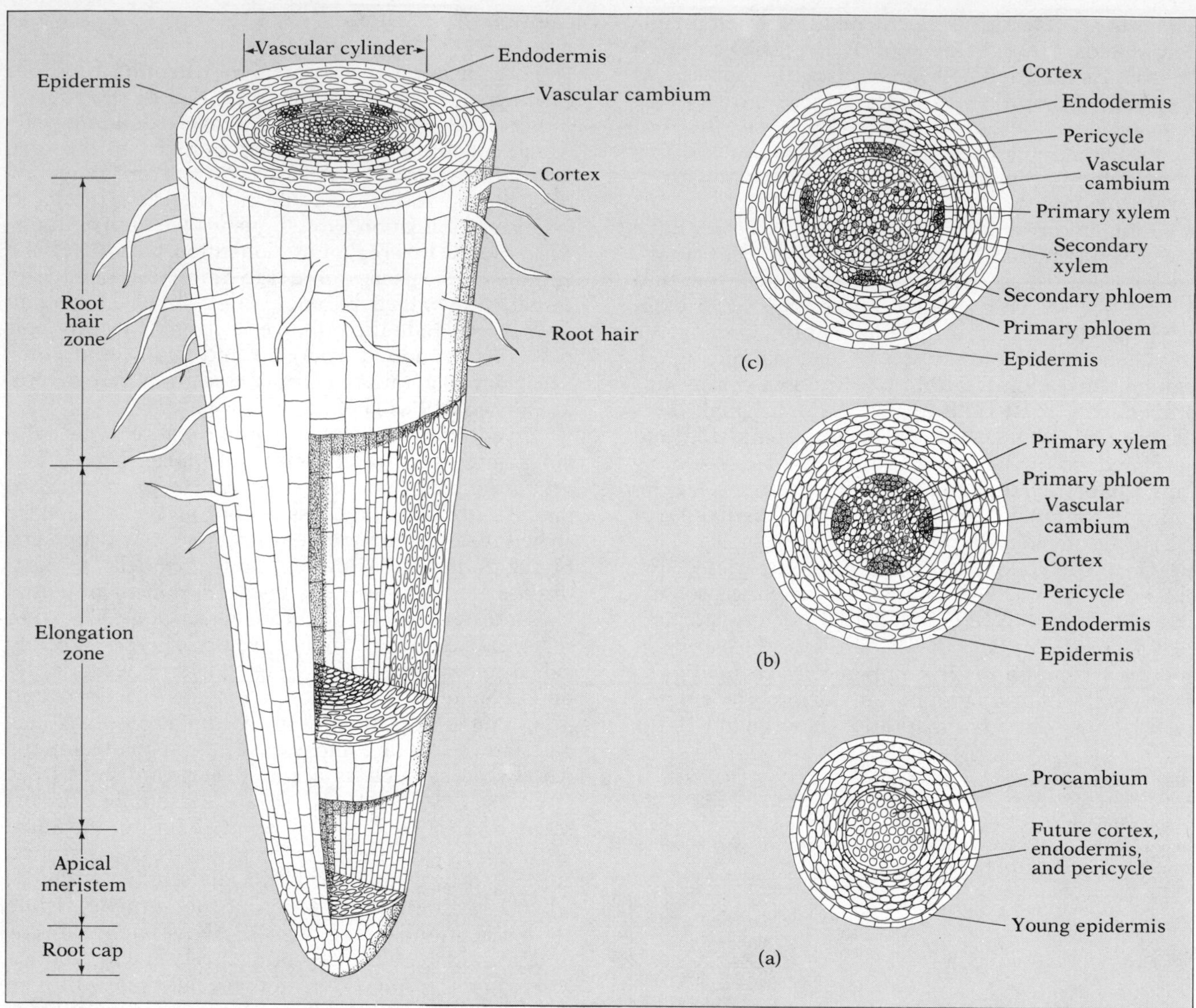

21.18 (a) Just above the apical meristem, root cells begin to show signs of differentiation. The outer cells form the epidermis, and within, an undifferentiated cylinder will soon form the *cortex, endodermis,* and *pericycle.* At the center, cells of the *procambium* are set aside to be used in producing the vascular system. (b) Further up, in the region of elongation, cells have grouped into tissues. Procambial cells have formed primary *xylem* and *phloem,* while tissues surrounding the emerging *stele* also differentiate. (c) At this level, evidence of secondary growth emerges. Primary xylem and phloem begin to form rings of *vascular tissue* as the second-generation xylem and phloem fill in the gaps. From here on, new rings of xylem and phloem will be produced by vascular cambium that has been set aside by the primary cambium.

It should be noted that the grasses don't follow these rules. We're not sure why, but it could be that grasses evolved under strong pressure from animal grazers. If grasses grew from their tips, it would be difficult for them to recover from grazing. But grasses survive grazing because they have a meristematic reserve below the leaves and stem, called the *basal meristem.* Thus, the loss of the stem does not interfere with growth, since the plant continually emerges from the basal meristem lying below.

Growth and Development Within the Stem

The developmental processes in the stem are similar to those in the root. In both cases, elongation is followed by the development of the vascular system. Keep in mind that the active tissue ahead must have continuous supplies to support growth. To imagine what's happening here, let's divide the rapidly growing stem into the four regions shown in Figure 21.21.

Region I is just below the meristem. It consists of a circle of procambium islands. At this stage, there has been no real differentiation. Region II is a little

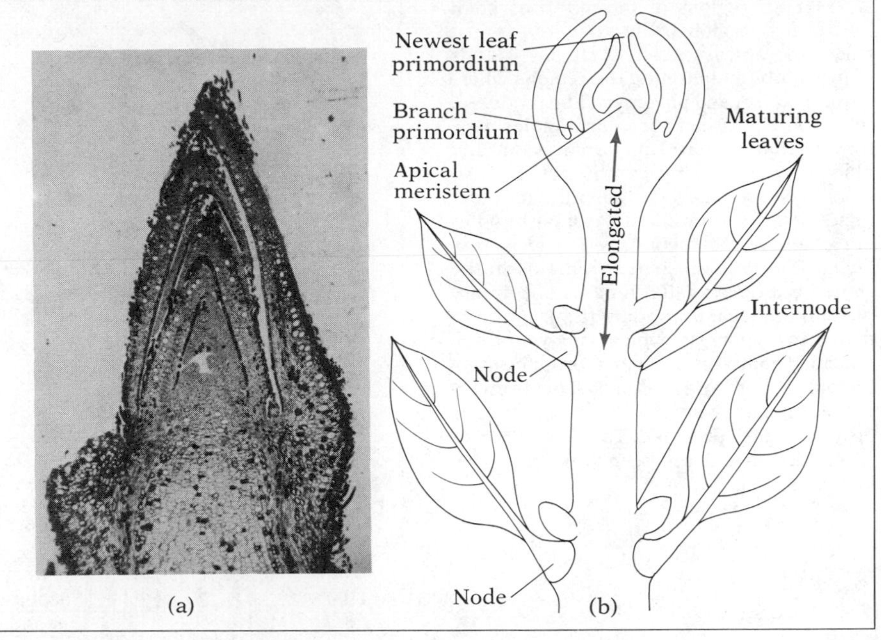

21.19 Stem growth and structure. (a) The photomicrograph reveals very clearly the initial regions of growth in the young stem shoot. The dome-shaped central region is the actual meristem from which new cells emerge. It consists of perpetually undifferentiated or embryonic cells, which, during growth periods, are in constant mitotic activity. Surrounding the meristem are two immature leaves. (b) By the time they resemble mature leaves, the meristem will have been pushed ahead, leaving them behind. Below the first leaves appear two mounds of simple tissue, the *leaf primordia,* or sites of future leaves. Below these are mounds of *branch primordia,* seen just above each leaf attachment.

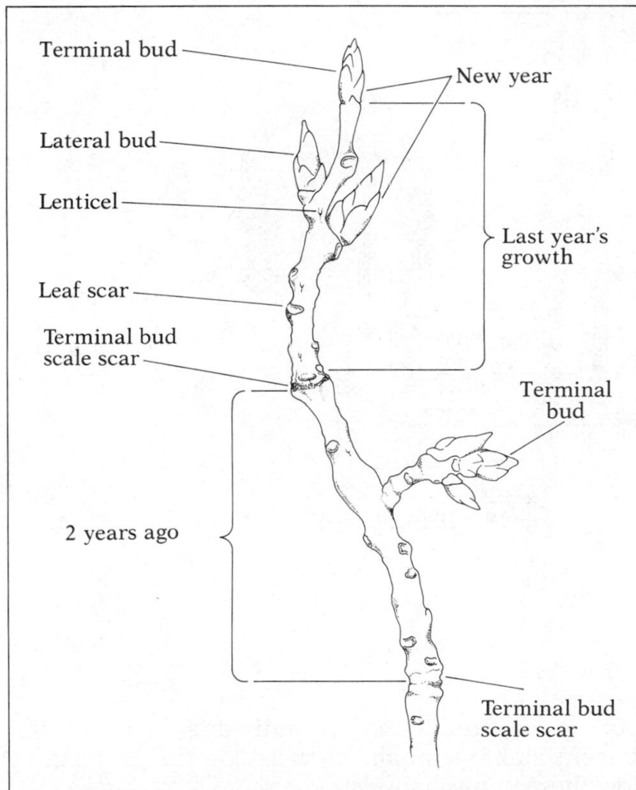

21.20 A look at this dormant deciduous woody stem reveals 2 years of growth history. It is tipped by a dormant terminal bud that will emerge as the next season's apical meristem. The previous season's growth extends down to the first terminal bud scale scar. Below this is a length of stem that represents growth from 2 years ago. It too is marked by a terminal bud scale scar. The young branch at the right will produce a similar growth record. The position of each leaf scar is a node, and the spaces between are internodes.

older. It reveals the same islands, but the procambium has undergone differentiation in which xylem and phloem cells have formed and have already become functional.

Both monocots and dicots have vascular bundles at this stage. The difference between them is that the vascular bundles of the dicots are arranged in a ring, as shown in Figure 18.12; in monocots, the vascular bundles are scattered throughout the cortex of the stem. Vertically, the monocot vascular bundles frequently interconnect with one another.

Monocots do not usually get beyond this stage. If you've ever observed the trunk of a palm tree in cross-section, you may have noticed the stringy strands of its scattered vascular bundles. In a living palm, the trunk is approximately the same diameter from the base to the crown.

In region III, the islands of vascular tissue start to change. The phloem and xylem begin to form the cylindrical tissue arrangement typical of older perennials. Soon the remaining phloem islands join, and only the primary xylem remains visible.

Region IV, further down the stem, has a different appearance. Here, we see the results of having lived three seasons. The vascular system seems quite different but the difference can be explained. It is due to secondary tissue. The islands of vascular tissue have vanished, and in their place we find a cylinder of tissue. The procambium, it seems, has grown from its position in the islands into a ring all the way around the stem. It is now called *cambium* or *vascular cambium.* It will divide continuously and provide a solid ring of xylem on its inner side and a solid ring of phloem on its outer border. The cambium will,

21.21 Four regions of the maturing stem are shown. Region I, the youngest, is just below the apical meristem. It has islands of procambium outlining the regions where primary xylem and phloem will be produced. In region II, distinct vascular bundles have formed that contain functional xylem and phloem separated by cambium. The vascular system changes radically in older region III. The prominent vascular bundles are changing as secondary tissue is produced. The primary xylem remains in the center, but only small remnants of primary phloem are seen. Secondary tissue predominates in region IV, which is 3 years old. Vascular cambium regularly produces secondary xylem and phloem. The primary phloem has long since been pushed out as growth in girth occurred. Three cylinders of secondary xylem are seen, marking each year's growth.

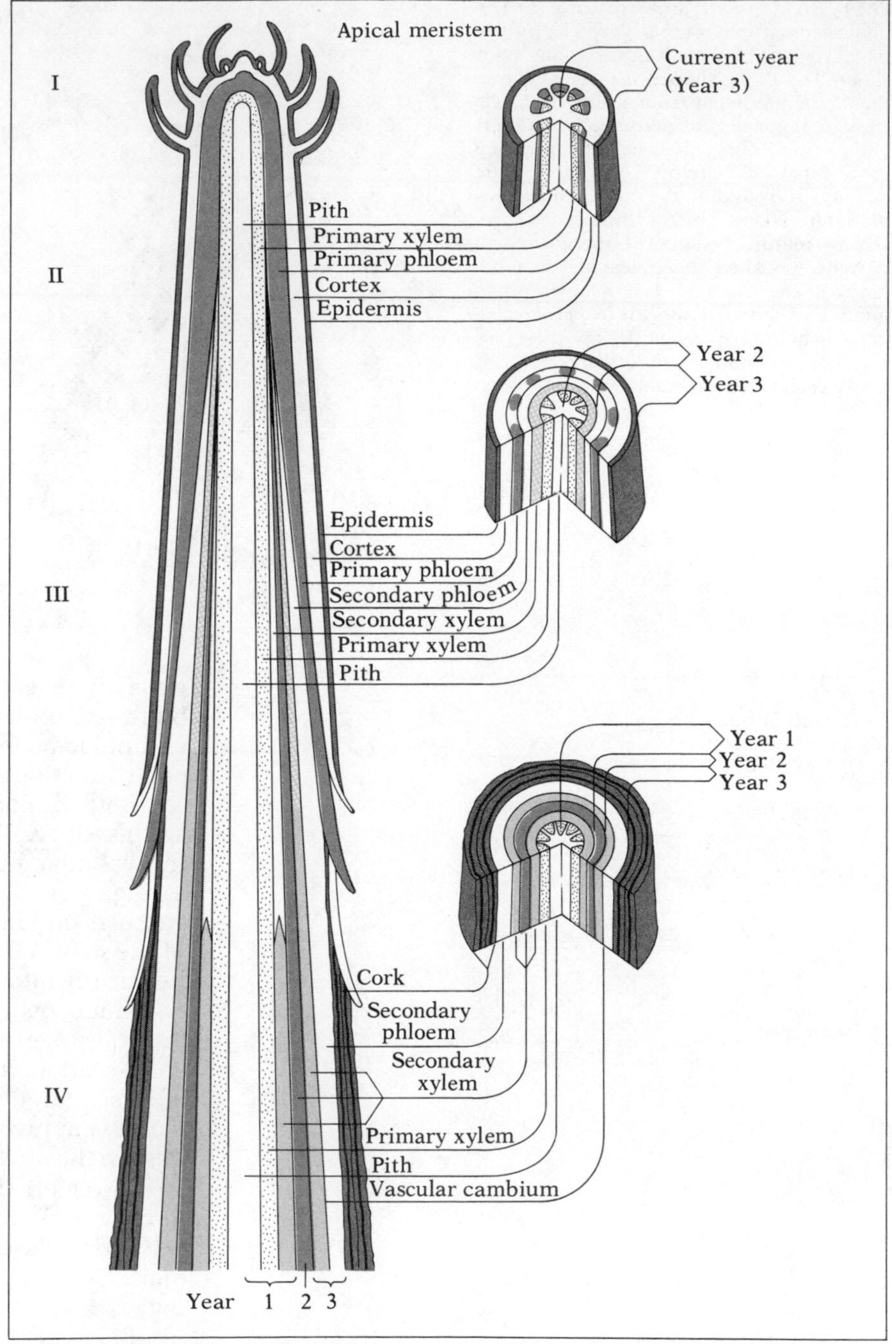

however, always retain a reserve of undifferentiated cells within its narrow band, no matter how many phloem and xylem cells it produces. This reserve must be maintained for continued growth in future seasons. Since our example was three years old we can see the results of the first two season's development as inner rings of tissue. This is entirely xylem. At the start of the present season's growth the phloem that had been produced previously was crowded outward by the new growth activity.

We won't linger on the woody stem's tissues since we discussed these in Chapter 18. We will, however, look at differentiation in individual cells of the phloem and xylem, and then tackle the problem of how these things happen.

The events begin with the division of a cambial cell. One of the daughters, under influences we don't understand, begins its differentiation into a xylem cell. The other daughter divides again. One cell from this division rests, but the other divides yet again. The two resulting cells begin their differentiation into a sieve member and companion cell, the basic elements of phloem.

Now the phloem and xylem lengthen and begin to take on the characteristics of mature cells. In the xylem this means a transition in the wall structure. Now, thickened rings, spirals, and pits emerge. Finally as the xylem matures, its end walls are destroyed or perforated and its cytoplasm is absorbed by neighboring cells. The result is the formation of long continuous tubes of nonliving material. The sieve-tube cell develops its pitted end walls, the sieve plates, uniting its cytoplasm with that of those above and below. They then lose their nuclei, apparently relying on the adjacent companion cells for nuclear functions.

The Problem of Differentiation

We can learn detail after detail about plant development until it seems that we surely must have the problems solved. But one haunting question prowls the attics of our minds: "How do plant cells differentiate?" This, of course, is the question we started with, so we might wonder, what do we really know? The answer is essentially the same for plants and animals: we really don't understand the basic phenomenon. As we have seen, however, certain observations give us some insight and tantalizing clues. We know, for example, that auxin in specific quantities stimulates cell elongation. In this role, it probably interacts with the gibberellins. Further, there is an important relationship between auxin and light. Cytokinins provide us with additional clues. It appears that hormones are definitely involved in the differentiation process. There are undoubtedly other interacting factors such as light and tissue interaction. Each cell, theory tells us, has the gene complement needed to produce an entire plant. Apparently, the successive masking and unmasking of genes as development proceeds is the key. Keep these ideas in mind as we look into some of the information gained from studies of development.

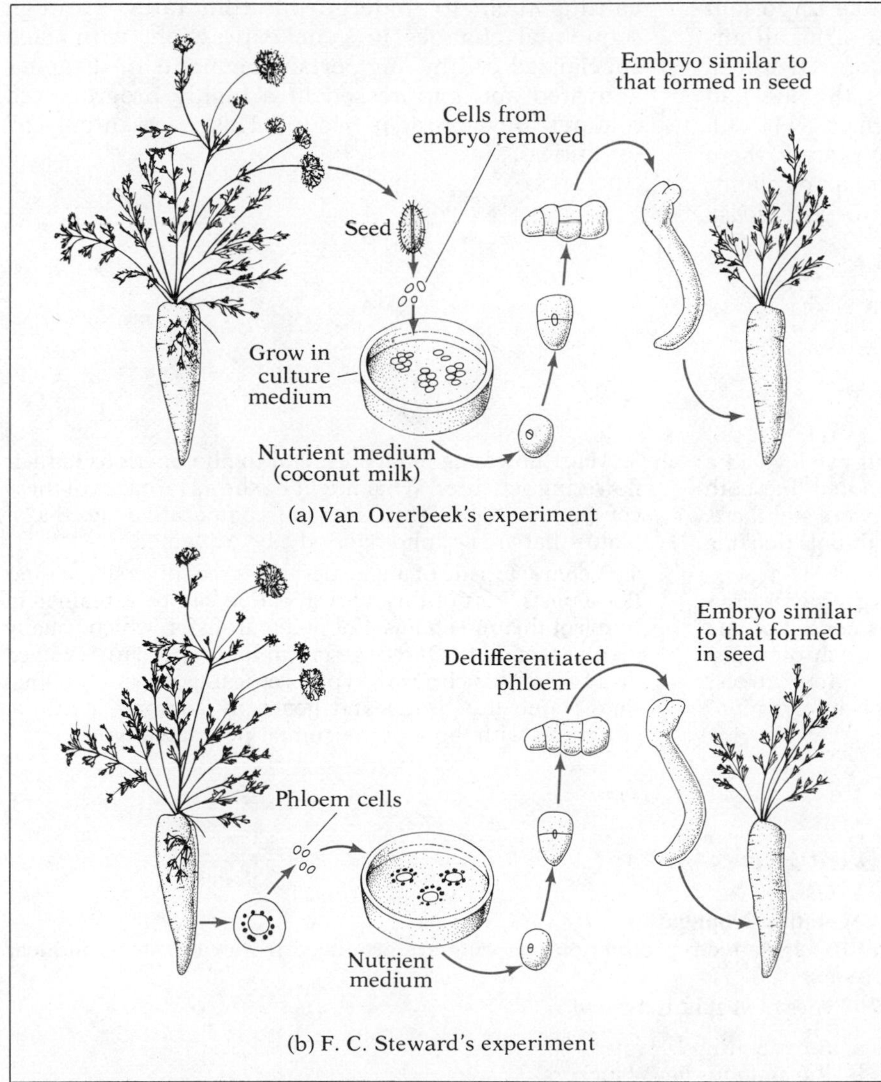

21.22 Tissue culture with the carrot. (a) Van Overbeek succeeded in culturing cells obtained from carrot embryos that had been grown from seeds. His nutritional medium contained coconut milk (endosperm). Isolated cells developed readily into mature carrots, proving that many cells in the embryo were *totipotent* (had the capability of the zygote). (b) Following these earlier leads, researchers developed even more penetrating techniques. In the late 1950s, F. C. Steward succeeded in removing mature carrot phloem and culturing this tissue through embryonic stages that eventually produced adult carrot plants. It now appears that few plant cells ever lose their totipotency.

Information from the Study of Plant Cells

In some ways plants make ideal subjects for studying cell growth and proliferation under controlled conditions. For one thing, some of them can regenerate from bits and pieces. Many plants can regenerate roots from stem cuttings, stems from bits of root, and even entire plants from leaves. This knowledge has been invaluable in agriculture through the ages, but for biologists it suggests clues to the puzzle of differentiation. It supports the idea, for example, that cells are *totipotent*—that is, they have all the genetic information needed to produce the entire organism from which they come. As this clue was recognized, however, it prompted the question: Can plant *cells* acting alone duplicate the regenerative feat we see in cuttings? This question has been partly answered through studies using tissue culture techniques.

Plant researchers turned to tissue culture methods in the 1930s and 1940s. A pioneer in these efforts, Johannes van Overbeek, finally discovered that a suitable medium was coconut milk. Besides important nutrients, coconut milk contains some critical hormones needed for plant growth. Van Overbeek succeeded in growing individual cells that he had separated out from young carrot embryos. He cultured them in his coconut milk and later planted them in soil. Lo! They produced normal adult carrot plants. The results indicated that cells in the embryo, at least, were totipotent.

It wasn't until the late 1950s, however, that mature tissue was first cultured in a similar manner by F. C. Steward at Cornell University. Phloem companion cells were successfully removed and grown in a nutrient medium. The mature cells grew into rootlike structures which, when planted, produced entire carrot plants (Figure 21.22). More recently, there has been excellent progress in cultering redwood trees and orchids.

All of this seems to indicate that, first of all, any cell does contain all the genetic information necessary to produce the plant. Second, when removed from their cell associations, they can be regressed to an undifferentiated state from whence they can proceed through normal embryonic development. How does this answer the question about differentiation? Looking at the observation from another perspective, we can hypothesize that the key to differentiation is in the interaction between cells. Neighboring cells can thus interact and be acted upon by hormones from remote regions of the plant, causing them to undergo different fates. Through sequential changes in gene expression, with each specialized cell having certain segments of its genes activated and suppressed in a highly programmed manner, development occurs. This is essentially all we know.

Application of Ideas

1. A young boy carved his initials at about eye-level in a small tree. Returning as a young man, he noted that both he and the tree had grown and the initials were still there. What position was he in when he read the initials (looking up, down, or straight ahead)? Explain your answer.
2. Dendrochronology is partly the study of growth rings in woody stems. What do the growth rings consist of and what kinds of information might a dendrochronologist gain by studying growth rings in living or fossil trees? What are some potential applications for this information?
3. Many flowering plants respond to photoperiods in their flowering activities. What might be the advantages of these complex systems? How can the comparative success of plants that are not photoperiodic be explained?
4. A characteristic of angiosperms is great diversity in their floral parts. Part of this diversification can be explained in terms of their mechanism of pollen transfer, which usually involves animals. One large group has apparently evolved away from this circus of types, returning to wind pollination. Name this group and generally compare its floral structures with those of the animal-pollinated types.

Key Words

adventitious bud
aleurone
androecium
anther
apomixis

Key Ideas

Vegetative Propagation

1. In vegetative reproduction offspring are produced from root, stem, and leaf tissue.
2. Types of vegetative growth:
   a. Stem cuttings
   b. *Adventitious buds* (suckers)

basal meristem
branch primordia
bud scale

callus tissue
calyx
carpel
coleoptile
corolla
cotyledon

dark clock
differentiation

embryo
embryo sac
endosperm
endosperm mother cell
epicotyl

filament
fruit

generative nucleus
germination
graft
gynoecium

hypocotyl

indeterminate plant
integument
internode

leaf primordia
leaf scar
legume
long-day plant

megagametophyte
megaspore
megaspore mother cell
megasporogenesis
micropyle
microspore
microspore mother cell
microsporogenesis

naphthalene acetic acid
nectar
node
nucellus

ovule

petal
photoperiod
photoperiodicity
phytochrome
pistil
pith
placenta
plumule
pollen sac

c. Leaves (*Bryophyllum*)
d. *Stolons* (horizontal stems)
e. *Tubers* (underground storage stems)
f. *Rhizomes* (underground stems)

Grafting

1. In *grafting,* older stem stock is used to receive cuttings from young shoots. Grafting is used to propagate seedless fruits and to overcome the long period required for seed growth.

2. When stems are grafted together, *callus tissue* forms first and eventually produces new tissue.

Apomixis: Asexual Reproduction with Seeds

1. *Apomixis* is seed production without union of gametes. Dandelions are known to carry out *apomixis,* thus the offspring is a clone. Most apomicts require *pollination* in order to produce the required triploid *endosperm.*

2. In other cases, *pollination* contributes nothing but an incentive to begin development of the *embryo.*

Essay 21.1 Asexual Reproduction: The Long Run

Asexual reproduction always results in identical offspring, both to each other and to the parent. This is ideal in agriculture, but not so in natural populations.

Sexual Reproduction in Flowering Plants

1. Angiosperms produce flowers, seeds, and *fruits. Fruit* has special botanical meaning. It is the ripened and mature ovary. Fruit is not necessarily sweet. It includes examples such as the bean pod, cereal grain, and tomato.

2. Flowers are composed of four regions whose parts occur in whorls. These are:

a. The *calyx:* a *whorl* of *petals* (outermost)
b. The *corolla:* a *whorl* of *petals*
c. The *androecium:* a *whorl* of *stamens,* each with a *filament* and an *anther*
d. The *gynoecium: pistil* composed of fused or separate *carpels;* each *pistil* has a *style* and a *stigma*

3. Flowers are often colorful and fragrant. Sugary nectar is produced in the *nectaries.* Such flowers are usually insect or bird pollinated. Wind-pollinated angiosperms produce inconspicuous flowers, without odor or color (grasses, other grains).

The Initiation of Flowering

1. A knowledge of the flowering mechanism is extremely useful for agricultural and horticultural reasons.

2. Plants follow seasonal and climatic regularity in flowering.

Photoperiodicity

1. Plants follow critical dark periods in their flowering activities. These vary with seasons and latitude.

2. Three categories have been recognized:

a. *Long-day plants* (critical shortness of night period)
b. *Short-day plants* (lengthening nights)
c. *Indeterminate* or "day neutral" (indifferent to light–dark periods)

3. Studies reveal that day periods can be varied in many ways without affecting the plant. Tampering with the dark periods, however, has an effect on flowering.

4. *Long-day plants,* for instance, can be made to flower by a short exposure to intense light during the night. *Short-day plants* can be inhibited by the same process.

5. Application of this knowledge has been used in controlling flowering in agriculture (chrysanthemums and sugarcane).

6. The cocklebur has been used for extensive flowering studies. They flower when the dark period is $8\frac{1}{2}$ hr. The leaves apparently contain the photoreceptor.

pollen tube
pollination
primary growth
procambium

radicle
receptacle
rhizome
root apical meristem

scutellum
secondary tissue
seed coat
sepal
shoot apical meristem
short-day plant
sperm nuclei
stamen
stigma
stolon
style
suspensor

totipotent
tube nucleus
tuber

vascular cambium
vegetative reproduction

whorl

Red and orange light are the most effective stimulants. A flash of far-red negates the middle-of-night flash, if it occurs in less than 35 min after red/orange exposure.

7. From these observations rose the idea of a pigment receptor. The pigment occurs in two forms, Pr, the red-absorbing form, and Pfr, the far-red absorber. In daylight, red light converts Pr to Pfr, but at night the pigments reverse to Pr. A nighttime flash of red immediately converts Pr to Pfr.

8. The receptor pigment is called *phytochrome.* It is also the photoreceptor for chloroplast orientation, phototropism in stems, and photoresponse in leaves. Its action in the dark clock is unknown.

Transmitting the Stimulus

1. Studies in the cocklebur have been carried out with six plants *grafted* together and one leaf at the end exposed to the appropriate *photoperiod.* The flowering response occurred in all six plants.

2. Transmission of information from the pigment receptor to the region of flowering has not been clarified, but a hormone presumptuously named florigen has been suggested.

Pineapples and Auxin

Auxin has been successfully used to control and synchronize flowering in pineapples.

Meanwhile, Back in the Ovary: Megasporogenesis

1. *Megasporogenesis* occurs in the *ovules.* Each *ovule* develops from placental tissue. Its structures include a stalk, *nucellus, integuments, micropyle,* and a *megaspore mother cell.*

2. The *megaspore mother cell* enters meiosis, producing four haploid daughters. One will survive to become the *megaspore.* It goes through three nuclear mitotic divisions, producing eight haploid nuclei. This is called the *megagametophyte* and is the female half of the gametophyte generation of the flowering plant.

3. The *megagametophyte* rearranges its nuclei into a seven-cell embryo sac. One becomes the egg cell. Another, a large. centrally located cell with two nuclei becomes the *endosperm mother cell.* The remaining cells will not participate in fertilization.

Similar Events in the Anthers: Microsporogenesis

1. Within the *anthers* are numerous *microspore mother cells.* After meiosis each will produce four haploid *microspores.* Each *microspore* will undergo mitosis, producing a *generative nucleus* and a *tube nucleus.*

2. Following this, a hard protective coat develops around the *microspore,* forming a pollen grain. When *pollination* occurs, the *generative nucleus* will divide mitotically, forming two *sperm nuclei.*

3. The *microgametophyte* represent the male half of the gametophyte generation in the flowering plant.

Pollination and Fertilization

1. Self-pollination is common in some species, but others develop various devices for avoiding it. Hybridization between species is more common in plants than in animals, but there are mechanisms that generally prevent this too.

2. Pollen on the *stigma* is stimulated to produce a *pollen tube.* As the *generative nucleus* follows the *tube nucleus* down the *pollen tube* it divides into two sperm nuclei.

3. The *pollen tube* penetrates the *micropyle* of the *ovule* and the two sperm enter the *embryo sac.* One fertilizes the egg cell, while the other fertilizes the binucleate *endosperm mother cell.*

4. The fertilized egg develops into the *embryo,* while the triploid *endosperm mother cell* produces the starchy *endosperm* of the seed.

5. Following fertilization, the ovary enlarges into the fruit.

Seed Development in Flowering Plants

1. Mature seeds consist of an *embryo,* a food supply, and protective *seed coats.*

2. In monocots, food is stored in undifferentiated *endosperm* tissue, the starchy bulk of the corn kernel.

3. *Seed coats* in wheat and corn are partly maternal diploid tissue and partly specialized *endosperm* tissue called the *aleurone.*

4. In dicots, the *endosperm* is absorbed into the two *cotyledons* or seed leaves, becoming part of the *embryo* itself.

5. In dicots, development of the *embryo* begins with the *endosperm* nucleus, undergoing numerous divisions, forming a multinucleate mass. Cell walls will form later. The zygote begins its development by forming a vertical row of cells. One cell enlarges to form the globular *suspensor.* At the other end a ball of cells forms what will become the embryo and the *cotyledons.* In dicots, two lines of cells advance and invade the *endosperm,* eventually absorbing all of it into two emerging *cotyledons.* In monocots, one line of cells advances, forming a single *cotyledon.* Most of the *endosperm* remains intact.

Further Seed Development in Dicots

1. In *legumes* (beans, etc.), the *cotyledons* continue, absorbing the *endosperm* and filling the *embryo sac.* The cellular advance turns back toward the base of the seed.

2. The *shoot apical meristem* and *root apical meristem* begin their development. The first forms the *epicotyl* and finally the *shoot.*

3. The second forms the *hypocotyl* from which the *radicle* grows.

4. Some cells and their descendants from each meristem will remain immature or undifferentiated for the life of the plant.

5. Midway between the apical meristems, a region of *procambium* forms. This will produce vascular tissue.

Further Seed Development in Monocots

1. The *grain* or kernel is not just a seed, but represents the entire fruit, encased in hardened coatings.

2. The single *cotyledon* forms the *scutellum,* a digestive organ that absorbs nutrients from the starchy endosperm to support the embryo and sprout at germination.

3. The *embryo* contains a *radicle* and an embryonic *shoot.*

4. The *shoot* apex in monocots is enclosed in the *coleoptile.* The *coleoptile* opens after the *shoot* emerges, freeing the first leaf.

Seed Dormancy

Following primary differentiation of the meristems, the seed enters a period of dormancy. Hardened, protective *seed coats* form, and the seeds become enclosed in the fruit.

Seed Dispersal

1. Distribution of offspring to prevent direct competition is a serious survival problem to nonmotile plants. Seeds are distributed by water, wind, and animals.

2. Coconuts are successful ocean travelers and have established themselves on islands throughout the world. Wind-borne seeds or fruit are commonly winged or plumed, while animal-borne seeds are sticky, barbed, or spined and cling to hair or feathers. Fleshy fruits are eaten by animals and the seeds often pass through the digestive system to be later deposited in the feces.

Germination and Primary Growth

1. Dormancy periods vary widely and some seeds require special *germination* conditions. Most require only water, oxygen, and favorable temperatures.

2. In grains, gibberellins are released from the *embryo* and stimulate the *aleurone* to produce α-amylase, the starch digester.

3. Following water intake, the *seed coats* split and the embryo emerges.

4. In beans, the *hypocotyl* enlarges and forms a bend that pushes its way through the soil, drawing the *cotyledons* behind it. It will later straighten and expose the young leaves to sunlight.

The Development of the Plant

When the seedling is exposed to sunlight, chloroplasts are produced. As the leaves develop and turn green, the food reserves in the *cotyledons* are exhausted and these bodies wither and fall off.

Growth and Development in the Root

1. *Root apical meristem* produces cells on either side. The outer cells become the epidermis, producing root hairs.

2. Above the meristem, rapidly elongating cells push the root tip into the soil. As the elongating cells complete growth, they begin to differentiate into the tissues of the vascular system.

Growth of the Stem

1. *Shoot meristem* produces cells below, which elongate, pushing the shoot tip upward. Upon completion of the elongation process, central cells form *pith,* cortex, and primary xylem and phloem.

2. *Apical meristems* also produce *leaf primordia* and *branch primordia,* which will give rise to leaves, flowers, and branches. They are left behind the growing stem in *nodes.* These are separated by growth regions known as *internodes.*

3. Many lateral buds remain dormant since they are inhibited by auxin from the uppermost apical meristem. Under certain conditions they will form branches.

4. Grasses are adapted to the grazing activities of herbivores. *Basal meristems* below the leaves and stems produce a continuous growth below the grazing level.

Growth and Development Within the Stem

1. The *vascular cambium* begins as islands of *procambium.* In older regions below the region of elongation, these islands differentiate into xylem and phloem bundles. In the monocots and many herbaceous dicots, this is the limit of vascular tissue. In the woody dicots, the bundles coalesce into rings of vascular tissue. The *vascular cambium,* located between the xylem and phloem, produces new areas of these tissues annually.

2. Most monocots reach the vascular bundle stage and produce no further vascular growth. They lack the *vascular cambium* of the dicots. Some produce new vascular growth and even increase in width, but through a different process.

3. Cambium cells produce xylem and phloem in the following manner:

   a. The cambium cell divides.
   b. The inner daughter begins to form a xylem cell.
   c. The other daughter divides again, producing two cells.
   d. One cell remains as cambium, but the other, outer cell divides. This division produces a sieve-tube member and a companion cell.
   e. Both xylem and phloem then mature into functional tissues.

The Problem of Differentiation

The plant hormones act as chemical messengers in the differentiation of plant tissues. They work by inducing and inhibiting genes in many complex interactions.

Information from the Study of Plant Cells

1. The concept of *totipotence* states that all cells have all of the genetic information needed to produce an entire organism. Experiments in plant tissue culture have supported this concept.

2. Van Overbeek succeeded in culturing carrots from cells separated from the *embryo.*

3. Steward succeeded in culturing mature phloem cells and producing entire carrots from these.

4. Van Overbeek's and Steward's results support the *totipotency* concept. They also indicate that mature cells can dedifferentiate (become unspecialized) and go on to produce the entire plant.

Review Questions

1. List five plant structures that are capable of *vegetative reproduction.* (pp. 557–558)

2. Distinguish between *stolons, tubers,* and *rhizomes.* (pp. 557–558)

3. List several advantages of *grafting* over growth from seeds. (p. 558)

4. Describe the process of *apomixis* and explain the genetic characteristics of offspring produced this way. (pp. 559–560)

5. Review the evolutionary disadvantages of asexual reproduction. (Essay 21.1)
6. Write a botanical definition of the term *fruit* and give examples that do not seem to fit the common definition. (p. 560)
7. Describe the organization of the *calyx, corolla, androecium,* and *gynoecium.* (pp. 560–561)
8. In what way are the color and fragrance of flowers an adaptive feature? (p. 561)
9. Describe the appearance of wind-pollinated flowers and give examples. (p. 561)
10. What is the evidence for the length of the night being more important than the length of day in flowering? (p. 562)
11. Define the term *photoperiodicity.* (p. 562)
12. Describe an experimental procedure that prevents flowering in *long-day* and in *short-day plants.* (p. 562)
13. Describe an experiment that indicates that the leaves of cockleburs contain the photoreceptors. (p. 563)
14. Relate the effect of red and far-red light on *phytochrome* to what happens during middle-of-the-night flashes. (p. 564)
15. What is florigen, and why haven't its characteristics been described? (p. 565)
16. Describe the structure of an *ovule.* (p. 565)
17. What exactly is the gametophyte generation (male and female) in the flowering plant? (p. 565)
18. Describe the events between *megaspore mother cell* meiosis and development of the *embryo sac.* (p. 565)
19. Describe the events between *microspore mother cell* meiosis and pollen development. (p. 566)
20. Review the events of pollination and fertilization. (pp. 567–568)
21. What structures are present in wheat or corn germ? (p. 569)
22. How does the *cotyledon* in a monocot differ from the one in a dicot? (p. 569)
23. List the major events in the early development of the dicot *embryo.* (p. 569)
24. How does development of the *endosperm* differ in dicots and monocots? (pp. 569, 574)
25. Why is it incorrect to refer to corn kernels as seeds? (p. 575)
26. How do wind-dispersed seeds differ from those that are animal dispersed? (p. 576)
27. What is the best evidence that fleshy, sweet *fruit* evolved along with the animal dispersal agent? (p. 577)
28. What are the usual requirements for *germination?* (p. 577)
29. Describe the emergence of a bean seedling, using the terms *radicle, plumule, hypocotyl,* and *cotyledons.* (p. 577)
30. What is the next step for root cells that have completed elongation? What kinds of cells form? (p. 579)
31. What is the role of the primordial region below the stem apical meristem? (p. 579)
32. What are the *nodes* and *internodes* of a woody stem? (p. 579)
33. Under what conditions do lateral branch buds produce branches? (pp. 579–580)
34. Describe an adaptation in grasses that permits them to survive grazing. (p. 580)
35. Describe the basic changes that occur in the vascular system of a woody stem, proceeding down from the tip. (pp. 580–583)
36. Outline the events in xylem and phloem production by the *vascular cambium.* (p. 583)
37. What does the term *totipotent* mean? (p. 584)
38. Explain how van Overbeek's experiment supports the concept of *totipotency.* (p. 584)
39. Describe Steward's experiment and explain what this added to van Overbeek's results. (p. 584)

Suggested Reading

Albersheim, P. 1975. "The Walls of Growing Plant Cells." *Scientific American,* April. How cellulose fibers are laid down in careful patterns.

Baker, H. G. 1963. "Evolutionary Mechanisms in Pollination Biology." *Science* 139:877. The always fascinating sex lives of plants.

Brill, W. J. 1977. "Biological Nitrogen Fixation." *Scientific American* 236:68.

Epstein, E. 1972. *Mineral Nutrition of Plants: Principles and Perspectives.* John Wiley, New York.

———. 1973. "Roots." *Scientific American,* August.

Esau, K. 1977. *Anatomy of Seed Plants,* 2d ed. Wiley, New York. The authoritative sourcebook by one of the world's most distinguished botanists.

Galston, W. W., and P. J. Davies. 1970. *Control Mechanisms in Plant Development.* Prentice-Hall, Englewood Cliffs, N.J.

Hardy, R. W. F., and V. D. Havelka. 1975. "Nitrogen Fixation Research: A Key to World Food?" *Science* 188:633. Basic research in plant biochemistry has impressive prospects for practical application in the world food crisis.

Heslop-Harrison, J. 1974. "Development, plant." In *The New Encyclopedia Brittanica,* 15th ed., 6:658. Thirteen pages densely packed with information.

Hutchinson, J. 1969. *Evolution and Phylogeny of Flowering Plants.* Academic Press, New York and London.

Jensen, W. A. 1973. "Fertilization in Flowering Plants." *Bioscience* 23:21.

Laetsch, W. M. 1979. *Plants: Basic Concepts in Botany.* Little, Brown and Company, Boston. This is a good, basic, and up-to-date botany text.

Lehner, R., and J. Lehner. 1962. *Folklore and Odysseys of Food and Medicinal Plants.* Tudor, New York. Did you know that crabgrass was deliberately introduced into the United States by Polish immigrants? In the 19th century Poland people made bread from crabgrass seed.

Rutter, A. J. 1972. *Transpiration.* Oxford University Press, Oxford.

Sigurbjörnsson, Björn. 1971. "Induced Mutations in Plants." *Scientific American,* January. The rapid progress in plant breeding that has led to the so-called "green revolution" has depended in part on deliberately caused mutations.

Sporne, K. R. 1971. *The Mysterious Origin of Flowering Plants.* Carolina Biological Supply Company, Burlington, N.C.

Steeves, T. A., and I. M. Sussex. 1972. *Patterns in Plant Development.* Prentice-Hall, Englewood Cliffs, N.J.

Wade, N. 1974. "Green Revolution (1): A Just Technology, Often Unjust in Use." *Science* 186:1093. All fundamental changes, even of the most promising kind, end up hurting someone. Peasants of the third world end up getting it in the neck, it seems, no matter what happens.

Wardlaw, I. F. 1974. "Phloem Transport: Physical, Chemical or Impossible?" *Annual Review of Plant Physiology* 25:515. What *does* make sap flow? Although the answer is still not known, at least some hypotheses can be ruled out. Unfortunately, it sometimes seems that all hypotheses can be ruled out.

Wareing, P. F., and I. D. J. Phillips. 1970. *The Control of Growth and Differentiation in Plants.* Pergamon, New York.

Went, F. W. 1963. *The Plants.* Life Nature Library. Time, Inc., New York. Although it is now getting along in years, this beautifully illustrated volume remains a superbly readable, informative, and intelligent nontextbook approach to botany.

Part Five
Animals

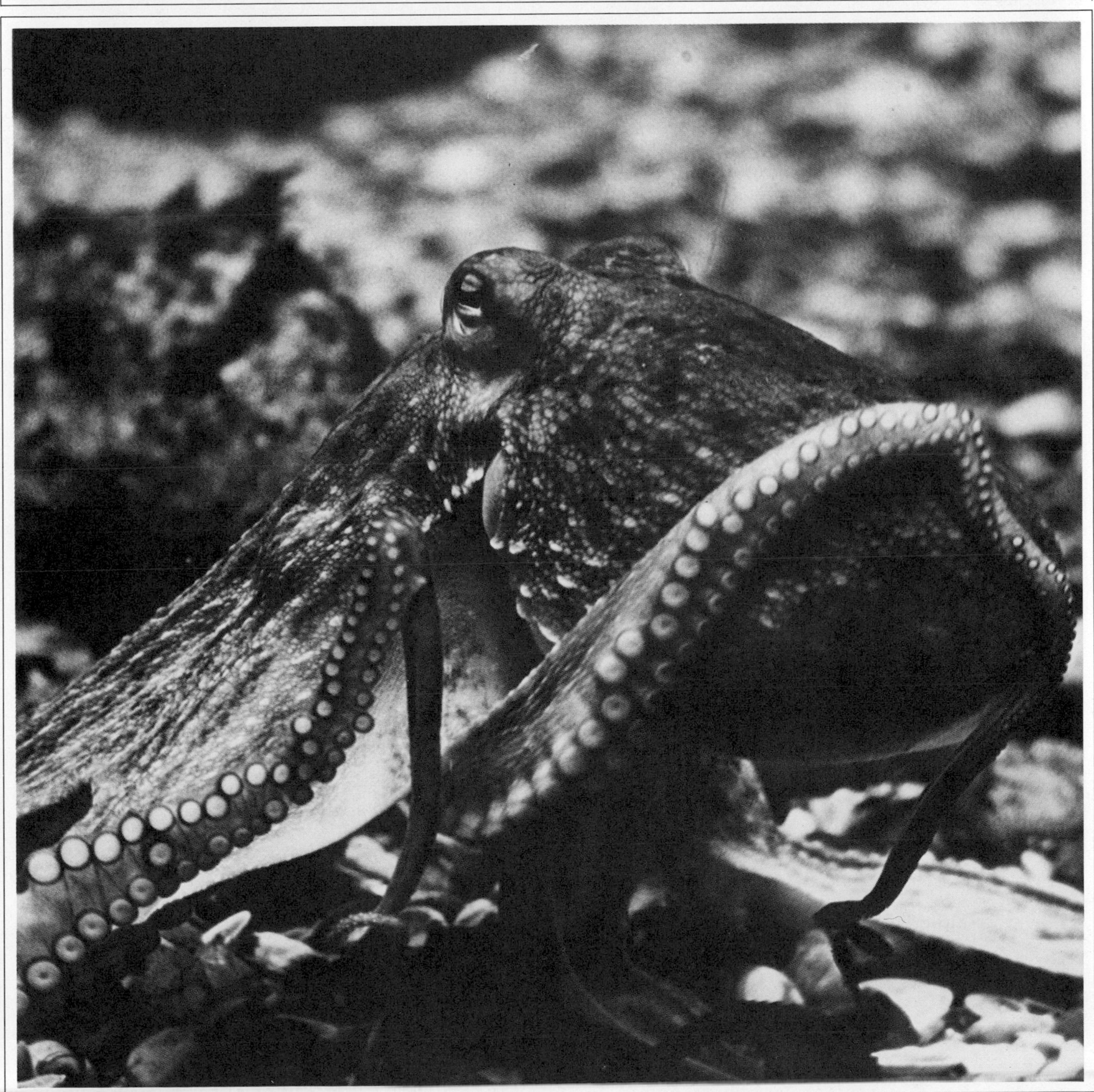

Chapter 22

Animal Diversity - The Nonchordates

If you were driving across the countryside of Kenya and passed a troop of baboons, would it be wise to stop, get out, and walk among them, or are they dangerous? Would you be afraid of wolves if you walked across Isle Royale in Michigan? Why don't people use watchcats instead of watchdogs? Can a horse be trusted to guard the pasture? Should you pick up a centipede? What signals of danger would you look for if you were to wade across a South American river? Painful nips? Loud splashes? Entwining pressure? High-pitched whines?

We live among other animals and so we have an idea of their characteristics. And we know they differ. A dog, we know, may obey out of what is apparently a real affection for us. A horse seemingly obeys because of what might happen if he doesn't. Dogs are smaller than horses and often harbor fleas that horses lack. Most people don't fear dogs, or horses, or fleas, but sometimes it is wiser to fear all of them, but for different reasons. Some people, we know, love dogs or horses, but not many care for fleas and it is even possible to love horses without loving dogs. These are all very different kinds of animals and each has its own fascination for us. You would even gain some respect for the flea if you knew more about its private life.

22.1 Origin and Relationships

What Is an Animal?

If someone asked you what an animal is, you could probably come up with a pretty good definition. But you would probably be wrong just the same. The reason is, animals are so varied that exceptions can be found to almost every rule. Most animals, one would think, move around and have mouths and eat things. If we add "and are multicellular" to that, we've pretty much covered the Animal kingdom. But some animals would virtually shout, "Not me." Nonetheless, animals do all have certain things in common.

All evidence indicates that the animals—the kingdom Metazoa—constitute a good, solid, monophyletic unit (providing we leave out the sponges, which are a special case). Because of their relatedness and basic similarities we can make some general statements about animals. Essentially, we will develop a rather longwinded definition of the group:

1. *Animals are multicellular and eucaryotic.* This criterion takes away the Animal kingdom membership privileges of the protozoa. As you know, biologists who follow the two-kingdom system don't go along with this ouster, and for some good reasons. One reason is that the protozoa are eucaryotic, and another is that most protozoa meet most of the other animal criteria. But then there are good reasons for the five-kingdom division, too, especially since we prefer to devise monophyletic groups when possible (see Figure 13.14).
2. *Animals have cell and tissue differentiation and well-organized systems that carry out different life functions.* Some are better organized than others, of course. But all animals (except perhaps a few types that have lost theirs) have nervous systems, guts, and reproductive systems—no matter how simple. Nearly all have muscles, and most have some form of excretory system. And, as we mentioned, nearly all animals have mouths and can eat things; a few (tapeworms, for instance) merely absorb nutrients through their body covering.
3. *Most animals reproduce sexually, with large, nonmotile eggs and small flagellated sperm.* The majority of species in every animal phylum is capable of sexual reproduction. However, each

phylum also contains *parthenogenetic* species that reproduce asexually. With few exceptions, animals are diploid organisms and all cells except the gametes are diploid. These, then, are the criteria of animals. But you have also undoubtedly noticed the many exceptions in our three-part definition.

Most people are familiar with at least a few animals from each of the major phyla. The most familiar phylum is that of the chordates, which includes people and sharks but also includes invertebrates such as sea squirts and salps. Everyone has seen insects, earthworms, and snails, which are representative arthropods, annelids, and mollusks, respectively. You probably know that your dog has roundworms (aschelminths); and if you live near a rocky coastline, you're likely to know about jellyfish (coelenterates), starfish (echinoderms), sponges (poriferans), and flatworms (platyhelminths), all of which abound in tidepools.

We've just named the nine major animal phyla. Each phylum has certain characteristics that set it apart from the others, and each has a range of diversity within it. There are at least as many rare animal phyla as the common ones listed above—all quite fascinating, if not outright weird. All the animal phyla can be found as fossils in rocks that are at least half a billion years old. The phyla themselves are quite distinct, and there is generally no question of who belongs to what phylum.

We still think that it is helpful in understanding animals to do what we can to reconstruct how the major phyla evolved from one another and from a series of common ancestors. Our plan in this chapter and the next is to describe the distinguishing characteristics of each major phylum and to take a look at the diversity within them. For instance, we'll consider what snails, octopuses, clams, nudibranchs, chitons, abalone, the chambered nautilus, and other mollusks have in common, and how they differ from one another. In passing, we'll comment on their interrelations with other organisms, where they live, and the different ways in which they cope with some of the problems all animals have. And we'll also examine some of the theories of the origins and interrelationships of the different phyla.

Metazoan Origins and Phylogeny

Be prepared for some confusion and, possibly, for working with some fascinating—if difficult—puzzles, because the origins and phylogenies of the major groups are often obscure. We can present some good evidence for quite contradictory hypotheses, and we'll introduce some different hypothetical family trees. The problem is, it's not possible for more than one of these trees to be "true," but it is possible for all of them to be false!

Part of the reason for all this obscurity is that animal evolution appears to have been very fast (as evolution goes) between the time of the first animal and the establishment of all the animal phyla. There's nothing unusual about this; it's an example of rapid adaptive radiation—something that happens again and again in evolution whenever some group gets together a new and particularly favorable array of adaptations. The first metazoan animal species, with its mouth, gut, nerves, multicellular organization, and muscles, may have been a crude thing by today's standards, but it was the crowning evolutionary achievement in the world of its own time and has left many, many descendants. Rapid speciation must have followed, with each group developing its own variations and basic body plan. This great surge of species appeared in the late Pre-Cambrian era. Many hopeful new phyla fell by the wayside. Paleontologists are forever digging up weird beasts from Pre-Cambrian strata. But by the onset of the Cambrian, all the major groups had established their identities—all except the vertebrates, which were the last major group on the scene.

At least, that's what we think happened. All we really know is that there are few fossils from the Pre-Cambrian, and that some of these clearly don't belong to any present phylum. Most of them are rather indistinct worms. The period between the first crude, rare, worm fossils and the rich, extensive, and diverse early Cambrian fossils was about 70 million years—practically overnight in evolutionary terms.

So let's look at the Animal kingdom, paying particular attention to how animals differ and how they are, in important ways, alike. We will undoubtedly come to have a greater appreciation for the term *adaptation* and, hopefully, we will not become too disenchanted with the field of systematics. Let's begin by noting, and then avoiding, the arguments about the animal family tree.

Ernst Haeckel, Jovan Hadzi, and the Animal Tree

Ernst Haeckel was a brilliant 19th century evolutionist who had some ideas about how animals evolved. His genius is reflected in the fact that he left us arguing among ourselves over his ideas and the argument goes on. The theory that he developed has been challenged again and again, but neither his supporters nor detractors have decisively won the argument even as Haeckel's very molecules are wafting on Atlantic breezes and gracing the leaves of Bavarian plants.

Haeckel thought that the animals must have sprung from some primitive flagellates. He tended to believe that the flagellate probably resembled the *Volvox*, since this flagellate seems to bridge the gap between single-celled and multicellular organisms. Haeckel even went a step further. He noted that in its asexual reproduction, the *Volvox* produces a bell of cells which then inverts itself, producing something resembling the early embryos of animals. The caved-in body would also resemble modern coelenterates (Figure 22.1).

Haeckel's opponents argue that there is no fossil evidence to support this theory. (Actually, this criticism holds for all theories of animal origin since invertebrate animals are already abundant in the *earliest* fossil records from the Cambrian, so we have almost no fossil data from rocks more than 500,000 years old.) In other words, we may be looking in vain to fossils to clarify the origin of animals. It is argued, also, that organisms similar to *Volvox* were more likely to be in the line leading to multicellular plants. *Volvox* itself is no longer seriously considered to be the ancestor of anything, but perhaps it serves as something of a living metaphor of what the earliest multicellular life might have been like.

Perhaps animals are descended from the shorter-armed cousins of the flagellates—the ciliates. This idea was proposed by Adam Sedgwick only 25 years after Haeckel suggested his theory, but it was brusquely shoved aside amidst all the shouting. However, in 1963, it was uncovered, dusted off, and revived by Jovan Hadzi of Yugoslavia, who made a few modifications of his own. He preferred ciliates partly because they resembled certain primitive flatworms. The ancestor, he said, was a multinucleate ciliate.

As evidence, Hadzi noted that:

1. Some multinucleate ciliates of today bear a striking similarity to contemporary, marine flatworms whose cellular organization is not complete (Figure 22.2).
2. Both are believed to have evolved in the primitive oceans.
3. Both are entirely heterotrophic, utilizing intracellular digestion of food materials and lacking gut development.
4. Their use of the cilia in locomotion is similar.
5. Sexual reproduction is animallike in both groups. Unlike the haploid *Volvox*, ciliates are diploid until conjugation occurs. Meiosis produces haploid nuclei, which are exchanged. Conjugation also bears a kind of resemblance to the simple reciprocal copulation of flatworms, which like ciliates are bisexual. Some of the simpler ciliates are multinucleate; Hadzi's idea, then, is that the original multicellular animal occurred when a multinucleate ciliate evolved true cells.

So, we still don't know who was more important among the hypothetical ancestor of all the animals, the coelenterate or the flatworm, although the latter appears to be most convincing. Probably neither idea

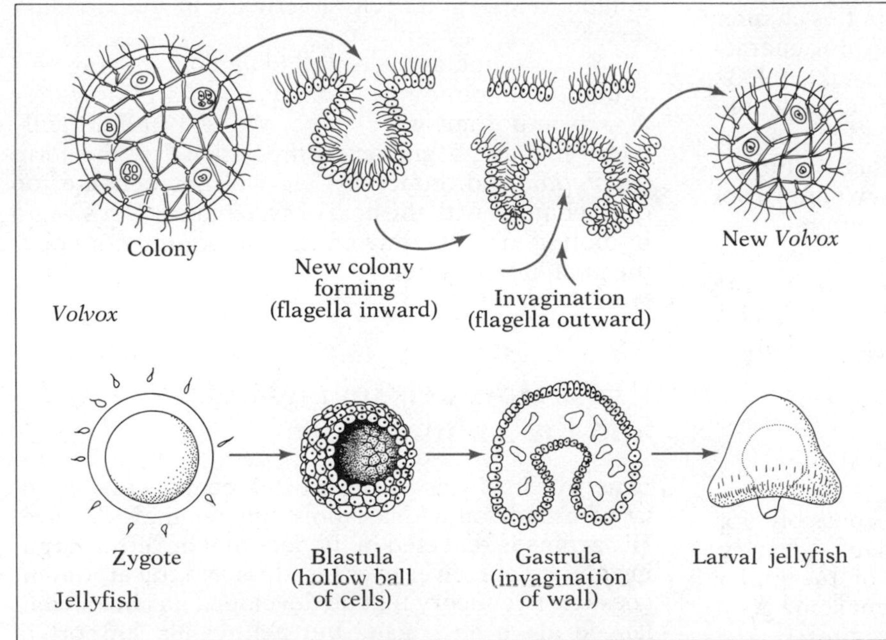

22.1 A comparison of *Volvox* with the coelenterate body plan. Haeckel suggested that the green algal flagellate *Volvox* was ancestral to the animal line. As evidence, he cited the events in their asexual reproduction in which a ball of flagellated cells forms from a colony, as shown here. The flagella are all turned inward until they evert, as the ball turns inside out. Haeckel likened the inversion process to the animal process of *gastrulation.* This process, shown in the coelenterate, includes the inpocketing (or invagination) of the blastula wall, forming a cavity that will be the basis for the adult form. Today, *Volvox* has been quietly discarded as a candidate for reasons discussed in the text, but it is easy to see why Haeckel was captivated by the idea.

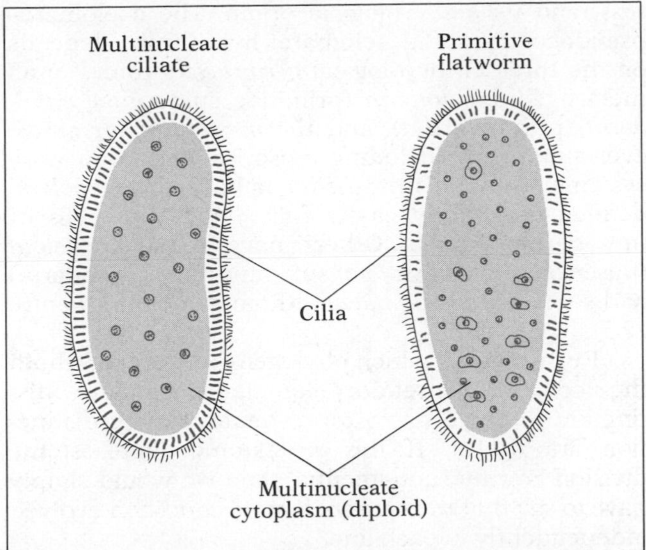

22.2 Ciliate hypothesis. An alternative hypothesis of animal ancestry involves the ciliates. Certain marine, multinucleate ciliates bear a resemblance to primitive, gutless marine flatworms. The comparison goes far beyond obvious physical features and includes methods of getting food, digestion, primitive sexual reproduction through conjugation, and, of course, habitat. Note that both are essentially multinucleate.

makes you swell with pride, but we can expect to focus in on the problem with a bit more confidence as biochemical taxonomy, now in its dawning years, develops into a mature science. At this point, the evidence from a single molecule, *cytochrome c*, indicates that animals are much more closely related to flagellate protozoans than to ciliate protozoans.

Trees and Trees

There is some agreement among systematists regarding the relationships of the various phyla and how the major branches divided. For example, most biologists believe that the sponges (Porifera) are so fundamentally different from any other kind of animal that they are in no evolutionary line to any other group.

Some scientists have proposed the groupings shown in Figure 22.3. Here, the sponges are isolated as a subkingdom, the animals with primitive saclike guts form another category, and all the rest of the animals are divided into two groups: the *protostomes* and *deuterostomes*. These are embryological designa-

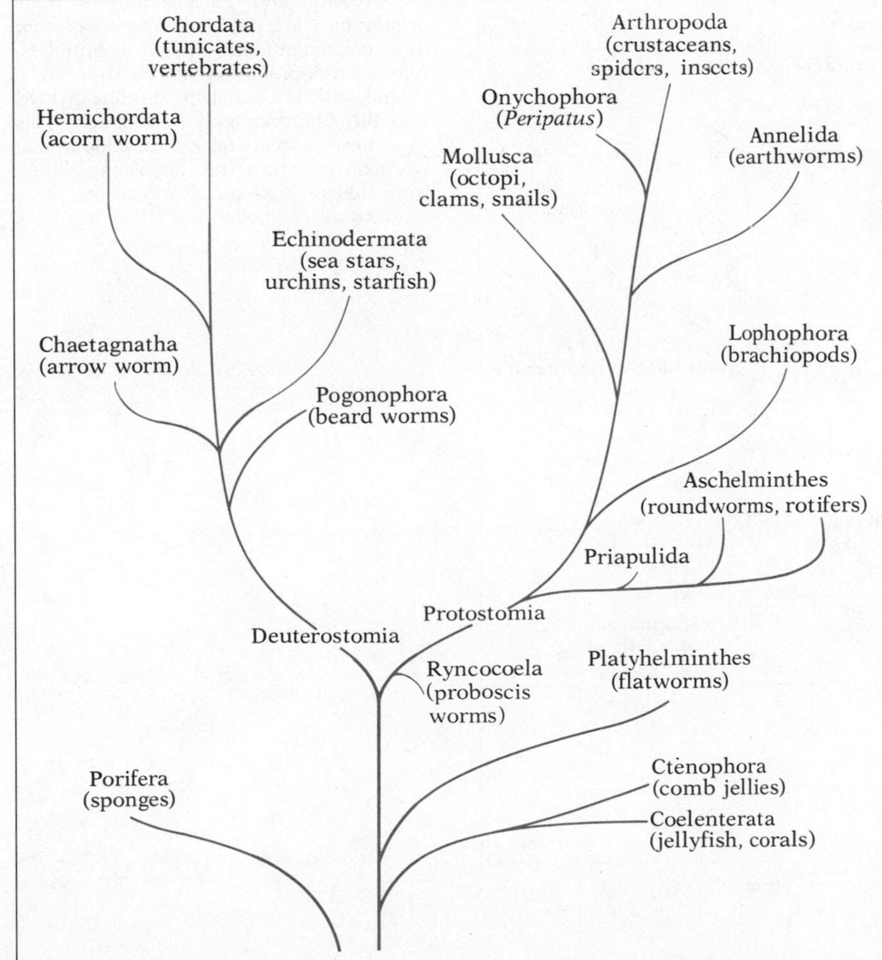

22.3 The protostome–deuterostome hypothesis. One of the preferred phylogenetic trees at present is formed according to the protostome–deuterostome hypothesis. According to the hypothesis, two great divisions arise early in the evolution of animals. The divisions are based on several basic differences in the embryonic development. These differences include cleavage patterns, origin of mesoderm, origin of the coelom, and the way the mouth and anus develop (from which the two names are derived). When all these factors are considered, the tree is divided into lines terminating in the phyla Chordata and Arthropoda. As you can see, greater diversity has occurred in the protostome line, which contains most of the familiar invertebrates.

tions that refer to an early process in which the mass of embryonic cells roll in, forming a *blastopore.* In most of the mollusks, annelids, and arthropods, this indentation will develop into a mouth. Later, another inpocketing will form the anus. Thus, this group is called the Protostomia ("first the mouth"). In echinoderms (such as starfish) and chordates (including the vertebrates), the blastopore generally forms the anus. The opening to the mouth forms later. Therefore, they are called the Deuterostomia ("second the mouth"). This sort of partitioning is favored by some, but not all, biologists.

A third possible scheme is presented in Figure 22.4. It indicates that, for example, the chordates and annelids are more closely related to one another (splitting off later in geologic time) than does the protostome–deuterostome classification. This tree shows the major division of animals to be based on the type of body cavity (or *coelom*) they possess, and its embryological origin. The coelomate–pseudocoelomate–acoelomate hypothesis depends on the three embryological *germ layers* found in all metazoa: the *ectoderm* (primitive epidermis), *endoderm* (primitive gut), and the *mesoderm* (primitive everything else, including muscular and circulatory systems). Animals in some phyla have a *true coelom,* defined as a body cavity fully lined with cells of mesodermal origin. Others have a *pseudocoelom* (false coelom; a cavity not surrounded by mesoderm) and some are *acoelomate* (without a coelom) (Figure 22.5).

Regardless of which phylogeny one accepts, both the coelomate–pseudocoelomate–acoelomate distinction and the protostome–deuterostome distinction are useful. If the protostome–deuterostome division is truly fundamental, then we would simply have to say that the true coelomate condition evolved independently in each line.

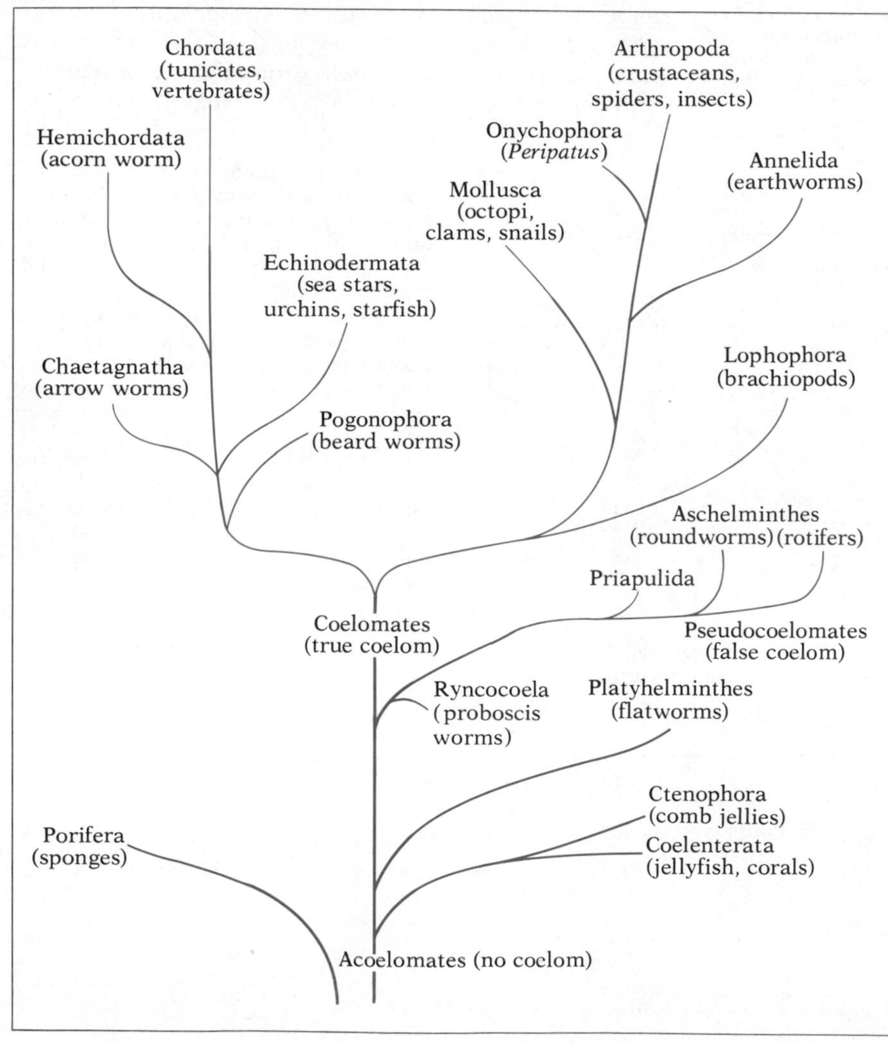

22.4 The coelomate–pseudocoelomate–acoelomate hypothesis. In this phylogenetic scheme, the emphasis is on the *coelom.* According to this hypothesis, the most primitive animals are *acoelomates,* with no body cavity. These include two important phyla, Coelenterata and Platyhelminthes. One evolutionary line produced a *pseudocoelom* (false coelom) and included Aschelminthes. A final development was the *coelomate* line, animals with a true coelom. This line divided into the Chordata and Arthropoda. This hypothesis emphasizes one developmental feature. Note that this phylogeny differs from that of Figure 22.3 only in the place at which the echinoderm–chordate line joins the rest of the tree.

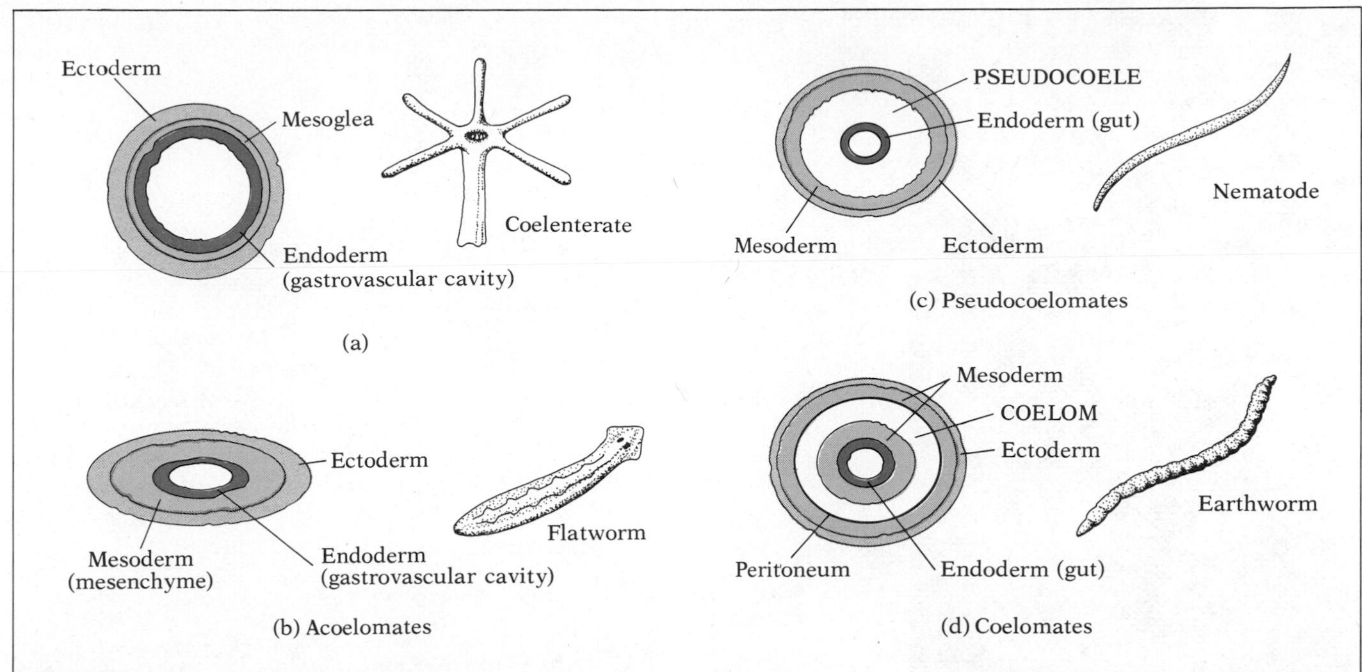

22.5 Development of a pseudocoelom and a coelom. (a) In coelenterates, this region is comprised of mesoglea, a jellylike matrix containing ameboid mesenchyme cells. (b) Acoelomates have a well-developed muscular mesoderm layer, but no coelom. (c) The pseudocoelomates have true cavities filled with fluids, but these cavities do not contain the specialized mesodermal linings present in the true coelom, and the gut lacks a covering of mesodermal tissue. (d) In the true coelom, the body cavity is generally hollow and is lined by the mesodermally derived *peritoneum.* The lining forms a supportive structure for the digestive system. Some organs, notably the kidney, are actually outside the peritoneum. Certain species of insects lose their coelom as they develop, replacing it with a body cavity called the *hemocoel,* in which blood circulates.

22.2 Animals Without a Coelom

Phylum Porifera: The Sponges

Sponges represent an offshoot from the main line of animal evolution. There is no doubt that sponges are animals, since they are multicellular heterotrophs, but their organization is extremely simple. Most sponges are marine, living in shallow ocean waters throughout the world, although a few live in fresh water. The adults are sessile, living quietly attached to the bottom in clear water, without apparent movement. In the simpler sponges, the body is vaselike, with a large central cavity opening (*osculum*) at the top (Figure 22.6). The thin body wall is riddled with porelike cells through which currents of water pass, created by flagellated collar cells, the *choanocytes,* lining the cavity. The incoming water brings with it fresh supplies of oxygen and food. Sponges feed on microscopic organisms and bits of organic matter that become trapped in mucous secretions of the collar cells. These trapped particles pass down the collar where they are phagocytized by the cytoplasm and digested in vacuoles. The water currents that brought in the particles then pass out through the osculum, laden with carbon dioxide and other cellular wastes.

In more complex sponges, the body wall is thicker with more distinct areas. The pores open into canal systems that widen frequently into chambers lined with collar cells before leading into a reduced central cavity.

Sponges maintain their body shape with spicules or fibers that are scattered in the body wall. These are produced by wandering ameboid cells (amebocytes), and this composition differs in each of the three classes of sponges. The calcareous sponges use calcium carbonate, while the beautiful glassy sponges use silicon dioxide (glass). A third material, *spongin,* is a fibrous protein found in proteinaceous sponges (previously used as bath sponges).

Sponges are usually *hermaphroditic,* with each individual producing both sperm and eggs. Sperm travel between sponges is aided by the currents created by the collar cells. The egg cells are often located just beneath the collar cells, where the sperm can easily penetrate. A fertilized egg becomes a swimming larva, which, after a brief period, settles to the bottom and begins to sit out its adult life. Some sponges can

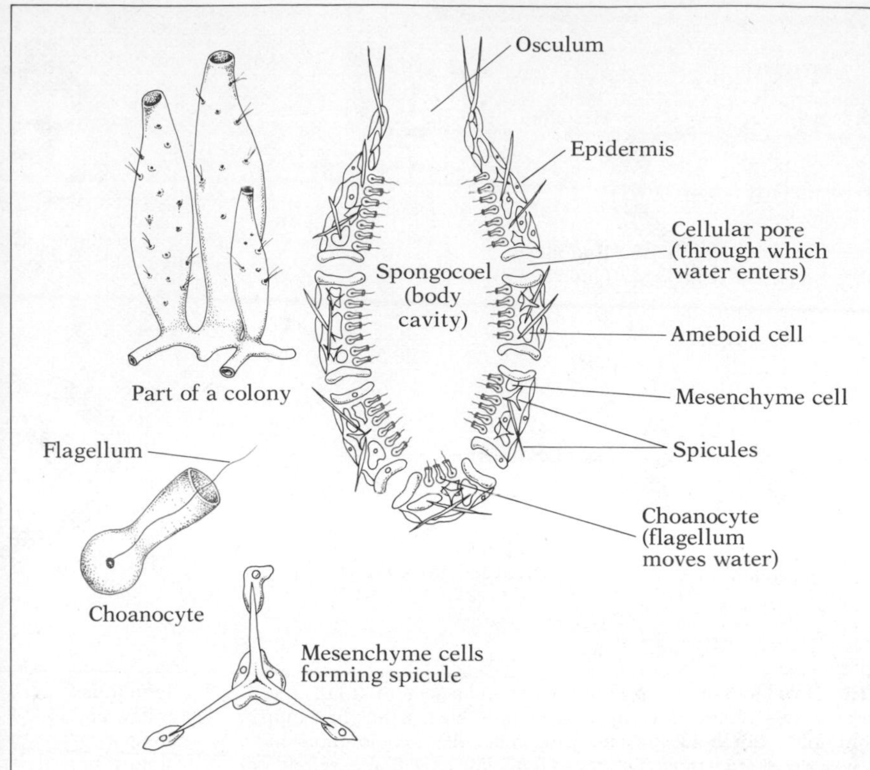

22.6 A simple calcareous sponge is essentially an organized accumulation of cells surrounding a central cavity. There are only a few cell types, the most specialized of which are the *choanocytes* (collar cells). Wandering ameboid cells perform some functions of complex systems in other animals. The structural framework is a system of *spicules* in the simpler sponges and a network of glass or protein fibers in those that are complex. The movement of flagella in the choanocytes provides the currents needed for feeding, exchanging gases, and releasing clouds of sperm cells during reproduction.

also reproduce asexually, releasing *gemmules*, which are clusters of motile, flagellated cells.

As adults, their ability to respond to the environment is extremely limited and is similar, in fact, to the colonial flagellates. There is no accumulation of nerve cells that could be considered a central nervous system. At best, sponges are organized at a tissue level, which means that functions are carried out by groups of similar cells with little coordination of activity among the tissues. The best evidence for this simple kind of organization is the ability of sponges to regenerate from small pieces of the individual and for the sponge body to reaggregate (crawl back together) when the cells are artificially separated. In more complex systems, this wouldn't be possible.

Phylum Coelenterata

The *coelenterates*, or Cnidaria, as they are also known, include the *hydra, jellyfish, corals,* and *anemones*. Most have tentacles armed with stinging cells called *cnidoblasts*. The coelenterates are very delicate and extremely thin-walled animals, composed essentially of an outer and inner layer of cells. Sandwiched between these layers is acellular, jellylike matter. A few wandering ameboid cells represent the only mesoderm that coelenterates can claim. The outer cells are mainly protective, while the inner cells form a saclike gut called the *gastrovascular cavity* because it serves both digestive and circulatory functions. A simple opening serves as both mouth and anus—or, to put it another way, coelenterates have a mouth but no anus and must spit out the undigested remnants of whatever they swallow.

Coelenterates are unusual among animals in that their bodies are organized along a radial principal rather than bilaterally as we see in most animals. Radial symmetry appears again in the echinoderms, but only in the adults. Unlike radially symmetrical sponges, however, coelenterates have nerves and muscles and are capable of coordinated movement. The nerves form a loose nerve net without a centralized brain. Both nerves and muscles are specialized epidermal cells.

The phylum contains two basic body patterns, called the *polyp* and the *medusa*. Polyps are sedentary, usually attached to rocks. The medusas can swim and are commonly known as jellyfish. In some coelenterate species, both polyp and medusa stages exist, appearing in an unusual alternation of generations. (See Figures 22.7 and 22.8.) Don't confuse this with alternation of generations in plants, however, since both polyp and medusa are diploid, true to their animal mode of life.

Class Hydrozoa

Most *hydrozoans* are found as colonies of polyps with specialization for feeding and reproduction among polyp members (Figure 22.7). The colonies grow asexually by simply budding more polyps. Sexual reproduction begins when minute medusae are produced. The free-swimming medusae are dioecious (of one sex or the other) and they release gametes which then join in fertilization. Hydrozoans, like all coelenterates, produce a planula larva, a swimming, ciliated form that eventually settles to the bottom, attaches, and there begins its strange metamorphosis into a colony of polyps.

An aberrant, but familiar hydrozoan is the tiny, free-living *Hydra*. *Hydra* and a closely related genus *Chlorhydra* are unusual in several ways, one being that they are freshwater animals in an almost exclusively marine phylum. There is no medusa stage and no planula. The *Hydra* polyp reproduces by budding new polyps. As in so many other minute freshwater forms, sexual reproduction occurs only under harsh conditions, and the zygote is encased in a toughened, resistant case. When conditions improve, the shelled zygote germinates into a tiny polyp. *Chlorohydra* is dependent for nutrition on an intracellular photosynthetic symbiont, a green alga.

Class Scyphozoa

In the scyphozoans, unlike the hydrozoan, jellyfish are the dominant generation. The polyps are entirely absent in some species and greatly reduced in the remainder. Most medusae have a bell-shaped body with the mouth centrally located on the underside, surrounded by tentacles. For the most part, they are

22.7 Life cycle of *Obelia*, a colonial marine hydrozoan. Each colony consists of a variable number of polyps interconnected by a common tubular cavity through which food is distributed. *Obelia* lives within a protective sheath of nonliving secreted material. The individuals within the colony are of two types: feeding polyps and reproductive polyps. The feeding polyps capture food in typical coelenterate fashion, using tentacles armed with nematocysts. The reproductive polyps form swimming medusae, which are released as they mature. This is an asexual process, differing from alternation of generations in plants in that there is no haploid gametophyte generation. In these animals, only gametes are haploid. When the small, free-swimming medusae mature, egg or sperm cells are produced meiotically and released into the water where fertilization occurs. The zygote develops into a planula, which is also free-swimming for a time. Eventually, this ciliated larva settles to the bottom of its aquatic environment, attaches to the surface, and begins to form a new colony. This life cycle differs from the typical scyphozoan cycle in that the polyp stage is large, long-lived, and colonial, and the medusa stage is small and brief.

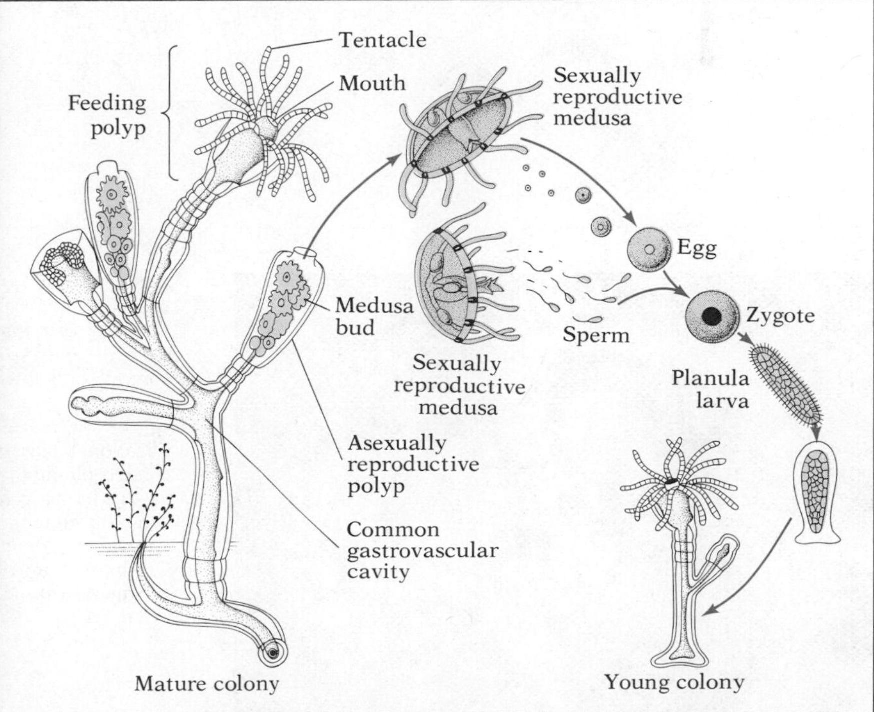

drifters, but some active swimming is possible by contraction of muscle fibers around the bell.

Sexes are separate in the jellyfish. Sperm, produced in the testes of males, swim out through the mouth and into the gastrovascular cavity of the female, penetrating the ovary where fertilization occurs. A swimming planula larva develops and settles to the ocean floor where a unique transition occurs. Usually hanging down from underside a rock ledge, the attached polyp begins to bud off miniature jellyfish in assembly line fashion (Figure 22.8). These mature into adults. Some of the Scyphozoa can form rather large structures, up to 2 m in diameter. These enormous jellyfish are indeed dramatic sights when seen drifting in clear, blue waters, far out at sea.

Some jellyfish, as you may know, can sting. The large Portuguese man-of-war (*Physalia*) can capture fishes of considerable size by stinging them senseless. A fish, perhaps seeking food or refuge in the dense "foliage" of the *Physalia*'s tentacles, is at once paralyzed by the venom of thousands of nematocysts as it brushes past the graceful, often brightly colored lures. The fish becomes totally ensnared and is then subjected to digestive enzymes that prepare its body for absorption into the animal's multiple feeding structures.

Class Anthozoa

The Anthozoa—the *corals* and *anemones*—are in some ways opposite to the Scyphozoa. They exist only as polyps. We mentioned earlier that polyps can reproduce asexually by simply budding new polyps. In their sexual reproduction, female and male polyps produce eggs or sperm separately and fertilization occurs in the water. The zygotes develop into swimming planula larvae that develop directly into more anthozoans (Figure 22.9).

The most familiar anthozoans in temperate waters are the heavy-bodied sea anemones. Corals are colonial anthozoans that secrete surrounding walls

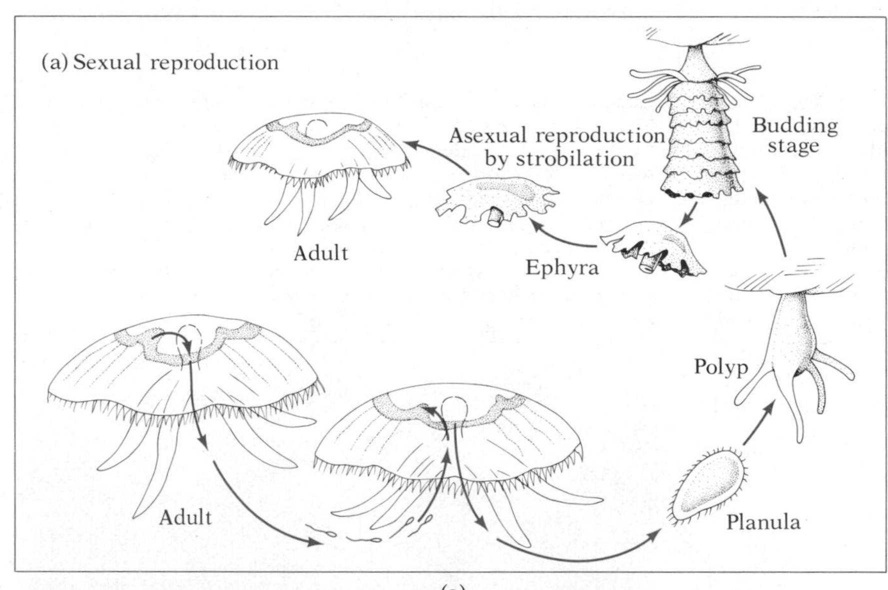

(a)

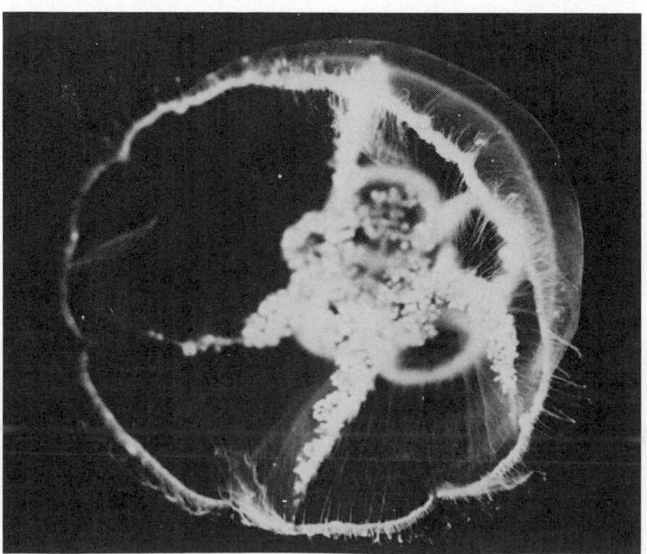

(b)

22.8 Jellyfish are free-swimming coelenterates with a highly reduced polyp stage. Most are relatively small, a few inches in diameter, but many are much larger. One North Atlantic species *Cyanea capillata* reaches 2 m in diameter, with tentacles trailing some 35 m below, and can weigh up to a (metric) ton.

(a) In its life cycle, the adult scyphozoan jellyfish produces egg or sperm cells. The sperm are shed into the water and fertilization is internal. The zygote is released and develops into ciliated planula larva that attaches as a *polyp* to the underside of an overhanging rock. There it feeds and develops into a budding stage, producing many new jellyfish through an asexual process known as *strobilation.* Each individual thus produced (an *ephyra*) will mature into an adult jellyfish. In the scyphozoan life cycle, the medusa stage (b) is dominant, and the polyp stage is small and brief.

(a)

(b)

22.9 Sea anemones and corals are members of the class, Anthozoa. (a) Corals are colonial, living in extensive limestone coverings that they secrete. Their secreted skeletons often produce extensive reefs and atolls. Corals are polyp forms, similar to the anemones, but generally much smaller. (b) Anemones are more solitary, although they are often seen in clusters, attached by their pedal disks to the substrate. They can move about, but most of the time remain attached to one place. In their feeding posture, they are extended, but when disturbed, they quickly retract into a short, rounded body.

of limestone (calcium carbonate). The cumulative effects of their secretions over millions of years have produced the barrier reefs that protect many islands in the Pacific, and often have produced the islands themselves (coral atolls). Although all corals can feed with tentacles, most tropical corals depend on photosynthetic, intracellular, symbiotic dinoflagellates for most of their caloric requirements.

Phylum Ctenophora

The ctenophores ("comb carriers") appear to be closely related to the coelenterates. They are also known as *comb jellies* or *sea walnuts* (Figure 22.10). Like the coelenterates, comb jellies are more or less radially symmetrical; they have a gastrovascular cavity primarily of two cell layers separated by a jellylike mesoglea. However, the comb jellies lack nematocysts. They are characterized by fused cilia that form eight rows or *combs*. The undulation of these combs push the animals along. The Ctenophora capture prey by the use of tentacles with "glue cells." Their sticky secretions capture plankton (small drifting invertebrates) and (to the dismay of oystermen), oyster larvae. The Ctenophora are well-known for their intriguing luminescence. One species, *Cestus veneris*, or Venus' girdle, has a ribbonlike glowing body nearly a meter long.

22.10 These delicate comb jellies are propelled through the water by rows of cilia found along their *combs*. The combs are the dense vertical ridges around the body wall. The central pouchlike object is the stomach or gastrovascular cavity. Comb jellies feed by snaring small prey in gluey secretions from their tentacles (not seen here). The tentacles can trail along behind the animal or may be retracted into sheaths.

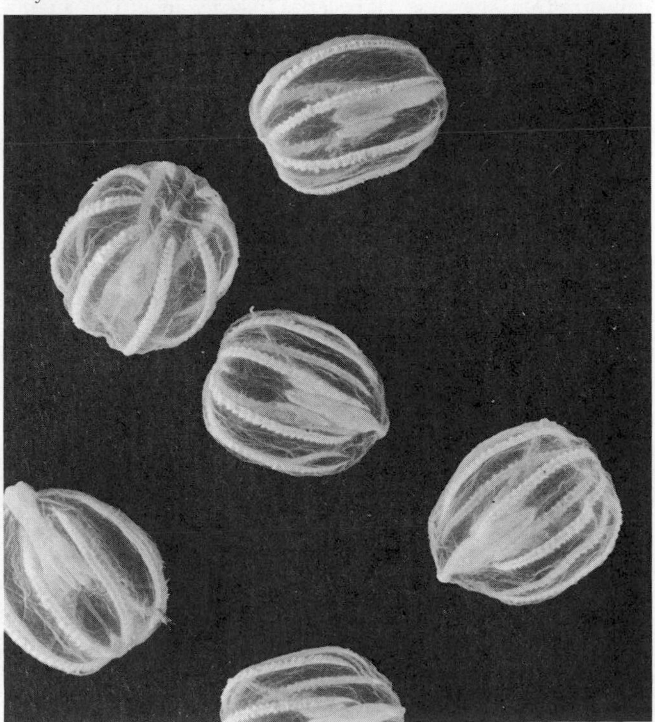

Phylum Platyhelminthes: The Flatworms

The Platyhelminthes are bilateral throughout life. The phylum includes one class of free-living types and two classes of parasites, the flukes and the tapeworms. Some are highly complex, with fully developed organs and organ systems. They have three well-developed germ layers: *ectoderm, endoderm,* and *mesoderm*. But in spite of the presence of mesoderm, the flatworms lack a coelom or any body cavity other than a coelenteratelike gastrovascular cavity.

As bilateral animals, flatworms have definite anterior and posterior ends orientating to their environment. It follows that if an animal has a right and left side, it is also likely to have a head and a tail end. Their nervous systems are not well-developed, but their nerves are concentrated at the anterior, or head, end, as with most animals. This, of course, means, in an evolutionary sense, that the leading end will be loaded with sensory apparatus and thus the animal can determine what kind of environment it is heading into. It wouldn't do to have the senses located at the other end.

Flatworms, like coelenterates, lack special respiratory and circulatory structures. Lacking a well-defined circulatory system, they depend upon diffusion for their oxygen needs, and their flattened structure is one way to keep all internal cells close to the surface. The digestive system is also incomplete (pouchlike, rather than tubular), with a highly branched cavity. But the flatworms do have gonads, an excretory system, crude eyes (in some species) and mesodermally derived muscle layers.

Class Turbellaria

The free-living flatworms are best represented by *Dugesia*, better known as *planaria* (Figure 22.11). These animals have been the subject of anatomical, developmental, and behavioral studies for many years. We will describe their digestive, reproductive, osmoregulatory, and nervous systems in more detail in the coming chapters. Planaria are important here because they represent a phylum that has arrived at an organ system level of development, yet they definitely retain some primitive characteristics, such as the gastrovascular cavity.

Most turbellarians are marine, but others, like planaria, are freshwater forms. Planarians move in a curious sort of crawl. As they travel along the bottom they secrete a layer of mucus underneath them and then beat their way through it with the cilia that cover their ventral surface. They are predators and scavengers, feeding through a protrusible, muscular pharynx, while pinning the unlucky victim in the slimy mucous layer below. Food enters the highly branched gastrovascular cavity where digestion begins. Bits of food are phagocytized by cells

22.11 (a) *Dugesia,* or planaria, as it is also known, is a free-living flatworm from the class Turbellaria. *Dugesia,* about half a centimeter long, is easily seen with the unaided eye. Its long and flattened body has a definite head which is marked by two light-sensitive areas (*ocelli*). Its underside is ciliated, and the cilia provide locomotion as it steers by muscular movement in the body. It feeds by the use of a protrusible pharynx. These are freshwater animals, found in quiet, clear water where they feed by scavenging on dead animals. (b) Planaria, though simple, are capable of a rather wide range of behavior.

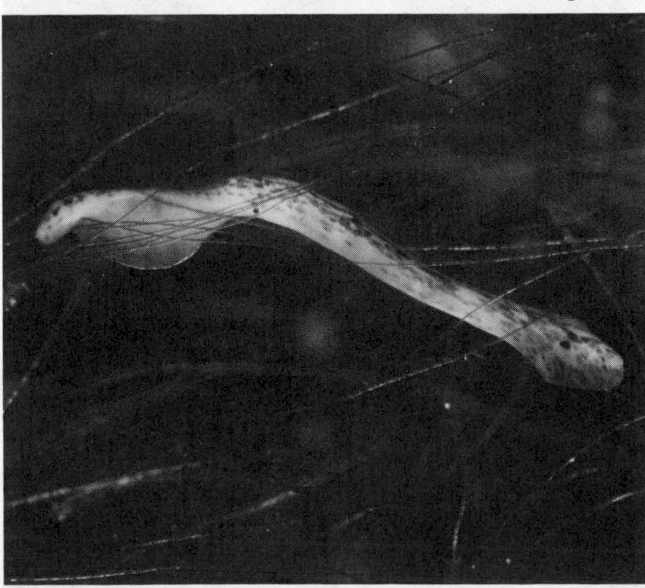

(a)

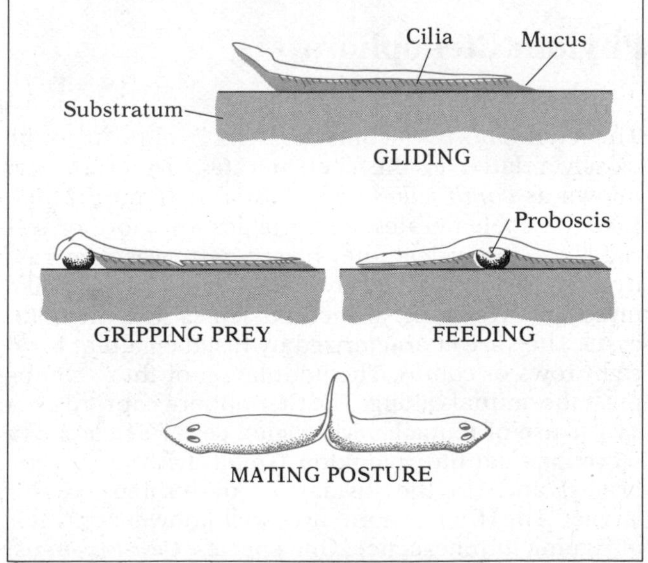

(b)

lining the cavity and digestion is completed intracellularly in vacuoles.

Some planaria can reproduce asexually by simply pinching in half just behind the pharynx, each half regenerating the missing structures. They have tremendous regenerative power and can recover from drastic experimental surgery as long as some critical portion of the body remains.

If some of their systems are primitive, others are quite advanced. One advancement is in the reproductive system. Each planarian has well-defined ovaries and testes. Since they are *hermaphrodites*, each has both a penis and a vagina, but they do not generally self-fertilize. Copulation is reciprocal, with simultaneous penetration and exchange of sperm.

Class Trematoda

The Trematoda, or *flukes*, are *parasites* with the features of the planaria. However, the gastrovascular cavity is divided into two parts and the flukes are not ciliated. They are equipped with large suckers, in front and on the bottom side, with which they attach to their host's tissues. Their epidermal layer of tissues is resistant to digestive fluids.

Flukes invade many kinds of animals and make their home in various parts of the host, including the gills, lungs, liver, and intestines of fish. One species has even found its niche in the urinary bladder of frogs. They also attack humans, you will be delighted to learn. One human fluke, common to Asia, is a giant over fifteen centimeters (six inches) long.

An important fluke in the Nile Valley of Egypt today is the human blood fluke, *Schistosoma mansoni*, which has recently become widespread in the area. The extensive irrigation canals fed by the Aswan Dam have allowed aquatic snails to flourish. The snails serve as an intermediate host to the *Schistosoma mansoni*, which causes the debilitating disease *schistosomiasis*. The parasite enters the body right through the skin when people wade in the water. It is estimated that well over 60% of the population in the delta is now infested with the blood fluke. Schistosomiasis is characterized by a gradual weakening of the host to the point that other diseases easily cause death. It is also a common disease in other tropical regions around the world.

Class Cestoda

The Cestoda, or *tapeworms*, are structurally different from other flatworms. They have no digestive cavity and are surrounded by a thick cuticle, which is resistant to digestive enzymes. This is all appropriate to their life style, since they live amidst powerful digestive juices.

Typically, the sexually mature tapeworm consists of an extremely small rounded *scolex*, or head, equipped with suckers and hooks for attaching to the host's intestinal wall. Extending from this is a long row of segments known as *proglottids*. (This type of segmentation is a form of asexual reproduction by budding, and is not to be confused with true segmentation as seen later in the annelids.) The proglottids are produced continuously from the neck region, and they mature, swelling with eggs, as they are moved back by the growing chain. Each proglottid contains both ovaries and testes (Figure 22.12), and

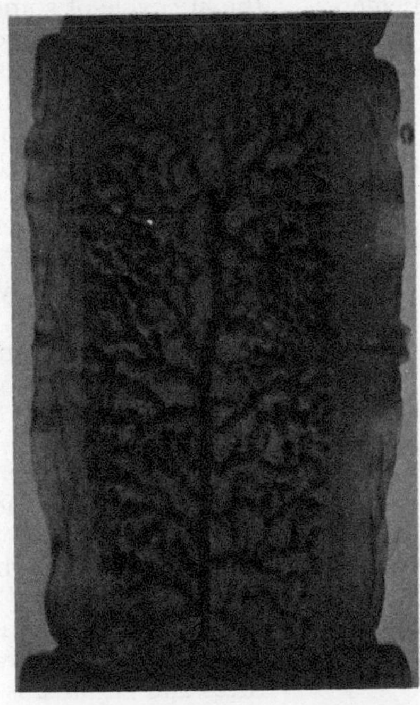
(a)

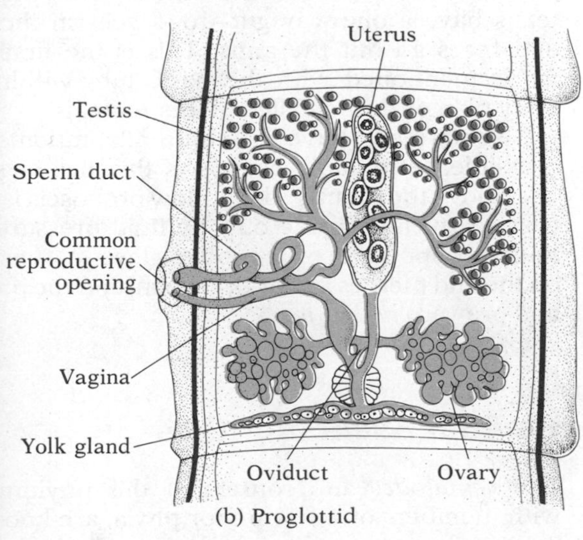

(b) Proglottid

22.12 Proglottids of the pork tapeworm *Taenia solium.* (a) The adult tapeworm consists of a scolex and a large number of proglottids. The proglottids are reproductive structures containing both ovaries and testes. Fertilization occurs in the host. It is internal, occurring either between different worms, or between proglottids of the same worm as they fold back and forth against each other. (b) The proglottid shown here is immature, but the ovary and testis, as well as the vagina and sperm duct, are easily visible. When the proglottid is fully mature, nearly all details are obliterated by the egg-filled uterus.

self-fertilization is common in tapeworms. When they reach the end of the tapeworm, mature proglottids containing thousands of fertilized eggs break away from the organism and pass out with the host's feces. The fertilized eggs must then lie about until they are ingested by the next host. Most tapeworms have one, or even two *intermediate hosts* (hosts of a different species in which the mature sexual forms of the parasite do not occur). If such an intermediate host is involved, an egg will be transformed into a tunneling creature that bores its way into the host's tissues, forms an encapsulated, protective cyst, develops into a bladderworm, and then lies dormant. Both pigs and cattle serve as intermediate hosts to human tapeworms. Uncooked pork or beef, then, may serve to carry the parasite to its primary host—humans—where the cyst will open and a tapeworm will then set up housekeeping. The cycle is complete (Figure 22.13).

Phylum Nemertinea: The Nemerteans, or Proboscis Worms

You quite likely have never seen a nemertean worm. Although they aren't really rare, they aren't headline-makers either. You may be relieved to know that members of this phylum won't burrow into your soft tissues: they are free-living marine predators. Nemerteans don't seem to be closely related to any other group. They have some characteristics in common with flatworms, other characteristics that are like those of most other animal phyla, and a few that are theirs alone. Like flatworms, nemerteans are acoelomate: their thin, ribbonlike bodies are solid with parenchymal tissue. But unlike flatworms, the nemerteans have a one-way gut—food goes in the mouth and feces go out the anus. This is the first animal we've mentioned with the basic tube-within-a-tube body plan of the so-called higher animals.

Nemerteans have a unique adaptation: a long, eversible, retractable proboscis that lies in a cavity just above the mouth. The hollow proboscis is rapidly everted (turned inside out) by fluid pressure, like a finger popped out of a rubber glove. A sharp stylet on the end pierces a prey organism, and the proboscis and prey are reeled in.

Phylum Aschelminthes

The *nematodes* and rotifers of this phylum, along with members of a few minor phyla, are known also as pseudocoelomate protostomes. The nematodes, rotifers, and several other groups are different enough that they are often placed in separate phyla of their own, but because of a number of structural similarities these pseudocoelomates can be organized under one phylum, Aschelminthes. We confess to going with the lumpers in this case. The pseudocoelom differs from the coelom in development rather than appearance. A true coelom is a body cavity between gut and body wall, lined with cells of mesodermal origin. A cavity exists in pseudocoelomate phyla, but its lining consists of cells of mesodermal and endodermal origin (instead of being entirely mesodermal). The gut, usually only one cell layer thick, lies free in the cavity, is not covered by a layer of mesoderm and hence lacks the muscular intestine found in coelomate animals. This seemingly minute point is important phylogenetically. You are probably getting the idea by our emphasis on embryonic details, that phylogeny cannot always be determined by comparing adult forms. The embryonic patterns are much more informative in most cases. Some of the developmental details from which we have derived our phylogeny are quite complex and beyond the level of our discussion, but we have included a few so you can see how scientists attempt to classify organisms.

Another important phylum characteristic of the Aschelminthes is the presence of a complete digestive tract (though it lacks muscular walls) with an anterior mouth and a posterior anus. Both rotifers and nematodes have syncytial tissues and tough external *cuticles*.

Class Nematoda

The nematodes are also called *roundworms* and they include some horrendous parasites. Simply reciting the life cycle of some of them can cause nightmares.

Roundworms are slender, cylindrical, and usually tapered at both ends. They are bilaterally symmetrical, although from outside appearances, they lack a distinctive head. Most are surrounded by a resistant but elastic cuticle. They move by flexing longitudinal muscles on one side or the other, producing an undulating motion. Sexes are separate. Nematodes are usually easy to recognize, since they all look like glistening, smooth worms, some simply being larger than others.

The parasitic roundworms have a muscular *pharynx* (throat) and strong lips. Thus, they suck in digested food in the host's intestine and blood and cell fluids from the intestinal wall.

Different nematodes thrive in nearly every conceivable moist or aquatic habitat, from the soil of flower gardens to the world's oceans. One has been found only in German beer mats (coasters). One researcher—amazed by the ubiquity of these beasts—said that if everything in the world were to disappear except nematodes, vague suggestions of all the structures and organisms of the earth would be represen-

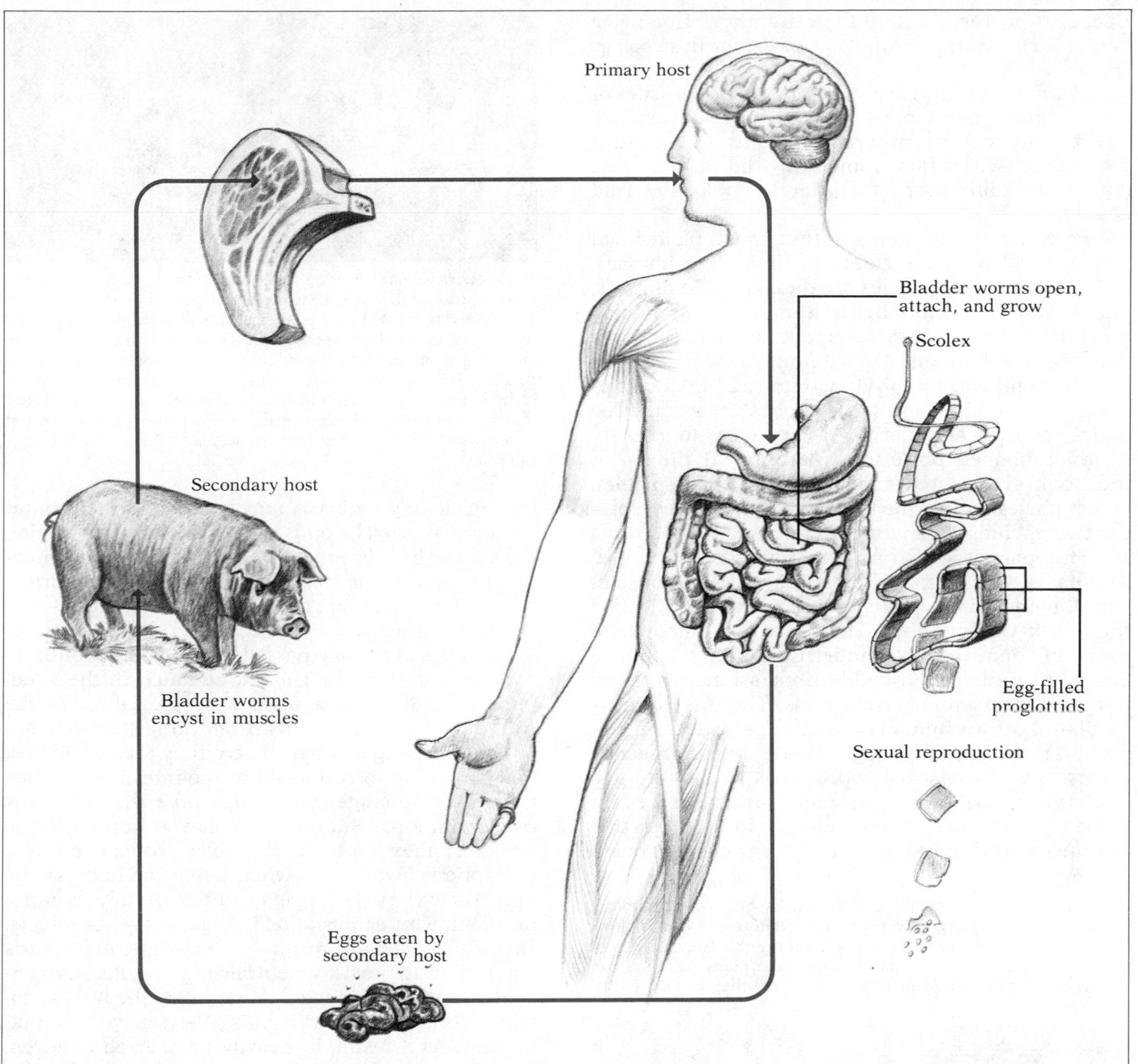

22.13 Pork tapeworm life cycle. Humans, the primary hosts of the pork tapeworm (*Taenia solium*), become infested when raw or partially cooked pork is eaten. The worms are encysted as *bladder worms* in the pig's muscles. Once in the human intestine, the cysts rupture, the head of the bladder worm everts (turns inside-out), and the juvenile worm attaches. Following this, the worms grow rapidly, producing numerous proglottids. Eventually, the mature proglottids (full of eggs) are shed with the host's feces. If they are eaten by another pig, the eggs will hatch in the intestine and find their way to the secondary host's muscles where they again encyst. It is also possible for the eggs to hatch in their human host and become encysted in muscle tissue there. This is usually not a serious problem, but worms have been known to encyst in the eyes and brain, creating serious vision problems and sometimes insanity. There is a beef tapeworm with a similar life cycle and similar effects on humans.

ted by suspended nematodes, and one would still be able to identify animals, ponds, rocks, plants, and buildings by the types of these tiny worms found where the earth once stood!

Most of the half million named species of nematodes are free-living, and many are predators in a microscopic food web. Winnowing their way among moist soil particles, they feed upon protozoans, rotifers, small earthworms, and each other. Some still their prey with a paralyzing saliva. Others make use of a tiny *blade* which impales the hapless prey while the sucking lips and pumping pharynx drain its juices.

Because of their astounding number, free-living nematodes are an essential part of the earth's ecology (Figure 22.14).

Many nematodes are very important parasites of plants. Plant nematodes may have a devastating effect on agriculture, and require constant control at great expense. In fact, sometimes the farmer must completely shift crops in order to make the land usable.

Parasitic roundworms infest most plants and probably all of the vertebrates, including humans. In fact, at least ten species are dangerous to humans, and perhaps fifty more live in humans without doing great harm. There are three types that are particularly dangerous to humans. *Ascaris lumbricoides* is a giant among members of its phylum (Figure 22.15). Females often exceed 20–35 cm (8–14 inches) in length. The males, shorter and narrower, are easy to identify by their hooked posterior extremity. At the tip of the hook are located a pair of bristlelike spicules, which it uses to hold the female's genital pore (reproductive opening) open during copulation.

The parasitologist Asa C. Chandler described *Ascaris* as "one of man's most faithful and constant companions since time immemorial . . . ." (We thought it was the dog!) The parasite is uncommon today in countries with modern plumbing, although cases of infestation in children occasionally pop up in areas of unsanitary conditions. There is no intermediate host, so humans pass the parasite along directly, not only infecting others, but reinfecting themselves. The *Ascaris* reproduces in the intestine and its resistant egg cases pass out with feces. In those places where people defecate in the open (not as unusual as it sounds), the tough egg cases containing developing embryos may survive on the ground for many years. The odds against any embryo finding the proper host is very great, so the *Ascaris* counters this by producing incredible numbers of offspring. A female may contain close to 30,000,000 eggs, which can be released at a rate of 200,000 a day! The microscopic eggs are easily passed along to the mouth by contaminated hands. The larvae hatch in the small intestine and begin a strange odyssey through the wall of the gut, to the heart and lungs, up and out of the breathing passage, where they are swallowed, finally coming to rest again in the intestine. As they travel, they change. So by this time they are now 2 or 3 mm long. But once they have returned to the intestine, they grow to immense proportions. Not everyone is aware of it when he or she harbors the pest. A few worms living in the intestine may go unnoticed, but where infestation is great, so are the effects. Dietary deficiencies result. For example, the *Ascaris* survives in its host by neutralizing the digestive enzyme trypsin. This can seriously affect the host's protein digestion, since trypsin is a key enzyme in that process. As a result, in heavily parasitized children, both growth and mental capacity are impaired.

You can also get worms from your pets. Americans love their dogs so much that they share everything with them. They romp with them on the lawn (which is the dog's toilet), let them "kiss the baby," and any number of other intimacies, unaware that dogs harbor a relative of *Ascaris* known as *Toxocara canis*, which causes dogs no great distress, but is ill-adapted to humans, and we cause each other great discomfort. Both parasite and host react to mutually unfamiliar proteins of each other. Consequently, the irritated worms roam through the human's liver and lungs, causing inflammation wherever it goes. Some-

22.14 Nematodes are common in most soils, but abound in moist, fertile places. Many are scavengers or predators of soil organisms and are considered free-living, but a great number are agricultural parasites. Because of their minute size, the plant parasites can readily move throughout plant tissues where they consume food reserves.

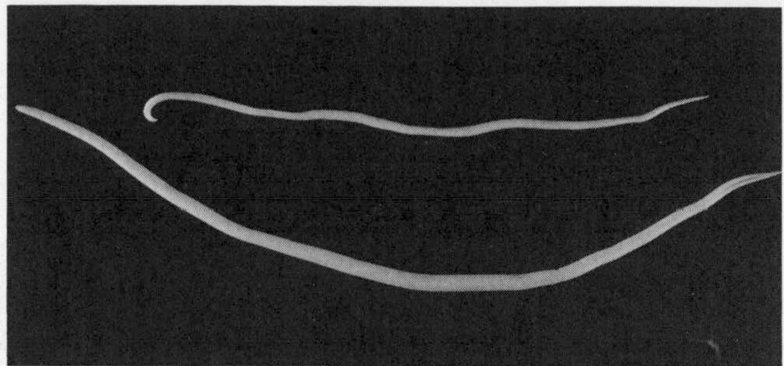

22.15 *Ascaris lumbricoides,* male and female. Female *ascarid worms* grow to 35 cm (14 inches) in length, while the males are shorter and more slender. Males can immediately be distinguished from females by the presence of a curled tail with fine bristles emerging from the tip. The adults are commonly found in the small intestine of infected people and hogs. The *Ascaris lumbricoides* of humans and hogs are identical in their physical features, but they are apparently physiologically distinct species. They do not develop well in the wrong host, so we can't blame the hogs for this one.

times they find their way to the brain and spinal cord or eyes where they may live for months, creating mysterious disease symptoms that are difficult for physicians to diagnose. Fortunately, the worms can't reproduce in humans, so the symptoms pass after a few months. There is no known method of killing the worms in humans. Instead, they are attacked by worming the dogs. Cat-lovers can't afford to be smug here because they have precisely the same problem with *Toxocara cati.*

Trichinella spiralis is a common inhabitant of swine. Sexually mature adults copulate in the hog intestine, whereupon the fertilized females burrow into the intestinal wall and remain there producing about 1500 live young. The young then begin to migrate, entering lymph channels and eventually finding their way into the voluntary muscles, tissue rich in blood and oxygen. The favored muscles include those of the diaphragm, ribs, tongue, eye, and larynx, although all voluntary muscles are susceptible. There, the worms mature and surround themselves with a protective cyst (Figure 22.16). The cycle ends there unless the hog is eaten by another hog or some other carnivore like ourselves. If we haven't first killed the worm by cooking it or freezing pork at $-15°C$ for 20 days, any infested pork or pork products (sausage or lunch meats) can lead to the worm traveling through our own lymph channels, to encyst in our muscles. It has been estimated that an ounce of heavily infected sausage can contain 100,000 or more encysted larvae which could produce at least 100 million venturesome offspring in their new host.

22.16 This light microscope photograph shows the *trichina* worm encysted in pork. If such meat is eaten without proper cooking, these worms will leave their cysts and venture through the body of the flesh-eater. Their favorite regions are the diaphragm, ribs, tongue, eye muscles, and larynx. The infestation is known as *trichinosis* and its severity depends upon how many worms have been ingested. The first rule of prevention is thorough cooking of all pork products. Apparently, deep-freezing at $-15°C$ for 20 days also kills the worms. Most cases of trichinosis can be traced to locally produced pork products, particularly from farms where uncooked garbage is (illegally) fed to hogs.

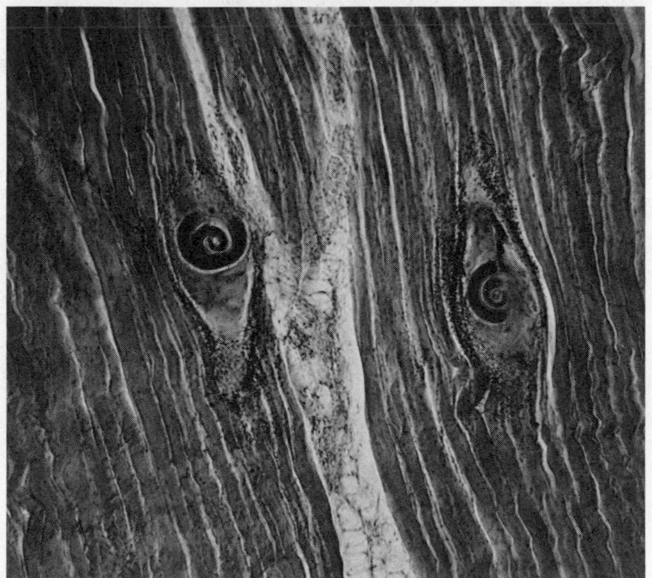

The symptoms of this infestation, called *trichinosis,* include muscle pain, fever, blood disorders, edema, and gastrointestinal disturbances. Most people survive the invasion, if it is not too severe. Medical treatment is purely symptomatic, which means making the patient comfortable. There is no way to kill the worms once they are in the human body. However, eventually they become entombed in hardened cysts and go with us to our graves.

Class Rotifera

At least the rotifers are not parasitic. In fact, their elaborate feeding system is unlike that of all other pseudocoelomates. They feed by sweeping microscopic organisms into their mouths through the action of double rings of cilia around their head area. Food is swept into a grinding gullet, or gizzard, which is unlike anything else in the animal world. The digestive system contains other specialized organs, including a stomach, two digestive glands, intestine, and anus (Figure 22.17).

The excretory system solves both the problem of nitrogen waste disposal and osmoregulation. Rotifers are equipped with flame cells like those of the flatworms. Beating cilia direct fluids along a system of ducts to a bladder, which subsequently empties through an opening to the outside. The system is quite advanced for such a "lowly" animal.

Female rotifers are much larger than the males and copulation proceeds in ways that should not surprise anyone. Males are scarce, however, and, in some species, they have never even been found. One reason may be that the male digestive system is so rudimentary and inadequate that, apparently, they mate and die off. The short interlude of the male's existence would seem to threaten the survival of these animals were it not for a surprising, compensating ability of the females. The females, it turns out, come in two kinds: those that mate with males and those that don't. Before you draw any unsubstantiated correlations to humans, we hasten to explain.

Rotifers reproduce sexually or asexually, with chaste females also producing offspring, but *parthenogenetically* (without fertilization). These offspring are all females. Apparently, environmental conditions determine whether these daughters will be of the sexual or asexual type. So, in the world of rotifers, males are reduced to the least common denominator.

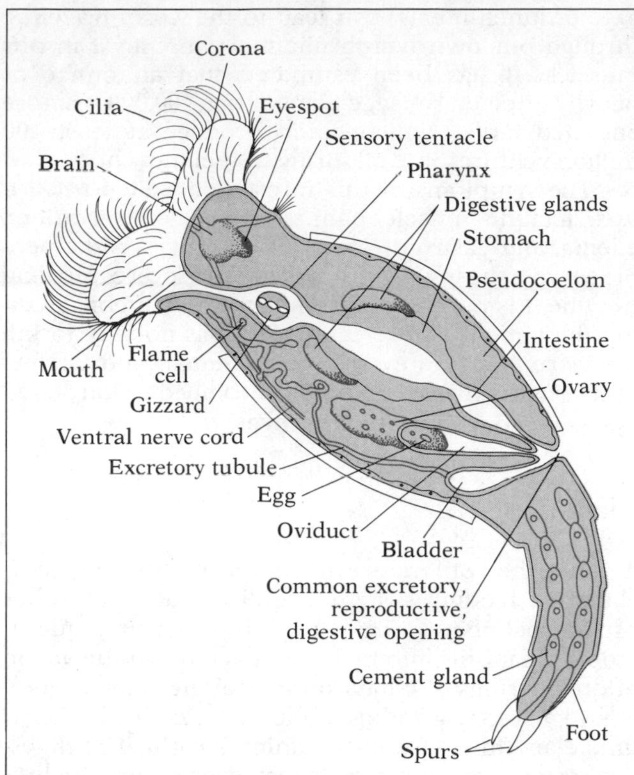

22.17 The rotifers are extremely common in fresh water. Although they are only about the size of many protists, the rotifers are far more complex. The digestive system is well-developed, with definite organ specialization. Unique to the rotifers is the gizzard, a set of grinding structures in the pharynx. The nervous, excretory, and reproductive systems of these animals are also quite complex. Many rotifers have cement glands at the base of the body that secrete a sticky substance, permitting the animal to fasten itself, temporarily, to an underwater surface.

All they contribute is an occasional input of genes, adding some genetic variability to an otherwise all-female population.

Other Aschelminthes

In addition to the nematodes and rotifers, a miscellany of other, obscure pseudocoelomate organisms are thrown into this phylum (when they are not given phyla of their own). Among them are several groups of microscopic marine worms and the bizarre "horsehair worms," (class Nematamorpha), which parasitize insects and are occasionally found swimming freely in small ponds, looking for all the world like lost strings from a violin bow.

There are only eight living species of *priapulids,* although this group is common in Cambrian fossil strata. Priapulids are more complex than other aschelminthes, and there is some question of whether they truly belong in this phylum. The adults have cellular rather than syncytial tissues and lack a cuticle; there is some argument, in fact, as to whether the priapulids are truly pseudocoelomate. But larval priapulids are almost indistinguishable from rotifers, suggesting that rotifers may have evolved from a priapulid line in which the adult stage was lost. Priapulids were named after the Roman god *Priapus,* noted for his exaggerated phallus (Figure 22.18).

22.18 An adult priapulid, a wormlike marine predator that may or may not belong to the phylum Aschelminthes. Larval priapulids are almost indistinguishable from adult rotifers.

22.3 Animals with a Coelom

The Lophophorate Phyla

You may have never heard of lophophores, but they are important in the scheme of things because they represent our first look at coelomate animals. Recall that in one system (Figure 22.4), the coelomate animals are separated from the other creatures.

Lophophores include three minor phyla: Ectoprocta, Phoronidea, and Brachiopoda (Figure 22.19). The ectoprocts, also known as bryozoans, are easily seen at low tide, forming branched, reddish fronds that look something like seaweed. These are actually colonies of ectoprocts, formed by asexual budding. Each individual bears tentacles, and feeds in the fashion of a coral, but there the similarity ends. The tentacles are attached to a semicircular ridge called the *lophophore*, and each tentacle is covered with a ciliated lining. Some feathery ectoprocts, dead and dried, are sprayed with green paint and sold in grocery stores as "living air ferns—requiring no care."

Brachiopods are shelled lophophores that superficially resemble a clam or scallop. However, the shell is fastened *dorsoventrally* (top to bottom), rather than *laterally* (right to left). The shell encloses two coiled and ciliated tentacles which comprise the lophophore.

The phoronids are wormlike lophophorates that live in tubes penetrating the oxygen-starved bottoms of estuaries and bays. They also bear the ciliated lophophore. Phoronids, brachiopods, and ectoprocts all produce a larva very similar to the *trochophore* larva of primitive annelids and mollusks.

While today's lophophorates are only minor phyla in terms of their species numbers, they were once more important. In fact, they are well-represented in the oldest fossils. Both brachiopods and ectoprocts are found in Cambrian deposits over 500 million years old. Some rock strata are composed entirely of their fossilized shells. The brachiopods have dwindled from about 3000 named species in the Ordovician fossil beds to about 200 named species alive today.

Phylum Annelida: The Segmented Worms

Who would have guessed that earthworms could be so important? Nonetheless, they are among the *annelids*, and annelids are important because they are segmented. Segmentation is perhaps the most profound change we've encountered after the bilateral body form, with the coelom bringing up a close third.

Class Oligochaeta: The Earthworms (Lumbricoides)

Earthworms are obviously segmented, as viewed from the outside, but it is important to know that repetition is seen in many of their internal structures as well. The internal organs are suspended in a segmented, fluid-filled coelom. Each segment contains elements of the circulatory, digestive, excretory, and nervous systems. The circulatory system is closed, which means that the blood is always enclosed in blood vessels. The system includes five pairs of aortic arches (hearts), comprised of pulsating vessels, along with arteries, veins, and capillaries. The blood of many annelids contains hemoglobin dissolved in the

22.19 The lophophores include three phyla that all have a structure known as the *lophophore*. It consists of a series of ciliated tentacles that are used in feeding and providing water currents for respiration. (a) One of the phyla, Ectoprocta, forms crusty colonies found on rocks or forms seaweedlike fronds. Close examination reveals tentacles thrusting in and out of the body. (b) Members of the second phylum, Brachiopoda, are often mistaken for clams, but the shell halves are arranged *dorsoventrally* rather than *laterally*. The dominating feature of brachiopod anatomy is the numerous tentacles of the lophophore. Some of the brachiopods live in tubelike burrows, attaching themselves to the ends of the burrows by a long contractile stalk. When disturbed, they draw themselves into the burrow. (c) The third phylum is Phoronidia. These wormlike creatures extend their lophophores from their burrows like a spirally coiled fan in order to feed.

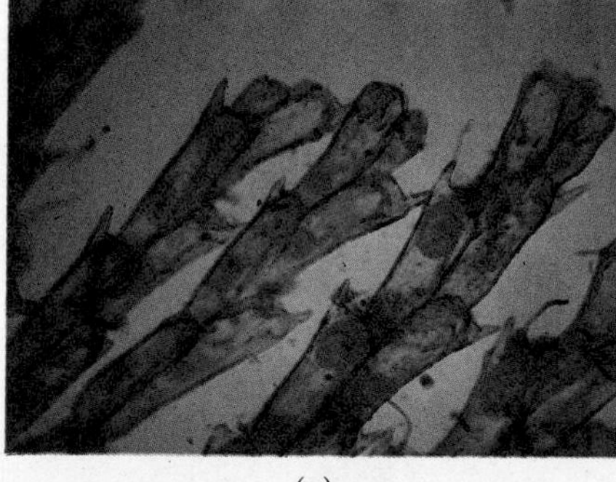

(a)

(b)

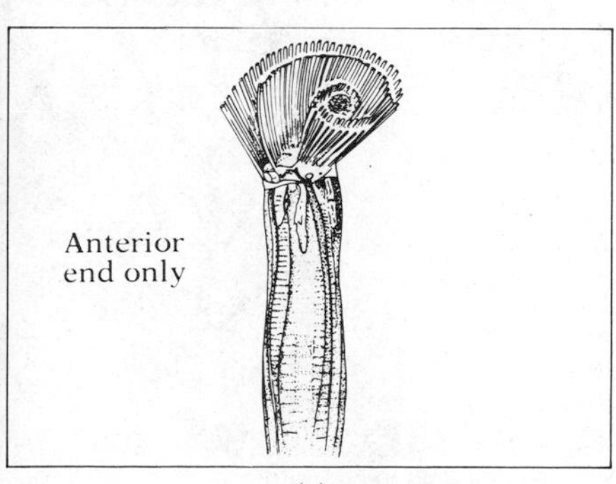

(c)

circulating fluid instead of as a component of blood cells. The excretory system is also highly developed in the terrestrial worms, consisting of paired nephridia in nearly all segments. The nephridia remove metabolic wastes from the coelomic fluids (Figure 22.20).

The earthworm has a complex reproductive system, with both male and female parts present in every individual, including sperm- and egg-producing structures and sperm receptacles. When earthworms copulate, they democratically exchange sperm, which are then stored until the proper time for fertilization. The *clitellum*, a smooth cylinder of tissue easily seen on earthworms, is the site of attachment in copulating worms. It later functions in secreting a mucous cocoon that will slide down the worm and surround the fertilized eggs until it is shrugged off, eggs inside.

The digestive system is complete (following the tube-within-a-tube plan seen in the roundworms). The tube is not simple, however, but is subdivided in the first twenty or so segments into swallowing, storing, and grinding regions. The intestine, a diges-

22.20 In cross-section the coelom of the earthworm is clearly visible. The various internal organs are suspended in the coelom and supported by partitions that divide each segment. The body cavity is a true coelom, lined by mesodermally derived tissue. The gut wall is muscular. Other structures seen in the drawing attest to the complexity of this annelid and demonstrate clearly the segmented plan. Nearly all segments bear paired *nephridia* (the excretory organs). At the anterior end, the complex complete (tubelike) digestive system is the axis around which a closed circulatory system, the paired reproductive structure, and the brain are found. The smooth region just beyond the cross-section is the *clitellum*, which secretes a mucous layer around the fertilized eggs as they are released.

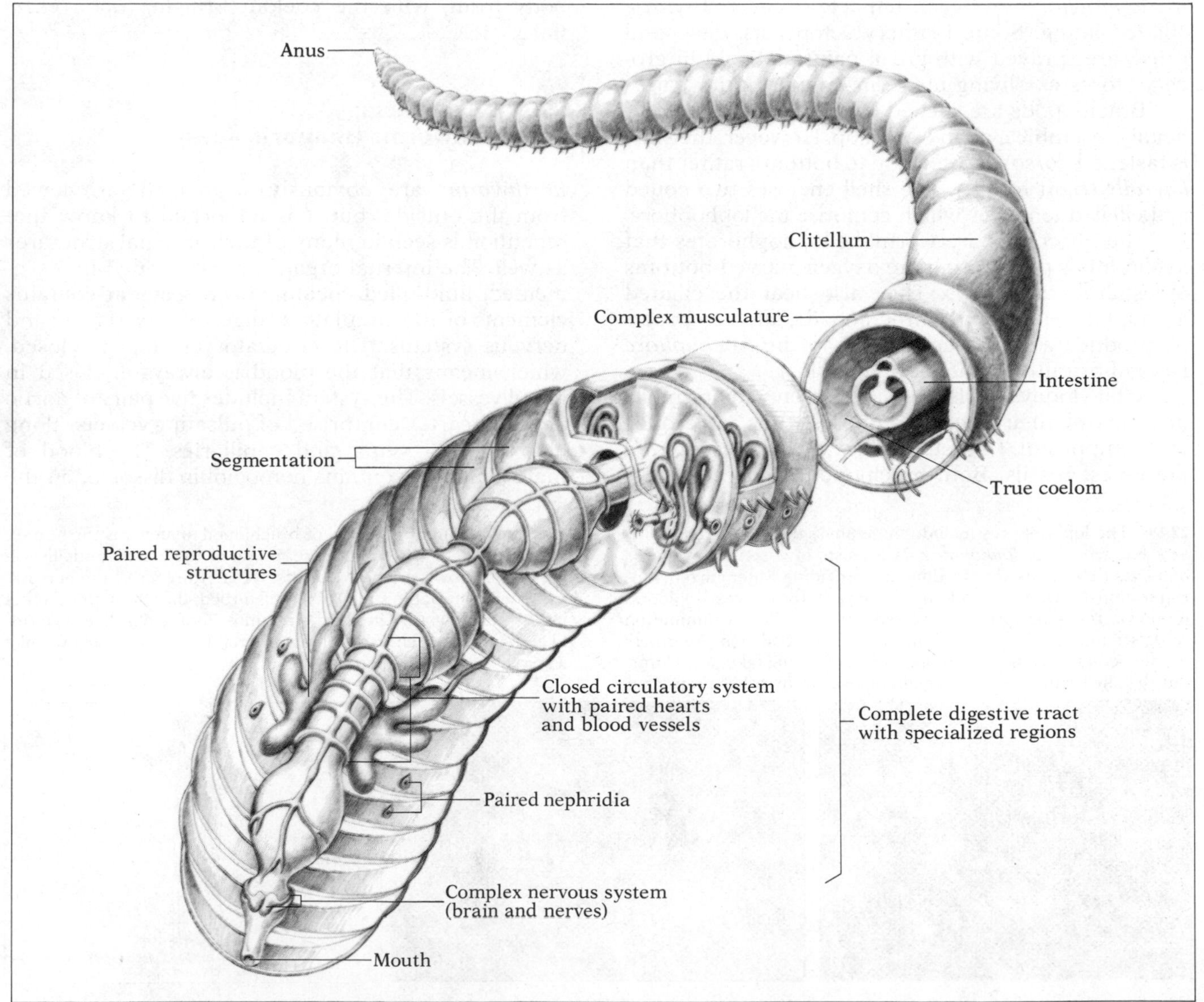

tive and absorptive structure, continues to the anus and is not segmented along its length. Earthworms feed on soil, absorbing its complement of decaying vegetation.

Darwin, in *The Formation of Vegetable Mould Through the Action of Worms* (1881), demonstrated that earthworms are a vital link in the chain of soil-building activities. They are surprisingly important in cultivating, aerating, and fertilizing soils. And they are, of course, part of the food webs in terrestrial ecosystems, having been known to sustain birds that are not late.

The nervous system consists of masses (ganglia) of nerve cells above and below the esophagus, and a ventral nerve cord that has paired ganglia and side branches reaching into each segment. The presence of large nerve cell masses in the head region is an example of cephalization and anterior orientation in the bilateral animals.

Burrowing and other movement are accomplished by two layers of muscle (circular and longitudinal) in the body wall, which can extend and contract the body, flexing against the fluid-filled body that maintains the earthworm's shape. Burrowing is assisted by numerous *setae,* bristles that can be inserted into the burrow to hold some segments fast while other parts of the body are extended or contracted.

Class Hirudinea: The Leeches

Leeches are mainly parasitic, but there are also some predatory or scavenger species. They are primarily found in fresh water, although there are marine species and some terrestrial species found on moist surfaces. The parasitic species are *ectoparasites*—that is, they attach themselves to the skin of many animals, including humans (Figure 22.21).

The segmented body of the leech is usually flattened with suckers at the anterior and posterior ends. Suction is applied by a muscular pharynx and some species have horny teeth in the anterior sucker that cut through the skin of the host organism, allowing the leech to ingest the host's blood. They also secrete a chemical called *hirudin,* which is an anticoagulant.

The body form of the leech is somewhat modified from the segmental condition. Although segmentation is still apparent in the nervous, reproductive, and excretory systems, the coelom itself is not divided. Like the earthworm, leeches are hermaphroditic but not self-fertilizing. They are also similar to earthworms in copulation, egg laying, and development. Their eggs hatch as little adults without going through a larval stage.

Leeches used to be a favorite means of treating bruises (especially black eyes) and were also used for bloodletting. Pharmacies in some countries still stock them for such medicinal purposes.

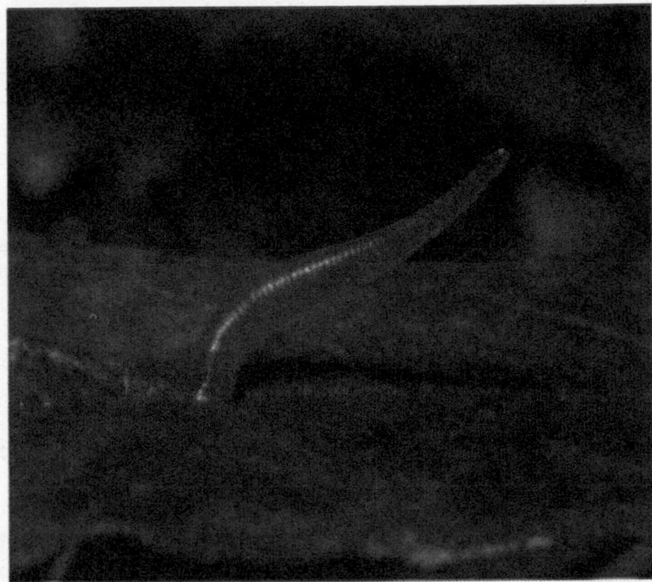

22.21 Leeches are armed with sharp, piercing mouthparts and suckers for holding onto their hosts and drawing blood. As a leech fastens onto the skin of its host, it secretes into the small wound an anticoagulant known as *hirudin.* Leeches feed voraciously once attached: they are able to ingest three times their body weight at one time. After feeding, they drop off and fast for long periods.

Class Polychaeta

This class includes a number of filter feeders (fan worms) as well as the predatory clam worm, *Nereis,* which is a marine carnivore (Figure 22.22). They make their homes in the sandy and muddy tidal flats.

22.22 *Nereis,* a marine polychaete worm, is an aquatic relative of the earthworm, but it is highly specialized for its marine existence. Its anterior has eyes, sensory projections, and a set of retractible grasping jaws which reveal it as a voracious predator. The body of *Nereis* is clearly segmented, and each segment below the head region is equipped with a pair of *parapods* which are used in locomotion (crawling, digging, and swimming) and in respiration.

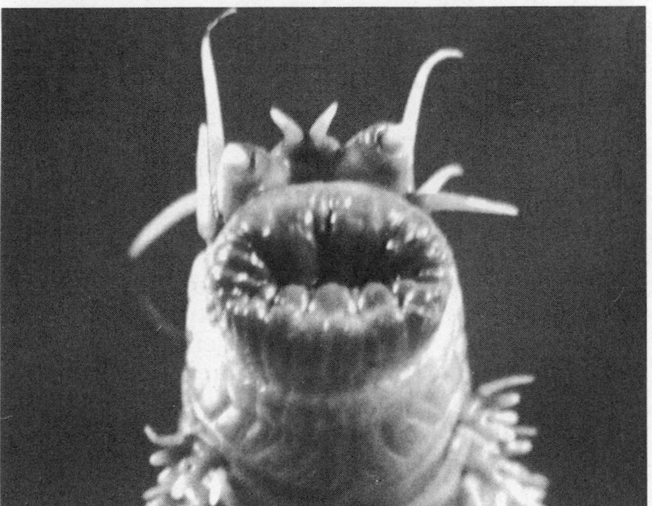

Although *Nereis* tends to burrow reclusively in the bottom, it can be roused to swim when startled or when mating. Unlike the earthworm and leech, the sexes are separate, but the clam worms don't actually copulate. Gametes are produced, strangely enough, in the coelomic lining. When the eggs are ready, the female swims near the surface of the water. Then her body literally bursts, releasing the eggs, which are then fertilized by the males. The zygotes develop into trochophore larvae.

Externally, the most striking features of the clam worm are the numerous pairs of fleshy *parapods* along its segments, and the sensory appendages at its anterior end. The parapods are muscular extensions of the body wall, used both in movement and in gas exchange. It is perhaps a bit more difficult to see the powerful jaws of this carnivore.

The marine polychaetes are a diverse group with many variations on the phyletic theme. In almost all cases, however, the segmentation is evident to some degree. Among the polychaetes are a number of tube-dwelling species. Some construct tubes of sand particles, cemented with mucus or calcium secretions. They are easy to see in their burrows because their feathery plumes (which are actually feeding and respiratory devices) constantly pop in and out. These colorful ciliated tentacles have inspired common names such as "fan worm," "feather worm," and "peacock worm."

The tube-building polychaetes have left important clues to the annelid history. Fossilized tubes have been found along with what seem to be tracks made by the ancient worms in Pre-Cambrian strata. The earliest tracks are random and dispersed, but those appearing later are aggregated, showing a possible change in the social behavior of the worm. These rocks, then, bear rare examples of fossilized behavior and are the earliest eucaryote fossils known. In addition, peculiar fossils in Cambrian strata have been identified as polychaete jaws.

Phylum Mollusca

The earth is burgeoning with *mollusks.* In fact, we already know of about 100,000 species, so that makes them behind only the nematodes and insects in sheer variety. There are six classes, four composed mostly of marine animals. An example from each class includes the chiton, clam, snail, octopus, toothshells, *Neopilina,* and a wormlike group, the Solenogastres. We will consider only the four larger classes to represent the phylum. Some mollusks are minute, living in their tiny inconspicuous shells, while the giant North Atlantic squid that swims confidently through the cold ocean waters reaches over 18 m (50 ft) in length.

The term *mollusk,* derived from Latin, means soft-bodied; but actually, this isn't very descriptive for an invertebrate. Aside from being soft-bodied, all mollusks have a muscular *foot* that contains sensory and motor systems and may be used for creeping, digging, holding on, and capturing prey. Some have external shells produced by secretions of a fleshy covering known as the *mantle,* which, by the way, is also present in nonshelled members of the phylum. So both snails and slugs have mantles. The *mantle cavity* (the space enclosed by the mantle) houses feathery respiratory *gills* in aquatic mollusks. In terrestrial mollusks, which lack gills, the lining of the mantle cavity is highly vascular and the mantle cavity itself serves as a lung.

Mollusks are coelomates in embryonic development, although this cavity is often reduced to an open region just around the heart. Except for the cephalopods the circulatory system is open, with blood leaving vessels to enter open spongelike sinuses before returning to the heart where it is picked up and returned to vessels. The digestive system is complete (with both a mouth and anus). Most classes have a rasping tonguelike structure called the *radula* with which they file their food into small particles. The enormous diversity among mollusks is suggested by comparing the development of the shell, foot, and mantle (Figure 22.23).

The fossil history of mollusks goes back to the Cambrian period. Interestingly, the earliest fossils include an ancestral form very similar to *Nautilus,* a living shelled cephalopod. In fact, it seems that the Cambrian fossils include at least twenty times the number of mollusk species that now exist. The presence of these and other complex invertebrates in the very oldest true fossil animal records demonstrates the total inadequacy of geological and paleontological evidence in tracing the dim origins of invertebrate phyla. We must assume that most of the important events in invertebrate history occurred in the millions of years preceding the Cambrian period.

The relationship of mollusks to annelids is suggested by two factors. First, many mollusks produce a trochophore larva, although many mollusks have eggs that hatch directly into miniature adults with no distinct larval stage. Second, some fossil mollusks were segmented and apparently had a well-developed coelom, similar to that of annelids. For many years this theoretical relationship was based primarily on fossil evidence of *Pilina,* mollusk shells with apparently segmental muscle attachments. But in 1952, a Danish research vessel dredged ten "living fossils" up from a depth of 3 km. They were the remnants of the class known as Monoplacaphora, long believed to be extinct. The specimens, dubbed *Neopilina,* are clearly segmented, and the coelom of the adult is well-developed and extensive.

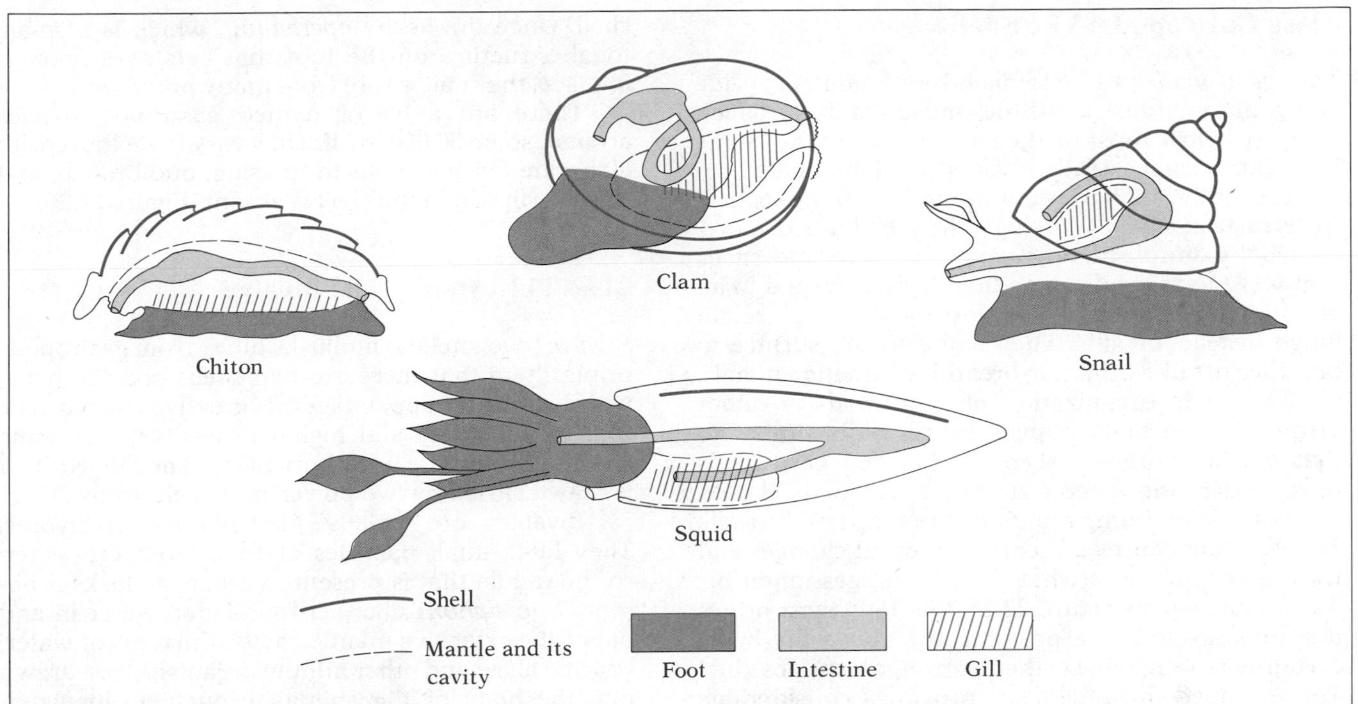

22.23 The basic body plan of the mollusks is best seen in the primitive chitons, where the *foot, mantle,* and *shell* are simple. The other classes shown here have each diverged from the basic plan in a different way in their evolution. The greatest change appears to have occurred in the cephalopods (the octopuses and squid), which have lost all or nearly all of the shell and modified the foot into grasping tentacles. Note the divergent evolution of the shell, mantle, gill, foot, mantle cavity, and intestine in these four classes.

Class Amphineura: The Chitons

Chitons are probably the least specialized mollusks. Lack of specialization, to many people, means primitive, that is, the chitons are thought to have retained many of the features of the ancestral mollusks. Their flattened bodies cover a long foot and are protected by a shell composed of eight plates (possibly reminiscent of segmentation). The gut, beginning with a radula-equipped mouth, extends the length of the body. Feathery gills extend along each side of the animal within the mantle cavity, which forms a groove entirely around the foot (Figure 22.24).

Chitons are slow-moving, rather dull creatures. They creep along over the rocky surfaces of tidal regions, grazing as they go by using their sharp radula to scrape the algae from the surfaces of the rocks. When disturbed, their defense is to grip the rocks with powerful foot muscles and flatten so that it is very difficult to remove them. These primitive beasts have been around a long time. Chitons very similar to today's show up in the fossil record in Ordovician deposits (500 million years old).

22.24 The primitive chitons reveal the basic mollusk body plan. Their shells consist of eight dorsal plates, and the foot is long and flat, somewhat similar to that of the gastropod mollusks (such as snails, slugs, and abalone). These simple and sluggish creatures may not be very interesting to watch, but they are sought by collectors, who denude tidepool rocks of chitons in order to make jewelry.

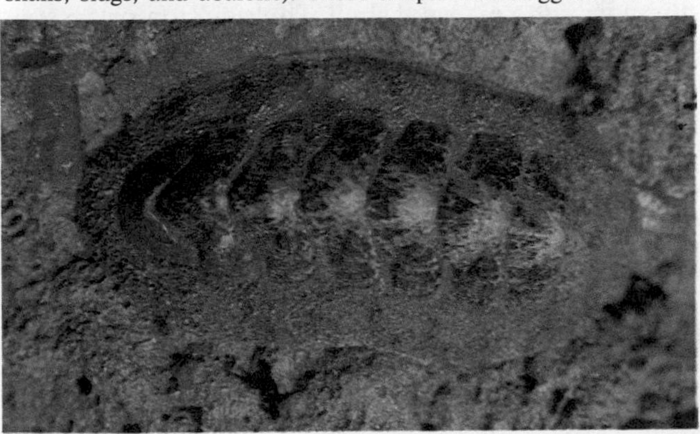

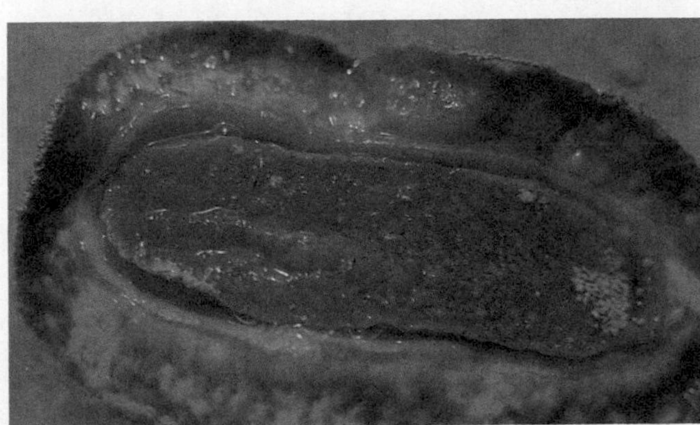

Class Gastropoda: The Snails

The *gastropods,* or "stomach-foot" animals, slide along on a large, extensible, muscular foot, which lies just below most of their digestive organs. Some have rather successfully invaded the land and a host of beautifully-shelled creatures move over the seabed. In terrestrial snails and slugs, the gills have been replaced by simple mantle-cavity lungs. Some snails that were terrestrial at one time have returned to an aquatic habitat, but their descendants still retain lungs instead of gills. They come to the surface to breathe just like other air-breathing aquatic animals.

The basic organization of gastropods is rather primitive. The body plan is superficially similar to that of the chitons, except that coiled gastropods have undergone *torsion* as they developed. The result is a twisted hump matching the shape of the shell. This twisting causes all sorts of internal changes and we aren't quite sure what the advantages might be. As you can see in Figure 22.25, this 180° twist brings the anus around to a position just above the head, certainly not the most esthetic arrangement possible! The humped arrangement provides considerable space for retraction of the entire head-foot into the

22.25 Torsion and its effects on the gastropod. In larval snails, the internal organs undergo a twisting as the animals develop. As you can see in the young ciliated larva, the digestive tract is simply U-shaped. As time passes, it completes a 180° twist that brings the anus up over the mouth. This twist is related to other developmental events as well. For example, organs on one side of the body are compressed and fail to develop. The nervous system contorts into a figure eight. Spiral development of the shell and body mass, which occurs in many snails, is a separate, later developmental effect.

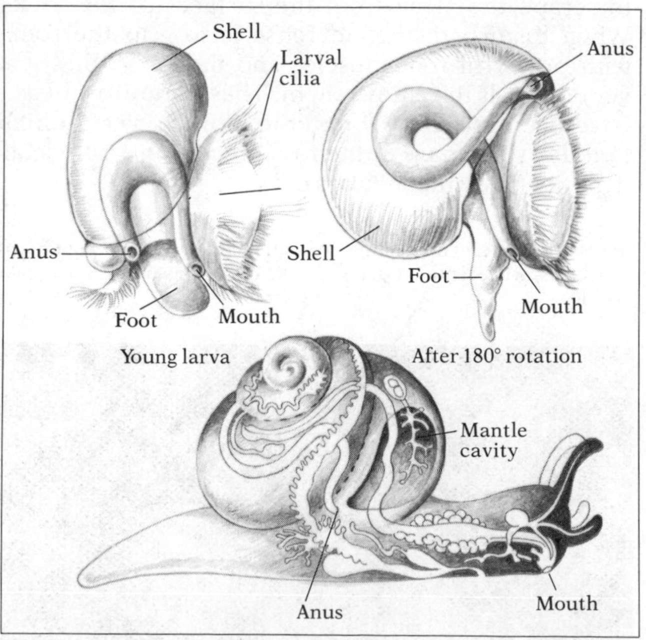

shell. Once the horny *operculum,* which is a small, tough structure on the foot that acts as a door, is in place, the snail is safe from many predators.

There are a lot of named gastropod species around, some 50,000 in all. They vary from those with elaborate twisted shells to the slug, nudibranch, and sea hare in which the shell is absent (Figure 22.26).

Class Pelecypoda: The Bivalves

Bivalve (two-shelled) mollusks differ from gastropods primarily in that there are two shells and the head, with its sensory appendages, is greatly reduced. The foot is still fleshy and highly extensible, and some species (clams) use it for burrowing. The hinged shell is drawn closed by two powerful muscle groups.

Bivalves are chiefly *filter-feeding* herbivores. They filter small particles of food from the water, so the radula that is present in other mollusks is absent. Two *siphons* channel food-laden water in and out of the extensive mantle cavity. Currents of water, bearing algae and other minute organisms, are drawn into the body of the animal through an incurrent siphon by the action of cilia that line the large gills. The food particles are then trapped in heavy mucous secretions and moved by liplike, ciliated *labial palps* which separate out grains of sand and permit the bits of food to move into the mouth. Of course, water flowing over the gills bears not only food, but oxygen.

The circulatory system of bivalves is typical of mollusks. The heart is located in its coelomic cavity. As you can see in Figure 22.27, the gut also passes through the cavity; in fact, it goes right through the heart! Mollusks have open circulatory systems. This means that blood leaves the vessels and moves through spongy openings (*sinuses*) before it returns to the heart.

Class Cephalopoda: The Squid and Octopus

The "head-foot" mollusks, the *squid* and *octopuses* are, in many ways, highly specialized creatures. They are active and voracious predators, feeding on fast-moving invertebrates and vertebrates. Almost all of them lack the external shell. Their circulatory system is closed, an advanced condition with the efficiency needed for large fast-moving animals. The gills are served by dense capillary beds with blood pumped by two pairs of hearts, one of which boosts blood through the gills. Water is moved over the gills by a pumping action of the muscular mantle, in contrast to the ciliary currents found in other marine mollusks. In cephalopods, the bulbous head is greatly enlarged and the mantle itself forms the external covering. The thick muscular molluscan foot is highly modified into tentacles—complete with suckers. The

22.26 There are 50,000 named species of gastropod mollusks. (a) These include a great number of marine, freshwater, and terrestrial snails. In addition, some of the gastropods have lost all but remnants of their shells, as seen in (b) the beautiful nudibranch and (c) the not so beautiful sea hare. (d) The abalone is a flattened creature that clings tenaciously to the rocks of the ocean bottom, but actually its shell is modified from a spiral form. (e) Most limpets have conical shells, but this deep-sea sulpher limpet has only a remnant shell and somewhat resembles the sea hare.

22.27 Anatomy of a clam. The clam, a bivalve mollusk, has highly specialized respiratory and digestive systems. In this diagram, one shell, one pair of gills, and part of the body wall have been cut away. The head is reduced, with none of the elaborate sensory features found in many other mollusks. The foot makes up most of the body. Clams are *filter-feeders;* they bring in a steady flow of water through the incurrent siphon, pass it through pores in the gills, and then send it back out through the excurrent siphon. As water passes through the gills, as shown by the arrows, rows of cilia sweep any particles forward to the mouth. Sensitive *labial palps* near the mouth decide what is edible. The mouth opens into a short gullet that leads into a stomach pouch. Food then moves into a lengthy intestine which is twisted back and forth through the foot. The intestine actually passes through the heart and back to the anus. The heart is a simple pumping chamber that forces blood out into an open circulatory system.

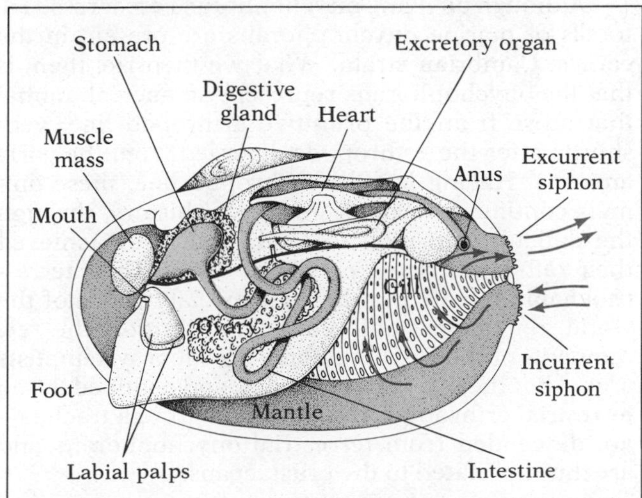

mouth is armed with a horny muscularized beak, used in ripping and tearing the prey.

Because of their hunting activities, the cephalopods are bound to attract the attention of other carnivores. Their defense is their speed and agility. Octopuses and squid move rapidly by forcefully ejecting water through their siphons in a jetlike action. The siphon can be turned in any direction. They may also quickly change colors or release a cloudy "ink" to foil predators. In addition, octopuses live in burrows and can squeeze their large bodies through and into tiny cracks and crevices. Cephalopods have highly developed senses, including eyes that are very similar to those of vertebrates (Figure 22.28). In addition, octopuses have a large, well-developed brain and are surprisingly intelligent invertebrates, as we will see in Chapter 31.

Phylum Onychophora

The Onychophora, which includes only about sixty-five to seventy named species, is so minor that it could well be ignored if it weren't for the fact that some researchers consider it an evolutionary link between the Annelida and Arthropoda. Actually, all three of these phyla are probably descended from annelid-like ancestors (Figure 22.29).

The best-known member of the phylum is the peculiar *Peripatus,* which lives among the damp leaves of the tropics, including Africa, Southeast Asia, New Zealand, Australia, and the West Indies. These

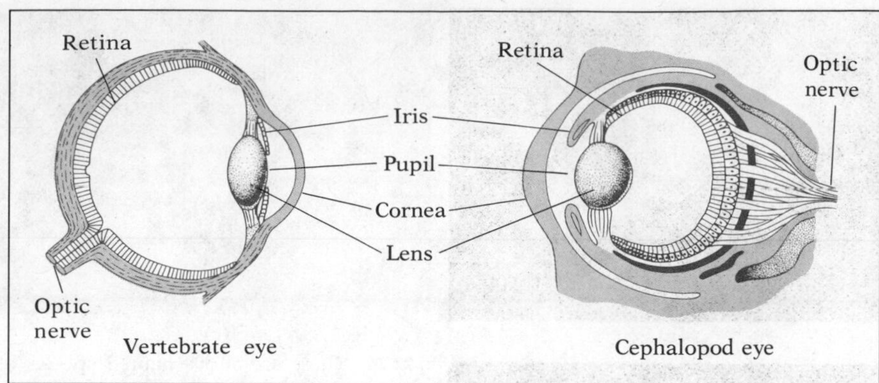

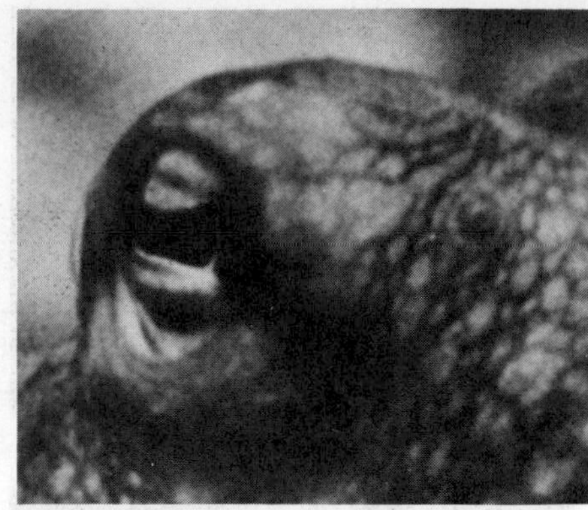

22.28 The eyes of the cephalopods (octopus and squid) are remarkably similar to those of vertebrates. The squid is believed to have binocular vision, like humans and a few other mammals, which gives it the ability to judge distances with great accuracy.

22.29 The phylum Onychophora has many characteristics of both the annelids and the arthropods. Its thin-walled, segmented body resembles the earthworm's and the serial duplication of internal structures such as the paired nephridia is also reminiscent of the annelids. On the other hand, onychophorans have jointed legs, tipped with claws, and they have an open circulatory system, both of which are arthropod characteristics. This mixture of traits suggests the evolutionary relationship in the diagram. The annelids are considered to be the ancestral type. The onychophorans branched off later, after the arthropod line began. But some systematists believe that millipedes, centipedes, and insects, all terrestrial organisms, descended from the terrestrial onychophorans and are unrelated to living marine arthropods.

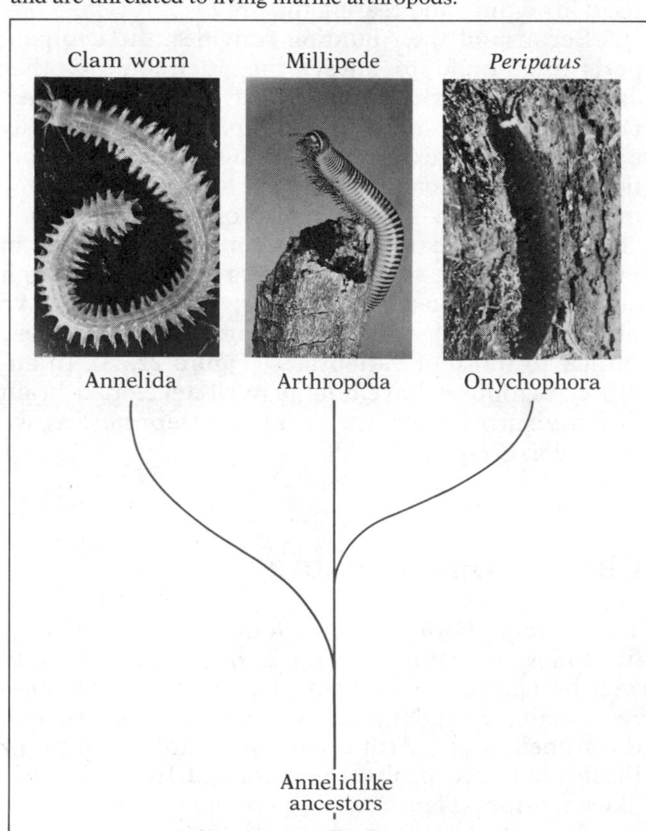

organisms are essentially nocturnal. If you want to see them in their native habitat, you will have to wander through the rain forests in the dead of night reaching into crevices, beneath fallen trees, and under bark and foliage. (We'll wait here for your report.)

Peripatus has a thin, soft cuticle covering a muscular body wall. The muscle layers run in both circular and longitudinal directions. It is a highly segmented animal, but the segmentation is largely confined to its appendages and internal structure; its body wall is fairly smooth. This segmentation reveals its annelid heritage. On the arthropod side, we see that its soft appendages are armed with tiny claws and small jaws. Its head even has *antennae.* It breathes through a series of holes (*spiracles*) along its body which open into tiny pits connected to a number of tubes known as *tracheae.* This arrangement permits air to reach the internal organs easily, but since there is no way of controlling the size of the opening, water loss is a hazard. This explains its preference for moist habitats.

Although all living onychophorans are terrestrial, fossils of marine onychophorans are present in the earliest Cambrian strata. What we theorize, then, is that the onychophorans represent an ancient animal that arose from the primitive arthropod line, very shortly after the arthropods diverged from the early annelids. Having left the arthropod line, these animals continued through geological history, through the difficult transition to land, until they achieved their rather restricted existence today. Their terrestrial habitat, scattered through special regions of the world, remained stable enough to support the few remnants of this phylum. Some dissident systematists argue that the insects, millipedes, and centipedes—all terrestrial arthropods that breathe through trachea—are descended from terrestrial onychophorans and are thus unrelated to the crustaceans.

Phylum Arthropoda

One way to give you some idea of what we're up against here is to imagine Noah trying to load his ark with animals. When he lined them up, two by two, he was confronted with eight arthropods for every nonarthropod species. (We won't include the nematodes.) If, by chasing, swatting, and kicking, he was able to get one pair aboard per minute, he would have spent 18 months, night and day simply loading his quota of crickets, crabs, lice, flies, centipedes, aphids, wasps, weevils, dragonflies, lobsters, ticks, and the like into the vessel.

In other words, there are a lot of species of arthropods. We don't know how many. Some say 800,000, others say a million, and still others estimate that many times that number have still never been identified. We must conclude that as a phylum, Arthropoda is the most successful on earth.

So what are these pervasive little creatures? *Arthropod* means jointed-leg, so they have jointed legs. And they are also segmented, but these segments are not simply repeating units suggestive of earthworms; they are highly specialized for different tasks. Arthropods have an *exoskeleton (exo,* outside) made of chitinous material, often hardened with calcium salts. All arthropods have open circulatory systems. Their nervous system consists of major *ganglia* (clumps of nerves) in the head and a *ventral nerve cord* bearing paired minor ganglia in most segments. Except for a few marine forms, arthropods are small because of the size restrictions placed on them by this plan, but size apparently doesn't seem to have much to do with evolutionary success. So at least we don't have to deal with eighty pound mosquitos.

Arthropods have successfully invaded most of the earth, establishing niches of often amazingly narrow specializations. (Consider the mayfly, a delicate creature that emerges without mouthparts with which to feed, and which must mate in the few precious hours of life allowed it.) Other arthropods are omnivores, herbivores, carnivores, scavengers, ectoparasites, and endoparasites. They live on, under, and above the surface of land and water, from deep ocean trenches to the highest mountain peaks. In many cases, because of their highly specialized niches, numerous species can exist side-by-side without seriously competing with each other. Sometimes one species' scraps are another's dinner.

Their success is due to other important factors as well. The reproductive capacity of many arthropods is phenomenal. In addition, many produce larvae that have an entirely different diet from the adults, thus expanding the niche and avoiding parent–offspring competition. Thanks to sexual reproduction, there is a great deal of genetic variation in most populations, so that they tend to overcome all sorts of environmental deterrents. (For example, pesticides; DDT-resistant mosquitos have evolved in only a few generations.)

Subphylum Chelicerata

Arthropods can be divided into two subphyla according to their feeding apparatus and appendages (Figure 22.30). The more ancient group, the Chelicerata, includes the arthropods that lack jaws. They have six pairs of appendages, four of which are legs. The first two pairs are sensory palps and *chelicerae* (fangs). None of the Chelicerata have antennae. They include horseshoe crabs, sea spiders, and class Arachnida: true spiders, daddy longlegs, ticks, mites, and scorpions.

Horseshoe crabs of today are nearly identical with fossils 200 million years old, except that during this time the mouth opening has migrated backward from an anterior position to down between the legs. Thus horseshoe crabs are primitive, with onc advanced feature.

So why aren't horseshoe crabs classified with crabs and other crustaceans? Partly because they possess *book gills* that strongly resemble the *book lungs* of spiders. Their eyes are also unlike those of other arthropods. In summary, they are quite different. The taxonomic difficulties in Arthropoda are such that many systematists believe it is a polyphyletic group, brought together by the common feature of jointed legs.

Class Arachnida

The *arachnids* don't win much affection with their habit of sucking the juices of other organisms. Spiders, equipped with hollow fangs and venom, are all carnivores. Typically, when a spider bites its prey, it injects venom which weakens the unfortunate animal. Then it injects digestive enzymes into its host's body to liquefy the tissues there, so it can suck the fluid out.

Spiders differ from other arthropods in another way. As mentioned, they respire through the use of a structure known as the book lung. Book lungs are sacs with slits that open to the outside. The sacs are lined with leafy folds—similar to the pages of a book. Blood passing through the thin "pages" exchanges gases with air in the sac.

Most spiders are equipped with silk-producing glands connected by ducts to external devices called *spinnerets* (Figure 22.31). Both the *web* and the *cocoon* of the spider are produced by this system. The silk of spiders (and insects) is a protein known as

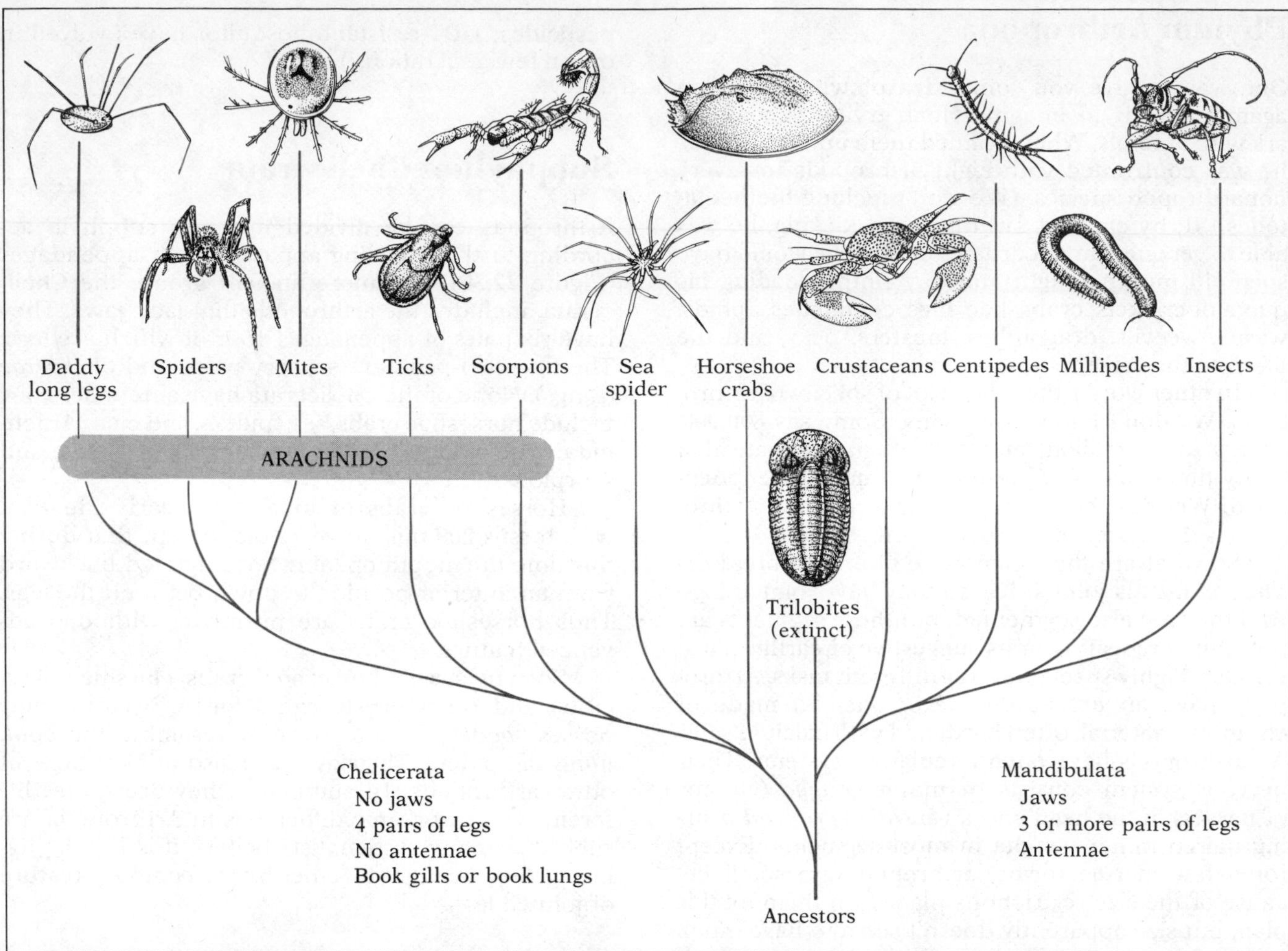

22.30 The phylum Arthropoda can be organized into two main subphyla on the basis of mouthparts and appendages. The Chelicerata, shown in the left branch, lack jaws, but produce clawlike *chelicerae* (which in the spiders are venomous fangs). In addition, they have four pairs of walking legs and other appendages that are generally related to feeding. The Mandibulata, branching off to the right, have jaws and three or more pairs of walking legs. Most have antennae. The subphylum Mandibulata includes the enormous class, Insecta.

fibroin, composed chiefly of the amino acids glycine, alanine, and tyrosine. Spiders as a group are known to produce seven different kinds of silk, each with its own function. Webs may have sticky parts and safe, dry parts along which the spider can run. It wouldn't do to become entangled in one's own web, no matter how many novels it spawned.

Subphylum Mandibulata

The subphylum Mandibulata consists of arthropods with *mandibles* (jaws). They also have antennae and various numbers of paired appendages, including three or more pairs of walking legs. There are four major classes and two rather minor ones. The major classes are the Crustacea (marine and freshwater crabs, lobsters, copepods, barnacles, sowbugs, and many others), Chilopoda (centipedes), Diplopoda (millipedes), and Insecta (flies, grasshoppers, bees, and so forth). Figure 22.32 illustrates the wide variety within this subphylum.

Class Crustacea

Crustaceans live in both marine and fresh water. Their size ranges from microscopic ostracods and copepods to giant crabs from the western North Pacific (see Figure 22.32). Some species in this diverse group clearly demonstrate their relationship to the primitive

22.31 Most spiders have silk glands associated with specialized external appendages known as *spinnerets.* The orb spider, a master web-maker, can produce several kinds of silk. The glands, as shown here, are located in the abdomen. They open into the spinnerets which emit threads of different diameters. The web in the photograph belongs to an orb spider. The spider can move about quickly on the web, but it must know where the sticky parts are, or it will become ensnared in its own trap.

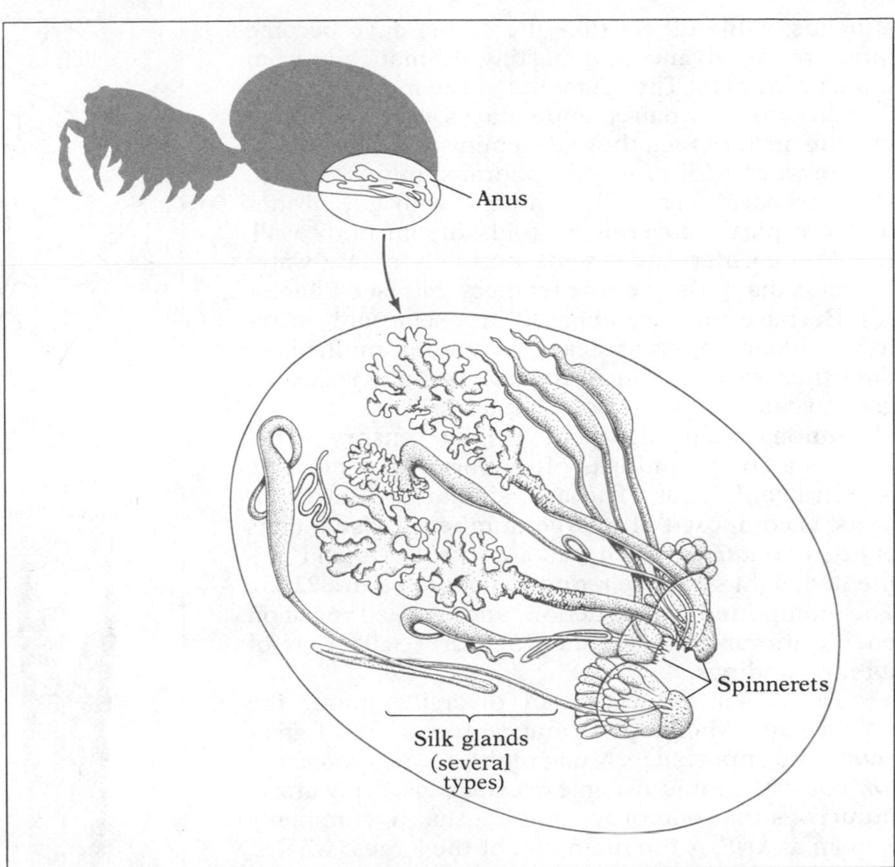

22.32 The Mandibulata are such a diverse group that it's difficult to find a representative example. These species can give you some idea about the diversity. The largest members of this subphylum are the giant crabs (a) that live in the waters off the coast of Japan. Keep in mind that these are crustaceans, as is the tiny freshwater *Daphnia,* shown magnified in photograph (b). The rhinoceros beetle (c) is antithetical to the delicate mayfly (d) though both are insects. The rock louse (e) lives along the rocky ocean shore.

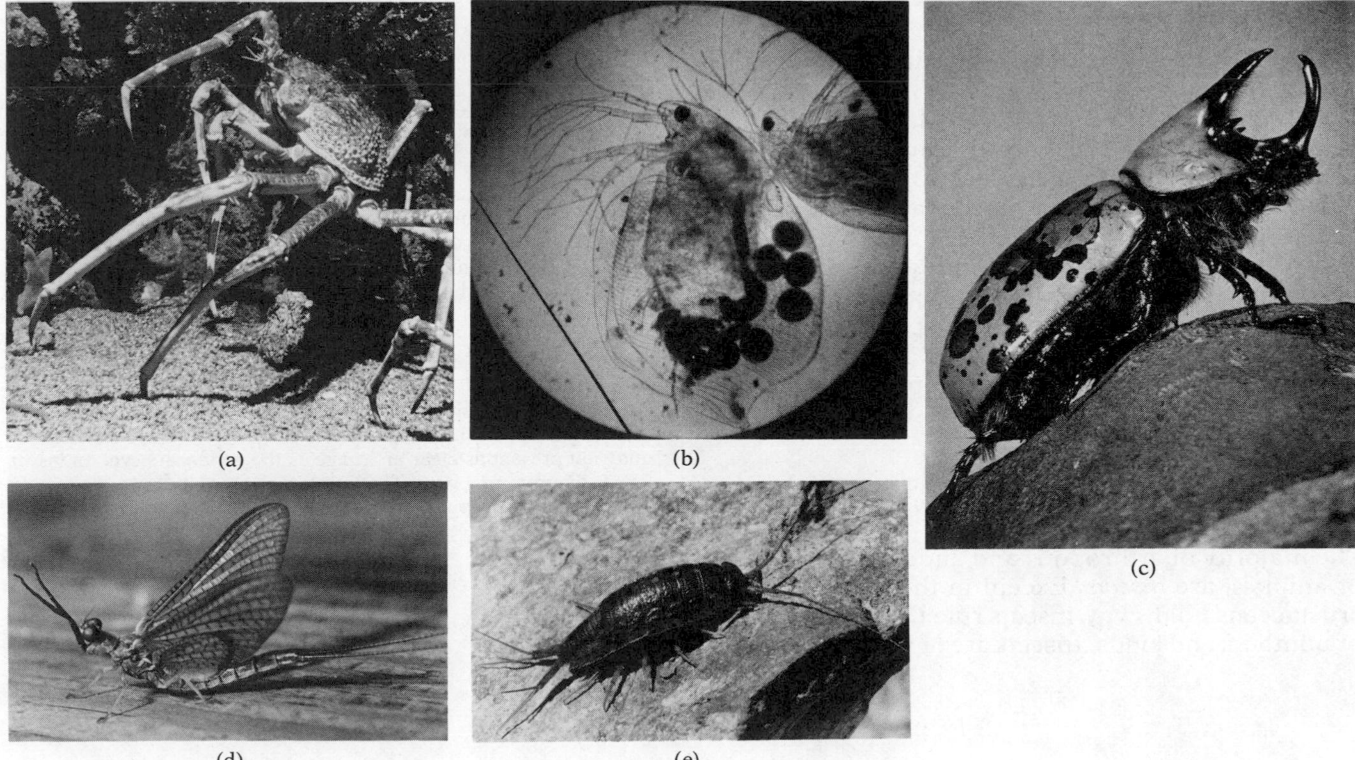

annelids, while others (like the crabs) have become much more advanced, departing dramatically from the ancient plan. The segments of the most primitive crustaceans bear paired appendages, each somewhat like the next. In fact, they are reminiscent of the polychaete worms (or of onychophorans) in this respect.

Crustaceans are gill breathers, with gill cavities that are partly covered by folds of the body wall. Active muscular movements normally create water currents that pass over the feathery gills (see Chapter 26). Because they are encased in a semirigid, secreted exoskeleton, crustaceans frequently molt, shedding their exoskeleton so that hormonally regulated growth can take place (see Chapter 27).

Among their highly developed sensory structures is a compound eye, often borne on a movable or retractable stalk. The eye, similar to that of insects, is composed of a large number of visual units called *ommatidia.* Each has a *corneal lens* and pigmented, light sensitive *retinal* cells (see Figure 22.33). The compound construction and stalked position enable the animal to obtain a nearly 360° view of its surroundings.

There is a great deal of diversity among the crustaceans. Many of the marine forms are of great economic importance. Much of the ocean's *zooplankton* consists of microscopic crustaceans. They are so numerous that one order, Euphausiacea, commonly known as *krill,* is the main diet of the largest whales. But beyond providing food for whales, crustaceans are an important part of the ocean's long food chain, since they are primary consumers and serve as food for other organisms only slightly larger, which serve as food for other organisms . . . .

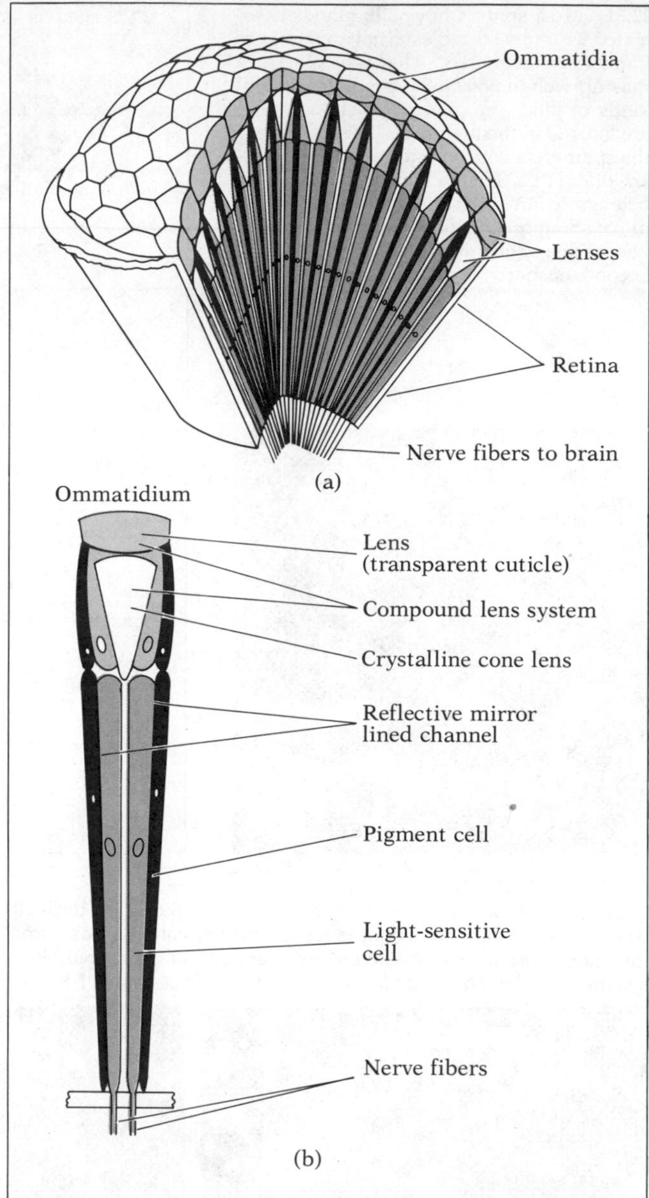

22.33 (a) The compound eye of insects and crustaceans. Crustaceans have compound eyes, as do the insects. The compound eye consists of many individual light-receiving units called *ommatidia.* (b) Each unit has a compound lens (a thickened cuticular lens and a crystalline cone lens), a mirror-lined channel and a cluster of long light-sensitive cells. The mirrored channel is an efficient light collector and is most abundant in crustaceans that live in dim light. The unit is surrounded by pigment cells that screen it and prevent light from diffusing from one unit to another. The eye is apparently well-adapted for detecting movement, but may not present as clear an image as the vertebrate eye. An insect eye is diagrammed here. The number of ommatidia varies greatly in different arthropods, from as few as 20 to as many as 28,000. The dragonfly has a great many since it must detect the movements of tiny flying prey.

Classes Chilopoda and Diplopoda

Don't confuse *centipedes* (Chilopoda) and *millipedes* (Diplopoda). Millipedes are harmless herbivores. Centipedes, on the other hand, are predators and not only have a more fearsome appearance, but, in fact, can be dangerous (see Figure 22.30). The centipede's first pair of legs are modified into perforated claws that are capable of injecting poison into prey. The sting of common centipedes can be painful, but probably no more life-threatening to humans than a wasp sting. The sting of larger tropical species can put an adult in bed with a fever for several days.

Class Insecta

The majority of arthropods and, indeed, the majority of animals, are *insects.* Except in the sea, where the crustaceans hold sway, insects rule the earth in terms of numbers and kinds. Insects are of such importance ecologically and economically that we literally could not have reached our own place in nature without understanding something about them. One reason is that we are continuously engaged in conflict with them and, even now, we don't always win.

The segmented body of insects has undergone considerable modification from the ancestral form, primarily through the fusion of the segments. Only in the abdominal region and in the embryo is it possible to see a clearly segmented body. Insects have three major body regions: *head, thorax*, and *abdomen* (Figure 22.34). The thorax gives rise to three pairs of legs, and, in some species, wings. The legs are modified in different species for running, jumping, or simply resting. The appendages on the abdomen have been modified to form the genitalia and anal structures. Many females have an ovipositor, a hollow appendage that is used to dig or pierce holes into which the eggs are laid.

Insects breath by admitting air into a complex tracheal system through external openings known as *spiracles*. The highly branched tracheae carry air through the body to individual cells where the gas exchange occurs. Some tracheae end in air sacs where surface area is greatly increased (see Chapter 26).

Of course, the head bears the major sense organs and the mouthparts. Consider, for example, the head of the grasshopper. It has one pair of sensory antennae, a pair of compound eyes, and three simple eyes (*ocelli*). Its mouthparts consist of a single upper lip (*labrum*), and a lower lip (*labium*) bearing sensory palps, and two laterally movable *mandibles* (jaws) that do most of the chewing, assisted by a pair of *maxilla* also bearing sensory palps. The entire apparatus is magnificently designed for biting and chewing. The sensory palps detect texture and flavor, the jaws rip off bits of vegetation, and the lips hold the food in place during chewing.

Mouthparts in other insects are often highly specialized to fit their specific feeding niche. Butterflies have their maxilla greatly extended into a flexible sucking device (neatly coiled up when not in use). The cicada has a piercing and sucking mouth, while the housefly has a large complex tonguelike apparatus for licking. (Insect mouthparts are illustrated in Chapter 25.)

Returning to our grasshopper, let's examine the locomotor structures. Grasshoppers are flying insects. They have two pairs of wings, although only one pair actually propels the insect. The other serves as a protective wing cover. Of course, many insects can fly, and some have considerable range. The desert locust (a close relative of grasshoppers), for example, can fly over 200 miles in a day.

Some insects can beat their wings at incredible rates. The midges, for example, have a recorded beat of 1000 cycles per second! Fast rates of wing beating cannot be explained by direct nerve action because nerves can't carry impulses that frequently. We're not sure how insects are able to accomplish this, but it appears to be due to the resonating of stretched muscles or thoracic shells. (See Chapter 24.)

The legs of insects vary with the animal's mode of life. The grasshopper, for example, has two pairs of legs of similar size for walking and grasping, while its third pair is greatly enlarged. The combination of extremely powerful muscles and a light exoskeleton produce a great mechanical advantage, so the grasshopper is able to jump great distances from a standing start.

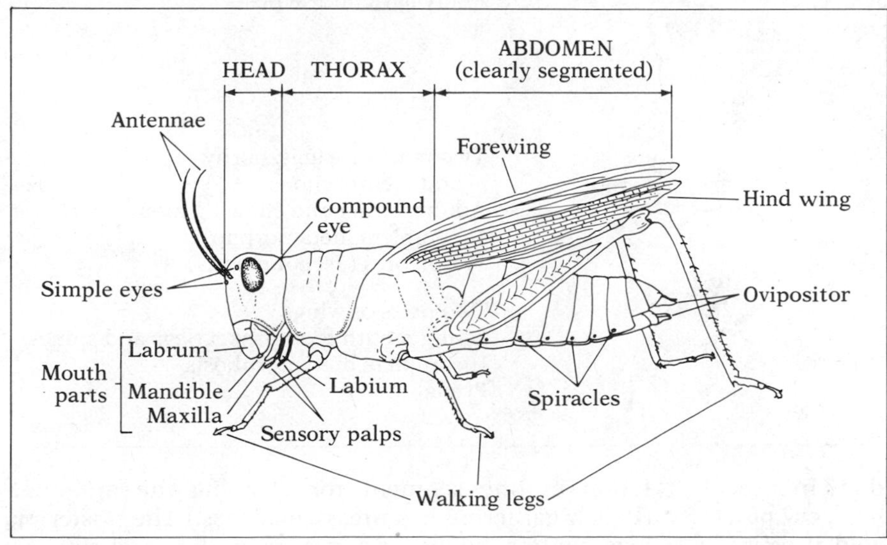

22.34 The insect body plan (represented by the grasshopper) consists of three main parts: *head, thorax*, and *abdomen*. Segmentation is most apparent in the abdomen and is obscured in the thorax and head because the segments there are fused. The head bears a pair of compound eyes and two or three simple eyes, in addition to a pair of sensory antennae. The mouthparts include the jaws and sensory *labrum* and *labium* (lips). The largest accumulation of nerves is located in the head, and comprises the insect brain. The thorax gives rise to three pairs of walking legs and two pairs of wings. The abdomen contains twelve hinged segments. A pair of *spiracles* penetrates each segment and admits air to the branching system of trachea that extend throughout the body. The last segments bear the reproductive structures and the anal opening. In the female, some of these segments are modified into *ovipositors* (structures for aiding the process of egg laying).

Table 22.1 Major Insect Orders[a]

Order (number of named species within order)	Examples	
Coleoptera (500,000)	Japanese beetle Stag beetle	Two pairs of wings, horny and membranous Heavy, armored exoskeleton Biting and chewing mouth parts Complete metamorphosis Herbivores, predatory carnivores, scavengers Serious agricultural pests
Lepidoptera (140,000)	Moths Butterflies	Two pairs of wings Hairy bodies Long coiled tongue for sucking Complete metamorphosis Serious agricultural pests
Hymenoptera (90,000)	Bees Wasps Ants	Two pairs of membranous wings Head mobile Well-developed eyes Chewing and sucking mouthparts Stinging Complete metamorphosis Many are social Important as pollinators and predators of other insects
Diptera (80,000)	Flies Mosquitos	One pair of wings and halteres Sucking, piercing, lapping mouthparts Complete metamorphosis Medically dangerous as carriers of diseases (malaria, yellow fever, sleeping sickness, encephalitis, filiariasis)
Hemiptera (55,000)	True bugs: Bed bug Assassin bug Homoptera: Aphids Scale insect Cicada Plant lice	Two pairs of wings, horny and membranous Piercing mouthparts Many agricultural pests
Orthoptera (30,000)	Roaches Grasshoppers Mantis Crickets	Two pairs of wings, horny and membranous Adults, biting and chewing mouth Incomplete metamorphosis Herbivores (except mantis)
Odonata (5000)	Dragonfly	Two pairs of wings Biting mouth food basket (legs and spines) Incomplete metamorphosis Predator

And there are a host of other variations in insect legs. The praying mantis, for example, has very powerful forelegs which snatch prey and hold it as it is ripped apart by the insect's jaws (Figure 22.35). Honey bees have walking legs modified for pollen collection. The first pair is equipped with a pollen brush for collecting pollen from the body and a circular indentation in one segment for cleaning the antenna. (Pollen gathering is a messy business.) The posterior legs contain hollow grooves, or pollen baskets, for compacting pollen (Figure 22.36).

Body lice are also highly specialized (see Table 22.1). They are able to get a good grip on your skin

Table 22.1 *(continued)*

Order (number of named species within order)	Examples	Characteristics
Neuroptera (4000)	Ant lion Dobson fly Lacewing	Two pairs of membranous wings Biting mouthparts Complete metamorphosis Silk cocoon
Anoplura (2400)	Lice	Wingless Sucking or biting mouthparts Small with flattened body, reduced eyes Legs with clawlike tarsi (for clinging to skin) Highly host-specific (humans have head, body, and pubic lice) Medically important (carry typhus and cause relapsing fever) Incomplete metamorphosis
Isoptera (2000)	Termites	Two pairs of wings, but some stages are wingless Chewing mouthparts Social, division of labor for reproduction, work, defense Incomplete metamorphosis
Siphonoptera (1200)	Fleas	Small, wingless, laterally compressed Piercing and sucking mouthparts Jumping legs Complete metamorphosis Medically important (carry bubonic plague, typhus)
Ephemeroptera (1000)	Mayfly	One pair of wings Vestigal mouthparts in adults (do not feed) Few days of adult life (reproduce and die) Incomplete metamorphosis
Dermaptera (1000)	Earwig	Two pairs of wings, leathery and membranous Biting mouthparts Large pincerlike cerci in males Incomplete metamorphosis

[a]The number of named species in each of the thirteen orders listed ranges from a very modest 1000 in the Dermaptera to a half million in the Coleoptera. In fact, by conservative estimate, it is accurate to say that one in four animals roaming the earth is a beetle. Most of the remainder of the animals are butterflies, moths, flies, mosquitos, bees, ants, and wasps (and their near relatives).

by using their grasping legs. These are equipped with sharp, curved tips that give the parasite six toe-holds while their biting and sucking mouthparts, buried deep in the tissues, draw blood. And then consider the sensitive legs of roaches that detect delicate air currents set up by anything that might be after them, such as an irate apartment dweller.

Of course, these are only a few examples of insect diversity. We discuss other specialized facets of their bodies and lives in other places, but consider the insect orders in Table 22.1 for a final reminder of what we're up against.

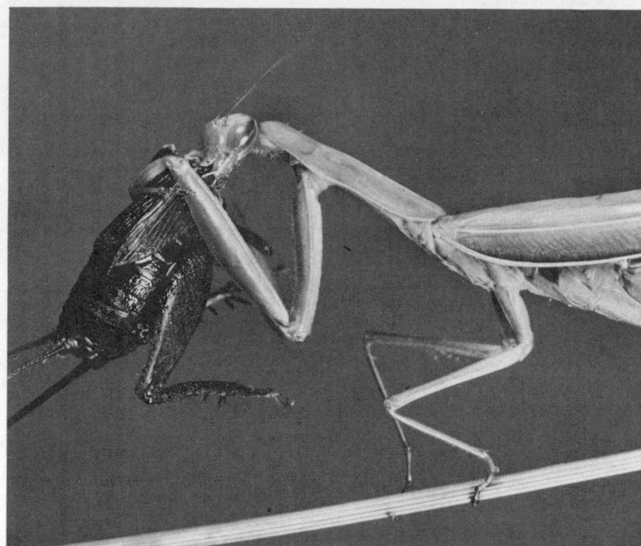

22.35 The praying mantis is a voracious predator. This mantis has torn its insect prey open with its powerful barbed forelegs and is feeding with its strong jaws.

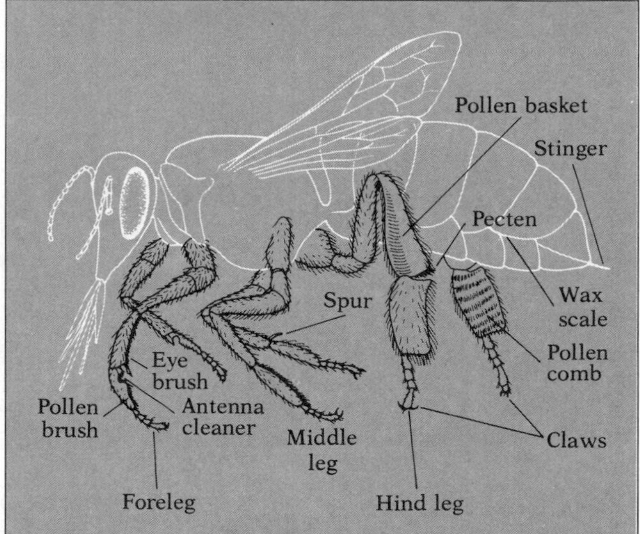

22.36 Specialized legs of the honeybee. The first and last pairs of legs on the honeybee are specialized for collecting pollen, which the bee uses for food. The first pair is used to collect pollen and to clean the eyes and antennae. Note the eye brush and antenna cleaner. The second pair of legs is relatively unspecialized, but does have spurs that the bee uses to remove wax "plates" from its abdomen (for use in hive building). The hind legs contain a pollen comb and a pollen basket. These utensils are used to remove pollen from the other legs, crush the pollen into compact masses, and carry it home. As the bee goes about its business, plants get pollinated because some pollen is lost, but most goes to form "bee bread," a food made of a combination of nectar and pollen.

Phylum Echinodermata: The Spiny-Skinned Animals

Your first look at *echinoderms* may lead you to believe you are on the wrong road to the vertebrates. Sea stars, brittle stars, urchins, sea cucumbers, and sand dollars (Figure 22.37) appear unlikely relatives to amphibians, fishes, mammals, and birds, not to mention humans. You won't find much in the comparative anatomy of echinoderms and vertebrates to instill confidence in the idea either. To see the relationship between the echinoderms and the vertebrates, we must look at their earliest embryonic development, and here the echinoderms are unveiled as deuterostomes. Their embryos indicate that they are related, not directly to vertebrates, but to their cousins, the soft-bodied hemichordates and urochordates, which in turn show clear relationships with the chordate phylum. We will review this relatedness in the next chapter.

Echinoderms are unusual in many respects. Their spiny, crusted covering is actually an endoskeleton, although it may seem to be on the outside. Echinoderms are covered by an epidermis, beneath which lie the calcareous plates that are secreted by the dermis (which is of mesodermal origin). (Recall that the shells of mollusks and the exoskeletons of arthropods are both secreted by epidermal cells that stem from ectoderm.)

The body of echinoderms is essentially a radial, five-part construction, a scheme found in no other animal. Echinoderms show no obvious segmentation, indicating they are not closely related to the annelid line.

One of the most unusual traits of the echinoderms is their water vascular system, which is composed of a series of canals which contain water, and by hydrostatic pressure help to evert their numerous muscular "tube feet." Each tube foot is equipped with a terminal sucker in most species.

Echinoderms are exclusively marine animals. Many species are found in the shallow waters of the continental shelf, and particularly just below the reaches of the lowest tides. However, one group, the brittle stars, have been found in the deepest trenches of the ocean.

The phylum consists of five classes: The Crinoidea (crinoids: sea lilies), Holothuroidea (sea cucumbers), Echinoidea (sea urchins and sand dollars), Asteroidea (sea stars), and Ophiuroidea (serpent stars and brittle stars). Each class differs considerably from the others, with perhaps the greatest departure being the soft-bodied, wormlike sea cucumbers, which have only scattered elements of an endoskeleton. The unusual modifications are adaptations for a burrowing mode of life.

Class Asteroidea: The Sea Stars

Let's take a look at one class of echinoderms—the sea stars (see Figure 22.37). Sea stars (starfish) are well-known for their voracious appetite when it comes to gourmet foods, such as oysters and clams. Obviously, they are the sworn enemy of oystermen. But these same oystermen may have inadvertently helped the spread of the sea stars. At one time, when they caught a starfish, they chopped it apart and vengefully kicked the pieces overboard. But they were unfamiliar with the regenerative powers of the starfish. The central disk merely grows new arms, and a single arm can form a new animal (Figure 22.38).

Stars are slow-moving predators, so their prey, obviously, are even slower-moving or immobile. Their ability to open an oyster shell is a testimony to their persistence. When a sea star finds an oyster or clam, the prey clamps its shell together tightly, a tactic that discourages most would-be predators, but not the starfish. It bends its body over the oyster and attaches its tube feet to the shell, and then begins to pull. Tiring is no problem since it uses tube feet in relays. Finally, the oyster can no longer hold itself shut, and it opens gradually—only a tiny bit, but it is enough. The star then protrudes its stomach out through its mouth. The soft stomach slips into the slightly opened shell, surrounds the oyster, and digests it in its own shell.

The tube feet are also used in movement, again through sequential application to a surface. Tube feet are located in double rows in the grooved underside (oral side, since the starfish mouth is on the bottom) of each arm (Figure 22.39). Each tube foot is attached to a central canal, and the canals all connect to a hollow ring canal (part of the water vascular system) in the central disk. Water enters and leaves the water vascular system through a ver-

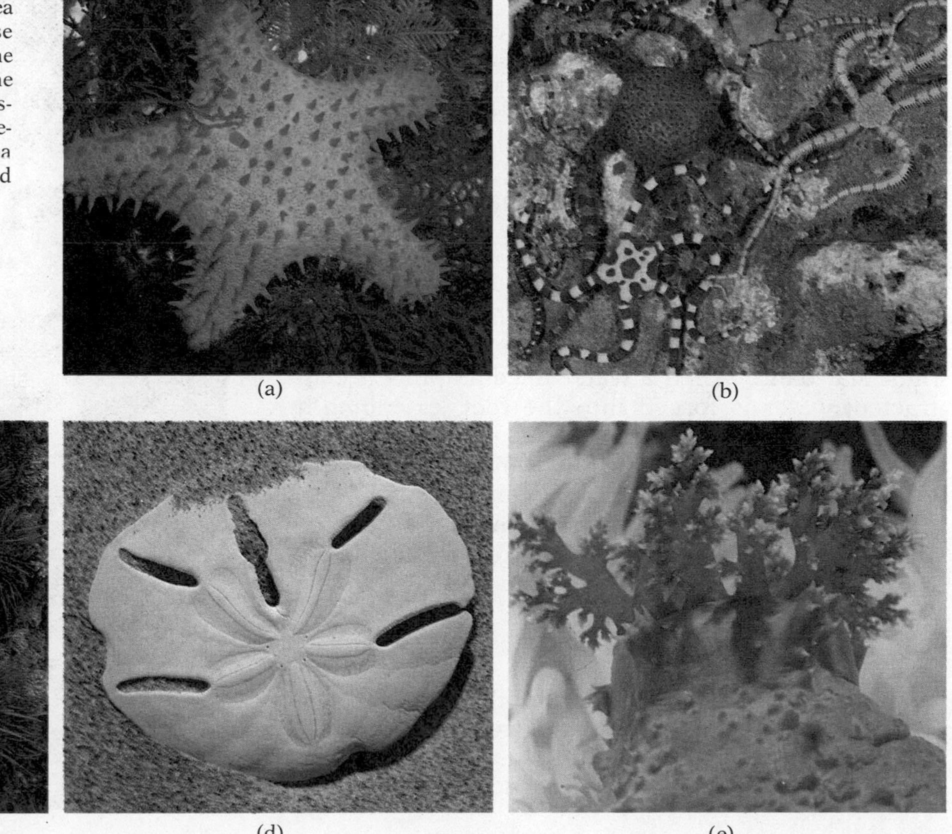

22.37 One of the most striking features of the echinoderms is their five-fold radial symmetry (this is easiest to see in the sea stars and the brittle stars). In addition, these marine animals have a spiny skin, with the longest spines found in the sea urchins. The echinoderms are unique in that they possess a water vascular system, used in movement and feeding. Shown here are (a) sea star, (b) brittle star, (c) sea urchin, (d) sand dollar (skeleton), and (e) sea cucumber.

(a) (b) (c) (d) (e)

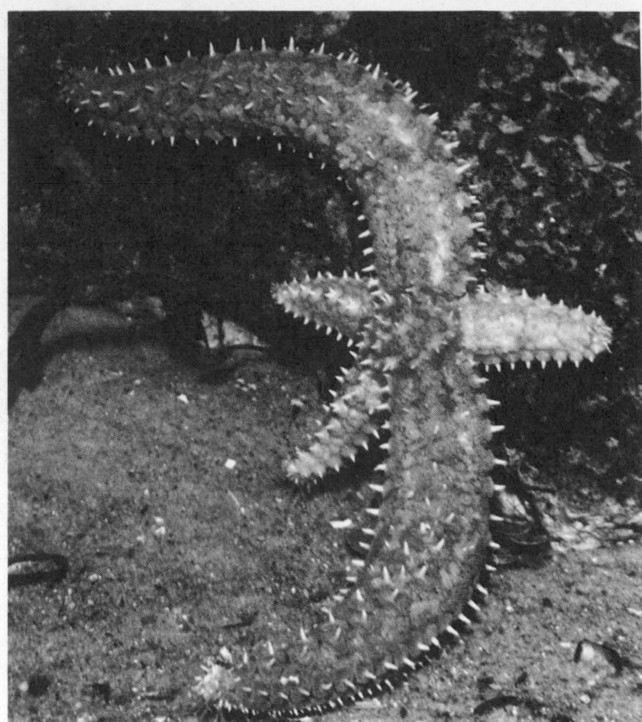

22.38 Sea stars are endowed with considerable regenerative ability. Most species can regenerate an entirely new body from any portion of the central disk. (One species, *Linckia,* can even regenerate a new body, complete with disk and arms, from very small pieces of any arm.) While their regenerative powers are well-known, the cellular details of their regeneration are not. However, it is generally believed that regeneration occurs through a dedifferentiation of cells and a return to the earlier embryological sequence. In some instances, mesenchyme cells (which have never really differentiated) can assume the task of building a new structure to replace a lost one.

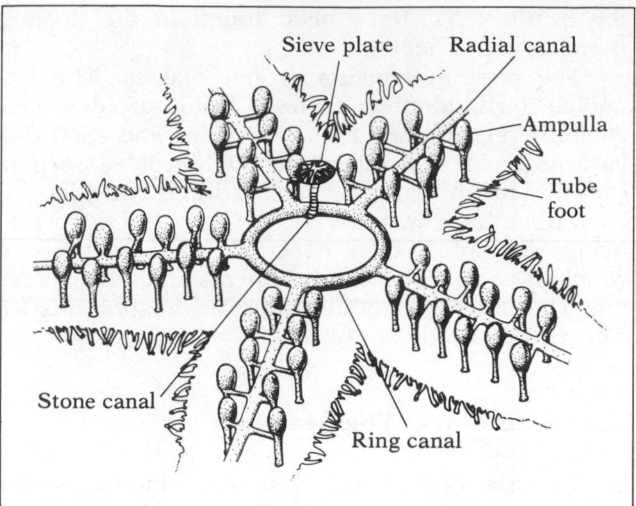

22.39 The water vascular system of echinoderms consists of a supply system composed of a *sieve plate,* a number of *canals,* and numerous *tube feet.* Water is used to fill the *ampullae.* These are, in turn, used to extend the tube feet. This is accomplished by contraction of the ampullae, which fill the tube feet with water, greatly lengthening them. Each tube foot ends in a sucker tip. When several of the tube feet contact a surface, they can contract by muscular action, forcing water out and sucking at the surface (water is permitted to reenter the ampulla and canal system by a series of valves). In this way, the tube feet, working in series, pull the sea star along the sea bottom. The tube feet are also used in opening clams.

tical tube (the *stone canal*), which has a sieved opening in the upper side of the star. Each tube foot contains longitudinal muscles that can contract, shortening the delicate structure. Above each foot is a rounded sac (the *ampulla*), which is similar to a squeeze-bulb. Each ampulla can be contracted by a muscular band. When a water-filled ampulla is contracted, water is forced into the foot, extending it hydrostatically. The sucker end of the foot attaches to a surface and the longitudinal muscles then contract, shortening the tube and causing suction on the undersurface. In this way, the animal gets its "footing." Thus, the tube feet, contracting in a highly coordinated fashion, can draw the animal over the rocky bottom of its habitat.

The starfish, like all echinoderms, lacks a centralized nervous system. The greatest concentration of nerves is in a ring around the mouth, with branches extending into the arms. Sensory receptors are limited in the starfish to one sensory tentacle and one light-sensitive region at the tip of each arm.

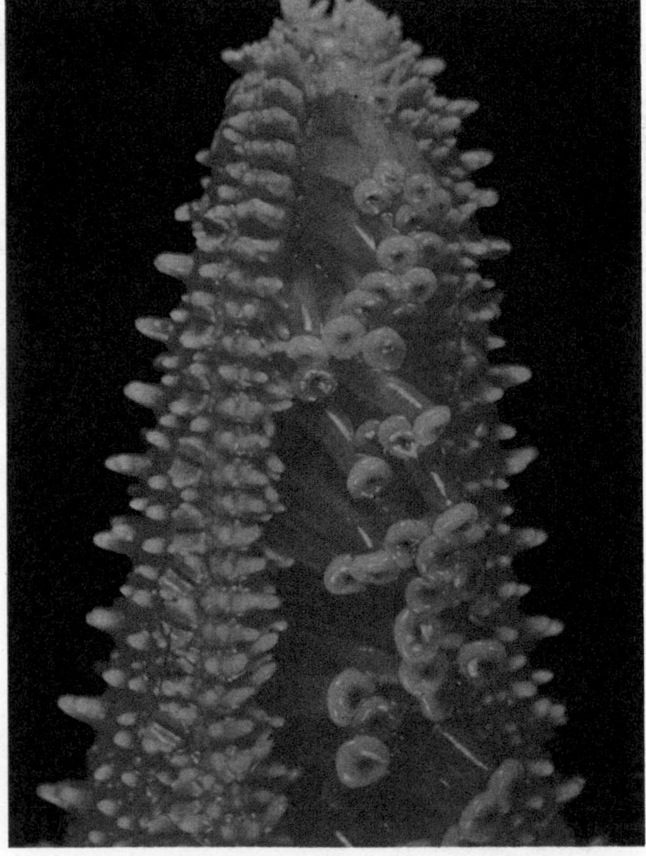

Echinoderms are either male or female, but it is hard to tell which is which. Sperm or eggs, produced in the well-defined testes or ovaries found in each arm, are shed into the water through tiny openings in the arms. Fertilization takes place in the water. The sea star zygote cleaves radially as opposed to the spiral cleavage of protostomes (Figure 22.40). The embryo differentiates into a free-swimming larva which is similar to other deuterostome larvae. The coelom also develops in typical deuterostome fashion. Thus, whereas adult sea stars don't resemble vertebrates at all, our ancestors were probably rather closely related, as suggested by echinoderm embryology.

The echinoderms, represented by fossil crinoids, are found in Cambrian deposits along with fossil Protostomia. These primitive echinoderms, called *eocrinoids*, were stalked animals somewhat different from their living descendants. The presence of deuterostomes in the earliest Cambrian strata means that the protostome–deuterostome division must have taken place somewhere in Pre-Cambrian times. Other classes of echinoderms appear first in the Ordovician, the next oldest period.

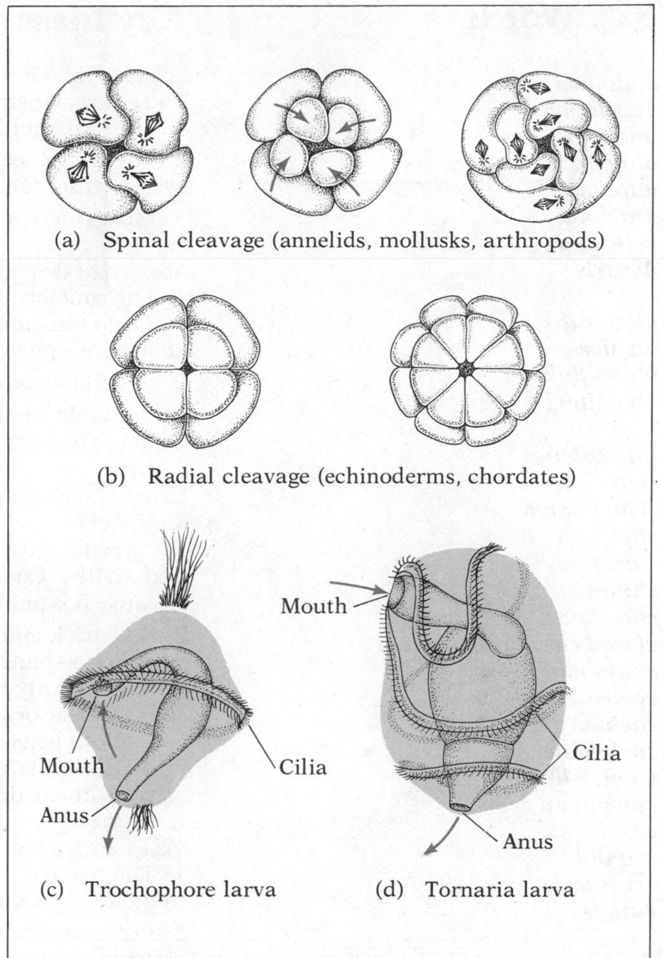

22.40 The pattern of development in echinoderms is unlike that of the annelids or mollusks. To begin with, the egg cleavage is not spiral (a), but radial (b). In addition, the larval stages do not resemble the trochophore larva (c). Instead, they resemble the tornaria larva (d) of the acorn worm, which is a primitive relative of the chordates (a hemichordate). For these and other reasons, the echinoderms are placed in the chordate line of evolution. Their radial symmetry is merely considered a secondary development. In addition, the larva itself is bilateral, as are the chordates.

Application of Ideas

1. Interestingly, one group of animals, the poriferans, and one group of plants, the bryophytes, each represent an aside from the main line of animal and plant evolution. Both emerged from protist ancestors, although very different ones, and each is doing well today but in restricted environments. Describe these two groups and how they relate to the main lines. In what way do they differ from the main-line organisms?

2. Discuss the idea that the coelom might have evolved twice in animals. What is the basis for this supposition and are there any alternative ideas?

3. Older concepts of animal evolution (still accepted by laymen) tended to place man at the pinnacle and treated other animals as though they represented failures or arrested stages on the way to human evolution. Present a strong argument against this point of view. Where does evolution lead? Are humans still evolving?

4. Insects are often regarded as man's chief competitor for supremacy on the earth. Comment on this idea by comparing our survival strategies with those of insects (as a group).

5. Using the mollusks as examples, explain the evolutionary terms: *primitive* and *advanced; specialized* and *generalized.* Why are the terms usually inappropriate when used to describe the entire organism?

6. Explain the importance of animal embryology to the development of evolutionary trees. Give examples. What problems would arise in systematics without embryological knowledge?

7. Radially symmetrical animals are all aquatic. Suggest reasons why this type of symmetry is not found in the terrestrial environment. Using your imagination, suggest a terrestrial radially symmetrical animal and explain how it might function on the land.

Key Words

abdomen
acoelomate
anemone
annelid
antenna
arachnid
arthropod
Ascaris

bilateral
bivalve
blood fluke
book lung

Cambrian
centipede
cephalopod
chelicera
chitin
chiton
chordate
closed circulatory system
cnidoblast
coelenteron
coelom
coelomate
comb jelly
compound eye
coral
crinoid
crustacean
cuticle

deuterostome

echinoderm
ectoderm
ectoparasite
ectoproct
endoderm
endoparasite
endoskeleton
eocrinoid
excretory system
exoskeleton

filter feeder
flame cell
flatworm
fluke

ganglion
gastropod
gastrotrich
gastrovascular cavity
gills
glue cells

hemichordate
hermaphroditic
horsehair worm
hydrozoan

Key Ideas

What Is an Animal?
The characteristics of animals are:
1. They are multicellular and eucaryotic.
2. They have cell and tissue differentiation. They often have well-organized systems that carry out life functions.
3. Most can reproduce sexually with motile sperm and stationary eggs.

Metazoan Origins and Phylogeny
1. The animals are organized into two groups, the parazoa (sponges) and the *metazoa* (all others). Sponges lack the organization of other animals. The metazoans are a monophyletic group.
2. Several schemes have been proposed for animal phylogeny.
3. Complex metazoans evolved in *Pre-Cambrian* times with nearly all phyla evolving in a period of about 70 million years.

Ernst Haeckel, Jovan Hadzi, and the Animal Tree
1. Haeckel selected the green alga, *Volvox,* as the ancestral type. He related the peculiar inversion of reproductive spheres to the form of *coelenterates* and to the gastrulation process in many embryos. Critics object to *Volvox* because it is photosynthetic and lives in fresh water.
2. Sedgwick and later Hadzi proposed the multinucleate marine ciliates of the protozoa as ancestral types for several reasons:
   a. Similarity to some coenocytic marine flatworms
   b. Marine origins
   c. Both heterotrophic
   d. Both ciliated
   e. Both are diploid and reproduce sexually in a similar manner

Trees and Trees
1. The two major phylogenetic trees emphasize different animal characteristics in their organization.
2. *Protostome–deuterostome:* The major divisions are made between these two embryological factors.
   a. *Protostomes* include *mollusks, annelids,* and *arthropods*. The embryonic mouth forms early as an inward movement of cells into the hollow blastula.
   b. *Deuterostomes* include *echinoderms* and *chordates*. The early invagination of cells produces an anus; the mouth appears later.
3. A second scheme divides the animals into *acoelomates, pseudocoelomates,* and *coelomates*. The *coelom* is the body cavity.
   a. The main trunk includes the *acoelomates,* animals with no *coelom* development at all (*coelenterates, flatworms,* and *comb jellies*).
   b. A first branch includes the *pseudocoelomates,* which have produced a cavity that is not a true *coelom* (roundworms and *rotifers*).
   c. The tree continues with the *coelomates* (all the higher phyla). It forms two main branches, one containing *echinoderms* and *chordates,* the other with the *mollusks, arthropods,* and *annelids.*

Phylum Porifera: The Sponges
1. The sponges apparently have a different origin in the protists than do the *metazoans*. Their semicolonial organization suggests a flagellate ancestor such as *Protospongia.*
2. Their level of organization is simple at the tissue level.
3. Sponges are sessile and mainly marine. Feeding and gas exchange is provided by collar cells. The skeleton consists of calcium, glass, or spongin. Eggs are fertilized internally and swimming larvae develop.

Phylum Coelenterata
1. The coelenterates include hydra, jellyfish, corals, and anemones. All have radial symmetry.

incurrent siphon

kinorynchid
krill

mandible
mantle
mantle cavity
mastax
maxilla
medusa
mesoderm
mesoglea
metazoa
millipede
mollusk
molt

nematocyst
nematode
nemertine
nervous system

ocellus
ommatidia
onychophoran
open circulatory system
operculum
organ system

parapod
parthenogenesis
peripatus
pharynx
planaria
planula
polyp
Pre-Cambrian
proglottid
protostome
pseudocoelomate

radial symmetry
radula
ring canal
rotifer

schistosomiasis
scolex
sea star
segmentation
sensory palp
spinneret
spiracle
stone canal

tape worm
tentacle
thorax
torsion
trachea
trochophore larva
tube foot

2. Characteristics include: *radial symmetry, cnidoblasts*, thin body wall of two cell layers, *coelenteron*, eight types of cells.

3. Two basic forms exist: the *polyp* and the *medusa*. Some species include both in an alternation of generations. Unlike plants, both generations are diploid.

4. The classes of *coelenterates* are arranged according to the dominance of *polyp* or *medusa* type.

Class Hydrozoa
The hydroids and *Hydra*. Except for the *Hydra*, a solitary freshwater polyp, *polyp* and *medusa* are equally important in most hydroids. The *polyp* is a colony with feeding and reproductive individuals. Hydroids reproduce by budding off *medusae*, which produce gametes. The larva settles to the bottom and develops into a new polyp colony.

Class Scyphozoa
The jellyfish. The *medusa* is the dominant form. These are the familiar jellyfish of coastal waters. Jellyfish feed by paralyzing prey with toxins from *cnidoblasts*. Sexes are separate and the polyp stage reduced or absent.

Class Anthozoa
1. The *corals* and *anemones* produce the *polyp* only. *Corals* are colonial, while anemones are mainly solitary. In their sexual reproduction a ciliated planula larva develops directly into the polyp.

2. *Coral* skeletons are significant in island and reef building.

Phylum Ctenophora
1. This phylum includes the *comb jellies*, which have many similarities to the *coelenterates*.

2. They lack stinging cells and capture prey with *tentacles* bearing *glue cells*.

3. The comb or combs are rows of fused cilia used in locomotion.

Phylum Platyhelminthes: The Flatworms
1. The *flatworms* are bilateral animals. They include both free-living and parasitic members. The body is derived from three germ layers, ectoderm, endoderm, and mesoderm.

2. They have anterior–posterior orientation with sensory structures at anterior end, and an organ–system level of organization.

3. The digestive system is pouchlike with one opening.

Class Turbellaria
1. Free-living, ciliated *planaria* is the best-known representative.

2. *Planarians* are *hermaphroditic* with a well-developed reproductive system utilizing internal fertilization.

3. The digestive system includes a muscular pharynx, but they retain the simple gastrovascular system. There is some centralization in the nervous system, and specialized flame cells aid in water regulation.

Class Trematoda
1. Trematoda includes the *flukes*. They are somewhat similar to *planaria*, but are not ciliated.

2. *Flukes* are common internal parasites with internal suckerlike mouths.

3. The human *blood fluke, Schistosoma mansoni*, is an important human parasite, common in Egypt and other tropical regions around the world. Infestation is called *schistosomiasis*.

Class Cestoda
1. This class includes the *tapeworms*. They lack digestive tracts, absorbing digested food from the intestine of their hosts.

2. The body consists of a *scolex* equipped with suckers and hooks, and a series of *proglottids*, which bud off the *scolex*. These contain testes and ovaries and, after fertilization, numerous fertilized eggs. The eggs pass out with the host's feces.

water vascular system

yolk gland

zooplankton

3. Some *tapeworms* have two hosts. In the alternate host, the *tapeworm* encysts in the muscles. It reaches the primary host again when the muscle is eaten.

Phylum Nemertinea: The Nemerteans, or Proboscis Worms
This phylum includes the proboscis worms. They are marine predators with an eversible pharynx and sharp lancet that pierces prey. The gut is complete with a separate mouth and anus, otherwise they are organized like flatworms.

Phylum Aschelminthes
1. This phylum includes round worms and *rotifers*. Taxonomically they are *pseudocoelomate protostomes*.
2. They have a complete (tube-within-a-tube) digestive tract with mouth and anus.

Class Nematoda
1. The roundworms are slender and cylindrical, with pointed ends. They move by flexing longitudinal muscles. Males and females are separate. They feed with a muscular *pharynx*.
2. Most soil nematodes are predators, but others are plant parasites. A few are animal parasites.
3. *Ascaris* parasitizes man and hogs. It completes its life cycle in one host. Each female produces millions of eggs.
4. Both dogs and cats harbor roundworms capable of infecting man.
5. *Trichinella spiralis*, a roundworm of pigs, is microscopic. The worms tend to burrow throughout the pig's body forming cysts, which will remain dormant until the flesh is eaten. Proper cooking or freezing techniques prevent infestation in humans.

Class Rotifera
1. The *rotifers* are free-living. They sweep in food with rings of cilia, grinding it with the *mastax*. The digestive tract is highly developed with specialized organs. The *excretory system* makes use of *flame cells*.
2. Females reproduce sexually or through *parthenogenesis*, depending on environmental conditions. Males are smaller and short-lived.

Other Aschelminthes
Other pseudocoelomates include the horsehair worms and priapulid worms. The latter are important because of their resemblance as embryos to rotifers. This suggest that rotifers may have evolved from priapulids, having lost the adult stage.

The Lopophorate Phyla
1. Three phyla, the Ectoprocta, Phoronidia, and Brachiopoda, exist. They are all coelomates.
2. Their name comes from a semicircular ridge, the *lophophore*, which bears nonstinging *tentacles*.
3. The phyla includes the plantlike colonial bryozoans, the wormlike 'phoronidians, which produce trochophorelike larva, and the clamlike brachiopod.

Phylum Annelida: The Segmented Worms
The key characteristic in this phylum is *segmentation*, a development that is seen to continue in the *arthropods*.

Class Oligochaeta: The Earthworms (Lumbricoides)
1. Earthworms are segmented inside and out. They have a *coelom* in which many organs are suspended. The digestive system is complex, with specialized organs, and it is *complete* with mouth and anus. The circulatory system is *closed* with arteries, veins, capillaries, and paired hearts.
2. Earthworms are *hermaphroditic*, with well-developed reproductive systems.
3. They are important ecologically as soil-cultivating agents.

Class Hirudinea: The Leeches

Leeches feed by drawing blood through their sucker mouths. They may be considered parasitic or carnivorous depending on your point of view.

Class Polychaeta

1. Polychaetes are marine, segmented worms, commonly represented by *Nereis.*
2. *Nereis* is a burrower, but it has sensory appendages and fleshy *parapods* used for swimming and respiration. It produces a *trochophore larva.*
3. Many polychaetes are tube dwellers, using feathery devices for feeding and respiration. Polychaetes are the oldest known animal fossils appearing in the Pre-Cambrian strata.

Phylum Mollusca

1. Mollusks comprise one of the largest animal phyla with 100,000 species. These are divided into seven classes, four of which are common.
2. They range in size from minute to over 50 ft in length.
3. Common characteristics include the *mantle,* foot, *gills,* and shell. These are modified in various ways in the seven classes.
4. Fossils of *cephalopods* go back to *Cambrian.*
5. The relationships of *mollusks* to *annelids* is seen in the larval forms in fossil *mollusks,* which are segmented with well-developed coeloms. (A surviving form, Monoplacaphora, has these ancient annelid traits.)

Class Amphineura: the Chitons

1. *Chitons* are the most primitive in body plan. Variations from this plan are considered specializations.
2. They have a long, flat foot, surrounded by a *mantle* and containing the *gills* within. Eight plates form the shell.
3. They feed by grazing on algae, using the rasping *radula.*
4. Their fossils extend back 500 million years.

Class Gastropoda: The Snails

1. Marine *gastropods* have *gills,* while terrestrial forms have lunglike *mantle cavities* for gas exchange.
2. Many have coiled shells and bodies that displace many organs during the developmental process.
3. The bodies in snaillike *gastropods* can be retracted and covered by an *operculum.*
4. In addition to snails, this class includes slugs, nudibranchs, limpets, sea hares, and abalones.

Class Pelecypoda: The Bivalves

1. The *bivalves* have two-part hinged shells, powerful adductor muscles, reduced sensory appendages, and a fleshy extensible foot. They are *filter feeders,* using siphons to bring in food particles.
2. *Gills* are used in respiration and an *open circulatory system* is present, as in most *mollusks.* The *coelom* is reduced to a chamber around the heart.

Class Cephalopoda: the Squid and Octopus

1. The head of *cephalopods* is greatly enlarged, covered by the *mantle.* The foot is modified into *tentacles* and the mouth contains a beak. The circulatory system is closed with two hearts and efficient gill circulation.
2. *Cephalopods* are fast-moving predators, equipped with highly developed eyes. Some have ink glands that can cloud the water. Octopuses and squid move rapidly by compressing the body and forcing water out through siphons (jet propulsion).
3. Octopuses are believed to be intelligent, with considerable learning ability.

Phylum Onychophora

1. The onychophorans, represented by *Peripatus,* suggest an interesting evolutionary link. They share characteristics of *annelids* and *arthropods.*

2. *Peripatus* is segmented, *cuticle*-covered like the annelids, but with appendages bearing claws. It also has jaws that are modified legs. It utilizes *spiracles* and *tracheae* in gas exchange, as do the insects, millipedes, and centipedes—a possible clue to the evolution of these groups.

Phylum Arthropoda

1. There are around one million species, representing the most diverse phylum.
2. *Arthropods* are segmented with jointed legs and exoskeletons.
3. They have open circulatory systems and gangliated nervous systems with a ventral nerve cord.
4. Their niches cover the earth with numerous narrow specializations.
5. They feed on nearly all energy sources, often with different foods for larval and adult stages.
6. *Arthropods* have an enormous reproductive capacity.

Subphylum Chelicerata

The jawless *arthropods* have six pairs of appendages, with four pairs of legs. They have sensory *palps* and *chelicera* fangs. The group includes spiders, horseshoe crabs, sea spiders, ticks, mites, and scorpions.

Class Arachnida

1. The arachnids include spiders, ticks, mites, and scorpions.
2. Spiders are carnivorous and venomous. The venom paralyzes and digests. Digested food is sucked out of the prey's body.
3. Spiders exchange gases via *book lungs,* which are heavily folded sacs.
4. Many spiders are equipped with silk glands and *spinnerets.*

Subphylum Mandibulata

This subphylum consists of *Arthropods* with jaws (*mandibles*).

Class Crustacea

1. These are marine and freshwater *gill* breathers. Their *exoskeletons* are commonly impregnated with calcium salts. Many *crustaceans molt* as they grow.
2. They are equipped with a compound eye consisting of units called *ommatidia.*
3. *Zooplankton,* part of the food chain of the sea, are principally *crustaceans.*

Classes Chilopoda and Diplopoda

Chilopoda includes poisonous and carnivorous *centipedes.* Diplopoda includes the herbivorous *millipedes.*

Class Insecta

1. The greatest number of animal species and the greatest diversity exists in the insects.
2. The general insect body plan is a three-part organization with *head, thorax,* and *abdomen.* Wings and legs arise from the *thorax.* Two pairs of wings are common along with three pairs of legs.
3. Mouthparts are often highly specialized to fit a specific feeding niche.
4. Insects use the tracheal breathing plan. Cells are never far from these branched passages.
5. Flying insects are numerous. The speed of wing movement in some is amplified by reverberating muscles.
6. The legs are also modified in a great number of specializations. Honey bees are an excellent example, with legs modified into pollen baskets, compactors, brushes, and wax extractors.

Phylum Echinodermata: The Spiny-Skinned Animals

1. *Echinoderms* include *sea stars,* brittle stars, sea urchins, sea cucumbers, and sand dollars. They have a *radially symmetrical* body as adults.
2. Their relationship to the *chordates* is established through the larval forms that are in the *deuterostome* group.
3. They have an *endoskeleton* covered with an epidermis and dermis. The endoskeleton consists of calcareous plates, secreted by the dermis.

4. Adult *echinoderms* have *radial symmetry* with a five-part division and they are unsegmented.
5. Their *water vascular system* is unique. It consists of a canal system and tube feet and works by hydrostatic pressure.
6. *Echinoderms* are organized in five classes:
   a. Crinoidea (sea lilies)
   b. Holothuroidea (sea cucumbers)
   c. Echinoidea (sea urchins and sand dollars)
   d. Asteroidea (sea stars)
   e. Ophiuroidea (serpent stars and brittle stars)

Class Asteroidea: The Sea Stars

1. Sea stars (starfish) are well known for feeding on oysters and clams.
2. The *water vascular system* with its *tube feet* are its means of locomotion and feeding.
3. In feeding on an oyster, the starfish surrounds it and attaches its *tube feet* in relays. When the oyster tires and opens, the starfish's stomach is everted over the soft parts of the oyster and digestion and absorption occur.
4. Each *tube foot* has longitudinal muscles that can contract the appendage. It is extended by water from the *water vascular system.*
5. In sexual reproduction, gametes are released into the water where fertilization occurs. The *dipleurula* larva, typical of *deuterostomes,* then develops.
6. Fossil *crinoids,* the *eocrinoids,* are present in *Cambrian* strata. This suggests that the *protostome–deuterostome* split occurred very early in animal evolution.

Review Questions

1. List several important characteristics of the animal kingdom. (p. 592)
2. What animal characteristics are shared with the protozoan protists? (p. 592)
3. What does the fossil record reveal about the length of time the major *phyla* took to emerge? (p. 593)
4. What was there about *Volvox* that captured Haeckel's imagination as he theorized it to be similar to the ancestor of all animals? (p. 594)
5. Review the five points made by Hadzi in his ciliate ancestor hypothesis. (p. 594)
6. Explain the differences between *protostome* and *deuterostome* embryos. (pp. 595–596)
7. Describe the differences between the *coelomate, pseudocoelomate,* and *acoelomate* condition. (p. 596)
8. Explain why sponges are not in the main scheme in the taxonomic trees. (p. 597)
9. List five characteristics of the *coelenterates.* (p. 598)
10. Describe the *alternation of generations* in a hydroid. How does this differ from alternation of generations in plants? (p. 599)
11. In what way are ctenophores similar to *coelenterates* and how are they different? (p. 601)
12. List five important characteristics of the *flatworms.* (p. 602)
13. What is a *complete digestive system* and in what phyla is it found? (p. 604)
14. List five characteristics of the *nematoda.* (p. 604)
15. Generally describe the abundance of *nematodes* and their distribution on the earth. (pp. 604–606)
16. Describe the activities of *Ascaris* in its hog or human host. (p. 606)
17. Describe sexual and asexual reproduction in *rotifers.* (pp. 607–608)
18. Describe the structure that the *lophophorate* phyla have in common. (p. 609)
19. List three examples of organ-system development in the earthworm. (pp. 609–611)
20. List five structures important to the classification of *mollusks.* (p. 612)
21. List the four major classes of *mollusks* and describe for each what has happened to the shell, *mantle,* and foot. (pp. 612–615)
22. What special feeding adaptation occurs in the pelecypods? (p. 614)
23. List four *cephalopod* adaptations for predation or defense. (pp. 614–615)
24. Where do the *Onychophora* fit in the evolutionary scheme? Explain. (pp. 615–616)
25. List four important characteristics of *arthropods.* (p. 617)
26. Describe three characteristics of spiders that make them different from most other *arthropods.* (p. 617)
27. How does the skeleton of *crustaceans* differ from that of other *arthropods?* (pp. 618, 620)
28. Describe the basic body plan of insects. (pp. 620–621)
29. List several examples of insect feeding specializations (pp. 621–623)
30. Describe the *echinoderm* skeleton. (p. 624)
31. What is the evolutionary link between *echinoderms* and *chordates?* (pp. 624–627)
32. Describe the action of the water *vascular system* as a *sea star* moves along a rock. (pp. 625–626)

Chapter 23

Animal Diversity - The Chordates

The hairy-nosed wombat cannot run very fast. Neither can its cousin, the naked-nosed wombat. But coyotes can, and cheetahs can outrun coyotes. None of these animals jump from the tops of trees, but flying squirrels and birds do. And a bird may go up instead of down. Both birds and chimpanzees build nests in trees, but only one gives milk and niether can hold its breath as long as a turtle. And most fish don't have to come up for air at all. Fish, however, do have certain things in common with turtles. In fact, all these animals have enough traits in common that they are placed in the same phylum: the *Chordata.* In addition, they are all in the same subphylum: the *Vertebrata;* that is, they all have backbones.

However, whereas all vertebrates are chordates, not all chordates are vertebrates. There are several groups of invertebrates—animals that never develop backbones—that are perfectly good chordates. The point is, chordates are a highly diverse group of animals that exploit their environments in very different ways, while sharing certain traits in common. Let's have a closer look at this fascinating phylum, now, paying particular attention to the vast spectrum of the group and their special qualities, while not losing sight of the critical ways they are the same. First, however, a glance at a different, relatively obscure phylum that is only distantly related to our own.

23.1 The Invertebrate Chordates

Phylum Hemichordata

The *hemichordates* ("half-chordates"), represented by the acorn worms, are marine animals. The acorn worms live in U-shaped burrows in the sand and mud along coastal areas. At the anterior end of the acorn worm is a conical or acorn-shaped *proboscis* (nose), and just behind it is a *collar* (Figure 23.1). Of particular importance are the rows of *gill slits* in the wall of the pharynx, behind the collar. These slits reveal the worm's affinity to the chordates. The apparatus acts as a kind of gill, and probably helps in feeding as well. When water is drawn in through the mouth, it is passed out through the gill slits, and oxygen is removed from the water by the tissue between the slits. The meager clue linking acorn worms to echinoderms is that the acorn worm larva resembles that of the echinoderms (Figure 22.40). In addition, of course, echinoderms, hemichordates, and chordates all are deuterostomes and form blastulas by radial cleavage. Keep in mind, however, that we are not implying that chordates descended from echinoderms. However, it does seem that in the distant past, echinoderms, chordates, and hemichordates had ancestral forms that were more closely related.

The name comes from the fact that hemichordates were assumed to have a very short, reduced *notochord* in the collar region. This structure appeared in drawings and diagrams for decades, but eventually the embarrassing truth emerged: acorn worms have absolutely no trace of a notochord.

Another hemichordate is much more mysterious. The *pterobranchs* are strange, tiny, budding colonial filter-feeders with leafy tentacles. They superficially resemble bryozoans more than anything else. In some species there is a single pair of gill slits. A collarlike region and a few other traits link them tenuously with the acorn worms. They are certainly among the most obscure of our relatives. Both groups of hemichordates have dorsal nerve cords that originate from the ectoderm that are sometimes hollow. But strangely enough, they have ventral nerve cords as well.

The larvae of hemichordates have a true coelom—actually, three pairs of coelomic cavities. This parti-

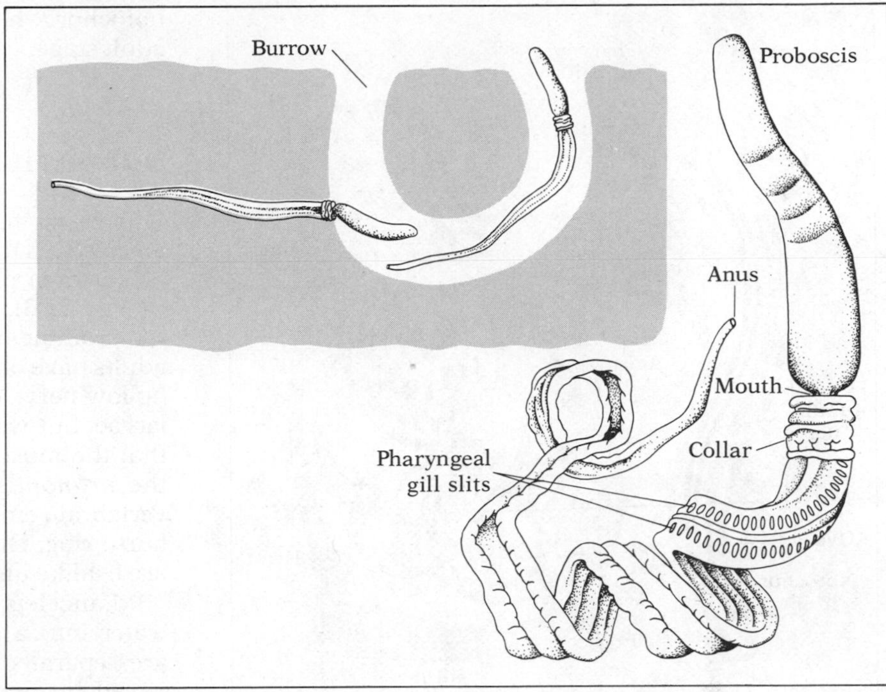

23.1 The acorn worm has a large number of pharyngeal *gill slits.* But there is little else to recommend this creature as a chordate. Its simple lifestyle places few evolutionary demands upon this animal. It burrows in mud flats, using its proboscis somewhat in the manner of an earthworm. As it tunnels along, water and organic debris enter the mouth, and the nutrients in the debris are sorted and digested by the gut, while the water passes out through the gill slits. As you might expect from its simple mode of living, there is very little development of the nervous system, which is simpler than that of the flatworm.

tioned coelom seems to be an ancient deuterostome character that is retained in some chordate larvae.

Phylum Chordata

The chordates are customarily divided into three subphyla: Urochordata, Cephalochordata, and, the Vertebrata. So far, we've been talking about this or that group being related to the chordates, but we haven't said what a chordate is. We have mentioned *gill slits,* however, and that's one of the structures present in all chordates. They all also have a *dorsal, hollow nerve cord.* (Our spinal cord is an example.) Most of the bilateral animals we've studied up to this point have a paired, solid *ventral nerve cord.* Finally, and perhaps most interestingly, chordates have (at least at some point in their development) a *notochord.* The notochord is a flexible, turgid rod running along the back of an animal and, when it is present, it serves as kind of a skeletal support. It consists of large cells, apparently under considerable hydrostatic pressure, encased in a tight covering of connective tissue; its rigidity is something like that of a tight-skinned Polish sausage. In nearly all living chordates the notochord exists only in the embryo or larva. In vertebrates it is replaced by the backbone early in the embryo's development.

A final defining characteristic of chordates is the postanal tail, which is present at least in some stage of development in all chordates. Whereas adults of one chordate, the tunicates, generally do not have such a tail, it is found in their larva, just as it is in our embryo.

Subphylum Urochordata

The most common *urochordates* are the tunicates (ascidians), or sea squirts. The name *tunicate* is derived from the tough covering (or tunic) that surrounds the soft, saclike body. Tunicates are revealed as chordates by their tadpolelike larvae, which bear notochords, a hollow, dorsal nerve cord (that enlarges anteriorly), gill slits, and a postanal tail (Figure 23.2). The larval stage only exists one or two days. But it is long enough so that we can call the tunicates chordates. The larva then attaches to the substrate and develops into a typical adult tunicate. Tunicates may live as solitary individuals, or may form colonies by asexual budding.

In its transition to the adult form it *loses* chordate characteristics as it develops. Its tail, notochord, and nerve cord are drawn into the body and absorbed, leaving only the enlarging gill sac or basket as a clue to its chordate relationship. They are filter-feeders, using their gill clefts as strainers. The adult circulatory system is open, consisting of little more than a bizarre heart that pumps blood first one way, then the other. To further confuse the biologist, tunicates secrete cellulose in producing the tunic; cellulose is rarely found in animals.

Other urochordates are free-swimming as adults, notably the bizarre, transparent *salps* that retain a simplified tunicatelike body plan. One urochordate group, the *larvacea,* retains the tadpole body plan,

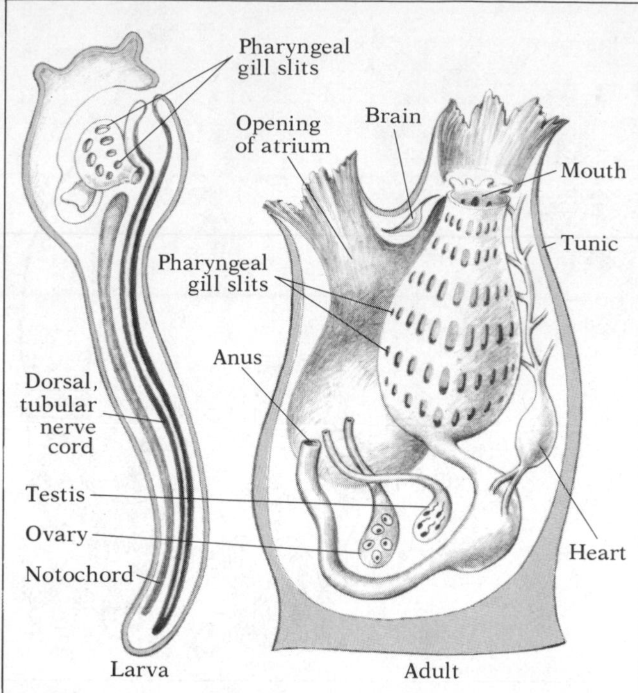

23.2 The tunicate and its larva. The tunicate as an adult hardly represents what we expect in a chordate. The adults commonly live in a colonial fashion on rocks and wharf pilings and are often found fouling boat hulls. The larval form of the tunicate places the animal securely in the chordate phylum and reminds us of the need for studying the entire life cycle of an organism when establishing phylogenetic relationships. As a larva, the tunicate has the pharyngeal *gill slits*, a lengthy *notochord*, and a *dorsal, tubular nerve cord*—all the hallmarks of the chordate phylum. As you can see, the body form of the adult is rather simple and there are no specialized sensory devices. But they maintain the pharyngeal gill slits throughout their adult life. Currents of water are produced by cilia in the gill structure drawing food particles into the digestive tract. Respiration occurs as water exits through the atrium, along with digestive wastes.

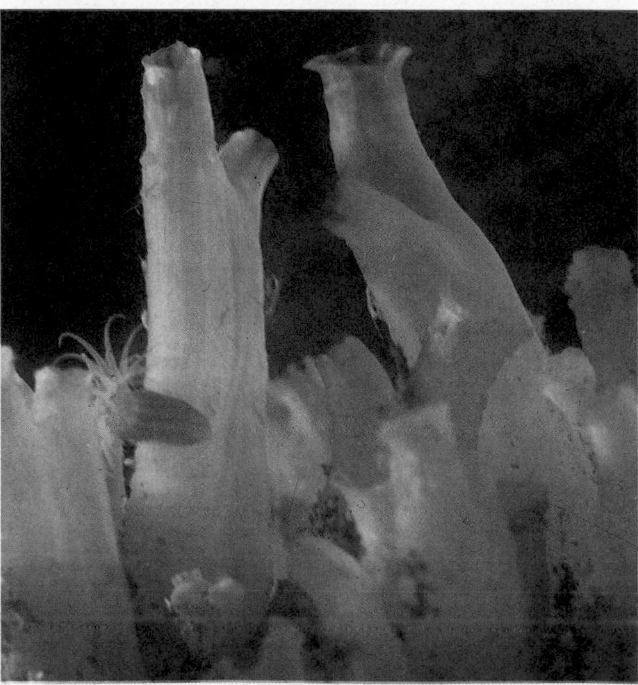

notochord, hollow, dorsal nerve cord and all, in the adult stage.

Subphylum Cephalochordata

It is easier to believe that the *cephalochordates,* or *lancelets,* as they are commonly known, are members of our own phylum. They look very much like fish. (Figure 23.3). As you can clearly see, the notochord is a prominent part of the adult anatomy. In addition, adults have obvious pharyngeal gill slits, and a dorsal, hollow nerve cord. They generally resemble tunicate larvae, but with a much larger gill apparatus. Notice that the musculature is a series of repeating units—the *myotomes*—a trait found in most fish and all vertebrate embryos, which aid in both swimming and burrowing. The lancelet's movements, not surprisingly, are fishlike undulations.

Lancelets are filter-feeders. Beating cilia draw water into a complex mouth, where food particles are separated out for digestion. An oral hood, projected beyond the mouth, bears twenty-two fleshy, sensory projections known as *cirri.* The mouth is surrounded by a delicate membrane, the *velum,* with twelve tentacles for sorting food. Water taken into the mouth passes through the gills, where O_2 and CO_2 are exchanged, enters an outer chamber (the *atrium*), and exits through an opening near the tail. A single, rather well-preserved cephalochordate fossil occurs in a rock from the early Cambrian.

23.3 This lancelet is *Amphioxus*, an inhabitant of sandy regions of the shore in most tropical and subtropical parts of the earth. The adult body is clearly similar to the vertebrate body plan, and its appearance is fishlike. The lancelet is a burrower without eyes. The notochord runs the length of the body and functions as an anchor for the paired rows of *myotomes* (muscle segments). In addition, there is a dorsal, tubular nerve cord and pharyngeal gill slits. Its filter-feeding apparatus is similar to that of the tunicate and is enclosed in an atrium. The gill basket strains food particles from water that is drawn into the mouth, passes through the gill slits, and finally exits through the posterior atrial opening. The anus opens separately, ahead of the tail (another chordate characteristic). In some ways, the lancelet resembles a tunicate larva; in other ways, a fish.

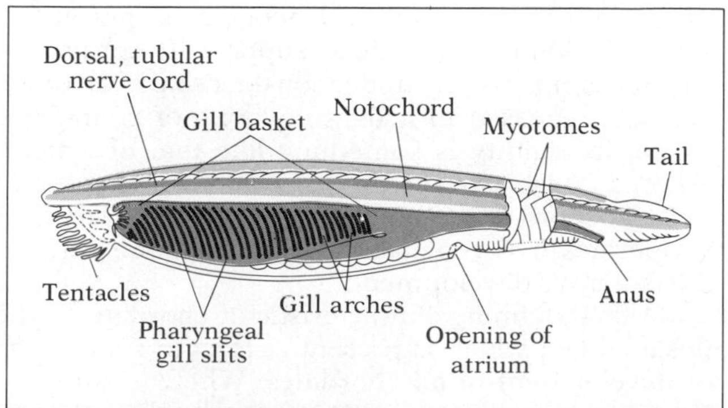

23.2 The Vertebrates

Subphylum Vertebrata: Animals with Backbones

Now, let's move on to the group that we are most familiar with. No more ventral nerve cords, or open circulatory systems, or bodies that can grow from arms, or anuses above the eyes. Finally, we find animals we know something about: the *vertebrates.*

Most vertebrates have several traits in common, in addition to the standard chordate characters. They include:

1. Backbone (vertebral column)
2. Cranial brain development
3. Closed circulatory system with a dorsal aorta and and a ventral heart
4. Gases exchange in gills or lungs
5. Two pairs of limbs
6. One pair of image-forming eyes
7. Excretory system consolidated in a paired kidney
8. Separate sexes

However, these are not all found in every member of the group and so can only be regarded as generalizations.

There are seven classes of living vertebrates, plus one that is extinct. They are:

1. Class Agnatha—jawless fishes; represented today by lampreys and hagfishes
2. Class Placodermi—extinct; first jawed fishes
3. Class Chondrichthyes—cartilaginous fishes (rays, sharks, and skates)
4. Class Osteichthyes—bony fishes
5. Class Amphibia—frogs, toads, and salamanders
6. Class Reptilia—turtles, snakes, lizards, crocodilians
7. Class Aves—birds
8. Class Mammalia—mammals

A great deal of vertebrate biology is covered in the chapters that follow, so here we will concentrate on an overview of the vertebrate groups.

Class Agnatha: Jawless Fishes

The earliest well-defined vertebrate fossils on earth are a group of agnathans called the *ostracoderms.* These were jawless, heavy, armored fish (Figure 23.4). Their fossils are found in the sediments of Ordovician, Silurian, and Devonian periods beginning half a billion years ago. The ostracoderms were fearsome-looking creatures with large heads and huge gills. They may be the first vertebrate fossils, but they were probably not the first gourmets. They most likely swam on the bottom, sucking up the sediments and filtering out any nutrients. Their fin structure tells us that they were sluggish swimmers, probably depending upon their armor and bottom-dwelling habits for protection against the large invertebrate predators that haunted the waters in those days.

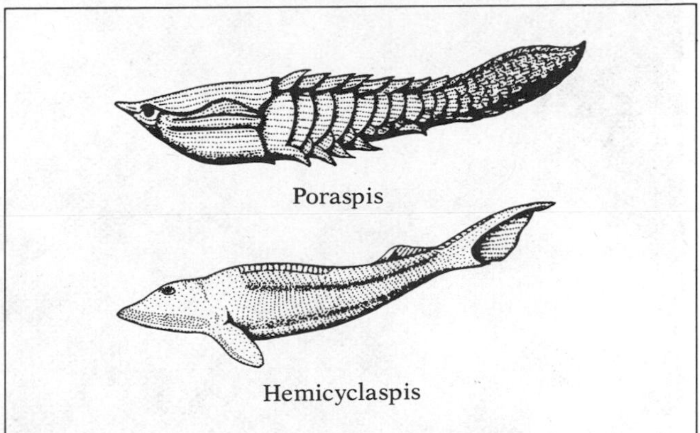

23.4 The ostracoderms were jawless fishes of the Ordovician period. *Hemicyclaspis* was large; its head contained the gill region and was enclosed in a hard bony shield with a single olfactory opening (nostril) on top. The mouth was a fleshy, sucking device. The heavy head may have been used to plow through the muddy bottom, where soft-bodied worms and mollusks could be found and sucked into the mouth. The *Poraspis* was more typically fish-like and was probably more adapted to swimming than bottom dwelling.

Today's agnathans, boneless remnants of the armored fishes (which died out at the end of the Devonian period), remain jawless. The lampreys and hagfish are members of the subclass Cyclostomata (round mouths). Their bodies are long and cylindrical and have simple median fins, better adapted for wriggling along the bottom than for fast active swimming seen in other fishes. The lamprey feeds with the aid of a rounded sucker mouth armed with horny spikes and a rasping tongue (Figure 23.5). The hagfishes, or slime hags as some species are called, lack the sucker mouth and feed by a rasping device, boring into the bodies of dead or dying animals. By our standards they aren't a particularly admirable group. Lampreys are parasitic, feeding by attaching to their host's body and, after filing away the skin, sucking out blood and body fluids. Incidentally, this method of feeding creates problems with gas exchange. In other fish, the mouth is used both for feeding and for the intake of oxygen-laden water that then passes over the gills and out through the gill clefts. This is the way the larval lamprey breathes, but in the adult, the gills are located in muscular pouches that pump water in and out their own openings, a feature that is absent in other fishes.

The lamprey life cycle includes a larva that somewhat resembles the adult lancelet. The larval period is very long. During this stage, the animal inhabits fresh

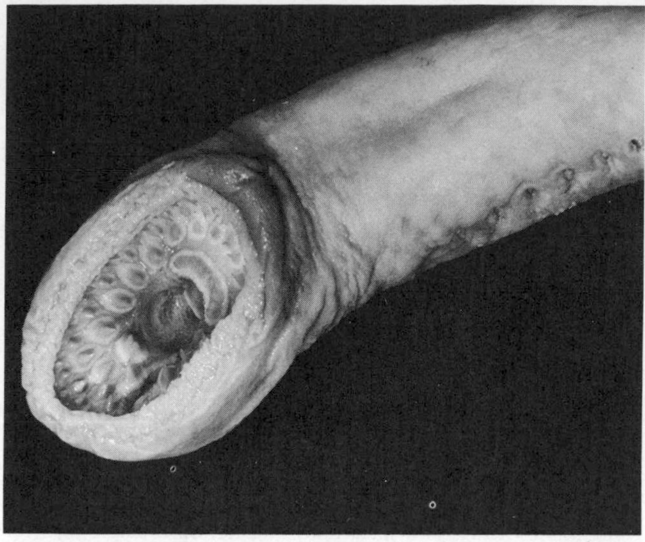

(a)

(b)

23.5 The cyclostomes. Today's jawless fishes are comprised of two groups, the lampreys (a) and the hagfish (b). The lamprey uses rasping and sucking mouthparts in feeding. Both groups lack a bony skeleton. A notochord persists in the adult along with a rudimentary cartilaginous backbone. Lack of bone was once considered to be a primitive condition, since cartilage is the developmental forerunner of bone in animals. But today many researchers believe that cartilaginous skeletons represent a secondary loss of bone instead. The adult lamprey is an ectoparasite, while the hagfish is a bottom scavenger known to attack weakened or injured fish.

water, but the adult form normally lives at sea until it is time to spawn. The lamprey has shown its physiological adaptability by surviving throughout life in the fresh waters of the Great Lakes, where it is a very successful parasite. Lampreys entered the lakes through the Welland Canal, which joins the St. Lawrence Seaway with Lake Erie. Since their unintended introduction, they now complete the larval stage of the life cycle in the surrounding streams and tributaries.

Class Placodermi: Extinct Jawed Fishes

The *placoderms* apparently descended from the ostracoderms, but no one can really be sure. They originated back in the Silurian period, their numbers swelling during the Devonian. Their line died out some 150 million years later. Clearly, they must be considered a successful group in terms of their tenure on earth.

Placoderms were another armored group, but they had hinged jaws, a hallmark of vertebrates (Figure 23.6). The presence of paired fins in the placoderms indicate they could swim a great deal better than the ostracoderms. In addition, the paired pectoral and pelvic fins of some were to give rise to the limbs of terrestrial animals. Hinged jaws and maneuverability enabled the primitive vertebrates to become predators, a role formerly dominated by invertebrates, surprisingly enough.

The placoderm jaw, an important evolutionary structure, appears to have been derived from the first pair of gill arches (the bony structure that supports the gills). (In later fish, additional gill arches became involved.) This, of course, required a considerable modification in both the position and strength of the gill arches. In addition, the musculature of the pharynx had to be rearranged. The upper and lower jaws of sharks and other fish form a hinge-like structure that is only loosely attached to the cranium (Figure 23.7).

From the placoderm line arose the two great classes of fishes we see today—the cartilaginous and bony fishes.

23.6 The placoderms are believed to have been ancestors to all living vertebrates other than cyclostomes. Some were probably predators. Note the teeth and hinged jaws of *Dunkleosteus*. Important advances in fin development undoubtedly helped with the predatory mode of life. The placoderms were armored, some with massive plates surrounding the head and other portions of the body. *Dunkleosteus* was 10 m long.

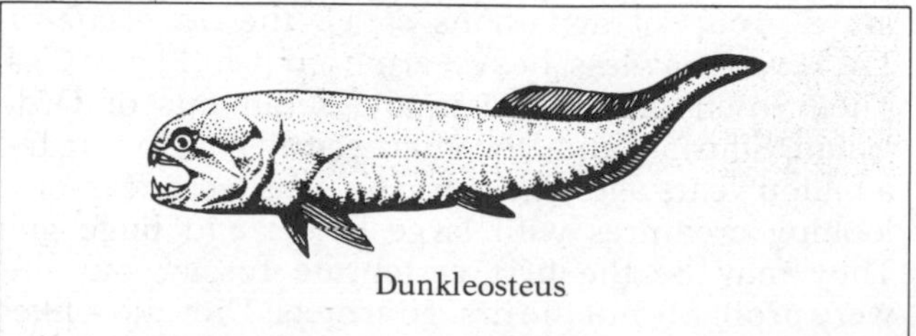

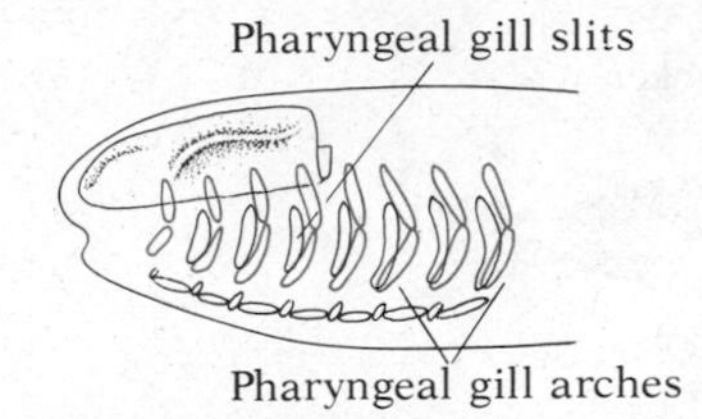

(a) Jawless ostracoderm (nine pairs of unspecialized gill arches)

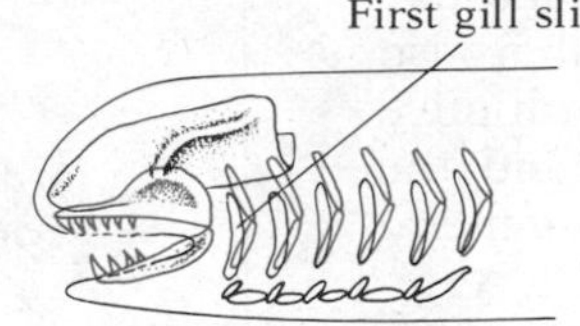

(b) Primitive jaw of placoderm (gill arches modified into weak jaws)

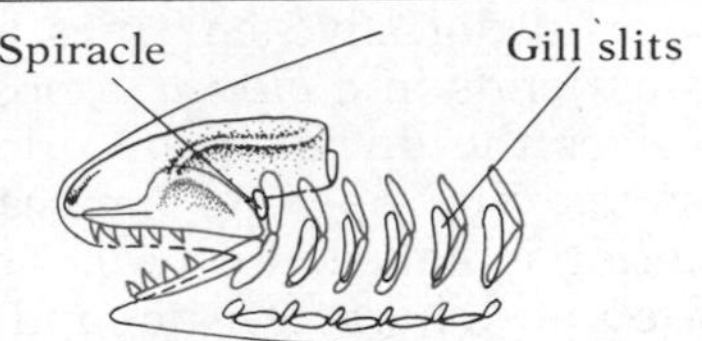

(c) Jaw of shark (gill arches modified into strong jaws and their supporting elements)

23.7 The jaws of today's vertebrates are believed to have evolved from the primitive gill arches. In this hypothetical series, the events unfold in a sequence from the ancient jawless fishes to a modern shark. (a) In ostracoderms, the gill arches are all similar and unspecialized. (b) In the placoderms, the first pair of gill arches has been modified into the primitive upper and lower jaws, accompanied by related muscle development. (c) In a modern shark, the first two pairs of gill arches are greatly modified into strong, hinged upper and lower jaws. The first gill slit has become the spiracle, a small, somewhat dorsal opening. In terrestrial vertebrates, the first gill slit becomes the middle-ear passage, and the bones of the second and third arch become the bones of the inner ear (not shown).

Class Chondrichthyes: Sharks, Rays, Skates, and Chimera

Chondrichthyes are the cartilaginous fishes—a large group of predators and scavengers. These modern predators have replaced protective armor and a heavy skeleton of their ancestors with a tough skin, light frame, and great speed. Their origin remains a puzzle, but their fossils first appear in the early Devonian period. By the Mesozoic era, such fossils dwindle as the numbers of bony fishes burgeon. But beginning in the Jurassic period, they again increased gradually, and since then they have held their own.

In addition to their cartilaginous skeleton, sharks and rays are unique in several other respects. Their body and tail shape is unlike that of most bony fishes (Figure 23.8). It is difficult to explain the advantages of either of these characteristics. In any case, note that the shark's tail is more similar to that of placoderms than to those of modern bony fishes.

The skin of sharks and rays is very rough, because it is embedded with miniature "teeth" known as *placoid scales.* Each consists of a *pulp cavity,* covered by a hard toothlike projection known as a *denticle.* Each is anchored into the dermis by a basal plate. Embryonically, these scales form in the same way teeth do. The teeth in the mouth are similar. They develop continuously with new teeth being formed and migrating forward to replace older or broken teeth throughout life. Also, the rays and several kinds of sharks have heavy, flattened "paved" teeth that are reminiscent of knurled nonslip surfaces on tool handles. These are used in crushing the shells and exoskeletons of their invertebrate prey. Since we have painted a picture of the shark as a large voracious predator we might mention that the largest sharks, the whale shark and the basking shark, are harmless filter-feeders that eat small zooplankton.

A unique feature in the shark's short digestive system is the spiral valve (Chapter 25), which consists of a spiralling flap of absorptive tissue which greatly increases the surface of the intestine. Sharks swallow large chunks of food and the spiral valve slows the movement of food, permitting more time for their

23.8 (a) The shark body can be a lesson in streamlining. Note the large fins and the primitive tail of a sort seen first in some of the ostracoderms and placoderms. (b) The rays, on the other hand, have undergone a great transition for their mode of life as bottom dwellers. The pectoral fins have expanded and have joined the head. Other fins and the tail are greatly reduced in most rays, although they have the same general shape as that seen in the shark. (c) The tuna, a modern bony fish, is also streamlined, but in a different configuration.

(a)

(b)

(c)

powerful digestive enzymes to work. The digestive system ends in a *cloaca,* a posterior cavity that also receives the ducts of the urinary and reproductive systems. The opening between the cloaca and the exterior is called the *vent.* The cloaca is a general vertebrate characteristic, and is found in all vertebrate classes. (Although you don't have a cloaca, the cloacal mammals are represented by the platypus.)

Sharks have keen senses and locate prey by smell, by certain patterns of vibrations (distress movements) that are picked up by the lateral line organs, and by sight. The *lateral line organ,* characteristic of all fish, is composed of canals running the length of the body, containing groups of sensory cells that respond to water movements. Other sensory canals on the head are sensitive to electrical disturbances or fields created by nearby animals although the importance of this isn't known.

It is generally believed that sharks must keep moving in order to pass oxygen-laden water into their mouths and over their gills. They lack the mouth muscles and valves bony fish have for propelling water over the gills. Many sharks will simply suffocate if restrained from moving. In sharks, as in other fish, the nostrils are not connected with respiratory passages and can be used only for smelling—not for breathing.

Fertilization is internal, by copulation. The shark penis consists of two slender, grooved appendages, the *claspers,* which are used to open the female's vent, whereupon the male ejaculates along the channel made by the paired claspers. The various systems in the sharks and rays will be revisited in the coming chapters.

Class Osteichthyes: The Bony Fishes

Like many modern animal groups, the bony fishes have dispersed into just about every conceivable underwater niche (and some above water). Obviously, they are a very diverse group (Figure 23.9), and, in fact, over 20,000 species have been identified. It should be made clear that bony fishes are not so closely related to sharks as a first glance might indicate. There are a number of important differences in these animals. For example, the fins of many bony fish are much more movable and finer in structure, permitting greater maneuverability. You've probably seen fish dart from one stationary position to another. Sharks can't do this and they can't make quick darting turns. Also, bony fish have an air bladder, which improves balance and permits the animal to remain stationary at varying depths. Sharks lack air bladders and swim continuously to maintain themselves at a desired depth. Sharks do store a lot of lipid, however, and this may help with their bouyancy problem.

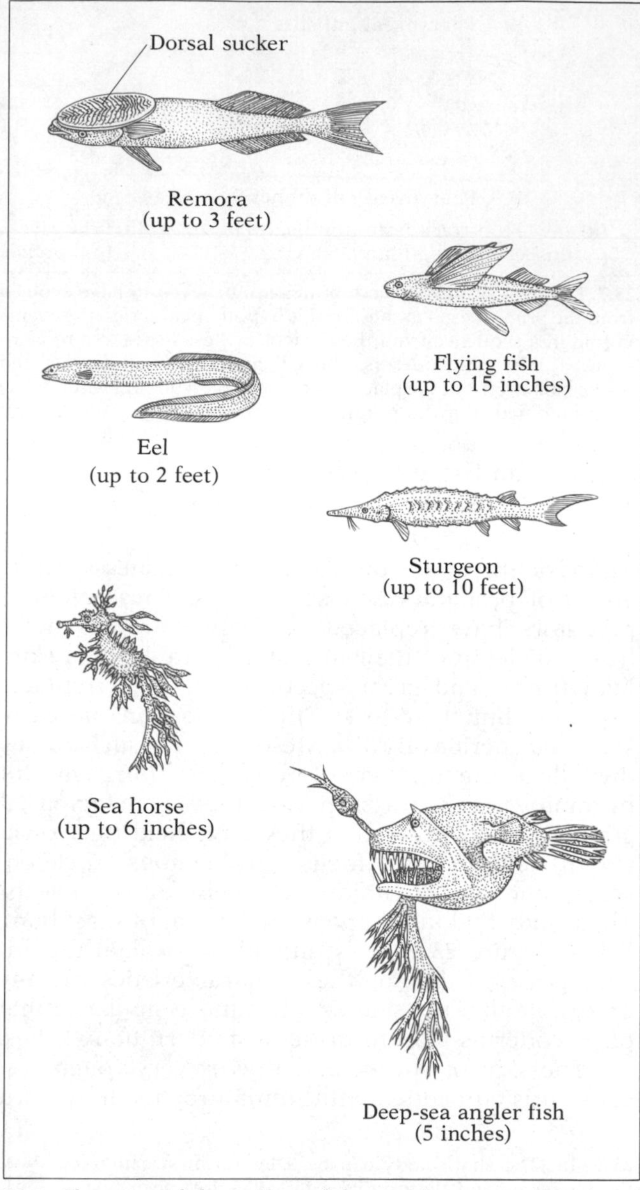

23.9 Class Osteichthyes. Bony fish come in a great diversity of shapes. Most of the structural differences are in fin and body shape, but all the rules are seemingly broken by the very odd sea horse, and even more dramatically by the creatures of the abysses such as the angler fish.

The bony fish produce flattened scales of several types, although they are absent in some fish. A few have toothlike scales similar to those of the sharks, but with surfaces of true enamel. The skin contains numerous mucous glands, which explains the slimy protective covering of a live fish. A faint line is usually obvious in the scales, running along the length of the body. This is the lateral line organ, which detects low frequency vibrations. Other sensory structures include the eyes, which probably focus best on nearby objects. Taste buds are scattered over the mouth, and balance centers are found in the inner ear. The respiratory structures include four gills on each side, located in a common *gill chamber* and protected by the bony operculum, a movable external flap. Each gill has a toothed raker that keeps food out of the gills. Unlike most sharks, bony fish pump water into the mouth and over the gills, using muscles in the walls of the gill chamber. The circulatory system of fishes includes a two-chambered heart and, of course, a closed system of vessels, like all vertebrates. Blood flows from the heart directly to the gills, where it exchanges CO_2 for O_2 before being circulated to the body and returned to the heart.

Reproduction in bony fish is varied. Most have external fertilization, some species releasing enormous numbers of gametes freely into the environment, while others seek sheltered places, and release far fewer gametes, perhaps into a nest, which either or both parents may then protect. Still other fish, live-bearers, copulate and retain the eggs in the body until the young emerge to fend for themselves. Stranger still are mouth breeders, some of which protect the embryos in the mouth until, and even after, hatching.

Many authorities believe that both bony fishes and cartilaginous fishes sprang from placoderm ancestors in the Devonian period. All bony fish are believed to have evolved from some kind of air-breathing, freshwater ancestor that had already developed lungs. These ancient lungs lost their respiratory function in most fish and became modified into the air bladder. A momentous event occurred when the bony fishes split into two groups: the ray-finned fish from which most of today's bony fishes evolved, and the lobe-finned fish. There were relatively few lobe-finned species but they had enormous evolutionary consequences (Figure 23.10). This line was to give rise to invaders of the land and thus to amphibians, reptiles, birds, and mammals. Those that did not invade the land include the lungfish (Dipnoi, six species) and the rare, deep-sea Coelacanth, *Latimeria* (one species). *Latimeria* belongs to a once-large, diverse group, the crossopterygians.

The lungfish are represented today by three genera that live in swamps of Australia, Africa, and South America (Figure 23.11). These strange fishes have highly vascularized lungs that connect to the pharynx. Dipnoi nostrils, unlike those of any other living fish, are connected internally with the mouth cavity. However, lungfish gulp air through their mouths. The Dipnoi survive in highly stagnated water by filling their lungs with oxygen gulped from the surface. Some species can even survive periods of pond drying by lying dormant in the mud. In spite of such adaptations, however, none of the Dipnoi can survive on land for very long; when all is said and done, they are fish.

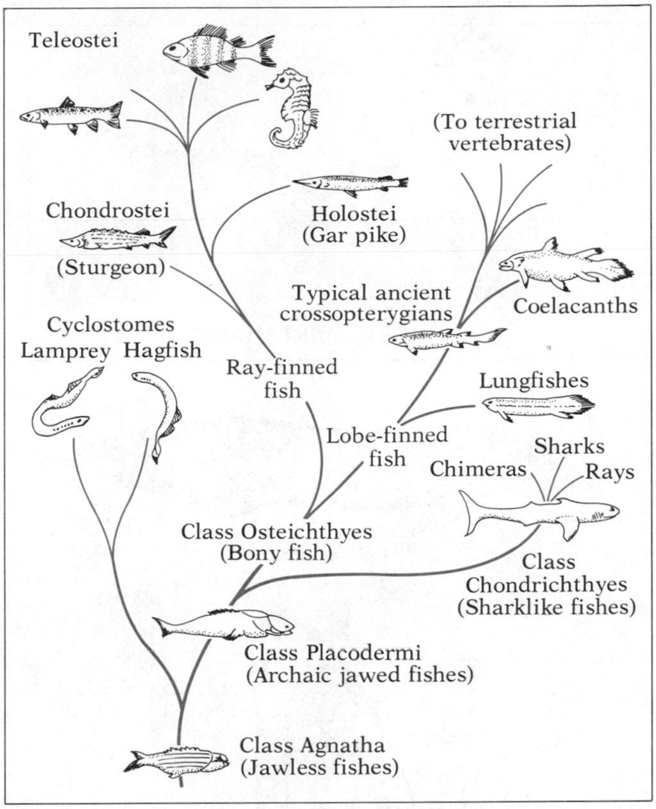

Teleostei
(To terrestrial vertebrates)
Chondrostei
(Sturgeon)
Holostei (Gar pike)
Typical ancient crossopterygians
Coelacanths
Cyclostomes
Lamprey Hagfish
Ray-finned fish
Lungfishes
Lobe-finned fish
Sharks
Chimeras
Rays
Class Osteichthyes (Bony fish)
Class Chondrichthyes (Sharklike fishes)
Class Placodermi (Archaic jawed fishes)
Class Agnatha (Jawless fishes)

23.10 A family tree of the fish. Included are three living classes (Agnatha, Chondrichthyes, and Osteichthyes) and one extinct class (Placodermi). The Osteichthyes, or bony fish, are divided into two subclasses: ray-finned fish and lobe-finned fish. Most living fish belong to one of the three orders of ray-finned fish, Teleostei. Only seven living species are lobe-finned fish, and of these, six are lungfish. The seventh is the coelacanth *Latimeria*. It was the lobe-finned fish, however, that gave rise to the terrestrial vertebrates.

The crossopterygians have a different, rather dramatic story. They were long thought to be extinct and scientists around the world were trying to learn what they could from their fossil bones, when one day a fishing boat plying the waters between Africa and Madagascar brought up a strange catch. It was a living crossopterygian of the coelacanth group. This one and the seventy or so that have been caught since were named *Latimeria* (Figure 23.12).

Latimeria, of course, is not ancestral to anything else. It does not even belong to the particular crossopterygian group from which land vertebrates evolved. It has several curious features, such as laying huge eggs and retaining a large notochord as an adult. The lungs in this deep-sea species are vestigial structures, filled with fat. It is a distant relative indeed, but it is, nonetheless, more closely related to us than it is to any other living fish.

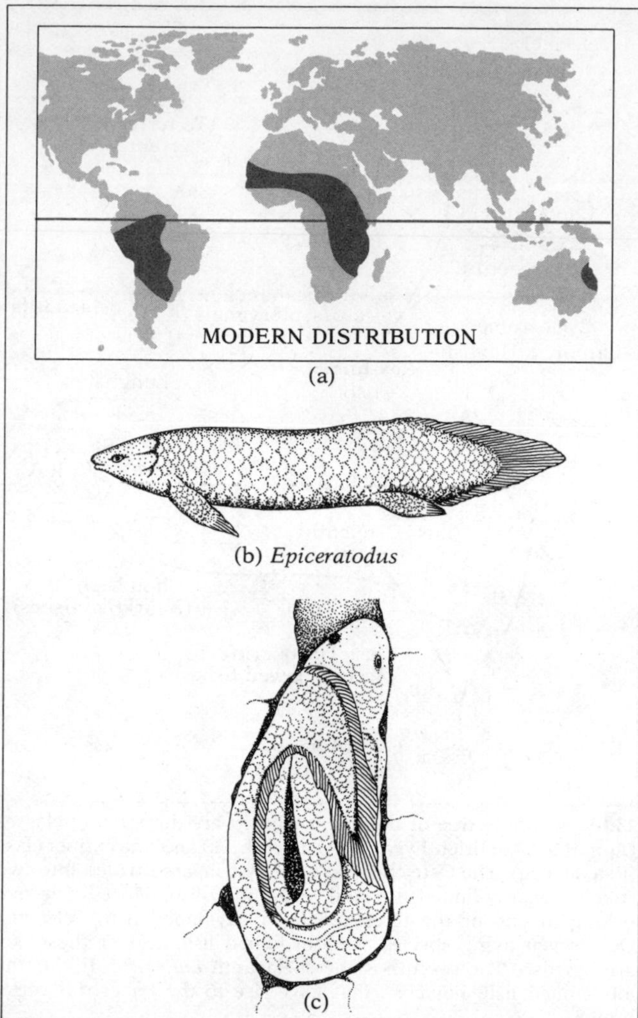

23.11 (a) The presence of lungfish on three continents long troubled evolutionists. How did they cross the ocean, or did they? But what was once believed to be a triple case of convergent evolution can now be explained in a simpler manner. The ancestral lungfishes evolved and were widespread before the continents drifted apart. The lungfishes are presently found in tropical regions where seasonal droughts are common. (b, c) All have the ability to withstand long periods of drought by breathing air or by closing themselves up in burrows where they survive by becoming dormant through the dry season.

23.12 The coelacanth, *Latimeria,* is not a fossil reconstruction, but a specimen recently caught by fishermen. The group to which it belongs, long believed extinct, is now known to exist off the Comoro Islands, which lie between the northern tip of Madagascar and the east African coast. Since their discovery in 1939, a number of *Latimeria* have been caught and subjected to intense study. The lungs of this deep-sea fish are vestigial and contain fatty tissue rather than air. Still, it is our closest living fishy relative.

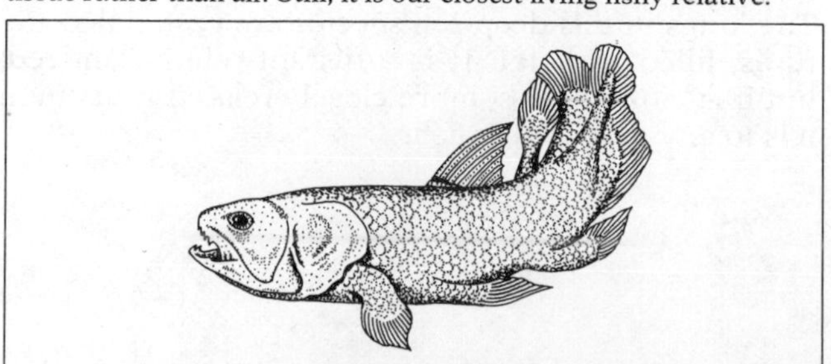

Class Amphibia: The Amphibians

Amphibians today are represented by three orders. These include Urodela, the salamanders; Anura, frogs and toads; and Apoda, the wormlike caecilians. Nearly all amphibians reproduce and develop in aquatic habitats, in spite of the fact that some are well-adapted for drier environments; a few even live in deserts. Both amphibians and reptiles are believed to have evolved from members of one ancient terrestrial group, *Stegocephalia,* found in Carboniferous and Permian strata. The skulls of stegocephalians were fused and had an amphibian–reptilian appearance (Figure 23.13). The three modern amphibian orders apparently branched off rather early from these Carboniferous old animals (although it is sometimes argued that the three orders may have evolved independently from different fish ancestors).

Note the differences in the skeletons of a lobe-finned fish and a fossil amphibian in Figure 23.14. Some of the more obvious skeletal features necessary

23.13 The amphibian line is believed to have arisen from certain lobe-finned fishes. These are then somehow connected to the labyrinthodonts (of the group Stegocephalia). The labyrinthodont shown here (*Diplovertebron*) may have been the primitive amphibian from which both reptiles and amphibians emerged, although the lineage remains obscure.

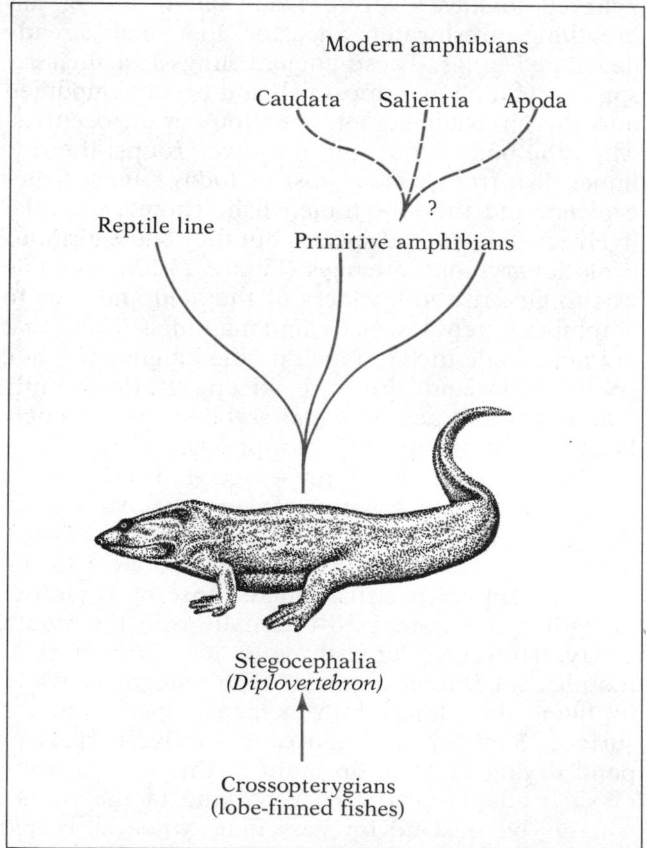

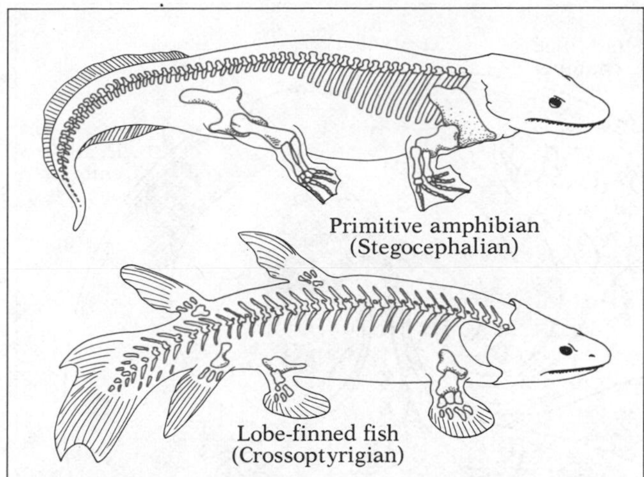

23.14 The legs of amphibians evolved from the bony, muscular fins of their crossopterygian ancestors. The lobe-finned fish probably began adapting to land by crawling from pond to pond or possibly by climbing out of the water to avoid aquatic predators.

for terrestrial life included greater development in the vertebral columns and the shoulder and hip girdles.

The modern salamander gives away its fishy ancestry as it quickly wriggles along, its long body undulating as the feet on either side are first brought together, then swung apart. The salamander's legs, by the way, are positioned at an awkward angle to the body, bearing the weight on flexed joints. In the evolution of other terrestrial vertebrates, particularly those to whom speed was important, the legs gradually came to be located under the body and projected downward rather than sideways. Locomotion came to depend on the muscles of the limbs and limb girdles, rather than on lateral undulations of the entire body.

Whereas salamanders drag their tails over the earth and are barely able to keep their bodies clear of the ground, the frogs have no such problem. First, they lack tails and, second, they hop. The caecilians, on the other hand, avoid the problem altogether, having lost their legs in favor of a wormlike burrowing.

Amphibians have moist, highly vascularized skin, which is an important organ of respiratory exchange in spite of the presence of lungs. Skin-breathing is only useful when an efficient circulatory system is present, and even then it is very limited. Frogs rely entirely on a skin exchange when they hibernate in mud, but their metabolic processes slow to a crawl at this time.

Amphibians have three-chambered hearts and go the fish one better by having a blood circuit through the lungs that is separated from that of the body in general. This arrangement means that oxygenated blood is returned to the heart for a second pump before entering the body circuit. The separation is imperfect, however, as is discussed in detail in Chapter 26.

Frogs and toads are external fertilizers. The male generally clasps the female and the eggs and sperm are released simultaneously. Fertilization in salamanders and caecilians is internal. Male salamanders release their sperm in small gelatinous packets called spermatophores. The females crawl over the packets and squat, picking them up in their cloacas, where fertilization occurs before the eggs are laid. Caecilians, on the other hand, actually copulate, the male introducing sperm directly into the female's cloaca.

Class Reptilia: The Reptiles

Finally, backboned animals conceived, lived, and died entirely on the dry earth—and these were the scaly creatures called *reptiles.* The new land dwellers gradually became more finely adapted to *terra firma,* until they now boast a wide variety of specializations for land. Sweeping changes in sexual reproduction and embryonic development were required. Reptiles retained the cloaca, but males developed a penis for copulation and internal fertilization, a prerequisite for full-time land dwelling. Since the placenta had not yet evolved (except possibly in a few shark species), egg laying was brought to the land. But the reptile (and bird) egg, exposed to the terrestrial environment, required its own changes if the embryo was to survive. The answer for reptiles was a porous, leathery case complete with a food supply, membranes for transport of food and oxygen in and waste out (Figure 23.15), and sufficient fluid to support and protect the embryo from drying out or injury. The nitrogen wastes from nucleic acid and protein metabolism could no longer be excreted as ammonia, which requires using large amounts of water; so the reptiles converted their waste to uric acid and other purines, which form relatively insoluble crystals. These solid nitrogenous wastes are also produced by many adult reptiles, birds, and terrestrial arthropods.

Adult reptiles developed a dry skin covered with protective scales and few mucus-secreting glands, thus they reduced water loss. They have well-developed lungs. Unlike the simple saclike lungs of amphibians, the reptile lungs are composed of many smaller compartments and thus have a spongy texture and a greatly increased surface area. The heart in most reptiles is three-chambered, but a partial septum divides the ventricles, except in the crocodilia where the septum is complete, forming a four-chambered heart. Although reptiles are called cold-blooded (poikilothermic), they have evolved a number of behavioral strategies for escaping both heat and cold (these are reviewed in Chapter 28).

Most of the reptiles are carnivorous, except for a few herbivorous lizards, tortoises, and turtles. Because their metabolism is slower, reptiles require considerably less food than do birds and mammals

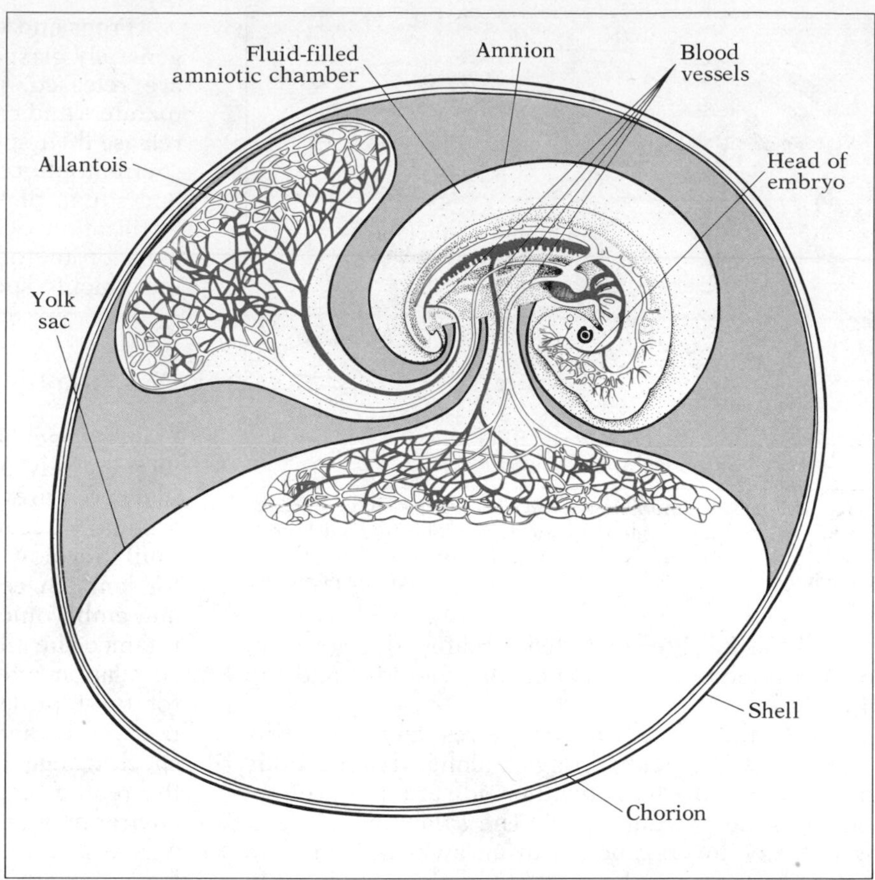

23.15 The reptile egg represents the complete transition from amphibian reproduction to the terrestrial form of life. The egg provides the necessary environment for the embryo's development, and is similar in some ways to the aquatic environment. The *amniotic cavity* formed by the *amnion* is fluid-filled, protecting the embryo. The *yolk sac* grows out over the food mass in the *yolk*, traversed by blood vessels which bring the vital food supply into the embryo. The *allantois*, an extension of the urinary bladder, collects the embryo's nitrogen wastes. In the case of birds and reptiles, this takes the form of crystals of uric acid and other purines, which are retained safely and left behind when the young animal hatches. The *chorion* assists the embryo in its exchange of respiratory gases by providing the necessary surface area. Finally, the egg case itself, leathery in the reptile, and calcified in the bird, protects the contents while permitting gas to be exchanged with the surroundings. This is the fertilized egg of a bird (chicken), which is not significantly different from that of a reptile.

of the same size. Reptiles locate whatever food they do eat by a variety of means, including vision, heat detection, olfaction, and hearing. Two species of lizards and many kinds of snakes are venomous. The poisonous pit vipers, commonly night hunters, rely on heat detectors in the two pits for sensing and locating prey. Pit vipers produce a hemotoxin that affects the circulatory system, causing blood cell destruction and hemorrhage; the venom of cobras and related species (such as the coral snake) is a complex protein neurotoxin which affects the nerves of the respiratory system, often causing death.

The earth doesn't harbor as many reptiles now as it once did. In fact, the Mesozoic era, which lasted over 160 million years, is known as the "Age of Reptiles." However, the reptiles actually orginated much earlier, branching off from the amphibian line in the early Pennsylvanian period, peaking during the Jurassic, and dwindling down to the comparatively few species of today.

Seymouria, an extinct amphibianlike reptile or reptilelike amphibian (nobody seems to be sure), is a good prospect for the ancestral type (Figurc 23.16). It was generally lizardlike, but its legs sprawled to the side, as do those of salamanders. From *Seymouria* or its close relatives, the cotylosaurs or "stem-reptiles" emerged in a fascinating array of species. The land, for a time free of nonreptilian competitors, was theirs. The opportunities were boundless. They branched out and went everywhere. There were giants and dwarfs. They climbed, hunted, hid, lurked, and threatened. As their numbers increased, competition and predation probably led some to return to the

23.16 The cotylosaurs were highly generalized amphibianlike reptiles that gave rise to the reptiles of the Mesozoic era. They have become known as the "stem-reptiles." *Seymouria,* an early member of this group, is found in lower Permian strata, some 220 million years old. About 60 cm (2 ft) long, *Seymouria* probably fed on insects and perhaps amphibians. Anatomically, it fits between amphibians and reptiles. That is, it shares many characteristics with each. From this group arose eight major reptile types, some of which would dominate the terrestrial realm for 200 million years.

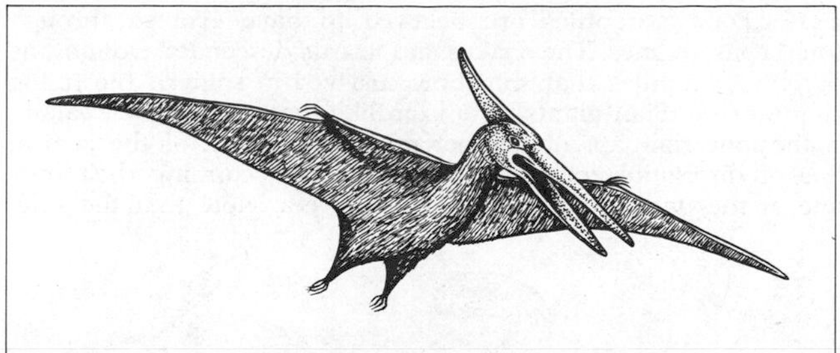

23.17 The pterosaurs represented an experiment in flight. Their fossils have left us with a host of questions. Some had a wingspan of 7–10 m, yet the wing was merely membranous skin, attached to an enormously elongated digit. There doesn't appear to be much evidence of large, strong wing muscles. Most paleontologists believe that they did more soaring and gliding than active flying. The problem is in trying to determine how they became air-borne. The legs seem too fragile for them to have made a running start. One idea is that they climbed trees or cliffs and then jumped. Most pterosaur fossils are in marine strata, indicating that the animals may have dived for fish from cliffs along the seashore. Pterosaurs, unlike other reptiles, had fur.

water from which they came. Others protected themselves with armor or sheer speed, striding around on their long hind legs. At least two lines of flying reptiles developed—one was the line that led to birds. The other, a much earlier and immediately successful group, were the pterosaurs of the Jurassic period through the end of the Cretaceous (Figure 23.17), the first flying vertebrates. They are not ancestral to anything alive today. Their bodies were usually about 1 m long. Some recently discovered fossils contained a surprise: pterosaurs, at least the smaller ones, were furry.

By the start of the Cretaceous period, some 135 million years ago, bigness was "in." This is the age of the great dramatic beasts that cross our minds when we think of dinosaurs. Even the pterosaurs produced a giant: *Pteranodon,* which was the size of a small airplane. The capstone of these great beasts was *Tyrannosaurus* ("tyrant lizard"; Figure 23.18), probably the largest terrestrial carnivore ever. But

23.18 *Tyrannosaurus,* although not the largest of the Mesozoic reptiles, was the largest predator and undoubtedly had things its way for a few million years. These dinosaurs were about 16 m long and stood 7 m high. Their prey were probably the even larger, lumbering herbivores. They shared the planet with a host of other peculiar beasts, including *Archeopteryx,* the earliest known bird, and a few small mammals.

larger dinosaurs stalked the earth. Great lumbering herds of plant-eating beasts devastated the taller trees, stripping them of their foliage. One called *Diplodocus* holds the record at 30 m, weighing in at 30 metric tons. The great herbivores were hardly equipped for a very complicated existence, since these giants had the tiniest of brains. Some even required a second, somewhat larger, mass of ganglia or a "brain" located at the base of their tail, above the hips. Presumably their bulk was too large to be controlled by a single nerve center.

The reptiles were hugely successful, as witnessed by their prominence on earth for so many millions of years. So why did they die off? Their demise came with the closing events of the Cretaceous period, and we have to admit that we really don't know what brought it about. We do know, however, that the delicate webs of life are highly vulnerable to change, and the loss of one member affects all. Perhaps a change in the plants started the process. We believe, for example, that their habitat was cooling at about this time. Some scientists blame the emergence of the early mammals, which might have preyed on the eggs of the giant reptiles. Also, any number of subtle changes could have acted in concert. In any case, it is clear that by the close of the Mesozoic era, only scattered remnants of the great reptilian class remained. Today, the largest of these are found in the tropics and the seas, represented by crocodiles, alligators, sea turtles, one giant lizard, and some very long snakes.

The modern reptiles are probably descended from several branches of the stem-reptiles. The turtles and side-necked turtles form one line of descent, the alligators and crocodiles another, the lone tuatara constitutes a third, and a fourth group is comprised of lizards and snakes (Figure 23.19). Snakes are believed to have evolved away from one branch of the lizard line about the start of Jurassic, mid-way through the reptile era. The four lines survived the rigors of the Cenozoic era and are represented today by about 6000 species.

Two other survivors of the ancient stem-reptiles made their way, almost unnoticed, into the Cenozoic era as well. These motley little creatures were to change the earth, and one descendant would build highways and devise elaborate justifications for all its behavior. But that was a long way off. The other group would grow feathers and fly. The time of the ancestral birds and mammals came with the vast geological changes that occurred with the beginning of the Cenozoic era.

Class Aves: The Birds

One would not tend to think that birds are any more closely related to reptiles than they are to, say, fish. However, beneath their modifications for flight, lie

23.19 Today's reptiles are believed to have evolved through four separate lines. The snakes and lizards descended from a line of smaller reptiles that somehow survived in spite of the ruling clique of reptilian giants. The lizardlike tuatara of New Zealand is the lone remnant of another group. Right out of the central line of dinosaurs emerged the crocodilians, forming their own line in the upper half of the Mesozoic era. Note that the side-

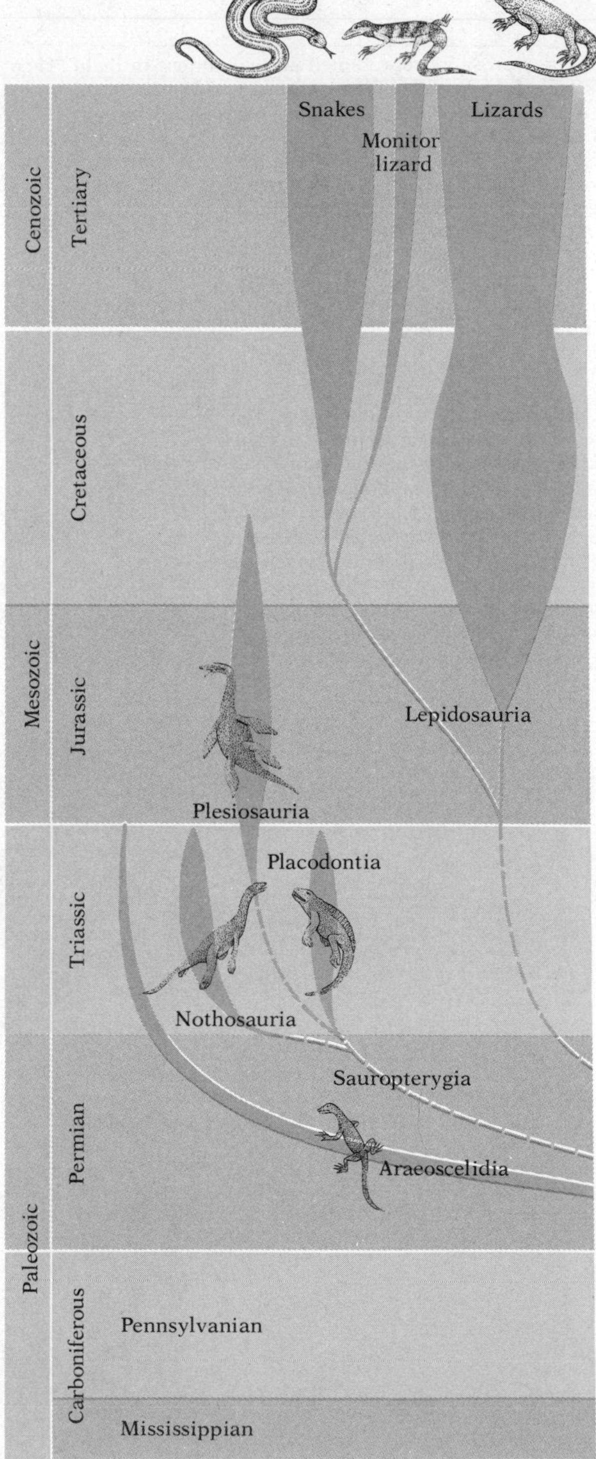

necked turtles (shelled reptiles with long, crooked necks that fold sideways) have had a long, separate history of their own. Both kinds of turtles, like the snakes and lizards, side-stepped the giants and emerged from ancient stock early in the Mesozoic. From this proposed lineage, it is apparent that although lizards and crocodiles are both reptiles, they are really not closely related. Many extinct groups are also shown here. The widths of the colored lines roughly indicate the relative numbers of species at any point in time; dashed lines indicate our best guesses in the absence of fossil data. Lines indicate the origins of the birds and mammals, from the archosaurs and therapsids respectively.

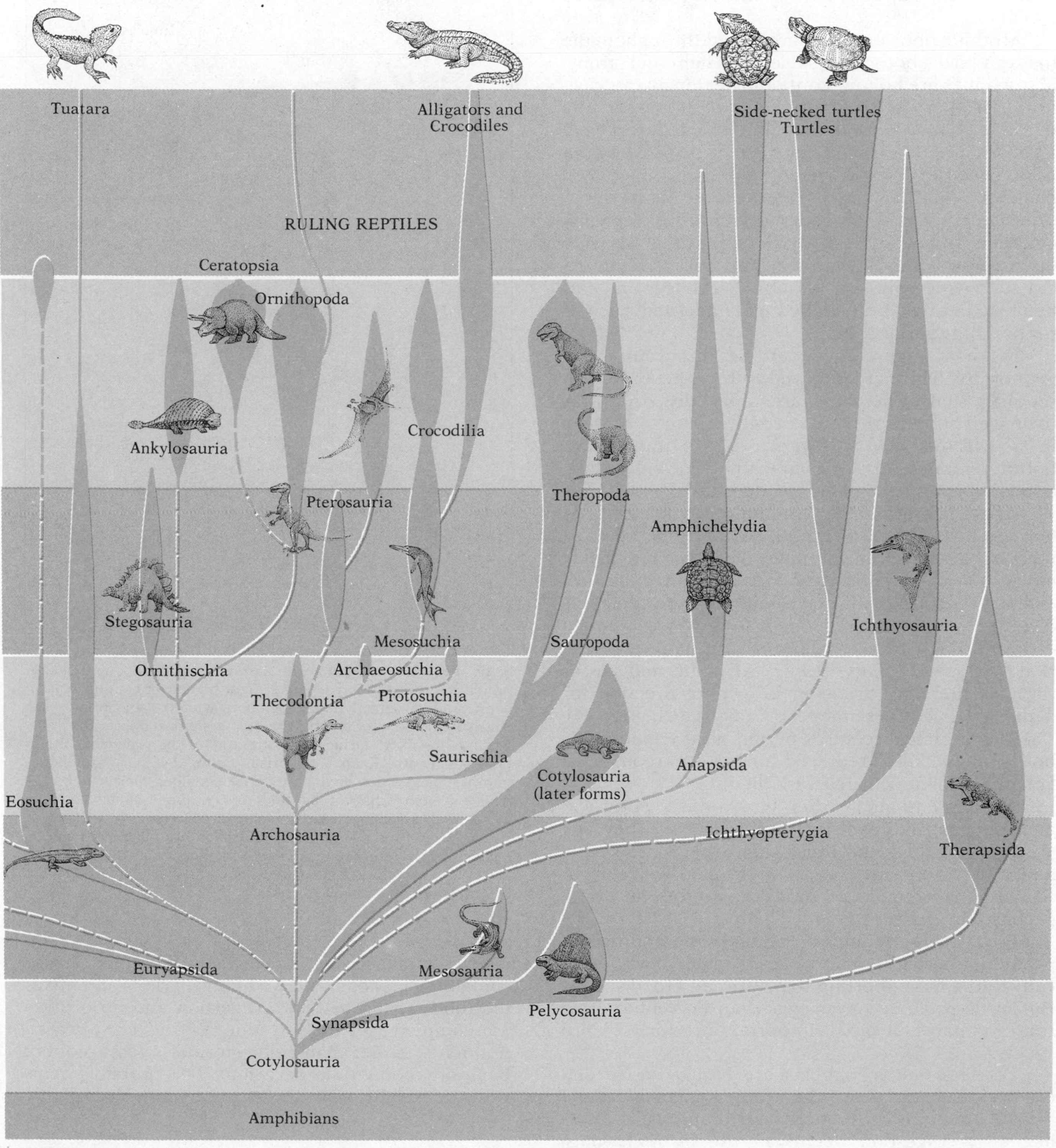

many reptilian traits. For example, the scales on the legs of birds are like the scales of reptiles. And the general skeletal framework of birds resembles that of one line of extinct reptiles. Of course there are notable differences between these ancient reptiles and modern birds. Many of the modern birds' changes, however, relate directly to their intense specialization for flight.

Modifications for flight are found throughout the modern bird's body. The skeleton is light and strong. Many bones are hollow, containing extensive air cavities crissed-crossed with bracings for strength. The reptilian jaw has been drastically lightened, and teeth have been replaced by a light, horny beak. The neck is long and flexible and the bones of the trunk (pelvis, backbone, and rib cage) have become fused into a semirigid unit. The breastbone (sternum) is greatly enlarged, and posseses a large keel to which the large flight (breast) muscles attach. The tail is greatly reduced, having only four vertebrae. The legs and particularly the feet are usually highly adapted, such as for perching and grasping.

Feathers (Figure 23.20), another important modification for flight, are believed to have evolved from reptilian scales. Feathers are unusually strong for their weight. Part of their strength is due to the way in which their barbs (branches) are interlocked. Softer, noninterlocking down feathers provide insulation in all young and in many adult birds.

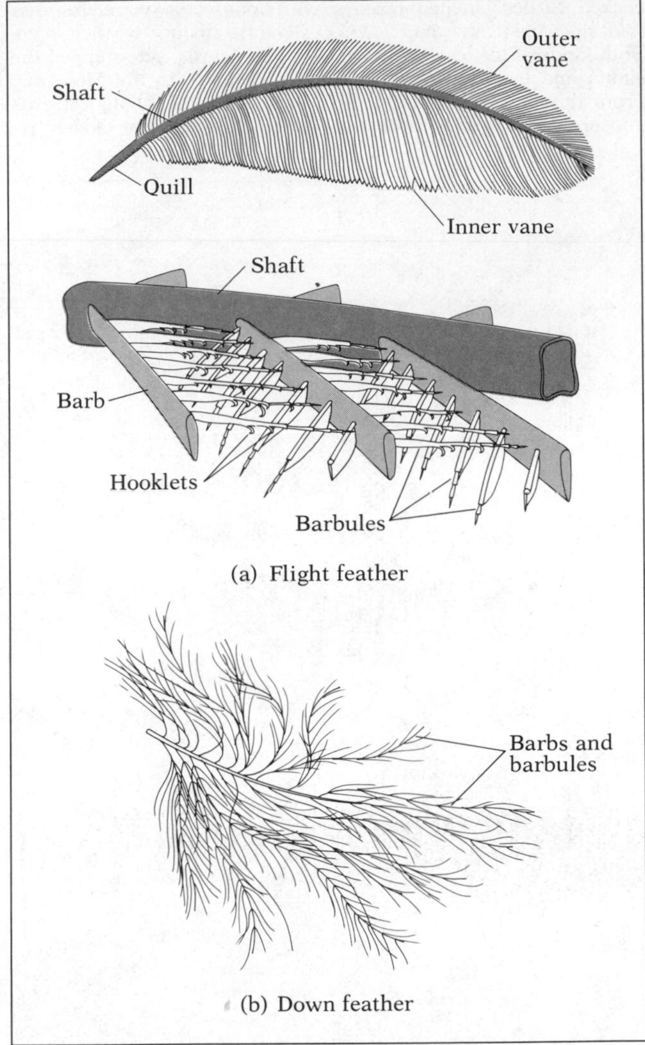

23.20 Feathers provide flight surfaces, insulation, and waterproofing for birds. (a) The larger flight feather or contour feather consists of a central shaft, beginning with the anchoring quill, which is the base of a long shaft. The shaft supports the vane, which is the flight surface. The structure of the vane lends it surprising strength. Each barb, arising from the shaft, subdivides into barbules, which have many hooklets along their length. The hooklets from adjacent barbules fasten the barbules together in a smooth, strong surface. Roosting birds frequently preen these surfaces, resetting the hooklets that have been disrupted during flight. (b) Delicate, plumelike down feathers form the insulating layer. They are less regular and softer and insulate best when ruffled. Down feathers lack hooklets.

Birds, like mammals, maintain a constant body temperature (homeothermy), which has allowed them to adapt to all climates. They have a highly efficient four-chambered heart and an extremely extensive respiratory surface, with numerous air sacs in addition to a complex, spongy lung. The movement of air through these structures is unique in vertebrates, since it follows a circular route and passes through the lung tissue in only one direction for a more efficient exchange of gasses (Chapter 26). These anatomical features permit a very high metabolic rate, which is required for the enormous muscular exertion associated with flight.

As we mentioned earlier, birds produce uric acid and other purines as their nitrogen waste. They lack a urinary bladder and these solid wastes are simply added to the digestive wastes in the cloaca. We might add that the female generally has only one ovary, perhaps further trimming for flight.

Birds have continued the reptilian tradition of producing large, resistant eggs, provided with basically the same supporting structures and food supply. But birds produce fewer eggs than do reptiles, as a rule, and parental care is important in nearly all bird species.

Feeding habits in birds are highly varied and often very specialized. When food is available many birds can temporarily store large quantities in a saclike *crop.* Digestion begins in the *proventriculus,* which secretes gastric juices. Next a muscular *gizzard* containing bits of swallowed gravel grinds and mixes the food, acting as teeth. Digestion and absorption are completed in the intestine. Wastes are released frequently rather than being stored for less frequent but more momentous occasions. This probably helps with the weight problem, and of course birds, being

23.21 Penguins represent the most extreme adaptation of birds to aquatic life. Their wings are reduced but extremely strong and aid in swimming. As in other flightless birds, the flightlessness doesn't indicate a primitive state, but a secondary loss as these birds evolved into new niches. The penguins must return to the land to reproduce.

highly mobile, don't have to worry very much about leaving droppings that predators might follow.

But not all modern birds fly. One group of birds, including the modern ostrich and emu, came to rely on size, strength, and fleetness of foot for its survival. Their tinier relative, the chicken-sized kiwi of New Zealand, evolved in the absence of serious predation. The strangest variant of all, the penguins (Figure 23.21), use their wings only as flippers, spending most of their lives far out at sea. The evolutionary transition was not complete, however. Penguins still return to land during the brief Antarctic summer to lay their eggs and rear their young.

There are approximately 8650 named species of birds, divided into about twenty-seven orders. In spite of such extensive speciation, however, the general plans of all birds are rather similar. Apparently the demands of flight have superimposed a set of design parameters that is more or less followed by all birds (Figure 23.22).

Class Mammalia: The Mammals

Mammals are hairy animals that give milk. They also maintain a constant body temperature, grow their embryos in the uterus, nourish and sustain their embryos by a placenta (in nearly all), and use a muscular diaphragm to help move air in and out of the lungs.

The hair or fur is produced continuously in follicles, associated with numerous oil glands that lubricate it. This insulating covering (and layers of fat) help the mammal adapt to the colder areas of the world. This adaptation is bolstered by relatively high metabolic activity that helps sustain constant internal body temperature. Unlike the reptiles and amphibians, mammals do not become sluggish when temperatures drop; but to maintain warmth they must

23.22 The general organization of birds is similar throughout the class, probably due to the demands of flight. However, there is marked diversity in the beaks and feet. These variations permit birds to explore many food sources and habitats. The greatest divergence from general organization is seen in the flightless birds, such as the emu. Some flightless birds, such as the ostrich, depend on their size and strength to discourage predators. Others, like the kiwi, are furtive and elusive.

Emu

Hornbill

Hummingbird

Hawk

Spoonbill

Woodpecker

eat considerably more than similarly sized reptiles. The furry covering also protects mammals from the hot sun. Homeothermy is not just a matter of adding greater or lesser body heat, but involves delicate heat-sensing centers in the brain and responsive mechanisms for making rapid adjustments.

Locomotion in the four-legged mammals has also changed, with the limbs more directly under the body. Rapid and complex responses to stimuli are made possible by increased brain size, particularly in the cerebrum, where a number of voluntary functions are coordinated and initiated.

Like the birds, mammals have a four-chambered heart and fully separated blood circuits to the lungs and body. The respiratory surface, while not nearly as extensive as that of the birds, is nevertheless considerable.

Mammals always fertilize internally, and in all but one minor group (monotremes) the cloaca has been discarded. A separate orifice, the anus, is present for defecation, and female mammals have also evolved separate urinary and reproductive openings.

The newest innovation is the placenta, which is a unique mammalian trait. The placenta is a matlike growth of highly vascularized tissue, produced by the embryo when it implants in the soft, receptive uterus. Blood vessels from the embryo form dense capillary beds, which compliment those of the uterine lining and all of the life-supportive functions—nutrition, respiratory exchange, metabolic waste removal—are provided. Marsupials do not have true placentas, and monotremes, like their reptilian ancestors, still lay eggs.

The mammals had their reptilian origin somewhere in the Permian period, branching off very early from a primitive stock. The first fossils we find in our line are the pelycosaurs, followed by the therapsids (Figure 23.19).

The therapsids were flesh eaters that developed the forelegs and musculature necessary for running. The angle of their elbows was quite different from that of the first terrestrial reptiles, whose elbows angled out to the side. Other indications of their mammalian association include tooth structure as well as skull and jaw features. Actually, there were quite a large number of these dim reptiles blinking at the early earth, preying and falling prey. They almost passed into oblivion with the rise of the great dinosaurs, but a few obscure, recalcitrant survivors managed to avoid the more powerful carnivores for the next 160 million years. (What would the earth be like today if they hadn't?) Their hazardous lives among the ruling reptiles may have placed a great premium on intelligence, coupled with keen senses. And, in fact, large brains are characteristic of today's mammals. Survival was further assured by the ability to bear young alive, rather than deserting the eggs to predators or having to defend a stationary nest. The advent of live birth is believed to have occurred in Mesozoic era. Another advancement was the development of milk, a kind of modified sweat, which provided the young with high-protein, high-calorie nutrition during the critical stages of early growth. In turn, the young would have tended to remain with at least one parent in order to be fed, and this continued association would have permitted the offspring to learn from the parent. Thus the tendency toward braininess would have been enhanced in order to exploit this opportunity.

Much of this is simply conjecture, however, fed by circumstantial evidence and *ad hoc* reasoning. There are very few mammal fossils from the Mesozoic era—just a few teeth and jaws. This fragmentary record indicates that the early mammals were not very inspiring beasts. For the most part, they were very small carnivores, probably feeding on insects, worms, and possibly the abundant, abandoned eggs of reptiles. There are those who would accuse the modern shrew (shown in Figure 23.23) of being similar to the Mesozoic placental mammal.

The Cenozoic era, however, is the "age of mammals." The great reptiles were soon to end their struggles. By the end of the Cretaceous period, there were essentially three types of mammals: (1) the egg-laying mammals (*monotremes*), (2) the pouched mammals (*marsupials*), and (3) mammals that had developed a *placenta* (Figure 23.24). The monotremes never became important and, today, are restricted to a few minor species. The marsupials reached prominence in the geographical isolation of their Australian continent and, for a while, in South America. But placental mammals became dominant in the rest of the world.

They all radiated from an explosion of species in the early Cenozoic era. By the start of the second Cenozoic epoch—the Eocene—which began 54 million years ago, the main lines of mammalian evolution had already become established. The Cenozoic era was a disruptive time, indeed, marked by drastically fluctuating climatic patterns.

The early Cenozoic mammals were therefore subjected to severe stress, and many didn't survive. The main lines of mammalian evolution, however, managed to remain intact. Today's mammals are grouped into three subclasses. Protheria, with the egg-laying monotremes (echidna and duckbill platypus); metatheria, the marsupials; and eutheria, which contains all the other orders. In the metatheria, a *pseudoplacenta* develops from the embryonic yolk sac. In the eutheria, the placenta develops from other membranes, the *chorion* and *allantois*. In all there are eighteen or nineteen living orders, depending on whether you consider the pinnipeds—walruses, seals, and sea lions—to be carnivores or in their own order (Figure 23.25 and Table 23.1).

23.23 Small, inconspicuous insectivores like the shrew shown here may have been ancestral to the placental mammalian line. As long as reptiles were abundant, these animals remained in the background, gradually gaining in intelligence, improving their reproductive strategies and their sensory structures. They probably lived in trees and burrows—spending most of their time in hiding—feeding at night on insects, or worms, reptile eggs, and smaller reptiles. In this subdued condition, they weathered the Mesozoic era to emerge as the dominant type in the Cenozoic.

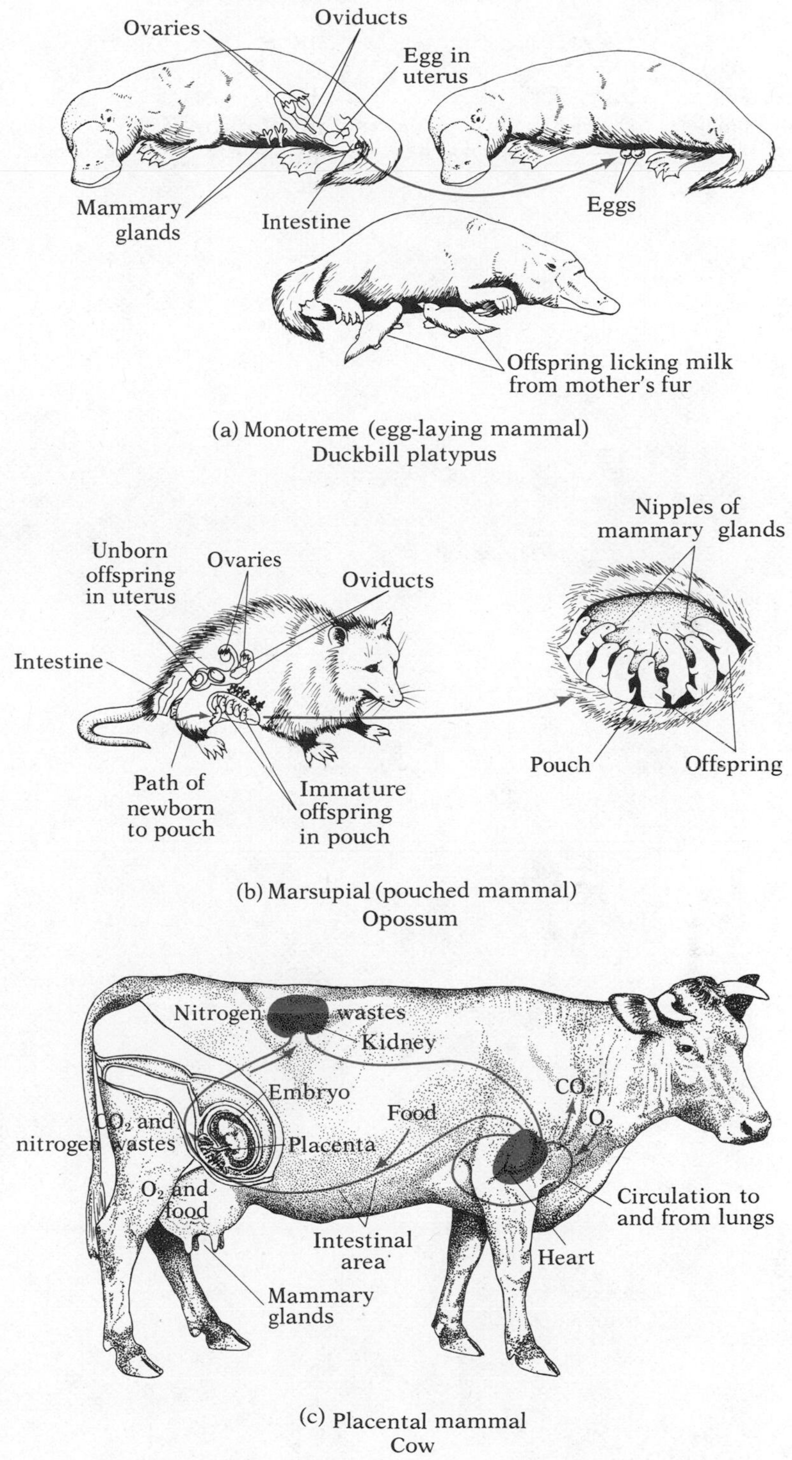

Ovaries
Oviducts
Egg in uterus
Mammary glands
Intestine
Eggs
Offspring licking milk from mother's fur

(a) Monotreme (egg-laying mammal)
Duckbill platypus

Unborn offspring in uterus
Ovaries
Oviducts
Nipples of mammary glands
Intestine
Pouch
Offspring
Path of newborn to pouch
Immature offspring in pouch

(b) Marsupial (pouched mammal)
Opossum

Nitrogen wastes
Kidney
Embryo
Food
CO_2
O_2
CO_2 and nitrogen wastes
Placenta
O_2 and food
Circulation to and from lungs
Intestinal area
Heart
Mammary glands

(c) Placental mammal
Cow

23.24 Three lines of mammals probably evolved from the ancestral therapsids. These were the primitive *monotremes*, the *marsupials*, and the *placental* mammals. Each differs in its manner of reproduction. The reproductive history is summarized in these three modern representatives. (a) Monotremes are hairy egg layers. Their young feed by simply lapping milk off the mother's fur. (b) The marsupials give birth to immature young, which climb to the pouch where they fasten on a nipple and remain there as they complete their development. (c) The placental mammal completes its embryonic development in the uterus, supported by membranes which form the placenta. It also nurses for a time after birth and, like the others, requires a period of time for growth and learning before it ventures out on its own. Such a long developmental period gives the complex brain time to develop. High intelligence, of course, has been important to the success of mammals.

23.25 Today's mammals occupy eighteen or nineteen orders and are dispersed into innumerable niches over the earth. They live below the earth's surface; on its surface; in trees; in oceans, streams, and lakes; and have even taken to the air as gliders and as true flyers. They feed on everything, from minute plankton to great trees, and, of course, each other. Mammalian roots reach back to the early Mesozoic era, but do not appear to branch out

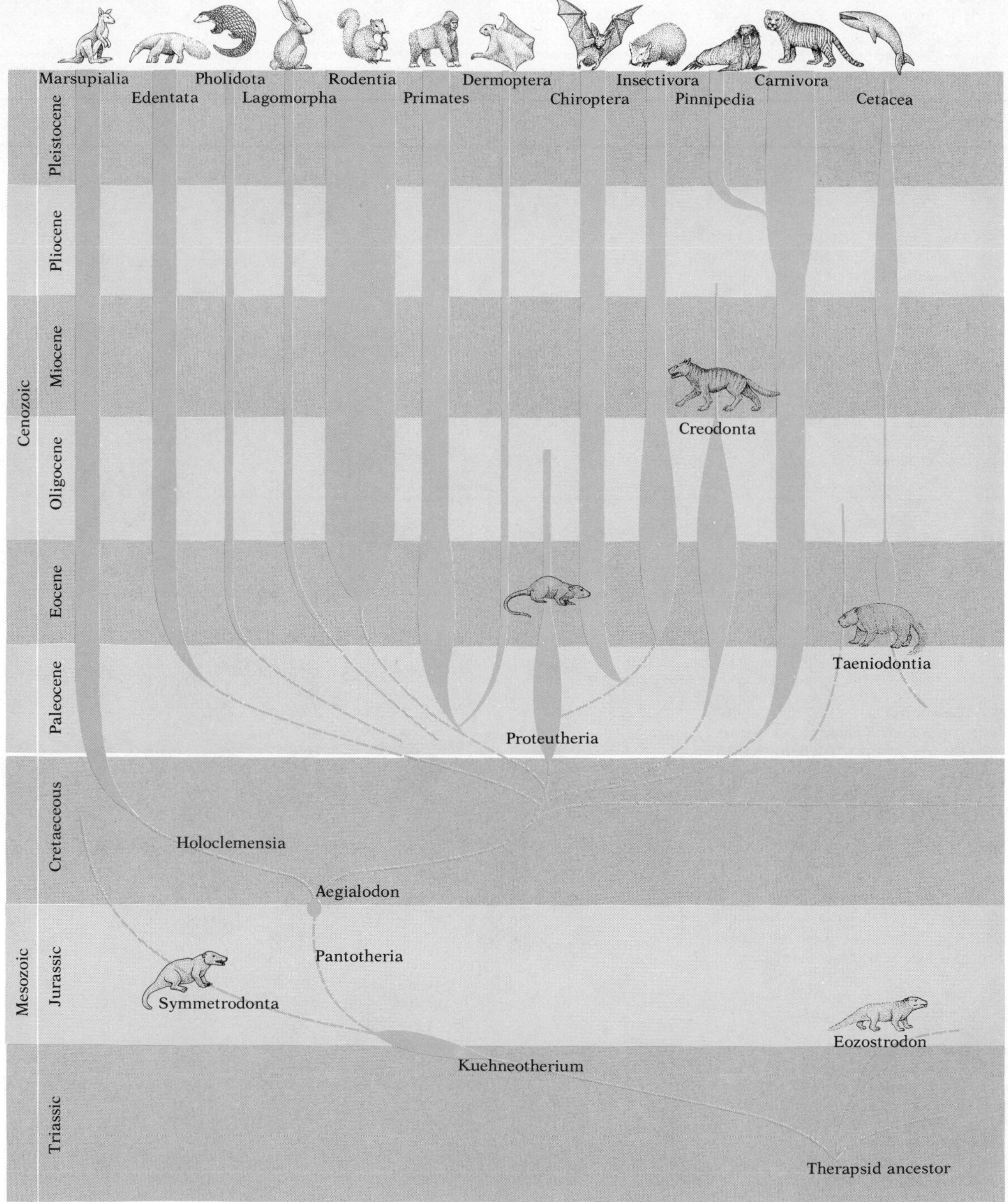

significantly until the Cenozoic, where they became the dominant vertebrates. The Cenozoic, an era of great climatic changes, saw the end of the ruling reptiles. A multitude of mammals arose and died off, but many persisted to give rise to today's forms. Dashed lines represent hypothetical relationships—note that nearly all mammalian relationships are hypothetical at this level.

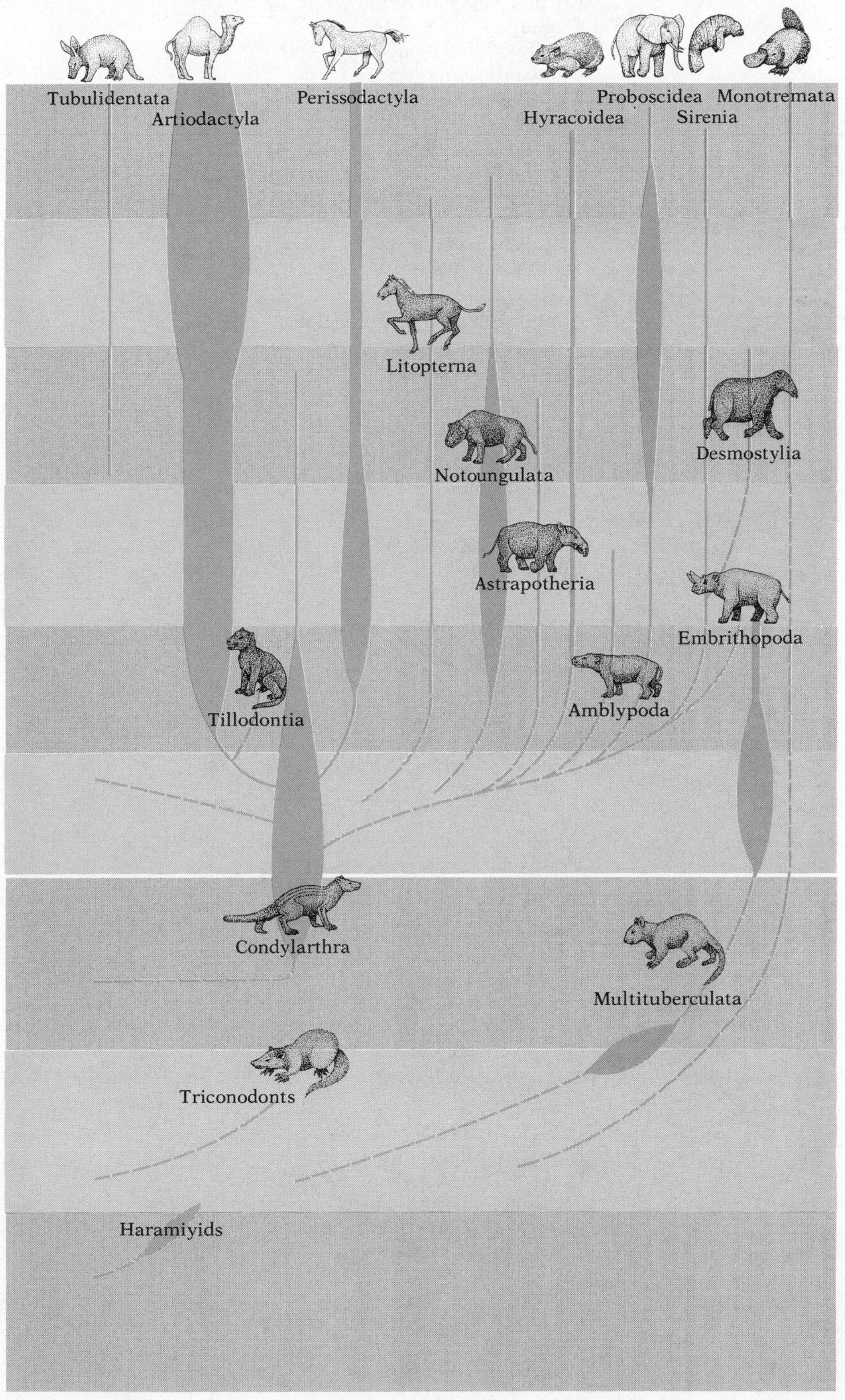

Table 23.1 Orders of Mammals

Order	Characteristics
Monotremata	
Duckbill platypus	Corpus callosum of brain absent Egg layers Adult with horny beak, no teeth Cloaca present Mammary glands without nipples
Marsupalia	
Kangaroo, opossum, koala	Corpus callosum of brain absent Pouched mammals (marsupium); pouch surrounds nipples on abdomen Pseudoplacenta develops from yolk sac Embryo begins development in uterus, finishes in pouch Fossils in Cretaceous Teeth relatively unspecialized
Insectivora	
Shrew, mole	Small, with long tapered snout Sharp, pointed teeth Fossils in Cretaceous
Dermaptera	
Flying lemur (only 2 species)	Squirrellike Wide, fur-covered patagium (flying membrane) Glide from trees Nocturnal
Chiroptera	
Bat	Capable of true flight Flight membrane supported by long forelimb and four long fingers; thumb modified into hook Sharp teeth Nocturnal Fossils in Paleocene
Primates	
Man, ape, monkey, lemur, loris	Highly movable head, eyes directed forward in most Many have hands with opposable thumb Hands and feet with five distinct digits Fossils in Paleocene
Edentata	
Sloth, anteater, armadillo	Toothless or molars only, no enamel, long snout Heavy claws Anteaters—no teeth, long neck, tapered mouth, long, protrusible and sticky tongue Sloths—arboreal, hang upside down Armadillos—teeth, horny covering divided by furrows Fossils in Paleocene
Pholidota	
Scaly anteater (pangolin)	Horny plates over body No teeth Long, slender tongue
Lagomorpha	
Hare, rabbit	Hind legs developed for jumping Many with long ears Lateral incisors behind medial incisors Teeth grow continuously, no canines Chewing with lateral motion only Fossils in Eocene to recent
Rodentia	
Squirrel, gopher, mouse, beaver, rat, porcupine, cavy	Small (mostly under 30 cm) Chisellike teeth, incisors grow continuously, no canines Chew in lateral and back-and-forth motion Widely distributed, most numerous order

Table 23.1 *(continued)*

Order	Characteristics
Cetacea	
Porpoise, whale	Fully aquatic, air breathers Young born underwater Streamlined body, no neck region Forelimbs modified into flippers, hind limbs absent Tail modified into vertical fluke
Carnivora	
Bear, cat family, dog family, hyena, raccoon, otter, weasel, mongoose	Feet with claws Small incisors, enlarged canines Carnivorous
Pinnipedia	
Seal, sea lion, walrus	Primarily aquatic, but rest, mate, and bear young on land Limbs modified into flippers, webbed toes Walruses bear enlarged canines (tusks)
Tubulidentata	
Aardvark	Stout body, long, piglike snout with slender, protrusible tongue No incisors or canines, teeth lack enamel
Proboscidea	
Elephant	Largest terrestrial mammal, large head and ears Nose and upper lip modified into flexible, prehensile trunk containing nasal passageways Upper incisors elongated into tusks
Hyracoidea	
Hyrax	Small (resembling guinea pigs) Forelimb with 4 toes, hind limb with 3; toes are hooved No canines
Sirenia	
Sea cow (manatee)	Forelimbs with paddles, hind limbs absent Tail has lateral flukes Heavy, blunt muzzle Herbivorous
Perissodactyla	
Horse, ass, zebra, tapir, rhinoceros	Hooved feet with odd numbers of toes Horse family has one functional digit on fore and hind limbs Tapirs have even numbers of toes (4) on forelimbs, odd number (3) on hind Herbivorous
Artiodactyla	
Pig, hippopotamus, camel, deer, giraffe, sheep, bison, cattle, antelope	Many are large, heavy animals Hooved, with even number of toes, usually 2 Antlers or horns on many, some with tusks Many are ruminants (4-chambered stomach) Many with teeth in upper jaw reduced

23.3 Primates and the Evolution of Man

The Primates

Modern primates, order Primates, are divided into two major suborders. The first—Prosimii—consists of tree shrews, tarsiers, lemurs, and the loris. These are the more "primitive" primates, though each has its own set of advanced features. The second suborder—Anthropoidea—includes monkeys, apes, and humans. This group is further subdivided into three superfamilies. Cebidae includes the New-World monkeys, those with the prehensile tails and nostrils set apart by a wide nasal septum (this group was formerly known as the platyrrhines). The second superfamily, also composed of monkeys, is the Cercopithecoidea, the Old-World group (formerly known as catarrhines). Their tails are never prehensile and are often short, and many have doglike snouts with the nostrils close together. The great apes (gibbons, orangutans, chimpanzees, and gorillas) and humans make up the third superfamily, Hominoidea (Figure 23.26).

23.26 New-World and Old-World monkeys. The common Spider monkey is a New-World monkey. Note the prehensile tail and its widespread nostrils, which tend to flare to the side. Old-World monkeys include the familiar baboon. These monkeys have a long snout and very large canine teeth. Note the foreward position and closeness of the nostrils.

23.27 *Plesiadapis,* a rodentlike tree mammal from the Paleocene epoch, is believed to be ancestral to the primate line. Its fossil remains have been found in North America and in Europe. *Plesiadapis* was rather small, about the size of a housecat—not very impressive, but it apparently was our ancestor. Its general body structure and long tail indicates that it was arboreal.

Primates can be traced through the fossil record as far back as the Paleocene epoch, some 65 million years. The earliest primates were unlike modern types in many ways. For example, they had long snouts and claws, and actually, they looked more like rodents. In fact, it is believed that they occupied very similar niches as modern rodents, some living in trees, but many scurrying about on the ground and some in burrows. It's a bit embarrassing to think of how many grubs our ancestors ate. But these little animals bore, unmistakably, teeth that resembled those of later fossil primates (Figure 23.27).

Interestingly, there was considerable primate activity in the New World during the Paleocene. Fossils are fairly common in the western part of the United States, but the record also tells us that primates began to die out in the more northern areas during the Eocene, becoming limited to the southerly regions.

It was once believed that because of geographic isolation, primates in the New and Old Worlds evolved independently into their modern groups in the period since the Eocene. This was, for a time, referred to as a "model case" of parallel evolution—that is, resemblance due to similar environmental pressure, not genetic relatedness. Fossil primates in both continents appeared to be changing together toward a monkey form. Now the entire idea has been reexamined in the light of new data on continental drift (see Figure 17.32). It now seems that there may well have been land bridges and narrow waterways between Africa and South America in the Eocene epoch, perhaps as recently as 50 million years ago. Protein sequence data supports a common monkey ancestry and a divergence time of about 60 million years ago; the resemblances between New-World and Old-World monkeys now appear to be simply due to the fact that they are rather closely related.

The early primates began to form groups that would be recognizable to us now by the Eocene. Fossils reveal a general shortening of the snout, a more forward location of the eyes, and a definite primate tooth structure. Toward the close of the Miocene epoch, we see the major divergences that would eventually lead to modern primates.

In Figure 23.28, you can see that the loris, tarsiers, and the prosimian lemurs of Madagascar branched from their Eocene ancestors early in the history of primates. The New-World monkeys did not diverge until early in the Oligocene epoch. A third branch split in the mid-Oligocene, but it only produced a group destined for a relatively speedy extinction during the Pliocene; it is represented by *Oreopithecus.*

The primate line leading to humans, shown in the center of Figure 23.28, has to its left, four fossil hominoids of uncertain affinity. Note that *Dryopithecus,* for example, is here placed in the main line just after the divergence of gibbons. The gibbons, them-

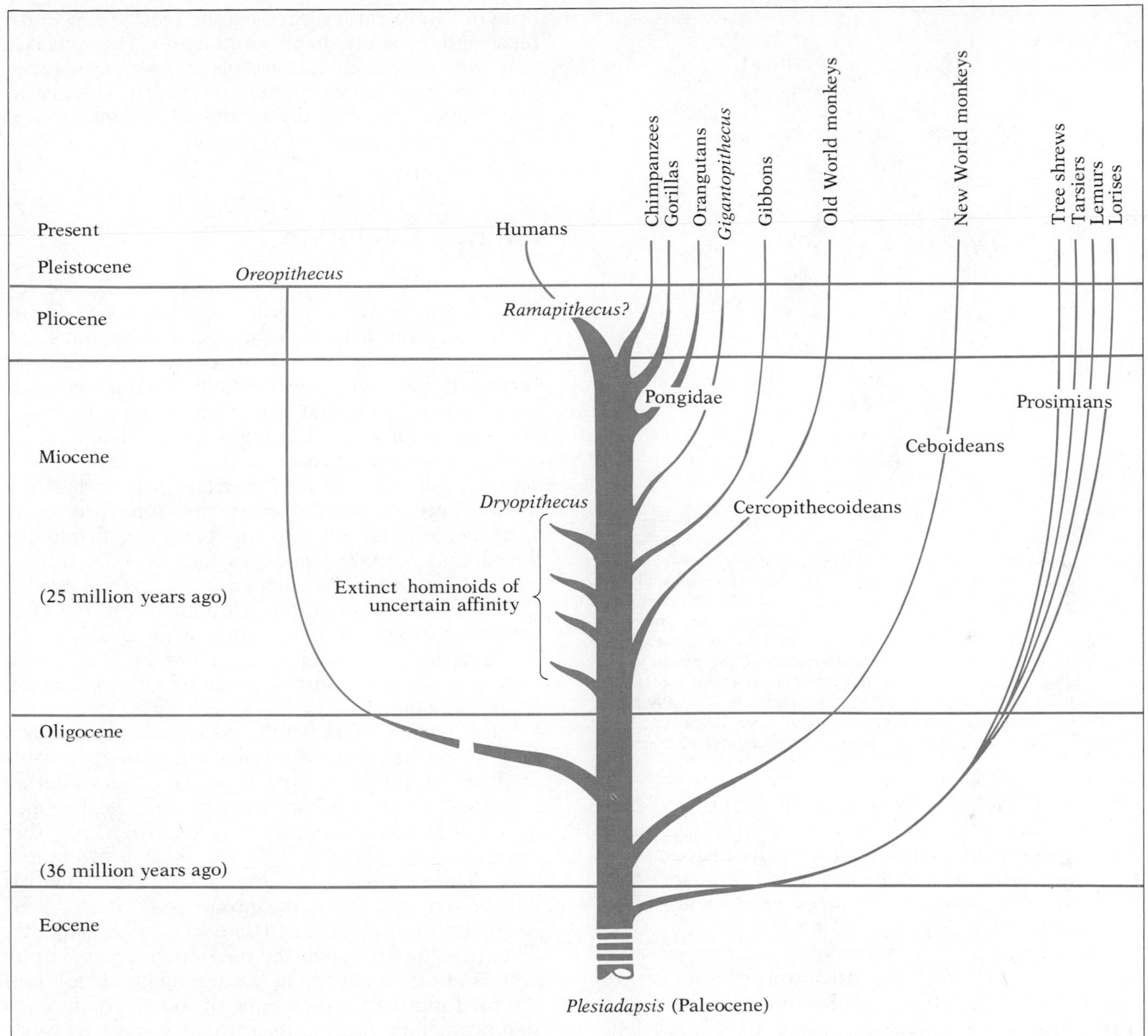

23.28 This tentative phylogeny of the primates places *Plesiadapsis* at the base. The earliest branch produced the prosimian line, followed next by the New-World monkeys. These two groups represent the greatest divergence in the living primates. Old-World monkeys diverged 10 million years later, followed by the great apes and *Gigantopithecus* (whose fossils were first found in Chinese drugstores). This largest primate survived almost into modern times (and, some say, still survives as the Yeti, or Abominable Snowman!). At the moment, *Ramapithecus* is placed just about at the division between the human line and that of the apes. *Oreopithecus,* another fossil ape or monkeylike creature is a second mystery, since is lived nearly to Pleistocene. It has been classified by some authorities as a hominid (human family), but the idea is not widely accepted among anthropologists.

selves, branched off not too long after the Old-World monkeys had diverged from the main stock. (Although the gibbon is an ape, structurally, it is not very closely related to orangutans, gorillas, chimpanzees, and humans.)

In the mid-Miocene, a branch of anthropoids represented by the *Gigantopithecus* made its debut, but it fell into extinction during the Pleistocene. (We should add that some serious-minded people believe it still roams the Himalayas as the notorious Yeti, the Abominable Snowman.) Finally, orangutans, gorillas, chimpanzees, and humans formed their distinct family lineages. The last divergence from the line, represented in the late Miocene by *Ramapithecus,* may have produced the human line (Figure 23.29).

How do primates differ from other mammals? This would seem to be an easy question, and it is unless one attempts to become too specific. Perhaps

23.29 *Ramapithecus.* The reconstruction of *Ramapithecus* was made from a fairly meager supply of fossil remains and may not reflect what the animal really looked like. Nevertheless, *Ramapithecus* is tentatively accepted as possibly being one of the first primates in the human line. The conclusion is based primarily on teeth, which constitute the bulk of the fossils found so far. Louis Leakey suggested that *Ramapithecus* was a tool user, based on a crushed animal skull and toollike volcanic rock found at one site. Fossils of *Ramapithecus* have been found in Africa, India, Pakistan, Greece, and possibly continental Europe and China.

this fact alone tells us something about the order; primates are often described as generalized mammals. This means that they have many unspecialized features, some even quite primitive. But, let's look at some more obvious features and decide about their status later.

The limbs and body of *most* primates are well-adapted for arboreal life. All have grasping or prehensile hands divided into clawless (nailed) fingers. Most have long arms, and some have prehensile (grasping) tails. These are all traits associated with living in trees, and the trees have made other demands as well. Primates have developed binocular, stereoscopic vision with eyes located in a forward position. If you are going to swing on limbs, you need to be able to judge distance. You also need good eye–hand coordination, and primates have that too.

The brain of primates has also responded to the special demands made on the beasts by increasing in size. The largest primate brains are in the apes and, of course, humans.

The other anatomical and physiological features of primates are similar to the plan of many mammals. Primates are often omnivores, so their dentition is unspecialized. With the exception of humans, the primates have rather large canine teeth, used in defense and (by some) to kill animal prey. Humans lack long canines and substitute tools such as rocks, sticks, guns, and nuclear weapons. An officer and a gentleman would never bite the enemy; his canines are too short.

Human Evolution

As is usual when humans try to assess themselves, their genes, or their heritage, the air becomes full of indignation and accusations. Things have become particularly volatile in recent years. But, of course, things were pretty volatile in Darwin's day, when the very idea that humans had an evolutionary ancestry at all was offensive to many. Even Alfred Wallace, the codiscoverer of the principle of natural selection, couldn't bring himself to believe that human beings had originated in the same way other species had. Darwin himself, however, firmly believed that humans and apes had evolved from a common ancestor, but there was no fossil evidence for this contention—none. Opponents of the idea of human evolution made much of the absence of a "missing link" and continued to make much of it long after many hominid (humanlike) fossils had been discovered.

The history of the study of hominid fossils is a long lesson in confusion. Fossils of extinct possible relatives of humans were found in a bewildering array and in no systematic order. The fossil groups included possible ancestors, extinct side lines that have no living progeny, and at least one prominent hoax. The problem of identifying species, which is difficult enough when the group is alive and well, is greatly multiplied when one has to consider individuals that are separated not only by space but by time, and especially when these individuals are represented only by fragments of bone. With each significant new find, anthropologists went to work. Often they revised everything they had previously said, and changed the entire structure of the presumed human ancestry. There is no reason to believe that there won't be similar drastic revisions in the future, as more and more fossil hominid groups are found. And always, when dealing with anything so personal as one's own ancestry, the interpretations of these old bones have been colored by prejudice, optimistic theories, and wishful thinking.

Species names were applied liberally, and theories abounded. In recent years there has been a tendency to simplify and organize the slowly emerging picture, and it is beginning to make some sense. As we go over the fragmentary evidence of a hypothetical phylogeny of human evolution, you might note one consistent pattern: reasonably well-defined species emerge rather suddenly, and persist almost unchanged for long periods before disappearing. We can assume that this constancy of form means that the species was well adapted to the niche in which it lived. Meanwhile, another group appears, presumably an offshoot of the first. But by the time it is leaving fossils, it is already adapted to a new and somewhat different environment. Both groups may coexist for a long time. This pattern is seen not only in human evolution, but throughout the fossil record of all living things. Now let's see what the picture looks like regarding humans—according to some authorities, at least.

The Divergence of the Human Line

In the early 1960s, biochemists began accumulating evidence that humans are more closely related to the chimpanzees and gorillas of Africa than to the orangutan of Asia. Furthermore, the steady, dependable changes in molecules have given us another kind of clock that appears to place the divergence of humans and African apes at not more than 10 million years ago. Anthropologists had assumed a much more ancient divergence, and were mostly hostile to the new ideas; however, in the absence of good fossil evidence to the contrary, a relatively recent emergence of the human line from the ape line has been accepted. The position of *Ramapithecus,* which appears to have some humanlike characteristics, is still debated.

The big problem is that there are no fossils joining *Ramapithecus,* who lived, say, 12 million years ago, and a rather large number of diverse hominid fossils that are about 2–3 million years old. Perhaps the emerging humanlike forms lived in forests for the intervening 9 million years; bones, we know, are not preserved in the moist, acid soil of forests. But then new fossils appear. About 2–3 million years ago, a number of humanlike forms lived in relatively dry open grasslands of Africa. They have been given a number of names, but when the dust settles three general groups can be recognized: *Australopithecus africanus, Australopithecus robustus,* and *Homo habilis.* It is not clear whether any of these forms was directly ancestral to *Homo sapiens,* but it is clear that the two *Australopithecus* genera persisted almost unchanged until at least 1 million years ago, long after the genus *Homo* was well established. Still, our own ancestors could have evolved from early offshoots of the *Australopithecus* line.

Australopithecus africanus

Australopithecus africanus, first discovered by Raymond Dart in 1924, were rather small-boned, lightbodied creatures, about 4 ft tall. They had humanlike teeth, with small incisors and canines, and walked on two legs. They were hunters, apparently killing baboons, gazelles, hares, birds, and even giraffes with clubs made from the long bones of their prey. (Near some *Australopithecus* bones, Dart found a baboon skull with an antelope leg bone jammed into it.) They also made stone tools—most surprising, since before this discovery it had been assumed that only true humans (*Homo sapiens*) were so sophisticated as to be able to make tools. The cranial capacity of *Australopithecus* fossils (expressed here in cubic centimeters) range from 450 to 650 cc, compared with 1200–1600 cc for modern adult humans. Some of this difference is simply due to their generally smaller size (even today, 4-ft-tall people don't have brains as large as 6-ft-tall people, though they are just as smart).

Australopithecus robustus

Australopithecus robustus, as the name suggests, was larger-boned although not taller. Most strikingly, *A. robustus* had huge jaws, greatly expanded cheekbones to accommodate overgrown jaw muscles, and very large and heavy teeth. These teeth, however, were more human than apelike in most respects; humans, for one thing, have a much thicker layer of enamel than do apes. Some individuals even had a saggital crest, a prominence on top of the head to which jaw muscles attached. The tooth-wear indicates that *A. robustus* were vegetarians. These creatures apparently did make tools.

Many physical anthropologists now question whether there really were two well-separated australopithecine species, pointing to an array of what seems to be intermediate forms. They prefer to refer to both types as *A. africanus,* distinguishing between them, when necessary, as *gracile* and *robust* types.

Homo habilis

Homo habilis is the name given by Louis Leakey to certain fossils from the Olduvai Gorge of Africa that seemed to be closer to the human form. Critics have argued that it is just another variant of *Australopithecus,* and that Leakey was unjustified in trying to place his fossils within the genus *Homo.* More recently, his son Richard and wife Mary found a hominid skull of unusual interest at Koobi Fora, east of Lake Turkana (formerly Lake Rudolph). Two different attempts at dating the strata gave widely divergent estimates: one laboratory said the skull was 1.6 million years old, older in fact than most *Australopithecus* fossils, and the other laboratory calculated that it was 2.5 million years old. But it appears to be much closer to the human line than is *Australopithecus.* For one thing, it has a greater cranial capacity (775 cc) than either *Australopithecus* or the earlier, disputed, *Homo habilis* finds. The Leakeys, apparently fed up with the sterile arguments that have raged over the assignment of species and genus names to hominid fossils, simply identified their unique find by its arbitrary field identification: fossil skull 1470. Another, similar skull, apparently that of a young female, has been whimsically named "Lucy." Were 1470 and Lucy *Homo habilis, Australopithecus,* or something else? Perhaps it doesn't matter: they were real, they were alive, they are possible human ancestors, or at least close relatives.

Homo erectus

The fossil beds of the eastern shore of Lake Turkana have now yielded other, more advanced fossils, that is, fossils closer to the modern human; these have been called *Homo erectus,* and are thought to be an extinct member of our own genus. The earliest African *Homo erectus* skulls are known to be more than 1.5 million years old. What is most interesting about them is that they appear to be of the same species as other *H. erectus* fossils—those that were once called "Java Ape Man" and the "Peking Man." But these were only regarded as about half a million years old. Some *H. erectus* fossils may even be less than 200,000 years old. In other words, *Homo erectus* flourished, relatively unchanged, for well over a million years. During its first 300,000 years they coexisted with other hominid species, including the australopithecines and possibly *Homo habilis.* In at least the last 100,000–200,000 years of their existence, they overlapped with yet another upstart species, *Homo sapiens.*

The first *H. erectus* discovered was in 1891 by Eugene Dubois, working in Java, who was deliberately searching for fossil evidence of human evolution. He was fortunate indeed. He found a very primitive-looking skull cap, along with a rather modern-looking thigh bone. He quickly recognized that both were more nearly human than apelike. The cranial capacity of his find was approximately 775 cc, and he called it *Pithecanthropus erectus;* it was popularly called the Java Ape Man, and that name persisted until it was changed to *H. erectus* in 1960. The small cranial capacity of Java Man seems confusing when considered in its lofty position as man's second most recent ancestor. We hasten to add, however, that most *H. erectus* fossils have a greater capacity and the range is great—775 to 1225 cc (Table 23.2). Not only did this find produce another definite candidate for the human ancestry, but it extended the probable geography of our forebears as well. *Homo erectus* has since been found in China, Europe, and southern Africa and of course East Africa. *Homo erectus* had a small brain, heavy brows, and strong jaws and teeth. From the neck down, however, they apparently were just like us.

In addition to their fossilized bones, *Homo erectus* left behind crude stone tools. The tools, consisting of hand axes and scrapers are designated *lower Paleolithic* or *lower Old Stone Age* (Figure 23.30).

Table 23.2 Cranial Capacity of Fossil and Modern Primates

Primate	Cranial Capacity[a]
Homo sapiens	1350–1450 cc
Homo erectus	750–1225 cc
Skull 1470	775 cc
Homo habilis	600–675 cc
Australopithecus robustus	450–550 cc
Australopithecus africanus	450–550 cc
Ramapithecus	—
Gorilla	500–600 cc
Chimpanzee	350–450 cc

[a]Cranial capacity is expressed in cubic centimeters.

There is evidence that these Stone Age people built shelters, lived by hunting small game, and gathered plant foods.

The era of *Homo erectus* apparently ended about 200,000 years ago, just after the first glaciers of the Pleistocene receded.

Homo erectus gave rise to our species, *H. sapiens,* but the oldest fossils that are presumed to be of this

23.30 *Homo erectus,* toolmaker. It was the discovery of simple stone tools near the Choukoutien, China, site that led to excavations and the discovery in 1927 of fossil teeth from which *Sinanthropus pekinensis* (Peking Man) was named. Since that time, many similar finds have produced a large collection of skull and jaw fragments as well as other bones. These have been assigned to the genus *Homo erectus.* The tools of *H. erectus* are fairly well-developed, revealing an advanced technology. These include chopping tools, hand axes, and scrapers, possibly for cleaning hides.

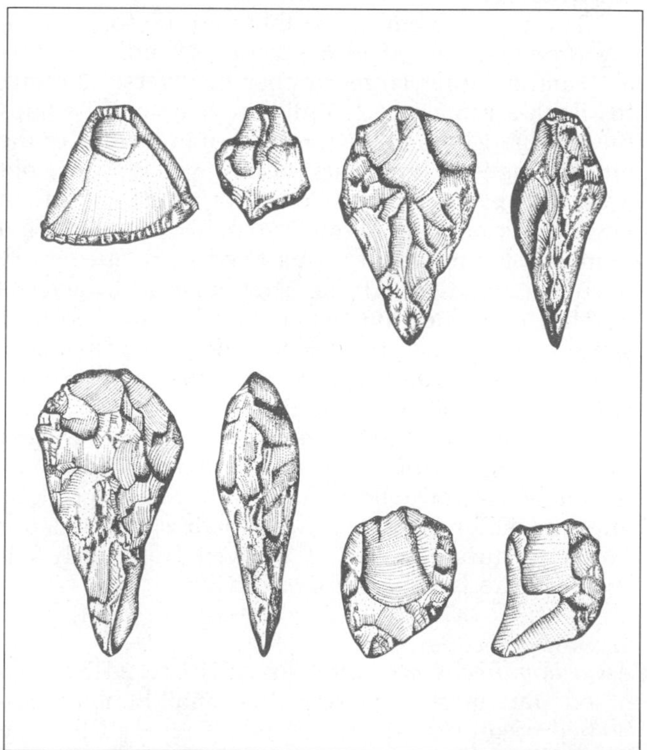

group are unfortunately highly suspect. The earliest probable *Homo sapiens* fossils are the fragmentary, ambiguous "Heidleberg man" (of which we have a jawbone) and the Vértess zöllös fossil (an occipital bone). Both of these are European and both are about 500,000 years old. Other, more nearly complete *Homo sapiens* skulls occur in strata formed during the next interglacial period, yielding "Swanscombe man" and "Steinheim man." It seemed for a time that *H. sapiens* had simply replaced *H. erectus.* Then another skull, found on the Solo river in Java, not far from where Dubois found the Java Ape Man, showed things were not quite that simple. This skull was clearly another *H. erectus.* The problem was, the "Solo man" appears to have been alive less than 70,000 years ago, well into the time of *H. sapiens.* Thus, not only were *Homo sapiens* and *Homo erectus* contemporary, they may have coexisted on the earth even when the relatively recent group of *H. sapiens* called Neanderthal lived. Let's now look at the Neanderthals.

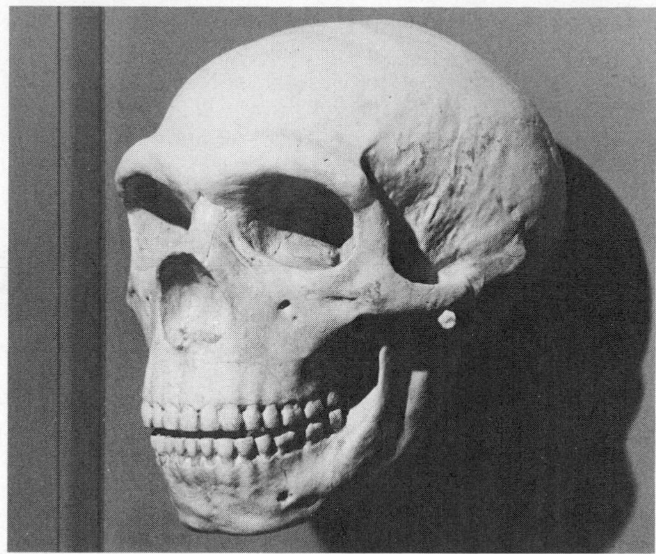

23.31 Neanderthal human. Fossil remains similar to this skull have been found in many parts of Europe, including a complete skeleton from La Chapelle-Aux-Saints, France. They have also been unearthed in Africa, Asia, and Malaysia. In spite of their resemblence to modern humans, they cannot be conclusively shown to have been a direct ancestor, and may represent a group that became extinct.

Homo sapiens: Neanderthals and Us

Even while Darwin was working at his country estate, workmen in a steep gorge in the Valley of Neander (Neanderthal) were pounding at something that turned out to be a skeleton, inadvertently smashing it to bits, but leaving enough for researchers to see evidence of a new and different kind of human (a similar skull had been unearthed at Gibralter a few years earlier, but had not created much of a stir).

Scientists are rarely at a loss for words, and an explanation was immediately forthcoming. Professor E. Mayer of Bonn examined the heavy-browed skull and proclaimed that the skull and bone fragments belonged to a Mongolian cossack chasing Napoleon's retreating troops through Prussia in 1814; an advanced case of rickets had caused him great pain and his furrowed brow had produced the great ridges; he was so distraught, he had crawled into a cave to rest, but, alas, had died there. This scenario has been rejected in favor of the idea that the bones were those of a member of an early form of human that became extinct—the *Neanderthal human* (Figure 23.31).

Homo sapiens neanderthalensis thrived around 70,000–40,000 years ago, and left a rich record of his fossil remains and some indication of his tools and culture. The Neanderthal fossils at first seemed to indicate exactly what anyone would expect of an ape-man. They were heavy-boned, heavy-browed (both males and females), with low foreheads, wide faces, and little or no chin. Heavy vertebral processes on the neck, and the angle of the head with the neck indicated that they walked in a crouch, their heads thrust forward. The heavy leg bones of the earlier-found fossils were quite bowed. They were, by our standards, not very attractive. Their brains appear actually to have been rather larger than those of modern humans, although their relatively sloping forehead has been (probably erroneously) cited as evidence that they may have had smaller frontal lobes and were less "sophisticated." There is no way to know whether they had a spoken language. However, we do know, from fossil pollens found with their remains, that Neanderthals sometimes buried their dead after first covering them with bouquets of flowers. Whether or not that's sophisticated, it certainly is human—touchingly so.

Neanderthal fossils have been found in Yugoslavia, Germany, Czechoslovakia, Uzbekistan in Central Asia, Kwantung Province of China, Iraq, Israel, North Africa, Italy, and Greece. But not all of them fit our dramatic descriptions of the first heavy, stooped skeletons. For a while anthropologists thought that there were two varieties of Neanderthal—the heavy, stooped *classical* Neanderthal and the much lighter-boned, fine-featured *progressive* Neanderthal. But more and more fossils appeared that seemed to fall somewhere in between. Eventually it was realized that there probably were not two distinct groups at all, but a single species that varied from place to place, from time to time, and even within a population (just as people vary today). The bowed legs of early finds were eventually diagnosed as being due to joint disease. In any case, since *Homo erectus* had an essentially modern skeleton, probably the later Neanderthals did not have an apelike body.

Today there is no agreement even on whether *Homo sapiens neanderthalensis* is even a legitimate taxonomic group; many anthropologists would rather drop the subspecies designation. At any rate, Neanderthal-type fossils are not found in rocks younger than about 40,000 years. With a few questionable, possibly transitional forms from 30,000 years ago onward, all hominid fossils are indistinguishable from modern human skeletons. This includes the bones of the once-celebrated Cro-Magnon Man, a European cave dweller who lived about 20,000 years ago.

What happened, and why did the Neanderthals disappear? One school has proposed that the classic Neanderthal and the line of modern humans diverged as long ago as 250,000 years, perhaps separated geographically by glaciation (Figure 23.32). For some reason, the Neanderthals flourished while the to-be-humans were scarce; then, about 40,000 years ago, the scarce *Homo sapiens sapiens* suddenly burgeoned, expanded their range, and Neanderthal went extinct (possibly being wiped out by warring *H. sapiens sapiens*).

A second theory minimizes the differences, considers Neanderthals to have been one highly variable species that included individuals much like ourselves, and places them in the mainstream of human evolution. Modern humans simply evolved from somewhat archaic Neanderthal ancestors. This theory says the Neanderthals didn't die out, but merely changed. Subscribers to this alternative believe that earlier anthropologists had simply put too much emphasis on some extremes of normal variation.

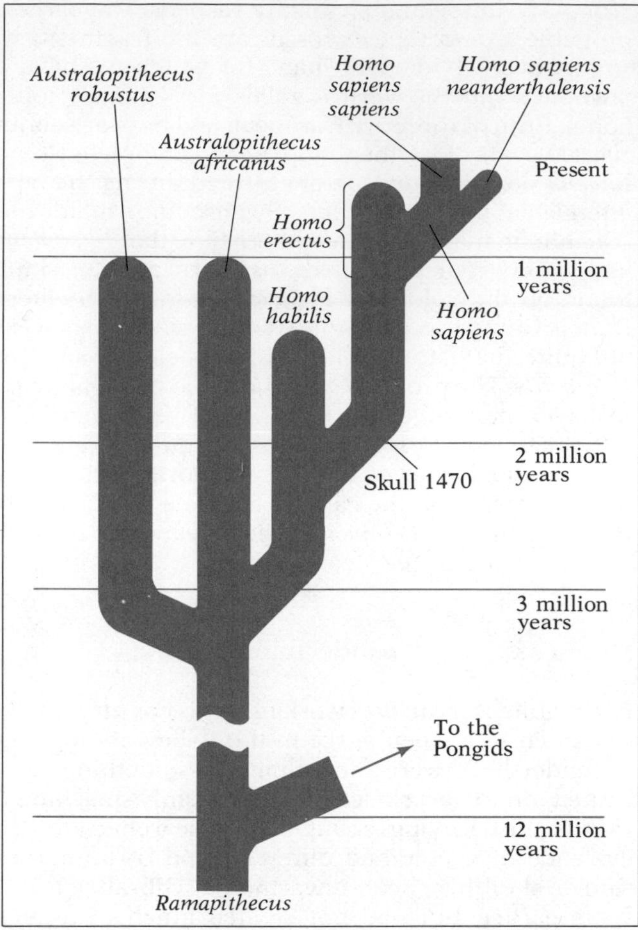

23.32 Fossil *Homo* and relatives. This scheme shows the tentative phylogenetic tree of humans beginning with *Ramapithecus*. There are no fossils from the time of *Ramapithecus* until the first proposed branching, some 7–8 million years later. This has presented serious obstacles to our forming the human family tree. About 6 million years ago, the line apparently diverged with the australopithecenes; both *Australopithecus africanus* and A. *robustus* headed down their own specialized paths and finally reached extinction. The other branch, *Homo habilis* (an apparent contemporary of the australopithecenes), represents the line toward modern humans. *Homo habilis* is followed 2 million years later by *H. erectus*. About 200,000 to 500,000 years ago, the line may have again diverged with *H. sapiens neanderthalensis*, branching off and becoming extinct about the time *H. sapiens sapiens* is first found in the fossil record. These last two may have been contemporaries and may have even met each other occasionally. We can only guess at the nature of such a meeting.

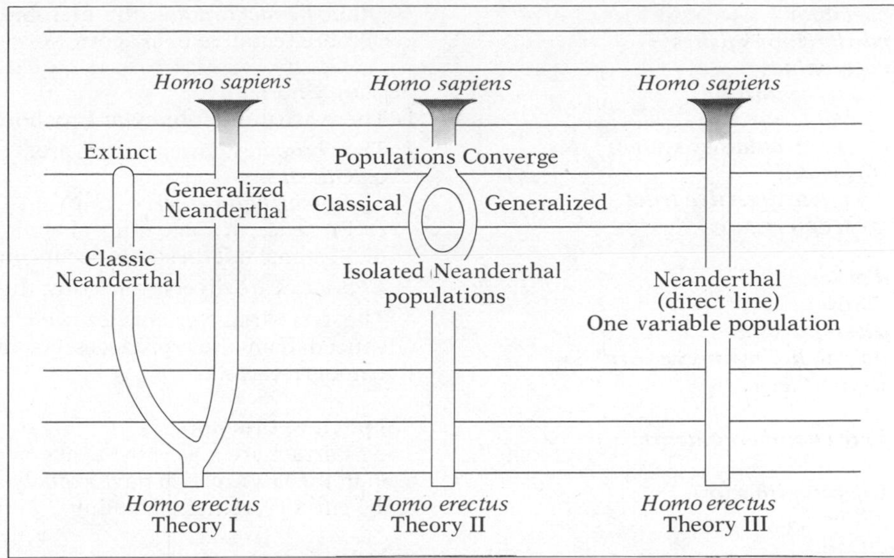

23.33 The manner in which the Neanderthal fits into the evolution of modern humans has not been clearly defined. The scheme depicts three theories that attempt to account for the fate of Neanderthal. The first assumes the population was divided by glaciers. The classical Neanderthal became extinct. The other line survived and their fossils resemble modern humans more closely. The second scheme is based on the possibility that classic Neanderthals and those less advanced were separated, but not long enough to become reproductively isolated. They merged during an interglacial period and from this merger arose modern humans. The third scheme simply considers the varying populations of Neanderthal to be simply that, a varying population that evolved into modern humans.

The third is an interesting compromise which suggests a divergence in the genus *Homo* about 75,000 years ago during, and possibly caused by, the Würm Glacial Period. The two isolated populations became morphologically distinct, subject to different selective pressures, but did *not* become so different that they couldn't interbreed. Some 40,000 years ago (the time of Neanderthal's extinction according to the first theory), the populations again converged. They then underwent a rather complete interbreeding, which produced modern humans. According to this idea, the Neanderthals merely became absorbed. Their descendants are among us. Their descendants, in fact, *are* us (Figure 23.33).

Application of Ideas

1. Compare a bony fish (a vertebrate) with the lancelet (a cephalochordate). In the comparison determine which characteristics in each represent primitive or advanced states.

2. The Cenozoic era is often referred to as the Age of Mammals. Did the mammals evolve in Cenozoic? When did the mammal line actually emerge and what were the earliest mammals like? What actually happened to the mammal line in Cenozoic?

3. Suggest three anatomical differences that clearly distinguish humans from other primates. Are the differences abrupt departures from the general primate characteristics? Comment on these differences and describe how they relate to human accomplishments.

4. Locomotion in the mammals can be visualized with an interesting demonstration using the hand. Pressing the hand flat on a surface with the forearm raised represents one type of locomotion. Next, lift just the palm of the hand with the fingers still in contact. Which animals does this represent? Now raise the palm and fingers with all finger tips touching. Which group walks this way? Next, raise the fingers upward until only three touch. Finally raise the hand still more until only the longest fingers touch. Which animals do the last two positions represent?

Key Words

agnathan
Anthropoid
Australopithecus africanus
Australopithecus boisei
Australopithecus robustus

binocular (stereoscopic) vision

Key Ideas

Phylum Hemichordata

1. These animals are wormlike and live in burrows.

2. Acorn worms reveal their relationship to the *chordate* line through their *gill slits* and larva, which like those of echinoderms and *chordates,* is a deuterostome.

caecilian
cartilaginous fishes
catarrhines
cephalochordate
chordate
closed circulatory system
coelacanth
complete digestive tract
crossopterygian

denticel
dinosaur
diversity
dorsal, hollow nerve cord
Dryopithecus

four-chambered heart

Gigantopithecus
gill arches
gill slit
glaciation

hemichordate
Homo erectus
Homo habilis
Homo sapiens neanderthalensis
Homo sapiens sapiens

ice ages

lancelet
lobe-finned fish
lungfish

marsupials
monotremes
myotome

Neanderthal human
New-World monkeys
notochord

Old-World monkeys
Old Stone Age
omnivore
Oreopithecus
ostracoderm

parallel evolution
"paved" teeth
pelycosaur
pharynx
Piltdown man
placental mammals
placoderm
placoid scales
platyrrhines
Pleisiadapis
prehensile
prosimian
pterodactyl
pulp cavity

3. Other *hemichordates*, the pterobranchs, share some *chordate* traits, but are even more tentative then acorn worms.

Phylum Chordata

1. There are three subphyla: Urochordata, Cephalochordata, and Vertebrata.
2. The *chordate* characteristics are:
   a. *Gill slits*
   b. *Dorsal, hollow nerve cord*
   c. Presence, at some time, of a *notochord*
   d. Postanal tail (in some, occurs only in larvae or embryo)
3. *Chordates* are diverse and solve their survival problems in a great many ways.
4. The terrestrial vertebrates were the last group to evolve. They are more advanced than their predecessors in the sense that they are the product of the modern earth.

Subphylum Urochordata

The *tunicates* are soft with saclike bodies. The relationship to other *chordates* is seen in the larva, which have *notochords, dorsal hollow nerve cords,* and *gill slits.* Only *gill slits* remain in adults, and they are used mainly in filter feeding.

Subphylum Cephalochordata

1. Cephalochordate fossils date back to early Cambrian times.
2. Lancelets are similar to vertebrates. The *notochord, dorsal, hollow nerve cord,* and the *gill slits* persist in the adult. Gills are used both in filter feeding and respiration.

Subphylum Vertebrata: Animals with Backbones

In addition to the standard *chordate* characteristics, vertebrates have several traits in common:
   a. Cranial brain
   b. *Vertebral column*
   c. Ventral heart and *closed circulatory system*
   d. *Gills* or lungs
   e. Two pairs of limbs
   f. Complete digestive tract
   g. *Kidney*
   h. Separate sexes

Class Agnatha: Jawless Fishes

1. The earliest vertebrate fossils are *Ostracoderms*, armored fish. The fossils are $\frac{1}{2}$ billion years old from Ordovician, Silurian, and Devonian deposits.
2. *Ostracoderms* had large heads and were probably bottom feeders.
3. Today's representatives in *Agnatha* are the cyclostomes, the lampreys and hagfish. They feed by a suckerlike rasping mouth. The larval lamprey lives in fresh water, while the adult lives in the sea. (In the Great Lakes they remain in fresh water.)

Class Placodermi: Extinct Jawed Fishes

1. *Placoderms* originated in Silurian, flourished in Devonian, and then died out. They had hinged jaws, paired fins, and were predators.
2. The jaw is believed to have derived from the *gill arches*. Some stages in this transition are seen in sharks.
3. *Placoderms* are ancestral to today's fishes.

Class Chondrichthyes: Sharks, Rays, Skates, and Chimera

1. The *cartilaginous fishes* are predators and scavengers.
2. They appeared first in Devonian but were replaced somewhat by bony fish by Mesozoic, increasing again in Jurassic.
3. The tail and fin shape of sharks are considered primitive today. Shark skin contains many *placoid scales*. The teeth are produced continuously in rows. They occur in a paved form in rays.

Ramapithecus

stegocephalian
stem-reptile

therapsid
tunicate

urochordate

ventral nerve cord
vertebral column

Class Osteichthyes: The Bony Fishes

1. Bony fishes have radiated into numerous niches.
2. They differ from sharks in many ways:
   a. Fins are much more movable, adding to manueverability.
   b. Air bladders add buoyancy and greater stability at varying depths.
   c. Active pumping of water over gills.
3. Bony fish have the lateral line organ, flattened scales, and a two-chambered heart. They are mostly external fertilizers, but some copulate and protect their young.
4. The bony fish line split into a ray-finned fish line that includes the teleosts of today, and the lobe-finned line that produced terrestrial animals. The remnants of the early split are the *lungfish.*
5. *Lungfish* live in Australia, Africa, and South America. The air bladder is vascularized and connects to the pharynx. This permits breathing and survival in stagnant water. They can survive long periods covered by mud. These traits are shared with amphibians.
6. The *crossopterygians* were believed to be extinct until the *lobe-finned coelacanth, Latimeria,* was found alive and well near Madagascar. *Lobe-fins* are believed to have preceded walking limbs.

Class Amphibia: The Amphibians

1. Three surviving orders are Urodela (salamanders), Anura (frogs and toads), and Apoda (wormlike *caecilians*).
2. Nearly all amphibians reproduce in water through external fertilization.
3. All have three-chambered hearts, most have lungs, but use skin respiration as well.
4. Amphibians appear first in late Devonian and are believed to have come from the *stegocephalian* line of early amphibians.
5. Terrestrial life required skeletal changes in the backbone and pectoral and pelvic girdles. The early terrestrial vertebrates probably moved like salamanders do today, with legs at an awkward, flexed angle. Frogs evolved a specialized form of locomotion, while *caecilians* move like snakes.

Class Reptilia: The Reptiles

1. Reptiles represent a final adaptation to dry land. Their eggs, like those of birds, do not require water. Their skins are scaly and protect against drying. Internal lungs and efficient circulatory systems help with the adaptation. Many are efficient predators with keen senses; some are venomous.
2. The Age of Reptiles lasted over 160 million years. Reptiles branched early from the amphibians (Pennsylvanian) and peaked in number during Jurassic, many dying out at the close of Cretaceous.
3. The ancestral split from amphibians is represented by *Seymouria*. It retains the salmanderlike stance. *Seymouria* was from a group known as cotylosaurs. These are called the *stem-reptiles.* Descendants of the *stem-reptiles* radiated out into numerous land niches, and some returned to water. At least two lines of flying reptiles evolved.
4. The large reptiles arrived in Cretaceous, with *Tyrannosaurus* representing the largest carnivores. *Diplodocus* was the largest herbivore.
5. The demise of reptiles accompanied the close of Cretaceous. The reasons are not known but there are guesses:
   a. Changing vegetation
   b. Changing climate
   c. Mammalian predators
6. Today's reptiles represent four branches from the *stem-reptiles*. These are the:
   a. Turtles and side-necked turtles
   b. Alligators and crocodiles
   c. Lizards and snakes
   d. Tuatara
7. Today's birds and mammals represent two additional branches from the *stem-reptiles.*

Class Aves: The Birds

1. Modern birds emerged from a line of birdlike reptiles.
2. Some of the modifications or departures from the ancestor include:
   a. Loss of tail vertebrae
   b. Reduction of wing bones and loss of wing claw
   c. Loss of teeth
   d. Development of hollow bones
3. Flight modifications are numerous, including hollow, strong bones, fusion of vertebrae, enlarged sternum with keel, feet for grasping, feathers, fusion in "hand" bones, highly efficient circulatory (four-chambered heart) and respiratory systems, and a constant body temperature. The eggs are similar to reptilian eggs and parental care is almost universal. Food is stored in a crop and ground in a gravel-filled gizzard.
4. Penguins, emus, ostriches, and other flightless birds developed different strategies than flight for survival.
5. Diversity in the 8650 species of birds is reflected mainly in beak and foot modifications. These relate to feeding and roosting habits.

Class Mammalia: The Mammals

1. Mammals produce milk, have hair, four-chambered hearts, are homeothermic, and most grow their embryos in a uterus, nourished through a placenta. Mammals are active and intelligent.
2. They probably originated in Permian, branching early from primitive stem-reptiles.
3. Mammals survived the reptile age as a small minority, relying on intelligence and keen senses. They protected their young, which were born live. They were probably very small predators, feeding on invertebrates, reptile eggs and young.
4. At the start of Cenozoic three groups of mammals emerged.
   a. *Monotremes* (egg layers)
   b. *Marsupials* (pouched)
   c. *Placental* (placenta)
5. By Eocene, the main lines leading to today's mammals had become established.
6. Today's mammals are organized into eighteen or nineteen orders, depending on whether you place the pinnipeds with the carnivores or in their own order.

The Primates

1. The modern primates, order Primates, are organized into two major suborders:
   a. Suborder Prosimiae: tree shrews, tarsiers, lemurs, lorises
   b. Suborder Anthropoidea: monkeys, apes, humans
2. Anthropoidea is subdivided into three superfamilies:
   a. Superfamily Ceboidea: New-World monkeys (prehensile tails, wide-set nostrils)
   b. Superfamily Cercopithecoidea: Old-World monkeys (tails not prehensile, often short, close-set nostrils)
   c. Superfamily Hominoidea: gibbons, orangs, chimps, gorillas, humans
3. Primates can be traced back 75 million years. Early primates were rodentlike insectivores. *Plesiadapsis* is a somewhat later example, but represents an early type.
4. Newer ideas based on continental drift hypothetically explain the distribution of *Old-* and *New-World monkeys.*
5. Modern primates emerged by the close of Miocene.
6. In the primate taxonomic tree, the earliest branches produced lorises, tarsiers, and lemurs. A later branch produced *New-World monkeys.* More recent branches produced *Old-World monkeys,* apes, and humans.
7. Primates have developed special characteristics:
   a. *Prehensile* hands (and tails in some)
   b. Nails rather than claws
   c. *Binocular, stereoscopic vision*
   d. Good hand–eye coordination
   e. Large eyes and brain

8. Primates have unspecialized dentition, generally with large canines for defense and killing prey.

Human Evolution
A characteristic of human and other species fossil analysis is that new species emerge suddenly and survive for long periods. Offshoots arise and coexist, both well adapted.

The Divergence of the Human Line
1. Biochemical evidence suggests that the human and ape line diverged no more than 10 million years ago.
2. Possible fossils of the human line date back to *Ramapithecus*, about 12 million years ago. A gap of 9 million years occurs before the next fossils.
3. The next fossils are *Australopithecus*, including *A. africanus*, *A. robustus*, and *Homo habilis*. The genus persisted until 1 million years ago.

Australopithecus africanus
Characteristics: small, 4 ft, humanlike teeth, upright, used simple tools, cranial capacity 650 cc.

Australopithecus robustus
Characteristics: small, but large-boned, heavy jaw, large teeth, large saggital crest, made tools.

Homo habilis
Homo habilis is controversial, believed by some to be a species of *Australopithecus*. The cranial capacity of one skull (1470) is 750 cc.

Homo erectus
1. *H. erectus* is decidedly human, but about 1.5 million years old (other names are Java Ape Man and Peking Man). The species lasted over 1 million years, coexisting with *Australopithecus* and *Homo sapiens*.
2. The cranial capacity of *H. erectus* ranges from 775 to 1225 cc. The species was widespread, found on three continents. They were generally similar to *Homo sapiens* and produced stone hand axes and scrapers.

Homo sapiens: Neanderthals and Us
1. *Neanderthal (Homo sapiens neanderthalensis)* fossils have been found throughout Europe, North Africa, and parts of Asia. Neanderthal left a record of toolmaking and other cultural artifacts.
2. They have been placed into two groups: classical (the heavier coarse features), and progressive (lighter, refined features). With many new finds it appears that *Neanderthal* is just a single highly varied type. Three theories of his fate are:
   a. *Neanderthal* may have coexisted with modern humans, but went extinct 40,000 years ago as *H. sapiens sapiens* expanded.
   b. Some variant of *Neanderthal* may have produced *H. sapiens sapiens.*
   c. *Neanderthal* and modern humans diverged about 75,000 years ago, then converged 40,000 years ago, and Neanderthal was absorbed by *H. sapiens sapiens.*

Review Questions

1. What structures in the acorn worm and its larva make it a candidate for the echinoderm–chordate line? (p. 634)
2. Describe the *notochord.* (p. 635)
3. List three characteristics of the *chordates*. (p. 635)
4. Why are *tunicates* believed to be related to *chordates?* (p. 635)
5. List three chordate characteristics of the lancelet. (p. 636)
6. List eight characteristics of vertebrates. (p. 637)
7. Are all of the eight characteristics exclusive to the vertebrates? Explain. (p. 637)
8. What were the earliest *Agnatha* and when did they live? (p. 637)
9. Describe one member of the modern *Agnatha*. (p. 637)

10. What were the advances of the *placoderms* over the *ostracoderms?* (p. 637–638)
11. Briefly describe the skeleton, skin, teeth, and fins of the modern *cartilaginous fish.* (pp. 639–640)
12. What is the function of the air bladder in bony fishes? (p. 640)
13. What evolutionary pathways were followed by the two divisions of the bony fishes? (pp. 640–641)
14. Describe the special adaptations of *lungfishes.* (p. 641)
15. List and give examples of the three types of amphibians. (p. 642)
16. What are some of the skeletal modifications required for terrestrial locomotion? (pp. 642–643)
17. Briefly review four adaptations made by reptiles for terrestrial life. (pp. 643–644)
18. When did the age of reptiles end and what are some of the suggested reasons for their demise? (p. 646)
19. Reptiles are believed to be paraphyletic. Describe the origins of the four groups of today. (p. 646)
20. In what ways have the birds of today become different from the ancestral reptiles? (pp. 646, 648)
21. What reptilian characteristics do birds have today? (p. 648)
22. What are two different adaptations of flightless birds that have helped them succeed? (p. 649)
23. List five important characteristics of the mammals. (pp. 649–650)
24. How might the early mammals have survived the long age of reptiles? (p. 650)
25. Describe the three groups of mammals that emerged into the Cenozoic. (pp. 650–651)
26. What conditions in the Cenozoic led to the extinction of many mammalian species? (pp. 650, 652–653)
27. List the divisions of the primates and add examples of the groups within each division. (pp. 655–656)
28. Describe the Paleocene ancestors of today's primates. (p. 656)
29. What recent theory has now displaced the older idea of parallel evolution in *New-* and *Old-World monkeys?* (p. 656)
30. What physical changes in the primates of Eocene brought in the modern primates? (p. 656)
31. In the primate taxonomic tree, which group branched off earliest? Which was second? Third? (p. 656 and Figure 23.28).
32. What is the evidence that suggests *Homo erectus* was a successful species? (pp. 660–661)
33. Why is *H. erectus* believed to have been an intelligent *hominid?* (p. 660)
34. What were some cultural traits of *Homo sapiens neanderthalensis?* (pp. 661–662)
35. Briefly review the three theories that attempt to establish the relationship between *Neanderthal* and *H. sapiens sapiens.* (pp. 662–663)

Chapter 24

Animal Support and Locomotion

Compared to other celestial bodies, the earth is a rather unspectacular place. It lacks many of the great mysteries of the universe, such as the black holes through which time and space can be stretched. Even as planets go, it's a rather unexciting place. Other planets are certainly larger and more dramatic. What must the great storms on Jupiter be like? Nonetheless, our small planet harbors life, and although our globe is insignificant by astronomical standards, it is large enough to present its animals with two specific sorts of problems.

First, because we are so small, it seems to have a rather large surface. Also, that surface is not homogeneous. It is quite different from one place to another. And it changes. This means that some areas are more suitable to life than others, and areas that are quite hospitable at one time may be quite unsuitable (even deadly) at another time. So many animals have had to develop the ability to travel from one place to another. Some travel easily, such as the golden plover, whose migratory distance would take it around the world each year as it moves between its wintering and breeding grounds. Other animals, however, such as the lowly chiton, simply creep slowly over their wave-washed rocks, scraping off morsels of slimy food as they go (Figure 24.1). Nonetheless, both animals must move or die.

24.1 Both the chiton and the golden plover utilize their skeletal and muscular structures in movement—but with quite different results. The chiton's travels, as it creeps along in its rocky marine world, are measured in centimeters. The golden plover, in contrast, is a world traveler capable of physical feats that stagger the imagination.

Earth also presents us with another basic problem. For some reason, celestial bodies with great mass tend to pull objects toward their surface. Not only do apples fall, but animals find it hard to jump. We are held fast. This means that there is a continuous tug on our bodies, seeking to flatten us against our great globe. But we resist with stiff skeletal systems that are able to withstand the persistent pull and, sometimes, with muscles that contract and defy that pull. Earth's gentle claim to our bodies has only recently been overcome as we have managed to yield a few people over to the gravity of another celestial body, but generally we are all subject to a certain tie to our mother earth, and many species have developed stiff structures that allow them to stand away, balanced, rather than collapsing in a great mass of quivering tissue. Our structures are often movable and serve as a place of attachment for those same muscles that help us not only to move from one place to another, but to behave in an adaptive manner wherever we are. So let's have a closer look at the structures of support and movement among the animals that live on this small planet we call home.

24.1 Invertebrate Form and Movement

We are aware that many animals have skeletons of one type or another. But some do not. There are many soft-bodied animals that have no visible means of support. However, some of these beasts are supported by a *hydrostatic* skeleton that maintains form and assists in the movement of muscles. A hydrostatic skeleton, as the name implies, involves fluids. These fluids exert pressure by being forced against a restraining outer wall. The best examples of hydrostatic systems are found in the saclike animals of the phylum *Coelenterata* and the worms of the phyla *Playthelminthes, Aschelminthes,* and *Annelida.*

Coelenterates, you will recall, are comparatively simple animals with saclike bodies that consist of two cell layers separated by a jellylike material. The sea anemone's body, for example, is a hollow sac filled with seawater (see Figure 22.9). The flow of seawater in and out is controlled by a valve at the opening to the sac. Muscle fibers in the body wall run both longitudinally and circularly, so the anemone has the ability to make itself short and thick or tall and thin. How does the anemone keep its shape and move its parts so precisely?

With its body cavity filled with water and the "mouth" closed, the contraction of circular muscle fibers and the relaxation of longitudinal fibers extend the body into a shape similar to a vase. Then, contraction of longitudinal fibers and relaxation of the circular fibers pulls the body down into a wider, flatter squat. Imagine squeezing a plastic bag filled with water and tied off at the top. Relaxation of both sets of muscle fibers slightly extends the body to a position consistent with its internal cell turgor (fluid pressure in the cells). Without pushing water around, the movements of an anemone would be greatly restricted and, of course, the animal would be less effective at catching its prey.

Hydrostatic skeletons can also work in animals in which the internal body fluids are used to extend muscles and maintain shape. For example, consider the earthworm (phylum Annelida). Since the earthworm is a segmented animal (the body essentially consists of a series of closed units), the fluid mechanism is a bit more complicated than that of the anemone. The complexity is not surprising, however, considering the earthworm's much greater range of movements and the fact that it is a terrestrial, burrowing animal, living a difficult life, indeed. The dense muscles in the earthworm are arranged in an outer circular layer and an inner layer that runs along its length. In this animal, though, each layer is organized into bundles rather than a simple sheet (Figure 24.2). Such complexity permits a wide range of movements, but, nevertheless, each segment moves according to the simple principle described for the anemone. Contraction of the circular muscles extends the body, while contraction of longitudinal muscles shortens and thickens the body. The fluid, located in the body cavity, or coelom, makes the inching motion possible. When coelomic fluid is experimentally removed, the worm's motion is impaired.

The roundworm (phylum Aschelminthes) has a simpler musculature than the annelid, in that only the longitudinal layer is present. However, it uses a similar principle in its movement. The body fluid is held in one long, unsegmented body cavity. It is under considerable pressure by the longitudinal

24.2 Body movement in the earthworm. The earthworm, an annelid, is capable of a wide range of movements. Like the anemone, the earthworm utilizes hydrostatic forces in maintaining its body shape and as a resistant base for muscle action. The muscles are arranged in longitudinal bundles and in a circular sheet. Without hydrostatic pressure, the earthworm's wide range of movement is seriously curtailed.

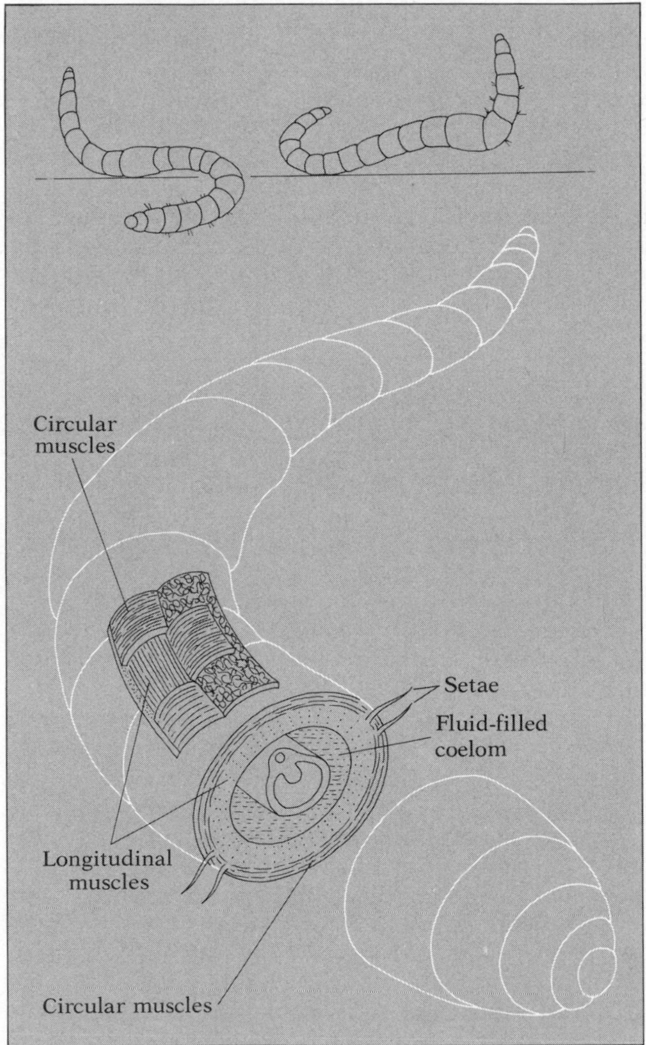

muscles and thus the worm has a very efficient hydrostatic skeleton. Roundworms are very firm, as you know if you've ever had the opportunity and courage to pick up one. Because it has only longitudinal muscles, the roundworm can't inch its way around. It can only whip back and forth. Forward motion is accomplished by the whipping action against the wave front it creates, a technique that works best in a thick liquid. Soil nematodes probably push against fine soil particles as they move through the moist earth. The muscles contract under their own power, but they can only extend by being stretched back out by hydrostatic pressure.

Exoskeletons

Now let's leave the concept of watery skeletons and talk about those that seem to make more sense: *hard* skeletons. These can lie outside the soft flesh or inside, but wherever they are, they serve in support and locomotion. *Exoskeletons* are generally on the outside of the organism and are formed from secretions of the epidermal cells. *Endoskeletons* are inside, surrounded by living tissue. The muscle attachments must vary accordingly, as we shall see.

Since exoskeletons are formed from secretions of the epidermal cells, they are derived from embryonic ectoderm. Hardened outer coverings are a characteristic of the phylum Arthropoda, the phylum that contains the most species including the insects, crustaceans, and other joint-legged types. The arthropod exoskeleton serves as protection and as a place for muscle attachment, just as the endoskeleton does, but it has the additional function of retarding water loss in terrestrial species. You may recall that its principal constituent is chitin, a complex carbohydrate. In crustaceans (lobsters, shrimp, crayfish, crabs, etc.), the chitin is hardened with calcium carbonate.

Undoubtedly, it is a great advantage to wear one's protection on the outside, but when that armor is made of a nonliving secretion, how does one grow? There is a solution. Periodically, the exoskeleton is shed, or *molted,* and a new one is formed. The molting process (also known as *ecdysis*) is controlled by the hormone *ecdysone* in insects and a similar hormone, *crustecdysone,* in the crustaceans. (We'll have more to say about this in Chapter 27.) In ecdysis the first order of business is to try to save some of the constituents of the old shell. After all, some of those minerals were hard to come by. Since the exoskeleton is a permanent material, it cannot be completely reabsorbed, but a few salts can be extracted before it splits and falls away from the epidermis.

During the period before a new shell is completed, the soft-bodied arthropod is highly vulnerable to its predators, and so molting animals are furtive beasts indeed. But a new exoskeleton is soon secreted by cells of the epidermis and allowed to harden. Just before this happens, the animal swells by taking in water. That water is excreted after the shell is laid down. Thus, when the animal returns to its normal size, its new suit of armor will, at first, be a bit too large.

Of course, an animal that has an exoskeleton will behave a bit differently from one that has an endoskeleton. It fact, muscle action in arthropods is more or less the opposite of that in vertebrates. Note in Figure 24.3 that a muscle that *extends* a limb in an arthropod is located about where one would be that would flex the limb in a vertebrate. The muscles of arthropods differ from those of vertebrates in another way. They are commonly attached to skeletal structures known as *apodemes.* These are ingrowths of the exoskeleton that serve as points of attachment for the muscles.

It seems that exoskeletons should be rather cumbersome, but you have most likely observed that insects are quick little beasts and are exceedingly agile. (One of your authors once saw a fly walk up a wall!) And we have all been impressed with the strength of insects, perhaps having watched an ant drag a dead roach many times its size. The strength and agility of insects, it turns out, is not due to a superior structure, but an inferior size. They can move so strongly and swiftly with an exoskeleton precisely because they are small. It has been calculated that an arthropod as large as a human would

24.3 The attachment of muscles in the arthropod limb (top) is the opposite of that in a vertebrate limb. In both, the muscles originate in the upper region and insert in the lower. The (a) muscles in both extend the limb, and the (b) muscles flex the limb. Note their opposite attachment with respect to the skeleton. Note also the *apodeme* shown in the insect exoskeleton.

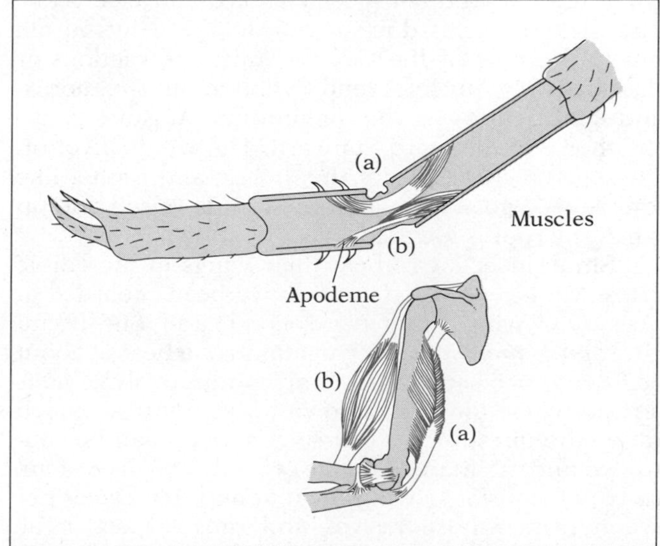

hardly be able to move at all. Thus, we don't have to be afraid of encountering science fiction's giant ants. We would be faster and stronger than they would be.

Insect Wings and Their Muscles

We have generalized about the arthropod skeleton and muscle attachments, noting that the skeleton is on the outside and the muscles are on the inside. An arthropod leg is a hollow, jointed tube, and its activating muscles and/or tendons extend down its length to move the extremity from the inside. The principle is illustrated in grasshoppers in which the great muscle that straightens out the hind leg for its powerful jump is located in the large *femur,* or third segment, and acts through a long thin tendon (a structure that generally attaches muscle to skeleton) that passes through the kneelike joint and inserts well down inside the second segment of the leg.

The movement of insect wings is somewhat different. Embryonically, the wing begins as a hollow organ with an exoskeleton, but in the adult it changes into a thin, chitinous sheet. There are no muscles in the wing, but there are powerful "wing muscles" in the insect thorax. They move the wings through a complex lever arrangement that depends on distorting the exoskeleton of the thorax itself (Figure 24.4).

There are minor muscles in the thorax that help fold the wings out of the way and spread them for flight, but the flight muscles themselves are divided into two groups of opposed muscles, the *transverse* thoracic muscles that raise the wings and the *longitudinal* thoracic muscles that bring them down again. In the wing beat, the transverse muscles contract while the longitudinal muscles relax, and vice versa. The flight muscles don't attach to the wings at all, but to the roof of the thorax. Thus, contractions of the transverse muscles tend to flatten out the thorax, and contractions of the longitudinal muscles cause the thorax roof to arch upward. The wings pivot on the skeletal side walls of the thorax, and, rather like long levers, go down when the roof is raised and up when the roof is lowered.

Small insects can beat their wings at incredible rates. Midges ("no-see-ums") have been recorded at over 1000 wing beats per second, and our friend *Drosophila melanogaster* can maintain a beat of about 300 beats per second for 2 hr continuously. This is a faster cycle than any known nerve impulse, which raises the question of how such activity can be controlled and maintained by the central nervous system. In blowflies, whose wings beat at only 140 cycles per second but whose nerves and muscles are large enough for neurophysiological examination, the neural impulses to beating wings have been timed at only about twelve per second, or one impulse for every eleven or twelve wing beats. Apparently,

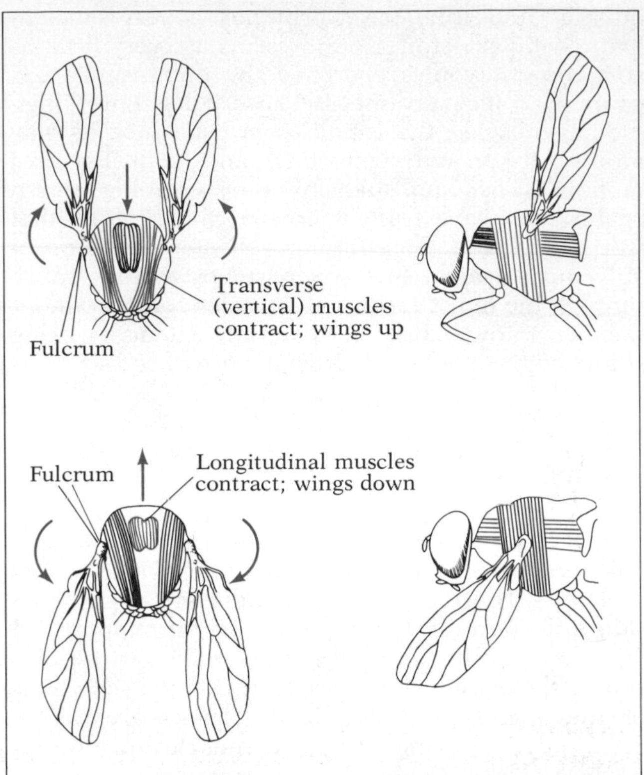

24.4 There are two sets of muscles in the thorax of most flying insects, called the *transverse* and *longitudinal* groups, although they are really arranged diagonally to each other. When they contract, they flex the thin shell which is the thorax, causing the attached wings to move. In some species, such as the mosquito, the wings move far faster than the neurons applying the muscles could fire. The mechanism involved was a great puzzle until it was found that the thorax, once flexed, would resonate, and each vibration would cause the wings to move, producing the well-known whine that has initiated many a midnight search for the tiny culprit.

the muscles themselves are capable of maintaining the cycle between neural impulses through some kind of resonating mechanism. In other words they may lengthen and shorten several times when stimulated by a single neural impulse.

In more primitive insects (such as the dragonflies) and those with slower wing beats (such as moths and butterflies), each wing beat is triggered by a separate nerve impulse.

Mollusk Shells—A Type of Exoskeleton

Mollusks, such as snails, clams, and octopuses, generally utilize both hydrostatic skeletons and hard exoskeletons. The exoskeletons are, of course, their *shells,* which usually cover only part of the animal. Clams can protect themselves with two shells, but when they are moving about, they use the hydrostatic pressure of their hollow, muscular foot. Snails dimly peer about on eyestalks raised by hydrostatic pressure.

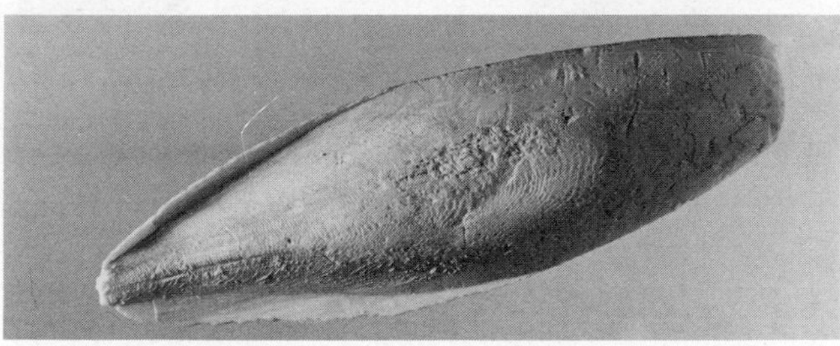

24.5 Cuttlefish and cuttlebone. The cuttlebone seen here is actually an internalized exoskeleton of the cuttlefish. This highly porous flotation device is located just behind the head and along the inside of the dorsal surface. Squid, close relatives of the cuttlefish, also have a cartilaginous structure, but it has only a skeletal function.

Chitons have a hard exoskeleton consisting of eight movable plates (see Figure 22.24). In a few species, the skin of the mantle grows up over the plates. Thus, the exoskeleton becomes an endoskeleton, although technically it is still an exoskeleton because it is secreted by ectodermal cells. Squids and some slugs have further internalized their exoskeletons into skeletal supporting rods—fully internal, but still derived from embryonic ectoderm.

One squid has found a new use for its internalized skeleton. This is the cuttlefish (Figure 24.5), and its skeleton is cuttlebone, the common beak sharpener for canaries. That isn't the new use we were referring to, however. The cuttlebone, in the live cuttlefish, is used as a flotation device. It has many hollow chambers supported by strong calcium carbonate walls. Question: How does the cuttlefish get air into its cuttlebone? Answer: It doesn't. Instead certain cells of the organism pump water out of the cuttlebone chambers, leaving only a vacuum behind—a most peculiar situation when you think about it.

Although the shells of mollusks are secreted by the epidermis, as are the exoskeletons of arthropods, unlike arthropods, mollusks don't shed their exoskeletons as they grow. Their shells grow only by increments at the edges, though they may be thickened with new layers on the inner surface. This puts a different kind of restraint on what sorts of shells they can have. It turns out that the problem is one of geometry. There are only a few rigid surfaces that can grow by increments and still retain the same shape (think about it). One is a special kind of spiral called the *logarithmic spiral;* a second is a *cone;* a third is a combination of the first two, or a *spiraled cone* (Figure 24.6). And cone shapes, it turns out,

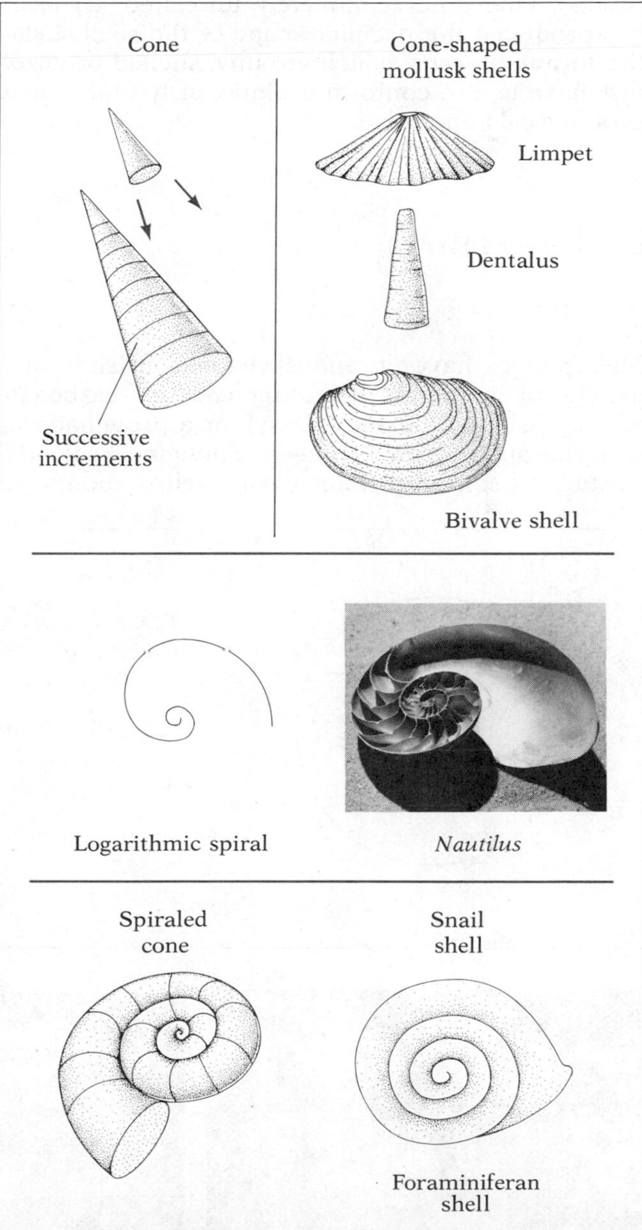

24.6 Exoskeleton growth in mollusks. Unlike the arthropods, the mollusks grow continuously without molting. This is accomplished through the addition of shell material. The mantle secretes the shell. Limpets and bivalves form a cone as they grow. The *Nautilus*, on the other hand, adds new shell material unevenly, forming a continuously enlarging logarithmically spiraled cone. As the *Nautilus* grows, the animal itself lies down cross ways behind it and lives in its most recent compartment. Note the similar problem of growth encountered by the snail and the foraminiferan, a protistan not related to mollusks at all.

are common in mollusks. In fact, limpet shells are simple cones. It's a bit harder to imagine, but the two valves (half-shells) of clams are modified or distorted cones, retaining their shape as they grow at the edge. The familiar snail shell is a spiraled cone. The mollusks have evolved the spiraled cone twice, in cephalopods (chambered nautilus) and in gastropods (snails). One other completely unrelated organism has produced the peculiar shape of the snail shell—the foraminiferans, which are tiny, shelled protozoa that have had to conform evolutionarily to the same geometrical imperative.

Endoskeletons

Sponges

The sponges have an endoskeleton consisting of a variety of materials including calcium carbonate (chalky), silicon dioxide (glassy), or a proteinaceous material aptly termed *spongin.* Spongin is soft and elastic and soaks up water like a, well, a sponge. In fact, sponges are commonly classified according to their skeletal materials, which are secreted by ameba-like cells (amebocytes). These cells roam through the loose gelatinous middle layer of the sponge body, known as *mesoglea* (Figure 24.7).

Echinoderms

The spiny-skinned echinoderms, such as starfish, also have endoskeletons. At first glance, the starfish appears to have an exoskeleton, but closer examination reveals that its skeleton actually lies below the living epidermis (Figure 24.8). *Echinoderm,* by the

24.7 Skeletal framework in three types of sponges. The skeletal elements of sponges, spicules and fibers, occur in a variety of shapes. (a) In the simple ascon sponge, spicules are composed of calcium carbonate and are scattered throughout the body wall. (b) In the much more elaborate glass sponges, spicules of silicon dioxide (akin to glass) are elaborate and interlacing. (c) In the familiar "bath" sponge, the skeletal elements are composed of *spongin,* a fibrous protein. The skeleton in all adult sponges acts strictly as a structural framework, since adult sponges are incapable of organized movement.

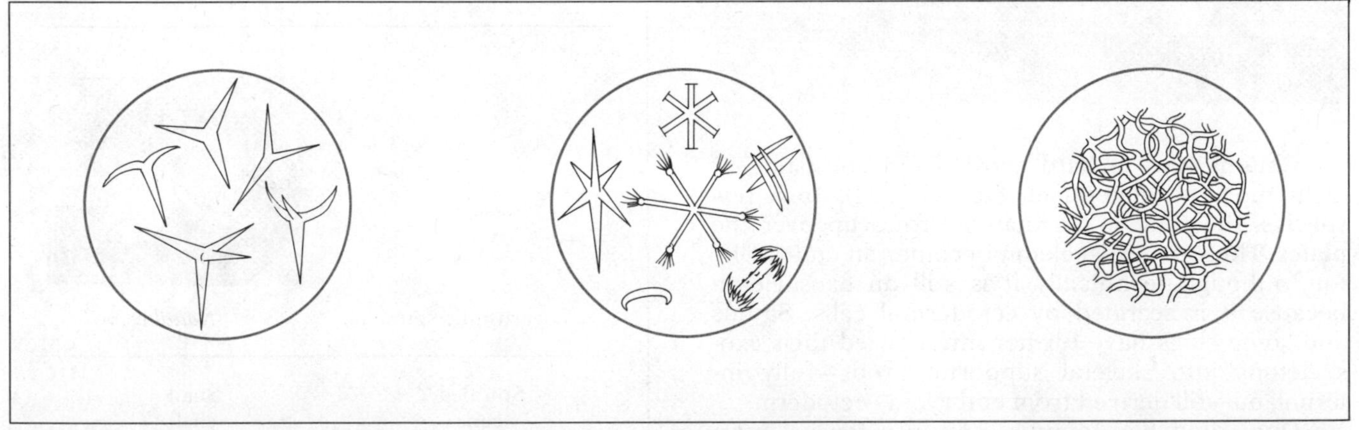

(a)

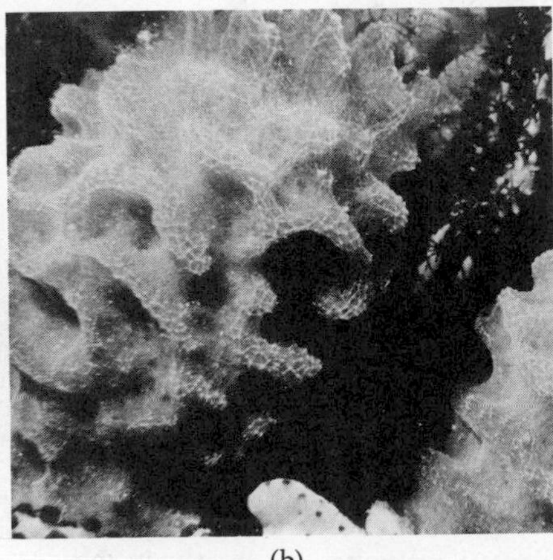

(b)

(c)

way, means "spiny skin." Under the epidermis, the thick dermis (of mesodermal origin) secretes a series of plates that fit closely against each other. The plates themselves are composed of many tightly interlocked little spined bodies, composed chiefly of calcium carbonate, a common material in marine animals. As the skeletal plates form, they erupt over the entire upper surface of the starfish, creating short, blunt spines. (In the sea urchin, these spines are very long, often very sharp, and movable.) By the way, the hinges between the plates give starfish the flexibility they need to wrap around an oyster and insistently pull on both halves of the shell until the hapless prey yields.

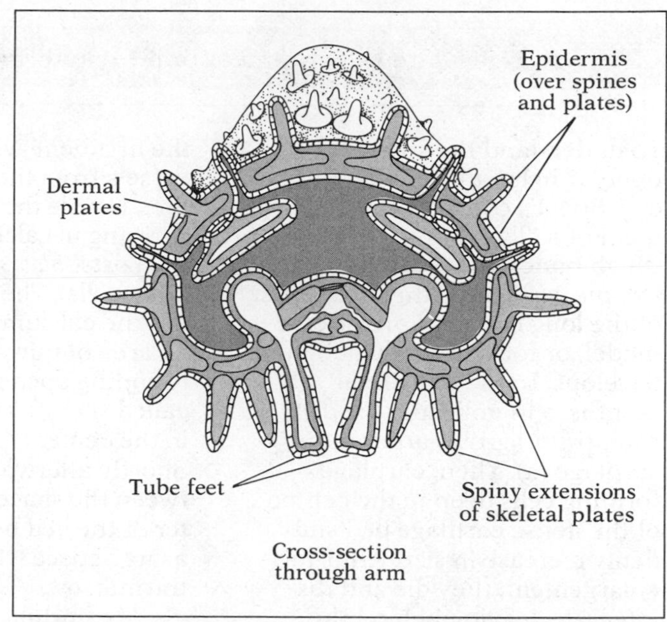

24.8 The endoskeleton of echinoderms. In the starfish, the skeleton consists of numerous plates composed of calcium carbonate. In fact, the familiar spines are actually overgrown plates. The crusty skeleton appears to be outside the animal, but it is actually an endoskeleton since it is covered with a thin epidermis and is secreted by mesodermal cells.

24.2 Vertebrate Form

Vertebrate Supporting Structures

Bone

When we think of skeletons, we, as vertebrates, don't usually think of shells, or spiny plates. We think of bone. Bones are the light, rocklike structures that hold us away from the earth's surface (keeping us from lying ignominiously in great heaps of protoplasm) and that shield our more delicate parts from hard environmental objects—which we sometimes encounter with considerable force.

The skeletal system of most vertebrates is composed of bone. The exceptions are the jawless fishes (lampreys and hagfishes) and the cartilaginous fishes (sharks, skates, rays, and chimera). The jawless fishes have a rather simple skeleton, similar to that of the *Amphioxus*. However, there is some skeletal development in the skull, and cartilaginous rods hold the fins upright.

All the other vertebrates (bony fishes, amphibians, reptiles, birds, and mammals) have skeletons composed primarily of bone, although some cartilage usually persists, as we shall see. The endoskeleton reaches its most extreme development in the backboned animals. The vertebrate skeleton may vary a bit from one individual to the next within a species but, in its essentials, it is remarkably conservative and doesn't vary. In fact, it is often used to classify animals and show relationships. Fortunately, systematists have fossil bones to work with rather than something as variable as blood vessels.

The principal functions of vertebrate skeletons include support, muscle attachment, protection, and one we haven't said much about yet—the production of blood cells.

Vertebrate skeletons are usually divided, for convenience, into *axial* and *appendicular* components. The axial skeleton includes the skull (*cranium*), the vertebral column, and the bones of the *thorax* (chest). The appendicular skeleton is particularly revealing when comparing locomotion in vertebrates. It consists of the limbs (always two pairs) with the two girdles, *pelvic* (hip) and *pectoral* (shoulder). All these, by the way, are absent in most snakes. Thus, there are no broad-shouldered snakes. Before getting into the details of all this, let's consider joints.

The various bones of the body meet at *joints*. Each kind of joint allows only a certain kind and degree of movement and some allow no movement at all (such as the joints between the bones of the skull—try flexing your skull). The joints of the three pelvic bones are only slightly movable, and only in certain people. In the human female, pelvic joints can move slightly so as to enlarge the birth canal when the need arises.

The more movable joints, though—for example, the elbow or hip—are both complex and elegant. The articulating (contacting) surfaces are covered with a thin layer of exceedingly smooth cartilage. Bones are held together by a tough *capsule* of ligament, which, despite its toughness, can be torn

Essay 24.1 The Development and Structure of Bone

To understand the structure of bone, it helps to see how it develops. So let's consider the development of a "long bone"—the *femur* (thigh bone). In vertebrates, *hyaline cartilage* is the forerunner of the long bones. It provides a model, or form, upon which bone develops. In the early fetal months, a membrane (the *perichondrium*) grows around the cartilage (a). Then, cartilage-forming cells deep in the center of the dense cartilage bed suddenly increase in size. After this enlargement, they die and disintegrate, leaving behind the cavities their bodies have created. These cavities become the primary ossification centers where the first bone will form (b). Blood vessels from the perichondrium then invade the cartilage bed, first carrying in calcium salts and then *osteoblasts* (embryonic bone cells). The osteoblasts will use the calcium salts to produce an area of thin, bony walls surrounding open pockets. This is called *spongy bone,* and it lies in the center of the cartilage bed. Shortly afterward, the walls between the spaces in the very center of the bed break down, leaving a large space which will become the marrow.

Meanwhile, at the periphery of the cartilage model, something else has been going on. Cells, again from the perichondrium, begin to lay down a collar of bone around the model. This bone is not thin-walled like the first bony wall. It is dense and compact, and as it accumulates, the developing femur grows in width (c).

As all this is happening, the embryonic femur is lengthening as the result of the development of two new bone-forming centers at either end (d). These are the *epiphyseal plates,* which are really areas of cartilage formation. But they yield to bone-forming cells on either side of each plate, which lay down material that causes the bone to grow longer. This lengthening takes place as long as growth continues. It usually stops

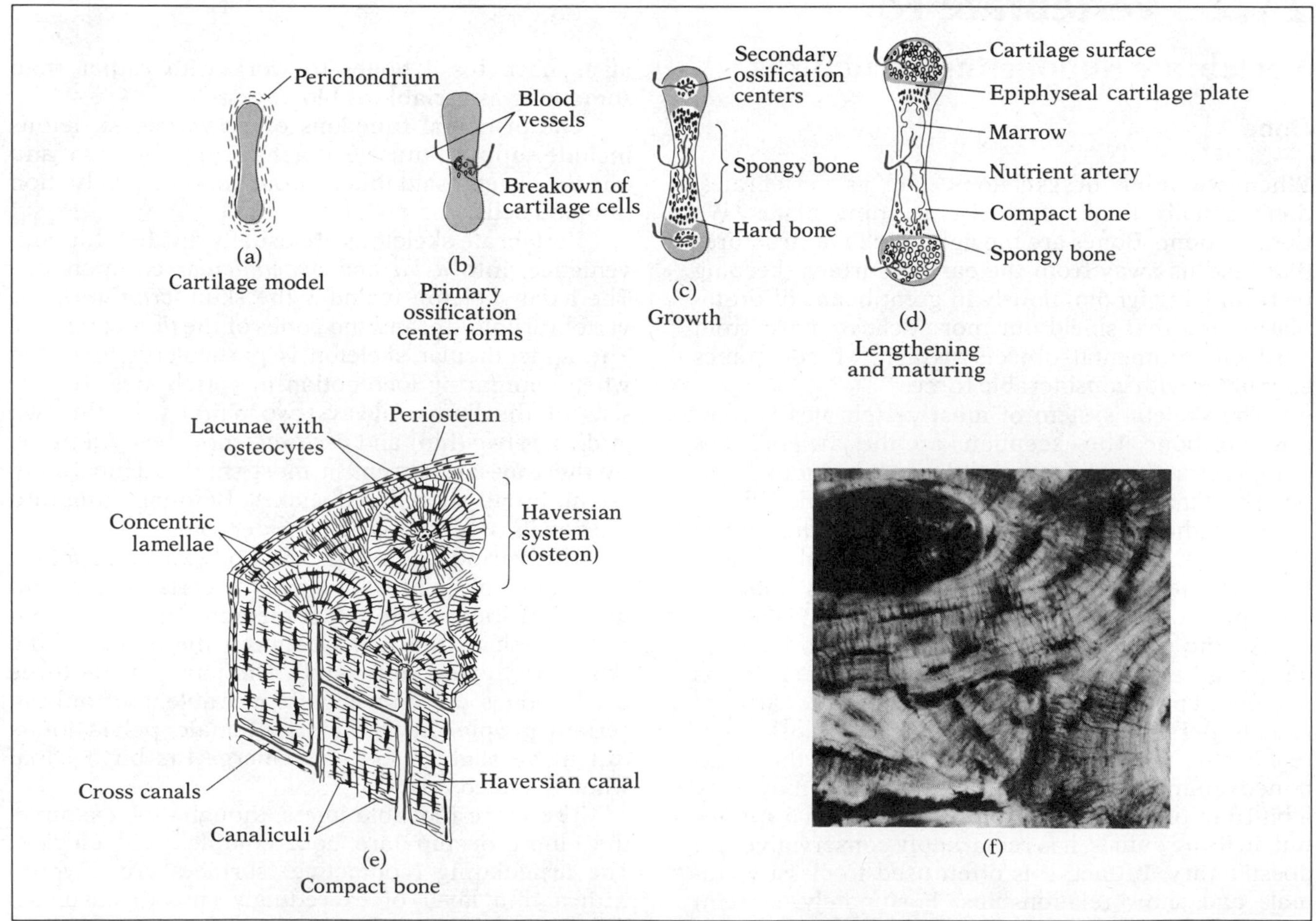

somewhere between the ages of 18 and 24 in humans, but possibly earlier or later. Finally, though, the epiphyseal plate itself ossifies and growth is stopped permanently. Exactly when this occurs depends on growth hormones and other factors, but ultimately upon the genetic constitution of the individual. So if you are taller or shorter than you would like, first blame your epiphyseal plates, and then your parents.

As ossification continues, the bone cells become entombed in minute, hardened cavities, and thus, in isolation, they enter a quiescent state. They are not completely isolated, however. They communicate with the outer world through a series of capillary canals, as we see in mature compact bone.

Cross-sections through mature compact bone reveal an intricate weblike structure called the *osteon* (or *Haversian system*) (e). An osteon is a unit of bone structure, and consists of a central (Haversian) canal that contains blood vessels, surrounded by concentric rings, or *lamellae,* of calcified bone. Within each ring are the osteocytes (bone cells), each in its own minute cavity, the *lacuna.* Reaching between the entombed bone cells are tiny little canals, the *canaliculi* (f). These also communicate with the Haversian canal. The bone cells exchange materials with the circulatory system by means of the fluids seeping through the canaliculi. So, in spite of their stony appearance, bones are living, changing systems.

If a bone is broken, the trauma arouses the sleeping osteocytes in the *periosteum,* and they spring into action, showing they have not lost their ability to produce bone, until the broken ends are fused.

(sprained). The entire joint is constantly lubricated by the secretions of a *synovial* membrane that lines the capsule. Synovial joints help explain the smoothness, great mobility, and speed of many vertebrates. And you can imagine that a hungry lion creaking and squeaking around in the grass is likely to come up short.

The movable joints are generally named according to the type and degree of their movement (Figure 24.9):

1. Gliding joints—permit movement back and forth, such as in the bones of the wrist and hand.
2. Pivotal joints—permit rotating movement in one direction, such as in the neck as the head rotates.
3. Ball-and-socket joints—permit a variety of motions, such as in the hip and shoulder.
4. Hinge joints—permit movement primarily in one plane, such as in the knee.

The elbow joint is a good example of a combination of hinge and pivotal functions.

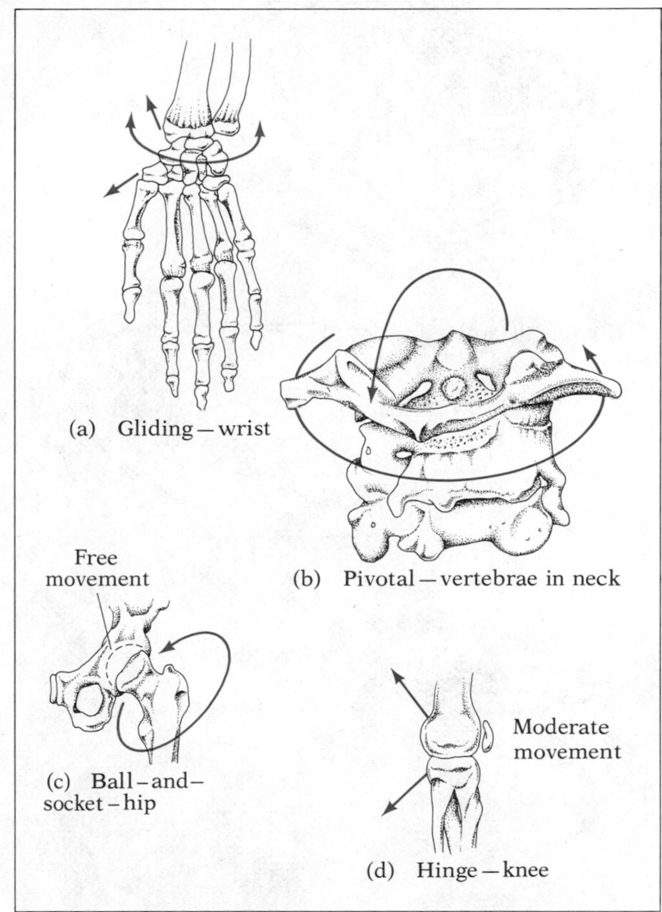

24.9 Joints in the human skeleton. The range of body movements in humans and other vertebrates is made possible by joints in the skeleton. Most movement occurs in (a) gliding, (b) pivotal, (c) ball-and-socket, and (d) hinge joints. Some joints permit very little movement.

The Axial Skeleton

The axial skeleton runs, more or less, along the axis of the body and is particularly fascinating for many evolutionary biologists because it includes the skull and jaw, structures that have traditionally yielded a great deal of information about the evolution of vertebrates. In addition, the axial skeleton includes the vertebral column and the rib cage. Let's have a closer look, now, at the skull.

The Skull

Skulls make a rather nice series, from simple to complex. In the fishes they consist of a loosely arranged series of individual plates. The jaws are simple and form a single-directional hinge, which means they can bite but they can't chew by moving the jaw sideways. In fish, both the top and bottom jaws move, a feature still found in some birds but one which has been lost in the mammals—our upper jaws are fused to our skull. As we compare the skulls of other vertebrates in the order in which we believe they progressed through their evolution, the trend is toward the fusion of parts, resulting in fewer bones (Figure 24.10). Jaws become more mobile in the mammals as the art of chewing develops. This mobility is made possible by the way the jaw articulates with the skull, coupled with a complex muscle arrangement. Amphibians and reptiles have more mobility in the jaws than do mammals, but, like birds, they swallow food whole. In fact, the jaws of snakes show a remarkable adaptation for swallowing whole prey: they unhinge and permit snakes to take prey that is often larger in diameter than the snakes themselves. They are able to do this because the lower jaw is hinged to the quadrate bone, which is loosely attached to the skull and can be pulled away from it (see Figure 25.10).

24.10 Comparison of skull structure in vertebrates. In the evolution of the vertebrate skull, bones have tended to become fused. We can trace this tendency by considering living species. The bony fish has many separate bones both in the cranium and the jaw. We see simpler skulls and jaws in the amphibian and reptile, and even more rigid skulls in birds and mammals. Note that the mammalian jaw, while comparatively simple, is very movable. Its great mobility permits chewing—a seemingly common act, but actually unusual among vertebrates.

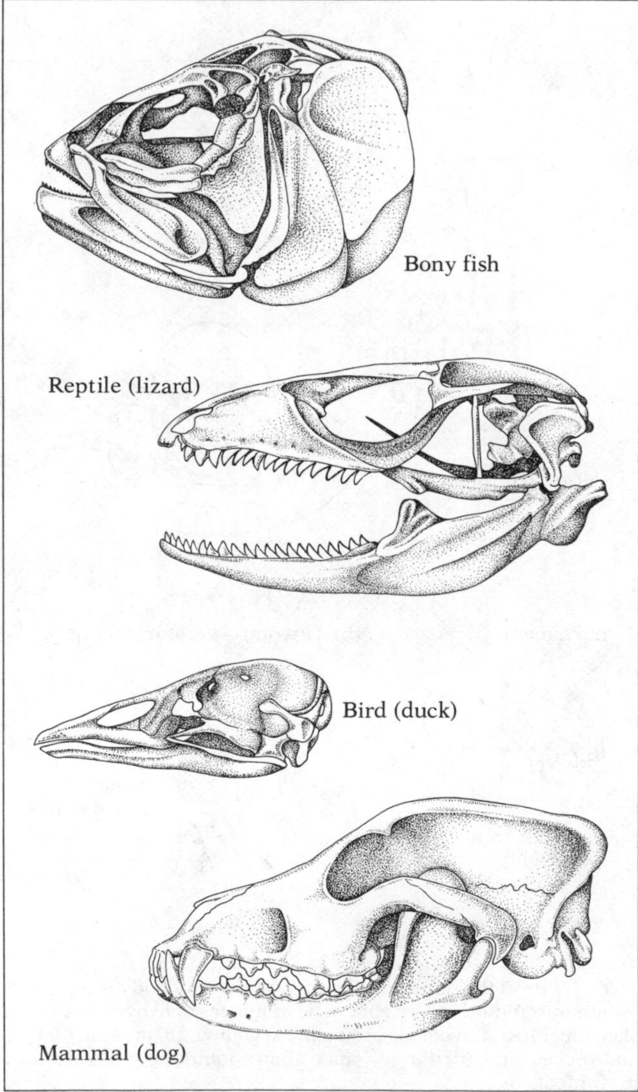

In vertebrates the skull not only protects the brain, but also houses a complex group of receptors. Many of the senses, you know, are "up front," such as are the organs of sight, hearing, balance, smell, and taste. The sensory structures are actually isolated from the brain itself, but the major nerve pathways enter the braincase via openings known as *foramina* (singular, *foramen*). The largest of these, the *foramen magnum,* is at the base of the skull and serves as an entrance for the spinal cord. These and other features of the vertebrate skull are shown in Figure 24.11.

The Vertebral Column

The backbone of vertebrates has had various roles in the evolution of animals. Keep in mind that very few animals walk upright, so the backbone in terrestrial animals has often had to support the great weight of the belly. On the other hand, aquatic species are buoyed by water and so their reduced weight places far less stress on the vertebral column. So, the backbone may serve various purposes, and we can expect it to be modified or specialized accordingly.

First, let's review some general features of the vertebral column (Figure 24.12). The backbone consists of a series of bones which form a movable axis for the body. Each bone is individually known as a *vertebra,* and their number varies greatly from one class of animal to another. The reptiles have the record, with as many as 400 (most of them very similar to each other). The birds, on the other hand, have comparatively few, partly because the posterior ones are fused. All mammals, from giraffes to shrews and whales, however, have exactly seven vertebrae in the neck, even though the remainder of the column varies.

Humans have thirty-three vertebrae, more or less, including the five that are fused to form the *sacrum* and the three to five that fuse to make the *coccyx* (tailbone). The remainder move and are separated by slippery disks of cartilage, which are compressible and can rupture. (Because they are compressible, you are shorter in the evening than in the morning, and you shrink as you grow older.) But these same disks give the vertebral column great strength and help it to absorb shock.

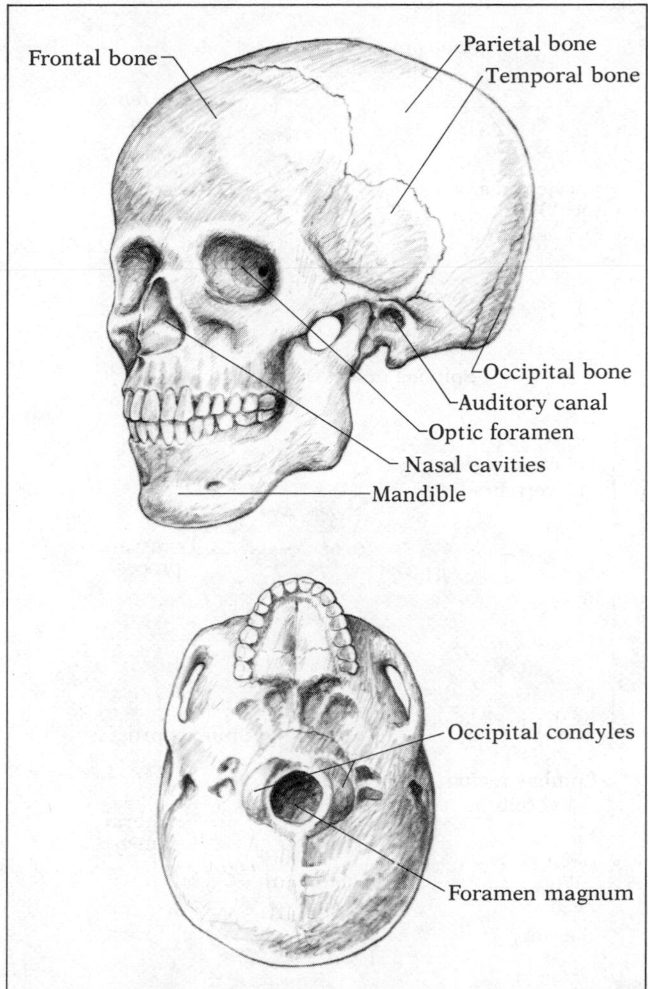

24.11 The human skull consists of twenty-eight individual bones, many of which fuse together in joints or sutures. At the base of the skull is a large opening known as the *foramen magnum*, through which the spinal cord passes into the brain. (Note the two smooth surfaces, the occipital condyles, which articulate with the first cervical vertebra permitting "yes" movements.) Other external openings include the *auditory canal, optic foramen, nasal cavities,* and various minor openings for blood vessels and nerves. Some of the major bones are named here. The primary functions of the skull are protecting the brain and housing the sensory structures of sight, smell, and hearing.

As we've said, the vertebral column has an important role in supporting the body, but, in addition, it protects the *spinal cord.* The spinal cord passes through the *neural canal,* which is formed by a series of vertebral foramena. In additon, some vertebrae buttress the ribs.

The vertebral column above the sacrum is divided into three regions. The neck area is called the *cervical* region, and it contains the seven vertebrae common to all mammals. The first and second cervical vertebrae, called the *atlas* and *axis,* respectively, support the skull and allow the yes/no motions of the head. Next, there are the twelve *thoracic* vertebrae. Each of these has well-developed lateral processes which articulate with the ribs. Because the ribs can move, the chest cavity can expand when you inhale and then contract when no one's looking. In addition, a central flattened spine rises from each of these vertebrae and serves as an area where the back muscles can attach. If you have a lot of muscles, they will stand higher than this row of spines and there will be a crease down your back. If you don't have well-developed muscles there, the spines will stand out.

A third region, between the thoracic region and the sacrum, consists of the five *lumbar* vertebrae. These are the largest, with a more massive central body and heavier spines. Their increase in size is to be expected since the vertebrae carry an increasing load as the column descends. The lumbar region is followed posteriorally by the sacrum, a triangular group of fused vertebrae wedged between the right and left hip bones, thus completing the pelvic girdle. If you are a runner, the constant pounding of the vertebral column exerts its force against the sacral wedge forcing the pelvic bone apart a bit and causing temporary lower back pain. If this happens you should slow down on the downhills.

Below the sacrum is the final region of the column, the coccyx. The coccyx is a vestigial tail, a gift from our ancestors. A tail is actually present in the human embryo, but it is usually resorbed before birth. However, some babies are born before the tail disappears, an event which, it would seem, could lead to some interesting philosophical discussions in some families. (Doctors routinely cut the tails off these days.)

Besides the skull and the vertebral column, the axial skeleton includes the ribs and the *sternum* (breastbone). The sternum is composed of three parts, the upper *manubrium,* middle *body,* and lower *xiphoid process.* Humans have twelve pairs of ribs. All articulate with the thoracic vertebrae, but not all connect to the sternum. A look at Figure 24.13 will show you that ribs number 8, 9, and 10 attach to the seventh rib through a common cartilage. Numbers 11 and 12 are not attached to the sternum at all, but are "floating ribs." The *rib cage* functions in breathing and in protecting the internal organs. The ribs and sternum also form blood cells in their marrow.

There are some examples of odd rib cages in the animal world. For example, amphibians lack rib development of any significance. Since they have no diaphragm or rib cage, they breathe by moving the muscles in the mouth and throat. Perhaps the most unusual variation of rib structure is in the turtle. Its ribs and vertebrae are solidly joined to the platelike bones of the *carapace* (upper shell). Birds also have an unusual thoracic structure. The sternum gives rise to an enormous *keel,* which has a large, flat vertical projection to which the powerful flight muscles are attached.

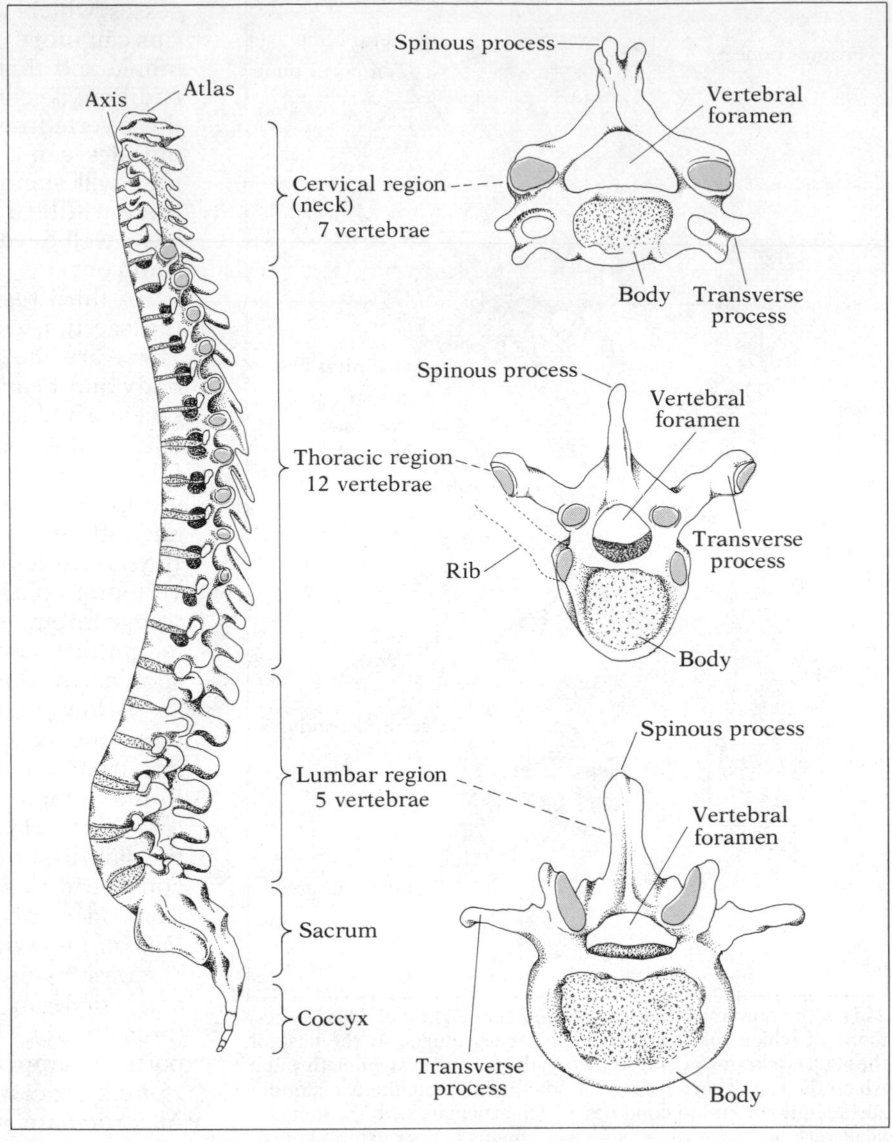

24.12 The vertebral column in humans, allowing for some variation in the *coccyx,* consists of about thirty-three bones. Note that there are seven *cervical* vertebrae, twelve *thoracic* vertebrae, five *lumbar* vertebrae, five (fused) *sacral* vertebrae, and three or four vertebrae in the *coccyx.* In lateral view the column is curved. Also notice surfaces (in color) that articulate with the ribs. In the inset, vertebrae from three regions are shown in detail. Each vertebra consists of a large disk (the body, normally separated by other disks of compressible cartilage), a vertebral foramen (for the spinal cord), and a number of processes, including the spinous processes which form a protective ridge along the back and serve as attachment for muscles. You can see that some processes articulate with other vertebrae.

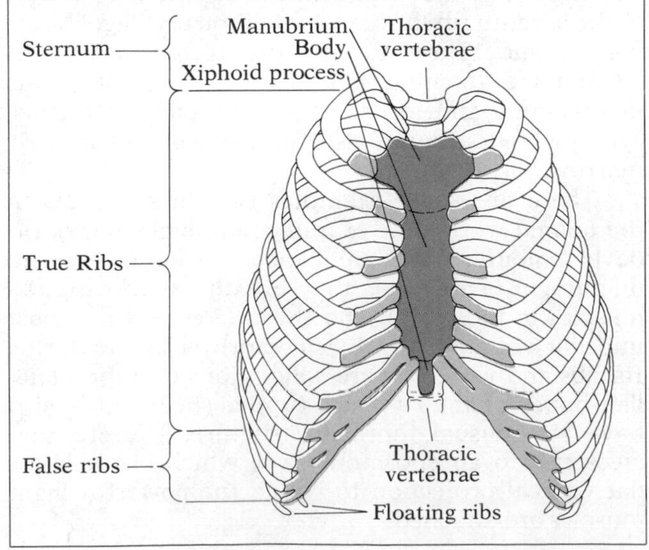

24.13 The human rib cage consists of the *sternum* (three bones), twelve pairs of ribs, and their related cartilages. Note that only the first seven pairs of ribs articulate directly with the sternum. The eighth, ninth, and tenth pairs ("false ribs") articulate with the cartilages of the seventh pair. The eleventh and twelfth pairs do not attach at all in the front, but are "floating ribs." Each rib articulates with one of the twelve thoracic vertebrae to complete the rib cage in the back.

The Appendicular Skeleton

Now let's consider the bones that are attached to the axial skeleton. These, in general, lie further away from the body's midline. Essentially, we will find two rings (girdles) of bone and limbs arising from them.

The Pectoral and Pelvic Girdles

If you have taken an art class, perhaps you learned to draw humans, not as stick figures, but as a head on the end of a backbone with two ovals along its length from which the arms and legs dangle. These ovals are the pectoral (shoulder) and pelvic (hip) girdles. By using just such drawings, we can show the transition of limb position from amphibian to mammal (Figure 24.14).

Humans, of course, are unusual. We (along with birds) are bipedal. Therefore, our pectoral girdles are relieved of their weight-supporting burden and are greatly reduced. Our pelvic girdles are also different even from the other primates, which are never quite comfortable when walking upright. As you can see in Figure 24.15, our upright posture has redistributed our weight considerably. The upper torso rests entirely on the *sacroiliac* joint (the joint between the *sacrum* and the *ilium*), with some help from certain back muscles that distribute the weight to the ilium.

24.14 Vertebrates arranged in order of their increasing adaptation to terrestrial life. Note in the stick figures the pectoral girdle and the position of the limbs with respect to the trunk. Under what conditions is the apparently awkward stance of the amphibian or reptile perfectly suitable?

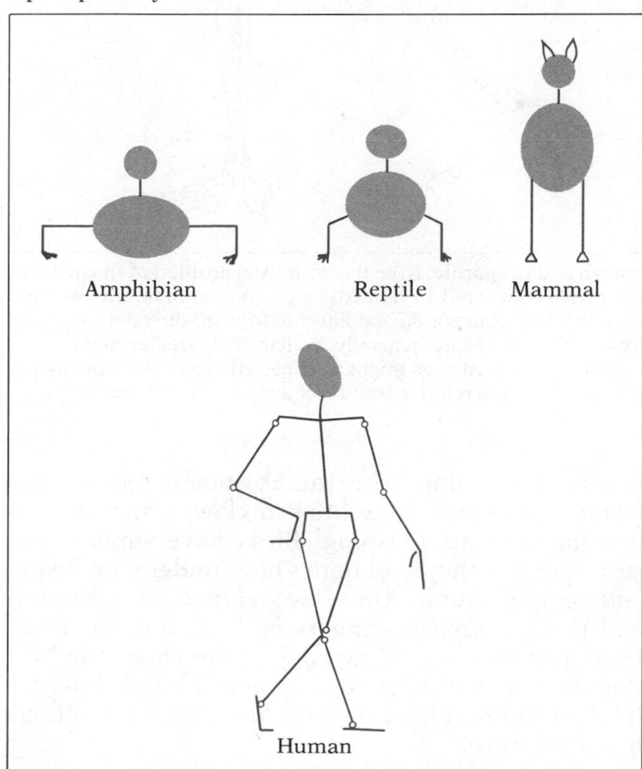

As a species, we are particularly prone to back troubles, whether we are athletes, heavy laborers, or sedentary office workers. Evolution obviously still has a few bugs to work out. Let's look at our own pelvic and pectoral anatomy in more detail.

The pelvic girdle is formed from six paired bones (and the sacrum), but they are not easily distinguishable in the adult skeleton because some of them are fused. For example, each half of the pelvis is formed from three fused bones, the *ilium* (hip), the *ischium* (the one you sit on), and the *pubis* (in front, where you would imagine). The two halves of the pelvis join ventrally but don't fuse. They remain slightly separable, and the juncture is called the *pubis symphysis.* Both sides join dorsally with the sacrum to form the sacroiliac joint, which creates many of our back problems. The joint transfers considerable weight from the vertebral column and upper torso to the pelvic girdle, as we have seen. Where the three pelvic bones join, they form the *acetabulum,* or hip socket, a deep depression that receives the spherical head of the femur, forming the ball-and-socket joint between the upper leg and hip. Note in Figure 24.16 that the pelvic bones of women are shaped somewhat differently from those of men. Specifically, the cavity of the pelvis in the female is wider than in the male, with less slope inside; and the angle of the pubic arch is considerably greater than in the male.

The pectoral girdle consists of two *scapulae* (shoulder blades) and two *clavicles* (collar bones). The shoulder blade is a triangular, flattened bone that provides a broad area for muscle attachment. The outer point of the triangle is thick and forms part of the shoulder joint. This corner of the scapula is concave and receives the rounded head of the upper arm, forming a ball-and-socket connection. In many movements, such as a bear-hug, the scapula glides over the rib cage. The clavicles are attached to the ventral (front) side of the shoulder joint. They join the manubrium of the sternum just above the first ribs and extend laterally to the scapulae at a point near the shoulder joint. You can *feel* your clavicles move as you shrug your shoulders.

Limbs

Whereas most arthropods have three or more pairs of legs, vertebrates have no more than two. Those with fewer than two are rather rare and include snakes, whales, dugongs, and kiwis. Even fish have paired pectoral and pelvic fins. Limbs, as you know, are primarily used for locomotion, except in most fishes. The fins simply serve to refine the powerful thrust-

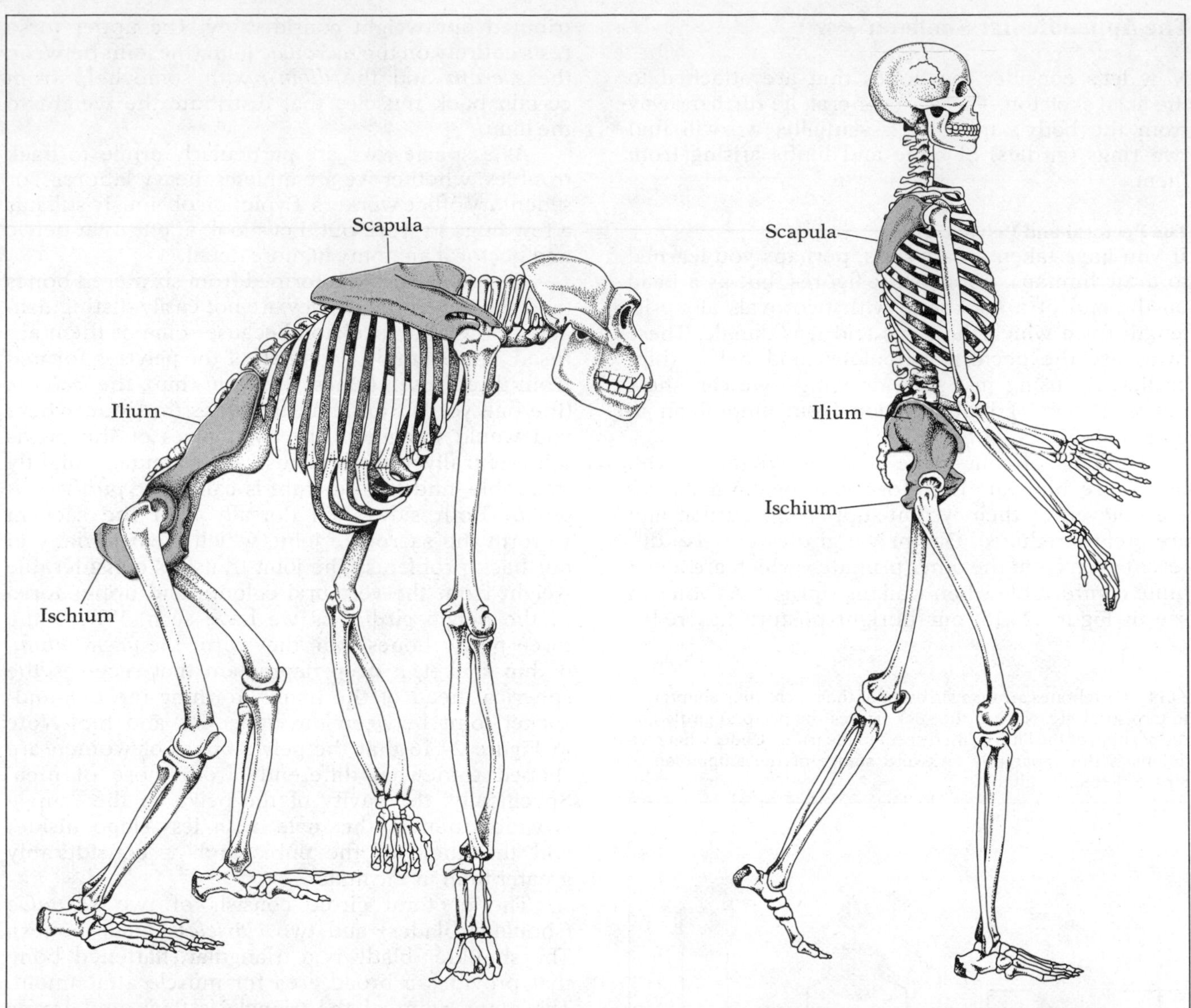

24.15 The appendicular skeleton of the human and the gorilla. Of all the primates, only humans are truly bipedal. This mode of transportation has caused humans to diverge skeletally from the other primates. All primates can stand erect, but only humans are good at it. The others are far more efficient when getting around on all fours. Compare the pectoral and pelvic girdles, position of skull, relative length of arms and legs, and the shape of the foot in the human and the gorilla. Note the horizontal position of the pelvis in the gorilla compared to its vertical position in man. The ischium is quite long in the gorilla and flares in forward direction. The pectoral girdles are more generally similar, with greater mass in the scapula of the gorilla, as might be expected in a heavy, brachiating primate. Note the relative lengths of arms and legs in each.

ing movement of the body and tail. But forgetting fish, snakes, and limbless lizards for now, the rest of the vertebrates do, in fact, use their limbs primarily for locomotion, as we see in Figure 24.17.

We can see in the flying vertebrates just how far some limbs have been modified for locomotion. In the two living groups, birds and bats, the forelimbs and especially the digits (fingers) have been greatly modified for flight. In the bats, the hands have been dramatically extended to form the frame-work of the skin-covered wing. The thumb, not a part of the framework, has a large hooked claw, which is used in climbing and grasping. Birds have similar arm structure, but the hand bones have undergone fusion and simplification. Only two *carpals* (wristbones) and three *digits* (the fingers or toes) remain; these have become the leading edge of the wing. The bird thumb, surprisingly enough, is still free and clawed—it forms a little spike called the *alula,* which functions in flight control.

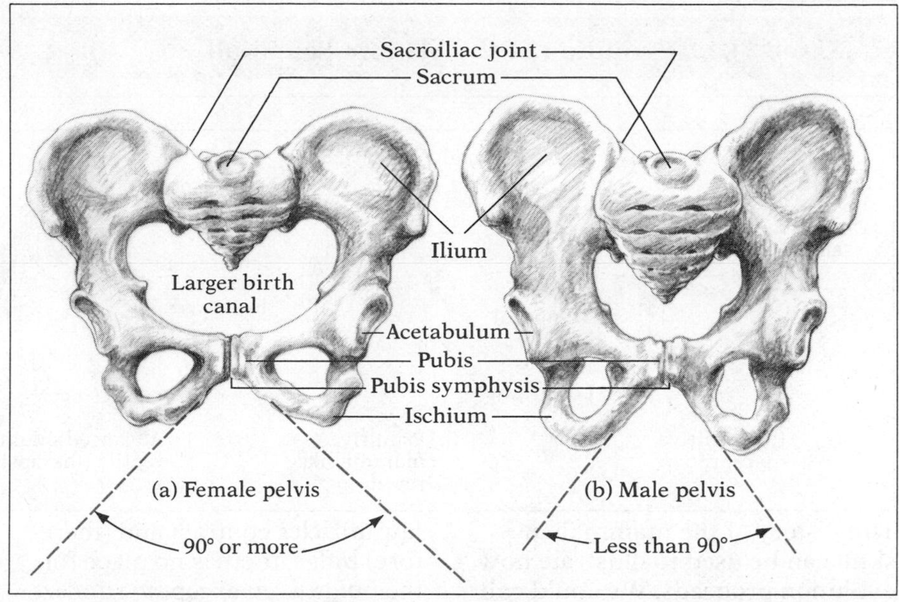

24.16 The human pelvis, male and female. Some of the differences between male and female pelvises can be detected at a glance. Note in particular the difference in the size of pelvic inlets. In addition, the pelvic arch may be quite different. In women, the arch generally forms an angle greater than 90°; in men it is generally less than 90°. The wider angle in women, of course, provides room for the birth canal. Note also the difference in shape and length of the sacrum and the flare of the ilium.

The Human Appendicular Skeleton

With the exception of the thumb and the arch of the foot, human appendages are rather generalized, so they make good examples for a detailed look at a primitive mammal. Our hands, with those famous opposable thumbs, are perhaps the most interesting part of our appendicular skeleton. (Does it seem strange to use humans as an example of a primitive condition? Our mouths are also primitive—as even cursory eavesdropping would substantiate.)

But what about the hand? Is it special? Anatomically, except for our magic thumb, the human hand is very generalized. This means it isn't special, but that it is similar to those of most primates and, in fact, very much like those of the most primitive land vertebrates: five splayed digits of roughly equal length. Our thumb, however, gives our hand a certain specialness. We can make tools and we can button a shirt. Chimpanzees, with their less opposable thumbs, have trouble with both. Chimpanzees can touch their thumbs to the other fingers; we can simply do it better. Of course, a chimpanzee will use its thumb for grasping, but generally only while its fingers are flexed under. Try buttoning your shirt this way. The

24.17 Specialized forelimbs of vertebrates are arranged here to reveal presumed pathways of evolution. As you can readily see, each region of the forelimb has undergone some modification for specialized methods of locomotion. These modifications include reduction and fusion, as shown dramatically in the foot bones of the horse and the wing of the bird. Reduction has occurred in the horse. Only the third toe remains and the heel is raised clear of the ground. The hand bones of the bird have undergone fusion with only a few of the carpals, metacarpals, and digits remaining. Also note the unusual feature in the hand bones of the bat. Both bird and bat wing have essentially the same function, but the supporting frameworks differ in a fundamental way. The feathers of birds provide both structural strength and flight surface, but the bat has greatly extended fingers to provide the structure, while the flight surface is a thin membrane. People are true bipeds, so the forearms and hands are not required for walking.

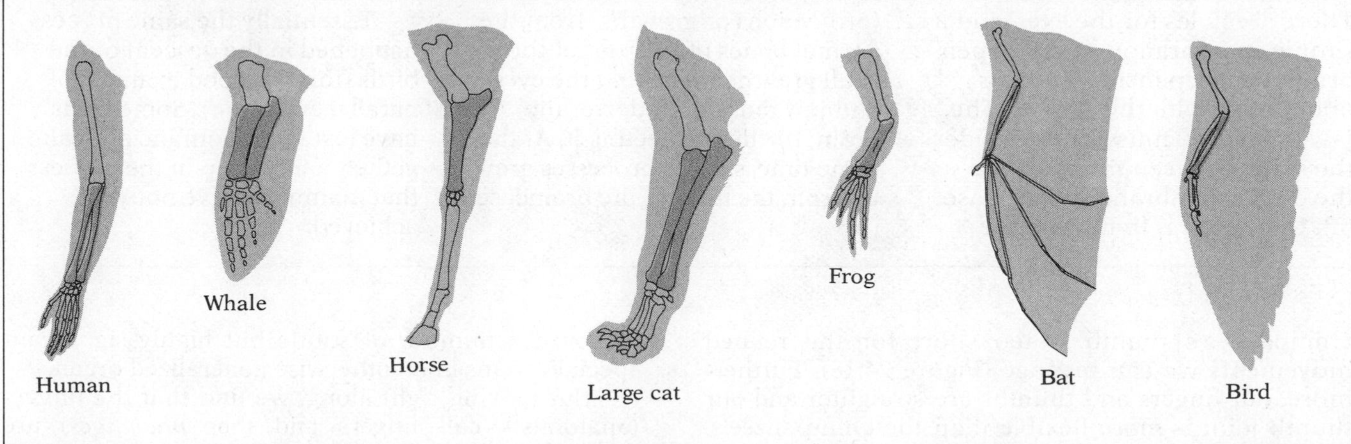

Essay 24.2 Evolution of the Mammalian Skull

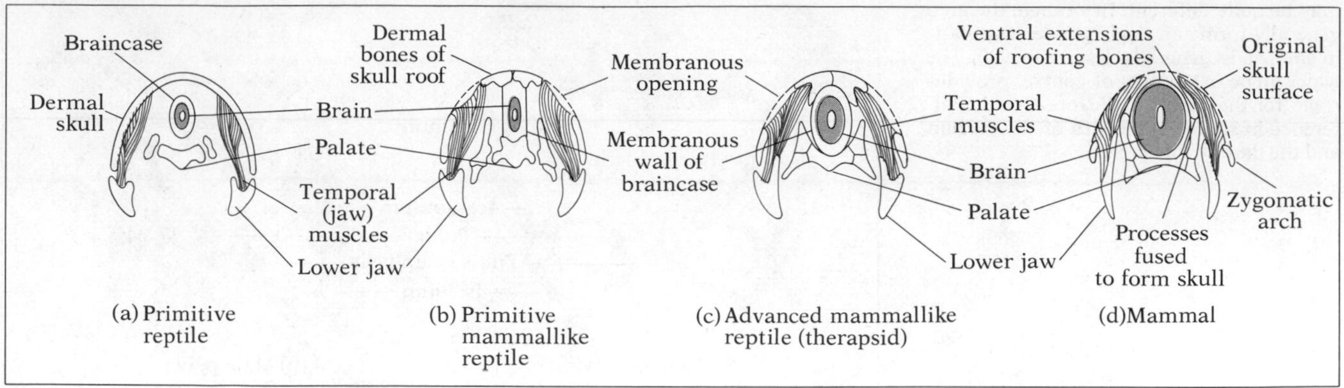

The history of the mammalian skull can be used to illustrate how evolution proceeds. We could call our lesson "How to repair a mistake, or you can't hardly get here from there." Originally most of the bones of the skull weren't part of the axial skeleton at all. The axial skeleton consisted of the vertebrae, the ribs, the gill arches (later, jaws), and a tiny *braincase,* suitable for a tiny brain. Most of the bones of the head were *dermal bone,* loose protective plates of bone developed in the lower layers of the skin, as we see in fish today.

When fish became terrestrial, the dermal bone plates fused and hardened, over evolutionary time, to become the skull of the ancient amphibians and early reptiles. Turtle skulls are more or less similar to these early reptilian skulls. There are holes for the eyes, and a simple, membranous, partly open braincase deep inside. There is one trouble with this design: The jaw (*temporal*) muscles are inside the skull, between the skull and the open, membranous braincase (a). One result is that when the jaw muscles contract and (therefore) bulge, there is no place for the bulge to go except to squeeze the brain. It would be much better to have the jaw muscles on the outside of the skull.

And evolution, as Darwin argued tirelessly, must proceed by small increments. It would seem that there could be no way to get the heavy jaw muscles from the inside of the skull to the outside, without having some maladapted monster halfway between.

But here is what happened: In the evolutionary descent of all reptiles other than the turtles, membranous openings in the side of the skull appeared. They were at first small but then grew larger. These openings gave the bulging jaw muscles some place to go, relieving pressure inside the skull. Meanwhile, flangelike *processes* (projections or growths) from the dermal bones of the roof of the skull grew down *behind* the eyes and jaw muscles and over the brain, further protecting it. At the same time, similar processes grew up from the floor of the braincase (b).

Eventually, the protective flanges, or side processes, that grew down over the brain fused, to create a new and more solid encasement for this delicate—and enlarging—organ. And the openings in the dermal skull grew larger and larger (c, d), until eventually all that was left of the original skull outside the jaw muscle was the thin, fragile, and useless *zygomatic arch* (cheekbone). It took many tens of millions of years to accomplish this by evolution and natural selection, with each incremental step along the way a slight improvement over the one before. But in the end, the jaw muscles were outside the skull where, logic seems to tell us, they belonged in the first place. Evolution had to create an entirely new skull on the inner side of the jaw muscles to accomplish this.

Essentially the same process happened in the descent of the birds (this is a good example of parallel evolution). Some birds have lost the zygomatic arch altogether, a final step in the process that mammals have not yet achieved.

---

chimpanzee's thumb is too short for the refined movements we can manage (Figure 24.18). Furthermore, our fingers and thumbs are straighter and our thumb joint is more flexible than the chimpanzee's. Finally, the muscles in our hands are different from those in the hands of other primates. So, we have achieved a number of subtle but highly significant specializations in an otherwise generalized organ.

But moving right along, we find that the fingers (anatomists call fingers and toes *phalanges*) are attached to the hand bones (called *metacarpals*). The joints are essentially hinges, but they are capable of

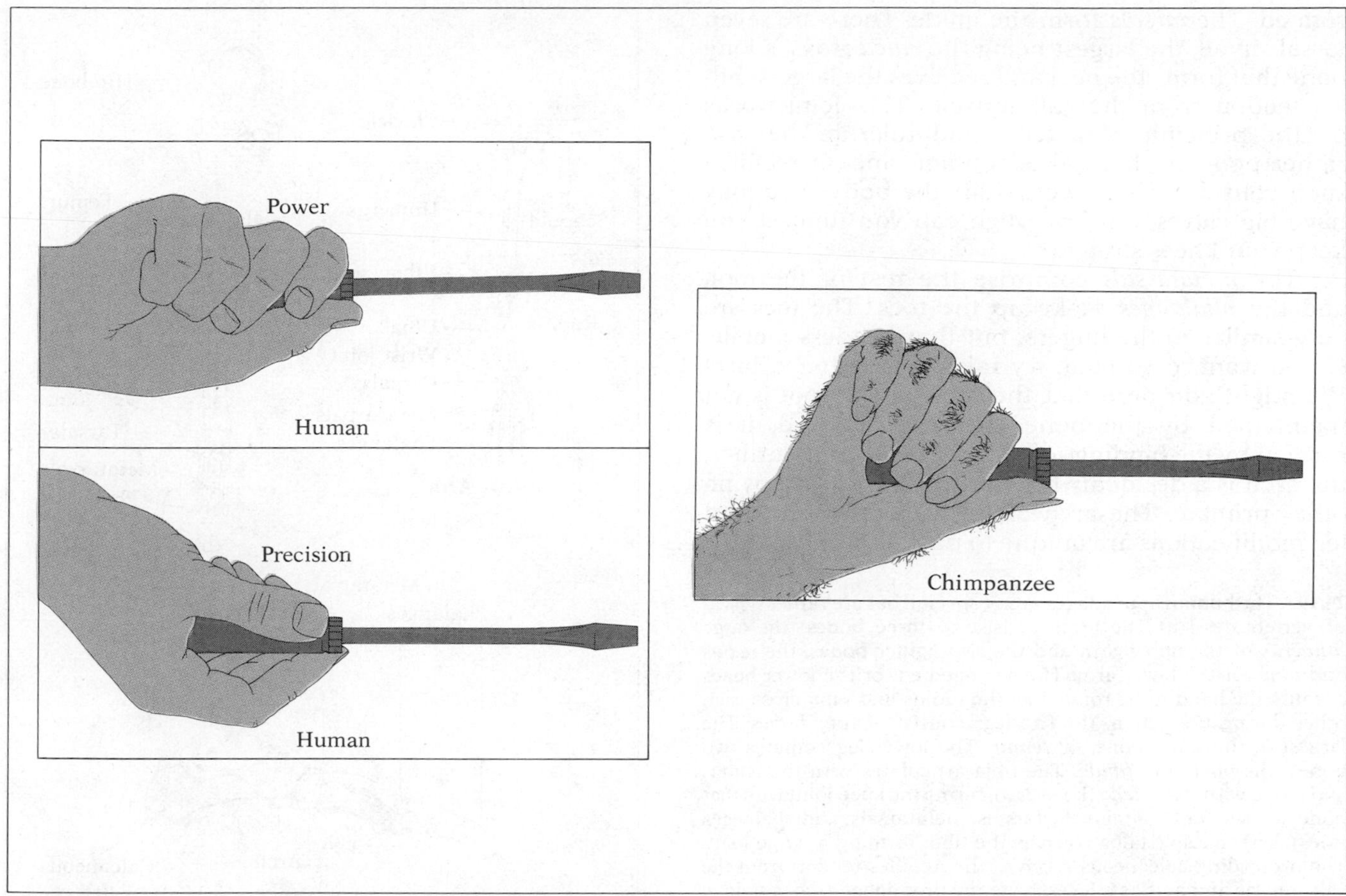

24.18 The chimpanzee and human hands. The secret of our manual dexterity resides in our remarkable thumb. No other animal boasts a bone that can duplicate its wide range of movements. Compare the position and musculature of the human and chimpanzee thumbs. Note the opposability of thumb to finger tips compared to that of the chimpanzee. With the exception of the thumb, however, human hands are quite unspecialized.

some lateral movement, as we see when the index finger is wagged back and forth as a warning. The metacarpals then join the *wrist,* which is composed of eight small bones, the *carpals.* These are all somewhat blocky or *cuboid* in shape and form gliding joints [Figure 24.19(a)]. Check the various movements we've described using your own hands—if no one is watching.

The mammalian forelimbs, such as our arms, are made up of three long bones: the large *humerus* of the upper arm and the *radius* and *ulna* of the lower arm. The elbow is the complex pivoting hinge joint where the bones join. The other ends of the radius and ulna form the wrist hinge. Notice in Figure 24.19(a) that the radius lies outside at the elbow and inside at the wrist. (It is easier to write on your radius than your ulna.) This twisting permits 180° rotation (which is how the radius got its name) as well as a hinge movement. At the elbow, the radius forms a rotating joint and the ulna forms a simple hinge. At the wrist, the opposite is true. Play around with a skeleton (if you have access to one) to verify this.

The mammalian leg, like the arm, consists of a single upper bone (*femur*) and two lower bones (*tibia* and *fibula*), as shown in Figure 24.19(b). Most other vertebrates have some modification of this plan. The upper and lower limbs meet at the knee joint, but actually, only the tibia articulates with the femur. The knee is more than a simple hinge. You can't rotate your lower leg as easily as you can your lower arm, but some degree of rotation is important. It has proven very difficult for surgeons to design an artificial knee joint that can manage these subtle movements. The fibula joins the tibia just below the knee. A fourth bone, the kneecap, or *patella,* which is chestnut-shaped, covers and protects the juncture of the tibia and femur to complete the knee joint. The two rounded heads of the femur rest rather precariously on the flattened head of the tibia. The presence of several strong ligaments (the ligamentous capsule) and the attached muscle tendons make the joint exceptionally strong. Nevertheless, knee injuries are common in the contact sports, usually because this kind of joint cannot withstand very much twisting.

At the foot, the tibia and fibula articulate with a large bone called the *talus,* forming a kind of hinge joint that permits the foot to be flexed, extended, and rotated. The *tarsals* form the ankle. There are seven tarsals in all, the largest being the *calcaneous,* a long bone that forms the heel and receives the large Achilles tendon from the calf muscle. This joint works on the principle of a lever and fulcrum, but has rather poor mechanical advantage since it requires such considerable power to lift the body. You may have big calves, but how high can you jump if you keep your knees straight?

The *metatarsals* comprise the rest of the foot, and the *phalanges* make up the toes. The toes are very similar to the fingers, but they are less mobile. If you want to go mad, try raising each toe in turn. We might add here that the arch of the foot is not maintained by the bone structure. Instead, it is formed by the bindings of its ligaments. And, further, the arch is a decidedly human feature, shared by no other primate. The arch along with other foot and leg modifications are unique to man, "the walker."

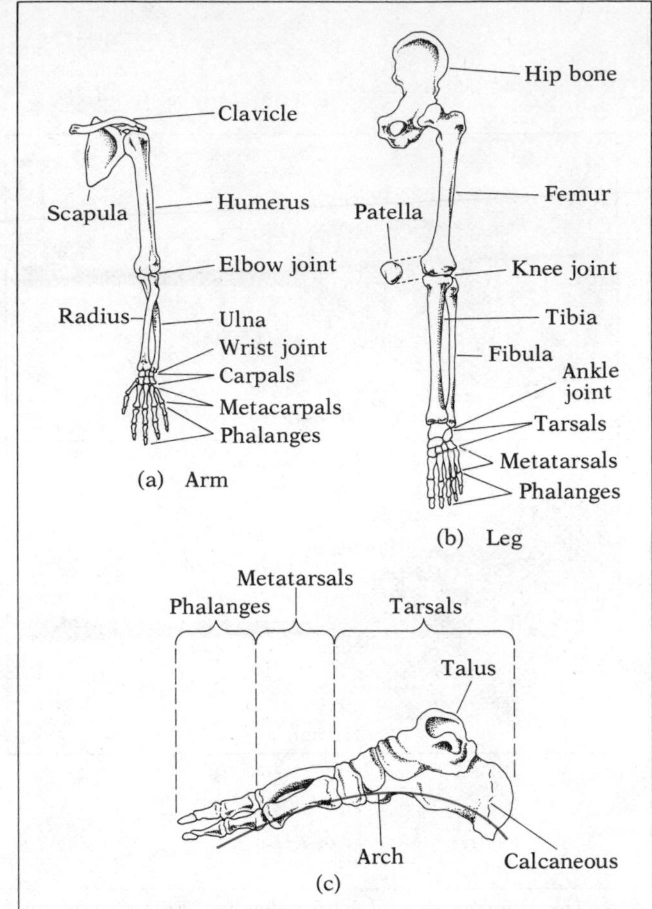

24.19 The human appendages aren't special, but are rather typical of vertebrates. (a) The arms consist of three bones: the large *humerus* of the upper arm and the two lighter bones, the *radius* and *ulna,* of the lower arm. The arrangement of the lower bones permits the hand to be rotated as the radius and ulna cross each other during the action. (b) The legs consist of four bones. The largest is the thigh bone, or *femur.* The lower leg contains two bones, the *tibia,* and *fibula.* The tibia articulates with the femur, and along with the *patella* (knee cap), forms the knee joint with that bone. (c) The foot contains the tarsals, metatarsals, and phalanges (toes). The massive talus receives the tibia forming a hinge joint. The protruding calcaneous receives the Achilles tendon from the calf muscle and acts as a lever. Note the prominent arch, a unique feature of our species.

24.3 Vertebrate Movement

Muscles and How They Move

That, then, covers the essentials of the human skeleton. But we need not fear the dancing bones of Halloween, because bones can't move unless muscles are attached to them. So let's take a look now at the contractile tissue that moves the vertebrate skeleton.

Types of Muscle

In a sense, the ability of a cell to alter its shape has altered the shape of our world. Cell contraction, after all, is no more than altering the cell shape, because as it contracts in one direction, the cell must lengthen in another. Some cells are specialized to contract—these are called muscle cells in most animals, including humans. Even the simpler species, however, use very similar biochemical changes in contracting their cells.

Remember that one type of cell movement depends on special whiplike organelles—the cilia and flagella. The function of these organelles depends on a biochemical system involving tubulin and microtubules. Another basic cell-movement system depends on the protein *actin,* which is biochemically similar wherever it is found throughout the eucaryote kingdoms. Recall that actin fibers cause movement in the cytoskeleton and cell membrane, utilizing the energy of ATP. They are also intimately involved along with other proteins in muscle contraction. The muscles of vertebrates and other metazoa behave in much the same way.

Vertebrate muscles can be divided into three types, according to their microscopic structure and the nerves that activate them. *Smooth muscle* has a smooth, glistening appearance, and it is also known as *involuntary muscle.* It is found in various internal organs: in the walls of the digestive tract, in the walls of blood vessels, at the base of every hair, in the iris of the eye, and in the uterus and other reproductive organs. You may have noticed that your hair rises involuntarily under eery conditions, or that you can't stop your stomach from growling on important dates, or that (as Da Vinci said) the penis seems to have a mind of its own. Smooth muscle is, logically enough, smooth in appearance, and it is generally not subject to voluntary control; hence its two names. It is controlled by the *autonomic* (involuntary) nervous system (see Chapter 31). The ultrastructure (that is, submicroscopic structure) of smooth muscle is rather different from that of voluntary muscle, as we'll see, although the same proteins are involved in the contracting mechanism.

A second type of muscle is *heart,* or *cardiac, muscle.* It too is under the involuntary control of the autonomic nervous system, although heart muscle also has control centers of its own. (Heart muscle cells in tissue culture will continue to beat.) Heart muscle is different from other muscle in its microscopic structure, although, again, the principal contractile proteins are the same.

The third type of muscle has three names: *skeletal, striated,* or *voluntary muscle.* It is controlled by the *somatic* (voluntary) nervous system. As for the names, it is *skeletal* because these are the muscles that move the skeleton; it is *striated* ("striped") when viewed under the microscope; and it is *voluntary* because it is *possible* to move the muscles at will. However, some skeletal muscles aren't attached to bones (your lip and anus muscles, for instance). And though voluntary muscles may be consciously controlled, they are not always under such control. After all, you can consciously control your breathing movements, within limits, but when was the last time you consciously remembered to take a breath? You certainly don't have to lie awake at night making yourself breathe. And you might be able to observe some other involuntary movement of voluntary muscles the next time you spill hot coffee in your lap. Still, voluntary muscles are directly controlled by the "conscious" levels of the central nervous system. Now, let's take a closer look at the three types of muscle.

Smooth Muscle

Smooth muscle cells are shaped like long spindles, more or less overlapping, and in sheets. A nucleus lies approximately at the center of each cell [Figure 24.20(a)]. Smooth muscles contract rather slowly, but they can remain contracted for extended periods of time. Unlike our voluntary muscles, they tend to lack *proprioceptors* (sensory receptors that tell us where our muscles are and what they are doing—are the toes on your right foot flexed right now?), so most of the time we can't tell, by feel, what our smooth muscles are doing. This seems to be one of the main reasons why we can't control them easily.

If we can find a way to monitor our smooth muscles, we may find that we have some control over them. For example, by staring at the pupil of your eye in a mirror and noting its contraction and dilation, it is possible to *make* it do what you want it to. It's a lot of hard work and the ego reward is scanty, so most people usually don't bother to learn the trick. But yogis can do it easily, as well as all sorts of other bizarre feats with their smooth muscles—like voluntarily shutting down circulation in one arm, or making an ear turn red. They have also boggled the minds of physiologists by bringing their hearts to a near standstill.

The inability of most of us to control smooth muscle contraction has caused us a great deal of stress. People who are tense may find that the ring of muscle that causes food to move from the stomach to the intestine does not relax. Thus, food just sits in their stomach, making them feel as if they had just swallowed a bowling ball. And many people suffer from sexual dysfunction. Males may gain an erection at an inopportune time, to say the least; and then at another time they may desperately want to, but to no avail. Sheer will isn't involved. Glandular secretions are also involuntary. Imagine skydiving. Do your palms feel damp? Can you control it?

24.20 The microscopic structure of three muscle types. (a) Smooth muscle lines the gut and is found in other places such as in blood vessels and the iris of the eye. Smooth muscle is involuntary and is commonly found in sheets, but individual cells are long and spindly, containing a single nucleus. (b) Cardiac muscle is found exclusively in the heart. It is also involuntary. The muscle fibers are cylindrical and striated like skeletal muscle, but unlike skeletal muscle, they branch and are interrupted by intermittent dense regions known as *intercalated* disks. (c) Skeletal muscle is voluntary and, like cardiac muscle, is striated. The muscles occur in bundles, and the individual fibers are long and multinucleate. Skeletal muscle, of course, moves the skeleton.

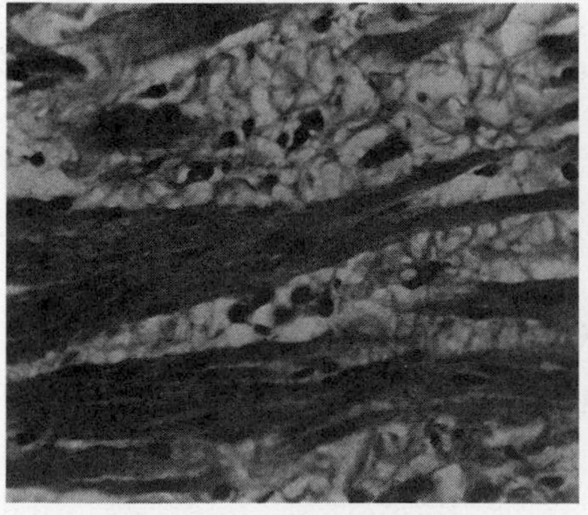
(a)

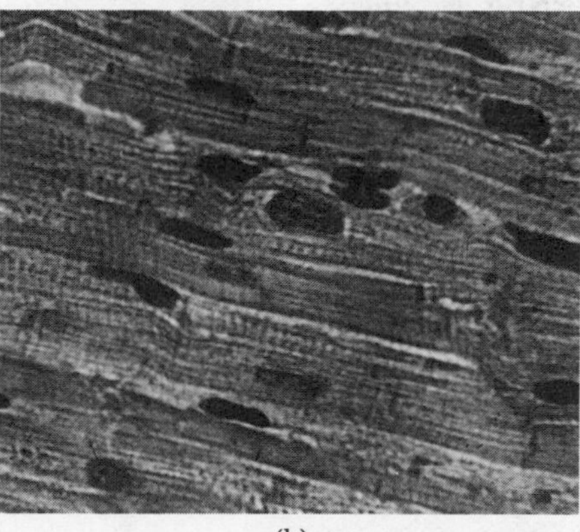
(b)

(c)

We might make one other interesting point here. If a hypnotist tells you that a wooden cane is a red-hot poker and lays it against your arm, a welt may appear where the cane touches you. The blister is caused by the localized relaxation of tiny circular muscles around arterioles. So, you see, your brain does have control over such functions; but just try raising a welt by sheer will power. You can't do it. The bottom line to all this is that we really don't understand the relationship between voluntary and involuntary muscle control.

Cardiac Muscle

Cardiac muscles have some things in common with both smooth muscles and skeletal muscles. For example, like smooth muscles, the heart is largely under control of the involuntary nervous system. Like voluntary muscles, the cardiac muscle cells are cylindrical and striated. And cardiac muscle differs from the other types in another way. The fibers are branched, unlike those of the voluntary or smooth muscles; but within those branching fibers the individual nucleated cells are separated by *intercalated disks,* which are areas where membranes are folded and tightly compressed and which easily transmit impulses. They are absent in other muscle cell types [Figure 24.20(b)].

The most impressive feature of cardiac muscle is its rhythmic beating. It begins early in embryological development, long before its nerve connections have formed. These first timorous flutterings send blood throughout the embryo in this direction and that, since even the circulatory system has not yet been completed.

Skeletal Muscle

Skeletal muscles are largely under voluntary control. All skeletal muscles are richly endowed with proprioceptors, which enable the individual to know where parts of the body are and what they are doing. Skeletal muscles can contract much more rapidly than can smooth muscles or even cardiac muscle, but they cannot stay contracted for very long.

The unbranching cylindrical muscle fibers are multinucleate, with the nuclei lying just beneath the cell surface [Figure 24.20(c)]. (Because of their multinucleate condition, we will be referring to *fibers* rather than cells in discussing skeletal muscles.) Each fiber is packed with precisely arranged contractile proteins. We will return to the details of their arrangement when we look at muscle ultrastructure. Fibers are bound into bundles known as *fasciculi* by a matrix of connective tissue called the *endomysium,* which is composed largely of an extracellular material. The fasciculi give meat its stringy appearance. Blood vessels run throughout the fascicular bundle, supplying oxygen and nutrients and carrying off wastes. Nerves also penetrate the bundle, their branches dividing ever more finely until they innervate each fiber. These nerves will carry the messages that cause the fibers to contract.

The whole mass of aligned bundles constitutes the *belly* of the muscle (Figure 24.21). It is enclosed in a tough casing of connective tissue, the *fascia,* which is continuous with the inner connective tissue sheaths and the endomysia. (*Fascia* is Latin for a bundle of sticks fastened together, emphasizing the strength that comes from binding. It may help to remember that *fascism* is a rather binding form of government.) At the ends of the muscle, the facia coalesces into denser collagenous tissue, foming the cordlike *tendons,* which may be broad and short or thin or long. Then the tendon is fastened to bone. When the muscle contracts, usually one end moves a great deal (the *insertion*) and one moves much less (the *origin*). As a general rule, the origin is closer to the midline of the body. Consider the biceps muscle in Figure 24.22. Where is the origin? The insertion?

24.21 Skeletal muscle has at least five levels of order or organization. The first three: the entire muscle, the *fascicular bundles,* and the fiber (or cell) are shown. Whole muscles are surrounded by a fascia, which is continuous with the tendons of origin (anchor) and insertion (part to be moved). The thicker midregion is commonly called the belly. Each muscle consists of a large number of strandlike fascicular bundles. Each bundle consists of many fibers (cells), which are bound together in a connective tissue matrix. Each multinucleate fiber has its own cell membrane, the sarcolemma. Within each fiber are two more visible levels of organization: the myofibrils (the contractile units) and the myofilaments (the contractile protein molecules).

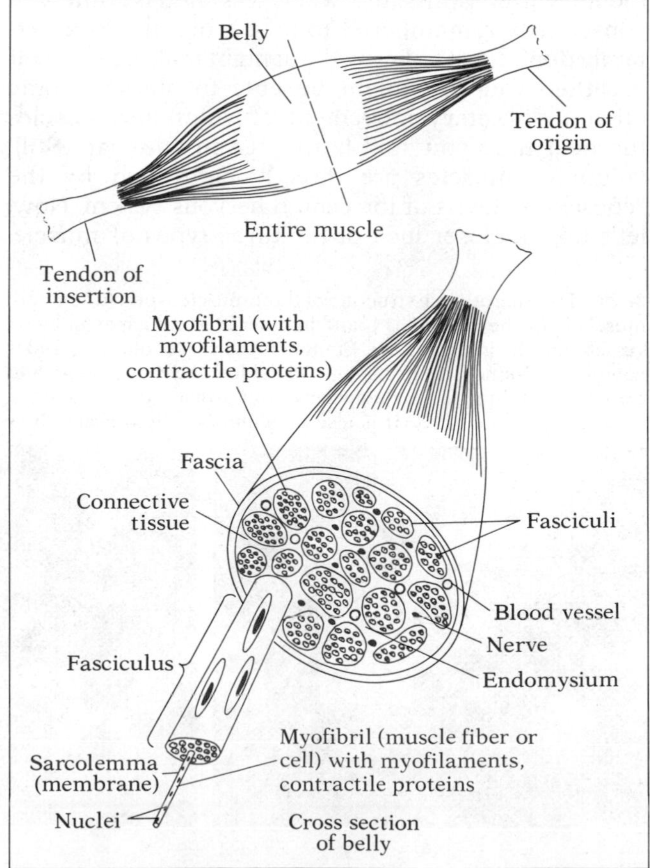

24.22 The externally visible muscles of the human illustrate the various types and arrangements and give us some idea about origins and insertions. Only the major muscles have been named. By finding tendons of origin and insertion, and the general orientation of a muscle, you can determine just what it does. Keep in mind that although muscles can contract forcefully, they cannot extend with force. For this reason, they work in opposing units known as *antagonists.* Note the marked difference in male and female muscle development.

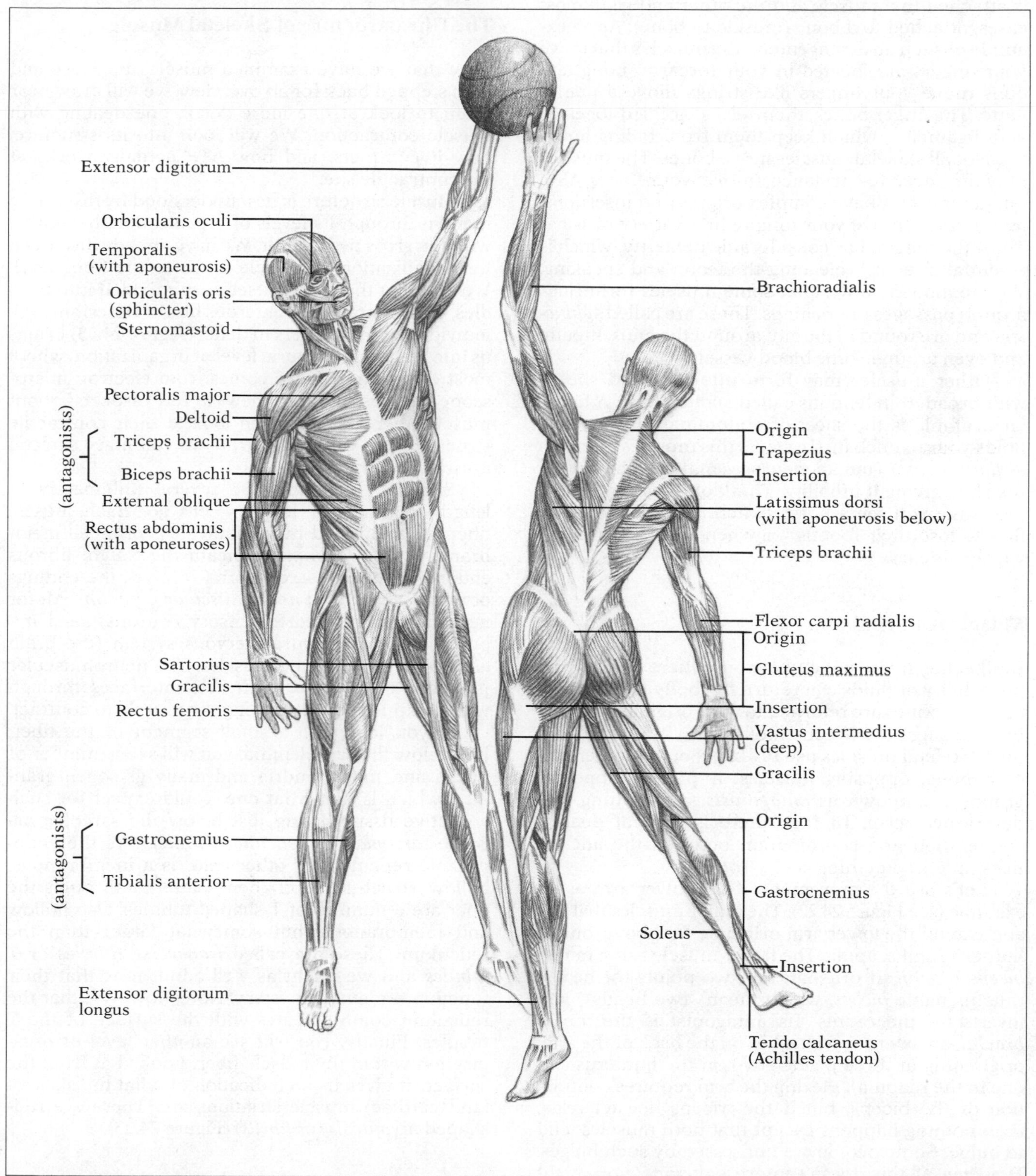

Tendons and *ligaments* are similar in structure and appearance, and are often confused. The difference is simple: both ends of a ligament are fastened to bone (bone to bone), whereas one end of a tendon is attached to a muscle and the other end is, in most cases, attached to a bone (muscle to bone). As an example of such an arrangement, the muscles that move your fingers are located in your forearm. Long tendons move your fingers like strings move a marionette. The finger bones, themselves, are tied together with ligaments, which keep them from falling apart.

Not all skeletal muscles move bones. The muscles of your face, for instance, move your face. Also, tongue muscles have complex origins and insertions, and you can move your tongue in a variety of ways. Thus, the tongue has considerable dexterity, which is essential in eating, cleaning the teeth, and speaking. We mentioned earlier that some muscles form rings around passages or openings. These are called *sphincters* and are found in the gut, around the anus, mouth, and even around some blood vessels.

Other muscles may form into flattened sheets, with broad, thin tendons called *aponeuroses.* A familiar example is the sheet of abdominal muscles that holds your paunch in. (In steers this muscles is known as *flank steak.*) Thin aponeuroses may attach beneath the skin, giving it mobility. Think of aponeuroses the next time you see a horse twitch its flanks to cause flies to lose their footing, or when you see someone wiggle his ears.

Muscle Antagonism

Unlike the hydrostatic skeleton, where the incompressibility of fluids can return the body to its former position, bones are returned to their original positions by opposing muscle action (or by gravity). Obviously, then, skeletal muscles usually have, on the other side of a bone, opposing muscles. A pair of opposing groups are known as *antagonists,* an ill-fitting, but descriptive term. In fact, there is a great deal of cooperation and coordination between the antagonists unless something goes wrong.

Let's use the movement of the lower arm as an example (see Figure 24.22). The major muscles that flex and extend the lower arm originate far above, on the humerus and scapula. The flexor muscle is the famed *biceps brachii.* It originates at two points (or heads; thus its name *biceps,* which mean "two heads") and inserts on the radius. Its antagonist is the *triceps brachii,* an extensor which lies on the back of the arm, originating at three places (two on the humerus and one in the scapula). Flexing the arm requires contraction of the biceps, but if the triceps doesn't relax, then nothing happens except that both muscles tend to bulge. Some people are impressed by such bulges. However, if this doesn't impress anyone, you might try falling to the ground and doing pushups. You can point out that now the biceps are relaxing and the triceps are contracting. If this doesn't impress anyone either, you may be asked to leave.

The Ultrastructure of Skeletal Muscle

Now that we have examined muscles up close and then stepped back for an overview, we will draw near again to look at one more detail, one dealing with muscle contraction. We will look into its structure, how it contracts, and how ATP actually works at the contractile site.

Muscle structure is best understood by dissecting it down through its levels of organization, beginning with the gross tissue itself. We have already discussed the organization of muscle down to the fiber level. We saw that the muscle itself is subdivided into bundles, each containing numerous fibers. Describing the individual muscle fibers in detail (Figure 24.23) brings us into the ultrastructural level of organization, where most of our knowledge comes from electron microscope studies. Let's review a few aspects about muscle fiber first and then look at their contractile structure. We will concentrate on voluntary skeletal muscle.

Skeletal fibers may be several millimeters in length—enormously long as cells go. Each muscle fiber is surrounded by its equivalent of a cell membrane, the *sarcolemma,* beneath the tough, fibrous endomysium. The sarcolemma receives the endings of motor neurons at *neuromuscular junctions.* Motor neurons (as opposed to sensory neurons) send impulses from the central nervous system (the brain and spinal cord) to the muscles. The neuromuscular junctions are the nerve–muscle interfaces through which impulses pass, causing the muscle to contract.

If you look into a small segment of the fiber, just below the sarcolemma, you will see a number of nuclei and mitochondria and many glycogen granules, which is just what one would expect for such an active tissue. Lying just below the sarcolemma is the *sarcoplasmic reticulum,* which, like the endoplasmic reticulum of other cells, is a membranous, hollow structure. And then scattered through the fiber are a number of T-shaped tubules, also hollow and membranous, but somewhat larger than the reticulum. These are called *transverse tubules* or *T tubules* and we might as well admit now that their function remains a mystery. Note, however, that the reticulum communicates with the surface of the T tubules. Finally, you will see another level of organization within the muscle fiber. Looked at from the surface, it gives us an indication of what the skeletal (and cardiac) muscle striations are. These are rod-shaped *myofibrils* (or *fibrils*) (Figure 24.23).

24.23 Each fiber constitutes a cell, surrounded by a cell membrane called a sarcolemma. Within the sarcolemma, some of the unique cellular organization of muscle is seen. Numerous nuclei are visible along with a membranous *sarcoplasmic reticulum,* a special system of *transverse,* or *T, tubules* and numerous mitochondria. Each fiber contains numerous *myofibrils* that contain the contractile units. And each myofibril has an even more minute level of organization. The myofibrils reveal the banding that gives skeletal muscle its other name, *striated* muscle. These same striations are visible in cardiac muscle as well.

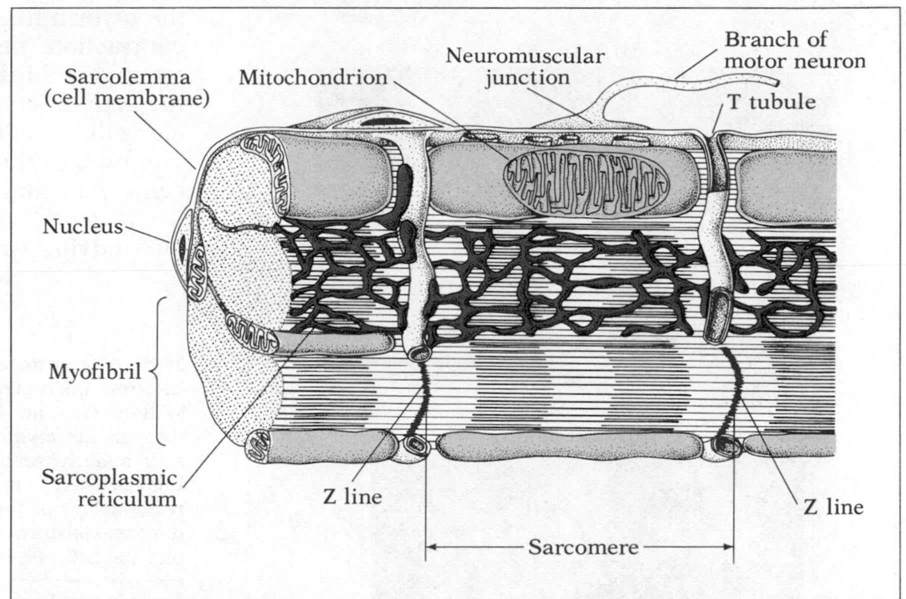

Isolating one of the myofibrils gives us a clearer view of the striations (Figure 24.24). The bandings, now seen in three dimensions, take on an entirely new aspect, but we will come to that in a moment. Let's look at the surface view first. In the figure, the region between the *Z lines* is the *sarcomere* or more simply, the *contractile unit.* Toward the center of each sarcomere is a broad region, the *A band,* with a smaller and lighter central strip, the *H zone.* The two darker parts of the A band consist of overlapping *myofilaments,* which are lengthy filamentous protein structures. The darker lines extending across the A band and running through the H zone are thick *myosin* filaments. These are overlapped by thinner filaments of *actin,* which begin at the Z lines and run part way into the A band from each side in resting muscle. That's why there is an H zone. When muscle contracts, the H zone tends to disappear, as we shall see. The *I bands* consist of actin filaments alone. The I bands don't include the overlapping actin–myosin regions.

When the muscle fiber is stimulated, the Z lines move closer together (the sarcomere is shortened). This changes the banding, as you can see in Figure

24.24 Closeup of the myofibril. The *sarcomere,* or *contractile unit,* is located between *Z lines.* The darkest band, containing a lighter zone at its center, is known as the *A band.* It consists of thick myosin filaments alternating with thin actin filaments. The actin filaments don't quite reach the center when the muscle is at rest, so the light *H zone* (*band*) in the center of the A band is myosin only. Note that the actin filaments reach all the way to the Z lines, and where they leave the A band they form the lighter I band. The cut end of the myofibril reveals that it is composed of *myofilaments* of actin and myosin. The cut has been made through the region of overlap and shows the arrangement of actin around myosin.

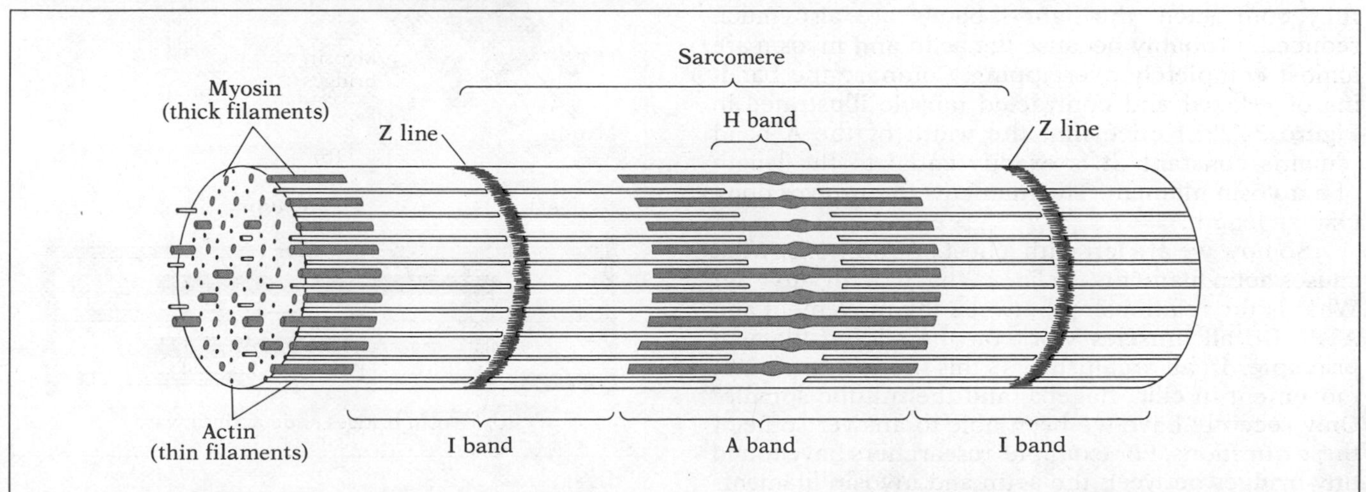

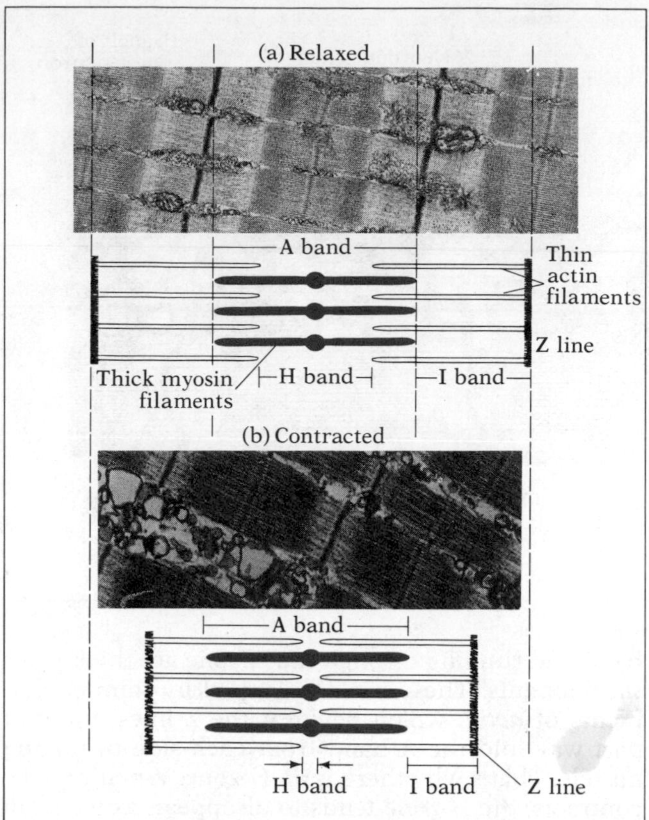

24.25 Two electron micrographs of muscle show what happens at contraction. (a) The muscle in a relaxed state: Note the distance between Z lines, the width of the I bands, and the density of the A band. (b) Now observe the muscle in a contracted state: The dashed lines have been included as guides to reveal changes in Z line distance and I band width. Note these changes and the increase in density in the A band. Has the A band width changed? What has happened in the H zone (band)? How would these changes affect an entire muscle?

24.25, producing a dark line where the H zone was. What actually happens is that actin myofilaments slide inward through the myosin. When the muscle is fully contracted, the light I bands are also much reduced, probably because the actin and myosin are almost completely overlapping. Compare the banding of relaxed and contracted muscle illustrated in Figure 24.25. Notice that the width of the A band remains constant—it is exactly equal to the length of a myosin filament. The filaments themselves don't change length.

So now we are left with a host of questions. What causes actin filaments to slide? What is actin, anyway? What is the relationship between the movement and ATP? Do all muscles work on the sliding filament principle? In all organisms? Is this the mechanism of movement in cilia, flagella, and the mitotic spindle? Only recently have we been able to answer some of these questions. For example, researchers have found tiny bridges between the actin and myosin filaments (Figure 24.26). These bridges apparently extend from the myosin filaments and fasten to the actin. During contraction, they act as tiny rachets that attach at an angle and then pull themselves perpendicular to the actin fibers; then they reattach at a point further along the actin filament and pull themselves up again. The bridges thus pull the actin rapidly in toward the center, causing the filaments to slide past each other.

Recently, muscle cytologists have concentrated on studying isolated actin and myosin. Using mag-

24.26 The arrangement of myofilaments and the cross-bridges. Electron micrographs reveal other features of the contractile system. (a) The electron photomicrograph of a cross-section through the myofibril shows the arrangement of myofilaments actin and myosin in respect to each other. The drawing shows the bridges between myosin and the surrounding actin. (b) Notice the relationships of the myofilaments and their bridges in the three-dimensional drawing. (c) During contraction, these bridges actively pull the actin fibers inward, thus shortening the contractile unit.

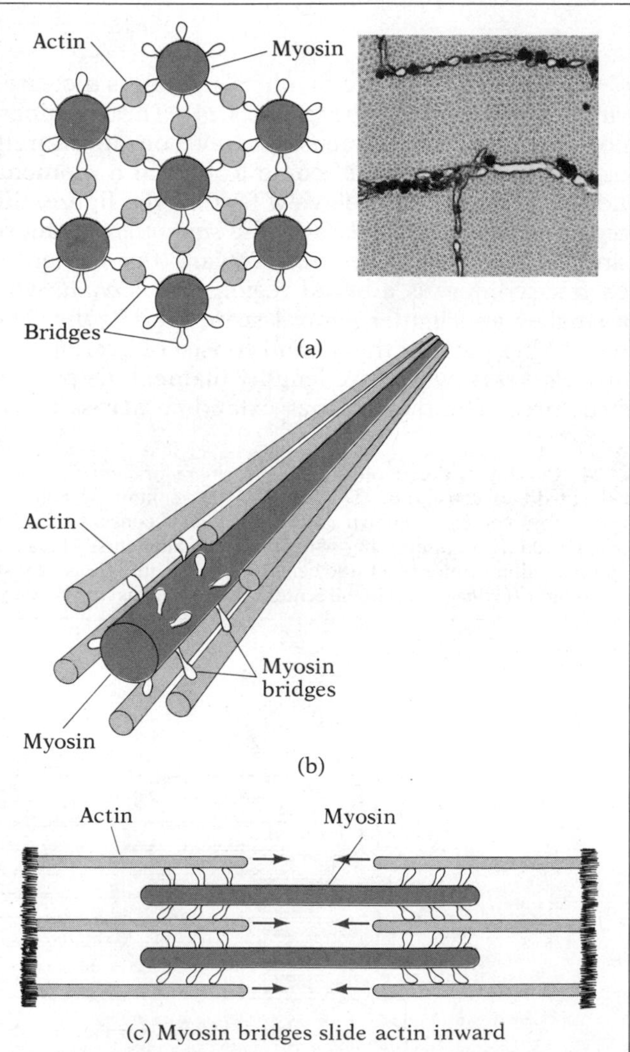

nifications of nearly half a million diameters, they have observed the filaments in enough detail to construct models of their structure. Myosin has been shown to be constructed of long, rodlike proteins spirally wound into the myofilaments. Where individual proteins end, they turn outward in a club-headed figure (Figure 24.27). These myosin heads, which occur in threes at specific distances along the filament, are believed to be the bridges. Actin filaments have been found to consist of three protein components. The first of these, actin, is globular, with its spheres arranged end to end in a helical strand. A second long and slender protein, *tropomyosin,* follows along the helix. At the end of each tropomyosin, the third protein, *troponin,* forms a globular tip.

Calcium and the Biochemical Mechanism of Muscle Contraction

Even the study of muscle chemistry has a long history that reaches back into the 19th century. In 1883, for example, it was determined that calcium ions (Ca^{2+}) are required for the contraction of frog muscles. A considerable body of biochemical and physiological information has emerged during this century. But it has only been in the past 20 or so years that we have been able to integrate this knowledge with what we have recently learned about muscle structure. Let's see what this synthesis has produced.

24.27 The molecular structure of actin and myosin. (a) The thicker myosin myofilament consists of long, slender proteins spirally wound into the filament. At the end of each rod there emerges a thick, club-shaped tip which is the myosin bridge. (b) The thinner actin myofilaments contain three kinds of protein. The major protein is globular actin. The individual actin spheres are arranged in a long, slender helical filament. *Tropomyosin,* a short filamentous protein, follows the curve of the helix for a distance, ending with a molecule of *troponin.* The tropomyosin–troponin–actin complex of the actin filament acts as a controlling mechanism for the myosin bridges.

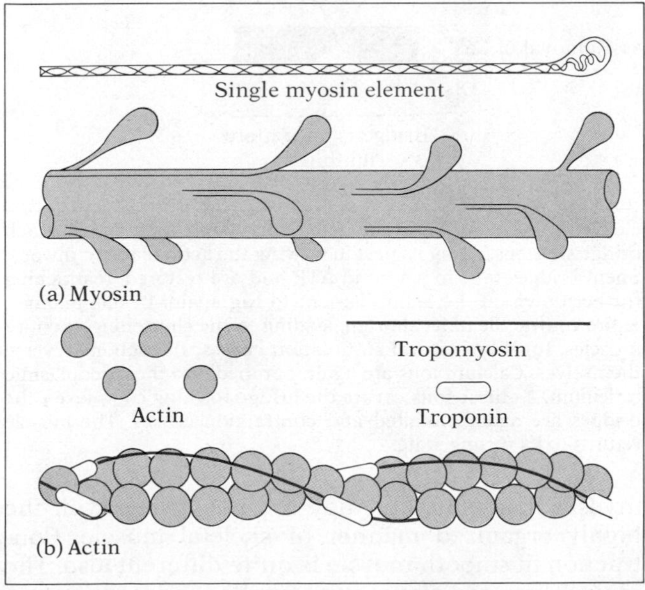

You recall that in a resting muscle filament, the Z lines are widely separated and that there is only a small amount of overlapping myosin and actin. In this position, only a few myosin bridges are near—but *not touching*—the slightly overlapping actin. Each bridge contains ATP, as we might expect. In the relaxed state, the troponin–tropomyosin–actin complex inhibits bridge formation with the myosin heads.

Muscle contraction begins when the sarcolemma receives an impulse from a motor neuron. The impulse changes the permeability of the sarcoplasmic reticulum throughout the fiber (and adjacent fibers). The reticulum, which is loaded with calcium, releases Ca^{2+} into the muscle cytoplasm. The transverse tubules seem to play a role here, but their exact function is unknown [Figure 24.28(a)].

First, the Ca^{2+} joins with the troponin–tropomyosin–actin complex. This breaks the inhibitor, thus permitting bridge formation. Second, the myosin bridges attach to the calcium–actin complex, triggering the enzymatic hydrolysis of ATP, thus forming ADP and P_i. Third, energy released from the breakdown of ATP causes the bridge to move, drawing the actin filament along [Figure 24.28(b)].

As the actin moves inward, more bridges contact the actin and new ATP reserves join the spent bridges, which are still attached. Adding ATP once again restores the bridges, permitting them to act again. The repeated action of the bridges brings the muscle into its fully shortened state. As long as calcium ions are present and ATP is available, the contracted state will hold.

Calcium ions, then, apparently are the controlling factor in muscle contraction. However, their presence depends on the permeability of the sarcoplasmic reticulum to calcium. This permeability is, in turn, under the influence of the depolarizing nerve impulse.

In the relaxation process, the actions reverse themselves [Figure 24.28(c)]. This occurs when the neuron that began the process ceases to stimulate the muscle. The membrane then becomes less permeable to Ca^{2+} ions, so they can no longer leave the sarcolemma. Further, in the absence of neural stimulation, the sarcolemma begins to *actively transport* Ca^{2+} back into the sarcoplasmic reticulum, where it will be stored. When Ca^{2+} is thus removed from the contractile units, the troponin–tropomyosin–actin complex again inhibits the myosin bridges from contracting the actin. The muscle is now at rest. Before it can contract again, it must be pulled out (extended) by some other force, such as the contraction of an opposing muscle.

What we have described here is a synthesis of ideas into a working model. But the model is by no means complete. There are two reasons for this: First,

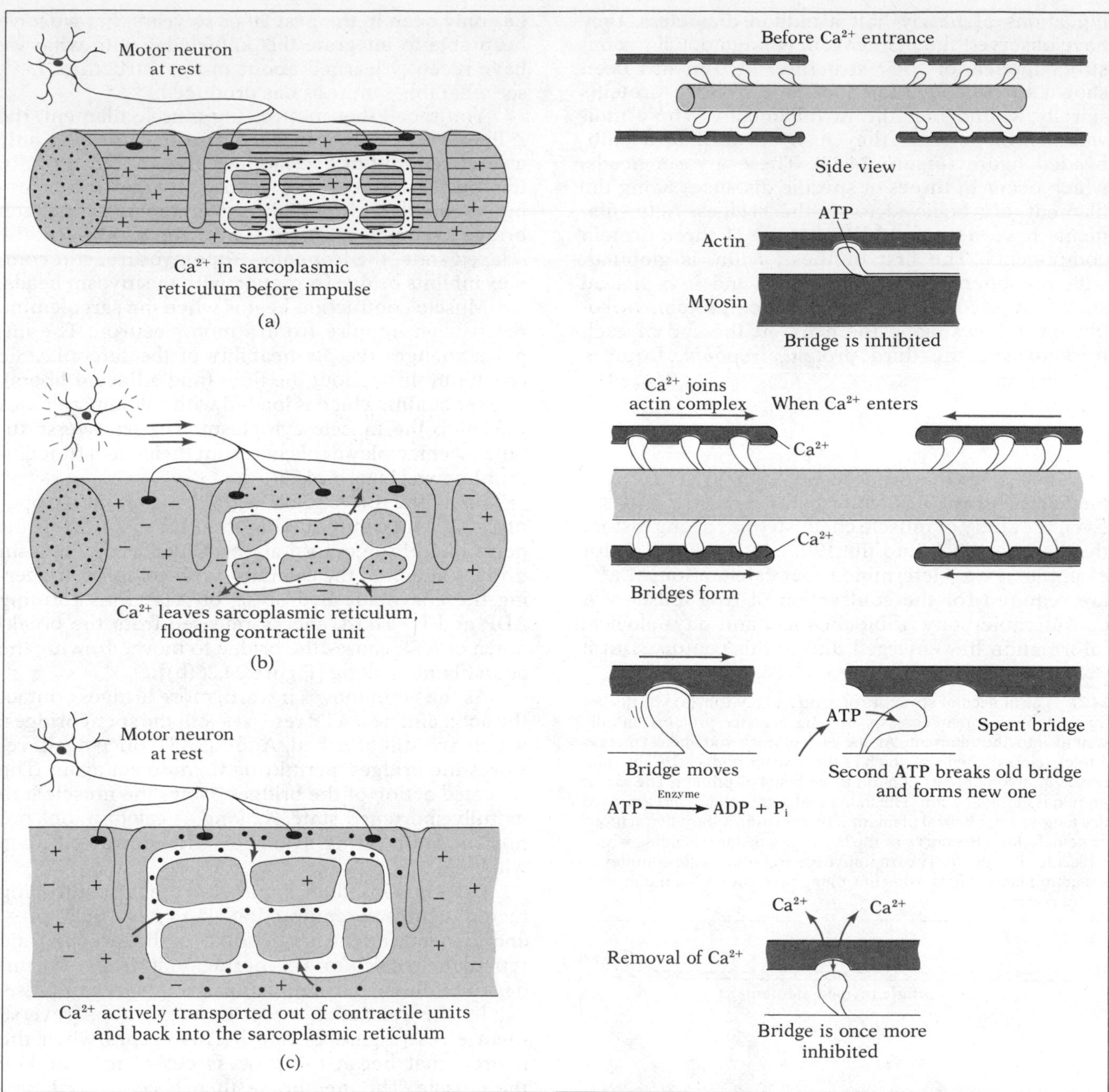

24.28 Muscles contract when a motor nerve transmits an impulse to the muscle surface. In the resting state (a), calcium ions are pumped into and stored in the sarcoplasmic reticulum and the muscle is polarized (positive charges outside, negative inside). In this state, myosin-actin bridge formation is inhibited. (b) When a motor neuron transmits an impulse and the neural impulse reaches the muscle surface, a depolarizing wave is created, stopping the calcium pumps momentarily, and calcium enters the contractile units. The Ca^{2+} removes the inhibition and the bridges form. Movement begins as ATP present in the bridges is enzymatically cleaved to ADP and P_i in the usual energy-releasing action. Each bridge shortens, doing its part in moving the actin filament inward. Spent bridges take in a second ATP and are restored, reattaching further down on the actin filament to tug again. In this manner, entire contractile units shorten, leading to the shortening of entire muscles. (c) When neural stimulation ceases, the actions reverse themselves. Calcium ions are again pumped into the sarcoplasmic reticulum. As these ions vacate the bridge-forming complexes, the bridges are again inhibited and contraction ceases. The muscle returns to its resting state.

we didn't tell you everything that is known. Second, not everything is known. Furthermore, the model cannot pretend to be universal. It may hold for skeletal and cardiac muscle, but smooth muscle may behave differently. Smooth muscle contains actin and myosin filaments, but they are not arranged in the highly organized manner of skeletal muscle. Contraction in smooth muscle is quite different also. The movements are slow, quite gentle, and continuous.

In this chapter, then, we have considered skeletal systems and the muscles that move them. As usual, many questions remain, particularly pertaining to muscle contraction. But let's continue now with our look at animals, keeping in mind that whereas the answers may be fascinating, so, in many cases, are the questions themselves.

Application of Ideas

1. The evolution of skeletons has occurred in two distinct directions, following the protostome and deuterostome lines. How do skeletons in these groups differ? Explain how one permitted only limited size in terrestrial animals while the other was far less limiting.

2. The bird skeleton probably has more specialized features than are seen in any other vertebrate. Describe several specific differences from other vertebrate skeletons. Note in your description of the use of the triangle in structure and comment on its appropriateness.

3. Studies of muscle contraction provide examples of how the scientific disciplines merge. Visual aspects of contraction have been traced right down to the biochemical level. Explain where visual studies leave off and biochemical studies begin in the contractile process. How much of a gap remains between the two?

4. Levers are classified according to the position of the fulcrum, the force applied, and the weight to be moved. First, second, and third class levers are defined as follows: fulcrum in the center, weight in the center, and force in the center. Illustrate these using simple drawings and then find an example of each in the joints and muscles of the body. Determine which offers the best mechanical advantage (the least amount of force needed to move a part the greatest distance).

5. The use of hydrostatic pressure in invertebrates is well known as is the use of turgor pressure by plants. Explain how these mechanisms are used and find similarities and differences. It would be interesting to include the Venus flytrap (Chapter 19) in your discussion.

Key Words

A band
acetabulum
Achilles tendon
actin
alula
antagonist
apodeme
aponeurosis
appendicular
articulate
atlas
autonomic nervous system
axial
axis

ball-and-socket joint
bipedal
bone
braincase

calcaneous
calcarious sponge
calcium carbonate
canaliculi
carapace
cardiac muscle
carpal
cartilage
cervical vertebrae
clavicle

Key Ideas

Introduction

1. One of the most obvious characteristics of animals is their ability to move from place to place.

2. Muscular and skeletal/muscular systems provide animals with motility and resistance to gravity.

3. Motility varies greatly among the animals. Some move only as larvae, others range only a few feet per day, while still others range hundreds, even thousands of miles.

Invertebrate Form and Movement

1. Soft-bodied invertebrates such as coelenterates, flatworms, roundworms, annelids, and others have contractile fibers or complex muscles, but utilize a *hydrostatic skeleton* for a base upon which the muscles act.

2. In the anemone, contractile fibers running in two directions provide the limited movement of shortening and extending the water-filled body.

3. Muscles, arranged in bundles in the earthworm, run in two directions. Contraction against the fluid-filled coelom provides a wide range of movements.

Exoskeletons

1. *Exoskeletons* surround the animal and are secreted by epidermal cells. The *exoskeleton* is characteristic of arthropods. The principal constituent is chitin, with the addition of calcium carbonate in the crustacean skeleton. The *exoskeleton* of terrestrial arthropods is suitable only if they remain small.

2. *Exoskeletons* provide a muscle base for movement, but also protect terrestrial animals from drying and in all cases help protect the soft parts from injury. Hardened appendages are also used in feeding and defense.

coccyx
compact bone
contractile unit
coxa
cranium
crustecdysone
cuttlebone
cuttlefish

degenerate
dermal bone
dermis
digit

ecdysis
ecdysone
ectoderm
endomysium
epidermal cell
epiphyseal plate
exoskeleton

fascia
fascicular bundle
femur
fibula
filament
foramina
foramen magnum

gliding joints

Haversian canal
Haversian system
hinge joints
humerus
hydrostatic skeleton
H zone

I band
ilium
insertion
intercalated disks
intervertebral disks
involuntary muscle
ischium

keel

lacuna
lamellae
ligament
logarithmic spiral
lumbar

marrow
mesoderm
mesoglea
metacarpals
metatarsals
molting
motor neurons

3. Many arthropods must molt or shed their *exoskeletons* in order to grow. This is followed by the secretion of a new skeleton.

4. Muscle action (*flexion* and *extension* of parts) in animals with *exoskeletons* is opposite to those with *endoskeletons*.

Insect Wings and Their Muscles
Movement in fast-beating insect wings exceeds the neural impulses to the muscle, working on a resonating or reverberating principle. Flying muscles are paired for raising and lowering the wing by acting on the shape of the thorax.

Mollusk Shells—A Type of Exoskeleton
1. Mollusks are often fully or partially enclosed in an ectodermally secreted shell. Movement is often by *hydrostatic* mechanisms. Squids and some slugs have internalized the shell, using it as a supporting rod.

2. The *cuttlefish* use the internalized shell as a flotation device. It pumps water out of the *cuttlebone*, creating a vacuum inside.

3. Mollusks do not *molt*. They secrete new shells as they grow. For geometric reasons the shells increase by forming a spiral or a cone or a combination of the two.

Endoskeletons

Sponges
1. Sponges produce skeletons of calcium carbonate, silicon dioxide, or proteinaceous *spongin*. Amebocytes (wandering ameboid cells) secrete the skeletal elements.

2. The sponge skeleton is used for support and form only. There is little movement if any in the sponge body.

Echinoderms
The echinoderm produces an *endoskeleton* that consists of calcified plates covered by skin. The plates are flexible enough in some to permit movement.

Vertebrate Supporting Structures

Bone
1. With the exception of the cartilaginous fishes (cylostomes and class Chondrichthyes), the vertebrates make use of an *endoskeleton* of bone.
2. Cartilaginous skeletons in sharks and rays are not considered primitive but represent a degenerate condition.

3. The skeleton of vertebrates has long been used to establish taxonomic schemes and relationships.

4. The principal functions of the skeletal system are support, locomotion, protection (skull and *rib* cage), and blood cell production.

5. The skeleton is divided into the *axial* (central) and *appendicular* (girdles and appendages) portions.

6. Movement is permitted by joints. Each has its own type of movement and some are immovable. Joints are usually cartilaginous surfaces, often in capsules, and always lubricated by fluids. They include *gliding*, *pivotal*, ball-and-socket, and hinge joints.

Essay 24.1 The Development and Structure of Bone

1. The *perichondrium* surrounds *cartilage* and its cells die and form cavities. Blood vessels penetrate, bringing in *osteoblasts*, which secrete bone. *Marrow* forms, surrounded by dense bone. Two lengthening centers form at the ends. *Cartilage* at the end fuses last.

2. In mature *compact bone*, the calcium is laid down in concentric *lamellae*. A unit of bone within the *lamellae* is called an *osteon* or *Haversian system*. It consists of isolated bone cells, *osteocytes*, in cavities called *lacunae*. Osteo-

motor units
muscle fiber
myofibril
myofilament
myosin
myosin bridge

nerve impulse
neural canal
neuromuscular

origin
ossification
osteoblast
osteon

patella
pectoral girdle
pelvic girdle
perichondrium
phalanges
pivotal joints
proprioceptive sensors
pubic symphysis
pubis

quadrate

radius
ribs

sacroiliac
sacrum
sarcolemma
sarcomere
sarcoplasmic reticulum
scapula
skeletal muscle
skull
smooth muscle
spicules
spinal cord
spongin
spongy bone
sternum
striated muscle
synovial fluid
synovial joints
synovial membrane

talus
tarsals
tendon
thoracic
thorax
tibia
transverse tubules (T tubules)
tropomyosin
troponin

ulna

vertebra (vertebrae)
vertebral column
voluntary muscle

cytes communicate with each other via *canaliculi*, which carry nerves and blood vessels throughout the system.

The Axial Skeleton

The Skull

The progress of skulls through evolution has shown a general fusion of bones. Jaws, at first absent, become simple biting structures, then chewing structures. The upper jaw became fused to the *cranium*.

Essay 24.2 Evolution of the Mammalian Skull

1. In primitive vertebrates, the braincase was internal, surrounding the brain. Bones of the cranium were *dermal bones*, plates developed in the skin. Jaw muscles were inside. The original outer bone gradually gave way, slowly being replaced by shelflike extensions that grew beneath the jaw muscles and behind the eyes.

2. Reptile jaws, snakes in particular, are the most mobile. A loosely hinged arrangement makes this possible.

3. The vertebrate *skull* houses the complex sense receptors (light, sound, taste, smell, balance) and receives the major nerve pathways through the *foramen magnum*.

The Vertebral Column

1. In most animals, the *vertebral column* forms the body axis and supports the weight of the internal organs. The massiveness varies with the task and the number of *bones* varies greatly from class to class.

2. The vertebrae are separated by *cartilaginous* disks, which give the column its mobility.

3. In addition to support, the *vertebral column* houses the *spinal cord* in its *neural canal* and its processes *articulate* with the *ribs*. The column is divided into five regions.
   a. *Cervical*: Seven neck *bones* in mammals. First two are the atlas and axis, which permit head movement.
   b. *Thoracic*: Twelve *bones* in humans. Articulate with the *ribs* and attachments for back muscles.
   c. Lumbar: Five *bones* in humans. Largest *vertebrae* with massive central spines and bodies.
   d. *Sacrum*: Five fused *vertebrae*. Forms part of *pelvic girdle*. Transfers weight from upper body to girdle and legs.
   e. *Coccyx*: Four (some variation) fused *vertebrae* make up "tailbone" (vestigial tail).

4. The *ribs* and *sternum* complete the *axial* skeleton. Humans have twelve pairs of *ribs*, seven of which extend from *vertebra* to *sternum*. The *rib* cage functions in breathing, protection, and red cell production.

5. In turtles, the *ribs* are fused to the upper shell (*carapace*). In birds, the *sternum* is continuous with the large *keel*, to which the flight muscles attach.

The Appendicular Skeleton

The Pectoral and Pelvic Girdles

1. The transition from amphibian to mammal required great changes in the *pectoral* and *pelvic girdles* as the legs became arranged to better support the weight.

2. The *pelvic girdle* is formed by the fusion of the paired *ilium, ischium,* and *pubis* bones. The *sacrum* forms the dorsal connection. The *acetabulum* receives the femur forming a *ball-and-socket joint.*

3. Our *pectoral girdle* consists of paired *scapulae* and *clavicles.*

Z lines
zygomatic arch

Limbs

1. Vertebrates are limited to two pairs of appendages. These are greatly reduced or absent in some.

2. In birds and bats, the forelimbs are greatly modified for flight. Bats have greatly extended forearms and fingers. In birds, the forearms form the leading edge of the wing, and the hands have undergone fusion and simplification.

The Human Appendicular Skeleton

1. Human appendages are generalized (except for the thumb and arch of the foot).

2. The thumb differs from that of other primates. It is longer and more flexible, permitting us to grasp objects without closing the hand.

3. The bones of the hand include the fingers (*phalanges*), hand (*metacarpals*), and wrist (*carpals*).

4. The forelimbs consist of a single *humerus* (upper arm) and the *radius* and *ulna* (lower arm). The latter forms the wrist joint. Together the *radius* and *ulna* permit a twisting rotation as they turn past each other.

5. The legs consist of the upper *femur* and lower *tibia* and *fibula*. Where they meet they form a *hinge joint*, which includes the *patella* or knee cap. The knee contains several strong ligaments, a capsule, and *tendons*.

6. The *tibia* and *fibula articulate* with the *talus* to form a *hinge joint* with the foot. The *tarsals* form the ankle. The large *calcaneous* forms the heel and receives the Achilles *tendon* from the calf muscle.

7. The feet consist of the *tarsals* in the ankle, *metatarsals* in the foot, and *phalanges* or toes. The arch is held in shape by *ligaments*.

Muscles and How They Move

Types of Muscle

There are three types of vertebrate muscle: Smooth, cardiac, and skeletal.

Smooth Muscle

Smooth muscle is found in the digestive tract, blood vessels, iris, and reproductive system. Its microscopic appearance is long and spindly. It is *involuntary*, controlled by the *autonomic nervous system.*

Cardiac Muscle

1. *Cardiac* or heart muscle is *involuntary*, controlled by the *autonomic nervous system.*

2. It forms *cylindrical* and *striated* branched fibers with *intercalated disks* between adjacent cells. Its contractions are rhythmic and intrinsic.

Skeletal Muscle

1. *Skeletal muscle* is *voluntary* and *striated*. It is found in the muscles that move the skeleton and maintain body shape. They are generally under conscious control and contain *proprioceptive sensors*, which inform the brain of their position and orientation.

2. *Skeletal muscle* is multinucleate, cylindrical, and prominently *striated* (striped).

3. *Bundles* of fibers are surrounded by a connective tissue called a *fascia*. The *fascia* form *tendons*, which attach to the skeleton. *Insertions* refer to *tendon* attachments to parts that move. *Origins* are *tendon* attachments to anchoring parts. *Ligaments* are used to hold *bone* to *bone* in *joints*.

4. Some *skeletal muscle* does not move *bones*, but moves the face and tongue, and forms flattened sheets as seen in the abdomen.

Muscle Antagonism

Each muscle or muscle group that moves a part has an opposing muscle or muscle group. Such pairs are called *antagonists*. The biceps brachii and the triceps brachii are examples.

The Ultrastructure of Skeletal Muscle

1. Muscle is best studied by descending the orders of organization.
   a. The organ: The belly of the muscle is subdivided into *bundles* of fibers and their sheaths.
   b. Bundles: *Fascicular bundles* contain blood vessels and nerves and consist of fibers (cells) in a matrix of connective tissue.
   c. Fibers: Muscle fibers are fused, multinucleate cells. Each fiber is surrounded by a *sarcolemma*, which receives neurons. Within the *sarcolemma* are *myofibrils*, nuclei, mitochondria, the *sarcoplasmic reticulum*, and *transverse (T) tubules*.
   d. The *myofibrils:* The *myofibrils* can be divided longitudinally by their *Z lines*. Between *Z lines* are the contractile units, which are identified by their banded patterns.
   e. The *myofilaments*: The *myofibrils are subdivided into myofilaments* of *actin* and *myosin*.
2. The banding of a *contractile unit* includes a central *A band*, with a lighter *H zone* at its center. To the right and left of the *A band* are two lighter *I bands*. The lighter and darker regions represent nonoverlapping and overlapping *actin* and *myosin myofilaments.*
3. When muscle contracts, the *Z lines* move closer together. The *H zone* and the *I bands* decrease in width. The *A band* remains the same width. The *actin* filaments from each side move together, sliding among the *myosin* filaments.
4. A *myosin myofilament* is composed of numerous protein strands with protruding, clubshaped heads. The heads interact with *actin,* forming movable bridges. In contraction, these fasten onto the *actin*, in a ratchetlike action pulling *actin* inward. The bridges do not form unless the muscle is stimulated to contract.
5. An *actin myofilament* consists of three kinds of protein: *actin, troponin*, and *tropomyosin*. The three interact to form the *troponin–tropomyosin–actin* bridge complexes. In the resting state the three inhibit bridge formation with *myosin*.

Calcium and the Biochemical Mechanism of Muscle Contraction

1. In the resting state, the *myosin* heads are close but not touching the *actin*. Each head or bridge contains ATP.
2. Muscle contraction occurs in several steps:
   a. The *sarcolemma* receives an impulse.
   b. The *sarcoplasmic reticulum* changes polarity and releases stored calcium ions into the contractile unit.
   c. Ca^{2+} joins the *tropin–tropomyosin–actin* complex and the *myosin bridges* attach.
   d. ATP is enzymatically hydrolyzed to ADP + P_i. The high-energy bond is used to bend the bridge.
   e. The bridges pull the *actin* inward. ATP is again added, permitting the bridge to relax.
   f. Bridges repeat their action throughout the contractile units as long as Ca^{2+} and ATP are present.
3. In muscle relaxtion the events are reversed:
   a. Neural stimulation stops.
   b. Calcium ions are pumped back into the *sarcoplasmic reticulum.*
   c. The *troponin–tropomyosin*-actin protein complex inhibits bridge formation.
   d. The contracted muscle can be elongated only by the pull of an opposing force.

Review Questions

1. Explain what a *hydrostatic skeleton* is and list several phyla where it is the principal type used. (p. 670)
2. How does the earthworm make use of its coelom in movement? (p. 670)
3. Account for the earthworm's wide range of movements. (p. 670)
4. List the basic differences between *endoskeletons* and *exoskeletons*. (p. 671)
5. What contributes to the heaviness of crustacean *endoskeletons* and what has living in water to do with its feasibility? (p. 671)

6. Explain how *ecdysis* works and how it enables crustaceans to grow. (p. 671)

7. Using your own biceps and triceps as an example, explain how they move your forearm and then explain how they would have to be arranged in an arthropod to do the same (a drawing is helpful). (p. 671)

8. Compare the insect *exoskeleton* to that of a crustacean and list any additional functions it serves on land. (pp. 671–672)

9. Describe the attachment of insect's wings and explain how the muscles move them up and down. (p. 672)

10. What is the apparent difficulty in explaining the insect's wing beat in terms of *neural impulses?* (p. 672)

11. Describe the special use of the skeleton in the *cuttlefish.* (p. 673)

12. Explain how the "geometric imperative" determines the shape of growing mollusk shells. (p. 673)

13. List the materials of sponge skeletons and explain how they function for these sedentary animals. (p. 674)

14. Fully describe the skeleton of echinoderms. (pp. 674–675)

15. Summarize the events in the development of *bones.* (Essay 24.1)

16. Describe the elements of an *osteon* or *Haversian system.* (Essay 24.1)

17. List four specific functions of the vertebrate skeleton. (p. 675)

18. List four types of joints and give examples of each. (pp. 675, 677)

19. What are the main parts of the *axial* skeleton? (p. 677)

20. Describe the general evolutionary trend in cranial and jaw structure in the vertebrates. (p. 678)

21. Review the main points in the transition of the vertebrate skull. (Essay 24.2)

22. What purposes do the *foramina* and in particular the *foramen magnum* serve? (p. 678)

23. Describe the special role of the *sacrum* and explain why problems might arise at its joint with the pelvis. (pp. 678–679)

24. List the three types of *ribs* and explain their differences in terms of attachments. (p. 679)

25. Describe different modifications in the *rib* cage and *sternum* of turtles, amphibians, and birds. (p. 679)

26. Compare the limb arrangement and walking stance in the amphibians, reptiles, mammals, and humans. (p. 681)

27. List the bones of the *pectoral girdle* in humans and explain their arrangement. (p. 681)

28. List the bones of the *pelvic girdle* in humans and describe their arrangement. (p. 681)

29. Describe two differences between the male and female pelvis of humans. (p. 681)

30. Compare the forearm and hand modifications for flying in the bats and birds. (pp. 681–682)

31. List the bones of the human upper appendages and describe the kinds of movement at the joints. (pp. 683–685)

32. Describe the subtle differences and capabilities between our hands and those of the chimpanzee. (pp. 683–685)

33. List the bones of our lower limbs and describe the kinds of movement at the joints. (p. 685)

34. List the three types of vertebrate muscle and describe their location. (pp. 686–688)

35. Explain the differences between the three types of muscle in the following way (pp. 686–688)
   a. Microscopic appearance
   b. Level of control
   c. Functions

36. *Cardiac* is the most unusual muscle type. Describe its three levels of control. (p. 688)

37. What is the function of the *endomysium?* The *fascia?* (p. 688)

38. Distinguish between *tendons* of *origin* and *tendons* of *insertion.* (p. 688)

39. How does the function of *tendons* differ from that of *ligaments?* (p. 690)

40. List some muscle actions that do not involve bone movement; there are many. (p. 690)

41. Using the biceps brachii and triceps brachii as a model, explain what is meant by muscle *antagonism.* (p. 690)

42. Describe the structures of a *muscle fiber.* (pp. 690–691)

43. Make a simple drawing of a *sarcomere,* the *contractile unit* of muscle. Add in all of the *lines, bands,* and *zones* and label each. (You should use: *I, A, H,* and *Z*) (54, 55) (pp. 691–692)

44. Describe the appearance of a *contractile unit* that has undergone contraction. (pp. 691–692)

45. Explain the ratchet theory of contraction. (pp. 692–693)

46. Describe the structure of *actin* and *myosin filaments.* (pp. 692–693)

47. Supply the information needed to explain the biochemistry of contraction (pp. 693–694)
   a. Neural action
   b. Changes in the *sarcolemma*
   c. Effect of Ca^{++} on the *troponin–tropomyosin*-actin complex
   d. Role of ATP
   e. Action in the *myosin bridge*
   f. Second ATP action

48. Now, using the organization from the last question, explain how muscle contraction is stopped. (pp. 693–694)

Chapter 25

Digestion and Nutrition

It's always a little risky to generalize about evolutionary trends, since evolutionary change runs in all directions, and at very different rates. Some traits change fast, some change slowly. If a species happens to harbor a lot of rapidly changing traits, then the species itself changes rapidly. By the same token, a great proportion of slowly changing traits will produce a species that is very similar to its distant ancestors. Even within the same animal, different organs may evolve at different rates and show different levels of specialization. For example, we humans have primitive mouths. Here, we must even yield before the lowly platypus with its highly specialized bill. But the gut! Ah, the gut. You can be proud. You have a very highly evolved gut—not as highly evolved as that of the cow, with its famous four stomachs, but it is still very efficient and rather complicated. So it is only fitting to look at how we ended up with such a nice gut. But first let's review some general processes of digestion.

25.1 Digestive Systems

Digestion refers to breaking down complex food molecules into their component parts so that they can be used as building blocks for other complex molecules. Sometimes digestion occurs outside the organism, such as when bacteria and fungi secrete enzymes into their environments and break food materials down into small molecules that can be absorbed by active transport through their cell membranes. *Extracellular digestion* and absorption is, in fact, the only way in which heterotrophic bacteria and fungi can obtain nourishment.

Most eucaryotes other than fungi and plants are capable of taking food particles into their cells, to be digested *intracellularly* in *digestive vacuoles.* You can see digestive vacuoles in large protozoa such as the *Paramecium* or in amebas (see Figure 15.15). Humans and other vertebrates practice both types of digestion. We have white blood cells that roam through our bodies engulfing bacteria and other foreign matter, but most of our digestive processes are extracellular. Remember, our gut is really a tube within a tube (the outer tube being the body wall). Thus, anything passing through our gut is, topographically, still outside our bodies. It enters our bodies through the gut lining. (So, it is proper, after all, to say, "I would like to get on the outside of a hamburger.")

Now, we'll review digestion in the simpler multicellular invertebrates and then we'll move on to the vertebrates, concentrating on humans.

Systems and Digestion in Invertebrates

The Simpler Invertebrates

Food getting in the loosely organized sponges is the responsibility of the flagellated *choanocytes,* also known as *collar cells* (see Figure 22.6). These highly specialized cells line the body canals, and their beating causes seawater to swirl and eddy throughout the body of the sponge. That water sometimes carries in tiny food particles which are trapped by the collar cells and passed through a layer of mucus into the base of the cell (Figure 25.1). The food is then moved from the collar cell to the cells beneath it, where it becomes engulged in digestive vacuoles. Digested food molecules then diffuse throughout all the cells of the sponge. It is believed that those peculiar, wandering cells called amebocytes also aid in the distribution processes by carrying food from one place to another.

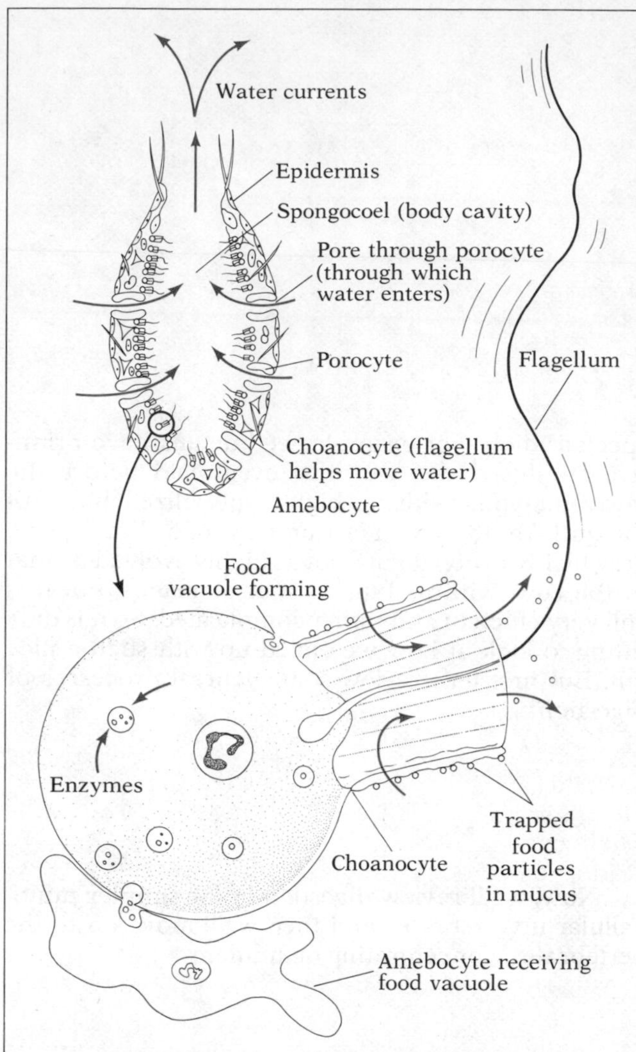

25.1 Intracellular digestion and food getting in sponges. Sponges obtain their food by filtering seawater. Water currents, created by flagellated *collar cells* (*choanocytes*), bring a constant flow of water into the body through pores surrounded by cells known as *porocytes.* The rhythmic movement of each flagellum brings bits of organic debris in contact with mucus on the collar. From there, food is carried to the choanocyte, is phagocytized, and ends up in a food vacuole. Some digestion occurs there, but other food particles are engulfed by amebocytes, which are believed to carry on digestion and distribution of food to other cells in the sponge.

So the sponge, a simple multicellular animal, carries on digestion without a well-defined digestive system.

The coelenterates have a somewhat more complex digestive system, comprised of a saclike gut in which digestion takes place. So, in these animals, digestion is first extracellular. However, after food is ingested and partially broken down, particles are engulfed by the lining of the gut where digestion continues, so, all in all, most coelenterate digestion is still intracellular (Figure 25.2).

Flatworms and comb jellies have similar blind gut cavities. Of course, indigestible material must eventually be ejected through the organism's mouth, which thus must also serve as its anus. It is not surprising that such an arrangement is considered primitive.

The saclike gut of the coelenterates and flatworms, called the *gastrovascular cavity,* is primitive in another way. It serves a dual purpose of digestion and distribution; that is, it serves as a sort of circulatory system as well as a digestive system. Coelenterates and flatworms have no circulatory systems as such. Food materials are transported throughout the body by tiny, fingerlike extensions of the gut. The gastrovascular cavity probably reaches its highest level of development in the flatworms, such as planaria. In fact, the planaria's remarkable feeding mechanism deserves some special attention, since it may have been the living forerunner of the vacuum cleaner (Figure 25.3).

The highly branched digestive structure of the flatworm begins with a long, protrusible *pharynx,* its feeding and swallowing organ. When the animal senses the presence of food (probably through a chemical stimulation), it everts its pharynx (like turning a glove inside out) and begins to suck at its prey. (Planarians feed on any animals, dead or alive, but luckily, the animal is only a few millimeters long, or water sports could become a bit eery.) If the flatworm is successful, the juices of the prey and a number of small particles enter its gastrovascular cavity. The larger particles are then phagocytized, and internal digestion begins. Since the cavity is a blind sac, the wastes are pumped back out through the pharynx, the same way they came in. Distribution of food is aided by the loose structure of the worm's middle layer of cells.

The problem of a digestive system in the parasitic relatives of planaria (class Trematoda) was discussed in Chapter 22.

More Complex Invertebrates

We will now move along to animals with more highly developed digestive systems. Although the sac gut works well enough—there are plenty of coelenterates and flatworms around to prove it—the obvious next step was the development of a second opening to the gut cavity. As we mentioned in Chapter 22, the progression from having a single opening to having two separate openings appears to have occurred in evolution at least twice. One group—the Protostomia—used the second opening to the gut as an anus, and the second group—the Deuterostomia—used the second opening as a mouth. Either way was an im-

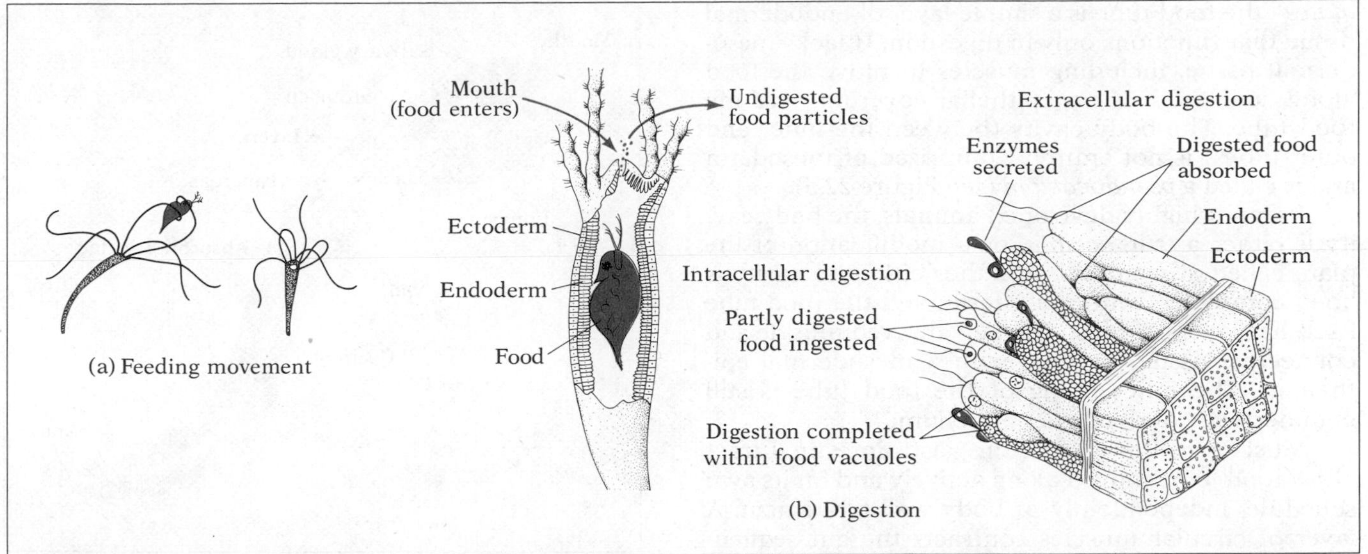

25.2 Hydra, a freshwater coelenterate, demonstrates the manner in which many coelenterates feed. (a) A minute crustacean is captured by paralyzing it with toxins from stinging cells in the tentacles. Then it is brought into the mouth. (b) The mouth opens into a *gastrovascular* cavity, where digestion begins. Glandular cells secrete powerful digestive enzymes into the sac, and other cells phagocytize small partially digested particles so that digestion is completed in vacuoles. We see, then, that digestion in hydra is both extracellular and intracellular.

25.3 Food getting and digestion in planaria. Planaria have a highly branched gastrovascular cavity. They obtain food by sucking with a protrusible *pharynx.* Bits of food are distributed throughout the *gastrovascular cavity,* where they are phagocytized and digested intracellularly. The gastrovascular cavity, therefore, acts more as a distribution system than as a digestive organ.

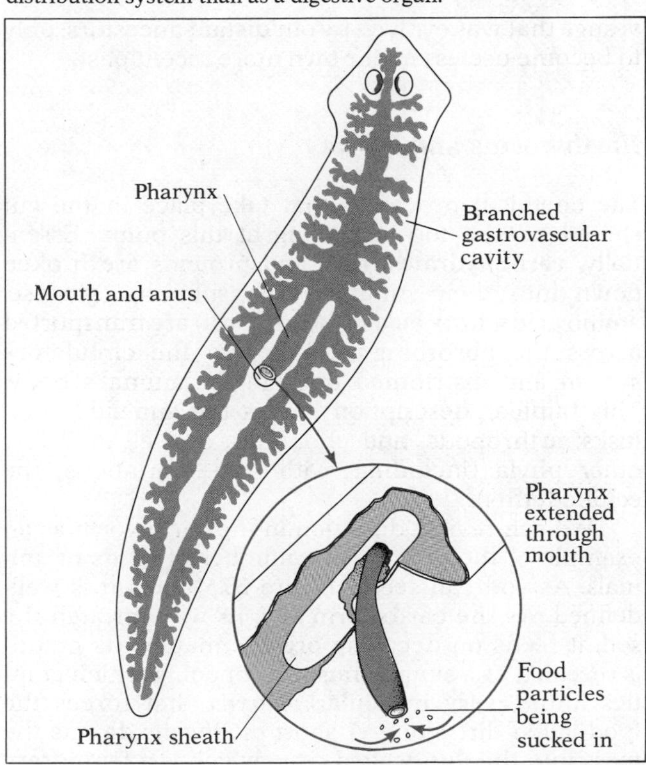

provement, and the great majority of the metazoa descend from these two lines. In addition to improved sanitation, the one-way gut made possible a shift from batch, or intermittent, processing to continuous processing of ingested food, and allowed for further specialization of the gastrointestinal tract.

Many metazoa still utilize a great deal of intracellular ingestion and digestion. The so-called "liver" of many invertebrates, such as mollusks and crustacea, actually consists of intestinal side passages lined with phagocytic cells, and some food is digested within these cells after having been broken down into small particles. The vertebrate gut usually doesn't bother with phagocytosis as a means of food ingestion, though dissolved fats are taken into cells as whole molecules. Generally, as we will see, vertebrates break down food into small molecules so that they can be transported across the gut epithelium directly into the circulatory system.

Pseudocoeloms, Coeloms, and Hemocoels

All groups of animals with one-way guts have a basic *tube within a tube* body plan. The outer tube is the body wall, which is typically muscular and rich in connective tissue. The inner tube is the food tube, or *gastrointestinal tract.* Between the inner tube and the outer tube, in most animal groups, is a fluid-filled cavity. The cavity allows for a certain independence of movement between the body wall and the gut.

In nematode worms, rotifers, and a number of similarly primitive forms—all known as *pseudocoelo-*

mates—the food tube is a simple layer of endodermal tissue that functions only in digestion. It lacks mesodermal tissue, including muscles to move the food along, and there is no epithelial covering over the food tube. The body cavity (between the inner and outer tubes) is not entirely comprised of mesoderm and is called a *pseudocoelom* (see Figure 22.5).

In more highly developed animals, the body cavity is either a true *coelom,* or a modification of the plan, called a *hemocoel.* In either case, the cavity is lined completely with mesoderm, and the food tube itself has layers of mesodermally derived muscle and connective tissue, covered with a mesodermal epithelium. The inner lining of the food tube is still endodermally derived gut epithelium.

Muscles in the wall of the gastrointestinal tract allow food to be moved along actively and on its own schedule, independently of body wall movement. A layer of circular muscles contracts the gut sequentially as if a constricting ring were moving along it, while segments of muscles aligned lengthways shorten, permitting the gut to increase in diameter. The regular, flowing alternation of circular and longitudinal muscle contractions creates *peristalsis,* which pumps the gut contents along toward the anus.

Of course, since food passes across the gut wall, it is advantageous to have as large a gut as possible. But this doesn't mean there is a tendency for more efficient guts to simply be longer. Different animal groups have evolved different ways to increase the inner surface area of the gut. One way, of course, is to lengthen and coil the gut; another way is to fold the epithelium thereby increasing the surface area. We'll look at examples of such folds later—in the *typhlosole* of the earthworm, the *spiral valve* of the shark, and our own *intestinal villi.*

Another development in various groups of animals is the outpocketing of the gut into blind sacs—which have subsequently developed all sorts of functions. We've already mentioned one such outpocketing—the "liver" of mollusks and crustacea. In fact, in embryonic development (and probably also in evolutionary development), the vertebrate liver begins as an outpocketing of the gut and remains attached to the cavity of the gut via the bile duct, as we shall see. But unlike the invertebrate liver, the vertebrate liver doesn't take in food materials directly. The pancreas and salivary glands are also gut outpocketings and remain attached through their ducts. In fact, the lungs are endodermal gut outpocketings (Figure 25.4).

In some species, blind outpocketings of guts serve as fermentation vats, such as the *ceca* (singular, *cecum*) of rodents, most mammals, and many fish. Our *appendix* protrudes from our own cecum as a vestige that was evolved by our distant ancestors, only to become useless in our own more recent past.

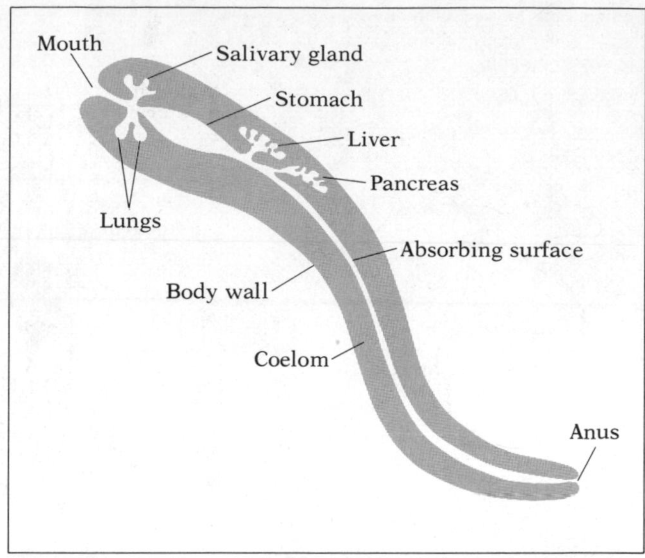

25.4 The tube within a tube plan. The simplified animal shown here typifies the tube within a tube plan of most animals. The outer tube is the body wall, while the inner tube represents the gut. Between the two tubes is the body cavity. Regions of the gut have become specialized in carrying out the various functions of digestion and absorption. Special outpocketings form accessory structures such as the liver and pancreas, branching off the inner tube (as does the respiratory system). Food passes through the tube where it is digested and some of the products are then absorbed. This body plan is representative of that found in many of the major phyla of animals—including our own.

Earthworms and Insects

The chemical processes that take place in the gut should not be too surprising at this point. Essentially, carbohydrates, fats, and proteins are broken down into their subunits. The subunits (glucose, amino acids, fatty acids, and glycerol) are transported across the absorbing surface into the circulatory system and distributed through the animal's body. This familiar description covers the annelids, mollusks, arthropods, and chordates, as well as a few other phyla (including, with some variations, the echinoderms).

We can review digestion in the earthworm as an example of the process in a number of kinds of animals. As you can see in Figure 25.5, the gut is well-defined. As the earthworm eats its way through the soil, it feeds on decaying organic matter. Its mouth is revealed as a simple muscular opening which empties into a very muscular *pharynx* that forces the food-laden dirt along. A short *esophagus* directs the mass into the thin-walled *crop,* which is a temporary storage organ. The ingested material is next moved

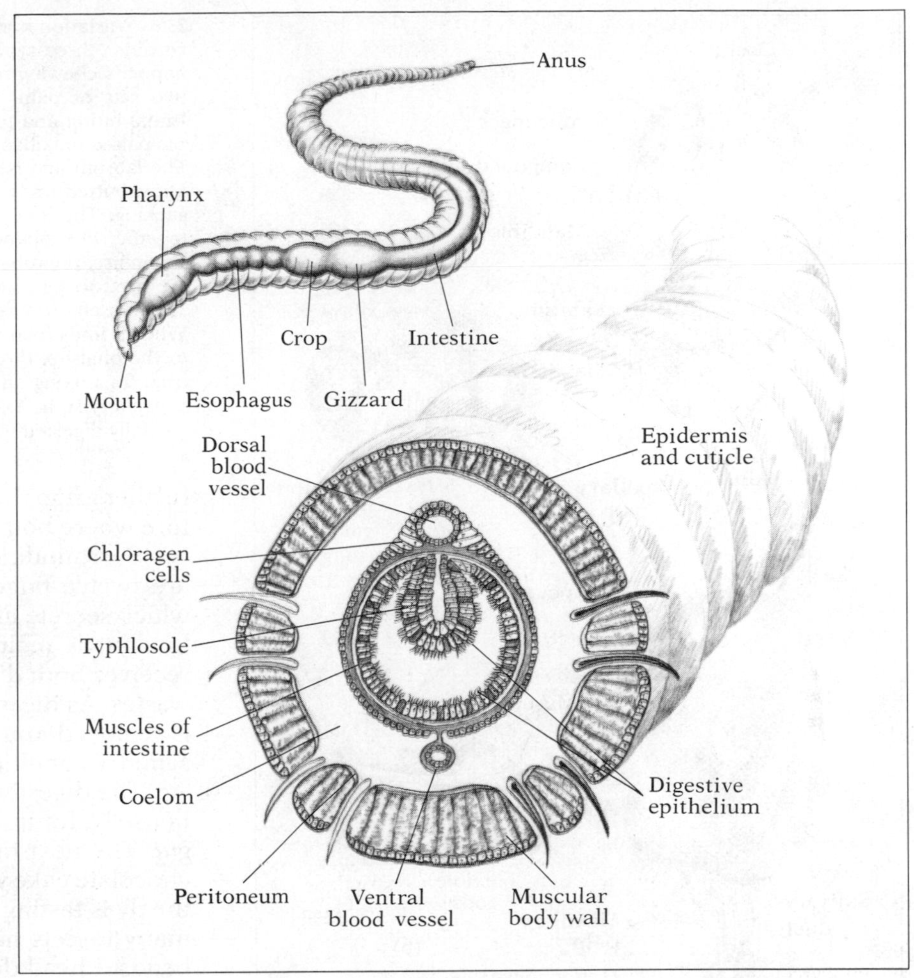

25.5 The digestive system of the earthworm is quite complex. The system is complete, which means food moves through a tube, with a mouth opening at one end and an anus at the other. The tube in the earthworm has several specialized regions. The *pharynx* aids in swallowing, while the *crop* and *gizzard* store and grind food, respectively. The *intestine* extends the length of the worm ending in the anus. It has a muscular wall and an absorbing and glandular lining, or epithelium. The absorptive area is greatly increased by an intestinal fold known as the *typhlosole.*

into the thick-walled, muscular *gizzard.* The gizzard is filled with tiny stones and its role is to grind the food. It is functionally similar to the gizzard of birds.

From the gizzard the food moves into the intestine, where it is digested by enzymes secreted by cells in the intestinal lining. Digestion and absorption occur along the entire length of the intestine. As you can see in Figure 25.5, a large fold in the earthworm's intestinal wall, the *typhlosole,* assists the absorption process by greatly increasing the surface area. From the absorptive surface, the food molecules pass into the circulatory system for distribution throughout the body. Although the earthworm doesn't have anything to compare with the vertebrate liver, groups of skin cells known as the *chloragen* cells perform some of the same functions. These include the conversion of glucose to glycogen, the deamination (removal of nitrogen) of amino acids, and the formation of urea. The accumulation of some of these products helps to give the worm its color. Strangely enough, roaming phagocytes pick up whole bits of soil passing through the earthworm's intestine and then move to the area of the skin, with the resulting earthy coloration causing the earthworm to present a major visual problem to the early bird.

It's difficult to generalize about insect feeding structures. In fact, the group serves as a lesson in digestive variation, one that illustrates the tremendous range of adaptations to the varied energy sources of the earth. It is difficult to think of anything that at least some insects don't eat. This versatility has been particularly troublesome since the dawn of agriculture, because growing food in large patches has made it possible for some species of insects to increase dramatically in numbers.

Biting and chewing mouthparts are very common in insects, particularly in predators and plant eaters (Figure 25.6). They can be clearly seen in the grasshopper, which has, essentially a five-part mouth. It tastes its food with its palps, pairs of sensory structures attached to the mouthparts. Biting and shearing is done with the large saw-toothed mandibles, while

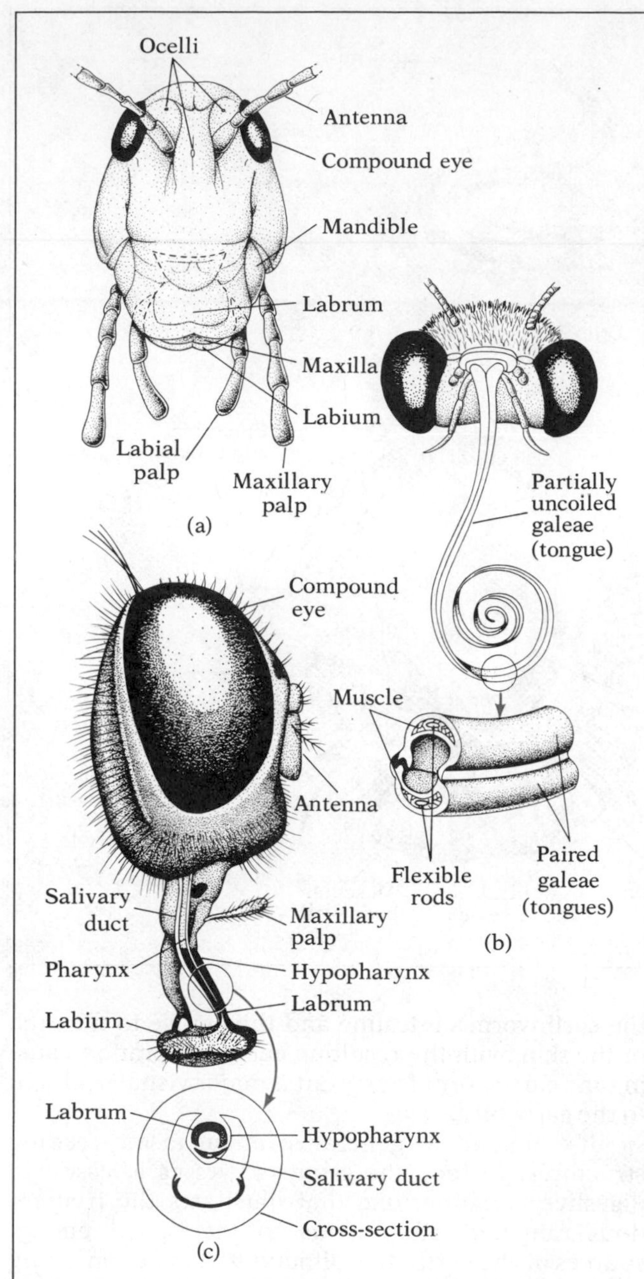

25.6 Variation in insect mouthparts is seemingly endless, so let's consider three types as representative examples. (a) The grasshopper's chewing mouth is nicely adapted for eating foliage. The two sets of palps contain taste receptors for identifying food. Initial biting and tearing is done with the large mandibles, while the paired maxillae do more delicate shredding and manipulating. The labrum and labium help in holding and positioning food. (b) Many butterflies feed on nectar, so their mouthparts are adapted to sucking. The long extensible paired *galeae,* or tongues, are held together like zip-lock bags, forming a long drinking tube. When not in use, the tube is kept in a tight coil by flexible rods. Muscles are used to pull against the rod to straighten it out for use. (c) The housefly has its own feeding strategy and related mouthparts. When it finds food, saliva is released along a duct running parallel to the pharynx through an enlarged spongelike labium, which is used for mixing and stirring. Its labrum forms part of a tubelike *hypopharynx,* an extension of the pharynx used for sucking up the partially digested food.

the smaller maxillae do the finer grinding. The upper and lower lips, the *labrum* and *labium,* assist by holding food so the shearing and grinding can occur efficiently. Saliva from distinct salivary glands in grasshoppers helps with the grinding and begins digestion.

The grasshopper's digestive system, similar to that of many insects, consists of three parts, the fore-, mid-, and hindgut (Figure 25.7). The foregut is divided into a pharynx, esophagus, and crop. The crop functions primarily in food storage. In some insects this region has a roughened surface which is used to further grind the food. The midgut is a digestive structure where both digestion and absorption occur.

Surrounding the junction of the fore- and midgut are twelve fingerlike accessory structures, the *ceca,* which secrete digestive enzymes into the midgut. The hindgut is mainly a water-absorbing structure that receives both digested food and cellular nitrogenous wastes. As digestive residues pass through, the water is removed and conserved. Thus the feces leave in a semidry condition.

The digestive systems of insects vary widely. The housefly, for instance, carries on digestion *outside the gut.* The next time you see one walking across your chocolate cake you might remember two things. First, the fly is tasting the icing with its feet—a place where many insects have taste receptors. Second, when it bends it head down, it is disgorging its enzymes into the surface and stirring it all around with its hairy labium. But, don't worry, it will suck up any mess it makes with its hollow, tubelike mouthparts.

Another variation is seen in the termite, which has a large expansion of the hindgut, housing a vast number of ciliated protists that digest the cellulose from the wood the termites eat. Microorganisms also inhabit the gut of bloodsucking tsetse flies and sucking lice, tucked away in special cells. It seems that as nutritious as blood might be—with its iron, protein, sugars, and vitamin B—it is not sufficient for the bloodsucker's diet. Yeasts and bacteria provide the missing nutrients as they form waste products.

Plant eaters, by the way, frequently have piercing and sucking mouthparts. Recall the aphids that suck up the phloem contents (Chapter 18) and some of the noctuid fruit-eating moths that tear the fruit first then suck up the exposed fluids. Of course, piercing and sucking mouthparts are found among the blood-eating and some predaceous species as well. When the ant lion sinks its huge curved mandibles and maxillae into the ant's vulnerable abdomen it doesn't chew. It uses channels formed between the two paired mouth parts to suck out the ant's body fluids.

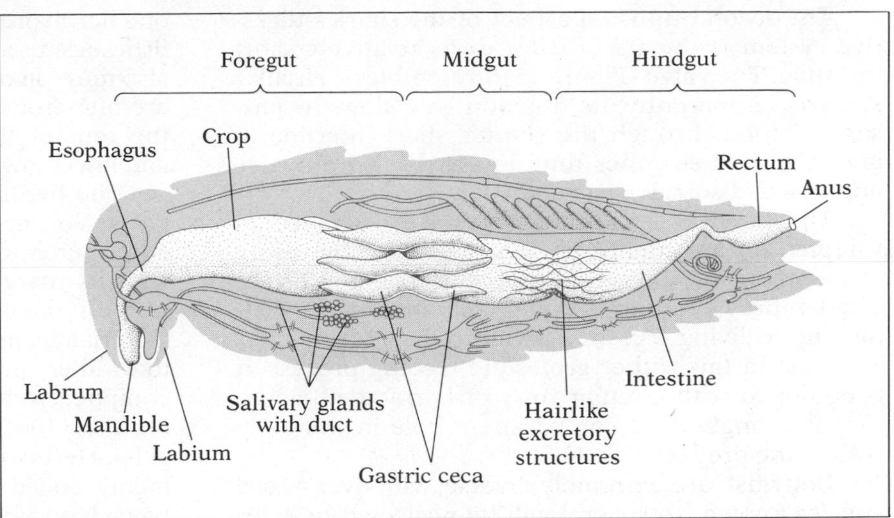

25.7 The grasshopper's digestive system is similar to that seen in many insects. The three major regions are the fore-, mid-, and hindgut. The foregut consists of the mouth, pharynx, esophagus (feeding and swallowing structures), and crop (storage). The last part of the foregut contains an abrasive lining for continued pulverizing of food. At the union of fore- and midgut, twelve finger-like gastric ceca secrete digestive enzymes into the midgut. The midgut is responsible for digestion and absorption. The hindgut receives the undigested residue and removes most of the water as the feces are produced. Wastes are temporarily held in the rectum and pass out through the anus.

Foraging and Digestive Structures in Fish, Amphibians, Reptiles, and Birds

The mouthparts of vertebrates, just as do those of insects, provide us with general clues as to what they eat. For example, if you find a large skull with all the teeth conical, you've probably found a fish eater. If the skull is small, the animal probably ate insects.

Most vertebrates are either plant-eating herbivores or predatory carnivores. Of course, there are the inevitable exceptions—for example, there are some vertebrate filter-feeders. But let's start with the aquatic vertebrates, including the sharks, the jawless fishes, and the bony fishes.

Fish

Sharks are not entirely carnivorous because they swallow and digest practically anything. One of your authors, while on an expedition, helped dissect a 12 ft tiger shark and found a polaroid negative that had been thrown overboard 6 days earlier and 200 miles away! It is disconcerting to find a picture of one's self in a shark's stomach.

However, sharks are basically carnivores, which means they eat animals. And we are animals, so we shouldn't be too surprised at the volumes of fact and fiction that have been accumulated about these relentless predators (probably more fiction than fact, but the fact is interesting enough). Two aspects of the shark's food-getting and digestive apparatus are especially noteworthy. To begin with, the teeth occur in rows, the largest of which are around the periphery of the mouth (Figure 25.8). As the teeth grow, the rows migrate outward so that the outermost ones fall out (and sink to the ocean floor to provide us with necklaces). They are continuously produced from the dermis of the mouth, which explains their increasing length. The system provides the animal with replacements for lost or broken teeth. The jaw itself moves only in one plane on a hinge joint. The sideways tearing action typical of shark attacks is primarily the result of its powerful thrashing body movements. Food is swallowed in great chunks.

There are exceptions to this arrangement, however. The sand shark and the rays have blocklike "pavement" teeth, an arrangement that is useful for crushing the shelled mollusks and crustaceans that they pick up from the bottom. Their teeth are not dangerous to humans and they avoid us when they can, thus the vengeance with which they are attacked by pier fishermen who act as if they are saving humanity from something terrible is not merited.

25.8 The fearsome mouth of the shark is lined with rows of teeth. The outer rows are the oldest and are often broken or lost as the shark feeds. As this happens, new teeth are rotated forward to replace them.

The second unusual aspect of the shark's digestive system is the *spiral valve* in its relatively short intestine. The valve (Figure 25.9) resembles a circular stairway. Apparently, its function is to slow the passage of food through the shark's short intestine to give digestive enzymes time to act. It also provides additional absorptive surface.

The lamprey and hagfish, as we discussed in Chapter 23, are totally jawless. They are the living remnants of an entire class of jawless fossil fishes. The lamprey's mouth is completely modified for attaching to living fish and draining the prey's fluids. To assist in this rather grotesque feeding process, it is equipped with circular rows of horny teeth and a rasplike tongue that can scrape a hole in the body wall of the prey (Figure 23.5).

Bony fish are extremely diverse, with over 40,000 species known, so several will be neglected here. In fact, we will actually only look at one carnivore and one herbivore. Like the sharks, the jaws of the bony fishes are used for biting or sucking only, since they also move in one plane (they cannot grind). The teeth are numerous, and in many species they grow from the roof of the mouth as well as from the upper and lower jaws.

25.9 The digestive system in sharks is abrupt and short. The resulting loss of absorbing surface is compensated by a special feature in the intestine called the *spiral valve.* Like the typhlosole in the earthworm, the spiral valve in sharks greatly increases its ability to absorb digested food in its brief transit through the system. Spiral valves are found in some primitive bony fish.

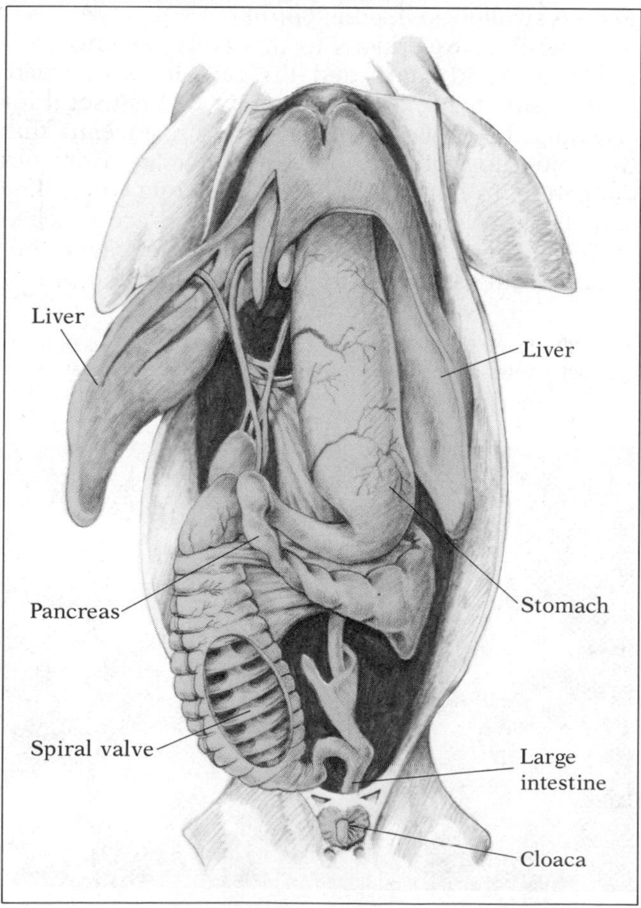

The herbivorous fishes have interesting mouthparts. You need not fear being gummed by an old carp, perch, or minnow, since they lack that ridge with its marginal teeth. Instead, they make use of teeth on the floor of the mouth to grind bits of plant material against leathery patches on the roof. Fishes that filter minute particles from the water are equipped with highly branched gill rakers for straining their food from the water that passes over the gills. Herbivorous fishes usually have very long, highly coiled intestines. The extended intestine is typical in animals with the problem of extracting nourishment from cellulose-enclosed plant cells. They need time to work on it.

Most fish, however, are predaceous. The digestive tracts of carnivorous fishes are short, since their food is more easily digested, but the absorptive surfaces are increased by the presence of blind pouches (*ceca*), which outpocket from the intestine. Their teeth are sharp and numerous. Some have highly specialized mouthparts. The Mediterranean parrot fishes, for example, have teeth fused into a beak that is excellent for feeding on corals.

Amphibians, Reptiles, and Birds

The amphibians have a digestive system similar to that of bony fishes, but with some important differences. For example, fish have rather immobile tongues, but the tongue in amphibians is highly movable. In the frog, in fact, the tongue is used to capture food. This feat is possible because the tongue is attached to the front of the mouth, permitting the sticky organ to be flicked far out and with considerable speed and accuracy. The prey is then scraped off against a peculiar patch of teeth on the roof of the mouth and swallowed whole—since amphibians can't chew.

The reptiles are a rather diverse group and boast a wide variety of techniques in detecting and catching prey. For example, some reptiles have many teeth and some show a trend toward fewer numbers but greater specialization. Crocodiles and alligators are unusual in that their teeth are seated in sockets in the bone, much like our own. Venomous snakes have developed grooved or hollow *fangs* with intricate injecting structures (Figure 25.10). All snakes, by the way, are carnivorous and, as we discussed in Chapter 24, have undergone a unique modification of the jaw for swallowing whole prey.

Snakes and lizards find food by the use of well-developed eyes and a specalized *olfactory* (odor sensing) *organ* at the roof of the mouth. This structure, called *Jacobson's organ,* detects molecules (odors) in the air which are brought to it by the moist, flicking tongue. So the darting reptile's tongue is actually its effort to sample the air and bring the molecules to its olfactory sensor.

Turtles don't have teeth. Instead, the jaws are edged with a sharp-edged horny ridge which is efficient for both the plant food of the box turtle and the carnivorous menu of the snapping turtle.

Bird's teeth are scarce as hens' teeth since all birds have bills rather than teeth. The bill or beak is a bony structure with a covering of keratin, a tough horny protein found in fingernails and hair. While the basic structure of the bill is similar in all birds, there is considerable variation in its shape and size. Each type strongly reflects the feeding habits of the bird (Figure 25.11).

In some instances the feeding habits are revealed in the feet as well. Raptorial birds, or birds of prey, typically have long, curved talons for grasping prey. Ground foraging species such as grouse and pheasants have heavy, strong feet for scratching into the soil.

In addition to bills and feet, some species have very specialized tongues. Woodpeckers have long barbed tongues for snagging insects, while sap-

25.10 Snakes do not chew their food, but simply swallow it whole. Their digestive system is specialized to deal with such food items. The mouth is extremely flexible because of the arrangement of bones in the head and jaw. The lower jaw is loosely attached to the quadrate bone. Even the bones of the palate move easily. In addition, the jaw can separate in the front and each side can move independently, helping to work the prey into the mouth. The esophagus and stomach can stretch considerably as can the body wall. There is no sternum, so the ribs can be moved freely as the prey passes through the esophagus.

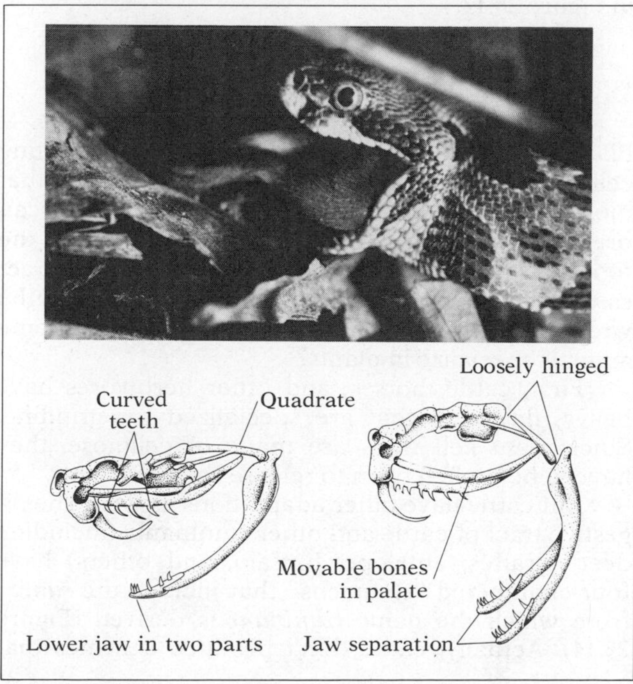

25.11 Although birds show little variation in many traits (for example, they all have feathers), there is considerable diversity in their feeding structures. Note the variations in their beaks, feet, and tongues. In many cases, the foot structure is only indirectly related to feeding, but in the birds of prey and ground foraging species, the foot structure is directly involved. Several species have highly specialized tongues for capturing insects and collecting fluids.

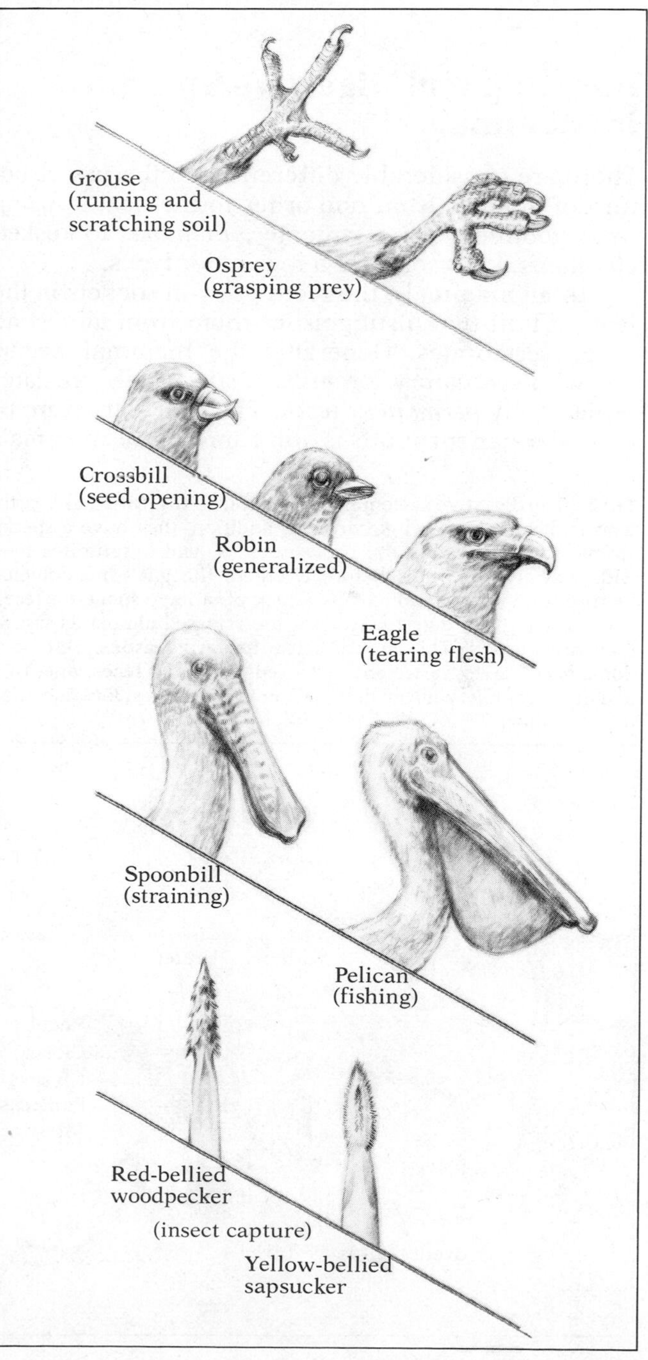

suckers have hairlike surfaces for absorbing sap. Hummingbirds have extremely long forked tongues that curl up to form a drinking trough for sucking up nectar.

The digestive system of herbivorous birds is somewhat unusual in that it includes a large *crop* for storing and moistening food, followed by a two-part stomach. The first portion—the very glandular *proventriculus*—secretes gastric juices. Food then passes into a very muscular *gizzard,* where it is pulverized with the aid of abrasive sand grains or small rocks (Figure 25.12).

Foraging and Digestive Structures in Mammals

There are considerable differences in the jaw structure of animals from one order to the next, ranging from toothless spiny anteater, *Echidna,* to tusked elephants. Let's consider a few major types.

In all mammals, the teeth grow in sockets in the jaws, a trait that distinguishes them from almost all other vertebrates. Generally, the mammal begins life with *temporary,* or *milk, teeth,* which are later replaced by *permanent teeth.* The basic structure of each permanent tooth is the same in all mammals (Figure 25.13). We don't know why many mammals have two sets of teeth or why they don't have more.

25.12 The digestive system in herbivorous birds. Birds, like earthworms, have *crops* and *gizzards.* In addition, they have a special stomach region called the *proventriculus,* which saturates food with digestive enzymes before it enters the gut. An additional feature (one related to flight) is a lack of storage space for feces. Both urinary and digestive wastes are released almost as fast as they are formed. This makes sense for two reasons. One, they don't need the extra weight of stored urine and feces, and, two, leaving droppings while in flight will not give the nest location away to a predator.

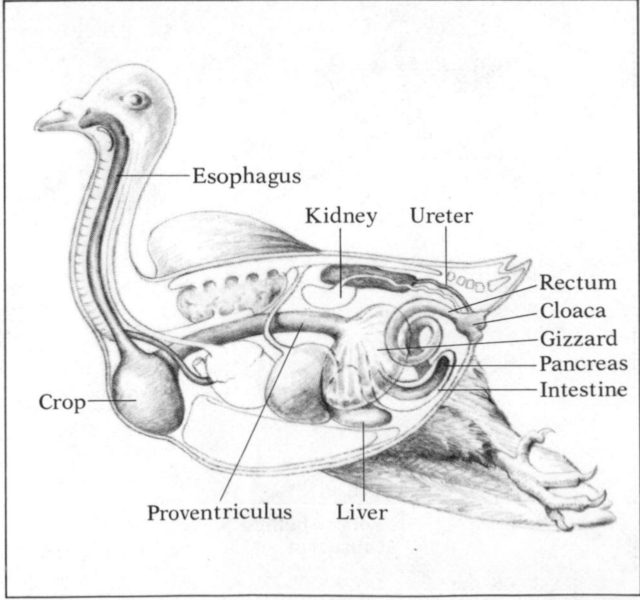

There are four types of teeth in mammals, each theoretically having its own specialized function. The *incisors* are chisel-shaped for cutting and are highly developed in the rodents. The second group—the *canines*—are used for tearing food and for defense. They are, of course, very prominent in carnivores and completely absent in some of the herbivores. Imagine a horse with fangs! The last two groups—the *premolars* and *molars*—are specialized for grinding. They are large and flat in the herbivores, and are used to break down the cellulose walls and fibers of plants. But they may have other uses. Dogs have large, sharp-edged premolars specialized for shearing bones.

The dentition in humans is actually rather generalized in structure, lacking the prominent specialized features of other mammals. Some zoologists regard human dentition as "primitive" because of the apparent lack of specialization. Their lack of specialization, of course, enables us to utilize a variety of foods. In any case, our teeth are specialized to a degree. For example, we have incisors and molars. Perhaps another human "specialization" is our canine teeth, which are much smaller than those of our primate ancestors. Human canine teeth still have very long roots, suitable for anchoring a much larger fang. Why have our fangs been reduced to *eye teeth?* Apparently, because we don't bite to kill prey now. We are, of course, adapted to a rather varied (general or unspecialized) diet of plant and animal food. Take a look now at the various dentition types shown in Figure 25.13.

Digestion in the Herbivores—Ruminants

Plant eaters have the formidable task of digesting cellulose. One problem faced by mammals is that most of them cannot produce enzymes that can break down cellulose. Any cellulose taken in with the food normally passes through the mammalian tract essentially unchanged. But can 90 million cattle be wrong? If cellulose is so hard to handle, why do some animals specialize in plants?

First, cattle, horses, and other herbivores have heavy, flat teeth that are specialized for grinding. Since plant cell walls are made of cellulose, they have to be broken open to release their nutrients.

But cattle have other adaptations as well. The digestive tract of cattle and other ruminants (including deer, giraffes, antelope, buffalo, and others) have four-chambered "stomachs" that include the *rumen* from which the name *ruminant* is derived (Figure 25.14). Actually, cattle don't produce enzymes that

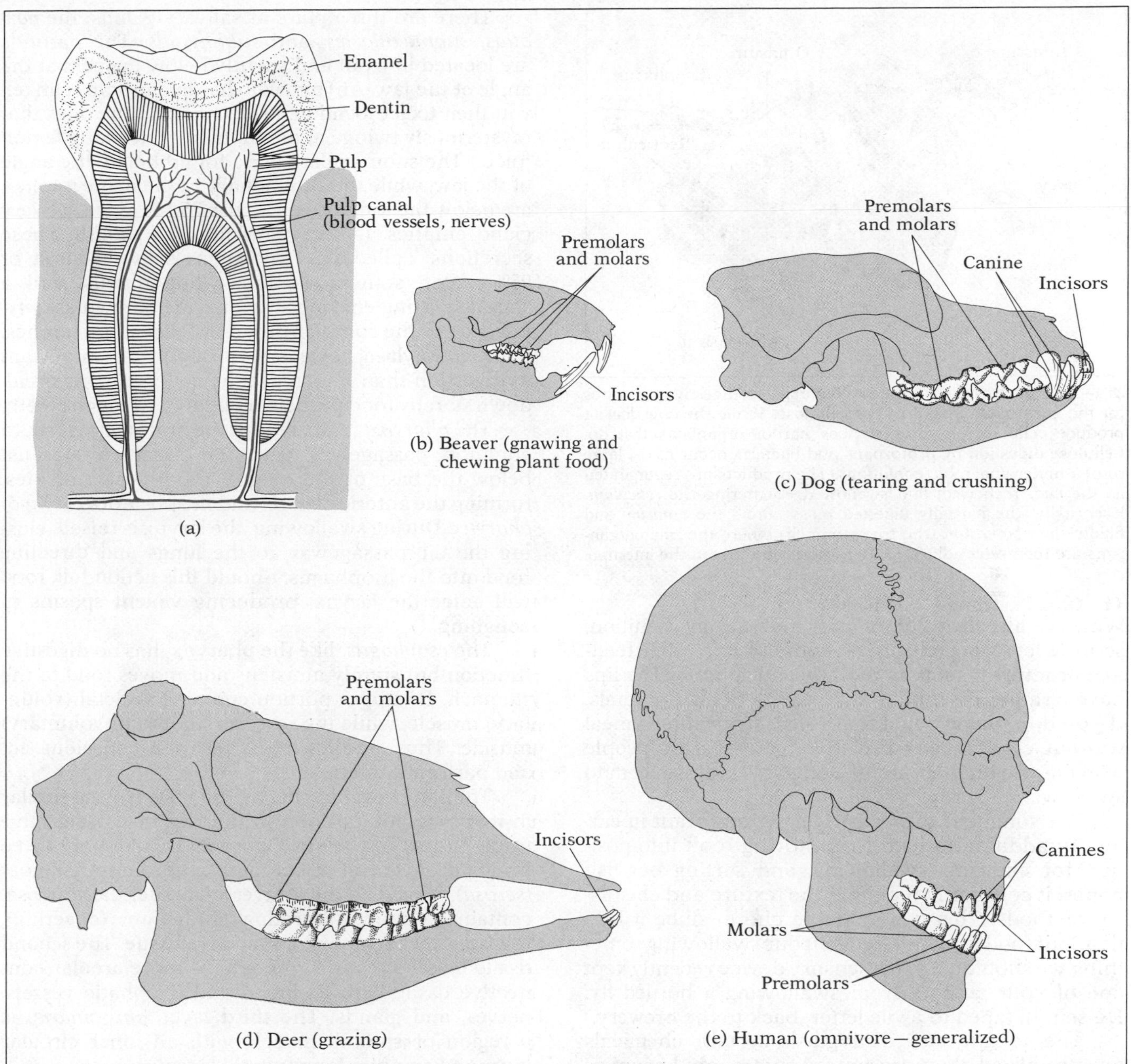

25.13 (a) The basic tooth structure is the same in most mammals (a few lack enamel). A hardened layer of *enamel* surrounds a softer *dentine* region. Within the dentine is the *pulp* cavity which is penetrated by blood vessels and nerve endings. But the arrangements and individual shapes of teeth vary considerably according to the feeding habits of the mammal. Note the differences in the gnawing and grinding teeth of a rodent (b), the tearing and crushing teeth of a canine (c), and the grinding and nipping teeth of a ruminant (d). Note the absence of upper incisors. The gum has a horny covering that meshes with the lower incisors for nipping foliage. (e) Human dentition is perhaps the most uniform, with little specialization.

can digest cellulose. Instead, they harbor certain protozoans and bacteria in their gastrointestinal tract. These organisms *can* break down cellulose through a fermentation process which forms products that can be digested and absorbed in the usual manner of mammals.

The Digestive System of Humans

The human digestive system (Figure 25.15) is a rather good representative of the mammalian system in many respects. So let's follow the path of food through the human tract, right to the very end.

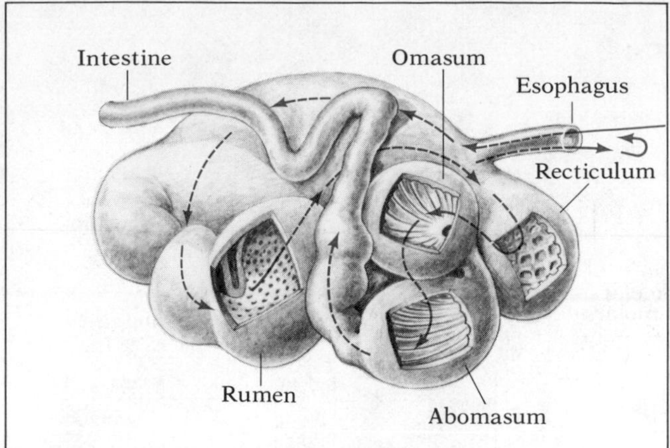

25.14 Ruminants such as the cow have special digestive structures for the enzymatic breakdown of cellulose. While the cow doesn't produce cellulase enzymes, it does harbor organisms that do. Cellulose digestion by protozoans and bacteria occur in the large rumen under anaerobic conditions. The products are regurgitated as the *cud,* rechewed, and swallowed, reentering the *reticulum.* Eventually, the partially digested mass enters the *omasum* and finally the *abomasum* (the true stomach) where the microorganisms are themselves digested. Digestion continues in the intestine.

The Oral Cavity and Esophagus

We have already taken a brief look at our dentition, so now let's concentrate on some of our other feeding structures, such as the lips and tongue. The lips have a rather essential role in eating in all mammals. If you don't know what that role is, try eating a meal without closing your lips. Even those crass people who chew with their mouths open must close them to swallow.

The tongue is also surprisingly important in eating. In addition to its role in moving food into position for chewing, swallowing, and sorting out fish bones, it constantly monitors the texture and chemistry of foods. This is a valuable chore, since it tells us about food in time to prevent our swallowing something we shouldn't. (This sensory device recently kept one of your authors from swallowing a bottled fly. He sent it, taped to a vile letter, back to the brewery.)

The tongue can distinguish certain chemicals by specialized chemoreceptors commonly known as *taste buds.* These receptors apparently have the ability to distinguish four basic tastes: salt, sour, sweet, and bitter. Each type of receptor is concentrated on specific parts of the tongue (Figure 25.16). No one can explain taste differences, since the receptor cells for each taste are morphologically identical. Of course, olfaction or smelling complicates the issue. Actually, a lot of what we think we taste, is really what we smell. You can prove this to yourself the next time you have a really bad cold. Your food will be rather tasteless. In addition to informing us of the flavor of food, by the way, the stimulation of taste buds also enhances the flow of *saliva.*

There are three pairs of salivary glands, the *parotids, submaxillaries,* and *sublinguals.* The parotids are located in front and slightly below the ears at the angle of the jaw. Anyone who has had mumps can tell you their exact location. (They are also the ones that mysteriously twinge at the thought of drinking lemon juice.) The submaxillaries are found below the angle of the jaw, while the sublinguals, as the name implies, are below the tongue (see Figure 25.15). Each salivary gland empties its secretions into the mouth. These secretions, collectively known as saliva, consist of 95% water, some ions, lubricating mucus, and a starch-splitting enzyme. The enzyme, known as *amylase,* begins the complex process of digesting starches. Salivary amylase, however, is probably less important in digestion than it is in oral hygiene. It helps break down starchy food particles caught between the teeth.

The *pharynx,* in the rear of the oral cavity, forms a common passageway with the nasal cavity. Just below the base of the tongue, the pharynx divides, forming the anterior *larynx* and the posterior *laryngopharynx.* During swallowing, the larynx is raised, closing the air passageway to the lungs and directing food into the esophagus. Should this action fail, food will enter the larynx, producing violent spasms of coughing.

The *esophagus,* like the pharynx, has no digestive function but simply moistens and moves food to the stomach. Its upper portion contains skeletal (voluntary) muscle, while the rest has smooth (involuntary) muscle. Thus, swallowing is partly a conscious act and partly automatic.

The lower esophagus, by the way, is very similar in its tissue composition to the rest of the digestive tract. As you can see in Figure 25.17, there are three principal layers of tissue and a fibrous outer coat (*serosa*). The innermost layer of tissue, the mucosa, contains a mucus-secreting epithelium (covering), followed by a region of connective tissue. The second tissue layer (*submucosa*) contains more areolar connective tissue with its blood and lymphatic vessels, nerves, and glands. The third layer (*muscularis*) is a region of smooth muscle with an inner circular layer and an outer longitudinal layer.

The esophagus serves two functions. It secretes mucus into the *lumen,* or cavity, of the digestive tract, and it moves food along by the organized contraction of the two smooth muscle groups. This action is peristaltic, similar to the movement in the gut. Food being swallowed is pressed into a *bolus* (clump) by the tongue and pharynx. As the bolus approaches the involuntary muscles of the esophagus, the circular layer just ahead of it relaxes as the muscles behind the bolus contract, pushing the food along. The muscles are coordinated by nerve endings that are activated by the presence of food. The autonomic

25.15 The digestive system in humans is not unusual, but is more or less representative of that of many vertebrates. It is the old tube within a tube system. The basic structure is similar from one end to the other with any differences mainly restricted to the linings. The *pharynx* and *esophagus* are primarily pasages with secretions that are limited to lubricating mucus. The *stomach* is chiefly secretory, producing hydrochloric acid and proteases. The *small intestine* is both secretory and absorptive, with villi and microvilli to greatly increase its absorbing surface. It also receives the digestive secretions of the *liver* and *pancreas.* The *colon,* or *large intestine,* is chiefly an organ of absorption, removing water from the digestive residue and compacting and storing wastes. It also provides a suitable environment for enormous populations of bacteria which synthesize products useful to their host.

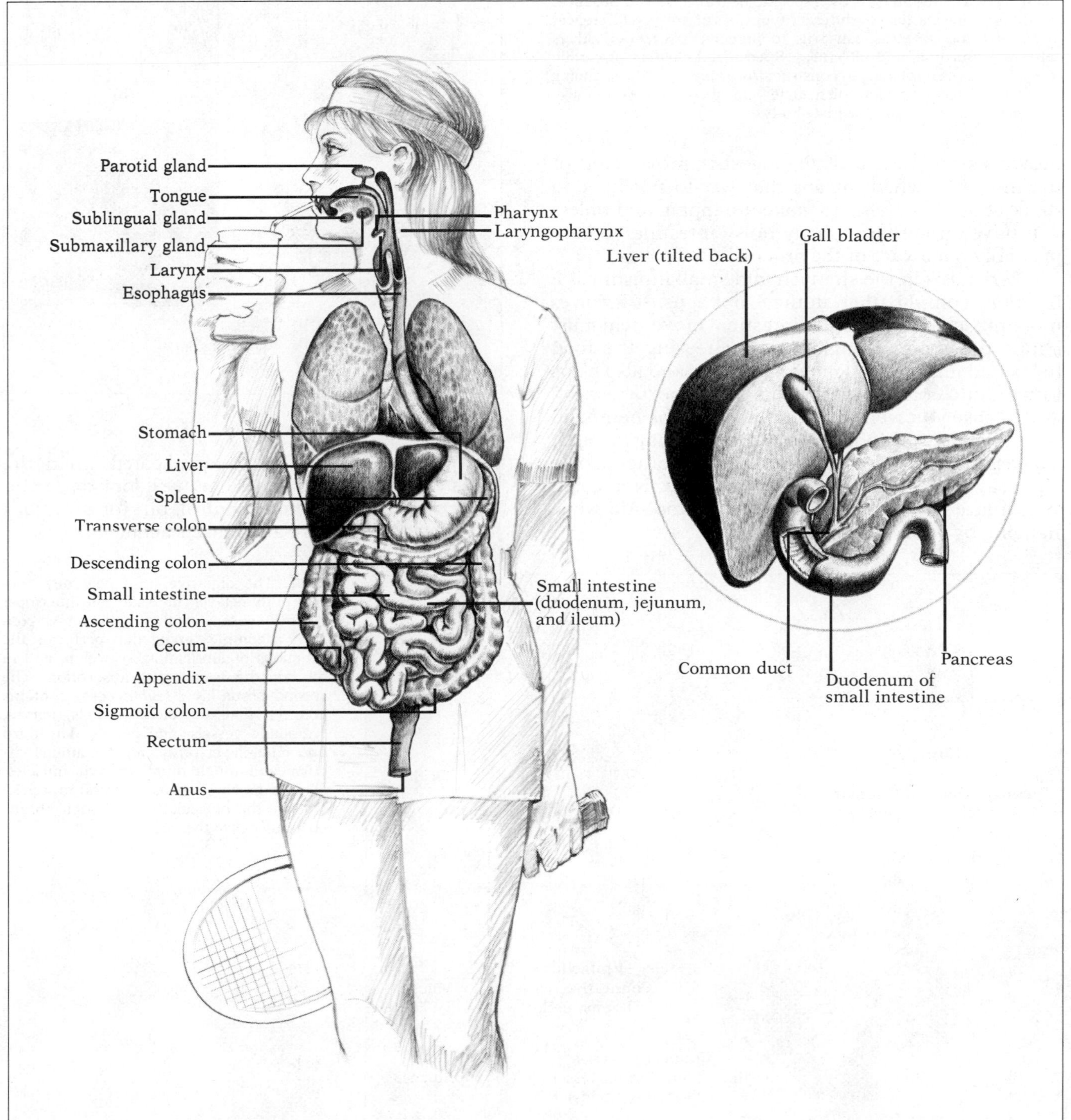

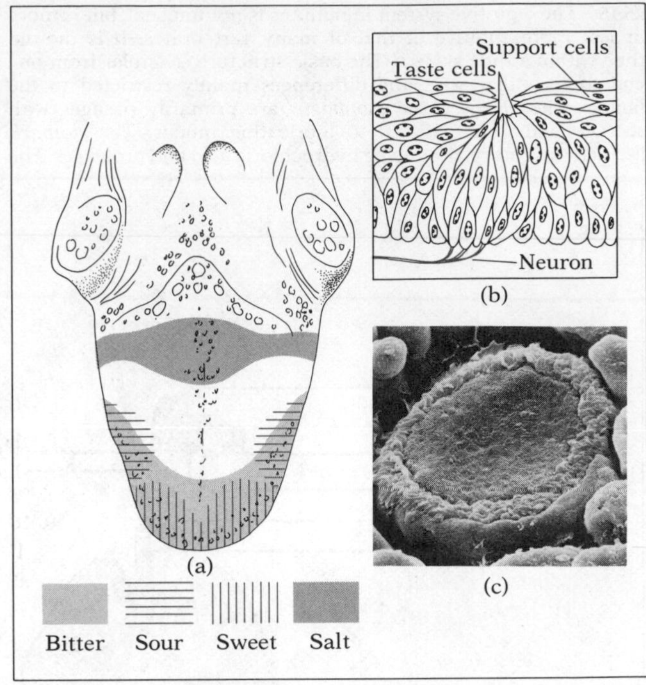

25.16 (a) While the tongue itself doesn't carry on digestion, it is an important organ for sensing the taste and texture of food to be swallowed. In addition, it is important in directing food to the teeth for chewing and to the pharynx for swallowing. (b) Taste buds do not contain receptors that are specialized to detect a certain taste but, instead, a single receptor responds to three or more tastes and is simply more responsive to one than the others. Each taste, we find, has its own specific neural code and produces a specific firing pattern in different types of receptors. Differences in these firing patterns cause us to perceive sweetness, others saltiness, sourness, and bitterness. Specialized clusters of certain receptors have enabled us to construct *taste maps.* (c) The scanning electron microscope view of a taste bud shows the columnlike projections that contain the taste buds.

nervous system controls the digestive movements of the intestine, which means that we do not have to think about peristalsis to make it happen, and unless you have an embarrassingly noisy intestine you are probably not aware of the process.

Peristalsis in the stomach and small intestine is a bit more complex than in the esophagus. It involves more muscle activity, since intestinal movement must churn the food. Wringing and squeezing the food helps it to mix with enzymes and water and also helps bring it into contact with the absorbing surfaces.

In case you were wondering, gravity is helpful in the movement of food through the esophagus, but it isn't essential. Otherwise, astronauts and other people who eat upside down wouldn't survive. Nor could you amaze your friends by chugging Kool-Aid while hanging by your knees.

The Stomach

Everything becomes more complicated when the bolus enters the stomach. Later, we'll look at chemical activity and hormonal control, but for now, let's just consider structure and muscle action.

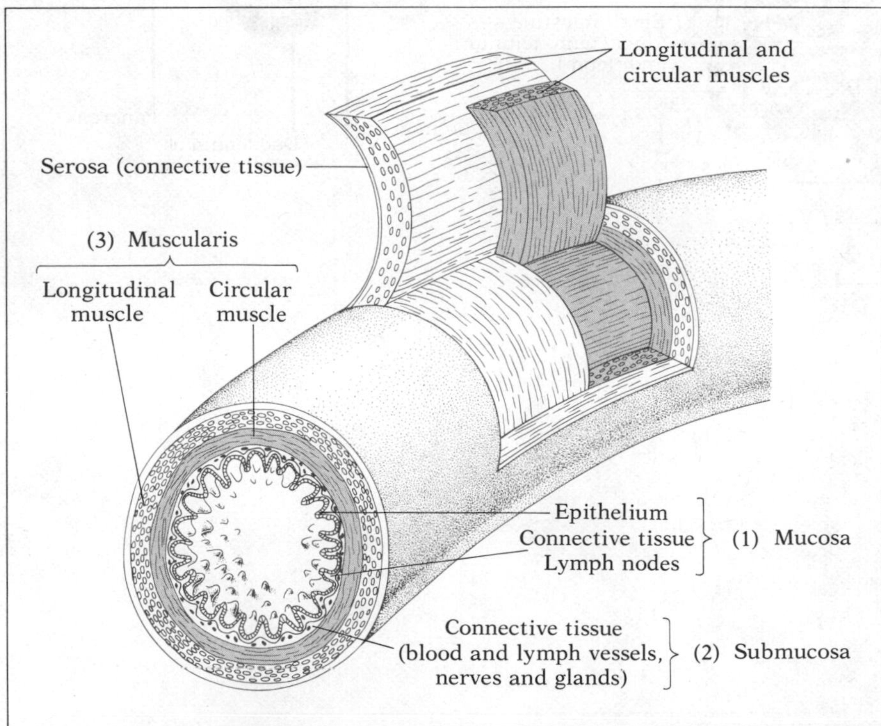

25.17 The digestive tract generally consists of three layers of tissue. The innermost layer (*mucosa*) surrounding the lumen contains a complex lining that performs the functions of lubrication, secretion, and in all but the esophagus, absorption. The second tissue layer (*submucosa*) contains connective tissue, blood vessels, nerves, lymphatic vessels, and glands. The third and outer layer (*muscularis*) contains circular and longitudinal smooth muscles. (The stomach also has diagonal muscles.) Outside the muscularis is a tough fibrous covering called the *serosa*.

Basically, the *stomach* operates in a similar manner in humans and other animals in that it temporarily stores food and then helps to digest it. (When empty, the stomach lining becomes highly folded.) Within reason, the stomach can accommodate a considerable amount of food without causing much pressure against its walls. This is because smooth muscle stretches easily. (Any resistance can cause cramps.) The average capacity is about 1 liter, although with repeated stretching, one's stomach can easily surpass this.

The stomach is closed off at either end by two muscular sphincters (Figure 25.18). These circular groups of muscles lock food in and permit the stomach to churn the food without forcing it back into the esophagus or into the intestine before it is ready. You may have noticed that the *upper* (*cardiac*) *sphincter* fails sometimes when the stomach is overfull or unbalanced somehow, and allows some acid juices to enter the esophagus, creating the familiar "heartburn." Unfortunately, the production of those acids is also related to emotional states.

The muscle layers of the stomach are more complex than those of the esophagus. Here, we encounter a third oblique layer of muscle just inside the circular layer. This layer, along with the circular and longitudinal layers, permits a twisting action that accompanies the usual wringing (by the circular muscles) and shortening (by the longitudinal muscles). Also, peristalsis in the stomach is nondirectional, which means that food is moved back and forth in an amazing sequence of contortions until the *lower* (*pyloric*) *sphincter* relaxes a bit, admitting a small amount of food to escape into the small intestine.

The stomach is lined with a *mucosa* (mucous-secreting tissue) that contains many long tubular glands which secrete the gastric juices. The glands may contain *chief cells,* which secrete the protein *pepsinogen,* and parietal cells, which secrete *hydrochloric acid* (HCl). The acid is required to activate pepsinogen, forming the digestive enzyme *pepsin.* Other glands secrete water, mucus, the enzyme *rennin* in young mammals (which helps to digest milk) and small quantities of *gastric lipase,* a fat-splitting enzyme. The presence of HCl in the stomach produces a very low pH of 1.6–2.4. This environment is hostile to almost everything and kills millions of potentially troublesome microorganisms that enter with food. The acidity of the stomach and the potency of its enzymes produces an environment that could (and sometimes does) endanger the lining itself. One reason we don't digest our stomachs is that a layer of *mucin,* an insoluble mucoprotein, forms a coating over the stomach lining.

There is another aspect of stomach activity that must be described. No discussion of the stomach is

25.18 The human stomach is a J-shaped pouch with a sphincter at either end. Its muscular layers run in three directions and can produce powerful writhing and wringing actions. The lining is complex, bearing cells that secrete water, mucus, acids, and enzymes. The primary enzyme is *pepsin,* which is activated by *hydrochloric acid* and performs well at a very low pH. The deadly acid environment of the stomach destroys potentially dangerous organisms we ingest with our food. The detailed inset reveals the cellular structure of the lining. Mucous-secreting cells abound at the tips of long tissue columns. At the deep crevices between these, *pepsinogen*-secreting *chief cells* and acid-secreting *parietal cells* release their products.

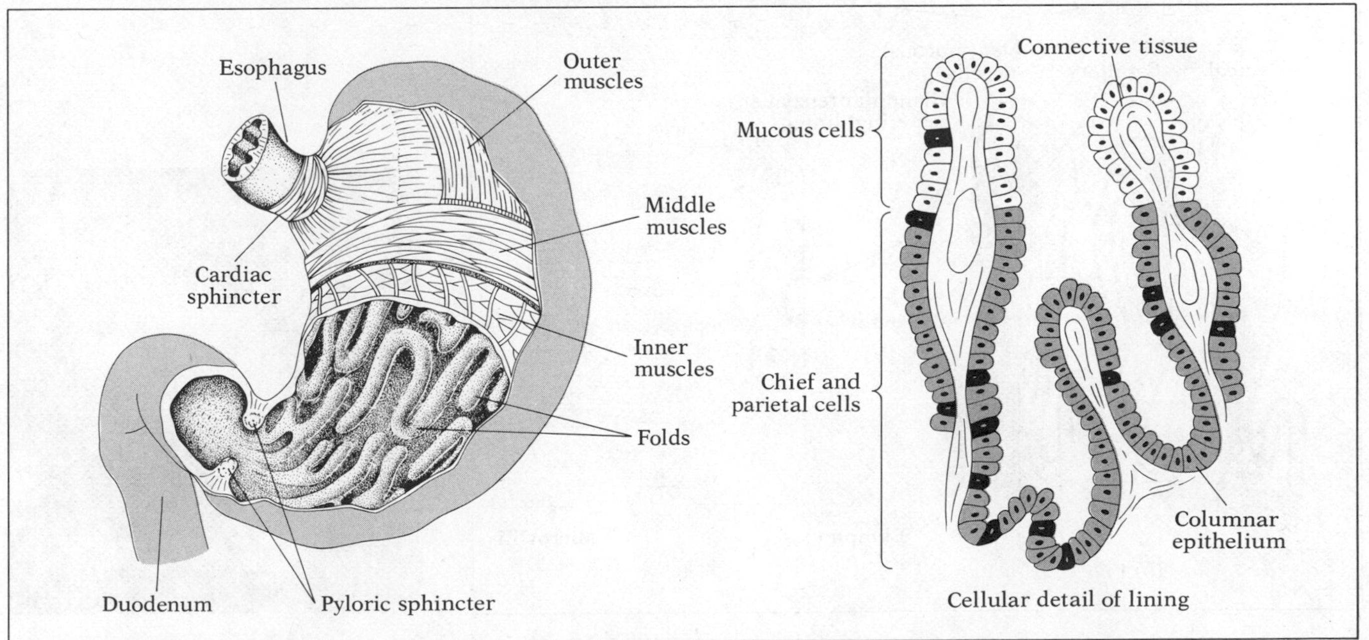

complete without it, and somebody is bound to ask. What causes you to vomit? Believe it or not, there is a "vomiting center" in the brain. It can be stimulated by a variety of influences (including stark fear). When the center is activated, air intake increases suddenly, the stomach contracts violently, the pyloric sphincter tightens, the cardiac sphincter relaxes, and the rest—well, the rest is history!

Small Intestine

But let's talk about something else. What happens if you *don't* become nauseous? Food, now in a highly liquefied state in the stomach, is released in spurts into the *small intestine*. It functions in the digestion and absorption of food and is about 20 ft long in humans, consisting of three regions: the *duodenum, jejunum,* and *ileum* (Figure 25.15).

The mucosa of the small intestine is uniquely adapted for its main function—the absorption of digested foods. Its inner surface has the appearance of a wrinkled bath towel. Closer examination of the surface reveals that it is made up of a mass of folds, each fold covered with tiny projections called *villi.* Each villus, as you see in Figure 25.19, is, in turn, covered with epithelial cells bristling with fine projections of the cell membrane known as *microvilli.* So what we have here is a highly efficient absorbing membrane. Because of these projections on its projections, the total surface area is enormous. Within each villus is a *capillary bed* and a *lacteal* (a lymphatic vessel). The villi are capable of rather vigorous movement, like millions of wiggling fingers, and help in mixing food and enzymes in the gut and also assist in absorption by creating pressures that tend to concentrate digested food in the many recesses of the lining.

The mucosa of the small intestine is also glandular, secreting enzymes and copious amounts of

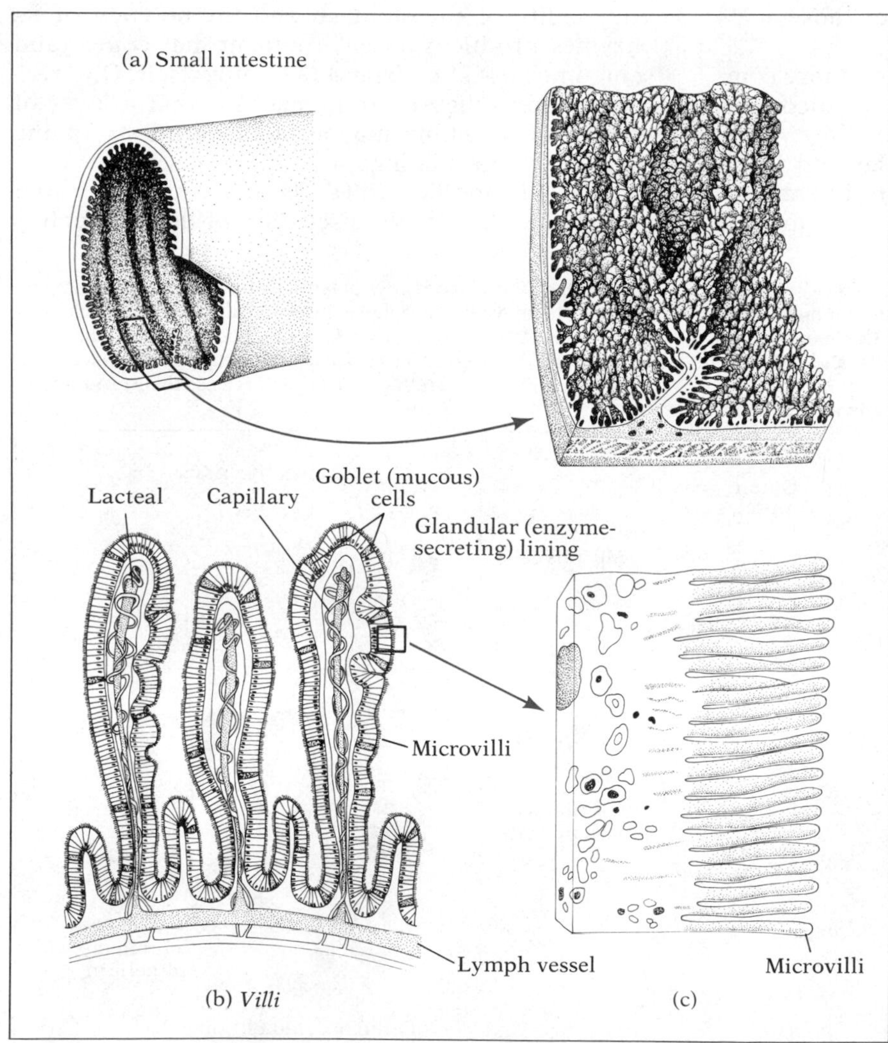

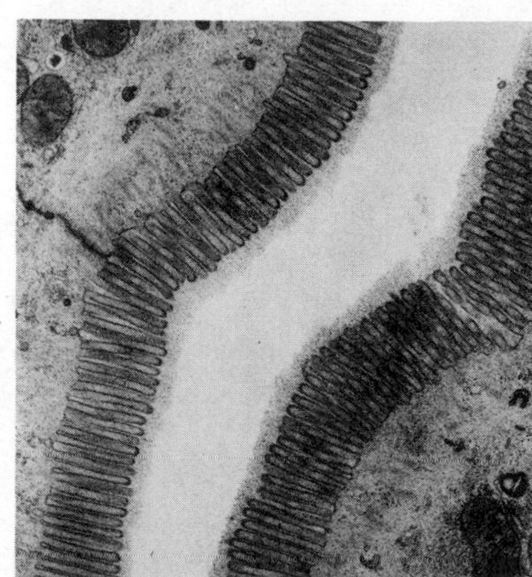

25.19 The inner surface or mucosa of the small intestine (a) is lined with intestinal villi (b). Each villus consists of a projection in the surface epithelium. Its outer cells secrete mucus and enzymes. Within the villus is a capillary network arising from blood vessels in the submucosa. At the center is a lacteal, a blind extension of the lymphatic system. Digested foods are absorbed into these two important structures. The surface cells of the villus have their membranes folded into numerous microvilli as seen in (c) and the electron micrograph. The folding provides an enormous surface. The villi move in an oscillating manner helping to mix the contents of the small intestine.

mucin, which helps to keep the intestine from digesting itself, just as it does the stomach. In addition, mucus is secreted by the epithelial cells and even more is secreted by numerous glands (Brunner's glands) in the small intestine.

Liver

The duodenum of the small intestine receives the secretions of two accessory organs, the liver and the pancreas. The *liver*, an organ of many functions, aids in digestion by secreting a slightly alkaline fat emulsifier known as *bile*. Bile is a complex substance containing salts, pigments, water, and a nucleoprotein. The bile salts are actually steroids, and are important in dissolving ingested fats. The bile pigments (red and green) represent products of hemoglobin destruction, since the liver along with the spleen is the red blood cell graveyard. These breakdown products of hemoglobin become part of the digestive wastes. The red and green, of course, produce brown and that is what gives feces its characteristic color.

The bile is stored in the *gall bladder* after it is produced in the liver. Its release, after a meal, is brought about by a hormone, *cholecystekinin—pancreozymin*—which causes the gall bladder to contract (and stimulates enzyme release by the pancreas). Bile reaches the duodenum via the *bile duct*, which is joined by highly alkaline fluids from the *pancreatic duct* just before it enters the intestine. So the bile carries off metabolic wastes and physically breaks down fats into minute droplets. In this condition, the fats are easier for *lipase* (the fat-digesting enzyme) to handle. If the bile duct becomes obstructed, serious digestive difficulties occur. To begin with, fatty meals cause severe abdominal cramps. If the duct remains obstructed, the affected person's feces become gray or white because of the absence of bile pigments and the presence of undigested fat. At the same time, the urine becomes dark, and the person's skin takes on a yellowish hue, particularly evident in the whites of the eyes. The condition is called *jaundice*, and is produced by a buildup of bile in the bloodstream.

Pancreas

In humans, the *pancreas* is a long glandular organ lying nestled in the first turn of the small intestine (Figure 25.15). It secretes *sodium bicarbonate* (which neutralizes excess stomach acid) and an entire battery of digestive enzymes that are involved in the breakdown of fats, carbohydrates, protein, and nucleic acids. The pancreatic enzymes, therefore, carry out much of the digestive process.

Large Intestine

The small intestine joins the *large intestine* (*colon*) on the right side of the body. Their union is just below the actual beginning of the colon, so a small portion of the colon forms the *cecum* (the pouch mentioned earlier), which has a hollow, fingerlike extension of its own—the *appendix*. At the union of the small and large intestines is a one-way valve (the *ileocecal valve*), which prevents a backflow of the food residues into the small intestine if peristalsis should temporarily create suction. It also slows the passage of food through the small intestine.

The large intestine consists of several parts. These are (in order): *cecum, ascending colon, transverse colon, descending colon,* and *sigmoid* (S-shaped) *colon* (Figure 25.15). The last portions of the digestive tract are the *rectum*, the *anal canal*, and the *anus*—where our story ends.

The primary function of the colon is to absorb water and minerals into the blood, and to prepare the feces to leave the digestive tract. There is very little, if any, digestive activity in the colon. The mucosa of the colon is glandular, but it secretes mucus only. This mucus is used in the production of feces and lubricates the lower portions of the colon.

The colon is a veritable hotbed of bacterial action, but one species, the familiar *Escherichia coli*, is predominant. Any time *E. coli* appears—for example, in water—it is a sure sign of contamination by feces and possibly other feces-borne organisms, such as the hepatitis virus. However, *E. coli* from humans is indistinguishable from *E. coli* from other mammals.

Strange as it may seem, the greater part of fecal material, by dry weight, is composed of dead *E. coli*. Bacterial action produces the rather remarkable odor of feces and the sometimes painful accumulation of gas. The gas, or *flatus*, is commonly the result of the body state rather than the food itself, as many people like to believe. The gases are products of bacterial metabolism and consist commonly of nitrogen, methane, and sometimes hydrogen sulfide.

Actually, people need *E. coli*. Unless a colon has a flourishing bacterial colony, problems can arise. Among these are deficiencies in vitamins, including vitamin K, biotin, and folic acid. It therefore seems that our relationship with our *E. coli* is, in fact, mutualistic. Unfortunately, *E. coli* can cause problems when it strays. These include skin, urinary tract, and even colon infections where the lining is eroded or damaged.

The final region of the digestive system is the rectum. Its structure is basically the same as that of the rest of the colon, except for the presence of folds in its walls. These are the *rectal valves*, which help support the weight of the feces so that gravity doesn't become the enemy of propriety. The anus is the external opening. The movement of wastes through the anal sphincter is usually under voluntary control, a fact for which we can all be grateful.

25.2 The Chemistry of Digestion

Chemical Digestion and Absorption

To understand the chemistry of digestion, it helps to review the structures of the molecules that are digested. You may recall from our previous discussions that proteins, carbohydrates, and fats are often very large, complex molecules composed of repeating molecular subunits. Except for some of the neutral fats, these macromolecules can't be absorbed through the lining of the gut. So in the digestive process, enzymes attack the linkages of the molecular subunits, breaking them down into their components. The bonds are broken through a process called *hydrolysis*. Essentially, water is added enzymatically to the linkages, disrupting them, and causing the molecular subunits to revert to their original form prior to the time when they were synthesized into "food." We discussed this synthetic process in Chapter 2, where we called it *dehydration linkage*.

Briefly, in digestion, carbohydrates are dismantled into simple sugars, fats into fatty acids and glycerol, DNA and RNA into free nucleotides, and proteins into their various amino acids. Figure 25.20 illustrates the general action of the digestive enzymes that aid these processes. The enzymes and their functions are listed in Table 25.1.

Carbohydrate Digestion

Starch digestion begins in the mouth, where *ptyalin* breaks some linkages, producing maltose and some larger fragments. Starch digestion is then temporarily stalled by the acidity of the stomach. It resumes, in full swing, in the small intestine, where *amylase* from the pancreas converts all the starch into maltose. But maltose is still too large to pass into the bloodstream. Finally, maltose is broken down (by *maltase*) into the monosaccharide, glucose, which can be absorbed. Sucrose (table sugar) and lactose (milk sugar), both disaccharides, are also split into monosaccharides by enzymes from the mucosa of the small intestine. Sucrose is composed of one molecule of glucose and one molecule of fructosc, while lactose consists of one molecule of glucose and one molecule of galactose. If sucrose or lactose are injected into the bloodstream, they will be excreted unchanged, since they can be broken down only in the gut. But if glucose, fructose, or galactose are injected, they will be used by the cells, since, after all, they are the common breakdown products of carbohydrate digestion.

The enzyme *lactase*, which breaks down milk sugar, is present in the intestines of all normal human babies and nearly all Caucasian adults. Negro and

Table 25.1
Digestive Enzymes and Their Functions

Tissue source (enzyme or substance)	Digestive function (substrate → product)
Mouth (salivary glands)	
Ptyalin (salivary amylase)	Starch → Polysaccharide fragments, maltose, some glucose
Stomach	
Pepsin	Protein → Peptides
Rennin	Soluble milk protein → Insoluble milk curd
Gastric lipase	Neutral fat → Fatty acids + Glycerol
Hydrochloric acid (HCl)[a]	Kills bacteria, activates pepsin, and aids hydrolytic cleavages
Pancreas	
Trypsin	Polypeptides → Dipeptides
Chymotrypsin	Polypeptides → Dipeptides
Pancreatic lipase	Fats → Fatty acids + Glycerol
Pancreatic amylase	Polysaccharides → Disaccharides
Nucleases	DNA, RNA → Nucleotides
Sodium bicarbonate ($NaHCO_3$)[a]	Neutralizes acid foods from stomach
Liver	
Bile salts[a]	Emulsifies fats
Small intestine	
Dipeptidase	Dipeptides → Amino acids
Maltase	Maltose → Glucose
Lactase	Lactose → Glucose + Galactose
Sucrase	Sucrose → Glucose + Fructose

[a]No enzymatic action.

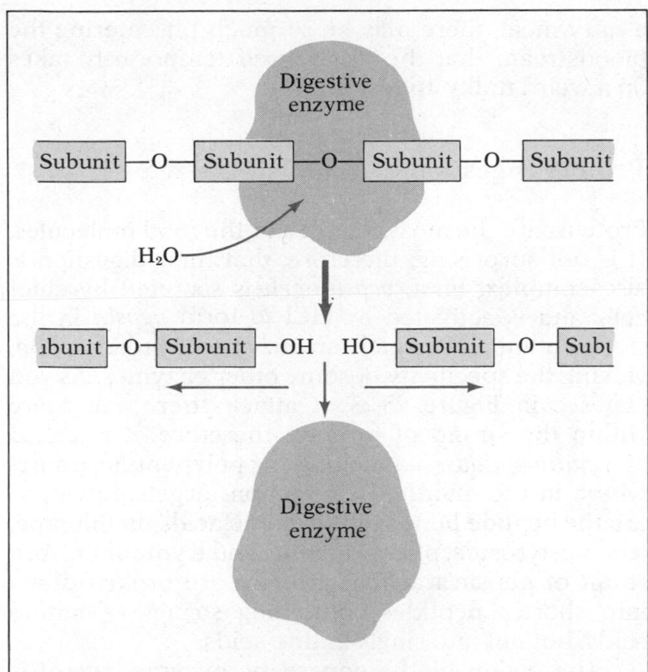

25.20 Digestion, or *hydrolysis*, is the opposite of dehydration synthesis. Hydrolyzing enzymes break the bonds between subunits of food molecules by the addition of water. In this manner, proteins are broken down into amino acids, starches into glucose, fats into fatty acids and glycerol, and nucleic acids into nucleotides. The smaller subunits can then be moved across absorbing membranes and transported in the circulatory and lymphatic systems.

Oriental adults and their older children usually lack intestinal lactase, and milk can cause excessive buildup of gas and also diarrhea. Cheese is more digestible, because its lactose has already been broken down. It may be that Caucasian adults continue to produce lactase by natural selection; milk has long been a major part of their ancestors' diet.

Figure 25.21 illustrates the digestion of the starches amylose and amylopectin. Note that the absorption of monosaccharides, at least the six-carbon molecules, such as glucose, is believed to be primarily by active transport. The absorption seems to be related to sodium transport, since if it stops, so does carbohydrate absorption. At any rate, glucose and fructose are transported into the capillaries of the villi in the small intestine, and from there directly

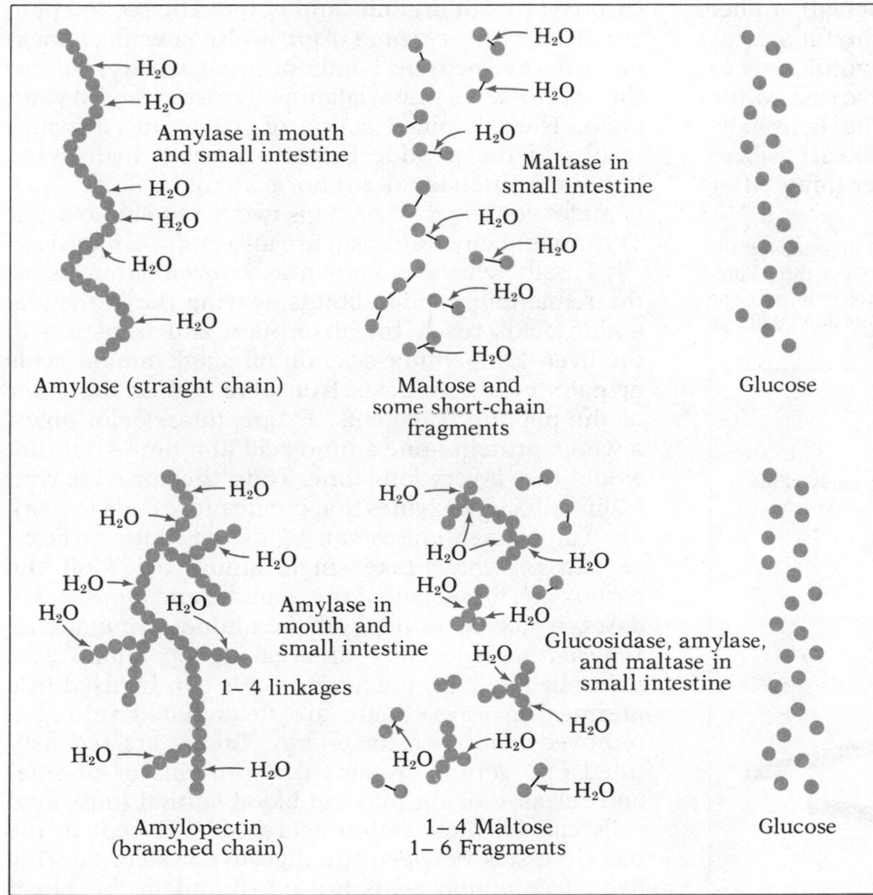

25.21 The chemistry of starch digestion. The starch of food consists of two types, amylose and amylopectin, generally in a 3:1 ratio. In its natural state, it is found in compact granules in the storage cells. Cooking starchy vegetables breaks up the granules, forming a pasty residue that greatly aids the digestive process by exposing more linkages to enzyme action. Amylose is readily digested by *amylase* in a series of steps which finally produce maltose (a double sugar) and some glucose. Maltose digestion occurs in the small intestine through the action of *maltase.* Amylopectin undergoes a more complicated digestive process because of its numerous branches. Amylase action proceeds along a side chain, breaking alternate 1–4 linkages. It cannot, however, break the 1–6 branching linkage. This requires the action of other enzymes. The result, then, of amylase action is a residue of maltose (1–4 linkages) and another residue of short branches (1–4 and 1–6 linkages).

to the liver where they are recombined to produce long chains of glycogen.

Fat Digestion

Fats, for the most part, reach the small intestine with little chemical change. This is a bit surprising, since we know that the stomach secretes *lipase.* Once in the small intestine, bile salts separate fats into tiny droplets. Some of these droplets are further broken down by lipase into fatty acids and glycerol (Figure 25.22), which cross the membrane and then are reconstituted within the villi. Strangely enough, the fat droplets can be used as they are, but they may be broken down in order to cross the intestinal membrane. It's like a couple holding hands, disengaging to pass through a revolving door, and then rejoining on the other side. Other fats (neutral fats—those bearing no ionic charges) can enter the villi intact, through the process of *pinocytosis* (see Chapter 4).

There are other interesting twists to the story of fat absorption. It seems that longer chains of fatty acids (those with more than twelve carbons) enter the lymphatic vessel of the villus (the lacteal), rather than the blood. On the other hand, if the fatty acids contain twelve or fewer carbons, they go directly to the blood capillaries and are quickly carried to the liver. The longer fatty acids follow the lymphatic system to the thoracic duct near the heart, where they enter the bloodstream. And another thing. After a fatty meal, there may be so much fat entering the bloodstream that the blood itself temporarily takes on a weird milky appearance.

25.22 The chemistry of fat digestion. Fats are digested by the enzyme *lipase* in the small intestine. During the process, three water molecules are utilized in separating each fat molecule into one molecule of glycerol and three fatty acids. Some fats don't undergo this process but are taken in whole by pinocytosis.

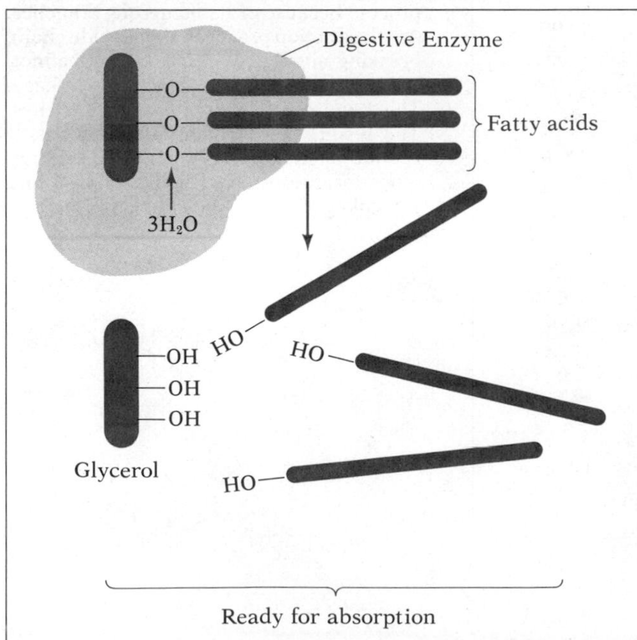

Protein Digestion

Proteins are the most complex of the food molecules. It is not surprising, therefore, that their digestion is also complex. First, *pepsinogen* is secreted by chief cells and is activated by HCl to form *pepsin* in the stomach lining. Pepsin is an enzyme of broad action, lacking the specificity of some other enzymes. As you can see in Figure 25.23, it attacks here and there within the strand of protein; therefore, it is called an *endopeptidase*—attacking a polypeptide somewhere in the middle. Its common targets, however, are the peptide bonds of the amino acids methionine, leucine, tyrosine, phenylalanine, and tryptophan. As a result of pepsin's action, proteins are broken down into shorter peptides containing strings of amino acids, but not into single amino acids.

The action of the pancreatic enzyme *trypsin* is more specific. It hydrolyzes only the bonds on the carboxyl side of arginine and lysine. The second pancreatic enzyme, *chymotrypsin,* is also specific, attacking only the peptide bonds on the carboxyl side of the amino acids phenylalanine, tyrosine, and tryptophan. The combined action of pancreatic enzymes results in the peptide fragments being hydrolyzed into some individual amino acids as well as small peptides consisting of perhaps two to ten amino acids. Trypsin and chymotrypsin are also endopeptidases.

Finally, enzymes known as *exopeptidases* cleave the remaining peptide bonds, leaving the individual amino acids ready for absorption and transport to the liver. Exopeptidases chop off single amino acids or pairs of amino acids from one end or the other of the peptide fragments. Exopeptidases can digest a whole protein—one amino acid at a time—but that would take a very long time. Thus, they operate with fragmentizing enzymes that create more ends to work on. The exopeptidases can be divided into *carboxypeptidases,* which take single amino acids off the carboxy terminal end of the peptide, and *aminopeptidases,* which take them off the amino terminal end. *Dipeptidases* specialize in breaking up amino acid pairs. In the liver, the amino acids can be used in a number of ways. Some are deaminated (nitrogen removed) and used for energy. Others are reconstituted into serum proteins (the proteins of plasma) and released again into the blood so that individual cells can fill their amino acid needs. Except in the blood vessels between the digestive system and the liver, free amino acids are not found in the blood in high concentrations.

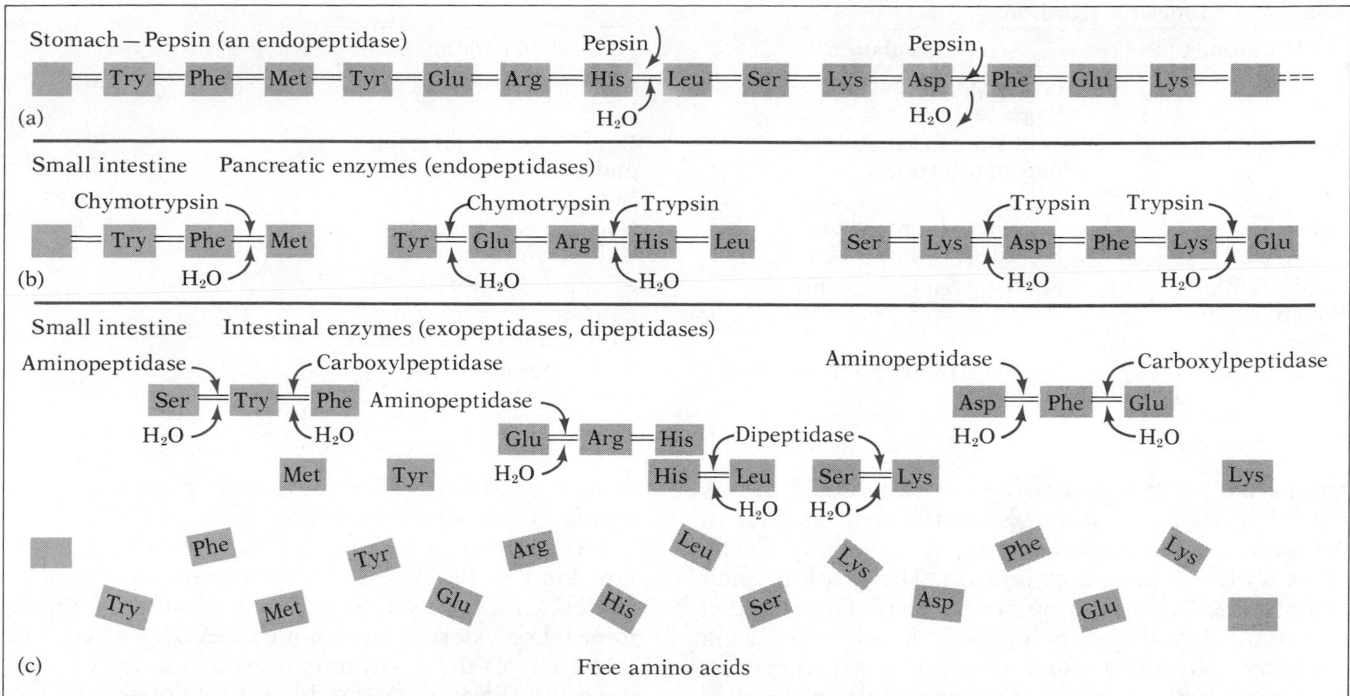

25.23 The chemistry of protein digestion. Protein digestion, because of the complexity of these enormous molecules, is a multi-step process. (a) It begins in the stomach with the action of pepsin. Pepsin (an *endopeptidase*) breaks the peptide bonds (utilizing water) of five specific amino acids, producing fragments of different lengths. (b) Chymotrypsin and trypsin from the pancreas (also endopeptidases) act next in the small intestine, again at specific sites. Some amino acids are released but mainly fragments of different lengths emerge. (c) Finally, the last battery of enzymes, from the intestinal cells (known as *exopeptidases* and *dipeptidases*) go to work. The first breaks off terminal amino acids and the second splits dipeptides. The result is a residue of single amino acids which are then absorbed.

Nucleic Acid Digestion

Almost everything we eat contains at least some nucleic acids. Nucleic acids, however, are large molecules, bonded in complex ways. Certain enzymes produced in the pancreas are capable of breaking such bonds. The highly acid environment of the stomach undoubtedly plays a part also. For example, *purines* (double ring nitrogen base of nucleotides) can be removed from DNA by acid hydrolysis at a pH of about 3, which is considerably less acid than the stomach juices. The enzymes that hydrolyze nucleic acids are called *nucleases,* and they are divided into two classes on the basis of how they attack the large molecules. Some nucleases break off the end of a nucleotide, while others attack bonds within the molecule. For example, *phosphodiesterase* (obtained from the venom of rattlesnakes) is a very potent nuclease that separates phosphates from sugars along the backbone of DNA molecules.

Nucleases break down nucleic acids into small units of either a single base or small chains of three or four bases. We know about the action of these enzymes because of the intense interest in DNA and RNA structure in the past years.

Integration and Control of the Digestive Process

The release of digestive juices, of course, must be very precisely timed. It turns out that such timing is under at least three types of control: mechanical, neural, and hormonal, with some interaction among the three types. For example, even the thought of a chocolate cake can evoke salivation. Saliva flow can also be stimulated by chewing tasteless substances like paraffin. So neural and mechanical stimulation can stimulate saliva flow. And, of course, acids such as lemon juice will also do the trick.

Some investigators report that the saliva itself may vary, depending on what evoked it. Chewing bread, for example, stimulates the rapid flow of a dilute saliva, while meat elicits the release of a dense, mucous saliva. This leaves the possibility of a more specific, recognition-based control system. On the other hand, sheer terror or anger can inhibit salivary secretion. Has your mouth ever gotten so dry you could hardly speak?

We find that, further along in the digestive tract, gastric secretions can be stimulated by our simply sensing the presence of food. A neural message is sent along a specific nerve (called the *vagus nerve*) from the brain to the stomach lining. In some cases,

Table 25.2 Digestive Hormones

Hormone	Stimulation	Function
Gastrin	Protein in stomach Vagus nerve	Stimulates release of gastric juice
Enterogastrone	Acid state in intestine Fats in intestine	Inhibits gastric juice release and slows stomach contractions
Secretin	Acid state in intestine Peptides in intestine	Stimulates release of sodium bicarbonate
Cholecystokinin–Pancreozymin	Food entering small intestine	Stimulates secretion of pancreatic enzymes and contraction of gall bladder
Enterocrinin	Stomach materials in intestine	Stimulates intestinal secretions

the gastric secretions can be overstimulated in this way. For example, a stressful life style can do it. Actually, over-secretion occurs as the body finally relaxes after a period of tension. The result of such massive secretions can be quite severe. For example, the stomach wall can be eroded, thereby producing an *ulcer*. Normally, such erosion is prevented by biological buffers, but, for some people, preventive medicine may be necessary. One treatment involves surgically severing the gastric branch of the vagus nerve. This relieves the problem by stopping the unneeded flow of gastric juices, but doesn't greatly impair digestive activity, since the secretions are normally stimulated in other ways as well.

The presence of food in the stomach mechanically evokes gastric secretions by stimulating sensory neurons in the stomach wall itself. Impulses reach the brain and again the brain sends down the word via the vagus nerve to resume the flow of gastric secretions.

A third mechanism is hormonal and independent of the vagus nerve. When food, particularly concentrated protein, is present in the stomach, the hormone *gastrin* is released from cells of the stomach wall into the bloodstream. Hormones, of course, have very specific target sites. These targets are tissues that are comprised of cells that are specifically sensitive to a hormone; they have specific means of binding to the hormones and react to them in very precise and regulated ways. In this case, the gastric glands are stimulated to release gastric juices.

The gastrin system again illustrates a very important kind of biological control system—the *negative feedback mechanism*. Simply put, a stimulus (in the internal or external environment) evokes a response that changes the environment in such a way as to reduce the stimulus. A simple example of this type of system is the thermostat on your household central heating system. The stimulus in this case is cold, which causes the thermostat to send a message to the furnace, which turns on, which changes the thermostat's environment and removes the stimulus. The system then shuts down until the house becomes cold again. In the case of gastrin, the stimulus is protein in the stomach, which causes receptive cells to produce gastrin, which causes other cells to produce gastric juices, which digest the protein enough to send it on to the small intestine, which removes the stimulus, which shuts the system down. Gastrin, like other natural hormones, is short-lived so that it doesn't hang around causing trouble after its message is no longer pertinent.

There are five major hormones involved in the integration and control of digestion. The example of gastrin illustrates the basic principle of their operation, although there are variations on this theme (Table 25.2).

25.3 Nutrition

What Happens to Nutrients?

The subject of nutrition is becoming increasingly fascinating to the general public as we learn more about it. Some of us are even obsessed with the subject, but at least a certain degree of interest is merited. One reason is that it might be useful personally. Who would have guessed that eggs can cause a vitamin deficiency? Also, we need a way to sort through the constant barrage of commercial advertising from the Madison Avenue crowd. A second reason is that food is a critical commodity, and it may come to play an even larger role in global politics. Now, more than ever, we need to be able to sort fact from fiction in order to use the world's food resources economically.

Unfortunately, the excitement about food doesn't necessarily imply an interest in scientifically established information about nutrition. Many food faddists are almost ignorant of the subject. Some do not even realize that organically grown food is no better for you than food grown with chemical fertilizers; rose hip vitamin C is no different from synthetic vitamin C; large doses of vitamins are not healthful and may be harmful; dietary protein in excess of body needs is simply converted to glucose and urea in the liver—and fat on the hips; and high-protein diets have no healthful value. Actually most Americans eat far more protein than they need, and tests with long-distance runners show that they have more stamina on a high-carbohydrate diet than on a high-protein diet. So, let's have a closer look at diet and see if the scientific findings have any relevance for us.

Carbohydrates

Carbohydrates are common sources of energy. As we discussed in Chapter 6, they are used in respiration, but let's look at other ways they can be used. To begin with, once glucose makes its way into the circulatory system, it is immediately removed and stored as glycogen by the liver. From there, it is meted out according to need. Thus, the blood glucose level remains rather constant, with a temporary 10–20% increase immediately after a high-carbohydrate meal. A second storage place for glycogen is in the muscles where, during glycolysis, the glucose is directly used for energy. Long-distance runners build additional glycogen reserves in muscles by continually depleting their reserves in long training runs, and letting them build back at even higher levels. The rate of glycogen breakdown and the release of glucose into the blood is under the control of a complex feedback system (Figure 25.24).

Glucose can be converted into fat, as people who eat a lot of carbohydrates can tell you. The conversion is accomplished in a respiratory pathway, where acetyl-CoA is converted to a fatty acid which, of course, can then become part of a fat molecule. The ability of the liver and muscles to store glycogen is sharply limited, so any excesses are stored as fat.

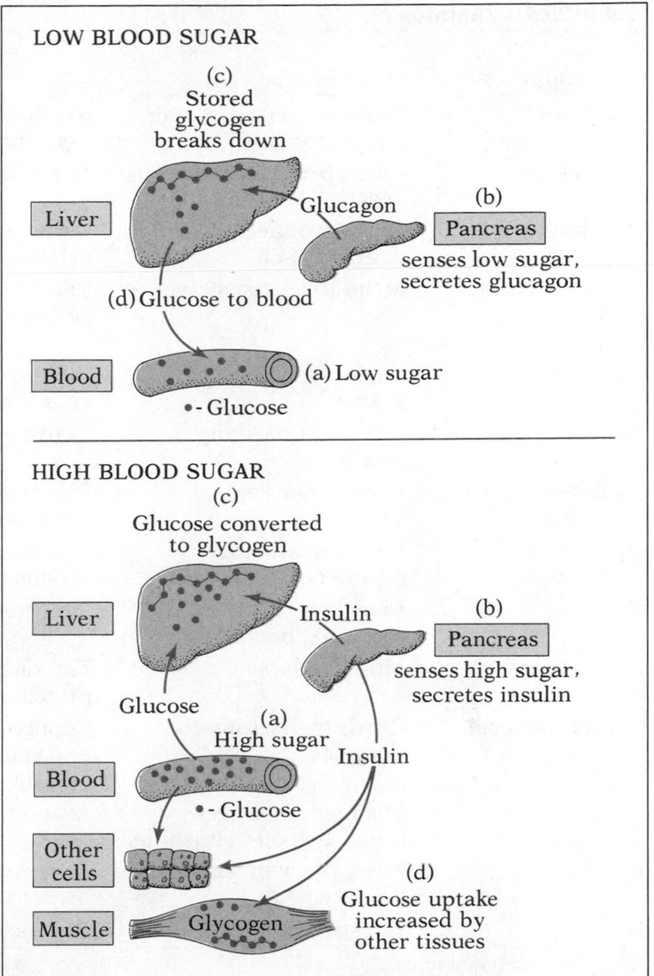

25.24 Control of blood glucose levels. As sugar reserves in the blood fall, the pancreas responds by releasing *glucagon* into the bloodstream. Its target, glycogen-storing cells of the liver, responds by breaking glycogen down into glucose. This elevates the blood sugar level as glucose leaves the liver cells. Alternatively, when sugar levels in the blood are high (after a big meal) the pancreas releases *insulin* into the blood. Insulin stimulates the uptake of glucose by cells in general and those of the liver and muscles in particular. In the liver and muscles, excess glucose is converted to glycogen. Under certain conditions, the liver can be stimulated to convert protein into glucose. The stimulating hormone is a family of glucocorticoids which are released in an interaction involving the adrenal gland and the pituitary.

Fats

Fats are not all bad. They remain an essential nutrient to humans as well as many other animals (although the quantity needed is open to debate). Unsaturated fats are necessary for membrane synthesis and various other tasks, but the vertebrate liver cannot make unsaturated fats from glucose; therefore, they must be consumed in food from plants. In addition to fat being an excellent energy source (with twice the energy value of carbohydrates, gram for gram), several essential vitamins, including vitamins A, D, E, and K, are found in fatty foods. Further, the fat under the skin is an important heat insulator.

Fats are not only synthesized from carbohydrates, but they can also be constructed from amino acids, which may also be broken down into acetyl-CoA. For this reason, high-protein weight-reducing diets may defeat their own purpose if the quantity is not carefully measured.

Table 25.3 Vitamins

Vitamin	Source	Function	Daily requirement	Result of deficiency
A, retinol	Fruits, vegetables, liver, dairy products	Synthesis of visual pigments	1500–5000 IU[a] 3 mg	Night blindness, crustiness about eyes
B_1, thiamine	Liver, peanuts, grains, yeast	Respiratory coenzyme	1–1.5 mg	Loss of appetite, beriberi, inflammation of nerves
B_2, riboflavin	Dairy products, liver, eggs, spinach	Oxidative chains in cell respiration	1.3–1.7 mg	Lesions in corners of mouth, skin disorders
Niacin, nicotinic acid	Meat, fowl, yeast, liver	Part of NAD and FAD cell respiration	12–20 mg	Skin problems, diarrhea, gum disease, mental disorders
Folic acid	Vegetables, eggs, liver, grains	Synthesis of blood cells	0.4 mg	Anemia, low white blood count, slow growth
B_6	Liver, grains, dairy products	Active transport	1.4–2.0 mg	Slow growth, skin problems, anemia
Pantothenic acid	Liver, eggs, yeast	Part of coenzyme A of cell respiration	Unknown	Reproductive problems, adrenal insufficiency
B_{12}	Liver, meat, dairy products, eggs	Red blood cell production	5–6 μg	Pernicious anemia
Biotin	Liver, yeast, intestinal bacteria	In coenzymes	Unknown	Skin problems, loss of hair and coordination
Choline	Most foods	Fat, carbohydrate, protein metabolism	Unknown	Fatty liver, kidney failure, metabolic disorders
C, ascorbic acid	Citrus fruits, tomatoes, potatoes	Connective tissues and matrix antioxidant	40–60 mg	Scurvy, poor bone growth, slows healing, (colds?)
D	Fortified milk, seafoods, fish oils, sunshine	Absorption of calcium	400 IU[a]	Rickets
E	Meat, dairy products, whole wheat	Uncertain	Unknown	Infertility, kidney problems
K	Intestinal bacteria	Blood-clotting factors	Unknown	Blood-clotting problems

[a]IU = International units.
mg = milligram ($\frac{1}{1000}$)
μg = microgram ($\frac{1}{1,000,000}$)

Protein

Protein is an essential nutrient. It is possible to store some protein in the liver and muscles, but there is a steady turnover of total body protein. We need protein the most, of course, when we are growing.

Amino acids are utilized in several ways. For example, they are the building blocks of protein; they are used to form nitrogen bases; they can be oxidized for energy, and they can be converted to fats and carbohydrates. In any case, however, the metabolism of amino acids invariably leads to leftover nitrogen wastes. These are first formed as molecules of poisonous ammonia which all organisms must immediately remove from their bodies, or somehow render harmless. We will discuss these processes later.

It is interesting that humans can convert some amino acids to others. In fact, we can produce twelve of the twenty amino acids we require for maintenance and growth. The remaining eight must be supplied, intact, in the diet. These are often referred to as the *essential amino acids*, a phrase that delights the advertising agencies, although it is something of a misnomer, since all twenty are biologically essential at some level. The best source of amino acids is probably animal protein, since many plant foods are deficient in the amino acids tryptophan, methionine, and/or lysine. So if you are contemplating a vegetarian diet, you will want to select foods with these amino acids present—soya protein is a good source, as are milk and cheese products.

Keep in mind that protein deficiency is one of the critical problems facing a large portion of the humans on earth today. The greatest sufferers are children, especially the unborn. Brain development, particularly, requires a great deal of protein, especially in the first 2 years when the brain reaches 90% of its adult size. A deficiency disease, *kwashiorkor*, has become commonplace in Africa, Latin America, India, and the Middle and Far East. The symptoms are numerous, but chiefly there is retardation of physical and mental growth, loss of hair pigment, general wasting of the body, and often, blindness related to vitamin A deficiency. Protein deficiency in the United States is usually limited to the very poor, the hopelessly ignorant, the old and helpless, dieting teenagers, alcoholics, and some food faddists.

Vitamins and Minerals

There are a variety of substances that are essential to metabolic processes. Most of these substances have been clearly identified by biochemists so that we know their molecular structure as well as their precise function. A few are only vaguely understood, however. All we can say about them is that if you don't have them, you will develop something-or-other, and it's usually bad. Most of the vitamins function in common enzymatic reactions as *coenzymes*. Some enzymes cannot function without coenzymes, so you can imagine the importance of vitamins—for example, in oxidation–reduction reactions.

Vitamins in our food are of two types: water-soluble or fat-soluble. In either case, however, they are transported across the villi of the small intestine directly into the blood for distribution. One, vitamin B_{12}, is so large that it must be transported with the help of a mucoprotein produced in the stomach. The process involves a carrier transport of some type, and probably pinocytosis as well, but at the moment we really don't know much more about it.

Biotin, another essential vitamin, binds with an egg-white protein and becomes useless. So eating too many raw eggs can theoretically result in clinical biotin deficiency. Some indirect evidence of this comes from dietary studies using dogs, in which it has been discovered that biotin deficiency is rather common in those that are fed raw eggs with their food. As any food enthusiast knows, good foods can actually interfere with each other and the vitamin balance is particularly easy to upset.

Let's look at the B vitamin, *niacin,* to see how one vitamin works. Another name for niacin is *nicotinic acid* or *nicotinamide.* But don't cast startled glances at someone's cigarette, because we are referring to a compound associated with respiration and photosynthesis, NAD^+ and $NADP^+$ (*nicotine* is something else). These are known as the *nicotinamide adenine dinucleotide coenzymes.* These coenzymes are very common in plant and animal tissues, and function in concert with enzymes called *dehydrogenases.* The coenzymes become reduced (accept electrons) when the enzyme oxidizes (takes electrons from) the substrate. Then the reduced coenzyme passes the electrons to some other compound. These functions are highly critical to both photosynthesis and respiration. Other vitamins act in a similar manner, and are listed in Table 25.3.

We already know that, in addition to the complex vitamins, animals require a variety of simple inor-

Table 25.4 Major Mineral Nutrients

Mineral	Food source	Function	Daily requirement	Result of deficiency
Calcium	Dairy foods, eggs	Growth of bones and teeth, blood clotting, muscle contraction, nerve action	1 g	Tetany, rickets, loss of bone minerals and muscle coordination
Cobalt	Common in foods, water	Vitamin B_{12}	1 mg	Anemia
Copper	Common in foods	Production of hemoglobin, enzyme action	2 mg	Anemia
Fluorine	Most water supplies	Prevents bacterial tooth decay	Unknown	Tooth decay, bone weakness
Iodine	Seafood, iodized salt	Thyroid hormone	0.25 mg	Hypothyroidism
Iron	Meat, eggs, spinach	Hemoglobin (oxygen transport)	10 mg (men) 18 mg (women)	Anemia, skin problems
Magnesium	Green vegetables	Enzyme function	350 mg	Dilated blood vessels, irregular heartbeat, loss of muscle coordination
Manganese	Liver, kidneys	Enzyme function	Unknown	Loss of fertility, menstrual irregularities
Phosphorus	Dairy foods, eggs, meat	Growth of bones and teeth, ATP nucleotides	1.3 g	Loss of bone minerals, metabolic disorders
Potassium	Most foods	Nerve and muscle activity	2–4 g	Muscle and nerve disorders
Sodium	Most foods, salt	pH balance, nerve and muscle activity, body fluid balance	0.5 g	Weakness, muscle cramps, diarrhea, dehydration
Zinc	Common in foods	Enzyme action	Unknown	Slow sexual development, loss of appetite, retarded growth

ganic ions or, more commonly, minerals. Minerals are also absorbed in considerable quantity from our food. Sodium, as we've mentioned, is rapidly absorbed, along with lesser amounts of calcium, potassium, and chloride. Iron absorption is under feedback control—the intestine absorbs iron only when tissue reserves are low. The minerals may be used in the formation of gross structures such as bone, or they may become active parts of functional molecules like hemoglobin. In addition, some of the ions are necessary in enzyme action. The specific roles of the minerals are listed in Table 25.4.

Application of Ideas

1. Select five distinct variations in mammalian dentition (including its absence) and relate these variations specifically to the food utilized by the animals concerned. Comment on how these variations might have arisen.

2. The feeding habits of animals are highly varied. Some are omnivores, feeding on a variety of vegetable and animal sources. Others are highly specialized carnivores or herbivores, feeding on one specific item. Describe an example of each and discuss the short- and long- run advantages and disadvantages of evolution in either direction.

3. Careful studies of herbivorous animals reveal that they rarely produce the enzyme cellulase, yet a considerable number get along fine on plant fibers. Account for this phenomenon and offer reasons why the enzyme hasn't evolved in the animals.

4. Among the terrestrial animals, socketed, enamel-covered teeth are primarily a characteristic of mammals as is the more mobile jaw. Of what particular advantage were these innovations to mammals as they experienced their rapid diversification in the terrestrial environment?

Key Words

aminopeptidase
amylase
anus
appendix
ascending colon
autonomic nervous system

bile
bile duct
bile salts
biotin

canine teeth
carboxypeptidase
cardiac sphincter
cecum
chemoreceptors
chief cells
cholecystekinin
chloragen cells
chymotrypsin
coelom
collar cell
colon
connective tissue
crop

deamination
dehydration linkage
dehydrogenase
descending colon
digestion
digestive vacuole

Key Ideas

Digestion

1. Digestion is the chemical breakdown of complex food molecules into their simpler component parts.

2. *Extracellular digestion* means outside of cells. Enzymes are released from cells to the food and then digested products are absorbed.

3. *Intracellular digestion* means inside cells, usually within vacuoles that have formed from the cell membrane during phagocytosis or pinocytosis.

4. Humans carry on both types, but food is technically outside the body when it is in the gut.

Digestion in Invertebrates

The Simpler Invertebrates

1. Sponges feed from phagocytized particles moving through their flagellated canal systems. *Digestion* is *intracellular.*

2. Coelenterates have a saclike gut where *digestion* begins. It is completed in vacuoles *intracellularly.*

3. Flatworms and comb jellies also use the saclike gut, with food entering and residues leaving through the mouth.

4. The saclike gut of coelenterates, flatworms, and others is called a *gastrovascular cavity.* In flatworms these are elaborate, highly branched distribution chambers.

5. Planaria, a free-living flatworm, uses a long, protrusible *pharynx* to suck up food. Food particles enter the cells by phagocytosis where *digestion* occurs intracellularly.

More Complex Invertebrates

1. The complete digestive system evolved twice, but differently in the protostomes and deuterostomes. These groups have the tube structure with both mouth and *anus.* Its development permitted specialization along the gut.

dipeptidase
duodenum

elastin
endoderm
endopeptidase
epithelium
esophagus
essential amino acids
exopeptidase
extracellular digestion

feces
filter feeder
flatus

gall bladder
gastric lipase
gastrin
gastrointestinal tract
gastrovascular cavity
gill rakers
gizzard

hemocoel
herbivorous
hydrochloric acid
hydrolysis

ileocecal valve
ileum
incisor
ingestion
intracellular digestion

Jacobson's organ
jejunum

kwashiorkor

lactase
lacteal
large intestine
laryngopharynx
larynx
lipase
liver
lymphatic vessel

maltase
mesoderm
microvillus
milk teeth
molar
mucous membrane

negative feedback mechanism
niacin
nuclease

olfaction
oral cavity

pancreas

2. The more complex invertebrates and vertebrates still use *intracellular digestion* to some degree. The *liver* of some, for example, still phagocytizes particles for further processing.

Pseudocoeloms, Coeloms, and Hemocoels

1. Animals with a one-way gut have the tube within a tube body plan. The outer tube or body wall is generally muscular. The inner tube or gut is suspended in a fluid-filled cavity.
2. In nematodes, the gut is simple and derived from *endoderm.* The cavity is a *pseudocoel* derived partially from *mesoderm.*
3. In many animals, the cavity is a *coelom* or *hemocoel.* The cavity is lined with *mesoderm,* and the gut has layers of *mesoderm*-derived muscle, connective tissue, and epithelium.
4. Muscles in the gut wall run in two directions and permit *peristalsis* to occur.
5. Animals have evolved different ways of increasing the surface area of the gut.
6. Special outpocketings forming blind sacs have permitted the development of the *liver, pancreas,* and *salivary glands.* Fermentation sacs or *ceca* have also formed as elaborations on the simple tube.

Earthworms and Insects

1. The earthworm serves as an example of *digestion* for a number of animals. Its gut contains six specialized regions, including a *pharynx, esophagus, crop, gizzard,* intestine, and *typhlosole,* a fold in the intestine wall for increasing surface area.
2. *Chlorogen cells* store glucose, deaminate amino acids, and accumulate wastes, while phagocytic cells transport soil to the skin, adding to the protective coloration.
3. Insects vary greatly in feeding habits, food, and mouthparts.
4. Mouthparts in grasshoppers include tasting, chewing, and holding structures and salivary glands. The gut includes the fore-, mid-, and hindgut for storage, digestion, and waste formation. Enzymes are secreted by *ceca.* Other insects use complex sucking and piercing mouthparts in their specialized diets, including plant juices, animal body fluid, and blood, etc.

Foraging and Digestive Structures in Fish, Amphibians, Reptiles, and Birds

Fish

1. Sharks are carnivores. Their teeth occur in rows with the outermost constantly replaced. The jaw is a simple hinge.
2. The shark's digestive area is increased by the *spiral valve,* which also slows food passage.
3. Lampreys are jawless, feeding by sucking fluids from prey.
4. Bony fish are very diverse. Teeth are numerous, and the jaw is a simple hinge.
5. Fish that remove particles from the water have highly branched *gill rakers.*
6. In general, *herbivorous* fish have long, highly coiled intestines, while those in *carnivorous* fishes are short and pouched.

Amphibians, Reptiles, and Birds

1. Amphibian mouths are fishlike, but the tongue is very mobile in some.
2. Reptiles show greater variation and specialization. Some have permanent, socketed teeth, venom-injecting fangs, olfactory organs, heat sensors, specialized jaws for swallowing whole prey, or fused horny beaks.
3. Birds have horny bills that vary with the species' feeding habits. The tongue is barbed in some and tubelike in others. The toes may bear long sharp talons, or may be broad and stubby for digging.
4. They also have specializations in the gut. The *crop* stores food, while *digestion* begins in the *proventriculus.* Food is pulverized by gravel in the gizzard.

pancreatic amylase
pancreatic duct
pancreatic lipase
parietal cell
parotid gland
pepsin
pepsinogen
peristalsis
pharynx
phosphodiesterase
premolar
proventriculus
pseudocoelom
pseudocoelomate
ptyalin (salivary amylase)
pyloric sphincter

rectal valve
rectum
rennin
rumen
ruminant

salivary gland
sigmoid colon
small intestine
sodium bicarbonate
sphincter
spiral valve
stomach
sublingual gland
submaxillary gland
sucrase

taste bud
transverse colon
typhlosole
trypsin

urea

vagus nerve
villus
vitamin

Foraging and Digestive Structures in Mammals
1. Teeth in mammals are all socketed. Generally milk teeth precede permanent teeth. The four types of teeth, the *incisors, canines, premolars* and *molars,* undergo great modification in the different mammals, while human teeth are considered very generalized.
2. Humans lack the large *canines* common in primates, a trait that is often used by anthropologists to place fossils in human or ape taxonomic lines.

Digestion in the Herbivores—Ruminants
Mammals do not have enzymes for digesting cellulose, but many have specialized teeth for eating plant food. Cattle are *ruminants* with four-chambered *stomachs* where protozoans and bacteria digest cellulose in a fermentation process. The cow absorbs the products of fermentation.

The Digestive System of Humans

The Oral Cavity and Esophagus
1. Feeding structures of the *oral cavity* are:
   a. Lips (aid in mastication and swallowing)
   b. Tongue (moves food for chewing; detects taste, texture)
   c. *Chemoreceptors* (taste and smell)
   d. *Salivary glands* (secrete saliva for lubrication and starch digestion)
   e. *Pharynx* (breathing and swallowing)
2. The *esophagus* directs food to the *stomach,* but has no digestive function. Its tissues are typical of the entire digestive tract. The three layers are:
   a. *Mucous membrane,* simple columnar epithelium, areolar connective tissue
   b. Connective tissue, blood and lymphatic vessels, nerves and glands
   c. Circular smooth muscle, longitudinal smooth muscle, fibrous outer coat
3. Food is moved by *peristalsis,* which includes coordinated contractions of muscle layers under the control of the *autonomic nervous system.*

The Stomach
1. The *stomach,* a J-shaped sack, stores and mixes food with enzymes.
2. The stomach has three muscle layers that work in a squeezing and twisting action.
3. The stomach lining (mucosa) is complex and glandular and includes:
   a. *Chief cells* (secrete *pepsinogen*)
   b. *Parietal cells* (secrete HCl)
   c. Mucus- and water-secreting glands
   d. *Rennin*-secreting cells
   e. *Gastric lipase*-secreting cells
4. The major enzyme, *pepsin,* starts protein *digestion.* Its precursor is *pepsinogen,* which is HCl-activated.
5. The acidic *stomach* environment kills microorganisms. The lining is protected by a coating of mucin.
6. Vomiting occurs when food irritants stimulate the medulla. Violent contractions occur along with a relaxation of the upper *cardiac sphincter.*

Small Intestine
1. The *small intestine* consists of three regions: the *duodenum, jejunum,* and *ileum.*
2. Its surface is highly folded and contains numerous *villi.* Each *villus* is covered with *microvilli,* minute folds in the cell membranes.
3. Each *villus* contains a *lacteal* and a capillary bed. Vigorous movements of the *villi* helps in mixing food with enzymes.

Liver
Bile is an important secretion of the liver that emulsifies fats. Its release from the *gall bladder* is controlled by the hormone, *cholecystekinin.*

Pancreas

The *pancreas* is a glandular accessory organ. It secretes *sodium bicarbonate* and digestive enzymes.

Large Intestine

1. *Small* and *large intestine* join, forming the *cecum* with its extension, the *appendix.* The *ileocecal valve* separates the two intestines.
2. The *large intestine* or *colon* consists of the *cecum, colon, rectum,* and *anus.* Its function is to absorb water and minerals and form the *feces,* which are stored in the *rectum.*
3. The *colon* contains enormous populations of bacteria. Bacterial corpses and food residues form the *feces. E. coli* is known to produce vitamin K, biotin, and folic acid, all nutritional requirements.

Chemical Digestion and Absorption

In *digestion,* the molecular subunits of food molecules are enzymatically cleaved by *hydrolysis.* The products of *digestion* are simple sugars, fatty acids, glycerol, amino acids, and nucleotides of RNA and DNA. (Table 25.1 summarizes the organs of digestion, the glands and their enzymes, the substrate and their products. We will not repeat these here except to mention special ideas.)

Carbohydrate Digestion

1. Starches are primarily digested in the *small intestine.*
2. *Lactase,* the enzyme of milk sugar digestion, is produced by babies and all Caucasian adults. Negro and Oriental adults often lack the enzyme. The evolutionary reason for this is not known except for the observations that reveal that Caucasian adults have drunk milk for thousands of years.
3. *Fructose* and *glucose* are absorbed by the *villi,* directly into the capillaries. They are deposited in the *liver* for *glycogen* storage.
4. Acidic mixtures from the *stomach* are neutralized by alkaline *bile* and *sodium bicarbonate* from the *pancreas.*
5. *Sodium bicarbonate* release is under the control of the hormone *secretin.*

Fat Digestion

1. Fats enter the *villi* in two ways. They enter as glycerol and fatty acids, are reconstituted in the *villus* cells, and then enter the *lacteal* and lymphatic system.
2. Neutral fats can also enter the *villi* and *lacteals* intact, through pinocytosis.
3. Long chain fatty acids enter the *lacteal,* while shorter ones enter the circulatory system.

Protein Digestion

1. Because of their complexity, proteins require a four-step process in their complete *digestion* to amino acids.
   a. *Pepsin,* an endopeptidase (stomach): nonspecific, produces peptides of several to many amino acids in length.
   b. *Trypsin,* an endopeptidase (*pancreas*): specific for bonds on carboxyl side of arginine and lysine.
   c. *Chymotrypsin,* an endopeptidase (*pancreas*): specific for bonds on carboxyl side of phenylalanine, tyrosine, and tryptophan; residues include amino acids, dipeptides, and tripeptides.
   d. *Exopeptidases* and *dipeptidases* (*duodenum*): produce the final amino acid products. The enzymes are *carboxypeptidases* or *aminopeptidases* working on either end of polypeptide chains.
2. Amino acids are transported to the *liver* where they are reconstituted into serum proteins for transport to the cells. Some are deaminated and used for energy or to produce other molecules.

Nucleic Acid Digestion

Nucleic acid *digestion* may begin in the acidic environment of the *stomach.* Two enzymes (*nucleases*) work at the ends or at the inner bonds. The products are short chains of nucleotides.

Integration and Control of the Digestive Process

1. Mechanical, *neural*, and hormonal mechanisms control *digestion*. Chewing evokes salivation. Thinking of food can also evoke enzyme secretion. The presence of food in the *stomach* mechanically stimulates gastric secretions. Neurons send impulses to the brain, and nerves returning to the *stomach* cause the release.

2. Hormonal mechanisms are independent of the others. For example, protein in the *stomach* stimulates the lining cells to release *gastrin*, which stimulates the gastric glands to release gastric juices. *Gastrin* action illustrates the *negative feedback system*, an important aspect of biological control.

3. Five major hormones are involved in *digestion* (see Table 25.2).

What Happens to Nutrients?

Carbohydrates

1. Carbohydrates are stored as glycogen in the *liver* and skeletal muscles.

2. Glucose is required for energy, but can be converted to fat when it reaches the acetyl-CoA step in respiration.

Fats

1. Fats are essential for energy, cell structure, and fat soluble *vitamins*.

2. Fats can be synthesized from carbohydrates and amino acids through the acetyl-CoA step.

Protein

1. Protein is essential for growth and turnover replacement.

2. Amino acids are used to construct proteins, purines and pyrimidines, and for energy.

3. Twelve amino acids can be produced in our bodies. The other eight are dietary requirements.

4. The best sources are animal protein since all twenty amino acids are commonly present.

5. Vegetarians may lack three amino acids in their diets unless they eat balanced protein sources such as soya protein, milk, or milk products.

6. Protein deficiency is critical in much of the world. Its presence is critical for normal fetal and childhood development. *Kwashiorkor* is a protein deficiency disease associated with mental and physical retardation.

Vitamins and Minerals

1. *Vitamins* function mainly as coenzymes in oxidation–reduction reactions. They are water soluble or fat soluble, crossing the *villi* directly into the blood stream.

2. Dietary imbalances can produce vitamin deficiencies.

3. The B vitamin, niacin, is nicotinic acid, a constituent of NAD and NADP. These are common coenzymes in animals and plants respectively. They function in dehydrogenase reactions along with enzymes.

4. In addition to *vitamins*, *minerals* are also highly important. (Tables 25.3 and 25.4 list the essential *vitamins* and *minerals* and their uses.)

5. They function in enzyme reactions, membrane activity, and parts of active molecules, and are important in fluid balance.

Review Questions

1. Explain the difference between *extracellular* and *intracellular digestion*. (p. 701)

2. Describe the manner in which sponges feed and the type of *digestion* used. (p. 701)

3. Describe the *gastrovascular cavity* in coelenterates and flatworms. (p. 702)

4. Describe the peculiar manner of feeding in *planaria.* (pp. 702–703)

5. Explain the idea, "two separate openings (mouth and *anus*) occurred in evolution at least twice." (p. 702)

6. Describe the tube within a tube body plan in the nematodes and include the *pseudocoelom* in your description. (pp. 703–704)

7. How does the tube within a tube body plan differ in *coelomate* animals? Include the term *mesoderm* in your answer. (pp. 703–704)

8. Explain the action of circular and logitudinal muscles in *peristalsis.* (p. 704)

9. List the digestive structures in the earthworm and briefly describe their functions. (pp. 704–705)

10. List the *"liver"* functions of the earthworm's *chloragen cells.* (p. 705)

11. What are the shark's specializations for feeding and *digestion?* (p. 707)

12. Describe the feeding structures in the jawless fishes. (p. 708)

13. List the special characteristics of fishes that (p. 708)
   a. Feed on particles in the water
   b. Are strictly *herbivorous*
   c. Are strictly *carnivorous*

14. List four feeding specializations in the reptiles. (pp. 708–709)

15. Describe four variations in birds' beaks and explain how each is utilized. (p. 709 and Figure 25.11)

16. List the specialized regions of the digestive tract in birds and explain their functions. (p. 710)

17. List the four groups of teeth in mammals and give examples of four animals that have emphasized the use of each group differently. (p. 710)

18. In what way have cattle solved the problem of digesting cellulose? (pp. 710–711)

19. Describe the layered structure of the *esophagus* and relate this to its functions. (p. 712)

20. Explain the muscle layers of the *stomach* and what kind of actions they produce. (p. 715)

21. List the specialized structures of the *stomach* mucosa and name their secretions. (p. 715)

22. With the aid of a simple drawing, describe the *villi* of the *small intestine* and summarize their functions. (p. 716)

23. What are the functions of the *liver* and *pancreas* in the digestive process? (p. 717)

24. Describe the content and function of *bile* and review its control. (p. 717)

25. List the structures of the *large intestine* in the order in which they appear and explain its general function. (p. 717)

26. List the activities of *E. coli* in the intestine. (p. 717)

27. Explain how the rate at which wastes pass through the *colon* is important in normal bowel movement. (p. 717)

28. Present a precise chemical definition of *digestion.* (p. 718)

29. List the enzymes responsible for carbohydrate *digestion* and their breakdown products. (Table 25.1 and p. 718)

30. How does the *small intestine* overcome the high pH of *stomach* contents it receives? (pp. 719–720)

31. Describe how fats are handled in the digestive system. (Table 25.1 and p. 720)

32. Describe the various ways fats are absorbed. (p. 720)

33. Explain what happens to protein molecules in the *stomach.* (p. 720)

34. Describe the manner in which the pancreatic enzymes *trypsin* and *chymotrypsin* alter peptides. (p. 720)

35. What is the role of *exopeptidases* in the final steps of protein *digestion?* (p. 720)

36. Distinguish between the terms *carboxypeptidase, aminopeptidase,* and *dipeptidase.* (p. 720)

37. What happens to amino acids before they are distributed to cells for use in protein synthesis? (p. 720)

38. Describe the way in which nucleic acids are digested. (p. 721)

39. List the three levels of control in *digestion.* (p. 721)

40. Using *gastrin* as an example, explain how *negative feedback* works in the hormonal control of *digestion.* (p. 722)

41. Besides their value as an energy source, how can the body dispose of carbohydrates? (p. 723)

42. What does acetyl-CoA of respiration have to do with utilizing glucose, fatty acids, and amino acids? (p. 723)

43. List three essential reasons for including fats in a complete diet. (p. 723)

44. Why is meat a better source of protein than most vegetables? (p. 724)

45. Describe the effects of *kwashiorkor,* the protein deficiency disease of undeveloped countries. (p. 724)

46. Explain how most *vitamins* are used in cells. (pp. 725–726)

Chapter 26

Respiration and Transport

This may not be the best of all possible worlds, but it's the only one we've got, and we must admit that the processes of life are essentially involved with overcoming the problems that our world has set for us. (An optimist might say that our task is to take advantage of the opportunities the planet provides.)

One thing our planet has provided is oxygen. And living things have devised ways to use that peculiar gas. Perhaps because we are all descended from the same kind of primordial life forms, most living things tend to use oxygen in the same way—as an acceptor of spent electrons.

Our planet has presented us with other problems (opportunities) too. And these may vary from place to place and from time to time. So, on this variable and changing earth, we living things have had to

26.1 (a) In skin breathers a simple exchange of O_2 and CO_2 occurs across the moist body wall. (b) External gills are a means of greatly increasing the surface area for gas exchange. (c) The gill becomes more elaborate in fish, with O_2-carrying water passing in one opening (mouth), moving over the gills where O_2 is exchanged for CO_2, and out through different openings. (d) In the insects, highly branched, thin-walled tracheae penetrate the fluid-filled body, exchanging gases throughout. (e) The vertebrate lung produces a great surface area in a relatively small space by the presence of numerous blind, thin-walled sacs where gases are exchanged. (f) The bird has a specialized respiratory system, having added air sacs to an already complex lung. The circular route of the air assures a much more complete exchange of gases.

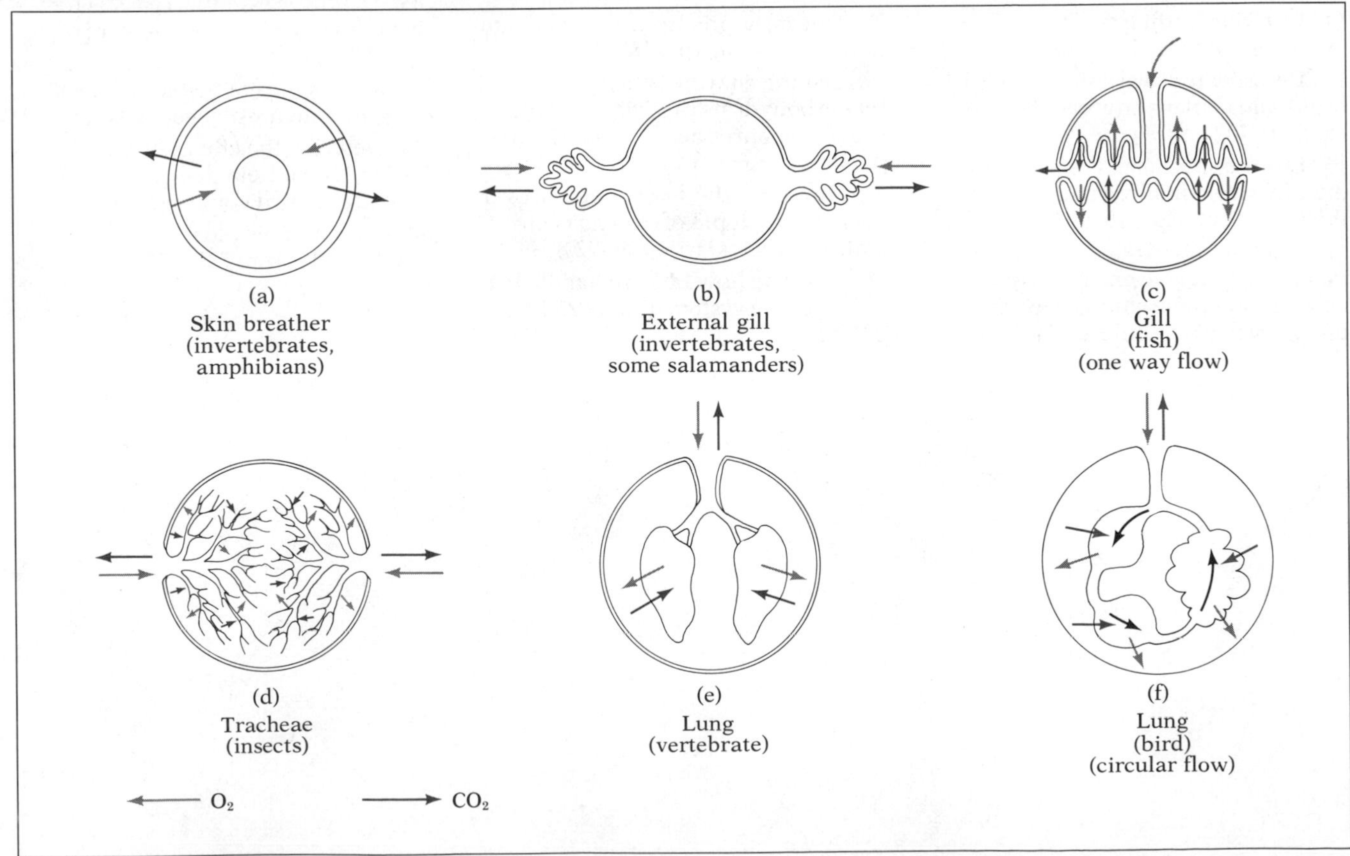

devise increasingly sophisticated ways of coping, and this means we have had to develop specialized systems. First, we became multicellular, and then we divided up our tasks among the cells. In time, groups of cells became very different from each other, each according to its chore. And so systems developed. One of those systems helped to get oxygen to the places where it was needed, and another system managed to develop ways to carry waste gases away from those places.

In this chapter, we'll take a look at these systems: the one that brings oxygen into the body and the one that carries it to those places where it can be used. We will also, of necessity, encounter waste gases and the ways various animals deal with them. Most of our discussion of gas exchange will relate to the more complex multicellular animals. These include most of the invertebrates and all the vertebrates—those animals that have developed complex organ systems and body masses too large and dense to rely on gases diffusing cell by cell through the body. In addition, most of the animals we will consider must live active lives in order to eat and avoid being eaten. This relentless pursuit requires muscular activity which creates a great demand for oxygen.

We must keep in mind that the exchange of gases requires structures with two basic characteristics. The first is a permeable surface area of sufficient size. The second is that this surface area be moist, since gases can't normally cross dry membranes. Surface area can be increased in a number of ways. Figure 26.1 will serve as an introduction to the evolutionary directions animals have taken to solve the problem. Let's see how these demands are satisfied in some of the animals that utilize a simple skin interface first, and then consider structures that are more complex and specific for gas exchange.

26.1 Evolution of Gas Exchange Surfaces

The Body Interface

We described the body form and feeding habits of some of the sponges in an earlier chapter. However, we should remind ourselves that not all sponges are small; in fact, the more complex sponges may reach an impressive size indeed. These animals, although sedentary, require a greater exchange of O_2 and CO_2 than their smaller relatives. But how do gases suffuse their great mass? The problem is solved by increasingly strong currents of oxygen laden water (created, you'll recall, by the flagellated collar cells) which constantly pass through the canals of the body (see Figure 25.1). Whereas smaller sponges need few canals, the larger ones increase their respiratory interface through an extensive canal system. Thus no cells are very far from the oxygen-carrying water.

Similarly, although some species of coelenterates (such as jellyfish) attain considerable size, they require no specialized respiratory structures. Their sacklike bodies consist essentially of two cell layers that are virtually always in contact with the external environment. Their outer body wall and extensive gastrovascular cavity provides an adequate surface.

The flatworm (planaria) is a dorsoventrally (back-to-belly) flattened animal with an extensive gastrovascular cavity (see Figure 25.3). While its body consists of several layers of cells, including dense muscle tissue, its external surface and highly branched gastrovascular cavity expose much of its body mass to its freshwater environment. It can, therefore, use cell-by-cell diffusion as a means of exchanging gases with the environment. Of course, as long as it relies on diffusion alone to transport gases, it can never grow very large—its mass would quickly outgrow its surface area. So if you ever see an 8 ft planaria coming toward you while you are swimming at the lake, ignore it. It doesn't exist. Marine flatworms, by the way, can get to be about 10 cm wide and perhaps 25 cm long—but only a few millimeters thick. They have solved their surface–volume problem by expanding in only two dimensions.

A few terrestrial animals, such as the earthworm and even some small salamanders, manage to use their moist skin as a respiratory surface. But these complex animals are very special cases. For example, the earthworm's habitat is really only semiterrestrial, since it restricts itself to moist soil. The lungless salamanders are also limited to damp places. They spend their days underground (or under logs), venturing to the deadly dry air only at night when they suffer less risk of drying out. The skin of earthworms and lungless salamanders, by the way, is kept moist by the secretion of a slime layer.

Although skin breathers may have no special structures to exchange gas with their environments, some have efficient circulatory systems for transporting gas once it is within their bodies. For example, both the earthworm and the lungless lizard have highly efficient, closed circulatory systems containing hemoglobin (see Figure 26.13). The salamander's hemoglobin is bound to special blood cells, as is our own, but the earthworm's hemoglobin is dissolved in the blood. Even so, it is still a far more efficient oxygen carrier than simple plasma. So skin breathing works for some complex terrestrial organisms—if an efficient circulatory system is present to distribute O_2 and remove CO_2, and they restrict themselves to a moist habitat. In summary, the earthworm and the lungless salamander share an interesting combination of behavior, anatomy, and physiology that permit their simple respiratory systems to work.

Expanding the Interface: Tracheae

Most of the terrestrial animals exchange gases across surfaces that are actually specialized infoldings of the gut or body surface. By internalizing their respiratory surfaces, land dwellers avoid the problem of excessive water loss. Furthermore, these internal pouches are usually highly folded, providing an increased area of exchange.

There are several variations in this internal arrangement, perhaps strangest in the arthropods, a group that does most things differently. The bodies of insects, for example, are riddled with a series of tiny, highly branched tubules called *tracheae* (Figure 26.2). These tubes open to the outside via tiny holes known as *spiracles*, situated along the insect's side. In some larger insects, air is pumped in and out the spiracles by a bellows action of the abdomen, the spiracles opening and closing in synchrony with the pumping action. In some insects, the branching tracheae terminate in elastic *air sacs* that expand and contract and aid in the exchange processes. In other species of insects, air simply diffuses through the tracheal system. Oxygen and carbon dioxide are exchanged through the incredibly fine branches of the trachea, which permeate the body and carry oxygen to the immediate vicinity of every active cell. The finer branches of the tracheae are filled with fluid through which the respiratory gases can diffuse. Although insects do have a heart and circulatory system, these systems appear to serve primarily to distribute food molecules and metabolic products; the respiratory functions are thus left to the tracheal system. The exoskeletons of insects, by the way, are waterproof, which helps to keep them from drying out. Thus, insects can go places and do things that earthworms and lungless salamanders can't.

26.2 The respiratory system in the grasshopper. Insects have developed an elaborate respiratory system for the exchange of gases. Pores known as *spiracles* line the body, opening into lengthy *tracheal tubes.* These branch through the body and some end in blind pouches known as *air sacs.* Gases pass through both the tracheae and the air sacs. The grasshopper assists the process, flexing its exoskeleton like a bellows. The extensive and branching tracheal tree permeates the body and negates the need for an efficient circulatory system.

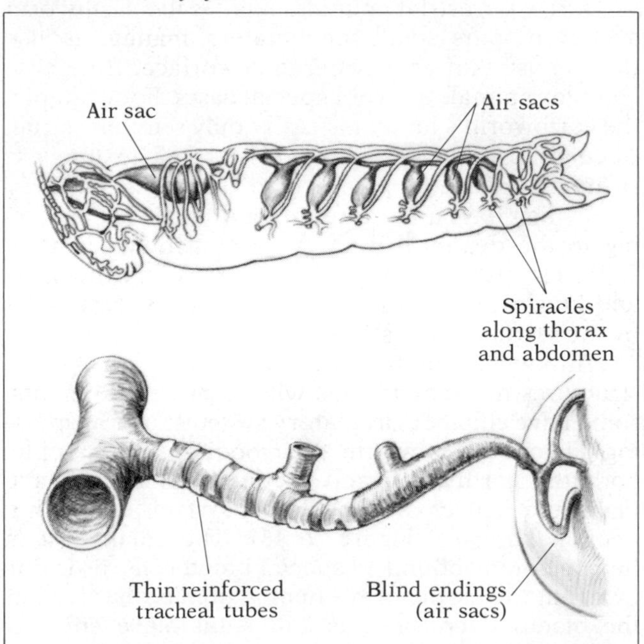

Expanding the Interface: Branchiae

The starfish, a typical echinoderm, has a different kind of respiratory apparatus from other invertebrates. You may recall from Chapter 24 that this animal has a skeleton of minute plates covered with dermal layers (see Figure 24.8). But scattered through the skeleton are a large number of pores through which a respiratory surface protrudes. These projections, the *dermal branchiae*, are actually outpocketings of the coelom (Figure 26.3). Since the cavity is fluid-filled, the respiratory wastes are continually excreted into the seawater through the walls of these projections. Of course, it helps if both the coelomic fluids and the seawater circulate a bit. This is accomplished by clusters of ciliated cells on both the inside and the outside surfaces of the dermal branchiae.

Complex Interfaces: Gills

Aquatic mollusks, arthropods, and chordates are more active than echinoderms, and their increased respiratory demands are met with a more complex structure than the dermal branchiae. Here we find the *gill*, which is a combination respiratory–circulatory structure. Gills are thin-walled, finely divided outpocketings with extensive capillary beds through which blood flows, bringing in CO_2 and carrying away

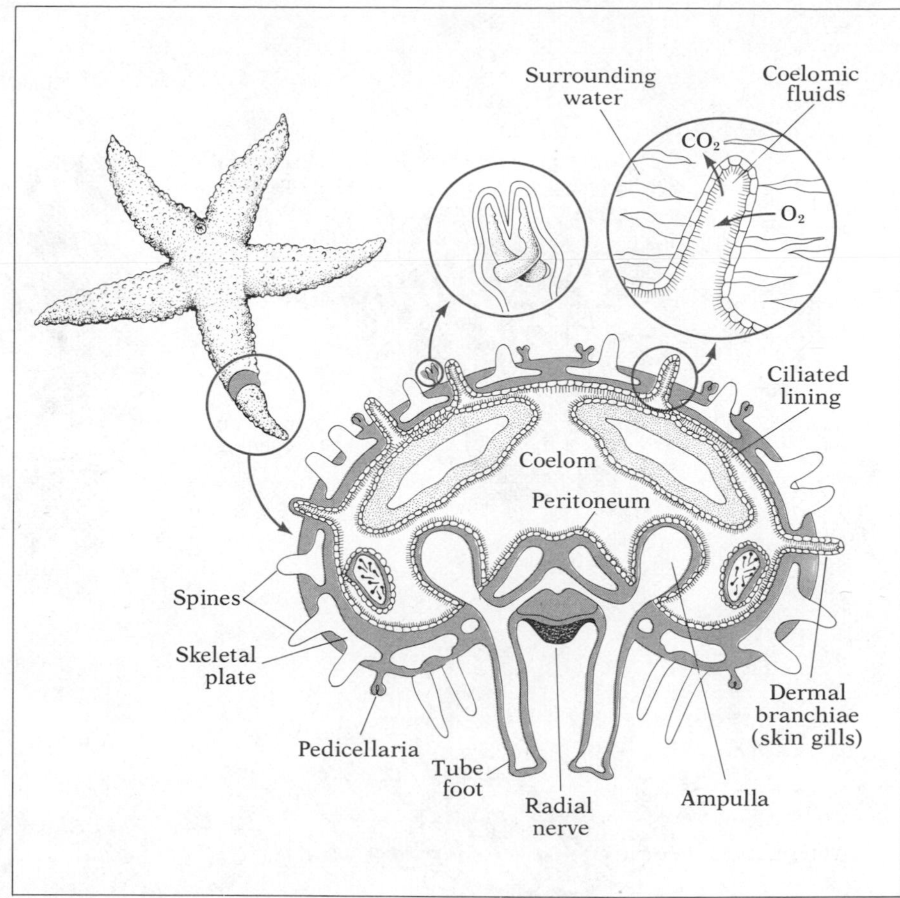

26.3 Respiratory exchange in the starfish. Echinoderms are protected by an endoskeleton of jointed plates equipped with crusty spines. However, to accommodate gas exchange, the animal must expose softer, membranous parts to the surrounding seawater. It does this partly through its ventral surface, which has softer areas, and partly through the use of *dermal branchiae* (skin gills). The dermal branchiae are membranous projections through the armor. The membranes are continuous with the fluid-filled lining of the body cavity. Thus, the dermal branchiae serve as an interface between the two fluid environments where CO_2 and O_2 can be exchanged by diffusion. Note the presence of tiny pincerlike structures, the *pedicellaria.* These minute claws grasp and drive off would-be encrusting or sucking organisms.

O_2. Capillaries will be described in detail later, but you should be aware that they are the smallest blood vessels, with walls only one cell thick.

The aquatic mollusks, with the exception of a few air breathers, rely on gills. In the bivalves (such as clams) the gill serves in both food filtering and respiration as was mentioned in Chapter 25. In bivalves, gill structure [Figure 26.4 (a)] consists of sheetlike folds that protrude into the mantle cavity. Water is drawn, by ciliary action, into the incurrent siphon. From there, it passes through microscopic pores in the gills, entering channels that direct it out through the excurrent siphon. As water passes through the channels, O_2 and CO_2 are exchanged through tiny blood vessels that permeate special partitions. Oxygenated blood is then returned directly to the heart for distribution to the body. Mollusks lack blood cells and the oxygen-carrying pigment, *hemocyanin* (a copper-containing protein) is carried along in solution.

The arthropod gills are varied, so we will concentrate on the lobster, a crustacean. The lobster gills are feathery extensions of the body wall, attached partly to the bases of the legs [Figure 26.4 (b)]. They are covered by a *carapace,* which is open at both ends. Water is drawn under the carapace, across the gills toward the head by the action of a paddlelike appendage near the jaws known as the *bailer.* Blood from an open cavity in the thorax floor passes through channels in the gills, exchanging O_2 and CO_2 as it goes. Following this vital circuit, it returns to the heart to be pumped out again.

Since the blood is not continuously enclosed in vessels, the system is an *open* one. In the open system, the blood is pumped from the heart into the vessels, but the vessels empty into channels and sinuses (hemocoels). The blood percolates through these sinuses and then returns to vessels that lead them to a cavity surrounding the heart. From there they are drawn into the heart and pumped out again into the incomplete vessels. For those of you who sail, this kind of heart is reminiscent of a sump or bilge pump, the submersible water pump that draws water from its surroundings and directs it into hoses. A *closed circulatory system* is one in which the blood is virtually always enclosed in vessels and not left free to percolate through tissues. Like the mollusks, the blood of arthropods contains hemocyanin.

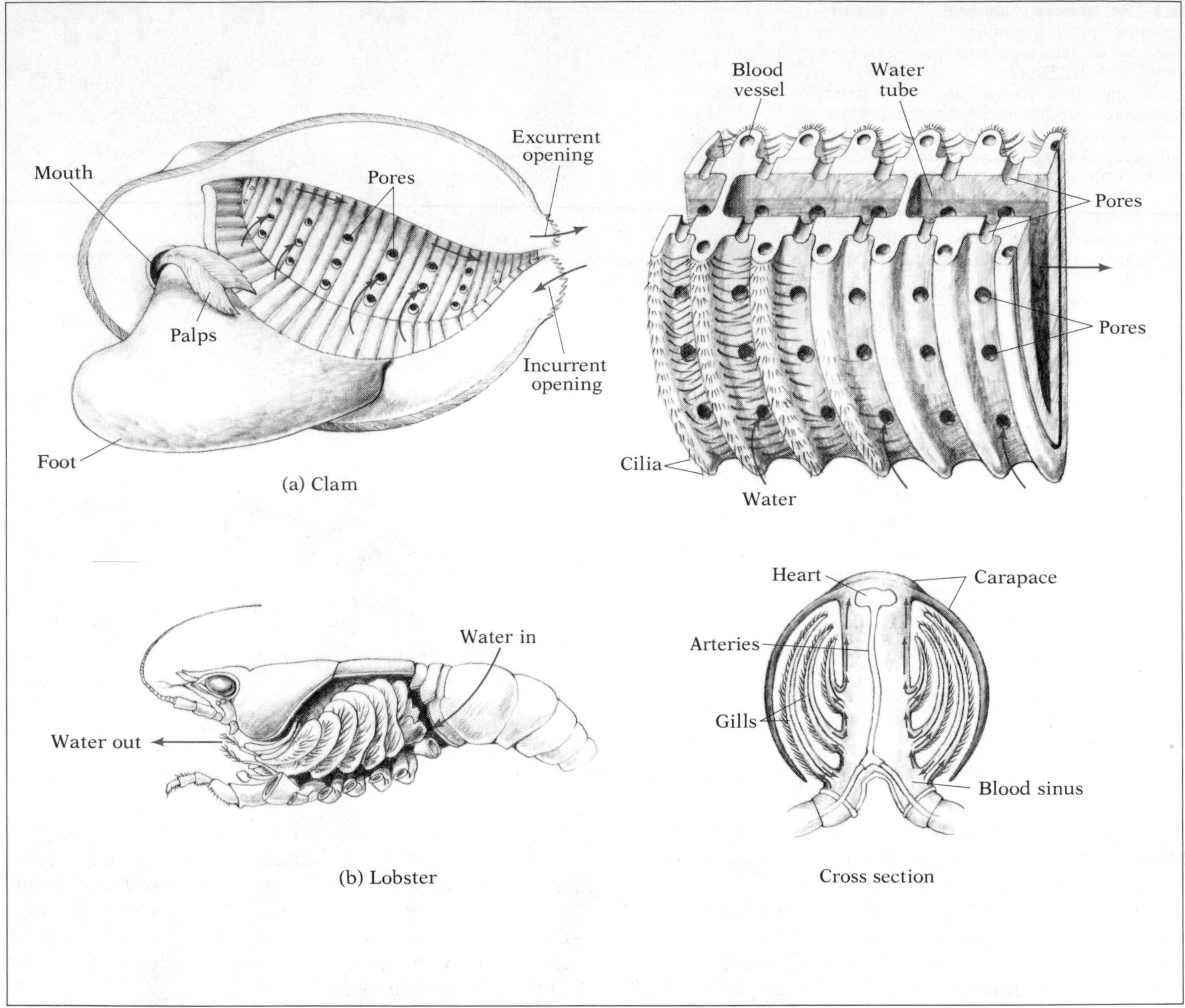

26.4 In each animal, the delicate gills are protected by a covering. The clam (a) and lobster (b) actively move water over their gill surfaces. In the clam, a stream of water created by ciliary action enters through an incurrent siphon, passes over the gills, and leaves by an excurrent siphon. Each gill contains numerous pores that direct water through passages leading to the excurrent opening. Blood vessels penetrate the gill, exchanging CO_2 for O_2 across its thin surfaces. The lobster makes use of a pair of specialized appendages (*bailers*) to draw currents of water under the carapace of its exoskeleton, across the gills, and out under the head. The lobster's gills are very similar to those of fishes. Each consists of a central supporting section that contains feathery filaments. Blood enters the gills from an open cavity below, flows into the gills, and exchanges CO_2 for O_2 in the thin-walled filaments. Oxygenated blood leaves the gills to flow through vessels directly to the heart.

Gill Baskets

Gill baskets are found in tunicates and lancelets, two of our chordate relatives. Tunicates have a fairly primitive, open circulatory system. The tunicate heart, in particular, has a curious feature: It periodically reverses direction, so that blood sloshes back and forth; running first one way, then the other. Tunicate blood doesn't contain hemoglobin, but instead has protein that contains the obscure element *vanadium*, which combines readily with oxygen. The tunicate's *gill basket*, a highly branched structure with an enormous surface, probably does function in gas exchange, but its principal task is that of straining microscopic food out of seawater.

In the lancelet, *Amphioxus*, another chordate (see Figure 23.2), the gill basket is very similar to that of the tunicate and also strains food out of water. But the gills quite clearly also provide the usual respiratory function of other invertebrate gills. The repiratory pigment is hemoglobin, which is not free in the plasma, but is contained in red blood cells. The general scheme of the *Amphioxus* circulatory system is almost exactly the same as that found in modern fish in early mammalian embryos, with one exception: *Amphioxus* is a heartless beast. Blood flows by peristaltic contractions of its larger arteries.

Gills in Fishes

Larval lampreys are structurally and functionally similar to *Amphioxus*, although they do have hearts. The gills of adult lampreys and almost all other fish have lost any food-straining function and are solely respiratory organs. In most fish, there are five pairs of gills.

Gills in fishes are typically rows of fingerlike rods from which feathery filaments arise on either side (Figure 26.5). Each supporting rod houses an artery that sends thin-walled capillaries into each thin-walled filament. Thus, O_2 and CO_2 can pass across the walls with relative ease. Blood oxygenated in this way then passes back into the supporting rod and joins a vein, which returns it to the body for distribution. While the origin and structure of gills varies, the functional principle is the same. The gills of fishes permit the blood to flow through innumerable tiny blood vessels that feed thin-walled capillaries imbedded in lamellae, over which water must pass. The gills themselves are comprised of concentric rows of supportive gill arches along which lie blood vessels that feed the fingerlike extensions that arise from the arches, the gill filaments. *Gill rakers* are stiffened fingerlike protrusions from the gill arches that keep swallowed food from passing out through the gill openings and provide some protection for the delicate gills. The gill filaments are fed by *efferent blood vessels* bearing deoxygenated blood. From the efferent vessels the blood flows into the capillaries where the gas exchange takes place and the freshly oxygenated blood then flows back into the system through the *afferent* vessels.

The exchange of gases in fish gills is made more efficient and complete in an interesting manner, utilizing a *countercurrent mechanism.* The capillaries of the gills are arranged in such a manner that the direction of blood flow opposes the direction of water movement. Since the blood crossing the capillary continually encounters a new supply of water with its full load of O_2, the O_2 gradient favors the movement of oxygen into the blood throughout. In other words, O_2 can continue to enter the circulatory system until the blood leaves the capillary. Carbon dioxide removal is similarly enhanced, but the gradient is opposite. The moving current of water never becomes saturated with CO_2, permitting a more complete exchange.

Countercurrent exchanges in organisms are established when substances (or heat) move in differ-

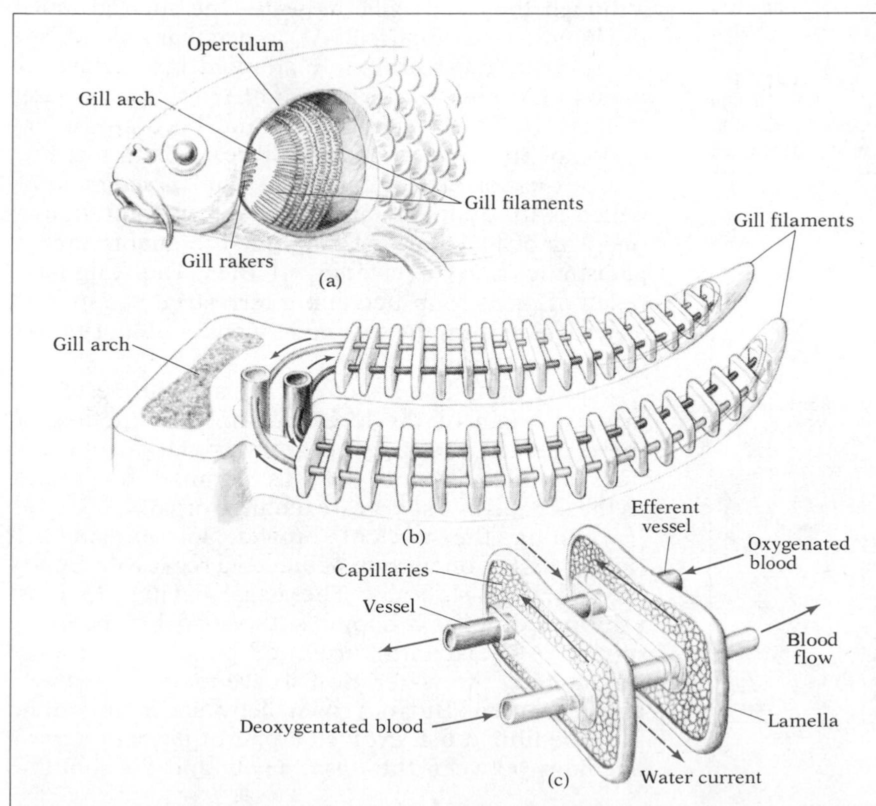

26.5 Gills are membranous structures through which CO_2 and O_2 pass. (a) They may be quite elaborate, with a supporting circulatory system as seen here in the fish gill. The operculum has been cut away to show the details. (b) The supporting structures of the fish gills are the cartilaginous gill arches. Inside the curve of the arch are the *rakers,* which act as a screen to prevent food from going out the gills. At the outer curve, two rows of gill filaments protrude from each arch. (c) The surface area of each filament is greatly increased by the presence of lamellae as shown. Each lamella has a rich supply of capillaries that branch from incoming vessels carrying deoxygenated blood. Only a few cells separate the capillaries from the water flowing over the lamella. Thus CO_2 and O_2 are readily exchanged between blood and environment. Note the arrows which describe the flow of water across the lamellae and the flow of blood in the capillary bed within. The opposite movement enhances the exchange of O_2 and CO_2 through a countercurrent exchange. The capillaries rejoin to form vessels which carry oxygenated blood out of the filament. From the gill it enters the general circulation of the fish.

ent directions, but in closely paralleled vessels or tubules and as long as an exchange can occur between the flowing materials. Animals have adapted to this phenomenon in gas exchange, heat regulation, and excretion.

Oxygen doesn't diffuse very rapidly through water, and the concentration of oxygen in seawater is rather low—only about 0.4% by weight. In order to maintain an adequate level of gas exchange, all active aquatic animals must keep water flowing almost constantly over their respiratory surfaces. Among the fish, sharks of the open sea manage this by constantly swimming forward with their mouths open. It adds to their sinister look, but they're only breathing. (Recently, divers claim to have found sharks motionless, apparently asleep. No one yet knows the implications of this finding. In any case, some bottom dwellers utilize a bellows mechanism, which works by contraction and expansion of the gill chambers and so the animals do not need to keep moving.)

Bony fish (sharks are not bony fish) keep the water passing over their gills by their mouth muscles acting together with a remarkable arrangement of valves (Figure 26.6). Water is taken through the mouth, as the gills are closed off to the outside by the protective *opercula*, or gill covers. The fish then closes its mouth, makes a sort of swallowing movement to contract its oral cavity, and opens its opercula to let the water flow out across the gills. And that's why fish so often stand motionless, gulping.

26.6 Breathing movements in fish. The movement of water over the gills is not a passive process. It is directed by active pumping that is controlled by valvelike structures. As the diagram indicates, water is drawn into the mouth cavity with the *opercula* closed. In a second action, water is forced back over the gills and out through the opened opercula. Movement back out through the mouth is prevented by closing of the oral valve. The cycle is repeated continuously as the fish satisfies its respiratory demands; the rate of pumping changes to accommodate varying demands.

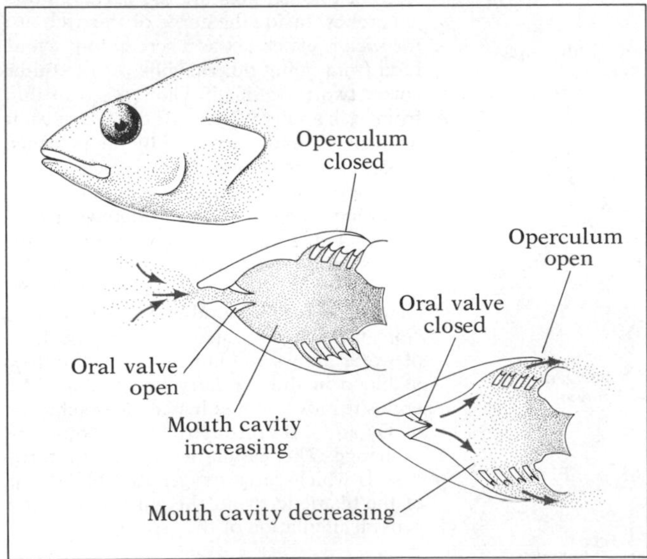

Some freshwater fish—for example, goldfish—can gulp air. Their oral cavities contain heavily vascularized tissues and chambers, and act sort of like lungs. Unlike the more exotic aquarium fish, goldfish can survive and even flourish in polluted water that is nearly devoid of oxygen, which is the main reason they are easily kept in fishbowls.

Complex Interfaces: Lungs

Evolution of the Vertebrate Lung

The vertebrate *lung* has had a curious evolutionary history, which apparently goes something like this: A long time ago, the ancestor of all modern bony fishes (that is, the ancestor of all living fishes except sharks, rays, chimera, and agnathans) was a freshwater fish, perhaps rather like the carp. This primitive fish also gulped air for its oxygen, and habitually swallowed it. A pocket of highly vascularized tissue became specialized in the pharynx of this ancestral fish as a place to hold its oxygen-laden air between swallows. Over evolutionary time, the air pocket grew and branched, and became the vertebrate lungs. Our lungs.

Some of the descendants of that early ancestral fish species retained the lungs as accessory organs, although they kept gills as well. One such group is the lungfish (see Figure 23.11), a peculiar assemblage of freshwater fish uniquely adapted for survival in ponds that tend to dry up. Another group was later to leave its freshwater habitat and become the ancestor of the terrestrial vertebrates. For this group, the possession of lungs was a vital *preadaptation*, which is to say it was an evolutionary adaptation to one way of life that just happened to enable the organism to also survive in another. The long-term result of one group becoming terrestrial is apparent as we watch humans staring in wonder at a primitive fish in an aquarium.

Other bony fish have kept the structure and embryonic origin of the lung, but not its function. In these species, the lung became a flotation device—the *swim bladder*—and the gills regained their status as the organism's sole gas-exchange organ.

Among the ancient lunged, lobe-finned fish (Figure 23.12), one group achieved a real evolutionary breakthrough—literally. The nasal cavities in most fish (including, as it happens, those of the one living species of coelacanth), are blind pouches used only for smelling the water. So fish are mouth breathers, as we've seen. But our own fishy ancestor among the lobe-finned fish evolved a pair of *internal nares*—openings between the nasal cavity and the mouth—

and for the first time, a fish could breathe through its nose. It's not clear that they actually did use their internal nares for breathing—lungfish also have internal nares, but gulp air through their mouths—still, this feature certainly proved to be yet another preadaptation. It helped greatly when certain descendants invaded the land. To this day, the internal nares of amphibia and most reptiles open directly into the mouth cavity, as did those of the fish ancestor, but, in any case, these animals can breathe with their mouths closed. In the evolution of mammals (and crocodilians, as it happens), a *palate* further separated the mouth cavity from the nasal cavity, pushing the internal nares back to the region of the pharynx. Thus we can chew and breathe at the same time.

The Vertebrate Lung

The lung is an inpocketing, branching tube that ends in a multitude of tiny air sacs (*alveoli*), where blood and air are separated by a thin, moist membrane. By muscular control of its breathing apparatus and the valves of its mouth and pharynx, the animal can control the amount of air that passes through its lungs. Of course, moisture loss is unavoidable; the exhaled air is saturated with water vapor.

The evolutionary pathway taken in changing from an aquatic to a terrestrial respiratory system is perhaps represented by today's amphibians. Nearly all amphibians have lungs (the lungless salamanders we mentioned earlier have secondarily lost their lungs along the way) and nearly all species exchange gases in two ways: through those rather simple, baglike lungs and through their moist, highly vascularized skin. The ability to breathe through the skin enables many amphibians to hibernate in mud through cold seasons, although their restricted oxygen supply means they must remain inactive. (It's probably hard to be very active while encased in mud anyway.) Not only are the lungs simple, but even breathing movements in amphibians are quite primitive. Since there is no diaphragm or well-developed rib cage, air is forced in and out of the lungs by the raising and lowering of the floor of the mouth. This coarse pumping action is aided by the presence of "check valves" in the nostrils and the *glottis* (the opening of the trachea) (Figure 26.7).

The amphibians also illustrate the transition from aquatic to terrestrial gas exchange in another, more dramatic way. Each individual makes this transition in its own lifetime. Larval forms—tadpoles and newts—breathe by gills. The tadpole gills are enclosed within a cavity, superficially much like the atrium that encloses *Amphioxus* gills. The newt's gills, on the other hand, are feathery extensions that stick out from

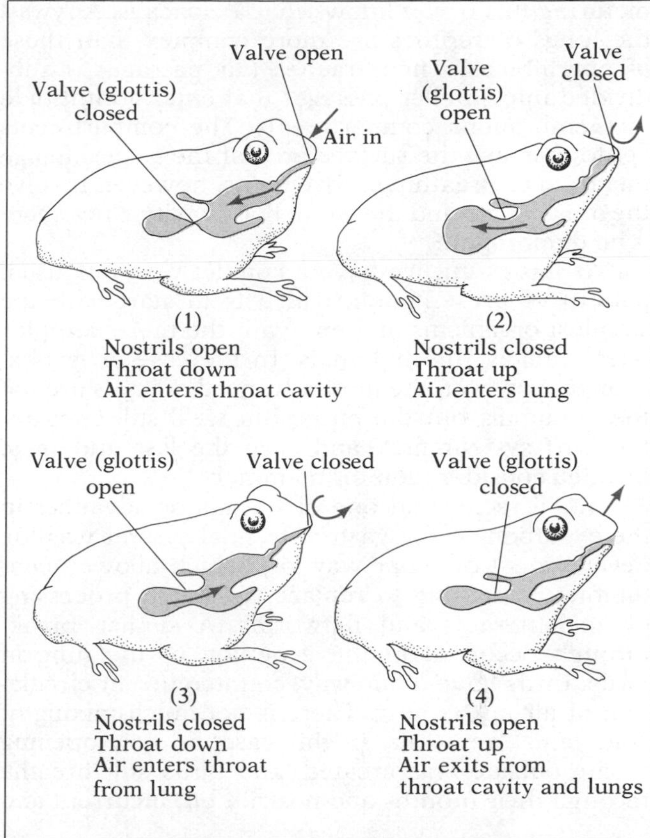

26.7 The frog breathes through lungs, but gases also pass across its moist, highly vascularized skin. Air is forced in and out by the bellows action of its throat and body wall, since there is no diaphragm. Note the four-stroke system using check valves in the nostrils and trachea. Like all vertebrates, the frog has a closed circulatory system with red blood cells that contain hemoglobin.

the sides of its throat. The gills are strictly for aquatic gas exchange, of course, and the skin is used as a gas-exchange surface both in the water and in the air. Larval newts don't retain the pharyngeal gill clefts of their fishy ancestors, however, so they strictly depend on water flowing past their bodies.

At the time of metamorphosis, the amphibia begin to swallow gulps of air, the lungs expand, and ultimately the gills are absorbed back into the body. Amphibians don't have waterproof integuments, thus they can breathe through their skin—or perhaps we should say that amphibians breathe through their skin, which doesn't allow them to have waterproof integuments. Unfortunately, one can't have it both ways, since each condition has its advantages.

Since the dry reptilian skin is impervious to air, reptiles, including the many aquatic species, are strictly lung breathers. Perhaps *strictly* is not the right word. The inevitable exception this time is found in some water-breathing turtles that use a large, heavily vascularized cloaca as a supplemental respiratory surface—a kind of water lung. You may have trouble picturing this if you know what a cloaca is. Anyway, the lungs of reptiles are more complex than those of amphibians. Their trachea (air passage) is subdivided into smaller passages that enter a multitude of membranous compartments. The compartments act to increase the surface area of the gas exchange membrane. Breathing movements, however, involve the muscles around the entire body cavity, since there is no diaphragm.

At this point, we have a conflict with our usual pattern of presentation, which is to start with the simplest organisms and end with the more complex ones, namely the mammals. In the case of gas exchange, however, the most advanced animals are not the mammals but the birds. But we'll still consider the bird system first, and save the last and most detailed consideration for mammals.

You'll recall that one of the major advances in the evolution of the gastrointestinal system was the development of a one-way gut, which allowed continuous processing to replace the batch processing of coelenterates and flatworms. A similar breakthrough occurred in the evolution of the lung in birds. Birds have a one-way, countercurrent circulation of air in the lung. There is not much mixing of "old" and "new" air. In this case, no new opening to the outside was created, and birds still breathe through their mouths and nostrils, but incurrent and excurrent air are separated deep within the respiratory cavity (an improvement over the efficiency of the mammalian lung, as we shall see).

It should not be surprisng that the respiratory system of birds is the most complex of all the vertebrates. After all, flight is incredibly demanding, and birds have astonishingly high metabolic rates; they use a great deal of oxygen in a very short time.

So how do they get it? Lets trace the pathway. What we will find is that the inhaled air passes in wide *bronchi* right through the lungs and into an extensive system of *air sacs* that penetrate throughout the body. The air sacs surround the gonads of at least some male birds, and even fill the cavities of some of the longer bones. Hollow bones, of course, mean a great saving in weight, and the system can serve as a cooling mechanism.

When the bird is not flying, the muscles of the thorax and abdomen are responsible for the breathing movements, but in flight, the rapid wing beats help move air.

Once the air has entered the air sacs, a series of check valves prevents it from leaving by the same path. Instead, it is now shunted back to the lungs, this time (and for the first time) passing through the alveoli and exchanging gases with the blood system (Figure 26.8). The one-way flow of air through the alveolar tissue assures that stale air can't accu-

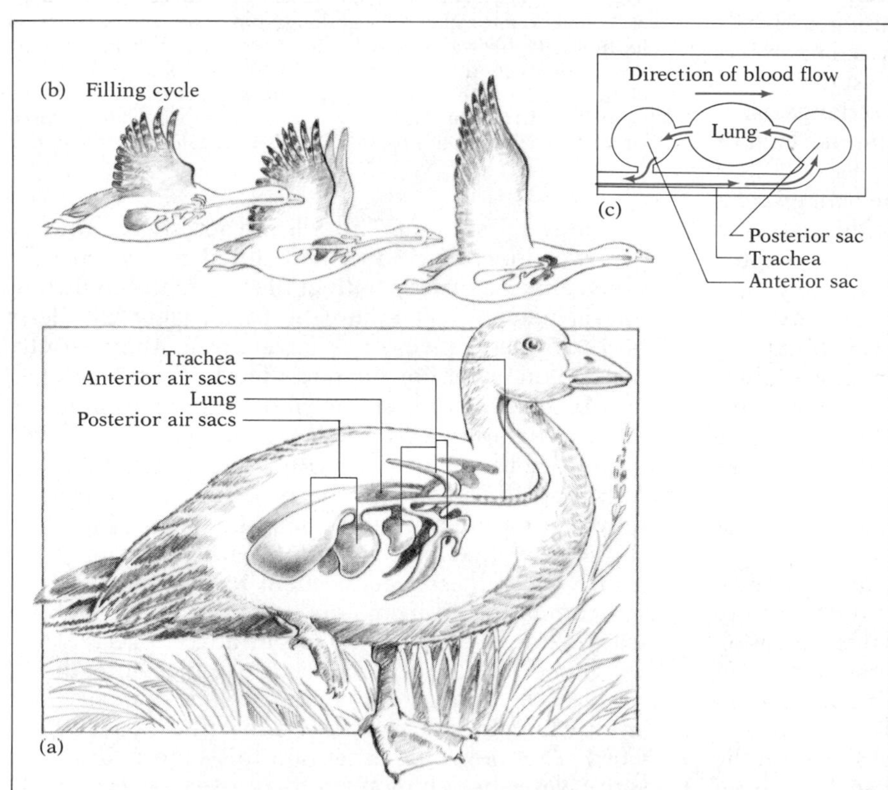

26.8 Consistent with the severe respiratory demands of flight, the respiratory system in birds is very well-developed and highly specialized. (a) Not only do birds have lungs, but numerous air sacs that extend into its long bones. The respiratory system has three regions, the posterior sacs, lung, and anterior sacs. (b) Air enters the posterior sacs, where CO_2 and O_2 are first exchanged. It then enters the lung where more exchange occurs, and finally the anterior sac for a final exchange. Then it is exhaled. (c) Interestingly, the blood flows opposite to the direction of air movement. This countercurrent assures a more efficient exchange of gases and makes it possible for the bird to be active at higher altitudes where oxygen is less plentiful. Bats, which have typical mammalian lungs, can hardly survive at altitudes where birds are active.

mulate there. In contrast, in the mammalian lung, a sizable proportion of the air in the lung at inhalation is still present at maximum exhalation, since the lung cavities can't be completely emptied. This remaining oxygen-depleted air must mix with the fresh air of the next incoming breath—not a particularly efficient system.

We now turn to a more detailed look at a respiratory system in the mammals, concentrating on ourselves.

The Anatomy of the Human Respiratory System

The human respiratory system is fairly typical of other mammals. In Figure 26.9, you can see that air first passes through the nasal passages (unless you are a chronic mouth breather). In addition to their role in directing the air flow, these passages are important in filtering, warming, and moistening the air prior to its entering the lungs. The nostril hairs act as an initial filter, trapping dust particles. A nose full of hair may not be esthetically pleasing, but your lungs undoubtedly appreciate it.

26.9 The human respiratory system is typical of mammals. It begins with nasal passages, which are separated from the mouth by the palate and function in warming and filtering air. Air then moves through the pharynx and larynx into the trachea. From the trachea it enters the lung proper through the two bronchi that branch and rebranch into bronchioles, finally ending in blind pouches (alveoli). Along with the extensive supporting circulatory system, the many passages and alveoli are the main components of the lungs. The lungs are in the thoracic cavity, which is bounded by the rib cage and lies above the dome-shaped, muscular diaphragm.

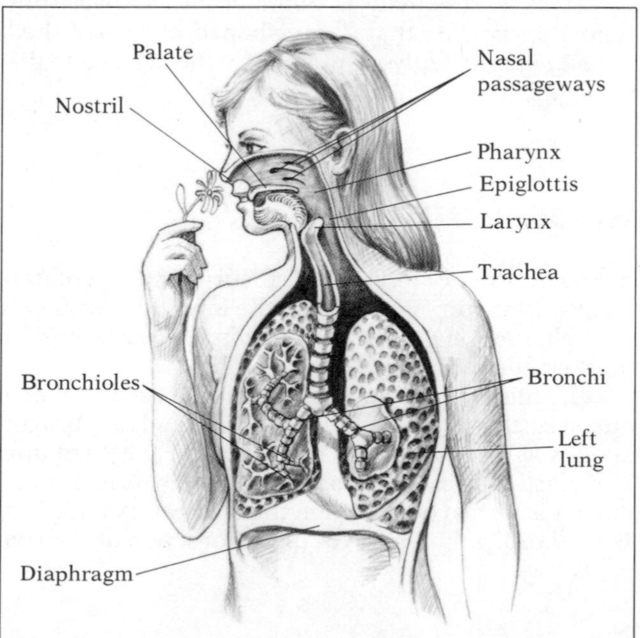

In all mammals, the nasal cavity is separated anteriorly from the mouth by the *hard palate*, a bony shelf in the roof of the mouth ending in a softer muscular region, the *soft palate*. The nasal cavity is lined with mucous membrane, which includes mucous glands, mucus-secreting *goblet cells*, and cells bearing cilia (Figure 26.10). These glands and cells

26.10 Microscopic structure of the respiratory passages. (a) Much of the surface of the respiratory passages is composed of a complex epithelium. Within this cellular lining are large numbers of goblet cells and ciliated cells that secrete mucus and sweep it, along with trapped dust and foreign materials, toward the throat. What happens then depends on your upbringing. (b) The scanning electron microscope view shows such a ciliated lining. In the normal, nonsmoker's trachea, the cilia can be visualized as a constantly moving column of mucus, which spills over into the pharynx. It is joined there by a downward moving stream of mucus from the upper passageways. The accumulated fluids are constantly swallowed, and numerous bacterial spores and other contaminants from the air therefore find themselves in the hostile acids of the stomach.

produce a film of mucus, which is constantly swept toward the throat by the ciliary action. In addition, the lining of the nose contains many dense capillary beds that warm the air before it enters the lungs. The system is very effective. Even on very cold days, air entering the lungs of gasping joggers is warmed almost to body temperature.

The inhaled air moves from the nasal passages into the pharynx and from there through the larynx into the trachea unless you are swallowing, whereupon the laryngeal opening is closed. You can't breathe while you swallow (of course, immediately everyone tries). The laryngeal opening itself contains the vocal mechanism. Below it, the *trachea* is a tube composed of C-shaped rings of stiff hyaline cartilage that hold the passage open. As the trachea enters the chest, it branches into right and left *primary bronchi* which branch again and again into the *bronchioles* that form the *respiratory tree*. The trachea and some of the other bronchial structures have a secreting and sweeping lining similar to that in the nasal passages, so the air is cleaned of dust and debris once again. (Probably the air going out is cleaner than the air coming in, so, in a sense, humans help to clean the earth's air.) Tiny particles of inhaled dust become trapped in the mucus, which is continually being moved up into the throat and swallowed.

The bronchioles, the tiniest branches of the respiratory tree, end in grapelike clusters of air spaces known as the *alveoli* (Figure 26.11). Each cluster is enclosed in a dense capillary bed, so here the atmosphere is only a membrane away from the blood. The clusters, of course, provide an enormous surface area. In fact, the total surface area of the approximately 300 million alveoli of the human lungs has been estimated at nearly 750 sq ft. So your lungs could pave a tennis court.

When viewed intact, the lungs appear large, roughly triangular in shape, and have a broad base.

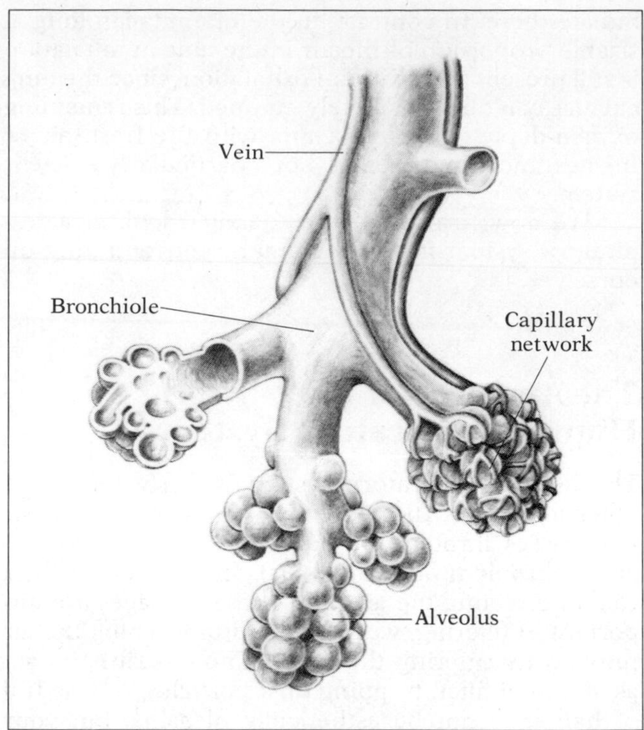

26.11 The alveoli and their circulatory supply. The bronchioles terminate in the alveoli. Each alveolus is a cluster of blind, thin-walled air sacs with dense capillary networks. The highly elastic air sacs provide the interface between environmental air and the circulatory system.

Two baglike membranes (the *pleurae*) enclose the lungs. The inner pleura is attached firmly to the spongy surface of the lung. The outer pleura forms the tough lining of the pleural cavity, which houses the lungs. (*Pleurisy* is an inflammation of the pleurae.)

The pleural cavity is bounded at its lower (posterior) portion by that dome-shaped muscular shelf, the *diaphragm*. Only mammals, by the way, have diaphragms.

26.2 Gas Exchange and Respiration Control

The Breathing Movements

Breathing, or *ventilation*, in humans involves the *intercostals* (the muscles of the rib cage) and the diaphragm. In the relaxed condition, the diaphragm rises into a dome shape and protrudes into the pleural cavity. Inhalation is accomplished by contracting (lowering and flattening) the diaphragm, contraction of the external intercostals, and lifting the rib cage (swinging the ribs upward). The change increases the volume of the pleural cavity, creating a partial vacuum. But why do the lungs fill? Strangely enough, *the weight of the atmosphere forces air down into the lungs, causing them to inflate.*

Exhalation, when passive (not forced), is produced by the relaxtion of the rib cage and diaphragm, actions which decrease the pleural cavity volume. The elastic lungs then contract to their former state, and air is forced out through the respiratory passages. Forced expiration involves the contraction of internal intercostal muscles which further depress the rib cage.

Partial Pressure

The amount of one particular kind of gas in a system can be described in terms of *partial pressure.* For instance, air is about 21% O_2; therefore, the partial pressure of O_2 at sea level, where the total atmospheric pressure is 1 atmosphere, is 0.21 atmosphere. On a mountain top, the air is still 21% O_2, but the total atmospheric pressure is much less and therefore, so is the partial pressure of O_2. So we breathe harder up there.

Consider another example. A diver at about 100 m breathes compressed air with a total pressure of about 10 atmospheres; the partial pressure of O_2 is 2.1 atmospheres, which is too high for normal physiological processes. For prolonged activity at such depths and pressures, divers may breathe a mixture of artificial air that is perhaps 98% inert helium and 2% O_2. At 10 atmospheres total pressure, that would give a partial pressure of O_2 of 0.20 atmosphere, approximately the same availability of oxygen as in normal air at sea level.

The availability of gases dissolved in a fluid can also be expressed in partial pressures. For instance, the partial pressure of O_2 at the surface of a stream may be, say, 0.20 atmosphere. This doesn't mean that the water is 20% O_2, or even that the gases dissolved in the water are 20% O_2; it just means that the water is in approximate equilibrium with air in which the partial pressure of O_2 is 0.20 atmosphere. At the bottom of a quiet lake, the partial pressure of O_2 may be much less, perhaps only 0.05 atmosphere or less. What that means is that the water, at whatever temperature and salinity it happens to have, has as much oxygen dissolved in it as if it were in equilibrium with 0.05 atmosphere of gaseous oxygen.

The concept of partial pressure will be important to our understanding of gas exchanges, because diffusion of dissolved gases always occurs from regions of higher partial pressure to regions of lower partial pressure. For instance, if we had an open container of a fluid that was rich in CO_2 and poor in O_2 (as measured by the partial pressures of these two gases), and placed it inside a closed container with a mixture of gases that was rich in O_2 and poor in CO_2, in due time the fluid and the gas would equilibrate, so that the partial pressure of O_2 would be the same in both the fluid and the gas, and the partial pressure of CO_2 would also be the same in both parts of our closed system. The actual amount—in contrast to the immediate availability—of O_2 and CO_2 in each part of the system would depend on the solubility of each gas in the fluid, which in turn would depend on such factors as temperature.

The Exchange of Gases in the Alveoli

We breathe in air rich in O_2 and breathe out air rich in CO_2. The exchange takes place on the moist inner surfaces of the alveoli, primarily through simple diffusion (Figure 26.12). The blood that enters the lung from the heart has previously been routed through the body tissues, where the constant depletion of O_2 in mitochondrial respiration has created a very low partial pressure of O_2. At the same time, the partial pressure of CO_2 is high in the body tissues, again due to metabolic activity. The blood has tended to equilibrate with the body tissues. Hemoglobin (the carrier of O_2 in the blood) is deoxygenated in the body tissues and picks up molecules of CO_2.

Now the blood itself, as it enters the lungs, has a low partial pressure of O_2 and a high partial pressure of CO_2. Some of the CO_2 is carried in the form of bicarbonate ions (HCO_3^-), some as dissolved CO_2, and some loosely bound to hemoglobin. In the arterioles of the alveoli, an enzyme (*carbonic anhydrase*) catalyzes the conversion of HCO_3^- to gaseous CO_2, further increasing the partial pressure of this gas. In the brief time that the blood and air are in near contact on opposite sides of the thin alveolar membrane, nearly complete equilibration takes place. The blood and air become almost equal in the partial pressures of O_2 and CO_2, as molecules of O_2 diffuse in and molecules of CO_2 diffuse out. Then the equilibrated air is exhaled and fresh air is inhaled, while simultaneously the blood flows continuously through the lung. Blood leaving the lung is relatively high in O_2 and low in CO_2.

One key to the efficiency of the system is the behavior of hemoglobin itself. Hemoglobin is highly specialized to associate and dissociate with oxygen. While an ordinary passive fluid (such as water) will absorb O_2 in rough proportion to its partial pressure, hemoglobin is more sensitive to O_2 availability. Whereas the affinity of H_2O for O_2 is simply dictated by partial pressures, hemoglobin has the remarkable ability to "change its mind" about how it feels about accepting oxygen. To put it a bit more scientifically, hemoglobin has a strong affinity for O_2 when the partial pressure is high, and loses its affinity for O_2 when the partial pressure is low. This allows the hemoglobin to give up all its O_2 in its journey through the body, and to become quickly saturated again in the lungs.

Because of hemoglobin's great affinity for O_2, it can carry sixty times as much O_2 as can an equal weight of water alone. So if we didn't have a specific O_2 carrier, our circulation would have to be sixty times as fast, or our activity would have to be greatly curtailed. There are plenty of organisms that do, in fact, manage without a circulating O_2 carrier, but they are either very small or sedentary, or both.

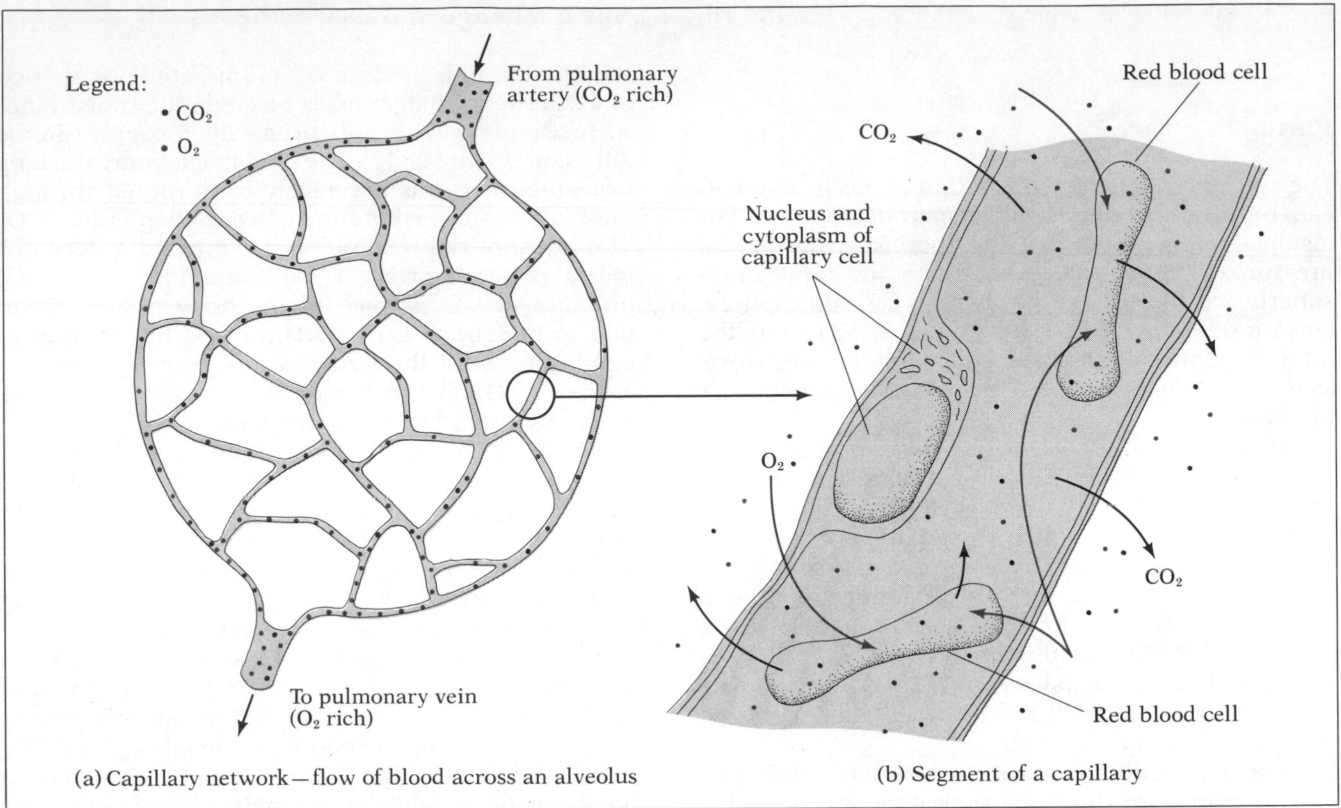

26.12 Exchange of gases across the alveolus. The net movement of molecules of CO_2 and O_2 is shown in the capillary network (a) and in a small segment of a capillary (b). In the capillary network, blood entering the alveolar region is rich in CO_2 and nearly depleted of O_2. But as it makes its way through the capillary bed, the situation reverses. Carbon dioxide diffuses out of the blood into the air sac while O_2 diffuses into the blood. Thus, blood leaving the air sac is rich in O_2 and has a low CO_2 content. In the enlarged view, note the thin membrane of the capillary wall and alveolar membrane separating the blood from the inhaled air. Red blood cells (drawn to scale) are instrumental in the exchange as they easily give up CO_2 and attach O_2 to their hemoglobin to form *oxyhemoglobin.*

So what, precisely, is this magic molecule and how does it work? A molecule of hemoglobin consists of four polypeptide chains and four heme groups (Figure 3.12), one heme group for each polypeptide chain. Each heme group contains an iron atom, and in the intact protein, the four iron–heme groups can bind reversibly to four molecules of O_2. This is not a covalent bond, and the iron doesn't change its oxidation state. The association and dissociation of O_2 and hemoglobin (Hb) can be written:

$$\text{Hb} + 4\,O_2 \rightleftharpoons \text{Hb} \cdot (4\,O_2)$$

The left side of the formula shows *deoxyhemoglobin* and the right side, *oxyhemoglobin*.

One more point about oxygenation of the blood: At sea level, and under normal conditions, blood passing over the alveoli is nearly saturated with O_2. That is, most of the hemoglobin molecules pick up their full allotment of four O_2 molecules. The partial pressure of O_2 in the lungs is about 0.16 atmospheres at sea level. Increasing the O_2 concentration in the lungs has very little effect on the amount that crosses into the lungs. There is a considerable safety margin in this relationship, because it also works the other way. Reducing the O_2 concentration doesn't easily change the amount of O_2 reaching the blood either. With 0.08 atmospheres partial pressure of O_2, the blood will still become 80% saturated. This essentially means that humans can live comfortably at varying altitudes, including high in mountains where O_2 partial pressure is low. Furthermore, at high altitudes, additional red blood cells are released by the spleen, and, in addition, more red blood cells are produced in the bone marrow, allowing for long-term adaptation. (Some long-distance runners train at high altitudes in order to manufacture more red blood cells and thus be able to get more O_2 into the blood than runners who train in the flatlands.)

Now let's look at a related phenomenon: CO_2 transport. First of all, the two gases are inversely related. It turns out that where partial pressures of CO_2 are high, O_2 has a lower affinity for hemoglobin. When CO_2 levels are low, hemoglobin holds its O_2

more tightly. In other words, we might consider the two gases to be in competition for hemoglobin. An important factor in how hemoglobin handles O_2 and CO_2, then, is the relative numbers of molecules of each gas present.

By the way, CO_2 doesn't actually combine with the heme groups the way O_2 does, but instead reacts with amine side groups of some of the amino acids in other parts of the protein. The combination of CO_2 and hemoglobin is called *carbaminohemoglobin,* and the effect of CO_2 on hemoglobin–oxygen dissociation is called the *Bohr effect:*

$$\mathrm{Hb\cdot(4\,O_2) + CO_2 \rightleftharpoons Hb\cdot CO_2 + 4\,O_2}$$

The adaptive significance of the Bohr effect is obvious if you think about it. The presence of CO_2 is an indicator that oxidative metabolism is occurring, and it means that the tissue could use some O_2. The most metabolically active regions produce the greatest quantities of CO_2, which in turn hastens the release of O_2 from the passing bloodstream. Interestingly, the Bohr effect varies with different animals. In mice and other very small animals with rapid circulation, the blood gives up its O_2 very readily, so that the hemoglobin returns to the lungs from its quick transit through the tiny mouse body in a deoxygenated state, ready to take on another load. In horses, humans, and other large mammals, the blood holds on more tightly to its O_2, not releasing it until it is deep into the relatively deoxygenated, carbon dioxide-laden tissues.

Most CO_2, you may recall, is not carried on the hemoglobin at all. About 80% of it is carried as bicarbonate ions (HCO_3^-) in the plasma, as we've seen. Carbonic anhydrase, the same enzyme that catalyzes the conversion of HCO_3^- to CO_2 and H_2O in the lungs, speeds the reaction in the opposite direction in the tissues:

$$\mathrm{CO_2 + H_2O \xrightleftharpoons{\text{Carbonic anhydrase}} \underset{\text{Carbonic acid}}{H_2CO_3} \rightleftharpoons HCO_3^- + H^+}$$

The release of hydrogen ions in this reaction means that blood becomes increasingly acid as it travels through respiring tissue. This higher acidity itself also affects the dissociation of hemoglobin, further inducing it to unload its O_2.

The Control of Respiration

The perpetual, incessant changes that are the hallmark of living things are usually only minor adjustments—fine tuning—a way of responding to shifting conditions within and without, as the body attempts to keep things constant. For example, the need for O_2 and the production of CO_2 are always changing in every part of the body—now a little more here, now a little less there. Considering all such adjustments, however, the body's task is enormous and its ability to respond is admirable, even dazzling. How can it handle such a complex chore as regulating O_2 and CO_2 levels in literally millions of places at once?

First, we should be aware that both the respiratory and the circulatory system must respond in a coordinated way to changing O_2 levels. After all, increasing the breathing rate would be of little value unless one also increased the rate of blood flow. Second, it is important to know that respiratory control is exceedingly complex, involving many types of sensors and a variety of complex interactions which are poorly understood. However, we do know about a few discrete centers that have to do with breathing and these are in the brain.

The major respiratory control centers are in the *pons* (a small sphere at the base of the brain) and the *medulla* (also known as the *brainstem*)—the part of the brain that tapers into the spinal cord. Actually, the medulla contains the control centers for several other vital functions, including the heartbeat. Three centers appear to be involved. The best understood is called the *rhythmicity center.* Its interaction with the other two is a bit of a mystery, but apparently the relationship is somehow critical. The rhythmicity center, itself, has two circuits of opposing neurons, one for inspiration and one for expiration. (So exhalation doesn't occur simply because inhalation stops. It has its own circuit.) Apparently, the coordination of these two circuits produces the rhythm of breathing. The breathing centers of the brain are integrated with stretch receptors in the lungs that help regulate the depth of breathing. The sequence of the action is:

1. The inspiratory circuit activates the muscles for inhalation and, at the same time, inhibits the expiratory circuit.
2. The lungs fill, activating stretch (sensory) receptors which fire a signal back to the rhythmicity center to inhibit the inspiratory circuit. In the meantime this circuit ceases to inhibit the expiratory circuit.
3. The expiratory circuit becomes active, inhibits the inspiration center, and causes exhalation.
4. When the lungs empty, the stretch receptors no longer inhibit the inspiratory circuit, and soon the inhibiting neurons of the expiration center fire. This frees the inspiration circuit all over again.

So there are two levels of control here. One is a cycling inhibition action between the two circuits, while the other is an overriding or reinforcing action from the respiratory organs themselves. The first produces a rhythm, while the second controls the depth of inhalation.

For many years, physiologists believed that breathing rate was influenced by the O_2 level in the blood, but now they are rather certain that the respiratory centers respond primarily to the CO_2 level. Actually, the centers probably respond to an increased acidity created when CO_2 forms H_2CO_3 in the blood. In essence, when the CO_2 level increases, the center is stimulated and breathing rate increases. More CO_2 is then exhaled, so the respiratory center, no longer so strongly stimulated by H_2CO_3, slows the breathing down (in yet another example of negative feedback control).

Of course, breathing control isn't quite that simple, since (as everyone knows) there is some influence from the higher control centers. We can consciously or voluntarily change our breathing rate. This type of control is limited, because the medulla's respiratory centers have an overriding capability. This means that the next time your little brother holds his breath in a tantrum, you needn't worry. Unless he has learned to control his involuntary mechanisms, his rising CO_2 level will overcome his deliberate intentions, and he will soon be breathing again. But maybe you weren't worried anyway.

On the other hand, if your little brother uses barbituates or other "downers," he may be in more serious trouble. Depressants can suppress the respiratory centers, and, in fact, most barbituate-related deaths occur this way. Alcohol is a strong respiratory depressant too, and more than a few little brothers have stopped breathing forever shortly after having demonstrated their manliness by chugging a fifth of something. You can imagine the combined effect of alcohol and downers.

Hyperventilating—deliberately breathing deeply—is the opposite of holding one's breath and, strangely enough, can be more dangerous. You might have tried it yourself and noted that hyperventilation can make you feel giddy and light-headed. The effect is paradoxical; it turns out that what you are experiencing is O_2 *starvation* of the brain. It works this way: the O_2 sensor is the *carotid body,* a group of nerves in the *carotid artery* of the neck, leading to the brain. In normal activities, the carotid body carefully monitors the partial pressure of O_2 in the blood, and dilates or constricts the artery in such a way as to assure a nearly constant O_2 supply to the brain. The distinctly abnormal practice of hyperventilation causes the mechanism to overreact, constricting the carotid artery too much and starving the brain. Unfortunately, the experience isn't particularly unpleasant and the situation doesn't automatically right itself. Deliberate hyperventilation before a dive is one of the principal causes of drowning.

In other circumstances, the inhalation of pure O_2 can have the same potentially lethal effect. Since the carotid body is also somewhat sensitive to CO_2 levels, hospital O_2 usually has about 5% CO_2 added to prevent such adverse reactions.

26.3 Evolution of Circulatory Systems

Transport in Multicellular Animals

So, as we see, it is important to living things to get good gas into the body and bad gas out. About the only way gas can go anywhere in a living body, of course, is for it first to be dissolved in a fluid, and, in all species, gases must cross wet membranes. As we've also seen, if the body is small enough—composed of only one cell—or, in multicellular animals, if it has a body shape that brings most cells very close to the environment, gases simply diffuse between the environment and the cells. But in larger and more complex animals, dissolved gases must be carried via transport systems into the deeper tissues. Let's quickly review a few of the ways that smaller multicellular animals move gas around; then we'll talk about real circulatory systems. Keep in mind that various things can be carried in these systems. In most species, nutrients and metabolic wastes are carried, as well as dissolved gases.

The now familiar gastrovascular cavity of coelenterates and flatworms functions in circulation as well as digestion and respiration. Movement of materials though the extensive body cavity are partly as a result of general body movement, although in some animals ciliated cells aid in the circulatory efforts. In animals such as the anemones and jellyfish, complex canal systems lined with ciliated cells create currents that distribute materials.

As we saw, in a *closed circulatory system,* the blood is enclosed within blood vessels throughout. An *open circulatory system* may have some blood vessels, but elsewhere the blood is free to percolate directly through the tissue. The annelids, mollusks, arthropods, and echinoderms are all invertebrates that have circulatory systems, whether open or closed.

As you may recall, the annelids have a remarkably well-developed closed system with distinct vessels, tubular hearts, and hemoglobin-rich blood. In addition, they have a fluid-filled coelom, described previously as

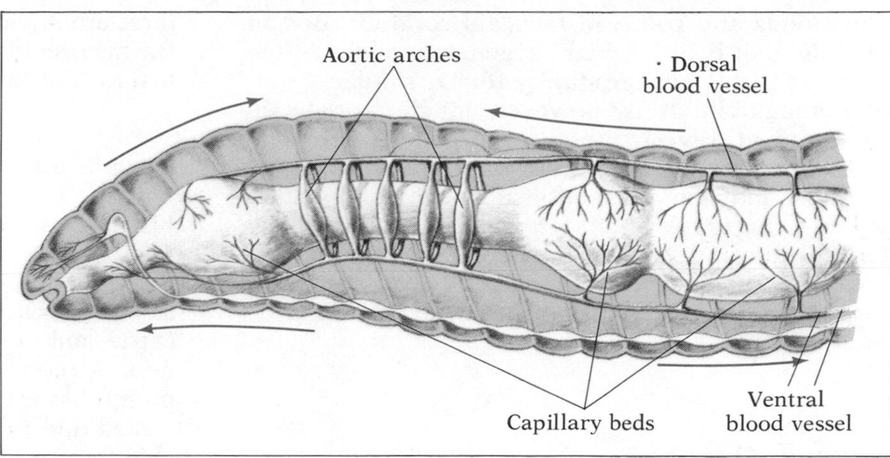

26.13 The circulatory system in the earthworm and other annelids is closed. In other words, blood is virtually always contained within vessels. Blood is pumped along by five pairs of *aortic arches*, complete with valves. Among the major vessels is the *dorsal vessel*, which contracts to push blood forward into the arches. The blood is then pumped down into the *ventral vessel*, which distributes it back toward the body and forward toward the head. The two large vessels are connected in each segment by extensive capillary beds that serve the digestive, excretory, reproductive, and respiratory regions. In the earthworm, the blood contains the respiratory pigment hemoglobin, but it is dissolved in the blood fluid, not bound within red blood cells.

a hydrostatic skeleton, which assists in the distribution of materials. In Figure 26.13, note the major vessels and the five pairs of *hearts*, or *aortic arches*. These aortic arches, contain valves that assure a one-way flow of blood, since the entire arch more or less contracts all at once.

Those joint-legged animals called arthropods give us particular problems because there are more than 800,000 species of them. Thus it becomes difficult to make generalizations. But at least we can say that probably all of them have open circulatory systems.

If you compare the circulatory systems of the crayfish and the grasshopper in Figure 26.14, you will see that the circulatory system is more complex in the gill-breathing crayfish. This is because its blood is involved in respiration as well as transport, whereas in the insect the two systems are independent. In addition to differences in blood vessels, the hearts of the two arthropods are quite different. Note that the grasshopper pumps blood with a heart that is no more than a long dorsal blood vessel with muscular thickenings along its posterior region. In the crayfish, the heart is larger, more central, and more complex.

Very few insects have respiratory pigments in the blood, since the blood plays only a minor role in respiration. In other arthropods, however (for example, the crustaceans, spiders, and scorpions), the blood contains *hemocyanin*, which, as we mentioned with regard to mollusks, binds O_2 to copper rather than iron. Thus, these lowly creatures were the original

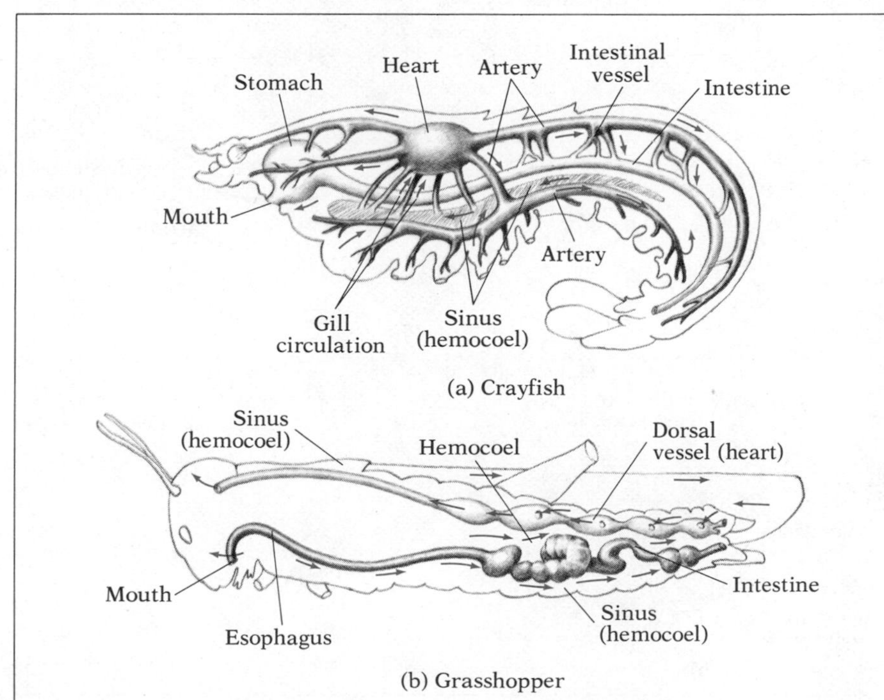

26.14 Both the crayfish and the grasshopper have open circulatory systems. In these systems, blood vessels empty out into coelomic cavities known as *sinuses* or hemocoels. (a) The system of vessels is more extensive in the crayfish, with the blood serving a respiratory function as it passes through the gills. Oxygenated blood from the gills flows directly to the sinus surrounding the heart, where it is again pumped out through a series of vessels, back to the sinuses. (b) The grasshopper has one vessel that is dorsal and contains several pumping regions or "hearts." Blood is collected by the hearts and pumped forward first to sinuses in the head and then through the body sinuses, eventually returning to the dorsal vessel. Movement of the sluggish fluid is assisted by body movements and diffusion.

bluebloods and some, in fact, did reside in Boston. The pigment is blue when oxygenated, and colorless when not, and has an affinity for O_2 similar to that of hemoglobin. It is, however, much more easily saturated at low O_2 pressures. The adaptive value of this phenomenon is interesting. The crayfish, for example, may live in places—such as highly polluted water—where O_2 content is low. Furthermore, it may burrow into muddy banks where O_2 levels are even lower. It can live in such places because it has a respiratory pigment that binds to O_2 even at very low pressures.

Circulation in the Vertebrates

All vertebrates have closed circulatory systems (if we ignore that blood does, in fact, percolate through the liver). Along with this system of vessels, they have an efficient, centrally located heart. The most efficient of these are found in the crocodiles, mammals, and birds. But in order to appreciate these marvelous organs, let's first look at circulation in some other vertebrates.

Circulation in Fishes

Fish are the most simple-hearted of all vertebrates. Their heart consists of a blood-receiving chamber, the *sinus venosus,* which is continuous with the *single atrium,* a *single ventricle,* and the *conus arteriosus* (Atria and ventricles are muscular pumping chambers. Atria are generally thin-walled chambers that pump blood into ventricles. Ventricles are thick-walled and pump blood into arteries with considerable force.) The fish heart can be considered a two-chambered heart, since the sinus venosus is really an enlargement of the *common cardinal vein* (the large vein that empties into the heart), while the *conus arteriosus* is the bulbous receiving end of the aorta, a large artery (Figure 26.15). Notice that blood

26.15 Like all vertebrates, fishes have closed circulatory systems. (a) The circuit made by the blood is simple, with blood following a circular path from heart to gills to body and back to heart. (b) The heart is considered to be essentially a two-chambered pumping device, with one atrium and one ventricle, although it actually has four major parts, with the *sinus venosus* and *conus arteriosus.* Blood entering the *sinus venosus* is passed to the *atrium.* As the atrium contracts, it empties its contents into the *single ventricle.* Ventricular contraction forces blood out into the *conus arteriosus* and then to the *ventral aorta.* The ventral aorta directs blood to the gills, where it is oxygenated before passing into the *dorsal aorta,* which directs blood to the head and body. From the capillary networks throughout the head and body, blood again returns to the heart along the cardinal veins.

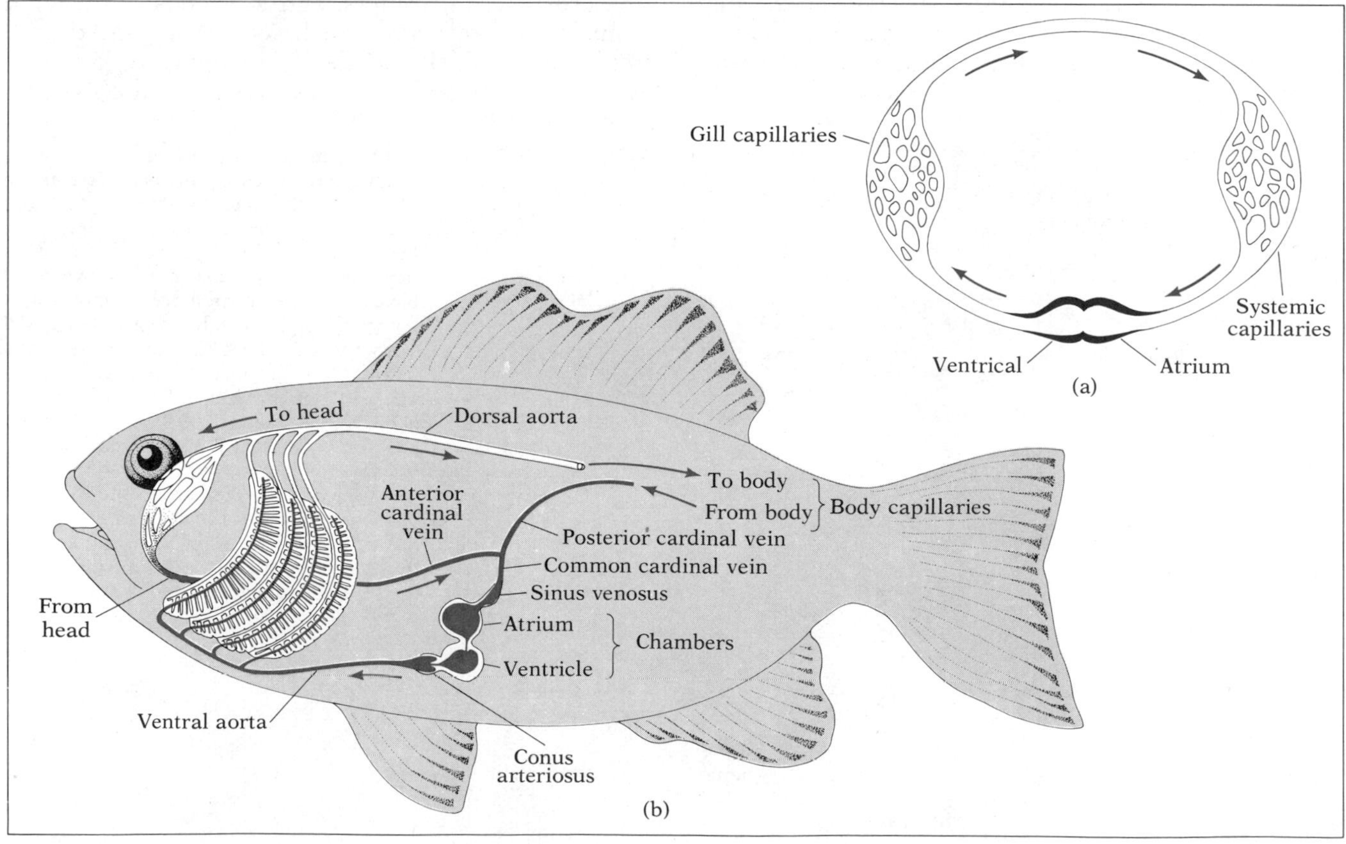

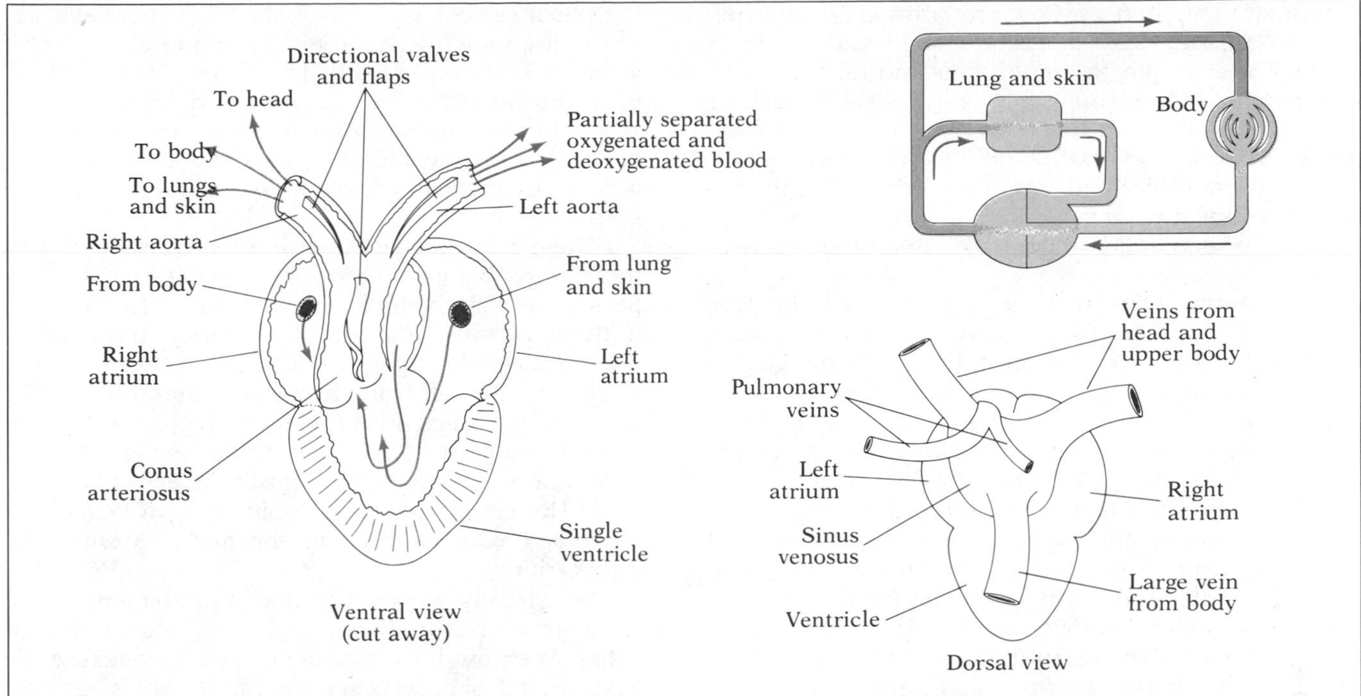

26.16 Circulation in the amphibian and most reptiles is almost a two-circuit system. These animals have a three-chambered heart, so there is some mixing of pulmonary and systemic blood. The three-chambered heart of the frog, shown here in both ventral (front) and dorsal (back) view, includes *two atria* and *one ventricle.* Deoxygenated blood arriving at the heart from the body enters the *right atrium* after being collected in the *sinus venosus* (dorsal view). The right atrium pumps its contents into the single ventricle. Simultaneously, the *left atrium,* which has received oxygenated blood from the lungs and skin, empties its contents into the single ventricle. Oxygenated and deoxygenated blood does mix. But an intricate system of guides helps to reduce the mixing. The ventricle then contracts, sending its contents into the conus arteriosus and then out into the divided aorta. Branches of the aorta carry blood into the lungs and skin through the pulmonary circuit and back to the heart again. Other branches supply blood to capillary networks in the head and body as seen in the schematic diagram. Blood returning through the veins reenters the heart through the sinus venosus once again.

pumped from the ventricle moves to the *conus arteriosus* and on to the aorta that carries it directly into the blood vessels and capillaries of the gills. From the gills, freshly oxygenated blood flows through the body, and from there returns to the heart. Blood returning to the heart moves sluggishly, since it has passed through two major capillary beds, one in the gills and the other in the body tissues. Most of the tremendous force it had when it left the heart has been lost to resistance in the capillaries, primarily those in the gills.

The circulatory system in cold-blooded animals may be involved in the conservation of body heat. Recent investigations of fast-swimming fishes are shedding some light on this question. For a long time, physiologists have wondered how cold-blooded fish, with their sluggish circulatory systems, could move so fast. Weren't their muscles at water temperature? And isn't that too cold to permit rapid contraction? The answer, it seems, is that the deeper, swimming muscles are not at water temperature. The fact is, their temperatures vary in different parts of their bodies, and the deeper muscle layers are warmer (see Chapter 28).

In summary, the flow of circulation in fish is: heart to gills (via the ventral aorta); through the gill arches; from the gills to the tissues (via the dorsal aorta); from the tissues to the heart (via the common cardinal vein). Note that in each cycle through the heart, all the blood flows through the gills and all the blood flows through the other body tissues. Oxygen is picked up by the gills and is released in the tissues. Therefore, the fish heart pumps only deoxygenated blood. Because of this, the fish heart muscle must receive its oxygenated blood supply indirectly, through a small artery of its own.

Circulation in Amphibians and Reptiles

In the evolution of animals that, even for a time, could survive on land, the development of the lungs and the deemphasis of the gills meant that the simple cycle of fish was no longer enough. The demands were greater and changes were in order. Now, some of the blood pumped from the heart was directed to the lungs and some was directed to the body tissues.

The tendency, then, was toward complexity and specialization. But the emerging system was far from ideal. For example, in amphibians and most reptiles, oxygenated blood from the lungs (and skin) reaching the heart was mixed with deoxygenated blood from other tissues. This rather inefficient system exists, today, in amphibians and most reptiles, although these groups have made some advances toward separating oxygenated and deoxygenated blood, as we'll see.

For now, let's consider the heart of the frog (Figure 26.16). The large *sinus venosus* is present as it was in the fish, but you will readily notice that there are *two atria.* Blood returning from the body enters the *right atrium* from the *sinus venosus.* From there, it passes into the *single ventricle.* In the ventricle, it is joined by blood from the *left atrium*—blood coming from the lungs. This system has produced a vexing question for biologists. Is the oxygenated blood from the lungs mixed with the deoxygenated blood from the body? This point has been argued for many years and the evidence indicates that it does mix, but only partially. In the amphibian heart, total mixing is prevented by flaps and partial valves that tend to direct the oxygenated and deoxygenated blood into the proper arteries. In the reptiles, there is even a partial *septum,* or partition, between the ventricles, which helps keep the two kinds of blood separated. If oxygenated and deoxygenated blood were thoroughly mixed in the ventricle, the system would be very inefficient, since a great deal of oxygenated blood would return to the lungs and, of course, deoxygenated blood would return to the body.

Now, to return to the circulatory pathways in these amphibians and reptiles, we can see that blood passes from the ventricle to the lungs, then returns to the left atrium and into the ventricle from where it is pumped throughout the body. In the body, the arteries branch off from the major routes into smaller *arterioles* which go on to divide into a greater number of capillary beds. It is in the capillaries that O_2, CO_2, food, and wastes are exchanged between blood and cells. The capillaries then rejoin to form veins, and the blood courses back to the heart to enter the right atrium again.

So, already we see a major improvement in the circulation of amphibians and reptiles over that of fishes. We now have a two-part system because we have added a *pulmonary circuit,* which is at least partly separated from the rest of circulation. As a result of the second circuit, blood pressure remains

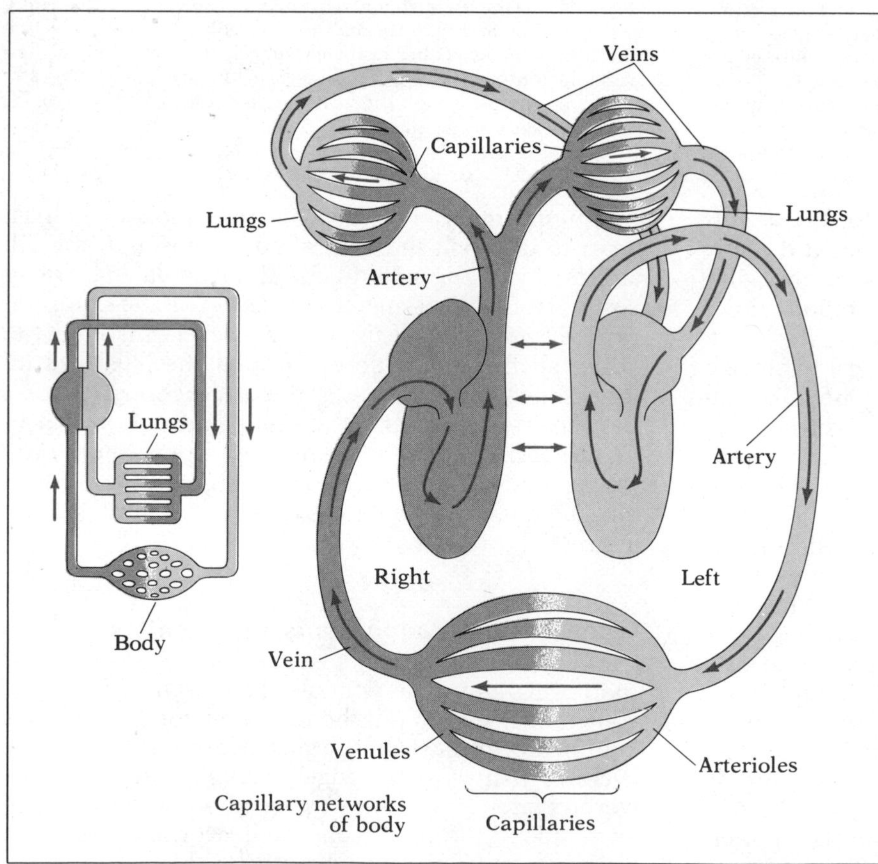

26.17 Mammals, birds, and crocodilians have four-chambered hearts. The four-chambered heart essentially allows two separate and distinct blood circuits. In fact, you can diagrammatically separate the right and left halves. This is shown more diagrammatically in the inset. The right side receives deoxygenated blood from the body and pumps it through the lungs in one pulmonary circuit. The left side receives blood from the lungs and pumps it through the body in a separate circuit. The two circuits actually unite in the capillary beds, where arteries from the systemic circuit branch into capillaries which rejoin to form veins, which return to the heart to be pumped through the pulmonary circuit.

strong as oxygenated blood is sent to the tissues. This is because a second boost is produced by the single ventricle when oxygenated blood returns to the heart from the lungs.

The Four-Chambered Heart

The evolution of the four-chambered heart was not a sudden thing. As predicted by the rules of Darwinian evolution, it had to occur by a succession of tiny steps, each one an improvement over the last. We can see some of these intermediate steps in living amphibians and reptiles, with their three-chambered hearts, which differ from each other only slightly, such as in the degree to which they have developed their partial septums between the ventricles.

The next step—achieved by birds, crocodiles, and mammals—is the division of the ventricle into two chambers: the *right ventricle* and the *left ventricle,* thereby producing, with the right and left atria, a four-chambered heart. In the four-chambered heart, once again all the blood is oxygenated (in the lungs) and all the blood is deoxygenated (in the body tissues) in each complete circuit. The crucial difference between this complete circuit and that of the fish is that a complete circuit in birds and mammals involves two trips through the heart, once through the left side and once through the right side (Figure 26.17).

26.4 Human Circulation and Blood

The Human Circulatory System

Humans, as with all good mammals, have a four-chambered heart and a closed circulatory system. The system consists of the heart, arteries, capillaries, and veins. The arteries are muscular vessels that always carry blood away from the heart. All human arteries, with the exception of the pulmonary arteries, carry oxygenated blood; hence, arterial blood is usually a bright red. Capillaries are thin-walled vessels that branch from small arteries, known as *arterioles.* Actually, capillaries are the functional parts of the circulatory system, since materials can only enter and leave the blood through these vessels. Small veins known as *venules* receive the blood from capillaries and flow into ever-larger veins; and these direct blood toward the heart. With the exception of the pulmonary veins, all veins carry deoxygenated blood.

Arteries tend to be round and thick-walled, with heavy layers of smooth muscle and elastic connective tissue. Constriction or dilation of arteries and arterioles controls the blood flow in specific tissues. Veins, on the other hand, tend to be thin-walled and flattened, with much larger openings in cross-section. They also generally lie nearer the surface, so most of the vessels you see under the skin are veins.

Circulation Through the Heart

In mammals, blood arrives at the right side of the heart from two major veins (Figures 26.18 and 26.19). Because of our upright posture, one is higher than the other so, they are named the *superior* and *inferior venae cavae,* but in other mammals they are known as the *anterior* and *posterior venae cavae.* The venae cavae empty into a thin-walled muscular chamber known as the *right atrium.* The right atrium pumps into a chamber below it, the muscular *right ventricle.* The ventricle assists the atrium by creating a suction as its walls relax. When the right ventricle contracts, it forces blood into the *pulmonary arteries* and on to the lungs.

When a chamber contracts, of course, blood could be forced either way. So there are one-way valves between the chambers and between the chambers and the vessels. Figure 26.20 shows the valves of the heart that assure the one-way flow of blood. The valve between the right atrium and ventricle is the *tricuspid* valve. It checks any backward flow created by the ventricular contraction. When the tricuspid is closed, it resembles a three-part parachute; the "lines" are a series of tendonous cords (*chordae tendineae*) that prevent the three flaps from collapsing backward like an umbrella in a strong wind.

As blood passes into the muscular pulmonary artery, its backflow is prevented by the *pulmonary semilunar valve,* a three-part curved flap that closes when the artery is full and expanded. The name semilunar refers to the half-moon shape of each of the three flaps. Arteries have complex walls of muscle and elastic connective tissue which, because they can contract, are able to maintain a nearly constant pressure between fillings.

Blood from the right and left branches of the pulmonary arteries finally weaves its way into the capillaries surrounding the alveoli of the lungs, where gases are exchanged. The capillaries then rejoin to form the pulmonary veins, which return oxygen-rich blood to the left atrium. Then, when the left atrium contracts, blood passes from there through the double-flapped *bicuspid* (or *mitral*) *valve* into the left ventricle. This valve is similar to the tricuspid of the right side, except that it has two parts.

The walls of the left ventricle are much thicker than those of the right, which probably gave rise to the myth that the heart is on the left side of the chest. (Actually, it is in the center, but tilted.) The left ventricle is large and muscular enough to force the oxygenated blood throughout all the tissues of the body and all the way back to the right side of the heart again—an enormous task! Contraction of the left ventricle sends blood through the *aortic semilunar valve*—and on into that great artery, the *aorta.* The aorta makes a U-turn to the left, giving rise to a number of side branches as it turns and still more as it passes down through the trunk. Finally, it splits into the iliac arteries, which branch into the legs.

Control of the Heart

The subject of the human heart has been overworked. Volumes have been devoted to describing its relentless and tireless activity. We have eulogized and venerated it, but in reality, it needs no romanticizing,

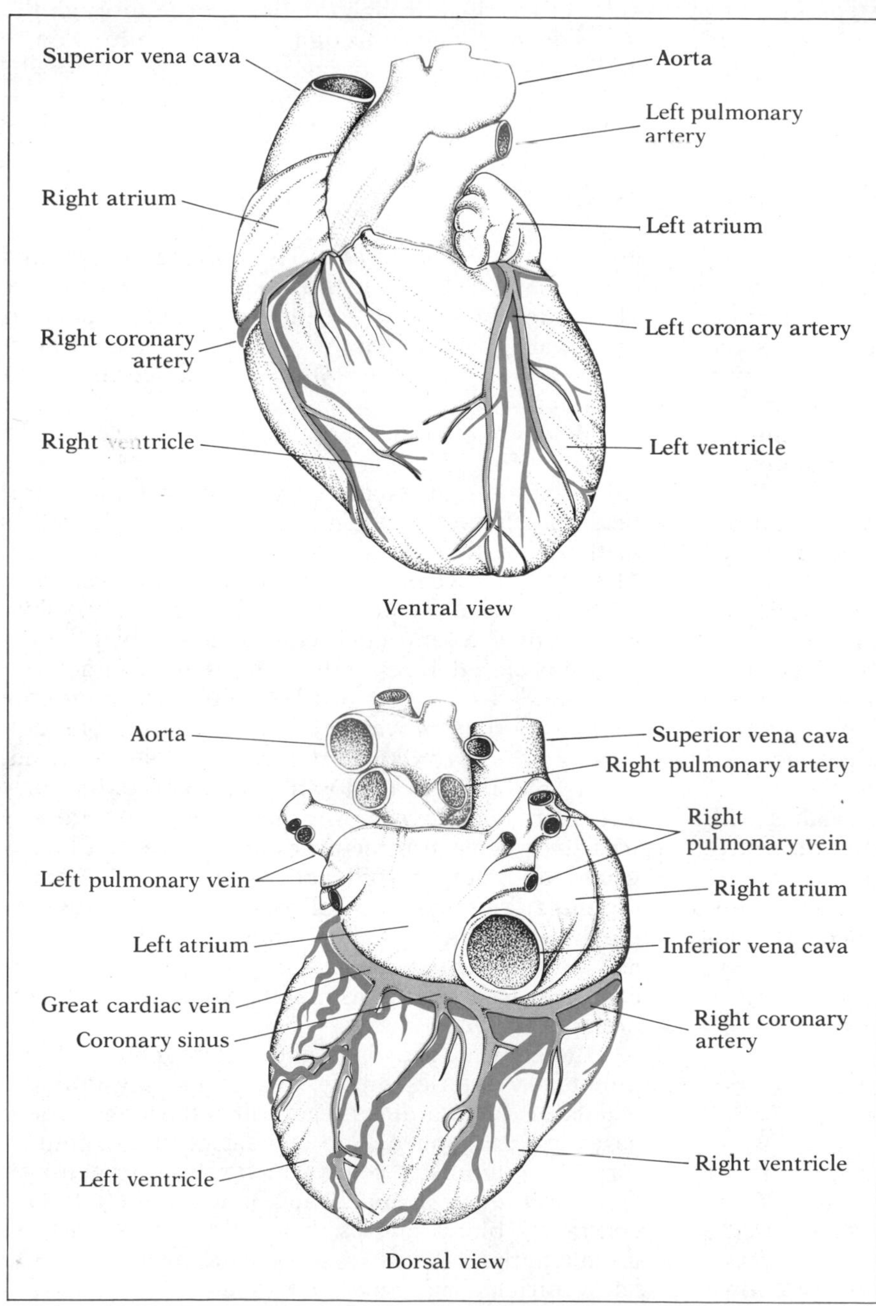

26.18 The human heart, rather typical of mammals, consists of four chambers. The large, muscular right and left ventricles make up the bulk of the heart. The right and left atria form muscular caps at the top of each ventricle. The large vessels near the atria are the superior and inferior venae cavae, which join at the right atrium, the pulmonary veins that enter the left atrium, and the thicker-walled pulmonary artery and the aorta that emerges from the left ventricle. The heart muscle has its own vessels, the coronary arteries and veins. The coronary arteries branch directly from the aorta, sending their branches deep into the heart muscle. The coronary veins emerge from heart muscle to form the coronary sinus, which empties into the right atrium. The coronary arteries are often involved in heart attacks where blockages stop the critical blood supply to areas of the heart muscle.

26.19 A cutaway diagram of the heart, illustrating its circulatory pathways. Blood enters the right and left atria almost simultaneously. The superior and inferior venae cavae bring in deoxygenated blood from the body, emptying into the right atrium. Oxygenated blood entering the left atrium comes from the pulmonary circuit via the pulmonary veins. Blood flows from the atria to the ventricles. The contraction of the ventricles sends blood from the right ventricle into the pulmonary arteries (to the lungs) and from the left ventricle into the aorta (to the head and body). Backflow into right and left atria is prevented by the *tricuspid* and *bicuspid valves.* Blood entering the pulmonary artery and aorta is prevented from flowing back into the ventricles by the *pulmonary semilunar* and *aortic semilunar valves,* respectively (a). The pumping action is summarized in (b).

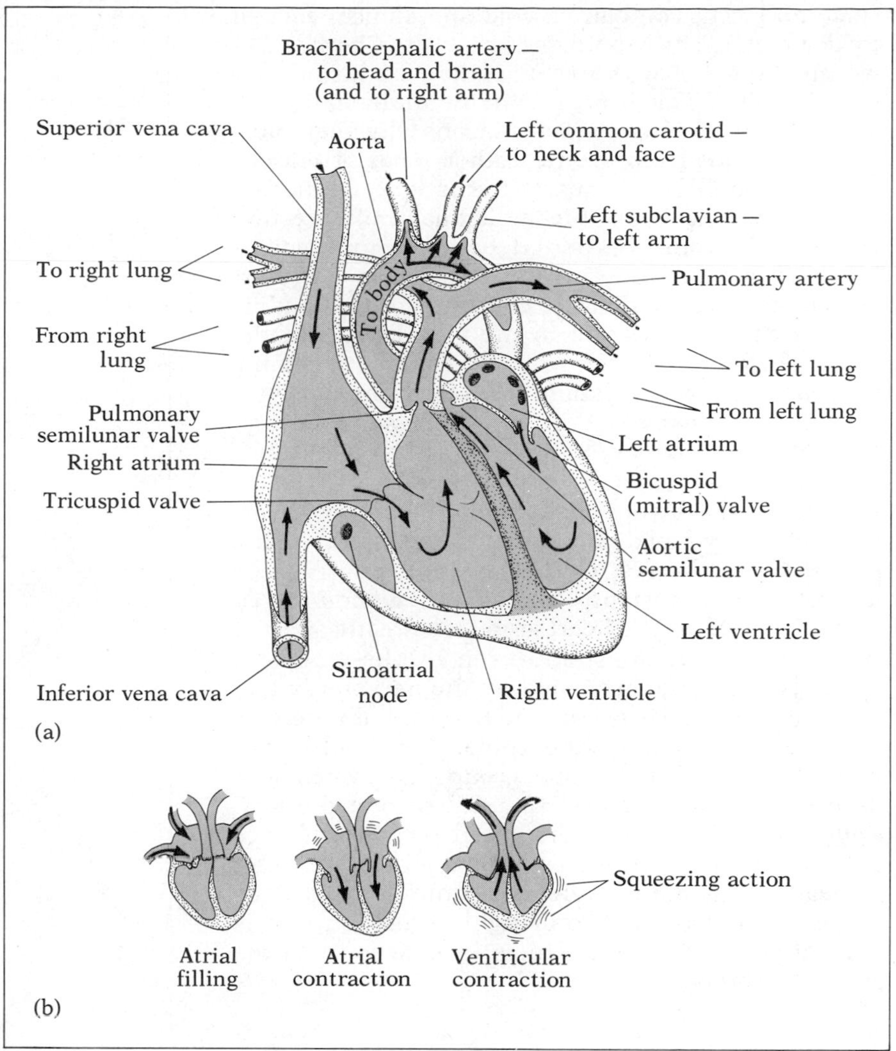

26.20 The valves of the heart. The larger valves are between the atria and ventricles. The *bicuspid (mitral) valve* has two flaps of tissue, while the *tricuspid* has three. The aortic and pulmonary valves have heavier and smaller flaps and lack the support of *chordae tendineae.* The stringlike chordae can be seen in the photograph. These attach to papillary muscles projecting from the base of the ventricles. When the ventricles contract, these cords are tightened, holding the thin flaps in place as blood surges against them.

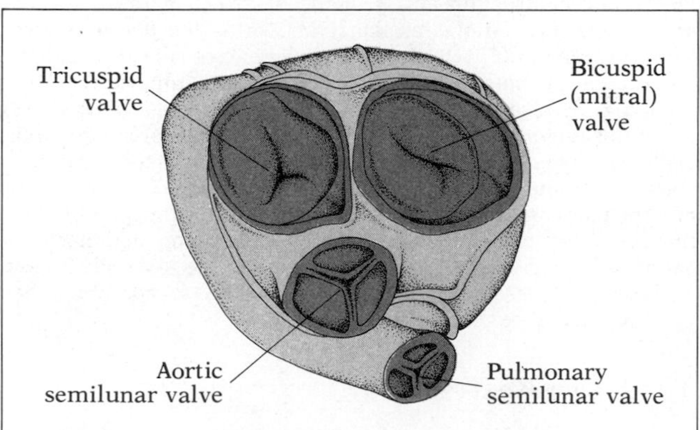

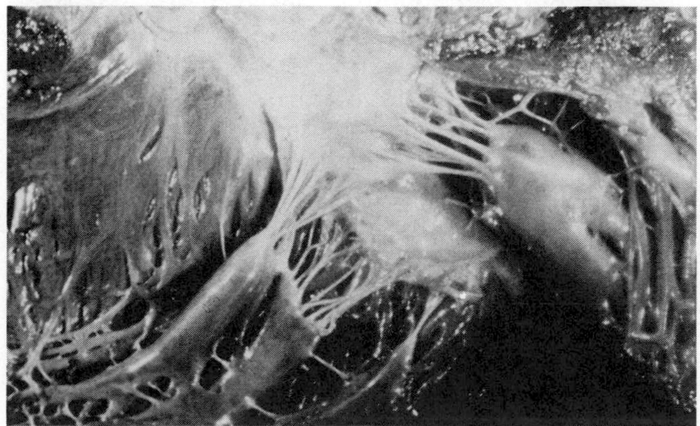

since no discussion can be cold and clinical enough to drain it of its wonder. Earlier (in Chapter 24), we mentioned that the heart contains unusual contractile tissue. This was an extreme understatement. There are three physiological reasons why heart muscle is different from the two other types of muscle. First, although heart muscle forms long, branching fibers and is striated, the individual cells are uninucleate. Second, the period between contractions is rather prolonged. And third—perhaps the most curious feature—is that the contraction itself and its rhythmicity are inherent in the muscle cells of the heart. The heart, in other words, is capable of beating without outside regulation. This is dramatically illustrated in transplanted hearts, which sometimes go on beating for many years with no nerve connections whatever.

We are not saying, though, that the normal heart is functionally isolated, beating merrily away, independently of the rest of the body. Obviously there are too many constantly changing demands placed upon the system. The heart is, in fact, greatly affected by external conditions, and even by the emotions. Heart rate can be fine-tuned by the nervous system and the endocrine system. The nerves that innervate the heart originate in the spinal cord and in the medulla of the brain. One group, the *sympathetic nerves*, accelerate the heart rate. The second, the *parasympathetic*, act to slow the heartbeat. We will have a close look at these nerves in Chapter 31. The point is, heart rate is partly under the control of two kinds of nerves and these permit tight regulation of this vital organ so that it operates in concert with the rest of the body.

The Pacemaker

One wonders, how does the heart direct itself if it is influenced by all these nerves? The origin of the heartbeat is in the right atrium, at a region known as the *sinoatrial node* (see Figure 26.19). More commonly, it is called the *pacemaker*. This is also the region influenced by the autonomic nerve fibers. The pacemaker transmits impulses to a second node—the *atrioventricular node*—and across the atrial walls. While the sinoatrial node causes atrial contraction, the atrioventricular node initiates the contraction of ventricles. Impulses from the atrioventricular node pass through a strand of specialized muscle in the *ventricular septum* known as the *bundle of His* (which might perturb feminists until it is pronounced: *Hiss*!). The bundle branches into right and left halves and travels to the pointed apex of the heart, where each ventricular contraction actually begins (Figure 26.21).

As a result of this arrangement, the atria contract simultaneously, filling the ventricles. There is enough delay in the impulse for the atria to complete their contraction before the ventricles begin. Then the ventricles contract simultaneously.

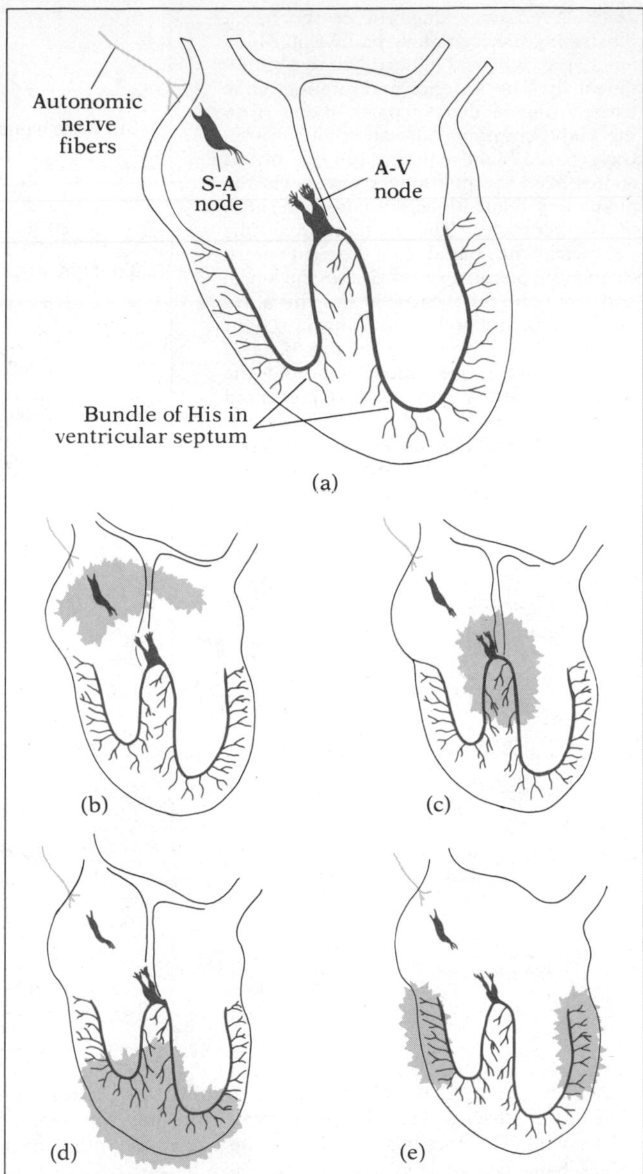

26.21 The contraction cycles of the heart originate in the *sinoatrial (S-A) node*, more familiarly known as the *pacemaker* (a). Impulses generated in the S-A node spread across the atria, causing contraction of the muscle walls (b). The same impulse reaches a second nodal area known as the *atrioventricular (A-V) node* (c). From this region, the impulse is regenerated and spread down the *ventricular septum* (d). The impulse spreading over the ventricular surfaces initiates the contraction of the ventricles, first at the tip and then moving rhythmically upward (e). The arrangement of the muscle fibers causes a squeezing and wringing action in the ventricles. Interestingly, conduction through the A-V node is slow. In fact, there is enough delay for the atria to complete their contraction before the ventricles begin.

The Heart Beat

There are two major heart sounds. You've heard them: *lub dub, lub dub*. The first is the sudden closing of the tricuspid and bicuspid valves as they shudder under the tremendous force of the contracting ventricles. The second sharp sound occurs as the valves in the aorta and pulmonary arteries snap shut by arterial pressure, as the ventricles relax. Then there is a brief pause. Then (hopefully) *lub dub*. The period of contraction in the heart, heralded by the first sound, is known as *systole*, while the longer period beginning with the second sound is *diastole*, during which the chambers of the heart fill up once again.

So, one wonders, just how much work does this remarkable organ actually do? Any answer, of course, can only be an estimate, since there will be great variation in any group of people. But, generally, it has been calculated that the resting heart rate is 72 contractions per minute, and that each contraction forces about 80 ml of blood into the aorta. This measurement is the *stroke volume*, comparable to engine displacement. So the volume pumped each minute is about 5–6 liters (about 11–13 pints) and is called the *cardiac output*. During activity, cardiac output can be increased up to 30–35 liters per minute—a remarkable reserve power! In case you have sensed an arithmetic problem here we hasten to explain that your heart doesn't have to increase its beat rate by six times to get a six-fold increase in cardiac output—not if you are in good condition. During vigorous activity the athlete's heart rate climbs rapidly, but so does the stroke volume. The nonathlete, on the other hand, has to keep up the cardiac output primarily through increased heartbeat alone with less increase in stroke volume. This is one reason an unconditioned person tires so quickly and takes so long to recover. In trained athletes, the heart rate may go up to 180 per minute, the difference is, theirs comes down faster (to about 120 in three minutes).

Arteries play a significant role in the maintenance of blood pressure because with their muscular layer they are able to contract, squeezing the blood, and thereby raising blood pressure (Figure 26.22). When they relax, the vessel opens and blood pressure is lowered. A common problem in older persons is *arteriosclerosis* or "hardening of the arteries." Arteriosclerosis is a general term that includes calcium deposits in the arterial connective tissue, general loss of arterial elasticity, and especially *atherosclerosis:* arterial lesions or thickenings (*plaques*) characterized by localized growth of smooth muscle cells and deposits of lipids, especially cholesterol. As vessels lose their elasticity, they fail to yield before the surge, creating serious burdening of the heart as resistance to flow increases. Normally elastic arteries can maintain pressure within their channels in spite of intermittent blood volume surges. Atherosclerotic plaques impede blood flow and further stress the arteries.

The aorta receives the full impact of the powerful surges of blood from the left ventricle. A great many body functions depend upon a rather constant pres-

26.22 Anatomy of an artery. Arteries are thick-walled, muscular vessels with considerable elasticity. (a) In cross-section they are shown to consist of three regions. The innermost *tunica intima* is a smooth endothelium surrounded by a connective tissue sheath. The central region is the *tunica media* which consists of a layer of smooth muscle arranged circularly around the artery. Outside the smooth muscle is an elastic membrane that gives the artery its resiliency. The outer region is the *tunica adventitia*, a sheath of loose connective tissue. Arteries depend a great deal on their elasticity to function properly in maintaining normal blood pressure. (b) In old age, for reasons of poor diet, lack of exercise, and other factors, hardening of the arterial wall can occur. The condition is known as arteriosclerosis.

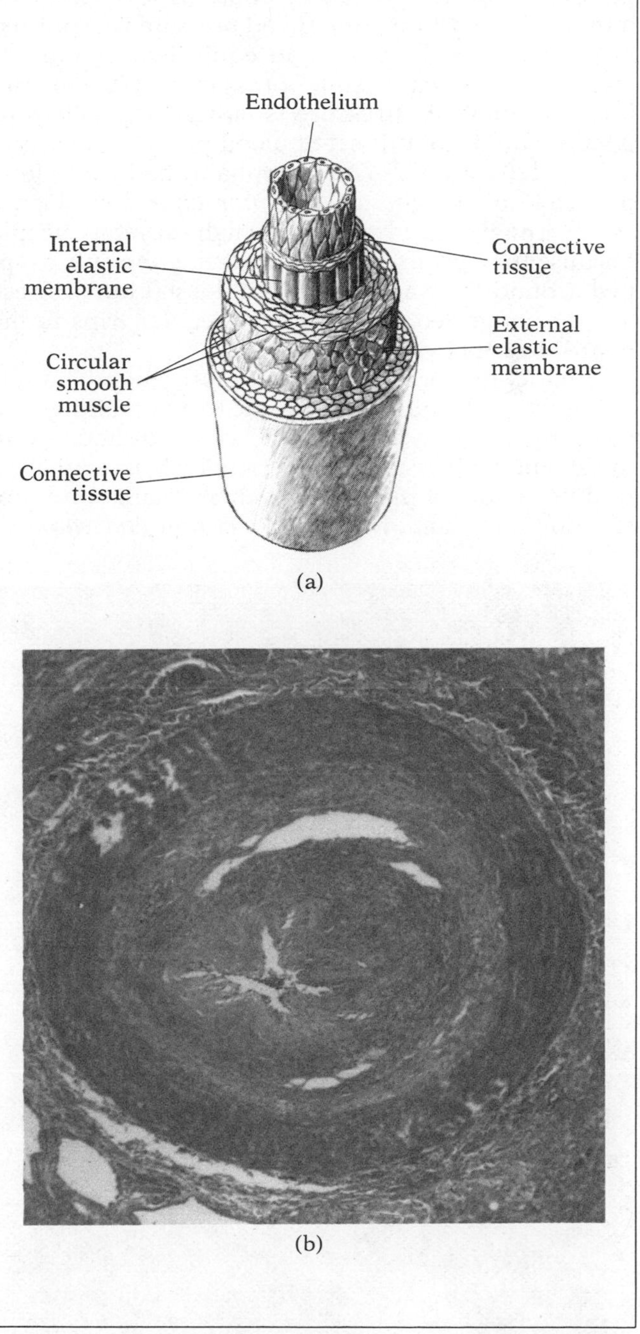

sure in the circulatory system, so the action of arterial walls is vital. The sudden swell of blood during systole (heart contraction) expands the elastic walls of the aorta, accommodating the increased volume. Backflow into the ventricle, as we explained earlier, is prevented by the aortic valve. As the blood moves onward through the arterial tree, the expanded aortic walls contract elastically, squeezing the rapidly decreasing blood volume. The speed of this recoil prevents a drastic pressure drop, but it does allow some lowering of blood pressure. For example, there is a pressure drop in diastole of about 35%, or of 40–60 mm of mercury (mm Hg). Blood pressure is expressed in terms of the height of an equivalent column of mercury. A typical systolic pressure is 120 mm Hg; a typical diastolic pressure is 80 mm Hg, and your doctor would call this set of blood pressure measurements "120 over 80." You've probably had your blood pressure taken with a *sphygmomanometer* (Figure 26.23). The device registers too high for many people because they get excited about having the cuff wrapped around their arm (even if it doesn't hurt). These are the same people who make regular trips to the dentist every 14 years.

In the smaller arteries (arterioles), smooth muscle lining the walls plays a vital role in maintaining and directing the flow of blood. There is little surge in these smaller vessels, however, which is actually a testimony to the precision of their regulation. Constriction and dilation (*vasoconstriction and vasodilation*) of the arterioles is partly under the control of the sympathetic fibers of the autonomic nervous system, which is augmented by hormonal influence. In addition, oxygen, carbon dioxide, hydrogen ions, and a variety of drugs can cause constriction or dilation of arterioles. There is some evidence that blood pressure can be consciously controlled as well, as is discussed in Chapter 31.

Circuits in the Human Circulatory System

We're not about to discuss all the pathways of the human circulatory system, but we would like to consider a few interesting circuits. A *circuit* can be loosely defined as an area where some distinct and major function occurs. Figure 26.24 shows the four major circuits: pulmonary, hepatic portal, renal, and general body (which includes everything else). Typically, these circuits pass through vital organs which are likely to change the blood.

Pulmonary Circuit

We don't want to describe the *pulmonary circuit* again and you don't want to hear it again, but as you know, the gases of respiration are exchanged here, and it constitutes a circuit. But since blood flows directly from the heart into the pulmonary circuit and back to the heart, it serves as a good example of what we mean.

(a)

26.23 (a) Blood pressure is commonly measured in the brachial artery with a device known as a *sphygmomanometer.* It measures two things. The first is *systolic pressure,* which is the arterial pressure during the contraction of the heart. The second is *diastolic pressure,* which is a measurement of arterial pressure between contractions. To make the measurements, a cuff is wrapped around the upper arm and inflated. At the same time, a stethoscope is inserted just under the cuff above the joint. As the cuff is inflated, the brachial artery is collapsed and no sound is heard. Pressure in the cuff is slowly reduced and when a beating sound is first heard the pressure on the dial is recorded (the systolic pressure). Then, as the pressure continues to be released, all sound ceases (the diastolic pressure). (b) In the graph, the peaks represent systolic pressure and the valleys diastolic pressure. The shaded area is the pressure range at which the arterial sounds are heard. The diagonal line represents pressure decreasing in the sphygmomanometer.

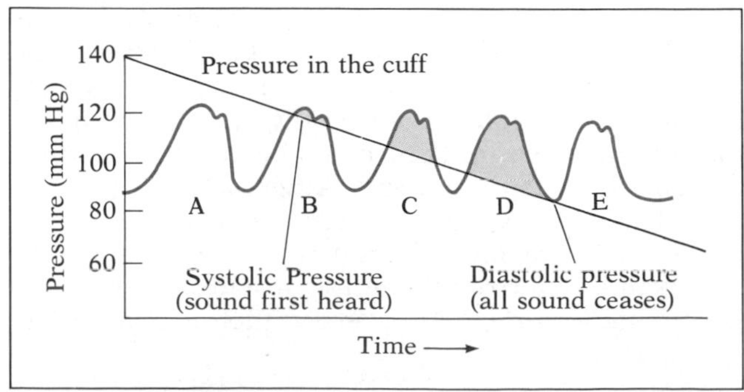

(b)

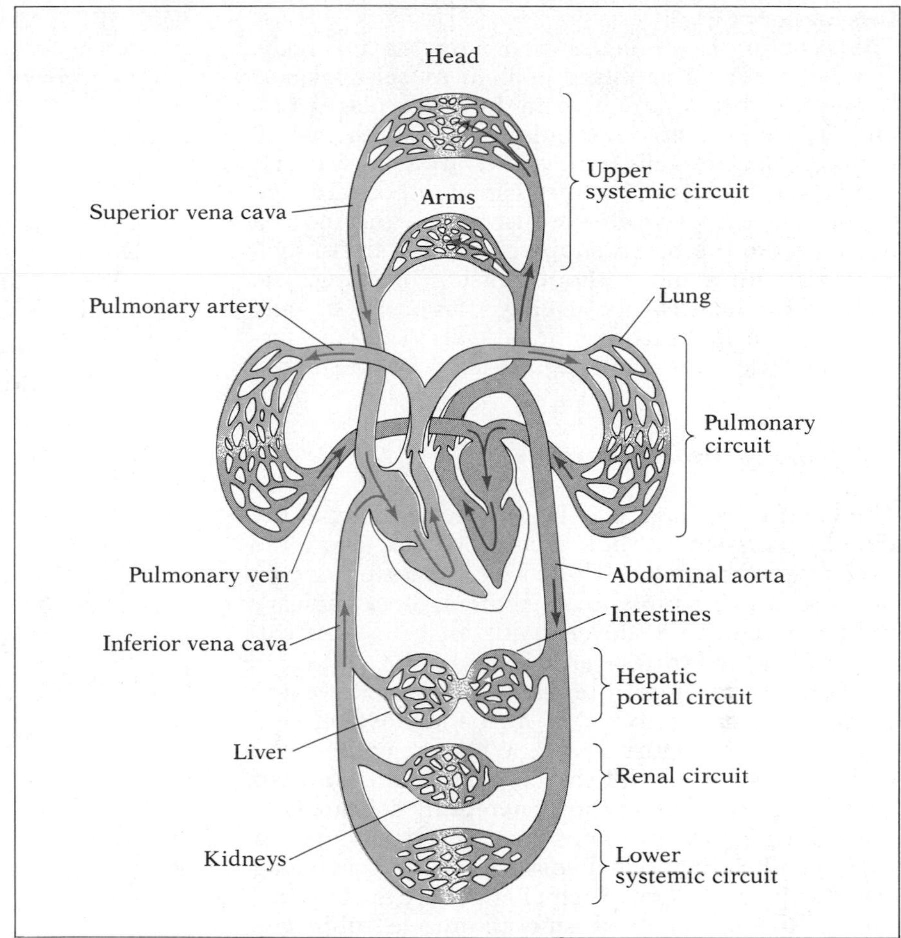

26.24 The four major circulatory circuits. Closest to the heart is the *pulmonary circuit* where CO_2 and O_2 are exchanged. Note the relationship of this circuit to the heart. The second circuit is the *hepatic portal circuit* and includes the digestive organs and the liver. Note that a major branch of the aorta goes directly to capillary beds in the digestive system, forms the portal vein and then another capillary bed in the liver, before returning to the heart. The third circuit is the *renal circuit,* which includes branches to and from the kidneys. After blood passes through the renal circuit, it returns immediately to the heart. Finally, there are the *upper* and *lower systemic circuits,* which simply designates blood flow to general areas of the body.

Hepatic Portal Circuit

In the *hepatic portal circuit,* vessels branching from the aorta spread between the intestinal membranes and the digestive organs. In these organs, digested foods are collected in small vessels that come together and carry blood directly to the liver where they branch again into a capillary bed. Actually, the liver contains not only capillaries, but many branched *sinusoids* (minute cavities) through which the blood passes. (As we said, it could be argued that this makes the human system an open one.) From there, the blood, cleansed by the liver of toxic materials, as well as much of its food, returns directly through the inferior vena cava to the heart. You will notice that this circuit is unusual in that the blood flows through capillary beds twice before returning to the heart. (In fact, a *portal system* is one connecting two capillary beds.) The liver, by the way, is also served by arteries, as you would expect in such an active organ.

Some liver diseases greatly impede the flow of the hepatic portal circuit. Persons with these conditions have a peculiar type of high blood pressure, one in which the high blood pressure is limited to the hepatic portal system, as blood entering the liver becomes slowed. Since the vessels of the esophagus also flow into the hepatic portal circuit, they may build up pressure and rupture, quickly causing death by hemorrhage. This problem can be prevented by surgery—specifically by grafting a vein that by-passes the liver, taking blood directly from the hepatic portal vein to the systemic circulation. The problem is, although the patients' lives are saved, they may live the remainder of their lives in a trancelike stupor that arises from having an excess of free amino acids in their general circulation. The liver closely regulates the amount of amino acids in the blood.

Renal Circuit

Further down, the aorta sends right and left branches (the *renal arteries*) into the kidneys where nitrogenous waste, urea, excess water, miscellaneous metabolic by-products, and some salts are removed from the blood. Blood returns from the kidneys via the *renal veins* to the inferior vena cava, and then to the heart.

General Body Circuit

The general circuit includes the rest of the body. Here we refer to capillary beds in muscles, glands, bones, the brain, and so on. In these places CO_2 and O_2 are exchanged, required substances for cell maintenance are delivered, wastes are removed, and so on. This requires immense capillary beds. In fact, capillaries are so profusely distributed that no cell lies far from the blood supply. As a test, try to stick yourself with a pin without causing bleeding. The walls of capillaries are usually composed of only one layer of flattened epithelial cells. They are fascinating little vessels and require a closer look.

The Role of the Capillaries

The capillaries, then, are the functional units of the circulatory system. They are the places where the exchanges take place. The role of every other part of the system is simply to assist them. Since capillary beds communicate closely with all cells, you can assume the total surface area is immense.

One of the more interesting traits of the circulatory system is its ability to open and close selected capillary beds (Figure 26.25). Whole regions of the body may suddenly blush red as they are suffused with blood traveling in these incredibly delicate little vessels, while, at the same time, blood may almost stop in other capillary beds—there is no incoming blood to push it along. Such changes are made possible by the presence of smooth-muscle sphincters located in the tiny arterioles just about where they divide into the capillaries. Thus, the vasoconstriction of arteriole sphincters in the digestive area, for example, will partially close down circulation in that system and will result in more blood being sent to the muscles and brain. This, in fact, is what happens in emergency situations. The routing of blood flow to systems of greatest demand is yet another example of an organism's ability to maintain a continuous harmony with its changing environment.

Not only is directing blood to specific regions important to changing vital functions, but it also can help to regulate the body's temperature. We discuss the process in detail in Chapter 28, but for now let's see how it works in capillaries.

The capillaries of the skin can be opened when internal temperatures are excessive, permitting the warm blood to be carried to the surface where the heat can be radiated away. You have probably seen people with a "flushed" appearance after exercise or on a hot day. Less obvious is the opposite effect, which occurs when we are exposed to extreme cold. The skin blanches as surface vessels are closed down and blood is shunted to deeper, warmer areas, preventing excessive heat radiation from the body. In extreme cases, the body can overdo it. The lack of

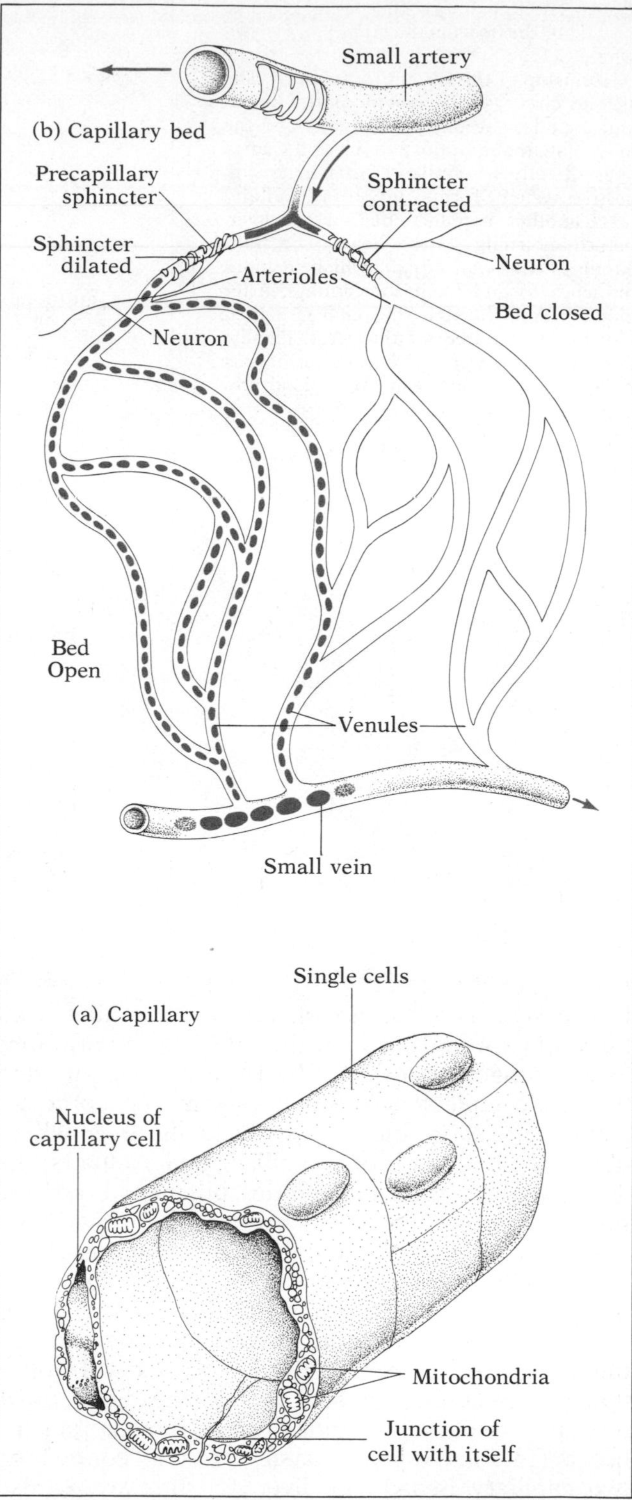

26.25 (a) Capillaries are composed principally of endothelial cells. Because their walls are only one cell thick, materials pass rapidly through them. (b) Capillary beds arise where arterioles branch. The capillaries, in their delicate profusion, serve tissues directly before they rejoin into thicker-walled venules. The beds can be partially shut down in various regions by action of muscular valves known as *precapillary sphincters* in the arteriolar walls. Thus, blood can be temporarily diverted to organs where the need is greatest. In this instance, the left side is open while the right side is shut down.

warm blood being pumped into the extremities permits cell fluids to freeze and *frostbite* can attack the fingertips, toes, ears, and even the nose (Figure 26.26). As grim as this sounds, the system is adaptive. None of these structures are absolutely necessary for survival, so they are the first to be sacrificed. The body's tendency, under these extreme conditions, is to maintain a constant internal temperature at all costs, and an ear is a small price.

As blood passes through a capillary bed, two things occur. You are aware that nutrients, oxygen, and other transported substances diffuse down the osmotic gradient into the cells, and that wastes and carbon dioxide, moving along an opposite gradient, enter the blood for transport to the liver, skin, excretory, and respiratory systems. But the capillaries are also actively involved in shifting water and ions across capillary walls. Electron micrographs of capillaries show that the flat, thin cells that make up the capillary wall are engaged in extensive pinocytosis (Figure 26.27). It is impossible to discern from the micrographs whether the pinocytic vesicles are picking up fluids from the bloodstream and depositing it in the tissue serum or vice versa, but it is likely that this kind of transport goes on in both directions most of the time.

The factors involved in the water and ion shift are complex. They involve constantly changing hydrostatic conditions. In general, what happens is that water and ions are lost at the arteriole end of the capillary bed, only to reenter at the venule end. Any excess water in the capillary bed is drained from the tissues through the lymphatic system (which we will consider shortly). Figure 26.28 illustrates some of the aspects of hydrostatic pressures in the capillary system. Notice that the blood is drained from capillary beds by the venous portion of the circulatory system. The joining of capillaries forms venules, which, in turn, join to form veins. The lowest blood pressure in the system occurs in the veins, where it actually approaches 0 mm Hg. The low pressure is due to the vast area of the capillaries and arterioles; with more surface area, there is greater frictional resistance against that surface. This means that although blood volume leaving the arteries is equal to the volume entering the veins, the force of its movement is greatly depleted. If you put your thumb over the water leaving a hose, the pressure in the hose (arteries) is considerable, the spray (capillaries) strikes the surface of the ground and the water then trickles down the gutter (veins). The high energy level is gone.

Because of the greatly reduced blood pressure in veins, the blood needs help in getting back to the heart. Many of the veins have one-way flap valves, which means that if the blood moves, it can only move one way. The walls of the veins, though much thinner than arteries, do contain some smooth muscles that can contract and help push the blood along (Figure 26.29). But the venous flow receives assistance in other ways as well. In the chest area, breathing

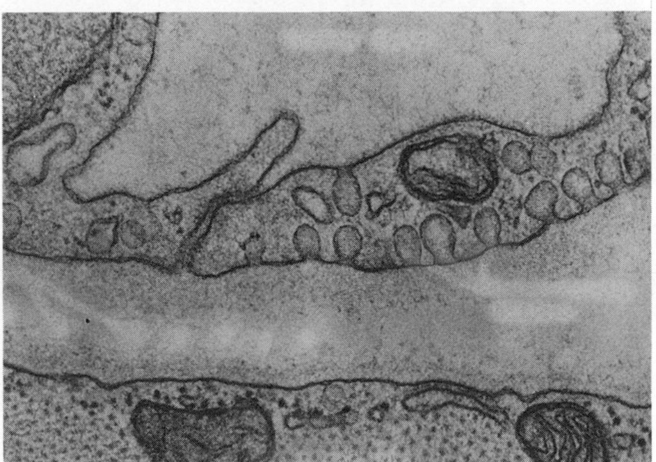

26.27 Pinocytic vesicles in the capillary wall. As the electron micrograph of a capillary wall clearly indicates, capillaries are far more than minute tubes. In addition to their more passive role in permitting substances to diffuse across, capillary cells are known to carry on a considerable amount of active transport, as is indicated by the numerous pinocytic vesicles in the capillary cells. Pinocytic vesicles are inpocketings in the cell membrane that close around substances, pinch off into vacuoles, and actively move these substances through the cytoplasm. In this instance, they may be transporting their contents to the lumen of the capillary for deposition into the bloodstream.

26.26 Frostbite! Frostbitten fingers, toes, ears, and noses are among the many risks humans face when exposed to subzero temperatures. When this happens, amputation of the destroyed tissue is inevitable. Mountain climbers who try Mt. Everest and survive, often reveal horror stories of frostbite where they have painlessly amputated their own fingers and toes—the only way to avoid gangrene. Frostbite occurs when capillary beds serving the extremities are shut down in defense against danger to more vital regions. The tissue, thus deprived of oxygen and other vital substances, simply freezes, dies, and cannot recover. The affected parts become black and decay begins.

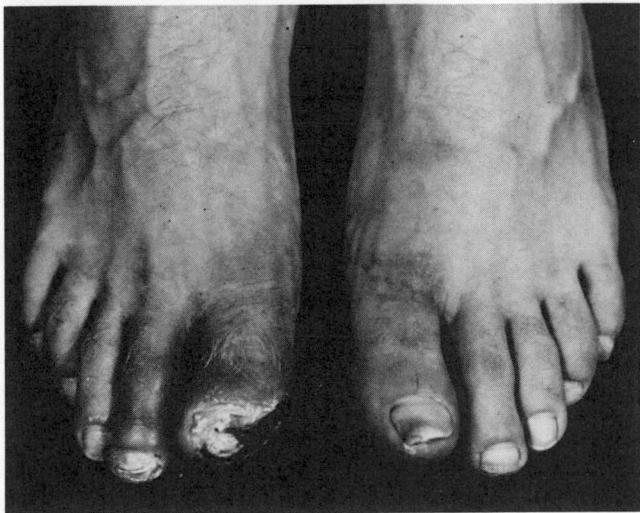

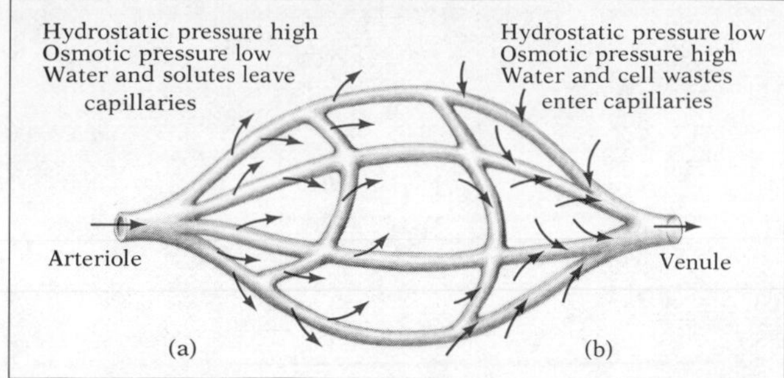

26.28 Pressure relationships in capillary beds. The movement of water and solutes in and out of the circulatory system occurs in the capillary beds. This movement, in turn, is controlled by pressure relationships in the bed. Two forces at work are hydrostatic pressure and osmotic pressure, and both operate inside and outside the capillary. (a) Blood entering a capillary bed from an arteriole is under a hydrostatic pressure that is greater than osmotic pressure, so water and solutes are forced out of the capillaries and into the surrounding tissue spaces and cells. (b) The situation reverses at the venule side of the capillary bed. Hydrostatic pressure is much less here because of frictional resistance in the greatly increased capillary surface area. Osmotic pressure has increased substantially because of the loss of water and solute, now concentrated outside the capillaries. Many of the solutes have by now been taken up by the cells. Most of the water follows the diffusion gradient back into the capillaries where they merge to form the venule. The remainder will be drained from the tissue spaces by the lymphatic system, and a balance is thus maintained. As water reenters the capillary, it is joined by cellular wastes and other metabolic products that have to be carried away.

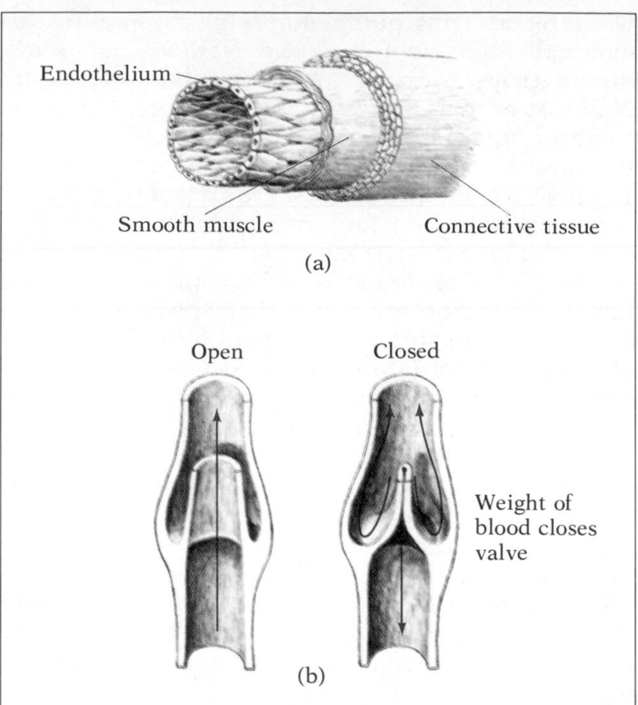

26.29 (a) Veins are commonly thinner-walled than arteries. They also tend to be larger. In cross-section, we see that a vein contains some of the same regions we saw in the artery in Figure 26.22. But, while the inner layer of tissue is similar, the middle tissue layer typically contains much less muscle and is not as elastic. The outer region is also similar to that of the artery. (b) The onward movement of blood in the veins is assisted by one-way valves. (c) The presence of these valves was first demonstrated by William Harvey in the 17th century, but the demonstration is elegantly simple and repeatable. The tourniquets used by Harvey are a bit dangerous, so simply grasp your arm with your hand, preferably the one on the other arm. And then let a partner trace the valves.

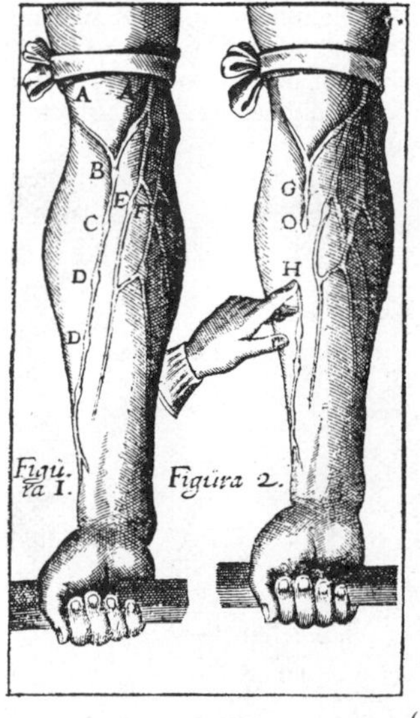

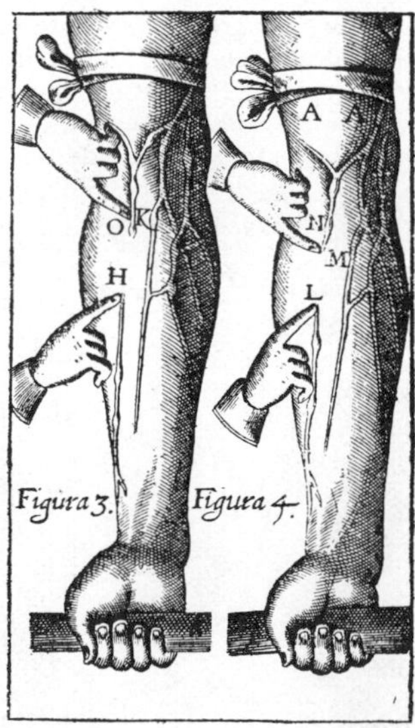

(c)

movements squeeze the walls of the vessels, sending the blood toward the heart. Breathing helps in two ways. During inspiration there is a negative pressure in the chest cavity, causing some blood to be drawn into veins there. At the same time, the flattening of the diaphragm presses on the organs below, squeezing the veins somewhat. The valves prevent backflow in both situations.

Muscular movement in the arms and legs also assists venous flow. A contracting muscle exerts a massaging or squeezing action against the veins. If you have been in one position for a long time you feel the need to "stretch." This is actually just a way of squeezing out old blood and letting fresh, invigorating blood enter the muscles. Soldiers forced to stand at attention for a long time may pass out because there is not enough muscular movement to force sufficient blood to the heart and on up to the brain. So we see, blood returning to the heart receives a boost from a number of sources.

The Blood and Lymph

By definition, blood is a tissue. That is, it is composed of several types of cells in more or less resident (unchanging) status. More precisely, it is a connective tissue, with *plasma* as its *matrix* (ground substance). Gentle centrifugation of blood easily divides it into three parts: plasma on top, a thin, clear band of *leukocytes* and *platelets* in the middle, and the heavy *erythrocytes* on the bottom. By volume, the red blood cells (erythrocytes) comprise about 45% of the whole blood in men. In women, the average percentage is about 43%. The upper layer, comprising the straw-colored plasma, consists of about 90% water. The remaining 10% is composed of a long list of substances; by weight, about 7% of plasma is protein, and the remaining 3% includes urea, amino acids, carbohydrates (mostly glucose), organic acids, fats, steroid hormones, and various inorganic ions.

The Red Blood Cell

Normally, there are about 5 million red cells per cubic millimeter of blood (Figure 26.30). (If you have a penchant for arithmetic, you might want to calculate the number per person. The average blood volume in a human is about 5 liters.) As we mentioned earlier in the chapter, red cells transport oxygen and carbon dioxide, an ability related to their hemoglobin content. Normally, the blood contains about 15 grams of hemoglobin for each 100 ml of whole blood.

Red and white blood cells are produced continuously in the *red bone marrow*. In adult humans, red cell formation occurs almost exclusively in the axial skeleton. Red cells have about a 4 month life expectancy. When they age, some rupture, spilling their hemoglobin into the blood. Most aging or damaged red cells are phagocytized by cells known as *macrophages* in the spleen and liver. The hemoglobin released during this destructive process is changed in the liver to *bilirubin*, which is subsequently excreted into the small intestine as a part of the bile. Should the bilirubin fail to be removed, it builds up in the blood, producing jaundice. *Hepatitis* can interfere with bilirubin excretion so that the waste builds up in the blood, turning the skin and the whites of the eyes yellowish.

26.30 Red blood cells (erythrocytes) make up about 45% of total blood volume in men and 43% in women. Individually, they are minute as animal cells go, about 6–8 μm in diameter. They form a biconcave disk in shape and lack a nucleus or other typical cell features. These are lost in their development. To say that the red cells are "numerous" is like saying that the defense budget is "costly." We each have about 25 billion at any one time.

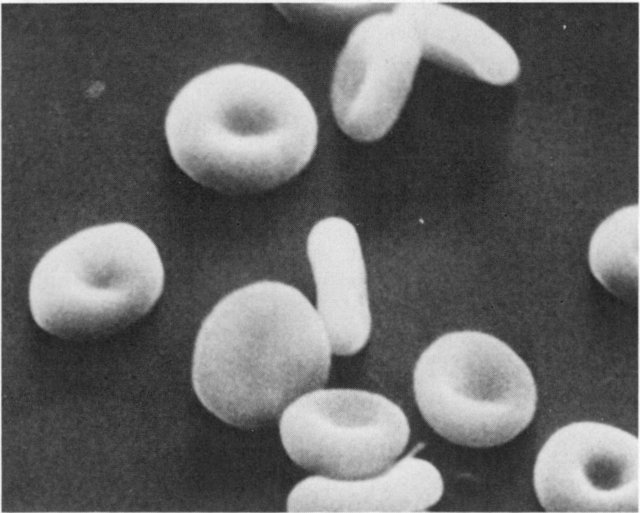

The slow, steady production of red cells can suddenly increase in an emergency in which blood is lost. Normally, however, a constant balance is maintained through a negative feedback system. When red cell numbers are deficient, oxygen supply to the tissues is lowered. The kidneys, in response, release a substance (*erthropoietin*) that stimulates the bone marrow, and soon there is a surge in red blood cell production. Increased red cell production eventually removes the stimulus, and thus the production slows. This is just another example of a self-regulated biological process. We have mentioned several of these, and will continue to point them out, since they are the key to understanding the complex functioning of living bodies.

The White Blood Cell

Unlike the red cells, white cells (leukocytes) do not discard their nuclei as they mature (Figure 26.31). Unfortunately, the names of the five types of white cells reflect their staining characteristics and tell us nothing of their functions. The white blood cells are active in the immune system, and their different roles will be discussed in Chapter 28.

Platelets

Blood platelets are extremely numerous, tiny structures in the blood that lack nuclei. Unlike erythrocytes, which are whole cells from which the nuclei have been lost, platelets are only fragments of cells. They are formed when a large white blood cell called a *platelet mother cell* (*megakaryocyte*) matures and fragments (without mitosis) to form the small, disk-shaped vesicles—a most peculiar process. Platelets, as we will see shortly, play a role in blood clot formation.

Plasma

So now that we have an idea of what the various little bodies are that float in our blood, let's see what it is that they float in. We mentioned earlier that plasma is 90% water, and that the other 10% is largely made up of various proteins. We will concern ourselves primarily with some important proteins of plasma, and then we will consider in detail how blood clots.

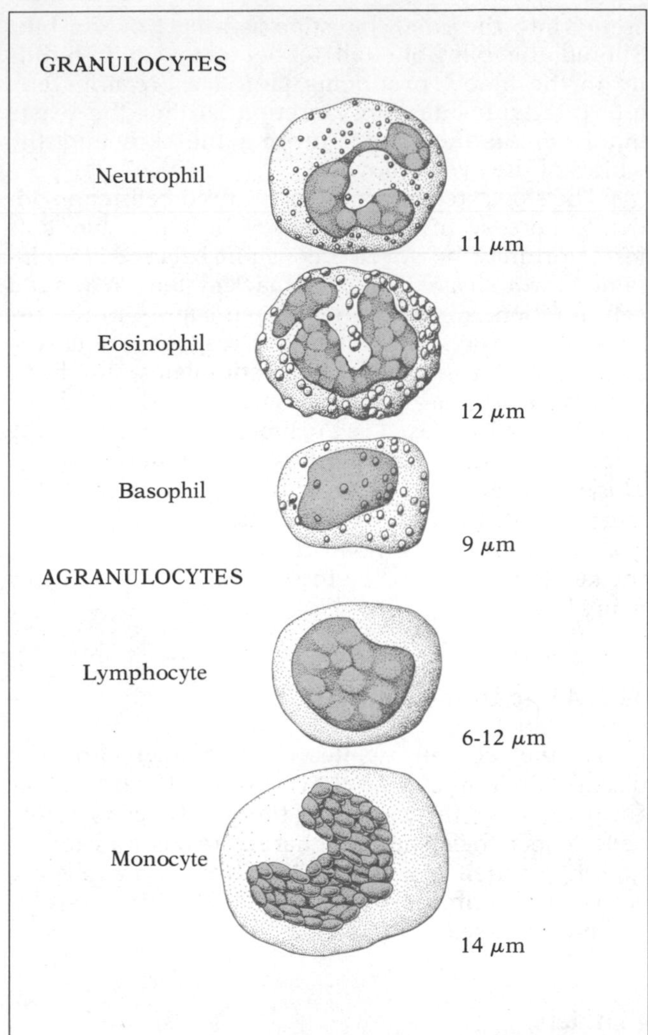

26.31 The five types of white cells (leukocytes) are grouped into two categories known as *granulocytes* and *agranulocytes*—terms which simply describe their staining characteristics. *Neutrophils* and *monocytes* are phagocytes; their role is to engulf foreign material, including bacteria and other phagocytes that have been destroyed at infection sites. *Lymphocytes* are undifferentiated; they can differentiate into other white cell types and even into red cells. They may also produce antibodies under certain conditions, as we'll see in Chapter 28. *Eosinophil* production increases in the presence of foreign proteins, but no one knows why. *Basophils* are known to secrete *heparin* (an anticoagulant), *histamine*, and *seratonin*. Neutrophils and lymphocytes far outnumber the others, making up about 95% of the white cells. (The measurements given next to each cell type are their diameters.)

Since proteins are too large to pass through most cell membranes, the overall protein content of the plasma is vital to maintaining a critical hydrostatic pressure in the vessels through osmotic action, particularly in the capillaries. This, in turn, is important to fluid balances throughout the body. Three of the critical plasma proteins are the *albumins*, *globulins*, and *fibrinogen*. The albumins are large proteins that bind miscellaneous impurities and toxins in the blood. They are also important in maintaining the osmotic potential of the blood, which in turn keeps plasma fluid from leaking out into the tissues. The globulins include the antibodies (see Chapter 28) and proteins involved in the transport of lipids and fat-soluble vitamins. Fibrinogen is the one that is most important in the blood clotting process. Clotting, obviously, is important to any organism that relies on circulating fluids. The world is full of puncturing objects that can pierce our epidermis and cause our fluids to leak out. Any such leakage must be minimized.

Clots can also form in diseased blood vessels such as the irregular inner surfaces caused by *atherosclerosis* (the localized growth of smooth muscle cells and the accumulation of fatty deposits in the arterial linings, producing roughened areas, plaques, and eventually occlusions). These hidden clots can cause great problems. For example, they can break free and later block important arterioles, causing heart attacks (by blocking vessels in the heart). Clots that break free and move through the vessels (*emboli*) can stop anywhere, blocking blood flow to large parts of the body.

So let's look at the major aspects of clotting, many of which remain obscure, while others lie safely beyond the scope of our presentation (Figure 26.32). But, in any case, there are at least fifteen substances involved. Two proteins are basic to the process: *prothrombin* and *fibrinogen*. In addition, platelets and damaged cells may play a role. Generally, the clotting process proceeds as follows:

1. A vessel is damaged.
2. Platelets and damaged cells release *thromboplastins* (enzymes).
3. In the presence of thromboplastins and calcium ions, *prothrombin* becomes *thrombin*.
4. Thrombin, a specific endopeptidase, breaks apart the large *fibrinogen* molecules, the smaller pieces forming *fibrin*.
5. Fibrin fibers, along with damaged platelets, form a network that solidifies, becoming a clot and stopping bleeding.
6. The clot contracts, pulling the wound together, further preventing bleeding and encouraging healing.

One of the many substances (called *cofactors*) involved in the clotting reaction is vitamin K, which is normally produced in ample amounts by bacteria in the gut. A synthetic antagonist of vitamin K, called *WARFARIN* (for the *W*isconsin *A*lumini *R*esearch *F*oundation, which financed its development and owns the patent), can prevent clotting. WARFARIN has been found to be valuable as medicine and as rat poison, depending on the dose and the recipient. As medicine, it is a blood thinner, decreasing the viscosity of blood and decreasing the likelihood of embolisms. As a rat poison, it causes death by massive spontaneous hemorrhage.

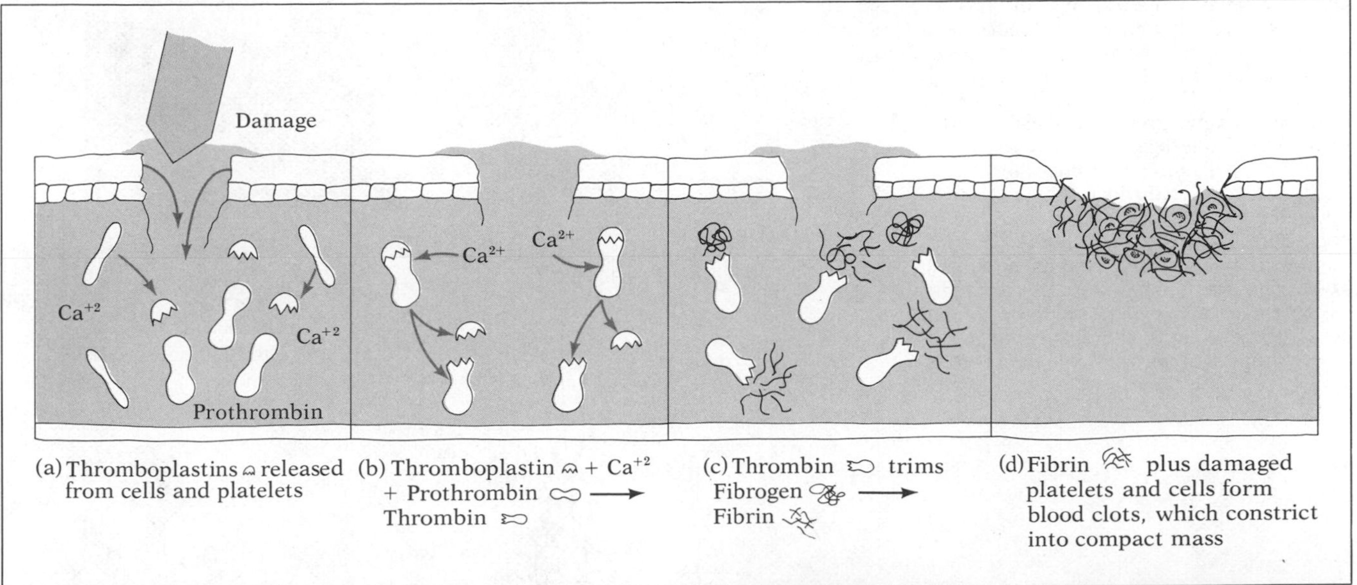
Damage
Ca^{+2}
Ca^{+2}
Prothrombin
Ca^{2+}
Ca^{2+}
(a) Thromboplastins released from cells and platelets
(b) Thromboplastin + Ca^{+2} + Prothrombin ⟶ Thrombin
(c) Thrombin trims Fibrogen ⟶ Fibrin
(d) Fibrin plus damaged platelets and cells form blood clots, which constrict into compact mass

26.32 The clotting of blood. When a blood vessel is ruptured, a complex series of events ensues which forms a clot. Basically, two changes occur. *Prothrombin*, a constituent of plasma, is converted to *thrombin*; and *fibrinogen* is converted to *fibrin*, the substance of a finished clot. As you can see, platelets and damaged cells release *thromboplastins*. When thromboplastins and calcium ions are present, prothrombin is converted to thrombin. Thrombin then reacts with fibrinogen to form fibrin, and fibrin along with cell elements become the clot.

Other cofactors are involved in two similar X-linked genetic diseases, *hemophilia* and *Christmas disease* (named after the family in which the disease was first described). The first is caused by a congenital lack of what is known as the *antihemophilic factor* (a globulin); the second is caused by a lack of the *Christmas factor* (a thromboplastin component). Both diseases are associated with poor blood clotting, which means that the affected individuals tend to bleed continuously internally, due to life's daily little traumas, and they show tragic susceptibility to cuts and bruises. The missing factors can be isolated from the whole blood of a donor now and used to treat clotting disorders.

The Lymphatic System

The lymphatic system is the "second circulatory system." And a rather simple one, at that. Basically, it drains tissue spaces and cavities of fluid that has leaked out of the capillaries because of high hydrostatic pressure, returning fluids to the bloodstream through ducts entering the subclavian veins. We've already mentioned its role in fat digestion and absorption in Chapter 25. Figure 26.33 illustrates the anatomy of this obscure system. As you can see, it consists of a large number of extremely thin-walled ducts, *lymph nodes*, and vessels. The system has no "heart," so the fluid is pushed along by the squeezing action of muscles and other organs. Check valves help keep the lymph from backing up so that the flow continues toward the heart.

Lymph itself is a watery substance that is often milky in color due to the presence of suspended fats. In addition, the lymph contains some proteins that have escaped through the capillary walls.

The lymph nodes, often called *lymph glands*, are scattered throughout the body. Their greatest concentrations, however, are in the neck, armpits, and groin (Figure 26.33). Each node is a compartmentalized mass of tissue that harbors multitudes of lympho-

26.33 (a) The lymphatic system consists of a number of lymph vessels, nodes, and ducts. The vessels begin as blind sacs in tissue spaces and join together forming larger vessels. The lymph vessels or ducts eventually empty into large veins near the heart. (b) Fluids are kept moving in one direction by a number of one-way valves, similar to those seen in veins. (c) Lymph nodes are located generally throughout the body but cluster in several regions including the groin, abdomen, armpits, neck, and head. The nodes are labyrinths through which the lymph passes en route to the ducts. Each node is about 1–2 mm in diameter and has incoming and outgoing vessels.

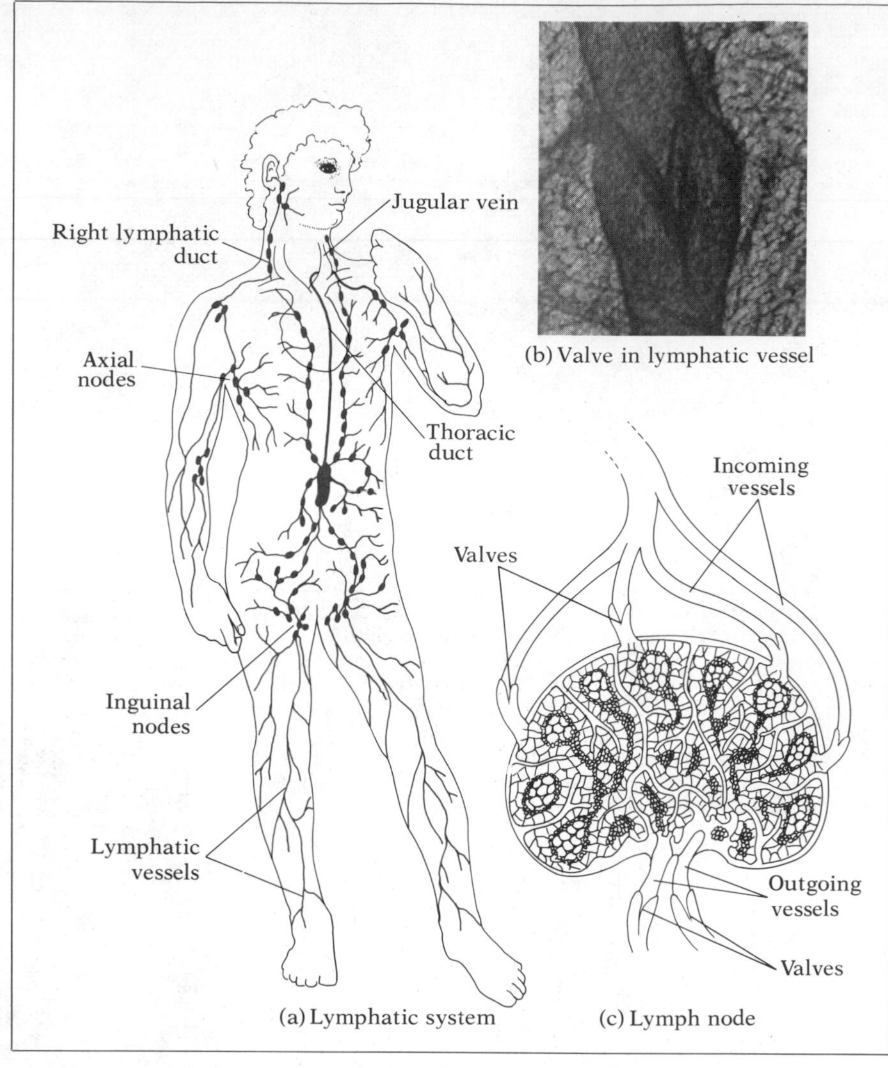

(b) Valve in lymphatic vessel

(a) Lymphatic system

(c) Lymph node

cytes. Foreign materials, bacteria, and viral particles that have managed to enter the tissue are swept into the nodes where they are attacked by resident lymphocytes. So we see that a main function of the lymphatic system is the removal of any foreign materials or invading microorganisms. The spleen and tonsils aid the lymphatic system in this function. When the lymphocytes in the nodes are under massive attack during an infection, the nodes enlarge, a telltale sign that something is wrong.

Application of Ideas

1. Tracing the embryological development of the four-chambered heart in humans reveals stages when the heart is two-chambered, then three, and finally, four. What does this suggest to you about the genetic basis for heart development in humans? How could your hypothesis be tested?

2. All aquatic mammals and birds are lung breathers just like their terrestrial relatives, no matter how well adapted they are to the aquatic environment in other ways. Given enough time, would it be probable for these animals to evolve a gill-like device? What related physiological problems would birds or mammals have to overcome to utilize gill breathing? Have any terrestrial vertebrates re-evolved gills?

3. The gravest danger in a coronary embolism (blockage) immediately follows its onset. This is true even when a small amount of heart muscle is affected. Why is a loss of function even in a small part of the heart muscle so significant to the proper functioning of the heart?

Key Words

albumin
air sac
alveolus (alveoli)
anterior (superior) venae cavae
aorta
aortic arch
arteriosclerosis
arteriole
artery
atherosclerosis
atrioventricular node
atrioventricular valve
atrium

bicarbonate ion
bicuspid valve
bilirubin
Bohr effect
bone marrow
bronchiole
bronchus (bronchi)
bundle of His

capillary
capillary bed
carbaminohemoglobin
carbonic acid
carbonic anhydrase
carotid artery
carotid body
cloaca
closed circulatory system
common cardinal veins
countercurrent multiplier

deoxyhemoglobin
dermal branchiae
diaphragm
diastole
dorsal aorta

embolism
erthropoietin
erythrocyte
expiration center

fibrin
fibrinogen

gastrovascular cavity
gill
gill arch
gill basket
gill filament
globulin
glottis
goblet cell

heme group
hemocoel

Key Ideas

Introduction

1. Nearly all of the organisms of the earth utilize oxygen to remove spent hydrogen, forming water at the end.

2. The more complex invertebrates and all of the vertebrates have to devise systems to provide transport of materials through bodies of increasing density and size. The important requirements are moist, thin surfaces that can be protected and a way of getting dissolved gases to and from these interfaces.

The Body Interface

1. In sponges, the same flagellated cells and canal systems that bring in food, bring in fresh supplies of water for an exchange of CO_2 with O_2.

2. The thin-walled saclike coelenterates require no special system, since all cells are close enough to water for an exchange of gases to readily occur through diffusion.

3. Flatworms are flat and their gastrovascular cavity is highly branched, so cell-by-cell diffusion of gases is sufficient for the necessary exchange.

4. Terrestrial life presents the problem of desiccation when large surfaces are exposed for the required gas exchange.

5. A few terrestrial animals exchange gases across the skin. Salamanders and earthworms can do this because of their damp environment, their moist, vascular skins and their efficient *hemoglobin*-carrying circulatory systems.

Expanding the Interface: Tracheae

1. Most terrestrial animals have developed extensive inpocketings for gas exchange surfaces.

2. Insects exchange gases across highly branched *tracheae,* which are everywhere, sometimes ending in sacs. They open to the air through *spiracles.*

Expanding the Interface: Branchiae

Echinoderms have a respiratory system consisting of *dermal branchiae* protruding through the skeletal plates. Coelomic fluid, moved by cilia, carries on the exchange of gases with the surrounding sea water.

Complex Interfaces: Gills

1. In the larger aquatic vertebrates, gas exchange occurs through complex *gills* of the respiratory system and transport is accomplished through circulatory systems.

2. *Gills* usually consist of numerous thin, fingerlike, structures, each containing a capillary network. The combined surface area is great enough to permit a rapid exchange of gases.

3. In the clam, gills also function in sorting food to be passed by cilia to the mouth. *Hemocyanin* is the O_2-carrying pigment. Currents enter and leave by ciliated movement.

4. Crayfish actively bail water across their *gills*. Blood is drawn from a cavity below, passes through the gills, and goes directly to the heart for redistribution.

Gill Baskets

Gill baskets serve both digestive and respiratory functions in tunicates and lancets. The first has an *open circulatory system* with vanadium used as the respiratory pigment. The second utilizes a *closed circulatory system* with *hemoglobin* in red blood cells.

Gills in Fishes

1. Because of the relative low O_2 content of seawater, fish must actively move water over their *gills*. Some sharks do this by swimming, while bony fish actively pump water across the *gills* with the mouth. The pumping action is coordinated with the opening and closing of the *opercula.* Some fish can gulp air and exchange gases across *vascularized* chambers in the mouth.

hemocyanin
hemoglobin
hepatic portal circuit
hyaline cartilage
hyperventilation

inferior (posterior) venae cavae
inspiration center
intercostal muscles
internal nare

jaundice

lacteal
leukocyte
lung
lymphatic system
lymph duct
lymph node
lymphocyte

medulla
mucus

open circulatory system
operculum
oxygen carrier
oxyhemoglobin

pacemaker
palate
parasympathetic nervous system
partial pressure
pharyngeal gill cleft
pharynx
plasma
platelet
pleura
pleural cavity
poikilothermal
posterior (inferior) venae cavae
primary bronchi
prothrombin
pulmonary circuit
pulmonary valve
pulmonary vein

renal artery
renal circuit
renal vein
respiratory center
respiratory tree

sinoatrial node
sinus
sinusoid
sinus venosus
skin breather
soft palate
sphincter
sphygmomanometer
spiracle
spleen

2. The *gills* are made more efficient by moving blood opposite to the flow of water, thereby setting up a countercurrent exchange. A more complete exchange of O_2 and CO_2 between the capillaries and water occurs.

Complex Interfaces: Lungs

Evolution of the Vertebrate Lung

1. The vertebrate *lung* apparently evolved from a primitive pocket of highly *vascularized* tissue, used by air-gulping fish. This group is represented by lungfish today. In other bony fish the primitive *lung* has evolved into a *swim bladder.*

2. Amphibians and most reptiles have nasal openings leading into the mouth in the primitive fashion. Crocodiles and mammals have formed a *palate* in the roof of the mouth.

The Vertebrate Lung

1. Frogs and toads have check valves in the nostrils and *trachea.* They have simple, saclike lungs and a highly *vascularized,* moist skin which also exchanges gases. Air is pumped in and out by throat muscles. Larval forms have temporary *gills.*

2. Reptiles have a dry skin and use lungs for breathing (except for some aquatic turtles that use the *cloaca* as well). The reptilian lung has a subdivided *trachea* and numerous compartments. There is no *diaphragm,* so breathing involves the body wall.

3. Bird *lungs* are the most highly evolved in the vertebrates. Once in the body, incurrent and excurrent air follow different pathways.

4. Inhaled air first passes the *lungs* to *air sacs* throughout the body. Inhaling and exhaling in flight is assisted by wing movements. Air leaves the sacs along a different path and then enters the lungs, exchanging more gases there. The lungs are almost completely ventilated as air passes back out through the *trachea.* The system allows countercurrent exchange of gases.

The Anatomy of the Human Respiratory System

1. Mouth and *pharynx:* nasal and oral passages are separated by the *palate.* The nasal passages filter, moisten, and warm air passing through.

2. *Larynx, trachea,* and *lung:* air passes through the *larynx* and vocal cords, enters the *trachea,* and passes into the right and left *bronchi.* The *trachea* and *bronchi* also contain ciliated and *mucus*-secreting cells.

3. *Bronchi* branch and rebranch into the *bronchioles,* which form the respiratory tree. The *bronchioles* end in *alveoli,* which are enclosed by capillaries of the *pulmonary circuit.*

The Breathing Movements

In inhalation, the dome-shaped *diaphragm* moves down while the *intercostal* muscles raise the rib cage. This creates a partial vacuum in the cavity and atmospheric pressure inflates the lungs. Exhalation is usually passive. The muscles relax and the elastic *lungs* deflate.

Partial Pressure

1. The amount of a particular kind of gas that is available to a system is called its *partial pressure.* At sea level where atmospheric pressure is 1 atmosphere, the *partial pressure* of oxygen is 0.21 (21% of air).

2. The diffusion of dissolved gases always occurs from regions of higher *partial pressure* to areas of lower *partial pressure* until equilibrium is reached.

The Exchange of Gases in the Alveoli

1. Carbon dioxide and O_2 are exchanged across the alveolar walls between the blood and air.

2. Blood entering the lungs has been in the tissues where *partial pressures* of CO_2 are high and O_2 low. As blood passes the cells it is equilibrated. *Hemoglobin* becomes deoxygenated and picks up CO_2.

3. Entering the *lungs,* blood has a low O_2 *partial pressure* and a high CO_2 *partial pressure.*Air in the *alveoli* is just the opposite. As blood passes, equilibration is reached. Carbon dioxide is lost and O_2 is gained.

stretch receptor
stroke volume
superior (anterior) venae cavae
swim bladder
syrinx
sympathetic nervous system
systole

thorax
thrombin
thromboplastin
trachea (tracheae)
tricuspid valve

vascularized
vasoconstriction
vasodilation
vein
ventilation
ventricle
venule
voice box

WARFARIN

4. Carbon dioxide is transported in blood in the form of dissolved CO_2, *bicarbonate ions,* and some loosely bound to *hemoglobin.* The enzyme *carbonic anhydrase* converts the bicarbonate ions to CO_2 gas.

5. *Hemoglobin* is the key to efficient respiratory exchange. It has a strong affinity for O_2 when *partial pressures* are high. This reverses when the *partial pressures* are low.

6. *Hemoglobin* contains four polypeptides, each with an iron-containing *heme* group. Each *heme* can bind reversibly with one oxygen molecule.

7. Humans can live comfortably at varying altitudes because *hemoglobin* is sufficiently saturated even when *partial pressure* of O_2 is low.

8. Carbon dioxide and O_2 appear to be in competition for hemoglobin. The binding of O_2 by hemoglobin is decreased in the presence of CO_2 (*Bohr* effect).

9. Because of the *Bohr effect, hemoglobin* gives up its O_2 in tissues that are producing the most CO_2, not just in tissues with low O_2 content.

10. Most of the CO_2 is transported as *bicarbonate ion. Carbonic anhydrase* works in the opposite direction in the tissues, speeding the formation of the ion.

The Control of Respiration

1. Both the respiratory and the circulatory system must respond in a coordinated way to changing O_2 requirements.

2. The major respiratory control centers are in the *medulla.* They include *inspiratory* and *expiratory centers.* Together they produce a breathing rhythm.

3. The two centers inhibit each other alternately and are, in turn, influenced by *stretch receptors* in the *lungs,* which determine the depth of inhalation. These controlling areas are influenced by other factors as well.

Transport in Multicellular Animals

1. The earthworm and other annelids have a *closed system* consisting of arteries, *capillaries,* and veins along with tubular hearts. Blood flow in the vessels is one way because of valves.

2. Arthropods have the *open system.* Blood, pumped by the heart, leaves its vessels entering *sinuses* known as *hemocoels.* It then reenters vessels and is directed to a cavity surrounding the heart, which draws it in to be pumped into vessels again.

3. Crustaceans are gillbreathers, so much greater vessel development is present, assuring rapid circulation through the *gill* structures.

4. Some crustaceans, spiders, and scorpions utilize a copper-containing protein pigment known as *hemocyanin* for carrying O_2.

Circulation in the Vertebrates

All vertebrates have *closed circulatory systems* and central hearts. Crocodiles, birds, and mammals have four-chambered hearts.

Circulation in Fishes

1. The fish heart is essentially two-chambered with an *atrium,* a *ventricle,* and an enlarged vein known as the *sinus venosus,* which acts as a blood-receiving chamber.

2. Blood received in the *sinus venosus* enters the *atrium,* is pumped to the *ventricle,* and then out to the *gill* capillaries first. From there it moves to the capillaries of the head and body before returning to the heart.

Circulation in Amphibians and Reptiles

1. In amphibians and reptiles the heart is three-chambered, with one *ventricle* and two *atria.* A large *sinus venosus* collects returning blood.

2. Blood returns to the heart from the body and lungs, entering right and left atria. From these it enters the single *ventricle,* which pumps it out to the body and *lungs* once again. Flaps and valves help separate oxygenated and deoxygenated blood.

The Four-Chambered Heart

In the four-chambered heart two separate circuits have evolved, finally separating the *pulmonary circuit* with its oxygenated blood from the deoxygenated blood returning from the body.

The Human Circulatory System

1. The human circulatory system, like that of other mammals (and of crocodilians and birds), consists of a four-chambered heart, *arteries, capillaries,* and *veins.*

2. *Arteries* with thick muscular walls carry blood away from the heart. *Veins* with thinner less muscular walls return it. With the exception of the pulmonary vessels, *arteries* carry oxygenated blood, and *veins* carry deoxygenated blood.

3. *Arteries* branch into *arterioles,* then into *capillary beds. Capillaries* are the functional parts where all exchanges occur. They rejoin to form *venules,* then veins.

Circulation Through the Heart

Blood circulates through the heart as follows:

1. Blood from the body and *lungs* enters right and left *atria,* respectively.
2. *Atria* contract as blood is forced into right and left *ventricles.*
3. Backflow into the *atria* is prevented by *valves.*
4. Contractions of the *ventricles* forces blood out through *pulmonary artery* and *aorta.* The valves prevent backflow.

Control of the Heart

1. Heart muscle itself is unlike other muscle types. There are two main physiological differences:
   a. The branching fibers are made up of a series of uninucleate cells.
   b. The repolarization period is prolonged.
   c. Rhythmic contraction is inherent in the muscle cells.
2. *Sympathetic* control originates in the *medulla* and is carried through the spinal nerves. These nerves can accelerate the heart rate.
3. *Parasympathetic* nerves originate directly in brain itself and slow the heart rate.

The Pacemaker

1. Heartbeat originates in the *sinoatrial node.* It transmits to the *atrioventricular node,* via the atrial walls. Impulses from the *atrioventricular node* reach the *ventricles* via strands of specialized muscle in the ventricular septum, called the *bundle of His.* Contractions in the atria are simultaneous as are those of the *ventricles.*

2. The first heart sound is the bicuspid and tricuspid atrioventricular valves snapping shut. This is followed by the sound of the closing of the semilunar valves that prevent backflow from the arteries into the ventricles.

3. The resting cardiac output is 5–6 liters per minute, while the active cardiac output may be 30–35 liters per minute.

4. Blood pressure is maintained by the elasticity of the large *arteries.* Pressure is highest during *systole* and lowest during *diastole.*

5. Flow through *capillary beds* is controlled by smooth muscle sphincters in the *arterioles.* These function in response to hormones, ions, and the nervous system.

Circuits in the Human Circulatory System

A *blood circuit* is a capillary region where a major function occurs, and the vessels entering and leaving it.

Pulmonary Circuit

This includes the pulmonary arteries and *veins* and the alveolar capillaries. Gases are exchanged.

Hepatic Portal Circuit

Two *capillary beds* function here. One absorbs digested foods from intestine to be carried to the liver. The second, in the liver, removes most food. The liver also removes toxic substances.

Renal Circuit
The *capillary beds* remove nitrogenous waste, excess water, and other substances, maintaining water and salt balances.

General Body Circuits
The brain, muscles, glands, bones, etc., all have *capillary beds* where gases are exchanged, nutrients carried in, and wastes removed.

The Role of the Capillaries
1. *Capillaries* are the functional units of the circulatory system. They carry on thermoregulation in the skin, exchange gases with the tissues, and transport foods, wastes, ions, and hormones.

2. Transport across *capillary* walls is through diffusion, osmosis, and active transport, including pinocytosis and exocytosis.

3. Ion and water transport involves hydrostatic conditions. They are lost at the arteriolar end of the *capillary bed* and reenter at the *venule* end. Excess water is drained via the *lymphatic system.*

4. Venus pressure is very low because of the tremendous combined frictional resistance of *capillary beds.* To continue the onward flow, *veins* have one-way valves. Muscular movement also helps in the onward movement.

The Blood and Lymph
Blood is a tissue. Its ground substance is *plasma.* It consists of 43–45% red cells, a small percentage of white cells, and *plasma. Plasma* is 90% water, 7% protein, and 3% urea, amino acids, carbohydrate, organic acids, fats, hormones, and ions.

The Red Blood Cell
Red cells, about 5 million per cubic millimeter, are produced and destroyed constantly in a 4 month cycle. The liver changes free *hemoglobin* to *bilirubin,* which is excreted. A low oxygen level stimulates the kidneys to release *erthropoietin,* which stimulates red cell production.

The White Blood Cell
White cells are nucleated. Their function is chiefly disease control.

Platelets
Platelets are cell fragments formed from *platelet mother cells.* They play a role in clotting.

Plasma
1. *Plasma* proteins are important in maintaining a proper hydrostatic pressure in the blood. This is related to general fluid balances throughout the body. Two *plasma* proteins, *prothrombin* and *fibrinogen,* are important to clotting.

2. Clot formation can occur within diseased vessels. When clots break free they are known as *embolisms.* They can block vital arteries, causing severe medical problems.

3. In hemophilia and Christmas disease, blood clotting factors are absent, so hemorrhaging can easily occur.

The Lymphatic System
1. The *lymphatic system* consists of ducts, vessels, and nodes. Its function is to return fluids to the blood, carry fats from the digestive system, and aid in defense against infection.

2. The nodes along with the *spleen* and tonsils help remove bacteria and isolate them for destruction.

3. The *lymphatic system* does not circulate fluids, but simply drains them for return to the blood.

4. *Lymph nodes* are scattered in the body, clustering in various places. When the *lymphocytes* are active, the nodes may become painfully swollen.

5. Any blockage of the *lymphatic* vessels or *ducts* results in fluid accumulation in that part served. Fluid accumulation is known as *lymphedema.*

Review Questions

1. What is it about the body structure of coelenterates and flatworms that enables them to get along without a respiratory or circulatory system? (p. 733)
2. Describe the general structure of a *gill.* (pp. 734–735)
3. Why is a circulatory system so essential in gilled animals? (p. 735)
4. What are *dermal branchiae* and in what organisms are they found? (p. 734)
5. What characteristics of gas exchange membranes present special problems to the terrestrial animal? (p. 733)
6. What solutions to terrestrial respiratory problems have the salamander and earthworm utilized? (p. 733)
7. What is a *lung?* (p. 738)
8. Describe the respiratory structures in an insect. Is the circulatory system involved? Explain. (p. 734)
9. List two functions of the *gill baskets* in lancelets and tunicates. (p. 736)
10. Describe the method of water circulation past the gills of sharks and bony fishes. (pp. 737–738)
11. Review the events in the evolution of the terrestrial *lung.* (pp. 738–739)
12. In what manner do bony fishes use the primitive *lung* today? (p. 738)
13. Describe the breathing action in the frog. (p. 739 and Figure 26.7)
14. Explain the one-way passage of air through the bird's *lungs* and its advantages. (p. 740)
15. Offer a physiological reason for the bird's complex respiratory system. (p. 740)
16. List the structures of the human respiratory system in the order in which they appear. (pp. 741–742)
17. Describe the epithelium of the *nasal cavity* and the *trachea* and how this tissue functions. (p. 742)
18. Describe the structure of the *alveoli,* including both respiratory and circulatory aspects. (p. 742)
19. Explain the action of the *diaphragm* and rib cage and the lung response in inhalation and expiration. (p. 742)
20. Explain what 0.10 *partial pressure* of O_2 in H_2O means. (p. 743)
21. Explain what happens in a closed system when the *partial pressure* of a gas is different in liquid than in the surrounding air. (p. 743)
22. Compare the *partial pressure* of O_2 and CO_2 in blood leaving the lung with that of blood arriving at the lung. (p. 743)
23. What is the action of *carbonic anhydrase* in blood arriving at the *lung?* (p. 743)
24. Describe the structure of hemoglobin and its relationship to oxygen-carrying capacity. (pp. 743–744)
25. Describe the safety margin in blood saturation with O_2 that permits us to live at sea level or at high elevations. (p. 744)
26. How does the *partial pressure* of CO_2 affect the binding of O_2 to *hemoglobin?* (pp. 744–745)
27. Using a chemical formula, describe the chemistry of CO_2 transport and the role of *carbonic anhydrase* in the reaction (there are two ways in which the enzyme works). (p. 745)
28. Review the role of the *inspiration* and *expiration centers* and the *stretch receptors* in controlling breathing. (p. 745)
29. How does the CO_2 level in the blood control the rate of breathing? (pp. 745–746)
30. Explain how the *carotid body* in the neck malfunctions during *hyperventilation.* (p. 746)
31. Describe the circulatory system in the earthworm. (Which kind is it?) (pp. 746–747)
32. Compare the circulatory system of an insect with that of a crustacean. (p. 747)
33. Define *countercurrent exchange* and explain how it operates in fish gills. (p. 748)
34. Describe the fish's heart and its general blood circulation. (pp. 748–749)
35. What is the apparent problem with oxygenated and deoxygenated blood in the three-chambered heart and how is it solved in amphibians and reptiles? (pp. 749–750)
36. Draw a diagram of a frog's three-chambered heart, label it, and indicate blood flow with arrows. (p. 750)
37. Explain the changes in the bird heart and offer a reason why this might be necessary in a flying animal. (p. 751)
38. Prepare a simple cutaway outline drawing of the four-chambered heart and label its parts. Use arrows to show blood flow. (pp. 750–751)
39. List the four heart valves and explain the action of each. (p. 751)
40. Compare the task of the right heart chambers with that of the left and explain how the left is adapted for its job. (pp. 751–752)
41. List three ways in which cardiac muscle is physiologically different from other muscle. (pp. 752, 754)
42. Trace the movement of a nervous impulse through the heart beginning with the *sinoatrial node.* Describe the heart's response to the impulse. (p. 754)
43. Describe the response of a healthy artery to *systole* and *diastole.* What does this response provide? (pp. 754–755)
44. Define *stroke volume* and compare it in the heart at rest with the heart at work. (p. 755)
45. What is *arteriosclerosis* and how does it effect the artery's response to *systole* and *diastole?* (p. 755)
46. Explain the kind of control possible in the *arterioles* and the adaptive advantage of this control. (pp. 756, 758)
47. List the four major blood circuits and describe a special change in the blood in each. (pp 756–758)
48. Describe the osmotic conditions and the movement of water and ions at either end of a *capillary bed.* (p. 759)
49. List the various factors that help venous blood return to the heart. (pp. 759–760)
50. List the main substances present in *plasma.* (pp. 760–761)
51. List the following information about red cells (p. 761):
    a. Number per cubic millimeter
    b. Lifespan of a red cell
    c. Place of origin
    d. Removal of dead cells and *hemoglobin.*
52. List the steps that occur in blood clotting, using the terms *prothrombin, fibrinogen, thromboplastin, thrombin,* and *fibrin.* (p. 762)
53. Explain the general functions of the *lymphatic system.* (pp. 763–764)
54. List the structures of the *lymphatic system* and their functions. (pp. 763–764)

Chapter 27

Chemical Messengers in Animals

Life. A simple term. We intuitively know what it is. And we seem to have some sort of great respect for it (possibly because of its alternative). We see life around us, we are alive, and many people work hard to keep other organisms alive (never mind for a moment that group of people who like to refer to themselves as "hunters" and who go to great time and expense to deprive free-ranging animals of their lives). The point is, we are familiar with life and have an intense interest in it, but no one knows what it is. Its definition, much less its nature, escapes us. One of the reasons, obviously, is its complexity.

Another reason is that some of life's processes are apparent and others are not. We see a bird wading and we say, "Aha! There's a bird wading." So some life wades. What we don't see, though, are the strange little chemicals silently leaking into the bird's bloodstream from a small clump of tissues deep inside its body. And we can't trace their paths as they course through that body, being ignored by all the tissues along the way, until they reach the places that are somehow sensitive to those molecules, and the cells in those places *do* react. Their reaction may be so important that they kept the bird alive. They may also have affected the bird's behavior so that it stalks its hunting ground more intensively, laying on fat that will sustain it on a long journey southward that will come soon. The bird probably has utterly no knowledge about any such flight to come. At the right time, though, it will just go. And even then much of its behavior will be dictated by these invisible little molecules as they interplay through its body in a dazzling and dissolving concert. So let's have a closer look now at these coordinating molecules that direct the symphony we call life.

27.1 The Actions of Hormones

What Are Hormones?

Hormones, the chemical messengers, are formed in one part of the body and travel to other parts of the body—sometimes distant, sometimes quite near—where they cause changes. They are essentially regulatory devices, ways to make the body behave in a coordinated fashion. One problem in defining a hormone is deciding how far it has to travel before it initiates its action. For example, secretions produced by certain neurons act locally, only activating the next neuron. Most biologists, therefore, do not consider such neurosecretions to be hormones in the strict sense. Hormones, of course, don't act alone. The nervous system, for example, has much the same function. In fact, it's difficult to consider the hormone system and the nervous system separately, because they are involved in the same functions and so dependent on one another. If a generalization has to be made, it is that the nervous system usually acts more quickly than the *endocrine* (hormone) *system.*

Hormones, as we know from personal experiences, can affect our behavior and get us in a lot of trouble. Furthermore, the influence is reciprocal, because our behavior, or at least our experiences, can also influence our hormone levels, as we will see in our discussions of behavior.

A fascinating organism—if it is an organism, rather than a population of organisms—is *Dictyostelum*, the cellular slime mold, a creepy organism in the literal sense. Its bizarre life cycle starts as a population of single-celled amebas which, upon some sort of signal, begin a drastic transformation. They first crawl together into a mass to form a multicellular, motile "slug" and finally a stalked, fully differentiated reproductive body. The slime mold is of interest here because it shows perhaps the most simple animal hormone interaction, and it thus helps us understand how cells can influence each other.

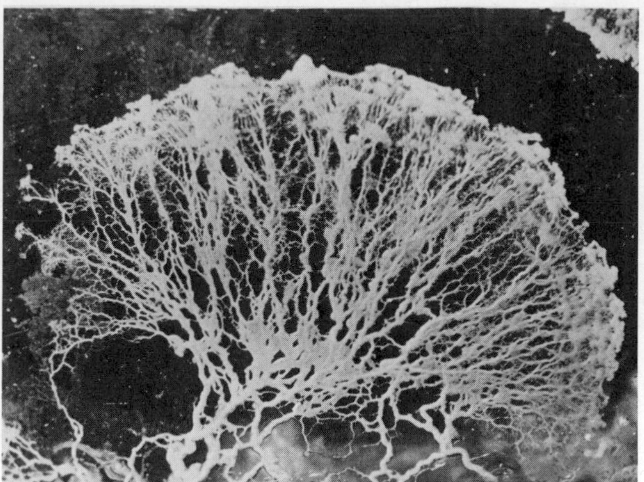

27.1 The slime mold *Dictyostelum.* One stage is marked by individual ameboid cells that carry on their activities independently. But, at some signal, the amebas begin to aggregate, marching in silent, flowing lines, and finally fusing into one multinucleate mass. The signal for massing is a chemical messenger known as *cyclic adenosine monophosphate (cAMP)* which is released by the individual amebas when conditions dictate.

Dictyostelum grows from a windblown spore into a free-living ameba, which moves in the classical ameboid fashion over soil or rotting wood and generates a clone of offspring by mitosis. Each of these show no allegiance to its parent and meanders on its own way, as would any other ameba. But when hard times befall it, it becomes more sociable. So in times of low food abundance or during a dry spell, the amebas seek each other out. It is interesting to see some that have found each other creeping along in little slimy rivulets, converging toward some central place (Figure 27.1) where they will crawl into a heap, undergo changes, and finally form a large multicellular slug. What calls them together? It is a chemical messenger—a hormone—that the cells begin to secrete when they are stimulated by changes in the environment. All each cell has to do is to move in the direction of greatest hormone concentration. As the cells aggregate, their additive effect increases the power of the chemical sirens, causing even more aggregation, which releases even more hormones, which . . . forms a positive feedback loop (an unusual system in biology since such loops are normally devastating to any life form, as we will see in our discussion of homeostasis).

The hormone that calls the cells together was fancifully named *acrasin*, after Acrasia, the mythological siren who lured seamen in Homeric legend. Later, the magical substance was discovered to be only a rather small molecule, *cyclic adenosine monophosphate* (*cAMP*), and so the mythological name was dropped. Cyclic AMP, after all, is just a simple nucleotide, adenine monophosphate, in which the single phosphate forms a bridge between the 3 and 5 carbons of the ribose sugar (Figure 27.2).

Presumably, if you were to spill cAMP on the ground, any nearby cellular slime molds would come running. We'll have more to say about cAMP later, because it turns out to be a key *intra*cellular messenger molecule that is involved in many of the hormone responses of higher organisms.

Hormones and the Invertebrates

We really don't know very much about hormones in invertebrate animals. However, the hormones of two classes of arthropods have been rather intensively studied: those of the crustaceans and insects. Because of our intimate association with the insects, much of the work that has been done has been critical to our own species.

The Molting Activity in Crustaceans

But let's begin on a less dramatic note. Let's consider molting in crayfish and other crustaceans, which we mentioned briefly in Chapter 24. If you recall, crus-

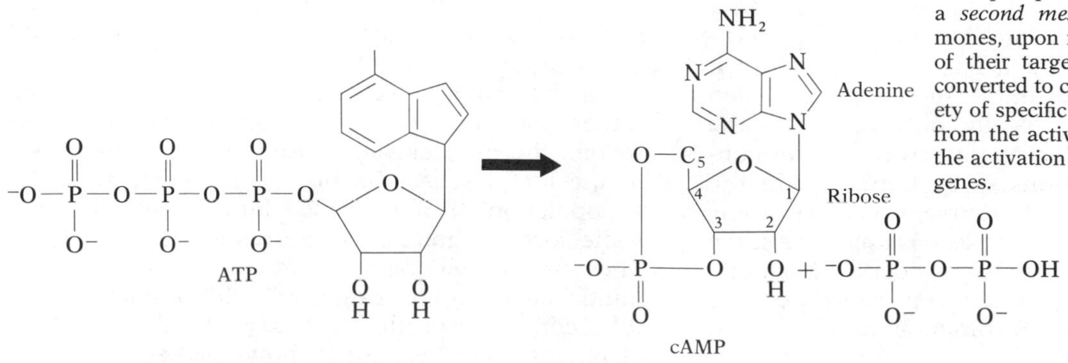

27.2 Cyclic AMP (cyclic 3,5-adenosine monophosphate, or simply cAMP) is called a *second messenger* because some hormones, upon reaching active sites on cells of their target organs, cause ATP to be converted to cAMP. Cyclic AMP has a variety of specific roles in the cell which range from the activation of enzyme systems to the activation and perhaps deactivation of genes.

taceans produce an exoskeleton from secretions of cells in the underlying epidermis. The problem is, physiologically, the animal can grow but the exoskeleton can't. So the exoskeleton must periodically be discarded to give the living organism within it space to expand. The process of shedding is known as *molting* (or *ecdysis*); it involves four distinct phases. In *preecdysis*, the crusty exoskeleton becomes thinner and more fragile as calcium and organic substances are reabsorbed into the body. The second stage, *ecdysis proper*, begins when the old exoskeleton splits down the back and the animal arches its back and pushes until its head emerges. Then, with its legs free of its old sleeves, it drags its body out of the old exoskeleton and moves away, leaving a ghostly shell of its former self. As soon as the confining shell is discarded, the animal quickly swells by absorbing water. In the third stage, *postecdysis*, the new shell is deposited on the swollen body and is hardened by being impregnated with calcium from the tissues (some of which was withdrawn from the old exoskeleton). Finally, there is a fourth stage called *interecdysis*. This is the time between moltings when the animal can go about its business without worrying about changing its armor. Unfortunately, many crustaceans grow so fast that they get little rest. Although by swelling up just before the new covering is laid down, and then shrinking a bit by losing water, the animal gives itself some growing room, the whole process repeats itself every few weeks. As the animal matures, however, interecdysis may lengthen to 5 or 6 months. Mature individuals of some species may even stop molting altogether.

Researchers began to suspect that hormones were involved in ecdysis when it was discovered that crayfish molted every 2 or 3 weeks if their eyestalks were removed. (One wonders why this surgery was done in the first place.) Then it was decided to try eyestalk implants. The operations were successful and, with the eyestalks restored, the crayfish returned to the normal molting cycle. Obviously, something in the eyestalk was regulating ecdysis. A search turned up the structure that was responsible and it was dramatically named the *X-organ* (Figure 27.3). But further surgical experimentation led to the discovery of another mysterious organ, located below the base of the eyestalk. Removal of this organ created an opposite effect: molting activity ceased altogether. The imaginative scientists named the second organ the *Y-organ*. They also decided that something from the two organs was controlling the process. They guessed that the "something" was hormonal.

However, proving it was another matter. Testing for hormone action is a rigorous business—one with very demanding rules. It is generally agreed that the role of any hormone can be described only after the following criteria are met:

1. Removal or damage to the suspected organ of production must cause a deficiency symptom.

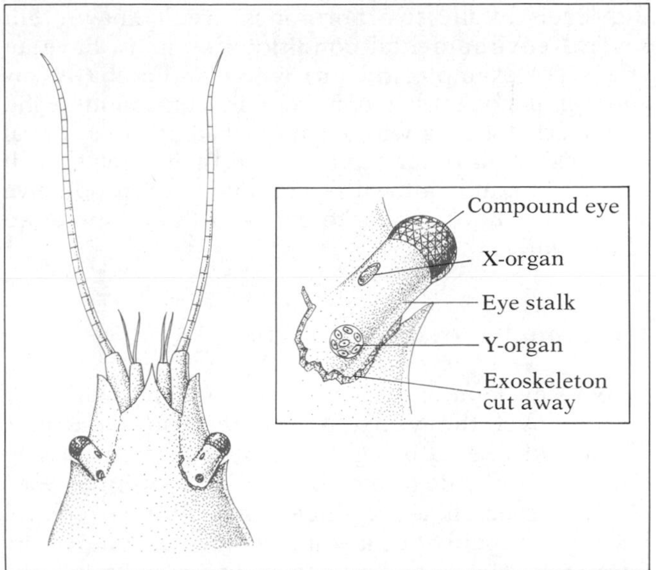

27.3 Hormonal control of *ecdysis*. The periodic shedding of the exoskeleton in crustaceans is under the control of two hormones. The molting hormone, *crustecdysone*, is secreted by the Y-organ, while molt-inhibiting hormone (*MIH*) is secreted by the nearby X-organ. Both of these endocrine structures are located at the base of the eyestalks. Light appears to be an important factor in controlling ecdysis.

2. The substance in question must be isolated and then refined and purified, if possible.
3. Reintroduction of the substance must alleviate the symptom.
4. Halting the artificial treatment must result in the reappearance of the symptoms.

At the moment, physiologists believe there are two hormones (or groups of hormones) involved in ecdysis. The Y-organ is believed to secrete the hormone *crustecdysone* (a variant of ecdysone found in insects), named for the crustacean. The X-organ, on the other hand, is believed to secrete a molt-inhibiting hormone known simply as *MIH*. These two hormones appear to be mutually interacting. That is, molting activity is determined by their relative concentrations. Increased ecdysone levels, it is believed, increase molting activity, while higher MIH levels suppress the activity. The factors controlling the levels of the two hormones aren't known, but several environmental conditions seem to have an effect. For example, the burrowing land crab *Geocarcinus* will not molt when kept in continuous light. Other crustaceans will not molt if kept in continual darkness. And other factors have their effects, such as the general health of the animal, its reproductive state, how much it gets to eat, and the temperature of the water.

Hormonal Activity in Insects

Insect development has received a great deal of attention over the years, partly because some people would like to bring much of it to a screeching halt. It is difficult to overemphasize the importance of this research area, since our ability to control insects—probably our leading ecological competitors (strangely enough)—depends on an exact knowledge of all aspects of their life. Know your enemy! Know, for one thing, that most insects are not enemies at all. For example, many of us don't like bugs, but we do like birds and birds depend on bugs. It is interesting that many ignorant people nonchalantly crush every insect they see, as if it were some religious imperative, while others almost never kill an insect (other than, perhaps, mosquitos and flies). Some Texans allow large spiders to share their homes in order to do battle with scorpions. The point is, our research into insect physiology can help the small beasts as well as hurt them.

Much of what we know about insect development has come from studies of a silkworm moth, *Halaphora cercropia*. As it grows, this moth undergoes a complete metamorphosis. This means that it grows through all the developmental stages of insects: *egg, larva, pupa,* and *adult* (Figure 27.4). The structural changes from one stage to another are strikingly dramatic. One wonders how a pupa could possibly develop from a larva. It's as if a chimpanzee metamorphosed into a toaster.

27.4 Metamorphosis in a silkworm moth, *Halaphora cercropia.* The silkworm moth passes through a complete metamorphosis in its development. The fertilized egg develops into a larva, which is an active feeding and food-storing stage. The larval stage is followed by pupation or diapause. The pupa remains dormant through winter, but in the spring undergoes important developmental changes within the cocoon. From this stage emerges the adult moth. Its primary activity is mating and reproduction. Each stage is directed by hormonal action.

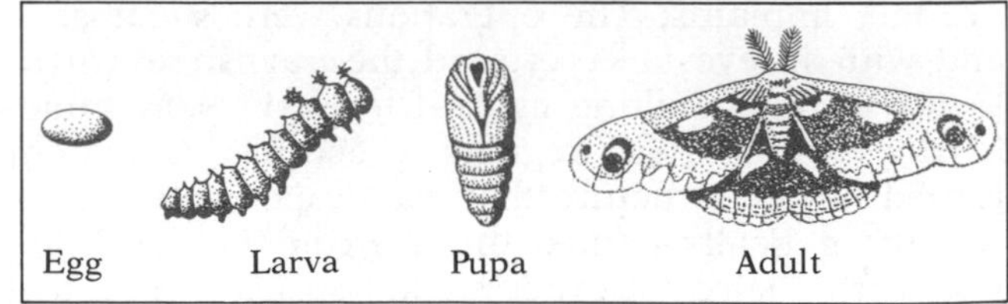

In its normal metamorphosis, the silkworm larva goes through several molts before entering pupation. It winters over in the pupal stage, emerging from the pupa case in the soft light of spring. It has carefully conserved its stored energy through the winter by existing in a very low metabolic state known as *diapause*.

Insect metamorphosis is known to be influenced by hormones from the brain itself and three pairs of endocrine structures. Two of these, the *corpus cardiacum*, and the *corpus allatum*, are extensions of the brain. The third are rather large structures known as the *prothoracic* or *ecdysial glands.*

During the larval period the corpus allatum is the dominating gland. Its secretion, *juvenile hormone* (JH), controls growth and molting as the larva progresses through several periods (Figure 27.5). As growth continues past the final molting, the secretion of JH begins to dwindle. As this happens (or perhaps the next event causes it to happen), secretory cells in the brain produce *brain hormone* (BH, also referred to as *prothoracotrophic hormone*). Brain hormone is transported to the corpus cardiacum from which it is secreted. Its target, the prothoracic gland, responds by secreting its hormone—*prothoracic gland hormone* (PGH, also known as *ecdysone*). When PGH is present, the larval stage ends abruptly and the pupal stage begins. The continued dominating effect of PGH eventually influences the adult emergence.

In an attempt to clarify the relationships among the insect growth hormones, physiologists have carried out some rather ingenious experiments. For example, when several corpora allata (plural of corpus allatum) are transplanted into larvae, the insects go on and molt, but they continue to molt into larger and larger larvae—they don't pupate! Since the corpora allata secrete juvenile hormone, the experiment illustrates the antagonistic role of the juvenile hormone over BH and PGH, and its ability, when abundant enough, to override the others.

In another series of experiments, a fine thread was tied behind the heads of larvae so that the flow of fluids from the head to other parts of the body was restricted. When a very young larva is treated this way, growth and molting fail. In older larvae, growth, molting, and pupation continue. These experiments indicate that the maturity hormones (BH and PGH) are released at a later time in the larval period, and that once they are distributed in the body, maturation will occur normally despite the subsequent restriction of circulation.

27.5 Metamorphosis in the insect is under the control of four secretory endocrine structures. These include parts of the brain, the *corpus cardiacum*, the *corpus allatum*, and the *prothoracic gland.* During the larval stages, growth and molting are under the control of *juvenile hormone* (JH), secreted by the corpus allatum. As long as this hormone dominates, the larval stage will persist. When growth is sufficient, the prothoracic gland, under the influence of *brain hormone* (BH), produces *prothoracic gland hormone* (PGH, or ecdysone) while the JH diminishes. A pupa is then formed, which is followed by emergence of the adult.

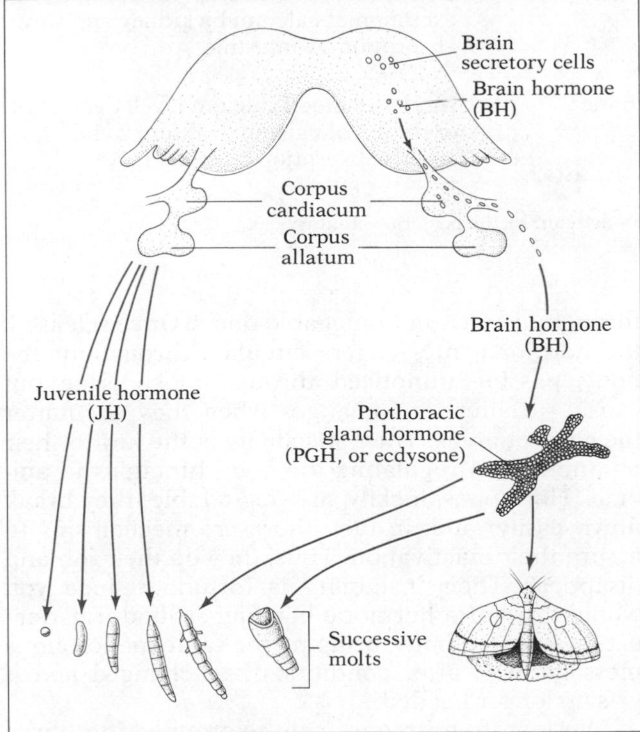

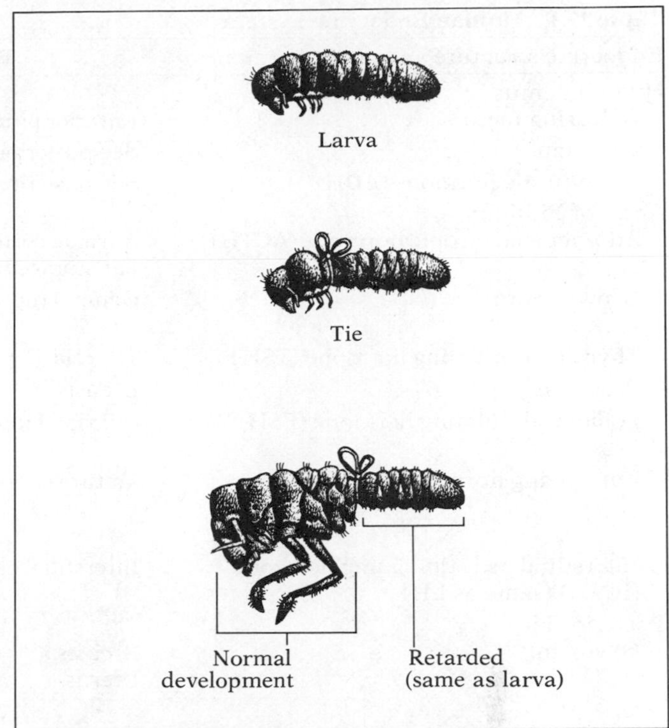

27.6 Experimental restriction of PGH. If prothoracic gland hormone (PGH) is restricted from reaching the posterior parts of an insect larva, those parts will remain in the larval stage while head and thorax undergo normal adult development. If the body is tied off just behind the thoracic gland, the head area will mature normally, but the rear area, not reached by hormones, will molt unchanged.

The third series of experiments produced the most dramatic results. Threads were tied around the larvae just behind the thoracic gland (Figure 27.6). The part of the insect anterior to the ligature pupated, while the posterior region remained in its larval state. Apparently, the difference was due to the failure of PGH to reach the hindpart of the insect, leaving it in a juvenile condition when it molted.

Because of such work we may soon be able to control insect populations by using their own growth hormones. When the technology is available, we will be able to control specific insect populations in much the same way we have selectively controlled plant growth. This would be a far more appealing method of dealing with the little creatures than the gigantic and risky pesticide programs now in use. (We should not be misled by the term *control*, however. The word is popular with many government agencies and seal killers, but it seems to assume some sort of innate wisdom on the part of the controller. Is this a safe assumption?)

Table 27.1 Human Endocrine System

Endocrine structure and major hormones	Target	Action
Hypothalamus		
Releasing factors	Anterior pituitary	Secretes hormones
Oxytocin[a]	See posterior pituitary	See posterior pituitary
Antidiuretic hormone (ADH)[a]	See posterior pituitary	See posterior pituitary
Anterior Pituitary		
Adrenocorticotropic hormone (ACTH)	Adrenal cortex Fat storage regions	Secretes steroid hormones Fatty acids released into blood
Growth hormone (GH)	General (no specific organs)	Stimulates growth; amino acid transport; slows carbohydrate utilization
Thyroid-stimulating hormone (TSH)	Thyroid	Secretes hormones
Prolactin	Breasts	Promotes milk production
Follicle-stimulating hormone (FSH)	Ovarian follicles	Stimulates growth of follicle and estrogen production
Luteinizing hormone (LH)	Mature ovarian follicle	Stimulates ovulation, conversion of follicle to corpus luteum, and production of progesterone
Interstitial cell stimulating hormone (ICSH) (same as LH)	Interstitial cells of testis	Stimulates sperm and testosterone production
Posterior Pituitary		
Oxytocin	Breasts Uterus	Stimulates release of milk Contraction of smooth muscle in childbirth and orgasm
Antidiuretic hormone (ADH)	Kidney	Increases water uptake in kidney (decreasing urine volume)
Thyroid		
Thyroxin Triiodothyronine	General (no specific organs)	Increases oxidation of carbohydrates; stimulates (with GH) growth and brain development
Calcitonin	Intestine, kidney, bone	Decreases blood calcium level; increases excretion of calcium by kidney and slows absorption by intestine
Parathyroid		
Parathormone	Intestine, kidney, bone	Increases blood calcium level; decreases excretion of calcium by kidney and speeds absorption by intestine

[a]The hypothalamic hormones, oxytocin and ADH, are released into the bloodstream by the posterior pituitary.

Hormones and the Vertebrates

As the vertebrates evolved, both the nervous systems and hormonal systems became increasingly important as a means of adjusting the animals to their environments. The hormones work in a rather similar fashion through many of the vertebrates, so we will again draw on that representative mammal with the primitive mouth and the untalented toes.

Anatomically, the human endocrine system consists of nine distinct glands plus a few other tissues scattered here and there in the body (Figure 27.7 and Table 27.1).

The endocrine glands are often referred to as the *ductless glands*, since their secretions seep directly from the gland into the blood without passing through distinct and pluggable ducts. Once released, the hormonal messengers circulate throughout the body, passing unnoticed through most tissue but causing tremendous changes when they encounter their target areas. Their specificity is the key to their unique role in regulating the body functions in animals. Hormones, luckily, are degradable (they break down easily) and, in fact, there are mechanisms to assure their inactivation. Thus, they do their job and disappear. Their instability is fortunate since you wouldn't want a hormone hanging around in a delicately adjusted body, bringing the same monotonous message even after conditions had changed and it was no longer needed.

Vertebrate hormones can be grouped into three major classes: *steroids, proteins* (including peptides),

Table 27.1 *(continued)*

Endocrine structure and major hormones	Target	Action
Islets of Langerhans		
Alpha cells: glucagon	Liver	Stimulates liver to convert glycogen to glucose; elevates glucose level in blood
Beta cells: insulin	Cell membranes	Facilitates transport of glucose into cells; lowers glucose level in blood
Adrenal Cortex		
Aldosterone	Kidneys	Increased recovery of sodium and excretion of potassium and hydrogen ions; uptake of chloride ion and water
Cortisol (hydrocortisone)	General (no specific organs)	Increases glucose, protein, fat metabolism; reduces inflammation
Androgens, Estrogens	General (many regions and organs)	Promotes secondary sex characteristics
Adrenal Medula		
Epinephrine, Norepinephrine	General (many regions and organs)	Increases heart rate and blood pressure; directs blood to muscles and brain; "fright or flight mechanism"
Ovaries		
Estrogen	General (many regions and organs)	Development of secondary sex characteristics; bone growth; sex drive
Progesterone (ovarian source replaced by placenta during pregnancy)	Uterus (lining)	Cyclic development of endometrium; maintenance of uterus during pregnancy
Testes		
Testosterone	General (many regions and organs)	Differentiation of male sex organs in embryo; secondary sex characteristics; bone growth; sex drive
Pineal Gland		
Melatonin	Uncertain in humans—perhaps hypothalamus and pituitary	May have some influence over hypothalamus or pituitary in cyclic activity
Thymus		
Thymosin	Lymphatic system	May stimulate development of lymphatic tissue
Damaged tissues		
Histamine	Capillaries of damaged area	Increases blood supply and capillary permeability; promotes clotting
Serotonin	Smooth muscles of blood vessels	Constriction—limiting blood loss

and *modified amino acids.* The steroids contain the usual four-ring structure:

```
                          CH3
                     CH2  |   CH2
              CH2        C        CH2
          CH3 |          |         |
     CH2  |   CH         CH ------ CH2
CH2      C        CH
|        |        |
CH2      C        CH2
     CH      CH2
```

Nonspecific steroid

This structure shows the carbon skeleton saturated with hydrogen, but don't take the diagram too literally since it doesn't represent any specific steroid. Actually, each steroid hormone has different side groups emerging from the carbons on the rings. The steroid hormones are all derived from cholesterol, which in turn is produced from the acetyl groups of acetyl CoA. A few of the more common steroid hormones are the sex hormones, *progesterone, testosterone,* and *estrogen*. Others include *aldosterone, corticosterone,* and *cortisol.*

Protein hormones are, of course, composed of chains of amino acids. The simplest are probably *antidiuretic hormone* (ADH) and *oxytocin,* which are short peptides of only nine amino acids each. Insulin consists of fifty-one amino acid molecules. Other protein or peptide hormones are *glucagon, follicle-stimulating hormone* (FSH), *luteinizing hormone* (LH), and *adrenocorticotrophic hormone* (ACTH). (See the top of page 778.)

We only know about a few modified amino acid hormones, but they are nevertheless important, since they include *epinephrine* and *norepinephrine* (also

Phe-Val-Asn-Gln-His-Leu-Cys-Gly-Ser-His$_{10}$-Leu-Val-Glu-Ala-Leu-Tyr-Leu-Val-Cys-Gly$_{20}$-Glu-Arg-Gly-Phe-Phe-Tyr-Thr-Pro-Lys-Ala$_{30}$-

S—S (Cys to Cys disulfide bridges)

-Gly-Ile-Val-Glu-Gln-Cys-Cys-Ala-Ser-Val$_{70}$-Cys-Ser-Leu-Tyr-Gln-Leu-Glu-Asn-Tyr-Cys$_{80}$-Asn

S—S

Protein hormone (insulin)

called *adrenalin* and *noradrenalin*), as well as *thyroxin, histamine,* and *serotonin.* Their basic structure consists of one or two carbon rings attached to a short carbon chain that contains a nitrogen group.

HO—(ring with I, I)—O—(ring with I, I)—CH_2—C(H)(NH_2)—COOH

Modified amino acid (thyroxin)

The Pituitary

Now, let's consider the various endocrine glands of the human body. Some will be familiar, but perhaps some won't. Some may even help to explain some of your less admirable behavior. We should begin our survey with the *pituitary* (or *hypophysis,* as it is also known). It was once thought to be the source of all mucus, but later, other scientists determined that this is not true at all—they said that it is, in fact, the seat of the soul. However, even more recently we have come up with yet other data.

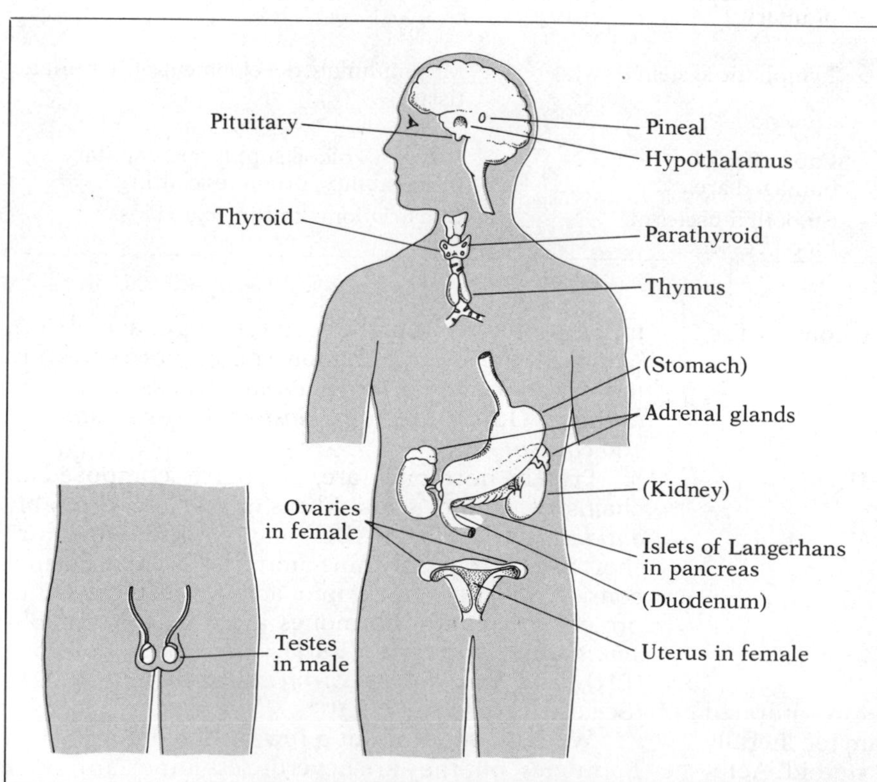

27.7 The human endocrine system consists of nine distinct ductless glands, and many minor (shown in parentheses), but important, specialized tissues (such as the hormone-secreting cells of the stomach, small intestine, and placenta). Endocrine glands secrete their hormones directly into the circulatory system. The secretions are released in very minute quantities, and are often controlled by delicate feedback mechanisms.

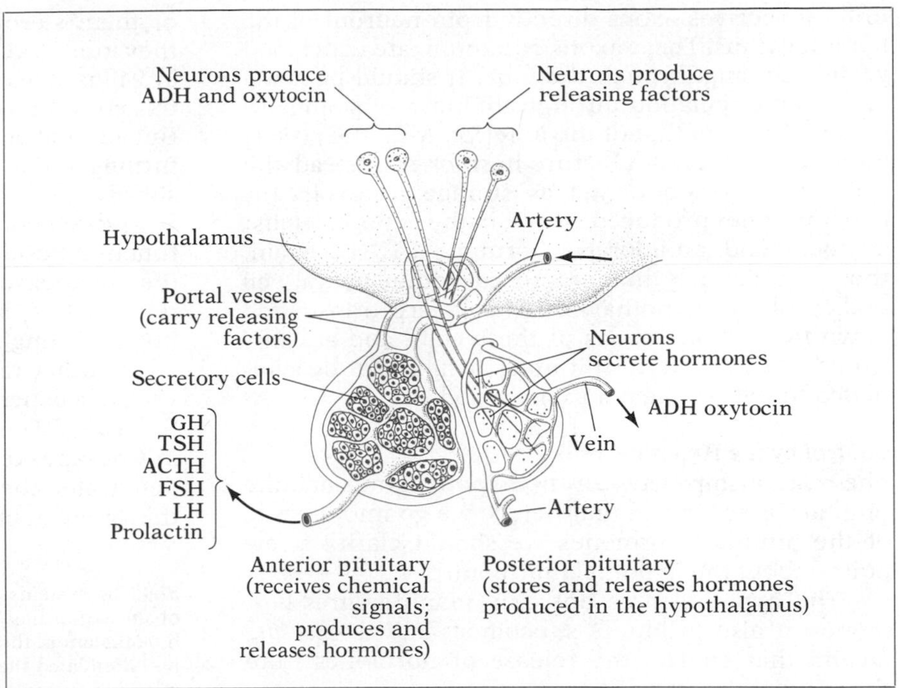

27.8 The bean-sized pituitary gland is really two quite distinct structures. The anterior portion is embryologically, structurally, and functionally distinct from the posterior. The anterior pituitary has its own capillary circuit and produces a number of hormones which it secretes directly into that capillary network. Several of its hormones stimulate other glands, a role that has earned it the name "master gland." The pituitary is intimately involved with a part of the brain called the hypothalamus, the secretions of which direct the activities of the anterior pituitary. The posterior pituitary receives hormones through the axons of neurons originating in the hypothalamus, stores them, and releases them upon demand.

The pituitary is a good place to start since we are now aware that it influences many other endocrine glands. As such, it has been called the "master gland," but if it does rule, it's a puppet government, as we shall see. In a rather democratic manner, it responds to its constituents through a system of checks and balances through that efficient biological process we call negative feedback control. In other words, the hormones of the pituitary activate other glands, which in turn send messages back that moderate the pituitary. (One is reminded of a popular old toy, a box with an "on-off" switch. When switched on, the lid would creak open, a hand slowly come out and switch the lever to "off," and quickly withdraw back into the box.) The pituitary is also quite responsive to signals from other parts of the brain, in particular the *hypothalamus*.

Because of such responsiveness, the pituitary is part of a direct link between the nervous system and the endocrine system. We pointed out earlier that most invertebrate hormones are actually the secretions of certain nerve cells rather than special glands. These *secretory neurons* manufacture substances that travel through the body and act as regulatory devices, just as do hormones. Humans, we should add, also utilize neural secretions, since the posterior part of the pituitary is actually part of the brain, and the anterior part responds to secretions from the brain. We will return to all this shortly.

The pituitary (Figure 27.8) is a tiny, bilobed structure 12–13 mm in diameter, or about the size of a kidney bean. If you point your finger directly between your eyes and stick your other finger straight into your ear, you will not only gain attention of the other people on the bus, but the lines will intersect at about the location of the pituitary.

The gland lies in a pocket of bone and is suspended by a stalk that emerges from the base of the brain, just below the hypothalamus, with which it is intimately related. The connections between the two structures run through the stalk that supports the pituitary.

The *front* (*anterior*) *lobe* of the pituitary is primarily glandular tissue and, as a matter of fact, secretes six important hormones. The anterior lobe develops in the embryo from an outpocketing of the primitive mouth. Of course, a gland is no good unless its hormone can reach the blood, and the anterior pituitary has a rich blood supply indeed. It consists of a portal system (beds of capillaries separated by a vein). The portal system originates as an artery that feeds the capillaries of the *median eminence* of the hypothalamus. The capillaries converge into a *portal vein* that in turn feeds the capillary bed of the anterior lobe, producing an artery–capillary–portal vein–capillary–vein arrangement. The capillaries of the median eminence are especially important because it is into these that the hypothalamus secretes neurohormones that stimulate the release of pituitary hormones. Thus, the hypothalamus is, in a sense, the "master" of the "master gland."

The *posterior lobe* of the pituitary originates in the embryo from an outpocketing of the great lobe which is the developing forebrain. Its downward growth meets the upward progress of the anterior pituitary until they join, finally forming the bilobed structure. Since the posterior lobe is really part of the brain it receives axons directly from neurons of the hypothalamus. These axons communicate with blood vessels passing through the lobe. It should be made clear that circulation through the posterior pituitary is separate from that of the anterior. Also, the posterior lobe doesn't manufacture hormones. Instead, the cells of the posterior act as storage reservoirs for two hormones produced above, in the hypothalamus: oxytocin and antidiuretic hormone (ADH). Again, these are actually manufactured within neural cell bodies of the hypothalamus itself, are transmitted down through the axons of these cells, and are only *stored* in the posterior pituitary, which can be stimulated to release them into the blood.

Control by the Hypothalamus

The relationship between the hypothalamus and the pituitary is so critical that before we go into the role of the pituitary hormones we should clarify a few points about the hypothalamic control.

The hypothalamus not only manufactures hormones, it also produces substances called *releasing factors* that control the release of hormones from the anterior pituitary. The releasing factors are manufactured by secretory neurons in the hypothalamus and carried to the anterior pituitary through a portal system. To date, only two releasing factors have been discovered and these release only three pituitary hormones. The thyrotrophic hormone is released by the *thyrotrophic hormone releasing factor* (TRF), while the luteinizing hormone and the follicle-stimulating hormone are both released by the *luteinizing hormone releasing factor* (LRF).

The known releasing factors are only short chains of amino acids, but their discoveries represent years of intense effort by researchers. For example, in isolating TRF (the first to be identified), 4 years of research and 2 million pig pituitaries were required for production of the first milligram! The reason the substance was so hard to purify is that releasing factors are so enormously powerful that only incredibly tiny amounts are produced to do the job. The problem is compounded because the molecules fall apart soon after they are released. In spite of such problems, the tedious search for other releasing factors goes on.

Hormones of the Anterior Pituitary

Now let's consider the hormones that are produced by the incredibly regular and busy little bean that rests so securely in the middle of our skulls. A good hormone to start with is ACTH, because it is produced according to the rules of our famed negative feedback system. The target of the adrenocorticotrophic hormone (ACTH) is the cells of the cortex (outer layer) of the adrenal gland, located on top of the kidney. When stimulated by ACTH, the adrenal cortex itself secretes an entire battery of hormones, most of them steroids. As their levels rise in the blood, they inhibit ACTH production in the pituitary. Figure 27.9 illustrates the manner in which it works. Actually, this may be one of the simpler feedback systems. But adrenal hormones have a greater role than simply turning themselves off. We'll see what that role is shortly.

Adrenocorticotrophic hormone also has another function besides activating the adrenals. It regulates the metabolism of fats. Under its influence, the body releases fatty acids into the bloodstream for redistribution through the organism.

Another rather fascinating role of ACTH was revealed in experiments in a population of overcrowded Sitka deer. The deer population confined on an island in Chesapeake Bay were provided with all the food and water they needed. They seemed healthy and they grew in number. But then something happened.

27.9 In systems with negative feedback control, the product of an action has a suppressing effect on that action. Here the hypothalamus, through a hypothetical releasing factor (ACTH-RF), has stimulated the pituitary to release ACTH. (To date the ACTH releasing factor has not been identified.) The target organ—the adrenal gland—in turn secretes hormones from its cortex (outer region). These hormones, in addition to instigating action at their target organs, reach the hypothalamus. When sufficient quantities have been produced, they inhibit the production of the releasing factor. The system slows or shuts down. Thus, the hormones of the ACTH target form a negative feedback loop.

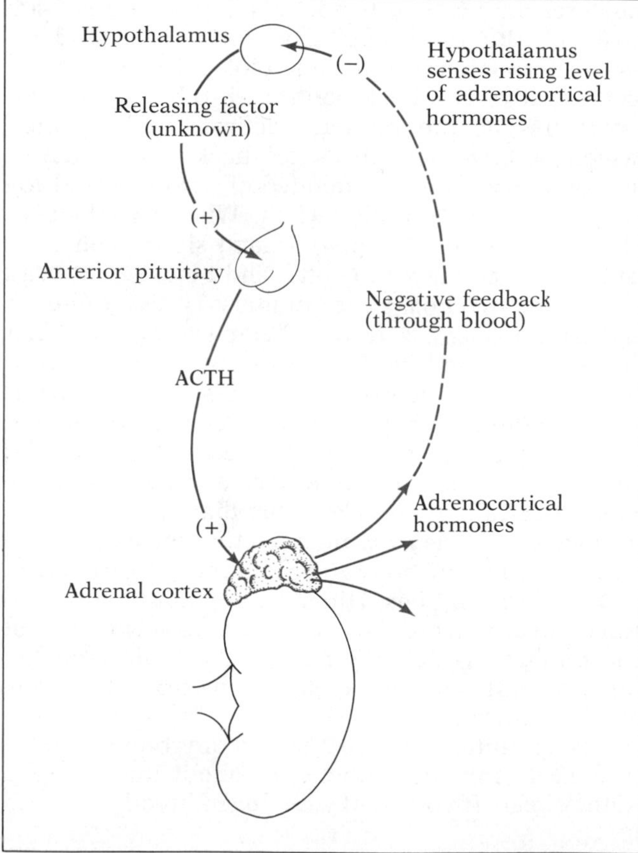

The animals apparently experienced what is called "physiological shock." One of the symptoms was high ACTH and adrenal steroid levels and this, in turn, drastically reduced the ability of the deer to reproduce. There was another strange effect. The antibody defenses of the deer were greatly suppressed, and diseases raged through the population. Furthermore, the animals began to die from hypoglycemic shock, a condition produced when hormonal imbalance reduces the level of blood sugar. The deer died off, apparently as a result of the stress produced by their own numbers.

This rather grisly story clearly illustrates the relationship between the nervous and endocrine systems. The deer, although well-fed and free of predators, were simply not able to cope with large numbers of their own species. They were barraged by each other's presence, perceiving each other through sight and smell. Somehow the information triggered abnormal levels of ACTH production and the bizarre chain of physiological events that followed. Obviously, interacting with each other so frequently upset the delicate endocrine system. Do you suppose there are upper limits for humans?

Another hormone, called *growth hormone* (GH), of course, stimulates growth. But GH is also responsible for other metabolic activities such as assisting amino acids across cell membranes, which results in an acceleration of protein synthesis within the cells. In addition, GH decreases carbohydrate utilization, thereby increasing blood glucose levels and stimulating the release of another hormone, insulin.

Proper levels of GH in the body are essential to normal growth. This becomes distressingly obvious when something goes wrong. Excessive secretion (such as might occur with a pituitary tumor) during the early years can produce *giantism* (*pituitary giants*). Many people with this condition range in height between 7 and 9 ft. Conversely, an underactive pituitary, with lower than normal GH levels, in young people results in *dwarfism* (pituitary dwarfs) [Figure 27.10 (a)]. People deficient in GH early in life keep infantile body proportions throughout their lives and are often called *midgets*. There are several thousand pituitary dwarfs in the United States today. Unfortunately, treatment at this time is not possible. Stimulating growth in even one pituitary dwarf would require a daily injection of all the GH that could be isolated from several human bodies. New promise looms on the horizon, however, with the possibility of cloning of a human GH gene and the manufacture of the protein in the bacteria, *E. coli*. It is likely that we can eventually eliminate the condition from the human population.

Should GH levels rise suddenly in an adult human, the resulting condition is known as *acromegaly*. In this abnormality, growth that had been stopped

(a)

27.10 Pituitary growth hormone (GH), a secretion of the anterior pituitary, influences metabolism, including growth in children. (a) Normally, GH stimulates growth until adulthood, but when its secretion is insufficient, the result can be growth retardation and the development of a *pituitary dwarf*. Oversecretion during adolescence can have the opposite effect, producing a *pituitary giant*. (b) Sudden increases in GH after maturity and when bone lengthening has ended, produces growth in certain body parts such as the face and hands. Here we see the result, a condition known as *acromegaly*.

(b)

on time mysteriously resumes, but it is restricted to areas where cartilage persists, such as the hands, feet, jaw, nose, tongue, and some internal organs [Figure 27.10 (b)].

Thyroid-stimulating hormone (TSH, also known as *thyrotropin*) is responsible for stimulating the thyroid to release its hormones, *thyroxin, triiodothyronine*, and *calcitonin*. These hormones regulate the metabolic rate of the body and calcium ion levels of the blood, but we will have a closer look at their functions when we consider the thyroid gland.

Prolactin promotes milk production in mammals, including humans. Toward the very end of pregnancy, the blood levels of prolactin increase to twenty-eight times that in nonpregnant females. But milk is still not produced, because it is inhibited by high estrogen and progesterone levels. When these fall off dramatically at birth, the milk flow begins, in part under the influence of *oxytocin* from the posterior pituitary. Continuing high levels of prolactin permit a woman to produce milk for very long periods, perhaps 3 years, after the birth of a child. Milk production stops, however, when the child stops nursing, since suckling directly stimulates continued secretion of the hormones.

Prolactin may also be involved in other activities which we know even less about. For example, it is present in males and in infants, but we can't imagine what it does in either case. Recent studies also reveal that prolactin rises in both males and females during sexual excitement, peaking at orgasm. Again, we don't know why.

Follicle-stimulating hormone (FSH) and *luteinizing hormone* (LH) are both involved in stimulating the production of gametes and sex hormones, and thus they are called *gonadotropins*. But since FSH and LH activity is discussed in Chapter 29, we will move along to the last two pituitary hormones—those stored in the posterior lobe.

Hormones of the Posterior Pituitary

The only two hormones we know of that are stored in the posterior pituitary are short peptides, *oxytocin* and *antidiuretic hormone* (ADH). Each is a chain of only nine amino acids, and they are remarkably similar, differing from each other in only two links of the chain:

Oxytocin: Cys-Tyr-Ile-Glu-Asp-Cys-Pro-Leu-Gly
ADH: Cys-Tyr-Phe-Glu-Asp-Cys-Pro-Arg-Gly

Both hormones are manufactured in the neural cell bodies of the hypothalamus and travel down to the posterior pituitary inside the axons of those cells. They are released by a variety of factors—some we know about and some we don't. Antidiuretic hormone, for instance, is released when the osmotic pressure of the blood falls below a certain level.

Oxytocin is released by, among other things, the stimulation of nipples and by sexual intercourse. Oxytocin functions with prolactin to stimulate milk secretion. Specifically, prolactin is involved in the actual manufacture of milk, and oxytocin contracts the smooth muscles around the milk ducts, forcing milk out of the nipples or at least allowing it to flow more freely. It is sometimes poetically called the "milk let-down hormone."

Oxytocin's role in bringing about the contractions of the uterine muscles during labor and after birth is described in Chapter 30. But once the baby is born, its suckling aids in producing oxytocin, which helps bring the distended uterus back down to its normal size. Oxytocin is sometimes administered by obstetricians to stimulate uterine contractions if labor has been unusually protracted, and sometimes so that the birth will fit the doctor's schedule. Apparently, the hormone increases the permeability of uterine myofibrils to calcium ions, an essential part of the contraction process.

Oxytocin, as was mentioned, also plays a role in sexual activity. Its levels in the bloodstream of women rise dramatically during sex play and orgasm. When oxytocin levels rise following orgasm, the uterus begins to contract rhythmically, which may help draw in the semen. We'll have more to say about this in Chapter 29.

Oxytocin levels also rise during menstruation, another time in which rhythmic uterine contractions occur. Sometimes these contractions are discernible as menstrual cramps. So far all the known activities of oxytocin have to do with female functions, although the hormone is found in males also. Perhaps it has other roles that remain to be discovered.

The other hormone of the posterior pituitary is ADH. The name *antidiuretic hormone* tells what the hormone does. A *diuretic* is something that increases the flow of urine, and an antidiuretic does the opposite. The target cells are the epithelial cells of the collecting ducts of the kidney, and the hormone is involved in water balance. Water conservation, you will recall, is one of the great problems of terrestrial life. The ADH level is influenced by the osmotic pressure of blood passing through the capillaries of the hypothalamus. If the blood is too "thick," the nerves fire and antiduretic hormone is released from storage in the posterior pituitary. A more complete discussion of this system is given in Chapter 28.

Incidentally, ADH was formerly known as *vasopressin* because it was shown to increase the blood pressure of laboratory animals. It was subsequently found not to have any such effect in humans, and so the name has been dropped. We occasionally will give you alternate names since you may run across them in your readings as you try to decipher what we were trying to say. In any case, so much for the pituitary hormones. Now let's talk about some strange molecules produced in the neck.

The Thyroid

The *thyroid gland* is shaped somewhat like a bow tie and is located in an appropriate place for one, in front of and slightly below the larynx (Figure 27.11). The thyroid weighs about 30 grams (1 oz). It produces two very similar hormones, thyroxin and triiodothyronine, both modified versions of the amino acid tyrosine, and calcitonin, a peptide composed of thirty-two amino acids. In the embryo, the thyroid develops as a ventral outpocketing of the throat that pinches off and embarks on its own developmental pathway. Its glandular epithelium, then, is endodermal. In evolution, the thyroid apparently derived from the *endostyle*, a structure still found in tunicates, *Amphioxus*, and the larva of the lamprey.

Thyroxin and Triiodothyronine

Thyroxin and triiodothyronine differ by one iodine atom only. (They carry four and three iodines, respectively.) In their production, two molecules of the amino acid tyrosine are joined and iodinated. (The structure of thyroxin is shown on page 778.) Both hormones apparently have the same action. Since thyroxin occurs in so much greater quantity we will focus on it in our discussion.

Thyroxin is important in bringing about metamorphosis in amphibians. Young tadpoles have very little thyroxin, but at an appropriate stage, thyroxin begins to surge within their bodies, causing dramatic changes. As the larva's tail and gills are resorbed, the legs and other adult structures emerge under the direction of thyroxin. The hormone is so potent that even if it is administered to newly hatched tadpoles, they will promptly turn into tiny "adult" frogs the size of flies.

27.11 The thyroid gland is a bilobed structure nearly surrounding the trachea at a point just below the larynx. Its hormones include *thyroxin, triiodothyronine,* and *calcitonin.* In general, these hormones increase the rate of body metabolism. The thyroid itself is controlled by the pituitary, which stimulates the thyroid hormone release through a negative feedback system. Note the position of the parathyroid glands on the posterior side.

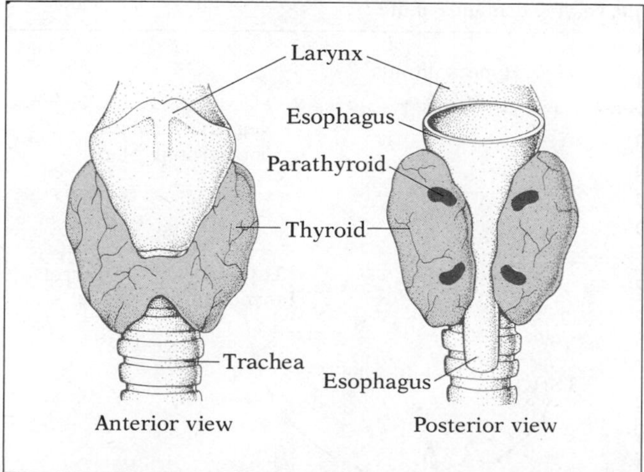

In humans, who do not undergo such drastic metamorphosis, thyroxin still plays a role. It influences the rate at which carbohydrates are oxidized in the cells. The *basal metabolism rate* is a measure of oxygen consumption for a given time and weight of an individual at rest. Although we have known this much for decades, just how thyroxin works remains a mystery. At the moment, our best guess is that it influences the synthesis of some of the intracellular enzymes of respiration.

Overactive and underactive thyroid glands produce *hyperthyroidism* and *hypothyroidism*, respectively. The symptoms of hyperthyroidism are nervousness, hyperactivity, insomnia, and weight loss. Skinny, active people are often accused of being hyperthyroid, but in fact most skinny, active people are perfectly normal—they're just skinny and active.

The thyroid is capable, under some conditions such as one caused by a tumor, of incredible growth. Its enlargement produces an *exophthalmic goiter* and can create an extreme hyperthyroid condition. The term *exophthalmic* ("out eyed") refers to a characteristic bulging of the eyes, which is part of the syndrome [Figure 27.12 (a)].

A goiter is any enlargement of the thyroid gland. A *colloid goiter* is unsightly [Figure 27.12 (b)], but it is actually a normal response to low levels of iodine in the diet. The overgrowth of the gland increases the efficiency of iodine utilization and retention, so that more nearly normal levels of thyroxin can be maintained. Before the introduction of iodized salt early in this century, such goiters were common in the iodine-poor parts of the United States, particularly in mountain regions. It is said that colloid goiters were once so nearly universal in some parts of the interior of Africa that persons without goiters were considered to be freakish and distinctly unattractive.

Hypothyroidism causes a general slowing of the metabolic rate. Hypothyroid individuals are usually overweight and physically and mentally sluggish. The condition sometimes occurs when a person is "run down" by a period of illness or other stress, and they subsequently fail to produce enough thyroxin. A chronically low basal metabolic rate can usually be treated successfully with thyroxin tablets.

In some cases, genetic abnormalities can cause a lack of thyroxin from birth. If the condition is untreated, the persons become *cretins. Cretinism* results in extreme mental and physical retardation [Figure 27.12 (c)]. Because the physical appearances may be similar, cretinism is sometimes confused with the chromosomal disorder, Down's syndrome, but the two conditions can easily be distinguished in the laboratory. At present, if the low thyroid condition is diagnosed early enough, cretinism can be effec-

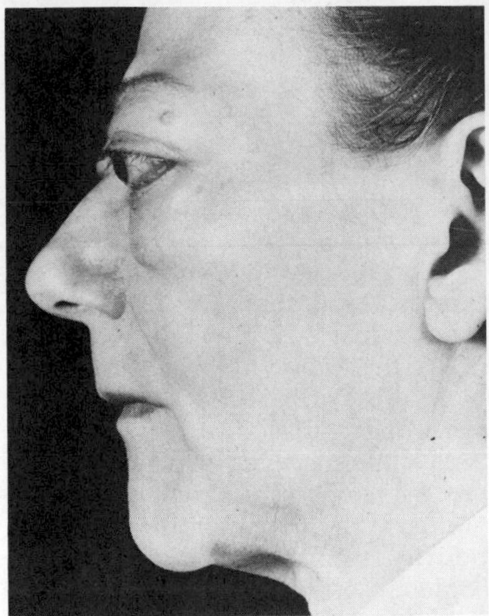

27.12 Thyroid-related abnormalities. A malfunctioning thyroid can cause a number of important clinical problems. (a) A hyperactive thyroid may produce *exophthalmic goiter.* The symptoms include bulging eyes, rapid metabolic rate, general nervousness, and a failure to gain weight irrespective of diet. (b) Many of these Peruvian villagers suffer from iodine deficiency, and their thyroids have become enlarged, resulting in a greater iodine recovery ability. The condition, known as *colloid goiter,* is rather rare today because of the availability of iodized salt.

tively prevented by the administration of a daily dose of thyroxin. The term *cretin,* by the way, is not a reflection on the good citizens of Crete, but is derived from *Christian.* In its original use, the term simply meant that such unfortunate individuals were, after all, fellow human beings.

Let's run over thyroxin control once again. The secretion of thyroxin is controlled by the thyroid-stimulating hormone (TSH) of the anterior pituitary. Secretion of TSH, in turn, is influenced by the release of a neurosecretion (releasing factor) from the hypothalamus into the pituitary portal circulation. The hypothalamus itself is sensitive to thyroxin levels, so the cycle operates according to a negative feedback loop (Figure 27.13). Thus, the immediate needs of the organism are met with short-term increases or decreases in thyroxin levels which change the basal metabolic rate.

Calcitonin

The other thyroid hormone, the peptide *calcitonin,* regulates the concentrations of calcium ions (Ca^{2+}) in the bloodstream. For many years it was believed that calcitonin was secreted by the parathyroid glands (discussed next), which are embedded in the tissue of the thyroid (see Figure 27.11). When it was finally determined that calcitonin is secreted by the thyroid itself, some people suggested that it should be renamed *thyrocalcitonin* so that everyone could show everyone else that they knew what they were talking about. The initial confusion was understandable, though, because the parathyroid glands do, in fact, secrete a hormone that regulates the concentration of calcium ions in the blood. So we'll hold off on considering how calcitonin works until we have

27.13 Negative feedback control of metabolism. The involvement of the thyroid in metabolic activity is governed by a negative feedback loop that involves the hypothalamus and the pituitary. When thyroid hormonal level in the blood is low, the hypothalamus responds by stimulating the pituitary via a specific releasing factor (TSH-RF). In response, the pituitary secretes TSH which, in turn, stimulates the thyroid to secrete its hormones. The body responds with an elevated metabolic rate. While this is happening, the hypothalamus senses the rise in the level of thyroid hormone in the blood and at some precise threshold, is itself inhibited, slowing its stimulating action on the pituitary. The cycle continues with the ebb and flow of messengers as the metabolic rate meets the varying demands of life.

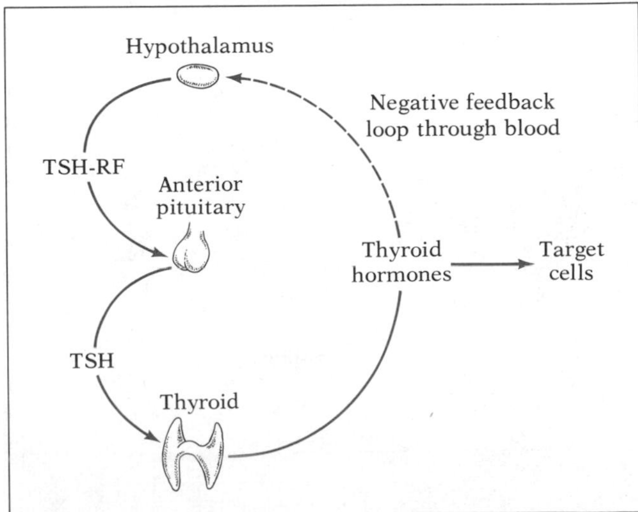

discussed the parathyroid hormone. Then we can consider these complementary calcium ion systems together.

The Parathyroid Glands and Parathormone

The *parathyroid glands*, four in number, are pea-sized bodies embedded in the tissue of the thyroid, two in each wing of the bow tie (see Figure 27.11). The parathyroids evaded scientific investigation for many years, although they had been found in, of all places, a rhinoceros, by a British anatomist in 1850. They were rediscovered when the thyroids of dogs were experimentally removed just to see what would happen. Unexpectedly, the dogs lost muscular control and appeared to be dying. In accordance with the rules of endocrinology mentioned earlier, thyroxin was administered, but it had no effect. The researchers wondered, then, what did the thyroid have to do with muscular control? Then the four tiny lumps were found embedded in the mass of excised tissue, and it was soon found that extracts of these lumps relieved the unexpected symptoms.

The vital hormone was purified and named *parathormone* or *parathyroid hormone* (PTH). Parathormone is a polypeptide of eighty-one amino acids, so it's large enough to be called a protein. It helps regulate the calcium levels of the blood, as does calcitonin. Actually, however, PTH and calcitonin have completely opposing actions all down the line: what one enhances, the other inhibits. Overall, the net effect of PTH is to increase the Ca^{2+} concentration in the blood, and the net effect of calcitonin is to decrease it (Figure 27.14). The parathyroid and thyroid cells are sensitive to blood calcium, releasing PTH when Ca^{2+} levels are lower than normal and calcitonin when Ca^{2+} levels are higher than normal. In a classic pair of negative feedback loops, a fine balance is struck between the two systems, and Ca^{2+} levels are kept at a constant 6 mg/ml. Such regulation is important, since many cells are so sensitive to Ca^{2+} that death can follow quickly if the level falls to 4 mg/ml or rises to 9 mg/ml.

These two opposing hormones have three known target tissues: the intestine, the kidney, and bone cells. In the intestine, PTH increases the absorption of Ca^{2+} and calcitonin does the opposite. Thus, dietary calcium is absorbed when it is needed and is passed through when it is not.

In the kidney, PTH inhibits the excretion of Ca^{2+} and calcitonin increases it. So, blood calcium is retained if it is needed and is excreted if it is present in excess.

The main buffer for the body's calcium, however, is the extensive reserve of calcium phosphate in the bones. Here, PTH increases the activity of *osteoclasts*,

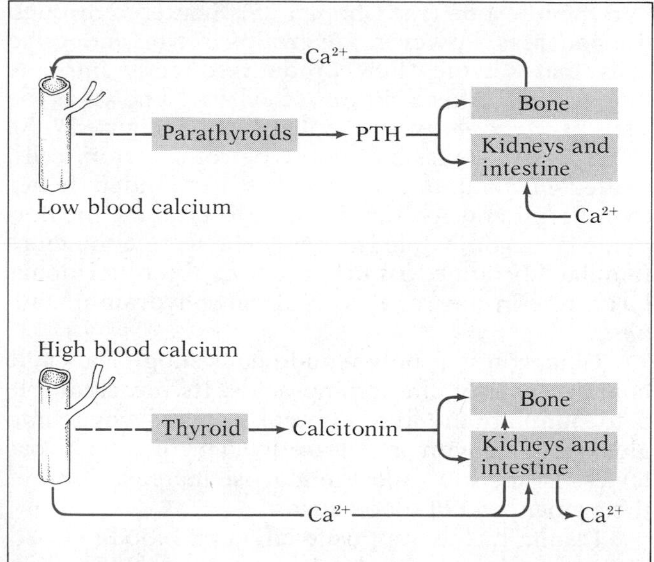

27.14 The parathyroids, four minute endocrine glands embedded in the thyroid gland, secrete *parathyroid hormone* (PTH), which regulates blood calcium levels. Its effect is opposite to that of *calcitonin*, which is secreted by the thyroid. Parathyroid hormone and calcitonin aren't under the influence of the pituitary; they respond directly to the calcium content of the blood. Secretion of PTH is increased when blood calcium drops. It brings that level up by withdrawing calcium from the bones, increasing intestinal uptake of calcium, and slowing the excretion of calcium by the kidneys. Calcitonin has the opposite effect. When blood calcium is high, it decreases the rate at which calcium leaves the bone, reduces intestinal uptake, and increases excretion of calcium through the kidneys.

peculiar bone-eating ameboid cells that dissolve calcium phosphate and release Ca^{2+} into the bloodstream. Calcitonin, on the other hand, increases the activity of *osteoblasts*, the bone-forming cells, thus reversing the action of the osteoclasts.

Abnormally low levels of PTH cause muscle convulsions and eventually brings about death. This can occur when the parathyroid glands are destroyed, as sometimes happens when the body's own immune system mistakenly begins to treat the parathyroid tissues as "foreign"—in what is known as an *autoimmune disease*. The opposite condition, abnormally high levels of PTH, can occur if the parathyroids should begin tumorous growth. High levels of PTH result in a severe decalcification of bone (*osteoporosis*), with the subsequent formation of fibrous cysts in the skeleton. If the condition is severe, death follows.

The Pancreas and the Islets of Langerhans

We've been talking about endocrine glands, but here we must diverge a bit because the pancreas is, for the most part, an *exocrine gland*. In other words, it has a duct. It manufactures digestive enzymes, which it then releases through the pancreatic duct into the intestine (see Chapter 25). Scattered through the pancreas, however, are groups of true endocrine cells that secrete their products directly into the bloodstream. These clumps of cells are known as the *islets of Langerhans*, or simply the *islets* (Figure 27.15). Each cluster consists of two types of secretory cells, named *alpha* and *beta* because it sounded better than Ralph and Arthur. Alpha cells produce the hormone *glucagon*, while the beta cells secrete the more familiar hormone, *insulin*. Each of these hormones has a role in the regulation of carbohydrate metabolism.

Glucagon is a polypeptide consisting of a single chain of twenty-nine amino acids. Its principal role is to stimulate the liver to break down glycogen into glucose. Glucagon is released when blood glucose levels fall too low, and the glucose increase restores the proper level of blood sugar.

Insulin has the opposite effect on blood glucose. That is, it decreases blood glucose levels. The two islet hormones complement each other in maintaining the delicate balance of glucose in the blood (Figure 27.16).

Insulin starts off as a protein that contains eighty-one amino acids. It is inactive in this form and must be subjected to cellular proteases before it can do its job. This process removes thirty of the amino acids from the middle of the sequence, producing the reactive insulin molecule, now composed of two short chains that have been cross-linked (page 778). Apparently, the long single-stranded inactive form is easier to store, since the storage organ could be disrupted by active insulin. Thus it is activated only a short time before it is secreted.

Insulin's specific role is to help glucose move across cell membranes. It is believed that target cells have very specific sites in their membranes—places at which insulin will bind. For some reason, once insulin attaches to the membrane at these sites, the membrane becomes more permeable to glucose.

Deficiency in the insulin system produces the familiar clinical symptoms of *hyperglycemia*, more commonly known as *diabetes* or *diabetes mellitus*. In some diabetics, the ultimate cause of the disease is an actual deficiency of insulin, which can be caused by insufficiency of insulin production or by increased levels of insulinase, an enzyme that destroys insulin. However, most adult diabetics have normal levels of insulin but a deficiency of receptor sites on the membranes of the target cells. In some cases, the target cells have enough receptor sites but simply fail to respond properly. Injected insulin helps, but high levels of injected insulin lead to further decreases in target cell receptivity, which requires yet more insulin—an unnatural "vicious cycle," or positive feedback loop.

The symptoms of diabetes include glucose in the urine, greatly increased urine volume, dehydration, constant thirst, excessive weight loss, exhaustion,

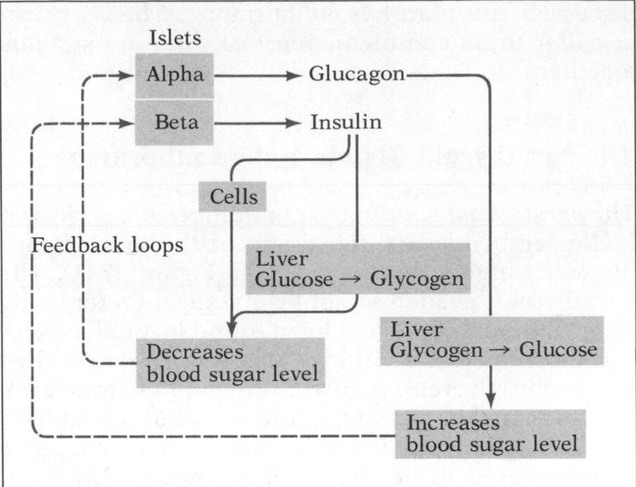

27.16 Glucagon and insulin control blood glucose levels independently of the pituitary and other endocrine glands, although the glands may influence these hormones indirectly. When blood glucose levels are low, the *alpha islet* cells are stimulated to secrete *glucagon*. The primary target organ of glucagon is the liver, where the hormone accelerates the rate at which liver cells convert glycogen to glucose, which then enters the blood. An elevated blood glucose level, on the other hand, activates the *beta islet* cells, which secrete *insulin*. Insulin's target is more general, since it influences the transport of glucose out of the blood into any metabolically active cell. It also affects the liver by increasing its rate of glycogen formation. While glucagon and insulin have opposite effects, they don't actually inhibit each other. Rather, they cooperatively maintain the critical blood glucose levels as the body demands are met.

27.15 The islets of Langerhans are groups of endocrine tissues embedded in the pancreas (an exocrine gland). Cells in the islets secrete insulin and glucagon, which are instrumental in regulating glucose levels in the blood. Note the differences in the islet cells and the surrounding pancreatic cells.

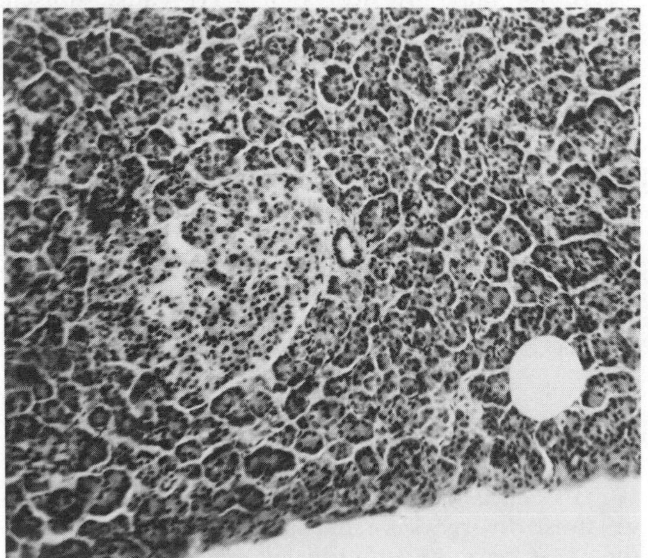

fatty liver, ulcerated skin, local infections, possible blindness, and many other problems. Some symptoms are the result of glucose starvation of the cells. Keep in mind that the glucose may be present but with insulin deficiency it can't cross the cell membranes. In addition, the diabetic has halitosis, exhaling chemicals called ketones, especially acetone, as those substances rise in the blood and cross the lung capillaries. Ketones are produced by an incomplete breakdown of fatty acids. Therefore, it becomes apparent, insulin is also involved in the breakdown of fatty acids.

Mild cases of some forms of diabetes in adults can be controlled by a well-regimented diet in which carbohydrates and fats are carefully regulated and somewhat more advanced forms can be controlled by insulin injections. Diabetes in young people is essentially a different disease and is likely to be very serious, indeed.

Cells may also be glucose-starved because of *hypoglycemia*, or low blood glucose. A pancreatic tumor may be at the root of this problem. As the pancreatic tissues grow out of control, too much insulin enters the blood. Any blood glucose is cleared away rapidly and sequestered as intracellular glycogen. The brain cells, which depend on glucose as a fuel, are the first to be affected. Low blood sugar in the brain results in dizziness, tremors, a violent temper, and sometimes blurred vision and fainting.

In the absence of pancreatic tumors, people can have milder (but sometimes serious) types of chronic hypoglycemia. In these people, our much-praised negative feedback system goes haywire and overreacts. An increase of blood glucose brings about an increase in insulin secretion, as you might suspect, and the insulin catalyzes the conversion of blood glucose into tissue glycogen, bringing the glucose level back down. Unfortunately, there is a time lag, and some peoples' bodies overreact before the situation can stabilize. The ingestion of a sweet roll at 9 a.m. may produce a case of severe hypoglycemia by 11 a.m. If the individual responds by eating, he or she will keep gaining weight. But failure to respond to the body's craving for sugar can bring on a tantrum.

This type of physiological hypoglycemia is sometimes a prelude to the more severe disease, diabetes, which may appear a few years later. How is it that hypoglycemia and hyperglycemia—opposite conditions—are actually different aspects of the same disease condition? We don't know, but both conditions involve the failure, one way or the other, of the same feedback control loop.

The Adrenal Glands

In most vertebrates, the *adrenal glands* are closely associated with the kidney. (In fact, in Latin *ad* means upon and *renal* refers to kidney, and in Greek, the same two roots are *epi-* and *nephron*, hence the name *epinephrine* for one of the hormones produced in the adrenals.) In humans the glands are perched atop each kidney, each adrenal being essentially two glands in terms of origin, structure, and function. There is an outer layer called the *cortex* and an inner one called the *medulla* (Figure 27.17).

Some of the adrenal cortical hormones are known by the peculiar name of *mineralocorticoids*, which reflects their role in mineral regulation. Each of these hormones has the basic steroid structure described earlier. The two best known are *aldosterone* and *cortisol*. Let's concentrate, for a moment, on aldosterone.

The target cells of aldosterone are the cells of the renal tubules of the kidney (Chapter 28). The hormone increases the kidney's recovery of sodium there, and increases the excretion of potassium and hydrogen ions in urine. It also helps chloride ions and water to get back into the circulatory system from the kidneys. As you may have already perceived, aldosterone and antidiuretic hormone (ADH) activity

27.17 The adrenal glands are roughly triangular structures situated on the top of the kidneys. The gland is divided into two secretory parts, an outer *cortex* and an inner *medulla*. Each produces its own battery of hormones. The cortex produces several hormones—all steroids—including aldosterone (which influences sodium excretion), hydrocortisone (which influences blood glucose levels and inhibits inflammatory responses), and several sex hormones. The medulla produces two hormones—both amines—epinephrine and norepinephrine.

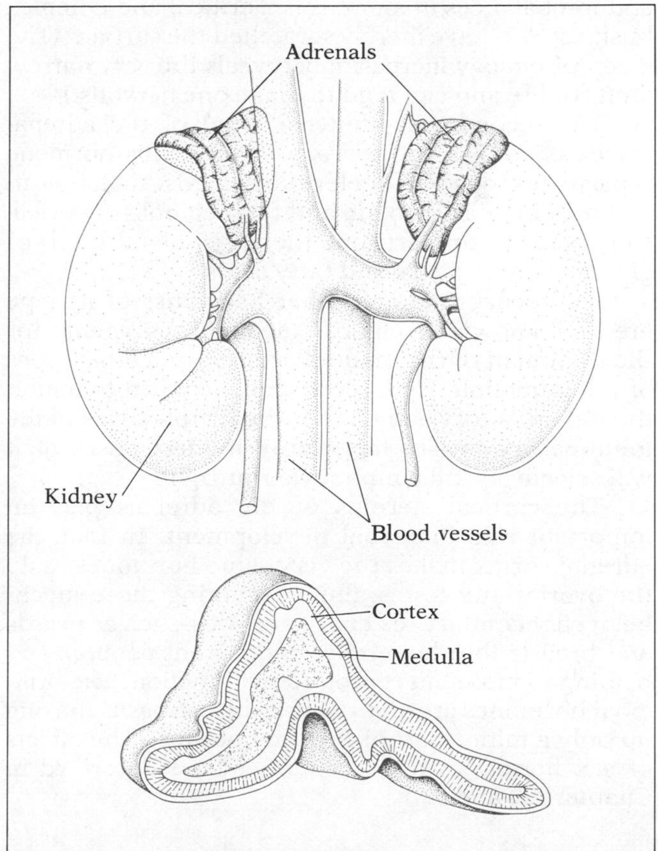

in the kidneys, operating together, closely regulate kidney action and keep the fluids of the body within tight limits. To illustrate, let's see what happens when aldosterone activity goes wrong.

Aldosterone hyperactivity simply causes sodium and water to be retained. When this happens, the urine increases in acidity, while virtually the entire body becomes more alkaline (*alkalosis*). The increase in alkalinity increases extracellular fluid, and causes a general body puffiness known as *edema* (or *dropsy*). The excessive fluid also elevates the blood pressure. In addition, potassium balances are disrupted, with a resultant loss of nerve function. Neurons can become overpolarized, resulting in muscle paralysis.

Aldosterone hypoactivity, on the other hand, means a general loss of sodium, chloride, and water. Potassium balances are again disrupted, and the body becomes acid. In this case, death follows within days.

The regulation of aldosterone secretion is very complex, but it seems to occur through the sodium content of blood being monitored by the adrenal cortex. A low sodium content results in increased aldosterone secretion, while high sodium creates the opposite effect. In addition, the rate of aldosterone secretion is influenced by the blood's content of potassium, its pH, the rate of blood flow through the kidney, and, as we indicated earlier, the secretion of ACTH in the pituitary. Of course, there may be additional factors involved in aldosterone control. As you can see by now, the management of water and ion balances in animals is a critical and complex business. We have merely scratched the surface. (The study of biology increasingly reveals the very narrow limits of life and can tend to make one nervous.)

The second well-known steroid of the adrenal cortex is cortisol (or *hydrocortisone*). This hormone is partly responsible for elevating glucose metabolism and may increase protein and fat metabolism as well. In general, hydrocortisone increases blood levels of glucose, amino acids, and fatty acids.

Hydrocortisone and other hormones of its type are well-known in clinical medicine as agents for the treatment of rheumatoid arthritis and other types of inflammation. Hydrocortisone is known to inhibit the inflammatory activities of the lymphocytes of the immune system—in fact, fairly modest doses of it will selectively kill lymphocytes outright.

The cortical steroids of the adrenals play an important role in sexual development. In fact, the adrenal cortex makes the very same hormones as do the ovaries and testes, thus promoting those much-beloved secondary sex characteristics, such as beards and breasts (but hopefully on different people), and helping to maintain reproductive function. The principal hormones are *androgens*, with *estrogens* making up only a minor part of the total output. (The effects of sex hormones on development are described in Chapters 29 and 30.)

We find we have encountered another biological puzzle here. Why are the same hormones produced in two separate regions of the body? Is this some sort of an evolutionary flaw, or is our ignorance showing again? It seems that each sex needs some amount of both hormones. For instance, the development of pubic hair is under the control of "male" sex hormones in *both* sexes.

The adrenal medulla produces two fascinating amine hormones: *epinephrine* and *norepinephrine*. (You may have heard them called *adrenalin* and *noradrenalin*. Adrenalin is actually an old trade name for *epinephrine*, coined by a pharmaceutical company.) These hormones cause very specific changes in the body, usually in response to stress. We will discuss this in more detail later, but for now just keep in mind that it reduces blood flow in the capillaries near the body surfaces (causing the skin to pale and reducing the extent of bleeding). The blood is instead diverted to the muscles which may need to perform some strenuous deed, like getting you out of there. At such times, digestive functions cease. Blood pressure increases and glucose is suddenly released into the bloodstream from the liver. All these responses, of course, are brought on by sudden fright or anger. They are often called the "flight or fight" responses. Note that, in general, the system readies the animal for emergency. The muscles receive the bulk of blood flow, glucose reserves are mobilized, and unneeded functions are shut down. We have all heard stories of people exhibiting superhuman feats of strength at these times (we've all heard of the little old lady lifting the car off her husband), or you may remember feeling the effects yourself.

So, one asks, just how does the adrenal medulla mediate all these respones? We haven't the foggiest. Part of the problem is, the data are confusing. For example, all these responses can be induced in animals that have had the adrenal medulla removed! What this seems to mean is that the adrenal medulla has a counterpart elsewhere. It is perhaps, quite simply, the sympathetic nervous system, since, in fact, all the neurons of the sympathetic nervous system secrete norepinephrine as a neurotransmitter (see Chapter 31). And so we assume that the endocrine system, in this case, is a redundant system, backing up a neural complex, presumably lowering the excitation thresholds of the sympathetic synapses. For now, we have no other explanation for the two systems carrying out the same function.

The Ovaries and Testes

The endocrine function of the ovary and testis is discussed in Chapter 29, so let's just briefly note some of the major points here. First, the two hormones of the ovary—*estrogen* and *progesterone*—and the principal hormone of the testis—*testosterone*—are all steroids. Estrogen, actually a whole family of molecules, includes two major variants, β-estradiol and *estrone,* the first apparently being more potent.

In addition to their reproductive function, the sex hormones have an important role in skeletal development. Their sudden increase in the blood at the onset of puberty stimulates the lengthening of the long bones, and causes a marked spurt of growth (interestingly, it also causes a simultaneous spurt in mental growth as measured by test performance). The onset of puberty is variable; some of us are precocious and some are definitely late bloomers. Toward the end of puberty, the sex hormones have the opposite effect. They promote the union of the diaphysis and epiphyses in the long bones (see Chapter 24), after which no further increase in height is possible. Thus late-maturing adolescents may grow to be unusually tall. Low levels of estrogen secretion may delay puberty and produce tall girls. The same relationship, by the way, holds for testosterone and boys. Remember that tough kid who matured way ahead of everyone else in your junior high school—the one who never grew very tall?

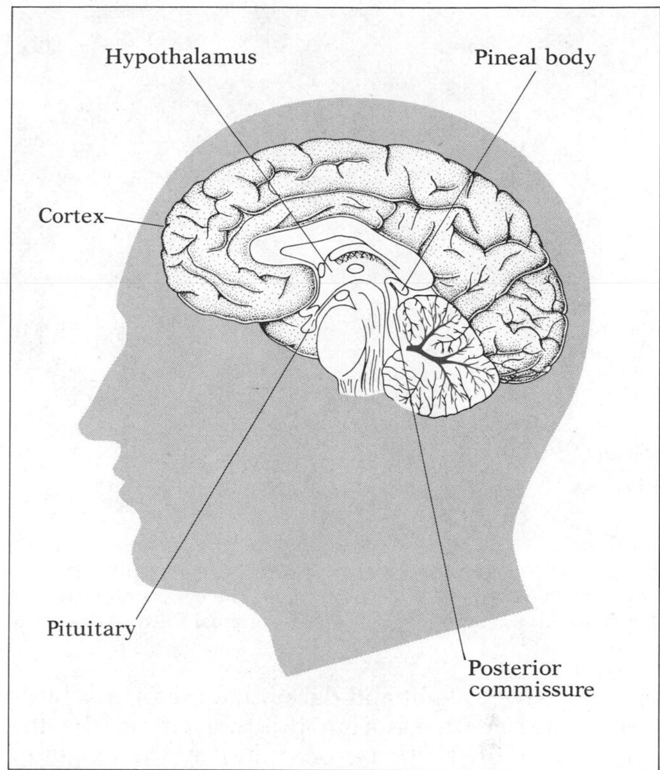

27.18 The *pineal body* remains a region of mystery. It is a minute stalked structure, connected to the *posterior commissure* in the brain. Although several substances that may have hormonal properties have been isolated from the pineal body, only one has been clearly identified as a hormone: *melatonin.* The function of this hormone is still obscure, but it may control the release of pituitary-stimulating releasing factors of the hypothalamus.

The Pineal Body

The *pineal body* is located just above the brainstem, connected by a stalk to a structure known as the *posterior commissure* (Figure 27.18). It's a strange gland and its function has been the subject of speculation for nearly 2000 years. Galen, the Roman anatomist, believed it served as a pathway for thought. Descartes, the French naturalist and philospher, believed it to be the "seat of the rational soul." Today, there are still serious reservations about the pineal's endocrine function, but many biologists are becoming convinced that is has such a role. The first strong evidence of an endocrine function resulted from investigations in the 1950s, when an active substance was extracted from the pineal body of cattle. When injected into frogs, the substance, now known as *melatonin,* caused the frogs to change color! A cattle extract caused frogs to change color? Now *there* was an interesting bit of news. People began to toy with the phenomenon and found out that the color changes resulted from the pigment granules clumping together. The phenomenon itself was not particularly surprising, since this is the usual way that many animals that change color change color. But cattle don't change color, so what was this peculiar hormone doing in their pineal?

People have continued to speculate about the pineal and one of the questions regarding it is, what is its evolutionary history? Where did it come from? The answer is a bit startling. It turns out that it may be the remains of some ancient and vestigial third eye!

Here is a case where an understanding of evolutionary origins was a real help in working out the physiology and function of an organ. In the fossil skulls of the earliest ostracoderms (jawless fishes), there are holes for three eyes. The third eye is in the middle, right on top of the skull. Only a remnant of the eye remains in modern fish (and humans, as well), but some amphibians and reptiles have retained a kind of recognizable eye in a rudimentary state. In the isolated New Zealand reptile *Sphenodon punctatus,* called the tuatara (Figure 27.19), the small, median third eye even has an easily recognizable lens and retina. Some very common American lizards also have a tiny, degenerate third eye on the top of their heads, complete with light receptors and an eyehole in the skull covered by a special translucent scale. This third eye can't form an image, but it is sensitive to light. It is, in fact, the light receptor by which the lizard sets its biological clock, behaving one way when days are long and another way when short, winter days draw on.

Birds have such a "third eye" also, and although it is enclosed within their translucent skulls, it, too,

27.19 The *Sphenodon punctatus*, a New Zealand reptile. It has a remnant third eye at the top of its skull. The eye is complete with lens and retina and may represent the original form of the pineal body.

is responsive to light and dark. Like the plants, birds determine whether spring has arrived not by the appearance of the first crocus, but by the length of the day—specifically, by the relative lengths of light and dark in a 24 hr cycle. And they measure the light with that third eye buried under their skulls.

We mammals don't have a third eye as such, and it has never been proven that our pineal body is light-sensitive at all. But the pineal body is certainly homologous with the third eye of the other vertebrates, and, light-sensitive or not, it apparently contains the timing mechanism that governs much of our activities.

Developmental Hormones, Mitogens, and Growth Factors

The endocrine glands and the hypothalamus aren't the only sources of hormones, as we've seen. Hormones are also produced in other tissues, such as the uterus, the stomach lining, and the placenta. In fact, we don't know where most hormones come from, and probably most hormones in the body haven't even been identified yet. A number of other hormones are suspected to exist on circumstantial evidence. There are good reasons to think that some cell movement, cellular interactions, and embryonic differentiation are influenced by hormones.

Many hormones were first identified as *growth factors* necessary for the maintenance of cells in laboratory tissue cultures. For example, it was found some time ago that mammalian tissue culture cells need blood serum in order to grow. Then it was discovered that the critical substances in the serum were hormones. Among those that have now been identified are *fibroblast growth factor, epidermal growth factor,* and *platelet-derived growth factor.* These are all polypeptide hormones, and the cells in the culture have specific receptors for each of them on their membranes. The action of all these hormones is the same: They are *mitogens,* which means that they induce G_1 cells to enter S phase and ultimately to undergo mitosis.

Perhaps the most interesting of these is the platelet-derived growth factor. As implied by the name, this hormone is released by blood cell platelets when they rupture, which is what they do at the site of a wound. The release of a mitogen at a specific site, then, has an obvious functional significance: It initiates local cell division and wound healing.

27.2 How Hormones Work

The Molecular Biology of Hormone Action

So far we have described the actions of hormones in terms of what they do—what adaptive changes they bring about in the organism. So now let's take a closer look at how they work. In other words, let's see how they act on their target cells.

The Second Messenger

It turns out that much of our information about the molecular biology of hormone action is very recent, and there are still a lot of unanswered questions. For example, it was only in 1971 that a Nobel Prize was awarded for the discovery of cAMP as a participant in hormone activity. Cyclic AMP (see Figure 27.2) is so important that it is now referred to as the *second messenger.* In light of this concept, we could refer to the other hormones we've been talking about as "first messengers," since it now appears that many of them bring their message only as far as the target cell membrane. Within the cell, the actual response is often triggered by cAMP. This intracellular second messenger is, in fact, the same molecule that gives the let's-get-together call to wandering slime mold cells.

The importance of cAMP came to light as scientists were trying to discover how epinephrine works by studying epinephrine activity in the liver. They found that, although most hormones operate by stimulating the transcription of specific kinds of mRNA in the target cells, epinephrine works on the cytoplasm.

You may recall that one of the effects of epinephrine is a sudden increase in the blood glucose level, which is accomplished through the enzymatic breakdown of glycogen reserves in the liver. The chain of events leading to the elevated glucose level is shown in Figure 27.20.

Epinephrine, released from the adrenal medulla, is carried by the blood to a target, such as the liver. It turns out that the liver cells have very specific receptor sites along their cell membranes. Once fixed to the membrane, epinephrine activates an enzyme known as *adenyl cyclase.* This enzyme is part of the membrane system itself and remains inactive until the epinephrine is in place. But when the enzyme is activated, it immediately converts cytoplasmic ATP to cAMP, starting a chain of events that leads to the release of glucose.

Apparently, at least nine other hormones stimulate the production of cAMP, utilizing it as a second messenger. The list includes glucagon, ACTH, LH, FSH, TSH, PTH, ADH, and calcitonin. But the question arises: How is the activity of so many hormones regulated if they all activate the same second messenger? The answer incorporates two important points. First, cAMP must remain in cells. If it circulated freely, it would presumably wreak havoc with body functions. Second, hormones have target cells and a target cell has a specific binding site for a specific hormone. Thus, no other hormone can bind to that site (although a cell may have binding sites for more than one hormone). Because of these factors, a number of hormones can utilize cAMP as a second messenger.

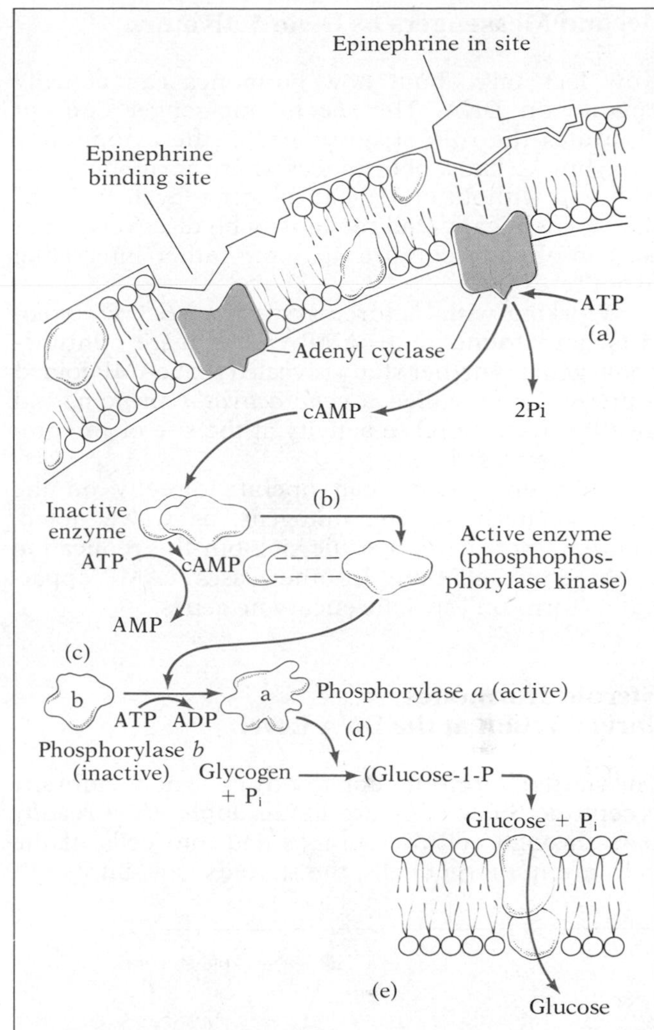

27.20 Epinephrine and the second messenger. Among its other roles, epinephrine (adrenalin) elevates blood sugar. The steps involved from the binding of epinephrine by a target cell to the release of glucose into the bloodstream are shown here. The cascading events begin with epinephrine binding to its specific receptor site on the membrane of a liver cell. (a) The enzyme *adrenyl cyclase,* which is also incorporated into the membrane, is activated by the receptor–hormone complex, and ATP is converted to cAMP within the cytoplasm. (b) Following this, the cAMP (the "second messenger") joins with an inactivated enzyme complex; activating the enzyme, with the aid of ATP, by removing a blocking polypeptide. (c) The activated enzyme *phosphophosphorylase kinase* converts another enzyme, *phosphorylase b,* to *phosphorylase a,* again with help from ATP. (d) *Phosphorylase a* then begins cleaving and phosphorylating the bonds between glucose molecules in glycogen. (e) Eventually, the phosphates are removed from glucose and the molecule crosses the liver cell membrane into the circulatory system. The blood sugar level has been increased.

Other Second Messengers?

At a time when laboratory after laboratory was reporting cAMP involvement in hormone after hormone, it began to appear that all polypeptide hormones might function this way. Alas, this turned out not to be the case. At least insulin and the mitogens involve other mechanisms. These polypeptide hormones interact with membrane receptors, all right, but then things get complex. At least one of the mitogens (the epidermal growth factor) is known to form a permanent covalent linkage with its membrane receptor. When a number of these receptors have been bound to the appropriate circulating hormone, all the receptors of that type cluster at one site on the surface of the cell. Then in one great, slow gulp, the patch of receptors and their hormones sink beneath the surface of the cell, engulfed in a vessel that pinches off from the cell membrane. Apparently, these hormones are then digested, and in some manner still unknown, the nucleus is stimulated to undergo mitosis (in the case of the mitogens). Other second messengers may be involved in this process, but so far they have not been identified.

Second Messengers as Gene Activators

Now let's talk about how hormones can actually operate on DNA. The second messenger concept illustrates the role of hormones at the cytoplasmic level, but we know of instances where hormones act—with and without cAMP—at the gene level. Much of the biochemical detail remains to be discovered, but we can already conjure up some rather interesting hypotheses.

Working with cell-free systems of *E. coli,* biologists have found that cAMP can activate nonfunctional genes. Further study revealed that cAMP joined a protein, now called a *gene activator* protein, and together they increase activity at the site of the previously repressed gene.

Hormones, then, can operate directly on the genes themselves. The mitogens, as we've noted, stimulate not just a specific site, but the replication of the entire genome. In other cases, cAMP apparently "turns on" specific eucaryote genes.

Steroid Hormones; Direct Action at the DNA Level

The steroid hormones apparently have no membrane receptors. Since they are lipid-soluble, they readily pass through cell membranes and into cells. If the cells are not target cells, the steroids can simply diffuse right out again. In appropriate target cells, however, the steroid hormones are caught and tightly bound by a *cytoplasmic binding protein.* Then the steroid-binding protein complex moves into the nucleus, where it joins with yet other proteins that are part of the chromatin complex. The chromosome-bound receptor might be thought of as a tiny organelle or cellular machine that has control of the genetic program. In any case, the complex of hormone, cytoplasmic receptor, and chromosomal receptor initiates gene activation, with the production of appropriate mRNA and protein (Figure 27.21).

The system was first worked out by Bert W. O'Malley and his co-workers at Vanderbilt University. It was known that estrogen and progesterone treatments could cause the oviducts of female baby chicks to produce egg-white albumin and other proteins. So these researchers labeled steroid hormones with radioactive tags and followed them through the sequence described above. At each step, the various intermediate compounds were isolated and identified. And at last something was known directly about how hormones turn on genes.

So, the continuous changes occurring within the living multicellular bodies are sometimes mediated by chemical messengers as the bodies constantly adjust and readjust to meet the demands of their specific worlds. One wonders why a complex system of messengers arose at all. Wouldn't it be simpler for cells that needed to change to initiate that change

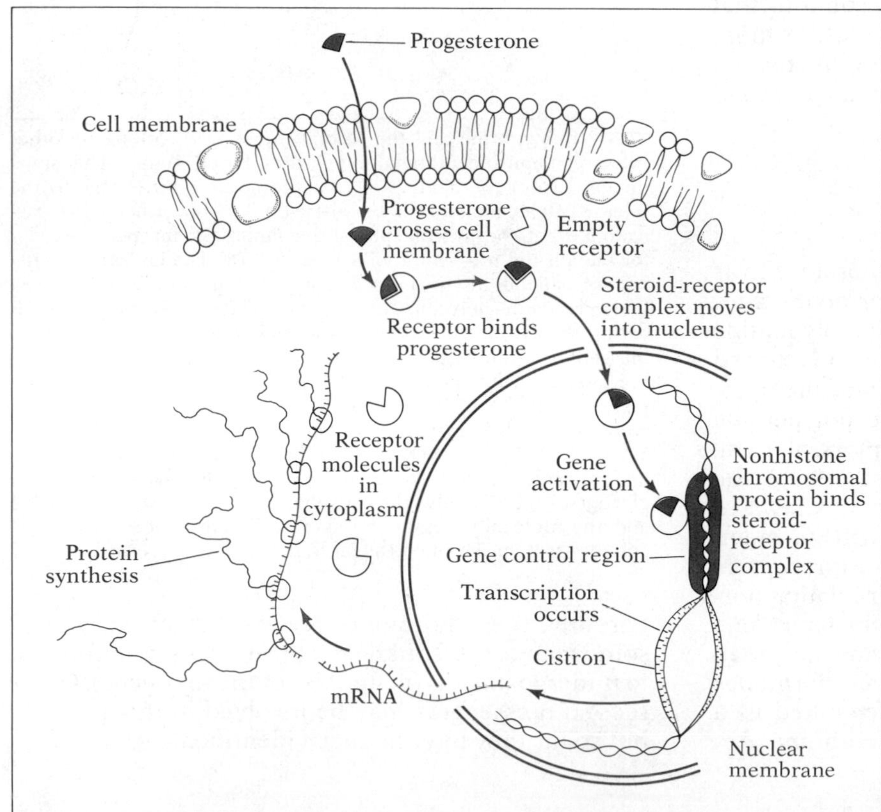

27.21 Hormonal action at the gene level. Estrogen and progesterone (two female sex hormones) are believed to act directly at the gene level. Here, progesterone, a lipid soluble steroid, readily passes through the membrane and joins a protein receptor molecule. Together they penetrate the nucleus where the hormone–protein complex activates a DNA segment, RNA is transcribed, and a protein is produced along ribosomes.

simply in response to environmental cues rather than to be directed by secretions from another part of the body—a part that was itself responding to the environment? Perhaps so, but nature doesn't always operate according to the rules of simplicity. It goes with what it has, and it may have, in fact, been simpler for existing tissues to respond to the environment by incidentally leaking fluids, fluids that caused other tissues to change. The kinds of changes that were adaptive were kept and refined. As time passed, the various systems would have interacted ever more precisely and become more regulated and coordinated, even as the target areas became increasingly sensitive. The result is an incredibly dizzying complex of chemical changes within living bodies as first this hormone, and then that, surges and dwindles, causing living things to change adaptively to a changing planet.

Application of Ideas

1. As two 19th century physiologists attempted to determine the function of the pancreas, they observed flies swarming around the urine of dogs from which the pancreas had been removed. This observation was an initial step in our understanding of the role of insulin in sugar metabolism. What might the next steps have been in isolating insulin and what rigorous rules should the investigators have followed in establishing its function?

2. Assuming that insect hormones could be synthesized in sufficient quantity, suggest a specific hormone that might be applied to insect control. Explain your choice, how it might work, and reasons why this method of insect control might be more ecologically suitable than the use of other chemical insecticides.

3. While the anterior pituitary and thyroid glands are known to control growth, malfunctions in any of the endocrine glands can also affect growth. Considering each of the endocrine glands, explain why this is so.

4. The controlling functions of the endocrine system often overlap those of the nervous systems. Site examples of this overlap and comment on the adaptive significance, if any.

Key Words

acetylcholine
acromegaly
adenyl cyclase
adrenal cortex
adrenal gland
adrenal medulla
adrenocorticotrophic hormone (ACTH)
aldosterone
alkalosis
androgen
anterior pituitary
antidiuretic hormone (ADH)

basal metabolism rate
biological clock
brain hormone (BH)

calcitonin
colloid goiter
complete metamorphosis
corpus allatum
corpus cardiacum
corpus luteum
cortisol
cretinism
crustecdysone
cyclic AMP (cAMP)

Key Ideas

What Are Hormones?

1. *Hormones* are chemical messengers produced in one place and causing changes in other, often remote places. They are regulatory molecules that help coordinate activities.

2. *Hormones* often work in concert with the nervous system. They generally work more slowly than nerves.

3. The slime molds have a massing stage when individuals join together. The chemical signal for massing is a molecule called *cyclic AMP.*

Hormones and the Invertebrates

Intensive studies of *neurosecretions* have been carried out in invertebrates such as crustaceans and insects.

The Molting Activity in Crustaceans

1. The process of *ecdysis* (molting) in crayfish has been studied in detail. The two controlling structures are called the *X-organ* and *Y-organ.* The *Y-organ* secretes *ecdysone* (or *crustecdysone*), and the *X-organ* secretes *molt-inhibiting hormone* (MIH). Control of molting is in the relative concentrations of the two, but light and other environmental factors are involved.

2. Testing for *hormone* action has demanding rules:
   a. Removal or damage of suspected organ produces a deficiency symptom.
   b. Substance in question isolated and if possible purified.
   c. Introduction of substance must alleviate symptom.
   d. Halting treatment must bring on symptoms again.

Hormonal Activity in Insects

1. Insect development is a vital subject in insect control.

2. The silkworm moth's development is a *complete metamorphosis* with four major stages: egg, larva, pupa, and adult. Development involves the brain,

diabetes
diapause
ductless gland

ecdysis
edema
endocrine system
epinephrine
estrogen
exocrine gland
exophthalmic goiter

first messenger

glucagon
growth hormone (GH)

hormone
hyperglycemia
hyperthyroidism
hypoglycemia
hypophysis
hypothalamus
hypothyroidism

insulin
islets of Langerhans

juvenile hormone (JH)

luteining hormone releasing factor (LRF)

melatonin
mineralocorticoids
mitogens
molt-inhibiting hormone (MIH)

negative feedback
neurosecretion
neurotransmitter
norepinephrine

osteoblast
osteoclast
osteoporosis
oxytocin

parathormone (parathyroid hormone)
parathyroid gland
pineal body
pituitary dwarf
pituitary giant
pituitary gland
postecdysis
posterior pituitary
progesterone
prolactin
prothoracic gland hormone (PGH)

releasing factor

corpus cardiacum, corpus allatum (brain extensions), and *prothoracic gland.* They interact as follows:

a. *Juvenile hormone (JH)* from the *corpus allatum* dominates the larval period, dwindling at the final molt.

b. *Brain hormone (BH)* is secreted by *corpus cadiacum.* Its target, the *prothoracic gland,* secretes *prothoracic gland hormone (PGH),* ending the larval stage and starting the pupal stage.

c. *PGH* then influences adult differentiation.

Hormones and the Vertebrates

1. The human *endocrine* system consists of nine distinct organs plus a number of other tissues.

2. Endocrine glands are ductless, releasing their products directly into the blood stream. The *hormone* products have *target cells* which are the only ones capable of identifying and responding to the *hormone. Hormones* work in extremely small quantities and are degraded rapidly.

3. Three classes of *hormones, steroids,* proteins, and modified amino acids are produced. All *steroids* have the four-ring structure, but each specific *hormone* has its own side groups. Protein *hormones* include from nine to eighty amino acids. Amino acid *hormones* consist of modified amino acids.

The Pituitary

1. The pituitary or *hypophysis,* as it is also known, influences the action of many other endocrine glands, but responds to these through a sensitive feedback control. It is also directly influenced by parts of the brain.

2. It lies in a pocket of bone beneath the brain and is intimately related to the *hypothalamus,* which communicates through the pituitary stalk.

3. The *anterior lobe* is glandular, producing six *hormones.* It has a rich capillary supply through which messengers from the *hypothalamus* arrive to stimulate *hormone* production and release.

4. The *posterior lobe* is a storage receptacle for secretions produced in the hypothalamus and carried in the axons of hypothalamic neurons.

Control by the Hypothalamus

Stimulation of the anterior pituitary by the *hypothalamus* is through releasing factors. There is probably one for each *hormone.*

Hormones of the Anterior Pituitary

1. The hormones of the *anterior pituitary* are summarized in Table 27.1.

2. The pituitary secretions are often controlled by *negative feedback* loops.

3. *Negative feedback* can be summarized this way:

a. The pituitary is stimulated to release a *hormone* by secretions of the *hypothalamus.*

b. The target, a distant endocrine gland, is then stimulated to release its *hormone.*

c. The second *hormone* does its job, but at the same time its level or quantity (or its effect) in the blood is sensed by the *hypothalamus* and the rising level inhibits the release of the first *hormone.*

4. Sometimes environmental stimuli evoke secretions. In the Sitka deer population it was noted that environmental stress due to overcrowding was related or caused by abnormally high levels of *ACTH.* Many physiological and behavioral functions were curtailed.

Hormones of the Posterior Pituitary

It is important to remember that the *posterior pituitary* does not produce its two *hormones.* They are produced by the *hypothalamus* and transmitted to the posterior pituitary through neurons.

The Thyroid

The *thyroid,* a bilobed gland, is located in the neck. It produces *thyroxin, triiodothyronine,* and *calcitonin.*

second messenger
steroid hormone

target cell
testosterone
thyrocalcitonin
thyroid stimulating hormone (TSH)
thyrotropic hormone releasing factor (ThRF)
thyroxin
triiodothyronine

vasopressin

X-organ

Y-organ

Thyroxin and Triiodothyronine
Thyroxin and *triiodothyronine* are modified amino acids (*tyrosine*). Their secretion is controlled by the *anterior pituitary* through *negative feedback.* (See Table 27.1 for a review.)

Calcitonin
Calcitonin is closely associated with the *parathyroid hormone, PTH.*

The Parathyroid Glands and Parathormone
1. Parathormone (PTH) is protein size, with eighty-one amino acids. The interplay of *PTH* and *calcitonin* is antagonistic, but cooperative. The first causes deposition of calcium in bone, while the second removes it. Together they form a delicate feedback loop with blood calcium levels at the center.
2. In the bones, *parathormone* stimulates *osteoclasts,* which convert bone calcium salts to mobile ions. *Calcitonin* increases the activity of *osteoblasts,* which deposit calcium as salts. (Review the parathyroid in Table 27.1.)

The Pancreas and the Islets of Langerhans
1. The *islets* consist of alpha cells that produce *glucagon* and beta cells that produce *insulin.* The two complement each other in maintaining the delicate blood glucose level.
2. *Diabetics* suffer from an insufficiency of *insulin* or an over production of insulinase. At other times, the problem is an insufficient number of receptor sites on the *target cells.*
3. *Hypoglycemia* can result from too much *insulin* being secreted. This causes an abnormal positive feedback system to work, with constantly increasing *insulin* levels causing an increasing amount of glucose to be removed and stored as *glycogen.* (Review the islets of Langerhans in Table 27.1.)

The Adrenal Glands
1. The *adrenal glands* lie across the top of the kidneys. The outer cortex and inner medulla are separate glands. The two best known cortical *hormones, aldosterone* and *cortisol,* are *steroids.*
2. For reasons that aren't clear, the *adrenal cortex* produces *androgens* and *estrogens,* which is redundant to the secretion by the *ovary* and *testis.*
3. The medullary secretions, *epinephrine* and *norepinephrine,* have the same effects as the sympathetic nervous system, which uses *norepinephrine* as a *neurotransmitter* (between neurons). The *adrenal gland* may simply represent a redundant system. (Review the *adrenal glands* in Table 27.1.)

The Ovaries and Testes
1. *Estrogen* and *progesterone* are the major *hormones* in females, while the major *hormone* of males is *testosterone.*
2. *Progesterone* is produced by the *corpus luteum* of the *ovary* during the menstrual cycle, but during pregnancy it is produced by the *placenta.*
3. Males experience two peaks in *testosterone* production. One is during sexual differentiation when the testes descend into the scrotal sac. The second occurs during puberty. (Review the *ovary, testis,* and *placenta* in Table 27.1.)

The Pineal Body
1. The pineal body is located in the brain. Its function in humans is not clear. In experimental animals, injections of pineal extracts cause pigment changes.
2. In primitive, jawless fishes the gland was a third eye, and today it is still found in *Sphenodon punctatus,* a reptile, acting as a light-sensitive organ. Birds may also respond to light through the *pineal body.* It may still act as some sort of biological timing mechanism in humans and other mammals, since that is its role in other animals.

Developmental Hormones, Mitogens, and Growth Factors
1. Other sources of *hormones* include the uterus, stomach lining, *placenta,* platelets, and numerous other regions as yet unidentified.

2. Tissue culture studies reveal a number of growth-stimulating factors present in blood serum that act as *mitogens* (mitosis stimulators). One *mitogen* is released by platelets and hastens the healing process.

The Molecular Biology of Hormone Action

The Second Messenger
1. The key to understanding the activity of *hormones* and other chemical messengers at the cellular level is *cyclic AMP.* Since its action is often a secondary response it is called the *second messenger.*
2. While many *hormones* stimulate the genes directly, the *hormone, epinephrine,* works in the cytoplasm. The chain of events is:
   a. *Epinephrine* arrives at a specific receptor site on the *target cell* membrane.
   b. Its presence activates *adrenyl cyclase,* an enzyme of the membrane, which converts cytoplasmic ATP to *cyclic AMP* (nine other *hormones* also do this).
   c. *Cyclic AMP* starts a chain of events that convert glycogen to glucose.
3. This brings up two vital points about *hormone* action:
   a. *Cyclic AMP* is restricted to the cytoplasm. Control would be lost if it circulated freely in the body of an animal.
   b. *Target cells* have very specific binding sites. No other *hormone* binds with that particular site, although a cell may have different binding sites for more than one *hormone.*

Other Second Messengers?
Mitogens accumulate in their receptor sites and both are carried in by phagocytosis. They are then modified by digestive processes. The link between this step and mitosis is still being sought.

Second Messengers as Gene Activators
Cyclic AMP is known to increase gene activity at normally repressed sites. It joins a gene activator protein and together they stimulate activity of previously repressed genes.

Steroid Hormones; Direct Action at the DNA Level
Lipid soluble *steroid hormones* pass through the cell membrane and join cytoplasmic binding proteins. The complex moves into the nucleus. There it joins a chromosomal receptor and initiates gene action leading to mRNA transcription and protein synthesis.

Review Questions

1. Write a comprehensive definition of the term *hormone.* (p. 771)
2. Develop a generalization that compares the action of the *endocrine system* with that of the nervous system. (p. 771)
3. Review the four steps in the process of *ecdysis,* summarizing each step. (p. 773)
4. Testing for *hormones* is described as a "rigorous business." List the four basic steps required in the identification of a new *hormone.* (p. 773)
5. List the stages of a *complete* metamorphosis in insects. (pp. 774–775)
6. What is the source and role of *JH, BH,* and *PGH* in the developing insect? (p. 775)
7. What observation tells us that the maturity *hormones* are released later in larval life? (p. 775)
8. Describe the experiment that rather conclusively proved the role of *PGH.* (p. 775)
9. List the nine distinct human *endocrine organs* or structures. (p. 776)
10. Explain the importance of specificity and biodegradability to *hormones* and *target* organs. (p. 776)
11. Describe the *pituitary* by providing information as follows (p. 779):
    a. Size, shape, and location
    b. Developmental origin of each lobe
    c. Each lobe's relationship to the *hypothalamus*
    d. Importance of its capillary network
    e. Function of each lobe (general)
12. What are *releasing factors* and what do they do? (p. 780)
13. Review the *target,* action, and symptoms of deficiency for each of the following *anterior pituitary hormones: ACTH, GH, TSH, prolactin,* and FSH. (pp. 780–782)
14. What are the two *hormones* of the posterior pituitary and what do they do? (p. 782)
15. How does the hypothalamus "know" when to release *ADH* and when to stop releasing it? (p. 782)

16. Construct a negative feedback diagram showing the way in which *ACTH* secretion is regulated. (p. 780)

17. Describe the chemical structure and role of *thyroxin.* (p. 783)

18. Compare the effects of a low *thyroxin* output on adults with that of children. (p. 783)

19. What is the supposed reason for the growth of a *colloid goiter* and how can it be prevented? (p. 783)

20. Draw a simple negative feedback diagram showing how *thyroxin* levels are controlled. (p. 784)

21. Describe the opposing roles of *calcitonin* and *parathyroid hormone.* (p. 785)

22. List the two specialized cells of the *islets of Langerhans* and name the *hormones* they produce. (p. 786)

23. Compare the general effects of the two *islet* hormones. (p. 786)

24. What goes on in the blood and cells of a person with *hyperglycemia?* (p. 786)

25. What is a positive feedback loop and how might this explain some forms of *diabetes?* (p. 786)

26. Define the term *hypoglycemia* and review two possible causes for the condition. (p. 787)

27. Name and locate the two endocrine regions of the *adrenal gland* and list the important *hormones* produced in each. (pp. 787–788)

28. How does the chemistry of the cortical *hormones* differ from those of the medulla? (pp. 787–788)

29. What general condition do *aldosterone* and *ADH* control? (p. 788)

30. Review the events that cause the condition known as *edema* or dropsy. (p. 788)

31. Why does it seem unusual to find the *androgens* and *estrogens* being produced and secreted by the *adrenal glands?* (p. 788)

32. Describe the physiological response we experience when *epinephrine* and *norepinephrine* are released. (p. 788)

33. List thc *ovarian hormones* and review their action. (p. 788–789)

34. What is the principal hormone of the testis? Review its action at the beginning and end of puberty. (p. 789)

35. Review the evolutionary history of the *pineal body* and summarize its hypothetical role in today's animals. (pp. 789–790)

36. What is the effect of a *mitogen* on the cell? (p. 790)

37. Describe the source and effect of platelet-derived *mitogens.* (p. 790)

38. List the events that occur when *epinephrine* reaches a target cell's membrane. (p. 791)

39. How do *hormones* identify their *target cells?* (p. 791)

40. Explain how a *mitogen* gets from a cell surface to the cytoplasm. (p. 791)

41. Explain why *steroids* are able to move readily into cells and how they operate once inside. (p. 792)

Chapter 28
Homeostasis

About 200 years ago, Charles Blagden, then secretary of the Royal Society of London, proved that he was one of the most persuasive people on earth. He managed to talk two friends into joining him in a small room in which the temperature had been raised to 126°C (260°F). They were, of course, aware that water boils at 100°C and so they were a bit reluctant. But Blagden prevailed, and the men, taking along a small dog and a steak, entered the room. They emerged 45 min later and were surprised to find that they were not only alive, but in good shape. The dog was fine too. But the steak was cooked!

Thus, Blagden demonstrated the remarkable abilities of living organisms to adjust to severe environmental conditions by regulating their internal processes. Of course, the delicate processes of life demand extremely constant and specific conditions and thus marked regulatory abilities should be expected. But, often, these abilities are startling, indeed.

Let's consider another example. Consider a young man who had thought that he was in good shape down in the city, but now he is finding it pretty tough going while backpacking high in the mountains. Like the dog in the hot room, he too is panting, but for a different reason. The level parts of the terrain are easy enough, but he and his companion are climbing now, and he finds that he has to stop every 100 m or so, just to catch his breath. His heart is racing, he feels light-headed and sick to his stomach, and he has a sharp pain in his right side. After each rest he feels much more comfortable as his heavy breathing subsides, so he resumes his climb. His companion, who has been in the mountains all summer and is having no problem, tells him that he'll get used to the altitude in a week or two. His companion is right. In fact, the very next morning he is already feeling a bit better, and in a couple of weeks he will hardly notice the altitude at all.

28.1 Feedback Systems and Temperature Regulation

The *Random House Dictionary* defines *homeostasis* as "the tendency of a system, especially the physiological system of higher animals, to maintain internal stability, owing to the coordinated response of its parts to any situation or stimulus tending to disturb its normal condition or function." Homeostasis, then, is one of the pervasive themes in biology, and we've already encountered it repeatedly. So now let's take a closer look at the concept.

Homeostasis can operate at a number of levels. In Blagden's experiment, the dog probably sought to leave the room—a behavioral response. Then he probably gave up and lay there panting, producing physiological changes. Regarding the novice mountain climber, he paused by the trailside to "catch his breath" in response to falling levels of oxygen in his blood. Neither the dog nor the man has much control over his panting; it's an almost completely automatic adjustment to the stimuli. The sharp pain in the climber's side is caused by the spontaneous muscular contraction of his spleen, which empties this organ of its reserve blood supply, while his nausea is due to his body's shunting of blood from his gut to his oxygen-starved brain and muscle tissue. On a different time scale, continued warm days would have caused the dog to shed his heavy fur, and continued exposure to high altitudes would have stimulated the climber to produce more red blood cells.

A Closer Look at Negative Feedback Loops

Negative feedback control, as we said time and again, simply means that a stimulus produces a reaction that ultimately reduces the stimulus. Such control can take a number of routes. A baby cries when it is hungry; the crying baby gets a reaction from its parent, who feeds it; the baby stops crying and is no longer hungry for a while. From the baby's point of view, the stimulus was the hunger, and its response was crying, which ultimately made the hunger go away. From the parent's point of view the stimulus was the crying and the response was feeding the baby, which made the crying go away.

Sometimes, the loop fails in one way or another. The baby may cry because, say, it is in pain from diaper rash, and is not at all hungry. If the parent tries to feed it, it cries harder, so the parent tries harder to feed it. So it . . . . This is *positive feedback,* where the stimulus evokes a response that further increases the stimulus instead of decreasing it. Thus, whereas negative feedback systems are stable, positive feedback loops are inherently unstable. Another more familiar *positive feedback loop* occurs in public address systems, when the microphone picks up a sound from the speaker, sends it to the amplifier, which amplifies it and returns it to the speaker, giving rise quickly to that full-volume high whine that sends us all up the walls.

Positive feedback loops are uncommon in nature and are usually associated with illness. High blood pressure, for instance, can damage arteries and arterioles, and the damaged vessel walls can become infused with lipid materials, scar tissue, and cellular growth. Thus, the size of the vessel opening is restricted, further increasing blood pressure and arterial disease. And then we all know of people who are depressed because they are overweight, and so they eat because they are depressed.

But let's take a closer look at the basics of the more usual, more functional, and more stable feedback system, the negative feedback system. The classic nonbiological example of negative feedback control is the thermostat on your house's heating system. Let's consider a house with two thermostats—one on the central heat and one on the air conditioner. The first is set at, say 65°F and the other at 80°F. If the temperature of the house goes below 65°F, the heater goes on. If the temperature goes above 80°F, the air conditioner goes on. If the system is a good one, the internal temperature stays within this 15°F range over a very wide range of outside temperatures.

Figure 28.1 shows the inside temperature of the house over a very wide range of outside temperatures. Note that there are limits to household homeostasis, however. On very, very cold winter nights, the central heat is on continuously, but it is still unable to keep the inside temperature high enough. Similarly, on the hottest summer days, the air conditioner

28.1 Household homeostasis. The typical home temperature-regulating thermostat provides an example of a negative feedback system. In this instance, two thermostats regulate the temperature of a home through alternate heating and cooling, keeping the temperature within a 15°F range. The thermostats sense the air temperature in the room and activate the furnace or air conditioner when temperatures fall below or above their settings. These same thermostats, like so many self-regulating systems in organisms, are subject to the effects they produce. In other words they shut down when air temperatures fall within the limits set. Again, like living mechanisms, they have their limits, or thresholds, at which they begin to fail. In the system shown here, the control mechanisms fail when the outside temperatures fall below 25°F or above 110°F.

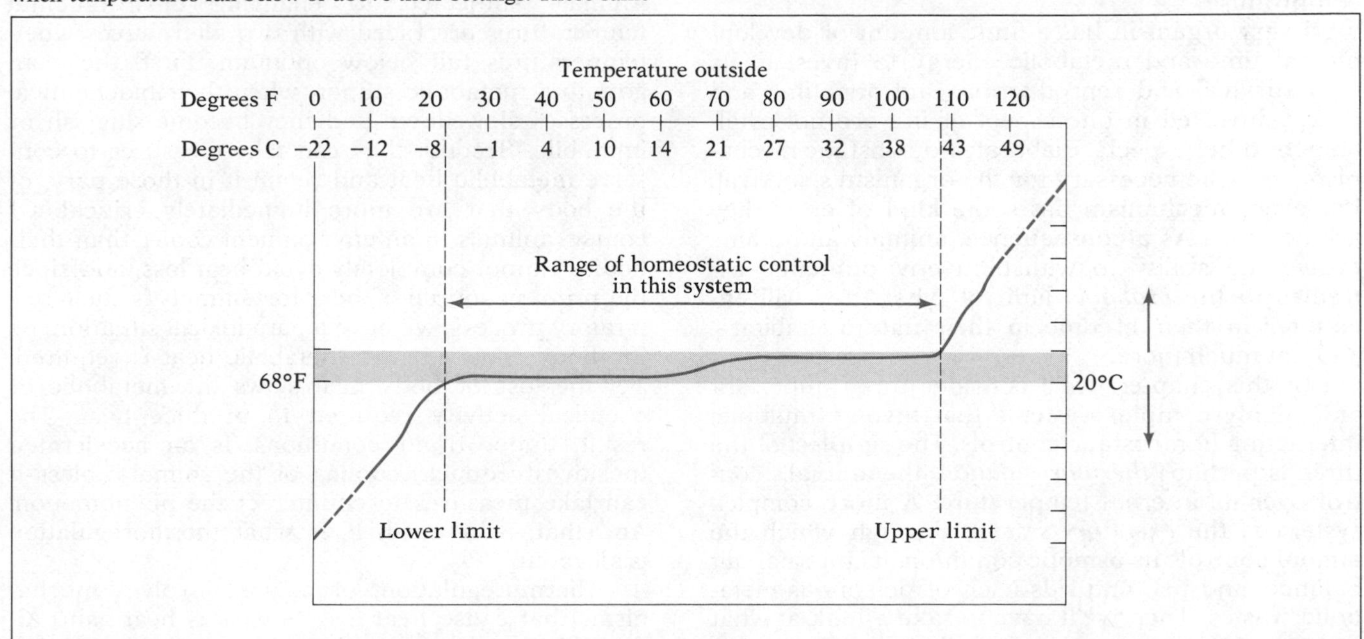

is on constantly, but at some point the inside temperature begins to rise. These stress points are the thresholds of environmental extremes at which homeostasis fails. To increase the range over which the inside temperature remains relatively constant, you would need to invest in a better furnace and a more expensive air conditioner.

Biological homeostatic systems have their limits, too. Many different organisms can keep their internal temperatures fairly constant, but some can do this over a much wider range than others. An Alaskan husky, for instance, can survive outdoors in the arctic night, and you can't. The husky has invested more of its resources in insulation and cold-resisting metabolism. A brine shrimp can thrive under extremely wide ranges of salinity in salt ponds, having invested in a complex homeostatic system that is specialized for such a varied environment. Brine shrimp can survive in fresh water and in seawater, but are seldom found in either habitat because there they cannot effectively compete with other species that have themselves specialized, in different ways, for these more constant environments.

The idea that every biological system has its limits is intuitive enough, and yet when many people encounter specific cases, they find them surprising and even somewhat troubling. As biologists, we are continually impressed with the marvelous and unexpected adaptations of living organisms to the extent that we begin to expect something like perfection. Then we are surprised when life, and living organisms, fall short of our expectations. Our expectations are usually not met because living things tend to compromise.

Every organism has a finite amount of developmental time and metabolic energy to invest in its own survival and reproduction, and any time and energy invested in one aspect of life are not available for other aspects. Elaborate homeostatic mechanisms may be necessary for the organism's survival, but every mechanism has some kind of cost—they are not free. As a consequence, animals and plants evolve the ability to withstand environmental extremes to the *probable* limit of what they will encounter in their lifetime in their natural habitat—and not much more.

In this chapter, we'll consider three important and highly complex systems that involve multiple, interacting homeostatic controls. The simplest of the three is perhaps *thermoregulation,* the animal's control over its internal temperature. A more complex system is the *excretory system,* through which the animal controls its osmotic condition, its internal salt balance, and pH, and rids itself of poisonous metabolic wastes. Then we'll have to take a look at what is perhaps the most complex of all biological systems this side of the brain: the immune system.

Thermoregulation

Because chemical reactions are generally influenced by temperature, an animal must develop a life style that permits it to survive over a range of internal temperatures, or it must develop ways to keep those temperatures relatively constant.

A great number of animals have some thermoregulatory capability, although the range may go from little control to complete control. Earlier, we introduced the terms *poikilothermic* and *homeothermic.* Traditionally, the latter refers to birds and mammals—the best thermoregulators—and the former to the rest of the animal kingdom—those that are "cold-blooded" and at the mercy of their surrounding temperatures. We would like to correct that point of view.

Many recent studies reveal that there is no neat dividing line between poikilothermic and homeothermic groups. At least there is no clear dividing line between taxonomic groups. We now know that thermoregulation is common in certain bony fishes, sharks, and some reptiles (probably the groups most intensively studied). As more of the earth's creatures are placed under close scrutiny, the list will probably grow. The most recently added are—of all things!—the *beetles.* Efficient metabolic thermoregulation may explain in part the extraordinary evolutionary success of this group.

Why Thermoregulate?

Animals subjected to changing environmental temperatures are faced with two alternatives when temperatures fall below optimum. First, they can go into a metabolic stupor, when their biochemical processes slow down and they become sluggish or immobile. Second, they can take measures to conserve metabolic heat and retain it in those parts of the body that are more immediately critical. Of course, animals in an environment cooler than their bodies cannot completely avoid heat loss, and since the primary source of heat for animals is their respiratory process, we have a paradoxical situation. As the body cools, greater metabolic heat is required. Yet the loss of body heat slows the metabolic or chemical activity required to produce heat. The result, under frigid conditions, is an accelerated (positive feedback) cooling of the animal, unless it can take measures to counteract the phenomenon. And that, in a nutshell, is what thermoregulation is all about.

Thermoregulation, of course, involves mechanisms that cause heat loss as well as heat gain. Although we've been talking mainly about heat losses, we hasten to point out that heat gain often represents ever greater difficulties. As a matter of fact, physiological damage created by excessive heat is often irreversible, whereas damage by low temperatures is more likely to be only temporary, except in extreme cases. We should also keep in mind that whether heat loss or heat gain is the goal, the mechanisms for achieving it may involve a number of techniques involving behavior as well as physiology. Combined strategies are very common. Where physiological mechanisms are not well-developed, behavioral strategies become the most essential. Let's note some specific examples.

The Countercurrent Heat-Exchange Mechanism, or How Tuna Swim Fast

Animals with closed circulatory systems have evolved a simple mechanism that prevents excessive cooling of blood as it circulates to the extremities. The mechanism, imposingly called *countercurrent multiplying,* borrows its name from the field of electronics. When in operation, it assists both in preventing heat loss throughout the body and retaining heat in specific and essential areas. Let's consider an example: If you immerse your hand in ice water for a period of time, the skin and fingertips will become chilled. Let's say the outer skin temperature drops from 36°C to 5°C. Thus, chilled blood returning through the veins could, hypothetically, begin to lower the net temperature of the entire body. But this doesn't happen—for two reasons. First, your body will automatically increase its heat output by changing its metabolic rate. Second, the blood returning through the veins is not chilled to 5°C. It is probably more in the neighborhood of 35–36°C—not very cold at all. But how could this be? It is due to a countercurrent exchange between veins and arteries, which works because many of these opposing vessels run closely parallel to each other, carrying blood in opposite directions. So, blood of different temperatures is passing in opposite directions. This, then, sets up the countercurrent heat exchange. The arterial blood warms the returning venous blood as heat moves from warmer areas to cooler areas (Figure 28.2).

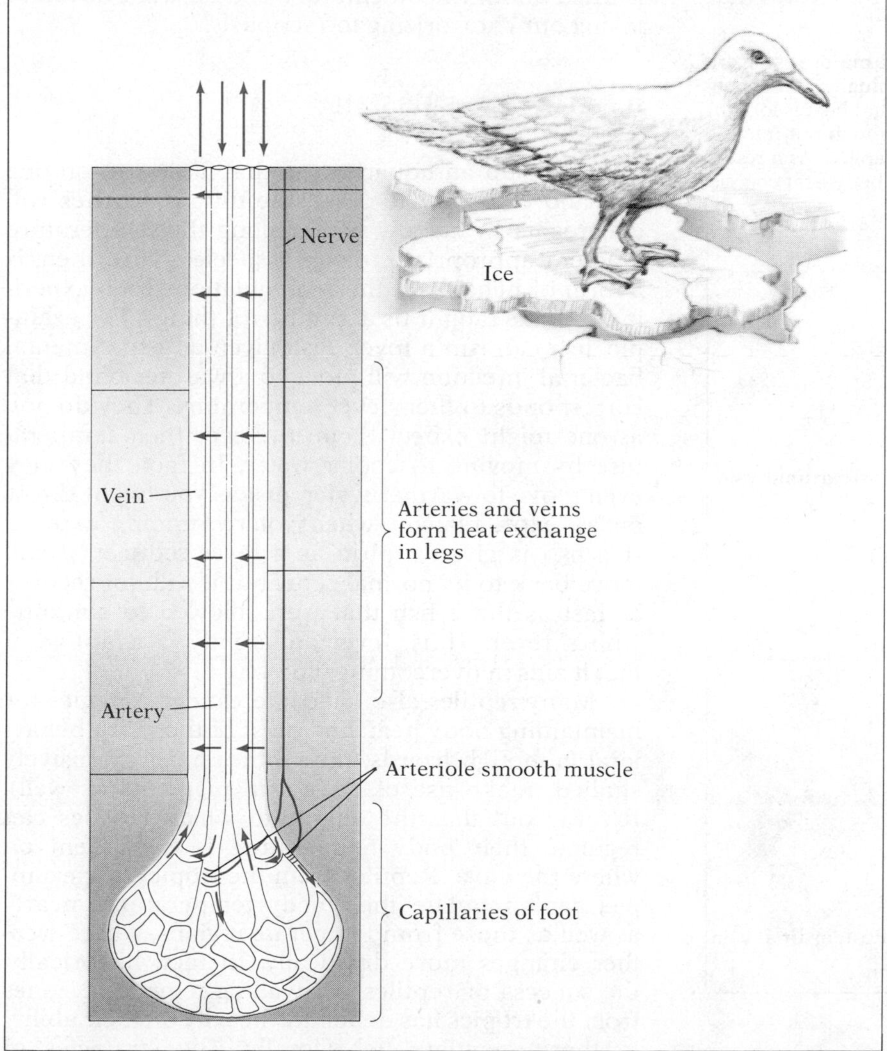

28.2 The gull walking on ice has no apparent problem with heat loss through its featherless legs and feet. Animals adapted to cold utilize a heat-exchanging countercurrent mechanism in vessels carrying blood to and from the extremities. This is seen in the diagrams of the veins and arteries of the legs and capillary beds in the feet. Blood entering the exchanger through arteries is at body temperature (38°C). As it passes through the legs its heat passes to cooler blood leaving the legs. The exchange goes on in those places where the arteries and veins lie close together and are parallel. As a result, temperature of the feet is similar to that of the environment, so not much body heat is lost there. In addition to this heat-conserving mechanism, smooth muscle in arterioles can shut down capillary circulation entirely, shunting the blood back to the body. This also helps prevent heat loss. These muscles are controlled by the autonomic nervous system.

As a specific case, let's see how this mechanism helps the fast-moving, "cold-blooded" predators of the sea. One of the problems with a cut-and-dried poikilothermic concept has been that it doesn't account for how fish can remain extremely active in the cold oceans. In fact, some of the fastest animals on earth are bony fishes and sharks. Among the fastest of the bony fishes is the bluefin tuna, which is often found in *very* cold water. Since there has been some research interest in the tuna, we are beginning to find some intriguing answers.

One thing we've found is that the tuna has a specialized circulatory arrangement. In Figure 28.3, compare the circulatory systems of slower- and faster-moving species of fish found in similar habitats. Notice that in the slower fish, the major vessels are centrally located, with branches serving the propulsion muscles and outer portions of the body. Unfortunately, the kind of system tends to cool the blood, and prevents heat from accumulating where needed.

On the other hand, look at the circulatory system of the bluefin tuna in Figure 28.3(b). Note that the major arteries and veins, always paired and parallel, are peripheral, with branches going inward. This arrangement, then, fosters a countercurrent exchange.

Furthermore, the tuna has another heat-controlling trick. Notice in Figure 28.4 that its swimming muscles are of two colors: dark lateral muscle, with the rest light. (You won't find the dark muscle in higher-priced cans of tuna; it's oily and has a strong flavor.) The darker muscle is a region of higher temperature—as much as 10°C higher than the skin temperature. One reason it is darker is because the muscles are surrounded by immense networks of blood vessels. This vascular layer, consisting of parallel arteries and veins, is known as the *rete mirabile* which translates into "wonderful net," and it truly is. The net, in fact, is a very dense countercurrent heat conserver. Because of this setup, the tuna has a comparatively warm group of swimming muscles. And that's why a tuna can swim fast.

At the moment, we don't know the extent to which other fishes thermoregulate. But what has been learned about the bluefin tuna should warn us about taking our categorizing too seriously.

28.3 (a) In fishes that do not thermoregulate, the major vessels are centrally located along the body axis. The branching vessels supply muscle beds in the body and tail in such a way that heat is lost. (b) In the warm-bodied fishes, the major vessels branch out into the muscles where countercurrent networks are arranged. As a result, these fishes are able to thermoregulate, keeping body heat concentrated in the swimming muscles.

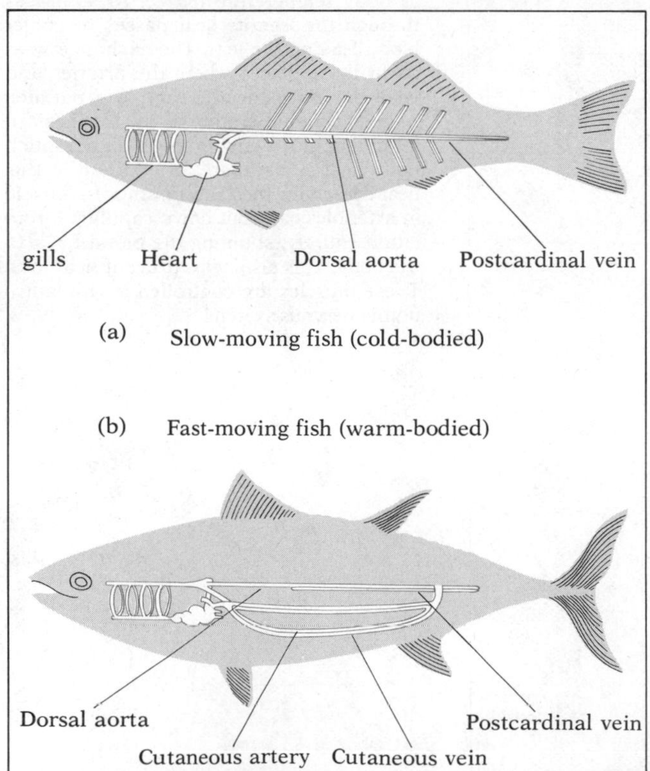

Regulating Behaviorally

If you set up an aquarium so that it's warm on one end and cold on the other, the different fishes will congregate in narrow bands along the temperature gradient appropriate for each species. This, then, is a form of behavioral thermoregulation. Such experiments have taught us a number of things. For example, fish can run a fever. Fish given an experimental bacterial infection will move to a warmer band that corresponds to their fever temperature. They do not, as one might expect them to, lower their temperature by moving to cooler water. In fact, they may even move to warmer water, just as you might throw on an extra blanket when you're running a fever. If a fish is given aspirin as a fever reducer, it will move back to its normal zone, but it will not recover as fast as those fish that were allowed to continue with a fever. Thus, fever apparently is adaptive in that it aids in overcoming illness.

Many reptiles also have interesting strategies for maintaining body heat, and most of these are behavioral (although lizards, the reptiles most extensively studied, make use of their skin pigments as well). It turns out that the degree to which reptiles can regulate their body temperature is dependent on where they live. Reptiles from the tropics, for example, can't regulate their body temperatures nearly as well as those from temperate regions, where weather changes more drastically. In fact, historically, the success of reptiles in invading temperate zones from the tropics has depended heavily on their ability to thermoregulate behaviorally. The strategies of

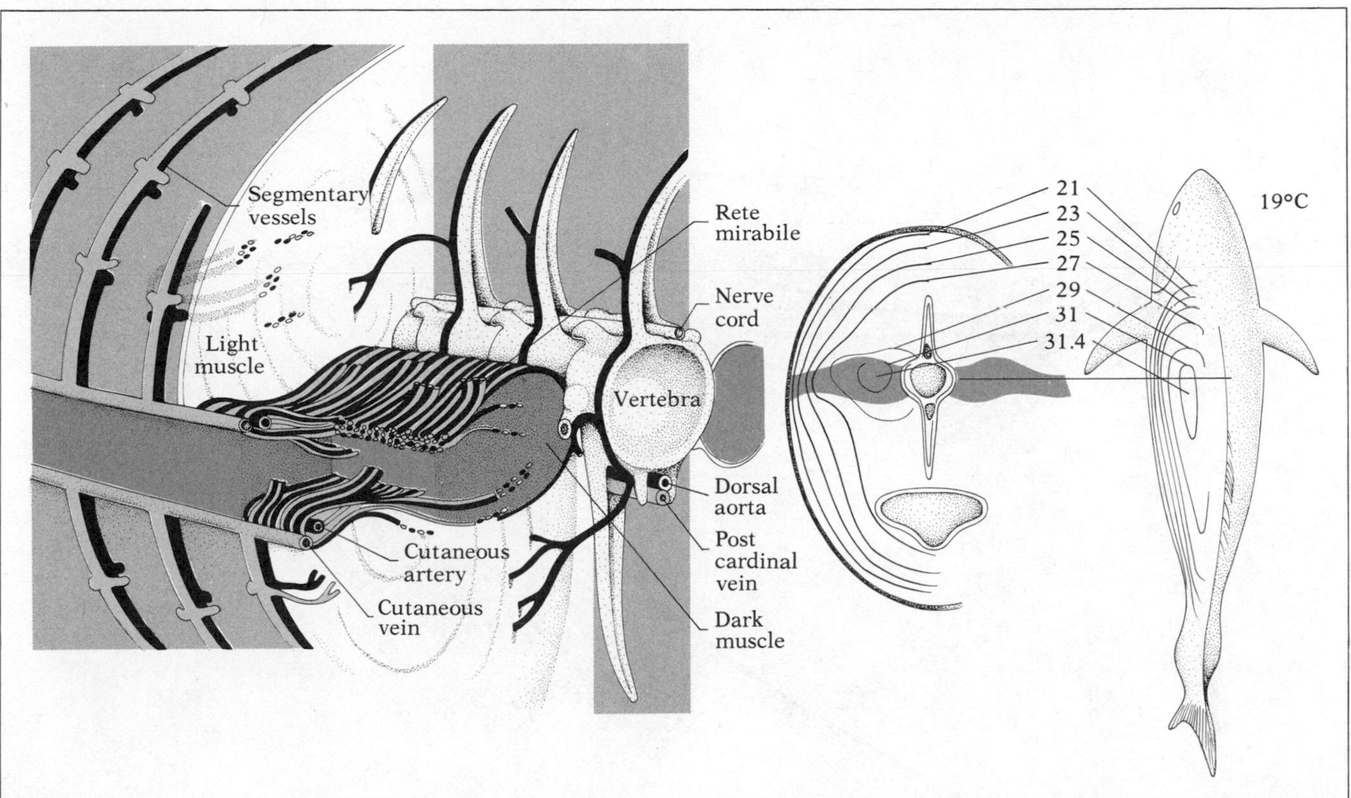

28.4 The countercurrent net of the tuna. Within the darker muscle regions, known as the *rete mirabile*, are found massive countercurrent exchangers. Veins and arteries arranged side-by-side set up a heat exchange that makes this one of the warmest regions in the tuna's body. Evidence of the effectiveness of this form of thermoregulation was produced by measuring temperatures all around the body at various distances below the skin. The results are plotted here in a *thermocline*, similar to what we see in weather maps. The tuna was immersed in water cooled to 19°C before measurements were begun. As you can see, the body temperature deep in the darker muscle region, where countercurrent exchange is greatest, is over 12°C warmer than the water temperature.

temperate zone lizards are different, but are apparently very successful since they are known to maintain remarkably constant body temperatures as they move about.

Both heating and cooling can be achieved through specific orientation to the sun's rays. Reptiles, for example, gain a great deal of body heat by basking in the sun, and the lizards have developed the technique into an art. They lose heat by simply avoiding the sun. By constantly adjusting its amount of exposure to the sun, the lizard keeps a rather constant body temperature (Figure 28.5).

Reptiles can improve their behavioral thermoregulation through color changes. The horned lizard (commonly called the "horny toad," although there is insufficient evidence to support the title) can alter its color to gain or lose the sun's heat. At low temperatures, the pigmented cells in its skin expand, darken its color and cause less light to be reflected and more to be absorbed. This ability promotes the efficiency of basking considerably (Figure 28.6). When body heat is high, the pigmented cells contract, producing a much lighter, heat-reflecting color.

There is rather convincing evidence of the presence of more complex thermoregulation in reptiles, although the precise mechanisms are not known. For example, it has been determined that on a warm day lizards of different sizes all reach a temperature peak at the same time. This seems to violate our understanding of heat transfer. Theoretically, the lighter and smaller reptiles should peak first.

Part of our problem with understanding such regulation is that most of our studies have been carried out in the laboratory under artificial conditions in which the animal is not able to exhibit its full behavioral repertoire.

As we mentioned earlier, there is currently some controversy over whether dinosaurs were warm-blooded or cold-blooded. Because there is no really sharp dividing line between the two categories, and because dinosaurs are scarce, the question may never be solved. It has been suggested, however, that the strange, highly vascularized backbone plates of the stegosaurs were well-adapted as heat radiators (Figure 28.7). Other dinosaurs had high dorsal fin structures that probably served both to capture solar heat

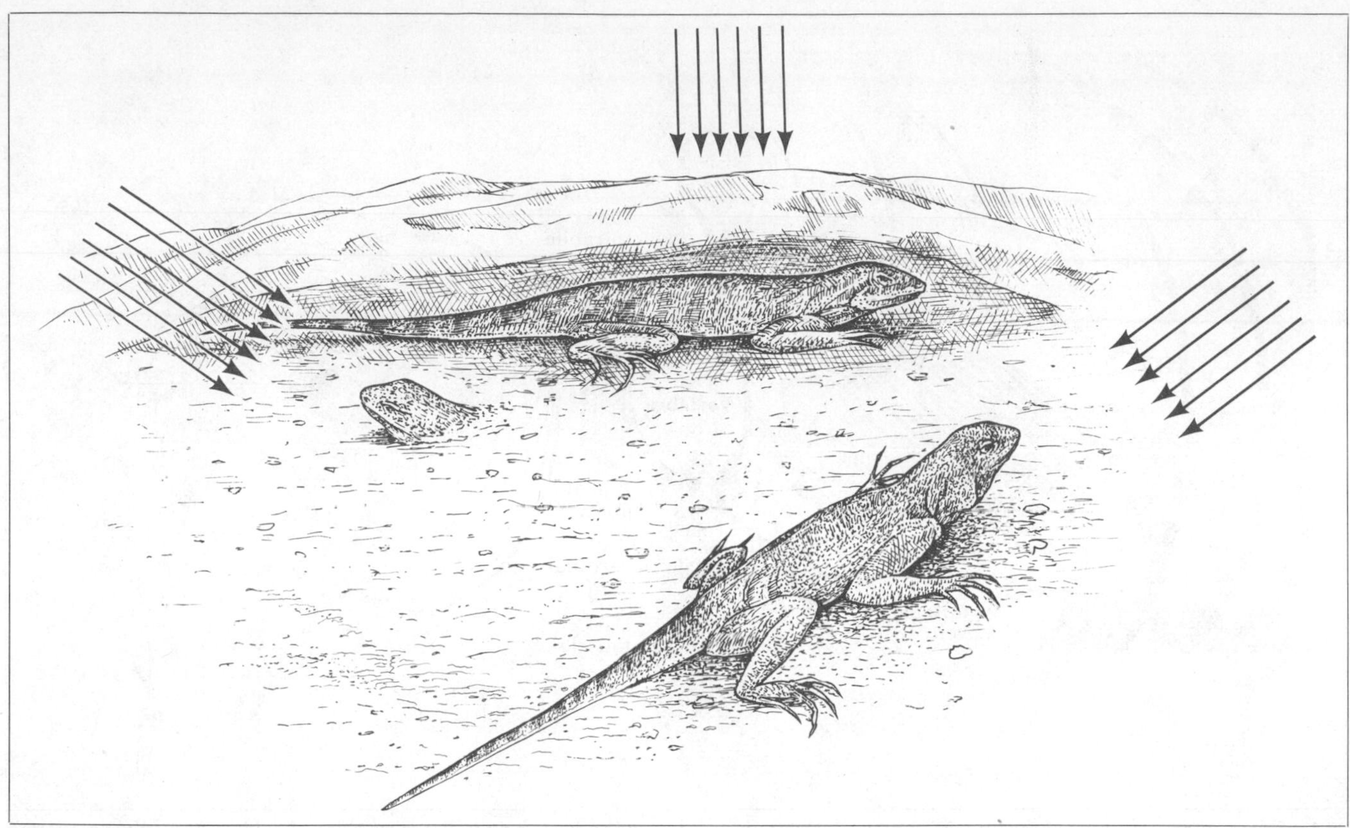

28.5 Lizards and other reptiles of the desert must develop strategies of thermoregulation to cope with drastic temperature changes. A lizard buries itself in the sand at night to conserve body heat. To become active in the morning, it must crawl out and absorb the sun's rays. First, the lizard exposes its head to the sun early in the day. By doing this, it slowly warms up, but doesn't expose its sluggish body to predators. During midday, when the lizard must escape from the heat, it finds shadows to linger in as it moves from one place to another. Later in the day, as cooling begins, the lizard basks in a position parallel to the sun's rays, absorbing the last remnants of the day's energy supply before digging in for the night. These are all behavioral mechanisms that assist the body in making physiological adjustments to changing temperatures.

28.6 Thermoregulation in the horned lizard. In addition to its behavioral repertoire, the horned lizard has developed a physiological mechanism for thermoregulating. Darker colors absorb more radiant energy than lighter ones. So the horned lizard (like many reptiles and amphibians and some fishes) is able to change its color through pigment migration—an automatic response. (a) When exposed to cooler temperatures in experimental situations, the lizard's color darkens. (b) At higher temperatures, it lightens considerably.

(a)

(b)

28.7 Thermoregulation in *Stegosaurus. Stegosaurus* had a prominent row of large plates covering its dorsal surface. These were formally considered to have been defensive armor that protected the stegosaurs against the giant carnivores of their day. While they may well have served this purpose, they are now thought to have been thermoregulating structures, containing capillary beds that helped in radiating body heat out of the animal and perhaps absorbing heat as part of a basking behavior. Exposing their large surface area to the early morning sun would have hastened the warm-up period many reptiles require to get themselves going each day.

and to radiate unwanted heat, depending on the animal's orientation to the sun.

There are many other examples of heat regulation in "cold-blooded" animals. But our point, in addition to introducing the subject and pouncing on some popular old ideas, is to illustrate the continuous diversity of form and function in living things. Each species, confronted with its own unique survival problems, is, over the long run, in a state of highly regulated, but constant, change.

Thermoregulation in the Homeotherms

Mammals and birds are most assuredly homeotherms. They have, indeed, made great inroads into maintaining a constant body temperature. Actually, we should be referring to *internal* body temperature, since the extremities and skin may be quite different from the temperature in the deeper tissues. Your skin temperature, for example, may vary widely and is usually cooler than the temperature of your internal organs.

Homeotherms adjust their temperatures in two ways. One involves varying the rate of metabolic heat production, and the other, the control of heat loss and gain through radiation, conduction, and convection. These two general mechanisms work in close harmony to maintain optimal internal temperatures.

Let's begin by looking at the outward manifestations of temperature control and then consider the more complex regulatory processes themselves. In particular, we'll focus on the processes common to humans.

Removal of Heat Energy

The most important avenues of heat escape in humans are the skin and respiratory passages, from which heat is lost through radiation, conduction, and convection. Heat, we know, radiates into the cooler surroundings. Radiation, therefore, is important in dissipating heat from the peripheral circulation, through the skin. In addition, heat escapes through conduction (heat passes by contact to a cooler solid, such as a tile wall) and convection (heat energy is carried away by the molecules of a cooler gas). The rate of heat escape by any means is greatly enhanced by the evaporation of moisture on the skin and in the respiratory passages. Since heated molecules move faster and more vigorously, they tend to break away and escape. Thus, the net effect is a cooling of the surface as less excited molecules are left behind. When air temperatures exceed body temperature, sweating is the only means by which the skin can be cooled. In situations where high air temperatures are combined with very high humidity, even this mechanism fails. At this time, humans may resort to behavioral strategies, ceasing all activity, bathing their bodies in cool water, or perhaps turning on an air conditioner. The cessation of activity is important, and humidity can be a far more important factor than you think. One of your authors (Wallace) recently suffered heat stroke in a marathon. He reports:

> The temperature was 75° and the humidity was 80%. All seemed well until the 24th mile. I was running at about 6:25 per mile when, suddenly, I felt remote, dizzy and "spacey." I slowed down, thinking I would quickly recoup. But I didn't recoup. My goal was anything under 2:50 and with 2 miles to go my time was 2:36. I tried to mentally subtract 36 from 50 to see how fast I would have to run the final 2 miles, but any mental calculations were impossible. I began to realize I was in trouble and when I felt my arm I found it dry and chilled. A wave of dismay swept over me because I knew continued stress could cause permanent kidney or brain damage, but I wanted to finish, so I slowed down even more. Then a strange thing happened. I began to fall asleep again and again while running. I managed to stay awake and trotted in at 2:55. Recovery, with water and mineral-laden drinks, was rapid. Strangely enough, a blanket was

needed to fend off what felt like a chill wind. Physiologists later told me that I had possibly altered my thermoregulatory mechanisms so that I had lost too much heat and that the sleepiness could have been the same sort described by people who are freezing. The point is, the thermoregulatory mechanisms are in delicate balance, and if shifted off balance, the results may not be merely drastic, but unexpected.

Conserving Heat

But let's talk about what happens when the system works. Generally, the circulatory system itself can help maintain temperature. For example, when internal temperatures rise, vessels close to the skin open up, producing the familiar flushed appearance, and increasing heat loss. The opposite occurs when the body cools. At very low external temperatures, blood leaves the skin, causing it to pale. Constriction of the muscular walls of the arterioles can minimize skin circulation and redirect the flow into deeper regions (Figure 28.8). For this reason, mountain climbers suffering from exposure will experience freezing in the fingers, toes, and ears before their internal temperatures begin to fall. As we discussed in Chapter 26, the vital organs will receive first priority in the ensuing battle for survival.

Actually, birds and mammals conserve heat by a number of means. Perhaps you've seen the little birds perched at your feeder on a cold winter day, sitting there with their feathers all fluffed up until they look like tennis balls with beaks. Mammals can also stand their hair on end, and the process, in both cases, is called *piloerection*. Insulating air is trapped under the hair or feathers and it buffers the animal from the environment (Figure 28.9).

While piloerection (goosebumps) in humans may not mean very much in terms of heat conservation, shivering does. Actually, shivering is a method of heat *production*. Since it involves the rapid contraction of muscles, the work itself produces heat. Interestingly, both shivering and piloerection can be produced without being stimulated by temperature. Did you ever see anything that made your hair stand on end?

Of course, these are not the only ways animals conserve heat. Woodland animals, such as skunks, may lay on great stores of fat that serve not only as food reserves, but also as a rather unvascularized buffer that protects the deeper tissues. Animals may also huddle together, as we see in great heaps of squealing and indignant pigs in winter farm lots, each trying for the center position. Also, the itinerants who sleep under the Paris bridges in winter tend to huddle together, the most revered occupying the more central positions.

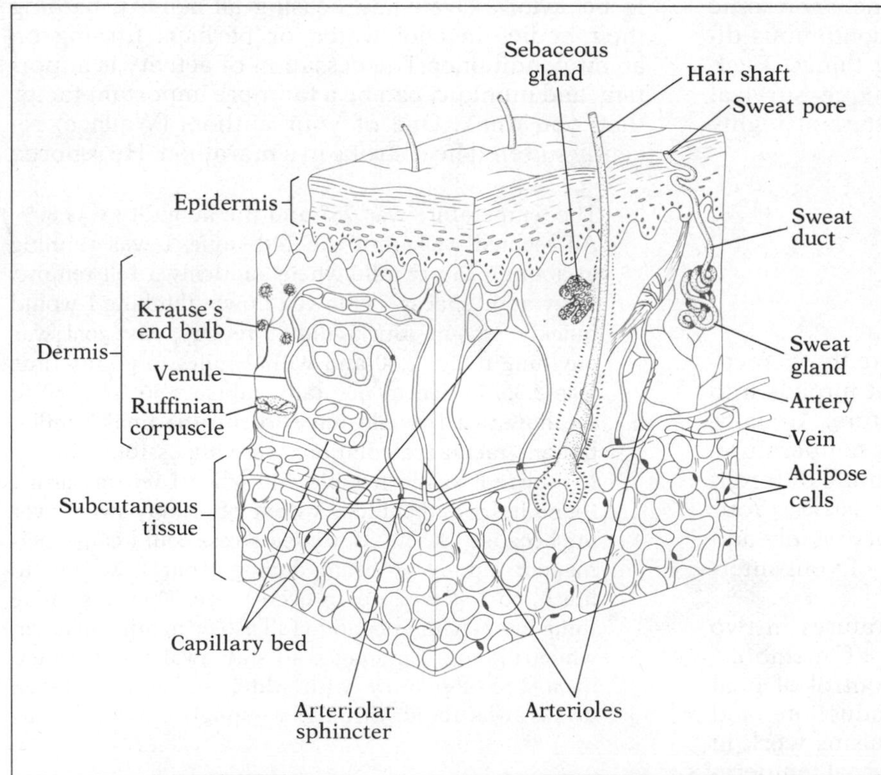

28.8 In addition to its other functions, the human skin is an excellent thermoregulatory organ. *Sweat glands* secrete water through ducts that empty into skin pores. When evaporating from the surface, water has a cooling effect. Sebaceous glands secrete oils that keep the hair and skin pliable and waterproof. The secretions empty along hair shafts. *Capillary beds* below the surface can bring blood near the skin for cooling or, alternatively, shut down, directing blood away to deeper regions in order to conserve heat. Temperature-sensing neural structures detect heat (*Krause's end bulbs*) and cold (*Ruffinian corpuscles*), passing this information along to the brain for the proper response. Insulation is further provided by layers of adipose tissue (fat) below the skin.

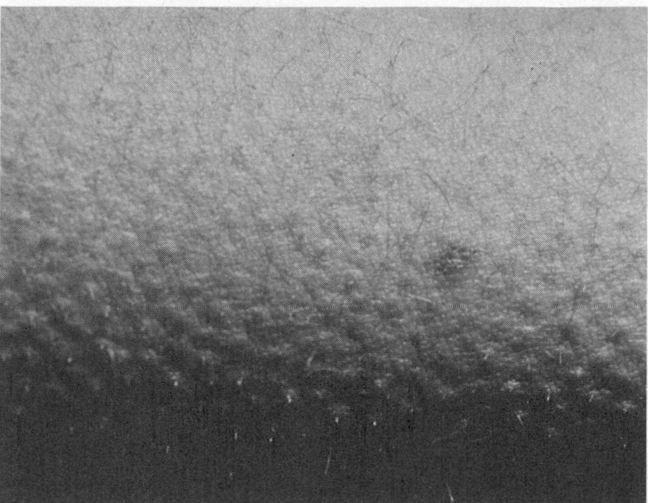

28.9 *Piloerection* may aid as an insulator in birds and mammals. The hair or feathers are raised above the skin by contraction of erectile muscles, producing air spaces that serve as insulation. Keep in mind that insulation merely serves in maintaining any temperature whether it is hot or cold. In humans, piloerection produces the familiar goosebumps and is often accompanied by shivering, which generates heat.

The Internal Source of Thermoregulation

So a living body can change in many ways to meet the challenge of unsuitable environmental temperatures. The question for now is: How are these responses regulated and integrated? Humans maintain an average internal temperature of about 37°C (98.6°F) within an amazingly narrow range. Surely some form of sophisticated internal control is required since we have so many independent ways to regulate heat. For example, consider just the responses of the skin: Piloerection, sweating, vasodilation, and vasoconstriction. Are these mediated by local receptors in the skin itself? Or are their initiation and coordination directed more centrally?

First, we need to know how the body measures temperature. Essentially, it is done two ways. One is through the sensory neurons of the skin. We react voluntarily to this information by doing whatever is necessary to bring the skin temperature back to its desired levels. We may move to a warmer place, take off some of our clothes, stomp our feet, or fan ourselves.

The second method of measuring heat is internal. The brain constantly monitors the temperature of the blood flowing through it. The responses initiated by the brain override all other controls; hence, the brain is the real thermostat of the body.

The heat-monitoring part of the brain is the hypothalamus (see Figure 27.7). Actually, it behaves very much like an ordinary home thermostat, but with far more precision. Physiologists have been able to detect neural activity in the hypothalamus when blood temperature varies by as little as 0.01°C. The sensitivity of the hypothalamus exceeds that of the best laboratory instruments, as far as we can tell.

The hypothalamus, while constantly measuring blood temperature, can elicit a variety of responses. For example, through its influence on the autonomic (sympathetic and parasympathetic) nervous system, it can increase the rate of heat production by stimulating the adrenal medulla, causing it to release norepinephrine into the bloodstream. This hormone increases the metabolic activity of cells, thus creating heat. The hypothalamus also stimulates thyroid activity through its pathways in the pituitary. It does this by stimulating the pituitary to release thyroid-stimulating hormone (TSH), which in turn causes the thyroid to increase its secretions. The result is an increase in metabolic activity and body heat. The resultant heat increase is detected in the hypothalamus, which then eases off on the pituitary (another example of negative feedback control).

The hypothalamus can also influence the autonomic nervous system to bring about the changes in the skin we described earlier.

And so, information arriving from peripheral and internal sensors results in a considerable variety of hypothalamic actions. As a homeostatic control structure, the hypothalamus keeps the organism in a finely tuned, responsive state, maintaining body temperatures at an optimal level. One might wonder, though, just how does the hypothalamus measure blood temperature; how does the thermostat work?

A Closer Look at How the Hypothalamus Works

It turns out that, after all is said and done, perhaps the hypothalamus doesn't measure heat! From what we have learned about nerve activity, we know that there are always chemical changes associated with any neural activity. This causes us to ask: Is it possible that the hypothalamus is actually measuring heat-related chemical changes in the blood rather than heat itself?

Let's look at some evidence from studies of fever. In a general sense, fever involves resetting the hypothalamal thermostat to a higher level. The agents of infection responsible for this change are collectively known as *pyrogens* (*pyro*, heat; *gen*, start). They may be toxins produced by bacteria agents, or they may be released by the body's own immune response system. In any case, the thermostat is altered chemically. When pyrogens are at work, there are measurable increases in blood calcium levels. This increase creates a calcium–sodium imbalance. In fact, we now know that the thermostat can be artificially altered by perfusing the brain with calcium solutions. This works so well that body temperature can be chemically lowered far enough to permit cryogenic (deep-cooled) surgery, duplicating the effect of submerging the body in ice water. The use of calcium removes the problem of having to lower body temperature by removing body heat; the thermostat is merely set to a lower level. This of course, is all a bit hypothetical. We really aren't sure that we're altering the thermostat itself. The important point is that the hypothalamus appears to function in heat regulation by measuring the level of calcium and/or sodium in the blood. At the moment, that's about all we know.

28.2 Osmoregulation and Excretion

Now let's look at another problem of regulation in animals. If temperature maintenance presents one classical problem for physiologists, this one certainly represents another. It is the problem of how animals regulate the water and ion concentrations in their cells. Most cells are about two-thirds water, but in some cases, "about" won't suffice. The amount of water in certain cells is critical indeed, as is the relative abundance of various ions in the cell fluids. The problem of maintaining the proper water and ion balance is called *osmoregulation*. And this necessarily leads us into the methods by which excesses of either are removed from the body—by the processes of *excretion*. In particular, *excretion* refers to the removal of metabolic wastes from the body—that is, the removal of nitrogen wastes that are released as amino acids are metabolized. So let's have a look, first, at how these wastes are produced.

Producing Nitrogen Wastes

Amino acids, you recall, are the constituents of proteins. Proteins must be broken down into individual amino acids before they can pass into the bloodstream. Those same amino acids can be used to build new proteins in the cells, or they may be used in a variety of other ways, such as by being converted to fatty acids or carbohydrates, or used to prepare amino acids for use as fuel in respiration. Some of these changes produce leftover fragments of nitrogen, and such nitrogen can be extremely poisonous.

As you can see in Figure 28.10, the amine group, $-NH_2$, is removed from the amino acid as NH_3, or *ammonia*. In the presence of protons from water, it readily forms ammonium ions (NH_4^+) and molecules of ammonia (NH_3). (Notice the role of NAD here—another example of the multiple use of these important molecules.) Ammonia is highly toxic, but many organisms can safely handle it if they live in water and if they are small enough to easily exchange materials with their environment. However, in many complex animals and especially those that live on land, the ammonia can't be removed fast enough to keep from poisoning the organism. In these cases, ammonia is changed to something more manageable.

When ammonia is altered, it usually winds up as one of two compounds. Some organisms, such as insects, reptiles, and birds, produce a solid, insoluble waste known as *uric acid*, or related compounds such as guanine. Others, such as many invertebrates, some fish, and mammals, produce the readily soluble compound, *urea*. The way that ammonia is handled is dictated by the need to conserve water. Some animals have no water problem and where water is readily available, soluble urea can be formed. Urea is relatively nontoxic, so it can be stored in the body temporarily, but it remains soluble. Its solubility means that large amounts of water must be used to flush it out of the system. In species that have water problems, on the other hand, it is best to produce insoluble uric acid and to excrete it as a pasty substance, suspended in very little water.

Not only is uric acid formation important for animals that must conserve water as adults, but also for bird and reptile embryos, which are confined within dry egg shells. So, birds and reptiles form uric acid which, because it is insoluble, can be handled somewhat independently of water. It is interesting that rather substantial amounts of dry wastes are left behind in the egg shells when birds and reptiles hatch. Also, many adult birds have, or can have, nearly dry droppings (except for some that sit in trees above outdoor concerts).

28.10 The deamination of alanine illustrates ammonia formation from the metabolism of amino acids. In the metabolic breakdown of alanine, pyruvic acid is formed, along with the waste product, ammonia. Pyruvic acid is then free to enter one of several pathways (such as the citric acid cycle). The waste ammonia must be rapidly excreted. Organisms have three choices: ammonia may be excreted directly, or it may be converted to uric acid, or to urea. Uric acid and urea are less toxic and can be eliminated more gradually. Humans produce urea in the deamination of amino acids. We do produce uric acid, but only in the metabolism of nucleic acids.

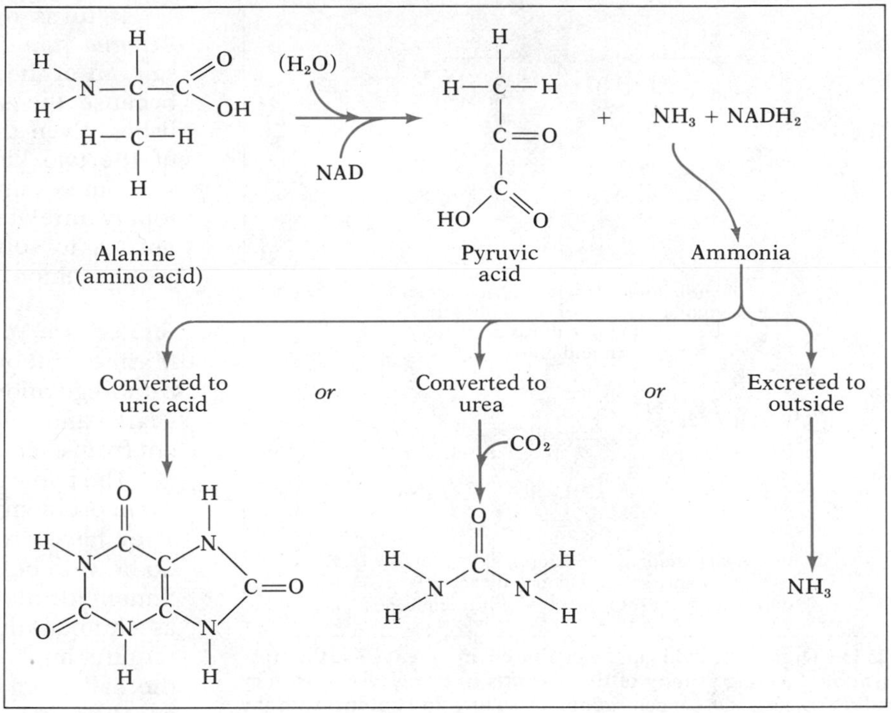

The Osmotic Environment

Each kind of environment presents its own osmoregulatory problems for animals. Over the eons, as the species attempted to explore new environments, they faced a multitude of water-regulatory problems. The results we find today are varied, but each organism has had to come up with a working mechanism in order to survive. Let's see how animals that live in salt water, fresh water, and on land have solved their problems.

The Marine Environment

Animals of the sea live immersed in a complex solution of ions, including sodium, potassium, calcium, magnesium, and chloride. When we talk about salt water, we are referring to the 3.5% sodium chloride content of the solution, but we can't neglect the others. They have their own specific effects on living systems and, together, they raise the osmotic pressure of seawater. To survive in this environment, organisms must solve two osmoregulatory problems. First, they are in a *hypertonic* medium—one in which the water concentration is higher inside the animals' cells than outside. The tendency, therefore, is for them to lose water by osmosis. Conversely, the seawater has more solutes (such as salts) in it than do their cells. Thus, the solutes also set up an osmotic gradient, but this time with the lower concentration toward the inside. This means, of course, that the ions will tend to enter the organism. So, marine dwellers must keep water in and salts out. To do this, they have evolved a variety of adaptive strategies.

The problem is somewhat restricted for deep-sea organisms because things don't change much in the briny depths and thus they have a constant osmotic environment, but near-shore, things can change drastically and animals that live in bays and estuaries have to put up with some rather marked changes in salinity. The invertebrates have solved their problems in two major ways. Those of the near-shore marine environment tend to be either *osmoconformers* or *osmoregulators.* The solute concentration within the body of a conformer changes according to the solute concentration of the seawater. The organism thus tends to stay *isotonic*; that is, it increases or decreases its solute concentration as the salinity around it changes. Actually many marine invertebrates are osmoconformers.

As an example, the limpet *Acmaea limatula* is a resident of the intertidal zone (Figure 28.11). As such, it commonly experiences changes in salinity as the tides change and as rainstorms dilute its surroundings. To adapt to these salinity changes, the limpet simply conforms, rather than fighting to maintain any particular internal salinity. Remarkably, its body fluids can remain isotonic in a salinity range of 1.5–5%. For some reason, these changes have little effect on the animal. A similar change in our body solutes would produce sudden death.

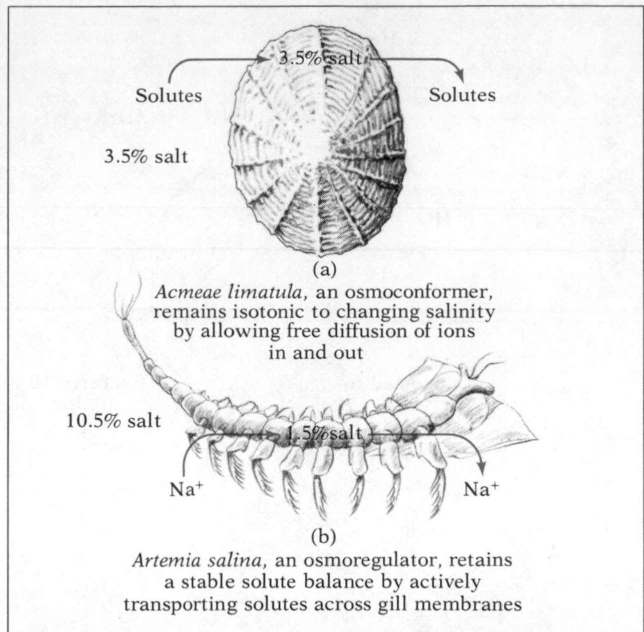

(a)
Acmeae limatula, an osmoconformer, remains isotonic to changing salinity by allowing free diffusion of ions in and out

(b)
Artemia salina, an osmoregulator, retains a stable solute balance by actively transporting solutes across gill membranes

28.11 Organisms living in the marine environment solve the problem of changing salinity of the seawater in two ways: either they adjust their body concentrations of solutes to conform with the concentration of the seawater (*osmoconforming*), or they actively get rid of the solutes as fast as they enter (*osmoregulating*). (a) *Acmeae limatula*, a limpet, is an osmoconformer. It permits the concentration of body solutes to increase and decrease with drastic changes in its environment. We aren't sure how it survives these osmotic changes, but not many organisms can. (b) *Artemia salina*, a brine shrimp, is an osmoregulator. It maintains a fairly constant solute concentration in its body fluids despite the fact that it can live in a 30% salt concentration. It accomplishes this by actively transporting excess solutes out through its gills. Of course, this means that it must devote a considerable portion of its energy reserves to osmoregulation.

On the other hand, many of the marine arthropods are osmoregulators. They maintain a relatively constant solute concentration within their bodies despite changes in salinity around them. Osmoregulation, though, involves work—active transport. One of the best-known regulators is the crustacean *Artemia salina* (Figure 28.11), which lives in salt ponds. It can maintain constant solute concentrations in its body when environmental salinities vary from less than 1% to more than 30%! We might mention that the solute concentrations of organisms are determined by finding the freezing point of their body fluids. Osmoconformers freeze at temperatures that also freeze the surrounding water. The body fluids of *Artemia salina* freeze at a temperature just below 0°C (the freezing point of *fresh water*). Samples of their usual salty environment, however, will not freeze until the temperature reaches about the range −14°C to −16°C. This indicates that the body fluids are considerably less salty, the mark of a marine osmoregulator.

It turns out that the osmoregulatory organ of *Artemia salina* is the gill, which actively transports salt out of the blood. This is particularly interesting, because the same mechanism is used by marine fishes. Even the tissues of the salt-excreting glands of the two kinds of animals are similar. This is a welcome example of convergent evolution—cases of totally unrelated animals evolving very similar mechanisms to solve the same problem are really not very common.

Strategies of Marine Vertebrates

Marine vertebrates have evolved several methods of osmoregulation. In particular, the bony fishes and sharks and rays have strategies that are quite different from each other (Figure 28.12).

The hypertonic marine environment presents the same problem to fish as it does to the invertebrates: They have problems with water loss and salt gain. So how do bony fishes solve the problem? They continually drink seawater. A terrestrial mammal (such as a human) that drank seawater would subsequently require huge quantities of fresh water to remove the salt from its body. The fish, however, has no problem since, as we mentioned earlier, it has special salt-secreting cells in the gills. With salt being removed from the body, seawater becomes perfectly suitable for drinking.

As an active vertebrate, a bony fish produces large quantities of nitrogen wastes, particularly if it is a carnivore. Thus, theoretically, it should require considerable fresh water for the removal of nitrogen. However, the fish doesn't use the excretory route of its terrestrial relatives. It has kidneys, but very little nitrogen waste is removed there. Instead, nitrogen wastes in the form of ammonia are excreted—again across the gill membranes.

So, what about sharks and rays? Their strategy for resisting water loss is to create an isotonic condition in the blood and tissue fluids by retaining the waste urea. (Interestingly, the ancient coelocanth uses the same strategy as sharks and rays.) This seems peculiar, but keep in mind that, as far as osmosis is concerned, it doesn't make much difference what is in solution, as long as the concentrations of *water* inside and outside the organism are equal. When they are unequal, there will be a net movement of water in or out in an attempt to reduce the gradient. In addition, sharks and rays have a mechanism for actively transporting salt out. In these animals, the site of transport is a special gland opening into the rectum.

Marine birds and marine reptiles also take in seawater with their food. The solute concentrations of their bodies, however, are about the same as those of terrestrial species. So what do they do with all those salts? Surprisingly, they actively excrete salt through special glands located near the eye and drain it through a duct into the nose. Their kidneys, in

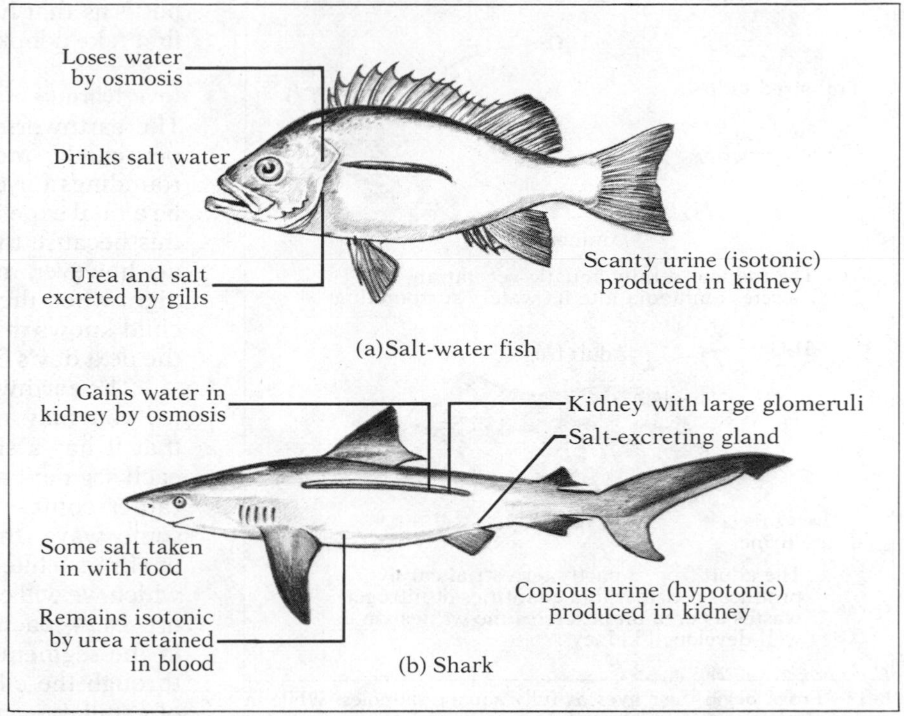

28.12 Osmoregulation in marine fishes and sharks. (a) The bony marine fish survives its salty environment by actively transporting salt and ammonia directly across the gills. It obtains its needed body water by drinking seawater. (b) The shark, however, uses a different strategy. It increases its solute concentration by maintaining a high urea concentration in the blood. In addition, it actively transports salt through special glands in the rectum. Does this make the shark an osmoconformer? An osmoregulator? Both?

fact, are not very efficient salt removers. Studies of gulls reveal that they swallow seawater with about 3.5% salt and excrete a urine that is only about 0.3% salt. The salt glands, however, produce a nasal secretion of 5% sodium chloride solution.

One last case: Marine mammals have highly efficient kidneys and can actively transport ions into their urinary collecting ducts. For them, then, the kidney is the salt-excreting gland. However, most marine mammals avoid drinking seawater and rely on relatively low-solute fluid from the body fluids of the fish they eat.

The Freshwater Environment

Just as salt water presents problems, fresh water, too, can be hazardous and even deadly to organisms that live in it. They tend to take in the fresh water through osmosis, and cells can swell and rupture. So their problem is how to keep the water out.

And fresh water presents another problem. It doesn't provide its denizens with the ions they need. In particular, sodium, potassium, and magnesium are in short supply, Thus, these particles tend to obey the laws of diffusion and leak out of the organisms into the environment. The organisms must expend energy to pump these errant molecules back in by active transport. Such transport may occur across the membranes of the skin, gills, or kidney tubules.

Of course, all the news is not bad. Freshwater organisms don't have much of a problem with handling nitrogenous wastes. In fact, many of them simply make ammonia and flush it out with the abundant water that is available.

Vertebrates of the Freshwater Environment

Earlier, we discussed the excretory mechanisms of such invertebrates as paramecia (contractile vacuoles) and planaria (flame cells), so, here, let's consider two groups of vertebrates: bony fishes and amphibians. Essentially, they solve their excess water problem by producing a copious, highly dilute urine. Their kidneys are well-adapted for removing water.

The excretory systems of fish and amphibians are rather highly developed. This is particularly true of amphibians. As they become more terrestrial with maturity, the changes include interesting shifts in the excretory processes (Figure 28.13). For example, a tadpole produces and excretes its nitrogen waste in the form of ammonia, much the way a fish does. This is possible because of its watery habitat and the fact that a tadpole is a vegetarian, existing on a diet light on proteins and heavy on carbohydrates. Adult frogs shift to the production of urea as the principal nitrogen waste. A frog must become something of a water conservationist, so the excretion of water-wasting ammonia is no longer feasible. (Even so, most frogs drink and excrete huge amounts of water—up to 30% of their body weight daily.)

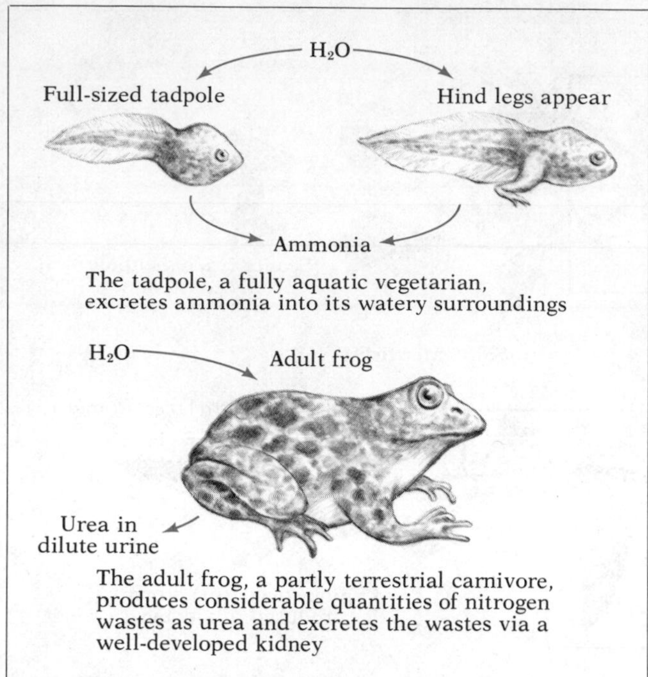

28.13 Frogs begin their lives as fully aquatic tadpoles. While in this developmental stage, the tadpole excretes copious amounts of water, with which it flushes out ammonia wastes. As an adult, however, the frog has a different problem, since it spends a considerable amount of time out of water. To manage its nitrogen waste problem, it switches to urea formation, which it excretes with an efficient kidney.

The Terrestrial Environment

In one important way, the terrestrial environment presents problems similar to those of the oceans: Animals have the problem of water loss. Terrestrial animals solve their water-retention problem through high water intake, highly efficient water-conserving excretory systems, water-tight skin, and behavioral patterns that help in the problem of exposure. Let's first take a look at a few land-dwelling invertebrates.

Invertebrates of the Terrestrial Environment

The earthworm is terrestrial, but actually lives in a perpetually moist environment. Leaving these surroundings for drier regions or the sunlit surface can be a fatal experience. Fortunately, they don't often do this because they are attracted to moist areas. They do, however, make frequent forays to the surface at night when the risk of desiccation is minimal. Every child knows this and collects the "nightcrawlers" for the next day's fishing.

The earthworm is a water conserver. One reason (as you may recall from our earlier discussion) is that it has a skeleton made of water. The fluids in each segment are constantly filtered and recycled by rather complex structures known as *nephridia.* (In some ways, the nephridium is reminiscent of the nephron, a filtering structure of the vertebrate kidney which we will consider shortly.) As you can see in Figure 28.14, each paired nephridium cleans the fluid in the segment just anterior to it. Any fluid moving through the ciliated tubule must pass a rich supply of capillaries in the tubule wall, where materials can be exchanged between the fluids and the blood. Water, minerals, and other essential materials are reabsorbed into the blood, while dissolved wastes pass to the outside through an opening known as the *nephridiopore.* So, what we have here is an example of osmoregulation and excretion being accomplished by the same organ, just as in the vertebrates. But before leaving the invertebrates, let's consider that enormous group called the arthropods.

The arthropods of dry land have developed some very distinct ways of solving their osmoregulatory and excretory problems. Mainly, they have concerned themselves with preventing water loss. This is accom-

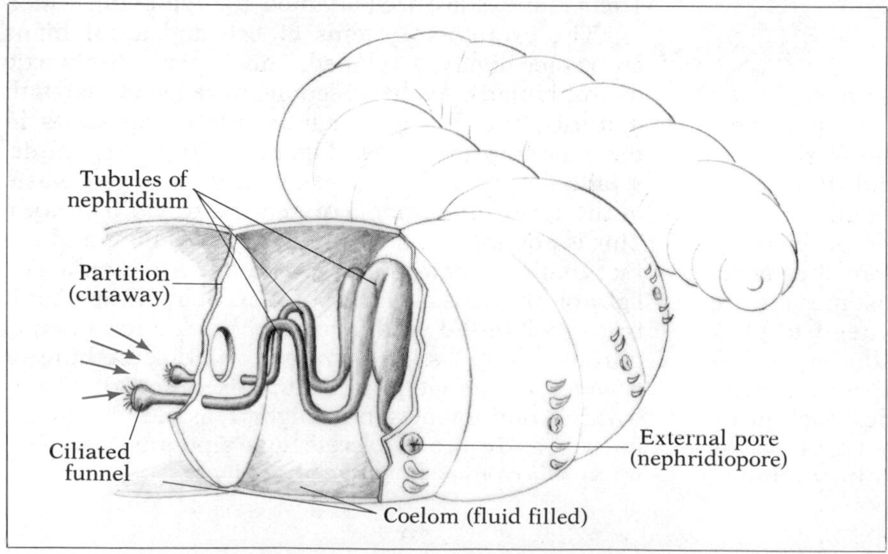

28.14 The earthworm lives in a moist environment, but it must conserve water nonetheless. Its water-conserving and nitrogen waste-eliminating structure is the *nephridium.* Each segment has a pair of nephridia. Body fluids enter the ciliated funnel, carrying nitrogen wastes and essential body solutes into its looped passage. As the fluids pass along, water and vital substances are selectively reabsorbed into the blood, leaving the nitrogen wastes, excess water, and excess salts to pass out through the *nephridiopore.*

plished through the presence of a water-resistant, waxy cuticle and the elimination of nitrogen wastes as the semisolid uric acid. (The respiratory arrangement is important in conserving body fluids as we saw in Chapter 26.) Let's consider other ways the insects handle their water and ion problems.

The main water source for insects is their food, although they also rely on "metabolic water" produced during cell respiration. Many insects, in fact, rely entirely on metabolic water and do not drink at all. The success of their water-conserving strategies is documented by the fact that some can live in the driest environments in the world, as long as sufficient food is available.

The excretory system in insects consists of many blind, hollow, tubular structures known as *Malpighian tubules*. They emerge at about where the mid- and hindgut join (Figure 28.15). The tubules extend into the coelomic fluids and here absorb body wastes. Inside the tubules, nitrogen wastes from the coelomic fluids are converted to insoluble uric acid crystals, which pass through the lumen of the Malpighian tubules directly into the gut. Once in the gut, the uric acid joins the digestive wastes, passing to the outside during defecation. This method of excretion conserves water in two ways. First, the nitrogen waste is a solid that doesn't require water for dilution. Second, fluids from the Malpighian tubules can be reabsorbed by cells in the lining of the gut. Thus the insects produce a dense, fairly dry, digestive and excretory waste.

Vertebrates of the Terrestrial Environment

Vertebrates that stalk the land are particularly at the mercy of water. They must constantly replace water that is lost to the air. Their most important means of saving water is the specialized kidney. As we said, two groups, the reptiles and the birds, typically conserve water through the production of uric acid, which can be eliminated in a semisolid state. The production of uric acid rather than urea is also necessary in the embryo within the egg. (Imagine the problem in the bird or reptile egg if ammonia gathered. Instead, though, uric acid crystals can accumulate harmlessly in special egg membranes.) Mammals, though, produce primarily urea (some uric acid is also produced), which requires a constant supply of water for its elimination. This is why a water-conserving kidney is so important. Since the most efficient kidney of this type belongs to the mammals, let's have a closer look at them—and at the human excretory system in particular.

28.15 The grasshopper resists water loss by producing a semisolid pasty digestive and excretory waste that contains insoluble uric acid crystals. Uric acid is gathered from the coelomic fluids by a large number of long, threadlike structures known as *Malpighian tubules.* The uric acid moves along the tubules, finally entering the intestine, where it joins the digestive wastes. This method of nitrogen waste excretion is an excellent means of conserving water, since uric acid can be excreted almost dry.

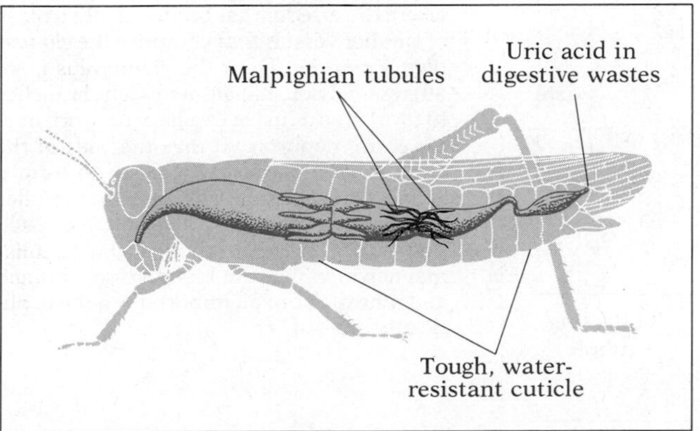

The Human Excretory System

Whereas humans are very wasteful of water (using about three gallons of highly purified water to flush a few ounces down the drain), their kidneys are not. The entire system is made more efficient by the fact that the kidneys are aided in balancing the body's water and minerals by the lungs and skin glands. But let's concentrate on the kidneys themselves, first seeing what they look like and then how they work.

To begin with, our system includes the *kidneys, ureters, bladder,* and *urethra* (Figure 28.16). Notice that each kidney receives a small branch from the descending aorta; this branch is the *renal artery.* The blood entering the kidney through this small artery is under extremely high pressure.

As expected, the excretory system removes excess water, excess salts, and the nitrogenous waste called urea. The volume and content of the urine thus produced depends ultimately on the quantity of water taken in as well as on the diet.

The kidney's highly regulated activity is under the control of hormones from the hypothalamus and adrenal glands which can alter the cells of the kidney.

Figure 28.17 illustrates the major features of the human kidney. The three regions of interest are the *cortex*, the *medulla*, and the *pyramids*. The cortex consists of the filtering units of the kidney—the *nephrons* and their related blood vessels. Each nephron consists of a spherical capsule and a long tubule. The tubules each form a very long loop that ventures down into the medulla area and returns to the cortex. Groups of nephrons empty into common collecting ducts that extend downward into the medulla, where they are joined by other ducts to form the pyramids. The pyramids empty into a hollow region known as the *renal pelvis*. This is the first place where pools of urine collect. But the urine passes from there to

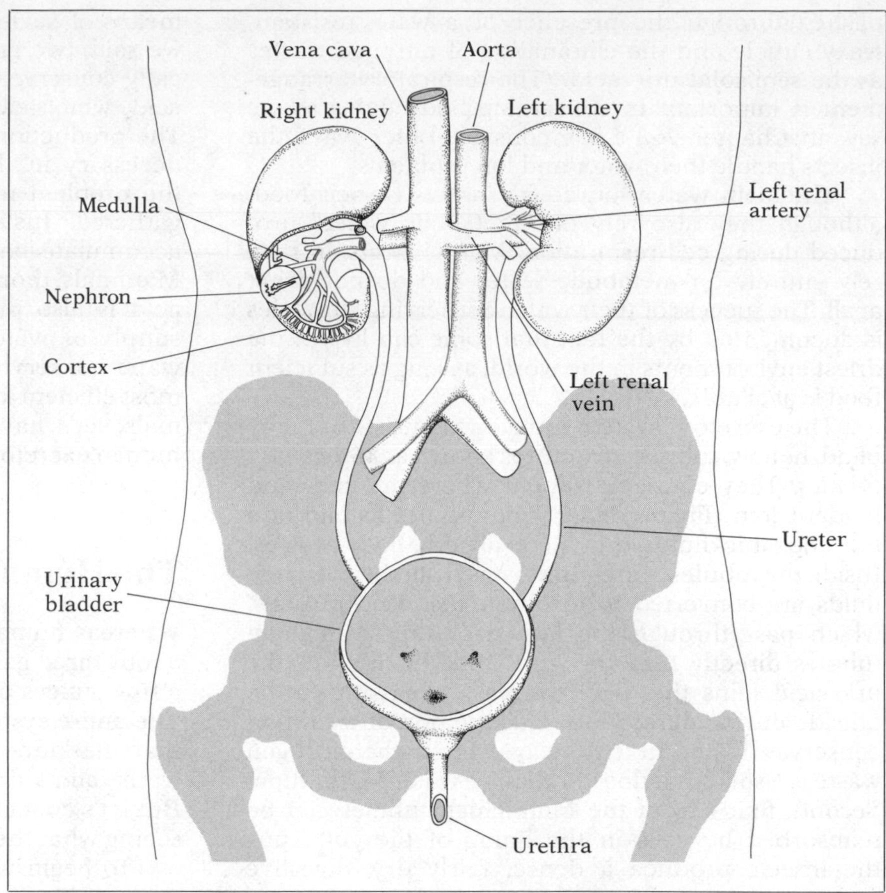

28.16 In humans, the excretory system consists of the paired *kidneys,* their blood vessels, the *ureters,* the *bladder,* and the *urethra.* The system has both an excretory and an osmoregulatory function. The blood vessels of the kidney, which form the *renal circuit,* consist of the paired *renal arteries,* a complex capillary network, and the paired *renal veins.*

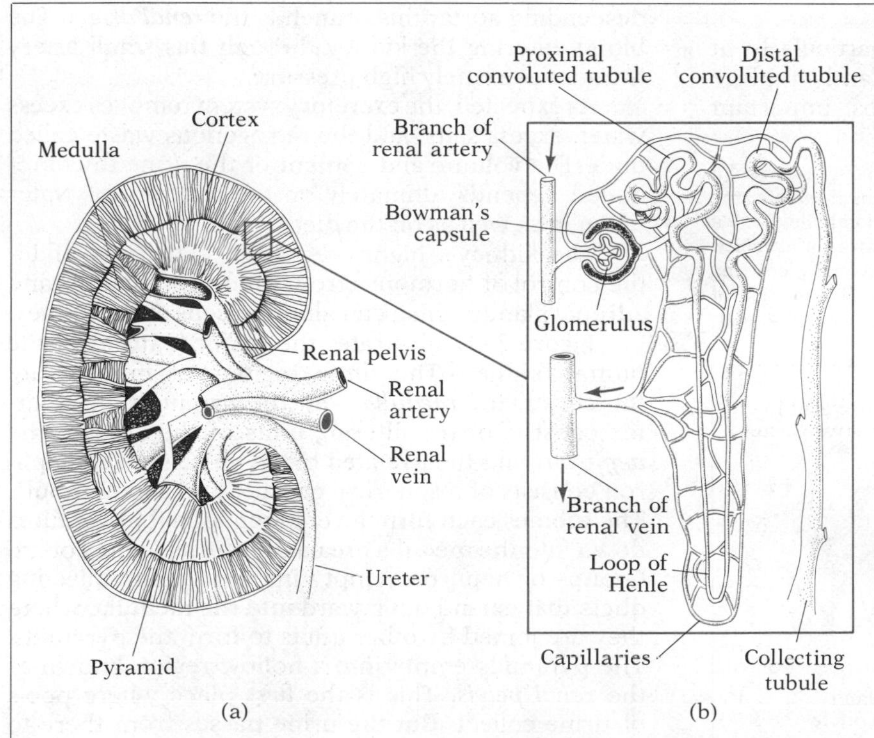

28.17 Anatomy and microanatomy of the kidney. (a) The kidney contains an outer *cortex,* and inner *medulla* (which contains the *pyramids*), and a collecting region known as the *renal pelvis.* (b) The functional units of the kidney are the *nephrons.* These are microscopic tubules that selectively remove urea, excess salt, and excess water from the circulatory system. Each nephron consists of five anatomical regions: Bowman's capsule, proximal convoluted tubule, loop of Henle, distal convoluted tubule, and collecting duct. Each has its own function. Note the blood supply to the nephron. Blood enters the nephron at the *Bowman's capsule,* where the arteriole has branched into a mass of smaller vessels that comprise the *glomerulus.* Emerging from the glomerulus is an afferent vessel that immediately branches to form an extensive capillary network over the entire nephron. At the other side of the network, the capillaries join up to form a venule that will eventually join other venules to form the *renal vein.* The shape of each region of the nephron and the peculiar manner in which the blood vessels branch and rebranch are all important to the nephric mechanism.

the bladder via the ureters. From the bladder, urine is passed to the outside through the *urethra,* which is much shorter in women than men. (This is one reason women suffer from bladder infections more often than men do—the germs don't have so far to travel.)

The nephrons, then, are the functional units, and there are about 1 million per kidney. Let's have a closer look at one (Figure 28.17). We begin with the *Bowman's capsule,* which is a hollow bulb or cup surrounding a ball of capillaries. The masses of capillary, each known as a *glomerulus,* arise from tiny branches of the renal artery. So blood, under high pressure, is carried into a Bowman's capsule, where it must remain for a while as it winds its way through the tortuous capillary ball. The blood finally emerges from the capsule and continues through vessels that form a fine network over the entire nephron before joining with those from other nephrons, finally coalescing into the renal vein. A long tube, or tubule, emerges from the capsule. The first region, the *proximal convoluted tubule,* winds a meandering route and then forms a long hairpin bend known as the *loop of Henle.* This loop is found in all water-conserving kidneys and is extremely prominent in the kidneys of desert-dwelling mammals. After forming the loop of Henle, the tube again begins to twist and contort into a convoluted tubule, this time called the *distal convoluted tubule.* It then joins the other nephrons to form the collecting ducts of the kidney.

The elements of the nephron, in the order of urine flow, are: (1) Bowman's capsule, containing the glomerulus; (2) the proximal convoluted tubule; (3) the descending limb of the loop of Henle; (4) the ascending limb of the loop of Henle; (5) the distal convoluted tubule; and (6) the collecting duct. The entire mechanism functions to remove wastes and toxic substances at all times, but also serves to remove excess water some of the time and excess salt some of the time.

Now, to see how all of this works, let's return to Bowman's capsule and the glomerulus, where blood is force-filtered (Figure 28.18). Normal arterial blood enters the glomerulus under pressure, and about one-fifth of its volume is force-filtered through the capillary walls and into the cavity of Bowman's capsule. This filtrate (in its primary, highly dilute state) flows into the upper regions of the nephron; and some rather viscid, concentrated, hypertonic blood is carried off by the venule that exits from the glomerulus. Osmotically, this blood is ready to pick up water from the dilute primary urine, and this is what it does. The venule forms a capillary network over the proximal convoluted tubule and recovers much of the water there by diffusion. In addition, valuable small molecules that have passed through the glomerular filter—such blood components as glucose and amino acids—are returned to the blood from the filtrate by active transport. Even proteins that escape into the nephron are recovered—through pinocytosis. Remaining in the urinary fluid are now urea, miscellaneous toxic substances (any small molecules not specifically reclaimed), some salt, and, still, much of the water.

It is in the loop of Henle and in the collecting ducts that most of the remaining water will be returned to the blood. This is an active process; yet biological systems usually cannot pump water directly. They must pump salt, and then water follows the salt. Afterward the salt can be recovered. Here's how the trick is done.

The descending limb of the loop of Henle, the ascending limb of the loop, and the collecting tubules form a curious S-shaped countercurrent multiplier, with all three portions lying more or less side-by-side. The glomeruli and the convoluted tubules lie in the cortex (outer part) of the kidney, while the loop and the collecting tubules extend into the medulla (inner part). Now, the tissue fluids of the medulla are relatively salty (hypertonic). As the dilute urine flows down the descending limb of the loop, it loses water and gains salt, both by passive diffusion. At the apex of the loop, the urine, still carrying with it the unwanted wastes, has lost much of its water, which was the object of the manuever. But it has picked up a lot of salt, and salt too must be conserved.

In the ascending limb of the loop of Henle, salt is actively transported from the nephron into the surrounding tissue. Actually, only chloride is pumped out, but sodium and potassium ions follow the chloride. This active transport accomplishes two things: it conserves the salt and it maintains the hypertonicity of the kidney medulla. In the ascending limb, as it climbs into the kidney cortex, and in the distal convoluted tubule, the urinary fluid tends to become isotonic with the circulating blood plasma of the surrounding capillary beds. But it has to traverse the salty medulla one more time, as it flows through the collecting ducts. As it again enters regions of increasing osmotic potential, the urinary fluid may lose yet more water—or not, depending on the needs of the organism. If the animal is conserving water, *antidiuretic hormone* (ADH) renders the walls of the collecting ducts and distal tubule permeable to water, which then leaves the nephron. If the organism has been drinking heavily, the walls become relatively impermeable to water, allowing excess water to flow on through the ducts and out of the system.

Note that salt in the nephron may be continuously recycled: into the descending limb by diffusion, then out of the ascending limb by active transport, only to enter the descending limb once again. Is the kidney just spinning its wheels? Absolutely not—the process maintains the osmotic gradient that assures a net flow of water out of the nephron and, eventually, back into the blood stream.

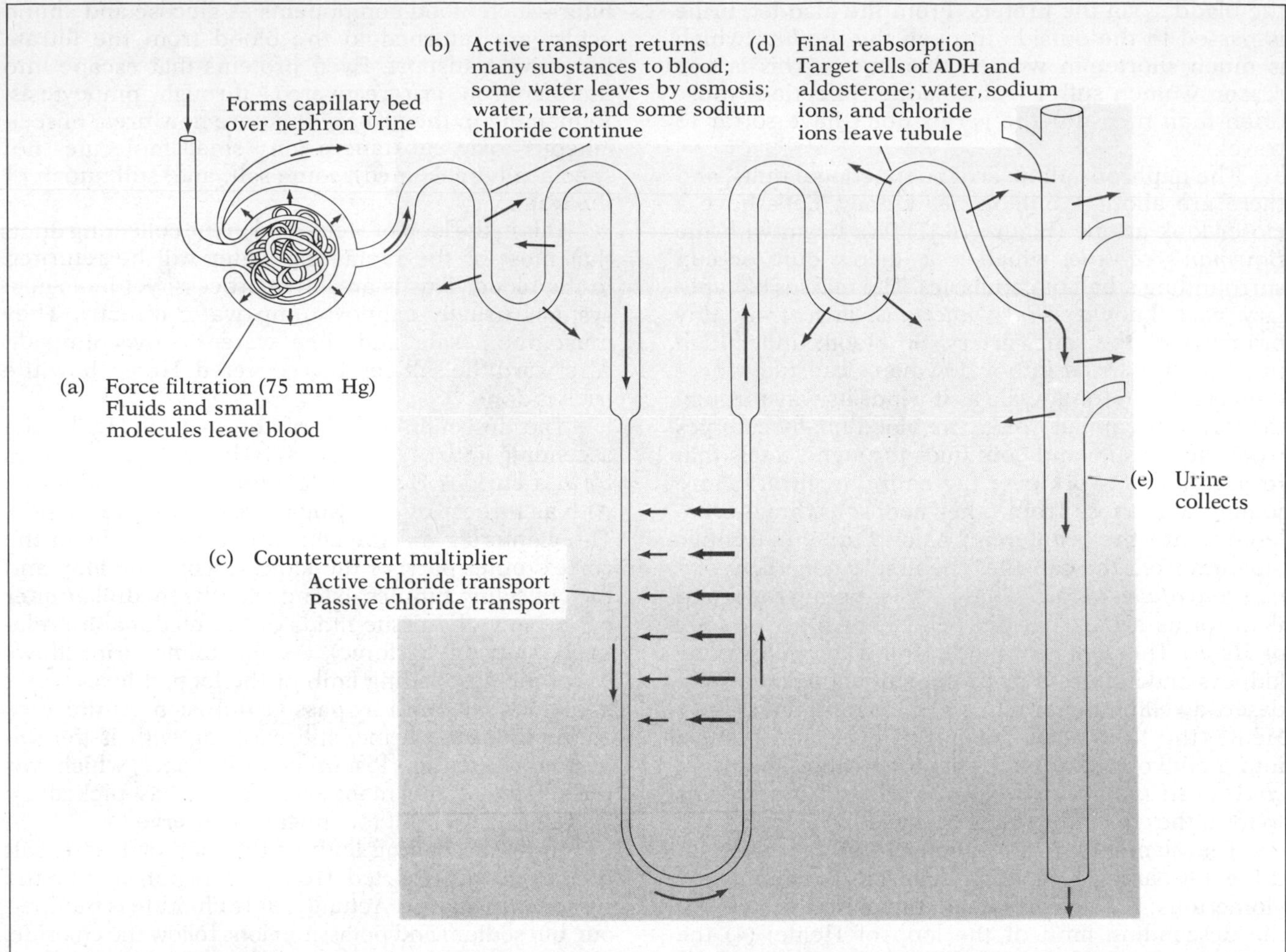

(b) Active transport returns many substances to blood; some water leaves by osmosis; water, urea, and some sodium chloride continue
(d) Final reabsorption Target cells of ADH and aldosterone; water, sodium ions, and chloride ions leave tubule
Forms capillary bed over nephron Urine
(a) Force filtration (75 mm Hg) Fluids and small molecules leave blood
(e) Urine collects
(c) Countercurrent multiplier
Active chloride transport
Passive chloride transport

28.18 The functioning nephron. (a) At the glomerulus, water, ions, amino acids, glucose, urea, and other molecules of low molecular weight leave the blood. (b) Most of the essential materials are reabsorbed (through active transport and diffusion) into the capillary network from the proximal convoluted tubule. (c) Passive movement of chloride into the descending limb of Henle's loop and its active transport out of the ascending limb of the loop create a countercurrent effect. When chloride leaves the tubule, sodium follows. Sodium chloride concentrates outside the nephron, creating an osmotic gradient that draws water from the tubule. (d) The final reabsorption of water and other essential materials occurs at the distal convoluted tubule and collecting ducts. (e) The filtrate—now urine—contains water, salt, urea, and miscellaneous toxins and other small molecules.

Control of Nephron Function

The hypothalamus, in addition to everything else it does, measures the osmotic pressure, or the concentration, of the blood passing through its capillaries. If the blood is hypotonic (with excess water), the kidney can begin to pump out water while retaining most of the sodium. If the blood begins to register as hypertonic (meaning the body is beginning to run short of water), the kidney is put on a water-rationing regime. In such cases, the hypothalamus secretes *antidiuretic hormone*, or ADH (see Chapter 27), into the posterior pituitary via nerve pathways. The hormone is then released into the blood. Its targets are the epithelial cells of the distal convoluted tubule and collecting ducts of the nephron.

The stimulation of the hypothalamus and the production of ADH has another effect. It creates the sensation of *thirst*. As the individual drinks, the blood becomes diluted, the hypothalamus stops sending out antidiuretic hormone, and the kidney increases its output of fluid. At the same time, the collecting ducts of the nephron become less permeable to water, so that the urine becomes more dilute and is produced in much greater quantities. It passes into the renal pelvis, through the ureters, and into the bladder. Eventually another sensation is produced, this time from the stretch receptors of the bladder. And a new behavior is initiated.

A similar mechanism operates to control the reabsorption of sodium chloride. In this case, the

Essay 28.1 Gout

"But that's hardly possible," explained the patient with a nervous smile. "You see, I'm a Lutheran minister." The physician had just given the diagnosis: the patient's great pain and swelling were due to a bad case of gout. To the incredulous minister, as with people through the ages, gout was associated with dissolute living and overindulgence of the flesh. But the doctor's diagnosis was correct, and even saints can get gout.

Gout is a fairly common disease of middle-aged men and tends to run in families. It seems to be a genetic defect in the metabolism of pyrimidines—the familiar thymine, cytosine and uracil of RNA and DNA. While humans discharge most of their metabolic wastes as urea, the pyrimidines are degraded to uric acid, which is actually a fairly simple pyrimidine itself. Uric acid is sufficiently soluble to pass out in the urine of most individuals, but for some reason, some people make much more uric acid than others, and it reaches high levels in the blood and tissues. When this happens, the excess uric acid tends to form sodium urate crystals and to accumulate in certain tissues where circulation is poor. The urate crystals are repeatedly phagocytized by macrophages, which form digestive vacuoles around them. Of course, they cannot digest these crystals; and in fact, the crystals form hydrogen bonds with the digestive vacuole membrane, causing it to break open and release its toxic enzymes into the cytoplasm. This kills the cell, which releases the urate crystal intact. Death of the macrophage also releases the digestive enzymes and other cell contents, which cause irritation and inflammation in the tissues. The result is gout.

The commonest symptoms are sudden pains in the knees or feet or, most especially, in the joint of the big toe. Pain is excruciating for most sufferers. Any movement or touch may set off howls of agony. The gouty patient sits disconsolate, his painfully inflamed foot swathed in layers of protective bandages, and simply waits—and waits—for the attack to go away.

But what about the popular association between gout and dissolute living? Is it without basis? Not entirely. Gout is exacerbated by anything in the diet that contains pyrimidines. The list includes sweetbreads, liver, tongue, an-

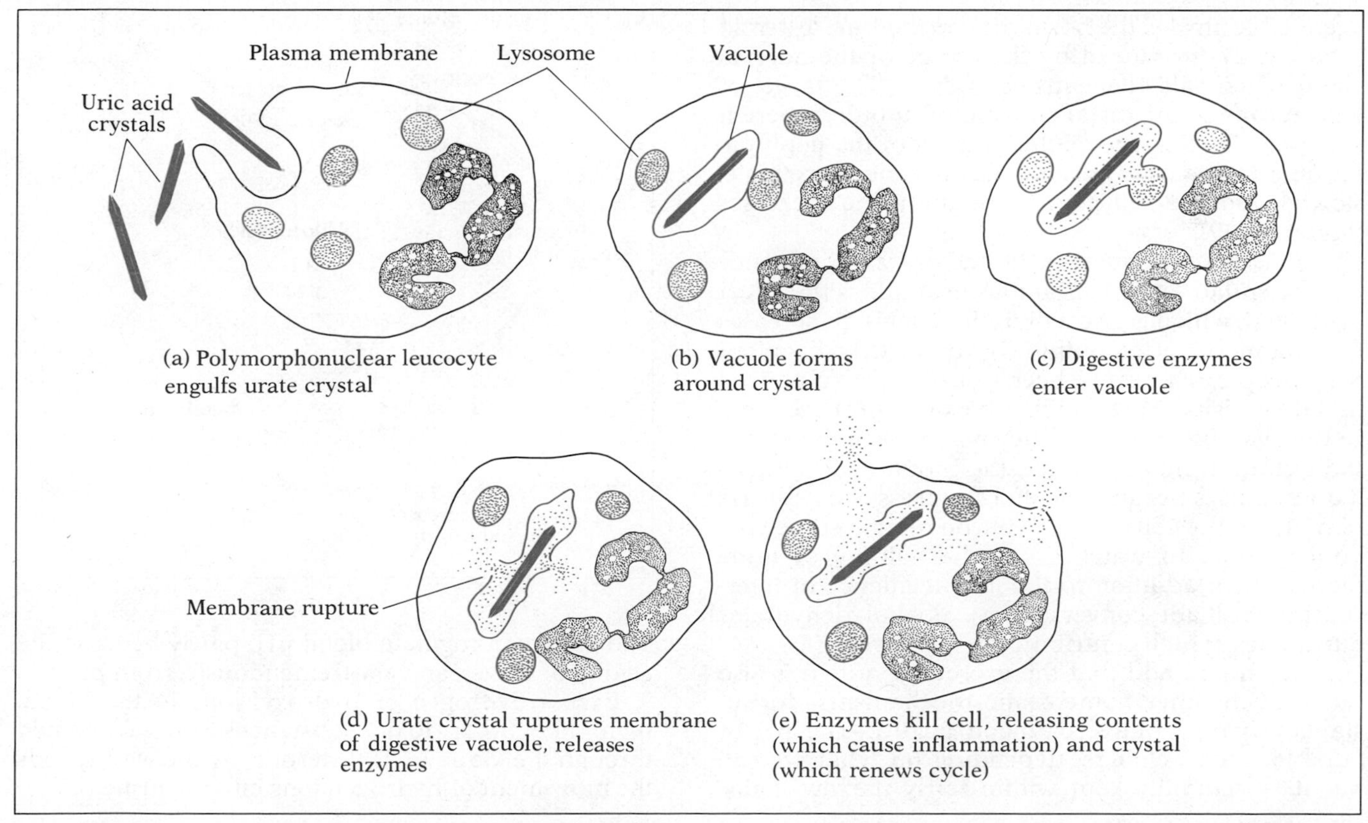

(a) Polymorphonuclear leucocyte engulfs urate crystal

(b) Vacuole forms around crystal

(c) Digestive enzymes enter vacuole

(d) Urate crystal ruptures membrane of digestive vacuole, releases enzymes

(e) Enzymes kill cell, releasing contents (which cause inflammation) and crystal (which renews cycle)

chovies, herring, veal, fowl, meat in general, caviar, wine, and beer. The wine and beer get their pyrimidines from yeast, which is a particularly rich source of RNA. In former times, when only the wealthy ate much meat, gout was a disease of rich men and alcoholics, and especially rich male alcoholics.

If you get gout, your doctor will put you on a diet that excludes meat, fowl, and fish, as well as wine and beer. You will be able to get your animal protein from eggs and milk, however, and will be perfectly free to indulge in distilled spirits.

If this account seems sexist, it's the sexism of nature. Women hardly ever get gout, although they may have the genetic constitution for it. An ancient herbal remedy, and still the medicine of choice for severe attacks, is the drug *colchicine*. This is a mitotic poison; it stops dividing cells in their tracks. It isn't clear just how this reduces the disease symptoms; perhaps it is a matter of holding down the proliferation of immune system cells.

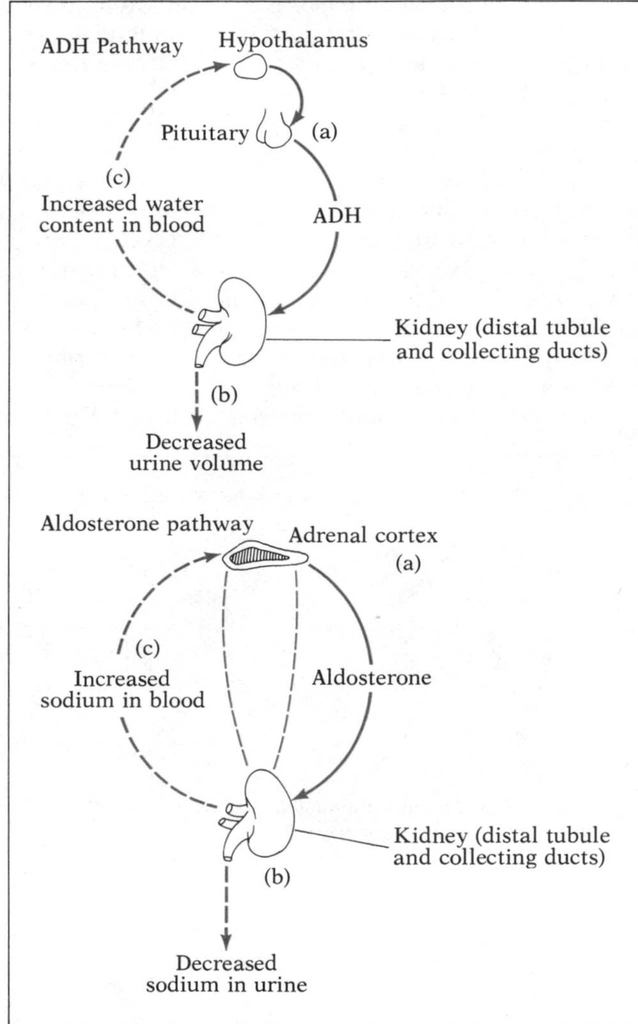

28.19 Two hormones—antidiuretic hormone (ADH) from the pituitary and aldosterone from the adrenal cortex—influence kidney function. When the hypothalamus senses a decrease in blood water content, its neurons to the posterior pituitary release ADH, which is then secreted into the blood (a). Its target is the distal convoluted tubule, where it renders the cell membranes more permeable to water, permitting more water to reenter the blood, thus decreasing urine volume (b). The increased water content of the blood then influences the hypothalamus, which slows the pituitary's release of ADH (c). On the other hand, aldosterone is released when cells in the adrenal cortex sense that sodium levels in the blood are lower than normal (a). The target cells are again those of the distal convoluted tubule, which respond by permitting more chloride to leave the tubule and reenter the blood (decreasing sodium and chloride ions in the urine) (b). The feedback loop is direct, with increased sodium chloride levels in the blood inhibiting the release of aldosterone (c).

agent of control is the hormone *aldosterone*, a steroid (Chapter 27) produced by the cortex of the adrenal gland when salt concentrations drop. The target of aldosterone is the distal convoluted tubule, where it increases the passage of chloride out of the nephron. When excess salt is present, aldosterone secretion is slowed and the salt content of the urine increases (Figure 28.19).

Of course, a number of factors can influence water regulation. One infamous example is the effect of alcohol, which is also a diuretic. That is, it increases the volume of urine output. If you've ever been out with people who drink beer, you may have noticed that their visits to the facilities become more frequent as the evening wears on. They may have noticed that the output seems to exceed the input. And perhaps it does. This is because alcohol suppresses the release of ADH, so the cells of the nephron remain somewhat impermeable to water. Thus, the volume of urine increases. In addition to the inconvenience of interrupting brilliant conversations, alcohol dehydrates our tissues, which contributes to hangovers.

We should add that the excretory system is also involved in other homeostatic mechanisms—for instance, in the control of blood acidity. The pH of our blood can change, depending on what we eat, but it is generally kept within fairly narrow limits. The body can regulate blood pH, partly because the acidity of urine can vary tremendously, from pH 4 to 9. Excess hydrogen or hydroxyl ions in the blood, then, like other toxic substances, are eliminated through the kidneys. Aldosterone, by the way, speeds the movement of hydrogen ions into the urine.

28.3 The Immune System

Homeostasis is the tendency of the body to maintain a constant internal environment in the face of insults from the external environment, and there is nothing quite so insulting as the invasion of foreign organisms. The invading microorganisms have defenses and offensives of their own, including devastating toxins and enzymes. However, some animals have a number of lines of defense against these invaders.

All animals have in common one very effective line of defense: ameboid cells that engulf and eat any foreign particle, including microorganisms. Some of the engulfing cells are fixed in place, for instance in the spleen and the lymph nodes, and simply engulf anything that passes their way. Other defensive cells roam about freely, as if they were on patrol, which, in a sense, they are. Human phagocytes come in two sizes: the huge *macrophages* and the somewhat smaller *neutrophils*. (The macrophages are sometimes called *polymorphonuclear leukocytes*.) The patrolling macrophages slip in and out of the bloodstream and wander everywhere, but they quickly congregate at the site of a wound or lesion, apparently drawn by a substance released from damaged platelet cells. If there is a sizable infection, arriving phagocytes will literally eat themselves to death, engulfing and digesting the resisting bacteria until the accumulation of toxic products kills them. The accumulation of dead phagocytes is called *pus*.

The neutrophils are involved in a more complex secondary defense; the *immune response*. The immune response actually requires the interaction of three different cell types: the neutrophils and two kinds of *lymphocytes*, the *B-cell lymphocytes* and the *T-cell lymphocytes*. The lymphocytes all look alike; they are small white blood cells that are filled with a large nucleus. The B and T stand for the sites in which these cells develop—*bone marrow* and the *thymus gland*, respectively.

The lymphocytes produce *antibodies*, which are protein molecules that have the amazing ability to recognize and to bind to "foreign" molecules. A foreign molecule that induces antibody formation is called an *antigen*. The B-cells produce circulating antibodies that drift freely in the blood, and the T-cells have antibodies that are firmly bound to the cell membranes. Before we delve into the mysteries of how the three cell types cooperate to produce the immune response and how antibodies are manufactured, let's consider what the antibodies look like and how they work.

There are actually a number of different kinds of antibodies, but the one most studied is *immunoglobulin G* (IgG), which is a four-part protein, consisting of two identical heavy (long) chains and two identical light (shorter) chains. The light chains have about 220 amino acids each and the heavy chains contain about 450, so the whole protein has about 1340 amino acids. The four chains are bound together by disulfide linkages (Figure 28.20). In some classes of immunoglobulins, small polymers of three or five subunits are formed.

Each of the chains is divided into two functional regions: a *constant region* and a *variable region*. The constant regions of both the light chains and the heavy chains are the same in thousands of different IgG antibodies, but the variable regions are specific to different antigens, so each antibody attacks only a certain kind of substance. The variable regions of the light chain and the heavy chain together form a cleft (*binding site*) that binds tightly to the surface of at least some part of the antigen. Since the molecule is symmetrical, with two identical specificity regions, two antigens can be bound simultaneously by the same antibody, which then forms a bridge between them. In turn, each antigen can be attacked by several antibodies, which thus cause the foreign molecules to coagulate into great clumps (Figure 28.21).

The antibodies can do a number of other things, partly because once an antibody–antigen complex is formed, the characteristics of the rest of the antibody molecule change. Thus, antibodies bound to the surface of a foreign cell induce rapid phagocytosis by phagocytes. The antigen–antibody complex also binds (and activates) a group of circulating blood plasma proteins that together are called *complement*, and the activated complement ruptures the cell membrane of the invading cell. In binding to the active sites of foreign enzymes and toxins, the antibodies can quickly inactivate them.

Antibody–antigen complexes also may bring about other, more indirect effects, sometimes employing hormones. Or at other times they may release *pyrogens*, which produce fever. And they may cause local inflammation at the site of an infection.

The membrane-bound antibodies of the T-cell lymphocytes can react with foreign antigens in much the same way as those produced by B-cells. But, in addition, the T-cells apparently have the power to lyse and to phagocytize. They can also attack the body's own cells when these cells have been infected with viruses, killing the infected cells in order to protect the organisms as a whole. The T-cell lymphocytes also recognize and destroy incipient cancer cells and are the body's first line of defense against spontaneous cancers.

Unfortunately, as we mentioned briefly in our discussion of the parathyroid glands, even this excel-

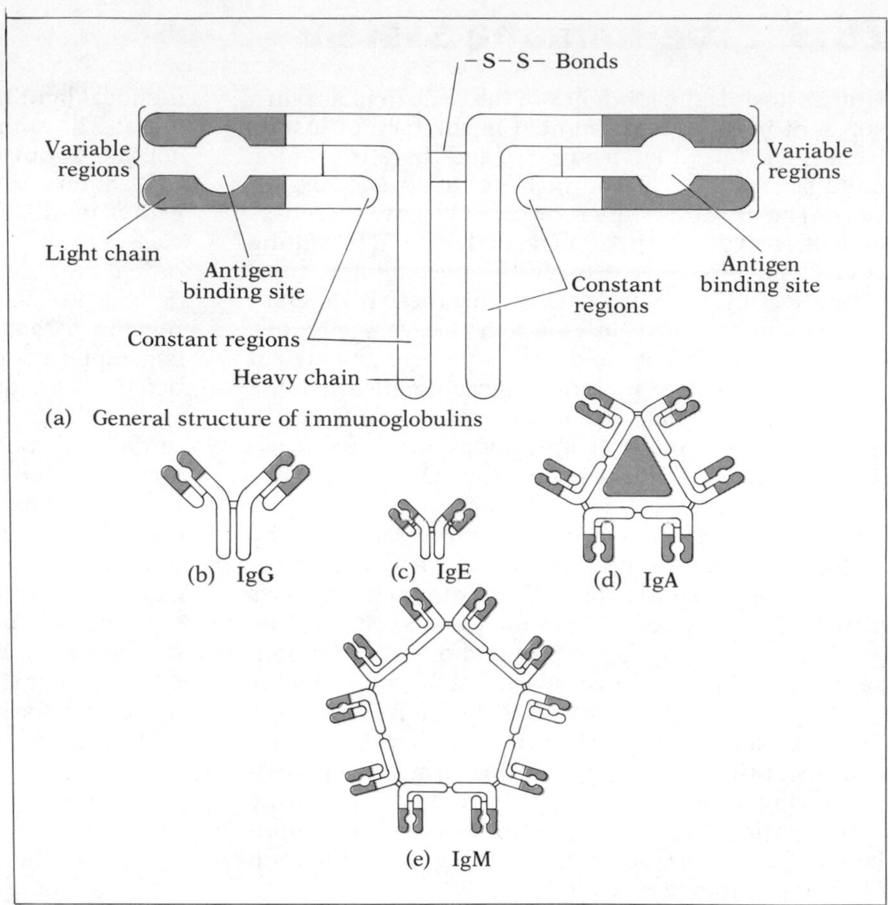

28.20 (a) Immunoglobulins are composed of two identical heavy chains and two identical light chains, connected by disulfide linkages. Each of the four chains has a constant region and a variable region. The variable regions enable the antibodies to become highly specific for a certain antigen. (b) The most abundant immunoglobulin is immunoglobulin G (IgG). It participates in secondary responses against infections. (c) IgE is very similar in structure to IgG. It is found in the respiratory passages and causes the familiar allergic response of hay-fever sufferers. (d) IgA consists of three immunoglobulin molecules that form a triangular structure. IgA also contains a transport polypeptide (dark triangle) that allows it to cross cell membranes. It is used in a first line of defense, stopping invaders before they gain a foothold in the body. IgA is present in mucus, mother's milk, saliva, and tears. (e) IgM, the largest of this group, is composed of five immunoglobulins arranged in a pentagon. It constitutes a first line of defense, since it is produced during the initial infection, although it is short-lived and is replaced in secondary responses by IgG. The light chains of the four immunoglobulin types may be the same, but each has a different type of heavy chain.

lent defense system sometimes goes awry. A line of lymphocytes may begin reacting to one of the organisms own proteins or tissues as though it were a foreign invader. Among the many known or suspected autoimmune diseases are arthritis, nephritis, rheumatic fever, various hormone disorders, and possibly diabetes and schizophrenia as well.

One curious thing: Apparently, each lymphocyte is fully committed to producing one, and only one, kind of antibody. Most of the circulating cells are terminally differentiated, and won't undergo any more mitoses. In the thymus gland and the bone marrow, however, cells that are already fixed for the production of a specific antibody will undergo repeated mitoses, producing clones of identically differentiated cells.

How Are Specific Antibodies Produced?

If you inject a rabbit with crocodile hemoglobin, the rabbit will make hundreds of different antibodies against crocodile hemoglobin. Then, if you inject alligator hemoglobin, the rabbit will make more antibodies that are specific for alligator hemoglobin. All mammals and birds, and apparently all reptiles (perhaps all vertebrates), have this ability to make antibodies against any large molecule (even completely artificial laboratory-synthesized molecules). So how does the immune system do such a trick? Can rabbits be *preprogrammed* to resist crocodile proteins? We do know one rather startling thing: the two ends of each antibody polypeptide—the constant region and the variable region—are coded for by three different genes from three different parts of the chromosome. During lymphocyte maturation, the DNA of the chromosome itself is permanently broken apart and rearranged to create a new gene. Since there are many ways to break and rearrange the chromosome, any one of a vast array of possible new antibody genes can be created in a given lymphocyte. You may recall that we've said that all the cells of the body have the same genetic information. Actually, a correction is in order. You can now see that each clone of differentiated lymphocytes has a pair of tailor-made genes that other cells don't have.

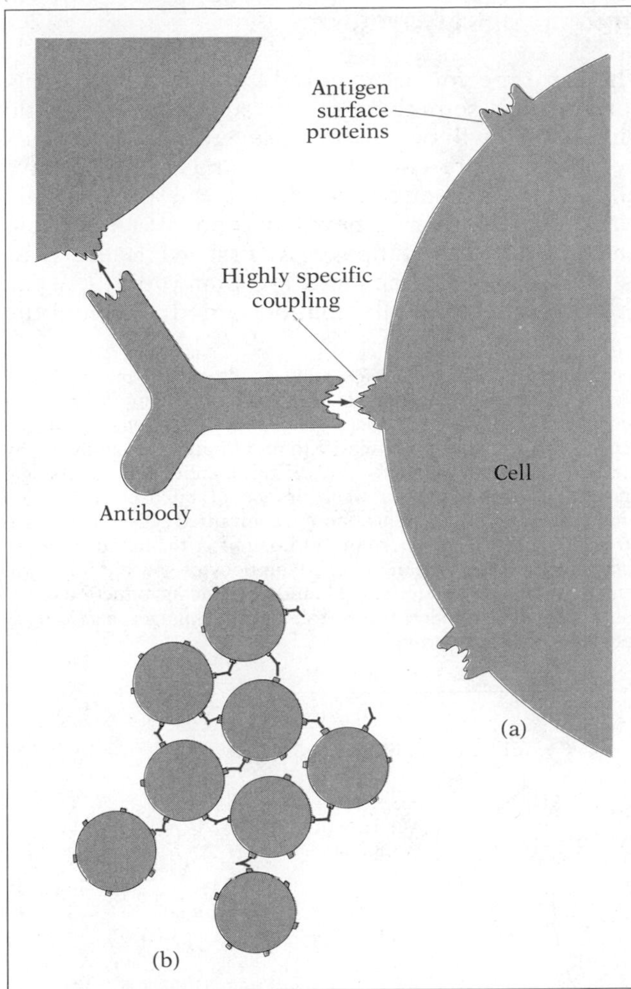

28.21 Clumping of cells by antibodies. (a) Each antibody has two identical antigen binding sites. Each cell has numerous surface antigens. The antibodies can form bridges between cells when the two binding sites interact with the antigens of different cells. (b) A network or clump is formed. Similarly, large antigen molecules can be precipitated from a solution by double-ended antibodies that bind them into a large, insoluble mass.

There is still a great controversy over how specific antibodies are produced (such as the ones against crocodile hemoglobin), but one leading notion is the *germ-line theory*. This simply states that all the possible variable regions of the antibody are coded in the DNA, just like the genes for every other protein. We have already found a few dozen "variable region genes" that direct both light-chain and heavy-chain variable regions, and there may well be thousands of such genes. Each variable region is made up by the splicing together of two different pieces of DNA, allowing for many permutations and combinations. Since each antibody molecule is formed from one heavy chain and one light chain, a thousand genes of each type would give a million possible combinations. And it is always possible that an antibody that is specific for crocodile hemoglobin might also react with other antigens, such as mayfly alcohol dehydrogenase or panda bear saliva.

The Clonal Selection Theory

Having the information coded in the DNA is, at best, only part of the answer, because there is still an enormous information retrieval problem. If the genes are there, how are they activated? Here, the leading explanation is the *clonal selection theory*.

According to this theory, the lymphocytes become specialized to react to specific antibodies early in embryonic development, long before they could have encountered the foreign antigens. Even at this stage, the embryo has billions of lymphocytes, and they are already busy manufacturing all the possible kinds of antibodies that the organisms will ever be able to produce.

Recognizing Self

Later in embryonic development, the theory goes, the lymphocytes begin to move through the bloodstream, getting to be familiar with their parent organism. During this stage, the antibodies are fixed to the surfaces of the lymphocytes. At this stage, the lymphocytes are extremely "touchy" and react strongly (perhaps overreact) if their one kind of surface antibody should happen to be triggered by some large molecule it might encounter. In fact, the lymphocytes seem to be so shocked at encountering antigens that they die. Actually, their death is beneficial to the organism since the only kind of macromolecule the lymphocytes are likely to encounter are of its own organism, and lymphocytes that react against the body's own molecules aren't wanted.

Evidence for this sensitive stage comes from experiments in which rat embryos or newborn guinea pigs are injected with a foreign antigen, whereupon these same rodents, when adult, are completely unable to make antibodies against the same antigen—their cells act as though they recognize the antigen as "self."

Clonal Selection

After their tour of the embryonic bloodstream, the lymphocytes end up in the thymus or bone marrow, where they live as a sort of military reserve unit. Now, their reaction to large molecules is different.

They are able to recognize "self" and to react against anything else. At this time, if their surface antibodies become bound to any macromolecule, the cells don't die—they divide. In fact, they undergo several rounds of replication, increasing the size of their clone. The small army continues to swell as it attacks the invaders. Ideally, the invading molecules are eliminated. If these cells don't encounter any more of this specific antigen, they will return into semiretirement; but they are ready for the next time. And the next time, there will be more of them (Figure 28.22).

Immunological Memory

The first time you are exposed to a specific antigen—for instance, something on the surface of a new flu virus—there will be a delay of several days to a week before there are enough B-cells and T-cells to have any measurable effect. So during this time, the flu virus has a wonderful new environment all to itself, and it goes on a rampage. At first, while the virus is multiplying, the chances are that you'll feel just fine. But, finally the incubation period ends and the

28.22 The *clonal selection hypothesis* holds that (a) the stem cells, which are precursors to all blood cells including the antibody-producing *lymphocytes,* contain the genetic information for making all possible antibodies. (b) At some point in embryonic development, rearrangements of DNA in the differentiating lymphocytes produce a pool of thousands (or perhaps hundreds of thousands) of different lymphocyte cell lines, each with a new, different immunoglobulin gene and the ability to produce one and only one unique immunoglobulin antibody. Such committed lymphocytes will carry their unique antibodies on their cell surfaces but will remain quiescent unless stimulated by an alerted macrophage. (c) In the primary immune response, patrolling macrophages phagocytize an introduced virus, bacterium, protein, or other macromolecule and recognize it as being a *foreign* antigen. (d) Alerted macrophages then make cell-to-cell contacts with numerous lymphocytes. (e) Any lymphocyte that produces an antibody that reacts with the foreign antigen will be stimulated to proliferate. (f) Usually, many different cell lines respond with different antibodies to a single antigen. It may require a week or two of intense lymphocyte proliferation before there are enough committed cells to neutralize an invading virus or bacterium. (g) Long after the initial response to a foreign antigen, many of the lymphocytes specific for it will remain in the lymphoid tissue. Should a second introduction of the same antigen again alert the patrolling macrophages, a *secondary response* will occur within hours.

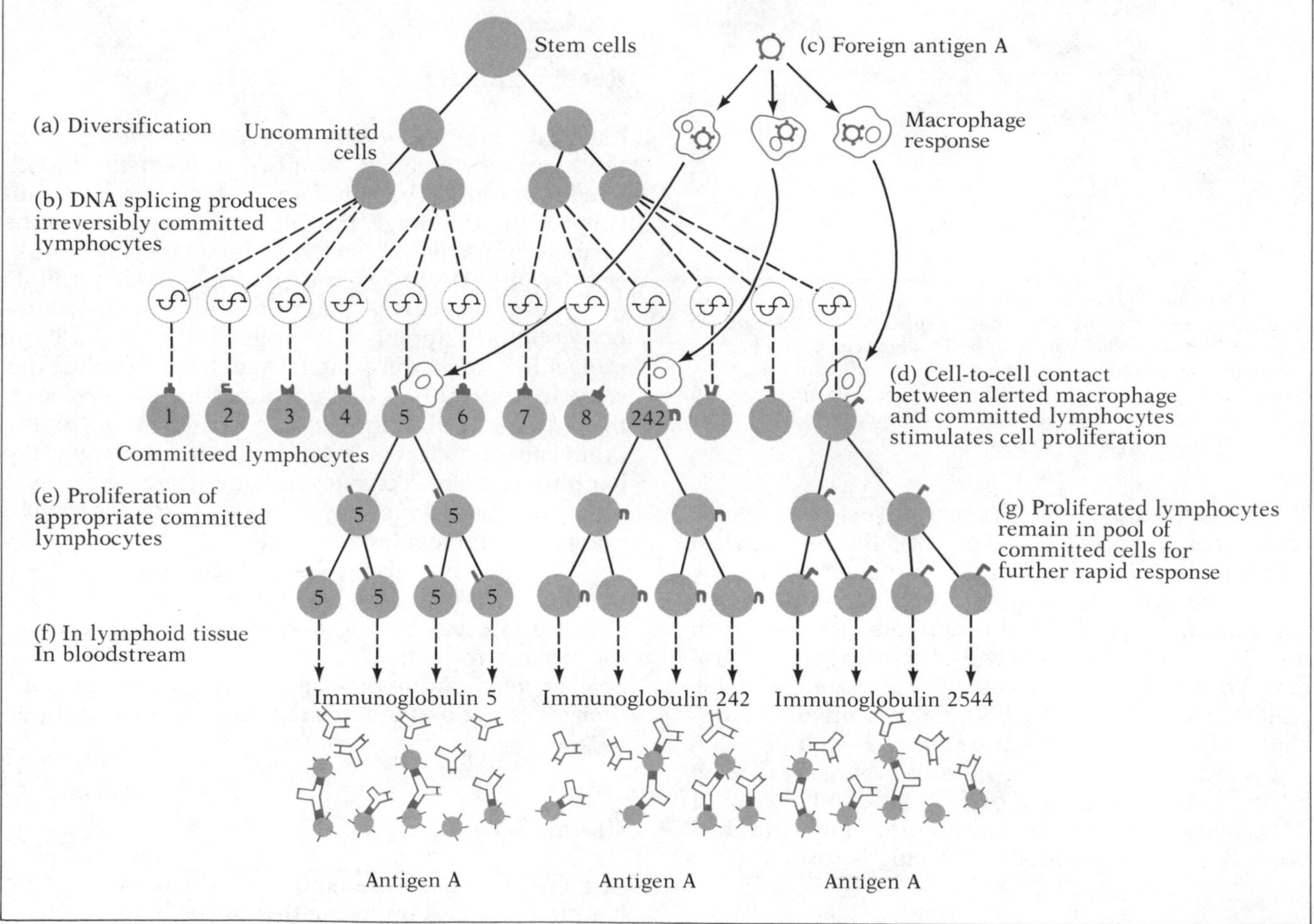

symptoms begin. Flu shouldn't be taken lightly, because the virus would almost certainly kill you if it weren't somehow promptly stopped.

And stopped it is, as the clones of specific antiviral lymphocytes grow, by successive doublings, into a large army. The flu viruses become coated with antibodies and they begin to clump together, to be engulfed by phagocytes. During this time, you will experience the fever, nausea, and inflammation that are the unpleasant signals of a successful immune response.

So the first time you will be sick. But the next time the same flu virus comes down your windpipe, it won't get to first base, because enough remnants of that army of lymphocytes and circulating antibodies will be waiting so that only a few rounds of replication will reconstitute the former army. The problem is, there are many kinds of flu viruses, and they evolve rapidly, so another flu virus might breach these defenses and make you sick again. But, in many cases, once you've had one specific infectious disease, you ordinarily won't be subject to that same disease again in this lifetime.

Some disease organisms are particularly insidious in that their surface antigens are sufficiently like those of your own body that the invaders are recognized as "self," which lets them slip by your defenses. Eventually, however, they will be detected and your immune system will come up with antibodies that will knock out these imposters, but unfortunately, the antibodies created may cross-react with your own tissues, causing a severe autoimmune reaction. The *Streptococcus* bacteria that cause "strep throat" are notorious for this. An infection in your throat creates antibodies that can attack your own tissue elsewhere, notably in the kidneys or heart valves, with serious and sometimes fatal results.

So we have seen that life is organization in the face of a disorganizing universe. Molecules must be brought together and caused to react in very specific ways if life is to exist. In order for such reactions to occur, conditions must be kept within rather rigid limits, and thus living things have developed innumerable mechanisms that help to ensure such regulation. Some regulation is obvious and easily measurable, but life works in subtle ways as well. There are multitudes of complex feedback loops, some of which we believe we understand. Such understanding, however, has left us prepared to acknowledge how much we really don't know about regulatory systems. So, in the laboratories, deserts, and oceans of our planet, the search goes on.

Application of Ideas

1. Many of the animal body functions are apparently automated through negative feedback control. What would be the alternative to this kind of control? What are the adaptive advantages of negative feedback control?

2. Homeostatic mechanisms involve far more than thermoregulation and osmoregulation. Describe examples of homeostasis in the digestive, respiratory, circulatory, and endocrine systems of animals.

3. Aside from the problem of cross-matching tissue types, the problem of kidney transplants is simple. All that is required is a place for the transplant and good arterial, venous, and ureter connections. Explain how the kidney can function properly with only the blood communicating with the rest of the body.

4. Is thermoregulation in birds and mammals more likely to fail under heated conditions or under cold conditions? Explain your answer.

Key Words

aldosterone
ammonia
antibody
antibody–antigen complex
antidiuretic
antidiuretic hormone (ADH)
antigen
autoimmune disease

basking
B-cell
behavioral thermoregulation

Key Ideas

Introduction

1. *Homeostasis* can be defined as the tendency of a physiological system to maintain internal stability.

2. *Homeostatic mechanisms* work on short-term and long-term bases.

A Closer Look at Negative Feedback Loops

1. In *negative feedback* a stimulus produces a reaction that ultimately reduces the stimulus.

2. In *positive feedback* the stimulus evokes a response that further increases the stimulus.

bladder
Bowman's capsule

clonal selection theory
collecting duct
complement
conduction
constant region
convection
cortex
countercurrent multiplier

distal convoluted tubule

excretion
excretory system

glomerulus

homeostasis
homeostatic mechanism
homeothermic
hypertonic
hypothalamus
hypotonic

immune response
immunoglobulin G
isotonic

kidney

loop of Henle
lymphocyte

macrophage
Malpighian tubule
medulla

negative feedback
nephridiopore
nephridium
nephron
neutrophil

osmoconformer
osmoregulation
osmoregulator
osmotic gradient

phagocyte
piloerection
poikilothermic
positive feedback
proximal convoluted tubule
pyramid
pyrogen

radiation
renal artery
renal pelvis
rete mirabile

3. In general, *positive feedback* is associated with illness or abnormalities in organisms.

4. *Homeostatic mechanisms* have their limits. These stress points, or *thresholds*, are reached in environmental extremes.

5. Some organisms specialize in coping with environmental extremes, which limits the competition from other species.

6. Most organisms evolve the ability to withstand extremes to the probable limit they will encounter.

Thermoregulation
Animals are categorized as *poikilothermic* (cold-blooded) or *homeothermic* (warm-blooded), but there are no real dividing lines, and many *poikilotherms* have physiological or behavioral mechanisms for retaining or losing heat.

Why Thermoregulate?
1. At lowered temperatures an organism can either go into a metabolic stupor or use *metabolic heat* to maintain itself at a constant level.

2. Heat loss can become a vicious circle. Loss of heat can slow the metabolic processes that produce heat.

3. *Thermoregulation* involves both warming and cooling.

The Countercurrent Heat-Exchange Mechanism, or How Tuna Swim Fast
1. *Countercurrent multipliers* assist in retaining heat in essential regions. The system operates when veins and arteries run closely parallel. Heat is exchanged between blood flowing in opposite directions. The result is cooler extremities, but no drastic change inside.

2. Fast-moving, coldwater fish use the *countercurrent exchange* system. The darker muscle contains an immense network of blood vessels, the *rete mirabile*, which is one large heat exchanger.

Regulating Behaviorally
1. Reptiles bask to gain heat and seek shadow to resist heat gain. It is all in orienting their bodies in the most strategic way. The horned lizard uses pigment changes to reflect or absorb heat.

2. The large vascular plates of *Stegosaurus* are believed to have acted as heat radiators and possibly as heat absorbers.

Thermoregulation in the Homeotherms
Homeotherms maintain their internal temperatures within a narrow range. They vary their *metabolic heat* production and control heat loss and gain in close harmony.

Removal of Heat Energy
1. Heat is lost through *radiation, conduction,* and *convection.* Evaporation enhances heat loss and can be considered a fourth mechanism.

2. The skin is important in *thermoregulation.* Blood can be shunted into the skin to lose body heat or away from the skin to retain heat.

Conserving Heat
Mammals conserve heat through behavior, countercurrent systems, *vasoconstriction, piloerection,* and insulating fur and fat. They actively produce heat by metabolic processes, sometimes enhanced by shivering.

The Internal Source of Thermoregulation
1. Humans measure temperature through sensory neurons in the skin. Voluntary responses to warmth or cold are numerous.

2. The hypothalamus is extremely sensitive to temperature changes. It responds through the endocrine system by instigating changes in the metabolic rate and through the sympathetic nervous system by opening or closing capillary beds in the skin.

salinity

T-cell
thermoregulation
tubule

urea
ureter
urethra
uric acid

variable region
vasoconstriction
vasodilation

A Closer Look at How the Hypothalamus Works

The *hypothalamus* may be measuring heat mediated changes. We know that it responds to chemicals known as *pyrogens*. These increase calcium in the blood, which changes the calcium–sodium balance. Temperature can be altered by perfusing the blood with calcium, so this may be what is really sensed by the *hypothalamus*.

Osmoregulation and Excretion

Osmoregulation is the management of water and ions in the body. The excretory system in many animals is involved in both *osmoregulation* and nitrogen waste elimination.

Producing Nitrogen Wastes

1. Nitrogen wastes are produced when amino acids are deaminized (amino group removed). The three nitrogen wastes produced by animals (and protists) are *ammonia, urea,* and *uric acid.*
2. *Ammonia* can be safely eliminated by animals as long it is rapidly removed.
3. Many invertebrates, some fish, and the mammals produce *urea,* while insects, reptiles, and birds produce *uric acid* and other purines.
4. The waste produced is related to the environmental restrictions. Where water is a premium item, *uric acid* is favored. Ridding the body of *urea* requires a constant water supply, but not as much as is required for *ammonia*.
5. Bird and reptile embryos store *uric acid* crystals in the egg during development. There is no way in which *ammonia* or *urea* could be safely removed.

The Osmotic Environment

Each environment presents its own specific *osmoregulatory* problems.

The Marine Environment

1. In the marine environment animals tend to lose water and gain salt.
2. Marine organisms *osmoregulate* in two major ways.
   a. The *osmoconformers*: Their solute concentration changes with the environment. They stay *isotonic.* Many marine invertebrates are *osmoconformers.*
   b. The *osmoregulators*: The solute concentration remains constant in spite of the environment. The regulator must resist water loss and salt entry. The gill is often involved in actively removing salt.

Strategies of Marine Vertebrates

1. Sharks and rays have different strategies than bony fishes.
2. Bony fishes constantly drink seawater, but actively transport salt out through their gills. Their nitrogen waste is *ammonia,* which also leaves via the gills.
3. Sharks and rays are *osmoconformers* in that they retain *urea* in their blood in an *isotonic* condition to seawater. They do however, actively transport salt out through a gland that opens into the rectum.
4. Marine birds actively excrete salt through glands near the eyes, while marine mammals use their *kidneys* to excrete salt.

The Freshwater Environment

1. Fresh water presents other problems in *osmoregulation.* Water tends to enter the body and salts or ions to leave. Active transport mechanisms are used to restore ions.
2. Nitrogen waste disposal is no problem for many since there is plenty of fresh water to dispose of *ammonia* or *urea.*

Vertebrates of the Freshwater Environment

1. Fish and amphibians have efficient, water-removing *kidneys.* Little water is reabsorbed as is the case in terrestrial animals.
2. The frog excretes *ammonia* as a tadpole and *urea* as an adult.

The Terrestrial Environment
Like marine animals, terrestrial animals must overcome water loss. They take in large amounts and have a water-conserving *excretory system.*

Invertebrates of the Terrestrial Environment
1. The earthworm utilizes the *nephridium* to remove wastes and conserve water. Each segment has a pair in its coelomic cavity. Fluid is swept into its ciliated funnel, moves along and has water and other vital substances reabsorbed, finally forming a waste that is passed out through the *nephridiopore.*
2. Terrestrial arthropods prevent water loss by growing a waxy cuticle and producing solid, *uric acid,* nitrogen waste.
3. Insects rely on metabolic water. Their excretory system consists of *Malpighian tubules,* which gather wastes from the coelom and pass *uric acid* crystals to the gut. Little fluid is needed for nitrogen waste disposal.

Vertebrates of the Terrestrial Environment
Humans and other terrestrial mammals use copious amounts of water to rid the body of *urea* and to maintain a water balance. The *kidney* is an efficient conserver of water and minerals.

The Human Excretory System
1. The excretory system consists of paired *kidneys, ureters,* the *bladder* and *urethra,* and the blood supply. Intake of water is usually balanced by water output in the urine, sweat, feces, and exhaled air.
2. Each *kidney* consists of the following:
   a. *Renal cortex, medulla,* and *pyramids*
   b. *Nephrons* containing a *capsule* and *tubule*
   c. Related blood vessels (arteries, capillaries, veins)
3. Urine is formed in the *nephrons* and enters the *pyramids,* which empty into the *renal pelvis* and then the *bladder* via the *ureters.* The *bladder* is emptied through the *urethra.*
4. The *nephron* begins with *Bowman's capsule,* which surrounds the vascular *glomerulus,* a capillary bed. The hollow capsule is continuous with the *proximal convoluted tubule,* the *loop of Henle,* the *distal convoluted tubule,* and the *collecting duct.* Each duct receives the *tubules* from many *nephrons.*
   a. *Bowman's capsule*: The *glomerulus,* a ball of capillaries with blood under considerable pressure, loses one-fifth of its *plasma* volume as water, *urea,* glucose, amino acids, salt, proteins, and vitamins.
   b. *Proximate convoluted tubule:* Reabsorbs by active transport much of the filtrate. Water, salt, and *urea* move along in the *tubule.*
   c. *Loop of Henle*: In the hypertonic cortex, water diffuses out of the descending limb of the loop, and salt diffuses in. In the ascending limb, chloride ions are actively pumped out again. Sodium ions follow the chloride ions out. A high concentration builds up outside the nephron. The chloride and sodium ion concentration around the loop causes water to leave the *tubule* osmotically.
   d. The *distal convoluted tubule*: Water leaves, moving down the gradient outside the *tubule.*
   e. *Collecting duct*: More water leaves the *nephron* here. The remainder is urine (water, *urea,* excess salts).
   f. Capillaries of the *nephron*: Blood leaving the *glomerulus* is dense. The vessel forms another capillary bed around the *nephron* and as fast as substances leave the *tubule* they diffuse or are actively transported into the blood.

Control of Nephron Function
1. Control is through hormones of the *hypothalamus,* anterior pituitary, and adrenal cortex.
2. Water balance:
   a. The *hypothalamus* senses blood density. When water is deficient, releasing factors travel to the pituitary, which responds by secreting *ADH.*

b. The target cells of *ADH* are in the *distal tubule* and *collecting ducts*. They respond by permitting more water to pass back into the blood.
c. When blood consistency is correct, this is sensed by the *hypothalamus* and and it stops its releasing factors. The system shuts down by *negative feedback*.

3. Sodium balance:
a. The adrenal cortex senses the sodium content of the blood and when it falls it releases *aldosterone*.
b. The target of *aldosterone* is the cells of the *distal tubule* and the *collecting ducts*. They respond by transporting more salt out of the *tubule* for reentry in the capillaries.
c. When the sodium level is up, the system shuts down.

4. Substances like alcohol and caffeine are diuretics: they depress *ADH* production and thus increase the urine output.

5. The *excretory system* also helps manage the blood pH by altering the acidity of the urine. A third effect of *aldosterone* is to enhance the loss of hydrogen ions.

The Immune System

1. The immune system fights invading disease organisms by destroying the invaders and by developing resistance.

2. Circulating *macrophages* and *neutrophils* move in and out of the circulatory system engulfing invading organisms. They are attracted by substances released by platelets.

3. *Neutrophils* are involved in the *immune response*, along with *B-cell lymphocytes* and *T-cell lymphocytes*.

4. *Lymphocytes* produce *antibodies* that are highly specific proteins. They bind to specific foreign bodies. The *B-cell antibodies* circulate, while those of the *T-cells* are bound to cell membranes.

5. *Antibodies* are induced by foreign proteins called *antigens*. One *antibody*, *immunoglobulin G*, is composed of two heavy and two light amino acid chains. Each chain has a constant and a variable region. The variable feature permits *antibodies* to be highly specific for an *antigen*.

6. When *antibodies* bind to *antigens*, they coagulate into a clump which can be engulfed by *phagocytes*.

7. *Antigen–antibody complexes* bind and activate blood plasma proteins known as *complement*. It ruptures the invading cells releasing its proteins and toxins, which are then inactivated by other *antibodies*.

8. *T-cell lymphocytes* can form *antibodies* as well, but in addition they can phagocytize infected body cells. At times this defense goes astray and *lymphocytes* can destroy healthy body cells. This is called an *autoimmune* disease.

9. *Lymphocytes* are specific, producing one kind of *antibody*. Thymus and bone marrow cells produce clones that are specific for producing one kind of *antibody*.

How Are Specific Antibodies Produced?

1. Animals can produce *antibodies* against large molecules such as proteins and polysaccharides.

2. *Antibody* production may be preprogrammed. It is known that the constant and variable regions of each *antibody* are coded by three different DNA segments.

3. The germ-line hypothesis explains how specific *antibodies* are produced. It states that all possible variable regions of *antibodies* are coded in DNA. (Some of these genes have been located.)

The Clonal Selection Theory

1. This produces an enormous information retrieval problem. We aren't sure how these specific, variable strand genes are activated. One answer is the *clonal selection theory*.

2. This theory states that *lymphocytes* become specialized in *antibody* production for all foreign *antigens*. They are just waiting for *antigens* to arrive.

Recognizing Self
The first confrontation of these specialized *lymphocytes* is with the body's own molecules. As each kind of molecule is encountered, the *lymphocyte* dies, eliminating undesired responses.

Clonal Selection
1. The differentiated *lymphocytes* move into thymus or bone marrow where they await the invasion of foreign molecules.
2. When they confront an *antigen*, they bind to it and divide several times. When the invader is destroyed, the specific *lymphocytes* have produced a large number of descendants to await the next invasion.

Immunological Memory
1. During the first exposure to an *antigen* there is a delay while *B-cells* and *T-cells* identify the *antigen* and produce *antibodies* and more cells with that capability. Once this is underway, the *antibody–antigen response* occurs with the invaders being engulfed and destroyed.
2. The second experience with the *antigen* is much faster since there are more differentiated *lymphocytes* to initiate the response.
3. Some invaders produce *antigens* that are similar to the host's. This can induce an *autoimmune* reaction. An example is strep throat where defending *lymphocytes* mistakenly destroy innocent tissues of the kidneys or heart.

Review Questions

1. List an example of a short-term and a long-term homeostatic response. (p. 798)
2. How do *negative feedback loops* differ from those that are positive? Which is prevalent in nature? (p. 798)
3. What is meant by the statement, "each (*homeostatic*) *mechanism* has some kind of cost"? (p. 800)
4. Define the terms *poikilothermic* and *homeothermic*. (p. 800)
5. What animals are considered to be the best thermoregulators? (p. 800)
6. Describe the vicious circle that occurs between metabolic activity and drastically lowered temperatures. (p. 800)
7. How might *countercurrent mechanisms* protect a bird wading in icy water? (p. 801)
8. Describe the *rete mirabile* and explain how it helps the tuna. (p. 802)
9. Describe *behavioral thermoregulation* in lizards. (p. 802)
10. What is the evidence that suggests that *Stegosaurus* was a pretty good *thermoregulator?* (p. 803)
11. What are the two general strategies of *thermoregulation* in mammals? (p. 805)
12. Compare heat *radiation, conduction,* and *convection* as colling devices and give an example of each in humans. (p. 805)
13. Describe the mechanisms in the skin that aid in both heat loss and heat conservation. (p. 806 and Figure 28.8)
14. What are the adaptive values of *piloerection* and of shivering? (p. 806)
15. What is the internal temperature control structure in humans and how sensitive is it? (p. 807)
16. Describe three mechanisms available to the *hypothalamus* to increase body heat. (p. 807)
17. What has the study of *pyrogens* told us about how the *hypothalamus* measures blood temperature? Describe some supporting evidence. (p. 808)
18. Define *osmoregulation*. (p. 808)
19. Describe the process in which nitrogen wastes are formed. (p. 808)
20. Under what conditions can organisms safely excrete *ammonia?* (p. 808)
21. List the three nitrogen wastes and add examples of organisms that produce these. (p. 808)
22. How is *uric acid* production a valuable adaptation in the embryos of reptiles and birds? (p. 808)
23. What special requirement is made of animals that excrete *urea?* (p. 808)
24. Contrast *osmoregulating* with *osmoconforming* and provide examples in marine organisms. (pp. 809–810)
25. How do marine animals handle the problem of salt intake? (p. 810)
26. How does the marine bony fish handle the problem of salt intake and nitrogen waste disposal? (p. 810)
27. Explain the unique way in which sharks and rays survive in their *hypertonic* environment. (p. 810)
28. Describe the problem freshwater animals have with water and inorganic ions. What is one solution? (p. 811)
29. Describe the arrangement, structure, and function of the *nephridia* in the earthworm. (p. 812 and Figure 28.14)
30. Where are the *Malpighian tubules* and how do they handle nitrogenous wastes? (p. 813)
31. List the *organs* of the human *excretory system,* including the major blood vessels, and give a function for each. (pp. 813–815)

32. Starting with the *glomerulus,* list the parts of the *nephron* and summarize the events in each part as urine is produced. (p. 815)

33. Of what value is the hairpin-shaped *loop of Henle?* Explain in detail. (p. 815)

34. What is the osmotic condition in the blood leaving the *glomerulus* and how does this help with the *nephron's* function? (p. 815)

35. Construct a simple diagram that shows how the water content of the blood is regulated. Begin with blood entering the *hypothalamus.* (p. 816)

36. What actually happens to the target cells when *ADH* is present? (p. 816)

37. What is the source of *aldosterone* and how does it effect its target cells? (p. 818)

38. What is the general role of the immune system? (p. 819)

39. Explain how the *macrophages* move about and what they do. (p. 819)

40. What kinds of white cells interact in an *immune response?* (p. 819)

41. Define the terms *antigen* and *antibody.* (p. 819)

42. Describe the structure of the *antibody immunoglobulin G* and explain how many different *antibodies* can be formed from this basic chemical structure. (p. 819)

43. Describe the events in a clumping reaction between *antibodies* and *antigens.* What is the value of this? (p. 819)

44. How does the *complement* help to neutralize toxins found inside a bacterium? (p. 819)

45. What are some of the capabilities of *T-cell lymphocytes?* (p. 819)

46. Describe how an *autoimmune disease* might get started. (p. 820)

47. How is the large number of highly specific *antibody*-producing cells explained in the germ-line hypothesis? (p. 821)

48. According to the *clonal selection theory,* what is the first job of the differentiated *lymphocytes?* (p. 821)

49. What is the evidence for the self-regulation step? (p. 821)

50. Describe the events that characterize the first encounter of a *lymphocyte* with its *antigen.* Contrast it with future encounters. (pp. 822–823)

51. Strep throat is known to be related to defective heart valves even though the bacterial infection doesn't occur there. Explain this according to the *autoimmune theory.* (p. 823)

Chapter 29

Animal Reproduction

As we've mentioned so many times before, the primary advantage of sex is that it permits combination of different types of chromosomes. Thus, sexual reproduction increases variation among offspring. This means our offspring are different from us. We may be very proud of ourselves and think we fit into our world very well indeed. We may even dominate parts of it. Everyone may even agree that we are splendid. However, we are only splendid in our world—the one we live in now. It would be quite presumptuous of us to assume that we would fit into every world with equal ease. And, as we know, even our world changes. Thus, if we were to make replicas of ourselves, each offspring bearing 100% of our genes, they would do well as long as the world remained as it is now, but they might be in a lot of trouble if it should change. In fact, they might not fit a different kind of world well at all and would die out.

But if we mix our genes randomly with someone else's, we will produce a motley crew of offspring—all of them different. Thus, should new conditions arise, at least some of them would be likely to survive and to propel our genes into yet other generations. This, then, is essentially the main argument in defense of sex. The theory is discussed further in Chapter 11.

Even with these advantages, our best theoreticians are still having trouble weighing them against the disadvantage of gene dilution. Obviously, our observations are not complete—or there is something amiss with the theory. But let's hold any judgments in abeyance for now and have a closer look at *how* animals reproduce.

29.1 Sex and Reproduction in Animals

Animals that practice sexual reproduction are faced with the very basic problem of how to get eggs and sperm together. You may think you know a way, but you may be surprised at how chancey an affair sex generally is. Under the best of circumstances, so many things can go wrong that one wonders how fertilization could *ever* occur. The chanciest conditions are found in those species that simply discharge eggs and sperm into the environment and then hope for the best. But that is precisely what many species do. The process is called *external fertilization* and is carried on by a number of aquatic invertebrates as well as some fishes and amphibians. In such cases, the eggs are usually discharged into the water and sperm are shed over them (Figure 29.1).

The Caribbean fireworm is about an inch long and spends its time burrowed in the bottom of bays in the West Indies. But on the fifth night after an August full moon, the worms crawl out of their burrows and swim to the surface of the warm tropical seas. In a remarkable timing feat, they reach the surface after sunset, but about an hour before the moon rises, in the blackness of night, so that nothing can interfere with what they are about to do. Then, for the only time in their lives, they become phosphorescent. The water sparkles and dances with greenish blinking lights. This is their time of reproduction. The males and females blink at different rates and thus they swim toward each other. The water seems to be alive with their shimmering bodies. Then, suddenly, in a tiny phosphorescent explosion, clouds of eggs and sperm fill the water, and the ruptured corpses of the fireworms sink slowly to the bottom, leaving their gametes in a game of chance.

A number of animals play this game of chance, from coelenterates, which depend on water currents to bear their gametes, to salmon, which return from the open sea to the very stream where they were born. There the splendid, colorful salmon mill about in the shallows, releasing eggs and sperm. In some

(a)

(b)

(c)

29.1 External fertilization is a common mode of reproduction in both marine and fresh waters. The risks of releasing gametes into the environment are lessened considerably where some kind of mating behavior is involved. While the more sessile, external fertilizing invertebrate must rely on water currents and huge numbers of gametes to assure fertilization (a), many of the vertebrates show intricate breeding patterns. The female fish must often be courted and induced into egg laying, whereupon the attentive male quickly releases his sperm over the egg mass (b). (c) Frogs and toads often provoke egg laying by clasping and squeezing the female, fertilizing the eggs as they are released. In the frogs and toads this follows a lengthy period of vocal courting.

species, the fish then die. In other species, they swim back to sea to make the journey another time.

Toads leave less to chance as males clasp females and, while pressing his opening against hers, he releases sperm as the eggs are laid. The toads, then, take the first step toward one-on-one (as it were) fertilization. A specific member of the opposite sex is selected and gametes are joined with those of that individual (although males may mate repeatedly). But couldn't there be a better way? Why risk having the sperm washed away in the hostile environment? Why not just put the sperm right where the eggs are held in the female? With the sperm safe in her body there would certainly be less risk of unsuccessful matings. Thus, internal fertilization arose, probably as a result of a constant *selection* for males that were able to get their sperm closer to the eggs that were ready to be fertilized. Finally, males would have developed various structures and behavioral patterns that actually placed the sperm inside the female.

Copulation and Viviparity

Let's see how some species go about getting the male's sperm to the female's eggs inside her body. In almost every case, it involves the process known as *copulation.* In spite of copulation being primarily a device of land dwellers, many aquatic organisms do it. Even the stationary barnacles, like other arthropods, rely exclusively on internal fertilization. In their case, this requires a long, roving and probing penis that can reach the barnacle next door, or even one several doors down (Figure 29.2). It doesn't matter which one, because the barnacle is hermaphroditic. Another aquatic example, the flatworm *Planaria,* is well-equipped for copulation (Figure 29.3). It is also hermaphroditic, so it doesn't have to find a planaria of the opposite sex; it finds another planaria. They simply exchange sperm. The sperm then moves up the lengthy oviducts to fertilize the eggs, which are waiting in the ovaries. As the fertilized eggs pass down the oviducts, they build up yolk stores as they go. An egg case forms around the eggs just before they are released.

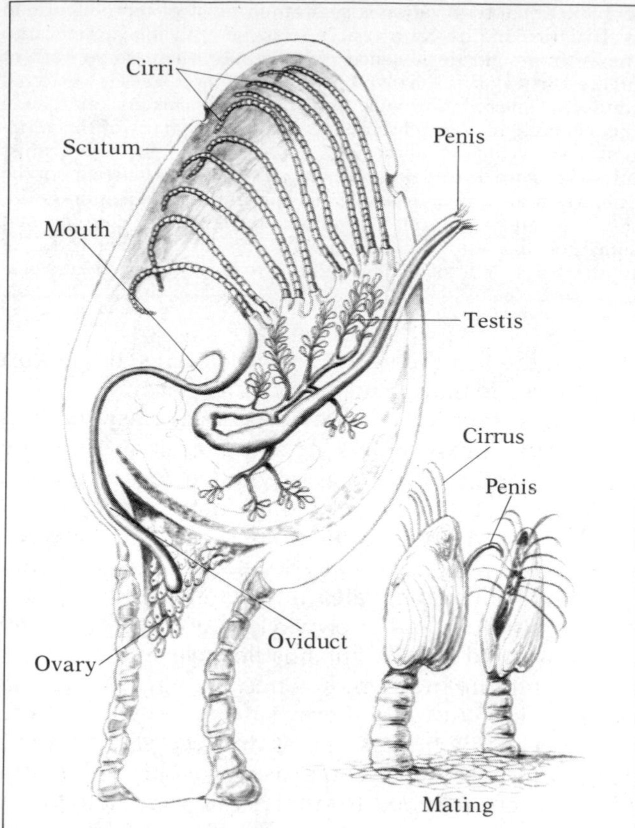

29.2 The barnacle *Lepas* is hermaphroditic but not self-fertilizing. Since is is a stationary animal and doesn't use external fertilization, the problem of transferring its sperm is solved by its long extensible penis. When it mates, *Lepas* sends its penis over to the neighboring barnacle (any one will do) and probing down into the *scutum,* releases its sperm. The sperm find their way into the nearby *oviduct,* where fertilization takes place.

A number of fishes also copulate as a mode of reproduction, and their offspring develop in a number of distinct ways. First, we should note that in external fertilization, the development occurs entirely outside the female's body and is called *oviparity.* In some fish that copulate, however, fertilized eggs are retained within the uterus while they mature. In fact, they hatch there and the female gives birth to live babies. You may have seen this in aquarium fishes such as guppies (Figure 29.4). This method of development is known as *ovoviviparity,* which refers to the housing of developing young in the uterus, with no apparent connection to the mother. This process may appear to be similar to *viviparity,* such as we see in the mammals, but the similarity is an illusion. In mammals, the young are in intimate association with the mother as they develop. They draw their sustenance from her and are not isolated in egg cases or shells. They, like the guppies, are born alive, but their development is drastically different from that of the fish.

The distinction between ovoviviparity and viviparity becomes less distinct in some of the sharks and rays. Generally, they practice internal fertilization. The male is equipped with external *claspers*—long, tapering modifications of the pelvic fins that can be erected (Figure 29.5). During sex, the erect claspers are held together and pushed into the female's cloacal opening (the common reproductive, excretory, and digestive waste passageway). The male then ejaculates between his grooved claspers, directly into her body.

Fertilized eggs in the sharks and rays become enclosed in leathery *egg cases.* In many species, these egg cases are maintained within the uterus, with the young emerging through the urogential opening, reminiscent of the mammals. In other species, the egg cases are dispensed with and young develop even more like the mammals. For example, at least three genera of sharks and rays may actually nourish their young in the uterus by exposing them to placentalike beds of capillaries—showing again that living things don't always fit into our neat categories.

29.3 *Planaria,* like most flatworms, are hermaphroditic, but they are not generally self-fertilizing. Sperm are produced in a series of testes on either side of the body. They travel down the paired sperm ducts to the penis. The eggs are produced in anterior ovaries. A lengthy oviduct with numerous yolk glands travels back to the genital chamber. Planaria are equipped with protusible penises, and copulating worms use them simultaneously. Each crowds its penis past the other through their respective *genital pores* and into the *copulatory sacs* where the sperm are released. The sperm then move up the oviducts to fertilize the eggs.

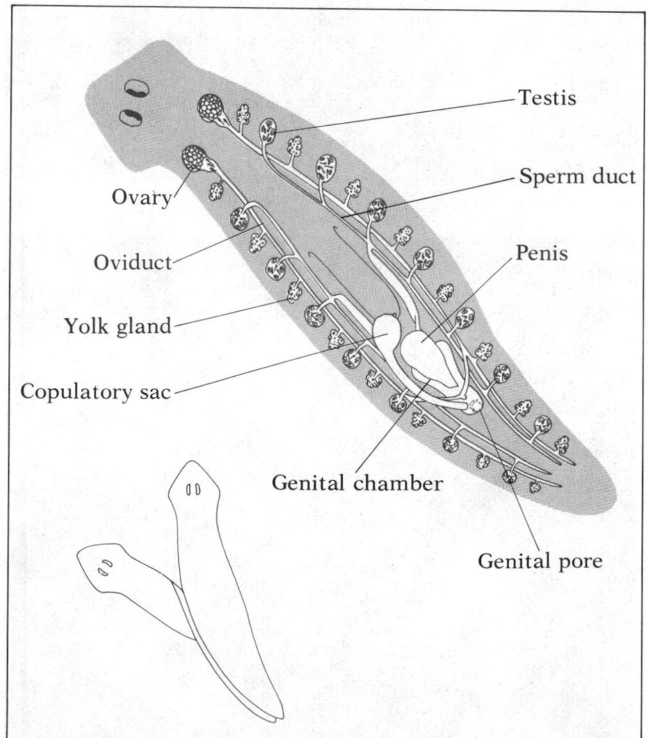

29.4 Anyone who has raised guppies (*Lebistes reticulatus*) knows about *ovoviviparity* from watching these live bearers. Ovoviviparity is not to be confused with *viviparity* as seen in the mammals. There is no placenta over which materials can be exchanged between the mother and her offspring. The mother simply provides safety by housing the eggs in her body as they develop. When ready, the young guppies emerge, just as other fishes emerge from eggs that have been less safely hatched outside the parent.

29.5 Male sharks can be readily distinguished from females by the presence of a pair of projections alongside the cloaca known as *claspers*. These are erectile structures used in copulation. They are inserted into the female cloaca and sperm runs along grooves in the claspers. With the exception of a few viviparous and oviparous species, most sharks are ovoviviparous, which means that the embryo develops in the uterus, but within eggs and not sustained by the mother. The leathery egg cases of oviparous sharks are often found washed up along beaches.

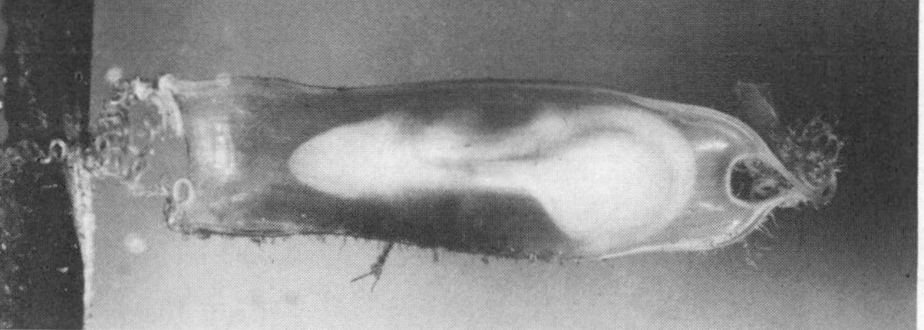

Reproduction and the Survival Principle

Before we go on to terrestrial animals, let's look at an important reproductive principle exhibited by the fishes. The sharks and rays don't produce many offspring, but to those they produce they offer a lot of protection (growing up in the uterus of a shark is a lot of protection). However, there is no evidence of maternal shark care after birth. The young must shift for themselves. On the other hand, the herring (a bony fish) produces about 50,000 eggs each season. To these they offer no protection. The eggs sink to the ocean floor and most of them perish. But some sift into concealed places where predators can't reach them. The eggs of turbot and cod are in an even more precarious position—they float. Thus, they are highly exposed in the open surface waters. In order to leave any offspring at all, these fishes must produce up to 9 million eggs per season. But whether a fish produces half a dozen eggs or 9 million, the average number of offspring surviving to adulthood is two for each pair of parents, in the long run. The point is, the more a parent of any species protects or cares for its offspring, the fewer offspring it produces (or can energetically *afford* to produce).

Reproduction in the Terrestrial Environment

The transition to land brought with it its own reproductive problems. After all, sperm cells couldn't be tossed out on the ground; they would dry out, and it's easier for a sperm to swim than to crawl. And how about the egg? The aquatic egg needed substantial modification to enable it to survive the rigors of a dry atmosphere and varying temperatures. Thus, copulation evolved, enabling sperm to be deposited safely in the moist environs of the female where the eggs are found. Let's look at some variations on this theme.

Reproduction in Terrestrial Arthropods

The insects have been described as one of nature's greatest success stories. One of the reasons for their success in the terrestrial environment is their tremendous reproductive capacity. Perhaps because their life spans are generally short, females are generally capable of producing large numbers of young. This ability can, in fact, set the stage for great fluctuations in numbers. Although predation and limited food supplies tend to suppress insect populations, they often surge out of control, notably in agricultural areas where predators may be few (often a result of human behavior) and where food is plentiful (always a result of human behavior).

All insects practice internal fertilization and they have developed an array of devices to accomplish this end (Figure 29.6). Males are commonly equipped with clasping organs for holding the female in position while the penis is introduced and sperm are released. Also, the penis is usually barbed, hooked, and braced in such a bizarre fashion that it can only fit into one kind of vagina—that of females of his own species. Students of evolutionary theory believe that this lock-and-key arrangement has evolved to prevent inappropriate matings between similar species.

Female insects are commonly equipped with an *ovipositor,* which, as the name implies, is used to deposit the eggs (Figure 29.7). In grasshoppers, the ovipositor is also used to dig holes in the ground where the eggs will be buried. Females of some species, such as the cicada and gall wasps, are even capable of boring into wood with the ovipositor and then filling the space with eggs.

29.6 Many male insects are equipped with accessory reproductive structures that are used in grasping the female and holding her in position for copulation. The genitalia are very complex, with barbs and hooks that bear little resemblance to anything we might think of as a penis. Each species has its own unique genitalia, long used by entomologists to identify species. When in the mating position seen in the photograph, the claspers assure the successful transfer of sperm.

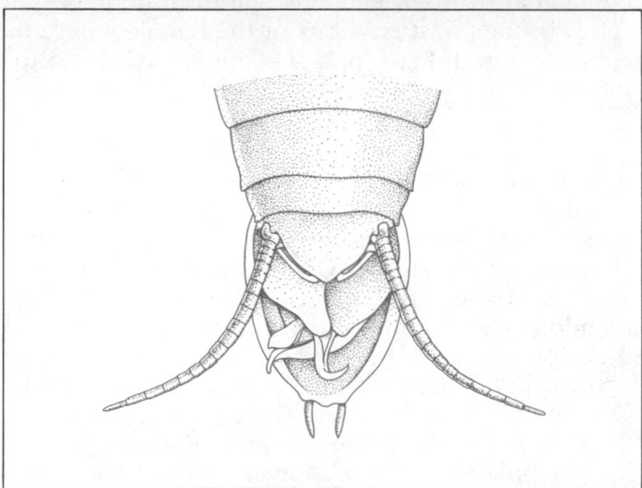

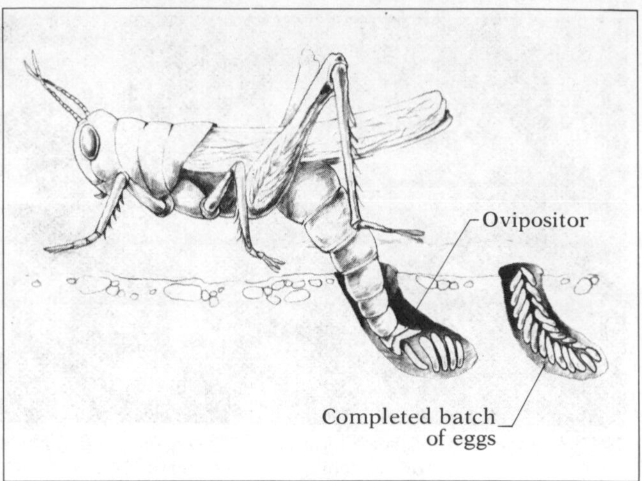

29.7 The ovipositor of the female grasshopper is a digging device used to make a hole in the earth for the eggs. She will bore about ten holes, depositing twenty or so eggs in each. Each group of eggs is enclosed in a sticky egg pod.

Many female insects mate only once in their lifetime and then store the sperm in a saclike structure called the *spermatheca.* They simply release the sperm when they are needed. Insect eggs are well-protected by a tough covering (the *chorion*), which forms before the eggs are laid. Each egg is fertilized through a small pore in its chorion (the *micropyle*) as it passes the spermatheca on its way through the vagina (Figure 29.8).

Mating in Spiders

Spiders are all carnivores and they are all venomous. Many species are solitary animals. And a venomous, carnivorous hermit must not be taken lightly. In particular, the habits of spiders add up to problems in mating, especially for the small, weak males who must initiate the process. In some species, they timorously approach the females at mating time, waving special appendages as though they were signal flags (Figure 29.9). The waving seems to mesmerize the female, at least long enough for the male to mate with her. The orb spider males signal their presence in an even more cautious manner. The male climbs on the female's intricate web and sort of strums out a signal that inhibits her feeding behavior. Males of other species even prepare a gift—usually a paralyzed insect, neatly wrapped in silk. Anything to distract the female or occupy her while he dashes in and accomplishes his task.

After all this trouble, most species of spiders don't actually copulate. Instead, the male may deposit his sperm on the ground and then scoop it up in a

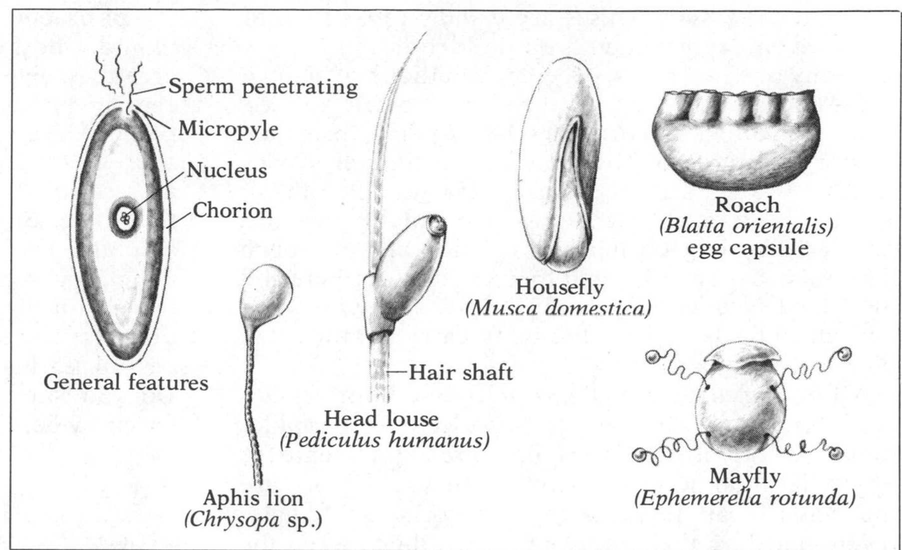

29.8 As you might expect, insect eggs occur in a great variety of shapes and sizes. Each egg shape or structural variation represents an adaptation to the particular circumstances in which the embryo develops. The general features shown here include a central yolky region harboring the nucleus, surrounded by a thin region of dense cytoplasm. The protective case (*chorion*) has a pore (*micropyle*) at one end through which a sperm can enter to achieve fertilization. Fertilization usually occurs as the eggs pass through the oviduct. At this time sperm emerge from their storage place (the *spermatheca*).

29.9 Mating in spiders. The smaller male spider must approach the large female with great caution. (a) He may use specific recognition signals so that she recognizes him as a mate and not as prey. (b) But even when all is well, male spiders don't actually copulate with the female. Instead, they quickly deposit their sperm in or near the female's genital tract using special bulblike pedipalps on the forelegs. If the male is fortunate, he will make good his retreat before the female attacks.

(a)

(b)

modified appendage. He then either draws near the female or climbs on her back and rakes his sperm-laden appendage over the plates covering her vagina. Then he clears out. He doesn't always make it, but in some spiders his chances are increased if he pinches the female hard as he leaves. She will then immediately tuck into a defensive ball, and before she can recover, he's gone.

Reproduction in Reptiles and Birds

As good terrestrial animals, reptiles and birds practice internal fertilization and are oviparous, with the exception of a few types of snakes that are ovoviviparous (and sea snakes, which have evolved a placentalike structure and can be said to be viviparous).

The reproductive system in reptiles is similar to those of the fishes and amphibians. There is a well-developed, but hidden, copulatory organ in the male. Actually, whereas turtles have only one penis, lizards and snakes have *two* penises (or, more correctly, two *hemipenises*), which can be everted (turned inside out, like the fingers of a rubber glove). However, only one is used at a time, depending on which side of the male the female's body is on, since lizards and snakes mate more or less side-by-side. (Actually, since each hemipenis is connected to only one testis, the males prefer to alternate sides, using the hemipenis on the side with the larger store of sperm.) Each hemipenis is covered with backwardly directed barbs and hooks and, whereas it goes in easily enough, it must be carefully withdrawn. The female is often larger than the male and if she should be startled in the act, she may take off, dragging the helpless male away by his organ.

Since the hemipenises are usually tucked out of sight, even experts have trouble identifying the sex of many of the reptiles, except when they are mating.

Male birds, with the exception of a few species, including ducks and ostriches, lack a penis. Insemination is accomplished by the male pushing his cloaca against the female's and quickly ejaculating into her duct, in a kind of cloacal kiss. He usually does this by first treading on her tail, causing her to crouch and raise her rear. His tail then cups under hers and the deed is done (Figure 29.10). By the way, some daredevil birds, such as the swift, carry out the act in mid-air.

Compared to aquatic vertebrates, both reptiles and birds produce few eggs, indeed. But reptiles tend to produce more eggs than birds and to care for them less (although crocodiles and alligators are particularly attentive as reptiles go, some females even guarding their nests, and then digging out the hatchlings and carrying them to the water in their jaws.) In general the reproductive success of reptiles depends in large part upon the female's successful concealment of the eggs. Sea turtles, for example, bury their eggs on sandy beaches where they are relatively safe from predators and are not exposed to the direct heat of the sun. But upon hatching, the young are completely on their own. As they scramble toward the sea, many fall prey to the voracious gulls wheeling above, and those that make it to the water face a battery of marine predators. Because of the high mortality rate of the young, then, the reproductive success of a female sea turtle often depends on her laying a lot of eggs each season and living through a lot of seasons. If she makes it through those first few years, her life span can be long because her environment (especially the aquatic one) is rather stable.

29.10 Most male birds lack a penis, so copulation is accomplished by the cloacas simply being pressed together. The act takes place in a flurry of wing flapping and apparent confusion, but it apparently works well. The success of a single barnyard rooster in inseminating a sizable flock of hens is legendary.

Birds employ a different reproductive strategy than do most reptiles. They are well-known for the care they give their offspring. Typically, only a few eggs are produced each season, and then one or both parents lavish them with warm attention. In some species, such as chickens, the young may hatch in a *precocial* state, ready to run around and feed on their own. But in other species, the young hatch in a completely helpless, semideveloped *altricial* state, doddering weakly about and demanding feeding and protection for many days (Figure 29.11). The reproductive success of birds depends, again, on the successful rearing of just a few well-protected offspring. Do you suppose parents of altricial or precocial species would lay more eggs? Do you think the con-

29.11 Birds can be divided into two groups according to the developmental state of the newly hatched young. (a) Altricial birds hatch out in a completely helpless state, often nearly featherless and requiring a lengthy period of constant care for survival. (b) Precocious young are more vigorous and independent. These birds are usually able to eat on their own, have feathers, and do not remain in any nest. But even precocious young are not entirely on their own; they are sheltered and protected from predators for some time after hatching.

(a)

(b)

tinued assistance of both parents is more necessary to precocial or altricial species? For which group do you suppose food is easier to find?

Reproduction in Mammals

Mammals give milk and have hair. They are almost all viviparous, and almost all viviparous animals are mammals. One reason for this is that we know that any viviparous animal probably has a placenta or something very much like it, and placentas are generally, but not always, associated with mammals. It is through the placenta that viviparous embryos receive their oxygen and nourishment, and discharge wastes and carbon dioxide. True placentas, we should add, are present only among the advanced mammals (*Eutheria*).

Placentas are not found in the monotremes and marsupials. You recall from Chapter 23 (Figure 23.24) that monotremes and marsupials are primitive subclasses of vertebrates. Monotremes include duck-billed platypuses and echidnas. The duckbill platypus is an unusual creature that not only has a bill like that of a duck but lays eggs; then to confuse the taxonomist, it has fur and gives milk. So it's a mammal. Interestingly, when the first stuffed specimen was sent from Australia to the British Museum, it was considered to be a crude hoax.

The marsupials, such as opossums and kangaroos, are also taxonomically distinct. They are mammals and the young even develop in the uterus (but not with a true placenta). Instead, a *pseudoplacenta* is formed, through which some nourishment is transferred from the mother's bloodstream to a capillary bed in the embryo's yolk sac. But a fully developed placenta is not necessary anyway, because the young don't stay in the uterus very long. While they are still hardly more than embryos—except for well-developed front limbs—they squirm out of the vagina almost unnoticed and, by using a motion similar to the Australian crawl, they make their way to the mother's pouch and attach to a nipple to begin their second stage of growth. The female kangaroo actually has two teats in her pouch, a big one and a little one. After the young kangaroo has grown a bit, it shifts from the small one to the larger one.

For a long time, no one could figure out how the infant made it to the pouch. The event remained clothed in mystery until one day a zookeeper just happened to be watching a kangaroo's vagina when not much else was happening. To his astonishment, a slimy little sluglike creature crawled out, and as he continued to watch, it began to crawl blindly to the pouch in a moist trail the mother had apparently laid down by licking her fur. (But recent studies reveal that the young can make the trip completely on their own.) The forelegs of the embryo are surprisingly well-developed, as are the supporting elements of the nervous system. The journey is flawless, according to all recorded observations.

The adaptive significance of this strange mode of development is, at the moment, anyone's guess. The kangaroo can conceive again while young are developing in the pouch, so it can support young in different stages of development. If it already has a young in its pouch, the development of the second embryo may be stopped, the second infant remaining in suspended animation until the pouch is free. Sometimes both teats are in use, with a new and an older offspring in the pouch at the same time. Whatever the ins and outs of marsupial reproduction, one point is very clear: it works. Kangaroos can reproduce prolifically if populations are not decimated for dog-food and exotic clothing. And opossums abound.

The rest of the mammals are placental. The pattern is similar in all of these, so we'll focus on humans, making comparisons with some other animals along the way.

29.2 Human Reproduction

The Human Reproductive System

Humans are reproductively unique in some ways, of course, and we will deal with some of our unusual behavior since it is biologically interesting and essential to our understanding of our own sexuality. But let's begin with anatomy, by examining our plumbing.

The Male Reproductive System

The reproductive anatomy of the male human is rather typical of mammalian systems in general (Figure 29.12). The external genitalia include the *penis* and the *testes* held in the *scrotum*. Shortly before birth, the testes descend from the abdomen, where they developed, up over the pubic bone and down into the scrotum. It's cooler in the scrotum and sperm won't develop at body temperature. In fact, unless the testes descend, sterility results. The scrotum is adapted for temperature control in an interesting way. When it's cold, muscle contractions draw the testes upward toward the body, making the scrotum tight and wrinkled. When it's hot, the opposite effect is created, with the testes sinking low

29.12 The genitalia of the human male include the *penis* and *testes.* The internal organs include the *vas deferens* and the accessory glands. The anatomical structures shown here are not unlike those seen in many mammals, although there are marked variations in the form of the penis from one species to another. The route of sperm travel is shown in color. The testis and *epididymus* consist of highly coiled tubules (inset). The coils within the testis, known as the *seminiferous tubules,* are the site of sperm production. Where the tubules join at the upper region, the epididymus begins, and it is here that the sperm are stored. The epididymus forms a long fold to the lower curve of the testis and returns to form the vas deferens. The vas deferens and related blood vessels are enclosed in a sheath known as the *spermatic cord.*

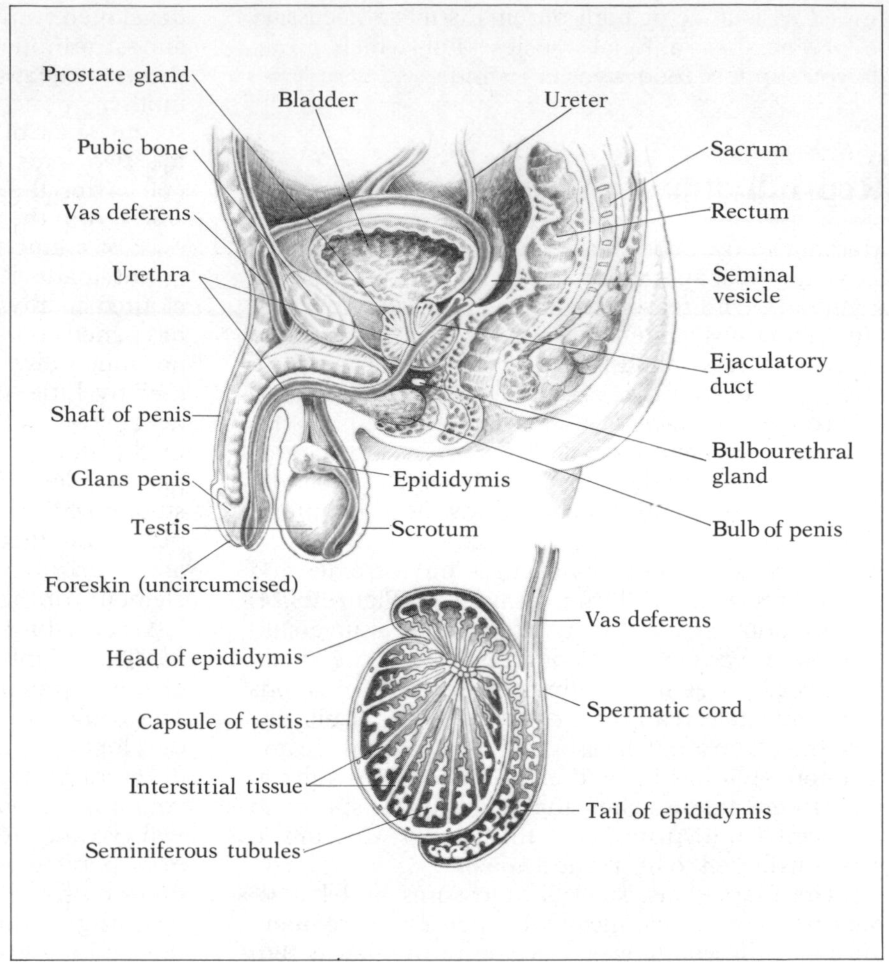

into the scrotum, producing a pendulous effect. One testicle, usually the left, hangs lower than the other.

The testis, which produces sperm and sex hormones, is an oval body consisting of two highly coiled series of tubules in a strong sheath. One set is a rather soft mass of coiled tubules known as the *epididymus.* The other is comprised of the highly coiled *seminiferous tubules,* along with their supporting connective tissue, interstitial endocrine gland cells, blood vessels, and nerves.

The *spermatic cord* from the epididymus carries the *vas deferens,* which is a tube for sperm transport, along with the blood vessels and nerves of the testes. The spermatic cords, one from each testis, travel upward, over the bones of the pubic arch, and then into the body cavity, retracing the path of the descending testes. The paired vas deferens continue to the underside of the bladder where they unite, joined by the duct of the *seminal vesicle.* Unlike the seminal vesicles in some invertebrates, these are not used to store sperm. Instead, they produce most of the volume of the *ejaculate* (*semen*). The fluids added from the seminal vesicles contain nutritive and buffering materials. The misleading name "seminal vesicle" is a heritage from early anatomists who named the structure before its function was understood.

After the vas deferens of the spermatic cord is joined by the ducts of the seminal vesicles, it unites with the *urethra* at about the place where it leaves the bladder. Surrounding this junction is the fleshy tissue of the *prostate gland.* It is undoubtedly one of the least perfected of the body's parts. It is said that if a man simply lives long enough, his prostate will go wrong in one way or another. (It is disconcertingly cancer-prone.) The functioning prostate produces alkaline secretions that raise the pH of the semen and produce its characteristic viscosity, color, and odor. Raising the pH of the semen has two important consequences. First, the sperm cells are activated. Sperm cells leaving the epididymus are immobile; they remain so until the alkaline environment arouses them. Second, the prostatic secretions are important in protecting the sperm cells from the decidedly acid environment of the female reproductive tract.

Just below the prostate, the urethra receives ducts from the last accessory structures, the *bulbourethral glands,* also known as *Cowper's glands.* Their function is not clear yet, but we know that they add mucous secretions to the semen. These secretions may precede ejaculation and therefore may have some lubricating value. Then the urethra passes through the penis. As you may have surmised, the urethra has the dual role of conducting both semen and urine—on different occasions, of course.

Erectile Tissue

The penises of mammals consist of very specialized tissue with an erectile capability. In some mammals, the *erection* may be due to spaces in the penis filling with blood, causing it to become turgid and grow in length and girth (as in men and horses), or it may be brought about by simply pulling the penis out from where it usually rests, already stiffed by cartilage or bone, deep within the body (as in dogs). In men, the penis is an erectile shaft tipped by an enlarged region, the *glans.* A vertical slit at the tip is the external opening of the urethra. Normally, the penis must be more or less turgid before it can be introduced into the female vagina.

The penis consists of three cylinders of very spongy tissue (Figure 29.13). Two of these are the paired *corpora cavernosa* and the third is the *corpus spongiosum,* terms that suggest rather well the functions of the structures. During sexual excitement, the blood flow into the spongy tissue increases, and venous flow back to the body is retarded. As the penis fills with blood, it becomes erect. The tough sheaths of connective tissue surrounding the spongy bodies resist the expansion, producing a certain firmness. The effect is somewhat like the hydrostatic skeleton produced in the earthworm, but we won't strain this analogy further.

29.13 The penis contains three regions of erectile tissue which are capable of retaining blood during an erection. The largest are the two *corpora cavernosa.* Note the large arteries at their midregions. The third region, the *corpus spongiosum,* is below these and contains the urethra. During erection, the spongy passages of these bodies become engorged with blood because its exit has been restricted. This enlarges, curves, and firms the penis in preparation for copulation.

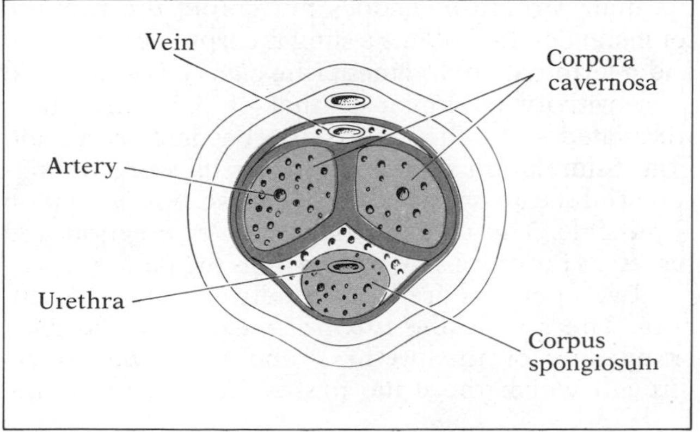

Loss of the erection is the result of restricted arterial blood flow and a subsequent drainage of the spongy tissue. Detumescence (softness) almost invariably follows ejaculation, no matter what you may have read in novels. Of course, other things besides ejaculation can have the same effect. Stepping on a cat's tail or putting one's head in ice water can do it. So can an unkind word. We are obviously omitting other important factors here, but we will return to the subject of copulation and ejaculation after we have discussed the female reproductive anatomy.

Sperm Production

Sperm are produced in the walls of the seminiferous tubules (Figure 29.14). The cells in the outermost region of the tubule continually divide by mitosis. Some mitotic daughter cells begin the series of meiotic divisions that will produce haploid gametes—the sperm.

Some of the outermost cells of the seminiferous tubules, just inside the tubule walls are referred to as *spermatogonia.* Although they have yet to enter the spermatogenic process, they are larger than surrounding or nonmeiotic cells of this region. When one of these cells enters meiosis, it becomes known as a *primary spermatocyte.* Then meiosis I separates homologous chromosomes into daughter cells known as *secondary spermatocytes.* Meiosis II follows, with the production of four *spermatids,* each with the haploid chromosome number (see Chapter 8). Each spermatid will then undergo the complete reorganization that is necessary to form mature *spermatozoa.* The spermatids at this time are apparently nourished by groups of specialized *Sertoli cells.*

In the mature spermatozoan, the chromatin will be condensed into a minute *sperm head.* The cytoplasm itself will differentiate into a midpiece and tailpiece, and excess cytoplasm will be sloughed off. The midpiece gives rise to a whiplike tailpiece and contains a peculiar spiral-shaped mitochondrion and a centriole. The tailpiece, a long, tapered flagellum, propels the sperm along. At the tip of the sperm head is the *acrosome,* a caplike structure that arises from a lysosome (Figure 29.15). (If you recall the role of a lysosome, you can probably guess how a sperm cell penetrates an egg.)

Sperm production begins at puberty and may continue throughout the lifetime of a male. Typically, however, it diminishes gradually as he grows older. There is no hard-and-fast rule, since some men have demonstrated their fertility in their sixties and seventies, and one 105 year old man became a father (he said).

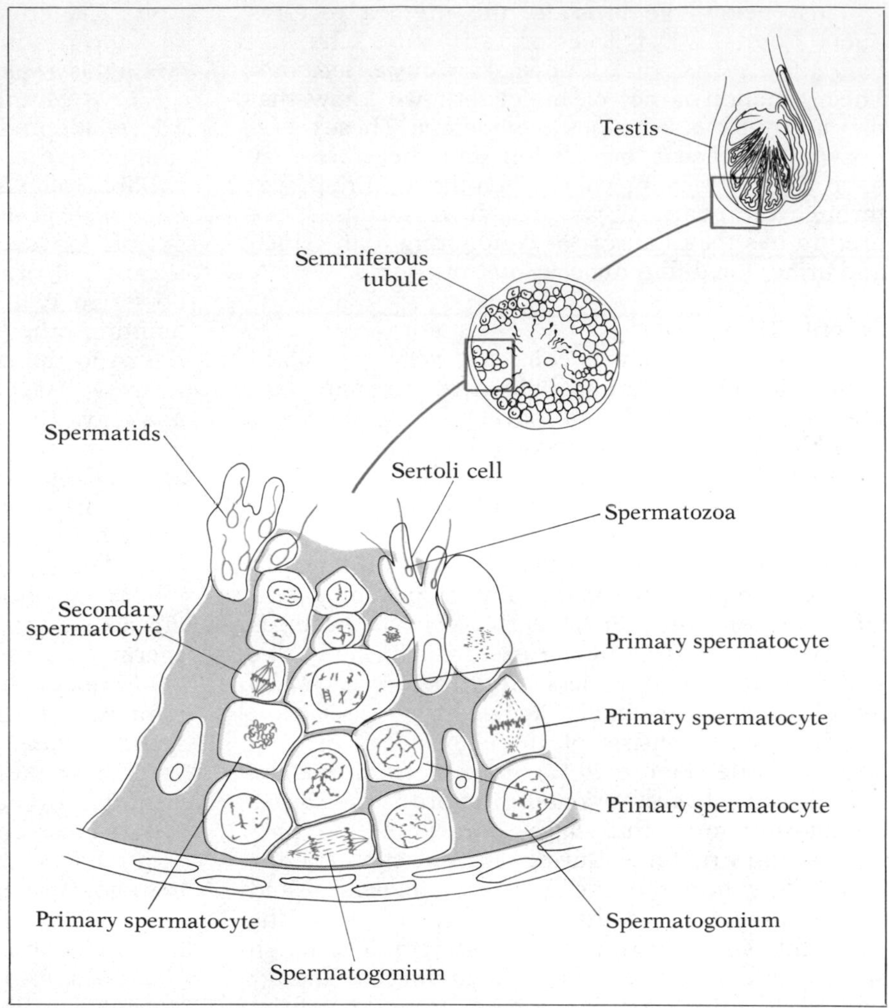

29.14 Spermatogenesis occurs in the walls of the seminiferous tubules of the testis, one of which is drawn out and enlarged here. Larger cells of the germinal layer, the *spermatogonia,* enter meiosis as *primary spermatocytes.* These are readily recognized by their large size and huge nuclei. After the first meiotic division occurs, the daughter cells are the *secondary spermatocytes.* These enter meiosis II, becoming *spermatids.* From this stage, nursed by *Sertoli cells,* each spermatid becomes greatly reduced in size, the flagellum develops, and the fully differentiated *spermatozoan emerges.* (The stages of meiosis are described in Chapter 8.)

Mature sperm cells, still in an immobile or inactive state, are swept along by ciliary action from the seminiferous tubules to the epididymus, where they are stored prior to ejaculation. Something important happens in the epididymus, but we don't know what it is. For some reason, unless the sperm travel through this tube, they are incapable of fertilization.

The Female Reproductive System

External Genitalia

The external genitalia of women (Figure 29.16) are collectively known as the *vulva,* the most prominent portion of which is a hair-covered mound of fatty tissue over the pubic arch called the *mons veneris.* Below the mons veneris lies the larger outer folds, or *labia majora* (the "major lips"), of the vulva. In an unaroused state, these folds cover a number of rather sensitive structures. Just inside the labia majora are the less prominent, thinner, inner folds, or *labia minora* (the "minor lips"). These join together at their upper portion to form a roof over a small prominence, the *clitoris.* This partially hooded, cylindrical organ is composed of highly sensitive, spongy erectile tissue. It is, in fact, complimentary to the penis and develops from the same embryonic structure as the glans of the penis. Although it is in no way a miniature of the male structure—it does not enfold the urethra, for instance—it contains a similar corpus cavernosum and a terminal enlargement (the *glans clitoridis*), and its sensitivity is similar to that of the penis. In a stimulated state, the clitoris also becomes erect and firm. Stimulation of the clitoris or its surroundings can produce an orgasm, but, of course, no ejaculation is possible. The clitoris both achieves erection and regresses far more slowly than does the penis.

Two openings are located between the labia minora. The first is the *urethral meatus* (passageway or opening of the urethra), about halfway down, through which the urine passes. The second is the

29.15 Structures in the human sperm cell include the *head, midpiece,* and *tail.* The head contains highly condensed haploid chromatin containing the twenty-three human chromosomes that have completed meiosis. At the tip is the enzyme-laden *acrosome,* which functions in egg penetration. The midpiece contains a centriole and a lengthy, spiraled mitochondrion. The tail is a flagellum, the propulsive structure of the sperm.

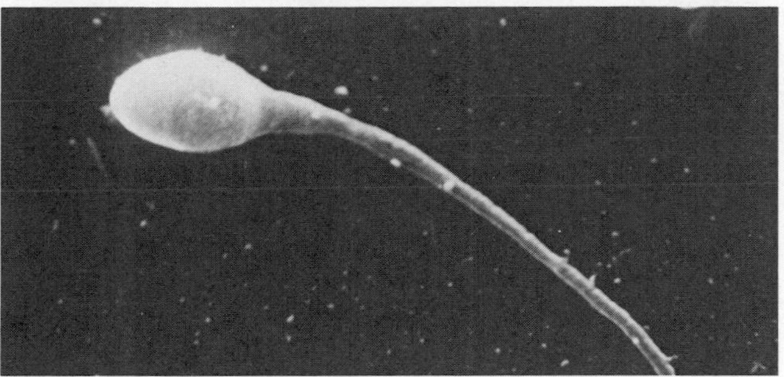

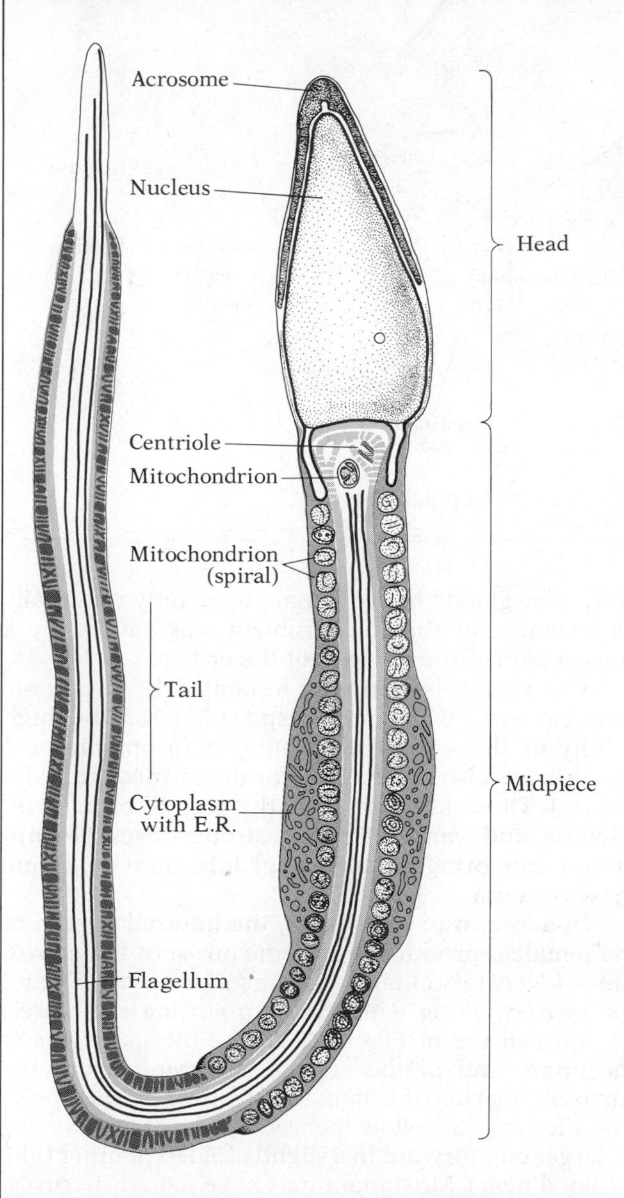

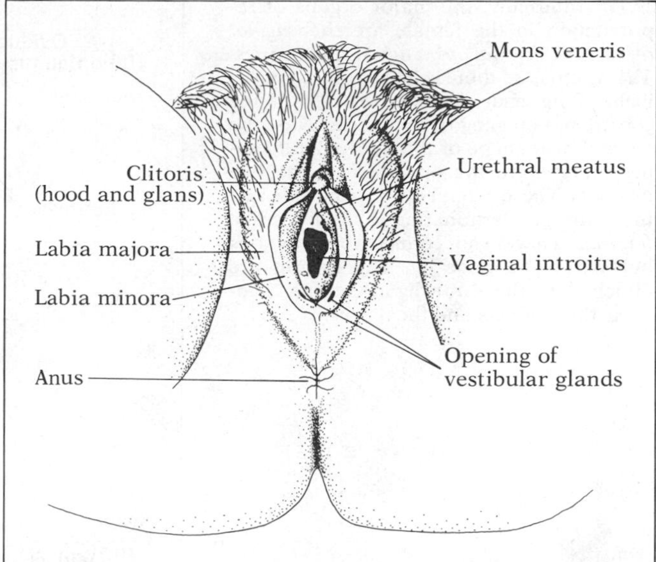

29.16 The external features of the female reproductive system include the *labia majora* and *minora,* the *clitoris,* introitus (opening) of the *vagina,* and the meatus (opening) of the urethra. The clitoris is homologous to the glans of the penis, containing a rich supply of sensory nerves and spongy erectile tissue. The clitoris appears to be the focal point of sexual stimulation during copulation, but there is some argument about this. Prior to sexual experience, the opening of the vagina is sometimes partially occluded by membranes known as the *hymen.*

introitus (opening of the *vagina*). In rodents and some other mammals, the urethral meatus is raised on a small *urinary papilla,* which presumably plays a minor role in protecting against infections.

In virgins, the opening of the vagina may be partially blocked by membranes known collectively as the *hymen.* Rupture of the hymen during the first intercourse may or may not cause bleeding. (Unfortunately, in some cultures, bleeding is considered a testimony of chastity on the wedding night in spite of the fact that the first intercourse does not always result in bleeding; hymens can be ruptured in a number of ways.)

Internal Genitalia

The *vagina* is a distendable, muscular tube about three inches long in its relaxed state (Figure 29.17). Its functions are to receive the erect penis during copulation and to act as a birth passageway during delivery of a baby. The vagina is marvelously adapted for both functions. Its highly folded walls consist of an inner layer of squamous epithelium, a middle muscular layer, and an outer layer of fibrous connective tissue. These layers, in addition to presenting a moistened, yet firm, stimulatory surface for the penis, are capable of great extension to accommodate the passage of the fetus during birth. Remember, the human vagina must accommodate a head about

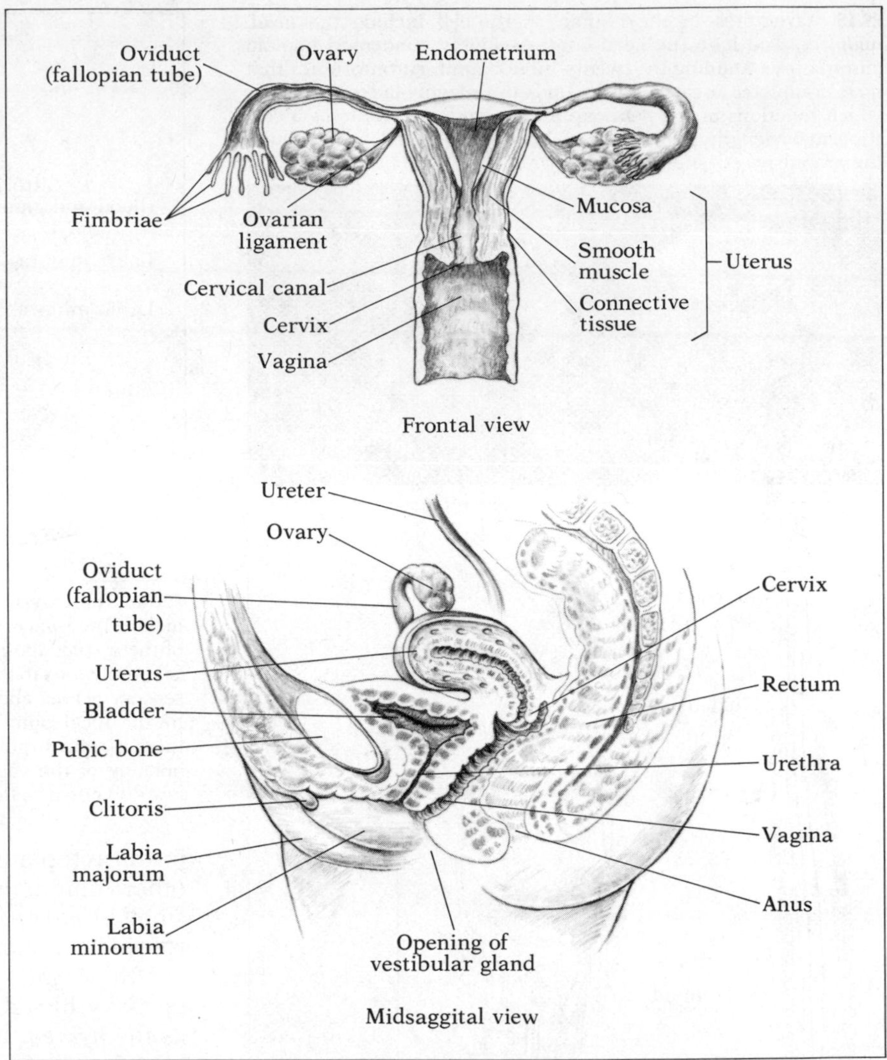

29.17 Internally, the major organs of reproduction in the female are the *vagina, uterus, oviducts* (*fallopian tubes*), and *ovaries*. The vagina, a tubular structure about 3 inches long, ends at the *cervix*. This is the constricted opening of the uterus. Note the general pear shape of the uterus. At its upper margin are the openings of the two oviducts. These extend out to each side and then curve in, terminating in the fingerlike *fimbriae*. They do not connect to the ovaries, but simply lie close by. The ovaries are attached to the same ligament that supports the oviducts and the uterus.

4 inches in diameter. What is more remarkable is the rapid recovery from this strained expansion to the original state.

Although the vaginal lining itself apparently produces most of the lubricating mucous secretions, some assistance is provided by the *vestibular glands*, or *glands of Bartholin*. These are located just beneath the surface of the lower margin of the vaginal opening, and correspond in structure and function to the bulbourethral glands in the male. These glands actively secrete fluids when the female is sexually aroused—much more quickly and actively for some than for others. Their importance in lubrication, a long-accepted idea, has been brought into question by the renowned sexual research of William Masters and Virginia Johnson. It was once believed that painful intercourse for the female was due to an insufficient lubrication from the Bartholin's gland. Masters and Johnson concluded that the secretions of the vestibular glands in the female were only minimally important, and that the problem was caused by a constriction of the muscles of the orifice.

The vagina is normally inhabited by propionic acid bacteria, which are responsible for the high acidity of the vaginal environment (propionic acid bacteria are also responsible for the tartness of Swiss cheese). The acidity prevents the growth of harmful bacteria and yeasts. In fact, strong doses of antibiotics can bring on a vaginal infection by killing these bacteria.

In addition to the vagina, the internal organs of the female reproductive system consist of the *uterus*, the *oviducts* (also known as the *fallopian tubes*), and the *ovaries*, along with their supporting structures. As you can see in Figure 29.17, the uterus begins at the upper end of the vagina. Its opening and the surrounding ring of firm tissue is known as the *cervix*. The uterus is a hollow, pear-shaped organ that tilts its larger end forward in a slightly folded manner (like a folded pear). Most mammals have paired uteruses;

the single uterus of humans and other primates (as well as bats) is an evolutionary specialization for species that customarily give birth to only one offspring at a time.

The walls of the uterus consist of three specialized layers. The inner layer is the versatile *mucosa.* It produces the soft, highly vascularized *endometrium* that will receive and support a zygote if fertilization occurs. (We will look at the cyclic production of the endometrium a little later in this chapter.) Below the mucosa lies the middle layer, a region of highly vascularized smooth muscle with fibers running in several directions. The muscle layer has an enormous capability to expand, as it must when accommodating an embryo. The outermost layer is connective tissue, which is continuous with the *broad ligament.* This latter structure forms two wings on either side of the uterus, supporting it as well as the ovaries and the oviducts.

The oviducts emerge from each side of the upper end of the uterus and extend for a few inches outward and then downward. They terminate in a cluster of fingerlike processes known as the *fimbriae.* Curiously, there is no direct connection between the oviduct and the ovary; eggs are shed into the body cavity and later picked up by the highly motile, ciliated fimbriae. The arrangement seems a bit peculiar, but it works. In fact, among the mammals, only the rodents directly release the ova into the reproductive tract. In humans, once the egg is in the oviduct, its passage is aided by the sweeping of the cilia that line the tube.

The ovaries are located about in the center of the fanlike broad ligament and are additionally supported by the *ovarian ligament.* They are oval-shaped and about an inch in length. They are made up of two main regions: an outer cortex and an inner medulla. The outer region is of primary interest to us. It contains the germinal epithelium where all the egg cells reside, arrested in metaphase I of meiosis during fetal life (Figure 29.18). The number of potential eggs in the ovaries is astounding, since so few can ever be used. Each ovary at puberty contains approximately 200,000 developing eggs, but nowhere near

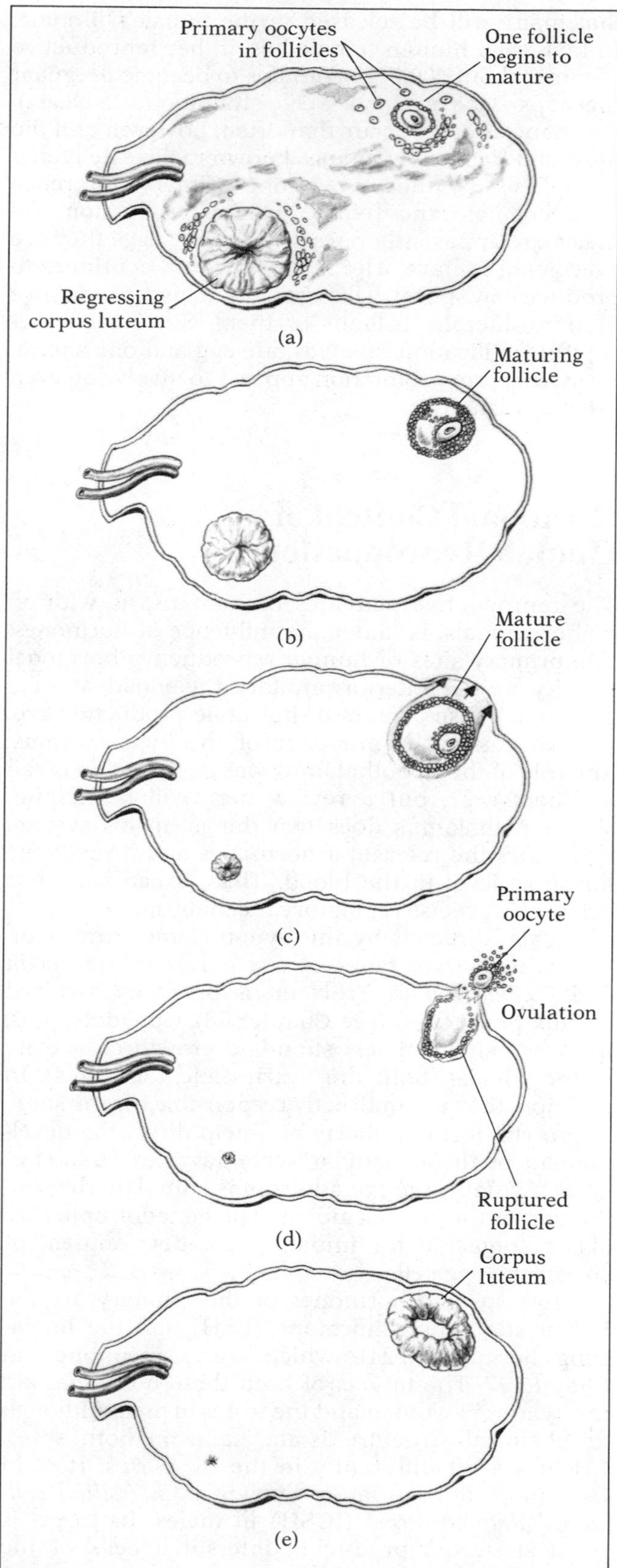

29.18 The ovary produces both sex hormones and egg cells, but unlike the testis, which produces a continual supply of sperm cells, the ovary contains all the egg cells it will ever produce when a baby girl is born. (a) These are primary oocytes, arrested in prophase I of meiosis I and surrounded by supporting follicle cells. Under the influence of pituitary hormones, a primary follicle will mature each month. The sequence of events in this maturation process is shown in this longitudinal section of a human ovary. (b) Note how the oocyte and its surrounding follicle cells enlarges as a follicle matures. (c) Upon maturation, which requires about 14 days, a shift of pituitary hormones occurs followed by ovulation. (d) The follicle ruptures and the oocyte bursts out carrying a covering of follicle cells with it. (e) The follicle then undergoes vast changes forming a corpus luteum, which for the next two weeks will secrete both estrogen and progesterone. Then it will regress and stop functioning as another follicle repeats the sequence.

that many will be released in the female's lifetime. In fact, each human female has, in her reproductive lifetime, about 400 opportunities to become pregnant (perhaps once every 4 weeks for about 35 years). Pregnancies can't occur that often, however, and the most children any woman is known to have delivered is 69. This, by the way, is one primary difference between male and female gamete production. Females are apparently born with all the eggs they are ever going to have. The male, however, continues to produce new sperm cells throughout his reproductive lifetime—literally trillions of them. Since each successful fertilization involves one egg and one sperm, massive sperm production appears to involve an even greater waste.

Hormonal Control of Human Reproduction

The reproductive activities of humans, as with all higher animals, is under the influence of hormones. The primary sites of human reproductive hormonal activity are the anterior pituitary, the gonad, and the placenta. These centers of hormone production are, in turn, under the influence of the hypothalamus. The role of the hypothalamus was described in detail in Chapter 27, but a review here will be helpful. The hypothalamus does two things in this system. It initiates the release of hormones and it reads the hormone level in the blood. Thus, it can act, once again, as a precise regulatory mechanism.

When directed by the hypothalamus, the pituitary secretes two types of sex hormones, *prolactin* and *gonadotropins*. Prolactin is primarily involved in milk production (see Chapter 28). Gonadotropins, as their name implies, stimulate growth or activity in the gonads (both the ovaries and the testes). In addition, they are indirectly responsible for the surge of growth during puberty and help direct the development of those familiar secondary sex characteristics. Actually, the gonadotropins stimulate the production of the sex hormones. The gonadotropins are also responsible for initiating the development of sperm and egg cells.

Two specific hormones of the pituitary are the follicle-stimulating hormone (FSH) and the luteinizing hormone (LH), which we've mentioned in Chapter 27. The targets of both these hormones are the ovaries in women and the testes in men. Although its chemical structure is the same in both sexes, LH acts a bit differently in the two sexes. It is, in fact, more appropriately known as *interstitial cell-stimulating hormone* (ICSH) in males. Its target is the testosterone-producing interstitial cells of the testes. Let's look further into the action of gonadotropins in the simpler system of males. Then we'll go on to hormones of females.

Male Hormonal Action

The specific target of FSH is the sperm-producing seminiferous tubules of the testes, and the hormone is rather continuously released throughout the life of a man. In addition to its role in sperm production, FSH synergistically acts with ICSH to induce the testes to produce sex hormones known as *androgens*, notably *testosterone*.

Androgen is a general term for all male sex hormones, but we are particularly interested in only testosterone. It is produced throughout the testes in the interstitial cells that lie outside the seminiferous tubules. Studies reported in 1972 indicate that the production of testosterone may rise and fall with increased and decreased sexual activity—or even with social interactions with the opposite sex. Testosterone levels fall in all-male crews of ships at sea, for instance. But even if a man's social life follows the pattern of flood and famine, his testosterone levels will be far more constant than the corresponding ovarian hormones in a woman, as we shall see.

Testosterone is vital to the development of the characteristics of "maleness." Although the appearance of the embryonic gonads is genetic, their subsequent development requires testosterone. If it is absent during the eighth or ninth month of development, the testes fail to descend into the scrotum. Later on, testosterone is essential for those perturbing events associated with puberty, such as voice changes, the growth of body hair, bone and muscle development, and the enlargement of the testes and penis. Abnormalities resulting from low testosterone levels in the developing male are readily corrected by injections of the hormone. Finally, the sex drive itself appears to be greatly influenced by the presence of testosterone. There is some evidence that testosterone is responsible for both pubic hair growth and sex drive in both men and women, although the precise nature of the relationships do not readily lend themselves to analysis.

High levels of testosterone inhibit the release of ICSH. The level is apparently detected by the hypothalamus, which then retards its stimulation of the pituitary. The pituitary, in response, slows its release of ICSH, thus finally lowering the production of testosterone by the interstitial cells. As you can see, a very delicate feedback mechanism regulates testosterone levels (Figure 29.19). The hormonal situation is not a simple negative feedback system, however, since a number of other factors may influence it. For example, we mentioned earlier that sexual and social activity influence hormonal levels.

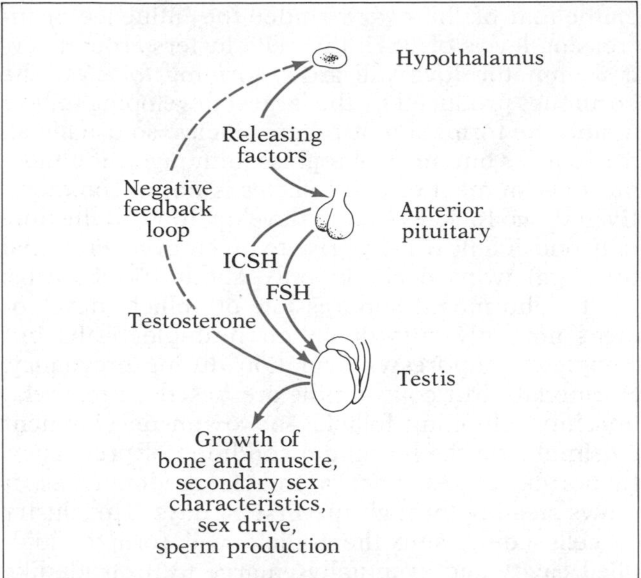

29.19 The ebb and flow of male reproductive hormones is much less extreme than that of females. The hypothalamus stimulates the anterior pituitary through its chemical releasing factors. The pituitary, in response, releases FSH and ICSH (which is the same as LH). The targets of FSH are the seminiferous tubules, which respond by increased sperm production. The target of ICSH is the interstitial cells surrounding the seminiferous tubules. Under the hormone's influence, the interstitial cells release testosterone, which stimulates spermatogenesis and a variety of other responses dealing with male sexuality. Testosterone levels are detected by the hypothalamus, and, when the level is high, it slows its stimulation of the pituitary, which responds by slowing its release of ICSH. Testosterone production, in turn, slackens. There are probably no pronounced peaks and valleys in the levels of these hormones, but rather, a number of minor oscillations.

Apparently, the frequency of ejaculation and/or sexual arousal has some influence on the hypothalamus, or perhaps directly on the pituitary, or the testes—or all three.

Now let's see if we can simplify some of this. Under the influence of the hypothalamus, the pituitary secretes the gonadotropins FSH and ICSH. These hormones stimulate sperm production and initiate the production and release of testosterone by the testes. Levels of testosterone are sensed by the hypothalamus, which responds by adjusting the release of gonadotropins. The result is a fairly steady level of hormone production. As you will soon see, all this is simple compared to what happens in the human female.

Female Hormonal Action

The onset of puberty in females occurs between 9 and 12 years of age. The beginning of fertility follows in 2–3 years. Both events are initiated by a rise in the pituitary gonadotropin, FSH (the follicle-stimulating hormone), which acts on the ovaries. The ovaries respond to FSH by producing estrogens. In turn, estrogens influence the growth of breasts and the nipples, broadening of the hips, and, in general, all the adult female contours. Less conspicuously, estrogen influences the growth of the uterus, the vaginal lining, the labia, and the clitoris. The appearance of pubic and axillary (underarm) hair, another signal of puberty, is initiated by the combined effects of estrogen and adrenocorticotrophic hormone (ACTH, from the pituitary) on the adrenal glands. In response to these hormones, the adrenals produce testosterone, which stimulates the growth of the coarser axillary and pubic hair. Interestingly, androgens are released in minute quantities throughout the life of women. If their production becomes abnormally high, the result can be increased body hair and even beard growth.

Puberty in the female marks the start of the *ovarian*, or *menstrual, cycle*, in which the gonadotropins and ovarian hormones rise and fall with some regularity. Other female vertebrates may experience somewhat analogous *estrous cycles* with a regular frequency or perhaps only one or two times each year. The remainder of the time they are sexually unresponsive and infertile. The human female is considered to be almost unique among mammals in always being sexually responsive (or at least in not having predictable cycles of responsiveness. Some other primate groups are sexually receptive during nonfertile periods, a point we will return to later.)

A mature ovum is released from one of the two ovaries about every 28 days. Of course, this is only an average figure, with normal ranges extending from 23 to 34 days. The cycles usually occur with considerable regularity, though it is not unusual for the menstrual cycle to vary unpredictably among women and for an individual woman at different times. Is it just a coincidence that the average period of the menstrual cycle is the same as the cycle of the full moon? The question has been asked many times, but we don't have an answer.

In discussing the menstrual cycle, it is common to place the essential event of ovulation (the release of an ovum) at midcycle, or the fourteenth day, and begin the discussion with the preceding events. The accompanying figures will help in understanding how the sequence is organized (Figures 29.20 and 29.21).

The first half of a cycle is known as the *proliferative*, or *preovulatory, phase*. During the first 3 or 4 days (again, there is some variation), the uterine lining is being shed in what is referred to as the *menses*, or *menstruation*. Since menstruation is cyclic, we will return to it, but first let's see what the hormones are doing.

In the proliferative phase, the pituitary, under the influence of the hypothalamus, releases more FSH. Its targets are the immature ova in the germinal

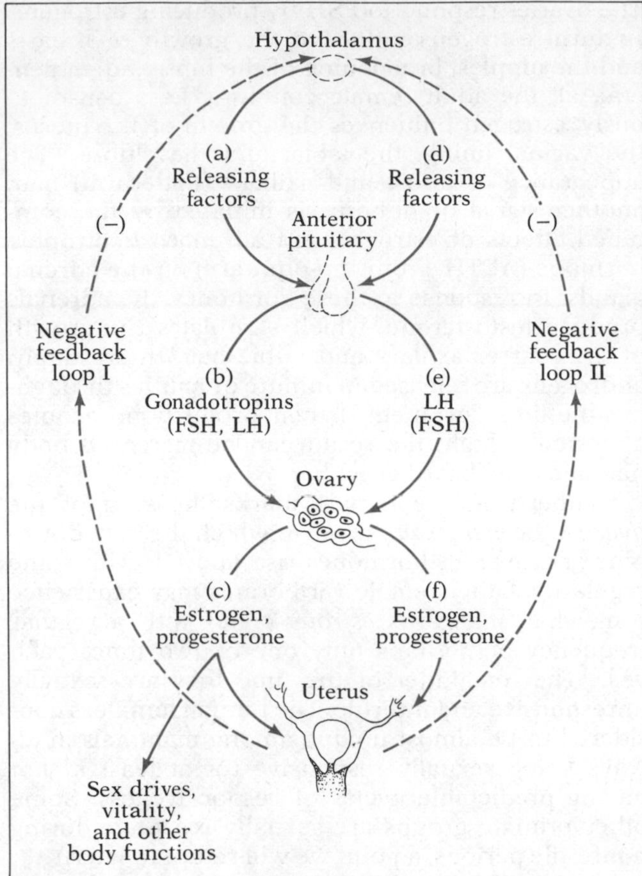

29.20 Hormonal interactions of the menstrual cycle. *Negative feedback loop I:* (a) During the first half of the menstrual cycle, releasing factors from the hypothalamus stimulate the release of FSH and LH from the pituitary. (b) FSH is the dominant hormone at this time. (c) The effect on the ovary is to stimulate the growth of a follicle and the production of estrogen. Estrogen, in turn, initiates cell division in the endometrium of the uterus. Toward the middle of the menstrual cycle, the rising level of estrogen acts as an inhibiting feedback mechanism on the hypothalamus. It responds by slowing the release of factors that stimulate FSH release from the pituitary. In addition, the hypothalamus stimulates the pituitary to produce more LH, which is released in a sudden surge. The drop in estrogen and rise in LH is believed to be the combination of events that triggers ovulation and the changing events of the second half of the cycle. *Negative feedback loop II:* (d) At midcycle, the hypothalamus switches its activity, now stimulating the pituitary to produce LH in a surge. (e) The change from FSH to LH dominance is accompanied by ovulation, marking the fourteenth day. (f) Under the influence of LH, the empty follicle in the ovary develops into the corpus luteum and produces both estrogen and progesterone. Under the combined influence of the two ovarian hormones, the uterus responds with more cell division and capillary growth in the endometrium. Toward the end of the cycle's second half, unless a fertilized egg has implanted, progesterone and estrogen levels in the blood reach a high and complete a negative feedback loop to the hypothalamus. In response, the hypothalamus slows its stimulation of the pituitary, and LH levels begin to dwindle. As a result, the corpus luteum fails and slows its production of estrogen and progesterone. Without their support, the endometrium breaks down and the menstruation begins. Meanwhile, the drop in estrogen and progesterone levels has removed the inhibition from the hypothalamus, and it once again stimulates the pituitary to release FSH as a new cycle begins.

epithelium of the ovary. Under the influence of increasing levels of FSH, the cell clusters around several immature ova will grow, forming *follicles*. The hormones produced by the largest developing follicle inhibit the formation of other follicles, so usually all the follicles but one will regress (although simultaneous development of two follicles is not uncommon). Even if two follicles should develop, most of the time only one, if any, will give rise to an embryo. Fraternal (two-egg) twins occur in only about 1% of births.

The hormonal suppression of follicle development not only cuts down on multiple births but completely suppresses ovulation during pregnancy. Hormonal birth control pills are based on this relationship, inhibiting follicle and ovum development by simulating the hormonal condition of pregnancy. In normal cycles, once a follicle predominates, it grows steadily through the first 14 days. The cluster of cells comprising the follicle will form a fluid-filled cavity and eventually enlarge to protrude like a blister from the ovary's surface. Sometime during the fourteenth day the ovum, in a layer of surrounding cells, will burst from the ovary and begin its journey to the oviduct (see Figure 29.18).

As it grows, the follicle secretes increasing amounts of estrogen whose principal target is the lining of the uterus. When the uterine lining is stimulated, it undergoes rapid cell division, building dense layers of cells over the entire surface, layers permeated by a profusion of blood vessels. The result is a highly vascularized lining known as the *endometrium*. The cervix, also in response to rising estrogen levels, secretes an alkaline mucus, changing the usually acid vaginal environment to one more hospitable to sperm cells.

Obviously, then, hormones tightly coordinate the events in the ovary with those in the uterus. One result is that at the time of ovulation the uterus is nearing a fully receptive state. Should the egg be fertilized within the oviduct, it will descend into a highly receptive environment.

As the fourteenth day of the cycle approaches, the hormone production changes. The rising blood estrogen level finally triggers action by the pituitary, instigating two changes. First, FSH secretion is inhibited, or at least dramatically lowered. Second, the pituitary rapidly increases its production of luteinizing hormone (the "LH surge").

The target of LH is the enlarged, fluid-filled follicle. Under LH influence, the fluid content rapidly increases and the follicle ruptures, releasing the ovum. The event is heralded by a slight, temporary elevation in body temperature (but only up to about 1°F (5/9°C) well within the range of normal daily fluctuations). Temperature elevation then, is used by those trying to establish the "fertile period" in the interest of conception, or, depending on one's point of view, contraception.

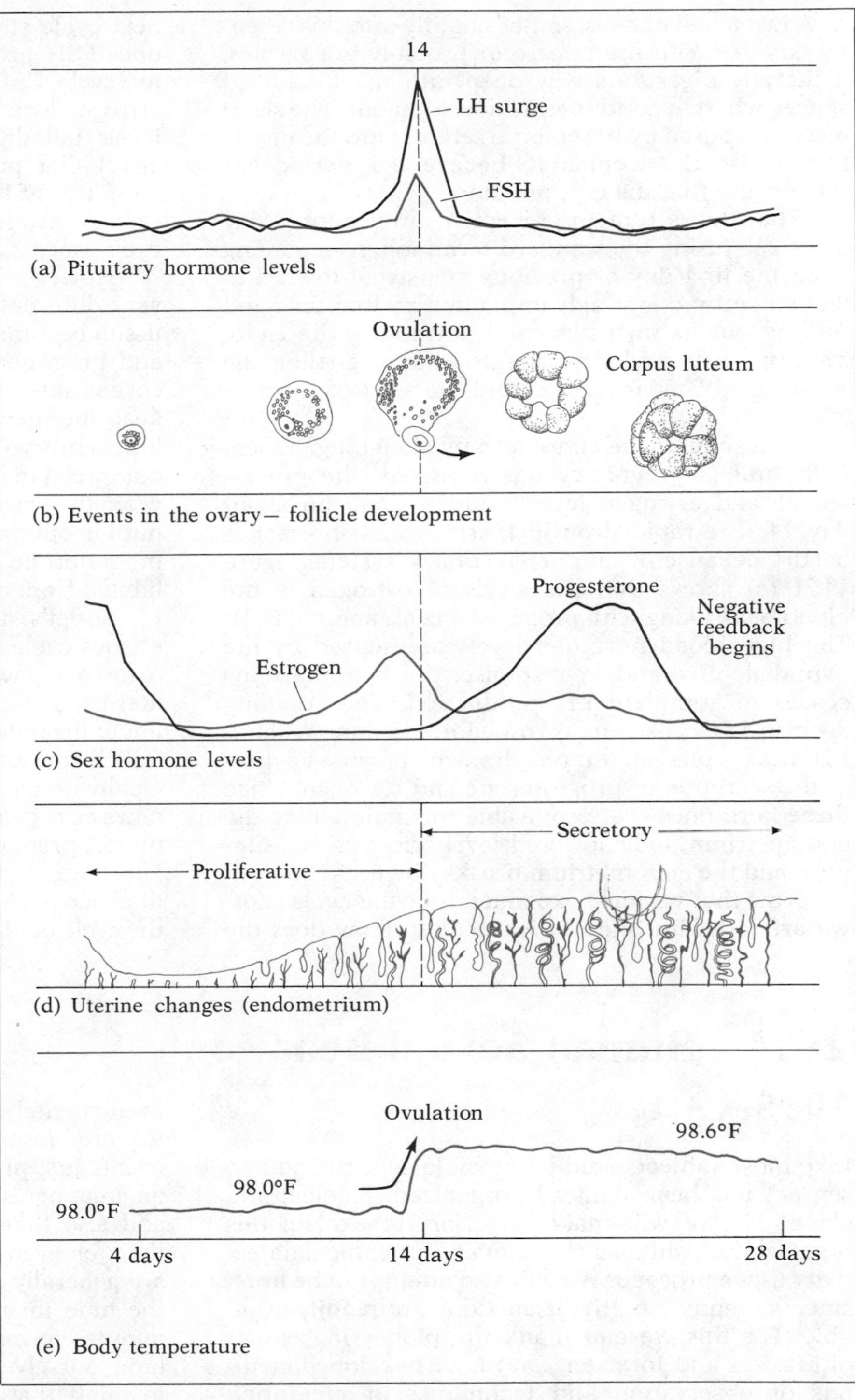

29.21 The interrelationships of the menstrual cycle. (a) The pituitary hormone levels through the 28 day period. Note the LH surge at midcycle. (b) The maturation of a follicle, with ovulation occurring on day 14. In the latter part of the cycle, the follicle becomes a corpus luteum, secreting estrogen and progesterone. (c) The rising and falling levels of these ovarian hormones. Note the temporary midcycle drop in estrogen and the rise of progesterone in the second half. (d) The uterus responds directly to estrogen and progesterone. In the second half of the cycle, the endometrium becomes highly vascularized. (e) Finally, a typical body temperature record through the 28 days. The subnormal temperature of 98°F (about 36.7°C) exists through the first half, but rises sharply at ovulation, to 98.6°F (37°C) through the remainder, gradually dropping back to 98°F as the ovarian hormones decrease.

Following ovulation, the *secretory,* or *postovulatory, phase* of the cycle begins. In the ovary, the vacated follicle, under the continued but somewhat decreased secretion of LH and FSH, forms a new structure, the *corpus luteum* (*corpus,* body; *luteum,* yellow). The corpus luteum is responsible for continued estrogen secretion and initiating the rise of a second hormone, *progesterone.* Progesterone helps to make the final preparations in the uterus for receiving the egg. And should pregnancy occur, it is responsible for maintaining the condition of the uterine lining until birth.

Progesterone and estrogen levels rise in the blood for about 6 days following ovulation. During this time, the endometrium of the uterus reaches its greatest development. Its loosely packed cell layers become fluid-filled sinuses, penetrated with glandular tissue that produces other secretions. The tiny embryo, now just a ball of cells, settles happily into this receptive tissue. With the release of proteolytic enzymes, it literally digests its way deep into the fluid-filled spaces where it continues its development. The slight bleeding caused by the embryo settling into the uterus may cause the woman to believe her period has started and that she isn't pregnant.

The uterus remains receptive until about "day 25 or 26" (using the standard terminology—counting from the first day of previous menstrual flow). Unless an embryo is snugly implanted by that time and putting out its own placental hormones, the endometrium will begin to slough away, signaling the beginning of menstruation and the start of the next 28 day cycle.

What causes the endometrium sloughing? Essentially, unless pregnancy has occurred, the progesterone and estrogen levels suddenly fall on about day 24. The rapid drop in these ovarian hormones occurs because of another feedback system. Figure 29.21 (c) shows that the levels of estrogen in the blood peak along with progesterone on about day 21. The high blood hormone levels are sensed by the hypothalamus, and in response, the hypothalamus ceases to stimulate LH production. The resulting drop in LH causes its own chain reaction. Without LH, the corpus luteum recedes, which causes a drop in the secretion of progesterone and estrogen. Since these hormones are responsible for maintaining the endometrium, their lowered level removes this support, and the endometrium breaks down.

Now that we have explained how the cycle ends, we are left with one more question. How does the next cycle start? What we are really asking is how does FSH production begin again? If you recall, at midcycle, FSH production slowed as a result of high estrogen levels. Now, at the end of the cycle, estrogen levels fall drastically. This removes the inhibitory effect that prevented FSH production. The signal "produce FSH" is again released, and its production starts. New follicles in the ovary respond and the cycle repeats.

However, if pregnancy has occurred, things are very different. The offspring's own extraembryonic tissue becomes an endocrine gland, putting out LH and progesterone. Back in the ovary, the withered corpus luteum continues to secrete hormones that keep the uterine lining hospitable for the voracious little embryo. It will remain at its task throughout pregnancy. The end of pregnancy is signaled when the extraembryonic tissue of the placenta stops putting out pregnancy-maintaining hormones. When progesterone levels drop, the uterus is no longer inhibited from contracting, and labor begins.

So let's summarize the important events of the estrous cycle. It involves the production of a mature ovum along with the simultaneous preparation of the uterus to receive it. The first half of the cycle is under the influence of FSH and estrogen which cause follicle and uterine growth. The second half is principally under the influence of LH, progesterone, and more estrogen. The effects are ovulation and the continued preparation of the uterus for implantation. Feedback mechanisms create the hormonal switch that occurs at midcycle and they finally terminate the cycle on day 28.

29.3 Human Sexual Behavior

The Sex Act

Like most subjects studied by biologists, the human sex act has been clinically organized, labeled, and classified. We will make an effort to explain this organization and also to maintain our clinical objectivity as we proceed. We will also attempt to be brief, since so many lengthy discussions are readily available. For this we can thank the pioneering efforts of Masters and Johnson. They have developed methods of observation and techniques of measuring many of the physiological aspects of human copulation that were heretofore only known anecdotally, if at all. They have also laid to rest many erroneous ideas about the subject.

The sex act (or *coitus, sexual intercourse, mating, copulation*) is somewhat arbitrarily divided into four phases in both males and females. These are *excitement, plateau, orgasm*, and *resolution*, in that order. We hasten to point out that these responses can occur in the absence of coitus, through manual or even mental stimulation. But we will assume that we are describing actual copulation—that is, the events just preceding and following the insertion of an erect penis into a receptive vagina. We had better add also that not all the events listed occur every time, or even most of the time. Further, the phases are generally continuous, not clearly separated, and the time involved may take anywhere from a few minutes to many minutes. Also, obviously we must limit ourselves to an idealized discussion, keeping in mind that when it comes to human sex, there are variations on variations.

Excitement

The initial events of sexual response in women are characterized by faster heartbeat, faster breathing, and a rise in blood pressure. This is accompanied by more outward manifestations such as firming of the nipples, spontaneous muscle contractions, and often what is known as the *sex flush*. This amounts to a reddening of the skin, particularly about the genitals, face, breasts, and abdomen, but not restricted to those areas. Changes also occur in the genital area, and include a swelling of the glans clitoridis (analogous to erection in the male), general moistening of the vaginal walls, a marked elevation and parting of the labia majora and a size increase in the labia minora. Insertion of the penis is possible as soon as the walls are sufficiently lubricated and the muscles of the vagina have relaxed. However, these changes do not necessarily signal psychological readiness of the woman.

In the male, excitement is characterized by somewhat similar events in circulation and respiration. His reactions, however, may occur more rapidly than hers. Genital changes include vasocongestion in the penis, bringing about an erection, accompanied by a general thickening and elevation of the scrotum. Lubricating fluid may also appear at the tip of the penis. While the penis is in this condition, let's assume that it has made its entrance.

Plateau

As copulation begins, in women, the events of the excitement phase continue, sometimes accompanied by vigorous pelvic thrusting. The swollen clitoris withdraws into its hood rather suddenly at this time, and is now extremely sensitive to touch. The labia become swollen, as does the outer portion of the vagina, especially the opening. The uterus may elevate, tilting backward somewhat. These events, as intensity grows, produce what has become known as the *orgasmic platform*—physiological and anatomical changes that may precede orgasm. Physical activity may peak now, with intense thrusting of the pelvis, although there is a great deal of variation here. Facial muscles characteristically relax, producing a slack appearance with an absence of expression.

In men, the intensity of the excitement phase has also increased steadily. The glans reaches its greatest size and the penis achieves its greatest curvature and rigidity. The testes increase in size and further elevate in the scrotum. Pelvic thrusting increases and secretions from the bulbourethral glands increase just prior to orgasm.

Orgasm

The term *climax*, often used to describe orgasm, is fitting. All the events in the first two phases, as their intensity has increased, come to a climax in our ideal copulation. In both men and women, the orgasm is a matter of complex muscular contractions, usually accompanied by intensely pleasurable sensations. These contractions are reflexive and involuntary, occurring in a steady series. They emanate in women from the orgasmic platform—the muscles and thickened tissue around the vaginal opening—joined to varying extent by those of the vagina and the uterus. During these contractions, the cervix is pulled upward, increasing the diameter of the surrounding inner vagina, and drawing semen upward. This event is known as *tenting*, and some authorities believe it permits semen to *pool* around the cervix. Contractions in the uterus are thought somehow to draw the semen into the cervix and uterus, but this idea remains to be substantiated. Uterine and cervical contractions are caused at least in part by the hormone *oxytocin*, which is released from the pituitary in large amounts at this time.

As the female orgasm continues, the general contractions occur in about 0.8 sec intervals, spreading through the lower pelvis. In an intense orgasm, many of the other muscles of the body may begin to contract spasmodically. The actual origin of an orgasm is a subject of controversy at this time, but the preponderance of opinion is that clitoral stimulation triggers this phase. Stimulation preceding and during orgasm is probably indirect, involving the clitoral hood, which is apparently moved rhythmically against the clitoris by the thrusting action of the penis and by shoving of the two pelvises. Thrusting also tugs against the labia minora which, as you can see in Figure 29.17, is continuous with the hood.

The duration of orgasm in women is highly variable. Several common reactions are shown in Figure 29.22. There may be one intense peak or several, or the peaks may be far more numerous but less intense. The ability of women to experience several orgasmic peaks in a single coitus is sharply contrasted to the usual single orgasmic peak to which men are limited.

Orgasm in men is synonymous with ejaculation. The event can, however, be divided into two distinct parts. The first is *emission*. Emission occurs with the rhythmic, peristaltic contractions in the vas deferens, prostate, and seminal vesicles. As a result, the elements of semen, spermatazoa, and prostatic fluids, are delivered to the bulbourethra prior to expulsion. Ejaculation, or part two, will immediately follow. This forceful expulsion of semen occurs as powerful striated muscles at the base of the penis contract, also in intervals of about 0.8 sec. The semen is forced through the urethra and out of the penis in several spurts, the first usually bearing more semen than the rest. Ejaculatory contractions usually continue for a short time (several seconds). Involuntary contractions of many muscles of the lower pelvis are usu-

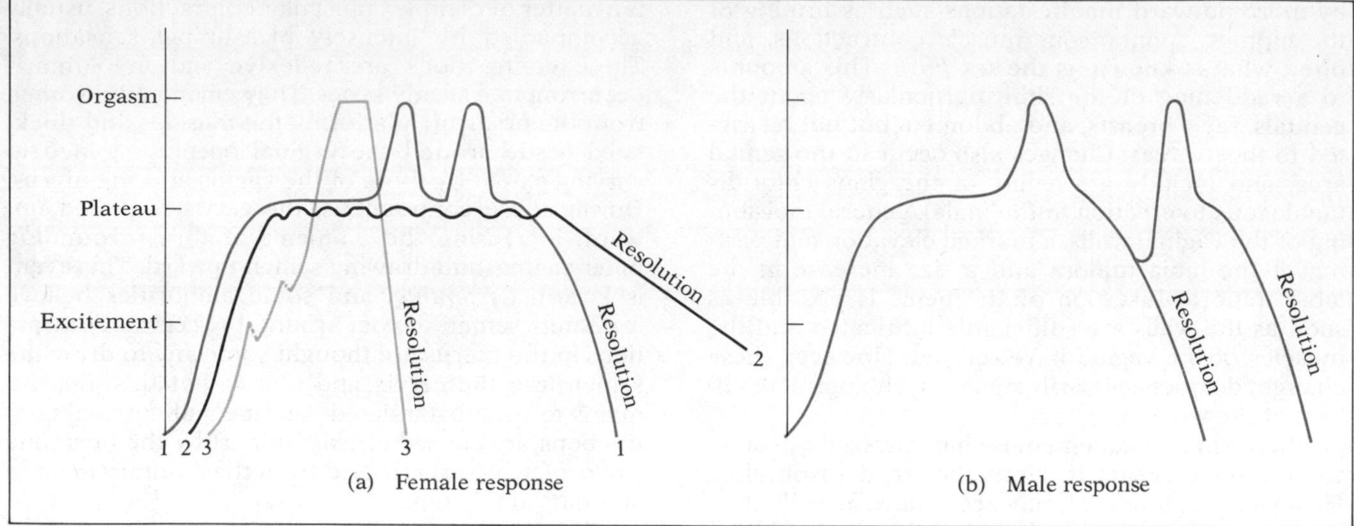

29.22 The orgasmic response. (a) In the highly variable female orgasm, we see three types. In ①, it occurs in several peaks of intense response. The number varies considerably in such multiple orgasms. In ②, there are a greater number of multiple events which are far less intense—just above the plateau level. In ③, the female orgasm is similar to that in males, with one intense period that lasts somewhat longer than the male's. Note the difference in resolution time in the three types of female orgasm. (b) Compared to this, the male orgasm is much simpler and more stereotyped, although as the graph indicates, orgasm can occur again in males after a period of resolution. The recovery time between orgasms in males is known to increase with age.

ally involved also. On the border between voluntary and involuntary movements are the intense, even violent pelvic thrustings and muscular contractions involving the whole body that commonly accompany this phase of the sex act.

Resolution

Following orgasm, both males and females retreat from the state of sexual excitement into what is known as *resolution*. This phase has been referred to as *afterglow*. Men lose erection rapidly, and enter a *refractory* period in which rearousal is nearly impossible—at least for some minutes and perhaps longer. Resolution in women is somewhat different, since some women can be aroused again almost immediately. In any case, the quiet, relaxed, and often tender resolution phase may have an intense pleasure of its own.

We remind you once again that this is an idealized account. In the variability of real-life situations, one often finds the idealized copulation to more of a fiction than a reality—though this may be one aspect of life where an imperfect reality is vastly preferable to a fictitious perfection. In reality, some of the stages—including orgasm—may be abbreviated or altogether lacking in one or both partners. Orgasm is often synchronous for some couples and is never synchronous for others, a fact that often turns out not to make very much difference to the people involved.

Masturbation

Masturbation, or *autoeroticism*, has been discouraged in almost every culture we know about. To this day, American children live in dread of getting caught at it, a disgrace indeed. The problem is, surveys indicate that about 90% of adults admit to masturbating regularly and the other 10% are suspect. It seems incongruous, then, that many people are still worried about whether masturbation is harmful, either physically or mentally. The answer is that it is not (unless you count fear and guilt). A controversial study by K. E. Money indicates that masturbation and early sexual experience (also taboo in Western culture) may even be beneficial. He argues that waiting until several years after the onset of puberty to have sexual stimulation may produce a dysfunctional system. He also thinks that early experience helps prepare the person's nervous system so that later the probability of sexual problems, such as frigidity in women, is reduced. Analogies can be made with other kinds of sensual deprivation. For example, if a kitten is kept in darkness during the few days that it normally learns to use its eyes, it will become permanently blind.

What Does Copulation Accomplish?

You might imagine a number of circumstances under which you might hear this question, but, for animals in general, the answer is obvious. The sex act, and the desire to engage in it, directly assures fertilization. Thus, we see animals either copulating or trying to, or at least thinking about it. But there are behavioral differences in the sexes, as you may noticed. Once they are sexually mature, males of many species are ready to engage in copulation at any time and as often as possible. Females, on the other hand, may be more selective, and are receptive only at certain times. In yet other species, sex is a seasonal thing, and neither males nor females are interested in copulation except at certain times—every spring, perhaps, or (for grunion) during the full moon. In such species, the males may not even be able to produce sperm at other times. Their testicles may regress in size or be pulled deep into the abdomen until the breeding season. Such males cannot be enticed at other times.

Humans appear to present a special case, as is so often true where behavior is concerned. Men produce sperm constantly and are usually willing to copulate whenever any suitable opportunity presents itself (social mores permitting). But the perpetual willingness of males is not unusual among mammals. Dogs and stallions, for example, are also always willing. On the other hand, women are unusual among mammals in not having pronounced cycles of sexual reproductivity. Women are able to copulate at any stage of their hormonal and ovulation cycle, and even during pregnancy (Figure 29.23). But most of the time, there is no ovum to be fertilized. In fact, the probability of pregnancy occurring as the result of one copulation in a woman at the peak of her fertile years—not using contraception—is less than 2% (one pregnancy in fifty matings).

29.23 Peaks and valleys of sexual activity through the menstrual cycle. The results of one study of female sexual response and orgasm through the 28 day menstrual cycle are very informative. If these data are truly representative of humans in general, we can make several observations about our sexual behavior. First, receptivity occurs throughout the cycle, but has peaks and valleys. The lowest frequency of orgasm occurred during the menstrual flow, although intercourse didn't drop off to any great degree. Second, an activity peak occurs just before ovulation and it is accompanied by an orgasmic peak. This is to be expected if fertilization is to be maximized. The second set of peaks is a little more difficult to account for since it is apparently too late in the cycle to accomplish fertilization.

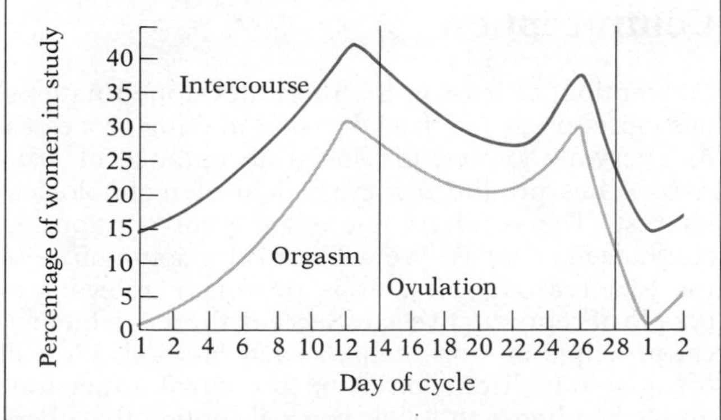

That's not very efficient. So why haven't women correlated their matings with their fertility? Could sex have a function other than reproduction in humans? A leading hypothesis is that the continual receptivity of human females is a result of *behavioral evolution*—evolution of behavior that has proven adaptive in our human ancestry. Men and women tend to form long-term bonds, an unusual trait in the mammal world. Even now, during a time of high divorce rates, people continue to marry, and more than half of all marriages in the United States end only in the death of one partner. So, the question is, why do we marry? Why do we have romantic notions about "love" between the sexes? Why do young people tend to pair off, always utterly convinced that they have found "that certain someone" with whom they will share love for the rest of their lives? When all goes awry, why does it take them so long to sever those frayed bonds and to get out of the marriage? Why do we tend to cling to each other so tenaciously? Part of the answer may have to do with sex. Sex definitely helps build ties. (This is one reason promiscuity is often discouraged; people aren't psychologically prepared to go around building bonds indiscriminantly.) Thus, it is argued, if a woman is always able (and often willing) to copulate, the eager male will tend to stay with her. She will reinforce this behavior by copulating with him, thus strengthening the bonds between them.

We tend to accept the notion of long-term bonds as "good." People who have lived together for 50 years are more venerated than those who have had fifty spouses. But why have we developed these values? Why do we love love? We must look to the evolutionary advantages of bonds.

In species in which fathers participate in supporting and/or caring for their offspring, it is highly adaptive for the survival of children—which is to say, for the reproductive fitness of both mothers and fathers—for the parents to stay together while children are growing up. Evolution is an opportunistic process, and apparently one of the ways that couples are encouraged to stay together was (and is) sex. Men whose mates were continually receptive had little need to wander off in search of other opportunities (and while they were gone, they ran the risk of being cuckolded). On the other hand, women who tended to be sexually oriented might like the idea of staying with a man. And from such simple beginnings, human bonds could have sprung. By being coerced to stay together, both parents would have been able to divide labor, cooperate, and, together, raise more children. Natural selection seized upon a behavioral drive that was already well-developed—the sex drive—and used it for a new purpose: cementing the pair bond. People who tended to form such bonds tended to leave more children like themselves. Thus, bond-building, and the things that lead to it, would have come to predominate social patterns in human culture.

Of course, other things keep people together too, under some mixture of social conventions and biological predispositions. "Falling in love" may well be another evolved behavior that has to do with creating a new pair bond, and any kind of sexual love (as compared with sexual desire) may be our subjective perception of our own drive for pairing behavior.

Human female sex hormones and behavior cycles are unusual among mammals in another way also. At about age 45 or 50, the cycle shuts down. Estrogen levels drop and ovulation ceases, although sexual receptivity normally doesn't cease. The event is called *menopause.* This apparently has evolved to maximize the reproductive potential of the woman. So let's engage in some evolutionary speculation about it.

How can a female increase her reproductive performance by becoming sterile? It would seem to have something to do with the lengthy childhood of the species and the fixed life span of humans. After all, reproductive performance depends on how many of one's offspring reach adulthood. Evolving women could leave more successful offspring by ceasing to produce babies at some point and to concentrate on caring for the children she already has or by assisting with grandchildren. If a primitive woman continued to become pregnant at a relatively advanced age, she wouldn't be likely to live long enough to see them to adulthood. Thus they would be left motherless and not likely to survive anyway. She had best shut down the system earlier and concentrate on whatever offspring she has, or on her grandchildren, since they, too, bear her genes. The sex bond is still useful, however, because the parents' reproductive role is not complete until the offspring are independent. And older people can continue to stay together and to care for their descendants, so perhaps sex, even in old age, is a method of reproduction.

We would be the first to admit that these views are not universally held, although we think that most biologists would go along with them. Other experts in the area hold that men and women form pair bonds because some higher being told them to; some say that the occurrence of sexual intercourse without the possibility of conception is unnatural and wrong. If, by unnatural, they mean that sex without chance of pregnancy is not practiced among the other creatures, they are mistaken. In fact, a number of species, particularly primates, engage in sex when there is no chance of reproduction. Their promiscuity is believed to be a reinforcing mechanism that helps hold the group together.

Fertilization

Fertilization occurs, almost invariably, in the upper region of the oviduct. This means that sperm cells have to move a considerable distance from where they arrive near the cervix (nearer in some cases than others). Actually, the sperms' movement is due more to the action of the oviduct's cilia than their own swimming. The journey up the oviduct (a distance of several inches) is a hazardous one, partly because of the acid environment of the vagina, the distance involved, and the random movement of fragile sperm cells. Indeed, of the hundreds of millions released at ejaculation, only about 5000 or 6000 ever reach the upper oviduct.

If there are several thousand sperm cells in the oviduct tubes when the ovum begins its journey, fertilization is almost assured. But what are the time limitations? A conservative estimate by gynecologists is that fertilization must occur within 48 hr of ovulation. Any time later means the sperm will reach a dead ovum. This means, essentially, that days 14–16 (or 2 days after body temperature elevation) is the *fertile period.* The big question then is: How long can a sperm cell live? And here is the gaping flaw in the rhythm system of birth control. We don't know! A conservative estimate, again, is 48 hr. Therefore, intercourse on days 13–16 present the best opportunities for fertilization. Keep in mind that these are just averages, and that people vary. In some cases, pregnancy has even resulted from intercourse during menstruation. You may recall the old joke, What do you call people who use the rhythm method? Parents. (The joke is funnier to some than others.) Let's see if there aren't some more dependable means of birth control.

Contraception

Conception, at least in the more developed nations, has increasingly involved decision-making processes. As everyone knows, the social acceptance of *birth control* has produced a great deal of technological interest. The result is, there are many options in *contraception* today. We will consider some of these for two reasons. First, it is obviously relevant to people of reproductive age. Second, there is a biological principle at work behind each method. We will organize our discussion of birth control under four major headings, although you will notice that there are instances when the groups overlap or are used in combination.

Natural Methods

The natural methods may be somewhat unnatural in their application. We call them this simply because they don't require chemical, mechanical, or surgical intervention. We will consider abstinence, coitus interruptus, douching, and the rhythm method.

For some reason, *abstinence* has never really become popular with most people, but it is rather dependable. Sociologists, by the way, tell us that abstinence among married people is more common that one might suspect.

Coitus interruptus has long been an extremely common method of contraception. It is simply the act of withdrawing the penis just prior to ejaculation. But it is risky. It is well-known that secretions from the bulbourethral glands, preceding ejaculation, are often accompanied by prematurely released sperm cells. In addition, some couples may engage in intercourse more than once, ignoring the fact that there are residual sperm cells still present in the male's urethra. Finally, there is the psychological frustration of a man's having to do exactly what he is least inclined to do, and some men aren't as dependable as others. The act places a great emotional strain on the woman as well as the man—and what's worse, it may deprive her of her own orgasm. Thus coitus interruptus is considered to rank only as "fair" in terms of dependability (Table 29.1).

Douching after intercourse was a fairly common method of contraception in past years. It involves flushing the vagina with plain water or various household or commercial solutions. Diluted vinegar has traditionally been the most common douche; its acid properties were intended to kill the sperm cells. Douching, regardless of the agent, has a 30–35% failure rate. (Failure rate in contraception studies refers to percentage of pregnancies per year of dependence on one method.) The reason it fails has to do with the time element. Sperm cells are known to pass safely through the cervix into the uterus only minutes after ejaculation. To improve its dependability, a woman would have to bolt from her bed as soon as ejaculation had occurred, rush to her

Table 29.1 Methods of Birth Control

Method	Mode of action	Effectiveness	Side effects or objectionable aspects
Coitus interruptus	Withdrawal of penis prior to ejaculation	Low to fair	Frustrating to many
Douche	Flush semen out (water, dilute vinegar, commercial preparations)	Very low	Possible irritation and infection
Rhythm	Abstinence during fertile part of cycle	Low to fair	Tension between partners, apprehension
Condoms	Prevents sperm from entering vagina (rubber sheath)	Fair to high	None (perhaps expense and inconvenience)
Diaphragm with spermicide	Prevents sperm from entering uterus; kills sperm	High	None (perhaps expense and inconvenience)
Cervical cap	Prevents sperm from entering uterus	High	None (difficulties with fitting and inserting for some)
Intrauterine device (IUD)	Unknown, but may interfere with implantation	High to highest	Loss of device; discomfort in some; perforation of uterus very rare, but 1/10,000
The pill	Hormones prevent ovulation	High to highest	Controversial; most evidence indicates much safer than pregnancy and birth
Sterilization	Vasectomy (cutting and tying vas deferens); tubal ligation (cutting and tying oviducts)	Highest	Possible psychological effects in some; risk of painful granuloma
Abortion			
Suction; Dilation and curettage (D&C)	First trimester, removal of embryo by suction device or currette		Some surgical risk; ethical objections common
Saline	After first trimester, injection of 20% salt solution into uterus, starts uterine contractions and expulsion of fetus		Increased risk (salt poisoning, infection, hemorrhage)

douche bag, and immediately flush the semen from her vagina. For some people this requirement might tend to erase some of the romance from one of life's tender moments. It also puts the burden entirely on the woman.

The so-called *rhythm method* is perhaps the most biologically interesting method of birth control. It is called the "most natural" method by its proponents, but its tools are papers, pencils, thermometers, and calendars. In effect, it relies heavily on the imposition of strict, clinical preparation if it is to be successful. Its success depends on a woman's precise knowledge of her menstrual cycle, including an accurate prediction of ovulation. Essentially, the practice involves abstinence during the period in which fertility is possible. Theoretically, this includes 4 days each month, 2 days on either side of ovulation. Usually, an additional day on each side of the 4 days is recommended for added insurance. A review of Figure 29.24 adds another dimension to the problem, however. As you can see, a peak in sexual receptiveness for many women occurs just prior to ovulation, a decidedly high-risk time, and a very difficult time to abstain. These problems add up to 15–35% failure rate, depending on one's data source.

29.24 The rhythm method of contraception is based on sound biological principles. It is logical to assume that there are definite periods of time during the menstrual cycle when conception is highly unlikely. The problem, as illustrated here, is in the variability among a large population of women and, sometimes, in our ignorance of some aspects of fertility. For example, consider the variability in the 28 day cycle. Ovulation on day 14 is probably a good average, but it is only an average. In addition, the survival period of spermatozoa in the female genital tract is estimated at 48 hr, but this information is not firm and the female environment undoubtedly varies. Finally, the rhythm system is based on the premise that people read thermometers well and keep accurate records. We must assume that they somehow compensate for variables that could affect the cycle (such as illness, stress, etc.). Note the first peak in receptiveness at about day 12, which corresponds with the start of the "most fertile period." The abstinence period is somewhat conservative, allowing extra time for sperm survival and bad record keeping.

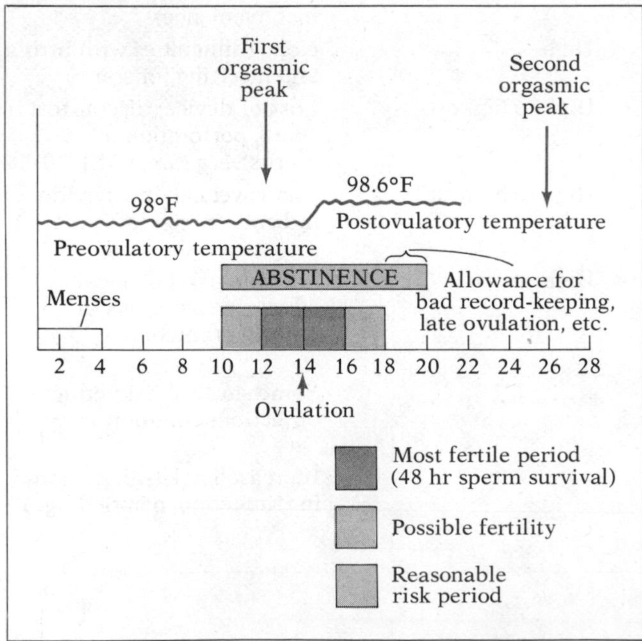

Mechanical Devices

We will consider mechanical devices to be any physical object used to block sperm travel or embryo implantation. These include the condom, diaphragm, cervical cap, and intrauterine device (IUD) (Figure 29.25). The first three are rubber sheaths designed to block sperm travel. We don't know how the fourth works.

Condoms are very thin rubber or animal-membrane sheaths designed to fit over the penis. They are commonly known as *rubbers* or *prophylactics,* and a host of other names you wouldn't believe. Condoms are probably the most common means of contraception, with nearly a billion sold annually in the United States alone. This attests to their easy availability rather than their dependability, since the failure rate for those using condoms is from 3 to 15%.

The *diaphragm* is a rubber dome with a thick rim. It is worn over the cervix during intercourse. Generally, the inner side is coated with a spermicidal jelly or cream before it is inserted. Since cervices come in a variety of sizes and shapes, selection and fitting require the service of a skilled practitioner, usually a physician. Data on the diaphragm's reliability are inconsistent, with failure rates varying from 5 to 20%.

Cervical caps were actually invented in the last century, and are much more popular in Europe than in the United States. They operate essentially in the same way as the diaphragm, but fit more firmly over the cervix and can be worn for days or even weeks. Their failure rate is about 8%.

The *IUD* is a widely used, very effective contraceptive device. It is also the least understood. It supposedly originated with Arab camel drivers who placed stones or apricot pits in the uteruses of their female camels to prevent them from becoming pregnant on long caravan marches. One can only wonder at how such a practice got started. Since the early 1930s, it has been known that a silver wire coil inserted into a woman's uterus is a very effective birth control device. The first study, involving 600 women, demonstrated a very low (1.6%) failure rate. However, interest in the IUD didn't really grow in the United States until the late 1960s, when the very promising reports of a 30 year study were published. Since that time, a variety of forms of the IUD has found use all over the world.

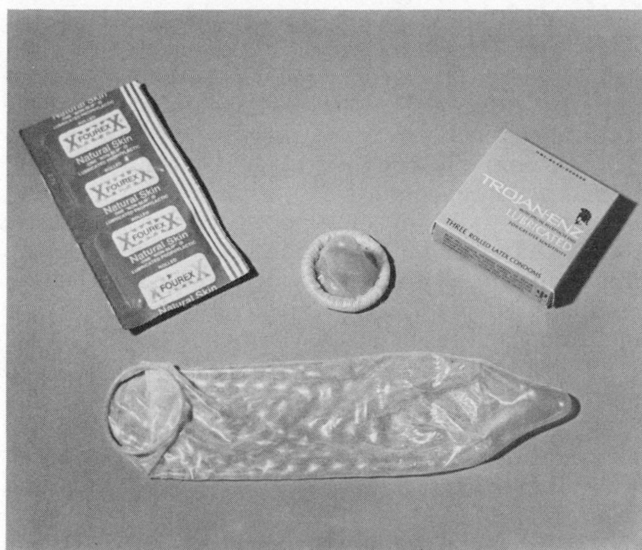

Condom

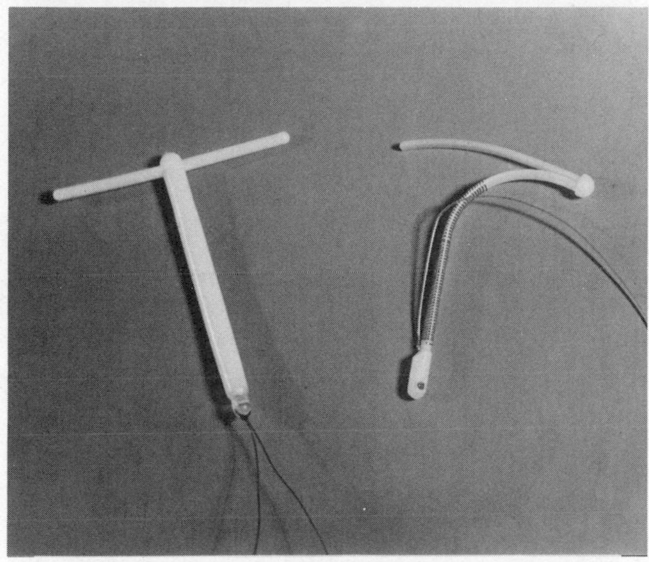

IUDs

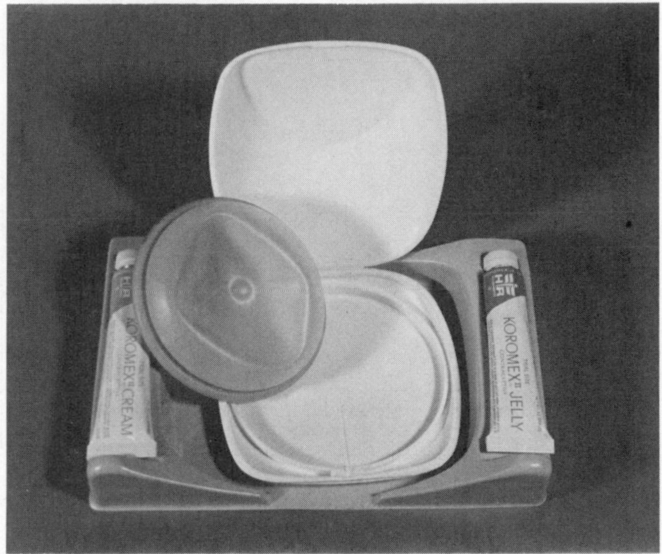

Diaphragm

29.25 Mechanical conception control devices. Most of the devices shown here block sperm entry into the uterus. The most reliable is probably the condom if failure rate is measured against its actual use—but many couples skip this step during presumably safe periods. The condom is also a fairly effective device for preventing the spread of venereal disease. The diaphragm is an effective contraceptive device if used properly. This means one should use it with spermicidal creams or jellies, carefully following prescribed directions. The cervical cap is similar in method to the diaphragm, but much smaller, covering only the opening of the cervix. Its application is much simpler, since it can remain in place for days to weeks, and doesn't require the spermicide. Each of the intrauterine devices (IUDs) is a manufacturer's variation on the same theme. The unusual shapes are the result of the manufacturer's attempt to produce a device with as much surface area as possible, yet not so much as to cause discomfort.

Although its mechanism of action is not fully understood, it appears that the IUD interferes with implantation of the embryo. Whatever its action, its failure rate today is a reported 1–3%, a very respectable range. It is used today as an alternative to the pill for women who do not wish to use the pill for one reason or another.

The primary advantage of the IUD is that it is convenient. Once it is inserted, it can more or less be forgotten, and fertility is restored when it is removed. Therefore, it is often an excellent choice for people interested in family planning. However, IUDs cause cramps and pain in almost a third of women receiving them, and frequently they must be removed. Some have also been associated with uterine infections.

Hormonal and Chemical Birth Control: The Pill

One of the most widely used contraceptives today is the birth control pill. At the present time, an estimated 50 million women around the world are using birth control pills (Figure 29.26).

The active substances in the pill are estrogen and progesterone. These are isolated from the urine of pregnant mares, or synthetic hormones may mimic the natural hormones. When taken according to the prescribed regimen, the pill is believed to prevent ovulation. Ovulation is normally suppressed during pregnancy, and it is thought that the pills simulate some crucial aspect of pregnancy.

In terms of their effectiveness, the two major types of birth control pills have been an overwhelm-

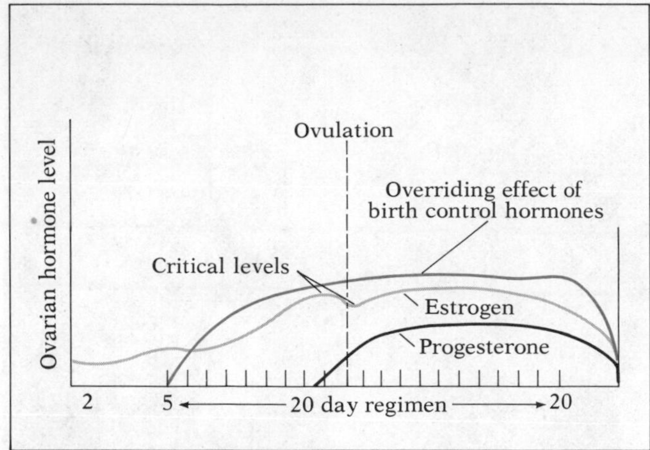

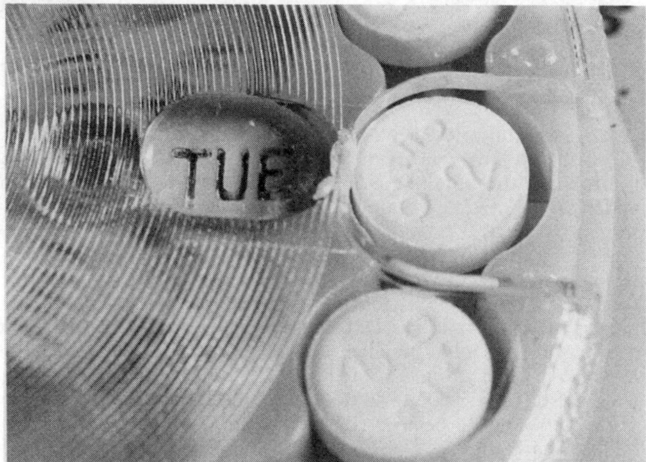

29.26 The effect of the 20 day birth control regimen is to override the normal ebb and flow of ovarian hormones, as seen here. It is believed that the delicate feedback mechanism between ovary and pituitary, which initiates ovulation is blocked, thus preventing ovulation. In addition, the presence of a steady high level of progesterone in combination pills is believed to simulate pregnancy, thus overriding the normal cycle. Actually, there is much to learn about the precise action of the synthetic hormones of birth control.

ing success. The combination pill is nearly 100% effective, and the sequential slightly lower because the hormone content is lower. Nearly all failures are the result of errors in their use.

The debate over possible side effects of long-term birth control pill usage is still going on. Since large-scale marketing didn't begin until 1960, it is unlikely that all the questions on side effects will be answered soon. By about 1990–1995, a substantial number of women will have spent most of their reproductive years on the pill, so, many questions will have been answered. We already know that birth control pills increase the risk of abnormal blood clotting and associated illness—just as does pregnancy—but the risk is still very small. Not using birth control is far more risky to one's health. Other alleged side effects are still problematical. One thing we can say with some assurance: The extensive use of the birth control pill has presented to medical science an experiment of unprecedented size and scope. It is unlikely that we will witness such an event again.

The possibility of chemically controlling fertility in males has repeatedly been considered. In spite of the fact that all the experiments in preventing male fertility have failed, the search goes on. Apparently, the regularity of male hormone production and the comparative simplicity of male reproductive physiology doesn't present an easy target.

In addition to fertility control chemicals, there are several other methods of chemical contraception. They are generally far less effective than hormonal control, however. For the most part, they consist of sperm-killing preparations such as jellies, foams, and suppositories, all used in the vagina. The most effective of these is undoubtedly the aerosol vaginal foam, since it can cover the entire cervix and vaginal lining with the spermicide (We won't even bother with such techniques as coke douches and Saran-wrap condoms. We hope you won't either.)

Surgical Intervention

We now find that surgical intervention in the reproductive process is gaining popularity throughout the Western world. Surgical intervention, of course, refers to sterilization and abortion. The abrupt change in attitudes toward these rather drastic measures seems to reflect a turnabout in values in a time when population increases appear to threaten the very core of human existence. We doubt very seriously whether conscious consideration of world problems are at the root of these changing values in the United States, however, although they may have had an indirect influence.

Male sterilization involves a minor operation known as a *vasectomy* [Figure 29.27(a)]. The procedure can be carried out in a doctor's office with the use of local anesthetic. Essentially, the operation consists of opening the scrotum in two places, and removing a small segment of the vas deferens. The ends are tied off and the scrotum is closed. As a result of the surgery, sperm travel is blocked just ahead of the epididymus. Sperm production continues, but they disintegrate and are cleared away by phagocytes. Sexual activity after vasectomy is completely normal as far as is known, but controlled studies have never been done with vasectomized humans. Vasectomized monkeys and rats show reduced sexual activity and a reduction in the size of the testicles.

Testicular cysts and *granulomas* (inflamed growths) of the vas deferens, caused by the excess trapped spermatozoa, have been reported both in

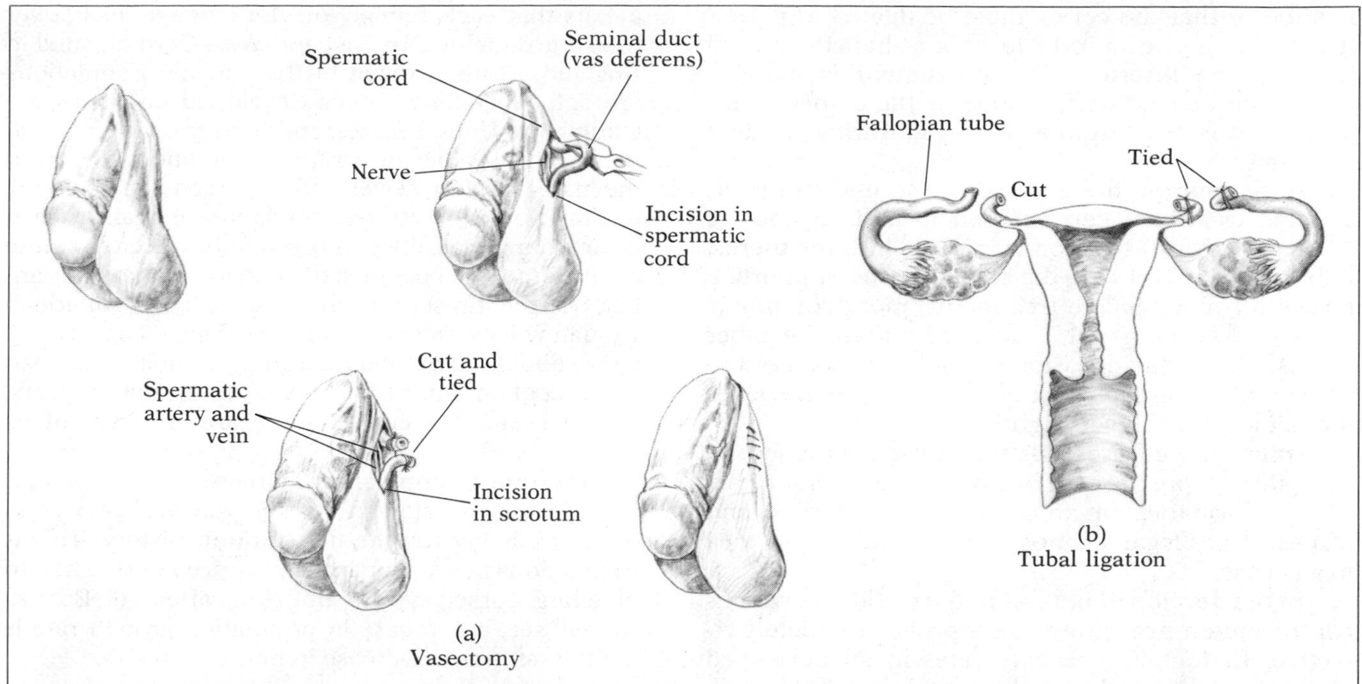

29.27 Both the vasectomy and the tubal ligation are common surgical methods of contraception. Both result in sterilization by preventing gametes from following their natural pathway. (a) *Vasectomy* involves a simple surgical procedure which is often done in a doctor's office. After an incision is made in the scrotum, the spermatic cord is located and cut open to reveal the vas deferens. The vas deferens is then cut and a section is removed. The ends are then tied tightly with suturing thread and subsequently grow closed, preventing sperm from reaching the urethra. This method doesn't interfere at all with normal sexual functions. The amount of semen ejaculated is not appreciably changed, only the spermatozoa are missing. These are absorbed by surrounding tissue. (b) *Tubal ligation* is more complex, surgically, than vasectomy, requiring abdominal surgery. More recently, it has been possible to reduce the size of the incision and use a viewing device and slender surgical instruments to complete the procedure. Tubal ligation consists of cutting and tying the oviducts, thus preventing the egg and sperm from meeting.

vasectomized humans and in experimental animals. It is known that vasectomized men ejaculate a smaller volume of semen and, of course, the sperm cells are absent. Many satisfied customers have reported that the sexual experiences of both partners are greatly enhanced by the removal of the fear of pregnancy. Others have reported unfavorable psychological effects. The operation is usually not recommended for any man not secure in his sexual identity.

Vasectomies are simple, but restoring fertility after a vasectomy is another matter. Restoration is a major operation, requiring full anesthesia and hospitalization. And only about half the operations are successful. Part of the problem is that the cut ends of the spermatic cord usually degenerate a bit and this makes them difficult to join again. Another difficulty is that most vasectomized men develop antibodies against their own sperm cells. In other words, one should have a firm conviction and good information before having a vasectomy. Barring surgical error, success rate as a contraceptive means is 100%, and it is a relatively simply operation.

Sterilization in women (*tubal ligation*) is similar in that the surgery involves cutting and tying the oviducts, thus preventing the sperm and egg from joining. The surgery can be done through small abdominal incisions or by working through the vaginal wall just below the cervix. An incision is used to locate, cut, and tie the fallopian tubes [Figure 29.27(b)]. There are no readily visible scars. The operation is essentially irreversible, and almost nothing is known about psychological effects.

Hysterectomy (surgical removal of the uterus) is an effective birth control measure, but is considered extreme. It is frequently performed, but for other reasons than contraception.

Abortion as a means of birth control is supposed to be more of an "if all else fails" procedure, although in most industrialized nations the abortion rate has come to exceed the birth rate! Abortion is defined as surgically removing, or causing the removal or discharge of, an embryo or fetus. Three major procedures are common. The most recently developed one is the *vacuum,* or suction, method. It is used in very early pregnancies, usually those less than 3 months along (in the first trimester). A slender tube attached to a vacuum pump pulls the embryonic mass from the uterus.

The second method is *dilation and curettage* (D&C). It is a bit more severe than the vacuum method in that the cervix must be dilated to permit the blade of the curette (a spoon-shaped surgical knife) to be inserted. The instrument is used to scrape the uterine wall, dislodging the embryo. This procedure is also usually employed during the first trimester.

If the embryo has entered the second trimester, a saline treatment can be used to induce labor. A 20% saline (NaCl) solution is injected into the uterus. Labor and delivery of the fetus and its supporting placenta follow. Saline treatments, more commonly known as "salting out," are riskier than the other means. There is an occasional problem of salt poisoning and also a greater risk of infection. No one takes the saline abortion very lightly.

Other surgical methods can be used in the second trimester of pregnancy, but none are without risk. The performance of an abortion after the second trimester is illegal in most places, and a bad idea in any case.

In the developed nations of the world, the various contraception procedures have proven extremely effective. In fact, the fertility rates in all developed countries have fallen below replacement, which means that each female, on the average, has fewer than one daughter. In East and West Germany and in England, deaths exceed births and the populations are actually falling. Other developed countries, including the United States, still have growing populations, but only because rapid population growth in the past has left fewer older persons than young persons; unless fertility rates increase again (which is far from impossible), in the next few decades these countries too will begin to decline in population number. At the present fertility rates, each American woman will average 0.8 daughter. There's hardly any worry about the human race going extinct because of contraception within the next several million years, however, simply because there are so many of us now.

The rate of population increase of the world as a whole has also slowed markedly in the last 5 years or so, probably for the first time in history. This is tremendous news: Apparently we are not doomed to breeding ourselves into oblivion, after all. But, as we will see, a decrease in population growth rate is not the same as a decrease in population size.

Application of Ideas

1. Hatching in the precocial state is considered by ornithologists to be primitive, and altrical hatching, advanced. Offer several reasons why this might be so. You might begin by considering the evolutionary origin of birds and the fact that precocious birds are generally ground foragers.
2. The pairing of human males and females has been described as bonding (i.e., the pleasure bond, the pair bond, etc.). Both cultural and genetic explanations of this bonding have been proposed, and these are often in conflict. Select one or the other of these points of view and offer arguments. Can either assertion be tested? Can both assertions be correct?
3. To date, most of the research and medical effort in chemical birth control has been aimed at women. Is this an attitudinal factor on the part of a male-dominated medical profession, or are there serious physiological problems that require this approach? You may have an opinion on the former idea, but explore the reproductive physiology of males and females as you formulate your answer.

Key Words

abortion
acrosome
altricial
androgen

birth control pill
broad ligament
bulbourethral glands

cervical caps
cervix

Key Ideas

Sexual Reproduction in Animals

In the aquatic environment, animals reproduce through both external and internal fertilization. External fertilizers shed their gametes into the environment. Many invertebrates, fish, and amphibians fertilize externally. Internal fertilization occurs in some aquatic animals.

Copulation and Viviparity

1. Some aquatic *copulators* include the sessile barnacle, the flatworm, planaria, the sharks, and some bony fishes.
2. Animals develop in three ways:
   a. *Oviparity:* Development outside female; food supply in egg
   b. *Ovoviviparity:* Internal fertilization and maintaining eggs within during development, but nourishment from the egg contents

chorion
claspers
clitoris
cloaca
coitus interruptus
condom
contraception
copulation
corpus cavernosum
corpus luteum
corpus spongiosum
Cowper's glands

diaphragm
dilation and curettage (D&C)

ejaculation
emission
endometrium
epididymus
erection
estrogen
external fertilization

fimbriae
follicle
follicle-stimulating hormone (FSH)

genital papilla
germinal epithelium
glans clitoridis
glans penis
gonadotropin

hemipenis
hermaphroditic
hymen
hysterectomy

labia majora
labia minora
LH surge
luteinizing hormone (LH)

marsupial
masturbation
meatus
menopause
menstrual cycle
menstruation
micropyle
midpiece
monotreme
mons veneris
mucosa

orgasm
orgasmic platform
ovary
oviduct (fallopian tube)
oviparity
ovipositor
ovoviviparity
ovulation
oxytocin

c. *Viviparity:* Internal fertilization and maternal nourishment of the embryo during development

Reproduction and the Survival Principle

1. Many reproductive strategies exist, but in the long run there must be trade-offs between energy spent on producing large numbers of offspring and energy spent assuring the survival of offspring.

2. The strategies range from massive egg production and little or no protection, to the other end of the spectrum with few eggs produced and great maternal or parental care and protection.

3. Both strategies involve risks, but in the long run they appear to equal out.

Reproduction in the Terrestrial Environment

Transition to land required *internal fertilization,* and in the *oviparous* animals an especially durable egg with adaptations that overcame the need for water.

Reproduction in Terrestrial Arthropods

1. One reason for the tremendous success of insects is their great reproductive capacity.

2. Insects have an array of often elaborate devices for *internal fertilization.* Males have clasping and holding structures and highly specialized, species specific, penises. Females often have *ovipositors* for digging and boring holes for the eggs.

3. Insects store sperm in a saclike *spermatheca.* Eggs are protected by a *chorion* and fertilized through a *micropyle.*

Mating in Spiders

Because they are solitary predators, mating requires complex behavior including recognition signals. Spiders do not copulate, but males usually rake sperm-laden appendages across the females' reproductive opening or on the abdominal surface.

Reproduction in the Reptiles and Birds

1. Birds and most reptiles are *oviparous* and their eggs are similar.

2. Lizards and snakes mate side by side using one of two barbed *hemipenises.* These are internal when not in use.

3. Reptiles have a common reproductive, excretory, and digestive channel, the *cloaca.* It receives the *hemipenis* during *copulation.*

4. In most bird species, males lack a *penis. Cloacas* are pressed together.

5. Reptiles produce more eggs than birds and many provide no maternal care. The strategy of sea turtles is to produce many eggs and bury them in the sand. Crocodiles and alligators produce fewer, but protect their nests and their young for a time.

6. Birds lay relatively few eggs and provide an extended period of care. Young hatch in either a *precocial* or an *altricial* state. The former hatches with feathers and ready to feed on its own, and the latter is initially naked and helpless.

Reproduction in the Mammals

Mammals are mainly *viviparous.* Most support their young through the *placenta.* Exceptions are the marsupials and monotremes. Marsupials rear their young in a pouch where they attach to a nipple. Monotremes are egg layers that nevertheless produce milk and provide intensive maternal care.

The Human Reproductive System

The Male Reproductive System

1. The external *genitalia* include the *penis, testes,* and *scrotum.*

2. The *testes* produce sperm and sex hormones. They consist of coiled seminiferous tubules forming ovoid bodies. The tubules emerge into the *epididymus.*

3. From each *epididymus* emerges *spermatic cords* containing blood vessels, nerves, and the *vas deferens.* The two *vas deferens* continue to the underside of the bladder where they are joined by the *seminal vesicle.* The latter produces

penis
placenta
polar body
precocial
primary oocyte
primary spermatocyte
progesterone
proliferative phase
prostate
pseudoplacenta

refractory period
resolution
rhythm method

salting out (saline abortion)
scrotum
secondary oocyte
secondary sex characteristics
secondary spermatocyte
semen
seminal vesicle
seminiferous tubule
Sertoli cells
sexual intercourse
sexual reproduction
spermatogonia
spermatheca
spermatic cord
spermatid
sperm head
squamous epithelium
sperm

tailpiece
tenting
testes
testosterone
tubal ligation

urethra
uterus

vagina
vas deferens
vasectomy
vestibular gland (glands of Bartholin)
viviparity
vulva

much of the *semen.* The *vas deferens* unites with the *urethra* in the *prostate gland.* Secretions from the *prostate* activate sperm and provide an alkaline environment. The *bulbourethral glands* add mucous secretions to the *semen.*

Erectile Tissue

All mammals have *erectile* mechanisms in the *penis.* In some, cartilage or bone may be involved, while *erectile tissue* alone is seen in others. Humans have three spongy regions of erectile tissue. *Erection* is produced by blood pressure.

Sperm Production

1. Sperm are produced meiotically in the erminal epithelium of the *seminiferous tubules.* Cells that enter meiosis are called *primary spermatocytes.* After meiosis I they are *secondary spermatocytes.* At its end they are called *spermatids* and these mature into *spermatozoa* (sperm).

2. Sperm structures include the minute *sperm head* (which contains the condensed chromatin and a caplike *acrosome*), the *midpiece* (which contains a *spiral mitochondrion* and a centriole), and the flagellum or *tailpiece.*

The Female Reproductive System

External Genitalia

The external *genitalia* or *vulva* consists of a prominent mound, the *mons veneris,* the *labia majora, labia minora,* hooded *clitoris,* and two orifices, the *meatus* of the *urethra* and the *introitus* of the *vagina.* The *clitoris* contains the *erectile glans clitoridis,* which enlarges during sexual excitement.

Internal Genitalia

1. The *vagina* is tubelike with highly folded walls capable of great extension during the birth process. The *vaginal* lining along with the *glands of Bartholin* (*vestibular glands*) produces lubricating mucus.

2. The other internal organs include the *uterus, oviducts,* (*fallopian tubes*), and the *ovaries.* The *uterus* opens at the *cervix.* (It is paired in most mammals.) Its lining consists of an active *mucosa* (which produces the *endometrium*).

3. The paired *oviducts* extend from the *uterus* to near the *ovaries,* forming fingerlike *fimbria* at their ends. These ciliated structures create currents that sweep the egg cell into the oviduct.

4. The *ovaries* are oval structures consisting of an outer cortex that contains the *germinal epithelium* with all the *oocytes.*

Hormonal Control of Human Reproduction

Under the influence of the *hypothalamus,* the *anterior pituitary* releases *luteinizing hormone (LH)* and *follicle-stimulating hormone (FSH).* The pituitary hormones stimulate the *ovaries* and *testes* to produce sex hormones and sex cells. (LH is identical to ICSH.)

Male Hormonal Action

1. *FSH* stimulates sperm production. *ICSH* induces the *testes* to produce *testosterone.* It is essential for sexual differentiation in male embryos, sexual development at puberty, and sex drive.

2. The production of *testosterone* is under negative feedback between the hypothalamus, pituitary, and testes.

Female Hormonal Action

1. The rise in *FSH* secretion initiates female *puberty* and fertility. The ovary responds to *FSH* by producing estrogens. Under their influence, the *secondary sex characteristics* begin to appear.

2. *Puberty* is accompanied with the start of the *menstrual cycle.* Ovarian cycles vary greatly in mammals, but in humans the cycle occurs about every 28 days. Human females are unique in that they are capable of sexual response throughout the ovarian cycle.

3. The 28 day cycle may be outlined as follows:
a. *Proliferative or preovulatory phase:* days 1–14
Days 1–4: Menses; menstrual flow
Days 1–14: Increasing levels of *FSH; follicle* development; one *follicle* enlarges; *follicle* secretes increasing amounts of *estrogen; uterine* lining (*endometrium*) responds by growth; *cervix* secretions become more alkaline
Day 14: FSH secretion inhibited; LH suddenly released; *follicle* bursts; *ovulation* occurs; body temperature slightly elevated
b. *Secretory or postovulatory phase:* Days 14–28
Days 14–20: LH and *FSH* stimulate vacated follicle to form *corpus luteum; corpus luteum* secretes *progesterone;* uterine lining (*endometrium*) increases in thickness and vascularity; fully receptive to a fertilized ovum.
Days 21–28: Some slowing in *estrogen* and *progesterone* secretion; slowdown increases rapidly through day 28 where the *endometrium* begins to slough away, signaling the end of the cycle

4. If an embryo has *implanted,* hormonal levels will not decline. Placenta produces *progesterone* and the lining is maintained.

5. Control of the pituitary and ovarian hormones includes two feedback events. One occurs at midcycle and instigates ovulation; the other occurs at the cycle's end, bringing on menstruation and a new cycle.

6. In pregnancy and under the influence of *birth control pills,* the hormonal interplay and feedback systems do not function. In pregnancy, the *corpus luteum* and the *placenta* produce *progesterone,* which remains at a high level until birth. *The pill* utilizes this overriding phenomenon also, by artificially maintaining a high level of hormones, thus inhibiting *ovulation.*

The Sex Act

In humans, *intercourse* is divided clinically into four phases: *excitement, plateau, orgasm,* and *resolution.* The first two show an increasing in certain events, including genital swelling, lubrication, blood pressure, pulse, and pelvic thrusting. These culminate in orgasm, which is physiologically somewhat different in the two sexes. Resolution is more definite in males, and orgasm more final.

Masturbation

Masturbation is almost universal in occurrence. Some studies indicate that early sexual experience improves later sexual functioning.

What Does Copulation Accomplish?

1. The obvious answer is that *copulation* assures *fertilization.*

2. In many mammals, *copulation* is impossible in off seasons. Our species' unique sexuality may have resulted from behavioral evolution. One answer is the theoretical "pair bond." Frequent sexual reward helps males form the pair bond with specific females. This is adaptive in terms of the long period of parental care by human offspring.

3. Studies of human female receptivity or frequency of *intercourse* and *orgasm* indicates that two peaks occur in each *menstrual cycle.*

4. In many primates, nonreproductive sexual activity is common. This may build bonds within social groups.

Fertilization

1. Human *oocytes* are arrested in prophase unless *fertilization* occurs, whereupon they complete meiosis. Fertilization occurs in the oviduct, whereupon the oocyte completes meiosis, usually producing two polar bodies. (Review meiosis in the human female in Chapter 8.)

2. At the time of *ejaculation,* hundreds of millions of *sperm* are released, but only a few thousand reach the *egg.* Studies indicate that *sperm* and egg must meet within 48 hr after *ovulation.*

Contraception

For a review of this section study Table 29.1.

Natural Methods
Natural methods refers to those without chemical or mechanical intervention. All of these methods have only limited reliability. The *rhythm method* relies on a number of variables that are difficult to control.

Mechanical Devices
When properly used, *condoms, diaphragms,* and *cervical caps* are reliable. Their failure is generally due to carelessness. The *IUD* is very effective, but many women cannot use them.

Hormonal and Chemical Birth Control: The Pill
Birth control hormones work by overriding the normal *menstrual cycle* and simulating pregnancy. They remain controversial because of undesirable side effects in some women. The medical risk associated with pregnancy is still greater than that of the pill.

Surgical Intervention
1. *Vasectomy* and *tubal ligation* are common permanent methods of birth control.
2. *Abortion,* the surgical removal of an embryo, involves three methods: *vacuum* or *suction, dilation and curettage (D&C),* and *salting out.* The latter is used for second trimester abortions and involves the most risk and discomfort.

Review Questions

1. Define and give an example of *external fertilization.* (pp. 830–831)
2. How have the amphibians improved on the *external fertilization* process? How might this affect the number of eggs and sperm necessary? (p. 831)
3. Describe *internal fertilization* in two aquatic invertebrates. (p. 831)
4. Distinguish between *oviparity, ovoviviparity,* and *viviparity,* and give an animal example for each. (p. 832)
5. In what way does the shark cross the line between *ovoviviparity* and *viviparity?* (p. 832)
6. Describe a common relationship between numbers of offspring and maternal care. Can any conclusion be reached as to which strategy is better? Support your answer. (p. 833)
7. Describe the insect reproductive structures, including those for mating, sperm storage, and depositing of eggs. (p. 834)
8. What are the functions of the *chorion* and the *micropyle* in insect eggs? (p. 834)
9. Describe mating behavior in the male orb spider. What is the male's problem when he attempts to mate? (p. 834)
10. Describe the reproductive anatomy and mating of lizards and snakes. (p. 835)
11. Describe birth and development in the marsupial. What is the evidence of its success? (p. 837)
12. List the structures found in the human *testes* and offer a function for each. (p. 838)
13. Trace the route of sperm travel from the *epididymus* through the *urethra.* List the essential structures along the route and describe the reproductive function of each. (p. 838)
14. List the *erectile tissues* of the human *penis* and explain their role in an erection. (p. 839)
15. Review the life history of the sperm cell beginning with *spermatogonia* and ending with the mature *spermatozoan.* (p. 839)
16. With the aid of a simple drawing of a sperm cell identify the following regions and their parts. Include the functions. (p. 839)
    a. *Head*
    b. *Midpiece*
    c. *Tailpiece*
17. List the structures of the external *genitalia* in the human female and add functions where they are specific. (pp. 840–841)
18. Describe the structure of the *vagina* and explain how it is adapted to its dual role. (p. 841)
19. Describe the structure of the *uterus* and its temporary lining. (pp. 842–843)
20. Describe the structure of the *oviducts* (fallopian tubes) and their curious relationship to the ovary. (p. 843)
21. How do the *ova* find their way into the *oviduct?* (p. 843)
22. Describe the main differences in gamete production in human males and females. (p. 844)
23. What is the role of the hypothalamus in the hormonal control of reproduction? (p. 844)
24. Describe the feedback control of *testosterone* production in the male, including the action of the pituitary. (p. 844)
25. Prepare a brief table of the reproductive hormones in females and include the following: name of hormone, structure that produces it, target organ(s), and effect. (p. 845)
26. Draw a simple graph showing the levels of *FSH, LH, estrogen,* and *progesterone* through the 28 day *menstrual cycle.* (pp. 845–847)

27. On a second graph, plot the events in the *endometrium* as it responds to hormonal levels in the 28 day cycle. (pp. 846, 848)
28. Describe the negative feedback loop that changes the events at midcycle. (p. 846)
29. What are the two effects of the *LH surge* at midcycle? (p. 846)
30. Describe changes in levels and sources of sex hormones in pregnancy. (pp. 847–848)
31. What events bring about the end of the *secretory phase* of the cycle? Describe the feedback loop. (p. 848)
32. In what manner does the end of one cycle automatically begin the next? (p. 848)
33. List the changes that occur in males and females during the *excitement* phase of *sexual intercourse.* (pp. 848–849)
34. What is an *orgasmic platform* and what are the accompanying anatomical changes? (p. 849)
35. List the anatomical and physiological events of *orgasm* in females. (p. 849)
36. Describe events in the two parts of the male *orgasm.* (pp. 849–850)
37. Review the argument of K. E. Money that favors childhood sexual experimentation and *masturbation.* (p. 850)
38. What is the apparent specialness of human sexuality as compared to that of other species? (p. 851)
39. How would a behavioral evolutionist explain human sexuality? (p. 851)
40. What are the apparent advantages of the bond? Is this as true today as when it evolved? (pp. 851–852)
41. Explain the theoretical adaptiveness of *menopause* in humans. (p. 852)
42. Using specific data, characterize sexual receptiveness in human females. (p. 851 and Figure 29.23)
43. Explain the problems associated with the *rhythm method* of birth control. (p. 854)
44. Comment on the effectiveness and acceptance of the four "natural methods" of birth control. (pp. 853–854)
45. List four mechanical methods of *contraception* and comment on the effectiveness of each. Why do most failures occur? (p. 854)
46. What is the strongest argument in favor of *birth control pills?* (p. 856)
47. What is a *vasectomy* and how does it usually affect sexual performance and *ejaculation?* (pp. 856–857)
48. Describe the surgery involved in *tubal ligation.* (p. 857)
49. List the three surgical procedures in *abortion* and compare their risks. (pp. 857–858)

Chapter 30

Development and Differentiation

One cell divides into two, the two into four. The four again divide and the divisions continue until eventually there is a great number of cells. All the cells have stemmed from that one primordial cell, but yet, many divisions later, far down the line, descendant cells may end up very different from one another. One descendant may become an elongated nerve cell, while another becomes a pigmented cell in the eye's complex retina, and yet another an unprepossessing liver cell with an impressive battery of biochemical tricks. But since they arise by mitotic divisions of a single fertilized egg, they all carry the same genetic information in their chromosomes.

How does this seeming miracle of development and differentiation occur? This, it turns out, is not a rhetorical question. We don't quite know. It would seem that there should be some ready answer, some astounding explanation. But if there is, it awaits our discovery. And perhaps it shall be discovered. After all, at one time, the basic mechanism of gene function was thought to be beyond our comprehension, but it turned out to be surprisingly straightforward, once the concept of the genetic code was grasped and the code itself was worked out. Of course, there may be no such easy explanation. Every year more is learned about the processes of development, but the questions keep getting more difficult, and the answers keep getting more complex.

Actually, all is not abysmal darkness, however. Parts of the puzzle have been worked out. The study of development has two aspects: the descriptive and the experimental. And the descriptive work—the story of what actually can be seen in a developing embryo—is already rather complete.

Descriptive embryology is just that—a record of what happens, without the how and why. Fortunately, what happens in development is interesting on its own account. And we must know about it if we hope to ever understand the underlying mechanisms. So, we will see that cell differentiation is only one aspect of development. Equally important, and equally baffling, are the organized movements of cells and tissues; the control of mitosis; the orientation of cells; programmed cell deaths; and, of course, the emergence of morphological shape and form. In order to understand these things, we must go beyond descriptive biology. We must experiment.

Embryologists have carefully described the development of a few select species, so that we now have a very solid base to work from. However, it may be unfortunate that the choice of experimental animals has been a matter of convenience and tradition and has been somewhat idiosyncratic. The attention of descriptive embryologists has been focused primarily on humans, pigs, chickens, frogs, sea urchins, and an occasional insect—quite a motley group. Experimental embryology has gravitated to frogs for reasons of practicality. Currently, the developmental biologist's favorite frog is the South African clawed frog, *Xenopus laevis*.

Our plan in this chapter will be to review developmental events in several species, offer a summary of observational data in each case, and from these attempt to present the important unifying ideas. For convenience, we begin with a description of the sperm and egg cells. We introduced these specialized cells earlier, but let's look at them in a bit more detail and in reference to their role in early development.

30.1 The Gametes and Fertilization

The Gametes

The Sperm

The sperm is one of the most highly specialized of cells. Its role is very straightforward. It will seek out and penetrate an egg cell, delivering its vital haploid nucleus. Its entire structure is devoted to this one mission. Most sperm cells consist of three main parts: the head, midpiece, and tail. Figure 30.1 shows a variety of sperm types.

The Egg

Animal egg cells are somewhat more difficult to describe because they differ so greatly from one species to the next. Whereas a human egg is about the size of a period, you couldn't hide an ostrich egg with this entire book. We can generalize a bit, however, by saying that each egg cell at the time of fertilization consists of a haploid nucleus surrounded by a region of cytoplasm containing yolky substances and all the usual cellular components. The egg cell membrane deserves special attention in our discussion since it plays an important role in fertilization.

30.1 These various types of sperm cells are all drawn to the same scale, indicating that the human sperm is rather small. Perhaps the most unusual sperm is that of the rat, with its sickle-shaped head. However, each sperm consists of essentially the same parts: head, midpiece, and tail.

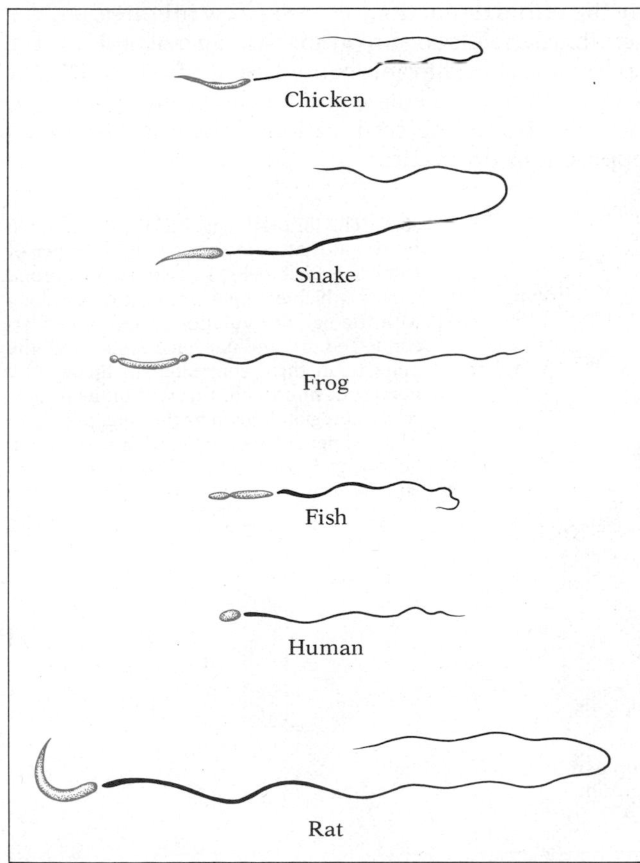

We can distinguish three types of egg cells on the basis of their yolk quantity and distribution. In birds and reptiles, the yolk reaches its greatest quantity, with the ostrich holding the record for living species. We shouldn't be fooled by the size of eggs, however, since the *blastodisc* (the region of the egg that becomes the embryo) is a relatively small portion of the contents in bird eggs. The quantity of yolk in the bird egg is, of course, related to the fact that most of development is completed within the confines of the shell, so the embryo lives with its life-support system. Amphibians, by the way, also have a relatively sizable yolk mass. Before the developing frog can hatch into the juvenile tadpole, able to find its own food, it requires a considerable amount of nourishment.

In both birds and amphibians, the yolk is concentrated in one region of the egg. After fertilization, this region will become the *vegetal pole*, named for its less active role in early development. The more actively dividing region is called the *animal pole*. Bird eggs, with their enormous yolk content highly concentrated in one region, are known as *telolecithal* (from *telo-*, "end" and *lecith-*, yolk). The yolk of amphibian eggs is not as great and is somewhat more moderately dispersed. Such eggs are termed *mesolecithal* (*meso-*, "middle").

Placental mammals produce eggs with a comparatively even distribution of yolk. These eggs are extremely small, barely visible to the naked eye, with enough nutritive content to support the embryo for only a very brief period. It is all these mammals require. After a few days in transit to the uterus, the embryo implants itself in the uterine wall and begins to develop the structures that will permit it to gain nourishment from its mother. Because the eggs do not need to provide their own nourishment, they have a meager, relatively evenly distributed yolk and are known as *isolecithal* (*iso-*, "same"). The eggs of many marine invertebrates, such as those of echinoderms and lancelets, are also isolecithal. They proceed very rapidly into a tiny but self-sustaining, free-swimming larval stage. Arthropods (such as insects) produce a different ovum known as a *centrolecithal* egg, with a central yolky region entirely surrounded by a thin layer of cytoplasm.

The egg cell membrane and its related structures have undergone extensive study by biologists in an effort to better understand the fertilization process.

Observations of the echinoderm egg reveal an interesting array of supplemental parts, such as the transparent substance surrounding the egg (known as the *jelly coat*), which covers a fibrous layer called the *vitelline membrane*. Just below the vitelline membrane is the egg proper and a region of granules and microvilli. As we shall soon see, these features are unique to egg cells and each plays an important role in fertilization.

Mammalian egg cells are somewhat different from echinoderm eggs. At ovulation, the egg emerges from the ovary surrounded by a dense covering of follicle cells known collectively as the *corona radiata*, the cells of which are held together by jellylike *hyaluronic acid*. Below the corona radiata lies a thick, glassy membrane that is secreted by the follicle cells. This is the *zona pellucida* (clear zone). Below this is the vitelline membrane of the egg proper (Figure 30.2). With these structures in mind, let's now examine the fertilization process itself.

Fertilization

Sperm reaching the vicinity of the egg, in the upper portion of the oviduct, release the enzyme *hyaluronidase*. This breaks down the corona radiata, allowing the sperm to contact the egg. A common cause of infertility, by the way, is a low sperm count. Even though only one sperm actually fuses with the egg, a great number of sperm are required to supply enough hyaluronidase.

As soon as a sperm cell contacts the egg surface, the acrosomal region begins to secrete a substance that enables the acrosomal membrane to fuse with the egg membrane. The released substance also causes competing sperm to clump together helplessly. When the secretions begin, an *acrosomal filament* may be seen projecting from the sperm head in some species, including echinoderms but not mammals. Finally, the sperm is securely attached, but at this point its nucleus has still not penetrated the egg cytoplasm.

The egg isn't passively waiting to be fertilized by some thrashing sperm, however. As soon as a sperm touches it, the egg's outer membrane undergoes several rapid changes. One of its first activities is to produce a barrier that will prevent additional sperm from penetrating it. In echinoderms, observers have seen the granules under the surface (*cortical granules*) rupture, releasing their contents into the *perivitelline* (around the vitelline) space. The barrier is formed by a series of physical and chemical events we don't fully understand, but we do know it takes the physical form of a *fertilization membrane* (Figure 30.3). This structure begins as a small blister at the point of penetration that enlarges in a rapidly expanding front around the entire egg—like a miniature tidal wave. Any additional sperm that have managed to penetrate to the vitelline region are detached by the cortical reaction. Within a few minutes, a clear semihardened halo surrounds the egg cell and further sperm attachment is prevented. So it's first come, first served. Similar events occur in the mammalian egg, but the rise of the fertilization membrane is not as apparent or dramatic.

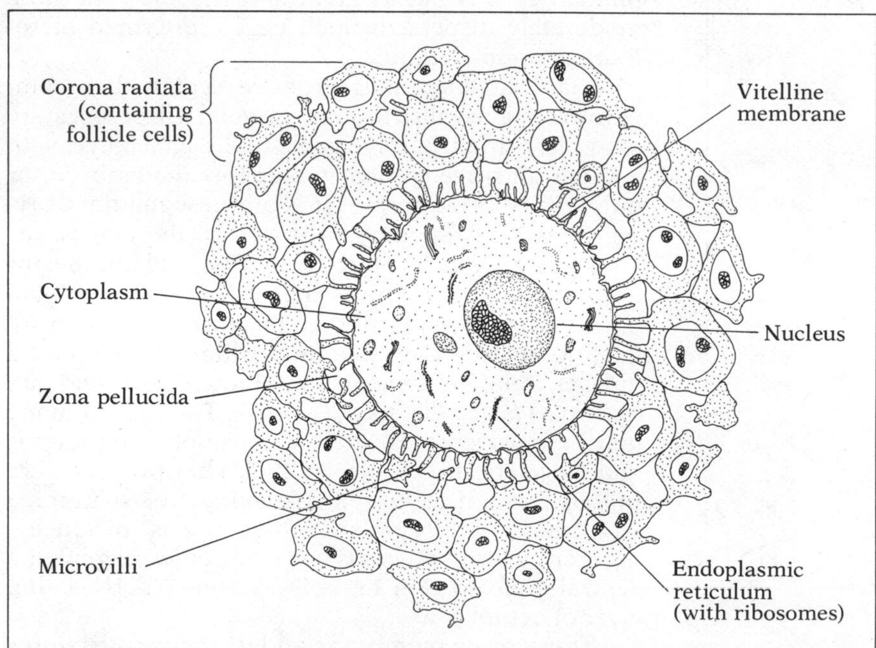

30.2 The human egg cell is surrounded by tissue that contains a large number of smaller follicle cells, known as the *corona radiata*, which emerges from the ovary along with the egg at ovulation. These are interconnected by cellular processes, and the mass is, in turn, connected to the egg by numerous microvilli. Just within the follicle cells is a region known as the *zona pellucida*, which is penetrated by the many microvilli. Following fertilization, the corona is shed and the entire surface changes significantly.

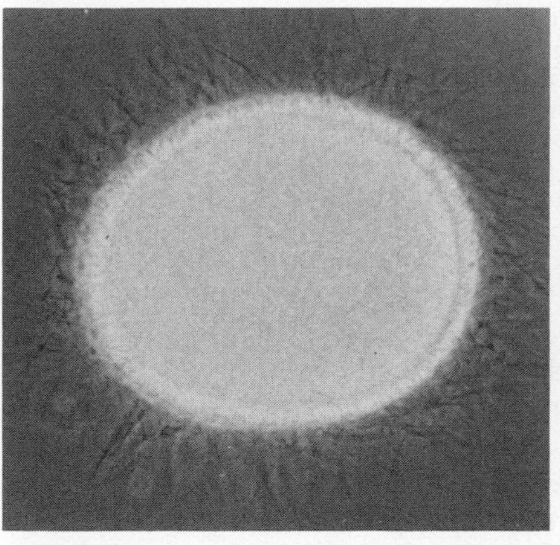
(a)

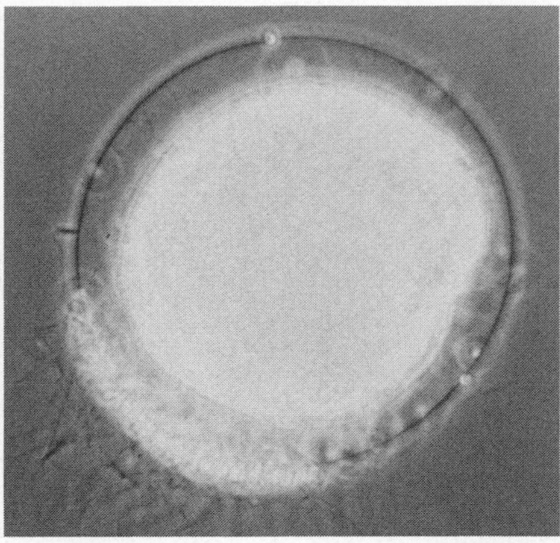
(b)

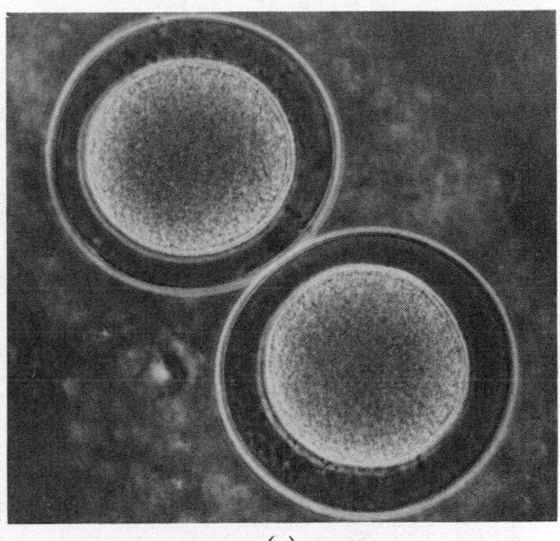
(c)

30.3 The fertilization process is easily observed in the sea urchin. Typically, the jelly coat will contain many sperm at the time of fertilization (a), but only one will actually penetrate the egg membrane. When one penetrates, cortical granules within rupture, creating a series of chemical events. The most visible evidence is the formation of a *fertilization membrane,* which will act as a barrier to other sperm (b). The fertilization membrane rises up from the egg surface at the point of penetration and spreads in a wave over the egg surface. Eventually, the membrane will surround the entire egg (c). Under ideal conditions, the egg will begin its first cleavage about an hour later.

Meanwhile, the attached sperm head becomes engulfed in an active mound of cytoplasm—the *fertilization cone*—which has risen up to surround it and to draw it into the female cell cytoplasm (Figure 30.4). The sperm head and neck region detach from the midpiece at this time. Soon the sperm nuclear membrane breaks down and releases the chromosomes into the egg cytoplasm, and the sperm centriole is also freed. Penetration is now complete, and the sperm nuclear substance begins to migrate to the egg nucleus. We'll now leave the sperm in this delicate position while we review the nuclear history of the egg as it prepares for the event.

In mammals, the egg nucleus (if you recall our previous discussion) is suspended in late meiotic prophase prior to ovulation. By the time of ovulation, though, meiosis I is completed and the first cell division occurs. Recall that this highly unequal division produces one large cell and a minute polar body. This may, in fact, be the end of the line. Unless fertilization occurs, there will be no second meiotic division. But sperm penetration signals the egg to complete meiosis. Now the chromatids separate, and a second unequal cleavage of cytoplasm occurs, forming a second polar body and a much larger haploid egg cell.

So, while the second meiotic event is occurring, the nuclear contents of the sperm become surrounded by a newly formed membrane, apparently formed by the egg cytoplasm. The sperm and egg haploid nuclei (now known as the male and female *pronuclei*, respectively) migrate through the thick cytoplasm toward each other and join in a tight union. At this point, fertilization is complete. The nucleus of the zygote—now diploid—will enter a period of DNA replication, from which the chromosomes will emerge in their doubled chromatid (G_2) condition, and the first mitosis can now begin.

The Future of a Zygote

Before getting into some of the details of development, let's first take stock of what we have so far and where we are going. In a multicellular organism, whether a plant or animal, we know that the newly fertilized egg cell, or zygote, is an independent entity with its own unique genotype. The zygote is one cell containing a full set of chromosomes, a bit of cytoplasm with its usual array of organelles, and some stored food. From this, barring some unforeseen accident, will emerge a complex organism whose very existence is seated in an incredible array of very tightly coordinated biochemical events. All the tissues, organs, and systems will be in place, cooperating as a functional whole. It's hard to believe all this springs from the simple zygote, but as one cell divides into two, two into four, four into eight, and so on, layers of cells are formed and these then are able to change and reposition themselves. Tissues interact, grow around and into each other, fusing, spreading, and shifting somehow until, finally, a recognizable form begins to emerge. Organs are vaguely silhouetted, then gradually refined and clarified. And, finally, they begin to function. Ultimately, a new individual is formed—formed from preexisting matter, yet with its own identity, and carrying with it a repertoire of genetic capability, now ready to be molded and redefined by the myriad experiences of its environment.

One is impressed by the enormity of the developmentalist's task. How can we begin to understand such a process? The questions about the process have been asked throughout recorded history. Early philosophers, questioning their own existence, worried about the "generation of form," and developed two points of view on development. One group concluded

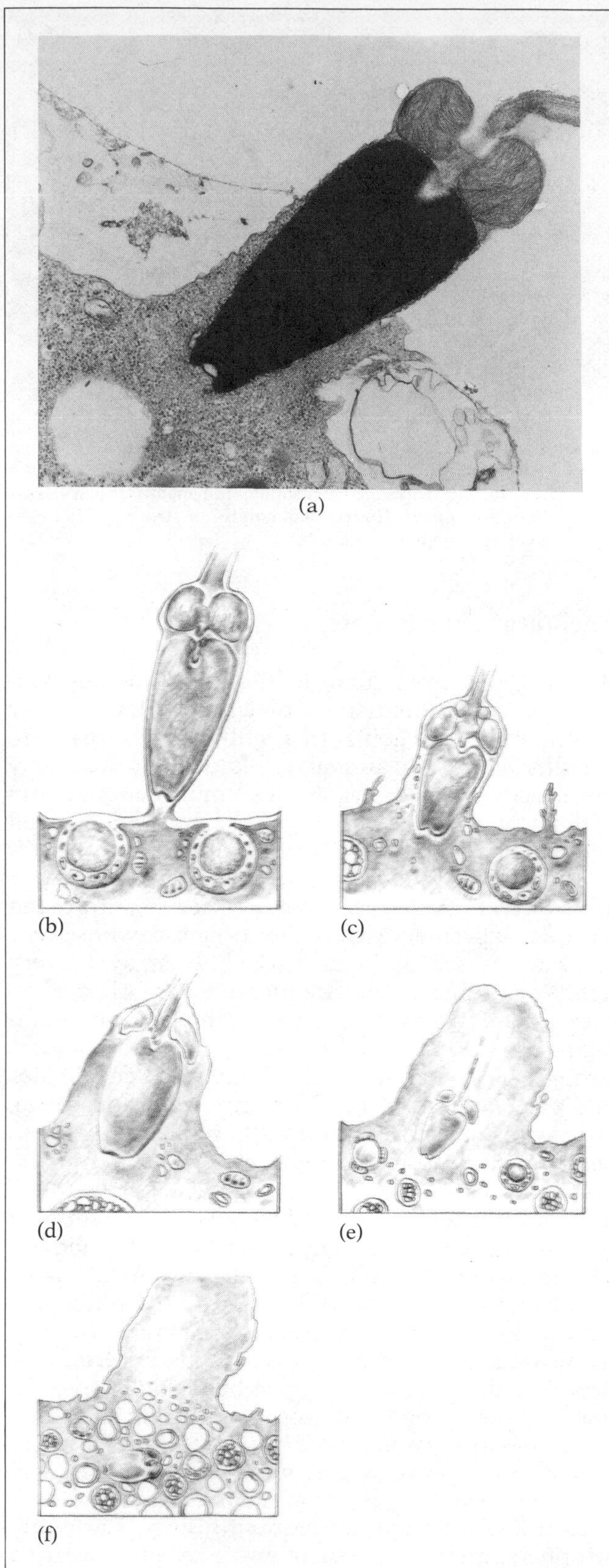

30.4 (a) The electron micrograph clearly points out that fertilization is a mutual effort. Not only has the egg surface risen up to surround the penetrating sperm head, but it is believed that the sperm is actively drawn into the egg interior. In (b), the acrosomal cap has touched the surface, spilling its contents. Cortical granules rupture to produce the egg reaction. The egg cytoplasm at (c) begins to rise and in (d) surrounds the sperm head. At (e), the sperm accessory structures, which are no longer needed, have begun to disintegrate, and at (f) the sperm nucleus is now clearly in the egg cytoplasm. It will enlarge and form the *pronucleus.*

that the finished form existed in the gamete and development was merely the expansion of this form. This point of view became known as the *preformation theory.* A 17th century microscopist reported seeing a miniature human form, which he called a *homunculus,* crouching in the head of a sperm cell. This report was widely circulated and quickly took hold in peoples' minds. Others calculated that within each of these miniatures were yet littler miniatures, and that all future humankind was packed into Adam's scrotum, like boxes within boxes. (We would intuitively reject such notions now, but consider how much of our "intuition" is really an *educated* guess based on information not available to earlier thinkers.)

But even the preformationists disagreed. One group—the *spermists*—believed that form was contributed by the sperm cell only, while the female provided a place for development to occur. Surprisingly, for a male-dominated era, there arose the *ovists,* with the opposing claim. It was the egg, they said, that held the little people. The famous early microscopist Leeuwenhoek put aside his gadget long enough to test the competing hypotheses with a rare, early genetic experiment. He mated a black male rabbit and a white female rabbit. The offspring were all black, "proving" the spermist position. And keep in mind that Leeuwenhoek was a great scientist of his time.

In opposition to the preformationists were the *epigenesists.* The epigenesis theory held that the egg and sperm merely joined to form an amorphous mass from which a definite form was molded (through processes they admitted they didn't understand). For the most part, the two explanations were based on simple observations and what the scientists of the day assumed to be logical deductions. The ideas proved very difficult to dislodge. In fact, it wasn't until the 19th century that further experiments were carried out. (Keep in mind the rarity of experimentation in the sciences at this time.)

Late in the 19th century, Wilhelm Roux devised an experiment that he believed tested the preformation concept. Using fertilized frogs' eggs that had cleaved once, he destroyed one cell with a hot needle (Figure 30.5). He reasoned that if form was already established in this very early embryo, then the resulting embryo would be badly deformed. Each time he carried out the procedure his results confirmed

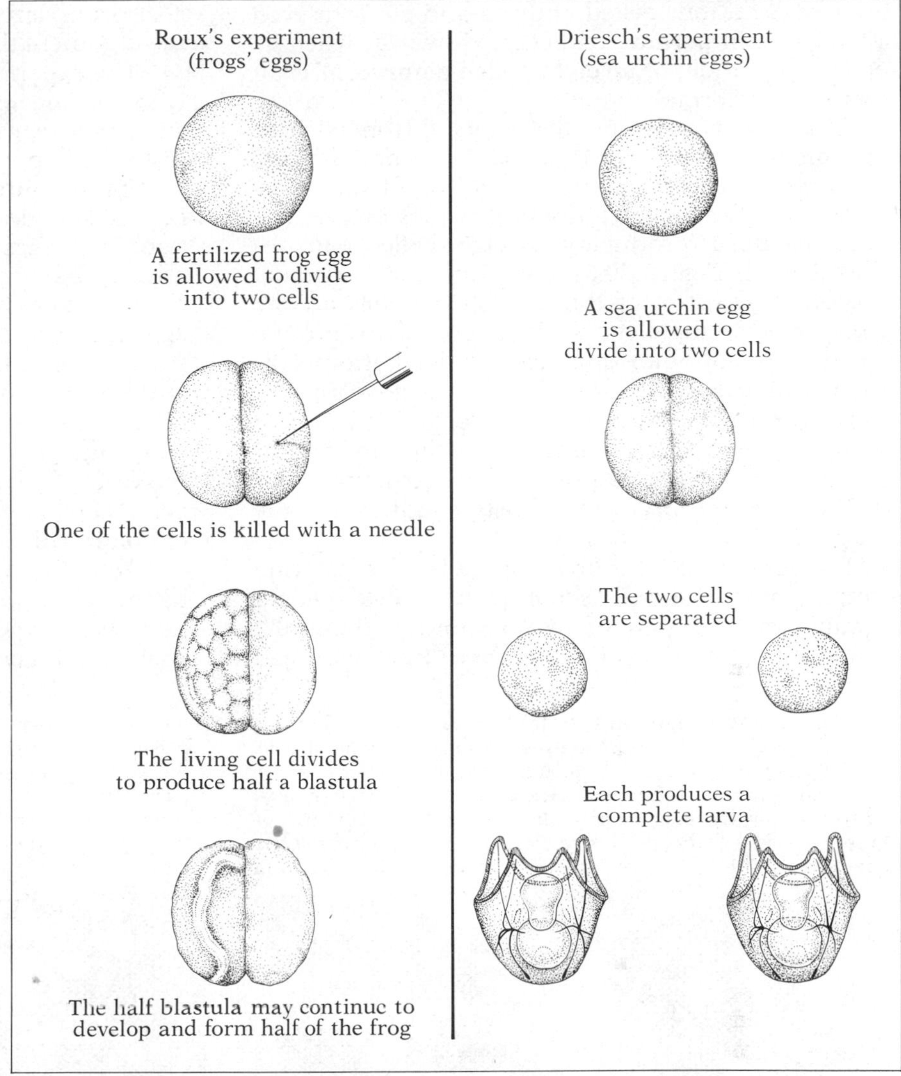
Roux's experiment
(frogs' eggs)
Driesch's experiment
(sea urchin eggs)
A fertilized frog egg
is allowed to divide
into two cells
A sea urchin egg
is allowed to
divide into two cells
One of the cells is killed with a needle
The two cells
are separated
The living cell divides
to produce half a blastula
Each produces a
complete larva
The half blastula may continue to
develop and form half of the frog

30.5 The experiments of Roux and Driesch. The preformationists and epigeneticists both claimed experimental evidence for their beliefs. Wilhelm Roux succeeded in killing one of the cells in a two-celled frog embryo. The resulting embryo proved to be abnormal. He concluded that this was evidence of preformation. On the other hand, Hans Driesch thought he had strong evidence for epigenesis when he succeeded in separating the two cells of an early sea urchin embryo. Each cell developed into a complete sea urchin larva. Actually, Roux would have produced a normal embryo if he had separated the two egg cells of the frogs' eggs. Apparently, the presence of a damaged cell caused the abnormalities.

this belief. Sometimes an embryo with a head region emerged, sometimes a tail region, but always an abnormal embryo. This was strong evidence for preformation.

But in a series of similar experiments involving sea urchin eggs, another experimenter, Hans Driesch, got different results. In this case, the two cells were separated and each undamaged cell generally produced a normal, whole embryo. From these contrary results, a lengthy debate arose in the scientific literature, and scientists rallied to one side or the other. The experiments, probably marking the beginning of experimental embryology, were incessantly repeated and elaborated on. Finally, the arguments died down when it was determined that either result could be produced simply by varying the experimental techniques. And so it was admitted that neither observation was very conclusive. When the damaged cell of the frog embryo was removed, the uninjured cell developed into a normal embryo. If it was allowed to remain attached to the healthy cell, a deformed embryo resulted. Apparently, the presence of the dead cell had an effect on the live one.

The early experimenters also found that half a multicellular embryo (at the early *blastula* stage) could go on to produce a normal animal. But the division had to be made in the early stages and along certain planes. A frog egg during early cleavage divisions could be experimentally separated into two parts, and the left and right halves would each develop into normal tadpoles. But if the top were separated from the bottom halves, they didn't develop normally.

It turned out that the timing was important. Driesch separated the cells of a two-celled sea urchin embryo by shaking it violently in a vial of seawater. He found that both cells developed into pint-sized, but otherwise normal, urchin embryos. Then he shook apart a four-celled embryo and got four even tinier, but still normal, embryos. However, when he separated the cells of an eight-celled embryo, all eight produced abnormal animals.

The most basic lesson that emerged from all these experiments was that the fertilized egg is not uniform, but is separated into areas with different functions. In the frog, the first two division planes are vertical while the third (producing an eight-celled embryo) is horizontal (Figure 30.6). The very first horizontal division makes it clear that not all the cells are the same, because the cleavage is displaced toward the top (the animal pole) thus leaving the bottom cells larger. Apparently, if an embryo is to develop normally, it must receive some cytoplasm from the top half of the egg and some from the bottom half of the egg. So the egg is not without structure after all. (Perhaps the preformationists could take some solace in that.)

Incidentally, if the first two cells of a cleaving human embryo should separate, the result would presumably be a pair of monozygotic ("identical") twins. But separation of the embryo into twins can also occur later. Does monozygotic twinning affect the argument that a human life begins at conception?

The experiments described above wouldn't work with all species, because some have more highly structured eggs. For example, in annelids and mollusks, and apparently insects as well, every definable part of the unfertilized egg has a strictly determined fate that cannot be affected by any kind of manipulation. Such eggs are called *mosaic eggs.* In these eggs, critical changes in the cytoplasm have occurred very early—changes which unwaveringly send part of the egg on its way to its developmental fate. Presumably, such changes occur during the long process of egg maturation. Think about this. If, in some species, differentiation is accomplished by cells interacting and influencing each other's development, how do the cells of mosaic embryos develop? They seem to act independently of each other. So what triggers or controls their changes? No one knows.

Actually, we have only a few clues as to what happens at the molecular level during egg maturation. We do know, though, that, in most animal species, the genes of the developing oocyte are very

30.6 The early cleavages. In both (a) the sea urchin and (b) the frog, the first and second cleavages transect the animal pole. The first four cells are nearly identical in yolk content. In the sea urchin, third and subsequent cleavages are consistently equal or nearly so up through the blastula stage. But this isn't true of the frog's embryo, where the third cleavage produces *micromeres* (small cells) in the upper region and *macromeres* (large cells) in the lower region. This unequal cell division continues into the blastula stage, with a mass of small rapidly dividing cells marking the animal pole and larger sluggishly dividing cells, the vegetal pole. The same pattern in the frog will continue as the embryo's form begins to emerge. The highly active animal pole represents the region of the egg that will produce most of the embryo's structure.

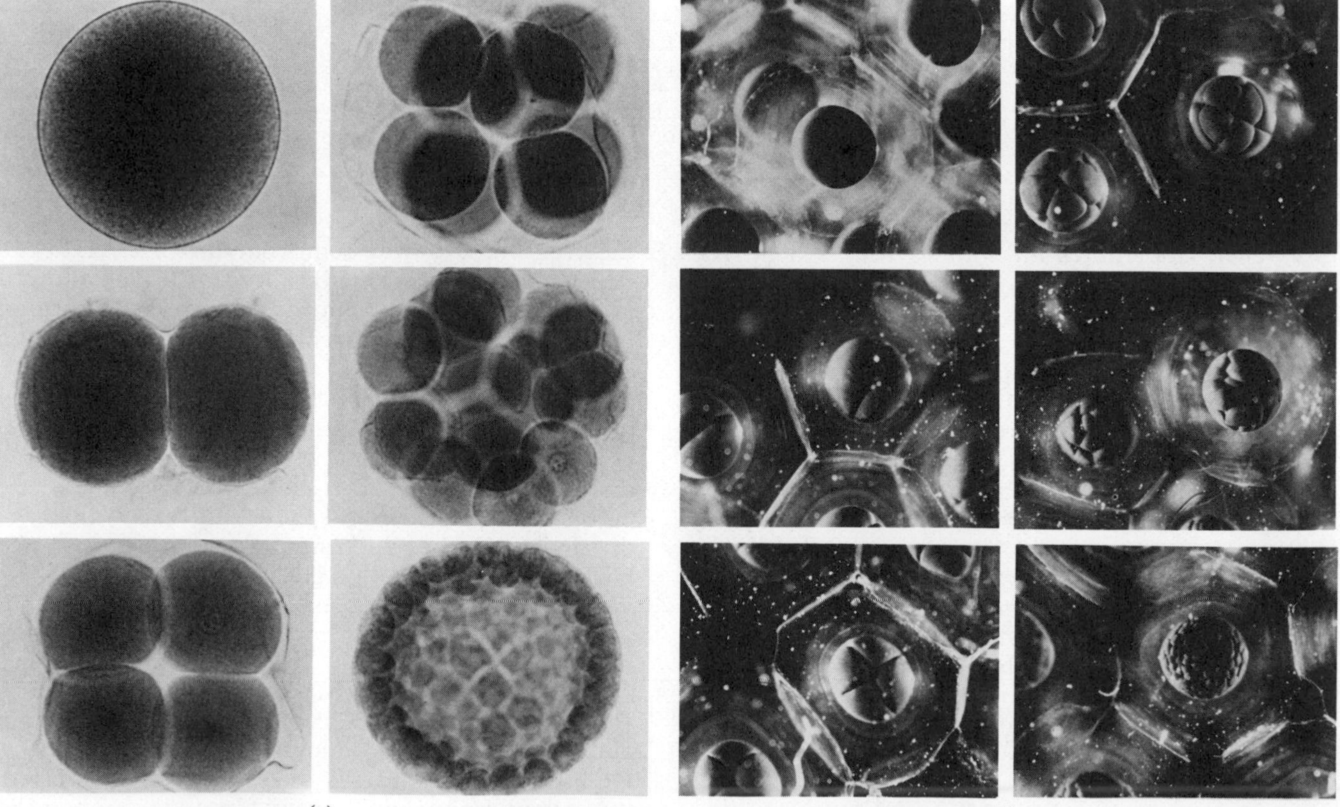

active. As the egg develops, the nucleus becomes unusually large and the chromosomes, too, seem to be active. They are called *lampbrush chromosomes* because they reminded early microscopists of the bristles extending from the core of a lamp brush or bottle brush. They have the same amount of DNA as normal chromosomes in any somatic cell, but the DNA is extended into long loops and is heavily coated with nuclear proteins and mRNA. The long loops are thought to be unwound, transcribing, active genes.

During egg maturation, individually recognizable loops of the lampbrush chromosomes extend, become very active, and later coil up again as other loops bloom along the chromosome, in a kind of programmed series. Messenger RNA has been found leaving the loops for the cytoplasm where it will set up templates for new proteins. Lampbrush chromosome activity occurs during the late prophase of meiosis, well before the chromosome number is halved, and, of course, long before fertilization. Thus, these active genes are of the mother's diploid genotype. (See Figure 8.20.)

None of this, unfortunately, explains how the egg cytoplasm becomes divided into regions with different developmental potentials. We can imagine that perhaps the different mRNAs have affinities for only a certain part of the cytoplasm, but that really doesn't tell us much. If you can recall our earlier groundwork discussions regarding gene action, perhaps you might like to toy with the idea.

The First Cleavage Divisions

The zygote divides shortly after fertilization. This is the first of many, many cell divisions that lay the foundations for later tissue specialization. Each division is preceded by DNA replication and the usual mitotic process. The time lapse between fertilization and first cleavage varies greatly among different animals, but in no case does the egg immediately respond after being violated by the nosing sperm; it almost always waits at least 30–40 min. As you might expect, zygotes with small quantities of yolk generally begin to respond sooner, but there are notable exceptions even to this simple rule. Fertilized human eggs, bearing little yolk, for example, require about 30 hours before the first cleavage.

Typically, a zygote will cleave into two equal halves. These are known as *blastomeres.* (A notable exception is found in the cleavage of the very yolky telolecithal eggs, which we will describe shortly.) The second cleavages usually occur more rapidly than the first and form at right angles to the plane of the first (Figure 30.6), dividing the zygote into four long segments. The plane and location of the third and subsequent cleavages are difficult to generalize about because of the wide range of variation, so we will have to select some specific examples.

Dividing sea urchin and human eggs continue to produce blastomeres of about equal size, and these, then, rearrange themselves until the embryo takes on the appearance of a hollow ball composed of small rounded cells. This is the blastula stage. As shown in Figure 30.6, in its third cleavage, the amphibian zygote divides horizontally, with the cleavage displaced toward the top (animal pole). This produces four smaller cells (known as *micromeres*) at the top and signals the start of a trend that will soon produce a rapidly dividing animal polc and a slower dividing vegetal pole (made up of larger yolk cells called *macromeres*).

Birds and reptiles have telolecithal eggs with large yolk reserves, and follow a very different pattern of early cleavage. In fact, their yolky area doesn't divide at all. Instead, cytoplasmic divisions are restricted to the blastodisc (Figure 30.7), which con-

30.7 Cleavages in the chicken zygote. The enormous quantity of yolk in the chicken egg produces special problems in cell division. As we see here, the problem is solved by cell divisions being restricted to a small patch of material at one side of the yolk—the *blastodisc.* Note that even here the first two cleavages are not complete. Cell division occurs more rapidly near the center of the blastodisc. There is no equivalent to the blastula stage of sea urchins and frogs in the bird embryo. Instead, a hollow region (the *blastocoel*) will form as the upper layer of cells rises away from the yolk.

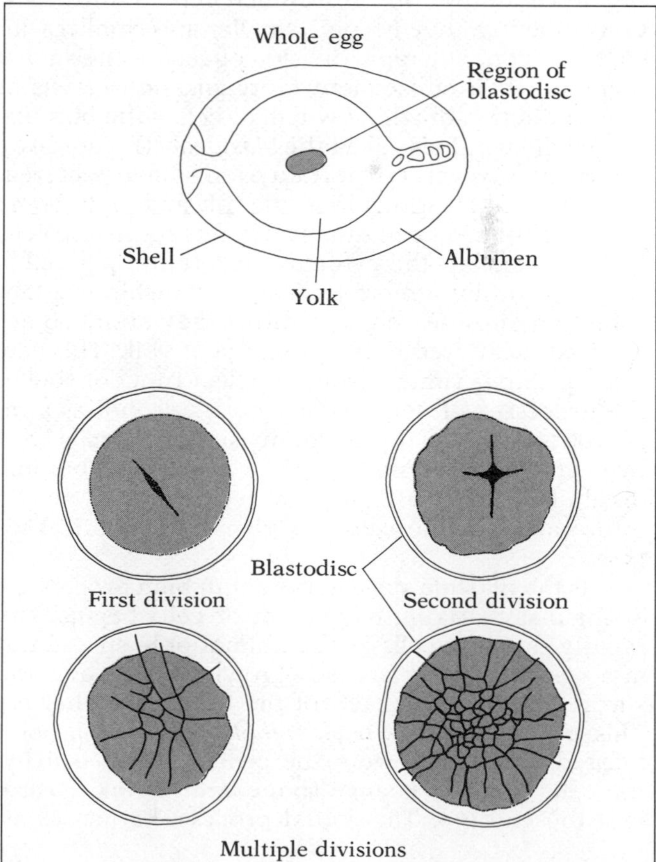

tains the nucleus and most of the cell's nonyolky cytoplasm. The first division forms a depression across the disk. The second forms a furrow at right angles to the first. Neither of these divisions completely divides the cytoplasm, but merely lays out a pattern for the next cleavages. Unequal cleavages of the first four cells produce a central cluster, with the most rapid rate of cell division occurring toward the center of the disk. Eventually, the cells pinch off over the yolk and become surrounded by membranes. All of this occurs in the blastodisc, which at this time, has become a tiny bit of gray tissue—the *blastoderm*—located on the yolk. (Before you go poking at your breakfast, you probably can't see any blastoderm on your eggs, since nearly all eggs sold in the stores are unfertilized.) The reason for the peculiar development of telolecithal eggs is simple. The yolk mass is just too large to divide each time mitosis occurs, so the bird or reptile embryo develops over only one portion of the yolk mass.

The Blastula and Gastrulation

But back to the frog embryo. At it continues its divisions, the greater activity at the animal pole becomes evident. At this stage of development, cells divide rapidly, but they don't grow much between divisions. Consequently, they become smaller and smaller with each division. The rapid division of cells at the animal pole and their subsequent reorganization results in the production of a hollow ball of cells—the blastula. The cavity within is called the *blastocoel* (Figure 30.8).

By now, several important events have occurred. First, the nuclear–cytoplasmic–plasma membrane relationship has been drastically altered. As the cells become smaller, the cytoplasmic volume per cell is reduced so that a more typical relationship is established among the cell structures; they aren't so abnormally scattered in the great sea of yolk. This new relationship is vital to proper nuclear function and to the processes of cellular transport. Second, as a result of so many cytoplasmic divisions, the cells have been able to differentiate—which is to say, to become dissimilar. Keep in mind, however, that the genetic information of the cells has remained identical *we think*.

As the blastula stage in the amphibian progresses, a singularly striking movement of cells begins. The rapidly dividing cells at the animal pole spread out in a wavelike front, actually growing down over the surface of the other cells of the blastula as they go. This process is known as *epiboly*. Then, at a point located about 160° from the center of this activity, the cells begin to change shape and to sink inward, into the embryo. This initial process, known as *invagination*, forms a small depression on the surface of the blastula (Figure 30.8). The cell deformation is caused by the activity of the microtubules and microfilaments of the cytoskeleton, similar to what occurs in ameboid movement. Then cells actually begin to migrate over the lip of the invagination and roll inward into the interior of the cell mass. The point of inward migration is known as the *blastopore*, and the rapidly dividing animal pole cells invade in a wave from above, continually turning inward over the top lip of the blastopore. Then a second wave of cells appears from the opposite direction—from the vegetal pole. These, too, begin to turn inward, into the embryo. Now the blastopore has formed a slit that looks like a down-curved, unhappy mouth. The upper layer of cells races over the surface to sink through the blastopore into the yolky mass within. Once inside, though, the layer turns upward again, retracing along the inner wall the route it had followed on the surface. This inward movement of cells, or *involution*, forms a second cavity known as the *archenteron* (or *primitive gut*). The archenteron will, in fact, become the gut of the developing animal.

The continued migration of cells from below—from the vegetal pole—extends the crescent-shaped down-turned mouth of the blastopore, until the ends of the crescent meet, forming a circular slit. Yolk-filled cells protruding from the crevice form the *yolk plug*. The original cavity (the *blastocoel*) becomes crowded to the rear and is finally obliterated as the archenteron enlarges. Meanwhile, *epiboly* (the proliferation of cells from above, growing down over the lower half of the embryo) continues. (Keep clearly in mind the differences between *epiboly*, *invagination*, and *involution*.)

As gastrulation proceeds, the reorientation and displacement of cells forms three major groups, or *germ layers* of cells (see Figure 30.8). The outer layer surrounding the embryo becomes the *ectoderm* (outer skin). The upper part of the ingrowing layer becomes the *mesoderm* (middle skin). Finally, the larger yolk-filled cells that have entered the blastopore from below form the *endoderm* (inner skin). Even before cleavage begins, the cytoplasm of the fertilized egg can be divided into areas that will eventually become the three germ layers. These areas can be identified by marking various areas of the embryo with dye and then watching where the dyed cells moved to. Such techniques made possible the construction of *fate maps* (see Figure 30.11). With the formation of three distinct germ layers, the embryo has now changed from a hollow ball of cells to a *gastrula*. These three germ layers hold great significance for the events that follow. As time passes, the importance of each layer will be clearly apparent. The formation of the gastrula and of the resulting

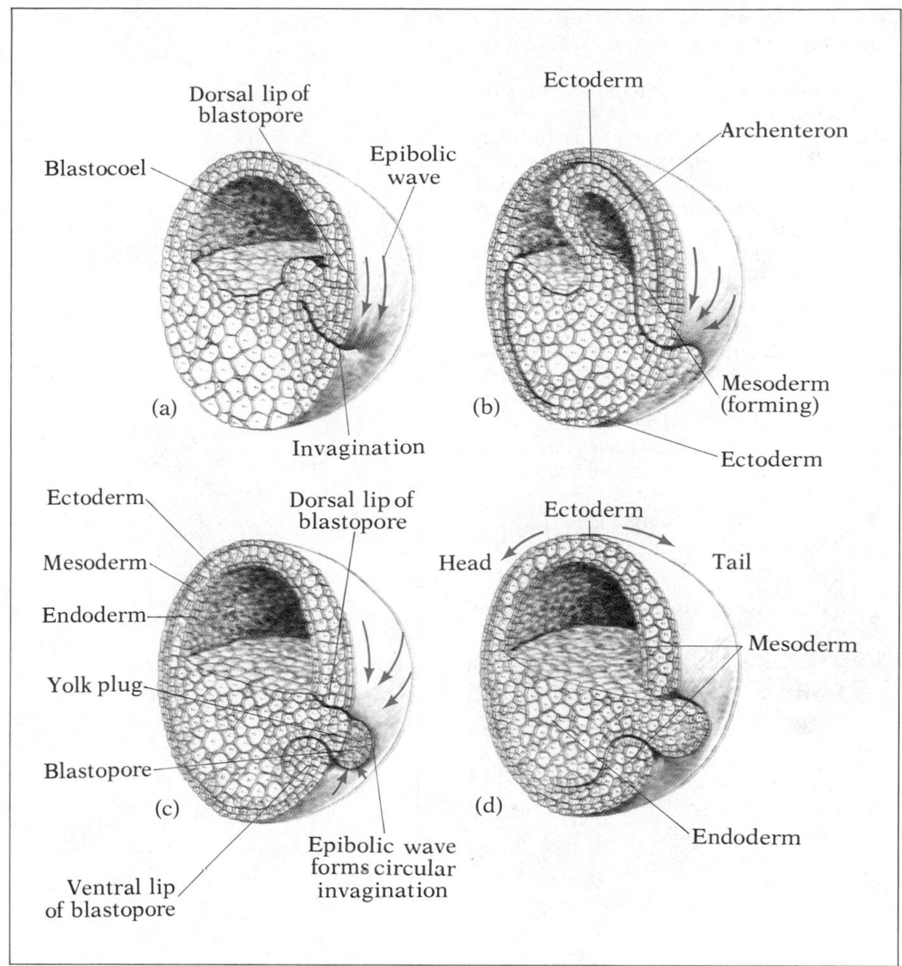

30.8 (a) Gastrulation in the frog embryo begins as an epibolic wave of rapidly dividing ectodermal cells. As the cells move around the embryo they begin to invaginate (turn in), forming the dorsal lip of an opening known as the blastopore. (b) As the ingrowth continues, the surface wave continues around the embryo, completing the blastopore. (c) It is temporarily filled with yolky cells forming a yolk plug. The ingrowth of cells forms a second cavity, the archenteron, squeezing the blastocoel back and eventually replacing it entirely. The ingrowth also produces an inner layer of cells, the mesoderm. With this development, the embryo contains three distinct cell layers, an outer ectoderm, middle mesoderm, and inner endoderm. The latter lines the archenteron and makes up the large region of yolky cells.

three germ layers is clear evidence of cellular differentiation. Now let's look at gastrulation in birds and mammals.

Gastrulation in Birds

Gastrulation proceeds somewhat differently in birds (and reptiles, which are very similar) than in amphibians. The most obvious differences lie in their cleavage patterns, and these, of course, are related to their yolk abundance. Other than that, the differences are not crucial, since the general development is the same. In the bird blastula the disk of actively dividing cells (two to three layers thick) splits and the upper layer rises, blisterlike, at its center to form a blastocoel (Figure 30.9). The lower cells (the floor of the blister) form a *primitive endoderm* and are joined by migrating cells from above to eventually complete the endodermal layer. The endodermal layer will grow over the yolk to form a *yolk sac,* in effect bringing the yolk inside the primitive gut lining.

Gastrulation first becomes apparent as the cells of the blastodisc begin to converge along a line that runs down the center of the disk. This line establishes the embryo's longitudinal axis. As the cells converge from both sides of the line, they roll under, forming an elongated depression called the *primitive streak.* Turning under on each side of the streak, the migrating cells form a middle (mesodermal) layer between the ectoderm and endoderm. The activity at the primitive streak is therefore homologous to the migration of cells around the blastopore in the amphibian embryo, and the primitive streak itself is really just an elongated blastopore.

Gastrulation in Mammals

Mammals have evolved along a different line. The mammalian egg is isolecithal, like the echinoderm egg. This means it has a meager yolk supply. The mammalian embryo will get nearly all its nourishment through maternal circulation, but it can't begin its parasitic existence without some preparation. So it temporarily sets aside gastrulation while it develops the tissues it will need to maintain itself.

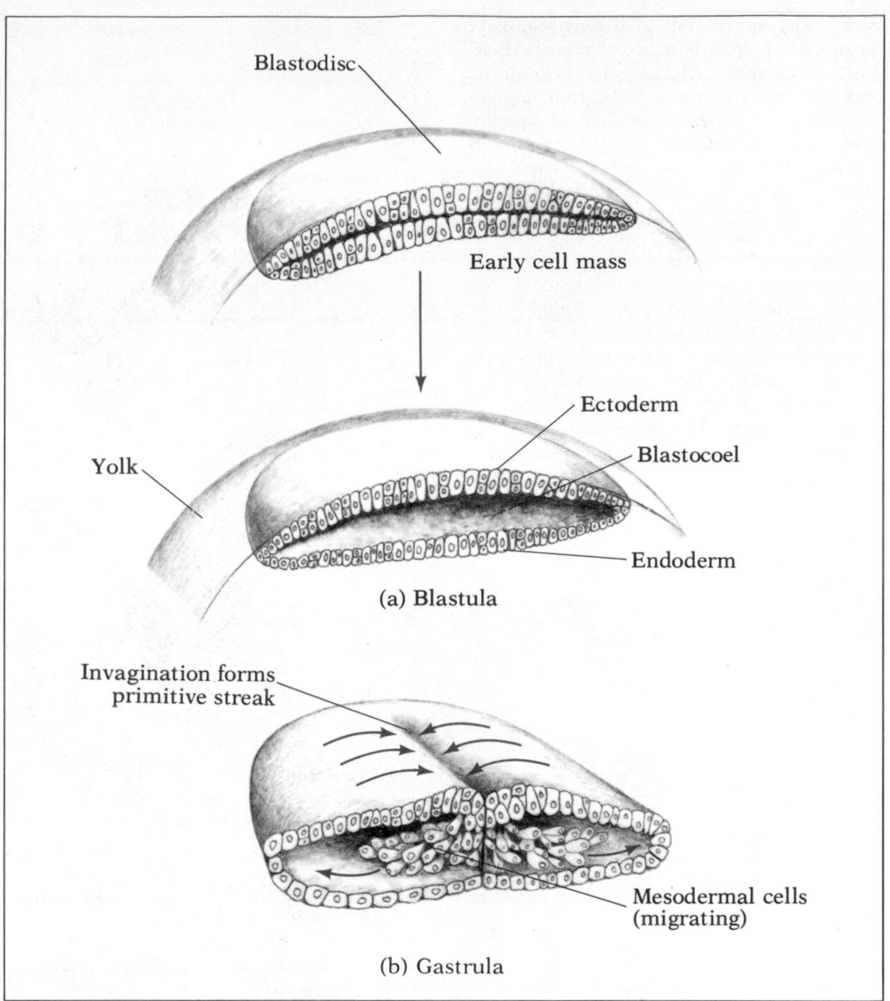

(a) Blastula

(b) Gastrula

30.9 (a) The blastula of birds forms as the upper layer of the blastodisc rises from the lower layer. The lower layer, along with cells that migrate in from above, forms the *endoderm.* The upper layer becomes the *ectoderm.* The space between is the *blastocoel.* (b) Gastrulation in the bird embryo is necessarily quite different from that of the frog. Cells from the upper surface form an ingrowth along a center line (the *primitive streak*). This will be the *embryonic axis.* As the cells turn under, they form a third layer, which at first consists of scattered cells. These will coalesce into islands of cells and will comprise the *mesoderm.* Like the amphibian, the gastrula of the bird finally consists of an outer ectoderm, a middle mesoderm, and an inner endoderm. Shown below is a surface view of the primitive streak stage of a chick embryo.

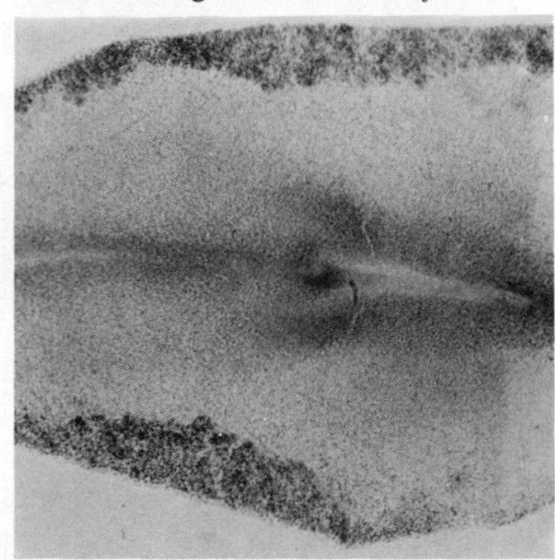

The equivalent of a blastula in the mammal is a stage known as the *blastocyst,* the stage that is reached at about the time of implantation. The blastocyst (sometimes known as the *signet ring* stage, because of what it looks like when a thin slice is cut out in cross-section) is a hollow sphere one cell in thickness nearly all the way around, except for an enlarged area known as the *inner cell mass,* which is several cells thick. The fluid-filled space within the sphere is referred to simply as the *blastocyst cavity.* The inner cell mass will become the embryo proper, while the thin outer sphere (the *trophoblast*) will become the extraembryonic membranes. These membranes are structures that are not part of the embryo but which are needed to sustain it, as we shall see when these special structures are discussed later in the chapter.

As preparations for gastrulation proceed, a layer of cells from the inner cell mass rises up to form the first of two membranes, the amnion, one of those supporting structures we will describe later. The remaining cells form the *embryonic disk,* essentially a two-layered plaque of cells. While it is still early in the history of the disk we can say that the upper layer constitutes the ectoderm and the lower, the endoderm. A second cavity, the *yolk sac,* forms as cells of the endoderm divide and migrate out to form its outlines. The stage is now set for gastrulation and the formation of a three-layered embryo. Before we proceed, however, let's consider the yolk sac in mammals. In birds and reptiles, as the yolk sac grows over the surface of the yolk, it develops blood cells and blood vessels that will bring yolk nutrients into the embryo. The mammalian yolk sac, on the other hand, is hollow (that is, it contains no yolk), but like the yolk sac of birds and reptiles, it is the site of the initial blood cell formation.

Gastrulation begins when ectodermal cells start dividing and migrating inward along the long axis of the elongate disk. The groove formed is called the *primitive streak,* as it was in the birds. As a matter of fact, what happens next is also familiar. Cells migrating in between ectoderm and endoderm produce the third layer, the loosely arranged mesoderm. Now that we have a three-layered embryo, gastrulation is complete (Figure 30.10).

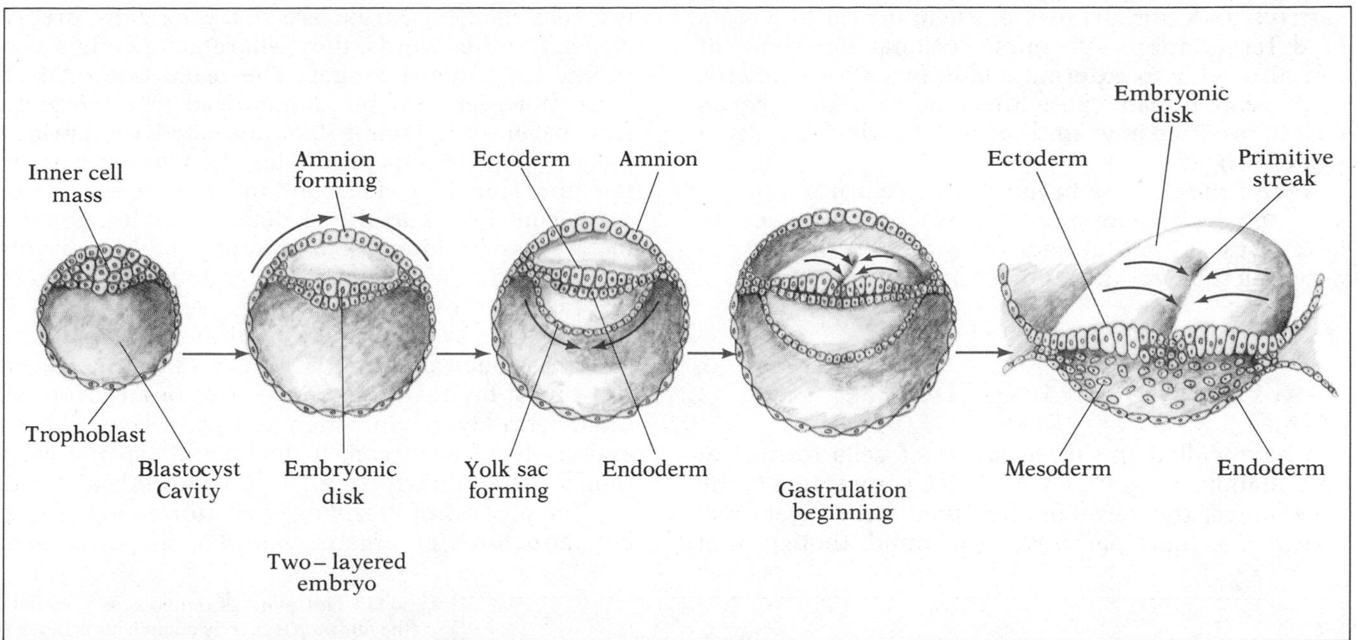

30.10 Gastrulation in mammals occurs after the blastocyst implants in the uterine lining, and supporting extraembryonic structures have begun their development. Cells of the inner cell mass rise up to form the amnion. This is followed by the formation of the yolk sac from cells at the lower portion of the embryonic disk. The two-layered disk matures into a layer of ectoderm and endoderm. Along the ectoderm, cell division increases along the long axis, forming the primitive streak as these new cells turn inward. The movement of cells inward between ectoderm and endoderm produces the third layer, the mesoderm. This completes gastrulation.

30.2 Morphogenesis: The Development of Form

Now, how do embryos differentiate? In other words, how do they change as they grow so that kidney cells become distinct from eye cells and arms and fingers develop from shapeless buds? One way to approach the problem is at the cellular level.

First we must emphasize again that, as far as we know, each cell has identical DNA.* It is this DNA that directs the production of all the enzymes to cell activity and the proteins used in building the parts of the cell. We also know that it is largely the cell's enzyme complement that make it different from other cells. It follows that control of DNA is crucial in the control of differentiation. This statement, however, only moves the question of how differentiation occurs back one step. What causes the DNA in one cell to behave differently from the DNA in another cell? Why are certain DNA segments active in one cell and totally inhibited in another?

*This statement sounds dogmatic, but we're considerably less sure now that at least one exception is known—namely the differentiated lymphocyte cell, in which there are known irreversible changes in the DNA.

Bringing the differentiation question down to the molecular level offers certain advantages. For example, we know something about gene or DNA activation in lower organisms from our studies of the operon. (See Chapter 12). We also know that hormones reaching target cells can activate certain genes of those cells. There is no reason to doubt that any number of chemical messengers from the cytoplasm of the cell, or neighboring cells, or even the surrounding environment might play a role in gene activation. In fact, this general idea is rather well entrenched in embryology today. However, we're stalled at the level of hypothesis. No embryological messengers other than the sex hormones have yet been identified. As various substances are tested, we learn that their activity is too complex to yield precise answers.

Regarding "environmental" explanations of cell differentiation, keep in mind that other cells may constitute a large part of any cell's environment. As the gastrulation process unfolds you can easily see that ectoderm is exposed on one side to the external environment, while the other side is exposed to an inner environment. Mesoderm and endoderm are totally internal and encounter quite different environments. So it is reasonable to assume that the differing environments of the three germinal layers of the gastrula undoubtedly play a role in the future events of differentiation. Of course, cellular environment can also refer to external conditions. (For example, cold weather can cause the hair cells in a hare's coat to produce new kinds of mRNA which results in white fur.)

With these ideas in mind, let's return to our description of development. As we proceed, we will look at experimental data from time to time to see how well the hypotheses hold up.

The Fate of the Germ Layers

We have called the three layers of cells formed at gastrulation the *germ layers*. Since *germ* refers to the beginnings, the term implies that these layers will give rise to new parts. Keep in mind, though, that the cells in the gastrula are not yet rigidly predisposed. In other words, they will retain a certain versatility for a while longer. The usual fate of each layer, however, can be summarized in a *fate map*. As you can see in Figure 30.11, the ectoderm is principally responsible for producing the nervous system, the outer skin, hair, nails, and most of the eye. Mesoderm contributes to the skeleton, muscles, gonads, blood vessels, kidneys, and many other typically internal structures; while the endoderm produces several types of internal linings, such as those of the digestive system, liver, respiratory system, and pancreas. Such a listing is a bit artificial, however, since most organs are comprised of tissues from all three germ layers. So, whereas a germ layer doesn't entirely form an organ, it does lay down the early foundation upon which other germ layers may build.

The process of *morphogenesis* (formation of specific structures or organs—*morpho,* shape; *genesis,*

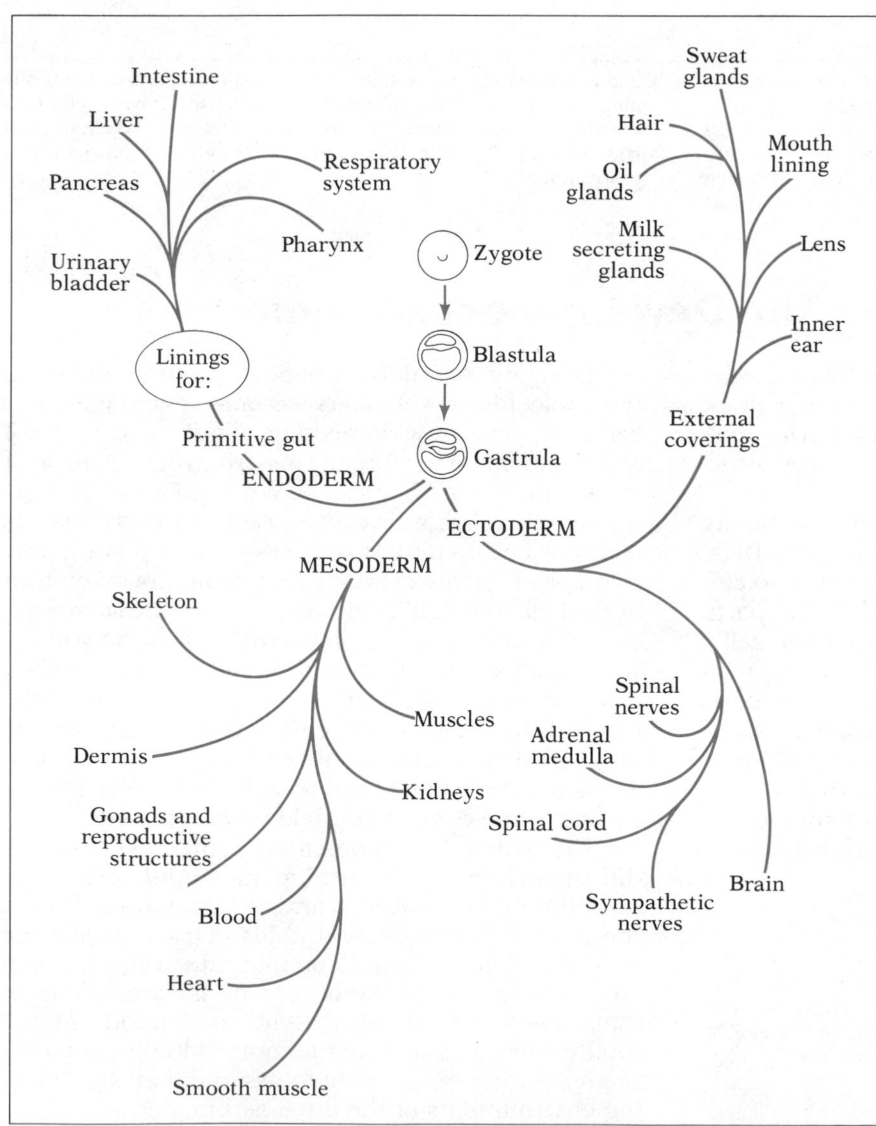

30.11 Not quite as ominous as it sounds, the *fate map* is merely an indication of what tissues each of the three germ layers will form. As the map indicates, endoderm produces the linings of many internal structures. Mesoderm contributes to the formation of muscles, skeleton, and blood vessels. Ectoderm produces many of the outer structures and is very active in the earliest phases of development, as we saw in gastrulation and will see in the neurulation process. It is important to note that no organ is entirely derived from any one germ layer, but rather, each organ represents the cooperative ingrowth of tissue from these derivatives. As the tissues meet during development and morphogenesis, they often have a profound effect on each other. One tissue seemingly will influence the developmental events of another as organ structures emerge from the formless embryo.

beginnings) begins to be apparent as we watch the early development of the central nervous system, the first system to develop in all vertebrates. The development of the nervous system is called *neurulation.* The construction of the neural framework involves three major processes, which are carried out in a highly coordinated manner. These are cell movement, cell differentiation, and growth. In essence, the vertebrate embryo produces a dorsal, hollow tube extending the length of the developing body. That tube will become the brain and spinal cord. Using Figure 30.12 as a guide, let's see how this central nervous system is actually formed.

In the frog, the first thing that forms is the *neural plate,* which extends from the dorsal lip of the blastopore, over the sphere of cells along a line anticipating the spinal column, toward what will be the head. As it proceeds along this line, it forms a depression—*neural groove*—along the body length. The ridges on either side of the valley, *neural folds,* are ectodermal thickenings. The widened portion marking the "northern" end of the valley (with the blastopore at the "southern" end) is the future head region. The neural ridges continue to thicken and gradually turn toward each other as they converge at the midline. Eventually, the tips of the folds fuse together to form

30.12 Neurulation in the frog embryo. Even when the frog embryo has completed gastrulation there is no respite in its frantic developmental activity. The gastrula, viewed from the outside, is at first a rather featureless ball of cells, giving no hint of what is happening inside. (a) The first indication of activity on the surface is the emergence of a thickened *neural plate* along the animal pole of the gastrula. The cells of this plate will form the primitive brain and spinal cord. (b) Next, a surge of mitotic activity produces a rising double ridge of tissue, tracing the outline of the neural plate. (c) The ridges, known as the *neural folds,* continue to rise up, until the two sides lie prominently on either side of the neural groove. (d) As the process continues, the ridges of the neural folds bend toward each other and come together, first at the center, then at what will be the tail of the embryo, and (e) finally joining at the incipient head. Thus, a hollow tube is formed which will be the *dorsal hollow nerve cord* of all vertebrates. The mesodermal notochord started the process by inducing the overlying ectoderm to form the neural plate. It will continue to lend some support, but will fade away as the mesoderm lying along its sides begins to form the vertebral column.

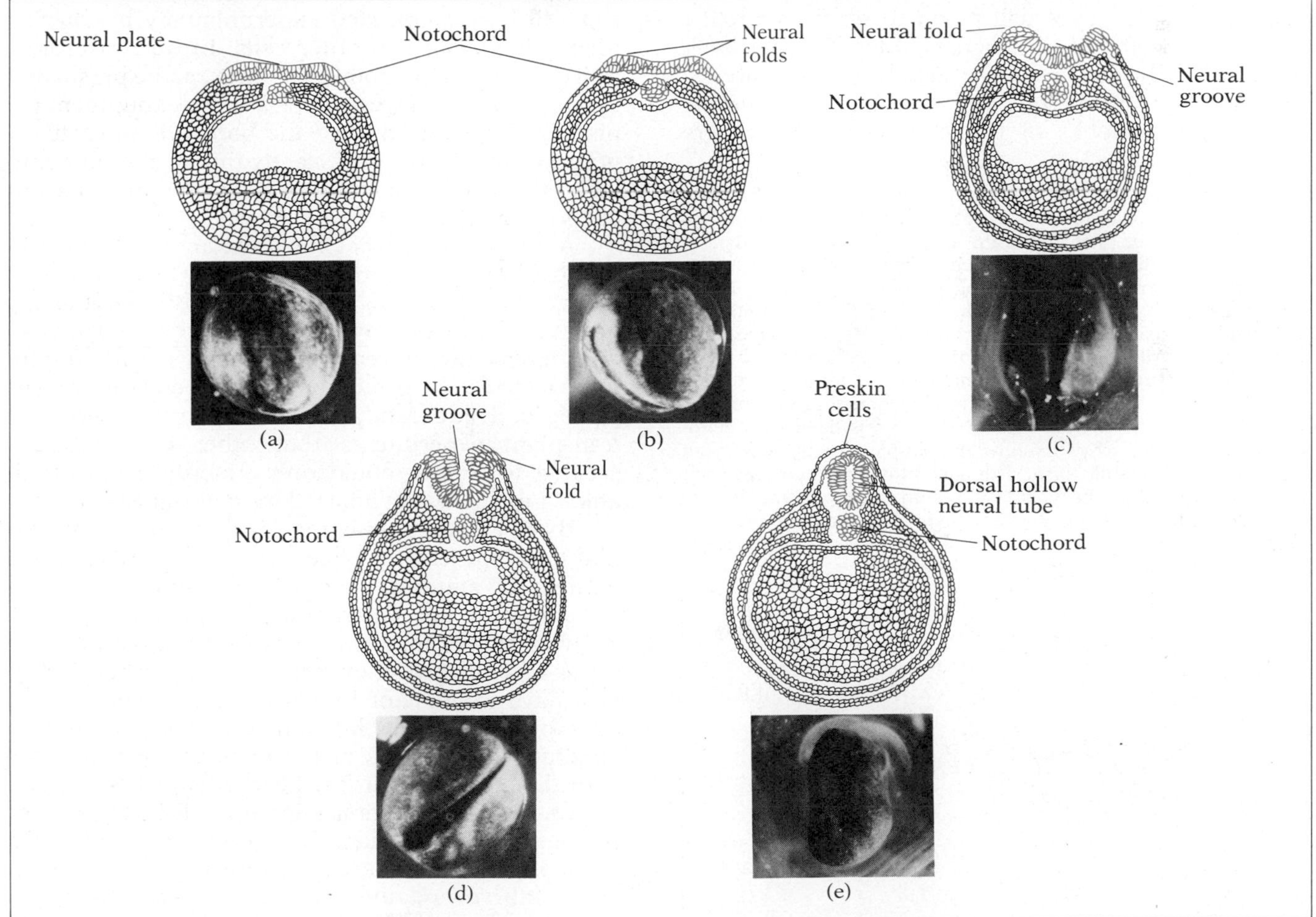

the hollow *neural tube.* This tube will form the central nervous system. Such a critical structure as the central nervous system, of course, can't lie exposed on the surface of the embryo, so it sinks beneath the surface, breaking its direct connection with the overlying ectoderm (Figure 30.12).

Meanwhile, below the neural ectoderm, the mesoderm begins to form the long, central, rodlike supporting structure called the *notochord.* (The notochord is so special that some authorities consider it to comprise a unique, fourth germ layer.) Vertebrates and all other chordates produce a notochord and rely on it for support in at least some stage of development. In most vertebrate embryos the notochord disappears as its supporting role is taken over by the developing spinal column. The notochords remain in the adults of only a few fish and some of the more primitive chordates.

As the nervous system and the notochord continue their development, the recognizable features of the embryo gradually emerge. Through subtle changes, groups of cells in the anterior region of the neural tube begin to produce the first brain tissue. Shortly, the eyes will begin to form as cuplike extensions of the brain (Figure 30.13). Further back along the embryonic axis, a series of mesodermal clusters on either side of the notochord form the *somites.* These paired blocks of tissue are the forerunners of the body muscles and the axial skeleton. Their appearance reminds us again of the segmented plan common to so many animals.

These, then, are some of the events of early vertebrate development. With this background, we'll consider some other processes associated with development, and some of the intriguing problems that remain with us.

30.13 As the frog continues to develop, the outline of the brain emerges. At first, the brain is nothing more than hollow bulges of the neural tube. The forebrain and midbrain are easily recognized, as are the cup-shaped rudimentary eyes and ear vesicles. Other changes are rapidly occurring along the notochord. Mesodermal tissue has grouped itself into the blocklike *somites,* which will send extensions out on either side of the axis. These will eventually contribute to muscles of the body wall and to vertebral structure. (Embryologists use the number of somites as a guide in identifying various stages of early development in vertebrates.) It is because of the somites that the coelom is lined with mesoderm. Do you see why? By now the simple gastrula has taken on clearly identifiable form, although many systems have not emerged. Note that the yolk has become reduced and is now centered in the developing gut region. The remainder of food reserves will last just long enough for the tadpole to complete its development, emerge, and begin feeding itself.

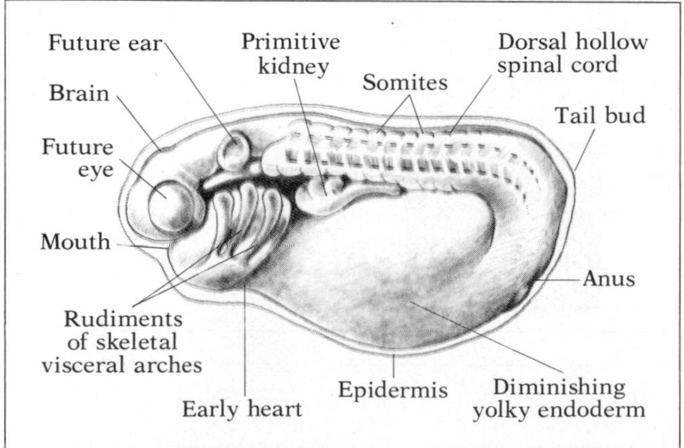

Embryonic Organizers and the Induction Theory

The gastrula provides an excellent subject for the study of development. As the stage just prior to morphogenesis, it presents us with a rather simple subject, but yet one poised and readied for a series of incredibly complex events. Interestingly, by the end of gastrulation, the cells of the ectoderm, mesoderm, and endoderm have become committed to producing specific tissue regions of the animals, but their fates have not been irrevocably sealed. They can still be manipulated experimentally by altering their relationships to other cells. In other words, a cell on its way to becoming muscle can be persuaded to become cartilage (such as by experimentally placing it in some part of the body where cartilage normally develops). It is exactly this kind of manipulation that has given embryologists several important clues to the nature of development.

A great deal of what we know about the interaction of the germ layers has emerged from the work of Hans Spemann. His many contributions to biology are almost overshadowed by his highly imaginative and technically clever manipulation of the amphibian gastrula. In the early 1920s, Spemann and his protégée, Hilde Mangold, carried out a series of transplant experiments that other embryologists, seeking to repeat, could only describe as "formidable." Let's see what amazed his colleagues.

By Spemann's time, many of the details of normal development had been clearly described for a number of animals, so he had a rather firm basis for designing his experiments. Essentially, what he did was take common information and carry out a series of uncommon experiments that others might well have thought of but would not have dared to attempt. Specifically, he wanted to know about the condition of the cells in the ectoderm just prior to neurulation. Had their fate been irrevocably sealed?

Choosing the region known as the dorsal lip of the blastopore, Spemann began by simply removing some of the ectoderm. Then, culturing both the damaged embryo and the severed ectoderm, he found that the embryo developed abnormally, and the piece of ectoderm simply formed a blob—an undifferentiated mass of cells. This observation verified what had been previously established—that the ectoderm was essential for normal development—but it also indicated that the ectoderm would not become nerve tissue by itself.

In a second series of experiments, Spemann transplanted ectodermal tissue here and there on recipient embryos. (Imagine transplanting tissue on a tiny frog embryo!) In each case, the ectoderm proved to be subservient to the surrounding tissue. That is, it took on the fate of its new neighbor cells. Thus, Spemann established the plasticity of ectodermal tissue; the fate of ectoderm at this stage had not become sealed.

In a third series of experiments, Spemann changed the procedure. Instead of removing ectoderm, he opened a flap near the dorsal lip and scooped out some of the mesoderm below. Replacing the flap, he observed that abnormal neural tissue developed in spite of the fact that the ectoderm was in place. (Remember, the neural apparatus is ectodermal.) The implications of this were intriguing. What was the role of the mesoderm in neurulation? Was the failure merely the result of harsh treatment? Spemann first did the logical thing. He repeated the experiment, but this time he replaced the scoop of mesoderm before closing the flap. A normal embryo developed. Harsh treatment apparently had nothing to do with the results.

In his next experiment, Spemann removed mesoderm from beneath the dorsal lip ectoderm and transplanted it to another embryo. He prepared the recipient embryo by opening a flap of ectoderm and removing a scoop of mesoderm on the side opposite the dorsal lip. This region, he knew, had nothing to do with neurulation (the development of the nervous system). Inserting the mesodermal transplant in the vacated cavity below the ectodermal flap on the ventral lip, he closed the flap (Figure 30.14). Development proceeded without hesitation. But much to his amazement, a second neural region emerged at the site of the transplant! In fact, if the embryo survived this treatement, it grew a second head! And so, the dorsal lip mesoderm took on a whole new significance.

Spemann's experiments reveal two major things. First, as we mentioned earlier, the ectodermal cells of the early gastrula are not yet differentiated, nor are they totally committed. We emphasize this discovery mainly because it led to a barrage of experiments that tested the commitment of embryonic tissues. We will summarize a few of these later on. Second, Spemann was able to conclude that germ layers in the embryo influence each other. The mesoderm, for example, influences, or organizes, the ectoderm above, a process now called *induction*. This discovery was the first real clue to the basis of differentiation.

In 1935 Spemann received the Nobel Prize for his work. It was followed by an era of experimental

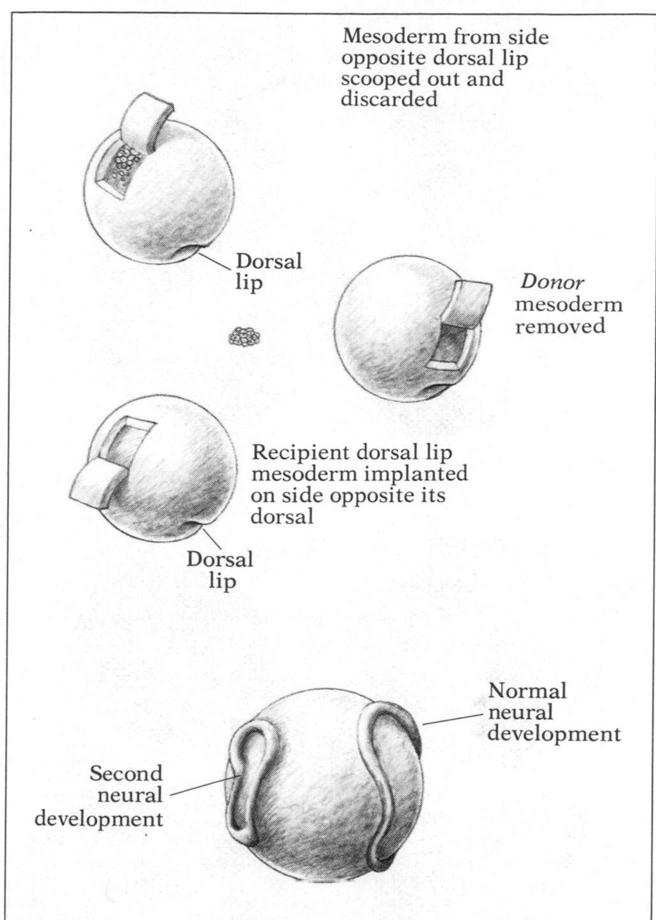

30.14 Hans Spemann developed a transplant technique which consisted of moving dorsal lip mesoderm to other regions of the gastrula. As shown here, one such procedure led to the production of an additional neural region on the "wrong" side of the embryo. From these and subsequent experiments Spemann developed his *induction theory* of development. He postulated that the germ layers influence, or induce, each other to develop specific structures.

embryology in which the limits on transplanting were the technical difficulties and the vulnerability of the embryo. Many examples of embryonic induction have been discovered since that time. One of the most noteworthy of these was related to the development of the lens in the eye. This added a new twist, since it involves a "triple-inductive" system, as described in Figure 30.15.

The discovery that mesoderm can influence the fate of ectoderm was undoubtedly important. But, as with most discoveries, the finding produced more questions than it answered. The obvious question is: How does the mesoderm influence the ectoderm? What does it really do? What is the relationship between the two layers? Is there some sort of chemical message transmitted across the cells? That last question, of course, leads us back to the molecular level. And if chemical messengers are involved in induction, then what kinds of molecules are they?

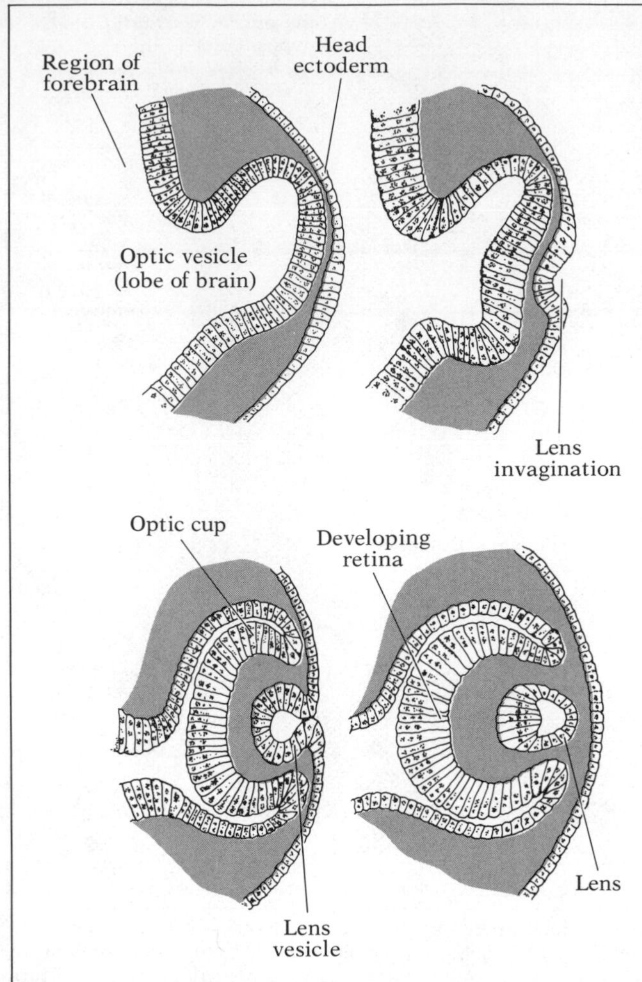

30.15 Early experiments on the development of the eye in frogs showed that there is a complex interplay among the embryonic tissues. As the brain begins its development, two bulges of tissue, the *optic vesicles,* begin their growth outward toward the surrounding ectoderm. When contact is made, the ectoderm responds by growing inward toward the vesicle, almost as if in response to a message. Meanwhile, the vesicle changes shape, forming the optic cup. Finally, ectodermal ingrowth pinches off and comes to rest in the curve of the optic cup. It will eventually form the lens. Embryologists now know that the optic vesicle induces the lens. But the interaction is rather complex. The developing lens, in turn, affects the optic cup. In fact, the development of the eye is a lesson in both induction and counterinduction.

Testing the assumption that a chemical messenger passes between mesoderm and ectoderm turned out to be relatively easy. (Most experiments are not nearly as difficult to perform as formulating the questions they will answer!) In this case, the procedure consisted of placing mesodermal tissue on a laboratory growth medium for a brief period and, after removing it, placing a segment of ectoderm on the same medium. In those first experiments, as the isolated ectoderm began to change, under the wide-eyed observations of the scientists, it produced crude structures strongly reminiscent of a neural apparatus. It was enough to encourage them to refine the experiment by placing mesoderm and ectoderm on either side of an agar block. The results were the same. Obviously, something passed through the agar.

But what is it? Experimenters tested the media and the agar block to determine what substance was being released from the mesoderm. This is the point at which we found ourselves in the early 1960s.

Unfortunately, it turned out that many substances pass between mesoderm and ectoderm. So people were left guessing. At first they guessed that a nucleic acid—perhaps an RNA—was the active substance. Later, proteins known as *histones* were suspected, but it was learned that all cells of an organism have the same five types of histones, and that, in any case, histones don't move between cells. The upshot is, we *still* don't know what the active substance is. We can only harbor suspicions. But we suspect a number of substances, including some inorganic ions. Actually, the possibilities are mind-boggling. With repeated failures to find "the inducer," researchers have become discouraged, and the search has waned, leaving us as confused as ever. It is suggested that there are many inducers acting alone or in concert, perhaps even changing from day-to-day as the embryo develops. Apparently, part of our problem is that our experimental techniques and/or our level of information have not made it possible for us to ask the right questions. It seems that we have probably learned all we can for now about embryonic induction through investigation at the tissue level. Today's researchers are devoting their time to investigations aimed primarily at the cellular and molecular level. They are using much simpler animals and plants, under more controllable experimental conditions. Eventually, their discoveries will undoubtedly bring scientists back to the amphibian embryo for another look.

The Cell Nucleus in Differentiation

The continued potency of ectodermal cells in the early gastrula established by Spemann was extended by other experimenters. One line of research reported in the late 1950s by J. B. Gurdon, R. W. Briggs, and T. J. King involved isolated nuclei of blastula cells. Through their ingenious experiments it was possible to test the potency of the nucleus as embryonic development proceeded.

The primary question asked was: When, if ever, does the nucleus lose any part of its capability to direct development? Or, put more simply: How does the nucleus change as cellular specialization proceeds?

Using microsurgical techniques even more delicate and sophisticated than Spemann's, Gurdon,

Briggs, and King succeeded in removing the nucleus from the mature egg cell. In its place, they transplanted nuclei from cells in more advanced stages. The frog's egg, then, provided a standard and undifferentiated cytoplasmic medium, with all the cytoplasmic capabilities of any fertilized egg (Figure 30.16).

Two important results were observed. First, when donor nuclei came from cells taken from a blastula, normal tadpoles developed. There was some variation in the success rate of nuclei from different regions of the blastula; but, in general, blastula nuclei were still young enough to produce normal tadpoles when placed into undifferentiated cytoplasm. Then it was discovered that when donor nuclei came from older embryos, most of the recipient embryos failed. The embryo would grow, but it would form hideous and deformed masses of tissue. The conclusion seemed clear at the time: As the embryo progresses in its development, irreversible changes occur in the nuclei of its cells. Furthermore, they begin to lose their potency during the gastrulation process.

The experiments were later to be refined (as if shifting nuclei from cell to cell were not already almost an art!). The problem was, since a variety of cell nuclei were used as donors, there was always some question of genetic differences. Perhaps it was wrong to assume that the genes of all cells from one organism are identical. Could some subtle differences cause an experimental embryo to fail? And were the changes in nuclei during development passed on to descendant cells? These questions arose because the abnormalities produced were often different from each other. The monsters were not alike; so, although the nuclei had been removed from closely associated cells, they apparently differed.

To answer these questions, a two-stage experiment was designed. The first stage involved the usual transplanting of older nuclei into *enucleated* (without nucleus) egg cells. In the second stage, nuclei from the blastula stage of the resulting embryos were used. These were transplanted into a second series of enucleated egg cells. Since the donor nuclei were all descendants of the original donor nucleus, they were presumed to be genetically identical. (Obviously, they were producing clones.) The experimental procedure was substantially more precise than before (Figure 30.17). The results were interesting. The monsters became more standardized (and everyone appreciates standard monsters). Also, it was interesting that when one cloned colony was compared with another, the usual differences appeared. So the basis of the specific abnormality was already present at the time of the first nuclear transplantation.

Additional refinements in technique produced another curious effect. The original conclusion, remember, had been that differentiation involved irreversible changes in the genetic material, but not until some stage in gastrulation. The idea was that age was a critical factor. However, the supposed critical stages—the points of no return—continued to be moved up as the transplanters became more expert with better and better techniques. These days, in fact, Gurdon can get as many as 90% of adult frog

30.16 Transplanting nuclei from various stages of development has answered some questions about development. Here, unfertilized frog eggs from the South African clawed frog (*Xenopus laevis*) are enucleated (that is, their nuclei are removed). This can be done by microsurgery with extremely fine-tipped glass needles or through selective destruction using ultraviolet radiation. Once the nucleus has been removed or destroyed, donor nuclei from various sources can be implanted with reasonable success. Using the gastrula or, in some cases, intestinal epithelium, cells are separated and disrupted. Nuclei are then selected out in micropipettes and implanted in the donor egg. Implanting a nucleus is, of course, the equivalent of fertilization, since the donor nucleus is diploid. The hybridized egg cell can then be cultured for more donor material or observed through its development.

Needle
Nucleus removed
Unfertilized frog's eggs with nuclei near surface
Cells dissociated from various sources
Nucleus implanted
Blastula—can be used for subsequent nuclear transplants or observed through development process
Egg cell with diploid donor nucleus

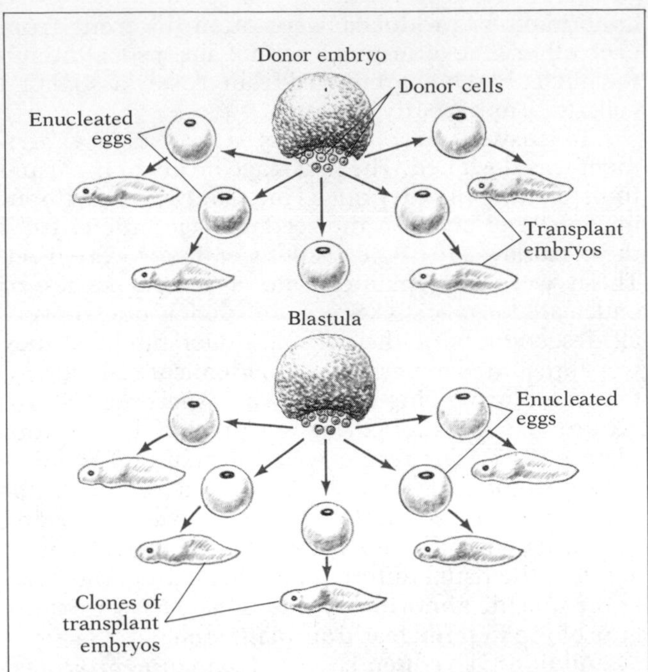

30.17 In later transplant experiments, Gurdon, Briggs, and King made an effort to produce genetically identical subjects that could then be used for transplants. The procedure included transplanting nuclei into enucleated eggs and allowing some to develop into embryos. But one line (shown at the center) was treated differently. When the recipient zygote developed into the blastula stage, its cells were used for nuclear transplanting into a number of unfertilized eggs. Each of the resulting tadpoles was a member of the resulting clone, which means that all were genetically identical. Using transplant nuclei from the clone group produced far different results in future experiments. Gurdon, Briggs, and King found that when nuclei from older embryonic stages were implanted into eggs, they could reliably develop normal embryos.

brain cell nuclei to produce normal embryos when transplanted into enucleate eggs. It is now believed that irreversible changes do *not* occur in most cell nuclei as the cell ages. So, whereas nuclei do undergo differentiation, previous indications of "irreversible nuclear changes" were, instead, apparently due to "irreversible nuclear damage" during the difficult transplantation process. Nuclear changes, then, are not irreversible—with the one known exception already mentioned (the immunoglobulin gene in differentiated lymphocytes).

In summary, it is clear that up through early gastrulation, there has not been much irreversible nuclear differentiation. During the gastrulation process, however, there is a sudden barrage of RNA transcription, with all kinds of new mRNA appearing in the cytoplasm. And neighboring cells may experience entirely different patterns of gene activation. Somehow, certain genes are activated while others are inhibited—perhaps forever (in normal development). Newly differentiated cells go their separate ways, passing their altered genetic state on to daughter cells.

Nerve Cells and Developmental Patterns

The longest cells of the vertebrate body are the *neurons,* or nerve cells. The *body* (nucleus and most of the cytoplasm) of the cell may be in or near the brain or spinal cord, while long extensions or processes travel to a distant part of the body where they branch into treelike endings. So when you stub your toe, the pain is transmitted to your spine through neurons about a meter long. A relay team of spinal neurons transmits the information to your brain, and you can then say whatever it is you say when you stub your toe.

In embryonic development, the neural processes follow along defined nerve pathways from their location in the brain or spinal cord. However, it appears to be largely a matter of chance which part of the body any given neuron in the nerve will eventually reach. Most of the neurons that grow out of the central nervous system, it appears, will not survive. Those which make connections with appropriate target tissues will live, while those which do not will perish during embryonic development.

Sprouting Factor and Antisprouting Factor

Once the growing neuron reaches its intended target area, it begins to sprout branches. Other neurons in the area will also branch, until the target area (say, your toe) has the proper density of neuron endings. How do the nerves know when the proper density has been reached? As occurs so often in development, regulation depends on negative feedback loops. In this case, two hormonelike substances are involved. The tissues receiving the neuron are continually producing a *sprouting factor.* At the same time, an *antisprouting factor,* manufactured back in the body of the nerve cell, is transmitted down through the process and released at the ends of the neuron branches. This substance inhibits further sprouting of that neuron, and also of any other neurons in the area. As more and more branches of the neurons sprout, a balance is reached between sprouting factor and antisprouting factor, and further branching ceases. But if some of the terminal processes in the area are killed or removed, the balance shifts and the remaining neurons will again put out new sprouts until the proper density of nerve endings is reestablished. If transmission of the antisprouting factor is experimentally blocked (for example, by colchicine, which inhibits cytoplasmic movement), all the neurons in the target area will sprout new branches.

How does the neuron know when its tip reached its own target area—or has failed to do so? The answer is complex and not fully developed. We do know that the target area itself has rather little effect on nerve growth patterns. This can be seen in experiments in which a patch of embryonic skin is removed prior to nerve growth and is replaced with a patch from another area. Regardless of the source of the transplanted skin, neurons reaching it will sprout and branch in a regular pattern, as if the original skin were still in place. But, as stated above, the eventual survival of the neuron depends on the nature of the tissue. The growing process, once it reaches its target area, apparently sends some unknown kind of message back to the body of the cell, informing it that contact has been made—*and with what kind of tissue*. The cell is then given the OK to continue living. If the appropriate contact is not made, the cell obligingly dies. After all, its only interest lies in the success of the organism as a whole.

An Experimental Investigation of Nerve Specificity in Development

Consider the classic experiment illustrated in Figure 30.18. A piece of epidermis is removed from the side of a developing frog embryo, at a time when its own future as back skin or belly skin has been irreversibly determined, but before the axons of the nervous system have yet reached their target tissues. The piece of epidermis is rotated 180° and replaced. When the frog finally becomes an adult, a patch of skin on one side is upside down. The frog has white belly skin on part of its back and green back skin on part of its belly. Now, it is known that frogs don't like to be touched with vinegar-soaked paper. Even if the frog is blindfolded, its feeble nervous system responds to where the hurt is, and the animal can scratch at it. When the experimenter puts acid-soaked filter paper on the white patch of belly skin on the frog's back, however, the animal keeps kicking itself in the stomach with its hind foot. If the experimenter then applies the stimulus to the green patch on the frog's belly, the poor animal reaches its hind foot up over its back, repeatedly hitting at the spot where that back skin ought to be. Which is to say, the central nervous system has received messages from neurons that correctly indicate exactly what target tissue had been innervated.

30.18 In this simple transplant experiment, a patch of skin is excised from a developing frog embryo and reversed. Later when the patch has grown in and its neuronal connections made, a simple test is carried out. The lower region of the patch is stimulated with an irritant. The frog responds by trying to wipe the area where the skin would have been had it not been reversed.

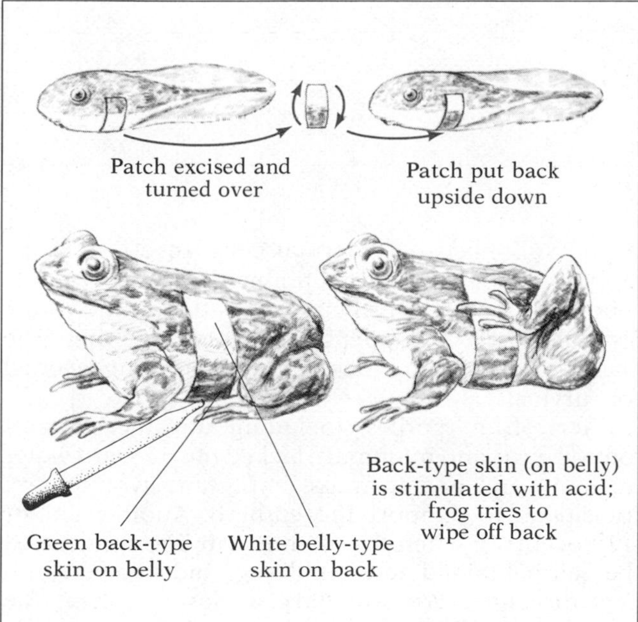
Patch excised and turned over
Patch put back upside down
Back-type skin (on belly) is stimulated with acid; frog tries to wipe off back
Green back-type skin on belly
White belly-type skin on back

If the frog is then killed and dissected, the visible nerve pathways are found to be just as they would have been if the ectoderm hadn't been reversed. The neurons going down the nerve pathway to the belly found themselves actually innervating back skin. Those innervating the patch area on the back also followed normal nerve pathways. In both cases, the developing neurons recognized the tissue they terminated in, and passed this information on. The tissue-selection hypothesis would hold that these particular neurons would have degenerated if the epidermis had not been rotated. But as luck would have it, some neurons targeted for belly skin actually made it to the misplaced belly skin, and other neurons—targeted for back skin—found their target tissue, even though it was in the wrong place. So when the back was touched with vinegar, belly neurons were activated, and the spinal column and brain deciphered the information accordingly.

How can the growing nerve cells know they have terminated in, say, belly skin? We don't know. Presumably each kind of target tissue has its identifying mix of chemical signals. We know even less about the programming and control of nerve cell interconnections in the brain, except that any such system must be phenomenally complex. Every brain neuron has connections—usually multiple connections—with the processes of many other neurons. The patterns are complex and surely not random. Does each neuron or class of neurons put out its own identifying set of chemical stimuli? We do know that mouse brains have many more kinds of mRNA than all other mouse tissues put together, which indicates that a substantial proportion of the genes and proteins of vertebrates are involved with brain function. And most humans would agree that human brains are probably even more complex than mouse brains. (For another experiment illustrating the complexity of neural pathways, see Figure 30.19.)

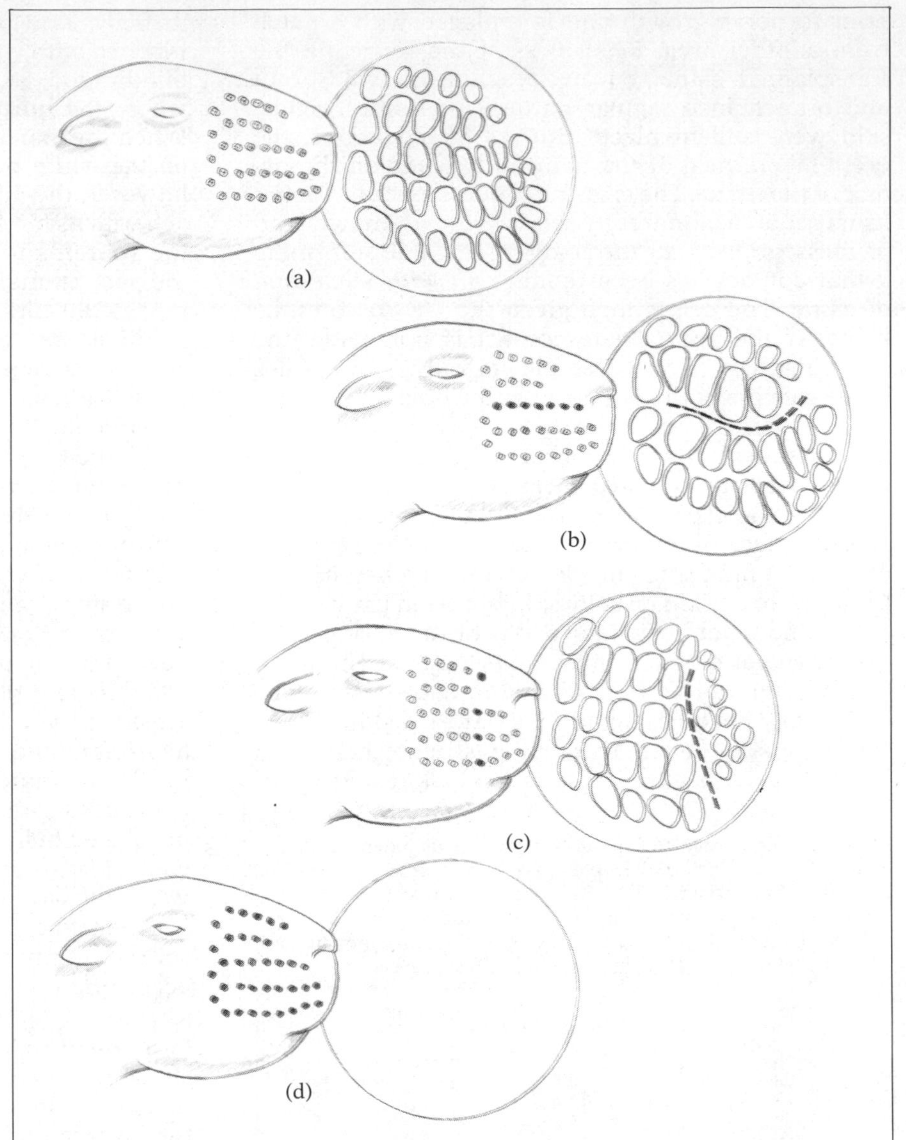

30.19 The target-cell hypothesis states that if a developing neuron does not make contact with its intended target tissue—which may be another specific neuron type—the neuron will die and disappear. (a) One of a mouse's most important sense organs is the whiskers on its snout. The touch receptor associated with each whisker sends its output to just one *barrel:* a specific, visible aggregate of neurons in the sensory cortex of the brain. The arrangement of the barrels is similar to the arrangement of the whiskers. (b,c) If one row of whiskers is destroyed shortly after birth, the corresponding row of barrels in the sensory cortex will later be found to be missing. (d) If all the whiskers are destroyed, all the barrels disappear. The fibers that innervate the whisker are not directly connected to the barrels, incidentally, but are linked by at least two synaptic relays. If the terminal neuron fails to find its intended whisker, an entire sequence of neurons must die.

Physical Support and Life Support of Vertebrate Embryos

In earlier chapters we emphasized some of the adaptations of animals making the transition to land life. Some of those changes, you recall, occurred in the developmental processes, particularly in the area of embryonic support and protection. Let's now have a closer look at some of these adaptations.

To begin with, the fishes and amphibians carry out their development in the water, so their problem of embryo support and protection is relatively simple. They don't require an elaborate system of membranes for respiration, excretion, and the maintenance of a proper water content. Food is stored intracellularly in the yolk-filled endodermal cells. In essence, the embryo of the fish and amphibian grows around its food supply, and to further simplify things, the young are able to feed independently upon hatching. The watery medium provides nearly everything needed for survival.

Terrestrial reptiles, including the ancestors of today's birds and mammals, lacked the buoyant water medium, and had to make important evolutionary transitions to support the embryo. Another major shift occurred when the mammalian line abandoned the self-contained terrestrial egg, and evolved in a new direction. We will have a closer look at the reptilian and bird method first and then proceed with the mammals.

The Self-Contained Egg

The zoologist Alfred S. Romer has suggested that the self-contained egg of reptiles developed while reptiles were still largely aquatic animals. The idea is that a protected egg, with its own food and water, could be laid on land and the hatchlings could then return to the water. The idea certainly has its appeal. Could there have been any safer place for a species to hide its eggs? After all, since the transition to land hadn't really begun, there were fewer predators on the land to devour the eggs whereas all manner of hungry beasts lurked in the water. But alas! Predators also evolve, so it probably wasn't long before they invaded the sanctuary of the dry land to see what else the earth held that was good to eat.

We can only guess about the structure of the primitive land egg, but we can make some reasonable inferences from what we see in such eggs today. Presently, both reptile and bird eggs consist essentially of four parts. The *shell* itself is composed of a soft leathery covering (reptiles) or a hardened calcareous material (birds). A heavy *yolk mass* constitutes most of the stored food. The fluid *albumen* (egg white) acts primarily as a water supply, but also contains some stored protein and some effective antibiotics. Finally, there is the *embryo*, which will develop four extraembryonic membranes. These membranes, which develop along with the embryo, consist of the *yolk sac*, the *chorion*, the *allantois*, and the *amnion*. Note the stages in the development of these membranes in Figure 30.20.

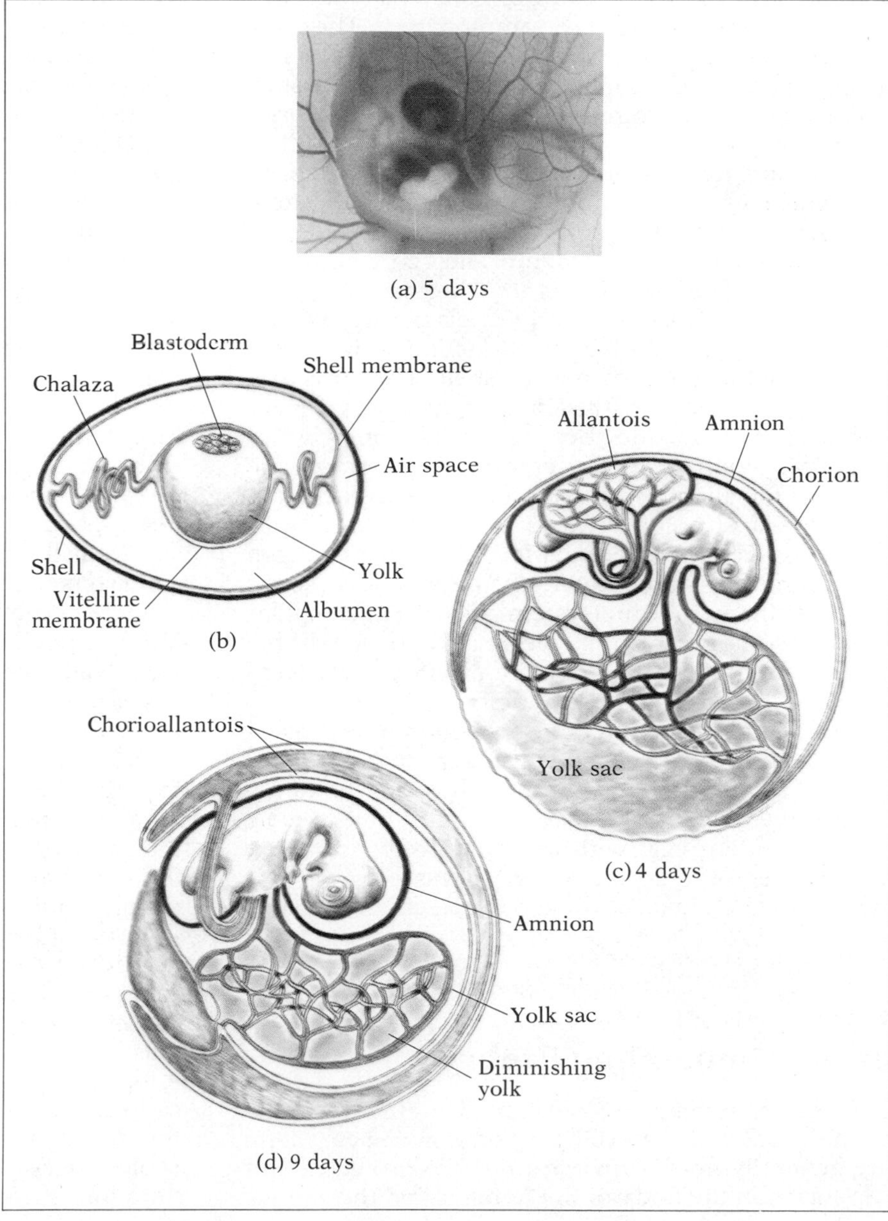

30.20 Extraembryonic membranes of the bird embryo. (a) The photograph shows an intact chicken embryo after about 5 days of development. Much of the supporting structure is not visible, but the *yolk sac*, with its vascular system, is easily seen, as is the fluid-filled *amnion* covering the embryo. The development of these all-important supporting structures can best be seen in a sequence of stages. In (b), the fertilized egg is shown before the blastoderm is well-developed. The familiar egg structure includes the porous *shell*, the *shell membrane*, a region of *albumen*, and the central *yolk* supported by the twisted *chalaza*. In the events that follow, the supporting membranes develop. (c) The yolk sac emerges early and is seen as a membrane spreading out over the yolk. Its development is accompanied by the development of a rudimentary heart and a vascular system that will carry nutrients into the embryo. The *allantois* begins as a simple sac that follows the curve of the *chorion* as it enlarges. It, too, will carry blood vessels, since its function is gas exchange and uric acid storage. The surrounding chorion is a simple sac that envelops the entire affair. The amnion forms an envelope of its own, directly surrounding the embryo and bathing it in a fluid medium. (d) As development continues, the yolk sac covers the entire but diminishing yolk, while the allantois has grown along the route of the chorion, fusing in one area to form the *chorioallantois*. The amnion enlarges to accommodate the growing embryo.

The yolk sac of birds and reptiles is the first extraembryonic membrane to develop. Actually, it is an extension of the gut. As it expands over the surface of the yolk, it becomes suffused with an extensive system of blood vessels. Food substances, absorbed into the blood as it circulates over the yolk's surface, are thus carried into the embryo. As food reserves diminish, the membrane folds inward upon itself and is eventually drawn into the abdomen of the embryo just prior to hatching, through what amounts to a navel.

The amnion and the chorion develop simultaneously. Figure 30.20 shows how they grow up over the embryo, the amnion joining over the tiny animal, forming a fluid-filled, protective sack which acts as a shock absorber and also keeps the body well-lubricated so that the growing parts don't fuse together. The amniotic fluid is kept in rather constant motion by continual contractions of the amnion itself. The chorion continues to grow around the entire egg contents, forming a continuous membrane just under the shell and later fusing with the allantois to form the *chorioallantois*.

The allantois begins as a peculiar little pouch emerging from the embryonic hindgut; it is an extension of the urinary bladder. But as it grows, it spreads out into a full-fledged membrane, coming to lie against the chorion and setting the stage for their fusion. The network of blood vessels permeating the fused chorioallantois has two functions. First, since it is pressed against the porous shell, it absorbs oxygen and carries it to the embryo; it also gives off waste carbon dioxide. Second, it transports water from the albumen to the embryo. In addition, the cavity of the allantois functions as a waste receptacle. The whitish, insoluble waste of reptiles and birds, consisting of uric acid, guanine, and other nitrogen compounds, is stored safely in the allantois. So we see that the membrane system of birds and reptiles is quite complex, but its role is only to sustain the embryo. When the young animal hatches, all these membranes remain with the shell.

We can see, then, that the transition of egg layers to the land involved considerable reproductive preparation. In a sense, the vertebrate brought its watery environment with it to the land. But the *cleidoic* or independent egg, with its supply of food, water, and extraembryonic membranes, was only one answer to the demands of the terrestrial environment.

Life Support of the Mammalian Embryo

The terrestrial mammals developed another response to the problems of developing on land. Their young are generally not left in eggs, but are protected and supported in the body of the female. And the various membranes have quite different functions from those of cleidoic eggs. The structues that support the embryo with food, water, and waste removal are collectively called the *placenta*—this is exclusive to the mammals. But not all mammals have placentas, and sharks and sea snakes may have very similar structures. Recall that the duckbill platypus and spiny anteaters (the most primitive mammals) still lay cleidoic eggs, while the marsupials have evolved a placenta-like organ—the *pseudoplacenta*—from the embryonic yolk sac.

Earlier, we described events in the mammalian embryo that led up to gastrulation, but we were primarily interested in the embryo. Now, in describing the development of the placenta, we begin on about the sixth or seventh day of life when the blastocyst settles into the soft lining of the uterus (Figure 30.21). Dissolving its way into the lining, the embryo begins at once to produce its extraembryonic membranes. Since there is so much variation in membrane development among the mammals, we will confine our discussion to humans.

Once the human blastocyst has reached the uterine lining, several things happen. The inner cell mass begins its transition into a three-layered embryo with its distinctive ectoderm, endoderm, and mesoderm. Simultaneously, the more external trophoblast secretes the digestive enzymes that enable the embryo to digest its way into its mother's uterus (see Figure 30.10). It must destroy a lot of blood vessels in so doing, so at first the embryo is nestled in a blood-filled sinus. For a brief time, this blood will suffice to supply food and oxygen and carry off wastes of the tiny embryonic mass. Recall that the mammalian embryo doesn't carry much food, so an outside source must be supplied at once.

Soon the trophoblast, along with cells from the embryonic mesoderm, produces a series of fingerlike branches, known as the *chorionic villi*, which extend into the thick endometrium of the receptive uterus. Each villus will be penetrated by embryonic blood vessels, so a large maternal–embryonic surface area is created which enables a ready exchange of materials. Actually, the villi are established well before they completely swing into action, and, in fact, blood will not begin to circulate through them until the twenty-first day of development. Keep in mind, though, that these villi are only the forerunners of the placenta.

After the many fingers of the chorionic villi probe deeply into the mother's tissues, additional membranes begin to develop. The amnion, which formed earlier through the hollowing out of the inner cell mass, becomes the second cavity in the blastocyst. A third cavity is produced when cells migrate out to form the hollow yolk sac. (We can probably add this to our list of vestigial structures—one of those evolutionary hangovers that had its roots in an earlier time but is no longer functional. In fact, there is no

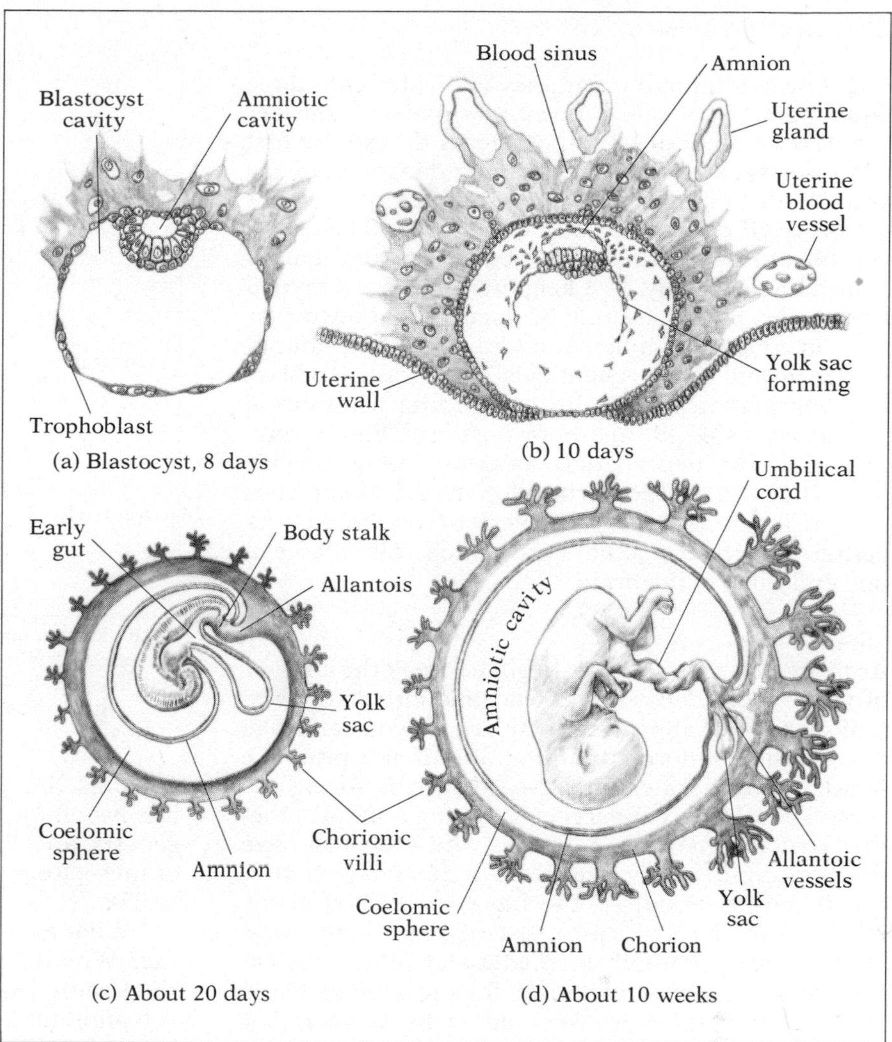

30.21 (a) In humans and other mammals, the process of implantation occurs when the embryonic blastocyst contacts the uterine lining and begins to sink below the surface, digesting the mother's uterine lining (endometrium). (b) The blastocyst adheres to the surface and finally digests its way in, becoming covered by endometrial cells. (c) At first, substances are absorbed and wastes and gases are exchanged in the surrounding maternal blood, but soon the fingerlike branches of the trophoblast and embryonic mesoderm begin to penetrate the lining. Note the formation of the *body stalk,* which connects the embryo to the developing placenta. Eventually, this structure will carry the blood vessels of the umbilical cord. Note also the presence of the yolk sac and the increase in the amnion. The larger sphere surrounding the embryo, yolk sac, and amnion is known as the coelomic sphere. The allantois, first seen here, is an extension of the gut into the coelomic cavity. In some mammals it collects urinary wastes, but in humans and other primates it leads the way for blood vessels as they grow toward the chorionic villi. (d) As events progress, the *chorionic villi* become more extensive, increasing the surface area for the vital exchanges. The embryo itself is growing, as is its amnion, and these begin to take up the space in the coelomic sphere. These blood vessels penetrate the body stalk, venturing into the fingerlike chorionic villi, where capillary networks will produce vast exchange interfaces to provide for the needs of the rapidly growing embryo.

yolk in this yolk sac. We do know that the yolk sac in humans produces the first embryonic blood cells, however, so it is good for something. However, even this function is replaced by the liver and spleen during the second month of development.)

About this time, the allantois begins to develop. Although it will not develop into the great extraembryonic structure it becomes in birds and reptiles, it will eventually form the urinary bladder. And, later on, the blood vessels that have grown into the allantois will become the major vessels of the placenta.

At this early stage of development, we see the human embryo, suspended by a *body stalk,* and surrounded by an amnion and dangling a yolk sac. All of this is, in turn, surrounded by the *coelomic sphere,* (earlier the blastocoel). The coelomic sphere, then, lies within the ring of chorionic villi. A close look at Figure 30.21(c) will show that, at first, there is not much indication that this is a developing human. Nothing at all will look familiar, in fact, until the embryo is about 8 weeks old.

30.3 Human Development and Birth

The Emergence of the Human Form

Finally, the mass of tissue begins to take on the appearance of the human form. For a while, it takes some imagination to see the resemblance, but as time goes by, the features are sharpened and brought into their proper relationships. Let's look at the process step-by-step.

The First Trimester

The 9 month human pregnancy is divided into three equal parts, each called a *trimester*. We will consider the development of various systems during the first trimester separately and there is a danger in giving you the idea that they develop sequentially, one after the other. Of course they don't, but it would be far too complex to try to consider processes, rather than organs, independently. Just keep in mind that a myriad of complex processes may be occurring at once, perhaps in concert with tightly coordinated eruptions of enzymes and other chemicals within the embryo.

Neurulation begins during the first trimester at about day 18 or 19, just before gastrulation is completed, as the neural folds arise on the embryonic disk. They will proceed to develop the central nervous system in the usual way of vertebrates, folding together to form the spinal cord which will enlarge at one end to form the brain.

Circulatory System

As those first neural ridges begin to break the contour of the human embryo, the heart and circulatory system make their appearance. By day 22, you can make out the first timorous palpitations of the primitive heart. In vertebrates, this great organ is formed as mesodermal cells converge, producing a simple tubelike structure. Within 4–5 days, the tube will have developed into a fully functional organ; it is primitive, but it moves blood. As the blood is pushed along, it follows the path of least resistance and, in this way, channels are eventually formed. Later, other cells will move in to line these channels, thus producing blood vessels. The major arteries and veins lie in rather prescribed places, but, because of chance, surface veins may lie anywhere. Compare the veins on the back of your hand to those of your neighbor's.

Developing blood vessels begin to penetrate the chorionic villi at this time. Then, within another 2 weeks (or about 40 days from fertilization), the tubular heart will have looped back on itself, paving the way for its four-chambered pattern, which, by now, is nearly completed (Figure 30.22). An opening between the atrial chambers will remain until some time after birth.

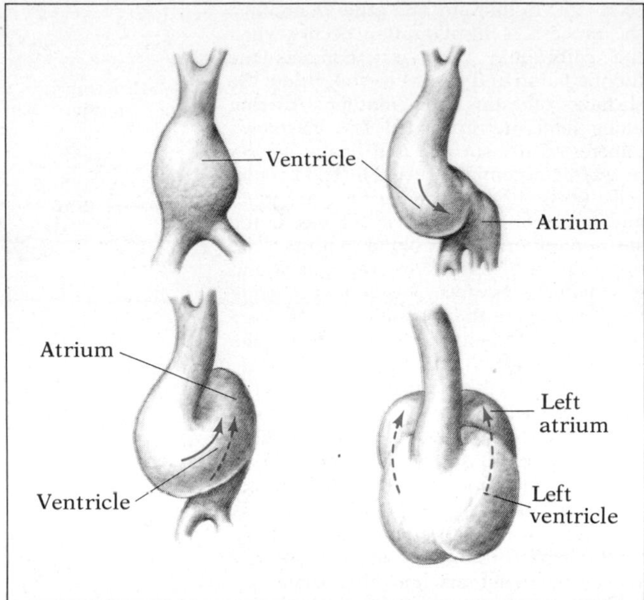

30.22 The human heart begins as a simple tubelike structure. By growing faster in the middle than at either end, the tube loops and thus the heart progresses from a single tube to a doubled one. In time, the atria and ventricles form.

Respiratory and Digestive Systems

Progress in the respiratory and digestive systems is fairly rapid, so that by 5–6 weeks their basic patterns are clearly established. Interestingly, the primitive gut in human embryos originates from an ingrowth of the yolk sac, rather than the other way around. First a few blind pouches form, then more and more outpocketings, until finally one can make out the indistinct outlines of the gut, liver, and pancreas. The trachea, bronchi, and lungs begin as a small outpocketing in the pharynx. Branching occurs, forming the two lung buds, which continue their morphogenesis, and the first individual lobes take form. All of these are primarily endodermal structures (Figure 30.23).

What about vestigial structures such as the yolk sac? Why do they remain with us? Perhaps we can shed some light on these problems by approaching yet another question. All mammals, including humans, produce embryos with a series of *gill arches* separated by *gill pouches* (Figure 30.24). They form gills in fish, but we don't need gills, so why do we have the gill arches? In a fish, the genetic instruction might read: "Produce gills." In mammals, however, these older instructions might have, superimposed upon them, the secondary instruction: "Modify gill arches into a jaw, a larynx, part of the middle ear, and the tongue bone structures." (Did you know about the tongue bone? It is also called the "hyoid.") No wonder the early embryologists have been accused of greeting each other with Haeckel's slogan, "Ontogeny recapitulates phylogeny!" Translation: "The development of the individual retraces the evolutionary development of the phylum." Every vertebrate embryo certainly appears to have a rough blueprint of its phylum incorporated in its DNA. And also, the evolutionarily older members generally appear to have less variation and refinement on the basic theme. But the point is, embryos of evolutionarily newer animals tend to begin their development along ancient lines; then with new newer, supplemental genetic instructions, they incorporate variations on the original plan. Nevertheless, the old patterns re-

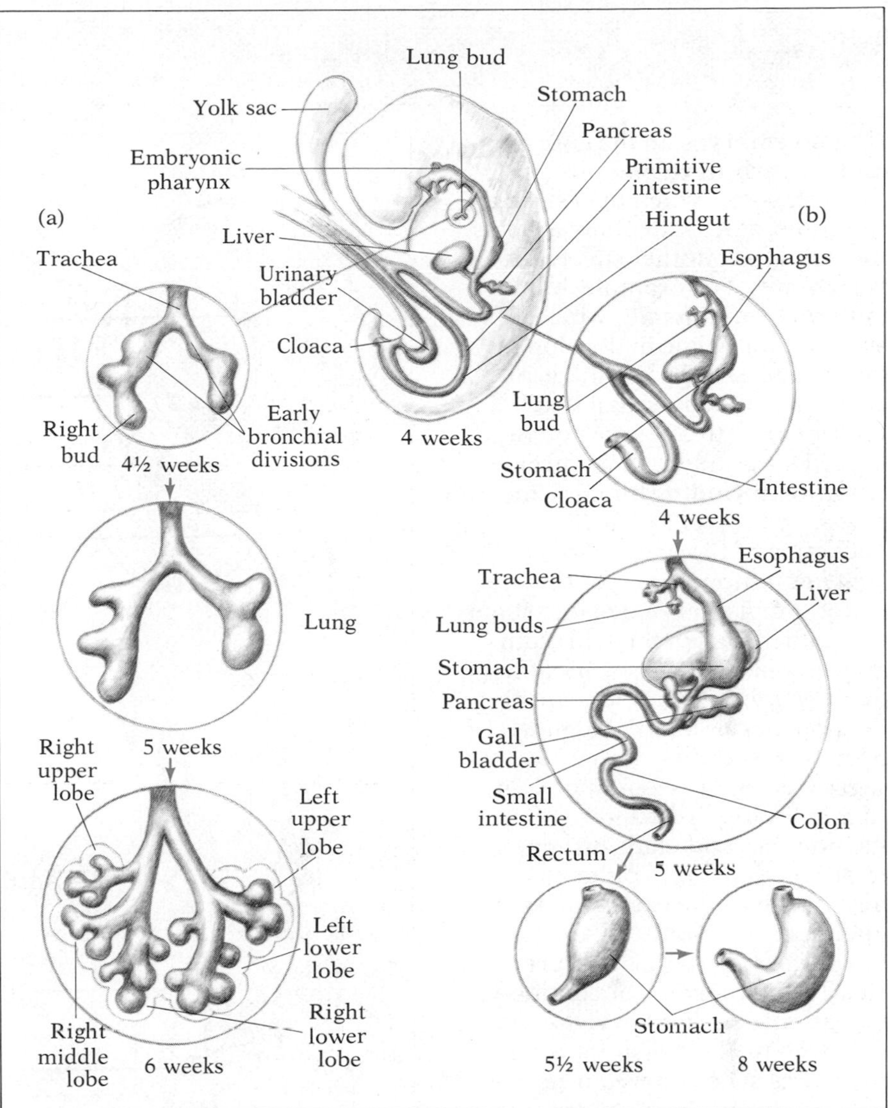

30.23 (a) The respiratory system makes its appearance in the third week as a simple outpocketing or lung bud from the embryonic pharynx. By the fourth week the lung bud has divided into right and left sides, which form the pattern for the two bronchi and lungs. By the fifth week of development, the lobes of the lung take on a crude form to be refined into a more definite shape one week later. (b) The digestive tube also takes on form during the fourth week. Its major regions, the esophagus, stomach, and intestine differentiate first. The tube ends in a cloaca, a temporary reminder of our chordate affiliation. By the fifth week, the tube has further differentiated, with the stomach enlarging and the small intestine, colon, and rectum clearly visible. In addition, the liver, gall bladder, and pancreas are clearly outlined. The small intestine will enter a coiling stage as development continues, and the stomach, seen at 5½ and 8 weeks, will assume its final form.

30.24 By the fifth week, the *gill arches* (numbers) are prominent. In a fish, these structures form the jaw and gill structure, but in humans they contribute to numerous structures in the head and neck, including the lower jaw, tongue (and its hyoid bone), larynx, middle ear, pharynx, and many others. The incorporation of the gill arches into these structures is completed early in development, so this fishlike stage lasts only a few weeks.

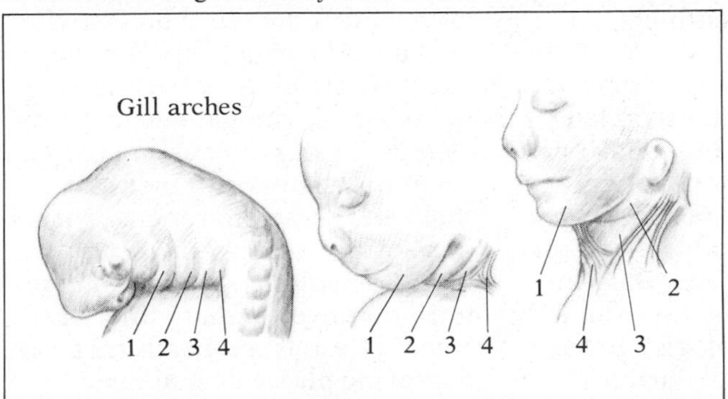

main. Recall that mammalian development begins with the newly evolved trophoblast, then ploddingly recapitulates reptilian development with its version of the blastodisc, even forming an empty yolk sac.

Nervous System

The nervous system continues its rapid progression in the first trimester. While it is the earliest system to begin development, the nervous system will not be completed until long after the birth process—perhaps not until the moment of death—since all learning, is in a sense, a developmental process. By the fourth week, the major regions of the brain and spinal cord are recognizable. When the first trimester ends, these are already well-defined. The still-smooth cortex now extends over much of the embryonic brain, and

Essay 30.1 The Theory of Recapitulation

Human embryos, in the course of their early development, go through stages when they strongly resemble, at least in many aspects, the embryos of other species of vertebrates. For example, human embryos develop *tails,* which persist until fairly late in development. (In a small percentage of cases, a rudimentary tail is still present at birth and must be removed surgically.) All vertebrate embryos, including human embryos, develop *gill clefts* (in-pocketing of the neck) and *gill pouches* (outpocketing of the endodermal pharynx) at one stage or another. In fish (and, presumably, in our fishy ancestors), the thin layer where the clefts meet the pouches breaks through to become respiratory water passages (see Figures 23.7 and 30.24). In mammals, birds, and reptiles, the pouches and clefts disappear or are modified into other structures, such as the ear canal. In mammals, the bones of some of the *gill arches*—bones that are clearly *homologous* (of common evolutionary origin) to those that will become the gill-supporting structures in fish—develop, instead, into the tiny bones of the inner ear.

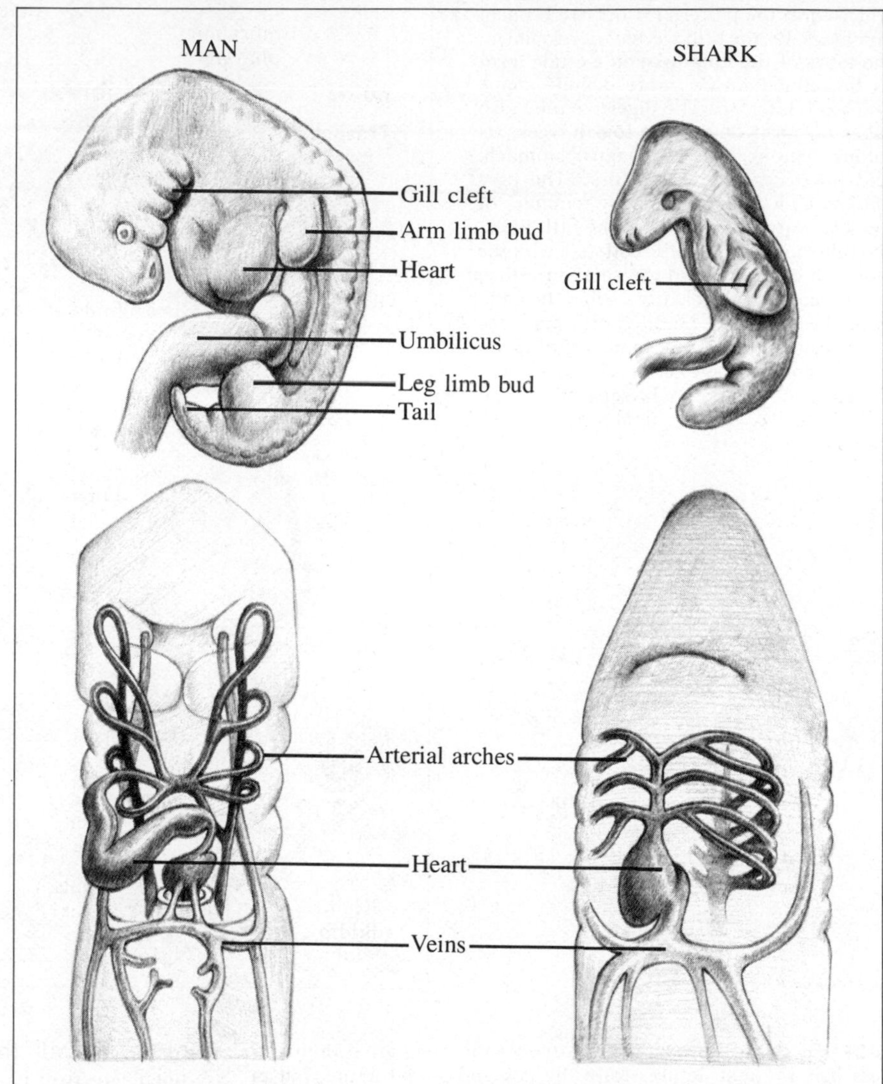

The circulatory system of the early human embryo resembles that of embryonic fish and, indeed, also resembles the circulatory system of adult fish. The heart is a tubular affair, and it feeds forward into five pairs of arterial gill arches. In fish, this basic plan is retained into the adult form; but most of the arterial gill arches are eventually lost or supplanted in the course of human embryological development.

In its earliest embryonic stage, the human organism may be considered a single-celled animal. At somewhat later stages, the embryo is progressively *Amphioxis*-like, fishlike, amphibianlike, and even —in some features of heart and kidney development—reptilian. Finally, however, it develops characteristics that are unmistakably mammalian and, eventually, unmistakably human.

The great similarity of embryonic stages of distantly related organisms was noted by Charles Darwin, who felt that the phenomenon indicated a common ancestry and that it supported his theory of evolution. In his old age, Darwin complained about the fact that a younger German biologist, Ernst Haeckel, was generally being credited with the idea although Darwin had thought of it first. In truth, Haeckel made much more of the phenomenon than Darwin ever did. Haeckel based a ponderous theoretical structure on what he called "the biogenetic law": *Ontogeny recapitulates phylogeny.* An impressive slogan that meant that each organism repeats, in its development, the entire series of morphological stages that characterized its evolutionary ancestry. Haeckel published beautiful steel-engraving plates that showed, quite impressively, just how similar early vertebrate embryos were.

Unfortunately, later studies showed that the engravings were much more similar than the actual embryos. Haeckel chose to ignore inconvenient exceptions to his "law," such as the facts that the early human trophoblast stage and the "higher" vertebrate blastodisc stage have no counterpart in "lower" forms. Instead, he looked to the much simpler gastrulas of echinoderms and annelids for clues to the earliest forms of multicellular life, and drew imaginative and far-reaching conclusions from them.

The biogenetic law is currently in comparative disrepute. It now seems evident that the embryos of higher organisms do not repeat the adult forms of their ancestors. Instead, the phenomenon makes more sense to us when we consider that many aspects of the *developmental mechanisms* of our fishy ancestors could have been inherited by us. Our embryos therefore resemble the embryos of our ancestors. In some living groups (such as many fish), the embryos subsequently undergo less extensive modification into the adult form than in other groups (such as the mammals). That seemed clear enough to Darwin, and it helps to explain why the early embryonic forms of distantly related contemporary species are often so much more similar than are the adult forms.

Another way of interpreting the patterns we see is to conclude (not unreasonably) that adaptive evolutionary innovations that occur late in embryonic development have always been less likely to be seriously disruptive than innovations that affect early embryos, and that novel changes have tended to be tacked on to what at one time had been the end of the embryonic process. But the generality can't be drawn out in too much detail, because even in early embryos there are many characteristics that appear to be of relatively recent evolutionary origin, and some may be found in only a single species.

the cerebellum and medulla have become distinct (Figure 30.25).

Limbs

The limbs of humans appear as rounded flaps during the fourth week (Figure 30.26). The arms and legs are distinguishable at 6 weeks, but fingers and toes require an additional week. The rudimentary hands and feet actually begin as simple paddles that take form through a kind of developmental programmed cell death, as the tissue between the fingers and toes is broken down and absorbed.

Excretory System

The urinary and excretory system of the human develops in a series of distinct stages. Beginning on day 25, three successive sets of paired tubules develop, with the final one forming on about day 30. This last one eventually becomes the embryonic kidney. The three sets of tubules are called the *pronephros, mesonephros,* and *metanephros.* The metanephros actually becomes the kidney in humans, but the two systems that precede it are necessary because the first helps lay down permanent tubules and the second actually acts as a kidney for a few days in human

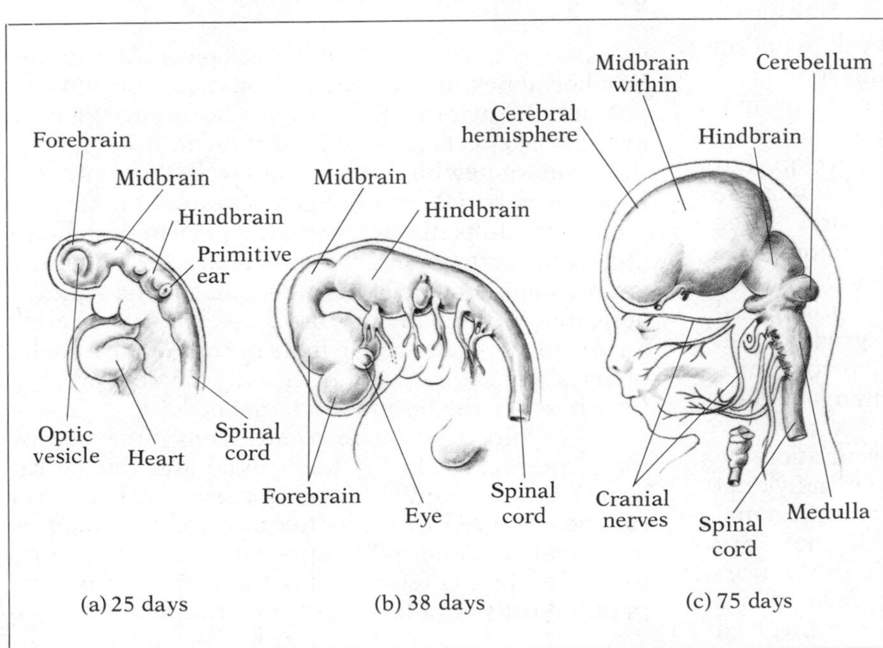

30.25 Following the neural stage, the regions of the brain and other sensory structures become prominent. (a) At about 25 days, the neural structures already include the forebrain, optical vesicles, midbrain, hindbrain, and primitive ear. In addition, some of the cranial nerves are apparent. (b) At 38 days, regions of the brain have differentiated and the eye is forming. (c) The regions commonly associated with the adult brain begin to form by the end of the first trimester. The primitive features of the brain become overgrown by the massive cerebrum.

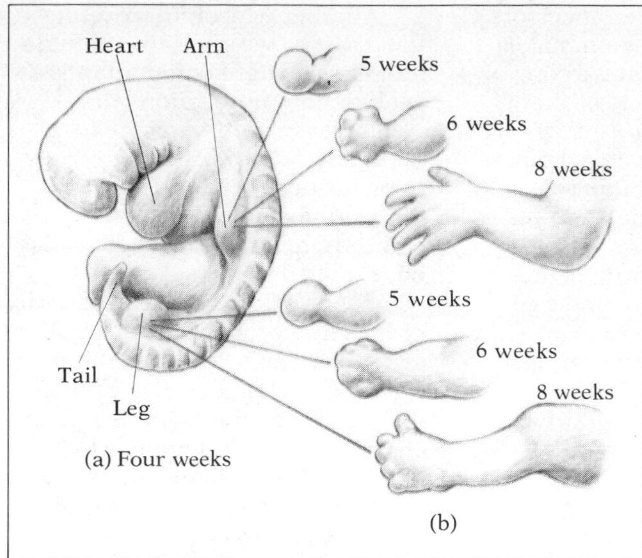

30.26 (a) By the fourth week the arms and legs appear as simple buds in the embryo. (b) By the fifth week they have grown significantly, forming paddlelike extensions. By 6 weeks, the finger and toe structure is visible, but a webbing remains. The tissue between fingers and toes will undergo a "programmed" death and by the eighth week separation will have become complete.

development. In some vertebrates, the pronephros or mesonephros become the final excretory organ. In humans, during the first 35 days, any wastes that are collected are deposited in the gut. Later when the bladder and urethra form their own passageway, the metabolic wastes of the embryo are carried into the amniotic cavity and removed by the mother's circulatory system. By the time of birth, about all the amniotic cavity contains is urine.

Reproductive System

The reproductive system begins to develop during the first few weeks, but even a trained observer can't determine the sex of the embryo (without a chromosome test) until the eighth week. It is true that before this time the genitals have begun to develop, but the genitals of the two sexes start off in much the same way. In the sexually "indifferent" state, three prominent parts of the reproductive anatomy are visible. These comprise the genital tubercle and include the *urogential groove, urogenital folds,* and *genital swellings* (Figure 30.27). It is no great step to build a female system from these structures, but a developing male must undergo great changes under the influence of male hormones.

The factors controlling sexual differentiation are complex. Certainly, the genetic sex of an individual is determined at fertilization by the chromosomal complement of the sperm cell. Sperm cells that carry X chromosomes produce females, while those bearing Y chromosomes produce males. However, this is only a beginning. A number of genes are involved in sexual development, and the control may stem from different levels (such as genetic, hormonal, or environmental). For example, it appears that the Y chromosome of mammals initiates the development of the testes and that the testes take over from there by producing two hormones. These determine the primary sexual characters.

One male hormone, not yet isolated and not well-understood, supresses the development of the oviduct, uterus, and primitive vagina, which would otherwise develop from the primitive *Müllerian duct,* an embryonic structure that is present in both male and female fetuses. (The vagina forms only partially from the Müllerian duct; its lining is an ingrowth of other tissue.)

The other testicular hormone is better known. It is the steroid *testosterone,* the principal mammalian androgen. (Recall that while an androgen is any male-determining hormone, testosterone is a specific chemical compound.) Testosterone is produced by the fetal testes and interacts with target tissues, such as the primitive and undifferentiated genitals, transforming them into the typical male apparatus. Perhaps most interestingly, it also changes the developing brain.

Two types of tissue arise in the rudimentary region of the gonads. One is known as *gonadal cortical tissue* and the other as *gonadal medullary tissue.* It is hypothesized that these tissues are antagonistic in the sense that one normally replaces the other as sexual development proceeds. It turns out that cortical dominance produces ovarian tissue, while medullary dominance results in the development of testes.

A brief summary of sexual differentiation is shown in Figure 30.27, where the fate of each of the four primary regions of the embryonic genitals is traced.

Prenatal Sex Hormones and the Development of Behavior

Sex hormones affect behavior, in case you haven't noticed. Behavioral effects of sex hormones include irreversible effects in early development. For example, the brains of newborn rats are not well-differentiated. If newborn female rats are given a dose of male hormones, as adults they will exhibit typical male fighting and even try to mount other females, even though the high levels of male hormones have long ago disappeared. If the baby rats are dissected shortly after an injection of radioactive male hormones, the nuclei of certain brain cells will be heavily labeled, proving that these are the hormone's target cells.

More direct evidence comes from dissection of the same regions in the adult male and female rat brains. Consistent differences between the two sexes can be observed in the patterns of nerve synapses in normal individuals. The application of testosterone to the developing female rat brain during the critical period shortly after birth results in malelike synapses.

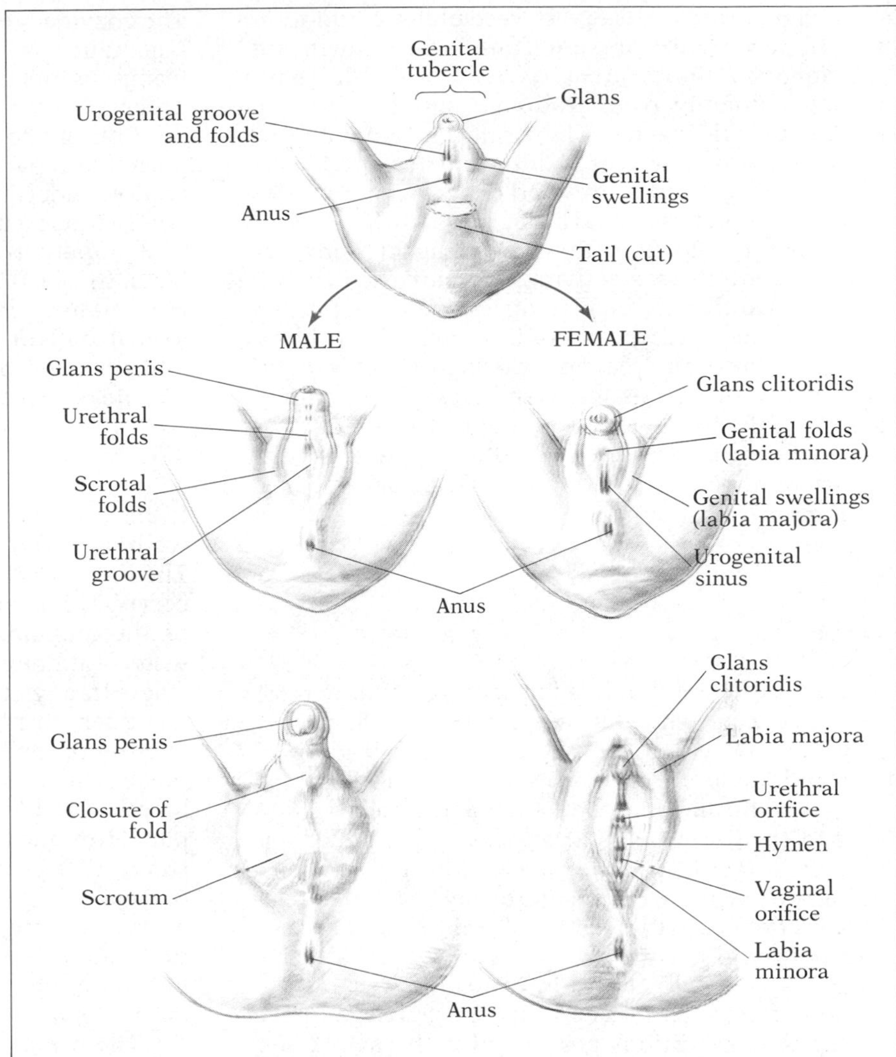

30.27 At 7 weeks, the embryo has produced a *genital tubercle* with a prominent *glans* and *urogenital folds* and *groove.* The tubercle is supported by a triangular *genital swelling.* Under the influence of male hormones from the embryonic gonad, these structures will produce typical male sex organs. In the absence of these hormones, female organs will emerge. In the genetic male (XY), the glans will enlarge to form the *glans penis,* with the shaft of erectile tissue emerging from the urethral folds. The urethral groove will close, while the genital swellings will differentiate into the *scrotal sac* to await the descent of the *testes* in the eighth month of life. In the genetic female (XX), the glans will become the *glans clitoridis,* and the urethral folds will remain open forming the *labia minora.* The genital swellings will form the *labia majora.*

In humans, the corresponding period of development comes during fetal life. Our evidence of irreversible sexual differentiation of the human brain is much less direct. According to John Money and Mary Jane Eberhardt, human females who have been (for some reason or other) exposed to unusually high levels of male hormones in utero show various behavioral and personality effects later on, such as exceptional "tomboyism," as well as a stated preference to enter the professions rather than become housewives.

The Second and Third Trimesters

A great deal of morphogenic activity marked the first trimester. So much, in fact, that by its close, most major systems have approached their final form. The second trimester is characterized by growth and refinement. Body length increases rapidly, catching up to the large head. Systems begin to approach a functional state. The fetus will breath, but only amniotic fluid enters and leaves the respiratory system. Secretory linings form in the digestive system, and enzymes are secreted. Bile from the liver joins these secretions. The fetus swallows, bringing in fluids and cellular debris that accumulates in the gut as dark, gelatinous *muconium.* The dermis and epidermis of the skin reach a differentiated state and are penetrated by sweat glands, sebaceous glands, hair follicles, and sensory neurons. If stimulated, the fetus will respond with feeble, uncoordinated movements at first, but within a few weeks these movements become more precise and brisk as the mother feels the first kick.

As the third trimester begins, the organ systems begin to be fully functional. A fetus may even survive if born now, but the chances of survival are very poor even under the most sophisticated medical supervision. At this time the fetus will measure about 35cm and weigh around 1000g. It resembles a full-term baby, but a very thin and emaciated one. Growth and refinement continue, but at a reduced rate. The fetus, no longer floating free, gradually fills the amniotic sac and uterine cavity. The brain and spinal cord develop rapidly. The cortex differentiates and begins to form the familiar fissures and convolutions. A layer of fat accumulates beneath the skin. In the skeletal framework *ossification,* the hardening of bone, increases rapidly. These activities increase the demand for protein and calcium, placing a substantial burden on the mother's reserves. As final prelude to birth, antibodies cross the placenta during the last month. These will provide the newborn baby with immunity against viral and bacterial infections for a month or two. As the third trimester ends, the fetus, now crowding the mother's abdomen, will weigh about 3200g and measure close to 50cm.

Birth

In spite of the fact that it is fascinating and that most of us have experienced it, we still don't really understand the birth process or the factors leading to it. Considerable current research is focused on the changes in the placenta that occur at the end of gestation and the hormonal shifts that bring them about. Since the level of progesterone is important in maintaining the placental tissue, there may be a direct relationship between the aging of this structure and lowered progesterone levels. Let's consider this briefly in the light of recent information.

Researchers interested in miscarriage have learned that conditions associated with vaginal and uterine infections can simulate those of the normal birth process. One wonders, what is the connection? High levels of the enzyme *phospholipase A2* are produced by certain common bacteria. Furthermore, this enzyme, when injected into the amnion of test animals, immediately instigated the birth process. It turns out that phospholipase A2 is an essential enzyme in the conversion of lipids into ubiquitous messengers known as *prostaglandins.* When prostaglandins are present in sufficient quantity, uterine contractions and the placental separation begin. In the normal course of pregnancy, the lipid precursors of prostaglandins are maintained in the amnion and the production of phospholipase A2 is inhibited by high progesterone levels. In other words, the trigger for labor exists all along. Then at the end of normal pregnancy, progesterone levels fall as the placenta, which produces the hormone, ages. Lowered progesterone levels spring the trigger and the genes that produce phospholipase A2 are no longer inhibited. The enzyme is produced, lipids are converted to prostaglandins, and labor begins. This new discovery brings us one step closer to understanding a vital process.

Shifting hormone levels, in addition to initiating placental separation, may also bring on oxytocin secretion, since this pituitary hormone starts uterine contractions early in the birth process. The fetus itself apparently has little to do with initiating its own birth. In fact, if an early fetus is removed from a pregnant monkey, leaving the placenta intact, the physiological state of pregnancy will continue for the normal gestation period and the monkey will deliver a nice healthy placenta, right on schedule.

The Stages of Birth

In accordance with our penchant for pigeonholing, we have divided the birth process into three stages. The first is *dilation,* which involves a softening of the cervix. This is sometimes accompanied by a rupturing of the amnion, often referred to as "breaking the water," although this usually happens in the second stage. In any case, a ruptured amnion is signaled by a sudden gush of amniotic fluid through the cervix and out of the vagina, a startling event that has traumatized many a cab driver. The period of dilation is highly variable, lasting from 2–16 hr. It is accompanied by those periodic contractions we call *labor pains.* These contractions increase in frequency through the first stage. When they begin to occur every 3–4 min, the fetus will *crown,* meaning that the head will begin its passage through the cervix and become visible. This is the start of stage two, the *expulsion of the fetus.*

The expulsion process may take just a few minutes or it may require hours. The mother may have to be given a spinal anesthetic at this time. The level of pain is highly variable, depending on emotional state, pain threshold, and preparedness of the mother. When mothers are asked about pain, they give a variety of answers from saying it is like pulling your upper lip over your head, to a certain joy with little pain. Apparently, few mothers swear off after one child because of the pain.

Culture may have a lot to do with childbirth. It is well-established that accounts of the pain and joy of childbirth differ dramatically from culture to culture. Birth is considered to be much more painful in Japan than in China, for instance. Our own culture has recently been undergoing a dramatic reevaluation of the childbirth experience. By all accounts, American childbirth is much less painful now that it was 30 years ago! In most cultures, midwives or physicians aid the expulsion process, but Papuan women prefer complete privacy, spurning any assistance. Also, different cultures prefer different positions for delivery (squatting, kneeling, flat on the back.) The

Essay 30.2 Steroid Hormones and Development

Hormone studies have provided important clues as to how differentiation occurs in general. Some of the best understood effects of hormones in development are those of the sex hormones. Perhaps this is because neither male hormones (androgens) nor female hormones (estrogens) are necessary for survival of the individual, so that the experimenter can safely manipulate their levels more or less at will. Sex hormones are steroids, a class of polycyclic (many-ringed) molecules related to lanolin, cholesterol, and vitamin D. There is direct evidence that steroid hormones can turn on specific genes.

Researchers have found that the circulating sex hormones are always present at very low concentrations, and they readily drift into and out of all kinds of cells. However, some cells contain specific receptors that bind the sex hormones, and the hormones quickly become concentrated in these target calls. In these cells, the hormone–receptor complex diffuses into the cell nucleus, where it joins with highly specific nuclear proteins which, after further changes, specifically bind to certain genes. After an injection of radioactive sex hormones, the chromosomes of target cells become highly radioactive, and the nuclei of target cells can be identified by autoradiography (see Essay 7.2 in Chapter 7). Furthermore, all the various intermediate complexes can be localized and isolated by biochemical techniques.

Among the target cells of the female hormone progesterone are the cells of the immature oviducts of female baby chicks. Baby chicks don't lay eggs and don't ordinarily produce egg white proteins, but the immature oviduct cells can be made to produce egg white proteins by giving them progesterone. In these chicks, the radioactive hormones are found bound tightly to the chromosomes of the target cells. Here, we have a case of a hormone activating a gene, or groups of genes.

Sex hormones are involved in the morphological development of embryos also—that is why baby boys aren't like baby girls. In such complex development, many different genes must be controlled by a single hormone. Curiously, male hormones do not appear to interact directly with the genetic material. In the target cells which are sensitive to male hormones, specific enzymes alter the circulating testosterone into an active form—either dihydroxy testosterone or, sometimes, estradiol. This means that estradiol, which is a female hormone, acts as a male hormone once it is within the nucleus of certain male target cells. If the testes are removed from a developing male rat or monkey embryo, it will develop phenotypically as a female. It seems that there are always both female and male hormones present in normal individuals of both sexes, and normal female or male development depends on the relative balance between the two. In the absence of either testes or ovaries, the female hormones that are produced in low amounts by the adrenal gland predominate (in normal females, the bulk of the female hormones are produced by the ovary). The fact that embryos without gonads develop as females has led some researchers to conclude that the female form is more "primitive" than the male. (It's difficult to interpret what is meant by "primitive" here—males have been around exactly as long as females!)

There is a curious, rare human genetic condition that dramatically illustrates the role of sex hormones in the development of sexual morphology. This is the *androgen-insensitivity syndrome,* which is due to the failure of a gene of the human X chromosome. In normal males, this gene makes the enzyme that alters testosterone into the form that is functional within the cell. When an XY individual has inherited, from her mother, an X chromosome with a nonfunctional form of this gene, her tissues will be unable to respond to male hormones. We say "her" because such individuals are indeed females, although normal XY individuals are male. Persons with the adrogen-insensitivity syndrome have only one readily observable abnormality—a completely bald pubic area. Internally, however, they lack ovaries, oviducts or a uterus; instead, they have testes in the labia majora. Such individuals are biochemically unlike normal females in having high levels of circulating testosterone, to which they are totally insensitive. Like normal XY males, and unlike normal XX females, they lack Barr bodies and do have the Y-linked H-Y cell surface antigen. They are, of course, sterile. Several investigators of this strange but informative phenomenon have noted that the affected individuals appear to have exaggerated female secondary characteristics, being unusually voluptuous in comparison with their own XX sisters. Although their adrenals produce only the same, relatively small amounts of estrogen normally found in males, the balance between male and female determinants is, functionally, 100% on the side of femaleness. The reason they lack pubic hair is that the development of pubic hair is a response to testosterone, in both males and females.

point is the human birth process appears to involve as much cultural anthropology as biology.

The final stage of birth is the separation of the placenta and its expulsion from the uterus as *afterbirth.* Other mammals, including herbivores, eat the placenta. Humans don't. But what happens to the distorted uterus once the placenta is gone? After all, it has been stretched to some sixteen times its former size. It turns out that it rapidly contracts to its original state. This contraction helps to stifle the bleeding at the site of placental attachment. A sluggish recovery may require the use of injected oxytocin as a stimulant to increase contraction of the smooth uterine muscle. But as we've mentioned before, the physical stimulus of the infant suckling on the mother's nipples accomplishes the natural release of oxytocin through complex nervous and hormonal interactions.

Miscarriage (Spontaneous Abortion)

Of course, not all pregnancies are terminated by delivery of a healthy offspring. The environment in which the fetus develops is optimal in terms of security and comfort; however, the route between conception and birth is fraught with risk. Part of the risk stems from the very complexity of living things, including the complex and uncertain process of meiosis, which often fails to maneuver the chromosomes properly. There is a phenomenal array of physical and chemical interactions going on within the body of a developing organism. If, at any point along the line, a component fails in its function or its timing, the embryo may be lost. If this happens very early, the embryo may be broken down—digested—and reabsorbed by the mother's body or expelled with a menstrual flow, usually without her ever knowing she was pregnant. If the embryo should fail at some period later in its development, the placenta dies. Without the continued production of progesterone, the uterus goes into rhythmic contractions and the fetus is spontaneously aborted as a *miscarriage.* About half of all miscarriages involve fetuses that are structurally, genetically, or physically abnormal.

Physiological Changes in the Newborn Baby

The systems of the fetus are in a state of readiness prior to birth. It must be prepared for a new and more threatening kind of existence, because it will be propelled into that new world with a startling speed. The vital exchange of gases has always been provided by the placenta. But no longer. So the infant's own respiratory system must function for the first time. And then there are the changes in the circulatory system (Figure 30.28). Oxygenated blood from the placental circulation had previously entered the fetus through the umbilical vein, proceeding directly into the vena cava and thence to the right side of the heart. A hole in the septum between the fetal atria, the *foramen ovale,* permitted the blood to flow from the right atrium to the left atrium, thus circumventing

30.28 This is the circulatory system of the fetus just before birth when the placenta is intact. Note the foramen ovale, a septum between right and left atria in the heart, the ductus arteriosis, connecting the pulmonary artery with the aorta and, of course, the umbilical circulation. First, the foramen ovale must close to prevent the mixing of oxygenated and deoxygenated atrial blood when the lungs become functional. Second, the ductus arteriosis must constrict, also to prevent the mixing of deoxygenated and oxygenated blood. The umbilical circulation is disrupted when the cord is tied and cut. Actually the umbilical vessels themselves will normally contract, closing off circulation and preventing substantial blood loss. Both the ductus arteriosis and the umbilical vessels and their connections to the permanent circulatory system will regress quickly after birth leaving little trace in the adult circulatory system. The foramen ovale is closed by flaps of tissue from each atrium, but will require about one year for permanent closure to occur.

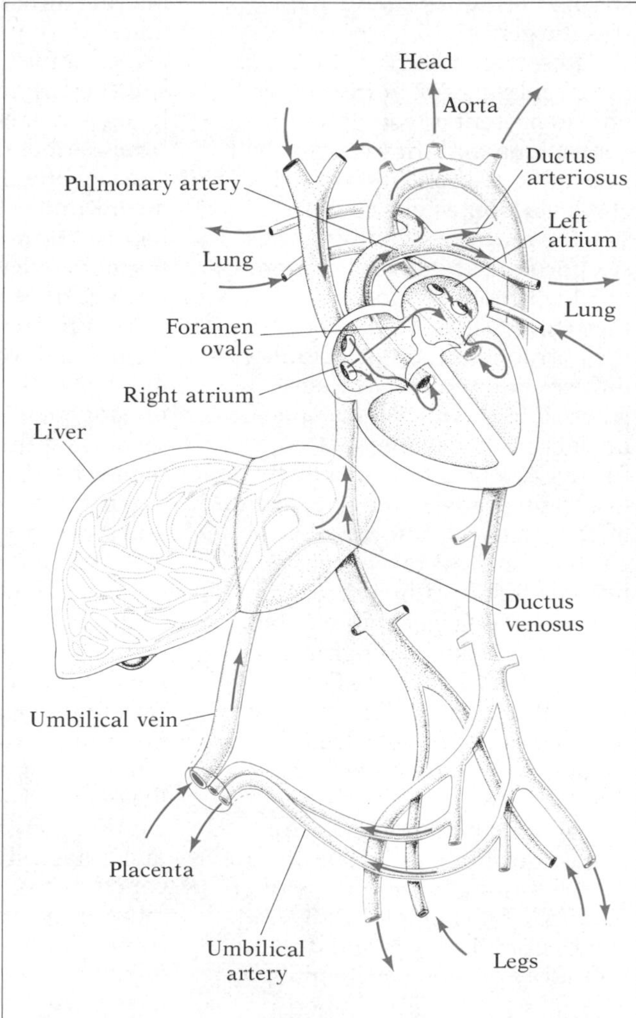

the route to the deflated and useless embryonic lungs. In addition, a connecting vessel between the pulmonary artery and aorta, the *ductus arteriosis,* permitted blood from the right ventricle to by-pass the lungs and go directly into arterial distribution.

For the most part, pulmonary circulation in the fetus is necessary only for tissue growth and maintenance. So, in effect, there are two short-circuits in the fetal circulation that aid in the rapid distribution of oxygen. For the newborn baby, however, these bypasses won't do. The structures must be changed immediately.

With the baby's first breaths, the ductus arteriosus constricts vigorously, closing that bypass. With the shortcut closed, the blood must move toward the rapidly expanding lungs. As a result of this increased pulmonary flow, the volume of blood returning directly to the left atrium suddenly increases. The foramen ovale is composed of two overlapping flaps of tissue which are forced closed when the atria contract. Actually, however, the foramen ovale will not seal itself completely until the baby is about a year old, but there is no rush since the flaps act as valves, preventing mixing of atrial blood until that time.

If the ductus arteriosus and/or the foramen ovale fail to close properly, the circulation will continue to bypass the lungs and the baby will turn "blue." The blue color is due to the dark deoxygenated blood showing through semitransparent skin. These "blue babies" require heart surgery.

The first breath of a newborn infant is taken in response to a sudden increase in blood CO_2 when the placental exchange is lost. The rise in CO_2 stimulates the respiratory center of the medulla with the usual result. So, you see, the baby will breathe with or without the traditional swat on its rear. In deference to tradition, however, we should add that the swat may stimulate movement, speeding up the respiratory process.

Some newer delivery procedures emphasize birth in a quiet, semidarkened room, no swats, and an immediate immersion of the newborn into a comfortable bath of lukewarm water. This purportedly convinces the infant that the world is a decent place, and hopefully makes a better person of him or her. At least it seems to make a better person of the obstetrician.

Lactation

Human mothers, as all good mammals, are prepared to nurse their young immediately (Figure 30.29). The breasts, under the influence of estrogen and progesterone, have enlarged during pregnancy, and with the delivery, *colostrum* secretion begins. Colostrum, a clear, yellowish fluid, differs from milk in that it contains more protein, vitamins, and minerals and less sugar and fat. It also contains the mother's antibodies which can be important to the infant in its first days. The secretion didn't begin earlier because, apparently, the milk-stimulating hormone prolactin is suppressed while the progesterone levels of pregnancy are high. Actually, though, two hormones are involved in the production of milk. The second hormone is oxytocin, which is released, along with prolactin, in response to suckling. Strangely enough, some women release milk from the nipples if they even hear a baby's cry. So, apparently, the pituitary receives sensory input from pathways through the auditory centers as well. Of course, a baby's cry elicits a strong rescue or aid response from most people, especially mothers.

In a few days, the colostrum is replaced by whiter, thicker milk. Human milk is sufficient in all required vitamins except vitamin D. Its caloric value is about 700 cal per liter. In comparison to cow's milk, human milk contains about one-third less casein protein, (making it easier to digest), much more milk sugar (lactose), and considerably less calcium and butterfat. It is less acid and, interestingly, is bacteria-free (which is not true of cow's milk, either in grocery stores or at the source).

There are additional benefits of breast feeding. Some studies reveal that breast-fed infants are less susceptible to disease, anemia, and vitamin deficiencies. Whether the disease aspect is related to some additional antibody supply from mother's milk, or the fact that cow's milk is contaminated with bacteria, is not known. In addition, psychologists tell us that there is a definite value in breast feeding in terms of the baby's feelings of security and well-being. The psychological effects on mothers are very variable. Some women receive great pleasure from their nursing young, a phenomenon that probably strengthens their attachment for their infants. Breast feeding immediately after delivery may also be of definite physiological value to the mother, as we've mentioned, since suckling stimulates the release of oxytocin which helps to contract the uterus and stop bleeding.

Bottle feeding is also considered by many to have its advantages. The jury is still out on the relationship, if any, between breast feeding and breast cancer. On the one hand, breast cancer viruses are known to be transmitted by mother's milk—in mice. On the other hand, groups of women who have had relatively little experience with breast feeding (such as nuns and other unmarried women) have extraordinarily high rates of breast cancer in their old age.

At one time, there was considerable cultural pressure on respectable American women to bottle feed their babies. At the present time, there appears to be an equally strong cultural pressure on women to breast feed their babies. People can get quite emotional on this issue. Fortunately, the babies themselves usually thrive on either regimen.

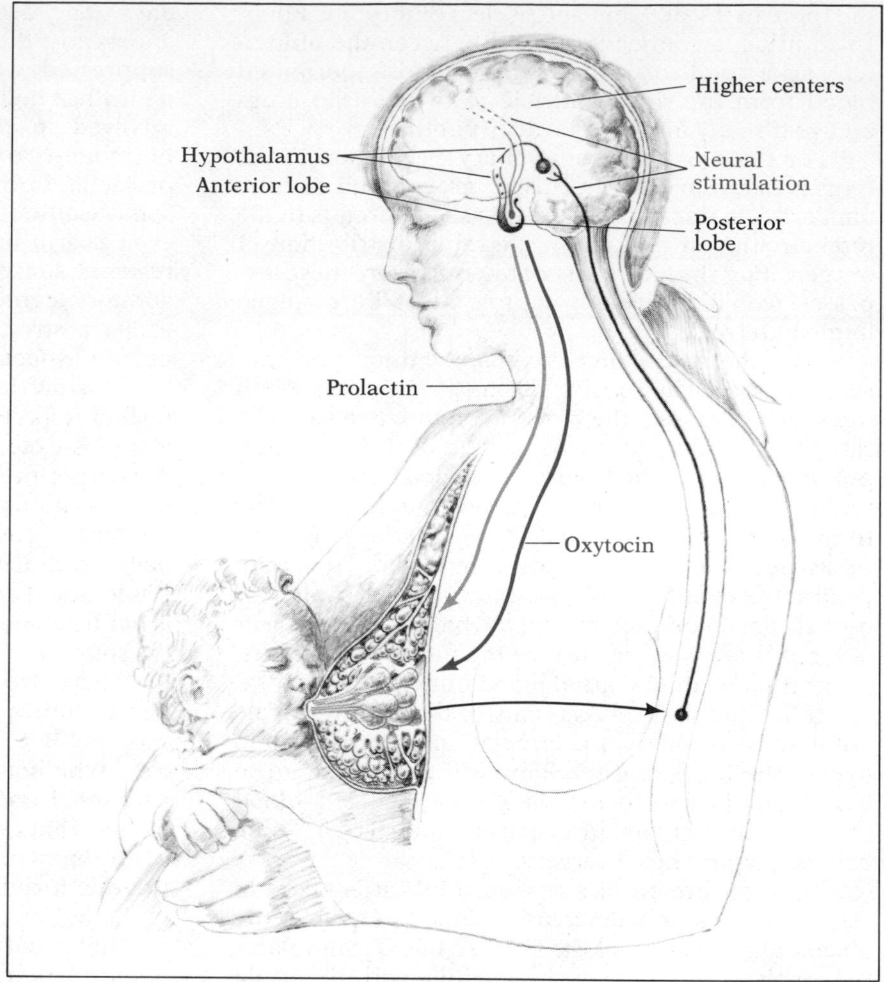

30.29 Suckling is believed to provide a neural stimulus which increases milk production and triggers its release (letdown). The stimulatory signal follows neural routes to the hypothalamus, where releasing factors stimulate the pituitary to release prolactin and oxytocin. Actually, in the case of prolactin, the suckling activity removes an inhibitory releasing factor from the pituitary, thereby permitting its release. If suckling is stopped for a long enough period, prolactin release is again inhibited and the breasts will slow down and eventually stop the production of milk.

Application of Ideas

1. Explain how the central dogma might be used to support either the preformation concept or epigenesis.
2. Fertilization has been avidly studied in echinoderms, primarily because of the easy availability and viability of gametes and embryos. Does studying fertilization in sea urchins and sand dollars really help us in understanding the process in vertebrates? In humans? Explain how information gained this way can be helpful and cite specific cautions to be applied.
3. The embryos of mammals go through stages where certain structures resemble those formed by other vertebrates, yet these structures are soon modified into tissues and organs not seen in other vertebrates. What is the significance of this interesting observation?
4. Some time during the evolution of mammals, the switch was made from the self-sustained land egg to placental support. Suggest what some of the intermediate stages might have been. What adaptive advantage might this change have offered the evolving mammal? Would the same advantages have applied to birds?

Key Words

acrosomal cap
acrosomal filament
albumen

Key Ideas

Introduction
Descriptive knowledge of development far exceeds experimental knowledge. Many fundamental questions about development remain unanswered.

amnion
animal pole
antisprouting factor
archenteron
axon

blastocoel
blastocyst
blastoderm
blastodisc
blastomere
blastopore
blastula
body stalk

casein
centrolecithal
chorioallantois
chorion
chorionic villus
cleavage
cleidoic
colostrum
corona radiata
cortical granules

development
differentiation
dilation
ductus arteriosis

ectoderm
embryology
embryonic disc
endometrium
epiboly
epigenesis
extraembryonic membranes

fate map
fertilization cone
fertilization membrane
fetus
foramen ovale

gastrulation
genital swelling
genital tubule
germ layers
gill arch
gill pouch

head
homunculus
hyaluronic acid
hyaluronidase
H-Y cell surface antigen

implantation
induction
inner cell mass
invagination
involution
isolecithal

The Gametes

The Sperm

The animal *sperm cell* contains an enzyme-filled *acrosome,* a compact haploid nucleus, a centriole, mitochondrion, and propulsive flagellum.

The Egg

1. Eggs are classified in terms of their yolk distribution as *isolecithal* (even, mammals and many marine invertebrates), *mesolecithal* (moderate, amphibians and mammals), *telolecithal* (uneven, birds and reptiles), and *centrolecithal* (insects).

2. The typical egg surface is very complex, containing a *zona pellucida, vitelline membrane,* and a number of *cortical granules* below the surface.

Fertilization

Fertilization is an active process for both egg and sperm. Six distinct events occur:

1. Release of *hyaluronidase*
2. Fusion of *acrosome* with egg surface
3. Rupture of *cortical granules*
4. Rise of *fertilization membrane*
5. Active engulfing of sperm by egg cytoplasm
6. Migration and fusion of *male* and *female pronuclei*

The Future of a Zygote

1. The *preformation* concept stated that human form was already present in miniature in the sperm or egg.

2. The *epigenetic* concept, championed by H. Driesch, opposing *preformation,* stated that form arose from an amorphous framework.

3. In their final analysis, the experiments of Roux and Driesch showed that different techniques and embryos produced different results. *Preformation* as originally conceived has, however, been disproven.

4. Experiments with young embryos has shown that there is considerable organization in the unfertilized egg.

5. Studies of *lampbrush chromosomes* reveals bursts of genetic activity in the unfertilized egg.

The First Cleavage Divisions

1. While the cell divisions in echinoderms and mammal embryos are roughly equal for a time, the early divisions in amphibians are very unequal. A more rapidly dividing region, the *animal pole* consists of small cells, while the opposing *vegetal pole* is a region of larger, sluggish, yolky cells.

2. Because of the great *yolk* mass, bird and reptile *zygotes* undergo surface divisions in a flattened region known as the *blastodisc.* The yolk itself is not involved.

3. In many vertebrates the early divisions produce a hollow ball of cells known as the *blastula.* Its cavity is known as the *blastocoel.*

The Blastula and Gastrulation

1. In amphibians *gastrulation* is an overgrowth of cells which spreads over the *blastula* and invades the *blastocoel.* The region of invasion is known as the *blastopore.*

2. When *gastrulation* is complete the ingrowth of cells produces a three-layered embryo, including an outer *ectoderm,* middle *mesoderm,* and inner *endoderm.* A new cavity, the *archenteron* or *primitive gut,* has also formed.

Gastrulation in Birds

Gastrulation in birds and reptiles occurs along a central axis in the *blastodisc* as cells push inward from each side. This inward migration supplies the cells that form a loose middle region of *mesoderm.* The cells below are the *endoderm.*

land egg

macromere
mesoderm
mesolecithal
mesonephros
metanephros
micromere
midpiece
miscarriage
monozygotic twins
morphogenesis
mosaic eggs
Müllerian duct

nerve synapse
neural groove
neural tube
neuron
neurulation
notochord

ontogeny

phospholipase A2
phylogeny
placenta
preformation theory
primitive gut
primitive streak
pronephros
pronuclei
prostaglandin
pseudoplacenta

somite
sprouting factor

telolecithal
totipotent
trimester
trophoblast
tubular heart

umbilical vessel
urogential fold
urogential groove

vegetal pole
vitelline membrane

yolk mass
yolk plug
yolk sac

zona pellucida
zygote

Gastrulation in Mammals

In mammals, the *blastocyst* (*blastula*) implants prior to *gastrulation. Ectoderm* and *endoderm* form first, then a *mesoderm* forms when exterior cells migrate into the interior.

Morphogenesis: The Development of Form

1. The basis for cellular *differentiation* apparently lies in factors that activate and deactivate genes at critical periods.
2. Activating factors include hormones, other chemical messengers, the cellular environment, and communication with neighboring cells.

The Fate of the Germ Layers

1. Each of the three *germ layers* of the *gastrula* is destined to produce specific tissues in the developing *embryo.*
2. *Ectoderm* produces skin, nervous system, hair, nails, and eyes, while *mesoderm* produces skeleton, muscles, gonads, kidneys, and vascular system. *Endoderm* produces linings of the gut and many internal organs.
3. The earliest system to develop is the nervous system. In vertebrates this begins with *neural groove,* which eventually forms the *neural tube.* Underlying this is the developing *notochord,* sometimes considered a fourth germ layer.

Embryonic Organizers and the Induction Theory

1. Spemann's experiments revealed that parts of the *ectoderm* are subservient to the underlying *mesoderm* and that this outer tissue has not yet fully differentiated in the *gastrula.*
2. In his third experiment, Spemann determined that the *ectoderm* is induced to form *neural* tissue by its underlying *mesoderm.* This process or influence is known as *induction.*
3. The chemical messengers associated with *induction* have never been clearly identified.

The Cell Nucleus in Differentiation

1. Gurdon, Briggs, and King tested the *totipotency* of the *blastula* nuclei by transplanting them into enucleated egg cells.
2. The nuclei of *blastula* cells are generally *totipotent,* while those of older embryos seemed to lose potency.
3. In later experiments with transplants, the source of transplant nuclei became *gastrulas* produced from an original transplant. Thus the new material provided numerous genetically identical nuclei. The rate of transplant failure dropped and the failures bore identical abnormalities.
4. Eventually the concept of *differentiated* nuclei became questionable, as older and older embryonic donor nuclei proved to be *totipotent.*
5. In the intact *embryo,* irreversible nuclear *differentiation* may or may not begin to occur in the *gastrula* where differential gene activation and deactivation is seen.

Nerve Cells and Developmental Patterns

Axons are sent from the developing brain or spinal column to all parts. Substances released by the neurons inhibit further neuron penetration. Neurons apparently identify the cells they invade. Neurons ending in the wrong tissue die off. Transplants substantiate the idea, since tissue placed in the wrong region still transmits to portions of the brain that associate with the former location. Returning messages go to the mislocated tissue as though it were in the right place.

Physical Support and Life Support of Vertebrate Embryos

In aquatic vertebrates (fish and amphibians) many of the requirements of the embryo are provided by the watery environment.

The Self-Contained Egg

In reptiles and birds, the environmental necessities of the embryo are provided by the *land egg* as follows:

1. Food: *yolk sac* and *yolk*
2. Water: *albumen*

3. Gas exchange: *chorioallantois*
4. Waste storage: *allantois*
5. Protection: *amnion* (fluid-filled)

Life Support of the Mammalian Embryo

1. In mammals, the life-supporting structure is the *placenta.*

2. When *implantation* into the uterine wall occurs, cells of the *trophoblast* digest out a cavity which fills with blood and nutrients.

3. After *implantation,* the *trophoblast* and *mesoderm* produce fingerlike *chorionic villi* that penetrate the uterine lining. Later, blood vessels will penetrate as the *placenta* develops and enlarges.

4. The *yolk sac* of mammals does not contain *yolk,* but does produce the first blood cells.

The Emergence of the Human Form

The First Trimester

1. The earliest events in human development include the emergence of the primitive nervous system followed shortly by the simple heart and first blood vessels.

2. The simple *tubular heart* loops back on itself, eventually developing into the four chambers required.

Respiratory and Digestive Systems

1. The respiratory system begins as an indentation in the early pharynx, deepens, branches, and forms the respiratory tree.

2. The *primitive gut* is produced by the *yolk sac,* which indicates a change in function for this vertebrate structure.

3. At one stage, the *gill arches,* seen in fish, are prominent, but these structures go on to form jaw, larynx, hyoid bone, and middle ear.

Nervous System

By the fourth week the brain and spinal cord have formed.

Limbs

The limbs begin as rounded flaps in the fourth week, continue into a paddlelike stage followed by elongation and separation of fingers and toes. The development of the legs and feet lag somewhat behind the arms and fingers.

Excretory System

The urinary and excretory system go through three distinct stages of development named for the tubules that form. These are the *pronephros, mesonephros,* and *metanephros,* the latter forming the final kidney and urinary structures.

Reproductive System

1. The reproductive system remains undifferentiated until the eighth week of life. Under the influence of sex hormones, each of the primitive structures will differentiate into male or female external genitalia.

2. Under the influence of testosterone, the primitive genital region will develop into the penis and scrotum. The testes developing in the abdomen will descend into the scrotum shortly before birth.

3. When testosterone is absent, the primitive genital region develops into the clitoris and labia majora and minora.

Prenatal Sex Hormones and the Development of Behavior

It has been noted in rats that the target cells of male hormones include cells of the cerebral cortex. Testosterone influences both nerve connections and behavior.

The Second and Third Trimesters

1. In the fourth month growth continues with muscle development, blood forming elements in the bones, general response to stimuli, and a general straightening of the back.

2. During the fifth month the *fetus* responds more specifically to stimuli. The digestive and respiratory systems are well developed but do not function. Bone growth is rapid and calcium demands heavy.

3. The sixth month is marked by greater movement and further refinement of structure. At the end of this month, the *fetus* may survive premature birth.

4. The *fetus* reaches full growth and growth rate declines. The demands on the mother are greatest at this time in terms of protein, iron, and calcium. In the latter part of this period antibodies cross the *placental* barrier.

Birth

Aging of the *placenta,* lowering progesterone levels, and increasing estrogen and oxytocin levels appear to be important factors in instigating the birth process. One hypothesis suggests that *prostaglandins* are released as the gene that produces the enzyme *phospholipase A2* is freed by lowered progesterone levels. *Phospholipase A2* produces *prostaglandins,* which stimulate contractions.

The Stages of Birth

1. The first stage of labor and birth is *dilation,* where the cervix softens and begins to open. This is accompanied by contractions that increase in both frequency and duration. The *amnion* may break at this time.

2. Stage two, expulsion, involves the passage through the cervix preceded by the *amnion's* bursting. Passage of the *fetus* may be rapid or prolonged.

3. Stage three involves the delivery of the *placenta.*

Miscarriage (Spontaneous Abortion)

Miscarriage is far more common than we realize. Many factors are involved in the failure of the embryo and or the *placenta.*

Physiological Changes in the Newborn Baby

1. At birth, all of the inactive systems whose functions were usurped by the *placenta* must become functional.

2. Changes in the circulatory system include closing of the *foramen ovale* of the heart, closure of the *ductus arteriosis,* and rerouting of blood from the *umbilical vessels.*

3. The lungs must function at once. Fluid must be cleared from passages, the respiratory centers become functional, and diaphragm, lungs, and rib muscles begin action.

Lactation

1. Declining progesterone levels remove inhibition of prolactin and the subsequent release of *colostrum* from the breasts. This will be replaced by milk in a few days.

2. Human milk differs from cow's milk in significant ways, including less casein, more sugar, and less calcium and butterfat. It is also sterile.

3. Suckling stimulates the release of oxytocin and prolactin from the pituitary. The first aids in needed uterine contraction and milk "letdown," while the second stimulates milk secretion.

Review Questions

1. What are the major structures of the human sperm? (p. 865)
2. Relate the terms *mesolecithal* and *telolecithal* to the mode of development in organisms they describe. (p. 865)
3. What is the apparent function of the *cortical granules* found near the surface of the egg cytoplasm? (p. 866)
4. Describe the active response of an egg to the attachment of a sperm cell. (pp. 866–867)
5. How did the techniques of Roux differ from those of Driesch? In what way did the differing techniques affect the outcome of their experiments? (pp. 868–870)
6. Distinguish *preformation* from *epigenesis.* In what way does our understanding of DNA lend some support to *both* hypotheses? (pp. 867–868)

7. Starting with the sperm and egg *pronuclei,* review the events that must occur in the *zygote* before the first mitotic division can occur. (p. 867)

8. Reviewing the early *cleavages* of the frog's egg, explain how *differentiation* may have already begun after the third *cleavage.* (p. 870)

9. Compare the patterns of cell division between frog and human. What is the obvious difference? (pp. 870–871)

10. What are *lampbrush chromosomes* and what do they tell us of the metabolic activity of an unfertilized egg? (p. 871)

11. Compare the early *cleavages* in the bird *zygote* with those of amphibians and mammals, and explain why they must differ. (pp. 871–872)

12. Compare the rate of cell division, size of cells, and the contents of cells in the *animal* and *vegetal pole* of the amphibian embryo. (p. 871)

13. Describe the *gastrulation* process in the frog using the terms *epiboly, invagination, blastopore, involution, archenteron,* and *yolk plug.* (p. 872)

14. List the *germ layers* found in the *gastrula* and indicate where they are in a simple cross-section drawing. (p. 872)

15. For each of the *germ layers,* list three structures (tissues) for which they are responsible. (p. 872)

16. Describe the way *gastrulation* proceeds in the bird. (p. 873)

17. What term is used to describe the human *blastula?* What becomes of its *inner cell mass* and *trophoblast?* (p. 874)

18. Describe *gastrulation* in the human. (p. 874)

19. Define the term *differentiation.* (p. 875)

20. Explain how our knowledge of DNA action and control helps in understanding *differentiation.* (p. 875)

21. Describe, in several steps, the process of *neurulation* in the frog. (p. 877)

22. What is the significance of the *notochord* to phylum Chordata? Which adult chordates retain this embryonic structure? (p. 878)

23. List the conclusions reached by Spemann in each of his three experiments. (p. 879)

24. Define the term *induction.* How many specific *inducer* substances have developmental biologists found to date? (pp. 879–880)

25. What was the fundamental question to which Gurdon, Briggs, and King addressed their experiments? (pp. 880–881)

26. Describe the general procedure and the improved procedure in the Gurdon, Briggs, King experiments. (p. 881)

27. What are the present conclusions of Gurdon, Briggs, and King about the embryonic nucleus? (p. 882)

28. List the requirements of the embryo which are satisfied by the land egg. Name the structures that are responsible. (p. 885)

29. Describe the process of *implantation* and the role of the *trophoblast* in producing the *chorionic villi.* (p. 886)

30. List the major events in the development of the human heart. (p. 888)

31. How are the basic genetic instructions that produce *gill arches* modified in the mammalian embryo? (p. 888)

32. Describe the role of "programmed death" in the development of the hands and feet. (p. 891)

33. Describe the reproductive system at 8 weeks of development. List the major structures. (p. 892)

34. Outline the fate of the following in producing male and female genitalia: *urogential groove, urogenital folds, genital swelling.* (p. 892)

35. What are the general roles of the X and Y chromosomes in determining sex? (p. 892)

36. Discuss the role of *cortical* and *medullary dominance* in the development of testis or ovary. (p. 892)

37. Summarize the effects of testosterone on rat behavior and nerve synapses. (p. 892)

38. Outline the major developments in the second *trimester* of human growth. (pp. 893–894)

39. List the increased demands placed on the mother during the third *trimester* of human development. (pp. 893–894)

40. Summarize the events in three stages of labor. (pp. 894, 896)

41. Describe three structural differences between the fetal circulatory system and that of the adult. (pp. 896–897)

42. Discuss the possible causes for a condition known as "blue baby." (p. 897)

43. Describe the hormonal control of milk production and the role of suckling. Describe a secondary effect of suckling that is important to the mother's recovery. (p. 897)

44. List several advantages of breast feeding. (p. 897)

Suggested Reading

Alexander, T. 1975. "A Revolution Called Plate Techtonics Has Given Us a Whole New Earth." *Smithsonian* 5:30.

Baker, M. A. 1979. "A Brain-Cooling System in Mammals." *Scientific American,* April. A countercurrent heat exchange keeps the brains of most mammals from overheating during heavy exertion. Primates have lost this system.

Balinsky, B. I. 1974. "Development, animal." In *The New Encyclopedia Brittanica,* 15th ed., 6:625. Although this article is almost entirely devoted to vertebrate embryology, it does that well.

Cairns, J. 1975. "The Cancer Problem." *Scientific American,* November. Cairns argues that almost all cancers could be prevented by eliminating cancer-associated factors from the environment.

Edelman, G. M. 1970. "The Structure and Function of Antibodies." *Scientific American,* August.

Epel, D. 1978. "The Program of Fertilization." *Scientific American.* Includes the most dramatic scanning electron micrographs of eggs and sperm that we've seen yet.

Garcia-Bellido, A., P. A. Lawrence, and G. Morata. 1979. "Compartments in Animal Development." *Scientific American,* July. How does a differentiating cell know where it is and what to do? In some animals it appears that differentiation is a stepwise procedure, with clones of cells becoming increasingly specialized as compartments within compartments.

Griffin, D. R., ed. 1974. *Animal Engineering.* Readings from *Scientific American.* W. H. Freeman, San Francisco. Engineering majors with an interest in biology or biologists with an engineering background will find themselves at home here.

Gurdon, J. B. 1968. "Transplanted Nuclei and Cell Differentiation." *Scientific American,* December. The technique of transplanting diploid nuclei into enucleated amphibian eggs and a still-controversial viewpoint of just what the resulting embryos tell us about the course of normal development.

Heimer, L. 1971. "Pathways in the Brain." *Scientific American,* January.

Hoyle, G. 1970. "How Is Muscle Turned On and Off?" *Scientific American,* April.

Johanson, D. C., and T. D. White. 1979. "A Systematic Assessment of Early African Hominids." *Science* 203:321. Since our ideas about early hominid relationships are rapidly changing, it pays to keep up with the news. Here the authors take a fresh look at some old bones from East Africa.

Katchadourian, H. 1974. *Human Sexuality: Sense and Nonsense.* W. H. Freeman, San Francisco.

Kimble, D. P. 1973. *Psychology as a Biological Science.* Goodyear Publishing Co., Santa Monica, Calif. Kimble lives up to his impudent title: biology students who have been baffled by the orientation of standard psychology courses will find some relief here.

Lazarides, E., and J. P. Revel. 1979. "The Molecular Basis of Cell Movement." *Scientific American,* March. Some superb photographs of the cytoskeleton and the incredible "cellular geodome," as revealed by immunoflorescence microscopy.

MacArthur, R. H. 1972. *Geographical Ecology: Patterns in the Distribution of Species.* Harper and Row, New York.

Mayr, E. 1963. *Animal Species and Evolution.* Harvard University Press, Cambridge, Mass. A truly great work uniting biogeography, natural history, evolution, and genetics. Here Mayr introduces and defends "the biological species concept."

Miller, W. H. 1974. "Photoreception." In *The New Encyclopedia Brittanica,* 15th ed., 14:353. In far more detail than found in most *Brittanica* articles, Miller gives a full account of the morphology, physiology, and biochemistry of eyes and other vertebrate and invertebrate photoreceptors.

Money, John, and Anke A. Ehrhardt. 1972. *Man and Woman, Boy and Girl: The Differentiation and Dimorphism of Gender Identity from Conception to Maturity.* Johns Hopkins University Press, Baltimore. From their studies of patients who have been misdiagnosed as to gender because of congenital abnormalities of the genetalia, the authors conclude that personal sexual identity is confirmed by 18 months of age. On the other hand, prenatal sex hormone levels are shown to have a decided influence on personality.

Morris, S. C., and H. B. Whittington. 1979. "The Animals of the Burgess Shale." *Scientific American,* July. An exciting look at an accidentally preserved nearshore community in the early Cambrian era, including primitive members of most major phyla, some members of mysterious and bizarre long-extinct phyla, a marine onychophoran, and one amphioxuslike chordate.

Morton, J. E. 1967. *Guts.* St. Martin's Press, New York.

Oppenheimer, J. H. 1979. "Thyroid Hormone Action at the Cellular Level." *Science* 203:971.

Peel, J., and M. Potts, 1969. *Textbook of Contraceptive Practices.* Cambridge University Press, New York. Slightly dated—the very newest techniques won't be found there—but still useful.

Petit, C., and L. Ehrman. 1969. "Sexual Selection in *Drosophila.*" *Evolutionary Biology* 3:177. Perhaps you thought those little flies wouldn't *care.* But they do.

Rugh, R., and L. B. Shettles. 1971. *From Conception to Birth: the Drama of Life's Beginnings.* Harper and Row, New York.

Simmons, J. A., M. B. Fenton, and M. J. O'Farrell. 1979. "Echolocation and Pursuit of Prey by Bats." *Science* 203:16.

Smith, C. A. 1963. "The First Breath." *Scientific American,* April. The dramatic changes that occur when fetal circulation and respiration suddenly shift to the adult patterns in mammals.

Storer, T. I., R. L. Usinger, R. C. Stebbins, and J. W. Nybakken. 1979. *General Zoology,* 6th ed. McGraw-Hill, New York. An updated—complete, but still somewhat old-fashioned in style—version of the classic 1943 zoology textbook.

Tanner, J. M. 1974. "Development, human." In *The New Encyclopedia Brittanica,* 15th ed., 6:650. Tanner's emphasis is not on embryology but on the growth and development of human beings between birth and adulthood.

Waddington, C. H. 1974. "Development, biological." In *The New Encyclopedia Brittanica,* 15th ed., 6:643. Waddington takes an overview of the *principles* of animal development.

Weisman, I. L., L. E. Hood, and W. B. Wood. 1978. *Essential Concepts in Immunology.* Benjamin, Menlo Park, Calif. There is no way to make the complexities of immunology simple. Advanced students will at least find this paperback a good place to get started.

Weissmann, G. 1975. "The Molecular Basis of Acute Gout." In Weissmann, G., and R. Clairborne, eds. *Cell Membranes.* HP Publishing Company, New York.

Wilson, A. C., S. S. Carlson, and T. J. White. 1977. "Biochemical Evolution." *Annual Review of Biochemistry* 46:573. Each kind of protein appears to evolve (change its amino acid sequence) at a steady rate, making it possible to use protein sequence divergence as an "evolutionary clock."

Woolacott, R. M., and R. L. Zimmer. 1979. *Biology of Bryozoans.* McGraw-Hill, New York.

Part Six

Behavior

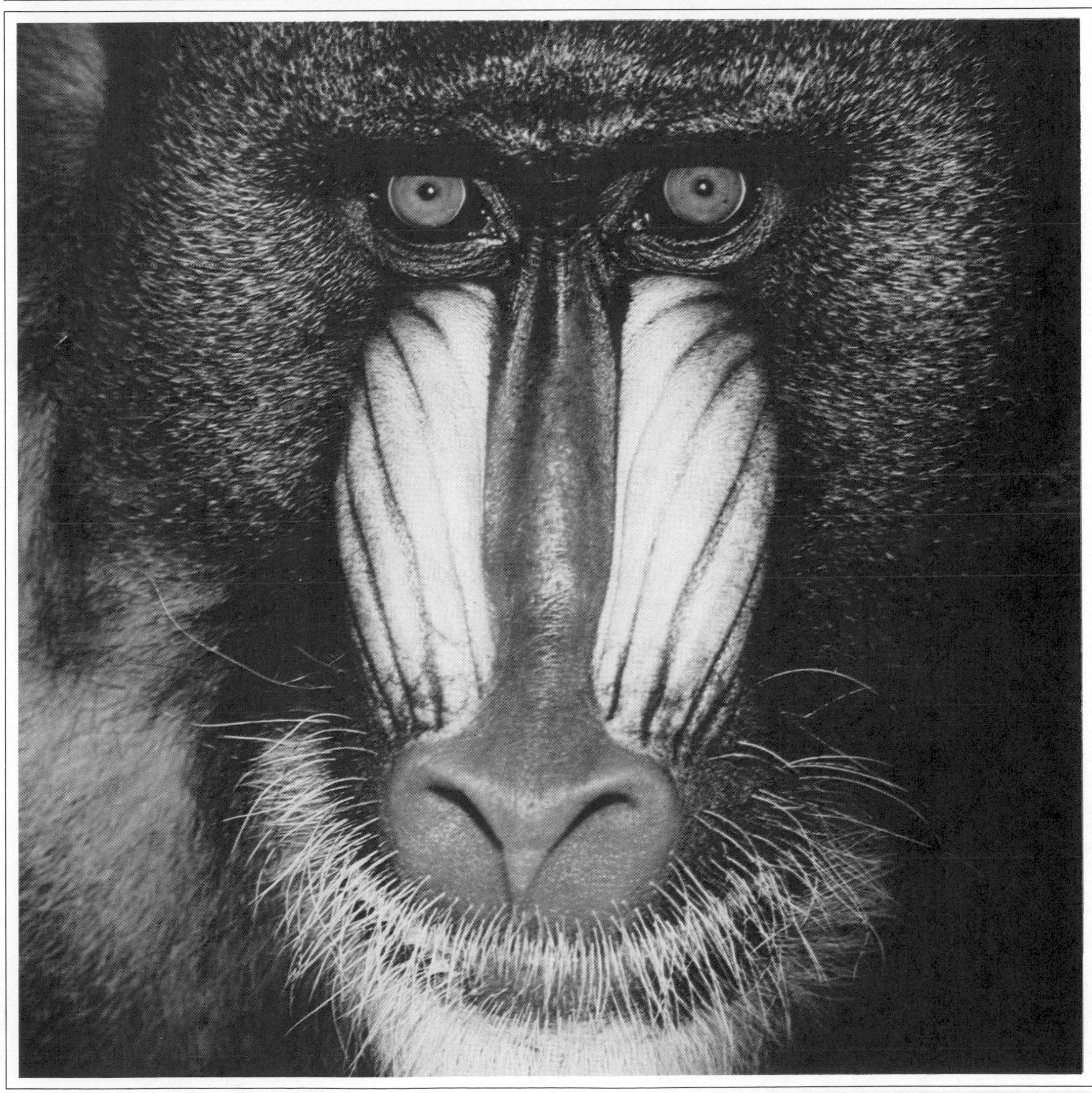

Chapter 31
The Nervous System

The earth is a variable, changing, and disruptive place. It is, of course, where we live and where life has evolved, but it is a highly dangerous place for living things just the same. Some places are better than others, we know, and some periods are less dangerous than others. In addition, some neighbors are essential, while others are innocuous or deadly. Obviously, then, an animal has to have some means of knowing the nature of its surroundings.

And then there is that other world. The inner one. Life processes can go on only within certain physical limits. Animals, especially complex ones such as mammals, are subject not only to external influences but to internal ones as well, and these must be closely regulated. The peculiar kinds of cells that give us information about our external and internal environments and then send signals to the body comprise the *nervous system.* It should be apparent that the nervous system operates with the hormonal system to help us adjust to our environment. Although both systems are regulatory, the nervous system works faster—sometimes almost instantly—and its effects are usually of much shorter duration.

31.1 Neurons and Nervous Systems

What Neurons Are

Neurons are nerve cells. Groups of neurons sometimes travel along the same channels and may form thick, white fibers called *nerves.* The special quality of neurons is that they are irritable—a physical, not an emotional, state. In other words, they can react to environmental stimuli, and in very specific ways. For example, some neurons are involved in the *detection* of environmental stimuli. These are *receptors.* Some detect general kinds of environmental conditions, while others register very specific stimuli, such as cold or pressure. The stimulation of any neural receptor usually results in other neurons being activated, such as those that carry the message to the spinal cord and brain, so some neurons are involved in *transmission.* Once the brain, or in some cases the spinal cord, has received the message, other neurons are involved in *decoding* it ("that was sound," "that was light"). And then other neurons may be brought into play as the brain or spinal cord transmits messages to various muscles—for example, to move the organism toward the sound or away from the light. What we've been talking about, of course, are vertebrate nervous systems. Not all animals have brains and spinal cords, as we shall see.

Neurons in most animals, however, basically behave very similarly, so let's consider in general terms how a neuron works.

How Neurons Work

First, neurons come in an incredible array of sizes and shapes, as you might guess, since some detect light, some color, others pressure, and so forth. A few representative types are shown in Figure 31.1. Basically, they all have a *dendrite* ("tree") region. ("Dendrite" comes from "tree"; this part of the neuron is usually highly branched.) The dendrite receives stimuli, either from the environment or from other neurons. The dendrites may or may not extend from the cell body (the part that harbors the nucleus). The cell body contains the nucleus and most of the usual cell organelles, such as the endoplasmic reticulum, mitochondria, and ribosomes. Clusters of cell bodies in the brain are called *nuclei,* while clusters outside are known as *ganglia.* You may recall that some invertebrates have no real brain development, but only ganglia. The *axon* is the "trunk" of the neuron. Axons may be very long. In humans, they range up to 3 ft long—such as those that run from the base of the

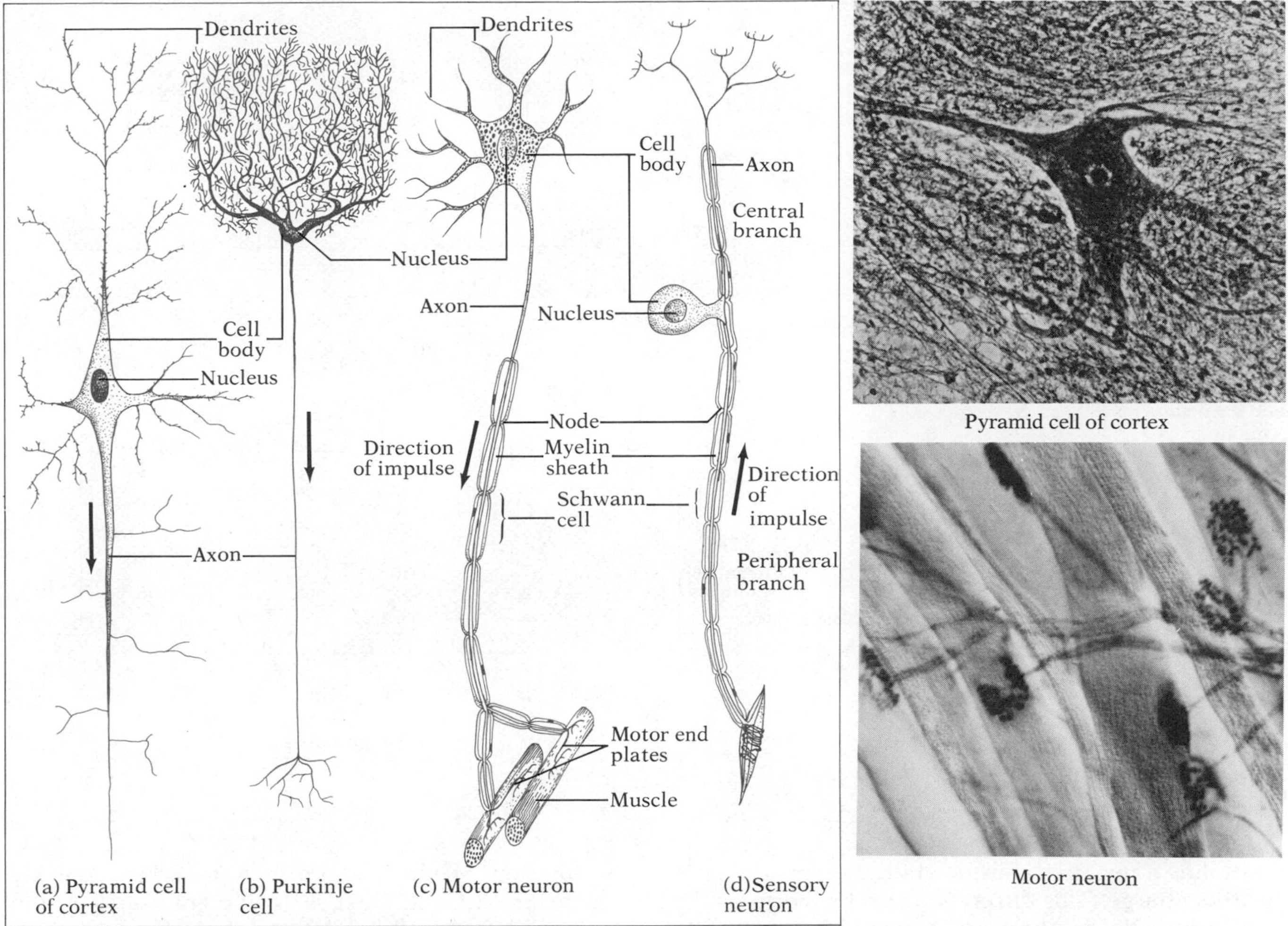

31.1 Four neurons found in human beings show the diversity of these cells. The neurons shown in (a) and (b) are cells from the cortex and cerebellum of the brain. Part (c) shows a *motor neuron* with axons that run from the nervous system to the effector (in this case, a muscle). A *sensory neuron* running from the receptor to the spine is shown in (d). Note that the sensory neuron has no true *dendrites.* The *nodes* in the *myelin sheath* are where one *Schwann cell* ends and another begins. A single nerve cell may be 9 ft long, such as those that run from the base of a giraffe's spine down its hind leg.

spine down the leg. Axons transmit impulses away from the cell body. The axon terminates in the *synaptic junction,* and this joins, or *synapses,* with other neurons; maybe one, maybe several. Ultimately the terminal of a chain of neurons synapses with cells of effector organs such as muscles or glands.

The neural impulse is transmitted both chemically and electrically. The impulse is possible because the peculiar cell membrane has the ability to pump certain molecules that bear electrical charges out of the cell while, at the same time, it allows other charged particles to move to the interior of the cell, thus setting up different concentrations of chemicals outside and inside. As a result of such shifts, the fluid outside the neuron bears the opposite electrical charge from that inside the cell.

The Resting Neuron

To examine the neural impulse more fully we need to take a closer look at the resting neuron, before the impulse begins. Neurophysiologists explain the situation in several ways, and while they aren't always in agreement, most agree that the familiar sodium pump (see Figure 4.12) is involved.

Specifically, neuronal membranes, using considerable amounts of ATP, pump sodium ions (Na^+) out as fast as they diffuse in. Potassium (K^+) is more mobile, easily moving in and out of the neuron. In addition to these two positive ions, neurons contain negative chloride ions (Cl^-) which are also highly mobile. Finally, there are large numbers of totally immobile negatively charged proteins within the neuron. These tend to attract the free potassium ions, thus preventing the K^+ from reaching an equilibrium inside and outside. Apparently, there is a kind of tug of war between the diffusion of K^+ out of the neuron and the attraction inward by the proteins. In addition, some scientists believe K^+ is actively pumped inward. The net result of the ionic gradients is a charge difference between the inside and outside of a resting neuron. A convenient simplification is to say that the

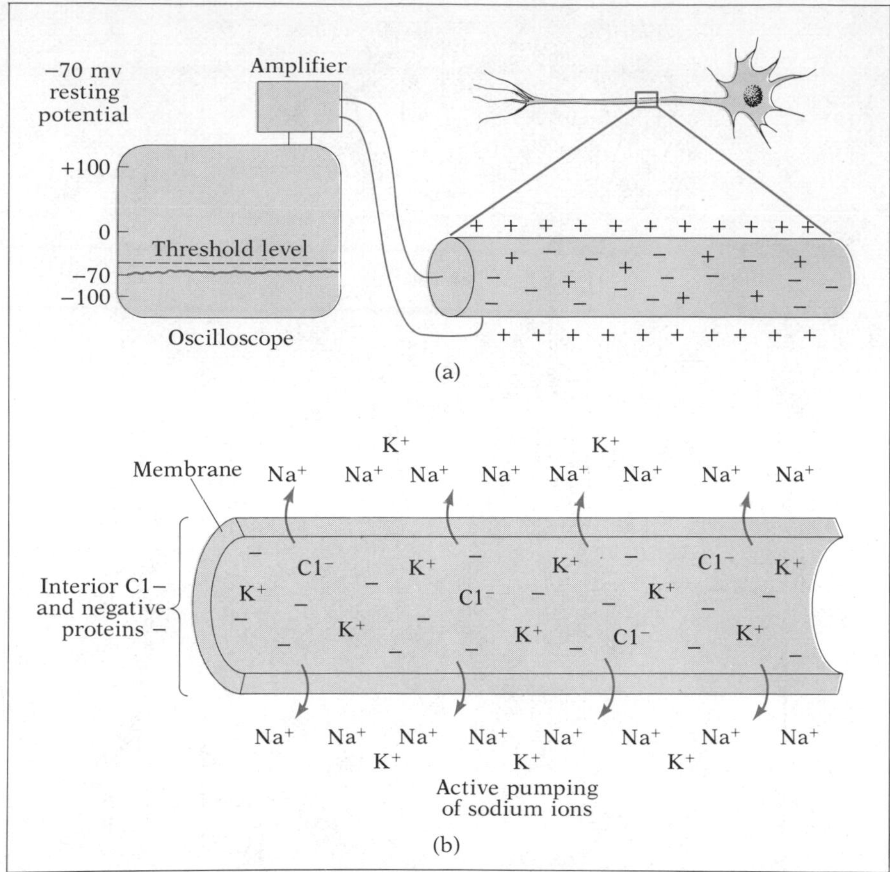

31.2 The resting neuron is *polarized,* with a greater positive charge on the outer membrane than within the neuron. The difference, or *resting potential,* is −70 mv. (a) In this view, a section of an axon has its electrical leads attached to an oscilloscope. The dotted line represents the change needed to bring the neuron to its *firing threshold.* The wavy line represents the measured resting potential. (b) This potential (voltage difference) between outside and inside the resting neuron may be explained in terms of the movement of ions. In the cutaway view, Na^+ is being pumped outside, where it accumulates, while K^+ moves in and out more freely, but is attracted to the negative proteins ⊖ and Cl^- within. All the activity shown produces a net positive charge outside the neuron, as long as the resting state is maintained.

inside has a negative charge while the outside has a positive charge. This difference is referred to as the *resting potential* and has been measured at −70 millivolts (mv). In other words, the resting neuron is *polarized* (Figure 31.2).

The Action Potential

If a stimulus of sufficient strength *(above the threshold)* is received all this changes immediately. First, at any point where a stimulus reaches a dendrite, the sodium pumps stop briefly (less than a millisecond) and Na^+ that has accumulated outside rushes in toward the negatively charged interior as some K^+ leaks out. The result is that the polarity changes from −70 mv overshooting to about +30 mv—a total change of 100 mv! If this event is traced on an oscilloscope (Figure 31.3), it produces what physiologists refer to as a *spike.* This whole *depolarizing* (and opposite repolarizing) event is called an *action potential.* Depolarization isn't a stationary event. It moves along the entire length of a neuron in a wavelike action. And like an ocean wave, it is followed by a trough.

Following depolarization, there is an almost immediate outpouring of K^+ as it diffuses through a very permeable membrane. The K^+ outflow is followed by restoration of the sodium pumps, which quickly move the Na^+ out. Then, K^+ will reenter the cell and its equilibrium will be restored, but before this can occur, as the oscilloscope tracing in Figure 31.3 tells us, the neuron is briefly hyperpolarized (in other words, the resting potential exceeds −70 mv). During the hyperpolarized state, the neuron can't react to any stimulus, no matter how strong.

In summary, then, a neural impulse or action potential occurs when the polarized, resting neuron allows a sudden influx of Na^+. The influx proceeds in a wavelike motion along the neuron, followed immediately by a return to the resting state. But how does a neural impulse move from one neuron to another and how does this impulse create a response by an *effector* such as a muscle or gland? To answer these questions we need to examine the activity at the terminal branches of an axon, at what is known as a *synaptic junction.*

The Synaptic Junction

Many axons terminate in fine branches, each ending in a bulbous swelling known as a *synaptic knob.* These knobs are arranged close to, but not quite in contact with, dendrites or cell bodies of another neuron or, in many cases, the surface of an effector such as a muscle or gland (Figure 31.4). The minute space between the synaptic knob and the next neuron or effector is known as the *synaptic cleft* and is usually about 0.02 μm across. Within each synaptic knob there are a large number of vesicles containing chemicals known as *neurotransmitters.* The neurotransmitters are the connecting link between one cell and another. When a depolarizing wave reaches the synaptic knobs, two things happen: First, the membrane surrounding the knob (the *presynaptic membrane*) freely admits calcium ions (Ca^{2+}) from the surrounding spaces. Then, Ca^{2+} activates a microtubular system within the knob and the microtubules attach to the vesicles and draw them against the presynaptic membrane. The vesicles then fuse with the membrane and spill their contents into the synaptic cleft. The neurotransmitters diffuse across the cleft, contacting receptor sites on the *postsynaptic membrane* of the next neuron or effector. If enough

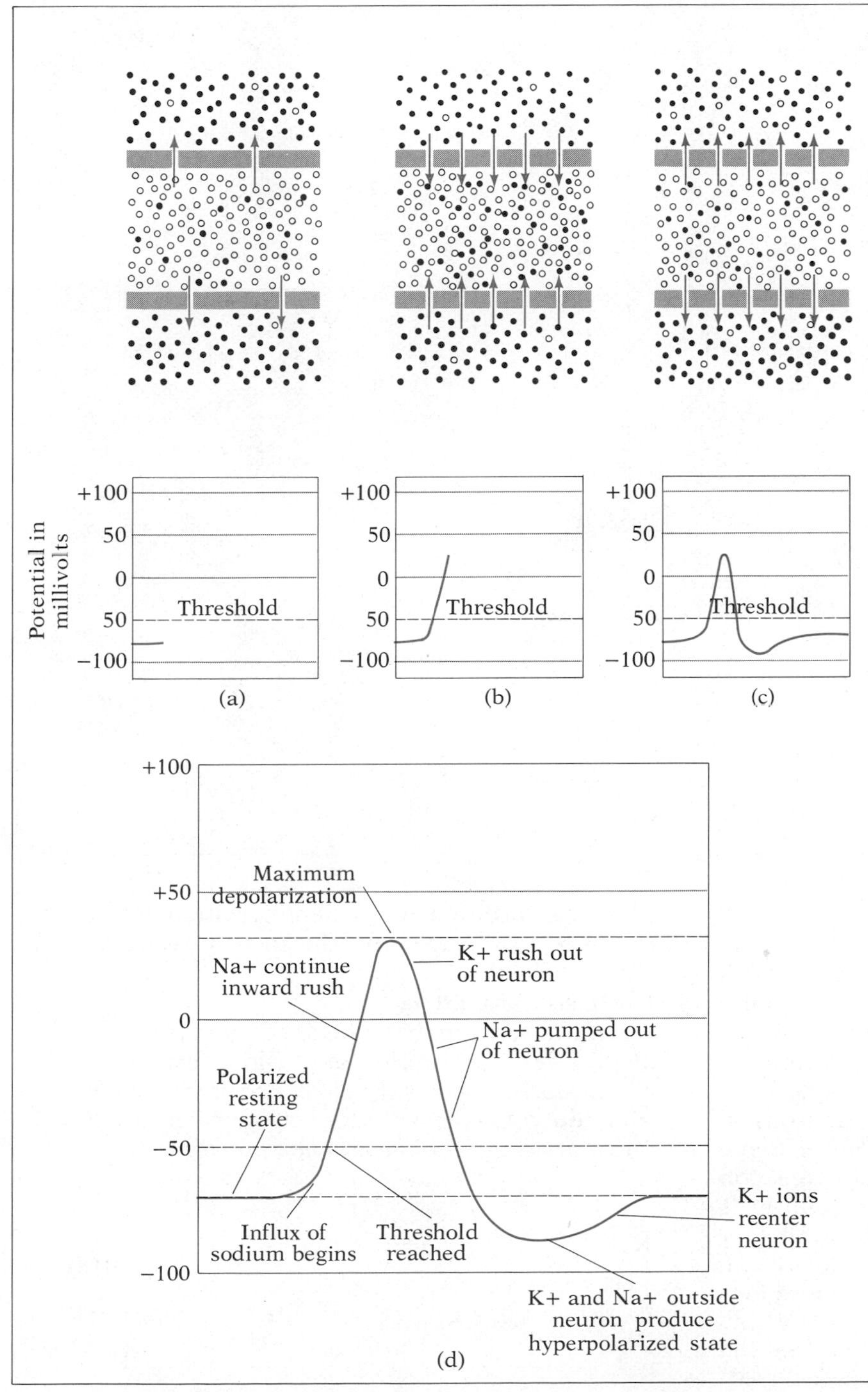

31.3 The *action potential.* In this sequence, the events in an action potential are shown in three steps although the entire sequence occurs in less than a millisecond. Sodium ions (Na^+) are represented by solid dots, while potassium ions (K^+) are the open circles. (Only Na^+ and K^+ are shown.) (a) The neuron is at rest, as revealed by both the oscilloscope tracing, which shows the electrical activity below the firing threshold, and the neuron, which is pumping Na^+ out. (b) A neural impulse begins as sodium pumps stop and Na^+ invades the neuron. (The resulting change in polarity is traced on the oscilloscope.) The inrushing positive charges bring the voltage potential up past the zero, or neutral, point to +30 mv. This is referred to as a *spike.* (c) Changes again occur when K^+ rushes out of the neuron, soon joined by Na^+ as the sodium pumps are restored. The outward movement of positive ions continues and the oscilloscope tracing drops below the resting potential to a temporary state of hyperpolarization (−85 mv). The resting state is finally reestablished when K^+ reenters the neuron. During its hyperpolarized state, a neuron cannot respond to stimuli regardless of the strength. The events surrounding an action potential are summarized at (d).

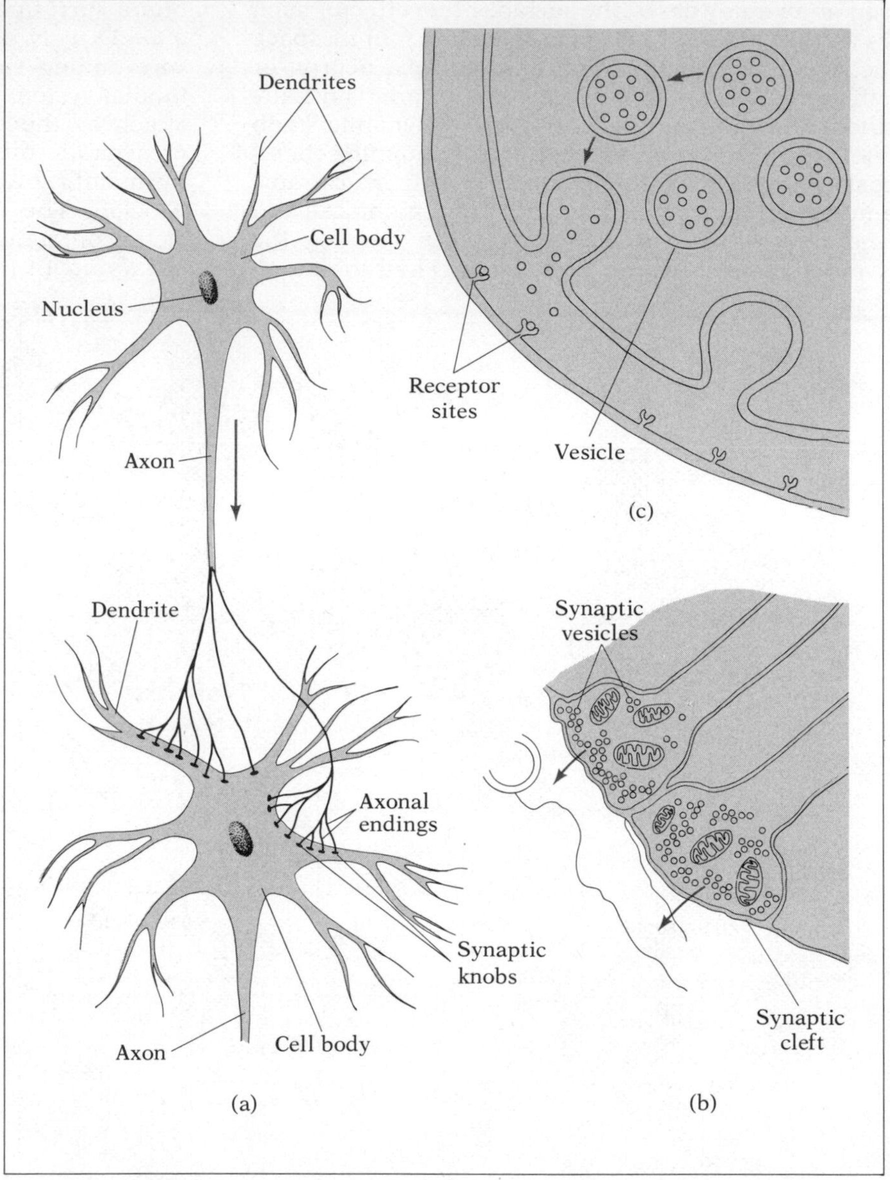

31.4 (a) The axons of nerve cells synapse with other neurons through delicate axonal endings that terminate in synaptic knobs. (b) Each knob harbors a number of mitochondria and numerous vesicles containing neurotransmitters. (c) As a depolarizing wave reaches the synaptic knob, the vesicles are drawn to the presynaptic cell membrane where fusion takes place. The neurotransmitter molecules are then free to diffuse across the synaptic cleft. The cleft is about 0.02μm across. As the drawing indicates, molecules of the neurotransmitter substance reach the postsynaptic membrane (the surface of the next cell) when numerous receptor sites are found. When enough receptor sites are filled, that neuron will respond by initiating its own depolarizing wave.

combining sites are filled, an action potential will be created. The sodium pumps of the postsynaptic neuron will stop and sodium will diffuse into the cell so that the cycle is repeated and the wave impulse is passed on. In the case of an effector, the response may be quite complex. The cell or tissue will respond in its own specific manner.

There is an alternative reaction by neurons in certain instances which is known as *inhibition.* Some neurons inhibit others—that is, they *prevent* an action potential from occurring. The exact mechanism is unknown, but one hypothesis suggests that inhibitory neurons cause the postsynaptic membrane to leak potassium. The added positive charges of these ions to an already polarized neuron creates a state of hyperpolarization. Thus, the hyperpolarized neuron will fail to react. The question of how a single neurotransmitter can produce either depolarization or hyperpolarization in a neuron remains to be answered.

The Neurotransmitters

Neurotransmitters have been of great interest to biologists for many years and a long list of possible chemicals has been suggested. For the moment, however, only two have been conclusively identified. These are *acetylcholine* and *norepinephrine.*

$$(CH_3)_3N^+-CH_2-CH_2-O-\overset{\overset{\displaystyle O}{\|}}{C}-CH_3$$

Acetylcholine

$$\text{(OH)}_2C_6H_3-\underset{\underset{\displaystyle H}{|}}{\overset{\overset{\displaystyle OH}{|}}{C}}-\underset{\underset{\displaystyle NH_2}{|}}{\overset{\overset{\displaystyle H}{|}}{C}}-H$$

Norepinephrine
(Noradrenalin)

Acetylcholine is found at neuromuscular junctions, in the brain, and at junctions in the internal organs. Norepinephrine (also called *noradrenalin*) is the neurotransmitter of the sympathetic nervous system, which will be discussed in some detail shortly. It is also secreted by the adrenal gland (see Chapter 27). Other suspected neurotransmitters are *epinephrine* (also called *adrenalin*), *serotonin*, and the amino acids, *glycine* and *glutamic acid*.

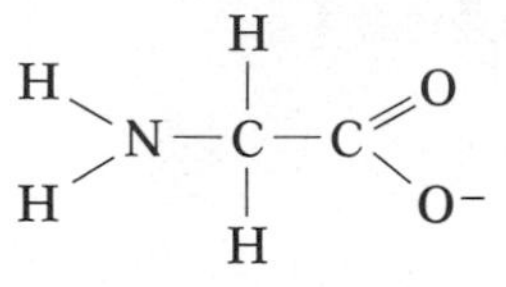

Glycine

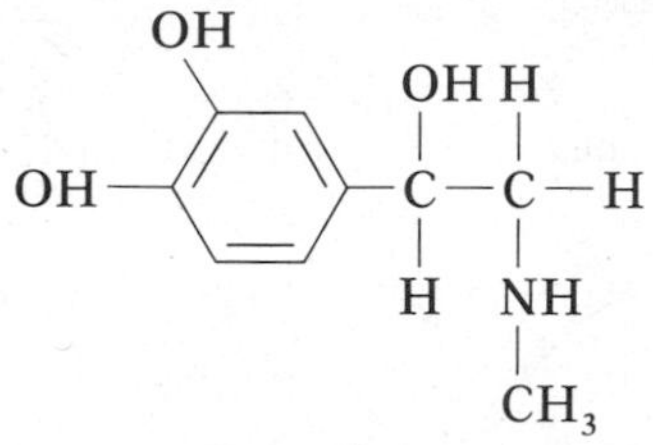

Epinephrine (Adrenalin)

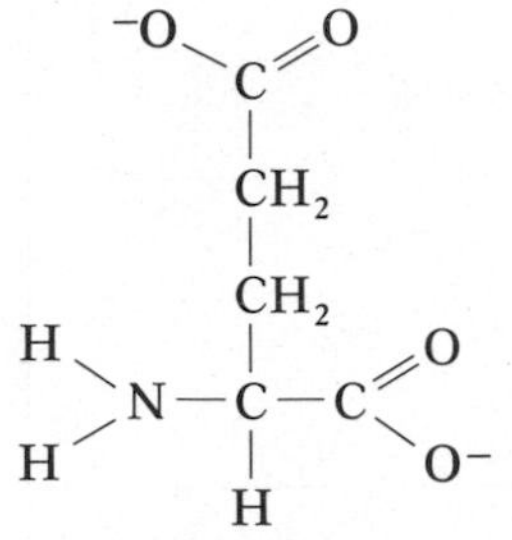

Glutamic acid

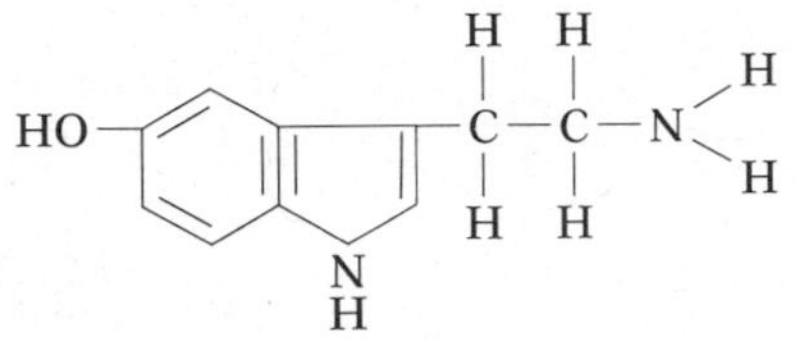

Serotonin

Our findings about what happens in the synaptic cleft answers some questions about neurons and raises some others. It explains, for example, how intraneural impulses flow in one direction, from dendrite to axon, and from the axon of one neuron to the dendrite of the next. Only the presynaptic knob can secrete the neurotransmitter substance, and only the postsynaptic membrane has receptor sites for activating the next neuron or effector. Therefore, once an action potential is set in motion, it can only proceed from neuron to neuron in the manner described. One question raised by the activity of neurotransmitters is: How are they shut off? To answer this question we turn once again to enzymes.

Acetylcholine, it turns out, doesn't last long in the synaptic cleft. No sooner does it appear, when an enzyme called *acetylcholinesterase* is secreted into the cleft and deactivates it. This breakdown is vitally important if there is to be any kind of control in the nervous system. Without the enzymatic event, neurons would constantly be stimulated by residual neurotransmitters. As a matter of fact, this is exactly what occurs in insects sprayed with a class of insecticides known as *organophosphates* (a group that includes malathion and parathion). The organophosphates were developed after World War II, using information gained in nerve gas research. They bind the active sites on the enzyme, thus destroying nervous control in the insect. (Occasionally, farm laborers spraying crops also fall victim to these highly toxic chemicals.)

Fast Neurons and the Schwann Cells

The propagation of an impulse just described is more typical of neurons that are naked or *nonmyelinated*. Compared to *myelinated* neurons the velocity of an impulse in these is slow, just poking along at a few meters per second.

The efficiency of neural transmission is related to two factors that are common considerations in electrical engineering. The first is resistance in the conductor or wire—which is a function of its diameter, among other things. Thicker conductors have less resistance than thin conductors, and so they carry current more efficiently. The second is insulation—which prevents the loss of current by a conductor to its surroundings.

The squid has solved its neural efficiency problem partly through the evolution of giant axonal fibers, up to 1 mm thick. As a result, the muscular activity of this fast-moving invertebrate is greatly enhanced. Whereas vertebrates lack giant axons, they have improved their neural efficiency by insulation.

The fast axons of vertebrates are *myelinated*. This means that they are wrapped by a whitish membranous sheath (see Figure 31.1). The sheaths consist of membranes from *Schwann cells*, which envelop the axons during embryonic development as described in Figure 31.5. The flattened Schwann cells are wrapped around the neuron so that each is several layers thick. The Schwann cells contain a great deal of fat and the resulting whitish color results in solid white regions in the spinal cord and brain. You may be wondering how this insulating layer might affect the neural impulse, since it would certainly interfere with the shift of ions that characterizes the process. The answer is found in the spacing of the sheath. As you can see in Figure 31.5, the sheath is not continuous, but is periodically interrupted by naked axonal spaces along its length. These naked regions, known as the *nodes of Ranvier*, are about 1 μm wide, while the individual Schwann cell wrappings are much larger, nearly 1 mm wide. A node, therefore, appears every millimeter along the axon. So how does the depolarizing wave travel along a myelinated axon? The answer is, very fast!

When a myelinated neuron is activated, the depolarizing wave (or action potential) moves along the axon by literally jumping from one node to another (Figure 31.6). This type of transmission, known as *saltatory propagation*, is up to twenty times faster than that seen in nonmyelinated axons. In addition, saltatory propagation provides a considerable reduction in ATP expenditure, since much of the axon is never depolarized. The fast-acting myelinated neurons innervate the skeletal muscles of many verbebrates. Myelinated motor neurons transmit their impulses at a velocity of 100 m per sec.

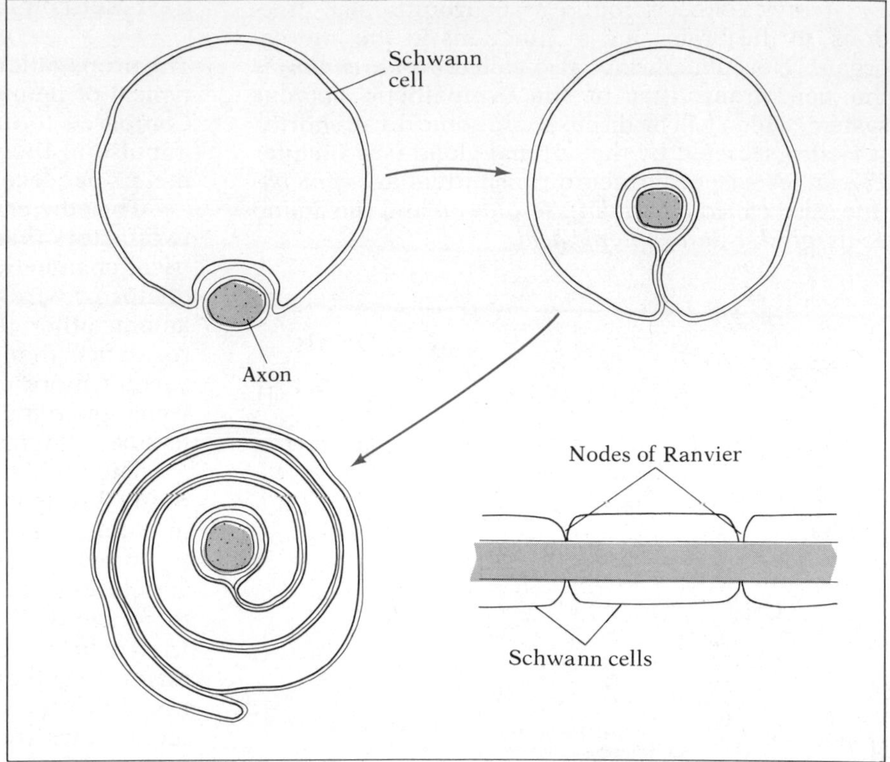

31.5 *Myelinated* neurons are covered by membranous sheaths formed by *Schwann cells.* These coverings are laid down during embryological development as Schwann cells wind themselves around and around the newly developed axonal fiber. Each Schwann cell is separated from the next by a naked region known as a *node of Ranvier.*

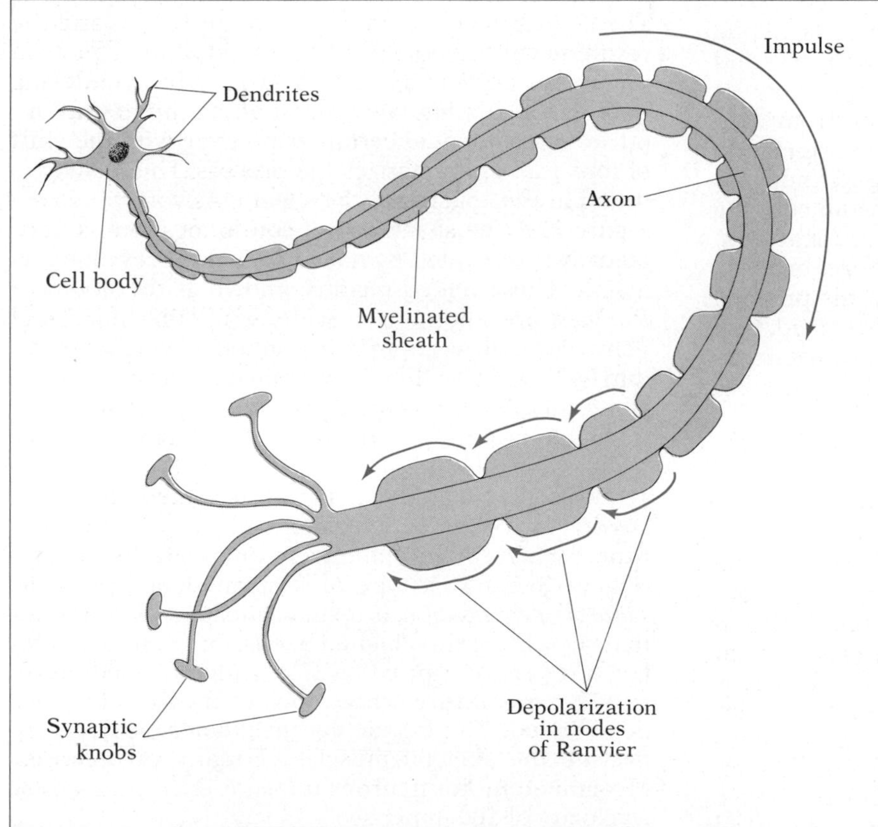

31.6 *Saltatory propagation.* In the myelinated neuron, the depolarizing wave doesn't involve the whole axon, but instead jumps from one node of Ranvier to another along its length. This type of impulse—saltatory propagation—is many times faster than impulses that move along nonmyelinated neurons. Saltatory propagation along myelinated neurons is commonly seen in motor neurons of the vertebrates, such as the one shown here.

Integration of Transmissions

This has been a brief and generalized account of neural transmission. But be not deceived. The actual complexities are staggering. It is important to know, however, that the scheme we have outlined here holds for all types of neurons. A retinal neuron that has been stimulated by light responds in the same way as does an auditory neuron that has reacted to sound.

Also, the sequence of events in a firing neuron remains exactly the same, no matter how strong the stimulus. The stimulus only has to be strong enough to initiate the action potential. A stronger stimulus will not produce a stronger neural response. The nervous system detects the intensity of the stimulus in other ways. For example, if you place your hand on a stove, how do you know the difference between a hot stove and a warm one? It turns out that a hot stove increases the *rate* of firing of neurons in your hand. Also, various heat-sensitive neurons in your hand have different thresholds. Some are easy to stimulate; others are harder. So you can also detect the level of the heat by the number of neurons that are triggered

One might ask, then, if all neurons fire the same way, how are we able to register so many sensations (salt, heat, light)? One reason is because different neurons use different pathways in getting to the brain, or to any other center where the messages are decoded. A series of neurons that register "salt" follow a different route from those that register "light," and they end up in different parts of the brain. Presumably, then, if a taste neuron were to be artificially synapsed with a light neuron, the taste would follow the route of the light neuron and we would be able to taste what we were seeing.

Actually, in spite of all the attention given the neural mechanism, the firing of individual neurons is probably not particularly very important to "behavior." It is the interplay of millions of neurons that results in the adaptive functioning of the nervous system. The brain (or its equivalent) is constantly monitoring, integrating, suppressing, ignoring, and coordinating the various signals coming from millions of neurons all over the body. How long has it been since you noticed the impulses continually being sent from your foot? Your ear? How many sounds have you ignored today?

And what determines whether a complex series of impulses will be sent to the various muscles of the body? If a man sitting alone notices an attractive woman watching him from across the room, the distant, pensive gaze he effects as he lights a cigarette may be purposeful. The slight nervous tapping of his foot under the table may not be. Notice the behavior of someone studying for a test when he or she is well-prepared, and then watch someone hastily cramming for an exam. Why the differences? What does clenching one's hair have to do with preparing for an exam? Why was *that* behavioral pattern generated?

This is the hard part. We know more about receptors and the mechanisms of neurons than we know about their integration. The problem is, we are dreadfully short on information (in spite of a vast body of data) regarding what goes on in the central nervous system. We will shortly have a look at its development in a range of species, but first let's think about the question.

The Central Nervous System

The *central nervous system* in higher animals is composed of the *brain* and *spinal cord.* This system was evolved very slowly, in a stepwise fashion. From animals with the coelenterate's netlike system with no brain, there probably arose animals with paired longitudinal cords such as we find in flatworms. In the flatworms, you recall, there is no real brain, but there are more nerves concentrated in the head region than in the rest of the body. Then, as time progressed the long nerves running the length of animals would have tended to become segmented, as we find in earthworms. The paired nerves would have fused into a single ventral cord with bulbous swellings (*ganglia*) in the head region. (The nervous system of insects is a modification of this scheme.)

In the vertebrates, the brain is highly developed and very large with respect to the spinal cord. Both are hollow and are derived from a plate of dorsal ectoderm that has sunk beneath the surface and has become encased in a protective column of bones or cartilage. The delicate brain is encased in a thick skull (some perhaps thicker than others). Now let's have a closer look at these changes.

Cephalization

We talk very casually about animals having brains. Almost anyone has some notion of what a brain is, but only the most romantic of souls ascribe brains to plants. Most of us would say that only animals have brains and that these organs are the seats of intelligence. The only problems are that not all animals have brains and we don't know what intelligence is. So to keep ourselves from going in circles, perhaps it would help to see if we can put the brain in some kind of evolutionary perspective.

First, we should be aware that brains are the result of an overdevelopment of one end of the nervous system. Since we're talking about ends, that means we're talking about animals that have longitudinal body plans. Starfish don't have ends. Neither do sea anemones. These animals are radially symmetrical and are just as likely to move off in one direction as another. But worms have ends and they tend to move along in the direction of their body axis. In the long course of evolutionary history, it seems that it would have become advantageous for animals with elongate bodies to tend to lead with the same end. Logic and evolutionary expediency thus seem to dictate that this leading end would be the site of a concentration of receptors so that the animal would be able to sense the nature of the environment into which it was moving before its whole body was exposed to that environment. It wouldn't do to back into an unfamiliar place. So *cephalization*, which is the development of an anterior end, set the stage for the formation of the brain by localizing neurons in the part of a longitudinal body that was likely to enter a new environment first. The brain was probably a subsequent specialization of a highly sensory and reactive area. (Do you suppose a localized brain would be as necessary for the denizens of a more benign planet?)

Another effect of an overdeveloped head region is that the brain develops relatively early in most kinds of embryos. Once it is functioning, the brain can help direct the development of the rest of the body. As we've already discussed, many of the hormones of growth and development originate in the brain tissue (especially, the hypothalamus and pituitary).

Let's now consider a few of the specializations in a selected sequence of animals, moving from those that don't have brains to those that do. We will arrange these species from simple to complex with the tacit assumption that we are tracing the evolutionary development of the nervous system. The assumption is a common one, but keep in mind that we don't really know what the ancestral species were, and even if we did, it isn't likely that their nervous systems would have fossilized. Put another way, who can say that flatworms evolved from primitive kinds of coelenterates? And even if they did, who knows what kinds of nervous systems those primitive coelenterates had? But with these reservations, let's see if we can trace the development of the brain.

Coelenterates

The most primitive true nervous system appears in the coelenterates, such as the hydra. This is a netlike system that was once thought to be continuous and unbroken, but now it seems that the net is composed of separate neurons that cross each other, giving the appearance of a net, but not actually touching (Figure 31.7). Actually, it is better if the system is not one continuous net, since a certain amount of facilitation and inhibition can occur if there are a number of synapses, making more complex behavior possible since an impulse can take alternate pathways.

Coelenterates may have two kinds of neurons: slow-conducting neurons (in which the impulse travels at about 0.5 cm/sec), and fast-conducting neurons (that transmit action potentials at about 2 m/sec). In the sea anemone, the fast, pulsing, twitching movements you may have seen are due to the action of the fast neurons. The slower, sustained contraction that results when you muster the courage to touch one in a tide pool is due to the activation of slower neurons.

Flatworms

The most primitive flatworms also have a netlike nervous system that is similar to that of the coelenterates, but others improve on that plan. The relatively advanced planarians have a longitudinal, bilaterally symmetrical body plan comprised of two nerve cords and a concentration of neurons in the head region (Figure 31.7). Here, in the anterior end, the rate of both cellular activity and sensitivity is higher than in the rest of the body. It is in the planarians that we first see *ganglia*, those clumps of nervous tissue in the head region that actually contain the cell bodies of the cells in that area.

Annelids

The annelids are segmented worms, such as the earthworm. In the earthworm, large ganglia surround the esophagus in the head of the worm (a common arrangement in invertebrates), and each segment along the length of the worm has a pair of fused ganglia emanating from the nerve cord (Figure 31.7). If the head of a flatworm is removed, it will simply grow a new one, but if an earthworm is beheaded, its repair is less complete, and its behavior is affected accordingly. For example, it will dig more slowly than a normal worm, but it will dig.

Mollusks

If you have ever owned a pet clam, you may have noticed that it really wasn't much company. Clams are notoriously short on personality and it may be because they are not very intelligent. But surprisingly enough, another mollusk is rather bright, perhaps smarter than some vertebrates. We're referring, now, to the octopus. It, like many other invertebrates, has a ganglionic mass of neural tissue around its esophagus, but in the case of the octopus, the mass is so differentiated that it may rightfully be called a *brain*. For example, a distinct and identifiable part of that com-

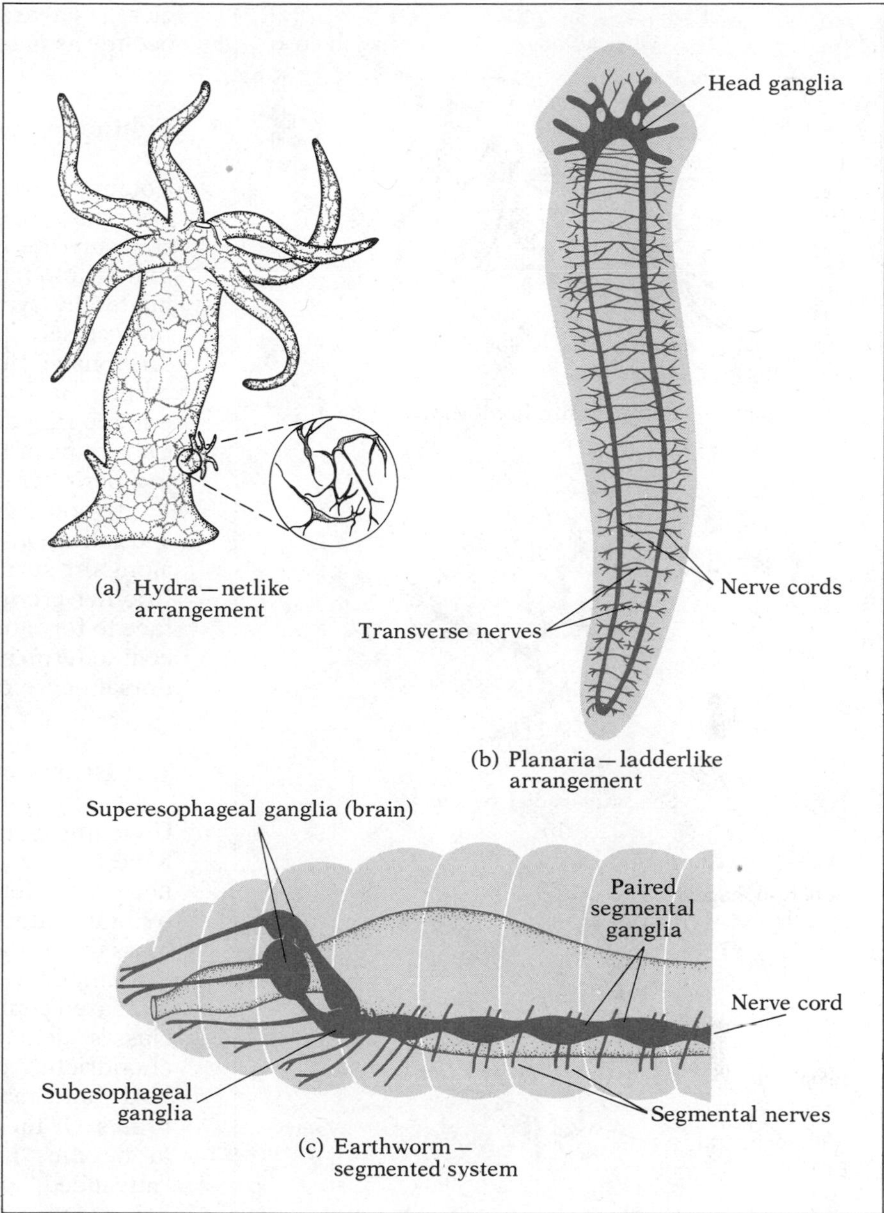

31.7 Representative invertebrate nervous systems. (a) The coelenterates, like this hydra, have a diffuse, rather uncentralized system. The netlike arrangement of nerve fibers is scattered over its body, with the greatest number near its mouth and tentacles. It was once believed to be a continuous net, but closer study (inset) revealed that many individual neurons simply lie in close proximity. (b) Planaria, the free-living flatworm, shows considerable cephalization in its nervous system. Many cell bodies of neurons are accumulated in the head region, forming ganglia. Close by are the two light-sensitive eye spots with many neurons grouped together, forming simple optic nerves. Two nerves emerge from the brain and run in parallel down the body. Branches from these, criss-cross the body, forming a ladderlike arrangement. (c) The earthworm, an active, free-living annelid, has a complex, segmented nervous system. The most prominent features include paired superesophageal and subesophageal ganglia, which are the largest groupings of cell bodies and are always found in the third and fourth segments. In addition, the earthworm has a ventral, solid nerve cord running the length of its body. Paired, but not fused ganglia appear in each segment, sending branching sensory and motor nerves around the body.

plex serves as a respiratory center; another part controls the animal's rapid color changes; other parts deal with associations and are hence important in learning; and other parts control eye movements (Figure 31.8). The octopus, in fact, has been an important research animal in the study of learning and memory.

Arthropods

Arthropods are interesting from an evolutionary point of view because they have taken the segmentation we first saw in annelids one step further—their segments are specialized. Not only are the body segments specialized, but so are the nerve ganglia with which they are associated. Thus, in a crayfish, the peculiar appendages that form the mouthparts are controlled by one set of ganglia; another set controls and coordinates the walking legs; and yet another set is associated with the tiny appendages under the abdomen.

The arthropod brain itself is composed of ganglia above and below the esophagus and each brain part has very specific roles (Figure 31.8). Interestingly, if an arthropod is of a species that just has a simple eye that registers only light, the visual centers will take up only about 3% of the brain. But in highly visual spe-

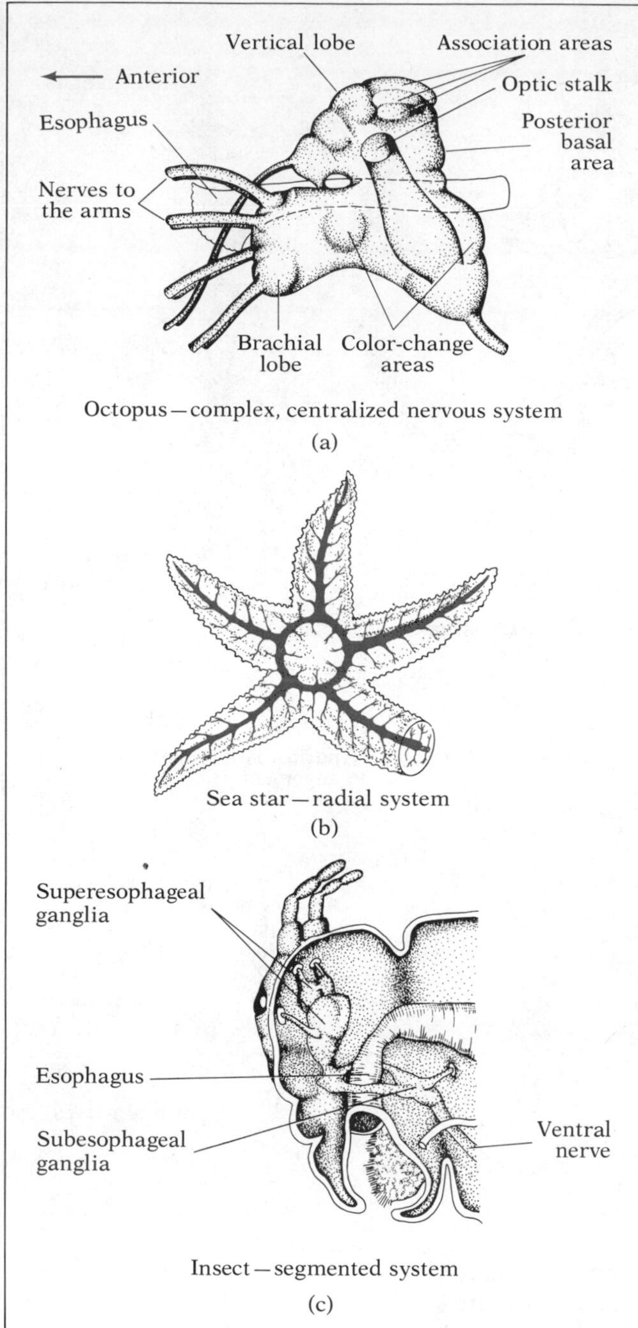

31.8 Nervous systems of cephalopod, echinoderm, and insect. (a) The octopus has a highly developed nervous system, complete with specialized regions related to keen eyesight, rapid movement, and intricate muscle control—all assets to such an active predator. In addition, the octopus is capable of learning tasks that are impossible for many of the supposedly advanced vertebrates. In other words, we can remove the quotation marks from the word *brain.* (b) As adult animals, the starfish and other echinoderms are radially symmetrical. The nervous system, of course, follows this organization. In the sea star, the system consists of a central ring with major nerves radiating into each arm. There is little cephalization in this animal. (c) The insect nervous system follows the segmental pattern seen first in the annelids. The structures common to both include superesophageal and subesophageal ganglia; a ventral, solid nerve cord; and ganglia serving most segments. Specialization is greater than that seen in the annelid or echinoderm.

cies, such as the housefly, the visual centers may occupy as much as 80% of the brain.

Echinoderms

You might wonder why we are now reverting to echinoderms, which are radially symmetrical, after stressing the importance of bilaterality. It turns out that echinoderms, such as sea urchins and starfish are bilaterally symmetrical as larvae, and at that stage, they closely resemble the larvae of hemichordates. They have simply reverted to radial symmetry as adults. Basically, their nervous system consists of a nerve ring around the mouth that gives rise to five (or groups of five) nerve trunks. These are connected to a peripheral nerve net (Figure 31.8).

In some groups, the five nerve trunks consist of specialized *neurectoderm* that forms shallow grooves along the surface of the arms, between the tube feet. In other groups, these grooves sink beneath the surface to form hollow nerve cords. Does that mean that echinoderm nerve trunks are homologous with the dorsal nerve cord of the chordates? We don't know.

The Brains of Vertebrates

Up to now you may have noticed that the nerve cords have been ventral. However, the vertebrate central nervous system is composed of a dorsal nerve and its bulbous anterior end—the large, complex, ganglionic mass that is the brain.

Among vertebrates, there is a tendency toward increased cephalization. That is, the more "advanced" classes generally have larger brains. Agnathans, chondrichthyes, bony fish, amphibians, reptiles, birds, and mammals are the living classes of vertebrates. Of these, birds and mammals are considered to be the "higher," or "advanced," classes. Here "advanced," with respect to any trait, simply means being farther from the presumed ancestral state. Our lowly cephalochordate relative, the lancelet, has a hollow spinal cord but no brain at all, a trait it presumably shares with our own Cambian ancestors.

In a relatively simple group—the chondrichthyes (fish with cartilaginous skeletons, such as sharks)—the thinking center of the brain (the *cerebrum*) is so small that if it is removed, the shark may show little or no change in behavior. It will swim right along without its brain and it may still even attack. A lot of a shark's brain, it turns out, is devoted to its sense of smell. The same is true of the bony fishes (the next highest group of vertebrates), but their brains are more differentiated and removal of the cerebrum leads to rather clear alterations in behavior.

There are basically three parts to any vertebrate brain: the *forebrain, midbrain,* and *hindbrain.* It is the forebrain that is associated with the cerebrum. In lower animals, the forebrain is reduced, and the posterior brain regions are disproportionately larger. The fish brain for example is dominated by the midbrain, but there is also a relatively large hindbrain. In many animals, the midbrain is important in analyzing visual and olfactory stimuli, but in humans it is greatly reduced and simply serves as a bridge between the forebrain and hindbrain. Compare the brains of the vertebrates (Figure 31.9). You can see that in higher vertebrates, the midbrain and hindbrain have yielded to the forebrain's cerebrum. And it is in mammals that we find the most complex development. There are more neurons, and they interact in more ways than in any other group of animals.

The Human Brain

The human brain is, humans believe, the most complex and highly developed brain on earth (in spite of the fact that we can't build a decent bird's nest). In humans, the *forebrain* consists of the *two cerebral hemispheres* and certain internal structures, such as the *thalamus* and *hypothalamus.* The *midbrain,* logically enough, is the area between the forebrain and hindbrain. It is not much different from the midbrain of most vertebrates, containing nerve pathways and receiving and integrating input from the eyes and ears, but with the help of other brain regions. The *hindbrain* consists of the *medulla,* the *cerebellum,* and the *pons* (Figure 31.10).

The Medulla

As a rough generality, the more subconscious, or mechanical, processes are directed by the more posterior parts of the brain. For example, the hindmost part, the medulla, is specialized as a control center for such functions as breathing and heartbeat. As we have already seen, it also functions as a connecting area between the spinal cord and the more anterior parts of the brain.

The Cerebellum and Pons

Above the medulla and somewhat further toward the back of the head is the cerebellum, which is concerned with balance, equilibrium, and coordination. Do you suppose there might be differences in this part of the brain between athletes and nonathletes? Do you think this "lower" center of the brain is subject to modification through learning? The pons, which is the portion of the brainstem just above the medulla, connects the cerebellum and the cerebral cortex, accenting the relationship between the cerebellar part of the hindbrain and more conscious centers of the forebrain.

The Thalamus and Reticular System

The thalamus and hypothalamus are located at the base of the forebrain. The thalamus is rather unpoetically called the "great relay station" of the brain

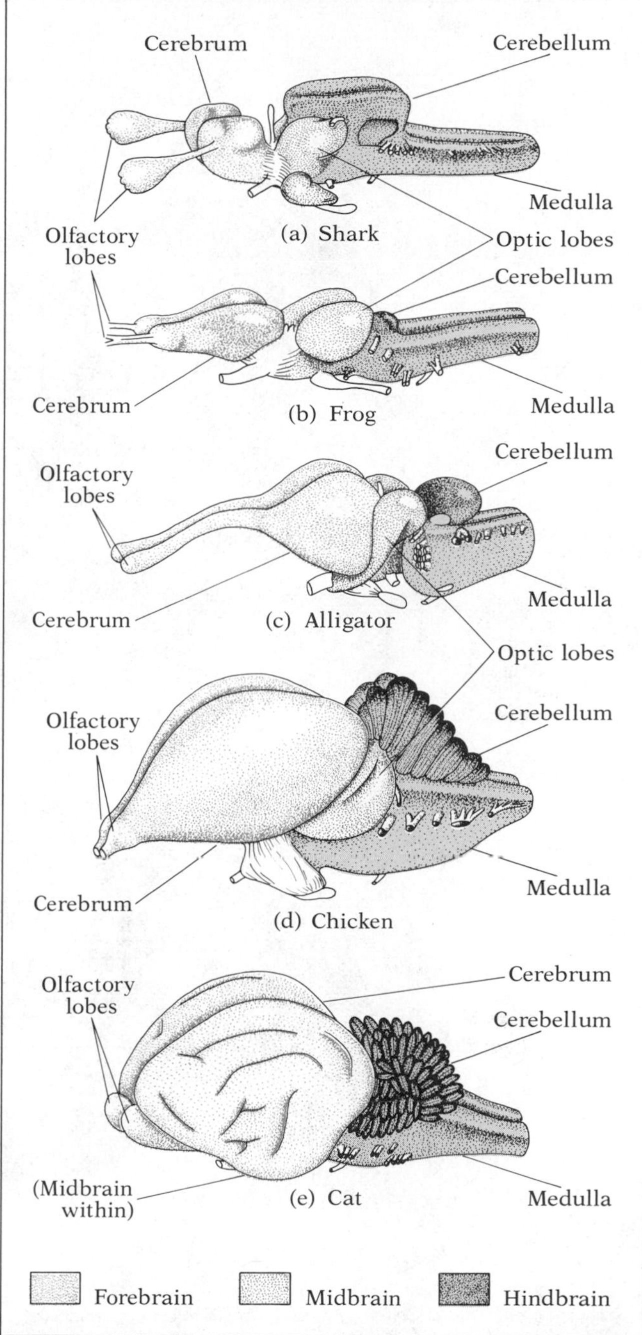

31.9 Comparing the anatomical structures of the brain in five classes of vertebrates (cartilaginous fish, amphibian, reptile, bird and mammal) reveals general evolutionary trends and specific trends in specialization. (a) Note the relatively large olfactory lobes in the shark's brain. It is obviously related to the importance of chemical detection to this predator. (b) The frog feeds by visual means as does the chicken. Note the size of their optic lobes. The trend toward increasing dominance of the cerebrum in vertebrate evolution is also obvious, beginning with the alligator (c) and becoming more pronounced in the bird (d). (e) The trend is greatest in the mammal, with increased convolutions in the cortex. Convolutions or foldings are a way of increasing cerebral size without greatly increasing cranial size.

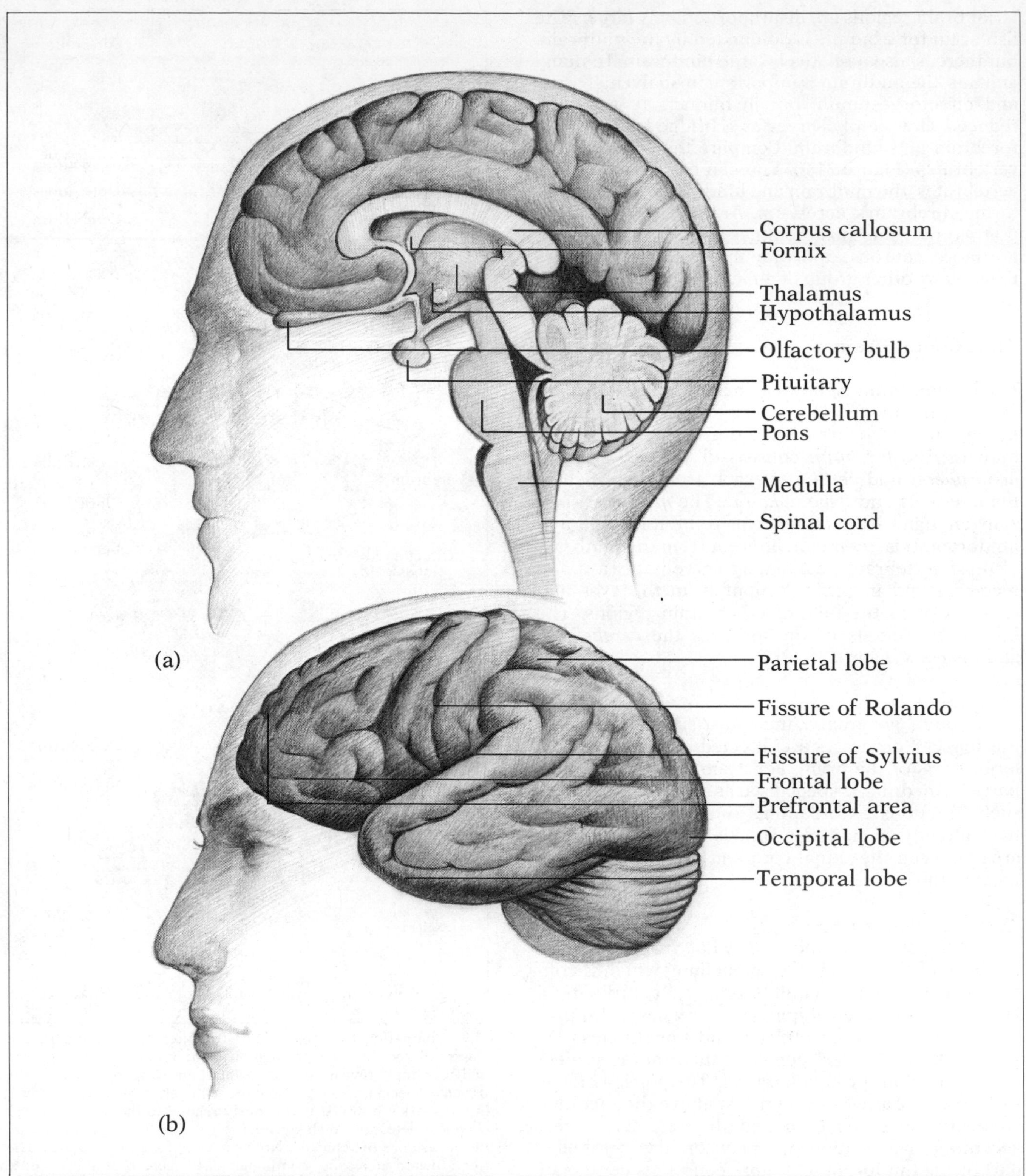

31.10 Most of what is seen in the midsagittal section (a) of the human brain is forebrain. This includes the two great cerebral hemispheres with their convoluted cortex and most of the structures directly below, including the corpus callosum, thalamus, hypothalamus, and olfactory bulb. The midbrain includes small featureless, but vital, pathways (not seen here). The hindbrain includes the medulla, pons, and prominent cerebellum. Most of the brain is covered by the cortex, the famed "gray matter."

(Figure 31.11). It consists of densely packed clusters of neurons, which presumably provide connections between the various parts of the brain—between the forebrain and the hindbrain, between different parts of the forebrain, and between parts of the sensory system and the cerebral cortex.

The thalamus contains a peculiar neural structure called the *reticular system,* an area of interconnected neurons that are almost feltlike in appearance. These neurons run throughout the thalamus and into the midbrain. The reticular system is still somewhat of a mystery, but several interesting facts are known about it. For example, it bugs your brain. Every pathway to and from the brain sends side branches to the reticular system as it passes through the thalamus. So all incoming and outgoing communications to the brain are tapped by this system. Also, these reticular neurons are rather unspecific. The same reticular neuron may respond to stimuli from the hand, foot, ear, or eye. We don't know the functions of this system, but it is believed that the reticular apparatus serves to activate the appropriate parts of the brain upon receiving a stimulus. In other words, it acts as an arousal system for certain brain areas. The more messages it intercepts, the more the brain is aroused. Thus, the reticular system may function importantly in sleep. You may have noticed it is much easier to fall asleep when you are lying on a soft bed in a quiet room with the lights off. Under these conditions, there are fewer incoming stimuli; as a result, the reticular system receives fewer messages, and the brain is lulled rather than aroused.

31.11 The reticular system of the human brain. The impulse originating at the lower right passes through the reticular system, a structure composed of untold millions of neurons. The smaller arrows indicate that, in this case, the entire cerebrum has been alerted, but a specific part (the shaded area) is the target of most of the impulses. It is likely, then, that this area of the cerebrum will be required to deal with whatever initiated the stimulus in the first place.

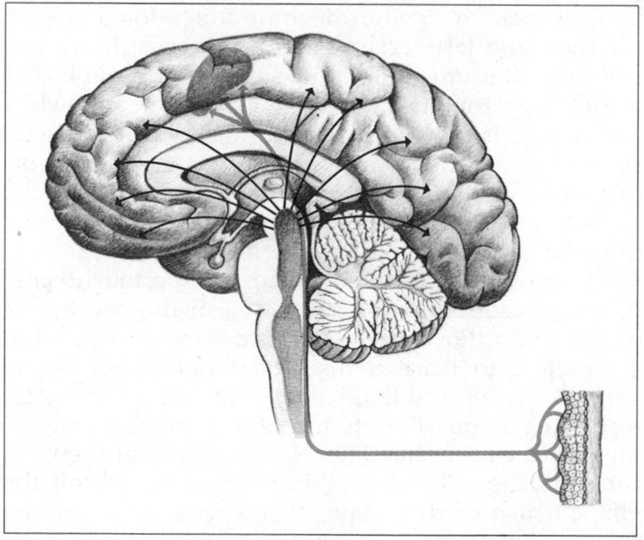

The reticular system may also regulate which impulses are allowed to register in your brain and which will be filtered out or diminished. When you are engrossed in a television program you may not notice that someone has entered the room. But when you are engaged in even more absorbing activities, it might take a *general stimulus* on the order of an earthquake to distract you, whereas the *specific* sound of just a turning doorknob would immediately attract your attention. Such filtering and selective repression of stimuli apparently take place in the reticular system.

The Hypothalamus

The hypothalamus is a small body densely packed with cells, which is extremely important in regulating the internal environment as well as certain general aspects of behavior. As we've discussed earlier, the hypothalamus helps control heart rate, blood pressure, and body temperature. It also plays a part in the regulation of the pituitary gland. And it controls such basic drives as hunger, thirst, sex, and rage. Experimental stimulation of various centers in the hypothalamus by means of electrodes can cause a cat to act hungry, angry, cold, hot, benign, or horny.

The Cerebrum

For many people the word *brain* conjures up an image of two large, deeply convoluted gray lobes. What they have in mind, of course, is the outside layer of the two cerebral hemispheres, the dominant physical aspect of the human brain. The cerebrum is present in all vertebrates, but it assumes particular importance in humans. In some animals, it may serve essentially as an elaborate refinement to implement behavior that could be performed to some degree without it. In more advanced animals, it takes on a more functional importance.

For example, if the cerebrum of a frog is removed, the frog will show relatively little change in behavior. If it is turned upside down, it will right itself; if it is touched with an irritant, it will scratch; it will even catch a fly. Also, sexual behaviors in frogs can occur without the use of the brain—but we'll try not to extrapolate from that. Rats are more dependent on their cerebral cortex. A rat that is surgically deprived of its cerebrum can visually distinguish only light and dark, although its patterns of body movement seem unimpaired. A cat with its cerebrum removed can meow and purr, swallow, and move to avoid pain, but its movements are sluggish and robotlike. A monkey whose cerebral cortex has been removed is severely paralyzed and can barely distinguish light and dark. The result in humans is total blindness and almost complete paralysis. Although such persons can breathe and swallow, they soon die.

Thus, it is apparent that the cerebrum is more than just the center of intelligence. It seems that, from an evolutionary standpoint, more and more of the functions of the lower brain have been transferred to the cerebrum. In this progession, even when the neural center of a certain function is not shifted to the cerebrum, a certain element of its control may be transferred there.

Generally, the degree to which neural control has been shifted to the cerebrum is reflected in the size of this structure. In other words, more intelligent animals have larger cerebral components in their brains. Increased dependence on the cerebrum may also be reflected by factors other than increased cerebral size. For example, the cerebrum may be more convoluted in more intelligent species. The deep convolutions of the human brain are lacking in the brain of the rat. Even the convoluted cerebrum of the highly touted and undoubtedly intelligent dolphin has fewer layers than the human cerebrum.

Hemispheres and Lobes

The human cerebrum consists of two hemispheres (the left and the right) and each of these is divided into four lobes. At the back is the *occipital lobe*, which receives and analyzes visual information. If this lobe is injured, black "holes" appear in the part of the visual field that is registered in that area.

The *temporal lobe* is at the side of the brain. It roughly resembles the thumb on a boxing glove, and it is bounded anteriorly by the *fissure of Sylvius*. The temporal lobe shares in the processing of visual information, but its main function is auditory reception.

The *frontal lobe* is right where you would expect to find it—at the front of the cerebrum, just behind the forehead. This is the part that people hit with the heel of the palm when they suddenly remember what they forgot. One part of the frontal lobe is the center for the regulation of precise voluntary movement. Another part functions importantly in the use of language, so that damage to this part results in speech impairment.

The area at the very front of the frontal lobe is called the *prefrontal area*. Whereas it was once believed that this area was the seat of the intellect, it is now apparent that its principal function is sorting out information and ordering stimuli. In other words, it places information and stimuli into their proper context. The gentle touch of a mate and the sight of a hand protruding from the bathtub drain might both serve as stimuli, but the two stimuli would be sorted differently by the prefrontal area. A few years ago parts of the frontal lobe were surgically removed in efforts to bring the behavior of certain aberrant individuals more into line with the norm. Fortunately, this practice has been largely discontinued.

The *parietal lobe* lies directly behind the frontal lobe and is separated from it by the *fissure of Rolando*. This lobe contains the sensory areas for the skin receptors and also the areas that detect bodily position. Even if you can't see your feet right now, you have some idea of where they are, thanks to receptors in the parietal lobe. Damage to the parietal lobe may cause numbness, but it can also cause a person to perceive his or her own body as wildly distorted. In addition, the person may suffer from an inability to perceive spatial relationships in the environment.

By probing the brain with electrodes, it has been possible to determine exactly which area of the cerebrum is involved in the body's various sensory and motor activities. We can see the results of such mapping in the rather grotesque Figure 31.12. The pictures are distorted not out of any appreciation of the macabre, but to demonstrate that the area of the cerebrum devoted to each body part is dependent not on the size of that part, but on the importance it has come to have in the natural history of the human being. Thus, we have the greatest number of *senses* in the face, hands, and genitals, but the greatest amount of control only in the face and hands.

Also note that the sensory and motor areas are not randomly scattered through the cortex, but that there is some order in their appearance. Thus, the index-finger control area lies near the thumb control area, and the elbow area lies closer to the finger area than does the shoulder area.

One important fact about the cerebral hemispheres is that although they are roughly equal in potential, the two hemispheres are not identical. Moreover, any chemical or structural differences between them are accentuated by learning, because the control for certain learned patterns takes place primarily in only one hemisphere. One example of the difference between the hemispheres is seen in handedness. (It is interesting that less cerebrated animals such as rats and parrots also show handedness.) Other functions, such as speech, IQ, and perception, are also more likely to be located in one hemisphere than the other. The primary speech center of right-handed people is located in the left hemisphere. But in left-handed people, damage to the right temporal lobe may result in poor performance on IQ tests, but perceptual-test performance is relatively unaffected.

The primary route of communication between the right and left cerebral hemispheres is the *corpus callosum*. It seems that if one side of the brain learns something—for instance, by covering one eye while learning a visual task—the information will be transferred to the other hemisphere. However, if the corpus callosum has been cut, the material has to be taught anew to the other hemisphere by uncovering the other eye and repeating the training program.

It should be pointed out that a certain degree of compensation is possible between the two halves of the brain. For example, if one hand is injured, it is possible to learn to use the other hand. So, in spite of the natural handedness you were born with, retraining is possible as muscles controlling movement of the other hand develop, and neural events in the opposite side of the brain begin to permit the efficient use of that hand. It has also been demonstrated that if one hemisphere of a child's brain is injured, the other side will take over some of its tasks.

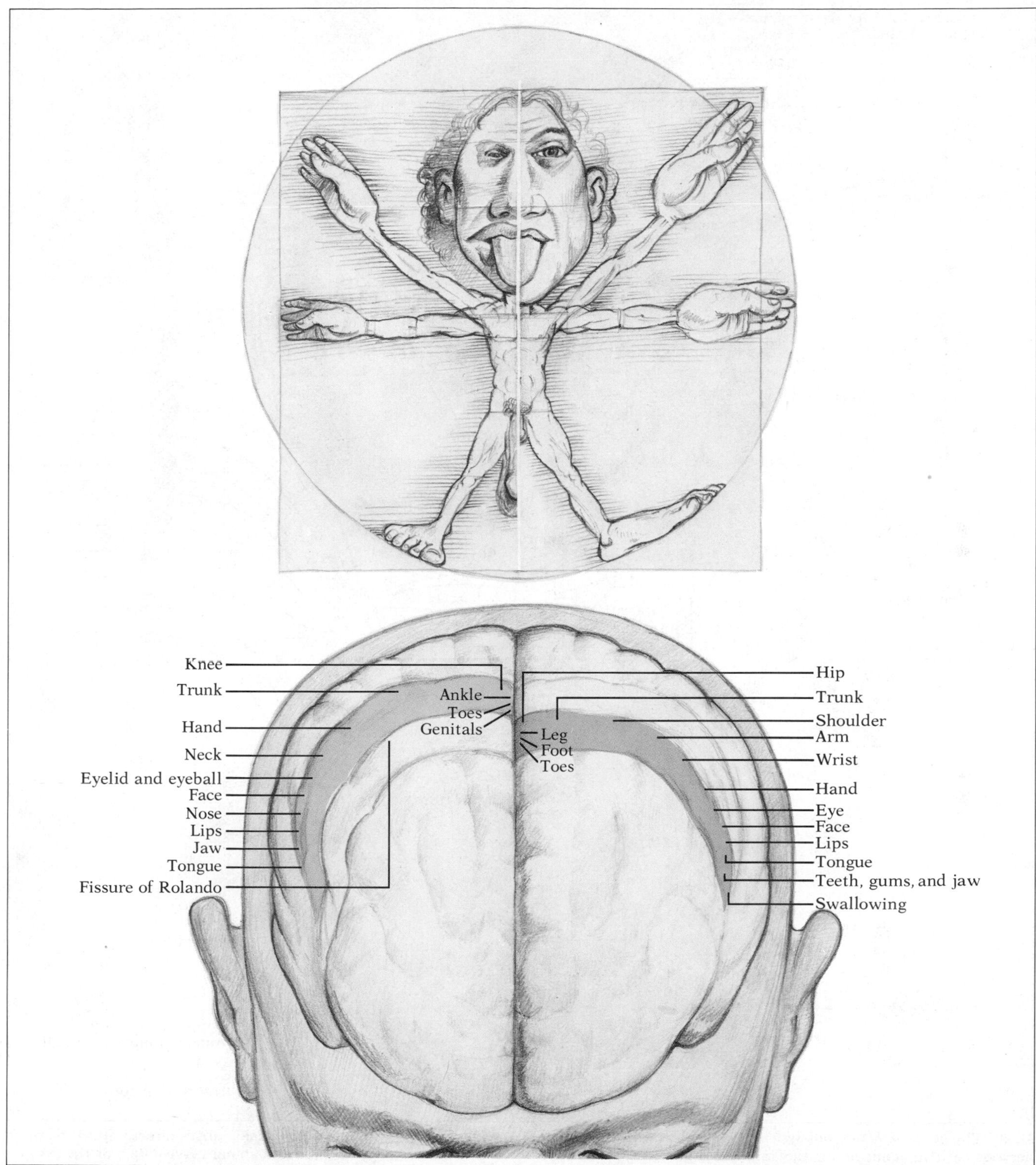

31.12 Function maps of the cerebral cortex showing sensory areas (left) and motor areas (right). The sensory section is taken posterior to the *fissure of Rolando,* the motor section anterior to it. Note that a large part of the human brain is devoted to face, hands, and genitalia. Also, the eyes and hands are given more motor than sensory space in the brain in spite of our great dependence on them for sensory information. Our lips and genitalia have a rather surprising amount of brain tissue allocated for their sensory input. Perhaps we have so much of the brain devoted to facial areas because the face is so important in the subtle processes of human communication.

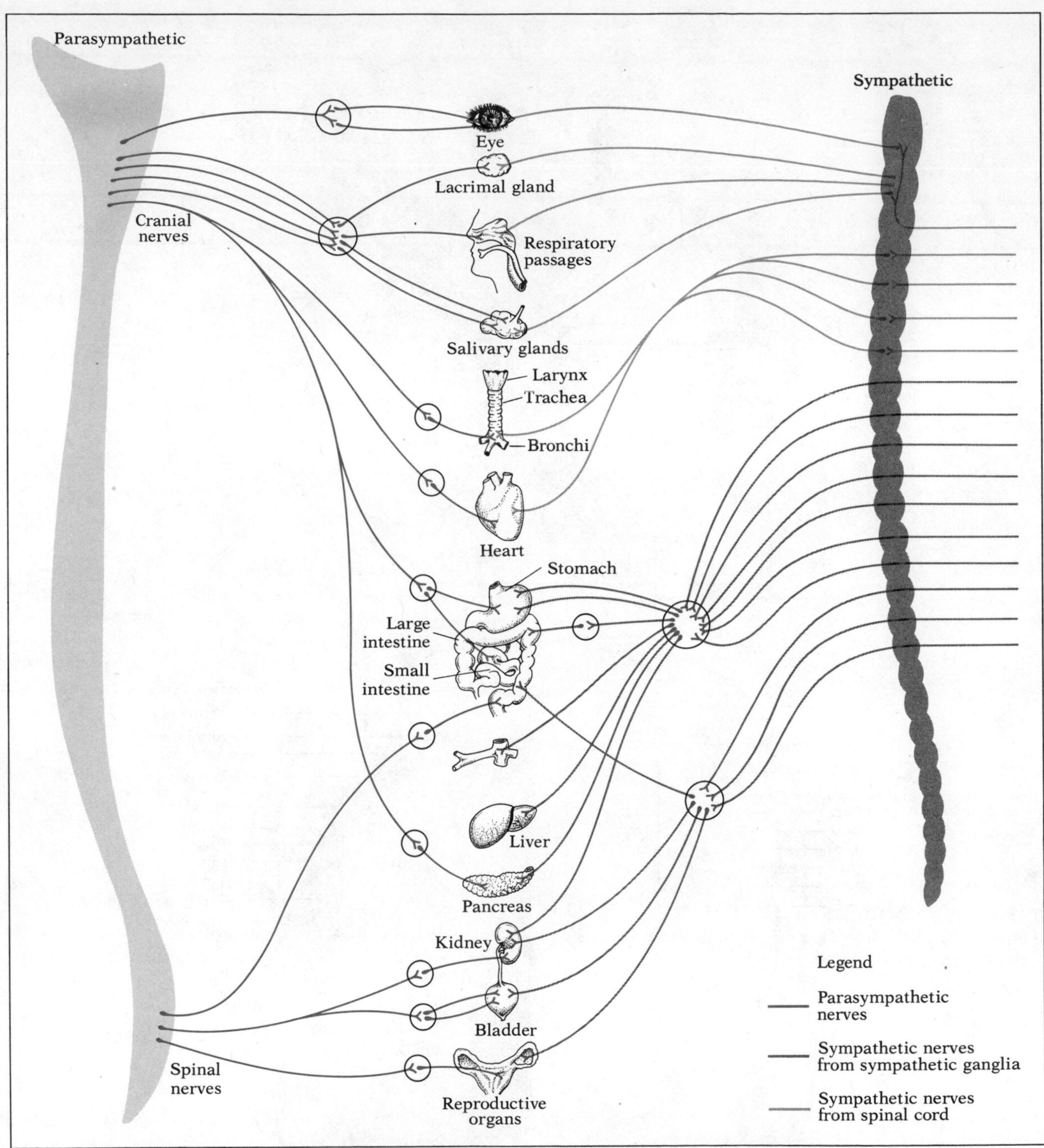

31.13 The *autonomic nervous system* is a branch of the peripheral nervous system, controlling the internal organs. It consists of two parts, the *sympathetic* division and the *parasympathetic* division. The first can generally be considered to speed up circulation and voluntary muscles; it acts primarily during stress. The second slows these processes down and is in control under normal conditions. The parasympathetic division arises directly from the brain and spinal cord as cranial and spinal nerves. Part of the sympathetic division arises from ganglia that run the length of the spinal cord. The remainder emerges directly from the spinal cord with their ganglia at some distance.

The Autonomic Nervous System

The *autonomic nervous system* is found only in the vertebrates. At its most complex, this system is composed of two divisions—the *sympathetic* and the *parasympathetic* systems—and these generally have opposing functions. For example, in humans, the sympathetic system functions in emergency situations. If a large, or even a medium-sized, bear rushes into the room where you are quietly reading, your heart rate will increase (carrying blood and oxygen to the running muscles), the pupils of your eyes will dilate (the better to see the bear), and blood will be shunted away from the digestive tract and peripheral vessels (going where it is needed and minimizing bleeding if the bear should beat you to the window). Actually, these are only a few of the changes that would be wrought by the sympathetic nervous system in response to the coming of the bear.

If the bear turns out to be only your roommate dressed in a bear suit, the parasympathetic system will take over, and, through an entirely different set of nerves, all the earlier responses will be reversed as your system returns to normal.

Autonomic systems of various complexities are found throughout the vertebrates. The relationship of the autonomic and central nervous systems is quite close. In fact, in some cases, autonomic centers may be located within the central nervous system. Also, commonly, autonomic fibers synapse with motor and sensory fibers going to and from the spinal cord. The cell bodies of sympathetic neurons are clustered in ganglia near the spinal cord, whereas the ganglia of the parasympathetic neurons are usually located near the organs they innervate (Figure 31.13). The interplay of opposing sympathetic and parasympathetic neurons results in bodily processes being kept within tight limits.

31.2 The Senses

Perceptual Systems

The question of perception is a fascinating one. If one were to ask a variety of animals in some afterlife what the earth was like, the answers would vary according to the animal that was asked. A fly might describe a world of swirling eddies of air, tasty surfaces, and shimmering mosaics of light. A dog might describe a gray world of thundering odors and constant sound. A tapeworm would speak of idyllic blackness and probably wouldn't have noticed much of anything. Humans would describe the world with great certainty and would talk at great length about colors and sounds, not saying too much about smell. The point is, we are capable of sampling only a small part of our world. It is far richer than we know, because much of it escapes us due to the limitations of our receptors. Let's consider receptors now, and see how it is that various species can exist side-by-side in different worlds.

Thermoreceptors

Most animals, whether cold- or warm-blooded, can only live within certain, rather restricted, temperature limits. Some smaller animals react to unfavorable heat conditions by simply dying or failing to reproduce, but many animals seek to regulate their temperature, either behaviorally or physiologically. It is especially important for these animals to be able to detect heat and change in temperature.

Heat detection is apparently a result of changes in the speed of temperature-dependent biochemical reactions. A number of types of receptors are temperature-dependent in their rate of firing, so most animals have never had to develop heat receptors per se.

Heat Detection in Invertebrates

Most invertebrates don't have thermoreceptors, but certain insects (such as locusts) have heat-sensitive cells scattered throughout their bodies. Cockroaches have heat receptors on their legs, which apparently help them to determine when they have found a good place to live—like your house.

Heat Detection in Vertebrates

Many vertebrates, on the other hand, do have specific heat receptors. Who has not heard of the deadly pit vipers? The pit, it turns out, is not where they live, but is an indentation between the eye and nostril. It is loaded with heat receptors and it tells the snake when it is facing a living thing that is generating metabolic heat (Figure 31.14).

There is some diagreement over whether mammals have specific receptors for detecting heat and cold, or whether free nerve endings register thermoreception. If the latter is true, it has been suggested that a single neuron can register both heat and cold. Is the feeling "cold" simply a variation of what we call "heat"? And how could a single neuron register both? One way is by firing in different patterns when stimulated by different temperatures. On the other hand, some physiologists believe that neurons that register warmth are apparently located deeper in the

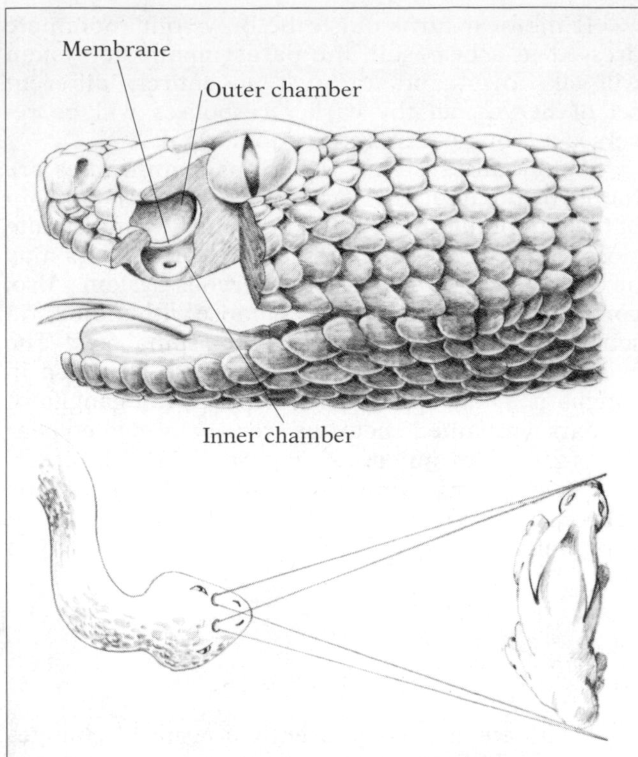

31.14 Heat sensors in the pit viper (rattlesnake, *Crotalus viridis*). Pit vipers detect their prey through a pair of heat-sensing devices located in the depressions near the eyes. Each pit consists of an outer chamber that ends in a thin membrane covering an inner chamber below. Extending over the membrane is a highly branched heat-sensing nerve. The recessed structure of the pits permits the snake to zero in on its prey, just as though it were using its eyes. By moving the head back and forth, it determines the exact location of the prey by the intensity of neural impulses from the membrane.

skin than are cold receptors. But there is another factor operating here. Whether the brain registers "warm" or "cold" also depends on the immediate previous experience of the receptors. If the skin has been very warm and touches something warm, but not as warm, the object will seem cool. If the skin has been cooler than the object, that object will register as warm.

Tactile Receptors

Tactile, or touch, receptors are extremely sensitive, fast-firing neurons that respond to any force which deforms or alters the shape of their cell membranes. In some cases, bristles or hairlike structures protrude from the surface of the cell membrane so that objects in the environment are touched by these dead *feelers* before the living cell reaches the objects.

Touch in Invertebrates

Bodies of invertebrates are often covered with tactile receptors from which bristles or hairs arise. Tactile sensitivity may be important for a variety of reasons. For example, web-building spiders have hairy legs that respond to vibrations set up by prey caught in the web. Cockroaches have tiny hairs protruding from their abdomen which, when stimulated by very light air currents, send them scurrying for safety. Many aquatic insects have bristles on their heads that are sensitive to water currents. R. S. Wilcox has accumulated a fascinating body of evidence that indicates how certain aquatic insects use water vibration in communicating with each other.

Touch sensitivity, in general, is common among a wide range of invertebrates and is usually employed in finding food, mating, and avoiding predators.

Touch in Vertebrates

Vertebrates have basically two kinds of tactile receptors. *Distance receptors,* such as the lateral line organs of some fish (Figure 31.16) are sensitive to water motion. These kinds of receptors can be used to detect predators or prey, and some fish use the information in complex ways to coordinate their schooling behavior.

And then there are *contact receptors.* These are of two types: pressure and touch. The pressure receptors are located deep in the skin and seem to utilize encapsulated nerve endings called *Pacinian corpuscles.* Light touch, on the other hand, is believed to be registered in *Meissner's corpuscles* near the surface of the skin. In most mammals, touch sensitivity is particularly great around the hairy areas, and genitals, as you may have noticed. The "whiskers" of many animals are especially sensitive to touch, but among primates, sensitivity to touch is greatest around the lips, eyes, and fingertips (Figure 31.17).

31.15 Touch receptors are commonly associated with the hair structures in spiders. When the hair is bent or moved, it activates a neuron which transmits its impulse to an associated ganglion for integration. Some sensory hairs are fine enough to be moved by air currents.

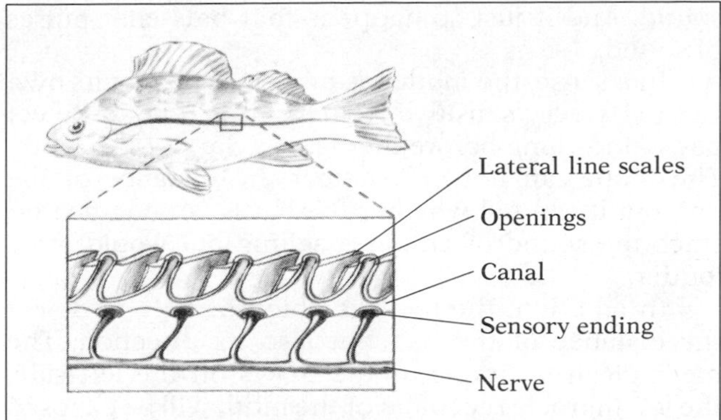

31.16 The lateral line organs of bony fishes and sharks are used to determine distant movements. As animals move through the water, they create disturbances that are picked up in the lateral line canal where sensory endings detect them, sending impulses to the cranial nerves and brain. The fish can then respond to this information. The details of one sensory ending are shown in the inset. The individual neural endings are enclosed in a gelatinous mass secreted by the neurons. Movement of water in the lateral line canal stimulates these neurons as the structure bends. Sharks have complex extensions of the lateral line organ around the head; these aid them in finding their way to disturbances that might indicate vulnerable prey.

Auditory Receptors

Audition, or hearing, is similar in some respects to lateral line reception in that both respond to distant stimuli. However, auditory receptors are sensitive to higher frequencies and they function both in air and water. You may have noticed how well you can hear sounds underwater. Beginning scuba divers may be surprised at how noisy their bubbles are.

Hearing in Invertebrates

Most invertebrates don't have specialized sound receptors, but many are sensitive to vibrations in the air, water, or soil in which they live. One notable exception among the invertebrates is the insects.

Insects have several kinds of sound receptors. Most species have sensory hairs that respond to low-frequency vibrations of air, but others have a specialized organ, located in a leg, that is sensitive to movements of whatever the insect is standing on. Yet others, including grasshoppers and crickets, have tympanal organs that respond to high-frequency vibrations, much like the human eardrum. The study of insect hearing has provided us with some good stories about scientific detective work, and one of the best ones is about the coadaptation of moth and bat hearing.

Noctuid moths are a favorite prey of certain bats. These bats fly swiftly and can turn on a dime to capture their prey on the wing. The bats locate the prey by emitting high-frequency sounds that bounce back from any structure in the environment. These echos provide the bat with information on the flying insect, and the bat simply intercepts the hapless moth. But just as bats have evolved ways of catch-

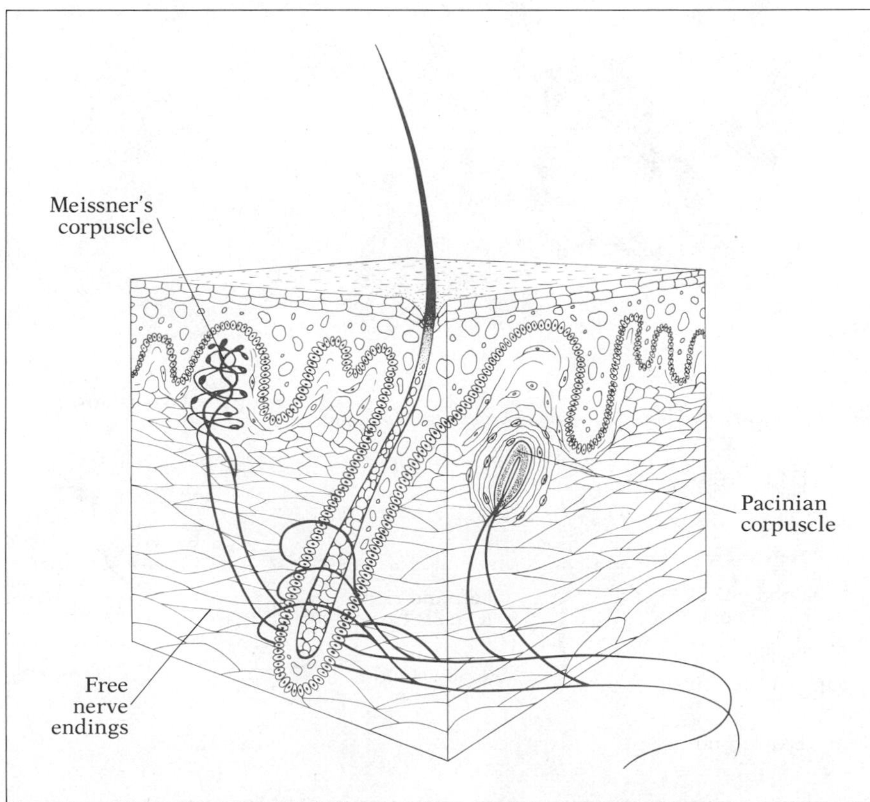

31.17 The specialized touch receptors are found in the skin of humans. *Meissner's corpuscles* are located close to the surface and register light touch. These are most numerous in the fingertips and around the lips. *Pacinian corpuscles* are in deeper skin locations and their complex end bulbs register pressure. Generalized sensory neurons (no distinct structures) surround the hair follicles of mammals and are stimulated by hair movement. (Try to move a single hair on your forearm without feeling it.)

ing moths, noctuid moths have developed ways of avoiding bats. When they hear the cry of a bat, they take evasive measures. As you might suspect when dealing with bats, this is easier said than done. It turns out, though, that the moth does it with a very simple hearing apparatus.

Kenneth Roeder found that noctuid moths have two *tympana,* one of either side of the thorax (the insect midsection), and that each tympanum has only two receptors. One, called the *A1 cell,* is sensitive to low-intensity sounds. The other, the *A2 cell,* responds only to loud sounds. Surprisingly, neither kind of receptor is very good at distinguishing frequencies (high versus low notes)—a sound of 20,000 Hertz (cycles per second) elicits the same neural action potential as one of 40,000 Hertz (a much higher sound).

As any sound becomes louder, however, the A1 cell fires more frequently and with a shorter lag time between receiving the stimulus and firing. Also, the A1 cell shows a greater firing frequency in response to pulses of sound than to continuous sounds; it fires increasingly slower if submitted to a continuous sound. And it just so happens that bats emit pulses of sound.

In a sense, the moth has beaten the bat at its own game. Its very sensitive A1 neuron is able to detect bat sounds long before the bat is aware of the moth. The moth can not only detect the distance of the bat, but it can tell whether the bat is coming nearer, since the sound of an approaching bat would grow louder.

In addition, the moth is able not only to detect the distance of the bat, but also its direction. The mechanism is simple. If the bat is on the left side, the left thoracic receptors of the moth will be exposed to the sounds while the receptors on the right will be shielded (Figure 31.18). Therefore, the left receptor fires sooner and more frequently than the right if the bat is on the left. If the bat is directly behind, both neurons will fire simultaneously. Thus, the moth can determine the distance and direction of the bat. But what about its altitude?

If the bat is above the moth, the bat sounds will be deflected by the upward beat of the moth's wings. But if the bat is beneath the moth, the wing beats

31.18 The relationship of sound pulses from a hunting bat and auditory neural firing in the hunted moth. (a) When a hunting bat, emitting its high-pitched sounds, approaches a noctuid moth from the side, the receptors on that side fire slightly sooner and more rapidly than those on the shielded side. (b) When the bat is behind the moth, the moth's receptors on both sides fire with a similar rapid pattern. (c) When the bat is above the moth, the moth's auditory receptors fire when its wings are up, but not when its wings cover the receptors on the down stroke.

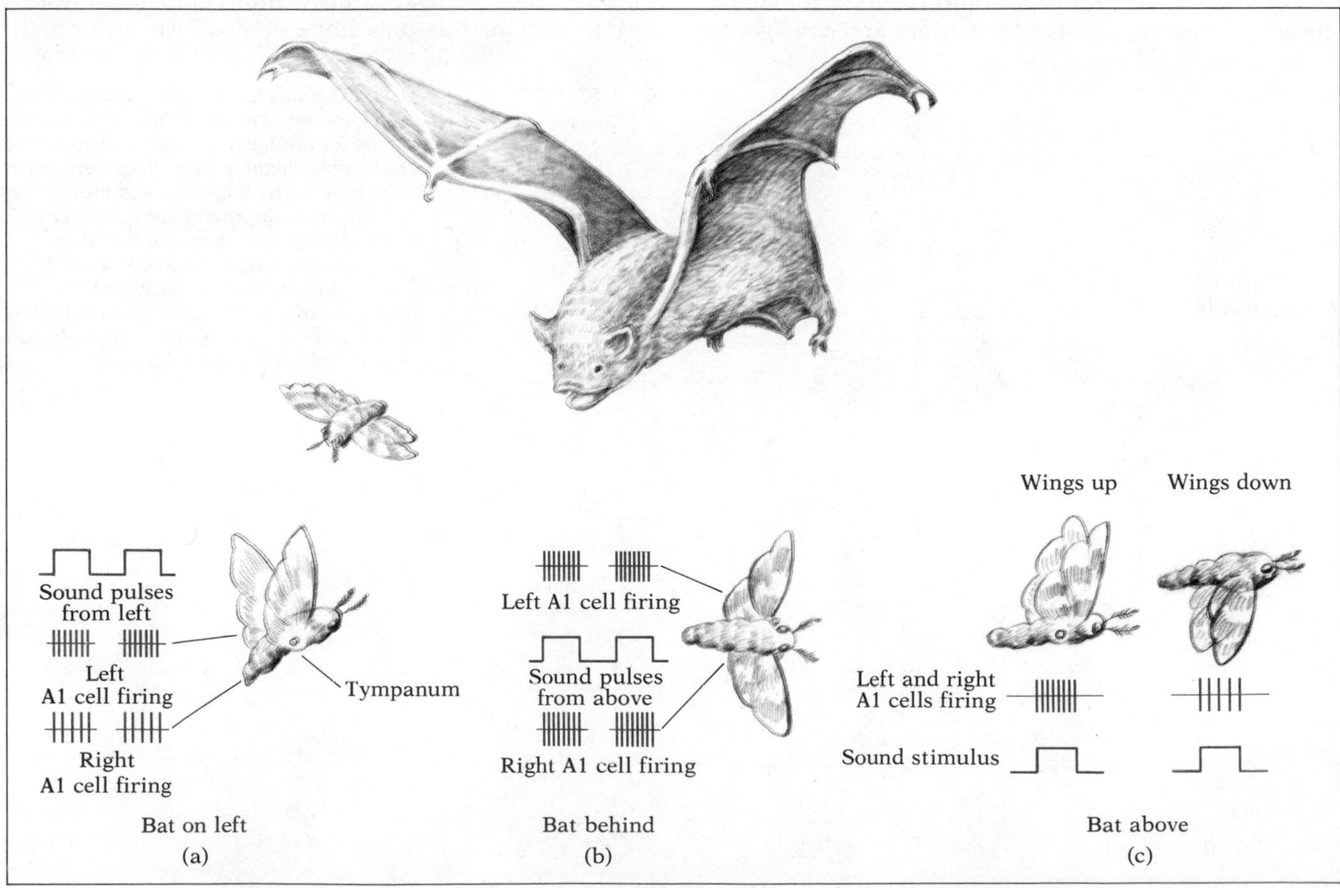

will have no effect on the pattern of neural firing. The moth, then, decodes the incoming data, probably in its thoracic ganglion (from which the auditory neurons emerge) so that it pretty well has the bat pinpointed.

But what does it do with this information? If the bat is some distance away, the moth simply turns and flies in the opposite direction, thus decreasing the likelihood of ever being detected. The moth probably turns until the A1 cell firing from each ear is equalized. When the bat changes direction, so does the moth.

Bats fly faster than moths, though, and if a bat should draw to within 2.5 m (8 ft) of the moth, the moth's number is up—at least if it tries to outrun the bat. So it doesn't. If the bat and moth are on a collision course, that is, if the moth is about to be caught, the sounds of the onrushing bat will become very loud. At this point, the A2 fiber begins to fire—the signal of imminent danger. These messages are relayed to the moth's brain, which then apparently shuts off the thoracic ganglion that had been coordinating the antidetection behavior. Now the jig is up and the moth changes tactics. Its wings begin to beat in peculiar, irregular patterns or not at all. The insect itself probably has no way of knowing where it is going as it begins a series of unpredictable loops, rolls, and dives. But it is also very difficult for the bat to plot a course to intercept the moth. Hopefully, the erratic course will take the moth to the ground, where it will be safe since the echos of the earth will mask its own echos.

The noctuid moth's evolutionary response to the hunting behavior of the bat serves as a beautiful example of the adaptive response of one organism to another. Also, it shows clearly that the sensory apparatus of any animal is not likely to respond to elements that are irrelevant to its well-being. It is not important for moths to be able to distinguish frequencies of sound, but it is important that they are sensitive to differences in sound volume. Anyone who tried to train a moth to respond to different sound frequencies could only conclude that moths are untrainable.

Hearing in Vertebrates

The sound receptors of most species of vertebrates are located in the *inner ear.* There is a lot of variation in how the vibrations reach the inner ear, however. Among many fishes, sound vibrations in water are conducted directly to the inner ear by vibrating water, but in some fishes (such as the minnows, catfishes, and suckers), a set of tiny bones—the *Weberian ossicles*—connect the swim bladder to the inner ear. Sound vibrates the air-filled bladder, which moves the bones and stimulates the receptors in the inner ear. The fishes that have such an apparatus can detect a much wider frequency of sound than those that lack it.

Land vertebrates have an *auditory canal* and one to three *middle ear* bones that transmit vibrations of the *tympanic membrane* (ear drum) to the inner ear. The tympanic vibrations usually stimulate auditory receptors which then carry the impulses to the brain.

Birds have a much more sensitive auditory system than lower vertebrates. Not only can birds hear well, but some can locate the source of a sound with astounding success. For example, the barn owl (*Tyto alba*) is able to pinpoint very small sounds. Apparently, it is able to do this because its ears are relatively far apart and the owl simply notes the difference in time it takes sound to reach the two ears. This would mean that the source of any sound reaching both ears simultaneously is located in the body's midplane.

In mammals, the auditory apparatus differs in three basic ways from that of other animals. First, there is an *external ear,* which most mammals are able to move so as to maximize sound input and locate its source. (Humans have largely lost this ability but those who *can* move their ears are in great demand for social events.) Second, the middle ear has three bones (as opposed to only one in other vertebrates). Third, the movement of the middle ear bones acting against the *oval window* sets up movements of fluids in a long *cochlea* that is coiled in the manner of a snail shell. The cochlea of other vertebrates is straight or looped. Frequency discrimination in the cochlea is possible by neurons attached to *hair cells* that rest on a *basilar membrane.* The sensory "hairs" of the hair cells are actually modified cilia that push against the relatively rigid *tectorial membrane* (Figure 31.19). The basilar membrane is relatively thick and stiff at its base near the tympanum, but at the other end the membrane is thin and flexible. This thinner end lies in what is the center of the coiled cochlea. High-pitched tones stimulate the hairs at the base of the membrane; low-pitched tones are registered at the apex. Mammals apparently detect loudness by the number of neurons stimulated and the number of impulses generated in each of these neurons.

Visual Receptors

Humans are highly visual animals and hence we seem to have a special appreciation for visual acuity in other animals. Some of those other animals see well indeed by our standards, some see poorly, and some see a world that would be alien to us.

Light receptors are sensitive to a particular part of the spectrum of electromagnetic energy—the part we call light. Its wavelength ranges from about 430 to 750 nm, but no animal can see more than a part of this range. Shorter wavelengths, such as X-rays, beta rays, and gamma rays, can't be detected by

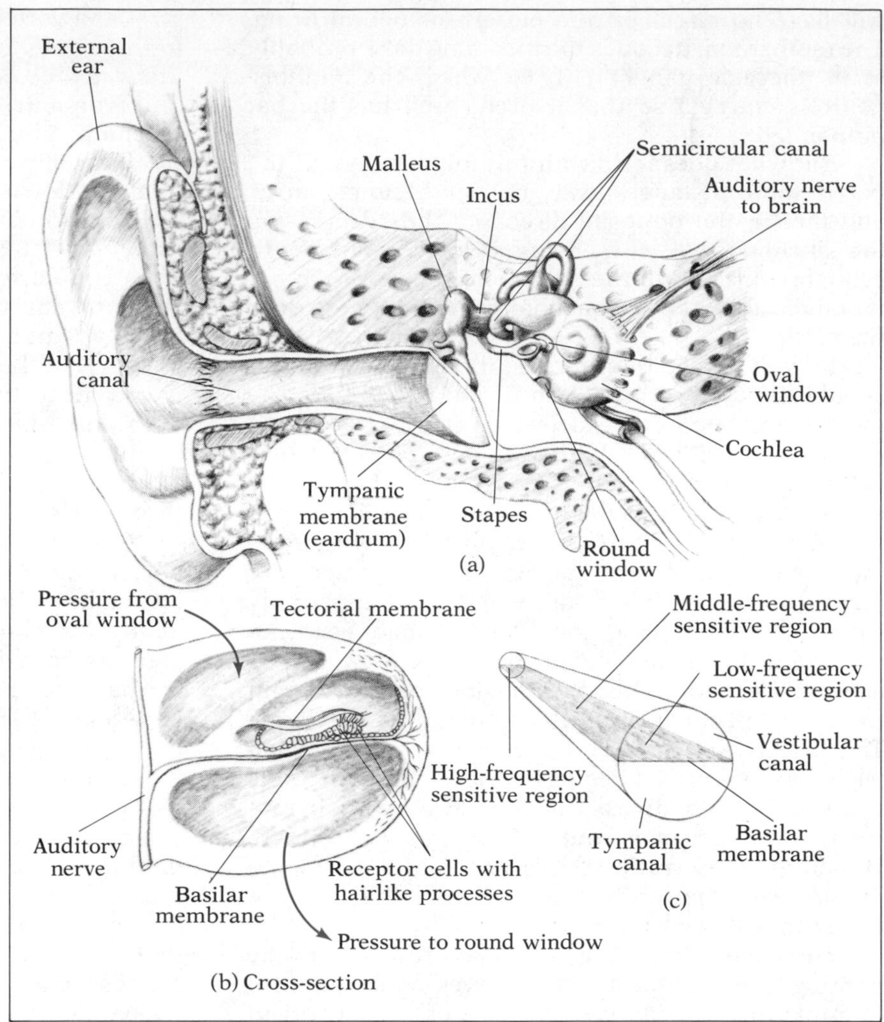

31.19 (a) When sound waves vibrate the human eardrum (tympanic membrane) it sets in motion three tiny leverlike bones, the *malleus, incus,* and *stapes.* The stapes is attached to an inner membrane of the oval window. Movement of the oval window sets fluids in motion within the snail-shaped cochlea. (b) The cochlea is divided by the basilar membrane. Sound waves move the *basilar membrane* and move its receptor cells, pushing their sensory hairs against the more rigid *tectorial membrane.* The distortion causes action potentials in the *auditory nerve.* (c) Here, a section of the cochlea is shown straightened out, showing the areas of sensitivity of various frequencies. Those near the oval window register the lowest sounds. Waves set in motion in the cochlea eventually reach the round window which is stretched outward, relieving any pressure. The looped structures at the base of the cochlea are the semicircular canals, which function in maintaining equilibrium.

any animal. Neither can the very long ones, such as radio waves.

Visual receptors contain *pigments* (colored substances), which absorb certain kinds of electromagnetic energy and transform that energy into a neural stimulus. In the light-sensitive organs of all animals—and also those of many light-sensitive algal protists—the initial light receptor is a visual pigment based on a specific carotenoid molecule, *retinal.* No animal can make retinal, however; it must be obtained from plant sources. We get ours in a slightly modified form as *vitamin A.* Retinal has a carbon ring and a number of light-absorbing conjugated double bonds. It exists in two forms, depending on the orientation of one of the internal double bonds: *cis-retinal* is the somewhat less stable, reactive form of the molecule and *trans-retinal* is the more stable, nonreactive form. Neither form of retinal is very efficient at absorbing light by itself. The actual light-sensitive pigment consists of one molecule of *cis*-retinal conjugated with a protein carrier called an *opsin.* Whereas the retinal is always the same, the opsin protein varies from species to species and even within a single animal.

In vertebrates the principal visual pigment is *visual purple* (it really is purple). Visual purple is present in large quantities in stacks of membranous disks in the distal segments of the *rods* of the vertebrate retina. (*Rods* and *cones* are specialized sensory neurons, and their light-sensitive parts are highly modified sensory cilia—see Figure 31.23). When visual purple absorbs a photon of light, the *cis*-retinal is transformed to the *trans-* form and dissociates from its opsin. The dissociated pigment loses its purple color and is said to be "bleached." The bleaching of visual purple alters the ion balance and membrane permeability of the rod in some way, either causing depolarization or hyperpolarization, depending on the type of rod. This in turn either causes or inhibits the propagation of an action potential and a nerve impulse. The *trans*-retinal enters the bloodstream and is reconverted into the reactive *cis-* form in the liver, in an enzyme-mediated, energy-requiring reaction. The *cis*-retinal then returns to the eye.

Rods are sensitive to all wavelengths of visible light, which, of course, is what makes visible light visible. The cones, however, respond only to specific colors. The visual pigments of the cones are based on the same retinal molecule, but different opsin proteins alter the absorption properties of the conjugated pigments. There are three kinds of cones in the eyes of humans and other primates: red-sensitive, green-sensitive, and blue-sensitive. The ranges of sensitivity overlap, and our perceptions of a multitude of colors derive from the balance of stimulation of the three populations of cones.

What about color? Many species are able to see color, that is, to discriminate between different wavelengths of light. The ability has developed in a wide range of animals including insects, fishes, reptiles, birds, and mammals, apparently having evolved several times.

Seeing in Invertebrates

The planarian flatworm has two eyespots that are shaded on opposite sides; thus, it has an improved directional system. It can tell which direction light is coming from according to which eye is being stimulated. Such ability is, of course, important in dark-seeking animals (Figure 31.20).

Among invertebrates, some of the sharpest-eyed creatures are the cephalopod mollusks such as the octopus and squid. The octopus has what seems, at first, to be a peculiar interest in the ends of what it sees. It is particularly attracted to the pointed end of any object, since, in the water, this is usually the head end and thus indicates which way the animal is likely to go. The octopus may not have learned to make this association; instead it may possess a nervous system that selectively filters incoming stimuli so that certain shapes can elicit particular behavioral patterns, as we will discuss later.

31.20 The two eyespots of planaria enable the animal to determine the direction of light. It will usually turn toward darkness. Light coming from the side is shielded from one photosensitive area, but is able to reach the other, stimulating the worm to behave appropriately. Here, we show one eyespot in detail. Light-sensitive cells are present in a cup which is pigmented on one side. The neurons of the eyespot join to form a simple optic nerve.

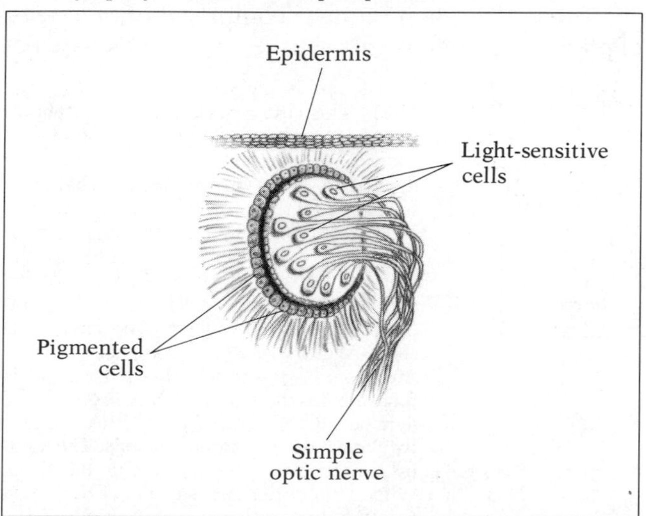

Arthropods, such as spiders, crayfish, and insects, have very good vision. Spiders, in fact, have eight eyes (two rows of four each), but most of these lack enough photoreceptors to form clear images. However, the jumping spiders (Salticidae) have a relatively large number of photoreceptors in their eyes. This is undoubtedly beneficial, since they leap from a distance onto their prey and it wouldn't do for a weak-eyed spider to leap onto a hungry bird.

Crustacea, such as crayfish, have two kinds of eyes: *median eyes* and *compound eyes*. The median eyes are simpler, lacking lenses, and are usually found only in the larval stages. The adult's compound eyes may rest on stalks and they are the first eyes we've considered that are especially sensitive to movement. Actually, the compound eyes are composed of numerous *ommatidia* (see Figure 22.33). The ommatidia can't move to follow an image, so each stares blankly in its own direction until something moves into its visual field. Then it fires an impulse to the central nervous system. If something moves across the animal's field of vision, each ommatidium is stimulated in turn, so even very slight movements are detected. The convex surface of such eyes may give a visual field of 180°, but the crayfish probably perceives the world as some sort of shimmering mosaic.

The praying mantis is a relentless killer of insects. Its vision is keen, partly because it has two types of ommatidia in its compound eyes. The more central ones are in opaque sleeves of pigment so that impulses can't "spark across." Thus, the stimulation of one can't trigger others. This means the praying mantis can detect such slight movements as an insect slowing crawling on the wall. Even the slowest movement will cause new ommatidia to fire, with no confusing effects caused by diffusion of stimuli (Figure 31.21). But since threshold intensity for each individual ommatidium must be reached in order for it to fire, these receptors function best in good light. The more peripheral ommatidia, on the other hand, have transparent sleeves, so light can pass from one ommatidium to another. This means that these areas yield fuzzy outlines, but they are highly effective in their own way because light can filter through one receptor to the next and dance about the thresholds of several receptors simultaneously.

Honeybees (*Apis*) see remarkably well, being particularly sensitive to form and color. Interestingly, certain forms are more preferred by bees than others. For example, bees seem to prefer images with many parts. In other words, a form that is strongly sub-

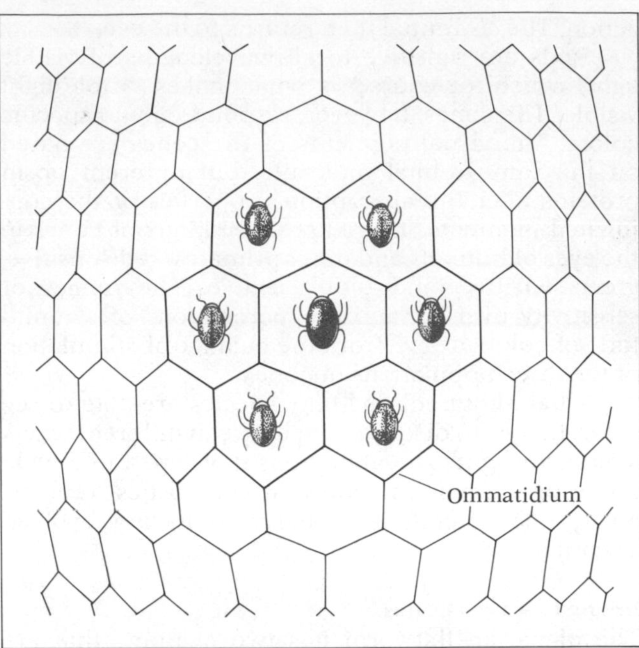

31.21 Compound eyes are common in insects as well as other arthropods. The eye consists of a large number of units known as *ommatidia*. Each ommatidium fires its impulse separately when an object crosses its field.

divided or has an uneven outline is particularly attractive. This may be because such forms produce numerous changes in the firing of the ommatidia and it is this flickering effect that attracts the bee. Flowers, of course, are subdivided and usually have uneven outlines.

In an experiment by the 1973 Nobelist Karl von Frisch, the color vision of humans and bees was compared. It was found that whereas we can't see ultraviolet light, bees can. In fact, it is the ultraviolet color of some flowers that attacts bees—flowers that appear entirely white to us. On the other hand, bees are blind to red light, which we see easily. You shouldn't be surprised to learn, then, that red flowers are not usually pollinated by bees, but by wind, birds, or moths. Thus, the color of some flowers and the sensory abilities of bees have evolved together to the mutual advantage of each kind of organism.

Seeing in Vertebrates

The vertebrate eye (Figure 31.22) is unexpectedly similar to that of cephalopods, such as the octopus. In each, light entering through the *pupil* is focused by a *lens* onto a sensitive layer of pigment at the back of the eye (the *retina*). The major difference between the two kinds of eyes is that their retinal layers face in opposite directions. In mammals, the photoreceptors lie separated from the lens by two layers of neurons, while in cephalopods, the light entering the eye falls directly on the photoreceptors.

The mammalian retina is comprised of two types of photoreceptors: highly sensitive *rods*, which detect

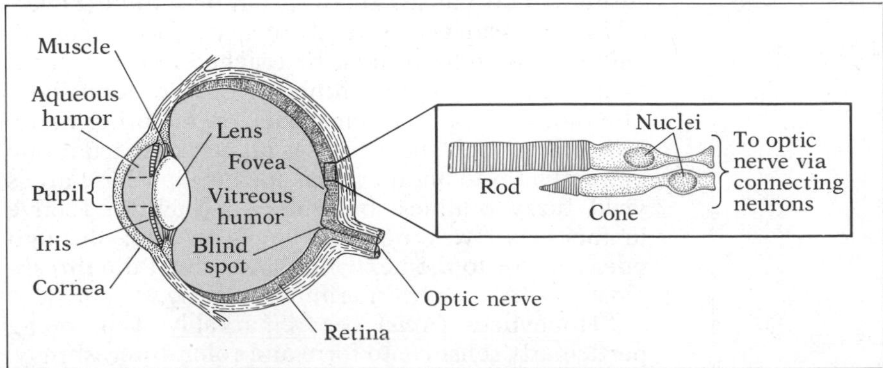

31.22 This cutaway view of the vertebrate (human) eye shows that it is a multilayered sphere. The open regions of the eyeball are fluid-filled, with the two main chambers separated by the *lens*. The outer chamber is enclosed by the tough, transparent cornea. Light entering the eye is regulated by the *iris*, which can change the diameter of the *pupil*. The transparent lens is pliable, and the muscles of the ciliary body change its shape slightly to compensate for distance. After passing through the lens, light is focused on the *retina*, which consists of light-sensitive *rods* and *cones* buried under two layers of associated neurons. The highest concentration of neurons is in the *fovea*. When stimulated, the neurons of the retina create impulses that are carried out of the eye, via the *optic nerve*, to the visual centers of the brain.

light, and *cones*, which are sensitive to color (Figure 31.23). The cones are independently innervated and much more sensitive than rods to fine detail in bright light. The rods tend to be interconnected in clusters, which increases their sensitivity to dim light, but lowers their sensitivity to detail. Since rods cannot distinguish color, "all cats are gray at night."

As you sit reading this fascinating account, out of the corner of your eye you might be distracted by a slight movement. You look up; it's only a fly walking along the wall. It's a good trick, but you've seen it before, and eagerly return to your text. You noticed the fly because the vertebrate eye is extremely sensitive to movement, although the movement-detecting mechanism is different from that employed by arthropods. The mammalian eye is sensitive to movement because of the interaction of three types of neurons in the optic nerve. Some neurons continue to fire as long as they are stimulated by light ("on" fibers). But others respond only when incoming light stops, or is turned off ("off" fibers). The third type fires either when a light appears or when it stops ("on–off" fibers). Remember that we're speaking of light in its broadest sense; light carries information regarding color, texture, and form in the vertebrate visual system. Because of a preponderance of "off" and "on–off" fibers in their eyes, vertebrates are highly sensitive to anything moving across their field of vision.

31.23 This scanning electron micrograph shows the structure of the rods (thinner bodies) and cones (thicker bodies) of the retina. Actually, the rods and cones constitute the deepest layer of the retina, furthest from the light. Rods and cones don't transmit their impulses directly to the optic nerve, but these impulses move through complex synaptic junctions in groups of neurons known as *bipolar* and *ganglion cells.*

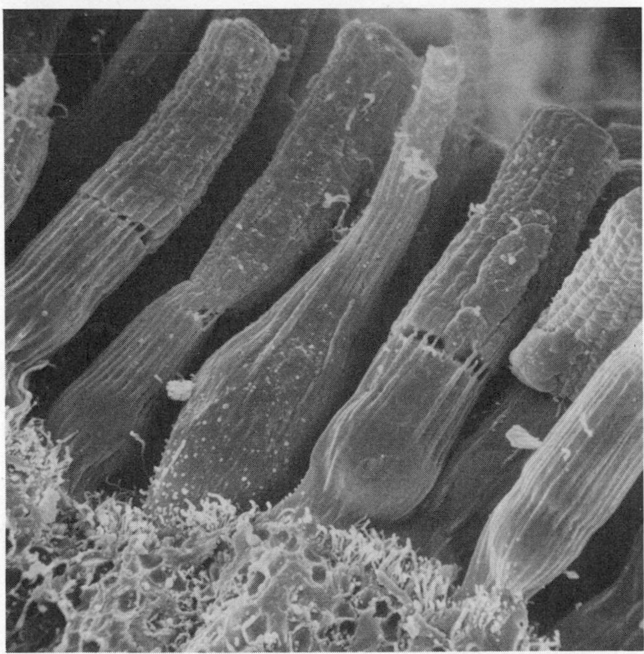

In most vertebrates and some invertebrates, there is a "visual alarm" reaction to anything that rapidly increases in size. If a small circle projected onto a wall should suddenly grow larger, very young and inexperienced animals will duck or try to run. There is obviously some mechanism in their perceptual system that triggers an evasive response to certain cues. Probably, through the long periods of the animals' evolution, rapidly enlarging objects would have meant that something was after them—and gaining. The response, then, could have become genetically "programmed" so that each animal didn't have to learn the proper response the hard way.

Many vertebrates also have an inborn ability to perceive depth. For example, many young land vertebrates will not cross a *visual cliff* (Figure 31.24), although they have had no experience with real cliffs. Apparently, the stimulus that triggers the behavioral response is associated with *parallax,* or the difference in apparent movement of nearby things as compared to those farther away. As the head is moved from side-to-side, nearer objects seem to shift more and it is this relative shifting of things in the field of vision that produces the perception of depth. The central nervous system seems to react to the parallax cue and sends out impulses to the animal to "stop" or "jump" or "back up."

Terrestrial life has presented vertebrate eyes with special problems. Amphibians spend part of their lives in the water and part on land, but their eyes are adapted to terrestrial environments. To protect their eyes from the drying atmosphere, they cover them with lids while special glands secrete moistening liquids. Of course, conservation of moisture is a problem with almost all terrestrial life, and the amphibian method of keeping the eyes wet isn't unusual among vertebrates. But the amphibian eye *is* unusual in some respects. For example, the frog is able to distort the shape of its eyeball and thus to adjust the distance from the lens to the retina. This ability not only aids in focusing, but in the perception of distance. Frogs, of course, are among those animals that are not able to judge distance by focusing their two eyes on the same object, since their eyes stare off in different directions. (Animals that can "fix" an object with both eyes (*stereoscopically*) apparently can extrapolate distance from the angle thus formed.) Since some frogs eat a lot of small insects (although some eat snakes or other frogs), their retina is especially adapted to detect movement.

Frogs, like humans, have two eyelids for each eye, but frogs, birds, most reptiles, and some mammals also have a third one, the *nictitating membrane,* which further moistens and protects the eye. In addition, many lizards and turtles have well-developed *tear glands.* They are able to move their eyes in their sockets more easily than the amphibians and can

31.24 The visual cliff. The infant need not have learned the dangers of cliffs, but it avoids them even when the cliff is harmlessly covered with glass. As it crawls out on the glass it suddenly notes the parallax—that is, part of the ground has less apparent motion as the infant moves. The infant has never seen a cliff before, but something about the cliff registers as danger and it may stop and try to back up or turn around.

adjust the shape of the lens instead of distorting the entire eyeball, as frogs do.

Snakes are also reptiles, but you can't make a snake blink, because they have no eyelids. Instead, their eyes are protected by transparent skin over the eyes. They have no tear glands and their eyes are relatively fixed in their sockets, and thus they stare in what humans take to be a most sinister fashion.

Birds have well-developed vision, although their eyes are structurally similar to those of reptiles. Their eyes are relatively large and are protected by nictitating membranes which they close during flight. High-flying birds are well-known for their visual acuity.

Not surprisingly, night-hunting birds have more rods than cones in their retinas, and day-flying birds have more color-registering cones. In addition, in some birds (such as hawks), the retina bears a horizontal streaklike fovea, which is particularly heavily laden with cones. This means that in order for a hawk to take a really good look at something it has noticed on the ground, it must tilt its head so that the image falls on the streak. However, the arrangement of the streak is ideal for other visual pursuits. For example, because of the placement of the streak, a hawk is able to scan the horizon carefully as it flies along.

Birds can see color; else why would the robin's breast be red? Color vision in birds seem to be facilitated by colored oil droplets within the receptors, and it turns out that each kind of receptor has its own color of oil. The result is that colored objects are strongly registered and sharply contrasted. Some receptors even have red droplets that act as haze filters, cutting out some of the wavelengths that produce glare. We find the greatest number of cells bearing red oil in early birds—those species that are most likely to be up and about when it is hazy.

Among mammals, vision is particularly good in primates and carnivores. As would be expected, nocturnal mammals have eyes that are highly sensitive to light so that they are able to discriminate vague forms in the darkened world in which they live. However, animals that are more active during the day have the sharpest vision, since they are more dependent upon their eyesight. Mammals, as most vertebrates, are very sensitive to movement, but, in some cases, only movements in a certain direction elicit neural impulses. For example, rabbits, with eyes located on the sides of their head, have certain ganglia that respond if an object is moving from the back toward the front, but not in the reverse direction. Thus, they are hard to creep up on from behind. Also, rabbits, have special ganglia in their retinas that respond only if a small or distant body moves into the receptor field and stops (as would a stealthy hunting predator), or if such a body moves out of the field of vision. Grazing rabbits often stop and raise their heads and remain perfectly still for a few moments. Thus, the rabbit's visual images remain stable and most cells stop firing. However, should something in the rabbit's visual field move slowly and then stop, the "predator-finding" ganglion will fire and the sneaky lynx will have been discovered while it is still a safe distance away.

Primates are unusual among mammals in that we can see color. It has been suggested that this is one way chimpanzees, for example, can distinguish ripe red or yellow fruit from unripe fruit. It might be added that color vision in primordial humans may have enabled them to make the switch from being hunter–gatherers to formidable and effective hunters. Ethologist George Schaller, in his study of the African lion, found that during the fawning season he could rather easily find newborn antelope on the African plains. The defense of the young antelope was to freeze (remain motionless). The effort is useless against apes or humans but rather effective against color-blind lions looking for movement or contrast. (Also, young animals often have very little scent.) Animals that are preyed upon by predators with color vision (such as birds) are often cryptically colored to enhance their camouflage (Figure 31.25).

31.25 Cryptic coloration. Animals that are preyed upon by predators with color vision must not only be cryptically marked but camouflaged by color as well. This toad (*Megophyrus nasuta*) is difficult to see among the dead leaves of its habitat.

The color- and detail-sensitive cones of the primate eye are especially concentrated in the circular *fovea*, or "center of vision." When you look straight at something, you are focusing its image on your fovea. To see a dim star, on the other hand, you may have to look slightly away, since the fovea is correspondingly poor in rods.

Chemoreceptors

Basically, there are three different kinds of chemoreception. The simplest type involves a general sensitivity of relatively unspecialized receptors to certain chemicals. Taste, or *gustation*, is a more specialized ability. It operates best when the receptors are in direct contact with the source of the chemical, a requirement which is not demanded of the third, and most sensitive type of chemoreception—smell, or *olfaction*. Olfactory neurons can react to very low concentrations of chemicals in air or water, and hence they can respond to molecules that have diffused or been swept from the stimulus source. Therefore, the source can be detected at a distance.

Chemical Detection in Invertebrates

Some invertebrates get along with no really specialized powers of chemical detection while others have the greatest powers of chemical detection yet discovered. Planarians, for example, have only generalized receptors located in pits on their bodies over which they move water with their beating cilia. At the other extreme, certain insects have taste and smell receptors of astounding sensitivity. Among insects, taste receptors may be found in the mouthparts, antennae, or forelegs—and, in some cases, the ovipositor (since many insects lay their eggs only on certain plants). However, insect olfactory receptors generally lie at the back of the minute trachea that branch throughout the insect body. As a molecule of a chemical diffuses in through the tubule, it becomes dissolved in the insect's body fluids which bathe the receptor.

Among the incredible stories of insect olfactory abilities we find the tale of the silk moth (*Bombyx*). The silk moth smells with its antennae and adult males have enormous antennae with thousands of sensory hairs (Figure 31.26). About 70% of the adult male receptors respond to only one complex molecule called *bombykol* and, astoundingly, only one molecule is necessary to generate an impulse. (A human receptor would need to contact several million of these molecules in order to detect anything at all.) A female in reproductive condition needs only to sit on a twig and release bombykol in order to attract a male. As the molecules drift downwind to finally touch

31.26 The antennae of the male *Bombyx* moth are remarkably large, complex, and sensitive. These devices are able to detect only a single molecule of sex attractant released by a receptive female. Each large bristle of the antenna has numerous hairs (*sensilla*) extending outward. Within each sensillum is a fluid-filled cavity containing sensory neurons. Chemical substances in the air enter these cavities through pores in the cuticle, stimulating the dendritic endings. Impulses thus created are transmitted to the brain.

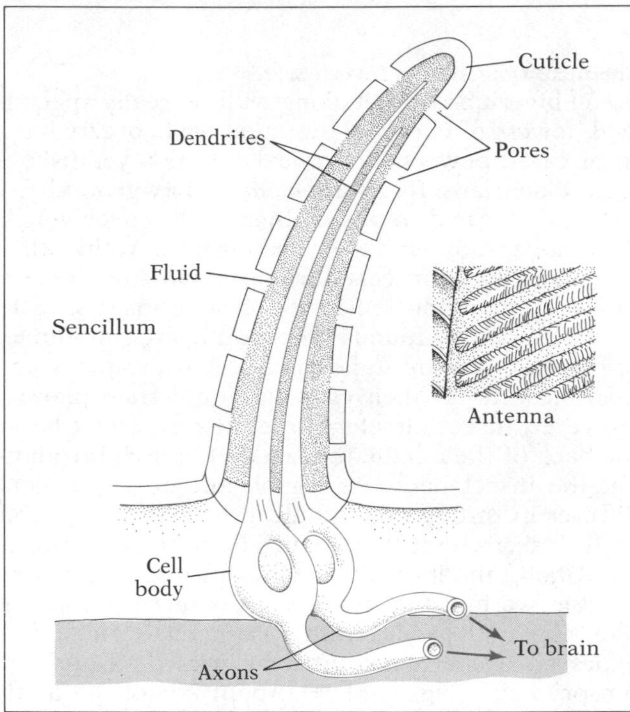

the antennae of a waiting male, they enter the tiny pores of a "hair" (*sensillum*) that houses his olfactory receptors. The molecules then dissolve in the fluid that moistens the receptor, and interact with its membrane. Upon perception of the molecule, the male becomes excited, flies into the air and heads upwind, a behavior which, sooner or later, should lead him to the waiting siren. They mate and life goes on.

The female, it should be noted, has only small antennae with much lower sensitivity. She is not so good at detecting males, but then, why should she be? Whereas the male uses his energy in seeking her out, she places her energy into egg formation. This is not to say that she is insensitive, however. She is very sensitive to other environmental cues, such as those that indicate a suitable site for egg laying.

Chemical Detection in Vertebrates

Vertebrates detect chemicals in a number of ways, employing all three major means (general sensitivity, gustation, olfaction). For example, many vertebrates have generalized chemical receptors scattered over the body surface. It has been argued that these are the same as pain receptors (acid on the skin is painful), but actually the two sensations have been separated experimentally. In one test, acid could still be detected on the skin after the pain receptors had been deadened by cocaine.

Olfaction in vertebrates is usually accomplished by moving chemical-laden water or air into a canal or sack that contains the chemical receptors. The chemical can't be detected, however, until it goes into solution. You may be aware that your nose is full of "a solution." Actually, the nasal passages in all terrestrial vertebrates are moistened by mucus since, in the entire group, this is where the olfactory receptors lie (Figure 31.27). On the other hand, you never see fishes or aquatic amphibians with a runny nose because they don't need to manufacture a nasal liquid. Any chemical they encounter is likely to already be in a solution. In these aquatic groups, chemical-laden water flows into a pouchlike nasal sac lined with a folded membrane that is covered with chemical receptors, actually modified cilia.

In land vertebrates, the best sense of smell is found among the mammals, especially the carnivores and rodents. It is obvious why good olfaction is important to a carnivore. Wolves and lions, for example, often locate their prey by smell. It is less clear why rodents have evolved such strong olfactory ability, since their food sources are not usually widely dispersed and do not usually run away. However, their sense of smell may not be so strongly associated with food finding. Many rodents discriminate among themselves on the basis of families (or tribes) and they may recognize each other individually by smell, so olfaction may be important to them in social interactions.

On the other hand, there are animals that aren't able to smell anything. For example, some toothed whales have no sense of smell at all. Apparently, whales are descended from terrestrial vertebrates that went back to the sea, and their descendants were just not able to readapt their noses to the aquatic environment.

Olfaction is rather poor in primates. Chimpanzees smell about like you do (no offense). Males often sniff the rear ends of female chimpanzees in heat, and they apparently detect something, but a key mating signal seems to be the red rear and soliciting behavior of the female.

The sense of taste is neurologically very similar to the sense of smell. In fact, in aquatic animals, the sense of taste is sometimes very difficult to distinguish from that of smell or some sort of general chemical detection. In water-dwelling vertebrates, receptors may be scattered over the body's surface, but they are often more concentrated in the head region. For example, the bullhead (or "catfish") has high concentrations of taste buds on its long, sensitive whiskers and around the mouth. Such sensitivity, of course, is important in an animal that feeds on the dark bottoms of murky rivers or lakes.

In all land-dwelling vertebrates, taste receptors are confined to the mouth area. And, as you know, there are four basic tastes: sweet, sour, salty, and bitter. In humans, these are located in specific areas of the tongue (sweet on the tip, sour on the sides, bitter on the back, and all parts of the tongue are sensitive to salt; see the taste map in Figure 25.16). Physiological studies indicate that the cells of any taste bud can respond to three or even four "tastes" but that they are more sensitive to some than to others. So, if both salty and sweet can cause a receptor to fire, how does the brain know what taste has been encountered? It turns out that each taste has its own neural code, producing distinctive firing patterns in the receptors.

You may have noticed that you have difficulty in trying to get your cat to eat a cube of sugar that your rat will nibble and your guinea pig might devour in a flash. The reason is cats can't taste sweets, while rats are moderately sensitive to them and guinea pigs are highly sensitive to the taste. The ecological differences among species are obviously important in determining their diets. For example, cats are almost exclusively flesh eaters and would therefore have little use for "sweet" taste buds. Humans are omnivorous; that is, we'll eat almost anything, plant or animal. So it has undoubtedly been important to us, through our long eons of development, to be able to distinguish among a variety of tastes and smells. Those things that smelled and tasted good were probably good for us. Something rotten (for example, bacteria-ridden) probably lost its appeal for us (but increased its value in the eyes of other animals, such as appreciative vultures who apparently aren't as susceptible to those bacteria).

31.27 The olfactory receptors in the human nose connect to the *olfactory bulb* of the brain. The receptors are able to distinguish a wider variety of stimuli than are the taste buds. The olfactory neurons are part of the nasal epithelium dispersed with other cells. Each neuron has numerous olfactory hairs that protrude from the epithelium. Each "hair" is a modified cilium.

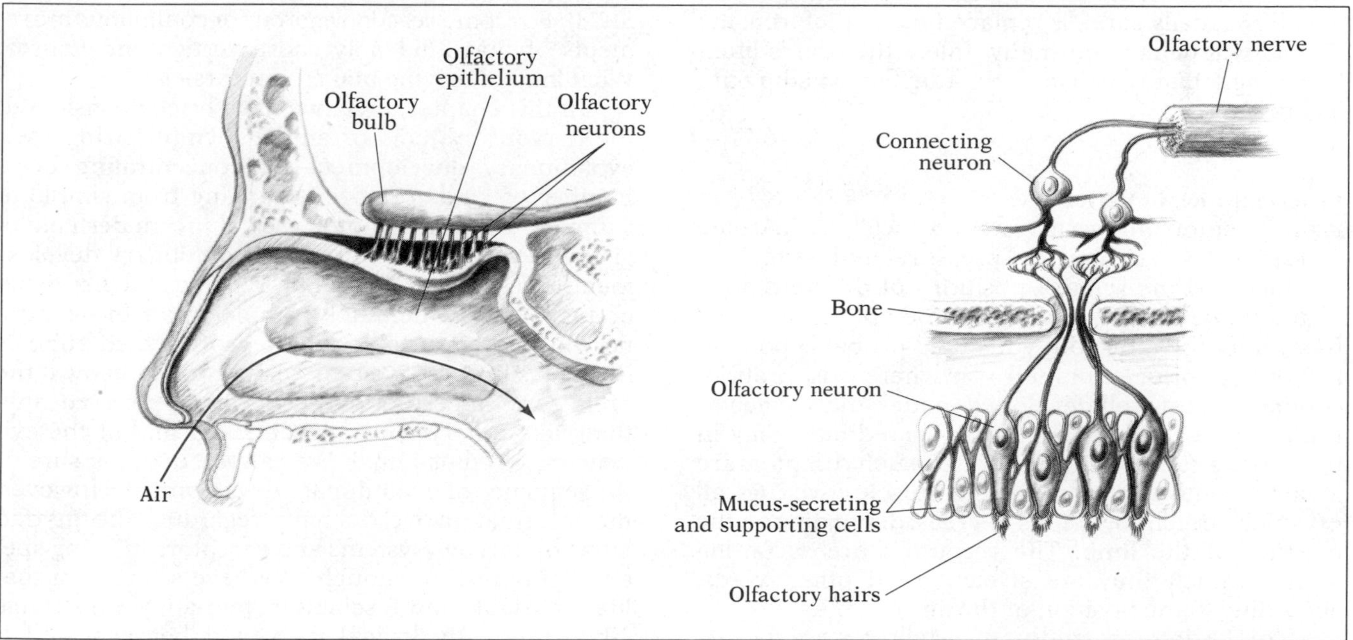

Proprioception

Proprioception is the ability to detect stimuli that arise within the body and tell the animal its position, or the position of one part of the body relative to another.

Proprioception in Invertebrates

It is particularly important for animals with many body parts to be able to coordinate those parts—to know what each is doing. Proprioception, then, is particularly important to the arthropods, which have several types of these sensors. They may be located peripherally, for example, at the base of "hairs," or deep within the muscles. Cockroaches—which are running animals—have hairs on the inside of their legs. These respond to flexion of the knees, so they are aware whether a leg is extended or flexed. Other arthropods—the crabs and lobsters—on the other hand, have proprioceptors located within the muscles themselves.

Another type of proprioception gives information regarding the position of the body with respect to the vertical—in other words, to gravity—and hence is important in equilibrium. In most invertebrates, this information is conveyed by *statocysts.* A statocyst is an opening, a chamber, that contains sensory neurons and a granule of some sort. In crayfish, as long as a granule is resting on the sensory hairs that line the bottom of the chamber, the animal will remain in an upright position. If the animal is turned over so that the granule falls away from the hairs, it will immediately right itself. If, on the other hand, the granule, which is usually sand, is replaced by a bit of iron, the animal will behave normally unless the iron is lifted by a magnet, in which case the crayfish will flip onto its back.

Proprioception in Vertebrates

Proprioceptors in vertebrates, as well as in some invertebrates, respond to pressure and stretching produced by the relative positions of different parts of the body. Thus, you may not be able to see your foot under the table, but you could probably point to it. Your proprioceptors tell you where it is. Proprioception is most well-developed in mammals—such as the primates, which move with incredible agility in their forest canopy. In vertebrates, such receptors are located in muscles, tendons, and muscle coverings, all of which stretch to various degrees, depending on the position of the limb. The receptors fire according to how much they are stretched—in other words, according to the position of the limb.

Equilibrium receptors in vertebrates are located in the inner ear. Usually, three fluid-filled loops, or *semicircular canals,* and two open chambers register

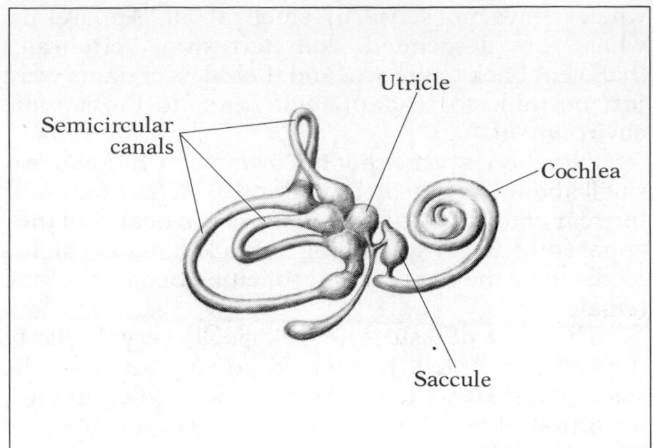

31.28 Semicircular canals. The middle ear is very sensitive to the body's position and movement. The *semicircular canals* lie at right angles to each other so that bodily movement in any direction shifts the fluid in at least one of them. It turns out, though, that humans are generally not particularly sensitive to accelerating motion, but very sensitive to vertical motion (and that's what usually makes you motion sick). The *saccule* and *utricle* are fluid-filled cavities in which crystals of calcium carbonate (called *otoliths*) are embedded in a jellylike matrix. Changes in the position of the otoliths send information to the brain regarding a charge in direction or a change with respect to gravity.

any movement. Each semicircular canal has a bulge at its base which is lined with sensory hairs. Shifting fluids move if the animal changes speed or direction. Since the three loops lie at right angles to each other, any sort of movement is detected. The two chambers —the *utricle* and *saccule*—contain granules or small bones and function very much like statocysts (Figure 31.28). For some reason, vigorous or continuous movements of that fluid may cause vertigo and nausea. What kind of driving makes you carsick?

In this chapter, then, we have briefly considered the nervous systems of animals, emphasizing their evolutionary development by concentrating on a number of kinds of species ranging from simple to complex. This is a common ruse of the modern biologist who wishes to emphasize evolutionary development. But, a word of caution. Whereas the nerve net of the hydra is definitely simpler than the central nervous system of a vertebrate, soft-bodied animals don't fossilize very well, so we really don't know if the prehistoric ancestors of vertebrates possessed anything like a nerve net. It seems logical, but the evidence is circumstantial. We cannot, then, be sure of the sequence of evolutionary development. However, the information we do have regarding the myriad kinds of nervous systems and receptors in living species is impressive enough. We have seen again that life is variable and fascinating, perhaps even strange —but eminently logical, as would be suspected if natural selection does indeed mold animals to fit their particular kind of world.

Application of Ideas

1. Use the terms *primitive* and *advanced* to describe the major parts of the human brain. Explain the basis for which you applied the terms.

2. The human brain has a tremendous information storage capacity we call memory. In view of this, and assuming that information is stored in molecules, which class of molecules is most likely involved? Explain why each of the other classes of molecules might not be appropriate.

3. Compare the action of pheromones with that of hormones. How is their action similar and in what fundamental way do they differ?

4. The central nervous system, particularly the conscious centers, exercise what might be called, "noisy control," with a barrage of sensory messages being received, integrated, and acted upon. The autonomic system, on the other hand, can be referred to as a "quiet" system, since its activity is generally below the conscious level. Elaborate on these ideas by considering how the two systems respond to a problem such as thermoregulation. Let's assume that the systems are busy adjusting to extreme cold first and then extreme heat in the surroundings.

5. Describe the presumed evolution of the central nervous system from a diffuse net to a longitudinal system with a head end. Why would cephalization have arisen to begin with?

6. What are the relative advantages and disadvantages of the human eye and the insect eye comprised of ommatidia?

7. Give examples of how specific senses have evolved that help an animal survive under the special conditions of its niche. Do animals have senses beyond what they are likely to be called on to utilize? Why do you suppose this is?

Key Words

A1 cell
A2 cell
acetylcholine
acetylcholinesterase
action potential
auditory receptor
audition
audltory canal
autonomic nervous system
axon

basilar membrane
bilateral symmetry
bombykol

cell body
central nervous system
cephalization
cerebellum
cis-*retinal*
cerebrum
chemoreceptor
cochlea
color vision
compound eye
cone cell
contact receptor
corpus callosum
cryptic coloration

dendrite
depolarization
distance receptor
dorsal nerve

effector
epinephrine

Key Ideas

Introduction

The *nervous system* consists of cells, tissues, and organs that give the animal clues about the external and internal environment and permit it to respond to stimuli.

What Neurons Are

1. The cellular unit of the nervous system is the *neuron* or *nerve cell. Neurons* join each other to form *nerves.*

2. *Neurons* specialize in detecting, processing, and responding to information.

How Neurons Work

1. Each neuron has a receiving end of *dendrites,* a *cell body,* and a transmitting *axon.*

2. *Neurons* join at *synapses,* which are minute spaces in the synaptic region.

3. *Nerve impulses* have both a chemical and an electrical nature.

4. Because of the differential movement of ions in and out of neurons, the two sides of the cell membrane often bear opposite electrical charges.

The Resting Neuron

1. Resting neurons are *polarized* with a *resting potential* of −70 mv across the cell membrane.

2. *Polarity* is maintained by active pumping of sodium ions (Na^+) to the outside. Potassium (K^+) tends to move into the *neuron* where negatively charged proteins and negative chloride (Cl^-) ions are found.

The Action Potential

1. When a *stimulus* reaches a threshold, a *depolarizing* event called an *action potential* flows down the *neuron.*

2. The *sodium pumps* stop and Na^+ rushes inward as some K^+ leaks out. The charge difference goes from −70 mv to +30 mv.

3. *Depolarizing waves* move along followed immediately by *repolarization.* This begins with K^+ flowing out, followed shortly by Na^+ as the *sodium pumps* resume. A short period of *hyperpolarization* ends as K^+ reenters the neuron and the *resting state* resumes.

4. *Impulses* move from one *neuron* to another or to *effector* organs across *synaptic junctions.*

equilibrium
external ear
eyespot

fissure of Rolando
fissure of Sylvius
forebrain
fovea
frontal lobe

ganglion
gustation

heat receptor
hemisphere
hindbrain
hyperpolarization
hypothalamus

inhibition
inner ear

lateral line organ
lobe

medulla
Meissner's corpuscle
midbrain
middle ear
myelinated

nerve
nerve cell
nerve cord
nerve impulse
nerve net
nervous system
neuron
neurotransmitter
nictitating membrane
nodes of Ranvier
norepinephrine

occipital lobe
"off" fibers
olfactory receptor
ommatidium
"on" fibers
"on–off" fibers
opsin
outer ear

Pacinian corpuscle
parallax
parasympathetic nervous system
perception
photoreceptor
pons
postsynaptic membrane
prefrontal area
pressure receptors
presynaptic membrane
proprioception

The Synaptic Junction

1. *Neurons* often end in *synaptic knobs* which come close to the next *neuron* or *effector.* The *synaptic cleft* is bridged by *neurotransmitters.* These are substances released from the *knobs* at the *presynaptic membrane,* crossing the cleft to receptors on the *postsynaptic membrane* and stimulating a second *depolarizing* wave there.

2. Some neurons are *inhibitory,* probably causing the next *neuron* or *effector* to leak potassium, thus reaching a state of *hyperpolarization.* This inhibits impulse generation.

The Neurotransmitters

1. *Neurotransmitters* include *acetylcholine, norepinephrine,* possibly *epinephrine,* and others. *Acetylcholine* is common, acting in muscles, brain, heart, and other effectors.

2. *Acetylcholine* is deactivated by the enzyme *acetylcholinesterase.* It is vital that *neurotransmitters* be short-lived so that control is possible.

Fast Neurons and the Schwann Cells

1. *Neurons* can increase the speed of impulses by being thick or by being *myelinated.* The giant *axon* of the squid is an example of the former.

2. *Myelin* is a white membranous sheath wrapped about *neurons* in 1 mm long segments. Nonmyelinated *nodes* (*nodes of Ranvier*) are minute spacings in the sheath.

3. Impulses jump from one *node* to another in *saltatory propagation. Depolarization* occurs only in the *nodes,* speeding the impulse and conserving ATP.

Integration of Transmissions

1. *Action potentials* are always of the same strength once threshold has been reached. Intensity can be increasing the number of *action potentials* and by getting more *neurons* involved.

2. Different sensations are recognized when specialized *sensory receptors* are stimulated. The *action potentials* are identical, but the pathways and *receptors* in the brain vary.

The Central Nervous System

1. The *central nervous system* consists of the *brain* and *spinal cord.*

2. Invertebrates have different degrees of complexity in their *nervous systems,* from simple *netlike* to more complex aggregations of *ganglia* along the body and in the head region.

Cephalization

Animals with longitudinal body plans have undergone *cephalization* with most *sensory* structures and brain development at the "head" end.

Coelenterates

Coelenterates have *netlike nervous systems* with many criss-cross *synapses.* They have slow and fast *neurons.*

Flatworms

Planaria has a bilateral body plan with two *nerve cords,* paired *ganglia,* and neural branches along the body. The largest *ganglia* are at the anterior end.

Annelids

The earthworm has large *ganglia* about the esophagus and a ventral *nerve cord* containing paired (fused) *ganglia* in each body segment.

Mollusks

The cephalopods, and especially the octopus, have a large *ganglionic mass* which is specialized enough to be a *brain.*

Arthropods

Arthropods are segmentedlike annelids, but the segments have become highly specialized. The specialized appendages are controlled by their own paired

radial symmetry
receptor
relay neuron
resting potential
retina
rod cell

saccule
saltatory propagation
Schwann cells
semicircular canals
sencillum
sensory
seratonin
sodium pump
spike
spinal cord
statocyst
stimulus
sympathetic nervous system
synapse
synaptic cleft
synaptic knob

tactile receptor
thalamus
temporal lobe
thermoreceptor
touch receptor
transmission
trans-*retinal*
tympanal organ
tympanic membrane
tympanum

utricle

visual cliff
visual pigment
visual purple
visual receptor

Weberian ossicle

ganglia. The *brain* consists of paired *ganglia* above and below the esophagus. Highly visual species have most of their *brain* devoted to vision.

Echinoderms
Adult echinoderms are *radially symmetrical* and have little *cephalization.* They have a nerve ring with nerve trunks radiating outward. The larvae, however, are bilaterally symmetrical.

The Brains of Vertebrates
1. Vertebrates differ from invertebrates by having a dorsal tubular *nerve cord* (*spinal cord*) and a much greater degree of *cephalization.*
2. Advanced vertebrates have greater *cephalization* than primitive vertebrates.
3. In the cartilaginous fishes (sharks, etc.) much of the *brain* is specialized for smell.
4. The vertebrate brain consists of *forebrain, midbrain,* and *hindbrain.* In lower vertebrates the first is greatly reduced in proportion to the others.

The Human Brain
In humans, the midbrain is reduced and the *forebrain* is most prominent.

The Medulla
The *medulla* controls many of the visceral functions, those of a mechanical or subconscious nature.

The Cerebellum and Pons
The *cerebellum* is a center for balance, equilibrium, and coordination. The *pons* contains pathways between *cerebellum* and *cortex.*

The Thalamus and Reticular System
The *thalamus* is a vital relay structure containing fibers and connections for parts of the *brain.* Within is the *reticular system* which contains branches from many parts of the *brain* and acts as an arousal mechanism. It may also filter out superfluous *stimuli* and permit specific *stimuli* to reach the conscious centers of the *brain.*

The Hypothalamus
The *hypothalamus* monitors and controls many internal functions.

The Cerebrum
1. The *cerebrum* or cerebral cortex becomes increasingly large and important in the higher vertebrates. In the primates, many lower brain functions are moderated by the *cerebrum.*
2. Expansion of the cortex is made possible by convolution (folding), which is most apparent in humans and other primates.

Hemispheres and Lobes
1. The two cerebral hemispheres each are subdivided into four *lobes.* The *temporal lobe* processes visual information. The *frontal lobe* regulates voluntary movement and speech, while its *prefrontal* area sorts and orders stimuli. The *parietal lobe* contains sensory areas for the skin *receptors* and detects body position.
2. Left and right *hemispheres* are equal in potential but develop functional differences as a result of learning. In right and left handedness, learning occurs in the opposite *hemisphere.*
3. Connecting pathways between *hemispheres* are in the *corpus callosum.* Split brain experiments reveal that learning is transferred from one hemisphere to the other.

The Autonomic Nervous System
1. The *autonomic nervous system* has two opposing parts, the *sympathetic* and *parasympathetic* systems.
2. The *sympathetic system* speeds up many functions, particularly those required in an emergency. The parasympathetic system counters these responses, bringing things back to rest.

3. The two divisions are intimately related to the *central nervous system. Sympathetic ganglia* are found along the *spinal cord. Parasympathetic ganglia* are found near the organs they serve. The two systems work in concert to maintain homeostasis.

Perceptual Systems
Animals have a varying perception of their surroundings, depending on their individual specializations for receiving *stimuli.*

Thermoreceptors
Animals respond to unfavorable temperatures through behavioral or physiological regulation.

Heat Detection in Invertebrates
A few invertebrates, such as locusts and cockroaches, have heat receptors.

Heat Detection in Vertebrates
Vertebrates have well developed heat *receptors.* The pits of pit vipers are one example. Mammals have receptors known as *Ruffinian corpuscles* and *Krause end bulbs* in the skin.

Tactile Receptors
Tactile receptors have extremely sensitive and fast-firing *neurons.*

Touch in Invertebrates
Invertebrates sense touch through bristles. Arthropods are often particularly well equipped with *sensory* bristles, which generate impulses when disturbed.

Touch in Vertebrates
Vertebrates have *distance receptors* and *touch receptors.* The *lateral line organ* of fishes is sensitive to water movement. *Contact receptors* respond to pressure and touch. *Pacinian corpuscles* are pressure receptors, deep in the skin, while *Meissner's corpuscles* are *touch receptors,* nearer the surface.

Auditory Receptors
Auditory receptors respond to distant *stimuli,* but at higher frequency than lateral line *distance receptors.*

Hearing in Invertebrates
1. Auditory receptors include the *tympanal organs,* which are similar to the eardrums.
2. Noctuid moths make use of their *tympani* to evade bats. The *tympani* have two kinds of *receptors,* the *A1 cell* for low-intensity, pulsing sounds and the *A2 cell* for loud sounds. The moths can detect distance, direction, and altitude of the predatory bat by the differential rate of firing.

Hearing in Vertebrates
1. Vertebrates have *sound receptors* in the *inner ear.* In some fishes, the sound communicates directly with the *inner ear,* but in others, *Weberian ossicles* connect the swim bladder to the *inner ear* and it acts as a vibrating chamber.
2. Land vertebrates have a three-part hearing organ. The *outer ear* opens into an *auditory canal* and *middle ear. The tympanum* connects to three *ossicles* in the *middle ear,* which in turn connect to the *inner ear.* From the *inner ear, auditory receptors* send impulses to the brain.
3. In mammals, the *middle ear* has three bones that transfer sound from the *tympanum* to the *oval window* of the *inner ear.* This sets up movement in fluids in thc long coiled *cochlea.*
4. *Neurons* in the *cochlea* detect frequencies of sound. The *basilar membrane* has different thicknesses that can distinguish frequencies.

Visual Receptors
1. Animals receive parts of the visible spectrum only. *Receptors* contain light sensitive pigments that transform light energy to neural *stimuli.* A reversible photochemical process is often utilized.

2. Differences in intensity are believed to be detected in one of three ways:
   a. Detection of number of photons.
   b. Relay *neurons* may fire at rates proportional to the number of *receptors* stimulating them.
   c. Decoder cells may measure frequency at which relay *neurons* fire.

Seeing in Invertebrates

1. *Euglena* and *Planaria* have directional, light-sensitive *eyespots.*

2. In the mollusks, the cephalopods have highly efficient and complex eyes.

3. Crayfish have *compound eyes.* They are composed of numerous units known as *ommatidia,* which detect objects crossing their field. Praying mantises have two types of *ommatidia.* The central portions contain isolated facets that only fire when they are individually stimulated. This helps detect slow movements of prey. The outer facets help in poor lighting, since light can be transferred from one facet to another.

4. Honey bees are sensitive to form and color and are attracted to strongly subdivided forms (flowers). They can see ultraviolet light, but not red light.

Seeing in Vertebrates

1. The vertebrate eye is similar to the cephalopod eye, but the *retinal* layers face in opposite directions.

2. Mammalian *retinas* have two types of *photoreceptors,* the *rods* and *cones.* The eyes are extremely sensitive to movement.

3. Sensitivity to movement is possible because of three types of *neurons.*
   a. *On fibers* respond to continuous light.
   b. *Off fibers* respond when light stops.
   c. *On–off fibers* respond when light appears or stops.

4. *Off fibers* and *on–off fibers* predominate, and so objects moving across the field are easily picked up.

5. Most vertebrates respond through a visual alarm reaction to objects that rapidly increase in size.

6. Vertebrates have good depth *perception.* This is related to *parallax,* the seeming movement of nearby objects as the head is moved.

7. Amphibians distort the shape of the eyeball in adjusting to distance. Frogs, lizards, birds, and some mammals have a third eyelid, the *nictitating membrane,* which protects and moistens the eye. Lizards have good eye movement, while snakes have none. Snake's eyes are fixed and covered with a transparent scale.

8. Vision in birds is highly developed, especially in high-flying predators and scavengers. The relative number of *rods* and *cones* is related to night and day hunting, respectively. In some birds, a horizontal streaklike *fovea,* is heavily invested with *cone* cells. The entire horizon can be scanned simultaneously.

9. *Color vision* in birds is facilitated by colored oil droplets within the *receptors,* each with its own colored oil; some have haze filters.

10. Mammals have excellent vision, especially primates and Carnivora. Rabbits are particularly sensitive to objects coming up from behind and objects that move and then stop.

11. Primates have good shape discrimination and *color vision.* Most mammals see only by contrast in intensity.

Chemoreceptors

Chemoreception occurs through general *receptors,* taste (*gustation*), and smell (*olfaction*).

Chemical Detection in Invertebrates

1. Insects have highly developed *chemoreceptors.* These are located on mouthparts, antennae, forelegs, and in some cases on the ovipositor.

2. The silk worm has highly specialized *chemoreceptors* in its antennae. Males respond to *bombykol,* which is released by females. One molecule can generate an impulse in the hairlike receptors.

Chemical Detection in Vertebrates

1. Some vertebrates have *chemoreceptors* on the body surface. The chemical must enter a pore or sac (such as the moist nasal surfaces).

2. The sense of smell is highly developed in rodents and carnivores. Rodents use it for social interaction, carnivores, for finding prey.

3. Primates have comparatively poorly developed *olfactory receptors.*

4. *Olfaction* in humans may be an important but neglected area of sexuality. The use of artificial scents to cover body odors probably eliminates its former role.

5. Taste and smell are neurologically similar. Taste *receptors* in fish are usually on the body.

6. In land dwellers, *taste receptors* are in the mouth. In humans, different parts of the tongue specialize in tasting four flavors: sweet, sour, bitter, and salty. Each taste apparently has its own neural code, which is interpreted by the *brain.*

Proprioception

Proprioception is the ability to detect the position of the body and its parts.

Proprioception in Invertebrates

Proprioceptors are located at the base of bristles, deep in the muscles or in specialized *statocysts.* The latter are minute chambers that contain *sensory neurons* and granules of sand. The movement of these granules against the *sensory hairs* informs the animal of its orientation.

Proprioception in Vertebrates

1. In vertebrates, *proprioceptors* are found in the muscles, tendons, and muscle coverings. The *receptors* respond to stretching.

2. Equilibrium is accomplished by *receptors* in the *inner ear.* These consist of three fluid-filled loops and two open chambers. Shifting fluid is registered by sensory hairs. The chambers (*utricle* and *saccule*) contain granules and function like the arthropod *statocysts.* Because of the arrangement of the *semicircular canals,* movement in any direction can be sensed.

Review Questions

1. Describe three general roles of *neurons.* (p. 906)
2. Prepare a simple drawing of a *motor neuron,* label its parts, and show the direction an *impulse* would travel. (pp. 906–907)
3. Describe the specific behavior of ions in a resting *neuron.* (pp. 907–908)
4. Describe the events that occur when a *neuron* fires, including the behavior of the ions described in Question 3. (p. 908)
5. Prepare a drawing of an *action potential* as seen on the oscilloscope and label the action in each part. (p. 908)
6. Using the terms *synaptic knob, synaptic cleft, presynaptic* and *postsynaptic membranes,* calcium, *neurotransmitter,* and *receptor sites,* explain the action at a *synapse.* (pp. 907–910)
7. How is the action of *neurotransmitters* controlled? (pp. 910–911)
8. Explain the one-direction flow of an *impulse* in terms of the *synapse.* (p. 911)
9. What determines whether a *neuron* will be "fast" or "slow"? (p. 911)
10. Explain how *depolarization* occurs along a *myelinated axon.* (p. 911)
11. In what ways is *saltatory propagation* more efficient? (p. 911)
12. If all *action potentials* are alike, how can the intensity of *neural impulses* be heightened (2 ways)? (p. 913)
13. What are *ganglia* and what is their role in animals without much *cephalization?* (pp. 913–914)
14. What is the advantage of *cephalization* to longitudinal (bilateral) body plans? (p. 914)
15. Describe the *nervous system* in the earthworm. Is *cephalization* complete? (p. 914)
16. What are two major differences between the *nervous systems* of arthropods and vertebrates? (p. 916)
17. List the three major regions of the *brain* and compare their importance in the lower vertebrates with the higher vertebrates. (pp. 916–917)
18. Where is the *medulla* and what are some of its functions? Can any vertebrate function without a *medulla?* (p. 917)
19. Explain how the *reticular system* of the *thalamus* works. (p. 919)
20. Using frogs, rats, monkeys, and humans as examples, describe the increasing importance of the *cerebrum* and the evidence for the idea. (p. 919)
21. List the lobes of the *cerebrum* and state a function in each. Include the *prefrontal area* of the *frontal lobe.* (p. 920)

22. Compare the role of the *sympathetic division* of the *autonomic nervous system* with that of the *parasympathetic division.* (p. 923)

23. Describe the location of the *ganglia* in both *sympathetic* and *parasympathetic divisions.* (p. 923)

24. Describe the pit vipers. (pp. 923–924)

25. Explain how a single *neuron* might register both heat and cold. (p. 923)

26. In general, what are the *touch receptors* in invertebrates and what are they used for? (p. 924)

27. Briefly describe how the noctuid moth uses its *tympanal organs* in avoiding moth-eating bats. (p. 925)

28. Explain the characteristics of hearing in barn owls that enables them to hunt so effectively. (p. 927)

29. Briefly describe how sound waves are converted to *neural impulses* in the mammalian ear. (p. 927)

30. What is the advantage of *shielded ommatidia* in the compound eye of the praying mantis? (p. 929)

31. In what important ways are the vertebrate and invertebrate (octopus) eye different? (p. 930)

32. List the three types of neural fibers in the optic nerve of mammals and explain how they function in perceiving moving objects. (p. 931)

33. Describe the *visual alarm reaction* in vertebrates and comment on its adaptiveness. (p. 931)

34. Define *parallax* and explain how it helps young vertebrates avoid danger. (p. 931)

35. What unusual aspects of vision are seen in the frog's eye? (p. 931)

36. List three special characteristics of the eyes of hunting birds. (p. 932)

37. What special kinds of movement are most easily detected by the rabbit and how does this help it? (pp. 932–933)

38. What is the proposed value of *color vision* to humans as hunters and gatherers? (p. 933)

39. What do *gustation* and *olfaction* have in common and how are they different? (p. 933)

40. Which groups of mammals have the most acute *olfactory* sense and where are these *receptors* located? (p. 933)

41. Of what value is an acute *olfactory* ability to mice and rats? (p. 934)

42. Do *neurons* distinguish flavors? How do they inform the *brain* of flavors? (p. 935)

43. Describe our threshold for salty, sweet, sour, and bitter flavors. How does distinguishing these tastes help? (p. 935)

44. What is the *statocyst* in invertebrates and how does it function? (p. 936)

45. How do we determine the position of our appendages without looking at them? (p. 936)

46. Describe the anatomy of our own *equilibrium receptors* and briefly explain how they work. (p. 936)

Chapter 32

The Mechanisms and Development of Behavior

The field of study loosely called *behavior* can't exactly be called rowdy, but it certainly hasn't been as sedate or circumspect as some other areas of science either. One reason is that when it began to be studied seriously in this century, research branched off in several different directions. As each branch developed more or less independently, and as researchers in each group became vaguely aware of what the others were up to, the air became filled with criticism. At first, the lines were rather sharply drawn. Some people considered behavior in general to be genetically determined; others were outraged and said that all behavior was a result of learning. Some believed that behavior should be studied under natural conditions; others believed it should be studied in the laboratory where the environment could be carefully controlled. And no one had much use for anyone else's ideas. It would be nice if we could say that the dissension is past history and that we are now beyond all that, but many basic disagreements remain to this day.

Much of the disagreement, of course, comes from the complexity of the problem. Here we are trying to study the behavior of other cultures and other species and we don't even understand our own, personal behavior. How often we surprise ourselves with our own behavior. We simply don't know what roams the cellars of our minds. Thus, how can we hope to understand ourselves by studying other species? But how often are our self-conceptions based on the behavior of bright-eyed and bewhiskered beasties with an intense interest in sunflower seeds? There are also those who say we can't even study other animals by studying other animals. What motivates a crayfish may not even exist in the world of the rat, and one strain of rat may behave differently from another strain.

For these reasons, some scientists tend to sniff and say that behavior is a "soft science"—a very pejorative term among scientists. These people, though, simply don't know enough about the field. Behavior has now come of age. It is attracting armies of hard-nosed young biologists who not only tromp around in muddied boots but also speak algebra and converse with computer terminals. The study of behavior does have its problems, of course, primarily because of our great gaps in information. We may be able to synthesize a moth attractant chemically, but we don't know what motivates moths, or what it feels like to be a moth, or what a moth senses and knows. Nonetheless, the only way to learn about behavior is to study it. The alternative is not to study it until we know enough about it to study it. Right now, we do know quite a bit about quite a lot of animal behavior. But let's begin with a little history.

32.1 Instinct and Learning

Ethology and Comparative Psychology

The study of animal behavior certainly isn't new. The ancients had theories to account for such events as migration and territorial behavior. But many great puzzles remained, so much so that more philosophy was created than was data. And so it remained for centuries. Then, in the 1930s, new dimensions were added to the discipline as a compelling theory of *instinct* was developed. The idea was crystallized in Europe, although it is difficult to trace to its precise origins. Its development is attributed primarily to the Austrian Konrad Lorenz and his co-workers, who set forth their theory in 1937, and again in 1950 (Figure 32.1). The Dutch zoologist Niko Tinbergen, at Oxford University, amended the scheme in 1942 and 1951, and a theory resulted which, in spite of its imperfec-

(a)

(b)

32.1 (a) Konrad Lorenz and gosling. Lorenz, a rather aloof Austrian, has time and again proven himself to be one of the most astute observers and synthesizers in the area of animal behavior in this century. Drawing upon the work of his predecessors and coupling it with his intimate knowledge of animals, he was, with Niko Tinbergen, primary in developing the modern *instinct* theory. He has been strongly criticized for his assertion that humans are instinctively aggressive.

(b) Niko Tinbergen, a humanistic and imaginative researcher, shared the Nobel Prize with Lorenz and Karl von Frisch, an impeccable experimenter who worked primarily with insects. Tinbergen, at Oxford, worked in close association with Lorenz in the early years of ethology and summarized the concept, with his own modifications, in his landmark book *The Study of Instinct.* Tinbergen, like Lorenz and von Frisch, entered retirement by continuing to work. Tinbergen was a hyperactive child who, at school, was allowed to dance on his desk periodically to let off steam. So in "retirement" he entered a new arena, stimulating the use of ethological methods in autism.

tions, demanded serious consideration. And then the arguments began.

Lorenz and Tinbergen, who shared the 1973 Nobel Prize in medicine and physiology with the German biologist Karl von Frisch, were the founders of an approach to behavior called *ethology,* which incorporates the instinct idea. Ethology traditionally concerns itself with *innate* behavioral patterns. Research is normally done in the field, under conditions as natural as possible, although ethologists have also relied on observations of the many tame or half-tame animals which can usually be found sharing their homes.

The strongest disagreements with the ethologists' instinct theory came from American *comparative psychologists,* led, in recent years, by B. F. Skinner. Their research has traditionally dealt with learning processes in laboratory animals, primarily the Norway rat. Although many of the arguments continue, there has been a steady closing of the gap between ethologists and comparative psychologists. Probably the final integration of the two approaches depends on further data from a third area of study, *neurophysiology,* which is concerned with how the nervous system works. Such information, however, is hard won and, at present, we just haven't filled in those nebulous gray areas. But let's take a look at the ethologist's concept of instinct and see what all the argument has been about.

Instinct

It was once believed that, whereas humans learn their behavior, other animals respond only to unalterable "instincts" that are indelibly stamped into their nervous systems at birth. Since mounting evidence over the years didn't bear out this sweeping generalization, the concept of instinct fell into some disrepute. In fact, many behavioral scientists declined to even use the word. The problem was compounded by the fact that while ethologists were seeking to clarify and dignify the concept and make it useful, the word *instinct* found its way into the public's vocabulary. Thus, while scientists were attempting to define the exact parameters of the instinct concept, others were using the term with wild abandon to explain everything from a baby's grip to swimming.

With all the confusion over the term itself, it is best to start with an explanation of the instinct concept. So let's first break it down and consider its components, according to the Lorenzian model.

Essay 32.1 History of the Instinct Idea

Instinct was referred to in Greek literature at least 2500 years ago, and, for some reason, the concept survived with almost no supporting evidence, as very few other ideas have been able to do. In the 1st century, the Stoic philosophers developed the idea that animal behavior takes place without reflection. This was the beginning of the movement to separate the behavior of humans from that of other animals. Humans were officially removed from their place on the natural scale of animals (where Aristotle had placed them at the top, just above the Indian elephant) by Albertus Magnus (Fat Albert) in the 13th century. St. Thomas Aquinas, a student of Albertus, supported the separation of humans and other animals based on the idea that we have "souls," divinely placed in the embryo sometime before birth. This marked the beginning of a religious explanation of instinct. The reasoning was: There is a life after death and the part of a person that continues to live is called the soul. The precise conditions, good or bad, under which the soul continues after death are determined by how the person behaved during life on earth. In order to make the "correct" decisions, reason is necessary. Animals don't share in the hereafter and don't need reason. Instead, they are guided by blind instinct.

This philosophical notion of instinct was rejected by many people. For example, Erasmus Darwin, Charles's grandfather, believed that behavior has no innate quality at all, but that it is entirely learned. Charles disagreed, and, in fact, was willing to treat behavioral characters just as morphological ones—species-specific and subject to natural selection. The instinct idea was to fall into some disrepute in the last part of the 19th century, partly because of the tendency of some researchers to name an action as instinctive and to let the matter drop. Such an attitude obviously would not encourage the search for verifying data. So it came to be that in this century the idea of instinct was scoffed at, and not considered valid subject matter for serious investigation.

Even by the end of the 19th century, however, some workers had not abandoned the instinct idea. C. O. Whitman, working on the behavior of pigeons at the University of Chicago, was accumulating evidence that in 1898 led him to declare, "Instincts and organs are to be studied from the common viewpoint of phyletic descent." Oskar Heinroth in 1910 described the behavioral similarities of related species and later noted that certain characters would appear in animals reared in isolation. An encompassing theory, however, was to be left to Heinroth's student, Konrad Lorenz.

Fixed Action Patterns and Orientation

It can't be denied that animals are born with certain behavioral repertoires. The first time a tern chick is given a small fish, it jerks its head and snaps in such a way that the fish lands head down so that it can be swallowed head first, with the spines safely flattened. Other birds build nests by using peculiar sideways swipes of their heads with which they jam twigs into the nest mass. And all dogs scratch their ears the same way, by moving their rear leg outside the foreleg. Such precise and identifiable patterns, which are innate and characteristic of a given species, are called *fixed action patterns.* When fixed action patterns are coupled with orienting movements, we speak of an *instinctive pattern.*

To illustrate the relationship between a fixed action pattern and its orientation, consider the fly-catching movements of a frog (Figure 32.2). For the fixed action pattern (the tongue flick) to be effective,

32.2 A leopard frog orienting to catch a fly. The tongue flick, a *fixed action pattern,* is not released until the central nervous system is stimulated by the sight of an insect in the proper position (close and in the midline). The orientation is obvious as the frog shifts its position, but once the tongue flick has started, no adjustment is possible; it will continue to its completion.

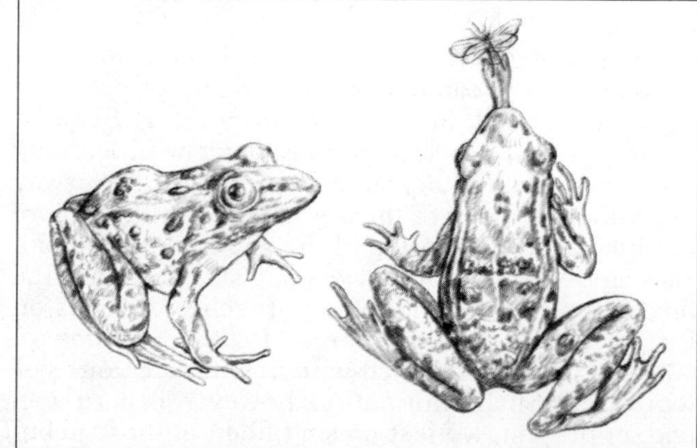

the frog must carefully orient itself with respect to the fly's position. The proper performance of the fixed action pattern results in the frog's gaining a fly. It is important to realize, however, that in most cases, once the fixed action pattern is initiated, it cannot be altered. If the fly should move after the tongue begins its motion, the frog will miss, since the fixed sequence of movements will be completed whether the fly is there or not. So the frog first *orients* and then *performs* the fixed action pattern.

Whereas the frog first orients and then flicks its tongue, with other instinctive patterns the orientation may be indistinguishable from the fixed action pattern itself. For example, if a goose sees an egg lying outside her nest, she will roll it back under her chin (Figure 32.3). She keeps the egg rolling in a straight line by lateral movements of the head. In this case, the fixed and orienting components are difficult to separate. However, they were distinguished in a simple, but classic, experiment. The orienting movements, it turns out, depend on the direction taken by the rolling egg. If it rolls to the right, the goose moves her head to the right as she continues to draw it toward her. If an experimenter removes the egg during the retrieval, the goose will continue to draw her head back (the fixed component)—but in a straight line, without the orienting side-to-side movements. Thus we see that the fixed action pattern, once initiated, is independent of any environmental cue, so that once it starts, it continues. The relationship between this fixed action pattern and its orientation component, then, is similar to that between a car's engine and its steering wheel. Once the engine is started, it continues to run, but each change of direction requires a new movement of the steering wheel and thus depends on the specific environmental circumstance.

32.3 In a classical experiment, Lorenz and Tinbergen showed that the behavior of a greylag goose rolling an egg back to her nest had two components: fixed and orienting. When they removed the egg while she was retrieving it, she continued the chin-tucking movements (the fixed component) but stopped making the slight side-to-side movements that kept the egg moving in a straight line (the orienting component).

It has been found that an animal may not be born with all its fixed action patterns. Many of these patterns develop as the animal matures, just as do morphological characteristics. For example, wing flapping or other patterns associated with flight normally don't appear until about the time a bird is actually ready to begin flying, although there may be some incipient flapping just prior to this time. Because some instinctive patterns are delayed, some researchers have assumed that they were learned. In addition, instinctive actions may improve with time, but it is important to realize that the improvement may or may not be a result of learning. In some cases, instinctive patterns are perfected by learning but in other cases instincts simply mature with age, like many physical structures. It is also interesting that the fixed action pattern and its orientation component do not always mature simultaneously. A baby mouse may scratch vigorously with its leg flailing thin air; only later will it apply the scratch to the itch.

Now let's consider another drama—one that might pull some of these ideas together. In this example, we will see very variable behavior (with strong learning components) becoming increasingly stereotyped until the final fixed action pattern is performed. The last orienting move is still variable, of course, but only within strict limits.

A peregrine falcon begins its hunt at daybreak. At first its behavior is highly variable; it may fly high or low, or bank to the left or the right. It is driven by hunger and would be equally pleased by the sight of a flitting sparrow or a scampering mouse. But suddenly it sees a flock of teal flying below. Its behavior now becomes less random and more predictable. It swoops toward them in what is now clearly "teal-hunting behavior" (Figure 32.4). So we see random searching until something triggers the swoop.

Almost all falcons perform this dive in just about the same way, although there are some variations. At the sight of the falcon, the teal close ranks for mutual protection. (After all, it is hard to pick a single individual out of a tightly packed group. A predator is likely to shift its attention from one individual to another, and in failing to concentrate on a single prey, it may come up empty-handed.) So first, the falcon makes a sham pass to scatter the flock, and the falcon is likely to pass through any part of the flock. The

32.4 The phenomenally swift and violent attack of a peregrine falcon. At this point, the bird's behavior is highly stereotyped, and if the teal makes a last minute change, the falcon will miss. This last stereotyped effort, however, was preceded by a series of modifiable acts, the first and most variable being the search for prey. Each step after that was performed in increasingly inflexible ways until the final strike. A successful strike may then initiate the next instinctive sequence—the one that results in swallowing. And that provides the greatest and longest-lasting relief.

terrified teal scatter. The falcon immediately beats its way upward, and after picking out an isolated target below, it begins a second dive.

If we were to watch this same falcon perform such a dive on several occasions, we might notice that the procedure is much the same every time. And if we were to watch several different falcons, we might notice that they all perform this action in much the same way. The greatest amount of variation is during the earliest part of such dives, since, at this stage, what the falcon does will be dictated in part by what the teal does. The falcon's options become fewer, and its behavior becomes more and more stereotyped, as it descends. Finally, there is a point at which the falcon's action is no longer variable. It now initiates the last part of the attack, the part that is no longer influenced by anything the teal does. At this instant the falcon's feet are clenched, and it may be traveling at over 150 miles an hour. Unless the teal makes a last split-second change in direction, it will be knocked from the sky, and the falcon's hunt will have been successful. The performance of this last, foot-clenched, fixed action pattern then triggers the *next* instinctive sequence. The falcon may follow the teal to the ground over any of several paths and approach the senseless teal a number of ways, perhaps fluttering or hopping. Then, with increasingly unvarying and stereotyped movements, it will grasp and pluck the teal, tear away its flesh with specific movements of its head, and swallow the meat. The sequence of muscle actions in swallowing is very precise and always the same. This is another fixed action pattern. Repeated swallowing actions seem to bring a measure of relief, so that the falcon doesn't resume hunting again for a while. Theoretically, it is the desire to perform swallowing movements that brought the falcon out to hunt in the first place. In other words, the falcon was searching for a situation that would enable it to release the fixed action pattern of swallowing. This probably seems a bit peculiar, but let's go on.

We can derive several points from this example. First, the state of the animal is significant. The falcon had not eaten for a time and was hungry. Hunger thus provided the impetus, drive, or motivation. In general, if an instinctive action pattern has not been performed for a time, there is an increasing likelihood of its appearance. Second, the earlier stages of this instinctive sequence were highly variable, but the behavior became more and more stereotyped until the final fixed action was accomplished. Third, the performance of this final action provided some relief. We have to assume this relief from the fact that once the falcon has performed this final action (swallowing), it does not immediately seek out conditions that will permit it to do so again. It has probably not escaped you that the performance of this fixed action pattern provides the animal with a biologically important element—in this case, food.

Let's take another case, one that tells us something about the desire to perform fixed action patterns. Courting behavior in birds is believed to be instinctive. In one experiment, Daniel Lehrman of Rutgers University found that when a male blond ring dove was isolated from females, it soon began to bow and coo to a stuffed model of a female—a model it had previously ignored. When the model was replaced by a rolled-up cloth, he began to court the cloth; and when this was removed, the sex-crazed bird directed his attention to a corner of the cage, where it could at least focus its gaze. It seems that the threshold for release of the fixed courtship pattern became increasingly lower as time went by without the sight of a live female dove. It is almost as though some specific energy for performing courting behavior were building up within the male ring dove. As the energy level increased, the *response threshold,* the minimum stimulus necessary to elicit a response, lowered to a point at which almost anything would stimulate the dove. Maybe you can relate all this to something in your own experience.

Hypothesizing the buildup of some mysterious "energy" seems less necessary to explain the hunting behavior of the falcon, since the reason the falcon hunts seems apparent: It is hungry. But then how do you explain hunting behavior in well-fed animals such as house cats? Perhaps the hunt is an instinctive action pattern in itself, so that its performance provides a measure of relief.

Appetitive and Consummatory Behavior

There are labels, of course, for the things we have been talking about. First of all, instinctive behavior is usually divided into two parts, the *appetitive* and *consummatory* stages. Perhaps the clearest examples of appetitive behavior are seen in feeding patterns, from which the name is probably derived. Appetitive behavior is usually variable, or nonstereotyped, and involves actually searching—for example, for food. Some stages of appetitive behavior may be more variable than others, as we saw in the increasingly stereotyped hunting behavior of the falcon. The specific objective of this appetitive behavior is the performance of consummatory behavior. Consummatory behavior is highly stereotyped and involves the performance of a fixed action pattern. In feeding behavior, swallowing is the consummatory pattern and involves a complex and highly coordinated sequence of contractions of throat muscles (Figure 32.5). These contractions are the components of a fixed action pattern. As you know, the performance of consummatory behavior is followed by a period of rest or

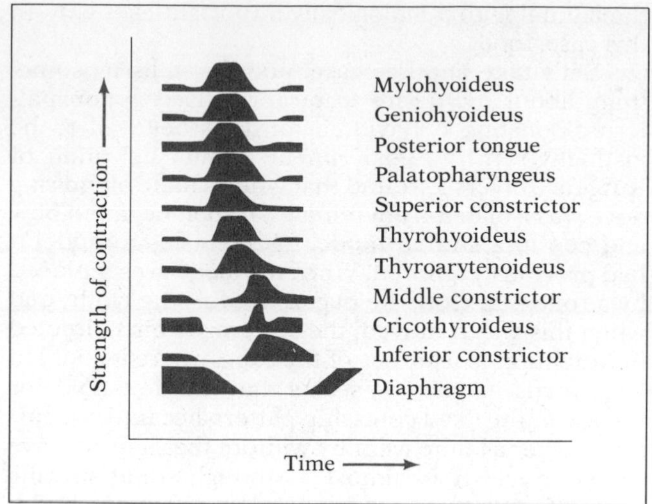

32.5 Diagrammatic summary of the sequence, timing, and intensity of the muscular contractions involved in swallowing in the dog. A fixed pattern such as swallowing, then, may have many components. In order for the pattern to be performed correctly, each component must function in a precise way. Such precision and complexity builds a certain conservatism into the overall pattern, since one part can't change drastically without affecting the other parts.

relief, during which the appetitive patterns can't be so easily initiated again.

Numerous experiments have demonstrated that the performance of consummatory behavior is, in fact, the actual goal of appetitive behavior. This means, in effect, that a hungry animal isn't looking for something to fill its stomach, but something to swallow. In an experimental demonstration, the food swallowed by a hungry dog was shunted away from the stomach by tubes leading to a collecting pan on the floor. As soon as the dog had swallowed the proper amount of food, it stopped eating, apparently satisfied, although its stomach was still empty. The performance of the consummatory swallowing patterns had apparently provided relief. The dog, however, began eating again sooner than it normally would have.

Releasers

Instinctive behavior is released by certain very specific signals from the environment. In the case of the falcon, the diving sequence was released by the sight of the teal flying below. The eating sequence was released by the sight of a downed bird. Courtship bowing is normally released in male ring doves by the sight of an adult female. Thus, environmental factors that evoke, or release, instinctive patterns are called *releasers*. The releaser itself may be only a small part of any appropriate situation. For example, fighting behavior may be released in territorial male European robins, not only by the sight of another male, but even by the sight of a tuft of red feathers at a certain height within their territories (Figure 32.6). Of course, such a response is usually adaptive because tufts of red feathers at that height are normally on the breast of a competitor. The point is that the instinctive act may be triggered by only *certain parts* of the total environmental situation.

The exact mechanism by which releasers work isn't known, but one theory is that there are certain neural centers, called *innate releasing mechanisms* (IRMs), which, when stimulated by impulses set up by the perception of a releaser, trigger a chain of neuromuscular events. And it is these events that comprise instinctive behavior. Many such IRMs may be involved in any complex pattern, so that specific releasers must be presented in sequence. Or, as we have learned, the performance of one consummatory act (such as a falcon's knocking a teal to the ground) may act as the releaser for the next part of the pattern (such as plucking the teal before eating it).

Action-Specific "Energy" and Vacuum Behavior

The energy that is believed to build up for each instinctive pattern is called *action-specific energy*. The continued buildup of this energy in the absence of a

32.6 A male European robin in breeding condition will attack a tuft of red feathers placed in his territory. Since red feathers are usually on the breast of a competitor, it is to his reproductive advantage to behave aggressively at the very sight of them. The phenomenon illustrates that releasers of instinctive behavior need to meet only certain criteria—in other words, need to represent only a specific part of the total situation. The model at left is more "birdlike," but lacks red coloration.

releaser may cause a progressive lowering of the threshold necessary to release a pattern, as we saw in the case of the male ring dove. Eventually, if no releaser is provided at all, the instinctive pattern may "go off" on its own. The performance of an instinctive pattern in the absence of a releaser is called *vacuum behavior.* Once, Konrad Lorenz noticed one of his tame starlings flying up from perches on the furniture and catching insects on the wing. He hadn't noticed any insects in his house, so he climbed up to see what the starling was catching. He found nothing. The well-fed starling had had no chance to release its insect-catching behavior, and so now the behavior was apparently being released in vacuo—in the absence of any stimulus.

The Behavioral Hierarchy

Niko Tinbergen proposed a *behavioral hierarchy* to define the relationships of the different levels of an instinctive behavioral pattern. According to this model, increasingly specific stimuli are needed to release the next lowest level of an instinctive pattern on a hierarchical scale. As an example, he described the pattern of breeding behavior in the male stickleback fish—a common, but fascinating, denizen of European ditches (Figure 32.7). The breeding behavior is triggered by the lengthening days of spring, and at that time the visual stimulus of a suitable territory may elicit either aggression or nest building. Nest building will commence unless another male appears;

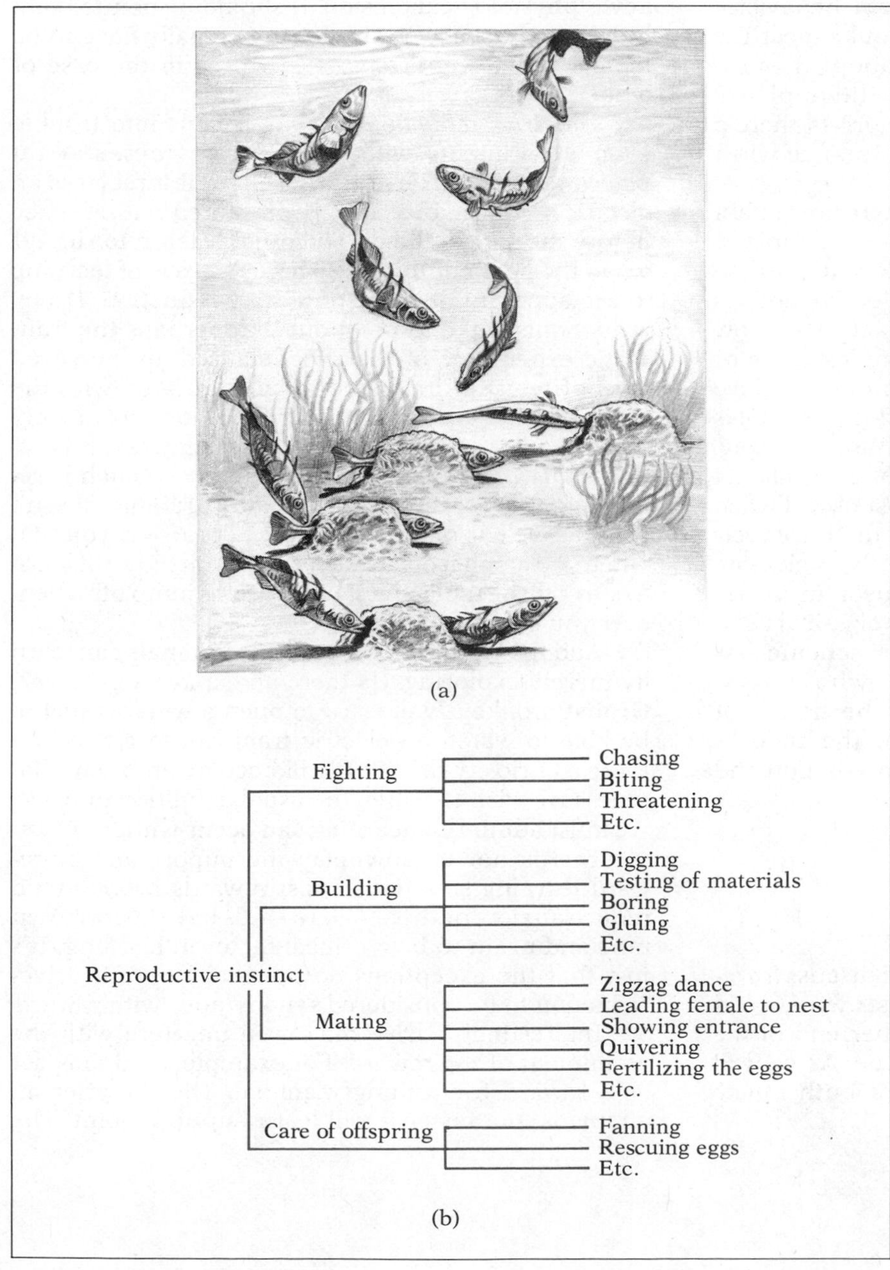

32.7 (a) The sequence in courtship and mating in the stickleback *Gasterosteus,* a fish found in European ditches and streams. The male is attracted by the sight of a female swollen with eggs. He is torn between chasing her and leading her to the nest he has built, and the result is a peculiar zigzag dance. Once she is maneuvered into his "tunnel of love," he pokes her tail with his snout and thus induces her to lay eggs. He then promptly chases her off, returns to fertilize the eggs, and then spends days fanning oxygen-laden water over them until they hatch. (b) The hierarchical organization of the reproductive instinct of the male stickleback, according to the model devised by Tinbergen in 1942. The overall reproductive instinct is composed of a number of different patterns. The pattern that is expressed is determined by what happens in the environment. These patterns are, in turn, composed of yet more specific ones, and their expression is also environmentally determined.

then aggression is released. The sight of a competitor will elicit any of a number of aggressive responses, depending on what that male does. If the intruder bites, the defender will bite back; if the intruder runs, the defender will chase, and so on.

According to Tinbergen's model, the "energy" for an instinct passes from higher centers to lower ones [from left to right in Figure 32.7 (b)]. This happens only when a block, or inhibitor, for a particular action is removed. Hence, the energy of the general reproductive instinct passes to a "fighting" center (an IRM) at the sight of another male. The sight of that male fleeing then removes a block at the next level of behavior, so that energy flows to the centers for chasing behavior.

It is thus apparent that the term *instinct* carries with it very specific implications. For a behavioral pattern to qualify as instinctive, it should meet the kinds of conditions we have talked about. For example, does the pattern show an appetitive phase? Does it ever appear as vacuum behavior? Is there a period immediately following its performance when the threshold for its release is raised?

You have undoubtedly noted that there are certain problems with the instinct model. For example, a strictly hierarchical organization makes it hard to account for the fact that different appetitive actions can lead to the same consummatory act. Also, how is it that an animal can perform a complex chain of fixed action patterns without intervening appetitive patterns? It is important to understand that *innate* is not synonymous with *instinctive.* Put another way, not all innate patterns are instinctive. The instinct concept described here is probably most useful when describing the behavior of insects and the lower vertebrates. But when not applied too rigidly, it gives an explanation for certain kinds of behavior in higher animals. What we have presented is the classical ethological interpretation of instinct. The scheme has been criticized by neurophysiologists, who, for example, can find no neural basis for the notion of action-specific energy. Perhaps, then, the model's greatest usefulness lies in behavioral prediction and as a springboard for further research.

Learning

Now let's leave the realm of instinct and discuss *learning.* There are many diehard behaviorists who would say that we are now moving to the other end of the behavioral spectrum. But this is not true. As we will see, adaptive behavior is comprised of both innate and learned components.

Reward and Reinforcement

Before we consider the various types of learning, it would be well to mention a much debated concept that enters into many types of learning. It is the concept of *reward,* or *reinforcement.*

In evolution, changes are preserved through their survival value. If their net effect helps the organism to survive, they are "good" and are preserved. But how does an individual ascertain which behavioral patterns are good? It has been said that if an act fulfills or relieves a body need, then it is good. That act is thereby reinforced so that it will probably be repeated when the same body need appears again. And then "good" has been defined as that which reduces a *drive* (such as the hunger drive); in other words, that which relieves physiological tension. It should be pointed out, however, that the reward doesn't actually have to be biologically advantageous (as we see in the case of addicting drugs).

The *drive-reduction hypothesis* runs into trouble from experiments with *avoidance responses.* A rat placed on an electric grid will run off if it receives an electrical shock. If a light is presented within 5 sec before the shock, the rat will quickly learn to run off when the light comes on. So here is a case of learning in the apparent absence of a drive-reduction. Drive-reductionists like to point out that perhaps the traumatic experience of the shock caused an increased level of tension that came to be associated with the light, and the rat ran in order to reduce its anxiety when the light came on. Therefore, they say, it is *reduced anxiety* (called a *secondary drive*), which is rewarding. (In some cases, avoidance training doesn't work as well as one might hope. If you swat your cat when you catch it on the kitchen table, it may not learn to stay off the table, but it will learn to jump off whenever you enter the room.)

And then, it has been found that animals can learn by merely exploring. (Is there an exploration drive?) Or that monkeys will learn to open a window just to be able to watch an electric train run in circles. (Is there a curiosity drive?) It will become apparent that unless we wish to widen the usual definition of drive, we must admit that learning can occur without it.

This is not to downplay the importance of reward learning situations. Most rewards probably do reduce anxiety or drive, and rewards have been shown time and again to be a critical factor in learning. It's just that the exceptions do not really permit drive-reduction to be considered synonymous with reward.

Interestingly, drive may even interfere with the attainment of the reward. For example, food may act as a reward for a hungry animal. The hungrier an animal is, the faster it will learn—up to a point. The animal may reach a level of hunger where its motivation is so high that it interferes with learning (Figure 32.8). In other words, the animal is so desperate for the reward that it can't put its mind to the problem of how to get the food. Its drive is so strong that it makes all the wrong moves and ends up with nothing. Perhaps you have experienced the same problem in a different context.

Another question is: At what level does reinforcement operate? Lorenz proposed that the mechanism which directly brings about the good, or adaptive, effect is the fixed action pattern—the consummatory act. (We have already discussed some problems arising with too strict an application of such an idea.) Lorenz said that the completion of an act (such as eating) has a reinforcing effect through *reafference,* or sensory feedback. As an example of such a learning–instinct interaction, consider nest building in crows. A crow, standing on a suitable nest site with nest material in its beak, performs a downward and sideward sweep of the head that forces the material against the substrate and later against the partially completed nest. When the twig meets resistance, the bird shoves harder, thrusting repeatedly, as would a man trying to push a pipe cleaner through a pipestem. When the twig is wedged in, its resistance is increased and the efforts of the bird are heightened until, when it sticks fast, the bird consummates its activities in an orgiastic maximum of effort. Then it loses interest in that twig. Some species of crows apparently possess no innate mechanism to guide their selection of nest material, so young birds will attempt to build with anything they can carry. Most of the unlikely material, which may include light bulbs or icicles, will not wedge firmly into the nest and so the birds are prevented from reaching the consummatory stimulus situation that leads to both relief and biological success. Such failure quickly extinguishes the bird's acceptance of inadequate material until, finally, the birds become true connoisseurs of twigs. In such situations, then, the reinforcer of a learning situation is the performance of an innate consummatory act.

32.8 When the trainee's motivation becomes too great, it may interfere with the efforts of the trainer.

It has also been shown that learning may function in placing innate patterns into their proper adaptive sequence through reinforcement mechanisms. Nest building in rats is accomplished by the performance of three motor patterns. First, the rat brings nest material to a selected site. Then, it stands in the center of the material and, turning to and fro, stacks the material into a roughly circular wall. Finally, it pats the wall with its paws, tamping it down and smoothing the inside. However, inexperienced rats, when offered paper strips, will get into a frenzy of all three activities—running about with nest material, patting, and building all at once. Each act is performed perfectly—but no nest results. The inexperienced rat, after carrying a few paper strips, will stand above them performing heaping movements in the air and tamping down a wall that doesn't exist. The failure of the pattern to provide a rewarding reaffirmation by its resistance to the rat's paws teaches the rat not to attempt heaping before the nest material is carried in, or patting before the wall is built.

Lorenz also pointed out that the structure and functional properties of the consummatory act were never fully understood until their teaching function was realized. He noted that there does not seem to be any case in which conditioning by reinforcement could be demonstrated in any behavioral system that does not include appetitive behavior. However, rats with electrodes implanted in their brains that are allowed to press a bar and send an impulse to their own pleasure centers will learn a task in order to be able to press the bar. On the face of it, this seems to refute Lorenz's notion that appetitive behavior must be involved, since there is not likely to be an appetitive pattern for bar-pressing or pleasure-center stimulation. However, we don't know what pleasure-center stimulation feels like. Does it feel like sexual relief? Eating sumptuously? We only know that the rat will frantically press the bar up to 7000 times per hour, until it collapses in total exhaustion. Also, since learning plays such an important part in appetitive behavior, can the goal sometimes be an acquired one? Must it always be innate? At this point we simply don't know enough about what appetitive action really is to be able to support or deny Lorenz's assumption.

Social Reinforcement in Humans

Not long ago, B. F. Skinner caused another stir—this time with his book, *Beyond Freedom and Dignity.* In it, Skinner surmised that the Utopian society might be reached through a system of positive rewards. He argued that all behavior is molded by reward and punishment, and that therefore "freedom" and "dignity" are illusions. This being the case, he said, it is better to have an open, planned, and rational system of behavioral control than a hidden and idiosyncratic one that is both inefficient and subject to the manipulations of unscrupulous persons. Those individuals who acted in accordance with some grand design would be rewarded. Aberrant individuals would go unrewarded and, hence, in time, almost all members of the society would be constrained to go along the prescribed route. Of course, a wide variety of humanists—from behavioral scientists to drunks—jumped on the idea with both feet.

It's worth noting that Joseph Stalin believed in a similar idea, and attempted to put into practice the rewards, punishment, and propaganda that would create "the new Soviet man." Karl Marx also stressed the effectiveness of social conditioning and the environment in determining the attitudes and behavior of people. This is why Pavlov's ideas on conditioning behavior have an almost sacred status in Russia, and it is also part of the reason that any notion of innate behavior patterns, especially in humans, is denounced with such vigor by contemporary Marxists.

It is probably erroneous to speak of "society's" expectations. The idea implies a "big brother," or some other consciously manipulating entity. However, it is more likely that the society or the "system" that molds us all—is actually nonexistent—at least in the sense of its being a body, aware of itself and its decisions. It may be what is often called—simply a *system*—not led, not controlled, but effective and self-perpetuating nonetheless, just as Skinner feared (Figure 32.9).

32.9 Throughout history people who do not fall in line with the values of the prevailing "system" have been treated with varying degrees of hostility. It has been assumed that even subtle disapproval, in the long run, can cause changes in behavior.

Four Types of Learning

The importance of learning for an animal varies widely from one species to the next. For example, there are probably very few clever tapeworms. But then, why should they be clever? They live in an environment that is soft, warm, moist, and filled with food. The matter of leaving offspring is also simplified; they merely lay thousands and thousands of eggs and leave the rest to chance—to sheer blind luck. In contrast, chimpanzees live in changeable and often dangerous environments, and they must learn to cope with a variety of complex conditions. They are long-lived and highly social, and the young mature slowly enough to give them time to accumulate information. We see, then, that learning is more important to some species than others. The way animals learn also differs from one species to the next, but we are only beginning to understand some of the differences.

Although learning may theoretically take place in a variety of ways, we will consider only *habituation, classical conditioning, operant conditioning,* and *autonomic learning.*

Habituation

Habituation is, in a sense, learning not to respond to a stimulus. In some cases, the first time a stimulus is presented, the response is immediate and vigorous. But if the stimulus is presented over and over again, the response to it gradually lessens and may disappear altogether. Habituation is not necessarily permanent, however. If the stimulus is withheld for a time after the animal has become habituated to it, the response may reappear when the stimulus is later presented again.

There are several ways in which habituation is important in the lives of animals. For example, a bird must learn not to waste energy by taking flight at the sight of every skittering leaf. A reef fish, holding a territory, may come to accept its neighbors, whereas it would immediately drive away a strange fish wandering through the area.

Habituation is often ignored in discussions of learning, perhaps because it seems so simple, but it may well be one of the more important learning phenomena in nature. For example, it may have something to do with consummatory satiation. It may also help to explain why animals continue to avoid predators (which they are likely to see only rarely), while ignoring more common, harmless species.

Classical Conditioning

Classical conditioning was first described through the well-known experiments of the Russian biologist Ivan Pavlov (Figure 32.10). In classical conditioning, a behavior that is normally released by a certain stimulus comes to be elicited by a substitute stimulus. For example, Pavlov found that a dog would normally salivate at the sight of food. He then experimentally presented a dog with a signal light 5 seconds before food was dropped onto a feeding tray. After every few trials, the light was presented without the food. Pavlov found that the number of drops of saliva elicited by the light alone was in direct proportion to the number of previous trials in which the light had been followed by food.

This conditioning process also worked in reverse. When the conditioned animal was presented over and over again with a light signal that was not followed by food, the salivary response to the light diminished until it finally was extinguished (Figure 32.11).

Pavlov's experiments demonstrated two other important properties of classical conditioning: *generalization* and *discrimination.* It was found that if a dog had been conditioned to salivate in response to a green light, it would also more-or-less respond to a blue or red light. In other words, the dog was able to generalize regarding the qualities of lighted bulbs. However, it was found that, with careful conditioning, the dog could be taught to respond only to light with certain properties. Dogs do not see color well, so they are more likely to respond to light-intensity differences. If food was consistently presented only with a green light, and never with a red or blue light, the dog would come to respond only to green lights, thus demonstrating clearly that it was able to discriminate differences among stimuli (discrimination) as well as the properties they had in common (generalization).

32.10 The Russian biologist Ivan Pavlov formulated the notion of *classical conditioning.* His findings, embraced by Soviet Communism, became an integral part of the Party's methodology in molding the "new Soviet man."

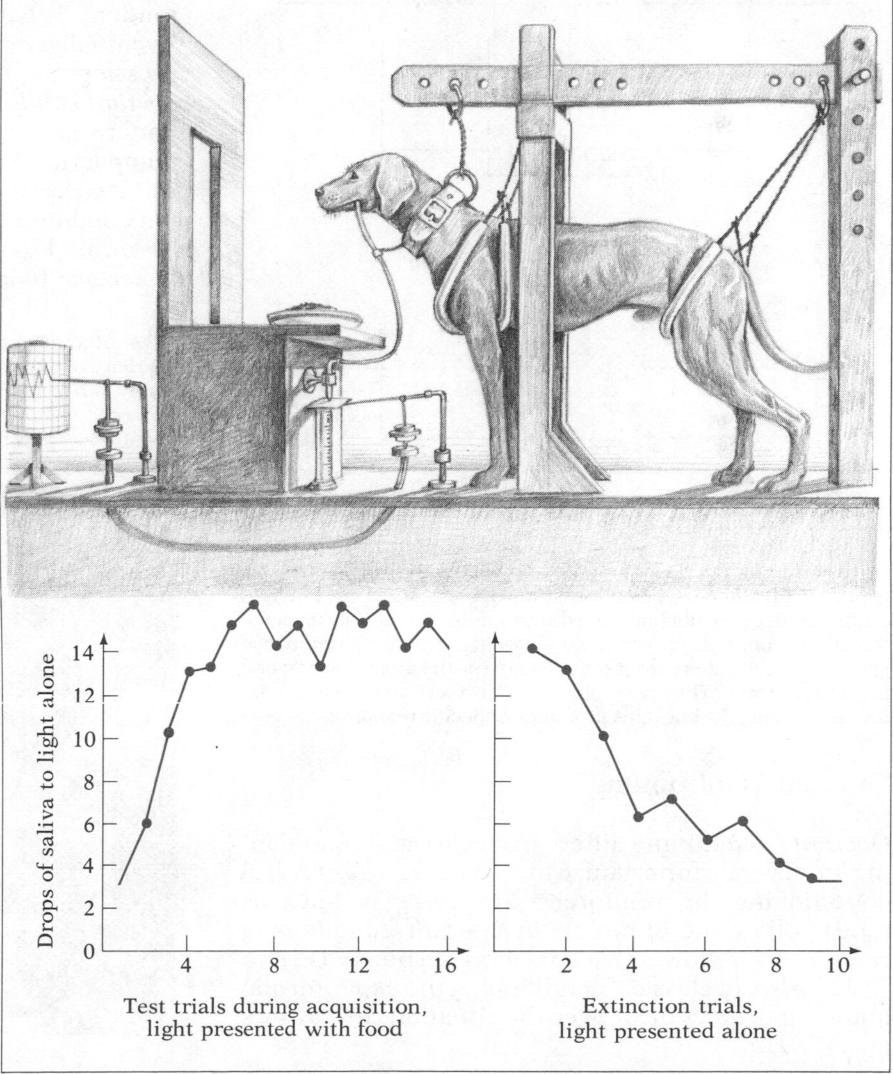

32.11 Pavlov's dog. This apparatus was devised to demonstrate classical conditioning. Upon presentation of a light, meat powder would be blown into the dog's mouth, causing him to salivate. Later, he came to salivate at the sight of a light alone. His salivation, then, was *conditional* on the light. Note in the first graph that the dog salivated at maximal levels after only eight trials. When the experiment was reversed and food no longer followed the light, the dog stopped salivating after only nine more trials.

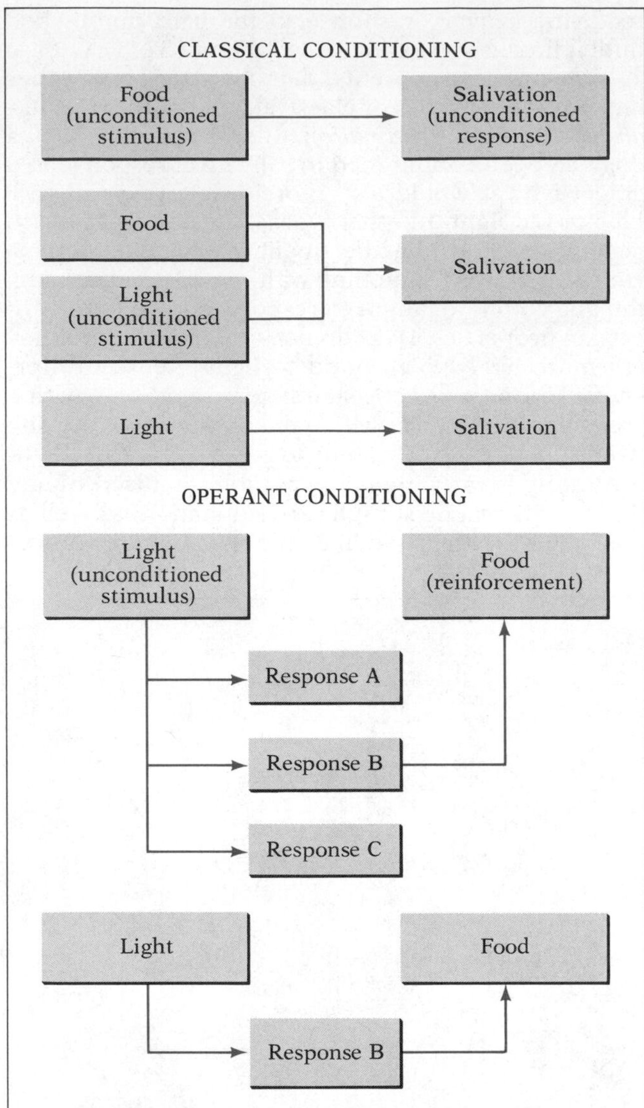

32.12 The organizational differences between classical conditioning and operant, or instrumental, conditioning. Note that in classical conditioning, an unconditioned stimulus becomes paired with a conditioned stimulus. The conditioned stimulus then becomes a substitute for the unconditioned stimulus to produce the unconditioned response. In operant conditioning, when a conditioned stimulus occurs, there is an opportunity for the animal to respond in various ways. However, only the "correct" response is reinforced. Finally, the stimulus produces a specific response.

Operant Conditioning

Operant conditioning differs from classical conditioning in several important ways. Whereas in classical conditioning the reinforcers (or rewards, such as food) follow the stimulus, in operant conditioning reinforcers follow the behavioral response (Figure 32.12). Also, in classical conditioning the experimental animal has no control over the situation. In Pavlov's experiment all the dog could do was wait for lights to go on and food to appear. There was nothing the dog could do one way or the other to make it happen. In operant conditioning the animal's own behavior determines whether or not the reward appears.

In the 1930s, Skinner developed an apparatus that made it possible to demonstrate operant conditioning. This device, now called a *Skinner box,* differed from earlier arrangements involving mazes and boxes from which the animal had to escape in order to reach the food.

Once inside the Skinner box, an animal has to press a small bar in order to receive a pellet of food from an automatic dispenser (Figure 32.13). When the experimental animal (usually a rat or a hamster) is first placed in the box, it ordinarily responds to hunger with random investigation of its surroundings. When it accidentally presses the bar, lo! a food pellet is delivered. The animal doesn't immediately show any signs of associating the two events, bar-pressing and food, but in time its searching behavior becomes less random. It begins to press the bar more frequently. Eventually, it spends most of its time just sitting and pressing the bar. Skinner called this sort of learning *operant conditioning.* It is based on the fact that if a pattern is rewarded, the probability of that pattern reappearing is increased.

It is possible to build "stimulus control" into operant conditioning situations. For example, a pigeon in a Skinner box may be required to peck a bar in order to receive food. If the apparatus is designed so that

32.13 Rat in a *Skinner box.* Boxes are designed to promote operant conditioning. For example, a hungry rat may move randomly, searching for food, until it accidentally presses a bar that delivers a food pellet. Each delivery means a greater probability that the rat will press the bar again, until finally the rat learns simply to press the bar each time it wants a pellet.

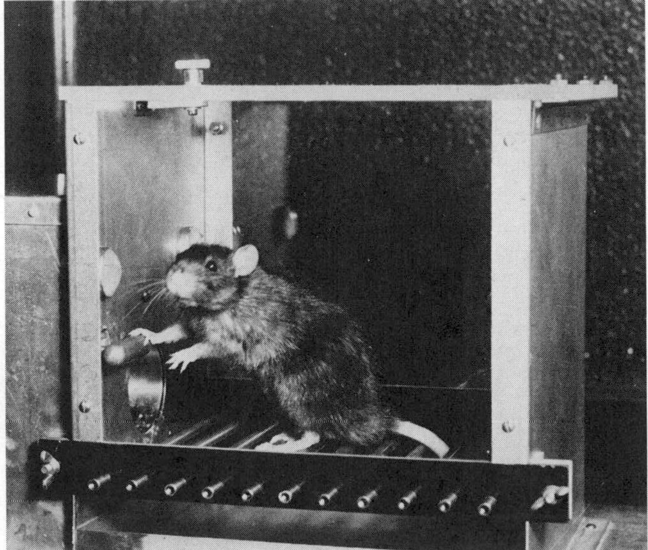

food is delivered only when the bar is pecked within a certain period after a sign that says "Peck" lights up, the pigeon will come to peck the bar mostly at that time. It will largely ignore the bar until the Peck sign is lighted. If the pigeon pecks after a "Don't peck" sign lights up, and that peck is not followed by food, the pigeon will come to ignore that sign through discrimination learning. (It should also be pointed out that pigeons can be trained to peck in response to the "Don't peck" sign—pigeons can't really read.)

The relative importance of each type of learning to animals in the wild isn't known at this point. However, it is likely that most behavior patterns arise in nature through the interaction of several types of learning.

The learning capabilities of an animal are largely related to its evolutionary heritage. For example, M. E. Bitterman of Bryn Mawr College demonstrated that there are strong species differences in *habit reversal* ability. In these experiments, once an animal had learned to discriminate between two choices for its reward, the reward was transferred to the opposite choice. Monkeys and rats showed the best performance, fish the worst, and turtles scored somewhere in the middle. The turtle scores were rather peculiar. They progressively improved in making spatial choices (for example, in differentiating high and low), but not in visual choices (such as in differentiating between a circle and a square). Thus, it appears that habit reversal requires a specific kind of learning ability that is present in mammals, but not in fish, and it exists in turtles only with respect to a certain class of problems.

Autonomic Learning

There is one other type of learning that should be mentioned—*autonomic learning.* The study of this type of learning is in its infancy, but it may have enormous implications. There have been problems in duplicating some of the original experiments, but the phenomenon has such tremendous potential that it is continuing to receive attention.

It has long been agreed that visceral responses (those under the control of the autonomic nervous system) could be classically conditioned, an example being Pavlov conditioning his dogs to salivate. However, even Pavlov didn't suspect that his dogs could be instrumentally conditioned to salivate *in order to receive* food. We have generally assumed that classical and instrumental conditioning operate by different neurophysiological mechanisms, and that visceral responses could only be learned through classical conditioning. However, there were those learning theorists all along who argued that the mechanisms are the same for both types of learning, and that if an arm could consciously be trained to bend, an iris could consciously be trained to contract.

About 1970, L. V. DiCara and N. E. Miller of Rockefeller University reported that they were able to produce, through instrumental training, increases and decreases in heart rate, blood pressure, intestinal contractions, control of blood vessel diameter, and even rate of urine formation.

Basically, their experiments involved using rats that were paralyzed with a curare derivative (the arrow poison of certain South American Indians). The effect of the poison was to allow the animals to remain fully conscious while unable to move skeletal muscles voluntarily. Miller and DiCara provided positive reinforcement to the rats by an electrode implanted in the "pleasure center" of the brain and negative reinforcement by a mildly unpleasant electric shock. The minor physiological variations that occur naturally throughout the body were the starting points for the experiments. For example, if they wished to train the animal to slow its heart rate, as the heart temporarily slowed naturally, the pleasure center would be stimulated; whenever the heart speeded up, the animal would receive a shock. They found that in some cases the heart rate soon remained slow. The reverse effect could also be achieved. When a rapid heart rate was rewarded, the heart rate increased.

Other recent experiments have shown that even brainwave patterns can be altered by training. Miller and his colleagues trained rats to raise or lower the voltage of their brainwaves. Then A. H. Black of McMasters University in Canada trained dogs to alter the activity of one kind of brainwave, the theta wave. Yogis, in practicing certain types of meditation, usually increase the number of certain brainwaves, particularly alpha waves (Figure 32.14). In this country, yoga novices are sometimes tested with simple electroencephalographs to determine whether they are properly controlling their thought patterns. And in recent years, the conscious control of alpha waves became another American fad among those trying to escape the rigors of modern life.

32.14 Yogis, and certain other practitioners of meditation, are able to change their brainwaves and produce primarily those associated with deep rest and relaxation.

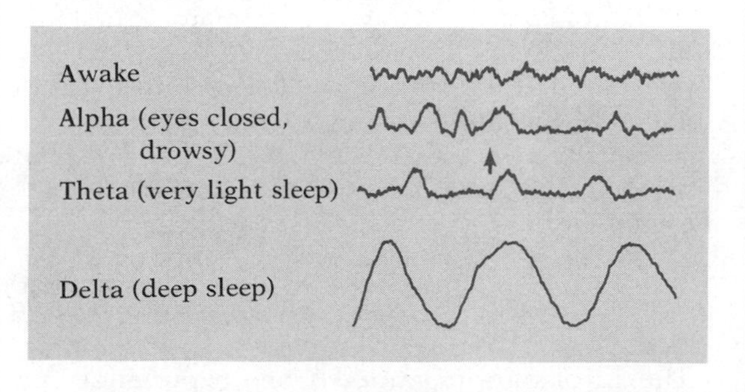

To rule out the possibility that such learning was really due to some sort of general changes in the autonomic nervous system, other tests were performed. For example, when heart rate and intestinal contraction were tested separately, it was found that changes could be brought about in one without altering the other. Also, it was found that visceral changes can be made in very specific areas without affecting other parts of the body. The blood vessels of one ear could even be trained to dilate more than those of the other ear.

The implications of autonomic learning in nature are enormous. Perhaps such learning plays a far more important role in animal physiology than has yet been imagined. For example, suppose gulls swimming in frigid North Atlantic waters learn to respond to unusually cold temperatures by constricting the vessels in their feet. This would reduce the amount of fluid flow in that area and thus prevent a dangerous lowering of the body temperature. Suppose an arctic fox learns to withdraw blood from its ears on the same basis. Such learning, of course, would function only as part of an overall adaptive pattern that involves morphological specialization as well. For example, the gull has circulatory shunts that permit blood to by-pass extremities under certain conditions, and the ear of the arctic fox is small with a minimal surface area exposed to the elements. We need only point out the implications of autonomic learning in habitat selection. This ability would permit any animal to exist in highly variable or changing habitats by increasing its flexibility through the expansion of the physiological tolerance of the animal.

What are the implications of such learning for humans? Aside from the benefits extolled by the practitioners of yoga and the bizarre feats accomplished by certain cultists (some of these phenomena may, it is believed, be based on autonomic conditioning), there are certain very practical therapeutic benefits of the process. Already, teams of doctors have shown some promising results in helping heart patients to control their heart rate, and in training epileptic patients to control a type of abnormal brainwave associated with this problem. This research, however, is in very preliminary stages.

On the other hand, visceral learning may be at the root of certain psychosomatic illness. If prolonged tension, for example, brings about headaches, the appearance of the discomfort may be "rewarded" by rest, a round of ale, or even (for some) a gratifying feeling of martyrdom. The appearance of the headache is repeatedly reinforced until finally a person unconsciously learns to bring about those internal changes (such as the constriction of certain blood vessels) in order to cause a headache which will be rewarded.

32.2 Memory

Memory and Learning

> Many years had elapsed during which nothing of Cambray had any existence for me, when one day in winter my mother offered me some tea . . . . I raised to my lips a spoonful of the tea in which I had soaked a morsel of cake. No sooner had the warm liquid, and the crumbs with it, touched my palate than a shudder ran through my whole body . . . . And suddenly the memory returns. The taste was that of the little crumb of madeleine which on Sunday mornings at Cambray my aunt Leonie used to give me, dipping it first in her own cup of real or of lime-flower tea . . . . Once I had recognized the taste, all the flowers in our garden and in M. Swann's park, and all the water lilies on the Vivonne and the good folk of the village and their little dwellings and the parish church and the whole of Cambray sprang into being.
>
> MARCEL PROUST
> *Remembrance of Things Past*

The information gained from experience must not only be stored but also must be available for recall in order for there to be any advantage to experience. The storage and retrieval process is what is known as *memory,* and its characteristics are important in any consideration of learning.

Learning is accompanied by changes in the central nervous system. There is a more or less permanent record of the learned thing, and in humans at any rate, it can sometimes be consciously recalled. At other times, it can be jarred into consciousness when an experience triggers a memory, perhaps involuntarily, as occurred in Proust's remembrances. Something in the neural apparatus changes. The physical change that is presumed to occur in the brain when something is learned is sometimes called the *memory scar, memory trace,* or *engram.*

It is suspected that virtually everything we encounter is learned—stored away in the brain. Wilder Penfield and his group working in Montreal used electrodes to probe the brain (which has no pain receptors) of conscious patients, and asked them to describe their sensations as various parts of the brain were stimulated. The results were startling. Some patients "heard" conversations that had taken place years before, some heard music or seemed to find themselves with old friends, long deceased. They seemed to almost "re-live" the experiences, rather than simply remember them. The powers of recall under such circumstances are phenomenal. A hypnotized bricklayer described markings on every brick that he had laid in building a wall many years before. His recollections were carefully recorded, then the wall was found and examined. The markings were there! (Our memory is not infallible, however. We also have a tendency to embellish partially recalled events!)

Although we may have entered every jot and tittle of our lives into our mind's ledger, we are not consciously able to recall very much of it. Sometimes, we can't even recall what we have tried to memorize—as you well know. But apparently the information is there, just the same. The fault lies with our recall techniques. The implications of this finding are enormous. If our research ever enables us to recall or "re-live" earlier events, we might "re-read" novels on long train trips and all our exams would be virtually open-book tests. Perhaps terminally ill and pain-racked patients could be stimulated to "re-live" happier, youthful days with loved ones, long dead, while life dwindled in a changed body.

The Advantage of Forgetting

There are all sorts of strange abilities associated with memory in humans. *Idiot savants,* a special class of retardates, have very low IQs, but some are able to accomplish incredible mathematical feats, such as multiplying two five-figure numbers in their heads. Others can immediately tell you the day of the week on which Christmas day fell in 1492—or any day in any year (although this is probably not a memory feat). A normally intelligent Russian man made his living giving stage performances as a *mnemonist.* He would sometimes memorize lines of fifty words. Once, in just a few moments he memorized the nonsense formula.

$$N \cdot \sqrt{d^2 \cdot \frac{85^3}{vx}} \cdot \sqrt[3]{\frac{276^2 \cdot 86x}{n^2v \cdot 264}} \cdot n^2b = SV \cdot \frac{1624}{32^2} \cdot r^2s$$

Fifteen years later, upon request, he repeated the entire formula without a single mistake.

And what about those people with "photographic memories"? They exist, and they are called *eidetikers.* Proof of their abilities has been demonstrated with *stereograms.* These are apparently randomized dot patterns on two different cards; when they are superimposed by use of a stereoscope, a three-dimensional image appears. One person was asked to look at a 10,000 dot pattern with her right eye for 1 min. After 10 sec, she viewed another "random" dot pattern. She then recalled the positions of the dots on the first card and conjured up the image of a "T."

So, if such abilities are possible for our species, why haven't they been selected for so that by now we can all, more or less, perform such feats? On the surface, the advantages seem enormous. The reason we can't is, in part, because there are serious drawbacks to remembering everything. Both the Russian mnemonist and the eidetiker could look at a barren tree, "recall" its leaves, and, when they looked away, be confused over whether the tree was leafy or not. The mnemonist could watch the hands on a clock and not notice they had moved, remembering (or "seeing") only where they had been. Also, what about all those insignificant events it is of no advantage to remember? The energetically expensive neural apparatus would be wasted retaining such information for recall (assuming such retention takes more energy than the storage of material we normally can't recall). And the mnemonist had trouble with discerning time lapse. He recalled everything so well that it seemed to him as if every event of his life had just occurred.

Obviously, the key here is that the reason we can't remember as well as the people in these examples is that we, as a species, have not found it necessary or useful. We generally don't need to recall every stone on the path to the place where we found food yesterday; we only need to remember the location of the path. In fact, remembering too much about our physical environment might cause us to relate to the environment on the basis of remembrances. Thus, any change in the environment might not be adjusted to as quickly as if we were never quite sure as to what to expect, because we had learned we couldn't rely too strongly on our memories.

The Consolidation Hypothesis and Octopus Memory

It now seems that there are two separate mechanisms by which information is stored, as evidenced in the differences between *long-term memory* and *short-term memory.* The existence of these two types of memory was first noticed in humans with brain concussions who were unable to recall what happened just before an accident, but could remember what happened much earlier. On the basis of such findings, two memory systems with information that could be affected independently were postulated. These two systems were conceptualized as long-term memory and short-term memory. Information stored in the long-term memory system is relatively permanent (subject to very slow decay). Once information is processed into the long-term system, it is not easily disrupted and therefore it may be recalled with minimum confusion after long periods of time. Short-term memory, on the other hand, is relatively easily disrupted and subject to rather rapid decay. (This is the reason cramming for exams is not a good idea. If the exam is postponed, all is lost.) Also, there is evidence that the short-term memory can be overloaded, whereas the long-term memory can't.

Apparently, short-term storage is necessary before long-term storage can occur. In a series of experiments, it was found that if an inedible object is presented to one arm of an octopus, over and over in rapid succession, the arm will come to reject it after learning that it is inedible. If the object is then immediately presented to other arms, they will accept it. But if some time is allowed to elapse before the second presentation, all the arms will reject it. This finding has been interpreted as meaning the information acquired by one arm is only locally adaptive and must filter from a hypothesized short-term center to a long-term system before a general adaptive (total rejection) pattern can appear. The experiment actually only shows that something filters from neural centers associated with one arm to general centers. The short- and long-term phenomena are postulated because the learning arm ends up with no more information than the other arms once the information has become integrated.

In another experiment, a mouse is put into an arena (enclosed area) containing only a box with a hole. The mouse will run into the hole and get a terrific shock to its feet. That mouse will never run into the same hole again as long as it lives. But if such a mouse is given electroconvulsive or chemical shock treatment within about an hour after the unpleasant incident, it will forget the whole experience, and helplessly run into the same hole the next day, and every day as long as it survives the experiment. But if the electroconvulsive treatment is delayed for 2 hrs, it is no longer effective, and the mouse cannot be made to forget its aversive experience. Mouse memory, then, is labile for up to about an hour, and then becomes fixed. We infer that transfer is made from short-term "holding" memory to permanent memory in that time.

Such observations have led to the development of a *consolidation hypothesis*, which states that memory is stored in two ways. First, any registered experience enters a short-term memory system, where it causes changes that last for about an hour; then the system returns to normal (forgets). But while events are still in the short-term system and subject to recall by memory, they are being shifted into the long-term memory system, where they form an engram. Actually, the engram theoretically need not be formed from the short-term memory system. It could be formed through some parallel mechanism. The transfer from short-term to long-term is called *consolidation*, and during this process, the memory is susceptible to all sorts of disturbances or even obliteration. Interestingly, it can be shown that consolidation does not occur if protein synthesis is temporarily blocked by selective poisons.

The Mammalian Memory

Some of the most surprising results in the area of mammalian memory have come from experiments involving what is called the *split-brain technique*. It would be nice, somehow, to be able to say that the term is a misnomer. But it isn't; brains are split. The technique involves separating the cerebral hemispheres by cutting them down the midline through the tract that links them—the tract called the *corpus callosum* (Figure 32.15).

A rather startling discovery in brain-damaged humans led to these experiments. The halves of the

32.15 The corpus callosum is a massive trace of white neuron fibers known as a *commissure*. These fibers link the two cerebral hemispheres together, thus permitting the relay of information from one side to the other.

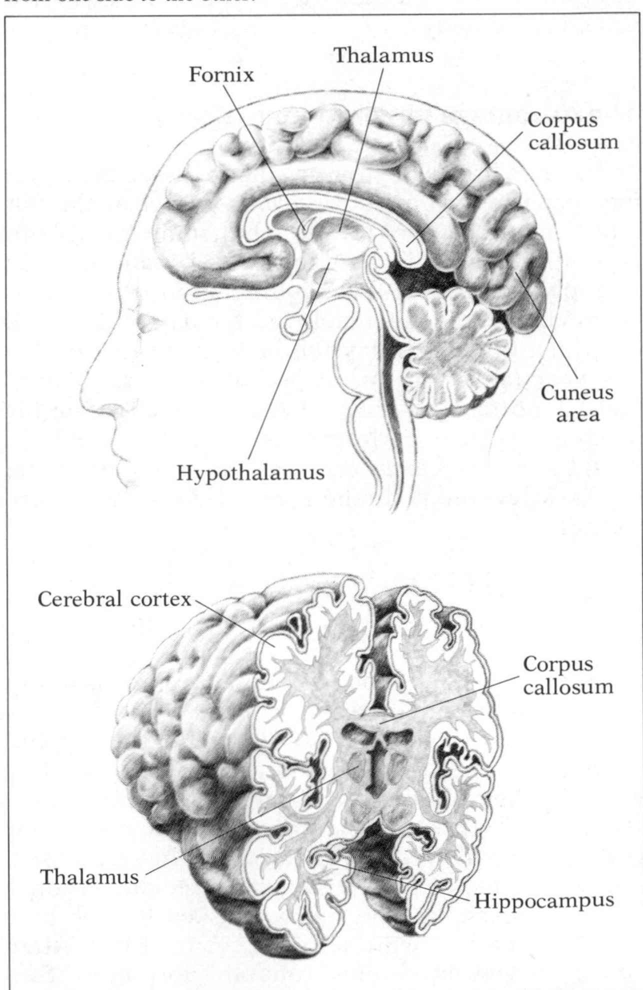

brain could be severed all the way back to the cerebellum with apparently no ill effect. It took a long series of careful experiments to determine the function of the corpus callosum and to solve the puzzle of how brains could be split without causing death. In fact, splitting brains by severing the corpus callosum is now a recognized treatment of certain kinds of epilepsy.

If we learn a visual discrimination with one eye covered, we can make the same discrimination with the other eye. Anatomically, the reason is because fibers from the right and left halves of each eye cross over the *optic chiasma* to the other side of the brain, as shown in Figure 32.16. Thus, the visual centers in both halves of the cerebrum receive information from both eyes. Now, if the optic chiasma is split, the right eye can still make the same discriminations as the left, and vice versa (Figure 32.17). However, if the corpus callosum is also split, then things learned by one eye remain unknown to the other eye. In fact, anything learned by one half of the brain is not transferred to the other half. This means, in effect, that mammals have two brains that can act independently. It is even possible to train both halves of the brain separately. A split-brain monkey can be trained to approach an object seen with one eye, and to withdraw from the same object if it is seen with the other eye. If that monkey sees the object with both eyes, however, usually one side of the brain will take over and the monkey will respond without hesitation by either approaching or withdrawing.

The split-brain research shows that one function of the corpus callosum may be to transfer engrams, or their precursors, from one cerebral hemisphere to the other, with the result that most information is stored bilaterally.

Since an animal can function normally with only one cerebral hemisphere, the other hemisphere can be experimented with rather drastically without endangering the health of the animal. For example, most of one hemisphere can be removed to see what the function of the remaining tissue might be. Through such techniques it was found that the various areas of the cortex are not of equal potential. Certain areas have abilities that cannot be assumed by other areas. For instance, in humans, language is localized in the left cerebral hemisphere, while spatial relations are controlled by the right. A split-brain subject simply can't draw even the simplest geometric figure with the right hand, which is controlled by the left hemisphere; but the subject can do very well with the left hand. In normal persons, apparently, spatial information is transferred to the right hand through the corpus callosum.

We should soon have answers to a series of important questions through split-brain experimentation. For example, how long does it take an engram to get through the corpus callosum? If only some of the corpus callosum's fibers are cut, does the engram become imperfect or just take longer to get through? What connections are necessary between cortical areas and subcortical areas in order for the former to store engrams?

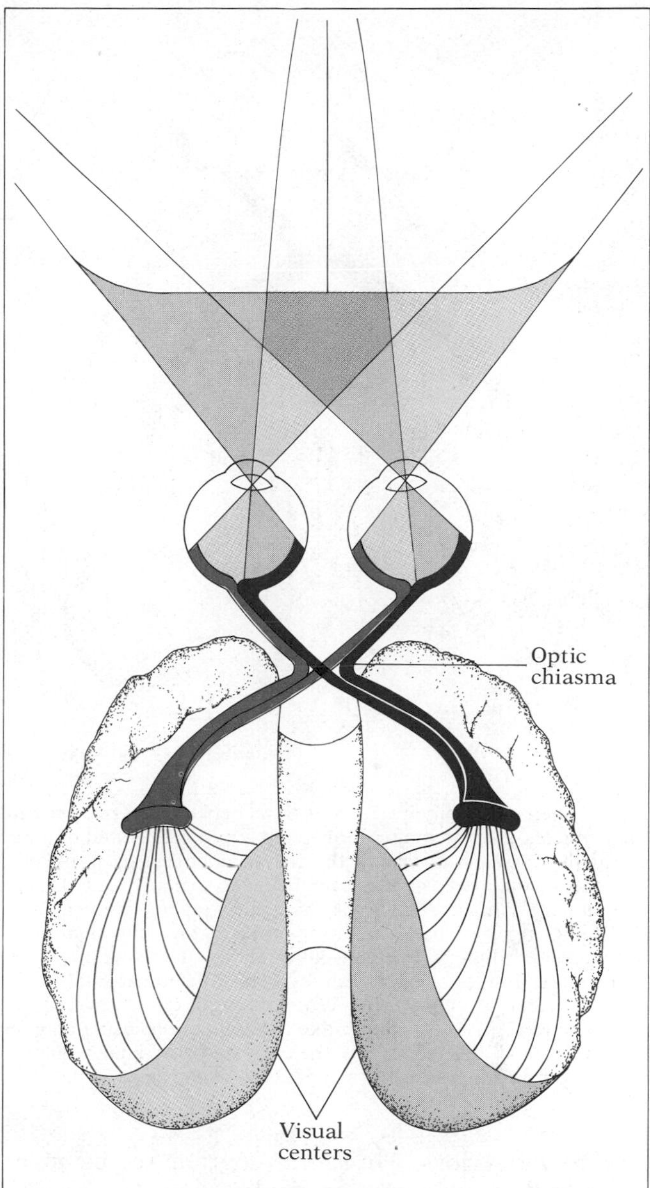

32.16 The optic chiasma is formed where tracts of visual nerves cross in the brain. The two visual centers of the brain each receive visual input from both eyes.

Experiments involving removal of parts of the cortical mass indicate that the engram doesn't seem to be localized in any particular part of the brain. For example, excluding the language center, it doesn't matter much *which part* of the cerebrum is excised, but *how much*. Fifty percent removal results in about

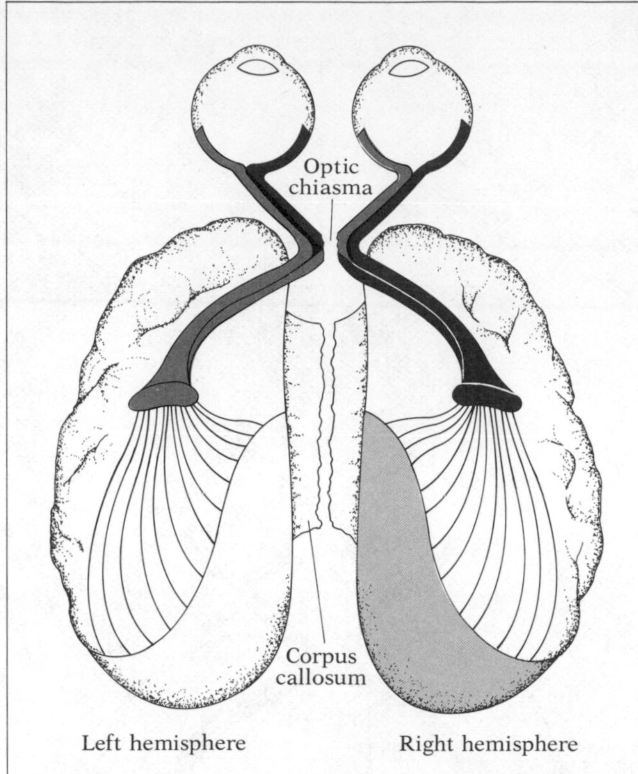

32.17 There are many interconnections between the two cerebral hemispheres, but the major connecting organ is the band of tissue called the *corpus callosum.* If the corpus callosum in monkeys is cut, the animals will continue to behave normally, but through experimentation, the halves of the brain can be revealed to be capable of functioning as separate and roughly equivalent units, although each has its own special qualities. If the *optic chiasma* is also cut, images seen by one eye can't be transferred to the opposite cerebral hemisphere. When the right eye is covered, the monkeys can learn tasks and make discriminations using only the left eye. But if the left eye is then covered, the animal can no longer perform the same tasks and has to be retrained.

50% memory loss. Thus, the engram seems to be diffused throughout the cortical tissue.

We still don't know whether consolidation takes place in the cortex itself, or if it occurs in some lower brain center and is then transferred to the cortex. But the lower brain centers undoubtedly play a role in consolidation. For example, the *hippocampus* is part of the limbic system and one hippocampic area lies along the lower inside edge of each cerebral hemisphere (Figure 32.15). The hippocampus connects directly to the hypothalamus. Damage to both lobes of the hippocampus results in a person being unable to recall anything that happened in the immediate past, but he or she has no problem in remembering what happened before the brain damage. Apparently, in the latter case, the person is drawing on a fully formed engram.

The hippocampus is also attached by broad tracts to the *reticular system,* that mysterious net that functions in arousing certain parts of the cortex. Damage to the reticulum also retards remembrance of recent events. Thus, it would seem that continuing *attention* is important in forming an engram. Attention is impaired by damage to either the hippocampus or reticulum.

Here we are speculating about how engrams are formed, when we don't even know whether engrams *exist,* at least not in the sense that we have been discussing them. There is a danger in hearing a name and constructing an imaginary working picture that accounts for what we've heard, and then assuming this imaginary thing exists. However, if we hope to ever gain some real understanding of behavior, we must keep in mind the level of supportive evidence for each assumption we encounter.

How Instinct and Learning Can Interact

Much of the argument in the field of behavior has been over one simple question: Is a certain behavior innate or learned? After many years of bickering, the question itself is no longer regarded as valid. The supposition that a particular behavior stems from one source or the other neglects the myriad ways in which behavioral components interact. In a sense, it's like asking whether the area of a triangle is due to its height or its base. A better question is, how do innate and learned patterns interact in the development of a particular behavior? (Notice that we are not equating innate and instinctive. Innate patterns, you recall, may not require releasers and may never be performed in vacuo. Instinctive patterns are a special kind of innate behavior.)

As an example of such interaction, consider the development of flight in birds. Flight is a largely innate pattern. Obviously, a bird must be able to fly pretty well on the first attempt, or it will crash to the ground as surely as would a launched mouse. It was once believed that the little fluttering hops of nestling songbirds were incipient flight movements, and that the birds were, in effect, learning to fly before they left the nest. But then someone performed the inevitable experiments. In one group, nestlings were allowed to flutter and hop up and down, while another group was reared in boxes that prevented any such movement. At the time in their lives when the young birds would have normally begun to fly, both groups were released. Surprise! The restricted birds flew just as well as the ones that had "practiced"!

Flight behavior, then, is obviously an innate, or unlearned, pattern. It is important to realize, though, that generally, young birds don't fly as well as adults.

Many innate patterns can be improved upon by learning through practice.

In other instances, the innate and learned components of a behavior can be more clearly differentiated. The process by which a red squirrel opens a hazelnut has been shown to be a rather complex behavior. The squirrels first cut a groove along the growth lines on one or both of the flat sides of the nut, where the shell is thinnest. Then they insert their incisors into the groove and break the shell open. In one study, ethologist Irenaüs Eibl-Eibesfeldt reared baby squirrels without any solid food. When they were finally presented with a hazelnut, they correctly performed the gnawing actions and even inserted their incisors into the grooves correctly. Thus, these patterns are apparently innate. The problem was that the young squirrel gnawed all over the surface of the nut and started several grooves without really completing any of them. Eventually they were usually able to break open the shell, but only after a great expenditure of energy. In time, however, their performance improved. After the squirrels had handled a few nuts, they began to make their grooves at the thinnest part of the shell, where gnawing was easier, so that the shell then opened more easily. Most of the squirrels soon became proficient at opening hazelnuts, as their innate gnawing pattern was modified through learning to produce the complex adaptive result.

The relative importance of innate and learned components of behavior may vary widely. For example, whereas reproductive behavior in stickleback fish may be largely innate, the nut opening in squirrels is perhaps less so. The song of songbirds may be largely learned, or almost completely innate, depending on the species.

Moreover, white-crowned sparrows, in order to learn their song properly, must hear the song of their species during a brief "sensitive period" when they are too young to sing at all. If they are not exposed to the song until after this period, they will produce abnormal songs, lacking the finer details.

So, we see that some behavioral patterns are almost completely innate, such as the reproductive behavior of the stickleback fish or the sucking response in young mammals, whereas others are almost entirely learned, such as songs in certain birds or the backstroke in humans. Behaviors which entail both instinct and learning may depend more on one element than another. All this suggests that it is better to ask how a certain behavior develops than to ask whether it is innate or learned.

Application of Ideas

1. Clearly distinguish between innate behavior and instinctive behavior.
2. What have been the main factors (or information) that have drawn ethology and comparative psychology together?
3. Choose what you consider to be an instinctive act and break it down into its appetitive and consummatory behaviors. (Remember that there may be several consummatory patterns involved in any complex behavior.)
4. How does our social system mold our behavior? How is it set up to administer subtle, as well as overt, rewards and punishments?
5. What is the evidence that engrams exist?

Key Words

action-specific energy
appetitive behavior
autonomic learning

behavioral hierarchy

classical conditioning
comparative psychology
consolidation
consummatory behavior
corpus callosum

Key Ideas

Ethology and Comparative Psychology

1. The study of animal behavior was principally philosophy until the theory of *instinct* was developed in the 1930s.
2. The founders of *ethology* were Konrad Lorenz, Niko Tinbergen, and Karl von Frisch. *Ethology* concerns itself with *innate* behavioral mechanisms, observed primarily in the field.
3. The field of *comparative psychology*, led by B. F. Skinner, studies behavior in the laboratory.
4. The two fields will probably find common ground in the fledgling science of *neurophysiology.*

discrimination
drive
drive reduction

eidetakers
engram
ethology

fixed action pattern

generalization

habit reversal
habituation
hippocampus

innate
innate releasing mechanisms (IRMs)
instinct
instinctive pattern

labile
learning
long-term memory

memory

neurophysiology

psychological tension

operant conditioning
orientation

reafference
recall
reduced anxiety
reinforcement
releaser
response threshold
reticular system
reward

short-term memory
Skinner box
split-brain
stimulus

vacuum behavior

Instinct

1. *Instinct* refers to certain classes of responses that are *innate* (inborn or present at birth). Overuse of the term caused it to lose its meaning.

2. The *instinct* concept can be broken down into components.

Fixed Action Patterns and Orientation

1. Precise identifiable behavior patterns that are *innate* and characteristic of a given species are called *fixed action patterns.* When coupled with *orienting movements* they are known as *instinctive patterns. Orienting* movements and *fixed action patterns* are not always clearly separated.

2. Delayed *instinctive patterns* may have learned aspects, but their delayed appearance does not automatically mean that learning is involved.

3. The falcon hunt has a pronounced variable *orienting* aspect: the search. It has a specific fixed action pattern: the dive. Its feeding pattern is also *fixed* and includes plucking and tearing in a predictable way. Its swallowing is *fixed.* It will not perform its hunt again for some time, having *released* the *fixed action pattern* of swallowing. Thus in the hunt, hunger, the drive, brought on the increasingly fixed patterns of behavior.

Appetitive and Consummatory Behavior

1. *Appetitive behavior* is usually variable and nonstereotyped, while *consummatory behavior* is *fixed. Consummatory behavior* may involve the execution of one or several *fixed action patterns.*

2. In feeding, the act of swallowing (rather than an increased blood sugar level or a full stomach) appears to be the *consummatory* aspect.

Releasers

Releasers are specific signals from the animal's environment. They bring on *instinctive* behavior. There may be neural centers called *innate releasing mechanisms* (IRMs) that trigger neuromuscular events. The performance of one *consummatory act* may provide the *releaser* for the next.

Action-Specific Energy and Vacuum Behavior

The buildup of energy for an *instinctive* pattern is known as *action-specific energy.* As it builds, the *threshold* for its *release* lowers. In the absence of a specific *releaser,* it may eventually go off on its own. This is called *vacuum behavior.*

The Behavioral Hierarchy

1. Tinbergen noted different levels of an *instinctive behavior pattern.* He arranged these in hierarchical scale. The classic story of stickleback mating behavior is an example. He claims that the energy for *instincts* passes from higher centers to lower. As inhibitors for each act are removed, behavior moves from one level down to another [see figure 32.7 (b)].

2. In summary, there are many unanswered questions about the *instinct* model. It is important to know that not all *innate patterns* are *instinctive.* There are subtle differences between the two terms.

Learning

Adaptive behavior is comprised of both *innate* and *learned* components.

Reward and Reinforcement

1. One theory states that some acts may not necessarily do the animal any biological good, but may simply relieve physiological or psychological tension. These are known as *drive-reduction* acts, and the theory's proponents claim that the animal is *rewarded* by *learning* to do what reduces anxiety.

2. Other observations indicate that animals learn by exploring and by being driven by curiosity. In other words *drive-reduction* is not synonymous with *reward.*

Social Reinforcement in Humans

1. B. F. Skinner surmised that society's problems could be solved through application of positive *reward* systems. Correct behavior would be *rewarded,*

while aberrant behavior would go *unrewarded* (and unpunished). Eventually people would go along with the prescribed route.

2. In human society the *rewards* are not always readily apparent, but may operate in small, subtle ways.

Four Types of Learning

The importance of *learning* is proportional to its need. Some animals exhibit little or no need, while others depend heavily on *learning* for survival. There are four ways animals learn: *habituation, classical conditioning, operant conditioning,* and *autonomic learning.*

Habituation

Habituation is *learning* not to respond to a *stimulus.* Animals learn to sort *stimuli* and to stop responding to those that are irrelevant to their survival.

Classical Conditioning

1. Pavlov's classical experiment demonstrated that *stimuli* can be substituted. Thus the dog learned to salivate at a signal that it had associated with food, even when the food was not present.

2. The dog also learned to *generalize,* so that a different colored signal light would provoke the same *response.* With further training, the dog would learn to discriminate colors and its *response* would become more specific.

Operant Conditioning

1. In *classical conditioning* the reward follows the *stimulus,* while in *operant conditioning* the *reward* follows the behavioral response. In the former, the animal's role is passive, while in the latter the animal is active.

2. *Operant conditioning* was developed using the Skinner box, a device that *rewarded* the subject for the correct response.

3. How these kinds of learning operate in the wild is not known, but all three probably occur.

Autonomic Learning

1. It was formerly believed that visceral or *autonomic learning* could only occur through *classical conditioning.* In 1970 scientists reported that certain visceral functions could be controlled by *operant conditioning.* Animals were *rewarded* or punished as they learned visceral control.

2. *Autonomic learning* may explain how animals survive in variable or changing habitats by *learning* to increase their physiological tolerance.

3. Physicians are interested in *autonomic learning* for its obvious applications to the treatment of chronic illnesses. It may be the root of psychosomatic illnesses where one unconsciously learns to bring on symptoms.

Memory and Learning

The physical change that is presumed to occur in the *brain* when something is learned is known as an *engram.* Everything we encounter may be stored in the brain. Studies by Wilder Penfield reveal that electrical stimulation can bring on *recall* of experiences long since forgotten.

The Advantage of Forgetting

There are serious disadvantages in total memory *recall.* There may be no advantage in terms of how we function in an ever-changing environment. Total *memory* may also be energetically prohibitive.

The Consolidation Hypothesis and Octopus Memory

1. Two separate mechanisms of information storage are known. These are *long-term memory* and *short-term memory.* The former is permanent (with slow decay). The latter is easily disrupted and can probably be overloaded.

2. From experiments it was concluded that information is first filtered through *short-term memory* centers and later finds its way to *long-term* centers. This transfer is called *consolidation.* The process will not occur if protein synthesis is temporarily blocked.

The Mammalian Memory

1. *Split-brain* experiments have provided information about *memory* in mammals. The technique involves cutting the *corpus callosum.* If done skillfully it does no apparent damage.

2. Normally information is transferred from one side of the brain to the other through the *corpus callosum.* Experiments with *split-brain* subjects reveal that *learning* in one side (through one eye) will be stored only on that side.

3. *Engrams* appear to be diffused throughout the cortex. When parts of the cortex are removed, the amount of *memory* loss is proportional to how much is removed, not where the removal takes place.

4. Lower brain centers apparently play a role in *consolidation.* Damage to one region, the *hippocampus* of the limbic system, results in loss of *memory* only in events since the damage. Long-term memory is not affected. Similar results occur when the *reticular system* is damaged.

How Instinct and Learning Can Interact

The question as to whether a particular behavior is *instinctive* or *learned* is no longer valid. It is better to ask how the two interact. Even bird flight, long believed to be *learned,* is *innate.* On the other hand, the behavior of squirrels opening nuts shows definite *learning* aspects as the animal gains experience.

Review Questions

1. Carefully define the term *ethology* and explain how it differs from *comparative psychology.* (p. 945)
2. Why did the term *instinct* fall into disrepute in the behavioral sciences? (p. 945)
3. Using the feeding of a frog as an example, break the *instinct* idea down into *orientation* and *fixed action patterns.* (pp. 946–947)
4. Outline the steps involved in the falcon's hunt and indicate which are *orientation movements* and which are *fixed action patterns.* (pp. 947–949)
5. Explain what happens to the *response threshold* when a *fixed action pattern* is delayed. (p. 949)
6. Carefully distinguish between *appetitive* and *consummatory behavior.* (p. 949)
7. What is the evidence that suggests that swallowing is a *consummatory* act? (p. 950)
8. Using the term *IRMs* (*innate releasing mechanisms*), explain how *releasers* might work. (p. 950)
9. Present an example of *vacuum behavior* and relate your example to the concept of *action-specific energy.* (pp. 950–951)
10. List the steps in the *behavioral hierarchy* in stickleback breeding and explain how one step might lead to another. (pp. 951–952)
11. List the three conditions that qualify an act as *instinctive.* (p. 952)
12. Explain what the *drive-reduction* hypothesis is and its drawbacks. (p. 952)
13. What kinds of learning do not seem to fit the *drive-reduction* hypothesis? (p. 952)
14. Explain how *learning* and *instinct* interact, using the example of nest building by crows. (p. 953)
15. What were Skinner's hopes for society through *operant conditioning* and positive rewards? (p. 954)
16. Explain how the "system of rewards and punishments" works in society to mold human behavior. (p. 954)
17. What is *habituation* and how is it of adaptive value to an animal? (p. 954)
18. Using Pavlov's experiment as an example, define *classical conditioning.* (p. 955)
19. How does *operant conditioning* differ from *classical conditioning?* (p. 956)
20. Explain how the *Skinner box* is used to demonstrate *operant conditioning.* (p. 956)
21. In what groups of animals is *habit reversal* best demonstrated? (p. 957)
22. What is *autonomic learning?* What is the evidence that suggests that *autonomic learning* may occur through *operant conditioning?* (p. 957)
23. Of what value could *autonomic learning* be to animals in the wild? (p. 958)
24. How could *autonomic learning* be valuable to people? (p. 958)
25. What is an *engram?* Review the evidence that suggests that *engrams* are rather permanent. (pp. 958–959)
26. Other than its theatrical value, would total *recall* be an advantage or a disadvantage in everyday life? Explain. (p. 959)
27. Describe the two types of *memory* and their characteristics. (pp. 959–960)
28. Explain the role of the *corpus callosum* in the functioning of the cerebral hemispheres. (pp. 960–961)
29. What happens to left- or right-eyed learning in *split-brain* subjects? (p. 961)
30. What evidence suggests that the *corpus callosum* transfers *engrams?* (p. 961)
31. Is there a special *memory* place in the cortex? Explain. (pp. 961–962)
32. What is the apparent role of the *hippocampus* and *reticular system* in *memory?* (p. 962)
33. What is wrong with asking the question, Is it *innate* or *learned?* when referring to a specific behavior? (pp. 962–963)

Chapter 33

The Adaptiveness of Behavior

Now that we've learned something about how behavior develops, let's turn our attention to the deceptively simple questions: What is behavior good for? How is it adaptive? We are no longer concerned with those niggling, academic questions of behavioral components, but with how whole behavioral patterns fit into the lives of animals. Behavior, like morphology, can be understood only by asking questions related to its evolution; that is: How has it come to be preserved by contributing to the reproductive success of an animal's ancestors? How does the performance of a behavioral act help the animal to adjust to a complex world, so that its chances of success are best?

We can begin by discussing navigation, a very complex behavioral phenomenon that has provided biologists with seemingly endless conceptual bones to gnaw. As one set of intellectual teeth has grown dull, another appears and attacks the bone from another direction. Yet the problematical bone has remained largely intact. Because of the resilience of the problem, we'll consider only some of the major concepts in the search for the precise mechanisms of navigation and orientation. Here the question is not one of "Why?"—the adaptedness of seasonal migration and of simply being able to find one's way around seems obvious enough—but the trickier question of "How?" The upshot of the story is, we don't really know how animals find their way around, but the search for mechanisms is a good introduction to the immense difficulties of animal behavior studies.

Then we'll shift to discussions of the intuitively apparent adaptedness of animal communication and aggression. One aspect of animal aggression is territorial defense—when an animal chases others away from a specific piece of real estate. Here, we will ask: Why? What specific advantage does this behavior give to the territorial animal? Finally, we'll turn to the fascinating field of social behavior, another area of animal behavior studies where philosophical and political passions often run high.

Navigation and Orientation

There are undoubtedly people in South America who wonder where their robins go every spring. Those robins, of course, are "our" birds; they merely vacation in the south each winter. If you've been keeping a careful eye on the robins in your neighborhood, you may have noticed that a specific tree is often inhabited by the same bird or pair of birds each year. You may have wondered, since these birds winter thousands of miles away, how they find their way back to that very tree each spring. Such questions have perplexed us for years. And the more we learn, the more strange the story becomes.

The migratory behavior of birds is truly astounding, but other animals may have similar, if less dramatic, adaptive patterns. The phlegmatic toad in your garden sits in a hole and gorges on bugs, but when the urge hits, the toad heads through grass jungles for the pond where it was born. In fact, thousands of toads suddenly begin to move together, hopping slowly and generally unnoticed. In contrast with the robins' 5000 mile journey, the toads may migrate perhaps 2 or 3 miles, but on a toad scale it is still an epic trek that may take weeks. We don't notice them much until they try to cross highways, where their flattened evidence becomes more apparent. But many make it, and after the males set up small territories around their ponds and begin chorusing, the females move toward specific males (usually the larger ones) and they mate. After mating and egg laying, the majority of the toads will make their way back to their own specific holes and drainpipes. How do they find their way?

Equally intriguing is the behavior of the homing pigeon (all pigeons are technically homing pigeons). A homing pigeon can be put into a dark box and transported to a distant location by a circuitous route. When it is released, it will often show up in its home loft that same evening or the next day. Some make it quickly, others never turn up—there is a great deal of

33.1 Two hundred fifty American soldiers in World War I owed their lives to this homing pigeon, which carried a distress message to its handler. Reinforcements were sent to rescue the beleaguered battalion.

variation in abilities (Figure 33.1). But again, the question is: How?

Orientation

One aspect of navigation is *orientation,* which simply means facing in the right direction. More is known about orientation than is known about other aspects of navigation. Much of the recent work with birds began in the 1950s, when a young German scientist, Gustav Kramer, provided some answers—and, in the process, raised new questions.

Kramer found that caged migratory birds became very restless at about the time they would normally have begun migration in the wild. Furthermore, he noticed that as they fluttered around in the cage, they often launched themselves in the direction of their normal migratory route. Kramer devised experiments with caged starlings and found that their orientation was, in fact, in the proper migratory direction—except when the sky was overcast. When they couldn't see the sun, there was no clear direction to their restless movements. Kramer surmised, therefore, that they were orienting by the sun. To test this idea, he blocked their view of the sun and used mirrors to change its apparent position. He found that under these circumstances the birds oriented with respect to the position of the new "sun." This seemed preposterous. How could a stupid bird navigate by the sun when we often lose our way with road maps? Obviously, more testing was in order.

So, in another set of experiments Kramer put identical food boxes around the cage, with food in only one of the boxes (Figure 33.2). The boxes were stationary, so that the one containing food was always at the same point of the compass. However, its position could be changed with respect to its surroundings by revolving either the inner cage containing the birds

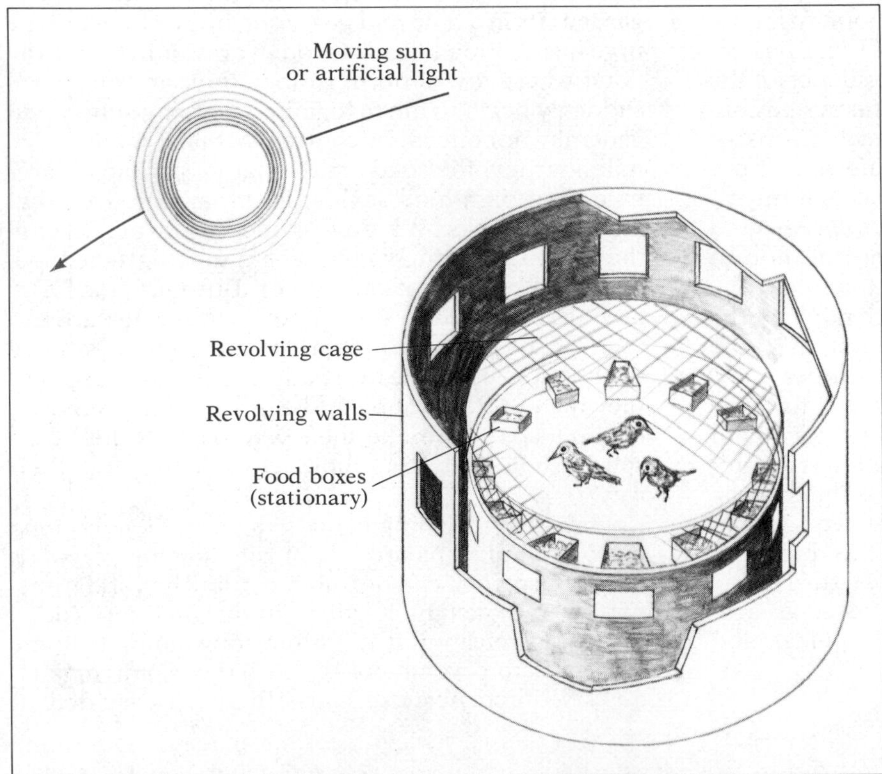

33.2 Kramer's orientation cage. The birds can see only the sky through the glass roof. The apparent direction of the sun can be shifted with mirrors.

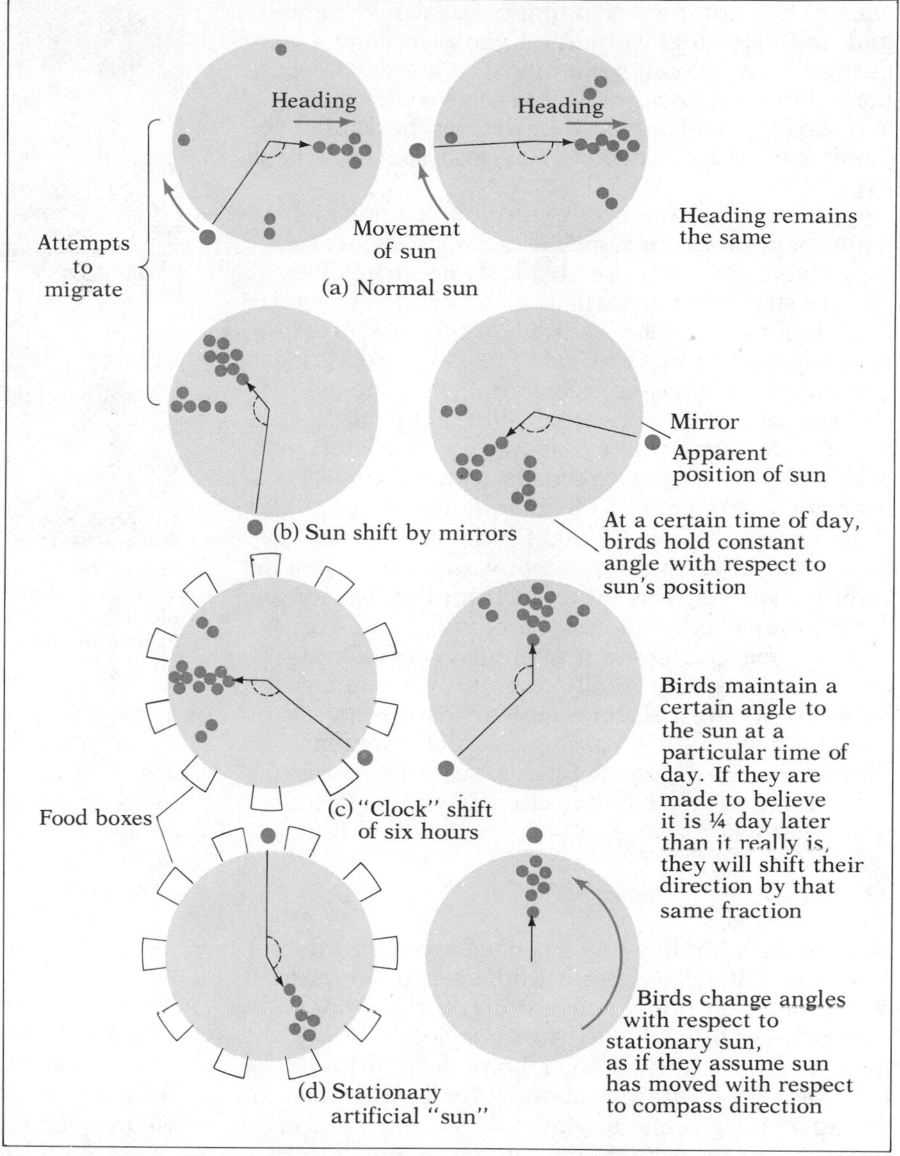
Heading
Heading
Attempts to migrate
Movement of sun
Heading remains the same
(a) Normal sun
Mirror
Apparent position of sun
(b) Sun shift by mirrors
At a certain time of day, birds hold constant angle with respect to sun's position
Birds maintain a certain angle to the sun at a particular time of day. If they are made to believe it is ¼ day later than it really is, they will shift their direction by that same fraction
Food boxes
(c) "Clock" shift of six hours
Birds change angles with respect to stationary sun, as if they assume sun has moved with respect to compass direction
(d) Stationary artificial "sun"

33.3 The analysis of sun-compass orientation in starlings tested in Kramer cages. The dots in (a) and (b) represent directions the birds flew as they tried to migrate. The results in (a) show that the birds maintain the same heading even as the sun changes position. In (b) orientation is shifted as the apparent position of the sun is changed by mirrors. In (c) the birds were subjected to artificial light–dark schedules that were off by 6 hr (¼ day). The birds shifted their directional behavior by 90° (¼ circle). In (d) an artificial sun was held stationary. The birds then changed their direction at about the same rate the real sun would have moved.

or the outer walls which served as the background. As long as the birds could see the sun through any window, no matter how their surroundings were altered, they went directly to the correct food box. It didn't matter whether the box appeared along the right wall or the left wall; the birds were not confused. Then, when the food was moved to a different box, they continued for a time to approach the old box until they gave up and searched around and finally discovered the new food box. Then they couldn't be fooled into approaching another box. On overcast days, however, the birds were completely disoriented and had great difficulty in locating their food. (Figure 33.3).

Such sun-compass orientation implies, of course, that starlings know the time of day and that they also know the normal course of the sun. In other words, incredible as it seems, they apparently know where the sun is supposed to be in the sky at any given time. It seems that part of this ability is innate. In one experiment, a starling reared without ever having seen the sun was able to orient itself fairly well, although not as well as birds that had been exposed to the sun earlier in their lives (this would appear to be another example of the interaction of innate behavior, maturation, and learning). Evidence for sun-compass orientation has also been found for a variety of other animals, including insects, fish, and reptiles.

The sun compass can't be the whole answer, of course. Possession of a compass may be necessary for navigation, but it isn't sufficient. Navigation involves starting at place A and finding place B. The difference would be made clear enough to you if you were blindfolded, driven out into the desert, and released with only a compass. Finding north would be easy; finding Bakersfield wouldn't. Even if you had a compass and a map, you would need to know your position on the map. And, as you sat on a rock, distraught and staring hopelessly at your devices, birds might be passing overhead on their way to a specific tree in Argentina.

As night drew on and you hadn't budged from your rock, other bird species might pass overhead. Since it is night, they are obviously not using the sun. As you strain to see their dim shapes through teary eyes, you might wonder: How do *they* navigate? Apparently, some birds navigate the way sailors used to —by the stars. In experiments somewhat analogous to the sun-compass work, night-migrating birds were brought, in cages, into a planetarium. A planetarium, of course, is essentially a theater with a domelike ceiling onto which can be projected simulated night skies. Sure enough, the fly-by-night birds oriented in their cages according to the sky that was projected on the ceiling, even if it was different from the sky outside. Again, they needed to know the time, just as the old sailors needed decent chronometers. In one experiment, birds that normally migrate north and south between Western Europe and Africa were shown a simulated sky as it would appear that moment over Siberia. The birds, behaving as though they were convinced that they were thousands of miles off course, oriented toward the "east."

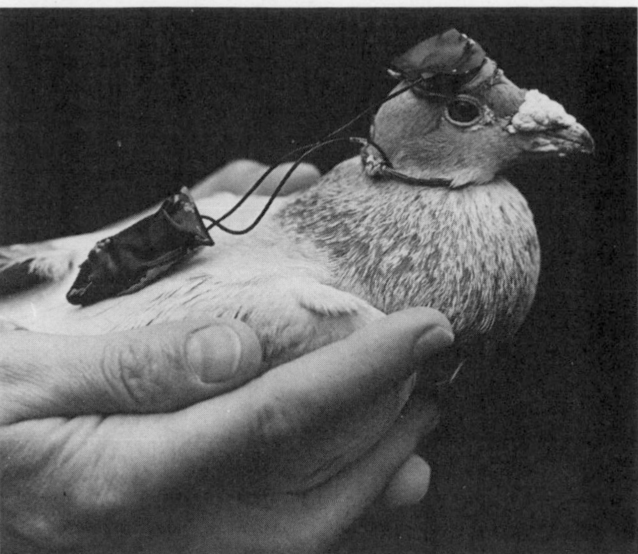

33.4 Pigeon with electrical coils on the head. Birds carrying coils with reversed polarity tended to home in opposite directions, indicating that pigeons can detect slight magnetic fields and that magnetism can influence their orientation.

Homing Pigeons

Studies with homing pigeons have proven equivocal, but instructive. Do pigeons utilize a sun compass? If so, how do they home successfully on cloudy days? In fact, pigeons fitted with frosted goggles find their way back to the vicinity of their home lofts, where they may be seen fluttering about directly overhead, or sitting on the ground near the loft, unable to see it. Do pigeons utilize a magnetic compass? Small electromagnets have been attached to the heads of homing pigeons (Figure 33.4) and by producing a magnetic field across the bird's head, but deflected from the normal magnetic field, the birds could be made to fly off course by the same degree that the artificial field deviated from normal. However, if the day were clear, and the sun were visible, some returned home anyway, but not as many as those birds who carried a similar, but nonfunctional, apparatus.

So, evidently, pigeons utilize a variety of clues—including the sun compass, magnetic force fields, and visual details. Researchers were foiled in trying to find *the* mechanism by blocking individual clues; the pigeon apparently has too many backup systems.

Navigation, as we've noted, requires a map as well as a compass. So do pigeons have built-in maps? W. T. Keeton at Cornell University concluded that they do, although pigeon maps apparently are very different from human maps. He noted that released pigeons do not fly for home in a straight line, but take indirect routes. For instance, pigeons released from one location due north of Cornell usually took off due east, and later made a series of sharply angled turns to take themselves home. Those pigeons that did fly straight south—there were a few—made the same turns, but ended up lost somewhere in western New York State.

It should be pointed out that the study of animal navigation and orientation is an extremely vigorous field at present, and new data are appearing almost daily. Any synthesis at this point would be extremely tenuous. However, these examples provide some idea of the value of innate characteristics in orientation, as well as the surprising sensitivity of some species to environmental cues.

Communication

If you should come upon a large dog eating a bone, walk right up and reach out as if to take the bone away. You may notice several changes in the dog's appearance. The hair on the back may rise, the lips may curl back to expose teeth, and the dog may utter peculiar guttural noises. This is communication. To see what the message is, grab the bone.

It would be a remarkable understatement to say that communication is an important phenomenon in the animal world. It is accomplished in many ways and with innumerable adaptive effects. Its ultimate function, however, is to increase reproductive success in those animals able to send and receive the signals.

Communication may be directly involved with reproductive success as a component of mating behavior or precopulatory displays. It may also indirectly increase reproductive output by helping the offspring to avoid danger when parents give warning cries, or by simply helping the reproducing animal to live better or longer so that it may successfully mate again. Remember, the charge to all living things is "reproduce as much as possible or your genes will be lost." Communication helps animals to carry out that reproductive imperative. Being able to receive and decode messages is adaptive, too. In a successful communication, both sender and receiver benefit. In the example above, if the message is properly understood, the dog gets to keep the bone and you get to keep your arm.

Let's consider a few general methods of communication, and in so doing perhaps we can learn something about the ways in which animals influence the behavior (or the probability of behavior) of other animals. That, after all, is what communication is all about.

33.5 An example of graded display. At left the bird, *Fringilla coelebs*, is showing a high-intensity threat. The posture at center is a medium-intensity threat and below is a low-intensity threat.

Visual Communication

Visual communication is particularly important among certain fish, lizards, birds, and insects—and among some primates as well. Visual messages may be communicated by a variety of means, such as through color, posture or shape, movement, or timing. As we mentioned in Chapter 32, the color red can release territorial behavior in European robins. The female quickly solicits the attention of the male bird by assuming a head-up posture while fluttering her drooping wings. Some male butterflies are attracted to the female by the way she flies. And as an example of communication by timing, fireflies are attracted to each other on the basis of their flash intervals, each species having its own frequency. (One predatory species flashes at another species' frequency and then eats whomever comes to call, so communication lines can be tapped.)

Because of the high information content possible with visual signals, subtle variations in the message can be conveyed by gradations in the display (Figure 33.5). Of course, graded displays are useful only to species that are sensitive and intelligent enough to be able to utilize such subtleties. Another advantage of the high information load in visual signals is that the same message may be conveyed by more than one means. Such redundancy may be used either to modify the message or to emphasize it in order to reduce the chance of error in interpretation.

A visual signal may be a permanent part of the animal, as in the elaborate coloration of the male pheasant and the striking facial markings of the male mandrill (Figure 33.6). These animals advertise their maleness at all times and are continually responded

33.6 A male mandrill has permanent markings that advertise his sex, but his signals of anger are temporary and dependent on his mood. It is probably often to his advantage to be related to as a male, but less frequently as an angry male.

to as males by members of their own species. On appropriate occasions they may emphasize their "machismo" by behavioral signals, such as the strutting of the pheasant and the glare of the mandrill. In other cases, the visual signal may be of a more temporary nature, such as the reddened rump areas in female chimpanzees and baboons during estrus. As a more short-term signal, a male baboon may expose his long canine teeth as a threat (Figure 33.7).

Short-term visual signals have the advantage that they can be started or stopped immediately. If a displaying bird suddenly spots a hawk and freezes, its position won't be given away by any lingering images in the environment. Also, in visual messages, the recipient is notified of the exact location of the sender. The recipient can then respond to him in terms of his precise location, as well as his general presence and behavioral state (aggressive, romantic, or whatever).

Visual signals also have certain disadvantages. For example, if a sender can't be seen, it can't communicate. And all sorts of things can block vision, from mountains and trees, to fog and big hats. Visual signals are generally useless at night or in dark places (except for light-producing species). Also, since visibility weakens with distance, long distances diminish the usefulness of such signals. As distance increases, the signal must become bolder and simpler; hence it can carry less information.

33.7 A male baboon displaying his large canine teeth as a threat signal. These males are extremely powerful and dangerous.

Sound Communication

Sound plays such an important part in communication in our own species that it may surprise you to learn that it is limited, for the most part, to arthropods and vertebrates. We're all familiar with the song of the cricket and the cicada, so you are probably aware that insect sounds are usually produced by some sort of friction—such as by rubbing the wings together or the legs against the wings. The important aspect of these sounds is cadence rather than pitch or tone, as is the case with most birds and mammals.

Vertebrates use a number of different forms of sound communication. Fish may produce sound by means of frictional devices in the head area or by manipulation of the air bladder. Land vertebrates, on the other hand, usually produce sounds by forcing air through vibrating membranes in the respiratory tract. Of course, they may communicate by sound in other ways. Rabbits thump the ground, gorillas pound their chests, and woodpeckers hammer hollow trees and drainpipes early on Sunday mornings.

Most of the lower vertebrates rely on signals other than sound, but some species of salamanders can squeak and whistle, and there is one that barks! Frogs and toads advertise territoriality and choose mates at least partly by sound. Most reptiles don't communicate by sound, but large territorial bull alligators can be heard roaring in more remote areas of the swamps in the southern United States. Darwin described the roaring and bellowing of mating tortoises when he visited the Galapagos Islands.

Sound may vary in pitch (low or high), in volume, and in tonal quality. The last is apparent as two people hum the same note, yet their voices remain distinguishable. It is possible to show the characteristics of sounds graphically by use of a *sound spectrogram.* In effect, this is a translation of sound into visual markings, which makes possible a more precise analysis of the sound.

The function of the message may dictate the characteristics of the sound. For example, Figure 33.8 shows a sound spectogram of the mobbing cries of several bird species. Such calls appear as brief, low-pitched *chuk* sounds, and their source is easy to pinpoint. When a bird hears the repetitious mobbing call of a member of its own species, it is able to locate the caller quickly and join it in driving away the object of concern—usually a hawk, crow, or other marauder. In contrast, when a hawk or other aerial predator is spotted flying overhead, the warning cry of songbirds is usually a high-pitched, extended *tweeeeee,* which is difficult to locate (Figure 33.9). The response of a bird hearing this call is quite different from its response to the mobbing call. The warning cry sends the listener heading for cover, often in a deep dive into protective foliage, from where it may also take up the plaintive, hard-to-locate cry.

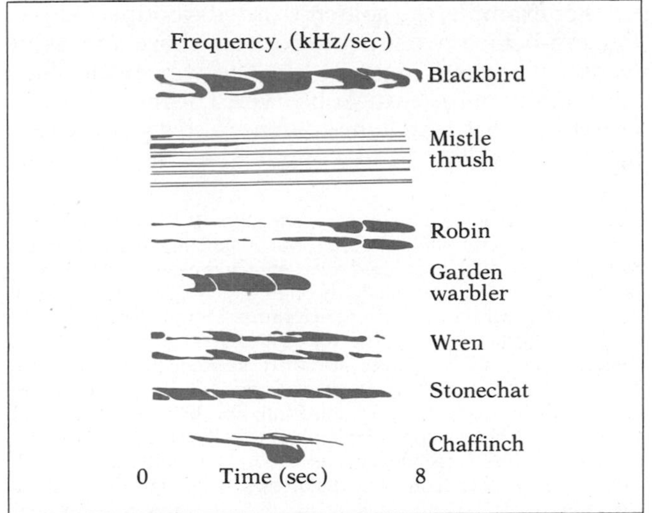

33.8 Sound spectrograms of the mobbing calls of several species of British birds while attacking an owl. The sounds have qualities that make them easy to locate. Such calls are low-pitched *chuk* sounds. Different species of birds have developed similar calls through convergent evolution. That is, their calls serve much the same purpose and were developed under relatively similar conditions; thus the qualities of the sounds came to be somewhat alike.

Sound signals have the advantage of a potentially high information load through subtle variations in frequency, volume, timing, and tonal quality. They are distinguishable at low levels, but they can also be made to carry over great distances by pumping more energy into them—that is, by becoming louder. In addition, sounds are transitory; they don't linger in the environment after they have been emitted. Thus, an animal can cut off a sound signal should its situation suddenly change—for example, with the appearance of a predator. A further advantage is that an animal doesn't ordinarily have to stop what it is doing to produce a sound. As everyone knows, sounds, unlike visual images, can go around or through many kinds of environmental objects.

33.9 Sound spectrograms of five species of British birds when a hawk flies over. Such calls are high-pitched, drawn out, and difficult to locate. They, too, have achieved their similarity through convergent evolution.

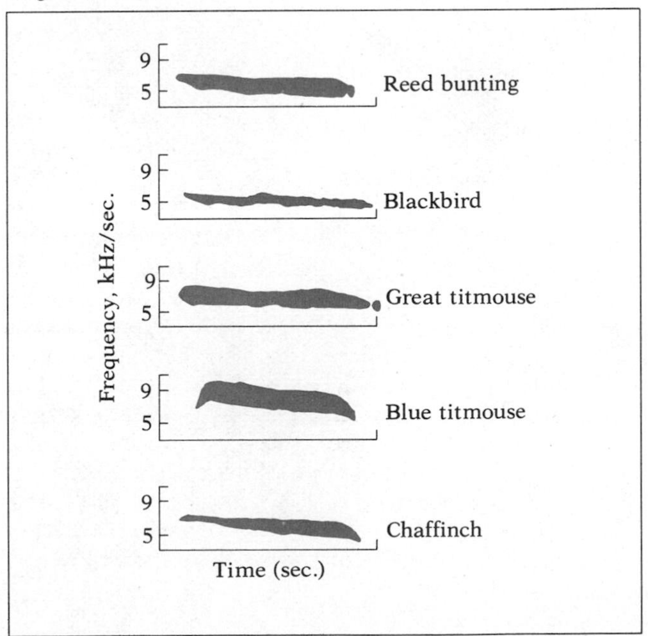

Sound communication also has its disadvantages. Sounds are rather useless in noisy environments. Thus, some sea birds that live on pounding, wave-beaten shorelines rely primarily on visual signaling. Also, sounds attenuate with distance. And the source of sound is sometimes difficult to locate, especially underwater.

Chemical Communication

You have probably seen ants rushing along single file as they sack your cupboard. You may also have taken perturbingly slow walks with dogs that stop to urinate on every bush. In both cases the animals are communicating by means of chemicals. The ants are following chemical trails laid down by their fellows, and the dog is advertising his presence.

Insects make extensive use of chemical signals, and certain chemicals elicit very rigid and stereotyped behavior. An *alarm chemical* is produced when an ant encounters some sort of threat, and as the chemical permeates the area, every ant of that species which detects the chemical reacts by rushing about in a very agitated manner. Female moths, we know, produce a *sex attractant* that attracts males from astonishing distances. You may recall our earlier discussion of one such substance called *bombykol;* a male moth can detect one molecule of this chemical (see Chapter 31).

Chemical signaling is so immediate and invariable that it can be compared to hormonal messages. But, whereas hormones work within an individual, these chemicals work between individuals and are called *pheromones* (from the Greek *pherein,* "to carry," and *horman,* "to excite").

It was at first believed that pheromones are found only among the insects, species with behavior that is largely genetically programmed. But then it was found that pheromones also exist in mammals. For example, if pregnant rats of certain species smell the urine of a strange male, some component of that urine will cause them to abort their fetuses and become sexually receptive again. And, as we have already seen, the females of most mammal species signal their sexual receptivity by some chemical means, with varying degrees of influence on the male, depending on the species.

Honeybees provide another interesting example of chemical communication. The queen produces a *queen substance* which, when fed to developing females, causes them to become sterile workers. When the queen dies, the inhibiting substance is no longer present, and new queens begin to develop (the first one to hatch may massacre the rest). At this time, the undeveloped ovaries of some adult workers also suddenly begin to mature. Their eggs will produce drones —royal consorts, one of which will mate with the new queen on her nuptial flight.

The molecular makeup of several chemical signals in the animal world has been carefully analyzed. It was found that the backbone of the molecule usually consists of five to twenty carbon atoms. The molecular weight of the chains ranges between 80 and 300. It is believed that molecules with at least five carbons are needed to provide the variation and specificity for different messages. The maximum molecular size is more likely to be set by other factors. For example, since these molecules are usually carried in the air, they must be small and light. Also, more energy is expended in the manufacture of larger molecules.

Chemical signals have the advantage of being extremely potent in very small amounts. Also, because of their persistence in the environment, the sender and receiver do not have to be precisely situated in time and space in order to communicate. And chemicals can move around many sorts of environmental obstacles.

The specificity of chemicals, however, limits their information load. Moreover, because chemicals do linger in the environment, they may advertise the signaler to arriving predators as well as to the intended recipient of the message.

Why Communicate?

Now that we have some idea of how communication can be accomplished, we might ask, why is it necessary to communicate at all? You can probably sit right there and think of a host of reasons, but let's see if we can't organize the ideas a bit.

Species Recognition

First, communication may permit one animal to know that another animal is of the same species. This may not seem too important—unless one is interested in reproduction. After all, if two animals of different species mate, they cannot normally produce healthy offspring. But a lot of species are very similar, and in the absence of some precise means of identification an animal might waste a lot of time and energy in trying to mate with a member of the wrong species. Therefore, species identification must somehow be quickly established.

For example, the golden-fronted woodpecker and the red-bellied woodpecker often share the same woods in Texas (Figure 33.10). The two species look very similar to the casual observer. The most conspicuous difference is that the golden-fronted woodpecker has a small yellow band along the base of its red cap,

33.10 (a) The golden-fronted woodpecker (*Centurus aurifrons*) looks a bit scruffy while molting, but is still remarkably similar to (b) the red-bellied woodpecker (*Centurus carolinus*). The range of the golden-fronted extends through dry areas from Honduras northward to central Texas, where it rather abruptly stops and that of the red-bellied begins. The red-bellied's range extends to the eastern coast, where it lives primarily in wooded habitats. The birds overlap in a narrow range in Texas, but in spite of strong similarities in appearance, vocalization, and behavior, they don't interbreed. Whereas they don't recognize each other as sexual partners, each does see the other as a competitor and they mutually exclude each other from their territories as though they were of the same species. Obviously, they use different cues in the recognition of mates and competitors.

(a)

(b)

which is absent on the red-bellied woodpecker. Also, the red-bellied woodpecker has slightly more white in its tail. The habits of the species are much alike and their calls are very similar, at least to the human ear. Nevertheless, each of these birds is able to recognize its own species. In any case, cross-mating has never been observed, nor have any hybrids. No one knows exactly what cue the birds use, but to them, apparently, the signals are clear.

Interestingly enough, whereas the two species ignore each other sexually, they have come to treat each other as competitors and will eject a member of the other species from a territory as quickly as they would a member of their own species. Apparently, the stimuli that release sexual and territorial behavior are quite different.

Individual Recognition

It may also be important for an individual to be able to recognize specific individuals within its own population. If you have ever watched gulls at the beach, you may have noticed that the adults all look alike. This is because you aren't a gull; all people probably look alike to them. If you should visit a gullery, the place where they nest, you might see thousands of "identical" gulls. Color banding and recording playback experiments, however, have shown that gulls can recognize their own mates on sight and are able to filter out the calls of their mates from the raucous din of the clamoring gulls wheeling above.

Individual mate recognition is important only in species that establish pairs. In species in which the sexes come together only for copulation, the mate-recognition problem simply becomes one of sexual identification.

Individual recognition is important for two primary reasons. First, one should be able to recognize one's own mate, especially when the rearing of offspring is done by both parents. Mate recognition ensures that pairs will attend to the nest that harbors their own offspring, thus increasing their chances of reproductive output. This is an essential factor in species that show a high level of coordination while rearing offspring. For example, some birds bring food to their mates while they are incubating eggs. In the African hunting dog, the male is more likely to regurgitate food to his own mate when she is nursing pups, than to any other females.

The second reason individual recognition is important is that it functions in the maintenance of dominance hierarchies. Once animals in a group know their rank with respect to the others, there will be less fighting over food or other commodities. In any group where rank is unknown, or not yet established, the incidence of fighting is likely to be high. Rank, however, can only be maintained when each animal can recognize the others individually so that it can respond to each according to its status.

Aggression

The old image of "Nature, red in tooth and claw" has recently been superseded by the popular notion that other animals get along with their own kind and humans are the only animals that kill each other. It might be a good idea to take a look at the animal kingdom to see what is actually going on.

First of all, we should note that whereas *aggressiveness* can be a mood, *aggression* is belligerent behavior, an act, which arises as a result of *competition.* A lioness may spring toward a group of wild pigs, single out one in the chase, catch it, and quickly smother it or break its neck. However, she is not acting out of aggression. She is being about as aggressive as you are toward a hamburger. However, if one of the big boars should turn and charge her, the lioness may momentarily show a component of aggression—fear. In fact, fear is an important component of aggression; it is simply at one end of the behavioral spectrum of aggressive behavior. At the other end is sheer, unbridled belligerence.

Fighting

Let's consider the most obvious form of aggression—fighting. You can discount the old films you have seen of leopards and pythons battling to the death. Such fights simply aren't likely to happen. What is a python likely to have that a leopard needs badly enough to risk its life for, and vice versa? Although such fighting might occur in the unlikely event that one should try to eat the other, fighting is much more likely to occur between two animals that are competing for the same commodity. Obviously, the strongest competitor is going to be one that utilizes the same habitat in the same way. And what sort of animal is most likely to do this? Obviously, a member of the same species. Moreover, if there is strong competition for mates, fighting is even more likely between members of the same sex within that species. And this is in fact where most fighting occurs—between members of the same sex of a given species.

Of course, fighting between species sometimes occurs. The golden-fronted and red-bellied woodpeckers exclude each other from their respective territories. And lions may attack and kill African cape dogs at the site of a kill. The lions don't eat the dogs; they just exclude them.

Animals may fight in a number of ways, but combatants of the same species usually manage to avoid injuring each other (Figure 33.11). There are several

33.11 Fighting male rattlesnakes (*Crotalus ruben*). Each male tries to push the other to the ground, but the deadly poisonous snakes never bite each other.

apparent benefits to such a system. First, no one is likely to get hurt. The competitor is permitted to continue its existence, it is true, but the possibility of having to compete again entails less risk than does serious fighting. Also, since animals are most likely to breed with the individuals around them, the opponent may be a relative that is carrying some of the same genes; hence, there is a reproductive advantage in sparing him. In fact, if the population doesn't tend to scatter much, the competitor might even be one's own mature offspring; it could also be a prospective mate. So it's best not to hurt each other—but notice that the motive is selfish, not benevolent.

When fighting occurs between potentially dangerous combatants, the fights are usually stylized and relatively harmless. For example, horned antelope may gore an attacking lion, but when they fight each other, the horns are never directed toward the exposed flank of the opponent (Figure 33.12). Such stylized fighting does, however, enable the combatants to establish which is the stronger animal. Once dominance is established, the loser is usually permitted to retreat.

On the other hand, all-out fighting may occur between animals that are unequipped to injure each other seriously, such as hornless female antelope (Figure 33.13), or between animals that are so fast that the loser can escape before serious injury, as in house cats.

We might ask ourselves, since it is useless to ask an antelope why they don't gore each other. An incurable romantic might assume that it is because they don't want to hurt one another. In all likelihood, what the two antelope "want" has little to do with it. The fact is, they can't hurt each other. When the system works, an antelope could no more gore an opponent than fly! In terms of the actual mechanism, it may be that the sight of an opponent's exposed flank acts as an inhibitor of butting behavior. Conversely, a facing view, under certain conditions, might serve as a release of very stereotyped fighting behavior.

But this begs the question. Regardless of what we may or may not surmise about animal motivation, such standardized behavior must have evolved because it was beneficial to the ancestors of the present combatants and so it has been retained. Thus, we return to the original question: How does the animal benefit by refusing to do serious injury to the opponent? One benefit is really not that subtle. The antelope might not gore its opponent because if it did so, it might get gored back. The situation is similar to that of two toughs in a barroom brawl. They slug and punch, and each is really trying to win the fight. On the other hand, each has a jackknife in his pocket—and both of them know it. Neither is willing to pull his knife, because then the other would be obliged to retaliate, and someone could get hurt that way. Either brawler would rather risk a loss in a fistfight than a draw in a knife fight.

On another level, two men might get into a screaming argument at a cocktail party but not resort to hit-

33.12 Male impalas fighting. Although the horns of these medium-sized antelope are formidable weapons, neither impala will attack the vulnerable flank of the other. Instead, a harmless pushing contest ensues as the tips of the ridged horns are engaged. These animals effectively employ horns and hoofs against species other than their own, but when confronted with a member of their own species, they are genetically constrained to behave in very circumscribed ways.

33.13 Hornless females of the Nilgau antelope have no inhibitions against attacking the flank of a competitor, but their butts are quite harmless. Keep in mind that "harmless" is used only in the immediate sense. The butt itself may not be dangerous, but if it establishes dominance so that the loser is deprived of commodities, the result can have far-reaching implications. An individual deprived of food, after all, is more likely to fall prey. If losing means that she is deprived of a mate, the result is genetic death even if she should have a long life. Her genes will die with her. If losing one fight meant that the animal would have absolutely no chance of breeding, or contributing her genes to the next generation, they might be expected to fight to the death. A loss for this species, however, usually just means a temporary setback, so it behooves the loser to accept it gracefully and attempt to breed another time. Interestingly, horned males of the same species almost never attack in this way, nor do horned females of other species.

ting—for fear of getting hit back. In the same vein, it was self-interest, not humanitarianism, that prevented either side from using poison gas in World War II; both sides had poison gas, and both sides knew it. With the atomic bomb it was different: The Americans had two, and the Japanese had none. The bomb could be used in 1945 without fear of retaliation, and so it was used.

Animal evolution has proceeded according to the same kind of logical accounting. J. Maynard Smith, of the University of Sussex, points out that the antlers of deer are used solely for intraspecific fights, and are constructed so as *not* to injure the opponent. The "points," or side branches, cause the antlers to lock in a head-to-head combat, so that the sharp tips do not actually reach the opponent's flesh. When they are seriously fighting a predator, they use their hooves with dazzling effect. Now, among deer are occasionally born males with antlers that will lack the side branches. When these males engage in what should be a ritual fight, their antlers slip through and gore their opponent. Does this give the mutant deer an advantage? Not at all! The gored opponent may be bigger and stronger and, once gored, becomes very angry. Ritual combat then becomes real combat, and the sharp-antlered deer doesn't always win. When a fight becomes serious, one only has to lose a single encounter and some other, branched-antlered stag will get the doe and father branch-antlered sons.

Maynard Smith has analyzed fighting strategies with mathematical models and computer simulations. What he strove to determine was the *evolutionarily stable strategy*—which he defined as an innate behavioral pattern that would outcompete all other behavioral patterns and would be stable against the invasion of a new mutant pattern of behavior. He asked: Is it best never to fight and always retreat? Then a born fighter will always win. So is it better always to fight? No, because there is too much risk of being beaten. Is it best to bluff consistently? No, your bluff will be called (it's better to bluff inconsistently and unpredictably). Maynard Smith found two behavioral "strategies" that proved to be the most stable in populations, depending on the circumstances. The first strategy was called *retaliator:* a retaliator engages only in ritual display and mock battle unless it is seriously attacked, whereupon it will retaliate with just as much seriousness. In a mathematical model, this behavior was always the most successful in the long run; and so it should not be surprising to see ritual display and mock battle so common in nature. There's always the threat of real injury to any animal that breaks the rules. The second strategy, that we will discuss shortly, is the *bourgeois* strategy. It employs quite a different set of rules, as we shall see.

Can we assume, then, that animals do not fight to the death with their own kind? As a general rule, they don't, but there are many exceptions. Animals do what is best for themselves, in the long run. Accidents may occur in normally harmless fighting, and this can lead to retaliation and escalation. There are also some species that normally engage in dangerous fighting. If a strange rat is placed in a cage with a group of established rats, the group may sniff at the newcomer carefully for a long time, but eventually they will attack it repeatedly until they kill it. If escape is impossible, male guinea pigs and mice often fight to the death. The males of a pride of lions may kill any strange male they find within their hunting area, and a pack of hyenas may kill any member of another pack that they can catch. Even gangs of male chimpanzees have been seen to ambush and kill isolated males from other troops.

Is Aggression Instinctive?

Our discussion of aggression so far suggests that aggressive behavior in many animals is largely innate. But there are many who contend that aggressive behavior, especially among mammals, is directly attributable to learning. Others, of course, stress that aggressive behavior has both learned and innate components. The argument has extreme significance for our species. If aggression is socially disruptive in an increasingly crowded world, then we must ask how it can be controlled. As a practical example, does television violence serve to release aggression, and hence dissipate it, or does it serve as a behavioral model for aggression? Since there is a tendency in most academic disciplines to focus on aggression as a cultural, or learned, phenomenon, let's consider the evidence that suggests an instinctive component.

In certain species of highly aggressive cichlid fish, the males must fight before they are able to mate. If a male's reproductive state is appropriate and a female is available, the male will frantically seek an opponent upon which to release his fighting behavior. Finding none, he will often attack and kill the female. He is then ready to mate, but of course by then it is too late. The behavior can be described in terms of Tinbergen's hierarchical model. In removing the inhibitory block for mating (by fighting), the male can move to the next level of reproductive behavior, but unfortunately he has destroyed the releaser of that behavioral level, the female. Among mammals, rats will learn mazes in order to be able to kill mice, and it has thus been inferred that the killing is a consummatory act that reduces tension.

Does an animal become aggressive when it is shielded from all opportunities to learn such behavior? Rats and mice have been reared in isolation, so that there was no opportunity for them to learn aggression. But when other members of the same species were introduced into their cages, the orphan rodents attacked, showing all the normal threat and fighting patterns. The evidence from all sides is essentially circumstantial, but we would do well to focus more research attention on the roots of aggression in these critical times.

Human Aggression

If aggressive behavior has innate components in other species, what about humans? Is there any evidence that members of our own species are genetically driven to behave shabbily toward each other? It is difficult, of course, to isolate behavioral components in a species as complex as our own. This is partly because we have the capacity to override both learned and innate behavioral tendencies by conscious decision. This ability has led some observers to assume that our urges—romantic, aggressive, or whatever—must be culturally based, and not innate. Hence, to correct any disruptive tendencies in our group, we need only to change our culture and/or our institutions.

However, if the behavior in question does have an innate basis, cultural stifling of it can only be cosmetic. An aggressive person may have learned that smiling and speaking in low tones reaps greater rewards than behaving aggressively, but if aggression is innate, then we can expect the person to be aggressive in small ways, or to direct aggression into harmless channels, or perhaps to show sporadic outbursts of violence when his or her guard is down. And, of course, smiles and soft voices can in themselves be the vehicle for extraordinary violence. Some of the most horribly bloodthirsty of programs have been proposed in congenial, reassuring, and even kindly tones by well-groomed, seemingly gentle and eminently "reasonable" souls, smiling into the camera. If aggression is an innate tendency in our species, we can expect it to permeate our social patterns. It may be temporarily suppressed, and can perhaps be rechanneled or disguised, but it will exert a subtle and pervasive influence in our daily lives.

In one rather interesting experiment, it was shown that aggression could be provoked in people and that the aggressive behavior, when released, caused some relief. First, an experimenter provoked a roomful of students to anger (that was the easy part), causing their blood pressures to rise. The students were then divided into two groups. Each group separately watched as the experimenter attemped to perform a series of simple tasks. One group was instructed to administer painful electrical shocks (or so they thought) to the experimenter whenever he made an error. The other group could only inform the experimenter of his errors by harmlessly flashing a light. It was found that the group that believed they were actually inflicting pain to the experimenter (who was in reality an actor) showed a rapid decrease in blood pressure, a physical sign of relief. The group that was not given an opportunity to release aggression maintained high blood pressures for a much longer time. In other experiments, blood pressure was found to drop quickly if the subject were allowed to hurl verbal insults at the experimenter. In both cases, the release of tension was short-term, as is the case with other types of instinctive patterns. It is important to note, however, that this experiment only indicated that aggressive behavior is a *response* to provocation, and did not address the question of whether such behavior is the result of a *drive.*

The arguments on both sides are much more complex than this brief summary might suggest. The important point, however, is to recognize that these alternative positions exist, and to attempt to structure our social programs as broadly as possible.

Territories

Aggression is the tool with which animals are able to hold *territories.* (Territories are defined as any defended area.) People first observed that animals held territories many hundreds of years ago, but our attention to the topic was not focused until 1920 when Eliot Howard wrote *Territory in Bird Life*. The book was based upon his observations of birds expelling each other from certain plots of land. Since one bird successfully defended one place and another bird was able to win in another place, it seemed as if the birds "owned" the land in the sense that a particular bird seemed to have certain rights when it was on a certain plot of land. In fact, when on that land, the bird was virtually undefeatable, and with apparent confidence would immediately drive out any intruder of the same species and sex. Of the same species and sex? Obviously, the bird was driving out competitors! Members of the same species would compete for commodities, such as food and nest sites, and members of the same sex would compete for mates. So here was a working hypothesis: Territories are a means of reducing competition.

Since Howard's time, territories in a vast range of species have been discovered. Many very different species, it turns out, have the same kind of territorial behavior. For example, some *Anolis* lizards (Figure 33.14) have territories very similar to those of many birds, and stickleback fish behave similarly. It is tempting, then, to look for encompassing underlying themes, broad generalizations to cover all the cases. Alas, all the cases don't conform to any useful general pattern. For example, a hummingbird may defend different bushes at different times of the day, taking a proprietory interest in each as its flowers open. Several tomcats may hold the same territory at different times of the night, each taking ownership in turn. Even closely related animals may show different territorial patterns. The Norway rat (*Rattus norvegicus*) chases strange rats only on special paths he has marked, whereas the black rat (*Rattus rattus*) defends the entire area crossed by his paths. Some animals, in a sense, carry their territories with them as *individual distances*. This is the radius that must not be violated by other members of the species. Starlings sitting on a wire sit evenly distributed (Figure 33.15), each a circumscribed distance from the others (perhaps having to do with how far its neighbor can reach with its bill).

In spite of such variation, however, we have been able to construct certain basic characteristics of territorial behavior. First, territories are rather specific places and do not, as a rule, include all the places where an animal may wander. The area in which an animal may be found at any given time is called its *home range*. Animals meeting within home ranges, but outside territories, normally don't fight. Baboons defend their sleeping trees and feeding grounds, but they meet peacefully at water holes.

To emphasize the variation in territorial behavior, let's consider a few examples from birds. In species that live as pairs, the territory is usually defended by the male against other males, although the female

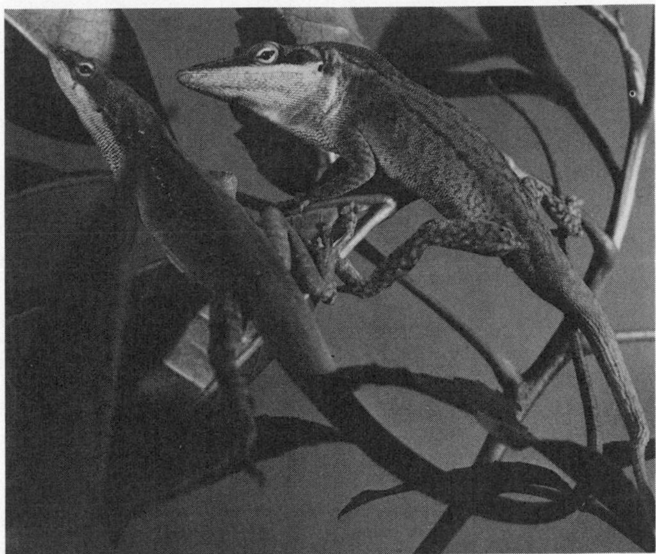

33.14 *Anolis* lizards are among the vast array of territorial species. The males may hold areas in which a number of females take residence, but all other males are excluded.

may join the fray. Whereas the male will not usually attack an intruding female, the resident female may attack her on sight. At certain times, such as in the winter, the sexes of some species may hold separate territories, but in spring they lower their defenses and come together to mate.

In some species, pairs may join together to defend a mutual territory in a community effort against members of other groups. In other species, pairs may

33.15 Starlings sit peacefully next to each other on a telephone line. They are regularly spaced according to their tolerance of each other's proximity.

live close together but may only defend small areas around their nests from other pairs. In yet other species, males may come together at the beginning of the mating season to form small temporary territories on a communal display ground.

These variations on a theme should underscore the difficulty in forming a sentence that begins, "Territories are to . . . ." Perhaps, then, it is best simply to list some of the adaptive features that have been attributed to territories by various people.

1. *Limitation of population and the assurance of space.* Territories obviously ensure space, but if space is limited and needed for breeding, they may also reduce the population size.
2. *Protection against predators, disease, and parasitism.* Disease and parasites obviously travel with more difficulty in a sparsely distributed population, but spacing also may be advantageous if it insures against a predator finding all members of the population just because it finds one.
3. *Selection of the most vigorous to breed.* A female should ideally choose the male who has successfully competed against other males. In this way she increases the probability of her offspring receiving "good" genes.
4. *Division of resources among dominant and subordinate individuals.* Once the resources have been distributed by the establishment of territories, less time will be spent in deciding who gets what. Losers have a better opportunity of finding food if they look elsewhere than if they challenge a territory holder.
5. *Stimulation of breeding behavior.* In some species males without territories cannot attract mates, and, in fact, males without territories may not even be able to develop sexually.
6. *Assurance of a food supply.* Territories often, perhaps usually, hold food.
7. *Provision of a mating and nesting place.* The territory may hold a nest site, or in the case of communal breeding grounds (*leks*) (Figure 33.16), a place to breed, since the male has signaled his vigor to other males and to females by possession of the territory.
8. *Increasing efficiency of habitat utilization of a familiar area.* Since an animal is likely to spend most of its time in or near its territory, it has reserved a place for itself with which it can become very familiar, learning where food and hiding places are most likely to be found.

J. Maynard Smith's theory of evolutionarily stable strategies (ESSs) has thrown some light on the adaptiveness of territorial behavior, and in particular on that aspect of territoriality that specifies that the "owner" of a territory always wins on an encounter within the confines of its turf. One stable strategy that came from the modeling studies was even better than that of retaliator, but it required a peculiar rule. If, in each encounter between adversaries, one would retreat immediately, then neither would waste time and energy fighting. For instance, this works in a dominance hierarchy (or *peck order*) and also works on the quite reasonable "rule" that the smaller of two potential combatants concedes immediately. But it also works if the determination of who concedes is completely arbitrary, say on the toss of a coin. Some of our traffic rules work this way. The first car to the intersection has the right of way; if two cars arrive simultaneously, the car on the right has the right of way and the car on the left concedes. As long as everyone knows the rules, it is advantageous to everyone to follow them.

In the case of territoriality, the rule is *stay off my turf.* The defender "knows" it (in an evolutionary sense, which means that it behaves as though it knows it); the intruder "knows" it; and both of them "know" that it is to their own advantage to follow the rule. According to the theory, it may be an arbitrary rule, but it works to the advantage of all parties, and has persisted and been stabilized by evolution. Maynard Smith called this bit of animal behavior the *bourgeois* strategy.

33.16 Males of the ruff (*Philomachus pugnax*) displaying on their *lek* (breeding ground). A female is at right. These birds are unusual in that the males are highly variable in plumage.

Cooperation

Cooperation seems much nicer than aggression, and most of the animal stories of our youth (our pre-Jack London youth, that is) involved animals that helped each other in some way. Certainly, cooperation is highly developed in some species, and its complexity and coordination in certain instances surpasses imagination. But again we have the problem of sifting fact from fiction—a difficult, if not impossible, task for those with a casual interest in the subject.

Cooperative behavior occurs both within species and between species. As examples of *interspecific* (between species) cooperation you have probably heard of the relationship of the rhinoceros and the tickbird. The little birds get free food while the rhinoceros rids itself of ticks and harbors a wary little lookout. The highest levels of cooperation, however, are most likely to be found between members of the same species.

Let's consider a few examples of *intraspecific* (within species) cooperative behavior. Porpoises are air-breathing mammals, much vaunted in the popular press for their intelligence. In fact, certain of their actions support the claim. Groups of porpoises will swim around a female who is in the throes of birth and will drive away any predatory sharks that might be attracted by the blood. They will also carry a wounded comrade to the surface so it can breathe. Their behavior in such cases is highly flexible, in that it is often adjusted to the subtleties of the environmental circumstances, rather than being a stereotyped behavior pattern. Such flexibility indicates that their behavior is not a blind response to innate genetic influences.

Group cooperation among mammals is probably most common in defensive and hunting behavior. For example, yaks of the Himalayas form a defensive circle around the young at the approach of danger, standing shoulder-to-shoulder with their massive horns directed outward. This defense is effective against all predators except humans, since the beasts often pitifully maintain this stance while they are felled one by one by rifles. Wolves, African cape dogs, jackals, and hyenas often hunt in packs and sometimes cooperate in bringing down their prey. In addition, they may bring food to members of the group that were unable to participate in the hunt.

Mammals such as the ones we have considered so far might be expected to show the highest levels of cooperative behavior as a result of their intelligence. It is a bit surprising, therefore, that among all animals social behavior and cooperation are most highly developed in the lowly insects (Figure 33.17). The complex and highly coordinated behavior patterns of insects are generally considered to be genetically programmed, highly stereotyped, and not much influenced by learning—except perhaps for very restricted classes of learning.

33.17 Bees around queen. These social insects show extremely high levels of cooperation, coordination, and self-sacrifice. The result is a hive in which the watchword is cold, sometimes brutal, efficiency.

In honeybee colonies, the queen lays the eggs and all other duties are performed by the workers, which are sterile females. Each worker has a specific job, but that job may change with time. For example, newly emerged workers prepare cells in the hive to receive eggs and food. After a day or so, their *brood glands* develop, and they begin to feed larvae. Later, they begin to accept nectar from field workers and pack pollen loads into cells. At about this time their wax glands develop, and they begin to build combs. Some of these "house bees" may become guards that patrol the area around the hive. Eventually, each bee becomes a field worker, or forager. She flies afield to collect nectar, pollen, or water, according to the needs of the hive. Apparently, these needs are indicated by the "eagerness" with which the field bees' different loads are accepted by the house bees.

If a large number of bees with a particular duty are removed from the hive, the normal sequence of duties can be altered. Young bees may shorten or omit certain duties and begin to fill in where they are needed. Other bees may revert to a previous job where they are now needed again.

The watchword in a beehive is *efficiency*. In the more "feminist" species, the drone (males) exist only as sex objects. Once the queen has been inseminated, the rest of the drones are quickly killed off by the workers. They are of no further use. The females themselves live only to work. They tend the queen, rear the young, and maintain and defend the hive. When their wings are so torn and battered that they can no longer fly, they either die or are killed by their sisters. But the hive goes on.

Altruism

Hopefully, you won't be shattered to learn that most of the Lassie stories aren't true. Consider what would happen to the genes of any dog that was given to rushing in front of speeding trains to save baby chickens. The reproductive advantages would be considerable to chickens, but dogs with those tendencies might be selected out of the population by the action of fast trains. Their reproductive activities would therefore be cut short. In contrast, the genes of a "chicken" dog that spent his energy, not in chivalrous deeds, but in seeking out estrous females, would be expected to increase in the population.

Altruism may be defined in a biological sense as an activity that benefits another organism but at the individual's own expense. Altruistic behavior may at first seem to have obvious advantages, but the question is a prickly one, especially when we consider how such altruism might have evolved.

It is easy to see how certain forms of altruism are maintained in any population. For example, pregnancy, in a sense, is altruistic. The prospective mother is swollen and slowed. Much of her energy goes to the maintenance of the developing fetus. At birth she is almost completely incapacitated and is in marked danger at this time. Pregnancy is clearly detrimental to her. So why do females so willingly take the risk? It may help to understand the enigma if we remember that the population at any time is composed entirely of the offspring of individuals who have made such a sacrifice. Thus, the females in the population may, to one degree or another, be predisposed to make such a sacrifice themselves. The tendency to have offspring or to do the things that result in having offspring can be innate (genetically based) or, in humans, transmitted culturally.

However, altruism on this basis doesn't explain why a bird may feed the young of another pair, or why an African hunting dog will regurgitate food to almost any puppy in the group. Why, also, would a bird that may have no offspring of its own give a warning cry at the approach of a hawk, alerting other birds at the risk of attracting the hawk's attention to itself?

To answer such questions we must look past the answers that first come to mind. It may seem cynical, but we must start with the premise that birds don't give a hoot about each other. A bird that issues a warning call isn't thinking, "I must save the others." At least, there is a simpler explanation of its behavior.

Keep in mind that the biologically "successful" individual is the one that maximizes its reproductive output. One way of accomplishing this is for the organism to behave in such a way as to leave as many individuals carrying its own type of genes as possible in the next generation. This would explain parental care. However—and this may not be so readily apparent—an individual can also leave its type of genes in the next generation by helping a relative's offspring to survive. An individual shares genes in common with a cousin, although fewer, of course, than with a son or a daughter. So it is reproductively beneficial for an individual to assist relatives, as long as the cost is not too high. In fact, there is theoretically a point at which an individual could increase the success of its genes by saving its nieces and nephews (provided there were enough of them), rather than its own offspring. From the standpoint of effective reproductive output, the organism would be better off leaving a hundred nephews than one son.

To illustrate, suppose a gene for altruism appears in a population (notice that this sets up the mechanism for the continuance of the behavior). As you can see from Figure 33.18, altruistic behavior would

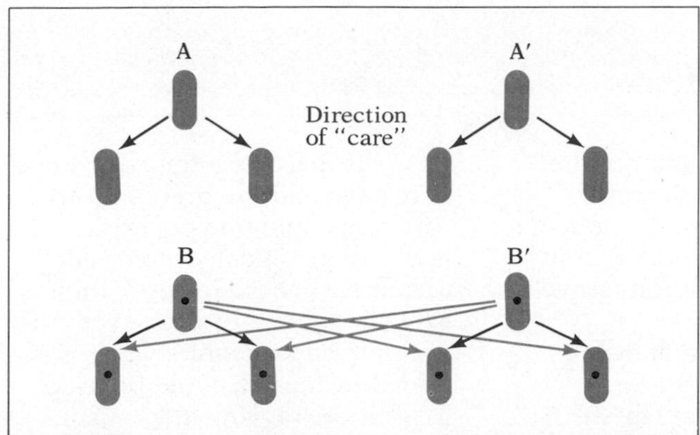

33.18 In this population, A and A′ are nonaltruists. They behave in such a way as to maximize their own reproductive success, but do nothing to benefit the offspring of other individuals. In another segment of the population (B and B′), a gene for altruism has appeared that results in individuals benefiting the offspring of others in some way. If can be seen that, assuming the altruistic behavior is only minimally disadvantageous to the altruist, generations springing from B and B′ are likely to increase in the population over those from A and A′. It should be apparent that the altruistic behavior is likely to be greatest where B and B′ are most strongly related, so the B shares the maximum number of genes in common with the offspring of B′ and vice versa. The idea is that B, for example, can increase its own reproductive success by caring for the offspring of a relative with whom it has some genes in common. After all, reproduction is simply a way of continuing one's own kinds of genes.

most likely be maintained only in groups in which the individuals are related (or have some kinds of genes in common). It might be expected, then, in nontransient populations (those that don't move around much) and those in which there is little mixing with outside groups—in other words, where there is a high probability that proximity indicates kinship.

Keep in mind that no conscious decision on the part of the altruist is necessary. It simply works out that when conditions are right, those individuals that behave altruistically increase their kinds of genes in the population, including the "altruism gene." Nonrelatives would be benefited by the behavior of altruists, of course, but there is an increased likelihood of individuals nearer an altruist to be related to it.

It has been determined mathematically that the probability of altruism increasing in a population depends on how closely the altruist and benefitee are related, as well as on the risk to the individual, of course. In other words, the advantages to the benefitee must increase as the kinship becomes more remote, or the behavior will disappear from the population. For instance, altruism toward brothers and sisters that results in a risk of death of the altruist will be selected for if the net genetic gain to the beneficiary is more than twice the loss to the altruist; for half-brothers, four times the loss; and so on. To put it another way, an altruistic animal would gain reproductively if it sacrificed its life for more than two brothers, but not for fewer, or for more than four half-brothers, but not for fewer, and so on. Therefore we can deduce that in highly related groups, such as a small troop of baboons, a male might fight a leopard to the death in defense of the troop. By the same token, a bird will give a warning cry when the chance of attracting a predator to itself is not too great, and when the average neighbor is not too distantly related.

This model, developed by J. B. S. Haldane in 1932 and expanded by W. D. Hamilton in 1963, helps to explain the extreme altruism shown by social insects such as honeybees. Since workers are sterile, their only hope of propagating their own genotype is to maximize the egg-laying output of the queen. In some species, the queen is inseminated only once (by a haploid male), so all the workers in a hive are sisters and have three-fourths of their genes in common. In such a system, then, almost any sacrifice is worth any net gain to the hive and to the queen.

In a brilliant essay, Robert Trivers expanded our understanding of the evolution of altruism by suggesting that most apparent altruism (outside of parental behavior) depends on the expectation of reciprocation—"I'll scratch your back if you'll scratch mine." *Reciprocal altruism* is an evolutionarily stable strategy with some complex rules. Help (altruistic acts) is given to others—even offered to strangers—when the cost or risk is not too great to the giver. The expectation is that some kind of help will be reciprocated at another time. If the expected reciprocation is not forthcoming, an evolutionarily derived emotion is experienced: *moral indignation.* The individual that fails to reciprocate is scorned, turned out of the social group, no longer aided. To get ahead, everyone again has to play by the rules—or appear to.

Trivers made an analogy with some stereotyped behavior between species of fish. Some fish specialize as cleaning symbionts: brightly colored, they tend obviously situated "stations," where other fish come to be cleaned. The cleaning fish pick off parasites and infected flesh, while the fish being cleaned—which might be a fierce predator—just holds still. Both fish are strongly inhibited from injuring the other, and both benefit from the relationship. So strong is the inhibition that the predator will usually starve before eating a cleaning fish. But sometimes the cleaner cheats and takes a chunk out of the bigger fish; then the game is over and the chase is on. In general, it pays to follow these innate rules of behavior and behavioral inhibition (Figure 33.19).

Some evidence of intraspecific reciprocal altruism has been reported in troops of social mammals,

Essay 33.1 Sociobiology

Sociobiology burst upon the academic scene in the mid-1970s, prompted and defined by a monumental book written by Edward O. Wilson at Harvard University. His book dealt almost entirely with other organisms, especially insects, with only the last twenty-seven pages dealing with humans. But it was enough. Wilson was attacked with a vigor usually reserved for the worst sorts of charlatans. But Wilson was no charlatan. So why the attack?

Wilson, in including humans in his discussion, had suggested what most biologists had always taken for granted—that human social behavior was molded in part by evolutionary processes. This, of course, meant that we were not *tabulae rasae,* blank slates waiting to be defined and molded solely by experience, by culture. Instead, suggested Wilson, we are born with certain tendencies and proclivities that have been shaped by our long processes of evolution. In a sense, he said that we are born with a certain genetically based foundation upon which we can build any number of social structures. However, as in building a house, those social structures can't exceed the limits of the foundation without grave risks.

The howls of denial began from down the hall. Some of Wilson's colleagues at Harvard were avowed Marxists and the notion that human social systems were not infinitely malleable and "perfectable" flew in the face of Marxist doctrine. Thus, "study groups" were set up and Wilson was denounced, and hounded, and even physically attacked whenever he tried to address the issue publicly. Because some of these detractors were among the best minds in biology, the academic community was rather perplexed. Why was Wilson being attacked? Was there something in his statements that had escaped the rest of us? When it dawned on the scientific community that the attacks were politically inspired, many people began to turn away from the noisemakers.

Wilson's detractors, of course, may have had the best of motives. After all, could one not argue that if we are genetically "compelled" to behave in certain ways, then isn't change impossible, even unnatural? The sociobiologist's position, they feared, could be used to promote racism and sexism. Perhaps they confused the attempted description of "what is" with "what ought to be." Sociobiology does not defend the status quo, it simply seeks to understand it. If changes are in order, we must know what prompts us to behave as we do. It does no good to pretend that people are, in fact, as we wish they were.

Obviously, there had been a grave misreading of sociobiology. The discipline is merely one more tool we can use in the examination of ourselves. If we are biologically constrained to behave in this way or that, we need to know it. Changes are indeed called for, but we need information first. After all, we didn't overcome gravity by simply denying its existence.

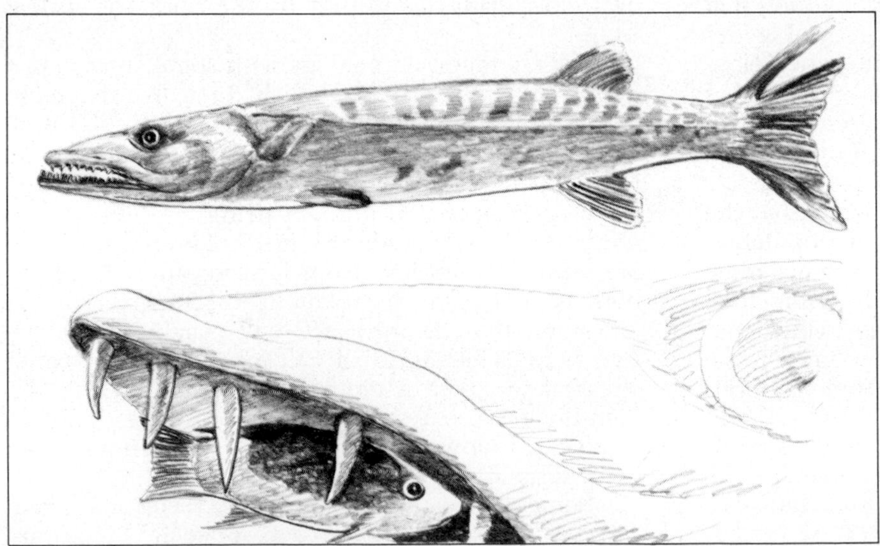

33.19 Cleaning symbiosis. A surprising number of marine organisms engage in cleaning. The Spanish hogfish (*Bodianus rufus*) would make a tasty morsel for the fearsome barracuda (*Sphyraena barracuda*), but the larger predator is genetically inhibited from attacking.

such as hunting dogs and higher primates, but the evidence is not very strong for such behavior in any animals but humans. Trivers implies, however, that reciprocal altruism is the key to human evolution. The complexity of such behavior, entailing as it does memory of past actions, the calculation of risk, the foreseeing of the probable consequences of present actions, the possibility of advantageous cheating, and the need to be able to detect such cheating, all require a level of intelligence that is beyond most organisms. To be a truly successful evolutionary strategy, in fact, reciprocal altruism requires a human intelligence. In the opinion of some anthropologists, it is exactly for the management of these elaborate social interactions that the human brain—and the conscious human mind—evolved.

The discussion of the development of altruistic behavior is admittedly somewhat esoteric. However, it seems important to consider even our most cherished and most despised behaviors in the context of the human as an evolutionary product.

Application of Ideas

1. Why is a biological clock (a precise sense of time) necessary for solar or stellar navigation?
2. Under what sort of competitive conditions (high versus low interspecific competition) would population recognition be particularly important?
3. What is the evidence that aggression is instinctive? Learned? Both?
4. What is an evolutionarily stable strategy? Can you think of another one in addition to those described by Maynard Smith?
5. Would you expect competition between males to be stronger under monogamous or polygamous conditions? Why?

Key Words

aggression
alarm chemical
altruism

chemical communication
cleaning symbiont
coloration
communication
competition

dominance hierarchy

evolutionarily stable strategy (ESS)

graded display

home range

individual distance
information load
inhibitory block
intraspecific cooperation
interspecific cooperation

navigation

orientation

pheromone
population recognition
precopulatory display

Key Ideas

Navigation and Orientation
Bird *navigation* is still filled with mysteries that biologists would like to unravel. Clues have come from several important experiments.

Orientation
1. Gustav Kramer's experiments with the migratory behavior of caged birds have provided strong evidence that some birds *orient* themselves to the sun's position. He made the following observations:
   a. In their cage movements, migratory birds *orient* themselves according to their usual migratory direction.
   b. Their *orienting movements* were clearest when the sun was out.
   c. When the normal sun direction was changed with mirrors they *oriented* themselves to the new "sun."
   d. When their surroundings were changed, they were still able to fly towards their food box as long as its compass direction was not changed and the sun was shining.
2. The *sun-orientation* studies indicate that some birds know the time of day and the normal course of the sun. The ability is innate since birds reared without sunlight also *orient* to the sun. Part of *sun orientation* involves maturation and learning.
3. Studies of night-migrating birds indicate that they are capable of *star navigation.* In experiments utilizing planetariums, birds showed the ability to correct their orientation when the "wrong" sky was shown.

Homing Pigeons
Pigeons utilize the *sun compass,* but they have backup methods using the earth's magnetic field and visual details on the ground when the sky is overcast.

reciprocal altruism
recognition
ritual combat
retaliator

short-term signal
sound communication
sound spectrograph
species recognition
star navigation
stylized fighting
sun compass
symbiont

territorial behavior
territory

visual communication
visual signal

Communication
The ultimate effect of communication is to increase reproductive success. It is an important part of mating behavior.

Visual Communication
1. *Visual communication* can occur through color, posture, movement, and timing of movements.

2. *Visual communication* carries a great information potential and a simple movement can become more complex by *graded displays.*

3. *Coloration* may be a permanent part of *visual signaling. Coloration* in male birds is an example. In other instances it may be temporary as in mating *coloration.*

4. *Short-term signals* are valuable to members of a species in locating each other and yet not giving predators a good fix on their position.

5. *Visual signals,* however, can be blocked by obstructions and except for light signals are ineffective at night.

Sound Communication
1. *Sound communication* is limited to arthropods and vertebrates. Arthropod sound is usually produced by friction (rubbing parts together). Vertebrates use a variety of devices including vocal sounds. Fish use the air bladder (swim bladder). *Sound signals* utilize pitch, volume, and tonal quality. Sound characteristics can be shown graphically in a *sound spectrograph.*

2. The quality of bird sounds can vary. Low-pitched sounds used in mobbing are easy to locate. High-pitched, extended sounds are used in predator alarm. These are difficult to locate.

3. The advantages of *sound communication* are high *information load,* distance, amplification, and transience. The disadvantages include problems in noisy environments, attenuation with distance, and difficulty of localization.

Chemical Communication
1. Chemical signaling is vital in social insects such as ants and bees.

2. Moths have extremely efficient chemical detectors (recall bombykol from Chapter 31).

3. The chemical substances used in *communication* are called *pheromones.* They are commonly used by insects, but are also important in mammals. Ants leave chemical trails and produce *alarm chemicals,* while rats rely on *chemical signals* in urine.

4. *Chemical signals* contain a carbon backbone with five to twenty carbon atoms. Their molecular weight ranges from 80 to 300, light enough to be carried in the air.

5. The advantages of *chemical signals* are their potency, their residual effect, and their ability to move around obstacles. A disadvantage is also found in the residual effect, which can alert a predator to the prey's presence. In addition there is a limited *information load.*

Why Communicate?

Species Recognition
Species recognition is vital for reproduction. Many species are very similar to others, and hybridization most likely produces sterile offspring if any. Similar species may ignore each other sexually, but become competitors when *territories* are involved.

Individual Recognition
1. In situations where maternal care is important, species must be able to recognize their mates and their offspring.

2. *Individual recognition* is not important where the sexes meet only for copulation, such as in spiders. Only sexual identification is important.

3. In social species, *individual recognition* is important in the establishment and maintenance of *dominance hierarchies.* In hierarchical populations there is a minimum of discordance and fighting.

Aggression
Aggression is belligerent behavior that arises from *competition.* Fear is an important component of *aggression.*

Fighting
1. Fighting between species is very unlikely simply because different species are not commonly in *competition.* The greatest *competition* occurs among members of the same species and, further, among those of the same sex.

2. In fighting within a species, there is usually very little injury. It is more adaptive to avoid injuring a competitor.

3. *Stylized fighting* is common between potentially dangerous combatants. Injury is rare and usually one emerges dominant while the other one retires from the scene.

4. The most vicious fighting may occur between two individuals that cannot really hurt each other. Exposed vital areas appear to act as inhibitors.

5. Fighting to the death or serious injury is nonadaptive to most animals.

6. Maynard Smith hypothesizes that engaging in ritual combat, but being willing to retaliate if injured, is an evolutionarily stable strategy.

7. Fighting to the death or to serious injury does occur in some species. The serious attacks are directed most commonly to outsiders.

Is Aggression Instinctive?
The origin of *aggression* is uncertain. Is it innate, learned, or both? Some animal behavior indicates a strong innate basis for *aggression.*

Human Aggression
1. Determining the source of human aggressive behavior is especially difficult because we can override both learned and innate behavior by conscious decision.

2. If *aggression* is innate in humans it can be expected to reveal itself in rechanneled or disguised ways.

Territories
1. Animals hold *territories* through aggressive behavior. The strongest attacks are against others of their species and in particular against others of their own sex.

2. *Territorial behavior* varies widely in different species. *Territorial* animals may only defend certain parts of their *home range* while sharing other parts.

3. Some of the proposed adaptive features of *territoriality* are:
   a. Insurance of space required
   b. Protection against predators, disease, and parasitism
   c. Selection of vigorous breeders
   d. Equitable division of resources
   e. Stimulation of breeding behavior
   f. Insurance of food supply
   g. Provision of mating and nesting places
   h. Increasing efficiency of habitat utilization

4. *Territorial behavior* is firmly set. Defenders appear to know that others must stay out, and intruders also appear to know they must give way. The behavior has become evolutionarily stabilized.

Cooperation
Cooperative behavior is most highly developed within species.

1. Porpoises protect females during birth process, and often aid injured members.
2. Himalayan yaks form defensive circles to resist predators.
3. Wolves and other similar predators hunt cooperatively.
4. Insects have developed the most complex and highly coordinated *cooperative behavior.* In bee colonies all aspects of colony maintenance require the attention of members specialized for that purpose.

Altruism

1. *Altruism* may be defined as an activity that benefits another organism at the altruist's expense.
2. Pregnancy in a sense is *altruistic* since it carries with it great risk and near incapacity at its end and during the birth process.
3. *Altruism* probably has its greatest adaptiveness where it maximizes the reproductive output of one's own genes. The genetic value of altruistic behavior is directly related to the closeness of kinship.
4. This hypothesis explains why extreme *altruism* is common in social insects such as honeybees. Because of their extremely close relatedness, any sacrifice by members directly benefits the genes that determine the behavior.
5. *Reciprocal altruism* is an evolutionarily stable strategy. The rules are strict; there must be reciprocal behavior.
6. In interspecific *altruism* (such as *cleaning symbiosis* in fishes), predators are strongly inhibited from injuring the cleaners unless the cleaners break the rules.
7. *Reciprocal altruism* is an integral part of human behavior, which may have an evolutionary basis. The successful management of *reciprocal altruism* requires intelligence not found in most animals.

Review Questions

1. Describe Kramer's experimental procedure and results where mirrors were used to test the *sun orientation* hypothesis. (p. 968)
2. What did Kramer discover from his second experiment utilizing the food box in changing surroundings? (pp. 968–969)
3. Can you describe the apparent accuracy with which starlings are able to use the sun as a compass? (pp. 968–969, Figure 33.3)
4. What *navigation* aids do night-migrating birds use and what is some of the evidence supporting the ideas? (p. 970)
5. Homing pigeons are excellent navigators. What kinds of environmental clues do they use and how are some of these redundant? (p. 970)
6. What are *graded displays* and how do they increase the information load of a *visual signal?* (p. 971)
7. List three examples of permanent *visual displays.* (p. 971)
8. What are some advantages of short-term *visual signals?* (p. 972)
9. List several disadvantages of *visual communication.* (p. 972)
10. In general, how do insects make sounds? (p. 972)
11. List three aspects of sound that help make it variable and potentially great in *information load.* (p. 973)
12. Using the mobbing and warning cries of birds as examples, show how bird sounds can be either easy or difficult to locate. (p. 972)
13. List several advantages of *sound communication.* (p. 973)
14. Describe two examples of *chemical communication* in insects. (p. 973)
15. Describe two examples of *pheromones* at work in mammals. (p. 973)
16. How are *pheromones* responsible for controlling the sex of developing bees? Of what advantage is this to the hive? (pp. 973–974)
17. What are the molecular requirements of a *pheromone?* (p. 974)
18. List several advantages in *chemical communication.* (p. 974)
19. Why is it important for individuals of one species to be able to recognize one another? (p. 974)
20. How is *individual recognition* important in *dominance hierarchies?* (p. 975)
21. Define the term *aggression* in its biological sense. (p. 975)
22. Is hunting by a predator *aggression* in the strict sense of the word? Explain. (p. 975)
23. Among what individuals is fighting most likely to occur and why is this so? (p. 975)
24. Why is it adaptive for fighting animals to avoid injuring or killing each other? (pp. 975–976)
25. Define and give an example of *stylized fighting.* (p. 976)
26. Present a theoretical reason why many animals do not attack exposed vital or weak areas while fighting. (p. 976)
27. Review Maynard Smith's concept of an *evolutionarily stable strategy.* What kind of fighting is most adaptive? (p. 977)
28. Are there any animals that fight to kill? Cite examples. (p. 977)
29. What are the three possible bases for *aggressive behavior?* (pp. 977–978)
30. Describe an example of *innate aggressive behavior.* (p. 978)
31. Why is it so difficult to determine the basis for human *aggressive behavior?* (p. 978)
32. List the main aspects of *territorial behavior* in birds. (p. 979)
33. List several adaptive features of *territoriality.* (p. 980)
34. What is the general rule about behavior in territories? (p. 980)
35. Briefly describe two examples of *cooperative behavior* in mammals. (p. 981)
36. In which group of animals has *cooperative* behavior reached its most intricate and coordinated level? Describe an example. (pp. 981–982)

37. Write a precise biological definition of *altruism.* (p. 982)

38. Explain the adaptive relationship between closeness of kin and *altruistic behavior.* (p. 983)

39. Explain how *altruism* becomes extreme in honeybees and what the adaptive significance is. (p. 983)

40. What is *reciprocal altruism* according to Trivers? What is the expectation for "rule-breakers"? (p. 983)

Suggested Reading

Barash, D. P. 1977. *Sociobiology and Behavior.* Elsevier, New York and Amsterdam.

Binkeley, S. 1979. "A Time-keeping Enzyme in the Pineal Gland." *Scientific American,* April. In at least some animals, the activity of *N-acetyltransferase* is the biological clock.

Brown, J. L. 1975. *The Evolution of Behavior.* W. W. Norton, New York.

Chomsky, N. 1972. *Language and Mind.* Harcourt, Brace, Jovanovich, New York. Chomsky argues that language is innate.

Christian, J. J. 1970. "Social Subordination, Population Density, and Mammalian Evolution." *Science* 168:84. Mammals have marked endocrine responses to crowding and to aggressive encounters.

Darwin, C. 1871. *The Descent of Man, and Selection in Relation to Sex.* Appleton, New York. Darwin has a great deal to say about the evolutionary importance of sexual behavior.

———. 1872. *The Expression of the Emotions in Man and Animals.* Appleton, New York. The first great work of comparative ethology.

Dawkins, R. 1976. *The Selfish Gene.* Oxford University Press, New York and Oxford. The selfish gene concept is applied to animal behavior in the second half of this popular paperback.

Ehrman, L., and P. A. Parsons. 1976. *The Genetics of Behavior.* Sinauer Associates, Sunderland, Mass. A textbook approach.

Fisher, J., and R. A. Hinde. 1948. "The Opening of Milk Bottles by Birds." *British Birds* 42:347. Birds learn a complex new item of behavior by watching other birds steal milk left on British stoops.

Gardner, R. A., and B. T. Gardner. 1971 "Two-way Communication with an Infant Chimpanzee." In *Behavior of Non-Human Primates,* ed. A. Schrier and F. Stollnitz, p. 117. Academic Press, New York.

Jansen, D. H. 1966. "Coevolution of Mutualisms between Ants and Acacias in Central America." *Evolution* 20:249. One of biology's best gee-whiz stories.

Johnsgard, P. A. 1967. "Dawn Rendezvous on the Lek." *Natural History* 76:16. A *lek* is a special place where males compete with one another for territory and prestige and wait for the arrival of sexually interested females.

Klopfer, P. H. and P. J. Hailman. 1967. *An Introduction to Animal Behavior: Ethology's First Century.* Prentice-Hall, Englewood Cliffs, N.J.

———. 1973. *Behavioral Aspects of Ecology.* Prentice-Hall, Englewood Cliffs, N.J.

Marler, P., and W. Hamilton. 1966. *Mechanisms of Animal Behavior.* Wiley, New York.

Maynard Smith, J. 1964. "Group Selection and Kin Selection." *Nature* 201:1145. The problem of group selection and the evolution of altruistic, group-directed behavior.

———. 1974. "The Theory of Games and the Evolution of Animal Conflicts." *Journal of Theoretical Biology* 47:209.

———. 1978. "The Evolution of Behavior." *Scientific American,* September. More on the key problem of altruism *and* on the logic of formalized, nonlethal aggressive conflicts.

Mech, L. D. 1970. *The Wolf: The Ecology and Behavior of an Endangered Species.* Natural History Press, Garden City, N.Y. Fascinating reading for all you wolf fans.

Milgram, S. 1974. *Obedience to Authority: An Experimental View.* Harper and Row, New York. People will do terrible things if they are told to with bland assurance: a full account of a famous and disturbing experiment in human behavior.

Money, J., and A. A. Eberhardt 1972. *Man and Woman, Boy and Girl: The Differentiation and Dimorphism of Gender Identity from Conception to Maturity.* Johns Hopkins University Press, Baltimore. Complex social and biological factors influence each person's concept of his or her own gender, but prenatal hormone levels have a profound influence on adolescent and adult behavior.

O'Keefe, J., and L. Nadel. 1978. *The Hippocampus as a Cognitive Map.* Clarendon (Oxford University Press), New York and Oxford. The mammalian brain stores "pictures" and "maps"—any two- or three-dimensional memory—in the hippocampus.

Oster, G. F., and E. O. Wilson. 1978. *Caste and Ecology in the Social Insects.* Princeton University Press, Princeton, N.J. Hedging bets against rare disasters demands very different allocations of scarce resources from what would be feasible if the world were predictable.

Robertson, D. R. 1972. "Social Control of Sex Reversal in a Coral-Reef Fish." *Science* 177:1007. Each population consists of one male and many females: when the male is removed, the females compete to determine which one will change sex.

Scott, J. P., and J. L. Fuller. 1965. *Genetics and the Social Behavior of the Dog.* University of Chicago Press, Chicago. Cocker spaniels and Basenjis behave very differently. Different aspects of these differences appear to undergo Mendelian segregation in the F_2 generation of a Basenji–Cocker cross.

Skinner, B. F. 1938. *The Behavior of Organisms: An Experimental Analysis.* Appleton-Century-Crofts, New York.

Trivers, R. L. 1971. "The Evolution of Reciprocal Altruism." *Quarterly Review of Biology* 46:35. Frequently behavior that appears to be altruistic is really selfish in the long run: it is insurance for similar help when it might be needed.

Wallace, R. A. 1979. *The Ecology and Evolution of Animal Behavior,* 2d ed. Goodyear Publishing Co., Santa Monica, Calif. The shorter version of Wallace's behavior book. The previous edition was the first text to bring together ecological and evolutionary aspects of animal behavior with a full treatment of the field of behavior studies.

———. 1979. *Animal Behavior: Its Development, Ecology, and Evolution.* Goodyear Publishing Co., Santa Monica, Calif. In this treatment, Wallace brings to the field of animal behavior the full implications of developmental neurobiology and endocrinology, always with a strong evolutionary viewpoint and a lively style.

———. 1979. *The Genesis Factor.* William Morrow, New York. Fearless Wallace shoots from the hip—but with deadly aim—in this enjoyable and thought-provoking popular book on the implications of sociobiology.

Wilson, E. O. 1971. *The Insect Societies.* Harvard University Press, Cambridge, Mass.

———. 1975. "Slavery in Ants." *Scientific American,* June. Some slave-making species have become so specialized that they are no longer capable of feeding themselves.

———. 1975. *Sociobiology, The New Synthesis.* Harvard University Press, Cambridge, Mass.

———. 1976. "Academic Vigilantism and the Political Significance of Sociobiology." *Bioscience* 26:183.

———. 1978. *On Human Nature.* Harvard University Press, Cambridge, Mass. The three books listed above form an "unplanned trilogy" in which Wilson successively considers insect societies, animal social behavior in general, and the application of evolutionary sociobiology to the study of human affairs. Wilson, who was an influential and widely respected biologist long before he coined the term "sociobiology," has been under heavy ideological fire for his views, which are in direct contradiction to those of Marx.

Part Seven

Ecology

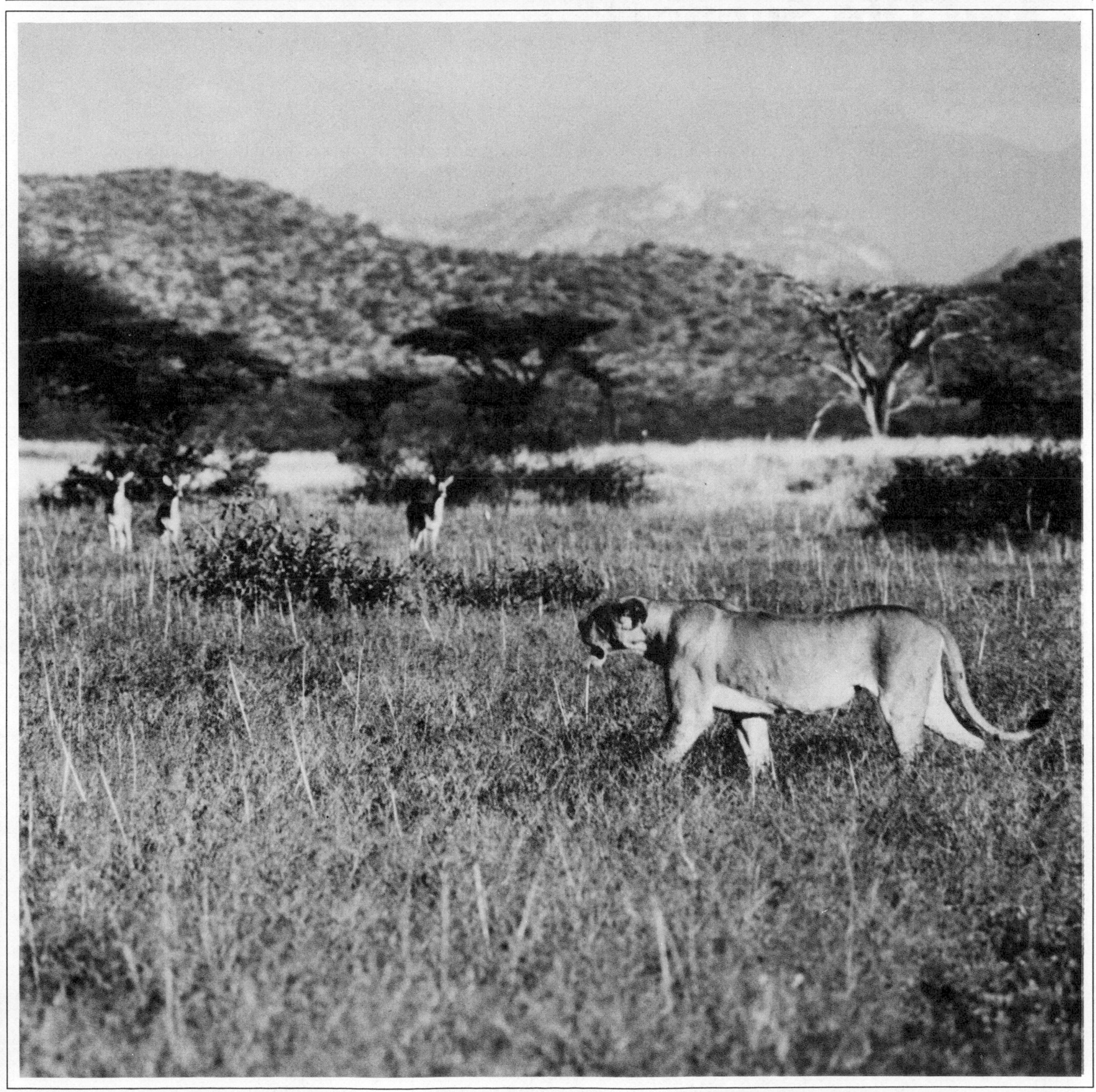

Chapter 34

Biomes, Communities, and Life-Support Systems

Ecology is the "study of" something—we know this from the *-ology* ending. The prefix *eco-* stems from the Greek word *oikos,* which means "house." So ecology is "the study of the house," and biologically, the house is the *environment.* The environment technically includes all of those external factors both nonliving and living that affect an organism. Ecology, then, is the study of the interaction between organisms and their environment, and the key word is *interaction.*

We saved the study of interaction until rather late in our discussion because a certain amount of basic information about organisms is necessary in order to understand their interactions. Ecology, then, is where the sciences come together. It is a relatively new field, probably emerging as much from what was called *natural history* as from anything else. Natural historians are those crusty old souls who tramp around in fields and who know where rabbits live. Unfortunately, natural history became unfashionable too soon, trampled in the rush to the altar of the molecule and the equation. It is being reborn as a valid topic for ecologists who find themselves a bit embarrassed by what they don't know about rabbits.

Today's ecologists have emerged from a long line of people who simply reported what they saw, but modern ecologists are armed with better powers of description—new ways of measuring and comparing. And they're asking questions. With the aid of computer technology, some ecologists now try to store the myriad data concerning physical and biological measurements of every conceivable aspect of the environment. From these stored measurements, computers can be programmed to solve problems dealing with interaction. Model environmental systems can be contrived, and variables tested for their impact on the model. (What happens if a predator dies out? What are the effects of too much rain?) Through such simulated studies, *systems ecologists* are learning to predict the eventual effect of both natural and artificial alterations of the environment. These efforts are in their embryonic stages, however, so don't look to the systems ecologists for answers yet. In fact, don't ever look to ecologists for unequivocal answers to anything. Ecology, as with all science, is at best able to provide us with *probabilities.* These are always based on the available data, but as you will soon see, interactions in living systems are incredibly complex, and attempts at their measurements can only be partially successful. In fact, our most modern descriptive methods will probably one day be regarded as unbelievably crude and naive. At present, we can only stroke our chins and pretend to be wise. After all our measuring and computing, we don't even know if competition is important in natural selection. We just don't have enough basic information. Predictions relying on incomplete data can't be considered very reliable.

Now that we have defined ecology and have some idea of what we're up against, let's look at ways in which the earth's environment and its organisms have been organized for study by ecologists. And let's begin by describing the *biosphere.*

34.1 The Biomes

The Biosphere

The biosphere is that thin veil over the earth wherein the wondrous properties of light and water are brought together to permit life. In addition to the land surface area, we must include the subterrestrial realm and the fresh and marine waters as well. In all, we're dealing with a lot of square footage but not much depth. Things can't live very far above or below the earth's surface. To be precise, the habitable regions of the earth lie within an amazingly thin layer of approximately 14 miles. This includes the highest mountains and the deepest ocean trenches. If the earth were the size of a basketball, the biosphere would be about the thickness of one coat of paint.

We are aware of the basic situation within the biosphere. It includes a certain amount of water, a rather narrow range of temperatures, sunlight, and carbon, hydrogen (in water vapor), oxygen, nitrogen, and various minerals. Our space probes have found no other place in the solar system where these conditions exist. A key difference between earth and other planets, then, lies in atmospheric composition. It is this atmosphere clinging to a tilted earth that produces climates and seasons. Without getting into a lengthy discourse on physical geography, we'll simply say that climatic patterns have a great effect on the distribution of life.

The Distribution of Life: Terrestrial Environment

With some imagination on your part, perhaps you can be persuaded to believe that life is distributed into *biomes* (Figure 34.1). A biome is an area which because of its unique conditions has a certain conveniently recognized array of plant life. Where dominant plant types are recognizable, it follows that the dominant animal types that live among these plants will also be recognized and labeled. (But this is truer in some biomes than others, since some are highly restrictive to animal life—allowing only certain kinds to exist in those places.) Actually, organizing the earth into biomes is somewhat artificial, since the boundaries are usually indistinct and local conditions within recognized biomes may produce well-organized, identifiable groups of organisms that don't fit our definitions. However, biomes remain a convenient way of organizing things for discussion. For example, grassland is a biome. Whether it occurs as prairie, pampas, savanna, or steppe, the dominant plants are grasses. If one knows the plant life of an area, certain assumptions can be made about the climate and the animals that will be found there. For example, in grasslands the animal life typically includes large mammalian herbivores, predatory mammalian carnivores, insects, and birds. So, describing biomes in terms of plants is not as narrow-minded as it might seem. Just keep in mind that any biome may

34.1 The *biomes.* Each of the biomes can be identified by its dominant form of plant life. The plant life is always the form that is best adapted to the climatic conditions, including precipitation, the availability of light, and, of course, temperature. Both latitude and altitude affect all these conditions.

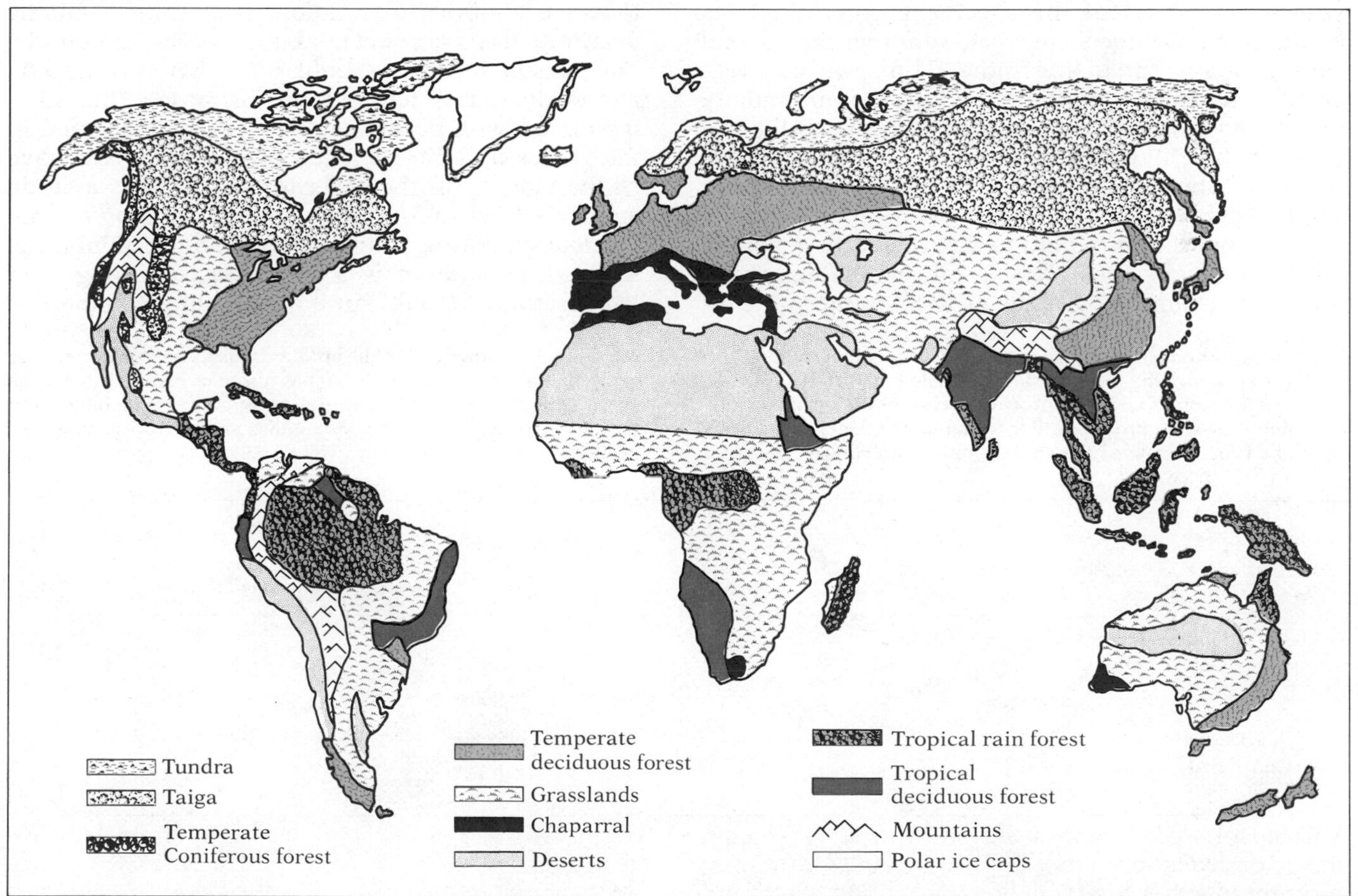

34.2 *Communities* are found within biomes and consist of associations of plants, animals, and microorganisms that can be identified as units. Mangrove communities form on low-lying tropical shorelines. The mangrove has adapted to brackish and salt water. Its stems and foliage are supported by dense prop roots. Specific organisms are associated with these plants.

be widely distributed since its existence is determined largely by climate.

Biomes, in turn may be subdivided into *communities* (Figure 34.2). Ecologically speaking, a community is a number of interacting populations of organisms. Several communities may exist within any biome. If you were to begin a walk across a biome, you might note that the specific groups of plants and animals change somewhat, although the general climate may remain the same. Thus, we can visit both a mesquite community and a sage community in a desert biome. We will return to how populations interact, but first let's make one more general observation about biomes and then describe them in some detail.

The distribution of biomes on the earth generally follows latitude lines. This is more obvious in the Northern Hemisphere than in the Southern Hemisphere. As the biomes change from one place to another, we find what is known as an *ecocline* (Figure 34.3). This is a gradual transition from one community type within the biome to another, and eventually into a new biome. The gradual transition existing across the United States from the Appalachian Mountains to the deserts is a good example. The moist forests of the Appalachians give way to drier oak–hickory forests, and then to forests consisting almost entirely of oak. The forest then yields to prairie, to a drier grassland biome, and finally to the communities of the desert. The borders of these biomes are usually indistinct, consisting of a general mixing of plants in each, seemingly engaged in an endless tug-of-war over boundaries. The principal determining factors in an ecocline are precipitation and temperature, as you might have guessed. (Look over the biome map in Figure 34.1.) Going from the equator northward we tend to move from tropical rain forest through desert, grassland, deciduous forest, then taiga (coniferous forest), and finally tundra and ice cap.

The arrangement of biomes along latitudinal lines is not entirely orderly for several reasons, one of which is the terrain, or topography. Mountain ranges, for example, interrupt the orderly distribution of biomes and are often responsible for the presence of deserts because they can block the movement of clouds. We're talking about mountains now in order to make another observation. Interestingly, biome distribution is a product of altitude as well as latitude. The reason for this should be rather obvious. As altitude increases, temperature decreases. Thus, high mountains with permanent snow or ice are found in the tropics, and these equatorial mountains may have all the biomes (in the order listed above) as altitude increases (Figure 34.4). Thus, we have an *altitudinal ecocline* similar to continental ones, but in miniature. Again, precipitation is a major factor, along with temperature. Actually, in the high altitudes, there is

34.3 A latitudinal *ecocline*. The profile shown is along the thirty-ninth parallel crossing the United States. The line transects several of the major biomes and is helpful in understanding the conditions by which they were produced. For example, the American deserts lie in the rain shadow of the western mountain ranges. Rain and snow fall on the western slopes only. Moisture-laden air from the Atlantic, at the other end of the profile, drops its burden as it moves west, the total rainfall diminishing as it moves inland. East of the Mississippi there is adequate rainfall to support a forest biome, but this dwindles near the Rocky Mountains, and the forests give way to prairie.

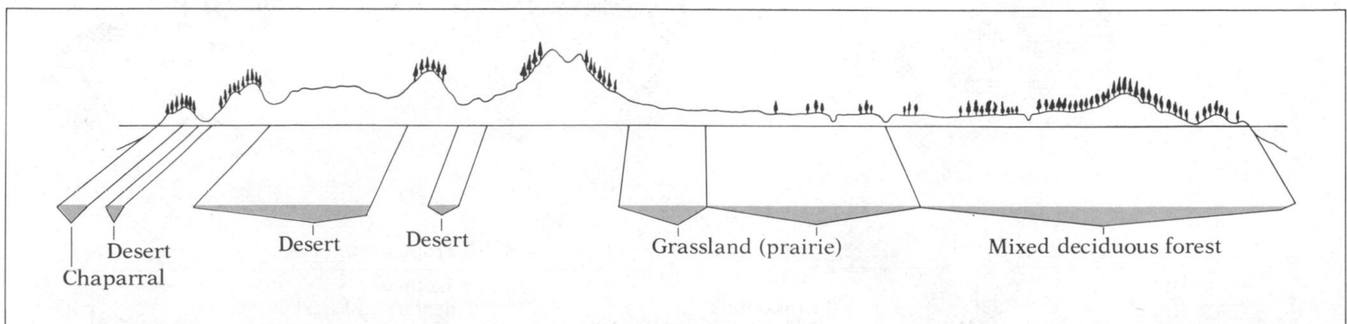

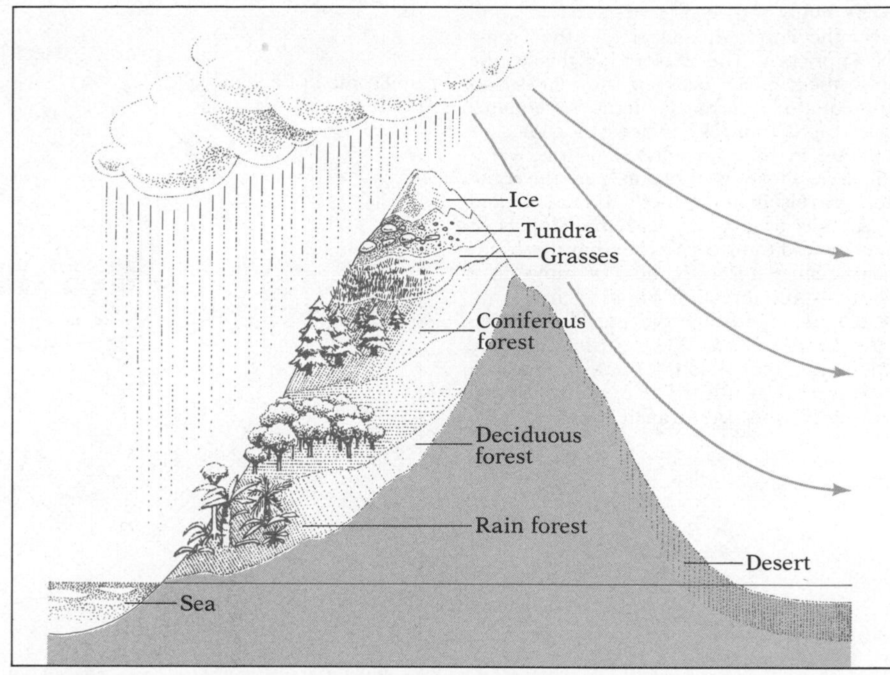

34.4 An altitudinal ecocline. The altitudinal variation provided by mountains often mimics latitudinal ecoclines. In this hypothetical model, a rain forest at sea level gives way to a deciduous forest at a higher elevation. Conifers replace the deciduous trees farther up, which yield to alpine meadows, and finally a rather typical tundra situation near the glacier. These transitions are a product of decreasing precipitation and temperatures. On the other side of the mountain, air masses, with little moisture remaining, rush down the slopes and across the barren terrain, leaving typical desert conditions behind as they pick up whatever moisture exists.

a relationship between temperature and precipitation. Since frozen water is unavailable to plants, there are plantless regions in high mountains. One of the best examples is Mount Ruwenzore in east central Africa. One can ascend from tropical rain forest through broad-leaved evergreens, to deciduous, coniferous, alpine meadow, and finally to the barren crags of the peaks. Once again we see that the factors determining the distribution of life are complex.

Biomes

Now let's consider the earth's biomes in more specific terms. Just keep in mind that the distinctions between them are not as tidy as we might lead you to believe here.

Compared to aquatic environments, the land is a turbulent and unstable place indeed. It changes seasonally, daily, and even hourly, and is disturbingly unpredictable. Land plants and animals, then, have special problems, but perhaps those with the greatest problems are those trying to survive in the desert.

The Desert Biome

Deserts are dry. By definition, they receive less than 25 cm (10 inches) of precious rain each year. You may be surprised to learn that deserts are not necessarily hot (even in the day) and some are found far from the tropics. The largest desert is the Sahara, which, as you can see on our biome map, covers nearly half of the African continent (and is getting bigger). Other large expanses of desert are found in Australia, Asia, western North America, and South America (in fact, every continent but Europe). Temperatures in these regions undergo dramatic day–night fluctuations, and may vary as much as 30°C (54°F) in a 24 hr period. One reason for such changes is that desert air isn't buffered by great amounts of stabilizing water vapor.

We have a pretty good idea of what causes deserts. Most of the earth's deserts are found near 30° north and south latitude. One explanation for this distribution has to do with air currents and heat at the equator. As equatorial air is heated, it rises, losing most of its moisture (in the tropical rain forests), and is replaced by cooler air moving in under it. This replacement produces revolving masses of air (called *air cells*) both north and south of the equator, extending to the 30° latitude line where the descending air at that end of the revolving cell is quite dry (Figure 34.5). These cells don't revolve exactly north and south, but are subject to the rotational forces of the earth (in a *Coriolis effect*), thus creating a diagonal flow (*trade winds*).

In addition, deserts generally occur near mountains, because the high terrain acts as a barrier to water-laden air currents. In California, for example, moisture-laden air from the Pacific Ocean encounters high mountains as it moves eastward. The air mass has to go upward, and in doing so it cools and condenses. Thus precipitation falls on the Pacific uplands and the Sierra Nevada Mountains. The region to the

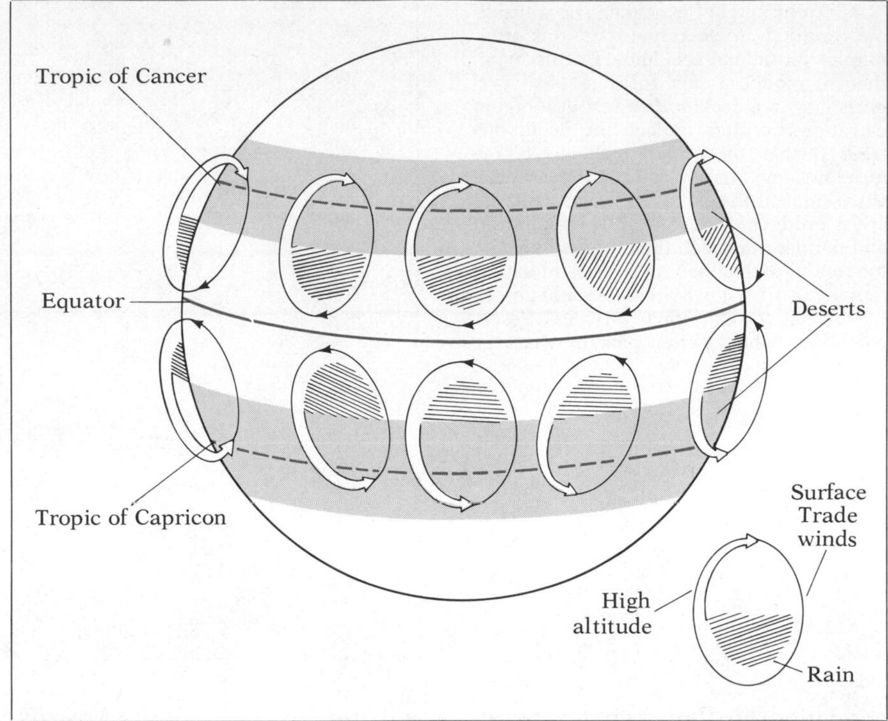

34.5 Many of the world's deserts are found near the Tropic of Cancer and the Tropic of Capricorn. The reason for this is the movement of air between these lines and the equator. Warm air along the equator rises rapidly and is replaced by cooler air moving in from the north. As the warm air rises, it loses its moisture near the equator, travels in a northerly direction, and then falls as dry air near the Tropics of Cancer and Capricorn. Note that these currents, called *air cells,* do not move in a north—south direction, but are thrown into a diagonal path by the earth's rotation (the *Coriolis effect*). The returning surface winds are known as the *trade winds* since they were once utilized by merchant sailing vessels in east—west transit.

lee of such a mountain range is said to lie in its *rain shadow*. Since moisture is lost as the air cools while ascending the western slope of this range, the air scurrying over the top and descending the eastern slope is very dry and fast-moving (see Figure 34.4). Making its way across the land, it gathers up whatever moisture there is, intensifying the dryness. And behold! the American deserts are born.

Conditions are reversed for the South American deserts, where prevailing winds carry their load of moisture from the Atlantic and cross the continent to the Andes, again dumping precipitation on the windward slopes so that the air moving down the other side toward the Pacific Ocean is dry. This causes the southern deserts to appear on the coast itself. The effectiveness of the Andean rain shadow is dramatized by the fact that the Atacama Desert in Chile has had *no* recorded rainfall! (Occasional local fogs are the only source of moisture.) Imagine an area where it *never* rains. The most northerly Peruvian deserts, however, almost always receive rainfall in December when sudden shifts in ocean currents and local winds bring torrential and destructive rains to the exposed coast.

Desert Plants

If you have never seen a desert, you might have fallen for the Hollywood image of lifeless regions of drifting sand studded with a few palm-covered oases. And if you looked hard enough, you could really find such places, but most deserts are alive with plants and animals. Since dry areas are called *xeric* (Greek for "dry"), plants adapted to deserts are called *xerophytes.* Their adaptations can take a number of routes, but they have one common imperative: Save water! Perennials, such as cactus, ocotillo, joshua tree, creosote bush, and palo verde (Figure 34.6), are adapted to living long periods on what little water they contain, while waiting for rain. The various cacti reduce water loss that might occur through leaves by having replaced leaves with spines. They are also covered with a thick, waxy cuticle over their photosynthetic stems, and in addition they can store water in the oversized cells of their tissues.

Compared to other plants, desert perennials tend to have fewer and more widely scattered guard cells. This helps in conserving water, but the price is a considerably slower growth rate, since CO_2 uptake and photosynthesis are retarded. Cacti generally close their stomata entirely during the day, and use stored energy from photosynthesis to fix CO_2 at night. Other desert perennials have deep root systems able to tap whatever water seeps into the porous soil, very small leathery leaves, and the ability to lie metabolically dormant, growing very little, for long periods.

If you're one of those people who can be made happier by the simple sight of a growing flower, you really should see what happens to annual plants in the desert. The rainy period may be short, perhaps confined to a single rain, so the plants have to act

34.6 Both of the North American desert plants shown here (the saguaro, top, and the Joshua tree, bottom) have adapted to limited water; these are called *xerophytes.* Their thorny epidermis discourages browsing and helps shield the green surfaces from the harsh, direct sunlight. Many xerophytes have pulpy water-storing tissues within, while others simply slow their metabolic processes, and become scraggy and leafless between rainy seasons.

quickly. After a spring rain (or *the* spring rain), the desert may become transformed as tiny flowers of all descriptions erupt into full bloom, a trembling and riotous offering of vibrant color and delicate forms. Their life cycle is short and they soon die. But their seeds will be highly resistant, lying in the sand until later rains propel them into their surge of growth and reproduction.

Desert communities are easily characterized according to their dominant plant species. Typically, the landscape is dominated by only one or two species, and these are spaced according to the availability of water.

Desert Animals

Just as plants have had to adapt to the rigors of the desert, so have the animals. Animals have the advantage in being able not only to adapt anatomically and physiologically, but in having more behavioral options.

Animal life in the desert consists primarily of arthropods, reptiles, birds, and mammals. Amphibians, scarce in the desert, are best represented by the fascinating spadefoot toad. The spadefoot toad has similar reproductive requirements to other amphibians. It must reproduce in water. It escapes heat and dryness by burrowing. It is also capable of *estivation,* a sort of reverse hibernation, remaining buried during long, hot, dry intervals. When sufficient rain occurs, the toads spring (hop?) into action. For them, time is short. Finding a temporary pond, they begin at once to call prospective mates in the briefest of courtships. Eggs, fertilized in the shallow ponds, hatch in a day or two. The young tadpoles complete their metamorphosis and emerge as adults in a few short weeks. Unlike relatives in moister environs, the spadefoot toad isn't a seasonal reproducer. Any heavy rainfall can bring on an amorous response. In fact, laboratory studies show that these amphibians can predict rain. Any sudden drop in barometric pressure sends the toad into a state of reproductive readiness. When laboratory researchers produced water for the toads without a drop in barometric pressure, the toads failed to reproduce.

Many desert animals beat the heat by simply staying out of the sun and becoming nocturnal. In fact, people visiting the desert may see only a few lizards, a bird or two, and a few insects, unless they want to traipse around at night with the snakes.

Mammals are largely represented by rodents, animals for which nocturnality has very clear advantages. They tend to lose water quickly because of rapid breathing and because small animals have a large surface area relative to their volume. Thus rodents spend the desert days deep in insulating burrows, venturing out only at night to forage for plant food. Of course, predators such as owls and rattlesnakes must follow their prey, so they too confine most of their activity to night or the cool of the evening.

The few daytime animals, such as long-legged and fast-moving lizards, are preyed upon by hawks and roadrunners, which are also adapted to daylight conditions. Even the daytime animals, however, restrict most of their activity to the morning and evening hours (Figure 34.7).

Perhaps we can best gain some insight into the adaptive strategies of desert animals by considering two mammals which inhabit different desert biomes: the kangaroo rat and the camel. The kangaroo rat (*Dipodomys deserti*) of the southern California desert is particularly interesting because it doesn't drink. It survives on the water content in its food supplemented by the metabolic water produced in cell respiration as electrons leave the electron transport system to be joined with oxygen. The kangaroo rat is also an incredible water miser. Its remarkably efficient kidney produces only highly concentrated urine, and very little at that (23% urea, 7% salt—compared to 6% urea and 2.2% salt in humans). The feces are also dry and crumbly. Most water loss, in fact, occurs through simple breathing. Even these special physiological features, however, wouldn't permit desert survival, were they not coupled with nocturnality. The rat spends its day in a humid, hair-lined burrow, venturing forth only in the cooler evening.

That great denizen of the desert, the camel, on the other hand, drinks. And drinks. In fact, it may drink up to one-third of its body weight at a time. Contrary to popular belief, it doesn't store water, and it sweats just as we do. So how does the camel survive for days without water in the dry desert? To begin with, it starts its day with a subnormal body temperature, so it takes a while to warm up.

(a)

(b)

34.7 Day and night desert food webs. Animals in the desert forage in two shifts. The day feeders are fewer in number but are much more active. These include birds such as the road runner (a), desert iguana (b), and many insects. But even these hardy foragers often remain in shadows at midday. At sundown the second shift begins as kangaroo rats (c) and other rodents start their browsing. This is, of course, an invitation to carnivores like the great-horned owl (d) and sidewinder that go quietly about their deadly business.

(c)

(d)

Once it does become warm, it doesn't immediately begin sweating, as we do. This cooling mechanism doesn't turn on until its body temperature reaches 41°C (106°F), a temperature that might kill us. Further, the camel's fur is just right for insulating (keeping its cool morning temperature in) without completely blocking heat escape. Finally, a camel can lose over 40% of its body water without great danger! Most mammals die after 20% loss (a human would behave mindlessly long before this). The camel doesn't spend its day in a burrow, but when resting it faces the sun, thus exposing a minimum of its flat body to the sun's direct rays.

Deserts, which seem so tough and unyielding, actually represent one of the more fragile biomes. Simple systems are always the most vulnerable, and deserts are essentially simple places, harboring few species compared to most other places. Because the desert is vulnerable, the impact of humans can be great (Figure 34.8).

The recent experience of the region to the south of the Sahara is sobering. People of the Sahel region long supported themselves and their small herds of cattle by hand-drawn water from a few scattered deep wells. In recent decades, gasoline pumps have been introduced, making water suddenly abundant. The result was a substantial increase in the human population and an explosive increase in the cattle population. The cattle, of course, have eaten most of the vegetation. A recent drought, not unlike the many droughts the region has previously survived, killed most of the cattle and many of the people. As a result of this destructive cycle, the Sahara itself has expanded by thousands of square miles, probably permanently.

34.8 Humans, supposedly intelligent creatures, may sometimes behave stupidly. We often foul, for almost no reason, the very earth that sustains us. In California, a compromise has been reached. The open desert is being divided by law into areas for off-road vehicle enthusiasts and natural reserves. Californians seem to have missed the point. The desert is so fragile that it may take centuries to recover from such abuse. It has recently been found that a brief exposure to the sound of a single motorcycle can permanently deafen lizards and small rodents.

The Grassland Biome

Grasslands seem like common places to us, so you may be surprised to learn they don't exist in Europe. You won't be surprised to learn they're also missing from Antarctica. In the Northern Hemisphere, they exist as huge inland plains, and include such places as the North American prairie and the Asian steppes. There are similarities between grassland and desert and, in fact, in these two regions the grassland gradually fades off into desert. The chief climatic difference between the two is precipitation; grasslands, of course, get more rain, but it is of a seasonal nature, not enough to support forests. An added factor that prevents forest penetration is fire, a natural and recurring phenomenon for grasslands.

In the Southern Hemisphere, grasslands are known under different names: the pampas in South America, and the veldt and savanna in Africa. The savanna is peculiar in that it is dotted with groves of trees. Grassland in Australia is very extensive and, in fact, occupies over half the continent (Figure 34.9).

We are compelled to say that the dominant plants of grassland are grasses. Fortunately, we may be able to regain your interest by noting that grasslands are very different from one place to the next. For example, on the American prairie (at least the part that is still identifiable), grasses east of the Mississippi may grow 10 ft tall while those in the west rarely get over a foot or two. Again, the difference is principally due to variation in rainfall.

Since rains are commonly seasonal in most grasslands, the plants have developed important strategies for drought survival. In lowland regions, root systems may penetrate a permanent water table as far as 3 m below the surface. More commonly, the grasses rely on a vast, diffuse root system. In addition, grasses readily become dormant, reviving when water is once again available. Some grasses produce underground stems (rhizomes) that remain alive after all the foliage has died. The rhizomes, as well as the above-ground stolons (horizontal stems), form the sods which also actively prevent the growth of trees where rainfall would otherwise permit it. This matlike growth prevented agriculture intrusion until improvements in plowing implements were accomplished. Hence the name, "sodbuster."

Grasslands are the habitats of many of the world's large herbivores. The herbivores may be relatively sparse, as are the pronghorns of the American West, or numerous, as are the wildebeest of the Serengheti plains of Africa. Normally these grazers don't destroy

Savanna

Pampas

34.9 The grassland biome. Unique communities exist in each geographic region where grassland is found. Each, however, has a common climatic feature—limited, seasonal rainfall. The African savanna differs from the other types of grassland in that its vast extent is dotted with groves of trees. Natural grasslands support enormous numbers of herbivores, some of which are among the largest of all mammals. These regions are slowly changing with the encroachment of civilization.

the grasses because the plants continue growing from their bases. Overgrazing, or grazing by herbivores that kill grass by cropping it too short (such as sheep may do) can irreparably destroy grassland. The mesquite- and cactus-covered wastelands area in Texas (55 million acres) was a stable grassland before the cattle barons subjected it to heavy grazing (Figure 34.10). Much of the Sahara and the deserts of the Middle East, in fact, have been created by humans. This, remember, was once called "the land of milk and honey."

Much of the earth's grassland has been disturbed by farming activities, as it is exploited for grain. Except in national parks and other refuges, the natural grazing populations have been replaced by domesticated grazers such as cattle, sheep, and goats. Where wildlife is protected and/or nurtured in the United States, herds of bison and pronghorn antelope are the conspicuous grazers. But the fields are full of unobtrusive grass eaters, such as jackrabbits, rodents, and prairie dogs, as well as insects and seed-eating birds. Obviously, we're not going to allow predators to get away with eating herbivores we have raised for profit, so large predators are restricted to protected areas. These animals include wolves, cougars, coyotes, and even foxes. Snakes and carnivorous birds range a bit wider, but they don't eat many animals from our domesticated herds. (Although some ranchers somehow justify shooting eagles, including our endangered national emblem. One peculiar and grisly exception is a parrot in New Zealand, the kea (*Nestor notabilis*), that swoops out of the forests, lands on a sheep's back, tears into its flesh, and eats the fat from around the kidneys of the living animal.)

34.10 This desert community was created by humans. The region near Cotulla, Texas, was once a marginal grassland. Overgrazing by herds of cattle was so disruptive to the grasses that they have never recovered. Today, it is an economic wasteland.

Occasionally, the balance between herbivorous insects and the birds that control their numbers is upset. Two insects, in particular, have periodically made history when this happened. The Rocky Mountain locust, after a phenomenal population explosion, left its lofty habitat for the Nebraska prairie in the 1820s. One swarm was measured to be 100 miles wide and 300 miles long. Its effect on wheat and other plants of the grassland was devastating. The Mormon cricket (a grasshopper) occasionally leaves its foothills in equally large numbers, but on foot, devouring every plant in sight. Their attack on the crops of Mormon settlers in Utah is legendary, as is their ultimate destruction by heroic California seagulls. The reason for these sudden population explosions and migrations are not known.

One final note: Grasslands, with their highly efficient primary producers, support complex food webs

34.11 The highly efficient producers of the grassland support an extensive food web, often including incredible numbers of large herbivores. These mammals, in turn, support a sizable predator population including the African lion and cheetah.

with larger populations of animals than any other terrestrial biome on earth (Figure 34.11).

Tropical Rain Forest

We find forests scattered over much of the world, but only in those places where the water supply is great enough to support enormous plants such as trees. In few places on earth is water more available than in the tropical rain forest. Typically, this biome receives about 2m (78 in) of rainfall per year, rather evenly distributed through the seasons. In fact, these forests don't show much seasonal change at all. The largest tropical rain forest is in the Amazon River Basin in South America. The second largest is in the wilds of the Indonesian Archipelago. And then there are those of the Congo Basin in Africa, parts of India, Burma, Central America, and the Philippines (Figure 34.12).

Tropical rain forests are best described as lush, with a great number of tree species growing to great heights. The floor is dark and wet, and the air is often cool, and laden with rich smells. Unlike what we find in the other biomes, no single kind of plant dominates the forest. Any tree is likely to be a different species from its neighbors, and trees of the same species are often miles apart. All are tall, however, forming a dense, continuous canopy overhead. They may range in height from 30–45 m. The trees are invaded by large numbers of vines, and both trees and vines together often support *epiphytes* ("plants of the air"), which have no roots of their own but are able to absorb water from the humid atmosphere.

Below the upper reaches of the dominant canopy, shorter plants form a subcanopy, so the darkened floor is rather barren and one is able to walk unhindered over cool, black soil (Figure 34.13). One would not take such a walk unadvisedly, however, since the warm air, humidity, and darkness make the floor ideal for fungal and bacterial activity. The forest floor is home to the scavenger and decomposer. Anything falling dead on the floor is quickly attacked by termites, beetles, fungi, and bacteria.

You are undoubtedly wondering by now where the "impenetrable jungle" is. The answer is, jungles are special parts of tropical forests. Essentially, they consist of a low-growing tangle of plants, which is, in fact, virtually impenetrable. Jungles arise where light reaches the forest floor. This includes steep slopes, disturbed areas such as clearings, deserted farms, and along river banks. The latter can best be described as a *wet jungle* and abounds with insects and reptiles. It is not a very friendly place, but actually it comprises only a limited part of the rain forest. The idea that jungles represent tropical rain forest grew from descripions by river travelers who weren't aware of the relatively clear forest floor, just a few hundred yards from the river bank.

34.12 Tropical rain forests receive enormous amounts of rain throughout the year. There is little seasonal change. No single plant species dominates the terrain, but a number of kinds form a dense canopy over the sodden earth. The humidity on the forest floor can be stifling.

34.13 In the struggle for sunlight, plants of the tropical rain forest form a multistoried canopy. The uppermost consists of trees that often reach 30–45 m above the floor. Below these, the foliage of a subcanopy layer is found, and there may exist an even lower subcanopy. The layered arrangement of trees is further complicated by climbing vines (the lianas), adding their foliage. The stratified foliage provides several types of animal habitats. Brightly colored birds abound in the leafy canopies, as well as reptiles, amphibians, and mammals. In fact, most of the vertebrates in such places live in the trees as opposed to the forest floor.

The tropical rain forest harbors an incredible number of animal species, matched by no other biome. Insects and birds are particularly abundant, and reptiles, small mammals, and amphibians are common. Many of the animals are arboreal (tree dwellers), because the ground is an exposed and dangerous place. Food webs are intricate, with many species of plants and animals interacting in complex ways.

Tropical seasonal forests occupy a considerable portion of the tropical biomes. These differ from true tropical rain forests in that rains are, of course, seasonal. A familiar example is the monsoon forest of Southeast Asia, although such forests exist in other tropical regions. In many instances, the trees here are deciduous, losing their leaves during the dry season. When the monsoons arrive with their heavy torrential rains, the forest takes on some of the characteristics of the tropical rain forest. Some of the more highly valued hardwood, such as Burmese teak, are found in the tropical seasonal forests.

The tropical rain forests of the world are rapidly disappearing. Most of the land covered by rain forest in 1900 is now agricultural. The Amazon Basin rain forest of Brazil, the last really large area of undisturbed forest, is now in the process of being cleared and exploited. The results may prove to be catastrophic, for several reasons.

The extremely heavy rainfall that gave rise to the rain forest in the first place is almost entirely composed of reprecipitated moisture from the steaming forest itself. Little additional moisture blows in from the ocean. Cutting down the trees, then greatly reduces the amount of moisture returned to the atmosphere within the Amazon Basin, and this will ultimately reduce the rainfall there by a substantial amount. The rain that does fall onto the cleared land will tend to run off into the river system and to the ocean.

Also, you may be surprised to learn that tropical soils are very poor and infertile. The nutrients formed by decomposition are immediately absorbed by the plants, so no reservoir of humus remains. In addition, the rains continually leach precious nutrients from the porous soil. This fact has become painfully clear to proponents of jungle agriculture. Once the native plants have been removed, the soil may rapidly change into a hard, water-resistant crust known as *laterite.* The enduring temples of Angkor Wat are made partly from laterite. In fact, the term means "brick" and it rather amply describes the reddened crust. Primitive people were able to farm tropical soils by using a *slash-and-burn* technique. They simply cleared the land by cutting and burning vegetation, grew one crop, and then moved on to repeat the process somewhere else, allowing the cleared area to recuperate before the soil hardened. Researchers are looking more closely at these primitive techniques, which they deplored until very recently. After all, all modern farming methods have failed miserably in these areas.

Chaparral or Mediterranean Scrub Forest

Our biome map indicates that the Mediterranean scrub forest or *chaparral* (as it is known in California) is rather insignificant among the forests of the world. It does, however, have unique characteristics and its own peculiar plant associations. Note, for example, that chaparral is exclusively coastal, found mainly along the Pacific coast of North America and the coastal hills of Chile, the Mediterranean, southernmost Africa, and southern Australia. This forest is unique in that it consists of broad-leaved evergreens in a region marked by seasonal drought.

Chaparral is produced primarily in areas with restricted rainfall. Seasonal rains occur in the winter, but growth is slow because of the sparseness of rain and its absence during the hot summer season. Here, we will concentrate on the chaparral of our own Pacific coast. Depending on the latitude, California's coastal chaparral receives between 10 and 30 inches of rainfall, almost all of it in the short winter rainy season. The seasonality of the rains means that the plants experience drought through most of the year. Plants and animals adapted to chaparral have adopted strategies similar to those of desert dwellers. In the chaparral, plants are chiefly represented by shrubs and a scattering of succulents, mostly with dwarfed, gnarled, and scrubby stems (Figure 34.14). Leaves are generally small, with very waxy and tough cuticles. Many plants become dormant after the rains, remaining that way through the dry summers. Along streams, the growth becomes more luxurious. Near seasonal creekbeds and on north-facing slopes (which have a reduced rate of evaporation in West Coast areas), the California live oak (*Quercus dumas*) grows to considerable height and girth.

34.14 The chaparral in southern California may lack the glamour and luster of many other forests, but its plants are tenacious and hardy. These scrubby plants resist an annual drought that would discourage most other plants. Many parts of the chaparral actually qualify as deserts in terms of total precipitation, but they are not called deserts because of their cool, moist marine air.

Chaparral has another problem—one related to drought. It is often seared by fast-moving brush fires. Controlling the fires is complicated by the rough, hilly terrain of the California coast. Fortunately, the chaparral readily recovers from fire damage, through a succession of growth stages. Little of the chaparral escapes the periodic fires, but the plants appear to have adapted to this violence. Some sprout quickly from burned stumps, and others scatter fire-resistant seeds over the burned ground. Ecologists have described the chaparral as a *fire-disclimax community.* This means that the chaparral virtually never reaches maturity, and any given stand is always in some stage of recovery. It shows, too, that fires were common before humans came on the scene.

The floor of chaparral may be heavily layered with leaf litter. Decomposition is slow because of the dryness. The accumulation of all those leaves (plants may shed leaves year-round) magnifies the fire problem. But when the leaves aren't going up in smoke, they are the habitat of large populations of insects and other invertebrates.

Other inhabitants of the chaparral include vertebrates such as mule deer, rabbits, bobcats, many rodents, lizards, snakes, and a considerable number of birds. Wrentits and towhees are typical residents, and large numbers of other birds pass through on their annual migrations.

The most consistent threat to chaparral is the steady encroachment by humans. It seems that no vertebrate is about to be allowed to go unshot, so the chaparral is full of "sportsmen." And builders are fond of organizing plots "with a mountain view" for those who can't afford "an ocean view." The changes produced by such activities can disrupt the chaparral's role as a natural watershed. Clearing, burning, and other such destructive activities produce a ravished landscape subject to severe erosion and mudslides. It is apparent that we need to reevaluate the importance of this coastal forest in terms of its usefulness as a watershed, even if some of us are unable to appreciate it for itself.

The Temperate Deciduous Forest

If you are from east of the Mississippi, the temperate deciduous forest may be your biome. It extends over much of the eastern United States, northward into southeast Canada. It is also found in the central and northern parts of Europe, including Great Britain and southern Norway and Sweden, and a long finger of the biome pushes into the center of the Soviet Union. In eastern Asia, deciduous forest occurs in China, the eastern Soviet Union, Korea, and Japan; while in the southern hemisphere, although it is much less conspicuous, it does exist in coastal Brazil, east Africa, along the eastern coast of Australia, and across most of New Zealand (Figure 34.15).

34.15 The deciduous forest changes its appearance at different times of the year. The lovely green hillside of summer will explode in a riot of colors when autumn arrives. Both are in sharp contrast to the starkness of winter. Animals here must adapt to the long cold winters. Many will hibernate, while others simply migrate. The remainder are faced with shortages of food and shelter for the winter months.

Temperate deciduous forests are characterized by changing seasons and leaf fall. With the onset of winter's freezing temperatures, water becomes unavailable for growth, so most plants go dormant or die. Winter also means reduced sunlight, a factor which alters animal behavior as well as plant behavior.

Rainfall in the deciduous forest is rather evenly distributed throughout the year, often averaging over 100 cm (39 inches). This is a lot of rain, enough to support a variety of trees. Typically, the larger trees such as oak, maple, beech, birch, and hickory form the canopy. Unlike the tropical rain forest, however, the forest floor may be covered with younger trees, shrubs, ground-hugging plants, and a variety of annuals. The annuals begin to grow early in the spring, often when it seems that winter is still here. They rather quickly produce their seeds and die off with the autumn frosts. The rich leaf litter and humus harbor an abundance of scavengers and reducers, especially in the warm summer months. Besides supporting a large soil population, the forest floor is like a giant soft sponge, soaking up rain and contributing to the luxurious forest growth.

There is a variety of animal life in deciduous forests. However, we have killed off most of the larger mammals, with the exception of those we have protected in order that hunters may have the pleasure of shooting them. We have also reduced the numbers of large mammals by cutting down their habitat. In fact, there is almost no virgin temperate forest left in the United States. The forests were once the home of deer, wolves, bears, foxes, and mountain lions. Today we have to settle for rabbits, squirrels, raccoons, opossums, and rodents. The largest predators are humans, owls, hawks, and occasional bobcats and badgers. In the northern deciduous forests, a few wolves have escaped having their skins trim cheap coats. Finally, as any camper knows, arthropods inhabit the forest in great numbers, primarily rummaging through the litter and foliage, and crawling into tents (Figure 34.16).

Winter in the deciduous forest is heralded by one of the most beautiful and moving events in nature: autumn. As the tubes in the leaf petioles begin to break and the green chlorophyll wanes, other colors come to dominate the woodland scene. The leaves almost seem to celebrate the cool days, as browns, reds, and yellows mix with the greens of conifers. It is too soon over. And the fading light of the shorter days, which changed the leaves, also warns the animals of the approach of winter. Some birds leave, others change color and give up territories. Some mammals, grown heavy with sleep and fat, find holes in which to hibernate. Others, unable to escape either behaviorally or physiologically, simply face the coming cold; some will not survive. Soon, the stark, lifeless days of winter settle in.

The deciduous biome produces trees with strong and flexible cell walls, so of course it has been exploited almost unmercifully by humans. In some cases, we have destroyed the trees because we wanted their cell walls to help shield us from the environment. In other cases, we simply wanted the trees out of the way because we had something else to put in that space. Often, native hardwoods have been cleared so that faster-growing, more marketable pines could be

34.16 Animals of the deciduous forest. Herbivores of the deciduous forest are numerous, including the gray squirrel, cardinal (center), and white tail deer (top), in places where they still remain. Carnivores here once included large cats, wolves, and bears, but in most of this disturbed biome these animals are now rare. Typical carnivores today include the raccoon (bottom), eastern garter snake, and the insect-eating redheaded woodpecker, although an occasional fox is still seen.

planted in orderly rows. In any case, we have removed so much forest that today almost all American hardwood forests are, at best, second-growth areas. As such, they are in transition, with many regions dominated by scrub oak and uninspiring shrubs.

The Taiga

The taiga is almost exclusively confined to the Northern Hemisphere. Great coniferous forests of pine, spruce, hemlock, and fir extend across the North American and Asian continents, with a narrow band reaching into Norway and Sweden and at higher elevations in many mountain ranges (Figure 34.17). The taiga is unmistakable. There is nothing else like it. It is subject to long, cold, moist winters and short summer growing seasons. It is difficult to generalize about rainfall because taiga is so extensive; some places get a great deal of rain, some don't. But in many parts, there is much less rainfall than in deciduous forests. The taiga is interrupted in places by extensive bogs, or *muskegs,* the remnants of large ponds. These wet regions are commonly dominated by spruce, but low-lying shrubs, mosses, and grasses form the spongy ground cover (Figure 34.18).

34.17 The taiga is an extensive biome, found almost exclusively in the Northern Hemisphere. Its dominant plant is the conifer, although communities of birch, willow, poplar, and alder are not unusual. Since the conditions of the taiga are duplicated in high mountains, similar communities are found here. The harsh climate includes a short growing season, limited rainfall, and a severe, long winter.

34.18 Where bogs and marshes form, the taiga is interrupted. These areas are often referred to as *muskegs.* They represent a perpetual tug-of-war between the aquatic and terrestrial environment, as plants continually invade the marshes, some failing, others becoming established.

At the northern reaches of the taiga, dense forest becomes scattered and scrubby, giving way in places to growth similar to that found in tundra. In the south, particularly where rainfall is sparse (as in Siberia), fingers of grassland penetrate deep into the taiga.

Whereas we can't generalize much about rainfall, we can say a few things about seasons and sunlight. First, the seasons can be quite peculiar. The usual short, wet growing season becomes distorted as one travels northward. The northernmost taiga survives days of only 6–8 hr long through much of the winter, and long 19 hr days in the summer.

The dominant plants of the taiga are conifers, but occasional communities of hardy, broad-leaved poplar, alder, willow, and birch may be found in disturbed places. Conifers have adapted to their cold environment by reducing their leaves to needles which are covered by a waxy secretion. This is an adaptation against water loss; ground water is frozen and unavailable much of the time. (One is reminded of the spines and waxy cuticles of desert plants.)

You may be aware that it is easy to walk through a coniferous forest because there is very little underbrush. One reason is because those soft needles that feel so good under your boots are murder on other plants. They make the soil notoriously acid and relieve the conifer of having to compete with many other species. The straight trunks and dark, unobstructed floor may lend a cathedral atmosphere to taiga forests, an atmosphere that some people find compelling and others forbidding.

The taiga harbors such large herbivores as moose, elk, and deer. It is the last refuge of the grizzly bear and black bear, and wolves still roam here, as do lynx and wolverines. While rabbits, porcupines, hares, and rodents abound, insect populations aren't as large as in deciduous forests, although there is an abundance of mosquitos and flies in boggy regions during the summer (Figure 34.19).

The taiga is a continuing target of lumbermen, but its vastness has protected its more distant regions. As timbering interests grind away at our huge forests (much of it on publicly owned land), they are careful to advertise that they often replant the areas. The problem is, mixed forests of genetically diverse and resilient plants are replaced by artificially developed and homogeneous trees that are fast-growing and can quickly be reharvested. We are aware of the dangers of genetic uniformity in any system. And where will the bear live while forest grows?

Coniferous forests aren't restricted to the climatic conditions described for the taiga. A number of temperate coniferous forests are to be found throughout

34.19 Herbivores of the taiga include the seed-eating ruffed grouse and large mammals such as the moose and caribou. The lynx and grizzly bear are still found in the taiga, protected somewhat from human intervention by the vastness of the territory.

Moose

Grouse

Lynx

the Northern Hemisphere. Often, different and distinct coniferous forests are seen where specific climatic and soil conditions prevail. The Olympic forest in western Washington is on its western slopes, a bona fide temperate rain forest with 500 cm (200 inches) of rain per year near the glacial peaks. Here, the Sitka spruce reaches its greatest size, approaching that of the famed California redwoods. Fingers of the western forests follow the Cascade Mountains into Oregon and California. Perhaps the most fascinating coniferous forests are the California coastal redwoods. Their gigantic potential is supported not by rain alone, but also by coastal fogs which help maintain moist conditions. Finally, on the western slopes of the Sierra Nevada Mountains, we find the magnificent forests of giant sequoias (*Sequoiadendion giganteum*). These awesome and splendid trees are among the world's oldest living organisms.

34.20 The tundra. In summer, the treeless tundra becomes a veritable marsh as the snow melts. With little runoff, the landscape becomes dotted with innumerable small ponds. Below the water, the soils of the tundra are perpetually frozen. The plants dotting the landscape include a number of dwarfed trees, grasses, and abundant reindeer moss.

The Tundra

The tundra is the northernmost land biome. It has no equivalent in the Southern Hemisphere, except in a few alpine meadows. Tundra is actually located in a narrow band between northern taiga and the Arctic ice extending from the tip of the Alaskan peninsula around the earth and back to the Bering Sea (Figure 34.20). Anyone traveling in the tundra may be struck by the absence of tall trees and shrubs. And in a climatic sense, the tundra may be equated with the desert. The annual precipitation is often less than 6 inches, and much of this occurs as snow. So, it's a dry place. But we must qualify that statement; you wouldn't believe it if you visited the tundra during its brief, damp summer. Anyway, it's dry because most of the year everything is frozen. When spring and summer finally arrive, only the upper few feet of soil thaw, leaving the frozen soil below in a state of *permafrost.* The surface thaw produces an unusual condition for a "desert." Ponds begin to form everywhere. Since there is no way for the water to percolate down, the plains of the tundra become a veritable bog.

The tundra receives less energy from the sun than any other biome. Because it is near the pole, any sunlight there strikes at an acute angle after diffusing diagonally through the atmosphere. The lack of sunlight drastically affects both the climate and the growing season. During the brief 6–8 week summer, plants must photosynthesize enough food to last through the rest of the year. And they must compress their reproductive period into this brief season, all of which means they can't put much of energy into growth. So the tundra is carpeted by low-lying plants. Another factor which dictates that plants hug the ground is the constant high winds of some tundra regions. A tall plant would simply be buffeted to death or even ripped away.

Because of the permafrost, the plants of the tundra can't form deep anchoring root systems; so they form shallow, diffuse roots that often become entangled with the roots of their neighbors to form a continuous mat. In the wet season, any disturbance on hilly slopes can break the entire root mat loose from the wet soil underneath and send a great mass of plant conglomerate sliding down, exposing the barren soil below. Recovery in the tundra is a very long, very slow process.

Lichens and mosses are common in the tundra (which may not surprise you), but so are willows and birches. These latter, however, are almost unrecognizable, being only dwarfed symbols of their more southerly relatives. They, with the lichens and mosses, are joined by grasses, rushes, sedges, and other annuals to complete the summer ground cover (Figure 34.21).

Animal life isn't rare in this peculiar northern biome. We should expect this, since consumers usually exist anyplace there are producers. In fact, year-round residents of the tundra include some rather large herbivores such as, in North America, the caribou and musk oxen, and in Europe and Asia, reindeer. Other browsing animals to be found are the ptarmigan, the snowshoe hare, and the ever-present and legendary lemmings. Actually, lemming populations are clear indicators of how good the season is for producers. In good years, their populations soar.

Lemmings, in turn, determine the success of a number of predators, including the arctic fox, lynx, snowy owl, weasel, and arctic wolf. Also, some migratory birds travel great distances to feed on the tiny rodents. The duration of their stay, as well as their reproductive success, will depend on the number of

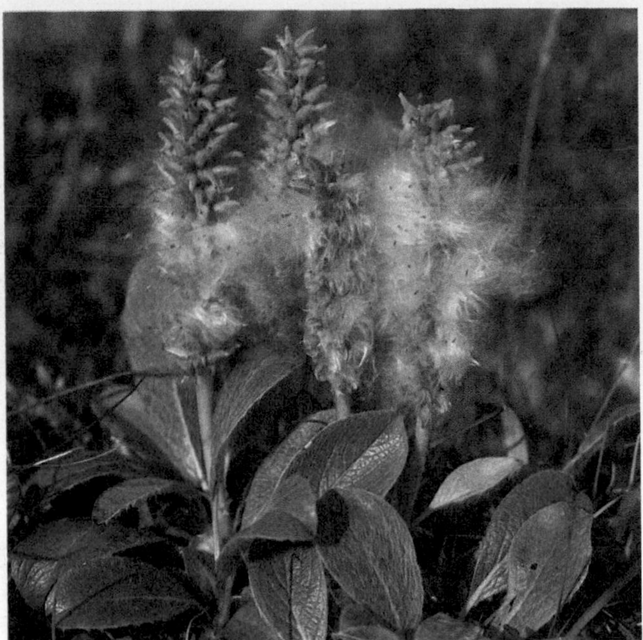

34.21 Plants of the tundra. The plants in the foreground are actually trees. The harsh tundra climate produces a stunted growth in woody plants. Willows that are 25 years old may only be 1 m high.

34.22 Because of the paucity of species in the tundra, the food webs may be comparatively simple (and, therefore, easy to disrupt). Common herbivores in the tundra include large animals such as musk ox, and small ones such as birds and lemmings. Unfortunately for the lemming, it is the major prey for several tundra carnivores, including the migratory snowy owl and the arctic fox.

Musk ox

Lemming

Snowy owl

lemmings. In years when lemmings are few, many of the predatory birds will migrate early while the permanent residents will survive by switching to other prey (Figure 34.22).

A number of waterfowl and shore birds also migrate to the tundra. When they arrive in the spring they must mate and rear their broods as quickly as possible, before the brief summer is over. Why do they travel, in some cases long distances, to breed in such a dismal and risky place? Perhaps because the long days permit extended feeding periods. And too, the winter was harsh, so resident competitors are few.

In the frantic race to complete their reproductive activities before summer ends, one also finds, remarkably, a variety of mosquitos. And tiny flies abound, swarming and darting in clouds over the sodden plain.

Winter comes early in the tundra and with the rapidly shortening days the migratory animals disappear. The caribou, for example, leave for the forested taiga where winter food is more plentiful. Those that remain prepare for survival by whatever strategy they have developed. Lemmings retreat to food-laden burrows, while ptarmigans tunnel into snow banks to emerge periodically on foraging expeditions. Since the larger herbivores and predators don't hibernate, they must rove the barren, windswept landscape to feed on mosses, lichens, and each other. Next summer's generation of insects avoids the issue by wintering over in immature stages.

Our own species, with the exception of a few Laplanders in northern Scandinavia and Finland, abhors the tundra. Even the hardy Eskimos prefer the Arctic coastline. For this reason, humans have had little effect on this most fragile ecosystem. This is fortunate because the links in the food chains here are few and important. The loss of a single predator or producer could be disastrous. This is one of the reasons that many people are greatly concerned about the possible ecological repercussions of the Alaska pipeline. This project, which was stalled only briefly by environmentalists, has now been completed; so we shall see. The pipeline carries hot oil across Alaska from the slopes to the port of Valdez. We have no idea what the effects of the construction and the oil spills will be on this fragile ecosystem. We can only hope. There have, in fact, already been oil spills, but so far they have been small ones. We know that recovery from damage is agonizingly slow and costly to the tundra communities. In places, there are wheel ruts left by a single wagon that passed over the soil 100 years ago.

North of the Tundra (or at higher altitudes), its vegetation gives way to barren and rocky soil similar to that of the Antarctic. Plant life is sparse and patchy. Eventually, these dry windswept plains end in the coastal ice floes and glaciers. Here the marine environment, in sharp contrast, is amazingly rich in life, even in its colder waters. Of course, the perpetual search for more petroleum threatens these rich waters as well.

The Water Environment

About three-quarters of the earth's surface is covered with water, and most of it is salty. But not only is the marine environment big, it's complex; it contains a vast array of communities. They fall into two major divisions: the shallow areas along continents, or the *neritic province,* and the deep-water *oceanic province.* (Figure 34.23). In the neritic province, the waters along the *continental shelf* are extremely variable. In these relatively shallow places, light can penetrate to the ocean bottom. The neritic zone is rich in many forms of life, particularly in the shallower *littoral region* (less than 15m [49 ft] deep), where wave and tidal action continually stir the waters and life-giving sunlight can penetrate. Part of this littoral region is the narrow but species-rich *intertidal zone.*

Perhaps the greatest mysteries lie in the dark waters of the oceanic province in the area called the *abyssal region.* Depths here can vary from 300 to nearly 11,000 m. The famed Marianas Trench is 10,680 m deep (over 6 miles). Little could be discovered about

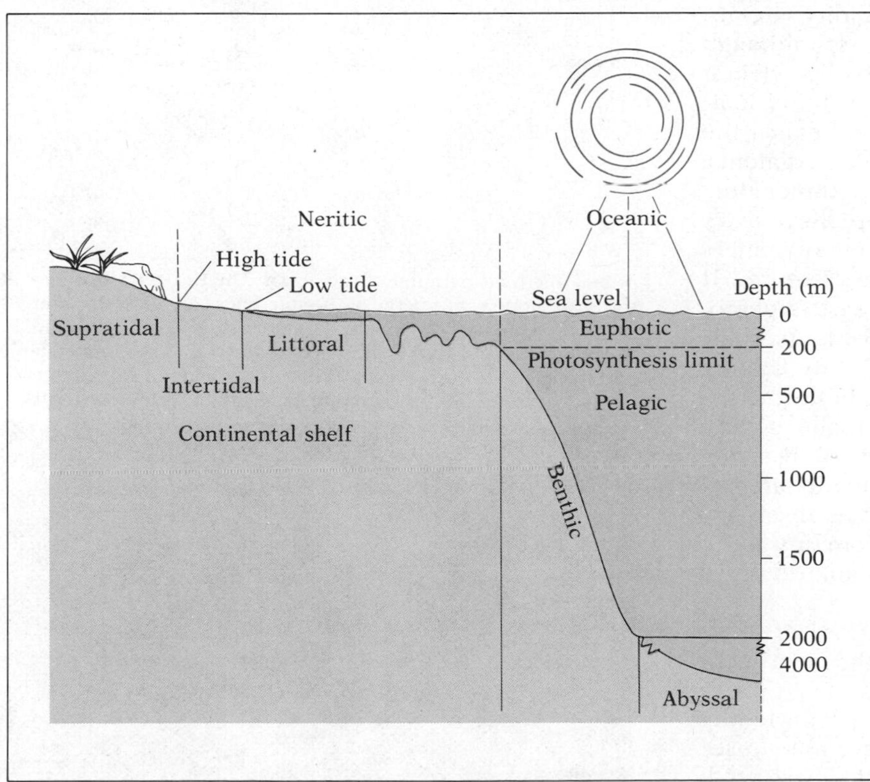

34.23 The oceans have been divided by oceanographers into ecologically distinct regions. The *neritic province* includes the *continental shelf,* which is composed of *littoral* and *intertidal zones.* These contain the highest density of living things. The *oceanic province* is comprised of the *pelagic zone* and the deep *abyssal region.* The extent that light can penetrate into the depths marks the *euphotic zone.*

conditions in the ocean abyss until the technology for its exploration was developed, and the first steps have now been taken. In 1960, the floor of the Marianas Trench was reached by Don Walsh and Jacques Piccard in the bathyscaph, *Trieste*. It was a momentous event, but our visits to the depths have been only brief and breathtaking glimpses, so we still know very little about what is going on down there. Aside from a few basic measurements, some remarkable photographs, and vivid impressions, we know little beyond the fact that it can be reached.

We also know that the ocean depths are a place of darkness, tremendous pressure, and chilling cold. Nevertheless, this formidable deep *abyssal region* supports a surprisingly large number of peculiar scavengers. Lacking a producer population, the benthic creatures rely on a continual rain of the remains of *pelagic* (surface-swimming) organisms settling to the ocean floor. Tethered cameras focused on bait, miles deep, have revealed the primitive hagfish, many species of bony fish, crustaceans, mollusks, echinoderms, and even an occasional shark.

Further above the ocean floor is the home of the pelagic organisms, those that swim or float in the open sea. Most of these are restricted to regions where light can penetrate. There are a number of fishes and other organisms in the vast region between the surface and the ocean floor, but they are difficult to catch and more difficult to photograph. Almost all wavelengths of light, depending on the turbidity, are absorbed by the upper 580–600 m of water, although some shorter wavelengths (blue) may be detected at 1000 m. The light is captured by a variety of tiny floating photosynthesizing plants, collectively called the *phytoplankton*. These are composed of algae, diatoms, and dinoflagellates (all microscopic). Phytoplankton, along with the larger algae (seaweeds), are the *primary producers* of the sea. This means that energy enters the marine ecosystem via these minute creatures. It has been estimated that 80–90% of the earth's photosynthetic activity is carried on by marine plants. Thus, the chain of life in the sea begins with tiny floating plants capturing the energy of the sun to build energy-containing compounds within their fragile bodies. But where there is energy, there is likely to be something trying to utilize it, and the tiny plants are fed upon by animals only a little larger than themselves—the *zooplankton*. A variety of animals, from tiny fish to the great baleen whales, feed upon plankton of all sorts (Figure 34.24).

The waters below the photosynthesizers, but above the ocean floor, harbor some of the most peculiar creatures on earth: the deep-sea fish (Figure 34.25). They are usually small and dark with big toothy mouths. Their sight is keen and they are able to see the faint light of luminescent organisms. These predators often produce their own light to signal each other or to attract prey. Some species cultivate lumi-

34.24 The food pyramid of the ocean. The tiny *phytoplankton* at the base capture the energy of the sun. These are eaten by animals larger than themselves, which are, in turn, eaten by larger animals. At the top are the largest carnivores of the sea. (There are no large herbivores in the open oceans as there are on land.) Note that each level is comprised of far fewer organisms than the one below it.

34.25 The deep-sea fishes are among the most bizarre of all creatures. They live where there is no sunlight and very few other animals. Thus, they must be equipped to capitalize on just about any living thing they might come across. *Chauliodus* uses the lengthy first ray of its dorsal fin as a lure, dangling in front of its toothy mouth.

nescent patches of bacteria beneath their eyes which become visible when the fish rolls down a specialized eye covering.

Coastal Communities

The ocean communities that are richest in life exist along the neritic province. The great fishing areas of the world are found in the relatively shallow areas above the continental shelves. In comparison, the vast, blue-water oceanic province is relatively devoid of life. The fish-laden neritic province is constantly exploited for its finny riches. We deliberately use the term *exploitation* because the fishing industry is systematically and efficiently depleting the resources here without adequate regard for the future. Our over-exploitation of the world fisheries is an established fact, recognized by our government as well as by international commissions. In fact, the United States has recently extended its traditional 12 mile limit to 200 miles in an effort to reserve more fish for itself. Other countries have also increased their limits.

Many of the world fisheries are being exploited at a rate that makes the likelihood of their recovery questionable. When increased fishing effort is not accompanied by increased catches, then it is fairly certain that the resource is in danger of depletion. Today's fishing technology is so sophisticated that even in areas where catches have diminished, the last remnants of once-abundant species are successfully being sought out and captured.

Coastal communities nearer the shore (*littoral zone*) are diverse, as anyone can discover by a simple drive along the coast. These areas include sandy beaches, rocky coasts, bays, estuaries, and tidal mud flats. In spite of the diversity, such places have one thing in common: They abound with a variety of life. The richness of life in the littoral zone is made possible by the availability of light, shelter (seaweeds, rocks, kelp beds), and nutrient runoff from rivers entering the sea. Many coastal waters are further enriched by *coastal upwellings* of mineral-rich water from the ocean floor.

Each region supports its own particular type of plant and animal community. The rocky tidal communities, for example, consist of a large variety of organisms, all adapted in some way for holding on or burrowing. Their big problem is to avoid being swept away or beached by the tremendous force of the waves. If you have dived in such places, you know about the nearly irresistible ebb and flow of the surging water.

Visitors to rocky tide pools are usually amazed at the variety of life found there (Figure 34.26). That life may include plants such as seaweeds, eel grass (which is not really grass), and microscopic phytoplankton. The animal life consists of tiny and tenacious mollusks, echinoderms, crustaceans, tunicates, and small fishes (often visible only when they dart from one tiny crevice to another). The more sedentary of these species are adapted to regular exposure to the wind and sun as the tides fall, baring them to the murderous atmosphere.

Estuaries, places where rivers enter the sea, are fascinating for humans and demanding for marine life because of the special problems of osmoregulation. The problems arise because of the constantly changing salinity, which varies with the changes in the tides and river flow. Estuaries commonly have silty or muddy bottoms, and their backwaters may form mud flats at low tide (Figure 34.27). The animals inhabiting the mud flats usually survive such drastic changes by burrowing to escape the drying sun. The mud flats are inhabited by mollusks such as clams

34.26 Rocky tide pools. The rocky coast is home for numerous marine animals. Each organism of the rocky tide pool is adapted in some way to withstand both the surging and pounding of waves and intermittent periods of exposure to the air at low tide. Some live in burrows; others cling tenaciously by bristles, sucker, or foot; while others come in and out with the tide.

and scallops, by sea cucumbers, and by arthropods, such as crabs. These quiet beasts are hunted relentlessly by numerous shore birds.

Estuaries are also regions of serious pollution, and many have been drastically altered by humans.

34.27 A mud flat, a saltwater area that is covered at high tide and exposed at low tide. The soil is often hard and tightly packed. This is a very difficult place to adapt to because it is subjected to repeated drying and soaking. Predators stalk these areas in search of exposed prey. Such flats may occur at estuaries, the river currents producing the prominent rills. The mud flat estuary harbors a variety of organisms if the river has not been poisoned upstream.

As long as rivers are used as dumping places for manufacturing wastes and sewage, the problem will persist. This is unfortunate, because bays and estuaries are rich sources of nutrients and are often the very places where marine animals reproduce. Thus, the young of many marine animals—even those that will eventually move to deeper waters—are frequently endangered by polluted conditions. In a very real sense, the ocean's chain of life begins in these shallow waters. You are also undoubtedly aware that, in addition to polluting the waters, "developers" constantly encroach on the estuaries by dredging and filling these natural breeding grounds to provide places to build expensive human dwellings.

Another shallow area that is among the richest of marine habitats is the coral reef, but you haven't seen one unless you have been to tropical regions where the ocean temperatures remain above 20°C. Corals, of course, are coelenterates that live in huge colonies, building heavy exoskeletons of calcium carbonate. As years pass, the mass of exoskeletons grows, with the newer members of the colonies growing on top of the old. Coral atolls, common in the South Pacific, were shown by Darwin to represent coral growth at the top of submerged volcanoes. Barrier reefs, on the other hand, are coral deposits that form along coastlines, usually a short distance from the beach. The largest and most spectacular of these is the Great Barrier Reef, which extends from 1200 miles along the east coast of Queensland, Australia.

Reefs, by their irregular growth, form natural refuges for marine animals and thus set the stage for complex food chains. The complexity is due to the diversity of food types and a multitude of hiding places which forces predators to be able to exploit a number of different kinds of food, specializing in none. Coral reefs usually shelter sponges, encrusting algae, and bryozoans, as well as mollusks such as the octopus. Fishes abound in reefs and sharks patrol the deeper waters alongside.

Productivity in the Oceans

Productivity (the rate at which organic matter is formed—measured in grams of carbon per unit of water per year) in the marine environment varies greatly from place to place. We have already mentioned where light reaches the shallow waters of the neritic province, life abounds, but that it may be rarer in deeper waters. A second factor that determines the numbers of organisms in the marine environment is the availability of nutrients. Interestingly, the regions of highest productivity are in the Antarctic and Arctic seas and in regions where cold ocean currents are common. This phenomenon has to do with the turbulent activity of cold layers of water. As surface waters cool, they become heavier, and their great masses

Table 34.1 Productivity of Organic Material

Area of ocean	Productivity ($gC/m^2/yr$)[a]	Efficiency (%)
Open (90%)	50	10
Coastal (9.9%)	100	15
Upwellings (0.1%)	300	20

[a]Measure is grams of carbon (organic material) per square meter per year.

sink into the depths. This action creates an upwelling, which brings nutrients from the bottom sediment to the surface waters. These nutrients support sudden increases in the plankton populations, which subsequently feed increasing fish populations. The important anchovy fishery off the coast of Peru is a good example, although all signs indicate that this vast fishery is being rapidly depleted by overfishing.

Upwellings also aid in the distribution of dissolved gases such as O_2 and CO_2. Productivity is summarized in Table 34.1, which shows that in the open sea (90% of the ocean area), the yearly productivity is only 50 grams of carbon (organic material) per square meter. Coastal regions produce twice that amount, while areas of upwellings produce six times that of the open sea.

The Fresh Waters

The freshwater environments include lakes, ponds, marshes, rivers, streams, and even puddles. We won't be able to describe each of these habitats in detail, but they are important and have marked similarities, so we will make some generalizations. Most freshwater bodies differ from the marine waters by the absence of appreciable amounts of minerals or salts. In fact, the solutes in freshwater habitats are generally lower in concentration than they are in the tissues of their inhabitants. The concentration of solutes presents interesting problems of adaptation, particularly for protists and animals. We described some survival strategies of these organisms in Chapter 28.

Freshwater Lakes and Ponds

The dividing line between lakes and ponds is vague. However, we might say that ponds are usually smaller, less permanent, and shallower. But once a body of water is classified as a lake, its properties can't be easily generalized. For example, the deeper lakes within the same climatic conditions are highly variable, but they do have some traits in common. Whereas in most ponds light is able to penetrate everywhere, enabling both producers and consumers to live at all levels, in the deeper lakes life may be confined to the shallower regions and near the surface. Besides light, there is another factor that determines the depths at which life can exist. Lake Tanganyika in Africa is 1450 m (almost a mile) deep, but only the top 60 m harbor life. This is not true, however, for the world's deepest lake, Baikal, in Sibera (1750 m deep). Here, animal life can be found even on the bottom. It turns out that the critical differences lie in the surrounding climates.

The climate over Lake Baikal is such that the surface tends to become cold, precipitating a turnover of its waters as surface layers cool and sink, driving deeper layers upward. It begins each autumn as the lake surface cools down with the approach of winter. Since water is most dense at 4°C, the surface water sinks down through warmer layers creating a rotation of the lake's waters. The colder water contains high quantities of dissolved oxygen. This means, of course, that animals which need oxygen can live at deeper levels. Life exists below the oxygenated level in all lakes, but it is restricted to anaerobic forms, chiefly bacteria. The anaerobes are supported by an abundance of organic nutrients which rain down from the aerobic life above.

The chief producers in ponds and lakes are algae, along with some flowering plants lining the shores. Consumers include protists, invertebrates, vertebrates, and their developing young (which may take different kinds of foods than the adults).

34.2 Ecosystems

Interaction in Ecosystems

So now we have organized life into climatic zones, or biomes. Our discussion has been descriptive, since our goal was simply to review the distribution of life on our small planet. Let's now look more closely at how organisms may interact with each other and with their physical environment. If we are able to isolate and identify a set of any such interactions, we can draw a line around what we have and call it an *ecosystem.* Of course, we have used the term before; it's a handy word, like *organism,* which biologists use when they don't want anything pinned on them.

Technically, an ecosystem is a conceptual unit of living organisms and all the environmental factors that effect them. As such, it is an operational term that can refer to the entire earth or a puddle in your backyard, depending on how broadly one wishes to define *interaction.* We are being a bit vague here, and you can probably see why. There are no isolated units. Everything interacts with everything else. But we choose to ignore this small embarrassment for the moment. Thus, we can select a system for a closer look. If we have selected our ecosystem properly, we will find it to be composed of four basic elements: the physical environment (known as an *abiotic* element); the living things (which are, of course, the *biotic* elements); energy (input and use); and, finally, the nutrients that cycle between the biotic and abiotic elements. Energy and nutrients represent what we will call *life-support systems* for reasons which will become obvious. So let's briefly consider these elements separately.

Abiotic Factors

When we introduced the marine biome, we mentioned several abiotic aspects of the oceans. For example, much of the ocean life depends on currents. Currents are abiotic, and so are temperature, moisture, light, and the local chemical composition of the earth (availability of essential minerals or presence of toxic or inhibiting substances). Some abiotic elements are influenced by the biotic elements, as you have probably guessed. One example that we have already considered is the rainfall in the Amazon Basin, which is related to the plant cover there. Life can only exist within a certain range of any of these factors. Beyond those limits, the abiotic element can become life-limiting; we then call these *limiting factors.*

Biotic Factors

Biotic factors of ecosystems are the living things, particularly as they are organized into communities. Communities are composed of interacting populations of different species. So one can break down ecosystems into communities and communities into their species. But communities (and hence ecosystems) can be described another way as well. One can rather impersonally measure their *biomass,* which is the combined organic content of the community. This calculation is especially useful when we are trying to determine the productivity of an ecosystem (as in Table 34.1), or when comparing one ecosystem with another. For example, if we can measure the energy and raw materials entering a community over a period of time and then measure the biomass of that community, we can learn something about its metabolic efficiency. We can also determine whether a community is growing or declining. This may sound rather academic and perhaps of no great interest—unless you eat. Farmers are interested in precisely these measurements, and, in fact, nations evaluate the state of their economy in a similar manner. As a matter of fact, the more you know about economics, the more you tend to appreciate ecology.

The biotic elements of an ecosystem can also be classified according to their activities. All the organisms fall into one of three categories: *producer, consumer,* or *reducer.* Obviously, producers are autotrophs. Most are photosynthetic, although there are bacteria called *chemotrophic autotrophs* that derive their energy, not from sunlight, but from inorganic substances such as sulfur and iron compounds. Entire complex ecosystems have been found in isolated patches of the ocean floor, where volcanic vents of hydrogen sulfide provide the entire energy input. These food chains extend from chemotrophic bacteria up through filter-feeders to arthropod predators. We, however, will concern ourselves with *phototrophs*—organisms that derive energy from light. In the sea, the most important producers, you will recall, are the phytoplankton, which account for 80–90% of the photosynthesis that takes place on the earth. On land, producers are represented by plants, algae, and cyanophytes.

Consumers, we might say, consume producers. No consumer can produce glucose from CO_2 and H_2O. Instead, consumers rely on the chemical bond energy and on materials manufactured by producers. Consumer populations therefore include all animals, the nonalgal protists, fungi, and bacteria. It may have occurred to you that we have now labeled everything on earth as a producer or consumer. So what is a reducer? A reducer is a special kind of consumer—one that eats the dead. Even their own. They break down amino acids, carbohydrates, and fats from plant and animal remains into inorganic nutrients which they then use. These nutrients can also be recycled into the producer population. This makes their activities ecologically vital and deserving of their own category. So we draw a rather fine but fairly distinct line between reducers and other consumers. Reducers generally feed by absorption. They secrete their digestive enzymes into their food and absorb nutrients into their cells. Bacteria and fungi are reducers. Since their food is dead organisms and sometimes waste material, we call these organisms *saprobes* (formerly, *saprophytes* from the days when bacteria and fungi were considered to be part of the plant kingdom).

But let's not stop here. We can devise even more categories. For example, we can further subdivide the consumers according to their *trophic levels,* or positions in food chains. Herbivores, for instance, are called *primary consumers,* while carnivores, which feed on herbivores, are *secondary consumers.* Less numerous, by far, are *tertiary* and *quaternary* consumers. Can you name a quaternary consumer? Perhaps you've eaten tuna, a large fish that feeds on smaller fish that in turn feeds on invertebrates. That would make the tuna a tertiary consumer, but what would it make you?

To represent these trophic levels, ecologists have devised simple structures called *Eltonian pyramids* (named after Charles Elton, a pioneering English ecologist who thought of the scheme). Actually, Eltonian pyramids can be constructed in several different ways, depending on what we want to compare. For example, in some instances, ecologists use sheer numbers of organisms, while in others they seek to show the distribution of stored energy and total biomass (Figure 34.28). Energy pyramids, while a bit simplistic, can be most instructive. They can point out, rather clearly, for example, the collapsing nature of food chains in an ecosystem. Pyramids show that the transfer of energy from one trophic level to another is far from efficient. (Cattlemen have known this for years. If they could get all the energy from their grassland into their cattle, they would be far richer.)

When we assign a community of organisms to specific trophic levels, we end up with a food web. We talk about food webs instead of food chains because the relationships are not simple and linear (Figure 34.29). Herbivores, for example, tend to eat several kinds of plants according to the plants' availability. Also, predators are often capable of switching prey, and some animals rummage through several trophic levels—humans are good examples. Now let's turn to the subject of energy itself and see how it wends its way through ecosystems.

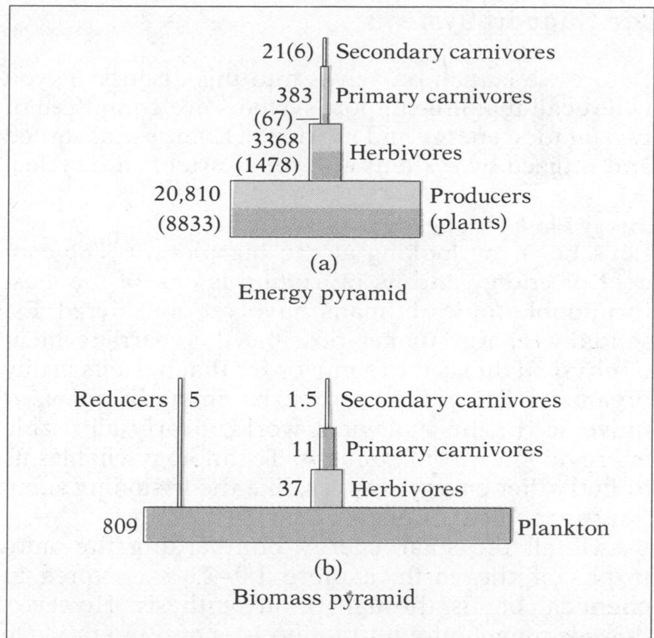

34.28 (a) An energy pyramid calculated for Silver Springs, Florida. The upper numbers for each tier designate the stored energy (expressed as kilocalories) for each square meter investigated over a year. Not all the stored energy at any level is available to the level above it, so the kilocalories that are available as food are shown in parentheses and as the shaded areas of each tier. Also, each tier is much smaller than the one below it because it is not completely efficient at recovering all the energy that *is* available. (b) The biomass pyramid is based on the total biomass at each trophic level. Its graph produces a different shape. The numbers represent grams of living material per square meter.

34.29 Even the simplest food webs are intricate. Herbivores are often preyed upon by several predators, and predators, in turn, may switch from one prey species to another, as one falls into short supply. Those carnivores that are not preyed upon make up the apex of the food web.

Life-Support Systems

Before we launch ourselves into this discussion, you will recall that life-support systems are comprised of two factors: energy and nutrients. Energy is absorbed and utilized by systems and the nutrients are cycled.

Energy Flow Through Ecosystems

Let's begin by looking at the big picture. The concept of energy and its utilization is one of the most formidable topics humans have yet considered. Essentially, energy makes possible the rearrangement of physical things into some order that permits highly organized living things to exist in a disorganized universe. In the biological world, nearly all usable energy comes from sunlight. Technology enables us to find other energy sources, like the fissioning atom. For the moment, though, consider the sun.

Of all the solar energy bombarding the autotrophs of the earth, a mere 1.0–2.0% is stored in chemical bonds through photosynthesis. However, this seemingly minute fraction is enough to produce 150–200 billion metric tons of dry organic material each year. About half of this stored energy is used at once by the plants themselves as they go about the business of being plants. The rest is stored in the plants' tissues for the waiting herbivores. But herbivores can only store about 10–20% of the plant material they eat. After all, they must metabolize much of what they eat to provide energy for running, eating, reproducing, and so on. Their activity helps to explain the 80–90% loss as energy moves between the two trophic levels. Also, animals are not 100% efficient in extracting the energy stored in plants, and much of what they do extract is dissipated. For instance, about 7% of the chemical energy ingested by dairy cattle is passed away as gas.

But back to contemplating matters on a grander scale. We can think of energy as entering an ecosystem, being stored by plants, and then being transferred through trophic levels, eventually leaving the ecosystem in its most degenerate form: Heat. Heat is a degraded form of energy as far as organisms are concerned. There is no way in which it can be used again. Thus, in referring to energy in ecosystems, we should describe it as a flow, rather than a cycle (Figure 34.30).

Nutrient Cycles

On the other hand, though, nutrients are generally cycled through ecosystems. Photosynthesis is simply a way of reorganizing chemicals that were already present, but in a low-energy form. When a consumer dies, its molecules do not disappear; they simply become disorganized until they are captured by another living thing that has the energy to utilize them in its own scheme of organization. You would probably be greatly embarrassed to learn where the molecules that comprise your body have been.

An absence of one or more of the essential nutrients (carbon, hydrogen, nitrogen, oxygen, phosphorus, sulfur, calcium, molybdenum, iron, magnesium, copper, zinc, boron, and cobalt) in an ecosystem can greatly curtail plant and animal life. So nutrients rank alongside climatic conditions and topography in determining the types of plant and animal communities that a region will support.

Measuring Ecosystem Productivity

Ecologists are forever trying to measure productivity in ecosystems. But it is virtually impossible to measure input and output directly in a real system. Undaunted, the ecologists have approached the problem indirectly.

One method, known as *harvesting*, involves measuring the caloric value per gram of living material. How much of any material is present? And how much

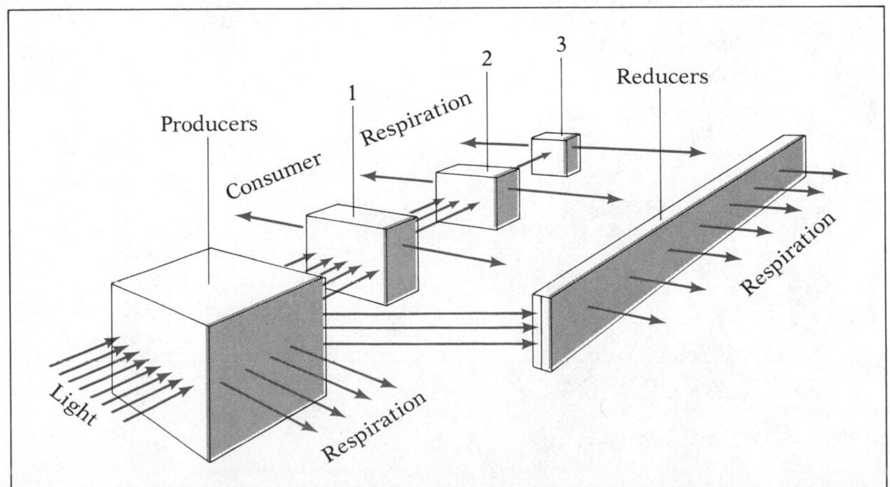

34.30 Energy entering an ecosystem is first captured as light by the green plants. The first energy loss occurs as these producers carry out respiration. Each succeeding level incorporates some of the energy of preceding levels, but loses as much heat as it, in turn, metabolizes that energy. In a balanced ecosystem, the light energy entering is equal to the energy that leaves as heat.

Essay 34.1 An Unusual Community

The Galapagos rift community, a bizarre assemblage of animals and bacteria, centers around the vents of sulfide hot springs in an area of active sea floor spreading some 380 miles from Darwin's islands. It was discovered in 1977 by geologists aboard the submarine *Alvin,* at 2500 m. The geothermal hot springs, spewing boiling-hot solutions of hydrogen sulfide and carbon dioxide, support an entire ecosystem based on autotrophic chemosynthesis. It is one of the most dense and productive communities on earth—near the vents, the mass of living tissue approaches 50 to 100 kg per m^2.

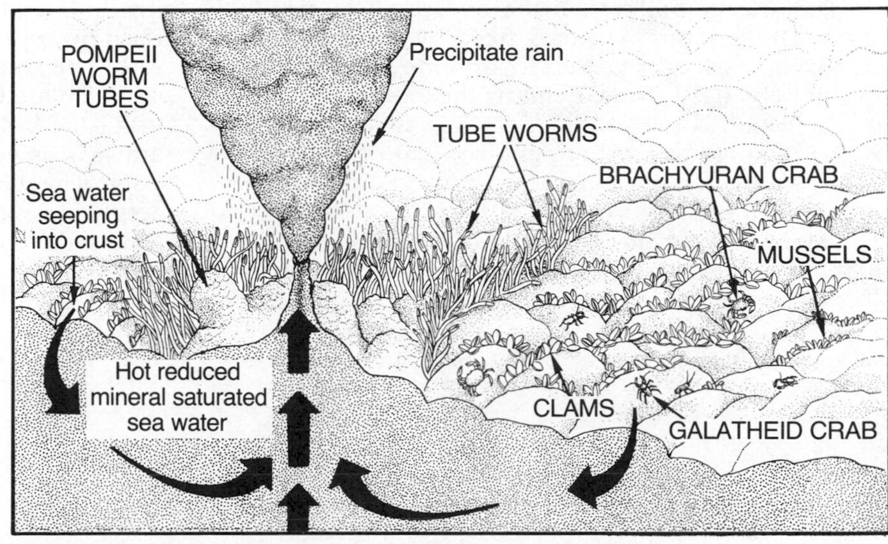

Prominent among these denizens are enormous, blood-red tube worms of a previously unknown group, apparently belonging to the pogonophores (an obscure phylum now thought to be related to annelids). Like all pogonophores, these *vestimentarians* lack all trace of a mouth, anus, or gut. They appear, in fact, to be chemosynthetic autotrophs, deriving all their energy and carbon needs directly from the oxidation of the hydrogen sulfide and the reduction of the carbon dioxide, probably with the help of intracellular bacterial symbionts. The redness of their flesh is due to heavy concentrations of oxygen-binding hemoglobin; oxygen from the surrounding cold seawater is needed both for the oxidation of H_2S and for the worm's own oxidative metabolism. The tube worms range up to nearly 3 m in length and are about the girth of a man's wrist.

The entire ecosystem, in fact, is based on hydrogen sulfide chemosynthesis and is entirely cut off from other, photosynthesis-based ecosystems. Apart from the tube worms, the primary producers are chemosynthetic bacteria, which swarm in enormous numbers where the carbon dioxide- and hydrogen sulfide-laden vent water (at 32°C to 330°C) mixes with the oxygenated abyssal seawater (at 2°C). At least 200 bacterial species proliferate rapidly near the vents and are visible to observers in *Alvin* as milky clouds. Filter-feeding crabs, clams, mussels, smaller worms (dubbed "spaghetti" by their geologist discoverers), and barnacles live off the bacteria. A little higher on the food chain, larger crabs and a variety of fish scavenge on clumps of bacteria as well as on animal remains. Whelks, leeches, limpets, and miscellaneous worms complete the community. Some species were previously known, and others, such as the tube worms, are new discoveries with uncertain affinities. Similar communities have been found now at at least one location on the East Pacific Rise Rift, and other deep-sea geothermal communities probably occur wherever there are appropriate sulfide springs.

The Galapagos rift community contrasts markedly with the rest of the deep-sea benthic biome, which is characterized by constant cold and a severely limited energy input. The usual benthic communities survive on what little organic material drifts down from the surface waters and are thus based ultimately on energy from distant photosynthesis. The animals of the cold water deep-sea bottom (there are no plants) are generally slow-moving and slow-growing, playing a variety of refrigerated waiting games: scavengers patrol listlessly for the chance arrival of a dead fish or a fecal pellet, and predators lie motionless in ambush for a wandering scavenger. In the hot springs communities, however, food is virtually unlimited, the temperature is not so uniformly cold, and both growth and metabolism are relatively rapid. The Galapagos rift clams, for instance, grow up to a third of a meter long at a rate of 4 cm per year, some 500 times faster than their smaller cold water relatives. The flesh of these clams, like that of the giant tube worms, is bright red with hemoglobin, indicating a high metabolic rate.

Rich and active as they are, the deep-sea rift communities are ephemeral. The hot springs eventually die down, like volcanoes, leaving behind ghostly communities of empty clam shells. As the earth's crust shifts and new hot springs form, immigrants from established or dying geothermal communities arrive to begin the unusual chemosynthetic ecosystem anew.

energy does it have? The required chemical analysis means the entire area must be sampled and parts of each organism harvested for analysis. The possibility of errors is obvious.

The energy budget can be expressed in simple terms. The amount of energy fixed by autotrophs in photosynthesis over a period of time is the gross productivity (GP). The GP minus the energy released in respiration is net productivity and is represented by the biomass or total organic content of the autotroph. These factors can be measured, with some difficulty, in various ecosystems.

In aquatic systems, it is possible to measure productivity using oxygen output as the indicator. Bottles containing the producers are arranged in dark and light situations. The oxygen produced in the lighted condition minus the oxygen utilized in the dark bottle gives us a measure of net productivity. Again, we must know what proportion of any system is comprised by any one kind of plant.

In one study, the productivity of an entire forest was calculated. The site was an oak–pine forest on Long Island, New York. The goal was to trace carbon pathways as well as to estimate productivity (Figure 34.31). The ecosystem, it was found, was storing energy beyond what was needed for respiration. Studies like these permit us to visualize total activity in ecosystems.

34.31 Productivity of a forest. The forest shown here is in a transitional state, just nearing its ecological climax. This means it is growing. How can we tell? The numbers represent grams of dry matter per square meter. The total amount of matter produced by the forest in this time period was 2650 grams per square meter. The

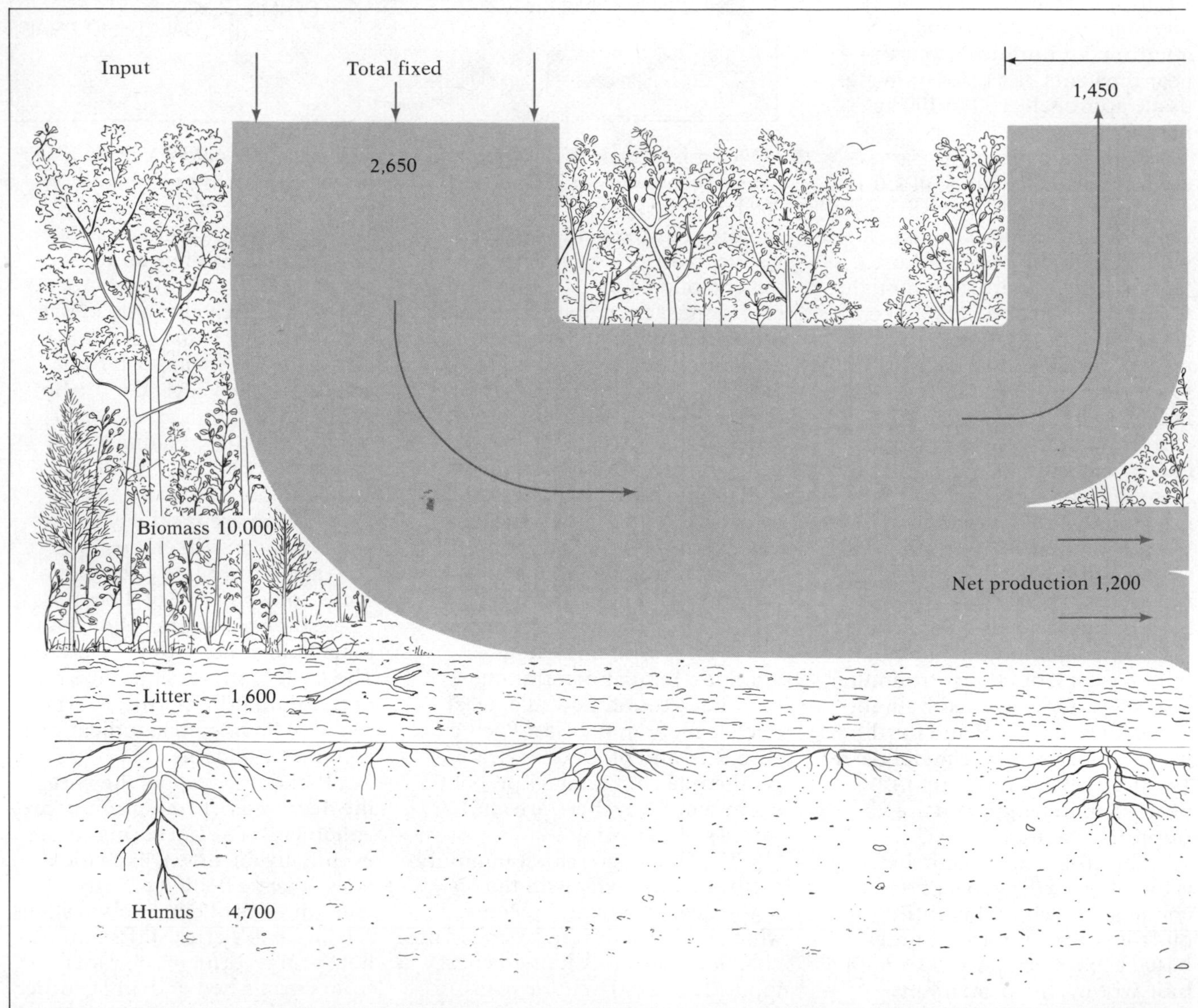

The Atmosphere

You are undoubtedly aware that the earth's atmosphere acts in several ways to shield living systems. It filters and reflects harmful radiation and slows the escape of heat from the earth's surface. Ultraviolet radiation striking the earth's surface is absorbed and released as longer waves or heat. The total heat of the atmosphere remains rather constant because of continuous heat loss into space. This constancy is important, but several human activities could alter the balance. For example, we are producing enormous amounts of heat by burning fossil fuels. Whether this heat is being added to the atmospheric burden faster than it can be dissipated is a good question. We don't know.

Another problem is the CO_2 released from burning fossil fuels and the clearing of forests. We have some answers here. We know that the CO_2 load in the atmosphere is increasing. The increase clearly correlates with the increasing use of fuels (Figure 34.32). But we don't know what the ultimate effect of increased atmospheric CO_2 will be.

What are the long-range effects of such changes on the earth's energy balance? We can only make educated guesses. A small increase in CO_2 hinders the escape of heat, since long-wave thermal radiation tends to be absorbed by CO_2 (the *greenhouse effect*).

total dry matter stored was 550 grams per square meter. The remainder was utilized in respiration by the organisms of the forest, releasing carbon dioxide and water. The 21% stored is trapped in humus or is reflected in new growth.

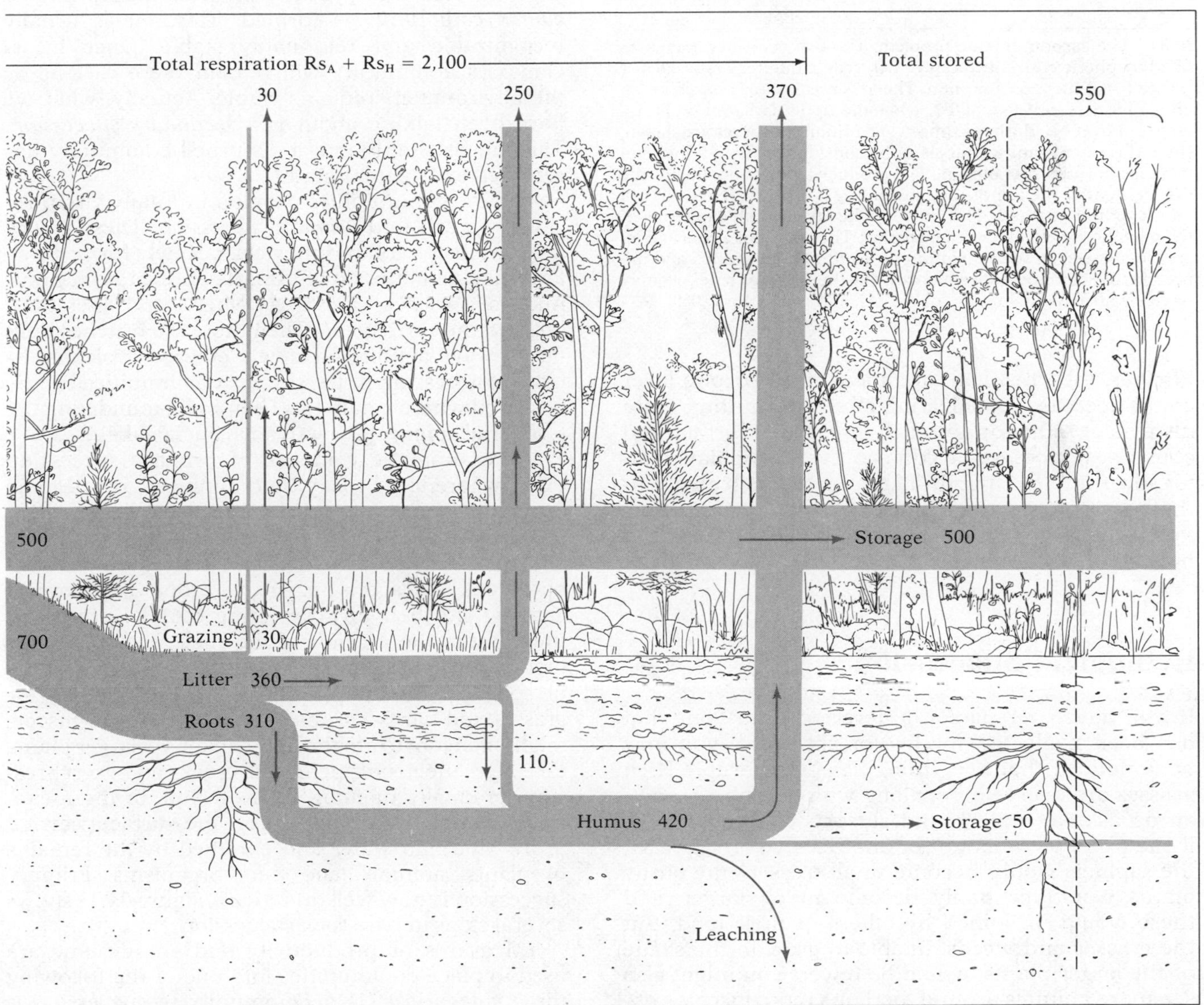

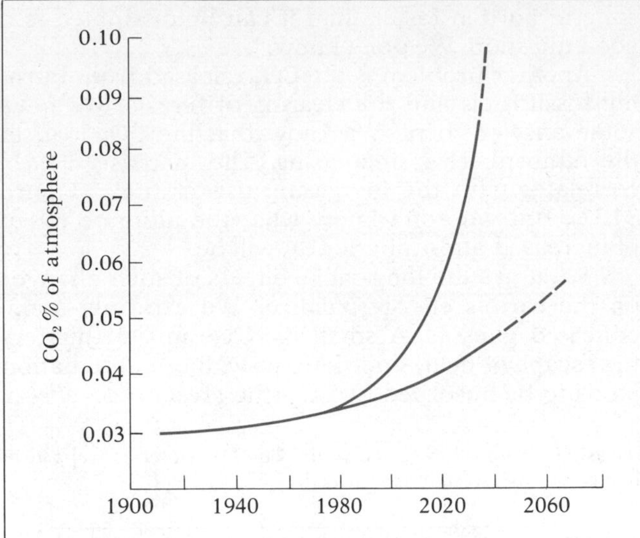

34.32 The carbon dioxide problem. The two projected increases in atmospheric concentrations of CO_2 reflect different estimates of future fossil fuel consumption. The lower estimate is based on an annually increased use of 2%, while the upper is based on an increase of over 4%. Both maximum and minimum curves are shown, since there are many variables to be considered. Even a minimal increase in fossil fuel burning will produce a near doubling of the CO_2 content of the atmosphere by mid-21st century. (How old will your children be then?) This may well be unacceptable in terms of the greenhouse effect described earlier. The effects of a fourfold increase, as seen in the maximum projection, hold serious and unpredictable climatic consequences. On the brighter side, we may have exhausted our supplies of oil, gas, and coal before 2050.

Further, the chemical alteration of our ozone layer could decrease its shielding effect, permitting more ultraviolet radiation to reach the earth's surface and causing an increase in skin cancers. Particulate matter, on the other hand, increases the reflectivity of the atmosphere, thus blocking solar energy—how much, we don't know. You would be astounded at how much we don't know.

Ecological Succession

If you have ever passed a deserted farm—one that has been neglected for five or six years—you may have noticed that everything has "grown up." High grasses are everywhere, along with scrubby bushes, an occasional sapling, and all sorts of weeds. Then, if the place were let go for another five or six years, the saplings would become small trees. Some bushy plants would probably become much larger, and these would be joined by others of their kind. But the grasses and weeds, unable to grow in the shade of the larger plants, would be fewer. You might also see more animals around, perhaps more insects, rabbits, and birds, and even an occasional deer from the neighboring woods. This is not the kind of place it used to be. The farm is undergoing *ecological succession.*

Succession usually doesn't mean the addition of the new, but a return to the old. The land is reverting to what it was before it was cleared. Eventually, it will closely resemble the surrounding forests (which may be successional themselves). During the period of succession, communities come and go. The activities of earlier communities often pave the way for successive ones, which in turn give way to others. The whole process is very gradual as some species are added and others are replaced. So we are forcing the issue a bit by talking about communities as identifiable entities. Who can say when one community has been replaced by another? But let's pretend we can. The process, then, continues until a *climax community* is formed. This one is usually recognizable and reasonably stable. Once in its climax, a community will remain there as long as the environment remains stable. Actually, what we have been talking about is a *secondary succession.* This means simply that a disturbed community has returned to its climax state.

Where communities develop in totally virgin regions, we can find *primary succession.* These places are obviously rare, but they exist. Newly formed deltas, rising up in the silt of estuaries, are virgin regions. And, in a way, so are areas where landslides have scraped the earth clean. Then there are beaches built up by wave action and newly emerged volcanic islands. Succession in freshwater communities can be observed where a dam has been built, inundating the terrestrial community with an artificial lake.

Less dramatic evidence of primary succession exists where large rocky outcroppings have become exposed. *Pioneer organisms* such as lichens and mosses are among the first inhabitants of the lifeless rock. Lichens, in fact, can exist quite well on bare rock surfaces. Because of their biochemical activity and their tiny, exploring rootlike processes, they hasten the erosion of the rock, and this is the first step in soil building. Tiny fissures become places of soil accumulation, presenting opportunities for grasses and hardy shrubs to take hold. A few insects might then move into the new shelter. As larger plants take hold, their roots penetrate cracks and exert pressure, gradually opening fissures. Soil, in the meantime, continues to build up on the surface, carried in by wind and water and enriched by the remains of plants, animals, and microorganisms. Primary succession is now well underway. Figure 34.33 shows several steps in bare rock succession.

Measures of productivity and energy flow are used to place communities into one of the following three categories: (1) A community in succession is storing energy in its biomass. That is, the biomass is increasing. The energy entering exceeds the energy

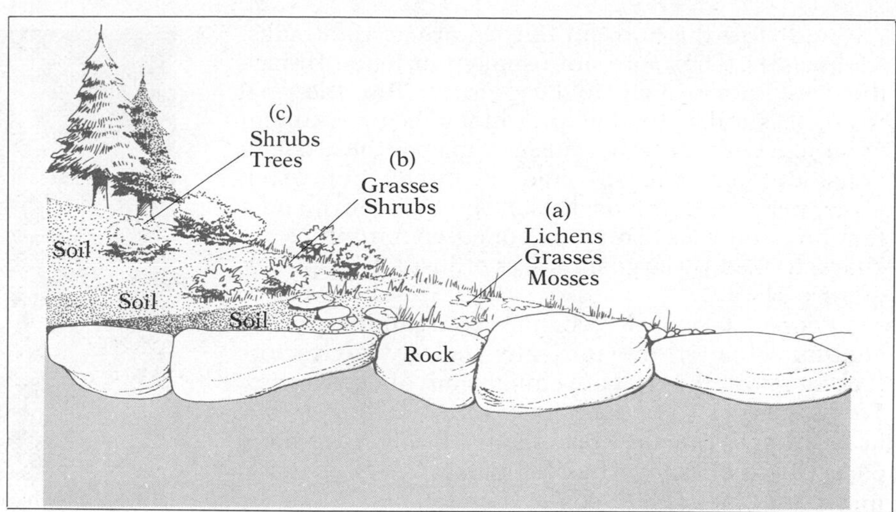

34.33 In primary succession, a rocky environment is gradually transformed into a forest community. (a) In the initial phase, pioneer lichens and mosses become established on the rocky surface, and help to erode the rocks with the assistance of natural weathering. Soil then begins to accumulate in crevices and low areas. (b) The pioneers are then displaced by grasses and annual plants as the soil continues to increase. More animal life and microorganisms become established, adding to the growing humus. (c) Finally, larger shrubs and trees become established as the soil cover becomes deeper and more complex. Food webs become more complex established as the community matures. It may take several hundred to thousands of years before the climax community is fully established.

released through respiration. (2) Once the climax state has been reached, the energy flowing in and out reaches equilibrium. (3) Finally, communities can simplify. A changing climate such as a drought, the release of unfavorable minerals through erosion, or any number of other factors can bring about a community's degradation. Forest fires are an extreme example. Degradation occurs when the energy released in respiration exceeds that gained in photosynthesis. The biomass of the community is therefore decreasing. The wearing out of a farm is another example.

Eutrophication in the Freshwater Community

Succession also occurs in aquatic systems and is called *eutrophication.* A eutrophic lake is one which is rich in nutrients and high in productivity. But this is a sign of aging. On the other hand, lakes with meager nutrients and low productivity are called *oligotrophic.* People generally prefer "clean" oligotrophic lakes.

Generally, freshwater bodies go from the oligotrophic to the eutrophic condition. Eventually, it is the fate of a lake to eutrophy to the point that it becomes so choked that it is invaded by organisms from the shore. The lake then gradually becomes dry land. So that is where lakes go.

Eutrophication is an extremely gradual process in lakes, but it occurs somewhat faster in ponds. One of the factors that determines the rate is the abundance of nutrients. Nutrients enter the pond or lake ecosystem from the surrounding terrestrial community via streams and runoff. Each particle of organic matter adds to the nutrient content. Some of the nutrients entering the pond or lake become part of the bottom sediment and are later brought up in turnovers (upwellings), thus becoming available to plants. As long as these processes are slow, eutrophication proceeds gradually.

Two nutrients strongly determine the rate of eutrophication because they are normally in such short supply. They are nitrogen and phosphorus. Recently, however, they have drastically increased in abundance. The abundance, of course, is not unrelated to human activity. Lakes and rivers are increasingly used as dumping places for treated and untreated sewage. In addition, there is an enormous runoff of fertilizers from farmland. Few natural waters in the world have escaped this influx. One of the richest sources of phosphate input has been from laundry detergents, which may contain 15–60% phosphate. The result of this nutrient pollution is a rapid increase in the aging of freshwater bodies.

Nutrient Pollution

It may seem strange to think about plant or algal nutrients as pollutants. And they're not—from the algae's point of view. For us humans, however, nutrients can turn something we like into something we don't like and something that kills fish, as well. High levels of nutrients have produced deadly algal blooms. The crowded algae take more oxygen from the water at night than they produce through photosynthesis in the daylight. By respiring at night, they deplete the oxygen in the water, and animal life perishes. Further, algal blooms can produce extreme crowding and competition for sunlight. Much of the bloom dies off and reducer populations flourish, rapidly depleting the oxygen as they carry on decomposition. The depletion brings on the death of many aerobic organisms, including invertebrates, vertebrates, plants, and protists. If the situation continues, the community becomes anaerobic and can be inhabited only by certain bacteria and a few other tolerant organisms.

Added to the nutrient burden are myriad industrial wastes. These are not usually nutrients. In fact, they are more likely to be poisons. But the final result is similar to that of sudden heavy nutrient pollution. Only a few organisms can bear these conditions. But one can still find the hardy blue-green algae, perhaps some protists such as diatoms, and a few invertebrates. These poisoned communities are characterized by large numbers of just a few species. (Figure 34.34).

People obviously change communities. They continually alter the intricate and myriad factors that maintain the delicate equilibrium of ecosystems. To put it simply, we just don't know yet how to live in harmony with the ecosystems of which we are a part. Our technology has far outstripped our basic understanding. While some of the claims made by self-styled environmentalists in the 1960s were outright exaggerations and some were based on faulty information, there is no doubt that we are dangerous to ourselves. But perhaps we are, at last, becoming aware of the kinds of changes we are capable of bringing about. As we turn to the scientific community for answers, ecologists will play an increasingly prominent role.

34.34 Eutrophication is a long, natural process, held in check by the availability of nutrients such as nitrogen and phosphorus. These are minimally available in most aquatic systems. When humans dump nitrogen- and phosphorus-rich wastes (sewage or agricultural and manufacturing wastes) into the water system, the process of eutrophication is greatly speeded. The result is not usually the same as natural eutrophication, but often results in the creation of an anaerobic, stagnant system like this one. In the earlier stages, there may be a great increase in the number of organisms, a sign of aging, but in the end, only a few species of tolerant worms, algae, and bottom fish remain.

Application of Ideas

1. Why are biomes defined in terms of plants? What alternatives might exist?
2. Why are there more species of animals and plants in the tropics than in temperate regions?
3. Birds in the tropics lay fewer eggs than birds in temperate regions. Why do you suppose this is?
4. What problems are involved with clearing tropical forests, especially regarding soil richness and oxygen production?
5. List the ways various animals have adapted to the desert biome. What general rules can be made for desert dwellers?

Key Words

abiotic
abyssal region
A horizon
altitudinal ecocline
arboreal

benthos
B horizon
biomass
biome
biosphere
biotic

canopy
chaparral
chemotropic autotroph

Key Ideas

Introduction

1. *Ecology* is the study of interaction between organisms and their environments.
2. Today's *ecologists* use the tools of the new technology in making their observations, analyzing their data, and recording information. *Systems ecology* utilizes model *ecosystems* for studying the impact of variables in the *environment.*

The Biosphere

1. The *biosphere* includes all of the regions of the earth that support life. The entire region is a thin layer some fourteen miles in thickness.
2. Conditions include adequate water, a certain narrow range of temperatures, sunlight, or other energy input, carbon, hydrogen, oxygen, nitrogen, and various minerals.

The Distribution of Life: Terrestrial Environment

1. Life is distributed into *biomes,* which support certain kinds of plant life. Climatic conditions within a biome are responsible for its characteristics.
2. *Biomes* are subdivided into *communities,* which are interacting *populations* of organisms.

climax community
community
consumer
continental shelf
Coriolis effect

desert

ecocline
ecological succession
ecology
ecosystem
El niño
Eltonian pyramid
energy pyramid
environment
epiphyte
estivation
eutrophication

fire-disclimax community
food chain
food web

grassland
greenhouse effect
gross productivity

intertidal zone

laterite
latitudinal ecocline
life-support system
limiting factor
litter
littoral zone
longitudinal ecocline

muskeg

neritic province
net productivity

oceanic province
oligotrophic

pampas
pelagic
permafrost
phototroph
phytoplankton
prairie
primary producer
producer
productivity
pioneer organism
population
primary consumer
primary succession

quaternary consumer

rain shadow
reducer

3. The distribution of *biomes* tends to follow latitude lines. Gradual transitions from one *community* type to another are called *ecoclines*.

Biomes
There are no sharp dividing lines between *biomes*. Transition is gradual.

The Desert Biome
1. *Deserts* receive less than 25 cm (10 in) of rain per year. Temperatures in *deserts* vary according to latitude and they are not necessarily hot places.
2. Some *deserts* are produced and maintained by high mountain ranges that block coastal precipitation.
3. Large deserts are found in all continents except Europe and Antarctica. All *deserts* experience dramatic day–night fluctuations in temperature.
4. *Deserts* are created along the 30° north and south latitude lines. They are in regions where air cells originating at the equator have lost their moisture.
5. Mountains cause *desert* conditions by forming *rain shadows*, as is clearly seen in the American deserts.

Desert Plants
1. Plants adapted to *desert* conditions are called *xerophytes*. Some become dormant, while others store water.
2. The annuals germinate after brief rains, quickly producing flowers and seeds before drought conditions return.
3. A few species, widely distributed, usually dominate a particular *desert*.

Desert Animals
1. Rodents survive by spending daytime in burrows and venturing out at night. Their predators follow the same routine. Some survive without water through behavioral and physiological strategies. The kangaroo rat rarely drinks, but lives off metabolic water conserved by a highly efficient kidney. The camel drinks huge quantities of water, survives dehydration, and withstands unusual changes in body temperature.
2. The *desert* is a relatively simple *biome* and can be easily disrupted through careless human acts. It can become productive if water is present, but its exploitation can produce irrevocable changes.

The Grassland Biome
1. *Grasslands* exist as huge inland plains in North America, Asia, Australia, South America, and Africa.
2. Grasses dominate this *biome*. Much of the *grassland* however is utilized for growing grains.
3. Rainfall is seasonal and limited. Grasses survive by producing deep diffuse root systems and by becoming dormant.
4. The animals of *grasslands* produce huge *populations* and include the largest herbivores. The grasses are well adapted to resist grazing.
5. The *food webs* of *grasslands* are complex and relatively stable. Occasionally insect populations explode out of control and devastate this *biome*. Normally *grasslands* support the largest populations of animals on earth.

Tropical Rain Forest
1. This forest is supported by heavy rainfall and a warm climate. These forests are found in South America, Africa, India, Burma, Central America, Indonesia, and the Philippines.
2. There are no dominant species of plants, but there are dominant types. The *canopy* is composed of very tall hardwoods. Vines abound as do *epiphytes*, which absorb water from the air.
3. The forest floor is dark and devoid of foliage. Scavenger and *reducer populations* abound.
4. The "jungle" refers to thick growth along rivers or where sunlight reaches the forest floor after disturbances.
5. The number of animal species is unequalled in other land biomes. Many are arboreal.

rocky tidal community

salinity
saprobe
savanna
secondary consumer
secondary succession
subcanopy
systems ecology

taiga
temperate deciduous forest
temperate rain forest
tertiary consumer
tide pool
tradewind
trophic level
tropical rain forest
tundra

upwelling

veldt

xerophyte

zooplankton

6. Tropical soils are relatively infertile and marginally useful because of the rapid cycling of nutrients. Agricultural interests and foresting have depleted much of this biome.

Chaparral or Mediterranean Scrub Forest

1. The *chaparral* is exclusively coastal in western North America, parts of the Mediterranean, South America, South Africa, and southern Australia.
2. Rainfall is seasonal and limited, but the coastal climate is somewhat moist.
3. The shrubs and trees have a dwarfed, scrubby appearance, with small leathery leaves.
4. The *chaparral* seldom reaches a *climax* state because of fire.
5. The forest floor contains a heavy *litter* because of slow decomposition, supporting large *populations* of insects and rodents. Large numbers of migratory birds are seasonal visitors.

The Temperate Deciduous Forest

1. This *biome* is found in all continents and like the *grassland* is greatly disturbed by human activities.
2. Seasonal change is characteristic and deciduous plants are common. Rainfall averages about 40 inches distributed throughout the year. Several species of trees dominate the various communities. The forest floor has heavy growth.
3. The soil is generally rich with a large soil population of scavengers and *reducers.*
4. Because of human activities, most of the larger herbivores and carnivores are gone.
5. Plants adapt to winter by losing their leaves and becoming dormant. Animals migrate, hibernate, or simply survive the winter as best they can.

The Taiga

1. The *taiga* is confined to the Northern Hemisphere. Rainfall varies but generally is less than the deciduous forest. It is often interrupted by bogs or *muskegs.*
2. Light can be a limiting factor to plant growth.
3. The dominant plants are conifers, with a few species of broadleafed deciduous trees. Conifers have fine, waxy leaves adapted for resisting water loss. Their decomposition in the litter produces an acidic soil condition.
4. Large herbivores and carnivores still survive in the *taiga.* Insect populations are large, but seasonal.
5. The *taiga* is important to people for its timber.
6. Conifers grow wherever the proper conditions exist, thus we have coniferous forests on the West Coast, and in many mountainous regions.

The Tundra

1. The *tundra* is the northernmost *biome,* forming a circle around the earth above the *taiga.*
2. Plants are small and ground hugging since water is scarce and frozen through most of the year. Roots cannot penetrate because of *permafrost.*
3. Light also limits plant growth, and is available in suitable intensity only in the brief summer season.
4. Lichens and mosses are common ground cover, along with dwarfed trees and grasses, rushes, and sedges.
5. The animal life includes some large herbivores along with numerous hares and lemmings. *Food webs* are usually abrupt, with small herbivores providing food for carnivore species. Migratory birds abound in summer, as do inscets.
6. Because of the simplicity of *food webs,* the *tundra* is highly vulnerable to human intervention.

The Water Environment

1. The marine environment is vast and complex, with many communities. Its most variable region is the *continental shelf.* The shelf includes the *neritic province* while the waters beyond are the *oceanic province.*

2. Shallower parts of the *neritic province* include the rich *littoral region* and *intertidal zone.*
3. Some *benthic* organisms, including fish and invertebrates, are found in the abyss.
4. The *oceanic province* contains *pelagic* organisms as far as light penetrates. These include phytoplankton, the *primary producers,* which carry on 80–90% of the earth's photosynthesis. *Phytoplankton* support *zooplankton* populations, forming the base of ocean food chains. Deep-sea fishes live as predators and scavengers in the perpetually dark depths.

Coastal Communities

1. The *neritic province* contains the world's fisheries.
2. Nearer the shore is the *littoral zone* with its diverse communities. These are rich in life, supported by nutrient input from rivers and coastal *upwellings.*
3. *Rocky tidal* organisms are adapted to wave surge and tidal forces. Many organisms in the *tide pools* must further adapt against the dry periods and changing salinity.
4. Estuaries are also rich in marine life, particularly larval forms, and are subject to the most serious pollution problems.
5. Tropical reefs form natural refuges and contain the richest flora and fauna of the oceans.

Productivity in the Oceans

Productivity in the marine environment varies with light and nutrients. Cold waters are the most productive, especially where *upwellings* occur. *Upwellings,* on the other hand, can produce disastrous red tides when *blooms* of dinoflagellates occur. They produce toxic wastes that result in giant fish kills.

The Fresh Waters

Freshwater bodies vary greatly in size and even in mineral content. Most lack minerals and produce osmoregulatory problems for their inhabitants (see Chapter 28).

Freshwater Lakes and Ponds

The extent of life in deep lakes is determined by the oxygen content of the water. If seasonal turnovers occur as water changes temperature, life can go on at the bottom of the deepest lakes. The *primary producers* of lakes are algae.

Interaction in Ecosystems

1. Technically an *ecosystem* is a conceptual unit of organisms and their *environment.* Its size has no upper or lower limits.
2. *Ecosystems* consist of four basic elements:
   a. *Abiotic* elements (physical environment)
   b. Energy that flows through
   c. *Biotic* elements (living things)
   d. Nutrients that cycle
3. Energy and nutrients represent life-support systems for an *ecosystem.*

Abiotic Factors

These include currents, temperature, moisture, light, chemical composition, and other nonliving entities. They are often referred to as *limiting factors.*

Biotic Factors

1. The living matter can be measured as the *biomass* or combined organic matter. Measuring this way is useful when studying the productivity, metabolic efficiency, or state of growth.
2. The *biotic* elements can be classified according to their activities, thus we have *producers, consumers,* and *reducers.*
3. *Producers* include photosynthetic and chemotrophic organisms. *Consumers* feed on *producers* and each other. *Reducers,* fungi and bacteria, are consumers that feed on the dead by absorption.

4. *Consumers* are classified as primary, secondary, tertiary, etc., according to *trophic* or feeding *levels.* The *trophic levels* can be represented quantitatively by *Eltonian pyramids.* These can convey numbers of individuals, *biomass,* or energy stored. The pyramidal shape indicates inefficiency in transfer. Feeding relationships can be shown in *food webs,* which are complex, and *food chains,* which are simple and linear.

Life-Support Systems
Life-support systems consist of energy and nutrients. The latter are cycled (see Chapter 19).

Energy Flow Through Ecosystems
1. Energy of *ecosystems* comes originally from sunlight. Only 1–2% is stored in photosynthesis. Half this amount is used by the plants in their activities.

2. Herbivores store about 10–20% of the energy they take in, using the rest in their activities. Eventually all incoming energy flows through an *ecosystem,* dissipating as heat.

Nutrient Cycles
Nutrients cycle through *ecosystems* in biogeochemical cycles (as discussed in Chapter 19). Cycling of nutrients is a vital activity of *reducers.*

Measuring Ecosystem Productivity
1. Productivity can be estimated by harvesting. This is done by determining the caloric value per gram of living material.

2. Productivity can be expressed as *gross productivity,* or if the energy of respiration is subtracted, the *net productivity.*

3. In aquatic systems, productivity can be measured by oxygen production in samples.

The Atmosphere
1. The atmosphere shields organisms from harmful radiation and retains heat. Energy absorbed by the earth is normally equal to energy leaving as heat, producing stable conditions.

2. Human activities may be inhibiting heat escape by constantly adding CO_2 from fossil fuel combustion. In addition we add other chemicals and particulate matter. The ultimate effects cannot be clearly predicted because of the conflicting effects of CO_2 and particulate matter, but a gradual increase in the atmospheric temperature is very possible.

Ecological Succession
1. *Ecological succession* is the gradual change from one type of *community* to another. It slows when a *climax* state is reached. *Climax communities* remain stable for long periods or until physical conditions change.

2. Where disturbances have altered a *community,* the trend is to return to the original type. This is called *secondary succession.*

3. The *succession* of *communities* from virgin regions is called *primary succession.* There are several recognizable steps. *Pioneer organisms* such as lichens and mosses grow first, modifying the soil and giving way to grasses, insects, etc. Each stage changes the soil in such a way as to make way for the next.

4. The status of a *community* can be determined by measuring net productivity. Systems that are storing energy are in *succession.* Where no *net productivity* is found a *community* has reached a *climax state.* If stored energy is decreasing, a *community* may be simplifying.

Eutrophication in the Freshwater Community
Succession in aquatic systems is called *eutrophication.* As it proceeds, lakes become increasingly productive. *Oligotrophic* lakes have low productivity and are stable. The trend in freshwater bodies is towards the *eutrophic* condition.

Nutrient Pollution
1. The rate of *eutrophication* is rapidly increased by nutrient input. In recent times, pollution has increased the rate of *eutrophication* in many aquatic systems.

2. Nutrient pollution, particularly with nitrogen and phosphorus, drastically alters aquatic *ecosystems.* Algal blooms can occur with a great increase in *reducer populations,* followed by oxygen depletion. Normally, nitrogen and phosphorus are limiting factors.

Review Questions

1. What is a *model ecosystem* and how does its study help? (p. 992)
2. Define the *biosphere* and describe its dimensions. (pp. 992–993)
3. List the necessary conditions for supporting life in the *biosphere.* (p. 993)
4. Explain the relationship between *biomes* and *communities.* (p. 994)
5. What is a *latitudinal ecocline.* Provide an example. (p. 994)
6. Describe the conditions that produce an *altitudinal ecocline.* (p. 994)
7. What are the special climatic conditions of the *desert biome*? (p. 995)
8. Summarize the geographic conditions that produce a *desert.* (p. 995)
9. Describe three ways in which plants have adapted to *desert* conditions. (p. 996)
10. Compare the physiological and behavioral adaptations to *desert* life of the kangaroo rat with those of the camel. (p. 998)
11. Why is it especially dangerous for humans to interfere with the *desert ecosystem*? (p. 999)
12. What are the physical conditions that produce a *grassland biome*? (p. 999)
13. What special adaptations do grasses have for resisting grazing and drought? (p. 999)
14. Describe humans' effect on the world's *grassland* today. (p. 1000)
15. Characterize the plant growth and its organization in the *tropical rain forest.* (p. 1001)
16. Describe the soils of the *tropical rain forest,* their fertility, and adaptability to agriculture. (p. 1002)
17. Describe the effects of agriculture on the *rain forests* of the Amazon Basin. (p. 1002)
18. List the geographical regions and special conditions of the *chaparral* (Mediterranean scrub forest). (p. 1003)
19. Characterize the plants of the *chaparral.* (p. 1003)
20. What are the most common animals of the *chaparral*? (p. 1003)
21. Where are the world's *deciduous forests* and what are their special climatic conditions? (pp. 1003–1004)
22. How do the *deciduous forest* animals cope with the changing seasons? (p. 1004)
23. What use do humans make of the *deciduous forests* and how has this affected plant and animal life? (p. 1004)
24. List the geographic regions where the *taiga* is found and state its special climatic conditions. (p. 1005)
25. Describe the plants of the *taiga* and their special climatic adaptations. (p. 1006)
26. What is the greatest threat to the *taiga*? What is wrong with reforestation? (p. 1006)
27. In what special way is the *tundra* similar to the *desert*? (p. 1007)
28. List several plants or plant types found in the *tundra.* (p. 1007)
29. Which carnivores are dependent on lemmings and how do the lemmings regulate the populations of these meat eaters? (pp. 1007–1008)
30. What strategies have the *tundra* animals developed for surviving the long winters? (p. 1008)
31. With the aid of a simple drawing, name and label the regions of the oceans. (p. 1009 and Figure 34.23)
32. What are the special conditions of the *abyss* and what forms of life, if any, are found at its bottom? (p. 1009)
33. What are the special characteristics of deep-sea fishes? (p. 1010)
34. List some reasons for the richness of life in the *littoral zone.* (p. 1011)
35. Why are estuaries so important to marine life and what have we done to threaten these regions? (p. 1011)
36. What is the importance of *upwellings* to marine life and how are they brought about? (p. 1011)
37. Compare ocean *productivity* in the open sea, coastal regions, and regions of *upwellings.* (pp. 1012–1013 and Table 34.1)
38. What conditions make it possible for vertebrates and other aerobes to live at the bottom of deep lakes? (p. 1013)
39. Carefully define the term *ecosystem* and characterize its dimensions. (pp. 1013–1014)
40. Make a list of the *abiotic factors* in an *ecosystem.* (p. 1014)
41. How is the term *biomass* useful in describing an *ecosystem*? (p. 1014)
42. List four *trophic levels* of the *consumers* and provide an example of each. (pp. 1014–1015)
43. Explain how an *Eltonian pyramid of energy* illustrates the collapsing nature of *food chains.* (p. 1015)
44. Why is a *food web* more descriptive of the interactions of the *producers* and *consumers* than a *food chain*? (p. 1015, Figure 34.29)
45. Using a simple diagram, illustrate the flow of energy through an *ecosystem.* Start with sunlight and *producers* and be sure to indicate how much energy is stored in each step. (p. 1016)
46. Describe the importance of *reducers* in the cycling of nutrients. (p. 1016, Figure 34.30)
47. Describe two ways in which the productivity of an *ecosystem* can be estimated. (pp. 1016, 1018)
48. Describe the role of the atmosphere in maintaining an acceptable range of temperatures for life on the earth. (pp. 1019–1020)
49. Distinguish between *primary* and *secondary succession.* (p. 1020)
50. List the steps that might be followed in *primary succession.* (p. 1020)
51. What kinds of productivity data would tell one that *succession* was going on? That a *climax community* was in place? That *ecosystem* deterioration was occurring? (p. 1020)
52. What is the apparent trend in many *oligotrophic* lakes? (p. 1021)
53. In what ways do human activities contribute to rapid *eutrophication*? (pp. 1021–1022)

Chapter 35

Populations

The numbers of animals in any area change from one time to the next. For example, wolves once inhabited most of the United States, but now they are rare over much of this area. Sometimes we know the reason for such population changes. Sometimes we don't. Wolves, for example, have simply fallen to the rifles of farmers and the rituals of America's manhood. But what about in Minnesota? Wolves still live there. Are their numbers changing? If so, why? If not, what is keeping them stable? How about other species? Why has the eastern bluebird become so rare? (The answer to that one is the entire species was nearly wiped out by a late winter storm several years ago and the species is only now recovering.) Why are vultures seen so infrequently in Arkansas? And how is it that beavers are making a comeback? The point is, the numbers of living things change, and usually we don't know why. Because population numbers reflect a myriad of environmental, evolutionary, and physiological effects (most of which are not very well understood), any question about populations tends to be a tough one. A host of researchers are, at this very moment, watching populations of tiny animals in little glass tanks —just trying to fit more small pieces into the great puzzle. Others are traipsing around with muddied boots or spending long hours watching a hole in the ground, waiting for some small face to appear. A great many people, in fact, are asking very basic questions about how populations change. So our goals here must remain modest. We can only describe, in general terms, what populations are, and a few of the factors that can make them change. But let's begin with a description of the *ecological niche*—what Darwin called an organism's "place in the economy of nature."

35.1 Population Growth and Reproductive Strategies

The Ecological Niche

It has been said that if the habitat is the organism's address, the niche is its profession. The niche, then, includes all the organism's interactions with its environment. And some niches are larger than others. For example, some animals are *broad-niched,* like crows. In other words, they interact with their environment in a number of ways. They walk along plowed fields eating newly planted corn, they fly into trees to sample berries, and they don't turn up their beaks at a dead squirrel along the roadside. They may eat the eggs of other birds and so they are attacked on sight by smaller, quicker, nesting birds. They retreat from these tiny tormentors, but they may turn around and mob an owl. Because they interact with their environment in a number of complex ways, they are called broad-niched. On the other hand, most microbes are *narrow-niched.* For example, such reducers as decomposing bacteria operate within very rigid and narrow limits. That rotting squirrel was not decomposing by the action of a single kind of bacteria. Instead, a variety of microbes were at work, each utilizing the waste products and by-products of others and each incapable of doing anything else. The decomposition process is complex, and each step is carried out by a microbial specialist. All specialists are regarded as having narrow niches.

The term *niche* is notoriously difficult to define; in fact, there is no record of any two biologists ever agreeing on a definition. A good working definition, however, and one that will elicit little argument, is this: The niche is the role a species plays in a community of interacting species. Its role includes where it lives, what it eats, what eats it, how it responds to stress, and what limits its population growth.

Population Changes

Some population changes are short-term or periodic and easily explainable. Grasshoppers, which are plentiful in summer, disappear in winter. That's easy to understand—they die. Some birds disappear from American forests in the winter as they migrate to more hospitable areas. These are familiar and rather predictable phenomena. We all expect grasshoppers and birds to be back in spring—unless DDT got them.

You may also have noticed that the numbers of many species don't seem to vary much from year to year. For example, the number of birds around your house has probably been about the same from one summer to the next. However, if you think back, you may also notice that some species you remember as having been fairly common just aren't around anymore.

And what about those sudden increases in numbers of certain animals? Where do all the locusts that periodically appear to ravage African crops come from? Certain populations seem to blossom for some mysterious reason, and we hear of "invasions" or "plagues." In spite of such changes as extinctions or plagues, however, most species seem to maintain rather stable numbers. They seem to be under some sort of control (Figure 35.1). In order to investigate this phenomenon, we must first understand something about how populations normally behave. The central questions are: What determines the numbers of a population? What causes them to fluctuate? And what stabilizing influences act on a population? We should keep in mind that, technically, a population is more than a group of organisms. It is a group of the same species which lives close enough together to be able to interact, for example, by breeding.

35.1 Stable populations. Note that this population of herons has oscillated but remained comparatively stable. The data were collected from a study of occupied nests in the area over a period of 30 years.

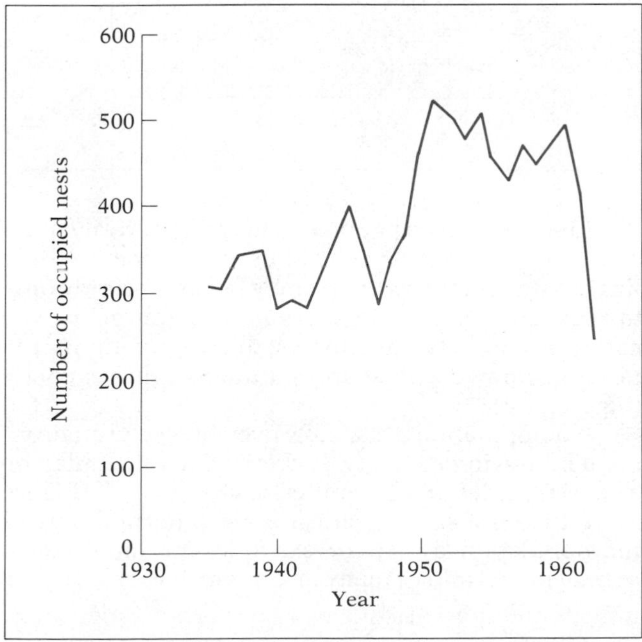

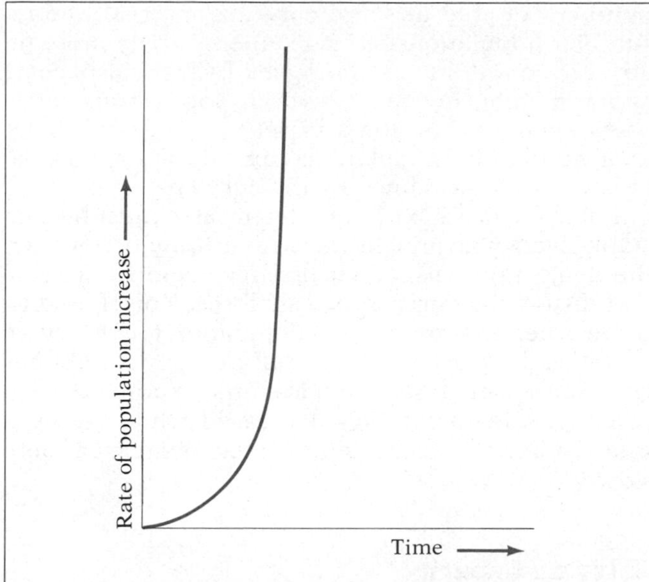

35.2 Theoretical growth. Time is plotted against population size. In such a case, there is nothing operating to restrict growth, so the population is achieving its *reproductive* (or *biotic*) *potential.* The characteristic curve is produced because larger populations increase more rapidly than smaller ones. This type of growth is called *exponential growth.*

Population Growth

If we were to place a few bacteria in a suitable medium, or a few goats on an uninhabited but hospitable island, the growth of their populations might be expected to show similar patterns, at least at first. Since we are assuming that their new environment provides them with the necessities of life, their growth will be limited primarily by the rate at which they can reproduce, rather than by environmental factors.

Note that the absolute numbers of individuals added to the population each generation increases with time. The more the population increases, the faster it increases. This is because as time passes, increasing numbers of individuals are producing offspring. If each individual in the original group leaves two offspring, or each pair leaves four offspring, the population will remain fairly small for several generations (one bacteria divides to leave two, these leave four, which leave eight—all low numbers). But once the population increases a bit, the numbers begin to skyrocket. (What would be the result of bacterial divisions after only ten generations? Twenty generations?) The reproductive capacity of any population, when it is unrestricted, is called its *biotic potential* (Figure 35.2). The theoretical growth curve for goats on the island would be the same as for the bacteria, except that the starting number would be two and the growth would proceed more slowly. Barring mishap, the shape of the curve would always be the same.

We see, then, that populations in ideal environments tend to *increase exponentially.* That is, if the numbers double at some constant interval, the result is a population that gains increasingly more (in absolute numbers) as time goes by. It will become apparent that, in the real world, populations rarely show such growth, for a variety of reasons. Under ideal conditions, populations can exhibit exponential growth (the type shown by the curve in Figure 35.2), but in such cases, birth and death rates must be constant. Even in an initially ideal environment, however, the death rate will sooner or later begin to increase and distort the curve, as we shall see. You should be forewarned that the human population growth curve is unusual. Over the long haul, our population has been increasing faster than a simple exponential, because the death rate has decreased while the birth rate has remained the same or has decreased more slowly.

Carrying Capacity

Because of limitations imposed by the environment, the full biotic potential of any species isn't likely to be reached. *Environmental resistance* is encountered when some essential commodity in the environment comes into short supply and thus reduces the population's rate of increase. The limit of biotic potential set by environmental resistance defines the *carrying capacity* of the environment. As a population approaches the carrying capacity of its environment, its numbers begin to level off (Figure 35.3).

35.3 The theoretical relationships between *biotic potential, environmental resistance,* and *carrying capacity.* The biotic potential of a population is rarely, if ever, achieved due to the problems of living in a real world (the problems are collectively called environmental resistance). Some individuals don't reproduce and some fail to reproduce maximally. The characteristic growth curve is produced because a small population grows slowly, since a few individuals can generate only a small number of offspring. Once the breeding population reaches a critical size, however, it skyrockets, to be slowed only by the effect of its own numbers. Theoretically, it should stabilize around the carrying capacity of the environment.

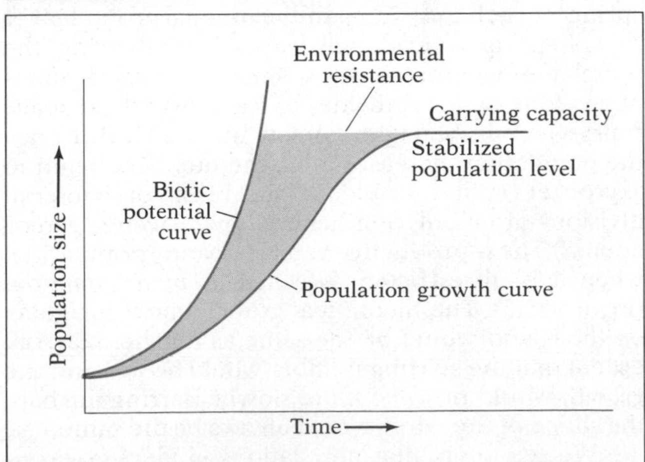

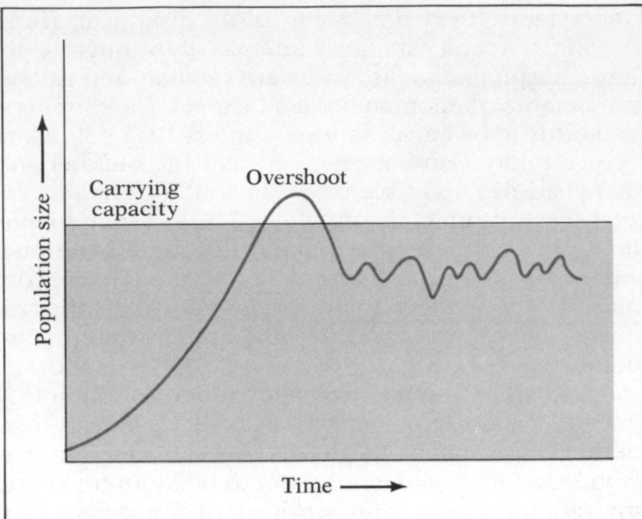

35.4 The type of population curve normally generated when a population reaches the environment's carrying capacity. Theoretically, a stable population fluctuates about the carrying capacity in a rather regular fashion.

35.5 If a population should drastically overshoot the environment's carrying capacity, or disrupt the environment and lower its carrying capacity, the population may *crash.*

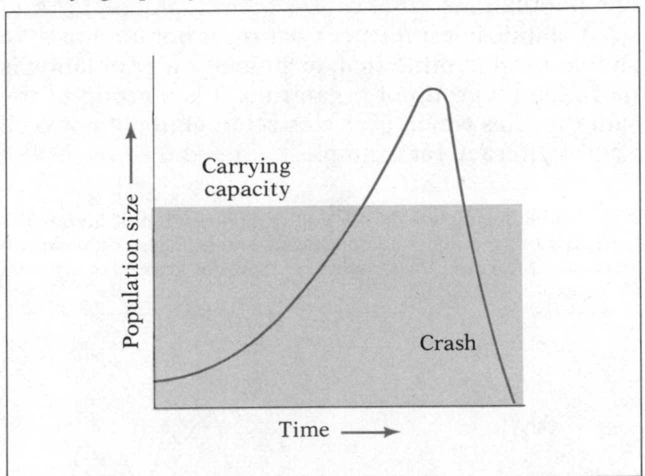

In some cases, a population may temporarily overshoot the carrying capacity of its environment. When this happens, its growth rate may fall off abruptly only to pick up again when the numbers fall below the carrying capacity. In more stable populations, the numbers may oscillate around the carrying capacity (Figure 35.4).

If a population drastically overshoots the environment's carrying capacity, however, the result may be a severe depletion of numbers called a *crash* (Figure 35.5). In some species, crashes are a normal part of the population cycle; for example, the numbers of annual plants and of many insects crash each autumn, because the carrying capacity of the environment has changed. When the environment improves again in the spring, their progeny begin to reproduce madly in the few months they have available.

Some crashes are more permanent. A bacterial population may rise rapidly, but if the population lives in a Petri dish on a laboratory shelf, or in a dead squirrel on the highway, it must eventually run out of food even as it befouls its environment with its own waste products. After it crashes, it cannot recover. As another example, if environmental resistance in the form of predators, disease, or food and water limitations is somehow reduced for a group of animals, its numbers may increase to the point that it eventually overshoots the carrying capacity and devastates its habitat. Oversized herds of elephants may eat all the young plants in their habitat, which, if left alone, would eventually have provided much more food—and would have continued to do so for generations afterward. The result of such devastation is a lowering of the carrying capacity of the habitat for long periods afterward. In fact, it is possible to lower the habitat's carrying capacity permanently. If stripping an area of its foliage should increase erosion, the habitat may never recover.

Reproduction Rate

There is an argument among biologists regarding whether anything is maintaining population numbers within certain limits—a control mechanism, or set of mechanisms, which only occasionally break down. Proponents argue that the relative stability of so many species indicates that some sort of stabilizing influence is operating. Any such influence, if it exists, could theoretically operate at two levels: (l) by affecting how many offspring are produced and (2) by affecting the longevity or mortality of organisms in the population.

How Many Offspring Are Produced?

The reproductive potential of most species, even the slowest-breeding ones, is incredibly high. Charles Darwin estimated that a single pair of elephants could leave over 19 million descendants in only 750 years. The fact that the entire world was not teeming with elephants indicated to him that some elephants were not reproducing, and that the ones that did reproduce were somehow "selected" by the environment.

Whereas elephants are very slow breeders, other species are more prolific. For example, the incredible reproductive potential of the housefly for a *single year* is shown in Table 35.1. But we are pleased to report that not all survive.

Table 35.1 Projected Populations of the Housefly (*Musca domestica*)[a]

Generation	Numbers if all survive
1	120
2	7,200
3	432,000
4	25,920,000
5	1,555,200,000
6	93,312,000,000
7	5,598,720,000,000

[a]In 1 year, about seven generations are produced. The numbers are based on each female laying 120 eggs per generation, each fly surviving just one generation, and half of these being females. Adapted from E. V. Kormondy, *Concepts of Ecology* (New Jersey: Prentice-Hall, Inc., 1969), p. 63.

What determines the rate at which an organism reproduces? Why haven't flies or elephants taken over the entire surface of the earth? It should be apparent that simple laws of energetics would not allow an elephant to become pregnant several times a year. There is just no way that a female could find enough food to produce that many offspring. And what would happen to the helpless baby elephants already born as new brothers and sisters appeared on the scene?

It may seem puzzling, then, to learn that each animal theoretically reproduces to its biological limit. Is it really true that all animals leave as many offspring as possible? You might intuitively agree when you recall that any generation is primarily made up of the offspring of the most successful reproducers. But some biologists have disagreed, saying animals deliberately adjust their birth rate in accordance with the population's death rate; as one goes up, so does the other. However, we will take the majority view that animals reproduce maximally. The next question we might ask regards method. How does an organism maximize its reproductive success?

Tapeworms

The tapeworm living so comfortably in the idyllic environs of your intestine is lucky to be there—although its luck is the inverse of yours. It is lucky because the life of a tapeworm is fraught with risk. There are incredible odds against its ever finding a host at all. In order to reach a human host, the egg must pass out with the feces of the previous human host. Then it lies around in the sun and rain until it is accidentally eaten by another animal such as a calf. Not only must it survive the elements, but it must also be found by a calf, so already the odds are against it. Once the egg has been eaten by the calf, it changes its appearance, moves through the intestinal wall of its host, and on through the bloodstream until it comes to lie in the muscle. Here it forms a protective capsule around itself and waits. It waits for a human to eat the beef without cooking it too well. When this happens, the worm attaches to the intestine of its human host, grows into an adult, and begins to lay eggs.

There is obviously a strong probability that not all these conditions will be met, and thus most eggs will not develop into adult tapeworms. So what is the tapeworm's answer to such a demanding life cycle? It has become capable of self-fertilization and it lays thousands and thousands of eggs. Its whole life is devoted to egg production. It exists only to lay eggs, eggs, EGGS! A few of these, hopefully, will wend their way, by chance, through the complex maze that is the tapeworm's life cycle. In countries with modern plumbing and sewage disposal, beef tapeworms aren't doing too well these days. Perhaps they will eventually be classed as an endangered species.

Chimpanzees

But what about other animals? Are they all in the numbers game? Let's consider the chimpanzee, an animal that has developed a different evolutionary strategy. When the female chimpanzee is sexually receptive, she may copulate with almost any male who signals his desire. Once she has become pregnant, she may not become sexually receptive again for years. Flo, the aging but sexy female studied by Jane Goodall, did not become sexually receptive again for 5 years after giving birth. During this time, she attended carefully to her baby.

In the first months, Flo carried her baby everywhere and diligently guarded it against danger. Later, the baby was permitted brief forays on its own, but it was not allowed to stray far from its mother's vigilant eye. During this time, the youngster would scurry back to Flo at any real or imagined sign of danger. As the young chimpanzee gradually became able to care for itself during those first few years, the association between mother and offspring relaxed, until finally Flo was free to mate again and rear another baby. So, among chimpanzees, the female does not maximize her reproductive success by giving birth to large numbers of offspring. Instead, few offspring are produced and these receive very careful attention until they become independent (Figure 35.6). Why has such a strategy evolved?

Chimpanzees have historically existed in a precarious and complex world. For centuries, before humans decimated the populations of African big cats in the name of sport and fashion, the chimpanzees were continually hunted by these vigorous and intelligent predators. And their lives were complicated by other factors as well. Chimpanzees feed on a number of kinds of food, including seasonal fruits, baby baboons and small monkeys, as well as termites, which they pick up by using tools made of broken twigs. So their diets are varied and complex. Also, they must move from place to place to exploit new food sources. Each night they must build a new nest of bent branches or twigs and leaves. Thus, they are constantly dealing with new and sometimes threatening situations.

35.6 Chimpanzees do not have thousands of offspring, leaving their survival to chance. A female has very few offspring in the normal course of her lifetime and, as they slowly mature, she very carefully tends to each one. The young are quite dependent for long periods as they continually acquire information about their environment. Chimpanzees may live for decades, so although their young mature slowly, the mothers may yet raise a fair number, one at a time.

To cope with such an environment, the chimpanzee needs not only high intelligence, but a system of extended parental care in order to give the young chimpanzee time to learn about its complex world. Extended parental care means that the parents can successfully rear only a limited number of young. So, in this species, selection has moved not toward maximum production of infants which are propelled into a hostile world to take their chances, but toward careful and efficient care of a few offspring. In other words, chimpanzees also rear as many young as possible, but in accordance with the demands of their particular kind of environment. So, we see that the number of offspring attempted by members of any species is likely to be set by natural selection and that the "strategy" of natural selection will depend upon a host of historical and environmental variables.

Birds

A quick look at some data from birds provides a clearer picture of how the dictates of a particular environment can influence the number of offspring attempted by a given species. For maximum reproductive success, birds do not lay as many eggs as possible, but the number of eggs that will enable them to rear as many young as possible. In most cases, this is limited to the number of young the parents can feed successfully.

Ornithologist David Lack found that females of the common swift (*Apus apus*) normally lay one to three eggs. Apparently, the number has not yet stabilized or, possibly, it is changing because of new ecological pressures. He found that in years when food was abundant, the number of young that reached feathering age was 1.9 for nests with two eggs, and 2.3 for nests with three eggs. In the nests in which Lack placed extra eggs to bring the total to four, only 1.4 young survived until feathering age. If the food supply were to remain constant over the years, one might expect the number of eggs laid by females of this population to stabilize. (Where?) But the environment is not constant, and Lack guessed that nests with two eggs might fledge the larger number of offspring in poorer years. Since the swift may not know in advance whether it will be a good year or a bad one, the most successful strategy is a mixed one—in effect "flipping a coin" to determine whether to lay two or three eggs. Or in terms of natural selection, the present swifts are necessarily descended from ancestors that laid two eggs in some years and three eggs in other years. The variability itself is adaptive.

In some cases, the increase in egg production because food is abundant may be accomplished by sequential multiple nestings. A pair of Jamaican woodpeckers reared three successive broods in one season when they were given extra food, though normally they might have been expected to rear one brood, or two at the most.

The number of offspring attempted by a species may also reflect the vulnerability of the young to predators. For example, certain species of gulls nest on accessible beaches where the young are in real danger of being found by prowling foxes and other marauders. These gulls often lay three eggs, although they can usually feed only two young successfully. The odds are that at least one of the young will be eaten, so apparently the three egg nest is an adaptation to high predation. On the other hand, the kittiwake, a species of gull that nests on steep cliffs, normally lays only two eggs. Since any marauding fox would be likely to break its neck trying to climb the cliffs, all the young are safe from predators (Figure 35.7).

There is less information on mammals, but wolves, for example, are reported to have no litters in years when food is scarce. In fact, during these periods they evidently do not mate. This may be because the hungry adults are more inclined to spend their time in foraging than in mating activities. It is possible that under extreme conditions some mammals might even be physically incapable of successful mating or of rearing young.

Humans

Human are subject to the same constraints. For many years it was observed that primitive people could space their children by deliberately extending the nursing period. Like Flo, the wild chimpanzee, the mothers would not become pregnant again until an older child was weaned, which might not be for 4–5

(a)

(b)

35.7 (a) The ground-nesting gulls, such as the mew gull, have been forced to assume a different reproductive strategy than (b) the cliff-nesting kittiwake. The nests of the ground-nesting species are vulnerable to marauding predators, so they compensate by laying more eggs than they will be likely to rear. The ground nesters, for example, lay three eggs, whereas they can probably rear only two young, but at least one of the eggs is likely to fall to a predator. The kittiwake, on the other hand, nests on cliffs safe from predators. Since all their young are likely to survive, they lay only two eggs.

years. If a young infant should die, however—a not uncommon occurrence among primitive peoples—the mother would usually become pregnant again almost immediately.

These reports by field anthropologists were not unchallenged, however, because of a bothersome discrepancy: the extended-nursing strategy clearly didn't work in more advanced societies (or what have been called "the overdeveloped countries"). Nursing mothers in Chicago and Frankfurt would usually become pregnant again right away, unless they took additional precautions. Apparently, extended nursing was not a very effective birth control measure, after all.

The discrepancy was eventually solved when it was determined that the effect of nursing on fertility was not entirely a hormonal effect, as had been assumed, but depended on the nutritional state of the mother. On a low-calorie diet—chronic hunger by our standards, but standard fare for primitive peoples—nursing mothers wouldn't become pregnant. Fertility seems to depend on some minimal level of available body fat, and on a low-calorie diet the additional burden of the production of milk lowers the mother's fat reserves below that threshold.

Since the reproductive strategy of any species is a response to its own evolutionary history and immediate environment, let's take a closer look at the factors that might have determined our own reproductive propensity. The well-nourished human female has the ability to produce one child each year for over 30 years. (One woman produced 69 children.) Does this mean that humans are prepared, either physiologically or psychologically, to rear thirty children? Or is our reproductive ability an evolutionary adjustment to a historically rigorous life in which most, or even all, of our offspring were not likely to survive? In other words, has the ability of the human female to reproduce prolifically evolved in response to a traditionally high death rate among children?

But what about now? As Table 35.2 and Figure 35.8 show, a child born in a modern overdeveloped country is very likely to live through its first critical years. In humans today, mortality rate decreases after the first year and is very low through the teens; then it increases gradually until about age 60, when it rises rather rapidly. This sort of curve is especially accurate for countries such as the United States, but doesn't necessarily apply to poorer countries.

The means by which the human species has increased the likelihood of survival are almost universally viewed as good and desirable. For example, we have specific medicines to combat various maladies which, in earlier days, would have proven fatal. We also have therapeutic and corrective devices to aid the sick. If someone in our midst is unable to provide for himself, or herself, that person will usually be cared for, however minimally. Our society often provides for those who, in harsher days, would have been selected against. A person doesn't have to be keen of wit and physically agile in order to cross a busy street. He or she simply waits for a light—a light that means it's safe to cross leisurely. The result of our social care has been a negation of many of the influences of natural selection. However, at the same time that we have reduced selection for swiftness, strength, and intelligence in our species, we have increased the level of

Table 35.2 Estimated Average Life Span in Human Populations

Population	Years
Neanderthal	29.4
Upper Paleolithic	32.4
Mesolithic	31.5
Neolithic Anatolia	38.2
Austrian Bronze Age	38
Classic Greece	35
Classic Rome	32
United States, 1900–1902	48
United States, 1950	70
United States, 1977	72

After E. S. Deevey, "The Probability of Death." Copyright © 1950 by Scientific American, Inc. All rights reserved.

35.8 Survivorship curves for five species are shown, each beginning with a population of 1000. The lines show the percent of life span reached as each population ages. In reality, the rotifer lives a maximum of 9 days compared to our 103 years, but as a population, rotifers do well. Their average life span is 6 days or 67% of their maximum, about the same as humans. The average black-tail deer fares poorly, completing only 6% of its total maximum life span. The average hydra completes 37% of its maximum life span, while the average gull completes only 24%. (Note that the vertical scale is logarithmic.)

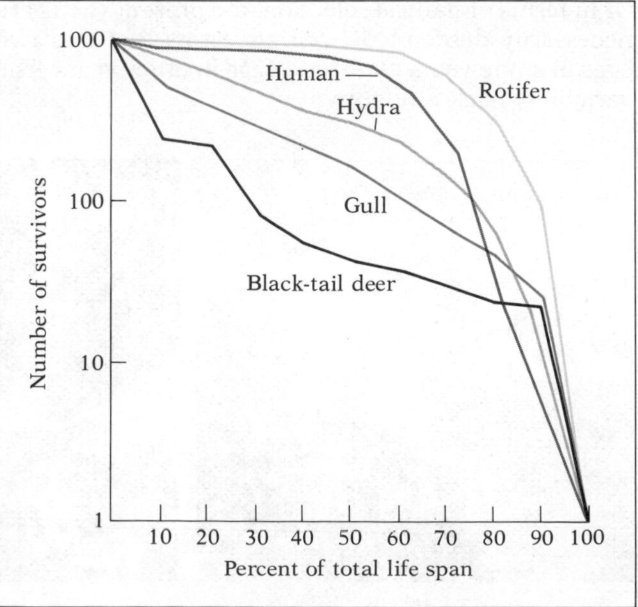

variation. Thus we find among us all sorts of interests, talents, tendencies, and appearances. The point is that, with its highly developed social programs, our society is attempting to ensure every individual that is born that he or she will live and reproduce. With our reproductive potential so high and so many of the natural curbs removed, we may be placing our species in an untenable position by our uncontrolled breeding. But the alternatives are harsh, indeed. What are the alternatives? Which ones would you personally be willing to endorse?

To sum up, reproductive rates are subject to natural selection, and the reproductive potential of humans has been established through the eons of our development. We have recently altered our environment so that the direction and strength of natural selection has been changed, but we are left with a reproductive capacity which better fits our earlier situation. If this is true, or even if it isn't, we might be led to ask ourselves a number of questions. Are large families with many children "natural"? Has the relaxation of natural selection resulted in our species increasing to the extent that we are physically or psychologically unable to deal effectively with the resulting swell of our own numbers? Do crowds make you uneasy? What might be the pressures on a social system in which parents are unable to rear each of their children carefully? Perhaps the most important questions involve the future. Where do we go from here? What can we expect? How much social change will be necessary to solve the problems resulting from an overbreeding population? How much social change can we tolerate? What are the odds we can pull it off? Those morose souls who spend time with such matters have imbued many of the rest of us with a deadening pessimism. But, looking on the brighter side, let's see how populations may be kept in check by death.

35.2 Interactions in Populations

Death

Have you ever wondered why death occurs at all? Why hasn't natural selection resulted in organisms that simply live forever? If "Life" is better than "Death," then why is there no marked selection for longevity? Why are there delicate insects that don't even have mouths with which to feed themselves, so that they must spend whatever energy their frail bodies possess in finding mates before their few precious hours of life are gone? Why do humans have so much trouble surviving past their "three score years and ten"? Why death?

People with all sorts of perspectives on life have ruminated on the meaning of death. Ideas have been fiddled with by philosophers and theologians, as well as scores of novelists, dramatists, poets, drunks, and others. Even aboriginal societies dealt with the meaning of death in one way or another—some very casually, not fearing it at all, but treating it as though it were simply the natural extension of life. Others have simply refused to accept death at all, maintaining a faith in an afterlife. So let's join the fray and consider death from a biological point of view.

Programmed Death

Many plants are *annuals*: their life cycles are intimately tied with the seasons, so that they all germinate in the early spring, flower in the late spring, grow and begin to set seed in the summer, and disperse seed and die in the fall. Actually, the final maturation of the seeds is often part of the death process: vital elements and energy are withdrawn from the leaves and moved through the phloem to the maturing seeds.

There are annuals among the animals, also. Certain Brazilian fish live in temporary ponds that exist only during the rainy season. Shortly before the ponds are due to dry up, the fish spawn and lay cystlike, drought-resistant eggs. The fish die when the ponds dwindle, but the eggs hatch the next year. Now, if these fish are netted and kept in an aquarium, they continue to follow their natural life cycle. After the females have laid their eggs and the males have spawned, the fish literally begin to fall apart. They cease eating, lose their scales, and soon are floating belly-up, in spite of the best care that can be given them. They have literally lost all purpose in living. After reproducing they, like the annual plants, are programmed for death.

Squid and octopuses have more or less the same one-shot life cycle. Some male octopuses can only mate once—their copulatory organ (a specialized arm) becomes detached in the process. All cephalopods, even the giant squids that do battle with sperm whales on the bottom of the sea, die shortly after their one massive reproductive effort. Death comes quietly and without a struggle. The animals don't merely starve, although they do stop eating. They just plain die.

Reproduction and death are under hormonal control in octopuses. If the cephalic gland of a mature octopus is removed, the animal won't engage in sexual activity or reproduce—and it also won't die;

that is, it won't die on schedule. The experimentally altered animal continues to feed and engage in normal activities. Sooner or later, however, death comes anyway. The altered animals begin to show signs of decrepitude that are never seen in normal octopuses. Tumors and infections, otherwise unknown in the species, eventually show up, and the long-lived but barren animals ultimately succumb.

Do humans have a similar kind of programmed death? There are some reasons for thinking that they do, to some extent at least, although the program is by no means as absolute as the octopus's. Usually between sixty and seventy the death rates suddenly begin a precipitous rise and continue to accelerate due to a host of apparently unrelated diseases. It seems that all kinds of systems wear out more or less at the same time. The time of this rapid aging may differ from individual to individual, which smoothes out the human death rate curves, but when you're old, you're old. It has been observed that people who have been saved from cancer very often soon die of heart disease or some other infirmity of age. In fact, although cancers claim about 25% of all humans, the combined human death rates for all causes increase so rapidly that if we were to eliminate cancer, say at noon today, we would only extend our average life expectancy by less than 3 years. So why do humans seem to wear out all at once?

For a suggestive answer we can turn to, of all people, Henry Ford. According to an apocryphal tale, Henry Ford once sent a team out to investigate the automobile junkyards of America, to find out how and why his Model T Fords wore out. In due time the investigators returned from what must have seemed to them like a fruitless mission indeed. Model T Fords, they found, wore out in hundreds of ways: the axles broke, the pistons seized, the radiators rusted, the planetary gears lost teeth, the bearings wore out, the springs cracked, and on and on. Henry pressed them: Was there nothing at all on their checksheet that did not show up as a major cause of automotive death? Well, yes, the investigators found that the Model T kingpins had stood up remarkably well. The kingpins of the junked cars generally had years of potential life remaining. "That's too bad," said Henry Ford, "we have been making the kingpins too strong." And he ordered designers to come up with cheaper, less robust kingpins for future production.

Just in case the biological implications of this fable aren't immediately obvious to you, we hasten to assure you that the great manufacturing genius was not being malevolent. Nor humanitarian, for that matter. He just wanted to build the best car for the cost, which is another way of saying the least cost for the car. Finite resources had to be allocated for a maximal return. We've made the same argument for the evolution of biological systems a number of times in this book. All organisms have finite resources of energy and time. The successful organisms in evolution allocate these resources in such a way as to maximize successful reproduction. Thus, as we have seen, a tapeworm puts its energies into gametes and a chimpanzee into child-rearing. One species of barnacle uses its finite resources to develop the metabolic devices necessary for survival high in the splash zone, and in less stressful environments it can't compete with another species that has put more of *its* energies into rapid physical growth.

What has this to do with longevity and death? Simply that investing in a physiology that provides longevity past the period of maximum reproduction is not a paying proposition. The energetic investment that would be necessary for prolonging life would get a better return if it were invested in faster development or more effective reproduction instead. More specifically, natural selection tends to maximize what population biologists call the "$l_x m_x$ curve" (el-ex-em-ex curve) of the species—the cumulative product of the probability of living to a certain age x, times the rate of reproduction at that age. But every increase in one component entails the expenditure of some time and energy, and hence some decrease in the other component. No species can reproduce at a high rate and at the same time survive infancy and live forever.

Different organisms, of course, have different strategies in the allocation of their metabolic resources. Mice reproduce early and prolifically, and die young. Because of high predation, most mice would die young in any case; but even in the protective custody of the laboratory, a 2 year old mouse is old and mice rarely live beyond three years. Prominent among the diseases of mousy old age is cancer; most 2 year old rats and mice become cancerous. However, a gene for increased resistance to cancer would not spread in a mouse population for two reasons: first, not enough mice would escape predation long enough for the gene to make very much of a positive difference; and second, a good cancer-suppressing mechanism would have metabolic costs that would reduce the energy available for rapid development and explosive reproduction.

Humans are among the longest-lived of all animal species. Our normal life span exceeds that of all other mammals and most vertebrates and invertebrates—possibly only giant tortoises and certain sea anenomes regularly outlive us. Correspondingly, we have very effective cancer-suppressing and disease-fighting mechanisms. As children, we may have seen our puppies and kittens grow up and die of old age before we ourselves reached maturity. Our bodies mature very slowly, but they last a long time. We have quite efficient defenses against the wear and tear of time, against infection and our own cancerous cells. We pay a price in physiological energy, in developmental time, and in our rather low rate of reproduction. However, after perhaps 15–20 years of maturation, another 25–30 years of peak reproduction, and another 20 years to see our youngest children reach their own maturity and independence, our bodies have had it. Some of us will live to be 100, but not many. There are no documented cases of any human being living beyond 115 years of age (well-publicized tales from the Soviet Caucasus to the contrary notwithstanding).

Like the well-engineered Model T, once the machine has served its allotted time, it wears out—all at once. Just as there is no need for immortal kingpins, there's no strong natural selection in favor of metabolic systems that protect against the depredations of postreproductive degeneration. From the relentless viewpoint of natural selection, there's just not enough of a payoff involved.

Death: The Long View

To help make this argument clearer, we should note two things: first, the world changes and, second, it is limited. Not only have continents moved, islands formed, ice ages come and gone, and weather patterns shifted, but humans have, in a very final sense, drastically altered the earth's surface in the past few thousand years. In a changing world, the organisms that live there must also change. This is what adaptation is all about—adjusting to new situations.

As a population is confronted by changing conditions, certain individuals can be expected to cope with them better than others. But the changes may well be too drastic for any individual, with its inherent limits, to be able to adjust. However, as new generations appear, their myriad genetic variations may provide material on which the new environment can have its molding effect through natural selection. Many in the new generations will fall by the wayside, their kinds of genes being separated out of the population. Individuals with other kinds and combinations of genes, though, may fit the new environment well, and these organisms will live and reproduce. The arrival of new and variable individuals is important in a changing world.

Now, in any system in which the commodities are limited, such as the planet earth, as the numbers of individuals increase, so will the competition for the available commodities, such as food, shelter, land, and natural resources. Consider, then, an organism that finishes its period of reproduction and then hangs around to compete with its offspring. Such an individual could be expected ultimately to leave fewer offspring than a parent which, after having reproduced, died, so as not to interfere with the success of its offspring.

David Blest has described certain moths with camoflaged markings in which the female, after laying her eggs, commits suicide. Suicide? Why? One theory is that if she were to continue to live and stay in the area where her offspring are hatching, she might eventually be caught by a bird; that bird would then have a better idea of what the moths around there look like and would likely search for more of them—including the female's offspring. So when the female has laid her eggs, she flies around frantically, using up her energy reserves and fraying her wings beyond recognition. Then she either dies of her own efforts, or, because of her conspicuous behavior, she is captured by a bird, which is then unable to find another tattered morsel like that one.

This explanation of the moth's behavior is entirely speculative, of course, but the consideration of programmed death as a means of continuing one's own genes provides food for thought. Keep in mind that the moth undoubtedly is not aware of the effects of her behavior. She doesn't try to save her young through self-sacrifice. Moths that behave in such a way, however, leave more offspring, each of which, perhaps, carries a "suicide gene."

We've tossed around some concepts fairly loosely here, as you are aware. However, these are legitimate questions for evolutionary biologists. Perhaps we are not in a position to uncover whole truths, but we can begin to accumulate certain kinds of evidence. Anyhow, this sort of approach makes as much sense as dealing with the concept in subjective, artistic, abstract, or mystical terms.

Controlling Populations Through Mortality

Just as the numbers that are added to any species are partly a function of environmental factors, so are the numbers that are allowed to remain in the population. What are some of the factors that tend to control populations by acting on the members themselves? What brings about death? Of course, many things can cause death, from being struck by lightning to falling off a cliff. Dogs occasionally catch squirrels, and all kinds of things tend to get run over. However, the population biologist can't afford to get side-tracked by such phenomena. The biologist wants to know how whole populations are affected by death. Dogs almost never catch squirrels when you get right down to it, and so they have very little effect on squirrel populations in spite of their constant threat. Actually, populations are changed most drastically by only one or two factors that influence death rate. Long-term studies of insects have shown that they are particularly vulnerable at certain times of the year, and during their vulnerable phase, their numbers are drastically lowered by only a few factors, such as cold weather or parasites.

We must also keep in mind that population numbers remain stable only when the birth rate and death rate are balanced. An imbalance in either direction will increase or decrease population numbers.

Abiotic Control

Just as seasonal weather changes can alter population numbers, so can irregular or unusual weather. Drought may kill many kinds of plants and animals. Many birds perish in some years because they begin their northward migration in the spring only to be caught by a late cold spell. Such physical factors that depress population numbers are called *abiotic controls.*

Density-Independent Mortality

It should be kept in mind that population-depressing influences such as severe weather are *density-independent* (population density is the number of individuals per unit of area). In a severe drought, the parching sun doesn't care how many corn plants are struggling in the field below. It kills them all. In an area saturated with DDT, most of the insects die—whether there are few or many. Their numbers mean nothing. Thus, mortality brought about by such means is independent of the density of the population.

Biotic Control

Now let's look at a much more complex phenomenon regarding how living organisms may influence each other's numbers. *Biotic population control* simply refers to living influences on populations. For example, the organism that causes bubonic plague can reduce populations. So can a tiger. We know how these work, but biotic influences can operate in more subtle ways as well. If a territorial bird drives a competitor into an area where there is less food, when winter comes, the underfed competitor may be more likely to succumb to the rigors of the season. Thus, the territory holder has indirectly brought about the reduction of the population.

Biotic controls on populations are likely to be *density-dependent,* which means that as population density rises, there will be increasing pressures on it acting to reduce that density. As the density falls, then, the pressures on it will lessen, thus permitting the numbers to increase again.

Density-Dependent Mortality

Generally speaking, abiotic factors do not really control population numbers in the usual sense of the word. They *affect* population numbers, but if one defines control to include some measure of regulation, then density-independent factors have no control at all. In other words, abiotic factors by themselves do not maintain population sizes within any kind of limits. For that, one needs density-dependent effects. Density-dependent effects do regulate, or limit, population sizes within a more or less defined range, because they generally involve some measure of negative feedback control, the familiar stabilizing effect seen so frequently in physiological systems.

In their interactions with biological populations, abiotic factors can have density-dependent effects and thus maintain population sizes within finite limits. One ecologist investigated population regulation as it occurs among the sowbugs in fields of grass near Berkeley, California. Sowbug populations increase greatly during the dry summer months, but when the rainy season comes, most of them drown. The only ones to survive were those that were able to find refuge in certain kinds of cracks and under rocks of a certain shape.

Those refuges, during rains, were jam packed with sowbugs. The ecologist hypothesized that the population size during the rainy season was determined solely by the volume of available places of refuge—the number of sowbugs to survive would equal exactly the number that could pack themselves into these refuges. The hypothesis was confirmed in laboratory experiments. So here we have an abiotic element—rain—with a density-dependent effect: The greater the density of sowbugs, the greater the proportion that will drown, once the population size exceeds the number of places of refuge.

That may not seem like a negative feedback system, but it is. It has the necessary feature: The more the population increases, the greater the death rate in the rainy season, and the more the population size is reduced.

Most density-dependent mortality is more directly under the influence of biotic factors—direct influences of the organisms on each other, or indirect influences on one another through their interactions with other organisms. The shape of the idealized population growth curve indicates that, as the population density increases, the growth rate (the absolute increase in numbers divided by the population size at the moment) declines, and eventually reaches zero. This is no abstraction; it really happens. It occurs because density-dependent effects increase mortality, or reduce fecundity, or both. Let's consider some important biological effects on population size, paying particular attention to when these effects are density-dependent and when they are not.

We'll consider seven categories of biotic population-controlling mechanisms, each of which can have density-dependent aspects:

1. Competition within species for finite resources
2. Self-poisoning through toxic products
3. Disease and parasitism

4. Competition between species for finite resources
5. Predation of one species on another
6. Mutualistic interactions
7. Emigration

The first two categories and the last involve only a single species and its environment. The other categories involve at least two species. In the case of disease, the second species might be a virus or a bacterium.

Competition Within a Species for a Finite Resource

The drowning of sowbugs in the rain seems to be an example of abiotic control but, as we have seen, it has a biological element as well. The sowbugs have to compete with one another for a limited resource—in this case, shelter. Since there is more than enough other commodities, such as dead grass for them to eat, shelter space is the *limiting resource* for this population.

Sometimes, however, populations are limited by the amount of food available. Certain poetic biologists have found that only so many maggots can survive in a cow dropping and the density of a population of flies may be limited to the density and size of cow piles in the area. When egg-laying adults find a cow pile, they have a remarkable ability to estimate whether there are already so many maggots or eggs present that their own offspring would be crowded out, and whether they should fly off and run the risk of not finding a fresher pile. Usually, though, limitations of food exert their influences much more bluntly. As competition for food intensifies, those who can't find enough weaken and die.

Territoriality and Dominance Hierarchies in Intraspecific Competition

In territorial species, some animals are excluded from certain areas held by other animals. If the habitat is variable, the result is that some animals will end up with "better" areas than others. If those areas hold more food, then in time of food shortage the most successful territory holders are more likely to survive. In some species of birds, the male must have a good territory in order to attract females. A splendid male holding an inferior territory will not find a mate. If there aren't enough good territories to go around, then not all individuals will be able to reproduce, and thus the population number is lowered (Figure 35.9).

In species that form hierarchies, or peck orders, certain animals have freer access to commodities than others. In such a system, each animal must be able to recognize others individually and to respond to them according to their rank. Hierarchies are established in various ways for different species, such as by fighting or in play. Hierarchies reduce conflict within groups by any animal acquiescing when confronted by a dominant individual. In times of shortage of critical commodities, then, the higher-ranking animals are more likely to survive as subordinates give way before them.

35.9 Among some birds, such as red-winged blackbirds, males may take territories of varying qualities. Females can assess territories and tend to choose the males with the best real estate. In some cases, a female will choose a male who already has a mate over a bachelor with an inferior territory.

In both territorial and hierarchical systems (which are not, by the way, mutually exclusive), when certain animals have freer access to commodities than other animals, harsh conditions may quickly bring about the deaths of the have-nots, thus increasing the likelihood of survival for the haves. This doesn't mean that the have-nots accept their role for the "good of the species." Quite the contrary. It is to their advantage to remain in the breeding population at all costs. However, a low-ranking individual may have better luck if it searches for unclaimed commodities than if it tries to challenge a dominant individual.

Self-Poisoning Through Toxic Wastes

In the growth curve of bacteria in a pure culture or in a rotting squirrel corpse, it might seem that food is again the limiting factor. Long after the growth curve has begun to fall off, however—even after growth has fallen to zero and the population has reached its plateau—a chemical analysis might show that there were ample foodstuffs available. The increased mortality and decreased rates of cell division, it turns out, have more to do with the accumulation of toxic waste products than lack of food. This is especially true in the pure culture, because in a rotting squirrel, where many species of decomposers are at work, one microbe's poison may be another microbe's food.

Some of the poisons, however, kill competitors of other species. For instance, bacteria may produce poisons that kill fungi which may be competitors for the food and fungi can produce substances, such as penicillin, that kill bacteria. Even so, these same poisons will also eventually affect the species that produce them. Microbes also produce poisonous metabolic breakdown products, such as ammonia, acids (such as carbon dioxide and lactate), and alcohol. These may poison the organisms that produce them. For example, yeast stop growing when their own waste product, ethanol, reaches 12% or so; which is the limit of the alcoholic content of unfortified wine. If carbon dioxide is not allowed to escape, the yeasts are poisoned even sooner. This is why homemade rootbeer gets very fizzy, but not very alcoholic. (Daniel Jansen, one of our more imaginative ecologists, has suggested that microorganisms make dead matter putrid and unpalatable so as to render themselves safe from most vertebrate scavengers.)

Not only microbes are subject to this kind of self-poisoning population control. Many plants secrete toxic substances into the soil, which reduces competition from other plant species but which also inhibits their own growth and seedling success.

Humans have, in recent centuries and especially in recent years, managed to escape most other biological checks on our population growth. Now we, like the yeast and bacteria, are faced with the possibility of polluting ourselves into extinction. Never mind that the poisons we release into the environment are made by mines, factories, and automobiles, and not directly by our simple primate bodies; the principle is the same. (Figure 35.10).

The awesome impact of the human species on the life-supporting world environment may soon bring the reality of human self-poisoning population control into sharp focus. For example, we are building, along with other countries such as Japan, huge tankers designed to carry incredible amounts of petroleum products (thirty times more than the traditional tankers). It is disquieting to learn that, in one recent 10-year period, there were about 5000 collisions of the smaller, more maneuverable, tankers. We have already witnessed the sinking of some of the most sophisticated supertankers afloat, such as the *Arco Cadiz.* The result was catastrophic, and other disasters have narrowly been averted by fortuitous winds. We might expect to see wholesale destruction of oxygen- and food-producing plankton over great areas of the ocean. Tankers have already ruptured in the Atlantic and released thousands of gallons of heavy fuel oil into one of the richest fishing areas of the world (Figure 35.11).

35.10 The discovery of vast dumps of discarded, highly toxic, chemical wastes in New Jersey and other places around the United States has struck terror in the people of nearby communities. Many of the companies responsible for discarding these materials in a lawful and safe manner have simply vanished under the protective cloak of bankruptcy.

There have also been recent reports indicating that in certain sections of the United States, industrial and automotive by-products emptied into the air are altering weather patterns. Recently, we have encountered the strange phenomenon of acid rains. Because of high levels of sulfur and nitrous oxides in the air. the rain contains startling concentrations of sulfuric and nitric acids. Sweden is particularly deluged with the reactive rain, although the Swedes are world lead-

35.11 This scene is all too often repeated in the oceans of the world. What is euphemistically called an "oil spill" can very well become an oil disaster for marine life. This is particularly true when refined or semirefined products are being transported. As the tankers get bigger, so do the accidents, yet we continue to build larger vessels.

ers in environmental affairs and have relatively little polluting industry. However, far away in industrial England and the Ruhr valley, tall smokestacks boil their poisonous fumes high into the atmosphere, to be carried safely away by the winds—safely, that is, to Sweden. We don't know what the results of continual acid rains will be, but we have already seen or can expect increased leaching of nutrients from the soil, lakes and streams turning acid, the corrosion of buildings and finally, metabolic changes in the organisms that drink the water.

We know that the quality of air throughout the entire United States has deteriorated sharply in recent years (although some people take solace in the fact that there has been a reduction of certain pollutants in some areas). We tend to sit in stoney complacency as televised reports warn our old people not to go outside and our children to refrain from running at recess, because the air is dangerous to breathe.

We humans have also figured out an entirely new way of poisoning our own environment: radioactivity. In the 1950s and 1960s, hundreds of nuclear bombs were exploded in the earth's atmosphere. (A presidential candidate, Adlai Stevenson, was derided and had his manhood questioned when he proposed a ban on such testing.) The atmosphere became seriously laden with radioactive fission products all over the world—enough to cause thousands of deaths by cancer, but not enough to have any measurable effects on population growth rates. The air is relatively clear of radiation now, in spite of continuing testing by the Chinese, but a nuclear war or power plant mishap (such as the one that almost occurred at Three-Mile Island in Pennsylvania in 1979) would change that in a hurry.

If we continue to befoul and alter our environment with such recklessness, we may, indeed, come to understand more clearly just what self-poisoning population control really means!

Disease and Parasitism

Infectious disease and parasitism have usually been considered as separate topics, but the distinction is rather arbitrary. Generally, we just say that infectious disease agents are viruses and bacteria, and parasites are larger organisms. Malaria is a disease caused by a parasite—in this case, a parasitic protozoan. And smallpox was a disease caused by a virus. So ecologically there are few really important distinctions between the effects of parasitism and those of other kinds of infections.

One general kind of distinction that can be made between parasites and other infectious disease agents, other than simple size, is that true parasitism is a long-term or permanent relationship between an individual parasite and its host; whereas most infectious diseases run a temporary course that ends when the host's defenses eliminate the infecting agent, or when the infecting agent eliminates the host. By this criterion, a ringworm fungus is not a parasite (you get over it), whereas *Herpes simplex,* the virus that causes recurrent cold sores, is a true parasite. (You've got it for life—sorry!)

Both parasitism and disease can certainly have substantial effects on populations. We will see in this chapter how the bubonic plague drastically reduced the human population. Less dramatic infectious diseases claim thousands of lives daily, and in the past the relative toll of infectious diseases in human populations was far greater. Infectious disease was probably the chief factor in human population control until about 1800. The great population explosion of the 19th and 20th centuries has not been due to a greater abundance of food but to the reduction of death from infectious diseases and parasites.

The density-dependence of disease as a population regulator is obvious. The more closely that susceptible individuals are packed together, the more opportunity there is for disease transmission. Also, the more individuals there are in a population, the greater will be the number of potential reservoirs in which more virulent mutant strains of the disease microorganism can develop.

In some cases, disease and predation may interact to control populations. For example, a 2 week old caribou fawn can already out-sprint a full-grown timber wolf, and healthy caribou are seldom prey to wolves. However, caribou are subject to a hoof disease that lames them before it affects other parts of the body, and it is these lamed animals that a wolf is likely to cull out of a herd. In areas where wolves were poisoned in order to protect the migrating caribou, this hoof disease spread unchecked, and in a few seasons decimated entire caribou herds.

The ecological relationship between a disease parasite and its host involves, of course, two populations. In most cases, the parasite is *host-specific;* it can infect only a single species, and thus its welfare is intimately tied to the welfare of its host. If the host species goes into a population decline, so does the parasite.

It has been said that it is "to the parasites' advantage" not to kill off its host species, at least if it is a host-specific parasite. Now, parasites are mindless things and don't necessarily know what is to their advantage and what is not, but we can note that we aren't likely to see many host-specific parasitic organisms that have driven their host species to extinction. A parasite that can permanently infect a long-lived, relatively healthy host has a selective advantage over one that destroys its host. Still, there are parasites that regularly kill their hosts, and others that weaken them so that they are nearly certain to die, as we've noted with the caribou hoof disease agent. Many parasites render their hosts permanently sterile, which of course will have an effect on population dynamics too.

There is a wide range of possible relationships between hosts and symbionts, with different effects on the population dynamics of both species. At one extreme is a mutualistic relationship. Here, one organism actually increases the survival and welfare of another. Next, are benign relationships, where the commensal symbiont depends on a host but doesn't measurably harm it. You won't be pleased to learn that virtually all humans harbor many commensals, especially on the skin. In addition to a considerable bacterial flora, we are *all* infected by a tiny arachnid mite that lives harmlessly in our sweat glands. It's there. Right now. As you read this.

Other symbionts, the parasites, may not be so benign. The bacteria that live on your teeth don't kill you, but they can make unpleasant holes in your teeth that may ultimately harm you. Intestinal worms usually don't kill their hosts either, but we've all seen some pretty sick puppies and sick puppies often die. Malaria can greatly reduce the vigor and especially the reproductive performance of infected individuals. Other parasites are more directly harmful. Some infectious diseases, like dog and cat distemper and human smallpox, regularly kill most of the individuals that are infected. Smallpox, the disease caused by a virus that survives now only in certain research laboratories, was once a quite effective population depressor. It nearly depressed American Indians right out of existence. In general, the less a parasite or infection harms the host, the better adapted to each other the parasite (and host) are said to be. In those cases where the life of the host is seriously impaired, the relationship is often a relatively new one in which the two species have not yet adjusted to each other very well. For instance, the American Indians apparently had not experienced the same natural selection that gave Europeans and their descendants a relatively high level of resistance to smallpox.

The infection of a new host species or strain by a parasite that had previously adapted to another host may sometimes have disastrous consequences. For example, in 1904, a fungal parasite of the Chinese chestnut was accidentally introduced into North America. In China, the parasite was held in check by a variety of control mechanisms that do not operate in North America. Sadly, the majestic American chestnut tree (Figure 35.12) proved defenseless to the fungus, and by the late 1940s, the great tree that had dominated our Appalachian forests was virtually extinct.

A special category of parasitic infection—very common among arthropod hosts but rare among mammals—is *parasitic castration.* Here, the parasite specifically infects the gonads, essentially taking over the reproductive energy of the hapless host. It's a remarkable strategy for the parasite; after all, a normal, uninfected animal sets aside as much energy as possible for the production of gametes. It is a rich source to be tapped. The parasitically castrated host survives to an old age, its health unimpaired but its reproductive fitness zero, and puts all its excess resources into the manufacture of more parasites.

35.12 The magnificent American chestnut tree once dominated our Appalachian forests, but now it is nearly extinct, having fallen to a newly introduced sac fungus from China. Such drastic effects of a parasite are usually associated with the parasite and host being new to each other and not having shared a long evolutionary history.

A final special category—also not one that applies to human populations—is the *host–parasitoid* relationship. A *parasitoid infection* is like a parasite, except that it always kills its host. The most common and best known parasitoids are the tiny wasps that parasitise the larvae or pupae of other insects. In many respects, a parasitoid–host relationship is more like a predator–prey relationship, with the important distinction that it is host-specific. The parasitoid wasp locates its victim—perhaps a caterpillar—and uses its long ovipositor to inject an egg or several eggs into the desperately squirming host. As the host develops, so does the parasitoid larvae within its tissues. The sickened host eventually dies at about the time that adult parasitoid wasps eat their way out of the body they have exploited, and fly off. Parasitoid wasps are very effective in population control, and have been used commercially in the biological control of insect pests of agricultural crops. Hordes of laboratory-raised wasps may be released in the fields or greenhouses. However, success in eliminating the host also eliminates the parasitic wasp, after which the inevitable new introduction of the pest species can result in an unchecked pest population explosion. In order to prevent this from happening, agriculturalists who use biological control find that they have to keep introducing the very pest that eats their crops, or provide some kind of sanctuary for them, just to prevent population fluctuations and local extinctions of the parasitoid that controls them.

There are other interspecific relationships that are the ecological equivalent of infection and parasitism, even if they go by different names. For instance, many plants are exploited by host-specific insect pests. The boll weevil is usually not thought of as being a parasite, but ecologically there is no difference between a cotton plant and a boll weevil on the one hand, and a human and a tapeworm on the other. The density-dependent population regulation is of the same kind.

Daniel Jansen found that, in tropical forests, almost all trees harbored one (and only one) host-specific insect *seed predator* (or, if you prefer, *fruit parasite*). Each tree had elaborate defenses against insects, defenses that were successful in keeping out all potential seed predators—except one. That insect had, of course, evolved defenses against the host plant's defenses, and in most cases was able to infest and destroy most of the host's seed output. Jansen found that the effect on population regulation is dramatic: The density of any one tree species remains extremely low, with the result that a single uniform area of forest might harbor hundreds of different tree species. Seedlings were never found to grow successfully under or near their parent tree; host-specific insect leaf predators (or host-specific fungal parasites) would get to them first. Only if an intact seed were to germinate far from other members of its species—and hence, far from any host-specific pests—did it have a reasonable chance of growing to maturity. This is one reason there are so many kinds of trees in the tropics and members of any one species are scattered so far apart.

What would happen if a tree were suddenly to evolve a new and entirely successful defense against its only insect parasite? Such sudden events are uncommon in evolution, because, as Darwin so often told us, evolution moves in imperceptible small steps. Thus, the parasite is usually able to keep up with its host species by evolutionary coadaptation. But if a tree species were to break free of this control, that species would presumably burgeon until it had achieved a high population density. The success would be short-lived, however, because if any insect species were to be able to make any inroads whatsoever, its effort would be greatly rewarded, and sooner or later a new, heavy infestation by some newly adapted insect pest would quickly bring the tree population density back to a low level.

Competition Between Members of Different Species

Out and out aggression, based on competition, between members of different species is relatively rare. Competition between species usually simply results in the quiet repression of one of the populations. Competition, by definition, usually results in harm coming to the loser of the race for any commodity that might be in short supply. This harm, in fact, may be so severe that the losing population completely dies out. The classic experiment demonstrating this principle was conducted by the Russian biologist, A. F. Gause. According to what is now called *Gause's law*, or *the principle of competitive exclusion*, two species cannot indefinitely live together and interact with the environment in the same way. The results of the key experiment that led him to this conclusion are shown in Figure 35.13. Of course, as we noted in Chapter 1, Darwin had long since come to the same conclusion; but he didn't have the experimental evidence.

Another experiment was done using two species of duckweed, *Lemna gibba* and *Lemna polyrhiza*. Grown alone in a tank, each did well, but when they

35.13 The graph is based on an experiment by A. F. Gause in which two species of *Paramecium*, *P. aurelia* and *P. caudatum*, were grown in pure culture conditions, feeding on a species of bacteria. Their food niches overlap; thus, the most efficient feeder, *P. aurelia*, won out. It may be that the food sources of these two species does not entirely overlap under natural conditions, so total extinction due to competition alone would be very rare.

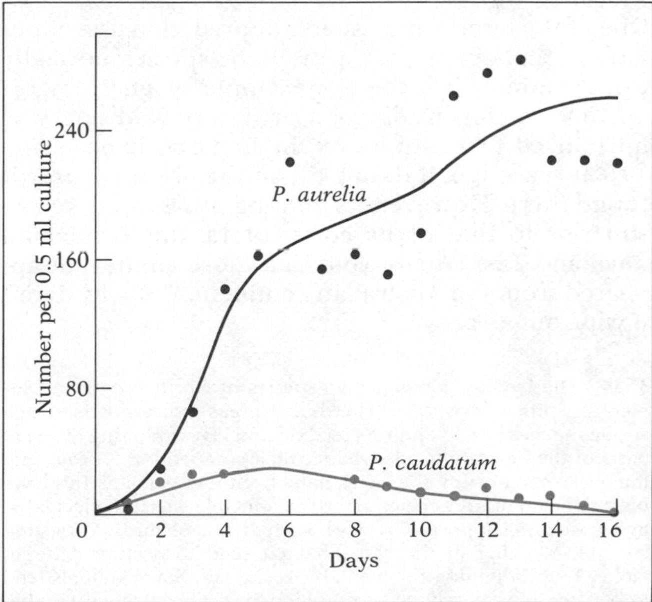

were grown in the same tank, *L. polyrhiza* died out. The reason is, *L. gibba* is a better competitor for light. It has air-filled sacs that cause it to float higher in the water, blocking light from its competitor.

Keep in mind that this and many other such experiments were done in the laboratory. It has been argued that competition in the wild rarely results in extermination, but rather in subdivision of the habitat, each species coming to live where it does best, as illustrated in a classical observation by the noted ecologist R. H. MacArthur (Figure 35.14).

One of the clearest examples of the effects of competition in the wild comes from J. H. Connell's studies of two species of barnacles in Scotland. Just before they change from free-swimming larvae into adults, they attach to a rock in the tidal zone and secrete a shell. One species, *Cthamalus stellatus,* lives in the higher part of the tidal zone, while another species, *Balanus balanoides,* lives further down. One can find very young members of *C. stellatus* in the lower depths, but they don't seem to survive there. No adults are ever found in that zone. It turns out that the faster-growing *B. balanoides* simply ousts *C. stellatus* by crowding it out or growing over it. *Cthamalus stellatus,* in fact, can grow quite nicely at that depth if *B. balanoides* is not there to harass it. The only place it can grow in the wild, though, is in the rough, harsh upper areas where pounding waves and periodic drying make it rough going. It is safe in these regions because *B. balanoides* simply can't cope with the conditions there. The two species obviously are quite different in their physiological requirements.

Predation

One of the interesting aspects of predation as a population regulator is that a predator species normally can't eliminate its prey. For example, when the vigorous and intelligent dingo, a predatory wild dog, was introduced to Australia by the first aborigines some 10,000 years ago, it didn't kill off the primitive prey it found there. However, its hunting prowess proved so superior to that of its competitors, the Tasmanian devil and Tasmanian wolf, that those animals disappeared from the Australian continent. But why didn't it wipe out its prey?

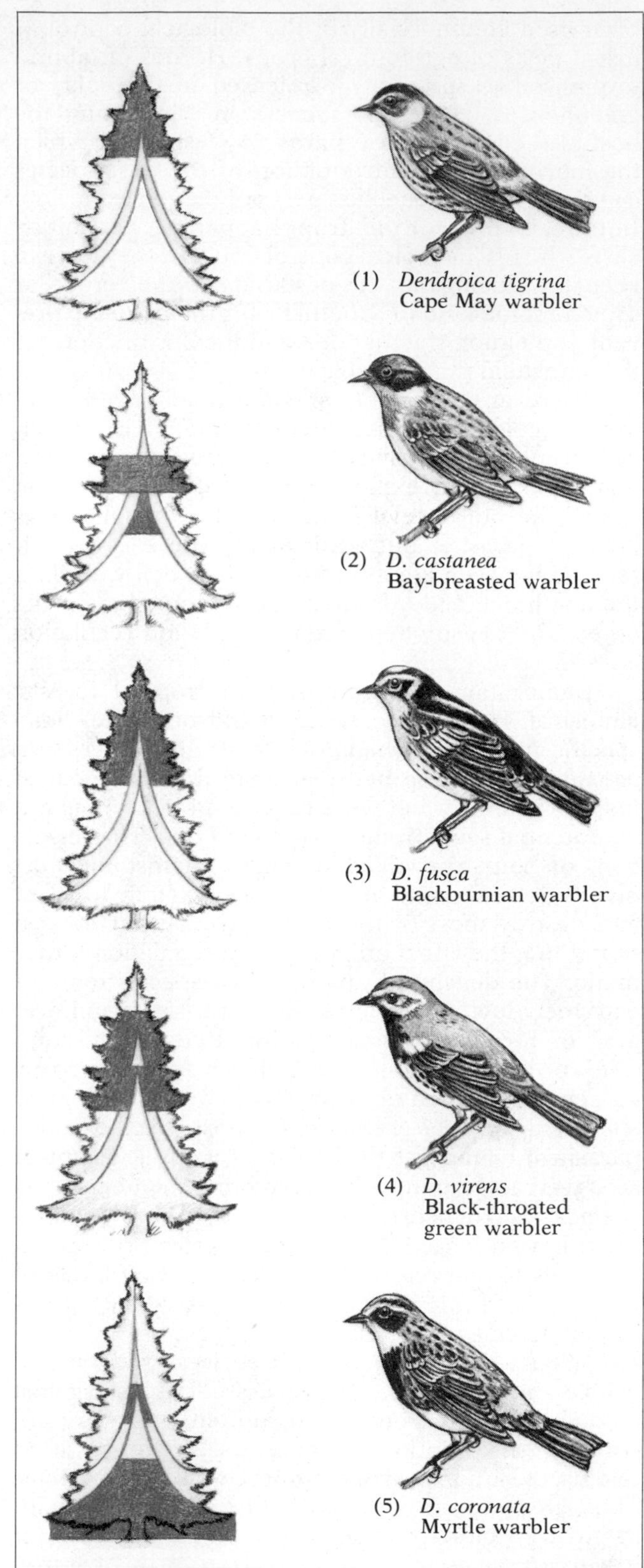

35.14 The feeding zones of five species of North American warblers in spruce trees vary. The darker areas indicate where each species spends at least half its feeding time. By exploiting different parts of the tree, the species reduce their competition for food and thus they can occupy the same habitat. Studies such as this have been done on many species since the principle was first elucidated by Gause and supported by work such as MacArthur's. Very similar species living in the same habitat tend to occupy different niches by subdividing the available resources. Since animals tend to behave so as to reduce competition, it has been suggested that competition must be an important factor in natural selection.

Essay 35.1 The Greenpeace Effort

In 1975, at the very time the fifteen nations of the International Whaling Commission were winding up their annual week-long discussions, the Russian whaling fleet was busy in the Pacific running down and harpooning some of the last of the earth's great whales. But this time the hunters were also the hunted. The Canadian Greenpeace Foundation vessel, *Phyllis Cormack,* after 58 days of fruitless searching and woefully short of supplies, had found the Russian fleet. They intended to physically put themselves between the whales and the harpoon and hoped that the Russians wouldn't risk an international incident by shooting anyway. The goal was twofold. First, they hoped to save a few whales. Second, they hoped to save a lot of whales—by focusing attention on the plight of the whales while the Whaling Commission was in session.

Their boat was older and slower than those of the Russian fleet, but when the Russians were spotted, they had already killed, floated, and flagged some whales, which meant that they couldn't leave the dead beasts. There were buyers at home and profit is, as always, the name of the game. The crew of the Greenpeace expedition scrambled for their small outboard Zodiacs for a closer look at the first whale they found. One of the men climbed aboard the floating carcass and cried, "My God! It's only a baby." And so it was. The infant whale was only 20 ft long. The crewman said the baby's skin "was warm and oily, the blood that flowed from a gaping wound hot on my hand. I . . . closed the eyelid. I felt lonely upon the ocean . . . ." But not for long. One of the Russian chase boats had turned and was bearing down on what they assumed to be poachers. A plea in Russian was to no avail—met only with vulgar retorts. As the Greenpeace crew followed, the Russians turned toward a pod of whales, gathered together as they sought now to escape their tormentors. The Zodiacs gave chase, caught the whales first, and rode practically on their backs to foil the harpooners. But the Russian harpooners fired anyway, a 250 lb harpoon with an exploding warhead sinking deep into an exhausted female that was too tired to dive again. The explosion shredded muscle, bone, and guts, but it didn't kill. She dived, but came up at once, thrashing in pain and screaming. The bull turned and rushed past the Zodiac like a great locomotive straight at the man who had fired the shot. The Russians were quite safe and were calmly reloading. As the great animal lunged near the bow, they fired again. The gallant bull made one dive under the hull and came up dead. The Canadians sat in their Zodiac and wept.

The Russians had made a profit. The IWC had heard of the effort. And millions of us were going about our business not really aware of how much the earth was changing.

Let's examine the question by reviewing the classical theory that accounts for the density-dependent effect of predators. According to the theory, under undisturbed conditions, prey numbers rise steadily, thus providing more and more food for predators. Then the predator numbers begin to rise. Their numbers do not rise immediately, however, since it takes time for the energy from food to be converted into successful reproductive efforts. When the predator numbers finally do rise, though, the large numbers of predators places increasing pressure on the prey. Then, as the prey begin to be killed off, the predators find themselves with less food, and so their own numbers fall off due to starvation or simply to less successful reproduction. Again there will be a lag time. Because the fluctuations of predator numbers lag behind those of the prey, the prey may be well on the road to recovery before the predator population begins to rise again (Figure 35.15). If the predator is able to hunt

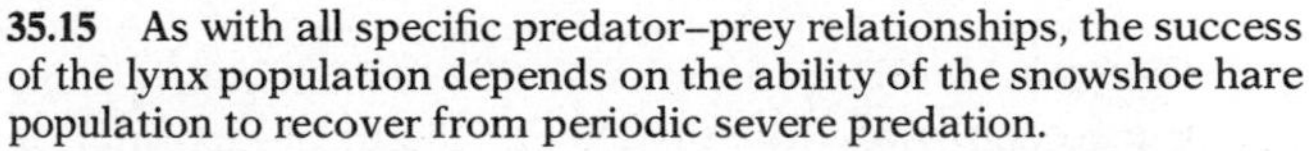
35.15 As with all specific predator–prey relationships, the success of the lynx population depends on the ability of the snowshoe hare population to recover from periodic severe predation.

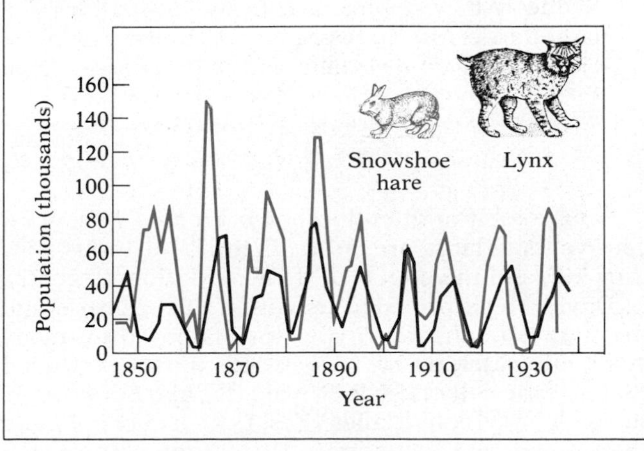

more than one prey type, the story is complicated by such factors as *prey switching* as the predators seek other, more available, food to maintain their numbers.

Our own species is providing us with clear examples of how density-dependent regulation can fail. The great whales have been hunted to the brink of oblivion over the past few decades as modern whaling methods have reduced personal risk while increasing profits. Although there is nothing that whales provide that can't be obtained elsewhere, the demand for whale products (and their price) hasn't diminished, especially in Japan. Thus, instead of the human predators relaxing their pressure and allowing the whale population to recover, whaling fleets continue to exert their depressing effect on populations of the great mammals (Figure 35.16). Then, as whales decrease in number, the price of whale products goes up, and the hunt becomes still more avid. If humans actually starved when they couldn't catch whales (which might once have been the case among the Eskimos) both populations might eventually stabilize (or cycle). But the current decline in whale numbers has had no effect on the growth of the human population.

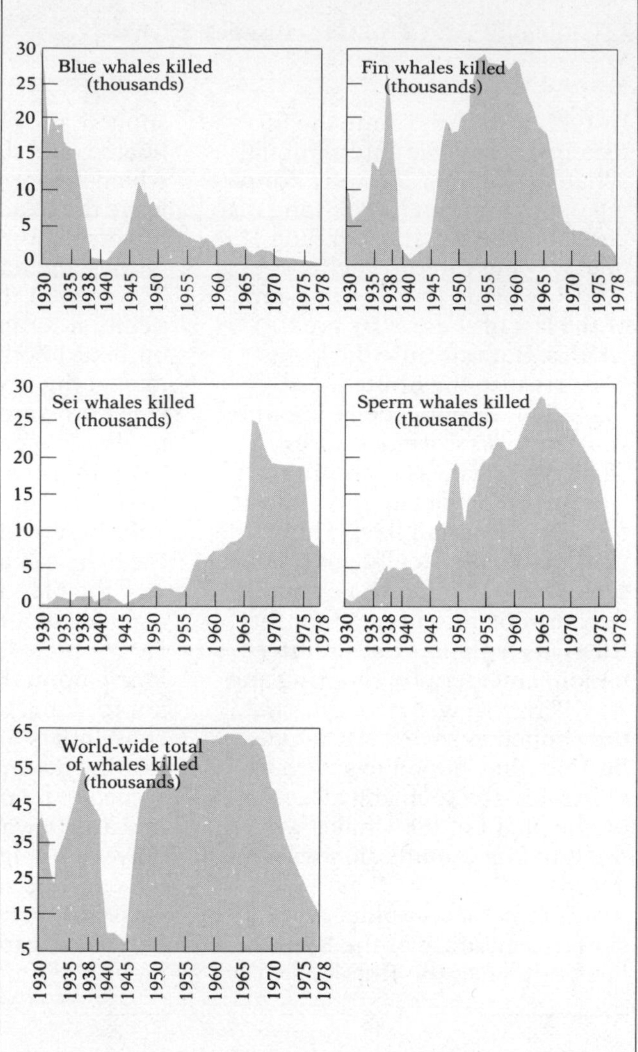

35.16 The numbers of some whales killed annually since 1930. Note that the slaughter stopped only when the civilized whale-hunting countries turned their weapons on each other, during World War II. The kill began dropping in the 1960s even as boats became faster and increasingly efficient electronic detection devices were employed. Blue whales have been legally protected since 1965, but ecologists point out that they probably will not recover because there are too few left for individuals to find mates. On a recent popular whale record (an LP you might like to hear), Roger Payne points out that all oceans are now rent with the rumblings of passing ships, making it impossible for the deep-voiced whales (like blues and finbacks) to communicate with each other across the vast ocean basins.

A Closeup of the Wolf as a Predator

At this point let's take a closer look at one of the most fascinating of our vanishing species—the wolf. In particular, we'll focus on the wolf as a predator. We can learn several things here. First, we can see the complexities in predation studies and how precisely the predator and prey are attuned. We can also learn something about the social system of another animal and how it is related to its role as a predator. Finally, you might like to know something more about wolves, those magnificient animals whose pelts keep turning up as trim on cheap coats in mediocre department stores.

> One day I watched a long line of wolves heading along the frozen shore line of Isle Royale in Lake Superior. Suddenly they stopped and faced upwind toward a large moose. After a few seconds the wolves assembled closely, wagged their tails, and touched noses. Then they started upwind single file toward the moose.
>
> L. DAVID MECH
> *The Wolf* (1970)

Quite a bit of attention has been paid in very recent years to large predators, their life histories, how they kill, and the effects of their predation. Recently, L. David Mech provided us with a rare glimpse into the nature of the wolf. His story is based on many years of watching wolves in the northern United States, especially on Isle Royale in Lake Superior. Is the wolf an effective killer? Yes. Does it ever kill more than it can eat? Sometimes. Does it attack humans? No. Is it always successful in its hunts? No. In fact, for every hundred attempts to kill moose on Isle Royale, only 7.8 attempts were successful.

One of the first things that should be mentioned about wolves is that they are continually on the hunt. They may attack a prey animal only hours or minutes after a successful kill, and when they are on the move, they are hunting. Their stomachs can hold up to 20 lb of food and it is apparently rapidly (or incompletely) digested.

Wolves probably locate prey most often by scent (although they may stumble across prey). Mech found in fifty-one hunts in which he could tell that the wolves were trailing their prey (moose) through the snow, forty-two cases involved direct scenting. They once detected a cow and her twins 1.5 miles away. He found that when the lead animals catch wind of the prey, they stop and all pack members stand alert with eyes, ears, and nose focused toward the prey. Then they may carry out a peculiar ceremony, standing nose-to-nose and wagging their tails. If they are in deep snow, they just pile up behind the leader, who then sets out straight for the prey. As they near their quarry, they quicken the pace and seem anxious to lunge ahead. But they hold themselves in check, alert and tails wagging. The restrained approach permits the wolves to approach their prey without sending it into flight. Of course, if they are scent tracking, they would be to the lee of the prey and their own scent could not alert the animal.

When a large prey such as a moose realizes its situation, it can do one of three things: go after the wolves, stand its ground, or flee. Only once did Mech observe a moose head straight for a pack of wolves that was stalking it. The wolves fled. The second best strategy seems to be to stand ground. As soon as wolves realize their quarry has spotted them but is not running, they stop stalking. If they proceed, they do so cautiously. There may be an advantage in this wariness. First of all, a moose is a strong and dangerous animal, easily capable of killing a wolf. If the animal doesn't run, then, it may be out of a spirit of confidence in its own prowess. Perhaps it has successfully dominated other moose—or other wolves—and hasn't developed the habit of retreat. Also, if it doesn't run, it isn't wasting its energy and can therefore probably put up a better fight. In fact, on Isle Royale, any moose was safe from attack as long as it stood its ground. But if it ran, it was chased and attacked.

Smaller or weaker animals such as deer don't have this option; they could certainly not fend off a wolf pack. But another factor may enter in here. A wolf may need the sight of a running animal in order to stimulate its closing rush. Mech believes the sight of a nonmoving animal inhibits the rush response. The wolves may still approach, but slowly. (You may have noticed that your dog will chase anything that runs from it. One of biologist John Fentress' hand-reared wolves would run from approaching sheep, and chase retreating ones. And a 5 month old wolf pup he owned always ignored horses until he saw one roll in the dust. The sight of a downed animal was just too much and he had to be restrained.)

So, if the prey runs, the wolves give chase. If the quarry is fast enough in its initial run, it usually gets away. Wolves are very perceptive about such matters. If the chase looks hopeless, they stop. But no one knows what kind of cues they use to determine whether to call off the chase. The chase may go on for miles, but it usually covers a very short distance. Of forty-one observed chases, thirty-two were called off within 0.5 mile. Mech saw only one that lasted 3 miles, although Lois Crisler, who also studies wolves, once saw one go on for 5 miles (the prey was a caribou).

Interestingly enough, the prey is also perceptive. It runs no farther than necessary. When the wolves give up, the prey (whether moose, deer, caribou, or Dall sheep) will stop and turn around to watch the wolves. Thus, it doesn't waste energy in running needlessly.

Wolves are widespread (though vanishing) animals, and confront a wide range of prey types. Since they are largely carnivorous, their best energy bet lies in bring down large prey. Perhaps, occasionally, a single wolf may be able to do this, but the wolf's social system permits the strength of numbers. This means they can hunt a number of kinds of large animals. Some of their prey are herding animals, some are solitary, some live in forests, others on tundra, and others on jagged, rocky slopes. However, wolves can and do hunt field mice, and apparently subsist on mice alone, if necessary. Obviously, some prey depend on strength and some on furtiveness or quickness and speed. Probably no group of wolves specializes in more than one or two of these types of animals. It would be interesting to see how wolves that rely primarily on deer would go about attacking a moose.

It is known that wolves attack different animals in different ways. For example, moose are usually bitten on the rump area, with one wolf sometimes clinging to the large, rubbery nose. If the kill is not at first successful, the wounded moose may stiffen and weaken so that it can be brought down some days later. On the other hand, caribou are usually attacked at the shoulder and neck area (Figure 35.17). Field mice are chased and jumped on, the wolf landing with all four feet on the hapless rodents.

35.17 The caribou has been singled out, separated from the herd and systematically attacked by wolves. Each type of prey is handled differently by the wolf pack. In this case, the prey is brought down by frontal attacks on the shoulder and neck.

It is important to realize that wolves are opportunists. They will eat not only mice, but insects, fish, rabbits, birds, or just about anything they can catch. They will also eat beavers and certain kinds of berries. They have even been known to attack powerful bison.

Wolves have also changed their diet to conform to the presence of humans. They raid garbage dumps, for example. Where humans have replaced wild game with domestic animals, the wolf's diet has changed in accordance. And, whereas wolves almost always dispose of all the remains of wild game, they sometimes eat only parts of domestic animals such as cattle and sheep, possibly because human activity drives them from such kills. This behavior has brought down the wrath of government agencies at the behest of lobbying ranchers. Wolves have traditionally brought bounties and their appetites make them easy to poison. An even more disagreeable habit of wolves from a modern human viewpoint is that they will kill and eat dogs and cats. And they may be cannibalistic under some circumstances—although there is no evidence of wolves killing other wolves for food (they may eat the carrion) or of members devouring even the corpse of a member of their own pack.

So, what kinds of animals do wolves take from among their prey populations? After all, their success rate is surprisingly low. In one pack on Isle Royale, of 131 moose they detected, only 6 were killed. And of the 131, 54 escaped without even being stalked by the wolves. Since so many attempts are made and since the success rate is so low, it seems that primarily disadvantaged prey are taken. The prey could be newborn, inexperienced, malformed, sick, old, wounded, parasitized, starving, crippled, or just plain stupid. Of course, the removal of most of these types would have a "sanitizing" effect on the group. A sick or parasitized animal would not be able to spread its condition to others. Malformed or stupid individuals would not be able to pass along their kinds of genes. Old, wounded, or crippled animals would not be able to interfere with group movement, in the case of herding animals, and would not compete for food or other commodities. Several studies have shown that wolves kill primarily animals less than 1 year old, or those that have lived at least half the usual life span for that species in the wild. In a sample of 93 deer killed by wolves in Minnesota, 59% of the adults were relatively old—at least 4.5 years old. (In contrast, most deer killed by hunters, only 20% were over 4.5 years old.) Why, then, do wolves take very young prey? It is hard to see how killing young animals could benefit the herd other than by simply reducing the herd's density. But we must keep in mind that wolves are not in the business of raising ungulates, so there is no reason to think that all their strategies should be beneficial to the prey population.

But what is the evidence? There is an accumulating body of data (much of it presently circumstantial) that wolves do kill a disproportionate number of diseased or starving animals. Mech states, "Because most of the figures were based merely on the examination of the bones or partly eaten carcasses of wolf kills, it is remarkable that such high rates of maladies were found. Such a situation strongly suggests that, if large numbers of intact carcasses could be examined, almost every animal killed by wolves could be classified as young, old, or otherwise inferior." Wolves would undoubtedly take any animal that they could. The problem is, they have no choice. Strong, healthy animals provide too great a risk.

Do wolves control the populations of their prey? For a number of reasons it is difficult to make a precise judgment, but consideration of the problem should provide further insight into the complexities of predator–prey interactions. First of all, it should be pointed out that, in many places where such studies have been made, there may have been unusually high numbers of ungulates. The law (and perhaps the "rites of manhood" among human males) have been harder on the wolves than on the ungulates. Wolf, coyote, lynx, and lion pelts are proudly displayed, and those wolf collars continue to be very popular among the suburban set. Whereas wolves have been hunted, ungulates have been "managed." The results are apparent in the overwhelming number of deer in New Jersey. Possibly, at one time, when ungulate numbers were lower and the number of wolves was higher, wolf activity may have been a major factor in the control of ungulates. But because of the artificial controls on wolves and their prey, it is difficult to know just what sort of population control was imposed by the predators under natural conditions.

Now, at first glance, it would seem that if a wolf killed a moose, it would be reducing the moose population. But it is possible that the wolf factor could be removed from a prey population (by killing all wolves) and that the numbers of prey would *not* increase—just as in the caribou example we mentioned earlier. Remember, the caribou's hoof disease was able to spread, and it destroyed whole herds of caribou. But if the disease is wiped out (by wolves catching the weak animals), more may starve or be caught by predators. This illustrates another aspect of population control: different mortality factors may come into play at different population levels. Disease and starvation, for example, are usually not important until the herd has built up to an unusually high level.

In discussing the relationship of predators to prey, one must consider three major variables: (1) density, the biotic potential and size of individual animals in the prey species; (2) the variation in predator–prey ratios for different prey species, and also for the same prey in different areas; and (3) the possible difference in the amount the predator does eat and what it might eat if food were readily available.

Wolves might have different effects on populations with different densities and reproductive potential. Deer may reach a density of thirty per square mile, while moose are much sparser. Therefore, more deer would have to be killed than moose in order for wolves to control deer populations. Deer may give birth when only 1 year old, and they often produce twins or triplets. But moose may not breed until they are 3–4 years old, and then they only produce one calf per year. Removing a moose, then, has a greater effect on the population than removing a deer.

Also, since a moose is four or five times as large as a deer, wolves would have to kill more deer than moose in order to feed themselves. A ratio of 30 moose per wolf is, in fact, roughly similar to 150 deer per wolf. This factor is partly compensated for by the fact that larger animals are usually sparser.

There are cases where wolves have been shown to control prey number. For example, on Isle Royale there were about 30 moose (averaging 800 lb each) per wolf (from 1959 through 1966). So there were about 24,000 lb of moose per wolf. The 210 square mile island supports about 600 moose in late winter, and about 225 calves are born each spring, but only about 85 survive to be recruited into the herd. There are about 23 wolves on the island and these kill about 140 calves and 83 adults each year for a total of 223 animals. Thus, the annual production of the moose herd is balanced by the average number of wolf kills. The implication is that the wolves are taking enough moose to control the herd.

As further evidence of the regulatory impact of the wolves, the moose have lived on the island since about the first of the century, but the wolves didn't arrive until 1949. In the early 1930s, A. Murie estimated the moose herd at 1000–3000 animals. They apparently overbrowsed their area and their numbers fell off drastically through the effects of starvation and disease only a few years later. The herd recovered, but began to starve again in the late 1940s. However, for decades since the wolves have arrived, moose numbers were lower and more stable than ever before, and the vegetation recovered. In recent years, the Isle Royale moose population has again been rapidly increasing, this time despite the presence of wolves.

The important point is that a predator–prey relationship is a dynamic one. If wolves do not exert effective population control at low wolf densities, or at high prey densities, the population of wolves can be expected to rise until the predator–prey relationship shifts to a controlling one. Predation is density-dependent in several ways. First, increases or decreases in one of the two populations will engender the opposite effect on the other. Second, predation becomes more efficient at higher densities, because the predator has to spend less time searching. Thus, it is not just the ratio of predator–prey populations that is crudely regulated, but also absolute numbers.

Care must also be taken in extrapolating from the wolf–moose figures to any other predator. For example, African lions are much larger than wolves (and larger animals normally need to feed less per unit weight than smaller animals). Also, lions live where it is warm and the energy provided by food is not as important in maintaining body temperature as it is for the wolves of colder areas. In addition, cats lack the endurance of wolves and stalk their prey very carefully, saving energy for a short rush. While wolves are always hunting, lions may nap about 22 hr a day. Different social systems may also make comparisons difficult. A male lion may expend no energy in the hunt, but merely walk over to the kill and take charge after someone else—perhaps a female lion or a hyena troop—brought down the prey. Thus, male lions may require fewer calories than if they actively hunted. In any case, perhaps this glimpse into the role of wolves as predators will give us an idea of how one system works and an appreciation of some of the complexities involved.

The "Balance of Nature"

The old phrase, "balance of nature," is in bad repute among biologists, but like many disreputable catch-phrases it may still contain a bit of truth. After all, we've been stressing the relative stability of species interactions and the density-dependent mechanisms that tend to regulate populations. Of course, there are also recurrent, sizable fluctuations in populations in nature—much more for some species than for others—but we also do see some measure of "balance." In fact, we see it most dramatically when human intervention introduces new factors that upset that balance.

We've noted that parasites generally don't exterminate their hosts and predators generally don't exterminate their prey. There is nothing magical about this, and neither does it demonstrate some beneficence on nature's part. It's more a matter of simple logic. We can assume that (nonhuman) predators have exterminated prey species many times in the past. When this has happened, however, neither the predator species nor the prey species have survived, and naturally the event has escaped human notice. Similarly, we don't often see host exterminations by lethal parasites or virulent diseases, except when human interventions or accidental introductions have set up novel conditions. Even without human intervention, mutant disease organisms sometimes cause massive die-offs, and sometimes species undergo population explosions that so disrupt their own environment as to be followed by local extinctions or near extinctions. Over geological time, well over 99% of the species that are known to have existed have, in fact, gone extinct for one reason or another. If there is a balance of nature—and it does seem that in some sense there is—it is a precarious balance without any long-term guarantees.

While most population biologists are concerned with mechanisms that tend to stabilize and regulate populations, this viewpoint has its critics. Why are we so concerned with stability? Some critics say that there is a large measure of wishful thinking involved—biologists want to believe that nature is "good" or that, one way or another, this must be the "best of all possible worlds." Some biologists, on the other hand, deny that populations are regulated at all. There are yet other critics who claim that stability is an obsession with some population biologists and that their need to believe in stabilizing forces has unconscious political roots. To these critics, belief in the "balance of nature" comes from a deep conservative plea for the sanctity of the status quo in a manifestly unjust world; and to conclude that, say, wolf predation actually benefits moose populations is a veiled and/or unconscious argument that exploitation actually benefits the masses.

Emigration

Of course, populations can be reduced if some animals just ship out. Biologically, this is called *emigration* and is distinct from *immigration* in that it refers to a one-way trip out of a population, while immigration considers the effect on a population of individuals arriving from somewhere else. The classic example of population reduction by emigration is the arctic lemming (Figure 35.18). Populations in northern Europe periodically begin to behave erratically and to emigrate in great numbers. They sometimes really do plunge into the sea and drown. The emigration is apparently a response to overcrowding. In the United States, certain species of squirrels begin periodic treks across inhospitable terrain, again driven by some unknown force. And then there are records of mass movements of East African antelope, which ended when the beasts surged into the surf of the Indian Ocean to drown.

From Biblical times we have heard of plagues of locusts (Figure 35.19), which have swarmed across the land, devouring entire crops that lay in their paths. It seems that when the population density of locusts reaches a certain critical level, the locusts begin to produce offspring with a different developmental hormone condition. These hormones cause longer wings, lighter bodies and darker colors, and it is this generation that emigrates in the dreaded swarms.

We don't know much about what causes emigrations, but we do know that the result is always the same: a reduction in the local population. Of course, since animals do not act out of altruism, we can assume that the emigrators derive some benefit from it all. They may just seek new areas where competition is lessened because of a sparser population. Since all habitats eventually change, old ones sometimes become unfavorable while new habitats emerge; therefore, all animals are descended from ancestors that have emigrated at one time or another. In the case of the antelope and lemmings, perhaps these species have historically gained something by emigration, but they simply haven't been programmed by evolution to deal with oceans.

35.18 The arctic lemming is among the most famous and least understood animals. The lemmings begin massive migrations when their numbers are high and, for some reason, some of them plunge into the sea—apparently, in futile attempts to reach new areas.

35.19 Locusts have periodically ravaged African crops as the insects move in great swarms over the land. Virtually all the plant material in their path is devoured.

If animals that normally undergo cycles of population growth and dispersal are not allowed to disperse—for instance by keeping them in cages or confining them to islands—the unnaturally high density of the population that results may be seen to cause social and behavioral breakdowns. Confined mice or rats in high population densities show abnormal interactions—either fighting constantly or completely ignoring one another, gorging on food to the point of obesity or wasting away even in the presence of abundant food, withdrawing from all activity, and so on. In particular, normal sexual and maternal functions break down, and mothers frequently abandon their litters. Death rates become very high in the absence of any obvious factor other than emotional (and endocrine) upsets. Apparently, similar effects can be seen at times of population explosions in the wild, for instance among lemmings and voles, and among deer confined on small islands.

35.3 Human Populations: Growth and Limits

Human Populations

Now, let's consider some points that relate specifically to the human population of the earth, something of the history of our species, and what our future looks like from here.

Early Humans

Somewhere between 1 and 2 million years ago one might have noticed among the different kinds of living things on the earth a small population of primates that differed somewhat, but not much, from other animals. Actually, the appearance of these animals, ambling in little bands over the vast African plains, was one of the most momentous events in the history of the planet. Who could have guessed that the entire earth, including the great mountains, endless oceans, vast plains, immense forests, and even the air itself would give way, change their form, before something as seemingly insignificant as that hunched figure sitting in the shade, picking at the soft parts of an insect.

We don't have a very firm idea of how many individuals comprised the human species in those early times. In fact, all the evidence before about 1650 A.D. is conjectural. We know, for example, that agriculture was almost unknown before about 8000 B.C., or possibly a bit earlier in Malaysia and the Middle East. We know that before that time humans lived by hunting and gathering their food. Considering the inefficiency of such methods, and given a land area of about 58 million square miles, it has been estimated that the earth could have supported no more than about 5 million people in 8000 B.C. That would approximate the densities found for the few hunting-and-gathering cultures that survived into the 20th century.

It is believed that these early human populations were separate, but interbreeding, groups, possibly divided into yet smaller bands. From studies of present-day hunters and gatherers, it has been calculated that the average size of such groups would have been about 500 men, women, and children, of whom about 200 would be of breeding age. Population size was probably relatively stable and mortality was highly density-dependent (keep in mind the food sources). It is also generally believed that the number and size of the groups approached the carrying capacity of the environment, as it does for other species.

At about 8000 B.C., however, that carrying capacity became elevated by the advent of agriculture. People began to cease roaming and began to grow and store their food and thus they added a relatively stable element to their food supply. The lean seasons were no longer marked by the deaths of so many children and old people, since surplus food could be stored. The result of the shift toward agriculture was a slow, steady increase in human populations.

How did the shift to agriculture raise the earth's carrying capacity? There are several considerations. The growing of cereal crops and the herding of domesticated animals appeared at about the same time, though not necessarily among the same peoples. By growing their own cereals, humans became primary consumers, moving some notches lower on the food pyramid. Most parts of most plants are inedible, but by concentrated agriculture, useless plants (now, for the first time, "weeds") could be removed, and only the most suitable plants allowed to grow. Because humans were still hunters, they could chase away or kill (and thus supplant) other primary consumers that were potential competitors for their food, such as deer or rabbits. By maintaining herds of domesticated animals, and keeping the wolves and mountain lions away, the Neolithic peoples supplanted wild carnivores and at the same time didn't have to waste their energies following and chasing game. Before agriculture, human populations had probably been limited by the availability of food. Now the populations began to grow, and other controlling factors became important. For instance, malaria doesn't affect nomadic tribes of hunters and gatherers, but until recently, it has been a major factor in the mortality of agricultural peoples.

All of this is not to say that the agricultural (or Neolithic) revolution has been the only factor that has changed the carrying capacity of the human environment. The advent of tool making also had its marked effect long before the advent of agriculture, and the industrial revolution had its effect long after. Finally, the scientific revolution, especially the germ theory of disease, greatly reduced the density-dependent mortality of communicable disease, and thus once again increased the carrying capacity of the environment.

Human Growth Rates

Not only has the human population increased steadily, with a few irregularities, since the introduction of agriculture 10,000 years ago, but so has the rate of increase. An originally low rate of increase steadily mounted until about 1970, by which time the annual world population growth rate was approximately 3%. To go from 5 million people on earth in 8000 B.C. (equivalent to the present population of Chicago) to 500 million in 1650 A.D. took between six and seven doublings, or an average doubling time of about 1500 years (for an annual growth rate of 0.047%). The population doubled again in 200 years—to 1 billion by 1850—for an average annual growth rate of 0.35%. By 1930—only 80 years later—there were 2 billion humans. A doubling time of 80 years is equivalent to an annual growth rate of 0.9%. By the mid-1970s, the population had doubled again, for an average annual growth rate during this period of about 1.6%, in spite of the fact that these years spanned the Great Depression and World War II. By 1970, the world growth rate was approaching 3% per year, which is equivalent to a doubling time of 23 years. (As a rule of thumb, to convert annual percentage rates into doubling time, divide the percentage into 70. Thus $70/3 = 23$. Seventy years is the doubling time for a 1% growth rate.) Doubling the population size either means that everyone gets less, or that the entire resources and production rates of the world must also double: twice as many schools, twice as many roads and automobiles, twice as much milk and underwear and cereal grain and bananas and thumbtacks and houses, just to keep even. That's no small task, and some parts of the world haven't kept up; they are poorer, on a per person basis, than ever before. This is not to mention what high population densities do to things like personal privacy, clean air, and quiet forests and open spaces.

Well, that's the bad news, and by 1970 the people of the world were waking up to the horrors of the "population explosion."

In the 1970s, and especially since 1973, the rate of world population growth has slowed for the first time in recorded history. Pessimists like to say that a slowing of the growth rate is like driving toward a cliff at 5 miles per hour instead of 10 miles per hour. But, in fact, the growth rate has slowed down dramatically, not only in the overdeveloped nations of the world, but in most other nations as well. Just in the nick of time, too—or possibly, not quite in the nick of time, since the world population will continue to grow for quite a while yet, and we have hardly the resources to support the 4 billion of us who are already here. But we have probably seen the last population doubling, and at current rates the world population should peak at about 6 billion some time in the mid-21st century. Our relief has to be tempered somewhat by the unpleasant fact that all population projections that have ever been made have been very *inaccurate,* usually in underestimating future population growth but recently in overestimating it. Why is this?

Population Changes Since the 17th Century

Somewhere around the 17th century the world's human population suddenly began to shoot upward. Such an abrupt rise has heralded population crashes in other species, and until we have some sort of evidence that humans are immune from natural laws, we might want to examine carefully the factors behind our skyrocketing numbers.

The sharp rise in the population of Europe between about 1650 and 1750 is attributed to innovations in agriculture and the beginning of the expoitation of the Western Hemisphere. The accompanying rise in Asian populations is more difficult to explain. India, for example, was in turmoil following the demise of the Mogul Empire—and yet India, too, experienced the same population boom. In China, after the fall of the Ming Dynasty in 1644, the Manchu emperors brought political stability and new agricultural policies, which may account for the increase in the Chinese population.

Not much is known about Africa in this period, but its population has been estimated at about 100 million in 1850. Between 1750 and 1850 the populations of Asia increased only about 50%, but in Europe the numbers of people swelled much more rapidly. The factors permitting such growth included newer agricultural techniques, better sanitation, and advances in medicine, such as the smallpox vaccine (Figure 35.20).

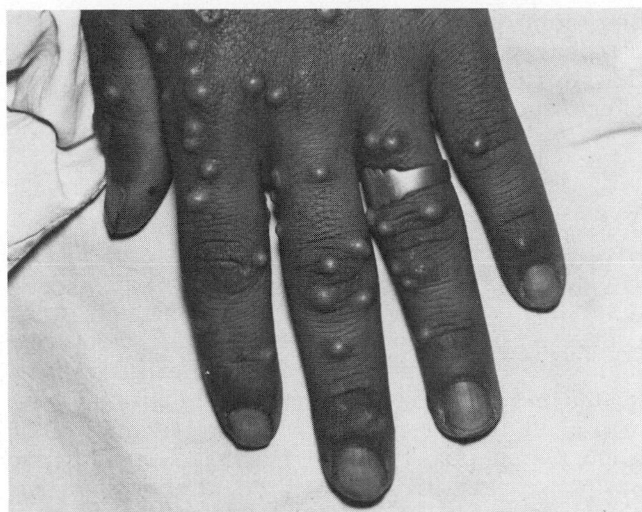

35.20 This is one of the last cases of smallpox on earth and marks one of the greatest large-scale medical achievements. Through medical intervention the human death rate has steadily declined. Along with other advances that promote longevity and survival of the young the effect has been a rapidly expanding world population.

After 1850

In the next 100 years the world population grew from over 1 billion in 1850 to about 2.5 billion by 1950. (Since the Korean War we have added another 1.5 billion to our numbers.) Between 1850 and 1950 the populations of Asia, Africa, and Europe approximately doubled, the population of Latin America increased about five times, and that of North America increased sixfold.

Between 1850 and 1900 the death rate continued to decline in the face of advances in industry, agriculture, and medicine. The lot of the European previous to this time had been incredibly bad. The cities were garbage-filled and overrun with rats. The stench was devastating. Life in the country was not much better, with the scattered hamlets being little more than isolated slums inhabited by people of numbing ignorance. But, by the beginning of the 20th century, the situation of these people improved as the likelihood of crop failure decreased. In addition, transportation advances meant that food could be imported to locally famished areas. Also, about this time the role of bacteria in infection and disease became known, through the work of Louis Pasteur and Florence Nightingale. Simple expedients like washing dishes (and hands), not spitting on the floor and keeping raw sewage out of the water supply saved more lives than all the efforts of modern medicine. Death rates in Europe dropped from about 24 to 20 per 1000 per year in those 50 years.

At about the turn of the century, a new and highly significant trend appeared in the world's populations. It seemed that, for some reason, industrialization was followed by a decreased birth rate. The phenomenon, which has been highly speculated upon, has never been entirely explained, but there are some good guesses at why it happened. First of all, in agricultural societies, children are a source of labor and a social security measure. They provide assistance before they leave home and they care for their aged parents later. In an industrial society, however, they are less likely to aid in production, but they maintain their status as "more mouths to feed." Also, mobility is important in industrial societies and mobility is reduced by the presence of children. Given these reasons, it is interesting that, in the early 20th century, as birth rates dropped off in the industrial cities they did the same in rural areas. One reason was the mechanization of farming, which reduced the need for labor. So it seems that industrialization tends to depress both death rate and birth rate.

About the time of World War II, many underdeveloped countries experienced a dramatic reduction in death rates. The cause was apparently the rapid import of public health programs, including rudimentary sanitation and new drugs, from the more developed countries. Among the most important changes was the control of malaria through the widespread use of DDT (Figure 35.21). Such importation of "death control" produced the most sweeping changes in human population dynamics in the history of the species. It is important to realize that the changes in death rate in these underdeveloped countries were not immediately accompanied by the social and institutional changes that marked the slower reduction in death rate in the more developed coun-

35.21 Large-scale use of DDT has been of great value in controlling certain diseases such as malaria and yellow fever. There have been marked side effects, however. For example, human populations rose dramatically, and the long-lived DDT entered practically every level of our food chain. It is seldom used now.

tries. The changes were much more rapid and were brought about from the outside.

Age Profiles

In the consideration of how any population changes, it is important to consider the ages of the people in the group. Figure 35.22 shows the *age profiles* of different kinds of populations. The shapes are very descriptive. For example, India is typical of many rapidly growing countries with high birth rates and declining death rates. Sweden, on the other hand, is representative of older developed countries which have had low birth and death rates for many generations.

The profile of India is produced by a high birth rate, combined with a sudden recent decline in the deaths of infants and children. There hasn't been enough time for those benefited by imported "death control" to have moved into the older age groups. Furthermore, the children of these young people will further inflate the youngest end of the age spectrum. However, this kind of growth obviously can't go on forever. A program of population control will have to be initiated, or environmental resistance in the form of famine or some other check will eventually alter the shape of the age pyramid. What would the age pyramid look like for a rapidly developing country 10 years after it had suddenly instigated strict birth-control programs? Judging from the age profile of Sweden, at about what age would you say that death-control programs begin to lose their effectiveness?

The Population Boom and Bust

The period of the Great Depression of the 1930s saw dramatic decrease in the birth rate, especially in the United States and other overdeveloped countries. Times were hard, and then the war came, with its deaths and disruptions. Married people found ways to put off having children and, more importantly, single people put off getting married. For many, the postponements became permanent. For instance, American women born in 1908 were 21 at the time of the crash and 37 by the end of World War II, and the opportunities for them were slim. Fully 35% never married, and unmarried women have fewer children, on the average, than married women. The 1930s and early 1940s were periods of low birth rates. Then came the postwar baby boom, and the birth rate skyrocketed. In the United States there were nearly twice as many babies born in 1947 as there had been in 1945, and the birth rate continued to grow for another 15 years. The rest of the world, especially the developing world, also was experiencing inflated growth rates at this time, although it was more a matter of dramat-

35.22 Age profiles break the population down according to sex (left and right halves) and age classes. They reveal the status of a population with great visual impact. The population of India, for example, is a broad pyramid, a definite sign of population growth out of control and increasing exponentially. Considering the plight of the average citizen, India's future may be bleak even if the growth rate suddenly slowed. Industrialization is having some influence on India's birth rate at the present time, but most birth-control programs have failed. The death rate has also dropped, adding another dimension to the problem.

Sweden's population growth is fully under control and has been for some time. The profile suggests other kinds of problems, however, with an "aging" population. Consider the social problems of a nation whose elderly consumers are rapidly increasing in number and the potential burden on the younger workers.

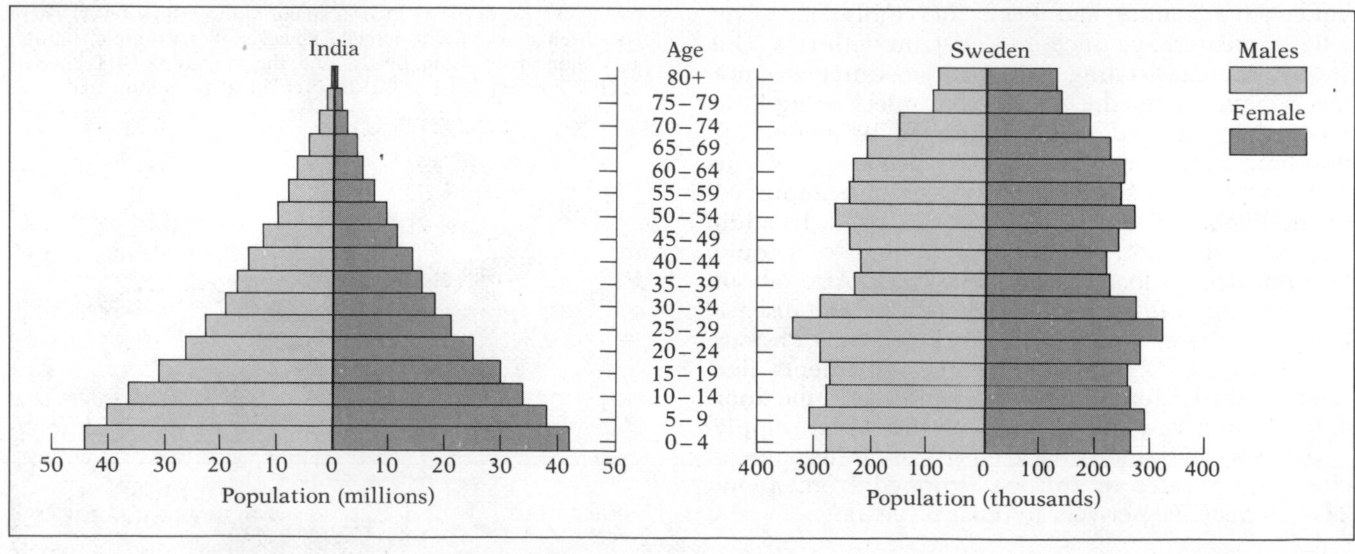

ically decreased infant mortality rates and the persistence of already high birth rates.

No one is really sure quite why the American birth rate zoomed to such high levels in the years between 1947 and 1965. In part, some of the people who had postponed reproduction in the 1930s and 1940s were in fact able to make up for lost time. Probably, a greater factor was fashion. In the postwar era, large families were looked on with great favor by the media and by people themselves, and individual couples responded cheerfully to this cultural imperative. The marriage rate soared, and the proportion of unmarried women among those born in 1932 fell to less than 7%. The great period of suburban building was upon us. One can only say that it must have seemed like a good idea at the time. Other possible reasons for the marriage and baby boom have been proposed, and there are probably factors that we will never understand. The fact of the baby boom of 1947–1965, however, cannot be denied.

Population biologists, notably Paul Ehrlich and Garrett Hardin, were among the first to sound warnings about the ultimate consequences of unbridled population growth. Simple arithmetic showed that the growth rates of the 1950s would soon outstrip the world of resources and space. By extrapolating then current population growth rates into the future, alarmists conjured up images of a sphere of human flesh expanding outward from the earth at the speed of light—and this, they said, was arithmetically possible in only a few thousand years.

The outlook wasn't promising. The huge crop of babies born in the first year of the baby boom, 1947, would reach 21 by 1968 and spend their most productive years in the 1970s. By all logic, another and more serious baby boom was imminent. Even if the birth rates fell substantially, the most optimistic projections showed more world population doublings and probable disaster.

As it turned out, the most optimistic projections weren't optimistic enough. Birth rates fell to unprecedented lows in the developed countries, and began to fall significantly in the rest of the world as well. Perhaps this is shown best by another statistic, the *replacement rate.* The replacement rate for any given year is the sum of the $l_x m_x$ curves for all women in that year—in other words, the number of births to women at age x divided by the total number of women who are that age. In a perfectly stable population, the replacement would be two—one daughter and one son, who would replace the parents that reproduced them. Over a long period of time, a replacement rate greater than two would mean an increasing population, and a replacement rate less than two would mean a shrinking population. In 1955, the peak year of the baby boom, the American replacement rate was 3.5. But by 1973 the American replacement rate fell to 1.7, and, so far, it has remained right about there.

But the American population is, in fact, still growing, because even with low rates there are many young people in the peak years of reproduction and few old people in the peak years of dying. But this is a temporary age profile effect. If the age-specific birth rates stay where they have been since 1973, the population of this country will eventually begin to decline in numbers.

In some developed countries with different age structures, populations have already begun to shrink. There are fewer English people and fewer West Germans today than there were a year ago today. Most of the countries of Eastern Europe are declining in numbers as well. With every year more European Russians die than are born, although the total population of the Soviet Union is growing because of the higher birth rates of the country's non-Caucasian peoples. In another generation the Soviet Union will be a predominantly Asian nation.

What happened? Apparently, changing fashion again played its role. To everyone's surprise, including that of the population biologists and demographers, the dire warnings that had been sounded were heard, and were paid attention to. Large families became looked on with less favor—sometimes to the embarrassment of those who had followed the fashion of the 1950s and were stuck with the results—and remaining single became more acceptable, too. The economic and social liberation of women, and the vast increase in the proportion of working women among both the married and unmarried must have been major factors. But the overriding cause of the "baby bust" has been the availability of effective contraception and abortion. The pill works, and millions of women take it. Other methods are available, and more importantly, information is available and the discussion of contraception is no longer taboo. Abortion has also been a major factor: the great fall of birth rates in 1973 coincided with the sudden availability of legal abortion in most of the United States for the first time. In California the birth rate dropped 10% in that one year.

The full impact of this transformation is difficult to underestimate. If birth control and abortion can halve the birth rate, then we must conclude that without these factors something like half the babies born were unplanned and, at least at the time of conception, unwanted. This is not to say that they were necessarily unwanted once they were born—people have a marvelous way of welcoming and embracing the inevitable, especially if it has big eyes and dimples.

The replacement rate has fallen well below the lowest point of the 1930s. The marriage rate has gone

Essay 35.2 Birth Rates and Death Rates, Increases and Doubling Times

Birth rates and death rates are commonly expressed as number born or died per 1000 people in a population. For example, if 200 babies are born in a year and the population contains 10,000 persons, the birth rate becomes 20. If the deaths total 180 in that population, the death rate is 18. The growth rate per thousand in this population is also stated, per thousand, and is determined by subtracting the death rate from the birth rate, in this case 20 − 18, or 2 per 1000. This is often stated as a percentage, so the growth rate becomes 0.2%, not a very high growth rate compared to most real situations. The estimated annual percent increase in the world for 1977 was 1.8% (18 per thousand per year). Converting percent increases to doubling times is often more meaningful. As we said, this can be done by dividing the percent increase into 70. The answer for the world in 1977 rounds off to 39 years, a very dangerous answer considering all it conveys.

Population Increase Converted to Doubling Time

Annual percent increase	Doubling Time (years)
0.5	140
0.8	88
1.0	70
1.8	39
2.0	35
3.0	23
4.0	18

If the present declining birth rate continues this won't really happen, but this is an important "if." The percent increase in the United States fell below 1.0% in 1977 and is reported to be between 0.6% and 0.7%, so we are stabilizing. Unfortunately several world regions hover between 2% and 3% (Africa, 2.3%; Asia, 2%; Latin American, 2.7%).

down, too. If the age-specific marriage rates of 1978 continue unchanged, fully one-third of all women will never marry. And nobody calls them old maids anymore. Statistics on men are similar, but are notoriously less reliable. Actually, more men than women never marry, and men that do marry tend to marry more times.

As kindergartens, and then first-grade classrooms, and then whole elementary schools were closed down in the 1970s because of falling enrollments, *demographers* (people who study human populations) first claimed that the downturn in the birth rate was only temporary. They thought the children of the baby boom hadn't come through with babies of their own to the extent they had predicted, but these people were only postponing their families. The demographers thought a new boom would come anyway, just a little behind schedule. This view was based on interviews in which most young people said that they wanted to marry eventually, and most young married people said that they wanted to have about three children. But the predicted, delayed boom still hasn't happened, and it looks like it never will. The children of the postwar baby boom, now entering their thirties, are coming to accept the fact that many of them will never marry after all, and earlier plans for large families begin to fade after the first few years of the reality of raising even one or two children. The clock is not going to roll back on the economic liberation of women or the availability of contraception, so we may never see another baby boom. Fashion, however, is unpredictable, so we can't really be sure.

Application of Ideas

1. Why do expanding populations so frequently describe an S-shaped curve on a graph?
2. The statement that organisms tend to leave as many offspring as possible has enormous implication for biology. What is the circumstantial evidence for the statement and what are some of its implications in the evolution of social behavior?
3. What sort of species would tend to have more young or leave more eggs, one that lived under competitive or noncompetitive situations?
4. Give several examples of density-dependent and density-independent mechanisms acting together to control populations. You may use hypothetical examples.
5. What are the important lessons regarding predators as population controllers that we have learned from studies of wolves?

Key Words

abiotic control
age profile
agricultural revolution

biotic potential
birth rate
broad-niched

competition
competitive exclusion
crash

death control
death rate
demographer
density-independent
dominance hierarchy
doubling time

ecological niche
emigration
environmental resistance
exponential growth

Gause's law
growth curve

habitat
host-specific

immigration
intraspecific

limiting resource
$l_x m_x$ *curve*

mortality
mortality rate
mutualistic

negative feedback system
neolithic revolution
niche

ornithologist

parasitic castration
parasitism
parasitoid
peck order
population
predation
prey switching

replacement rate

survivorship curve

territoriality

Key Ideas

The Ecological Niche

1. The *niche* is the role a species plays in a community of interacting species.
2. The *habitat* is the specific place or kind of place in which an organism is found.

Population Changes

Populations are groups of the same species that interact with each other as a breeding unit. Their size often varies in a regular and cyclic way.

Population Growth

1. The growth in absolute numbers of *populations* entering new *habitats* is slow at first, but as more numbers are added, more are able to reproduce, and the absolute growth increases.
2. The maximum rate at which a *population* can increase is its *biotic potential.*
3. In the ideal environment, *populations* increase *exponentially.*
4. Human *population* growth history is unusual among animals in that the *death rate* has decreased.

Carrying Capacity

1. *Populations* do not reach their *biotic potential* because of *environmental resistance.* The limit set by *environmental resistance* is the *carrying capacity.*
2. When a population exceeds the normal *carrying capacity,* a *crash* can occur. This can be temporary or permanent, depending on the environmental damage, but the *carrying capacity* is often reduced, sometimes permanently.

Reproduction Rate

Population stabilizing factors are believed to regulate both *birth rate* and longevity.

How Many Offspring Are Produced?

Organisms tend to reproduce at a biological maximum. There are many factors that determine what is biologically possible.

Tapeworms

Survival of individual tapeworm offspring is so limited that the tapeworm must produce many thousands of eggs.

Chimpanzees

1. Female chimpanzees produce few offspring and have long periods of maternal care during which they do not reproduce.
2. The chimpanzee's *niche* is complex, requiring intelligence and a long period of learning.
3. Although they reproduce maximally, relatively few offspring are produced.

Birds

1. Reproductive success in birds is provided by the bird's laying the number of eggs that can be successfully reared. Birds may produce more eggs when *predation* pressure is strong, and fewer when it is not.
2. Other factors determining the number of eggs are the time available for rearing and the food supply.

Humans

1. In primitive tribes constantly confronted with marginal food supplies, nursing mothers may have been unable to produce offspring.
2. Where food is adequate, women can produce offspring yearly. This potential may have been important at a time when infant *mortality* was very high and the reproductive years limited.
3. In recent times, the *death rate* has declined, while the *birth rate* has remained relatively unchanged. This has led to the rapid growth of human *population.*
4. The selective forces that at one time produced the high *death rate* are minimized by human intervention.

Death

Programmed Death
1. Many plants are annuals, living one year or season. Some animals are similar, living through one reproductive cycle.
2. The life span of humans appears to be programmed also, since many diseases progress rapidly in old age. Since humans are living longer, diseases of the aged are much more in evidence. Specific death rates suggest that we wear out all at once.
3. The animal's physiology provides for a period of life that is long enough for maximum reproduction and little more.

Death: The Long View
Death, in the long view, is adaptive. Without it, parents would remain to compete with offspring for the basic commodities.

Controlling Populations Through Mortality

Abiotic Control
Abiotic controls are physical factors such as drought or flooding.

Density-Independent Mortality
Factors such as drought, flooding, frost, etc., kill members of a *population* without respect to density.

Biotic Control
Biotic factors are living influences. Most are *density-dependent. Density-dependent* factors exert pressure on a *population* as its numbers increase.

Density-Dependent Mortality
1. *Density-dependent* factors act as *population* controls or regulators by increasing *mortality* and/or reducing fertility.
2. *Abiotic factors* can be *density-dependent* in their effects.

Territoriality and Dominance Hierarchies in Intraspecific Competition
1. Population size in *territorial species* is strongly controlled by the availability of space.
2. Where *dominance hierarchies* are established, the critical commodities go preferentially to the dominant individuals.

Self-Poisoning Through Toxic Wastes
1. An important aspect of *population* control in microorganisms is the toxic waste they produce. The waste of one species often becomes the food of the next.
2. Plant species also produce toxic products that tend to inhibit both their own and other species from growing nearby.
3. Humans also produce toxic wastes both natural and industrial, which constantly threaten the environment. Some important examples include oil spills, acid rains, automobile pollutants, and radiation.

Disease and Parasitism
1. Infectious diseases and parasitism are excellent examples of density-dependent controls, since they spread more readily in dense *populations.*
2. Disease and *predation* interact when the weakened animal is picked off by the predator.
3. Parasites and hosts evolve together. In time the host is not killed by the infestation. True *parasitism* is generally a long-term relationship. Short-term relationships are generally more harmful to the host.
4. The chestnut fungus is an example of a newly introduced parasite. The new host has virtually no defense.
5. *Parasitic castration,* infection of the gonads, has an obvious effect on the host population.

6. In *parasitoid* relationships the parasite always kills its host. An example is the *parasitoid* wasp, which lays its eggs in a caterpillar.

7. In tropical forests, the population of any one species is often kept in check by highly host-specific "seed predators." This leads to a diversity of species in these forests.

Competition Between Members of Different Species
Competition between different species is usually minimal because they rarely coexist if they have totally overlapping *niches.* (This is *Gause's law.*)

Predation
1. Predator species do not kill all of a prey species, but are important regulators. The two *populations* tend to fluctuate together because the abundance of prey also controls the predator population.

2. The co-regulation between predator and prey often fails when humans are predators.

A Closeup of the Wolf as a Predator
1. The success rate of wolves preying on large animals is essentially low, usually confined to young, aged, or debilitated animals. Large, healthy animals apparently offer too much risk.

2. The few studies of *population* regulation by wolves indicates that where *populations* are small, predation by wolves is a regulating factor. *Predation* has a stabilizing influence, and death by starvation and disease is less common. Where predators are eliminated, herds of grazing animals may increase to a point where they drastically exceed the *carrying capacity.*

The "Balance of Nature"
1. Whether or not a balance exists is speculative. Undoubtedly there have been many situations where a predator or parasite has killed off a species. Ninety-nine percent of all known species produced on the earth are extinct.

2. Some population biologists doubt the existence of regulating factors.

Emigration
1. *Emigration* may be a response to overcrowding. In some instances the migrating animals perish in a hostile environment.

2. Locust plagues may result from *populations* reaching a critical density, followed by mass *emigration.* Hormonal factors are apparently involved in producing special characteristics in the *emigrating populations.*

3. In laboratory *populations* where *emigration* is prevented, high *population* densities produce behavioral abnormalities.

Human Populations

Early Humans
1. Prior to the beginnings of agriculture, the earth probably supported about 5 million people; close to *carrying capacity* for hunting and gathering.

2. About 8000 B.C., with the origin of agriculture, the human *population* started to increase. Agriculture greatly increases the human *carrying capacity* of the environment.

3. Tool making also stimulated *population* growth by making hunting more effiicient.

Human Growth Rates
1. Both population size and its growth rate increased since the start of agriculture. By 1970 the growth rate was 3% per year.

2. *Population doubling times* were as follows:

8000 B.C. to 1650 A.D.	5–6 doublings (1608–1930 years)
1650 to 1850	1 doubling (200 years)
1850 to 1930	1 doubling (80 years)
1930 to mid-1970	1 doubling (45 years)

3. Anticipated *doubling time* in 1970 was 23 years.

4. Since 1973, the world *population* growth rate has slowed for the first time in recorded history. World *population* should peak at 6 billion in the mid-21st century.

Population Changes Since the 17th Century

1. Between 1650 and 1750 human *population* rapidly increased. Some of the reasons include European expansion and agricultural and political stability in Asia.

2. Between 1750 and 1850 the Asian *population* increased by 50%, while that in Europe was even greater. Some of the reasons were better agricultural techniques, sanitation, and advances in medicine, all of which tended to depress the *death rate.*

After 1850

1. Between 1850 and 1950 population went from 1 to 2.5 billion. Asia, Africa, and Europe doubled their populations. Latin America increased fivefold, while North America increased sixfold.

2. Industrialization, for reasons not known, depresses the *birth rate.*

3. The *death rate,* highest in undeveloped countries, has been decreasing since World War II because of improved public health, sanitation, and malarial control.

Age Profiles

Age profiles are important visual indicators of how populations will change. Pyramidal shapes indicate rapid growth.

The Population Boom and Bust

1. The *birth rate,* depressed in the 1930s and 1940s, rose drastically after World War II. However, a second boom 20 years later did not occur, and the *birth rate* in developed countries plunged.

2. The reasons for the new decline were:
   a. Changing attitudes about large families
   b. Economic and social liberation of women
   c. Increase of women in working force
   d. Birth control
   e. Delayed marriage

Review Questions

1. Carefully define the term *niche* and give an example. (p. 1028)

2. Draw an S-shaped *population growth curve* and explain its three main parts. (p. 1029)

3. Explain the relationship between *biotic potential* and *environmental resistance.* (pp. 1029–1030)

4. Using an analogy, explain what happens to the growth rate when *populations* increase *exponentially.* (pp. 1029–1030)

5. Under what conditions does a *population crash* become permanent? (p. 1031)

6. What are the two ways *population regulation mechanisms* work? (p. 1031)

7. What are some of the chimpanzee's problems with successful reproduction and how does it cope? (p. 1032)

8. Under what conditions does nursing the young act to reduce fertility in humans? (p. 1034)

9. What evidence suggests that human longevity is well programmed? (p. 1036)

10. Why is longevity past the reproductive age often nonadaptive? (p. 1037)

11. Explain how the "suicide gene" of the moth assures its offspring of greater success. (p. 1037)

12. List an example of an *abiotic control* that is *density-independent.* (p. 1038)

13. What makes a *population* control *density-dependent?* (p. 1038)

14. Explain how *population* is so closely controlled in *territorial species.* (p. 1039)

15. In what serious way are humans similar to bacteria in their effect on the environment? (p. 1040)

16. Explain how high *population* density affects the role of disease organisms and parasites as agents of control. (p. 1041)

17. Describe the manner in which *parasitoid* wasps control some insect populations. (pp. 1042–1043)

18. What has been the effect of seed or fruit predators on the distribution of species in tropical rain forests? (p. 1043)

19. How does *Gause's law* explain how *interspecific competition* is kept at a minimum? (p. 1043)

20. Explain how predator and prey *populations* regulate each other. (p. 1045)

21. What is L. David Mech's conclusion about the efficiency of the wolf hunt? (p. 1046)

22. What members of the prey *population* are most often killed by hunting wolf packs? (p. 1048)

23. Present a theoretical explanation for animal *emigration.* (p. 1050)

24. What is the estimated *carrying capacity* of the earth for humans prior to the dawn of agriculture and why was it so low? (p. 1051)

25. List the doubling times of human populations from 8000 B.C. to the 1970s. (p. 1052)

26. What reasons are suggested for *population* increases between 1650 and 1750? (p. 1052)

27. List the innovations that promoted *population* growth between 1850 and 1900. (p. 1053)

28. How has industrialization generally affected *birth rates* in recent history? What are the explanations? (p. 1053)

29. Compare the *age profiles* of Sweden and India and summarize the futures of these two populations if trends were to continue unchanged. (p. 1054)

30. Can we expect a *population* decline in the United States very soon? Explain. (p. 1055)

Suggested Reading

Borgstrom, G. 1976. "Never Before Has Humankind Had to Face the Problem of Feeding So Many People with So Little Food." *Smithsonian* 7:70.

Calhoun, J. R. 1962. "Population Density and Social Pathology." *Scientific American,* February. Rats kept in crowded conditions mistreat their babies and lose interest in sex.

Connell, J. H. 1978. "Diversity in Tropical Rain Forests and Coral Reefs." *Science* 199:1302. What makes some habitats so species-rich? Connell suggests that unpredictable natural disasters, such as typhoons, prevent them from ever approaching species equilibrium.

Connell, J. H., D. Mertz, and W. Murdoch, eds. 1970. *Readings in Ecology and Ecological Genetics.* Harper and Row, New York.

Corliss, J. B., et al. 1979. "Submarine Thermal Springs on the Galápagos Rift." *Science* 203:1073. *Alvin* takes a look at a novel ecosystem built on sulfur-oxidizing chemosynthetic bacteria.

Ehrlich, P. R., and A. H. Ehrlich. 1979. "What Happened to the Population Bomb?" *Human Nature,* January. Everyone expected an unprecedented population explosion in the United States in the 1970s, as the children of the 1945–1960 baby boom came of age. The bomb's failure to explode offers some hope for humanity.

Gressit, J. L. 1977. "Symbiosis Runs Wild in a Lilliputian 'Forest' Growing on the Backs of High-living Weevils in New Guinea." *Smithsonian* 7:135. With a title like that, what more can we say?

Horn, H. H. 1975. "Forest Succession." *Scientific American,* May. The succession of trees in a New Jersey forest depends primarily on soil moisture and the geometry of leaves.

Hutchinson, G. E. 1959. "Homage to Santa Rosalia, or Why Are There So Many Kinds of Animals?" *American Naturalist* 93:145. A masterpiece by the century's most prestigious ecologist.

Jennings, P. R. 1974. "Rice Breeding and World Food Production." *Science* 186:1085.

Murdoch, W. W., ed. 1975. *Environment,* 2d ed. Sinauer, Stamford, Conn. Basic readings.

Murdoch, W. W., and A. Oaten. 1975. "Population and Food: Metaphors and the Reality." *Bioscience* 25:561.

National Academy of Sciences (U.S.). 1975. *Underexploited Tropical Plants with Promising Economic Value.* Instead of continuing to grow the standard "world crops" in new tropical nations, why not take a closer look at the native plants?

Ricklefs, R. E. 1978. *Ecology,* 2d ed. Chiron Press, Newton, Mass. Our favorite all-around ecology textbook, which also has an excellent coverage of the basics of population genetics.

Schaller, G. B. 1972. *The Serengeti Lion: A Study of Predator–Prey Relations.* University of Chicago Press, Chicago.

A *Scientific American* book. 1970. *The Biosphere.* W. H. Freeman, San Francisco.

Shepard, P., and O. McKinley, eds. 1969. *The Subversive Science: Essays Toward an Ecology of Man.* Houghton Mifflin, Boston.

Terborgh, J. 1974. "Preservation of Natural Diversity, the Problem of Extinction-prone Species." *Bioscience* 24:715. How much responsibility should we accept for endangered species—and at what expense?

Wilson, E. O., ed. 1974. *Ecology, Evolution and Population Biology.* Readings from *Scientific American.* W. H. Freeman, San Francisco.

Wilson, E. O., and W. H. Bossert. 1971. *A Primer of Population Biology.* Sinauer, Sunderland, Mass. (paperback).

APPENDIX A

CLASSIFICATION OF ORGANISMS

The classification scheme presented here has been adopted, at the kingdom level, from the five-kingdom classification scheme proposed by R. H. Whittaker in 1969 (*Science* 163(10): 150–160). We have modified the scheme below that level in various ways as we attempted to clarify relationships among the organisms. Most of the taxa mentioned here are also described in the text, particularly in Chapters 14–17 and 22–23.

Absent in this listing are the viruses. Some authors include them in the Kingdom Monera, while others have invented Kingdom Virus. It is our opinion that there is too little known about viral taxonomy or systematics to propose a serious taxonomic category.

The Procaryotes

Membrane-bounded organelles and nucleus absent. Simple, circular chromosome of DNA only. Cell division by fission. Limited sexual recombination (plasmid transfer). Solid, rotating flagella where present.

KINGDOM I MONERA: Includes bacteria and cyanophytes. Generally single-celled, some forming aggregates or colonies. Cell specialization in some.

SUBKINGDOM SCHIZOPHYTA (Schizomycetes): bacteria. Single-celled or cells in chains or clusters. Some genetic recombination Mainly heterotrophic but some chemotrophs and phototrophs. Motile forms with rotating flagella.

Phylum Eubacteriacea (true bacteria): bacillus and coccus cells. Formed single or in chains. Latter single diplococcus, staphylococcus, sarcinoid, or streptococcus. Mostly saprobes, but some medically and ecologically important parasites. Aerobic and anaerobic. Phototrophs and chemotrophs. Examples are *Escherichia coli, Claustridium tetani, Diplococcus pneumoniae.*

Phylum Myxobacteriae: slime or gliding bacteria. Cells produce slime for gliding movement. Some cell specialization. Fungallike fruiting bodies.

Phylum Chlamydobacteriae: mycelium bacteria. Cells form fungallike, mycelial filaments within a sheath. Flagellated cells.

Phylum Spirochaetae: spirochaetes. Cork-screw shaped cells. Some are the largest of the kingdom. Flagellated. Some important parasites. An example is *Treponema pallidum,* the agent of syphilis.

Phylum Mycoplasmae: Smallest cells known, parasitic. Form myceliallike colonies.

SUBKINGDOM CYANOPHYTA

Phylum Cyanophyta: blue-green algae. Photosynthetic. Single-celled and colonial. Photosynthetic pigments on complex internal lamellae with chlorophyll *a*, beta carotenes, phycocyanins and phycoerythrins. Some with specialized cells for nitrogen fixation. Examples are *Nostoc* and *Oscillatoria.*

The Eucaryotes

Membrane-bounded organelles including nucleus, complex chromosomes with protein, mitotic division, hollow microtubular flagella where present, sexual reproduction (unless noted).

KINGDOM II PROTISTA: Includes some algae and all protozoans. Primarily single-celled or colonial. Animallike heterotrophs and photosynthetic autotrophs. Sexual reproduction by meiosis and fertilization (unless otherwise noted)

Phylum Pyrrophyta: dinoflagellates. Chiefly single-celled with flagella. Photosynthetic. Cell walls of chitin. Primitive fission and unusual single division meiosis. Organisms of red tide and mussel poison included. Examples are *Gonyaulax* and *Gymnodinium.*

Phylum Euglenophyta: *Euglena.* Chiefly single-celled, photosynthetic with light sensitive "eye spot." Flagellated. Some feed by absorption. No known sexual recombination. Examples are *Euglena gracilis* and *Phacus.*

Phylum Chrysophyta: yellow-green and golden brown algae, diatoms. Single celled and colonial. Photosynthetic. Cell walls of pectin and silicon dioxide. Form large marine sediments of diatomaceous earth.

Phylum Mastigophora: flagellated protozoans. Chiefly single-celled heterotrophs. Feed by engulfing or absorbing. Sexual reproduction rare. Includes important parasitic trypanosomes such as *Trypanosoma gambiense* (African sleeping sickness).

Phylum Sarcodina: the amebas. Single cells, amoeboid movement. Feed by phagocytosis and absorption. Marine forms include siliceous Radiolaria and calcareous Foraminifera. Important in sediment formation. Includes agent of amoeboid dysentery, *Endamoeba hystolytica.*

Phylum Sporozoa: Single cells, chiefly nonmotile, parasitic spore formers. Life cycle may include two hosts. Includes agent of malaria, *Plasmodium vivax.*

Phylum Ciliophora: ciliates. Mostly single-celled. Movement by coordinated cilia in rows or fused into cirri. Complex organelles, including oral groove in some with feeding by phagocytosis. Contractile vacuoles for water regulation. Sexual reproduction by conjugation. Examples are *Paramecium sp.* and *Euplotes.*

KINGDOM III FUNGI: All heterotrophs with saprophytic and parasitic nutrition. Feed by extracellular digestion and simple absorption. Mainly multicellular, often with complex organization. Cells often form a mycelium with cell walls of cellulose or chitin. Body primarily haploid with typically brief diploid state following fertilization. Many with no known sexual reproduction. Asexual reproduction includes mitotic spore formation.

SUBKINGDOM GYMNOMYCOTA: slime molds. *Acellular slime molds* have a large, multinucleate plasmodium. Feeding by phagocytosis. Haploid asexual spores formed in aerial sporangiophores. May also divide, forming haploid, flagellated swarm cells that fuse, producing a diploid state. *Cellular slime molds* have a single-celled plasmodium. Acellular reproduction by mitosis and division. Fusing of individuals is followed by production of sporangia and spores.

SUBKINGDOM DIMASTIGOMYCOTA: Oomycetes, water molds. Mycelium with nonseptate cells. Saprophytes and parasites. Flagellated, asexual zoospores. Sexual stages include large oogonia and fingerlike antheridia. Important parasites include downy mildew and potato late blight.

SUBKINGDOM EUMYCOTA: true fungi. Fungi that lack amoeboid and flagellated movement (with the exception noted below).

Phylum Chytridomycetes: chytrids. Unicellular or with limited mycelium. Flagellated spores and gametes with highly resistant sporangia. Saprophytes and parasites.

Phylum Zygomycetes: bread molds. Extensive nonseptate mycelium. Asexual reproduction by mitotic spore formation, sexual, by conjugation and formation, sexual by conjugation and formation of diploid zygospores. Examples are *Rhizopus* and *Neurospora.*

Phylum Ascomycetes: sac fungi. Extensive, septate mycelium. Asexual reproduction by conidia and conidiospores, sexual reproduction by saclike asci and ascospores. Asci are produced by conjugation without nuclear fusion, resulting in a dicaryotic condition, persisting until fusion, followed by spore formation. Very brief diploid state is followed by meiosis and ascospore formation. Includes yeast, *Claviceps,* and lichens.

Phylum Basidiomycetes: club fungi. Extensive septate mycelium, often producing large fruiting basidiocarp. Saprophytes with some parasites. Sexual reproduction by conjugation and persistent dicaryotic condition until fusion, followed by meiosis and spore formation. Includes mushrooms, shelf and bracket fungi, and wheat rust.

Phylum Deuteromycetes: Fungi Imperfecti. Convenience group that includes fungi with no known sexual state. Examples are *Trichophyton* (athletes foot) and *Dactylaria* (nematode-snaring fungi).

KINGDOM IV PLANTAE: the plants. Primarily nonmotile, multicellular organisms with dense cellulose cell walls. Photosynthetic with chlorophyll *a* and carotenoids. Most with specialized tissues and organs. Asexual and sexual reproduction with an alternation of generations in which the haploid and diploid states are often clearly separated in the more primitive species.

SUBKINGDOM RHODOPHYCOPHYTA: red algae **Division (phylum) Rhodophyta:** Coastal seaweeds that contain the pigments phycocyanin, phycoerythrin, chlorophyll *a,* and carotenoids. Separate sporophyte and gametophyte generations, but with varying dominance in different species. Flagella and cilia absent. Storage carbohydrate is floridean or carageen. Examples are irish moss and the edible *Porphyra.*

SUBKINGDOM PHAEOPHYCOPHYTA: brown algae **Division (phylum) Phaeophyta:** Coastal seaweeds and kelps that contain the pigments fucoxanthin, chlorophyll *a* and *c,* and carotenoids. Storage carbohydrates are laminarin and mannitol. Separate sporophyte and gametophyte generations and flagellated sperm. Considerable tissue specialization with holdfasts, stipes, blades, and bladderlike floats. Examples are *Macrocystis, Nereocystis,* and *Sargassum.*

SUBKINGDOM EUCHLOROPHYTA: green algae and the land plants. Members contain chlorophyll *a* and *b* and carotenoids, storing carbohydrates as the starches amylose and amylopectin. (Includes all remaining plants.)

Division (phylum) Charophyta: stoneworts. Ancient, complex freshwater alga with apical growth. Calcareous cell walls. *Chara.*

Division (phylum) Chlorophyta: green algae. Diverse plants, chiefly aquatic in both marine and fresh waters. Single-celled, colonial, and multicellular organization. Some flagellated. Varied dominance in gametophyte and sporophyte. Multicellular forms share ancestry with land plants. Examples are *Chlamydomonas, Ulva,* and *Volvox.*

Division (phylum) Bryophyta: mosses, liverworts, hornworts. Multicellular terrestrial plants with tissue specialization, but lacking true vascular systems. Water required for fertilization.

CLASS MUSCI: mosses. Widely distributed terrestrial and semiaquatic forms. Dominant, photosynthetic haploid gametophyte, small and upright or hanging, that produce archegonia and/or antheridia. Following fertilization, the dependent diploid sporophyte emerges from the archegonia and produces spores meiotically, restoring the haploid gametophyte generation. Examples are sphagnum (peat) moss, pigeonwheat moss, and cord mosses.

CLASS HEPATICAE: liverworts. Terrestrial plants of limited distribution. The dominant, photosynthetic gametophyte forms a flattened ground-hugging thallus. Some reproduce asexually by gemmae cups. Antheridia and archegonia are borne on stalks. Dependent diploid sporophyte emerges to produce haploid aerial spores through meiosis. Examples are *Marchantia* and *Riccia.*

CLASS ANTHOCEROTAE: hornworts. Terrestrial plants of very limited distribution. Dominant haploid gametophyte is small and leafy, producing antheridia and archegonia on the leafy surface. Hornlike diploid sporophytes are often photosynthetic and independent, producing haploid aerial spores through meiosis. *Anthoceros.*

Division (phylum) Tracheophyta: vascular plants. Sporophytes contain vascular tissue with true roots, stems, and leaves. Sporophytes are dominant with greatly reduced gametophytes. Water requirement for fertilization in more primitive species.

SUBDIVISION (SUBPHYLUM) PSILOPHYTA: The four surviving species are of very limited distribution. Rhizoids are nonvascular. Vascular stems bear scalelike leaves along highly branched stems. Haploid aerial spores are produced in sporangia situated along the stems. Spores develop into greatly reduced gametophytes from which the sporophyte emerges after fertilization. Water required for sperm travel. *Psilotum nudum.*

SUBDIVISION (SUBPHYLUM) LYCOPHYTA: club mosses. Widely distributed terrestrial plant. Vascular roots, stems, and leaves. Dominant upright, leafy sporophyte that produces haploid spores in terminal strobili. Spores produce minute independent gametophyte from which the new sporophyte emerges. Water required for sperm travel. Examples are *Lycopodium* and *Salaginella* (club mosses or ground pines).

SUBDIVISION (SUBPHYLUM) SPHENOPHYTA: horsetails. Limited number and distribution. Dominant sporophyte grows in branched vegetative and single reproductive stems. Leaves are reduced and nonphotosynthetic. Sporangia are borne at the tip of reproductive shoot, producing haploid aerial spores that develop into minute gametophytes requiring water for sperm travel. *Equisetum.*

SUBDIVISION (SUBPHYLUM) PTEROPHYTA: ferns. Very broad distribution. Dominant sporophyte includes very leafy aerial parts, often treelike, produced from underground rhizomes with small, hairy roots. Haploid spores are produced in sori borne on the leaves. Spores produce a highly reduced, independent gametophyte that requires water for sperm travel. Examples are tree ferns, sword ferns, bracken ferns, and staghorn.

SUBDIVISION (SUBPHYLUM) SPERMOPHYTA: seed plants. Gametophyte becomes incorporated into dominant sporophyte, aerial spore gives way to pollen development, and water requirement for sperm travel limited to a few types.

GYMNOSPERMS: Nonflowering seed plants. Naked seeds, not enclosed in fruit.

CLASS GINKOOPSIDA: *Ginkgo.* Sole surviving species, *Ginkgo biloba,* may no longer exist in the wild. Large sporophyte (tree) with bilobed fan-shaped leaves. Dioecious with pollen trees and seed trees (in which ovules develop). Swimming sperm utilize fluids produced by the plant.

CLASS CYCADOPSIDA: cycads. Tropical distribution. Palmlike appearance with thick partly subterranean stem. Produces numerous divided leaves. Dioecious with seeds borne in large seed cones and pollen in pollen cones. Flagellated sperm travel through pollen tubes to fertilize the female gamete. Examples are *Cycas* (Sago palm), *Zamia,* a North American species.

CLASS PINOPSIDA: (also known as Division Coniferophyta): conifers. Widely distributed in northern and mountainous areas with considerable penetration into temperate regions. Commonly monoecious, with individual trees bearing pollen and seed cones. Gametophyte generation is reduced to relatively small masses of cells housed in the cones. Seeds, without fruit, are formed. Examples are pines, spruce, hemlock, cedar, larch, fir, juniper.

THE ANGIOSPERMS: Flowering plants. Seeds enclosed in fruit.

CLASS DICOTYLEDONAE: the dicots. Includes both herbaceous and woody plants, annuals, biannuals, and perennials. Most have two cotyledons (seed leaves), net-veined leaves, floral parts in fours, fives, or multiples of these, vascular cambium with secondary growth, and vascular bundles in single or coincentric rings. Examples are the magnolia, rose, cucumber, bean, pea, carrot, oak, willow, and maple.

CLASS MONOCOTYLEDONAE: the monocots. Primarily herbaceous plants, often annual, but many perennials. Single cotyledon (seed leaf), parallel-veined leaves, floral parts in threes or sixes, or multiples of these. Vascular cambium rare, and vascular bundles commonly scattered in the stem. Examples are the lily, iris, palms, orchids, and the grasses (grains).

KINGDOM V ANIMALIA: the animals. Multicellular, heterotrophic eucaryotes with specialized cells, tissues, and in most, organ systems. Most are capable of rapid response to external stimuli through specialized sensory and motor cells and muscle tissue. Chiefly diploid with sexual reproduction through meiosis. Motile sperm and stationary eggs produced. Asexual reproduction through fission, budding, regeneration, and parthenogenesis.

SUBKINGDOM PARAZOA: Primarily tissue-level organization with evolutionary origin in flagellate line.

Phylum Porifera: sponges. Nonmotile as adults, solitary or colonial. Filter feeding with phagocytosis. Extraordinary regenerative powers through reaggregation. Asexual reproduction through budding. Sexual reproduction with sperm released externally to swim to internal egg. Swimming larval stage.

CLASS CALCARIA: calcareous sponges. Simpler sponges with calcareous spicules (skeletal elements). *Leucosolenia, Grantia.*

CLASS HEXACTINELLIDA: glass sponges. Spicules of silicon dioxide. Venus flower basket, glass rope sponge.

CLASS DEMOSPONGIAE: horny sponge. Complex sponge with numerous canals and pumping chambers. Skeleton of protein fibers. Bath sponge.

SUBKINGDOM METAZOA: Tissue and organ system levels of organization. Probably evolved from ciliate protists.

RADIATE PHYLA: Radial Symmetry.

Phylum Coelenterata: Radially symmetrical, thin-walled body, which consists of essentially two cell layers that enclose a saclike digestive cavity, the coelenteron. All carnivores with tentacles and stinging or snaring cnidoblasts. Both extracellular and intracellular digestion. Exist as polyps, medusae, or an alternating cycle of the two.

CLASS HYDROZOA: hydroids. Colonial, stationary feeding, polyps, and reproductive polyps. Asexual reproduction by budding new polyps. Sexual generation separate as medusae, which produce haploid sperm or egg. Sperm released externally to enter egg-bearing medusae. Examples are marine *Obelia* and freshwater *Hydra* (which lacks the medusae stage).

CLASS SCYPHOZOA: jellyfish. Primarily medusoid with gas-filled bell. Sexual reproduction includes an external sperm release and fertilization in the egg-bearing individual. Larval stage attaches to substratum and produces numerous medusae by budding. Examples are *Aurelia,* Portugese-Man-O-War.

CLASS ANTHOZOA: corals and anemones. Primarily polyps, with highly colonial corals and both colonial and solitary anemones. Asexual reproduction by budding of new polyps. Sexual reproduction by external release of sperm and fertilization in egg-bearing individuals. Larva develop into polyps directly. Examples are *Metridium, Actinia* (beadlet anemone), sea fan coral, brain coral.

Phylum Ctenophora: comb jellies. Radially symmetrical, thin-bodied animals, similar to coelenterates but with glue cells. Eight rows of combs (fused cilia). Luminescent. All marine. Examples are comb jellies and Venus' girdle.

THE BILATERAL PHYLA: Animals with bilateral symmetry.

ACOELOMATES: Animals with no coelomic development.

Phylum Platyhelminthes: flatworms. Dorsoventrally flattened worms found in marine, freshwater, and terrestrial environments. Some with complex organ systems developing from three embryonic cell layers. Highly branched but saclike digestive system.

CLASS TURBELLARIA: planarians. Free-living carnivores with extensive organ-system development. Hermaphroditic with copulatory organs for internal fertilization. Reproduce asexually by whole body fission. *Dugesia* (planaria).

CLASS TREMATODA: flukes. Parasitic with oral suckers for attachment and feeding. Life cycle may require two or more hosts. Examples are lung fluke and liver fluke.

CLASS CESTODA: tapeworms. Parasitic with minute head that bears a scolex for attachment. Reproductive structures in serial proglottids, budding continuously from scolex. Following fertilization, the egg-filled proglottids break away. Life cycle requires more than one host. Examples are beef and pork tapeworm.

Phylum Nemertina: proboscis worms. Thin ribbonlike worms with a complete, one-way digestive tract. Protrucible pharynx with piercing stylet. Marine.

PROTOSTOMATE—PSEUDO-COELOMATE PHYLA: Animals with mouth development first in the embryo and anus later. False coelom, not surrounded exclusively by mesodermally derived tissue.

Phylum Aschelminthes: roundworms and rotifers. Many with cylindrical bodies. Complete digestive tract.

CLASS NEMATODA: roundworms. Free-living soil nematodes and intestinal parasites. Long smooth body with longitudinal muscles. Separate sexes. Parasites may require more than one host. Most vertebrates infested. Examples are vinegar eels, *Ascaris,* hookworm, lung worms, pin worms.

CLASS ROTIFERA: rotifers. Free living. Cilia for movement. Complete digestive tract with complex grinding organ. Separate sexes, but parthenogenetic reproduction common.

PROTOSTOME—COELOMATE PHYLA: Protostomate phyla with a true coelom, surrounded by mesodermally derived tissue.

Phylum Ectoprocta: Phoronidia, Brachiopoda: the lophophorate phyla. Ciliated tentacles attached to ridgelike lophophore. Examples are moss animals, lampshells, and the wormlike *Phoronis.*

Phylum Annelida: segmented worms. Highly segmented worms with well-developed coelom, complete and specialized digestive tract, and closed circulatory system.

CLASS OLIGOCHAETA: earthworms. Burrowing, free-living soil scavengers. Complex muscular body wall with setae for burrowing and anchoring. Some concentration of ganglia in head region, but no external sensory appendages.

CLASS POLYCHAETES: marine worms. Free-living marine carnivores. Similar to earthworm but with sensory appendages, mandibles, and fleshy parapods for respiration and swimming. Examples are clamworms, tube worms, and fan worms.

CLASS HIRUDINEA: leeches. Ectoparasites with limited segmentation. Sucker mouth and muscular pharynx for blood-sucking.

Phylum Mollusca: the mollusks. Reduced coelom and segmentation. Foot, shell, and mantle are basic but highly varied. Open circulatory system in most.

CLASS AMPHINURA: chitons. Primitive body plan with long muscular foot, simple mantle, multiple gills, and shell in eight separate plates. Mouth with radula. Marine.

CLASS GASTROPODA: single-shell mollusks. Terrestrial. Large muscular foot, shell often coiled with body parts displaced. Foot often retractible. Shell reduced or absent in some. Marine and freshwater. Radula used in feeding by most. Examples are snails, whelks, limpets, sea hares, abalone, and slugs.

CLASS PELECYPODA: bivalve mollusks. Large, muscular, and retractible foot for digging. Hinged valves (shells) with strong retractor muscles. Mainly filter feeders. Marine and freshwater. Examples are clams, mussels, scallops, and oysters.

CLASS CEPHALOPODA: squid and octopus. Enlarged head with well-developed eyes and brain. Foot modified into tentacles that bear muscular suckers. Fast-moving predators with muscular beak and ink glands. Closed circulatory system. Complex mantle for jet propulsion. Marine. Examples are octopus, squid, and the chambered nautilus.

Phylum Onycophora: Peripatus. Segmented, wormlike body. but with jointed appendages that bear claws. Tracheal respiratory system.

Phylum Arthropoda: Animals with paired jointed appendages and an ectodermally secreted, chitinous exoskeleton. Segmented, but with great variation in segments. Universally distributed with greatest number of species of any phylum.

SUBPHYLUM CHELICERATA: Arthropods with six pairs of appendages, including four pairs of legs, paired sensory palps and chelicerae (fangs) rather than mandibles. Most have two body regions.

CLASS MEROSTOMATA: horseshoe crabs. Aquatic, with book gills. Mouth in cephalothorax.

CLASS ARACHNIDA: spiders and relatives. Terrestrial, mostly venomous carnivores. Book lungs. Examples are spiders, ticks, mites, scorpions, whipscorpions, harvestman, and daddy-longlegs.

CLASS PYCNOGONIDA: sea spiders. Marine. Short, slender body, seven pairs of appendages (four to six pairs of walking legs), sucking mouth parts.

SUBPHYLUM MANDIBULATA: Most with three pairs of walking appendages. Jaws or mandibles are present, along with antennas and compound eyes.

CLASS CRUSTACEA: Primarily aquatic, with two pairs of antennas, great diversity in body organization, with general body shortening and great specialization in the appendages. Many with a hardened exoskeleton containing calcium and a carapace covering the anterior regions. Includes lobsters, crabs, crayfish, shrimp, barnacles, water fleas, isopods, copepods, and many others.

CLASS CHILOPODA: centipedes. Terrestrial. Clearly segmented body with two regions. Flattened dorsoventrally. One pair of legs per segment, with venomous duct in first pair. Examples are *Scutigera* (house centipede) and *Scolopendra* (giant centipede).

CLASS DIPLOPODA: millipedes. Terrestrial. Body generally long and cylindrical with two clearly segmented body regions. Two pairs of legs on most segments.

CLASS INSECTA: insects. Largest number of species in any taxon. Inhabit all environments. Body in three regions with head, thorax, and abdomen. One pair of antennae, highly varied mouth parts, three pairs of walking legs, and two pairs of wings in most flying species. Body segmentation highly modified through specialization. Approximately twenty-five orders of insects. (The thirteen largest orders are reviewed in Table 22.1, pages 622–623.) Examples are beetles, moths, bees, ants, flies, bugs, grasshoppers, lice, termites, fleas, dragonflies, and many others.

DEUTEROSTOME—COELOMATE PHYLA: Animals with a true coelom, but following deuterostome development, with the anus forming in the embryo first, and the mouth forming later.

Phylum Echinodermata: spiny skinned animals. Animals with a five-part radial symmetry as adults, bilateral as larvae. They produce an endoskeleton of spiny calcareous plates. All have a water vascular system with tube feet in most. Fertilization is external. Marine.

CLASS CRINOIDIA: sea lilies. Cup-shaped body, with branched arms, no spines or pedicellaria, and tube feet without suckers.

CLASS HOLOTHUROIDIA: sea cucumbers. Long cylindrical body with microscopic plates. Arms, spines, and pedicellaria absent. *Cucumaria.*

CLASS ECHINOIDEA: urchins and dollars. Round and flattened or spherical body without arms. Rigid endoskeleton with movable spines. Pedicellaria in some. Tube feet with suckers. Examples are *Echinus* (an edible sea urchin), *Spatangus, Echinocardium,* and Paracentrotus (boring urchin).

CLASS ASTEROIDEA: sea stars. Arms prominent. Short spines. Pedicellaria. Prominent ambulacral groove on underside, with rows of tube feet bearing suckers. Examples are the common *Pisaster* and *Echinaster* (the thorny star).

CLASS OPHIUROIDEA: serpent and brittle stars. Arms prominent but flexible in horizontal plane only. Movable spines, tube feet without suckers, and no pedicellaria.

Phylum Hemichordata: acorn worms. Marine worm that forms a U-shaped burrow. Acorn-shaped proboscis with collar. Gill slits along the pharynx. Larva resembles that of the echinoderms. A second group, the pterobranchs, resemble bryozoans but have the collar and gill slits. The dorsal nerve cord is present in both.

Phylum Chordata: the chordates, including the vertebrates. Chordates have gill slits, a notochord, and post anal tail at some stage in their lifetime. They also have a hollow dorsal nerve cord.

SUBPHYLUM UROCHORDATA: tunicates and salps. Marine filter feeders. Adult with soft, saclike body with gill slits and open circulatory system. Some with cellulose in the surrounding tunic. Chordate characteristics primarily in the larva. *Ciona.*

SUBPHYLUM CEPHALOCHORDATA: lancelet. Marine filter feeders. Fishlike body form. Notochord persists in adults. Prominent gill slits. Muscles of body in serial myotomes. Sensory cirri and mouth with small tentacles. *Amphioxus.*

SUBPHYLUM VERTEBRATA: the vertebrates. Vertebral column of bone or cartilage. Increasing cephalization (brain development). Ventral heart, dorsal aorta. Two pairs of limbs. Sexes separate.

CLASS AGNATHA: jawless fishes. Includes the extinct Ostracodermi and the Cyclostoma. Cyclostomes have a cartilaginous skeleton, eellike body, and a round mouth, with a rasping device in some. Examples are the lamprey and hagfish.

CLASS CHONDRICHTHYES: cartilaginous jawed fishes. Mostly marine. Cartilaginous endoskeleton. Placoid scales form tough coarse skin covering. Gill openings form slits. Internal fertilization and ovoviviparous development. Examples are sharks, rays, chimera.

CLASS OSTEICHTHYES: bony fishes. Fishes in which most cartilage is replaced by hard bone.

Subclass Sarcopterygii: lobe-finned fishes. Fleshy paired fins.

Order Crossopterygii: ancient fishes. One known survivor, *Latimeria,* a coelacanth. Terrestrial vertebrates are believed to have evolved from the crossopterygians.

Order Dipnoi: lung fishes. Functional nostrils. Breathe air by gulping it into lungs. Limited habitat. *Epiceratodus.*

Subclass Actinopterygii: ray-finned fishes. Main line of bony fishes. Consists of three orders: Chondrostei, Holostei, and Teleostei. The first is represented by sturgeons and paddlefishes today. The second order is represented by gar pikes and the bowfin. The third order contains the great majority of bony fishes with about 20,000 species representing the largest single vertebrate group. Examples are perch, bass, trout, mackerel, eel, sunfish, tuna.

CLASS AMPHIBIA: the amphibians. Both freshwater and terrestrial. Moist, vascular skin without scales. Most are external fertilizers, requiring water for development from aquatic, gilled larva to adult form. Three orders include Anura (frogs and toads), Urodela (salamanders), and Apoda (caecilians).

CLASS REPTILIA: the reptiles. Mostly terrestrial species with some having returned to fresh water and marine environments. Dry skin with scales. Air breathers throughout life. Most are oviparous with some ovoviviparity. Fertilization is internal through copulation. Young develop in the self-contained egg. Four orders include Chelonia (turtles), Squamata (snakes and lizards), Crocodilia (crocodiles and alligators), and Rhynchocephalia (*Sphenodon* or tuatara).

CLASS AVES: birds. Primarily terrestrial, flying species. Skeleton modified for lightness by reduction and fusion, loss of teeth, and hollow bone structure. Body covering of feathers for flight surfaces and insulation. Homeothermic, with four-chambered heart and highly efficient respiratory system. Most variation occurs in beaks and feet, which often reflect feeding niche. Fertilization is internal, but development is oviparous, utilizing the self-contained egg. Examples are flightless birds such as the ostrich, kiwi, and penguin, along with numerous flying species.

CLASS MAMMALIA: the mammals. Primarily terrestrial, but with numerous marine and freshwater aquatic species (all air breathers). Mammals have a body covering of hair, a glandular skin, and produce milk in mammary glands. The skull has undergone considerable fusion of bones. The specialized teeth are socketed and undergo only one replacement. All are homeothermic, with four-chambered hearts. Fertilization is internal through copulation and most are placental.

Subclass Protheria: egg layers. The few species of egg layers live in highly restricted ranges. Milk is secreted through pores rather than nipples. Birdlike cloaca is present. Includes the duckbill platypus and spiny anteater, both included in Order Monotremata.

Subclass Metatheria: marsupials. Pouched mammals, primarily from Australia and New Zealand. Internal fertilization through copulation, but only a rudimentary placenta (pseudoplacenta) forms. Young born immature and complete their development in the marsupium, where they fasten to a nipple to nurse. Examples are kangaroos, koala, wombat, and the widely distributed opossum, and others organized into Order Marsupalia.

Subclass Eutheria: placental mammals. Mammals whose young complete their prenatal development in the uterus, nourished by a placenta. (The seventeen orders of placental mammals are reviewed in Table 23.1, pages 654–655.)

APPENDIX B

A Biological Lexicon

Being a list of Greek and Latin prefixes, suffixes, and word roots commonly used in biological terms, alphabetized by the most common combining form in English; with assorted and miscellaneous examples illustrating each usage.

a-, an- [Gk. *an-*, not, without, lacking]: anaerobic, atom, abiotic, anonymous, anaphrodisiac, apnea, anorexia, apathy, anesthesia, aseptic, asexual, ataxia, anemia

acro- [Gk. *akros,* highest]: acrocentric, acrophobia, acrocephalic, acromegaly, acrosome

ad- [L. *ad-*, toward, to]: adhesion, admixture, adopt, adrenalin, adventitious, adsorption, advanced

allo- [Gk. *allos,* other]: allele, allopatric, allozyme, allosteric, allotetraploid, allelopathy

ammo- [from ammonium salts produced from camel dung near the Libyan shrine to the god Ammos]: ammonia, ammonium, amine, amide, amino group, amino acid, aminopeptidase, antihistamine

amphi- [Gk. *amphi-*, two, both, both sides of]: amphibian, amphidiploid, Amphineura, amphipod, ampholyte

ana- [Gk. *ana-*, up, up against]: anaphase, analogy, analysis, anatomy

andro- [Gk. *andros,* an old man]: androecium, androgen, androgynous, polyandry

anti- [Gk. *anti-*, against, opposite, opposed to]: antibiotic, antibody, antigen, antidiuretic hormone, antihistamine, antiseptic, antipathy, antiparallel, antihistamine

apo- [Gk., *apo-*, different]: apoenzyme, apomixis, apodeme, aponeurosis

archeo- [Gk. *archaios,* beginning]: archegonium, archenteron, archaic, Archeozoic, menarche

arthro- [Gk. *arthron,* a joint]: arthropod, arthritis, arthrodire, condylarthra

auto- [Gk. *auto-*, self, same]: autoimmune, autotroph, autosome, autonomic, autoeroticism, autopolyploid, autonomous, autotomy

auxo- [Gk. *aux,* to grow or increase]: auxin, auxospore, auxotroph, auxillary

bi-, bin- [L. *bis,* twice; *bini,* two-by-two]: binary fission, binocular vision, biennial, bifurcating, bicarbonate, bilateral, binomial, bipolar

bio- [Gk. *bios,* life]: biology, biomass, biome, biosphere, biosynthesis, biotic, amphibian, abiotic, antibiotic, biotin

blasto-, -blast [Gk. *blastos,* sprout; now "pertaining to the embryo"]: blastoderm, blastodisc, blastopore, blastula, trophoblast, osteoblast, neuroblast, neuroblastoma

brachi- [Gk. *brachion,* arm]: brachiation, brachiopod

brachy- [Gk. *brachys,* short]: brachydactyly, brachycardia, brachycephalic, brachyuran

branchio-, -branch [Gk. *branchia,* a gill]: branchiopod, elasmobranch, branchial arch, pseudobranch, opistobranch, lamellibranch, nudibranch

broncho- [Gk. *bronchos,* windpipe]: bronchus, bronchi, bronchiole, bronchitis

carcino- [Gk. *karkin,* a crab, cancer]: carcinogen, carcinoma

cardio- [Gk. *kardia,* heart]: cardiac, cardiology, brachycardia, tachycardia, myocardium, electrocardiogram

cephalo- [Gk. *kephale,* head]: cephalic index, cephalization, cephalochordate, cephalopod, cephalothorax, encephalitis, encephalogram, brachycephaly, diencephalon

chloro- [Gk. *chloros,* green]: chlorobacterium, chlorogen cell, chloroxybacterium, chlorophyll, chloroplast, chlorosis, chlorine

chole- [Gk. *chole,* bile; Gk. cholecyst, gall bladder]: cholesterol, cholera, acetylcholine, cholinesterase, cholecystectomy, cholecystokinin, hypercholestrolemia

chondro- [Gk. *chondros,* cartilage]: achondroplasia, chondrocranium, chondrodysplasia, chondroitin, chondroma, chondroblast

chromo- [Gk. *chroma,* color]: chromosome, chromoplast, chromatophore, chromatin, hemochromotosis

coelo-, -coel [Gk. *koilos,* hollow, cavity]: coelacanth, coelenteron, coelenterate, coelom, hydrocoel, encephalocoel, acoel, enterocoelous, pseudocoelom

com-, con-, col-, cor-, co- [L. *cum,* with, together]: coenzyme, cohesion, commensal, compete, compound, conjugation, convergence, covalent

cranio- [Gk. *kranios,* L. *cranium,* skull]: cranial, craniotomy, cranium

cuti- [L. *cutis,* skin]: cutaneous, cuticle, cutin

cyclo-, -cycle [Gk. *kyklos,* circle, ring, cycle]: cyclosis, cyclostome, pericycle

cyto-, -cyte [Gk. *kytos,* vessel or container; now, "cell"]: cytoplasm, cytology, cytochrome, cytokinesis, polycythemia, erythrocyte, leucocyte, amebocyte, cytosine, cytostome

de- [L. *de-*, "away, off"; deprivation, removal, separation, negation]: deciduous, decomposer, deoxyribose, decarboxylase, dehydration, dephosphorylation

derm-, dermato- [Gk. *derma*, skin]: dermatitis, dermis, epidermis, dermal, ectoderm, endoderm, mesoderm, blastoderm

di- [Gk. *dis*, twice]: dicaryon, dicotyledon, dioxide, dipole, dipleurula, diencephalon, diatom

dia- [Gk. through, passing through, thorough, thoroughly]: diabetes, diagnosis, dialysis, diapause, diaphragm, diastole

diplo- [Gk. *diploos*, two-fold]: diplotene, diploid, diploblastic

eco- [Gk. *oikos*, house, home]: ecology, androecium, gynoecium, ecosphere, ecosystem, ecotone, ecotype, economy

ecto- [Gk. *ektos*, outside]: ectoderm, ectoplasm, ectomorph, ectoparasite, ectopia, ectoskeleton, ectoproct

en- [Gk., L. in, into]: encephalon, encephalitis, entropy, encapsulate, energy, engram, environment, enzyme, diencephalon

endo- [Gk. *endon*, within]: endocardium, endocrine, endoderm, endodermis, endogenous, endometrium, endoparasite, endoskeleton, endothelium, endotoxin, entoproct, endergonic, endoplasmic reticulum

entero- [Gk. *enteron*, intestine]: enteron, enteropneust, archenteron, enterocoelous, enteritis, enteric, entameba

epi- [Gk. *epi*, on, upon, over]: epiblast, epiboly, epicardium, epicanthal fold, epicotyl, epidemic, epidermis, epididymis, epigastric, epiglottis, epilepsy, epinephrine, epiphyte, epithelium

eu- [Gk. *eus*, good; *eu*, well; now "true"]: eubacterium, eucalyptus, eucaryote, euchromatin, euphoria, euplacental, eusocial, euthanasia, eutherial

ex-, exo-, ec-, e- [Gk., L. out, out of, from, beyond]: emission, ejaculation, evagination, evolution, excretion, exergonic, exhale, exocrine, exocytosis, exogenous, exon, exoskeleton, experiment, extrude, appendectomy, tonsillectomy

extra- [L. outside of, beyond]: extracellular, extraembryonic, extravert, extrapolate, extrauterine

-fer [L. *ferre*, to bear]: fertile, fertilization, afferent, efferent, Carboniferous, conifer, rotifer

galacto- [Gk. *galakt-*, milk]: galactose, galactic [Milky Way]

gam-, gameto- [Gk. gamos, marriage; now usually in reference to gametes (sex cells)]: gamete, gametangia, gamont, bigamy, polygamy, gametogenesis, isogamete, isogamous, heterogamy

gastro- [Gk. *gaster*, stomach]: gastric, gastrula, gastritis, gastrectomy, gastrin, gastrovascular cavity, gastropod, *Drosophila melanogaster*, gastrotrich

gen- [Gk. *gen*, born, produced by; Gk. *genos*, race, kind; L. *genus*, *generare*, to beget]: gene, genetics, polygenic, genotype, geneology, generic, genus, gender, general, genital, genocide, generate, regenerate, glycogen, florigen, pyrogen, fibrinogen, androgen, estrogen, heterogenous, endogenous, exogenous

gluco-, glyco- [Gk. *glykys*, sweet; now pertaining to sugar]: glucose, glycogen, glycolysis, glucouronidase, glycoside, glycine, glycerol, glyceride, phosphoglycerate, glycol, glycolipid, glycoprotein

gyn-, gyno-, gyneco- [Gk. *gyne*, woman]: gynecology, gynecomastia, gynandromorph, gynoecium, polygyny, androgynous, misogyny, epigyny, hypogyny

hemo-, hemato-, -hemia, -emia [Gk. *haima*, blood]: hematocrit, hematology, hematoma, hematopoisis, heme, hemocoel, hemodialysis, hemoglobin, hemophilia, hemorrhage, hemorrhoid, hemostat, hemotoxin, hemorrhagic septicemia, anemia, leukemia, polycythemia, toxemia, hypercholesterolemia

hepato- [Gk. *hepar, hepat-*, liver]: heparin, hepatitis, hepatoma, hepatic portal system

hetero- [Gk. *heteros*, other, different]: heterochromatin, heterosexual, heterogeneous, heterogamous, heterograft, heterologous, heteromorphic, heterozygote

histo- [Gk. *histos*, web of a loom, tissue; now pertaining to biological tissues]: histology, histone, histocompatibility, histamine, antihistamine, histidine, histolytic

homo-, homeo- [Gk. *homos*, same; Gk. *homios*, similar]: homeostasis, homeothermy, homogeneous, homogenized, homograft, homosexual, homologous, homologue, homology, Homoptera, homozygote

hydro- [Gk. *hydor*, water; now, confusingly, pertaining either to water or to hydrogen]: hydrogen, hydrate, dehydration, hydraulic, carbohydrate, hydrocephalus, hydrolytic, hydrometer, hydrophobia, hydride, hydrocarbon, hydrochloric acid, hydroxide

hyper- [Gk. *hyper*, over, above, more than]: hyperacidity, hypercholesterolemia, hyperglycemia, hyperkinetic, hypersensitive, hypertension, hyperthermia, hyperthyroid, hypertonic, hyperventilation

hypo- [Gk. *hypo*, under, below, beneath, less than]: hypobranchial, hypochlorite, hypochondria, hypocotyl, hypodermic needle, hypoglossal nerve, hypoglycemia, hypogyny, hypophysis, hypotension, hypothalamus, hypothermia, hypothesis, hypothyroid, hypotonic, hypoxia

in-, im- [L. *in*, in]: imprinting, inbreeding, instinct, insulin, intelligence

inter- [L. *inter*, between, among, together, during]: interaction, interbreed, intercellular, intercostal muscles, interface, interior, intermolecular, interoceptive, interphase, interpopulation, interspecific, interstitial

intra-, intro- [L. *intra*, within]: intracellular, intracranial, intramolecular, intrauterine, intravenous, introduced, introitus, intromittent organ, introversion

-itis [L., Gk. *-itis*, inflammation of]: arthritis, bronchitis, dermatitis, enteritis, gastritis, neuritis

leuko-, leuco- [Gk. *leukos*, white]: leukocyte, leukemia, leucosis, leukoplast, leukorrhea, leukopenia, leukocytosis, leukoderma

-logue, -logy [Gk. *-logos*, word, language, type of speech]: analogue (analogy), homologue (homology), dialogue, monologue, travelogue

-logy [Gk. *-logia*, the study of, from *logos*, word]: anthropology, biology, cytology, embryology, histology

-lysis, lys-, lyso-, -lyze, -lyte [Gk. *lysis,* a loosening, dissolution]: lyse, lysin, lysis, lysogeny, lysozyme, paralysis, hydrolysis, phosphorolysis, analysis, analyze, catalysis, catalytic, electrolyte

macro- [Gk. *makro-,* now "great," "large"]: macrocyte, macrogamete, macrogametophyte, macromere, macromolecule, macromutation, macronucleus, macroscopic, macrospore

mega-, megalo-, -megaly [Gk. *megas,* large, great, powerful]: megagamete, megalecithal, megalith, megalocephalic, megalomaniac, megaspore, Megatherium, megaton [atom bomb], acromegaly, cardiomegaly

-mere, -mer, mero- [Gk. *meros,* part]: blastomere, centromere, chromomere, macromere, micromere, dimer, isomer, isomerization, polymer, meroblastic, merodiploid, merozoite

meso-, mes- [Gk. *mesos,* middle, in the middle]: mesencephalon, mesentery, mesic, Mesoamerican, mesoderm, mesoglea, mesolecithal, mesomorph, mesonephros, mesophyll, meson, Mesopotamia, mesothelium, Mesozoic

meta-, met- [Gk. *meta,* after, beyond; now often denoting change]: metabolism, metacarpal, metamerism, metamorphic, metamorphosis, metanephros, metaphase, metastasis, metatarsal, metatherian, Metazoa, metencephalon, methemoglobin

micro- [Gk. *mikros,* small]: microbalance, microbe, microbiology, microcephalic, microclimate, micrococcus, microgamete, microgametophyte, microliter, micrometer, micronucleus, micropyle, microscope, microspore, microtome

myo- [Gk. *mys,* mouse, muscle]: myasthenia gravis, myocardial infarction, myocardium, myocyte, myoglobin, myoma, myosin, myotome

neuro- [Gk. *neuron,* nerve, sinew, tendon]: neurasthenia, neurenctoderm, neuritis, neuroanatomy, neuroblast, neurofibril, neuroma, neuron, neurosis, neurotoxin, neurotransmitter, neurula, aponeurosis

oligo- [Gk. *oligos,* few, little]: oligocarpous, Oligocene, oligochaete, oligocythemia, oligopeptide, oligosaccharide, oligotrophic

-oma [Gk. *-oma,* tumor, swelling]: adenoma, carcinoma, glaucoma, hematoma, lipoma, neuroblastoma, sarcoma

oo- [Gk. *oion,* egg]: oogenesis, oogonium, oophyte, oospore, ootid

-osis [Gk. *-osis,* a state of being, condition]: arteriosclerosis, cirrhosis, halitosis, hemochromatosis, metamorphosis, tuberculosis, neurosis

osteo-, oss- [Gk. *osteon,* bone; L. *os, ossa,* bone]: os penis, ossicle, ossification, ossified, ostectomy, Osteichthys, osteitis, osteoblast, osteoclast, osteomyelitis, osteopath, periosteum, teleost

para- [Gk. *para-,* alongside of, beside, beyond]: paradigm, paralysis, paramedic, parameter, parapatric, paraphyletic, parasite, parasympatric, parathyroid, Parazoa, antiparallel

patho-, -pathy, -path [Gk. *pathos,* suffering; now often disease or the treatment of disease]: pathogen, pathology, pathological, osteopath, psychopath, homeopathy

peri- [Gk. *peri,* around]: perianth, pericardial, pericarp, pericycle, periderm, perigyny, perimeter, perinium, periosteum, photoperiod, peripheral, peristome, peritoneum

phago-, -phage [Gk. *phagein,* to eat]: phagocyte, phagocytosis, phage, bacteriophage

plasm-, -plasm, -plast, -plasty [Gk. *plasm,* something molded or formed; Gk. *plassein,* to form or mold]: plasma, plasma membrane, plasmid, plasmasol, plasma thromboplastin, plasminogen, plasmodium, plasmolysis, germ plasm, cytoplasm, nucleoplasm, neoplasm; plastid, chloroplast, chromoplast, leucoplast, protoplast; dermoplasty, stomatoplasty, plastic surgery

-pod [Gk. *pod,* foot]: amphipod, Apoda, arthropod, cephalopod, decapod crustacean, gastropod, isopod, pseudopod, podiatrist

poly- [Gk. *poly-, polys,* many]: polyandry, polybasic, polychaete, polyclad flatworm, polycotyledonous, polydactyly, polycythemia, polygamy, polygenic inheritance, polygyny, polymer, polymorphism, polypeptide, polyploid, polysome

-rrhea [Gk. *rhoia,* flow]: amenorrhea, diarrhea, gonorrhea, leukorrhea, rheostat

septi-, -sepsis, -septic [Gk. *septicos,* rotten, infected]: septic, septicemia, aseptic, antiseptic

-some, somat- [Gk. *soma,* body; Gk. *somat-,* of the body]: somatic cell, somatoplasm, psychosomatic, centrosome, acrosome, chromosome, ribosome, polysome

-stat, -stasis, stato- [Gk. *stasis,* stand]: statocyst, statolith, metastasis, stasipatric, thermostat, venistasis, electrostatic, hydrostatic

stoma-, stomato-, -stome [Gk. *stoma,* mouth, opening]: stoma, stomatitis, stomatoplasty, stomodeum, cyclostome, cytostome, deuterostome, protostome, colostomy

sym-, syn- [Gk. *syn,* with, together]: symbiont, symbiosis, symmetry, sympathetic nervous system, pubic symphysis, symposium ["drinking together"—an important scientific term], symptom, synapsis, synaptinemal complex, synchrony, syncytium, syndactyly, syndrome, synergist, synovial fluid, synsacrum, synthesis, synthetic

taxo-, -taxis [Gk. *taxis,* to arrange, put in order; now often referring to ordered movement]: taxonomy, taxon, syntax, ataxia, chemotaxis, geotaxis, phototaxis

tomo-, -tome, -tomy [Gk. *tome,* a cutting; Gk. *tomos,* slice]: tomography, microtome, atom [you can't cut it], anatomy, dichotomy, craniotomy, lobotomy, phlebotomy, appendectomy, tonsillectomy

trich-, tricho-, -trich [Gk. *tricho-,* combining form of *thrix,* hair]: trichina, trichinosis, trichocyst, trichome, trichomonas, trichonymph, gastrotrich

tropho-, -troph, -trophy [Gk. *trophe,* nutrition]: trophic level, trophoblast, atrophy, hypertrophy, autotroph, auxotroph, heterotroph

trop-, tropo-, -tropy, -tropism [Gk. *tropos,* to turn, to turn toward]: tropism, tropical, troposphere, entropy, geotropism, phototropism

ur-, -uria [Gk. *ouron,* urine]: uracil, uranalysis, urea, uremia, ureter, urethra, uric acid, urine, uremia, albuminuria, alcaptonuria, cystinuria, glycosuria, phenylketonuria

uro-, -uran [Gk. *oura,* tail]: urochordate, uropod, uropygium, anuran, macruran, brachyuran

GLOSSARY

A band one of the bands of striated muscle, corresponding to the fixed length of the myosin filament.
abdomen 1. in mammals, the portion of the body between the diaphragm and the pelvis; the cavity of this region. 2. in other vertebrates, the corresponding part of the body, containing the stomach, intestines, liver, and reproductive organs. 3. in arthropods, the posterior section of the body, behind the thorax or cephalothorax.
abiotic 1. characterized by the absence of life. 2. nonbiological; factors independent of living organisms.
abiotic control control of population numbers by nonbiological factors, such as weather.
ABO blood group system a genetically controlled polymorphic cell surface polysaccharide antigen.
abortion the spontaneous or *induced* expulsion of the human fetus before it is viable.
abscisic acid a plant hormone that causes leaf abscission.
abscission the normal separation, through an abscission zone, of flowers, fruit, and especially leaves from plants.
abscission zone also *abscission layer,* a layer of specialized, cutinized parenchyma cells at the base of a leaf petiole, fruit stalk, or branch, in which normal separation occurs.
absorption 1. the being taken up by capillary, osmotic, chemical, or solvent action. 2. interception of light or sound waves.
absorption spectrum a graph indicating the relative light absorption of a molecule as a function of the wave length of light.
abyssal region the lowest depths of the ocean, especially the bottom waters; also, *abysmal region.*
accessory pigment a pigment other than chlorophyll that absorbs energy from light and is capable of transferring that energy to chlorophyll for photosynthesis.
acellular not composed of cells; not divisible into smaller cellular units; also, *noncellular.*
Acetabularium a siphonous marine green alga, also, *mermaid's cup.*
acetabulum the socket in the hip bone that receives the head of the thigh bone.
acetic acid a colorless, pungent, water-miscible liquid, CH_3COOH, the essential constituent of vinegar. The corresponding anion, CH_3COO^-, is acetate.
acetyl CoA a key intermediate in metabolism, consisting of an acetyl group covalently bonded to coenzyme A.
acetyl group acetic acid (acetate) in an ester linkage with R—OH to form R—O—CO—CH_3.
acetylcholine a neurotransmitter released and hydrolyzed in synaptic transmission and in the initiation of muscle contraction; biochemically, it is an acetic acid ester of choline.
acetylcholinesterase a membrane-bound enzyme that hydrolyzes acetylcholine in the course of synaptic nerve impulse transmission.
achondroplasia a genetic pathology, usually due to a dominant allele, in which the conversion of cartilage into bone is defective, resulting in extremely short appendages, dwarfism, and facial anomalies.
acid a compound capable of neutralizing alkalis and of lowering the pH of aqueous solutions. Acids are proton donors that yield hydronium ions in water solutions.
acoelomate lacking a coelom.
acromegaly a pathology of adults caused by abnormally high levels of growth hormone, as from a pituitary tumor; characterized by enlargement of bones of the face, head, hands, and feet.
acrosomal filament a filament produced by rupture of the acrosome of the sperm of echinoderms, which penetrates the egg and initiates fertilization.
acrosome an organelle in the tip of a spermatozoan, which ruptures in fertilization with the release of enzymes and (in echinoderms) of an acrosomal filament.
actin a cytoplasmic protein known in both globular and fibrous forms. See *actin filament.*
actin filament also *microfilament,* a cytoplasmic protein fiber found in the cytoskeleton and in muscle contractile units, associated with cellular movement.
actinomycin an antibiotic derived from actinomyces filamentous bacteria.
action-specific energy (ASE), a hypothetical endogenous tension that increases in the nervous system until it is discharged, either on its own or due to the perception of an appropriate stimulus. The discharge produces motor pattens that result in a particular behavior.
action spectrum a graph relating light-induced biological activity (e.g., carbon fixation or oxygen release in photosynthesis) per photon as a function of wave length.
active site that part of an enzyme directly involved in specific enzymatic activity.
active transport energy-requiring transport of a substance across the cell membrane usually against the concentration gradient.
adaptation 1. an adjusting to conditions. 2. a change that improves function. 3. any alteration in the structure or function of an organism or any of its parts that results from natural selection and by which the organism becomes better able to survive and multiply in its environment.
additive law in mathematical probability theory, the relation that the probability of the occurrence of one of two or more mutally exclusive events is equal to the sum of the individual probabilites.
adenine a purine, one of the nitrogenous bases found in both DNA and RNA as well as in several coenzymes.
adenosine a nucleoside formed of adenine and ribose covalently linked.
adenyl cyclase an enzyme, usually incorporated into the cell membrane, that is capable of transforming ATP into cyclic AMP and pyrophosphate.
adhesion the attraction between two dissimilar substances; the molecular force of attraction in the area of contact between unlike bodies, or between a solid and an adhering liquid.
ADP adenosine diphosphate, a compound of adenine, ribose, and two phosphate groups.
adrenal also *adrenal gland,* one of a pair of endocrine (ductless) glands located above the kidneys, each consisting of two functionally and embryologically distinct parts, a central *medulla* and an outer *cortex.*
adrenal cortex the outer portion of the adrenal gland, mesodermal in origin and producing cortisol, estrogens, androgens, and various other steroid hormones.
adrenal medulla the inner portion of the adrenal gland, derived embryologically from neural crest cells and producing epinephrine and norepinephrine (adrenalin and noradrenalin) and its principle hormonal products.
adrenocorticotropic hormone (ACTH), a pituitary hormone that stimulates the production of cortisol and other hormones of the adrenal cortex.
advanced relatively later in evolutionary origin or state, as opposed to *primitive.* See also *derived.*
adventitious in plants, 1. originating from mature nonmeristematic tissue, especially if such development is not ordinarily expected. 2. growth of a root or bud in an unusual place; *adventitious root,* a secondary root growing from stem tissue; *adventitious bud,* a bud growing directly from a root, or from callus tissue of a cut stem.
aerobe an organism that utilizes oxygen in cellular respiration.
aerobic 1. growing in the presence of oxygen. 2. utilizing oxygen in respiration.
African sleeping sickness a disease of cattle and humans caused by a parasitic protozoan, transmitted by the tsetse fly and eventually affecting the central nervous system.
agar also *agar-agar,* a polysaccharide produced by red algae; also, a gel made from this material, used as a moist semi-solid base for the experimental growth of microorganisms.
age profile see *population pyramid.*
agglutination the clumping of bacteria, erythrocytes, or other cells by an antibody or plant lectin.
aggregation a number of independent organisms grouped together either in a casual, temporary way or for more permanent mutual reasons.
aggression hostility, attack, or threat, especially unprovoked, usually against a competitor or potential competitor.
agnathan a member of the vertebrate class *Agnatha* (see Appendix A).
agricultural revolution also *green revolution,* the introduction of advanced agricultural technology and improved strains into the third world economy.
A horizon the outer, usually dark-colored and light-textured layer of soil, rich in humus.

air sac in birds and insects, any of numerous extensions of the respiratory system as an air-filled, membraneous sacs into various body parts.
alarm chemicals pheromones released as alarm signals.
albinism a marked and abnormal, genetically based deficiency in pigmentation.
albino a human being, plant, or animal lacking or abnormally deficient in pigmentation.
albumen 1. egg white. 2. also *ovalbumin,* the principle protein of egg white.
albumin 1. any of a class of clear, water-soluble plant or animal proteins. 2. *serum albumin,* a clear, water-soluble constituent of blood plasma, thought to serve detoxifying and osmotic functions. 3. *albumen* (egg white protein).
alcohol 1. ethyl alcohol, C_2H_5OH. 2. any of a class of organic compounds analogous to ethyl alcohol in having a $-CH_2OH$ group and lacking aldehyde or carboxyl groups.
aldehyde any organic compound with the reactive group $-CHO$.
aldosterone a steroid hormone produced by the adrenal cortex, involved in controlling sodium and potassium reabsorption by the kidney tubule epithelium.
aleurone 1. also *aleurone layer,* a single, outermost layer of cells in the endosperm of cereals. 2. pigmented protein granules found in this layer.
alga (pl. *algae*), a somewhat ambiguous term originally referring to any oxygen-evolving photosynthetic organism other than a "higher plant." Now, 1. any photosynthetic member of the kingdom *Protista.* 2. a member of the plant divisions *Rhodophyta* (red alga), *Phaeophyta* (brown alga), or *Chlorophyta* (green alga). 3. a cyanophyte (blue-green alga) (obsolete usage).
alkaloid any organic, nitrogenous, alkaline, water-soluble, bitter-tasting compound produced by plants, usually with pharmacological effects; e.g., morphine, nicotine, quinine, caffeine.
alkalosis an alkaline condition in the entire body characterized by abnormally high levels of bicarbonate in the blood.
alkaptonuria a genetic pathology caused by the lack of an essential enzyme and characterized by the excretion of homogentisic acid.
allantois one of the extraembryonic membrances. In birds and reptiles it serves as a repository for nitrogenous wastes of the embryo.
allele a particular form of a gene at a locus.
allele frequency the proportion (relative number) of alleles of a given type at a specific gene locus.
Allen's rule the generalization that members of a warm-blooded taxon tend to have longer, thinner extremities in warmer regional and tropical latitudes and shorter, thicker extremities in colder regions and high latitudes.
allometry the science that deals with empirical laws governing the relative growth and sizes of parts in relation to the growth and sizes of whole organisms.
allopatric living in different geographical regions; separated by geographical barriers.
allosteric site on certain enzymes, a secondary binding site for small metobolites involved in the regulation of enzymatic activity.
allostery the control of enzyme activity through the binding of small metabolites at one or more secondary (allosteric) binding sites.
allotetraploidy adj., derived from the hybridization of two distinct species and carrying the full diploid chromosome complements of both parental species; n., an allotetraploid organism. Also, *amphidiploid.*
alpha amylase a starch-digesting enzyme.
alpha glucose glucose in one of its two possible ring configurations.
altitudinal ecocline see ecocline.
altricial of birds, helpless at hatching and requiring parental feeding and care. Compare *precocial.*
altruism behavior that is directly beneficial to others at some cost or risk to the altruistic individual.
alula 1. a small feathered, clawed spike (the thumb) at the leading edge of the wing that functions in the flight control of birds. 2. the feathers growing from the first digit of the wing.
alveolus (pl. *alveoli*), the air cells of the vertebrate lung, formed by the terminal dilation of tiny air passages.
amanitin a deadly poison produced by mushrooms of the genus *Amanita,* which includes the death angel and the death cap.
ameba also *amoeba,* 1. any protozoan of the large genus *Amoeba,* characterized by lobose pseudopods and the lack of permanent organelles or supporting structures; 2. any ameboid protozoan, such as the ameboid stage of a flagellate or sporozoan.
amebic dysentery a contagious disease caused by an enteric ameba.
ameboid also *amoeboid,* like an ameba in moving or changing shape by protoplasmic flow.
ameboid movement movement or change in the shape of a cell by cytoskeleton deformation, the formation of pseudopods, and protoplasmic flow.
amino acid also *alpha-amino acid.* 1. any organic molecule of the general formula R—CH (NH_2)COOH, having both acidic and basic properties; 2. any of the 20 subunits found as normal constituents of polypeptides.
amino group the univalent group $-NH_2$, often ionized as $-NH_3^+$.
aminopeptidase a peptidase (proteolytic enzyme) that attacks peptide bonds adjacent to the amino end of a polypeptide chain.
ammonia 1. a colorless, pungent, poisonous, high soluble gas, NH_3. 2. this compound dissolved in water; ammonium hydroxide.
ammonification the production of ammonia by soil organisms, particularly by the reduction of nitrates or nitrites.
ammonium ion the univalent ion NH_4^+. Ammonia (NH_3) reacts with water to form an ammonium ion and a hydroxyl ion.
amnion the innermost of the extraembryonic membranes of reptiles, birds, and mammals, and thus the sac in which the embryo itself is suspended.
amoeba see *ameba.*
AMP *adenosine monophosphate,* a molecule consisting of adenine, ribose and one phosphate group.
ampulla (pl. *ampullae*) a dilated portion of a canal or duct, such as the ampulae of the semicircular canals of the inner ear, and the ampullae of the echinoderm water vascular system, which are muscular sacs located above the tube feet.
amylase any enzyme that digests starch.
amylopectin a form of plant starch consisting of branched chains of alpha glucose subunits.
amylose a form of plant starch consisting of unbranched chains of alpha glucose subunits.
anaerobe an organism that does not require free oxygen to live. *facultative anerobe,* an organism that can utilize free oxygen when available but does not require it (e.g., yeast). *obligate anerobe,* an organism that cannot tolerate free oxygen.
anaerobic 1. living or functioning in the absence of oxygen. 2. *anaerobic respiration,* respiration not requiring or utilizing oxygen; glycolysis.
analog also *analogue,* 1. an analogous relationship. 2. an artificial biochemical sharing some properties (e.g. enzyme binding) with a natural biochemical.
analogous in comparative morphology, similar in form or function but derived from different evolutionary or embryonic precursors (e.g., the wings of insects and birds).
analogy similarity in function, and perhaps appearance, but stemming from different evolutionary origins.
anaphase the stage of mitosis or meiosis in which the centromeres divide and separate and the two daughter chromosomes of each chromosome travel to opposite poles of the spindle; and in which the spindle elongates while the centromeric spindle fibers shorten.
anaphase I of meiosis the stage at which the chromosomal tetrads part and the homologous centromeres of each pair travel to opposite poles of the spindle.
anaphase II of meiosis the stage of which the centromeres divide and daughter centromeres separate, moving to opposite poles of the spindle.
androecium a whorl of stamens; the stamens of a flower taken together.
androgen any substance that promotes masculine characteristics; male sex hormone.
anemone *sea anemone,* a sedentary coelenterate having a largish, cylindrical body and one or more whorls of tentacles.
angiosperm a plant in which the seeds are enclosed in an ovary; a flowering plant; a monocot or dicot.
animal a member of the animal kingdom, consisting solely of multicellular forms, almost entirely diploid except for the gamete stage.
animal pole 1. the part of the ovum having the most cytoplasm and containing the nucleus. 2. in the early embryo, the comparable region and the region in which the greatest amount of cell division occurs. Compare *vegetal pole.*
anion a negatively charged ion.
annelid a worm of the phylum *Annelida,* including polychaetes, earthworms, and leeches.
annual 1. yearly; *annual growth ring,* the concentric rings seen in the cross-section of a woody stem, each ring corresponding to one year's growth (rapid spring and summer growth and slower autumn and winter growth). 2. a plant that completes its life cycle within a year and dies a programmed death; see *perennial.*
annulus (pl. *annuli*), ("little ring"), 1. a pore in the nuclear membrane. 2. the wall of a sporangium, within a sorus (ferns).
antagonist one of a pair of skeletal muscles (or groups of muscles) the actions of which oppose one another.
antenna (pl. *antennae*) 1. one of the paired, jointed, movable sensory appendages of the heads of insects or other arthropods. 2. *light antenna,* a cluster of photosynthetic pigments (chlorophyll and accessory pigments) which receives energy from photons and transfers that energy to a single photocenter.
anterior (adj.) toward the front or head end of an organism.
anterior pituitary see *pituitary, anterior.*
anterior vena cava see *vena cava, anterior.*
anther the pollen-producing organ of the stamen.
antheridium (pl. *antheridia*) in lower plants and fungi, a male reproductive organ containing motile sperm.
antheridogen a hormone inducing the formation of antheridia.
anthocyanin a class of water-soluble plant pigments, including most of those that give blue and red flowers their colors.
anthropoid 1. resembling a human. 2. *anthropoid ape,* any tailless ape of the family *Pongidae,* comprising the gorillas, chimpanzees, orangutans, gibbons and simiang.
antibiotic any of a large number of substances, produced by various microorganisms and fungi, capable of inhibiting or killing bacteria and usually not harmful to higher organisms; e.g., *penicillin, streptomycin.*
antibiotic resistance the ability of an organism to survive in the presence of an antibiotic. Antibiotic resistance is sometimes transmitted from one bacterium to another by means of a plasmid vector.
antibody a protein molecule that can recognize and bind to a foreign molecule. See also *immunoglobulin.*
antibody-antigen complex a molecular complex consisting of an antibody bound noncovalently to one or more specific antigens.
anticodon a region of a transfer RNA molecule consisting of three sequential nucleotides capable of antiparallel Watson-Crick pairing with the three sequential mucleotides of a codon.
anticodon loop the region of a transfer RNA molecule containing the anticodon; one of the three or four loops of a transfer RNA molecule.
antidiuretic hormone (ADH) also *vasopressin,* a polypeptide hormone secreted by the posterior pituitary, the action of which is to increase the resorption of fluid from the kidney filtrate.

antigen 1. any large molecule, such as a cell-surface protein or carbohydrate, that stimulates the production of specific antibodies, or that binds specifically with such antibodies. 2. any antibody-specific site on such a molecule.
anti-gibberellin a plant hormone that counterpacts the action of gibberellin.
antihemophilic factor a blood factor necessary for normal clotting, congenitally lacking in hemophilics. Antihemophilic factor can be isolated from normal blood and administered to hemophilics.
antiparallel (adj.) running parallel but in the opposite direction; specifically, the two strands of DNA, which form parallel interwound helices in which the 5′-to-3′ direction in one strand is the 3′-to-5′ direction in the other.
antiserum (pl. *antisera*), blood serum containing antibodies specific to some particular antigen.
antisprouting factor a local hormone produced by the axons of nerves that inhibits the growth of other nerve axons, in a developmental negative feedback control loop.
anus the opening of the lower end of the alimentary canal (gut), through which solid wastes are voided.
aorta in vertebrates, the principle or largest artery, conveying blood away from the heart.
aortic arch in embryology, one of a series of five paired, curved blood vessels that arise in the embryo from the ventral aorta, pass through the branchial arches, and unite to form the dorsal aorta.
apical pertaining to the apex or tip, as of a shoot or root.
apical dominance the uppermost growing tip of a plant stem that hormonally inhibits the upward growth or formation of other branches.
apical meristem the meristem (actively dividing cells) of the growing tip of a root or shoot.
apodeme an ingrowth of the arthropod exoskeleton that serves as the point of attachment for a muscle.
apoenzyme the protein part of an enzyme containing a bound coenzyme.
apomixis the production of seeds without the union of gametes.
aponeurosis a flattened sheet of dense connective tissue covering certain muscles.
appendicular skeleton the bones of the limbs (two pairs), along with the pelvis and pectoral girdles.
appendix also *vermiform appendix*, a hollow, fingerlike, blind extension of the cecum, having no known function in humans.
appetitive behavior a variable, nonstereotyped part of instinctive behavior involving searching (for food, water, a mate) for the opportunity to perform.
aqueous pertaining to water; dissolved in water; watery.
arabinose a five-carbon sugar first isolated from gum arabic.
arachnid any arthropod of the class *Arachnida*, comprised of spiders, scorpions, mites, ticks, and others.
arboreal tree-dwelling.
archegonium the female reproductive organs of ferns and bryophytes.
archenteron in embryology, the primitive digestive cavity of the gastrula.
areolar connective tissue fibrous connective tissue with the fibers forming a loose net or meshwork.
arginine one of the twenty principle amino acids coded by the genetic code.
arteriole a small artery.
arteriosclerosis inelasticity and thickening of the arterial walls.
artery a vessel carrying blood away from the heart and toward a capillary bed.
arthropod any member of the animal phylum *Arthropoda*, consisting of segmented invertebrates having exoskeletons with jointed legs (see Appendix A).
articulate (v.), to form a joint (with).
artificial selection the deliberate selection for breeding by humans of domesticated animals or plants on the basis of desired characteristics.
ascaris any nematode (roundworm) of the family *Ascaridae*.
ascending colon first region of the large intestine (see Figure 25.15).
Aschelminthes a pseudocoelomate animal phylum that includes nematode worms, ectoprocts, rotifers, gastrotrichs, priapulids, horsehair worms and others.
ascocarp a cup-like or sac-like body in the Ascomycota in which asci are produced.
ascogonium female reproductive organ of ascomycetes; the portion of the ascocarp which receives the antheridial nuclei in fertilization, and from which dicaryotic hyphae emerge.
ascomycete a sac fungus.
ascospore a haploid spore produced after sexual reproduction within the ascus of a sac fungus.
ascus (pl. *asci*), in ascomycetes, the sac in which meiosis occurs and in which four or eight ascospores are subsequently formed.
aseptic free of microorganisms.
asexual reproduction reproduction not involving the union of genetic material.
assay a test or determination; also, *bioassay*.
assay system in biology, a closed system in which the amount or character of growth or other biological activity is an indicator of the presence or activity of a micronutrient, toxin, hormone, etc.
aster one of a pair of structures, formed in mitosis and meiosis of all animal cells and many other eucaryotic cells, each consisting of *astral rays* (of microtubules and microfilaments) radiating from a centriole.
astral ray one of the many visible linear structures radiating from a centriole to form an aster, presumed to consist of microtubules, microfilaments, or both.
atherosclerosis a form of arteriosclerosis characterized by fat deposits in the inner lining of the arterial wall.
athlete's foot a contagious disease of the feet caused by a fungus that thrives on moist surfaces.
atlas the first cervical vertebra which supports the head.
atom the smallest indivisible unit of an element still retaining the element's characteristics.
atomic mass (also, *atomic weight*), the average mass of the atoms of an element, given in daltons; the exact mass of a specific isotope.
atomic number the number of protons in an atomic nucleus.
ATP, adenosine triphosphate a ubiquitous small molecule involved in many biological energy exchange reactions, consisting of the nitrogenous base adenine, the sugar ribose, and three phosphate residues.
atrioventricular bundle see *bundle of His*.
atrioventricular node a small mass of specialized muscle fibers at the base of the wall between the atria of the heart, conducting impulses from the bundle of His.
atrioventricular valve a valve between an atrium and the corresponding ventricle.
atrium either of the two upper chambers of the heart, each of which receives blood from veins and in turn forces it into the corresponding ventricle.
audition the sense of hearing.
auditory canal the open, bony canal from the outer ear to the eardrum.
auditory receptor 1. a neuron or organ specialized for the reception of sound.
Aureomycin an artificial antibiotic, chlortetracycline (trade name).
Australopithecus an extinct genus of hominid primates, including *A. africanus, A. boisei*, and *A. robustus*.
autoimmune disease a disease in which the organism's immune system attacks and destroys one or more of the organism's own tissues.
autonomic learning any learned response mediated by the autonomic nervous system.
autonomic nervous system also *involuntary nervous system*, the system of nerves and ganglion which innervates blood vessels, heart, smooth muscles, and glands and controls their involuntary functions. See *sympathetic* and the *parasympathetic nervous systems*.
autoradiograph a picture revealing the presence of radioactive material in a thin object or section, produced by *autoradiography* and usually superimposed on a photographic image of the object.
autoradiography the production of pictures revealing the presence of radioactive material in a thin object or section, in which a film or photographic emulsion is laid directly on the object to be tested, is exposed to radiation for a period of time, and is developed photographically.
autosome any chromosome other than the sex (X and Y) chromosome.
autotetraploid adj. having four homologues of each chromosome type, derived from chromosomal duplication without cell division in a diploid organism; n., an autotetraploid organism.
autotroph 1. a microorganism capable of using carbon dioxide as its only source of carbon (and thus receiving its energy from sources other than organic compounds). 2. a microorganism capable of growth on minimal medium (as opposed to *auxotroph)*.
auxin a class of natural or artificial substances that act as the principle growth hormone in plants. The naturally occuring auxin is primarily, or solely, *indoleacetic acid*.
auxospore the zygote of a diatom, consisting of a naked unflagellated cell capable of considerable growth before becoming encapsulated.
auxotroph a mutant organism requiring at least one growth factor not normally required of its species, and thus unable to grow on minimal medium.
avidin an egg-white protein that binds and inactivates the vitamin *biotin*.
axial pertaining to the axis. Axial fruit and flowers grow close to the stem as opposed to terminal fruit and flowers, which grow on the tips of branches.
axial skeleton in vertebrates, the skull, vertebral column, and bones of the chest, opposed to appendicular skeleton.
axil the angle between a petiole or branch and the supporting stem.
axis 1. the second cervical vertebra. 2. in plants, the longitudinal support; the stem and root.
axon the (usually) long extension of a neuron that conducts nerve impulses away from the body of the cell.

bacillus 1. an aerobic, rod-shaped, spore-producing bacterium of the genus *Bacillus*. .2 any rod-shaped bacterium.
back cross a cross between an individual of the first filial generation (i.e., an individual produced by the cross between two true-breeding strains), and an individual of either parental strain; see *test cross*.
bacteria see *bacterium*.
bacteriochlorophyll a purple photosynthetic pigment utilized by various anaerobic photosynthetic bacteria.
bacteriophage a virus that infects and lyses bacteria.
bacterium (pl. *bacteria*), any of numerous procaryotic organisms; any procaryote other than a cyanophyte.
ball-and-socket joint a joint allowing maximal rotation and flexion, consisting of a ball-like termination on one part, held within a concave, spherical socket on the other; e.g., the hip joint.
ball-and-stick model 1. a solid model of a molecule, in which atomic centers are represented by colored balls and covalent bonds are represented by sticks. 2. a drawing of such a model.
bark the portion of a stem outside of the wood (xylem), consisting of cambium, phloem, cortex, epidermis, cork cambium, and cork; everything from the vascular cambium outward.
Barr body a dark-staining feature of the nucleus of the cells of female mammals, representing the condensed X chromosome.
basal body a structure found beneath each eucaryotic flagellum or cilium, consisting of a circle of nine short triplets of microtubules.
basal meristem a meristematic reserve in grasses that is below the leaves and stem, and thus is protected from grazing.
basal metabolism rate a measure of oxygen consumption per minute per kilogram of an individual at rest.

base 1. (a) a compound that reacts with an acid to form a salt; (b) the oxide or hydroxide of a metal; (c) a substance that releases hydroxyl ions when dissolved in water. 2. a nitrogenous base (purine or pyrimidine) of nucleic acids.
base deletion a mutation in DNA in which a single nucleotide or nucleotide pair is removed and the phosphate-sugar backbone is rejoined.
base insertion mutation in DNA in which a single nucleotide is inserted into the chain.
base substitution a mutation in DNA in which one nucleotide is replaced by another, or modified into another.
basement membrane in animals, a sheet of collagen that underlies and supports the cells of a tissue; also *basal membrane, basilar membrane.*
basidiocarp the spore-producing organ in the *Basidiomycetes;* a mushroom.
basidiospore an aerial spore produced by meiosis in *Basidiomycetes.*
basidium *(pl. basidia*), the meiotic cell of *Basidiomycetes* that produces basidiospores by budding.
basilar membrane in the vertebrate ear, a membrane that conducts sound waves; also *basement membrane, basal membrane.*
basking to expose one's body to warmth, as the warmth of the sun; a frequent component of *behavioral thermoregulation.*
B cell a thymus cell that matures in the bone marrow (mammals) or bursa of fabricius (birds), and later circulates in the blood; involved in the immune response, especially in the production of free antibodies.
behavior observable activity of an organism; anything an organism does that involves action and/or response to stimulation.
behavioral phenotype the phenotype with respect to behavior.
behavioral thermoregulation regulating one's internal body temperature by behavioral means, such as *basking,* seeking shade, and taking cold showers.
bell-shaped curve see *normal distribution.*
belly 1. the abdomen. 2. of a muscle, the fleshy, bulging portion.
benthic referring to the benthos, or bottom-dwelling community of organisms.
benthos the aggregation of organisms living on or at the bottom of a body of water.
Bergmann's rule the generalization that members of a warm-blooded taxon tend to be larger in colder regions and higher latitudes and smaller in warmer regions and tropical latitudes.
beta glucose glucose in one of its two possible ring configurations.
beta glycoside a glycoside in which the glucose subunit is in its beta configuration.
beta hemoglobin the beta chain of adult hemoglobin; see under *hemoglobin.*
beta linkage a glycosidic linkage in which gulcose is fixed in its beta form; e.g.cellulose.
B horizon the soil layer immediately beneath the A horizon, from which it receives humus; usually less weathered and more compact than the A horizon.
bicarbonate ion HCO_3^-.
bicuspid valve also *mitral valve,* the two-flapped valve between the left atrium and the left ventricle of the heart.
bilateral symmetry having left and right sides that are approximate mirror images; having a single plane of symmetry.
bile a bitter, highly pigmented, alkaline, fat-emulsifying liquid secreted by the liver, containing bile salts and bile pigments.
bile duct the duct that carries bile from the cystic duct of the gall bladder to the small intestine.
bile pigments breakdown products of hemoglobin (bilirubin and biliverdin).
bile salts cholesterol and other steroid compounds secreted by the liver and essential for the emulsifying, digestion and absorption of dietary fats.
bilirubin a reddish-yellow pigment occuring in bile, blood, and urine; a breakdown product of hemoglobin.
binary fission fission (splitting) into two organisms of approximately the same size; cell division (asexual reproduction) in procaryotes.
binocular having or involving two eyes; *binocular vision,* stereoscopic vision, the ability to perceive depth through the integration of two overlapping fields of vision.
binomial system the tradition, introduced by Linnaeus, that each organism is given a taxonomic name in Latin consisting of a generic term and a specific term.
binucleate having two nuclei.
bioassay see *assay.*
biochemical pathway a series of enzymatic steps by which an organic molecule is progressively modified.
biochemistry the science dealing with the chemistry of living organisms.
biodegradable capable of being rendered harmless upon exposure to the elements and organisms of the soil or water.
biogeochemical cycle the pathway of elements (e.g., carbon, nitrogen) or compounds (water) as they are taken up and released by organisms into the physical environment.
biogeography the study of the geographical distribution of living things.
biological clock an innate mechanism by which living organisms are able to perceive the lapse of time.
biological species concept see *species.*
biomass 1. the total weight of living organisms, usually expressed as dry weight, per unit area or volume in a particular habitat. 2. the weight of organic material produced in a unit time period under specified conditions.
biome a complex of ecological communities characterized by a distinctive type of vegetation, as determined by the climate.
biometry also biometrics, 1. the application of mathematical statistics to biology. 2. the application of mathematical statistics to the study of polygenic inheritance. *biometrician,* one who practices or studies biometry.
biosphere the entire part of the earth's land, soil, waters and atmosphere in which living organisms are found.
biotic 1. pertaining to life. 2. ecological factors due to the interactions of living organisms, as opposed to abiotic factors such as climate.
biotic potential the maximum growth rate of a population when it is unrestricted by environmental resistance.
biotin a water-soluble vitamin essential to most organisms, produced in adequate amounts for vertebrates by intestinal flora; an essential growth factor for *Neurospora;* complexed and inactivated by the egg-white protein *avidin.*
bipedal literally two-footed; walking on two feet, as birds, humans, kangaroos and some dinosaurs.
biradial having two planes of symmetry at right angles, as ctenophores.
birth control the deliberate regulation of the number of one's offspring born especially by preventing or lessening the frequency of conception.
birth control pill an oral steroid contraceptive that inhibits ovulation, fertilization or implantation, causing temporary infertility in women.
birth rate crude birth rate, the number of births per year per 1000 individuals of all ages in the population. Other birth rates may be expressed in terms of births per year per number of women of a specific age range, the expected lifetime reproduction per female and so on.
Biston betularia the peppered moth, the subject of several key studies of rapid evolutionary adaptation to altered conditions. See *industrial melanism.*
bivalve a mollusk with two shells (valves) hinged together; a pelecypod.
bladder 1. any membraneous sac serving as a receptacle for fluid or gas. 2. in brown algae, the hollow spheres that act as floats to keep the algae at the surface of the water. 3. *gall bladder,* a muscular sac-like organ that is the temporary receptacle of bile. 4. *urinary bladder,* the storage organ for urine in many vertebrates.
blade 1. *leaf blade,* broad part of a leaf, as distinguished from petiole or midrib. 2. any broad, thin part of the thallus (body) of a red, green or brown alga.
blastocoel the cavity of a blastula.
blastocyst the blastula of a mammal.
blastoderm the single upper layer of cells of a blastula in embryos that develop in telolecithal eggs (fish, birds, and reptiles).
blastodisk the small disc of cytoplasm on the yolk of a bird or reptile egg, containing the egg nucleus, that becomes the early embryo.
blastomere any cell produced during cleavage, through the blastula stage.
blastopore in a gastrula, the opening of the archenteron produced by the inward growth of ectoderm during gastrulation.
blastula the early embryonic stage, consisting of a single layer of cells that form a hollow ball enclosing a central cavity, the blastocoel.
blending inheritance a theory of inheritance, which states that parental characters blend to produce an intermediate character in the offspring.
blood a circulating fluid in animals that helps distribute gases, nutrients, etc., and often collects cellular wastes.
blood fluke see *schistosome.*
blood type also *blood group,* one of a group of categories into which an individual can be categorized depending on his or her blood cell surface antigens.
blue mold also called *green mold,* any fungus of the genus *Penicillium,* which has bluish-green spores.
body stalk in embryology, the connection between the early embryo and the extraembryonic tissues; later the body stalk becomes the umbilical cord.
Bohr effect the competitive effect of carbon dioxide in reducing the affinity of hemoglobin for oxygen.
bolus a clump of chewed food; blood clot within a vessel.
bombykol the sex attractant produced by the female silkworm moth.
bond see *chemical bond, covalent bond, hydrogen bond, salt linkage, van der Waals forces.*
bone the hard connective tissue forming the skeleton of most vertebrates, consisting principally of a collagen matrix impregnated with calcium phosphate.
bone compact, see *compact bone;* bone, spongy, see *spongy bone.*
bone marrow see *marrow.*
book lung the respiratory organ of a spider, scorpion or other terrestrial arachnid, consisting of thin, membranous structures arranged like the leaves of a book; *book gill,* the similar structure in horseshoe crabs.
botanist a scientist who studies plants.
botany the scientific study of plants.
bottleneck *population bottleneck,* a relatively short period of time during which the size of a population becomes unusually small, resulting in a random change in gene frequencies.
bound (adj.), 1. attached to; e.g., a *bound* ribosome, an antibody *bound* to an antigen. 2. bounded by, as in a *membrane-bound* vessicle; see also *membrane-bound, membrane-bounded.*
bound ribosome a ribosome attached to the surface of a membrane, as of the endoplasmic reticulum. See also *endoplasmic reticulum, rough.*
bounded (adj.), limited by: e.g., a *membrane-bounded vessicle.*
Bowman's capsule one of numerous double-walled membraneous capsules in the nephron, each surrounding a glomerulus (ball of capillaries).
brain 1. in vertebrates, the anterior enlargement of the central nervous system, encased in the cranium. 2. in invertebrates, any anterior concentration of neurons more or less corresponding in function to the vertebrate brain.
braincase the primitive bony covering of the vertebrate brain as found in most fish and in the embryos of other vertebrates.
branch primordium see *primordium.*
branchial arch see *gill arch.*
bread mold *Rhizopus nigrans,* a zygomycete.
broad ligament one of a pair of supporting ligaments of the uterus, bearing the ovaries.
broad-niched an organism able to live under a wide variety of conditions.

bronchiole a small branch of a bronchus, part of the respiratory tree of the lungs.
bronchus (pl. *bronchi*), either of the two main branches of the trachea.
brown alga any alga of the *Phaeophyceae*, usually brown due to fucoxanthin pigment; e.g., kelp.
Brownian motion also *Brownian movement*, the irregular motion of microscopic particles in a liquid or gas, caused by random thermal agitation of molecules in the medium.
bryophyte a moss, liverwort, or hornwort of the Division *Bryophyta* (see Appendix A).
bud a small protuberance on the stem of a plant, containing meristematic tissue and covered with overlapping rudimentary foliage.
bud scale an often hairy, waxy, or resinous scale enclosing an immature bud.
budding 1. asexual cell reproduction with unequal cytoplasmic division, as of yeasts. 2. similar asexual reproduction of a multicellular animal, as in hydras.
buffer a solution of chemical compounds capable of neutralizing both acids and bases, and thus able to maintain an equilibrium pH.
bulbourethral glands (Cowper's glands), a pair of small glands that secrete a mucous substance into the urethra in males, especially on sexual arousal.
bundle of His, also atrioventricular bundle, a bundle of specialized muscle fibers that conduct impulses from the right atrium to the ventricles.
bundle sheath one of more layers of compactly arranged parenchyma cells that enclose the small veins in a leaf.
bundle sheath cell a cell of the bundle sheath; in C_4 plants, the site of Calvin cycle (C_3) photosynthesis.
bursa a pouch, sac or vesicle, especially a sac containing synovial fluid to allow the smooth movement of tendons or joints.
bursa of fabricius in birds, a discrete abdominal organ in which B cell thymocytes mature.

caecilian also *blind worm* an amphibian of the order *Apoda*, all of which are legless, wormlike, burrowing and nearly blind.
calcaneous the largest of the tarsal bones, forming the heel.
calcareous containing or composed of calcium carbonate.
calcareous sponge, a sponge with a skeleton of calcium carbonate spicules.
calcitonin one of the two major hormones of the thyroid gland, whose major action is to inhibit the release of calcium from the bone.
calcium salt a salt in which calcium is the metallic element; in nature, most often calcium carbonate or calcium phosphate.
callus tissue an undifferentiated tissue that forms over wounds in plants.
calorie 1. also *small calorie* (calorie proper), the amount of heat (or equivalent chemical energy) needed to raise the temperature of one gram of water by 1°C. 2. also *large calorie, kilocalorie,* the heat needed to raise the temperature of a kilogram of water by 1°C; 1,000 small calories.
Calvin cycle the cycle in C_3 photosynthesis in which $NADPH_2$ and ATP reduce carbon dioxide to glyceraldehyde phosphate (PGAL).
calyx the outermost whorl of floral parts (the sepals), usually green and leaf-like.
cambium undifferentiated meristematic tissue in a plant. See *cork cambium, primary growth, secondary growth, procambium, vascular cambium.*
Cambrian see the geological timetable inside the front cover.
canaliculus (pl. *canaliculi*), a tiny canal, as in bone, where canals communicate between osteocytes.
canine tooth in many mammals, any of four pointed, often elongate teeth used for tearing in feeding and in defense.
canopy the upper leafy area of a tree or especially of a forest.
capillary the smallest blood vessels; the fine channel between the arteriole and venule.
capillary action the tendency of a liquid to rise in a small tube due to adhesion to its inner surfaces and cohesion among water molecules.
capillary bed tissue rich in capillaries; a body of capillaries taken together.
capsule 1. a type of fruit which becomes dry and hard before rupturing to release seeds. 2. in animals, any membraneous sac or covering.
carapace a hard, shield-like covering of the dorsal surface of an animal, as a lobster, turtle, armadillo, etc.
carbaminohemoglobin deoxygenated hemoglobin complexed with carbon dioxide.
carbohydrate a class of organic compounds with multiple hydroxyl side groups and an aldehyde or ketone group; sugars, starches, cellulose, and chitin; empirical formula $(CH_2O)n$.
carbon dioxide a colorless gas of the formula CO_2, readily dissolved in water and capable of reacting with water to form a hydroxyl ion (OH^-) and a bicarbonate ion (HCO_3^+).
carbonate rock rock rich in calcium carbonate, formed by deposits of marine shells; included are limestone and marble.
carbon-4 plant (C_4 plant) any plant in which the first product of CO_2 fixation is a four-carbon compound.
carbonic acid the acid formed when CO_2 dissolves in water; H_2CO_3.
carbonic anyhdrase an enzyme, plentiful in lung tissue, that catalyses the conversion of carbonic acid to carbon dioxide gas and water.
Carboniferous see the geological timetable in the inside front cover.
carboxyhemoglobin see *carbaminohemoglobin.*
carboxylase also *decarboxylase,* any enzyme that catalyzes the release of carbon dioxide from a carboxyl group.
carboxypeptidase a protease that cleaves peptide bonds near the carboxy terminal (*C*-terminal) end of a polypeptide.
cardiac pertaining to the heart; near or toward the heart.
cardiac muscle specialized muscle of the heart that is both striated and involuntary.
cardiac sphincter the ring of muscle that closes the passageway between the lower esophagus and the stomach.
cardinal vein also *cardinal sinus,* any of four longitudinal veins of the vertebrate embryo and certain fish, running anteriorly and posteriorly along the vertebral column and discharging into the right and left common cardinal veins, or ducts of Cuvier.
carnivorous (adj.), flesh eating.
carotene a red or orange hydrocarbon, $C_{40}H_{56}$, found in most plants as an accessory photosynthetic pigment.
carotenoid any of a group of red, yellow and orange plant pigments chemically and functionally similar to carotene.
carotid artery either of a pair of large arteries on each side of the neck and head.
carotid body a mass of cells and nerve endings on either carotid artery, that senses blood oxygen levels and blood pressure and responds by affecting the rate of breathing and the heart beat.
carpal also *carpale,* any wrist bone.
carpel in flowers, a simple pistil, or a single member of a compound pistil; one sector or chamber of a compound fruit.
carrageen a polysaccharide produced by Irish moss, a red algae; used as a thickening agent.
carrier a heterozygote for a mutant or rare recessive allele.
carrier molecule a protein that binds a small molecule in facilitated diffusion or active transport.
carrying capacity a property of the environment defined as the size of a population that can be maintained indefinitely.
cartilage a firm, elastic, flexible, translucent type of connective tissue; in development, a precursor of bone formation.
cartilagenous fishes fish having skeletons of cartilage, comprising sharks, rays and chimeras; one of the eight vertebrate classes.
casein the principle protein component of milk.
casparian strip a waxy strip on cell walls of endoderm that serves as a barrier to the conduction of moisture in roots and stems.
Castle–Hardy–Weinberg law also *Hardy–Weinberg law, binomial law;* a statement of the mathematical expectations of genotype frequencies given allele frequencies in a population of random mating diploid individuals.
catalase an enzyme that decomposes hydrogen peroxide to molecular oxygen and water, in the reaction $2\ HOOH \rightarrow 2\ H_2O + O_2$.
catalyst an agent that causes or accelerates a chemical reaction, while not being permanently altered itself.
catarrhines (obsolete) a primate suborder comprised of the Old-World monkeys, apes, and humans.
cation a positively charged ion.
CCA stem a sequence of three nucleotides added enzymatically to the 3′ end of all tRNA molecules.
cecum in vertebrates, a blind pouch or diverticulum of the intestine, at the juncture of the small and large intestines.
cell 1. the structural unit of plant and animal life, consisting of cytoplasm and a nucleus, enclosed in a semipermeable membrane. 2. any similar organization, as that of a protist or procaryotic organism.
cell division also *cytokinesis,* the division of a cell into two daughter cells.
cell-free system a preparation of cytoplasmic constituents extracted from cells by the disruption of the cell membrane.
cell membrane the semipermeable membrane enclosing the cell.
cell plate in plant cell division, the forming cell membrane between nascent daughter cells.
cell sap the fluid that fills large vacuoles in plant cells; primarily water with various substances in solution or suspension.
cell streaming see *cyclosis.*
cell-surface antigen any constituent of the cell surface capable of eliciting a specific antibody response.
cellulase an enzyme that digests cellulose.
cellulose an inert, insoluble carbohydrate, a principle constituent of plant cell walls, consisting of unbranched chains of beta glucose.
cell wall the semirigid extracellular encasement of a plant, fungal, algal, cyanophyte, or bacterial cell that gives it a definite shape.
centipede a predaceous, nocturnal terrestrial arthropod of the class *Chilopoda* (see Appendix A).
central cylinder the supporting structure of moss plant, which some botanists believe to have specialized conductive cells.
central dogma the proposition that all biological information is encoded in DNA, transmitted by DNA relication, transcribed into RNA, and translated into protein; together with several exceptions for certain misbehaving viruses. The term was coined by Francis Crick.
central nervous system the brain and spinal cord.
centriole one of a pair of organelles in animal, protist, fungi and lower plant cells, apparently involved in organizing the spindle and asters in mitosis and meiosis; each centriole in turn consisting of a pair of short cylinders of nine triplet microtubules. Centrioles are absent (or invisible) in cells of seed plants.
centrolecithal of an egg, having a central yolk entirely surrounded by cytoplasm, as in insects.
centromere also *kinetochore,* also *spindle fiber attachement,* the specialized region of a chromosome to which spindle fibers are attached, visible under the light microscope as a light-staining body, and under the electron microscope as a pair of shield-like plates. A centromere is considered to be a single object until the actual physical separation of daughter centromeres in anaphase.
centromeric pertaining to the centromere.
centromeric heterochromatin a heterochromatin region adjacent to the centromere.
centromeric spindle fiber any of a group of microtubules attached to each centromere and proceeding to a spindle pole in mitosis or meiosis.
cephalic pertaining to the head.
cephalization the evolutionary tendency to concentrate neural and sensory functions in an anterior end.

cephalochordate ("head cord animal"), a lancelet (*Amphioxus*) of a chordate subphylum in which the notochord extends through what would be the head if it had one, which it doesn't.
cephalopod a mollusk of the class *Cephalopoda* (see Appendix A).
cerebellum a portion of the brain serving to coordinate voluntary movement, posture, and balance; it is located behind the cerebrum.
cerebrum the anterior, dorsal portion of the vertebrate brain, the largest portion in humans, consisting of two *cerebral hemispheres* and controlling many localized functions, among them voluntary movement, perception, speech, memory and thought.
cervical 1. pertaining to the neck. 2. pertaining to the cervix of the uterus.
cervical cap a birth control device consisting of a small cap that encloses the cervix.
cervical vertebra any vertebra of the neck region.
cervix the opening to the uterus and the surrounding ring of firm, muscular tissue.
CF1 particle an ultramicroscopic structure on the outer surface of the thylakoid, the site of chemiosmotic phosphorylation in photosynthesis.
C4 plant a plant in which the initial product of CO_2 fixation is a four-carbon compound.
chain reaction 1. a reaction which produces a product necessary for the continuation of the reaction. 2. any series of events in which each event causes the next.
chain termination codon any of the three codons of the genetic code causing the termination of polypeptide synthesis; in RNA these are UGA, UAG, and UAA.
chain termination mutation a base substitution mutation in which the new codon created is a chain termination codon.
chaparral a vegetation type common in California, characterized by a dense growth of low, evergreen shrubs and trees.
character a classifiable feature, trait or characteristic of an individual organism; a specific component of a phenotype.
character state a specific state of a morphological character, defined in terms of presence or absence.
charging enzyme any of a group of specific enzymes that covalently attach amino acids to their appropriate tRNA's.
chelicera (pl. *chelicerae*), one of the first pair of appendages in the *Chelicerata;* a spider's fang.
chemical bond any of several forms of attraction between atoms in a molecule.
chemical bond energy the potential energy invested in the formation of a chemical bond or that released upon its dissolution. See also *covalent bond, hydrogen bond, salt linkage, van der Waals forces.*
chemical communication the transmission of information between individuals by the use of phenomones.
chemical reaction the reciprocal action of chemical agents on one another; chemical change.
chemiosmosis the process in mitochondria, chloroplasts and aerobic bacteria in which an electron transport system utilizes the energy of photosynthesis or oxidation to pump hydrogen ions across a membrane, resulting in a proton concentration gradient that can be utilized to produce ATP.
chemiosmotic phosphorylation the production of ATP utilizing the energy of protons passing across a membrane and through F_1 of CF_1 particles.
chemoreceptor a neural receptor sensitive to a specific chemical or class of chemicals.
chemosynthesis the synthesis of organic compounds with energy derived from inorganic chemical reactions.
chemosynthetic (adj.), utilizing chemosynthesis.
chemotroph an organism that lives on the energy of inorganic chemical reactions; *chemotrophic,* adj.
chew to grind with the teeth, especially utilizing sideways motion.
chief cell a stomach cell that secretes pepsinogen.
chitin a structural carbohydrate that is the principle organic component of arthropod exoskeletons.
chiton a marine mollusk of the class *Amphineura* (see Appendix A).
chlamydobacterium a filamentous sheathed bacterium.
chlorenchyma chloroplast-containing parenchymal tissue in leaves.
chlorobacterium a green sulfur bacterium.
chlorophyll a green photosynthetic pigment found in chloroplasts, cyanophytes and chloroxybacteria. It occurs in several forms, *chlorophyll a, b,* and *c.*
chloroplast a plastid containing chlorophyll.
chlorosis an abnormally pale or yellow condition of plants caused by lack of chlorophyll, due to disease or mineral deficiency.
chloroxybacterium a recently discovered photosynthetic procaryote utilizing chlorophylls *a* and *b,* and apparently more closely related to the ancestor of the plant chloroplast than are cyanophytes; a symbiont of certain tunicates.
choanocyte also *collar cell,* a type of cell in all sponges and certain protists in which a single flagellum is surrounded at its base by a screen of fused cilia (collar), that filters food from the water current created by the flagellum.
cholecystekinin a hormone that stimulates the contraction of the gall bladder.
cholesterol a common sterol occurring in all animal fats, a vital component of cell membranes, an important constituent of bile for fat absorption, a precursor of vitamin D, and too much of which is not good for you.
chondroitin the principle structural protein of cartilage.
chordate a member of the phylum *Chordata,* including the vertebrates, tunicates, and lancelets, all of which have a notochord and a dorsal tubular nerve cord at some point in development (see Appendix A).
chorioallantois a highly vascular extraembryonic membrane of birds, reptiles and some mammals, formed by the fusion of the chorion and the allantois.
chorion 1. the tough covering on insect eggs; 2. the outermost of the extraembryonic membranes of birds, reptiles and mammals, contributing to the formation of the placenta in placental mammals.
chorionic villi small finger-like processes of the chorion of the early mammalian embryo, especially before the formation of the placenta.
chromatid in a G2 chromosome, one of the two identical strands of the chromosome following replication and prior to cell division.
chromatin the substance of chromosomes, a molecular complex consisting of DNA, histones, nonhistone chromosomal proteins, and usually some RNA of unknown function.
chromomere one of the beadlike clumps or granules arranged in a linear array on a chromonema, especially when visible in partially condensed prophase chromosomes. Chromomeres may be more or less representative of functional genetic units (genes).
chromonema (pl. *chromonemata*), one long DNA strand with its associated histones and proteins; the G_1 chromosome or the G_2 chromatid (literally, "colored thread").
chromoplast a plastid containing primarily pigments other than chlorophyll, as certain flower pigments.
chromoprotein a colored protein; a protenaceous pigment.
chromosome 1. in eucaryotes, an independent nuclear body carrying genetic information in a specific linear order, and consisting of one linear DNA molecule (in G_1) or two DNA molecules (in G_2), one centromere, and associated proteins. 2. in procaryotes, an analogous circular DNA molecule bound to the cell membrane. 3. the analogous DNA or genetic RNA molecule of a virus. A eucaryote chromosome is considered to be a single entity until the physical separation of centromeres in anaphase.
chromosome puff in polytene chromosomes, an enlargement of one band associated with transcriptional activity (mRNA production).
chrysophyte a yellow-green alga.
chymotrypsin a proteolytic enzyme produced by the pancreas.
chytrid an aquatic fungus.
cilia (sing. *cilium*), fine, hairlike, motile organelles found in groups on the surface of some cells; shorter and more numerous than flagella, but similar in structure, they exhibit coordinated oarlike movement.
ciliary basal body see *basal body.*
ciliate any protist of the class *Ciliata,* having cilia on part or all of the body (see Appendix A).
circulatory system the vascular system, consisting of blood-forming organs or tissues, vessels, the heart, and blood.
cis-dominant being expressed in a heterozygous state, but only with regard to control of other gene loci on the same chromosome.
cistron 1. also *structural gene,* a sequence of DNA specifying the sequence of a polypextide chain. 2. any continuous genetic unit in which different recessive mutant lesions fail to complement one another in a double heterozygote.
citric acid cycle also *Krebs cycle, tricarboxylic acid cycle,* a cyclic series of chemical transformation in the mitochondrion by which pyruvate is degraded to carbon dioxide, NAD, and FAD are reduced to $NADH_2$ and $FADH_2$, and ATP is generated.
citrulline an amino acid precursor in the biosynthesis of arginine; not found in proteins.
clasper one of a pair of specialized grooved caudal fins used by male sharks and rays as a penis.
class a major taxonomic grouping intermediate between *phylum* (or *division*) and *order.*
classical conditioning see *conditioning, classical.*
clavicle also *collar bone,* one of a pair of bones of the pectoral girdle, articulating with the sternum and scapula.
cleaning symbiont a usually conspicuous organism that interacts mutualistically with others by removing ectoparasites, necrotic tissue, etc., receiving nourishment and inhibiting agressive behavior of the individual being cleaned.
cleavage 1. the total or partial division of a zygot into smaller cells (blastomeres).
cleavage furrow the initial depression indicating cytoplasmic cleavage.
cleidoic egg the egg of birds and reptiles, protected by a shell or membrane and independent of the environment because of extraembryonic membranes for food reserves, moisture reserve, and sequestering of toxic wastes.
climate the weather conditions of a region taken as a group; the prevailing weather conditions; as temperature, humidity, precipitations, seasonality, wind, and cloudiness.
climax 1. that stage in ecological succession of a plant or animal community that is stable and self-perpetuating. 2. an orgasm.
cline 1. a simple geographic gradient, or regular change, in a given character. 2. a regular change in allele frequency over geographic space.
clitellum a thickened, ring-like glandular portion of the body wall of earthworms, which secretes mucus to form a cocoon for eggs.
clitoris in female mammals, an erectile, erotically sensitive organ of the vulva, homologous embryologically to most of the penis.
cloaca the common cavity into which the intestinal, urinary, and reproductive canals open in vertebrates other than placental and marsupial mammals.
clonal selection theory in immunology, the proposition that all potential antibody specificity is present in differentiated cells early in development and that the specific immune response consists of inducing appropriately differentiated cells to proliferate clonally.
clone 1. a group of genetically identical organisms derived from a single individual by asexual reproduction. 2. a group of identically differentiated cells derived mitotically from a single differentiated cell.
cloning, gene see *gene cloning.*
closed circulatory system a system in which blood is enclosed within arteries, veins, and capillaries throughout, and is not in direct contact with cells other than those lining these vessels.

closed system 1. in thermodynamics, a system in which matter and energy neither enter nor leave. 2. see *closed circulatory system.*
club fungus (pl. *club fungi*), a basidiomycete, as mushroom.
club moss any plant of the genus *Lycopodium*; ground pine.
cniderian see *coelenterate.*
cnidoblast in coelenterates, a specialized cell containing a stinging or snaring nematocyst.
coat 1. *seed coat,* the outer protective covering of a seed, comprised of one or more layers of tissue derived from the parent sporophyte. 2. *viral coat,* the outer protective protein jacket of a virus.
coccus (pl. *cocci*), any spherical bacterium (principally eubacteria), a condition found in many distantly related groups.
coccyx, the tailbone or lowest portion of the vertebral column in humans and tailless apes, consisting usually of four fused vertebrae.
cochlea in mammals, a spiral cavity of the inner ear containing fluid, vibrating membranes, and sound-sensitive neural receptors.
code in general usage, "a system of communication in which arbitrarily chosen words, letters or symbols are assigned definite meanings"; see also *genetic code.*
codominance the individual expression of both alleles in a heterozygote; see *dominance interactions.*
codon also *code group,* 1. a series of three nucleotides in mRNA specifying a specific amino acid (or chain termination) in protein synthesis. 2. the colinear, complementary series of three nucleotides or nucleotide pairs in the DNA from which mRNA codon is transcribed.
coelacanth a lobe-finned crossopterigian fish, *Latimeria chalumnae,* related to the fishy ancestors of land vertebrates; thought to be extinct since the Cretaceous until found alive in 1938.
coelenterate one of the Coelenterata, a phylum of radically symmetrical animals lacking a true body cavity or anus; included are sea anenomes, corals, hydroids, and jellyfishes.
coelenteron the saclike cavity of coelenterates, which carries out ingestion, digestion, phagocytosis and absorption of food, and also has circulatory functions; also, gastrovascular cavity.
coelom *true coelom,* a principal body cavity, or one of several such cavities between the body wall and gut, entirely lined with mesodermal epithelium; compare *pseudocoelom.*
coelomate having a true coelom.
coenocyte an organism made up of a multinucleate mass of cytoplasm without subdivision into cells, as certain algae and fungi.
coenocytic consisting of a multinucleate mass of cytoplasm without subdivision into cells.
coenzyme a small organic molecule required for an enzymatic reaction.
coenzyme A also *CoA,* a coenzyme of the Krebs cycle; *acetyl CoA, succinyl CoA,* coenzyme A covalently bonded to an acetyl or succinyl group.
cofactor any organic or inorganic substance, especially an ion, that is required for the function of an enzyme.
cohesion the attraction between the molecules of a single substance.
coitus the act of sexual intercourse, especially between people (Latin, coming together).
coitus interruptus coitus that is intentionally interrupted by withdrawal before ejaculation.
coleoptile in the embryos and early growth of grasses, a specialized tubular leaf structure, completely enclosing and protecting the pulmule during emergence.
colinearity *principle of colinearity,* the finding that corresponding parts of a structural gene, mRNA, and polypeptide occur in the same linear order.
collar cell see *choanocyte.*
collecting duct also *collecting tubule,* the part of a nephron that collects fluids from distal convoluted tubules and discharges it into the renal pelvis.
collenchyma in plants, a strengthening tissue, a modified parenchyma consisting of elongated cells with greatly thickened cellulose walls.
colloid goiter see *goiter, colloid.*
colon in mammals, the large intestine from the cecum to the rectum; see Figure 25.15
colonial 1. generally existing in colonies. 2. *colonial organism,* an organism of semindependent parts, derived by asexual reproduction.
colony 1. a group of animals or plants of the same kind living in a close semidependent association. 2. an aggregation of bacteria growing together as the descendents of a single individual, usually on a culture plate.
color blindness the inability to distinguish colors. *red-green color blindness,* the inability to distinguish certain shades of red from corresponding shades of green.
color vision the ability to discern colors.
coloration appearance as to color and patterns of color.
colostrum clear, yellowish milk secreted a few days before and after giving birth; colostrum is rich in maternal antibodies.
columnar epithelium epithelium consisting of one (*simple columnar*) or more (*stratified columnar*) layers of elongated, cylindrical cells.
comb jelly see *ctenophore.*
combination pill a birth control pill that combines two or more female hormones.
common cardinal vein *duct of Cuvier*; see *cardinal vein.*
common descent descent of two or more species (or individuals) from a common ancestor; e.g., the similarity in blood chemistry of apes and humans is due to common descent.
communication in its most general sense, any activity that functions to alter the behavior of another animal.
community an assemblage of interacting plants and animals forming an identifiable group within a biome, as in salt marsh or sage desert community.
compact bone dense, hard bone with spaces of microscopic size.
companion cell in plants, a nucleated cell adjacent to a sieve tube member, believed to assist it in its functions.
comparative anatomy the science of comparing the anatomy of animals, tracing homologies and drawing evolutionary and phylogenetic inferences; see also *allometry.*
comparative psychology the study of mental processes (behavior) in animals from a comparative point of view, usually in a laboratory environment.
competition 1. seeking to gain what another is seeking to gain at the same time; a common struggle for the same object. 2. in ecology, the utilization by two or more individuals or species of the same limiting resource.
competitive exclusion principle see *Gause's law.*
complement (n.) 1. a group of blood proteins that ineract with antibody-antigen complexes to destroy foreign cells. 2. (v.t.) the production of a normal phenotype by two recessive mutations in a double heterozygote; see *complementation test.*
complementation test a test in which two recessive mutants are brought together as a double heterozygote; failure to complement (produce a normal phenotype) indicates that the mutants are allelic or contain overlapping deletions.
complementation group a set of recessive mutants that appear to be allelic with one another in that all double heterozygotes show a recessive phenotype.
complete flower see *flower.*
complete digestive tract a tubular digestive tract with an anal as well as an oral opening.
complete metamorphosis of insects, development includes distinct larval, pupal, and adult stages.
complex (n.), two or more molecules held together, usually by weak and reversible bonds, as *antibody–antigen complex, enzyme–substrate complex.*
compound eye an arthropod eye consisting of many simple eyes closely crowded together, each with an individual lens and a restricted field of vision, so that a mosaic image is formed.
compound microscope an optical instrument for forming magnified images of small objects, consisting of an *objective lens* that can be brought close to the object being examined, aligned with an *ocular lens* mounted at the other end of a body tube of fixed length.
compound nucleus in certain siphonous algae, an aggregate of numerous nuclei within the coenocytic cytoplasm.
compound in chemistry, a pure substance consisting of two or more elements in a fixed ratio; consisting of a single molecular type.
compound lens a lens made of several simple lenses glued or mounted together on the same axis.
concentration gradient 1. a slow, consistent decrease in the concentration of a substance along a line in space. 2. for any spatial difference in concentration, the direction away from the region of greater concentration.
condensation of chromosomes, the coiling and supercoiling that transforms diffuse chromatin into a compact, discrete body in mitosis.
conditioned response an involuntary response that becomes associated with an arbitrary, previously unrelated stimulus through repeated presentation of the arbitrary stimulus simultaneously with a stimulus normally yielding the response.
conditioning, classical the process by which a conditional response is learned and elicited; compare *operant conditioning.*
condom a thin sheath of rubber or animal membrane worn over the penis during sexual intercourse to prevent conception or venereal infection (after Dr. Condom, an 18th-century English physician); also, *rubber, prophylactic.*
conservative replication replication of a molecule in which the original molecule remains intact. Compare *semi-conservative replication.*
consolidation the transfer of engrams from short-term memory to long-term memory.
consolidation hypothesis the presumption that memory is first stored in short-term centers from where it is rapidly lost unless it is transferred to long-term centers where its loss is more gradual.
constant region the *C*-terminal portion of an immunoglobulin (antibody) light or heavy chain, not involved in antigen-specific binding and coded by a constant-region cistron.
constitutive always present independent of environmental conditions.
constitutive heterochromatin fixed heterochromatic chromosomes or regions of chromosome, generally lacking in specific genetic functions. *Constitutive mutant* a mutant that constantly produces a normally inducible or repressible enzyme regardless of the cellular environment.
consummatory behavior the satisfying, stereotyped behavior that completes an instinctive act. Usually preceeded by *appetitive behavior.*
consumer one that consumes; in ecology, an animal that feeds on plants (primary consumer) or other animals (secondary consumer).
continental drift the slow movement of continents relative to one another on the surface of the earth; also, *plate techtonics.*
continental shelf the sea bottom of shallow oceanic waters adjacent to the shores of a continent; considered to be a submerged part of the continent.
continuous distribution see *distribution.*
continuous spindle fibers two partially overlapping spindle fibers (not actually continuous), each originating near a pole and extending across the chromosomes without attaching.
contraception the prevention of pregnancy, usually by preventing the union of gametes or by preventing implantation of the blastocyst; see also *birth control.*
contractile capable of contracting (shortening, drawing in, or becoming smaller).
contractile unit (of muscle), see *sarcomere.*
contractile vacuole also *water vacuole* an organelle of many fresh water protists that maintains the osmotic equilibrium of the cell by bailing excess water.

control a standard of comparison in a scientific experiment; a replicate of the experiment in which a possibly crucial factor being studied is omitted.
controlled experiment an experiment involving several replicates, each differing by a single variable factor.
convection the transfer of heat by the movement of a circulating fluid or gas; compare *radiation, conduction.*
conduction 1. the transfer or movement of fluid, heat, ions, impulses, etc., through or along a channel or medium; e.g., the conduction of impulses through the central nervous system. 2. the transfer of heat through a solid, as opposed to *convection* or *radiation.*
conducting tissue see *vascular tissue.*
cone 1. one of a class of conical photoreceptors in retina that detect color, consisting of a highly modified cilium with specialized proteinaceous pigments for detecting red, green, or blue wavelengths. 2. a male or female reproductive structure of conifers, consisting of a cluster of scale-like modified leaves and either pollen or ovules, or the seed.
configuration in chemistry, the structure of a molecule with reference to the spatial relations of its atoms and bonds.
conidium (pl. *conidia*), an asexual spore borne on the tip of a fungal hypha; also *conidiospore.*
conidiophore a specialized branch of the mycelium bearing conidia.
conifer an evergreen gymnosperm of the order *Coniferales,* bearing ovules and pollen in cones; included are spruce, fir, pine, cedar, and juniper.
conjugated protein a compound of one or more polypeptides with one or more nonprotein substance; e.g., hemoglobin, which consists of four polypeptide chains and four heme groups.
conjugation in ciliates, a temporary cytoplasmic union in pairs, accompanied by meiosis and the exchange of haploid nuclei.
connective tissue in animals, any tissue of mesodermal origin having abundant extracellular substance or interlacing fibers, usually including collagen; in which the cells do not aggregate to form sheets or solid masses.
consummatory behavior a part of instinctive behavior, usually following a successful search, that involves highly stereotyped (invariable) behavior and the performance of a fixed action pattern.
convergent evolution the independent evolution of similar structures in distantly related organisms; often found in organisms that have adopted similar ecological niches, as marsupial moles and placental moles.
copulation sexual union or sexual intercourse in animals involving internal fertilization.
coral 1. a colonial anthozoan that secretes a calcareous skeleton. 2. the skeleton itself.
cord in anatomy, a ropelike structure; as *spinal cord, umbilical cord.*
coriolus effect the deflection of a moving body of water or air caused by the rotation of the earth, resulting in clockwise motion in the Northern hemisphere and counterclockwise motion in the Southern hemisphere, as seen in tornados, ocean currents, and prevailing winds.
cork 1. in plants, secondary tissue, produced by the *cork cambium,* consisting of cells that become heavily suberized and die at maturity, resistant to the passage of moisture and gasses; the outer layer of bark. 2. such a tissue from the cork oak, used to seal wine bottles.
cork cambium also *phellogen,* in plants, the outermost meristematic layer of the stem of woody plants, from which the outer layer of bark is produced.
corn smut a fungal disease of corn; see *smut.*
corolla in flowers, the whorl of petals surrounding the carpels or carpel.
corona radiate ("radiating crown"), an aggregation of follicle cells surrounding the mammalian egg at ovulation.
corpus callosum a broad, white neural tract that connects the cerebral hemispheres and correlates their activities.
corpus cavernosum (pl. *corpora cavernosa*), in the penis or clitoris, a mass or erectile tissue with large interspaces capable of being distended with blood.
corpus luteum (pl. *corpora lutea*), ("yellow body"), a temporary yellow endocrine body on the surface of the ovary, consisting of secretory cells filling a follicle after ovulation; regressing quickly if the ovum is not fertilized or persisting throughout pregnancy.
corpus spongiosum a mass of erectile tissue in the penis or clitoris.
cortex 1. (*zoology*) the outer layer or rind of an organ, as *adrenal cortex, kidney cortex.* 2. (*botany*) the portion of stem between the epidermis and the vascular tissue.
cortical (adj.), referring to the cortex.
cortical dominance increase of the cortex at the expense of the medulla. In the embryonic development of the vertebrate gonad, the early undifferentiated gonad consists of an inner *medulla* and an outer *cortex*; in females, cortical dominance results in the formation of an ovary, while medullary dominance results in the formation of a testis in males.
cortical granules membranous vesicles beneath the surface of unfertilized echinoderm eggs; at fertilization these rupture to form the *fertilization membrane.*
cortisol also *hydrocortisone,* a steroid hormone of the adrenal cortex that inhibits immune responses, lessening inflamation, and is also active in carbohydrate and protein metabolism, response to stress and other processes.
CO_2 carbon dioxide.
cotyledon also *seed leaf,* a food-storing structure in dicot seeds, sometimes emerging as seed leaves; food-digesting organ in most monocot seeds, remains in the soil.
countercurrent multiplier an arrangement such that two fluids (or gases) flow in opposite directions on either side of a membrane, increasing the efficiency of heat exchange, solute transfer, etc.
covalent bond a relatively strong chemical bond in which an electron pair is shared by two atoms, simultaneously filling the outer electron shells of both.
Cowper's gland also *bulbourethral gland,* either of a pair of small exocrine glands discharging lubricating mucus into the male urethra at times of sexual arousal; after William Cowper, 1666–1709.
coxa (pl. *coxae*), 1. also *innominate bone, hip bone;* either half of the pelvis; created by the fusion of the ilium, ischium and pubis.
CPHL, chlorophyl CPHL (ox), oxidized chlorophyll. CPHL (re), reduced chlorophyll.
cranium 1. the skull. 2. the part of the skull enclosing the brain.
crash population crash, a sudden die-off or diminution in numbers, particularly after the carrying capacity of the environment has been exceeded and resources have been depleted.
creatine a modified amino acid found in vertebrate muscles. creatine phosphate, also *phosphocreatine,* phosphorylated creatine in which the high-energy phosphate bond serves as an energy store for the regeneration of ATP in rapid muscle metabolism.
creationism the doctrine that all things, especially plant and animal species, were created fairly recently and substantially as they now exist, by an omnipotent creator and not by gradual evolution.
cretin a person affected by cretinism (human being; derived from *Christian*).
cretinism a recessive genetic abnormality that results in the inability to produce thyroxine; affected persons are extremely retarded physically and mentally unless thyroxine is administered from early infancy.
Crick strand either of the two strands of DNA, the other being designated the *Watson strand.*
crinoid a type of sessile, stalked echinoderm.
crista (pl. *cristae*), a shelf-like fold of the inner mitochondrial membrane into the central cavity of the mitochondrion.
crop also *craw,* 1. an enlargement of the gullet of many birds, which serves as a temporary storage organ. 2. an analogous organ in certain insects and earthworms.
crossing over 1. the exchange of chromatid segments by enzymatic breakage and reunion during meiotic prophase. 2. a specific instance of such an exchange; a crossover.
crossopterygian any of a group of lobe-finned fish, including the ancestors of terrestrial vertebrates; the group is now extinct except for *Latimeria.*
crossover 1. a crossing over. 2. the result of crossing over. 3. the place on a cromatid where crossing over has occurred.
cross-reacting material (CRM) a mutant gene product that has lost its biological function but not all of its antigenic determinants, and thus responds to antibodies against the functional gene product.
crustacean any arthropod of the class *Crustacea* (see Appendix A).
crustecdysone the molting hormone of crustaceans.
cryptic coloration coloration that serves to conceal or to render inconspicuous.
crystal a solid having a regular internal arrangement of molecules.
ctenophore also *comb jelly,* any marine swimming invertebrate of the phylum *Ctenophora,* with a gelatinous body and eight rows of plates (combs) of fused cilia.
***C*-terminal end** also *carboxy-terminal end,* the end of a polypeptide chain in which the final amino acid has its primary amino group fixed in a peptide linkage and its primary carboxyl group free; the last part of the chain to be synthesized.
C_3 plant a plant in which the initial product of carbon dioxide fixation is a three-carbon compound of the Calvin cycle. *C_3 pathway,* the Calvin cycle.
CTP cytosine triphosphate.
cuticle a tough, often waterproof, non-living covering, usualy secreted by epidermal cells.
cuttlebone the skeleton of a cuttlefish, filled with thin, hollow chambers and used as a flotation device.
cuttlefish any of several swimming cephalopods, especially of the genus *Sepia.*
cyanobacterium (pl. *cyanobacteria*), a cyanophyte.
cyanophyte also *cyanobacterium,* formerly *blue-green alga,* any of a large group of blue-green photosynthetic procaryotes having as photopigments chlorophyll *a,* phycocyanin and phycoerythrin; utilizing photosystems I and II, and producing oxygen as a photosynthetic waste product.
cycad a plant of an order of palm-like gymnosperms, known from the Triassic to the present.
cyclic AMP (cAMP) adenosine monophosphate in which the phosphate is linked between the 3′ and 5′ carbons of the ribose group; serves as an intracellular gene regulator under a variety of circumstances.
cyclosis also *cell streaming,* the movement of cytoplasm within the cell, often in more or less regular, circular pathways.
cyst 1. a bladder, sac or vissicle. 2. a closed sac, containing fluid, embedded in a tissue. 3. a heavy protective covering of a dormant animal or protist. 4. an encysted organism.
cytochrome 1. any of a group of iron heme enzymes or carrier proteins in the oxidative and photosynthetic electron transport chains. 2. any iron heme protein other than hemoglobin or myoglobin.
cytochrome *c* the most readily isolated of the cytochromes of the mitochondrial electron transport system and one which has been extensively used in studies of phylogeny and biochemical evolution.
cytochrome P450 an oxygen carrier of the cell membrane, involved in facilitated diffusion of oxygen and also in detoxification of poisons, especially in lung tissue.
cytokinesis cytoplasmic cell division; actual division of the cell into two daughter cells.
cytokinin a plant cell hormone, mitogen, and plant tissue culture growth factor, which interacts with other plant hormones in the control of cell differentiation.
cytological map a map locating genes on the physical chromosome by means other than recombination mapping.
cytology the scientific study of cells.
cytoplasm the semisolid, protein-rich matrix of a cell exclusive of the cell membrane, the nucleus or other large inclusions.

cytopyge the fixed position on the surface of a protist from which wastes are discharged by exocytosis; the "anus" of a ciliate protozoan.
cytosine a pyrimadine, one of the four nucleotide bases of DNA, and also one of the four bases of RNA.
cytoskeleton the internal structure of animal cells, composed of microtubules and actin microfilaments; it controls the size, shape and movement of the cell.
cytostome the mouth of a protist.

dalton a unit of atomic and molecular mass, equivalent to one twelfth the mass of an atom of carbon 12.
D and C see *dilation* and *curettage.*
Danielli model a historically important (and largely validated) conception of the cell membrane as a phospholipid bilayer stabilized by proteins with channels for the passage of water-soluble compounds.
dark clock an unknown mechanism by which plants are able to measure a period of darkness with fair accuracy, and thus produce flowers at the appropriate season.
dark reaction see *light-independent reaction.*
Darwinian fitness see under *fitness.*
dATP, deoxyadenosine triphosphate a phosphorylated deoxynucleotide consisting of adenine, deoxyribose, and three phosphate groups.
daughter cell either of the two cells created when one cell divides.
DDT, dichloro-diphenyl-trichloro-ethane a colorless, odorless, highly persistent insecticide, relatively harmless to humans and other mammals but toxic to birds and, of course, insects.
deamination the removal of an amino group.
death control the reduction of a population growth rate by an increase in the death rate, as opposed to achieving the same end by a reduction in the birth rate.
death rate 1. in human populations, the number of deaths per 1000 population per year. 2. any analogous rate.
decarboxylase see *carboxylase.*
deciduous plant a perennial plant that seasonally drops its leaves.
decomposer also *reducer,* an organism that breaks down organic wastes and the remains of dead organisms into simpler compounds, such as carbon dioxide, ammonia and water.
decomposition the action of decomposers.
dedifferention the return of a differentiated cell to an undifferentiated, plastic state.
degenerate (adj.) in evolution, having become relatively simplified; having lost structures; having function or structures become vestigial; *degenerate code,* any code in which more than one symbol is used to signify the same thing; with regard to the genetic code, the fact that more than one codon is used to code genetically for individual amino acids. See *code, genetic code.*
dehydrated (adj.) having lost water or moisture; desiccated
dehydration linkage a covalent bond formed between two compounds by the removal of one oxygen atom and two hydrogen atoms.
deletion 1. the removal of any segment of a chromosome or gene. 2. the site of such a removal after chromosome healing, considered as a mutation. See also *base deletion.* 3. the deleted chromosome.
deletion map a genetic map constructed from recombination tests between various deletion chromosomes.
demography 1. the statistical study of human population (or other) with reference to growth, birth, and death rates, size, density, and migration.
denaturation the alteration of a protein so as to destroy its properties, through heating or chemical treatment (protein denaturation may be reversible or irreversible).
dendrite an extension of a neuron that conducts impulses toward the cell body and axon.
dendrograph a device for recording fluctuations in the diameter of a tree trunk.
denitrification the conversion of nitrates and nitrites to nitrogen gas; *dentrifying bacteria,* common soil and manure bacteria that are responsible for denitrification.
density-dependent effects factors affecting population parameters (growth, birth and death rates) in different ways depending on competition and population density, e.g., food availability, nesting site availability, disease, and predation.
density-independent effects those factors affecting population parameters independently of population size, as temperature, salinity, and meteorites.
denticle 1. a small tooth. 2. a small pointed projection, developmentally homologous with teeth, found in shark skin.
deoxyhemoglobin hemoglobin not carrying oxygen.
deoxyribonucleotide a nucleotide in which the sugar portion is deoxyribose rather than ribose.
deoxyribose a five-carbon sugar identical to ribose except that the 2′ hydroxyl group is replaced by a hydrogen.
depressant a "downer," any psychoactive drug that lowers spirit or mood, or decreases neurological or physiological activity.
derived a character that has undergone an evolutionary change in the group being considered; the opposite of *primitive.* See also *advanced.*
dermal (adj.) pertaining to the skin or dermis; *dermal bone,* bone originating in evolution and embryology as flat plates beneath the skin, as certain bones of the cranium.
dermal branchiae "skin gills" of echnoderms, consisting of ciliated outpocketings of the coelom and body wall.
dermis in animals, the inner mesodermally-derived layer of the skin, beneath the ectodermally-derived epidermis.
descending colon a part of the large intestine that descends to the rectum; see Figure 25.15.
descent derivation from an ancestor or ancestors.
desert a region characterized by scanty rainfall (especially less than 25 cm annually).
desiccate (v.t.) to dry up, cause to dry up, dehydrate; desiccated, (adj.) dried up, dehydrated.
desmosome an ultrastructural organelle joining the cytoplasm of adjacent cells in plants or animals. See *gap junctions.*
deuterostome a bilateral animal with radial cleavage and a mouth that does not arise, developmentally or phylogenetically, from the blastopore; compare *protostome.*
development 1. the whole process of growth and differentiation by which a zygote, spore or embryo is transformed into a functioning organism. 2. any part of this process, as the development of the kidney. 3. evolution.
Devonian see geological timetable inside the front cover.
dextrose see *glucose.*
diabetes mellitus a genetic disease of carbohydrate metabolism characterized by abnormally high levels of glucose in the blood and urine, and the inadequate secretion or utilization of insulin.
diakinesis final stage of meiotic prophase, in which the chromosomes are maximally condensed and homologs are attached only at their ends.
diapause in insects, a period of dormancy, interrupting developmental activity of the pupa or other stage, frequently occurring during hibernation or estivation.
diaphragm in mammals, a dome-shaped, muscularized body partition separating the chest and abdominal cavities, involved in breathing movement.
diastole the period of expansion and dilation of the heart during which it fills with blood; the period between forceful contractions (systole) of the heart. See *systole.*
diatom an acellular or colonial yellow-green photosynthetic protist, having a silicon-impregnated cell wall in two parts.
diatomaceous earth geological deposits consisting largely of the cell walls of diatoms.
dicot a flowering plant of the angiosperm class *Dicotyledonae,* characterized by producing seeds with two cotyledons; compare *monocot.*
dicotyledon see *dicot* and *cotyledon.*
differential *concentration differential,* the difference in concentration of a solute in two fluids, especially the fluids on either side of a semipermeable membrane.
differential *pH differential,* the difference in pH of two fluids, especially the fluids on either side of a semipermeable membrane.
differentiation in development, the process whereby a cell or cell line becomes morphologically, developmentally, or physiologically specialized. *terminal differentiation,* differentiation to a form in which the cell will normally not undergo further cell division. See *dedifferentiation.*
diffuse widely spread or scattered. *diffuse chromatin,* chromosomes or portions of chromosomes that are not condensed.
diffuse root system also *fibrous root system,* a root system of a plant in which there are many roots all about the same size.
diffusion the random movement of molecules of a gas or solute under thermal agitation, resulting in a net movement from regions of higher initial concentration to regions of lower initial concentration.
digestion the process of making food absorbable by breaking it down into simpler chemical compounds, chiefly through the action of enzymes, acid, and emulsifiers in digestive secretions; see *extracellular* and *intracellular digestion.*
digestive vacuole in animals and protists, intracellular vacuoles that arise by the phagocytosis of solid food materials and in which digestive processes occur, nutrient molecules being absorbed and undigested materials eventually being expelled by exocytosis.
digit in terrestrial vertebrates, any of the terminal divisions of the limbs; a finger, thumb or toe.
dihybrid an organism that is heterozygous at two different loci. *dihybrid cross,* a cross between two genotypically identical dihybrids.
dilation an enlargement or widening; in the human birth process, the initial stage, in which the body of the cervix softens and its opening widens.
dilation and curettage (D and C) a surgical operation in which the cervix is forcefully dilated and the uterine mucosa scraped with a curette; a common method of induced abortion as well as a procedure for removing cysts and polyps.
dimer a molecle consisting of two similar subunits.
dinocaryote a dinoflagellate, considered as being other than a eucaryote; the term emphasizing the fact that cell and nuclear organization in the dinoflagellates is different from that of either procaryotes or eucaryotes.
dinoflagellate a flagellated, photosynthetic, marine protist of the *Dinoflagellata,* a group considered by various authorities to be an order, class, division, kingdom, or even higher group (*dinocaryotes*), on a level with procaryotes and eucaryotes.
dinosaur 1. any of a taxon of large, extinct reptiles, widely distributed from the Triassic to the Mesozoic, including the largest known land animals. 2. any large, extinct reptile.
dioecious 1. in flowering plants, having separate sexes in the sporophyte generation. 2. in algae plants and bryophytes, having separate sexes in the gametophyte generation; that is, producing only eggs or sperm in any one thallus.
dipeptide two amino acids united by a peptide linkage; a common intermediate product of digestion. *dipeptidase* an enzyme that hydrolyzes dipeptides but not longer polypeptides.
diplococcus (pl. *diplococci*), 1. a coccus (spherical) in which the cells are arranged in colonies of two. 2. a bacterium of the genus *Diplococcus,* which includes several serious human pathogens, causing pneumonia and gonorrhea.
diploid having a double set of genes and chromosomes—one set from each parent.

diplont an organism that is diploid throughout its life cycle except for the haploid gamete stage, mitosis occurring only in diploid cells.
diplotene also *diplonema,* the stage in meiotic prophase during which chiasmata become visible and separation of homologs begins.
directional selection selection favoring one extreme of a continuous phenotypic distribution.
disaccharide a carbohydrate consisting of two simple sugar subunits.
discrete distribution see *distribution.*
discrimination the ability to perceive the difference between similar stimuli.
dispersive replication replication in which the original material is dispersed among all descendents.
dissociate to come apart into discrete units; *dissociation constant,* of a chemical reaction, a constant depending on the equilibrium between the combined and dissociated forms of a chemical or chemicals.
distal convoluted tubule a portion of the nephron between the loop of Henle and the collecting duct; see Figures 28.17 and 28.18.
distal away from the center of the body, heart, or other reference point.
distance receptor a sensory receptor of stimuli arising at a distance, such as sight and sound.
distribution in statistics, a set of values or measurements of a population, each measurement being associated with one individual. *continuous distribution,* the distribution of a character that varies continuously or by small increments, such as heights and bristle number. *discrete distribution,* the distribution of a character that exists in a finite number of distinct character states; see also polymorphism.
diverticulum a blind sac branching off a cavity or canal.
diversity 1. variety; variability. 2. the range of types in a major taxon: *plant diversity.* 3. in ecology, a measure of the number of species coexisting in a community.
division a major primary category or taxon of the plant, fungal and sometimes moneran and protist kingdoms, equivalent to *phylum.*
DNA, deoxyribonucleic acid the genetic material of all organisms (except RNA viruses); in eucaryotes DNA is confined to the nucleus, mitochondria and plastids.
DNA polymerase any of several enzymes or enzyme complexes that catalyze the replication of DNA.
Dollo's law a generalization in biology; characters lost in the course of evolution are never regained in their original form.
dominance the phenotypic expression of only one of the two alleles in a heterozygote. See *dominance interaction.*
dominance, behavioral dominance a behavioral relationship between two animals where the *subordinate* individual withdraws or behaves submissively toward the *dominant* individual in any conflict, potential conflict, or interaction.
dominance, cortical see *cortical dominance.*
dominance hierarchy 1. also *pecking order,* behavioral interactions established in a troop, flock or other species group, in which every individual is dominant to those lower on the order and submissive to those above. 2. in genetics, the relationship among a group of multiple alleles in which each ordered series behaves as a dominant to those below it and as a recessive to those above it.
dominance interaction the interaction between two alleles of a given locus in the development and expression of a specific phenotypic trait. One allele of the pair is *dominant* if its typical phenotype is expressed in both the homozygote and the heterozygote; in this case the other allele, which is expressed only when it is homozygous, is said to be recessive.
If the phenotype of the heterozygote is half-way between the homozygous phenotypes, both alleles are said to be *semidominant;* if the phenotype of the heterozygote exceeds those of both homozygotes, both alleles are said to be *overdominant.* If each of two alleles has a unique phenotypic expression and both are independently expressed in the heterozygote, both are said to be *codominant.* For other dominance interactions consult the text.
dominance, medullary see *medullary dominance.*
dominant 1. a phenotypic trait that is always expressed when a certain allele is present. 2. an allele that expresses a given phenotypic trait whether homozygous or heterozygous. See *dominance interaction.*
dormancy the temporary suspension of growth, development, or other biological activity; suspended animation, *dormant,* in a state of dormancy.
dorsal toward the back or (usually) upper surface of a bilateral animal.
dorsal aorta see under *aorta.*
dorsal hollow nerve cord a defining characteristic of chordates, arising in development as a flat dorsal plate of neurectoderm that rounds up and sinks beneath the surface to form, in vertebrates, the *brain* and *spinal cord* (*central nervous system*).
dorsoventrally from top to bottom, or back to front, in humans.
double bond a bond between two atoms consisting of two covalent bonds, with a total of four electrons shared.
double helix the configuration of the native DNA molecule, which consists of two antiparallel strands wound helically around each other.
doubling time the time it takes a growing population to double in numbers.
Down's syndrome see *trisomy 21.*
downy mildew 1. a fungal parasite of plant leaves. 2. the disease caused by the fungus.
DPGA, 1, 3 diphosphoglycerate an intermediate metobolite in glycolysis.
trisomy 21 also *Down's syndrome,* a congenital pathological condition in humans, owing to the presence of an extra chromosome 21, either as a free trisomic or translocated to another chromosome. It is characterized by low intelligence, facial and body abnormalities, and various physiological defects.
drift *random drift,* also *genetic drift,* 1. the chance fluctuation of allele frequencies from generation to generation in a finite population. 2. the long-term consequences of such fluctuations, such as the loss or fixation of selectively neutral alleles.
drive an urgent, basic or instinctual need pressing for satisfaction; a physiological or psychological tension, lack or imbalance that the organism actively seeks to redress, quench or fulfull.
drive-reduction hypothesis the idea that most behavior aims at (or is rewarded by) the reduction of specific innate drives.
dropsy see *edema.*
Drosophilia melanogaster the common fruit fly.
drumstick a small projection (not unlike a chicken leg) of the nucleus of the polymorphonuclear leucocytes of human females, attributed to the condensed second X chromosome.
Dryopithecus an extinct ape, possibly ancestral both to humans and modern apes.
duct of Cuvier see *cardinal vein.*
ductless gland see *endocrine gland.*
ductus arteriosis in the mammalian fetus, a short broad vessel conducting blood from the pulmonary artery to the aorta, thus bypassing the fetal lungs.
duodenum the first region of small intestine, just posterior to the stomach, receiving the common bile duct.
dwarf 1. any abnormally short person. 2. also *achondroplastic dwarf,* a genetic abnormality, an abnormally short person with markedly atypical body proportions. 3. any plant or animal unusually small for its species or type. *pituitary dwarf,* a dwarf whose small stature is due to a lack of growth hormone during development.

ecdysis also *molting,* 1. in arthropods, the shedding of the outer, non cellular cuticle. 2. this shedding together with associated changes in size, shape or function. *ecdysiast,* a strip-tease dancer.
ecdysone the molting hormone of insects.
echinoderm any organism of the marine, coelomate, deuterostome phylum *Echinodermata* (see Appendix A).
ecocline a gradual transition from one community type to another within a biome.
ecological niche see *niche.*
ecological succession see *succession.*
ecologist a scientist who studies ecology.
ecology the branch of biology dealing with the relationships between organisms and their environment.
ecosystem the biotic and abiotic factors of an ecological community considered together.
ecotype in plants, a morphological variety found in, and presumably adapted to, specific regions within the species range, but interfertile with other ecotypes of the species.
ectoderm in animal development, the outermost of the three primary germ layers of an embryo: the source of all neural tissue, sense organs, the outer cellular layer of the skin and associated organs of the skin; also, any of the tissues derived from embryonic ectoderm; also, the outer cellular layer of a coelenterate.
ectoparasite a parasite that feeds on or attaches to the surface tissue of the host; fleas, lice, ticks and athlete's foot fungus are human ectoparasites.
ectoproct a lophophorate, coelomate, sessile, colonial bryozoan ("moss animal") distinguished from the outwardly similar but unrelated pseudocoelomate entoproct by having the anus outside of the whorl of tentacles.
edema also *drospy,* an abnormal accumulation of fluid in connective tissues, causing a general puffiness.
effector also *effector organ,* a bodily organ actively used in behavior, especially in communication—distinguished from *receptor;* e.g., the human eyebrows and associated muscles constitute an effector organ, and a rather effective one at that.
efficiency *thermal efficiency,* in a metabolic pathway, the proportion of potential chemical energy retained in usable form and not dissipated as heat or entropy.
egg cell the larger gamete of an anisogametic organism; in plants, the one haploid cell of the embryo sac that will participate as the female gamete; in animals, the functional product of meiosis in females.
eidetiker a person with a photographic memory.
ejaculate (v.i.), 1. to eject semen by involuntary muscular contractions of the vas deferens and urethra in an orgasm. 2. (n.) the amount of semen ejaculated in one orgasm. *ejaculation,* the act of ejaculating.
elastin an extracellular structural protein that forms long, elastic fibers in connective tissue.
electromagnetic radiation waves of energy propagated by simultaneous electric and magnetic oscillations, including radio waves, microwaves, infrared, visible light, ultraviolet radiation, X-rays and gamma rays. *electromagnetic spectrum,* the range of electromagnetic radiation from low-energy, low frequency radio waves to high-energy, high frequency gamma rays.
electron one of the three common constituents of an atom, a *lepton* with a mass of 1/1837 of that of a proton and an electrostatic charge of $^-1$.
electron acceptor a molecule that accepts one or more electrons in an oxidation-reduction reaction (and thus becomes reduced).
electron carrier a molecule that behaves cyclically as an electron acceptor and an electron donor.
electron donor a molecle that loses one or more electrons to an electron acceptor in an oxidation-reduction reaction (and thereby becomes oxidized).
electron microscope a device for creating magnified images of small specimens through bombarding it with an electron beam and by subsequent magnetic focusing.

electron orbit also *electron orbital,* 1. the state of an electron as determined by its energy as it moves within an atom. 2. the space within which an electron pair moves within an atom.
electron shell 1. the space occupied by the orbits of a group of electrons of approximately equal energy; 2. an energy level of a group of electrons in an atom.
electron transport chain also *electron transport system,* a series of cytochromes and other proteins, bound within a membrane or thylakoid, mitochondrion, or procaryote cell, that passes electrons and/or hydrogen atoms in a series of oxidation-reduction reactions that result in the net movement of hydrogen ions across the membrane.
electrostatic (adj.) pertaining to the attraction or repulsion of electric charges independent of their motion.
element a substance that cannot be separated into simpler substances by purely chemical means.
elephantiasis enlargement of extremities due to blockage of lymphatic channels by parasitic nematodes.
elongation lengthening; in plant development, the stretching in one direction of stem or root cells under turgor.
Eltonian pyramid also *pyramid of numbers,* a concept in ecology that, because of thermodynamic inefficiency, organisms forming the base of a food chain are numerically abundant and comprise a large total biomass, while organisms of each succeeding level of the chain are successively less abundant and of smaller total biomass, and the top predator is always numerically rare and of relatively small total biomass.
embolism 1. the sudden blockage of a blood vessel by a dislodged blood clot, air bubble or other obstruction. 2. also *embolus,* the clot or obstruction itself.
embryo an animal or seed plant sporophyte in an early state of development. In animals this stage begins with cleavage, includes the laying down of germ layers, organ systems and basic tissue types, and grades imperceptibly into the *fetal* stage; in seeds the embryo stage lasts from fertilization to germination, the "mature embryo" comprising a rudimentary plant with plumule, radicle and cotyledons.
embryo sac in flowering plants, the mature megagametophyte after division into six haploid cells and one binucleate cell, enclosed in a common cell wall.
embryology the scientific study of early development in plants and animals.
embryonic disk 1. also *embryonic shield,* the part of the inner cell mass from which a mammalia embryo develops. 2. *blastodisk.*
emigrate to move out of an area; the moving out of an area by former residents.
emission the discharge of semen; radiation of energy.
empirical based on observation rather than theory. *empirical law,* a statement that describes a relationship but offers no explanation of its mechanism.
endergonic (adj.) of a biochemical reaction or half-reaction, requiring work or the expenditure of energy; moving from a state of lower potential energy to one of higher potential energy; compare *exergonic.*
endocrine gland also *ductless gland,* a discrete gland that secretes hormones into the blood system.
endocrine system the endocrine glands taken together, and their hormonal actions and interactions.
endocytosis the process of taking food or solutes into the cell by engulfment (a form of active transport); see also *phagocytosis, pinocytosis;* compare *exocytosis.*
endoderm also *entoderm* 1. the innermost of the three primary germ layers of a metazoan embryo and the source of the gut epithelium and its embryonic outpocketings (in vertebrates, the liver, pancreas, lung). 2. any tissue derived from endoderm. 3. in coelenterates, the epithelium of the gastrovascular cavity.
endodermis a single layer of cells around the stele of vascular plant roots that forms a moisture barrier in that the lateral cell walls are closely oppressed and waterproofed in bands, forming the Casparian strip.
endometrium in mammals, the mucous membrane tissue lining the cavity of the uterus, responds cyclically to ovarian hormones by thickening as preparation for implantation of the blastocyst.
endomysium the delicate connective tissue sheath surrounding the individual muscle fibers.
endoparasite a parasite that lives in the gut, internal organs or tissues of its host; opposed to *ectoparasite.*
endopeptidase a proteolytic enzyme that cleaves peptide bonds in the middle portions of polypeptides.
endoplasmic reticulum (ER) internal membranes of the cell, usually a site of synthesis; *rough endoplasmic reticulum,* ER with bound ribosomes, the site of synthesis of noncytoplasmic proteins; *smooth endoplasmic reticulum,* ER without bound ribosomes, usually the site of synthesis of non-protein materials.
endoskeleton a mesodermally derived supporting skeleton inside the organism, surrounded by living tissue, as in vertebrates and echinoderms.
endosperm a nutritive tissue of seeds, formed around the embryo in the embryo sac by the proliferation of the (usually) triploid endosperm nucleus to form a starch-rich mass of thin-walled cells; the endosperm may persist until germination or be resorbed by the cotyledon(s) during seed maturation.
endosperm mother cell a large, central, usually diploid or binucleate cell formed in the megagametophyte by fusion of haploid nuclei; when fertilized in double fertilization, forms the *endosperm nucleus.*
endospore an asexual resistant spore formed within a bacterial cell; see *spore.*
endosymbiont a mutualistic symbiont that resides within the cells of its symbiotic partner.
energy a fundamental concept of physics, either being associated with physical dodies (potential energy, kinetic energy) or with electromagnetic radiation; under some conditions interconvertable with mass by the relation $E=mc^2$ or with entropy but not capable of being created or destroyed.
energy pyramid the Eltonian pyramid with regard to chemical energy rather than numbers or biomass; see *Eltonian pyramid.*
energy state also *energy level,* a quantum state of an electron in an atom, a form of potential energy as the electron changes from a higher energy state to a lower one; a measure of the free energy in a system.
energy store in a molecule, a bond capable of absorbing, concentrating, and storing potential energy.
engram (also *memory scar, memory trace*) the physical change that is presumed to occur in the brain when something is learned.
enteric relating to the gut.
entoderm see *endoderm.*
entoproct a pseudocoelomate, sessile, colonial bryozoan ("moss animal") distinguished from the outwardly similar but unrelated *ectoprocts* by having the anus inside of its whorl of tentacles.
entropy in thermodynamics, the amount of energy in a closed system that is not avalable for doing work; also defined as a measure of the randomness or disorder of such a system. *negative entropy,* free energy in the form of organization. See *energy, free energy.*
enucleate without a nucleus.
environment the surrounding conditions, influences, or forces that influence or modify an organism, population or community.
environmental resistance the sum of environmental factors (e.g., limited resources, drought, disease, predation) that restrict the growth of a population below its biotic potential (maximum possible population size).
environmental variance that part of the total phenotypic variance of a population that is not due to genetic differences between individuals.
enzyme a protein that catalyzes chemical reactions.
enzyme-substrate complex the unit formed by an enzyme bound by non-covalent bonds to its substrate.
eocrinoid the original ancestor of all crinoids; the oldest crinoid.
epiboly in gastrulation, the growth of proliferating micromeres (ectoderm) down over the surface of the embryo.
epicotyl in the embryo of a seed plant, the part of the stem above the attachment of the cotyledon(s); forms the epicotyl hook in the growth of some seeds.
epidermal pertaining to the epidermis, arising from the epidermis.
epidermis 1. in plants, the outer protective cell layer in leaves and in root and stem primary growth, one cell thick and made waterproof with an outer layer of cutin; replaced in secondary growth by *periderm.* 2. in animals, the outer epithelial layer, derived from ectoderm; lacking innervation or vascularization in vertebrates.
epididymis in mammals, an elongated soft mass lying alongside each testis, consisting of convoluted tubules; the site of sperm maturation.
epigenesis development considered as involving gradual diversification of an initially homogenous zygote (or spore)—contrast with *preformation.*
epinephrine also *adrenaline,* a neurohormone and systemic hormone with numerous effects, including raising of blood pressure, increasing heart beat, constricting blood flow to internal organs and skin and increasing blood flow to muscles, together with general arousal and excitement; produced by the adrenal medulla and by nerve synapses of the autonomic nervous system.
epiphysis in vertebrates, a part of a bone that ossifies separately and later becomes fused to the rest. *epiphyseal cartilage,* also *epiphyseal plate,* a plate of cartilage separating the body of a long bone from its terminal epiphysis, being the site of bone growth and elongation.
epiphyte a plant that grows nonparasitically on another plant or sometimes an object.
episome a separable genetic unit in a procaryote; see *plasmid.*
epistasis the masking of a trait ordinarily determined by one gene locus by the action of a gene or genes at another locus.
epithelial (adj.) of cells or tissue, covering a surface or lining a cavity.
epithelium (pl. *epithelia*) a tissue consisting of tightly adjoining cells that cover a surface or line a canal or cavity, and that serves to enclose and protect.
equilibrium (pl. *equilibria*), a state of rest or balance due to the equal action of opposing forces; in chemistry, the condition existing when a chemical reaction and its reverse reaction proceed at equal rates, so that there is no net change in the system.
equilibrium frequency in population genetics, the idealized allele frequency at which selection, mutation and other forces balance, so that there is no expected net change; chance fluctuations may cause oscillations around this idealized equilibrium.
ER see *endoplasmic reticulum.*
era one of the major divisions of geological time; see the geological timetable inside the front cover.
erection 1. the becoming stiff and firm (turgid), by dilation under blood pressure, of a penis, clitoris, or nipple. 2. the stiffened, turgid state of such an organ.
ergot a toxic fungal parasite of rye, from which various pharmaceuticals legal and illegal, are prepared.
erythrocyte also *red blood cell,* a hemoglobin-filled, oxygen-carrying, circulating blood cell; enucleate in mammals, but having a physiologically inactive, condensed nucleus in other vertebrates.
erythropoesis the production and differentiation of blood cells, especially erythrocytes.

erthropoietin a hormone that stimulates the production and differentiation of erythrocytes.
esophagus the anterior part of the digestive tract; in mammals it is muscularized and leads from the pharynx to the stomach.
essential nutrient a food item that must be included in the diet for healthful functioning; *essential amino acids,* an amino acid that an animal cannot manufacture from simpler food items.
ester one of a class of chemical compounds formed by a dehydration linkage between a hydroxyl group and a carboxyl group: $R_1-CH_2OH + HOOCH-R_2 \rightarrow R_1-CH_2-O-CO-R_2+H_2O$. *ester linkage,* the dehydration linkage forming an ester.
estivation a state of dormancy or torpor induced by heat or dryness; summer dormancy.
estrogen 1. any female hormone or feminizing hormone 2. *estradiol, estriol* or *estrone,* steriod female hormones produced primarily in vertebrate ovaries but also in the adrenal cortex, able to promote estrus and the development of secondary sexual characteristics.
estrous cycle the sequence of endocrine and other reproductive system changes that occur from one estrus period to the next.
estrus also *heat,* in the females of most mammalian species, the regularly recurring state of sexual excitement around the time of ovulation; the only time during which the female will accept the male and is capable of conceiving; the term is not applicable to people.
ethanol also *ethyl alcohol,* a two-carbon alcohol, CH_3CH_2OH, a common byproduct of anerobic metabolism in yeast and (as it happens) goldfish.
ethology the scientific study of animal behavior as it occurs in the organism's natural environment, or among free-ranging animals in the ethologist's environment.
ethylene a colorless, sweet-smelling unsaturated hydrocarbon gas, $CH_2=CH_2$, emitted by ripening fruit and inducing further ripening; also produced from petroleum and used in the manufacture of synthetic ethanol, polystyrene and polyethylene plastic.
eubacterium a "true bacterium," a bacterium of the largest bacterial phylum, Eubacteriacea (see Appendix A).
eucaryote a nucleated organism.
euchromatin chromatin other than heterochromatin.
euglena any of a number of nonsexual photosynthetic flagellate protists of the genus *Euglena,* whose photosynthetic pigments are similar to those of chlorophytes (green algae and higher plants).
euglenoid a *Euglena* or any of several related organisms.
eutrophic (adj.) a rapidly changing lake or other body of fresh water, rich in nutrients but periodically depleted of oxygen.
eutrophication the process whereby a lake matures and becomes eutrophic, either through evaporation and silting, or artificially by pollution.
evolution 1. any gradual process of formation, growth, or change. 2. descent with modification. 3. long term change and speciation (division into discrete species) of biological entities. 4. the continuous genetic adaptation of organisms or populations through mutation, hybridization, random drift and natural selection.
evolutionary stable strategy (ESS) an innate behavioral strategy (of conflict, etc.) that confers greater fitness on individuals than can any alternative behavior that might arise by mutation or recombination.
excise to cut out.
excretion the removal of metabolic wastes from the body.
excretory system an organ system involved in the removal of metabolic wastes; e.g., the kidney, urinary bladder and associated ducts and valves.
exergonic (adj.) of a biochemical reaction or half-reaction, producing work, heat or other energy; moving from a higher energy state to a lower energy state; as opposed to *endergonic.*
exocrine gland also *ducted gland,* gland that releases its secretions externally, into the gut, or anywhere other than the blood stream; usually exocrine glands are connected to ducts, but glandular secretory cells of the gut epithelium are considered exocrine.
exocytosis the process of expelling material from the cell through the cell membrane by the fusion of vacuoles, secretion granules, etc. with the cell membrane, and their subsequent eversion.
exopeptidase a proteolytic enzyme that hydrolyzes peptide linkages near the two ends of a polypeptide; see *carboxypeptidase* and *aminopeptidase;* compare *dipeptidase, endopeptidase.*
exophthalmic goiter an enlargement of the thyroid, usually due to a tumor or metastasizing cancer, that is associated with abnormally elevated levels of thyroxine (which, among other things, causes the eyes to bulge out).
exoskeleton an external skeleton or supportive covering, as in arthropods and armadillos; arthropod exoskeletons are formed by secretions of epidermal cells, largely chitin, proteins and calcium carbonate.
expected value also *expectation,* in mathematical probability, the mean value of all possible outcomes of an event, being the summed products of the probability of each outcome and the value assigned to that outcome.
expiration center a hypothesized brain center governing expiration (release of air from the lungs).
exponential growth also *geometric growth,* population growth in which the population size increases by a fixed proportion in each time period, successive values forming an exponential series, $P_t=P_0r^t$ where P_0 is the initial population size, t is the time, and r is the Malthusian parameter.
extend (v.t.), of a limb or body part, to bring outward, away from the body; opposed to *flex.*
external ear all parts of the ear external to the eardrum, comprising the external ear canal, the external auditory meatus, and the pinna (pl. *pinnae*—what you wriggle when you wriggle your ears).
external fertilization fertilization outside the body, as in echinoderms and most bony fish.
extinction a coming to an end or dying out of a species or other taxon (group of related organisms).
extracellular digestion digestion outside of cells (usually in the gut).
extraembryonic outside of the embryo proper. *extraembryonic coelom,* the space between the chorion and amnion, which in the early embryo is continuous with the body coelom. *extraembryonic membrane,* any of several membranes (*amion, chorion, allantois, yolk sac*) produced by the zygote but not part of the embryo proper.
extra-Y syndrome see *XYY syndrome.*
eyespot 1. a simple visual organ in many invertebrates, consisting of a pigmented cup with light-sensitive receptors. 2. a simple pigmented visual organelle in various flagellate protists, including euglenoids and *Chlamydomonas.*

facilitated diffusion diffusion of specific molecules across a cell membrane that is facilitated by reversible association with carrier molecules able to traverse the membrane. Facilitated diffusion differs from active transport in that no energy is expended and net movement follows the concentration gradient.
facultative having alternative responses according to conditions.
facultative anaerobe an organism such as yeast that can utilize oxygen when it is available but is also able to live anaerobically.
facultative heterochromatin chromatin that is condensed only under specific conditions, and functions as euchromatin under other conditions. See also *heterochromatin.*
FAD, $FADH_2$ flavine adenine dinucleotide, a hydrogen carrier in metabolism. FAD is the oxidized form and $FADH_2$ is the reduced form.
family in classification, a grouping smaller than *order* but larger than *genus.*
fascia (pl. *fasciae* or *fascias*), a heavy sheet of connective tissue covering or binding together muscles or other internal structure of the body, often connecting with ligaments or tendons.
fat 1. *neutral fat,* also *triglyceride,* any of the solid, semi-solid or liquid, lipid-soluble compound consisting of glycerol bound by ester linkages to three fatty acid molecules. 2. also *adipose tissue,* parts of an animal consisting largely of cells distended with triglycerides.
fate map a plan of a zygote, blastula or early embryo indicating the normal developmental fates of the cellular descendents of embryonic ectoderm, mesoderm, and endoderm.
fatty acid an organic acid consisting of a linear hydrocarbon "tail" and one terminal carboxyl group.
feces bodily waste discharged through the anus.
feedback of any system, the return to the input of a part of the output, which serves to regulate or affect the functioning of the system. *negative feedback:* the output damps the activity of the system, reducing the flow of output. *positive feedback:* the output stimulates the activity of the system, further increasing the flow of output.
femur 1. also *thigh bone,* the upper bone of the hind leg of a vertebrate. 2. in insect legs, the third segment from the body, often enlarged.
fermentation 1. anaerobic breakdown of glucose by yeast to form alcohol and carbon dioxide. 2. any controlled enzymatic transformation of an organic substrate by microorganisms.
fertility factor also F^+ episome, in *E. coli,* a plasmid or episome, transmitted by conjugation to uninfected hosts that sometimes incorporates into the bacterial chromosome and results in the transfer of genetic information between bacteria.
fertility rate the number of births per 1,000 women from 15 to 44 years of age; a clearer indicator of reproductive activity in a population than the birth rate.
fertilization the process of the union of gametes or gamete nuclei.
fertilization cone a cone-shaped mound of egg cytoplasm that, upon activation of the egg by a fertilizing sperm, rises to engulf the sperm.
fertilization membrane in echinoderms, a membrane formed upon egg activation by the rupture of cortical granules, preventing further sperm penetration.
fetus in vertebrates, an unborn or unhatched individual past the embryo stage; in humans, a developing, unborn individual past the first 12 weeks of pregnancy.
fiber see *muscle fiber.*
fibrin an insoluble fibrous protein forming blood clots and also contributing to the viscosity of blood; see *fibrinogen.*
fibrinogen a globular blood protein (*globulin*) that is converted into fibrin by the action of thrombin as part of the normal blood clotting process.
fibrinopeptide either of two small peptides released from fibrinogen in its conversion to fibrin by the enzymatic action of thrombin.
fibroin the insoluble fibrous protein of silk.
fibrous protein a protein in which the molecules tend to align in linear arrays, in contrast to globular proteins.
fibrous root system also *diffuse root system,* the roots collectively of a plant in which there are many equivalent roots rather than a prominent central root.
fibula in land vertebrates, the smaller of the two bones of the lower hind leg; fused with the *tibia* in birds and many mammals.
fiddlehead a yong unfurling frond of a fern, resembling the head of a violin.
filament 1. a long fiber. 2. see *muscle filament.* 3. in flowers, the slender stalk of the stamen, on which the anther is situated.
filial generation in Mendelian genetics, 1. *first filial generation,* F_1, the offspring of a cross between two homozygotes for different alleles (or individuals of two true-breeding strains). 2. subsequent filial generations (F_2, F_3 etc.) The progeny of selfing, of inbreeding or of crosses between individuals of a given filial generation.
filter-feeder an animal that obtains its food by filtering minute organisms from a current of water.
fimbria (pl. *fimbriae*), in vertebrates, a fringe of ciliated tissue surrounding the opening of the oviduct into the peritoneal cavity.

fire-disclimax community an ecological community in continual transition because of recurring fires (e.g., chaparral).
five-prime end (*5′ end*), of a nucleic acid, the end at which the 5′ carbon of ribose or deoxyribose is free to form a phosphate linkage; the first end of the chain in the direction of synthesis.
first messenger a hormone, as distinguished from a *second messenger.*
first law of thermodynamics the physical law that states that energy cannot be created or destroyed; later amended to allow for the interconversion of matter and energy.
fission the division of an organism into two (binary fission) or more organisms, as a process of asexual reproduction.
fissure of Rolando also *central sulcus,* the deep groove separating the frontal lobe from the parietal lobe of the cerebral cortex.
Fissure of Sylvius also lateral fissure, a deep groove in the human cerebral cortex separating the temporal lobe from the frontal and parietal lobes.
fit adapted or suited, e.g., "fit to be a policeman." In biology, adapted to the environment and thus well able to survive and reproduce.
fitness 1. the state of being adapted or suited (e.g., to the environment. 2. also *relative fitness, Darwinian fitness,* the relative expectation of surviving and reproducing of an individual or of a specific genotype, compared with that of the general population or of a standard genotype. 3. one half the expected number of offspring of a diploid genotype.
fixation 1. *nitrogen fixation,* the energy-consuming transformation of nitrogen gas to soluble nitrogen compounds. 2. the preparation of specimens for microscopic viewing by treatment that precipitates proteins and leaves spatial relationships intact; *to fix,* v.t., to so treat a specimen.
fixed action pattern a precise and identifiable set of movements, innate and characteristic of a given species; see also *instinctive pattern.*
flagellate protozoan a unicellular nonphotosynthetic protist of the group Mastigophora, propelled by one or more flagella.
flagellum (pl. *flagella*) 1. a long, whiplike, motile eucaryote cell organelle, projecting from the cell surface but enclosed in a membrane continuous with the cell membrane; longer and fewer in number than cilia, it propels the cell by undulations. 2. an analagous, non-homologous organelle of bacteria, consisting of a solid, helical protein fiber that passes through the cell wall and propels the cell by rotating like a propeller.
flame cell in flatworms, rotifers and lancelets, a cup-shaped cell with a tuft of cilia in its depression, the flame-like beating of which creates suction that draws waste materials into an excretory duct.
flatus gas expelled through the anus.
flatworm a platyhelminth, especially a turbellarian such as the common planaria.
flex (v.t.), to move muscles so as to close the angle of a joint; the opposite of *extend.*
florigen a hypothetical flower-inducing hormone.
flower the part of a seed plant that bears the reproductive organs, especially around the time of reproduction or during the time in which some or all of its parts are conspicuously white or brightly colored. *complete flower,* a flower having all four sets of floral organs; pistils, stamens, petals, sepals. *incomplete flower,* a flower lacking one or more of the four sets of floral organs. *regular flower,* a radially symmetrical flower. *irregular flower,* a bilaterally symmetrical flower. *peloric flower,* an abnormally regular flower in a species in which flowers are normally irregular.
fluid mosaic model also *Singer model,* a description of the cell membrane as a phospholipid bilayer stabilized by specifically oriented proteins, with some proteins extending through to both surfaces and other proteins specific for the inner or outer surfaces.
fluke a parasitic trematode flatworm; see also *Schistosome.*
follicle 1. *hair follicle,* a deep, narrow-mouthed depression from which a hair shaft grows. 2. *ovarian follicle* or *Graafian follicle,* a membraneous cavity or vesicle in a mammalian ovary, enclosing the maturing egg and associated cells (follicle cells); after ovulation, the follicle changes to become a *corpus luteum.*
follicle-stimulating hormone (FSH) a peptide hormone of the mammalian anterior pituitary, which stimulates the growth of ovarian follicles; the same peptide activates sperm production in males.
food chain a sequence of organisms in an ecological community, each of which is food for the next higher organism, from the primary producer to the top predator.
food vacuole see *digestive vacuole.*
food web also *food cycle,* a group of interacting food chains; all of the feeding relations of a community taken together; the flow of chemical energy between organisms.
F-one particle usually **F1 particle,** an ultramicroscopic mitochondrial organelle attached to the inner surface of the crista; the site of chemiosmotic phosphorylation.
foramen (pl. *foramina*), an opening or perforation.
foramen magnum the large opening at the base of the skull through which the spinal cord passes into the brain.
foramen of Monro the opening of a lateral ventricle into the third ventricle of the brain.
foramen ovale in the fetal mammalian heart, an opening in the septum between the atria; it normally closes at the time of birth.
foraminiferan a shelled marine sarcodine protist with one or more nuclei in a single cell membrane; the carbonate shell having numerous minute openings (foramina) through which slender branching pseudopods are extended.
fork see *replication fork.*
forebrain 1. the anterior of the three primary divisions of the vertebrate brain. 2. the parts of the brain developed from the embryonic forebrain.
fossil any remains, impression, or trace of an animal or plant of a formal geological age, such as a mineralized skeleton, a footprint, or a frozen mammoth.
fossil fuel coal, petroleum, or natural gas.
founder effect the population genetic effect of the chance assortment of genes carried by the successful founder or by a few founders of a subsequently large population.
four-chambered heart the heart of mammals, birds and crocodilians (and probably of dinosaurs), in which the atrium and the ventricle are both separated into left and right chambers by the evolutionary development of septa.
fovea (pl. *foveae*), the most sensitive part of the retina; a small central depression rich in cones and devoid of rods.
F plus eposome usually written **F+ episome,** see *fertility factor.*
frame shift mutation a small insertion or deletion in a structural gene such that all mRNA codons downstream are misread in translation.
free energy energy theoretically available to be used for work.
free radical a highly reactive atom or molecular unit with an unpaired electron.
freeze-fracture a technique for the preparation of electron microscope specimens, in which frozen cells are broken under conditions in which the membrane bilayer parts.
frequency see *allele frequency, genotype frequency.*
frequency (of electromagnetic radiation), the number of wave oscillations per second, which is inversely proportional to the wave length and directly proportional to photon energy.
frequency dependent selection natural selection in which a phenotype is more fit than average when it is numerically rare, and is less fit than average when its relative number surpasses some equilibrium frequency value.
frontal lobe an anterior division of the cerebral hemisphere, believed to be a site of higher cognition.
fructose a six-carbon ketose sugar, occuring free in honey, and combined with glucose to form the disaccharide *sucrose.*
fruit the seed-bearing ovary of a flowering plant.
fruiting body in fungi, slime molds, and myxobacteria, a structure specialized for the production of spores.
fucoxanthin a brown caretenoid pigment characteristic of brown algae.
fulcrum the support about which a lever moves.
functional group in biochemistry, a side group with a characteristic chemical behavior or function.
fungi imperfecti a wastebasket category consisting of fungi whose life cycle is largely unknown, or which lack sexuality, so that assignment into other, more discrete categories is impossible or has not been done.
fungus (pl. *fungi*), an organism of one of Whittaker's five kingdoms, Kingdom *Fungi,* which includes yeasts, mushrooms, molds, mildews, part of the lichen symbiosis, rusts, smuts, sac fungi, puffballs, water molds, and sometimes the slime molds.

galactose a hexose sugar, one of the two subunits of lactose, the disaccharide of milk sugar.
galactoside a compound of galactose; e.g., lactose. *galactosidase,* an enzyme that dissociates galactose from a galactoside.
galactoside permease in *E. coli,* a membrane protein involved in the facilitated diffusion of galactosides into the cell.
gall bladder, see under *bladder.*
gametangium (pl. *gametangia*), in algae and fungi, the structure (cell or organ) in which gametes are formed.
gametocyte also *meiocyte,* the cell that undergoes meiosis to form gametes; a spermatocyte or oocyte.
gametophyte in plants with alternation of generations, the haploid form, in which gametes are produced.
gamma radiation electromagnetic radiation of very high energy per photon, beyond X-radiation.
gamont the gametocyte of a protozoan.
ganglion a mass of nerve tissue containing the cell bodies of neurons.
gap junction dense structures that physically connect cell membranes of adjacent cells along with channels for cell-to-cell transport.
gastric lipase a fat-digesting enzyme secreted by the stomach lining.
gastrin a hormone, produced by the stomach lining, that induces the secretion of gastric juice (and may be identical with histamine).
gastrointestinal tract 1. the stomach and intestines as a functional unit. 2. the entire passageway from the mouth to and including the anus.
gastropod a mollusk of the large and varied class Gastropoda, present in Cambrian fossils; included are snails, slugs, nudibranchs, pterobranchs, opistobranchs, limpets, keyhold limpets, abalone and others.
gastiotrichs an Aschelminth class related to the rotifers and similar to them in body plan.
gastrovascular cavity also *coelenteron,* the cavity of coelenterates, ctenophores and flatworms, which opens to the outside only via the mouth, and serves the functions of a digestive cavity, crude circulatory system, and coelom.
gastrula an early metazoan embryo consisting of a hollow, two-layered cup with an inner cavity (archenteron) opening out through a blastopore.
gastrulation the process usually involving invagination, involution, and epipiboly, whereby a blastula becomes a gastrula.
Gause's law the ecological principle that no two genetically isolated kinds of organisms can coexist (exist at the same time and place) while occupying the same ecological niche; the principle that each species of a group of coexisting species must be specialized in some way so as to occupy a unique niche. Also, the *competitive exclusion principle.*
gel electrophoresis an analytical procedure in biochemistry, whereby a mixture of biomolecules enters a gel and separates by the differential migration of its constituents in an imposed electrical field.
gemma (pl. *gemmae*), in liverworts, a multicellular bud that becomes detached from the parent thallus in a form of asexual reproduction.

gemma cup a cup-shaped organ of certain liverworts, containing gemmae, which are dislodged by raindrops.
gemmule in Darwin's discredited theory of pangenesis, minute particles produced by body parts and transmitted somehow to the gametes to participate in heredity and the developmental processes of the offspring.
gene (variously defined), 1. the unit of heredity. 2. the unit of heredity transmitted in the chromosome and that, through the interaction with other genes and gene products, controls the development of hereditary character. 3. a continuous length of DNA with a single genetic function. 4. the unit of transcription of DNA. 5. a *cistron* or *structural gene,* that is, the sequence of DNA coding for a single polypeptide sequence.
gene cloning recent techniques whereby pieces of DNA from any source are spliced into plasmid DNA, cultured in growing bacteria, purified and recovered in quantity. also *gene splicing, DNA cloning, recombinant DNA.*
gene flow the exchange of genes between populations through migration, pollen dispersal, chance encounters and the like.
gene frequency see *allele frequency.*
gene pool all the genetic information of a population considered collectively.
gene splicing the use of recombinant DNA techniques to form covalent bonds between DNA of different sources. See *gene cloning.*
generalization the process whereby a response is made to stimuli similar to, but not identical with, a reference stimulus.
generative nucleus one of the haploid nuclei of a pollen grain; when successful pollination occurs, the generative nucleus divides to form two sperm nuclei.
genetic background also *background genotype,* the sum total of the genetic constitution of a organism apart from the specific locus under consideration.
genetic code the relationship by which the sixty-four possible codons (sequences of three nucleotides in DNA or RNA) each specify an amino acid or chain termination in protein synthesis.
genetic drift see *drift.*
genetic dwarf a dwarf whose condition is due to genetic rather than environmental causes.
genetic map a chart of the sequence of and relative distances between specific genes along a chromosome, based on recombination frequencies observed between gene loci.
genetic marker a standard mutant allele used in genetic tests in which it serves only to trace the fate or distribution of a chromosome, chromosome region, or gene locus, especially in recombination studies.
genetic variability a broad term indicating the presence of different genetic constitutions in a population or populations.
genetic variance a specific mathematical value with regard to a single trait in a single population, specifically the amount of the total phenotypic variance that is attributable to genetic differences between individuals; see also *variance.*
genetics 1. the science of heredity, dealing with the resemblances and differences of related organisms resulting from the interaction of their genes and the environment. 2. the study of the structure, function and transmission of genes.
genital swelling in mammalian embryos, a paired swelling beneath the *genital tubercle;* it will form the scrotum in males and the *labia majora* in females.
genital tubercle in the sexually undifferentiated mammalian embryo, three small swellings, the glans and two urogenital folds; in the male these will form the head and shaft of the penis, and in the female they will form the *glans clitoridis* and the *labia majora.*
genitals also *genitalia,* 1. the organs of the reproductive system. 2. also *external genitalia,* the external reproductive organs; the penis and scrotum in human males and the vulva in females.
genome the full haploid complement of genetic information of a diploid organism, usually viewed as a property of the species.
genotype 1. the genetic constitution of an organism with respect to the gene locus or loci under consideration. 2. also *total genotype,* the sum total of genetic information of an organism.
genotype frequency the proportion of individuals in a population having a specified genotype.
genus (pl. *genera*), a group of similar and related species; a taxon smaller than *family* and larger than *species.*
geological era one of five divisions of the earth's history (Cenozoic, Mesozoic, Paleozoic, Proterozoic, and Archeozoic); see the geological timetable inside the front cover.
geology the scientific study of the physical history of the earth, the rocks, and soil of which it is composed, and the physical changes that the earth has undergone or is undergoing.
geotaxis also *positive geotaxis,* the tendency to move downward. *negative geotaxis,* the tendency to move upward.
geotropism also *positive geotropism,* the tendency of a growing root tip to turn downward. *negative geotropism,* the tendency of a growing shoot to turn upward.
germ layer any of the three layers of cells formed at gastrulation, the partially differentiated precursors of specific body tissues and organs; see *ectoderm, endoderm* and *mesoderm.*
germ line *germ line cell,* the gametes and any cell or tissue that will or might give rise to gametocytes, or that otherwise contains the physical precursor of DNA that will be transmitted of offspring. *germ-line DNA,* the DNA of germ line cells.
germ plasm 1. an amorphous substance which contained all hereditary information according to Weissmann's theory. 2. the physical embodiment of the genetic information of an organism, especially as transmitted in gametes.
germinal adj. 1. pertaining to the germ line. 2. pertaining to reproduction. 3. giving rise to embryonic structures. 4. *germinal layer,* the innermost layer of the epithelium, containing mitotic cells.
germinal epithelium 1. the epithelium of the gonads, which through mitosis gives rise to gametocytes. 2. also *germinal layer,* the innermost layer of an epithelium, containing mitotic cells.
germination the process whereby a seed and embryo ends dormancy, the stored materials of the seed being digested, the seed coat ruptured, and the plumule and hypocotyl begin to grow; spore germination, the end of dormancy in a spore.
giant chromosome see *polytene chromosome, lampbrush chromosome.*
giantism 1. also *gigantism,* growth to abnormal or unusual size. 2. see *pituitary giant.*
gibberellin any of a family of plant growth hormones that control cell elongation, bud development, differentiation, and other growth effects.
Gigantopithecus a genus of giant fossil primates from the Pleistocene, possibly of a group ancestral both to humans and to anthropoid apes.
gill arches also *branchial arches* in fish, a row of bony or cartilagenous curved bars extending vertically between the gill slits on either side of the pharynx, supporting the gills. 2. in land vertebrate embryos, a row of corresponding, homologous, rudimentary ridges that give rise to jaw, tongue and ear bones.
gill basket also *branchial basket,* the cartilagenous structure supporting the gills in urochordates, cephalochordates, and larval lampresys, serving as a sieve for filter feeding.
gill clefts also *gill slits, branchial clefts,* 1. in fish, the openings between the gills. 2. in vertebrate embryos, the corresponding and homologous grooves in the neck region, between the branchial arches.
gill filament one of the many feathery, leaf-like or filamentous structures making up the working surfaces of the gill.
gill pouch also *branchial pouch,* one of the hollowed-out cavities corresponding to the gill clefts in cyclostomes and some sharks.
gill raker in fish, one of the bony processes on the inner side of a gill arch, serving to prevent solids from passing out through the gill clefts.
gill slit see *gill cleft.*
gill 1. an organ for obtaining oxygen from water. 1a. any of the highly vascular, filamentous processes of the pharynx of fishes, between the gill clefts. 1b. any of the analagous but not homologous organs of invertebrates. 2. the thin, laminar structures on the underside of a mushroom (basidocarp), bearing the spore-forming basidia.
girdle see *pectora girdle; pelvic girdle.*
gizzard 1. in birds, a thick-walled, musclar enlargement of the alimentary canal, filled with small stones and used to grind seeds or other food. 2. any analogous structure in invertebrates, as in earthworms, rotifers and gastrotrichs.
glaciation the formation of ice sheets, such as the great ice sheets that covered Europe and northern North America four times during the Pleistocene (last 1 million years). The last such glaciation ended about 10,000 years ago.
glans (L. *glans,* acorn), 1. also *glans penis,* the conical, vasular, highly innervated body forming the end of the penis. 2. also *glans clitoridis,* a homologous body forming the tip of the clitoris.
gliding joint a skeletal joint in which the articulating surfaces glide over one another without twisting.
globulin and of a large class of globular proteins occurring in plant and animal tissues and in blood.
Gloger's rule a generalization in biogeography: within a geographic range of a species or genus of warm-blooded animals, the degree of pigmentation increases as the mean environmental temperature increases.
glomerulus (pl. *glomeruli*), the mass or tuft of capillaries within a Bowman's capsule.
glucagon a polypeptide hormone secreted by the pancreatic islets of Langerhans, whose action increases the blood glucose level by stimulating the breakdown of glycogen in the liver.
glottis the opening between the vocal chords in the larynx.
glucose also *dextrose, blood sugar, corn sugar, grape sugar,* a six-carbon sugar, occuring in an open chain form or either of two ring forms; the subunit of which the polysaccharides starch, glycogen, and cellulose are composed, and a constituent of most other polysaccharides and disacchrides.
glucose-1-phosphate glucose covalently bonded to a phosphate group; an intermediate in the synthesis and breakdown of glucose as well as in the synthesis and breakdown of most polysaccharides.
glucuronic acid an acid aldehyde carbohydrate related to glucose, one of the principle constituents of cartilage and the carbohydrate group of several mucoproteins.
glucosamine an amino derivative of glucose, a constituent of chitin.
glue cell a glandular, thread-bearing cell, found only in ctenophores; used to capture prey by adhesion.
glycerol formerly *glycerin,* an organic compound composed of a three-carbon backbone with an alcohol (hydroxyl) group on each carbon; a component of neutral fats and of phospholipids. Pure glycerol is a sweet, sticky and disgusting liquid.
glycogen also *animal starch,* a highly branched polysaccharide consisting of alpha glucose subunits; a carbohydrate storage material in the liver, muscle and other animal tissues.
glycolipid a compound with lipid and carbohydrate subunits; e.g., cerebrosides and gangliosides.
glycolysis the enzymatic, anaerobic breakdown of glucose in cells, yielding ATP (from ADP), and either lactic acid, or alcohol and CO_2, or pyruvate and $NADH_2$ (from NAD).
glycoprotein a compound containing polypeptide and carbohydrate subunits.

G_1 also *G_1 phase, gap one,* the period in the eucaryotic cell cycle between the end of mitosis and the onset of DNA replication. During G_1 each chromosome consists of a single chromonema.
goblet cell a goblet-shaped, mucus-secreting epithelial cell; found in the lining of the nasal cavity, bronchi and elsewhere.
goiter a visible enlargement of the thyroid gland on the front and sides of the neck. *Colloid goiter,* a goiter resulting from iodine deficiency, associated with normal or subnormal levels of thyroid hormone. See *exophthalmic goiter.*
Golgi body also *Golgi complex, Golgi apparatus,* a cytoplasmic, membranous, subcellular structure found especially in secretory cells, and which is involved in the packaging of cell products from the endoplasmic reticulum into secretion granules.
gonad in animals, the primary sex gland, with endocrine functions, and the site of meiosis; an ovary or testis.
gonadotropin a hormone that stimuates growth or activity in the gonads; in vertebrates, a specific peptide hormone of the anterior pituitary.
graded display a communication pattern in which there exist intermediate states such as a behavioral continuum from low-to-high levels of motivation.
gradient a continuous change in concentration (or of any other quantity such as temperature, partial pressure, or allele frequency) over space. *with the gradient* (in the movement of solutes), in the direction from higher to lower concentration. *against the gradient,* in the direction from lower to higher concentration, i.e., in the direction opposite to that of expected net diffusion.
graft 1. the act of uniting two plants in such a way that they grow together. 2. the resulting growth. 3. v.t., to make a graft.
grafting an artificial method of propagation in which the bud, stem, or shoot of a plant is inserted into a cut in a recipient stem such that the cambium of both plants comes into contact and will eventually grow together.
grain also *cereal grain,* the hard fruit of a cereal grass, such as wheat, rye, corn, rice, or barley, consisting of a single seed in a tough covering.
gram stain a solution of iodine and iodide used to stain certain large classes of bacteria. *gram negative bacterium,* a bacterium that cannot be stained by gram stain. *gram positive bacterium,* a bacterium that can be stained with gram stain.
granum (pl. *grana*), a stack of thylakoids disks in a chloroplast; seen as a series of minute, multiple green bodies containing most of the chlorophyll of the plant.
green alga (pl. *green algae*), a unicellular or multicellular alga of the Plant Kingdom division Chlorophyta; an alga in which the green color of chlorophyll is not masked by other pigments.
green mold see *blue mold.*
greenhouse effect the warming principle of greenhouses, in which high-energy solar rays enter easily, while less energetic heat waves are not radiated outward; now especially applied to the analogous effect of increasing atmospheric concentrations of carbon dioxide through the burning of fossil fuels and forest biomass.
gross productivity the amount of biochemical energy captured by photosynthesis in a particular area per unit time.
growth curve a graph of population numbers per unit time, especially in a period of rapid initial growth leading into populations size stabilization or to a population crash.
growth factor a hormone, usually a mitogen, necessary for the proliferation of cells in tissue culture.
growth hormone (GH) 1. a polypeptide hormone of the anterior pituitary that regulates growth in vertebrates. 2. any hormone that regulates growth, e.g., auxin or gibberelin in plants.
G_2 also *G_2 phase, gap two,* the period in the eucaryotic cell cycle between the end of S phase and the onset of mitotic interphase. During G_2 each chromosome consists of a pair of chromatids.
guanine one of the nitrogenous bases of RNA and DNA.
guanosine a nucleoside of guanine; guanine linked to ribose.
gullet 1. the esophagus. 2. an invagination in the surface of certain protozoan protists, especially when used for the intake of food.
guard cell in leaf and stem epidermis, either of a pair of cresent-shaped cells that surround each stoma, and whose change in turgor regulate the size of the stomatal opening.
gustation the sense of taste; the act of tasting.
guttation the normal exuding of moisture from the tip of a leaf or stem, presumably due to root pressure.
gymnosperm a seed plant of the paraphyletic group Gymnospermae, producing naked seeds and lacking fruit or flowers; included are conifers, cycads, and ginkgos.
gynoecium in a flower, the whorl or carpels taken as a group; the pistil or pistils.

habit reversal a test involving the ease with which an animal learns to respond in a manner opposite to that to which he has previously been trained.
habitat the place where a plant or animal species naturally lives and grows.
habituation learning not to respond to environmental stimuli that may have no relevance to the organism.
hadron any of a class of heavy subatomic particles, including protons and neutrons.
half life 1. of a radioisotope, the time it takes for half of the atoms in a sample to undergo spontaneous decay. 2. the time it takes for half of the amount of an introduced substance (such as a drug) to be eliminated by a natural system, either by excretion or by metabolic breakdown.
haploid having a single set of genes and chromosomes; compare *diploid, polyploid.*
haplont an organism that is haploid throughout its life cycle except for the diploid zygote stage; mitosis occurring only in haploid cells. compare *diplont.*
haptoglobin a blood protein having a specific affinity for extracellular hemoglobin; a molecular hemoglobin scavenger.
Hardy–Weinberg law see *Castle–Hardy–Weinberg law.*
Haversian canal in bone, a small canal through which a blood vessel runs.
Haversian system a Haversian canal together with its surrounding, concentrically arranged layers of bone, canaliculi, lacuna, and osteocytes.
H band in striated muscle, the variable light-staining band in the center of each sarcomere, representing the distance between the ends of two sets of actin microfilaments.
head in all vertebrates and insects, some other arthropods, and many mollusks and worms, the anterior division of the body that contains the eyes, mouth, other sensory organs, and brain.
heart a hollow, pulsating muscular enlargement of a circulatory vessel, which by its rhythmic contraction functions in pumping blood.
heat of activation the energy input needed before an exergonic chemical reaction can proceed.
heat receptor a sense organ that perceives warmth or radiant heat.
heliozoan also *sun animacule,* a protist of a group in the phylum Sarcodina, consisting of free-living, freshwater forms looking rather like tiny suns, with multiple thin, stiff, radiating pseudopods.
heme group an iron-containing, oxygen-binding porphyrin ring present in all hemoglobin chains and in myoglobin; see *porphyrin group.*
hemicellulose any of a number of complex polysaccharides found in wood and other plant cell walls, having as subunits not only glucose but also uronic acids, xylose, galactose and various pentose sugars.
hemichordate an animal of the deuterostome phylum *Hemichordata* (see Appendix A).
hemipenis either of the paired copulatory organs of a male lizard or snake.
hemisphere *cerebral hemisphere,* either the right or left half of the cerebrum.
hemizygous a term used to describe the condition of X-linked genes in males, which are neither homozygous nor heterozygous.
hemocoel a body cavity, especially in arthropods and mollusks, formed by expansion of parts of the blood vascular system.
hemochromatosis a genetic disease in human males, characterized by widespread deposition of iron in the tissues, resulting in bronzing of the skin, cirrhosis of the liver and pancreas, diabetes, and loss of armpit hair; being a disease of males, it affects neither females nor children.
hemocyanin a copper-containing respiratory pigment occurring in solution in the blood plasma of various arthropods and mollusks.
hemoglobin an iron-containing respiratory pigment consisting of one or more polypeptide chains, each associated with a heme group.
hemophilia a genetic tendency to uncontrolled bleeding in humans, due to the lack of a necessary constituent of the blood clotting process; caused by recessive alleles at either of two sex-linked loci, it affects males almost exclusively.
hepatic portal vein in all vertebrates, a large blood vessel that collects blood from the capillaries and venules of the esophagus, stomach and intestine and carries it to the liver, where it once more divides into capillaries. *hepatic portal system,* the hepatic portal vein and associated blood vessels.
herbivorous plant-eating. *herbivore,* an animal that feeds only on plants.
heredity the transmission of genetic characters from parents to offspring, and the effects of this transmission. See also *genetics.*
heritability symbolized h^2, a measure of the tendency of variation in a continuous trait to be inherited; technically it is the additive genetic variance divided by the total phenotypic variance of a specified trait in a specified population.
hermaphrodite an animal or plant that is equipped with both male and female reproductive organs.
heterocaryon fungus cell or mycelium that contains two genetically unlike haploid nuclei as a result of sexual fusion; the heterocaryotic state persists until nuclear fusion occurs just prior to meiosis.
heterochromtin chromatin that is relatively heavily condensed at times other than mitosis or meiosis, or that condenses early in prophase, and is genetically largely inert. *centric heterochromatin,* heterochromatin surrounding the centromere region. *constitutive heterochromatin,* genetic material that always behaves as heterochromatin. *facultative heterochromatin,* genetic material that behaves as heterochromatin only under certain conditions, e.g., the heterochromatic nucleus of bird erythrocytes or the condensed extra X of mammalian females.
heterocyst in filamentous cyanophytes, a relatively large, transparent, thick-walled cell specialized for nitrogen fixation.
heterogeneous nuclear RNA (HnRNA) nuclear RNA of miscellaneous known and unknown functions, including the very large initial transcripts from which mRNA are tailored.
heteromorphic chromosomes homologous chromosomes that can be distinguished from one another by some visible difference, as a variable knob or constriction; e.g., X and Y chromosomes.
heterotroph a microorganism which requires organic compounds as an energy or carbon source.
heterozygous having two different alleles at a specific gene locus on homologous chromosomes of a diploid organism.
high energy bond a molecular bond that releases a large amount of energy when it is broken (that is, replaced with a different type of bond).
high frequency of recombination strain (HFr strain) a strain of bacteria derived from an individual in which the F^+ episome has integrated into the host chromosome; characterized by a high frequency of genetic recombination when mixed with a fertility negative strain.
hilum (pl. *hila*), a scar left on a seed, marking the point of separation from the ovary; a bean's belly button.

hindbrain 1. the posterior of the three primary divisions of the embryonic vertebrate brain. 2. the parts of the adult brain derived from the embryonic hindbrain, including the *cerebellum, pons,* and *medulla oblongata.*
hinge joint a joint that moves in one plane, like a door hinge.
hip bone see *coxa.*
hippocampus 1. a long, curved ridge of grey and white matter on the inner surface of each lateral ventricle of the brain, involved in memory consolidation, visual memory and other functions. 2. a sea horse.
hirudin an anticoagulant secreted by leeches.
histocompatibility antigen any of several extremely active cell-surface antigens involved in normal cell-to-cell interactions in the immune response, and highly variable in all mammalian populations; these antigens are the principle cause of rejection of tissue and organ transplants between individuals.
histology the scientific study of tissues.
histone one of a class of small, highly basic proteins that complex with the nuclear DNA of higher eucaryotes, forming nucleosomes that presumably protect the DNA from degradation; one histone (histone I) appears to have a role in chromatin condensation and thus in one level of gene control.
HnRNA see *heterogeneous nuclear RNA.*
holdfast a rhizoidal base of a seaweed, serving to anchor the algal thallus to the ocean floor or rock substratum.
homeostasis the tendency toward maintaining a stable internal environment in the body of a higher animal through interacting physiological processes involving negative feedback control.
homeostatic mechanism any mechanism involved in maintaining a stable internal equilibrium, especially one involving self-correcting negative feedback.
homeothermic also *warm-blooded,* maintaining a stable, high internal body temperature independently of the environment, compare *poikilothermic.*
home range the area to which an animal confines its activities; compare *territory.*
hominids a primate group comprised of humans, and related extinct forms.
hominoid a primate group comprised of humans, apes and related extinct forms.
homologous 1. similar because of a common evolutionary origin. 2. derived by independent evolutionary modification from a corresponding body part of a common ancestor. 3. derived from a common embryological precursor; e.g., the *glans clitoridis* is homologous with the *glans penis.* 4. *homologous chromosomes,* chromosomes which, because of common descent, have the same kinds of genes in the same order and which pair in meiosis.
homology similarity due to common descent.
homologue also *homologous chromosome,* either of the two members of each pair of chromosomes in a diploid cell; see *homologous.*
homozygous having the same allele at a given locus in both homologous chromosomes of a diploid organism.
homunculus ("little man"), in preformationist theory, the tiny but fully organized human organism present in the sperm or egg.
hormone a chemical messenger transmitted in body fluids or sap from one part of the organism to another, that produces a specific effect on target cells often remote from its point of origin, and that functions to regulate physiology, growth, differentiation, or behavior.
hornwort a nonvascular terrestrial plant allied to mosses and liverworts.
horsehair worm the free-living adult stage of an aschelminth worm parasitic on insects.
horsetail a vascular plant of the ancient group *Equisetum,* with many primitive characters.
host a living organism that harbors or sustains a parasite, pathogen or other symbiont.
host-specific a parasite or pathogen that can infect only a single host species.
h^2, see *heritability.*
humerus the long bone that forms the upper portion of the vertebrate forelimb, e.g., the bone of the human upper arm.
humus partially decayed organic materials in the soil.
Huntington's disease also *Huntington's chorea,* a lethal hereditary disease of the nervous system developing in adult life and attributable to a dominant allele.
H-Y cell-surface antigen also *H-YA* an antigen found on the surface of all cells of all male mammals, but not in female mammals; and also on the cells of all female birds, but not male birds. It is believed to be controlled by a Y-linked gene in mammals, and to be responsible for the differentiation of the mammalian testis.
hyaline transparent.
hyaline cartilage translucent, bluish-white connective tissue consisting of cells embedded in a homogeneous matrix.
hyaluronic acid a viscous mucopolysaccharide occurring chiefly in connective tissues, but also as a cell-binding agent in other tissues, e.g., it holds together the cells of the *corona radiata.*
hyaluronidase an enzyme that dissolves hyaluronic acid and thus disrupts connective tissue; produced by invading bacteria, and also by mammalian spermatozoa.
hybrid the offspring of two animals or plants of different races, breeds, varieties, species, or genera.
hydration shell the layer or layers of water surrounding and loosely bound to an ion in solution.
hydrocarbon 1. a chemical compound consisting of hydrogen and carbon only, often in chains or rings: e.g., gasoline, benzone, parafin; 2. a portion of an organic molecule consisting of hydrogen and carbon only, e.g., the hydrocarbon portion of a fatty acid.
hydrochloric acid HCl, or a balance of the ions H^+ and Cl^-.
hydrogen bond a weak, non-covalent, usually intramolecular electrostatic attraction between the negatively polar hydrogen of a side group and a positively polar oxygen of another side group.
hydrogen ion 1. a proton. 2. as a convenient fiction, a proton in solution and bound to a water molecule; actually a *hydronium ion* (H_3O^+).
hydrolysis see *hydrolytic cleavage.*
hydrolytic cleavage also *hydrolysis,* the reaction of a compound with water such that the compound is split into two parts by the breaking of a covalent bond; and water is added in the place of the bond, an $-OH$ group going to one subunit and an $-H$ group going to the other.
hydronium ion a symmetrical charged ion, H_3O^+, that can dissociate into water and a hydrogen ion; most hydrogen ions in solution are actually in the form of hydronium ions.
hydrophobic with regard to a molecule or side group, tending to dissolve readily in organic solvents but not in water; resisting wetting; not containing polar groups or subgroups.
hydrostatic pressure the pressure exerted in all directions within a liquid at rest.
hydrostatic skeleton a supporting and locomotory mechanism involving a fluid confined in a space within layers of muscle and connective tissue; movement occurring when muscle contractions increase or decrease the hydrostatic pressure of the fluids.
hydroxyl group $-OH$, consisting of an oxygen and hydrogen covalently bonded to the remainder of the molecule; a constituent of alcohols, sugars, glycols, phenols and other compounds.
hydroxyl ions also *hydroxide ion,* the univalent anion OH^-, one of the dissociation products of water in the reaction $2H_2O \rightarrow H_3O^+ + OH^-$.
hydroxyl radical the highly reactive free radical $\cdot OH$, containing an unpaired electron; a product of the action of ionizing radiation on water or aqueous systems, and a principle intermediate in radiation damage.
hymen also *maidenhead,* fold of thin mucous membrane closing the orifice of the vagina, especially in virgins.
hyperglycemia an abnormal excess of glucose in the blood.
hyperosmotic see *hypertonic.*
hyperthyroidism a condition caused by an excess of circulating thyroid hormone; symptoms in humans include nervousness, sleeplessness, hyperactivity, weight loss, and, if prolonged, a bulging of the eyeballs.
hypertonic (adj.), having a higher osmotic potential (e.g., higher solute concentration) than the cytoplasm of a living cell (or other reference solution).
hyperventilate to breathe deeply deliberately, purging the body of CO_2.
hypha (pl. *hyphae*), one of the individual filaments that make up a fungal mycelium.
hypocotyl the part of a plant embryo below the point of attachment of the cotyledon.
hypoglycemia 1. an abnormally low level of blood glucose. 2. a medical condition involving recurrent episodes of low blood sugar, due to the overproduction of insulin.
hypophysis see *pituitary gland.*
hypothalamus a portion of the floor of the midbrain, containing vital autonomic regulatory centers, and closely associated functionally with the pituitary gland.
hypothesis a proposition set forth as an explanation for a specified group of phenomena, either asserted merely as a provisional conjecture to guide investigation (e.g., working hypothesis) or accepted as highly probably in the light of established facts; see also *null hypothesis, theory.*
hypothesis tested see *null hypothesis.*
hypothyroidism a condition caused by abnormally low levels of circulating thyroid hormone; symptoms include physical and mental sluggishness and weight gain.
hypotonic (adj.), having a lower osmotic potential (e.g., a lower concentration of solutes) than the cytoplasm of a living cell (or other reference solution).
hysterectomy the surgical removal of the uterus.
H zone see *H band.*

IAA indoleacetic acid; see *auxin.*
I band one of the striations of striated muscle, of variable width, corresponding to the distance between the ends of the myosin filaments of adjacent units.
ice ages the periods of recurrent glaciation of the last million years; the last such glaciation, ending about 10,000 years ago.
ileocecal valve a sphincter separating the large intestine from the small intestine.
ileum a region of small intestine.
ilium the hip bone, one of the three pairs of fused bones forming the pelvis.
imbibition the taking up (absorption) of a fluid (often water) by a hygroscopic gel, colloid, or fibrous matrix; such as the ultramicroscopic spaces in a plant cell wall.
immigration moving into a given area or population and taking up residence.
immune response the entire array of physiological and developmental responses involving specific protective actions against a foreign substance; including phagocytosis, the production of antibodies, complement fixation, lysis, agglutionation and inflammation.
immunoglobin a protein antibody produced by thymocytes in response to specific foreign substances, consisting of constant regions and two specific antigen binding sites; or a polymer of such molecules.

implantation the act of attachment of the mammalian embryo (blastocyst) to the uterine endometrium.
impulse neural impulse, 1. a wave of excitement (transitory membrane depolarization) transmitted through a neuron; 2. the same wave as it travels a nerve pathway, including its transmission across synapses.
inborn error of metabolism a genetic defect in which an individual lacks one of the enzymes of a biochemical pathway.
incisors in humans and most mammals, the four upper and four lower front teeth, which are chisel-shaped for biting and cutting; in rodents the medial incisors are greatly exaggerated and the lateral incisors are absent.
incomplete flower see under *flower.*
incomplete penetrance see *penetrance.*
indeterminate plant one that shows no photoperiodicity in its flowering.
individual distance the space around an individual within which intrusion by another individual will elicit flight or attack.
indoleacetic acid also *IAA,* the naturally-occurring form of auxin, the plant growth hormone; see *auxin.*
induced mutation a change in DNA caused by ionizing radiation, ultraviolet light or mutagenic chemicals.
inducer 1. in molecular genetics, a small molecule that triggers the activity of an inducible enzyme. 2. in embryology, a substance that stimulates the differentiation of cells or the development of a particular structure.
inducible enzyme an enzyme produced by an organism only when induced by an appropriate stimulus. *inducible operon,* the entire operon involved in the control of a battery of inducible enzymes.
induction the process by which the fate of embryonic cells of tissue is determined, especially as due to tissue interactions.
indusium in ferns, a covering over the sorus (spore-forming organ).
inferior situated below or beneath; e.g., *inferior vena cava,* inferior ovary in a flower (epigyny).
information load a measure of the level of information that can be communicated in a single display.
ingestion swallowing.
inhibition the restraining, diminishing, or preventing of activity.
inhibitor a substance that interferes with a chemical or biological process.
inhibitory block according to the classical definition of instinct, the neurological inhibitors of behavior that are selectively removed by the perception of the appropriate releaser.
initiator codon a codon (sequence of three mRNA nucleotides) that initiates translation of a polypeptide; specifically, AUG, which also codes for methionine and N-formyl methionine.
innate inborn, not due to learning.
innate behavioral pattern a genetically programmed behavior pattern.
innate releasing mechanism (IRM) a hypothetical center in the central nervous system that, when stimulated by the proper environmental cue, activates neural pathways that result in an *instinctive behavior pattern.*
inner cell mass in a mammalian blastocyst, the portion that is destined to become the embryo proper.
inner ear the portion of the ear enclosed within the temporal bone, consisting of a complex fluid-filled labyrinth and the associated auditory nerve; included are the three *semicircular canals,* the *cochlea,* and the *round* and *oval windows;* compare *external ear, middle ear.*
innoculum a small quantity of living cells or organisms introduced into a suitable medium for growth.
innominate bone (*"nameless bone"*). see *coxa.*
inorganic molecule one not containing carbon, generally one found occurring outside of living organisms.
insectivore 1. an eater of insects. 2. a member of the mammalian order *Insectivora,* all of which eat insects: living insectivores are the moles, shrews, and hedgehogs.
insectivorous habitually eating insects; *insectivorous plant,* one with special adaptations for trapping and digesting insects for their nitrogen and mineral content.
insertion 1. in genetics, the addition of extra genetic material into the middle of a chromosome, gene or other DNA sequence; *base insertion,* the insertion of a single base pair into a DNA sequence. also *muscle insertion,* 2. in anatomy, the distal attachment of a tendon or muscle; compare *origin.*
inspiration center a center in the brain responsible for control of the musclar actions involved in the intake of air in breathing.
instar see *interecdysis.*
instinct innate behavior involving appetitive and consummatory phases, the latter usually in response to an environmental releaser but sometimes occurring as vacuum behavior.
instinctive pattern *a fixed-action pattern* that is coupled with orienting movements.
insulin a polypeptide hormone secreted by the islets of Langerhans of the pancreas, whose principle action is to facilitate the transport of glucose into cells and stimulate the conversion of glucose to glycogen, but having many other functions; it is an essential growth factor in mammalian cell culture.
integument a covering or envelope; in the flower, the covering that encloses the nucellus of an ovule, and later forms part of the seed coat; in animals, the skin, exoskeleton, tunic, cuticle or other covering.
interbreeding 1. commonly breeding together. 2. breeding together; hybridizing.
intercalated disk in heart muscle, the highly convoluted double cell membrane between two adjacent cells in a heart muscle fiber.
intercostal (adj.), between the ribs. *intercostal muscles,* the muscles of the rib cage, involved in breathing.
intercdysis also *intermolt, instar,* in arthropods, the period between successive molts.
intermediate optimum of a continuously distributed measurable phenotype, the value with the highest reproductive fitness when selection is against higher and lower values.
internal nares see *nares.*
internode the interval between two nodes, see *nodes.*
interphase all of the cell cycle between successive mitoses; incuded are G_1, S phase, and G_2.
interstitial situated within, but not characteristic of the principle structure of a particular organ or tissue, in reference to isolated cells or of fibrous tissue; e.g., the interstitial cells of the testis, which have an endocrine function.
interdital zone the zone between the mean high tide line and the mean low tide line, intermittently covered by the sea or exposed to air.
intervertebral disk one of the tough, elastic, fibrous disks situated between the centra of adjacent vertebrae.
intracellular digestion digestion within cells, in digestive vacuoles following phagocytosis.
intraspecific with a species; between members of the same species.
introgression backcrossing of a hybrid with one or the other parental types. Common means of diversification in flowering plants.
introitus the external opening of the vagina.
invagination the folding inward of a surface or tissue to make a cavity; specifically, the formation of a gastrula by the inward folding of part of the wall of the blastula.
in vitro ("in glass"), in a test tube or other artificial environment; compare *in vivo.*
in vivo in the living body of a plant or animal; compare *in vitro.*
involuntary muscle see *smooth muscle.*
involution an inward roll or curve, in gastrulation, the movement of cells toward the blastopore, over the blastopore lip, and into the archenteron.
ion any electrostatically charged atom or molecule.
ionic bond a chemical attraction between ions of opposite charge.
ionization 1. Thedissociation of a molecule into oppositely charged ions in solution. 2. the creation of ions by the energy of ionizing radiation.
ionizing radiation energetic radiation that produces ions in air and free hydroxyl radicals in water, the latter being responsible for induced mutation and tissue damage; included are X-rays, gamma rays, and streams of charged particles from the breakdown of radioactive isotopes.
ischium one of the bones of the pelvis, specifically the one you sit on. See also *coxa.*
islets of Langerhans clumps of endocrine cells in the pancreas, which secrete *insulin* and *glucagon.*
isolecithal of an egg, having a small amount of yolk that is evenly dispersed in the cytoplasm.
isogametes gametes that are identical in size and appearance, such as the (+) and (−) isogametes of *Chlamydomonas.*
isoosmotic see *isotonic.*
isotonic having the same osmotic potential (e.g., the same concentration of solutes) as the cytoplasm of a living cell, or of some other reference fluid.
isotope a particular form of an element in terms of the number of neutrons in the nucleus; *radioactive isotope,* also *radioisotope,* an unstable isotope that spontaneously breaks down with the release of ionizing radiation.

Jacobson's organ a pouch-like olfactory organ in the roof of the mouth of most mammals and other vertebrates; in lizards and snakes it receives the ends of the forked tongue; absent in primates.
jaundice a yellowish pigmentation of the skin and other tissues caused by the deposition of bile pigments, following bile duct obstruction, liver disease, or the excessive breakdown of red blood cells.
jejunum a region of small intestine.
jelly coat a layer of transparent substance surrounding the egg of an echinoderm, fish or amphibian.
juvenile hormone (JH), an insect hormone that prevents the differentiation of adult characters; secreted by the corpora allata of larval and subadult insects.

karyotype 1. a mounted display of enlarged microphotographs of all the stained chromosomes of an individual, arranged in order of decreasing size. 2. the total chromosome constitution of an individual. 3. the chromosomal makeup of a species.
keel the enlarged breastbone to which the flight muscles of birds are attached.
kelp any of various large brown algal seaweeds.
keratin a cysteine-rich, fibrous, insoluble, intracellular structural protein making up most of the substance of the dead cells of hair, horn, nails, claws, feathers and the outer epidermis.
kidney 1. in vertebrates, one of a pair of ducted excretory organs situated in the body cavity beneath the dorsal peritoneum, serving to excrete nitrogenous wastes and to regulate the balance of body ions and fluids. 2. any analogous organ in invertebrate metazoans.
kilocalorie see *calorie.*
kinetic energy the energy of motion.
kinetics the pattern of a chemical reaction over time, including rates and changes in rates.
kinetochore see *centromere.*
kingdom one of the primary divisions of life forms; to the traditional plant and animal kingdoms Whittaker has added fungal, protist, and moneran (procaryote) kingdoms.
kinorhynchid a minute worm of the class *Kinoryncha,* usually placed with the Aschelminthes but sometimes considered to be an arthropod or annelid.

Klinefelter's syndrome in humans, an abnormal condition of males, caused by the chromosomal condition XXY; characterized by increased stature, moderate retardation, small testes, low testosterone and secondary effects of low testosterone.
Krebs cycle see *citric acid cycle.*
krill planktonic crustaceans that constitute the principal food of baleen (toothless) whales.
kwashiorkor a syndrome of severe protein deficiency in human infants and children, including failure to grow, deficiency of melanin pigment, edema, degeneration of the liver, anemia, and retardation.

labia majora (sing. *labium majorum*), in human females, the outer, fatty, often hairy pair of folds bounding the vulva.
labia minora (sing. *labium minorum*), in the human female, the inner, thinner, highly vascular pair of folds surrounding the introitus, enfolding and extending downward from the clitoris.
labial palps a pair of leaf-like, fleshy appendages on either side of the mouth of a bivalve mollusk which remove food from the gills in feeding.
labile readily undergoing chemical breakdown, dissociation, or denaturation.
labium 1. see *labia majora, labia minora.* 2. the lower lip of a snapdragon or similar flower. 3. the single lower lip of an insect or crustacean.
labrum the single upper lip of an insect or crustacean.
lac operon in *E. coli,* a complex of structural genes and controlling genes induced by lactose or other beta galactosides.
lactase a digestive enzyme that hydrolyzes lactose to glucose and galactose; absent in most non-Caucasian adults.
lacteal a lymphatic vessel of the intestinal villi.
lactose also *milk sugar,* a disaccharide consisting of glucose and galactose subunits.
lacuna (pl. *lacunas, lacunae*), a minute cavity in bone or cartilage that holds an osteocyte or chondrocyte.
lamella (pl. *lamellae*), 1. one of the bony concentric layers (ring-like in cross section) that surround a Haversian canal in bone. 2. one of the thin plates composing the gills of a bivalve mollusk. 3. any thin, plate-like structure.
lamellibranch a bivalve mollusk.
laminarian a large brown algal kelp.
lancelet also *Amphioxus,* any of the genus of living cephalochordates, *Branchiostoma;* elongate, fish-like, translucent marine animals of one to ten centimeters in length.
lanolin a greasy sterol found in untreated wool and expensive cosmetics.
large intestine also colon or bowel, a division of the alimentary canal, primarily functioning in the resorption of water.
laryngopharynx the lower part of the pharynx at the region of the larynx.
larynx in terrestrial vertebrates, the expanded part of the respiratory passage just below the glottis and at the top of the trachea; in mammals it contains the vocal chords and constitutes a resonating *voice box* or *Adam's apple.*
late blight the fungal disease caused by *Phytophthora infestans,* responsible for the Irish potato famine of the 1840s.
latency 1. of a pathogenic infection, the state or period of living and developing in a host without producing symptoms. 2. the age, from about five to puberty, in which children are assumed by their parents to have no interest in sex. 3. Darwin's term for the gross phenomenon of recessivity.
lateral line organ in nearly all fish, a sensory organ considered to be responsive to water currents and vibrations; thought to be distantly homologous with the inner ear.
laterally from side to side.
laterite a hard, water-resistant, crusty soil, largely aluminum oxides leached of mineral nutrients, common in tropical rainforest regions.
law of alternate segregation also *Mendel's First law,* the observation, based on the regularity of meiosis, that heterozygous alleles separate in gamete formation, each allele going to approximately half the gametes produced.
law of independent assortment also *Mendel's second law,* the observation, based on the regularity of meiosis, that genetic elements on different chromosomes behave independently in the production of gametes.
leaf in vascular plants, a lateral outgrowth from stem functioning primarily in photosynthesis, arising in regular succession from the apical meristem, consisting typically of flattened blade jointed to the stem by a petiole.
leaf generation vegetative plant reproduction in which new shoots emerge from the margins of a leaf.
leaf primordium see *primordium.*
leaf scale a flattened, photosynthetic, leaflike structure in bryophytes.
leaf scar the mark left on a stem by a fallen leaf.
learning in animal behavior, the process of acquiring a persistent change in a behavioral response as a result of experience.
lecithin a choline phospholipid widely distributed in membranes and a major component of egg yolk.
lectin in plants, a substance that combines with a specific antigen; antibody-like function but, unlike antibodies,not induced by foreign antigens.
lens any clear, biconvex organelle that functions to focus or collect light.
lenticel a pore in stem of plant for the passage of gases between the atmosphere and stem tissues.
leptotene (adj. and n), also *leptonema* (n.), the first stage of meiotic prophase, at which time the still unsynapsed chromosomes appear as elongated, thin threads.
leucine one of the twenty amino acids of protein structure.
leucoplast a colorless plastid; see also *plastid.*
leukocyte also *leucocyte,* a vertebrate white blood cell; see page 762.
LH see *luteinizing hormone.*
lethal–recessive dominant an allele that has a prominent visible effect as a heterozygote, and is lethal as a homozygote; in humans such an allele is also called a *rare dominant* since it may never occur as a homozygote.
lichen a combination of a fungus and an alga growing in a symbiotic relationship.
ligament a tough, flexible, but inelastic band of connective tissue that connects bones, or that supports an organ in place; compare *tendon.*
ligase an enzyme that heals nicks (single-strand breaks) in DNA.
light antenna see *antenna.*
light-independent reaction also *dark reaction,* that part of photosynthesis not immediately involved in chemiosmosis, specifically the fixation of CO_2 into carbohydrate from the $NADPH_2$ and ATP produced by the light reaction (Calvin cycle).
light reaction that part of photosynthesis directly dependent on the capture of photons; specifically the photolysis of water, the thylakoid electron transport system, and the chemiosmotic synthesis of ATP and $NADPH_2$.
lignin an amorphous substance that gives wood its rigidity.
limiting factor any factor in the environment which, by its presence or relative absence, limits the growth of a population; e.g., disease, predation, food supply, mineral nutrient.
limiting resource a resource, such as food, shelter, or mineral nutrient, that by its scarcity limits the growth of a population or determines the carrying capacity of the environment.
linkage group a group of gene loci shown to be linked by recombination tests; ultimately, a chromosome.
linked (adj.), of two gene loci, not segregating independently.
linoleic acid an essential polyunsaturated fatty acid, necessary for cell membrane function.
lipase any fat-digesting enzyme.
lipid an organic molecule that tends to be more soluble in non-polar solvents (such as gasoline) than in polar solvents (such as water).
lipoprotein a conjugated protein with a lipid subgroup.
litter also *leaf litter, duff,* the uppermost slightly decayed layer of organic material on a forest floor.
littoral in reference to the seashore. *littoral zone,* 1. a costal region including both the land along the coast, the water along the shore and the intertidal. 2. see *intertidal.*
liver 1. in vertebrates, a large, glandular, highly vascular organ that serves many metabolic functions including detoxification, the production of blood proteins, food storage, the biochemical alteration of food molecules, and the production of bile. 2. in mollusks, a digestive organ opening into the gut, in which intracellular digestion of food by phagocytosis occurs.
liverwort a kind of ground-hugging bryophyte.
loam a highly fertile soil defined ideally as a mixture of three-quarters silt and sand and one-quarter clay.
lobe a curved or rounded projection or division, as of the liver or lung.
lobe-finned fish a fish of the subclass Sarcopterygii, including lung fish and crossopterygians (see Appendix A).
locus (pl. *loci*), also *gene locus,* specific place on a chromosome where a gene is located.
logarithmic spiral a form of spiralling curve that keeps its overall shape constant as it increases in length.
long-day plant a plant that begins flowering at some specific time before the summer solstice when day length exceeds night, flowering being triggered by a critical, established period of darkness.
long-term memory 1. learning that persists more than a few hours, the memory trace of which is physically located in a different part of the brain than short-term memory. 2. the part of the brain and the general neural function with which such persistent memory traces are associated.
loop of Henle the hairpin loop of the vertebrate nephron, between the proximal and distal convoluted tubules, which leaves the cortex and descends into the medulla, then loops back to the cortex; it is involved in water resorption.
lophophore in brachiopods, ectoprocts and phoronids, a filter-feeding organ consisting of a spiral or horseshoe-shaped ridge surrounding the mouth, and bearing tentacles.
love (variously defined), in humans, the emotional experience of the development and maintenance of any of several kinds of strong interpersonal bonds, including incipient or consummated pair bonding; interpreted subjectively as a feeling of attraction, desire, affection, sympathetic interest, enthusiasm, and devotion, and often associated with a mixture of euphoria and anxiety.
lumbar pertaining to the lower trunk; *lumbar region* of the spine, the vertebrae between the thorax and the sacrum.
lumen the cavity or channel of a hollow tubular organ or organelle.
lumper a taxonomist who prefers fewer, larger groupings; compare *splitter.*
lung in land vertebrates, one of a pair of compound, saccular organs that function in the exchange of gases between the atmosphere and the bloodstream. 2. any of several analogous organs in invertebrates, e.g., *book lung.*
lungfish a lobe-finned, partially air-breathing fish of the order Dipnoi.
luteinizing hormone (LH) a vertebrate polypeptide hormone of the anterior pituitary that in the female stimulates ovulation and the development and maintenance of the corpus luteum; the same polypeptide functions in the male as the *interstitial cell stimulating hormone.*
luteining hormone releasing factor (LHRF), see *releasing factor.*
$l_x m_x$ **curve,** a graphic representation of the reproductive strategy of a species, in which l_x is the probability of surviving to age x, and m_x is the probability of reproducing during ages x to $x + dx$.
lycophyte also *ground pine, club moss,* any of several small terrestrial vascular plants constituting the plant subdivision *Lycophyta.*
lymph duct see *thoracic duct.*

lymph node a rounded, encapsulated mass of lymphoid tissue through which lymph ducts drain, consisting of a fibrous mesh containing numerous lymphocytes and phagocytes.
lymphatic system the system of lymphatic vessels, lymph nodes, lymphocytes, the thoracic duct and the thymus, which together serve to drain body tissues of excess fluids and to combat infections.
lymphatic vessel also *lymphatic*, a thin-walled vessel that conveys lymph, originating as an open intercellular cleft and draining eventually into the *thoracic duct*.
lymphedema edema due to faulty lymphatic drainage; see also *edema*.
lymphocyte any of several varieties of similar-appearing leukocytes involved in the production of antibodies and in other aspects of the immune response; see *B cell, T cell*.
lysate the material released from a cell upon lysis.
lyse (v.t.), to destroy a cell by rupturing the cell membrane.
lysis (n.), the destruction or lysing of a cell by rupture of the cell membrane.
lysogeny the incorporation of a bacteriophage chromosome into the bacterial host chromosome, together with mechanisms preventing further infection and lysis; in later cell generations the incorporated DNA may excise, replicate and eventually cause cell lysis.
lysosome a small membrane-bounded cytoplasmic organelle, generally containing strong digestive enzymes or other cytotoxic materials.
lysozyme an enzyme that lyses bacteria by dissolving the bacterial cell wall; a natural constituent of eggwhite and tears.

macromere a relatively large, yolk-filled blastomere of the lower half (vegetal pole) of a blastula.
macromolecule any large biological polymer, as a protein, carbohydrate or nucleic acid.
macronucleus the larger of the two nuclei of *Paramecium* and certain other ciliate protozoans, carrying somatic line DNA; compare *micronucleus*.
macronutrient a plant nutrient required in relatively substantial quantities, as nitrogen, phosphorus, potassium sulfer, magnesium, and calcium; compare *micronutrient*.
macrophage a large phagocyte.
malaria an acute or chronic disease caused by parasitic infection by sporozoan parasites (*Plasmodium*) of the red blood cells; transmitted by *Anopheles* mosquitos.
malate a negatively charged, a four-carbon intermediate of the Krebs cycle; as *malic acid* it is responsible for the tartness of green apples.
malic acid the acid form of *malate*.
malpighian tubule any of numerous blind, hollow, tubular structures that empty into the insect midgut and function as a nitrogenous excretory system.
maltase an enzyme that hydrolyzes maltose.
maltose a disaccharide consisting of two glucose subunits in an alpha linkage; a digestion product of starch.
mammal any vertebrate of the class *Mammalia* (see Appendix A).
mammary gland also *milk gland, mamma, mammary*, the organ that, in female mammals, secretes milk for the nourishment of the young; a breast or udder.
mandible 1. the vertebrate lower jaw or jaw bone. 2. either the upper or lower segment of a bird's beak. 3. either of two laterally paired anterior mouth appendages of an arthropod, which form strong biting jaws. 4. any invertebrate biting mouthpart.
mannose a hexose sugar.
mantle a fleshy covering of mollusks that secretes material to form the external shell.
mantle cavity in mollusks, a cavity between the mantle and the body proper, in which the respiratory organs lie; a highly vascularized mantle cavity serves as a lung in pulmonate snails.
mark, release, and recapture experiment any of a number of related techniques in which animals (e,g, birds, fish) are captured, tagged or banded, released into the wild and later recaptured; used to determine migration, mortality, population size and other demographic statistics.
marrow *bone marrow*, the soft tissue in the interior cavity of a bone; *blood marrow*, vascularized bone marrow in which blood cells of all types are produced; *fatty marrow*, bone marrow consisting of adipose tissue.
marsupial a mammal of the subclass Metatheria, usually with a pouch (*marsupium*) in the female; included are the kangaroo, wombat, koala, Tasmanian devil, opossum, and wallaby.
mass action on *law of mass action*, in reversible chemical reactions, the rate and net direction of the reaction depends on the relative concentration of reactants and products. Reactions proceed from high to low.
mastax also *gizzard*, the grinding gullet of rotifers and gastrotrichs.
masturbation manipulation of one's sexual organs for erotic gratification.
materialism the philosophical theory that regards matter and its motions as constituting the universe, and regards all phenomena, including those of mind, as due to material agencies; the denial of supernatural forces; the denial that life constitutes a special, nonmaterial state or agency. contrast with *vitalism;* e.g., biological activity at all levels can be explained potentially in terms of physical laws.
mating type in fungi and various protists, a grouping of organisms incapable of sexual reproduction with one another but capable of such reproduction with members of one or more other groups or mating types.
maturation the process of coming to full development.
maxilla (pl. *maxillae*), 1. the upper jaw of a vertebrate, *anterior* to the mandible. 2. in insects, one of the first or second pair of mouth parts *posterior* to the mandibles.
mean 1. also *sample mean, arithmetic mean, average*, the sum of a finite number of numerical values divided by the number of valued summed; symbolized $\bar{X}$ 2. also *true mean, population mean*, the idealized limit of the arithmetic mean as the number of values summed becomes infinite; symbolized μ.
meatus 1. the opening of any passageway to the exterior. 2. *urinary meatus*, the opening of the urethra into the vulva in human females. 3. *penile meatus*, the opening of the urethra at the end of the penis in males. 4. *external auditory meatus*, the opening of the external ear canal.
mechanism 1. the agency or means by which an effect is produced or a purpose is accomplished; 2. the theory that everything in the universe is produced by matter in motion through definable physical laws; *materialism*.
medial also *median* (adj.), in or toward the middle or midline.
median eye 1. a simple photoreceptive organ, usually lacking a lens, found in lancelets and certain reptiles; apparently used in responses to photoperiodicity. 2. also *pineal body*, a homologous organ found in birds and various other vertebrates. 3. also *parietal eye*, a similar dorsal outgrowth of the brain of various vertebrates. 4. a midline ocellus in insects.
medulla 1. the inner portion of a gland or organ; compare *cortex*. 2. also *medulla oblongata*, a part of the brainstem developed from the posterior portion of the hindbrain and tapering into the spinal cord.
medullary dominance see *cortical dominance*.
medusa the motile, free–swimming jelly fish body type of coelenterate.
megagametophyte the female gametophyte produced from a megaspore in flowering plants; the *embryo sac* consisting of eight haploid nuclei or cells.
megaspore a single large cell with one haploid nucleus formed after meiosis of a megaspore mother cell in plants, and from which the megagametophyte develops by mitosis.
megaspore mother cell in the ovule, a large diploid cell that will give rise to the megaspore by meiosis and the degeneration of three of the four haploid nuclei.
megasporogenesis the formation of megaspores.
meiosis also *reduction division*, in all sexually reproducing eucaryotes, the process in which a diploid cell or cell nucleus is transformed into four haploid cells or nuclei through one round of chromosome replication and two rounds of nuclear division, the first of which involves the unique pairing of chromosome homologues.
meiospore a haploid spore produced by meiosis.
meiotic interphase the period, which may be prolonged or so brief as to be virtually nonexistent, between telophase I of meiosis and prophase II of meiosis, during which there is no DNA replication.
Meissner's corpuscle in mammals, a small elipsoidal touch-responsive neural end organ.
melanin the characteristic animal surface pigmentation, a class of polymers of tyrosine or dopa, giving black, brown, orange, or red coloration depending on chemical composition and pigment granule size; also found in plants.
melanism an unusual development of a black or nearly black color in a species not generally so pigmented.
melatonin a hormone produced in the mammalian pineal body that has a role in sexual development and maturation through its effect on the hypothalamus; it also causes color changes in frogs, and appears to be related to photoperiodicity responses.
membrane 1. any thin, soft, pliable sheet, as of connective tissue. 2. *cell membrane*, also *plasma membrane*, the semipermeable, lipid bilayer, proteinaceous sac enclosing the cytoplasm. 3. any internal cell structure of a similar construction, such as the nuclear membrane, thylakoid membrane and others; see also *fluid-mosaic model*.
membrane-bound 1. bound to a membrane, e.g., *membrane-bound ribosomes*. 2. also *membrane-bounded*, having an outer boundary consisting of a membrane, e.g., a lysosome is a membrane-bounded organelle.
membrane-bounded having an outer boundary consisting of a membrane.
memory 1. the ability to retain a learned response or to recognize a stimulus previously encountered. 2. the part or function of the brain in which learned responses are maintained. See *engram, long-term memory, short-term memory*.
memory trace also *memory scar*, see *engram*.
menarche in human females, the onset of menstruation; the first menses, taken as the beginning of puberty.
menopause in human females, 1. the cessation of menstruation, usually occurring between the ages of 45 and 50. 2. the whole group of physical, physiological and behavioral occurrences and changes associated with the cessation of menstruation.
menstrual cycle also *ovarian cycle*, the entire cycle of hormonal and physiological events and changes involving the growth of the uterine mucosa, ovulation, and the subsequent breakdown and discharge of the uterine mucosa in menses; normally occurring at three to five week periods in nonpregnant women.
menstruation also *menses*, in nonpregnant females of the human species only, the periodic discharge of blood, secretions, and tissue debris resulting from the normal, temporary breakdown of the uterine mucosa in the absence of implantation following ovulation.
menses see *menstruation, menstrual cycle*.
meristem also *meristematic tissue*, a plant tissue consisting of small undifferentiated cells that give rise by cell division both to additional meristematic cells and to cells that undergo terminal differentiation; all postembryonic plant growth depends on meristem.
meristematic pertaining to meristem.
merozoite a stage of *Plasmodium* that reproduces asexually in a mammalian red blood cell, releasing the poisons that cause malarial symptoms.
mesoderm the middle of the three primary germ layers of the gastrula, giving rise in development to the skeletal, muscular, vascular, renal, and connective tissues, and to the inner layer of the skin and to the epithelium of the coelom (*peritoneum*).

mesoglea the loose, gelatinous middle layer of the bodies of sponges and coelenterates, between the outer ectoderm and the inner endoderm.
mesolecithal of an egg, having the yolk displaced primarily toward the vegetal pole but with cleavage complete, as in amphibia.
mesonephros 1. the adult kidney of an amphibian or fish. 2. the second of three pairs of kidneys formed in reptile, bird and mammalian embryos, superceded in later development by the *metanephros.*
mesophyll the chloroplast-rich parenchyma of a leaf, between the upper and lower epidermal cells.
mesosome in many bacteria, an inward extension of the cell membrane that forms a spherical membranous network; function not resolved.
Mesozoic era see the geological timetable inside the front cover.
messenger RNA *mRNA,* 1. in procaryotes, RNA directly transcribed from an operon or structural gene, containing one or more contiguous regions (cistrons) specifying a polypeptide sequence. 2. in eucaryotes, RNA transcribed from a structural gene, tailored and usually capped and polyadenylated in the nucleus and transported to the cytoplasm; containing a single contiguous region specifying a polypeptide sequence as well as leader and follower sequences.
metabolic defect see *inborn error of metabolism.*
metabolism the chemical changes and processes of living cells, including but not limited to respiration, the synthesis of biochemicals, and the breakdown of wastes; see also *basal metabolism, biochemical pathway.*
metabolite 1. a metabolic waste, especially one that is toxic. 2. an intermediate in a biochemical pathway.
metacarpals one of the usually five bones of the hand or forefoot, between the carpals and digits.
metanephros in the vertebrate embryo, the third and most posterior of the paired kidney structures to develop, and the adult kidney of reptiles, birds, and mammals; see *mesonephrose, pronephros.*
metaphase the stage of mitosis or meiosis in which the centromeres of the chromosomes are brought to a well-defined plane in the middle of the mitotic spindle prior to separation in anaphase.
metaphase plate the equatorial plane of the mitotic spindle on which the centromeres are oriented in mitosis.
metaphyte a multicellular plant, especially as contrasted with algal protists considered as plants; equivalent to Whittaker's Kingdom Plantae.
metatarsal one of the usually five bones of the foot or hindfoot, between tarsals and toes, forming the instep in humans and part of the rear leg in ungulates and birds.
metazoa in Whittaker's five-kingdom scheme all animals other than sponges.
methemoglobin (*met-hemoglobin*), the hemoglobin polypeptides from which the heme group has been detached. *methemoglobinemia,* a pathological condition in which methemoglobin is present in the blood (a symptom of nitrite poisoning).
methionine one of the twenty amino acids of polypeptide structure, and the one normally initiating translation, in eucaryotes as methionine and in procaryotes as *N*-formyl methionine.
mica a mineral that forms thin, clear, waterproof crystals.
micelle a loose structure consisting of highly associated particles, commonly lipids, in a colloidal solution.
microbe also *microorganism,* any very small organism, such as a bacterium, yeast or protist.
microbial genetics the scientific study of inheritance and DNA function in bacteria, bacteriophage, yeast, fungi, or protists.
microfibril one of the submicroscopic elongated bundles of cellulose in a plant cell wall.
microfilament 1. a submicroscopic filament of the cytoskeleton, involved in cell movement and shape; the principle component is fibers of the protein *actin.* 2. see *myofilament.*
microfossil a microscopic fossil; a fossilized microbe.
micromere in cleavage and in the blastula, a relatively small cell of the animal pole; compare *macromere.*
micrometer 1. symbol μm, one millionth of a meter. 2. an instrument for making very precise or very small measurements.
micronucleus (pl. *micronuclei*), in ciliate protists, the smaller of the two nuclei and the one carrying germ-line DNA; compare *macronucleus.*
micronutrient also *trace element,* an element necessary for plant growth but needed only in vanishingly small quantities.
microorganism also *microbe,* any organism too small to be seen readily without the aid of a microscope; such as bacterium, protist, or yeast.
micropyle 1. in insect eggs, a depression or differentiated area through which the sperm enters. 2. in seed plants, a minute opening in the integument of an ovule through which the pollen tube enters.
microspore in seed plants, one of the four haploid cells formed from meiosis of the microspore mother cell, which undergoes mitosis and differentiation to form a pollen grain.
microspore mother cell in the anther of a flowering plant, the diploid cells that will undergo meiosis to form the haploid microspores.
microsporogenesis the formation of microspores.
microtubule a cytoplasmic hollow tubule composed of spherical molecules of tubulin, found in the cytoskeleton, the spindle, centrioles, basal bodies, cilia and flagella.
microvilli (sing. *microvillus*), also *brush border,* tiny finger-like outpocketings of the cell membrane of various epithelial secretory or absorbing cells, such as those of kidney tubule epithelium and the intestinal epithelium.
midbrain the middle of the three divisions of the vertebrate embryonic brain; the adult structures derived from the embryonic midbrain.
middle ear in mammals, the part of the ear between the eardrum and the oval window, consisting of a chain of three ear ossicles in an air-filled chamber communicating with the pharynx by way of the eustacian tube.
middle lamella a layer of cementing material between adjacent plant cell walls.
midpiece in a spermatazoan, the segment between the head and the tail, containing one or more mitochondria.
midrib the large central vein of a dicot leaf.
milk teeth the temporary, deciduous first teeth in a mammal, replaced by permanent teeth during growth and maturation.
millipede also *myriopod,* nonpoisonous, herbivourous terrestrial arthropod of the class *Diplopoda,* with a long, cylindrical, segmented body, hard integument, and two pairs of legs per segment.
mineral nutrient an inorganic compound, element or ion needed for normal plant growth.
minimal medium broth or agar gel consisting of glucose, mineral salts, water, and glycine or other nitrogen source; the least complex medium capable of sustaining the growth of a specific microorganism.
miscarriage see under *abortion.*
missense mutation a base change in a structural gene that alters the coding property of the codon in which it appears, producing an abnormal polypeptide with one amino acid difference.
mitochondrion (pl. *mitochondria*), thread-like, self-replicating, membrane-bounded organelle found in every eucaryotic cell, functioning in oxidative chemiosmotic phosphorylation, apparently derived in evolution from a bacterial endosymbiont.
mitogen a hormone that stimulates cell division.
mitosis cell division or nuclear division in eucaryotes, involving chromosome consensation, spindle formation, precise alignment of centromeres, and the regular segregation of daughter chromosomes to daughter nuclei.
mitotic apparatus the centrioles, spindle, spindle fibers and asters; in plants, the spindle and spindle fibers.
mnemonic pertaining to or aiding memory.
modifier gene a gene or allele that modifies the expression of genotypes at another gene locus.
molar also *molar tooth,* a cheek tooth adapted for grinding; in human adults, the three most posterior pairs of teeth in each jaw.
mole 1. *gram molecule,* the quantity of a chemical substance that has a mass in grams numerically equal to its moleclar mass in daltons; 6.023×10^{23} molecules of a substance. 2. a congential spot or small skin growth. 3. a fossorial (digging) insectivore.
molecular biology a branch of biology concerned with the ultimate physiochemical organization of living matter; the study of biological systems using biochemical methods.
molecular mass also, *molecular weight,* the mass of a molecule expressed in daltons.
molecule a unit of a chemical substance, consisting of atoms bound to one another by covalent bonds.
mollusk any animal of the large phylum *Mollusca* (see Appendix A).
molt (v.i.), to cast off an outer covering in a periodic process of growth or renewal.
molt (n.) 1. the process of molting; see *ecdysis.* 2. the cast-off covering. 3. the period of a molt; an intermolt in interecdysis.
molting hormone a hormone that initiates or causes molting.
monera the procaryotes, comprised of the bacteria, cyanophytes and chloroxybacteria.
mongolism an archaic term for *trisomy 21.*
monocotyledon see *monocot.*
monocot a flowering plant of the angiosperm class *Monocotyledonae,* characterized by producing seeds with one cotyledon; included are palms, grasses, orchids, lilies, irises, and others.
monoecious 1. of a gametophyte, producing both egg and sperm in the same thallus. 2. of a seed plant sporophyte, producing both ovules and pollen. Compare *dioecious.*
monomer one of the subunits or potential subunits of a polymer.
monophyletic (adj.), of a taxonomic group, deriving entirely from a single ancestral species that possessed the defining characters of the group, while not having given rise to organisms outside of the group.
monosaccharide a sugar not composed of smaller sugar subunits (e.g., glucose, fructose).
monotreme a platypus, echidna, or extinct egg-laying mammal of the order *Monotremata,* subclass *Protheria.*
monozygotic twins also *identical twins,* twins derived from a single fertilized egg.
mons veneris in women, a rounded, usually hairy bulge of fatty tissue over the pubic symphysis and above the vulva.
morph one of the particular forms of an organism that exists in two or more distinct forms in a single population.
morphogenesis the development of form and structure.
mortality death, considered as an aspect of population dynamics.
mortality rate also *death rate,* the number of deaths per unit time occurring among a specified number of individuals (usually per 1000) in a given area or population.
mosaic egg an egg in which the developmental fates of different parts of the cytoplasm or surface are fairly rigidly determined prior to fertilization.
motor end plate the terminal branching of the axon of a motor neuron as it forms multiple synapses with a muscle fiber.
motor neuron a neuron that innervates muscle fibers, and impulses from which cause the muscle fibers to contract.
motor unit a motor neuron together with the muscle fibers it innervates, which contract as a unit.
M phase mitosis as part of the cell cycle (including prophase, metaphase, anaphase, telophase and, if it occurs, cytokinesis).
mRNA *messenger ribonucleic acid,* see *messenger RNA.*

mucin any of a class of mucoproteins that bind water and form thick, slimy, viscid fluids in various secretions; e.g., *mucin* secreted by the stomach lining protects it from the action of gastric juices.
mucopolysaccharide a polysaccharide that binds water to form a thick, gelatinous material that serves to cement animal cells together; a constituent of mucoproteins; e.g., *chondroitin, hyaluronic acid, heparin.*
mucoprotein a combination of mucopolysaccharide and a polypeptide to form *mucin.*
mucosa also *uterine mucosa, endometrium,* the highly glandular mucous membrane lining of the uterus; inner lining of the gut.
mucous (adj.), covered with or secreting mucus. *mucous membrane,* 1. any epithelium rich in mucus-secreting cells. 2. the epithelium that lines those passages and cavities of the body that communicate directly with the exterior such as the mouth, nasal cavity, and genital tract.
mucus a viscid, slippery secretion rich in mucins, that is secreted by mucous membranes and that serves to moisten and protect such membranes.
Mullerian ducts a pair of ducts parallel to the Wolffian ducts in vertebrate embryos, giving rise to the oviducts and uterus in female development and disappearing in males.
multicellular (adj.), consisting of a number of specialized cells that cooperatively carry out the functions of life.
multinucleate (adj.), having many nuclei within a common cytoplasm.
multiple alleles the alleles of a gene locus when there are more than two alternatives in a population.
multiplicative law in mathematical probability theory, the statement that the probability that all of a group of independent events will occur is equal to the product of their individual probabilities.
muscle fiber 1. one of the multinucleate cells of striated muscle, which takes the form of a long, unbranched cylinder. 2. cardiac muscle fiber, one of the long, branched, cylindrical subunits of heart muscle, consisting of individual cells joined end to end by intercalated disks.
muscular dystrophy an incurable hereditary disease characterized by the progressive wasting of muscle tissue and eventual death. *Duchenne-type muscular dystrophy,* a common sex-linked variety affecting primarily pre-teen boys.
muscularis the smooth muscle layer of the wall of a hollow contractile organ such as the uterus, gall bladder or urinary bladder.
mushroom a basidiocarp, especially but not necessarily an edible one.
muskeg a North American sphagnum bog, particularly in the taiga.
mutagen a chemical or physical agent that causes mutations.
mutagenize (v.t.), to treat with mutagens; to cause mutations to occur in.
mutant 1. a new or abnormal type of organism produced by a mutation. 2. a mutated gene. 3. an individual that bears and is affected by a mutated gene. 4. (adj.), mutated.
mutate 1. (v.t.), to alter, cause a change, or cause a mutation or DNA change to occur in; to mutagenize. 2. (v.i.), to change in state or genetic condition, to become altered, to undergo a mutation.
mutation 1. any change in DNA. 2. a change in DNA that is not immediately and properly repaired. 3. any abnormal, heritable change in genetic material. 4. the act or process of mutating. 5. a sudden departure from the parental type caused by a change in genetic material.
mutation rate the rate at which new mutations occur, generally in terms of mutations per locus per gamete per generation.
mutualism a mutually beneficial association between different kinds of organisms; symbiosis in which both partners gain fitness; compare *symbiosis, parasitism.*
mutually exclusive events possible events or outcomes such that the actual occurrence of one precludes the possibility of the other.
mycelium (pl. *mycelia*), 1. the mass of interwoven hyphae that forms the vegetative body of a fungus. 2. the analagous filaments formed by certain filamentous bacteria.
mycoplasma a tiny, nonmotile, wall-less bacterium of irregular shape, occurring as an intracellular parasite of animals and plants.
-mycota also *-mycetes,* a taxonomic suffix indicating that something is a fungus or, in the case of certain bacteria groups, is like a fungus.
myelin a soft, white, somewhat fatty material that forms a sheath around certain nerve axons, consisting of the phospholipid and proteins of the cell membranes of Schwann cells. *myelin sheath,* the sheath around many mature neuronal axons, consisting of many layers of Schwann cell membranes.
myelinated having a myelin sheath.
myeloid (adj.), pertaining to the bone marrow, e.g., *myeloid luekemia.*
myofibril a tubular subunit of muscle fiber structure, consisting of many sarcomeres end-to-end.
myofilament the highly organized microfilaments of striated muscle. *actin myofilaments,* thin filaments of the protein actin bound to the Z-line of a sarcomere; *myosin myofilament,* thicker filaments of the protein myosin interspersed among the actin myofilaments in a regular hexagonal array.
myosin a protein involved in cell movement, and structure; especially in muscle cells; see also *myofibril, sarcomere, cytoskeleton.*
myosin bridge the more or less globular head of the generally fibrous protein myosin, which forms a movable bridge between a myosin myofilament and an actin myofilament.
myotome 1. the portion of a vertebrate embryonic somite from which skeletal musculature is developed. 2. one of the muscular segments of the body wall of a fish or lancelet.
myxameba the free-living, unicellular, ameboid stage of a slime mold.
myxobacterium a slime or gliding bacterium, many species of which are capable of forming spores in stalked fruiting bodies.
myxomycetes see *slime mold.*

NAD also NAD^+, nicotine adenine dinucleotide, a hydrogen carrier in respiration. $NADH_2$, also $NADH + H^+$, the reduced form of NAD. *NADP, $NADPH_2$,* nicotine adenine dinucleotide phosphate, a similar hydrogen carrier in photosynthesis.
naphthalene-acetic acid also *NAA,* an artifical auxin.
nares (sing. *naris*), 1. *external nares,* also *nostrils,* the paired external openings of the nasal cavity. 2. *internal nares,* the paired openings of the nasal cavity into the pharynx.
nascent in the process of being born. *nacent RNA,* RNA in the process of being transcribed.
natural selection the differential survival and reproduction in nature of organisms having different heritable characteristics, resulting in the perpetuation of those characteristics and/or organisms that are best adapted to a specific environment. Also, *survival of the fittest.*
navigation the ability to locate or maintain reference to a particular place without the use of landmarks.
Neanderthal human an extinct, highly variable type or race of humans (*Homo sapiens neanderthalensis*) of the middle Paleolithic.
nectar a sweet liquid secreted by the nectaries of a flower or plant; the material used by bees in the production of honey.
nectary a plant gland that secretes a sweet fluid (nectar), most often at the base of a flower, that functions as a reinforcing reward for the visits of pollinating animals.
negative control in a genetic control system, control such that transcription occurs at all times except when prevented by an inhibitor molecule or an inhibitor complex.
negative feedback see under *feedback.*
negative entropy potential energy in the form of nonrandom organization.
negative geotaxis *negative geotropism,* negative phototropism; see *geotaxis, geotropism, phototropism.*
nematocyst one of the minute stinging cells of the coelenterates, consisting of a hollow thread coiled within a capsule, and an external hair trigger.
nematode any elongate, cylindrical, tapered roundworm of the Aschelminthe class *Nematoda* (see Appendix A).
nemertine also *nemertean, nemertinean, proboscis worm,* a member of a phylum of acoelomate worms with a complete gut, a ribbon-like, solid body, and a unique eversible proboscis.
neolithic revolution the human attainment of agriculture, commerce, improved stone tools and city-building that occurred world-wide at about the end of the last ice age, some 10,000 years ago.
nephridiopore the exterior orifice of a nephridium.
nephridium an excretory organ primitive to many coelomate phyla and still found in annelids, brachipods, mollusks and some arthropods, occurring paired in each body segment in segmented animals, and typically consisting of a ciliated funnel (*nephrostome*) draining the coclom through a convoluted, glandular duct to the exterior.
nephron a single excretory unit of a kidney, consisting of a glomerulus, Bowman's capsule, proximal convoluted tubule, loop of Henle, distal convoluted tubule, and collecting duct discharging into the renal pelvis, see also Figures 28.17 and 28.18.
nerve (n.), 1. a filamentous band of nerve cell axons and protective and supporting tissue, that connect parts of the nervous system with other parts of the body. 2. (adj.), pertaining to the nerve or nervous system, e.g., *nerve cell, nerve net, nerve fiber.*
nerve cell see *neuron.*
nerve cord 1. *dorsal tubular nerve cord,* the spinal cord of chordates. 2. *ventral nerve cord,* the solid, paired, segmentally ganglionated, longitudinal central nerve of many invertebrates.
nerve fiber an axon or dendrite, especially when myelinated.
nerve impulse see under *impulse.*
nerve synapse see under *synapse.*
neritic province the coastal sea from the low-tide line to a depth of 100 fathoms, generally waters of the continental shelf.
nervous system the brain, spinal cord, nerves, ganglia and the neural parts of receptor organs, considered as an integrated whole.
net productivity in ecology, the production of energy or biomass by plants, being the gross productivity less the energy used by the plants in their own activities.
neural arch the open ring formed by vertebral processes rising from the centrum and closing above the spinal cord.
neural canal 1. the canal formed by the series of vertebral neural arches, through which the spinal cord passes. 2. the lumen (cavity) of the spinal cord.
neural cleft see *synaptic cleft, synapse.*
neural crest the ridge of a neural fold, migrating cells of which will give rise to spinal ganglia, the adrenal medulla, the autonomic nervous system, and melanocytes (pigment cells).
neural folds in early vertebrate embryology, a pair of longitudinal ridges that arise from the neural plate on either side of the neural groove, which fold over and fuse to give rise to the *neural tube.*
neural groove in early vertebrate embryology, the linear depression in the neural plate between the neural folds that will invaginate to form the neural tube.
neural plate in the vertebrate gastrula, a thickened plate of ectoderm along the dorsal midline that gives rise to the neural tube, neural crests and ultimately to the nervous system.
neural tube 1. in early vertebrate embryology, the hollow dorsal tube formed by the fusion of the neural folds over the neural groove. 2. the spinal cord.
neurectoderm also *neuroectoderm,* ectoderm destined to give rise to neural tissue, sometimes considered to be a fourth germ layer; coextensive with the neural plate in vertebrate embryos.

neuromuscular junction the synapse between a neural motor end plate and a muscle fiber.
neuron also *nerve cell,* a cell specialized for the transmission of nerve impulses, consisting of one or more branched *dendrites,* a *nerve cell body* in which the nucleus resides, and a terminally branched *axon.*
neurophysiology the physiology of nerve cells, neural receptors, and other aspects of the nervous system.
neurosecretion 1. a secretion of a nerve cell, especially hormonal secretions into the general circulation, usually produced in the nerve cell body, transmitted along the axon, and released at the terminus of the axon. 2. the process of secretion by neurons.
Neurospora a genus of sac fungus often used in genetic studies.
neurotransmitter a short-lived, hormone-like chemical (e.g., *acetylcholine, norepinephrine*) released from the terminus of an axon into a synaptic cleft, where it stimulates a dendrite in the transmission of a nerve impulse from one cell to another; see also *synapse, synaptic cleft, synaptic knob.*
neurula an early vertebrate embryo slightly older than a gastula, in which the neural tube has formed.
neurulation the development of a neurula from a gastrula; equivalently, the development of the neural plate, neural folds and neural tube.
neutral mutation also *selectively neutral mutation,* a mutational change in DNA that has no measurable effect on the fitness of an organism, and which may sometimes be incorporated into the genome of a species by genetic drift.
neutralist a population geneticist or evolutionary theorist who argues that a substantial proportion of protein and DNA changes in evolution have been due to selectively neutral mutations fixed by random drift; compare *selectionist.*
neutrophil the most common mammalian phagocytic leukocyte.
neutron one of the two common constituents of an atomic nucleus, a hadron having no charge and no effect on chemical reactions.
New-World monkeys the primate superfamily Cebidea native to South America, with wide-set nostrils and prehensile tails.
niche "the place of an animal or plant in the polity (organized community) of nature" (Darwin); the position or function of an organism in a community of plants and animals. It has been said that if a *habitat* is an organism's address, a *niche* is its occupation. *Niche* has also been defined as the totality of the adaptations, specializations, tolerance limits, functions, biological interactions and behavior of a species.
nick 1. (n.), a single-strand break in DNA. 2. (v.t.), to cause a single-strand break in DNA by the hydrolysis of a phosphate linkage.
nicotine a poisonous, addictive, weakly basic liquid alkaloid found most commonly in tobacco, in which it is the principal active ingredient; used sometimes in solution as a potent insecticide but more often in smoke as a mood-altering drug.
nicotinic acid a five-carbon nitrogenous acid, a constituent of NAD and NADP in respiration and photosynthesis; a vitamin (dietary requirement) of animals.
nictitating membrane in many vertebrates, including many mammals and most birds, a thin translucent membrane that can be folded into the inner angle of the eye or drawn across the front of the eyeball; vestigial in humans.
nitrate the anion NO_3^-.
nitrification the chemical oxidation of ammonium salts into nitrites and nitrates by the action of soil bacteria.
nitrite the anion NO_2^-.
nitrogen fixation the chemical change of nitrogen from the stable, generally unavailable form of atmospheric nitrogen ($N\equiv N$) to soluble and more readily utilized forms such as ammonia, nitrate, or nitrite; usually by the action of cyanophytes or nitrogen-fixing bacteria, but sometimes by lightning, automobile engines, or synthetic fertilizer factories.
nitrogenase an enzyme that catalyzes nitrogen fixation utilizing cell energy.
nitrogenous base 1. a purine or pyrimadine used in the synthesis of nucleic acids; specifically, adenine, cytosine, guanine, thymine or uracil. 2. any related heterocyclic compound, such as xanthine or uric acid.
nitrous oxide also *laughing gas,* N_2O, a colorless gas with anesthetic and intoxicating properties; used also in aerosol whipped cream.
noble element also *noble gas,* an element in which the uncharged atoms have all electron shells filled and all electrons paired and which consequently are chemically unreactive.
node 1. in a plant stem, any point at which one or more leaves emerge. 2. in a growing stem tip, a region of potential leaf growth, containing meristematic tissue. 3. see *node of Ranvier.* 4. see *lymph node.*
node of Ranvier a constriction in a myelin sheath, corresponding to a gap between successive Schwann cells. See *saltatory propagation.*
nodule 1. any small lump. 2. see *root nodule.*
noncellular see *acellular.*
non-Darwinian evolution decent with modification attributable to genetic drift rather than natural selection; the fixation of selectively neutral mutations in evolutionary lineages.
nondisjunction 1. the failure of a pair of homologous chromosomes to segregate to opposite poles in anaphase I of meiosis. 2. the failure of a pair of daughter chromosomes or daughter centromeres to segregate to opposite poles in mitotic anaphase or in anaphase II of meiosis.
nongenetic variance that portion of total phenotypic variance in a population not attributed to genetic differences between individuals.
nonhistone chromosomal protein a large class of proteins other than histones associated with the chromosomes, constituting about one-third of the substance of chromatin in animals and plants, highly variable from tissue to tissue and known in at least some cases to be involved in gene control.
noninducible control mutant a mutation in a mappable controlling locus that prevents a normally inducible protein from being induced.
nonpolar of a molecule or part of a molecule, not forming hydrogen bonds with water, uncharged.
nonpolar attraction see *van der Waals forces.*
nonpurple sulfur bacterium a nonphotosynthetic bacterium that utilizes sulfur or sulfur compounds in chemosynthesis.
nonrecombinant chromosome a chromosome derived from a chromatid that has gone through meiosis without having undergone crossing over, or without exhibiting net recombination between gene markers.
nonseptate of a fungal hypha, not having septa; not being separated into individual cells.
nontranscribed strand the DNA strand of a cistron opposite the strand that is transcribed into RNA.
norepinephrine also *noradrenalin,* a compound that serves as a synaptic neurotransmitter, and also as a systemic hormone where it acts as a vasoconstrictor.
normal distribution also *normal curve, Gaussinan distribution, bell-shaped curve,* the idealized, symmetrical distribution taken by a population of values centering around a mean, when departures from the mean are due to the change occurrence of a large number of individually small independent effects; often approached in real populations. *log normal distribution,* the asymmetric distribution of values around a mean, when departures from the mean are due to many independent factors with multiplicative effects; also often approached in real populations, and not always readily distinguishable from the normal distribution.
notochord in all chordates at some point in development, a turgid, flexible rod running along the back beneath the nerve cord and serving as a skeletal support; replaced during development by the vertebral column in most vertebrates but persistent in adult coelocanths, cyclostomes, lancelets and larvacean urochordates.
***N*-terminal end** also *amino-terminal end,* the end of a polypeptide chain in which the end amino acid has its primary carboxyl group fixed in a peptide linkage and its primary amino group free (or formylated); the first part of the chain to be synthesized.
nucellus in flowers, a mass of thin-walled parenchymal cells that composes most of the ovule, encloses the embryo sac, and is enclosed by one or more integuments.
nuclear membrane in all eucaryotes, the bounding membrane of the nucleus, separating the cytoplasm from the nucleoplasm.
nuclear pore a structure in the nuclear membrane that appears as a hole in electron micrographs although probably filled with a protein mass in life; through which mRNA, nuclear proteins and other large molecules are presumably able to pass.
nuclease any enzyme that hydrolyzes a nucleic acid; see also *endonuclease, exonuclease.*
nucleic acid either DNA or RNA, DNA being a double polymer of deoxynucleotides and RNA being a polymer of nucleotides.
nucleolus (pl. *nucleoli*) a dark-staining body of RNA and protein found within the interphase nucleus of a cell, the site of synthesis and storage of ribosomes and ribosomal materials; disperses during mitosis and is reconstituted following nuclear membrane reorganization.
nucleoplasm the viscous fluid matrix of the nucleus, contrasted with *cytoplasm.*
nucleoside 1. a compound consisting of a nitrogenous base linked to the 1′ carbon of ribose. 2. also *deoxynucleoside,* a compound consisting of a nitrogenous base linked to the 1′ carbon of deoxyribose. compare *nucleotide.*
nucleotide 1. a compound consisting of a nitrogenous base and a phosphate group linked to the 1′ and 5′ carbons of ribose respectively; the repeating subunit of RNA. 2. also *deoxynucleotide,* a compound consisting of a nitrogenous base and a phosphate group linked to the 1′ and 5′ carbons of deoxyribose respectively, the repeating subunit of DNA.
nucleotide triphosphate also *nucleoside triphosphate, triphosphonucleotide,* a nucleoside with a chain of three phosphate groups linked to the 5′ carbon of ribose or deoxyribose.
nucleus in all eucaryote cells, a prominent, usually spherical or elipsoidal membrane-bounded sac containing the chromosomes and providing physical separation between transcription and translation.
null hypothesis the hypothesis that one variable has no effect on another, or that two variables are independent or unrelated: e.g., the hypothesis that there is no effect of vitamin C on cold symptoms. Also, *hypothesis tested,* since in general null hypotheses can be disproven if false, but specific counter-hypotheses cannot be proven to be true.
numerical taxonomy a mathematical approach to taxonomy in which phenotypes are reduced to character states classes as either present or absent, no assumptions are made about biological species or evolutionary history, and individuals are clustered into hierarchies of groupings by rigorous statistical procedures, usually with the aid of computers.
nutrient (n.), 1. any substance required for growth and maintenance of an organism. 2. a chemical element or inorganic compound needed for normal growth of a plant, see also *macronutrient, micronutrient, mineral nutrient.* 3. (adj.) furnishing nourishment, e.g., *nutrient broth,* a broth in which microorganisms may readily be grown.

obligate restricted to a particular condition of life; e.g., obligate aerobe, obligate anerobe, obligate parasite, obligate symbiont.
occipital condyle a rounded portion of the occipital bone at the base of the skull, which articulates with the first vertebra.
occipital lobe the posterior lobe of each cerebral hemisphere.
oceanic province (n.), the open sea as distinguished from the neritic province; *oceanic* (adj.), pertaining to the open sea.

ocellus (pl. *ocelli*), a minute simple eye or eyespot of any organism. 2. one of the elements of an arthropod compound eye; an *ommatidium.* 3. one of the three simple eyes that form a triangle dorsally between the compound eyes.
"off" fibers nerve fibers of the retina that discharge only when a light stimulus ceases.
oil a common term for any fat or viscous lipid that is liquid at ambient temperatures.
Okazaki fragment in DNA synthesis, the original form of a newly synthesized single strand, being a polynucleotide of some 200-300 bases.
Old Stone Age see *Paleolithic.*
Old-World monkey a primate of the superfamily *Cercopithecoidea,* found only in Asia, Africa and Gibralter, characterized by close-set nostrils on a somewhat dog-like snout; included are baboons, Barbary ape, rhesus, and macaques.
oleic acid a liquid unsaturated fatty acid of eighteen carbons, with a double bond between the 9th and 10th carbons.
olfaction 1. the sense of smell. 2. the process of smelling.
olfactory organ a sensory organ that responds to chemicals or mixes of chemicals, especially to trace quantities dispersed in air or water.
olfactory receptor the odor-sensitive cells of the mucous membrane of an olfactory organ.
oligotrophic of a lake, rich in dissolved oxygen and poor in plant nutrients and algal growth, with clear water and no marked stratification.
ommatidium (pl. *ommatidia*), one of the elements of a compound eye, consisting of a corneal lens, crystalline cone, rhabdome, light-sensitive retinula, and sheathing pigment cells.
omnivorous (adj.), feeding on both animal and plant material; literally, eating everything. *omnivore* (n.), an omnivorous animal e.g., pigs, people.
"on" fibers nerve fibers of the retina that fire when a light stimulus begins.
"on-off" fibers nerve fibers of the retina that respond both to the commencement and the cessation of a light stimulus.
ontogeny the development of an individual organism.
onychophoran *Peripatus* an animal of the phylum *Onychophora* (see Appendix A).
oocyst the zygote of the sporozoan *Plasmodium* which divides asexually to produce sporozoites.
oocyte an egg cell before maturation; *primary oocyte,* a diploid cell precursor of an egg, before meiosis; *secondary oocyte,* an egg cell after the formation of the first polar body.
oogonium (pl. *oogonia*), the large, spherical, unicellular female sex organ of water molds and egg-producing algae in which egg cells are produced.
oomycetes a fungal group that includes the water molds, egg molds, white rusts and downy mildews.
open circulatory system a circulatory system in which the arterioles end openly into intercellular space, allowing blood to percolate directly through nonvascular tissues.
operant conditioning see *conditioning, operant.*
operational taxonomic unit (OTU) a group of related organisms sharing a certain set of character states, as determined by the methods of numerical taxonomy, which may or may not correspond to species, genus or other defined taxonomic group.
operator locus in an operon, the binding site of an inhibitor protein or inhibitor complex.
operculum a body process functioning as a lid or cover, e.g., (a) the horny plate on the foot of certain gastropods, which serves to close off the opening of the shell; (b) the covering flap of a moss spore capsule; (c) the skin-covered bony plates that cover the gills of a fish.
operon in bacteria, the functional genetic unit of a set of inducible or repressible enzymes, including the inhibitor locus and a contiguous stretch of DNA that includes the promoter, operator, and usually the cistrons of several structural genes.
optimum intermediate, see *intermediate optimum.*
oral cavity also *mouth cavity,* the cavity between the mouth and the pharynx.
oral-facial-digital syndrome a genetic condition of human females, caused by an allele that is dominant with regard to abnormalities of the mouth, face and fingers, and is lethal in the hemizygous state.
orbit 1. see *electron orbit.* 2. in vertebrates, the bony cavity of the eye socket.
order a taxonomic level between class and family.
Oreopithecus an extinct hominoid primate.
organelle a functionally and morphologically specialized part of a cell.
organ system a group of functionally interrelated organs.
organic compound a chemical compound containing carbon.
organism 1. a form of life composed of mutually dependent parts that maintain various vital processes. 2. any form of life. 3. an individual plant, animal or microorganism.
orgasm in humans, the climax of sexual excitement typically occurring toward the end of coitus, usually accompanied in men by ejaculation and in women by rhythmic contractions of the cervix.
orientation the directing of bodily position according to the location of a particular stimulus.
origin 1. evolutionary ancestry. 2. the proximal skeletal attachment of a muscle or tendon; compare *insertion.*
ornithine an intermediate compound in the biosynthesis of arginine.
ornithologist a scientist who studies birds.
osmoconformer an aquatic organism that does not regulate the osmotic potential of its body tissues, but allows it to fluctuate with that of the environment.
osmoregulation in aquatic organisms, the homeostatic regulation of the osmotic potential of body fluids.
osmoregulator an aquatic organism that maintains a constant internal osmotic potential in spite of fluctuations in the salinity of its environment.
osmosis the tendency of water to diffuse through a semipermeable membrane in the net direction from lower to higher solute concentration.
osmotic gradient the difference between two fluids in terms of osmotic potential.
osmotic potential also *water potential,* a property of a solution equal to the osmotic pressure that would occur if it were rigidly confined and separated from pure water by a semipermeable membrane at a standard temperature; generally a function of the molar concentration of solutes.
osmotic pressure the actual hydrostatic pressure that builds up in a confined fluid because of osmosis.
ossification the process of bone formation and especially of the replacement of cartilage by bone.
Osteichthyes a vertebrate class consisting of the bony fish, including both lobe-finned and ray-finned fish (see Appendix A).
osteoblast a bone-forming cell.
osteoclast a bone-destroying ameboid cell that dissolves calcium phosphate and releases Ca^{2+} into the bloodstream.
osteocyte a bone cell isolated in a lacuna of bone tissue.
osteon see *Haversian system.*
osteoporosis a condition involving the decalcification of bone, producing bone porosity and fragility.
ostracoderm any of a group of extinct jawless fish.
OTU see *operational taxonomic unit.*
outer ear see *external ear.*
ovary 1. in animals, the (usually paired) organ in which oogenesis occurs and in which eggs mature. 2. in flowering plants, the enlarged, rounded base of a pistil, consisting of a carpel or several united carpels, in which ovules mature and megasporogenesis occurs.
overdominance see *dominance interactions.*
oviduct a tube, usually paired, for the passage of eggs from the ovary toward the exterior or to a uterus, often modified for the secretion of a shell or protective membrane; in humans it is sometimes known as a *Fallopian tube.*
oviparous (adj.), producing eggs that develop and hatch outside of the mother's body; compare *ovoviviparous, viviparous.*
oviparity (n.), the condition of being *oviparous.*
ovipositor an insect organ specialized for the depositing of eggs and often for boring holes in which eggs may be deposited.
ovoviviparity the condition of being ovoviparous.
ovoviviparous producing eggs that are fertilized internally, develop within the mother's body but without any direct connection with the maternal circulation; the young being released shortly before or after hatching. Compare *viviparous, oviparous.*
ovulate to release one or more eggs from an ovary.
ovulation the release of one or more eggs from an ovary.
ovule 1. in animals, a small egg; an egg in the process of growth and maturation. 2. in seed plants, a rounded outgrowth of the ovary, consisting of the embryo sac surrounded by maternal tissue including a stalk, the nucellus, and one or more integruments.
oxaloacetate the anion of *oxaloacetic acid,* a four-carbon intermediate in the citric acid cycle.
oxidation 1. the loss of electrons from an element or compound. 2. also dehydrogenation, the loss of hydrogens from a compound. 3. the addition of oxygen to an element or compound. 4. a multi-step process in which oxygen is added or hydrogens or electrons are removed (e.g., the oxidation of glucose).
oxidation-reduction reaction a chemical reaction in which one reactant is oxidized and another is reduced.
oxidative phosphorylation the production of ATP from ADP and phosphate in a process consuming oxygen, as by mitochondria or aerobic bacteria; see also *chemiosmosis, chemiosmotic, phosphorylation, oxidation.*
oxidative respiration the breakdown of biochemicals to produce cellular energy, utilizing oxygen as the final electron acceptor; see *oxidative phosphorylation, respiration.*
oxygen carrier a molecule specialized for the transport or storage of oxygen, e.g., *hemoglobin, cytochrome P450, myoglobin.*
oxygen debt a state of oxygen depletion after extreme physical exertion; measured by the amount of oxygen required to restore the system to its original state by the resynthesis of glycogen from lactate.
oxyhemoglobin hemoglobin carrying four oxygen molecules; the bright red arterial form of hemoglobin.
oxytocin a polypeptide hormone of the posterior pituitary that stimulates the contraction of the uterus and the release of milk.

pacemaker also *sinoatrial node,* the portion of the heart in which the impulse for the heartbeat originates and is regulated.
pachyderm 1. an elephant. 2. any of an artificial assemblage of large, thick-skinned animals, including elephants, rhinoceroses and hippopotamuses.
pachytene a phase of meosis in which crossing over occurs and in which the synapsed chromosomes appear as solid, thickened threads; between zygotene and diplotene.
Pacinian corpuscles oval pressure receptors containing the termini of sensory nerves, especially in the skin of hands and feet.
palate in mammals, the roof of the mouth cavity, separating the mouth cavity from the nasal cavity; consisting of an anterior, bony *hard palate* and a posterior fleshy *soft palate.*
paleobiology the branch of paleontology dealing with the origin, growth, biological functioning and ecological relationships of fossil organisms.
Paleolithic also *old stone age,* a period in human cultural evolution characterized by rough or chipped stone implements; the time of Neanderthal humans.

paleontology the scientific study of the forms of life existing in former geological periods, as represented by fossil animals, plants and micro-organisms.
Paleozoic see the geological timetable inside the front cover.
palmitic acid a saturated 16-carbon fatty acid.
palp see *labial palps.*
pampas (sing. *pampa*), the extensive, grassy plains of southern South America.
pancreas a large digestive and endocrine gland of vertebrates, which secretes various digestive enzymes into the duodenum by way of the *pancreatic duct,* and which also contains endocrine tissues in the form of interspersed *islets of Langerhans,* responsible for the production of the hormones *insulin* and *glucagon.*
pancreatic amylase a starch-digesting enzyme secreted by the pancreas. *pancreatic lipase,* a fat-digesting enzyme of the pancreas. *pancreatic proteases,* the protein-digesting enzymes *trypsin, chymotrypsin* and *elastin. pancreatic hormones,* insulin and glucagon.
pangenesis Darwin's theory that every organ of the adult body produced gemmules of information which were transported to the gonads and into the gametes.
parallax the apparent displacement of objects seen from different points of view, especially from the slightly different placement of the eyes in binocular vision.
parallel evolution similar evolutionary change occurring simultaneously in separated lines of descent.
parapatric living in separate but adjacent geographic regions not separated by geographical barriers; compare *allopatric, sympatric.*
paraphyletic of a taxonomic group, sharing defining characters because of common descent from a single ancestral species having those characters, but having also given rise to other organisms not now included in the group.
parapod also *parapodium,* one of the short, unsegmented, paired, leg-like or fin-like locomotive organs borne on either side of each body segment in *Nereis* and certain other polycheate worms.
parasite an organism living in or on another living organism from which it obtains its organic sustenance to the detriment of its host; see also *endoparasite, ectoparasite;* compare *symbiont, parasitoid, mutualism.*
parasitic castration a common form of parasitism in which the parasite invades the host's reproductive system, rendering it sterile; sometimes involving hormonal changes that lead to phenotypic sex reversal.
parasitism a relationship in which an organism of one kind (the parasite) lives in or on an organism of another kind (the host) at the expense of which it obtains food and shelter, causing some degree of damage but usually not killing the host directly; compare *symbiosis, mutualism.*
parasitoid an organism that invades the body of a host organism and eventually but regularly causes its death; specifically certain insects that lay one or more eggs in a host organism, the resulting larvae developing within and eventually killing the host at about the time that larval development is complete.
parasympathetic nervous system of the two divisions of the vertebrate autonomic nervous system, the one that utilizes acetylcholine as a neurotransmitter, that increases the activity of smooth muscle and digestive glands, slows the heart, and dilates blood vessels; compare *sympathetic nervous system.*
parathormone also *parathyroid hormone,* the internal secretion of the parathyroid glands, involved in maintaining normal calcium balance.
parathyroid glands four small endocrine glands embedded in or adjacent to the thyroid gland and involved in the regulation of calcium ion levels in the blood.
Parazoa a monophyletic group consisting of the sponges, usually included with the metazoa in the Animal Kingdom but not directly related to any other animal groups.
parenchyma in higher plants, a tissue consisting of thin-walled living cells that remain capable of cell division even when mature, and function in photosynthesis or food storage.
parietal bones a pair of membraneous bones of the roof of the skull.
parietal cells large cells of the stomach lining that secrete hydrochloric acid.
parietal eye a median light-sensitive organ of ancient vertebrates, or its vestige in certain contemporary vertebrates; analogous to the pineal eye but apparently not homologous with it.
parietal lobe the middle division of each cerebral hemisphere.
parotid glands a pair of ducted salivary glands situated below the ears.
parthenogenesis asexual reproduction in which gametes develop without fertilization, either with or without having undergone meiosis; e.g., male hymenopterans are produced parthenogenetically from haploid eggs, and *Daphnia* of either sex may be produced parthenogenetically from unreduced diploid eggs.
partial dominance see *dominance interactions.*
partial pressure 1. the independent pressure exerted by each gas in a mixture of gases. *total pressure* is equal to the sum of the partial pressures ($P_{total} = P_1 + P_2 + P_3 + \ldots$). 2. of a gas in solution, the partial pressure that would yield the observed concentration under equilibrium conditions.
particulate inheritance inheritance consisting of discrete, separable units of information, that is, of genes; compare *blending inheritance.*
passive transport movement of fluids, solutes or other materials without the expenditure of energy, e.g., by diffusion, especially across a membrane.
pasteurization heating of wine or milk to a certain temperature for a given amount of time in order to kill pathogenic or otherwise undesirable bacteria without destroying the quality of the wine or milk.
patella the kneecap.
pathway see *biochemical pathway.*
paved teeth also *pavement teeth,* teeth that form a continuous, flat, roughened surface, and that function in crushing or grinding.
pecking order, also *peck order,* in many gregarious animals, a specific linear order of relative dominance and submissiveness; see *dominance hierarchy.*
pecten 1. any comb-like organ. 2. a comb-like, pigmented, highly vascularized projection from the retina into the vitreous humor of the eyes of most birds; function unknown.
pectin any of a group of complex methylated polysaccharides occurring in plant tissues, especially of fruits and succulent leaves, that bind water and sugar to make viscous solutions or gels.
pectoral girdle the bones or cartilage supporting and articulating with the vertebrate forelimb; in most vertebrates not connected with the axial skeleton, but in humans connecting to the sternum by way of the clavicle and consisting of the clavicle, coracoid process, and scapula.
pelagic (adj.), living in the open sea; *pelagic zone,* the region of the open sea beyond the littoral and neritic zones and extending from the surface to the depth of light penetration; compare *abyssal, benthic, neritic, littoral.*
pellicle 1. the semirigid, proteinaceous integument of many protists. 2. the plasma membrane and associated cytoskeleton of an animal cell.
peloric of a flower, having radial symmetry abnormally in a species in which flowers are normally irregular (bilaterally symmetrical); see also *flower.*
pelvic girdle the bones or cartilage supporting and articulating with the vertebrate hind limbs; in humans consisting of the fused bones of the *pelvis.*
pelvis 1. the bones of the pelvic girdle, consisting of the *sacrum, coccyx* and the paired and fused *ischium, ilium* and *pubis;* see also *coxa.* 2. *renal pelvis,* the main cavity of the kidney, into which the nephrons discharge urine.
pelycosaur a large primitive reptile of the Permian, thought to be allied with the line of descent leading to mammals, most remarkable for the extreme development of the dorsal vertebral process suggesting a sail-like structure probably used in behavioral thermoregulation.
penicillin a group of closely related powerful antibiotics produced by fungi of the genus *Penicillium,* that function by disrupting the systhesis of bacterial cell walls.
penis 1. the copulatory organ of the male in any species in which internal fertilization is achieved by the insertion of a male body part into the female genital tract. 2. the erectile intromittant organ of male mammals, which also serves as a channel for the discharge of urine.
peppered moth see *Biston betularia.*
pepsin a proteolytic enzyme secreted as *pepsinogen* by glands of the stomach lining, that is active only at low pH and which acts primarily to reduce complex proteins to simply polypeptides.
pepsinogen the initial, inactive form of pepsin as occurring in and secreted by gastric glands, that readily converts to pepsin in an acid medium.
peptidase a proteolytic enzyme that hydrolyzes peptide bonds; see *aminopeptidase, carboxypeptidase, dipeptidase, endopeptidase, exopeptidase.*
peptide a chain of two or more amino acids linked by peptide bonds, too short to be coagulated by heat or precipitated by saturated ammonium sulfate; most often seen as a partial digestion product of a protein or polypeptide.
peptide bond also *peptide linkage,* the dehydration linkage formed between the carboxyl group of one amino acid and the amino group of another: $R_1\text{—COOH} + NH_2\text{—}R_2 \rightarrow R_1\text{—CO—NH—}R_2 + H_20$.
perception response or responsiveness to sensory stimuli, e.g., *color perception, depth perception,* perception of movement.
perennial (adj.), 1. continuing or lasting for several years. 2. (n.), a plant that lives for an indefinite number of years, as compared with *annual* or *biennial.*
perfect flower see under *flower.*
pericardium 1. the membraneous sac surrounding the vertebrate heart. 2. the cavity or hemocoel containing the arthropod heart.
pericarp the ripened and variously modified wall of a plant ovary (fruit) such as a pea pod, seed capsule, berry, or nutshell.
perichondrium the fibrous connective tissue covering a cartilage.
pericycle a layer of parenchyma or sclerenchyma that sheathes the stele of the root, and is associated with the formation of vascular cambium and lateral roots.
periderm in plants, a protective layer of secondary tissue derived from epidermal cells and consisting of cork, cork cambium and underlying parenchyma.
perineum the area between the anus and the exteria genitalia, especially in the female.
Peripatus 1. a genus of onychophorans. 2. *peripatus,* the common name of any onychophoran.
peristalsis successive waves of involuntary contractions passing along the walls of the esophagus, intestine or other hollow muscularized tube, forcing the contents onward.
peristome 1. the fringe of teeth around the opening of a moss spore capsule. 2. the lip of a spiral shell. 3. the area surrounding the mouth of an invertebrate.
peritoneum the smooth, transparent membrane lining the abdominal cavity of a mammal. *peritoneal cavity,* the principle body cavity (abdominal coelom) of a mammal.
permafrost the permanently frozen layer of soil and/or subsoil in arctic and subarctic regions.
permanent tissues plant tissues that have become specialized; in *simple permanent tissue,* one type of cell predominates; in *complex permanent tissue,* two or more types of cell are in close association.
permeable allowing materials to pass through. *permeability,* the degree to which materials are able to pass through a substance, membrane, or barrier.
permease an enzyme-like carrier that functions in facilitated transport of a specific substrate across a cell membrane.

Permian see the geological timetable inside the front cover.
peroxisome a cytoplasmic organelle involved in the detoxification of peroxides.
petal one of the usually white or brightly colored leaf-like elements of the corolla of a flower.
petiole the small stalk that supports the blade of a leaf; a *leafstalk.*
Petri dish a small shallow dish of thin glass on plastic with a loosely fitting overlapping cover, used for plate cultures in bacteriology.
PGAL *phosphoglyceraldehyde,* a small molecule intermediate in glycolysis.
pH the negative logarithm of the hydronium ion concentration of a solution and a common measure of the acidity or alkalinity of a liquid; pH values of less than 7 indicating acidity, and values greater than 7 indicating alkalinity.
pH differential the difference in acidity between two solutions.
phage also *bacteriophage,* a virus that infects and lyses bacteria.
phagoctye 1. any leucocyte that engulfs particulate matter. 2. any cell that characteristically engulfs foreign matter, e.g., cells of the reticuloendothelial system.
phagocytosis taking solid materials into the cell by engulfment and the subsequent pinching off of the cell membrane to form a digestive vacuole.
phalanges (sing. *phalanx*), in land vertebrates, the bones of the fingers and toes.
pharyngeal gill cleft see *gill cleft.*
pharyngeal pouch in vertebrate embryos, any of the series of outpocketings of the ectoderm of the pharynx opposite the gill clefts, which may or may not break through; in mammals, one of the pairs of pharyngeal pouches become the eustachian tubes and the cavities of the middle ears.
pharynx 1. in fishes, the portion of the alimentary canal in which the gills and gill slits are located. 2. in land vertebrates, the corresponding region, posterior to the oral and nasal cavities and anterior to the esophagus. 3. an analogous region in the alimentary canals of various invertebrates, including some in which it is eversible and toothed.
phenotype frequency the proportion of individuals in a population falling into specific phenotypic category.
phenotypic variance the total measurable variance of a character in a population; see *variance.*
phenylalanine one of the 20 amino acids of protein structure.
phlebotomy deliberate bleeding; the letting of blood.
phloem a complex vascular tissue of higher plants consisting of sieve tubes, companion cells, and phloem fibers, and functioning in transport of sugars and nutrients.
phloem fiber a fiber cell associated with phloem, with great strength and pliability, e.g., the linen fibers of flax.
phloem ray the part of a vascular ray located in the phloem; see *vascular ray.*
phosphate linkage also *phosphate diester,* a phosphate that is covalently linked to two organic residues and thus serves as a bridge between them: $R_1—PO_4—R_2$.
phosphate group phosphate linked by a dehydration linkage to an organic molecule: $R—HPO_4^-$.
phosphate ion any of the four ionization steps of phosphate: H_3PO_4, $H_2PO_4^-$, HPO_4^{2-} and PO_4^{3-}; the monobasic and diabasic forms being the more common in physiological solutions.
phosphodiesterase a phosphatase, common in snake venom, that hydrolyzes phosphate links of nucleic acids.
phosphoenolpyruvate also *phosphopyruvate,* an intermediate in carbohydrate metabolism.
phospholipid also *phosphatide,* any of a class of phosphate-esterified lipids, including *lecithins, cephalins, sphingomyelins;* a major component of cell membranes.
phosphorolase an enzyme that catalyzes phosphorolysis.
phosphorolysis also *phosphorolytic cleavage,* a reversible reaction analogous to hydrolysis, in which a dehydration linkage is broken by the addition of phosphate: $R_1—O—R_2 + HPO_4^{2-} \rightarrow R_1PO_4^{2-} + HO—R_2$.
phosphorylation the addition of a phosphate group to a compound, e.g., the addition of phosphate to ADP to produce ATP and water.
photoevent the unitary event of the light-dependent reaction of photosynthesis, conceived as the pathway followed by two electrons released by the photolysis of water.
photolysis of water oxidation of water through the removal of hydrogen by highly oxidizing elements of photosystem 680, which in turn are regenerated with the energy of captured photons; the net reaction of a complex series of events being $2H_20 \rightarrow 4H^+ + 4e^- + O_2\uparrow$.
photon a quantum of eletromagnetic radiant energy.
photoperiod the relative lengths of alternating periods of lightness and darkness, which change with the season especially in temperate climates.
photoperiodism the response of an organism to photoperiods, involving sensitivity to the onset of light or darkness and a capacity to measure time.
photoreceptor a receptor of light stimuli.
photorespiration a puzzling phenomenon in which abundant light energy is captured by photosynthesis but little or no net carbon dioxide fixation occurs; common in C_3 plants in bright sunlight on hot days.
photosynthesis the organized capture of light energy and its transformation into usable chemical energy in the synthesis of organic compounds.
photosynthetic pigment any pigment involved in the capture of light; see *chlorophyll, bacteriochlorophyll, accessory pigments.*
photosystem also *photosynthetic unit, pigment system,* an organized array of proteins and pigments bound within the thylakoid membrane, consisting of a light antenna, a photocenter, and an associated electron transport system.
photosystem 680 also *photosystem II, P680,* the first of the two photosystems in the electron pathway of photosynthesis in cyanophytes and all photosynthetic eucaryotes, and the one involving the photolysis of water; it probably occurred later in evolution than photosystem 700.
photosystem 700 also *photosystem I, P700,* the second in the two photosystems in the electron pathway of photosynthesis in cyanophytes and chloroplasts, and the one involving the reduction of NADP to $NADPH_2$; it may be evolutionarily more ancient than photosystem 680.
phototaxis a tendency to move toward light. *negative phototaxis,* a tendency to move away from light.
phototroph a photosynthetic organism; compare *autotroph, heterotroph, chemotroph.*
phototropism 1. the turning toward light of a growing plant stem. 2. phototaxis. 3. *negative phototropism,* the turning away from light, as of a root tip.
phragmoplast in plants, the cytoplasmic organelle initiating the formation of a cell plate following mitosis.
phycobilin a photosynthetic accessary pigment of cyanophytes and of red algal chloroplasts (which are apparently derived from symbiotic cyanophytes).
phycocyanin a blue-green photosynthetic pigment of cyanophytes (blue-green algae) and red algal chloroplasts.
phycoerythrin a red photosynthetic pigment of cyanophytes and red algal chloroplasts.
phylogenetic tree a graphical representation of the interrelations and evolutionary history of a group of organism, indicating the relative order of successive divisions of the line of descent, coincident with past speciation events.
phylogeny 1. the evolutionary history of an organism or group of organism. 2. a phylogenetic tree.
phylum 1. a major taxonomic unit of related, similar classes of animals, e.g. Phylum Annelida. 2. a division of the plant kingdom.
physiological characteristic to an organism's healthy functioning, as contrasted with *pathological;* e.g., *physiological pH,* the range of pH found in the tissues of a healthy organism.
physiology 1. a branch of biology dealing with the processes, activities, and phenomena of individual living organisms, organs, tissues and cells. 2. the normal functioning of an organism.
phytochrome a red-light-sensitive protein complex of certain plant cell membranes, involved in many light-induced phenomena including phototropism, photoperiodicity, and others.
phytoplankton photosynthesizing planktonic organisms.
Pi inorganic phosphate.
pigment any chemical substance that absorbs light, whether or not its normal function involves light absorption: e.g., chlorophyll, cytochrome c, hemoglobin, melanin.
piloerection the lifting up of hair by tiny involuntary muscles; bristling.
Piltdown man an historic fraud that no longer belongs in biology textbook glossaries.
pilus (pl. *pili*), 1. a hollow tube extended from a plasmid-plus bacterium to a plasmid-minus bacterium through which DNA is transferred. 2. any hair-like projection from the surface of a bacterium.
pineal body also *pineal organ, pineal gland,* a small body arising from the roof of the third ventricle in all vertebrates, forming a small red cone in most but forming an eye-like photoreceptive organ in larval lampreys, tuatara and some lizards; directly sensitive to light in reptiles and birds but not in mammals, although it appears to be the center of photoperiod responses in all vertebrates; it secretes the hormone *melatonin.*
pinocytosis taking dissolved molecular food materials, such as proteins, into the cell by adhering them to the cell membrane and invaginating portions of the cell membrane to form digestive vacuoles; a form of active transport.
pioneer organism 1. an organism that successfully establishes residence and produces offspring in an area not previously inhabited by its kind. 2. a type of organism specialized for the initial invasion of a disturbed area, such as a landslide or burned-out region.
pistil in flowering plants, a unit comprised of one or more ovaries, a style and a stigma; see also *gynocecium.*
pistillate having pistils but no stamens.
pith thin-walled parenchymous tissue in the central strand of the primary growth of a stem; the dead remains of such tissue at the center of a woody stem.
pituitary also *hypophysis,* a small double gland lying just below the brain and intimately associated with the hypothalamus in all vertebrates, consisting of an anterior lobe and a functionally distinct posterior lobe; *anterior pituitary,* also *adenohypophysis,* a glandular body communicating with the hypothalamus only by way of a small vascular portal system; secretes growth hormone, prolactin, melanocyte-stimulating hormone, thyrotrophic hormone, ACTH, FSH, and LH; *posterior pituitary,* also *neurohypophysis,* not actually a gland but the enlarged, gland-like termini of axons of cell bodies in the hypothalamus, the hormones being synthesized in the cell bodies and translocated to the posterior pituitary within the axons, to be stored pending release; among these are *oxytocin* and *antidiuretic hormone.*
pituitary dwarf a small individual resulting from the failure of the anterior pituitary to secrete growth hormone.
pituitary giant an abnormally large individual whose excessive growth is due to excessive secretions of growth hormone, usually because of tumerous overgrowth of the anterior pituitary.
pivital joint also *pivot joint,* 1. a joint that permits rotation. 2. a joint that permits only rotation.
placebo a substance having no pharmacological effect but administered as a control in testing experimentally or clinically the efficacy of a biologically active preparation.

placenta 1. in mammals other than monotremes and marsupials, the organ formed by the union of the uterine mucosa with the extraembryonic membranes of the fetus, which provides for the nourishment of the fetus, the elimination of waste products, and the exchange of dissolved gases. 2. in flowering plants, the part of the ovary that bears and nourishes the ovule.

placental mammals also *eutheria,* mammals that form chorioallantoic placentas; apparently a monophyletic group originating about 65 million years ago, and including all living mammals other than marsupials and monotremes.

placoderm a fish of the extinct vertebrate class *Placodermi,* fossil members of which were armored with bony plates, and had primitive jaw, fin and tail structure.

placoid scale see *denticle* (2).

planaria also *planarian,* any free-living flatworm of Class Turbellaria, also *Planaria,* a genus of planaria.

planula the early, ciliated, free-swimming larva of a coelenterate.

plaque 1. a pathological deposit of lipid, fibrous material and often calcium salts in the inner wall of a blood vessel. 2. a film of bacterial polysaccharide, mucus and detritus harboring dental bacteria.

plasma the fluid matrix of blood tissue, 90% water and 10% various other substances; distinguished from *blood serum* by the presence of fibrinogen and the absnece of certain platelet-derived hormones.

plasmalemma see *plasma membrane.*

plasma membrane the external semipermeable limiting layer of the cytoplasm; see also *membrane, fluid-mosaic model.*

plasmid in bacteria, a small ring of DNA that occurs in addition to the main bacterial chromosome and is transferred from host to host by direct contact through pili. *plasmid plus,* a bacterial strain carrying a plasmid and immune to further transfer. *plasmid minus,* a bacterial strain not harboring a plasmid and susceptible to plasmid infection.

plasmodium (pl. *plasmodia*), 1. a motile, multinucleate mass of protoplasm produced by the fusion of uninucleate slime mold ameboid cells. 2. also *Plasmodium,* the malarial parasite.

plastid any of several forms of a self-replicating, semiautonomous plant cell organelle, primarily as a *chloroplast* specialized for photosynthesis, a *chromoplast* specialized for pigmentation, or a *leucoplast* specialized for starch storage; believed to be functional specializations of the genetically identical entity, derived by hereditary symbiosis of a formerly free-living green, photosynthetic, starch-forming procaryote.

plasticity the capacity of an organism for adjustments to changes in the environment; *genetic plasticity,* the capacity of a population to respond rapidly to changes in the environment through short-term natural selection.

plastiquinone a lipid-soluble hydrogen carrier that moves within the membrane in an electron transport chain.

plate (v.t.), to spread a thin suspension of living microorganisms over the surface of an agar gel, so as to be able to isolate and count colonies that result from subsequent growth.

plate tectonics the movement of great land and ocean floor masses (plates) on the surface of the earth relative to one another, occurring largely in the Cenozoic era; see also *continental drift.*

platelets minute, cell-like but enucleate, fragile, membrane-bounded cytoplasmic disks present in the vertebrate blood as the result of the programmed fragmentation of thrombocytes; rupture of platelets releases factors that initiate blood clotting, produce or enhance pain, and induce the proliferation of fibroblasts.

Platyrrhines (obsolete), the New-World Monkeys.

pleura (pl. *pleurae*), in mammals, the tough, clear, serous connective tissue membrane covering a lung and lining the cavity in which the lung lies.

pleural cavity in mammals, either of two divisions of the coelom constituting the thoracic cavities, harboring the lungs and lined with the pleurae.

plumule 1. the apex of the plant embryo, consisting of immature leaves and an epicotyl that forms the primary stem. 2. a down feather. 3. a feathery extension of certain air-borne fruit, such as that of the dandelion.

poikilothermic also *cold-blooded,* not able to maintain a constant body temperature by physiological means; generally having a body temperature only slightly above that of the environment.

polar of a chemical or a part of a molecule, capable of forming hydrogen bonds with water and with other polar molecules, charged portion.

polar body either of two small cells produced by meiosis of an egg cell: *first polar body,* the recipient of the chromosomes going to one pole in anaphase I; *second polar body,* the recipient of the chromosomes going to one pole in anaphase II of the egg. Neither polar body divides again.

pollen grain one of the microscopic particles making up pollen, each being the microgametophyte of a seed plant containing a generative nucleus and a tube nucleus.

pollen sac one of two or four chambers in an anther in which pollen develops and is held.

pollen tube a tube that extends from a germinating pollen grain and grows down through style to the embryo sac, into which it releases sperm nuclei.

pollination the transfer of pollen from a stamen to a stigma, preceding fertilization of a flowering plant.

poly-A tail *polyadenylic acid tail,* a string of about 200 adenosine nucleotides added enzymatically to the 3′ end of HnRNA transcripts in the nucleus during the maturation of mRNA; thought to protect mRNA from degradation by cytoplasmic exonucleases.

polyadenylization the enzymatic addition of a poly-A tail to an RNA transcript.

polymer a molecule made up of a string of more or less identical subunits.

polymerase an enzyme that causes polymerization; e.g., *RNA polymerase, DNA polymerase.*

polydactyly the genetically derived state of having extra fingers or toes.

polygenic inheritance inheritance involving many interacting variable genes, each having a small effect on a specific trait; also, *quantitative inheritance.*

polygenic trait also *quantitative trait, metric trait,* a trait in which variation in a population is expressed in continuous increments about a mean, and is attributable to the action and interaction of many variable gene loci and usually also to multiple variable effects of the environment; see also *distribution, continuous.*

polymorphism 1. the existence within a population of two or more discrete, genetically determined forms other than variations in sex or maturity and apart from rare mutant forms; e.g., black and spotted leopards; see also *distribution, discrete.* 2. the existence within a population of two or more alleles at a locus, at allele frequencies greater than some arbitrary value such as 1% or 5%; e.g., **ABO** and **MN** blood group polymorphisms.

polymorphonuclear leukocyte a large white blood cell with permanently condensed, highly lobed nucleus.

polyp the typical sessile form of a coelenterate; compare *medusa.*

polypeptide a continuous string of amino acids in peptide linkage, longer than a peptide; compare *protein.*

polyphyletic of a taxonomic group, having two or more ancestral lines of origin; having defining characteristics in common because of convergent evolution rather than because of common descent; having a last common ancestor not belonging to the group or sharing its defining characters; e.g., *Pachyderma,* a discarded group originally containing elephants, rhinoceroses, hippopotamuses, pigs and horses; also the *animal kingdom* inasmuch as metazoa and parazoa are derived from different protist ancestors.

polyploid having more than two complete sets of chromosomes; *autopolyploid,* having three or more complete sets of a specific complement of chromosomes; *allopolyploid,* having two or more full diploid sets of chromosomes derived from different ancestral diploid species.

polysaccharide a polymer of sugar subunits.

polysome also *polyribosome,* a small group of ribosomes held together by their common attachment to a messenger RNA molecule.

polytene chromosome in certain tissues of certain insects, enlarged chromosomes created by chromatin replication without chromosome division, see also *salivary chromosome.*

polyunsaturated of a fatty acid, having more than one carbon-carbon double bond.

pons a broad mass of nerve fibers running across the ventral surface of the mammalian brain.

population 1. (demography) the total number of persons inhabiting a given geographical or political area. 2. (genetics) an aggregate of individuals of one species, interbreeding or closely related through interbreeding and recent common descent, and evolving as a unit. 3. (ecology), the assemblage of plants and/or animals living in a given area; or all of the individuals of one species in a given area. 4. (statistics) a finite or infinite aggregation of individuals under study.

population bottleneck see *bottleneck.*

population genetics the scientific study of genetic variation within populations, of the genetic correlation between related individuals in a population, and of the genetic basis of evolutionary change.

population of variants a slogan definition of species, emphasizing a departure from typological thinking and recognizing the problem of individual variation.

population pyramid a graph of a population in which the relative number of each age group of each sex is represented by a horizontal bar; also, *age profile.*

population recognition the ability to recognize members of one's own particular population, or subgroup of a species; see also *recognition.*

porphyrin group also *porphrin ring,* a complex organic compound consisting of a flat circle of nitrogenous rings, usually with iron or another metal complexed in the center; a constituent of hemoglobin, myoglobin, chlorophyll, and the cytochromes.

positive control of a gene or operon, control such that a gene remains inactive until stimulated by an appropriate controlling protein, hormone or other signal.

positive feedback see *feedback.*

positive geotropism see *geotropism.*

positive phototropism see *phototropism.*

posterior 1. in most bilateral organisms, away from the head end of an organism, the opposite of *anterior;* 2. in human anatomy, in view of the upright position of the organism at least in the daytime, *anterior* and *posterior* are often used in place of *ventral* and *dorsal* to mean toward the front and toward the back respectively, while *superior* and *inferior* are used in place of *anterior* and *posterior* meaning toward the head end and toward the tail end respectively.

posterior pituitary the posterior lobe of the pituitary, a stalked extension of neural tissue extending from the hypothalamus.

posterior vena cava in humans *inferior vena cava,* either of two large viens draining the body posterior to the heart; see *vena cava.*

postsynaptic membrane the receptive surface of a dendrite or cell body adjacent to a synapse; see also *synapse.*

post–transcriptional modification the entire series of enzymatic changes made in RNA in the nucleus after transcription, including *capping, polyadenylation,* and *tailoring.*

potential energy energy stored in chemical bonds, in nonrandom organization, in elastic bodies, in elevated weight or any other static form in which it can theoretically be transformed into another form or into work.

potentially interbreeding of two populations, capable of interbreeding except for the presence of geographical barriers.

powdery mildew 1. a fungal plant disease characterized by the presence of powdery conidia. 2. the fungus that causes the disease.

prairie the originally extensive tract of level or rolling treeless grassland of the Mississippi–Missouri valley, characterized by deep fertile soil and tall, coarse perennial grasses.
Pre-Cambrian era see the geological timetable inside the front cover.
precocial capable of walking and a high degree of independent activity from hatching or birth; compare *altricial.*
precursor a chemical substance from which another substance is formed, especially in a specific step of a biochemical pathway.
predation 1. the act of catching and eating. 2. being caught and eaten; e.g., *subject to predation.* 3. a mode of life in which food is primarily obtained by killing and eating other animals.
predator an animal that habitually preys on other animals, a carnivorous animal.
preformation theory a now discredited theory that every germ cell contains the organism fully formed and that development consists of increases in size of the parts to form the adult whole.
prehensile adapted for seizing, grasping, or wrapping around, as the tail of a New World monkey or the upper lip of a rhinoceros.
premeiotic condensation condensation followed by decondensation of the chromosomes of a meiocyte prior to the onset of meiotic prophase; function unknown.
premeiotic interphase the interphase following the last mitotic division of a meiocyte, ending with the onset of meiotic prophase.
pressure receptor a sensory receptor sensitive to the pressure of touch.
presynaptic membrane the membrane of an axon or of the synaptic knob of an axon in the region of a synaptic cleft, into which it secretes neurotransmitters in the course of the transmission of a nerve impulse; see *synapse, synaptic cleft.*
prey 1. (n.), an animal hunted or seized for food by a predator. 2. (v.t.), (followed by *upon*), to hunt or seize for food.
prey switching behavior of a predator in response to the relative densities of two prey species, in which the predator ceases hunting one prey and begins to hunt the other.
primary consumer a herbivore; see *consumer.*
primary growth of a plant, the initial growth or elongation of a stem or root, resulting primarily in an increase in length and the addition of leaves, buds and branches. *primary growth pattern,* the distinctive patten of xylem, phloem and other tissues in primary growth; compare *secondary growth, secondary growth pattern.*
primary mutation an alteration in DNA subject to repair by repair enzymes, as distinguished from mutations that cannot be repaired.
primary oocyte a diploid cell of egg-producing potential prior to its first meiotic division.
primary phloem phloem developed from apical meristem, that is, the phloem of primary growth.
primary producer in ecology, plants, algae or other photosynthetic (or chemosynthetic) organisms that produce the food of a food chain.
primary productivity see *gross productivity.*
primary spermatocyte a diploid meiotic cell of sperm-producing potential prior to its first meiotic division.
primary structure the linear sequence of amino acids in a polypeptide, or of nucleotides in a nucleic acid.
primary succession the succession of vegetational states that occurs as an area changes from bare earth to a climax community.
primary xylem xylem in primary growth, which is produced by apical meristem rather than vascular cambium.
primitive a character or character state, characteristic of the original condition of the group under consideration; ancestral; not derived; of or like the earliest state within the group considered; old; compare *advanced, derived.*
primitive gut see *archenteron.*
primordium the rudiment or commencement of a part or organ, often appearing as a mass or lump of undifferentiated cells; e.g., *leaf primordium, branch primordium, liver primordium.*
probability 1. the relative likelihood of the occurrence of a specific outcome or event based on the proportion of its occurrence among the number of trials or opportunities for its occurrence, viewed as indefintely extended. 2. the field of mathematics dealing with probability relationships.
probability cloud a graphic representation of a three-dimensional electron orbital or group of orbitals, in which the relative probability that an electron will occupy a specific location at a given instant is represented by shading or stippling.
probability curve a curve indicating the relative distribution of possible outcomes, called the *probability density function;* see also *distribution, normal curve.*
procambium the part of a meristem that gives rise to cambium.
procaryote also *prokaryote,* any organism of the kingdom Monera, having no nucleus and a single circular chromosome of nearly naked DNA; a *bacterium, cyanophyte, archebacterium* or *chloroxybacterium.*
process in anatomy, a small projection of the main mass of a structure or organism; *bone process,* a small projection of a bone, which may serve as a muscle origin or insertion.
producer see *primary producer.*
product something produced; *enzyme product,* one of the resulting compounds of an enzymatic reaction; *gene product,* either RNA directly transcribed from a gene or, more commonly, a polypeptide translated from the mRNA of an active gene.
productivity see *gross productivity, net productivity.*
progeny testing determination of an organism's genotype by crossing it with a known recessive homozygous individual and observing the resulting offspring.
progesterone a steroid hormone produced by the placenta and corpus luteum, that functions in maintaining the uterine mucosa and the pregnant state; a frequent constituent of birth control pills.
proglottid any segment of a mature tapeworm formed by strobilation, containing male and female organs and being shed when full of mature fertilized eggs.
prolactin a hormone of the anterior pituitary that in mammals induces lactation and helps maintain the corpus leuteum; in pigeons and related birds, stimulates the crop gland to produce a milk-like substance fed to nestlings.
proliferation rapid and repeated production of new cells by a succession of cell divisions.
pronucleus the nucleus of either gamete after the entry of a sperm into an egg and before nuclear fusion; *male pronucleus,* the sperm nucleus in the egg cytoplasm; *female pronucleus,* the egg nucleus after sperm entry.
prophylactic 1. (adj.), preventing or contributing to the prevention of disease. 2. (n.) any drug or other substance used as a disease preventitive measure. 3. a condom.
proplastid a minute cytoplasmic body from which a mature plastid is formed; an immature or unspecialized plastid.
proprioception the sum of neural mechanisms involved in sensing, integrating and responding to stimuli from within the body, including the integration of the movement of body parts, balance, and stance.
proprioceptive sensor see *proprioceptor.*
proprioceptor a sensory receptor that is located in internal tissues (as in skeletal, smooth and heart muscle, tendons, carotid body) and responds to conditions of body positions, muscle tension, or internal chemistry.
prosimian a primate of the suborder *Prosimii,* including tarsiers, lemurs, lorises, and pottos.
prostaglandin any of a group of hormone-like substances derived from long-chain fatty acids and produced by most animal tissues by a variety of stimuli including endocrine, neural, and mechanical stimuli as well as inflammation and oxygen deprivation; present in tissues, sometimes transported by the blood stream but more often found in other body fluids, e.g., prostaglandins in semen, which stimulate uterine contractions; appear to act by adenyl cyclase activation.
prostate gland a pale, firm, partly muscular and partly glandular organ that surrounds and connects with the base of the urethra in male mammals; its viscid, opalescent secretions is a major component of semen.
prosthetic group a part of a functional protein not part of a polypeptide, such as a heme group, coenzyme, carbohydrate or lipid.
protein a naturally occurring functional macromolecule consisting of one or more polypeptides held together by hydrogen bonds and van der Waals forces and often sulfhydryl linkages, as well frequently including one or more prosthetic groups.
prothallium also *prothallus,* the fern gametocyte, typically a small, flat, green thallus with rhizoids, bearing numerous antheridia and/or archegonia.
prothoracic gland an endocrine gland located in the prothorax of an insect, the source of the molting hormone *ecdysone.*
protist any organism of the Kingdom Protista, an extremely varied paraphyletic group consisting of most of the unicellular eucaryotes (see Appendix A).
protoeucaryote a hypothetical predatory microorganism, capable of the engulfment of prey, that by the successive engulfment of a protomitochondrion, a protochloroplast and perhaps a protocilium evolved into the ancestral eucaryotic cell, perhaps a billion years ago.
proton 1. one of the two hadrons composing the atomic nucleus in ordinary matter, having an electrostatic charge of +1, and a mass 1837 times that of an electron, and by its numbers determining the chemical properties of the atom. 2. a hydrogen ion.
proton pump an active transport system utilizing energy to move hydrogen ions from one side of a membrane to the other against a concentration gradient, as in chemiosmosis.
protonema (pl. *protonemata*), the thread-like and usually transitory stage of a moss or liverwort gametophyte as it emerges from a germinating spore; indistinguishable from a filamentous green alga in structure and appearance.
protoplasm the inner living substance of cells, including cytoplasm and nucleoplasm.
protoplast 1. a plant or bacterial cell artificially deprived of its cell wall. 2. a plant or bacterial cell, considered apart from its cell wall.
protostome an animal in which the mouth derives (developmentally or phylogenetically) from the blastopore; compare *deuterostome.*
protozoan 1. any of a large group of protists. 2. any nonphotosynthetic protist.
proximal nearer the point of attachment or origin; nearer the center of the body; compare *distal.*
proximal convoluted tubule see *nephron;* see Figure 28.17 and 28.18.
pruning cutting off of the growing tips of branches in order to produce a better shaped or more fruitful growth.
pseudocoelom also *pseudocoel,* "false coelom"; the body cavity in nematodes, rotifers and certain other invertebrates, between the body wall and the intestine, which is not entirely lined with mesodermal epithelium; in which the gut is entirely endodermal and not muscularized.
pseudoplacenta 1. also *yolk-sac placenta,* the placenta-like structure of marsupials. 2. any placenta-like structure, as in sea snakes, some cartilaginous fishes and a few insects.
pseudopod also *pseudopodium,* any temporary protrusion of the protoplasm of a cell serving as an organ of locomotion or engulfment; often having a fairly definite filamentous form, and sometimes fusing with others to form a network.
P700 see *photosystem 700.*
psilophytes 1. the earliest known vascular plants; simple dichotomously branching plants of the Silurian and Devonian. 2. see *psilopsid.*
psilopsid a simple plant of the tracheophyte subdivision *Psilopsida,* vascular plants lacking roots, cambium, leaves and leaf traces, and have a reduced, subterranean, nonphotosynthetic gametophyte.
P680 see *photosystem 680*

psychological tension an emotional stress that prompts activity to release the tension, or that manifests itself in other ways.
pterosaur also *pterodactyl,* any of numerous extinct flying reptiles of the lower Jurassic to the close of the Mesozoic.
ptyalin also *salivary amylase,* a starch-digesting enzyme found in the saliva of humans and of certain other mammals.
pubic bone also *pubis,* one of the constituent bones of the coxa or innominate bone.
pubic symphysis in mammals, the usually semirigid fibrous articulation of the two pubic bones in the midline above the genitalia; subject to considerable hormone-induced loosening toward the end of pregnancy.
pubis see *pubic bone.*
puff a temporary enlargement of a band of a polytene chromosome indicating gene transcriptional activity.
pulmonary artery a branching artery that conveys venous (i.e., deoxygenated) blood from the right ventricle to the lungs, consisting of a common pulmonary artery, right and left pulmonary arteries, and further subdivisions.
pulmonary circuit the passage of venous blood from the right side of the heart through the pulmonary arteries to the capillaries of the lung where it is oxygenated and from which it returns by way of the pulmonary veins to the left atrium of the heart.
pulmonary valve a three-cusped valve of the heart separating the cavity of the right ventricle from the common pulmonary artery.
pulmonary vein in birds and mammals the only vein that carries arterial (oxygenated) blood, returning it from the lungs to the left atrium.
pulp cavity the central cavity of a tooth, containing the nerves and blood vessels comprising *dental pulp.*
pupa a metamorphic insect in a specific quiescent intermolt between the larval and adult stages, during which time extensive body transformations occur, the pupa generally being enclosed in a hardened pupal case or cocoon.
pupil the contractile aperture in the iris of the eye.
purine a heterocyclic nitrogenous base consisting of conjoined 5-membered and 6-membered rings; e.g., adenine, guanine, uric acid.
purple nonsulfur bacterium a photosynthetic bacterium utilizing bacteriochlorophyll as a photosynthetic pigment and utilizing organic compounds from its environment as hydrogen sources.
purple sulfur bacterium a photosynthetic bacterium utilizing bacteriochlorophyll as a photosynthetic pigment and hydrogen sulfide as a hydrogen source.
pus a thick, yellowish-white accumulation of dead phagocytes, dead or living bacteria, and tissue debris in a fluid exudate.
pyloric sphincter also *pyloric valve,* a ring of muscle capable of closing off the opening between the stomach and the small intestine.
pylorus the opening between the stomach and the small intestine in a vertebrate.
pyramid 1. see *population pyramid.* 2. also *pyramid of numbers:* see *Eltonian pyramid.* 3. any of various anatomical features shaped like the Egyptian pyramids, including certain structures of the brain and kidney.
pyrimidine a nitrogenous base consisting of a six-membered heterocyclic ring; notably cytosine, uracil and thymine, constituents of nucleic acids.
pyrogen any fever-producing substance or hormone.
pyrophosphate an inorganic ion consisting of two phosphate groups joined by a phosphate-phosphate high energy bond: $H_2P_2O_7^{2-}$.
pyrrophyte *Pyrrophyta,* a division of the protist Kingdom comprised of the dinoflagellates; see also *dinoflagellate, dinocaryote.*
pyruvate the anion of pyruvic acid, the final organic product of glycolysis in animal cells and the initial substrate of the citric acid cycle.

quadrate also *quadrate bone,* the bony element of the skull articulating with the lower jaw in most vertebrates other than mammals.
quantasome one of the chlorophyll-containing spheroids that make up the central region of a thylakoid, possibly the functional unit of structure in photosynthesis.
quantum (pl. *quanta*), one of the very small increments or parcels into which many forms of energy are subdivided; directly proportional to the wave frequency of a photon, electron, or other particle.
quantum mechanics also *quantum theory,* an extensive branch of physics based on Planck's concept of radiant and other energy being subdivided into discrete quanta.
quark in current particle theory, an ultimate constituent of matter, each hadron being composed of three quarks.
quaternary of the fourth order, e.g., *quaternary consumer. quaternary structure,* protein structure on the level of the concatenation of polypeptide subunits.

race 1. a breeding stock or variety. 2. any of a variety of taxonomic groupings below the level of species, especially a geographic variety with a distinctive phenotype blending without isolation into other geographic varieties. 3. one of the three, five or forty genetically distinctive human groups according to various authorities.
racism the belief that psychological, cultural and mental traits and capacities are determined by biological race, that races differ decisively from one another without substantial overlap, that one race is inherently superior to the others, and that this superiority implies a right of domination.
radial having parts arranged like rays eminating from a center.
radial canal 1. in a sponge, one of the numerous choanocyte-lined canals radiating from the central cavity. 2. in a jellyfish, one of the gastro vascular canals extending from the central cavity to the marginal circular canal. 3. in an echinoderm water vascular system, a tube extending outward from the ring canal.
radial symmetry the condition of having similar parts regularly arranged as radii from a central axis, as in a starfish.
radiation 1. the transfer of heat or other energy as particles or waves; compare *convection, conduction.* 2. heat or other energy transmitted as particles or waves; see also *ionizing radiation.*
radiation, electromagnetic see *electromagnetic radiation.*
radiation, ionizing see *ionizing radiation.*
radical 1. also *moiety, residue,* a part or functional group of a molecule. 2. see *free radical.*
radicle the lower portion of the axis of a plant embryo, especially the part that will become the root.
radioactive isotope see *isotope.*
radioactive tracer a radioactive element or a compound containing a radioactive element, that is put into a biological system and transformed or translocated, its eventual fate being determined by physiochemical means.
radiolarian any marine ameboid protozoan of the order *Radiolaria* (see Appendix A).
radiolarian ooze a siliceous mud of deep sea bottoms, composed largely of the skeletal remains of radiolarians.
radioisotope see *isotope.*
radius one of the two bones of the forearm of land vertebrates, rotating about the *ulna* and articulating with the wrist.
radula in all mollusks other than bivalves, a toothed, chitinous band that slides backward and forward over a protrusible prominance from the floor of the mouth, scraping and tearing food and bringing it into the mouth.
rain shadow a region of low rainfall on the prevailing lee side of a mountain or mountain range.
Ramapithecus a genus of upper Pliocene apes, represented by a few fossil fragments, thought by some to be in the line of descent leading to humans and not to the living great apes.
random drift see *drift.*
rare dominant 1. an allele of low frequency leading to a definable abnormal phenotype in the heterozygote, and unknown or lethal in the homozygote. 2. a lethal-recessive dominant.
reaction center also *photocenter,* the part of a photosystem in which light-activated chlorophyll *a* transfers an electron to the electron transport system.
reaction see *chemical reaction.*
reafference the presumed reward associated with the performance of a particular movement when an animal is performing an adaptive or instinctive pattern; behavior as its own reward.
reagent 1. a chemical used in preparative, experimental, or diagnostic work because of the chemical reactions it undergoes. 2. any laboratory chemical.
recall 1. to bring something remembered into the consciousness. 2. the ability to reconstruct the details of something remembered.
receptacle the end of a floral stalk, which forms the base on which the flower parts are borne.
receptor 1. a specialized neuron involved in the detection of environmental stimuli. 2. any sense organ or group of sensitive cells that transmit impulses.
recessive (adj.), of an allele, not expressed in a heterozygote; see *dominance interaction.*
recessive-lethal dominant an allele that has a pathological or visible effect as a heterozygote and is lethal as a homozygote.
recessivity the failure of one allele of a heterozygote to have any discernable effect on a specific phenotypic trait; see *dominance interaction.*
reciprocal altruism behavior that appears to be altruistic in individual instances but in fact functions to increase the fitness of the individual insofar as it increases the likelihood that the individual will be the recipient of beneficial behavior at another time.
recognition 1. the ability to respond appropriately to the sight, sound or smell of another organism or a situation, whether on the basis of genetic propensities or of individual experience. 2. an act of appropriate response.
recombinant chromosome a chromosome emerging from meiosis with a combination of alleles not present on the chromosomes entering meiosis.
recombinant DNA a recent general term for the laboratory manipulation of DNA in which DNA molecules or fragments from various sources are severed and combined enzymatically and reinserted into living organisms; see also *gene cloning.*
recombination frequency the proportion of gametes bearing recombinant chromosomes for two specific gene markers that were heterozygous in the parent.
recombination map a representation of a linear series of gene loci of a chromosome or linkage group, the *map distance* between loci being derived from recombination frequencies and proportional to the expected number of crossovers per gamete occurring in the corresponding region.
rectal valve any of several folds of the rectal wall.
rectum the terminal part of the intestine, used for the temporary storage of feces.
red alga any alga of the Plant Kingdom division *Rhodophyta* (see Appendix A).
red blood cell see *erythrocyte.*
red tide seawater discolored by a dinoflagellate bloom in a density fatal to many forms of life.
reducer in ecology, a fungus or microorganism that breaks down plant or animal matter into small molecules.

reducing power the relative amount of energy released by a substance for each electron or hydrogen transferred from it in an oxidation-reduction reaction; the relative ability of one substance to transfer electrons or hydrogen atoms to another.
reduction of a substance, the addition of electrons or hydrogen atoms.
reduction division the first division of meiosis.
reductionist in science, one who attempts to divide phenomena and mechanisms to their most elemental parts, isolating as far as possible the effects of individual factors and testing these effects in separate controlled experiments.
redundancy having information present in more than one copy.
refraction the deflection of a ray of light as it passes obliquely from one medium into another of a different density.
refractive index also *index of refraction,* the ratio of the speed of light passing through a substance to the speed of light in a vacuum.
renal circuit 1. in mammals, the circuit of the branching *renal arteries* arising from the abdominal aorta and continuing through the glomerulus, the capillary beds of the nephron, and the renal veins to return to the vena cava. 2. in most other vertebrates, a renal portal circuit that originates in the tissues of the tail or hind limbs, coalesceses into a *renal portal vein* to the gloreruli, capillary beds and renal veins.
renal pelvis the cavity of the kidney into which the collecting ducts empty.
rennin an enzyme produced by the stomach linings of young mammals, the action of which is to coagulate milk.
repair enzyme any of several different complexes of enzymes that recognize improper base pairing in DNA, excise a region of one of the strands, and rebuild the DNA according to the rules of Watson-Crick pairing; or that otherwise repair mutational damage, including double-strand chromosome breaks.
replacement rate the birth rate (in terms of the lifetime production of offspring per woman) which, at age structure equilibrium, results in a stable population size; e.g., about 2.1 infants per adult woman.
replica plating a screening technique in which an array of colonies on a nutrient-supplemented agar plate are transferred as a group to one or more additional plates of nutrient-deficient agar to test for nutritional mutants.
replication fork the point at which unwinding proteins separate the two DNA strands in the course of DNA replication.
repressible operon an operon governing a synthetic pathway, and which is generally active but which can be inactivated by the presence of its normal metabolic product in the medium.
repressor-inducer complex in an inducible operon, the repressor protein bound to and temporarily inactivated by the small molecule that induces operon activity; see also *operon.*
repressor molecule in bacterial operons, a small metabolite that combines with an inactive repressor protein to form an active complex that binds with an operator to prevent transcription of an operon.
repressor protein in bacterial operons, a protein that binds the operator and prevents transcription either when bound to an inducing molecule (inducible operon) or when not bound (repressible operon).
reproductive isolation 1. the state of a population in which there is no mating between members of the group and members of other groups, and no immigration or emigration of individuals. 2. the state of a population or species in which successful matings outside of the group is biologically impossible because of physical mating barriers, hybrid inviability, or hybrid sterility.
reproduction see *sexual reproduction, asexual reproduction.*
resolution 1. the ability to distinguish detail in stimuli, specifically the ability to distinguish two closely spaced stimuli from a single stimulus. 2. in optics, the minimum distance between two points at which they still appear to be separate entities. 3. in sexual activity, a retreat from the excitement states, generally following orgasm.
resonate to vibrate sympathetically with the source of sound or other oscillation.
respiration 1. breathing. 2. the physical and chemical process by which an organism supplies oxygen to its tissues and removes carbon dioxide. 3. also *aerobic respiration,* any energy-yielding oxidative reaction in living matter. 4. the metabolic transformation of food or food storage molecules yielding energy, including *anaerobic respiration* (e.g., glycolysis) as well as *aerobic respiration* (e.g., citric acid cycle and electron transport).
respiratory center a region in the medulla oblongata that regulates breathing movements.
resting potential the charge difference across the membrane of a neuron or muscle fiber while not transmitting an impulse.
restriction enzyme in bacteria, an enzyme that recognizes and serves a specific, short DNA sequence, thus protecting the cell from all but a few highly adapted, host-specific viruses; such enzymes have proven useful for experimental DNA manipulation.
retaliator an animal that displays and threatens but does not attack others of its kind unless attacked first, but always fights back if attacked.
rete mirabile a small but very dense network of blood vessels in a counter-current heat exchange.
reticuloendothelial system a diffuse system of cells, including reticulum cells, the endothelial cells of capillaries, spleen, ducts and cavities, and all noncirculating phagocytes, functioning to cleanse the body and blood of debris.
reticulum cell also *reticular cell,* one of the branched, net-like cells of the reticuloendothelial system, forming intricate interstital networks ramifying through other tissues.
reverse transcriptase also *RNA-dependent DNA polymerase,* an enzyme of certain RNA viruses that copies RNA sequences into single-stranded and double-stranded DNA sequences in a minor reversal of the central dogma.
R-group 1. a chemistry shorthand where R stands for the bulk of a molecule. 2. a similar usage where R stands for the variable part of an amino acid.
Rh blood group system also *rhesus factor, Rh factor,* a polymorphic blood cell antigen system consisting of three variable antigenic sites controlled by eight alleles at a single locus.
Rh negative (adj.), homozygous for the absence of the most antigenic of the three Rh antigen sites; not responding to anti-Rh antibodies.
Rh positive homozygous or heterozygous for the most antigenic of the three Rh antigen sites; responding with blood cell agglutination to anti-Rh antibodies.
rhizoid 1. a rootlike structure that serves to anchor the gametophyte of a fern or byophyte to the soil and to absorb water and mineral nutrients. 2. a portion of a fungal mycelium that penetrates its food medium.
rhizome an underground horizontal shoot specialized for food storage.
ribose a five-carbon aldose sugar, a constituent of many nucleosides and nucleotides.
ribosomal RNA also *rRNA,* the RNA that forms the matrix of ribosome structure, consisting of a large, intermediate and two relatively small sequences.
ribosome an ultramicroscopic cytoplasmic organelle or very large molecule consisting of a matrix of RNA and numerous specific proteins, binding a messenger RNA molecule, a charged tRNA and a growing polypeptide chain during protein synthesis.
rib one of the numerous curved bony or cartilaginous rods that stiffen the lateral walls of most vertebrates and articulate with the vertebral column.
ring canal 1. also *circumoral canal,* a circular tube of the water vascular system that surrounds the esophagus in echinoderms. 2. a circular, tubular extension of the gastrovascular cavity in the edge of the umbrella of a jellyfish; see also *radial canal.*
ritual combat intraspecific combat following a rigidly defined format that precludes serious or permanent damage to either participant.
RNA also *ribonucleic acid,* a single-stranded nucleic acid macromolecule consisting of adenine, guanine, cytosine and uracil on a backbone of repeating ribose and phosphate units; divided functionally into rRNA (ribosomal RNA), mRNA (messenger RNA), tRNA (transfer RNA) HnRNA (heterogeneous nuclear RNA) and viral RNA.
RNA-dependent DNA polymerase also *reverse transcriptase,* a viral enzyme that catalyzes the copying of an RNA sequence into single-stranded or double-stranded DNA.
RNA polymerase the enzyme or enzyme complex catalyzing transcription.
RNA synthetase see RNA polymerase.
rRNA see *ribosomal RNA.*
rod one of the numerous long, rod-shaped sensory bodies in the vertebrate retina, containing many membrane layers bearing visual pigments, responsive to faint light but not to detail or to variations in color; compare *cone.*
root the portion of a seed plant, originating from the radicle, that functions as an organ of absorption, anchorage and sometimes food storage, and differs from the stem in lacking nodes, buds, and leaves.
root apical meristem see *apical meristem.*
root cap a protective mass of parenchmal cells that covers the root apical meristem.
root hair one of the many tiny tubular outgrowths of root epidermal cells, especially just behind the root apex, that function in absorption.
root meristem see *apical meristem.*
root nodule one of the multiple swellings on the roots of a leguminous plant, developed in response to a symbiotic nitrogen fixing bacterium and harboring and nourishing colonies of such bacteria.
root pressure the active force by which water rises into the stems from the roots, the mechanism is unknown.
rotifer a microscopic pseudocoelomate animal of the Aschelminthe class *Rotifera* (see Appendix A).
rough endoplasmic reticulum see under *endoplasmic reticulum.*
rumen the large first compartment of the stomach of cattle and other ruminants, used as a temporary repository for food which is partly broken down by bacterial and protozoan symbionts and later regurgitated for rumination.
ruminant a cud-chewing artiodactyl, including cattle, sheep, deer, giraffes, and camels.
ruminate 1. to chew a cud; to chew again food that has been chewed slightly, swallowed, and regurgitated. 2. to think it over.

S phase (synthesis phase), one of the phases of the cell cycle, occurring in interphase between G_1 and G_2; the period of chromosome replication synthesis.
saccule 1. a small sac. 2. the smaller of two chambers of the membranous labyrinth of the ear; compare *utricle.*
sac fungus a fungus of the phylum *Ascomycetes,* with sexual spores formed in sac-like *asci.*
sacroiliac 1. the juncture of the sacrum of the vertebral column and the iliac bones of the pelvis. 2. the fibrous cartilage between the sacrum and the iliac.
sacrum the part of the vertebral column that connects with the pelvis; in people it consists of five fused vertebrae with transverse processes fused into a solid bony mass on either side.
saggital (adj.), 1. the median plane of the body of a bilateral animal. 2. any plane parallel to the median plane.
saline 1. also *saline solution, physiological saline,* a solution of sodium, potassium and sometimes magnesium salts isotonic with physiological fluids. 2. (adj.), salty.
salinity the concentration of salt in a solution; saltiness.

saliva 1. in land vertebrates, a viscous, colorless, mucoid fluid secreted into the mouth by ducted salivary glands. 2. in invertebrates, any analogous secretion into or from the mouth.
salivary chromosome also *polytene chromosome,* any of the gigantic interphase chromosomes of the salivary glands of larval dipterans (e.g., *Drosophila*), each consisting of 1024 chromonemata aligned side by side, with thousands of prominently stainable bands corresponding with condensed and folded genes and occasional puffs corresponding to unfolded, actively transcribing genes.
salivary gland 1. in land vertebrates, any of several glands secreting saliva into the mouth; in mammals these are the paired *parotid, sublingual* and *submaxillary glands;* in snakes the salivary glands are modified into venom glands. 2. in invertebrates, any analogous gland ducting secretions into or from the mouth.
salt linkage an electrostatic bond formed between a negatively charged ion or side group and a positively charged ion or side group.
saltatory proceeding by leaps and bounds rather than by gradual transitions. *saltatory evolution,* also *macroevolution, saltation,* evolution by sudden variation or by periods of active change with intervening periods of morphological stability. *saltatory propagation,* the skipping movement of an impulse from one node of a myelinated neuron to another.
salting out also *saline abortion,* a method of abortion used in the second trimester, in which the fetus is killed with an injection of a concentrated salt solution into the amniotic cavity.
sap the watery fluid transported by phloem that circulates dissolved gases, sugars, other organic compounds, and mineral nutrients from one part of the plant to another.
saprobe an organism that reduces dead plant and animal matter.
saprophyte 1. a plant saprobe. 2. any saprobe; see *saprobe.*
saprophytic living on dead matter.
sarcodine an ameboid protozoan of the phylum *Sarcodina,* including amebas, forams, heliozoans, and radiolarians (see Appendix A).
sarcolemma the membraneous sheath enclosing a muscle fibril.
sarcomere the contractile unit of striated muscle bounded by Z-line partitions; consisting of regular hexagonal arrays of actin filaments bound to the Z-line partitions and parallel myosin filaments regularly interspersed between them.
sarcoplasmic reticulum membranous, hollow tubules in the cytoplasm of a muscle fiber, similar to the *endoplasmic reticulum* of other cells.
saturated having accepted as many hydrogens as possible. *saturated fat,* a triglyceride lacking carbon-carbon double bonds.
savanna a tropical or subtropical grassland with scattered trees or shrubs, usually maintained by such human activities as burning and by foraging for firewood.
scale insect any of various homopteran insects in which adult females become scale-like and permanently attached to a host plant and adult males are mouthless, winged and short-lived; a destructive agricultural pest.
scanning electron microscope (SEM), a device for visualizing microscopic objects by scanning them with a moving beam of electrons, recording impulses from scattered electrons and displaying the image by means of the syncronized scan of an electron beam in a cathode ray (television) tube.
scapula in land vertebrates, either of a pair of large, flat, triangular bones that slide dorsally above the rib cage and bear opposing muscles of the upper arm and the socket articulating with the humerus; the scapula forms the bone of a chuck roast.
schistosome also *blood fluke,* 1. any of several parasitic trematode flatworms, the primary host of which is a human, the secondary host being a fresh-water snail. 2. any related trematode parasite.
schistosomiasis any of several diseases caused by various schistosomes parasitic on humans; *Schistosoma haematobium,* infecting the veins of the urinary bladder; *Schistosoma japonicum,* infecting the portal and mesenteric veins; *Schistosoma mansoni,* also infecting the mesenteric veins.
Schwann cell also *neura lemma cell,* one of the cells that constitute the myelin sheath, each cell being greatly flattened so as to consist almost entirely of cell membrane, and being repeatedly wrapped around an axon so as to form a sheath many layers thick.
scientific method (variously defined), 1. any method used by scientists to investigate phenomena. 2. any of several more or less formal schemes of research methodology, in which a problem is identified, relevant data are gathered, a hypothesis and its null alternative are formulated from these data, and the null hypothesis is empirically tested for possible rejection.
sclera the heavy, white connective tissue enclosing most of the eyeball; the white of the eye.
sclerenchyma a protective or supporting plant tissue composed of cells with greatly thickened, lignified, and often mineralized cell walls.
scolex the hook-bearing head of a larval or adult tapeworm, from which the proglottids are produced by strobilation.
scrotum the external pouch of skin that contains the testes in most adult male mammals.
scutellum in monocots, the cotyledon in its specialized form as a digestive and absorptive organ.
sea anemone see *anemone.*
sea star also *starfish,* any echinoderm of the class *Asteroidea* (see Appendix A).
seaweed a multicellular marine alga, of the red, green or brown algal groups.
second law of thermodynamics the statement in physics that all processes occurring spontaneously within a closed system result in an increase of total entropy.
second messenger an intracellular chemical compound transferring a hormonal message from the cell membrane to the nucleus or cytoplasm.
second consumer a carnivore; see *consumer.*
secondary growth growth in dicot plants that results from the activity of secondary meristem, producing primarily an increase in diameter of stem or root; compare *primary growth.*
secondary growth pattern the arrangement of xylem, cambium, phloem, cork cambium, cork and other tissues in dicot secondary growth, in which the various tissues form concentric sheaths that appear as rings in cross-section; compare *primary growth pattern.*
secondary oocyte a developing egg cell after the first polar body has been produced and before the second polar body appears, after meiosis I.
secondary sex characteristic structural and psychological changes that occur at puberty, or seasonally in seasonal breeders, usually in one sex only, and is not directly part of reproduction; in humans, enlarged breasts and hips in women, beard growth, enlarged larynx and increased muscle development in men, pubic and axillary hair in both sexes; breeding plumage in adult male birds.
secondary spermatocyte a sperm-forming cell after the first meiotic division and before the second division.
secondary structure of a macromolecule, the pattern of folding of adjacent residues, usually in the form of helices or sheets.
secondary succession succession occurring in abandoned croplands.
secondary tissue see *secondary growth.*
seed the fertilized and ripened ovule of a seed plant, comprised of an embryo (miniature plant) including one, two or more cotyledons, and usually a supply of food in a protective seed coat; capable of germinating under proper conditions and developing into a plant.
seed coat see *coat.*
seed plant a plant of the tracheophyte subdivision *Spermophyta,* consisting of gymnosperms and angiosperms.
segmentation 1. the condition of being divided into segments, originally repetitions of nearly identical parts (as still seen in some annelids, chilopods, diplopods, and *Peripatus*), but frequently followed in evolution by the specialization of different segments, as seen in most arthropods, some annelids, and vertebrates. 2. the serial repetition of body parts, such as ganglia or somites.
self-fertilization fertilization of ovules or eggs effected by pollen or sperm from the same individual.
self-pollination the transfer of pollen from the anther to the stigma of the same flower.
SEM see *scanning electron microscope.*
semen 1. in mammals, a viscous white fluid produced in the male reproductive tract and released by ejaculation, consisting of spermatozoa suspended in the secretions of accessory glands, principally the prostate and seminal vesicle. 2. any fluid containing sperm and released by male animals.
semicircular canals in vertebrates, a group of three loop-shaped tubular portions of the membranous labyrinth of the inner ear. serves to sense balance, orientation and movement.
semiconservative replication replication of a DNA molecule in which the original molecule divides into two complementary parts, both halves being preserved while each half promotes the synthesis of a new complement to itself.
seminal vesicle 1. in various invertebrates, a pouch in the male reproductive tract serving for the temporary storage of sperm. 2. in male mammals, paired outpocketings of the vas deferens once believed to be sperm storage organs and now known to be exocrine glands producing much of the fluid substance of semen.
seminiferous tubule any of the coiled, thread-like tubules that make up the bulk of a testis, and are lined with germinal epithelium from which sperm are produced.
semiparasite also *hemiparasite,* a plant that parasitizes another plant for moisture and elevation, but conducts photosynthesis for its own energy needs; e.g., mistletoe.
semipermeable membrane a membrane that allows some substances to pass freely through it but which restricts or prevents the passage of other substances.
senescence 1. aging. 2. changes, especially deterioration, associated with increasing age.
sensory 1. pertaining to the senses. 2. receptive of stimuli. 3. conveying nerve impulses from a sense organ to the central nervous system.
sensory cell a peripheral nerve cell that is the primary receptor of a sensory stimulus.
sensory hair an innervated arthropod hair that serves as a sense organ.
sensory palp also *palpus,* a segmented process on either side of the mouth of an arthropod, occurring in insects in pairs on the maxillae and on the labium, having tactile and gustatory functions.
sepals green, leaf-like parts of flowers that surround the flower bud before it opens and later forms a whorl (the *calyx*) beneath and outside the whorl of petals.
septate divided by septa
septum (pl. *septa*), any dividing wall or membrane, such as those dividing the cavities of a compound fruit; the transverse partitions dividing the chambers of a chambered nautilus, the transverse partitions dividing the coelom of an annelid, the cross-walls of the hyphae of septate fungus, the partition dividing the left and right chambers of the heart, etc.
sequester to remove from circulation; to separate out without destroying.
serial dilution a technique for measuring the density of suspended bodies (e.g., bacteria, viruses) in a fluid by progressively, repeatedly and accurately diluting the fluid until the density is small enough for accurate counts to be made.
serosa 1. see *chorion.* 2. also *serous membrane,* any thin, clear membrane consisting of a single layer of flattened, moist cells covering a connective tissue base, lining a body cavity or covering an internal organ.
serotonin a vasoconstricting hormone released from damaged tissues.
Sertoli cell also *nurse cell,* one of the elongated cells of the tubules of the testis to which spermatids become attached while they mature.

serum (pl. *sera*), 1. the clear fluid portion of vertebrate blood remaining after clotting, devoid of cells, platelets and fibrinogen, and containing various hormones and factors released from ruptured platelets. 2. such a fluid containing or enriched for specific antibodies.
sex chromosome an X or Y chromosome, or any chromosome involved in the determination of sex.
sexduction the transfer of bacterial host genes from an F^+ bacterium to an F^- bacterium following its fortuitous incorporation into the genome of an F^+ episome.
sex-influenced trait a congenital, gene-influenced or chronic abnormality that can occur in either sex but is more common in one than the other; e.g., breast cancer, ulcers, pyloric stenosis.
sex-limited trait a variable trait that affects members of one sex only; e.g., prostate cancer, pattern baldness, endometriosis.
sexual intercourse also *coitus, copulation,* the entry of the erect penis into the vagina.
sexual recombination genetic recombination as the result of meiosis and biparental reproduction.
sexual reproduction reproduction involving the union of gametes; biparental reproduction.
shelf fungus also *bracket fungus,* a basidiomycete that forms a shelf-like basidiocarp on the sides of infected tree trunks.
shell 1. see *electron shell.* 2. a hard, rigid usually calcaneous covering of an animal. 3. the calcaneous covering of a bird egg. 4. by extension, the leathery covering of a reptile or monotreme egg. 5. the hard, rigid outer covering of a fruit or seed, e.g., walnut shell, sunflower seed shell.
shoot apical meristem see under *apical meristem.*
short-day plant a plant that begins flowering after the summer solstice and in which flowering is triggered by periods of dark longer than some innately determined minimum.
short-term memory see under *memory.*
sickle-cell an abnormally crescent-shaped erythrocyte. *sickle-cell anemia,* a severe recessive condition attributable to homozygosity for an allele producing sickle-cell hemoglobin. *sickle-cell hemoglobin,* symbol Hb^S, adult hemoglobin in which the beta chain has valine rather than glutamic acid in the sixth position, resulting in the formation of sickle-shaped crystals. *sickle-cell trait,* the heterozygous condition for the allele specifying Hb^S which entails an increased risk of mild anemia but also an increased resistance to malaria.
sieve plate 1. partition or cell wall between sieve tube members, bearing multiple perforations allowing for the continuity of sieve tube cytoplasm. 2. also *madreporite,* the perforated calcareous disk through which the echinoderm water vascular system communicates with the exterior water.
sieve tube a thin-walled tube consisting of an end-to-end series of enucleate, living cells joined by sieve plates, forming the channels through which sap flows in phloem.
sieve-tube element also *sieve-tube member,* a thin-walled phloem cell having no nucleus at maturity and forming a continuous cytoplasm with other such cells to form *sieve tubes,* functioning in the transport of organic solutes, hormones, and mineral nutrients.
sigmoid S-shaped; *sigmoid colon,* an S-shaped part of the large intestine (Figure 25.15); *sigmoid curve,* an S-shaped graphed function, as of the cumulative normal distribution, population growth in a finite environment, or allele frequency under directional selection.
silage plant biomass (fodder) converted into succulent winter feed for livestock through anaerobic fermentation in an airtight chamber (silo).
silica silicon dioxide.
siliceous containing silica; *siliceous sponge,* a sponge having a skeleton of silica spicules.
silicon one of the most abundant elements in the earth's crust, a major component of glass, found in certain animal spicules, diatom shells, nettle spines, the outer cell walls of grasses and elsewhere.
Singer model see *fluid-mosaic model.*
sinoatrial node also *pacemaker,* a small mass of tissue made up of Purkinje fibers, ganglion cells, and nerve fibers, embedded in the musculature of the right atrium of a land vertebrate, representing the vestige of the sinus venosus of ancestral forms, serving as a source of regular contraction impulses to the heart, and transmitting the impulse to the auricles, the atrioventricular node, the bundle of His, and thus to the ventricles.
sinus a cavity that forms part of an animal body, e.g., any of the several air-filled, mucous-membrane lined cavities of the skull; a hemocoel; any large dilation of a blood vessel.
sinus venosus in fish and in the embryos of land vertebrates, a distinct chamber of the heart formed by the union of the large systemic veins, opening into the single auricle and being the site of origin of the heart beat impulse; developing in ontogeny and phylogeny into the *sinoatrial node.*
sinusoid a minute, endothelium-lined space or passage for blood in the tissues of an organ, such as the liver or spleen.
siphonous alga coenocytic green algae, an evolutionary line of green algae characterized by a multinucleote cytoplasm, without the usual division by cell walls or membranes, e.g., *Codium.*
site a specific place; a part of a large molecule; e.g., the *active site* of an enzyme.
skeletal muscle also *voluntary muscle, striated muscle,* muscle attached to the skeleton or in some cases to the dermis, under direct control of the central nervous system, characterized by distinct striations at the subcellular level, with multinucleate unbranched muscle fibers; compare *smooth muscle, cardiac muscle.*
skin-breather any organism in which a significant proportion of the exchange of respiratory gasses occurs through a vascularized moist skin; e.g., most amphibia.
Skinner box a device for investigating operant conditioning, being a small cage provided with one or more levers and an automatic feeder that provides a food pellet when a lever is depressed under appropriate conditions designated by the experimenter; after B. F. Skinner, b. 1904.
skull the skeleton of the head of a vertebrate, being a cartilaginous braincase in cyclostomes, sharks and the embryos of all forms; in most vertebrates the cartilage is replaced by bone and the structure made more complete by its union with other bones developed in dermal membranes; consisting of the cranium, the bony capsules of the nose, ear, and eye, and the jaws and teeth.
slime mold an organism of the fungal subkingdom *Gymnomycota,* possibly not related to other fungi, consisting of an amoeboid, saprophytic *plasmodium* that on maturity coalesces into structured fruiting bodies; two subgroups, the *acellular slime molds* and the *cellular slime molds,* though sharing many features may also be only distantly related to one another.
small intestine in vertebrates, the region of the alimentary canal and the cecum; the region in which most food absorption occurs.
smooth endoplasmic reticulum see *endoplasmic reticulum.*
smooth muscle also *involuntary muscle,* the muscle tissue of the glands, viscera, iris, piloerection and other involuntary functions, consisting of masses of uninucleate, unstriated, spindle-shaped cells occurring usually in thin sheets.
smut 1. a fungal disease of plants characterized by black, powdery masses of spores. 2. the fungus that causes the disease.
sodium bicarbonate the salt $Na^+ + HCO_3^-$
sodium pump a membrane active transport mechanism that utilizes energy to move sodium ions from one side of the membrane to the other.
soft palate in the oval cavity, the membranous, non-bony posterior extension of the hard palate.
solute something dissolved in solution.
somatic cells cells of the body; any cells other than germinal cells, especially of terminally differentiated tissues.
somatic-line DNA the genetic material of somatic cells, destined to perish with the organism, the macronuclear substance in protists such as *Paramecium.*
somatic nervous system the voluntary nervous system, as distinguished from the visceral (autonomic) nervous system.
somite 1. in the early vertebrate embryo, one of a longitudinal series of paired blocks of tissue that are the forerunners of body muscles and the axial skeleton. 2. one of the segments of any segmented animal.
sorus (pl. *sori*), one of the clusters of sporangia on the underside of a fern frond.
sound spectrograph a device for visually displaying the elements of a complex sound; *sound spectrogram,* one such display.
space-filling model a solid model of a molecule, in which the outer orbitals of atoms are represented proportionately by large intersecting spheres.
speciation 1. the division of a species into two or more species. 2. the process or processes whereby new species are formed.
species (sing.; pl., *species*) (variously defined), 1. a class of things or individuals having some common characteristics; a distinct type. 2. the major subdivision of a genus or subgenus, regarded as the basic category of biological classification and composed of related individuals that resemble one another through recent common ancestry and that share a single ecological niche. 3. in sexual organisms, a group whose members are at least potentially able to breed with one another but are unable to breed with members of any other group; a reproductively isolated group. 4. according to the "biological species concept," a biologically integrated group of individuals, tied together by genetic exchange and common adaptations, being the basic unit of evolutionary change.
species recognition the ability to recognize individuals as being members of one's own species.
specific name the second part of a formal scientific name, indicating species within a genus, always italicized and never capitalized.
spectrum see *absorption spectrum, action spectrum, electromagnetic spectrum, sound spectrograph.*
sperm 1. a male gamete. 2. a spermatozoan. 3. a spermatozoid. 4. (adj.), pertaining to a male gamete or male gamete function, as in *sperm head, sperm nucleus, sperm cell.*
sperm nucleus 1. the nucleus of a sperm. 2. either of the two nuclei that arise from the generative nucleus of a pollen tube and function in the double fertilization characteristic of seed plants.
spermatheca a sac connected with the female reproductive tract of most insects, many other invertebrates, and some vertebrates, that receives and stores spermatozoa after insemination and retains them, often for a long time, until egg maturation and fertilization.
spermatic cord in mammals, the structure that suspends the testis in the scrotum, consisting of a tough fibrous coat containing an extension of the coelom, the vas deferens, the nerves and blood vessels supplying the testis, and a retractory muscle, the last being vestigial in humans.
spermatic tubule see *seminal tubule.*
spermatid one of the four haploid cells formed by meiosis of a spermatocyte, prior to maturation into a spermatozoan.
spermatocyte a cell giving rise to spermatozoa; see also *primary spermatocyte, secondary spermatocyte.*
spermatogonium (pl. *spermatogonia*), a diploid testicular cell from which spermatocytes and ultimately sperm are produced; a germ line cell of an adult male.
spermatozoan (pl. *spermatozoa*) also *spermatozoon, sperm, sperm cell,* a motile male gamete of an animal, produced in great numbers in the male gonad, consisting of a sperm head containing the nucleus and an acrosome, a midpiece and a single flagellum (sperm tail).

spermatozoid also *sperm,* a motile male gamete of a plant other than an angiosperm or conifer (e.g., green or brown alga, moss, liverwort, fern, cycad, ginkgo), produced in an antheridium, that swims freely by means of two or more anterior flagella.
sphincter a ring of muscle surrounding a bodily opening or channel and able to close it off; e.g., *oral sphincter, anal sphincter.*
sphygmomanometer an instrument used to measure arterial blood pressure.
spicule 1. one of the many small to minute calcareous or siliceous pointed bodies embedded in and serving to stiffen and support the tissues of various invertebrates, including sponges, radiolorians, and sea cucumbers. 2. any small, stiff, spike-like or needle-like body part.
spike an action potential, in a traveling nerve impulse, the period of most extreme depolarization.
spinal column see *vertebral column.*
spinal cord see *dorsal hollow nerve cord.*
spindle a cytoplasmic body present in mitosis and meiosis, in shape resembling the spindle of a primitive loom, and composed of tubulin microtubules and actin microfilaments; serving to segregate daughter chromosomes in anaphase.
spindle apparatus a functional unit consisting of the continous and centromeric fibers of the spindle, the centromeres, the centrioles (or the analogous *polar ground substance*) and, when present, the asters.
spindle fiber one of the microtubule filaments constituting the mitotic spindle. *centromeric spindle fiber,* one of the several microtubules attaching firmly to a centromere and extending to a spindle pole. *continuous spindle fiber,* a microtubule extending from one pole, interlacing with continuous spindle fibers of the other pole, and ending blindly somewhat short of the other spindle pole.
spindle fiber attachment see *centromere, kinetochore.*
spinneret 1. any organ for producing threads of silk from the secretion of a silk gland. 2. one of the six nipple-like, jointed, processes near the end of a spider's abdomen, each bearing numerous minute orifices of silk gland ducts.
spiracle 1. the blowhole of a cetacean. 2. one of the external openings of an insect tracheal system. 3. one of the pair of openings on the upper back part of the head of an elasmobranch or sturgeon, communicating with the pharynx, derived from gill clefts but serving as incurrent respiratory openings in rays.
spiral valve a helical fold of the intestinal wall in the short intestine of a shark, which serves to slow the passage of food and provides additional absorptive surface.
Spirillum a genus of large spirochaete bacteria.
spirochaete a corkscrew-shaped procaryote of the phylum *Spirochaetae.*
spleen a discrete organ of the reticuloendothelial system, consisting of lymphoid, reticular and endothelial tissues with blood supply and red blood cells circulating freely in intercellular spaces; all enclosed in a fibroelastic, muscularized capsule; functions include the scavenging of debris and the maintenance of blood volume.
split-brain experiments and observations various experiments performed on individuals in which the corpus callosum, a major fiber tract between the hemispheres, is surgically cut; inferences from such experiments include the localization of various neural functions in either the right or left cerebral hemisphere.
spontaneous mutation a chemical change in DNA not induced by external agents; compare *induced mutation, mutation, mutagen.*
splitter a taxonomist who perfers to divide life forms into more named taxons with fewer members in each, or who at the extreme tends to regard every identifiable variant as a significant, nameable, natural unit; compare *lumper.*
SPONCH an mnemonic acronym of the six most common elements of living matter, in the order of decreasing atomic mass: Sulfur, Phosphorus, Oxygen, Nitrogen, Carbon, and Hydrogen.
sponge any aquatic organism of the animal phylum *Porifera* (see Appendix A).
spongin a tough, insoluble protein that makes up the skeleton of a class of sponges including the once commercially important bath sponge.
spongy bone bone with a network of thin, hard walls surrounding open, non-bony pockets.
sporangium a case, cell, or organ in which asexual spores are borne, in algae, fungi, bryophytes and ferns.
spore a minute unicellular reproductive or resistant body, specialized for dispersal, for surviving unfavorable environmental conditions, and for germinating to produce a new vegetative individual when conditions improve.
sporophyte in plants having an alternation of generations, a diploid individual capable of producing haploid spores by meiosis; the prominent form of ferns and seed plants; compare *gametophyte.*
sporozoan a parasitic protozoan of the phylum *Sporozoa* (see Appendix A).
sporozoite a small, motile and elongate infective stage of a sporozoan, a product of sexual fusion that initiates a new aseuxal cycle when transmitted to a new host.
sprouting factor a local hormone produced by tissues that induces nerve axon sprouting; compare *antisprouting factor.*
squamous epithelim stratified epithelium that consists, at least in its outer layers, of small, flattened, scale-like cells, generated from an underlying germinal layer.
stabilizing selection selection against both extremes of a continous phenotype, favoring an intermediate optimum.
stain a reagent that preferentially colors or darkens certain cell structures or tissue types, used in the preparation of sectioned material for microscopy.
stamen the male reproductive structure of the flower, consisting of a pollen-bearing anther and the filament on which it is borne.
staminate having stamens but no pistils; an exclusively male flower.
Staphylococcus a genus of nonmotile, clumped, spherical, gram-positive, aerobic eubacteria, frequently parasitizing the skin or mucous membranes, sometimes causing boils and other serious infections.
starfish see *sea star.*
star navigation the ability to navigate using the visible pattern of stars and a clock or biological clock.
starch a polysaccharide of glucose subunits in alpha glycoside linkages, produced by cyanophytes and by plastids in protists and plants. See *amylose, amylopectin;* compare *glycogen, cellulose.*
statistics 1. the science that deals with the classification, analysis and interpretation of numerical data, and that uses mathematical probability theory to seek order from such data. 2. the use of mathematical probability theory in hypothesis testing, specifically in determining whether or not observed correlations or departures from expectation can reasonably be attributed to chance; and in determining confidence limits on estimations of parameters deduced from data. 3. the numerical data.
statocyst a cell containing one or more statoliths. 2. a sense organ of equilibrium, orientation and movement consisting of a fluid-filled chamber containing a statolith; widely distributed among invertebrate animals.
statolith 1. the calcalcareous body of a statocyst which, by its greater specific gravity, tends to move downward or against the direction of acceleration. 2. a similar body in the inner ear of a fish or amphibian. 3. a small stone taken into the statocyst of certain invertebrates, such as the lobster. 4. dense crystals or vacuolar inclusions that are believed to function in gravity–sensing in plant cells.
steady state a controlled condition of physiological stability in an open system, differing from an *equilibrium* in that it is a function of input and output as well as internal interactions.
stegocephalian an extinct amphibian of the fossil order *Stegocephalia,* with a bony, reptilian or salamander-like body form, well-developed limbs and large size; included are all fossil amphibia occurring before the Jurassic.
stele the cylindrical central portion of the axis of a vascular plant, including pith, xylem, and phloem, surrounded by a pericycle.
stem reptile also *cotylosaur,* a Permian reptile of the group believed to be ancestral to all reptiles.
stereoscopic vision the ability to perceive depth through the integration of two overlapping fields of vision.
sternum a median ventral bone or cartilage in land vertebrates, connecting with the ribs, shoulder gridle or both.
steroid any of a class of lipid-soluble compounds on four interlocking saturated hydrocarbon rings; included are all *sterols* (alchoholic steroids), e.g., *cholesterol, estradiol, testosterone, cortisol.*
steroid hormone a class of hormones consisting of the steroid molecule with various side group substitutions, believed to freely pass across the cell membrane and, once bound by a specific carrier protein, interact directly with the chromatin in gene control; included are the vertebrate sex hormones.
sterol any steroid with an alcohol side group.
stigma the top, slightly enlarged and often sticky end of the style, on which pollen adhere and germinate.
stigmatic fluid a fluid secreted by the stigma that adheres pollen and triggers pollen germination.
stimulus an aspect of the environment that influences the activity of a living organism or part of an organism, especially through a sense organ.
stimulus generalization the tendency to accept a variety of stimuli that have traits in common with a previously rewarded stimulus.
stipe the stemlike structure in red or brown algae that supports the blades or blade.
stolon 1. in plants, a horizontal branch that produces new plants from buds; also *runner.* 1. a long horizontal fungal hypha that spreads across the surface and periodically sends down a mycelium into the substrate and a group of conidia into the air.
stoma (pl. *stomata*), one of the minute openings in the epidermis of leaves, stems and other plant organs, allowing the diffusion of gases in and out of intercellular spaces, formed by the concave walls of two cells; see also *guard cell.*
stomach 1. a muscular dilation of the alimentary canal in vertebrates, between the esophagus and the duodenum, that functions in temporary storage, preliminary digestion, sterilization and physical breakdown of ingested food.
stone canal a calcified tube of the echinoderm water vascular system, leading from an external sieve plate (madreporite) to the ring canal.
streaming see *cyclosis.*
Streptococcus a genus of nonmotile, chiefly parasitic, gram-positive, spherical bacteria that tend to occur in long chains; included are pathogens of humans as well as important agents in the manufacture of dairy products.
Streptomyces a genus of actinomycete bacterium that forms mycelia and produces chains of conidia from aerial hyphae.
streptomycin a powerful antibiotic produced by a soil actinomycete, *Streptomycetes grisens,* that functions by interferring with ribosomal function.
stretch receptor a proprioceptive sensory receptor that is stimulated by stretching, as in a tendon, muscle, or bladder wall.
striated muscle see *skeletal muscle.*
strobilation asexual reproduction by transverse division of the body into segments which break free as independent organisms; occurring in certain coelenterates and flatworms (tapeworms).
strobilus (pl. *strobili*), a conelike aggregation of sporophylls in a club moss or horsetail.

stroke volume the amount of blood passing through the heart per heartbeat.
stroma 1. the matrix or supporting connective tissue of an organ. 2. the cytoskeleton. 3. the proteinaceous matrix surrounding the thylakoids of a chloroplast.
stromatolite a macroscopic geological structure of layered domes of deposited material attributed to the presence of shallow water photosynthetic procaryotes.
strong nuclear force one of the basic forces in physics, binding hadrons within a nucleus.
structural formula a two-dimensional representation of the topological relationships of atoms and bonds in a molecule; sometimes including information on spatial relationships as well.
style the stalk of the pistil in a flower, connecting the stigma with the ovary.
stylized fighting see *ritual combat.*
suberin a complex fatty substance of cork cell walls and other waterproofed cell walls.
sublingual under the tongue; see *salivary gland.*
submucosa the layer of aveolar connective tissue beneath a mucous membrane.
subspecies a more or less clearly defined, morphologically distinct, named geographic variety of a species; when named, the name forms a third part of the scientific name (*genus, species, subspecies*).
substitution 1. see *base substitution.* 2. *amino acid substitution,* a base substitution that changes the identity of one of the amino acids of a protein.
substrate 1. the base on which an organism lives. 2. a substance acted upon by an enzyme. 3. a nutrient source or medium.
substrate-level phosphorylation the production in one or more steps of ATP from ADP, P_i, and an appropriate organic substrate; the capture of high-energy phosphate bonds directly from metabolic transformations; compare *chemiosmostic phosphorylation.*
succession also *ecological succession,* 1. the process of change in a biological population of an area as the available competing organisms respond to and in turn alter the environment. 2. the sequence of identifiable ecological stages or communities occurring over time in the progress of bare ground to a climax community.
sucker a shoot that is produced by an adventitious bud on the root of a plant.
sucrase also *invertase,* an enzyme that breaks sucrose into *invert sugar,* an equimolar mixture of glucose and fructose.
sucrose also *table sugar,* a sweet, 12-carbon disaccharide consisting of glucose and fructose subunits.
sulfhydryl group the side group R-SH. *sulfhydryl bridge,* a covalent link formed by the oxidation of two sulfhydryl groups: $R_1—SH + HS—R_2 \rightarrow R_1—S—S—R_2 + H_2$.
sulfur bacterium 1. *purple sulfur bacterium,* a photosynthetic bacterium utilizing H_2S as a source of hydrogen for photosynthesis; 2. *nonpurple sulfur bacterium,* a nonphotosynthetic chemotroph utilizing sulfur compounds as a source of energy.
sun compass an innate mechanism utilizing the angle of the sun and the time of day to compute direction for navigation.
supercoiling bending of a helical linear structure into larger orders of helices, as occurs in the mitotic condensation of chromosomes.
superior higher; toward the top; in human anatomy, in view of the unusual posture of the organism, *superior* is often used in places where *anterior* would be appropriate for other vertebrates.
superior vena cava the anterior vena cava in humans; see *vena cava.*
supplemented medium a growth medium providing metabolites not normally required by the organism, and thus able to sustain the growth of certain nutritional mutants that would not be recovered on minimal medium.
surface tension a condition that exists on the free surface of water or other liquid by reason of intermolecular forces unsymmetrically disposed around individual surface molecules, tending to make the surface layer behave in some respects as an elastic membrane.
survivorship curve a graph with age on the X axis and zero to one on the Y axis, the monotonic curve presenting the probability of surviving to age X for each X.
suspensor in a developing seed, a group or chain of large cells, swollen with water, produced by the zygote and serving to anchor the embryo in place and to push it into contact with the endosperm.
swarm cell also *swarm spore,* any of various minute, motile, sexual or asexual spores produced in large numbers by certain algae, fungi and protists.
swim bladder also *air bladder,* a gas-filled sac giving controlled bouyancy to most bony fish; homologous with the lungs of land vertebrates and lungfish and probably derived from the primitive lung of an ancestral air-breathing fish.
symbiont a symbiotic organism.
symbiosis 1. the living together in intimate association of two dissimilar organisms. 2. mutualism. *parasitic symbiosis,* also *parasitism,* symbiosis in which one organism gains fitness at some expense to the other. *commensalism,* symbiosis in which one organism gains at no cost to the other. *mutualistic symbiosis,* also *mutualism,* symbiosis in which both individuals gain fitness.
symbiosis hypothesis also *serial endosymbiosis hypothesis,* three interrelated hypotheses attributable to Lynn Margulis to the effect that the eucaryotic cell evolved from the mutualistic endosymbiosis of various procaryotic organisms, one of which gave rise basically to the cytoplasm, nucleus and motile membranes; a second to mitochondria; a third to chloroplasts and other plastids; and a fourth to cilia, eucaryotic flagella, basal bodies, centrioles, the spindle and all other microtubule structures.
symbiotic (adj.), living together; see *symbiosis.*
sympathetic nervous system one of the two divisions of the vertebrate autonomic nervous system, that utilizes norepinephrine as a neurotransmitter, decreases glandular secretions, decreases smooth muscle tone and activity, increases heart beat, causes piloerection, contracts blood vessels and contributes to a state of arousal; see *autonomic nervous system;* compare *parasympathetic nervous system.*
sympatric occurring in the same geographical area.
symphysis see *pubic symphysis.*
synapse 1. (n.), the place at which nerve impulses pass from the axon or the synaptic knobs of an axon of one neuron to the dendrite or cell body of another. 2. (v.i.), of homologous chromosomes, to pair together, chromomere by chromomere, in zygotene of meiosis.
synaptic cleft the minute space between synaptic knob of one neuron and the dendrite or cell body of another, into which neurotransmitters are released in the transmission of nerve impulses between cells.
synaptic knob one of the multiple bulbous swellings at ends of the branching terminus of an axon, containing neurotransmitters in secretion granules and forming one side of the synaptic cleft.
synaptinemal complex a complex structure composed of protein and RNA, the material of which is first formed between sister strands in leptotene of meiotic prophase and combines with a like structure to form the complex and accompish the specific zipper-like pairing of homologous chromosomes in zygotene.
syncytium 1. a multinucleate mass of protoplasm resulting from the fusion of cells, as in the plasmodium of a slime mold. 2. also *coenocyte,* a multinucleate mass created by repeated nuclear mitosis without cytoplasmic cell division.
synovial capsule a completely enclosed capsule containing *synovial fluid,* formed by the smooth cartilages covering the articular surface of the bones of a synovial joint and a surrounding *capsular ligament.*
synovial fluid also *synovia,* a transparent, viscid lubricating fluid secreted by the synovial membranes of joints, bursae, and tendon sheaths.
synthesist in science, one who attempts to deduce broad general principles from widely disparate observations, or who draws upon such observations to substantiate or refute such principles; one who reinterprets established relationships in the support of a new general idea or *synthesis.*
systole the period of heart contraction particularly of the ventricles; compare *diastole.*
systolic pressure the highest arterial blood pressure of the cardiac cycle.
syrinx a voice-generating structure in birds formed by modifications of the lower part of the trachea and/or the adjacent bronchi.
systematics 1. the science of classifying organisms or groups of organisms on the basis of their evolutionary relatedness. 2. the science of determining the evolutionary relatedness of groups of organisms, particularly at higher taxonomic levels.
systematist a biologist who tries to determine the evolutionary relationships of organisms.

tactile receptor, a sensory receptor responsive to light touch; compare *pressure receptor.*
taiga moist, subarctic forest biome of Europe and North America dominated by spruces and firs.
tailpiece the portion of a bacteriophage between the head and the end plate.
talus also *astragalus, anklebone,* the principle bone of the human ankle.
tap root also *taproot,* 1. a root having a prominent central portion giving off smaller lateral roots in succession. 2. the central root of such a system, especially when it grows vertically and deep.
tapeworm a parasitic flatworm of the class *Cestoda.* (see Appendix A).
tapetum, *tapetum lucidum,* a highly reflective layer of the choriod of the eye of nocturnal mammals (e.g., cat's eyes), consisting of layers of flattened cells covered by a zone of doubly refracting crystals.
target cell a cell acted upon by a specific hormone, generally containing or bearing specific hormone receptor proteins not found in other cells.
tarsal bone one of the smaller bones of the ankle, between the talus and the metatarsals.
tarsus 1. the tarsal bones taken together. 2. the vertebrate ankle. 3. the distal portion of an arthropod leg.
taste bud a sensory receptor sensitive to taste, found chiefly in the epithelium of the tongue, being a conical mass of cells made up of supporting cells and *taste cells,* which terminate in modified sensory cilia that project into a pore-like space in the overlying epithelium.
tautology a statement that excludes no logical possibilities.
taxon 1. any taxonomic category, such as *species, genus, subspecies, phylum* etc. 2. any named group of related organisms.
taxonomist an individual skilled in identifying, describing and classifying organisms, usually specializing in a particular group.
taxonomy the science dealing with the identification, naming, and classification of organisms.
Tay-Sachs disease a disease due to homozygosity for a rare recessive autosomal allele, characterized by inclusion bodies in neurons, juvenile idiocy and early death. The recessive allele and the disease unusually common in persons of Eastern European Jewish descent.
T-cell a lymphocyte of a variety that matures in the thymus, produces immunoglobulins bound tightly to its exterior surface, and interacts in complex ways with other types of cells in the immune system.
telolecithal of an egg, having a large yolk, with active cytoplasm confined to a small disk on the upper surface, as in reptiles, birds and monotremes.
telomere an invisible but inferred body occurring at either end of all eucaryotic chromosomes.

telophase the stage of mitosis or meiosis in which new nuclear membranes form around each group of daughter chromosomes, the nucleoli appear and the chromosomes decondense; at which time the cell membrane and cytoplasm of the cell usually divide to form two daughter cells.
TEM acronym for *transmission electron microscope;* see *electron microscope.*
temperate (adj.), 1. having a mild but usually seasonal climate. 2. in the temperate zone.
temperate deciduous forest a forest biome of the temperate zone, usually in areas of significant winter snowfall, in which the dominant tree species and most other trees are deciduous and are bare in winter months.
temperate rain forest a cool woodland of the temperate zone in an area of heavy rainfall but little or no snow, usually including many different kinds of trees as well as a single dominant tree species.
temperate zone the region of the earth between the Tropic of Cancer and the Arctic Circle together with the region between the Tropic of Capricorn and the Antarctic Circle.
temporal bone a compound bone of the side of the human skull.
temporal lobe a large lobe on the lateral portion of each cerebral hemisphere.
temporal muscle a large muscle originating on the temporal bone, passing beneath the zygomatic arch and inserting into the lower jaw; a principle muscle of the jaw.
tendon a tough dense cord of fibrous connective tissue that is attached at one end to a muscle and at the other to the skeleton, and transmits the force exerted by the muscle in moving the skeleton.
tendril a slender, fast-growing tip of a climbing plant that grows in spirals around an external supporting tree, bush, or trellis.
tensile strength the ability to be pulled without breaking.
tentacle an elongate, flexible, fleshy, sometimes branched process borne by animals chiefly in groups or pairs on the head or surrounding the mouth, and serving as a tactile and/or prehensile organ.
tenting the expansion of the internal vagina (cul-de-sac) and elevation of the uterus on approaching orgasm.
terminal differentiation cell differentiation in which the cell remains in G_1 and does not normally undergo subsequent cell division.
terminalization in meiotic prophase, the movement of chiasmata away from the centromeres and toward the telomeres.
terminator codon (also *termination codon, stop codon, nonsense codon*), one of the codons UAA, UAG, or UGA in mRNA, or the corresponding codons in DNA, signalling the end of polypeptide transcription.
territorial (adj.), exhibiting territorial behavior.
territorial behavior also *territoriality,* a species-specific pattern of behavior associated with the defense of a territory, in most territorial species by the male, and often including territorial marking, singing or other displays, threats, attacks on trespassing conspecifics, and patrolling by the defending animal, as well as avoidance and submissive behavior by others when in or around a defended territory.
territory the area defended by a territorial animal.
tertiary consumer a consumer at the third level; see *consumer.*
tertiary structure of a protein, the pattern of folding of a polypeptide upon itself, which is generally quite specific for each protein type.
test cross 1. a cross of an individual of unknown genotype with a homozygous recessive to determine zygosity by progeny testing. 2. the cross of a known double heterozygote with a double homozygote to determine linkage relationships.
testis (pl. *testes*), the male gonad, in which spermatozoa are produced by meiosis.
testosterone a steroid hormone, the principle androgen of vertebrates.
tetrad 1. a group of four. 2. a group of four equal cells, such as microspores or spermatids, produced by meiosis. 3. also *bivalent,* the synapsed homologues in pachytene through diakinesis, considered as a group of four synapsed chromatids.
tetrahedron a perfect geometric solid with four equal sides, four equal edges, and four corners.
tetramer a molecule composed of four subunits, e.g., hemoglobin, a protein which is composed of two alpha chains and two beta chains.
tetrapod also *land vertebrate,* any vertebrate of the classes Amphibia, Reptilia, Mammalia, and Aves, including some that don't have four feet (e.g., birds, people, whales, dugongs).
tetraploid having four complete genomes in each nucleus; *autotetraploid,* having four genomes (i.e., twice the diploid number) of one normally diploid species; *allotetraploid,* also *amphidiploid,* having two genomes (i.e., the usual diploid number) of each of two different, normally diploid species.
T-even phage one general type of *E. coli* bacteriophage, with a double-stranded DNA genome and a complex body consisting of a head, tailpiece, end plate and attachment fibers; arbitrarily designated T_2, T_4, T_6, etc.
thalamus a large subdivision of the diencephalon, consisting of a mass of nuclei in each lateral wall of the centrally-located third ventricle of the brain.
thalassemia also *Cooley's anemia,* a severe congenital anemia due to homozygosity for a variant allele of the beta hemoglobin locus, producing little or no adult hemoglobin. *alpha thalassemia,* an analagous partial deficiency of the hemoglobin alpha chain.
thallus a plant body of a multicellular alga, that does not grow from an apical meristem, shows no differentiation into distinct tissues, lacks stems, leaves, or roots; my be simple or branched, consisting of filaments, thin plates or solid bodies of cells, and may include fairly complex blades, bladders and holdfasts.
theory 1. a coherent group of general propositions used as principles of explanation for a class of phenomena: e.g., *Newton's theory of gravitation, Darwin's theory of the origin of species.* 2. a proposed explanation whose status is still conjectural, in contrast to well-established propositions that are often regarded as facts. *Theory* and *hypothesis* are terms often used colloquially to mean an untested idea or opinion. A *theory* properly is a more or less verified explanation accounting for a body of known facts or phenomena, whereas a *hypothesis* is a conjecture put forth as a possible explanation of a specific phenomenon or relationship that serves as a basis for argument or experimentation.
therapsid also *mammal-like reptile,* any Permian or Triassic reptile of the extinct order *Therapsida,* with mammal-like upright locomotion rather than crawling or bipedal locomotion.
thermal (adj.), pertaining to heat.
thermal energy 1. energy in the form of heat. 2. the heat equivalent of some other form of energy.
thermal efficiency also *thermodynamic efficiency,* in a chemical reaction or series of reactions, the proportion of the original potential chemical bond energy of the substrate still retained in the potential bond energy of the product and *not* dissipated as thermal energy or entropy.
thermodynamics 1. the branch of physics that deals with the interconversions of energy as heat, potential energy, kinetic energy, radiant energy, entropy and work. 2. the processes and phenomena of energy interconversions.
thermoreceptor a sensory receptor that responds to temperature or to changes in temperature.
thermoregulation 1. an animal's control over its internal temperature. 2. the physiological mechanisms that maintain a body at a particular temperature in an environment with a fluctuating temperature. *behavioral thermoregulation,* maintaining a body temperature within acceptable limits by behavioral means, as by basking and by seeking out appropriate microenvironments.
thiamine also vitamin B_1, an essential water-soluble vitamin for most animals, the deficiency of which is responsible for the disease beriberi; utilized as a major portion of the coenzyme A of the citric acid cycle and many other pathways.
thoracic (adj.), pertaining to the thorax.
thoracic region the portion of vertebral column that articulates with ribs, between the cervical and lumbar regions.
thorax 1. in mammals, the part of the body anterior to the diaphragm and posterior to the neck, containing the lungs and heart. 2. the middle of the three parts of an insect body, bearing the legs and wings.
three-prime end (3′ end) of a nucleic acid, the end or direction toward the 3′ carbon of ribose or deoxyribose subunits, as opposed to the 5′ end; replication, transcription and translation proceed from the 5′ end toward the 3′ end.
threonine one of the twenty amino acids of polypeptide synthesis.
thrombin in the blood clotting reactions, a proteolytic enzyme that catalyzes the conversion of fibrinogen to fibrin by the removal of two short peptide segments, and in turn is produced from *prothrombin* by the action of *thromboplastin.*
thromboplastin a complex phosphoprotein, released from damaged platelets, that catalyzes the conversion of the inactive serum component *prothrombin* into the active enzyme *thrombin* in the initiation of blood clotting.
thylakoid a membraneous structure of chloroplasts and cyanophytes consisting of a *thylakoid membrane* containing chlorophyll, accessory pigments, photocenters and the photosynthetic electron transport chain; an inner lumen that becomes acidic during active photosynthesis; and CF_1 particles attached to the outer sides of the thylakoid membrane, which are the sites of chemiosomtic phosphorylation.
thymidine a nucleoside consisting of the pyrimidine *thymine* and a deoxyribose residue.
thymidylic acid a thymine deoxynucleotide.
thymine a pyrimidine, one of the four nitrogenous bases of DNA and the only nitrogenous base specific for DNA.
thyrocalcitonin another name for *calcitonin,* emphasizing the belated discovery that the hormone is secreted by the thyroid and not the parathyroid as had been assumed; see *calcitonin.*
thyroid also *thyroid gland,* a large endocrine in the lower neck region of all vertebrates, arising as a ventral median outpocketing of the pharynx and believed to be homologous with the endostyle of cephalochordates, lamprey larvae, and urochordates; secretes *throxine* and *calcitonin.*
thyroid-stimulating hormone (TSH) also *thyrotropin,* a peptide hormone of the anterior pituitary, the action of which is to stimulate the release of thyroxin from the thyroid.
thyrotropic hormone releasing factor (THRF) a small peptide produced in the hypothalamus and carried by the pituitary portal vein to the anterior lobe of the pituitary, where it stimulates the release of TSH.
thyroxin also *thyroxine,* also *thyroid hormone,* a specialized amino acid with four iodine residues, $C_{15}H_{11}I_4NO_4$, a hormone with an effect of increasing general body metabolism and the stimulation of growth and also serves as a stimulus to metamorphosis in amphibia.
tibia 1. the larger of the two bones of the vertebrate lower hind leg, between the femur and the talus. 2. an analogous segment in an insect leg.
tide pool a pool of sea water left behind by the receding tide and covered at high tide.
toadstool a mushroom especially if poisonous, distasteful, or suspicious.
torsion an event of gastropod development in which the upper part of the body is rotated 180 degrees, bringing the anus to a position above the head.
totipotent of a cell, able to undergo development and differentiation along any of the lines inherently possible for the species; undifferentiated.

touch receptor see *tactile receptor.*
toxin a poisonous substance produced by a biological organism.
trace element an element necessary for growth and development but only in extremely small amounts.
tracer see *radioactive tracer.*
trachea (pl. *tracheae*), 1. in land vertebrates, the main trunk of the air passage between the lungs and the larynx, usually stiffened with rings of cartilage; bifurcating to form the bronchi; lined with ciliated, mucus-secreting columnar epithelium. 2. one of the air-conveying chitinous tubules comprising the respiratory system of an insect, millipede, centipede, wood louse or onychophoran, and in the insects and onychophorans constituting a ramifying network including air sacs and small fluid-filled passages penetrating to all parts of the body.
tracheid a long tubular xylem cell or its lignified, empty cell wall, which functions in support and the conduction of water, distinguished from *xylem vessels* by having tapered, closed ends communicating with other tracheids through pits.
tracheophyte a vascular plant.
trade wind a wind that blows almost constantly from the northeast in a latitudinal belt between the horse latitudes and the doldrums, or from the southeast between the doldrums and the southern horse latitudes, as a result of equatorial warming and the coriolus effect.
transduction the transfer of a host DNA fragment from one bacterium to another by a viral particle.
transcribed strand the DNA strand of a structural gene that is physically involved in transcription.
transcription the process of RNA synthesis as the RNA nucleotide sequence is directed by specific base pairing with the nucleotide sequence of the transcribed strand of a DNA cistron.
transfer RNA (tRNA), any of a class of relatively short ribonucleotides with a common secondary and tertiary structure consisting of three or four loops and a double-stranded stem, with numerous modified bases in addition to the four normally found in RNA; specific varieties become covalently linked to specific amino acids by specific enzymes, then function in transcription to recognize appropriate mRNA codons and to transfer the amino acid to the growing polypeptide chain.
transferrin an iron-carrying molecule of the blood serum.
transformation in a bacterium, the direct incorporation of a DNA fragment from its medium into its own chromosome.
translation polypeptide synthesis as it is directed by an mRNA on a ribosome; the transfer of linear information from a nucleotide sequence to an amino acid sequence according to the strictures of the genetic code.
translocation 1. the step in protein synthesis in which a transfer RNA molecule that is covalently attached to the growing polypeptide chain is moved (translocated) from one ribosomal tRNA attachment site to the other. 2. a chromosome rearrangement in which a terminal segment of one chromosome replaces a terminal segment of another, nonhomologous chromosome. *reciprocal translocation,* a chromosome rearrangement in which terminal segments of two nonhomologous chromosomes are exchanged. 3. the directed movement of materials from one part of an organism to another part, especially the directed movement of solutes through phloem.
transmission electron microscope (TEM), see *electron microscope.*
transpiration 1. the evaporation of water vapor from leaves, especially through the stomata. 2. the physical effects of such evaporation taken together. *transpiration pull,* the pulling of water up through the xylem of a plant utilizing the energy of evaporation and the tensile strength of water.
transverse tubule also *T tubule,* a specialized system of tubules in a muscle fiber, transmitting the contractile impulse from the sarcolemma to the sarcomeres.
Triassic period see geological timetable inside the front cover.
tricuspid valve the valve between the right atrium and the right ventricle, consisting of three triangular membranous flaps.
triglyceride, a compound of glycerol with each of its three hydroxyl side groups bound in an ester linkage to a fatty acid or other residue.
triiodothyronine a hormonally active substance similar to thyroxin but lacking one of the four iodines of thyroxin, usually made synthetically.
trimester one of three three-month periods of the nine months of human gestation, which thus is divided into first, second, and third trimesters.
triradial also *triradiate,* having three radiating branches; compare *radial, pentaradial.*
trisomy the condition in which an otherwise diploid individual has a third homologue of one chromosome.
trisomy 21 also *Down's syndrome,* originally *mongolian idiocy,* a severe human congential pathology attributable to the presence of three rather than two homologues of chromosome 21.
trisomy X syndrome also *XXXfemale,* a mild congenital human abnormality attributable to the presence of three rather than two X chromosomes, characterized by moderately increased stature, moderately decreased mental functioning, and frequently by amenorrhea and sterility.
tRNA see *transfer RNA.*
tRNA charging enzyme see *charging enzyme.*
trochophore also *trochophore larva,* a rather specific body form of a free-swimming ciliated larva of marine invertebrates of several phyla, including annelids, mollusks and rotifers, having an oval, bilaterally symmetrical body with an equatorial band of cilia, a complete gut, an apical sensory tuft and often nephridia and a second band of cilia behind the mouth.
trophic relating to nutrition. *trophic level,* a level in a food pyramid.
trophoblast the outer ectodermal layer of the blastocysts of many mammals that digests the portion of the uterine endometrium into which it comes in contact, providing nutrition and anchorage for the early embryo and eventually forming part of the placenta.
tropical rain forest a tropical woodland biome that has an annual rainfall of at least 250 cm and often much more, typically restricted to lowland areas, characterized by a mix of many species of tall broad-leaved evergreen trees forming a continuous canopy, with vines and woody epiphytes, and a dark, nearly bare forest floor.
tropism a turning toward or away from a stimulus, usually accomplished by differential growth; see *geotropism, phototropism.*
tropomyosin a low-molecular-weight filamentous protein that accompanies the globular protein actin in making up actin microfilaments; see p. 693.
troponin a protein of low molecular weight that binds calcium in muscle contraction, a specific variant of the ubiquitous calcium-binding molecule *calmodulin;* see page 693.
true bacteria see *eubacterium.*
true-breeding (adj.), of an organism or strain, when mated with individuals like itself producing offspring like itself; specifically, homozygous for all relevant loci.
true fungus an organism of the fungal subkingdom Eumycota, lacking flagella and amoeboid movement; included are sac fungi, club fungi, zygomycetes (bread molds), chytrids, and yeasts.
trypanosome a parasitic, flagellated protozoan of the genus *Trypanosoma,* infecting mammals and transmitted by insect vectors, responsible for *chagas disease, sleeping sickness, nagana,* and other serious and unpleasant diseases.
trypsin a powerful proteolytic enzyme secreted by the pancreas in the inactive form *trypsinogen* and activated in the intestine.
tryptophan one of the twenty amino acids of polypeptide synthesis.
tsetse fly also *tsetse,* a biting fly of Africa that is the carrier of the trypanosome responsible for sleeping sickness.
T tubule see transverse tubule.
TTP *thymidine triphosphate,* a deoxynucleotide that is one of the precursors of DNA synthesis.
tubal ligation sterilization of a female mammal by cutting and tying the oviducts.
tube foot one of the numerous specialized, hollow, extensible, flexible organs of locomotion and grasping in echinoderms formed as extensions of the water vascular system and each usually bearing a terminal sucker.
tube nucleus in seed plants, one of the two haploid nuclei of a pollen grain, which controls the subsequent growth and enzyme production of the pollen tube upon germination.
tuber a thick stem that grows underground and is capable of producing offshoots; e.g., potato, Jerusalem artichokes.
tubule any slender, elongated channel in an anatomical structure; *kidney tubule,* the tubular part of a nephron, being all of the nephron except Bowman's capsule; see also *proximal convoluted tubule, distal convoluted tubule.*
tubulin the protein that comprises the spherical subunit of microtubules.
tundra a biome, characterized by level or gently undulating treeless plains of the arctic and subarctic, supporting a dense growth of mosses and lichens and dwarf herbs and shrubs, seasonally covered with snow and underlain with permafrost.
tunicate 1. any animal of the subphylum *Urochordata.* (see Appendix A).
turgid stiffened and distended by hydrostatic pressure from within.
turgor the normal state of turgidity and tension in living plants cells, created by osmosis. *turgor pressure,* the actual hydrostatic pressure developed by the fluid in a turgid plant cell.
Turner's syndrome also *XO syndrome,* a congential human abnormality due to the presence of a single X chromosome and no other sex chromosome; resulting in immature female development, short stature, sterility, certain anatomical peculiarities and normal mental functioning except for a marked deficiency in certain narrowly delimited spatial tasks.
two, four D usually **2,4 D** *dichlorophenoxyacetic acid,* a synthetic hormone that has the agriculturally useful effect of killing dicots (e.g., dandelions) and not harming monocots (e.g., lawns).
tympanic membrane also *tympanum, eardrum,* a thin clear, tense double membrane of connective tissue covered with living epithelium, dividing the middle ear from the external auditory canal, vibrating sympathetically with received sound and transmitting the impulses to the inner ear by way of the ossicles of the middle ear; deep within a bony recess in mammals, on the surface in frogs and toads, and intermediate in birds and reptiles.
tympanum 1. see *tympanic membrane.* 2. an analogous thin tense membrane of an organ of hearing in an insect. 3. the membrane of a sound-producing organ that acts as a resonator.
type specimen a preserved individual maintained in a museum and designated as the type of a species or subspecies for the identification and comparison of other specimens, and serving as the final criterion of the characteristics of the group.
typhlosole 1. a longitudinal fold in the inner intestinal wall of the earthworm. 2. a similar fold projecting into the gut of certain pelecypod mollusks and starfish.
tyrosine one of the twenty amino acids of polypeptide synthesis.

ulcer 1. a slow-healing break in the skin or mucous membrane, with a loss of surface tissue, inflammation, necrosis, disintegration of epithelial tissue, and sometimes infection. 2. such an area of tissue damage in the lining of the stomach or duodenum, exacerbated by self-digestion by acidity and gastric juices.

ulna one of the two bones of the forearm or the corresponding part of the forelimb of land vertebrates, which forms an articulation with the humerus at the elbow; secondarily fused with the radius and greatly reduced in many groups.
ultraviolet light also *ultraviolet radiation, u.v.*, electromagnetic radiation having a shorter wavelength than visible light and longer than X-rays, including wavelengths normally invisible to humans but visible to bees and hummingbirds; interacts destructively with DNA and thus is mutagenic, carcinogenic, and destructive of skin tissues.
umbilical cord in placental mammals, a vascular cord connecting the fetus with the placenta, containing two *umbilical arteries*, an *umbilical vein* and the blind remnant of the yolk sac, in a connective tissue matrix.
umbilical vesicle also *umbilical vessel*, the vestigial, yolkless yolk sac of the embryo of a placental mammal, having the form of a fluid-filled pouch with a transitory connection with the gut.
umbilicus 1. a small depression in the middle of the abdomen marking the place where the umbilical cord was attached. 2. see *hilum*.
ungulate a mammal that has hoofs, especially of the orders *Perissodactyla* and *Artiodactyla*.
unicellular also *acellular*, 1. consisting of a single cell. 2. having a single cell nucleus.
unsaturated capable of accepting hydrogens; *unsaturated fat*, a triglyceride with one or more carbon-carbon double bonds.
upwelling the coming to the surface of water from the depths of the ocean, which is associated with the introduction of mineral nutrients and a consequent richness in productivity and biomass.
uracil a pyrimidine, one of the four nitrogenous bases of RNA and the only one specific to RNA.
urea a highly soluble nitrogenous compound, the principle nitrogenous waste of the urine of mammals; retained in high concentrations in the blood of elasmobranchs and the coelocanth as a means of osmoregulation.
ureter either of the pair of ducts that carry urine from the kidney to the bladder in mammals, or from the kidney the cloaca in other vertebrates; compare *urethra*.
urethra the canal that carries urine from the mammalian bladder to the exterior, and in males serves also for the transmission of semen.
uric acid a relatively insoluble pyrimidine, the principle nitrogenous excretion product of reptiles, birds and insects; excreted in small quantities as a product of nucleic acid breakdown in mammals; sometimes crystalizing in the tissues of men causing a painful inflammation (*gout*).
uridine the nucleoside of uracil.
urinary bladder see *bladder*.
Urochordata one of the chordate subphyla (see Appendix A).
urogential folds a pair of early prominences in the undifferentiated stage of the development of the mammalian external genitalia; form the shaft of the penis in males and the *labia minora* in females.
urogenital groove in the undifferentiated stage of the embryonic development of the mammalian genitalia, a longitudinal depression between the urogenital folds; becomes the *penile urethra* in males and the *vestibule* between the labia minora in females.
"use and disuse" the idea, shared by Lamarck, Erasmus Darwin and Charles Darwin, that use or disuse of a body part caused hereditary changes in the direction of hypertrophy or atrophy respectively.
uterine mucosa see *endometrium*.
uterus 1. in female mammals, a muscular, vascularized, mucous-membrane-lined organ for containing and nourishing the developing young prior to birth and for expelling them at the time of birth, consisting of enlargements of the paired oviducts or of a single median enlargement of the fused oviducts. 2. an enlarged section of the oviduct of various vertebrates and invertebrates modified to serve as a place of development of the young or of eggs.
utricle the chamber of the membranous labyrinth of the middle ear into which the semicircular canals open.
uv light *uv radiation*, see *ultraviolet light*.

vacuole a general term for any fluid-filled membrane-bounded body within the ctyoplasm of a cell, in higher plants occupying most at the volume of most differentiated living cells, in protozoa performing numerous functions; see *contractile vacuole, digestive vacuole*.
vacuum behavior instinctive actions that are released spontaneously, without the stimulus of a releaser, presumably due to the build-up of action-specific energy that finally demands to be discharged.
vagina 1. an expandable canal that leads from the cervix of the uterus of a female mammal to the external orifice (*introitus*), serving to receive the penis in copulation and as a *birth canal* in parturition. 2. any canal of similar function in other animals.
vagus nerve either of the tenth pair of cranial nerves, with sensory and autonomic motor fibers innervating the heart and viscera.
vanadium an element that occurs widely in small amounts, concentrated as the chief oxygen carrier in the blood of tunicates.
van der Waals forces also *nonpolar attraction*, relatively weak attractive forces acting between nonpolar atoms and molecules, arising because of the electric polarization induced in each particle by the other, effective at relatively great molecular distances, serving to bind together lipid-soluble portions of molecules and important in the stability of membranes, the specific folding of proteins, and the attachment of proteins to or within membranes.
variability *genetic variability*, the general qualitative term for the presence of genetic differences between individuals in a population.
variable age of onset of a genetic pathology, not being observed at birth but becoming manifest at some variable time in later development.
variable expressivity of a genetic pathology, having different degrees of abnormality in different individuals with the same or similar genotype.
variable region the portion of an immunoglobulin polypeptide concerned with the binding of an antigen, which varies from antigen to antigen.
variance a specific quantitative measure of the variability of a continous trait having the property that the contributing factors in such variability can be ascribed individual variances that sum to the total variance: specifically, the *total variance* of a continuous trait, also termed the *phenotypic variance*, symbolized V_p, is defined as the *mean squared deviation* or $V_p = \Sigma(X_i - \overline{X})^2 / (N-1)$ for a population of N individuals with phenotypes X_i and phenotypic mean $\overline{X}$. *additive genetic variance*, symbolized V_a, is the portion of the total variance due to the average effects of the different alleles carried by individuals in the population. *dominance genetic variance*, symbolized V_d, is the portion of the total variance due to specific dominance interactions. *environmental variance*, symbolized V_e, is the proportion of the total variance due to nongenetic effects, and if desired can further be broken down into different nongenetic sources of phenotypic variability. The variance of a normally distributed trait is equal to the square of the *standard deviation* of that trait in a specified population.
variety 1. a species biotype or ecotype. 2. a subspecies. 3. a particular breed or genetic type of a domesticated plant or animal. 4. any group of organisms below the species level having some genetic characteristic or group of characteristics in common.
vascular (adj.), pertaining to vessels or tissues of conduction, as phloem and xylem in plants or the blood vessels in animals.
vascular bundle a unit of the vascular system of plants, a strand of xylem associated with a strand of phloem and usually a sheath of fibrous sclerenchyma, in the primary growth pattern and in leaves and petioles.
vascular cambium the cylinder of lateral meristem that produces xylem on its inner side and phloem on its outer side in secondary growth, contributing thus to growth in circumference.
vascular plant a plant with xylem and phloem; a tracheophyte.
vascular system 1. the circulatory system of an animal. 2. the xylem and phloem of a vascular plant.
vascularized having many blood vessels, especially capillaries.
vas deferens (pl. *vasa deferentia*), the male genital duct or spermatic duct, especially of a higher vertebrate, in men being one of two small, thick-walled, muscular tubes, about 65 cm. long, which run in the spermatic cords from the testes through the inguinal canal, joining the duct of the seminal vesicle to form the ejaculatory duct.
vasectomy sterilization of the male by severance and ligature of the vasa deferentia (see vas deferens).
vasoconstriction contraction of sphinctors in arterioles resulting in reduced blood flow.
vasodilation relaxation of arteriolar sphincters, allowing increased blood flow.
vasopressin an old term for *antidiuretic hormone* (ADH).
vector an organism that transmits a parasite, virus, bacterium or other pathogen from one host to another.
vegetal pole the lower, more yolk-filled end of a zygote or early cleavage blastual, determining the ventral side in development.
vegetative reproduction asexual reproduction in plants, in which offspring are produced from portions of the roots, stems, or leaves of the parent.
vein 1. a vessel returning blood toward the heart at relatively lower velocity and pressures; such blood is usually relatively deoxygenated, but in the case of the *pulmonary veins* it is freshly oxygenated. 2. a vascular bundle in a leaf or petiole.
veldt eastern and southern African grassland, usually level, often with scattered trees or shrubs; where the lions live (see savanna).
vena cava (pl. *venae cavae*), any one of three large veins by which blood is returned to the right atrium of land vertebrates, which replace the cardinal veins and ducts of Cuvier of fishes and embryos. *anterior venae cavae* (in humans, *superior venae cavae*), the paired major collecting veins of the head and forelimbs; *posterior vena cava* (in humans, *inferior vena cava*), the major vein running alongside the dorsal aorta and returning blood from the posterior part of the body and from the viscera.
ventilation breathing.
ventral (adj.), in bilateral animals, toward the belly; downward, opposite the back; compare *dorsal*. In human anatomy, in view of the peculiar upright stance of the organism, *ventral* is frequently replaced by *anterior* (toward the front).
ventral aorta see *aorta*.
ventral nerve cord a common feature of many invertebrate phyla, the main longitudinal nerve cord of the body, being solid, paired, and ventral, with a series of ganglionic masses.
ventricle a cavity of a body part or organ: 1. one of the large muscular chambers of the heart. 2. one of the systems of communicating cavities of the brain, consisting of two *lateral ventricles* and a median *third ventricle*.
venule a small vein.
vertebra (pl. *vertebrae*), one of the bony or cartilaginous elements that together make up the vertebrate spinal column, in land vertebrates being a complex bone consisting of a cylindrical centrum articulating with adjacent centra by a fibrous cartilagenous *intervertebral disk*, a dorsal *neural arch* providing a passage for the spinal cord, and various small spines, articular processes, and muscle attachments.
vertebral column also *spinal column*, the articulated series of vertebrae connected by ligaments and separated by intervertebral disks that in vertebrates forms the supporting axis of the body, and of the tail in most forms.

vesicle a small cavity, fluid-filled sac, blister, cyst, vacuole, etc.
vessel 1. a tube or canal in which a body fluid is contained or circulated within the interior, in contrast to a *duct* that empties to the exterior or into the alimentary canal; e.g., *blood vessel, lymph vessel.* 2. a conducting tube in a dicot formed in the xylem by the end-to-end fusion of a series of cells (*vessel elements*) followed by the loss of adjacent endwalls and of cell cytoplasm; compare *tracheid.*
vestibular gland also glands of Bartholin, a pair of oval, ducted glands lying on either side of the lower part of the vagina, secreting a lubricating mucus especially on appropriate occasions.
villus (pl. *villi*), a small, slender, vascular, fingerlike process: 1. one of the minute processes that cover and give a velvety appearance to the inner surface of the small intestine, that contain blood vessels and a lacteal, are covered with microvilli and serve in the absorption of nutrients. 2. one of the branching processes of the surface of the implanted blastodermic vesicle of most mammals.
viral RNA the genetic material of certain viruses that utilize RNA rather than DNA in transmission and replication.
virulence 1. disease-producing capability. 2. the capacity of an infective organism to overcome host defenses.
virus a noncellular organism transmitted as DNA or RNA enclosed in a membrane or protein coat, often together with one or a few enzymes; replicating only within a host cell, utilizing host ribosomes and enzymes of synthesis.
visible light electromagnetic wavelengths longer than about 400 nm and shorter than about 750 nm, which can serve as visual stimuli to most photoreceptive organisms.
visual cliff 1. a dropoff placed beneath a plate of glass. 2. a flat pattern creating the optical illusion of a steep dropoff.
visual receptor a sense receptor stimulated by light or by patterns of light.
vitalism a doctrine in biology that ascribes the functions of a living organism to a *vital principle* on *vital force* distinct from chemical and physical forces; the doctrine is no longer taken seriously by practicing biologists.
vitamin any organic substance that is essential to the nutrition of an organism, usually by supplying part of a coenzyme not made by the organism (see Table 25.3).
vitelline membrane 1. see *fertilization membrane.* 2. the cell membrane of an egg.
viviparity the condition of producing live young.
viviparous (adj.), producing live young from within the body of the female, following development of the young within a uterus in intimate association with the tissues of the mother; compare *ovoviviparous, oviparous.*
voice box see *larynx.*
voluntary muscle see *skeletal muscle.*
volvocine pertaining to the colonial green algal flagellate *Volvox.*
vulva the external genitalia of a woman.

water mold an aquatic fungus of the oomycetes.
water vacuole see *contractile vacuole.*
water vascular system a system of vessels in echinoderms, containing sea water, used in the movement of tentacles and tube feet and possibly in excretion and respiration.
Watson strand either of the two strands of DNA, the other being designated the *Crick strand.*
wave a disturbance or variation that transfers itself and energy progressively from point to point in a medium or in space in such a way that each particle or element influences the adjacent particles or elements, and that may be in the form of an elastic deformation or of a variation in pressure (e.g., sound wave), or of electric and/or magnetic intensity (e.g., electromagnetic wave); see also *electromagnetic radiation.*
wave length the physical distance from one point of maximum intensity to the next in a propagated wave, especially of electromagnetic radiation in a vacuum; inversely proportional to *frequency,* the number of waves passing a point per second, and also inversely proportional to the energy per photon.
wax 1. a dense, hard, lipid-soluble ester of a long-chain alcohol and a fatty acid. 2. beeswax, one such compound. 3. a natural mixture of such monohydroxy esters and other lipid substances.
W chromosome one of the sex chromosomes of birds and butterflies such that the chromosome constitution ZW results in female development; sometimes considered to be a *Y chromosome;* carries the H-Y antigen gene.
Weberian ossicles a set of tiny bones that connects the swim bladder to the inner ear in some fish.
wheat rust 1. any of three distinct fungal diseases of wheat. 2. a rust fungus responsible for such a disease.
whorl 1. a group of parts repeated in a circle. 2. any of the four basic radially repeated groups of flower parts; see *gynoecium, androecium, corolla, calyx.* 3. a fingerprint in which the ridges form concentric circles.
wild type 1. of a phenotype, normal or typical in appearance; not mutant, rare or abnormal. 2. of an allele, normal and common in the population; not mutant; giving rise to a typical phenotype when homozygous.
wilt (v.i.), to loss turgor, especially because of an inadequate supply of water to a plant or plant part.
wood the hard, fibrous xylem of secondary growth, especially that of the central stem of a tree, consisting of lignified cellulose cell walls.
woody containing wood or wood fibers; consisting largely of secondary growth with hard lignified tissues.
work in physics, the transfer of energy resulting in the movement of an object formerly at rest.

X chromosome 1. one of the two heteromorphic sex chromosomes of mammals, flies and certain other insects, such that the possession of two X chromosomes without a Y chromosome results in the development of a normal female, while the possession of one X and one Y chromosome results in a normal male and other combinations are rare and abnormal. 2. the sex chromosome of grasshoppers and certain other insects such that the possession of two X chromosomes results in female development while the possession of only one results in male development. 3. see *Z chromosome.*
X-linked also *sex-linked,* the condition, common to many genes of functions unrelated to sex, of being present on the X chromosome; thus males have only one copy rather than the normal diploid two and recessive X-linked alleles are always expressed in males, thus radically affecting patterns of inheritance.
xanthophyll any of the bright yellow caretenoid pigments of flowers, leaves, or fruit.
xeric (adj.), of an environment, low in moisture available for plant life; dry.
xeroderma pigmentosum a rare autosomal recessive condition of humans in which the DNA repair system specific for u.v.-induced damage is defunct, resulting in a pathological sensitivity to sunlight that usually leads to multiple skin cancers.
xerophyte a plant adapted to dry areas.
X-organ an endocrine gland, found in the eyestalk of crayfish, that produces *molt-inhibiting hormone;* compare *Y-organ.*
X-ray a highly energetic form of electromagnetic radiation with wavelengths of .01 nm to 10 nm, between gamma radiation and u.v. radiation.
X-ray crystallography a procedure of directing X rays at a crystal of a protein or other large molecule, recording the pattern produced as the rays are bent by the regular, repeating molecular structures within the crystal, and using this pattern to reconstruct the 3-dimensional stucture of the molecule.
XXY syndrome see *Klinefelter's Syndrome.*
xylem one of the two complex tissues in the vascular system of vascular plants, consisting of the dead cell walls of vessels, tracheids or both, often together with sclerenchyma and parenchyma cells; functioning chiefly in the conduction of water and in support, but sometimes also in food storage; the substance of wood in secondary growth as well as a part of the vascular bundles of primary growth; see *tracheid, vessel, lignin;* compare *phloem.*
xylose a five-carbon sugar obtained chiefly from wood.
XYY syndrome also called extra-Y syndrome, an abnormal genetic condition of males due to nondisjunction, characterized by markedly excessive height, a tendency to severe acne and mild mental retardation that in itself tends to get affected individuals into trouble of various kinds.

Y chromosome 1. one of the two heteromorphic sex chromosomes of mammals, flies and certain other insects such that XY individuals are male and XX individuals are female; in mammals carrying the primary male sex determinant for the H-Y antigen, a determinant for increased height, and few other genes; in general nearly devoid of genes not concerned with maleness; compare *X chromosome.* 2. see *W chromosome.*
yeast a unicellular sac fungus, especially *Saccharomyces cerevisiae,* which is used in the production of bread, beer, wine, and raised doughnuts.
yellow-green alga any photosynthetic protist of the phylum *Chrysophyta* (see Appendix A).
yolk the material stored in an ovum that supplies food material to the developing embryo, consisting chiefly of *vitellin* (a lecithin-containing phosphoprotein), nucleoproteins, other proteins, lecithin (a phospholipid) and cholesterol.
yolk gland also *vitellarium,* 1. a modified part of the ovary of flatworms and rotifers that produces yolk-filled cells that nourish the true eggs. 2. the organ of an insect in which the egg cells grow to mature size.
yolk mass the mass of yolk in an egg cytoplasm; see *yolk.*
yolk plug a plug-like mass of yolk-filled endoderm cells left protruding from the blastopore of an amphibian embryo after the cresentic blastopore enlarges to form a complete circle.
yolk sac 1. one of the extraembryonic membranes of a bird, reptile or mammal, that in birds and reptiles grow over and encloses the yolk mass and in placental mammals encloses a fluid-filled space; the first site of blood cell and circulatory system formation. 2. an analogous structure in elasmobranchs and cephalochordates.
Y-organ one of the endocrine glands of a crayfish eyestalk, secreting *crustecdysone,* the crustacean molt-inducing hormone.

Z chromosome one of the two heteromorphic sex chromosomes of birds and butterflies, such that ZZ individuals are male and ZW individuals are female; sometimes considered to be an X chromosome; compare *W chromosome, X chromosome, Y chromosome.*
zeatin a plant cytokinin hormone and growth factor extracted from corn.
Z line in striated muscle, the partition between adjacent contractile units to which actin filaments are anchored.
zona pellucida a thick, transparent, elastic membrane or envelope secreted around an ovum by follicle cells.
zooplankton animal life drifting at or near the surface of the open sea.
zoospore 1. any independently motile spore. 2. the flagellated gamete of a foraminiferan.

zooxanthela intracellular, photosynthetic, symbiotic dinoflagellates common in radiolarians, corals, marine flatworms and various other invertebrates.

Z protein an enzyme in the thylakoid that is involved in a complex series of reactions resulting in the oxidation of water, releasing molecular oxygen and hydrogen ions into the thylakoid lumen and passing electrons to oxidized chlorophyll a.

Z scheme a graphic presentation of the oxidation-reduction reactions occurring in the light reaction and electron transport chain in photosynthesis.

Zwitterion an ion with both positively charged and negatively charged molecular side groups; e.g., a free amino acid, $^{+}NH_3$—HCR—COO^{-}.

zygomatic arch the arch of bone that extends along the front and side of the skull below the orbit, passing outside of the jaw muscle; in humans also called the *cheekbone.*

zygospore a diploid fungal or algal spore formed by the union of two similar sexual cells, having a thickened and usually ornamented wall, that serves as a resistant resting spore.

zygote a cell formed by the union of two gametes; a fertilized egg.

zygotene the stage in meiotic prophase in which homologous chromosomes synapse; see *synapse, synaptinemal complex.*

ILLUSTRATION ACKNOWLEDGMENTS

All photographs have been supplied by Photo Researchers (PR)/National Audubon Society Collection, 60 E. 56th Street, New York, New York 10022.

Cover Dr. Kakefuda, Courtesy of NHI
Page viii Russ Kinne
Page xi Russ Kinne
Page xiv Eric V. Gravé
Page xv Kjell B. Sandved
Page xvii Soames Summerhays
Page xxii Cosmos Blank
Page xxiii Karl H. and Stephen Maslowski
Page 1 P. Schulz, Biology Media

Chapter 1
1.1 Jen and Des Bartlett
1.5 (left) Roy Pinney; (center) Robert McLeod; (right) Zoological Society of San Diego
1.7 (left) Al Lowry; (right) Sonja Bullaty
1.8 A. Francesconi
1.11 (left) A. W. Ambler; (right) Herman Schlenker
1.12 Courtesy of the New York Public Library

Chapter 2
2.13 (left) Harold W. Hoffman; (right) Peter B. Kaplan
2.14 Dr. William J. Jahoda
2.20 Michael P. Gadomski
2.26 Hugh Spencer
Page 46 (top) Farrell Grehan; (left) Cosmos Blank; (center) Cosmos Blank; (bottom) Karl W. Kenyon
Page 47 (top) Stephen Dalton; (center) M. I. Walker; (bottom, left) Michael Tweedie; (bottom, right) Michael Tweedie
2.28 Courtesy of the New York Public Library

Chapter 3
3.4 Eric V. Gravé
3.8 Omikron
3.9 Walter Dawn
3.17 (top) Jeanne White; (bottom) Joe Munroe
3.19 Marshall Sklar

Chapter 4
4.4 (top, left) Bruce Roberts; (top, center) M. I. Walker; (top, right) Hugh Spencer; (bottom, left) Eric V. Gravé; (bottom, center) R. Knauft, Biology Media; (bottom, right) Eric V. Gravé
Page 91 (left) Walter Dawn; (center) Omikron; (right, top) Eric V. Gravé; (right, bottom) Dr. William M. Harlow
Page 92 Lee D. Simon, Omikron
4.7 (a) Omikron; (b) Omikron
Page 98 Biology Media
Page 99 Daniel Branton
4.8 J. F. Gennaro, Jr.
4.9 Daniel Branton
4.17 (a) Marshall Sklar; (b) Daniel Branton; (c) K. R. Porter, University of Colorado
4.19 Warren Rosenberg
4.20 P. Schulz, Biology Media
4.22 Marshall Sklar
4.23 (a) W. A. Jensen, Biology Media; (b) Biology Media
4.24 Marshall Sklar
4.25 Biology Media
4.27 (left) Omikron; (center) Benjamin Bouck, originally published in *J. Cell Biol*, 50: 373
4.28 E. Visil, Biology Media

Chapter 5
5.1 Kent and Donna Dannen
5.20 (top, left) Courtesy of O. M. Scott Company; (top, right) Porterfield-Chickering; (bottom, left) R. C. Hermes

Chapter 6
6.7 J. Paul Kirouac
6.8 Bradley Smith
Page 168 Bettye Lane
Page 181 Dr. Joseph Gall, Omikron

Chapter 7
Page 189 Courtesy Dr. Kakefuda, NHI
Page 194 (top) Lee D. Simon, Omikron; (bottom) Lee D. Simon, Omikron;
Page 195 (top, left) E. Kellenberger, Omikron; (top, left) E. Kellenberger, Omikron
7.10 Rosalind Franklin
7.13 O. L. Miller, Jr., and B. R. Beatty, *J. Cell Physiol.*, Vol. 74 (1969)
7.14 (d) J. F. Gennaro; (e) Alexander Rich
7.22 Alexander Rich

Chapter 8
8.1 (top, left) Courtesy Harry S. Truman Library; (top, center) Hare Studios, Courtesy Harry S. Truman Library; (top, right) Strauss-Peyton, Courtesy Harry S. Truman Library; (bottom, left) St. Louis Globe Democrat, Courtesy Harry S. Truman Library; (bottom, right) United Press International
8-5a Carolina Biological Supply Company
8-5b (1–4) Biology Media; (5) Carolina Biological Supply Company; (6–9) Biology Media
8.7 W. Engler, Armed Forces Institute of Pathology (Courtesy of G. F. Bahr)
8.8 J. Pickett-Heaps
8.9 Carolina Biological Supply Company
8.10 (a) R. Knauft, Biology Media; (b) George Whitely; (c) R. Knauft, Biology Media; (d) Carolina Biological Supply Company
8.11 (left) Peter Hepler; (right) Peter Hepler
8.13 Jean-Marie Lucciani, Courtesy Pasteur Institute of Paris
8.14 Jean-Marie Lucciani, Courtesy Pasteur Institute of Paris
8.15 Courtesy James Walters
8.17 Courtesy James Walters
8.18 (a) Courtesy James Walters; (b) Jean-Marie Lucciani, Courtesy Pasteur Institute of Paris; (c, d) Courtesy James Walters
8.20 Dr. Joseph Gall, Omikron
Page 256 Bruce Roberts
8.22 Landrum B. Shettles, M.D.

Chapter 9
9.1 Courtesy Burpee Seed Company
9.2 Leslie Holzer
9.11 Toni Angermayer
Page 284 (left) Omikron; (right) Nigel Calder, Omikron
9.15 Eric V. Gravé

Chapter 10
10.2 Carolina Biological Supply Company
10.5 Russ Kinne
Page 302 Courtesy New York Public Library
10.18 Courtesy Mrs. Marjorie Guthrie, president, Committee to Combat Huntington's Disease
10.19 Albert F. Blakeslee, *Journal of Heredity*, Vol. V, No. 11, p. 512 (1914)
10.14 Hans Laufer
10.16 (top) Keystone Press Agency; (bottom) Wide World Photos

Chapter 11
Page 327 M. W. F. Tweedie
11.5 Ray Ellis
11.6 George Holton
11.16 (left) Tom McHugh, Courtesy Wildlife Unlimited; (right) Mark N. Boulton
11.17 Lynn McLaren

Chapter 12
12.2 Stadler, Omikron
12.10 Drs. N. Yamamoto and T. F. Anderson
12.13 Drs. T. F. Anderson, F. Jacob, and E. Wollman
Page 385 Eric V. Gravé

Chapter 13
13.1 (top, left) Harry Engels; (top, right) Tom McHugh, Courtesy Pt. Defiance Zoo, Tacoma; (right, center) Leonard Lee Rue III; (right, bottom) Tom McHugh, Courtesy Healesville Sanctuary, Australia; (bottom, center, right) Mark N. Boulton; (bottom, center, left) M. Philip Kahl, Jr.; (bottom, left) Russ Kinne; (center) Jeanne White
13.2 (left) Carleton Ray; (right) Tom McHugh, Courtesy Chicago Zoological Park
13.3 (left) Joseph T. Collins; (right, bottom) George Porter; (right, top) George Porter; (center, top) Norman R. Lightfoot
13.5 (center) Karl H. Maslowski; (right) Irvin L. Oakes; (left) Russ Kinne
13.10 (bottom, left) Leonard Lee Rue III; (bottom, right) Paolo Koch

Chapter 14
14.3 Omikron
14.4 C. Robinow, Omikron
14.5 Eric V. Gravé
14.6 (top, left) Omikron; (top, center) Omikron; (top, right) Eric V. Gravé; (bottom, left) Omikron; (bottom, center) Eric V. Gravé; (bottom, right) Eric V. Gravé
14.7 Omikron
14.9 (left) Courtesy Lederle Laboratories, Omikron; (right) Eric V. Gravé
14.11 Courtesy Prof. J. William Schopf, U.C.L.A.
14.12 (top, left) R. Knauft, Biology Media; (top, right) Hugh Spencer; (bottom, left) Eric V. Gravé
14.13 Biology Media
14.15 K. R. Porter
14.18 Tom McHugh

Chapter 15
15.1 Russ Kinne
15.6 Russ Kinne
15.7 Omikron
15.8 Eric V. Gravé
15.10 Eric V. Gravé
15.11 Eric V. Gravé
15.12 N. E. Beck, Jr.
15.14 (top) M. I. Walker; (top, middle) Michael Abbey; (bottom, center, left) Eric V. Gravé; (bottom, middle, right) Eric V. Gravé; (bottom, left) Eric V. Gravé

Chapter 16
16.1 W. V. Crick
16.2 Eric V. Gravé
16.3 (left) P. W. Grace; (right) Robert Knauft, Biology Media
16.4 P. W. Grace
16.6 Eric V. Gravé
16.7 Pfizer, Inc.
16.11 (left) Barry Lopez; (right) Ken Brate
16.12 Biology Media
16.13 (a) Hugh Spencer; (b) Grant M. Haist; (c) Hugh Spencer; (d) H. E. Stork
16.15 (a) Eric V. Gravé; (b) New Jersey Experimental Station, Omikron
Page 463 Grant Haist

Chapter 17
17.1 Tom McHugh
17.5 (a) Ron Curbow; (b) Avis Reeves
17.7 M. I. Walker
17.9 (a) Richard Chesher; (b) Anne Hubbard
17.10 Edna Bennett
17.14 Carolina Biological Supply Company
17.15 Glenn Foss
17.16 Carolina Biological Supply Company
17.17 (left) Noble Proctor; (right) Carolina Biological Supply Company
17.18 Ken Brate
17.19 Courtesy American Museum of Natural History
17.20 Grant Haist
17.21 Russ Kinne
17.25 Courtesy New York Public Library © 1931 Field Museum, Chicago
17.26 A. J. Belling, Dept. Biology, New York University
17.27 (left) Russ Kinne; (right) A–Z Collections
17.28 Ken Brate
17.29 S. J. Krasemann
17.31 Courtesy American Museum of Natural History
17.33 (a) M. W. F. Tweedie; (b) M. W. F. Tweedie; (c) Hermann Eisenbeiss; (f) Kjell B. Sandved
17.35 Courtesy American Museum of Natural History
17.40 V. Claerhout

Chapter 18
18.1 Joe Munroe
18.6 Hugh Spencer
18.9 George Whitely
18.12 (left) M. I. Walker; (right) Omikron
18.17 Robert Boardman
18.18 Omikron

Chapter 19
19.2 Charles R. Belinky
19.5 U.S.D.A. Photo, Omikron
19.7 A. B. Joyce
19.8 J. Paul Kirouac
19.12 Kenneth W. Fink
19.13 (a) Jerome Wexler; (b) Hal H. Harrison; (c) Alvin E. Staffan

Chapter 20
20.1 Max and Kit Hunn
20.5 Farrell Grehan
20.6 (a) Pierre Berger; (b) Bettina Ciccione
20.8 Omikron
20.12 Omikron

Chapter 21
21.4 Kjell B. Sandved
21.5 (a) Peter B. Kaplan; (b) M. E. Warren; (c) James Simon
Page 570 (top, left) Leonard Lee Rue IV; (top, right) Leonard Lee Rue III; (bottom, left) Jewel Craig; (bottom, center) Michael P. Gadomski; (bottom, right) Jane Kinne
Page 571 (left) Farrell Grehan; (top, right) Russ Kinne; (bottom, right) Dan Guravich
Page 572 (top) Yeager and Kay; (bottom) Jerome Wexler
Page 573 (top, left) Robert Bornemann; (top, right) George Carva; (bottom, left) Ken Brate; (bottom, right) Leonard Lee Rue III
21.19 R. Knauft, Biology Media
Page 591 Russ Kinne

Chapter 22
22.6 (left) Nancy Sefton
22.7 (left) Thomas W. Martin
22.8 Carleton C. Ray
22.9 (a) Soames Summerhays; (b) Bucky and Avis Reeves
22.10 Jack Dermid
22.11 N. E. Beck, Jr.
22.12 Eric V. Gravé
22.14 W. F., Mai, Omikron
22.15 Eric V. Gravé
22.16 Russ Kinne
22.18 Tom McHugh
22.19 (a) Russ Kinne; (b) Russ Kinne
22.21 Russ Kinne
22.22 Eric V. Gravé
22.24 (left) Mark N. Boulton; (right) Bruce Hunter
22.26 (a) Allan Power; (b) Anthony Mercieca; (c) Jen and Des Bartlett; (d) Anthony Mercieca; (e) Anthony Mercieca
22.28 Russ Kinne
22.29 (left) Walter Dawn; (center) Robert Hermes; (right) Carolina Biological Supply Company
22.31 Jerome Wexler
22.32 (a) Tom McHugh, Courtesy of Steinhart Aquarium; (b) G. T. Hillman; (c) Karl H. Maslowski; (d) N. E. Beck, Jr.; (e) Tom McHugh
22.35 Robert C. Hermes
22.36 Treat Davidson
22.37 (a) Soames Summerhays; (b) Kjell B. Sandved; (c) Al Lowry; (d) Dan Sudia; (e) Tom McHugh, Courtesy of Steinhart Aquarium
22.38 Russ Kinne
22.39 George G. Lower

Chapter 23
23.2 Tom McHugh, Courtesy of Taronga Zoo, Sydney
23.5 (a) Karl H. Maslowski; (b) Tom McHugh, Courtesy of Steinhart Aquarium
23.8 (a) Soames Summerhays; (b) Tom McHugh, Courtesy of Steinhart Aquarium; (c) NOAA, Omikron
23.16 Lynwood M. Chace
23.21 Robert W. Hernandez
23.32 (left) Tom McHugh; (center, left) Jen and Des Bartlett; (center, right, top) Anthony Mercieca; (top, right) Charles G. Summers, Jr.; (center, right, bottom) Allan D. Cruickshank; (right, bottom) Dan Sudia
23.23 A. W. Ambler
23.31 Tom McHugh, Courtesy Field Museum, Chicago

Chapter 24
24.1 (top) Kjell B. Sandved; (bottom) Louis Quitt
24.5 (top) Tom McHugh; (bottom) Russ Kinne
24.6 Russ Kinne
24.7 (a) Nancy Sefton; (b) John C. Deitz; (c) Carleton Ray
Page 676 Michael Abbey
24.20 (a) Marshall Sklar; (b) Omikron; (c) Omikron
24.25 (top) G. F. Gauthier, Omikron; (bottom) Biology Media
24.26 Omikron

Chapter 25
25.8 Russ Kinne
25.10 Ken M. Highfill
25.16 Graziadai, Omikron
25.18 Omikron

Chapter 26
26.10 Ellen R. Dirksen, Courtesy U.C.L.A., Dept. of Anatomy, School of Medicine
26.20 Courtesy of the Armed Forces Institute of Pathology
26.22 Russ Kinne
26.23 Stephen L. Feldman
26.26 Courtesy of the Armed Forces Institute of Pathology (negative CA-44271-C)
26.27 D. W. Fawcett
26.29 Courtesy New York Public Library
26.30 Omikron
26.32 Emil Bernstein and Eila Kairinen, Gillette Research Institute
26.33 Eric V. Gravé

Chapter 27
27.1 Hugh Spencer
27.3 Cosmos Blank
27.4 Stephen Dalton
27.10 (a) UPI; (b) World Wide Photos
27.12 (a) Omikron; (b) Paul Almasy, Omikron
27.15 N.Y.U. College of Medicine, Omikron
27.19 Arthur W. Ambler

Chapter 28
28.2, 28.3, 28.4 From "Fishes with Warm Bodies," by Francis C. Carey. Copyright © 1973 by Scientific American, Inc. All rights reserved.
28.5 From "How Reptiles Regulate Their Body Temperature," by Charles M. Bogert. Copyright © 1959 by Scientific American, Inc. All rights reserved.
28.6 (a) William Belknap, Jr.; (b) Allan D. Cruickshank
28.7 Field Museum
28.9 (left) Russ Kinne; (right) Russ Kinne

Chapter 29
29.1 (a) Nancy Sefton; (b) Tom McHugh; (c) Ruth H. Smiley
29.4 A. Cosmos Blank
29.5 (top) Richard Ellis; (bottom) Russ Kinne
29.6 Ken M. Highfill
29.9 Tom McHugh
29.10 Mary M. Thacher
29.11 (a) Allan D. Cruickshank; (b) A. Cosmos Blank
29.15 Dr. Gerald P. Schatten, Science Photo Library
29.25 (top, left) Russ Kinne; (top, right) Russ Kinne; (bottom, left) Russ Kinne
29.26 Robert de Gast

Chapter 30
30.3 (a) V. D. Vacquier, *Exptl. Cell Res.* 82 (1973); (b) V. D. Vacquier, *Exptl. Cell Res.* 82 (1973); (c) R. Knauft, Biology Media
30.4 Dr. Everett Anderson, Science Photo Library
30.6 Carolina Biological Supply Company
30.9 Carolina Biological Supply Company
30.12 Carolina Biological Supply Company
30.16 Russ Kinne
30.19 From "The Development of the Brain," by W. Maxwell Cowan. Copyright © 1979 by Scientific American, Inc. All rights reserved.
30.20 (a) Jerome Wexler
Page 905 Toni Angermayer

Chapter 31
31.1 (top, right) Robert Knauft, Biology Media; (bottom, right) Eric V. Gravé
31.15 Gordon S. Smith
31.21 Stephen Dalton
31.23 Omikron
31.25 M. W. F. Tweedie
31.26 Jerome Wexler

Chapter 32
32.1 (a) Nina Leen, *Life Magazine;* (b) World Wide Photos
32.9 Van Bucher
32.10 Van Bucher
32.13 Will Rapport

Chapter 33
33.1 UPI
33.4 Charles Walcott
33.6 Toni Angermayer
33.7 Toni Angermayer
33.10 (a) Allan D. Cruickshank; (b) A. A. Francesconi
33.11 W. B. Allen, Jr.
33.12 Leonard Lee Rue III
33.14 Jen and Des Bartlett
33.15 Mary M. Thacher
33.17 Colin G. Butler, F.R.P.S.
Page 991 M. P. Kahl

Chapter 34
34.2 Mark N. Boulton
34.6 (top) Russ Kinne; (bottom) Verna R. Johnston
34.7 (a) Allan D. Cruickshank; (b) Robert H. Wright; (c) Woodrow Goodpaster; (d) George M. Bradt
34.8 Bill Eppridge, *Life Magazine*
34.9 (left) M. P. Kahl; (right) François Gohier
34.10 Russ Kinne
34.11 Helen Cruickshank
34.12 Townsend P. Dickinson
34.14 Tom McHugh
34.15 (left) Van Bucher; (center) F. B. Grunzweig; (right) Michael Philip Manheim
34.16 (top) Karl H. and Stephen Maslowski; (center) Kirtley Perkins; (bottom) Karl Maslowski
34.17 Tracy Knauer
34.18 Charlie Ott
34.19 (left) Charlie Ott; (top, right) Hugh M. Halliday; (bottom, right) Charlie Ott
34.20 Charlie Ott
34.21 Leonard Lee Rue III
34.22 (top) Stephen J. Krasemann; (center) Karl W. Kenyon; (bottom) Karl H. and Stephen Maslowski
34.24 George G. Lower
34.26 Tom McHugh
34.27 Jen and Des Bartlett
34.31 From "The Energy Cycle of the Biosphere," by George M. Woodwell. Copyright © 1970 by Scientific American, Inc. All rights reserved.
34.34 J. Paul Kirouac

Chapter 35
35.6 Tom McHugh
35.7 (a) Leonard Lee Rue III; (b) Robert Bright
35.9 Karl H. Maslowski
35.10 Georg Gerster, Rapho Div., PR
35.11 De Sazo, Rapho Div., PR
35.12 Leonard Lee Rue III
35.17 Tom McHugh, Courtesy Wildlife Unlimited
35.18 Karl W. Kenyon
35.19 Gianni Tortoli
35.20 Courtesy of the Armed Forces Institute of Pathology (Neg. CA #44271-C)
35.21 Joe Munroe

INDEX

A

A band of sarcomere, *691, 692*
A1 cell, *926,* 927
A2 cell, *926,* 927
aardvark, 655
abiotic factors, 1014
  in mortality, 1038
ABO blood group system, 282, *283*
abomasum, 712
abominable snowman, 657
abortion, 853, 857, 858
  spontaneous, 894
  spontaneous, due to chromosomal anomalies, 256
  responsible for declining birth rate, 1055
abscisic acid, 552, 553
abscission, 553
abscission zone, 553
absorption spectrum, *see* spectrum, absorption
abstinence, 853
abyssal region, *1009,* 1017
acellular organism, 91
acellular slime mold, *see* slime mold
Acetabularia, 91, 470, *471*
acetabulum, 681
acetaldehyde, in fermentation, 164
acetylcholine, *910,* 911
  structural formula, *910*
acetylcholinesterase, 911
acetyl CoA, 168, *169*
  conversion to fat, 723
acetylene, 39
Achilles tendon, 686
achondroplastic dwarf, 280, 281
acid, 40
acid balance, control by kidney, 818
acid rain, 1040, 1041
acoel flatworm, 594, *595*
acoelomates, *596,* 597–604
acorn, 577
acorn worm, 634, *635*
  allied with echinoderms, 627
acquired immunity, 821, *822*
acrasin, 772
acromegaly, *781,* 782
acrosomal filament, 866
acrosome, 839, *841,* 866, 868
ACTH, 75, 776, 779, 780, 791; *see also* adrenocorticotropic hormone
actin, 686
  division furrow in cytokinesis, 240, *241*
  sarcomere, *691–693*
  spindle, 234, 237
  see also cytoskeleton
Actinomycetes, 418
Actinopterigii, *641*
action potential, 908, *909*
action-specific energy, 950, 951
  no evident neural basis, 952
action spectrum, *see* spectrum, action
active site, 69, 70
active transport, 95, 100, 104, 105
  of ADP into mitochondrion, *172,* 175
  of auxin, 549
  calcium ions and sarcoplasmic reticulum, 693, *694*
  of chloride from kidney tubule, 815, *816*
  of ions by root hairs, 538
  pinocytotic vesicles in capillary wall, *759*
  of water, 815
  of water into roots, *509,* 515; *see also* root pressure
Adam, all humanity in his scrotum, 868
adaptation, 593
  a characteristic of life, *46*
  see also evolution; natural selection; homeostasis
adaptation to land, *see* transition to land
adaptedness of behavior, chapter, 967–985
adaptive radiation, 18
  of animal phyla, 593
  of Darwin's finches, 5, *6,* 7
  of placental mammals, 652, *653*
additive law, 271, *272*
Adelie penguin, *328*
adenine, structural formula, 184
adenosine triphosphate, see ATP
adenyl cyclase, *791*
adhesion, 516
  of water, 37
adolescence and sex hormones, 789
ADP, 126
adrenal cortex, 727, *787,* 788, 895; *see also* adrenal gland
adrenal gland, *787,* 788
  control by pituitary, 780
  hormones listed, 777
  medulla, ectodermal origin, *876*
  sex hormones produced in cortex, 788, 895
adrenalin, *see* epinephrine
adrenocorticotropic hormone, 776, 779, *780*
  stimulates cAMP in target cells, 791
  39 amino acids, 75
advanced trait, brain size in vertebrates, 916, *917*
adventitious bud, 557, *558*
adventitious root, *506,* 507
afferent blood vessel, 737
African hunting dog, regurgitates food, 975, 982
African sleeping sickness, *433,* 440
Afrikaaners, 338
afterbirth, 896
agar, 63, 465
age profile, *1054*
age pyramid, *1054*
agglutination, 281–283, *821*
  ABO system, 283
  by antibodies, 282, 821
  MN system, 281
aggregate fruit, 572, *573*
aggression, 975–980
  fight to death, 977
  human, 978
  innate, 315, 977, 978
  ritualized combat, *976,* 977
  stickleback, *951, 952*
aggressive mimicry, 971
aging, 223, *224*
  hardening of the arteries, *755*
  human, 1036
  Harry Truman, *224*
  see also programmed death
agnatha, *637, 638,* 639, *641*
agranulocyte, 762
agriculture, beginnings of, 1051
  slash-and-burn, 1002
air bladder, 640, 641; *see also* swim bladder
air conditioner, example of homeostasis, 799
air currents cause deserts, 995, *996*
air fern, 609
air quality, 1041
air sac, 648
  alveoli of lungs, 742
  bird, 732, *740*
  insect tracheal system, *734*
alanine, structural formula, *66*
Alaskan pipeline, 1009
Albertus Magnus, scala natura, 946
albinism, 278, 279, 343, 351, *352*
albumen, egg white, 885
albumin, blood, 762
alcohol
  defined, 76
  depresses breathing centers, 746
  ethyl, *see* ethyl alcohol; ethanol
  suppresses ADH, 818
alcoholism, 724
aldehyde group, of simple sugar, 56
aldosterone, 777, 787
  control of kidney, *818*
  regulates blood acidity, 818
aleurone, *550,* 569, 577
alga, 428
  in competing kingdom schemes, 402, *403*
  lakes and ponds, 1013
  yellow-green and golden-brown, 430
  see also brown alga; green alga; red alga; cyanophyte; eutrophication
algin, 63, 466
alimentary canal, 703, 701–726
  human, 712–717
alkaloids, 533
alkalosis, 788
allantois
  bird and reptile egg, 643, *644, 885*
  develops into mammalian placenta, 650
  mammalian embryo, *887*
allele, 271, 321
  alternate, 273
  dominant and recessive, 278
allele frequency, 321, 323
  changes under artificial selection, 331
  equilibrium between mutation and selection, 338, *339*
allelic exclusion, 820
allelic interactions, 278–282, 328, 329, 356
  see also dominance relationships
allelopathy, 1040
Allen, Joel A., 337
Allen's rule, 336, *337*
alligator, brain, *917*
allopolyploidy, 390
allosteric site, 73, *74*
allotetraploidy, 389, 390
alpha amylase, 550, 577
alpha helix, *68*
alpha islet cell, 777, 786
alpha ray, 196
alternate alleles, 273; *see also* allele
alternate segregation, 297; *see also* Mendel, first law
alternation of generations, 464, 465, *466,* 473
  coelenterates, 599
  ferns, 481
  haploid spores, 353, 355
  horsetails, 481
  seed plants, 488
  see also life cycle
altitude sickness, 798
altricial and precocial birds, *836*
altruism, 982–985
  defined, 982
  evolution of, 20
  reciprocal, 983, 984
alula, 682, *683*
alveolus, 739, *741–744*
  gas exchange, *744*
Alvin, submersible, 1017
Amanita, 454
Amazon River basin, 1001, 1002, 1014
amber, *see* terminator codon
ameba, 433, *434, 435*
ameboid cell, of sponges, *598*
American Indian
  nearly extincted by smallpox, 1042
  superior vision, 340

amino acid, 64
abbreviations, 206
dietary requirements, 75, 720
structural formulas, *66, 67*
amino acid code, 206; *see also* genetic code
amino acid substitution, 216
amino group, 48
of amino acid, *64*
amino sugar, in bacterial cell wall, 411
aminoacyl-tRNA complex, 210
aminopeptidase, 720, 721
Amish, 338
ammonia, 41
as a nitrogenous waste, 808, *809,* 811
ammonification, *531*
ammonium ion, 41
atomic diagram, *44*
amnion, *885,* 886
bird and reptile egg, *644*
mammalian, 874, *875,* 886, *887*
ruptures at birth, 894
amniotic cavity, 892
AMP, 126
amphibia, *642, 643, 831*
absence on oceanic islands, 11
caecilian, *395,* 643
excretory strategies, 811, *812*
gastrulation, 872, *873*
migrating toad, 976
mouth and digestive system, 708
neurulation, *877*
phylogeny, *395*
respiratory system, *739*
sound communication, 972
spadefoot toad of desert, 977
tail bud stage, *878*
transplantation of nuclei, 880–882
Xenopus, 864, *881*
see also caecilian; frog; salamander; toad
Amphioxus, 636, see also lancelet; cephalochordate
amplexus, 643, *831*
ampulla of tube foot, *626,* 735
amylase, 712, 718
amylopectin, *60, 719*
amylose, 59, *719*
anaerobic organisms, 163, *165*
bacteria, 416
lake bottom, 1013
anaerobic respiration, 159–167, 454
in muscle, 15, *166,* 168
analogy, 399
distinguished from homology, 396
preposterous, 240
anaphase
in cell cycle, *229*
mitotic, 240
anaphase I of meiosis, *248*
anaphase II of meiosis, 250, *251*
Andes, sea shells on, 4
androecium, defined, *560*
androgen insensitivity syndrome, 895
androgens, 777, 788, 892, 893
produced by adrenal cortex, 788
anenome, 600, *601*
outlives humans, 1036
angiosperm, 486, 488–496
origins, 486, 491
see also flowering plants
Angkor Wat, 1002
angler fish, *640*
animal, defined, 592
diversity, 592–663
animal behavior, *see* behavior
animal kingdom, 592–663
characteristics, 404
evolutionary relationships with other kingdoms, 402, *403*
fossil record of major phyla, 593
origins and phylogeny, 558, 593
animal pole, 865, 871
animal reproduction, 830–858
see also reproduction; sexual reproduction; copulation; reproductive system; asexual reproduction
annelid, 609–612
hydrostatic skeleton, *670*
see also earthworm
annual animals, 1035
annual plant
desert flowers, 996, 997
programmed death, 1035
annulus, nuclear pore, 224
fern, 481, *482*
Anolis lizard, territoriality, *979*
Anopheles mosquito, 435–437
ant pheromones, 973
Antarctic Sea, high productivity, 1012
anteater, 654
antelope, females fight, 976, 977
antenna, insect, *706,* 934
antenna, light, 141–143
anterior vena cava, 751, *752, 753,* 763
see also vena cava
anther, *560,* 567
antheridium
of ascomycete, 451
of water mold, 447, *448*
antheridogen, 482
anthocyanin, 117
Anthozoa, 600, 601, *601*
anthrax, 415
Anthropoidea, 655
anthropology, 658–663, 1034
antibiotic resistance factor (plasmid), 360
antibiotics
cause of vaginal yeast infection, 842
dangers of overuse, 424, 454, 842
evolution of antibiotic resistance, 324
produced by chlamydobacteria, 418
resistance carried by plasmid, 360
antibody, 282, 283, 819
in colostrum, 897
crosses placenta, 894
defined, 282
membrane-bound, 819
specificity, 820–823
anticodon, 208
anticodon loop of tRNA, 208
antidiuretic hormone, 726, 779, 780, 815, 816
amino acid sequence compared with oxytocin, 75, 782
feedback loop, 818
interacts with aldosterone, 787
stimulates cAMP in target cells, 791
antigen, 819
defined, 282
antigen binding site, immunoglobulin, 819, *820*
antihemophilic factor, 301
antiparallel, 186, *187*
antisera
MN system, 281
anti-Rh prevents erythroblastosis, *286, 287*
antisprouting factor, 882
antlers, function to prevent serious injury, 977
anuran, 643, *see also* frog; toad; amphibia
anus, human, *713,* 717
anxiety and drive reduction, 952
aorta, human, *752, 753*
aortic arch, earthworm, *747*
aphid, *518,* 622
apical dominance, 547, *548,* 579
apical meristem, 546, 547
root, 597, *508, 509,* 579
shoot, 579, *581, 582*
apodeme, *671*
apomixis, 559, 560
aponeurosis, 690
Appalachian Mountains ecocline, *994*
appendicular skeleton, 681–686
defined, 675
appendix, 704, 717
appetitive behavior, 949, 950, 953
Aquinas, St. Thomas, and instinct, 946
arachnids, 617, *618, 619*
arch of foot, *686*
archegonium, 473, *474, 477,* 479, 481
archenteron, amphibian embryo, 872, *873*
Archeopteryx, *645*
archosaur, *646, 647*
arctic fox, 1007
arctic sea
high productivity, 1012
arctic tundra, *993,* 1007–1009
arginine
and ionic bonds, 65
structural formula, *67*
synthetic pathway, *355*
argon, atomic diagram, 30
Aristotle, scala natura, 946
arm, of a chromosome, 234, 240
armadillo, 4, 654
extinct giant, 4
arousal, and reticular system, *919*
Artemia, 800, *810*
arterial arches, *890*
arterioles, 750, 751
arterioles, of capillary bed, *758*
arteriosclerosis, *79, 755*
artery, *755,* 756
anatomy, *755*
defined, 751
arthritis
and alcaptonuira, 351
an autoimmune disease, 820
treatment with cortisol, 788
Arthrodira, 638
arthropod, 617–623, *624*
compound eye, *620*
exoskeleton, *671, 672*
origins, *616*
polyphyletic group, *616,* 617
reproduction in terrestrial forms, 833, *834, 835*
sound in communication, 972
articulation, 675, *677*
artificial selection
decline in fitness, 329, 330
dog breeds, 329
leg length of chickens, *329*
geotaxis in flies, 330, *331, 332*
mouse weight, 329, *330*
Artiodactyla, *653,* 655, 710
Ascaris lumbricoides, 606
ascending colon, *713,* 717
Ascension Island, 9
Aschelminthes, 604–608
movements, 670, 671
musculature, 670, 671
see also nematode; rotifer
ascocarp, *452*
Ascomycetes, 450–454
ascorbic acid, 724
ascus (pl. asci), *451, 452*
of yeast, *453*
asexual organisms, population genetics of, 323, *324*
asexual reproduction
contrasted with sexual reproduction, *325*
evolutionary significance, 559
by mitosis, 242, 243
in plants, 557–560
planaria, 603
rotifers, 607
by strobilation, 480, *600*
Ashkenazi Jewish, 338
asparagine, structural formula, *66*
aspartic acid
and ionic bonding, 64
structural formula, *67*
Aspergillus, 451
assortment, independent, 273–275
aster, 117, *238, 239*
formed by centriole, 237
lacking in flowering plants, 238
whitefish blastula, *231*
asteroid, collides with earth, 646
Asteroidea, *625, 626*
astral ray, *238, 239; see also* aster
atherosclerosis, 79, *755,* 762
athlete's foot, 445, 456
atlas, *677,* 679
atmosphere, 1019, 1020
origin of oxygen, 536
atom
defined, 24
structure of, 26
atom bomb, 977
fallout and test ban, 1041
atomic mass, 26
atomic number, 26
atomic weight, 26
ATP
balance sheet in respiration, 174, 175
cycle, *127,* 128
muscle contraction, 693, *694*
net production in respiration of glucose, *175*
reserves in muscle, *166*
structural formula, *127*
synthesis in CF1 particle, 145
as universal molecule of energy transfer, 126
atrioventricular node, *754*
atrium, *636*
amphibian, *749*
fish, *748*
human heart, 751, *752, 753*
reptile, *749*
attachment sites, ribosome, 205, *206,* 211, *212, 213*
auditory canal, 927, *928*
skull, *679*
auditory nerve, *928*
auditory reception, temporal lobe, 920
Australia, tablelands of, 532
Australopithecus, 659
phylogony, 665
autoimmunity, 819, 820
parathyroid glands, 785
autolysis, *109*
automobile exhaust, source of nitrogen oxides, 532
autonomic learning, 687, *957,* 958
autonomic nervous system, 922, 923
control of digestion, 721, 722
control of heartbeat, 754
diagram, *922*
peristalsis, 713, 714
smooth muscle, 687
and thermoregulation, 807
autoradiography, *196, 197*
and localization of RNA transcription, 202
autosome
define, 256

autotrophic bacteria, 416
  Galapagos rift community, 1017
autotrophic worm, 1017
auxin, 546–549, 583
  control of flowering, 561, 565
  discovery of, 545, 546
  structural formula, 546
auxospore, *431*
A-V node, *754*
avidin, 73, 725
Avogadro's number, 128
avoidance response, 952
Awramik, Stanley, discovers oldest fossil, 413
axial skeleton, 677–682
  defined, 675
axis, *677*, 679
axon, 882, 883, 906, *907*
  giant squid axon, 911
axon tree, 882
aye-aye, *398*

B

B cell lymphocyte, 819
baboon, male threat display, *972*
  territoriality, 979
baby boom, 1054, 1055
bacillus, *409*, 415, *416*
back cross, *276*, 277
backbone, 678–680
  carbon, 39, *41*
  of DNA, 184
bacteria
  chemosynthetic, 136, 416, 1017
  counting techniques, *362*
  exponential growth, 1029
  inhibit own growth, 1039, 1040
  major shapes, 413
  oldest fossils, 413, *419*
  phylogeny of major groups, 413
  in plant kingdom, 402
  screening for mutants, 336, 356, *357*
  "villains," 415
bacterial genetics, 356, 357, *358*
bacteriochloropyll, 417
bacteriophage, *92*, 357–361
  activity in host, 358
  described, 193, *194*, *195*, 200
  and Hershey–Chase experiment, 193
  life cycle, 358, *359*
bacterium, *90*
bailer, 735, *736*
Baja California, 423
Bakker, R. T., 489, 491
balance
  semicircular canals, *928*, *936*
  sense of, 641
balance of nature, 1049, 1050
baldness, 310
ball-and-socket joint, *677*, 681
ball-and-stick model, 33
bananas, artificially ripened, 552
barberry, 455, 456
barbiturates, 746
bare rock succession, 1020, *1021*
bark, 510, 516; *see also* cork
barley, 506
  grown on sea water, 506, 515
barnacle
  copulation, 831, *832*
  habitat specialization and competition, 1036, 1044
barometric pressure, sensed by spadefoot toad, 997
Barr body, *226*
  absent in adrogen insensitivity syndrome females, 895
  in chromosomal anomalies, 257
barracuda, in cleaning symbiosis, *984*
barrels and whiskers, *884*
barrier reef, 601, 1012
Bartholin's glands, 842
basal body, 117, 118, 437
  and centriole, 237, 259
  replication, 190
basal meristem, 580
basal metabolism, 783
base, 40
base pairing, in DNA, *186*
base substitution, defined, 216
basidiocarp, *454*
Basidiomycetes, *454*, *455*, 456
basidiospore, 454
basidium (pl. basidia), 454, *455*
basilar membrane, inner ear, 927, *928*
basophil, *762*
bat, *652*, *653*, 654
  captures moth, 925, *926*, 927
  on oceanic islands, 11
  single uterus, 843
  skeleton of forelimb, *683*
bath sponge, 597
Bauer, 315
Baur, crosses snapdragons, 290
Beadle, George, 352–356
Beagle, 2, 4, 5, 9, 10
beak, variety of structure, *709*
bean plant, Mendel's partly successful experiment, 288, 289
bean seed, *569*, *575*, *578*
beaver, 654
  skull, *711*
bee, 20, 489
  specialization of legs, *624*
  worker cooling hive, *47*
bee bread, 624
beeswax, 79, *624*
beetle, 622
  homeothermic, 800
behavior, 20, 22, 944–984
  adaptedness of, 967–985
  aggression, 975–980
  altrusion, 983–985
  autonomic learning, *957*, 958
  avoidance, 952
  classical conditioning, *955*, *956*
  communication, 970–975
  consummatory, 953
  cooperation, *981*, 982
  criminal, and low IQ, 257
  development of innate patterns, 963
  difficulties of scientific study, 944
  drive reduction, 952, 953
  ethology, 945
  evolution of, 967
  evolutionarily stable strategy, 977, 980, 983, 985
  exploration, 952
  fixed action pattern and orientation, 946–949
  fixed component, 947
  habit reversal, 957
  habituation, 954
  hierarchy, *951*, 952
  human aggression, 978
  hunting and attack pattern of peregrine falcon, 947–949
  innate contrasted with instinctive, 962
  innate sex differences, 892, 893
  learning, 952–958
  mechanisms and development, 944–963
  motivation, 952, *953*
  navigation, 967–970
  operant conditioning, *956*, 957
  pheromones, 772, 973, 974
  pleasure center, 953
  releasers, *950*
  ritualized combat, *976*, 977
  sociobiology, 984
  stereotyped, *948*, 949
  vacuum, 951
behavioral genetics, 330–332
behavioral phenotype, 331
behavioral thermoregulation, 802, 803, *804*, *805*
belly, of a muscle, *688*
Benchuga, 11
benthic community, 1010, 1011
  compared with Galapagos rift community, 1017
benthic region, *1009*, 1010
benzene, 41, *42*
Bergmann, 337
Bergmann's rule, 335, *336*, 337
bermuda grass, *149*, 558
berry, 572, *573*
Bessey, C. E., 492, *495*
beta cells, 777, 786
beta estradiol, 789; *see also* estrogen
beta hemoglobin
  multiple alleles, 282
beta islet cell, 777, 786
beta ray, 196
beta-galactosidase, 366
beta-glycoside, 62
bicarbonate ion, *43*
  in blood, 743, 745
biceps, 688, *689*, 690
bicuspid valve, 751, *753*, 754
bilateral symmetry
  flatworms, 602
  larval echinoderms, 627
bile, 717
bile acid, 79
bile duct, 704, 717
bile pigments, 717
bile salts, 718, 720
bilirubin, 761
binding site, of enzyme, *70*
Binet, Alfred, I.Q. tests, 313
binocular vision, 658, 931
  of primates, 654
  squid, *616*
binomial system of nomenclature, *392*
bioassay, for gibberellin, 550
biochemical pathway, and genetics, 351–356
biochemical taxonomy, 595
  humans and chimps, 659
biogenetic law, 890, 891
biogeochemical cycle, 530–536; *see also* nitrogen cycle, carbon cycle, nutrients
biogeography, 4, 5, 18
  absence of mammals on islands, 11
  oceanic islands, 15
biological clock, third eye and pineal body, 789; *see also* photoperiod
biological control, 1043
biological species concept, 386–390
bioluminescence, 1010, 1011
biomass, 1014, *1015*, *1018*, *1019*
  Galapagos rift community, 1017
  pyramid, 1015
  and succession, 1020, 1021
biome, 535, 992–1013
  aquatic, 1009–1013
  chaparral, *1003*
  coastal communities, 1011, *1012*
  desert, 995–999
  fresh water, 535, 1013
  grassland, 999–1001
  marine, 535, 1010, *1011*
  taiga, 1005–1007
  temperate deciduous forest, 1003, *1004*, *1005*
  tropical rain forest, *1001*, *1002*
  tundra, 1007–1009
  world map, *993*
biometricians, and Mendelians, 313
biometry, 313, 390
biosphere, 136, 992, 993
biotic factors, 1014, 1015
  in population mortality, 1038–1050
biotic potential, *1029*, *1030*
biotin, 717, 724, 725
  binding by avidin, 73
bipedalism, 681, *682*, 683
bird
  air sacs, 648
  brain, 913
  cleavage divisions, *871*
  egg, *885*, 886
  nucleated erythrocytes, 226, 227
  reproductive strategy, *1033*
  respiratory system, *732*, *740*
  salt gland in nose, 810, 811
  similarity to reptiles, 648
  skeleton, 648
  sound communication, 972, *973*
  territoriality, 979, *980*
  third eye, 789
  variety in beaks and feet, *709*
  wing skeleton, 682, *683*
bird song, learned component, 963
birth, 894–897
  perceived pain and culture, 894
  physiological changes in newborn, *896*, 897
  stages, 894, 895
birth canal, 841
birth control, 852–858; *see also* contraception
birth control pill, 853, 855, *856*
birth rate, 858
  and doubling times, 1056
  since 1930, 1054–1056
birth weight, selection for intermediate optimum, 323, *333*
Bishop, J., estimates human gene number, 309
bison, 1000
Biston betularia, 47, *327*
Bitterman, M. E., habit reversal, 957
bivalves, 614, *615*
  circulatory system, 735, *736*
Black. A. H., brain wave control, 957
Black, Joseph, 528
black sheep, *276*, 279
black-tail deer, survivorship curve, 1034
Blacks, American, superior vision, 340
blackberry, 572, *573*
bladder, 813, *814*, 815, *876*
  float of brown alga, 467, *468*
bladder, urinary, endodermal origin, *876*
bladderworm, *605*
blade
  of brown alga, 467
  of green alga, 467
  of leaf, 512
  of red alga, 465
  of soil nematode, 605
Blagden, Charles, demonstrates thermal homeostasis, 798
blastocoel, 871, 872, *873*, *874*
blastocyst, 848, 874, *875*, 886, *887*
blastoderm, *872*
  bird egg, *885*
blastodisc, 865, *871*, 872, *874*, 891
blastomere, 868–872
blastopore
  amphibian embryo, 872, 873
  fate of, 596
blastula, 872
  early experimental work, 869, *968*
blending inheritance, 15
  assumed by Darwin, 265
Blest, David, programmed death in a moth, 1037
blood, 760–764
  composition in men and women, 761
  plasma composition, 761–763
  as a tissue, 760

blood circuits, 756–758
blood clotting, 762, *763*
blood fluke, 603
blood group systems, 282–287
  ABO, 282, *283*
  MN, 281
  Rh, 283, *286, 287*
blood pressure, *756,* 757
  in capillary beds, 759, *760*
  lowered by release of aggressive behavior, 978
blood sugar, 57, *723; see also* glucose
blood vessels, 746–764; see also circulatory system
blubber, 78
blue baby, 897
blue-green alga, *409; see also* cyanophyte
body, of neuron, 882, 883
body plan, insects, 621
body stalk, *887*
body temperature, camel, 998, 999
Boers, 338
bog, *1005*
Bohr atom, 29–33
Bohr effect, 745
boll weevil, 1043
bolus, 712
bombykol, 933, *934,* 973
Bombyx, 933, *934*
bond angle
  of carbon backbone, 41
  of methane, 33
  of water, 36
bond cleavage, *70*
bone, 675–686
  compact and spongy, *676*
  dermal bone of skull, *684*
  development, 676
  growth, *676,* 677
  growth stimulated by sex hormones, *789*
  hollow bones of birds, *740*
bone marrow
  red, 761
  site of B-cell maturation, 819
bony fish, 640, 641, *642*
  freshwater ancestry, 738
book gill, 617
book lung, 17
baron, plant micronutrient, 537
bottle feeding, 897
bottleneck, 338, *339*
botulism, 373
Bound for Glory, 313
bound ribosome, 206, 215, *216*
bourgeois strategy, 977, 980
Bowman's capsule, *814,* 815, *816*
Boysen-Jensen, P., 545
brachiocephalic artery, *753*
brachiopid, *609*
brain
  development, 889, *891*
  early evolution, 913, 914
  frontal section, 960
  high phospholipid content, 78
  human, 917–920, *921*
  human embryo, *891*
  human, evolved for complex social interactions, 985
  integration of neural impulses, 913
  memory store, 958, 959
  motor and sensory areas, 920, *921*
  pattern of interconnections, 883
  pleasure center, 953
  representative vertebrate types, *917*
  sex-specific morphology, 892, 893
  specialization of right and left hemispheres, 920
  speech center, 920
  split-brain experiments, 960–962
  vertebrate, 916, 917
brain hormone, 774, *775*
brain stem, control of respiration, 745
brain waves, *957*
braincase, *684*
bran, 569
  laxative action, 62
branch primordium, 579, *581*
branchiae, echinoderm, 734, *735*
Branchiostoma (lancelet), *636,* 675, 736, 737, 916
bread mold, *449,* 450; *see also* zygomycetes
breaking water, 894
breast cancer, viruses in mouse milk, 897
breast feeding, 987
breasts, enlarge during pregnancy, 897
breathing, 742
  control, 745
  involuntary control, 687
  see also respiratory system
breeding, fighting bulls, 265
Briggs, R. W., transplants frog nuclei, 880–882
brine shrimp
  cost of specialization, 800
  osmoregulator, *810*
brittle star, *625*
broad ligament, *842,* 843
broad-niched organisms, 1028
bronchi, *741,* 742
bronchiole, *741,* 742
Brontosaurus, 489, *490*
brood gland, honeybee, 982
Brotherton, Robert, 516
Brown, nucleus and Brownian movement, 223
brown alga, 466, *467, 468*
  life cycle, *468*
brown teeth, dominant, 301, *303*
Bruce effect, 973
Brunner's glands, 717
brush border, *89, 94, 716*
bryophytes, 472–478
  water transport, *504,* 505
bryozoa, *see* ectoproct, entoproct
bubble, in DNA replication, *189*
bubonic plague, 623, 1041
bud, 579, *581*
  lateral, *581*
  scale, 579
  terminal, *581*
budding, of yeast, 452, *453*
bug, *622*
bulb of penis, *838*
bulbourethral gland, *838,* 839
bullfighting, 265
bundle of His, 754
bundle sheath, 513, *514*
bundle-sheath cells, in C4 plants, 148, *149, 150*
buoyancy, sharks and bony fish, 640
burr clover, *576,* 577
buttercup, 492, *493*
butterfly
  mouthparts, 621
  see also Lepidoptera

C

C3 pathway
  inefficient use of CO_2, 148
C3 plants, 145–147
C4 pathway, 148–151, 399
  efficient use of, CO_2, 150, 151
  overall formula, 150
  theoretical efficiency, 150
C4 photosynthesis, *see* C4 pathway
C4 plants, 148–151, 399
cactus, *398,* 996, 997
  guard cell function, 520, 521
  saguaro, 507
caecilian, *395,* 643
caffeine, 533
Cairns Figure, replicating DNA, *189*
calcaneous, *686*
calcareous sponge, 597
calcitonin, 776, 784, 785
  stimulates cAMP in target cells, 791
calcium
  biogeochemical cycle, *534,* 535
  in diet, 725
  ion regulation by calcitonin and parathormone, 785
  and muscle contraction, 693, *694*
  and thermoregulation, 808
  in transmission of neural impulse, 909
calcium carbonate, in arthropod skeleton, 63
calcium phosphate, 48
calla lily, *570*
callus tissue, *551,* 557, 558
calorie, 128
Calvin cycle, 146–148
  overall formula, 148
  in phylogeny, 414
Calvin, Melvin, 140
calyx, *560*
cambium, *513*
cambium, vascular, *580*
Cambrian era, 405, 593, 608, 609
  brachiopods and ectoprocts, common, 609
  crinoid fossils, 627
  diversity of mollusks, 612
  fossil cephalachordate, 636
  onychophoran fossil, 616
  priapulids common, 608
camel, 655, 998
cAMP, *see* cyclic AMP
canaliculus, *676,* 677
cancer
  anaerobic metabolism, 167
  closing of cancer virus, 373, 376
  and life expectancy, 1036
  in mice, 1036
  phagocytosis of cancerous cells, 819
  somatic mutation, 187
  suppression, effective in human species, 1036
canine teeth, 710, *711*
  in threat display, *972*
Canis, 386, *387; see also* dog
canopy, rain forest, *1001, 1002*
cap, of mRNA, 207, 211
Cape Verde Islands, similarity to Galapagos, 5
capillary, 751, *758–760*
  pinocytic vesicles, *759*
  structure, *758*
capillary action, 37, *38,* 516
capillary bed, 750, *758*
  hepatic portal system, *757*
  water and ion flow, 759, *760*
caproic acid, structural formula, *76*
capsule
  of a joint, 675, 677
  seed pod, 560, *572,* 577
carapace
  lobster, 735, *736*
  turtle, 679
carbaminohemoglobin, 745
carbohydrates, 57–63
  digestion, 723
carbon
  biogeochemical cycle, *535, 536*
  bond angles, 32–34
  compounds of, 39
carbon backbone, 39, *41*
carbon cycle, 538
carbon dioxide (CO_2), 43, 146, *535,* 538, 743, *744,* 746, 1019, *1020*
  accumulation in atmosphere, *535,* 538, 1020
  in air, 43
  controls breathing rate, 746
  fixation in photosyntesis, 146
  greenhouse effect, 1019, *1020*
  released from lung, 743, *744*
  transport, 744, 745
carbonic acid, 43
carbonic anhydrase, 743, 745
carboniferous, 480, 482, 485
carboxyl group, 48
  of amino acid, *64*
carboxypeptidase, 720, *721*
cardiac muscle, *687,* 688, 754
cardiac output, 755
cardiac sphincter, *715*
cardinal, *1005*
cardinal vein, *752*
  fish, *748*
caribou, 1007
  attacked by wolves, *1047*
  hoof disease and wolf predation, 1041
carnivores, *652, 653,* 655
carotene, 114, 421
carotenoids, 137, *139*
carotid artery, *753*
carotid body, 746
carpal, 682, 685, *686*
  of bird wing, 682
carpel, *560*
  of primitive flower, 492, *493*
carrageen, 63, 465
Carribean Fireworm, 830
carrier pigeon, *968*
carrot, in cell culture, *583,* 584
carrying capacity, *1030,* 1031
  human environment, 1052
cartilage, *91,* 675
  of joint, 675, *676*
cartilaginous fish, 639, 640, 641; *see also* chondrichthyes, shark
cartilaginous skeleton
  cyclostomes, 637, 638
  of sharks, rays, chimeras, 639
Cascade range, 1007
Casparian strip, 507, 508, *509*
Castle, W. E., 341
Castle–Hardy–Weinberg, 313, 314, 340–344
  defined, 341
  underlying assumptions, 344
cat
  brain, *917*
  can't taste sweets, 935
  manx allele, 281, 284
  short life span, 1036
  territoriality, 979
catalase, in peroxisomes, 113, *114*
catalyst, 50, 51
  platinum, 50
  reversibility in theory and fact, 70, 71
cattle, digestive system, 710, *711, 712*
cavy, 654
Cebidae, 655, *656, 657*
cecum
  grasshopper, 706, *707*
  human, 704, *713,* 717
celestial navigation, 970

cell, 84–120, 223–259
animal, table of characteristics, 87
animal, "typical," *89*
cycle, 227, 228, *229*
defined, 86
diversity, *90*
in multicellular organism, 457
origin of, 85
plant, table of characteristics, 87
plant, "typical," *88*
procaryote, *90*
procaryote, table of characteristics, 87
stages of cycle, 228, *229*
terminal differentiation, 227, 228
cell culture
carrot, *583*, 584
plant, 551
cell differentiation, 223, 456, 864, 875
characteristic of animal kingdom, 592
irreversible changes in nuclei, 881, 882
morphogenesis, 875–884
myxobacteria, 417
nerve cell density and connections, 882–884
neurons, 882, *883*, *884*
and plant hormones, 551
in plants, 583
root tip, *508*
terminal, 227, 228
Volvox, 470, *471*
see also morphogenesis, gene control
cell diversity, 86–91
cell division, *241*
animals, 240, *241*
contrasted with mitosis, 229
plants, 241, *242*
table of events, 234
cell elongation, 545, 546
cell membrane, 39, 77, 78, 85, 86, *92*, 95
Danielli model, *100*
fluid mosaic model, *101*, 102
phospholipids in structure, 78
rich in unsaturated fats, 77
and van der Waals forces, 39
cell movement, in development, 864
cell nucleus, *see* nucleus
cell plate, *242*
cell recognition, 102
and binding proteins, 74
neurons, 882, *883*, *884*
cell sap, 117
cell size, 87, 92–95
cell streaming, 92
cell surface, antigens, 102
blood groups, 282–287
histocompatibility loci, 287
cell theory, 85, 91, 92
exceptions, 40
cell wall, 86, *88*, *242*, 404, 464
cell elongation, *579*
of fungi, 445
organization of cellulose fibers, *62*
plant, 95
procaryote, *410*
cellulase, 62
cellulose, 61–63, 511
compared with starch, 61, 62
digestion in ruminants, 710, 711, *712*
in fungal cell walls, 445
orientation of fibrils, *579*
in tunicate tunic, 635
Cenozoic, 486, 491, 650
adaptive radiation of placental mammals, *652*, *653*
centipede, 620
central dogma, 182, *183*, 322
exceptions to, 190, 208
preformation and epigenesis, 898
central nervous system, 913–921
human brain, 917–920, *921*
and memory, *958*
see also nervous system, brain
Central Valley, Caifornia, 532
centric heterochromation, 225, 234, 307
centriole, 89, *117*, *238*, *239*
amorphous or absent in flowering plants, 237
and basal body, 259
cycads and ginkgos, 258, 259
replication, 237
role in spindle formation, 237, *238*
centrolecithal egg, 865
centromere, 234, *246*, 254
centromere number is chromosome number, 234
described, 234
flanked by constitutive heterochromatin, 225
in meiotic cells (table), 254
centromeric spindle fiber, *237*, 238, *239*
centrum, *680*
cephalic gland, one-shot mating and programmed death, 1035, 1036
cephalization, 602, 913, 914
vertebrates, 916, *917*
cephalochordate, *636; see also* lancelet
cephalopod, *613*, 614
programmed death, 1035, 1036
cerci, *623*
Cercopithecoidea, 655, 656, *657*
cerebellum
human, 917, *918*
human embryo, *891*
various vertebrates, *917*
cerebral cortex, 650, 916, *917*, *918*, 919, 920
human embryology, *891*, 894
in split-brain experiments, *960–962*
in various vertebrates, *917*
cerebral hemisphere, *see* cerebral cortex
cerebrum, 650, *891*, 894, 916, *917*, *918*, 919, 920, 960–962
cervical cap, 853, 854
cervix, *842*
secretes alkaline mucus, 846
cesium chloride density gradient, *198*, *199*
Cestoda, 592, *603*, 604, *605*
cetaceans, 392, *652*, *653*, 655, 935, 1045, 1046
CF1 particle, *130*, *142*
Chaetagnatha, *596*
Chagas' disease, 11
chain reaction, 50
chain termination mutation, 216; *see also* terminator codon
chalaza, *885*
chambered nautilus, *673*, 674
Chandler, Asa C., 606
chaparral, *1003*
world map, *993*
character, use in taxonomy, 396, 397
character state, 391, 396, 397
Charaphyta, 467
Chargaff, E., 201
Chargaff's rule, 186, 201
charging enzyme, 207, 208, *210*
Chase, Margaret, 193, 200, 357
Chatham Island, 5; *see also* Galapagos
cheekbone (zygomatic arch), *684*
chelicera, 617
Chelicerata, 617, *619*, *620*
chemical bond, 29–33
chemical communication, 973, 974; *see also* pheromone
chemical energy, potential, 50
chemical forces, 29, 30
chemical messengers, 771–793
in embryology, 879, 880
chemical reaction, 30
endergonic, 50, 71, *72*, 152, *159*
exergonic, 50, 71, *72*, 152, *159*
irreversible, 50
reversible, 50
chemical wastes, 1012, 1021, *1022*, *1040*
chemiosmosis, 22, 125, 129–145, 160, *172*
in aerobic bacteria, 130, 131
in chloroplast, 125, *129*, 130, *142*, 145
Mitchell hypothesis, 22, 129, 133, 134
in mitochondrial membrane, schematic, *172*
in mitochondrion, 130–132, 160
mitochondrion compared with chloroplast, 173
in procaryote phylogeny, *414*
and rotation of bacterial flagellum, 411, *412*
chemiosmotic gradient, 145
chemiosmotic phosphorylation, *see* chemiosmosis
chemistry, importance to biology, 51
chemoreceptors, 933–935
insect, 933, *934*
planaria, 933
chemosynthetic bacteria, 136, 416, 1017
chemosynthetic giant tube worms, 1017
chemotrophy, 136, 416
Galapagos rift community, 1017
cherry, 570
chestnut blight, *1042*
chewing, 678
human, 713
chiasma (chiasmata), *247*, *248*, *249*, 250
related to recombination per chromatid, 306
chicken
brain, *917*
comb types, 289
dominant white leghorn, 279, 288
embryonic development, *885*
selected for long legs, 329
chicken fat, 39
chief cells, *715*
childbirth, 894–897
Chilopoda, 620
chimera, 639, *641*
chimpanzee, 398, *657*
chromosome number, 227
gangs of males kill males of other troops, 977
grasp, 683, 684, *685*
reproductive strategy, 954, *1032*
smells like you, 935
Chiroptera, 11, *652*, *653*, 654, *683*, 925, *926*, 927
chitin, 61, 63
in fungal cell walls, 445
structural formula of, 63
chiton, *613*, *669*, 673
chlamydobacteria, 417
Chlamydomonas, 469, *469*
life cycle, 469
chloragen cell, 705
Chlorhydra, 599
chloride
active transport from kidney tubule, 815, *816*
plant micronutrient, 537
chlorobacterium, *414*, 423, 465
chlorophyll, 114, 137–139, *143*
in phylogeny, *414*
chlorophyll *a*, 137–139
absorption spectrum, *138*, *139*
structural formula, *138*
in Z scheme, *143*
chlorophyll *b*, 137–139
spectrum, *139*
structural formula, *138*
Chlorophyta, 467, 469–472
chloroplast, *88*, 114, *115*, 140
and guard cell function, 521
of red alga, 465
symbiotic origin, *414*, 422, 423
chlorosis, 537
choanocyte, 597, *598*, 701, *702*
cholecystekinin, 717, 722
cholesterol, 79
bile salts, 718, 720
source of steroids, 777
structural formula, *79*
choline, 724
Chondrichthyes, *639*, *641*
brain, 916, *917*
modes of reproduction, 832, *833*
chondrodysplastic dwarf, 280, 281
chondroitin, structure of, 63
Chondrostei, *641*
chordae tendineae, 751, *753*
Chordata, *596*
chordate, 593, 596, 634–663
chorioalantois, *885*, 886
chorion
bird, *885*
covering insect egg, 834, *835*
develops into placenta, 650
mammalian, 886, *887*
chorionic villi, 886, *887*
Christmas disease, 763
chromatid
defined, 233
in G_2, 228
in meiosis, 247, 248, *249*
chromatin, *225*
condensation, *see* chromosome, condensation
diffuse regions indicate transcription, 226
net, 226
chromatin net, *88*, *89*
chromomere, 246, *246*
chromonema (chromonemata), 233
chromoplast, 114
chromoprotein, 67
chromosomal rearrangement, 216
chromosome, 109
anomalies, *256*, 257
arms, 234, 240
composition, *225*
condensation, function in mitosis, 233
condensation, gene control, 226
condensation, mitotic prophase, *230*
condensation, premeiotic, 245
homologues, 243, *244*, 245
levels of coiling, *235*
nondisjunction, *256*, 257
number, table of various organisms, 227
chromosome mapping, 304–309
using kitchen blender, *365*
chrysanthemum, 44, 494, *562*
chrysophytes, 430–432
chymotrypsin, 718
chytrid, 449
cicada, 972
mouthparts 621

cichlid fish, must fight before it can mate, 978
cilia (cilium), 89, 117, *118, 119*
  of ciliates, 437, 438
  modified for hearing and vision, 118, 119
  modified for olfaction, 934, *935*
  respiratory passages, *741*, 742
ciliate, 437–440, 457
  as metazoan ancestor, 594, *595*
Ciliophora, 437–440
circuit, blood, 756–758
  defined, 756
circular muscles, 670
circulation, 746–764; *see also* circulatory system
circulatory system, 746–764
  amphibian, *749*, 750
  annelid, 746, *747*
  bivalves, 614
  blood clotting, 762, *763*
  capillaries, 758–760
  capillary bed, *758*
  changes at time of birth, *896*, 897
  clam, 735, *736*
  closed vs. open, 735, 746
  crayfish, *747*
  earthworm, 609, *610*
  evolution, 746–748
  fish, 641, *748*, 749
  fish gills, *737, 738*
  four-chambered heart, 643, 648, 650, *750*, 751
  grasshopper, *747*
  heart, 641–650, 751–756
  heartbeat, 754–756
  hepatic portal circuit, *757*
  human, 751–760
  human embryo resembles adult shark, *890*
  human embryonic development, 888
  lancelet, 736
  lobster, 735, *736*
  lungs, 738–742
  lymphatic system, 763, *764*
  major circuits, 756–758, *757*
  mammals, 650
  mammals, birds, and crocodiles diagrammed, 750
  mollusk, 612
  open, 735
  reptile, 739, 740, *749*, 751
  tunicate, 736
  valves in veins, 759, *760*
  vertebrate, 748–760
circumesophageal ganglion, 914, *915, 916*
cirri, of lancelet, *636*
cis-dominance, 367
cisternal space, of endoplasmic reticulum, *215*
cistron, 207, 215, 216
  of lac operon, 366
citrate, 169; *see also* citric acid cycle
citric acid cycle, *159*, 160, 169–172
  phylogeny of procaryotes, *414*
  schematic diagram, *170*
citrulline, 355
clam, *615*
  circulatory system, 735, *736*
clam worm, *611*
clasper, 640, 832, *833*
classical conditioning, *955, 956*
  contrasted with operant conditioning, *956*
classification, *392*
  kingdoms, 402, *403*
clavicle, 681
clay, in soil, *528, 529*
cleaning symbiosis, 983, *984*
cleavage, *241*, 870–872
  of bond by hydrolyzing enzyme, 70
  chicken egg, *871*
  human zygote, 871
  ray, *870*
  sea urchin, *870*
cleidoic egg, *644, 885, 886*
  of monotremes, 886
climate (biome world map), *993; see also* biome
climax (orgasm), 840, 849, *850*
climax community, 1020
climbing plants, *544*
cline, 334–337
clitellum, *610*
clitoris, 840, *841*
  embroyology, *893*
  erection in excitement phase, 849
  role in orgasm, 840, 849
cloaca
  in human embryo, *889*
  monotreme, 650
  reptile, 643
  shark, 640
cloacal kiss, *836*
cloacal lung, 739, 740
clonal selection hypothesis, 821, *822*
clone
  evolution in, 323, 326
  by mitosis from single cell, 243
  recombinant DNA, 372, 373, *374, 375*
Clostridium, *414*, 415
clotting, blood, 762, *763*
club fungi, 454–456
club moss, *479*, 480
clumping, cells by antibodies, 281–283, *821*
clutch size, *1033*
Clydesdale horse, *11*
Cnideria, 598; *see* coelenterate
Cnidoblast, 598
coadaptation
  bacterium and phage, 358
  bat and moth, 925, *926*, 927
  giraffes and trees, *333*
  host and parasite, 1042
  pines and scale insects, 326
  tropical trees and insect seed predators, 1043
coastal communities, *1009*, 1011
coastal upwelling, 1011
cobalt
  animal micronutrient, 537
  in vitamin B_{12}, 27, 725
coccyx, 678, 679, *680*
cochlea, 927, *928*
  sensory hairs, 118
cockroach, tactile receptors, 924
cod, 833
Codium, 467, 470, *471*
codominance, 281, 284, *285*
codon, 205, *206*, 207, 217
coelacanth, *641, 642*
  urea in blood for osmoregulation, 810
coelenterate, 458, 598–601
  hydrostatic skeleton, 670
  intracellular digestion, 702, *703*
  muscle fibers, 670
  nerve net, 598, 913, 914, *915*
coelom, 703, 704
  development, *597*
  earthworm, *610*
  hemichordate, 634, 635
coelomate–pseudocoelomate–acoelomate hypothesis, *596*
coelomates, *596, 597*
coelomic fluid, in dermal branchiae, 734, *735*
coelomic sphere, *887*
coenocyte, of water mold, 447
coenzyme A, 168, 169
coenzymes, 724
  as prosthetic group, 67
coevolution, *see* coadaptation
cofactors, in blood clotting, 762
cohesion of water, *38*, 516, 517
coitus, *see* copulation; sexual behavior, human
coitus interruptus, 853
colchicine, treatment for gout, 818
  in karyotyping, 236
Coleoptera, *622*
coleoptile, *575*
  phototropism, 545–547
colinearity, principle of, 183
collagen, 74, 75
  and animal origins, 458
collar bone, 681
collar cell, 597, *598*, 701, *702*
colloid goiter, 783, *784*
colon, *713*, 717
colonial organization, of hydrozoans, *599*
colonial structure, Portuguese man-of-war, 600
color blindness, 301, *302*, 303
  test of, 301
color change
  frogs and melatonin, 789
  horned lizard, 804
  in thermoregulation, 802, 803
color vision, 929, 932, 933
  lions, 933
  primates, 933
coloration
  as visual communication, *971*
  protective, 14, *327, 933; see also* protective coloration
colostrum, 897
comb jelly, *601; see also* Ctenophora
combat, 975–978
  ritualized, 976, *976, 977*
commensalism, 1042; *see also* symbionts and symbiosis
common cardinal vein, *748, 752*
communication, 970–975
  chemical, 973, 974
  sound, 972–974
  visual, *971, 972*
community, 1014
  benthic, 1010, 1011
  Galapagos rift, 1017
  mangrove, 994
  see also biome
compact bone, *676*, 678
companion cell, *511, 513*, 582
comparative anatomy, 399
compass
  magnetic, *970*
  sun, *968, 969*
competition
  and aggression, 975
  between species, *1043*, 1044
  limiting resources, 1039
  reduced by habitat specialization, *1044*
  territoriality, 979
  by toxic products, 1040
competitive exclusion, 15, 17, *1043*
  and derived species, 7
  of Tasmanian devil by dingo, 1044
  barnacles, 1044
complement, 819
complete and incomplete flowers, *570*
complete metamorphosis, *622, 623*, *774, 775*
Compositae, 494, *495*
compound, defined, 26
compound eye, *620*, 929, *930*
conception, as beginning of human life, 870
condensation
  chromosome, 226, *230*, 233, 245
  and gene activity, 226
  see also chromosome, condensation
condensed X chromosome, *226*
conditioned response, 955
conditioning, classical, *955, 956*
condom, 853, 854, *855*
Condylarthra, *653*
cone
  color blindness, 301
  as modified cilium, 118
  mollusk shell shape, *673*
  retinal, 928–933
congenital night blindness, 281
conidiospore, of ascomycetes, *451*
conifer, 484–487
  apical dominance, 348
  dominant plant of taiga, 1006
coniferous forest, 1005–1007; *see also* taiga
conjugated protein, 67
conjugation, of *Paramecium*, *439*, 440
Connell, J. H., competing barnacles, 1036, 1044
conservation of energy, 131; *see also* thermodynamics
conservative replication, *198, 199*
conservative synthesis of RNA, 203
consolidation hypothesis, 959, 960
constant region, immunoglobulin, 819, *820*
constitutive heterochromatin, 225
constitutive mutant, 366
consumer, 1014, *1015*
  and energy flow in ecosystem, *1016*
consummatory behavior, 949, *950*, 953
  killing mice, 978
contact receptors, 924, *925*
continental drift, 488, *489*
  New-World and Old-World monkeys, 656
continental shelf, *1009*
continuity of the germ plasm, 321, 322, *323*
continuous spindle fiber, *238, 239*
continuous traits, 312–315, 333
contraception, 852–858
  coitus interruptus, 853
  and declining birthrate, 1055
  effect of nursing among primitive peoples, 1033, 1034
  table of methods, 853
contractile vacuole, 429, *430*, 439
  heliozoan, 434
  Paramecium, 438
contraction, of muscle unit, 693, *694*
control, in scientific method, 9
control gene
  eucaryotes, 371
  negative and positive control, 370
  operon, 366–371
control mutants
  at lac operon, 366, 367
control proteins, 279, 371
conus arteriosus, *748, 749*
convergent evolution, 395, *397, 398*
  of coeloms, 596
  of fungal groups, 457
  mara and rabbit, 3
cooling, 805
cooperation, *981*, 982
copper
  dietary need, 725
  hemocyanin, 27
  plant nutrient, 537

copulation, 831, *832*
birds, *836*
caecilians, 643
flatworms, *602*, 603
human, 848–852
nematodes, 606
painful intercourse caused by muscular constriction, 842
octopus, 1035
reptiles, 835, 836
sharks, 640
copulatory structure, insect, *834*
coral, 600, *601*, 1012
coral atoll, 601
explained by Darwin, 1012
coral reef, 1012
coral snake, 644
Coriolus effect, 995, *996*
cork, 84, *510*
corn (*Zea mays*), *149*, 507
ear and silk, *571*
kernel, *569*, *575*, *578*
corn smut, *454*
corn sugar, 57
corolla, *560*
corona radiata, *866*
coronary artery, *752*
coronary sinus, *752*
corpus allatum, 774, *775*
corpus callosum, *960–962*
absent in monotremes and marsupials, 654
human, *918*, 920
corpus cardiacum, 774, *775*
corpus cavernosum, *838*, *839*
of clitoris, 840
corpus luteum, 776, *843*, 847
corpus spongiosum, *839*
cortex
adrenal, *787*
cerebral, 650, 916, *917*, *918*, 919, 920, *960–962*
of gonad, 892
kidney, 813, *814*
of ovary, 843
of plant embryo, *574*
root tip, 507, *508*, *509*, 579, *580*
cortical dominance, 892
cortical granules, echinoderm egg, 866, *867*, *868*
corticotropin (ACTH), 75
cortisol, 777, 787
antiinflammatory, 788
cost-benefit ratio, 1036
cotton, *495*
CO_2, *see* carbon dioxide
cotyledon, 492, *569*, *574*, *575*, 577, *578*
gymnosperm, *484*
cotylosaur, *644*, *647*
countercurrent exchange
fish gills, *737*, 738
heat exchange and extremities, *801*
in kidney, 815, *816*
temperature in tuna, *802*, *803*
coupled reaction, 71
courtship
blond ring dove, 949
stickleback, *951*, 952
covalent bond, 31–33
energy states of, 49
cow
digestive system, 710, 711, *712*
pile, limiting resource for maggots, 1039
placental system, *651*
Cowper's gland (bulbourethral gland), *838*, 839
coyote, 386, *387*
crab, giant, *619*
crab grass, 558
cramming for exams, 960
cranial capacity
Australopithecus, 659
Homo erectus, 660
modern humans, 659
Neanderthal, 662
skull 1470, 659
cranial nerves, *922*
cranium, 675–678, *679*, *684*, *711*
crayfish
circulatory system, 747
fooled with an iron filing, 936
molting hormones, *773*
nervous system, 915
creatin phosphate, structural formula, 165
creation account, 3, 5
creationism, 405
Cretaceous, 491
Cretaceous–Cenozoic discontinuity, 646
Crick, Francis, 182, 186, 188, 190, 198, 201
central dogma, 182
DNA replication, 190
Crick strand, 190, *191*
in meiosis, 249
cricket, 972
criminal behavior, and low IQ, 257
crippling a microbe, 376
Crisler, Lois, tracks wolf chasing caribous, 1047
criss-cross inheritance, *299*, 300
crista of mitochondrion, *116*
Crithidia, 428
crocodilian, *646*, *647*, *750*, 751
crocus, 501
Cro-Magnon Man, 662
crop, 648
birds, *710*
earthworm, 704, *705*
grasshopper, 706, *707*
crop rotation, 531
crossing over, 244, 248, *249*, 304, *306*
occurs in pachytene, 248, *249*
transformation, *359*
two-strand double, *306*
crossopterygian, *641*
skeleton, *643*
crossover, *249*
cross-reacting material, 366
crow, nest-building behavior, 953
crowding, 780, 781
human populations, 1035, 1051
crowning, 894
Crustacea, principle entry, 618, *619*, 620
crustacean liver, 703
crustacean molting hormones, 772–774
crustecdysome, 671, 772–774
crying, response in mothers, 897
cryptic coloration
Biston betularia (peppered moth), 47, *327*
and color-blind lion, 933
leaf toad, *933*
crystal
ice, 36, *37*
NaCl, 31
crystalline cone, *620*
ctenidia, 612
Ctenophora, 601, *601*
C-terminal end of a polypeptide, 64, 211
cuneus, 960
cup fungus, *452*
curare, 957
Curtis, Roy, III, 376
cuticle
insects, *813*
of rotifers and nematodes, 604
thick in xerophytes, 996
cutin, 95
cuttings, 557, *558*
cuttle bone, *673*
cuttle fish, *673*
cuttle fish and cuttle bone, *673*
cyanide, 39
cyanobacterium, *see* cyanophyte
cyanophyte, *90*, *409*, 419
cellular structure, *420*
cycad, 485, *486*
flagellated gamete, 258
cycle
carbon, 538
cell, 227, 228, *229*
nitrogen, 530–534
nutrient, 1016
cyclic AMP, 790, *791*
in *E. coli*, 792
as slime mold pheromone, *447*, *772*
structural formula, 772
cyclic polypeptides, and central dogma, 190
cyclohexane, 41, *42*
cyclosis, *519*
cyclostome, 637, *638*, 641, 675, 708, 737
cypress, *520*
cysteine, 537
structural formula, *66*
sulfhydryl bridge, 65
cytochrome *c*, 173, 423, 456
flagellates, ciliates, and metazoa, 595
mitochondrial membrane, *172*
phylogeny, 19, *20*
cytochrome *f*, in Z scheme, *143*
cytochrome oxidase, *172*
cytochrome P450, 104
cytokinesis
contrasted with mitosis, 229
see also cell division
cytokinins, 551
cytological map, 307
cytology, 97
cytoplasm, as sole site of protein translation, 202
cytoplasmic binding protein of steroid hormones, *792*
cytopyge, *438*, 439
cytosine, structural formula, 184
cytoskeleton, 86, *89*, 102
cyclosis, 519
involved in invagination, 872
cytostome, *438*, 439

D

D loop of tRNA, 208, *209*
dalton, 26
Dan, S. K., spindle apparatus, 238
Danielli, J. F., 100, 101
model 100–102
Daphnia, *619*
reproductive strategy, 258
dark clock, 564; *see also* photoperiodicity
dark reaction, 140; *see also* light-independent reaction
Dart, Raymond, discovery of *Australopithecus*, 659
Darwin, Charles, 2–5, 7–13, 15–22, 25, 297, 312, 321, 322, 328, 329, 333, 337, 351, 388, 390
anticipates Haeckel, 890
bellowing of mating tortoises, 972
breeds snapdragons, *266*, 268
competitive exclusion principle, 1043
Coral Islands, 1012
and criticisms of *Origin*, 266
denies reality of species concept, 16, 17
early journals, 10, 11
earthworms, 611
ecological niche, 1028
experimental work, 11, 12
fails genetics, 265–268
finches, 5, *6*, 7
health fails, 11
ignorance of biochemistry, 25
ignores Mendel's work, 278
instinct and evolution of behavioral characters, 946
international nomenclature commission, 392
later publications, 21
Origin of Species, 11, *12*, 13, 337
pangenesis, 322
plant hormones, 544, 545
proposes root tip hormone, 549
quoted on tree of life, *393*
selectively neutral variation, 337
South America, 2–7
voyage of the *Beagle*, 10
world teeming with elephants, 1031
Darwin, Erasmus, 8
rejects instinct, 946
Darwin, Francis, 545, 549
Darwin, Robert, 2, 3
Darwin's finches, 5, *6*, 7
dATP, structural formula, 184
daughter cell, *252*
in cell cycle, 227, 228, *229*
DDT, 435, 437, 440, 1053
resistance, 617
de Duve, Christian, and lysosomes, 112, 113
de Vries, Hugo, 519
deamination, 720, *809*
death
abiotic factors, 1038
biotic factors in populations, 1038–1050
evolution of, 258
"the long view," 1037
mortality rate, *1034*
as population control, 1037–1046
at the population level, 1035–1044
programmed death in a moth, 1037
reduction in death rate induces population explosion, 1041
why death?, 1035–1037
see also programmed death
death angel, *454*
death rate, historical, 1052, 1053
deciduous plants, 538, 553
deciduous teeth, 710
decoding neurons, 906
decomposer, 410, 445, 454, 530, *1014–1016*, 1028
decomposition, by Basidiomycetes, *454*
dedifferentiation
animal cells in regeneration, 228
plant cells in culture, *582*, 584
deep sea
fish, *1011*
Galapagos rift community, 1017
deer
black-tail, 1034
sitka, 780, 781
skull and dentition, 711
white tail, *1005*

degeneracy, of genetic code, 206, 207
dehydration linkage
  sucrose, 57
  triglycerides, 76
dehydrogenases, 725
deletion, at DNA level, 216
deletion mapping, 307
Delwiche, C. C., 532
demographers, 1056
demographic transition, 1053–1056
demography; *see* human population
denaturation, of protein, 65
dendrite, 906, *907*
dendrochronology, 584
dendrograph, *517*, 518
denitrification, 417, *531*, 532
density-dependent mortality, 1038, 1039
  disease, 1041, 1052
  predation, *1045*, 1046
  tropical trees and insect seed predators, 1043
  wolves and moose, 1049
dental pulp, *711*
dentalus, *673*
denticle, 639
dentin, *711*
dentition, *711*
deoxyhemoglobin, 744
deoxynucleotide, 183
deoxyribose, 184
  structural formula, 183, 203
dephosporylation linkage, 61
  of starch, 71
depolarization, neuron, 908, *909*
deprivation experiment; *see* mineral deficiency, plant
depth perception, 931, *932*
dermal bone, *684*
dermal branchiae, 734, *735*
dermis, *806*
  echinoderm, *675*
  mesodermal origin, *876*
Dermoptera, *623*, *652*, *653*, 654
Descartes, R., calls pineal body the seat of the soul, 789
desert, 995–999
  animals, 997–999
  annual wildflowers, 34, 996, *997*
  definition, 995
  plants, 996, *997*
  and rain shadow, *994*, 996
  world map, 993
desiccation, 473, 475
  bryophyte, 504
detergent, 39
  properties of phospholipid, 78
detumescence, 839
deuterostome, *595*, 596, 635, 702
development, 864–898
  echinoderm, *627*
  human embryo, 887–896
  of innate behavior, 962
  of instinctive patterns, 947
deviation from mean, 314
Devonian, 480, 482
dextrose, 57; *see also* glucose
diabetes
  in Jewish populations, 338
  possibly an autoimmune disease, 820
  syndrome and insulin control, 786, 787
diakinesis, *247*
diapause, 774
diaphragm
  contraceptive, 853, 854, *855*
  human, *741*, 742
  lacking in reptiles, 740
  mammalian characteristic, 649
diaphysis, 789
diastole, 755, *756*
diatom, *47*, 430–432
  life cycle, *431*
  in polluted waters, 1022
diatomaceous earth, 431, 432
DiCara, L. V., autonomic learning, 957
dicaryon, 451
  Ascomycetes, *452*
  Basidiomycetes, *455*
Dickerson, Richard, 413
dicot, 491, *495*
  compared with monocots, *492*
  paraphyletic group, 395, 396
  primary growth pattern, *513*
  vascular bundles, 581
Dicotyledonae; *see* dicots
Dictyostelum, 402, 771, *772*; *see also* slime mold
diet, 80
dietary needs, 722–726; *see also* nutrition, animal
differentiation, 223, 456, 864, 875, 882
  terminal, 227, 228
  see also cell differentiation
diffuse root system, *506*
diffusion, 95–104, 505
  defined, 95
  facilitated, 100, *104*
  gradient, *103*, *105*
  in phloem, 518
  simple, *103*
diffusion gradient, in capillary bed, *760*
digestion
  chemistry, 718–726
  in plants, 539, 540
  starch in grain, *550*
  see also digestive system
digestive enzymes
  pancreatic, 717
  table, 718
digestive system, chapter, 701–726
  amphibian, 708
  bird, *709*, 710
  chemistry of digestion, 718–726
  coelenterate, 702, *703*
  cyclostome, 708
  earthworm, 704, *705*
  embryonic development of, human, 888, *889*
  fat digestion, *720*
  fish, *707*, *708*
  flatworm, 702, *703*
  grasshopper, *706*, *707*
  hormones (table), 722
  housefly, *706*
  human, 711–718
  insect, 705, *706*, *707*
  integration and control, 721
  intracellular and extracellular digestion, 701
  invertebrates, 701–706
  large intestine, *713*, 717
  liver and pancreas, *713*, 717
  mammal, 710, *711*, 712
  nucleic acid digestion, 721
  protein digestion, 720, *721*
  reptile, 708, *709*
  ruminant, 710, *711*, *712*
  small intestine, *713*, 716
  sponge, 701, *702*
  vitamins (table), 724
digestive vacuole, *434*, 701
digit, 682, *683*
dignity and freedom, 954
dihybrid ratios, 287–289, 356
dihydroxytestosterone, 895
dilation, stage of birth process, 894
dilation and curretage (D&C), 853
dimer, 65, 68
dingo, 386, *387*, 1044
Dinichthys, 638
dinoflagellate, *429*, 430
  as photosynthetic symbionts (zooxanthelae), 423
  unusual meiosis, 430
dinosaur, 486, 489, 490, *645*, *646*, *647*
  sudden extinction, killed by asteroid, 646
  thermoregulation, *805*
  possibly warm-blooded, *19*
dipeptidase, 718, 720, *721*
dipleurula larva, *627*
diploblastic organization, 597, *598*
diplococcus, 415, *416*
Diplodocus, *646*
diploidy, 243
  characteristic of animal kingdom, 593
diplont, defined, 469
Diplopoda, 620; *see also* millipede
diplotene, *247*
Dipnoi, 641; *see also* lungfish
Diptera, 622
directional selection, *333*
disaccharide, 57
disclimax community, 1003
discrimination, Pavlovian, 955
disease and population control, 1041, *1042*
dispersal, *1050*, 1051
dispersive replication, *198*, *199*
display, on lek, 980; *see also* aggression
distal convoluted tubule, *see* nephron
distance receptors, 924, 925
distemper, 1042
diurnal animals of desert, 997, *998*
divergence, 17, 18, 397; *see also* adaptive radiation
diversity
  animal, 592–663
  plants, 464–497
  tree species in a tropical rain forest, 1043
division, of plant kingdom, 393, 464
division furrow, 240, *241*
DNA, 183
  bonding, 184, *185*
  and "central dogma," 182
  cloning, 372–377
  component of chromatin, *225*
  condensation, 226, 230, 233; *see also* chromosome
  double helix diagrammed, *187*
  function in nucleus, 226
  history of discovery, 192, 193, 200
  identical in nearly every cell, 875
  junk, 216
  length in human cell, 186, 187
  library of cloned human DNA, 372
  amounts in meiotic cells (table), 254
  mitochondrial, 115, 202, 217
  noncoding, 216
  plastid DNA, 114
  DNA polymerase, 190, *191*
  recombinant DNA, 372–377
  DNA repair enzymes, 187, *188*, 190
  replication, *189*, 190, 191, *198*, *199*, 322
  replication in cell cycle, *229*
  DNA sequencing, 372
  stability of, 187
  synthesis, addition of nucleotides, *185*
  transcriptional activity in cell cycle, *229*
  Watson-Crick pairing, 186, 188, 190, *191*, 208, 248
dodder, *539*
dodo, 577
dog, 329, 386, *387*, 592
  African hunting dog regurgitates food, 975, 982
  brain wave control learned, 957
  chases anything that runs away, 1047
  chromosome number, 227
  communication with human, 970
  consummatory behavior in swallowing, 950
  domestic, 329
  genus *Canis*, 386, *387*
  Pavlov's, 955
  recognizes Darwin, 10
  seldom catches squirrels, 1037
  short life span, 1036
  skull, *711*
  spreads parasites, 606
  survives arctic nights, 800
  survives sauna, 798
  swallowing pattern analyzed, *950*
  urinating on tree, nonreciprocal, 533
  varieties of domestic dog, 17
  see also wolf; dingo
dogma, central, *see* central dogma
Dolichotis patagonium (mara), *4*
Dollo, Louis, 399
Dollo's law, 399
dominance
  causes, 278, 279
  defined, 278
  discovered by Darwin, 266
  discovered by Mendel, 268, 269
  and muscular dystrophy, 301
dominance hierarchy, 368
  behavioral, 980, 1039
  and individual recognition, 975
dominance interactions, *see* dominance relationships
dominance relationships, 278–282, 328, 329, 356
  codominance, 281
  lethal-recessive dominant alleles, 280, *281*, 285
  multiple alleles, 282
  overdominance, 282, 285
  partial dominance, 279, *280*, 284, 285
  in a partially diploid bacterium, 366–368
  sickle-cell hemoglobin, *284*, *285*
  simple dominance, 278, 279, 284, 285
dominant allele, defined, 278
dominant brown teeth, 301, *303*
dominant plant type, 993
dormancy
  bacterial spores, 412
  frogs, 643, 739
  lungfish, *642*
  plant, 553
  seed, 575, 576
  spadefoot toad, 997
dorsal aorta, fish, 748; *see also* aorta
dorsal blood vessel, earthworm, *747*
dorsal hollow nerve cord, 635, 877
  chordate characteristic, 635
  frog embryo, *877*
  hemichordates, 634
  lancelet, *636*
  may be homologous with echinoderm neurectoderm, 916
dorsal lip of blastopore, 872, *873*
  experimental manipulation, 878, *879*
  dorsal lip mesoderm, *879*
double bond, *32*
double fertilization, 565, 567, *568*
double helix, 186, *235*
  and DNA repair, 188

double Y syndrome, 257
doubling time, human populations, 1052, 1056
douche, 853, 854
Dover, white cliffs of, 434
down feather, *648*
downer (depressant), 746
Down's syndrome, *256*, 257
downy mildew, 447, *448*
dragonfly, 620, *622*
wingbeat, 672
Driesch, Hans, separates blastomeres, 869
drive
aggression, 977, 978
hunger, 949
drive reduction hypothesis, 944, 952, 953
drone bee, 982
Drosophila melanogaster
chromosome map, *307*
early mutation studies, 192
length of sperm, 118
life cycle, 298
metaphase chromosomes, *300*
mutants of, 298
selection for geotaxis, 330, *331*, *332*
used by Morgan, 297–300
wing beat, 672
drought
in chaparral, 1003
in Sahel, 999
drupe, 572, *573*
Dryopithecus, 656, *657*
duckbill platypus, 393, 640, *651*, 654, 886; *see also* monotreme
duckweed, competitive exclusion, 1043, 1044
ductless gland, 776; *see also* endocrine system
ductus arteriosis, *896*, 897
Dugesia, *602*
dugong, 393, 681
dumping of wastes, 1012, 1021, *1022*, *1040*
Dunkers, 338
Dunkleostus, *638*
duodenum, *713*, 715, 716
dwarf
achondroplastic, 280, 281
pituitary, 338, 376, *781*

E

E. coli, *see Escherichia coli*
eagle, 1000
ear, 925–927
external, 927
homologies of middle ear bones, 399, 890, 927, *928*
human embryology of, *891*
moth, *926*
vertebrate, 927, *928*
ear canal, origin from gill cleft, 890
ear drum, 927, *928*
ear ossicles, 399, 927, *928*
derived from gill arches, 399, 890
earthquake, experienced by Darwin, 4
earthworm, 609, *610*
circulatory system, *747*
digestive system, 704, *705*
excretory system, *812*
hydrostatic skeleton, *670*
nervous system, *915*
see also annelid
East Pacific Rise rift community, 1017
eastern bluebird, decimated by winter storm, 1028
eastern deciduous forest, *see* temperate deciduous forest
Eberhardt, Mary Jane, prenatal sex hormones, 893
ecdysis, 671, 772–775
ecdysone, 671
induces chromosome puffing, *308*
echidna, 710
echinoderm, *596*, 624, *625*
affinity with chordates, 624, *627*, 634
egg, 865, 866, *867*
endoskeleton, *674*, *675*
gills (dermal branchiae), 734, *735*
nervous system, *916*
see also sea star, sea urchin
echinoderm–chordate line, 624, *627*, 634
ecocline
altitudinal, 994, *995*
latitudinal, *994*
ecological niche, 15, 331, 480, 617, 659, 1028, 1043
ecological succession, 1020–1022
ecology, chapter, 992–1056
abiotic factors, 1014, 1038, 1039
biomes, 992–1013; *see also* biome
biotic factors, 1014, 1015, 1038, 1039
competition, 1039
competitive exclusion, 7, 15, 17, *1043*, 1044
death, 1035–1044; *see also* death; programmed death
density-dependent mortality, 1038, 1039
diversity in tropical rain forest, 1043
ecology, defined, 992
ecological niche, 15, 331, 480, 617, 659, 1028, 1043
ecosystems, 1013–1022
Galapagos rift community, 1017
hoof disease, caribou, and wolf predation, 1041
limiting resources, 1038, 1039
population control by disease and parasitism, 1041–1043
populations, 1028–1056; *see also* populations
predator-prey relationships, 19, *1045*, *1046*
succession, 1020–1022
systems ecology, 992
ecosystems, 1013–1022
defined, 1013
Galapagos rift community, 1017
see also biome
ectoderm, 596, 872–876
amphibian gastrula, 872, *873*
mammalian embryonic disk, 874, *875*
and phylogeny, 596
tissues derived from, *876*
ectoparasite, leech, 611
ectoproct, *609*
edentates, *652*, *653*, 654
eel grass, 1011
efferent blood vessel, 737
efficiency of energy flow in ecosystem, 1016
egg, 865–870
amphibian, 865
arthropod, 865
bird and reptile, 643, *644*, 865
cell membrane, 865, 866
echinoderm, 865, 866, *867*
insect, *835*
mammalian, 865, *866*, 873
mosaic, 870
oocyte, 245, *843*, 870, 871
ovule, 369, *566*, 574
egg mold, 445
egg white, *885*
and biotin deficiency, 725
Ehrlich, Paul, alerts U.S. to the population bomb, 1055
Eibl-Eibesfeldt, Iraneaus, innate patterns in squirrels, 963
eidetiker, 959
Einstein, Albert, 21
ejaculation, 838, 849
elastin, 74, 75
elator, *477*
electroconvulsive shock clears short-term memory, 960
electromagnetic spectrum, *96*, 137
electron, 26
electron accepter, 31
electron carrier, 135, 136, 143, 170–172, 808, 809; *see also* electron transport
electron density cloud, *28*
electron donor, 31
electron orbital, *28*
electron pairing, 28, 29
electron shell, 28–33
electron spin, 29
electron transport
in chloroplast, 142–145
in mitochondrial membrane, *159*, 160, 172–174
in mitochondrial membrane, schematic, *172*
electrostatic force, 26, 29, 31
in protein structure, 65, 68
element
defined, 25
table of elements with roles in biology, *27*
elephant, *653*, 655
population crash, 1031
on railroad track, 368, 370
world teeming with elephants, 1031
elongation
of growing polypeptide, 211–214
of root and stem cells, *579*
Elton, Charles, Eltonian pyramid, *1015*
Eltonian pyramid, *1015*
embolus and embolism, 762
embryo, plant, 568, *569*, 574
embryo sac, 488, 565, 569, *574*
embryology (animal), 864–898
and phylogeny, 604
birth process, 894, 896
brain features dependent on target innervation, *884*
chicken, *885*
descriptive compared with experimental, 864
experimental, 868, *869*, 878–884
extraembryonic membranes, 884–887
eye, *878*, *880*
frog, *878*
gastrulation, 872–874
human, 887–896
human circulatory system, 888
human digestive system, 888, *889*
human excretory system, 891, 892
human limbs, 891, *892*, 894
human nervous system, 889, *891*
human reproductive system, 892, *893*
human respiratory system, 888, *889*
mammal, 886, 887
messengers and inducers, 875
morphogenesis, 875–884
mosaic egg, 870
neurulation, frog, *877*
nervous system, 882, *883*, *884*, 889, *891*
ossification in third trimester, 894
programmed death of cells, 228
and steriod hormones, 892, 893, 895
tissue interactions, *879*
transplanted nuclei, 880–882
embryonic disk, 874, *875*
emigration, *1050*, 1051
emission, 849
empirical laws, 313
emu, *649*
enamel, *711*
on scales of holostean fish, 640
end product, and enzyme inhibition, 73
endergonic reaction, 50, 71, *72*, 152, *159*
endocrine system, chapter, 771–793
cephalic gland of octopus, 1035, 1036
compared with nervous system, 771
control of human reproduction, 844–848
interaction with nervous system, 781
major human endocrine glands, 777, *779*
parathyroid, 776, 783, *785*, 791
pineal body, 777, *789*, 790
table of human hormones, 776–777
see also hormone; pituitary; steroid; sex hormones
endocytosis, 105, 434, 701, 720, 721, 725, 791, *817*
endoderm
amphibian gastrula, 872, *873*
bird gastrula, 873, *874*
fate, *876–878*
mammalian embryonic disk, 874, *875*
origin of glandular viscera, 704
and phylogeny, 596
adult tissues derived from, *876*
endodermis, root, 507, *508*, *509*, 579, *580*
endometrium, 843, 846, *847*, 848, 886
endomysium, *688*
endonuclease, 721
endopeptidase, 720, 721
endoplasmic reticulum (ER), *88*, *89*, *110*, *111*, 422
rough, *215*
smooth, 111
endoskeleton
echinoderm, 624, *674*, *675*
sponge, 597, *598*, *674*, 675
vertebrate, *see* skeletal system
endosperm, 567, 569, *574*, *575*
endosperm mother cell, 565
endosymbiont, 421–423, 599, 601; *see also* symbiosis; sequential endosymbiosis theory
endothelium, of capillary, *758*
endothermic reaction, 50, 71, 72, 152, 159
energetic tendency, 29
energy
action-specific, 950–952
chemical, 50
flow through ecosystem, *1015*, *1016*, 1021
free, 126, 131, *143*, 158, 173, *174*, 505, 516, 517
kinetic, 49, 126
potential, 50, 126
yield from glycolysis, *162*, 163
Z scheme, *143*
energy budget, ecosystem, *1018*, *1019*
energy of activation, 49, *50*
energy pyramid, *1015*
energy state of a mixture, 49
energy store, 162, *163*
England, declining population, 1055
Engler, Adolph, 492

English sparrow, 337
engram, 958, 960–962
  long- and short-term memory, 959, 960
  transferred through corpus callosum, 961
Ensatina escholtzi, ring-species salamander, 388, *389*
enterocrinin, 722
enterogastone, 722
Entomeba histolytica 433, *434*
entropy, 126, 131; *see also* thermodynamics
environment, 992
  interaction in gene expression, 309, 312
  pollution, 1040, 1041
environmental resistance, *1030,* 1031
environmentalists, 1022
enzyme, 69–73
  allostery, 73, *74*
  as a catalyst, 51, 69
  control of enzyme activity, 72, 73
  hydrolyzing activity, *70*
  kinetics, 71–73
  regulatory, 73
  reversibility of action in theory and fact, 70, 71
  enzyme-substrate complex, *69, 70*
eocrinoid, 627
eosinophil, *762*
Ephemeroptera, 623
Ephrussi, Boris, one gene one enzyme, 353
ephyra, 600
epiboly, 872, *873*
epicotyl, *574, 575*
epicotyl hook, *522,* 577
epidermal growth factor, 791
epidermis
  echinoderm, *675*
  leaf, *90, 521*
  root, *90, 580*
  root tip, 507
epididymus, *838*
  sperm maturation, 840
epigenesis, 868
epiglottis, *741*
epigyny, *570*
epinephrine, 777, 787, 788
  cell membrane receptor, *791*
  mechanism of action, 790, 791
  stress response, 788
  structural formula, *911*
epiphyseal cartilage, *676,* 677, 789
epiphyseal plate, *676*
epiphysis, 789
epiphyte, 1001
episome, 360
  F+, 360–364
  F+ lac, 367
epithelium
  ciliated, *90*
  defined, 245
  intestinal, *94*
equational division, 250; *see also* meiosis; anaphase II
equilibrium, 504, 516
  frequency of morphs under frequency-dependent selection, 334, *335*
  mutation with selection, 338, *339*
  partial pressures of atmospheric with dissolved gases, 743
  of population size, 1029
  semicircular canals, 641, 928, *936*
Equisetum, 480, *481*
erectile tissue, *839*
erection, 839, 840; *see also* sexaul behavior, human
ergot, 450
erosion and carrying capacity, 1031
erythroblastosis fetalis, *286, 287*
erythrocyte, 760, *761*
  nucleated in birds, 226
  nucleated in fish, 222
  metabolized in liver, 717
erythropoietin, 761
Escherichia coli, 413
  chromosome map, *365*
  chromosome replication, *189*
  genetics of, 356–366
  Hershey-Chase experiment, 193
  makes vitamin K in human colon, 717
  operon, 366–371
  specially weakened laboratory strains, 376
Eskimo, 337
esophagus, 712, *713, 715*
essential amino acids, 724
ester linkage, 76
estivation
  lungfish, *642*
  spadefoot toad, 997
estrogen, 371, 777, 788, 789, 846–848, 895
  activities in different tissues, 371
  and menstrual cycle, *846, 847,* 848
  in development, 895
  induces activity in chick oviduct cells, 371
  produced also by adrenal cortex, 895
  produced by corpus luteum, *843*
  see also sex hormones
estrone, 788, 789
estrous cycle, 845
estuary, 1011
ethanol; ethyl alcohol, 159, *164,* 454, 1040
ethology, 945
ethylene, *552*
eubacteria, 410, *414,* 415
eucaryote
  cell, 86
  characteristics (table), 87
  movable membranes, 87, 241, 421
  origins, *421, 422,* 423
euchromatin, 225
  compared with heterochromatin, 226
Euglena, 402, 423, 428, *430,* 440
  distantly related to slime mold, 456
euglenoid, *see Euglena*
Eumycota, 445, 449–456
euphotic region, 1009
Euplotes, 437, 438
eutheria, 650
eutrophication, 474, 534, 1013, 1021, *1022*
Evans, Dale, 279
evaporation, *504, 505,* 506
  cooling function, *806*
  free energy of, 506, 516, 517
evasive action
  moth pursued by bat, 925–927
  teal pursued by falcon, 949
evolution, 8, 321–349
  and adaptedness, 10
  a characteristic of life, *47*
  of human brain and mind, 985
  in sexual vs. asexual organisms, *325*
  molecular, 19, *20, 400*
  mammalian skull and zygomatic arch, *684*
  rate of, 15, 338, 397
  similarity due to common descent, 4
  testing the hypothesis, 18, 19, 396
  see also natural selection
evolutionarily stable strategy (ESS), 977, 980
  reciprocal altruism, 983, 985
  retaliation, 977
  territoriality, 980
evolutionary strategy of sexual recombination, 258, 326, 622, 830, 850–852
excision
  of bacteriophage DNA from host chromosome, 358, *359*
  of intervening sequences from HnRNA, 207
excitement phase, sexual, 848, 849, *850*
excretion, 808–818; *see also* excretory system
excretory system, 533, 808–818
  amphibians, 811
  annelid, *610*
  in bird egg, 886
  earthworm, *812*
  fish, *811*
  frog, *812*
  grasshopper, *813*
  human, 813–818
  in human embryology, 891, 892
  kangaroo rat, 998
  mesodermal origin, *876*
  plants, 117
  see also kidney
exergonic reaction, 50, 71, *72,* 152, *159*
exhalation, 742, 745
exocrine cell, *113*
exocytosis, 105, *106*
exon, 207
exonuclease, 721
exopeptidase, 720, 721
exophthalmic goiter, 783, *784*
exoskeleton, 617, 671–674
  arthropod, formed of chitin, 63
  crustacean, 620
  mollusk, 672–674
exothermic reaction, 50, 71, *72,* 152, 159
expected value, 314
experiment, and scientific method, 8, *9*
experimental embryology, 868, *869,* 878–884
exploitation, by commercial fisheries, 1011
exploration, 952
exponential growth, 1029
  bacteria, 1029
  elephants, 1031
  houseflies, *1031*
expulsion of fetus, 894
expulsion of placenta, 896
external ear, 927
external fertilization, 830, *831*
external genitalia, 837–841
  embryology, *893*
  female, 840, *841*
  male, 837–839
external gill, 739
extinction, 1049, 1050
  by competition, 4, 7
  of a conditioned response, *955*
  of dinosaurs by asteroid, 646
extraembryonic membranes, 874, *875,* 884–887
  endocrine function, 848
  mammalian, 886, 887
extra Y syndrome, 257
eye, 927–933
  embrology, *878, 889, 891*
  human, diagrammed, *930*
  insect and crustacean, *620,* 929, 930
  vertebrate compared with cephalopod, *616,* 929
eye tooth, 710
eyespot
  Chlamydomonas, 469
  algal protist, 928
  Euglena, 430
  planaria, *929*
eyestalk, X- and Y-organs of crayfish, *773*

F

F1 particle, 116, 131, *132, 172,* 175
F+ episome, 360–366
  responsible for bacterial maleness, 364
  transfer, *361,* 363
F+ lac episome, 367
facial muscles, 690
facilitated diffusion, 100, *104*
factor, Mendel's term for allele, 271
facultative anaerobe, 416
facultative heterochromatin, 226
FAD (flavine adenine dinucleotide)
  in citric acid cycle, *170,* 171, *172*
  in Z scheme, *143*
fairy ring, *444*
falcon, hunting and attack pattern, 947–949
Falconer, D. S., artificial selection in mice, 329
Fallopian tube, *842,* 892
far red light, 562, *563*
fascia, *688*
fasciculus (fascicular bundle), *688*
fast neuron, 911, *912*
fat, 75
  body needs, 723
  cell, *85*
  chemistry of digestion, 720
  emulsified by bile, 717
  from carbohydrates, 723
  saturated, 77
  as thermal insulation, *806*
  under ACTH control, 780
  unsaturated, *77,* 723
fate map, 872, *876*
fatty acid, 75, *76, 77,* 720; *see also* fat
fear, a component of aggression, 975
feather, *648*
feces
  bird, 710
  gray when bile duct is obstructed, 717
  human, mostly dead *E. coli,* 717
  insect, 706
fecundity: woman with 69 children, 1034
feedback
  and reafferance, 953
  see also negative feedback, positive feedback
feelers, 924, *925*
female pronucleus, 254, *255,* 867
femur, 685, *686*
  insect, grasshopper, 672
Fentress, John, whose tame wolf runs from advancing sheep, 1047
fermentation, 159, *164,* 454
  in gut of ruminant, 711, *712*
fern, *481, 482, 483*
ferredoxin, in Z scheme, *143*
fertile period (human), 846, 852
fertility, and nutritional state, 1034
fertility factor, 360–366
fertility rate, 858
fertilization, 254, *255, 258,* 866–871
  human, 852
  human ovum, *258*

fertilization cone, 867, *868*
fertilization membrane, 866, *867*
fertilizer, chemical, 531, 532
fetus
  circulation, *896*, 897
  development, 887–896
Feulgen, Robert, DNA-specific stain, 192
fever, adaptedness of, 808
fiber, muscle, 91, 688; *see also* muscular system
fibril (myofibril), 690, *691, 692*
fibrin, 762, *763*
fibrinogen, 762, *763*
fibroin, 618
fibrous protein, 64
fibula, 685, *686*
fiddlehead, 481
Field, John, 519
fighting, 975–978
  cichlid fish must fight before it can mate, 978
  evolutionarily stable strategies, 977
  by human males, 976, 977
  lethal fights do occur between males of a species, 977
  rattlesnakes, *976*
  ritualized, *976*, 977
fighting center, 952
filament
  of a fish gill, *737*
  of a stamen, *560*
filter feeders
  basking shark, 639
  bivalve mollusks, 614, *615*
  Galapagos rift crabs, 1017
  lancelet, 636
  larval lamprey, 737
  lophophorates, 609
  ostracoderm, 637
  polychaete, 611
  pterobranch, 634
  sponge, 597, 598
fimbriae, *842*, 843
fin, 640, 641, 681, 682
finches, Darwin's, 5, *6*, 7
finger, embryological development, *892*
fingernails, a primate characteristic, 658
fire, in chaparral, 1003
fire disclimax community, 1003
firefly, communication in, 971
fireworm, carefully timed reproduction, 830
first trimester, 888–893
fish, 637–642
  advantage of rare red morph, 334, *335*
  annual life cycle of Brazilian fish, 1035
  behavioral thermoregulation and fever, 802
  deep sea, *1011*
  digestive system, *707, 708*
  gills, 732, *737, 738*
  nucleated erythrocytes, 222, 277
  osmoregulation, 810, *811*
  phylogenetic tree, *641*
  sound communication, 972
Fisher, Ronald A., 275, 314
  finds Mendel's results suspicious, 275
fisheries, coastal, 1011
fissure of Rolando, 920, 921
fissure of Sylvius, 920
fitness, 326
  original meaning, 333
  relative, 324
Fitzroy, Robert, captain of *Beagle*, 3, 5, 7
five-kingdom scheme, 402, *403*, 444, 592; *see also* kingdom
5S RNA, 205
fixation, of a favored allele, 324
fixed action pattern, 946–949
flagella, *see* flagellum
flagellate, *432, 433*
flagellin, 411
flagellum, *89*, 90, 117–119, 404, 410, *411, 412*
  and central dogma, 190
  bacterial, *90, 118*, 404, 410, *411*
  diagram of bacterial motor, *412*
  eucaryote, 117, *118, 119*
flame cell, rotifer, 607, *608*
flatus, 717
flatworm, *602, 603*, 604
  as ancestor to metazoa, 594, *595*
  digestive system, 702, *703*
  ladderlike nervous system, 913, 914, *915*
  respiratory system, 733
flavine adenine dinucleotide (FAD), *143, 170, 172*
flea, 592, *623*
Fleming, Walther, describes cell nucleus, 224
flight muscles, insect, *672*
flightless birds, 488, *649*, 681
floridean starch, 465
florigen, 564, 565
florists, manipulate photoperiod, 562, 563
flotation
  bladder of brown alga, 467, *468*
  cuttlebone, *673*
  swim bladder, 640, 641, 738
flower, *492, 493*, 560–573
  diagram, *560*
  fragrance, 561
  perfect, 493
varieties, *570–573*
flowering, initiation of, 561–565
flowering plant, 486, 488–496
  centriole absent or amorphous, 237, 238
  origins, 491
flu virus, rapid evolution, 823
fluid mosaic model of cell membrane, *101*, 102
fluke, 603
fluorine
  dietary need, 725
  in tooth enamel, 27
  reacts with noble elements, 30
flying lemur, *652, 653*, 654
folic acid, 717, 724
follicle, *843*
  cells, *866*
  corpus luteum, 776, *843*, 847
  endocrine function, 846
follicle-stimulating hormone (FSH), 779
  cycle in males (as ICSH), *845*
  and menstrual cycle, 845, *846, 847*
  stimulates cAMP synthesis in target cells, 791
  target tissues in testis, 844
food chain, 1015
  of Galapagos rift community, 1017
  see also food web
food faddists, 723
food pyramid, *1010, 1015*
food vacuole, *434*, 701
  Paramecium, *438*
food web, 611, *1015*
  and earthworms, 611
  Galapagos rift community, 1017
  grassland, 1000, 1001
  nematodes, 605
  tundra, 1008
foot
  bones, *686*
  mollusk, 612
  variety of structures in birds, *709*
foramen magnum, 678, *679*
foramen ovale, *896*, 897
Foraminifera, 434, *435, 673*, 674
  spiralled shell, *673*, 674
Ford, Henry, programmed death of Model T, 1036
forearm, homology and divergence, 399, *400, 683*
forebrain, 916, *917*
  and development of eye, *880*
  in human embryo, *891*
foreskin, *838*
forest
  coniferous, 1005–1007
  fire, 1003, 1021
  Mediterranean scrub, *1003*
  monsoon, 1002
  productivity, *1018, 1019*
  succession, *1021*
  taiga, 1005–1007
  temperate deciduous, 1003–1005
  tropical rain, *1001, 1002*
  tropical seasonal, 1002
forgetting, 959
fossil
  animal phyla, 593
  comparative anatomy, 399
  oldest procaryotic and eucaryotic, 409, 413
fossil fuels, 535, 538
  and greenhouse effect, 1019, *1020*
founder effect, 338, *339*
four-chambered heart, 643, 648, *750*, 751
fovea, 933
  of hawk, 932
foxtail, *576*, 577
fragrance, of flowers, 561
frameshift mutation, 217
Franklin, Rosalind, X-ray diffraction of DNA, 201
free energy, 126, 131, *143*, 158
  of evaporation, 505, 516, 517
  of glucose, 158
  loss in electron transport, 173, *174*
  Z scheme, *143*
freedom and dignity, 954
freeze-etching, *99, 101*
freeze-fracture, *99, 101*
frequency, 323
  allele frequency, 321, 323, 331, 338, 339, *339*
  genotype, 323
  of morphs, 334, *335*
frequency-dependent selection, 334, *335*
fresh water
  environment, 1013
  eutrophication, 1021, *1022*
  osmoregulation, 811
Fritos, 59
frog, *395*, 643
  brain, *917*
  catching flies, *946*, 947
  cleavage divisions, *870*
  color change by melatonin, 789
  distribution, *388*
  excretion and osmoregulation, 811, 812, *812*
  responding to stimulus, *46*
  skin graft and innervation, *883*
  Xenopus, 864, *881*
  see also amphibian; toad
frontal bone, *679*
frontal lobe, 920
frostbite, *759*, 806
fructose
  fructose-1, 6 diphosphate, *147*, 160, *161*
  in photosynthesis, *147*
  structural formula, *57*
fruit
  defined, 560
  variety, photoessay, *572, 573*
fruit fly, 298; *see also Drosophila melanogaster*
fruiting body
  myxobacterium, 417
  slime mold, 446, 447
FSH, *see* follicle-stimulating hormone
fungus (fungi), 444–456
  characteristics of kingdom, 404
  and chestnut blight, 1042
  evolutionary relationships with other kingdoms, 402, *403*, 456
  fungi imperfecti, 456
  as a polyphyletic group, 395
furrow
  cleavage furrow, *870*, 872
  division furrow in cytokinesis, 240, *241*
fucoxanthin, 466
functional side groups (table), 48

G

G_1, G_2, in cell cycle, 228, 229, 234
galactosamine, 63
galactose, structural formula, 59
Galapagos Islands, 5
  Darwin's finches, 5, *6*, 7
  tortoises, 5
Galapagos rift deep sea community, 1017
galeae, *706*
Galen, 789
Galileo, 84
gall bladder, 717, 722
Galton, Francis and the study of metric traits, 265, 312–314
gametangium, of bread mold, 449, *450*
gamete, *865, 866*
  fertilization, 254, *255, 258*, 866–871
  ginkgos and cycads, 258
  ovum, *258*
  sperm, 245, 567, 568, 839, *840, 841*, 866
  sperm nucleus (pollen tube), 567, *568*
  see also life cycle
gametocyte, 245, 251
  oocyte, 245, 251, *843*, 870, 871
  spermatocyte, 245, *248*, 839, *840*
gametophyte, 466
  fern, 482, *483*
  moss, 474
  seed plant, 488
  see also life cycle
gamma ray, 196
gamont, *436*, 437
ganglion (ganglia), 906, 914, 915
  earthworm and flatworm, 914, *915*
  spinal, *922*
gangrene, caused by ergot poisoning, 450
gap junction, 108, *109*, 514
garden snail
  and cellulase, 62
  survival in seawater, 12
Garrod, A. E., inborn errors of metabolism, 352, 353, 356

gas exchange, 732–746, *744*
in alveolus, 743, *744*
in leaf, 514
in plants, 519–522
partial pressure, 743
see also respiratory system
gastric juice, 721, 722
acidity of, 41
gastric lipase, 715, 718
gastrin, 722
gastrointestinal tract, 703; *see also* digestive system
gastropod, 614, *615*
snail, 12, 62, 614, *615*, 672, *673*, *674*
gastrovascular cavity, *597*, 598, 702, *703*
ctenophore, *601*
flatworm, 602
gastrulation, 872, *873*, *874*
birds, 873, *874*
compared with *Volvox* development, *594*
mammals, 873, 874, *875*
Gause, A. F., competitive exclusion principle, 15, 1043
Gause's law (competitive exclusion), 7, 15, 17, *1043*, 1044
gemma (gemmae), 476, *477*
gemmule
of sponge, 598
in theory of pangenesis, 322, *323*
gene
composition of, early theories, 182
controlling gene, 370, 371
environmental interaction in expression, 309
factor, Mendel's term, 271
modifying, 309
Muller confirms chemical nature of, 192
structural, 207, 215, 216, 366
gene activation
by second messenger, 792
by steroid hormones, *792*, 793, 895
see also gene control
gene cloning, 372–377, 781
gene control
and chromosome condensation, 226
control proteins, 279
in differentiation, 875
during gastrulation, 882
hormone induction in chick oviduct, 371
operon, 366–371
and polytene chromosome puffs, *308*, 309
repressible enzymes, 370, *371*
steroid hormones, *371*, *792*, 793, 895
and transplanted nuclei, 880–882
gene expression, 309–312
gene flow, 334–337
gene frequency, 321, 323, 331, 338, *339*
gene inserts, 207, 372, 820
gene interactions, 287–289, *309*
chicken combs, 289
and response to artificial selection, 330
gene locus, 273
gene number
in *Drosophila*, 308
in humans, 309
gene pool, 321
gene protein activator, 792
gene rearrangement in antibodies, 820
gene splicing, 372–377
gene substitution, 331
generalization, behavioral (Pavlov), 955
genetic background, 310
genetic code, 205, *206*, 207
in mitochondria, 217
genetic deterioration under civilization, 340
genetic drift, 337, 338, *339*
genetic mapping, 304–309
chromosome mapping, *306*, *307*
kitchen blender, 365, *365*
map distance and recombination frequency, 306
genetic variability, *see* variability, genetic
genetics, classical, 265–290
prior to 1950, 182
problem set, 289, 290
genital swelling, 892, *893*
genital tubercle, *893*
genitalia
embryonic development, 892, *893*
female external, 840, *841*
male, 837–839
penis, 687, 836–839, *893*
see also reproductive system
genotype, 277
genotype frequency, 323
Geocapsa, *420*
geographic cline, 334, *336*, *337*
geographic isolation, *388*
geographic variation in humans, *337*
geography and natural selection, 334–337
geology, Lyell's *Principles*, 4, 5
geotaxis, 330, *331*, *332*
geothermal hot spring, 1017
geotropism, 546–548
geraniol, 489
germ layers
amphibian gastrula, 872, *873*
and coelom, *597*
fate, 876–878
and phylogeny, 596
germ line, 244, 321, *322*, 440, 821
germ plasm, 321, 322, *323*
germ theory of disease, 410, 1053
germinal epithelium, 245
ovary, 843
germination, 577, *578*
desert annuals, 997
GH, *see* growth hormone
giant
giant axon, of squid, 911
pituitary giant, *781*
giant chromosome, 306, *307*, *308*, 309
giant squid, 612, 1035
gibberellin, 549, *550*, 577
gibbon, 656, *657*
Gigantopithecus, *657*
gill
external, *732*
fish, *732*
fish, anatomy and function, *737*
invertebrate, 734, 735, *736*
mollusk, 612
mushroom, *454*, *455*
salt excretion, 810, *811*
gill arch, *737*, *889*
amphibian embryo, *878*
ear ossicles, 399, 890
and origin of jaw, *639*
in embryology, 888, *889*
gill basket, 634–636, 736
gill chamber, 641
gill cleft, *890*
lamprey, 637
gill pouch, *890*
in embryology, 888, *889*
gill raker, *737*
gill slit
hemichordate, 634, *635*, 736
tunicate, 635, *636*, 736
Ginkgo, 258, *485*, 488
flagellated sperm, 258
giraffe, 655
and directional selection, *333*
and germ plasm theory, 322
and stabilizing selection, *334*
girdling, 516
gizzard, *710*
bird, 648
earthworm, *705*
rotifer, 607, *608*
gland of Bartholin, 842
glans clitoridis, 840, *841*
glans penis, *838*, 839
gliding bacterium, 410, *417*
gliding joint, *677*
globulin, 64, *762*
Gloger's rule, 336, 337
glomerulus, *814*, 815, *816*
glottis, 739, *739*
glucagon, *723*, 777, *786*, 787
stimulates cAMP synthesis in target cells, 791
glucocorticoids, *723*
glucosamine, in chitin, *63*
glucose, 56, 57
alpha and beta forms, *62*
blood level, *723*
open chain form, *57*
structural formula, *57*
synthesis in Calvin cycle, *147*, 148
glucouronic acid, 63
glutamic acid
and ionic bonding, 65
structural formula, *67*, *911*
glutamine, structural formula, *66*
Glutton Club, 2
glycerol (glycerine), *720*
structural formula, *75*, *76*, *77*
glycogen, 60, *61*, 723
breakdown mediated by epinephrine, *791*
conversion to glucose, 71
metabolism, 166, *167*
glycolysis, *159*, 160, *161*
compared with fermentation, *164*
energy yield, *162*, 163
muscle, 15, *166*, 168
schematic diagram, *160*
goblet cell, *716*, *741*
going to seed, 561
goiter, 783, *784*
golden plover, *669*
goldfish, breathes air, 738
Golgi body, 88, 89, 111, *112*, *113*
forms cell plate, 242
Golgi, Camillo, finds Golgi body, 111
Golgi complex, 88, 89, 111, *112*, *113*, 242
gonad
embryology, human, 892
mesodermal origin, *876*
the site of meiosis, 244
see also ovary, testis
gonadal cortex, tissue, 892
gonadal medula, tissue, 892
gonadotropins, 782, 844, 845
follicle-stimulating hormone, 779, 791, 844–847
interstitial-cell stimulating hormone, 776, *845*
luteinizing hormone, 779, 791, *846*, *847*
Gondwanaland, *488*
gonorrhea, 415
Goodall, Jane, reproduction of female chimpanzees, *1032*
goose, greylag egg roll, *947*
goosebumps, 806, *807*
gorilla, 657
skeleton compared with human, *682*
gout, *817*, 818
grasshopper, 706, *707*, *734*, *834*
growth hormone (GH), 75, 376, *776*, 779, 781
grunion, *831*, 851
GTP (guanosine triphosphate) in protein translation, *213*, 214
guanine, structural formula, 184
guard cell, 514, 520–522
fewer in xerophytes, 996
potassium dependent action, 537
guinea pig, 3, 4
males fight to death, 977
gull
individual recognition, 975
reproductive strategies of two species, *1033*
saves Mormons, 1000
gullet
Euglena, 430
Paramecium, *438*
rotifer, 607, *608*
Gunflint chert fossils, 413
guppy, 832, *833*
Gurdon, J. B., transplants frog nuclei, 880–882
gut, 701–726; *see also* digestive system
Guthrie, Arlo, Joady, and Nora Lee, 312
Guthrie, Woody, 311, 312
guttation, 515
Gymnomycota, 446, 447
slime mold, 395, 445–458, 771, 772
gymnosperm, 483–486, *487*
apical dominance, 348
life cycle, *487*
as a paraphyletic group, 396
gynoecium, *560*, 561

H

H zone, *691*, *692*
Haber, Fritz, 532
habit reversal, 957
habitat specialization
barnacles, 1044
brine shrimp, 800, 810
warblers in spruce, *1044*
habituation, 954
hadron, 26
Hadzi, Joran, ciliate theory of metazoan ancestry, 594
Haeckel, Ernst, recapitulation, 593, 594, 888, 890, 891
hagfish, 637, *638*
mouth and digestive system, 708
Hagiwara, crosses morning glories, 289
hair
ectodermal origin, *876*
a mammalian characteristic, 649
hair cell, of inner ear, 927, *928*
Haldane, J. B. S., 315, 327
and kin selection, *983*
Hales, Stephen, 514
half-life, of a radioactive isotope, 196
haltere, 622
Hamilton, W. D., kin selection, *983*
hand, 683, 684, *685*
hangover, and dehydration, 818
haploid cell, 243
haplont, defined, 469
haptoglobin, 73
hard palate, 741
Hardin, Garrett, alerts world to overpopulation, 1055
Hardy, G. H., 340, 341
Hardy–Weinberg law (Castle–Hardy–Weinberg law), 313, 314, 340–344

Harlech Castle, Wales, centuries of horse urine, 413
Harvey, William, demonstrates valves of veins, *760*
Haversian canal; Haversian system, *676*, 677
hawk, *647*
Hayes, William, 643
health foods, 723
hearing, 925–927, *928*
heart
  amphibian three-chambered, 643, *744*
  beat, 754–756
  bird four-chambered, 648
  changes at birth, *896*, 897
  control of beat, 752, 754–756
  crayfish, *747*
  crocodilian four-chambered, 643
  earthworm five pairs, 747
  embryonic development, human, *888*, *892*
  fish, 641, *748*
  four-chambered heart, 643, 648, 650, *750*, 751
  grasshopper, *747*
  human, 751–756
  mesodermal origin, *876*
  output, 755
  reptile three-chambered, 643, *749*
heart muscle, *see* cardiac muscle
heart transplant, 254
heartburn, 715
heat, and thermodynamics, 126
heat conservation, countercurrent exchange, 801, *810*
heat of activation, 50, *50*
heat receptors, 913, 923, *924*
heat stroke, Wallace, 805, 806
heavy chain of immunoglobulin, 819, *820*
hedges, 548
Heidleberg man, 661
height, human, 257, 313
  pituitary giant and dwarf, *781*
  and sex chromosomes, 257
Heinroth, Oskar, instinct theory, 946
heliozoa, *434*
helium, 28, *29*
heme group, structural formula, *67*
hemicellulose, 63
Hemichordata, *596*, 634, *635*
  acorn worm, 634, *635*, *637*
  pterobranch, 634
hemiparasite, *539*
hemipenis, 835
Hemiptera, 622
hemizygosity, 300
hemochromotosis, 312
hemocoel, 597, 704, *747*
hemocyanin, 735, 747
hemoglobin, 743, *744*
  annelid, 609, 733
  binds carbon dioxide, 744, 745
  Bohr effect, 745
  of chemosynthetic giant red pogonophores, 1017
  Constant Spring mutation in termination codon, 217
  DNA sequences of humans and rabbits, 372
  earthworm, 747
  function, 743–745
  and heme group, 67
  human and carp sequences compared, 338
  human and horse compared, 337
  lancelet, 737
  multiple alleles, 282
  metabolic breakdown, 761
  sickle-cell allele, *284*, *285*
  thalassemia allele, 282
  tetrameric form, 65
hemophilia, 301, 763
  Christmas disease, 763
hemotoxin, 644
henbane, *562*
Henking, H., discovers X chromosome, 300
Henslow, John, the man who walked with Darwin, 2, 3, 9
heparin, *762*
hepatic artery, *757*
hepatic portal circuit, *757*
hepatitis, 717, 761
herbaceous plants, 496
herbicides, 553
  effect on embryo, *9*
heritability, 314, *315*
heron, stable population, 1028
Herpes simplex virus, 1041
herring, 833
Hershey, Alfred, and the genetic role of DNA, 193, 200, 357
Hershey–Chase experiment, 193, *200*
heterocercal tail, 639
heterochromatin, 225, 226, 234
  centric, 225
  centric, condenses early, 234
  compared with euchromatin, 226
  constitutive, 225
  facultative, 226
  reduced recombination within, 307
heterocyst, 419, *420*
heterogeneous nuclear RNA (HnRNA), 205, 227
heteromorphic chromosomes, 297
heterotroph, 422
  fungi, 444
heterozygosity, 268, 278
Hfr (High frequency of recombination episome), 365
hibernation, amphibians, 643, 739
high energy bond, 126
  in glycolysis, 162
  see also ATP
hindbrain, 916, *917*
  human embryo, *891*
Hindenberg zeppelin, 48, *49*
hinge joint, 677
hip bones, 681
hip joint, 677
hippocampus, *960*, 962
hippopotamus, 655
Hirsch, Jerry, and geotaxis maze, 330, *332*
hirudin, 611
histamine, *762*, 777, 778
histidine
  and ionic bonding, 65
  structural formula, *67*
histocompatibility antigens, 287
histology, 97
histone
  in nucleosomes, *225*
  protective function, *187*
historicity of evolution, 5
Hitler, Adolf, 401
HnRNA (heterogeneous nuclear RNA), 205, 227
holdfast
  brown alga, 467
  red alga, *465*
Holmes, Sherlock, 193
holostei, *641*
Holy Fire, 450
home range, 979
homeostasis, chapter, 798–823
  a characteristic of life, *47*
  defined, 798
  see also negative feedback
homeothermy, 648–650, 800, 805–808
  birds, 648
  fast-swimming fish, 749
  mammals, 649, 650
  dinosaurs, 19, *19*
Hominidae, 655, *656*, *657*
hominid phylogeny, *662*
homing pigeon, 967, *968*, *970*
Homo erectus, 660, 661, *662*
Homo habilis, 659, *662*
Homo sapiens, 658–663
  earliest fossils, 658, 661
  genetic future, 340
  rapid recent evolution, 398
  see also human; human population
Homo sapiens neanderthalensis, *661*, *662*
Homo sapiens sapiens, *662*
homologous chromosomes, 243, *244*, 245
  pairing in meiosis, 245, *246*
homologue, 243
homology, 396, 399
  bones of gill arches, 399, *639*, 890
  chordate and echinoderm central nervous system, 916
  distinguished from analogy, 396
  vertebrate forelimb, 399, *400*, *683*
Homoptera, *622*
homozygosity, of true-breeding strains, 268
homunculus, 868, *921*
honeybee
  cooperative behavior, *981*
  kin selection, 983
  pheromones, 973, 974
  visual acuity, 929, 930
Hooke, Robert
  observes and names cell, 84, 85
  girdles a tree, 516
Hooker, John Dalton, friend of Darwin, 12
hop plant, 544
hormones (animal), 771–793
  birth control pill, 853, 855, *856*
  cAMP stimulated in target cells, 791
  control of gene activity by steroids, 371, *792*
  control of human reproduction, 844–848
  digestive hormones (table), 722
  distant action, 771
  ecdysone, *308*, 671
  epinephrine, 777, 787, 788, 790, *791*, *911*
  estrogen, 371, 777–789, *843*, 846–848, 895
  follicle-stimulating hormone (FSH), 779, 845–847
  formal proof of existence, 773
  human hormones (table), 776, 777
  invertebrates, 772–776
  juvenile, 774, *775*
  and locust migration, 1050
  luteinizing, 776, 779
  major classes of, 776, 777
  mechanisms of action, *308*, 371, 790–793
  menstrual cycle, 845–847
  modified amino acids, 777, 778
  overcrowding upsets balance, 1051
  pathways involved in response to nipple stimulation, *898*
  peptide hormones, 75
  and programmed death in cephalopods, 1035, 1036
  rapid degredation of, 776
  steroid hormones, 79, 776, 777, *792*
  steroids and development, 892, 893, 895
  target cells in brain, 892
  vertebrates, 776–790
hormones (plant), 544–554
  auxin, 545–549, 561, 565, 583
  gibberellins, 549, *550*, 577
  induction of flowering, 564, 565
hornbill, *649*
horned lizard, temperature regulation and color, 804
hornwort, 477, *478*
horny toad, 804
horse, modern breeds, *11*
horsehair worm, 608
horseshoe crab, 397, 617
horsetail, 480, *481*
host specificity, 1041
  of insect seed predators, 1043
hot spring, Galapagos rift community, 1017
house, example of homeostatic control, *793*
housefly
  egg, *835*
  exponential growth potential, 1031
  mouthparts, 621
housekeeping genes, 366
Howard, Eliot, *Territory in Bird Life*, 979
human, 658–663
  aggression, 978
  chromosome number, 227, 243
  early fossil finds, 658, 661
  embryonic development, 887–896
  evolution and origins, 657–663
  evolution of reciprocal altruism, 985
  genetic future, 340
  phylogeny, *662*
  longest-lived of all animals, almost, 1036
  programmed death of, 1036
  survivorship curve, *1034*
human populations, 1051–1056
  accelerating growth rate, 1030
  age profiles, *1054*
  and agriculture, 1051, 1052
  decelerating growth rate since 1973, 1052, 1055
  doubling time calculations, 1056
  fashion in birth rates, 1055, 1056
  growth rates, 858, 1030, 1052, 1055, 1056
  prognosis, 1035–1056
humerus, 685, *686*
hummingbird, 561, *649*
  long tongue, 710
  migration across Gulf, 25
  territoriality, 979
humus, *1018*, *1019*
  defined, 529
hunters, 771
Huntington's disease (Huntington's chorea), 310–312
Hutterites, 338
Huxley, Thomas, the bulldog of Darwinism, 18, *22*
hyaline cartilage, *676*
hyaluronic acid, 63, 866
  structure of, 63
hyaluronidase, 866
hybrid DNA, and Meselson–Stahl experiment, *199*
hybrid swarm, 389
hybridization, 386, *387*
  as a criterion of species, 17, *18*, 386, *387*
  flowering plants, 388–390, 488, 489
  liger, *18*
  and species recognition, *974*, 975
  wolf and coyote, *387*
Hydra, 599
  survivorship curve, *1034*

hydration shell, 36, *38*
hydrocarbon, 39, *41, 42*
hydrochloric acid (HCl), 40, 715, 718
  secreted by stomach parietal cells, *715*, 718
hydrocortisone (cortisol), 777, 787, 788
hydrogen
  combines with oxygen gas, 48, *49*
  ion, 40
  molecular, *32*
  molecular, spontaneous dissociation of, 50
hydrogen bond, 31, 36
  in protein structure, 65, 68
hydrogen sulfide
  example of diffusion, *103*
  source of energy in Galapagos rift community, 1017
hydrolysis; hydrolytic cleavage; 70, 127, 718, *719*
  of ATP, 127, 128
  enzymatic, *70*
  fats, *720*
  proteins, 720, *721*
  sucrose, 58
hydronium ion, 40
hydrophilic amino acid, 65
hydrophobia, 39
hydrophobic bonding, 39; *see also* van der Waals forces
hydrostatic pressure, 504, 505
hydrostatic skeleton, 670–672
  snail eyestalk, 672
hydrotropism, 522
hydroxide ion, 40
hydroxyl group, 48
hydroxyproline, 458
Hydrozoa, *599*
hymen, 841
  embryonic development, *893*
Hymenoptera, *622*
hyoid bone, 888, *889*
hyperpolarization, neuron membrane, 908, *909*
hypertonic solution, 107, *108*
hyperventilation, 746
hypha (hyphae), 91, 353, 444, *445, 449*
hypnotism, 688
hypocotyl, *574, 575*, 577, *578*
hypoglycemia, 786, 787
  and stress, 781
hypogyny, *570*
hypopharynx of insects, *706*
hypophysis (pituitary), 776–782, 844–848; *see also* pituitary
hypothalamus, 919
  hormones listed, 776
  human, 917, *918*, 919
  and menstrual cycle, 845–848
  oxytocin, prolactin, suckling, and releasing factors, *898*
  and pituitary, *779*
  and reproductive system, 844
  and body temperature, 807
hypothesis testing
  of evolution, 18, 19, 396
  Mendel, 274, 275
  and scientific method, 8, 9
hypothyroidism 783; *see also* cretinism
hypotonic solution, 107, 108
Hyracoidea, 653, 655
hysterectomy, 857

I

IAA, indoleacetic acid, *see* auxin
I band, *691, 692*
ice, 36, *37*
ICSH (interstitial cell stimulating hormone), 776, *845*
identical twins, 870
idiot savant, 959
iguana, *398*
  desert, 997, *998*
  Galapagos, 5
ileocecal valve, 717
ileum, *713*, 716
iliac arteries, 752
illium, 681, *682, 683*
imino group, 48
immortality
  potential of germ-line cells, 244
  potential of single-celled organisms, 258
immune response, 73, 74, 819–823
  primary and secondary response, *822*, 823
immune system, 73, 74, 819–823
  antibodies cross placenta, 894
  antibodies in colostrum, 897
immunoglobulin, 819, *820*
  IgG, IgA, IgE, IgM, *820*
immunological memory, 822, 823
impala, ritual fighting, 976
implantation, 848
impulse, neural, 907
inbibition, 38
Inborn Errors of Metabolism, 351
inbreeding depression, unavoidable in artificial selection, 330
incisor, 710, *711*
inclusion bodies, *88*
incomplete penetrance, 310–312
incubation period, 822, 823
incurrent siphon, 735, *736*
incus, *928*
independent assortment, 273–275, 297
  diagrammed, *275*
  in haploid spores, 354, *355*
  law restated, 274
  see also Mendel, second law
India, age profile, *1054*
Indian corn, 569, *571*
Indians, American, 340, 1042
indifferent stage, 892, *893*
individual distance, 974, *979*
individual recognition, 975
indoleacetic acid (IAA), 546; *see also* auxin
inducer locus, 367, *369, 371*
inducer molecule, 367, *369*
inducible enzymes, 366
inducing substance, 880
induction, embryonic, 878–880
  by dorsal lip mesoderm, *879*
  of lens, *880*
industrial melanism, *47, 327*
industrial waste, 1012, 1021, *1022*, 1040
infant mortality, and birth weight, *333*
infection, role of lymphatic system, 764; *see also* disease; parasitism
inferior vena cava, 757; *see also* posterior vena cava; vena cava
infertility, 866
influenza, primary and secondary responses, 822, 823
informational macromolecules, 79, 192
inhalation, 742, 745
inheritance of acquired characters, 401
inhibitor, color allele, *276*, 279, 286
inhibition, in neural pathways, 910
initial velocity, 71, *73*
initiation, of protein translation, 210
initiation complex, 210–211
initiator codon, 206, *207*
initiator tRNA, 210
innate, compared with instinctive, 952, 962
innate behavioral patterns, 945
innate releasing mechanism, *950*
inner cell mass, 874, *875*
inner ear, 927, *928*
  ectodermal origin, *876*
  of fish, 641
inorganic phosphate, *see* phosphate ion
insect, 620–623
  all cells of adult terminally differentiated, 228
  anatomy, *621*
  chemoreceptors, 933, *934*
  compound eye, *620*
  cuticle, 79
  digestive system, 705, *706, 707*
  eggs, 835
  flight muscles, *672*
  hormonal control of molting, *774, 775*
  mouthparts, 705, *706*
  nervous system, *916*
  orders defined and illustrated, *622, 623*
  organs of hearing, 925, *926*
  reproduction, *834*
  repiratory system, *732*
  sound communication, 972
  wings and wing muscles, *672*
insect pollination, 489, *490*, 561
  by yucca moth, *496*
insecticide, 911
insectivore, *651–653*, 654
  common ancestor of placental mammals, 337
insectivorous plants, *539*, 544
insertion, *688, 689*
  at DNA level, 216
inspiration, 745
  and blood flow, 759, 760
inspiratory circuit, 745
instinct, 944–952
  history of concept, 946
  instinctive aggression, 977, 978
  interacts with learning, 962, 963
  problems with concept, 945
  specific implications of term, 952
instinctive behavior
  compared with innate behavior, 952, 962
instinctive pattern, 946–949
  development, 947
insulin, 75, *723*, 777
  and cloned DNA, 373
  mechanism of action in target cells, 791
integration
  of bacteriophage DNA, 195
  of F+ into *E. coli* chromosome, 364
integration of neural impulses, 913
intelligence, 313
  required for successful reciprocal altruism, 985
  spurt in adolescence due to sex hormones, 789
  wild chimpanzee, 1032
interaction, ecosystem, 1014, 1015
interbreeding, 386–388
intercalated disks, *687*, 688
intercostal muscles, 742
interecdysis, 773
intermediate host, of tapeworm, 604, *605*
internal nares, 738, 739
international commissions, 392, 1045
internode, of plant stem, 579, *581*
interphase
  in cell cycle, 228, *229*
  meiotic, *250*
  premeiotic, 245
interstitial cell stimulating hormone, 776, *845*
interstitial cells of testis, *838*
intertidal zone, *1009*, 1011, *1012*
  barnacle habitats, 1044
interuterine device, 853, 854, *855*
intervening sequence, 207, 372, 820
intestine
  cow, *712*
  endodermal origin, *876*
  human, *713*, 717
  long in herbivorous fish, 708
intima, arterial, *755*
introgression, 389, 390
introitus, *841, 893*
intron, 207, 372
  antibody gene, 820
  intervening sequence, 207
intuition, 868
invagination, 872, *873*
  in *Volvox* and gastrula, *594*
  lens vesicle, *880*
involuntary muscle, 686–688; *see also* smooth muscle
involuntary nervous system; *see also* autonomic nervous system
involution, 872, *873*
iodine
  dietary need, 725
  in thyroxine, 27
  radioactive, 197
iodized salt, 783, 784
ion, defined, 31
ionic bond, 31
ionization, 30, 31
  of side groups, 48
ionizing radiation, 217
I.Q. tests, 313
iridium, cosmic origin of, 647
iris, *495*, 686
Irish moss, 465
IRM (innate releasing mechanism), *950*
iron
  dietary need, 725
  plant need, 537
iron absorption, abnormal, 312
irregular flower, 493
  snapdragon, *266*, 268
ischium, 681, *682, 683*
island, biogeography, 5–7, 11, 1049
Isle Royale
  predator-prey cycles, 1049
  wolves, 1046–1049
islets of Langerhans, 777, *786*
isogamete, defined, 469
isolecithal egg, 865, 873
isoleucine, structural formula, *66*
Isoptera, *see* termites
isotonic solution, 107, *108*
isotope, 27
  radioactive, 27
Israel, 506, 515
ivy, 507

J

jackal, 386, *387*
jackrabbit, 1000
Jacob, Francios, and the operon, 365–371
Jacobson's organ, 709
Jamaican woodpecker, extra brood, 1033
Jansen, Daniel
  host-specific seed predators, 1043
  protective role of putrification by microbes, 1040
jaundice, 717, 761

Java ape man, 660; *see also Homo erectus*
jaw, 677–678
origin in placoderms, 638, *639*
shark, *707*
vertebrate classes compared, *678*
jawless fish, 637–639, *641*, 675, 708, 737
jejunum, *713*, 716
jelly coat, 866
jellyfish, 599, 600; *see also* coelenterate
Jewish diseases, 338
Johnson, Virginia, 842, 848
joints, 675, *677*
capsule, 675
types, *677*
vertebrate skeleton, 675, *677*
Joshua tree, 996, *997*
Judd, Burke, one vital gene one chromosomal band, 308
jungle, 1001
junk DNA, 216
juvenile hormone, 774, *775*

K

kangaroo rat, 997, *998*
kangaroo reproduction, 837
karyotype, *237*
karyotyping, *236, 237*
kea, carnivorous parrot of New Zealand, 1000
keel, bird sternum, 648, 679
Keeton, William K., pigeon navigation, *970*
kelp, 467, *468*
Kenya, 592
keratin, 74, 75
and sulfhydryl bridges, 67
in bird beak, 709
ketosis, 787
Kettlewell, H. B. D., natural selection in peppered moth, 326
Khrushchev, Nikita, supports Lysenko, 401
kidney, 813–818
action of aldosterone, *787*
embryological development, 891, 892
kangaroo rat, 998
mesodermal origin, *876*
of marine fish, *811*
of marine mammals, 811
regulates blood acidity, 818
releases erythropoietin, 761
tubule epithelium, *94*
see also excretory system
kidney tubule, 813–818
Kilmer, Joyce, *Trees*, 505
kilocalorie, 128
kin selection, 982, *983*
kinetic energy, 49, 126
kinetics, enzyme, 71, *73*
kinetochore, *see* centromere
King, J. L., human gene number, 309
King, T. J., transplants frog nuclei, 880–882
kingdoms, 391
various kingdom schemes, 402, *403*
kingpin, 1036
kittiwake, 1033
kiwi, 649, 681
Klinefelter's syndrome, 257
knee, *677*, 685, *686*
Kramer, Gustav, orientation studies, 968
Krause's end-bulbs, *806*
Krebs, H. A., Krebs cycle, 169
Krebs cycle (citric acid cycle), *159*, 160, 169–172, *414*
krill, 620
kwashiorkor, 724

L

labia (labium minorum and labium majorum), 840, *841*
embryology, *893*
labial palp (insect), *706, 707*
Labiatae, *495*
labium (labia), insect mouthpart, *706, 707*
labor, induction of
drop in hormone levels, 848
prostaglandins, 894
labor pains, 894
labrinthodont, *642*
labrum, insect mouthpart, *706, 707*
lac operon, 366–369
Lack, David, optimal clutch size, 1033
lactase, 718, 719
absent in non-Caucasion adults, 58
lactate (lactic acid) in glycolysis, 165, *166*
lactation, in humans, 897
lacteal, *716*
lactogen, 75
lactose, 58
and lac operon, 366
structural formula, *59*
lacuna (lacunae) of bone, *676*, 677
Lagomorpha, *652, 653*, 654
lake, 1013
eutrophication, 1021, *1022*
Lake Baikal, 1013
Lake Tahoe, 467
Lake Tanganyika, 1013
Lamarck, Jean Baptiste, 401
lamella (lamellae)
bone, *676*, 677
chloroplast, 140
fish gill, *737*
lampbrush chromosome, *254*, 871
lamprey, 637, *638*, 641
lacks bone, 675
larva similar to adult lancelet, 737
mouth and digestive system, 708
lancelet, *636*
adult similar to larval lamprey, 737
brainless creature, 916
cartilaginous skeleton, 675
heartless creature, 736
hemoglobin, 737
myotomes, *636*
notochord, *636*
resembles vertebrate embryo, 890
land crab, 774
lanolin, 79
lanosterol, 79
Lapland, 337, 1009
Lapps, 337
lard, *77*
large intestine, human, *713*, 717
larvacean, 635
laryngopharynx, 712
larynx, 712, *713, 741*, 742
and thyroid gland, *783*
human, 712, *713*
Lassie, altruism, 982
latency, Darwin's term for recessivity, 268
lateral line organ, 640, 924, *925*
laterite, 1002
Latimeria, *641, 642*
Laurasia, *488*
Lavoisier, Antoine, 527
lawn, of bacteria, 358, *359*
laxative, natural occurrence in fruits, 577
leaf primordium, 579, *581*
leaf scale, 473
leaf scar, *581*
leaf toad, *933*
leaf vascular system and venation, *492*, 512–514
Leakey, Louis, and early hominids, 658
names *Homo habilis*, 659
Leakey, Mary and Richard, and skull 1470, 659, 660
learning, 944–958
autonomic, *957*, 958
bird song, 963
classical conditioning, *955, 956*
flying in birds, 962
and habituation, 954
interaction with instinct, 962, 963
and memory, 958, 959
operant conditioning, *956*, 957
lecithin, 78
in mayonnaise, 39
Lederberg, Esther, 357
Lederberg, Joshua, 356, 357, 360, 361
leech, *611*
Leuwenhoek, Anton van
observes cell nucleus, 223
rabbit genetics, 868
left and right hemispheres of the brain, 920
legume, *495, 572*
and fixation of nitrogen in root nodules, *45*, 532
Lehrman, Daniel, isolated male ring dove, 949
lek, *980*
lemming migration, 1007, *1008, 1050*
lemur, 655
lens, *930*
ectodermal origin of, *876, 880*
induced by optic cup, *880*
lens vesicle, *880*
lenticel, 519, 520, *581*
Leonardo da Vinci, 535, 687
leopard, 975
Lepidoptera, *622*
mouthparts, 621
wingbeat, 672
see also Bombyx; Biston betularia; moth
leptotene, *246*
lethal-recessive dominant allele, 280, *281*, 285
leucine, structural formula, *66*
leucoplast, *88*, 114, 421
leukocyte, 226, 701, 760, 761, *762*, 819
and drumstick, *226*
induced to divide in vitro, 228, *236*
a terminally differentiated cell, 227
varieties of in humans, *762*
Levin, Donald A., 390
LH, *see* luteinizing hormone
liana, *1002*
library of DNA, 372
lice (louse), *619*, 622, *623, 835*
lichen, 423, 451, *453*
as a pioneer organism, 1020
and pollution, 327
tundra, 1007
licorice, and subunit disassembly, 240
life
characteristics of, *46, 47*
conditions for, 992, 993
does it begin at conception?, 870
life cycle
acellular slime mold, *446*
angiosperm, *493*
Ascaris, 606
brown alga, *468*
cellular slime mold, *447*
Chlamydomonas, *469*
conifer, *487*
Drosophila melanogaster, *298*
fern, *483*
gymnosperm, *487*
in different kindgoms, 404
liverwort, *477*
moss, *476*
mushroom, *455*
pork tapeworm, *605*
red alga, *466*
Rhizopus, *450*
Scyphozoan, *600*
slime mold, *466, 477*
Ulothrix, *472*
Volvox, *471*
yeast, *453*
life expectancy and deterioration in old age, 1036
life history strategies, *see* reproductive strategy
life span, *1034*
human maximum of 115 years, 1037
octopus and squid, 1035, 1036
programmed death, 1035–1037
ligament
contrasted with tendon, 690
of knee, 685
ligation and insect hormones, *775*
liger, 18
light antenna, 141–143
light chain of immunoglobulin, 819, *820*
light reactions, 136, 140, *141*
summary of, 145, 151
light receptor, 927–933; *see also* visual receptor
light-independent reactions, 145–148
summary of, *141*, 145, 151
ligase
and DNA repair, 188
and DNA replication, 190, *191*
and recombinant DNA, 373, *374, 375*
lignin, *62*, 63, 95
and origin of terrestrial plants, 458
lily, capsule of, 572
limb, vertebrate, 681–682, 683
embryology, 891, *892*
vertebrate groups compared, 399, *400, 683*
limb girdle, *643*, 675, 681, *682, 683*
limiting factors; limiting resources, 1014
food for primitive humans, 1052
cow pile for maggots, 1039
refuge for sowbugs, 1038, 1039
limits of biological systems, *799*, 800; *see also* threshold; homeostasis
limpet, *673*, 674
osmoconforming, 809, *810*
linkage, 304–307
absent in Mendel's work, 274, *275*
in bacteria, 361
sex linkage, 299–303
Linnaeus, Carolus, binomial nomenclature, 392
linoleic acid, structural formula, *77*
linolenic acid, structural formual, *76*
linseed oil, *77*
lion, 975, *1000*
kill strange male in territory, 977
prey lack protective coloration, 933
lipase, 715, 717, 718
lipid, 75–79
lipid cell, *85*

lips, role in eating, 712
litter, forest, *1018, 1019*
littoral region; littoral zone; *1009,* 1011
live oak, 1003
liver
  endodermal origin in development, *876*
  human, *713,* 717
  human, circulatory system of, *757*
  invertebrates, 703
  origin in vertebrate phylogeny, 704
liverwort, 475, *476*
  life cycle, *477*
living fossil, 397
lizard, *646, 647*
  behavioral thermoregulation, *804*
  desert, 997, *998*
  monitor, *395,* 646
  retains third eye, 789
loam, *528*
lobe-finned fish, *641,* 738
lobes of the brain, 920
lobotomy, prefrontal, 920
lobster, 63
  circulatory system, 735, *736*
  crustecdysone, 620
locomotion, 669–695
  development in therapsids, 650
  evolution in terrestrial vertebrates, 643
  insects, 621
  mammals, 650
  vertebrate forelimb, 682, *683*
locus, gene, 273
locust, periodic plagues, 1029, *1050*
  Rocky Mountain locust and Mormon cricket, 1000
logarithmic spiral, *673*
Lompoc, California, locale of *The Bank Dick,* 432
London, Jack, 981
long-day plant, 562
  photoperiodicity, 562–565
longevity, 1036
  life span, *1034,* 1035–1037
longitudinal muscles, 670
long-term memory, 959, 960
loop of Henle, *814, 815*
  nephron, 813–818
lophophorates, *596, 609*
Lorenz, Konrad
  and gosling, *945*
  instinct theory, 944, *945*
  reafference and consummatory act, 953
  vacuum behavior, 951
loris, 655
lotus, 576
louse, *619,* 622, *623*
  egg (nit), *835*
love, 851, 852
low energy state, 32, 49, *70, 72*
LSD (lysergic acid diethylamide), 450
Lucciani, Jean-Marie, photomicrographs of human cells in meiosis, *246–248*
"Lucy," fossil hominid, 660
lumbar region; lumbar vertebrae, 679, *680*
lumber industry, 1006
luminescence, ctenophores, 601
lumpers and splitters, *394*
lung, *732,* 738–742
  of bird, 648, *740*
  bud, 888, *889*
  evolutionary origin, 641, 738
  human, *741, 742*
  mammalian, 650
  origin by outpocketing, 641, 704, 888, *889*
  reptile, 643
  see also respiratory system
lung bud, 888, *889*
lungfish, 488, *641,* 738
  distribution map, *642*
lungless salamander, 733
luteinizing hormone (LH), 776, 779
  and menstrual cycle, *846, 847*
  stimulates cAMP synthesis in target cells, 791
l_xm_x curve, 1036
lycophyte (club moss), *479,* 480
Lyell, Charles, *Principles of Geology,* 4, 5, 10, 390
lymph, 763
lymph node, 763, *764*
lymph vessel, small intestine, *716*
lymphatic system, 763, *764*
lymphocyte, *762*
  B-cells and T-cells, 819
  clonal selection hypothesis, 821, *822*
  killed by cortisol, 788
lynx, 1006, 1007
  cycles with snowshoe hare, *1045*
Lysenko, T. D., 397, 401
lysergic acid diethylamide, *see* LSD
lysine
  and ionic bonding, 65
  structural formula, *67*
lysis, of host by bacteriophage, *195,* 358, 359
lysogeny, *195,* 358, *359*
lysosomal diseases, 113
lysosome, *89, 817*
  defined, *112*

M

M phase, 228, *229*
MacArthur, Robert H., subdivided warbler habitat, *1044*
MacDougal, D. T., 517, 518
macromere, 871
macromolecule, 39
macronucleus of ciliate, *438, 439,* 440
macronutrients, plant, 530
macrophage, 819
  and clonal selection, *822*
  phagocytizes aged erythrocytes, 761
madreporite (echinoderm sieve plate), *626*
magnesium
  dietary need (animal), 725
  in enzyme activity, 27
  needed for photosynthesis, 25
  as a plant nutrient, 537
magnetic compass, pigeon, *970*
magnolia, 492
maize (Indian corn), *149,* 507, *571*
  kernal, *569, 575, 578*
  smut, *454*
malaria, 434, *435, 436,* 1041
  confined to agricultural peoples, 1052
  controlled by DDT, *1053*
malate (malic acid), in C4 pathway, 148, 149, *150*
malathion, 911
male pronucleus, 254, *255,* 258, 867
maleness, communicated visually, *971,* 972
malleus, *928*
  ear ossicles, 399, 927, 980
Malpighian tubules, *707, 813*
maltase, 718
Malthus, Thomas, treatise on population, 10, 13
maltose, 147, 718
mammals, 649–655
  defined, 837
  mammalian orders, *652, 653,* 654, 655
  origin in Permian, 650
  phylogeny, *395, 652, 653*
mammallike reptiles, *647,* 650, *684*
mammary glands
  ectodermal origin, *876*
  opossum, 651
  placental mammal, 651
  platypus, 651
manatee, *653,* 655
mandible
  arthropod, 618
  human, *679*
  insect, *621, 706, 707*
Mandibulata, 618–624
mandrill, anger display, *971*
manganese
  dietary need, 725
  in enzyme activity, 27
  as a plant nutrient, 537
Mangold, Hilde, and gastrula transplants, 878
mannose, 59
mantis, 622, *624*
  copulatory structure, *834*
  vision, 929
mantle, mollusk, 612
manubrium, 679, *680,* 681
Manx cat, 281, 284
map
  chromosome, 304–307
  pigeon homing map, 970
  world biomes, *993*
map distance, compared with recombination frequency, 306
mara, *4*
marathon race, *168,* 805, 806
Marchantia, 476, *477*
Margulis, Lynn, 114, 115, 413, 421–424
  and five kingdom scheme, 402, *403*
Marianas trench, 1009, 1010
marijuana, 315
marine mammals, and osmoregulation, 811
mark-and-recapture experiments, 9
  of peppered moths, 327
marker genes, in bacteria, 363
marriage rate, 1055, 1056
marrow, *676*
Mars, 35
marsh gas, 32
marsupial, 650, *652, 653,* 654
  reproduction, 836
  yolk sac pseudoplacenta, 886
marsupium, *651,* 654, 837
Marx, Karl, and Marxism
  efficacy of environment, 401
  social reinforcement, 954
  and sociobiology, 984
Mary Lyon hypothesis, 226
mass action, 71
  and glycolysis, 162
mastax (rotifer gizzard), 607, *608*
master gland, *see* pituitary
Masters, William, physiology of sexual response, 842, 848
Mastigophora, 428, *432, 433*
masturbation, 850
materialism, 322
maternal and paternal chromosomes, 243, 249, 289
mathematical biology
  Darwin unreceptive, 278
  Mendel, 267, 269
  models, 272, 273
  numerical taxonomy, 390, 391
  systems ecology, 992, *1018, 1019*
mating types
  in Ascomycetes, *452*
  in *Paramecium,* 440
  in yeasts, *453*
  in Zygomycetes (bread mold), 449, *450*
maturation, of annual plants, 1035
Mauritius, Island of, 577
Mayer, E., misidentifies Neanderthal skeleton, 661
mayfly, 617, *619*
Maynard-Smith, John
  adaptedness of territoriality, 980
  evolutionarily stable strategies, 977
mayonnaise, 39
Mayr, Ernst, biological species concept, 386–390
maze, geotactic, 331, *332*
Mazia, Daniel, isolates spindle apparatus, 238
McPhee, 315
meadowlark, *398*
Mech, L. David, and the wolves of Isle Royale, 1046–1049
median eminence, 779
medical science and human mortality, *1053*
Mediterranean scrub forest, *1003*
medulla
  adrenal, *787*
  brain, 745, 891, *917, 918*
  kidney, 813, *814*
  gonad, 892
  ovary, 843
medulla oblongata, 745, 891, *917, 918*
  human embryo, 891
medullary dominance, 892
medusa, 598, *599,* 600
megagametophyte, 565
megakaryocyte, 761
megaspore mother cell, 565
megasporogenesis, 565, *566*
meiosis, 243–258
  compared with mitosis, *252, 253*
  detailed diagram, *253*
  of diatom, 431
  of dinoflagellate, 430
  evolutionary function of, 258, 326, 622, 830
  in flowering plants, 565, *566*
  in human ovaries and testes, photomicrographs, *246–248*
  in human ovary, 245, *843*
  in human testis, *840*
  and independent assortment, *275*
  nondisjunction, *256,* 257
  prophase prolonged in females, 245, 251, 254, *255*
  in yeast, induced, 454
  see also life cycle; sexual reproduction
meiospore
  Chlamydomonas, 469
  yeast, *453*
meiotic interphase, *250*
Meissner's corpuscle, 924, *925*
melanotropin, 75
melatonin, 777, 789, 790
membrane
  cell, 39, 77, 78, 85, 86, *92,* 95, *100, 101,* 102
  nuclear, *99, 110,* 224, 430
membrane potential, neuron, 907, *908*
membrane receptor, *791*
memory, 958–962
memory scar, memory trace, 958, 960–962
menarche, 845

Mendel, Gregor, 16, 22, 182, 192, 266–277, 297, 299, 312–314, 321, 322, 351
attacked by Lysenkoists, 401
bean experiment, and gene interactions, 288, 289, 314
biography and portrait, 266, *267*
data criticized by Fisher as unlikely, 275
first law, 271–273
Mendelism rediscovered and confirmed, 297
pea experiment diagrammed, *270*
second law (independent assortment), 273–275
test cross, 276, 277
Mendelian genetics, 265–290
Mendelian ratios, 269, *270*, 271, *272*
modified by gene interaction, 287–289
Mendelians and biometricians, 313
menopause, 852
menses, 845
menstrual cramps, 782
menstrual cycle, 845–848
and frequencies of intercourse and orgasm, *851*
menstruation, 845
meristem
apical, 546, 547
apical, root, 507, *508*, *509*, 578
apical, shoot, 579, *581*, *582*
basal, 580
leaf primordium, 579, *581*
merozoite, *436*
merozygote (partial diploid), 366
mesenchyme, *597*
of sponge, *598*
mesoderm
amphibian gastrula, 872, *873*
and phylogeny, 596
dorsal lip mesoderm acts as an inducer, *879*
fate, 876–878
mammalian gastrula, 874, *875*
tissues derived from, *876*
mesoglea, sponge, 674
mesolecithal egg, 865
mesonephros, 891
mesophyll, 513, *514*
in C4 pathway, 148, *149*, *150*
mesosome, *411*
Mesozoic, *483*, *486*, *487*, 491
messenger RNA, 205, *207*
constituent of nucleus, 226, 227
hypothesized by Jacob and Monod, 368, 370
of lampbrush chromosomes, 871
number of sequences of mRNA in *Drosophila* and human cells, 309
only one cistron in eucaryote mRNA, 370
passes through nuclear pore, 224
polyadenylation, *207*
transcription in chromosome puffs, *308*, 309
metabolism, inborn errors of, 351–356
metacarpal, 684, *686*
metamorphosis
amphibian, 739
amphibian, controlled by thyroxin, 783
insect, *622*, *623*, *774*, *775*
metanephros, 891
metaphase, in cell cycle, 229
metaphase I of meiosis, *248*
metaphase II of meiosis, 250, *251*
metaphase plate, 234, 238
metatarsal, *686*
metatheria, 650, *652*, *653*, 654, 836
metazoa, 402, *403*, 458, 592
origins and phylogeny, 558, 593
animal kingdom, 592–663
methane, 32–34
methanogenic bacteria, 402, *403*, 410, 413, 416
methemoglobinemia, 533
methionine, 537
structural formula, *66*
methyl group, 48
metric system, 93, *94*; *see also* the inside back cover
mew gull, *1033*
micelle, *85*
Mickey Mouse, *see* Mouse, Mickey
microfibril, cellulose, *62*
microfilament, actin
cytokinesis, 240, *241*
cytoskeleton, *89*, 102
sarcomere, 691–694
water vacuole (contractile vacuole), *106*, *438*, 439
micromere, 871
micronucleus, *438*, *439*, 440
micronutrients, plants, 530, *532*, 537
micropyle, 488
insect egg, 834, *835*
seed, *574*, *575*
microscope, 96–99, 297
microspore, 565, 566, *567*
microspore mother cell, 565, 566, *567*
microsporogenesis, 566, *567*
microtubule, 86, *119*, 424
assembly, 237
in cytoskeleton, *89*, 102, 119
in plant cytokinesis, 242, *243*
and serial endosymbiosis hypothesis, *414*, 424
in spindle, 234, *237*
structure, *237*
subunit disassembly hypothesis, *239*, 240
microvillus (microvilli), *89*, *94*, *716*
midbrain, 916, *917*
human embryo, *891*
middle lamella, 95, 234, *242*
midge, 621, 672
midget, 338, *781*
achondroplastic dwarf, 280, 281
and growth hormone, 376
midnight flash experiments, *562*
midpiece of sperm, *841*
Miescher, Johann Friedrich, discovery of DNA, 22, 192
migraine, 450
migration, 771, *1050*, 1051
adaptedness of seasonal migration, 967
establishment of species, 5
gene flow, 334–337
migratory birds, 967
hummingbird, 25
plover, *669*
and sun compass, *968*, *969*
of tundra, 1007, *1008*
MIH, *see* molt inhibiting hormone
mildew, 447, *448*
milk, 650, *651*
indigestible by non-Caucasian adults, 718, 719
mammary glands, 651, *876*
milk let-down, 897, *898*
milk teeth, 710
milkshake, thickened with carrageenin, 465
Miller, N. E., autonomic learning, 957
millipede, *616*, 620
mimicry
leaf toad, *933*
orchid pollination, *490*
mineral deficiency, plant, *528*, *529*, 530, 532
mineral nutrients, plant, 527–538
and eutrophication, 1021, *1022*
macronutrients, 530
micronutrients, 530, *532*, 537
transport, 538
mineral nutrition, animal
human requirements, table, 725
recycled in molting, 671
mineralocorticoids, 787
minimal medium, 357, 361
mint, 494, *495*
miscarriage, 256, 894
missense mutation (amino acid substitution), 216
mistletoe, 539
Mitchell, Peter, chemiosmosis, 22, 129, 133, 134
mite, commensal on human skin, 1042
mitochondrial DNA, 115, 202, 217
mitochondrion, 86, 88, 89, 115, *116*, 404
and acetyl CoA, 169
DNA, 115, 202, 217
electron transport in membrane, *172*, 173
origin by endosymbiosis, 414, 421, *424*
turnover, 113
mitogens, 791, 792
and target cells, 791
mitosis, 228–240, *252*
in cell cycle, 228, *229*
compared with meiosis, *252*, *253*
in dinoflagellates, 430
onion root tip, *232*, *233*
mitotic apparatus, 237–240
mitral valve, 751, *753*
mixed strategy, 1033
MN blood group system, 281
mnemnonist, 959
mobbing, 972, *973*
model, of biological phenomena, 272, 273; *see also* mathematical biology
Model T Ford, programmed death, 1036
modification for flight, 682
modified bases, tRNA, 207, 208, *209*
modifier genes, 309
molar concentration, defined, 40
mole, 40, 128
molecular evolution, 337, 338
molecular mass, 26
molecule, defined, 26
mollusk, 593, *596*, 612–615
cephalopod, *613*, 614, 615, *616*, 929, 960, 1035
clam, *615*, 735, *736*
exoskeletons, 672, *673*, 674
eye compared with vertebrate eye, *616*
giant squid, 612, 1035
liver, 703
nervous system, 914, 915, *916*
programmed death, 1035
respiratory system, 734, 735, *736*
snail, 12, 62, 614, *615*, 672, *673*, *674*
molt inhibiting hormone (MIH), *773*, *774*
molting, 671, 772–775
crustecdysone, 620
ecdysone, *308*, 671
hormonal control, 671, 772–775
molybdenum, 530, *532*
plant nutrient, 537
Monera (procaryote kingdom), 409–427; *see also* procaryote
Money, John
beneficial effects of early sex experience, 850
prenatal sex hormones, 893
mongolism, mongolian idiocy, *see* trisomy-21
monitor lizard, *395*, *646*
monocistronic mRNA, 370
monocot, 491, *495*
a monophyletic group, 395, 396
compared with dicot, *492*
primary growth pattern, *513*
vascular bundles, *513*, 581
monocyte, 762
Monod, Jacques, and the operon, 366–368, 370, 371
monophyletic groups, 394, *395*
Monoplacophora, 612, *613*, *669*, 673
monosaccharide, 56–58
monotreme, *395*, 650, *652*, *653*, 654, 837
cleidoic egg, 886
monozygotic twins, 870
mons veneris, 840, *841*
monsoon forest, 1002
moose, Isle Royale, *1006*, 1047, 1049
moral indignation, 983
Morgan, Thomas Hunt, fruit flies and sex linkage, 297–300, 315, 316
attacked by Lysenkoists, 401
Mormon cricket, 1000
morning glory, Hagiwara's experiment, 289, 290
morphine, 533
morphogenesis, 867, 875–884
and germ layers, 876–878
human prenatal, 887–896
morphology, prime interest of early geneticists, 351
mortality, 258, 1035–1050; *see also* death; programmed death
mortality rate, *1034*
mosaic egg, 870
mosquito
larva, *35*
reason for existing, 18
of tundra, 1008
see also malaria
moss, *474*, 475, 520
life cycle, *476*
pioneer organism, 1020
water transport in, 504
moth
avoids bat, 925, *926*, 927
peppered (*Biston betularia*), *47*, *327*
programmed death, 1037
silkworm (*Bombyx*) and sex attractant, 933, *934*, 973
motherhood, chimpanzee style, *1032*
motility in different kingdoms, 404
motivation, 952, *953*
motor area of brain, 920, *921*
motor end plate, *907*
motor neuron, 690, *691*, *907*
saltatory propagation, 911, *912*
motorcyclists destroy desert, *999*
Mount Ruwenzore, 995
mountain
altitudinal ecocline, 994, *995*
rain shadow, 994–996
mouse
artificial selection experiments, 329, *330*
innervation of whiskers, *884*
jumped by wolf, 1047
Mickey, 36
patterns of coat color inheritance, 287, *288*
reproductive strategy, 1036
mouth, a metazoan characteristic, 592
mouth-breeding fish, 641
Moyle, Jennifer, experimental validation of chemiosmosis hypothesis, 133, 134
mRNA (messenger RNA), 205, *207*, 224, 226, 227, 309, 368, 370, 871; *see also* messenger RNA; tailoring; transcription

mucin, 715
muconium, 893
mucopolysaccharide, 63
mucoprotein, 715
mucosa
  colon, 717
  esophagus, 712, *714*
  small intestine, *716*
  stomach, 715
  uterus, 843, 846, 847, *848*
mucus, produced by respiratory passages, *741*
  goblet cell, *716, 741*
mudflat, 1011, *1012*
Muller, H. J., chemical nature of the gene, 192
Müllerian duct, 892
multicellularity, 243, 456–459
  and fungi, 444
  defined, 457
  in different kingdoms, 404, 458
  in green algae, 470, 475
  in plants, 457, 458
multinucleate cell, 241
multiple fruit, 572, *573*
multiplicative law, 271, *272*
mumps, 712
Munch, Ernest, 518
Murie, A., moose of Isle Royale, 1049
Mus musculus (mouse), 287, *288*, 329, *330, 884*, 1047
muscle, 686–695; *see also* muscular system
muscle antagonism, *689*, 690
muscle fiber, *91, 688, 690, 691*
  ultrastructure, 691–694
muscular dystrophy, 301
  variable age of onset, 310
muscular system, 686–695
  antagonists, *671, 689*, 690
  arthropod joint compared with vertebrate joint, *671*
  cell types, *90*, 686, *687*
  contractile unit, 691–694
  fasciculus, *688*
  fibers, *91, 688, 690, 691*
  gross anatomy of a muscle, *688*
  human, whole body, *689*
  of hydrostatic skeleton, *670*
  insect flight muscles, *672*
  insertions and origins, *688, 689*
  mesodermal origin, *876*
  myofibril, 690–692
  sarcomere, 691–694
  skeletal muscle, *91*, 688–690
  smooth muscle, *90*, 686–688, 694, 714, 715, 755–760, 843
  ultrastructure, 690–695
muscularis, 712, *714*, 843
mushroom, *444*, 454
  edible, *454, 455*
  life cycle, *455*
musk ox, 1007, *1008*
muskeg, *1005*
mustard, *495*
mutagen, 216
mutation, 16, 187, 188, 192, 216, 217, 321
  defined, 321
  and DNA repair, 187
  equilibrium balance with selection, 338, *339*
  and genetic drift, 338, *339*
  primary, 216
  and protein translation, 217
  screening for bacterial mutants, 356, *357*
  screening for *Neurospora* mutants, *354*
  as a vital part of adaptive evolution, 331
  within clones, 324
mutation rate, *339*, 340
  and temperature, 192
  in bacteria, 324, 357
  used to estimate gene number in flies and humans, 308, 309
mutualism, 1042
mycelial bacteria, 410, 417
mycelium, 91, 444, *445*
  of bread mold, *449*
mycetes, 445
mycologist, 455
mycoplasma, *409*, 410, 418, *419*
mycota, 445
myelin sheath, *907*, 911, *912*
myofibril, 690–692
myopia, 340
myosin bridge, *692, 694*
myosin filament, 691–693
myotome, of lancelet, *636*
myxameba, 446
myxobacterium, 410, *417*
Myxotricha paradoxa, 423, *424*

N

N-terminal end of a polypeptide, 64, 211, 215
NAD, $NADH_2$, 725
  in citric acid cycle, *170*, 171, *172*
  NAD^+ reduced to NADH + H^+, 135, 136
  role in nitrogen waste cycle, 808, *809*
  structural formula, 135
NADP, 725
  $NADP^+$ reduced to NADPH + H^+, 135
  structural formula, 135
  in Z scheme of photosynthesis, 143
narcotic, as a plant defense, 533
nares, internal, 738, 739
narrow-niched organisms, 1028
nasal cavity, fish, 738, 934
nasal epithelium, 934, *935*
nasal passages, *741*, 934
nasal sac, 934
natural history, 992
natural selection, 11, 13–17, *327*
  adaptedness of variability itself, 1036
  and adult intestinal lactase, 719
  and allocation of finite biological resources, 1036
  and behavior, 20, 952
  under civilization, 1034
  competition leads to specialization, 1044
  and Darwin's finches, 7
  in development of nerve cell pathways, 882, *883, 884*
  difficulties with the theory, 15
  and diseases of old age, 1036, 1037
  for dispersal, 1050, 1051
  and equilibrium frequencies of harmful mutations, 338, *339*
  experimental test of, 19, 396
  favoring a measure of altruistic behavior toward kin, 982, 983
  and female receptiveness, 851
  giraffe necks, *333, 334*
  group selection, 20
  the imperfectness of, 20, 1036
  kin selection, 982, *983*
  pines and scale insect coevolution, 326
  and reproductive potential, 1031
  and resistance to smallpox, 1042
  for ritualization of combat, 977
  and sexual reproduction, 258, 326, 622, 830, 850–852
  tracking a fluctuating environment, 333
Nature (journal), 313, 314
nautilus, *673*, 674
navigation, 967–970
  celestial, 970
  homing pigeon, *968, 970*
Neanderthal, *661*, 662, *663*
  life span estimated, 1034
nearsightedness, 340
nectar guide, *490*
nectary, 561
negative feedback control systems, 798–800
  density-dependent mortality in population control, 1038
  digestive enzymes, 722
  enzyme activity, 72, 73
  and hormones, 778, 779
  household thermostat example, *793*
  insulin, glucagon, and blood sugar, *723*, 777, *786*, 787
  iron absorption, 726
  kidney resorption, *818*
  limits of any system, *799*, 800
  and menstrual cycle, *846*
  neuron growth, sprouting and antisprouting factors, 882
  testosterone and ICSH, 844, *845*
  thyroid and TSH, *784*
negative gene control, 370
nematocyst, 600
nematode, 604–607, 703
  musculature, 670, 671
  trapped by a fungus, *456*
nemertean worm, 604
Neolithic revolution, 1051, 1052
neon, 28, *29*
Neopilina, 612
nephridiopore, *812*
nephridium
  earthworm, *610, 812*
  onychophoran, 616
nephritis, 820
nephron, 813–816
  control of function, *816*, 818
  tubule epithelium, *94*
  see also excretory system; kidney
Nereis, *611*
neritic province, *1009*
nerve, 906; *see also* nervous system
nerve cell, 87, *90*, 882–884, 906–913; *see also* neuron; nervous system
nerve gas, 911
nerve impulse, 908, *909*
nerve net, 598, 913, 914, *915*
nerve pathways, 882, *883, 884*
nerve relay, 882, 883
nervous system, 882–884, 906–936
  arthropod, 915, *916*
  autonomic, *922*, 923
  coelenterate, 598, 913, 914, *915*
  of earthworm, 914, *915*
  ectodermal origin, *876*
  embryology, *877*, 878, 882, *883, 884*, 899, *891*
  and endocrine system, 771, 781
  engram, 958, 960–962
  flatworm, 914, *915*
  impulse generation, 908, *909*
  insect, *916*
  mollusk, 914
  motor and sensory areas of the brain, 920, *921*
  nerve pathways, 882, *883, 884*
  octopus, 914, 915, *916*
  sea star, *916*
  sensory, 923–936
  sprouting factor and anti-sprouting factor, 882
  synapse, 907–911
  target tissues, proper and improper connections, 883
  whiskers and barrels, *884*
nest building, 953
nest predation, *1033*
neural arch, neural canal, 679, *680*
neural crest, neural fold, neural groove, neural plate, *877*, 878
neural impulse, 908, *909*
neurectoderm, echinoderm, *916*
neurohormones, 779
neuromuscular junction, 690, *691*, 693
neuron, 87, *90*, 882–884, 906–913, *907*
  described, 882
  development, 882, *883, 884*
  representative types, *907*
  terminal differentiation, 228
Neuroptera, *623*
neurosecretion, 771, *779*
Neurospora crassa, *353, 354*, 449
  metabolic pathways, *353*
  screening for mutants, *354*
neurotoxin, 644
neurotransmitter, 909, *910, 911*
neurula, *877, 878*
neurulation, 877
  frog, *877*
  human, 888
  induced by dorsal lip mesoderm, *879*
neutralists and selectionists, 337
neutron, 26
neutrophil, *762*, 819
New-World monkeys, 655, *656, 657*
New Zealand
  extinction of native species, 4
  carnivorous parrot, 1000
newt, 739
Newton, Sir Isaac, 21
N-formyl amino acid, 211
niche, 15, 331, 480, 617, 659, 1028, 1043
  broad and narrow niches, 1028
  and fitness, 331
  each hominid species well adapted at the time, 659
  niche overlap, 1043
niche specialization
  arthropods, 617
  barnacles, 1036, 1044
  birds, 5, *6*, 7, *709*
  brine shrimp, 800
Nicholas II, 302
nicotinamide, 725
nicotinamide adenine dinucleotide (NAD), 135, 136, 170–172, 725, 808, 809
nicotinamide adenine dinucleotide phosphate (NADP), 135, 143, 725
nicotinic acid, 135, 724
nictitating membrane, 931, 932
Nightingale, Florence, promotes germ theory of disease, 410, 1053
nipple, stimulation releases oxytocin, 782, 897, 898
nitrate ion, atomic diagram, *44*
nitrogen, organic compounds of, 41, *44*, 45
nitrogen base, 183, *184*, 201
nitrogen cycle, 530–534
nitrogen fixation, 44, 415, 531–534, *531*
  by heterotrophs, *420*
  by legumes, *45*
nitrogen wastes, 808, *809; see also* urea; uric acid; excretory system
nitrogenase, 537
nitrogenous base, 183, *184*, 201
Noah, 5, 617
noble elements, 30
noctuid moth, 925, *926*, 927
nocturnal animals of desert, 997, *998*
nocturnal birds, retina, 932
node, 506
  of plant stem, 579, *581*
  node of Ranvier, 911, *912*

nondisjunction, *256*, 257
nonhistone chromosomal proteins, 225
  in chick oviduct, 371
  steroid receptor, 371, *792*
noninducible mutant, 367
nonpolar attraction (van der Waals forces), 32, 39, 65, 68
nonpolar solvents, 39
nonpurple sulfur bacteria, 1017
nonrecombinant chromosome, *305*
nontranscribed strand of DNA, 203, *204*
noradrenaline (norepinephrine), 777, 807, 910, 911
norepinephrine, 777, 910, 911
  as a neurotransmitter, *910*, 911
  structural formula, *910*
  and thermoregulation, 807
normal distribution, *313*
North Pole, Australia, 413
no-see-um, 672
Nostoc, *420*
nostril, *741*
notochord
  absent in acorn worm, 634
  a chordate characteristic, 635
  frog embryo, *877*, *878*
  of lancelet, *636*
  persists in adult lamprey, 638
  retained in other adult fish, 878
N-terminal end of a polypeptide, 64, 211, 215
nu body (nucleosome), *225*
nuclear membrane, *99*, *110*, 224
  of dinoflagellate, 430
  pores, *99*, *110*, 224
nuclear transplantation, 880–882
nuclease, 718, 721
nucleic acid, 183
  digestion, 721
  see also DNA; RNA
nucleoid, *90*, *411*
nucleolus, *88*, *89*, 110
  disperses in prophase, 234
  site of rRNA synthesis, 225
nucleoplasm, 224
nucleoprotein, 67
nucleosome, *225*
nucleotide bases, 183, *184*
  of DNA, 183
  ratios (Chargaff's rule), 201
nucleus, 109, *110*, 223–225, 404
  of atom, *26*
  cluster of cell bodies in brain, 906
  in differentiation, 880–882
  discovery of, 223
  sole site of DNA replication and RNA transcription in eucaryotes, 202
  structure, 224, 225
  transplanted, 880–882
nucellus, 565, *574*
numerical taxonomy, 390, 391
nuns, high incidence of breast cancer, 897
nurse cell (Sertoli cell), 839, *840*
nursing
  chimpanzee, *1032*
  as a contraceptive practice, 1033, 1034
  releases oxytocin and prolactin, 782, 897, *898*
nut, 560
nutrient cycle, 1016
nutrients, mineral, 527–538
nutrition, animal, 722–726
  and fertility of nursing mothers, 1034
  mineral requirements, table, 725
  vitamins, table, 724

O

oak
  California live oak, 1003
  deciduous, 1004
  hybridization and variation in *Quercus*, 389
  scrub, 1005
Obelia, *599*
obligate anaerobe, 416
occipital condyles, *674*
occipital lobe, 920
ocean floor, 1010, 1011
oceanic province, *1009*
ocellus (ocelli)
  flatworm, *602*
  grasshopper, *706*
  insect, *621*
octet rule of electron shells, 29
octopus, 614, 615, *616*
  brain, 914, 915, *916*
  consolidation of learning, 960
  eye, *616*, 929
  mates only once and dies, 1035
Odonata, 622
off fibers, 930
off-road vehicles, 999
oil (biological), 75, 77
oil (petroleum)
  Alaskan pipeline, 1009
  spills, *1040*
Okazaki fragment, 190, *191*
Oken, Lorenza, 85
Old-World monkeys, 655, 656, *657*
Olduvai Gorge, 659
olfaction, 712, 933–935
  snakes and lizards, 709
  vertebrates, 934, *935*
olfactory bulb, *935*
  and human brain, *918*
olfactory hair as modified cilium, 118
olfactory lobe, various vertebrate brains, *917*
oligochaete, 609
  earthworm, 609, *610*, *670*, 704, *705*, 747, *812*, *815*
Olympic National Forest, Washington, 1007
O'Malley, Bert W., steroid hormone function, *792*
omasum, 712
ommatidium (ommatidia), *620*, 929, *930*
omnivore, 935
on fibers, 930
one gene, one enzyme hypothesis, 352–356
one polytene band, one gene hypothesis, 308
ontogeny recapitulates phylogeny, *890*
Onychophora, *596*, 615, *616*
oocyte, 245, 251, *843*
  genetically active, 870, 871
  lampbrush chromosomes, *254*, 871
  meiosis photomicrographs, *246–248*
oogonium, of water mold, 447, *448*
Oomycetes, 447, *448*
open circulatory system, 735, *736*
operant conditioning, *956*, 957
operational taxonomic unit (OTU), 391
operator locus, 367, *369*, *371*
operculum, *738*
  gastropod, 614
operon, 366–371
opossum, *398*, *651*
opposable thumb, 654
opsin, 928
optic chiasma, *961*, *962*
optic cup, *880*
optic nerve, 931
optic vesicle, *880*, *891*
optimal clutch size, *1033*
oral cavity, 712, *713*
oral groove, *438*
oral hygiene, 712
oral surface of an echinoderm, 625
oral valve of a fish, *738*
oral-facial-digital syndrome, 303
orangutan, *657*
orb spider, *619*
orbital, electron, *28*
orchid, *490*, 494, *495*
  mimicry, *490*
  tissue culture, 584
Oreopithecus, 656, *657*
organ systems, a characteristic of the animal kingdom, 592
organ transplant, 287
organelle, 109
organic chemistry, 39, *42*
organization, a characteristic of life, *47*
organizers, embryonic, 878–880
organophosphate insecticides, 911
orgasm, 849, *850*
  by stimulation of the clitoris, 840, 849
orgasmic platform, 849
orientation, 946–949, *968–970*
origin (of muscle), *688*, *689*
origin
  cells, 85
  of chloroplast, mitochondrion and eucaryote cell, 114, 115, 413, 421–424
Origin of Species, 11, *12*, 13, 15–22, 266, 337
ornithine, *355*
Ornithischian dinosaurs, *491*
Orthoptera, 622
Oscillatoria, 419, *420*
osculum, 597, *598*
osmoconformers, 809, *810*
osmoregulation, 808–818
osmoregulators, 809, *810*
osmosis, 86, 105–107, *504*, *505*
  in capillary bed, *760*
  loss of body fluid in seawater, 809
  schematic diagram of, *107*
osmotic potential, 108, 504
  of blood, 762
  in leaf cell, 516
ossification, *676*, 677
Osteichthyes, 640–642
osteoblast, 676, 785
osteoclast, 785
osteocyte, *676*
osteon, *676*, 677
osteoporosis, and parathyroid tumor, 785
ostracoderm, *637*
  prominent third eye, 789
ostrich, 488, 649, 836
otolith, 936
OTU, *see* operational taxonomic unit
oval window, 927, *928*
ovarian cycle, *843*, 845
  menstrual cycle, 845–848, *851*
ovarian ligament, 843
ovary
  endocrine functions, 788, 789
  human embryology, 892
  of a flower, 560, 565
overcrowding, pathological effects, 780, 781, 1035, 1051
overdeveloped nations, 1052
overdominance, 282, 285, 329
  pigeon transferrins, 282
  sickle-cell hemoglobin, 285
overfishing, 1011, 1013
  whales, *1046*
overgrazing, 999, *1000*
overwatering, 520
oviduct, *842*
  experimental induction of eggwhite genes, 371
  formed from Müllerian duct, 892
ovipositor, *621*, *834*
  of a parasitoid wasp, 1042
ovists and spermists, 868
ovoviviparity, 832, *833*
ovulate cone, 486, *487*
ovulation, 254, *843*, 845, 846
ovule, 369, *566*, *574*
owl
  desert, 997, *998*
  sound detection, 927
oxaloacetate
  in C4 pathway, 148, *150*
  in photosynthesis, 146
oxidation, 132, 134
  of iron, 132, *134*
  of water, 144
oxidative phosphorylation, oxidative respiration, 130–132, *172*, 173, *414*; *see also* chemiosmosis
oxygen
  biogeochemical cycle, 535, 536
  concentration in seawater, 738
  molecular, *32*
  origin of oxidizing atmosphere, 536
  partial pressure, 743
  toxic effect, 746
oxygen debt, 166, *168*
oxygen respiration, origin of, *414*
oxyhemoglobin, 67, 744
oxytocin, 776, 779, 780
  amino acid sequence, 75, 782
  and lactation, 782, 897, *898*
  and sexual activity, 782
  released by stimulation of nipple, 782, 897, *898*
  and uterine contractions, 896

P

P680, *141–143*, 144
P700, *141–143*, 144
Paal, A., 545
pacemaker, *754*
pachytene, *246*
Pacinian corpuscle, 924, *925*
pair bonding
  in humans, 851, 852, 858
  and mate recognition, 975
Palade, George, endoplasmic reticulum, 111
palate, 739, *741*
  evolution of, *684*
paleocene, 488
Paleozoic era, 405, *480*
palmitic acid, *76*
palomino horse, 279, *280*, 345
palp, labial palp of insect, *706*, *707*
palynology, 484
pampas, 999, *1000*
Panama, Isthmus land bridge, 4
pancreas
  endocrine function of islets, 785–787
  endodermal origin, *876*
  in digestive system, *713*, 717
  origin by outpocketing of gut, 704
  secretion stimulated by pancreozymin, 717
pancreatic enzymes, 717, 718, 720, *721*
pancreatic lipase, 718
pancreatic proteases, 720, *721*
pancreozymin (cholecystokinin), 717, 722
Pangea, *488*
pangenesis, 322, *323*
pangolin, *652*, *653*, 654

panting, 798
pantothenic acid, 724
paracoccus, *414*
parallax, 931, *932*
parallel evolution, 684
Paramecium, 91, 437, 438, 439
  competitive exclusion of one species by another, *1043*
  digestive vacuole, 701
paraphyletic groups, 394, *395*
  dicots, 395, 396
  Homo erectus, 660–662
  insectivores, 337, 654
  lizards, 395
  protists, 428
  reptiles, *395*, 644–647
parapod, 611, 612
parasite
  chestnut blight fungus, *1042*
  defined, 1041
  fungal, 444
  plant, *539*
parasitic castration, 1042
parasitism
  and mortality, 1041, 1042
  and population control, 1041–1043
parasympathetic nervous system, *922*, 923
parathion, 911
parathormone (PTH), 776, 785, 791
parathyroid gland, 776, *783, 785*
parathyroid hormone (PTH), 776, 785
  stimulates cAMP production in target cells, 791
parazoa, 402, *403*, 458, 592, 595, 597, 598, *674*, 675, 701, *702*
parenchyma, *511–513*
parental care
  birds, 648, *836, 1033*
  chimpanzees, *1032*
parietal bone, *679*
parietal cells, *715*
parietal lobe, 920
Paris, Oscar, density-dependent mortality in Berkeley sowbugs, 1038
parotid salivary gland, 712
parrot, carnivorous kea of New Zealand, 1000
parrot fish, 708
parthenogenesis
  facultative, 258
  rotifers, 607
partial diploid, 366, 367
partial dominance, 279, *280*, 284, 285
partial pressure, 743
parturition (birth) 894–897
passive transport, 95–104; *see also* diffusion
Pasteur, Louis, germ theory of disease, 410, 1053
Patagonian hare (mara), *4*
patella, 685, *686*
paternal and maternal chromosomes, 243, 249, 289
pathogenic bacteria, 410
Pauling, Linus, 356
pavement teeth, 707
Pavlov, Ivan, classical conditioning, 954, *955*
pea, 267, 268, *269, 572*
  advantages for genetic work, 267, 268
  alternate traits listed, 267, *269*
  Mendel's experiments diagrammed, *270*
peach, 572, *573*
Pearson, Karl, and biometry, 313
peck order, pecking order, 980, 1039
pecten, of bee leg, *624*
pectin, 63, 95, 242
  in diatom shell, 430
pectoral fin, 681
  of ray, 639
pectoral girdle, *643*, 675, 681, *682, 683*
pedicellaria, *735*
pedigree
  color blindness, 303
  hemophilia in Royal Family, *302*
  X-linked dominant brown teeth, 301, *303*
Peking man, 660, 661, *662*
pelagic region, *1009*, 1010
pellicle, *438*, 439
peloric snapdragon, *266*, 268
pelvic fin, 681
pelvic girdle, *643*, 675, 681, *682, 683*
pelvis, 681, *682, 683*
  of kidney, 813, *814*
pelycosaur, *647*, 650
penetrance, 310, *311*
Penfield, Wilder, physiology of memory, 958, 959
penguin, *328, 649*
penicillin
  ecological role, 1040
  resistance to, 324
  use in screening for bacterial mutants, 356, 357
Penicillium, 451
penis
  duck and ostrich, 836
  embryology, *893*
  human, 837, *838, 839*
  insect, *834*
  involuntary control, 687
peppered moth, *47, 327*
pepsin, *715*, 718, 720, *721*
  optimal pH, 72
pepsinogen, 715, 720
peptide, 75
peptide bond, 64, 211
  in polypeptide elongation, *213*
peptide hormones, 75, 777, 782
peregrine falcon, 947–949
perennial plants, 576
perfect flower, 493
pericarp, *569*
perichondrium, *676*
pericycle, *509*, 580
  of root tip, 579
perigyny, *570*
periosteum, *676*
Peripatus, 596, 615, *616*
peripheral nervous system
  autonomic, *922*, 923
  developing nerve pathways, 882, *883, 884*
periplasmic space, 86
Perissodactyla, *563*, 655
peristalsis, 704, 712, 713
peritoneum, *597*
  echinoderm, *735*
perivitelline space, 866, *867*
permafrost, 1007
permease, *104*
  galactoside permease, 366
Permian, 482
pernicious anemia, 724
peroxide, 113
peroxisomes, *88*, 113, *114*
perpetual motion, 152
perspiration, 805, *806*
petal, *492, 493, 560*
petiole, 512
Petri dish, 430
  replica plating, *357*
  serial dilution, 361, *362*
PGAL, *see* phosphoglyceraldehyde
  in Calvin cycle, *146, 147*
PGH, *see* prothoracic gland hormone
pH, 40, 41
  of blood in runners, *168*
  control of blood acidity by kidney, 818
  and enzyme activity, 72
Phaeophyta (brown algae), 466, *467, 468*
phage (bacteriophage), *92*, 193, *194, 195*, 200, 357–361, *359*
phagocyte, 761, 819, *822*
phagocytosis, 105, 434, 701, *702*
  of uric acid crystals, *817*
phalanges, 684, *686*
pharynx
  acorn worm, 634, *635*
  and origin of jaw, 638, *639*
  earthworm, 704, *705*
  human, 712, *713, 741*
  lancelet, *636*
  leech, 611
  nematode, 604
  planaria, 602, 702, *703*
phenotype, 279
  and continuity of the germ plasm, 322, *323*
phenotype frequency, 323
phenylalanine, structural formula, *66*
phenylketonuria, 351, *352*
phenylthiourea (phenythiocarbamide, PTC), 345
pheromone, 973, 974
  bombykol sex attractant, 933, *934*, 973
  cyclic AMP of cellular slime mold, 447
  and insect pollination, 489
phloem, 505, *510, 511, 513, 518, 519*
  origin from cambium, 582
  primary, 507, *508, 509*
  primary, root, *580*
  primary, stem, *582*
  rate of nutrient flow, 519
  secondary, root, 580
  secondary, stem, *582*
phloem cell, lacks nucleus, 228
phloem parenchyma, 511
phloem ray, *510, 511*
Pholidota (pangolins), *652, 653*, 654
phoronids, *609*
phosphate group, 48
phosphate ion, *45*
  atomic diagram, *45*
  high energy bond, 126, 162
  as a plant nutrient, 534
phosphatyl, 48
phosphodiesterase, 721
phosphoenolpyruvate, in C4 pathway, 148, *150*
phosphoglyceraldehyde (PGAL), *146, 147*, 161, 162
phospholipase A2, and spontaneous abortion, 894
phospholipid, 48, 75, *78*, 79, 85
  structural formula, *78*
phosphophosphorylase, *791*
phosphoric acid, *45; see also* phosphate ion
phosphorolysis (phosphorolytic cleavage)
  of glycogen, *61, 167*
  of starch, 60, *167*
  of sucrose, 69
phosphorus, *45*
  biogeochemical cycle, *534*, 535
  dietary requirement, 725
  in hormone action, 791
  transport measured by autoradiography, *197*
phosphorylation, *128*
  by ATP, 127
  oxidative, 130–132, *172*, 173, *414*
  photosynthetic, *129*, 130, *142*, 145
  substrate-level, 160, 162
photoevent, 140, 141, 145
photographic memory, 959
photolysis of water, 140, 141
  abiotic, in upper atmosphere, 536
photon, in photosynthesis, 137
photoperiodicity
  and the initiation of flowering, 562–565
  and the vertebrate third eye, 789, 790
photoreception, photoreceptor
  Euglena, 430
  guard cell, 521, *522*
  in leaves, 562, *564*
  in shoot tip, 547
  see also eye; visual receptors
photorespiration, 148
photosynthesis, 125–157
  analysis of pathways, 140
  and carbon cycle, *536*
  compared with respiration, 158, *159*
  energy source of ecosystems, *1016*
  overview of chemistry, 140
  by phytoplankton, 1010
  summary, 151
photosynthetic autotroph, 416, 1014
photosynthetic lamella, 86, 114, *130*, 140, 144, 145, 410, *420*
photosynthetic phosphorylation, *129*, 130, *142*, 145
photosynthetic pigments
  carotenoids, 114, 137, 139, 421
  chlorophyll, 114, 137, *138, 139*, 143, *414*
  phycobilins, 421, 423
  phycocyanin, phycoerythrin, and red algae, 465
photosynthetic unit, 140
photosystem I (photosystem 680), *141, 142, 143*, 144
photosystem II (photosystem 700 or P700), *141, 142, 143*, 145
phototaxis, *430*
  of *Chlamydomonas*, 469
phototroph, 416, 1014
phototropism, of shoot tip, 545–547
phragmoplast, *242*
phycobilins, 421, 423
phycocyanin, 465
phycoerythrin, 465
phylogenetic tree, *7, 20*, 399, 400
  of all life, *403, 414*
  angiospersm, *495*
  of Animal Kingdom, deuterostome–protostome hypothesis, *595*
  of Animal Kingdom, coelomate–acoelomate hypothesis, *596*
  arthropods, *618*
  based on cytochrome *c, 20, 400*
  Darwin's finches, 7
  Darwin's tree of life, *393*
  fish, *641*
  formal construction of, 399, 400
  fungi, 456
  hominids, *662*
  lumpers and splitters, *394*
  of mammals, *652–653*
  primates, *399, 657*
  procaryotes, *414*
  reptiles, *646, 647*
  tetrapod vertebrates, *395*
phylogeny, biochemical, 464
Physalia, 600
phytochrome, 522, 547
  and photoperiodicity, 564
phytoplankton, 429, *1010*, 1014, *1015*
Phytophthora, 448
Piccard, Jaques, explores Marianas trench, 1010
pig, 655
pigeon
  egg-white transferrin, 282
  homing, 967, *968, 970*
  in Skinner box, 956, 957

pigment, 67
  flower, *490*
  visual, 928, 929
  see also photosynthetic pigments
pigment systems I and II, *see* photosystems I and II
piloerection, *806, 807*
pilus (pili), 360, *361, 411*
pine, and scale insect, 326, 622
pine nut, 486
pineal body, 777, *789,* 790
pineapple, *565,* 572, *573*
pinnipeds, 650, *652, 653,* 655
pinocytosis, 105, 725
  and hormone function, 791
  through capillary wall, *759*
  intestinal villi, 720
pioneer organism, 1020
pistil, *560,* 561
pit viper, 644, 923, *924*
pitcher plant, *539*
pith, *513*
pituitary, 778–782, *789*
  anterior lobe, 780–782
  controlled by hypothalamus, *779, 780*
  controlling reproductive system, 844
  diagram, *779*
  effect on ovary, *843*
  hormones listed, 776
  hormones of posterior lobe, 782
  and menstrual cycle, 845, *846*
  oxytocin, prolactin, and suckling, 782, 897, *898*
  pituitary portal system, *779*
  posterior lobe, *779,* 780
  role in thermoregulation, 807
pituitary dwarf and pituitary giant, *781*
pivot joint, *677*
place in the economy of nature, 1028
placebo, and scientific method, *9*
placenta, 649, 650, *651,* 886
  at birth, 894
  endocrine function, 848
  expulsion, 896
  a mammalian characteristic, 649
  of a plant ovary, 565
  oxygen transport across, 104
placoderm, *638, 641*
placoid scale, 639
plague, 1041
planaria, *602,* 702, *703*
  chemoreceptors, 933
  copulation, 831, *832*
  eyes, *929*
  nervous system, 913, 914, *915*
Planck, Max, 133
plankton, 429, *1010,* 1014, *1015*
plant, 464–497
  cell, 87, *88*
  cell wall, *242*
  vascular system, 503–522
plant kingdom, 464–497
  characteristics, 404, 464
  divisions, 494
  evolutionary relationships to other kingdoms, 402, *403,* 457, 458
  principal subkingdoms and divisions, 464
plant nutrition, 527–528
planula larva, 599, 600
plaque
  in arteries, *755*
  in counting bacteriophage, 358, *359*
plasma, 760–763
plasma membrane, 39, 77, 78, 85, 86, 92, 95, *100,* 410
plasmalemma, 410
plasmid, bacterial, 360, *361*
  and gene cloning, 372–377
  responsible for bacterial sex, 360, *361,* 363
plasmodium of acellular slime mold, *446*
Plasmodium vivax, malarial parasite, 435–437
plasticity in development, 879
plastid, *88,* 114, *115,* 421
  DNA of, 202
  chloroplast, *88,* 114, *115,* 140, 422–424, 465, 521
  chromoplast, 114
  leucoplast, *88,* 114, 421
  lysosomelike, 113
  origin by endosymbiosis, *414,* 422–424
  proplastid, 114
plastocyanin (PC) in Z scheme, *143*
plastoquinone (PQ), *143,* 144
plate tectonics, 488, *489*
plateau phase of sexual arousal, 849, *850*
platelet, 760–762
platelet mother cell, 761
Plato, 328
Platyhelminthes, 602–604, 913, 914, *915*
  fluke, 603
  metazoan ancestor, 594, *595*
  planaria, *602,* 702, *703,* 831, *832, 929,* 933
platypus, 393, 640, *651,* 654, 886
platyrrhines (New-World monkeys), 655, *656, 657*
pleasure center, 953, 957
Plesiadapsis, 656, 657
pleura (pleurae), 742
pleural cavity, *741,* 742
pleurisy, 742
plover, *669*
plumule, *574, 575,* 577
Pneumonococcus 192, 193
Pogonophore, *596,* 1017
poikilothermy, 643, 800
Poinsettia, 562
poison gas warfare, 977
polar body, 254, *255, 822,* 867
  function, 254, 258
polarity, molecular, 36
polarization, nerve cell membrane, 907–909
pollen, 483, *484,* 488, 566, *567*
pollen basket, 622, *624*
pollen brush, 622, *624*
pollen sac, 566, *567*
pollen tube, 488, 489, 567, *568*
pollination, 488, 489, *490,* 567, *568*
  floral generalists and specialists, 496
  by insects, 489, *490,* 561
  by yucca moth, *496*
pollinium, *490*
pollution, 327, *1040,* 1041
  of air, 1041
  by nitrates and phosphates, 474, 532, 533, *534,* 1013, 1021, *1022*
  oil spills, *1040*
  wastes, 1012, 1021, *1022*
poly-A tail, polyadenylation, *207*
polydactyly, 310, 311
polygenic inheritance, 312–315, 329–334
  and artificial selection, 329
  with an intermediate optimum, 332, *334*
  for geotactic response, 331, *332*
  and Mendel's bean experiment, 289
polymer, 56
polymorphism, 329, 334
polymorphonuclear leukocyte
  drumstick, *226*
  leukocytes, *236,* 760, *762,* 819
polyp, 598, *599*
polypeptide, 64
  elongation, 211–214
  termination, 214
  see also protein translation
polyphyletic group, 394, *395*
  plant kingdom, 464
polyploid species complex, 390
polyploidy, 489
  in animals, 390
  in plants, 389, 390, 489
polyribosome, 214, *215*
polysaccharide, 56–61
  structural, 61
  synthesis, 61
Polysiphonia, 465, *466*
polysome, 214, *215*
polytene chromosome, 306–309
polyunsaturated fat, 77
pond, 1013
Pongidae, *657*
pons, 745, 917, *918*
population
  defined, 340
  as the unit of evolution, 321
population explosion, 1052
population genetics, 313, 314, 323–344
  and artificial selection, 331
  of asexual organisms, 323, *324*
  balance between mutation and selection, 338, *339*
  Castle–Hardy–Weinberg distribution, 313, 314, 340–344
  genetic drift, 338, *339*
  and genetic variability, 13, 16, 326, 328–340
  hypothetical models, 344
  polygenic inheritance, 329–334
  of sexual organisms compared with asexual, *325*
population growth, human
  anomalies in growth curve, 1030
  doubling time, 1052, 1056
  effect of the germ theory of disease, 410, 1053
  rate, 858, 1030, 1052, 1056
  see also populations, human; human populations
population pyramid, *1054*
  Eltonian pyramid, *1015*
population size and rate of evolution, 15
populations, 1028–1056
  abiotic factors in mortality, 1038
  biotic factors in mortality, 1038–1050
  changes in numbers, 1028–1031
  control by disease and parasitism, 410, 1041–1043, 1053
  control by emigration, *1050,* 1051
  control through mortality, 1037–1046
  control through self-poisoning, 1039–1041
  control of ungulates by wolf predation, 1048
  crash, *1030,* 1031
  death, 1035–1044
  density-dependent control, 1041
  fluctuating numbers, 1028
  growth and reproductive strategies, 1028–1035
  human populations, 1051–1056; *see also* human populations entry
  human growth rate, 858, 1030, 1052, 1056
  may not be regulated at all, 1050
  stability and instability, 1049, 1050
porcupine, 654
Porifera (sponges), 597, 598; *see also* sponge
pork and trichinosis, *607*
pork tapeworm, 603, *605*
porpoise, 655
  cooperative behavior, 981
portal system, portal vein
  hepatic, *757*
  pituitary, *779*
  renal, 757
Porter, Keith, endoplasmic reticulum, 111
Portuguese man-of-war, 600
positive feedback, 538
  examples, 799
  and slime mold aggregation, 772
positive gene control, 370
postanal tail, a chordate characteristic, 635, *636*
postecdysis, 773
posterior commissure, *789*
posterior pituitary, 779–782; *see also* pituitary
posterior vena cava, 751, *752, 753*
postovulatory phase, of menstrual cycle, 847
postsynaptic membrane, 909, *910*
posttranscriptional modification
  capping and polyadenylation, *207*
  introns, 205, 207, 372
  tRNA tailoring, 207
potassium
  dietary need, 725
  and guard cell function, *522*
  natural radioactivity, 196
  and neuron function, 907–909
  plant requirement, 536, 537
potato famine, 445, 448
potato starch, *59*
potential energy, 126
  chemical, 50
pouch (marsupium), *651,* 654, 837
PPLO (mycoplasma), *409,* 410, 418, *419*
prairie, 999
prairie dog, 1000
praying mantis, 622, *624*
preadaptation, 738
Pre-Cambrian era, 593
  worm tracks, 612
precipitation by antibodies, 282, 821
precocial and altricial birds, *836*
predation, 1044–1050
  data from wolf kills, 1048
  and frequency-dependent selection, 334, *335*
  lynx and hare, *1045*
  wolf as predator, 1046–1049
predator-prey cycle, *1045*
prediction, and scientific method, 8, *9*
preecdysis, 773
preformation hypothesis, 868
prefrontal area, 920
pregnancy
  as an altruistic act, 982
  and the ovarian cycle, 848
  see also menstrual cycle; corpus luteum; sexual reproduction
prehensile tail, 655, *656,* 658
prehistoric humans, life span, 1034
premeiotic condensation, 245
premeiotic interphase, 245
premolar, 710, *711*
preovulatory phase of ovarian cycle, 845
prepotence, *266,* 268
pressure receptors, 924, *925*
presynaptic membrane, 909, *910*

prey switching, 1015, 1046
Priestley, Joseph, 527
primary consumer, 1014
primary gametocyte, 245, *255*
primary growth pattern, 512, *513, 582*
primary oocyte, 245, *255*
primary producer, 1010
primary spermatocyte, 245
primary structure
  of DNA, 372
  of a polypeptide or protein, 64
  of macromolecules in general, 60
  of starch, 60
  of tRNA, 208, *209*
primary succession, 1020, *1021*
primates, *652, 653,* 654–658
  good color vision, 932, 933
  early fossil record in Paleocene, 656
  phylogenetic tree, *399,* 657, *662*
  phylogeny, 657
primitive and advanced characters, 492, 494, 495
  horseshoe crab, 617
  human dentition, 710
  penguin flippers, *649*
primitive gut, 872, *873,* 888
primitive peoples, 1034
primitive streak
  bird, 873, *874*
  mammal, 874, *875*
probability theory
  the additive law, 271, *272*
  mixed strategy in clutch size, 1033
  the multiplicative law, 271, *272*
  normal distribution, *313*
proboscis
  acorn worm, 634, *635*
  nemertean, 604
proboscis worm, 604
Proboscoida (elephants), *653,* 655; *see also* elephant
procambium, *508, 574,* 575, 579, *582*
  of root, 580
  transition to vascular cambium, *581*
procaryotes, 402–404, 409–427
  characteristics of group, 404; table, 87
  cell, *86*
  evolutionary relationships, 402, *403*
  relative size, *409*
  see also bacterium; cyanophyte
producer, 1014–1016
product, of enzymatic reaction, 69
productivity
  defined, 1012
  measuring productivity, 1016, *1018, 1019*
  oceanic, 1012, 1013
progeny test, 268, *270, 276*
progesterone, *777,* 788, 789
  falling level initiates birth process, 894
  production by corpus luteum, 843, 847
  suppresses prolactin, 897
proglottid, *603, 605*
programmed death, 1035–1037
  and antibodies, 821
  at the cell level, 228, 882
  cephalopods, 1035, 1036
  of humans, 1036
  in morphogenesis of fingers and toes, *892*
  of most nerve cells, 882, *883, 884*
  of a postreproductive moth, 1037
prolactin, 75, 776, 779, 782
  action of, 782
  released by stimulation of nipple, *898*
  suppressed by progesterone, 897
proliferative phase of ovarian cycle, 845
proline, structural formula, *66*
promitochondrion, *421*
pronephros, 891
pronghorn antelope, 999, 1000
pronucleus (pronuclei), 254, 255, 867
prophase
  in mitotic cell cycle, *229, 230, 231,* 233
  meiotic, 245–247
  meiotic prophase lasts up to 50 years, 245, 251, 254, *255*
  onion root tip, *232*
  whitefish blastula, *230*
propionic acid bacteria, 842
proplastid, 114
proprioception, 936
proprioceptor, 936
  lacking in smooth muscle, 687
prosimian, 655, 656, *657*
prostaglandins, and birth process, 894
prostate, *838*
prosthesis, 67
prosthetic group, 67
proteases, 720, *721*
protective coloration, 14
  Biston betularia, 327
  lacking in prey of color-blind lions, 933
  leaf toad, *933*
protein, 64–68
  binding, 73, 74
  conjugated, 67
  deficiency, 724
  denaturation of, 65
  dietary need, 723, 724
  enzymatic cleavage, 720, *721*
  structural proteins, 74, 75
  synthesis, 210–217, 960
  tertiary structure, 39
protein synthesis (translation), 210–217
  and central dogma, 182
  elongation of polypeptide chain, 211–214
  necessary for memory consolidation, 960
prothallium, 482, *483*
Protheria (monotremes), *395,* 650, *652, 653,* 654, 837, 886
prothoracic gland and prothoracic gland hormone (PGH), 774, *775*
prothrombin, 762, *763*
protist, Protista, 428–433
  characteristics of kingdom, 404
  defining traits, 428
  evolutionary relationships, 402, *403*
  protozoa, 428, 432, 592
  see also alga; ameba; ciliate; *Paramecium;* sporozoa
protoeucaryote, *414, 421*
proton, 26
proton pump, 145, 173
protonema, 475
protoplast, 410
  of a carrot, *583,* 584
  of *E. coli,* 373
Protospongium, 432
protostome, *595,* 702
  protostome–deuterostome dichotomy, 627
Protostomia, 596
protozoan, 428, 432, 592
  agent in Chagas' disease, 11
  see also ameba; ciliate; Mastigophora; *Paramecium;* protist; sporozoan
Proust, Marcel, *Remembrance of Things Past,* 958
proventriculus, 648, *710*
proximal convoluted tubule, *see* nephron
pruning, 548, *548*
Przewalski's horse, *11*
pseudocoel, *579,* 604, 703, 704
  development of, *597*
  and phylogeny, 596
pseudocoelomate phyla, *596, 597,* 604–608, 703, 704
pseudoplacenta, 650, 836, 886
pseudopod, *89,* 428, 433, 434
Psilotum, 479
psychosomatic illness and autonomic learning, 958
ptarmigan, seasonal coloration, 14
PTC (phenylthiocarbamide; phenylthiourea), 345
pterobranch, 634
pterodactyl, *645*
pterophyte, *481–483*
pterosaur, *645*
PTH, *see* parathormone; parathyroid hormone
ptyalin, 712, 718
puberty
  and sex hormones, 789
  females, 845
  and stature, *676,* 677, 789
pubic bone, 681, *682, 683, 838*
pubic hair, 788, 845, 895
  and androgen insensitivity syndrome, 895
pubic symphysis, 681, *682, 683*
public health, 410, 1053, *1053*
puff, chromosome, *308,* 309
pulmonary artery, 751, *752, 753,* 756, 757
pulmonary circuit, 756, *757*
  amphibian, 750
  mammals and birds, *750,* 751
  at parturition, *896,* 897
pulmonate snails, 614, *615*
pulp, dental, *711*
pumpkin, *495*
punctuated equilibria, 338
  and human evolution, 658, 659
Punnett, Reginald Crandall, *272,* 340, 341
  Punnett square, *272, 273, 280, 288,* 340, 341
pupa, *774*
  insect metamorphosis, *622, 623, 774, 775*
pupil, eye, *930*
purine, *184*
  in digestion, 721
Purkinje cell, *907*
purple nonsulfur bacterium, 417
purple sulfur bacterium, *414,* 417
pus, 819
  DNA isolated from, 192
putricene, 531, 1040
pyloric sphincter, 310, 340, *715*
pyloric stenosis, 310, 340
pyramid
  age profile, *1054*
  cell, *907*
  Eltonian, *1015*
  energy pyramid, *1015*
  food pyramid, *1010, 1015*
  of kidney, 813, *814*
pyrimidine, *184*
  gout, *817,* 818
pyrogens, 808, 819
pyrophosphate, 127
Pyrrophyta, *429*
  dinoflagellates, 423, *429,* 430
Pyrsonympha, 424
pyruvate, 163, *164*
  in C4 pathway, *150*
pyruvic acid, *see* pyruvate
python, 975

Q

Q (photosystem 680), *143*
Q (ubiquinone), *172,* 173
Q_{10}, 72
quadrate bone, 678
quantasome, *115,* 140
quantitative inheritance, 312–315, 329–334
  and Mendel's bean experiment, 289
quark, 26
quaternary structure, 65
queen substance, 974
Queen Victoria, 302
Quercus, 389, 1003

R

R group, 64, 65, *66, 67*
rabbit, 654
  absent from Patagonia, 3, 4
raccoon, 1005
racism, 984
radial cleavage, *627*
radial nerve of echinoderms, *735*
radial symmetry, echinoderm, 624, *625*
radicle, *574, 575*
radioactive isotope, 27, 196
  in photosynthetic pathway, 152
radioisotope, 27, 152, 196
radiolarian, 434, *435*
radius (arm bone), 685, *686*
radula, 612
rain, 38
rain forest, tropical, *1001, 1002*
rain shadow, *994, 995,* 996
Ramapithecus, 657, 658, 659
Rana pipiens, 388
random drift, 337, 338, *339*
random mating, *342; see also* Castle–Hardy–Weinberg distribution
Ranunculus, 492, *493*
Ranvier, node of, 911, *912*
Rapoport, I. A., Soviet geneticist, 401
raptorial birds, *709,* 947–949
rare dominant allele, 280, *281,* 285
Rasputin, Grigor Efimovich, 302
rat
  kangaroo, *998*
  lethal fights, 977
  nest building, 953
  in Skinner box, *956*
  territoriality, 979
rat poison, 762
rat's nest, 953
rattites, 649
  ostrich, 488, 649, 836
rattlesnake
  fighting without biting, *976*
  heat receptor, *924*
  venom, 721
Raven, Peter, predicts chloroxybacterium, 422
ray, *639, 641*
ray-finned fish, *641*
reading-frame shift, 217
reafference, 953
rearrangement, chromosomal, 216
recapitulation theory, *890*
receptacle, of a flower, *493, 560*
receptor cell, 906, 913
receptor molecule, in chick oviduct cells, 371
recessive allele, defined, 278
recessive lethal, 307, 308
recessive trait, 278, 326
  discovered by Darwin, 268
  discovered by Mendel, 268, 269
reciprocal altruism, 983, 985
reciprocal test cross, *299,* 342

recognition
individual, 975
species, 974, 975
recombinant chromosomes, *305*
recombinant DNA, 372–377
recombination, 304–307
bacterial, 357–366
characteristic of sexual organisms, 325
and chromosome mapping, 305–307
evolutionary function, 258, 326, 622, 830, 850–852
pine trees and scale insects, 326
see also crossing over
recombination frequency, compared with map distance, 306
rectal gland, 810, *811*
rectal valves, 717
rectum, human, *713*, 717
red alga, *465*, 466
life cycle, *466*
red blood cell, 222, 226, *284*, 717, 760, *761*
sickle-cell anemia, 216, 282, *284*, *285*, 356
red snow, 430
red tide, *429*
reducers, 410, 445, 454, 530, *1014–1016*, 1028
reduction division, 248
reductionists and synthesists, 21, 22, 422, 545
redundancy
of DNA structure, 187, 191
in spindle function, 240
in visual communication, 971
red-winged blackbird, territoriality and mating, *1039*
redwood tree, *503*, 515, 516, 1007
chromosome number, 227
from cell culture, 584
regeneration, 243
and cell dedifferentiation, 228
of sponges, 598
of starfish, 625, *626*
regression to the mean, 314, *333*
regular flower, 493
peloric (regular) snapdragon, *266*, 268
regulation of body temperature, 800–808
regulatory enzymes, 73
reindeer, 1007
reindeer moss, 452, *453*, *1007*
reinforcement, 952, 953
social, *954*
rejection, of organ transplant, 287
relapsing fever, 623
relative fitness, 324
releaser, *950*
releasing factor
in protein synthesis, *214*
behavioral, 776, *779*, *780*
releasing protein, *214*
relief, in consummatory behavior, *948*, 949, 950
renal circuit; renal artery; renal vein, *757*, 813, 814
rennin, 715, 718
repair enzymes, 187, *188*, 190
replacement rate, 1055
replica plating technique, *357*
replication of DNA, 189–191, *198*, *199*, 229, 322
repication fork, 189, 190, *191*
repressible enzymes, 370, *371*
repressor molecule, 368, *369*, 370, *371*
reproduction, animal, 830–858
amphibia, 643, *831*
arthropods, 833, *834*, *835*
barnacles, 831, *832*
a characteristic of life, *46*
external fertilization, 643, 831, 830, *831*
fish, 832, *833*
mammals, 837
monotremes and marsupials, 837
planaria, 831, *832*
reptiles and birds, 835, *836*
seasonality, 851
spiders, 834, *835*
see also sexual reproduction
reproduction, human 837–856
embryonic development, 892, *893*
female reproductive system, 840–848, *841–843*
fertilization, 852
hormonal action in females, 845–848
hormonal action in males, 844–845
male reproductive system, 837–839, *840*, *841*
mesodermal origin of reproductive organs, *876*
sexual behavior, 848–852; *see also* sexual behavior, human
reproduction, plant, 557–584
angiosperms, *493*
brown algae, *468*
conifer, 487
fern, 483
liverwort, *477*
mushroom, *455*
see also life cycles
reproductive isolation, 334–337, 386
in polyploid complexes, 390
reproductive polyp, *599*
reproductive potential, 1031–1034
reproductive rate, 1031–1034
reproductive strategy, 603, 305, 1031–1043
allocation of resources, 1036
altruism, 982, 983
birds, *1033*
chimpanzee, *1032*
Daphnia, 258
and genetic recombination, 258, 326, 622, 830, 850–852
humans, 1033–1035
Model T Ford, 1036
mouse, 1036
parasite, 1041, 1042
parasitoid wasp, 1042, 1043
tapeworm, 1031, 1032
reproductive success
altruists and nonaltruists, 982, 983
and communication, 970, 971
reproductive system
embryonic origin, *876*, 892, *893*
see reproduction, human; sexual reproduction, human; reproduction, animal
reptile, 643–649
behavioral thermoregulation, 802, 803, *804*, *805*
brain, *917*
circulatory system, 739, 740
egg, *885*, 886
mouth and digestive system, 709
origins, 644, 645
phylogeny, *395*, *646–647*
reproduction, 835, *836*
resolution, sexual, *850*
resolving power, 96
respiration (breathing), 643, 650, 733–746; *see* respiratory system
respiration (molecular), 159–173
aerobic, 168–173
anaerobic, 159–167
and carbon cycle, *536*
citric acid cycle, *159*, 160, 169–172, *414*
defined, 158
and energy flow in ecosystem, *1016*
fermentation, 159, *164*, 454
fermentation in gut of ruminant, 711, 712
glycolysis, *159*, 160, *161*, *162*, 163
muscle, 15, *166*, 168
oxidative (chemiosmotic) phosphorylation, 130–132, *172*, 173, *414*
and photosynthesis, compared, 158, *159*
respiratory system, 733–746
amphibian, 643, *739*
bird, 648, *740*
breathing movements, 742
changes at birth in mammals, *896*, 897
coelenterates, 733
control of respiratory movements, 745, 746
different systems compared, *732*
echinoderm, *734*, 735
embryonic development, 888, *889*
endodermal origin, *876*, 888
epithelium of passages, *741*
evolution and variety, *732*, 733–742
fish, 732, 737, *738*
gas exchange, 742–745
grasshopper, 734
human, *741*, *742*
inspiration and expiration, 745
lobster, 735, *736*
lungless salamander, 733
lungs, *732*, 738–742
mammals, 650, 738–742
mollusk, 734, 735, *736*
planarian, 732
sponge, 733
respiratory tree, 742
response threshold, 949
resting potential of a neuron, 907, *908*
restriction enzyme, 357, 358, 372, 373, *374*, *375*
retaliation; retaliator, 976, 977
rete mirabile, 802, *803*
reticular system of the brain, *919*, 962
reticulum, *712*
retina
development of, *880*
ectodermal origin, 878, *880*
human eye, *930*
rods and cones, 928–933
retinal, 928, 929
reverse transcriptase, 217
and central dogma, 183
reward, 952, 953
Rh blood group system, 283, *286*, *287*
rheumatic fever, 820
rhinoceros, 20, *21*
rhinoceros beetle, *619*
Rhipidura, 334, *336*
rhizoid, *504*
of bread mold, *449*
of moss, 473, *474*
of *Psilotum*, *479*
rhizome, *558*
of grasses, 999
Rhizopus (bread mold), *449*, 450
Rhodophyta (red alga), *465*, 466
rhythm method, 853, *854*
rhythmicity center, 745
rib cage, 679, *680*, 742
riboflavin, 724
ribose, structural formula, 203
ribosomal RNA (rRNA), 205
ribosome, 88, 89, 110, 205, 206, 211, *212*, *213*
bound (rough ER), 206, *216*
bound, in procaryote, 215
mitochondrial, 206
movement along mRNA, *213*
procaryotes, 206, 215
subunits, 206
ribulose, 1, 5 diphosphate (Ru-DP), *146*, 147, 148
rickets, 724
right and left hemispheres of the brain, 920
ring dove, 949
ringworm, 1041
Rise and Fall of the Third Reich, 64
ritualized combat; ritualized fighting, *976*, 977
RNA (ribonucleic acid), 203
nuclear, 226, 227
RNA polymerase, *204*
and the operon, 368
RNA replication, 183
RNA synthesis; RNA transcription, 202–205, 322
in cell cycle, 229
and central dogma, 182
and lampbrush chromosome loops, 871
posttranscriptional modification, *207*
RNA-dependent DNA synthesis, 183, 217
roach, 623, *835*
road runner, 997, *998*
robin
attacks red feather, *950*
migration, 967
red breast, 971
rod (retinal), 928–933
absent in fovea, 933
lacking in congenital night blindness, 281
as modified cilium 118
scanning electron micrograph, *931*
rodent, *652*, *653*, 654
desert, 997, *998*
mouse, 287, *288*, 329, *330*, *884*, 1036, 1047
rat, 953, *956*, 977, 979, *998*
Roeder, Kenneth, and moth ears, *926*, 927
Rogers, Roy, and palomino horse, 279
Romer, Alfred S., and development of the land egg, 885
root, 505–509
types, *506*, 507
root apical meristem, 578
root beer, homemade, 1040
root cap, 507, *508*, 578, 579, *580*
root hair, 507, *508*, *509*, 578–580
and gas exchange, 520
root nodules, *45*
root pressure, *515*
root system, *506*, 507
desert plants, 996
grasslands, 999
tundra, 1007
Rosaceae, *495*
rose hip, 723
rotifer, 607, *608*, 703
survivorship curve, *1034*
rough endoplasmic reticulum, 206, *215*, *216*
roundworm (nematode), *456*, 604–607, 703
parasite of dogs, 593
Roux, Wilhelm, killed frog blastomere, 868, *869*
Royal Society, 10
rRNA, 205

Ru-DP (ribulose diphosphate)
  in Calvin cycle, *146,* 147
  synthesis from PGAL, 148
ruff, on lek, *980*
Ruffinian corpuscle, *806*
rumen, 710, *712*
ruminant, 655
  digestive system, 710, *711, 712*
runner, 557, *558*
  marathon, *168,* 805, 806
  training at high altitudes, 744
running lights, 301
rust, 454–456, 459
Ryncocoela, 596

S

S phase, 228, *229,* 234
saccule, *936*
  of Golgi complex, *112*
sacroiliac, 681, *683*
sacrum, 678, 679, *680,* 681, *682, 683*
sage, *495*
saggital crest, of *Australopithecus robustus,* 659
Sahara Desert, *993,* 995, 999
  created by human activities, 1000
Sahel region, 999
St. Anthony's fire, 450
salamander, 388, *389, 395,* 733, 972
  Ensatina ring species, 388, *389*
  lungless, 733
  regeneration, 228
  sound communication, 972
saline abortion, 853, 858
saliva, stimulated by taste, 712
salivary amylase, 712, 718
salivary chromosome, 306, *307; see also* polytene chromosome
salivary gland
  control, 721
  human, 712, *713*
  insect, 706, *707*
  origin by outpocketing, 704
Salkowski, E. and H., 545
salmon, 830, 831
salp, 635
salt, 31
salt balance
  aldosterone and ADH, 787, 788
  in plants, 538
salt gland, 810, *811*
salt linkage, 31
salt tolerance, 506, 515
saltatory propagation of neural impulse, 911, *912*
salting out, 858
San Francisco Bay, 532
sand dollar, 624, *625*
sanitation, 410
  and population growth, 1053
sap, *518,* 519
saprobe, 1014
  fungal, 444
saprophyte, *see* saprobe
Saran Wrap, 856
Sarcina, 415, *416*
Sarcodina, 433, *434; see also* ameba
sarcolemma, 690, *691,* 693
sarcomere, 691–694
  mechanism of action, 693, *694*
sarcoplasmic reticulum, 690, *691,* 693, *694*
Sarcopterigii, *see* lobe-finned fish
Sargasso Sea, 466, *467*
sargassum, 466, *467*
saturated fat, 77
saturation, organic compounds in general, 41
savanna, 999, *1000*
scala natura, 946
scale insect, 326, 622
scale scar, *581*
scales, birds, and reptiles compared, 648
scaly anteater, 654
scanning electron microscope (SEM), *99*
scapula, 681, *682*
scar tissue, 243
Schaller, George, color vision in lions, 933
schistosomiasis, 603
schizophrenia, possibly an autoimmune disease, 820
Schleiden, Matthias Jacob
  cell theory, 85
  observes nucleus, 223
Schopf, William J., oldest fossil, 413
Schwann, Theodor
  cell theory, 85
  cell nucleus, 223
Schwann cell, *907,* 911, *912*
scientific method, *9*
  reductionisits and synthesists, 21, 22
  "saving" hypotheses, 370
scrotum, 837, *838*
  embryology, *893*
scurvy, 724
scutellum, monocot cotyledon, 569, *575*
Scyphozoa, 599, *600*
  life cycle, *600*
sea anenome, *see* anenome
sea cucumber, 624, *625,* 1012
sea horse, *640*
sea lettuce, 467, *472*
sea lily, *see* crinoid
sea lion, 655
sea otter, *46*
sea snake, 835
sea star, *625, 626, 675,* 734, *735, 916, 1012*
  endoskeleton, *675*
  nervous system, *916*
  respiration, 734, *735*
  in tidepool, *1012*
sea turtle, 836
sea urchin, 343, 634, *625,* 675
  cleavage divisions, *870*
  egg, 866, *867*
seal, 655
seaweed, *see* red alga; brown alga; green alga
second law of thermodynamics, *see* thermodynamics, second law
second messenger, 790, *791*
second polar body, *see* polar body
second trimester, 893
secondary consumer, 1014
secondary growth
  monocot vs. dicot, 495
  of root, 507
  of stem, 510, *582*
secondary oocyte, 254, *255; see also* oocyte; gametocyte
secondary response, *822,* 823
secondary sex characteristics
  corticosteriods, 788
  development in female, 845
secondary structure
  protein, 64, 65
  starch, *59,* 60
  tRNA, 208, *209*
secret of life, 25
secretin, 722
secretion granule, *89,* 113
secretory neurons, *779*
secretory phase, of menstrual cycle, 847
Sedgwick, Adam, 9, 594
seed, 483, *484,* 488
  development, 568, 569, *574,* 575, 576
  monocot and dicot compared, *569, 575*
seed capsule, 560, *572,* 577
seed dispersal, 557, *576,* 577
  Darwin's experiments, 11, 12
seed fern, *480*
seed leaves, 569; *see* cotyledon
seed plant, origins, 482–484
segmentation
  annelids, 609
  arthropods, 617
  crustacea, 620
  nervous system, 915, *916*
  Onychophora, 616
segregation of alternate alleles, 271, 272; *see also* Mendel, second law
Seisura, 334, *336*
selection
  artificial, *see* artificial selection
  in bacterial recombination experiments, 363
  natural, *see* natural selection
selection coefficient, 339
selectionist, 337
selenium, cofactor in enzyme activity, 27
self and not-self, 821, *822*
self-fertilization
  ferns, 482
  pea, 267
  tapeworms, 604
self-incompatibility, ferns, 482
self-poisoning
  of humans by pollution, *1040,* 1041
  limits population growth, 1039, *1040*
semen, 838, 849
semicircular canals, 928, *936*
  sensory hairs as modified cilia, 119
semiconservative replication, *191, 198, 199*
semilunar valve, 751, 752, *753*
seminal vesicle, *838*
seminiferous tubule, *838*
semipermeable membrane, 106; *see also* membrane; osmosis
senescence, in plants, 551; *see also* aging
sense organs, 923–936
  chemoreceptors, 933–935
  hearing, 925–927, *928*
  olfaction, vertebrates, 934, *935*
  proprioception, 936
  tactile, *924, 925*
  temperature, 923, *924*
  visual, 927–933; *see also* eye
senses, 920, *921*
sensillum, *934*
sensitivity, a characteristic of life, *46*
sensory area of brain, 920, *921*
sensory deprivation, 850
sensory hair
  hearing in insects, 925
  inner ear, 927, *928*
  proprioception in arthropods, 936
  semicircular canal, 936
  silkworm antenna, *934*
  touch, *924,* 925
sensory neuron, *907*
sensory receptors
  acid on skin, 934
  echinoderm, 626
  temperature, *806,* 807
sepal, *493, 560*
sequential endosymbiosis theory, 114, 115, *414, 421, 422*
sequoia, 1007; *see also* redwood
Serengeti Plain, 999
serial dilution, 361, *362*
serial endosymbiosis, hypothesis, 114, 115, 414, 421, *422*
serine, structural formula, *66*
serosa, of esophagus, 712, *714*
serotonin, *762, 777,* 778
  structural formula and possible role as neurotransmitter, 911
Sertoli cell, 839, *840*
seta (setae), earthworm, *610,* 611
7S RNA, 205
sewage amd eutrophication, 1021, *1022*
sex
  bacterial, 360–366
  broad definition, 358
  evolutionary function, 258, 326, 622, 830, 850–852
  origin of, 326
  see also sexual reproduction; sexual behavior; reproductive system; sex hormones
sex act, *see* copulation; sexual behavior
sex attractant, silkworm moth, 933, *934*
sex chromosome anomalies, 257; *see also* X chromosome; Y chromosome
sex determination
  flies and humans, 301
  turtles, 316
  X and Y chromosomes, 892
sex flush, 849
sex hormones
  balance between estrogens and androgens, 895
  control of reproduction, 844–848
  cycle in males, *845*
  and the developing brain, 892, 893
  and menstrual cycle, 845–848
  produced by adrenal cortex, 788
  and skeletal development, 789
see also estrogen; hormone; testosterone; secondary sex characteristics; reproductive system; sexual reproduction
sex linkage
  in bugs, 315
  Drosophila, 299, 300
  humans, 300–303
  oral-facial-digital syndrome, 303
sex ratio, by age, *1054*
sexduction, 366, 367
sex-influenced trait, 310
sex-limited trait, 310, 312
  gout, 818
  hemochromatosis, 312
sexual behavior, human, 848–852
  beneficial effects of early experience, 850
  erection, emission, ejaculation, 849
  function and evolutionary implications, 850–852
  menstrual cycle and sexual behavior, *851*
  more rapid response in males, 849
  orgasm, 849, *850*
  orgasmic platform, 849
  plateau phase, 849, *850*
  resolution, *850*
  variability, 848
sexual differentiation and dimorphism
  embryological, 892, *893*
  brain and behavior, 892, 893
sexual intercourse, *see* copulation; sexual behavior, human
sexual receptiveness, 845, *851*
  delayed in nursing female chimpanzees, *1032*
  signalled by pheromones, 973

sexual recombination, *see* recombination
sexual reproduction, 830–858
amphibian, 643, *831*
angiosperm, *493*
bony fish, 641
brown alga, *468*
characteristic of animal kingdom, 592
chimpanzee, 1031
Chlamydomonas, 469
clam worm (polychaete,) 612
conifer, 487
contrasted with asexual reproduction, *325*
different kingdoms, 404
earthworm, 610
echinoderm, 627
evolutionary advantages of sex, 258, 326, 622, 830
fern, 483
fireworm (polychaete), 830
gymnosperm, 487
hormonal control in females, 845–848
hormonal control in males, 844, 845
human, 837–858; *see also* reproduction, human; sexual behavior, human
jellyfish, 600
liverwort, *477*
Obelia, 599
mammals, *650*
moss, *476*
mushroom *455*
nematodes, 606
planaria, 603
red alga, *466*
Rhizopus fungus, *450*
Scyphozoa, 600
shark, 832, *833*
slime mold, *446*
spadefoot toad, 997
sponge, 597
stickleback courtship, *951,* 952
strategies of, 1031–1035
tapeworm, 603, *605,* 1031
toad, 967, 997
Ulothrix, 472
yeast, *453*
see also reproduction; reproductive strategy; copulation; fertilization; sexual behavior, human
sexual responsiveness, 845, *851,* 973, 1032
sexual selection, and territoriality, *980*
Seymouria, 644
shadow casting, *98*
shark, *639, 640, 641*
brain, 916, *917*
claspers, 832, *833*
constant swimming, 738
digestive system and teeth, *707, 708*
modes of reproduction, 832, *833*
osmoregulation, *811*
sheathed bacteria, 413, 417, *418, 419; see also* chlamydobacteria
sheep, recessive black trait, *276*
shelf fungus, 454
shell
bird and reptile egg, 885
electron, 28–33
hydration shell, 36, *38*
lobster, 735, *736*
mollusk, 672–674
turtle, 679
see also exoskeleton
shivering, 806, 807
short-day plant, 562; *see also* photoperiod
short-term memory, 959, 960
shoulder blade, 681
shrew, 392
Shull and Bauer, 315
Siamese cat, 309, 310
sickle-cell anemia (sickle-cell hemoglobin), 216, 282, 356
dominance relationships, *284, 285*
side groups (table), 48
side-necked turtle, *647*
Sierra Nevada, 1007
sieve plate, *511*
echinoderm, *626*
sieve tube, *511, 513, 518*
sieve-tube member
lacks nucleus, 228
maturation, 583
origin from cambium, 582
see also phloem
sigmoid colon, *713,* 717
sigmoid curve, 324, 330
signet ring stage, 874; *see also* blastocyst
silicaceous sponge, 597
silk gland, spider, 617, 618, *619*
silkworm, 774
moth chemoreceptor, 933, *934,* 973
Singer, S. J., cell membrane model, *101,* 102
sinoatrial node, *753, 754*
sinus venosus
amphibian, *749,* 750
fish, *748*
sinusoid, liver, 757
siphon, clam, 614, *615*
siphonous alga, 471, *472*
Sirenia, *653,* 655, 681
sister strands in meiosis, *249,* 250
sitka deer, overcrowding affects hormones, 780, 781
skeletal muscle, *91,* 688–695; *see also* muscular system
skeletal system, 669–686
amphibian and crossopterygian compared, *643*
appendicular, 675, 681–686
arthropod joint compared with vertebrate, *671*
axial, 675, 677–682
coelenterate, 670
development of bone, *676,* 677
earthworm, 670
echinoderm, *674, 675*
embryology of limbs, 891, *892*
evolution of skull, *684*
exoskeleton, 671–674
forelimb homologies, *683*
gorilla, complete, compared with human, *682*
hollow bones of birds, *740*
human, complete, *682*
human skull, *674*
hydrostatic, 670–672
invertebrate, 670–675
joints, *671,* 675, *677*
mesodermal origin, *876*
mollusk shells, 672–674
ossification in third trimester, 894
sponge, *674*
vertebral column, 678–680
vertebrate, 675–686
skin, *806,* 876
color, 337
echinoderm, *675*
thermoregulation, *806*
skin breather, 643, *732, 739*
lungless salamander, 733
Skinner, B. F.
comparative psychology and behaviorism, 945
develops Skinner box, *956,* 957
and social reinforcement, 954
Skinner box, *956,* 957
skull, 675, 677, *678, 679*
bird, reptile, fish, dog, *678*
evolution of zygomatic arch, *684*
human, *679*
various mammals, *711*
vertebrate classes compared, *678*
skull 1470, 659, 660, *662*
skunk cabbage, 496
slash-and-burn agriculture, 1002
sleeping sickness, *433,* 440
sliding filaments
sarcomere, 693, *694*
spindle fibers, *239*
slime bacterium, 410, *417*
slime mold, 395, 402, 445–458
acellular, *91*
hormonally induced aggregation, 771, *772*
life cycle, *446, 447,* 772
phylogenetic origins, 456
slug of cellular slime mold, *447*
sloth, 654
extinct giant ground sloth, 4
tree, 4
small intestine, *713, 716*
smallpox, 373, 1041, 1042
last case, 1053
smell, *see* olfaction; pheromone
Smith and Haldane, 315
smog, 1041
smoking, 741
smooth muscle, 686–688
arterial, *755*
contactile mechanisms are unknown, 694
sphincters in arterioles control blood flow, *758*
three smooth muscle layers in stomach, *715*
two layers in esophagus, *714*
of uterus, 843
in walls of veins, 759, *760*
smut, *454*
snail, 614, *615*
cellulase, 62
eyestalk, 672
shell, *673, 674*
survival in seawater, 12
snake, 644, *646, 647*
hollow fangs, 708
jaw bones, 678, *709*
venom, 644
snapdragon
Baur's experiment, 290
partial dominance of flower color, 279, *280*
peloric and irregular, *266,* 268
snare fungus, 456
snowy owl, 1007, *1008*
soap, 76
social behavior
altruism, 983–985
honeybees, *981,* 982
and olfaction, 934
social insects, 20
social reinforcement, *954*
wolves, 1046–1049
sociobiology, 984
altruism, 20, 982–985
sex activity and bonding, 851, 852
wolves, 1046–1049
sodium
dietary need, 725
required for animals but not for plants, 25, *26*
see also excretory system
sodium bicarbonate, 43, 718
secreted by pancreas, 717
sodium pump, 104, *105*
of neuron, 907
soft palate, 741
soil, *528, 529*
action of earthworms, 611
formation, 1020, *1021*
tropical rain forest laterite, 1002
Solenogastres, 612
Solo man, 661
solution, 31
somatic DNA, *438; see also* terminal differentiation; immortality; germ line
somatic tissue, 244
somite, frog embryo, *878*
sorus (sori) of fern, 481, *482*
soul, 946
sound communication, 972–974
sound receptor, 925–927, *928*
sound spectrogram, 972, *973*
South African clawed frog, 864, 881; *see also Xenopus*
South America, deserts, 996
Soviet Union, and T. D. Lysenko, 401
sowbug, density-dependent mortality, 1038
space probe, 993
space-filling model, *33*
of glucose, *58*
spacer region
between ribosomal RNA genes, *205*
in DNA, 207, 216
spadefoot toad, 997
Spanish hogfish, *984*
sparrow
English, 337
white-crowned, 963
spatial abilities, localized in brain, 961
specialization
barnacle habitat, 1036, 1044
bird beaks and feet, *6, 709*
cost-benefit ratio for, 800
teeth, 710, *711*
see also differentiation
speciation, 16, *17*
adaptive radiation, 593, *652, 653*
and Darwin's finches, 7
faster rate in flowering plants, 488
and geographic isolation, 7, 14, 15
by species fusion, 389
sudden appearance of new species in fossil record, 338
species concept, 386–390
considered unchanging before Darwin, 4, 16
Darwin denies reality of species concept outright, 16, 17
and OTU, 391
in plants, 388, 389
population of variants, 390
type specimens, 390
species number, 386
species recognition, *974,* 975
specific heat of melting, 36
spectrum
absorption, 137, *139*
action, *139*
electromagnetic, *96,* 137
sound spectrogram, 972, *973*
visible, *96*
speech center of brain, 920, 961
Spemann, Hans, experimental embryology, 878, *879*
sperm (spermatozoan, spermatozoa), 865, 866
flagellated in cycads and ginkgos, 258
head, 839, *841*
human, 839, *840, 841*
nucleus, 567, *568*
production in human male, 245, 839, *840*

sperm count, 866
spermatagonia, 839, *840*
spermatheca, 834
spermatic cord, *838*
spermatic tubule, 245
spermatid, 839, *840*
spermatocyte, 245, 839, *840*
  metaphase I, *248*
spermatozoa, 839, *840, 841; see also* sperm
spermicides, 856
spermiogenesis, 839, *840*
spermists and ovists, 868
Spermophyta, 482–484; *see also* seed plant; plant kingdom
Sphenodon, 646, 647, 789, *790*
Sphenophyta, 480, *481*
sphincter, 690
  anal, 717
  controls circulation in capillary bed, *758*
  cardiac and pyloric of stomach, *715*
sphygmomanometer, *756*
spicule
  nematode sex organ, 606
  sponge, 597, *598, 674*
spider, 617, 618, *619*
  book lung, 617
  mating, 834, *835*
  silk gland, 617, 618, *619*
  touch receptors, *924*
spike of nerve impulse, 908, *909*
spinal cord
  in neural canal, 679
  see also dorsal hollow nerve cord
spinal ganglia, *922*
spindle, 117, 230, *231,* 234, 237–240
  centromeric fibers, 234
  spindle apparatus, 237–240
  theories of function, *239,* 240
  tubulin proteins formed in G_1, 234
spindle fiber
  centromeric, 234, *237,* 238, *239*
  continuous, 238, *239*
  in meiosis, 248
spindle fiber attachment, *see* centromere
spinneret, spider, 617, *619*
spiny anteater, 710; *see also* monotreme
spiracle
  insect, *621, 734*
  Peripatus (Onychophoran), 616
  shark, *639*
spiral valve, 639, *708*
spirochaete, 409, 410, *414, 418*
  ancestor of eucaryote flagella and cilia, *422*
spleen, contracts to regulate blood supply, 798
split brain experiments, 920, *960–962*
SPONCH (Sulfur, Potassium, Oxygen, Nitrogen, Carbon, Hydrogen), 25, 26
sponge, 402, *403,* 592, 595, 597, 598
  choanocyte, 597, *598,* 701, *702*
  intracellular digestion, 701, *702*
  phylogenetic origin, 458
  skeletal structure, *674,* 675
spongin, 597, 674
spongocoel, *598*
spongy bone, *676*
spontaneous abortion, 256, 894, 896
spoonbill, *649*
sporangiophore
  acellular slime mold, 446
  bread mold, *449*
sporangium
  bread mold, *449*
  fungus, *445*
  slime mold, *446*
  water mold, 447, *448*
spore and spore formation
  bacteria, 411, *412*
  fern, 481, *482*
  fungi, *445,* 457
  myxobacterium, 417
  seed plant, 488
  spore formation in different kingdoms, 404
  sporulation, 481
  water mold, 447
sporophyll, 485
sporophyte, 466
  of fern, 481
  of hornwort, *478*
  of moss, *474*
  see also life cycle
sporozoa, 434, *435*
sporozoite, *436*
sporulation, *see* spore and spore formation
sprain, 675, 677
sprouting factor, 882
spruce, 486; *see also* taiga
squid, 614, 615
  cuttlefish and cuttlebone, *673*
  giant axon, 911
  giant squid, 612, 1035
squirrel
  Grand Canyon species divergence, 17
  innate behavior, 963
stability
  equilibria, 334, *335,* 504, 516, 1029
  of population size, 1029, 1049, 1050
  steady state, 504, *505*
stabilizing selection, 332, 333, *334*
  of birth weight, 332, *333*
Stalin, Josef, 401
  social reinforcement, 954
stapes, *928*
Staphylococcus, 415
star navigation, 970
starch, *59*
  metabolism, 166, *167*
  in plastids, 114
  synthesis, 71, 147
starfish, 624, *625, 626, 675,* 734, *735, 916, 1012; see also* sea star
starling
  individual distance on power line, *979*
  vacuum behavior, 951
statocyst, 936
statolith, 936
  of plant cell, 548, 549
stature
  and sex chromosomes, 257
  and sex hormones, 676, 789
steady state, 504, *505*
stearic acid, structural formula, *77*
Stegocephalia, 642, 643
stegosaur, thermoregulation, 803, *805*
stele, 507, *508, 509, 580*
stem, *510–512*
  primary growth, 579, *581*
  see also primary growth pattern; secondary growth pattern
stem cell, 227
stem reptile, *644, 647*
stereoscopic vision, 658, 931
  primates, 654
  squid, *616*
stereotyped behavior, 981
  falcon hunting, *948,* 949
  fighting, *976,* 977
sternum, 679, *680*
steroid, 75, 79
  bile salts, 717
steroid hormones, 79, 776, 777
  and development, 892, 893, 895
  mechanism of action in target cells, 71, *792*
  see also sex hormones; cortisol; testosterone; estrogen
sterol, *see* steroid
Stevenson, Adlai, derided for proposing test ban, 1041
Steward, F. C., *583,* 584
stickleback, behavioral hierarchy in courtship and defense, *951,* 952
sticky DNA, 373
Stine, G. J., frequency of sex chromosome anomalies, 257
stipe, of brown alga, *468*
stolon
  bread mold, *449*
  flowering plant, 557, *558*
stoma (stomata), 514, 520–522
  of C4 plants, 151
  of leaf, potassium dependence, and function, 537
stomach
  cow, *712*
  starfish, eversible, 625
  human, 712–716
Stone Age tools, 660
stone canal, *626*
stonewort, 467
storage diseases, 113
storage vacuole, *89,* 113
strategies of aggressive behavior, 977
strategies, reproductive, 1031–1036, 1042, 1043; *see also* reproductive strategy
strawberry, 557
streaming, cell, *519; see also* cyclosis
Streptobacillus, 416
Streptococcus, 415, *416*
streptomycin, use in bacterial genetics, 363
stress
  and epinephrine, 788
  and hormone imbalance, 781
stretch receptors, 936
  of bladder, 816
  in breathing, 745
striated muscle, 687–692; *see also* muscular system, skeletal
striation, of striated muscle, *691*
strobilation, 480, *600*
strobilus, 480
stroke volume, 755
stroma of chloroplast, 114, *115, 130,* 140
structural formulas, 32, 33, *58*
structural gene, 207, 215, 216, 366
structural protein, 74, 75
structure, primary, *see* primary structure
structure, secondary, *see* secondary structure
struggle for existence, 13
sturgeon, *641*
Sturtevant, A. H., conceives of recombination mapping, 306
style, *560,* 561
stylet, of nemertean worm, 604
stylized fighting, *976,* 977
subcanopy, 1001, *1002*
suberin, 95, 507, *509*
subesophageal ganglion, *916*
sublingual salivary gland, 712
submaxillary salivary gland, 712
submucosa, of esophagus, 712, *714*
subspecies, 334, 335
  geographical distribution, *388*
substrate, 69
  concentration, 71, *73*
substrate-level phosphorylation, 160, 162
subunit disassembly hypothesis, *239,* 240
succession ecological, 1020–1022
succulent, 996
sucker (fish), 72
sucker (plant), 557
suckling
  stimulates release of oxytocin, 782, 897, *898*
  as a contraceptive, 1032–1034
sucrose, 56
  structural formula, *57*
sugar, 56–59
  concentration in plant tissues, 518
sugar cane, *149*
  efficiency of C4 photosynthesis, 150
suicide, of a postreproductive moth, 1037
sulfate, as a plant nutrient, 537
sulfhydryl bond, sulfhydryl bridge, 48, 65
  in protein structure, 65, 68
sulfhydryl group, 48, 65
sulfur
  compounds of, 48
  plant requirement, 537
sun compass, *968, 969*
sunburn, and skin color, 337
sundew, *539*
sunflower, 507
sunflower seed, 560
superesophageal ganglion, *916*
superhelix (supercoil), *225,* 230, 233, *235*
superior vena cava, 751, *752,* 763; *see also* vena cava
support and locomotion, animal (chapter), 669–686
suppressor genes, 309
surface tension, *38*
surface–volume effects, 87, 92, 671, 672
  and Bergmann's rule, 336
  and cleavage divisions, 872
  and flatworms, 733
  gut villi and microvilli, *716*
  insect feats of strength, 671, 672
survival of the fittest, 13, 16, 332; *see also* natural selection
survival value, 321
survivorship curve, *1034*
suspensor, of plant embryo, 569, *574*
Sutton, Walter, chromosome theory of inheritance, 297, 300
Svedberg unit, 206
swallowing
  detailed sequence of muscle contractions, *950*
  consummatory behavior, 949
swarm cells, of acellular slime mold, 446
sweat glands and sweating, 805, *806*
  ectodermal origin, *876*
swift, optimal clutch size, 1033
swim bladder, 640, 641, 738
  as an organ of hearing, 927
Swiss cheese, 842
switching, by predators, 1015, 1046
symbionts and symbiosis
  Chlorhydra and green alga, 599
  cleaning, 983, *984*
  corals and dinoflagellate zooxanthellae, 601
  human skin mite, 1042
  lichens, 451
  origin of red algal chloroplast, 465
  photosynthetic, 422, 423, 429, 465, 599, 601
  protozoa in cattle and termite guts, 62
  termite hindgut, 62, 423, *424,* 706
  tsetse fly gut, 706
  varieties of symbiotic interaction, 1042
  Yucca and yucca moth, *496*
  see also parasite; parasitism

symbiosis hypothesis, 114, 115, 414, 421–423, 465
sympathetic nervous system, *922*, 923
 and circulating epinephrine, 788
synapse, neural, 907–911
synaptic cleft; synaptic junction; synaptic knob, 908, 909, *910*
synaptinemal complex, 245, 246, *249*
syncytium
 Aschelminthes, 604
 defined, 445
synonymous base substitution, 216
synovial joint, 675–677
synovial membrane, 677
synthesists, 21, 22, 422
synthetic reaction, 128
syphilis, 415, 418
systematics, *7*, *20*, 386–408
 compared with taxonomy, 394
 is it science?, 396
 see also phylogenetic tree; phylogeny
systemic circulation, *757*, 758
systems ecology, 992
systole, 755, *756*
Szent-Györgi, Albert, life slips through his fingers, 25, 51

T

T-cell lymphocyte, 819
T tubule, 690, *691*
tabula rasa, 984
tactile receptors, *924*, *925*
taiga, *993*, 1005–1007
tail
 a chordate characteristic, 635, *636*
 human embryo, 679, 890, *892*, *893*
tail bud stage, *878*
tailoring, 205, 207
 and gene inserts, 372
talus, 685, *686*
tanker, *1040*
tapeworm, 592, 603–605
 reproductive strategy, 1031, 1032
taproot, *506*, 507
tardigrade, 475
target cells
 binding site and adenyl cyclase, *791*
 brain targets of sex hormones, 892
 listed, 776, 777
 of sex hormones, 892, 895
target tissue, of nerve cell growth, 882, *883*, *884*
tarsals, *686*
tarsier, 655
tarsus, of body louse, 623
Tasmanian devil, excluded from Australia by dingo, 1044
taste bud, 712, *714*, *935*
 on fish, 640
taster allele for PTC, 345
Tatum, Edward, molecular genetics, 352–356, 360, 361
taxon, defined, 393
taxonomy, 386–408
 of bacteria, 415, 419
 biochemical, 595
 compared with systematics, 394
 is it a science?, 396
 King Philip, 392
Tay-Sachs disease, 113, 338
teak, 1002
teal, hunted by peregrine falcon, 947–949
tear gland, 931, 932
tectorial membrane, 927, *928*
teeth (tooth)
 deciduous and permanent, 710
 dental pulp, *711*
 fluorine requirement, 725
 as key to mammalian orders, 654, 655
 lacking in birds, 709
 mammalian, 710, *711*
 shark and ray, 639, *707*
 vertebrate classes compared, *678*
 X-linked dominant brown teeth, 301, *303*
teleost, *641*
television and violence, 678
telolecithal, egg, 865
 cleavage, *871*, 872
telophase, 229, *250*
temperate deciduous forest, 1003, *1004*, *1005*
 world map, *993*
temperature
 and camel, 998, 999
 effect on chemical reactions and on enzyme activity, 72
temperature receptors, *80*, 923, *924*
temperature regulation, 800–808; *see also* thermoregulation
temporal bone, *679*
temporal lobe, 920
temporal muscle, *684*
tendon, *688*
 compared with ligament, 690
 insect, 672
tendril, 544
tensile strength, of water, 516
tenting, 849
terminal differentiation, 227, 228
 reversibility, 228
terminator codon, 206, *213*, 214
 mutation to, 216
termite, *623*
 gut symbionts, 423, *424*, 706
terrestrial life, *see* transition to land
terrestrial vertebrates, phylogeny, *395*
territoriality, 978–980
 adaptive features attributed by various authorities, 980
 bourgeois strategy, 980
 and limiting resources, *1039*
tertiary consumer, 1015
tertiary structure
 of a protein, 65
 of tRNA, 208, *209*
test, of a sand dollar, *625*
test cross, 276, 277
 progeny test, 268, *270*, *276*
 and recombination frequency, *304*, *305*
testicle, *see* testis
testicular feminization, 895
testis (testes), 837, *838*
 endocrine function, 777, 788, 789
 embryonic development, 892
testosterone, 777, 789
 action in development, 844, 892, 895
 in brain development, 892, 893
 converted to estradiol in some target cells, 895
 cycle in males, *845*
 and pubic hair in both sexes, 844, 895
 and sex drive in both sexes, 844
 and skeletal development, 789
 and social environment, 844, *845*
tetrad, of meiotic chromatids, 247
tetrahedron, 33, 36
tetramer, 65
thalamus, 917, *918*
thalassemia, 282
thallus, 470, 476
therapsid, *647*, 650
 skull, 684, *684*
thermal energy
 of a collision, 50
 of hydrogen and oxygen combined, 37
thermodynamics, 49, 141
 and ATP synthesis, 175
 ecosystem, *1016*
 entropy, 126, 131
 evaporation and transpiration, 506, 516, 517
 free energy, 126, 131, *143*, 158, 173, *174*, 505, 516, 517
 laws of, 131
 second law stated, 126
thermoreception and thermoreceptors, *806*, 807, 923, *924*
thermoregulation, 649, 650, 800–808
 behavioral, 802, 803, *804*, *805*
 behavioral, in humans, 802, 805
 control by calcium ions, 808
 and fever, 802
 in fish, *802*, *803*
 mammals, 649, 650
 reptiles, 643
thermostat, example of negative feedback inhibition, 722, *799*
thiamin, 724
thickeners in milkshakes, 63, 465, 466
third eye, 789, 790
third trimester, 893, 894
thirst, 816
thoracic duct, 720, *764*
thoracic region, 679, *680*
thorax
 human, 679, *680*, *741*
 insect, *672*
 vertebrate, 675
threat, *971*
 display, 977
 male baboon, *972*
 mandrill, *971*
Three-Mile Island, 1041
3-phosphoglycerate (3PG), 146
threonine, structural formala, *66*
threshold
 behavioral response, 949
 firing threshold of a neuron, *908*, *909*
 as a limit of negative feedback control, *799*, 800
 nutritional state and fertility, 1034
thrombin, 762, *763*
thromboplastin, 763
thumb, 654, 683, *684*, *685*, 1032
thylakoid, 86, 114, *130*, 140, *145*, 410, *420*
 and chemiosmosis, *130*
 of a cyanophyte, 86, *420*
 membrane, 410, *420*
 and Z scheme, 144
thymine
 DNA damage and repair, 188
 structural formula, 184, 203
thymosin, 777
thymus
 endocrine function, 777
 site of T-cell lymphocyte development, 819
thyrocalcitonin (calcitonin), 776, 784, 785, 791
thyroid gland, *783*, *784*
 accumulation of radioactive iodine, 197
 hormones listed, 776
 tumors, 197
thyroid-stimulating hormone (TSH), 776, 779, 782, 784
 stimulates cAMP synthesis in target cells, 791
thyrotropin, 776, 779, 782, 784, 791
thyroxin, 27, 776, 783, 784
 and metamorphosis in amphibia, 783
 structural formula, 778
tibia, 685, *686*
tide pool, 1011, *1012*
tiger lily, *570*
tiglon, *18*
Tinbergen, Niko
 behavioral hierarchy, *951*, 952
 ethology, 944, *945*
 fighting behavior, 978
tissue culture
 carrot, *583*, 584
 and karyotyping, *236*
tissue damage, 777
tissue-level organization, sponges, 598
toad
 external fertilization, *831*
 migration and navigation, 967
 spadefoot, 997
tobacco, *495*
Tolkien, J. R., 475
tomboyism, 893
tongue, 690
 bone (hyoid), 888, *889*
 cyclostomes, 708
 human, role in eating, 712, *713*, *714*
 hummingbird, 710
 specializations in birds, *709*
tool use, 658, 659
 by chimpanzees, 1032
 by *Homo erectus*, *660*
tooth, *see* teeth
tornaria larva, *627*
torsion, gastropod, *614*
tortoise
 Galapagos, 5
 outlives humans, 1036
totipotency
 of a carrot cell, *583*, 584
 of transplanted nuclei, 880–882
touch receptors, *924*, *925*
toxic waste
 dumping, 1012, 1021, *1022*, *1040*
 self-poisoning limits population growth, 1039, *1040*
Toxocara canis, 606, 607
trace elements, 530, *532*, 537
trachea (tracheae)
 human, *741*, 742
 insect, *732*, *734*
 Peripatus, 616
tracheid, 511, *512*, *513*
tracheophytes, 478–496, 504
trade winds, 995, *996*
trade-offs, 1036
transcribed strand of DNA, 203, *204*
transcription of RNA, 202–205, 322
 in cell cycle, 229
 and central dogma, 182
 of introns, 372
 on lampbrush chromosome loops, 871
 posttranscriptional modification, *207*, 372
 in puffs on polytene chromosomes, 308, *309*
 reverse transcription, 183, 217
transduction, *360*
transfer RNA (tRNA), 205, 207, *208*, *209*, *210*, 226, 227
 charging enzymes, 207, 208, *210*
 elongation of growing polypeptide, 211–214
 tertiary structure, 208, *209*
transferrin, 73
 pigeon eggwhite, 282

transformation, 192, *193*, 358, *359*
transfusion of blood, 283
transition to land, 458, 472, 473, 738, 739
  amphibia, 642, 643
  eyes, 931
  lignin and terrestrial plants, 458
  limb girdles, *681*
  Onychophora, 616
  and reproduction, 833–837
  the reptilian egg, *644*, 884–886
  reptiles, 643, 644
  stance in land vertebrates, *681*
  water conservation and excretory system, *812, 813*
translation (protein synthesis), 210–217
  and central dogma, 182
  elongation of a polypeptide chain, 211–214
translocation
  of nascent polypeptides across membranes, *215*
  in protein synthesis, *213*
  rearranged chromosomes, 216
transpiration, 514–517
transplantation
  amphibian gastrula, 878–884
  cell nucleus, 880–882
  frog epidermis, *883*
  of organs, 287
transport across cell membranes
  active, 95, 100, 104, 105, *172*, 175, 538, 549, 693, *694*, *759*, 815, *816*
  diffusion, 95–104
  facilitated diffusion, 100, *104*
transverse colon, *713*, 717
transverse tubule, 690, *691*
tree of life, *393; see also* phylogenetic tree
tree shrew, *651*, 655
trematode, 603
Treponema pallidum, 415, 418
tricarboxylic acid cycle, *159*, 160, 169–172, *414*
triceps brachii, *689*, 690
trichina, *607*
Trichinella, *607*
trichinosis, *607*
Trichomonas vaginalis, *432*
Trichonympha campanala, *432*
tricuspid valve, 751, *753*, 754
Trigger, 279
triglyceride, 75–77
triiodothyronine, 776, 783
trimester
  first, 888–893
  second, 893
  third, 893, 894
triple bond, 39
trisomy-21 syndrome, *256*, 257
  confused with cretinism, 783
  and maternal age, 257
trisomy-X female, 257
Trivers, Robert, reciprocal altruism, 983, 985
tRNA, 205–214, 226, 227; *see also* transfer RNA
trochophore larva, 609, *627*
trophic level, 1014, *1015*
trophoblast, 874, *875*, 886, *887*
  violates biogenetic law, 891
tropical deciduous forest, 1002
  world map, *993*
tropical rain forest, *1001*, *1002*
  diversity of tree species, 1043
  rapidly disappearing, 1002
  world map, *993*
tropical seasonal forest, *993*, 1002
tropisms, plant, 546–548
tropomyosin, *693*
troponin, *693*
true-breeding strain
  peas, 267, 268
  snapdragons, 266
Truman, Harry, *224*
Trypanosoma gambiense, *433*
trypsin, 718, 720, *721*
  neutralized by roundworms, 606
  pH optimum, 72
tryptophan, structural formula, *66*
tryptophan synthetase operon, 370, *371*
tsetse fly, *433*
  gut symbionts, 706
TSH, *see* thyroid-stimulating hormone
tuatara, *646*, *647*, *789*, *790*
tubal ligation, 853, *857*
tube foot, 624, 625, *626*, *675*, *735*
tube nucleus, 565–567
tube-within-a-tube body plan, 604, 701, 703, *704*
  earthworm, 610
tube worm, 611, 612
  Galapagos rift community, 1017
tuber, *558*
tubule, kidney, *94*, 813–818
Tubulidentata, *652*, *653*, 655
tubulin, *119*, 424
  and origin of eucaryote cell, *414*, 422–424
  in spindle, 234, *237*
  subunit disassembly hypothesis, *239*, 240
tulip, *495*, *570*
tumbleweed, 577
tumor, thyroid, 197, 783, *784*
tuna, *639*, 1014, 1015
  thermoregulation, *802*, *803*
tundra, 1007–1009
  world map, *993*
tunica adventitia, tunica intima, and tunica media, *755*
tunicate, 635, *636*
  host of chloroxybacterium, *423*
Turbellaria, 602, 603; *see also* planaria
turbot, 833
turgidity, of penis, 839
turgor, *107*
  and cell elongation, *579*
  dependent on potassium, 536
  guard cell function, 520–522
  see also osmosis
Turner's syndrome, 257
turtle, *647*
  carapace, 679
  cloacal lung, 739, 740
  horny beak, 709
  learning behavior, 957
  modified rib cage, 679
  primitive skull, 684
  reproduction in sea turtle, 836
  sex determination, 316
  side-necked, *647*
twins
  fraternal (dizygotic), 846
  identical (monozygotic), 870
two, four dichlorophenoxyacetic acid (2, 4D), 553
tympanic membrane, 927, *928*
tympanum
  mammalian, 927, *928*
  noctuid moth, *926*
type specimen, 390
typhlosole, *705*
typhus, 623
Tyrannosaurus rex, *645*
tyrosine, structural formula, *66*

U

ulcer, 310, 715, 722
ulna, 685, *686*
Ulothrix, 470, 472
ultracentrifuge, 102
ultraviolet pigment, 561
Ulva, 467, 470, *472*
umbilical cord, *896*
unconditioned response, 955, *956*
underdeveloped nations, 1053
unicellular organisms, 91, 428–433; *see also* protist
unsaturated fats and fatty acids, 41, *77*, 723
unselected loci, 363
unwinding proteins
  and DNA replication, 190, *191*
  and RNA transcription, *204*
upwelling, 1011
  lake, 1013, 1021
  and productivity, 1013
uracil, structural formula, 203
urea, 808, *809*, 813
  and protein denaturation, 65
  in shark, ray, and coelocanth blood, 810, *811*
ureter, 813, *814*, 815
urethra, 813, *814*, 815, *838*, *839*
urethral meatus, 840, *841*, *893*
uric acid, 808, *809*, *813*
  crystals, in gout, *817*, 818
  excretion, 643, *644*, 648
  stored in bird and reptile egg, 886
urine, 813–818
  produced by fetus, 892
  see also excretory system
urochordate, 635, *636*
urogenital fold, 892, *893*
urogenital groove, 892, *893*
urogenital sinus, *893*
uterine mucosa (endometrium), 843, 846, *847*, 848, 886
uterus, *842*, 843
  contraction induced by oxytocin, 782, 896
  contraction after parturition, 896
  formed from Müllerian duct, 892
utricle, *936*

V

vacuole, 87, *88*
  plant cell, *88*, 116
  storage, *113*
  water (contractile), 105, *106*, 429, *430*, 434, 438, 439
vacuum behavior, 950, 951
vagina, 841, *842*
  bacterial flora protects, 454, 842
  constriction is cause of painful intercourse, 842
  formed from Müllerian duct, 892
  infection by yeasts, 454
vaginal jelly, 856
vaginal orifice (introitus), *841*, *893*
vagus nerve, 721, 722
valine, structural formula, *66*
valve
  heart, 751–756
  rectal, 717
  spiral, 639, *708*
  of veins, 759, *760*
van der Waals forces, 32, 39
  in protein structure, 65, 68
van Helmont, Jan Baptist, 527
van Overbeek, Johannes, *583*, 584
vanadium, 27, 736
variable age of onset, 312
variable expressivity, 309, *310*
variable region of immunoglobulin, 819, *820*
variability, genetic
  adaptedness of variability itself, 1033
  advantage of hidden variation, 326
  and artificial selection, 329, 330
  equilibrium between mutation and selection, *338*, 339
  frequency-dependent selection, 334, *335*
  hidden, 327–329
  for maladaptive conditions, 340
  mechanisms maintaining diversity, 328
  at the molecular level, 337
  and natural selection, 13, 16, 328
  origin of, 16
  overdominance, 282, 285, 329
  and selectively neutral variation, 337
  stabilizes under selection, *334*
vas deferens, *838*
  cut in vasectomy, *857*
vascular bundle, *492*, 495, 512, *513*, *580*, 581
  air channel, 520
  of dicot, *582*
vascular cambium 507, *508*, *509*, 581
  absent in monocots, 496
  production of xylem and phloem cells, 582
vascular plants, 478–496
vascular system, animals, 746–764; *see also* circulatory system
vascular system, plants, 503–522
  club moss, 480
  of leaf, *492*, 512–514
  see also phloem; transpiration; vascular bundle; xylem
vasectomy, 853, 856, *857*
vasoconstriction, 756, *758*
vasodilation, 756
vasopressin, former name for antidiuretic hormone, 782
vegetable oil, 75, 77
vegetal pole, 868
vegetative propagation, 557–560
vegetarianism, 724
vein
  defined, 751
  of a leaf, 512, *514*
  valves, 759, *760*
  see also circulatory system, 746–764
veldt, 999, *1001*
velum, 636
vena cava, 751, *752*, *753*
vent, 640
ventilation (breathing), 742–746
  inspiration and expiration, 745, 746
ventral aorta, *748*
ventral blood vessel, earthworm, *747*
ventral nerve cord, 611, 617, 635, *707*, 913, *914*, *915*, *916*
ventricle
  amphibian, *749*
  fish, *748*
  human heart, 751, *752*, *753*
  reptile, *749*
ventricular septum, *754*
venules, 751
venus fly trap, *539*
vermiform appendix, 704, 717
vernalization, 401
vertebra, 678, *680*
vertebral column, 678–680
vertebral disk, 678
vertebral foramen, 679, *680*
vertebrate subphylum, 637–655
  characteristics defined, 637
vertigo, 936
vesicle, of Golgi complex, *112*
vessel (xylem) and vessel element, 511, *512*

vestibular gland, 842
vestigial organs
  in embryo, 890
  gill arches, 888
  vermiform appendix, 704, 717
  yolk sac, 886, 887
Vestimentaria, 1017
vibrissae, 924, *925*
  innervation experiment, *884*
Victoria I, Queen, 302
villi (s. villus)
  chorionic, 886, *887*
  intestine, *716*
  microvilli, *89, 94, 716*
vine, *544, 1002*
violence and human behavior, 978
Virchow, Rudolf, 85
virulence, 192
virus
  activity in bacterium, 358
  bacteriophage, *92*, 193, *194, 195*, 200, 357–361
  and cell theory, 92
  Herpes simplex, 1041
  not in any kingdom, 402
  phagocytosis of infected cells, 819
visceral arch, *878*
visceral nervous system, *see* autonomic nervous system
visible light, 927, 928
vision, 927–933
  color blindness, 301, *302*, 303
  color vision, 932, 933
  depth perception, 931, *932*
  detection of movement, 931
  on and off fibers, 931
  squid, *616*
  stereoscopic, *616*, 654, 658, 931
visual acuity
  American Blacks and American Indians, 340
  honeybees, 929, 930
  mantis, 929
  squid, 616, 929
visual pigments, 928, 929
visual purple, 928
visual receptors, 927–933
  algae, 469, 430, 928
  arthropod compound eye, *620*, 929, *930*
  Euglena, *430*
  honeybee, 929, 930
  human eye diagram, *930*
  planaria, *929*
  plants, 521, *522*, 547, 562, *564*
  rods and cones, 928–933
  visual pigments, 928, 929
vital genes, mapped to chromosome bands, 307, 308
vitamins, 67, 724–726
  fat-soluble, 723
  table of, 724
vitamin A, 724
  and retinal, 928
vitamin B_1, 724
vitamin B_2, 724
vitamin B_6, 135, 724
vitamin B_{12}, 537, 724
vitamin C, 723, 724
  evolutionary loss in primates, 533
vitamin D, 79, 337, 724
vitamin E, 724
vitamin K, 717, 724, 763
vitelline membrane, 866, *867*
  of bird egg, *885*
viviparity, 832
  of sea snake, 835
Vogt, 316
voluntary muscle, *91*, 688–695; *see also* muscular system
vomiting center, 716
von Frisch, Karl
  color vision of bees, 930
  helps found ethology, 945
von Linne, Karl (Linnaeus), binomial nomenclature, 392
Volvox, 469, *470, 471*
  Haeckel's theories, *594*
Vorticella, *437*
vulva, 840, *841*
  embryology, 892, *893*

W

wading bird, 771
wall, the, *168*
Wallace, Alfred Russel, 13, 22
  and human evolution, 658
Wallace, Robert A., runs the marathon, 805, 806
walnut, 560
walrus, 655
Walsh, Don, explores the Marianas trench, 1010
Walters, James, grasshopper meiotic cells, *248, 250, 251*
WARFARIN, 763
Waring, Fred, invents the blender, 365
Waring blender, 365
warning cry, 972, *973*
  as an altruistic act, 982
waste
  deposited in plant vacuoles, 117
  dumping, 1012, 1021, *1022, 1040*
  see also excretory system, 808–818
water, 34–38
  atomic diagram, *36*
  cohesion, 516, 517
  content of cells, 35
  dissociation of, 40
  as ice, 36, *37*
  needs of plants, 514
  properties of, 34–38, 516, 517
  as a solvent, 36
water bug, *46*
water cycle, *35*
water flea, *619*
  evolutionary strategy, 258
water mold, 445, 447, *448*
water potential, 505
water strider, *38*
water vacuole (contractile vacuole), 105, *106*, 429, *430*, 434, 439
  heliozoan, 434
  Paramecium, 438
water vapor, 36, *37*
water vascular system, 624–626
watermelon, 572, *573*
Watson, James R., structure of DNA, 182, 186, 188, 190, 198, 201
Watson–Crick pairing of nucleic acid bases, 186
  codon and anticodon, 208
  and DNA repair, 188
  and DNA replication, *190, 191*
  in meiosis, 248
Watson strand of DNA, 190, *191*
wax, 75, 79
weather, and abiotic control of populations, 1038
web, spider, 617, 618, *619*
Weberian ossicles, 927
Wedgwood, Emma, 10
Wedgwood, Josiah, 3, 312
Weismann, August, continuity of the germ plasm, 264, 321, 322
Went, F. W., and auxin, 545
wet jungle, 1001
whale, 392, *652, 653*, 655
  baleen, 1010
  historical kills, *1046*
  killed by Russians, 1045
  toothed whales don't smell, 935
wheat, 390, *495*
  fungal infection of, 454–456, 459
wheat germ, 569
wheat rust, 454–456, 459
whiskers
  of catfish, 935
  mouse, innervation experiment, *884*
  tactile receptor, 924, *925*
white blood cell (leukocyte), *226*, 227, 228, *236*, 760, 761, *762*, 819
white leghorn, dominant color inhibitor, 279, 288
Whitman, C. O., pigeon instincts, 946
Whittaker, R. H., 402
  five-kingdom scheme, 402, *403*
  see also Appendix A
whorl, of a flower, 560
Wilcox, R. S., communication between water insects, 924
wildebeest, 999, *1001*
Wilkins, Maurice, X-ray diffraction of DNA, 201
willow, 527
Wilson, E. O., *Sociobiology*, 984
wilting, 515
wind pollination, 561
winemaking, *165*
wing
  bird and bat compared, *683*
  insect flight muscles, *672*
Woese, Carl, methanogenic bacteria, *402*, 404
wolf, 17, 386, *387*, 1046–1049
  arctic, 1007
  cooperative hunting behavior, 981
  and natural selection (Darwin's example), 14
  in northern deciduous forest, 1004
  population, 1028
  portrait of wolf as predator, 1046–1049
  preys on diseased caribou preferentially, 1041
  tame wolf runs from advancing sheep, 1047
  won't mate in bad years, 1033
Wollman, Elie, 365
women's liberation, 1055, 1056
wood, 62, 63, 503, *512*
woodpecker, *649, 709, 974*, 975
  competition and fighting between close species, 975
  given extra food, sets extra clutch, 1033
  species recognition, *974*
work, end of, 1144
world map of biomes, *993*
writs, *677*, 684, 685, *686*

X

X (photosystem 700), *143*
X chromosome
  Barr body, *226*
  drumstick, *226*
  grasshopper, 297
  factor for height, 257
  functions in embryo, 892
  heterochromatization, 300
  in metaphase I, *248*
  nondisjunction, 256, 257
X linkage, 299–303, 315; *see also* sex linkage
X-organ, *773*
X-ray chrystallography, 201, *202*
xanthophyll, 114
Xenopus, 864, *881*
xeric habitat, 996
xeroderma pigmentosum, 188, 190
xerophyte, 996, *997*
xiphoid process, 679, *680*
XO, in humans and flies, 257, 300
XXX female, 257
XXY, 257, 300
xylem, *90*, 503, 505, *510–513*
  maturation, 583
  origin from cambium, 582
  parenchyma, 511, *512*
  primary, 507, *508, 509, 582*
  primary, root, *580*
  secondary, root, 580
  secondary, stem, *582*
  terminally differentiated tissue, *229*
XYY, 257

Y

Y chromosome
  Drosophila, *300*
  has factor for height, 257
  human metaphase I, *248*
  near absence of genes, 300
  nondisjunction, 257
yak, cooperative defense, 981
yeast, 452–454
  as an ascomycete, 450
  fermentation, 159, *164*, 454, 1040
  infection of vagina, 154
  life cycle, *453*
  self-poisoning, 1040
yellow fever, *1053*
yoga, 687
yolk gland, planaria, *832*
yolk mass, *885*
yolk plug, 872, *873*
yolk sac
  bird, *644*, 873, *885*, 886
  develops into pseudoplacenta, 650
  mammalian embryo, 874, *875*, 886, *887, 889*
Y-organ, *773*
Yucca and yucca moth, *496*
Yule, G. V., 340, 341

Z

Z diagram, Z scheme, *143*, 144, 173
Z line, 691–693
Z protein, *143*, 144
zeatin, 550, 551
zebra, *1001*
zinc
  in alcohol dehydrogenase, 27
  dietary need, 725
  plant micronutrient, 537
zona pellucida, *866*
zooplankton, 620, *1010*
zoospore, *471, 472*
  of water mold, 447, *448*
zwitterion, 64
zygomatic arch, *684*
zygomycetes, *449*, 450
zygospore
  of bread mold, 449, *450*
  of *Chlamydomonas*, 469
zygote
  fate of, 867–871
  in germ line, 244
  human, *258*
zygotene, 245, *246*

The Metric System

Fahrenheit–Celsius Conversion

Centimeters–Inches Conversion

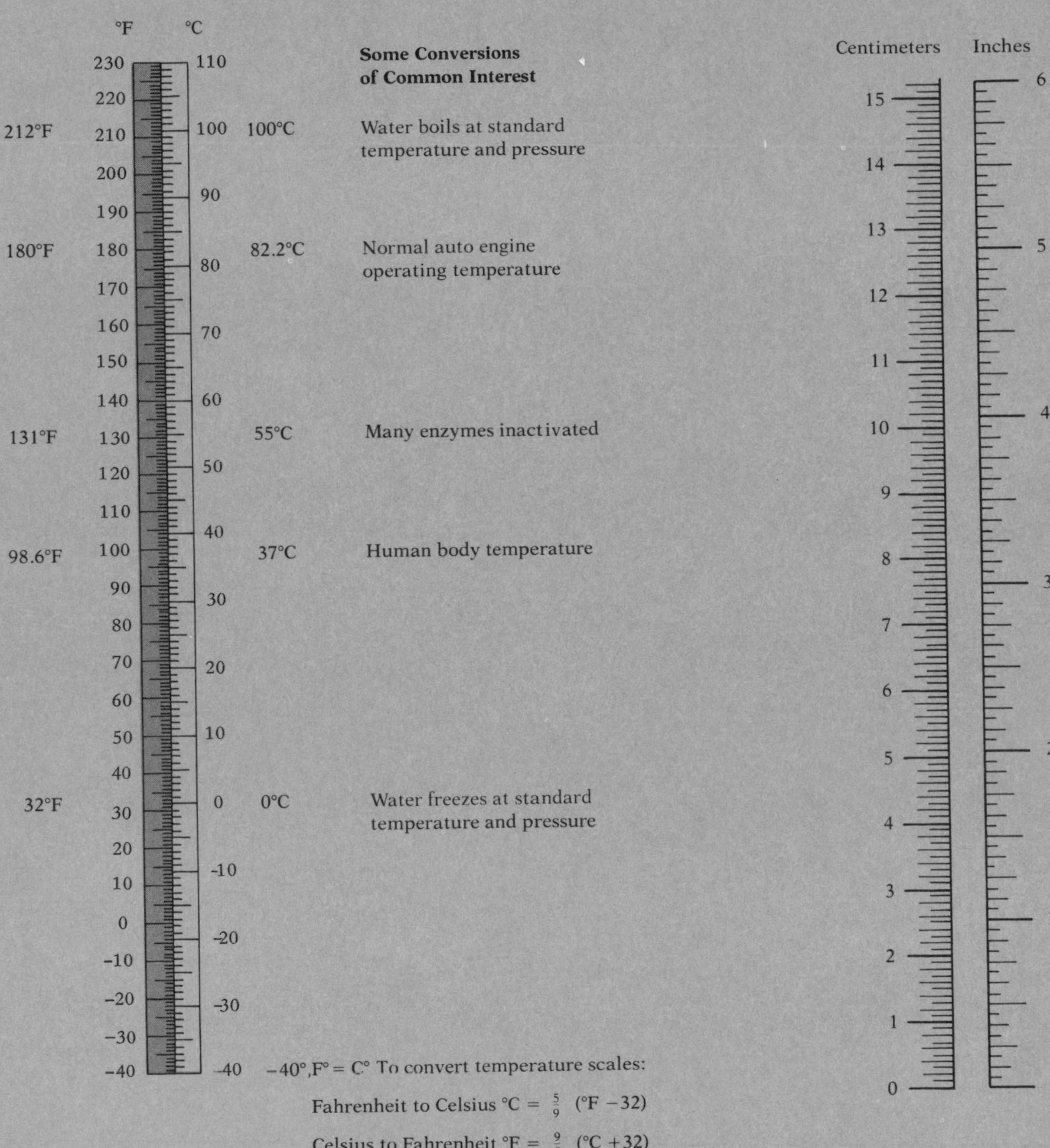

Some Conversions of Common Interest

Water boils at standard temperature and pressure

Normal auto engine operating temperature

Many enzymes inactivated

Human body temperature

Water freezes at standard temperature and pressure

−40°,F° = C° To convert temperature scales:

Fahrenheit to Celsius °C = $\frac{5}{9}$ (°F −32)

Celsius to Fahrenheit °F = $\frac{9}{5}$ (°C +32)